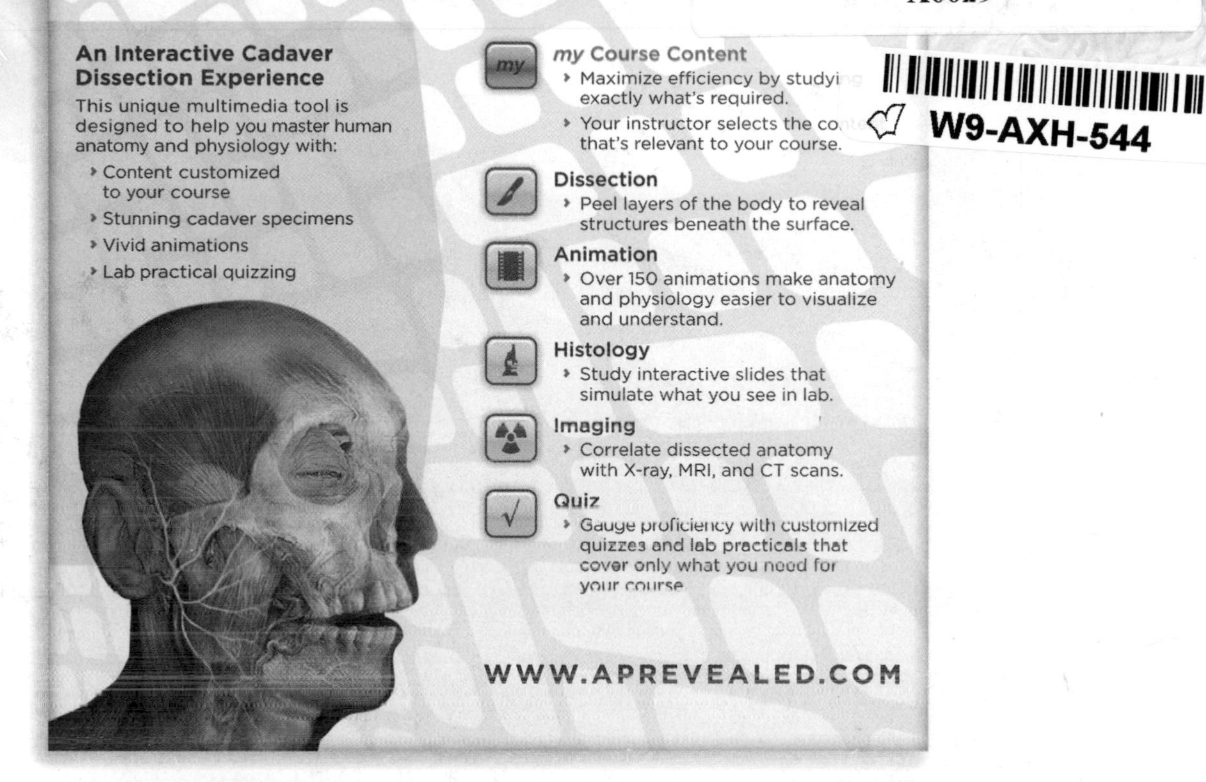
Anatomy & Physiology | REVEAL
An Interactive Cadaver Dissection Experience
This unique multimedia tool is designed to help you master human anatomy and physiology with:
Content customized to your course
Stunning cadaver specimens
Vivid animations
Lab practical quizzing
my Course Content
Maximize efficiency by studyi
exactly what's required.
Your instructor selects the co
that's relevant to your course.
Dissection
Peel layers of the body to reveal structures beneath the surface.
Animation
Over 150 animations make anatomy and physiology easier to visualize and understand.
Histology
Study interactive slides that simulate what you see in lab.
Imaging
Correlate dissected anatomy with X-ray, MRI, and CT scans.
Quiz
Gauge proficiency with customized quizzes and lab practicals that cover only what you need for your course.
WWW.APREVEALED.COM

Full Textbook Integration! APR

Icons throughout the book indicate specific McGraw-Hill Anatomy & Physiology | REVEALED® content that corresponds to the text and figures.

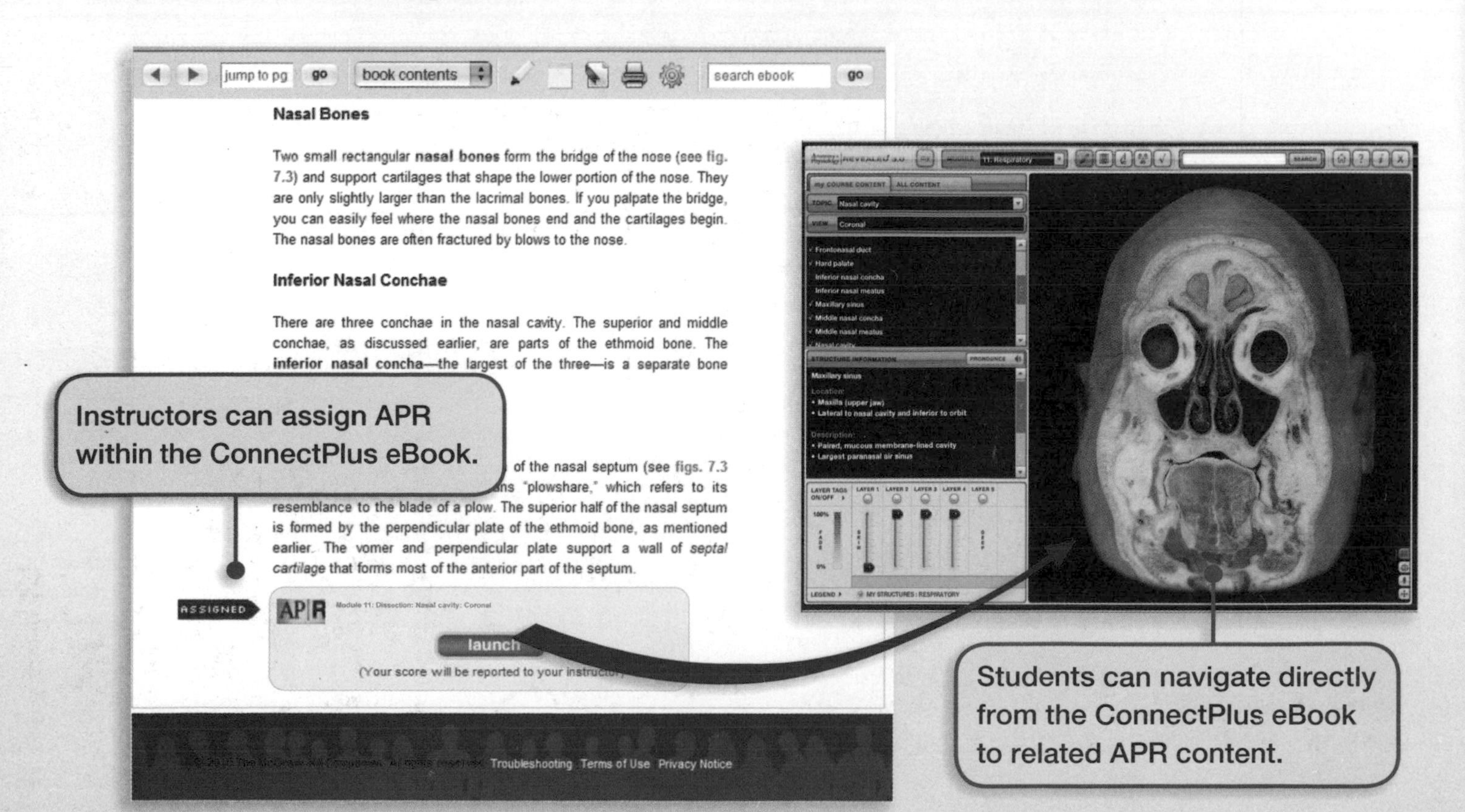
jump to pg
go
book contents
search ebook
go
Nasal Bones
Two small rectangular nasal bones form the bridge of the nose (see fig. 7.3) and support cartilages that shape the lower portion of the nose. They are only slightly larger than the lacrimal bones. If you palpate the bridge, you can easily feel where the nasal bones end and the cartilages begin. The nasal bones are often fractured by blows to the nose.
Inferior Nasal Conchae
There are three conchae in the nasal cavity. The superior and middle conchae, as discussed earlier, are parts of the ethmoid bone. The inferior nasal concha—the largest of the three—is a separate bone
Instructors can assign APR within the ConnectPlus eBook.
of the nasal septum (see figs. 7.3
"plowshare," which refers to its resemblance to the blade of a plow. The superior half of the nasal septum is formed by the perpendicular plate of the ethmoid bone, as mentioned earlier. The vomer and perpendicular plate support a wall of septal cartilage that forms most of the anterior part of the septum.
ASSIGNED
APR
launch
(Your score will be reported to your instructor)
Troubleshooting Terms of Use Privacy Notice
Students can navigate directly from the ConnectPlus eBook to related APR content.

THIRTEENTH EDITION

Human Physiology

Stuart Ira Fox

Pierce College

The McGraw·Hill Companies

HUMAN PHYSIOLOGY, THIRTEENTH EDITION

Published by McGraw-Hill, a business unit of The McGraw-Hill Companies, Inc., 1221 Avenue of the Americas, New York, NY 10020.

Some ancillaries, including electronic and print components, may not be available to customers outside the United States.

This book is printed on acid-free paper.

4 5 6 7 8 9 0 DOW/DOW 10 9 8 7 6 5 4

ISBN 978-0-07-340362-5
MHID 0-07-340362-8

Senior Vice President, Products & Markets: *Kurt L. Strand*
Vice President, General Manager, Products & Markets: *Marty Lange*
Vice President, Content Production & Technology Services: *Kimberly Meriwether David*
Director: *James F. Connely*
Brand Manager: *Marija Magner*
Director of Development: *Rose Koos*
Senior Development Editor: *Fran Simon*
Director of Digital Content Development: *Barbekka Hurtt, Ph.D.*
Marketing Manager: *Chris Loewenberg*
Project Manager: *Mary Jane Lampe*
Senior Buyer: *Sandy Ludovissy*
Senior Designer: *Laurie B. Janssen*
Cover Illustration: *© 2012 William B. Westwood, all rights reserved.*
Senior Photo Research Coordinator: *John C. Leland*
Photo Research: *David Tietz/Editorial Image, LLC*
Media Project Manager: *Laura L. Bies*
Typeface: *10/12 Utopia Std*
Compositor: *Electronic Publishing Services Inc., NYC*
Printer: *R. R. Donnelley*

Library of Congress Cataloging-in-Publication Data

Fox, Stuart Ira.
Human physiology / Stuart Ira Fox.—13th ed.
p. cm.
Includes index.
ISBN 978-0-07-340362-5—ISBN 0-07-340362-8 (hard copy : alk. paper)
1. Human physiology—Textbooks. I. Title.
QP34.5.F68 2013
612—dc23

2012003894

www.mhhe.com

Brief Contents

About the Author

Stuart Ira Fox earned a Ph.D. in human physiology from the Department of Physiology, School of Medicine, at the University of Southern California, after earning degrees at the University of California at Los Angeles (UCLA); California State University, Los Angeles; and UC Santa Barbara. He has spent most of his professional life teaching at Los Angeles City College; California State University, Northridge; and Pierce College, where he has won numerous teaching awards, including several Golden Apples. Stuart has authored thirty-eight editions of seven textbooks, which are used worldwide and have been translated into several languages. When not engaged in professional activities, he likes to hike, fly fish, and cross-country ski in the Sierra Nevada Mountains.

I wrote the first edition of *Human Physiology* to provide my students with a readable textbook to support the lecture material and help them understand physiology concepts they would need later in their health curricula and professions. This approach turned out to have wide appeal, which afforded me the opportunity to refine and update the text with each new edition. Writing new editions is a challenging educational experience, and an activity I find immensely enjoyable. Although changes have occurred in the scientific understanding and applications of physiological concepts, the students using this thirteenth edition have the same needs as those who used the first, and so my writing goals have remained the same. I am thankful for the privilege of being able to serve students and their instructors through these thirteen editions of *Human Physiology*.

—Stuart Ira Fox

To my wife, Ellen;
and to Laura, Eric, Kayleigh, and Jacob Van Gilder;
for all the important reasons.

Preface

FEATURES

The Cover

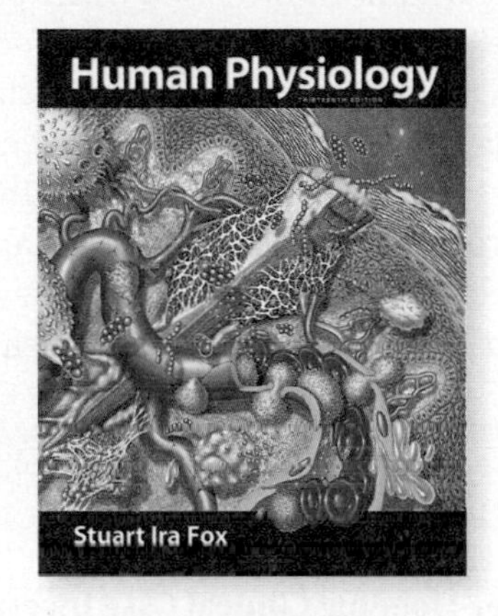

The cover illustration by William B. Westwood depicts a battle waged on multiple fronts in response to an invasion. A splinter has pierced the skin and allowed bacteria to enter the underlying tissue. This has enlisted numerous defense mechansims to plug the torn tissue and blood vessels and combat the bacteria. When the physiological events are viewed, the common experience of getting a splinter becomes rightly appreciated as high drama.

The splinter has pierced the epidermis and dermis of the skin, as well as blood vessels in the dermis. This has initiated a cascade of reactions involving blood platelets (small tan spheres) and clotting factors that produce a web of fibrin proteins forming a clot in the wound. Bacteria (staphylococcus are shown as purple grape-like clusters, streptococcus as short chains of orange-brown ovals) invade the body through this breach in the epidermal barrier. Neutrophils (the numerous large purple cells covered with bumps representing microvilli) leave small blood vessels and enter the sourrounding tissues along with fluid, which causes swelling in the area. Neutrophils can "eat" bacteria by phagocytosis, as shown in this illustration by two neutrophils engulfing staphylococci bacteria. Neutrophils also trap bacteria in secreted web-like NETS (neutrophil extracellular traps) and release enzymes and other molecules that kill bacteria.

Macrophages (green cells) are also phagocytic and catch bacteria in their filamentous pseudopod extensions. Lymphocytes (another type of white blood cell here colored blue-green; one is shown in the 12:00 o'clock position) secrete antibodies, which are depicted as yellow Y-shaped protein molecules. Antibodies initiate events that activate complement proteins (small turquoise spheres) and that stimulate mast cells (gray cells) to release histamine (small gray eggs) and other mediators of inflammation. Antibodies promote phagocytosis by neutrophils and macrophages, and complement proteins help antibodies destroy bacterial cells.

All this is but a snapshot of ongoing events that normally result in eradication of the bacteria and healing of the wound. The common experience of getting a splinter and recovering from it presents a fine example of homeostasis, of how physiological mechanisms act to re-establish a healthy state after normal conditions are perturbed.

What Sets This Book Apart?

The study of human physiology provides the scientific foundation for the field of medicine and all other professions related to human health and physical performance. The scope of topics included in a human physiology course is therefore wide-ranging, yet each topic must be covered in sufficient detail to provide a firm basis for future expansion and application.

> "The rigor of this course, however, need not diminish the student's initial fascination with how the body works. On the contrary, a basic understanding of physiological mechanisms can instill a deeper appreciation for the complexity and beauty of the human body and motivate students to continue learning more."
>
> —Stuart Fox

Human Physiology, thirteenth edition, is written for the undergraduate introductory human physiology course. Based on the author's extensive experience with teaching this course, the framework of the textbook is designed to provide basic biology and chemistry (chapters 2–5) before delving into more complex physiological processes. This approach is appreciated by both instructors and students; specific references in later chapters direct readers back to the foundational material as needed, presenting a self-contained study of human physiology.

In addition to not presupposing student's preparedness, this popular textbook is known for its clear and approachable writing style, detailed realistic art, and unsurpassed clinical information.

GUIDED TOUR

What Makes This Text a Market Leader?

Clinical Applications—No Other Human Physiology Text Has More!

The framework of this textbook is based on integrating clinically germane information with knowledge of the body's physiological processes. Examples of this abound throughout the book. For example, in a clinical setting we record electrical activity from the body: this includes action potentials (chapter 7, section 7.2); EEG (chapter 8, section 8.2); and ECG (chapter 13, section 13.5). We also record mechanical force in muscle contractions (chapter 12, section 12.3). We note blood plasma measurements of many chemicals to assess internal body conditions. These include measurements of blood glucose (chapter 1, section 1.2) and the oral glucose tolerance test (chapter 19, section 19.4); and measurements of the blood cholesterol profile (chapter 13, section 13.7). These are just a few of many examples the author includes that focus on the connections between the study of physiology and our health industry.

Clinical Investigation

Tom is a 77-year-old man brought to the hospital because of severe chest pain. He also complained that he had difficulty urinating and "got the runs" whenever he ate ice cream.

Some of the new terms and concepts you will encounter include:

- Isoenzymes
- Creatine phosphokinase and

◀ **Chapter-Opening Clinical Case Investigations, Clues, and Summaries** are diagnostic case studies found in each chapter. Clues are given throughout and the case is finally resolved at the end of the chapter.

Clinical Investigation CLUES

Laboratory tests showed elevated levels of acid phosphatase and creatine kinase in Tom's plasma.

- Which laboratory test might be related to Tom's difficulty in urinating?
- Given Tom's chest pain, what condition might his elevated creatine phosphokinase indicate?

Clinical Investigation SUMMARY

[illegible]da's great fatigue following workouts is partially related [illegible]e depletion of her glycogen reserves and extensive uti[illegible]on of anaerobic metabolism (with consequent produc[illegible]of lactic acid) for energy. Production of large amounts [illegible]ctic acid during exercise causes her need for extra [illegible]en to metabolize the lactic acid following exercise (the [illegible]en debt)—hence, her gasping and panting. Eating [illegible]e carbohydrates would help Brenda maintain the gly[illegible]n stores in her liver and muscles, and training more [illegible]ually could increase the ability of her muscles to obtain more of their energy through aerobic metabolism, so that she would experience less pain and fatigue.

The pain in her arms and shoulders is probably the result of lactic acid production by the exercised skeletal muscles. However, the intense pain in her left pectoral region could be angina pectoris, caused by anaerobic metabolism of the heart. If this is the case, it would indicate that the heart became ischemic because blood flow was inadequate for the demands placed upon it. Blood tests for particular enzymes released by damaged heart muscle (chapter 4) and an electrocardiogram (ECG) should be performed.

See the additional chapter 5 Clinical Investigation on *Metabolic Disease* in the Connect site for this text at www.mhhe.com/Fox13

▶ **Clinical Investigations** are enhanced with even more clinical assessments available on Connect. These Clinical Investigations are written by the author and are specific to each chapter. They will offer the students great insight into that specific chapter.

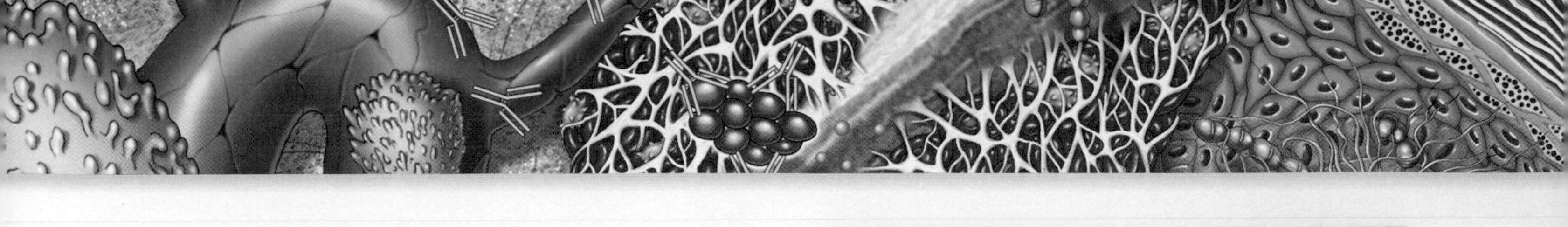

▶ **Clinical Application Boxes** are in-depth boxed essays that explore relevant topics of clinical interest and are placed at key points in the chapter to support the surrounding material. Subjects covered include pathologies, current research, pharmacology, and a variety of clinical diseases.

CLINICAL APPLICATION

When tissues become damaged as a result of diseases, some of the dead cells disintegrate and release their enzymes into the blood. Most of these enzymes are not normally active in the blood for lack of their specific substrates, but their enzymatic activity can be measured in a test tube by the addition of the appropriate substrates to samples of plasma. Such measurements are clinically useful because abnormally high plasma concentrations of particular enzymes are characteristic of certain diseases (table 4.1).

FITNESS APPLICATION

The ingestion of excessive calories increases fat production. The rise in blood glucose that follows carbohydrate-rich meals stimulates insulin secretion, and this hormone, in turn, promotes the entry of blood glucose into adipose cells. Increased availability of glucose within adipose cells, under conditions of high insulin secretion, promotes the conversion of glucose to fat (see figs. 5.12 and 5.13). The lowering of insulin secretion, conversely, promotes the breakdown of fat. This is exploited for weight reduction by low-carbohydrate diets. There has been concern that too much consumption of sugar, either as sucrose or as *high fructose corn syrup* (which contains about equal amounts of glucose and fructose), could promote obesity and the **metabolic syndrome**—a combination of central obesity, insulin resistance, type 2 diabetes mellitus, and hypertension (chapter 19, section 19.2).

◀ **Fitness Application Boxes** are readings that explore physiological principles as applied to well-being, sports medicine, exercise physiology, and aging. They are also placed at relevant points in the text to highlight concepts just covered in the chapter.

New Features!

▶ **Learning Outcomes** are now numbered for easy referencing in digital material!

LEARNING OUTCOMES

After studying this section, you should be able to:

2. Describe the aerobic cell respiration of glucose through the citric acid cycle.

3. Describe the electron transport system and oxidative phosphorylation, explaining the role of oxygen in this process.

▶ Learning Outcome numbers are now tied directly to **Checkpoint numbers!**

CHECKPOINT

2a. Compare the fate of pyruvate in aerobic and anaerobic cell respiration.

2b. Draw a simplified citric acid cycle and indicate the high-energy products.

3a. Explain how NADH and $FADH_2$ contribute to oxidative phosphorylation.

3b. Explain how ATP is produced in oxidative phosphorylation.

What Makes This Text a Market Leader?

Writing Style—Easygoing, Logical, and Concise

The words in *Human Physiology*, thirteenth edition, read as if the author is explaining concepts to you in a one-on-one conversation, pausing now and then to check and make sure you understand what he is saying. Each major section begins with a short overview of the information to follow. Numerous **comparisons** ("Unlike the life of an organism, which can be viewed as a linear progression from birth to death, the life of a cell follows a cyclical pattern"), **examples** ("A callus on the hand, for example, involves thickening of the skin by hyperplasia due to frequent abrasion"), **reminders** ("Recall that each member of a homologous pair came from a different parent"), and **analogies** ("In addition to this 'shuffling of the deck' of chromosomes...") lend the author's style a comfortable grace that enables readers to easily flow from one topic to the next.

Exceptional Art—Designed from the Student's Point of View

What better way to support such unparalleled writing than with high-quality art? Large, bright illustrations demonstrate the physiological processes of the human body beautifully in a variety of ways.

▶ **Stepped-out art** clearly depicts various stages or movements with numbered explanations.

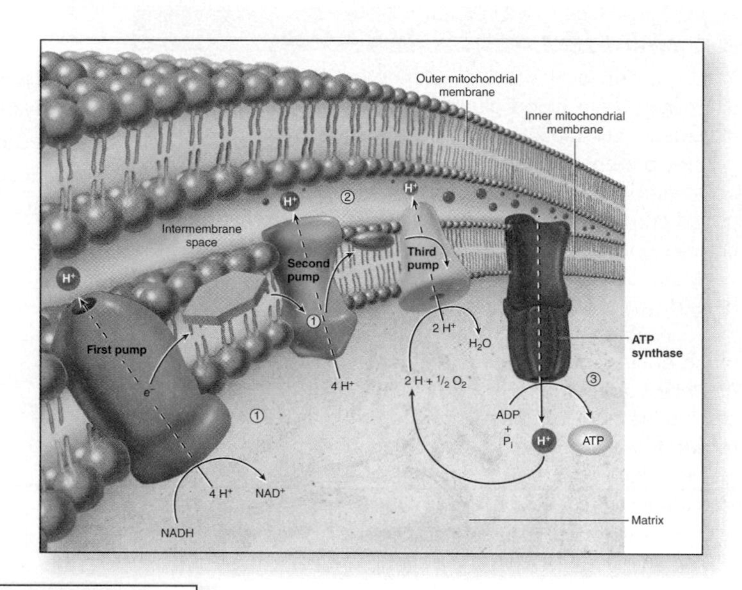

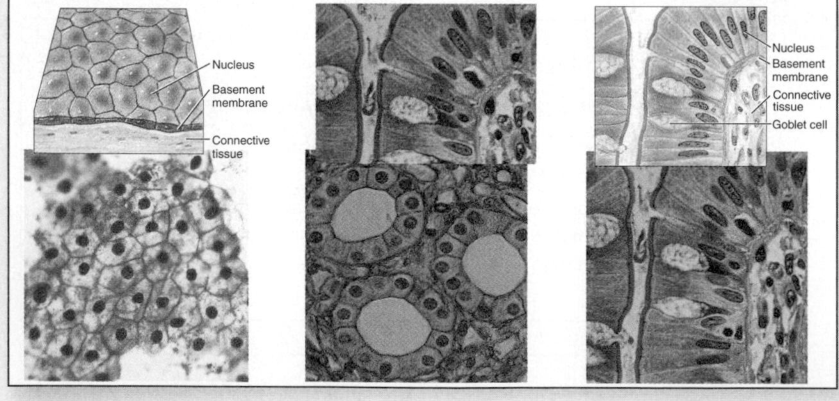

◀ **Labeled photos placed side by side with illustrations** allow diagrammatic detail and realistic application.

▶ **Macro-to-micro art** helps students put context around detailed concepts.

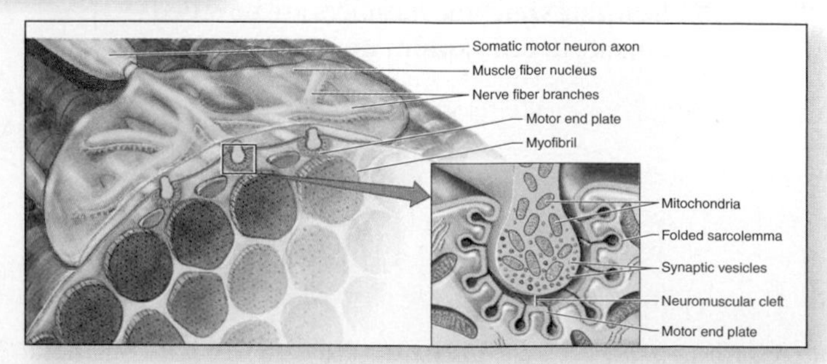

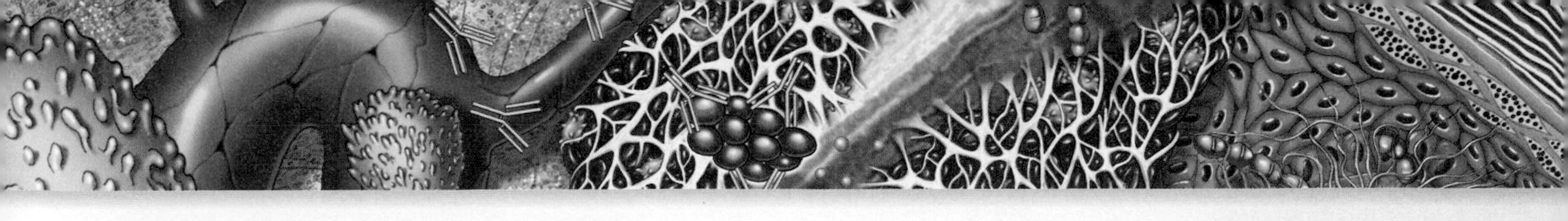

Thirteenth Edition Changes

What's New?

Human Physiology, thirteenth edition, incorporates a number of new and recently modified physiological concepts. This may surprise people who are unfamiliar with the subject; indeed, the author sometimes is asked if the field really changes much from one edition to the next. It does; that's one of the reasons physiology is so much fun to study. Stuart has tried to impart this sense of excitement and fun in the book by indicating, in a manner appropriate for this level of student, where knowledge is new and where gaps in our knowledge remain.

The list that follows indicates only the larger areas of text and figure revisions and updates. It doesn't indicate instances where passages were rewritten to improve the clarity or accuracy of the existing material, or smaller changes made in response to information from recently published journals and from the reviewers of the previous edition.

GLOBAL CHANGES:

- Case Investigation renamed Clinical Investigation.
- Learning Outcomes now numbered consecutively in each chapter.
- Figures that are referenced in the Test Your Quantitative Ability section of the Review Activities are now so indicated.
- Clinical Investigation Summary boxes now specify further Clinical Investigation available on Connect for this text.
- Figure colors brightened in all of the chapters.

MAJOR CHANGES IN CHAPTERS

Chapter 1: The Study of Body Function

- A new Clinical Investigation was added to chapter 1 with instructions on how to proceed and where to find additional Clinical Investigations in Connect.
- Expanded discussion of Goodpasture's syndrome.

Chapter 2: Chemical Composition of the Body

- Expanded discussion of stereoisomers.
- New figure 2.13 of stereoisomers.

Chapter 3: Cell Structure and Genetic Control

- Expanded discussion of the centrosome and basal body.
- Updated discussion of autophagy and added information on autophagosome.
- Discussion of mitochondrial DNA expanded and updated.
- Description of HapMap Project added to Clinical Application box.
- Explanation of microRNA (miRNA) expanded and updated.
- Expanded and updated description of epigenetic inheritance.

Chapter 4: Enzymes and Energy

- Figures 4.1 and 4.17 modified and several legends modified.

Chapter 5: Cell Respiration and Metabolism

- Overview of metabolism revised for enhanced clarity.
- Chapter reorganized to keep aerobic respiration of glucose in one section.
- New section added: "Interconversion of Glucose, Lactic Acid, and Glycogen."
- Table of ATP yields modified with footnotes for added clarity.
- New description of metabolic syndrome added.
- New Clinical Application box added on liver enzyme tests.

Chapter 6: Interactions Between Cells and the Extracellular Environment

- Discussion of diabetes mellitus in Clinical Application box revised for clarity.
- Updated and expanded discussion of the GLUT family of glucose transporters.
- New information added on membrane transporters for fatty acids.
- Updated and expanded discussion of paracellular transport through epithelial membranes.
- Updated and expanded description of ORT solution for rehydration therapy

Chapter 7: The Nervous System: Neurons and Synapses

- Expanded description of neuron structure.
- Updated and expanded discussion of microglia and their functions.
- Updated and expanded discussion of astrocytes and their functions.
- Updated discussion of gliotransmitters.
- Updated discussion of action potential initiation with new description of the back propagation of action potentials.
- Discussion of serotonin transporters added and action of SSRI drugs updated.
- Revised and updated description of the inhibitory effects of GABA and glycine.
- Revised description of spatial and temporal summation.
- Updated discussion of excitotoxicity.

Chapter 8: The Central Nervous System

- New discussion added on the structure and function of choroid plexuses.
- Discussion of mirror neurons updated and expanded.
- Expanded and updated discussion of PET scans.
- Updated and expanded discussion of the effects of sleep phases on memory.
- Discussion of Parkinson's disease updated and expanded.
- Updated and expanded description of memory consolidation.
- New section on Alzheimer's disease added with updated and expanded information.
- New discussion added on the role of CREB in synaptic plasticity.
- Discussion of neurogenesis updated and expanded.

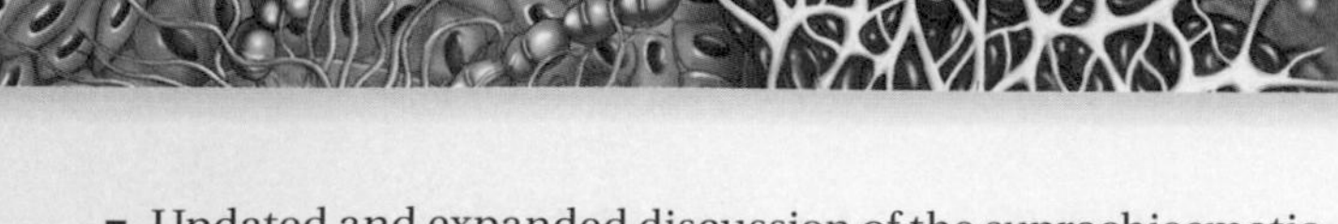

- Updated and expanded discussion of the suprachiasmatic nucleus and circadian rhythms.
- Discussion of the nucleus accumbens in the mesolimbic system and its relationship to drug abuse updated and expanded.
- Updated and expanded information on orexin neurons and narcolepsy.

Chapter 9: The Autonomic Nervous System

- Updated and expanded discussion of the mass activation of the sympathetic system.
- Updated and expanded discussion of adrenergic receptors.
- New table 9.5 on adrenergic and cholinergic receptor agonists and antagonists.
- Updated table 9.7 on adrenergic and cholinergic effects.
- Neural regulation of micturition updated.

Chapter 10: Sensory Physiology

- Updated and expanded discussion of pain and itch reception.
- Neural pathways of somatic sensation updated and expanded.
- Neural pathways of taste and smell updated and expanded.
- Updated and expanded discussion of function of stereocilia of inner ear.
- Neural pathways of hearing updated and expanded. New discussion of sound localization added.
- New discussion of corneal stem cells added.
- Updated discussion of the structure and function of the lens.
- Discussion of retinal pigment epithelium updated.
- Discussion of the physiology of cones updated.
- Updated discussion of macular degeneration.

Chapter 11: Endocrine Glands: Secretion and Action of Hormones

- Mechanism of action of methylxanthines expanded and updated.
- Description of the regulation of prolactin secretion updated and expanded.
- Effects of glucocorticoids on metabolism updated.
- Updated discussion of the effects of stress hormones on brain function and memory.
- Discussion of Graves' disease updated.
- Updated and expanded description of the melatonin and circadian rhythms.

Chapter 12: Muscle: Mechanisms of Contraction and Neural Control

- Updated and expanded discussion of titin.
- Discussion of the events occurring during twitch and summation updated and expanded.
- New figure 12.18 on twitch and summation.
- Updated and expanded discussion of the length–tension relationship.
- Updated and expanded discussion of glucose uptake and skeletal muscle metabolism.
- Discussion of the role of phosphocreatine in skeletal muscle metabolism updated and expanded.
- Description of muscle fiber types and their functions expanded and updated.
- Updated and expanded discussion of muscle fatigue.
- Description of the effects of exercise training on muscles and health updated and expanded.
- Expanded discussion of obscurin and nebulin.
- Reorganized section on smooth muscles.
- Updated discussion of the regulation of smooth muscle contraction and relaxation.

Chapter 13: Blood, Heart, and Circulation

- Updated discussion of hematopoietic stem cells.
- Information on blood clotting and clot prevention updated.
- Discussion of heart murmurs updated.
- Discussion of the cardiac cycle expanded to include blood pressure variations and the dicrotic notch.
- Figure 13.17 revised to show arterial pressure changes during the cardiac cycle.
- Updated and expanded description of excitation–contraction coupling in myocardial cells.
- New Clinical Application box added on aneurisms.
- Information on angiogenesis updated.
- New discussion of deep vein thrombosis added.
- Updated and expanded description of cholesterol carriers and atherosclerosis.
- Description of myocardial infarction updated and expanded.

Chapter 14: Cardiac Output, Blood Flow, and Blood Pressure

- Explanation of the mechanism of the Frank–Starling relationship updated and expanded.
- Description of the lymphatic return of interstitial fluid updated.
- New description of the sympathetic system regulation of blood volume added.
- Updated and expanded description of myocardial ATP turnover.
- Description of the cardiovascular adjustments during exercise updated and expanded.
- Discussion of the regulation of cerebral blood flow updated and expanded.
- Discussion of hypertension updated.
- Updated and expanded discussion of antihypertensive drugs.

Chapter 15: The Immune System

- Updated and expanded discussion of pathogen recognition receptors.
- New discussion added on danger-associated molecular patterns (DAMPS).
- New discussion added on cytokines and chemokines.
- Updated and expanded description of mast cells.

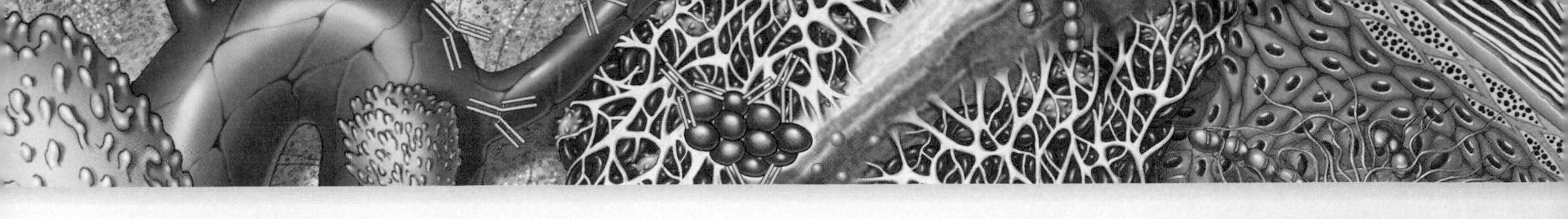

- New discussion added on neutrophil extracellular traps (NETS).
- Description of the events in a local inflammation updated and expanded.
- Discussion of HIV and AIDS updated and expanded.
- Updated description of interleukins.
- Discussion of adjuvants updated and expanded.
- Updated and expanded discussion of the clinical uses of monoclonal antibodies.
- Discussion of nuclear factor-κB updated and expanded.
- Updated and expanded discussion of natural killer (NK) cells.
- Discussion of the immunotherapy for cancer updated and expanded.
- Updated and expanded discussion of rheumatoid arthritis and systemic lupus erythematosus (SLE).

Chapter 16: Respiratory Physiology

- Updated and expanded discussion of asthma.
- Revised description of partial pressure measurements.
- New discussion of cor pulmonale.
- New discussion of phrenic motor nuclei and their function in the regulation of breathing.
- Updated discussion of the effect of partial pressure of oxygen in the regulation of breathing.
- Discussion of obstructive sleep apnea expanded.
- Updated discussion of β-thalassemia.
- Discussion of volatile and nonvolatile acids expanded.
- Updated discussion of polycythemia at high altitudes.

Chapter 17: Physiology of the Kidneys

- Description of kidney stones updated and revised.
- Updated discussion of polycystic kidney disease.
- Description of tubuloglomerular feedback updated and expanded.
- Updated and expanded description of organic anion and cation transporters.
- New description added of sulfate and phosphate reabsorption.
- Updated and expanded discussion of the juxtaglomerular apparatus.
- Discussion of the macula densa updated.
- Clinical Application box on Addison's disease and Conn's syndrome updated.
- Description of the generation of bicarbonate and ammonia by the renal tubules updated.

Chapter 18: The Digestive System

- Discussion of the gastrin stimulation of stomach acid secretion updated.
- Updated discussion of peptic ulcers.
- Updated and expanded description of the roles of Paneth cells.
- Explanation of the roles of the intestinal microbiota expanded and updated.
- New clinical applications box added on inflammatory bowel disease (IBD).
- Updated and expanded discussion of fluid and electrolyte absorption by the intestine.
- New discussion added on emulsification by bile acids.
- Updated table 18.6 on the regulation of gastric secretion.
- Enteric nervous system discussion updated and expanded.
- Updated discussion of the intestinal absorption of glucose and amino acids.
- New figure 18.37 showing effect of protein-bound triglycerides on the turbidity of the plasma.

Chapter 19: Regulation of Metabolism

- Updated and expanded information on essential fatty acids.
- New clinical application box on Wernicke-Korsakoff syndrome and beriberi.
- Discussion of hepatic steatosis and related conditions added.
- Updated and expanded discussion of the hypothalamic regulation of hunger.
- Updated discussion of leptin regulation of hunger and metabolism.
- Updated discussion of the role of insulin in hunger regulation.
- Discussion of diet-induced thermogenesis updated.
- Description of the hormonal responses to fasting updated and expanded.
- Description of the effects of fatty acids during fasting updated.
- Updated and expanded discussion of type 1 diabetes mellitus.
- Updated and expanded discussion of type 2 diabetes mellitus.
- Description of reactive hypoglycemia updated and expanded.
- New discussion added on glycated hemoglobin (hgb A1c) measurements.
- Updated and expanded discussion of drugs used to treat diabetes mellitus.
- Discussion of the hormonal regulation of bone physiology updated and expanded.
- Updated and expanded description of vitamin D function.

Chapter 20: Reproduction

- Discussion of genomic imprinting updated.
- Updated description of sex determination.
- New discussion added on the effect of testosterone on brain function.
- Description of sperm structure updated and expanded.
- Updated and expanded discussion of events at fertilization.
- New discussion added on monozygotic and dizygotic twins.
- Updated description of the fate of sperm mitochondria.
- Discussion of pluripotency updated.
- Updated and expanded discussion of iPS cells and adult stem cells.
- New discussion added on noninvasive fetal sex determination.
- Updated information added on placental steroid metabolism.
- Description of mammary gland structure and function updated and expanded.

Engaging Presentation Materials for Lecture and Lab

▶ **McGraw-Hill ConnectPlus® Anatomy & Physiology** is a web-based assignment and assessment platform that gives students the means to better connect with their coursework, with their instructors, and with the important concepts that they will need to know for success now and in the future. With Connect® Anatomy & Physiology, instructors can deliver assignments, quizzes, and tests easily online. Students can practice important skills at their own pace and on their own schedule. With ConnectPlus Anatomy & Physiology, students also get 24/7 online access to an eBook—an online edition of the text—to aid them in successfully completing their work, wherever and whenever they choose. (www.mhhe.com/Fox13)

NEW! **ConnectPlus®** with **LearnSmart™** makes teaching easier and learning smarter.

NEW! All content in Connect® is correlated to **HAPS Learning Outcomes!**

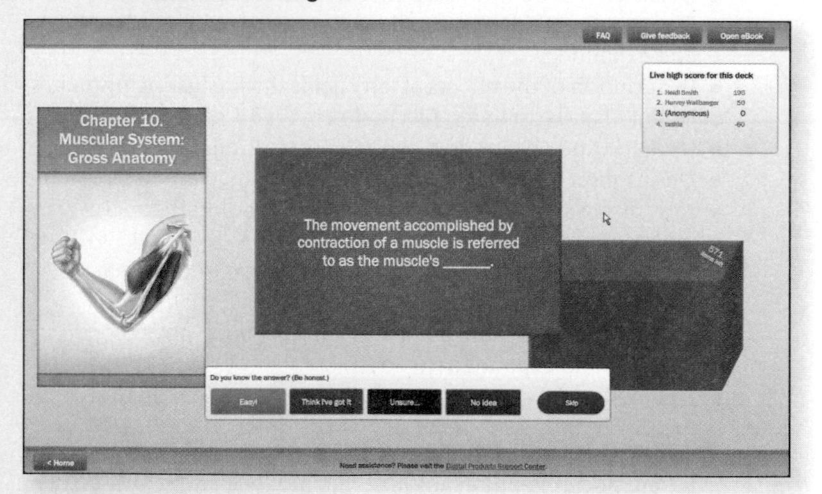

◀ **McGraw-Hill LearnSmart™** is an online diagnostic learning system that determines the level of student knowledge, and provides the student with suitable content for the Anatomy & Physiology course. Students learn faster and study more effectively. As a student works within the system, LearnSmart develops a personal learning path adapted to what the student has learned and retained. LearnSmart is able to recommend additional study resources to help the student master topics. This innovative and outstanding study tool also has features for instructors where they can see exactly what students have accomplished, and a built in assessment tool for graded assignments. You can access LearnSmart through ConnectPlus.

McGraw-Hill Higher Education and Blackboard® have teamed up!

What does this mean for you?

- Life simplified. Now, all McGraw-Hill content (text, tools, & homework) can be accessed directly from within your Blackboard course. All with one sign-on.
- Deep integration. McGraw-Hill's content and content engines are seamlessly woven within your Blackboard course.
- No more manual synching! Connect® assignments within Blackboard automatically (and instantly) feed grades directly to your Blackboard grade center. No more keeping track of two gradebooks!
- A solution for everyone. Even if your institution is not currently using Blackboard, we have a solution for you. Ask your McGraw-Hill representative for details.

NEW! Anatomy & Physiology Revealed 3.0 icons appear on figures that have a corresponding image in APR 3.0.

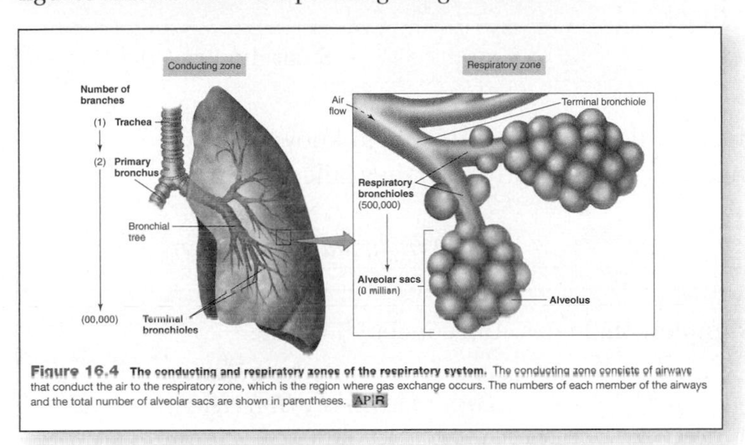

Figure 16.4 The conducting and respiratory zones of the respiratory system. The conducting zone consists of airways that conduct the air to the respiratory zone, which is the region where gas exchange occurs. The numbers of each member of the airways and the total number of alveolar sacs are shown in parentheses. APR

Text Website www.mhhe.com/Fox13

You will find the following items on the text website:

- **Art** Full-color digital files of all illustrations in the book and unlabeled versions of the same artwork can be readily incorporated into lecture presentations, exams, or custom-made classroom materials.
- **Photos** Digital files of all photographs from the text can be reproduced for multiple classroom uses.
- **Tables** Every table that appears in the text is available to instructors in electronic form.
- **Animations** Numerous full-color animations illustrating physiological processes are provided. Harness the visual impact of processes in motion by importing these files into classroom presentations or online course materials.
- **Lecture PPTs** Three different sets of PPTs are now available for instructors, including one with embedded animations. Rather build your own? No problem! All McGraw-Hill art is at your disposal with an easy-to-use search engine.

McGraw-Hill's Presentation Tools

Presentation Materials for Lecture and Lab—incorporate customized lectures, visually enhanced tests and quizzes, compelling course websites, or attractive printed support materials.

- A complete set of animation embedded PowerPoint slides are now available.
- Along with our online digital library containing photos, artwork, and animations, we now also offer **FlexArt**. FlexArt allows the instructor to customize artwork.
- Computerized test bank is powered by McGraw-Hill's flexible electronic testing program EZ Test Online.

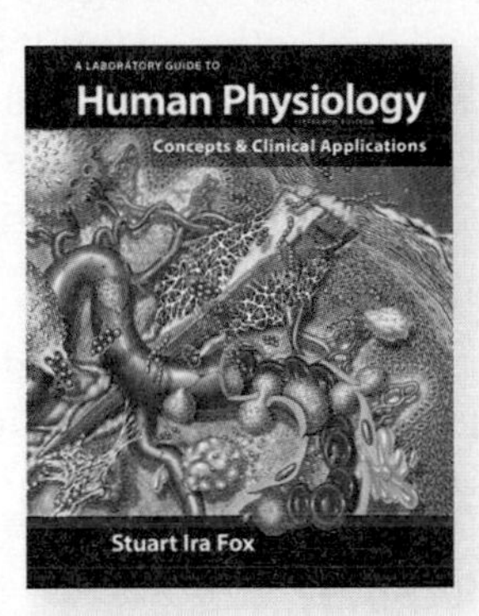

***Fox's Laboratory Guide to Human Physiology: Concepts and Clinical Applications*, 13th edition,** contains time- and student-tested laboratory exercises that support most of the subjects covered in a human physiology course. It functions as a standalone manual, but is particularly well suited for use with Fox's *Human Physiology, 13th edition* textbook. There are cross-references to relevant text information at the start of each exercise and cross-references in the figure legends to corresponding textbook figures. Also, the pedagogical hierarchy of questions in the Laboratory Report is the same as in the Review Activities of the textbook.

NEW! for the 13th edition: this Laboratory Guide now has clinical investigations at the start of most exercises! This is similar to the way that each chapter in *Human Physiology* begins with a clinical investigation. These provide added incentive for students to learn the concepts of the laboratory exercises and see how they apply in practice to the health professions. Questions relating to these clinical investigations have been added to the Laboratory Report of each exercise.

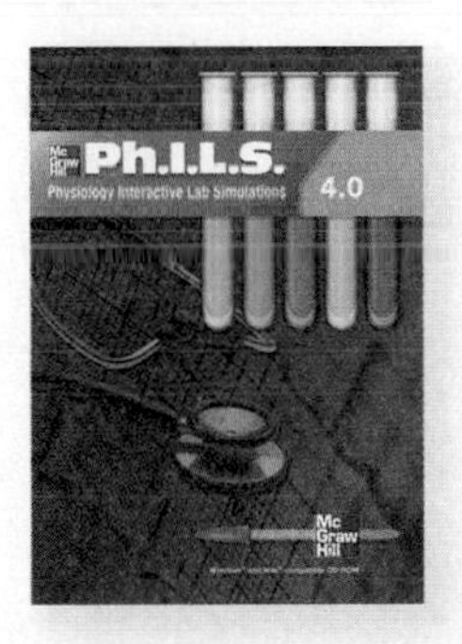

NEW! Ph.I.L.S. 4.0 has been updated! Users have requested and we are providing ***five new exercises*** (Respiratory Quotient, Weight & Contraction, Insulin and Glucose Tolerance, Blood Typing, and Anti-Diuretic Hormone). **Ph.I.L.S. 4.0** is the perfect way to reinforce key physiology concepts with powerful lab experiments. Created by Dr. Phil Stephens at Villanova University, this program offers ***42 laboratory simulations*** that may be used to supplement or substitute for wet labs. All 42 labs are self-contained experiments—no lengthy instruction manual required. Users can adjust variables, view outcomes, make predictions, draw conclusions, and print lab reports. This easy-to-use software offers the flexibility to change the parameters of the lab experiment. There is no limit!

NEW! Tegrity App—The Tegrity App lets students stream the recordings on-demand to their iPod Touch®, iPhone, iPad®, or Android™ device. This app is free for students taking any course in which the Tegrity service is being used.

Acknowledgments

The thirteenth edition of *Human Physiology* is the result of extensive analysis of new research in the field of physiology and evaluation of input from instructors who have thoroughly reviewed chapters. I am grateful to these colleagues and have used their constructive feedback to update and enhance the features and strengths of this textbook.

—Stuart Ira Fox

The "Learning Outcomes" is a strong point in this text. It allows the students to know exactly what they should have learned and will force them to go back and study what they failed to learn.

—Nick Ritucci
Wright State University

I give the author an "A" overall for explanation, examples, analogies, and concept building.

—Barbara Davis
Eastern Kentucky University

I think that the author is right on target with the explanations and examples. A perfect example is in chapter 2, the structure of atom including mass number and atomic number is explained in two short and concise paragraphs with one table and one figure as emphasis.

—Sheryl Ribbing
Shawnee Community College

Fox's text is at just the right level of sophomore and junior undergraduates. I have reviewed a large number of texts, and Fox's is among the best.

—Robert E. Farrell, Jr.
Penn State University

Reviewers

Paige J. Baugher, *Pacific University*
Gerrit J. Bouma, *Colorado State University*
Barbara Davis, *Eastern Kentucky University*
Robert E. Farrell, Jr., *Penn State University*
Cindy L. Hansen, *Community College of Rhode Island*
Kelly Johnson, *University of Kansas*
Susannah Nelson Longenbaker, *Columbus State Community College*
Royal A. McGraw, *University of Georgia*
Randy Mogg, *Columbus State Community College*
Richard G. Mynark, *Indiana University School of Medicine*
Jon S. Powell, *Southwest Colorado Community College*
Sheryl Ribbing, *Shawnee Community College*
Nick Ritucci, *Wright State University*
Katharina Rodriguez, *Pasadena City College*
Amber D. Ruskell, *Southeastern Community College*
Merideth Sellars, *Columbus State Community College*
Joseph Shostell, *Penn State University–Uniontown*

Contents

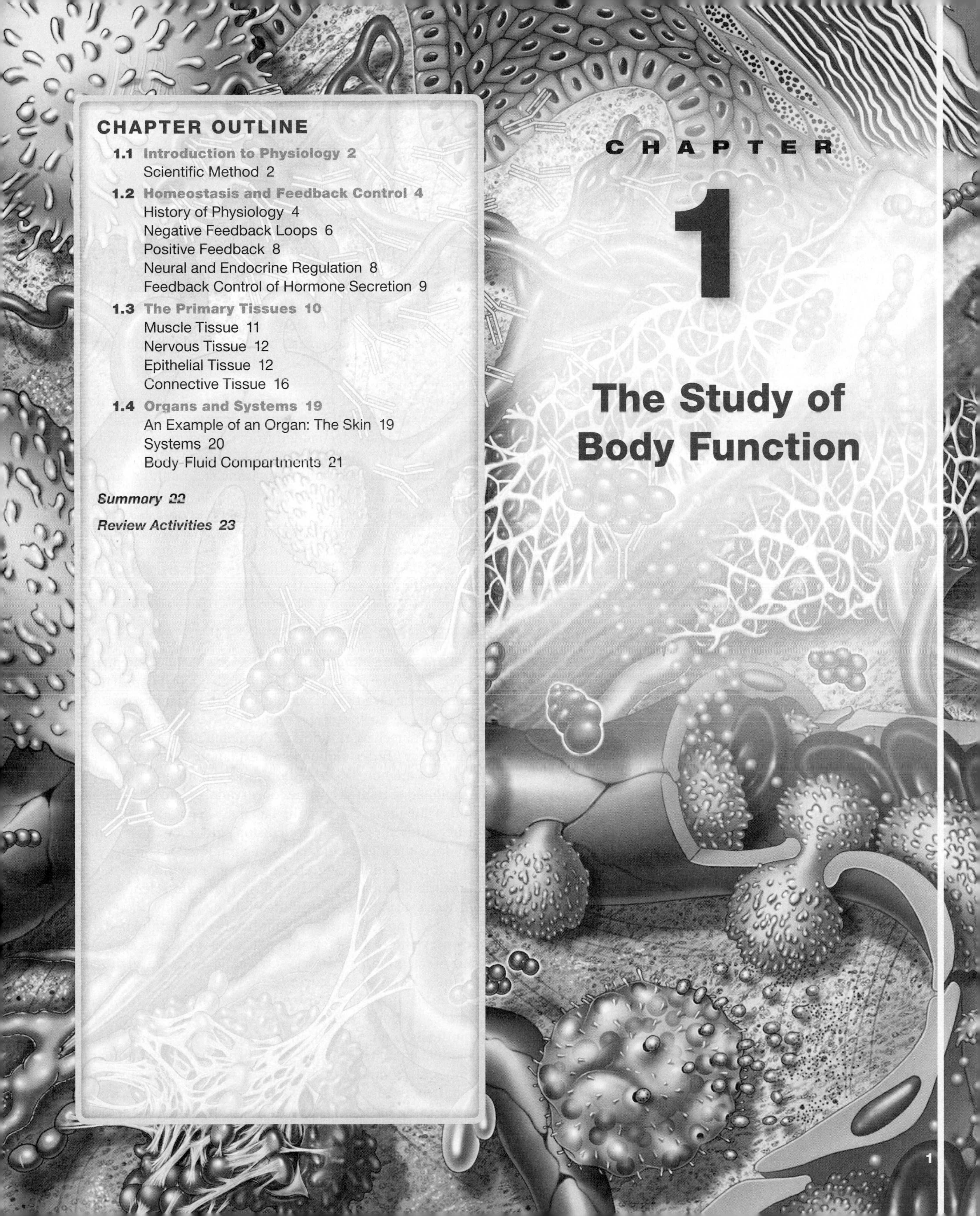

CHAPTER

1

The Study of Body Function

CHAPTER OUTLINE

Clinical Investigation

As you learn the material in chapter 1, you can begin to see how your new knowledge can be applied to health issues of interest to you now and of importance to you in your intended health-related career. This can add zest to your studies and enhance your motivation to truly understand physiological concepts, rather than to just memorize material for examinations.

In each later chapter, this box contains a medical mystery that you can solve using clues provided in the chapter. The "Clinical Investigation Clues" are located immediately following the relevant text explanations for each clue. At the end of each chapter is a "Clinical Investigation Summary" that you can use to check your ability to solve the medical mystery using the information provided in the chapter.

1.1 INTRODUCTION TO PHYSIOLOGY

Human physiology is the study of how the human body functions, with emphasis on specific cause-and-effect mechanisms. Knowledge of these mechanisms has been obtained experimentally through applications of the scientific method.

LEARNING OUTCOMES

After studying this section, you should be able to:

1. Describe the scientific study of human physiology.
2. Describe the characteristics of the scientific method.

Physiology (from the Greek *physis* = nature; *logos* = study) is the study of biological function—of how the body works, from molecular mechanisms within cells to the actions of tissues, organs, and systems, and how the organism as a whole accomplishes particular tasks essential for life. In the study of physiology, the emphasis is on mechanisms—with questions that begin with the word *how* and answers that involve cause-and-effect sequences. These sequences can be woven into larger and larger stories that include descriptions of the structures involved (anatomy) and that overlap with the sciences of chemistry and physics.

The separate facts and relationships of these cause-and-effect sequences are derived empirically from experimental evidence. Explanations that seem logical are not necessarily true; they are only as valid as the data on which they are based, and they can change as new techniques are developed and further experiments are performed. The ultimate objective of physiological research is to understand the normal functioning of cells, organs, and systems. A related science—*pathophysiology*—is concerned with how physiological processes are altered in disease or injury.

Pathophysiology and the study of normal physiology complement one another. For example, a standard technique for investigating the functioning of an organ is to observe what happens when the organ is surgically removed from an experimental animal or when its function is altered in a specific way. This study is often aided by "experiments of nature"—diseases—that involve specific damage to the functioning of an organ. The study of disease processes has thus aided our understanding of normal functioning, and the study of normal physiology has provided much of the scientific basis of modern medicine. This relationship is recognized by the Nobel Prize committee, whose members award prizes in the category "Physiology or Medicine."

The physiology of invertebrates and of different vertebrate groups is studied in the science of *comparative physiology*. Much of the knowledge gained from comparative physiology has benefited the study of human physiology. This is because animals, including humans, are more alike than they are different. This is especially true when comparing humans with other mammals. The small differences in physiology between humans and other mammals can be of crucial importance in the development of pharmaceutical drugs (discussed later in this section), but these differences are relatively slight in the overall study of physiology.

Scientific Method

All of the information in this text has been gained by people applying the **scientific method.** Although many different techniques are involved when people apply the scientific method, all share three attributes: (1) confidence that the natural world, including ourselves, is ultimately explainable in terms we can understand; (2) descriptions and explanations of the natural world that are honestly based on observations and that could be modified or refuted by other observations; and (3) humility, or the willingness to accept the fact that we could be wrong. If further study should yield conclusions that refuted all or part of an idea, the idea would have to be modified accordingly. In short, the scientific method is based on a confidence in our rational ability, honesty, and humility. Practicing scientists may not always display these attributes, but the validity of the large body of scientific knowledge that has been accumulated—as shown by the technological applications and the predictive value of scientific hypotheses—are ample testimony to the fact that the scientific method works.

The scientific method involves specific steps. After certain observations regarding the natural world are made, a **hypothesis** is formulated. In order for this hypothesis to be scientific, it must be capable of being refuted by experiments or other observations of the natural world. For example, one might hypothesize that people who exercise regularly have a lower resting pulse rate than other people. Experiments are conducted, or other observations are made, and the results are analyzed. Conclusions are then drawn as to whether the new data either refute or support the hypothesis. If the hypothesis survives such testing, it might be incorporated into a more general **theory.** Scientific theories are thus not simply conjectures; they are statements about the natural world that incorporate a number of proven hypotheses. They serve as a logical framework by which these hypotheses can be interrelated and provide the basis for predictions that may as yet be untested.

The hypothesis in the preceding example is scientific because it is *testable;* the pulse rates of 100 athletes and 100 sedentary people could be measured, for example, to see if there were statistically significant differences. If there were, the statement that athletes, on the average, have lower resting pulse rates than other people would be justified *based on these data.* One must still be open to the fact that this conclusion could be wrong. Before the discovery could become generally accepted as fact, other scientists would have to consistently replicate the results. Scientific theories are based on *reproducible* data.

It is quite possible that when others attempt to replicate the experiment, their results will be slightly different. They may then construct scientific hypotheses that the differences in resting pulse rate also depend on other factors, such as the nature of the exercise performed. When scientists attempt to test these hypotheses, they will likely encounter new problems requiring new explanatory hypotheses, which then must be tested by additional experiments.

In this way, a large body of highly specialized information is gradually accumulated, and a more generalized explanation (a scientific theory) can be formulated. This explanation will almost always be different from preconceived notions. People who follow the scientific method will then appropriately modify their concepts, realizing that their new ideas will probably have to be changed again in the future as additional experiments are performed.

Use of Measurements, Controls, and Statistics

Suppose you wanted to test the hypothesis that a regular exercise program causes people to have a lower resting heart rate. First, you would have to decide on the nature of the exercise program. Then, you would have to decide how the heart rate (or pulse rate) would be measured. This is a typical problem in physiology research because the testing of most physiological hypotheses requires quantitative **measurements.**

The group that is subject to the testing condition—in this case, exercise—is called the **experimental group.** A measurement of the heart rate for this group would be meaningful only if it is compared to that of another group, known as the **control group.** How shall this control group be chosen? Perhaps the subjects could serve as their own controls—that is, a person's resting heart rate could be measured before and after the exercise regimen. If this isn't possible, a control group could be other people who do not follow the exercise program. The choice of control groups is often a controversial aspect of physiology studies. In this example, did the people in the control group really refrain from *any* exercise? Were they comparable to the people in the experimental group with regard to age, sex, ethnicity, body weight, health status, and so on? You can see how difficult it could be in practice to get a control group that could satisfy any potential criticism.

Another possible criticism could be bias in the way that the scientists perform the measurements. This bias could be completely unintentional; scientists are human, after all, and they may have invested months or years in this project. To prevent such bias, the person doing the measurements often does not know if a subject is part of the experimental or the control group. This is known as a *blind measurement.*

Now suppose the data are in and it looks like the experimental group indeed has a lower average resting heart rate than the control group. But there is overlap—some people in the control group have measurements that are lower than some people in the experimental group. Is the difference in the average measurements of the groups due to a real physiological difference, or is it due to chance variations in the measurements? Scientists attempt to test the *null hypothesis* (the hypothesis that the difference is due to chance) by employing the mathematical tools of **statistics.** If the statistical results so warrant, the null hypothesis can be rejected and the experimental hypothesis can be deemed to be supported by this study.

The statistical test chosen will depend upon the design of the experiment, and it can also be a source of contention among scientists in evaluating the validity of the results. Because of the nature of the scientific method, "proof" in science is always provisional. Some other researchers, employing the scientific method in a different way (with different measuring techniques, experimental procedures, choice of control groups, statistical tests, and so on), may later obtain different results. The scientific method is thus an ongoing enterprise.

The results of the scientific enterprise are written up as research articles, and these must be reviewed by other scientists who work in the same field before they can be published in **peer-reviewed journals.** More often than not, the reviewers will suggest that certain changes be made in the articles before they can be accepted for publication.

Examples of such peer-reviewed journals that publish articles in many scientific fields include *Science* (www.sciencemag.org/), *Nature* (www.nature.com/nature/), and *Proceedings of the National Academy of Sciences* (www.pnas.org/). Review articles on physiology can be found in *Annual Review of Physiology* (physiol.annualreviews.org/), *Physiological Reviews* (physrev.physiology.org/), and *Physiology* (physiologyonline.physiology.org). Medical research journals, such as the *New England Journal of Medicine* (content.nejm.org/) and *Nature Medicine* (www.nature.com/nm/), also publish articles of physiological interest. There are also many specialty journals in areas of physiology such as neurophysiology, endocrinology, and cardiovascular physiology.

Students who wish to look online for scientific articles published in peer-reviewed journals that relate to a particular subject can do so at the National Library of Medicine website, *PubMed* (www.ncbi.nlm.nih.gov/entrez/query.fcgi).

Development of Pharmaceutical Drugs

The development of new pharmaceutical drugs can serve as an example of how the scientific method is used in physiology and its health applications. The process usually starts with basic physiological research, often at cellular and molecular levels. Perhaps a new family of drugs is developed using cells in tissue culture (*in vitro,* or outside the body). For example, cell physiologists studying membrane transport may discover that a particular family of compounds blocks membrane channels for calcium ions (Ca^{2+}). Because of their knowledge of physiology, other scientists may predict that a drug of this nature might be useful in the treatment of hypertension (high blood pressure). This drug may then be tried in animal experiments.

If a drug is effective at extremely low concentrations *in vitro* (in cells cultured outside of the body), there is a chance that it may work *in vivo* (in the body) at concentrations low enough not to be toxic (poisonous). This possibility must be thoroughly tested utilizing experimental animals, primarily rats and mice. More than 90% of drugs tested in experimental animals are too toxic for further development. Only in those rare cases when the toxicity is low enough may development progress to human/clinical trials.

Biomedical research is often aided by **animal models** of particular diseases. These are strains of laboratory rats and mice that are genetically susceptible to particular diseases that resemble human diseases. Research utilizing laboratory animals typically takes several years and always precedes human (clinical) trials of promising drugs. It should be noted that this length of time does not include all of the years of "basic" physiological research (involving laboratory animals) that provided the scientific foundation for the specific medical application.

In **phase I clinical trials,** the drug is tested on healthy human volunteers. This is done to test its toxicity in humans and to study how the drug is "handled" by the body: how it is metabolized, how rapidly it is removed from the blood by the liver and kidneys, how it can be most effectively administered, and so on. If significant toxic effects are not observed, the drug can proceed to the next stage. In **phase II clinical trials,** the drug is tested on the target human population (for example, those with hypertension). Only in those exceptional cases where the drug seems to be effective but has minimal toxicity does testing move to the next phase. **Phase III trials** occur in many research centers across the country to maximize the number of test participants. At this point, the test population must include a sufficient number of subjects of both sexes, as well as people of different ethnic groups. In addition, people are tested who have other health problems besides the one that the drug is intended to benefit. For example, those who have diabetes in addition to hypertension would be included in this phase. If the drug passes phase III trials, it goes to the Food and Drug Administration (FDA) for approval. **Phase IV trials** test other potential uses of the drug.

Less than 10% of the tested drugs make it all the way through clinical trials to eventually become approved and marketed. This low success rate does not count those that fail after approval because of unexpected toxicity, nor does it take into account the great amount of drugs that fail earlier in research before clinical trials begin. Notice the crucial role of basic research, using experimental animals, in this process. Virtually every prescription drug on the market owes its existence to such research.

CHECKPOINTS

1. How has the study of physiology aided, and been aided by, the study of diseases?
2a. Describe the steps involved in the scientific method. What would qualify a statement as unscientific?
2b. Describe the different types of trials a new drug must undergo before it is "ready for market."

1.2 HOMEOSTASIS AND FEEDBACK CONTROL

The regulatory mechanisms of the body can be understood in terms of a single shared function: that of maintaining constancy of the internal environment. A state of relative constancy of the internal environment is known as homeostasis, maintained by negative feedback loops.

LEARNING OUTCOMES

After studying this section, you should be able to:

3. Define homeostasis, and identify the components of negative feedback loops.
4. Explain the role of antagonistic effectors in maintaining homeostasis, and the nature of positive feedback loops.
5. Give examples of how negative feedback loops involving the nervous and endocrine systems help to maintain homeostasis.

History of Physiology

The Greek philosopher Aristotle (384–322 B.C.) speculated on the function of the human body, but another ancient Greek, Erasistratus (304–250? B.C.), is considered to be the first to study physiology because he attempted to apply physical laws to understand human function. Galen (A.D. 130–201) wrote widely on the subject and was considered the supreme authority until the Renaissance. Physiology became a fully experimental science with the revolutionary work of the English physician William Harvey (1578–1657), who demonstrated that the heart pumps blood through a closed system of vessels.

However, the originator of modern physiology is the French physiologist Claude Bernard (1813–1878), who observed that the *milieu intérieur* (internal environment) remains remarkably constant despite changing conditions in the external environment. In a book entitled *The Wisdom of the Body*, published in 1932, the American physiologist Walter Cannon (1871–1945) coined the term **homeostasis** to describe this internal constancy. Cannon further suggested that the many mechanisms of physiological regulation have but one purpose—the maintenance of internal constancy.

Most of our present knowledge of human physiology has been gained in the twentieth century. However, new knowledge in the twenty-first century is being added at an ever more rapid pace, fueled in more recent decades by the revolutionary growth of molecular genetics and its associated biotechnologies, and by the availability of more powerful computers and other equipment. A very brief history of twentieth- and twenty-first-century physiology, limited by space to only two citations per decade, is provided in table 1.1.

Most of the citations in table 1.1 indicate the winners of Nobel prizes. The **Nobel Prize in Physiology or Medicine** (a single prize category) was first awarded in 1901 to Emil Adolf von Behring, a pioneer in immunology who coined the term *antibody* and whose many other discoveries included the use of serum (containing antibodies) to treat diphtheria.

Table 1.1 | History of Twentieth- and Twenty-First-Century Physiology (two citations per decade)

1900	Karl Landsteiner discovers the A, B, and O blood groups.
1904	Ivan Pavlov wins the Nobel Prize for his work on the physiology of digestion.
1910	Sir Henry Dale describes properties of histamine.
1918	Earnest Starling describes how the force of the heart's contraction relates to the amount of blood in it.
1921	John Langley describes the functions of the autonomic nervous system.
1923	Sir Frederick Banting, Charles Best, and John Macleod win the Nobel Prize for the discovery of insulin.
1932	Sir Charles Sherrington and Lord Edgar Adrian win the Nobel Prize for discoveries related to the functions of neurons.
1936	Sir Henry Dale and Otto Loewi win the Nobel Prize for the discovery of acetylcholine in synaptic transmission.
1939–47	Albert von Szent-Györgyi explains the role of ATP and contributes to the understanding of actin and myosin in muscle contraction.
1949	Hans Selye discovers the common physiological responses to stress.
1953	Sir Hans Krebs wins the Nobel Prize for his discovery of the citric acid cycle.
1954	Hugh Huxley, Jean Hanson, R. Niedergerde, and Andrew Huxley propose the sliding filament theory of muscle contraction.
1962	Francis Crick, James Watson, and Maurice Wilkins win the Nobel Prize for determining the structure of DNA.
1963	Sir John Eccles, Sir Alan Hodgkin, and Sir Andrew Huxley win the Nobel Prize for their discoveries relating to the nerve impulse.
1971	Earl Sutherland wins the Nobel Prize for his discovery of the mechanism of hormone action.
1977	Roger Guillemin and Andrew Schally win the Nobel Prize for discoveries of the brain's production of peptide hormone.
1981	Roger Sperry wins the Nobel Prize for his discoveries regarding the specializations of the right and left cerebral hemispheres.
1986	Stanley Cohen and Rita Levi-Montalcini win the Nobel Prize for their discoveries of growth factors regulating the nervous system.
1994	Alfred Gilman and Martin Rodbell win the Nobel Prize for their discovery of the functions of G-proteins in signal transduction in cells.
1998	Robert Furchgott, Louis Ignarro, and Ferid Murad win the Nobel Prize for discovering the role of nitric oxide as a signaling molecule in the cardiovascular system.
2004	Linda B. Buck and Richard Axel win the Nobel Prize for their discoveries of odorant receptors and the organization of the olfactory system.
2006	Andrew Z. Fine and Craig C. Mello win the Noble Prize for their discovery of RNA interference by short, double-stranded RNA molecules.

Many scientists who might deserve a Nobel Prize never receive one, and the prizes are given for particular achievements and not others (Einstein didn't win his Nobel Prize in Physics for relativity, for example) and are often awarded many years after the discoveries were made. Nevertheless, the awarding of the Nobel Prize in Physiology or Medicine each year is a celebrated event in the biomedical community, and the awards can be a useful yardstick for tracking the course of physiological research over time.

Negative Feedback Loops

The concept of homeostasis has been of immense value in the study of physiology because it allows diverse regulatory mechanisms to be understood in terms of their "why" as well as their "how." The concept of homeostasis also provides a major foundation for medical diagnostic procedures. When a particular measurement of the internal environment, such as a blood measurement (table 1.2), deviates significantly from the normal range of values, it can be concluded that homeostasis is not being maintained and that the person is sick. A number of such measurements, combined with clinical observations, may allow the particular defective mechanism to be identified.

In order for internal constancy to be maintained, changes in the body must stimulate **sensors** that can send information to an **integrating center.** This allows the integrating center to detect changes from a **set point.** The set point is analogous to the temperature set on a house thermostat. In a similar manner, there is a set point for body temperature, blood glucose concentration, the tension on a tendon, and so on. The integrating center is often a particular region of the brain or spinal cord, but it can also be a group of cells in an endocrine gland. A number of different sensors may send information to a particular integrating center, which can then integrate this information and direct the responses of **effectors**—generally muscles or glands. The integrating center may cause increases or decreases in effector action to counter the deviations from the set point and defend homeostasis.

The thermostat of a house can serve as a simple example. Suppose you set the thermostat at a set point of 70° F. If the temperature in the house rises sufficiently above the set point, a sensor connected to an integrating center within the thermostat will detect that deviation and turn on the air conditioner (the effector in this example). The air conditioner will turn off when the room temperature falls and the thermostat no longer detects a deviation from the set-point temperature. However, this simple example gives a wrong impression: the effectors in the body are generally increased or decreased in activity, *not* just turned on or off. Because of this, negative feedback control in the body works far more efficiently than does a house thermostat.

If the body temperature exceeds the set point of 37° C, sensors in a part of the brain detect this deviation and, acting via an integrating center (also in the brain), stimulate activities of effectors (including sweat glands) that lower the temperature. For another example, if the blood glucose concentration falls below normal, the effectors act to increase the blood glucose. One can think of the effectors as "defending" the set points against deviations. Because the activity of the effectors is influenced by the effects they produce, and because this regulation is in a negative, or reverse, direction, this type of control system is known as a **negative feedback loop** (fig. 1.1). (Notice that in figure 1.1 and in all subsequent figures, negative feedback is indicated by a dashed line and a negative sign.)

Table 1.2 | Approximate Normal Ranges for Measurements of Some Fasting Blood Values

Measurement	Normal Range
Arterial pH	7.35–7.45
Bicarbonate	24–28 mEq/L
Sodium	135–145 mEq/L
Calcium	4.5–5.5 mEq/L
Oxygen content	17.2–22.0 ml/100 ml
Urea	12–35 mg/100 ml
Amino acids	3.3–5.1 mg/100 ml
Protein	6.5–8.0 g/100 ml
Total lipids	400–800 mg/100 ml
Glucose	75–110 mg/100 ml

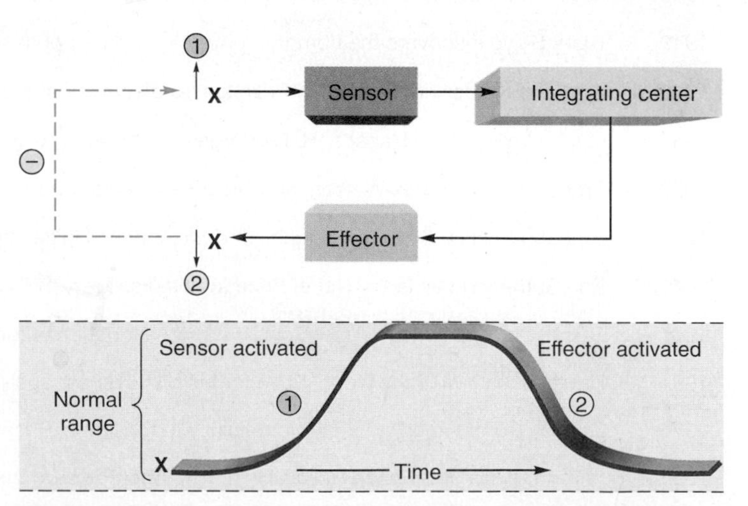

Figure 1.1 A rise in some factor of the internal environment (↑X) is detected by a sensor. This information is relayed to an integrating center, which causes an effector to produce a change (1) in the opposite direction (↓X). The initial deviation is thus reversed (2), completing a negative feedback loop (shown by the dashed arrow and negative sign). The numbers indicate the sequence of changes.

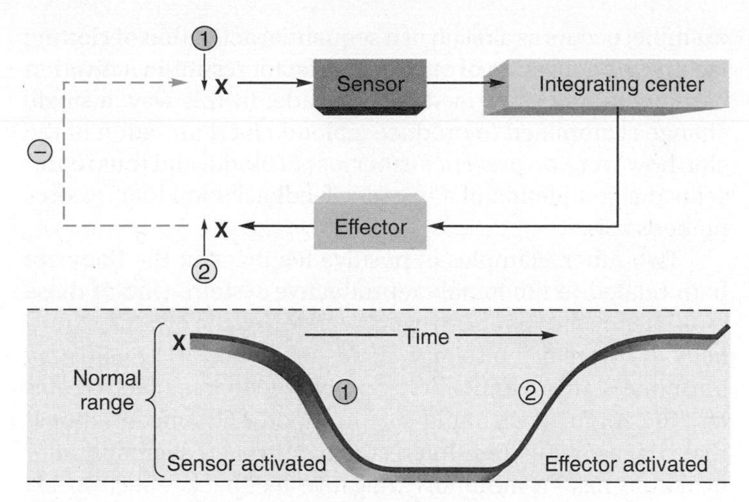

Figure 1.2 A fall in some factor of the internal environment (↓X) is detected by a sensor. (Compare this negative feedback loop with that shown in figure 1.1.)

The nature of the negative feedback loop can be understood by again referring to the analogy of the thermostat and air conditioner. After the air conditioner has been on for some time, the room temperature may fall significantly below the set point of the thermostat. When this occurs, the air conditioner will be turned off. The effector (air conditioner) is turned on by a high temperature and, when activated, produces a negative change (lowering of the temperature) that ultimately causes the effector to be turned off. In this way, constancy is maintained.

It is important to realize that these negative feedback loops are continuous, ongoing processes. Thus, a particular nerve fiber that is part of an effector mechanism may always display some activity, and a particular hormone that is part of another effector mechanism may always be present in the blood. The nerve activity and hormone concentration may decrease in response to deviations of the internal environment in one direction (fig. 1.1), or they may increase in response to deviations in the opposite direction (fig. 1.2). Changes from the normal range in either direction are thus compensated for by reverse changes in effector activity.

Because negative feedback loops respond after deviations from the set point have stimulated sensors, the internal environment is never absolutely constant. Homeostasis is best conceived as a state of **dynamic constancy** in which conditions are stabilized above and below the set point. These conditions can be measured quantitatively, in degrees Celsius for body temperature, for example, or in milligrams per deciliter (one-tenth of a liter) for blood glucose. The set point can be taken as the average value within the normal range of measurements (fig. 1.3).

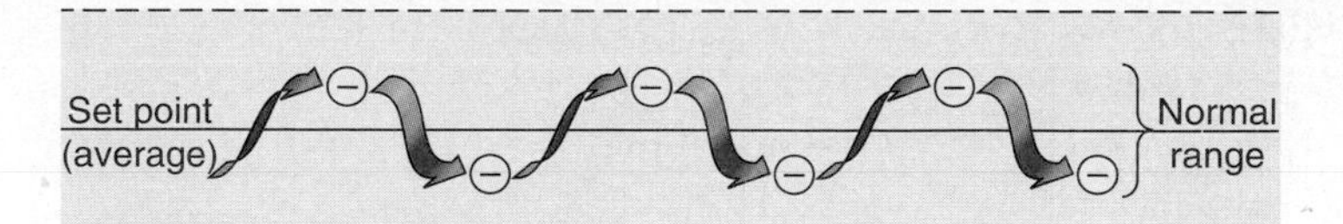

Figure 1.3 Negative feedback loops maintain a state of dynamic constancy within the internal environment. The completion of the negative feedback loop is indicated by negative signs.

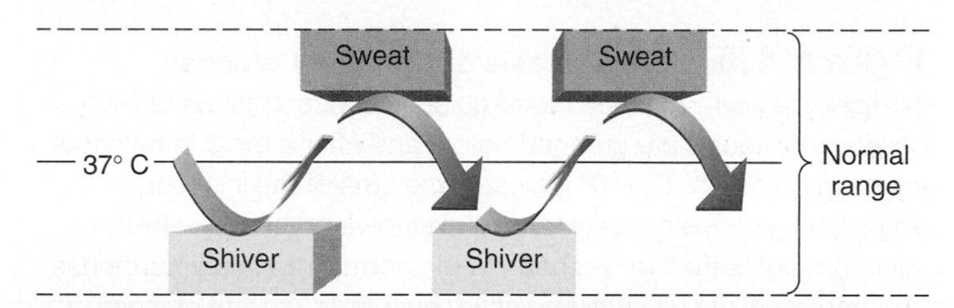

Figure 1.4 How body temperature is maintained within the normal range. The body temperature normally has a set point of 37° C. This is maintained, in part, by two antagonistic mechanisms—shivering and sweating. Shivering is induced when the body temperature falls too low, and it gradually subsides as the temperature rises. Sweating occurs when the body temperature is too high, and it diminishes as the temperature falls. Most aspects of the internal environment are regulated by the antagonistic actions of different effector mechanisms.

See the *Test Your Quantitative Ability* section of the Review Activities at the end of this chapter.

Antagonistic Effectors

Most factors in the internal environment are controlled by several effectors, which often have antagonistic actions. Control by antagonistic effectors is sometimes described as "push-pull," where the increasing activity of one effector is accompanied by decreasing activity of an antagonistic effector. This affords a finer degree of control than could be achieved by simply switching one effector on and off.

Room temperature can be maintained, for example, by simply turning an air conditioner on and off, or by just turning a heater on and off. A much more stable temperature, however, can be achieved if the air conditioner and heater are both controlled by a thermostat. Then the heater is turned on when the air conditioner is turned off, and vice versa. Normal body temperature is maintained about a set point of 37° C by the antagonistic effects of sweating, shivering, and other mechanisms (fig. 1.4).

The blood concentrations of glucose, calcium, and other substances are regulated by negative feedback loops involving hormones that promote opposite effects. Insulin, for example, lowers blood glucose, and other hormones raise the blood glucose concentration. The heart rate, similarly, is controlled by nerve fibers that produce opposite effects: stimulation of one group of nerve fibers increases heart rate; stimulation of another group slows the heart rate.

Quantitative Measurements

In order to study physiological mechanisms, scientists must measure specific values and mathematically determine such statistics as their normal range, their averages, and their deviations from the average (which can represent the set point).

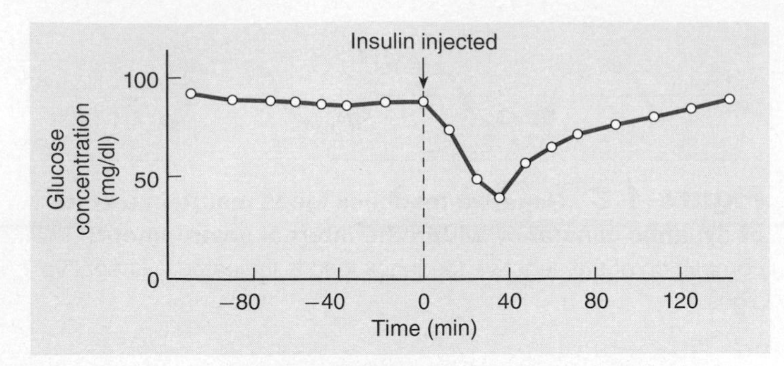

Figure 1.5 Homeostasis of the blood glucose concentration. Average blood glucose concentrations of five healthy individuals are graphed before and after a rapid intravenous injection of insulin. The "0" indicates the time of the injection. The blood glucose concentration is first lowered by the insulin injection, but is then raised back to the normal range (by hormones antagonistic to insulin that stimulate the liver to secrete glucose into the blood). Homeostasis of blood glucose is maintained by the antagonistic actions of insulin and several other hormones.

For these and other reasons, quantitative measurements are basic to the science of physiology. One example of this, and of the actions of antagonistic mechanisms in maintaining homeostasis, is shown in figure 1.5. Blood glucose concentrations were measured in five healthy people before and after an injection of insulin, a hormone that acts to lower the blood glucose concentration. A graph of the data reveals that the blood glucose concentration decreased rapidly but was brought back up to normal levels within 80 minutes after the injection. This demonstrates that negative feedback mechanisms acted to restore homeostasis in this experiment. These mechanisms involve the action of hormones whose effects are antagonistic to that of insulin—that is, they promote the secretion of glucose from the liver (see chapter 19).

Positive Feedback

Constancy of the internal environment is maintained by effectors that act to compensate for the change that served as the stimulus for their activation; in short, by negative feedback loops. A thermostat, for example, maintains a constant temperature by increasing heat production when it is cold and decreasing heat production when it is warm. The opposite occurs during **positive feedback**—in this case, the action of effectors *amplifies* those changes that stimulated the effectors. A thermostat that works by positive feedback, for example, would increase heat production in response to a rise in temperature.

It is clear that homeostasis must ultimately be maintained by negative rather than by positive feedback mechanisms. The effectiveness of some negative feedback loops, however, is increased by positive feedback mechanisms that amplify the actions of a negative feedback response. Blood clotting, for example, occurs as a result of a sequential activation of clotting factors; the activation of one clotting factor results in activation of many in a positive feedback cascade. In this way, a single change is amplified to produce a blood clot. Formation of the clot, however, can prevent further loss of blood, and thus represents the completion of a negative feedback loop that restores homeostasis.

Two other examples of positive feedback in the body are both related to the female reproductive system. One of these examples occurs when estrogen, secreted by the ovaries, stimulates the women's pituitary gland to secrete LH (luteinizing hormone). This stimulatory, positive feedback effect creates an "LH surge" (very rapid rise in blood LH concentrations) that triggers ovulation. Interestingly, estrogen secretion after ovulation has an inhibitory, negative feedback, effect on LH secretion (this is the physiological basis for the birth control pill, discussed in chapter 20). Another example of positive feedback is contraction of the uterus during childbirth (parturition). Contraction of the uterus is stimulated by the pituitary hormone oxytocin, and the secretion of oxytocin is increased by sensory feedback from contractions of the uterus during labor. The strength of uterine contractions during labor is thus increased through positive feedback. The mechanisms involved in labor are discussed in more detail in chapter 20 (see fig. 20.50).

Neural and Endocrine Regulation

Homeostasis is maintained by two general categories of regulatory mechanisms: (1) those that are **intrinsic,** or "built into" the organs being regulated (such as molecules produced in the walls of blood vessels that cause vessel dilation or constriction); and (2) those that are **extrinsic,** as in regulation of an organ by the nervous and endocrine systems. The endocrine system functions closely with the nervous system in regulating and integrating body processes and maintaining homeostasis. The nervous system controls the secretion of many endocrine glands, and some hormones in turn affect the function of the nervous system. Together, the nervous and endocrine systems regulate the activities of most of the other systems of the body.

Regulation by the endocrine system is achieved by the secretion of chemical regulators called **hormones** into the blood, which carries the hormones to all organs in the body. Only specific organs can respond to a particular hormone, however; these are known as the **target organs** of that hormone.

Nerve fibers are said to *innervate* the organs that they regulate. When stimulated, these fibers produce electrochemical nerve impulses that are conducted from the origin of the fiber to its terminals in the target organ innervated by the fiber. These target organs can be muscles or glands that may function as effectors in the maintenance of homeostasis.

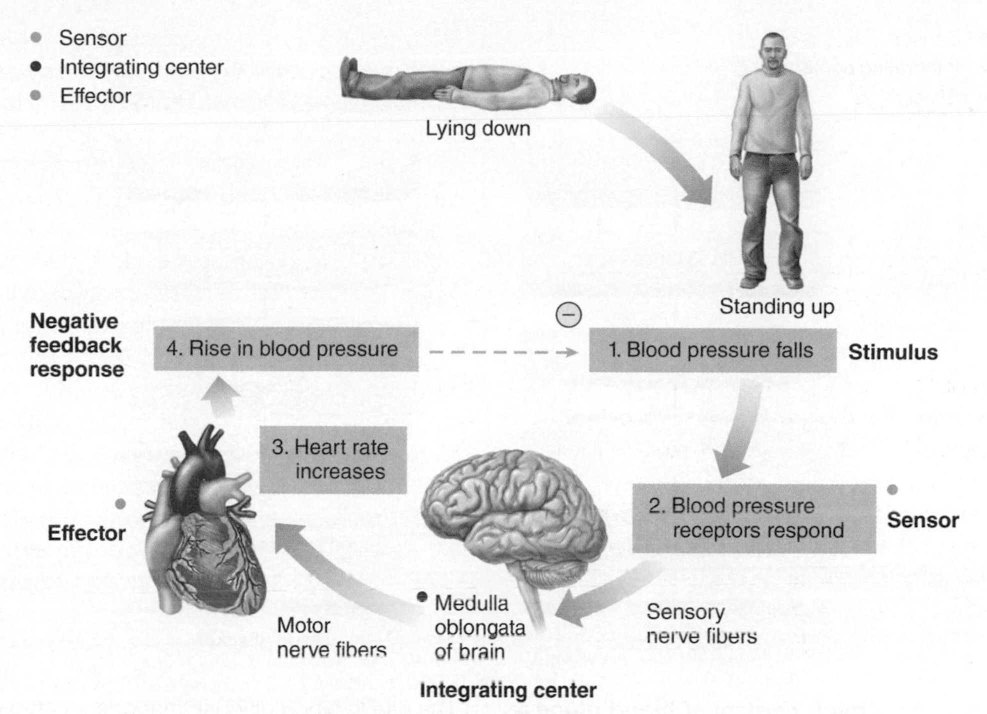

Figure 1.6 Negative feedback control of blood pressure. Blood pressure influences the activity of sensory neurons from the blood pressure receptors (sensors); a rise in pressure increases the firing rate, and a fall in pressure decreases the firing rate of nerve impulses. When a person stands up from a lying-down position, the blood pressure momentarily falls. The resulting decreased firing rate of nerve impulses in sensory neurons affects the medulla oblongata of the brain (the integrating center). This causes the motor nerves to the heart (effector) to increase the heart rate, helping to raise the blood pressure.

For example, we have negative feedback loops that help maintain homeostasis of arterial blood pressure, in part by adjusting the heart rate. If everything else is equal, blood pressure is lowered by a decreased heart rate and raised by an increased heart rate. This is accomplished by regulating the activity of the autonomic nervous system, as will be discussed in later chapters. Thus, a fall in blood pressure produced daily as we go from a lying to a standing position—is compensated by a faster heart rate (fig. 1.6). As a consequence of this negative feedback loop, our heart rate varies as we go through our day, speeding up and slowing down, so that we can maintain homeostasis of blood pressure and keep it within normal limits.

Feedback Control of Hormone Secretion

The nature of the endocrine glands, the interaction of the nervous and endocrine systems, and the actions of hormones will be discussed in detail in later chapters. For now, it is sufficient to describe the regulation of hormone secretion very broadly, because it so superbly illustrates the principles of homeostasis and negative feedback regulation.

Hormones are secreted in response to specific chemical stimuli. A rise in the plasma glucose concentration, for example, stimulates insulin secretion from structures in the pancreas known as the *pancreatic islets,* or *islets of Langerhans.* Hormones are also secreted in response to nerve stimulation and stimulation by other hormones.

The secretion of a hormone can be inhibited by its own effects in a negative feedback manner. Insulin, as previously described, produces a lowering of blood glucose. Because a rise in blood glucose stimulates insulin secretion, a lowering of blood glucose caused by insulin's action inhibits further insulin secretion. This closed-loop control system is called **negative feedback inhibition** (fig. 1.7*a*).

Homeostasis of blood glucose is too important—the brain uses blood glucose as its primary source of energy—to entrust to the regulation of only one hormone, insulin. So, when blood glucose falls during fasting, several mechanisms prevent it from falling too far (fig. 1.7*b*). First, insulin secretion decreases, preventing muscle, liver, and adipose cells from taking too much glucose from the blood. Second, the secretion of a hormone antagonistic to insulin, called *glucagon,* increases. Glucagon stimulates processes in the liver (breakdown of a stored, starchlike molecule called glycogen; chapter 2, section 2.2) that cause it to secrete glucose into the blood. Through these and other antagonistic negative feedback mechanisms, the blood glucose is maintained within a homeostatic range.

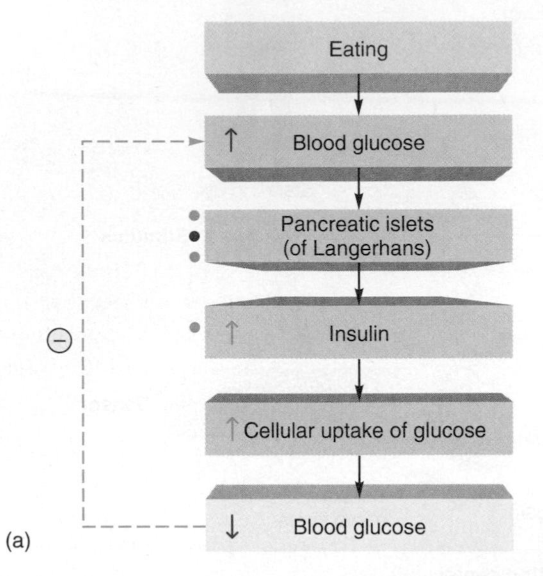

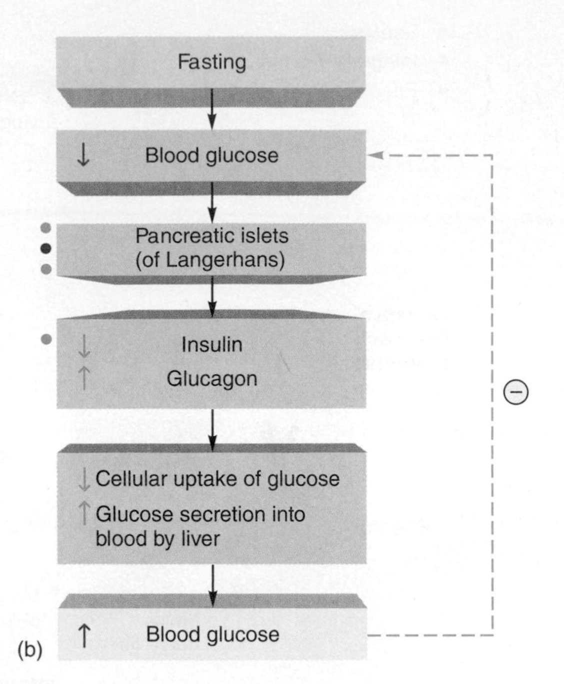

Figure 1.7 Negative feedback control of blood glucose. (*a*) The rise in blood glucose that occurs after eating carbohydrates is corrected by the action of insulin, which is secreted in increasing amounts at that time. (*b*) During fasting, when blood glucose falls, insulin secretion is inhibited and the secretion of an antagonistic hormone, glucagon, is increased. This stimulates the liver to secrete glucose into the blood, helping to prevent blood glucose from continuing to fall. In this way, blood glucose concentrations are maintained within a homeostatic range following eating and during fasting.

Clinical Investigation CLUES

In later chapters, Clinical Investigation Clues are placed where some of the preceding physiological concepts can be used to solve the Clinical Investigation presented at the beginning of the chapters. This can be a good motivator to re-read the relevant sections and see how they relate to biomedical applications. There may be several of these "Clues" in a chapter and by the last one you should be able to solve the medical mystery of the Clinical Investigation.

CHECKPOINTS

3a. Define *homeostasis* and describe how this concept can be used to explain physiological control mechanisms.

3b. Define *negative feedback* and explain how it contributes to homeostasis. Illustrate this concept by drawing and labeling a negative feedback loop.

4. Describe *positive feedback* and explain how this process functions in the body.

5. Explain how the secretion of a hormone is controlled by negative feedback inhibition. Use the control of insulin secretion as an example.

1.3 THE PRIMARY TISSUES

The organs of the body are composed of four different primary tissues, each of which has its own characteristic structure and function. The activities and interactions of these tissues determine the physiology of the organs.

LEARNING OUTCOMES

After studying this section, you should be able to:

6. Distinguish the primary tissues and their subtypes.

7. Relate the structure of the primary tissues to their functions.

Although physiology is the study of function, it is difficult to properly understand the function of the body without some knowledge of its anatomy, particularly at a microscopic level. Microscopic anatomy constitutes a field of study known as *histology.* The anatomy and histology of specific organs will be discussed together with their functions in later chapters. In this section, the common "fabric" of all organs is described.

Cells are the basic units of structure and function in the body. Cells that have similar functions are grouped into categories called *tissues.* The entire body is composed of only four major types of tissues. These **primary tissues** are (1) muscle, (2) nervous, (3) epithelial, and (4) connective tissues. Groupings of these four primary tissues into anatomical and functional units are called **organs.** Organs, in turn, may be grouped together by common functions into **systems.** The systems of the body act in a coordinated fashion to maintain the entire organism.

Muscle Tissue

Muscle tissue is specialized for contraction. There are three types of muscle tissue: **skeletal, cardiac,** and **smooth.** Skeletal muscle is often called *voluntary muscle* because its contraction is consciously controlled. Both skeletal and cardiac muscles are **striated;** they have striations, or stripes, that extend across the width of the muscle cell (figs. 1.8 and 1.9). These striations are produced by a characteristic arrangement of contractile proteins, and for this reason skeletal and cardiac muscle have similar mechanisms of contraction. Smooth muscle (fig. 1.10) lacks these striations and has a different mechanism of contraction.

Skeletal Muscle

Skeletal muscles are generally attached to bones at both ends by means of tendons; hence, contraction produces movements of the skeleton. There are exceptions to this pattern, however. The tongue, superior portion of the esophagus, anal sphincter, and diaphragm are also composed of skeletal muscle, but they do not cause movements of the skeleton.

Beginning at about the fourth week of embryonic development, separate cells called *myoblasts* fuse together to form **skeletal muscle fibers,** or **myofibers** (from the Greek *myos* = muscle). Although myofibers are often referred to as skeletal muscle cells, each is actually a *syncytium,* or multinucleate mass formed from the union of separate cells. Despite their unique origin and structure, each myofiber contains mitochondria and other organelles (described in chapter 3) common to all cells.

The muscle fibers within a skeletal muscle are arranged in bundles, and within these bundles the fibers extend in parallel from one end of the bundle to the other. The parallel arrangement of muscle fibers (fig. 1.8) allows each fiber to be controlled individually: one can thus contract fewer or more muscle fibers and, in this way, vary the strength of contraction of the whole muscle. The ability to vary, or "grade," the strength of skeletal muscle contraction is needed for precise control of skeletal movements.

Cardiac Muscle

Although cardiac muscle is striated, it differs markedly from skeletal muscle in appearance. Cardiac muscle is found only in the heart where the **myocardial cells** are short, branched, and intimately interconnected to form a continuous fabric. Special areas of contact between adjacent cells stain darkly to show *intercalated discs* (fig. 1.9), which are characteristic of heart muscle.

The intercalated discs couple myocardial cells together mechanically and electrically. Unlike skeletal muscles, therefore, the heart cannot produce a graded contraction by varying

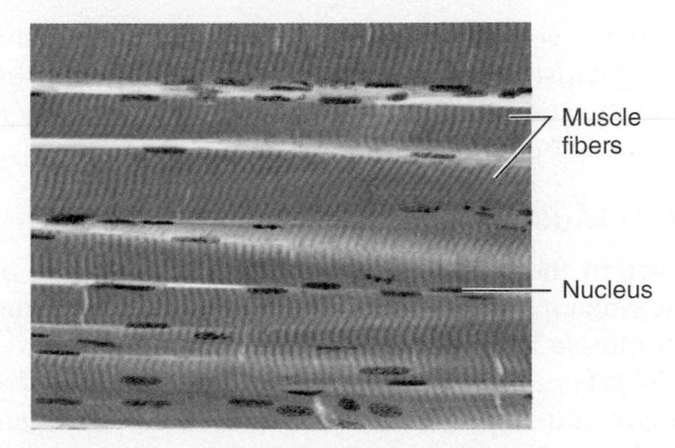

Figure 1.8 Skeletal muscle fibers showing the characteristic light and dark cross striations. Because of this feature, skeletal muscle is also called striated muscle. AP|R

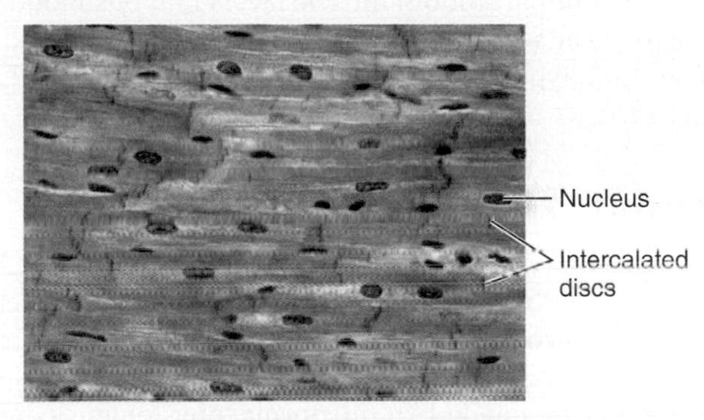

Figure 1.9 Human cardiac muscle. Notice the striated appearance and dark-staining intercalated discs. AP|R

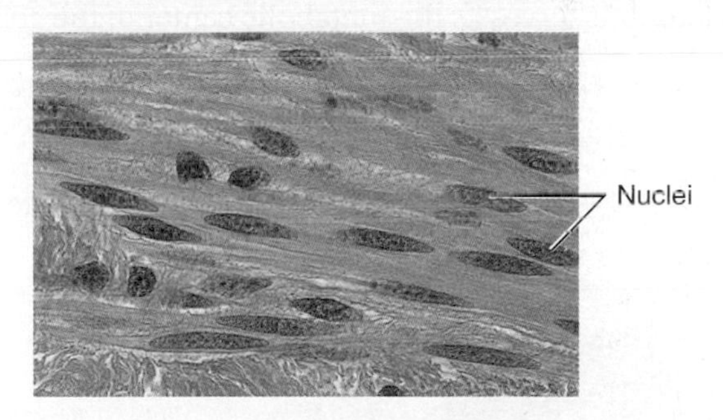

Figure 1.10 A photomicrograph of smooth muscle cells. Notice that these cells contain single, centrally located nuclei and lack striations. AP|R

the number of cells stimulated to contract. Because of the way the heart is constructed, the stimulation of one myocardial cell results in the stimulation of all other cells in the mass and a "wholehearted" contraction.

Smooth Muscle

As implied by the name, smooth muscle cells (fig. 1.10) do not have the striations characteristic of skeletal and cardiac muscle. Smooth muscle is found in the digestive tract, blood vessels, bronchioles (small air passages in the lungs), and the ducts of the urinary and reproductive systems. Circular arrangements of smooth muscle in these organs produce constriction of the *lumen* (cavity) when the muscle cells contract. The digestive tract also contains longitudinally arranged layers of smooth muscle. *Peristalsis* is the coordinated wavelike contractions of the circular and longitudinal smooth muscle layers that push food from the oral to the anal end of the digestive tract.

The three types of muscle tissue are discussed further in chapter 12.

Nervous Tissue

Nervous tissue consists of nerve cells, or **neurons,** which are specialized for the generation and conduction of electrical events, and **neuroglial** (or **glial**) **cells.** Neuroglial cells provide the neurons with structural support and perform a variety of functions that are needed for the normal physiology of the nervous system.

Each neuron consists of three parts: (1) a *cell body,* (2) *dendrites,* and (3) an *axon* (fig. 1.11). The cell body contains the nucleus and serves as the metabolic center of the cell. The dendrites (literally, "branches") are highly branched cytoplasmic extensions of the cell body that receive input from other neurons or from receptor cells. The axon is a single cytoplasmic extension of the cell body that can be quite long (up to a few feet in length). It is specialized for conducting nerve impulses from the cell body to another neuron or to an effector (muscle or gland) cell.

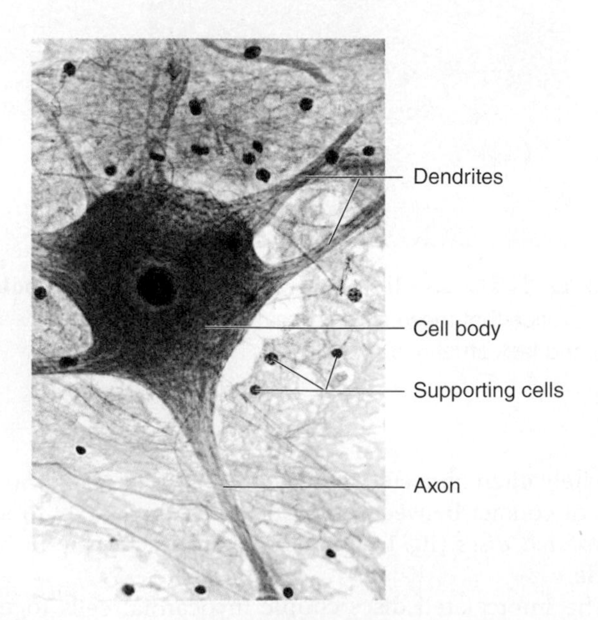

Figure 1.11 A photomicrograph of nerve tissue. A single neuron and numerous smaller supporting cells can be seen. APIR

The neuroglial cells do not conduct impulses but instead serve to bind neurons together, modify the extracellular environment of the nervous system, and influence the nourishment and electrical activity of neurons. In recent years, neuroglial cells have been shown to cooperate with neurons in chemical neurotransmission (chapter 7), and to have many other roles in the normal physiology (as well as disease processes) of the brain and spinal cord. Neuroglial cells are about five times more abundant than neurons in the nervous system and, unlike neurons, maintain a limited ability to divide by mitosis throughout life.

Neurons and neuroglial cells are discussed in detail in chapter 7.

Epithelial Tissue

Epithelial tissue consists of cells that form **membranes,** which cover and line the body surfaces, and of **glands,** which are derived from these membranes. There are two categories of glands. *Exocrine glands* (from the Greek *exo* = outside) secrete chemicals through a duct that leads to the outside of a membrane, and thus to the outside of a body surface. *Endocrine glands* (from the Greek *endon* = within) secrete chemicals called *hormones* into the blood. Endocrine glands are discussed in chapter 11.

Epithelial Membranes

Epithelial membranes are classified according to the number of their layers and the shape of the cells in the upper layer (table 1.3). Epithelial cells that are flattened in shape are **squamous;** those that are as wide as they are tall are **cuboidal;** and those that are taller than they are wide are **columnar** (fig. 1.12*a–c*). Those epithelial membranes that are only one cell layer thick are known as **simple membranes;** those that are composed of a number of layers are **stratified membranes.**

Epithelial membranes cover all body surfaces and line the cavity (lumen) of every hollow organ. Thus, epithelial membranes provide a barrier between the external environment and the internal environment of the body. Stratified epithelial membranes are specialized to provide protection. Simple epithelial membranes, in contrast, provide little protection; instead, they are specialized for transport of substances between the internal and external environments. In order for a substance to get into the body, it must pass through an epithelial membrane, and simple epithelia are specialized for this function. For example, a simple squamous epithelium

Table 1.3 | Summary of Epithelial Membranes

Type	Structure and Function	Location
Simple Epithelia	Single layer of cells; function varies with type	Covering visceral organs; linings of body cavities, tubes, and ducts
Simple squamous epithelium	Single layer of flattened, tightly bound cells; diffusion and filtration	Capillary walls; pulmonary alveoli of lungs; covering visceral organs; linings of body cavities
Simple cuboidal epithelium	Single layer of cube-shaped cells; excretion, secretion, or absorption	Surface of ovaries; linings of kidney tubules, salivary ducts, and pancreatic ducts
Simple columnar epithelium	Single layer of nonciliated, tall, column-shaped cells; protection, secretion, and absorption	Lining of most of digestive tract
Simple ciliated columnar epithelium	Single layer of ciliated, column-shaped cells; transportive role through ciliary motion	Lining of uterine tubes
Pseudostratified ciliated columnar epithelium	Single layer of ciliated, irregularly shaped cells; many goblet cells; protection, secretion, ciliary movement	Lining of respiratory passageways
Stratified Epithelia	Two or more layers of cells; function varies with type	Epidermal layer of skin; linings of body openings, ducts, and urinary bladder
Stratified squamous epithelium (keratinized)	Numerous layers containing keratin, with outer layers flattened and dead; protection	Epidermis of skin
Stratified squamous epithelium (nonkeratinized)	Numerous layers lacking keratin, with outer layers moistened and alive; protection and pliability	Linings of oral and nasal cavities, vagina, and anal canal
Stratified cuboidal epithelium	Usually two layers of cube-shaped cells; strengthening of luminal walls	Large ducts of sweat glands, salivary glands, and pancreas
Transitional epithelium	Numerous layers of rounded, nonkeratinized cells; distension	Walls of ureters, part of urethra, and urinary bladder

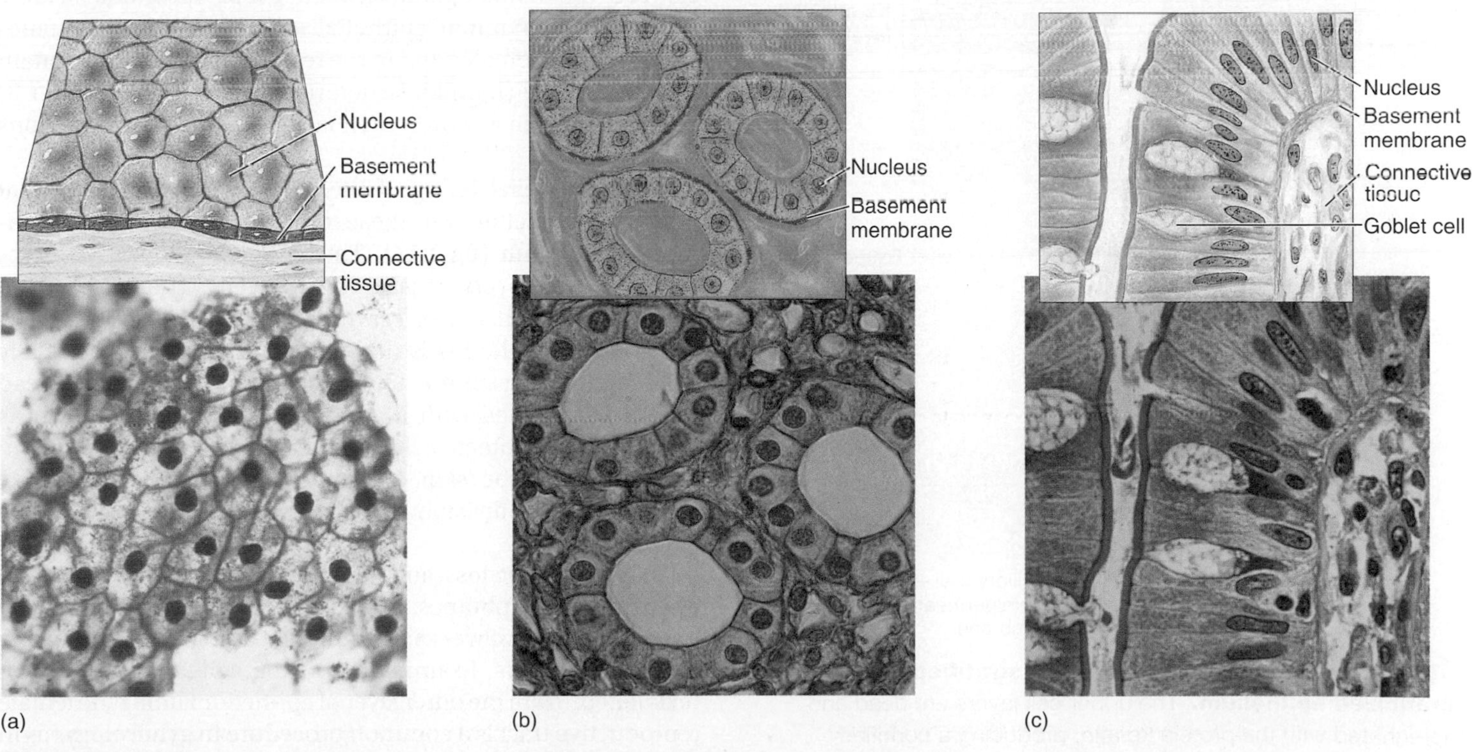

Figure 1.12 Different types of simple epithelial membranes. (*a*) Simple squamous, (*b*) simple cuboidal, and (*c*) simple columnar epithelial membranes. The tissue beneath each membrane is connective tissue.

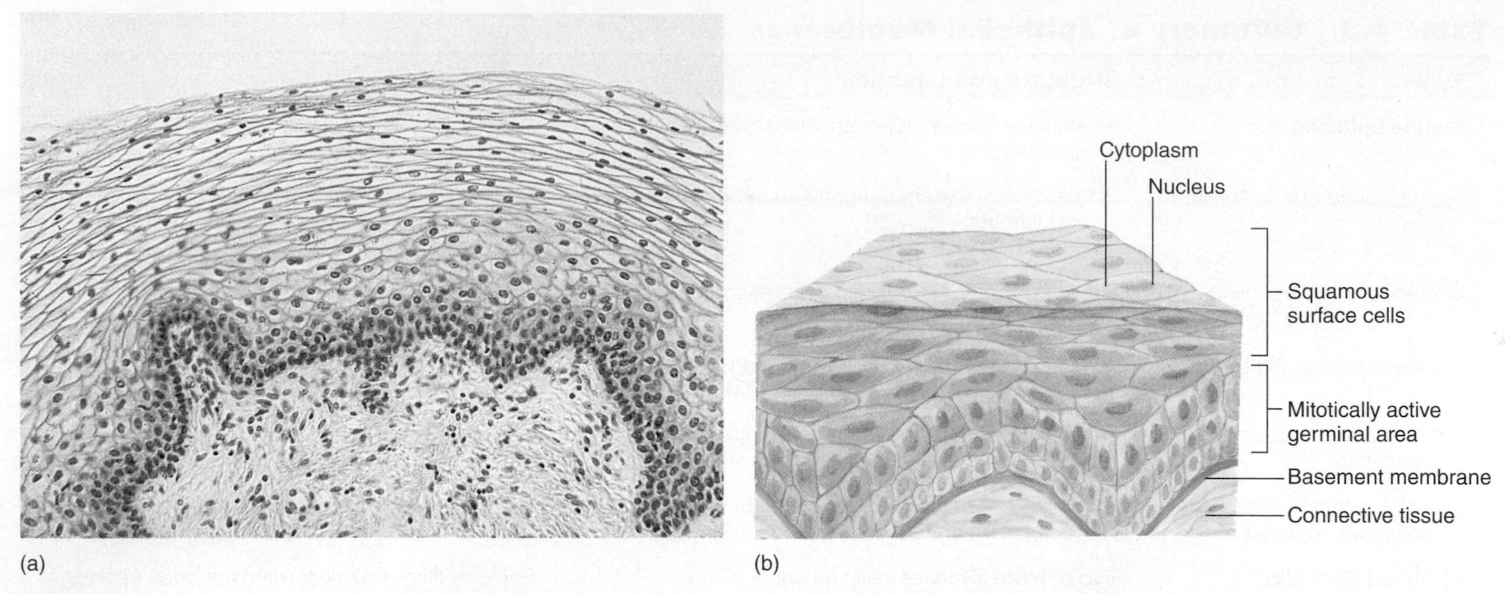

Figure 1.13 A stratified squamous nonkeratinized epithelial membrane. This is a photomicrograph (*a*) and illustration (*b*) of the epithelial lining of the vagina. APIR

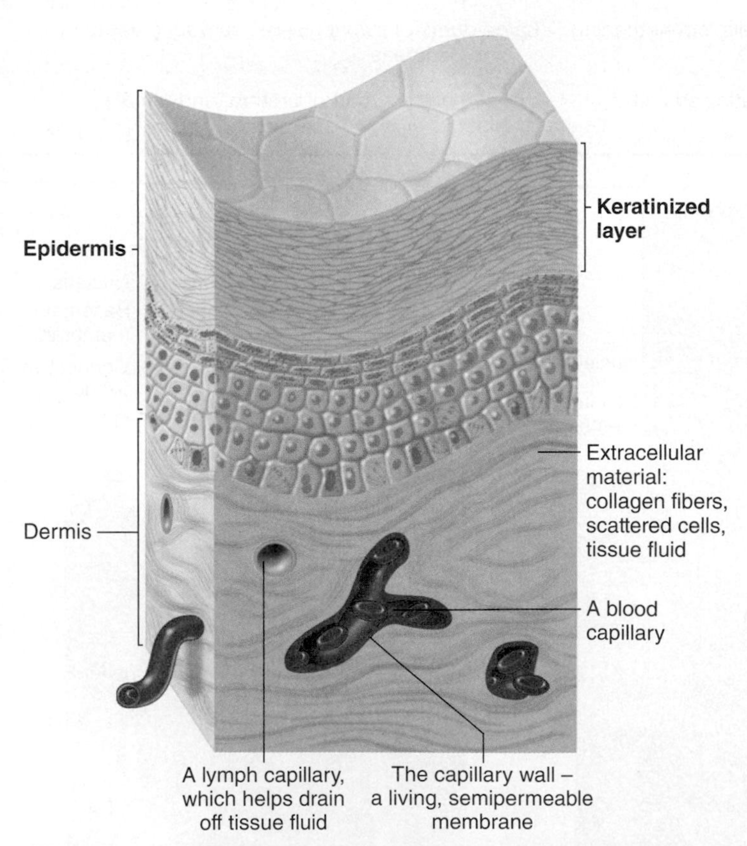

Figure 1.14 The epidermis is a stratified, squamous, keratinized epithelium. The upper cell layers are dead and impregnated with the protein keratin, producing a cornified epithelial membrane, which is supported by layers of living cells. The epidermis is nourished by blood vessels located in the loose connective tissue of the dermis. APIR

in the lungs allows the rapid passage of oxygen and carbon dioxide between the air (external environment) and blood (internal environment). A simple columnar epithelium in the small intestine, as another example, allows digestion products to pass from the intestinal lumen (external environment) to the blood (internal environment).

Dispersed among the columnar epithelial cells are specialized unicellular glands called *goblet cells* that secrete mucus. The columnar epithelial cells in the uterine (fallopian) tubes of females and in the respiratory passages contain numerous *cilia* (hairlike structures, described in chapter 3) that can move in a coordinated fashion and aid the functions of these organs.

The epithelial lining of the esophagus and vagina that provides protection for these organs is a stratified squamous epithelium (fig. 1.13). This is a *nonkeratinized* membrane, and all layers consist of living cells. The *epidermis* of the skin, by contrast, is *keratinized,* or *cornified* (fig. 1.14). Because the epidermis is dry and exposed to the potentially desiccating effects of the air, the surface is covered with dead cells that are filled with a water-resistant protein known as *keratin.* This protective layer is constantly flaked off from the surface of the skin and therefore must be constantly replaced by the division of cells in the deeper layers of the epidermis.

The constant loss and renewal of cells is characteristic of epithelial membranes. The entire epidermis is completely replaced every two weeks; the stomach lining is renewed every two to three days. Examination of the cells that are lost, or "exfoliated," from the outer layer of epithelium lining the female reproductive tract is a common procedure in gynecology (as in the Pap smear).

In order to form a strong membrane that is effective as a barrier at the body surfaces, epithelial cells are very closely packed and are joined together by structures collectively called **junctional complexes** (chapter 6; see fig. 6.22). There is no room for blood vessels between adjacent epithelial cells. The epithelium must therefore receive nourishment from the tissue beneath, which has large intercellular spaces that can accommodate blood vessels and nerves. This underlying tissue is called *connective tissue.* Epithelial membranes are attached to the underlying connective tissue by a layer of proteins and polysaccharides known as the **basement membrane.** This layer can be observed only under the microscope using specialized staining techniques.

Basement membranes are believed to induce a polarity to the cells of epithelial membranes; that is, the top (apical) portion of epithelial cells has different structural and functional components than the bottom (basal) portion. This is important in many physiological processes. For example, substances are transported in specific directions across simple epithelial membranes (discussed in chapter 6; see fig. 6.21). In stratified membranes, only the basal (bottom) layer of cells is on the basement membrane, and it is these cells that undergo mitosis to form new epithelial cells to replace those lost from the top. Scientists recently demonstrated that when these basal cells divide, one of the daughter cells is attached to the basement membrane (renewing the basal cell population), while the other is not. The daughter cell that is "unstuck" from the basement membrane differentiates and migrates upward in the stratified epithelium.

CLINICAL APPLICATION

Basement membranes consist primarily of a structural protein known as *collagen* (see fig. 1.17), together with assorted other types of proteins. The specific type of collagen in basement membranes is known as *collagen IV,* a large protein assembled from six different polypeptide chains coded by six different genes. (The structure of proteins is described in chapter 2, and the genetic coding of protein structure in chapter 3.)

Alport's syndrome is a genetic disorder of the collagen subunits. This leads to their degradation and can cause a variety of problems, including kidney failure.

Goodpasture's syndrome occurs when the person's own immune system makes antibodies that attack antigens (molecules that stimulate an immune response) in the basement membranes of the glomeruli (filtering units) of the kidneys. This results in an autoimmune inflammation of the glomeruli, or *glomerulonephritis,* which can produce acute renal failure and other problems.

Exocrine Glands

Exocrine glands are derived from cells of epithelial membranes. The secretions of these cells are passed to the outside of the epithelial membranes (and hence to the surface of the body) through *ducts.* This is in contrast to *endocrine glands,* which lack ducts and which therefore secrete into capillaries within the body (fig. 1.15). The structure of endocrine glands will be described in chapter 11.

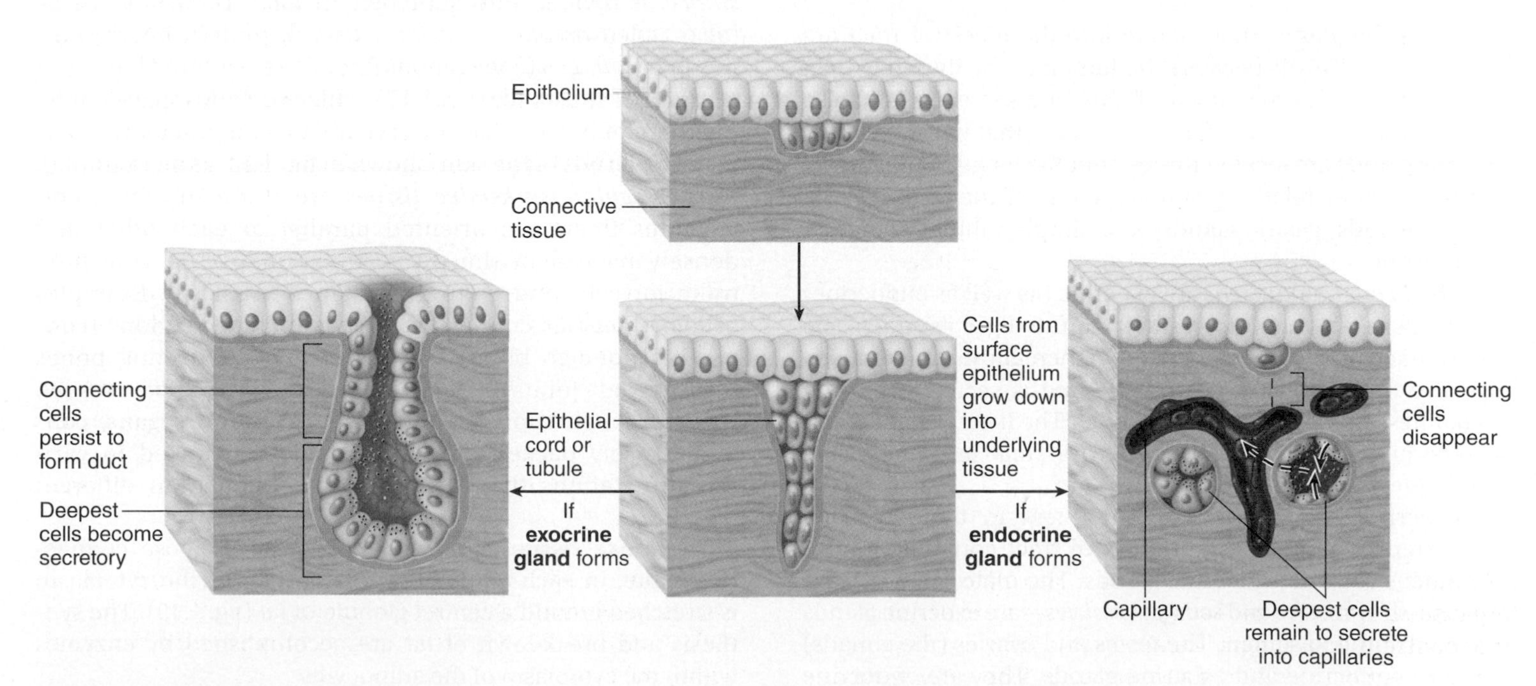

Figure 1.15 The formation of exocrine and endocrine glands from epithelial membranes. Note that exocrine glands retain a duct that can carry their secretion to the surface of the epithelial membrane, whereas endocrine glands are ductless.

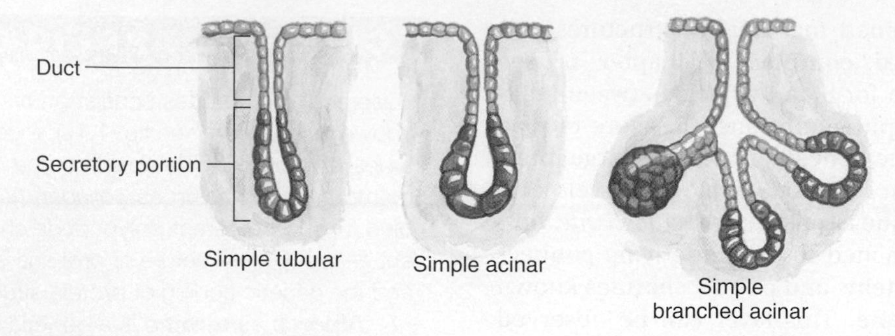

Figure 1.16 The structure of exocrine glands. Exocrine glands may be simple invaginations of epithelial membranes, or they may be more complex derivatives. AP|R

The secretory units of exocrine glands may be simple tubes, or they may be modified to form clusters of units around branched ducts (fig. 1.16). These clusters, or **acini,** are often surrounded by tentacle-like extensions of *myoepithelial cells* that contract and squeeze the secretions through the ducts. The rate of secretion and the action of myoepithelial cells are subject to neural and endocrine regulation.

Examples of exocrine glands in the skin include the lacrimal (tear) glands, sebaceous glands (which secrete oily sebum into hair follicles), and sweat glands. There are two types of sweat glands. The more numerous, the *eccrine* (or *merocrine*) *sweat glands,* secrete a dilute salt solution that serves in thermoregulation (evaporation cools the skin). The *apocrine sweat glands,* located in the axillae (underarms) and pubic region, secrete a protein-rich fluid. This provides nourishment for bacteria that produce the characteristic odor of this type of sweat.

All of the glands that secrete into the digestive tract are also exocrine. This is because the lumen of the digestive tract is a part of the external environment, and secretions of these glands go to the outside of the membrane that lines this tract. Mucous glands are located throughout the length of the digestive tract. Other relatively simple glands of the tract include salivary glands, gastric glands, and simple tubular glands in the intestine.

The *liver* and *pancreas* are exocrine (as well as endocrine) glands, derived embryologically from the digestive tract. The exocrine secretion of the pancreas—pancreatic juice—contains digestive enzymes and bicarbonate and is secreted into the small intestine via the pancreatic duct. The liver produces and secretes bile (an emulsifier of fat) into the small intestine via the gallbladder and bile duct.

Exocrine glands are also prominent in the reproductive system. The female reproductive tract contains numerous mucus-secreting exocrine glands. The male accessory sex organs—the *prostate* and *seminal vesicles*—are exocrine glands that contribute to semen. The testes and ovaries (the gonads) are both endocrine and exocrine glands. They are endocrine because they secrete sex steroid hormones into the blood; they are exocrine because they release gametes (ova and sperm) into the reproductive tracts.

Connective Tissue

Connective tissue is characterized by large amounts of extracellular material between the different types of connective tissue cells. The extracellular material, called the connective tissue *matrix,* varies in the four primary types of connective tissues: (1) connective tissue proper; (2) cartilage; (3) bone; and (4) blood. **Blood** is classified as a type of connective tissue because about half its volume is an extracellular fluid, the blood plasma (chapter 13, section 13.1).

Connective tissue proper, in which the matrix consists of protein fibers and a proteinaceous, gel-like *ground substance,* is divided into subtypes. In *loose connective tissue* (also called *areolar connective tissue*), protein fibers composed of *collagen* (collagenous fibers) are scattered loosely in the ground substance (fig. 1.17), which provides space for the presence of blood vessels, nerve fibers, and other structures (see the dermis of the skin, shown in fig. 1.14, as an example). *Dense regular connective tissues* are those in which collagenous fibers are oriented parallel to each other and densely packed in the extracellular matrix, leaving little room for cells and ground substance (fig. 1.18). Examples of dense regular connective tissues include tendons (connecting bone to bone) and ligaments (connecting bones together at joints). *Dense irregular connective tissues,* forming tough capsules and sheaths around organs, contain densely packed collagenous fibers arranged in various orientations that resist forces applied from different directions.

Adipose tissue is a specialized type of loose connective tissue. In each adipose cell, or *adipocyte,* the cytoplasm is stretched around a central globule of fat (fig. 1.19). The synthesis and breakdown of fat are accomplished by enzymes within the cytoplasm of the adipocytes.

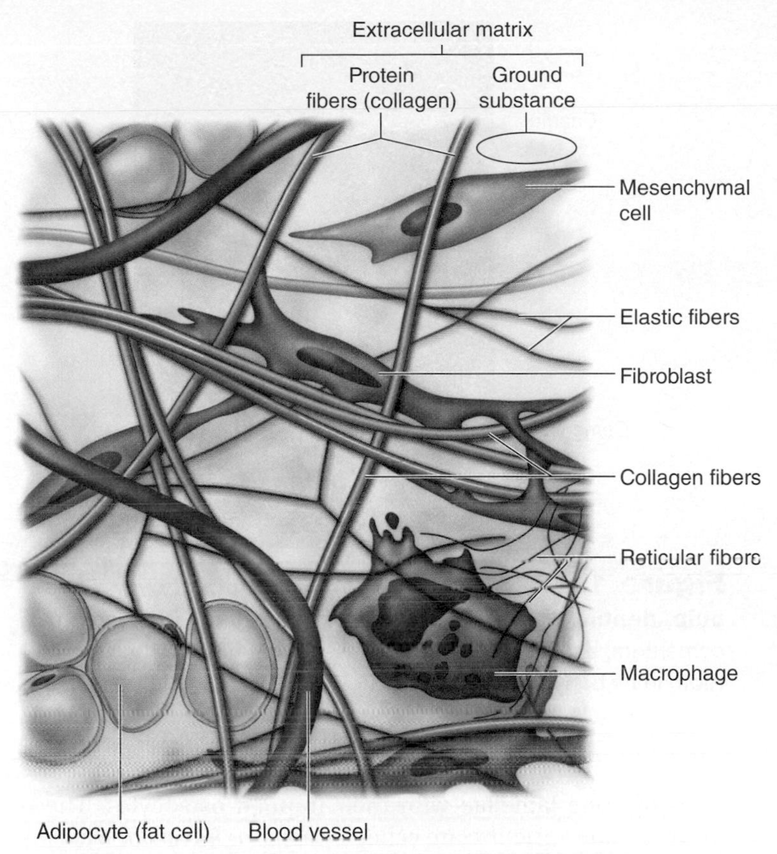

Figure 1.17 Loose connective tissue. This illustration shows the cells and protein fibers characteristic of connective tissue proper. The ground substance is the extracellular background material, against which the different protein fibers can be seen. The macrophage is a phagocytic connective tissue cell, which can be derived from monocytes (a type of white blood cell). AP|R

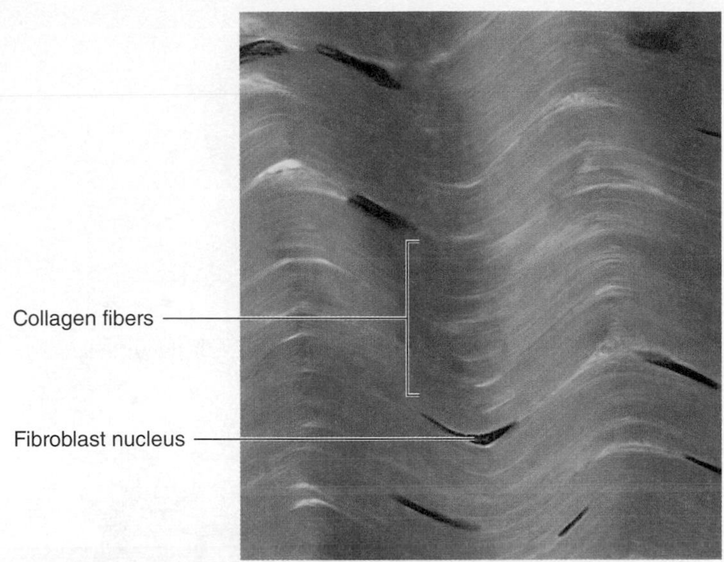

Figure 1.18 Dense regular connective tissue. In this photomicrograph, the collagen fibers in a tendon are packaged densely into parallel groups. The ground substance is in the tiny spaces between the collagen fibers. AP|R

Cartilage consists of cells, called *chondrocytes,* surrounded by a semisolid ground substance that imparts elastic properties to the tissue. Cartilage is a type of supportive and protective tissue commonly called "gristle." It forms the precursor to many bones that develop in the fetus and persists at the articular (joint) surfaces on the bones at all movable joints in adults.

Bone is produced as concentric layers, or *lamellae,* of calcified material laid around blood vessels. The bone-forming cells, or *osteoblasts,* surrounded by their calcified products, become trapped within cavities called *lacunae.* The trapped cells, which are now called *osteocytes,* remain alive because they are nourished by "lifelines" of cytoplasm that extend from the cells to the blood vessels in *canaliculi* (little canals). The blood vessels lie within central canals, surrounded by concentric

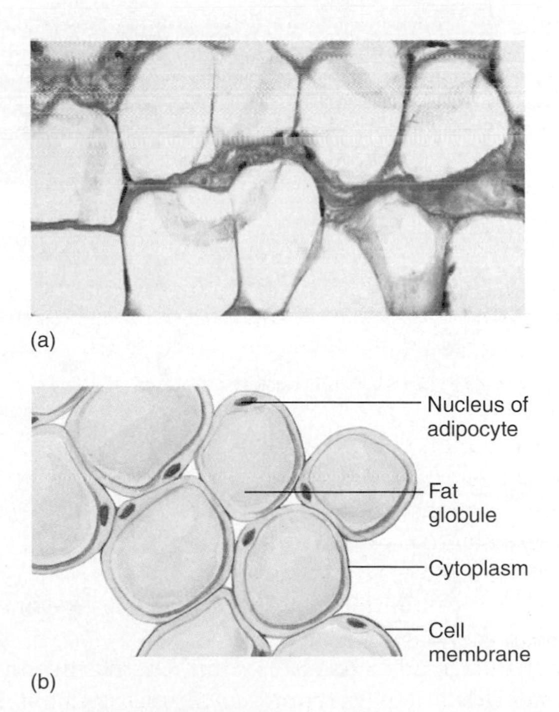

Figure 1.19 Adipose tissue. Each adipocyte contains a large, central globule of fat surrounded by the cytoplasm of the adipocyte. (*a*) Photomicrograph and (*b*) illustration of adipose tissue. AP|R

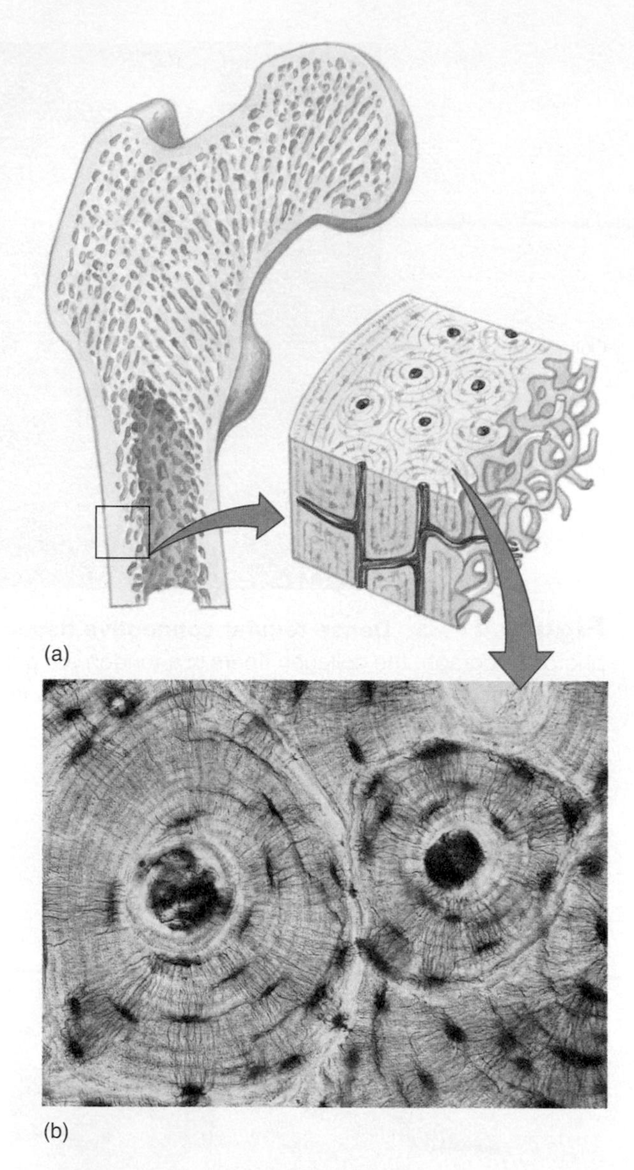

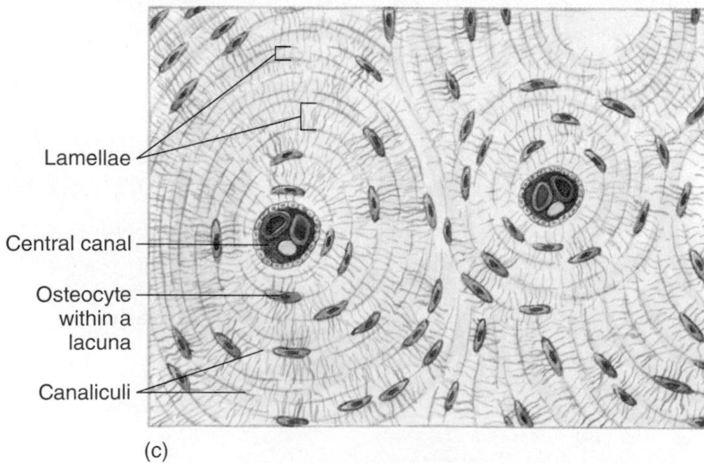

Figure 1.20 The structure of bone. (*a*) A diagram of a long bone, (*b*) a photomicrograph showing osteons (haversian systems), and (*c*) a diagram of osteons. Within each central canal, an artery (red), a vein (blue), and a nerve (yellow) is illustrated. APR

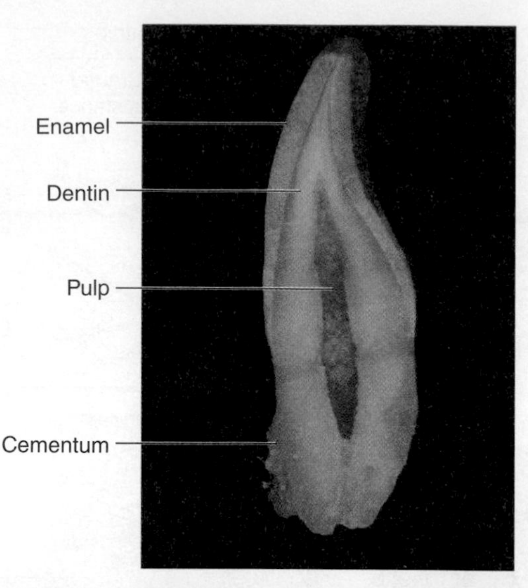

Figure 1.21 A cross section of a tooth showing pulp, dentin, and enamel. The root of the tooth is covered by cementum, a calcified connective tissue that helps to anchor the tooth in its bony socket.

rings of bone lamellae with their trapped osteocytes. These units of bone structure are called *osteons,* or *haversian systems* (fig. 1.20).

The *dentin* of a tooth (fig. 1.21) is similar in composition to bone, but the cells that form this calcified tissue are located in the pulp (composed of loose connective tissue). These cells send cytoplasmic extensions, called *dentinal tubules,* into the dentin. Dentin, like bone, is thus a living tissue that can be remodeled in response to stresses. The cells that form the outer *enamel* of a tooth, by contrast, are lost as the tooth erupts. Enamel is a highly calcified material, harder than bone or dentin, that cannot be regenerated; artificial "fillings" are therefore required to patch holes in the enamel.

CHECKPOINTS

6a. List the four primary tissues and describe the distinguishing features of each type.

6b. Compare and contrast the three types of muscle tissue.

6c. Describe the different types of epithelial membranes and state their locations in the body.

7a. Explain why exocrine and endocrine glands are considered epithelial tissues and distinguish between these two types of glands.

7b. Describe the different types of connective tissues and explain how they differ from one another in their content of extracellular material.

1.4 ORGANS AND SYSTEMS

Organs are composed of two or more primary tissues that serve the different functions of the organ. The skin is an organ that has numerous functions provided by its constituent tissues.

LEARNING OUTCOMES

After studying this section, you should be able to:

8. Use the skin as an example to describe how the different primary tissues compose organs.

9. Identify the body fluid compartments.

An **organ** is a structure composed of at least two, and usually all four, primary tissues. The largest organ in the body, in terms of surface area, is the skin (fig. 1.22). In this section, the numerous functions of the skin serve to illustrate how primary tissues cooperate in the service of organ physiology.

An Example of an Organ: The Skin

The cornified *epidermis* protects the skin against water loss and against invasion by disease-causing organisms. Invaginations of the epithelium into the underlying connective tissue *dermis* create the exocrine glands of the skin. These include hair follicles (which produce the hair), sweat glands, and sebaceous glands. The secretion of sweat glands cools the body by evaporation and produces odors that, at least in lower animals, serve as sexual attractants. Sebaceous glands secrete oily sebum into hair follicles, which transport the sebum to the surface of the skin. Sebum lubricates the cornified surface of the skin, helping to prevent it from drying and cracking.

The skin is nourished by blood vessels within the dermis. In addition to blood vessels, the dermis contains wandering white blood cells and other types of cells that protect against invading disease-causing organisms. It also contains nerve fibers and adipose (fat) cells; however, most of the adipose cells are grouped together to form the *hypodermis* (a layer

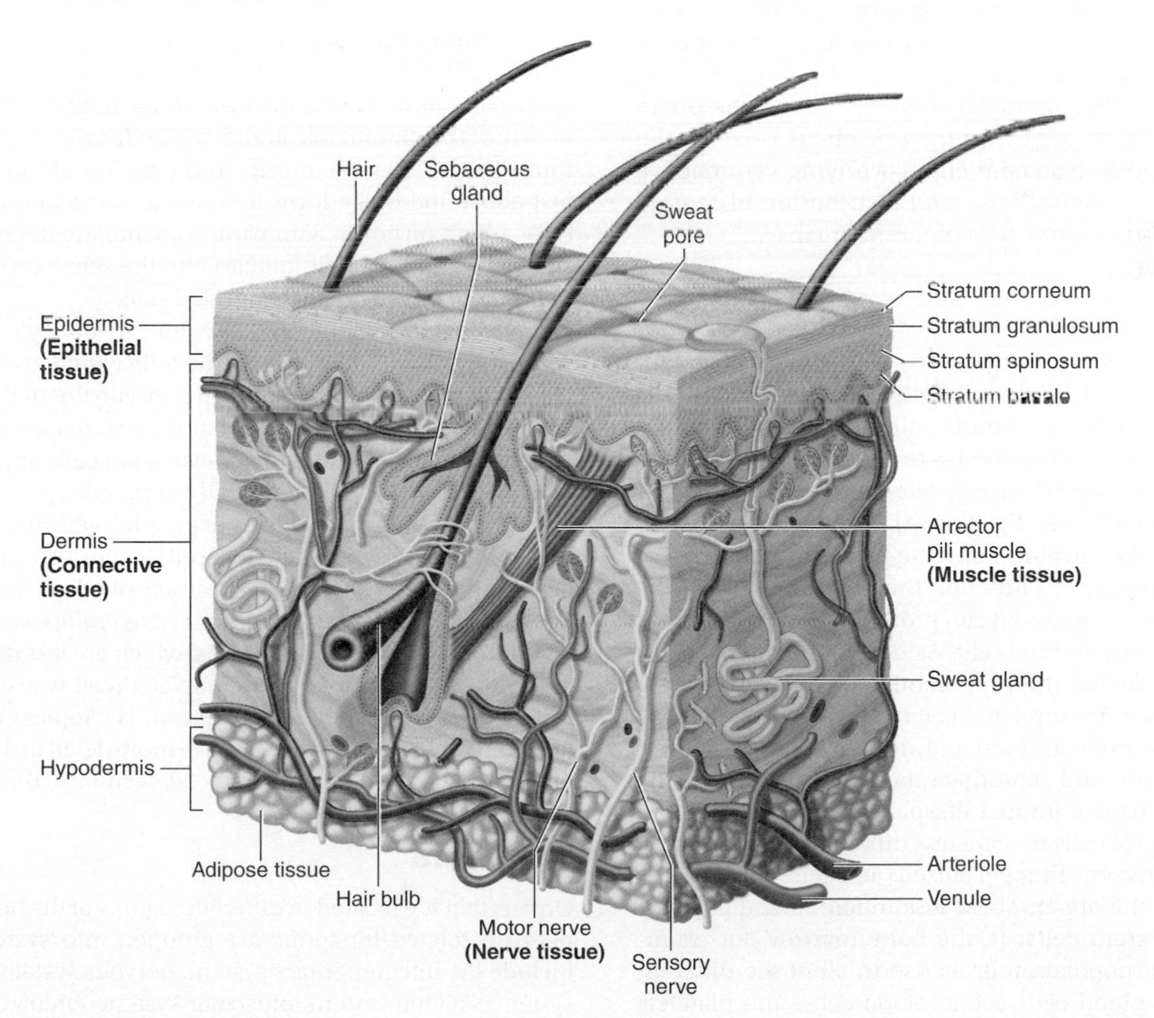

Figure 1.22 A diagram of the skin. The skin is an organ that contains all four types of primary tissues. AP|R

beneath the dermis). Although adipose cells are a type of connective tissue, masses of fat deposits throughout the body—such as subcutaneous fat—are referred to as *adipose tissue.*

Sensory nerve endings within the dermis mediate the cutaneous sensations of touch, pressure, heat, cold, and pain. Motor nerve fibers in the skin stimulate effector organs, resulting in, for example, the secretions of exocrine glands and contractions of the arrector pili muscles, which attach to hair follicles and surrounding connective tissue (producing goose bumps). The degree of constriction or dilation of cutaneous blood vessels—and therefore the rate of blood flow—is also regulated by motor nerve fibers.

The epidermis itself is a dynamic structure that can respond to environmental stimuli. The rate of its cell division—and consequently the thickness of the cornified layer—increases under the stimulus of constant abrasion. This produces calluses. The skin also protects itself against the dangers of ultraviolet light by increasing its production of *melanin* pigment, which absorbs ultraviolet light while producing a tan. In addition, the skin is an endocrine gland; it synthesizes and secretes vitamin D (derived from cholesterol under the influence of ultraviolet light), which functions as a hormone.

The architecture of most organs is similar to that of the skin. Most are covered by an epithelium that lies immediately over a connective tissue layer. The connective tissue contains blood vessels, nerve endings, scattered cells for fighting infection, and possibly glandular tissue as well. If the organ is hollow—as with the digestive tract or blood vessels—the lumen is also lined with an epithelium overlying a connective tissue layer. The presence, type, and distribution of muscle tissue and nervous tissue vary in different organs.

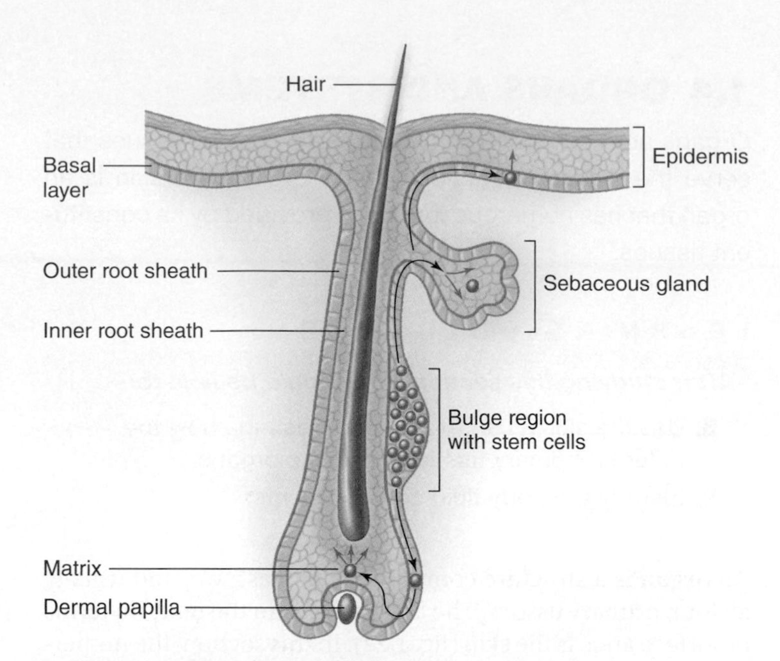

Figure 1.23 The bulge region of the hair follicle with stem cells. Stem cells in this region migrate to form the differentiated cells of the hair follicle, sebaceous gland, and epidermis. AP|R

Stem Cells

The different tissues of an organ are composed of cells that are highly specialized, or *differentiated.* The process of differentiation begins during embryonic development, when the fertilized egg, or *zygote,* divides to produce three embryonic tissue layers, or *germ layers: ectoderm, mesoderm,* and *endoderm* (chapter 20; see fig. 20.45*a*). During the course of embryonic and fetal development, the three germ layers give rise to the four primary tissues and their subtypes.

The zygote is *totipotent*—it can produce all of the different specialized cell types in the body. As development proceeds, the cells become increasingly differentiated (specialized) and lose the ability to form unrelated cell types. Some specialized cells—such as neurons and striated muscle cells—lose even the ability to divide and reproduce themselves. Because the specialized cells have a limited lifespan, many organs retain small populations of cells that are less differentiated and more able to divide to become the specialized (and generally related) cell types within the organ. These less-differentiated cells are known as **adult stem cells.** In the bone marrow, for example, the stem cell population gives rise to all of the different blood cells—red blood cells, white blood cells, and platelets (chapter 13). Similarly, there are stem cells in the brain (chapter 8), skeletal muscles (chapter 12), and intestine (chapter 18).

Scientists have recently discovered that there are also stem cells in the bulge region of the hair follicle (fig. 1.23). These stem cells form keratinocytes, which migrate down to the matrix of the hair follicle and divide to form the hair shaft and root sheath. Other stem cells in the region of the hair follicle just above the bulge form new sebaceous gland cells, which have a high turnover. Skin wounds stimulate the migration of stem cells from the hair follicles into the skin between follicles to promote healing of the wounded skin.

The bulge region also contains melanocyte stem cells, which migrate to the matrix of the follicle and give the hair its color. Scientists have now shown that graying of the hair with age is caused by loss of the melanocyte stem cells in the bulge of the hair follicles. The melanocyte stem cells appeared to be present in most of the hair follicles of people aged 20 to 30 and absent from most hair follicles of people aged 70 to 90.

As demonstrated by the stem cells in the bulge of the hair follicle, adult stem cells can form a variety of related cell types; the adult stem cells are therefore described as *multipotent.* This is different from **embryonic stem cells,** which are less differentiated and more capable of forming unrelated cell types; embryonic stem cells are described as *pluripotent.* The topics of embryonic and adult stem cells are discussed in more detail in the context of embryonic development (chapter 20, section 20.6).

Systems

Organs that are located in different regions of the body and that perform related functions are grouped into **systems.** These include the integumentary system, nervous system, endocrine system, skeletal system, muscular system, circulatory system, immune system, respiratory system, urinary system, digestive system, and reproductive system (table 1.4). By means

Table 1.4 | Organ Systems of the Body

System	Major Organs	Primary Functions
Integumentary	Skin, hair, nails	Protection, thermoregulation
Nervous	Brain, spinal cord, nerves	Regulation of other body systems
Endocrine	Hormone-secreting glands, such as the pituitary, thyroid, and adrenals	Secretion of regulatory molecules called hormones
Skeletal	Bones, cartilages	Movement and support
Muscular	Skeletal muscles	Movements of the skeleton
Circulatory	Heart, blood vessels, lymphatic vessels	Movement of blood and lymph
Immune	Bone marrow, lymphoid organs	Defense of the body against invading pathogens
Respiratory	Lungs, airways	Gas exchange
Urinary	Kidneys, ureters, urethra	Regulation of blood volume and composition
Digestive	Mouth, stomach, intestine, liver, gallbladder, pancreas	Breakdown of food into molecules that enter the body
Reproductive	Gonads, external genitalia, associated glands and ducts	Continuation of the human species

of numerous regulatory mechanisms, these systems work together to maintain the life and health of the entire organism.

Body-Fluid Compartments

Tissues, organs, and systems can all be divided into two major parts, or compartments. The **intracellular compartment** is that part inside the cells; the **extracellular compartment** is that part outside the cells. Both compartments consist primarily of water—they are said to be *aqueous.* About 65% of the total body water is in the intracellular compartment, while about 35% is in the extracellular compartment. The two compartments are separated by the cell membrane surrounding each cell (chapter 3, section 3.1).

The extracellular compartment is subdivided into two parts. One part is the *blood plasma,* the fluid portion of the blood. The other is the fluid that bathes the cells within the organs of the body. This is called *tissue fluid,* or *interstitial fluid.* In most parts of the body, blood plasma and tissue fluid communicate freely through blood capillaries. The kidneys regulate the volume and composition of the blood plasma, and thus, indirectly, the fluid volume and composition of the entire extracellular compartment.

There is also selective communication between the intracellular and extracellular compartments through the movement of molecules and ions through the cell membrane, as described in chapter 6. This is how cells obtain the molecules they need for life and how they eliminate waste products.

CHECKPOINTS

8a. State the location of each type of primary tissue in the skin.

8b. Describe the functions of nervous, muscle, and connective tissue in the skin.

8c. Describe the functions of the epidermis and explain why this tissue is called "dynamic."

9. Distinguish between the intracellular and extracellular compartments and explain their significance.

Clinical Investigation SUMMARY

At the end of each subsequent chapter is a Clinical Investigation Summary that presents solutions to the medical mystery presented in the Clinical Investigation at the beginning of the chapter. Don't cheat yourself—wait until you have tried to solve the mystery yourself based on your studies and the Clinical Investigation Clues—before you read the "Summary" at the end.

Would you like to solve a medical mystery now? Try the additional Clinical Investigation entitled *Negative Feedback Control of Blood Glucose* based on chapter 1 material. This is located in the **Connect** site for this text at www.mhhe.comn/Fox13. This should be fun, and you will soon realize that the more you play with physiological concepts the better you will understand them. The Summary boxes at the end of each subsequent chapter present additional Clinical Investigations on the website that relate to those chapters.

SUMMARY

1.1 Introduction to Physiology 2

A. Physiology is the study of how cells, tissues, and organs function.
  1. In the study of physiology, cause-and-effect sequences are emphasized.
  2. Knowledge of physiological mechanisms is deduced from data obtained experimentally.

B. The science of physiology shares knowledge with the related sciences of pathophysiology and comparative physiology.
  1. Pathophysiology is concerned with the functions of diseased or injured body systems and is based on knowledge of how normal systems function, which is the focus of physiology.
  2. Comparative physiology is concerned with the physiology of animals other than humans and shares much information with human physiology.

C. All of the information in this book has been gained by applications of the scientific method. This method has three essential characteristics:
  1. It is assumed that the subject under study can ultimately be explained in terms we can understand.
  2. Descriptions and explanations are honestly based on observations of the natural world and can be changed as warranted by new observations.
  3. Humility is an important characteristic of the scientific method; the scientist must be willing to change his or her theories when warranted by the weight of the evidence.

1.2 Homeostasis and Feedback Control 4

A. Homeostasis refers to the dynamic constancy of the internal environment.
  1. Homeostasis is maintained by mechanisms that act through negative feedback loops.
    a. A negative feedback loop requires (1) a sensor that can detect a change in the internal environment and (2) an effector that can be activated by the sensor.
    b. In a negative feedback loop, the effector acts to cause changes in the internal environment that compensate for the initial deviations that were detected by the sensor.
  2. Positive feedback loops serve to amplify changes and may be part of the action of an overall negative feedback mechanism.
  3. The nervous and endocrine systems provide extrinsic regulation of other body systems and act to maintain homeostasis.
  4. The secretion of hormones is stimulated by specific chemicals and is inhibited by negative feedback mechanisms.

B. Effectors act antagonistically to defend the set point against deviations in any direction.

1.3 The Primary Tissues 10

A. The body is composed of four types of primary tissues: muscle, nervous, epithelial, and connective tissues.
  1. There are three types of muscle tissue: skeletal, cardiac, and smooth muscle.
    a. Skeletal and cardiac muscle are striated.
    b. Smooth muscle is found in the walls of the internal organs.
  2. Nervous tissue is composed of neurons and neuroglial cells.
    a. Neurons are specialized for the generation and conduction of electrical impulses.
    b. Neuroglial cells provide the neurons with anatomical and functional support.
  3. Epithelial tissue includes membranes and glands.
    a. Epithelial membranes cover and line the body surfaces, and their cells are tightly joined by junctional complexes.
    b. Epithelial membranes may be simple or stratified, and their cells may be squamous, cuboidal, or columnar.
    c. Exocrine glands, which secrete into ducts, and endocrine glands, which lack ducts and secrete hormones into the blood, are derived from epithelial membranes.
  4. Connective tissue is characterized by large intercellular spaces that contain extracellular material.
    a. Connective tissue proper is categorized into subtypes, including loose, dense fibrous, adipose, and others.
    b. Cartilage, bone, and blood are classified as connective tissues because their cells are widely spaced with abundant extracellular material between them.

1.4 Organs and Systems 19

A. Organs are units of structure and function that are composed of at least two, and usually all four, of the primary types of tissues.
  1. The skin is a good example of an organ.
    a. The epidermis is a stratified squamous keratinized epithelium that protects underlying structures and produces vitamin D.
    b. The dermis is an example of loose connective tissue.
    c. Hair follicles, sweat glands, and sebaceous glands are exocrine glands located within the dermis.
    d. Sensory and motor nerve fibers enter the spaces within the dermis to innervate sensory organs and smooth muscles.
    e. The arrector pili muscles that attach to the hair follicles are composed of smooth muscle.
  2. Organs that are located in different regions of the body and that perform related functions are grouped into systems. These include, among others, the circulatory system, digestive system, and endocrine system.
  3. Many organs contain adult stem cells, which are able to differentiate into a number of related cell types.
    a. Because of their limited flexibility, adult stem cells are described as multipotent, rather than as totipotent or pluripotent.
    b. For example, the bulge region of a hair follicle contains stem cells that can become keratinocytes, epithelial cells, and melanocytes; the loss of the melanocyte stem cells causes graying of the hair.

B. The fluids of the body are divided into two major compartments.
  1. The intracellular compartment refers to the fluid within cells.
  2. The extracellular compartment refers to the fluid outside of cells; extracellular fluid is subdivided into plasma (the fluid portion of the blood) and tissue (interstitial) fluid.

REVIEW ACTIVITIES

Test Your Knowledge

1. Glands are derived from
   a. nervous tissue.
   b. connective tissue.
   c. muscle tissue.
   d. epithelial tissue.
2. Cells joined tightly together are characteristic of
   a. nervous tissue.
   b. connective tissue.
   c. muscle tissue.
   d. epithelial tissue.
3. Cells are separated by large extracellular spaces in
   a. nervous tissue.
   b. connective tissue.
   c. muscle tissue.
   d. epithelial tissue.
4. Blood vessels and nerves are usually located within
   a. nervous tissue.
   b. connective tissue.
   c. muscle tissue.
   d. epithelial tissue.
5. Most organs are composed of
   a. epithelial tissue.
   b. muscle tissue.
   c. connective tissue.
   d. all of these.
6. Sweat is secreted by exocrine glands. This means that
   a. it is produced by endocrine cells.
   b. it is a hormone.
   c. it is secreted into a duct.
   d. it is produced outside the body.
7. Which of these statements about homeostasis is *true?*
   a. The internal environment is maintained absolutely constant.
   b. Negative feedback mechanisms act to correct deviations from a normal range within the internal environment.
   c. Homeostasis is maintained by turning effectors on and off.
   d. All of these are true.
8. In a negative feedback loop, the effector organ produces changes that are
   a. in the same direction as the change produced by the initial stimulus.
   b. opposite in direction to the change produced by the initial stimulus.
   c. unrelated to the initial stimulus.
9. A hormone called parathyroid hormone acts to help raise the blood calcium concentration. According to the principles of negative feedback, an effective stimulus for parathyroid hormone secretion would be
   a. a fall in blood calcium.
   b. a rise in blood calcium.
10. Which of these consists of dense parallel arrangements of collagen fibers?
    a. skeletal muscle tissue
    b. nervous tissue
    c. tendons
    d. dermis of the skin
11. The act of breathing raises the blood oxygen level, lowers the blood carbon dioxide concentration, and raises the blood pH. According to the principles of negative feedback, sensors that regulate breathing should respond to
    a. a rise in blood oxygen.
    b. a rise in blood pH.
    c. a rise in blood carbon dioxide concentration.
    d. all of these.
12. Adult stem cells, such as those in the bone marrow, brain, or hair follicles, can best be described as ____________, whereas embryonic stem cells are described as ____________.
    a. totipotent; pluripotent
    b. pluripotent; multipotent
    c. multipotent; pluripotent
    d. totipotent; multipotent

Test Your Understanding

13. Describe the structure of the various epithelial membranes and explain how their structures relate to their functions.
14. Compare bone, blood, and the dermis of the skin in terms of their similarities. What are the major structural differences between these tissues?
15. Describe the role of antagonistic negative feedback processes in the maintenance of homeostasis.
16. Using insulin as an example, explain how the secretion of a hormone is controlled by the effects of that hormone's actions.
17. Describe the steps in the development of pharmaceutical drugs and evaluate the role of animal research in this process.
18. Why is Claude Bernard considered the father of modern physiology? Why is the concept he introduced so important in physiology and medicine?

Test Your Analytical Ability

19. What do you think would happen if most of your physiological regulatory mechanisms were to operate by positive feedback rather than by negative feedback? Would life even be possible?
20. Examine figure 1.5 and determine when the compensatory physiological responses began to act, and how many minutes they required to restore the initial set point of blood glucose concentration. Comment on the importance of quantitative measurements in physiology.
21. Why are interactions between the body-fluid compartments essential for sustaining life?
22. Suppose a person has collapsed due to a rapid drop in blood pressure. What would you expect to find regarding the rate and strength of this person's pulse? Explain how this illustrates the principle of negative feedback regulation.
23. Give examples of adult stem cells and explain their abilities and limitations. Why are adult stem cells needed in the body?

Test Your Quantitative Ability

Suppose body temperature varies between 36.6° C and 37.7° C over a period of a few hours (see fig. 1.4).

24. Calculate the set point as the average value.
25. Calculate the range of values (lowest to highest).
26. Calculate the sensitivity of the negative feedback loop; this is the deviation from the set point to the lowest (or highest) value.

CHAPTER

2

Chemical Composition of the Body

CHAPTER OUTLINE

Clinical Investigation

George decides it is immoral to eat plants or animals, and so resolves to eat only artificial food he obtains from his chemistry lab: D-amino acids and L-sugars. After several days, he feels very weak and seeks medical attention.

Some of the new terms and concepts you will encounter include:

- Stereoisomers (D- and L-amino acids and sugars)
- Ketone bodies and ketonuria

2.1 ATOMS, IONS, AND CHEMICAL BONDS

The study of physiology requires some familiarity with the basic concepts and terminology of chemistry. A knowledge of atomic and molecular structure, the nature of chemical bonds, and the nature of pH and associated concepts provides the foundation for much of human physiology.

LEARNING OUTCOMES

After studying this section, you should be able to:

1. Describe the structure of an atom and ion and the nature of covalent, ionic, and hydrogen bonds.
2. Explain the meaning of the terms **polar** and **nonpolar; hydrophilic** and **hydrophobic.**
3. Define **acid** and **base** and explain the pH scale.
4. Identify the characteristics of organic molecules.

The structures and physiological processes of the body are based, to a large degree, on the properties and interactions of atoms, ions, and molecules. Water is the major constituent of the body and accounts for 60% to 70% of the total weight of an average adult. Of this amount, two-thirds is contained within the body cells, or in the *intracellular compartment;* the remainder is contained in the *extracellular compartment*, a term that refers to the blood and tissue fluids. Dissolved in this water are many organic molecules (carbon-containing molecules such as carbohydrates, lipids, proteins, and nucleic acids), as well as inorganic molecules and ions (atoms with a net charge). Before describing the structure and function of organic molecules within the body, it would be useful to consider some basic chemical concepts, terminology, and symbols.

Atoms

Atoms are the smallest units of the chemical elements. They are much too small to be seen individually, even with the most powerful electron microscope. Through the efforts of generations of scientists, however, atomic structure is now well understood. At the center of an atom is its **nucleus.** The nucleus contains two types of particles—**protons,** which bear a positive charge, and **neutrons,** which carry no charge (are neutral). The mass of a proton is equal to the mass of a neutron, and the sum of the protons and neutrons in an atom is the **mass number** of the atom. For example, an atom of carbon, which contains 6 protons and 6 neutrons, has an atomic mass of 12 (table 2.1). Note that the mass of electrons is not considered when calculating the atomic mass, because it is insignificantly small compared to the mass of protons and neutrons.

The number of protons in an atom is given as its **atomic number.** Carbon has 6 protons and thus has an atomic number of 6. Outside the positively charged nucleus are negatively charged subatomic particles called **electrons.** Because the number of electrons in an atom is equal to the number of protons, atoms have a net charge of zero.

Although it is often convenient to think of electrons as orbiting the nucleus like planets orbiting the sun, this simplified model of atomic structure is no longer believed to be correct. A given electron can occupy any position in a certain volume of space called the *orbital* of the electron. The orbitals form a "shell," or energy level, beyond which the electron usually does not pass.

There are potentially several such shells surrounding a nucleus, with each successive shell being farther from the nucleus. The first shell, closest to the nucleus, can contain only 2 electrons. If an atom has more than 2 electrons (as do all

Table 2.1 | Atoms Commonly Present in Organic Molecules

Atom	Symbol	Atomic Number	Atomic Mass	Electrons in Shell 1	Electrons in Shell 2	Electrons in Shell 3	Number of Chemical Bonds
Hydrogen	H	1	1	1	0	0	1
Carbon	C	6	12	2	4	0	4
Nitrogen	N	7	14	2	5	0	3
Oxygen	O	8	16	2	6	0	2
Sulfur	S	16	32	2	8	6	2

atoms except hydrogen and helium), the additional electrons must occupy shells that are more distant from the nucleus. The second shell can contain a maximum of 8 electrons, and higher shells can contain still more electrons that possess more energy the farther they are from the nucleus. Most elements of biological significance (other than hydrogen), however, require 8 electrons to complete the outermost shell. The shells are filled from the innermost outward. Carbon, with 6 electrons, has 2 electrons in its first shell and 4 electrons in its second shell (fig. 2.1).

It is always the electrons in the outermost shell, if this shell is incomplete, that participate in chemical reactions and form chemical bonds. These outermost electrons are known as the **valence electrons** of the atom.

Isotopes

A particular atom with a given number of protons in its nucleus may exist in several forms that differ from one another in their number of neutrons. The atomic number of these forms is thus the same, but their atomic mass is different. These different forms are called **isotopes.** All of the isotopic forms of a given atom are included in the term **chemical element.** The element hydrogen, for example, has three isotopes. The most common of these has a nucleus consisting of only 1 proton. Another isotope of hydrogen (called *deuterium*) has 1 proton and 1 neutron in the nucleus, whereas the third isotope (*tritium*) has 1 proton and 2 neutrons. Tritium is a radioactive isotope that is commonly used in physiological research and in many clinical laboratory procedures.

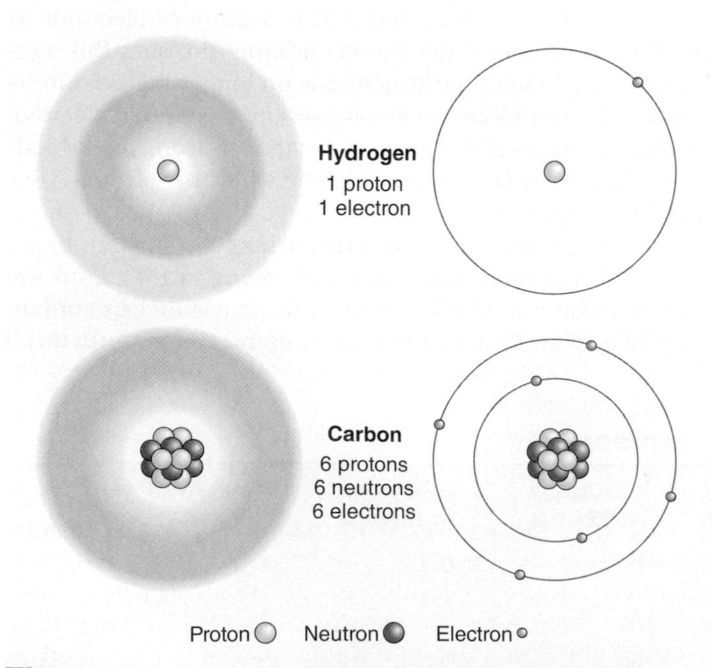

Figure 2.1 Diagrams of the hydrogen and carbon atoms. On the left, the electron shells are represented by shaded spheres indicating probable positions of the electrons. On the right, the shells are represented by concentric circles. AP|R

Chemical Bonds, Molecules, and Ionic Compounds

Molecules are formed through interaction of the valence electrons between two or more atoms. These interactions, such as the sharing of electrons, produce **chemical bonds** (fig. 2.2). The number of bonds that each atom can have is determined by the number of electrons needed to complete the outermost shell. Hydrogen, for example, must obtain only 1 more electron—and can thus form only one chemical bond—to complete the first shell of 2 electrons. Carbon, by contrast, must obtain 4 more electrons—and can thus form four chemical bonds—to complete the second shell of 8 electrons (fig. 2.3, *left*).

Covalent Bonds

Covalent bonds result when atoms share their valence electrons. Covalent bonds that are formed between identical atoms, as in oxygen gas (O_2) and hydrogen gas (H_2), are the strongest because their electrons are equally shared. Because the electrons are equally distributed between the 2 atoms, these molecules are said to be **nonpolar** and the bonds between them are nonpolar covalent bonds. Such bonds are also important in living organisms. The unique nature of carbon atoms and the organic molecules formed through covalent bonds between carbon atoms provides the chemical foundation of life.

When covalent bonds are formed between two different atoms, the electrons may be pulled more toward one atom than the other. The end of the molecule toward which the electrons are pulled is electrically negative compared to the other end. Such a molecule is said to be **polar** (has a positive and negative "pole"). Atoms of oxygen, nitrogen, and phosphorus have a particularly strong tendency to pull electrons

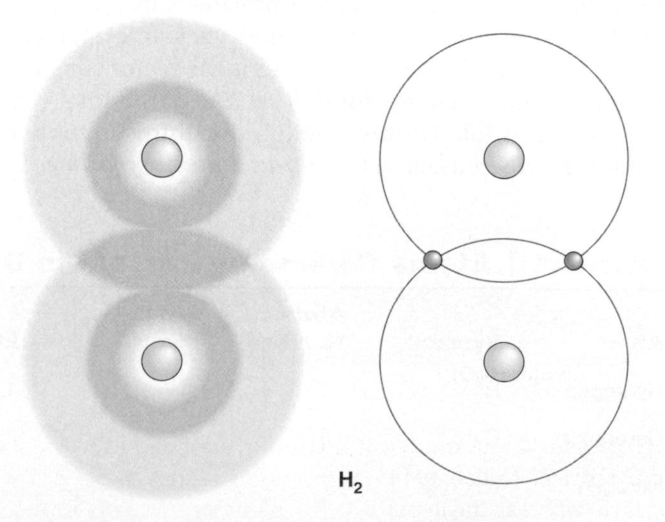

Figure 2.2 A hydrogen molecule showing the covalent bonds between hydrogen atoms. These bonds are formed by the equal sharing of electrons. AP|R

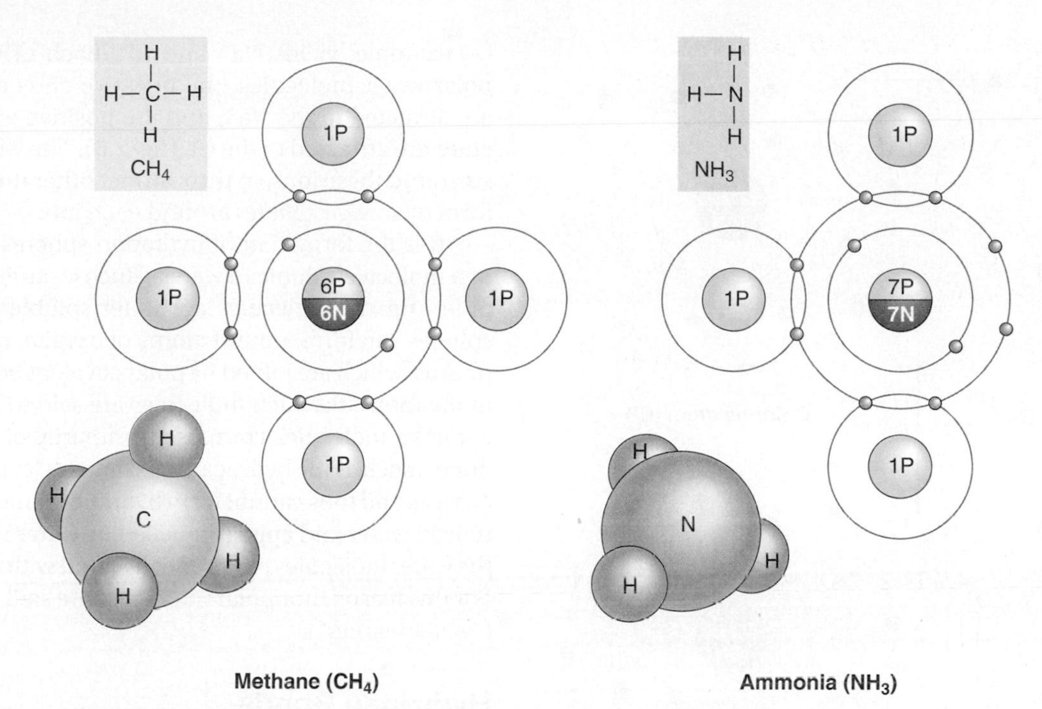

Figure 2.3 The molecules methane and ammonia represented in three different ways. Notice that a bond between 2 atoms consists of a pair of shared electrons (the electrons from the outer shell of each atom). AP|R

toward themselves when they bond with other atoms; thus, they tend to form polar molecules.

Water is the most abundant molecule in the body and serves as the solvent for body fluids. Water is a good solvent because it is polar; the oxygen atom pulls electrons from the 2 hydrogens toward its side of the water molecule, so that the oxygen side is more negatively charged than the hydrogen side of the molecule (fig. 2.4). The significance of the polar nature of water in its function as a solvent is discussed in the next section.

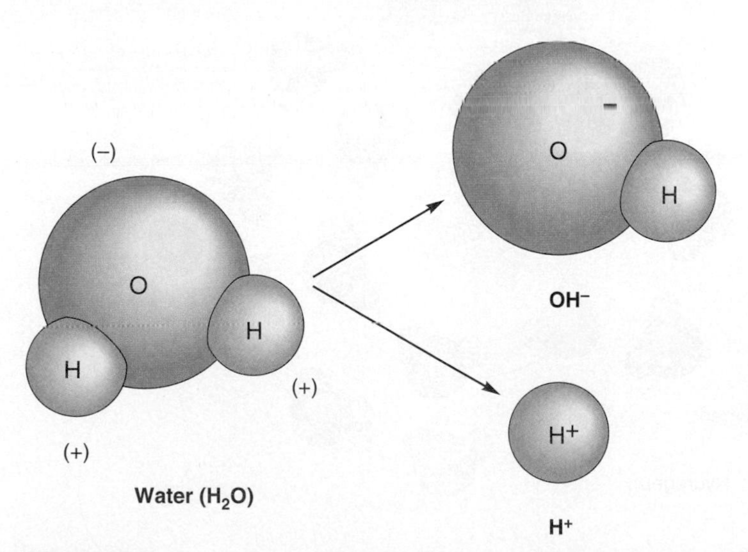

Figure 2.4 A model of a water molecule showing its polar nature. Notice that the oxygen side of the molecule is negative, whereas the hydrogen side is positive. Polar covalent bonds are weaker than nonpolar covalent bonds. As a result, some water molecules ionize to form a hydroxide ion (OH^-) and a hydrogen ion (H^+). AP|R

Ionic Bonds

Ionic bonds result when one or more valence electrons from one atom are completely transferred to a second atom. Thus, the electrons are not shared at all. The first atom loses electrons so that its number of electrons becomes smaller than its number of protons; it becomes positively charged. Atoms or molecules that have positive or negative charges are called **ions.** Positively charged ions are called *cations* because they move toward the negative pole, or cathode, in an electric field. The second atom now has more electrons than it has protons and becomes a negatively charged ion, or *anion* (so called because it moves toward the positive pole, or anode, in an electric field). The cation and anion then attract each other to form an **ionic compound.**

Common table salt, sodium chloride (NaCl), is an example of an ionic compound. Sodium, with a total of 11 electrons, has 2 in its first shell, 8 in its second shell, and only 1 in its third shell. Chlorine, conversely, is 1 electron short of completing its outer shell of 8 electrons. The lone electron in sodium's outer shell is attracted to chlorine's outer shell. This creates a chloride ion (represented as Cl^-) and a sodium ion (Na^+). Although table salt is shown as NaCl, it is actually composed of Na^+Cl^- (fig. 2.5).

Ionic compounds are held together by the attraction of opposite charges, and these compounds easily dissociate (separate) when they are dissolved in water. Dissociation of NaCl,

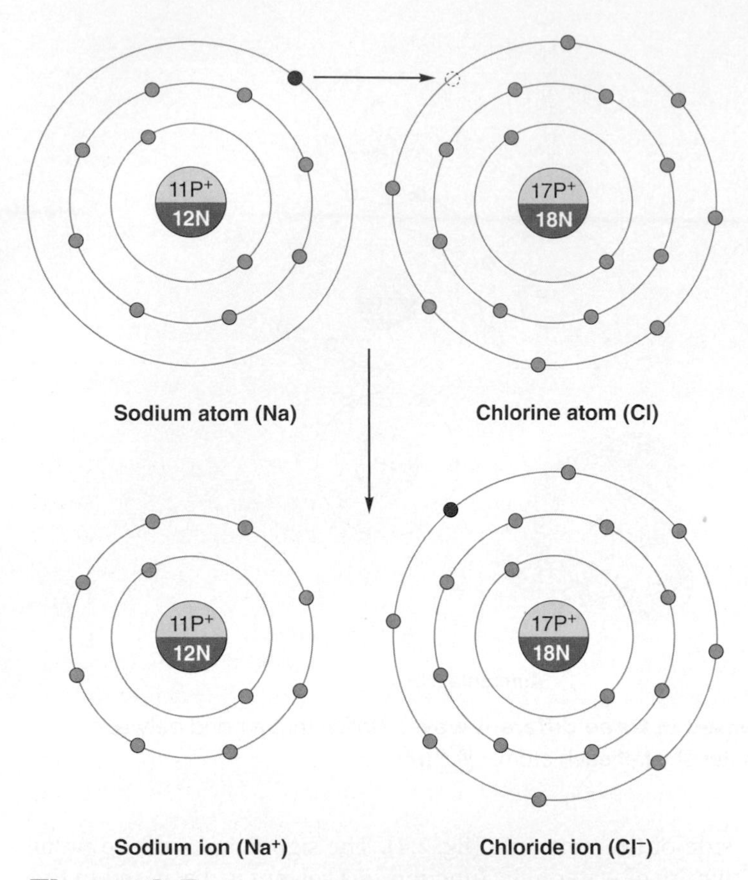

Figure 2.5 The reaction of sodium with chlorine to produce sodium and chloride ions. The positive sodium and negative chloride ions attract each other, producing the ionic compound sodium chloride (NaCl). AP|R

for example, yields Na^+ and Cl^-. Each of these ions attracts polar water molecules; the negative ends of water molecules are attracted to the Na^+, and the positive ends of water molecules are attracted to the Cl^- (fig. 2.6). The water molecules that surround these ions, in turn, attract other molecules of water to form *hydration spheres* around each ion.

It is the formation of hydration spheres that makes an ion or a molecule soluble in water. Glucose, amino acids, and many other organic molecules are water soluble because hydration spheres can form around atoms of oxygen, nitrogen, and phosphorus, which are joined by polar covalent bonds to other atoms in the molecule. Such molecules are said to be **hydrophilic.** By contrast, molecules composed primarily of nonpolar covalent bonds, such as the hydrocarbon chains of fat molecules, have few charges and thus cannot form hydration spheres. They are insoluble in water and appear repelled by water molecules (because the water molecules preferentially bond with each other; fig. 2.7). For this reason, nonpolar molecules are said to be **hydrophobic** ("water fearing").

Hydrogen Bonds

When a hydrogen atom forms a polar covalent bond with an atom of oxygen or nitrogen, the hydrogen gains a slight positive charge as its electron is pulled toward the other atom. This other atom is thus described as being *electronegative.* Because the hydrogen has a slight positive charge, it will have a weak attraction for a second electronegative atom (oxygen or nitrogen) that may be located near it. This weak attraction is called a **hydrogen bond.** Hydrogen bonds are usually shown with dashed or dotted lines

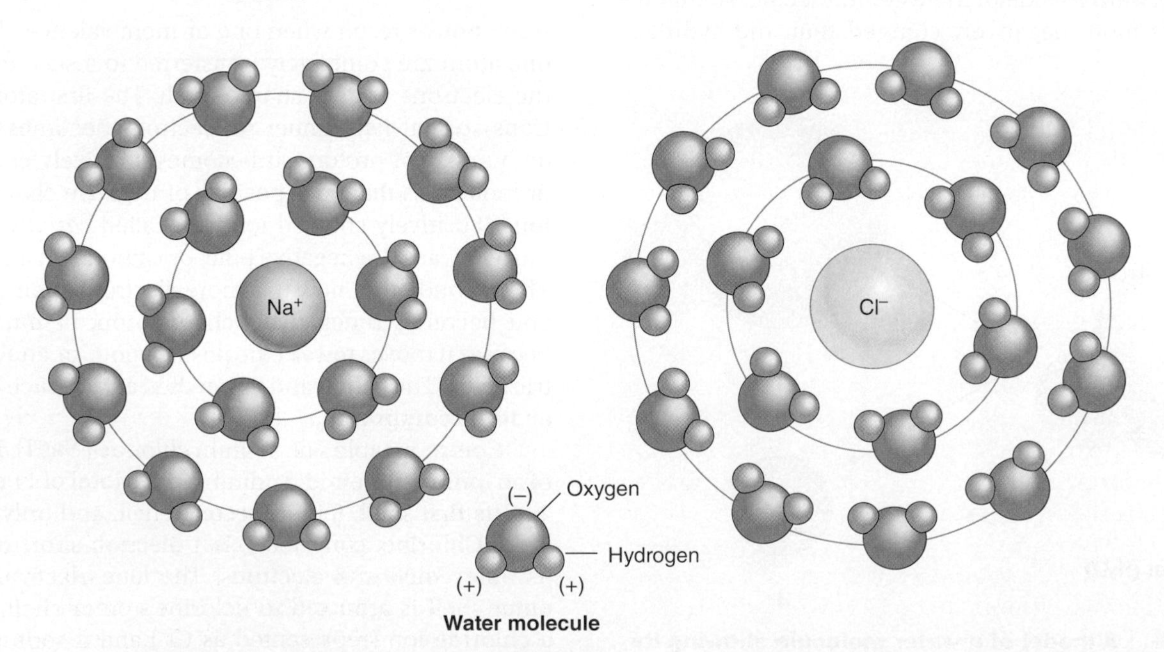

Figure 2.6 How NaCl dissolves in water. The negatively charged oxygen-ends of water molecules are attracted to the positively charged Na^+, whereas the positively charged hydrogen-ends of water molecules are attracted to the negatively charged Cl^-. Other water molecules are attracted to this first concentric layer of water, forming hydration spheres around the sodium and chloride ions. AP|R

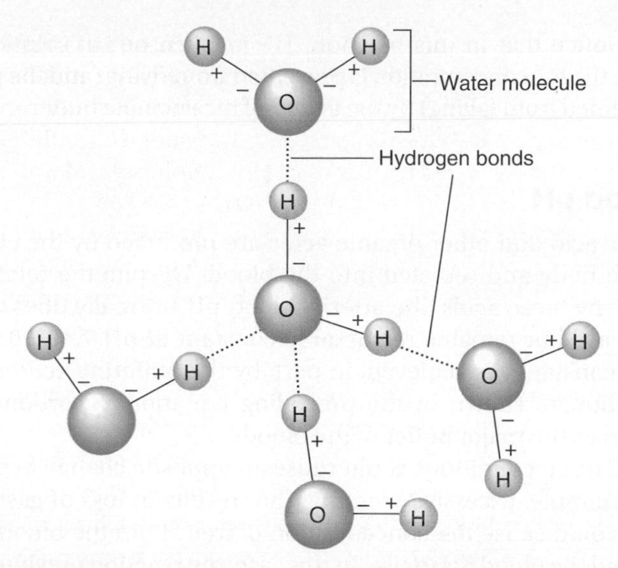

Figure 2.7 Hydrogen bonds between water molecules. The oxygen atoms of water molecules are weakly joined together by the attraction of the negatively charged oxygen for the positively charged hydrogen. These weak bonds are called hydrogen bonds. AP|R

(fig. 2.7) to distinguish them from strong covalent bonds, which are shown with solid lines.

Although each hydrogen bond is relatively weak, the sum of their attractive forces is largely responsible for the folding and bending of long organic molecules such as proteins and for the holding together of the two strands of a DNA molecule (described in section 2.4). Hydrogen bonds can also be formed between adjacent water molecules (fig. 2.7). The hydrogen bonding between water molecules is responsible for many of the biologically important properties of water, including its *surface tension* and its ability to be pulled as a column through narrow channels in a process called *capillary action.*

Acids, Bases, and the pH Scale

The bonds in water molecules joining hydrogen and oxygen atoms together are, as previously discussed, polar covalent bonds. Although these bonds are strong, a small proportion of them break as the electron from the hydrogen atom is completely transferred to oxygen. When this occurs, the water molecule ionizes to form a *hydroxide ion* (OH^-) and a hydrogen ion (H^+), which is simply a free proton (see fig. 2.4). A proton released in this way does not remain free for long, however, because it is attracted to the electrons of oxygen atoms in water molecules. This forms a *hydronium ion,* shown by the formula H_3O^+. For the sake of clarity in the following discussion, however, H^+ will be used to represent the ion resulting from the ionization of water.

Ionization of water molecules produces equal amounts of OH^- and H^+. Only a small proportion of water molecules ionize,

Table 2.2 | Common Acids and Bases

Acid	Symbol	Base	Symbol
Hydrochloric acid	HCl	Sodium hydroxide	NaOH
Phosphoric acid	H_3PO_4	Potassium hydroxide	KOH
Nitric acid	HNO_3	Calcium hydroxide	$Ca(OH)_2$
Sulfuric acid	H_2SO_4	Ammonium hydroxide	NH_4OH
Carbonic acid	H_2CO_3		

so the concentrations of H^+ and OH^- are each equal to only 10^{-7} molar (the term *molar* is a unit of concentration, described in chapter 6; for hydrogen, 1 molar equals 1 gram per liter). A solution with 10^{-7} molar hydrogen ion, which is produced by the ionization of water molecules in which the H^+ and OH^- concentrations are equal, is said to be **neutral.**

A solution that has a higher H^+ concentration than that of water is called *acidic;* one with a lower H^+ concentration is called *basic,* or *alkaline.* An **acid** is defined as a molecule that can release protons (H^+) into a solution; it is a "proton donor." A **base** can be a molecule such as ammonia (NH_3) that can combine with H^+ (to form NH_4^+, ammonium ion). More commonly, it is a molecule such as NaOH that can ionize to produce a negatively charged ion (hydroxide, OH^-), which, in turn, can combine with H^+ (to form H_2O, water). A base thus removes H^+ from solution; it is a "proton acceptor," thereby lowering the H^+ concentration of the solution. Examples of common acids and bases are shown in table 2.2.

pH

The H^+ concentration of a solution is usually indicated in pH units on a pH scale that runs from 0 to 14. The pH value is equal to the logarithm of 1 over the H^+ concentration:

$$pH = \log \frac{1}{[H^+]}$$

where $[H^+]$ = molar H^+ concentration. This can also be expressed as $pH = -\log [H^+]$.

Pure water has a H^+ concentration of 10^{-7} molar at 25°C, and thus has a pH of 7 (neutral). Because of the logarithmic relationship, a solution with 10 times the hydrogen ion concentration (10^{-6} M) has a pH of 6, whereas a solution with one-tenth the H^+ concentration (10^{-8} M) has a pH of 8. The pH value is easier to write than the molar H^+ concentration, but it is admittedly confusing because it is *inversely related* to the H^+ concentration. That is, a solution with a higher H^+ concentration has a lower pH value, and one with a lower H^+ concentration has a higher pH value. A strong acid with a high H^+ concentration of 10^{-2} molar, for example, has a pH of 2, whereas a solution with only 10^{-10} molar H^+ has a pH of 10. **Acidic solutions,** therefore, have a pH of less than 7 (that

Table 2.3 | The pH Scale

	H^+ Concentration (Molar)*	pH	OH^- Concentration (Molar)*
Acids	1.0	0	10^{-14}
	0.1	1	10^{-13}
	0.01	2	10^{-12}
	0.001	3	10^{-11}
	0.0001	4	10^{-10}
	10^{-5}	5	10^{-9}
	10^{-6}	6	10^{-8}
Neutral	10^{-7}	7	10^{-7}
Bases	10^{-8}	8	10^{-6}
	10^{-9}	9	10^{-5}
	10^{-10}	10	0.0001
	10^{-11}	11	0.001
	10^{-12}	12	0.01
	10^{-13}	13	0.1
	10^{-14}	14	1.0

*Molar concentration is the number of moles of a solute dissolved in one liter. One mole is the atomic or molecular weight of the solute in grams. Since hydrogen has an atomic weight of one, one molar hydrogen is one gram of hydrogen per liter of solution.

of pure water), whereas **basic (alkaline) solutions** have a pH between 7 and 14 (table 2.3).

Buffers

A **buffer** is a system of molecules and ions that acts to prevent changes in H^+ concentration and thus serves to stabilize the pH of a solution. In blood plasma, for example, the pH is stabilized by the following reversible reaction involving the bicarbonate ion (HCO_3^-) and carbonic acid (H_2CO_3):

$$HCO_3^- + H^+ \rightleftarrows H_2CO_3$$

The double arrows indicate that the reaction could go either to the right or to the left; the net direction depends on the concentration of molecules and ions on each side. If an acid (such as lactic acid) should release H^+ into the solution, for example, the increased concentration of H^+ would drive the equilibrium to the right and the following reaction would be promoted:

$$HCO_3^- + H^+ \rightarrow H_2CO_3$$

Notice that in this reaction, H^+ is taken out of solution. Thus, the H^+ concentration is prevented from rising (and the pH prevented from falling) by the action of bicarbonate buffer.

Blood pH

Lactic acid and other organic acids are produced by the cells of the body and secreted into the blood. Despite the release of H^+ by these acids, the arterial blood pH normally does not decrease but remains remarkably constant at pH 7.40 ± 0.05. This constancy is achieved, in part, by the buffering action of bicarbonate shown in the preceding equation. Bicarbonate serves as the major buffer of the blood.

Certain conditions could cause an opposite change in pH. For example, excessive vomiting that results in loss of gastric acid could cause the concentration of free H^+ in the blood to fall and the blood pH to rise. In this case, the reaction previously described could be reversed:

$$H_2CO_3 \rightarrow H^+ + HCO_3^-$$

The dissociation of carbonic acid yields free H^+, which helps to prevent an increase in pH. Bicarbonate ions and carbonic acid thus act as a *buffer pair* to prevent either decreases or increases in pH, respectively. This buffering action normally maintains the blood pH within the narrow range of 7.35 to 7.45.

If the arterial blood pH falls below 7.35, the condition is called *acidosis*. A blood pH of 7.20, for example, represents significant acidosis. Notice that acidotic blood need not be acidic (have a pH less than 7.00). An increase in blood pH above 7.45, conversely, is known as *alkalosis*. Acidosis and alkalosis are normally prevented by the action of the bicarbonate/carbonic acid buffer pair and by the functions of the lungs and kidneys. Regulation of blood pH is discussed in more detail in chapters 16 and 17.

Organic Molecules

Organic molecules are those molecules that contain the atoms carbon and hydrogen. Because the carbon atom has 4 electrons in its outer shell, it must share 4 additional electrons by covalently bonding with other atoms to fill its outer shell with 8 electrons. The unique bonding requirements of carbon enable it to join with other carbon atoms to form chains and rings while still allowing the carbon atoms to bond with hydrogen and other atoms.

Most organic molecules in the body contain hydrocarbon chains and rings, as well as other atoms bonded to carbon. Two adjacent carbon atoms in a chain or ring may share one or two pairs of electrons. If the 2 carbon atoms share one pair of electrons, they are said to have a *single covalent bond;* this leaves each carbon atom free to bond with as many as 3 other atoms. If the 2 carbon atoms share two pairs of

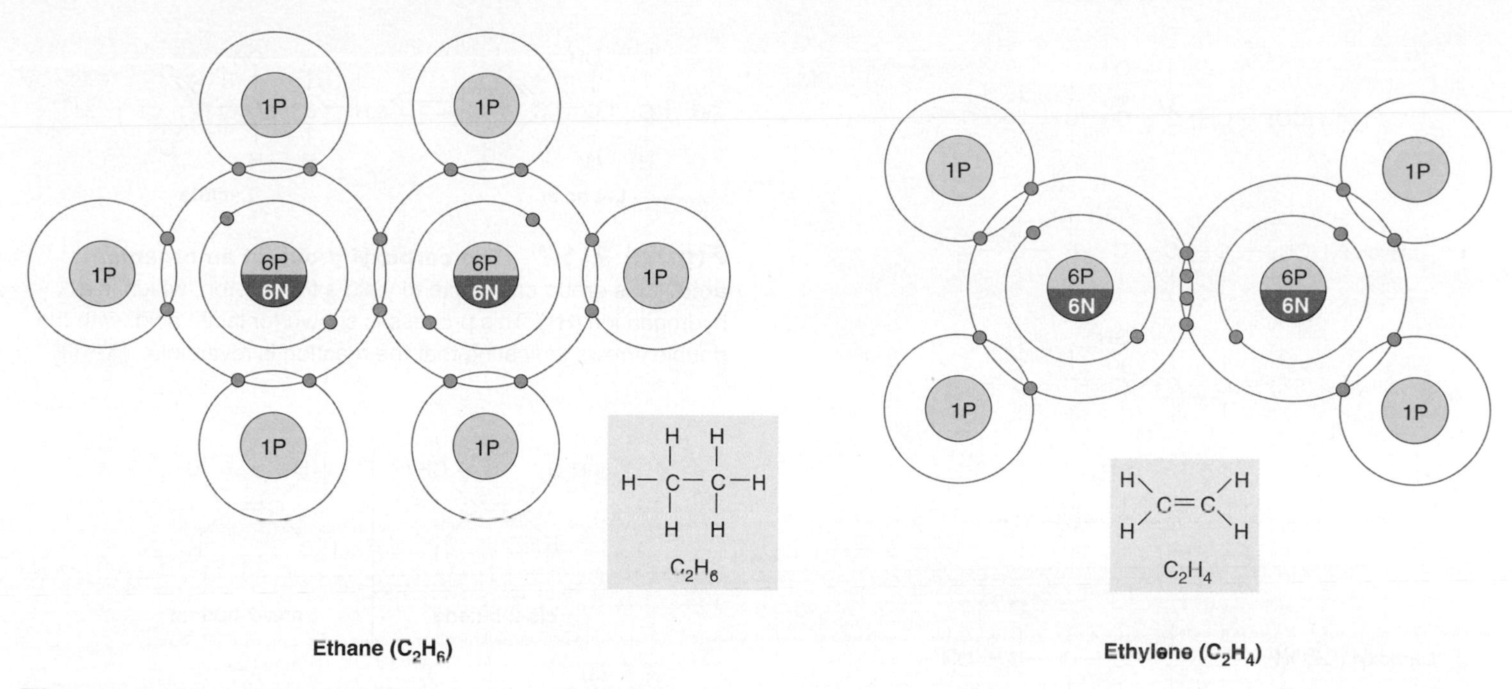

Figure 2.8 Single and double covalent bonds. Two carbon atoms may be joined by a single covalent bond (*left*) or a double covalent bond (*right*). In both cases, each carbon atom shares four pairs of electrons (has four bonds) to complete the 8 electrons required to fill its outer shell. AP|R

electrons, they have a *double covalent bond,* and each carbon atom can bond with a maximum of only 2 additional atoms (fig. 2.8).

The ends of some hydrocarbons are joined together to form rings. In the shorthand structural formulas for these molecules, the carbon atoms are not shown but are understood to be located at the corners of the ring. Some of these cyclic molecules have a double bond between 2 adjacent carbon atoms. Benzene and related molecules are shown as a six-sided ring with alternating double bonds. Such compounds are called **aromatic.** Because all of the carbons in an aromatic ring are equivalent, double bonds can be shown between any 2 adjacent carbons in the ring (fig. 2.9), or even as a circle within the hexagonal structure of carbons.

The hydrocarbon chain or ring of many organic molecules provides a relatively inactive molecular "backbone" to which more reactive groups of atoms are attached. Known as *functional groups* of the molecule, these reactive groups usually contain atoms of oxygen, nitrogen, phosphorus, or sulfur. They are largely responsible for the unique chemical properties of the molecule (fig. 2.10).

Classes of organic molecules can be named according to their functional groups. **Ketones,** for example, have a *carbonyl group* within the carbon chain. An organic molecule is an **alcohol** if it has a *hydroxyl group* bound to a hydrocarbon chain. All **organic acids** (acetic acid, citric acids, lactic acid, and others) have a *carboxyl group* (fig. 2.11).

A carboxyl group can be abbreviated COOH. This group is an acid because it can donate its proton (H^+) to the solution. Ionization of the OH part of COOH forms COO^- and H^+

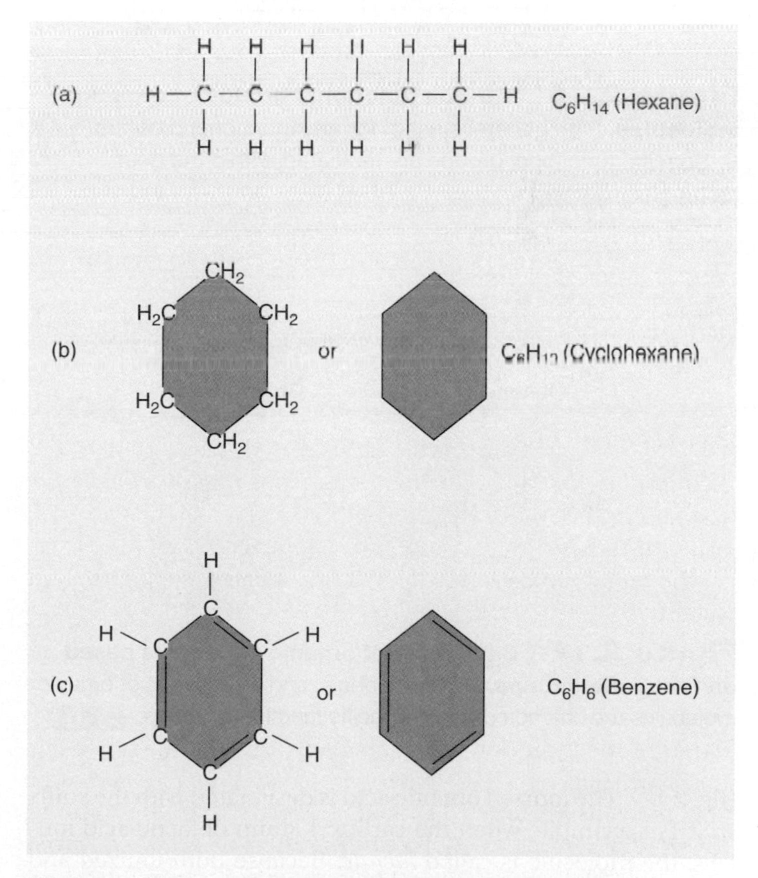

Figure 2.9 Different shapes of hydrocarbon molecules. Hydrocarbon molecules can be (*a*) linear or (*b*) cyclic or have (*c*) aromatic rings. AP|R

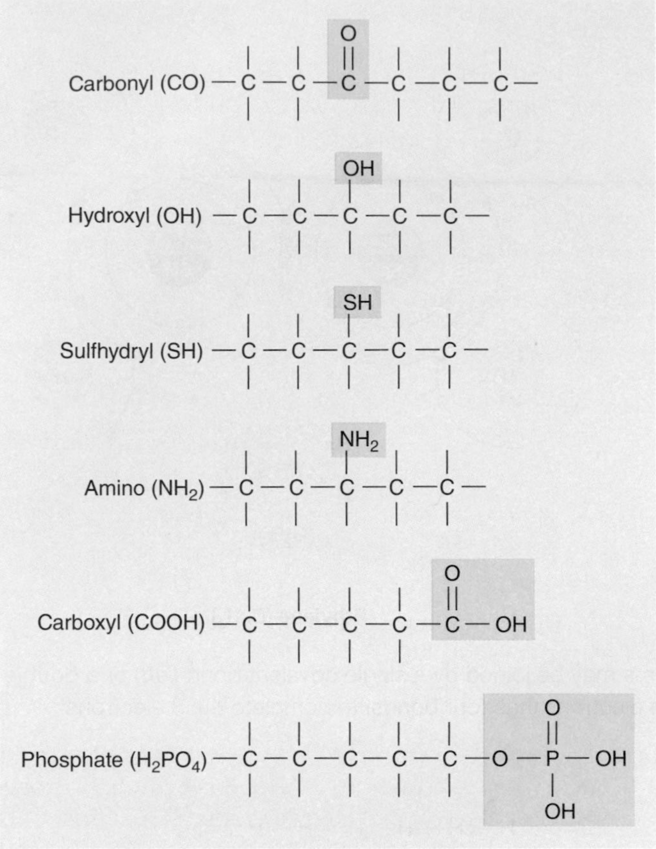

Figure 2.10 Various functional groups of organic molecules. The general symbol for any functional group is *R*. Specific functional groups are indicated within the gold rectangles. APIR

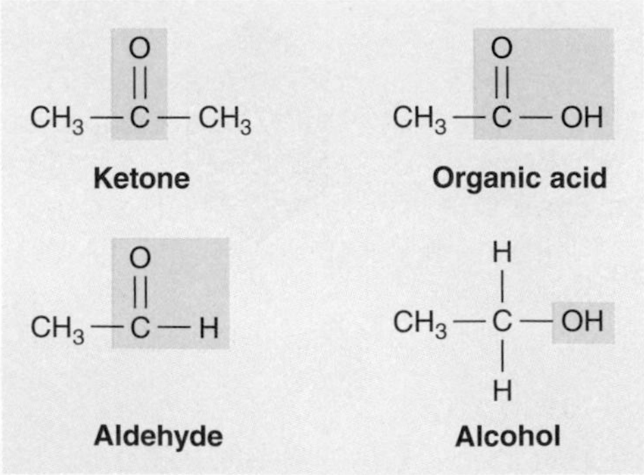

Figure 2.11 Categories of organic molecules based on functional groups. Acids, alcohols, and other types of organic molecules are characterized by specific functional groups. APIR

(fig. 2.12). The ionized organic acid is designated with the suffix *-ate*. For example, when the carboxyl group of lactic acid ionizes, the molecule is called *lactate*. Because both ionized and unionized forms of the molecule exist together in a solution (the proportion of each depends on the pH of the solution), one can correctly refer to the molecule as either lactic acid or lactate.

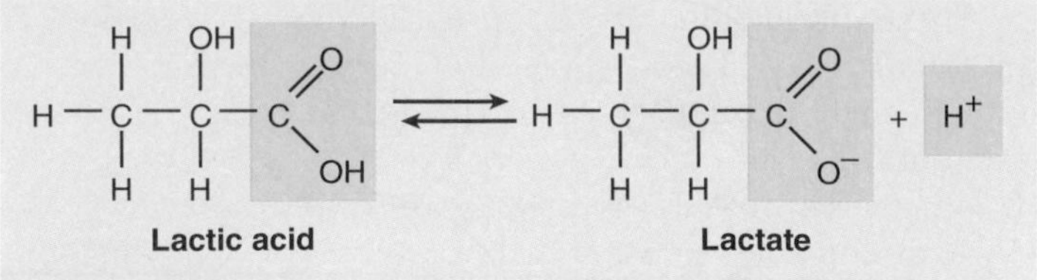

Figure 2.12 The carboxyl group of an organic acid. This group can ionize to yield a free proton, which is a hydrogen ion (H^+). This process is shown for lactic acid, with the double arrows indicating that the reaction is reversible. APIR

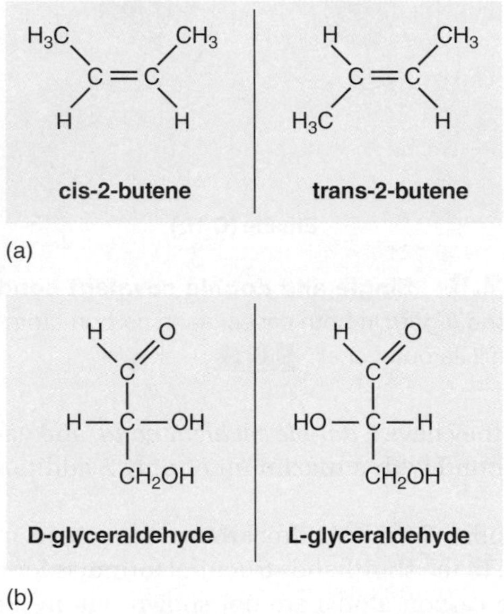

Figure 2.13 Stereoisomers. (*a*) Stereoisomers include molecules such as butene, which can have its methyl groups (CH_3) on either the same side (*cis*) or the opposite side (*trans*) of the molecule. (*b*) Other stereoisomers are enantiomers (optical isomers), which are mirror images of each other. Using the molecule glyceraldehyde as a reference, these can be designated as either *D* or *L* isomers. APIR

Stereoisomers

Stereoisomers are molecules that have the same atoms in the same sequence, but differ from each other in the way their atoms are arranged three-dimensionally in space. Stereoisomers include (1) those that are designated as *cis* (Latin for "on this side") and *trans* (meaning "across"), where two functional groups are located either on the same side (*cis*) or across from each other (*trans*) in the molecule (fig. 2.13*a*); and (2) *enantiomers* (also called *optical isomers*) that are mirror images of each other (fig. 2.13*b*). Enantiomers are like a left- and right-hand glove; if the palms are both facing in the same direction, they cannot be superimposed on each other. By convention, one enantiomer is often designated the *D*-isomer (for *dextro*, or right-handed) and the other is designated the *L*-isomer (for *levo*, or left-handed).

These subtle differences in structure are extremely important biologically. They ensure that enzymes—which interact with such molecules in a stereo-specific way in chemical reactions—cannot combine with the "wrong" stereoisomer. The enzymes of all cells (human and others) can combine only with L-amino acids and D-sugars, for example. The opposite stereoisomers (D-amino acids and L-sugars) cannot be used by any enzyme in metabolism.

CLINICAL APPLICATION

Severe birth defects often resulted when pregnant women used the sedative **thalidomide** in the early 1960s to alleviate morning sickness. The drug contains a mixture of both right-handed (D) and left-handed (L) forms. This tragic circumstance emphasizes the clinical importance of stereoisomers. It has since been learned that the L-stereoisomer is a potent tranquilizer, but the right handed version causes disruption of fetal development and the resulting birth defects. Interestingly, thalidomide is now being used in the treatment of people with AIDS, leprosy, and *cachexia* (prolonged ill health and malnutrition).

Clinical Investigation CLUES

George ate only D-amino acids and L-sugars that he obtained from the chemistry lab.

- What are these, and how do they relate to the amino acids and sugars normally found in food?
- What would be his nutritional status as a result of this diet?

CHECKPOINTS

1. List the components of an atom and explain how they are organized. Explain why different atoms are able to form characteristic numbers of chemical bonds.
2. Describe the nature of nonpolar and polar covalent bonds, ionic bonds, and hydrogen bonds. Why are ions and polar molecules soluble in water?

3a. Define the terms *acidic, basic, acid,* and *base.* Also define *pH* and describe the relationship between pH and the H^+ concentration of a solution.

3b. Using chemical equations, explain how bicarbonate ion and carbonic acid function as a buffer pair.

4. Explain how carbon atoms can bond with each other and with atoms of hydrogen, oxygen, and nitrogen.

2.2 CARBOHYDRATES AND LIPIDS

Carbohydrates are a class of organic molecules that includes monosaccharides, disaccharides, and polysaccharides. All of these molecules are based on a characteristic ratio of carbon, hydrogen, and oxygen atoms. Lipids constitute a category of diverse organic molecules that share the physical property of being nonpolar, and thus insoluble in water.

LEARNING OUTCOMES

After studying this section, you should be able to:

5. **Identify the different types of carbohydrates and lipids, and give examples of each type.**
6. **Explain how dehydration synthesis and hydrolysis reactions occur in carbohydrates and triglycerides.**
7. **Describe the nature of phospholipids and prostaglandins.**

Carbohydrates and lipids are similar in many ways. Both groups of molecules consist primarily of the atoms carbon, hydrogen, and oxygen, and both serve as major sources of energy in the body (accounting for most of the calories consumed in food). Carbohydrates and lipids differ, however, in some important aspects of their chemical structures and physical properties. Such differences significantly affect the functions of these molecules in the body.

Carbohydrates

Carbohydrates are organic molecules that contain carbon, hydrogen, and oxygen in the ratio described by their name—*carbo* (carbon) and *hydrate* (water, H_2O). The general formula for a carbohydrate molecule is thus $C_nH_{2n}O_n$; the molecule contains twice as many hydrogen atoms as carbon or oxygen atoms (the number of each is indicated by the subscript n).

Monosaccharides, Disaccharides, and Polysaccharides

Carbohydrates include simple sugars, or **monosaccharides,** and longer molecules that contain a number of monosaccharides joined together. The suffix *-ose* denotes a sugar molecule; the term *hexose,* for example, refers to a six-carbon monosaccharide with the formula $C_6H_{12}O_6$. This formula is adequate for some purposes, but it does not distinguish between related hexose sugars, which are *structural isomers* of each other. The structural isomers glucose, galactose, and fructose, for example, are monosaccharides that have the same ratio of atoms arranged in slightly different ways (fig. 2.14).

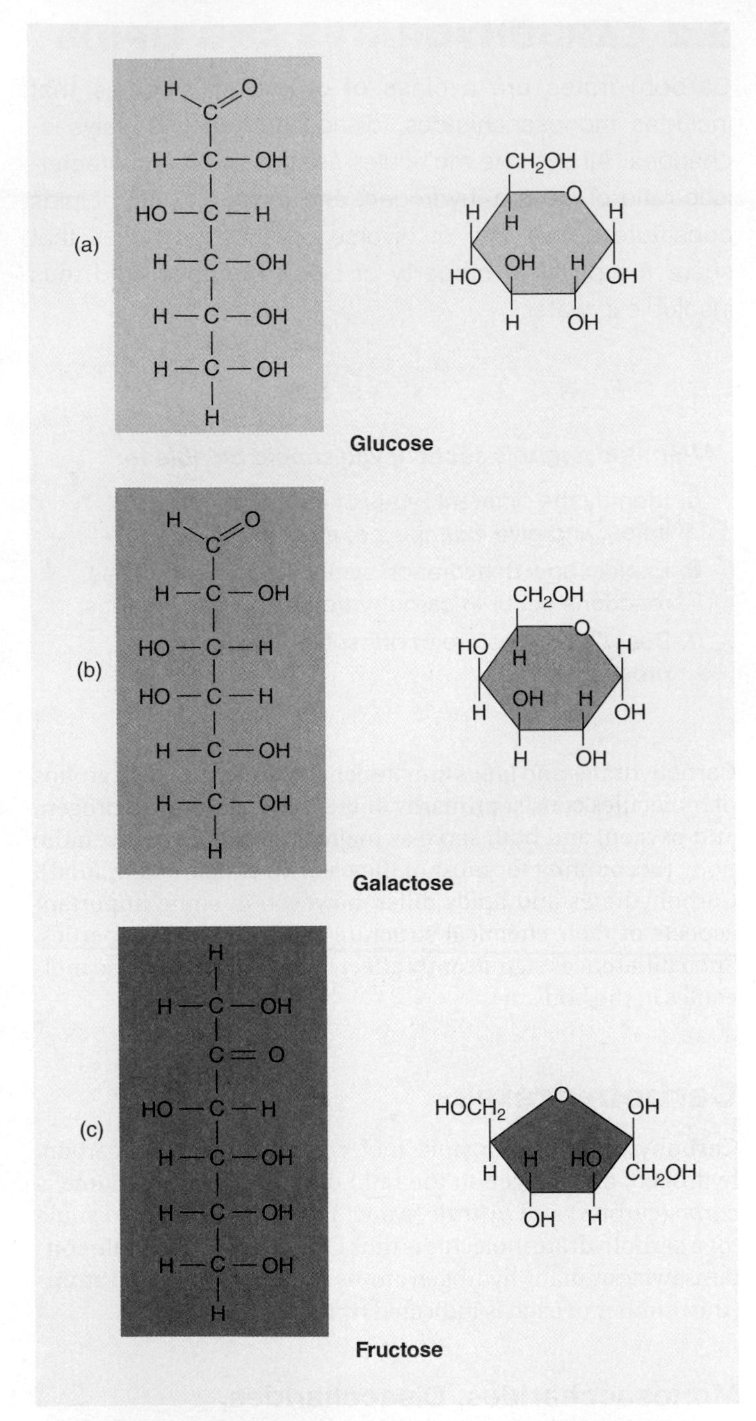

Figure 2.14 Structural formulas for three hexose sugars. These are (*a*) glucose, (*b*) galactose, and (*c*) fructose. All three have the same ratio of atoms—$C_6H_{12}O_6$. The representations on the left more clearly show the atoms in each molecule, while the ring structures on the right more accurately reflect the way these atoms are arranged. **AP|R**

See the *Test Your Quantitative Ability* section of the Review Activities at the end of this chapter.

Two monosaccharides can be joined covalently to form a **disaccharide,** or double sugar. Common disaccharides include table sugar, or *sucrose* (composed of glucose and fructose); milk sugar, or *lactose* (composed of glucose and galactose); and malt sugar, or *maltose* (composed of two glucose molecules). When numerous monosaccharides are joined together, the resulting molecule is called a **polysaccharide.**

The major polysaccharides are chains of repeating glucose subunits. *Starch* is a plant product formed by the bonding together of thousands of glucose subunits into long chains, and **glycogen** (sometimes called animal starch) is similar, but more highly branched (fig. 2.15). Animals have the enzymes to digest the bonds (chemically called alpha-1,4 glycosidic bonds) between adjacent glucose subunits of these polysaccharides. *Cellulose* (produced by plants) is also a polysaccharide of glucose, but the bonds joining its glucose subunits are oriented differently (forming beta-1,4 glycosidic bonds) than those in starch or glycogen. Because of this, our digestive enzymes cannot hydrolyze cellulose into its glucose subunits. However, animals such as cows, horses, and sheep—which eat grasses—can digest cellulose because they have symbiotic bacteria with the necessary enzymes in their digestive tracts. *Chitin* (poly-N-acetylglucosamine) is a polysaccharide similar to cellulose (with beta-1,4 glycosidic bonds) but with amine-containing groups in the glucose subunits. Chitin forms the exoskeleton of arthropods such as insects and crustaceans.

Many cells store carbohydrates for use as an energy source, as described in chapter 5. If many thousands of separate monosaccharide molecules were stored in a cell, however, their high concentration would draw an excessive amount of water into the cell, damaging or even killing it. The net movement of water through membranes is called osmosis, and is discussed in chapter 6. Cells that store carbohydrates for energy minimize this osmotic damage by instead joining the glucose molecules together to form the polysaccharides starch or glycogen. Because there are fewer of these larger molecules, less water is drawn into the cell by osmosis (see chapter 6).

Dehydration Synthesis and Hydrolysis

In the formation of disaccharides and polysaccharides, the separate subunits (monosaccharides) are bonded together covalently by a type of reaction called **dehydration synthesis,** or **condensation.** In this reaction, which requires the participation of specific enzymes (chapter 4), a hydrogen atom is removed from one monosaccharide and a hydroxyl group (OH) is removed from another. As a covalent bond is formed between the two monosaccharides, water (H_2O) is produced. Dehydration synthesis reactions are illustrated in figure 2.16.

When a person eats disaccharides or polysaccharides, or when the stored glycogen in the liver and muscles is to be used by tissue cells, the covalent bonds that join monosaccharides to form disaccharides and polysaccharides must be broken. These *digestion reactions* occur by means of **hydrolysis.** Hydrolysis (from the Greek *hydro* = water; *lysis* = break) is the reverse of dehydration synthesis. When a covalent bond

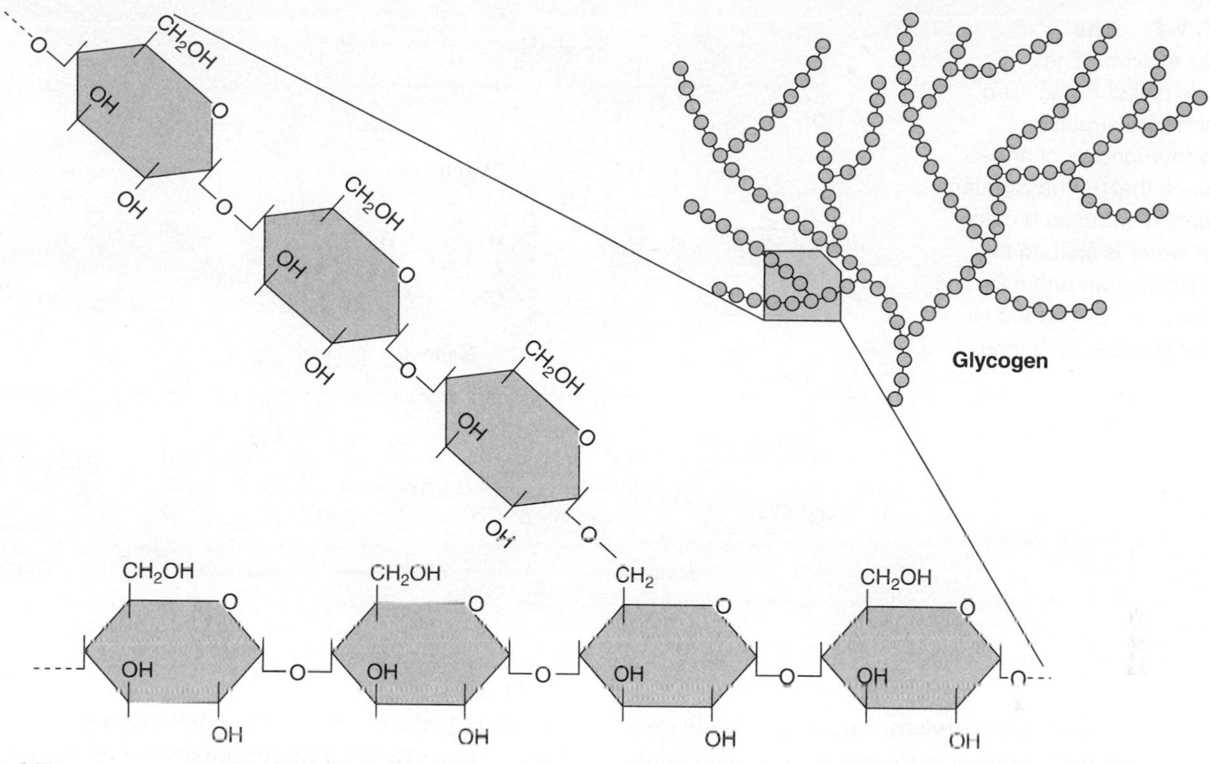

Figure 2.15 The structure of glycogen. Glycogen is a polysaccharide composed of glucose subunits joined together to form a large, highly branched molecule. AP|R

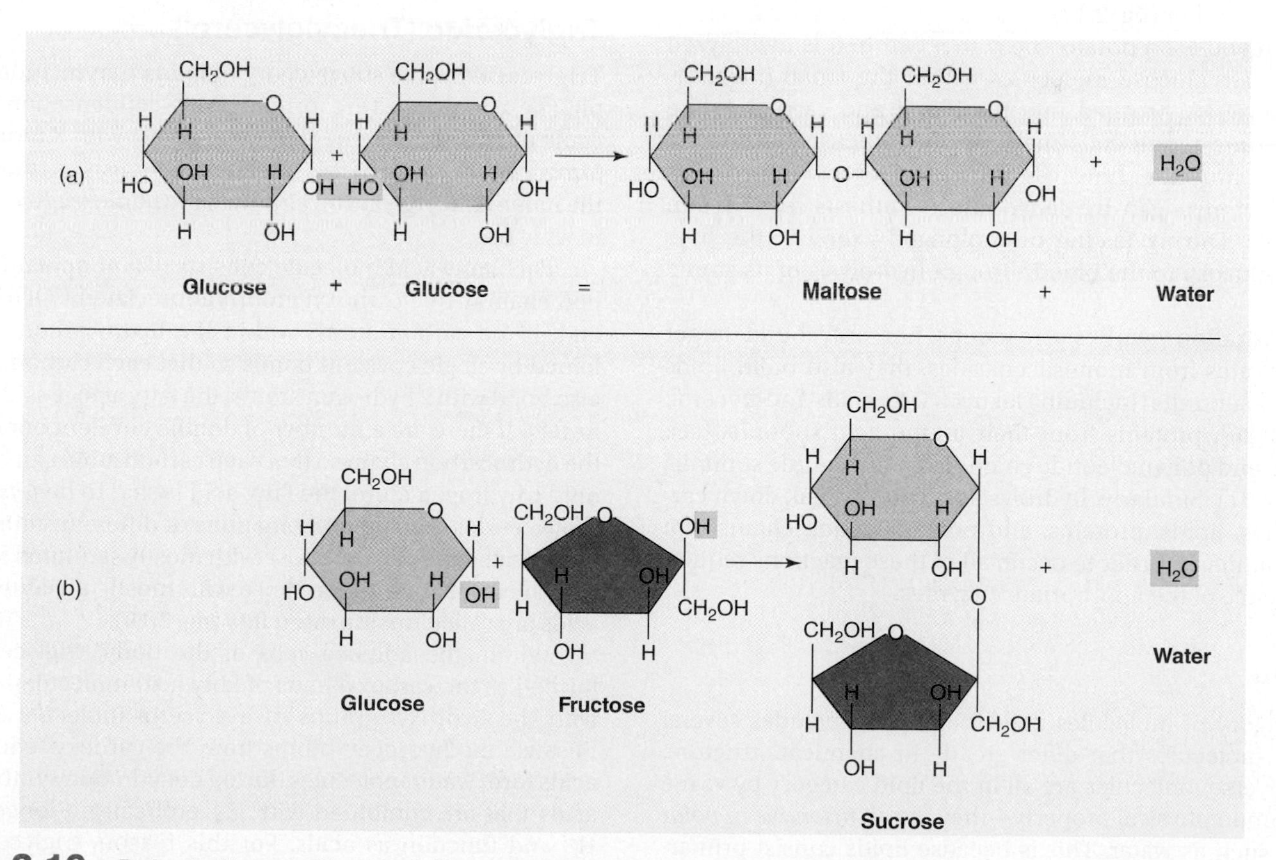

Figure 2.16 Dehydration synthesis of disaccharides. The two disaccharides formed here are (*a*) maltose and (*b*) sucrose (table sugar). Notice that a molecule of water is produced as the disaccharides are formed.

See *Test Your Quantitative Ability* section of the Review Activities at the end of this chapter.

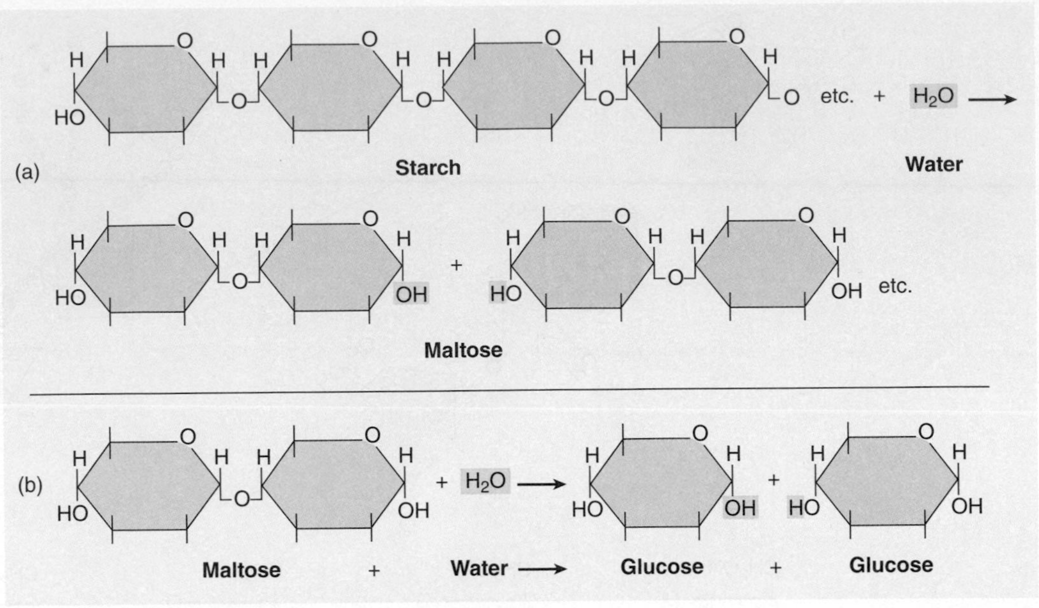

Figure 2.17 The hydrolysis of starch. The polysaccharide is first hydrolyzed into (*a*) disaccharides (maltose) and then into (*b*) monosaccharides (glucose). Notice that as the covalent bond between the subunits breaks, a molecule of water is split. In this way, the hydrogen atom and hydroxyl group from the water are added to the ends of the released subunits.

joining two monosaccharides is broken, a water molecule provides the atoms needed to complete their structure. The water molecule is split, and the resulting hydrogen atom is added to one of the free glucose molecules as the hydroxyl group is added to the other (fig. 2.17).

When you eat a potato, the starch within it is hydrolyzed into separate glucose molecules within the small intestine. This glucose is absorbed into the blood and carried to the tissues. Some tissue cells may use this glucose for energy. Liver and muscles, however, can store excess glucose in the form of glycogen by dehydration synthesis reactions in these cells. During fasting or prolonged exercise, the liver can add glucose to the blood through hydrolysis of its stored glycogen.

Dehydration synthesis reactions not only build larger carbohydrates from monosaccharides, they also build lipids from their subunits (including fat from fatty acids and glycerol; see fig. 2.20), proteins from their amino acid subunits (see fig. 2.27), and polynucleotide chains from nucleotide subunits (see fig. 2.31). Similarly, hydrolysis reactions break down carbohydrates, lipids, proteins, and polynucleotide chains into their subunits. In order to occur, all of these reactions require the presence of the appropriate enzymes.

Lipids

The category of molecules known as **lipids** includes several types of molecules that differ greatly in chemical structure. These diverse molecules are all in the lipid category by virtue of a common physical property—they are all *insoluble in polar solvents* such as water. This is because lipids consist primarily of hydrocarbon chains and rings, which are nonpolar and therefore hydrophobic. Although lipids are insoluble in water, they can be dissolved in nonpolar solvents such as ether, benzene, and related compounds.

Triglyceride (Triacylglycerol)

Triglyceride is the subcategory of lipids that includes fat and oil. These molecules are formed by the condensation of 1 molecule of *glycerol* (a three-carbon alcohol) with 3 molecules of *fatty acids.* Because of this structure, chemists currently prefer the name **triacylglycerol,** although the name *triglyceride* is still in wide use.

Each fatty acid molecule consists of a nonpolar hydrocarbon chain with a carboxyl group (abbreviated COOH) on one end. If the carbon atoms within the hydrocarbon chain are joined by single covalent bonds so that each carbon atom can also bond with 2 hydrogen atoms, the fatty acid is said to be *saturated.* If there are a number of double covalent bonds within the hydrocarbon chain so that each carbon atom can bond with only 1 hydrogen atom, the fatty acid is said to be *unsaturated.* Triglycerides contain combinations of different saturated and unsaturated fatty acids. Those with mostly saturated fatty acids are called **saturated fats;** those with mostly unsaturated fatty acids are called **unsaturated fats** (fig. 2.18).

Within the adipose cells of the body, triglycerides are formed as the carboxyl ends of fatty acid molecules condense with the hydroxyl groups of a glycerol molecule (fig. 2.20). Because the hydrogen atoms from the carboxyl ends of fatty acids form water molecules during dehydration synthesis, fatty acids that are combined with glycerol can no longer release H^+ and function as acids. For this reason, triglycerides are described as *neutral fats.*

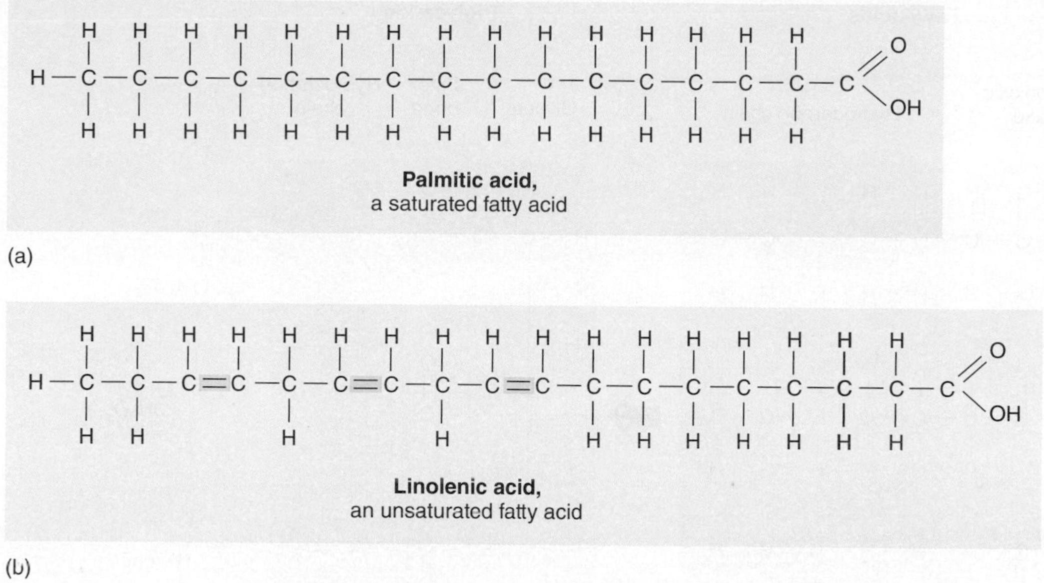

Palmitic acid,
a saturated fatty acid

(a)

Linolenic acid,
an unsaturated fatty acid

(b)

Figure 2.18 Structural formulas for fatty acids. (*a*) The formula for saturated fatty acids and (*b*) the formula for unsaturated fatty acids. Double bonds, which are points of unsaturation, are highlighted in yellow.

FITNESS APPLICATION

The saturated fat content (expressed as a percentage of total fat) for some food items is as follows: canola, or rapeseed, oil (6%); olive oil (14%); margarine (17%); chicken fat (31%); palm oil (51%); beef fat (52%); butter fat (66%); and coconut oil (77%). Health authorities recommend that a person's total fat intake not exceed 30% of the total energy intake per day, and that saturated fat contribute less than 10% of the daily energy intake.

Animal fats, which are solid at room temperatures, are more saturated than vegetable oils, because the hardness of the triglyceride is determined partly by the degree of saturation. *Trans fats,* which are also solid at room temperature, are produced artificially by partially hydrogenating vegetable oils (this is how margarine is made). This results in **trans fatty acids,** in which the single hydrogen atom bonded to each carbon atom is located on the opposite side of the double bond between carbons, and the carbon atoms form a straight chain. By contrast, in most naturally occurring unsaturated fatty acids the hydrogen atoms are on the same side as the double bond (forming *cis fatty acids*), and their carbon atoms bend at the double bonds to produce a sawtoothed pattern (fig. 2.19). Trans fats are used in almost all commercially prepared fried and baked foods. Saturated fat and *trans* fatty acids have been shown to raise LDL cholesterol (the "bad" cholesterol), lower HDL cholesterol (the "good" cholesterol), and thereby to increase the risk of coronary heart disease. The Food and Drug Administration (FDA) now requires all manufacturers to list trans fats on their food labels.

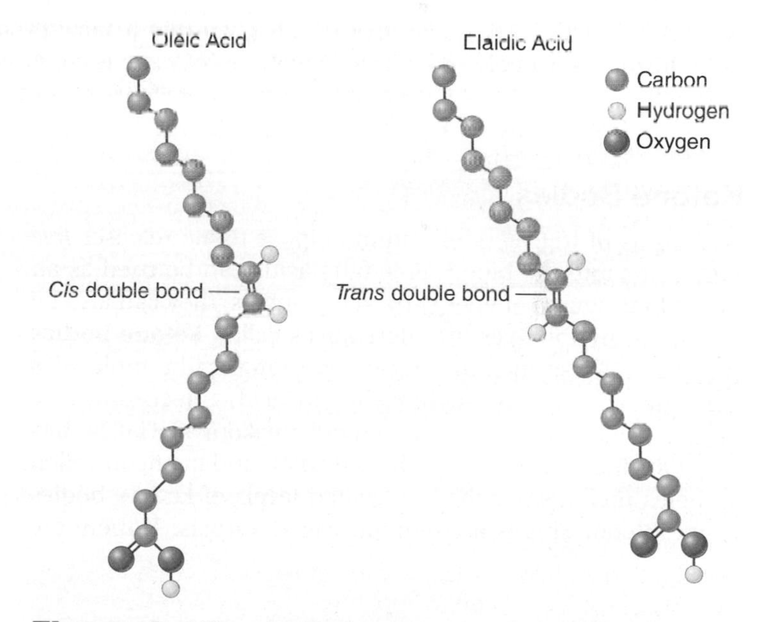

Figure 2.19 The structure of cis and trans fatty acids. Oleic acid is a naturally occurring fatty acid with one double bond. Notice that both hydrogen atoms (yellow) on the carbons that share this double bond are on the same side of the molecule—this is called the *cis* configuration. The cis configuration makes this naturally occurring fatty acid bend. The fatty acid on the right is the same size and also has one double bond, but its hydrogens here are on opposite sides of the molecule, known as the *trans* configuration. This makes the fatty acid stay straight, more like a saturated fatty acid. Note that only these hydrogens and the ones on the carboxyl groups (*bottom*) are shown. Those carbons that are joined by single bonds are also each bonded to 2 hydrogen atoms, but those hydrogens are not illustrated.

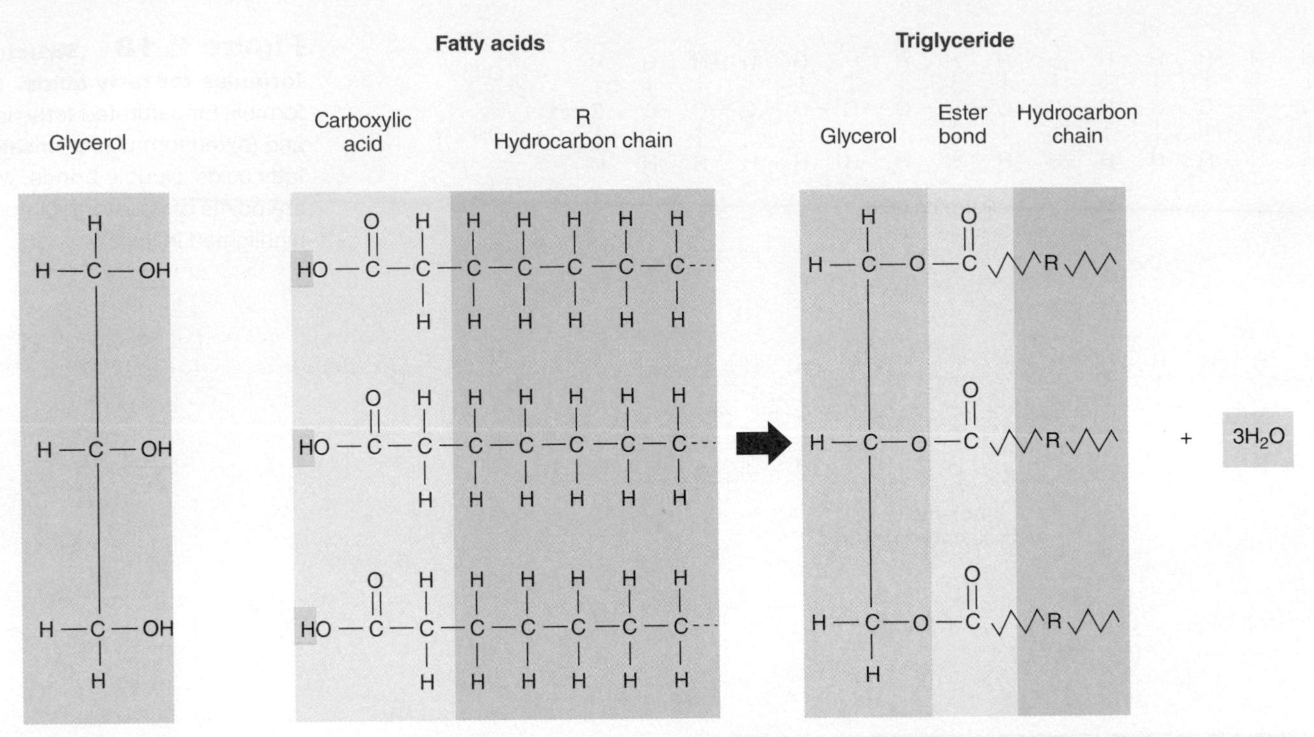

Figure 2.20 The formation of a triglyceride (triacylglycerol) molecule from glycerol and three fatty acids by dehydration synthesis reactions. A molecule of water is produced as an ester bond forms between each fatty acid and the glycerol. Sawtooth lines represent hydrocarbon chains, generally 16 to 22 carbons long, which are symbolized by an *R*.

Ketone Bodies

Hydrolysis of triglycerides within adipose tissue releases *free fatty acids* into the blood. Free fatty acids can be used as an immediate source of energy by many organs; they can also be converted by the liver into derivatives called **ketone bodies** (fig. 2.21). These include four-carbon-long acidic molecules (acetoacetic acid and β-hydroxybutyric acid) and acetone (the solvent in nailpolish remover). A rapid breakdown of fat, as may occur during strict low-carbohydrate diets and in uncontrolled diabetes mellitus, results in elevated levels of ketone bodies in the blood. This is a condition called **ketosis.** If there are sufficient amounts of ketone bodies in the blood to lower the blood pH, the condition is called **ketoacidosis.** Severe ketoacidosis, which may occur in diabetes mellitus, can lead to coma and death.

Acetoacetic acid → Acetone $+ CO_2$

Figure 2.21 Ketone bodies. Acetoacetic acid, an acidic ketone body, can spontaneously decarboxylate (lose carbon dioxide) to form acetone. Acetone is a volatile ketone body that escapes in the exhaled breath, thereby lending a "fruity" smell to the breath of people with ketosis (elevated blood ketone bodies).

Phospholipids

The group of lipids known as **phospholipids** includes a number of different categories of lipids, all of which contain a phosphate group. The most common type of phospholipid molecule is one in which the three-carbon alcohol molecule glycerol is attached to two fatty acid molecules; the third carbon atom of the glycerol is attached to a phosphate group, and the phosphate group, in turn, is bound to other molecules. If the phosphate group is attached to a nitrogen-containing choline molecule, the phospholipid thus formed is known as **lecithin**

Clinical Investigation CLUES 

George has his urine tested in the laboratory, where they discover that he has ketonuria (elevated levels of ketone bodies in the urine).

- What are ketone bodies, and how do they originate?
- What benefit does George's body derive from the ketone bodies?

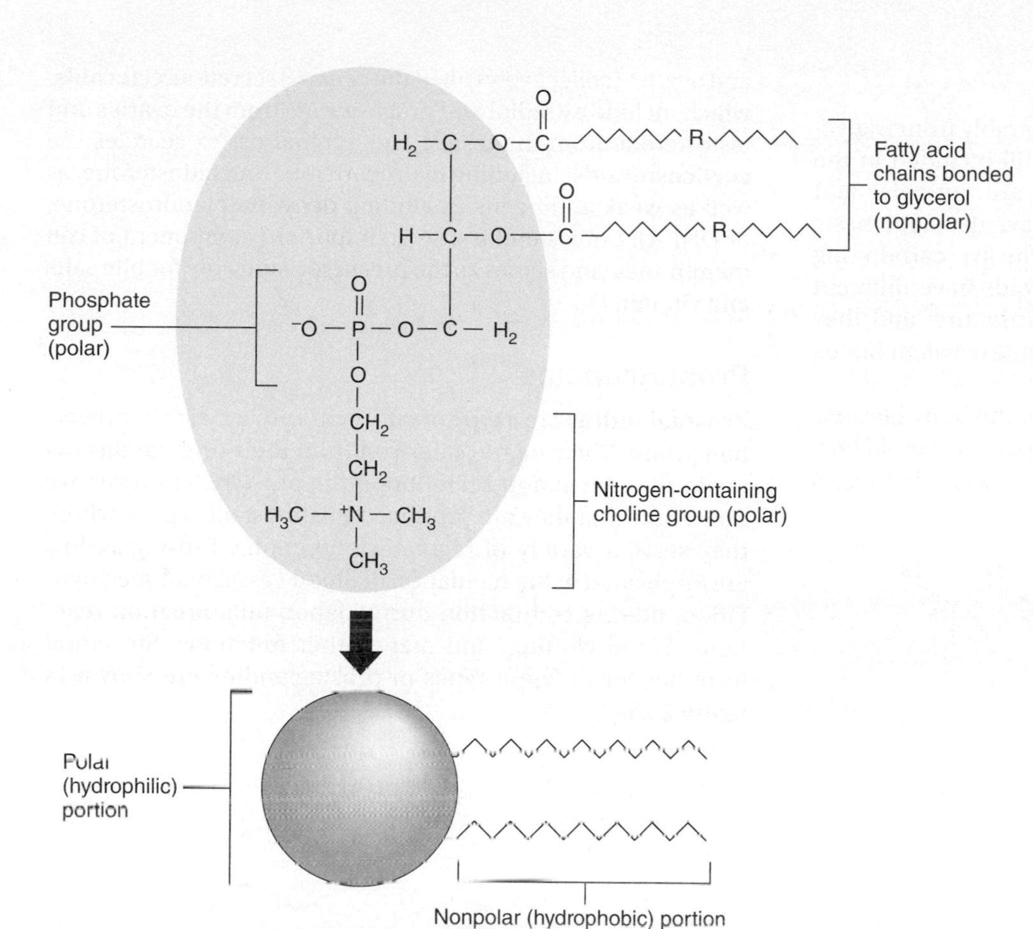

Figure 2.22 The structure of lecithin. Lecithin is also called phosphatidylcholine, where choline is the nitrogen-containing portion of the molecule. (Interestingly, choline is also part of an important neurotransmitter known as acetylcholine, discussed in chapter 7.) The detailed structure of the phospholipid (*top*) is usually shown in simplified form (*bottom*), where the circle represents the polar portion and the sawtoothed lines the nonpolar portion of the molecule.

(or *phosphatidylcholine*). Figure 2.22 shows a simple way of illustrating the structure of a phospholipid—the parts of the molecule capable of ionizing (and thus becoming charged) are shown as a circle, whereas the nonpolar parts of the molecule are represented by sawtooth lines. Molecules that are part polar and part nonpolar, such as phospholipids and bile acids (which are derived from cholesterol), are described as *amphipathic* molecules.

Phospholipids are the major component of cell membranes; their amphipathic nature allows them to form a double layer with their polar portions facing water on each side of the membrane (chapter 3). When phospholipids are mixed in water, they tend to group together so that their polar parts face the surrounding water molecules (fig. 2.23). Such aggregates of molecules are called **micelles.** Bile acids (which are not phospholipids, but are amphipathic molecules derived from cholesterol) form similar micelles in the small intestine (chapter 18, section 18.5). The amphipathic nature of phospholipids (part polar, part nonpolar) allows them to alter the interaction of water molecules and thus to decrease the surface tension of water. This function of phospholipids makes them **surfactants** (surface-active agents). The surfactant effect of phospholipids prevents the lungs from collapsing due to surface tension forces (chapter 16, section 16.2).

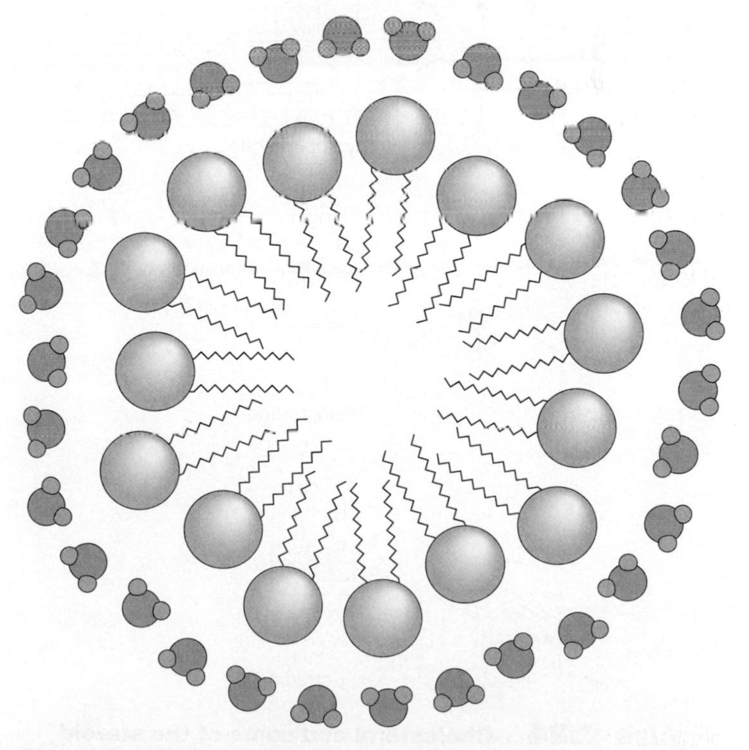

Figure 2.23 The formation of a micelle structure by phospholipids such as lecithin. The hydrophilic outer layer of the micelle faces the aqueous environment. AP|R

Steroids

In terms of structure, **steroids** differ considerably from triglycerides or phospholipids, yet steroids are still included in the lipid category of molecules because they are nonpolar and insoluble in water. All steroid molecules have the same basic structure: three six-carbon rings joined to one five-carbon ring (fig. 2.24). However, different kinds of steroids have different functional groups attached to this basic structure, and they vary in the number and position of the double covalent bonds between the carbon atoms in the rings.

Cholesterol is an important molecule in the body because it serves as the precursor (parent molecule) for the steroid hormones produced by the gonads and adrenal cortex. The testes and ovaries (collectively called the *gonads*) secrete **sex steroids,** which include estradiol and progesterone from the ovaries and testosterone from the testes. The adrenal cortex secretes the **corticosteroids,** including hydrocortisone and aldosterone, as well as weak androgens (including dehydroepiandrosterone, or DHEA). Cholesterol is also an important component of cell membranes, and serves as the precursor molecule for bile salts and vitamin D_3.

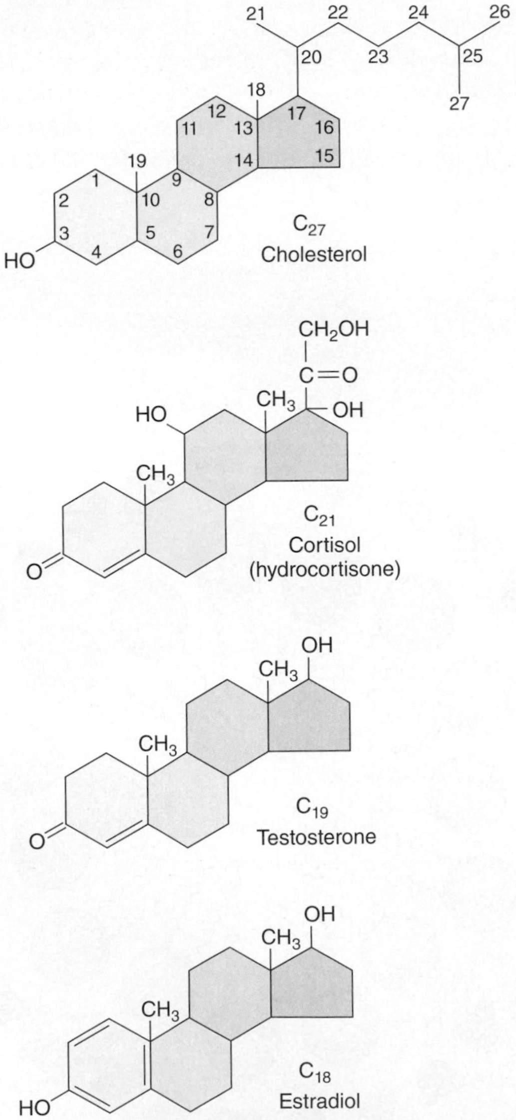

C_{27} Cholesterol

C_{21} Cortisol (hydrocortisone)

C_{19} Testosterone

C_{18} Estradiol

Figure 2.24 Cholesterol and some of the steroid hormones derived from cholesterol. The steroid hormones are secreted by the gonads and the adrenal cortex. The carbon atoms in cholesterol are indicated by the numbers.

Prostaglandins

Prostaglandins are a type of fatty acid with a cyclic hydrocarbon group. Their name is derived from their original discovery in the semen as a secretion of the prostate. However, we now know that they are produced in almost all organs where they serve a variety of regulatory functions. Prostaglandins are implicated in the regulation of blood vessel diameter, ovulation, uterine contraction during labor, inflammation reactions, blood clotting, and many other functions. Structural formulas for different types of prostaglandins are shown in figure 2.25.

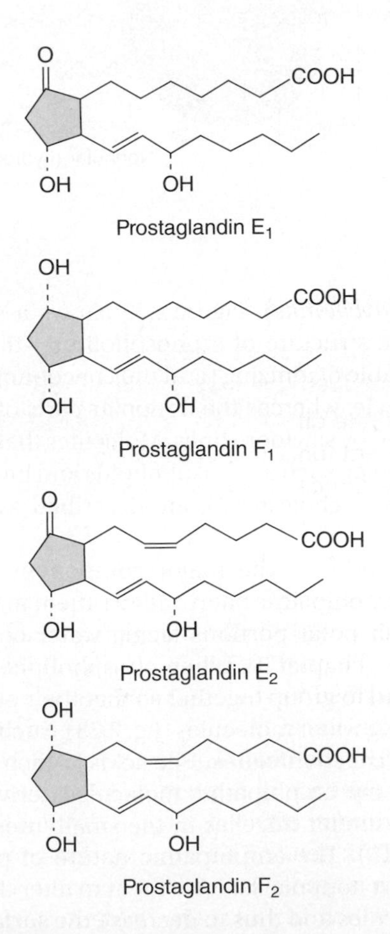

Figure 2.25 Structural formulas for various prostaglandins. Prostaglandins are a family of regulatory compounds derived from a membrane lipid known as arachidonic acid.

CHECKPOINTS

5a. Describe the structural characteristic of all carbohydrates, and distinguish between monosaccharides, disaccharides, and polysaccharides.
5b. Describe the characteristics of a lipid, and discuss the different subcategories of lipids.
6. Explain, in terms of dehydration synthesis and hydrolysis reactions, how disaccharides and monosaccharides can be interconverted and how triglycerides can be formed and broken down.
7. Relate the functions of phospholipids to their structure, and explain the significance of the prostaglandins.

2.3 PROTEINS

Proteins are large molecules composed of amino acid subunits. There are about 20 different types of amino acids that can be used in constructing a given protein, so the variety of protein structures is immense. This variety allows each type of protein to perform very specific functions.

LEARNING OUTCOMES

After studying this section, you should be able to:

8. Describe amino acids and explain how peptide bonds between them are formed and broken.
9. Describe the different orders of protein structure, the different functions of proteins, and how protein structure grants specificity of function.

The enormous diversity of protein structure results from the fact that there are 20 different building blocks—the *amino acids*—that can be used to form a protein. These amino acids, as will be described in the next section, are joined together to form a chain. Because of chemical interactions between the amino acids, the chain can twist and fold in a specific manner. The sequence of amino acids in a protein, and thus the specific structure of the protein, is determined by genetic information. This genetic information for protein synthesis is contained in another category of organic molecules, the *nucleic acids,* which includes the macromolecules DNA and RNA. The structure of nucleic acids is described in the next section, and the mechanisms by which the genetic information they encode directs protein synthesis are described in chapter 3.

Structure of Proteins

Proteins consist of long chains of subunits called **amino acids.** As the name implies, each amino acid contains an *amino group* (NH_2) on one end of the molecule and a *carboxyl group* (COOH) on another end. There are about 20 different amino acids, each with a distinct structure and chemical properties, that are used to build proteins. The differences between the amino acids are due to differences in their *functional groups. R* is the abbreviation for the functional group in the general formula for an amino acid (fig. 2.26). The *R* symbol actually stands for the word *residue,* but it can be thought of as indicating the "*rest* of the molecule."

When amino acids are joined together by dehydration synthesis, the hydrogen from the amino end of one amino acid combines with the hydroxyl group in the carboxyl end of another amino acid. As a covalent bond is formed between the two amino acids, water is produced (fig. 2.27). The bond between adjacent amino acids is called a **peptide bond,** and the compound formed is called a *peptide.* Two amino acids bound together are called a *dipeptide;* three, a *tripeptide.* When

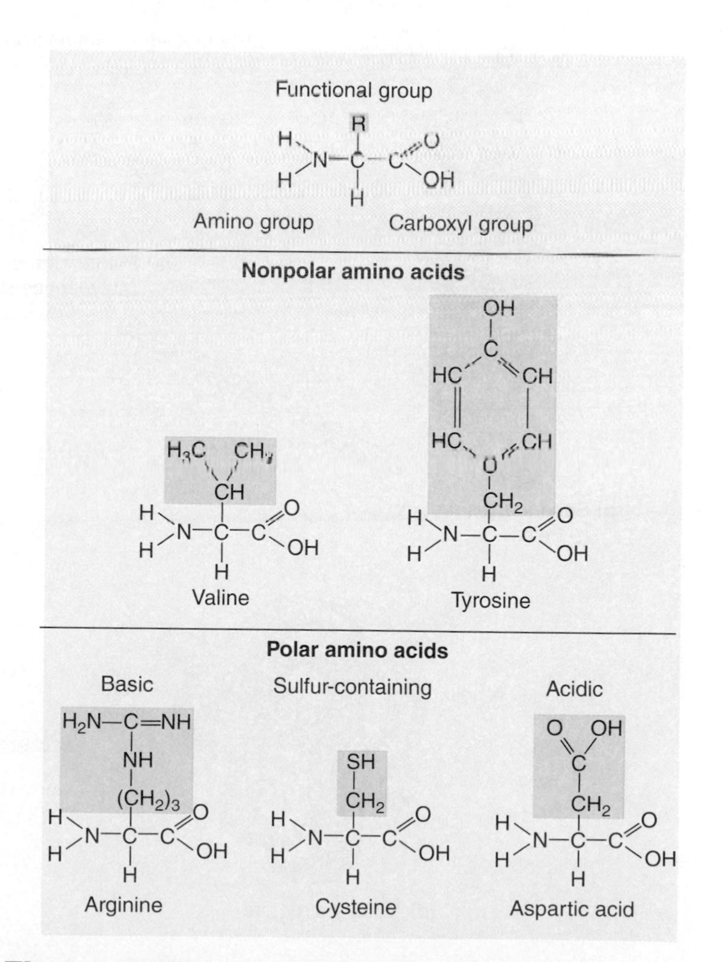

Figure 2.26 Representative amino acids. The figure depicts different types of functional (R) groups. Each amino acid differs from other amino acids in the number and arrangement of atoms in its functional groups.

Figure 2.27 The formation of peptide bonds by dehydration synthesis reactions. Water molecules are split off as the peptide bonds (highlighted in red) are produced between the amino acids.

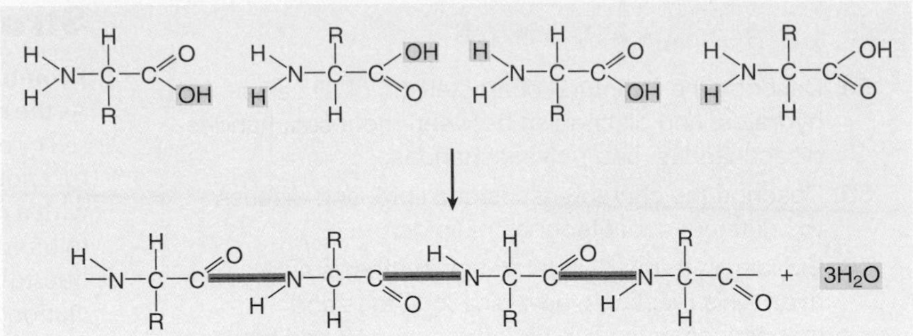

numerous amino acids are joined in this way, a chain of amino acids, or a **polypeptide,** is produced.

The lengths of polypeptide chains vary widely. A hormone called *thyrotropin-releasing hormone,* for example, is only three amino acids long, whereas myosin, a muscle protein, contains about 4,500 amino acids. When the length of a polypeptide chain becomes very long (containing more than about 100 amino acids), the molecule is called a *protein.*

The structure of a protein can be described at four different levels. The first level of structure describes the sequence of amino acids in the particular protein; this is the **primary structure** of the protein. Each type of protein has a different primary structure. All of the billions of *copies* of a given type of protein in a person have the same structure, however, because the structure of a given protein is coded by the person's genes. The primary structure of a protein is illustrated in figure 2.28*a*.

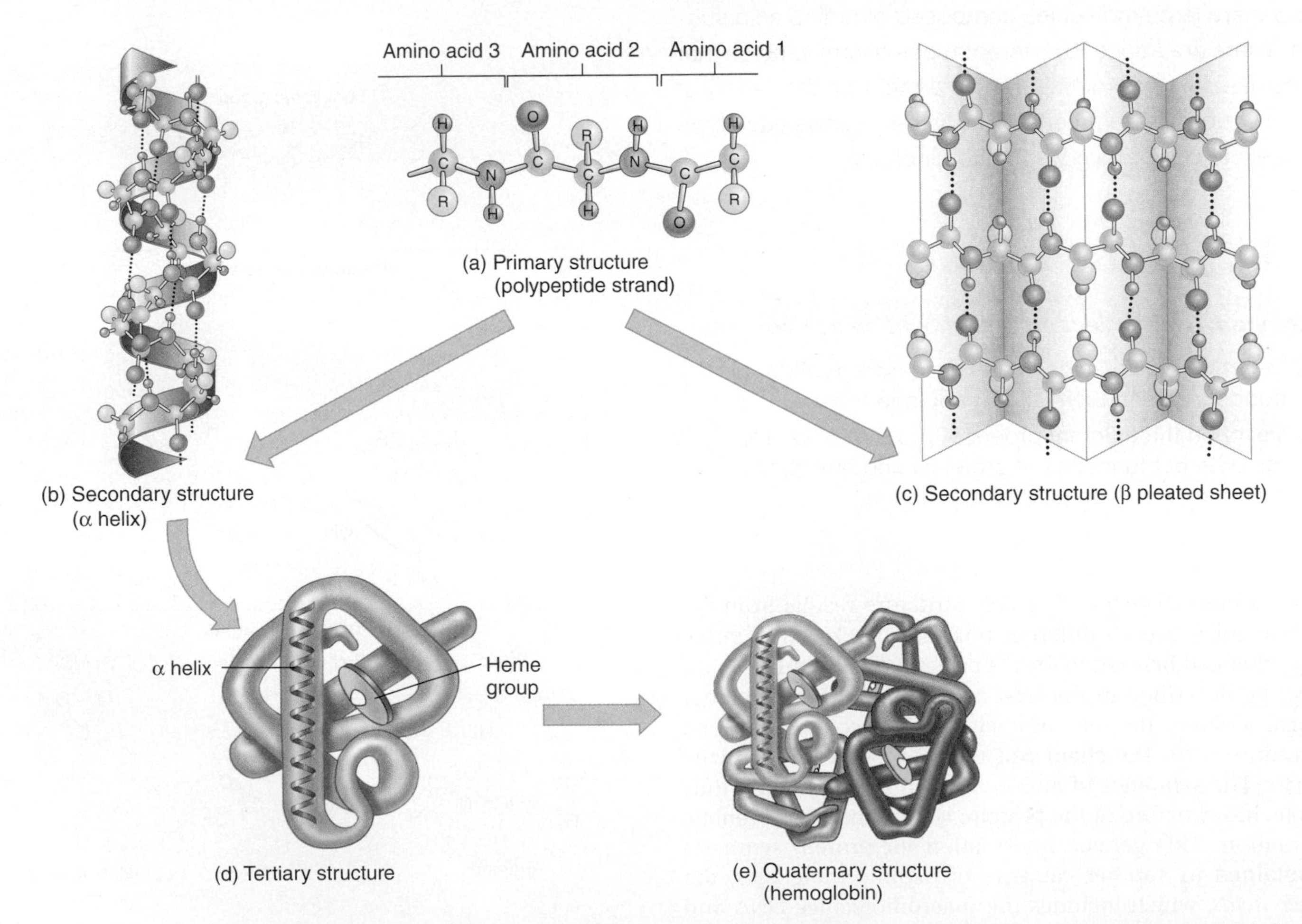

Figure 2.28 The structure of proteins. The primary structure (*a*) is the sequence of amino acids in the polypeptide chain. The secondary structure is the conformation of the chain created by hydrogen bonding between amino acids; this can be either an alpha helix (*b*) or a beta pleated sheet (*c*). The tertiary structure (*d*) is the three-dimensional structure of the protein. The formation of a protein by the bonding together of two or more polypeptide chains is the quaternary structure (*e*) of the protein. Hemoglobin, the protein in red blood cells that carries oxygen, is used here as an example. AP|R

Weak hydrogen bonds may form between the hydrogen atom of an amino group and an oxygen atom from a different amino acid nearby. These weak bonds cause the polypeptide chain to assume a particular shape, known as the **secondary structure** of the protein (fig. 2.28*b,c*). This can be the shape of an *alpha* (α) *helix,* or alternatively, the shape of what is called a *beta* (β) *pleated sheet.*

Most polypeptide chains bend and fold upon themselves to produce complex three-dimensional shapes called the **tertiary structure** of the protein (fig. 2.28*d*). Each type of protein has its own characteristic tertiary structure. This is because the folding and bending of the polypeptide chain is produced by chemical interactions between particular amino acids located in different regions of the chain.

Most of the tertiary structure of proteins is formed and stabilized by weak chemical interactions between the functional groups of amino acids located some distance apart along the polypeptide chain. In terms of their strengths, these weak interactions are relatively stronger for ionic bonds, weaker for hydrogen bonds, and weakest for van der Waals forces (fig. 2.29). The natures of ionic bonds and hydrogen bonds have been previously discussed. *Van der Waals forces* are weak forces between electrically neutral molecules that come very close together. These forces occur because, even in electrically neutral molecules, the electrons are not always evenly distributed but can at some instants be found at one end of the molecule.

Because most of the tertiary structure is stabilized by weak bonds, this structure can easily be disrupted by high temperature or by changes in pH. Changes in the tertiary structure of proteins that occur by these means are referred to as *denaturation* of the proteins. The tertiary structure of some proteins, however, is made more stable by strong covalent bonds between sulfur atoms (called *disulfide bonds* and abbreviated S—S) in the functional group of an amino acid known as cysteine (fig. 2.29).

Denatured proteins retain their primary structure (the peptide bonds are not broken) but have altered chemical properties. Cooking a pot roast, for example, alters the texture of the meat proteins—it doesn't result in an amino acid soup. Denaturation is most dramatically demonstrated by frying an egg. Egg albumin proteins are soluble in their native state in which they form the clear, viscous fluid of a raw egg. When denatured by cooking, these proteins change shape, cross-bond with each other, and by this means form an insoluble white precipitate—the egg white.

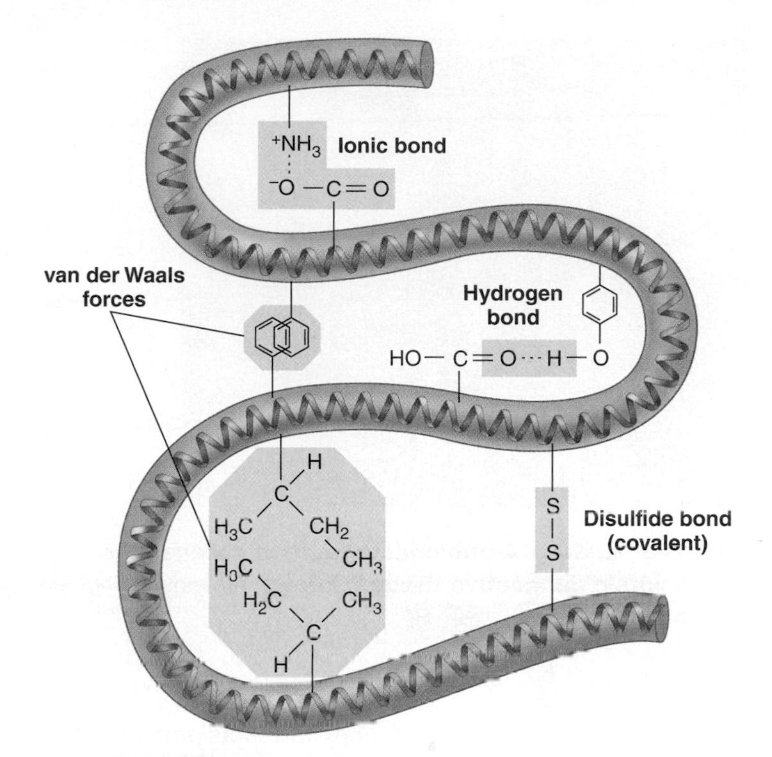

Figure 2.29 The bonds responsible for the tertiary structure of a protein. The tertiary structure of a protein is held in place by a variety of bonds. These include relatively weak bonds, such as hydrogen bonds, ionic bonds, and van der Waals (hydrophobic) forces, as well as the strong covalent disulfide bonds. AP|R

Hemoglobin and insulin are composed of a number of polypeptide chains covalently bonded together. This is the **quaternary structure** of these molecules. Insulin, for example, is composed of two polypeptide chains—one that is 21 amino acids long, the other that is 30 amino acids long. Hemoglobin (the protein in red blood cells that carries oxygen) is composed of four separate polypeptide chains (see fig. 2.28*e*). The composition of various body proteins is shown in table 2.4.

Many proteins in the body are normally found combined, or *conjugated,* with other types of molecules. **Glycoproteins** are proteins conjugated with carbohydrates. Examples of such

Table 2.4 | Composition of Selected Proteins Found in the Body

Protein	Number of Polypeptide Chains	Nonprotein Component	Function
Hemoglobin	4	Heme pigment	Carries oxygen in the blood
Myoglobin	1	Heme pigment	Stores oxygen in muscle
Insulin	2	None	Hormonal regulation of metabolism
Blood group proteins	1	Carbohydrate	Produces blood types
Lipoproteins	1	Lipids	Transports lipids in blood

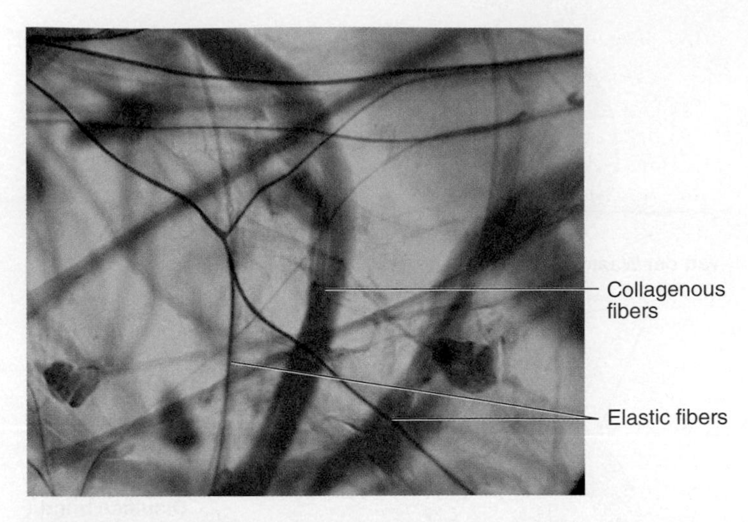

Figure 2.30 A photomicrograph of collagenous fibers within connective tissue. Collagen proteins strengthen the connective tissues. AP|R

molecules include certain hormones and some proteins found in the cell membrane. **Lipoproteins** are proteins conjugated with lipids. These are found in cell membranes and in the plasma (the fluid portion of the blood). Proteins may also be conjugated with pigment molecules. These include hemoglobin, which transports oxygen in red blood cells, and the cytochromes, which are needed for oxygen utilization and energy production within cells.

Functions of Proteins

Because of their tremendous structural diversity, proteins can serve a wider variety of functions than any other type of molecule in the body. Many proteins, for example, contribute significantly to the structure of different tissues and in this way play a passive role in the functions of these tissues. Examples of such *structural proteins* include collagen (fig. 2.30) and keratin. Collagen is a fibrous protein that provides tensile strength to connective tissues, such as tendons and ligaments. Keratin is found in the outer layer of dead cells in the epidermis where it prevents water loss through the skin.

Many proteins play a more active role in the body where specificity of structure and function is required. *Enzymes* and *antibodies,* for example, are proteins—no other type of molecule could provide the vast array of different structures needed for their tremendously varied functions. As another example, proteins in cell membranes may serve as *receptors* for specific regulatory molecules (such as hormones) and as *carriers* for transport of specific molecules across the membrane. Proteins provide the diversity of shape and chemical properties required by these functions.

CHECKPOINTS

8a. Write the general formula for an amino acid, and describe how amino acids differ from one another.

8b. Describe and account for the different levels of protein structure.

9. Describe the different categories of protein function in the body, and explain why proteins can serve functions that are so diverse.

2.4 NUCLEIC ACIDS

Nucleic acids include the macromolecules DNA and RNA, which are critically important in genetic regulation, and the subunits from which these molecules are formed. These subunits are known as nucleotides.

LEARNING OUTCOMES

After studying this section, you should be able to:

10. Describe the structure of nucleotides and distinguish between the structure of DNA and RNA.

11. Explain the law of complementary base pairing, and describe how that occurs between the two strands of DNA.

Nucleotides are the subunits of nucleic acids, bonded together in dehydration synthesis reactions to form long polynucleotide chains. Each nucleotide, however, is itself composed of three smaller subunits: a five-carbon (*pentose*) sugar, a phosphate group attached to one end of the sugar, and a *nitrogenous base* attached to the other end of the sugar (fig. 2.31). The nitrogenous bases are nitrogen-containing molecules of two kinds: pyrimidines and purines. The *pyrimidines* contain a single ring of carbon and nitrogen, whereas the *purines* have two such rings.

Deoxyribonucleic Acid

The structure of **DNA (deoxyribonucleic acid)** serves as the basis for the genetic code. For this reason, it might seem logical that DNA should have an extremely complex structure. DNA is indeed larger than any other molecule in the cell, but its structure is actually simpler than that of most proteins. This simplicity of structure deceived some early investigators into believing that the protein content of chromosomes, rather than their DNA content, provided the basis for the genetic code.

Sugar molecules in the nucleotides of DNA are a type of pentose (five-carbon) sugar called **deoxyribose.** Each

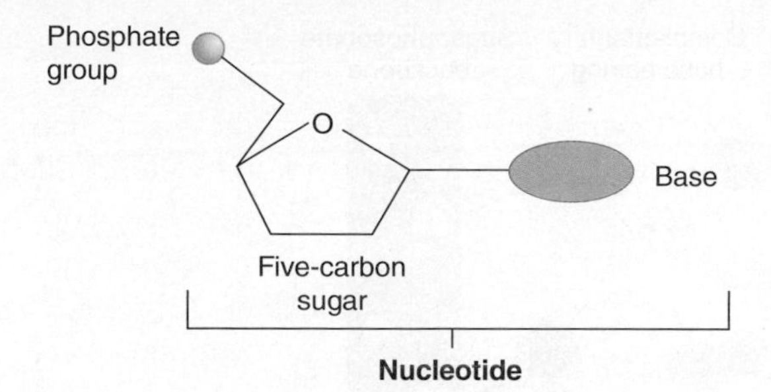

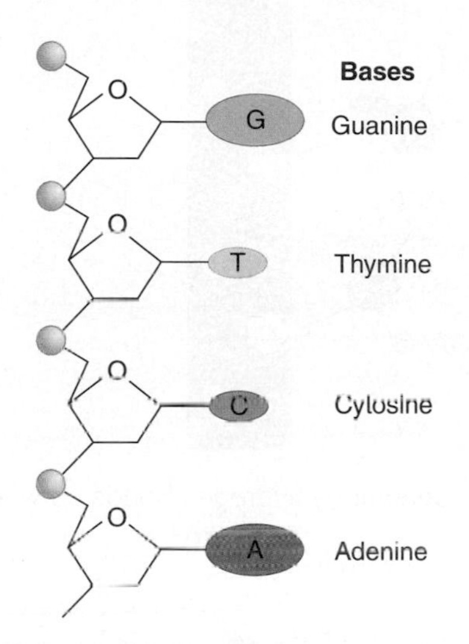

Figure 2.31 The structure of nucleic acids. The components of a single nucleotide are shown above, and the structure of a polynucleotide is shown below. The polynucleotide was formed by dehydration reactions between nucleotides that join the nucleotides together by sugar-phosphate bonds. APIR

deoxyribose can be covalently bonded to one of four possible bases. These bases include the two purines (**guanine** and **adenine**) and the two pyrimidines (**cytosine** and **thymine**) (fig. 2.32). There are thus four different types of nucleotides that can be used to produce the long DNA chains. If you remember that there are about 20 different amino acids used to produce proteins, you can now understand why many scientists were deceived into thinking that genes were composed of proteins rather than nucleic acids.

When nucleotides combine to form a chain, the phosphate group of one condenses with the deoxyribose sugar of another nucleotide. This forms a sugar-phosphate chain as water is removed in dehydration synthesis. Because the nitrogenous bases are attached to the sugar molecules, the sugar-phosphate chain looks like a "backbone" from which the bases project. Each of these bases can form hydrogen bonds with other bases, which are in turn joined to a different chain of nucleotides. Such hydrogen bonding between bases thus produces a *double-stranded* DNA molecule; the two strands are like a staircase, with the paired bases as steps (fig. 2.33).

Actually, the two chains of DNA twist about each other to form a **double helix,** so that the molecule resembles a spiral staircase (fig. 2.33). It has been shown that the number of purine bases in DNA is equal to the number of pyrimidine bases. The reason for this is explained by the **law of complementary base pairing:** *adenine can pair only with thymine* (through two hydrogen bonds), whereas *guanine can pair only with cytosine* (through three hydrogen bonds). With

Figure 2.32 The four nitrogenous bases in deoxyribonucleic acid (DNA). Notice that hydrogen bonds can form between guanine and cytosine and between thymine and adenine. APIR

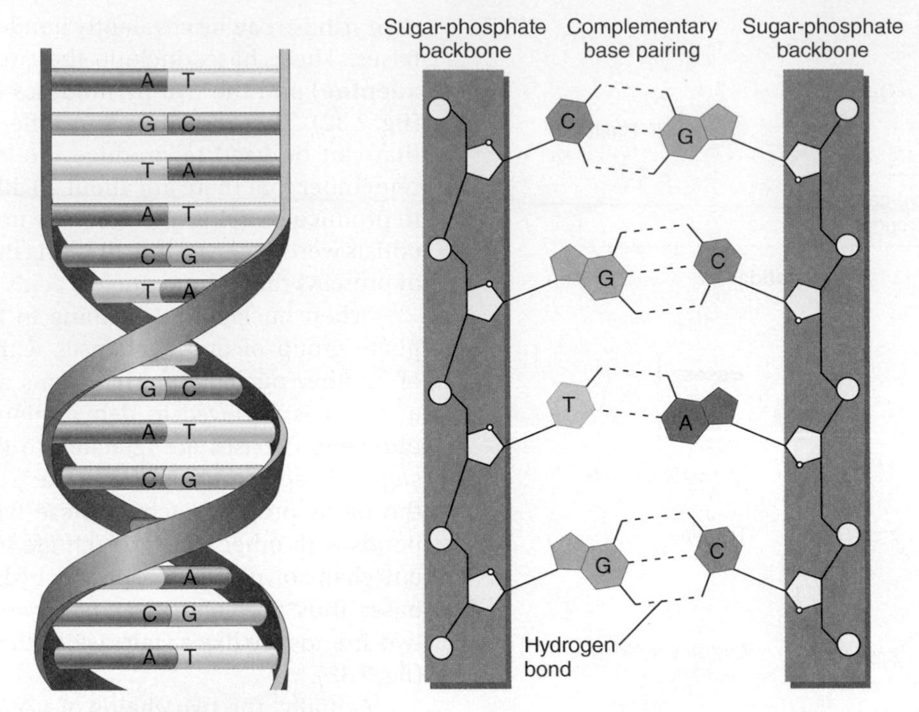

Figure 2.33 The double-helix structure of DNA. The two strands are held together by hydrogen bonds between complementary bases in each strand. APR

knowledge of this rule, we could predict the base sequence of one DNA strand if we knew the sequence of bases in the complementary strand.

Although we can be certain which base is opposite a given base in DNA, we cannot predict which bases will be above or below that particular pair within a single polynucleotide chain. Although there are only four bases, the number of possible base sequences along a stretch of several thousand nucleotides (the length of most genes) is almost infinite. To gain perspective, it is useful to realize that the total human *genome* (all of the genes in a cell) consists of over 3 billion base pairs that would extend over a meter if the DNA molecules were unraveled and stretched out.

Yet even with this amazing variety of possible base sequences, almost all of the billions of copies of a particular gene in a person are identical. The mechanisms by which identical DNA copies are made and distributed to the daughter cells when a cell divides will be described in chapter 3.

Ribonucleic Acid

DNA can direct the activities of the cell only by means of another type of nucleic acid—**RNA (ribonucleic acid).** Like DNA, RNA consists of long chains of nucleotides joined together by sugar-phosphate bonds. Nucleotides in RNA, however, differ from those in DNA (fig. 2.34) in three ways: (1) a **ribonucleotide** contains the sugar **ribose** (instead of deoxyribose), (2) the base **uracil** is found in place of thymine, and (3) RNA is composed of a single polynucleotide strand (it is not double-stranded like DNA).

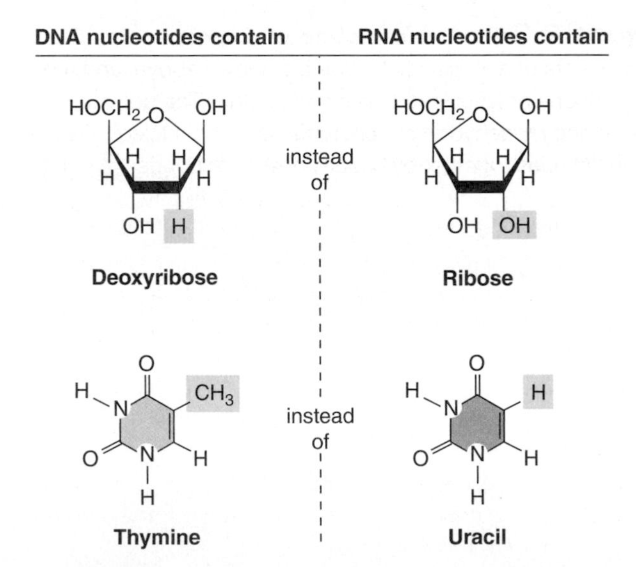

Figure 2.34 Differences between the nucleotides and sugars in DNA and RNA. DNA has deoxyribose and thymine; RNA has ribose and uracil. The other three bases are the same in DNA and RNA. APR

There are three major types of RNA molecules that function in the cytoplasm of cells: *messenger RNA (mRNA), transfer RNA (tRNA),* and *ribosomal RNA (rRNA).* All three types are made within the cell nucleus by using information contained in DNA as a guide. The functions of RNA are described in chapter 3.

In addition to their participation in genetic regulation as part of RNA, purine-containing nucleotides are used for other purposes as well. These include roles as energy carriers (ATP and GTP); regulation of cellular events (cyclic AMP, or cAMP); and coenzymes (nicotinamide adenine dinucleotide, or NAD; and flavine adenine dinucleotide, or FAD). These are discussed in chapters 4, 5, and 6. Purines (ATP and adenosine) are even used as neurotransmitters by some neurons (chapter 7, section 7.6).

CHECKPOINTS

10a. What are nucleotides, and of what are they composed?

10b. List the types of RNA, and explain how the structure of RNA differs from the structure of DNA.

11. Describe the structure of DNA, and explain the law of complementary base pairing.

Clinical Investigation SUMMARY

Because our enzymes can recognize only L-amino acids and D-sugars, the opposite stereoisomers that George was eating could not be used by his body. He was weak because he was literally starving. The ketonuria also may have contributed to his malaise. Because he was starving, his stored fat was being rapidly hydrolyzed into glycerol and fatty acids for use as energy sources. The excessive release of fatty acids from his adipose tissue resulted in the excessive production of ketone bodies by his liver; hence, his ketonuria.

See the additional chapter 2 Clinical Investigation on *High Cholesterol* in the Connect site for this text at www.mhhe.com/Fox13.

SUMMARY

2.1 Atoms, Ions, and Chemical Bonds 25

A. Covalent bonds are formed by atoms that share electrons. They are the strongest type of chemical bond.

1. Electrons are equally shared in nonpolar covalent bonds and unequally shared in polar covalent bonds.
2. Atoms of oxygen, nitrogen, and phosphorus strongly attract electrons and become electrically negative compared to the other atoms sharing electrons with them.

B. Ionic bonds are formed by atoms that transfer electrons. These weak bonds join atoms together in an ionic compound.

1. If one atom in this compound takes an electron from another atom, it gains a net negative charge and the other atom becomes positively charged.
2. Ionic bonds easily break when the ionic compound is dissolved in water. Dissociation of the ionic compound yields charged atoms called ions.

C. When hydrogen bonds with an electronegative atom, it gains a slight positive charge and is weakly attracted to another electronegative atom. This weak attraction is a hydrogen bond.

D. Acids donate hydrogen ions to solution, whereas bases lower the hydrogen ion concentration of a solution.

1. The pH scale is a negative function of the logarithm of the hydrogen ion concentration.
2. In a neutral solution, the concentration of H^+ is equal to the concentration of OH^-, and the pH is 7.
3. Acids raise the H^+ concentration and thus lower the pH below 7; bases lower the H^+ concentration and thus raise the pH above 7.

E. Organic molecules contain atoms of carbon and hydrogen joined together by covalent bonds. Atoms of nitrogen, oxygen, phosphorus, or sulfur may be present as specific functional groups in the organic molecule.

2.2 Carbohydrates and Lipids 33

A. Carbohydrates contain carbon, hydrogen, and oxygen, usually in a ratio of 1:2:1.

1. Carbohydrates consist of simple sugars (monosaccharides), disaccharides, and polysaccharides (such as glycogen).
2. Covalent bonds between monosaccharides are formed by dehydration synthesis, or condensation. Bonds are broken by hydrolysis reactions.

B. Lipids are organic molecules that are insoluble in polar solvents such as water.

1. Triglycerides (fat and oil) consist of three fatty acid molecules joined to a molecule of glycerol.
2. Ketone bodies are smaller derivatives of fatty acids.
3. Phospholipids (such as lecithin) are phosphate-containing lipids that have a hydrophilic polar group. The rest of the molecule is hydrophobic.
4. Steroids (including the hormones of the adrenal cortex and gonads) are lipids with a characteristic four-ring structure.

5. Prostaglandins are a family of cyclic fatty acids that serve a variety of regulatory functions.

2.3 Proteins 41

A. Proteins are composed of long chains of amino acids bound together by covalent peptide bonds.
   1. Each amino acid contains an amino group, a carboxyl group, and a functional group. Differences in the functional groups give each of the more than 20 different amino acids an individual identity.
   2. The polypeptide chain may be twisted into a helix (secondary structure) and bent and folded to form the tertiary structure of the protein.
   3. Proteins that are composed of two or more polypeptide chains are said to have a quaternary structure.
   4. Proteins may be combined with carbohydrates, lipids, or other molecules.
   5. Because they are so diverse structurally, proteins serve a wider variety of specific functions than any other type of molecule.

2.4 Nucleic Acids 44

A. DNA is composed of four nucleotides, each of which contains the sugar deoxyribose.
   1. Two of the bases contain the purines adenine and guanine; two contain the pyrimidines cytosine and thymine.
   2. DNA consists of two polynucleotide chains joined together by hydrogen bonds between their bases.
   3. Hydrogen bonds can only form between the bases adenine and thymine, and between the bases guanine and cytosine.
   4. This complementary base pairing is critical for DNA synthesis and for genetic expression.

B. RNA consists of four nucleotides, each of which contains the sugar ribose.
   1. The nucleotide bases are adenine, guanine, cytosine, and uracil (in place of the DNA base thymine).
   2. RNA consists of only a single polynucleotide chain.
   3. There are different types of RNA, which have different functions in genetic expression.

REVIEW ACTIVITIES

Test Your Knowledge

1. Which of these statements about atoms is *true?*
   a. They have more protons than electrons.
   b. They have more electrons than protons.
   c. They are electrically neutral.
   d. They have as many neutrons as they have electrons.
2. The bond between oxygen and hydrogen in a water molecule is
   a. a hydrogen bond.
   b. a polar covalent bond.
   c. a nonpolar covalent bond.
   d. an ionic bond.
3. Which of these is a nonpolar covalent bond?
   a. bond between two carbons
   b. bond between sodium and chloride
   c. bond between two water molecules
   d. bond between nitrogen and hydrogen
4. Solution A has a pH of 2, and solution B has a pH of 10. Which of these statements about these solutions is *true?*
   a. Solution A has a higher H^+ concentration than solution B.
   b. Solution B is basic.
   c. Solution A is acidic.
   d. All of these are true.
5. Glucose is
   a. a disaccharide. c. a monosaccharide.
   b. a polysaccharide. d. a phospholipid.
6. Digestion reactions occur by means of
   a. dehydration synthesis. b. hydrolysis.
7. Carbohydrates are stored in the liver and muscles in the form of
   a. glucose. c. glycogen.
   b. triglycerides. d. cholesterol.
8. Lecithin is
   a. a carbohydrate. c. a steroid.
   b. a protein. d. a phospholipid.
9. Which of these lipids have regulatory roles in the body?
   a. steroids d. both *a* and *b*
   b. prostaglandins e. both *b* and *c*
   c. triglycerides
10. The tertiary structure of a protein is *directly* determined by
    a. genes.
    b. the primary structure of the protein.
    c. enzymes that "mold" the shape of the protein.
    d. the position of peptide bonds.
11. The type of bond formed between two molecules of water is
    a. a hydrolytic bond.
    b. a polar covalent bond.
    c. a nonpolar covalent bond.
    d. a hydrogen bond.
12. The carbon-to-nitrogen bond that joins amino acids together is called
    a. a glycosidic bond. c. a hydrogen bond.
    b. a peptide bond. d. a double bond.

13. The RNA nucleotide base that pairs with adenine in DNA is
    a. thymine.
    b. uracil.
    c. guanine.
    d. cytosine.
14. If four bases in one DNA strand are A (adenine), G (guanine), C (cytosine), and T (thymine), the complementary bases in the RNA strand made from this region are
    a. T,C,G,A.
    b. C,G,A,U.
    c. A,G,C,U.
    d. U,C,G,A.

Test Your Understanding

15. Compare and contrast nonpolar covalent bonds, polar covalent bonds, and ionic bonds.
16. Define *acid* and *base* and explain how acids and bases influence the pH of a solution.
17. Explain, in terms of dehydration synthesis and hydrolysis reactions, the relationships between starch in an ingested potato, liver glycogen, and blood glucose.
18. "All fats are lipids, but not all lipids are fats." Explain why this is an accurate statement.
19. What are the similarities and differences between a fat and an oil? Comment on the physiological and clinical significance of the degree of saturation of fatty acid chains.
20. Explain how one DNA molecule serves as a template for the formation of another DNA molecule and why DNA synthesis is said to be semiconservative.

Test Your Analytical Ability

21. Explain the relationship between the primary structure of a protein and its secondary and tertiary structures. What do you think would happen to the tertiary structure if some amino acids were substituted for others in the primary structure? What physiological significance might this have?
22. Suppose you try to discover a hormone by homogenizing an organ in a fluid, filtering the fluid to eliminate the solid material, and then injecting the extract into an animal to see the effect. If an aqueous (water) extract does not work, but one using benzene as the solvent does have an effect, what might you conclude about the chemical nature of the hormone? Explain.
23. From the ingredients listed on a food wrapper, it would appear that the food contains high amounts of fat. Yet on the front of the package is the large slogan, "Cholesterol Free!" In what sense is this slogan chemically correct? In what way is it misleading?
24. A butter substitute says "Nonhydrogenated, zero trans fats" on the label. Explain the meaning of these terms and their relationship to health.
25. When you cook a pot roast, you don't end up with an amino acid soup. Explain why this is true, in terms of the strengths of the different types of bonds in a protein.

Test Your Quantitative Ability

The molecular weight is the sum of the atomic weights (mass numbers) of its atoms. Use table 2.1 to perform the following calculations.

26. Calculate the molecular weight of water (H_2O) and glucose ($C_6H_{12}O_6$).
27. Given that fructose is a structural isomer of glucose (see fig. 2.14), what is its molecular weight?
28. Review the dehydration synthesis of sucrose in figure 2.16*b* and calculate the molecular weight of sucrose.
29. Account for the difference between the molecular weight of sucrose and the sum of the molecular weights of glucose and fructose.

CHAPTER

3

Cell Structure and Genetic Control

CHAPTER OUTLINE

Clinical Investigation

Timothy is only 18 years old, but he appears to have liver disease. A liver biopsy is performed and reveals microscopic abnormalities as well as an abnormal chemical test. Timothy admits that he has a history of drug abuse, but claims that he is now in recovery.

Some of the new terms and concepts you will encounter include:

- Glycogen granules and glycogen hydrolysis
- Smooth endoplasmic reticulum and lysosomes

3.1 PLASMA MEMBRANE AND ASSOCIATED STRUCTURES

The cell is the basic unit of structure and function in the body. Many of the functions of cells are performed by particular subcellular structures known as organelles. The plasma (cell) membrane allows selective communication between the intracellular and extracellular compartments and aids cellular movement.

LEARNING OUTCOMES

After studying this section, you should be able to:

1. **Describe the structure of the plasma membrane, cilia, and flagella.**
2. **Describe amoeboid movement, phagocytosis, pinocytosis, receptor-mediated endocytosis, and exocytosis.**

Cells look so small and simple when viewed with the ordinary (light) microscope that it is difficult to think of each one as a living entity unto itself. Equally amazing is the fact that the physiology of our organs and systems derives from the complex functions of the cells of which they are composed. Complexity of function demands complexity of structure, even at the subcellular level.

As the basic functional unit of the body, each cell is a highly organized molecular factory. Cells come in a wide variety of shapes and sizes. This great diversity, which is also apparent in the subcellular structures within different cells, reflects the diversity of function of different cells in the body. All cells, however, share certain characteristics; for example, they are all surrounded by a plasma membrane, and most of them possess the structures listed in table 3.1. Thus, although no single cell can be considered "typical," the general structure of cells can be indicated by a single illustration (fig. 3.1).

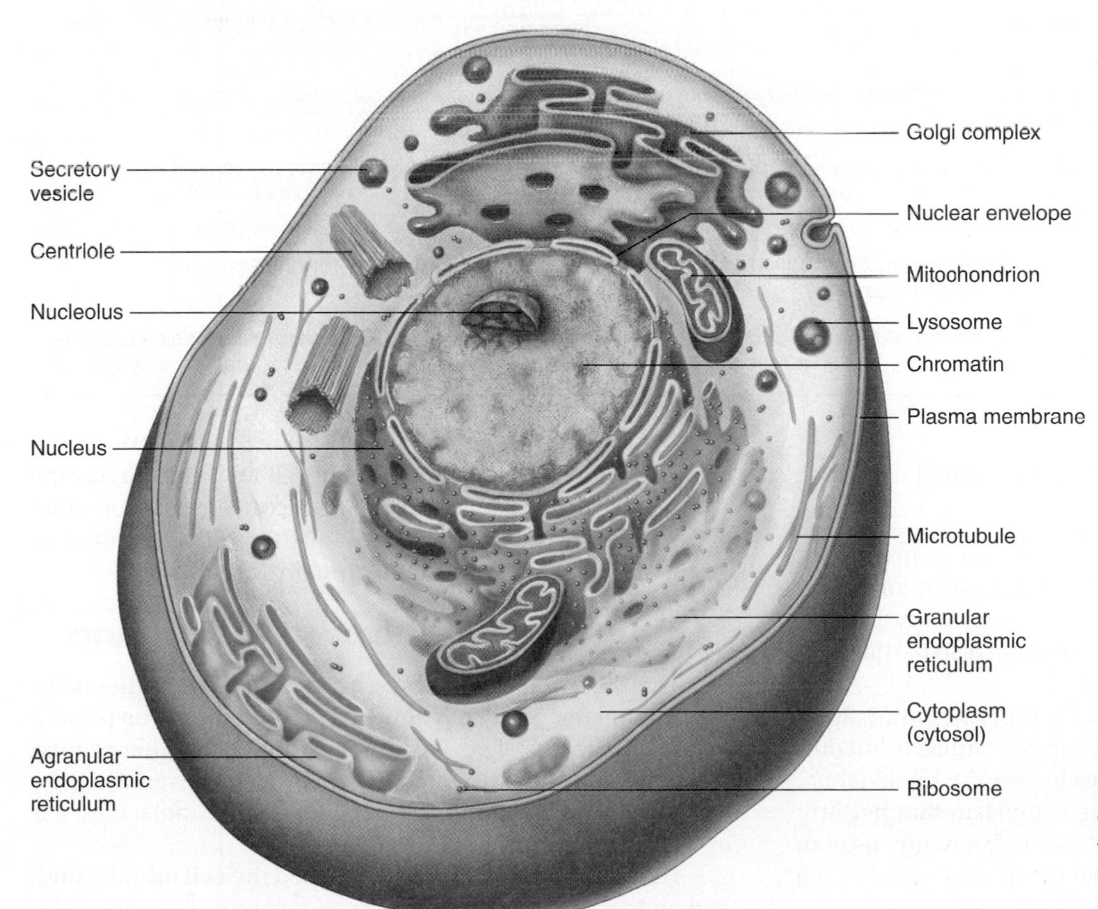

Figure 3.1 A generalized human cell showing the principal organelles. Because most cells of the body are highly specialized, they have structures that differ from those shown here. AP|R

Table 3.1 | Cellular Components: Structure and Function

Component	Structure	Function
Plasma (cell) membrane	Membrane composed of double layer of phospholipids in which proteins are embedded	Gives form to cell and controls passage of materials into and out of cell
Cytoplasm	Fluid, jellylike substance between the cell membrane and the nucleus in which organelles are suspended	Serves as matrix substance in which chemical reactions occur
Endoplasmic reticulum	System of interconnected membrane-forming canals and tubules	Agranular (smooth) endoplasmic reticulum metabolizes nonpolar compounds and stores Ca^{2+} in striated muscle cells, granular (rough) endoplasmic reticulum assists in protein synthesis
Ribosomes	Granular particles composed of protein and RNA	Synthesize proteins
Golgi complex	Cluster of flattened membranous sacs	Synthesizes carbohydrates and packages molecules for secretion, secretes lipids and glycoproteins
Mitochondria	Membranous sacs with folded inner partitions	Release energy from food molecules and transform energy into usable ATP
Lysosomes	Membranous sacs	Digest foreign molecules and worn and damaged organelles
Peroxisomes	Spherical membranous vesicles	Contain enzymes that detoxify harmful molecules and break down hydrogen peroxide
Centrosome	Nonmembranous mass of two rodlike centrioles	Helps to organize spindle fibers and distribute chromosomes during mitosis
Vacuoles	Membranous sacs	Store and release various substances within the cytoplasm
Microfilaments and microtubules	Thin, hollow tubes	Support cytoplasm and transport materials within the cytoplasm
Cilia and flagella	Minute cytoplasmic projections that extend from the cell surface	Move particles along cell surface or move the cell
Nuclear envelope	Double-layered membrane that surrounds the nucleus, composed of protein and lipid molecules	Supports nucleus and controls passage of materials between nucleus and cytoplasm
Nucleolus	Dense nonmembranous mass composed of protein and RNA molecules	Produces ribosomal RNA for ribosomes
Chromatin	Fibrous strands composed of protein and DNA	Contains genetic code that determines which proteins (including enzymes) will be manufactured by the cell

For descriptive purposes, a cell can be divided into three principal parts:

1. **Plasma (cell) membrane.** The selectively permeable plasma membrane surrounds the cell, gives it form, and separates the cell's internal structures from the extracellular environment. The plasma membrane also participates in intercellular communication.
2. **Cytoplasm and organelles.** The cytoplasm is the aqueous content of a cell inside the plasma membrane but outside the nucleus. Organelles (excluding the nucleus) are subcellular structures within the cytoplasm that perform specific functions. The term **cytosol** is frequently used to describe the fluid portion of the cytoplasm—that is, the part that cannot be removed by centrifugation.
3. **Nucleus.** The nucleus is a large, generally spheroid body within a cell. The largest of the organelles, it contains the DNA, or genetic material, of the cell and thus directs the cell's activities. The nucleus also contains one or more *nucleoli.* Nucleoli are centers for the production of ribosomes, which are the sites of protein synthesis.

Structure of the Plasma Membrane

Because the intracellular and extracellular environments (or "compartments") are both aqueous, a barrier must be present to prevent the loss of enzymes, nucleotides, and other cellular molecules that are water-soluble. This barrier surrounding the cell cannot itself be composed of water-soluble molecules; it is instead composed of lipids.

The **plasma membrane** (also called the **cell membrane**), and indeed all of the membranes surrounding organelles within the cell, are composed primarily of phospholipids and proteins. Phospholipids, described in chapter 2, are polar (and hydrophilic) in the region that contains the phosphate group and nonpolar (and hydrophobic) throughout the rest of the molecule. Since the environment on each side of the membrane is aqueous, the hydrophobic parts of the molecules "huddle together" in the center of the membrane, leaving the polar parts exposed to water on both surfaces. This results in the formation of a double layer of phospholipids in the cell membrane.

The hydrophobic middle of the membrane restricts the passage of water and water-soluble molecules and ions. Certain of these polar compounds, however, do pass through the membrane. The specialized functions and selective transport properties of the membrane are primarily due to its protein content. Membrane proteins are described as peripheral or integral. *Peripheral proteins* are only partially embedded in one face of the membrane, whereas *integral proteins* span the membrane from one side to the other. Because the membrane is not solid—phospholipids and proteins are free to move laterally—the proteins within the phospholipid "sea" are not uniformly distributed. Rather, they present a constantly changing mosaic pattern, an arrangement known as the **fluid-mosaic model** of membrane structure (fig. 3.2).

Scientists now recognize that the fluid-mosaic model of the plasma membrane is somewhat misleading, in that the membrane is not as uniform in structure as implied by figure 3.2. The proteins in the plasma membrane can be localized according to their function, so that their distribution is patchy rather than uniform. Thus, proteins in some regions are much more crowded together in the plasma membrane than is indicated in figure 3.2. This can be extremely important, as when the membrane proteins serve as receptors for neurotransmitter chemicals released by nerve fibers at the synapse (chapter 7).

The proteins found in the plasma membrane serve a variety of functions, including structural support, transport of molecules across the membrane, and enzymatic control of chemical reactions at the cell surface. Some proteins function as receptors for hormones and other regulatory molecules that arrive at the outer surface of the membrane. Receptor proteins are usually

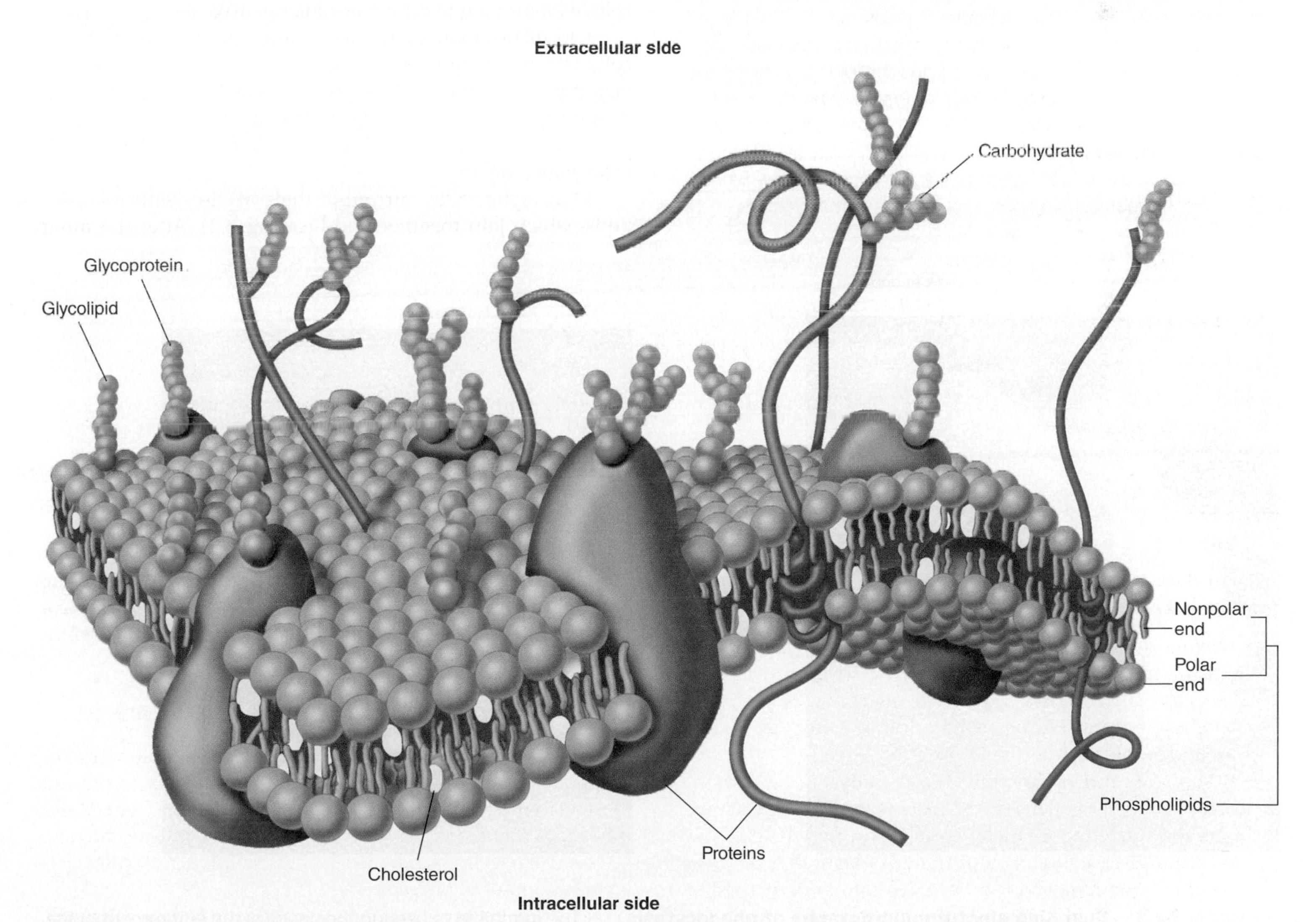

Figure 3.2 The fluid-mosaic model of the plasma membrane. The membrane consists of a double layer of phospholipids, with the polar regions (shown by spheres) oriented outward and the nonpolar hydrocarbons (wavy tails) oriented toward the center. Proteins may completely or partially span the membrane. Carbohydrates are attached to the outer surface. AP|R

specific for one particular messenger, much like an enzyme that is specific for a single substrate. Other cellular proteins serve as "markers" (antigens) that identify the tissue type of an individual.

In addition to lipids and proteins, the plasma membrane also contains carbohydrates, which are primarily attached to the outer surface of the membrane as glycoproteins and glycolipids. Certain glycolipids on the plasma membrane of red blood cells serve as antigens that determine the blood type. Other carbohydrates on the plasma membrane have numerous negative charges and, as a result, affect the interaction of regulatory molecules with the membrane. The negative charges at the surface also affect interactions between cells—they help keep red blood cells apart, for example. Stripping the carbohydrates from the outer red blood cell surface results in their more rapid destruction by the liver, spleen, and bone marrow.

CLINICAL APPLICATION

The plasma membrane contains cholesterol, which accounts for 20% to 25% of the total lipid content of the membrane. The cells in the body with the highest content of cholesterol are the Schwann cells, which form insulating layers by wrapping around certain nerve fibers (chapter 7, section 7.1). Their high cholesterol content is believed to be important in this insulating function. The ratio of cholesterol to phospholipids also helps to determine the flexibility of a plasma membrane. When there is an inherited defect in this ratio, the flexibility of the cell may be reduced. This could result, for example, in the inability of red blood cells to flex at the middle when passing through narrow blood channels, thereby causing occlusion of these small vessels.

Phagocytosis

Most of the movement of molecules and ions between the intracellular and extracellular compartments involves passage through the plasma membrane (chapter 6). However, the plasma membrane also participates in the **bulk transport** of larger portions of the extracellular environment. Bulk transport includes the processes of *phagocytosis* and *endocytosis.*

White blood cells known as *neutrophils,* and connective tissue cells called *macrophages* (literally, "big eaters"), are able to perform **amoeboid movement** (move like an amoeba, a single-celled animal). This involves extending parts of their cytoplasm to form *pseudopods* (false feet), which pull the cell through the *extracellular matrix*—generally, an extracellular gel of proteins and carbohydrates. This process depends on the bonding of proteins called *integrins,* which span the plasma membrane of these cells, with proteins in the extracellular matrix.

Cells that exhibit amoeboid motion—as well as certain liver cells, which are not mobile—use pseudopods to surround and engulf particles of organic matter (such as bacteria). This process is a type of cellular "eating" called **phagocytosis.** It serves to protect the body from invading microorganisms and to remove extracellular debris.

Phagocytic cells surround their victim with pseudopods, which join together and fuse (fig. 3.3). After the inner

Pseudopod

(a)

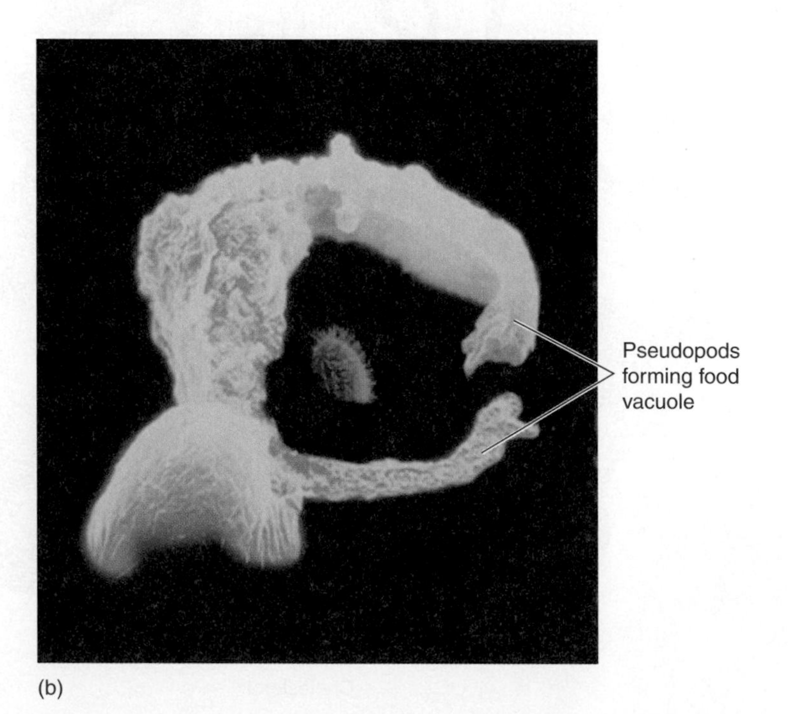

(b)

Figure 3.3 Scanning electron micrographs of phagocytosis. (*a*) The formation of pseudopods and (*b*) the entrapment of the prey within a food vacuole. APIR

membrane of the pseudopods has become a continuous membrane surrounding the ingested particle, it pinches off from the plasma membrane. The ingested particle is now contained in an organelle called a *food vacuole* within the cell. The food vacuole will subsequently fuse with an organelle called a lysosome (described later), and the particle will be digested by lysosomal enzymes.

Phagocytosis, largely by neutrophils and macrophages, is an important immune process that defends the body and promotes inflammation. Phagocytosis by macrophages is also needed for the removal of senescent (aged) cells and those that die by *apoptosis* (cell suicide, described later in this chapter). Phagocytes recognize "eat me" signals—primarily phosphatidylserine—on the plasma membrane surface of dying cells. Apoptosis is a normal, ongoing activity in the body and is not accompanied by inflammation.

Endocytosis

Endocytosis is a process in which the plasma membrane furrows inward, instead of extending outward with pseudopods. One form of endocytosis, **pinocytosis,** is a nonspecific process performed by many cells. The plasma membrane invaginates to produce a deep, narrow furrow. The membrane near the surface of this furrow then fuses, and a small vesicle containing the extracellular fluid is pinched off and enters the cell. Pinocytosis allows a cell to engulf large molecules such as proteins, as well as any other molecules that may be present in the extracellular fluid.

Another type of endocytosis involves a smaller area of plasma membrane, and it occurs only in response to specific molecules in the extracellular environment. Because the extracellular molecules must bind to very specific *receptor proteins* in the plasma membrane, this process is known as **receptor-mediated endocytosis.**

In receptor-mediated endocytosis, the interaction of specific molecules in the extracellular fluid with specific membrane receptor proteins causes the membrane to invaginate, fuse, and pinch off to form a vesicle (fig. 3.4). Vesicles formed in this way contain extracellular fluid and molecules that could not have passed by other means into the cell. Cholesterol attached to specific proteins, for example, is taken up into artery cells by receptor-mediated endocytosis. This is in part responsible for atherosclerosis (chapter 13, section 13.7). Hepatitis, polio, and AIDS viruses also exploit the process of receptor-mediated endocytosis to invade cells.

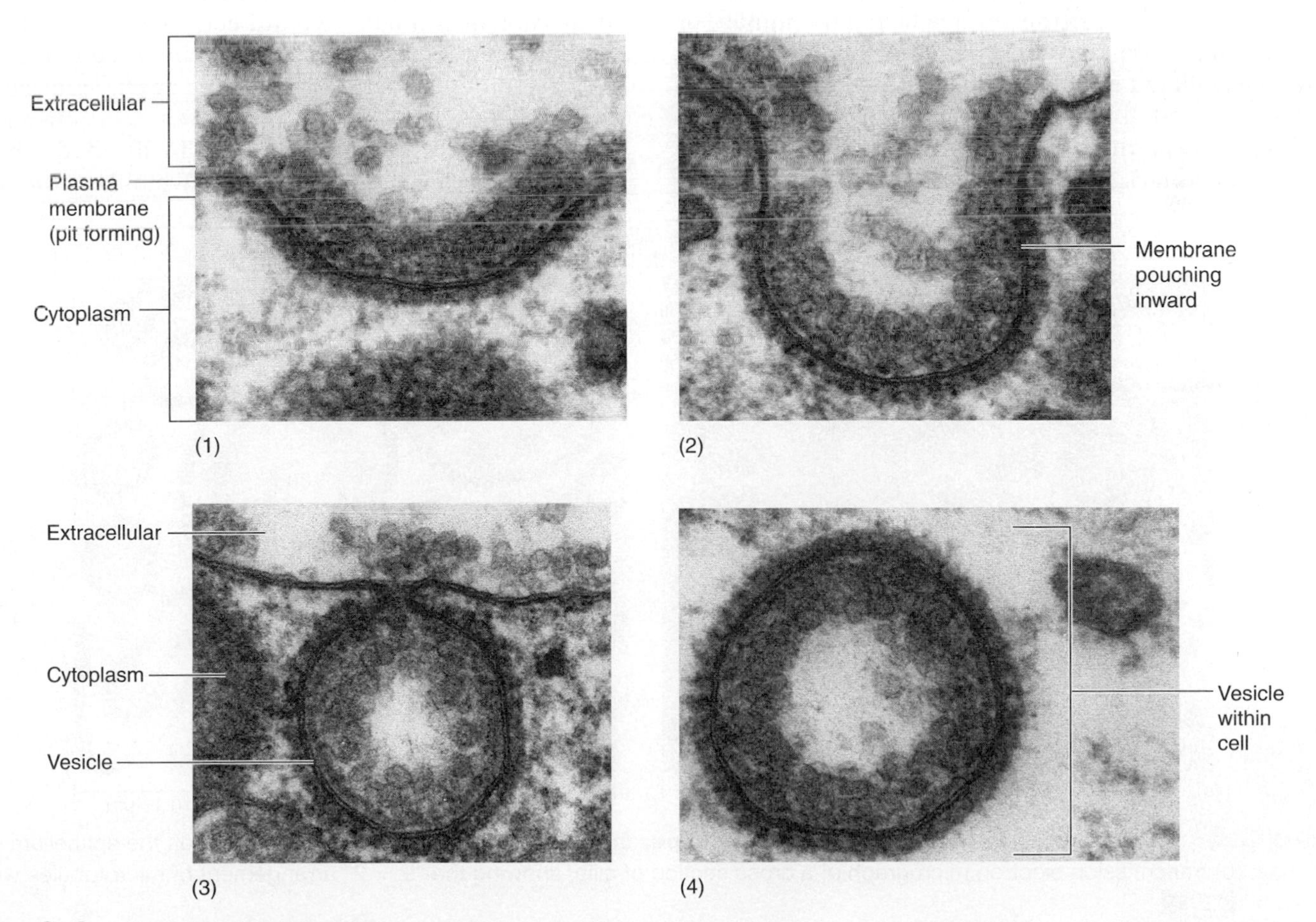

Figure 3.4 Receptor-mediated endocytosis. In stages 1 through 4 shown here, specific bonding of extracellular particles with membrane receptor proteins results in the formation of endocytotic vesicles. AP|R

Exocytosis

Exocytosis is a process by which cellular products are secreted into the extracellular environment. Proteins and other molecules produced within the cell that are destined for export (secretion) are packaged within vesicles by an organelle known as the Golgi complex. In the process of exocytosis, these secretory vesicles fuse with the plasma membrane and release their contents into the extracellular environment (see fig. 3.12). Nerve endings, for example, release their chemical neurotransmitters in this manner (chapter 7, section 7.3).

When the vesicle containing the secretory products of the cell fuses with the plasma membrane during exocytosis, the total surface area of the plasma membrane is increased. This process replaces material that was lost from the plasma membrane during endocytosis.

Cilia and Flagella

Cilia are tiny hairlike structures that project from the surface of a cell into the extracellular fluid. *Motile cilia* (those able to move) can beat like rowers in a boat, stroking in unison. Such motile cilia are found in only particular locations in the human body, where they project from the apical surface of epithelial cells (the surface facing the lumen, or cavity) that are stationary and line certain hollow organs. For example, ciliated epithelial cells are found in the respiratory system and the female reproductive tract. In the respiratory airways, the cilia transport strands of mucus to the pharynx (throat), where the mucus can be swallowed or expectorated. In the female reproductive tract, the beating of cilia on the epithelial lining of the uterine tube draws the ovum (egg) into the tube and moves it toward the uterus.

Almost every cell in the body has a single, nonmotile *primary cilium.* The functions of the primary cilia in most organs of the body are not presently understood, but primary cilia are believed to serve sensory functions. For example they are modified to form part of the photoreceptors in the retina of the eyes (chapter 10) and are believed to detect fluid movement within the tubules of the kidneys (chapter 17).

Cilia are composed of *microtubules* (thin cylinders formed from proteins) and are surrounded by a specialized part of the plasma membrane. There are 9 pairs of microtubules arranged around the circumference of the cilium; in motile cilia, there is also a pair of microtubules in the center, producing an arrangement described as "9 + 2" (fig. 3.5). The nonmotile primary cilium lacks the central pair of microtubules, and so is described as having a "9 + 0" arrangement.

Within the cell cytoplasm at the base of each cilium is a pair of structures called *centrioles,* composed of microtubules and oriented at right angles to each other (see fig. 3.28). The pair together is called a *centrosome.* The centriole that points along the axis of the cilium is also known as the *basal body,* and this structure is required to form the microtubules of the cilium. Centrosomes are also involved in the process of pulling duplicated chromosomes apart, as discussed in section 3.5.

Sperm cells are the only cells in the body that have **flagella.** The flagellum is a single, whiplike structure that

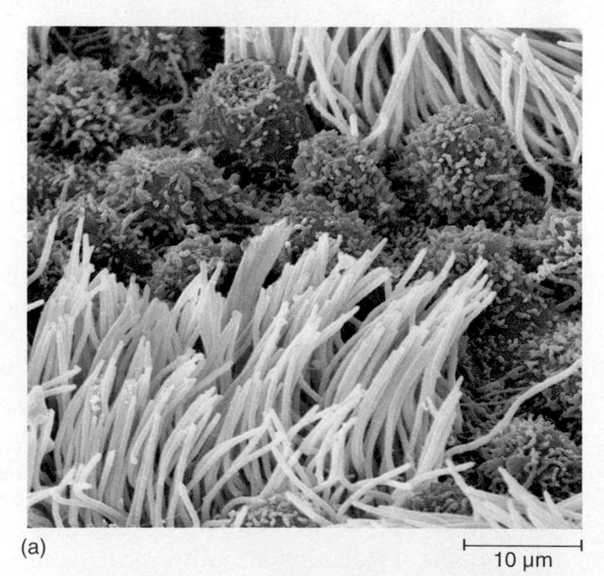

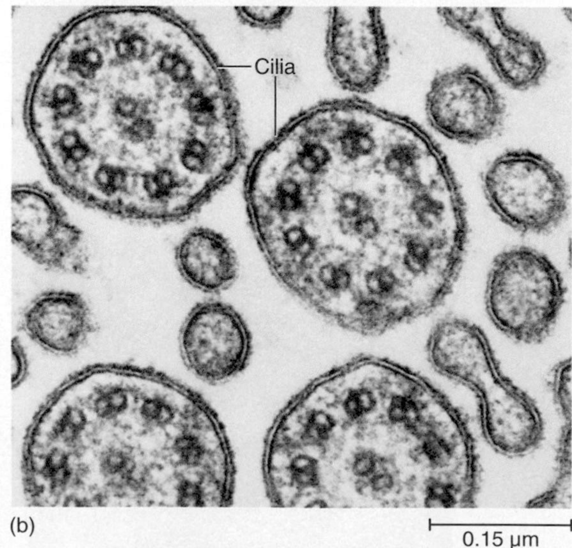

Figure 3.5 Cilia, as seen with the electron microscope. (*a*) Scanning electron micrograph of cilia on the epithelium lining the trachea; (*b*) transmission electron micrograph of a cross section of cilia, showing the "9 + 2" arrangement of microtubules within each cilium. APIR

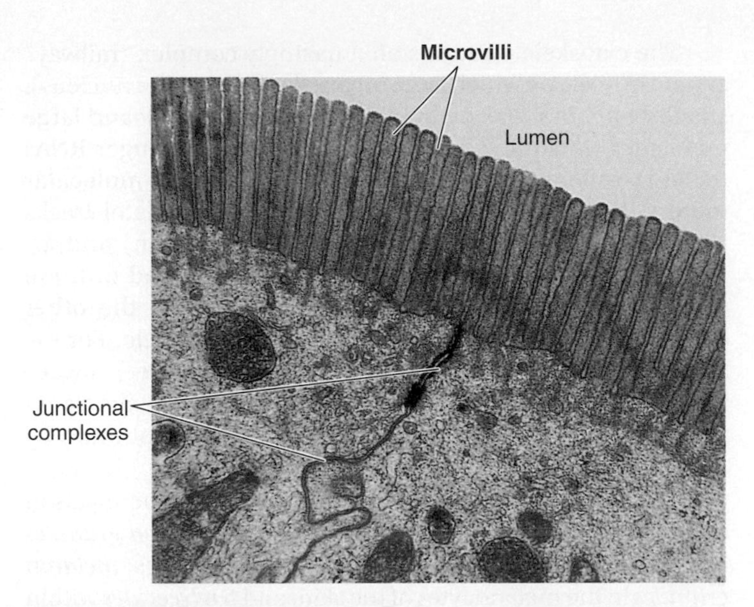

Figure 3.6 Microvilli in the small intestine. Microvilli are seen in this colorized electron micrograph, which shows two adjacent cells joined together by junctional complexes. AP|R

propels the sperm through its environment. Like the motile cilia, a flagellum is composed of microtubules with a "9 + 2" arrangement. The subject of sperm motility by means of flagella is considered with the reproductive system in chapter 20.

Microvilli

In areas of the body that are specialized for rapid diffusion, the surface area of the cell membranes may be increased by numerous folds called **microvilli.** The rapid passage of the products of digestion across the epithelial membranes in the intestine, for example, is aided by these structural adaptations. The surface area of the apical membranes (the part facing the lumen) in the intestine is increased by the numerous tiny fingerlike projections (fig. 3.6). Similar microvilli are found in the epithelia of the kidney tubules, which must reabsorb various molecules that are filtered out of the blood.

 CHECKPOINT

1a. Describe the structure of the plasma membrane.

1b. Describe the structure and function of cilia, flagella, and microvilli.

2a. Describe the different ways that cells can engulf materials in the extracellular fluid.

2b. Explain the process of exocytosis.

3.2 CYTOPLASM AND ITS ORGANELLES

Many of the functions of a cell are performed by structures called organelles. Among these are the lysosomes, which contain digestive enzymes, and the mitochondria, where most of the cellular energy is produced. Other organelles participate in the synthesis and secretion of cellular products.

LEARNING OUTCOMES

After studying this section, you should be able to:

3. Describe the structure and function of the cytoskeleton, lysosomes, peroxisomes, mitochondria, and ribosomes.

4. Describe the structure and functions of the endoplasmic reticulum and Golgi complex, and explain how they interact.

Cytoplasm and Cytoskeleton

The material within a cell (exclusive of that within the nucleus) is known as **cytoplasm.** Cytoplasm contains structures called **organelles** that are visible under the microscope, and the fluidlike **cytosol** that surrounds the organelles. When viewed in a microscope without special techniques, the cytoplasm appears to be uniform and unstructured. However, the cytosol is not a homogeneous solution; it is, rather, a highly organized structure in which protein fibers—in the form of *microtubules* and *microfilaments*—are arranged in a complex latticework surrounding the membrane-bound organelles. Using fluorescence microscopy, these structures can be visualized with the aid of antibodies against their protein components (fig. 3.7). The interconnected microfilaments and microtubules are believed to provide structural organization for cytoplasmic enzymes and support for various organelles.

The latticework of microfilaments and microtubules is said to function as a **cytoskeleton** (fig. 3.8). The structure of this "skeleton" is not rigid; it is capable of quite rapid movement and reorganization. Contractile proteins—including actin and myosin, which are responsible for muscle contraction—are associated with the microfilaments and microtubules in most cells. These structures aid in amoeboid movement, for example, so that the cytoskeleton is also the cell's "musculature." Microtubules, as another example, form the *spindle apparatus* that pulls chromosomes away from each other in cell division. Microtubules also form the central parts of cilia and flagella and contribute to the structure and movements of these projections from the cells.

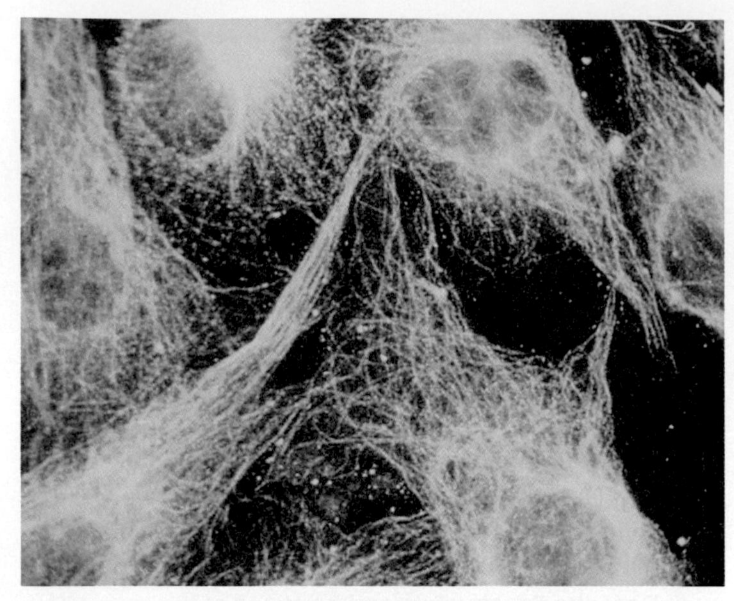

Figure 3.7 An immunofluorescence photograph of microtubules. The microtubules in this photograph are visualized with the aid of fluorescent antibodies against tubulin, the major protein component of the microtubules. AP|R

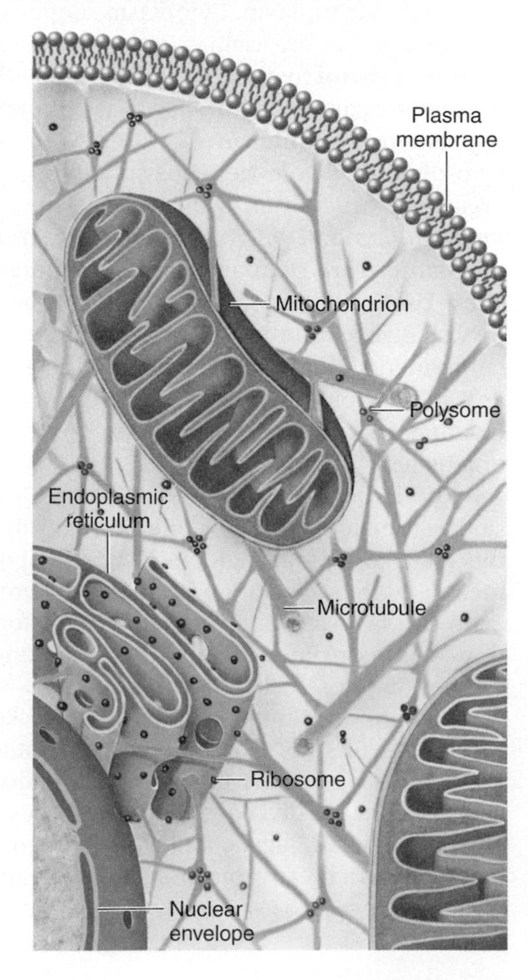

Figure 3.8 The formation of the cytoskeleton by microtubules. Microtubules are also important in the motility (movement) of the cell and movement of materials within the cell. AP|R

The cytoskeleton forms an amazingly complex "railway" system in a cell, on which large organelles (such as the nucleus), smaller membranous organelles (such as vesicles), and large molecules (including certain proteins and messenger RNA) travel to different and specific destinations. The molecular motors that move this cargo along their cytoskeletal tracks are the proteins *myosin* (along filaments of actin) and *kinesins* and *dyneins* (along microtubules). One end of these molecular motors attaches to their cargo while the other end moves along the microfilament or microtubule. For example, vesicles are moved in an axon (nerve fiber) toward its terminal by kinesin, while other vesicles can be transported in the opposite direction along the microtubule by dynein.

The cytoplasm of some cells contains stored chemicals in aggregates called **inclusions.** Examples are *glycogen granules* in the liver, striated muscles, and some other tissues; *melanin granules* in the melanocytes of the skin; and *triglycerides* within adipose cells.

Lysosomes

After a phagocytic cell has engulfed the proteins, polysaccharides, and lipids present in a particle of "food" (such as a bacterium), these molecules are still kept isolated from the cytoplasm by the membranes surrounding the food vacuole. The large molecules of proteins, polysaccharides, and lipids must first be digested into their smaller subunits (including amino acids, monosaccharides, and fatty acids) before they can cross the vacuole membrane and enter the cytoplasm.

The digestive enzymes of a cell are isolated from the cytoplasm and concentrated within membrane-bound organelles called **lysosomes,** which contain more than 60 different enzymes. A *primary lysosome* is one that contains only digestive enzymes within an environment that is more acidic than the surrounding cytoplasm.

A primary lysosome may fuse with a food vacuole to form a *secondary lysosome* that now contains the engulfed extracellular material. The digestion of structures and molecules within a vacuole by the enzymes within lysosomes is a process known as **autophagy.** Extracellular material digested by this process includes potentially disease-causing bacteria. Similarly, a membrane forms to become a vacuole around viruses and other pathogens that have entered the cell; such a vacuole is called an *autophagosome.* The membrane of the autophagosome can then fuse with the lysosomal membrane so that lysosomal enzymes can degrade the viruses. In these and other ways, autophagy by lysosomes contributes to immunity.

In addition, autophagosomes may contain damaged organelles such as peroxisomes, mitochondria (discussed shortly), and others. These fuse with primary lysosomes so autophagy can protect the cell from the toxic effects of these damaged organelles. In a similar manner, potentially toxic aggregations of proteins within the cytoplasm may be contained within autophagosomes and digested by lysosomes. Additionally, digestion by lysosomal enzymes is needed for the proper turnover of glycogen and certain lipids; lack of a particular enzyme can result in the undue accumulation of these molecules in the cell (described in the next Clinical Applications box). A lysosome that contains undigested wastes is called a *residual body.* Residual bodies may eliminate their wastes by exocytosis, or the wastes may accumulate within the cell as the cell ages.

Lysosomes have also been called "suicide bags" because a break in their membranes would release their digestive enzymes and thus destroy the cell. This happens normally in *programmed cell death* (or *apoptosis*), described in section 3.5. An example is the loss of tissues that must accompany embryonic development, when earlier structures (such as gill pouches) are remodeled or replaced as the embryo matures.

CLINICAL APPLICATION

Most, if not all, molecules in the cell have a limited life span. They are continuously destroyed and must be continuously replaced. Glycogen and some complex lipids in the brain, for example, are normally digested at a particular rate by lysosomes. If a person, because of some genetic defect, does not have the proper amount of these lysosomal enzymes, the resulting abnormal accumulation of glycogen and lipids could destroy the tissues. Examples of such defects include **Pompe disease, Tay-Sach's disease,** and **Gaucher's disease.** These are examples of the 50 known *lysosomal storage diseases.* Each is caused by a different defective enzyme produced by a defect in a single gene.

Clinical Investigation CLUES

Timothy has large amounts of glycogen granules in his liver cells, and many are seen to be intact within his secondary lysosomes.

- What are lysosomes, and why should they contain glycogen granules?
- What kind of inherited disorder could account for these observations?

Peroxisomes

Peroxisomes are membrane-enclosed organelles containing several specific enzymes that promote oxidative reactions. Although peroxisomes are present in most cells, they are particularly large and active in the liver.

Peroxisomes contain enzymes that remove hydrogen from particular organic molecules. Removal of hydrogen oxidizes the molecules and the enzymes that promote these reactions are called *oxidases* (chapter 4, section 4.3). The hydrogen is transferred to molecular oxygen (O_2), forming *hydrogen peroxide* (H_2O_2). Peroxisomes are important in the metabolism of amino acids and lipids and the production of bile acids. Peroxisomes also oxidize toxic molecules, such as formaldehyde and alcohol. For example, much of the ethanol (alcohol) ingested in drinks is oxidized to acetaldehyde by liver peroxisomes.

The enzyme *catalase* within the peroxisomes prevents the excessive accumulation of hydrogen peroxide by catalyzing the reaction $2H_2O_2 \rightarrow 2\,H_2O + O_2$. Catalase is one of the fastest acting enzymes known (see chapter 4), and it is this reaction that produces the characteristic fizzing when hydrogen peroxide is poured on a wound.

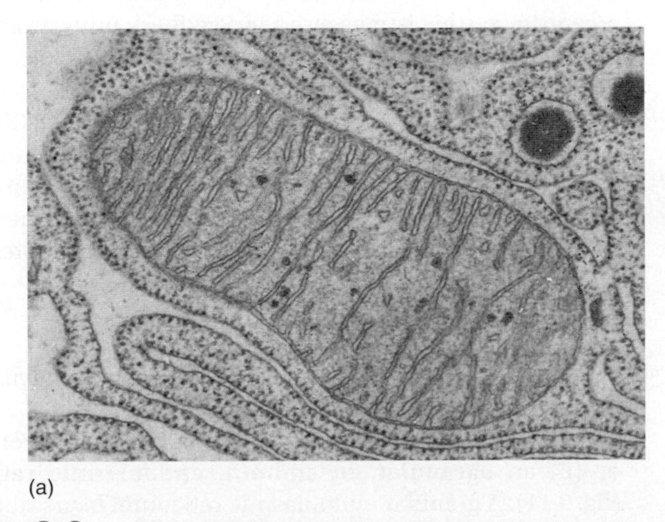

(a)

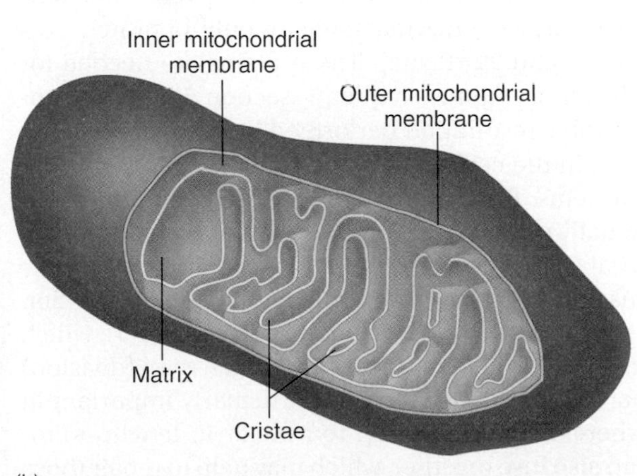

(b)

Figure 3.9 The structure of a mitochondrion. (*a*) An electron micrograph of a mitochondrion. The outer mitochondrial membrane and the infoldings of the inner membrane—the cristae—are clearly seen. The fluid in the center is the matrix. (*b*) A diagram of the structure of a mitochondrion. AP|R

Mitochondria

All cells in the body, with the exception of mature red blood cells, have from a hundred to a few thousand organelles called **mitochondria** (singular, **mitochondrion**). Mitochondria serve as sites for the production of most of the energy of cells (chapter 5, section 5.2).

Mitochondria vary in size and shape, but all have the same basic structure (fig. 3.9). Each mitochondrion is surrounded by an inner and outer membrane, separated by a narrow intermembranous space. The outer mitochondrial membrane is smooth, but the inner membrane is characterized by many folds, called *cristae,* which project like shelves into the central area (or *matrix*) of the mitochondrion. The cristae and the matrix compartmentalize the space within the mitochondrion and have different roles in the generation of cellular energy. The structure and functions of mitochondria will be described in more detail in the context of cellular metabolism in chapter 5.

Mitochondria can migrate through the cytoplasm of a cell and are able to reproduce themselves. Indeed, mitochondria contain their own DNA. All of the mitochondria in a person's body are derived from those inherited from the mother's fertilized egg cell. Thus, all of a person's mitochondrial genes are inherited from the mother. Mitochondrial DNA is more primitive (consisting of a circular, relatively small, double-stranded molecule) than that found within the cell nucleus. For this and other reasons, many scientists believe that mitochondria evolved from separate organisms, related to bacteria, that invaded the ancestors of animal cells and remained in a state of symbiosis.

This symbiosis might not always benefit the host; for example, mitochondria produce superoxide radicals that can provoke an oxidative stress (chapters 5 and 19), and some scientists believe that accumulations of mutations in mitochondrial DNA may contribute to aging. Mutations in mitochondrial DNA occur at a rate at least ten times faster than in nuclear DNA (probably due to the superoxide radicals), and there are more than 150 mutations of mitochondrial DNA presently known to contribute to different human diseases. However, mitochondrial DNA has only 37 genes that code for only 13 proteins (as well as 2 rRNAs and 22 tRNAs). The proteins are needed for oxidative phosphoryration (chapter 5, section 5.2), an essential part of aerobic respiration performed by mitochondria. By contrast, DNA in the cell nucleus codes for about 1,500 mitochondrial proteins. Because of this, many mitochondrial diseases are actually produced by mutations in nuclear rather than mitochondrial DNA.

Neurons obtain energy solely from aerobic cell respiration (a process that requires oxygen, described in chapter 5), which occurs in mitochondria. Thus, mitochondrial fission (division) and transport over long distances is particularly important in neurons, where axons can be up to 1 meter in length. Mitochondria can also fuse together, which may help to repair those damaged by "reactive oxygen species" generated within mitochondria (chapters 5 and 19).

Although mitochondria are needed for aerobic cell respiration and are thus essential for the life of the cell, the production of reactive oxygen species by mitochondria can kill the cell. Also, when mitochondria become damaged they can harm their host cells through the leakage of toxic mitochondrial molecules into the cytoplasm. The cell protects itself by enclosing the damaged mitochondria within an autophagosome, which can then fuse with a lysosome and be digested by the process of autophagy (discussed earlier).

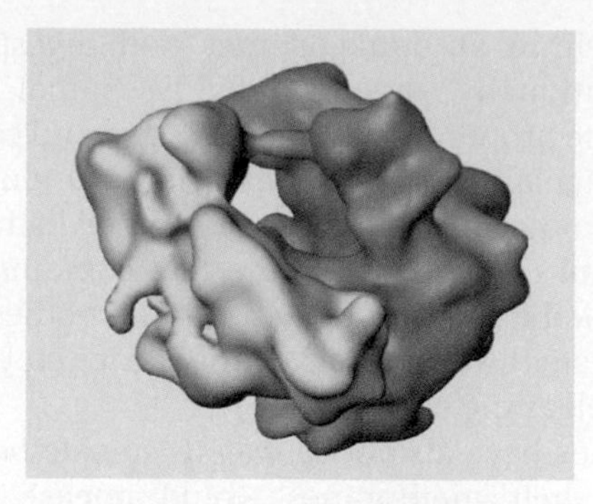

Figure 3.10 A ribosome is composed of two subunits. This is a model of the structure of a ribosome, showing the smaller (lighter) and larger (darker) subunits. The space between the two subunits accommodates a molecule of transfer RNA, needed to bring amino acids to the growing polypeptide chain. AP|R

Ribosomes

Ribosomes are often called the "protein factories" of the cell because it is here that proteins are produced according to the genetic information contained in messenger RNA (discussed in section 3.4). The ribosomes are quite tiny, about 25 nanometers in size, and can be found both free in the cytoplasm and located on the surface of an organelle called the endoplasmic reticulum (discussed next).

Each ribosome consists of two subunits (fig. 3.10), which are designated 30S and 50S after their sedimentation rate in a centrifuge (this is measured in Svedberg units, from which the "S" is derived). Each of the subunits is composed of both ribosomal RNA and proteins. Contrary to earlier expectations of most scientists, it now appears that the ribosomal RNA molecules serve as enzymes (called *ribozymes*) for many of the reactions in the ribosomes that are required for protein synthesis. Protein synthesis is covered in section 3.4, and the general subject of enzymes and catalysis is discussed in chapter 4.

Endoplasmic Reticulum

Most cells contain a system of membranes known as the **endoplasmic reticulum,** or **ER.** The ER may be either of two types: (1) a **granular,** or **rough, endoplasmic reticulum** or (2) an **agranular,** or **smooth, endoplasmic reticulum** (fig. 3.11). A granular endoplasmic reticulum bears ribosomes on its surface, whereas an agranular endoplasmic reticulum does not. The agranular endoplasmic reticulum serves a variety of purposes in different cells; it provides a site for enzyme

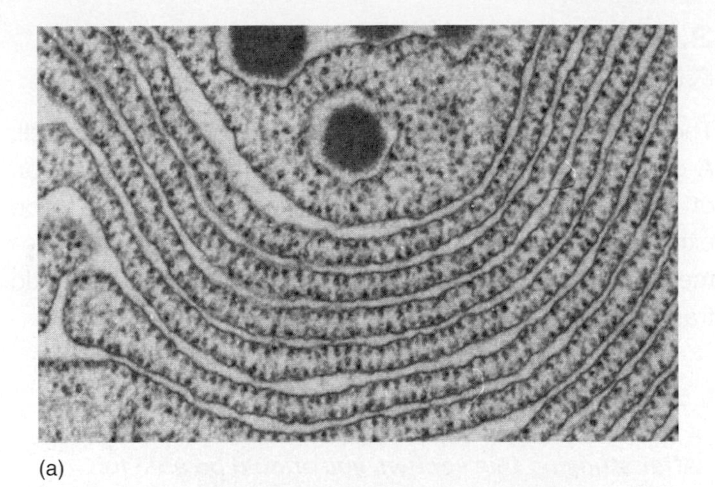

(a)

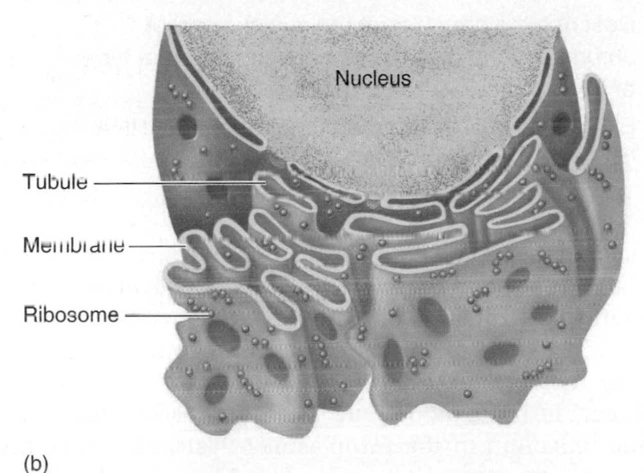

(b)

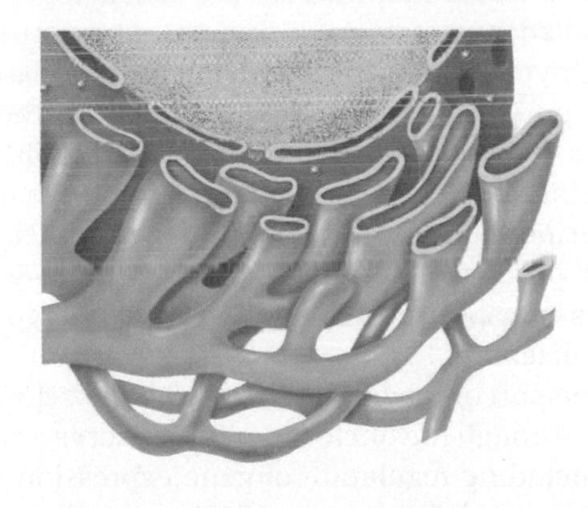

(c)

Figure 3.11 The endoplasmic reticulum. (*a*) An electron micrograph of a granular endoplasmic reticulum (about 100,000×). The granular endoplasmic reticulum (*b*) has ribosomes attached to its surface, whereas the agranular endoplasmic reticulum (*c*) lacks ribosomes. AP|R

reactions in steroid hormone production and inactivation, for example, and a site for the storage of Ca^{2+} in striated muscle cells. The granular endoplasmic reticulum is abundant in cells that are active in protein synthesis and secretion, such as those of many exocrine and endocrine glands.

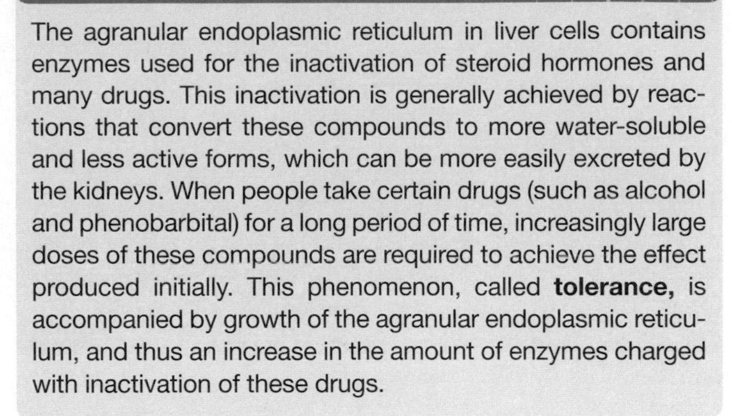

CLINICAL APPLICATION

The agranular endoplasmic reticulum in liver cells contains enzymes used for the inactivation of steroid hormones and many drugs. This inactivation is generally achieved by reactions that convert these compounds to more water-soluble and less active forms, which can be more easily excreted by the kidneys. When people take certain drugs (such as alcohol and phenobarbital) for a long period of time, increasingly large doses of these compounds are required to achieve the effect produced initially. This phenomenon, called **tolerance,** is accompanied by growth of the agranular endoplasmic reticulum, and thus an increase in the amount of enzymes charged with inactivation of these drugs.

Clinical Investigation CLUES

Timothy's liver cells show an unusually extensive smooth endoplasmic reticulum.

- What is a smooth endoplasmic reticulum, and why would it be unusually extensive in Timothy's liver cells?

Golgi Complex

The **Golgi complex,** also called the **Golgi apparatus,** consists of a stack of several flattened sacs (fig. 3.12). This is something like a stack of pancakes, but the Golgi sac "pancakes" are hollow, with cavities called *cisternae* within each sac. One side of the stack faces the endoplasmic reticulum and serves as a site of entry for vesicles from the endoplasmic reticulum that contain cellular products. The other side of the stack faces the plasma membrane, and the cellular products somehow get transferred to that side. This may be because the products are passed from one sac to the next, probably in vesicles, until reaching the sac facing the plasma membrane. Alternatively, the sac that receives the products from the endoplasmic reticulum may move through the stack until reaching the other side.

By whichever mechanism the cell product is moved through the Golgi complex, it becomes chemically modified and then, in the sac facing the plasma membrane, is packaged into vesicles that bud off the sac. Depending on the nature of the cell product, the vesicles that leave the Golgi complex may become lysosomes, or secretory vesicles (in which the product is released from the cell by exocytosis), or may serve other functions.

The reverse of exocytosis is endocytosis, as previously described; the membranous vesicle formed by that process is an *endosome.* Some cellular proteins that were released by exocytosis are recycled by a pathway that is essentially the reverse of the one depicted in figure 3.12. This reverse pathway is called **retrograde transport,** because proteins within the extracellular fluid are brought into the cell and then taken to the Golgi apparatus and the endoplasmic reticulum. Some toxins, such as the cholera toxin, and proteins from viruses (including components of HIV) rely on retrograde transport for their ability to infect cells.

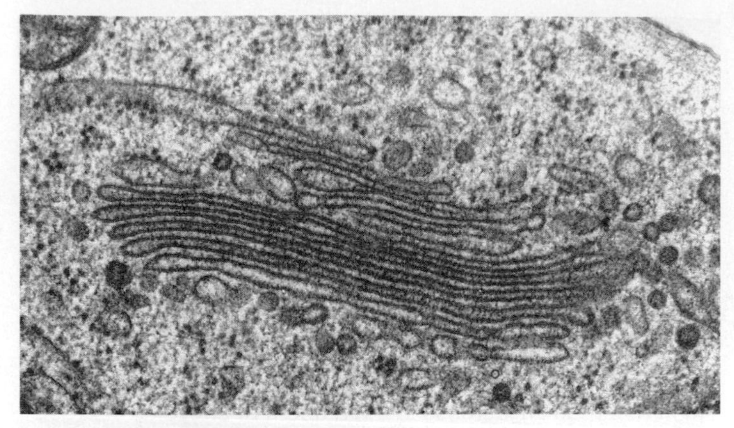

(a)

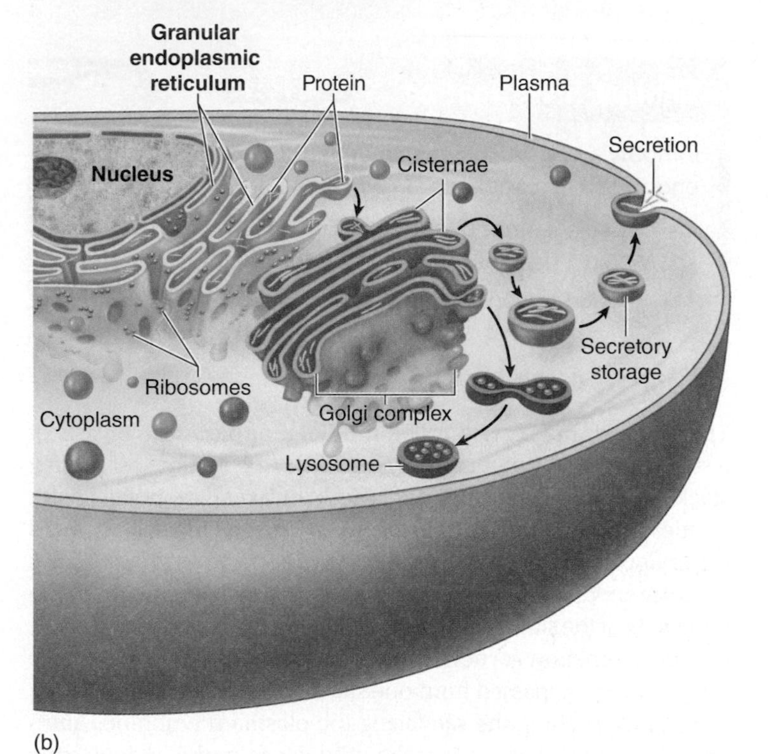

(b)

Figure 3.12 The Golgi complex. (*a*) An electron micrograph of a Golgi complex. Notice the formation of vesicles at the ends of some of the flattened sacs. (*b*) An illustration of the processing of proteins by the granular endoplasmic reticulum and Golgi complex. AP|R

CHECKPOINT

3a. Explain why microtubules and microfilaments can be thought of as the skeleton and musculature of a cell.

3b. Describe the functions of lysosomes and peroxisomes.

3c. Describe the structure and functions of mitochondria.

3d. Explain how mitochondria can provide a genetic inheritance derived only from the mother.

3e. Describe the structure and function of ribosomes.

4. Distinguish the two types of endoplasmic reticulum and explain the relationship between the endoplasmic reticulum and the Golgi complex.

3.3 CELL NUCLEUS AND GENE EXPRESSION

The nucleus is the organelle that contains the DNA of a cell. A gene is a length of DNA that codes for the production of a specific polypeptide chain. In order for genes to be expressed, they must first direct the production of complementary RNA molecules. That process is called genetic transcription.

LEARNING OUTCOMES

After studying this section, you should be able to:

5. Describe the structure of the nucleus and of chromatin, and distinguish between different types of RNA.

6. Explain how DNA directs the synthesis of RNA in genetic transcription.

Most cells in the body have a single **nucleus** (fig. 3.13). Exceptions include skeletal muscle cells, which have many nuclei, and mature red blood cells, which have none. The nucleus is enclosed by two membranes—an inner membrane and an outer membrane—that together are called the **nuclear envelope.** The outer membrane is continuous with the endoplasmic reticulum in the cytoplasm. At various points, the inner and outer membranes are fused together by structures called *nuclear pore complexes.* These structures function as rivets, holding the two membranes together. Each nuclear pore complex has a central opening, the *nuclear pore* (fig. 3.13), surrounded by interconnected rings and columns of proteins. Small molecules may pass through the complexes by diffusion, but movement of protein and RNA through the nuclear pores is a selective, energy-requiring process that requires transport proteins to ferry their cargo into and out of the nucleus.

Transport of specific proteins from the cytoplasm into the nucleus through the nuclear pores may serve a variety of functions, including regulation of gene expression by hormones (see chapter 11). Transport of RNA out of the nucleus, where it is formed, is required for gene expression. As described in this section, genes are regions of the DNA within the nucleus. Each gene contains the code for the production of a particular type of RNA called messenger RNA (mRNA). As an mRNA molecule is transported through the nuclear pore, it becomes associated with ribosomes that are either free in the cytoplasm or associated with the granular endoplasmic reticulum. The mRNA then provides the code for the production of a specific type of protein.

The primary structure of the protein (its amino acid sequence) is determined by the sequence of bases in mRNA. The base sequence of mRNA has been previously determined by the sequence of bases in the region of the DNA (the gene) that codes for the mRNA. **Genetic expression** therefore occurs

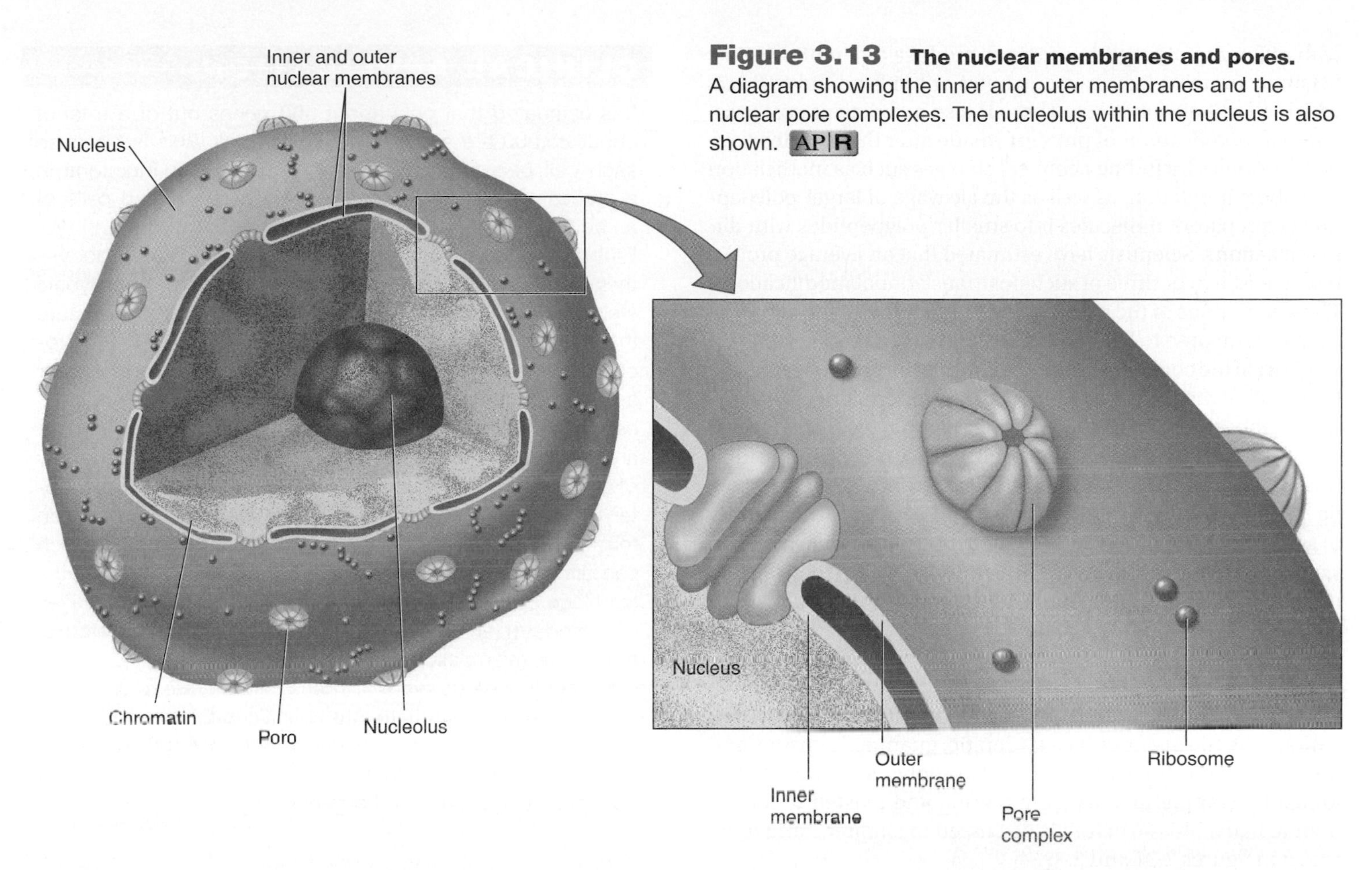

Figure 3.13 The nuclear membranes and pores. A diagram showing the inner and outer membranes and the nuclear pore complexes. The nucleolus within the nucleus is also shown. AP|R

in two stages: first **genetic transcription** (synthesis of RNA) and then **genetic translation** (synthesis of protein).

Each nucleus contains one or more dark areas (fig. 3.13). These regions, which are not surrounded by membranes, are called **nucleoli.** The DNA within the nucleoli contains the genes that code for the production of ribosomal RNA (rRNA).

Genome and Proteome

The term **genome** can refer to all of the genes in a particular individual or all of the genes in a particular species. From information gained by the Human Genome Project, scientists currently believe that a person has approximately 25,000 different genes. **Genes** are regions of DNA that code (through RNA) for polypeptide chains. Until recently it was believed that one gene coded for one protein, or at least one polypeptide chain (recall that some proteins consist of two or more polypeptide chains; see fig. 2.27*e,* for example). However, each cell produces well over 100,000 different proteins, so the number of proteins greatly exceeds the number of genes.

The term **proteome** has been coined to refer to all of the proteins produced by the genome. This concept is complicated because, in a given cell, some portion of the genome is inactive. There are proteins produced by a neuron that are not produced by a liver cell, and vice versa. Further, a given cell will produce different proteins at different times, as a result of signaling by hormones and other regulators.

So, how does a gene produce more than one protein? This is not yet completely understood. Part of the answer may include the following: (1) a given RNA coded by a gene may be cut and spliced together in different ways as described shortly (see fig. 3.17); (2) a particular polypeptide chain may associate

CLINICAL APPLICATION

The **Human Genome Project** began in 1990 as an international effort to sequence the human genome. Success was announced in 2001 when two groups published their sequences. It soon became apparent that human DNA is 99.6% similar among people; a mere 0.4% is responsible for human genetic variation. Because the human genome contains about 6 billion base pairs, this difference between one person and another still amounts to approximately 24 million base pairs. Scientists were also surprised to discover that humans have less than 25,000 genes (segments that code for polypeptide chains), rather than the 100,000 genes previously believed.

In 2005 the International Haplotype Project, or **HapMap Project,** provided a genome map of common single base pair variations, also known as *single-nucleotide polymorphisms* (abbreviated *SNPs*), among different people. Scientists hope to use this information to determine different predispositions to complex diseases and to devise personalized medical treatments. However, the application of genomic knowledge to medicine is still in its infancy.

with different polypeptide chains to produce different proteins; (3) many proteins have carbohydrates or lipids bound to them, which alter their functions. There is also a variety of *posttranslational modifications* of proteins (made after the proteins have been formed), including chemical changes such as methylation and phosphorylation, as well as the cleavage of larger polypeptide chain parent molecules into smaller polypeptides with different actions. Scientists have estimated that an average protein has at least two or three of such posttranslational modifications. These variations of the polypeptide products of a gene allow the human proteome to be many times larger than the genome.

Part of the challenge of understanding the proteome is identifying all of the proteins. This is a huge undertaking, involving many laboratories and biotechnology companies. The function of a protein, however, depends not only on its composition but also on its three-dimensional, or tertiary, structure (see fig. 2.27*d*) and on how it interacts with other proteins. The study of genomics, proteomics, and related disciplines will challenge scientists into the foreseeable future and, it is hoped, will yield important medical applications in the coming years.

CLINICAL APPLICATION

It is estimated that only about 300 genes out of a total of about 25,000 are active in any given cell. This is because each cell becomes specialized for particular functions in a process called *differentiation.* The differentiated cells of an adult are derived, or "stem from," those of the embryo. **Embryonic stem cells** can become any cell in the body—they are said to be *pluripotent.* The chromatin in embryonic stem cells is mostly euchromatin, with an open structure that permits its genes to be expressed. As development proceeds, more condensed regions of heterochromatin appear as genes become silenced during differentiation. *Adult stem cells* can differentiate into a range of specific cell types, but are not normally pluripotent. For example, the bone marrow of an adult contains such stem cells (also described in chapter 13). These include **hematopoietic stem cells,** which can form the blood cells, and **mesenchymal stem cells,** which can differentiate into osteocytes (bone cells), chondrocytes (cartilage cells), adipocytes (fat cells), and other derivatives of mesoderm (an embryonic germ layer; chapter 20). **Neural stem cells** (also described in chapter 8) have been identified in the adult nervous system. These can migrate to particular locations and differentiate into specific neuron and glial cell types in these locations. Many scientists hope that stem cells grown in tissue culture might someday be used to grow transplantable tissues and organs.

Chromatin

DNA is composed of four different nucleotide subunits that contain the nitrogenous bases adenine, guanine, cytosine, and thymine. These nucleotides form two polynucleotide chains, joined by complementary base pairing and twisted to form a double helix. This structure is discussed in chapter 2 and illustrated in figures 2.31 and 2.32.

The DNA within the cell nucleus is combined with protein to form **chromatin,** the threadlike material that makes up the chromosomes. Much of the protein content of chromatin is of a type known as *histones.* Histone proteins are positively charged and organized to form spools, about which the negatively charged strands of DNA are wound. Each spool consists of two turns of DNA, comprising 146 base pairs, wound around a core of histone proteins. This spooling creates particles known as **nucleosomes** (fig. 3.14).

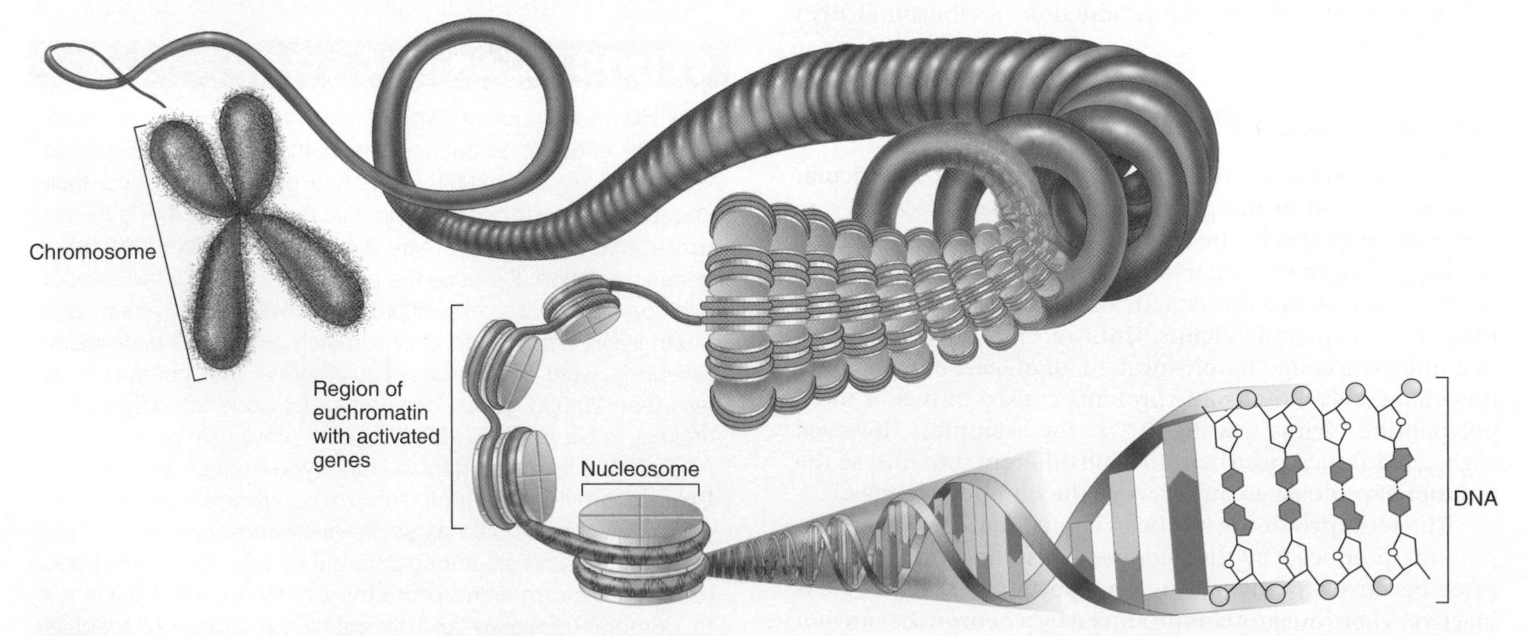

Figure 3.14 The structure of chromatin. Part of the DNA is wound around complexes of histone proteins, forming particles known as nucleosomes. AP|R

Chromatin that is active in genetic transcription (RNA synthesis) is in a relatively extended form known as **euchromatin.** By contrast, **heterochromatin** is highly condensed and forms blotchy-looking areas in the nucleus. The condensed heterochromatin contains genes that are permanently inactivated.

In the euchromatin, genes may be activated or repressed at different times. This is believed to be accomplished by chemical changes in the histones. Such changes include acetylation (the addition of two-carbon-long chemical groups), which turns on genetic transcription, and deacetylation (the removal of those groups), which stops the gene from being transcribed. The acetylation of histone proteins produces a less condensed, more open configuration of the chromatin in specific locations (fig. 3.15), allowing the DNA to be "read" by transcription factors (those that promote RNA synthesis, described next).

RNA Synthesis

Each gene is a stretch of DNA that is several thousand nucleotide pairs long. The DNA in a human cell contains over 3 billion base pairs—enough to code for at least 3 million proteins. Because the average human cell contains fewer proteins than this (30,000 to 150,000 different proteins), it follows that only a fraction of the DNA in each cell is used to code for proteins. Some of the DNA may be inactive or redundant, and some serves to regulate those regions that do code for proteins.

In order for the genetic code to be translated into the synthesis of specific proteins, the DNA code first must be copied onto a strand of RNA. This is accomplished by *DNA-directed RNA synthesis*—the process of **genetic transcription.**

There are base sequences for "start" and "stop," and regions of DNA that function as *promoters* of gene transcription. Many regulatory molecules, such as some hormones, act

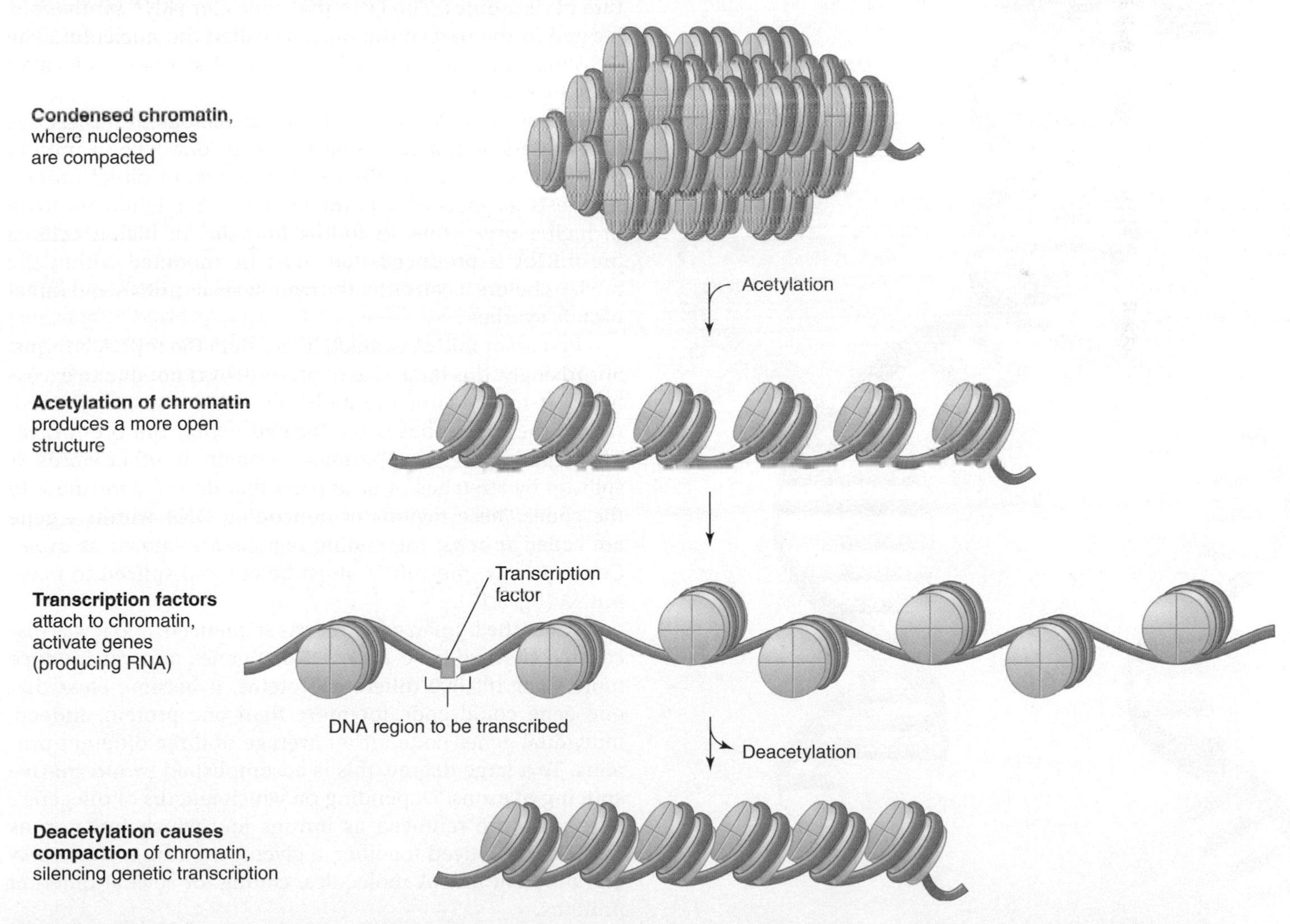
Condensed chromatin, where nucleosomes are compacted
Acetylation
Acetylation of chromatin produces a more open structure
Transcription factor
Transcription factors attach to chromatin, activate genes (producing RNA)
DNA region to be transcribed
Deacetylation
Deacetylation causes compaction of chromatin, silencing genetic transcription

Figure 3.15 Chromatin structure affects gene expression. The ability of DNA to be transcribed into messenger RNA is affected by the structure of the chromatin. The genes are silenced when the chromatin is condensed. Acetylation (addition of two-carbon groups) produces a more open chromatin structure that can be activated by transcription factors, producing mRNA. Deacetylation (removal of the acetyl groups) silences genetic transcription. AP|R

as **transcription factors** by binding to the promoter region of a specific gene and stimulating genetic transcription. Transcription (RNA synthesis) requires the enzyme **RNA polymerase,** which engages with a promoter region to transcribe an individual gene. This enzyme has a globular structure with a large central cavity; when it breaks the hydrogen bonds between DNA strands, the separated strands are forced apart within this cavity. The freed bases can then pair (by hydrogen bonding) with complementary RNA nucleotide bases present in the nucleoplasm.

This pairing of bases, like that which occurs in DNA replication (described in a later section), follows the law of complementary base pairing: *guanine bonds with cytosine* (and vice versa), and *adenine bonds with uracil* (because uracil in RNA is equivalent to thymine in DNA). Unlike DNA replication, however, only *one* of the two freed strands of DNA serves as a guide for RNA synthesis (fig. 3.16). Once an RNA molecule has been produced, it detaches from the DNA strand on which it was formed. This process can continue indefinitely, producing many thousands of RNA copies of the DNA strand that is being transcribed. When the gene is no longer to be transcribed, the separated DNA strands can then go back together again.

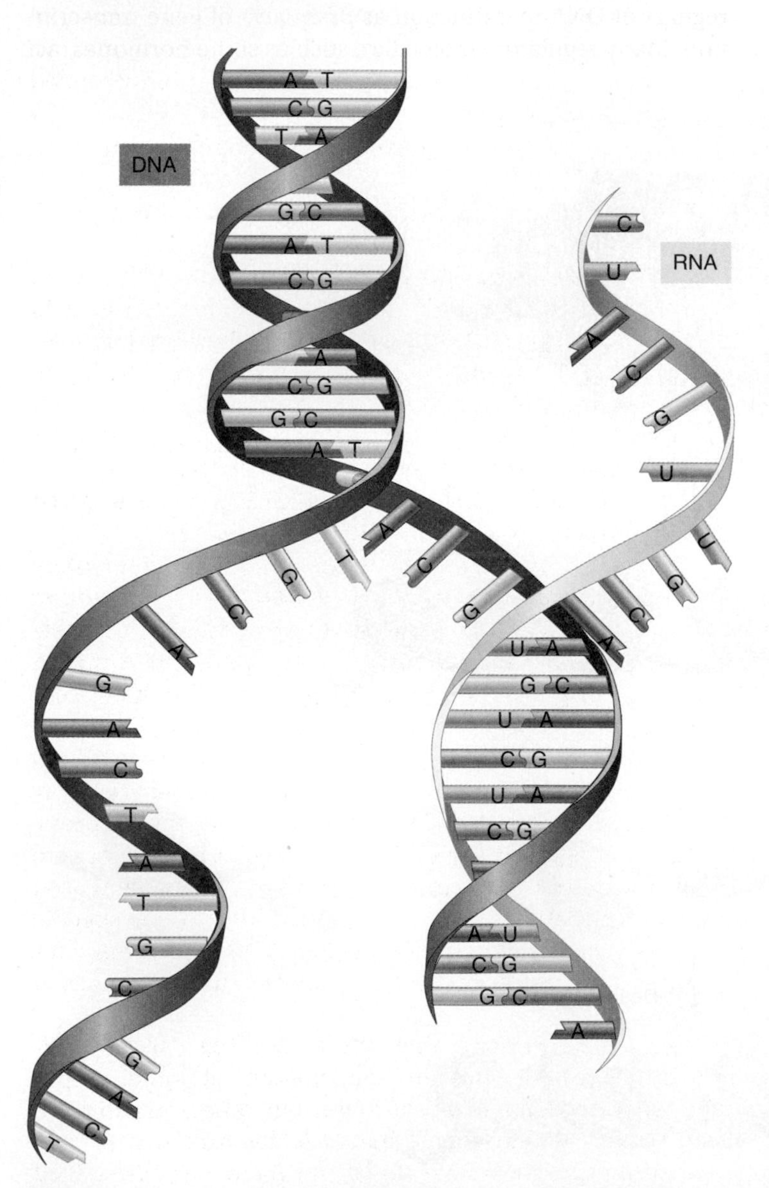

Figure 3.16 RNA synthesis (transcription). Notice that only one of the two DNA strands is used to form a single-stranded molecule of RNA. AP|R

Types of RNA

There are four types of RNA required for gene expression: (1) **precursor messenger RNA (pre-mRNA),** which is altered within the nucleus to form mRNA; (2) **messenger RNA (mRNA),** which contains the code for the synthesis of specific proteins; (3) **transfer RNA (tRNA),** which is needed for decoding the genetic message contained in mRNA; and (4) **ribosomal RNA (rRNA),** which forms part of the structure of ribosomes. The DNA that codes for rRNA synthesis is located in the part of the nucleus called the nucleolus. The DNA that codes for pre-mRNA and tRNA synthesis is located elsewhere in the nucleus.

In bacteria, where the molecular biology of the gene is best understood, a gene that codes for one type of protein produces an mRNA molecule that begins to direct protein synthesis as soon as it is transcribed. This is not the case in higher organisms, including humans. In higher cells, a pre-mRNA is produced that must be modified within the nucleus before it can enter the cytoplasm as mRNA and direct protein synthesis.

Precursor mRNA is much larger than the mRNA it forms. Surprisingly, this large size of pre-mRNA is not due to excess bases at the ends of the molecule that must be trimmed; rather, the excess bases are located *within* the pre-mRNA. The genetic code for a particular protein, in other words, is split up by stretches of base pairs that do not contribute to the code. These regions of noncoding DNA within a gene are called *introns;* the coding regions are known as *exons.* Consequently, pre-mRNA must be cut and spliced to make mRNA (fig. 3.17).

When the human genome was sequenced, and it was discovered that we have about 25,000 genes and yet produce more than 100,000 different proteins, it became clear that one gene could code for more than one protein. Indeed, individual genes code for an average of three different proteins. To a large degree, this is accomplished by **alternative splicing** of exons. Depending on which lengths of the gene's base pairs are removed as introns and which function as exons to be spliced together, a given gene can produce several different mRNA molecules, coding for several different proteins.

An estimated 92% to 94% of human genes undergo alternative splicing of exons, with most of the variation occurring between different tissues. The average gene contains eight

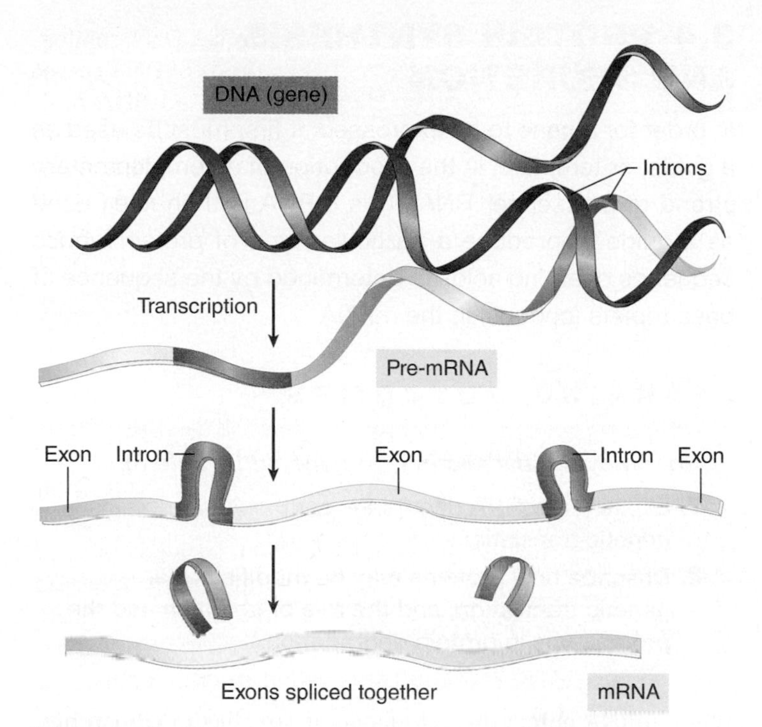

Figure 3.17 The processing of pre-mRNA into mRNA. Noncoding regions of the genes, called introns, produce excess bases within the pre-mRNA. These excess bases are removed, and the coding regions of mRNA are spliced together. Exons can be spliced together in different sequences to produce different mRNAs, and thus different proteins. AP|R

exons, although the number can be much larger—the gene for the protein "titin" contains 234 exons! Splicing together these exons in different ways could produce many variations of the protein product. The human proteome is thus much larger than the genome, allowing tremendous flexibility for different functions.

Introns are cut out of the pre-mRNA, and the ends of the exons are spliced, by macromolecules called *snRNPs* (pronounced "snurps"), producing the functional mRNA that leaves the nucleus and enters the cytoplasm. SnRNPs stands for *small nuclear ribonucleoproteins.* These are small, ribosome-like aggregates of RNA and protein that form a body called a *spliceosome* that splices the exons together.

Do the introns—removed from pre-mRNA in the formation of mRNA—have a functional significance? And, since less than 2% of the DNA codes for proteins, what about all of the other DNA located between the protein-coding genes? Is it all "junk"? Scientists once thought so, but evidence suggests that RNA molecules can themselves have important regulatory functions in the cell. For example, in some cases the RNA transcribed from regions of DNA that don't code for proteins helps regulate the expression of regions that do. This indicates that a description of the genome, and even of the proteome, may not provide a complete understanding of all of the ways that DNA regulates the cell.

RNA Interference

The 2006 Nobel Prize in Physiology or Medicine was awarded for the discovery of **RNA interference (RNA_i)**, a regulatory process performed by RNA molecules. In this process, certain RNA molecules that don't code for proteins may prevent specific mRNA molecules from being expressed (translated). RNA interference is mediated by two very similar types of RNA. One type is formed from longer double-stranded RNA molecules that leave the nucleus and are processed in the cytoplasm by an enzyme (called *Dicer*) into short (21 to 25 nucleotides long) double-stranded RNA molecules called **short interfering RNA,** or **siRNA.** The double-stranded RNA is formed from either the transcription of a segment of two complementary DNA strands, or from double-stranded RNA produced by a virus inside the host cell. In this, RNA interference is a mechanism to help combat the viral infection.

The other type of short RNA that participates in RNA interference is formed from longer RNA strands that fold into hairpin loops that resemble double-stranded RNA. These are processed by an enzyme in the nucleus and then Dicer in the cytoplasm into short (about 22 nucleotides long) double-stranded RNA molecules known as **microRNA (miRNA).** One of the two strands from the siRNA and miRNA then enter a protein particle called the *RNA-induced silencing complex (RISC),* so that this single-stranded RNA can pair by complementary base bonding to specific mRNA molecules targeted for interference.

There can be a range in the degree of complementary base pairings between one siRNA or miRNA and a number of different mRNAs. An siRNA can be perfectly complementary to a particular mRNA, forming an siRNA-mRNA duplex. In this case, the RISC will prevent the mRNA from being translated by causing destruction of the mRNA. As a result, a single siRNA can silence one particular mRNA. In contrast to siRNA, most miRNA molecules are only partially complementary to the mRNA molecules that they repress. The mechanism of this repression is complex and appears to involve both impaired translation of the mRNA and increased mRNA degradation. As a result, the synthesis of the specific proteins is reduced but not abolished. This repression contributes to the proper control of many essential processes in the cell.

Scientists estimate that there are 700–1,000 different miRNA molecules in the human genome. Separate genes code for many of these, but others are derived from introns within genes that code for proteins. In that case, when the pre-mRNA is cut and the exons are spliced together to make an mRNA (fig. 3.17), an intron removed in the process is processed into an miRNA that regulates the mRNA.

However, one miRNA can regulate the expression of more than one mRNA. This is possible because one miRNA can be incompletely complementary to a number of different mRNA molecules (from different genes), causing them to be silenced. In this way, a single miRNA may silence as many as an estimated 200 different mRNA molecules. Scientists currently estimate that at least 30% of human genes are regulated by miRNAs.

Scientists have discovered a few hundred different miRNA molecules in humans and have generated libraries of miRNAs to silence the expression of many genes. This can help in the study of normal genetic regulation and may lead to medical applications. For example, an miRNA that inhibits expression of a *tumor suppressor gene* can promote cancer, whereas a different miRNA that represses an *oncogene* (which promotes cancer) could have the opposite effect. In general, tumor cells produce fewer miRNA molecules than normal cells, and changes in the miRNA profile of metastatic cancer might be used to determine the origin, aggressiveness, and most effective treatment of the cancer. A particular miRNA that suppresses the expression of *cyclin proteins,* needed for progression through the cell cycle (discussed in section 3.5), was recently found to be abnormally lowered in mouse liver cancer cells; the introduction of this miRNA into the tumor cells inhibited their proliferation and the growth of this cancer.

In the future, RNA interference may be used medically to suppress the expression of specific genes, either abnormal genes of the patient or the genes of infectious viruses. At the time of this writing, the use of an siRNA to treat age-related macular degeneration (a major cause of blindness) is in phase III clinical trials, and others are in development to treat this same disease as well as the respiratory syncytial virus, high blood cholesterol, Huntington's disease, hepatitis C, solid tumors, AIDS lymphoma, and other conditions. Alternatively, drugs in development to treat hepatitis C and other conditions are designed to block the ability of specific miRNA molecules to inhibit genetic expression. Although these drugs may prove effective, their safety is a continuing concern.

CHECKPOINT

5. Describe the appearance and composition of chromatin and the structure of nucleosomes. Comment on the significance of histone proteins.

6a. Explain how RNA is produced within the nucleus according to the information contained in DNA.

6b. Explain how precursor mRNA is modified to produce mRNA.

3.4 PROTEIN SYNTHESIS AND SECRETION

In order for a gene to be expressed, it first must be used as a guide, or template, in the production of a complementary strand of messenger RNA. This mRNA is then itself used as a guide to produce a particular type of protein whose sequence of amino acids is determined by the sequence of base triplets (codons) in the mRNA.

LEARNING OUTCOMES

After studying this section, you should be able to:

7. **Explain how RNA directs the synthesis of proteins in genetic translation.**
8. **Describe how proteins may be modified after genetic translation, and the role of ubiquitin and the proteasome in protein degradation.**

When mRNA enters the cytoplasm, it attaches to **ribosomes,** which appear in the electron microscope as numerous small particles. A ribosome is composed of 4 molecules of ribosomal RNA and 82 proteins, arranged to form two subunits of unequal size. The mRNA passes through a number of ribosomes to form a "string-of-pearls" structure called a *polyribosome* (or *polysome,* for short), as shown in figure 3.18. The association of mRNA with ribosomes is needed for the process of **genetic translation**—the production of specific proteins according to the code contained in the mRNA base sequence.

Each mRNA molecule contains several hundred or more nucleotides, arranged in the sequence determined by complementary base pairing with DNA during transcription (RNA synthesis). Every three bases, or *base triplet,* is a code word—called a **codon**—for a specific amino acid. Sample codons and their amino acid "translations" are listed in table 3.2 and illustrated in figure 3.19. As mRNA moves through the ribosome, the sequence of codons is translated into a sequence of specific amino acids within a growing polypeptide chain.

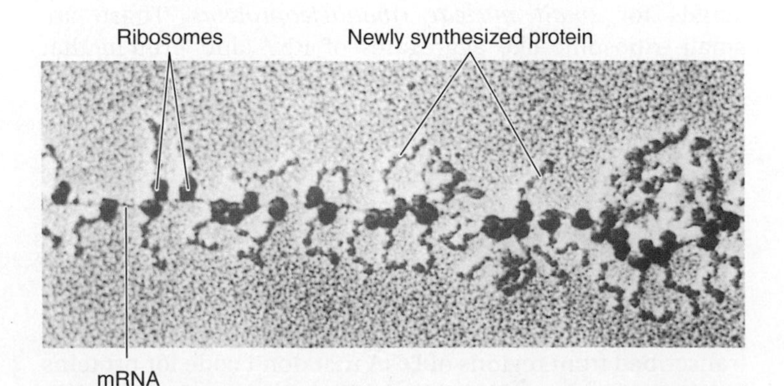

Figure 3.18 An electron micrograph of a polyribosome. An RNA strand joins the ribosomes together. AP|R

Table 3.2 | Selected DNA Base Triplets and mRNA Codons*

DNA Triplet	RNA Codon	Amino Acid
TAC	AUG	"Start" (Methionine)
ATC	UAG	"Stop"
AAA	UUU	Phenylalanine
AGG	UCC	Serine
ACA	UGU	Cysteine
GGG	CCC	Proline
GAA	CUU	Leucine
GCT	CGA	Arginine
TTT	AAA	Lysine
TGC	ACG	Threonine
CCG	GGC	Glycine
CTC	GAG	Glutamic acid

*In most cases there is actually more than one codon for each of the different amino acids, although only one codon per amino acid is shown in this table. Also, there are three different "stop" codons, for a total of 64 different codons.

CLINICAL APPLICATION

Huntington's disease is a progressive neurological disease of the brain, particularly of the basal ganglia (chapter 8; see fig. 8.11), causing a variety of crippling physical and psychological conditions. It's a genetic disease, inherited as a dominant trait on chromosome number 4. The defective gene, termed *huntingtin,* has a characteristic "stutter" where the base triplet CAG can be repeated from 40 to as many as 250 times. This causes the amino acid glutamine, coded by CAG, to be repeated in the protein product of the gene. For unknown reasons, this defective protein causes neural degeneration. In a similar manner **fragile X syndrome,** the most common genetic cause of mental retardation, is produced when there are 200 or more repeats of CGG in a gene known as *FMR1*.

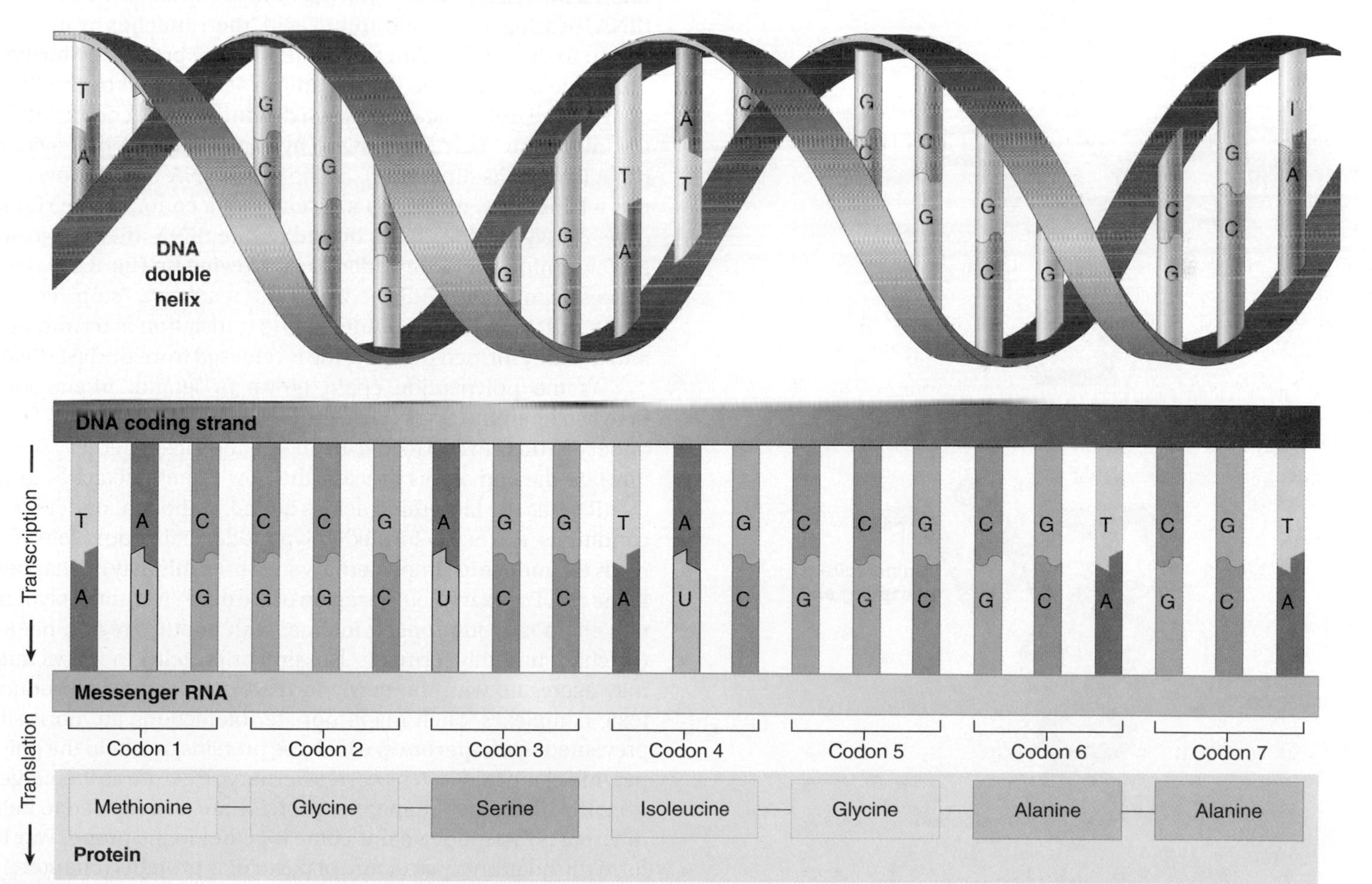

Figure 3.19 Transcription and translation. The genetic code is first transcribed into base triplets (codons) in mRNA and then translated into a specific sequence of amino acids in a polypeptide. AP|R

See the *Test Your Quantitative Ability* section of the Review Activities at the end of this chapter.

Transfer RNA

Translation of the codons is accomplished by tRNA and particular enzymes. Each tRNA molecule, like mRNA and rRNA, is single-stranded. Although tRNA is single-stranded, it bends in on itself to form a cloverleaf structure (fig. 3.20*a*), which is further twisted into an upside down "L" shape (fig. 3.20*b*). One end of the "L" contains the **anticodon**—three nucleotides that are complementary to a specific codon in mRNA.

Enzymes in the cell cytoplasm called *aminoacyl-tRNA synthetase enzymes* join specific amino acids to the ends of tRNA, so that a tRNA with a given anticodon can bind to only one specific amino acid. There are 61 different codons for the 20 different amino acids (and 3 that code for "stop"), so there must be different tRNA molecules and synthetase enzymes specific for each codon and amino acid. Each synthetase enzyme recognizes its amino acid and joins it to the tRNA that bears a specific anticodon. The cytoplasm of a cell thus contains tRNA molecules that are each bonded to a specific amino acid, and each of these tRNA molecules is capable of bonding with a specific codon in mRNA via its anticodon base triplet.

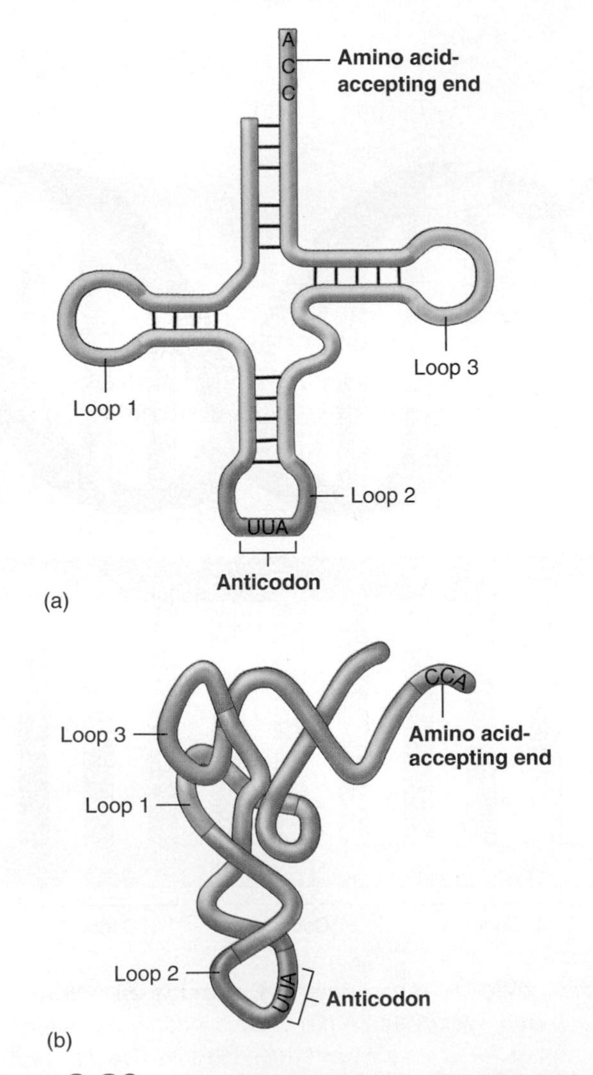

Figure 3.20 The structure of transfer RNA (tRNA). (*a*) A simplified cloverleaf representation and (*b*) the three-dimensional structure of tRNA. AP|R

Formation of a Polypeptide

The anticodons of tRNA bind to the codons of mRNA as the mRNA moves through the ribosome. Because each tRNA molecule carries a specific amino acid, the joining together of these amino acids by peptide bonds creates a polypeptide whose amino acid sequence has been determined by the sequence of codons in mRNA.

Two tRNA molecules containing anticodons specific to the first and second mRNA codons enter a ribosome, each carrying its own specific amino acid. After anticodon-codon binding between the tRNA and mRNA, the first amino acid detaches from its tRNA and bonds to the second amino acid, forming a dipeptide attached to the second tRNA. While this occurs, the mRNA moves down a distance of one codon within the ribosome, allowing the first tRNA (now minus its amino acid) to detach from the mRNA. The second tRNA with its dipeptide thereby moves down one position in the ribosome. A third tRNA, bearing its specific amino acid, then attaches by its anticodon to the third codon of the mRNA. The previously formed dipeptide is now moved to the amino acid carried by the third tRNA as the mRNA again moves a distance of one codon within the ribosome. This is followed by the release of the second tRNA (minus its dipeptide), as the third tRNA, which now carries a tripeptide, moves up a distance of a codon in the ribosome. A polypeptide chain, bound to one tRNA, thereby grows as new amino acids are added to its growing tip (fig. 3.21). This process continues until the ribosome reaches a "stop" codon in the mRNA, at which point genetic translation is terminated and the fully formed polypeptide is released from the last tRNA.

As the polypeptide chain grows in length, interactions between its amino acids cause the chain to twist into a helix (secondary structure) and to fold and bend upon itself (tertiary structure). At the end of this process, the new protein detaches from the tRNA as the last amino acid is added. Although, under ideal conditions, the newly formed polypeptide chain could fold correctly to produce its proper tertiary structure, this may not happen in the cell. For example, one region of the newly forming polypeptide chain may improperly interact with another region before the chain has fully formed. Also, similar proteins in the vicinity may aggregate with the newly formed polypeptide to produce toxic complexes. Such inappropriate interactions are normally prevented by **chaperones,** which are proteins that help the polypeptide chain fold into its correct tertiary structure as it emerges from the ribosome. Chaperone proteins are also needed to help different polypeptide chains come together in the proper way to form the quaternary structure of particular proteins (chapter 2).

Many proteins are further modified after they are formed; these modifications occur in the rough endoplasmic reticulum and Golgi complex.

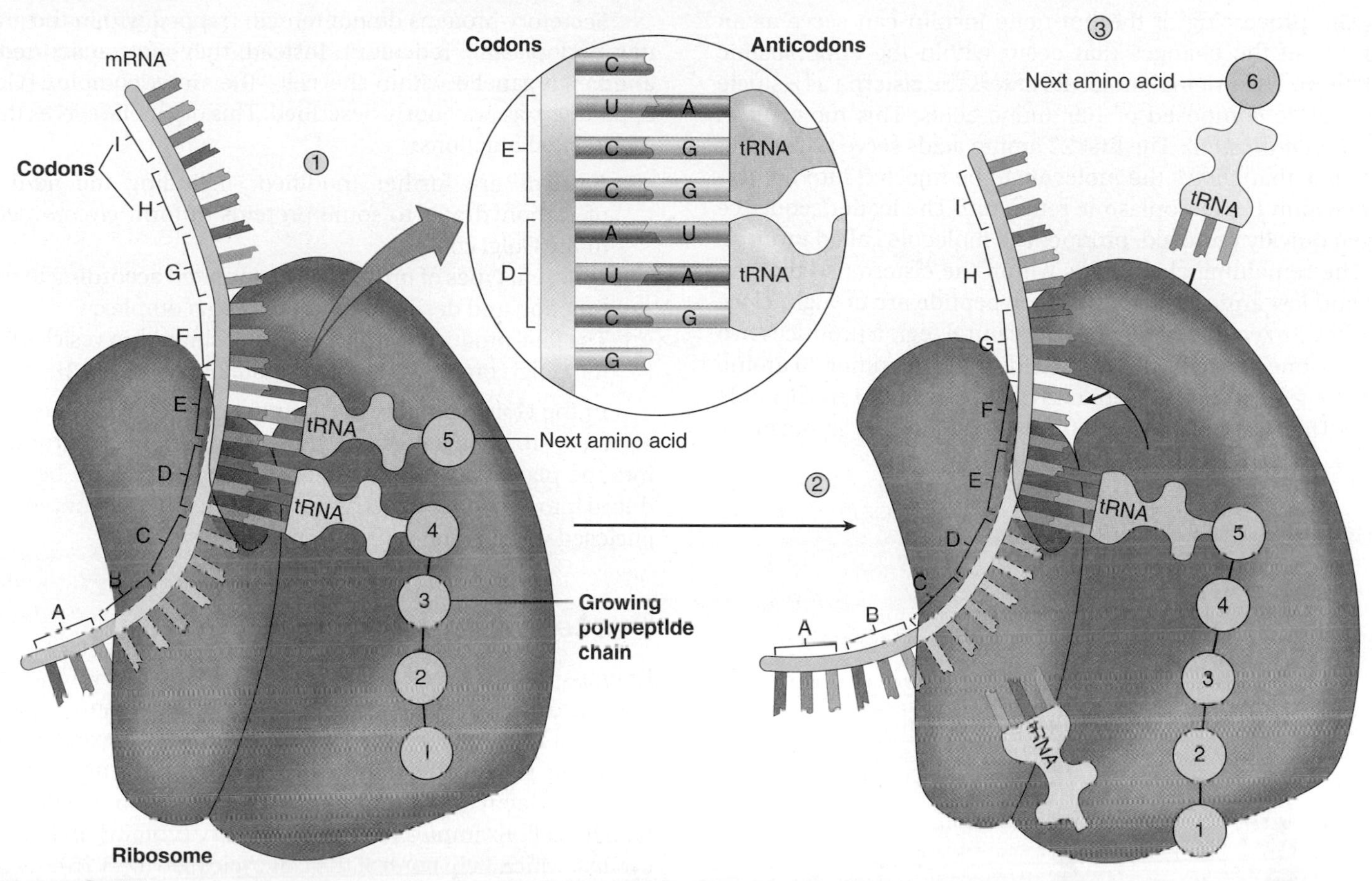

Figure 3.21 The translation of messenger RNA (mRNA). (1) The anticodon of an aminoacyl-tRNA bonds with a codon on the mRNA, so that the specific amino acid it carries can form a peptide bond with the last amino acid of a growing polypeptide. (2) The tRNA that brought the next-to-last amino acid dissociates from the mRNA, so that the growing polypeptide is attached to only the last tRNA. (3) Another tRNA carrying another amino acid will bond to the next codon in the mRNA, so that this amino acid will be at the new growing end of the polypeptide. AP|R

Functions of the Endoplasmic Reticulum and Golgi Complex

Proteins that are to be used within the cell are likely to be produced by polyribosomes that float freely in the cytoplasm, unattached to other organelles. If the protein is to be secreted by the cell, however, it is made by mRNA-ribosome complexes that are located on the granular endoplasmic reticulum. The membranes of this system enclose fluid-filled spaces called *cisternae,* into which the newly formed proteins may enter. Once in the cisternae, the structure of these proteins is modified in specific ways.

When proteins destined for secretion are produced, the first 30 or so amino acids are primarily hydrophobic. This *leader sequence* is attracted to the lipid component of the membranes of the endoplasmic reticulum. As the polypeptide chain elongates, it is "injected" into the cisterna within the endoplasmic reticulum. The leader sequence is, in a sense, an "address" that directs secretory proteins into the endoplasmic reticulum. Once the proteins are in the cisterna, the leader sequence is enzymatically removed so that the protein cannot reenter the cytoplasm (fig. 3.22).

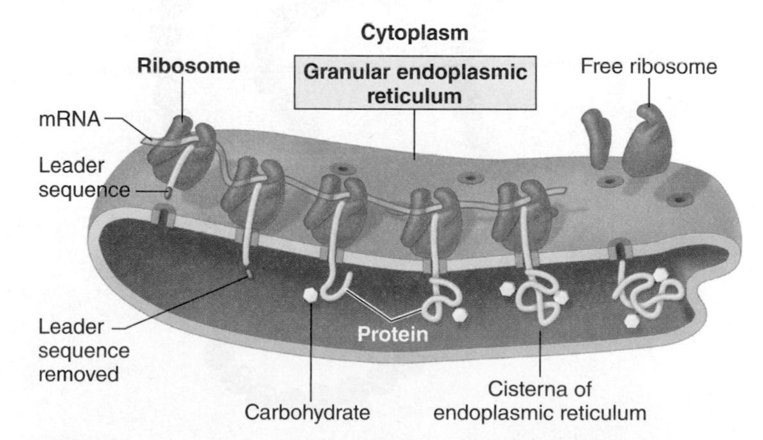

Figure 3.22 How secretory proteins enter the endoplasmic reticulum. A protein destined for secretion begins with a leader sequence that enables it to be inserted into the cisterna (cavity) of the endoplasmic reticulum. Once it has been inserted, the leader sequence is removed and carbohydrate is added to the protein. AP|R

The processing of the hormone insulin can serve as an example of the changes that occur within the endoplasmic reticulum. The original molecule enters the cisterna as a single polypeptide composed of 109 amino acids. This molecule is called *preproinsulin.* The first 23 amino acids serve as a leader sequence that allows the molecule to be injected into the cisterna within the endoplasmic reticulum. The leader sequence is then quickly removed, producing a molecule called *proinsulin.* The remaining chain folds within the cisterna so that the first and last amino acids in the polypeptide are brought close together. Enzymatic removal of the central region produces two chains—one of them 21 amino acids long, the other 30 amino acids long—that are subsequently joined together by disulfide bonds (fig. 3.23). This is the form of insulin that is normally secreted from the cell.

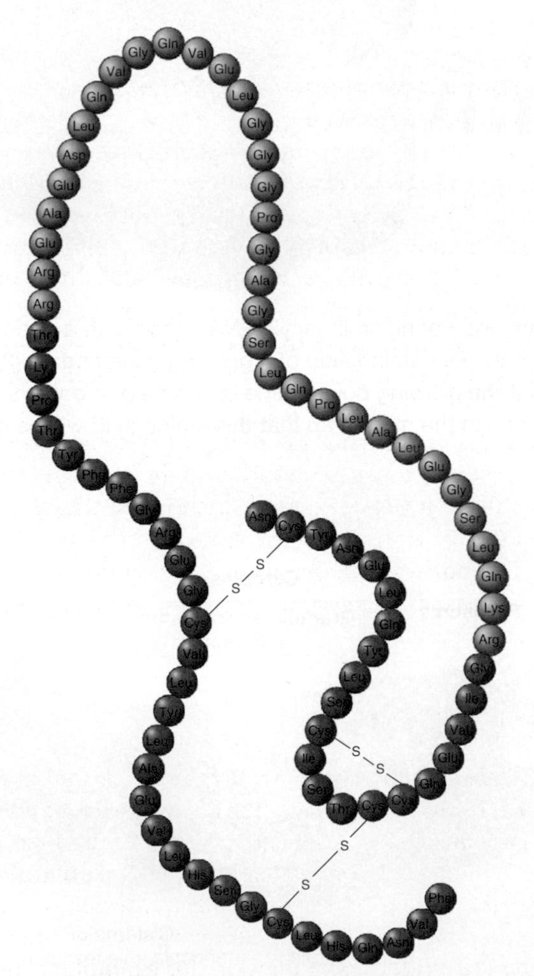

Figure 3.23 The conversion of proinsulin into insulin. The long polypeptide chain called proinsulin is converted into the active hormone insulin by enzymatic removal of a length of amino acids (shown in green). The insulin molecule produced in this way consists of two polypeptide chains (red circles) joined by disulfide bonds. AP|R

Secretory proteins do not remain trapped within the granular endoplasmic reticulum. Instead, they are transported to another organelle within the cell—the Golgi complex (Golgi apparatus), as previously described. This organelle serves three interrelated functions:

1. Proteins are further modified (including the addition of carbohydrates to some proteins to form *glycoproteins*) in the Golgi complex.
2. Different types of proteins are separated according to their function and destination in the Golgi complex.
3. The final products are packaged and shipped in vesicles from the Golgi complex to their destinations (see fig. 3.12).

In the Golgi complex, for example, proteins that are to be secreted are separated from those that will be incorporated into the plasma membrane and from those that will be introduced into lysosomes. Each is packaged in different membrane-enclosed vesicles and sent to its proper destination.

Protein Degradation

Proteins within a cell have numerous regulatory functions. Many proteins are enzymes, which increase the rate of specific chemical reactions (chapter 4). This can have diverse effects, including gene activation and inactivation. Other proteins modify the activity of particular enzymes, and so help to regulate the cell. Examples of such regulatory proteins include the cyclins, which help control the cell cycle (see fig. 3.25).

Because proteins have so many important functions, the processes of genetic transcription and translation have to be physiologically regulated. Hormones and other chemical signals can turn specific genes on or off, regulating protein synthesis. However, for critically important proteins, tighter control is required. Regulatory proteins are rapidly degraded (hydrolyzed, or digested), quickly ending their effects so that other proteins can produce new actions. This affords a much tighter control of specific regulatory proteins than would be possible if they persisted longer and only their synthesis was regulated.

Protease enzymes (those that digest proteins) located in the lysosomes digest many types of cellular proteins. In recent years, however, scientists learned that critical regulatory proteins are also degraded outside of lysosomes in a process that requires cellular energy (ATP). In this process, the regulatory proteins to be destroyed are first tagged by binding to molecules of **ubiquitin** (Latin for "everywhere"), a short polypeptide composed of 76 amino acids. Ubiquitin bonds to one or more lysine amino acids in the targeted cell protein, in a complex process that requires many enzymes and is subject to regulation. This tagging with ubiquitin is required for the proteins to be degraded by the **proteasome,** a large protease enzyme complex. Degradation of ubiquitin-tagged proteins within proteasomes eliminates defective proteins (for example, incorrectly folded proteins produced in the endoplasmic reticulum) and promotes cell regulation.

For example, the stepwise progression through the cell cycle requires the stepwise degradation of particular cyclin proteins.

The ubiquitin-proteasome system is the major route by which regulatory proteins in the cytoplasm are degraded. In addition, tagging with ubiquitin helps remove selected plasma membrane proteins (receptor proteins, for example; chapter 6, section 6.5). In that process, the membrane containing the selected proteins invaginates to form a vesicle that is digested within lysosomes. Ubiquination is even believed to tag organelles such as mitochondria for selective destruction by lysosomes in the process of autophagy (section 3.2).

CHECKPOINT

7a. Explain how mRNA, rRNA, and tRNA function during the process of protein synthesis.

7b. Describe the granular endoplasmic reticulum, and explain how the processing of secretory proteins differs from the processing of proteins that remain within the cell.

8. Describe posttranslational changes and other functions of the Golgi complex and the roles of ubiquitin and the proteasome.

3.5 DNA SYNTHESIS AND CELL DIVISION

When a cell is going to divide, each strand of the DNA within its nucleus acts as a template for the formation of a new complementary strand. Organs grow and repair themselves through a type of cell division known as mitosis. The two daughter cells produced by mitosis both contain the same genetic information as the parent cell. Gametes contain only half the number of chromosomes as their parent cell and are formed by a type of cell division called meiosis.

LEARNING OUTCOMES

After studying this section, you should be able to:

9. Explain the semiconservative replication of DNA in DNA synthesis.

10. Describe the cell cycle and identify some factors that affect it, and explain the significance of apoptosis.

11. Identify the phases of mitosis and meiosis, and distinguish between them.

Genetic information is required for the life of the cell and for the cell to be able to perform its functions in the body. Each cell obtains this genetic information from its parent cell through the process of DNA replication and cell division. DNA is the only type of molecule in the body capable of replicating itself, and mechanisms exist within the dividing cell to ensure that the duplicate copies of DNA will be properly distributed to the daughter cells.

DNA Replication

When a cell is going to divide, each DNA molecule replicates itself, and each of the identical DNA copies thus produced is distributed to the two daughter cells. Replication of DNA requires the action of a complex composed of many enzymes and proteins. As this complex moves along the DNA molecule, certain enzymes (*DNA helicases*) break the weak hydrogen bonds between complementary bases to produce two free strands at a fork in the double-stranded molecule. As a result, the bases of each of the two freed DNA strands can bond with new complementary bases (which are part of nucleotides) that are available in the surrounding environment.

According to the rules of complementary base pairing, the bases of each original strand will bond with the appropriate free nucleotides—adenine bases pair with thymine-containing nucleotides, guanine bases pair with cytosine-containing nucleotides. Enzymes called **DNA polymerases** join the nucleotides together to form a second polynucleotide chain in each DNA that is complementary to the first DNA strand. In this way, two new molecules of DNA, each containing two complementary strands, are formed. Thus, two new double-helix DNA molecules are produced that contain the same base sequence as the parent molecule (fig. 3.24).

When DNA replicates, therefore, each copy is composed of one new strand and one strand from the original DNA molecule. Replication is said to be **semiconservative** (half of the original DNA is "conserved" in each of the new DNA molecules). Through this mechanism, the sequence of bases in DNA—the basis of the genetic code—is preserved from one cell generation to the next.

The Cell Cycle

Unlike the life of an organism, which can be viewed as a linear progression from birth to death, the life of a cell follows a cyclical pattern. Each cell is produced as a part of its "parent" cell; when the daughter cell divides, it in turn becomes two new cells. In a sense, then, each cell is potentially immortal as long as its progeny can continue to divide. Some cells in the body divide frequently; the epidermis of the skin, for example, is renewed approximately every two weeks, and the stomach lining is renewed every two or three days. Other cells, such as striated muscle cells in the adult, do not divide at all. All cells in the body, of course, live only as long as the person lives (some cells live longer than others, but eventually all cells die when vital functions cease).

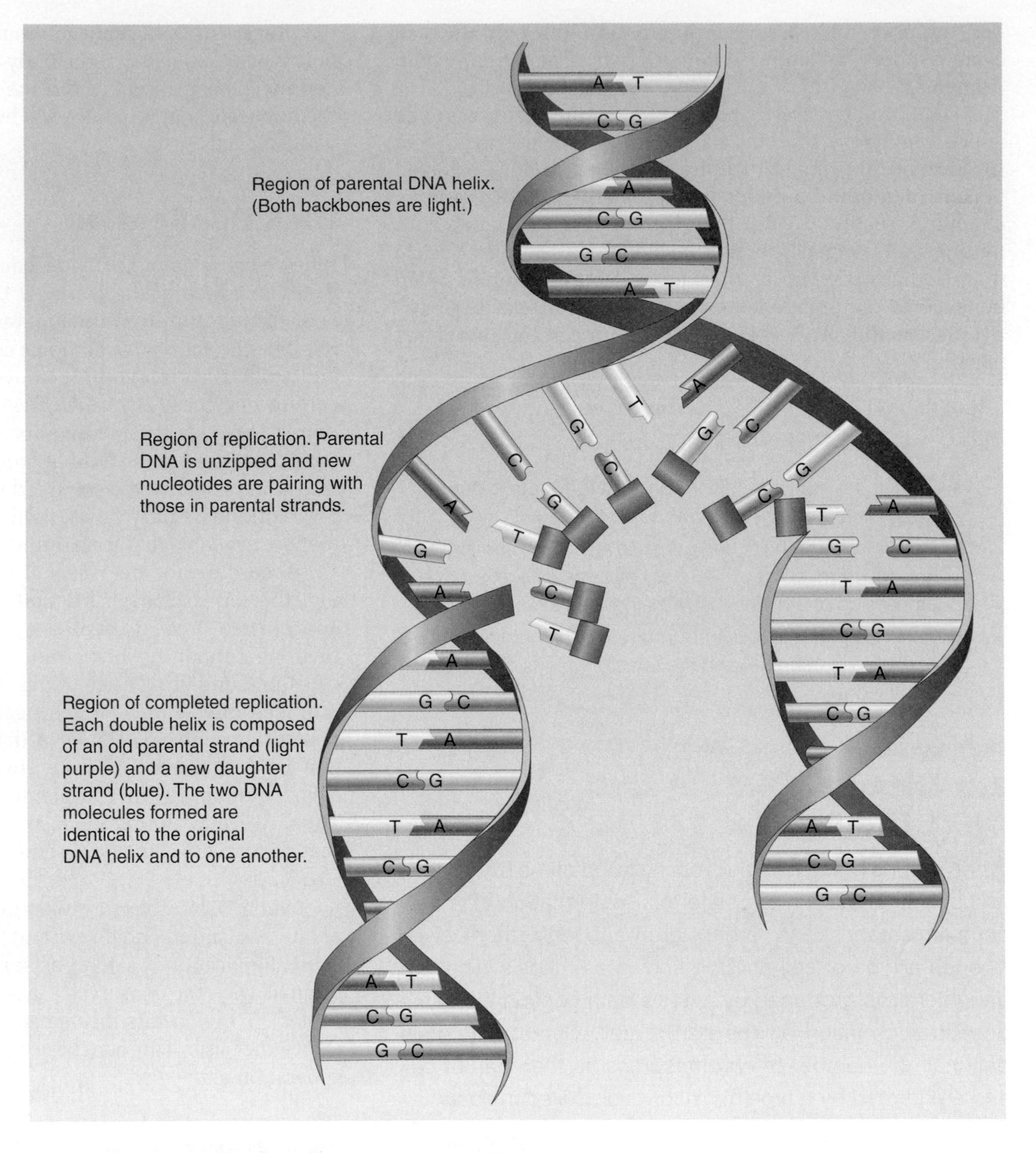

Region of parental DNA helix. (Both backbones are light.)
Region of replication. Parental DNA is unzipped and new nucleotides are pairing with those in parental strands.
Region of completed replication. Each double helix is composed of an old parental strand (light purple) and a new daughter strand (blue). The two DNA molecules formed are identical to the original DNA helix and to one another.

Figure 3.24 The replication of DNA. Each new double helix is composed of one old and one new strand. The base sequence of each of the new molecules is identical to that of the parent DNA because of complementary base pairing. APR

The nondividing cell is in a part of its life cycle known as interphase (fig. 3.25), which is subdivided into G_1, S, and G_2 phases, as will be described shortly. The chromosomes are in their extended form, and their genes actively direct the synthesis of RNA. Through their direction of RNA synthesis, genes control the metabolism of the cell. The cell may be growing during this time, and this part of interphase is known as the *G_1 phase* (*G* stands for *gap*). Although sometimes described as "resting," cells in the G_1 phase perform the physiological functions characteristic of the tissue in which they are found. The DNA of resting cells in the G_1 phase thus produces mRNA and proteins as previously described.

If a cell is going to divide, it replicates its DNA in a part of interphase known as the *S phase* (*S* stands for *synthesis*). Once DNA has replicated in the S phase, the chromatin condenses in the *G_2 phase* to form short, thick structures by the end of G_2. Though condensed, the chromosomes are not yet in their more familiar, visible form in the ordinary (light) microscope; these will first make their appearance at prophase of mitosis (fig. 3.26).

Cyclins and p53

A group of proteins known as the **cyclins**—so called because they accumulate prior to mitosis and then are rapidly destroyed during cell division—promote different phases of the cell cycle. During the G_1 phase of the cycle, for example, an increase in the concentration of *cyclin D* proteins within the cell acts to move the cell quickly through this phase. Cyclin D proteins do this by activating a group of otherwise inactive enzymes known as *cyclin-dependent kinases.*

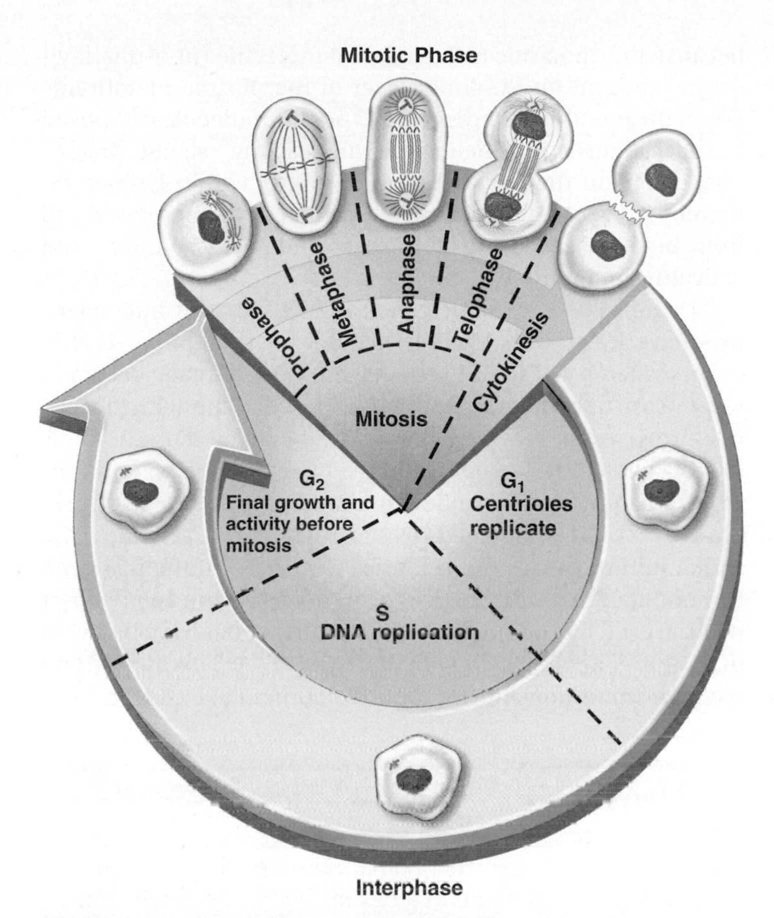

Figure 3.25 The life cycle of a cell. The different stages of mitotic division are shown; it should be noted, however, that not all cells undergo mitosis. AP|R

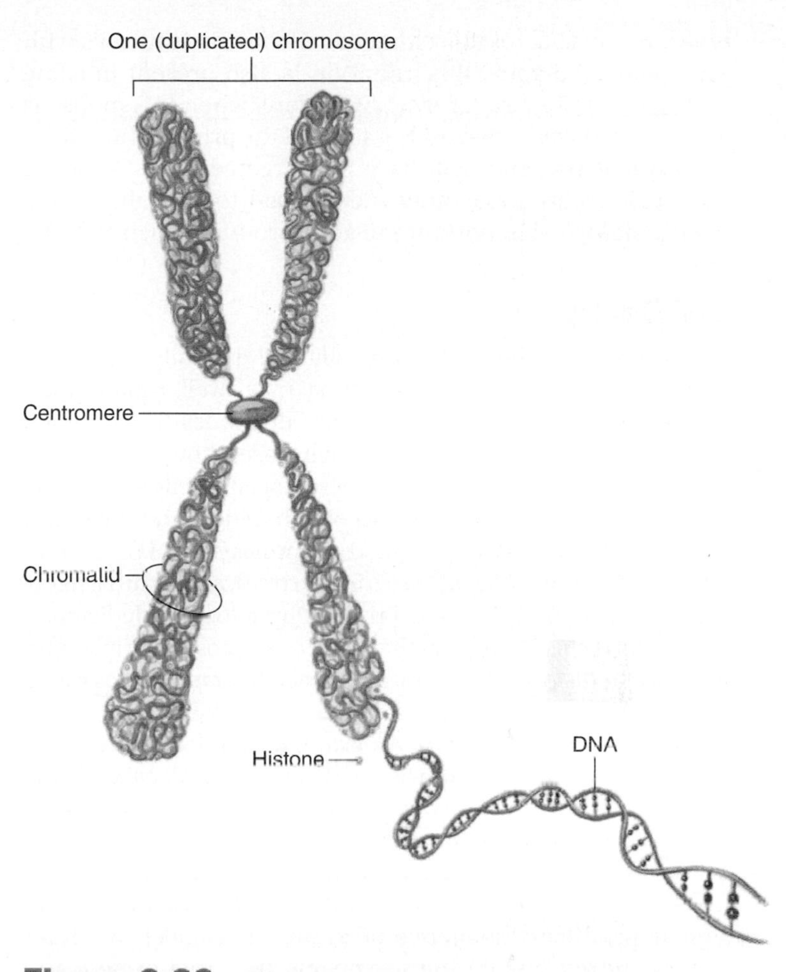

Figure 3.26 The structure of a chromosome after DNA replication. At this stage, a chromosome consists of two identical strands, or chromatids. AP|R

Overactivity of a gene that codes for a cyclin D might be predicted to cause uncontrolled cell division, as occurs in a cancer. Indeed, overexpression of the gene for cyclin D1 has been shown to occur in some cancers, including those of the breast and esophagus. Genes that contribute to cancer are called **oncogenes.** Oncogenes are altered forms of normal *proto-oncogenes,* which code for proteins that control cell division and *apoptosis* (cell suicide, discussed shortly). Conversion of proto-oncogenes to active oncogenes occurs because of genetic mutations and chromosome rearrangements (including translocations and inversions of particular chromosomal segments in different cancers).

Whereas oncogenes promote cancer, other genes—called **tumor suppressor genes**—inhibit its development. One very important tumor suppressor gene is known as **p53.** This name refers to the protein coded by the gene, which has a molecular weight of 53,000. The p53 is a *transcription factor:* a protein that can bind to DNA and activate or repress a large number of genes. When there is damage to DNA, p53 acts to stall cell division, mainly at the G_1 to S checkpoint of the cell cycle. Depending on the situation, p53 could help repair DNA while the cell cycle is arrested, or it could help promote *apoptosis* (cell death) so that the damaged DNA isn't replicated and passed on to daughter cells.

Through these and other mechanisms, the normal p53 gene protects against cancer caused by damage to DNA through radiation, toxic chemicals, or other cellular stresses. The ability of p53 to suppress cancer by causing cell cycle arrest or apoptosis is reduced in over 50% of all cancers. In about half of these cases, the gene for p53 is mutated, usually by a "point mutation" that causes a change in a single amino acid in the p53 protein. In the other half of cancers associated with inadequate p53 action, the p53 is normal but its ability to function is reduced. This can occur if there are defects in other proteins needed for p53 to function. Scientists are currently investigating molecules that might restore p53 activity as potential treatments for cancer.

Experimental mice with their gene for p53 "knocked out" all developed tumors. The 2007 Nobel Prize in Physiology or Medicine was awarded to the scientists who developed **knockout mice**—strains of mice in which a specific, targeted gene has been inactivated. This is done using mouse embryonic stem cells (chapter 20, section 20.6), which can be grown *in vitro.* A defective copy of the gene is made and introduced into the embryonic stem cells, which are then put into a normal (wild-type) embryo. The mouse that develops from this embryo is a *chimera,* or mixture of the normal and mutant types. Because all of this chimera's tissues contain cells with the inactivated gene, this mutation is also present in some of its gametes (sperm or ova). Therefore, when this mouse is mated with a wild-type mouse, some of the progeny (and their subsequent progeny) will have the targeted gene "knocked out." This technique is now widely used to help determine the physiological importance of gene products, such as p53.

Cell Death

Cell death occurs both pathologically and naturally. Pathologically, cells deprived of a blood supply may swell, rupture their membranes, and burst. Such cellular death, leading to tissue death, is known as **necrosis.** In certain cases, however, a different pattern is observed. Instead of swelling, the cells shrink. The membranes remain intact but become bubbled, and the nuclei condense. This process was named **apoptosis** (from a Greek term describing the shedding of leaves from a tree), and its discoverers were awarded the 2002 Nobel Prize in Physiology or Medicine.

There are two pathways that lead to apoptosis: extrinsic and intrinsic. In the extrinsic pathway, extracellular molecules called *death ligands* bind to receptor proteins on the plasma membrane called *death receptors.* An example of a death receptor is one known as FAS; the death ligand that binds to it is called FASL.

In the intrinsic pathway, apoptosis occurs in response to intracellular signals. This may be triggered by DNA damage, for example, or by reactive oxygen species that cause oxidative stress (discussed in chapters 5 and 19). Cellular stress signals produce a sequence of events that make the outer mitochondrial membrane permeable to cytochrome *c* and some other mitochondrial molecules, which leak into the cytoplasm and participate in the next phase of apoptosis.

The intrinsic and extrinsic pathways of apoptosis both result in the activation of a group of previously inactive cytoplasmic enzymes known as **caspases.** Caspases have been called the "executioners" of the cell, activating processes that lead to fragmentation of the DNA and death of the cell. Apoptosis is a normal, physiological process that also helps the body rid itself of cancerous cells with damaged DNA.

Apoptosis occurs normally as part of programmed cell death—a process described previously in the section on lysosomes. Programmed cell death is the physiological process responsible for the remodeling of tissues during embryonic development and for tissue turnover in the adult body. As mentioned earlier, the epithelial cells lining the digestive tract are programmed to die two to three days after they are produced, and epidermal cells of the skin live only for about two weeks until they die and become completely cornified. Apoptosis is also important in the functioning of the immune system. A neutrophil (a type of white blood cell), for example, is programmed to die by apoptosis 24 hours after its creation in the bone marrow. A killer T lymphocyte (another type of white blood cell) destroys targeted cells by triggering their apoptosis.

When a cell is dying by apoptosis, it releases chemicals that attract phagocytic macrophages (section 3.1). Macrophages recognize and eat dead cells, sparing the healthy cells. This is because the apoptotic cell displays a molecule (phosphatidylserine) present on the inner layer of the plasma membrane. When the membrane is disrupted and this molecule is exposed to macrophages, it functions as an "eat me" signal. Macrophages engulf the dead cell and digest it within lysosomes, thereby preventing the interior contents of the apoptotic cell from being released into the extracellular environment and activating an immune response.

Using mice with their gene for p53 knocked out, scientists have learned that p53 is needed for the apoptosis that occurs when a cell's DNA is damaged. DNA damage occurs in response to ultraviolet light in cells exposed to sunlight; tobacco (all forms); cancer-causing chemicals, including those in foods (such as heterocyclic amines in overcooked meats); and ionizing radiation (as from radioactive radon gas produced by uranium decay). The damaged DNA, if not repaired, activates p53, which in turn causes the cell to be destroyed. If the p53 gene has mutated to an ineffective form, however, the cell will not be destroyed by apoptosis as it should be; instead it will divide to produce daughter cells with damaged DNA. This may be one mechanism responsible for the development of a cancer.

CLINICAL APPLICATION

There are three forms of **skin cancer**—squamous cell carcinoma, basal cell carcinoma, and melanoma, depending on the type of epidermal cell involved—all of which are promoted by the damaging effects of the ultraviolet portion of sunlight. Ultraviolet light promotes a characteristic type of DNA mutation in which either of two pyrimidines (cytosine or thymine) is affected. In squamous cell and basal cell carcinoma (but not melanoma), the cancer is believed to involve mutations that affect the p53 gene, among others. Whereas cells with normal p53 genes may die by apoptosis when their DNA is damaged, and are thus prevented from replicating themselves and perpetuating the damaged DNA, those damaged cells with a mutated p53 gene survive and divide to produce the cancer.

Mitosis

At the end of the G_2 phase of the cell cycle, which is generally shorter than G_1, each chromosome consists of two strands called **chromatids** that are joined together by a *centromere* (fig. 3.26). The two chromatids within a chromosome contain identical DNA base sequences because each is produced by the semiconservative replication of DNA. Each chromatid, therefore, contains a complete double-helix DNA molecule that is a copy of the single DNA molecule existing prior to replication. Each chromatid will become a separate chromosome once mitotic cell division has been completed.

The G_2 phase completes interphase. The cell next proceeds through the various stages of cell division, or **mitosis.** This is the *M phase* of the cell cycle. Mitosis is subdivided into four stages: *prophase, metaphase, anaphase,* and *telophase* (fig. 3.27).

(a) **Interphase**
- The chromosomes are in an extended form and seen as chromatin in the electron microscope.
- The nucleus is visible.

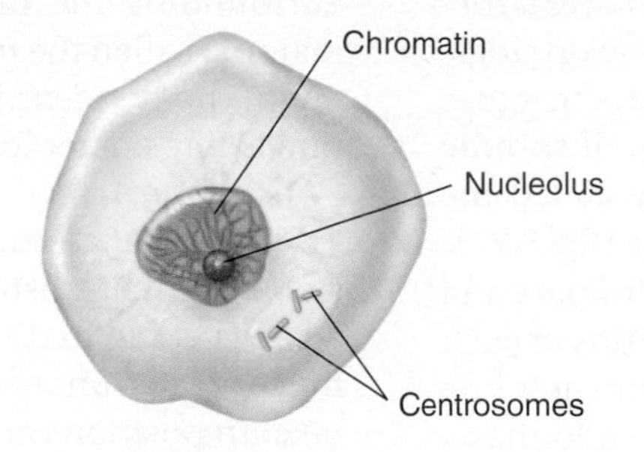

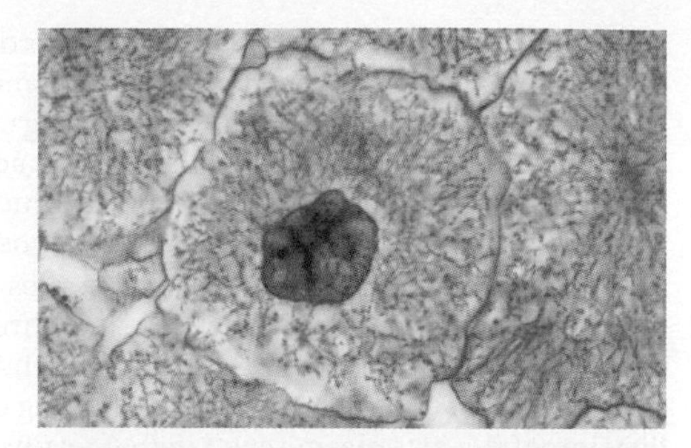

(b) **Prophase**
- The chromosomes are seen to consist of two chromatids joined by a centromere.
- The centrioles move apart toward opposite poles of the cell.
- Spindle fibers are produced and extend from each centrosome.
- The nuclear membrane starts to disappear.
- The nucleolus is no longer visible.

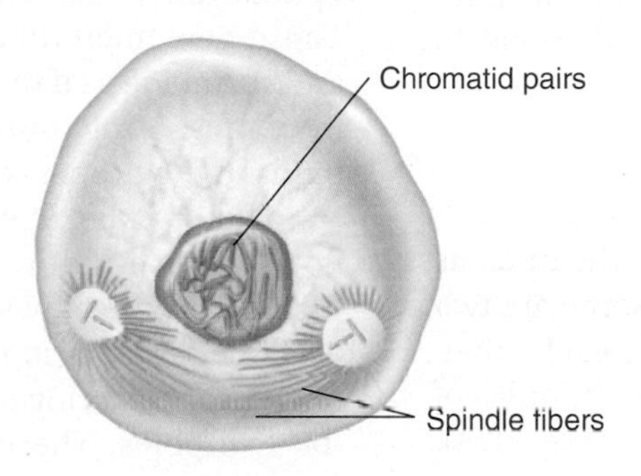

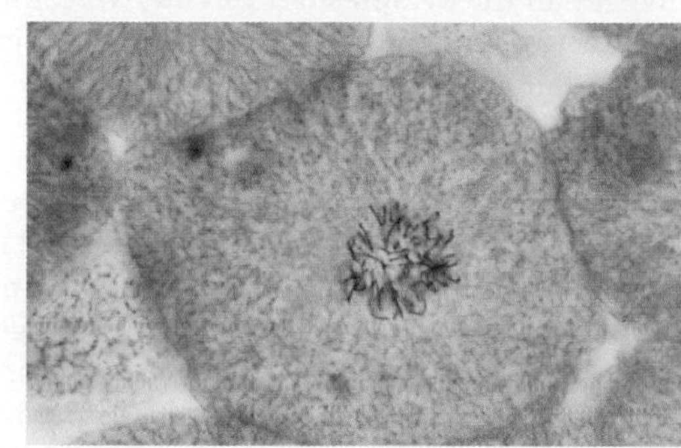

(c) **Metaphase**
- The chromosomes are lined up at the equator of the cell.
- The spindle fibers from each centriole are attached to the centromeres of the chromosomes.
- The nuclear membrane has disappeared.

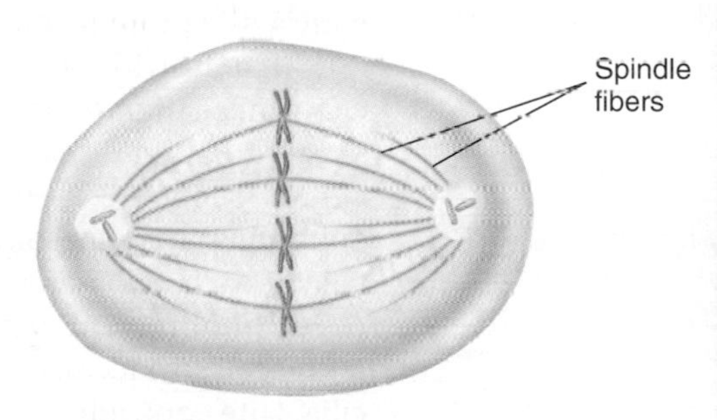

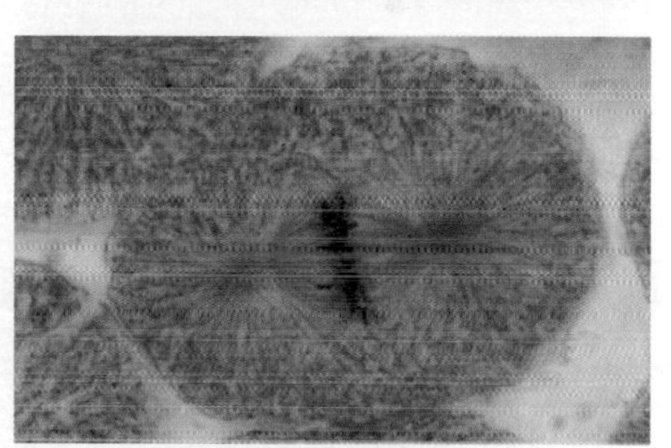

(d) **Anaphase**
- The centromeres split, and the sister chromatids separate as each is pulled to an opposite pole.

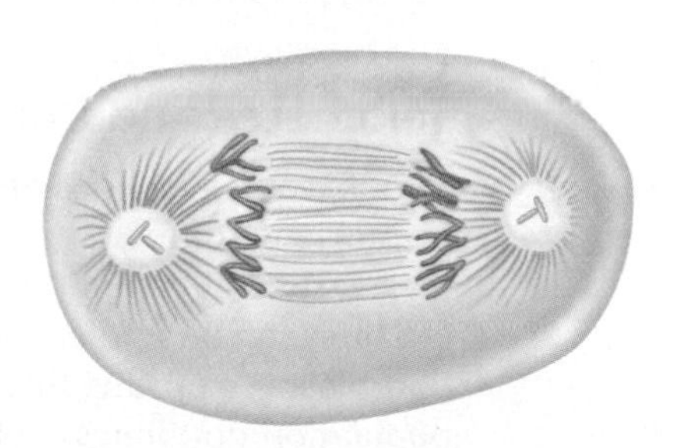

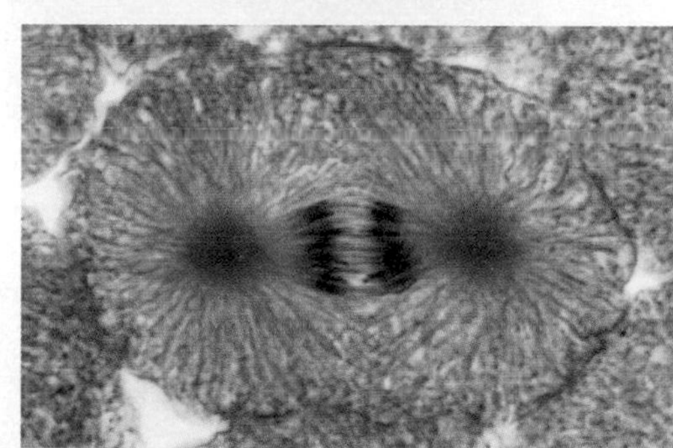

(e) **Telophase**
- The chromosomes become longer, thinner, and less distinct.
- New nuclear membranes form.
- The nucleolus reappears.
- Cell division is nearly complete.

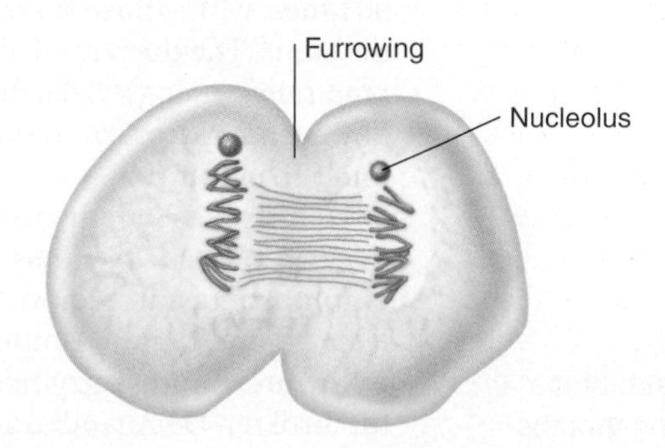

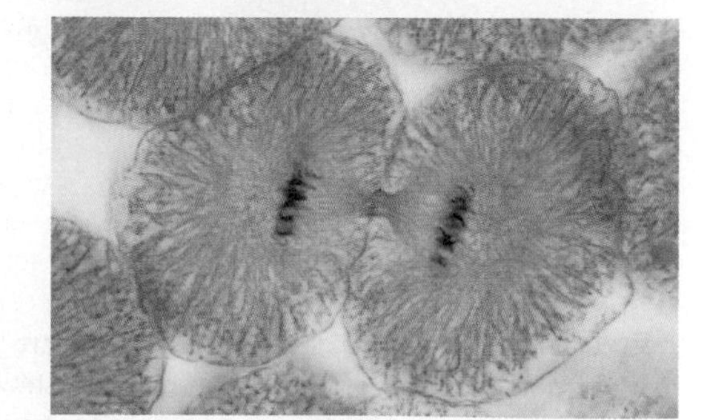

Figure 3.27 **The stages of mitosis.** The events that occur in each stage are indicated in the figure. APIR

In prophase, chromosomes become visible as distinctive structures. In metaphase of mitosis, the chromosomes line up single file along the equator of the cell. This aligning of chromosomes at the equator is believed to result from the action of **spindle fibers,** which are attached to a protein structure called the *kinetochore* at the centromere of each chromosome (fig. 3.27).

Anaphase begins when the centromeres split apart and the spindle fibers shorten, pulling the two chromatids in each chromosome to opposite poles. Each pole therefore gets one copy of each of the 46 chromosomes. During early telophase, division of the cytoplasm (*cytokinesis*) results in the production of two daughter cells that are genetically identical to each other and to the original parent cell.

Role of the Centrosome

All animal cells have a **centrosome,** located near the nucleus in a nondividing cell. At the center of the centrosome are two **centrioles,** which are positioned at right angles to each other. Each centriole is composed of nine evenly spaced bundles of microtubules, with three microtubules per bundle (fig. 3.28).

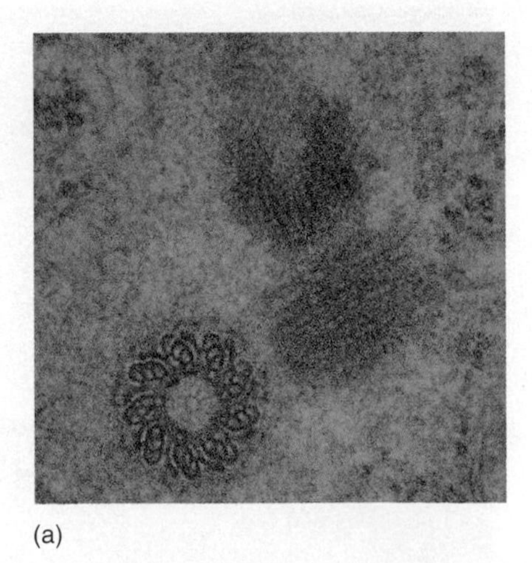

(a)

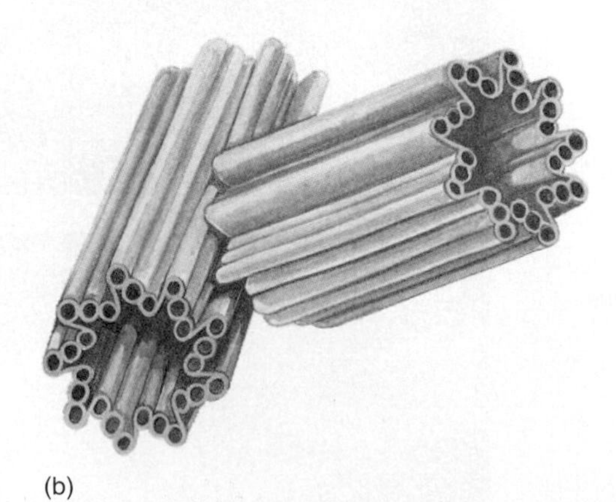

(b)

Figure 3.28 **The centrioles.** (*a*) A micrograph of the two centrioles in a centrosome. (*b*) A diagram showing that the centrioles are positioned at right angles to each other. AP|R

Surrounding the two centrioles is an amorphous mass of material called the *pericentriolar material.* Microtubules grow out of the pericentriolar material, which is believed to function as the center for the organization of microtubules in the cytoskeleton.

Through a mechanism that is still incompletely understood, the centrosome replicates itself during interphase if a cell is going to divide. The two identical centrosomes then move away from each other during prophase of mitosis and take up positions at opposite poles of the cell by metaphase. At this time, the centrosomes produce new microtubules. These new microtubules are very dynamic, rapidly growing and shrinking as if they were "feeling out" randomly for chromosomes. A microtubule becomes stabilized when it finally binds to the proper region of a chromosome.

The spindle fibers pull the chromosomes to opposite poles of the cell during anaphase, so that at telophase, when the cell pinches inward, two identical daughter cells will be produced. This also requires the centrosomes, which somehow organize a ring of contractile filaments halfway between the two poles. These filaments are attached to the plasma membrane, and when they contract the cell is pinched in two. The filaments consist of actin and myosin proteins, the same contractile proteins present in muscle.

In nondividing cells, the centrosome (containing two centrioles) migrates towards the outer portion of the cell cytoplasm and organizes the production of a nonmotile primary cilium (section 3.1). In some epithelial tissues, such as the epithelium that lines the respiratory passages, the apical surface of each cell contains hundreds of beating cilia. In those cases, hundreds of centrosomes are produced and migrate to the cell surface where they become the basal bodies of the cilia. One centriole from each centrosome pair helps to form the microtubules of the cilia.

Telomeres and Cell Division

Certain types of cells can be removed from the body and grown in nutrient solutions (outside the body, or *in vitro*). Under these artificial conditions, the potential longevity of different cell lines can be studied. Normal connective tissue cells (called fibroblasts) stop dividing *in vitro* after a certain number of population doublings. Cells from a newborn will divide 80 to 90 times, while those from a 70-year-old will stop after 20 to 30 divisions. The decreased ability to divide is thus an indicator of *senescence* (aging). Cells that become transformed into cancer, however, apparently do not age and continue dividing indefinitely in culture.

This senescent decrease in the ability of cells to replicate may be related to a loss of DNA sequences at the ends of chromosomes, in regions called **telomeres** (from the Greek *telos* = end). The telomeres serve as caps on the ends of DNA, preventing enzymes from mistaking the normal ends for broken DNA and doing damage by trying to "repair" them.

The telomeres are not fully copied by DNA polymerase, so that a chromosome loses 50 to 100 base pairs in its telomeres each time the chromosome replicates. Cell division may ultimately stop when there is too much loss of DNA in its telomeres, and the cell eventually dies because of damage sustained in the course of aging. There is evidence that telomere damage may contribute to the decline in organ function and the increased risk of disease with age. In part, this may be because the damaged telomeres activate p53, which induces cell cycle arrest, senescence, and apoptosis, as previously described.

However, stem cells that can divide indefinitely—germinal stem cells (which give rise to ova and sperm), hematopoietic stem cells in the bone marrow (which give rise to blood cells), and others—have an enzyme called **telomerase,** which duplicates the telomere DNA. Most cancer cells also produce telomerase, which may be responsible for the ability of cancer cells to divide indefinitely. Telomerase consists of an RNA portion containing nucleotide bases complementary to the telomere DNA, and a protein portion that acts as a *reverse transcriptase* enzyme, producing telomere DNA using the RNA as a template. Because of the significance of telomeres and telomerase in physiology, cancer, and senescence, the 2009 Nobel Prize in Physiology or Medicine was awarded to three scientists who were instrumental in their discovery.

Hypertrophy and Hyperplasia

The growth of an individual from a fertilized egg into an adult involves an increase in the number of cells and an increase in the size of cells. Growth that is due to an increase in cell number results from an increased rate of mitotic cell division and is termed **hyperplasia.** Growth of a tissue or organ due to an increase in cell size is termed **hypertrophy.**

Most growth is due to hyperplasia. A callus on the palm of the hand, for example, involves thickening of the skin by hyperplasia due to frequent abrasion. An increase in skeletal muscle size as a result of exercise, by contrast, is produced by hypertrophy.

CLINICAL APPLICATION

Skeletal muscle and cardiac (heart) muscle can grow only by hypertrophy. When growth occurs in skeletal muscles in response to an increased workload—during weight training, for example—it is called **compensatory hypertrophy.** The heart muscle may also demonstrate compensatory hypertrophy when its workload increases because of hypertension (high blood pressure). The opposite of hypertrophy is **atrophy,** the wasting or decrease in size of a cell, tissue, or organ. This may result from the disuse of skeletal muscles, as occurs in prolonged bed rest, various diseases, or advanced age.

Meiosis

When a cell is going to divide, either by mitosis or meiosis, the DNA is replicated (forming chromatids) and the chromosomes become shorter and thicker, as previously described. At this point the cell has 46 chromosomes, each of which consists of two duplicate chromatids.

The short, thick chromosomes seen at the end of the G_2 phase can be matched as pairs, the members of each pair appearing to be structurally identical. These matched chromosomes are called **homologous chromosomes.** One member of each homologous pair is derived from a chromosome inherited from the father, and the other member is a copy of one of the chromosomes inherited from the mother. Homologous chromosomes do not have identical DNA base sequences; one member of the pair may code for blue eyes, for example, and the other for brown eyes. There are 22 homologous pairs of *autosomal chromosomes* and one pair of *sex chromosomes,* described as X and Y. Females have two X chromosomes, whereas males have one X and one Y chromosome (fig. 3.29).

Meiosis (fig. 3.30) includes two sequences of cell division and occurs only in the gonads (testes and ovaries), where it is used only in the production of gametes—sperm and ova. (Gamete production is described in detail in chapter 20.)

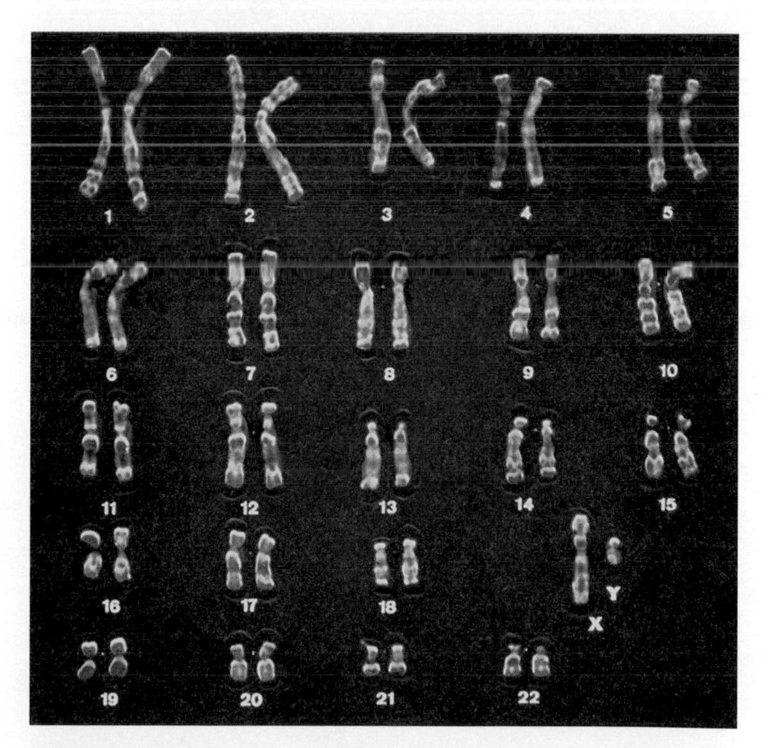

Figure 3.29 A karyotype, in which chromosomes are arranged in homologous pairs. A false-color light micrograph of chromosomes from a male arranged in numbered homologous pairs, from the largest to the smallest. AP|R

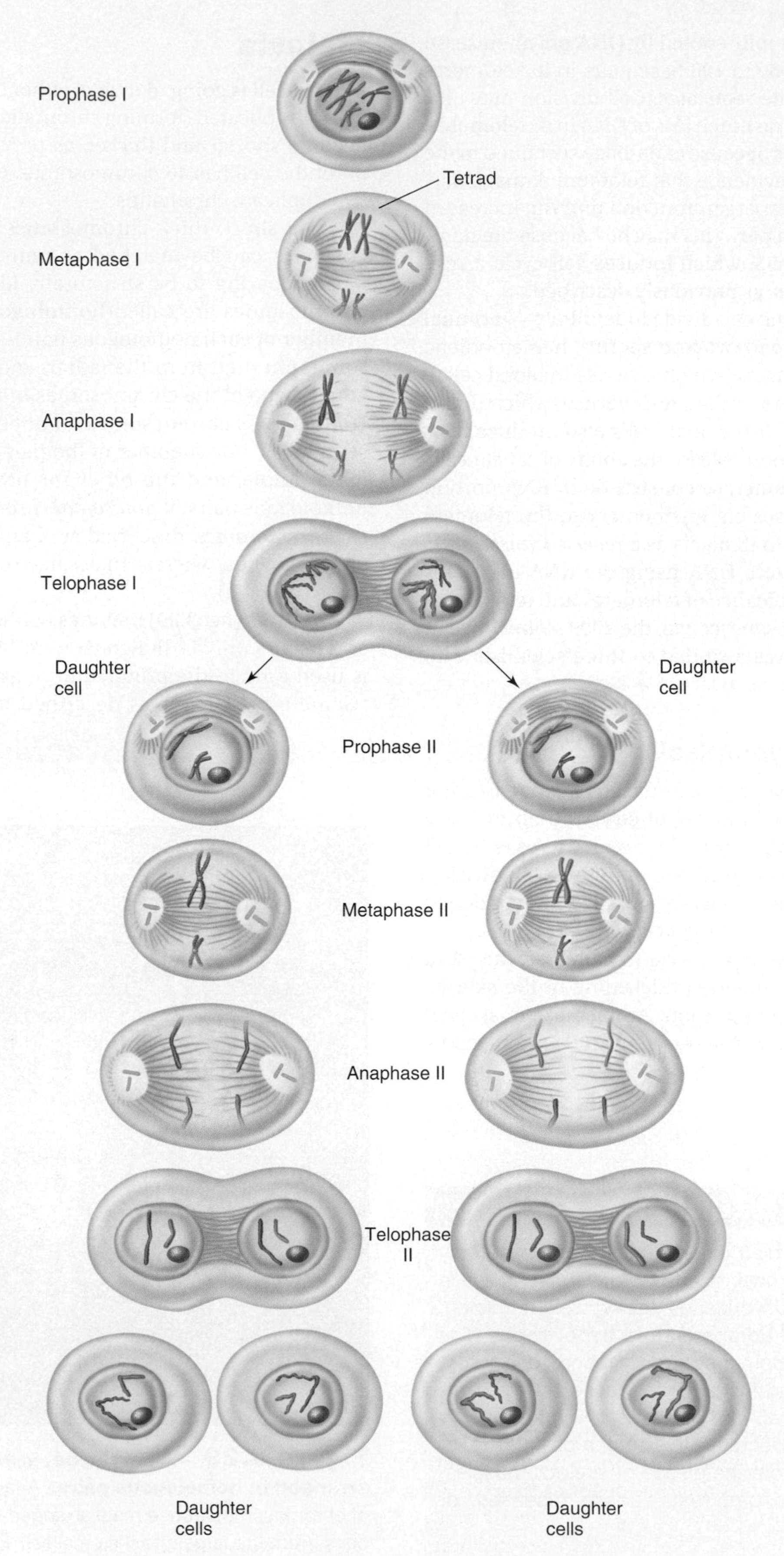

Figure 3.30 Meiosis, or reduction division. In the first meiotic division, the homologous chromosomes of a diploid parent cell are separated into two haploid daughter cells. Each of these chromosomes contains duplicate strands, or chromatids. In the second meiotic division, these chromosomes are distributed to two new haploid daughter cells. AP|R

In the first division of meiosis, the homologous chromosomes line up side by side, rather than single file, along the equator of the cell. The spindle fibers then pull one member of a homologous pair to one pole of the cell, and the other member of the pair to the other pole. Each of the two daughter cells thus acquires only one chromosome from each of the 23 homologous pairs contained in the parent. The daughter cells, in other words, contain 23 rather than 46 chromosomes. For this reason, meiosis (from the Greek *meion* = less) is also known as **reduction division.**

At the end of this cell division, each daughter cell contains 23 chromosomes—but *each of these consists of two chromatids.* (Since the two chromatids per chromosome are identical, this does not make 46 chromosomes; there are still only 23 *different* chromosomes per cell at this point.) The chromatids are separated by a second meiotic division. Each of the daughter cells from the first cell division itself divides, with the duplicate chromatids going to each of two new daughter cells. A grand total of four daughter cells can thus be produced from the meiotic cell division of one parent cell. This occurs in the testes, where one parent cell produces four sperm cells. In the ovaries, one parent cell also produces four daughter cells, but three of these die and only one progresses to become a mature egg cell (as will be described in chapter 20).

The stages of meiosis are subdivided according to whether they occur in the first or the second meiotic cell division. These stages are designated as prophase I, metaphase I, anaphase I, telophase I; and then prophase II, metaphase II, anaphase II, and telophase II (table 3.3 and fig. 3.30).

The reduction of the chromosome number from 46 to 23 is obviously necessary for sexual reproduction, where the sex cells join and add their content of chromosomes together to produce a new individual. The significance of meiosis, however, goes beyond the reduction of chromosome number. At metaphase I, the pairs of homologous chromosomes can line up with either member facing a given pole of the cell. (Recall that each member of a homologous pair came from a different parent.) Maternal and paternal members of homologous pairs are thus randomly shuffled. Hence, when the first meiotic division occurs, each daughter cell will obtain a complement of 23 chromosomes that are randomly derived from the maternal or paternal contribution to the homologous pairs of chromosomes of the parent cell.

In addition to this "shuffling of the deck" of chromosomes, exchanges of parts of homologous chromosomes can occur at prophase I. That is, pieces of one chromosome of a homologous pair can be exchanged with the other homologous chromosome in a process called *crossing-over* (fig. 3.31). These events together result in **genetic recombination** and ensure that the gametes produced by meiosis are genetically unique. This provides additional genetic diversity for organisms that reproduce sexually, and genetic diversity is needed to promote survival of species over evolutionary time.

Table 3.3 | Stages of Meiosis

Stage	Events
First Meiotic Division	
Prophase I	Chromosomes appear double-stranded.
	Each strand, called a chromatid, contains duplicate DNA joined together by a structure known as a centromere.
	Homologous chromosomes pair up side by side.
Metaphase I	Homologous chromosome pairs line up at equator.
	Spindle apparatus is complete.
Anaphase I	Homologous chromosomes separate; the two members of a homologous pair move to opposite poles.
Telophase I	Cytoplasm divides to produce two haploid cells.
Second Meiotic Division	
Prophase II	Chromosomes appear, each containing two chromatids.
Metaphase II	Chromosomes line up single file along equator as spindle formation is completed.
Anaphase II	Centromeres split and chromatids move to opposite poles.
Telophase II	Cytoplasm divides to produce two haploid cells from each of the haploid cells formed at telophase I.

Epigenetic Inheritance

Genetic inheritance is determined by the sequence of DNA base pairs in the chromosomes. However, as previously discussed, not all of these genes are active in each cell of the body. Some genes are switched from active to inactive and back again, as required by a particular cell; activity of these genes is subject to physiological regulation. Other genes may be silenced in all the cells in a tissue, or even in all of the cells in the body. Such long-term gene silencing occurs either in the gametes (and so is inherited) or in early embryonic development. Because the silencing of these genes is carried forward to the daughter cells through mitotic or meiotic cell division, without a change in the DNA base sequence, this is called **epigenetic inheritance.**

Epigenetic inheritance occurs by a variety of mechanisms, including (1) posttranslational modifications of histone proteins (the basic proteins that regulate the degree of compaction of the chromatin; see fig. 3.15); and (2) methylation (the addition of single-carbon methyl groups) of cytosine bases in DNA (specifically cytosines that precede guanines). DNA methylation is generally associated with decreased genetic transcription and gene silencing. Through these means, only one allele (gene) of a pair (from the maternal or paternal chromosomes)

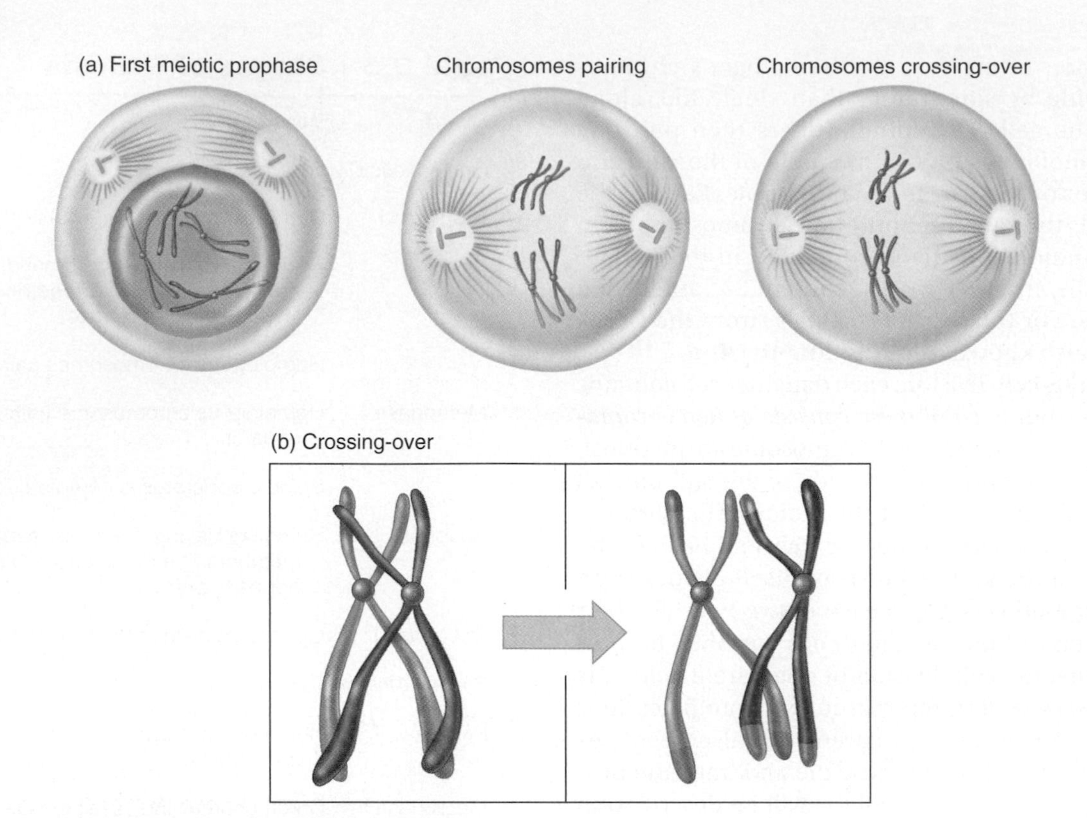

Figure 3.31 Crossing-over. (*a*) Genetic variation results from the crossing-over of tetrads, which occurs during the first meiotic prophase. (*b*) A diagram depicting the recombination of chromosomes that occurs as a result of crossing-over.

may be expressed, and only one X chromosome of the two Xs in a female is active. Acetylation (the addition of a two-carbon acetyl groups) to lysine amino acids in histone proteins has the opposite effect, increasing gene activity (genetic transcription). These and other epigenetic changes can be induced by environmental factors. Because of this, even identical twins can have differences in gene expression.

Problems with epigenetic inheritance are known to contribute to a number of diseases, including cancer, fragile X syndrome, and systemic lupus erythematosus. For example, because methylation of cytosine bases is an epigenetic mechanism for long-term gene silencing, it may not be surprising that cancers show a global (widespread) reduction in DNA methylation. This is associated with activation of genes and instability of chromosome structure in cells that have become transformed in a tumor. However, not all genes are activated; many cancers have inactivated tumor suppressor genes, as well as a generally reduced expression of microRNA (miRNA) genes.

CHECKPOINT

9. Draw a simple diagram of the semiconservative replication of DNA using stick figures and two colors.

10a. Describe the cell cycle using the proper symbols to indicate the different stages of the cycle.

10b. Define apoptosis and explain its significance.

11a. List the phases of mitosis and briefly describe the events in each phase.

11b. Distinguish between mitosis and meiosis, describe the phases of meiosis, and explain its functional significance.

Clinical Investigation SUMMARY

Timothy's past drug abuse could have resulted in the development of an extensive smooth endoplasmic reticulum, which contains many of the enzymes required to metabolize drugs. Liver disease could have been caused by the drug abuse, but there is an alternative explanation. The low amount of the enzyme that breaks down glycogen signals the presence of glycogen storage disease, a genetic condition in which a key lysosomal enzyme is lacking. This enzymatic evidence is supported by the observations of large amounts of glycogen granules and the lack of partially digested glycogen granules within secondary lysosomes. (In reality, such a genetic condition would more likely be diagnosed in early childhood.)

See additional chapter 3 Clinical Investigations on *Mitochondrial Disease, Macular Degeneration, and Breast Cancer Treatment* in the Connect site for this text at www.mhhe.com/Fox13.

Interactions

    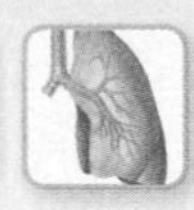 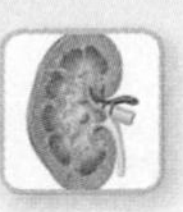  

HPer Links of Basic Cell Concepts to the Body Systems

Nervous System

- Regeneration of neurons is regulated by several different chemicals (p. 170)
- Different forms (alleles) of a gene produce different forms of receptors for particular neurotransmitter chemicals (p. 192)
- Microglia, located in the brain and spinal cord, are cells that transport themselves by amoeboid movement (p. 166)
- The insulating material around nerve fibers, called a myelin sheath, is derived from the cell membrane of certain cells in the nervous system (p. 167)
- Cytoplasmic transport processes are important for the movement of neurotransmitters and other substances within neurons (p. 164)

Endocrine System

- Many hormones act on their target cells by regulating gene expression (p. 323)
- Other hormones bind to receptor proteins located on the outer surface of the cell membrane of the target cells (p. 326)
- The endoplasmic reticulum of some cells stores Ca^{2+}, which is released in response to hormone action (p. 328)
- Chemical regulators called prostaglandins are derived from a type of lipid associated with the cell membrane (p. 352)
- Liver and adipose cells store glycogen and triglycerides, respectively, which can be mobilized for energy needs by the action of particular hormones (p. 675)
- The sex of an individual is determined by the presence of a particular region of DNA in the Y chromosome (p. 703)

Muscular System

- Muscle cells have cytoplasmic proteins called actin and myosin that are needed for contraction (p. 365)
- The endoplasmic reticulum of skeletal muscle fibers stores Ca^{2+}, which is needed for muscle contraction (p. 372)

Circulatory System

- Blood cells are formed in the bone marrow (p. 409)
- Mature red blood cells lack nuclei and mitochondria (p. 408)
- The different white blood cells are distinguished by the shape of their nuclei and the presence of cytoplasmic granules (p. 408)

Immune System

- The carbohydrates outside the cell membrane of many bacteria help to target these cells for immune attack (p. 495)
- Some white blood cells and tissue macrophages destroy bacteria by phagocytosis (p. 495)
- When a B lymphocyte is stimulated by a foreign molecule (antigen), its endoplasmic reticulum becomes more developed and produces more antibody proteins (p. 503)
- Apoptosis is responsible for the destruction of T lymphocytes after an infection has been cleared (p. 497)

Respiratory System

- The air sacs (alveoli) of the lungs are composed of cells that are very thin, minimizing the separation between air and blood (p. 533)
- The epithelial cells lining the airways of the conducting zone have cilia that move mucus (p. 536)

Urinary System

- Parts of the renal tubules have microvilli that increase the rate of reabsorption (p. 586)
- Some regions of the renal tubules have water channels; these are produced by the Golgi complex and inserted by means of vesicles into the cell membrane (p. 596)

Digestive System

- The mucosa of the digestive tract has unicellular glands called goblet cells that secrete mucus (p. 622)
- The cells of the small intestine have microvilli that increase the rate of absorption (p. 629)
- The liver contains phagocytic cells (p. 635)

Reproductive System

- Males have an X and a Y chromosome, whereas females have two X chromosomes per diploid cell (p. 702)
- Gametes are produced by meiotic cell division (p. 79)
- Follicles degenerate (undergo atresia) in the ovaries by means of apoptosis (p. 724)
- Sperm cells are motile through the action of flagella (p. 716)
- The uterine tubes are lined with cilia that help to move the ovulated egg toward the uterus (p. 722)

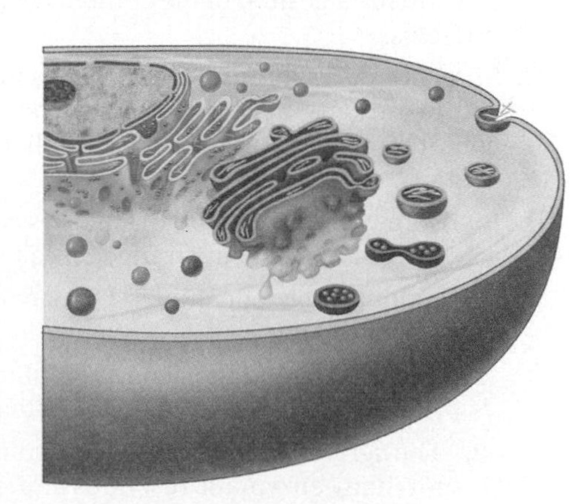

Visit this book's website at **www.mhhe.com/Fox13** for:
- Chapter quizzes, interactive learning exercises, and other study tools
- Additional clinical investigations
- Access to LearnSmart—An adaptive diagnostic tool that constantly assesses student knowledge of course material
- Ph.I.L.S. 4.0—physiology interactive lab simulations that may be used to supplement or substitute for wet labs

SUMMARY

3.1 Plasma Membrane and Associated Structures 51

A. The structure of the plasma membrane is described by a fluid-mosaic model.
  1. The membrane is composed predominantly of a double layer of phospholipids.
  2. The membrane also contains proteins, most of which span its entire width.

B. Some cells move by extending pseudopods; cilia and flagella protrude from the cell membrane of some specialized cells.

C. In the process of endocytosis, invaginations of the plasma membrane allow the cells to take up molecules from the external environment.
  1. In phagocytosis, the cell extends pseudopods that eventually fuse together to create a food vacuole; pinocytosis involves the formation of a narrow furrow in the membrane, which eventually fuses.
  2. Receptor-mediated endocytosis requires the interaction of a specific molecule in the extracellular environment with a specific receptor protein in the cell membrane.
  3. Exocytosis, the reverse of endocytosis, is a process that allows the cell to secrete its products.

3.2 Cytoplasm and Its Organelles 57

A. Microfilaments and microtubules produce a cytoskeleton that aids movements of organelles within a cell.

B. Lysosomes contain digestive enzymes and are responsible for the elimination of structures and molecules within the cell and for digestion of the contents of phagocytic food vacuoles.

C. Mitochondria serve as the major sites for energy production within the cell. They have an outer membrane with a smooth contour and an inner membrane with infoldings called cristae.

D. Ribosomes are small protein factories composed of ribosomal RNA and protein arranged into two subunits.

E. The endoplasmic reticulum is a system of membranous tubules in the cell.
  1. The granular endoplasmic reticulum is covered with ribosomes and is involved in protein synthesis.
  2. The agranular endoplasmic reticulum provides a site for many enzymatic reactions and, in skeletal muscles, serves to store Ca^{2+}.

F. The Golgi complex is a series of membranous sacs that receive products from the endoplasmic reticulum, modify those products, and release the products within vesicles.

3.3 Cell Nucleus and Gene Expression 62

A. The cell nucleus is surrounded by a double-layered nuclear envelope. At some points, the two layers are fused by nuclear pore complexes that allow for the passage of molecules.

B. Genetic expression occurs in two stages: transcription (RNA synthesis) and translation (protein synthesis).
  1. The DNA in the nucleus is combined with proteins to form the threadlike material known as chromatin.
  2. In chromatin, DNA is wound around regulatory proteins known as histones to form particles called nucleosomes.
  3. Chromatin that is active in directing RNA synthesis is euchromatin; the highly condensed, inactive chromatin is heterochromatin.

C. RNA is single-stranded. Four types are produced within the nucleus: ribosomal RNA, transfer RNA, precursor messenger RNA, and messenger RNA.

D. Active euchromatin directs the synthesis of RNA in a process called transcription.
  1. The enzyme RNA polymerase causes separation of the two strands of DNA along the region of the DNA that constitutes a gene.
  2. One of the two separated strands of DNA serves as a template for the production of RNA. This occurs by complementary base pairing between the DNA bases and ribonucleotide bases.

E. The human genome is now known to contain approximately 25,000 genes, while the human proteome consists of about 100,000 proteins.
  1. A gene is transcribed into pre-mRNA, which is then cut and spliced in alternative ways to produce a number of different mRNA molecules that code for different proteins.
  2. The RNA nucleotide sequences that are spliced together to make mRNA are called exons; the RNA nucleotides between them that are removed are known as introns.
  3. Some RNA molecules, known as short interfering RNA (siRNA), participate in silencing the expression of mRNA molecules that contain base sequences that are at least partially complementary to the siRNA.

3.4 Protein Synthesis and Secretion 68

A. Messenger RNA leaves the nucleus and attaches to the ribosomes.

B. Each transfer RNA, with a specific base triplet in its anticodon, binds to a specific amino acid.

1. As the mRNA moves through the ribosomes, complementary base pairing between tRNA anticodons and mRNA codons occurs.
2. As each successive tRNA molecule binds to its complementary codon, the amino acid it carries is added to the end of a growing polypeptide chain.

C. Proteins destined for secretion are produced in ribosomes located on the granular endoplasmic reticulum and enter the cisternae of this organelle.

D. Secretory proteins move from the granular endoplasmic reticulum to the Golgi complex.

1. The Golgi complex modifies the proteins it contains, separates different proteins, and packages them in vesicles.
2. Secretory vesicles from the Golgi complex fuse with the plasma membrane and release their products by exocytosis.

E. The concentration of regulatory proteins is controlled by their degradation as well as by their synthesis through genetic expression.

1. Regulatory proteins targeted for destruction are tagged by binding to a polypeptide known as ubiquitin.
2. The proteasome, an organelle consisting of several protease enzymes (those that digest proteins), then degrades the regulatory proteins that are bound to ubiquitin.

3.5 DNA Synthesis and Cell Division 73

A. Replication of DNA is semiconservative; each DNA strand serves as a template for the production of a new strand.

1. The strands of the original DNA molecule gradually separate along their entire length and, through complementary base pairing, form a new complementary strand.
2. In this way, each DNA molecule consists of one old and one new strand.

B. During the G_1 phase of the cell cycle, the DNA directs the synthesis of RNA, and hence that of proteins.

C. During the S phase of the cycle, DNA directs the synthesis of new DNA and replicates itself.

D. After a brief time gap (G_2), the cell begins mitosis (the M stage of the cycle).

1. Mitosis consists of the following phases: interphase, prophase, metaphase, anaphase, and telophase.
2. In mitosis, the homologous chromosomes line up single file and are pulled by spindle fibers to opposite poles.
3. This results in the production of two daughter cells, each containing 46 chromosomes, just like the parent cell.

E. Proteins known as cyclins, the expression of which can be altered in cancer, regulate the progression through the cell cycle.

F. Apoptosis is a regulated process of cell suicide, which can be triggered by external molecules ("death ligands") or by molecules released by mitochondria into the cytoplasm.

G. Meiosis is a special type of cell division that results in the production of gametes in the gonads.

1. The homologous chromosomes line up side by side, so that only one of each pair is pulled to each pole.
2. This results in the production of two daughter cells, each containing only 23 chromosomes, which are duplicated.
3. The duplicate chromatids are separated into two new daughter cells during the second meiotic cell division.

H. Epigenetic inheritance refers to the inheritance of gene silencing from the gametes or early embryo that is carried forward by cell division into all cells of the body.

REVIEW ACTIVITIES

Test Your Knowledge

1. According to the fluid-mosaic model of the plasma membrane,
   a. protein and phospholipids form a regular, repeating structure.
   b. the membrane is a rigid structure.
   c. phospholipids form a double layer, with the polar parts facing each other.
   d. proteins are free to move within a double layer of phospholipids.
2. After the DNA molecule has replicated itself, the duplicate strands are called
   a. homologous chromosomes.
   b. chromatids.
   c. centromeres.
   d. spindle fibers.
3. Nerve and skeletal muscle cells in the adult, which do not divide, remain in the
   a. G_1 phase.
   b. S phase.
   c. G_2 phase.
   d. M phase.
4. The phase of mitosis in which the chromosomes line up at the equator of the cell is called
   a. interphase.
   b. prophase.
   c. metaphase.
   d. anaphase.
   e. telophase.

5. The phase of mitosis in which the chromatids separate is called
   a. interphase.
   b. prophase.
   c. metaphase.
   d. anaphase.
   e. telophase.
6. Chemical modifications of histone proteins are believed to directly influence
   a. genetic transcription.
   b. genetic translation.
   c. both transcription and translation.
   d. posttranslational changes in the newly synthesized proteins.
7. Which of these statements about RNA is *true?*
   a. It is made in the nucleus.
   b. It is double-stranded.
   c. It contains the sugar deoxyribose.
   d. It is a complementary copy of the entire DNA molecule.
8. Which of these statements about mRNA is *false?*
   a. It is produced as a larger pre-mRNA.
   b. It forms associations with ribosomes.
   c. Its base triplets are called anticodons.
   d. It codes for the synthesis of specific proteins.
9. The organelle that combines proteins with carbohydrates and packages them within vesicles for secretion is
   a. the Golgi complex.
   b. the granular endoplasmic reticulum.
   c. the agranular endoplasmic reticulum.
   d. the ribosome.
10. The organelle that contains digestive enzymes is
    a. the mitochondrion.
    b. the lysosome.
    c. the endoplasmic reticulum.
    d. the Golgi complex.
11. Which of these descriptions of rRNA is *true?*
    a. It is single-stranded.
    b. It catalyzes steps in protein synthesis.
    c. It forms part of the structure of both subunits of a ribosome.
    d. It is produced in the nucleolus.
    e. All of these are true.
12. Which of these statements about tRNA is *true?*
    a. It is made in the nucleus.
    b. It is looped back on itself.
    c. It contains the anticodon.
    d. There are over 20 different types.
    e. All of these are true.
13. The step in protein synthesis during which tRNA, rRNA, and mRNA are all active is known as
    a. transcription.
    b. translation.
    c. replication.
    d. RNA polymerization.
14. The anticodons are located in
    a. tRNA.
    b. rRNA.
    c. mRNA.
    d. ribosomes.
    e. endoplasmic reticulum.
15. Alternative splicing of exons results in
    a. posttranslational modifications of proteins.
    b. the production of different mRNA molecules from a common precursor RNA molecule.
    c. the production of siRNA and RNA silencing.
    d. the production of a genome that is larger than the proteome.
16. The molecule that tags regulatory proteins for destruction by the proteasome is
    a. ubiquitin.
    b. chaperone.
    c. microRNA.
    d. cyclin.

Test Your Understanding

17. Give some specific examples that illustrate the dynamic nature of the plasma membrane.
18. Describe the structure of nucleosomes, and explain the role of histone proteins in chromatin structure and function.
19. What is the genetic code, and how does it affect the structure and function of the body?
20. Why may tRNA be considered the "interpreter" of the genetic code?

21. Compare the processing of cellular proteins with that of proteins secreted by a cell.
22. Define the terms *genome* and *proteome,* and explain how they are related.
23. Explain the interrelationship between the endoplasmic reticulum and the Golgi complex. What becomes of vesicles released from the Golgi complex?
24. Explain the functions of centrioles in nondividing and dividing cells.
25. Describe the phases of the cell cycle, and explain how this cycle may be regulated.
26. Distinguish between oncogenes and tumor suppressor genes, and give examples of how such genes may function.
27. Define *apoptosis* and explain the physiological significance of this process.
28. Describe what is meant by *epigenetic inheritance,* and explain its significance.

Test Your Analytical Ability

29. Discuss the role of chromatin proteins in regulating gene expression. How does the three-dimensional structure of the chromatin affect genetic regulation? How do hormones influence genetic regulation?
30. Explain how p53 functions as a tumor suppressor gene. How can mutations in p53 lead to cancer, and how might gene therapy or other drug interventions inhibit the growth of a tumor?
31. Release of lysosomal enzymes from white blood cells during a local immune attack can contribute to the symptoms of inflammation. Suppose, to alleviate inflammation, you develop a drug that destroys all lysosomes. Would this drug have negative side effects? Explain.
32. Antibiotics can have different mechanisms of action. An antibiotic called puromycin blocks genetic translation. One called actinomycin D blocks genetic transcription. These drugs can be used to determine how regulatory molecules, such as hormones, work. For example, if a hormone's effects on a tissue were blocked immediately by puromycin but not by actinomycin D, what would that tell you about the mechanism of action of the hormone?
33. Explain how it is possible for the human proteome to consist of over 100,000 proteins while the human genome consists only of about 25,000 genes.
34. Explain RNA interference (RNA_i) by siRNA and miRNA in the regulation of gene expression.
35. Describe the function and significance of ubiquitin and the proteasome in the regulation of gene expression.

Test Your Quantitative Ability

Review figure 3.19 and answer the following questions about a protein that is composed of 600 amino acids.

36. How many mRNA bases are needed to code for this protein?
37. If the gene coding for this protein contains two introns, how many exons does it contain?
38. If the exons are of equal length, how many bases are in each exon?

CHAPTER

4

Enzymes and Energy

REFRESH YOUR MEMORY

Before you begin this chapter, you may want to review these concepts from previous chapters:

CHAPTER OUTLINE

Clinical Investigation

Tom is a 77-year-old man brought to the hospital because of severe chest pain. He also complained that he had difficulty urinating and "got the runs" whenever he ate ice cream.

Some of the new terms and concepts you will encounter include:

- Isoenzymes
- Creatine phosphokinase and acid phosphatase

4.1 ENZYMES AS CATALYSTS

Enzymes are biological catalysts that increase the rate of chemical reactions. Most enzymes are proteins, and their catalytic action results from their complex structure. The great diversity of protein structure allows different enzymes to be very specific in the reactions they catalyze.

LEARNING OUTCOMES

After studying this section, you should be able to:

1. Explain the properties of a catalyst and how enzymes function as catalysts.
2. Describe how enzymes are named.

The ability of yeast cells to make alcohol from glucose (a process called *fermentation*) had been known since antiquity, yet even as late as the mid-nineteenth century no scientist had been able to duplicate this process in the absence of living yeast. Also, a vast array of chemical reactions occurred in yeast and other living cells at body temperature that could not be duplicated in the chemistry laboratory without adding substantial amounts of heat energy. These observations led many mid-nineteenth-century scientists to believe that chemical reactions in living cells were aided by a "vital force" that operated beyond the laws of the physical world. This *vitalist concept* was squashed along with the yeast cells when a pioneering biochemist, Eduard Buchner, demonstrated that juice obtained from yeast could ferment glucose to alcohol. The yeast juice was not alive—evidently some chemicals in the cells were responsible for fermentation. Buchner didn't know what these chemicals were, so he simply named them **enzymes** (Greek for "in yeast").

Chemically, enzymes are a subclass of proteins. The only known exceptions are the few special cases in which RNA demonstrates enzymatic activity; in these cases they are called *ribozymes.* Ribozymes function as enzymes in reactions involving remodeling of the RNA molecules themselves, and in the formation of a growing polypeptide in ribosomes.

Functionally, enzymes (and ribozymes) are biological **catalysts.** A catalyst is a chemical that (1) increases the rate of a reaction, (2) is not itself changed at the end of the reaction, and (3) does not change the nature of the reaction or its final result. The same reaction would have occurred to the same degree in the absence of the catalyst, but it would have progressed at a much slower rate.

In order for a given reaction to occur, the reactants must have sufficient energy. The amount of energy required for a reaction to proceed is called the **activation energy.** By analogy, a match will not burn and release heat energy unless it is first "activated" by striking the match or by placing it in a flame.

In a large population of molecules, only a small fraction will possess sufficient energy for a reaction. Adding heat will raise the energy level of all the reactant molecules, thus increasing the percentage of the population that has the activation energy. Heat makes reactions go faster, but it also produces undesirable side effects in cells. Catalysts make reactions go faster at lower temperatures by lowering the activation energy required, thus ensuring that a larger percentage of the population of reactant molecules will have sufficient energy to participate in the reaction (fig. 4.1).

Because a small fraction of the reactants will have the activation energy required for a reaction even in the absence of a catalyst, the reaction could theoretically occur spontaneously at a slow rate. This rate, however, would be much too slow for the needs of a cell. So, from a biological standpoint, the presence or absence of a specific enzyme catalyst acts as a switch—the reaction will occur if the enzyme is present and will not occur if the enzyme is absent.

Mechanism of Enzyme Action

The ability of enzymes to lower the activation energy of a reaction is a result of their structure. Enzymes are large proteins with complex, highly ordered, three-dimensional shapes produced by physical and chemical interactions between their amino acid subunits. Each type of enzyme has a characteristic three-dimensional shape, or *conformation,* with ridges, grooves, and pockets lined with specific amino acids. The particular pockets that are active in catalyzing a reaction are called the *active sites* of the enzyme.

The reactant molecules, which are called the **substrates** of the enzyme, have specific shapes that allow them to fit into the active sites. The enzyme can thus be thought of as a lock into which only a specifically shaped key—the substrate—can fit. This **lock-and-key model** of enzyme activity is illustrated in figure 4.2.

In some cases, the fit between an enzyme and its substrate may not be perfect at first. A perfect fit may be induced, however, as the substrate gradually slips into the active site. This induced fit, together with temporary bonds that form between the substrate and the amino acids lining the active sites of the enzyme, weakens the existing bonds within the substrate molecules and allows them to be more easily broken. New bonds are more easily

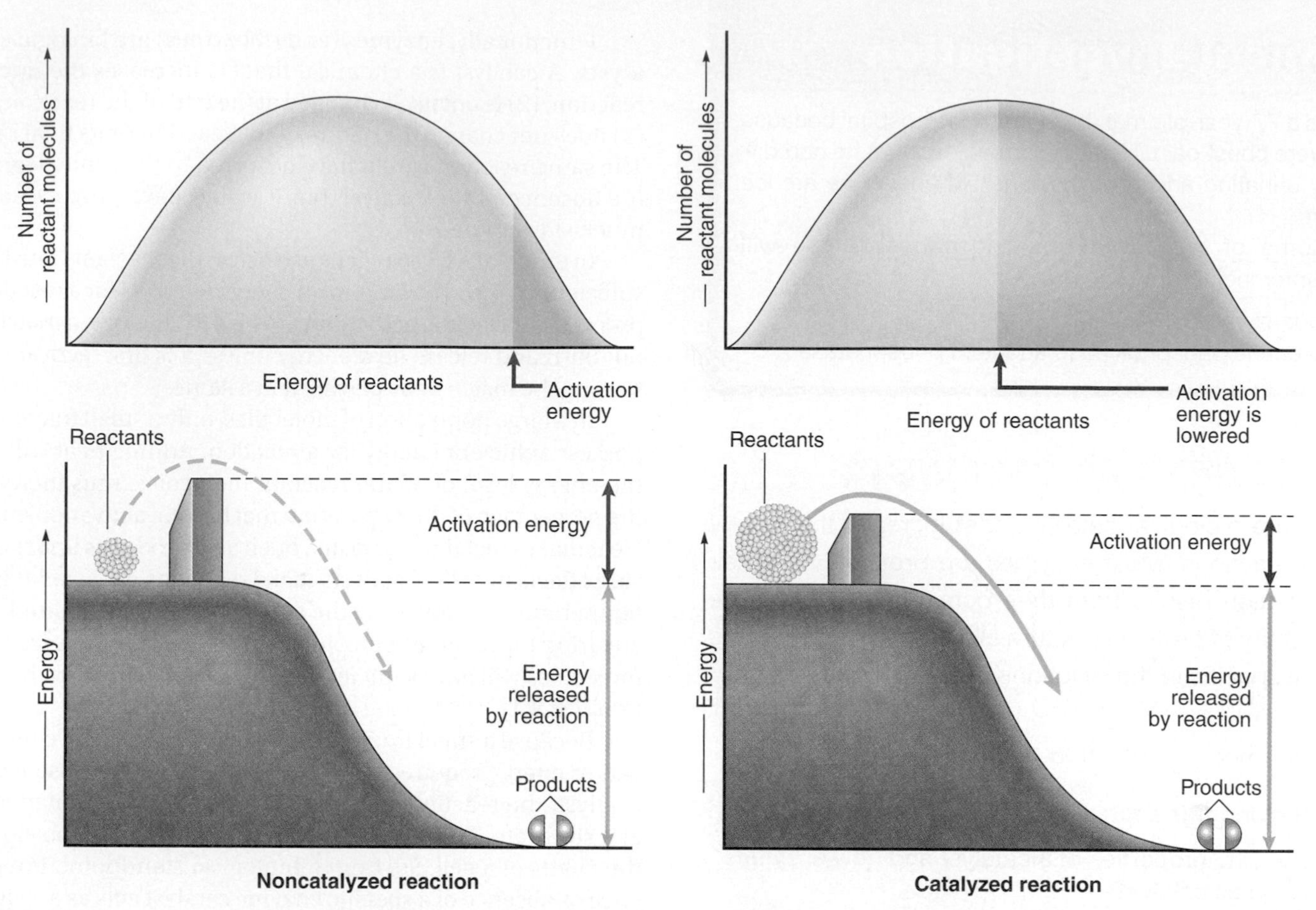

Figure 4.1 A comparison of noncatalyzed and catalyzed reactions. The upper figures compare the proportion of reactant molecules that have sufficient activation energy to participate in the reaction (blue = insufficient energy; green = sufficient energy). This proportion is increased in the enzyme-catalyzed reaction because enzymes lower the activation energy required for the reaction (shown as a barrier on top of an energy "hill" in the lower figures). Reactants that can overcome this barrier are able to participate in the reaction, as shown by arrows pointing to the bottom of the energy hill. AP|R

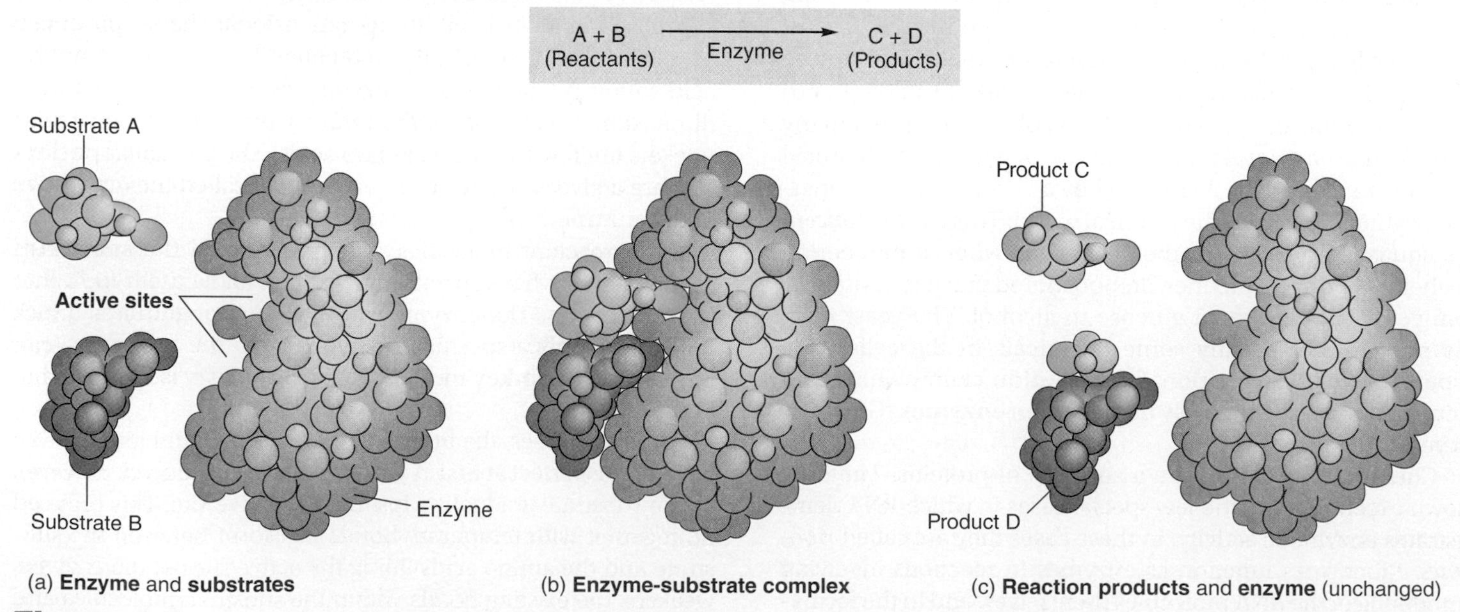

Figure 4.2 The lock-and-key model of enzyme action. (*a*) Substrates A and B fit into active sites in the enzyme, forming (*b*) an enzyme-substrate complex. This complex then (*c*) dissociates, releasing the products of the reaction and the free enzyme. AP|R

formed as substrates are brought close together in the proper orientation. This model of enzyme activity, in which the enzyme undergoes a slight structural change to better fit the substrate, is called the **induced-fit model.** This has been likened to putting on a thin leather glove. As your hand enters it, the glove is induced to fit the contours of your hand. The *enzyme-substrate complex,* formed temporarily in the course of the reaction, then dissociates to yield **products** and the free unaltered enzyme.

Because enzymes are very specific as to their substrates and activity, the concentration of a specific enzyme in a sample of fluid can be measured relatively easily. This is usually done by measuring the rate of conversion of the enzyme's substrates into products under specified conditions. The presence of an enzyme in a sample can thus be detected by the job it does, and its concentration can be measured by how rapidly it performs its job.

CLINICAL APPLICATION

When tissues become damaged as a result of diseases, some of the dead cells disintegrate and release their enzymes into the blood. Most of these enzymes are not normally active in the blood for lack of their specific substrates, but their enzymatic activity can be measured in a test tube by the addition of the appropriate substrates to samples of plasma. Such measurements are clinically useful because abnormally high plasma concentrations of particular enzymes are characteristic of certain diseases (table 4.1).

Clinical Investigation CLUES

Laboratory tests showed elevated levels of acid phosphatase and creatine kinase in Tom's plasma.

- Which laboratory test might be related to Tom's difficulty in urinating?
- Given Tom's chest pain, what condition might his elevated creatine phosphokinase indicate?

Naming of Enzymes

In the past, enzymes were given names that were somewhat arbitrary. The modern system for naming enzymes, established by an international committee, is more orderly and informative. With the exception of some older enzyme names (such as *pepsin, trypsin,* and *renin*), all enzyme names end with the suffix *-ase* (table 4.2), and classes of enzymes are named according to their activity, or "job category." *Hydrolases,* for example, promote hydrolysis reactions. Other enzyme categories include *phosphatases,* which catalyze the removal of phosphate groups; *synthases* and *synthetases,* which catalyze dehydration synthesis reactions; *dehydrogenases,* which remove hydrogen atoms from their substrates; and *kinases,* which add a phosphate group to (phosphorylate) particular molecules. Enzymes called *isomerases* rearrange atoms within their substrate molecules to form structural isomers, such as glucose and fructose (chapter 2; see fig. 2.13).

The names of many enzymes specify both the substrate of the enzyme and the job category of the enzyme. Lactic acid dehydrogenase, for example, removes hydrogens from lactic acid. Enzymes that do exactly the same job (that catalyze the same reaction) in different organs have the same name, since the name describes the activity of the enzyme. Different organs, however, may make slightly different "models" of the enzyme that differ in one or a few amino acids. These different models of the same enzyme are called **isoenzymes.** The differences in structure do not affect the active sites (otherwise the enzymes would not catalyze the same reaction), but they do alter the structure of the enzymes at other locations so that the different isoenzymatic forms can be separated by standard biochemical procedures. These techniques are useful in the diagnosis of diseases.

Table 4.1 | Examples of the Diagnostic Value of Some Enzymes Found in Plasma

Enzyme	Diseases Associated with Abnormal Plasma Enzyme Concentrations
Alkaline phosphatase	Obstructive jaundice, Paget's disease (osteitis deformans), carcinoma of bone
Acid phosphatase	Benign hypertrophy of prostate, cancer of prostate
Amylase	Pancreatitis, perforated peptic ulcer
Aldolase	Muscular dystrophy
Creatine kinase (or creatine phosphokinase-CPK)	Muscular dystrophy, myocardial infarction
Lactate dehydrogenase (LDH)	Myocardial infarction, liver disease, renal disease, pernicious anemia
Transaminases (AST and ALT)	Myocardial infarction, hepatitis, muscular dystrophy

Table 4.2 | Selected Enzymes and the Reactions They Catalyze

Enzyme	Reaction Catalyzed
Catalase	$2\ H_2O_2 \rightarrow 2\ H_2O + O_2$
Carbonic anhydrase	$H_2CO_3 \rightarrow H_2O + CO_2$
Amylase	starch + $H_2O \rightarrow$ maltose
Lactate dehydrogenase	lactic acid $\rightarrow$ pyruvic acid + NADH + H^+
Ribonuclease	RNA + $H_2O \rightarrow$ ribonucleotides

CLINICAL APPLICATION

Different organs, when they are diseased, may liberate different isoenzymatic forms of an enzyme that can be measured in a clinical laboratory. For example, the enzyme **creatine phosphokinase,** abbreviated either **CPK** or **CK,** exists in three isoenzymatic forms. These forms are identified by two letters that indicate two components of this enzyme. One form is identified as MM and is liberated from diseased skeletal muscle; the second is BB, released by a damaged brain; and the third is MB, released from a diseased heart. Clinical tests utilizing antibodies that can bind to the M and B components are now available to specifically measure the level of the MB form in the blood when heart disease is suspected.

Clinical Investigation CLUES

Tom had an elevated plasma level of the MB isoform of creatine phosphokinase.

- What isoenzymatic forms of creatine phosphokinase are there?
- What condition does his elevated MB isoform of creatine phosphokinase suggest?

CHECKPOINT

1. Use the lock-and-key model to explain how enzymes function as catalysts.
2. Explain how enzymes are named, and the nature of isoenzymes.

4.2 CONTROL OF ENZYME ACTIVITY

The rate of an enzyme-catalyzed reaction depends on the concentration of the enzyme and the pH and temperature of the solution. Genetic control of enzyme concentration, for example, affects the rate of progress along particular metabolic pathways and thus regulates cellular metabolism.

LEARNING OUTCOMES

After studying this section, you should be able to:

3. Describe the effects of pH and temperature on enzyme-catalyzed reactions, and the nature of cofactors and coenzymes.
4. Explain the law of mass action in reversible reactions.
5. Describe a metabolic pathway and how it is affected by end-product inhibition and inborn errors of metabolism.

The activity of an enzyme, as measured by the rate at which its substrates are converted to products, is influenced by such factors as (1) the temperature and pH of the solution; (2) the concentration of cofactors and coenzymes, which are needed by many enzymes as "helpers" for their catalytic activity; (3) the concentration of enzyme and substrate molecules in the solution; and (4) the stimulatory and inhibitory effects of some products of enzyme action on the activity of the enzymes that helped to form these products.

Effects of Temperature and pH

An increase in temperature will increase the rate of nonenzyme-catalyzed reactions. A similar relationship between temperature and reaction rate occurs in enzyme-catalyzed reactions. At a temperature of 0° C the reaction rate is immeasurably slow. As the temperature is raised above 0° C the reaction rate increases, but only up to a point. At a few degrees above body temperature (which is 37° C) the reaction rate reaches a plateau; further increases in temperature actually *decrease* the rate of the reaction (fig. 4.3). This decrease is due to the altered tertiary structure of enzymes at higher temperatures.

A similar relationship is observed when the rate of an enzymatic reaction is measured at different pH values. Each enzyme characteristically exhibits peak activity in a very narrow pH range, which is the **pH optimum** for the enzyme. If the pH is changed so that it is no longer within the enzyme's optimum range, the reaction rate will decrease (fig. 4.4). This decreased enzyme activity is due to changes in the conformation of the enzyme and in the charges of the R groups of the amino acids lining the active sites.

The pH optimum of an enzyme usually reflects the pH of the body fluid in which the enzyme is found. The acidic pH optimum of the protein-digesting enzyme *pepsin,* for

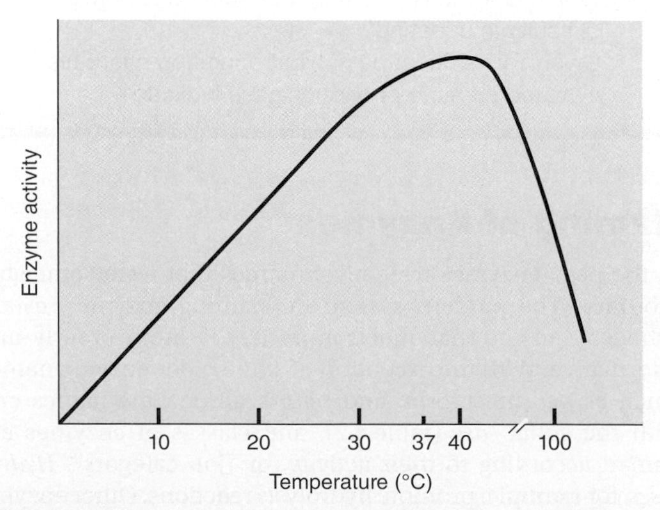

Figure 4.3 The effect of temperature on enzyme activity. This effect is measured by the rate of the enzyme-catalyzed reaction under standardized conditions as the temperature of the reaction is varied. AP|R

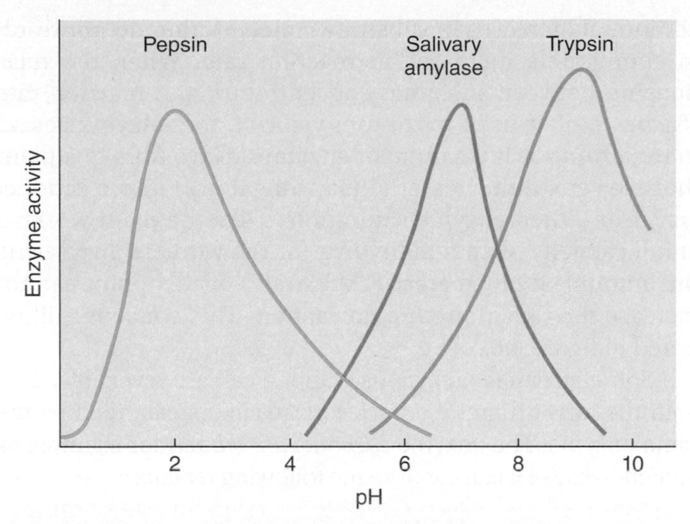

Figure 4.4 The effect of pH on the activity of three digestive enzymes. Salivary amylase is found in saliva, which has a pH close to neutral; pepsin is found in acidic gastric juice, and trypsin is found in alkaline pancreatic juice. APIR

See the *Test Your Quantitative Ability* section of the Review Activities at the end of this chapter.

Table 4.3 | pH Optima of Selected Enzymes

Enzyme	Reaction Catalyzed	pH Optimum
Pepsin (stomach)	Digestion of protein	2.0
Acid phosphatase (prostate)	Removal of phosphate group	5.5
Salivary amylase (saliva)	Digestion of starch	6.8
Lipase (pancreatic juice)	Digestion of fat	7.0
Alkaline phosphatase (bone)	Removal of phosphate group	9.0
Trypsin (pancreatic juice)	Digestion of protein	9.5
Monoamine oxidase (nerve endings)	Removal of amine group from norepinephrine	9.8

example, allows it to be active in the strong hydrochloric acid of gastric juice (fig. 4.4). Similarly, the neutral pH optimum of *salivary amylase* and the alkaline pH optimum of *trypsin* in pancreatic juice allow these enzymes to digest starch and protein, respectively, in other parts of the digestive tract.

CLINICAL APPLICATION

Although the pH of other body fluids shows less variation than that of the fluids of the digestive tract, the pH optima of different enzymes found throughout the body do show significant differences (table 4.3). Some of these differences can be exploited for diagnostic purposes. Disease of the prostate, for example, may be associated with elevated blood levels of a prostatic phosphatase with an acidic pH optimum (descriptively called *acid phosphatase*). Bone disease, on the other hand, may be associated with elevated blood levels of *alkaline phosphatase,* which has a higher pH optimum than the similar enzyme released from the diseased prostate.

Cofactors and Coenzymes

Many enzymes are completely inactive when isolated in a pure state. Evidently some of the ions and smaller organic molecules that are removed in the purification procedure play an essential role in enzyme activity. These ions and smaller organic molecules needed for the activity of specific enzymes are called *cofactors* and *coenzymes.*

Cofactors include metal ions such as Ca^{2+}, Mg^{2+}, Mn^{2+}, Cu^{2+}, Zn^{2+}, and selenium. Some enzymes with a cofactor requirement do not have a properly shaped active site in the absence of the cofactor. In these enzymes, the attachment of cofactors causes a conformational change in the protein that allows it to combine with its substrate. The cofactors of other enzymes participate in the temporary bonds between the enzyme and its substrate when the enzyme-substrate complex is formed (fig. 4.5).

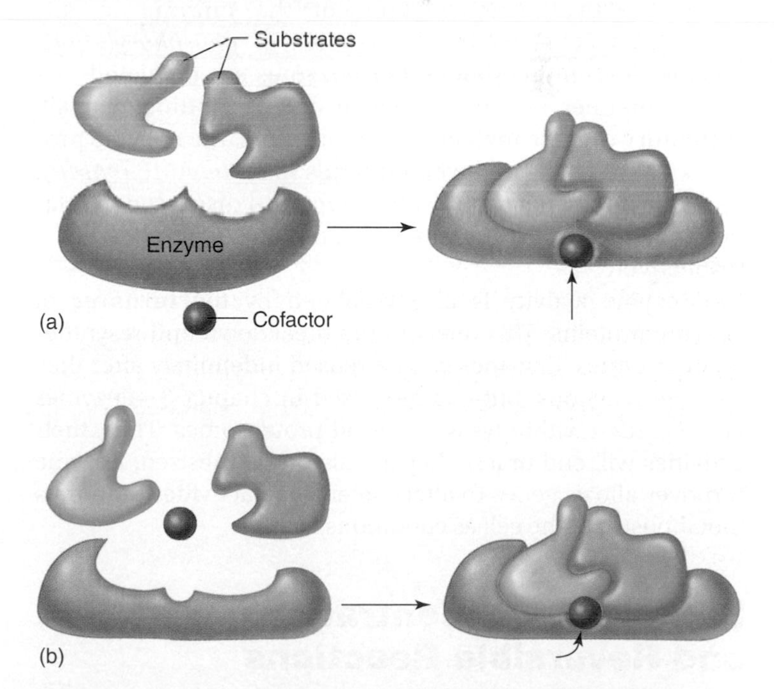

Figure 4.5 The roles of cofactors in enzyme function. In (*a*) the cofactor changes the conformation of the active site, allowing for a better fit between the enzyme and its substrates. In (*b*) the cofactor participates in the temporary bonding between the active site and the substrates.

Coenzymes are organic molecules, derived from water-soluble vitamins such as niacin and riboflavin, that are needed for the function of particular enzymes. Coenzymes participate in enzyme-catalyzed reactions by transporting hydrogen atoms and small molecules from one enzyme to another. Examples of the actions of cofactors and coenzymes in specific reactions will be given in the context of their roles in cellular metabolism in section 4.3.

Enzyme Activation

There are a number of important cases in which enzymes are produced as inactive forms. In the cells of the pancreas, for example, many digestive enzymes are produced as inactive *zymogens,* which are activated after they are secreted into the intestine. Activation of zymogens in the intestinal lumen (cavity) protects the pancreatic cells from self-digestion.

In liver cells, as another example, the enzyme that catalyzes the hydrolysis of stored glycogen is inactive when it is produced, and must later be activated by the addition of a phosphate group. A different enzyme, called a *protein kinase,* catalyzes the addition of the phosphate group to that enzyme. This enzyme activation occurs between meals (in a fasting state), when the breakdown of glycogen to glucose allows the liver to secrete glucose into the blood. After a carbohydrate meal, when glucose enters the blood from the intestine, the liver enzyme that hydrolyzes glycogen is inactivated by the removal of its phosphate group (by yet a different enzyme). This allows glycogen breakdown in the liver to be replaced by glycogen synthesis.

The activation/inactivation of the enzyme in this example is achieved by the process of *phosphorylation/dephosphorylation.* Many other enzymes are regulated in a similar manner, but some are activated by binding to small, regulatory organic molecules. For example, the enzyme protein kinase is activated when it binds to *cyclic AMP (cAMP),* a second-messenger molecule (chapter 6) discussed in relation to neural and endocrine regulation in chapters 7 and 11, respectively.

Enzyme activity is also regulated by the **turnover** of enzyme proteins. This refers to the breakdown and resynthesis of enzymes. Enzymes can be reused indefinitely after they catalyze reactions, but—as discussed in chapter 3—enzymes are degraded within lysosomes and proteosomes. Thus, their activities will end unless they are also resynthesized. Enzyme turnover allows genes to alter the enzyme activities (and thus metabolism) of the cell as conditions change.

Substrate Concentration and Reversible Reactions

At a given level of enzyme concentration, the rate of product formation will increase as the substrate concentration increases. Eventually, however, a point will be reached where additional increases in substrate concentration do not result in comparable increases in reaction rate. When the relationship between substrate concentration and reaction rate reaches a plateau of maximum velocity, the enzyme is said to be *saturated.* If we think of enzymes as workers in a plant that converts a raw material (say, metal ore) into a product (say, iron), then enzyme saturation is like the plant working at full capacity, with no idle time for the workers. Increasing the amount of raw material (substrate) at this point cannot increase the rate of product formation. This concept is illustrated in figure 4.6.

Some enzymatic reactions within a cell are reversible, with both the forward and the backward reactions catalyzed by the same enzyme. The enzyme *carbonic anhydrase,* for example, is named because it can catalyze the following reaction:

$$H_2CO_3 \rightarrow H_2O + CO_2$$

The same enzyme, however, can also catalyze the reverse reaction:

$$H_2O + CO_2 \rightarrow H_2CO_3$$

The two reactions can be more conveniently illustrated by a single equation with double arrows:

$$H_2O + CO_2 \rightleftarrows H_2CO_3$$

The direction of the reversible reaction depends on the relative concentrations of the molecules to the left and right of the arrows. If the concentration of CO_2 is very high (as it is in the tissues), the reaction will be driven to the right. If the concentration of CO_2 is low and that of H_2CO_3 is high (as it is in the lungs), the reaction will be driven to the left. The principle that reversible reactions will be driven from the side of the equation where the concentration is higher to the side where the concentration is lower is known as the **law of mass action.**

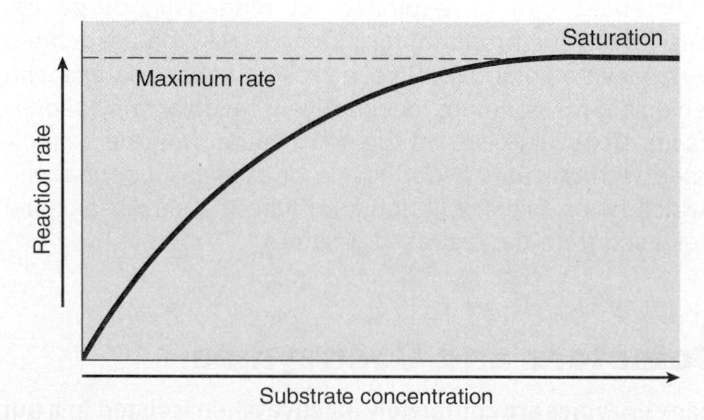

Figure 4.6 The effect of substrate concentration on the rate of an enzyme-catalyzed reaction. When the reaction rate is at a maximum, the enzyme is said to be saturated.

Although some enzymatic reactions are not directly reversible, the net effects of the reactions can be reversed by the action of different enzymes. Some of the enzymes that convert glucose to pyruvic acid, for example, are different from those that reverse the pathway and produce glucose from pyruvic acid. Likewise, the formation and breakdown of glycogen (a polymer of glucose; see fig. 2.14) are catalyzed by different enzymes.

Metabolic Pathways

The many thousands of different types of enzymatic reactions within a cell do not occur independently of each other. They are, rather, all linked together by intricate webs of interrelationships, the total pattern of which constitutes cellular metabolism. A sequence of enzymatic reactions that begins with an *initial substrate,* progresses through a number of *intermediates,* and ends with a *final product* is known as a **metabolic pathway.**

The enzymes in a metabolic pathway cooperate in a manner analogous to workers on an assembly line, where each contributes a small part to the final product. In this process, the product of one enzyme in the line becomes the substrate of the next enzyme, and so on (fig. 4.7).

Few metabolic pathways are completely linear. Most are branched so that one intermediate at the branch point can serve as a substrate for two different enzymes. Two different products can thus be formed that serve as intermediates of two pathways (fig. 4.8). Generally, certain key enzymes in these pathways are subject to regulation, so that the direction taken by the metabolic pathways can be changed at different times by the activation or inhibition of these enzymes.

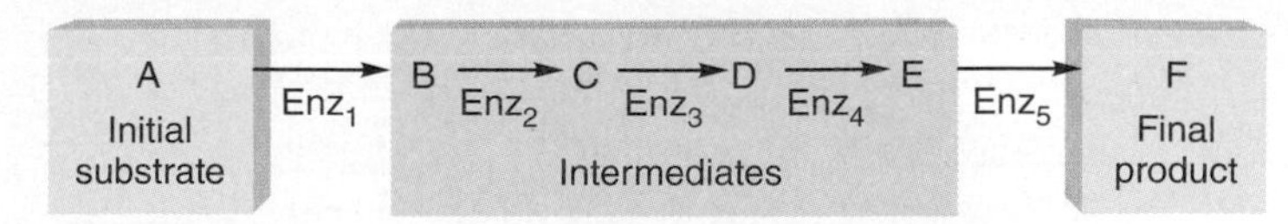

Figure 4.7 The general pattern of a metabolic pathway. In metabolic pathways, the product of one enzyme becomes the substrate of the next.

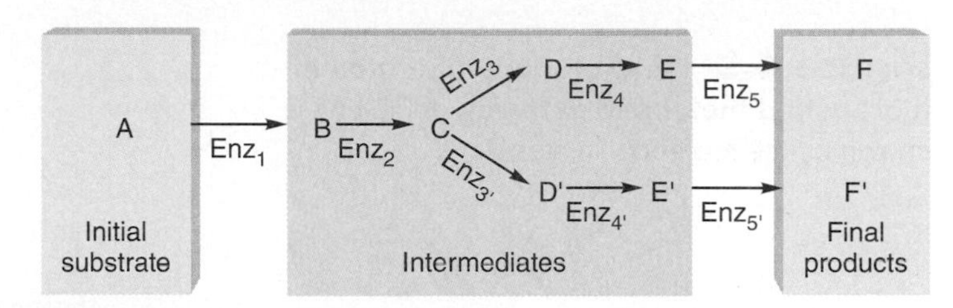

Figure 4.8 A branched metabolic pathway. Two or more different enzymes can work on the same substrate at the branch point of the pathway, catalyzing two or more different reactions.

CLINICAL APPLICATION

Severe combined immunodeficiency disease (SCID), which can be caused by a deficiency in *adenosine deaminase (ADA),* is a fatal disease inherited as an autosomal recessive trait. Because of this enzyme deficiency, the pathways of purine metabolism are disrupted and toxic metabolites accumulate to cause failure of the immune system (the "boy in the bubble" disease). Beginning in the 1980s, scientists attempted to cure this disease by **gene therapy,** using viruses to deliver the gene (for ADA) to the patients' hematopoietic stem cells (those from the bone marrow that form blood cells; chapter 13, section 13.2). Techniques for delivering this gene using viruses as vectors (carriers) have improved over the years, and a recent report demonstrated the relative safety and effectiveness of ADA gene therapy for SCID. Recent successes in gene therapy for *Leber's congenital amaurosis* (a form of congenital blindness) and *X-linked adrenoleukodystrophy* (a congenital brain disorder) have also been reported. Hopefully, the safety and effectiveness of gene therapies will continue to improve so that other inherited diseases can also be treated.

End-Product Inhibition

The activities of enzymes at the branch points of metabolic pathways are often regulated by a process called **end-product inhibition,** which is a form of negative feedback. In this process, one of the final products of a divergent pathway inhibits the activity of the branch-point enzyme that began the path toward the production of this inhibitor. This inhibition prevents that final product from accumulating excessively and results in a shift toward the final product of the alternate pathway (fig. 4.9).

The mechanism by which a final product inhibits an earlier enzymatic step in its pathway is known as **allosteric inhibition.** The allosteric inhibitor combines with a part of the enzyme at a location other than the active site. This causes the active site to change shape so that it can no longer combine properly with its substrate.

Inborn Errors of Metabolism

Because each different polypeptide in the body is coded by a different gene (chapter 3), each enzyme protein that participates in a metabolic pathway is coded by a different gene. An inherited defect in one of these genes may result in a disease known as an **inborn error of metabolism.** In this type of disease, the quantity of intermediates formed *prior* to the defective enzymatic step *increases,* and the quantity of intermediates and final products formed *after* the defective step *decreases.* Diseases may result from deficiencies of the normal end product or from excessive accumulation of intermediates formed prior to the defective step. If the defective enzyme is active at a step

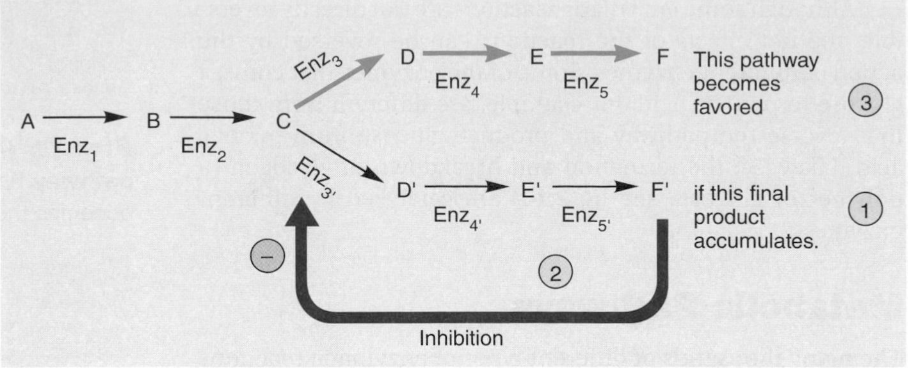

A → B → C
Enz_1 Enz_2
Enz_3 D → E → F This pathway becomes favored (3)
Enz_4 Enz_5
$Enz_{3'}$ D' → E' → F' if this final product accumulates. (1)
$Enz_{4'}$ $Enz_{5'}$
(–) (2) Inhibition

Figure 4.9 End-product inhibition in a branched metabolic pathway. Inhibition is shown by the red arrow in step 2.

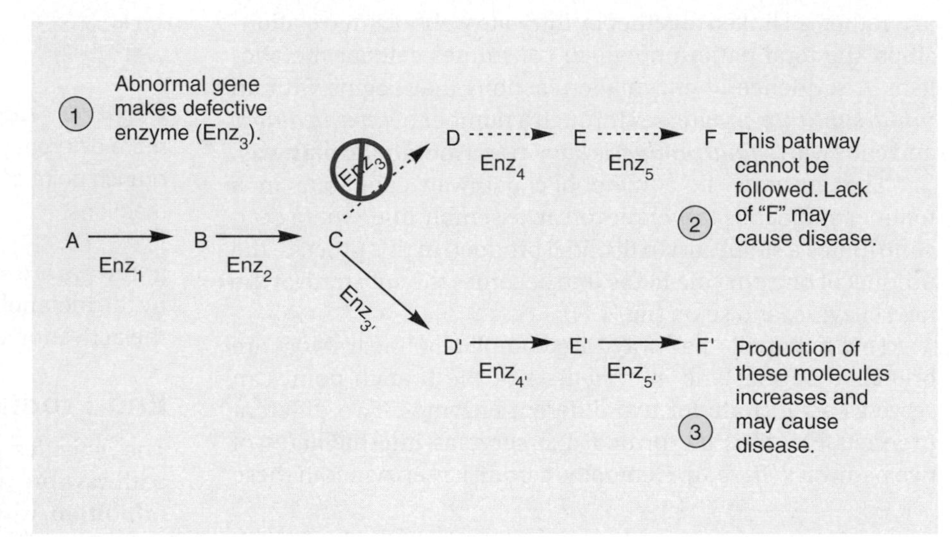

Figure 4.10 The effects of an inborn error of metabolism on a branched metabolic pathway. The defective gene produces a defective enzyme, indicated here by a line through its symbol.

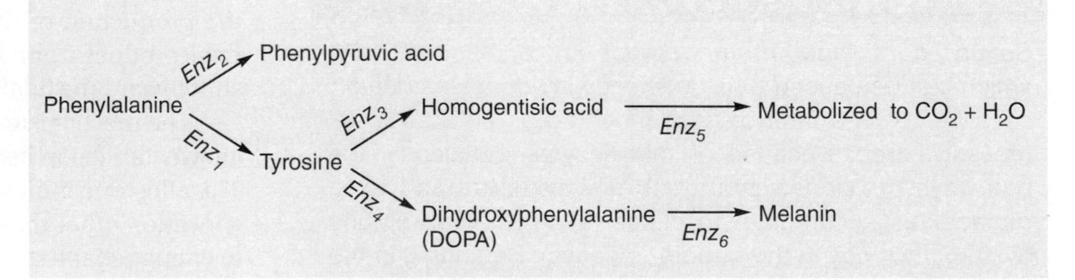

Figure 4.11 Metabolic pathways for the degradation of the amino acid phenylalanine. Defective *enzyme*$_1$ produces phenylketonuria (PKU), defective *enzyme*$_5$ produces alcaptonuria (not a clinically significant condition), and defective *enzyme*$_6$ produces albinism.

that follows a branch point in a pathway, the intermediates and final products of the alternate pathway will increase (fig. 4.10). An abnormal increase in the production of these products can be the cause of some metabolic diseases.

One of the conversion products of phenylalanine is a molecule called DOPA, an acronym for dihydroxyphenylalanine. DOPA is a precursor of the pigment molecule *melanin,* which gives skin, eyes, and hair their normal coloration. The condition of *albinism* results from an inherited defect in the enzyme that catalyzes the formation of melanin from DOPA (fig. 4.11). Besides albinism and PKU (described in the next Clincial Applications box), there are many other inborn errors of amino acid metabolism, as well as errors in carbohydrate and lipid metabolism. Some of these are described in table 4.4.

CLINICAL APPLICATION

The branched metabolic pathway that begins with phenylalanine as the initial substrate is subject to a number of inborn errors of metabolism (fig. 4.11). When the enzyme that converts this amino acid to the amino acid tyrosine is defective, the final product of a divergent pathway accumulates and can be detected in the blood and urine. This disease—**phenylketonuria (PKU)**—can result in severe mental retardation and a shortened life span. PKU occurs often enough (although no inborn error of metabolism is common) to warrant the testing of all newborn babies for the defect. If the disease is detected early, brain damage can be prevented by placing the child on an artificial diet low in the amino acid phenylalanine.

Table 4.4 | Examples of Inborn Errors in the Metabolism of Amino Acids, Carbohydrates, and Lipids

Metabolic Defect	Disease	Abnormality	Clinical Result
Amino acid metabolism	Phenylketonuria (PKU)	Increase in phenylpyruvic acid	Mental retardation, epilepsy
	Albinism	Lack of melanin	Susceptibility to skin cancer
	Maple-syrup disease	Increase in leucine, isoleucine, and valine	Degeneration of brain, early death
	Homocystinuria	Accumulation of homocystine	Mental retardation, eye problems
Carbohydrate metabolism	Lactose intolerance	Lactose not utilized	Diarrhea
	Glucose 6-phosphatase deficiency (Gierke's disease)	Accumulation of glycogen in liver	Liver enlargement, hypoglycemia
	Glycogen phosphorylase deficiency	Accumulation of glycogen in muscle	Muscle fatigue and pain
Lipid metabolism	Gaucher's disease	Lipid accumulation (glucocerebroside)	Liver and spleen enlargement, brain degeneration
	Tay-Sachs disease	Lipid accumulation (ganglioside G_{M2})	Brain degeneration, death by age five
	Hypercholestremia	High blood cholesterol	Atherosclerosis of coronary and large arteries

CHECKPOINT

3. Draw graphs to represent the effects of changes in temperature, pH, and enzyme and substrate concentration on the rate of enzymatic reactions. Explain the mechanisms responsible for the effects you have graphed.
4. Describe a reversible reaction and explain how the law of mass action affects this reaction.

5a. Using arrows and letters of the alphabet, draw a flowchart of a metabolic pathway with one branch point.

5b. Define *end-product inhibition* and use your diagram of a branched metabolic pathway to explain how this process will affect the concentrations of different intermediates.

5c. Because of an inborn error of metabolism, suppose that the enzyme that catalyzed the third reaction in your pathway (see no. 5a) was defective. Describe the effects this would have on the concentrations of the intermediates in your pathway.

4.3 BIOENERGETICS

Living organisms require the constant expenditure of energy to maintain their complex structures and processes. Central to life processes are chemical reactions that are coupled, so that the energy released by one reaction is incorporated into the products of another reaction.

LEARNING OUTCOMES

After studying this section, you should be able to:

6. **Distinguish between endergonic and exergonic reactions, and explain how ATP functions as a universal energy carrier.**
7. **Distinguish between oxidation and reduction reactions, and explain the functions of NAD and FAD.**

Energy may be defined as the ability to do work (exert a force that acts over a distance). **Bioenergetics** refers to the flow of energy in living systems. Organisms maintain their highly ordered structure and life-sustaining activities through the constant expenditure of energy obtained ultimately from the environment. The energy flow in living systems obeys the first and second laws of a branch of physics known as *thermodynamics.*

According to the **first law of thermodynamics,** energy can be transformed (changed from one form to another), but it can neither be created nor destroyed. This is sometimes called the *law of conservation of energy.* For example, the mechanical energy of a waterfall can be transformed into the electrical energy produced by a hydroelectric plant; the chemical bond energy in gasoline can be transformed into the mechanical energy of turning gears; and (in a hybrid car), mechanical energy can be transformed into electrical energy. Figure 4.12 shows a more biological example; indeed, this is the energy transformation upon which all animal and plant life depends: the transformation of light energy into the chemical bond energy in glucose molecules.

However, in all energy transformations, you can never get out what you put in; the transformation is never 100% efficient

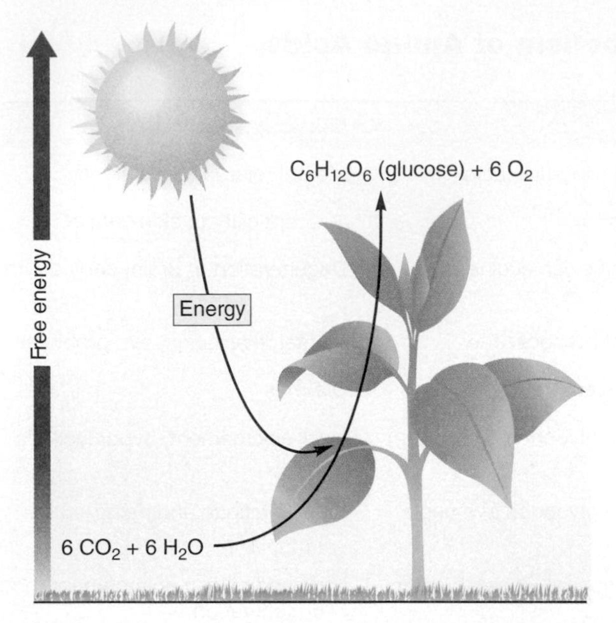

Figure 4.12 A simplified diagram of photosynthesis. Some of the sun's radiant energy is captured by plants and used to produce glucose from carbon dioxide and water. As the product of this endergonic reaction, glucose has more free energy than the initial reactants.

(that's why a perpetual motion machine is impossible in principle). The total energy is conserved in these transformations (first law of thermodynamics), but a proportion of the energy is lost as heat. Therefore, the amount of energy in an "organized" form—the energy available to do work—decreases in every energy transformation. *Entropy* is the degree of disorganization of a system's total energy. The **second law of thermodynamics** states that the amount of entropy increases in every energy transformation. Because only energy in an organized state—called *free energy*—is available to do work, this means that the free energy of a system decreases as its entropy increases. A hybrid car transforms chemical bond energy in gasoline to the mechanical energy of turning gears, which is then transformed into electrical energy that can later be used to turn gears. But the second law dictates that the process cannot simply be reversed and continued indefinitely; more gasoline will have to be burned. The second law also explains why plants require the continued input of light energy, and why we need the continued input of the chemical bond energy in food molecules.

The chemical bonding of atoms into molecules obeys the laws of thermodynamics. Six separate molecules of carbon dioxide and 6 separate molecules of water is a more disorganized state than 1 molecule of glucose ($C_6H_{12}O_6$). To go from a more disorganized state (higher entropy) to a more organized state (lower entropy) that has more free energy requires the addition of energy from an outside source. Thus, plants require the input of light energy from the sun to produce glucose from carbon dioxide and water in the process of *photosynthesis* (fig. 4.12). Because light energy was required to form the bonds of glucose, a portion of that energy (never 100%, according to the second law) must still be present in the chemical bonds of glucose (first law). It also follows that when the chemical bonds of glucose are broken, converting the glucose back into carbon dioxide and water, energy must be released. This energy indirectly powers all of the energy-requiring processes of our bodies.

Endergonic and Exergonic Reactions

Chemical reactions that require an input of energy are known as **endergonic reactions.** Because energy is added to make these reactions "go," the products of endergonic reactions must contain more free energy than the reactants. A portion of the energy added, in other words, is contained within the product molecules. This follows from the fact that energy cannot be created or destroyed (first law of thermodynamics) and from the fact that a more-organized state of matter contains more free energy, or less entropy, than a less-organized state (second law of thermodynamics).

That glucose contains more free energy than carbon dioxide and water can easily be proven by combusting glucose to CO_2 and H_2O. This reaction releases energy in the form of heat. Reactions that convert molecules with more free energy to molecules with less—and, therefore, that release energy as they proceed—are called **exergonic reactions.**

As illustrated in figure 4.13, the total amount of energy released by a molecule in a combustion reaction can be released in smaller portions by enzymatically controlled exergonic reactions within cells. This allows the cells to use the energy to "drive" other processes, as described in the next section. The energy obtained by the body from the cellular oxidation of a molecule is the same as the amount released when the molecule is combusted, so the energy in food molecules can conveniently be measured by the heat released when the molecules are combusted.

Heat is measured in units called *calories.* One calorie is defined as the amount of heat required to raise the temperature of 1 cubic centimeter of water 1 degree on the Celsius scale. The caloric value of food is usually indicated in *kilocalories* (1 kilocalorie = 1,000 calories), which are often called large calories and spelled with a capital C.

Coupled Reactions: ATP

In order to remain alive, a cell must maintain its highly organized, low-entropy state at the expense of free energy in its environment. Accordingly, the cell contains many enzymes that catalyze exergonic reactions using substrates that come ultimately from the environment. The energy released by these exergonic reactions is used to drive the energy-requiring processes (endergonic reactions) in the cell. Because cells cannot use heat energy to drive energy-requiring processes, the chemical-bond energy that is released in exergonic reactions must be directly transferred to chemical-bond energy in the

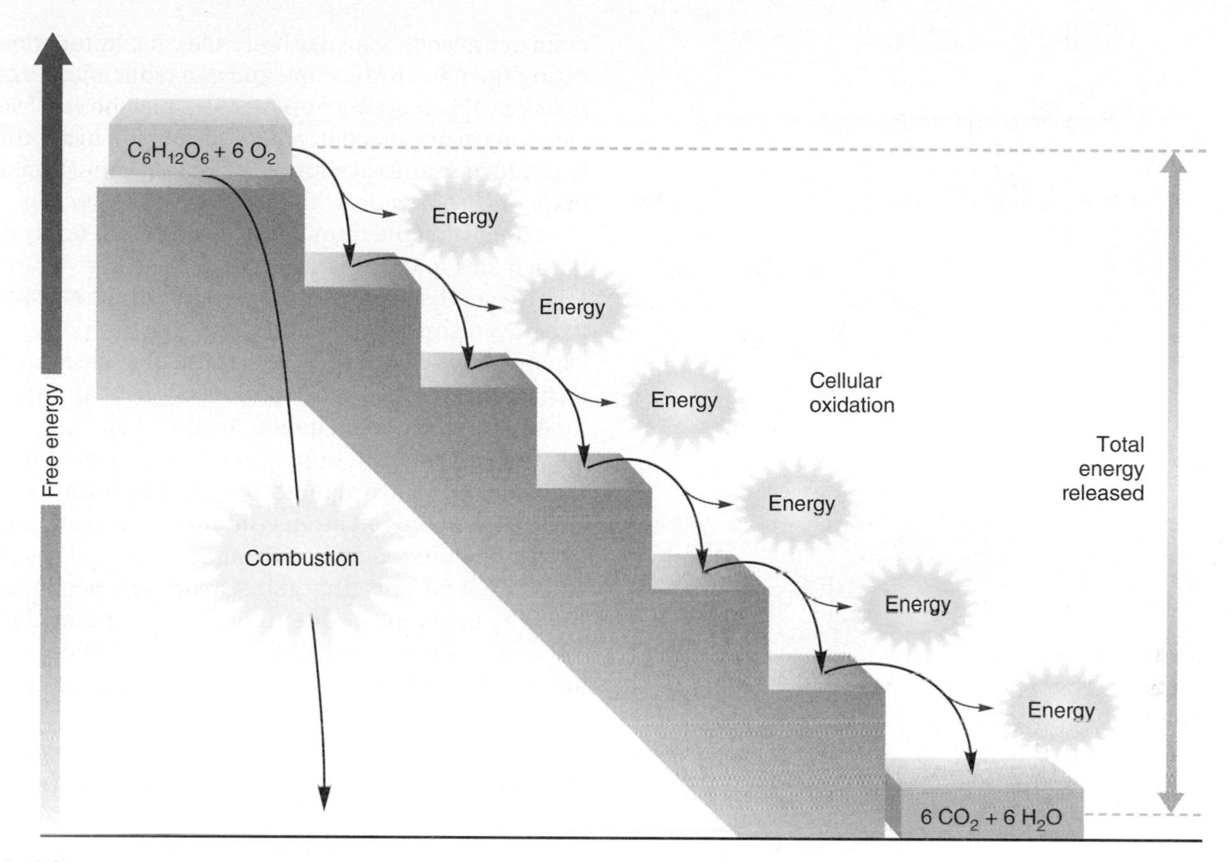

Figure 4.13 A comparison of combustion and cell respiration. Because glucose contains more energy than six separate molecules each of carbon dioxide and water, the combustion of glucose is an exergonic process. The same amount of energy is released when glucose is broken down stepwise within the cell. Each step represents an intermediate compound in the aerobic respiration of glucose.

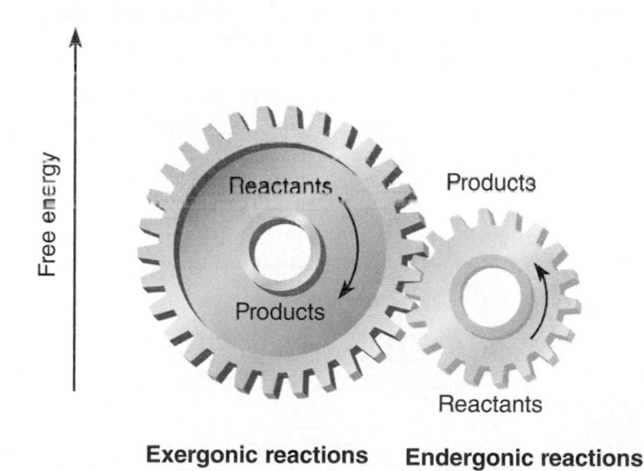

Figure 4.14 A model of the coupling of exergonic and endergonic reactions. The reactants of the exergonic reaction (represented by the larger gear) have more free energy than the products of the endergonic reaction because the coupling is not 100% efficient—some energy is lost as heat.

products of endergonic reactions. Energy-liberating reactions are thus *coupled* to energy-requiring reactions. This relationship is like that of two meshed gears; the turning of one (the energy-releasing exergonic gear) causes turning of the other (the energy-requiring endergonic gear). This relationship is illustrated in figure 4.14.

The energy released by most exergonic reactions in the cell is used, either directly or indirectly, to drive one particular endergonic reaction (fig. 4.15): the formation of **adenosine triphosphate (ATP)** from adenosine diphosphate (ADP) and inorganic phosphate (abbreviated P_i).

The formation of ATP requires the input of a fairly large amount of energy. Because this energy must be conserved (first law of thermodynamics), the bond produced by joining P_i to ADP must contain a part of this energy. Thus, when enzymes reverse this reaction and convert ATP to ADP and P_i, a large amount of energy is released. Energy released from the breakdown of ATP is used to power the energy-requiring processes in all cells. As the **universal energy carrier,** ATP serves to more efficiently couple the energy released by the breakdown of food molecules to the energy required by the diverse endergonic processes in the cell (fig. 4.16).

Coupled Reactions: Oxidation-Reduction

When an atom or a molecule gains electrons, it is said to become **reduced;** when it loses electrons, it is said to become **oxidized.**

Reduction and oxidation are always coupled reactions: an atom or a molecule cannot become oxidized unless it donates

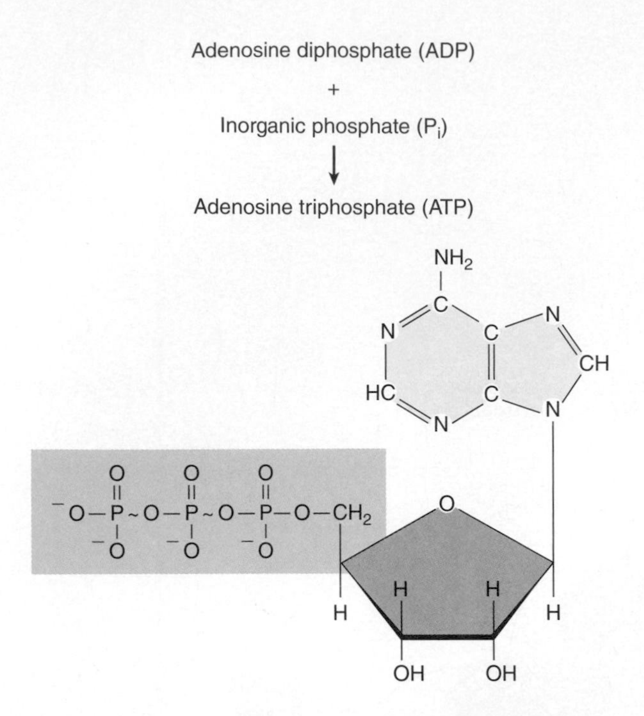

Figure 4.15 The formation and structure of adenosine triphosphate (ATP). ATP is the universal energy carrier of the cell. High-energy bonds are indicated by a squiggle (~). When the bond to the last phosphate is broken, energy is released and ATP is converted back into ADP and P_i.

electrons to another, which therefore becomes reduced. The atom or molecule that donates electrons *to* another is a **reducing agent,** and the one that accepts electrons *from* another is an **oxidizing agent.** It is important to understand that a particular atom or molecule can play both roles; it may function as an oxidizing agent in one reaction and as a reducing agent in another reaction. When atoms or molecules play both roles, they gain electrons in one reaction and pass them on in another reaction to produce a series of coupled oxidation-reduction reactions—like a bucket brigade, with electrons in the buckets.

Notice that the term *oxidation* does not imply that oxygen participates in the reaction. This term is derived from the fact that oxygen has a great tendency to accept electrons; that is, to act as a strong oxidizing agent. This property of oxygen is exploited by cells; oxygen acts as the final electron acceptor in a chain of oxidation-reduction reactions that provides energy for ATP production (chapter 5, section 5.2).

Oxidation-reduction reactions in cells often involve the transfer of hydrogen atoms rather than free electrons. Because a hydrogen atom contains 1 electron (and 1 proton in the nucleus), a molecule that loses hydrogen becomes oxidized, and one that gains hydrogen becomes reduced. In many oxidation-reduction reactions, pairs of electrons—either as free electrons or as a pair of hydrogen atoms—are transferred from the reducing agent to the oxidizing agent.

Two molecules that serve important roles in the transfer of hydrogens are **nicotinamide adenine dinucleotide (NAD),** which is derived from the vitamin niacin (vitamin B_3), and **flavin adenine dinucleotide (FAD),** which is derived from the vitamin riboflavin (vitamin B_2). These molecules (fig. 4.17) are coenzymes that function as *hydrogen carriers* because they accept hydrogens (becoming reduced) in one enzyme reaction and donate hydrogens (becoming oxidized) in a different enzyme reaction (fig. 4.18). The oxidized forms of these molecules are written simply as NAD (or NAD^+) and FAD.

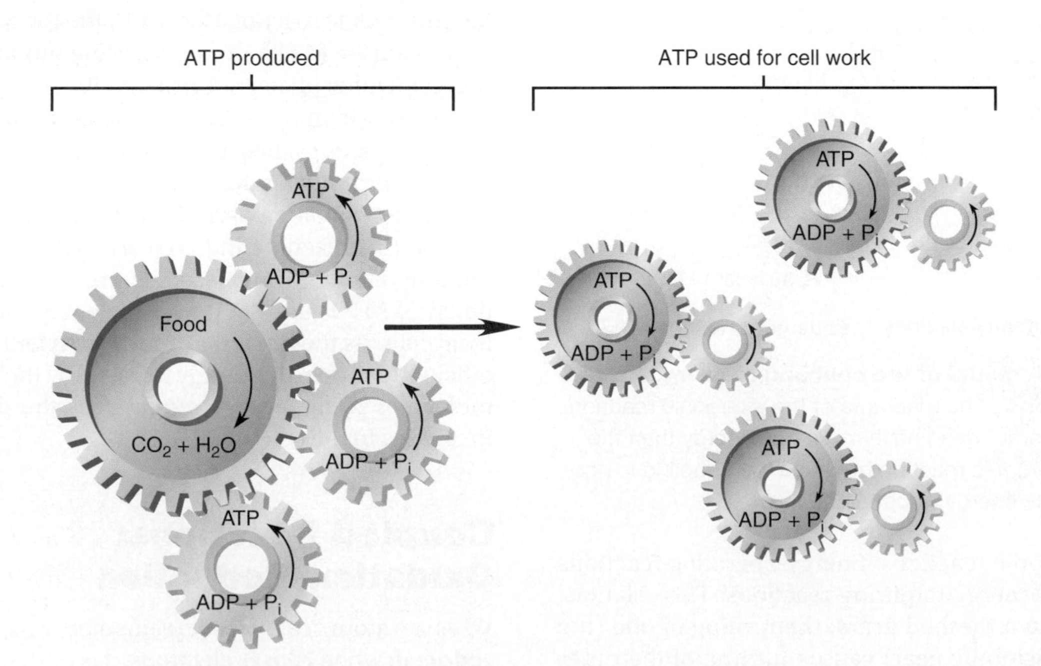

Figure 4.16 A model of ATP as the universal energy carrier of the cell. Exergonic reactions are shown as blue gears with arrows going down (these reactions produce a decrease in free energy); endergonic reactions are shown as green gears with arrows going up (these reactions produce an increase in free energy).

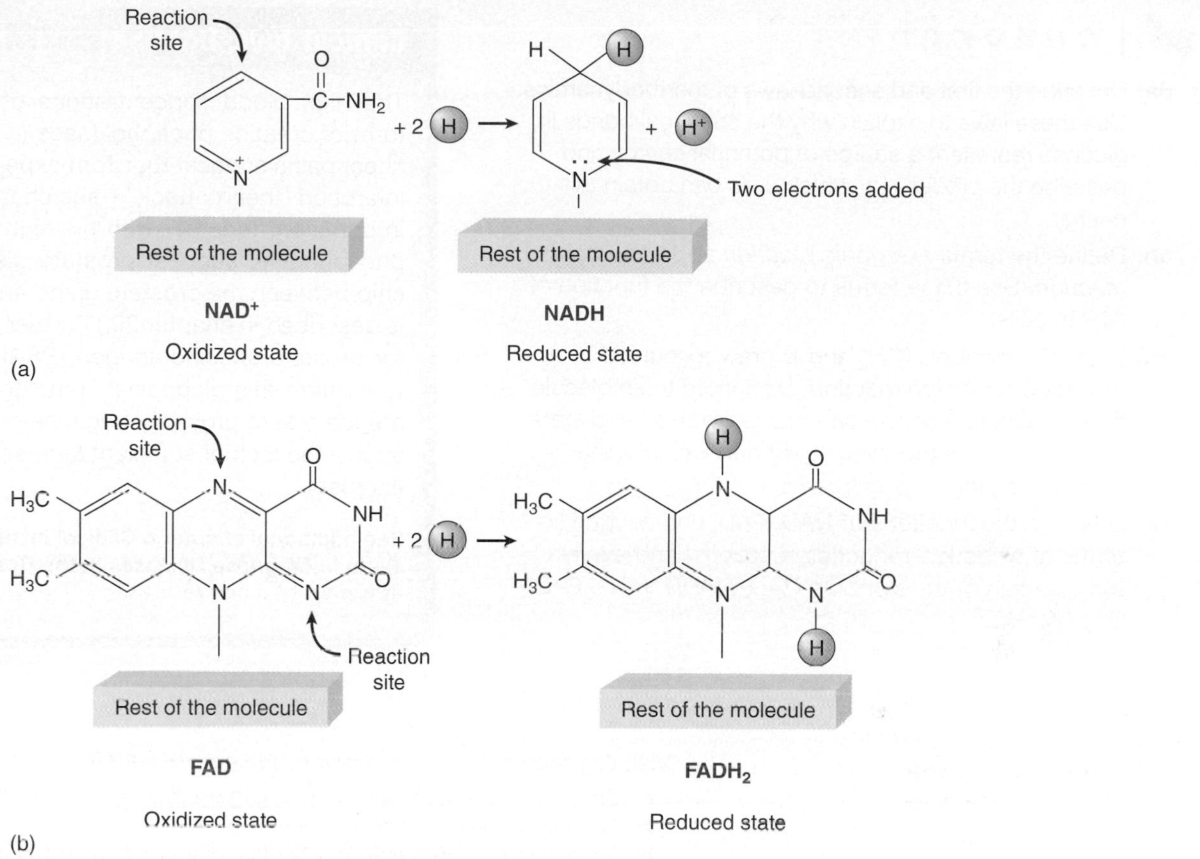

Reaction site
O
C–NH2
+ 2 H →
N+
Rest of the molecule
NAD+
Oxidized state
(a)
H H
+ H+
N
Two electrons added
Rest of the molecule
NADH
Reduced state

Reaction site
H3C
H3C
N
NH
O
N
N
O
+ 2 H →
Reaction site
Rest of the molecule
FAD
Oxidized state
(b)
H
H3C
H3C
N
NH
N
N
O
H
Rest of the molecule
FADH2
Reduced state

Figure 4.17 Structural formulas for NAD^+, NADH, FAD, and $FADH_2$. (*a*) When NAD^+ reacts with two hydrogen atoms, it binds to one of them and accepts the electron from the other. This is shown by two dots above the nitrogen ($\ddot{N}$) in the formula for NADH. (*b*) When FAD reacts with two hydrogen atoms to form $FADH_2$, it binds each of them to a nitrogen atom at the reaction sites. AP|R

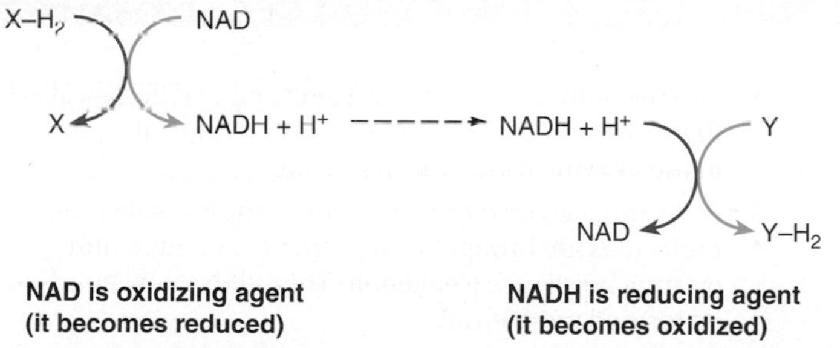

X–H2
NAD
X
NADH + H+ - - - - - → NADH + H+
Y
NAD
Y–H2
NAD is oxidizing agent (it becomes reduced)
NADH is reducing agent (it becomes oxidized)

Figure 4.18 The action of NAD. NAD is a coenzyme that transfers pairs of hydrogen atoms from one molecule to another. In the first reaction, NAD is reduced (acts as an oxidizing agent); in the second reaction, NADH is oxidized (acts as a reducing agent). Oxidation reactions are shown by red arrows, reduction reactions by blue arrows. AP|R

Each FAD can accept two electrons and can bind two protons. Therefore, the reduced form of FAD is combined with the equivalent of two hydrogen atoms and may be written as $FADH_2$. Each NAD can also accept two electrons but can bind only one proton (see fig. 4.17). The reduced form of NAD is therefore indicated by NADH + H^+ (the H^+ represents a free proton). When the reduced forms of these two coenzymes participate in an oxidation-reduction reaction, they transfer two hydrogen atoms to the oxidizing agent (fig. 4.18). The ability of FAD and NAD to transfer protons and electrons in this way is particularly important in metabolic reactions that provide energy (ATP) for the cells, as described in chapter 5.

CLINICAL APPLICATION

Production of the coenzymes NAD and FAD is the major reason that we need the vitamins niacin and riboflavin in our diet. As described in chapter 5, NAD and FAD are required to transfer hydrogen atoms in the chemical reactions that provide energy for the body. Niacin and riboflavin do not themselves provide the energy, although this is often claimed in misleading advertisements for health foods. Nor can eating extra amounts of niacin and riboflavin provide extra energy. Once the cells have obtained sufficient NAD and FAD, the excess amounts of these vitamins are simply eliminated in the urine.

CHECKPOINT

6a. Describe the first and second laws of thermodynamics. Use these laws to explain why the chemical bonds in glucose represent a source of potential energy and describe the process by which cells can obtain this energy.

6b. Define the terms *exergonic reaction* and *endergonic reaction.* Use these terms to describe the function of ATP in cells.

7a. Using the symbols $X\text{-}H_2$ and Y, draw a coupled oxidation-reduction reaction. Designate the molecule that is reduced and the one that is oxidized and state which one is the reducing agent and which is the oxidizing agent.

7b. Describe the functions of NAD, FAD, and oxygen (in terms of oxidation-reduction reactions) and explain the meaning of the symbols *NAD*, $NADH + H^+$, *FAD*, and $FADH_2$.

Clinical Investigation SUMMARY

The high blood concentrations of the MB isoenzyme form of creatine phosphokinase (CPK) following severe chest pains suggest that Tom experienced a myocardial infarction ("heart attack"—see chapter 13). His difficulty in urination, together with his high blood levels of acid phosphatase, suggest prostate disease. (The relationship between the prostate gland and the urinary system is described in chapter 20.) Further tests—including one for *prostate-specific antigen (PSA)*—can be performed to confirm this diagnosis. Tom "got the runs" when he ate ice cream probably because he has lactose intolerance—the lack of sufficient lactase to digest milk sugar (lactose).

See additional chapter 4 Clinical Investigation on *Enzyme Tests to Diagnose Diseases* in the Connect site for this text at www.mhhe.com/Fox13.

Visit this book's website at **www.mhhe.com/Fox13** for:

- Chapter quizzes, interactive learning exercises, and other study tools
- Additional clinical investigations
- Access to LearnSmart—An adaptive diagnostic tool that constantly assesses student knowledge of course material
- Ph.I.L.S. 4.0—physiology interactive lab simulations that may be used to supplement or substitute for wet labs

SUMMARY

4.1 Enzymes as Catalysts 89

A. Enzymes are biological catalysts.
  1. Catalysts increase the rate of chemical reactions.
    a. A catalyst is not altered by the reaction.
    b. A catalyst does not change the final result of a reaction.
  2. Catalysts lower the activation energy of chemical reactions.
    a. The activation energy is the amount of energy needed by the reactant molecules to participate in a reaction.
    b. In the absence of a catalyst, only a small proportion of the reactants possess the activation energy to participate.
    c. By lowering the activation energy, enzymes allow a larger proportion of the reactants to participate in the reaction, thus increasing the reaction rate.

B. Most enzymes are proteins.
  1. Protein enzymes have specific three-dimensional shapes that are determined by the amino acid sequence and, ultimately, by the genes.
  2. The reactants in an enzyme-catalyzed reaction—called the substrates of the enzyme—fit into a specific pocket in the enzyme called the active site.
  3. By forming an enzyme-substrate complex, substrate molecules are brought into proper orientation and existing bonds are weakened. This allows new bonds to be formed more easily.

4.2 Control of Enzyme Activity 92

A. The activity of an enzyme is affected by a variety of factors.
  1. The rate of enzyme-catalyzed reactions increases with increasing temperature, up to a maximum rate.
    a. This is because increasing the temperature increases the energy in the total population of reactant molecules, thus increasing the proportion of reactants that have the activation energy.
    b. At a few degrees above body temperature, however, most enzymes start to denature, which decreases the rate of the reactions that they catalyze.
  2. Each enzyme has optimal activity at a characteristic pH—called the pH optimum for that enzyme.

a. Deviations from the pH optimum will decrease the reaction rate because the pH affects the shape of the enzyme and charges within the active site.
b. The pH optima of different enzymes can vary widely—pepsin has a pH optimum of 2, for example, while trypsin is most active at a pH of 9.

3. Many enzymes require metal ions in order to be active. These ions are therefore said to be cofactors for the enzymes.

4. Many enzymes require smaller organic molecules for activity. These smaller organic molecules are called coenzymes.
   a. Coenzymes are derived from water-soluble vitamins.
   b. Coenzymes transport hydrogen atoms and small substrate molecules from one enzyme to another.

5. Some enzymes are produced as inactive forms that are later activated within the cell.
   a. Activation may be achieved by phosphorylation of the enzyme, in which case the enzyme can later be inactivated by dephosphorylation.
   b. Phosphorylation of enzymes is catalyzed by an enzyme called protein kinase.
   c. Protein kinase itself may be inactive and require the binding of a second messenger called cyclic AMP in order to become activated.

6. The rate of enzymatic reactions increases when either the substrate concentration or the enzyme concentration is increased.
   a. If the enzyme concentration remains constant, the rate of the reaction increases as the substrate concentration is raised, up to a maximum rate.
   b. When the rate of the reaction does not increase upon further addition of substrate, the enzyme is said to be saturated.

B. Metabolic pathways involve a number of enzyme-catalyzed reactions.
   1. A number of enzymes usually cooperate to convert an initial substrate to a final product by way of several intermediates.
   2. Metabolic pathways are produced by multienzyme systems in which the product of one enzyme becomes the substrate of the next.
   3. If an enzyme is defective due to an abnormal gene, the intermediates that are formed following the step catalyzed by the defective enzyme will decrease, and the intermediates that are formed prior to the defective step will accumulate.
      a. Diseases that result from defective enzymes are called inborn errors of metabolism.
      b. Accumulation of intermediates often results in damage to the organ in which the defective enzyme is found.
   4. Many metabolic pathways are branched, so that one intermediate can serve as the substrate for two different enzymes.
   5. The activity of a particular pathway can be regulated by end-product inhibition.
      a. In end-product inhibition, one of the products of the pathway inhibits the activity of a key enzyme.
      b. This is an example of allosteric inhibition, in which the product combines with its specific site on the enzyme, changing the conformation of the active site.

4.3 Bioenergetics 97

A. The flow of energy in the cell is called bioenergetics.
   1. According to the first law of thermodynamics, energy can neither be created nor destroyed but only transformed from one form to another.
   2. According to the second law of thermodynamics, all energy transformation reactions result in an increase in entropy (disorder).
      a. As a result of the increase in entropy, there is a decrease in free (usable) energy.
      b. Atoms that are organized into large organic molecules contain more free energy than the more disorganized, smaller molecules.
   3. In order to produce glucose from carbon dioxide and water, energy must be added.
      a. Plants use energy from the sun for this conversion, in a process called photosynthesis.
      b. Reactions that require the input of energy to produce molecules with more free energy than the reactants are called endergonic reactions.
   4. The combustion of glucose to carbon dioxide and water releases energy in the form of heat.
      a. A reaction that releases energy, thus forming products that contain less free energy than the reactants, is called an exergonic reaction.
      b. The same total amount of energy is released when glucose is converted into carbon dioxide and water within cells, even though this process occurs in many small steps.
   5. The exergonic reactions that convert food molecules into carbon dioxide and water in cells are coupled to endergonic reactions that form adenosine triphosphate (ATP).
      a. Some of the chemical-bond energy in glucose is therefore transferred to the "high energy" bonds of ATP.
      b. The breakdown of ATP into adenosine diphosphate (ADP) and inorganic phosphate results in the liberation of energy.
      c. The energy liberated by the breakdown of ATP is used to power all of the energy-requiring processes of the cell. ATP is thus the "universal energy carrier" of the cell.

B. Oxidation-reduction reactions are coupled and usually involve the transfer of hydrogen atoms.
   1. A molecule is said to be oxidized when it loses electrons; it is said to be reduced when it gains electrons.
   2. A reducing agent is thus an electron donor; an oxidizing agent is an electron acceptor.

3. Although oxygen is the final electron acceptor in the cell, other molecules can act as oxidizing agents.
4. A single molecule can be an electron acceptor in one reaction and an electron donor in another.
   a. NAD and FAD can become reduced by accepting electrons from hydrogen atoms removed from other molecules.
   b. NADH + H^+, and $FADH_2$, in turn, donate these electrons to other molecules in other locations within the cells.
   c. Oxygen is the final electron acceptor (oxidizing agent) in a chain of oxidation-reduction reactions that provide energy for ATP production.

REVIEW ACTIVITIES

Test Your Knowledge

1. Which of these statements about enzymes is *true?*
   a. Most proteins are enzymes.
   b. Most enzymes are proteins.
   c. Enzymes are changed by the reactions they catalyze.
   d. The active sites of enzymes have little specificity for substrates.
2. Which of these statements about enzyme-catalyzed reactions is *true?*
   a. The rate of reaction is independent of temperature.
   b. The rate of all enzyme-catalyzed reactions is decreased when the pH is lowered from 7 to 2.
   c. The rate of reaction is independent of substrate concentration.
   d. Under given conditions of substrate concentration, pH, and temperature, the rate of product formation varies directly with enzyme concentration up to a maximum, at which point the rate cannot be increased further.
3. Which of these statements about lactate dehydrogenase is *true?*
   a. It is a protein.
   b. It oxidizes lactic acid.
   c. It reduces another molecule (pyruvic acid).
   d. All of these are true.
4. In a metabolic pathway,
   a. the product of one enzyme becomes the substrate of the next.
   b. the substrate of one enzyme becomes the product of the next.
5. In an inborn error of metabolism,
   a. a genetic change results in the production of a defective enzyme.
   b. intermediates produced prior to the defective step accumulate.
   c. alternate pathways are taken by intermediates at branch points that precede the defective step.
   d. All of these are true.
6. Which of these represents an endergonic reaction?
   a. ADP + $P_i \rightarrow$ ATP
   b. ATP $\rightarrow$ ADP + P_i
   c. glucose + $O_2 \rightarrow CO_2 + H_2O$
   d. $CO_2 + H_2O \rightarrow$ glucose
   e. both *a* and *d*
   f. both *b* and *c*
7. Which of these statements about ATP is *true?*
   a. The bond joining ADP and the third phosphate is a high-energy bond.
   b. The formation of ATP is coupled to energy-liberating reactions.
   c. The conversion of ATP to ADP and P_i provides energy for biosynthesis, cell movement, and other cellular processes that require energy.
   d. ATP is the "universal energy carrier" of cells.
   e. All of these are true.
8. When oxygen is combined with 2 hydrogens to make water,
   a. oxygen is reduced.
   b. the molecule that donated the hydrogens becomes oxidized.
   c. oxygen acts as a reducing agent.
   d. both *a* and *b* apply.
   e. both *a* and *c* apply.
9. Enzymes increase the rate of chemical reactions by
   a. increasing the body temperature.
   b. decreasing the blood pH.
   c. increasing the affinity of reactant molecules for each other.
   d. decreasing the activation energy of the reactants.
10. According to the law of mass action, which of these conditions will drive the reaction $A + B \rightleftarrows C$ to the right?
    a. an increase in the concentration of A and B
    b. a decrease in the concentration of C
    c. an increase in the concentration of enzyme
    d. both *a* and *b*
    e. both *b* and *c*

Test Your Understanding

11. Explain the relationship between an enzyme's chemical structure and the function of the enzyme, and describe how both structure and function may be altered in various ways.

12. Explain how the rate of enzymatic reactions may be regulated by the relative concentrations of substrates and products.
13. Explain how end-product inhibition represents a form of negative feedback regulation.
14. Using the first and second laws of thermodynamics, explain how ATP is formed and how it serves as the universal energy carrier.
15. The coenzymes NAD and FAD can "shuttle" hydrogens from one reaction to another. How does this process serve to couple oxidation and reduction reactions?
16. Using albinism and phenylketonuria as examples, explain what is meant by inborn errors of metabolism.
17. Why do we need to eat food containing niacin and riboflavin? How do these vitamins function in body metabolism?

Test Your Analytical Ability

18. Metabolic pathways can be likened to intersecting railroad tracks, with enzymes as the switches. Discuss this analogy.
19. A student, learning that someone has an elevated blood level of lactate dehydrogenase (LDH), wonders how the enzyme got into this person's blood and worries about whether it will digest the blood. What explanation can you give to allay the student's fears?
20. Suppose you come across a bottle of enzyme tablets at your local health food store. The clerk tells you this enzyme will help your digestion, but you notice that it is derived from a plant. What concerns might you have regarding the effectiveness of these tablets?
21. Describe the energy transformations that occur when sunlight falls on a field of grass, the grass is eaten by a herbivore (such as a deer), and the herbivore is eaten by a carnivore (such as a cougar). Use the laws of thermodynamics to explain why there is more grass than deer, and more deer than cougars (in terms of their total biomass).
22. Use the reversible reactions involving the formation and breakdown of carbonic acid, and the law of mass action, to explain what would happen to the blood pH of a person who is hypoventilating (breathing inadequately) or hyperventilating (breathing excessively).

Test Your Quantitative Ability

Use the graph here and in figure 4.4 to answer the following questions:

23. What are the pH optima of these three enzymes?
24. What two pH values produce half-maximal activity of salivary amylase?
25. What two pH values produce half-maximal activity of pepsin?
26. At what pH is the activity of pepsin and salivary amylase equal?
27. Gastric juice normally has a pH of 2. Given this, what happens to the activity of salivary amylase, a starch-digesting enzyme in saliva, when it arrives in the stomach?

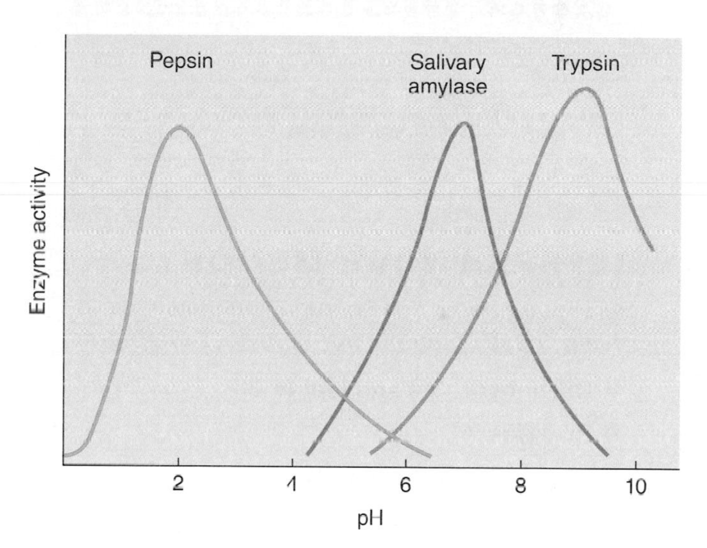

CHAPTER

5

Cell Respiration and Metabolism

REFRESH YOUR MEMORY

Before you begin this chapter, you may want to review these concepts from previous chapters:

CHAPTER OUTLINE

Clinical Investigation

Brenda, a college student, was training to make the swim team. Compared to her teammates, she experienced great fatigue during and following her workouts, and spent longer gasping for air after her workouts. Her coach suggested that she train more gradually and eat more carbohydrates than was her habit. Brenda also complained of intense muscle pain in her arms and shoulders that started with her training. Following a particularly intense workout, she experienced severe pain in her left pectoral region and sought medical aid.

Some of the new terms and concepts you will encounter include:

- Lactic acid fermentation and myocardial ischemia
- Glycogenesis and glycogenolysis

5.1 GLYCOLYSIS AND THE LACTIC ACID PATHWAY

In cellular respiration, energy is released by the stepwise breakdown of glucose and other molecules, and some of this energy is used to produce ATP. The complete combustion of glucose requires the presence of oxygen and yields about 30 ATP for each molecule of glucose. However, some energy can be obtained in the absence of oxygen by the pathway that leads to the production of lactic acid.

LEARNING OUTCOMES

After studying this section, you should be able to:

1. **Describe the metabolic pathway of glycolysis, how lactic acid is produced, and the physiological significance of the lactic acid pathway.**

All of the reactions in the body that involve energy transformation are collectively termed **metabolism.** Metabolism may be divided into two categories: *anabolism* and *catabolism.* Catabolic reactions release energy, usually by the breakdown of larger organic molecules into smaller molecules. Anabolic reactions require the input of energy and include the synthesis of large energy-storage molecules, including glycogen, fat, and protein.

The catabolic reactions that break down glucose, fatty acids, and amino acids serve as the primary sources of energy for the synthesis of ATP. For example, some of the chemical-bond energy in glucose is transferred to the chemical-bond energy in ATP. Because energy transfers can never be 100% efficient (according to the second law of thermodynamics, discussed in chapter 4), some of the chemical-bond energy from glucose is lost as heat.

This energy transfer involves oxidation-reduction reactions. Oxidation of a molecule occurs when the molecule loses electrons (chapter 4, section 4.3). This must be coupled to the reduction of another atom or molecule, which accepts the electrons. In the breakdown of glucose and other molecules for energy, some of the electrons initially present in these molecules are transferred to intermediate carriers and then to a *final electron acceptor.* When a molecule is completely broken down to carbon dioxide and water within an animal cell, the final electron acceptor is always an atom of oxygen. Because of the involvement of oxygen, the metabolic pathway that converts molecules such as glucose or fatty acid to carbon dioxide and water (transferring some of the energy to ATP) is called **aerobic cell respiration.** The oxygen for this process is obtained from the blood. The blood, in turn, obtains oxygen from air in the lungs through the process of breathing, or ventilation, as described in chapter 16. Ventilation also serves the important function of eliminating the carbon dioxide produced by aerobic cell respiration.

Unlike the process of burning, or combustion, which quickly releases the energy content of molecules as heat (and which can be measured as kilocalories—see chapter 4), the conversion of glucose to carbon dioxide and water within the cells occurs in small, enzymatically catalyzed steps. Oxygen is used only at the last step. Because a small amount of the chemical-bond energy of glucose is released at early steps in the metabolic pathway, some tissue cells can obtain energy for ATP production in the temporary absence of oxygen. This process is described in the next two sections.

Figure 5.1 presents an overview of the processes by which glucose can be obtained by the body cells and used for energy. Plasma glucose is derived from the digestion of food or from the breakdown of liver glycogen. The aerobic cell respiration of glucose occurs in three successive steps: (1) *glycolysis* (a metabolic pathway that takes place in the cytoplasm); (2) the *citric acid cycle* (a metabolic pathway that occurs in the mitochondrial matrix); and (3) *electron transport* (a process that occurs in the mitochondrial cristae). When glucose is metabolized anaerobically (without oxygen), glycolysis produces pyruvic acid that is converted into lactic acid. Although anaerobic metabolism is important (it provides energy for exercising skeletal muscles at certain times, for example), most body cells obtain energy by aerobic cell respiration.

Glycolysis

The breakdown of glucose for energy begins with a metabolic pathway in the cytoplasm known as **glycolysis.** This term is derived from the Greek (*glykys* = sweet, *lysis* = a loosening), and it refers to the cleavage of sugar. Glycolysis is the metabolic pathway by which glucose—a six-carbon (hexose) sugar (see fig. 2.13)—is converted into 2 molecules of pyruvic acid, or pyruvate. Even though each pyruvic acid molecule is roughly half the size of a glucose, glycolysis is *not* simply the breaking in half of glucose. Glycolysis is a metabolic pathway involving many enzymatically controlled steps.

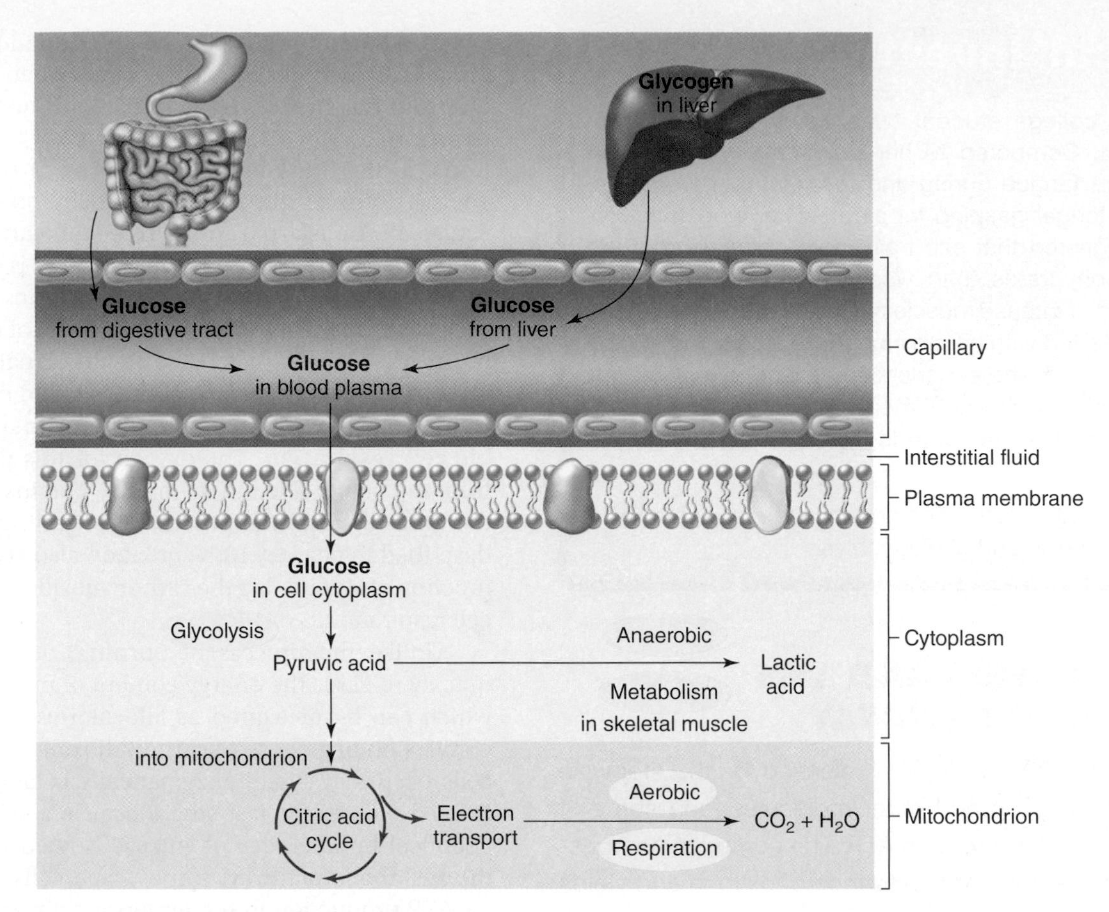

Figure 5.1 Overview of energy metabolism using blood glucose. The blood glucose may be obtained from food via the digestive tract, or the liver may produce it from stored glycogen. Plasma glucose enters the cytoplasm of cells, where it can be used for energy by either anaerobic metabolism or aerobic cell respiration. In this schematic diagram, the size of the plasma membrane is greatly exaggerated compared to the size of the other structures and the interstitial (extracellular tissue) fluid. APR

Each pyruvic acid molecule contains 3 carbons, 3 oxygens, and 4 hydrogens (see fig. 5.4). The number of carbon and oxygen atoms in 1 molecule of glucose—$C_6H_{12}O_6$—can thus be accounted for in the 2 pyruvic acid molecules. Because the 2 pyruvic acids together account for only 8 hydrogens, however, it is clear that 4 hydrogen atoms are removed from the intermediates in glycolysis. Each pair of these hydrogen atoms is used to reduce a molecule of NAD. In this process, each pair of hydrogen atoms donates 2 electrons to NAD, thereby reducing it. The reduced NAD binds 1 proton from the hydrogen atoms, leaving 1 proton unbound as H^+ (see chapter 4, fig. 4.17). Starting from 1 glucose molecule, therefore, glycolysis results in the production of 2 molecules of NADH and 2 H^+. The H^+ will follow the NADH in subsequent reactions, so for simplicity we can refer to reduced NAD simply as NADH.

Glycolysis is exergonic, and a portion of the energy that is released is used to drive the endergonic reaction $ADP + P_i \rightarrow ATP$. At the end of the glycolytic pathway, there is a net gain of 2 ATP molecules per glucose molecule, as indicated in the overall equation for glycolysis:

$$\text{Glucose} + 2\,\text{NAD} + 2\,\text{ADP} + 2\,P_i \rightarrow$$
$$2\,\text{pyruvic acid} + 2\,\text{NADH} + 2\,\text{ATP}$$

Although the overall equation for glycolysis is exergonic, glucose must be "activated" at the beginning of the pathway before energy can be obtained. This activation requires the addition of two phosphate groups derived from 2 molecules of ATP. Energy from the reaction $ATP \rightarrow ADP + P_i$ is therefore consumed at the beginning of glycolysis. This is shown as an "up-staircase" in figure 5.2. Notice that the P_i is not shown in these reactions in figure 5.2; this is because the phosphate is not released, but instead is added to the intermediate molecules of glycolysis. The addition of a phosphate group is known as *phosphorylation.* Besides being essential for glycolysis, the phosphorylation of glucose (to glucose 6-phosphate) has an important side benefit: it traps the glucose within the cell. This is because *phosphorylated organic molecules cannot cross plasma membranes.*

At later steps in glycolysis, 4 molecules of ATP are produced (and 2 molecules of NAD are reduced) as energy is liberated (the "down-staircase" in fig. 5.2). The 2 molecules of ATP used in the beginning, therefore, represent an energy investment; the net gain of 2 ATP and 2 NADH molecules by the end of the pathway represents an energy profit. The overall equation for glycolysis obscures the fact that this is a metabolic pathway consisting of nine separate

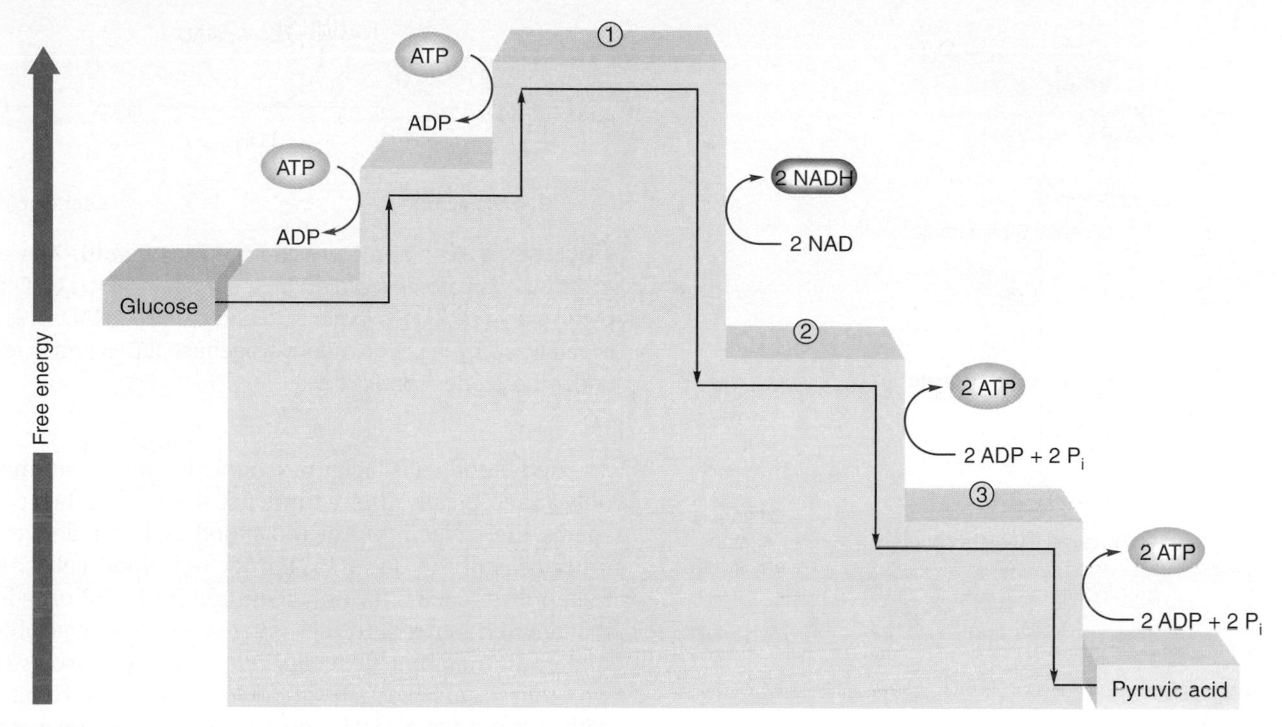

Figure 5.2 The energy expenditure and gain in glycolysis. Notice that there is a "net profit" of 2 ATP and 2 NADH for every molecule of glucose that enters the glycolytic pathway. Molecules listed by number are (*1*) fructose 1,6-biphosphate, (*2*) 1,3-biphosphoglyceric acid, and (*3*) 3-phosphoglyceric acid (see fig. 5.3). AP|R

steps. The individual steps in this pathway are shown in figure 5.3.

In figure 5.3, glucose is phosphorylated to glucose 6-phosphate using ATP at step 1, and then is converted into its isomer, fructose 6-phosphate, in step 2. Another ATP is used to form fructose 1,6-biphosphate at step 3. Notice that the six-carbon-long molecule is split into 2 separate three-carbon-long molecules at step 4. At step 5, two pairs of hydrogens are removed and used to reduce 2 NAD to 2 NADH + H^+. These reduced coenzymes are important products of glycolysis. Then, at step 6, a phosphate group is removed from each 1,3-biphosphoglyceric acid, forming 2 ATP and 2 molecules of 3-phosphoglyceric acid. Steps 7 and 8 are isomerizations. Then, at step 9, the last phosphate group is removed from each intermediate; this forms another 2 ATP (for a net gain of 2 ATP), and 2 molecules of pyruvic acid.

Lactic Acid Pathway

In order for glycolysis to continue, there must be adequate amounts of NAD available to accept hydrogen atoms. Therefore, the NADH produced in glycolysis must become oxidized by donating its electrons to another molecule. (In aerobic respiration this other molecule is located in the mitochondria and ultimately passes its electrons to oxygen.)

When oxygen is not available in sufficient amounts, the NADH (+ H^+) produced in glycolysis is oxidized in the cytoplasm by donating its electrons to pyruvic acid. This results in the re-formation of NAD and the addition of 2 hydrogen atoms to pyruvic acid, which is thus reduced. This addition of 2 hydrogen atoms to pyruvic acid produces lactic acid (fig. 5.4). Most lactic acid dissociates to form the *lactate* anion and H^+ at normal cellular pH (chapter 2; see fig. 2.12).

The metabolic pathway by which glucose is converted into lactic acid is a type of **anaerobic metabolism,** in the sense that the term *anaerobic* means that oxygen is not used in the process. Many biologists prefer the name **lactic acid fermentation** for this pathway because of its similarity to the way that yeast cells ferment glucose into ethyl alcohol (ethanol). In both lactic acid and ethanol production, the last electron acceptor is an organic molecule. This contrasts with aerobic respiration, in which the last electron acceptor is an atom of oxygen. Biologists reserve the term *anaerobic respiration* for pathways (in some microorganisms) that use atoms other than oxygen (such as sulfur) as the last electron acceptor. In this text, the terms *lactic acid pathway, anaerobic metabolism,* and *lactic acid fermentation* will be used interchangeably to describe the pathway by which glucose is converted into lactic acid.

The lactic acid pathway yields a net gain of two ATP molecules (produced by glycolysis) per glucose molecule. A cell can survive without oxygen as long as it can produce sufficient energy for its needs in this way and as long as lactic acid concentrations do not become excessive. Some tissues are better adapted to anaerobic conditions than others—skeletal muscles survive longer than cardiac muscle, which in turn survives under anaerobic conditions longer than the brain.

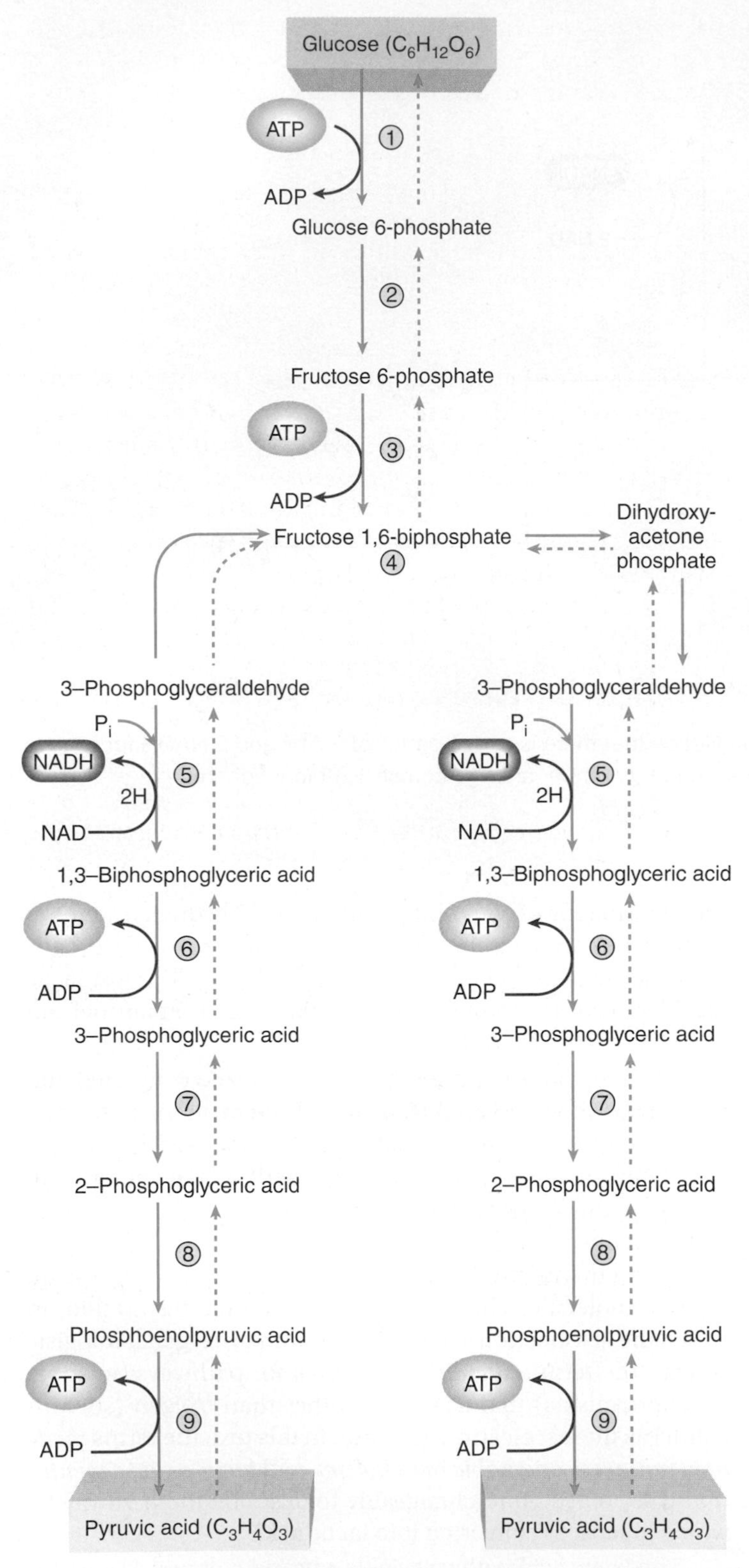

Figure 5.3 **Glycolysis.** In glycolysis, 1 glucose is converted into 2 pyruvic acids in nine separate steps. In addition to 2 pyruvic acids, the products of glycolysis include 2 NADH and 4 ATP. Because 2 ATP were used at the beginning, however, the net gain is 2 ATP per glucose. Dashed arrows indicate reverse reactions that may occur under other conditions. AP|R

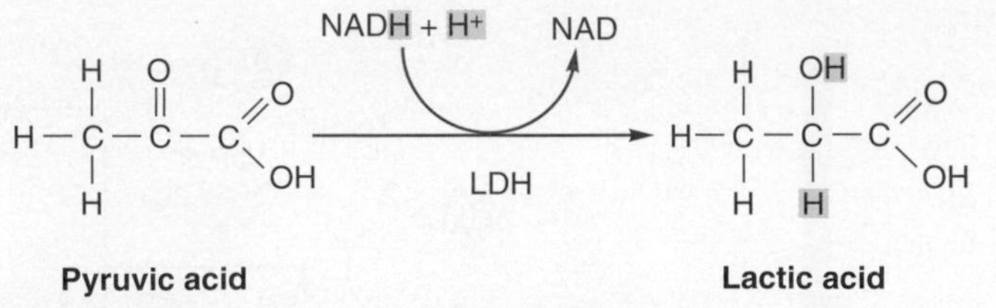

NADH + H+ → NAD (LDH)

Pyruvic acid → Lactic acid

Figure 5.4 **The formation of lactic acid.** The addition of 2 hydrogen atoms (colored boxes) from reduced NAD to pyruvic acid produces lactic acid and oxidized NAD. This reaction is catalyzed by lactic acid dehydrogenase (LDH) and is reversible under the proper conditions.

Red blood cells lack mitochondria and so can metabolize only anaerobically; they cannot use the oxygen they carry. This spares the oxygen within red blood cells for delivery to the other cells of the body. Except for red blood cells, anaerobic metabolism occurs for only a limited period of time in tissues that have energy requirements in excess of their aerobic ability. Anaerobic metabolism occurs in the skeletal muscles and heart when the *ratio of oxygen supply to oxygen need* (related to the concentration of NADH) falls below a critical level. Anaerobic metabolism is, in a sense, an emergency procedure that provides some ATP until the emergency (oxygen deficiency) has passed.

It should be noted, though, that there is no real "emergency" in the case of skeletal muscles, where lactic acid fermentation is a normal, daily occurrence that does not harm muscle tissue or the individual. Excessive lactic acid production by muscles, however, is associated with pain and muscle fatigue. (The metabolism of skeletal muscles is discussed in chapter 12, section 12.4.) In contrast to skeletal muscles, the heart normally respires only aerobically. If anaerobic conditions do occur in the heart, a potentially dangerous situation may be present.

CLINICAL APPLICATION

Ischemia refers to inadequate blood flow to an organ, such that the rate of oxygen delivery is insufficient to maintain aerobic respiration. Inadequate blood flow to the heart, or *myocardial ischemia,* may occur if the coronary blood flow is occluded by atherosclerosis, a blood clot, or by an artery spasm. People with myocardial ischemia often experience *angina pectoris*—severe pain in the chest and left (or sometimes, right) arm area. This pain is associated with increased blood levels of lactic acid which are produced by the ischemic heart muscle. If the ischemia is prolonged, the cells may die and produce an area called an *infarct.* The degree of ischemia and angina can be decreased by vasodilator drugs such as nitroglycerin, which improve blood flow to the heart and also decrease the work of the heart by dilating peripheral blood vessels.

Clinical Investigation CLUES

Brenda experienced muscle pain and fatigue, and during an intense workout felt a severe pain in her left pectoral region.

- What metabolic pathway is associated with muscle pain and fatigue?
- What serious medical condition could the pain in her left pectoral region indicate, and what produces this condition?

CHECKPOINT

1a. Define the term *glycolysis* in terms of its initial substrates and products. Explain why there is a net gain of 2 molecules of ATP in this process.

1b. What are the initial substrates and final products of anaerobic metabolism?

1c. Describe the physiological functions of lactic acid fermentation. In which tissue(s) is anaerobic metabolism normal? In which tissue is it abnormal?

5.2 AEROBIC RESPIRATION

In the aerobic respiration of glucose, pyruvic acid is formed by glycolysis and then converted into acetyl coenzyme A. This begins a cyclic metabolic pathway called the citric acid (Krebs) cycle. As a result of these pathways, a large amount of reduced NAD and FAD (NADH and $FADH_2$) is generated. These reduced coenzymes provide electrons for a process that drives the formation of ATP.

LEARNING OUTCOMES

After studying this section, you should be able to:

2. **Describe the aerobic cell respiration of glucose through the citric acid cycle.**

3. **Describe the electron transport system and oxidative phosphorylation, explaining the role of oxygen in this process.**

The aerobic respiration of glucose ($C_6H_{12}O_6$) is given in the following overall equation:

$$C_6H_{12}O_6 + O_2 \rightarrow 6\ CO_2 + 6\ H_2O$$

Aerobic respiration is equivalent to combustion in terms of its final products (CO_2 and H_2O) and in terms of the total amount of energy liberated. In aerobic respiration, however, the energy is released in small, enzymatically controlled oxidation reactions, and a portion (38% to 40%) of the energy released is captured in the high-energy bonds of ATP.

The aerobic respiration of glucose begins with glycolysis. Glycolysis in both anaerobic metabolism and aerobic respiration results in the production of 2 molecules of pyruvic acid, 2 ATP, and 2 NADH + H^+ per glucose molecule. In aerobic respiration, however, the electrons in NADH are *not* donated to pyruvic acid and lactic acid is not formed, as happens in the lactic acid pathway. Instead, the pyruvic acids will move to a different cellular location and undergo a different reaction; the NADH produced by glycolysis will eventually be oxidized, but that occurs later in the story.

In aerobic respiration, pyruvic acid leaves the cell cytoplasm and enters the interior (the matrix) of mitochondria. Once pyruvic acid is inside a mitochondrion, carbon dioxide is enzymatically removed from each three-carbon-long pyruvic acid to form a two-carbon-long organic acid—acetic acid. The enzyme that catalyzes this reaction combines the acetic acid with a coenzyme (derived from the vitamin pantothenic acid) called *coenzyme A*. The combination thus produced is called **acetyl coenzyme A,** abbreviated **acetyl CoA** (fig. 5.5).

Glycolysis converts 1 glucose molecule into 2 molecules of pyruvic acid. Since each pyruvic acid molecule is converted into 1 molecule of acetyl CoA and 1 CO_2, 2 molecules of acetyl CoA and 2 molecules of CO_2 are derived from each glucose. These acetyl CoA molecules serve as substrates for mitochondrial enzymes in the aerobic pathway, while the carbon dioxide is carried by the blood to the lungs for elimination. It is important to note that the oxygen in CO_2 is derived from pyruvic acid, not from oxygen gas.

Citric Acid Cycle

Once acetyl CoA has been formed, the acetic acid subunit (2 carbons long) combines with oxaloacetic acid (4 carbons long) to form a molecule of citric acid (6 carbons long). Coenzyme A acts only as a transporter of acetic acid from one enzyme to another (similar to the transport of hydrogen by

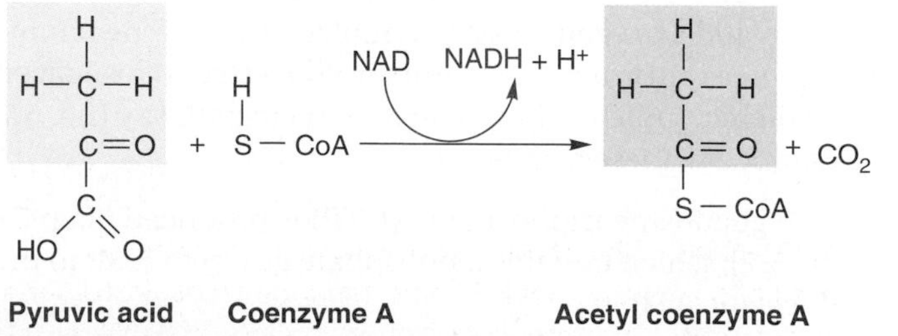

Pyruvic acid + Coenzyme A (NAD → NADH + H^+) → Acetyl coenzyme A + CO_2

Figure 5.5 The formation of acetyl coenzyme A in aerobic respiration. Notice that NAD is reduced to NADH in this process.

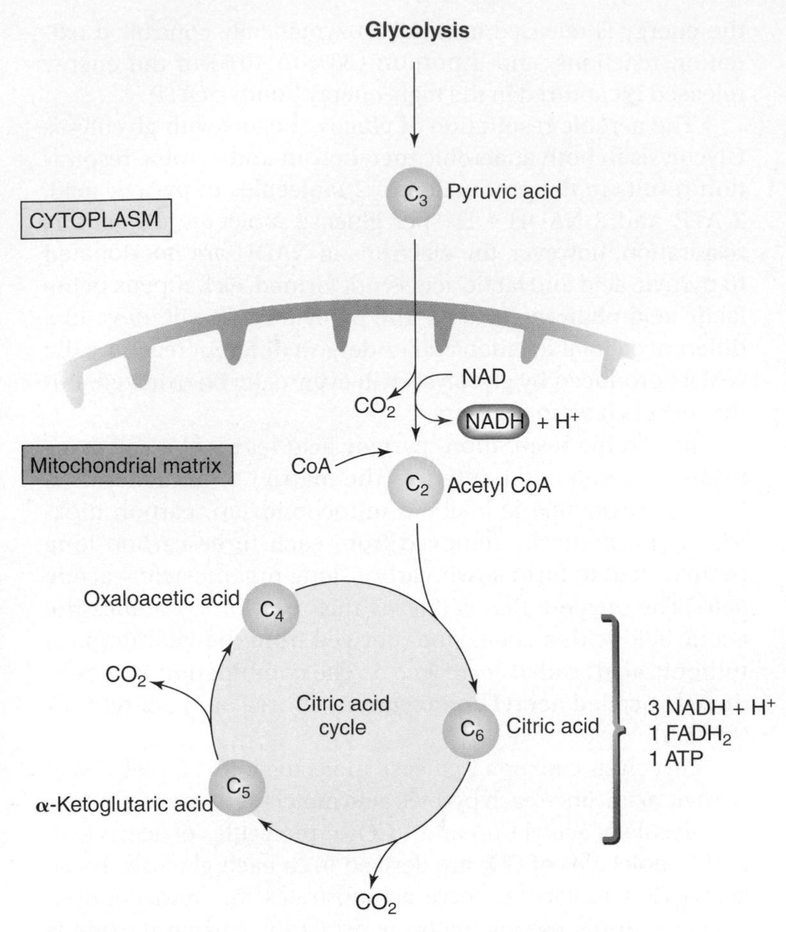

Figure 5.6 A simplified diagram of the citric acid cycle. This diagram shows how the original four-carbon-long oxaloacetic acid is regenerated at the end of the cyclic pathway. Only the numbers of carbon atoms in the citric acid cycle intermediates are shown; the numbers of hydrogens and oxygens are not accounted for in this simplified scheme. AP|R

NAD). The formation of citric acid begins a cyclic metabolic pathway known as the **citric acid cycle,** or **TCA cycle** (for tricarboxylic acid; citric acid has three carboxylic acid groups). It is often also called the **Krebs cycle,** after its principal discoverer, Sir Hans Krebs. A simplified illustration of this pathway is shown in figure 5.6.

Through a series of reactions involving the elimination of 2 carbons and 4 oxygens (as 2 CO_2 molecules) and the removal of hydrogens, citric acid is eventually converted to oxaloacetic acid, which completes the cyclic metabolic pathway (fig. 5.7). In this process, these events occur:

1. One guanosine triphosphate (GTP) is produced (step 5 of fig. 5.7), which donates a phosphate group to ADP to produce one ATP.
2. Three molecules of NAD are reduced to NADH (steps 4, 5, and 8 of fig. 5.7).
3. One molecule of FAD is reduced to $FADH_2$ (step 6).

The production of NADH and $FADH_2$ by each "turn" of the citric acid cycle is far more significant, in terms of energy production, than the single GTP (converted to ATP) produced directly by the cycle. This is because NADH and $FADH_2$ eventually donate their electrons to an energy-transferring process that results in the formation of a large number of ATP.

Electron Transport and Oxidative Phosphorylation

Built into the foldings, or cristae, of the inner mitochondrial membrane are a series of molecules that serve as an **electron-transport system** during aerobic respiration. This electron-transport chain of molecules consists of a protein containing *flavin mononucleotide* (abbreviated *FMN* and derived from the vitamin riboflavin), *coenzyme Q,* and a group of iron-containing pigments called *cytochromes.* The last of these cytochromes is cytochrome a_3, which donates electrons to oxygen in the final oxidation-reduction reaction (as will be described shortly). These molecules of the electron-transport system are fixed in position within the inner mitochondrial membrane in such a way that they can pick up electrons from NADH and $FADH_2$ and transport them in a definite sequence and direction.

In aerobic respiration, NADH and $FADH_2$ become oxidized by transferring their pairs of electrons to the electron-transport system of the cristae. It should be noted that the protons (H^+) are not transported together with the electrons; their fate will be described a little later. The oxidized forms of NAD and FAD are thus regenerated and can continue to "shuttle" electrons from the citric acid cycle to the electron-transport chain. The first molecule of the electron-transport chain becomes reduced when it accepts the electron pair from NADH. When the cytochromes receive a pair of electrons, 2 ferric ions (Fe^{3+}) become reduced to 2 ferrous ions (Fe^{2+}).

The electron-transport chain thus acts as an oxidizing agent for NAD and FAD. Each element in the chain, however, also functions as a reducing agent; one reduced cytochrome transfers its electron pair to the next cytochrome in the chain (fig. 5.8). In this way, the iron ions in each cytochrome alternately become reduced (from Fe^{3+} to Fe^{2+}) and oxidized (from Fe^{2+} to Fe^{3+}). This is an exergonic process, and the energy derived is used to phosphorylate ADP to ATP. The production of ATP through the coupling of the electron-transport system with the phosphorylation of ADP is appropriately termed **oxidative phosphorylation.**

The coupling is not 100% efficient between the energy released by electron transport (the "oxidative" part of oxidative phosphorylation) and the energy incorporated into the chemical bonds of ATP (the "phosphorylation" part of the term). This difference in energy escapes the body as heat. Metabolic heat production is needed to maintain our internal body temperature.

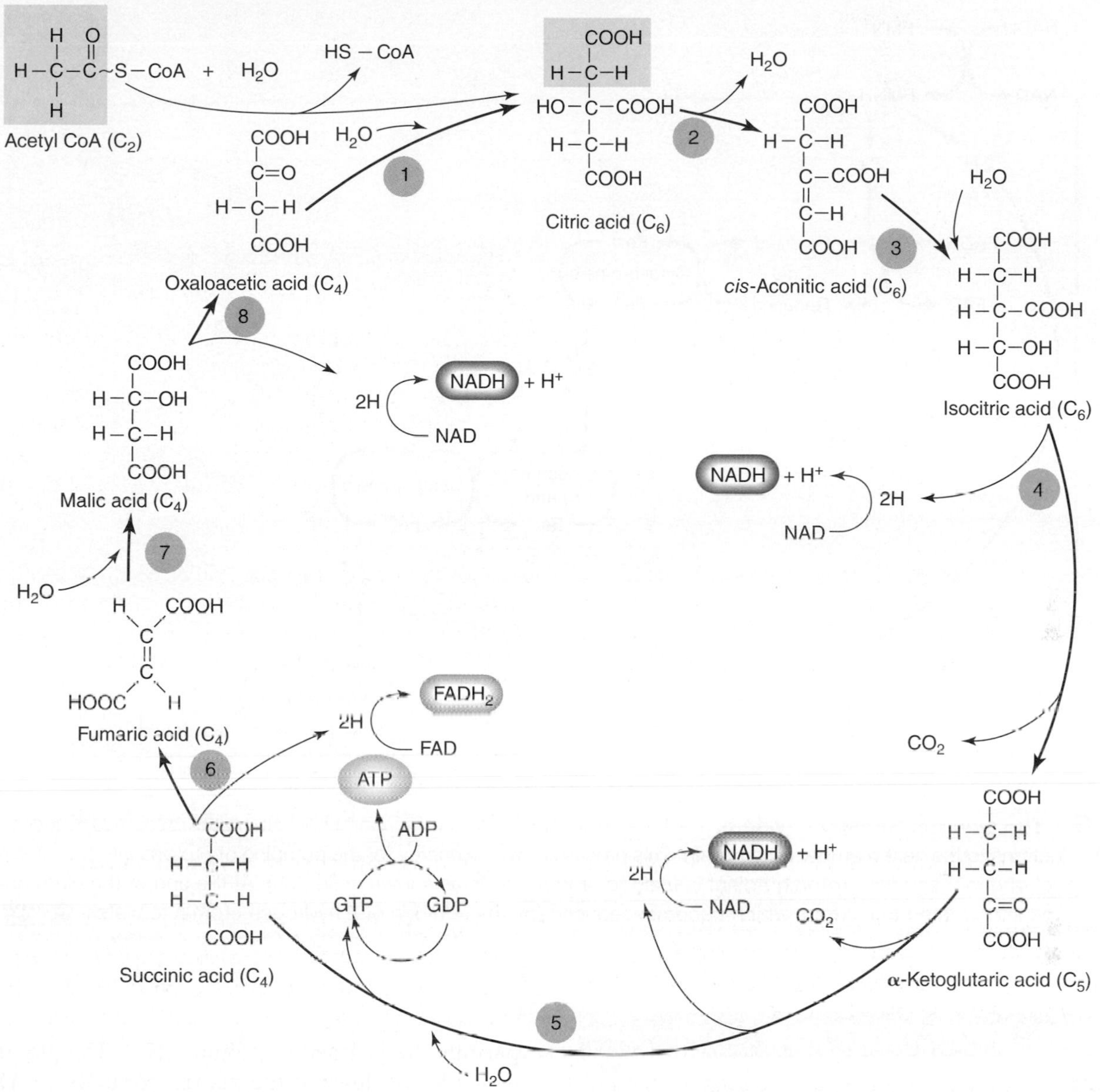

Figure 5.7 The complete citric acid cycle. Notice that, for each "turn" of the cycle, 1 ATP, 3 NADH, and 1 $FADH_2$ are produced. AP|R

Coupling of Electron Transport to ATP Production

According to the **chemiosmotic theory,** the electron-transport system, powered by the transport of electrons, pumps protons (H^+) from the mitochondrial matrix into the space between the inner and outer mitochondrial membranes. The electron-transport system is grouped into three complexes that serve as **proton pumps** (fig. 5.9). The first pump (the NADH-coenzyme Q reductase complex) transports 4 H^+ from the matrix to the intermembrane space for every pair of electrons moved along the electron-transport system. The second pump (the cytochrome c reductase complex) also transports 4 protons into the intermembrane space, and the third pump (the cytochrome c oxidase complex) transports 2 protons into the intermembrane space. As a result, there is a higher concentration of H^+ in the intermembrane space than in the matrix, favoring the diffusion of H^+ back out into the matrix. The inner mitochondrial membrane, however, does not permit diffusion of H^+, except through structures called *respiratory assemblies.*

The respiratory assemblies consist of a group of proteins that form a "stem" and a globular subunit. The stem contains a channel through the inner mitochondrial membrane that

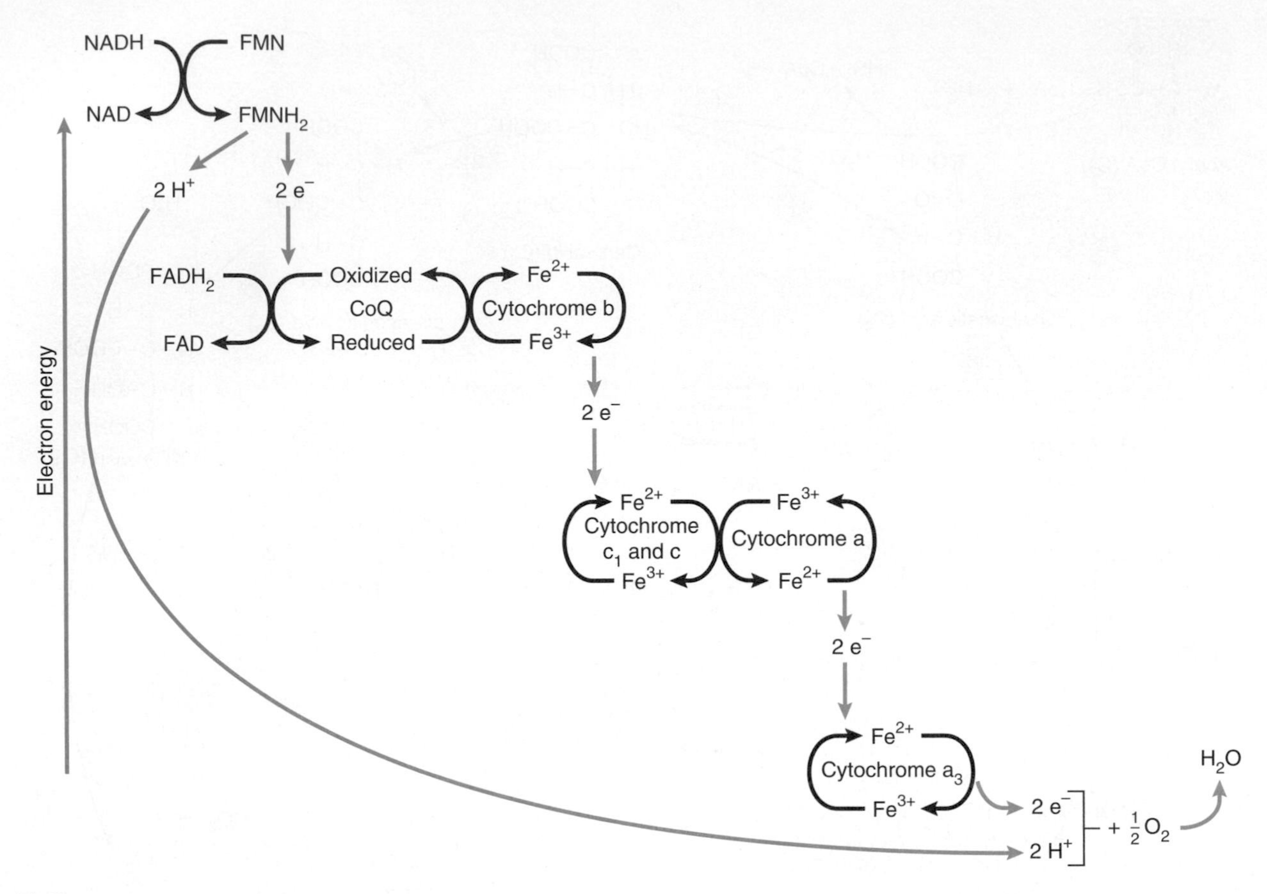

Figure 5.8 **The electron transport system.** Each element in the electron-transport chain alternately becomes reduced and oxidized as it transports electrons to the next member of the chain. This process provides energy for the pumping of protons into the intermembranous space of the mitochondrion, and the proton gradient is used to produce ATP (as shown in fig. 5.9). At the end of the electron-transport chain, the electrons are donated to oxygen, which becomes reduced (by the addition of 2 hydrogen atoms) to water. AP|R

CLINICAL APPLICATION

Free radicals are molecules with unpaired electrons, in contrast to molecules that are not free radicals because they have two electrons per orbital. A *superoxide radical* is an oxygen molecule with an extra, unpaired electron. These can be generated in mitochondria through the leakage of electrons from the electron-transport system. Superoxide radicals have some known physiological functions; for example, they are produced in phagocytic white blood cells where they are needed for the destruction of bacteria. However, the production of the superoxide, hydroxyl, and nitric oxide free radicals and other molecules classified as *reactive oxygen species (ROS)* contributes to many diseases, including atherosclerosis (hardening of the arteries—chapter 13, section 13.7). The mitochondrial production of ROS is also central to current theories of aging. Accordingly, reactive oxygen species have been described as exerting an *oxidative stress* on the body. **Antioxidants** are molecules that scavenge free radicals and protect the body from reactive oxygen species.

permits the passage of protons (H^+). The globular subunit, which protrudes into the matrix, contains an **ATP synthase** enzyme that catalyzes the reaction $ADP + P_i \rightarrow ATP$ when it is activated by the diffusion of protons through the respiratory assemblies and into the matrix (fig. 5.9). In this way, phosphorylation (the addition of phosphate to ADP) is coupled to oxidation (the transport of electrons) in oxidative phosphorylation.

Function of Oxygen

If the last cytochrome remained in a reduced state, it would be unable to accept more electrons. Electron transport would then progress only to the next-to-last cytochrome. This process would continue until all of the elements of the electron-transport chain remained in the reduced state. At this point, the electron-transport system would stop functioning and no ATP could be produced in the mitochondria. With the electron-transport system incapacitated, NADH and

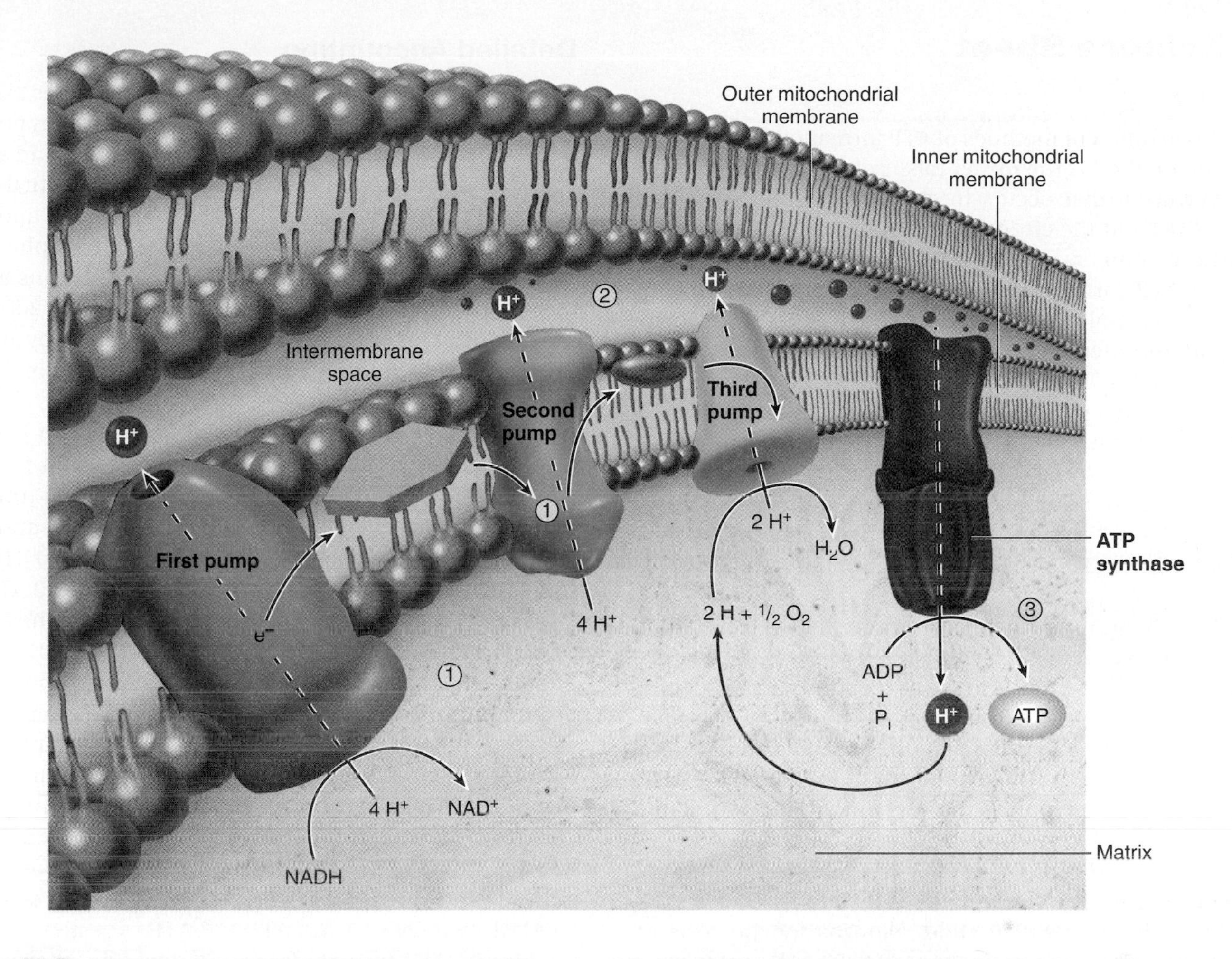

Figure 5.9 The steps of oxidative phosphorylation. (*1*) Molecules of the electron-transport system function to pump H^+ from the matrix to the intermembrane space. (*2*) This results in a steep H^+ gradient between the intermembrane space and the cytoplasm of the cell. (*3*) The diffusion of H^+ through ATP synthase results in the production of ATP. AP|R

$FADH_2$ could not become oxidized by donating their electrons to the chain and, through inhibition of citric acid cycle enzymes, no more NADH and $FADH_2$ could be produced in the mitochondria. The citric acid cycle would stop and only anaerobic metabolism could occur.

Oxygen, from the air we breathe, allows electron transport to continue by functioning as the **final electron acceptor** of the electron-transport chain. This oxidizes cytochrome a_3, allowing electron transport and oxidative phosphorylation to continue. At the very last step of aerobic respiration, therefore, oxygen becomes reduced by the 2 electrons that were passed to the chain from NADH and $FADH_2$. This reduced oxygen binds 2 protons, and a molecule of water is formed. Because the oxygen atom is part of a molecule of oxygen gas (O_2), this last reaction can be shown as follows:

$$O_2 + 4\,e^- + 4\,H^+ \rightarrow 2\,H_2O$$

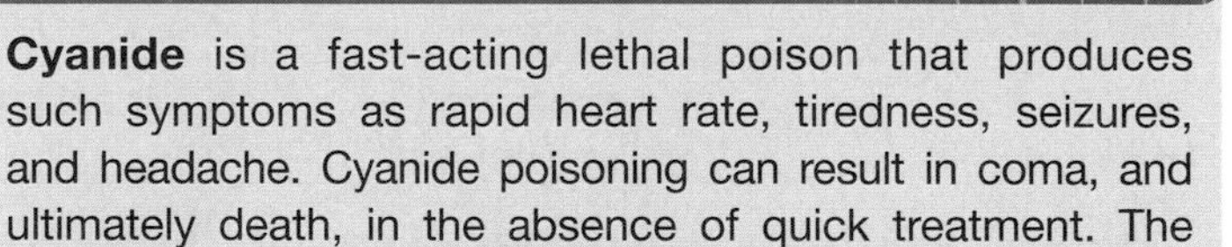

Cyanide is a fast-acting lethal poison that produces such symptoms as rapid heart rate, tiredness, seizures, and headache. Cyanide poisoning can result in coma, and ultimately death, in the absence of quick treatment. The reason that cyanide is so deadly is that it has one very specific action: it blocks the transfer of electrons from cytochrome a_3 to oxygen. The effects are thus the same as would occur if oxygen were completely removed—aerobic cell respiration and the production of ATP by oxidative phosphorylation comes to a halt.

ATP Balance Sheet

Overview

There are two different methods of ATP formation in cell respiration. One method is the **direct** (also called **substrate-level**) **phosphorylation** that occurs in glycolysis (producing a net gain of 2 ATP) and the citric acid cycle (producing 1 ATP per cycle). These numbers are certain and constant. In the second method of ATP formation, **oxidative phosphorylation,** the numbers of ATP molecules produced vary under different conditions and for different kinds of cells. For many years, it was believed that 1 NADH yielded 3 ATP and that 1 $FADH_2$ yielded 2 ATP by oxidative phosphorylation. This gave a grand total of 36 to 38 molecules of ATP per glucose through cell respiration (table 5.1). Newer biochemical information, however, suggests that these numbers may be overestimates, because, of the 36 to 38 ATP produced per glucose in the mitochondrion, only 30 to 32 ATP actually enter the cytoplasm of the cell.

Roughly 3 protons must pass through the respiratory assemblies and activate ATP synthase to produce 1 ATP. However, the newly formed ATP is in the mitochondrial matrix and must be moved into the cytoplasm; this transport also uses the proton gradient and costs 1 more proton. The ATP and H^+ are transported into the cytoplasm in exchange for ADP and P_i, which are transported into the mitochondrion. Thus, it effectively takes 4 protons to produce 1 ATP that enters the cytoplasm.

To summarize: The **theoretical ATP yield** is 36 to 38 ATP per glucose. The **actual ATP yield,** allowing for the costs of transport into the cytoplasm, is about 30 to 32 ATP per glucose. The details of how these numbers are obtained are described in the following section.

Detailed Accounting

Each NADH formed in the mitochondrion donates 2 electrons to the electron transport system at the first proton pump (see fig. 5.9). The electrons are then passed to the second and third proton pumps, activating each of them in turn until the 2 electrons are ultimately passed to oxygen. The first and second pumps transport 4 protons each, and the third pump transports 2 protons, for a total of 10. Dividing 10 protons by the 4 it takes to produce an ATP (in the cytoplasm) gives 2.5 ATP that are produced for every pair of electrons donated by an NADH. (There is no such thing as half an ATP; the decimal fraction simply indicates an average.)

Three molecules of NADH are formed with each citric acid cycle, and 1 NADH is also produced when pyruvate is converted into acetyl CoA (see fig. 5.5). Starting from 1 glucose, two citric acid cycles (producing 6 NADH) and 2 pyruvates converted to acetyl CoA (producing 2 NADH) yield 8 NADH. Multiplying by 2.5 ATP per NADH gives 20 ATP.

Electrons from $FADH_2$ are donated later in the electron-transport system than those donated by NADH; consequently, these electrons activate only the second and third proton pumps. Since the first proton pump is bypassed, the electrons passed from $FADH_2$ result in the pumping of only 6 protons (4 by the second pump and 2 by the third pump). Because 1 ATP is produced for every 4 protons pumped, electrons derived from $FADH_2$ result in the formation of $6 \div 4 = 1.5$ ATP. Each citric acid cycle produces 1 $FADH_2$ and we get two citric acid cycles from 1 glucose, so there are 2 $FADH_2$ that give 2×1.5 ATP = 3 ATP.

The 23 ATP subtotal from oxidative phosphorylation we have at this point includes only the NADH and $FADH_2$ produced in the mitochondrion. Remember that glycolysis, which

Table 5.1 | ATP Yield per Glucose in Aerobic Respiration

Phases of Respiration	ATP Made Directly	Reduced Coenzymes	ATP Made by Oxidative Phosphorylation: Theoretical Yield*	ATP Made by Oxidative Phosphorylation: Actual Yield**
Glucose to pyruvate (in cytoplasm)	**2 ATP** (net gain)	2 NADH, but usually goes into mitochondria as 2 $FADH_2$	**If from $FADH_2$:** 2 ATP (× 2) = 4 ATP or if stays NADH: 3 ATP (× 2) = 6 ATP	**If from $FADH_2$:** 1.5 ATP (× 2) = 3 ATP or if stays NADH: 2.5 ATP (× 2) = 5 ATP
Pyruvate to acetyl CoA (× 2)	None	1 NADH (× 2) = 2 NADH	3 ATP (× 2) = 6 ATP	2.5 ATP (× 2) = 5 ATP
Citric acid cycle (× 2)	1 ATP (× 2) = **2 ATP**	3 NADH (× 2) = 6 NADH 1 $FADH_2$ (× 2) = 2 $FADH_2$	3 ATP (× 6) = 18 ATP 2 ATP (× 2) = 4 ATP	2.5 ATP (× 6) = 15 ATP 1.5 ATP (× 2) = 3 ATP
Total ATP	**4 ATP**		**32 (or 34) ATP**	**26 (or 28) ATP**

* The *theoretical yield* is the number of ATP produced by oxidative phosphorylation inside the mitochondria.
**The *actual yield* takes into account the energy cost of transporting ATP out of the mitochondria and into the cytoplasm.

occurs in the cytoplasm, also produces 2 NADH. These cytoplasmic NADH cannot directly enter the mitochondrion, but there is a process by which their electrons can be "shuttled" in. The net effect of the most common shuttle is that a molecule of NADH in the cytoplasm is translated into a molecule of $FADH_2$ in the mitochondrion. The 2 NADH produced in glycolysis, therefore, usually become 2 $FADH_2$ and yield 2×1.5 ATP = 3 ATP by oxidative phosphorylation. (An alternative pathway, where the cytoplasmic NADH is transformed into mitochondrial NADH and produces 2×2.5 ATP = 5 ATP, is less common; however, this is the dominant pathway in the liver and heart, which are metabolically highly active.)

We now have a total of 26 ATP (or, less commonly, 28 ATP) produced by oxidative phosphorylation from glucose. We can add the 2 ATP made by direct (substrate-level) phosphorylation in glycolysis and the 2 ATP made directly by the two citric acid cycles to give a grand total of 30 ATP (or, less commonly, 32 ATP) produced by the aerobic respiration of glucose (see table 5.1).

CHECKPOINT

2a. Compare the fate of pyruvate in aerobic and anaerobic cell respiration.

2b. Draw a simplified citric acid cycle and indicate the high-energy products.

3a. Explain how NADH and $FADH_2$ contribute to oxidative phosphorylation.

3b. Explain how ATP is produced in oxidative phosphorylation.

5.3 INTERCONVERSION OF GLUCOSE, LACTIC ACID, AND GLYCOGEN

Glucose can be stored as glycogen. In the liver, stored glycogen can be hydrolyzed and used to form free glucose, which the liver can secrete into the blood. Lactic acid produced by exercising skeletal muscles can travel to the liver and be converted into glucose.

LEARNING OUTCOMES

After studying this section, you should be able to:

4. Explain how glucose and glycogen can be interconverted and why only the liver can use its stored glycogen to produce free glucose for secretion.

5. Define the term *gluconeogenesis* and explain the Cori cycle.

Cells cannot accumulate very many separate glucose molecules, because an abundance of these would exert an osmotic pressure (chapter 6) that would draw a dangerous amount of water into the cells. Instead, many organs, particularly the liver, skeletal muscles, and heart store carbohydrates in the form of glycogen.

Glycogenesis and Glycogenolysis

The formation of glycogen from glucose is called **glycogenesis** (table 5.2). In this process, glucose is converted to glucose 6-phosphate by utilizing the terminal phosphate group of ATP. Glucose 6-phosphate is then converted into its isomer, glucose 1-phosphate. Finally, the enzyme *glycogen synthase* removes these phosphate groups as it polymerizes glucose to form glycogen.

The reverse reactions are similar. The enzyme *glycogen phosphorylase* catalyzes the breakdown of glycogen to glucose 1-phosphate. (The phosphates are derived from inorganic phosphate, not from ATP, so glycogen breakdown does not require metabolic energy.) Glucose 1-phosphate is then converted to glucose 6-phosphate. The conversion of glycogen to glucose 6-phosphate is called **glycogenolysis.** In most tissues, glucose 6-phosphate can then be broken down for energy (through glycolysis) or used to resynthesize glycogen. Only in the liver, for reasons that will now be explained, can the glucose 6-phosphate also be used to produce free glucose for secretion into the blood.

Table 5.2 | Common Terms for Some Metabolic Processes in the Body

Term	Process
Glycolysis	Conversion of glucose into two molecules of pyruvic acid
Glycogenesis	The production of glycogen, mostly in skeletal muscles and the liver
Glycogenolysis	Hydrolysis (breakdown) of glycogen; yields glucose 6-phosphate for glycolysis, or (in the liver only) free glucose that can be secreted into the blood
Gluconeogenesis	The production of glucose from noncarbohydrate molecules, including lactic acid and amino acids, primarily in the liver
Lipogenesis	The formation of triglycerides (fat), primarily in adipose tissue
Lipolysis	Hydrolysis (breakdown) of triglycerides, primarily in adipose tissue
Ketogenesis	The formation of ketone bodies, which are four-carbon-long organic acids, from fatty acids; occurs in the liver

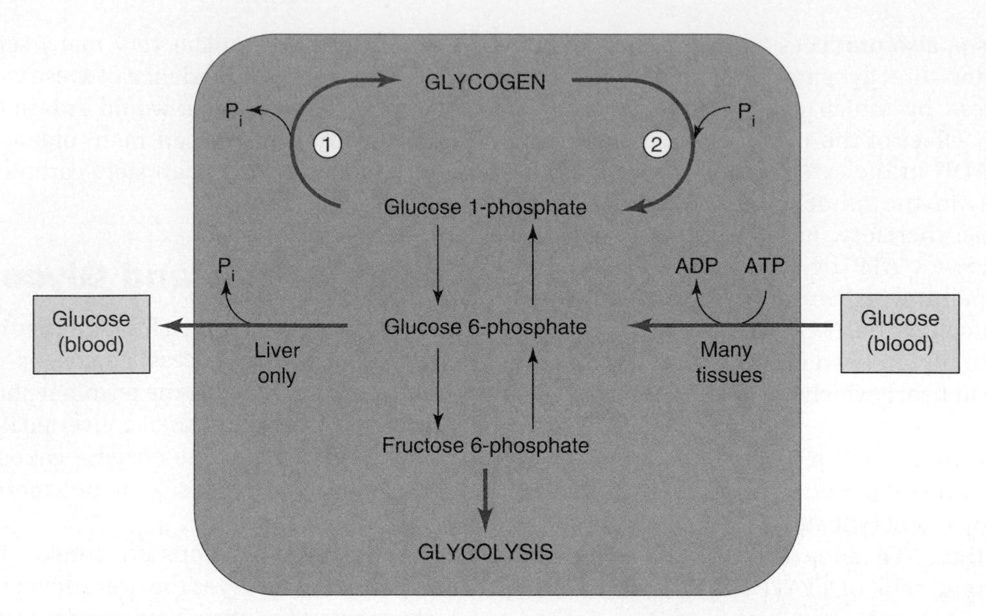

Figure 5.10 Glycogenesis and glycogenolysis. Blood glucose entering tissue cells is phosphorylated to glucose 6-phosphate. This intermediate can be metabolized for energy in glycolysis, or it can be converted to glycogen (*1*) in a process called *glycogenesis.* Glycogen represents a storage form of carbohydrates that can be used as a source for new glucose 6-phosphate (*2*) in a process called *glycogenolysis.* The liver contains an enzyme that can remove the phosphate from glucose 6-phosphate; liver glycogen thus serves as a source for new blood glucose.

As mentioned earlier, organic molecules with phosphate groups cannot cross plasma membranes. Because the glucose derived from glycogen is in the form of glucose 1-phosphate and then glucose 6-phosphate, it cannot leak out of the cell. Similarly, glucose that enters the cell from the blood is "trapped" within the cell by conversion to glucose 6-phosphate. Skeletal muscles, which have large amounts of glycogen, can generate glucose 6-phosphate for their own glycolytic needs, but they cannot secrete glucose into the blood because they lack the ability to remove the phosphate group.

Unlike skeletal muscles, the liver contains an enzyme—known as *glucose 6-phosphatase*—that can remove the phosphate groups and produce free glucose (fig. 5.10). This free glucose can then be transported through the plasma membrane. Thus, the liver can secrete glucose into the blood, whereas skeletal muscles cannot. Liver glycogen can thereby supply blood glucose for use by other organs, including exercising skeletal muscles that may have depleted much of their own stored glycogen during exercise.

Clinical Investigation CLUES

Brenda's coach advised her to eat more carbohydrates during her training.

- What will happen to the extra carbohydrates she eats?
- What benefit might be derived from such "carbohydrate loading"?

Cori Cycle

In humans and other mammals, much of the lactic acid produced in anaerobic metabolism is later eliminated by aerobic respiration of the lactic acid to carbon dioxide and water. However, some of the lactic acid produced by exercising skeletal muscles is delivered by the blood to the liver. Within the liver cells under these conditions, the enzyme *lactic acid dehydrogenase (LDH)* converts lactic acid to pyruvic acid. This is the reverse of the step of the lactic acid pathway shown in figure 5.4, and in the process NAD is reduced to NADH + H^+. Unlike most other organs, the liver contains the enzymes needed to take pyruvic acid molecules and convert them to glucose 6-phosphate, a process that is essentially the reverse of glycolysis.

Glucose 6-phosphate in liver cells can then be used as an intermediate for glycogen synthesis, or it can be converted to free glucose that is secreted into the blood. The conversion of noncarbohydrate molecules (not just lactic acid, but also amino acids and glycerol) through pyruvic acid to glucose is an extremely important process called **gluconeogenesis.** The significance of this process in conditions of fasting will be discussed together with amino acid metabolism (section 5.4).

During exercise, some of the lactic acid produced by skeletal muscles may be transformed through gluconeogenesis in the liver to blood glucose. This new glucose can serve as an energy source during exercise and can be used after exercise to help replenish the depleted muscle glycogen. This two-way traffic between skeletal muscles and the liver is called the **Cori cycle** (fig. 5.11). Through the Cori cycle, gluconeogenesis in the liver allows depleted skeletal muscle glycogen to be restored within 48 hours.

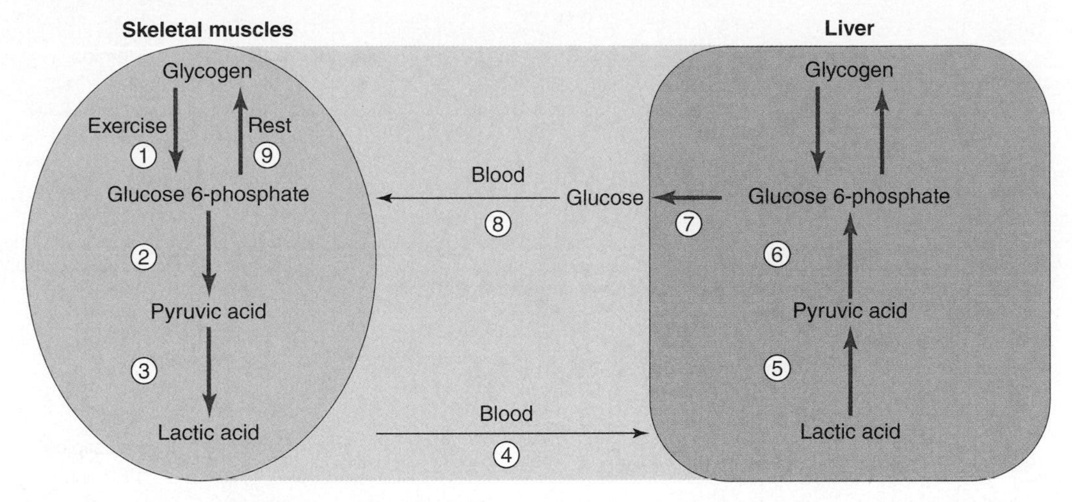

Figure 5.11 **The Cori cycle.** During exercise, muscle glycogen serves as a source of glucose 6-phosphate for the lactic acid pathway *(steps 1 through 3).* This lactic acid is carried by the blood *(step 4)* to the liver, where it is converted back to glucose 6-phosphate *(steps 5 and 6).* This is next converted into free glucose *(step 7),* which can be carried by the blood *(step 8)* back to the skeletal muscles. During rest, this glucose can be used to restore muscle glycogen *(step 9).*

4. Describe glycogenesis and glycogenolysis, and explain why the liver, but not skeletal muscles, can use stored glycogen to replenish blood glucose.
5. Explain the significance of gluconeogenesis and the Cori cycle in the homeostasis of blood glucose.

5.4 METABOLISM OF LIPIDS AND PROTEINS

Triglycerides can be hydrolyzed into glycerol and fatty acids. The latter are of particular importance because they can be converted into numerous molecules of acetyl CoA that can enter citric acid cycles and generate a large amount of ATP. Amino acids derived from proteins also may be used for energy. This involves the removal of the amine group and the conversion of the remaining molecule into either pyruvic acid or one of the Krebs cycle molecules.

LEARNING OUTCOMES

After studying this section, you should be able to:

6. **Describe how triglycerides can be used in aerobic cell respiration, and the nature of ketone bodies.**
7. **Describe how amino acids can be metabolized for energy, and explain how proteins, fats, and carbohydrates can be interconverted.**

Energy can be derived by the cellular respiration of lipids and proteins using the same aerobic pathway previously described for the metabolism of pyruvic acid. Indeed, some organs preferentially use molecules other than glucose as an energy source. Pyruvic acid and the Krebs cycle acids also serve as common intermediates in the interconversion of glucose, lipids, and amino acids.

When food energy is taken into the body faster than it is consumed, the concentration of ATP within body cells rises. Cells, however, do not store extra energy in the form of extra ATP. When cellular ATP concentrations rise because more energy (from food) is available than can be immediately used, ATP production is inhibited and glucose is instead converted into glycogen and fat (fig. 5.12).

Lipid Metabolism

When glucose is going to be converted into fat, glycolysis occurs and pyruvic acid is converted into acetyl CoA. Some of the glycolytic intermediates—phosphoglyceraldehyde and dihydroxyacetone phosphate—do not complete their conversion to pyruvic acid, however, and acetyl CoA does not enter a citric acid cycle. The acetic acid subunits of these acetyl CoA molecules can instead be used to produce a variety of lipids, including cholesterol (used in the synthesis of bile acids and steroid hormones), ketone bodies, and fatty acids (fig. 5.13). Acetyl CoA may thus be considered a branch point from which a number of different possible metabolic pathways may progress.

In the formation of fatty acids, a number of acetic acid (two-carbon) subunits are joined together to form the fatty acid chain. Six acetyl CoA molecules, for example, will produce a fatty acid that is 12 carbons long. When three of these fatty acids condense

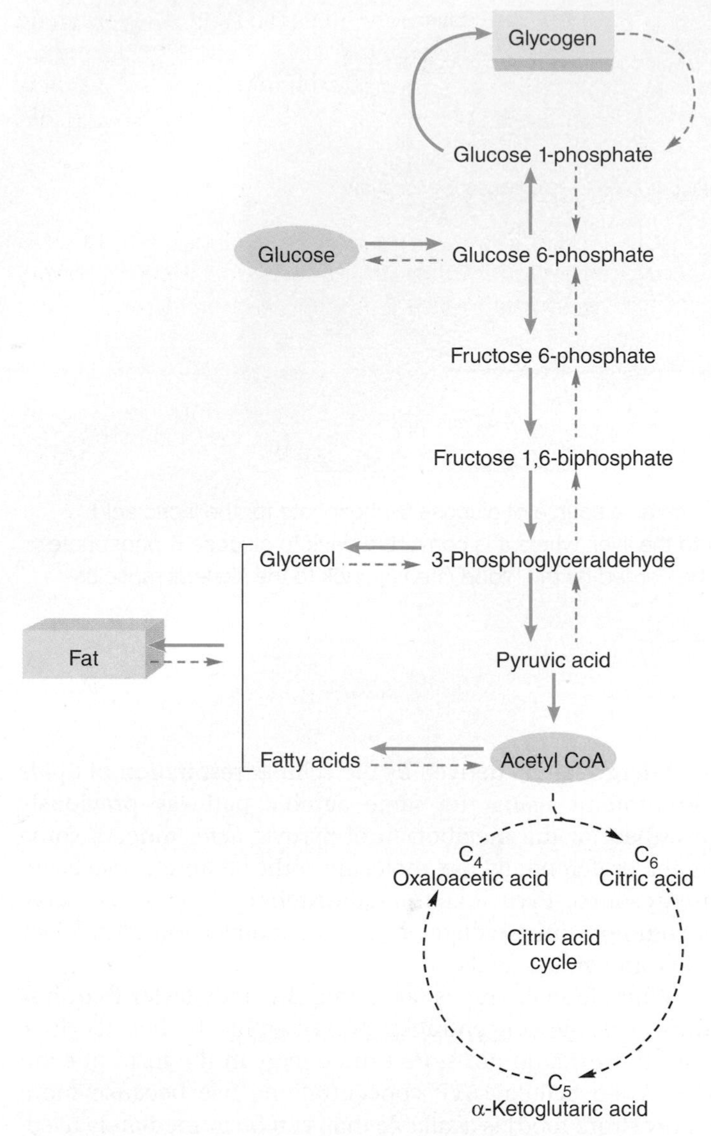

Figure 5.12 The conversion of glucose into glycogen and fat. This occurs as a result of inhibition of respiratory enzymes when the cell has adequate amounts of ATP. Favored pathways are indicated by blue arrows. AP|R

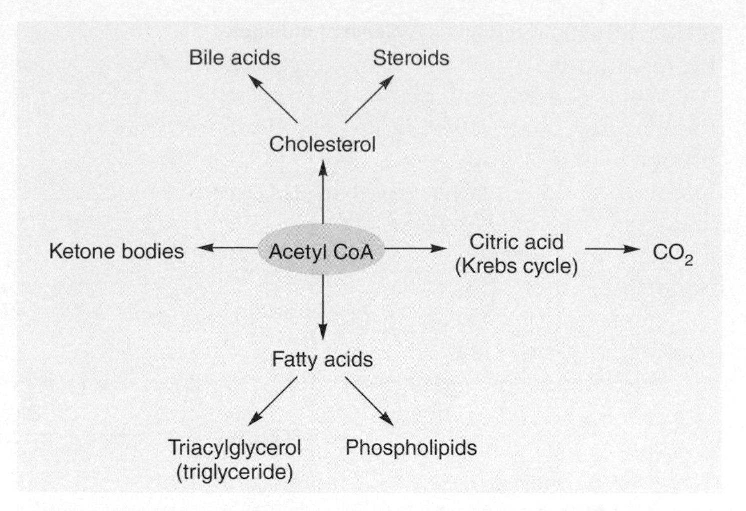

Figure 5.13 Divergent metabolic pathways for acetyl coenzyme A. Acetyl CoA is a common substrate that can be used to produce a number of chemically related products.

with 1 glycerol (derived from phosphoglyceraldehyde), a *triglyceride* (also called *triacylglycerol*) molecule is produced. The formation of fat, or **lipogenesis,** occurs primarily in adipose tissue and in the liver when the concentration of blood glucose is elevated following a meal.

Fat stored in adipose cells (*adipocytes*) of *white adipose tissue* (or *white fat*) serves as the major form of energy storage in the body. One gram of fat provides 9 kilocalories of energy, compared to 4 kilocalories for a gram of carbohydrates or protein. In a non-obese 70-kilogram (155-pound) man, 80% to 85% of the body's energy is stored as fat, which amounts to about 140,000 kilocalories. Stored glycogen, by contrast, accounts for less than 2,000 kilocalories, most of which (about 350 g) is stored in skeletal muscles and is available for use only by the muscles. The liver contains between 80 and 90 grams of glycogen, which can be converted to glucose and used by other organs. Protein accounts for 15% to 20% of the stored calories in the body, but protein usually is not used extensively as an energy source because that would involve the loss of muscle mass.

FITNESS APPLICATION

The ingestion of excessive calories increases fat production. The rise in blood glucose that follows carbohydrate-rich meals stimulates insulin secretion, and this hormone, in turn, promotes the entry of blood glucose into adipose cells. Increased availability of glucose within adipose cells, under conditions of high insulin secretion, promotes the conversion of glucose to fat (see figs. 5.12 and 5.13). The lowering of insulin secretion, conversely, promotes the breakdown of fat. This is exploited for weight reduction by low-carbohydrate diets. There has been concern that too much consumption of sugar, either as sucrose or as *high fructose corn syrup* (which contains about equal amounts of glucose and fructose), could promote obesity and the **metabolic syndrome**—a combination of central obesity, insulin resistance, type 2 diabetes mellitus, and hypertension (chapter 19, section 19.2).

White Adipose Tissue

White adipose tissue, or **white fat,** is where most of the triglycerides in the body are stored. When fat stored in adipose tissue is going to be used as an energy source, *lipase* enzymes hydrolyze triglycerides into glycerol and free fatty acids in a process called **lipolysis.** These molecules (primarily the free fatty acids) serve as *blood-borne energy carriers* that can be used by the liver, skeletal muscles, and other organs for aerobic respiration.

When adipocytes (adipose cells) hydrolyze triglycerides, the glycerol leaves through certain protein channels in the plasma membrane and enters the blood. The glycerol released into the blood is mostly taken up by the liver, which converts it into glucose through gluconeogenesis. By this means, glycerol released from adipocytes during exercise or fasting can be an important source of liver glucose.

However, the most significant energy carriers provided by lipolysis are the free fatty acids. Most fatty acids consist of a long hydrocarbon chain with a carboxyl group (COOH) at one end. In a process known as **β-oxidation** (β is the Greek letter *beta*), enzymes remove two-carbon acetic acid molecules from the acid end of a fatty acid chain. This results in the formation of acetyl CoA, as the third carbon from the end becomes oxidized to produce a new carboxyl group. The fatty acid chain is thus decreased in length by 2 carbons. The process of oxidation continues until the entire fatty acid molecule is converted to acetyl CoA (fig. 5.14).

A sixteen-carbon-long fatty acid, for example, yields 8 acetyl CoA molecules. Each of these can enter a citric acid cycle and produce 10 ATP per turn of the cycle, producing $8 \times 10 = 80$ ATP. In addition, each time an acetyl CoA molecule is formed and the end carbon of the fatty acid chain is oxidized, 1 NADH and 1 $FADH_2$ are produced. Oxidative phosphorylation produces 2.5 ATP per NADH and 1.5 ATP per $FADH_2$. For a sixteen-carbon-long fatty acid, these 4 ATP molecules would be formed seven times (producing $4 \times 7 = 28$ ATP). Not counting the single ATP used to start β-oxidation (fig. 5.14), this fatty acid could yield a grand total of $28 + 80$, or 108 ATP molecules!

Clinical Investigation CLUES

Brenda's coach advised her to exercise more gradually.

- If her muscles obtain a higher proportion of their energy from fatty acids under these conditions, what benefits would that have?
- How would this help to reduce her muscle fatigue?

Brown Adipose Tissue

Brown adipose tissue, or **brown fat,** develops from different cells than does white adipose tissue and has a different primary function: it is the major site for *thermogenesis* (heat production)

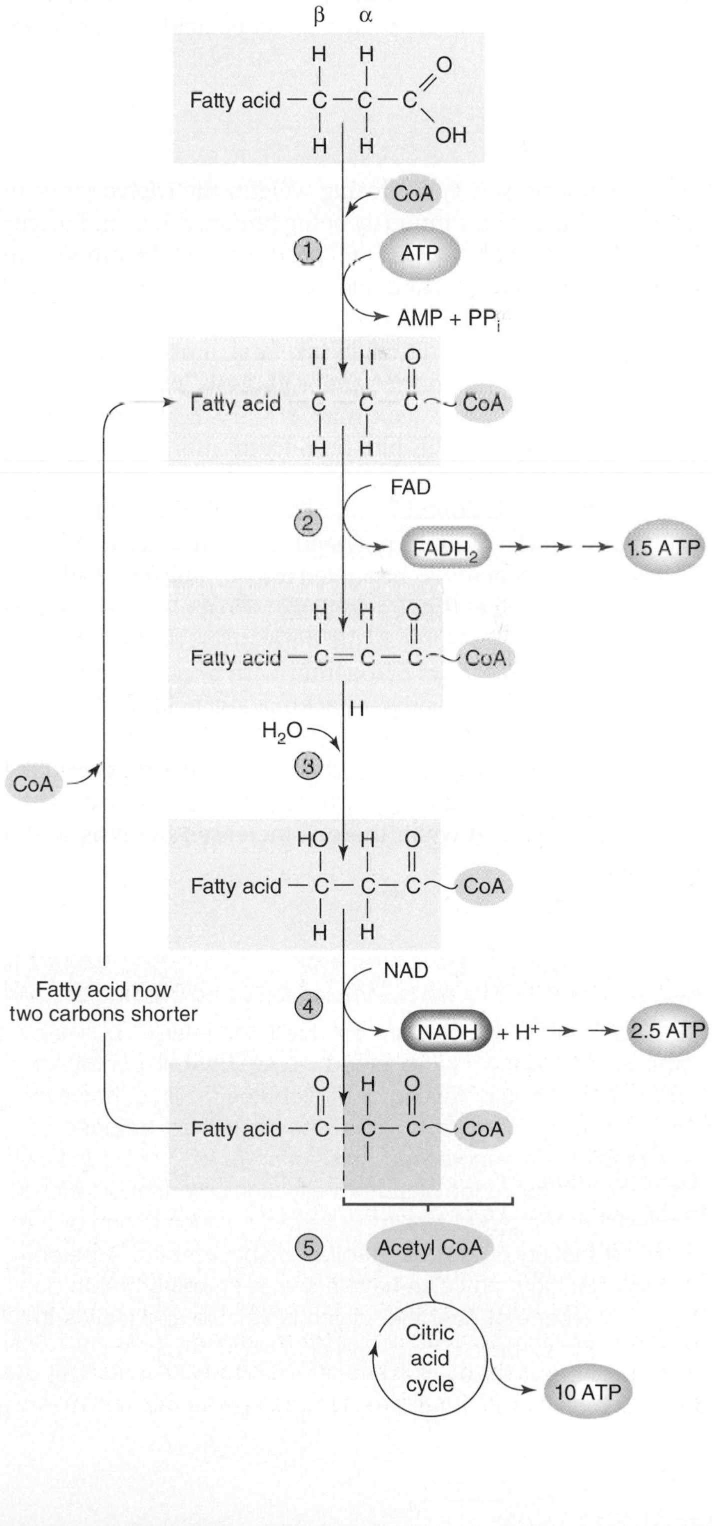

Figure 5.14 Beta-oxidation of a fatty acid. After the attachment of coenzyme A to the carboxyl group (*step 1*), a pair of hydrogens is removed from the fatty acid and used to reduce 1 molecule of FAD (*step 2*). When this electron pair is donated to the cytochrome chain, 1.5 ATP are produced. The addition of a hydroxyl group from water (*step 3*), followed by the oxidation of the β-carbon (*step 4*), results in the production of 2.5 ATP from the electron pair donated by NADH. The bond between the α and β carbons in the fatty acid is broken (*step 5*), releasing acetyl coenzyme A and a fatty acid chain that is 2 carbons shorter than the original. With the addition of a new coenzyme A to the shorter fatty acid, the process begins again (*step 2*). Acetyl CoA enters a citric acid cycle and generates 1 ATP directly and 9 ATP from the oxidative phosphorylation of 3 NADH and 1 $FADH_2$ obtained from the cycle.

See the *Test Your Quantitative Ability* section of the Review Activities at the end of this chapter.

in the newborn, who have a greater rate of heat loss and less muscle mass (for shivering) than do adults. Adults also have limited deposits of brown fat (mostly in the supraclavicular area of the ventral neck) that may contribute to calorie expenditure and heat production (chapter 19, section 19.2). In response to norepinephrine from sympathetic nerves (chapter 9), brown fat produces a unique *uncoupling protein* called **UCP1.** This protein uncouples oxidative phosphorylation by allowing H^+ to leak out of the inner mitochondrial membrane. As a result, less H^+ is available to drive ATP synthase activity so that less ATP is made by the electron transport system. Lower ATP concentrations exert less inhibition of the electron transport system, allowing an increased oxidation of fatty acids to generate more heat.

Ketone Bodies

Even when a person is not losing weight, the triglycerides in adipose tissue are continuously being broken down and resynthesized. New triglycerides are produced while others are hydrolyzed into glycerol and fatty acids. This turnover ensures that the blood will normally contain a sufficient level of fatty acids for aerobic respiration by skeletal muscles, the liver, and other organs. When the rate of lipolysis exceeds the rate of fatty acid utilization—as it may in starvation, dieting, and in diabetes mellitus—the blood concentration of fatty acids increases.

If the liver cells contain sufficient amounts of ATP so that further production of ATP is not needed, some of the acetyl CoA derived from fatty acids is channeled into an alternate pathway. This pathway involves the conversion of two molecules of acetyl CoA into four-carbon-long acidic derivatives, *acetoacetic acid* and *β-hydroxybutyric acid.* Together with *acetone,* which is a three-carbon-long derivative of acetoacetic acid, these products are known as **ketone bodies** (see chapter 2, fig. 2.19). The three ketone bodies are water-soluble molecules that circulate in the blood plasma, and their production from fatty acids by the liver is increased when there is increased lipolysis in the white adipose tissue.

CLINICAL APPLICATION

Ketone bodies, which can be used for energy by many organs, are found in the blood under normal conditions. Under conditions of fasting or of diabetes mellitus, however, the increased liberation of free fatty acids from adipose tissue results in the increased production of ketone bodies by the liver. The secretion of abnormally high amounts of ketone bodies into the blood produces **ketosis,** which is one of the signs of fasting or an uncontrolled diabetic state. A person in this condition may also have a sweet-smelling breath due to the presence of acetone, which is volatile and leaves the blood in the exhaled air.

Amino Acid Metabolism

Nitrogen is ingested primarily as proteins, enters the body as amino acids, and is excreted mainly as urea in the urine. In childhood, the amount of nitrogen excreted may be less than the amount ingested because amino acids are incorporated into proteins during growth. Growing children are thus said to be in a state of *positive nitrogen balance.* People who are starving or suffering from prolonged wasting diseases, by contrast, are in a state of *negative nitrogen balance;* they excrete more nitrogen than they ingest because they are breaking down their tissue proteins.

Healthy adults maintain a state of nitrogen balance in which the amount of nitrogen excreted is equal to the amount ingested. This does not imply that the amino acids ingested are unnecessary; on the contrary, they are needed to replace the protein that is "turned over" each day. When more amino acids are ingested than are needed to replace proteins, the excess amino acids are not stored as additional protein (one cannot build muscles simply by eating large amounts of protein). Rather, the amine groups can be removed, and the "carbon skeletons" of the organic acids that are left can be used for energy or converted to carbohydrate and fat.

Transamination

An adequate amount of all 20 amino acids is required to build proteins for growth and to replace the proteins that are turned over. However, only 8 of these (9 in children) cannot be produced by the body and must be obtained in the diet. These are the **essential amino acids** (table 5.3). The remaining amino acids are "nonessential" only in the sense that the body can

Table 5.3 | The Essential and Nonessential Amino Acids

Essential Amino Acids	Nonessential Amino Acids
Lysine	Aspartic acid
Tryptophan	Glutamic acid
Phenylalanine	Proline
Threonine	Glycine
Valine	Serine
Methionine	Alanine
Leucine	Cysteine
Isoleucine	Arginine
Histidine (in children)	Asparagine
	Glutamine
	Tyrosine

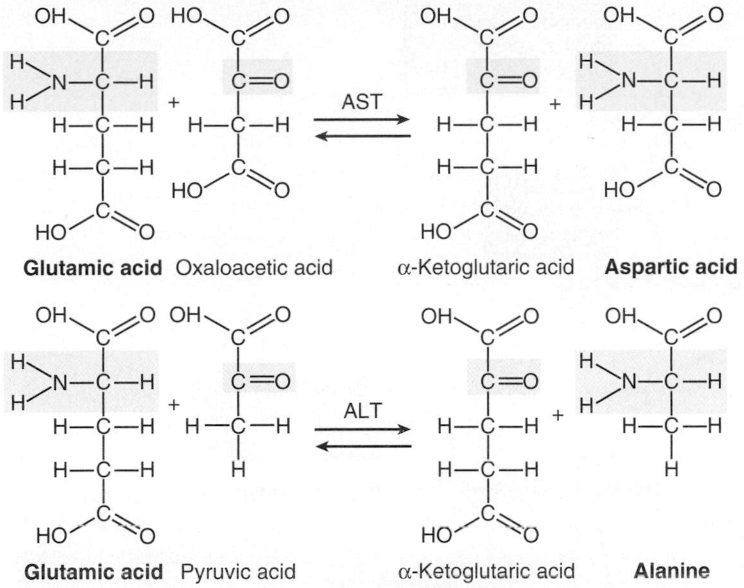

Glutamic acid + Oxaloacetic acid ⇌ (AST) α-Ketoglutaric acid + Aspartic acid

Glutamic acid + Pyruvic acid ⇌ (ALT) α-Ketoglutaric acid + Alanine

Figure 5.15 Two important transamination reactions. The areas shaded in blue indicate the parts of the molecules that are changed. (AST = aspartate transaminase; ALT = alanine transaminase. The amino acids are identified in boldface.)

produce them if provided with a sufficient amount of carbohydrates and the essential amino acids.

Pyruvic acid and the citric acid cycle acids are collectively termed *keto acids* because they have a ketone group; these should not be confused with the ketone bodies (derived from acetyl CoA) discussed in the previous section. Keto acids can be converted to amino acids by the addition of an amine (NH_2) group. This amine group is usually obtained by "cannibalizing" another amino acid; in this process, a new amino acid is formed as the one that was cannibalized is converted to a new keto acid. This type of reaction, in which the amine group is transferred from one amino acid to form another, is called **transamination** (fig. 5.15).

CLINICAL APPLICATION

Measurements of the AST and ALT enzyme levels in blood plasma are commonly referred to as **liver enzyme tests,** although their correlation with liver disease is imperfect. For example, these levels may be greatly elevated in hepatitis A and B, but they may be normal in hepatitis C. Toxic chemicals that damage liver cells, such as acetaminophen (Tylenol), can also greatly elevate the blood levels of AST and ALT. These enzyme levels may be moderately increased in people with a fatty liver, which can be caused by alcohol abuse. However, the enzyme levels may be somewhat elevated at times in healthy people, and damaged cardiac and skeletal muscles may also be a source of elevated plasma levels of these enzymes (particularly of AST).

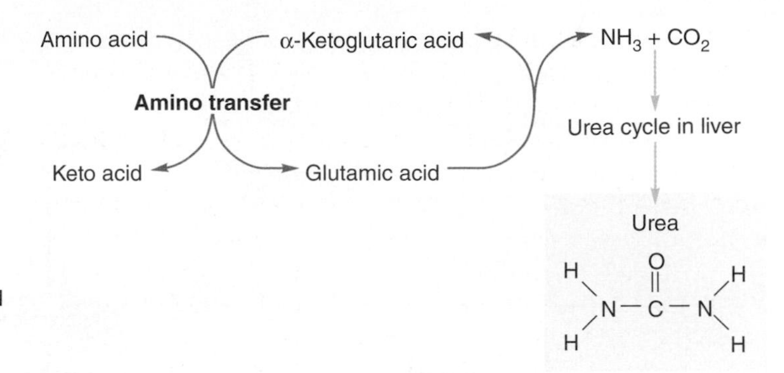

Figure 5.16 Oxidative deamination. Glutamic acid is converted to α-ketoglutaric acid as it donates its amine group to the metabolic pathway that results in the formation of urea.

Each transamination reaction is catalyzed by a specific enzyme (a transaminase) that requires vitamin B_6 (pyridoxine) as a coenzyme. The amine group from glutamic acid, for example, may be transferred to either pyruvic acid or oxaloacetic acid. The former reaction is catalyzed by the enzyme *alanine transaminase (ALT)*; the latter reaction is catalyzed by *aspartate transaminase (AST).* These enzyme names reflect the fact that the addition of an amine group to pyruvic acid produces the amino acid alanine; the addition of an amine group to oxaloacetic acid produces the amino acid known as aspartic acid (fig. 5.15).

Oxidative Deamination

As shown in figure 5.16, glutamic acid can be formed through transamination by the combination of an amine group with α-ketoglutaric acid. Glutamic acid is also produced in the liver from the ammonia that is generated by intestinal bacteria and carried to the liver in the hepatic portal vein. Because free ammonia is very toxic, its removal from the blood and incorporation into glutamic acid is an important function of a healthy liver.

If there are more amino acids than are needed for protein synthesis, the amine group from glutamic acid may be removed and excreted as *urea* in the urine (fig. 5.16). The metabolic pathway that removes amine groups from amino acids—leaving a keto acid and ammonia (which is converted to urea)—is known as **oxidative deamination.**

A number of amino acids can be converted into glutamic acid by transamination. Since glutamic acid can donate amine groups to urea (through deamination), it serves as a channel through which other amino acids can be used to produce keto acids (pyruvic acid and Krebs cycle acids). These keto acids may then be used in the citric acid cycle as a source of energy (fig. 5.17).

Depending upon which amino acid is deaminated, the keto acid left over may be either pyruvic acid or one of the Krebs cycle acids. These can be respired for energy, converted to fat, or

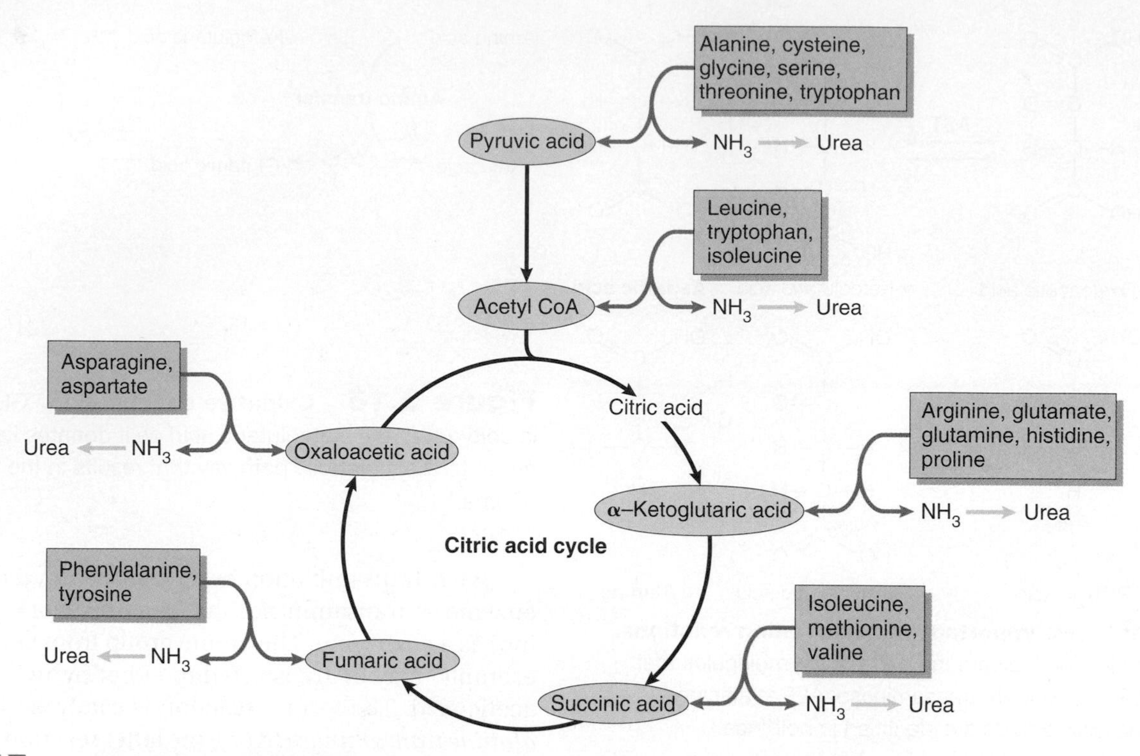

Figure 5.17 **Pathways by which amino acids can be catabolized for energy.** These pathways are indirect for some amino acids, which first must be transaminated into other amino acids before being converted into keto acids by deamination. AP|R

converted to glucose. In the last case, the amino acids are eventually changed to pyruvic acid, which is used to form glucose. This process—the formation of glucose from amino acids or other noncarbohydrate molecules—is called *gluconeogenesis,* as mentioned previously in connection with the Cori cycle.

The main substrates for gluconeogenesis are the three-carbon-long molecules of alanine (an amino acid), lactic acid, and glycerol. This illustrates the interrelationship between amino acids, carbohydrates, and fat, as shown in figure 5.18.

Recent experiments in humans have suggested that, even in the early stages of fasting, most of the glucose secreted by the liver is derived through gluconeogenesis. Findings indicate that hydrolysis of liver glycogen (glycogenolysis) contributes only 36% of the glucose secreted during the early stages of a fast. At 42 hours of fasting, all of the glucose secreted by the liver is produced by gluconeogenesis.

Uses of Different Energy Sources

The blood serves as a common trough from which all the cells in the body are fed. If all cells used the same energy source, such as glucose, this source would quickly be depleted and cellular starvation would occur. Normally, however, the blood contains a variety of energy sources from which to draw: glucose and ketone bodies that come from the liver, fatty acids from adipose tissue, and lactic acid and amino acids from muscles. Some organs preferentially use one energy source more than the others, so that each energy source is "spared" for organs with strict energy needs.

The brain uses blood glucose as its major energy source. Under fasting conditions, blood glucose is supplied primarily by the liver through glycogenolysis and gluconeogenesis. In addition, the blood glucose concentration is maintained because many organs spare glucose by using fatty acids, ketone bodies, and lactic acid as energy sources (table 5.4). During severe starvation, the brain also gains some ability to metabolize ketone bodies for energy.

Table 5.4 | Relative Importance of Different Molecules in the Blood with Respect to the Energy Requirements of Different Organs

Organ	Glucose	Fatty Acids	Ketone Bodies	Lactic Acid
Brain	+++	–	+	–
Skeletal muscles (resting)	+	+++	+	–
Liver	+	+++	++	+
Heart	+	++	+	+

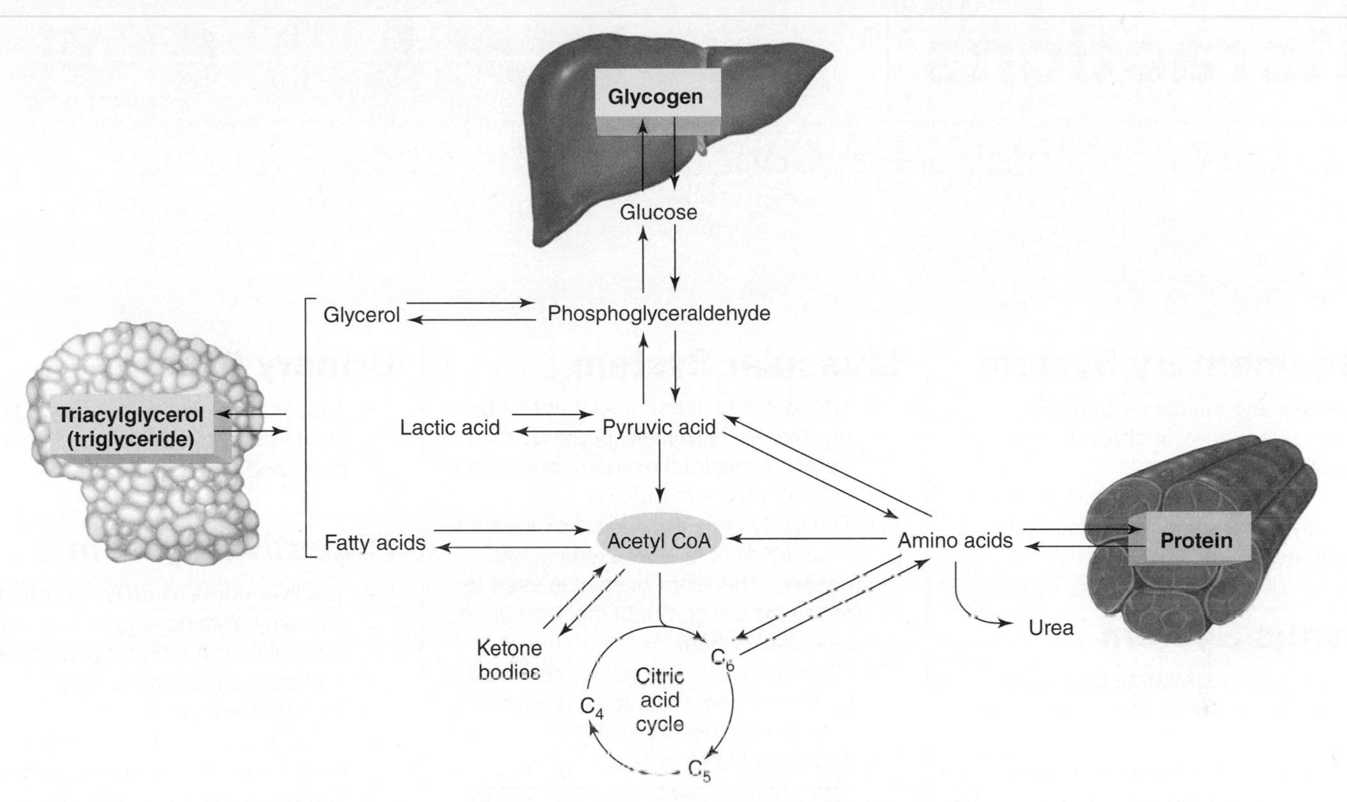

Figure 5.18 The interconversion of glycogen, fat, and protein. These simplified metabolic pathways show how glycogen, fat, and protein can be interconverted. Note that while most reactions are reversible, the reaction from pyruvic acid to acetyl CoA is not. This is because a CO_2 is removed in the process. (Only plants, in a phase of photosynthesis called the dark reaction, can use CO_2 to produce glucose.) AP|R

As mentioned earlier, lactic acid produced anaerobically during exercise can be used for energy following the cessation of exercise. The lactic acid, under aerobic conditions, is reconverted to pyruvic acid, which then enters the aerobic respiratory pathway. The extra oxygen required to metabolize lactic acid contributes to the *oxygen debt* following exercise (chapter 12, section 12.4).

Clinical Investigation CLUES

Brenda gasped and panted for air longer after her workouts than her teammates did.

- What is the extra oxygen needed after exercise called, and what functions does it serve?
- How would a more gradual training program help Brenda to gasp and pant less after her workouts?

CHECKPOINT

6a. Construct a flowchart to show the metabolic pathway by which glucose can be converted to fat. Indicate only the major intermediates involved (not all of the steps of glycolysis).

6b. Define the terms *lipolysis* and *β-oxidation* and explain, in general terms, how fat can be used for energy.

7a. Describe transamination and deamination and explain their functional significance.

7b. List five blood-borne energy carriers and explain, in general terms, how these are used as sources of energy.

Interactions

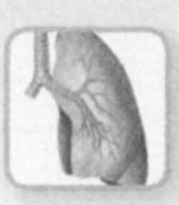

HPer Links of Metabolism Concepts to the Body Systems

Integumentary System

- The skin synthesizes vitamin D from a derivative of cholesterol (p. 693)
- The metabolic rate of the skin varies greatly, depending upon ambient temperature (p. 474)

Nervous System

- The aerobic respiration of glucose serves most of the energy needs of the brain (p. 124)
- Regions of the brain with a faster metabolic rate, resulting from increased brain activity, receive a more abundant blood supply than regions with a slower metabolic rate (p. 473)

Endocrine System

- Hormones that bind to receptors in the plasma membrane of their target cells activate enzymes in the target cell cytoplasm (p. 326)
- Hormones that bind to nuclear receptors in their target cells alter the target cell metabolism by regulating gene expression (p. 323)
- Hormonal secretions from adipose cells regulate hunger and metabolism (p. 669)
- Anabolism and catabolism are regulated by a number of hormones (p. 675)
- Insulin stimulates the synthesis of glycogen and fat (p. 347)
- The adrenal hormones stimulate the breakdown of glycogen, fat, and protein (p. 685)
- Thyroxine stimulates the production of a protein that uncouples oxidative phosphorylation. This helps to increase the body's metabolic rate (p. 687)
- Growth hormone stimulates protein synthesis (p. 688)

Muscular System

- The intensity of exercise that can be performed aerobically depends on a person's maximal oxygen uptake and lactate threshold (p. 382)
- The body consumes extra oxygen for a period of time after exercise has ceased. This extra oxygen is used to repay the oxygen debt incurred during exercise (p. 380)
- Glycogenolysis and gluconeogenesis by the liver help to supply glucose for exercising muscles (p. 379)
- Trained athletes obtain a higher proportion of skeletal muscle energy from the aerobic respiration of fatty acids than do nonathletes (p. 383)
- Muscle fatigue is associated with anaerobic metabolism and the production of lactic acid (p. 382)
- The proportion of energy derived from carbohydrates or lipids by exercising skeletal muscles depends on the intensity of the exercise (p. 378)

Circulatory System

- Metabolic acidosis may result from excessive production of either ketone bodies or lactic acid (p. 568)
- The metabolic rate of skeletal muscles determines the degree of blood vessel dilation, and thus the rate of blood flow to the organ (p. 470)
- Atherosclerosis of coronary arteries can force a region of the heart to metabolize anaerobically and produce lactic acid. This is associated with angina pectoris (p. 439)

Respiratory System

- Ventilation oxygenates the blood going to the cells for aerobic cell respiration and removes the carbon dioxide produced by the cells (p. 533)
- Breathing is regulated primarily by the effects of carbon dioxide produced by aerobic cell respiration (p. 555)

Urinary System

- The kidneys eliminate urea and other waste products of metabolism from the blood plasma (p. 598)

Digestive System

- The liver contains enzymes needed for many metabolic reactions involved in regulating the blood glucose and lipid concentrations (p. 637)
- The pancreas produces many enzymes needed for the digestion of food in the small intestine (p. 642)
- The digestion and absorption of carbohydrates, lipids, and proteins provides the body with the substrates used in cell metabolism (p. 649)
- Vitamins A and D help to regulate metabolism through the activation of nuclear receptors, which bind to regions of DNA (p. 325)

Reproductive System

- The sperm do not contribute mitochondria to the fertilized oocyte (p. 735)
- The endometrium contains glycogen that nourishes the developing embryo (p. 737)

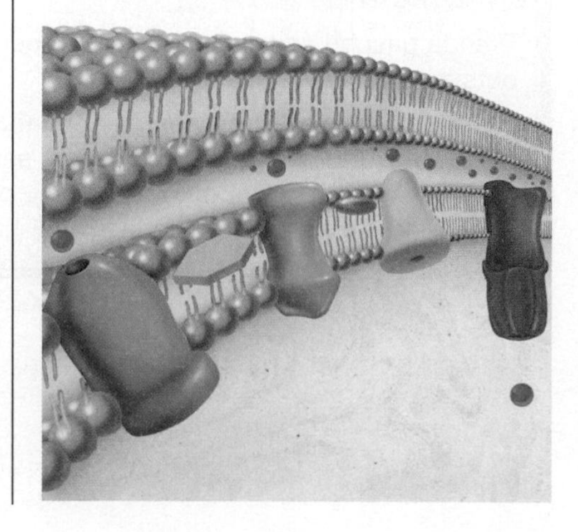

Clinical Investigation SUMMARY

Brenda's great fatigue following workouts is partially related to the depletion of her glycogen reserves and extensive utilization of anaerobic metabolism (with consequent production of lactic acid) for energy. Production of large amounts of lactic acid during exercise causes her need for extra oxygen to metabolize the lactic acid following exercise (the oxygen debt)—hence, her gasping and panting. Eating more carbohydrates would help Brenda maintain the glycogen stores in her liver and muscles, and training more gradually could increase the ability of her muscles to obtain more of their energy through aerobic metabolism, so that she would experience less pain and fatigue.

The pain in her arms and shoulders is probably the result of lactic acid production by the exercised skeletal muscles. However, the intense pain in her left pectoral region could be angina pectoris, caused by anaerobic metabolism of the heart. If this is the case, it would indicate that the heart became ischemic because blood flow was inadequate for the demands placed upon it. Blood tests for particular enzymes released by damaged heart muscle (chapter 4) and an electrocardiogram (ECG) should be performed.

See the additional chapter 5 Clinical Investigation on *Metabolic Disease* in the Connect site for this text at www.mhhe.com/Fox13

SUMMARY

5.1 Glycolysis and the Lactic Acid Pathway 107

A. *Glycolysis* refers to the conversion of glucose to two molecules of pyruvic acid.

1. In the process, two molecules of ATP are consumed and 4 molecules of ATP are formed. Thus, there is a net gain of two ATP.
2. In the steps of glycolysis, two pairs of hydrogens are released. Electrons from these hydrogens reduce two molecules of NAD.

B. When metabolism is anaerobic, reduced NAD is oxidized by pyruvic acid, which accepts two hydrogen atoms and is thereby reduced to lactic acid.

1. Skeletal muscles produce lactic acid during exercise.
2. Heart muscle undergoes lactic acid fermentation for just a short time, under conditions of ischemia.

5.2 Aerobic Respiration 111

A. The citric acid cycle begins when coenzyme A donates acetic acid to an enzyme that adds it to oxaloacetic acid to form citric acid.

1. Acetyl CoA is formed from pyruvic acid by the removal of carbon dioxide and 2 hydrogens.
2. The formation of citric acid begins a cyclic pathway that ultimately forms a new molecule of oxaloacetic acid.
3. As the citric acid cycle progresses, 1 molecule of ATP is formed, and 3 molecules of NAD and 1 of FAD are reduced by hydrogens from the citric acid cycle.

B. Reduced NAD and FAD donate their electrons to an electron-transport chain of molecules located in the cristae.

1. The electrons from NAD and FAD are passed from one cytochrome of the electron-transport chain to the next in a series of coupled oxidation-reduction reactions.
2. As each cytochrome ion gains an electron, it becomes reduced; as it passes the electron to the next cytochrome, it becomes oxidized.
3. The last cytochrome becomes oxidized by donating its electron to oxygen, which functions as the final electron acceptor.
4. When 1 oxygen atom accepts 2 electrons and 2 protons, it becomes reduced to form water.
5. The energy provided by electron transport is used to form ATP from ADP and P_i in the process known as oxidative phosphorylation.

C. Thirty to 32 molecules of ATP are produced by the aerobic respiration of 1 glucose molecule. Of these, 2 are produced in the cytoplasm by glycolysis and the remainder are produced in the mitochondria.

D. The formation of glycogen from glucose is called glycogenesis; the breakdown of glycogen is called glycogenolysis.

1. Glycogenolysis yields glucose 6-phosphate, which can enter the pathway of glycolysis.
2. The liver contains an enzyme (which skeletal muscles do not) that can produce free glucose from glucose 6-phosphate. Thus, the liver can secrete glucose derived from glycogen.

E. Carbohydrate metabolism is influenced by the availability of oxygen and by a negative feedback effect of ATP on glycolysis and the citric acid cycle.

5.3 Interconversion of Glucose, Lactic Acid, and Glycogen 117

A. Glycogen can be converted into free glucose by the liver because the liver has glucose 6-phosphatase; this allows the liver to secrete glucose into the blood.

B. Lactic acid produced by skeletal muscles during anaerobic metabolism can travel through the blood to the liver, where it is converted into glucose that can travel back to the muscles to replenish their glycogen in the Cori cycle.

5.4 Metabolism of Lipids and Proteins 119

A. In lipolysis, triglycerides yield glycerol and fatty acids.

1. Glycerol can be converted to phosphoglyceraldehyde and used for energy.
2. In the process of β-oxidation of fatty acids, a number of acetyl CoA molecules are produced.
3. Processes that operate in the reverse direction can convert glucose to triglycerides.

B. Amino acids derived from the hydrolysis of proteins can serve as sources of energy.

1. Through transamination, a particular amino acid and a particular keto acid (pyruvic acid or one of the Krebs cycle acids) can serve as substrates to form a new amino acid and a new keto acid.
2. In oxidative deamination, amino acids are converted into keto acids as their amino group is incorporated into urea.

C. Each organ uses certain blood-borne energy carriers as its preferred energy source.

1. The brain has an almost absolute requirement for blood glucose as its energy source.
2. During exercise, the needs of skeletal muscles for blood glucose can be met by glycogenolysis and by gluconeogenesis in the liver.

REVIEW ACTIVITIES

Test Your Knowledge

1. The net gain of ATP per glucose molecule in lactic acid fermentation is ______; the net gain in aerobic respiration is generally ______.
   a. 2;4 c. 30;2
   b. 2;30 d. 24;38
2. In anaerobic metabolism, the oxidizing agent for NADH (that is, the molecule that removes electrons from NADH) is
   a. pyruvic acid. c. citric acid.
   b. lactic acid. d. oxygen.
3. When skeletal muscles lack sufficient oxygen, there is an increased blood concentration of
   a. pyruvic acid. c. lactic acid.
   b. glucose. d. ATP.
4. The conversion of lactic acid to pyruvic acid occurs
   a. in anaerobic respiration.
   b. in the heart, where lactic acid is aerobically respired.
   c. in the liver, where lactic acid can be converted to glucose.
   d. in both *a* and *b*.
   e. in both *b* and *c*.
5. Which of these statements about the oxygen in the air we breathe is *true?*
   a. It functions as the final electron acceptor of the electron-transport chain.
   b. It combines with hydrogen to form water.
   c. It combines with carbon to form CO_2.
   d. Both *a* and *b* are true.
   e. Both *a* and *c* are true.
6. In terms of the number of ATP molecules directly produced, the major energy-yielding process in the cell is
   a. glycolysis. c. oxidative phosphorylation.
   b. the citric acid cycle. d. gluconeogenesis.
7. Ketone bodies are derived from
   a. fatty acids. c. glucose.
   b. glycerol. d. amino acids.
8. The conversion of glycogen to glucose 6-phosphate occurs in
   a. the liver.
   b. skeletal muscles.
   c. both *a* and *b*.
9. The conversion of glucose 6-phosphate to free glucose, which can be secreted into the blood, occurs in
   a. the liver.
   b. skeletal muscles.
   c. both *a* and *b*.

10. The formation of glucose from pyruvic acid derived from lactic acid, amino acids, or glycerol is called
    a. glycogenesis.
    b. glycogenolysis.
    c. glycolysis.
    d. gluconeogenesis.
11. Which of these organs has an almost absolute requirement for blood glucose as its energy source?
    a. liver
    b. brain
    c. skeletal muscles
    d. heart
12. When amino acids are used as an energy source,
    a. oxidative deamination occurs.
    b. pyruvic acid or one of the Krebs cycle acids (keto acids) is formed.
    c. urea is produced.
    d. all of these occur.
13. Intermediates formed during fatty acid metabolism can enter the citric acid cycle as
    a. keto acids.
    b. acetyl CoA.
    c. Krebs cycle acids.
    d. pyruvic acid.

Test Your Understanding

14. State the advantages and disadvantages of the lactic acid pathway.
15. What purpose is served by the formation of lactic acid during anaerobic metabolism? How is this accomplished during aerobic respiration?
16. Describe the effect of cyanide on oxidative phosphorylation and on the citric acid cycle. Why is cyanide deadly?
17. Describe the metabolic pathway by which glucose can be converted into fat. How can end-product inhibition by ATP favor this pathway?
18. Describe the metabolic pathway by which fat can be used as a source of energy and explain why the metabolism of fatty acids can yield more ATP than the metabolism of glucose.
19. Explain how energy is obtained from the metabolism of amino acids. Why does a starving person have a high concentration of urea in the blood?
20. Explain why the liver is the only organ able to secrete glucose into the blood. What are the different molecular sources and metabolic pathways that the liver uses to obtain glucose?
21. Why is the production of lactic acid termed a "fermentation" pathway?
22. Explain the function of brown fat. What does its mechanism imply about the effect of ATP concentrations on the rate of cell respiration?
23. What three molecules serve as the major substrates for gluconeogenesis? Describe the situations in which each one would be involved in this process. Why can't fatty acids be used as a substrate for gluconeogenesis? (*Hint:* Count the carbons in acetyl CoA and pyruvic acid.)

Test Your Analytical Ability

24. A friend, wanting to lose weight, eliminates all fat from her diet. How would this help her to lose weight? Could she possibly gain weight on this diet? How? Discuss the health consequences of such a diet.
25. Suppose a drug is developed that promotes the channeling of H^+ out of the intermembrane space into the matrix of the mitochondria of adipose cells. How could this drug affect the production of ATP, body temperature, and body weight?
26. For many years, the total number of molecules of ATP produced for each molecule of glucose in aerobic respiration was given as 38. Later, it was estimated to be closer to 36, and now it is believed to be closer to 30. What factors must be considered in estimating the yield of ATP molecules? Why are the recent numbers considered to be approximate values?
27. People who are starving have very thin arms and legs. Because they're not eating, no glucose is coming in from the gastrointestinal tract, yet the brain must still be getting glucose from the blood to keep them alive. Explain the relationship between these observations, and the particular metabolic pathways involved.
28. Suppose you eat a chicken sandwich. Trace the fate of the chicken protein and the bread starch from your intestine to your liver and muscles. Using this information, evaluate the statement "You are what you eat."

Test Your Quantitative Ability

Answer the following questions regarding a twenty-carbon-long fatty acid (refer to figure 5.14).

29. How many acetyl CoA molecules can be produced during the complete β-oxidation of this fatty acid?
30. How many ATP will be broken down in the complete β-oxidation of this fatty acid?
31. How many ATP will be produced by direct (substrate-level) phosphorylation when this fatty acid is completely metabolized?
32. How many NADH and $FADH_2$ will be produced when this fatty acid is completely metabolized?
33. How many ATP will be made by oxidative phosphorylation when this fatty acid is completely metabolized for energy?

CHAPTER

6

Interactions Between Cells and the Extracellular Environment

REFRESH YOUR MEMORY

Before you begin this chapter, you may want to review these concepts from previous chapters:

CHAPTER OUTLINE

Clinical Investigation

Jessica, a physiology student, says she drinks water constantly and yet is constantly thirsty. During a physiology lab exercise involving tests of her urine, she discovers that her urine contains measurable amounts of glucose. Alarmed, she goes to her physician, who tests her blood and urine and obtains an electrocardiogram (ECG).

Some of the new terms and concepts you will encounter include:

- Hyperglycemia, glycosuria, and hyperkalemia.
- Osmolality and osmotic pressure.

6.1 EXTRACELLULAR ENVIRONMENT

The extracellular environment surrounding cells consists of a fluid compartment in which molecules are dissolved, and a matrix of polysaccharides and proteins that give form to the tissues. Interactions between the intracellular and extracellular environment occur across the plasma membrane.

LEARNING OUTCOMES

After studying this section, you should be able to:

1. Describe the intracellular and extracellular compartments of the body.
2. Identify the components of passive transport, and distinguish passive from active transport.

The extracellular environment includes all constituents of the body located outside of the cells. The cells of our body must receive nourishment from, and release their waste products into, the extracellular environment. Further, the different cells of a tissue, the cells of different tissues within an organ, and the cells of different organs interact with each other through chemical regulators secreted into the extracellular environment.

Body Fluids

The water content of the body is divided into two compartments. Approximately 67% of the total body water is contained within cells, in the **intracellular compartment.** The remaining 33% of the total body water is found in the **extracellular compartment.** About 20% of this extracellular fluid is contained within the vessels of the cardiovascular system, where it constitutes the fluid portion of the blood, or **blood plasma.**

The blood transports oxygen from the lungs to the body cells, and carbon dioxide from the body cells to the lungs. It also transports nutrients derived from food in the intestine to the body cells; other nutrients between organs (such as glucose from the liver to the brain, or lactic acid from muscles to the liver); metabolic wastes from the body cells to the liver and kidneys for elimination in the bile and urine, respectively; and regulatory molecules called hormones from endocrine glands to the cells of their target organs.

The remaining 80% of the extracellular fluid is located outside of the vascular system, and makes up the **tissue fluid,** also called **interstitial fluid.** This fluid is contained in the gel-like extracellular matrix, as described in the next section. Body fluid distribution is illustrated in chapter 14, figure 14.8, in conjunction with a discussion of the cardiovascular system. This is because the interstitial fluid is formed continuously from blood plasma, and it continuously returns to the blood plasma through mechanisms described in chapter 14 (see fig. 14.9). Oxygen, nutrients, and regulatory molecules traveling in the blood must first pass into the interstitial fluid before reaching the body cells; waste products and hormone secretions from the cells must first pass into the interstitial fluid before reaching the blood plasma (fig. 6.1).

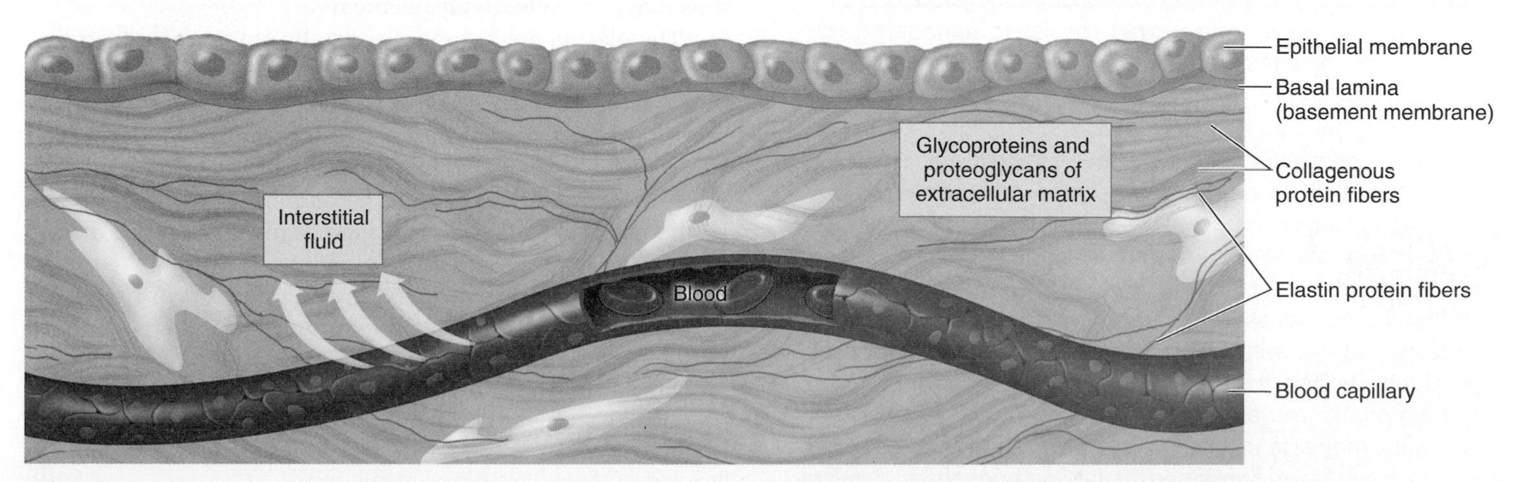

Figure 6.1 The extracellular environment. The extracellular environment contains interstitial tissue fluid within a matrix of glycoproteins and proteoglycans. Interstitial fluid is derived from blood plasma that filters through the pores (not shown) between the cells of the capillary walls, and delivers nutrients and regulatory molecules to the tissue cells. The extracellular environment is supported by collagen and elastin protein fibers, which also form the basal lamina below epithelial membranes. AP|R

Extracellular Matrix

The cells that compose the organs of our body are embedded within the extracellular material of connective tissues (fig. 6.1). This material is called the **extracellular matrix,** and it consists of the protein fibers *collagen* and *elastin* (see chapter 2, fig. 2.29), as well as gel-like *ground substance.* The interstitial fluid referred to previously exists primarily in the hydrated gel of the ground substance.

Although the ground substance seemingly lacks form (is amorphous) when viewed under a microscope, it is actually a highly functional, complex organization of molecules chemically linked to the extracellular protein fibers of collagen and elastin, as well as to the carbohydrates that cover the outside surface of the cell's plasma membrane (see chapter 3, fig. 3.2). The gel is composed of *glycoproteins* (proteins with numerous side chains of sugars) and molecules called *proteoglycans.* These molecules (formerly called *mucopolysaccharides*) are composed primarily of polysaccharides and have a high content of bound water molecules.

The collagen and elastin fibers have been likened to the reinforcing iron bars in concrete—they provide structural strength to the connective tissues. One type of collagen (collagen IV; there are about 15 different types known) contributes to the *basal lamina* (or *basement membrane*) underlying epithelial membranes (see chapter 1, fig. 1.12). By forming chemical bonds between the carbohydrates on the outside surface of the plasma membrane of the epithelial cells, and the glycoproteins and proteoglycans of the matrix in the connective tissues, the basal lamina helps to wed the epithelium to its underlying connective tissues (fig. 6.1).

CLINICAL APPLICATION

There is an important family of enzymes that can break down extracellular matrix proteins. These enzymes are called **matrix metalloproteinases (MMPs)** because of their need for a zinc ion cofactor. MMPs are required for tissue remodeling (for example, during embryonic development and wound healing), and for migration of phagocytic cells and other white blood cells during the fight against infection. MMPs are secreted as inactive enzymes and then activated extracellularly. However, they can contribute to disease processes if they are produced or activated inappropriately.

For example, MMPs are involved in the ability of a cancerous tumor to metastasize, or invade different locations. The MMPs are peptidases that break down extracellular matrix proteins and cell adhesion proteins, allowing the tumor cells to migrate across the basement membrane and into blood and lymphatic vessels. The destruction of cartilage protein in arthritis may also involve the action of these enzymes, and MMPs have been implicated in the pathogenesis of such neural diseases as multiple sclerosis, Alzheimer's disease, and others.

The surface of the cell's plasma membrane contains glycoproteins that affect the interactions between the cell and its extracellular environment. **Integrins** are a class of glycoproteins that extend from the cytoskeleton within a cell, through its plasma membrane, and into the extracellular matrix. By binding to components within the matrix, they serve as a sort of "glue" (or adhesion molecule) between cells and the extracellular matrix. Moreover, by physically joining the intracellular to the extracellular compartments, they serve to relay signals between these two compartments (or integrate these two compartments—hence the origin of the term *integrin*).

Through these interactions, integrins help to impart a polarity to the cell, so that one side is distinguished structurally and functionally from another (apical side from basal side, for example). They affect cell adhesion in a tissue and the ability of certain cells to be motile, and they affect the ability of cells to proliferate in their tissues. Extracellular matrix proteins and proteoglycans also bind to secreted regulatory chemicals, particularly various growth factors, and help to deliver these to integrins and receptor proteins at the cell surface (receptor proteins are discussed in section 6.5).

Categories of Transport Across the Plasma Membrane

Because the extracellular fluid is either blood plasma or derived from blood plasma, the term *plasma membrane* is used for describing the membrane around cells that separates the intracellular from the extracellular compartments. Molecules that move from the blood to the interstitial fluid, or molecules that move within the interstitial fluid between different cells, must eventually come into contact with the plasma membrane surrounding the cells. Some of these molecules may be able to penetrate the membrane, while others may not. Similarly, some intracellular molecules can penetrate, or "permeate," the plasma membrane and some cannot. The plasma membrane is thus said to be **selectively permeable.**

The plasma membrane is generally not permeable to proteins, nucleic acids, and other molecules needed for the structure and function of the cell. It is, however, permeable to many other molecules, permitting the two-way traffic of nutrients and wastes needed to sustain metabolism. The plasma membrane is also selectively permeable to certain ions; this permits electrochemical currents across the membrane used for production of impulses in nerve and muscle cells.

The mechanisms involved in the transport of molecules and ions through the plasma membrane can be categorized in different ways. One way is to group the different transport processes into those that require membrane *protein carriers* (described in section 6.3) and those that do not utilize membrane carriers.

1. **Carrier-mediated transport**
   a. *Facilitated diffusion*
   b. *Active transport*

2. Non-carrier-mediated transport

  a. *Simple diffusion* (diffusion that is not carrier-mediated) of lipid-soluble molecules through the phospholipid layers of the plasma membrane

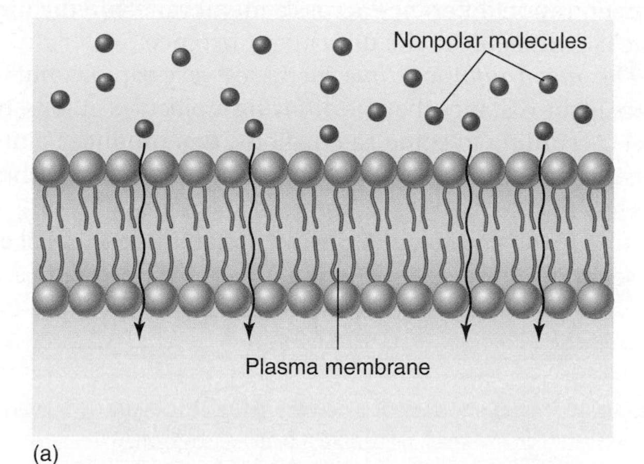

(a)

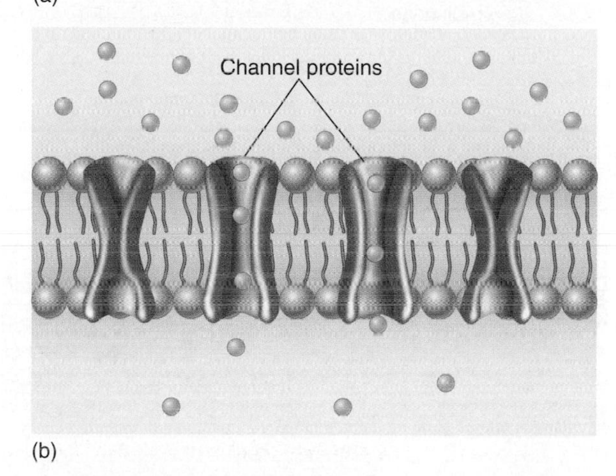

(b)

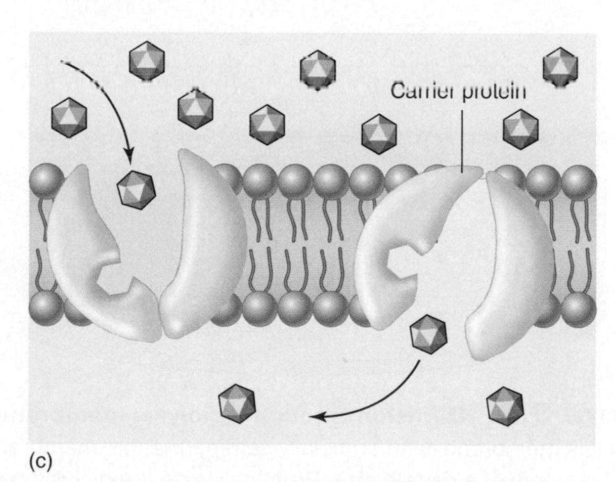

(c)

Figure 6.2 Three types of passive transport. (*a*) Nonpolar molecules can move by simple diffusion through the double phospholipid layers of the plasma membrane. (*b*) Inorganic ions and water molecules can move by simple diffusion through protein channels in the plasma membrane. (*c*) Small organic molecules, such as glucose, can move by facilitated diffusion through the plasma membrane using carrier proteins. AP|R

  b. Simple diffusion of ions through membrane channel proteins in the plasma membrane

  c. Simple diffusion of water molecules (osmosis) through aquaporin (water) channels in the plasma membrane

Membrane transport processes may also be categorized by their energy requirements. **Passive transport** is the net movement of molecules and ions across a membrane from higher to lower concentration (down a concentration gradient); it does not require metabolic energy. Passive transport includes all of the non-carrier-mediated diffusion processes (the simple diffusion of lipid-soluble molecules, ions, and water), plus the carrier-mediated facilitated diffusion (fig. 6.2). **Active transport** is the net movement of molecules and ions across a membrane from the region of lower to the region of higher concentrations. Because active transport occurs against the concentration gradient, it requires the expenditure of metabolic energy (ATP) that powers specific carrier proteins, which are often called **pumps**.

CHECKPOINT

1a. Describe the distribution of fluid in the body.

1b. Describe the composition of the extracellular matrix and explain the importance of the matrix metalloproteinases.

2. List the subcategories of passive transport and distinguish between passive transport and active transport.

6.2 DIFFUSION AND OSMOSIS

Net diffusion of a molecule or ion through a membrane always occurs in the direction of its lower concentration. Nonpolar molecules can penetrate the phospholipid barrier of the plasma membrane, and small inorganic ions can pass through protein channels in the plasma membrane. The net diffusion of water through a membrane is known as osmosis.

LEARNING OUTCOMES

After studying this section, you should be able to:

3. Define diffusion and describe the factors that influence the rate of diffusion.

4. Define osmosis, describe the conditions required for it to occur, and explain how osmosis relates to osmolality and osmotic pressure.

5. Explain the nature and significance of hypotonic, isotonic, and hypertonic solutions.

6. Explain how homeostasis of plasma osmolality is maintained.

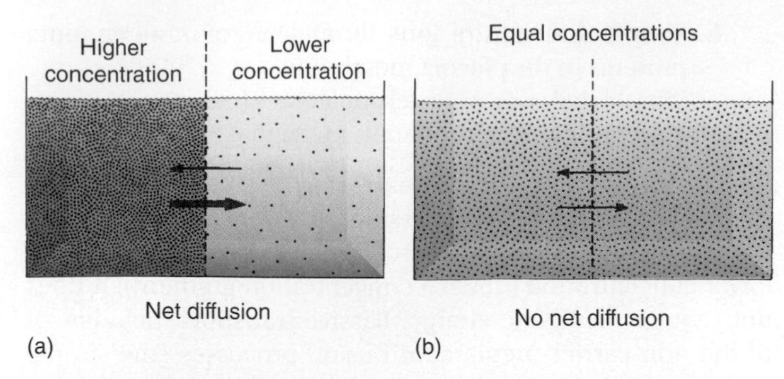

Figure 6.3 **Diffusion of a solute.** (*a*) Net diffusion occurs when there is a concentration difference (or concentration gradient) between two regions of a solution, provided that the membrane separating these regions is permeable to the diffusing substance. (*b*) Diffusion tends to equalize the concentrations of these regions, and thus to eliminate the concentration differences. AP|R

A *solution* consists of the *solvent,* water, and *solute* molecules that are dissolved in the water. The molecules of a solution (solvent and solute) are in a constant state of random motion as a result of their thermal (heat) energy. If there is a *concentration difference,* or *concentration gradient,* between two regions of a solution, this random motion tends to eliminate the concentration difference as the molecules become more diffusely spread out (fig. 6.3). Hence, this random molecular motion is known as **diffusion.** In terms of the second law of thermodynamics, the concentration difference represents an unstable state of high organization (low entropy) that changes to produce a uniformly distributed solution with maximum disorganization (high entropy).

As a result of random molecular motion, molecules in the part of the solution with a higher concentration will enter the area of lower concentration. Molecules will also move in the opposite direction, but not as frequently. As a result, there will be a *net movement* from the region of higher to the region of lower concentration until the concentration difference no longer exists. This net movement is called **net diffusion.** Net diffusion is a physical process that occurs whenever there is a concentration difference across a membrane and the membrane is permeable to the diffusing substance.

The *mean diffusion time* increases very rapidly with the square of the distance that the diffusing molecules or ions must travel. According to some calculations, this produces a mean diffusion time of (a) 10^{-7} sec. to cross a plasma membrane (10 nm); (b) 1.6×10^{-6} sec. to cross a synapse (40 nm); and (c) 1 to 2×10^{-3} sec. to cross the two squamous epithelial cells that separate air from blood in the lungs (1–2 μm). Notice that

CLINICAL APPLICATION

In the kidneys, blood is filtered through pores in capillary walls to produce a filtrate that will become urine. Wastes and other dissolved molecules can pass through the pores, but blood cells and proteins are held back. Then, the molecules needed by the body are reabsorbed from the filtrate back into the blood by transport processes. Wastes generally remain in the filtrate and are thus excreted in the urine. When the kidneys fail to perform this function, the wastes must be removed from the blood artificially by means of **dialysis** (fig. 6.4). In this process, waste molecules are removed from the blood by having them diffuse through an artificial porous membrane. The wastes pass into a solution (called a *dialysate*) surrounding the dialysis membrane. Molecules needed by the body, however, are kept in the blood by including them in the dialysate. This prevents their net diffusion by abolishing their concentration gradients.

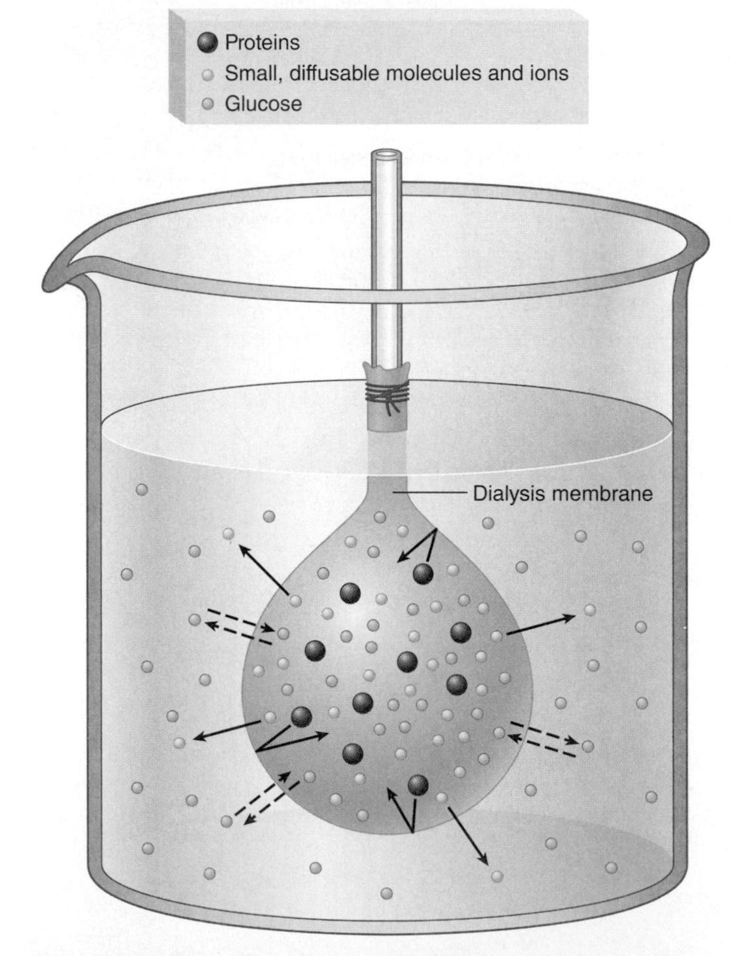

Figure 6.4 **Diffusion through a dialysis membrane.** A dialysis membrane is an artificially semipermeable membrane with tiny pores of a certain size. Proteins inside the dialysis bag are too large to get through the pores (bent arrows), but the small, diffusible molecules and ions are able to fit through the pores and diffuse (solid, straight arrows) from higher to lower concentration out of the bag and into the surrounding fluid. Glucose can also fit through the pores, but because it is present at the same concentration outside of the bag, there is no net diffusion (double dashed arrows). AP|R

diffusion occurs quickly across such small distances. However, with distances much beyond about 100 μm, the mean diffusion time becomes too long for effective exchange of molecules and ions by diffusion, which is why cells in most body organs are within 100 μm of a blood capillary, and why neurons have special transport mechanisms to move molecules along axons (which can be as long as a meter in length).

Diffusion Through the Plasma Membrane

Because the plasma (cell) membrane consists primarily of a double layer of phospholipids, molecules that are nonpolar, and thus lipid-soluble, can easily pass from one side of the membrane to the other. The plasma membrane, in other words, does not present a barrier to the diffusion of nonpolar molecules such as oxygen gas (O_2) or steroid hormones. Small molecules that have polar covalent bonds, but which are uncharged, such as CO_2 (as well as ethanol and urea), are also able to penetrate the phospholipid bilayer. Net diffusion of these molecules can thus easily occur between the intracellular and extracellular compartments when concentration gradients exist.

The oxygen concentration is relatively high, for example, in the extracellular fluid because oxygen is carried from the lungs to the body tissues by the blood. Because oxygen is combined with hydrogen to form water in aerobic cell respiration, the oxygen concentration within the cells is lower than in the extracellular fluid. The concentration gradient for carbon dioxide is in the opposite direction because cells produce CO_2. *Gas exchange* thus occurs by diffusion between the cells and their extracellular environments (fig. 6.5). Gas exchange by diffusion also occurs in the lungs (chapter 16), where the concentration gradient for oxygen produces net diffusion from air to blood, and the concentration gradient for carbon dioxide produces net diffusion from blood to air. In all cases, the direction of net diffusion is from higher to lower concentration.

Although water is not lipid-soluble, water molecules can diffuse through the plasma membrane to a limited degree because of their small size and lack of net charge. In most membranes, however, the passage of water is greatly aided by specific water channels (called *aquaporins*) that are inserted into the membrane in response to physiological regulation. This is the case in the kidneys, where aquaporins aid water retention by promoting the net diffusion of water out of microscopic tubules into the blood (chapter 17). The net diffusion of water molecules (the solvent) across the membrane is known as *osmosis.* Because osmosis is the simple diffusion of solvent instead of solute, a unique terminology (discussed shortly) is used to describe it.

Larger polar molecules, such as glucose, cannot pass through the double layer of phospholipid molecules and thus require special *carrier proteins* in the membrane for transport. Carrier proteins will be discussed separately in section 6.3. The phospholipid portion of the membrane is similarly impermeable to charged inorganic ions, such as Na^+ and K^+. However, tiny **ion channels** through the membrane permit passage of these ions. The ion channels are provided by some of the proteins that span the thickness of the membrane (fig. 6.6).

Some ion channels are always open, so that diffusion of the ion through the plasma membrane is an ongoing process. Many ion channels, however, are **gated**—they have structures

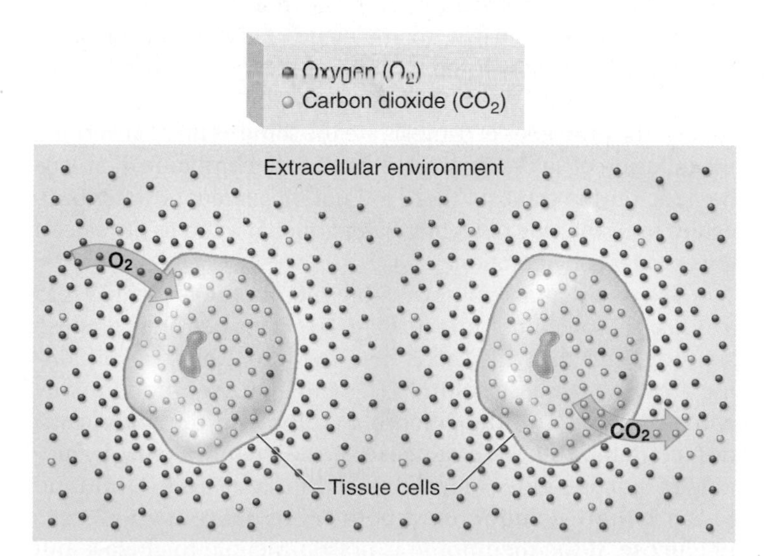

Figure 6.5 Gas exchange occurs by diffusion. The colored spheres, which represent oxygen and carbon dioxide molecules, indicate relative concentrations inside the cell and in the extracellular environment. Gas exchange between the intracellular and extracellular compartments thus occurs by diffusion. AP|R

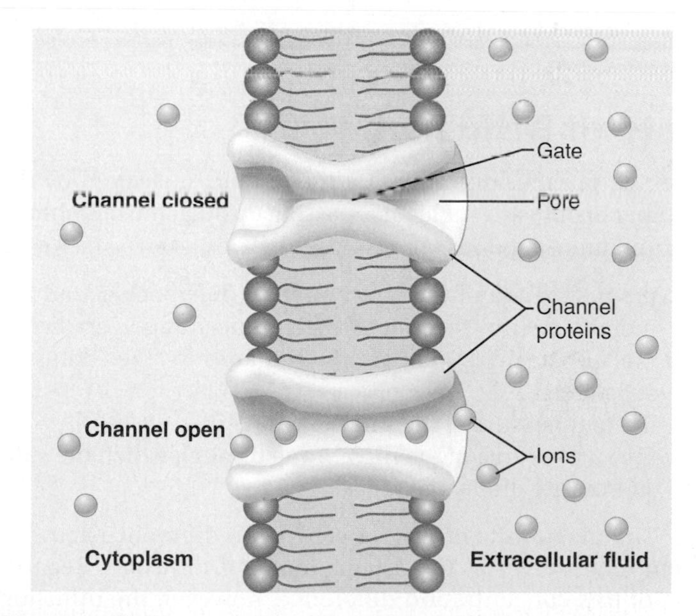

Figure 6.6 Ions pass through membrane channels. These channels are composed of integral proteins that span the thickness of the membrane. Although some channels are always open, many others have structures known as "gates" that can open or close the channel. This figure depicts a generalized ion channel; most, however, are relatively selective—they allow only particular ions to pass. AP|R

("gates") that can open or close the channel (fig. 6.6). In this way, particular physiological stimuli (such as binding of the channel to a specific chemical regulator) can open an otherwise closed channel. In the production of nerve and muscle impulses, specific channels for Na^+ and others for K^+ open and close in response to changes in membrane voltage (discussed in chapter 7, section 7.2).

CLINICAL APPLICATION

Cystic fibrosis occurs about once in every 2,500 births in the Caucasian population. As a result of a genetic defect, abnormal NaCl and water movement occurs across wet epithelial membranes. Where such membranes line the pancreatic ductules and small respiratory airways, they produce a dense, viscous mucus that cannot be properly cleared, which may lead to pancreatic and pulmonary disorders. The genetic defect involves a particular glycoprotein that forms chloride (Cl^-) channels in the apical membrane of the epithelial cells. This protein, known as *CFTR* (for cystic fibrosis transmembrane conductance regulator), is formed in the usual manner in the endoplasmic reticulum. It does not move into the Golgi complex for processing, however, and therefore, it doesn't get correctly processed and inserted into vesicles that would introduce it into the plasma membrane (chapter 3). The most prevalent gene mutation that affects CFTR function has been identified and cloned, and therapies based on this information—such as one that increases the time that CFTR channels remain open—are beginning to be developed.

Rate of Diffusion

The rate of diffusion through a membrane, measured by the number of diffusing molecules passing through the membrane per unit time, depends on

1. the magnitude of the concentration difference across the membrane (the "steepness" of the concentration gradient),
2. the permeability of the membrane to the diffusing substances,
3. the temperature of the solution, and
4. the surface area of the membrane through which the substances are diffusing.

The magnitude of the concentration difference across a membrane serves as the driving force for diffusion. Regardless of this concentration difference, however, the diffusion of a substance across a membrane will not occur if the membrane is not permeable to that substance. With a given concentration difference, the speed at which a substance diffuses through a membrane will depend on how permeable the membrane is to it. In a resting neuron, for example, the plasma membrane is about 20 times more permeable to potassium (K^+) than to sodium (Na^+); consequently, K^+ diffuses much more rapidly than does Na^+. Changes in the protein structure of the membrane channels, however, can change the permeability of the membrane. This occurs during the production of a nerve impulse (chapter 7, section 7.2), when specific stimulation opens Na^+ channels temporarily and allows a faster diffusion rate for Na^+ than for K^+.

In areas of the body that are specialized for rapid diffusion, the surface area of the plasma membranes may be increased by numerous folds. The rapid passage of the products of digestion across the epithelial membranes in the small intestine, for example, is aided by tiny fingerlike projections called *microvilli* (chapter 3, section 3.1). Similar microvilli are found in the kidney tubule epithelium, which must reabsorb various molecules that are filtered out of the blood.

Osmosis

Osmosis is the net diffusion of water (the solvent) across the membrane. For osmosis to occur, the membrane must be *selectively permeable;* that is, it must be more permeable to water molecules than to at least one species of solute. There are thus two requirements for osmosis: (1) there must be a difference in the concentration of a solute on the two sides of a selectively permeable membrane; and (2) the membrane must be relatively impermeable to the solute. Solutes that cannot freely pass through the membrane can promote the osmotic movement of water and are said to be **osmotically active.**

Like the diffusion of solute molecules, the diffusion of water occurs when the water is more concentrated on one side of the membrane than on the other side; that is, when one solution is more dilute than the other (fig. 6.7). The more dilute solution has a higher concentration of water molecules and a lower concentration of solute. Although the terminology associated with osmosis can be awkward (because we are describing water instead of solute), the principles of osmosis are the same as those governing the diffusion of solute molecules through a membrane. Remember that, during osmosis, there is a net movement of water molecules from the side of higher water concentration to the side of lower water concentration.

Imagine that a semipermeable membrane is formed into a spherical sac containing a 360 g/L (grams per liter) glucose solution, and that this sac is inserted into a beaker containing a 180 g/L glucose solution (fig. 6.8). One solution initially contains 180 g/L of the glucose solution and the other solution contains 360 g/L of glucose. If the membrane is permeable to glucose, glucose will diffuse from the 360 g/L solution to the 180 g/L solution until both solutions contain 270 g/L of glucose. If the membrane is not permeable to glucose but is permeable to water, the same result (270 g/L solutions on both sides of the membrane) will be achieved by the diffusion of water. As water diffuses from the 180 g/L solution to

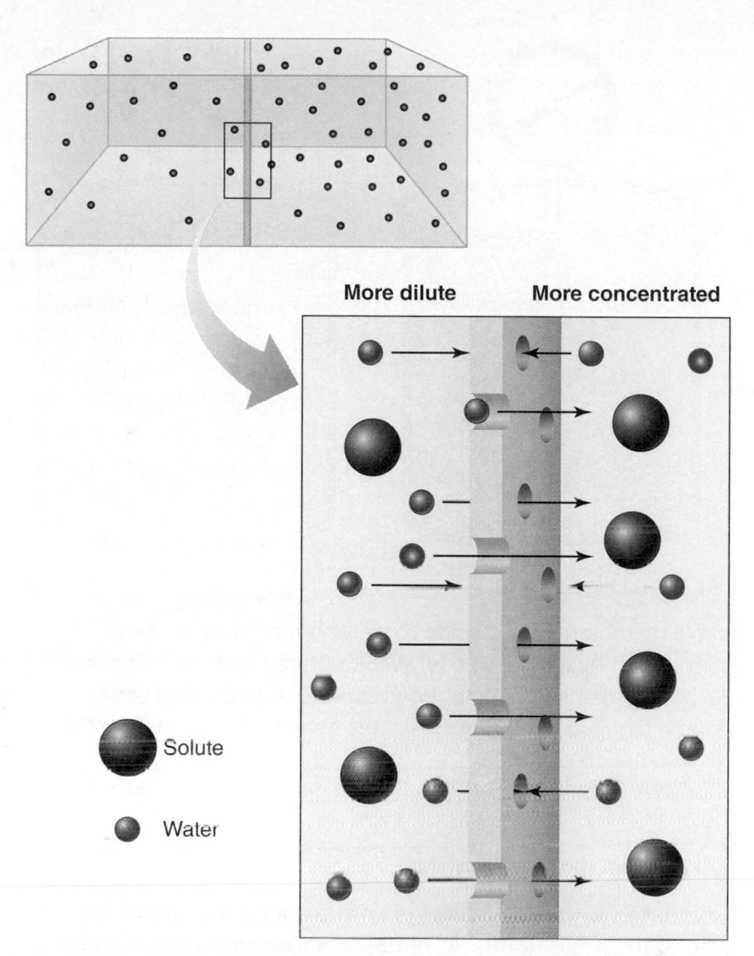

Figure 6.7 A model of osmosis. The diagram illustrates the net movement of water from the solution of lesser solute concentration (higher water concentration) to the solution of greater solute concentration (lower water concentration). APIR

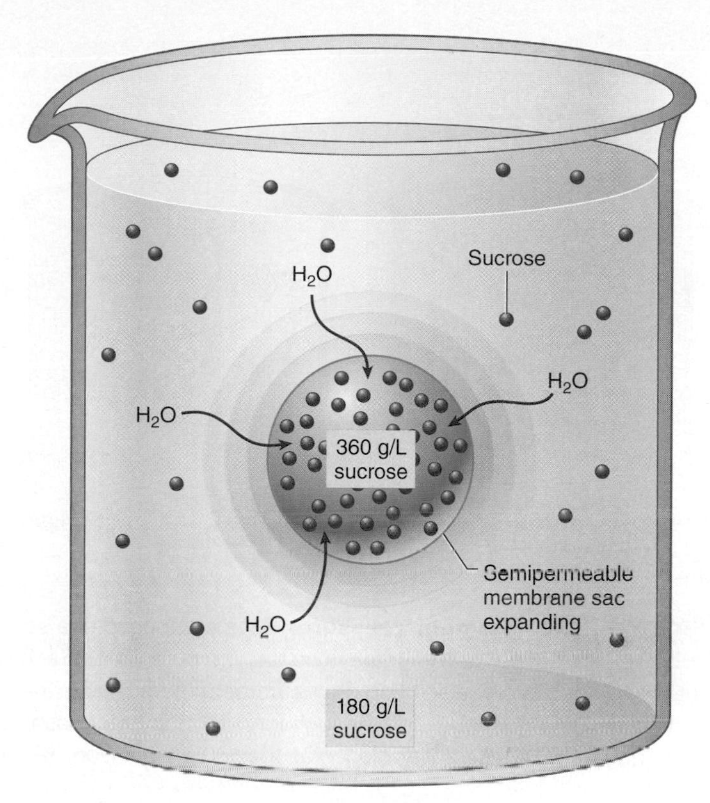

Figure 6.8 The effects of osmosis. A membranous sac composed of a semipermeable membrane that is permeable to water but not to the solute (sucrose) is immersed in a beaker. The solution in the sac contains twice the solute concentration as the solution surrounding it in the beaker. Because sucrose cannot diffuse through the membrane, water moves by osmosis into the sac. If the bag is able to expand without resistance, it will continue to take in water until both solutions have the same concentration (270 g/L sucrose). APIR

the 360 g/L solution (from the higher to the lower water concentration), the former solution becomes more concentrated while the latter becomes more dilute. This is accompanied by volume changes (assuming the sac can expand freely), as illustrated in figure 6.8. The net movement of water (osmosis) ceases when the concentrations become equal on both sides of the membrane.

Plasma membranes are permeable to water and so behave in a similar manner. Specific proteins present in the plasma membranes serve as water channels, known as **aquaporins,** which permit osmosis. In some cells, the plasma membrane always has aquaporin channels; in others, the aquaporin channels are inserted into the plasma membrane in response to regulatory molecules. Such regulation is especially important in the functioning of the kidneys (chapter 17, section 17.3), which are the major organs regulating total body water balance. Other organs notable for aquaporin channels in the plasma membrane of particular cells include the lungs, eyes, salivary glands, and brain.

Osmotic Pressure

Osmosis could be prevented by an opposing force. Imagine two beakers of pure water, each with a semipermeable membrane sac; one sac contains a 180 g/L glucose solution, the other a 360 g/L glucose solution. Each sac is surrounded by a rigid box (fig. 6.9*a*). As water enters each sac by osmosis, the sac expands until it presses against the surrounding box. As each sac presses tightly against the box, the box exerts a pressure against the sac that can prevent the further osmosis of water into the sac (fig. 6.9*b*). The pressure needed to just stop osmosis is the **osmotic pressure** of the solution. Plant cells have such rigid boxes, cell walls composed of cellulose, around them; animal (including human) cells lack cell walls, and so animal cells would burst if placed in pure water.

Because osmotic pressure is a measure of the force required to stop osmosis, it indicates how strongly a solution "draws" water by osmosis. Thus, the greater the solute concentration of a solution, the greater its osmotic pressure.

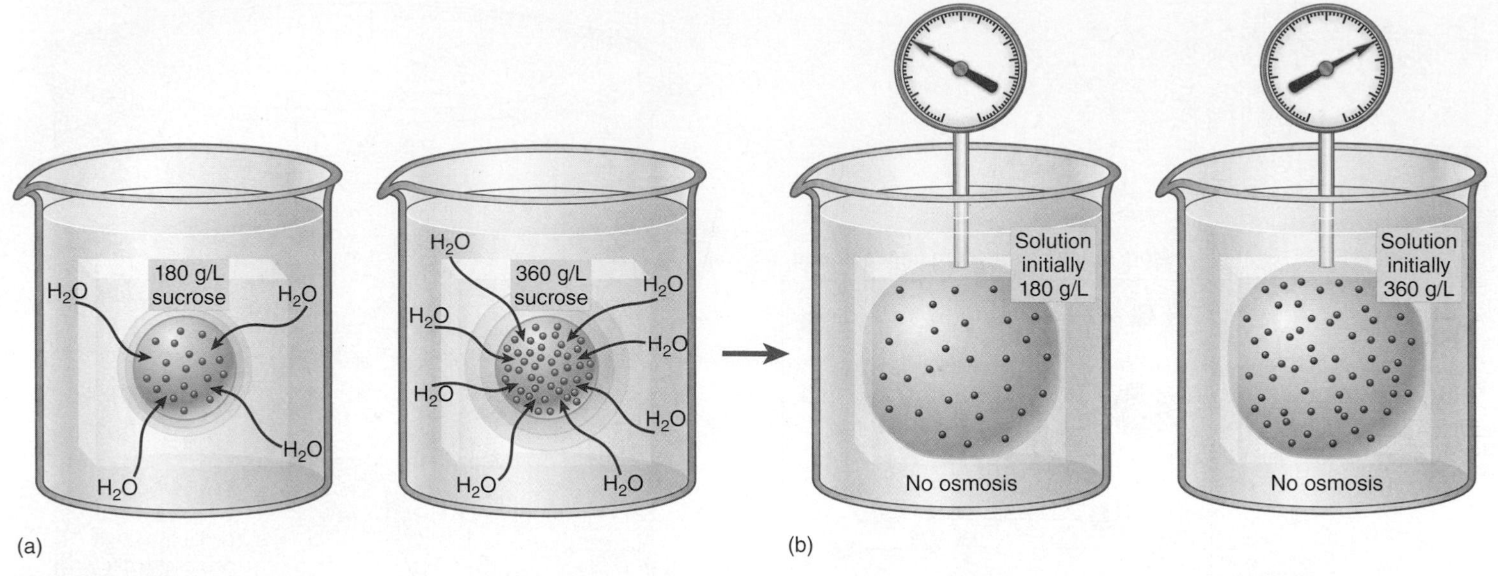

Figure 6.9 Osmotic pressure. Sacs composed of a semipermeable membrane, permeable to water but not to the solute (sucrose), are suspended in beakers containing pure water. Each sac is surrounded by a rigid box. (*a*) Water enters each sac by osmosis, but the 360 g/L sucrose solution draws water in more rapidly than the 180 g/L sucrose solution. (*b*) Each sac expands until it presses against its surrounding box with enough force to stop further osmosis. The force required to stop osmosis, the osmotic pressure, is twice as great for the 360 g/L sucrose solution as the 180 g/L solution. AP|R

Pure water has an osmotic pressure of zero, and a 360 g/L glucose solution has twice the osmotic pressure of a 180 g/L glucose solution (fig. 6.9*b*).

CLINICAL APPLICATION

Water returns from tissue fluid to blood capillaries because the protein concentration of blood plasma is higher than the protein concentration of tissue fluid. Plasma proteins, in contrast to other plasma solutes, cannot freely pass from the capillaries into the tissue fluid. Therefore, plasma proteins are *osmotically active.* If a person has an abnormally low concentration of plasma proteins, excessive accumulation of fluid in the tissues—a condition called **edema**—will result. This may occur, for example, when a damaged liver (as in *cirrhosis*) is unable to produce sufficient amounts of albumin, the major protein in the blood plasma.

Clinical Investigation CLUES

Urine does not normally contain glucose, yet Jessica has glycosuria (glucose in the urine).

- What would the presence of an extra solute, glucose, have on the osmotic pressure of the urine?
- How might this cause more water to be excreted in the urine, leading to frequent urination?

Molarity and Molality

Glucose is a monosaccharide with a molecular weight of 180 (the sum of its atomic weights). Sucrose is a disaccharide of glucose and fructose, which have molecular weights of 180 each. When glucose and fructose join together by dehydration synthesis to form sucrose, a molecule of water (molecular weight = 18) is split off. Therefore, sucrose has a molecular weight of 342 (180 + 180 − 18). Because the molecular weights of sucrose and glucose are in a ratio of 342/180, it follows that 342 grams of sucrose must contain the same number of molecules as 180 grams of glucose.

Notice that an amount of any compound equal to its molecular weight in grams must contain the same number of molecules as an amount of any other compound equal to its molecular weight in grams. This unit of weight, a *mole,* always contains 6.02×10^{23} molecules **(Avogadro's number).** One mole of solute dissolved in water to make 1 liter of solution is described as a **one-molar solution** (abbreviated 1.0 *M*). Although this unit of measurement is commonly used in chemistry, it is not completely desirable in discussions of osmosis because the exact ratio of solute to water is not specified. For example, more water is needed to make a 1.0 *M* NaCl solution (where a mole of NaCl weighs 58.5 grams) than is needed to make a 1.0 *M* glucose solution, since 180 grams of glucose takes up more volume than 58.5 grams of salt.

Because the ratio of solute to water molecules is of critical importance in osmosis, a more desirable measurement of concentration is **molality.** In a 1-molal solution (abbreviated

1.0 *m*), 1 mole of solute (180 grams of glucose, for example) is dissolved in 1 kilogram of water (equal to 1 liter at 4° C). A 1.0 *m* NaCl solution and a 1.0 *m* glucose solution therefore both contain a mole of solute dissolved in exactly the same amount of water (fig. 6.10).

Osmolality

If 180 grams of glucose and 180 grams of fructose were dissolved in the same kilogram of water, the osmotic pressure of the solution would be the same as that of a 360 g/L glucose solution. Osmotic pressure depends on the ratio of solute to solvent, *not* on the chemical nature of the solute molecules. The expression for the total molality of a solution is **osmolality**

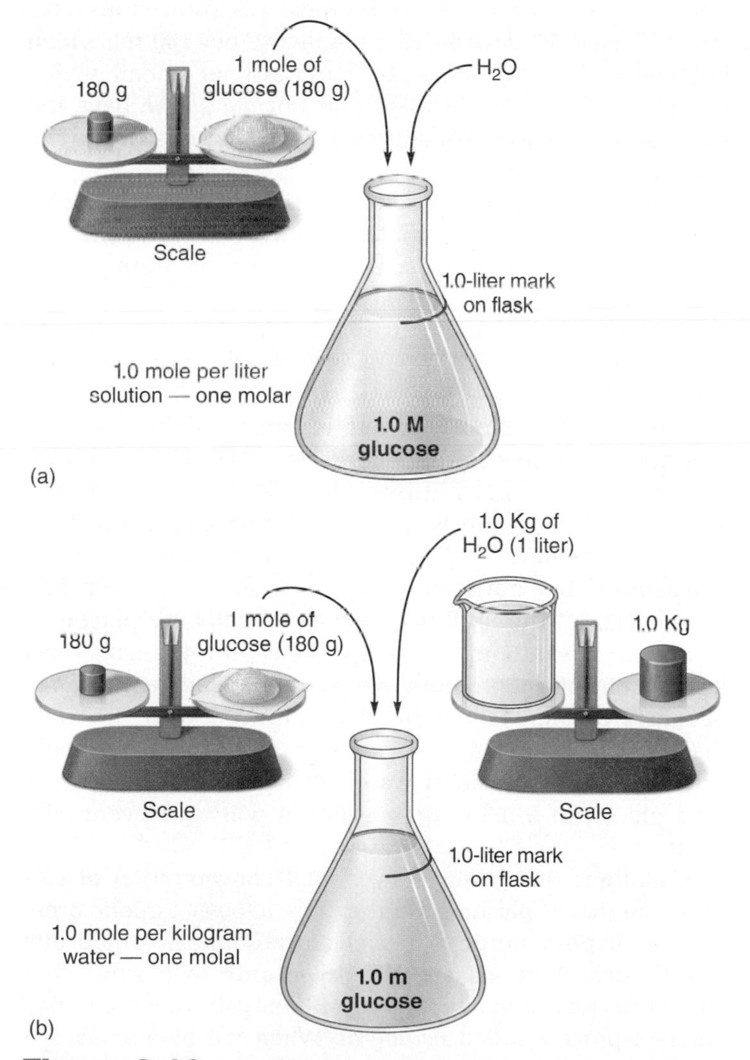

Figure 6.10 Molar and molal solutions. The diagrams illustrate the difference between (*a*) a one-molar (1.0 *M*) and (*b*) a one-molal (1.0 *m*) glucose solution.

See the *Test Your Quantitative Ability* section of the Review Activities at the end of this chapter.

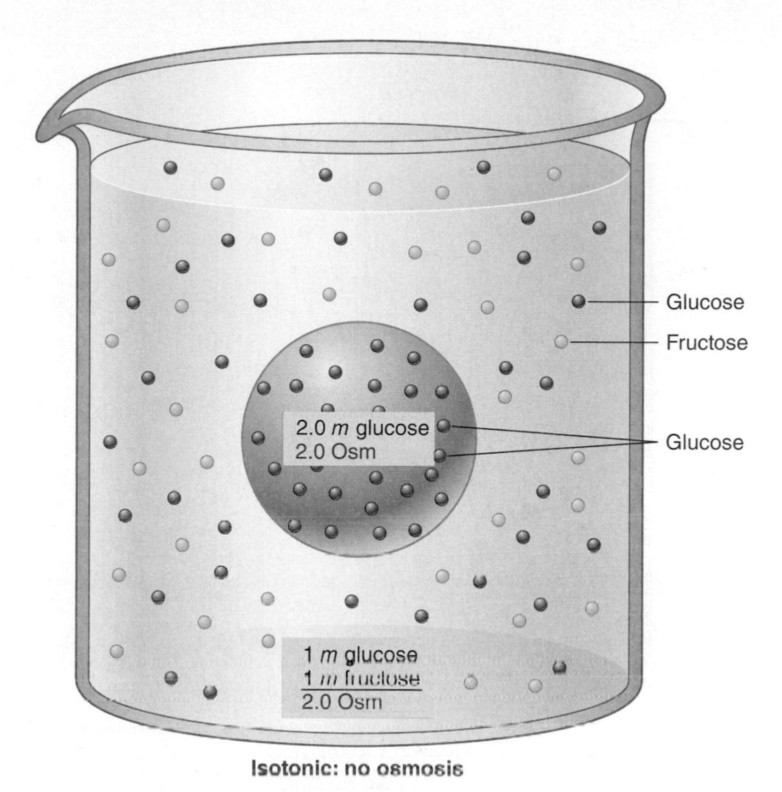

Figure 6.11 The osmolality of a solution. The osmolality (Osm) is equal to the sum of the molalities of each solute in the solution. If a selectively permeable membrane separates two solutions with equal osmolalities, no osmosis will occur. AP|R

(Osm). Thus, the solution of 1.0 *m* glucose plus 1.0 *m* fructose has a total molality, or *osmolality,* of 2.0 osmol/L (abbreviated 2.0 Osm). This osmolality is the same as that of the 360 g/L glucose solution, which has a concentration of 2.0 *m* and 2.0 Osm (fig. 6.11).

Unlike glucose, fructose, and sucrose, electrolytes such as NaCl ionize when they dissolve in water. One molecule of NaCl dissolved in water yields two ions (Na^+ and Cl^-); 1 mole of NaCl ionizes to form 1 mole of Na^+ and 1 mole of Cl^-. Thus, a 1.0 *m* NaCl solution has a total concentration of 2.0 Osm. The effect of this ionization on osmosis is illustrated in fig. 6.12.

Measurement of Osmolality

Plasma and other biological fluids contain many organic molecules and electrolytes. The osmolality of such complex solutions can only be estimated by calculations. Fortunately, however, there is a relatively simple method for measuring osmolality. This method is based on the fact that the freezing point of a solution, like its osmotic pressure, is affected by the total concentration of the solution and not by the chemical nature of the solute.

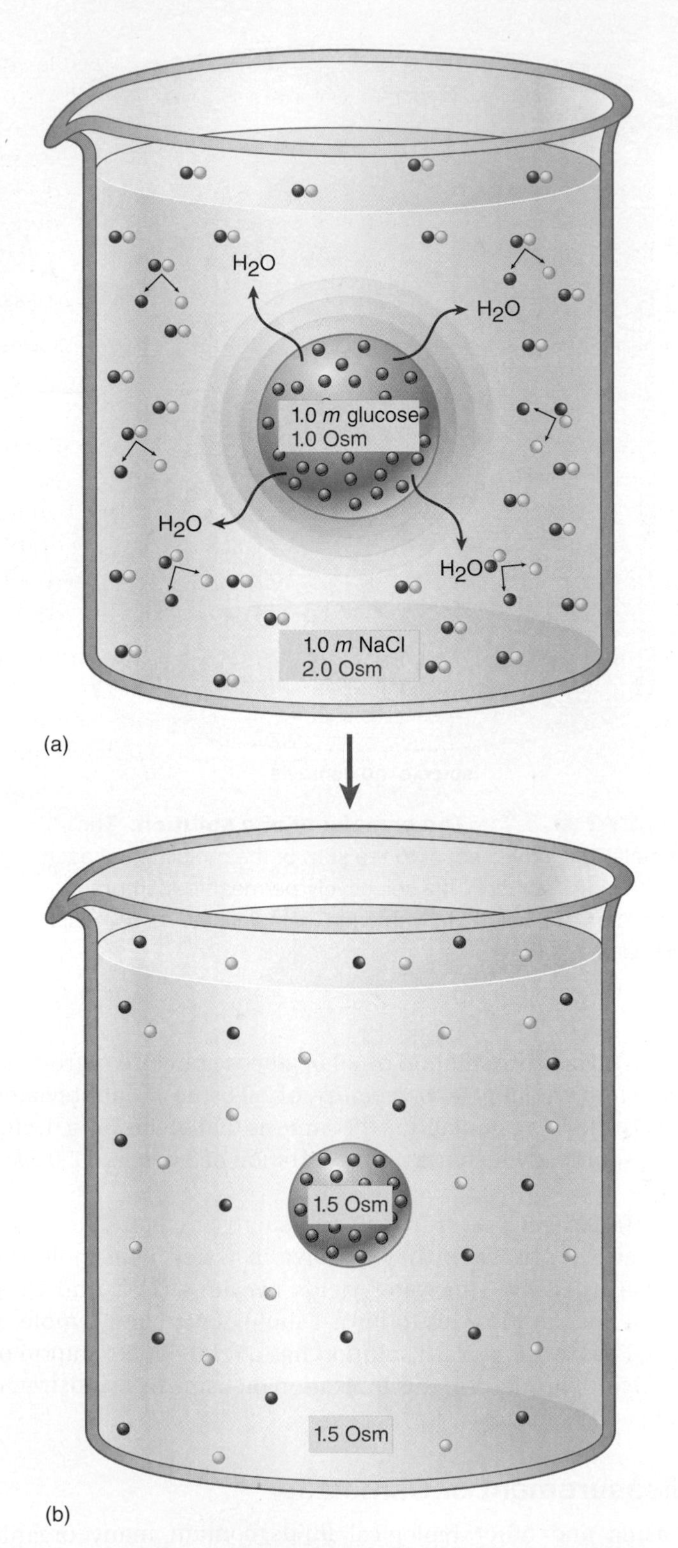

Figure 6.12 The effect of ionization on the osmotic pressure. (*a*) If a selectively permeable membrane (permeable to water but not to glucose, Na^+, or Cl^-) separates a 1.0 *m* glucose solution from a 1.0 *m* NaCl solution, water will move by osmosis into the NaCl solution. This is because a 1.0 *m* NaCl solution has a total solute concentration of 2.0 Osm, since NaCl can ionize to yield one-molal Na^+ plus one-molal Cl^-. (*b*) After osmosis, the total concentration, or osmolality, of the two solutions is equal. AP|R

One mole of solute per liter depresses the freezing point of water by $-1.86°$ C. Accordingly, a 1.0 *m* glucose solution freezes at a temperature of $-1.86°$ C, and a 1.0 *m* NaCl solution freezes at a temperature of $2 \times (-1.86) = -3.72°$ C because of ionization. Thus, the *freezing-point depression* is a measure of the osmolality. Since plasma freezes at about $-0.56°$ C, its osmolality is equal to $0.56 \div 1.86 = 0.3$ Osm, which is more commonly indicated as 300 milliosmolal (or 300 mOsm).

Tonicity

A 0.3 *m* glucose solution, which is 0.3 Osm, or 300 milliosmolal (300 mOsm), has the same osmolality and osmotic pressure as plasma. The same is true of a 0.15 *m* NaCl solution, which ionizes to produce a total concentration of 300 mOsm. Both of these solutions are used clinically as intravenous infusions, labeled *5% dextrose* (5 g of glucose per 100 ml, which is 0.3 *m*) and *normal saline* (0.9 g of NaCl per 100 ml, which is 0.15 *m*). Since 5% dextrose and normal saline have the same osmolality as plasma, they are said to be **isosmotic** to plasma.

The term **tonicity** is used to describe the effect of a solution on the osmotic movement of water. For example, if an isosmotic glucose or saline solution is separated from plasma by a membrane that is permeable to water, but not to glucose or NaCl, osmosis will not occur. In this case, the solution is said to be **isotonic** (from the Greek *isos* = equal; *tonos* = tension) to plasma.

Red blood cells placed in an isotonic solution will neither gain nor lose water. It should be noted that a solution may be isosmotic but not isotonic; such is the case whenever the solute in the isosmotic solution can freely penetrate the membrane. A 0.3 *m* urea solution, for example, is isosmotic but not isotonic because the cell membrane is permeable to urea. When red blood cells are placed in a 0.3 *m* urea solution, the urea diffuses into the cells until its concentration on both sides of the cell membranes becomes equal. Meanwhile, the solutes within the cells that cannot exit—and which are therefore osmotically active—cause osmosis of water into the cells. Red blood cells placed in a 0.3 *m* urea solution will thus eventually burst.

Solutions that have a lower total concentration of solutes than that of plasma, and therefore a lower osmotic pressure, are **hypo-osmotic** to plasma. If the solute is osmotically active, such solutions are also **hypotonic** to plasma. Red blood cells placed in hypotonic solutions gain water and may burst—a process called *hemolysis.* When red blood cells are placed in a **hypertonic** solution (such as seawater), which contains osmotically active solutes at a higher osmolality and osmotic pressure than plasma, they shrink because of the osmosis of water out of the cells. This process is called *crenation* (from the Medieval Latin *crena* = notch) because the cell surface takes on a scalloped appearance (fig. 6.13).

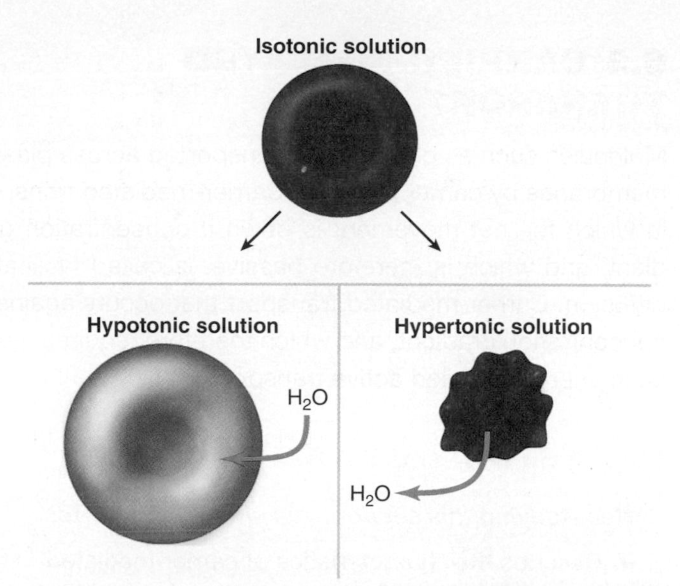

Figure 6.13 Red blood cells in isotonic, hypotonic, and hypertonic solutions. In each case, the external solution has an equal, lower, or higher osmotic pressure, respectively, than the intracellular fluid. As a result, water moves by osmosis into the red blood cells placed in hypotonic solutions, causing them to swell and even to burst. Similarly, water moves out of red blood cells placed in a hypertonic solution, causing them to shrink and become crenated.

CLINICAL APPLICATION

Intravenous fluids must be isotonic to blood in order to maintain the correct osmotic pressure and prevent cells from either expanding or shrinking from the gain or loss of water. Common fluids used for this purpose are *normal saline* and *5% dextrose,* which, as previously described, have about the same osmolality as normal plasma (approximately 300 mOsm). Another isotonic solution frequently used in hospitals is *Ringer's lactate.* This solution contains glucose and lactic acid in addition to a number of different salts. In contrast to isotonic solutions, *hypertonic solutions* of *mannitol* (an osmotically active solute) are given intravenously to promote osmosis and thereby reduce the swelling in **cerebral edema,** a significant cause of mortality in people with brain trauma or stroke.

Regulation of Blood Osmolality

A relatively constant osmolality of extracellular fluid must be maintained. This is mostly because neurons could be damaged by swelling or shrinkage of the brain within the skull, and because neural activity is altered by changes in the concentrations of ions (chapter 7). A variety of mechanisms defend the homeostasis of plasma osmolality, usually preventing it from changing by more than 1% to 3%. For example, dehydration due to strenuous exercise can increase plasma osmolality by 10 mOsm or more; ingestion of salt likewise increases plasma osmolality, whereas plasma osmolality is lowered by drinking water.

When a person becomes dehydrated, the blood becomes more concentrated and the total blood volume is reduced. The increased plasma osmolality and osmotic pressure stimulate **osmoreceptors,** which are neurons mainly located in a part of the brain called the hypothalamus (chapter 8, section 8.3). During dehydration, water leaves the osmoreceptor neurons because of the increased osmolality of the extracellular fluid. This causes the osmoreceptors to shrink, which mechanically stimulates them to increase their production of nerve impulses.

As a result of increased osmoreceptor stimulation, the person becomes thirsty and, if water is available, drinks. Along with increased water intake, a person who is dehydrated excretes a lower volume of urine. This occurs as a result of the following sequence of events (fig. 6.14):

1. Increased plasma osmolality stimulates osmoreceptors in the hypothalamus of the brain.
2. The osmoreceptors in the hypothalamus then stimulate a tract of axons that terminate in the posterior pituitary;

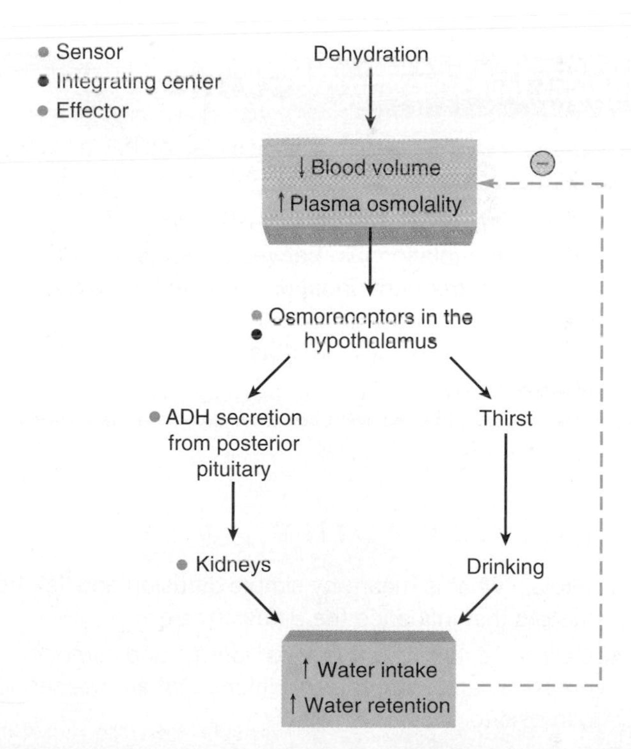

Figure 6.14 Homeostasis of plasma concentration. An increase in plasma osmolality (increased concentration and osmotic pressure) due to dehydration stimulates thirst and increased ADH secretion. These effects cause the person to drink more and urinate less. The blood volume, as a result, is increased while the plasma osmolality is decreased. These effects help to bring the blood volume back to the normal range and complete the negative feedback loop (indicated by a negative sign).

this causes the posterior pituitary to release **antidiuretic hormone (ADH),** also known as **vasopressin,** into the blood.

3. ADH acts on the kidneys to promote water retention, so that a lower volume of more concentrated urine is excreted.

A person who is dehydrated, therefore, drinks more and urinates less. This represents a negative feedback loop (fig. 6.14), which acts to maintain homeostasis of the plasma concentration (osmolality) and, in the process, helps to maintain a proper blood volume.

A person with a normal blood volume who eats salty food will also get thirsty, and more ADH will be released from the posterior pituitary. By drinking more and excreting less water in the urine, the salt from the food will become diluted to restore the normal blood concentration, but at a higher blood volume. The opposite occurs in salt deprivation. With a lower plasma osmolality, the osmoreceptors are not stimulated as much, and the posterior pituitary releases less ADH. Consequently, more water is excreted in the urine to again restore the proper range of plasma concentration, but at a lower blood volume. Low blood volume and pressure as a result of prolonged salt deprivation can be fatal (chapter 14, section 14.7).

Clinical Investigation CLUES

Laboratory tests reveal that Jessica's plasma has a higher than normal osmolality.

- What is the normal osmolality of plasma?
- What is the relationship between Jessica's glycosuria, frequent urination, and her high plasma osmolality?
- How does this relate to her statement that she is always thirsty?

CHECKPOINT

3. Explain what is meant by simple diffusion and list the factors that influence the diffusion rate.
4. Define the terms *osmosis, osmolality,* and *osmotic pressure,* and state the conditions that are needed for osmosis to occur.
5. Define the terms *isotonic, hypotonic,* and *hypertonic,* and explain why hospitals use 5% dextrose and normal saline as intravenous infusions.
6. Explain how the body detects changes in the osmolality of plasma and describe the regulatory mechanisms by which a proper range of plasma osmolality is maintained.

6.3 CARRIER-MEDIATED TRANSPORT

Molecules such as glucose are transported across plasma membranes by carrier proteins. Carrier-mediated transport in which the net movement is down a concentration gradient, and which is therefore passive, is called facilitated diffusion. Carrier-mediated transport that occurs against a concentration gradient, and which therefore requires metabolic energy, is called active transport.

LEARNING OUTCOMES

After studying this section, you should be able to:

7. Describe the characteristics of carrier-mediated transport, and distinguish between simple diffusion, facilitated diffusion, and active transport.
8. Explain the action and significance of the Ca^{2+} pump and the Na^+/K^+ pumps.

In order to sustain metabolism, cells must take up glucose, amino acids, and other organic molecules from the extracellular environment. Molecules such as these, however, are too large and polar to pass through the lipid barrier of the plasma membrane by a process of simple diffusion. The transport of such molecules is mediated by **carrier proteins** within the membrane. Although the action of carrier proteins cannot be directly observed, carrier-mediated transport can be inferred by characteristics it shares with enzyme activity. The common characteristics of enzymes and carrier proteins are (1) *specificity,* (2) *competition,* and (3) *saturation.*

Like enzyme proteins, carrier proteins interact only with specific molecules. Glucose carriers, for example, can interact only with glucose and not with closely related monosaccharides. As a further example of specificity, particular carriers for amino acids transport some types of amino acids but not others. Two amino acids that are transported by the same carrier compete with each other, so that the rate of transport for each is lower when they are present together than it would be if each were present alone (fig. 6.15).

As the concentration of a transported molecule is increased, its rate of transport will also be increased—but only up to a **transport maximum (T_m).** Concentrations greater than the transport maximum do not produce further increase in the transport rate, indicating that the carrier transport is saturated (fig. 6.15).

As an example of saturation, imagine a bus stop that is serviced once an hour by a bus that can hold a maximum of 40 people (its "transport maximum"). If there are 10 people waiting at the bus stop, 10 will be transported each hour. If 20 people are waiting, 20 will be transported each hour. This linear relationship will hold up to a maximum of 40 people; if there are 80 people at the bus stop, the transport rate will still be 40 per hour.

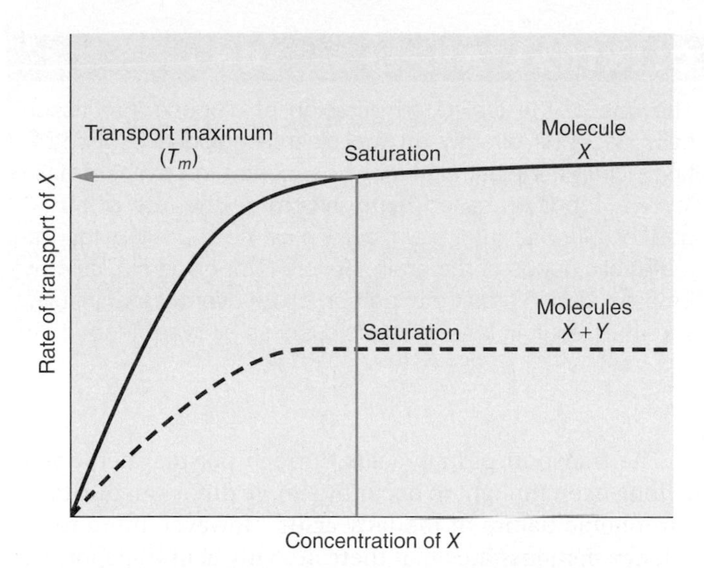

Figure 6.15 Characteristics of carrier-mediated transport. Carrier-mediated transport displays the characteristics of saturation (illustrated by the transport maximum) and competition. Since molecules X and Y compete for the same carrier, the rate of transport of each is lower when they are both present than when either is present alone. AP|R

CLINICAL APPLICATION

The microscopic tubules of the kidneys contain different transport carrier proteins that transport specific molecules from a blood filtrate (destined to become urine) within those tubules back into the blood. This transport is called *reabsorption.* Glucose, for example, is normally completely reabsorbed by carrier proteins so that urine is normally free of glucose. However, if the person has *hyperglycemia* (abnormally high plasma glucose concentration), the transport maximum of the glucose carriers can be exceeded. The untransported glucose in the filtrate continues along the tubules and is excreted in the urine, a condition called *glycosuria.* In the disease **diabetes mellitus,** glycosuria results from hyperglycemia caused by the inadequate secretion and/or action of the hormone insulin.

Facilitated Diffusion

The transport of glucose from the blood across plasma membranes occurs by **facilitated diffusion.** Facilitated diffusion, like simple diffusion, is powered by the thermal energy of the diffusing molecules and involves net transport from the side of higher to the side of lower concentration. ATP is not required for either facilitated or simple diffusion.

Unlike simple diffusion of nonpolar molecules, water, and inorganic ions through a membrane, the diffusion of glucose through the plasma membrane displays the properties of carrier-mediated transport: specificity, competition, and saturation. The diffusion of glucose through a plasma membrane must therefore

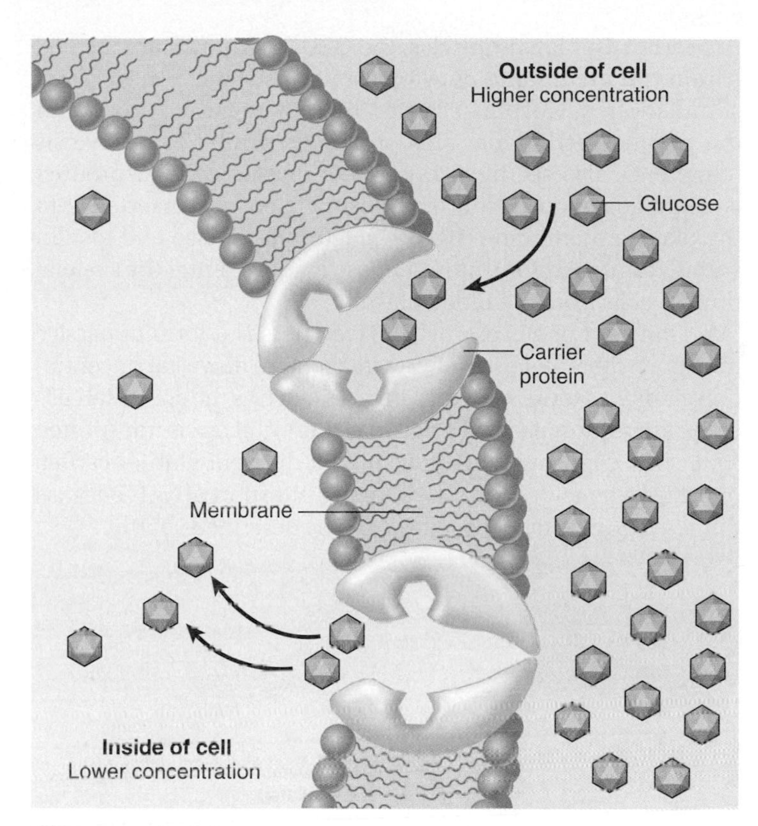

Figure 6.16 A model of the facilitated diffusion of glucose. A carrier—with characteristics of specificity and saturation—is required for this transport, which occurs from the blood into cells such as muscle, liver, and fat cells. This is passive transport because the net movement is to the region of lower concentrations, and ATP is not required. AP|R

be mediated by carrier proteins. In the conceptual model shown in figure 6.16, the carrier protein has a site that can bind specifically to glucose, and such binding causes a conformational change in the carrier so that a pathway is formed through the membrane. As a result, glucose is allowed to diffuse down its concentration gradient into the cell.

Like the isoenzymes described in chapter 4, carrier proteins that do the same job may exist in various tissues in slightly different forms. The transport carriers for the facilitative diffusion of glucose are designated with the letters **GLUT,** followed by a number for the isoform. For example, the *GLUT3* isoform is the major glucose transporter in neurons, but *GLUT1* is also present in the central nervous system and is increased under certain conditions. The pancreatic beta cells, which secrete insulin, and the hepatocytes of the liver produce *GLUT2.* The GLUT2 transporters allow an exceptionally high rate of glucose transport into these cells from the external environment. *GLUT4* is present in adipose tissue and skeletal muscles, and the insertion of GLUT4 carriers into the plasma membrane of adipocytes and skeletal muscle fibers is regulated by exercise and insulin. Because of this, GLUT4 is important for muscle physiology (chapter 12) and for glucose homeostasis in health and diabetes (chapter 19).

In unstimulated muscles, the GLUT4 proteins are located within the membrane of cytoplasmic vesicles. Exercise—and stimulation by insulin—causes these vesicles to fuse with the plasma membrane. This process is similar to exocytosis (chapter 3; also see fig. 6.23), except that no cellular product is secreted. Instead, the transport carriers are inserted into the plasma membrane (fig. 6.17). During exercise and insulin stimulation, therefore, more glucose is able to enter the skeletal muscle cells from the blood plasma.

Transport of glucose by GLUT carriers is a form of passive transport where glucose is always transported down its concentration gradient. However, in certain cases (such as the epithelial cells of the kidney tubules and small intestine), glucose is transported against its concentration gradient by a different kind of carrier, one that is dependent on simultaneous transport of Na^+. Because this is a type of active transport, it will be described shortly.

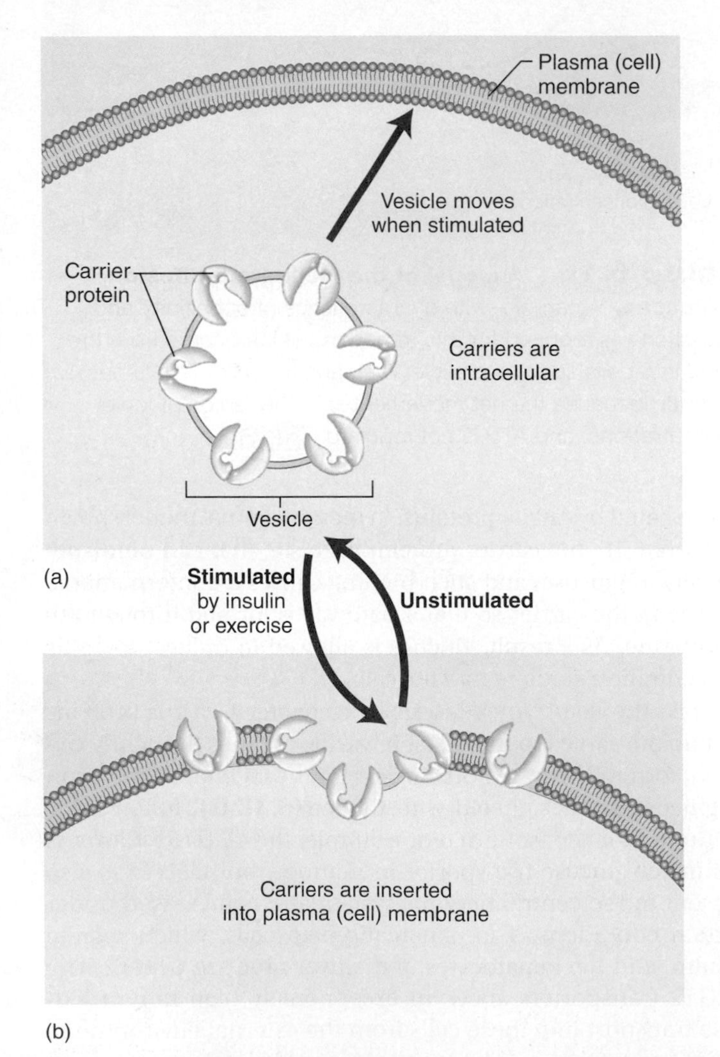

Figure 6.17 The insertion of carrier proteins into the plasma (cell) membrane. *(a)* In the unstimulated state, carrier proteins (such as those for glucose) may be located in the membrane of intracellular vesicles. *(b)* In response to stimulation, the vesicle fuses with the plasma membrane and the carriers are thereby inserted into the membrane.

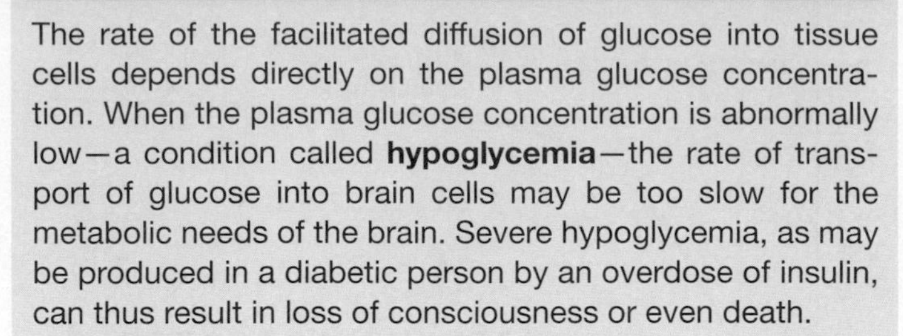

CLINICAL APPLICATION

The rate of the facilitated diffusion of glucose into tissue cells depends directly on the plasma glucose concentration. When the plasma glucose concentration is abnormally low—a condition called **hypoglycemia**—the rate of transport of glucose into brain cells may be too slow for the metabolic needs of the brain. Severe hypoglycemia, as may be produced in a diabetic person by an overdose of insulin, can thus result in loss of consciousness or even death.

The transport of fatty acids through plasma membranes has long been thought to occur by simple diffusion, due to the hydrophobic nature of the fatty acids. However, more recent evidence demonstrates that there are fatty acid transport carriers analogous to the GLUT carriers. These carrier proteins facilitate the diffusion of fatty acids out of adipocytes and into the blood, and out of the blood and into organs. Similar to glucose transport by GLUT4, the uptake of fatty acids into skeletal muscle fibers by facilitative diffusion carriers is increased by exercise and insulin.

Active Transport

Some aspects of cell transport cannot be explained by simple or facilitated diffusion. The epithelial linings of the small intestine and kidney tubules, for example, move glucose from the side of lower to the side of higher concentration—from the space within the tube (*lumen*) to the blood. Similarly, all cells extrude Ca^{2+} into the extracellular environment (fig. 6.18) and, by this means, maintain an intracellular Ca^{2+} concentration that is 1,000 to 10,000 times lower than the extracellular Ca^{2+} concentration.

Active transport is the movement of molecules and ions against their concentration gradients, from lower to higher concentrations. This transport requires the expenditure of cellular energy obtained from ATP; if a cell is poisoned with cyanide (which inhibits oxidative phosphorylation; see chapter 5, fig. 5.11), active transport will stop. Passive transport, by contrast, can continue even if metabolic poisons kill the cell by preventing the formation of ATP. Because active transport involves the transport of ions and molecules "uphill" (against their concentration gradients), and because it uses metabolic energy, the primary active transport carriers are referred to as *pumps.*

Primary active transport occurs when the hydrolysis of ATP is directly responsible for the function of the carriers, which are proteins that span the thickness of the membrane. Pumps of this type—including the Ca^{2+} pump (fig. 6.18), the proton (H^+) pump (responsible for the acidity of the stomach's gastric juice), and the Na^+/K^+ pump (fig. 6.19)—are also ATPase enzymes, and their pumping action is controlled by the addition and removal of phosphate groups obtained from ATP.

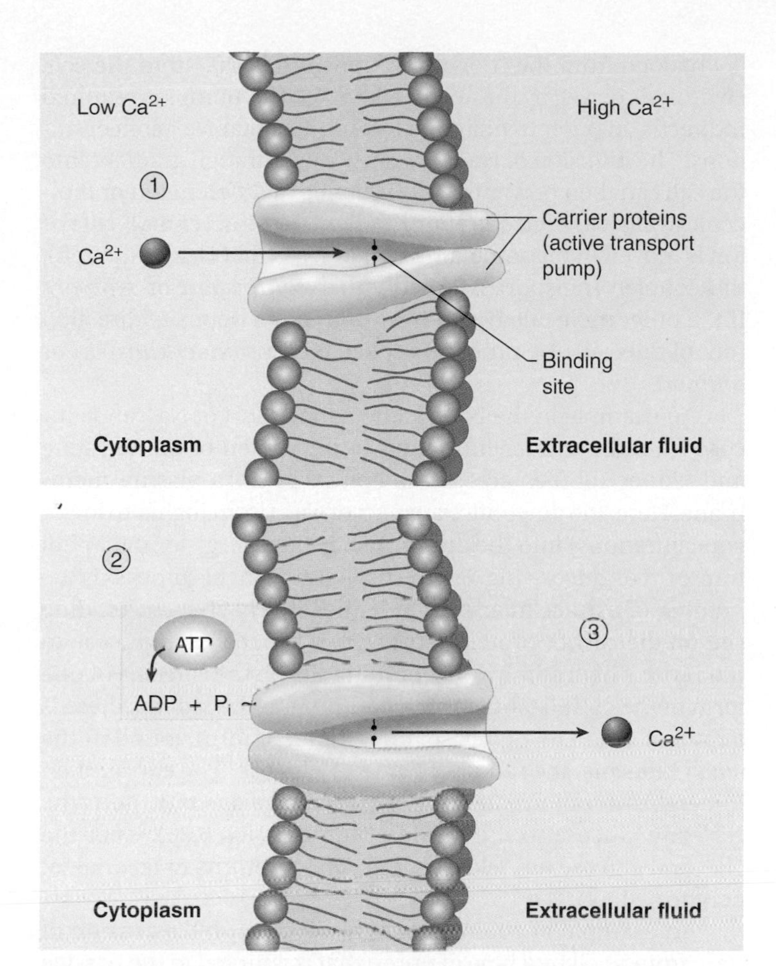

Figure 6.18 An active transport pump. This carrier protein transports Ca^{2+} from a lower concentration inside the cell to a higher concentration outside of the cell, and is thus known as a Ca^{2+} pump. (1) Ca^{2+} within the cell binds to sites in the carrier protein. (2) ATP is hydrolyzed into ADP and phosphate (P_i), and the phosphate is added to the carrier protein; this phosphorylation causes a hingelike motion of the carrier. (3) The hingelike motion of the carrier protein allows Ca^{2+} to be released into the extracellular fluid.

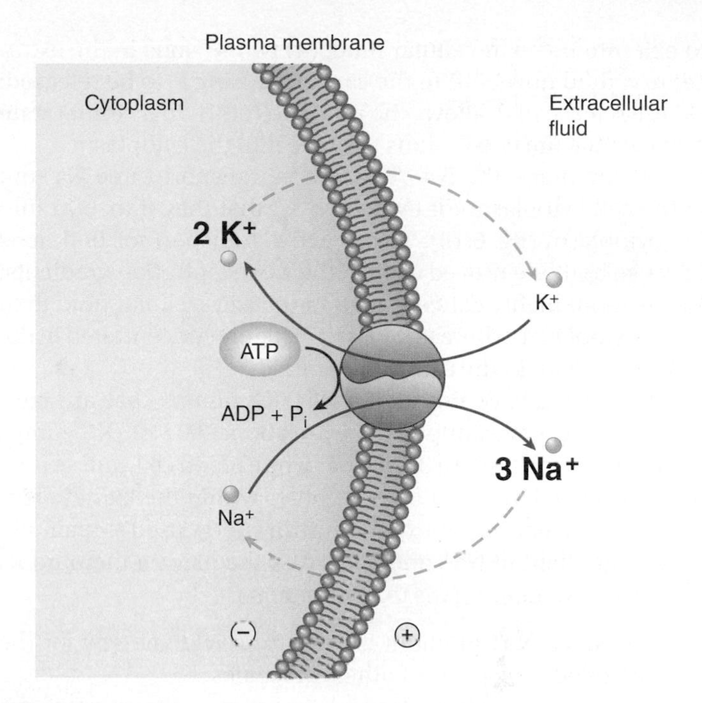

Figure 6.19 The exchange of intracellular Na^+ for K^+ by the Na^+/K^+ pump. The active transport carrier itself is an ATPase that breaks down ATP for energy. Dashed arrows indicate the direction of passive transport (diffusion); solid arrows indicate the direction of active transport. Because 3 Na^+ are pumped out for every 2 K^+ pumped in, the action of the Na^+/K^+ (ATPase) pumps help to produce a difference in charge, or potential difference, across the membrane. APIR

The Ca^{2+} Pump

Ca^{2+} pumps are located in the plasma membrane of all cells, and in the membrane of the endoplasmic reticulum (chapter 3) of striated muscle cells and others. Active transport by these pumps removes Ca^{2+} from the cytoplasm by pumping it into the extracellular fluid or the cisternae of the endoplasmic reticulum. Because of the concentration gradient thus created, when ion channels for Ca^{2+} are opened in the plasma membrane or endoplasmic reticulum, Ca^{2+} will diffuse rapidly down its concentration gradient into the cytoplasm. This sudden rise in cytoplasmic Ca^{2+} serves as a signal for diverse processes, including the release of neurotransmitters from axon terminals (chapter 7, section 7.3) and muscle contraction (chapter 12, section 12.2).

Figure 6.18 presents a simplified model of the Ca^{2+} pump. Notice that there is a binding site that is accessible to Ca^{2+} from the cytoplasm, and that the pump is activated by phosphorylation, using the P_i derived from ATP. Newer studies of these pumps have revealed the following: (1) binding of a cytoplasmic Ca^{2+} to an amino acid site in the pump activates the ATPase, causing the hydrolysis of ATP into ADP and P_i, which are bound to the pump; (2) both of the exits for Ca^{2+} are now momentarily blocked; (3) the ADP is released, producing a shape change in the protein that opens a passageway for Ca^{2+} to the extracellular fluid (or cisterna of the endoplasmic reticulum), so that Ca^{2+} can move to the other side of the membrane; (4) the P_i group is released from the pump, allowing the carrier to return to its initial state where cytoplasmic Ca^{2+} once again has access to the binding site.

The Sodium-Potassium Pump

A very important primary active transport carrier found in all body cells is the **Na^+/K^+ pump.** Like the Ca^{2+} pumps previously described, the Na^+/K^+ pumps are also ATPase enzymes. The Na^+/K^+ pump cycle occurs as follows: (1) three Na^+ ions in the cytoplasm move partway into the pump and bind to three amino acid sites; (2) this activates the ATPase, hydrolyzing ATP into ADP and P_i and causing both exits to be momentarily blocked; (3) the ADP is released, producing a shape change in the carrier that opens a passageway for the three Na^+ ions to exit into the extracellular fluid; (4) two K^+ ions in the extracellular fluid now bind to the carrier, causing P_i to be released; (5) the release of P_i allows the pump to return to its initial state and permits the two K^+ ions to move into the cytoplasm.

In summary, the Na^+/K^+ pumps transport three Na^+ out of the cell cytoplasm for every two K^+ that they transport into the cytoplasm (fig. 6.19). This is active transport for both ions because both are moved against their concentration gradients: Na^+ is more highly concentrated in the extracellular fluid than in the cytoplasm, whereas K^+ is more highly concentrated in the cytoplasm than in the extracellular fluid.

Most cells have numerous Na^+/K^+ pumps that are constantly active. For example, there are about 200 Na^+/K^+ pumps per red blood cell, about 35,000 per white blood cell, and several million per cell in a part of the tubules within the kidney. This represents an enormous expenditure of energy used to maintain a steep gradient of Na^+ and K^+ across the plasma membrane. This steep gradient serves three functions:

1. The steep Na^+ gradient is used to provide energy for the "coupled transport" of other molecules.
2. The gradients for Na^+ and K^+ concentrations across the plasma membranes of nerve and muscle cells are used to produce electrochemical impulses needed for functions of the nerve and muscles, including the heart muscle.
3. The active extrusion of Na^+ is important for osmotic reasons; if the pumps stop, the increased Na^+ concentrations within cells promote the osmotic inflow of water, damaging the cells.

Secondary Active Transport (Coupled Transport)

In **secondary active transport,** or **coupled transport,** the energy needed for the "uphill" movement of a molecule or ion is obtained from the "downhill" transport of Na^+ into the cell. Hydrolysis of ATP by the action of the Na^+/K^+ pumps is required indirectly, in order to maintain low intracellular Na^+ concentrations. The diffusion of Na^+ down its concentration gradient into the cell can then power the movement of a different ion or molecule against its concentration gradient. If the other molecule or ion is moved in the same direction as Na^+ (that is, into the cell), the coupled transport is called either *cotransport* or *symport.* If the other molecule or ion is moved in the opposite direction (out of the cell), the process is called either *countertransport* or *antiport.*

An example in the body is the cotransport of Na^+ and glucose from the extracellular fluid in the lumen of the intestine and kidney tubules across the epithelial cell's plasma membrane. Here, the downhill transport of Na^+ (from higher to lower concentrations) into the cell furnishes the energy for the uphill transport of glucose (fig. 6.19). The first step in this process is the binding of extracellular Na^+ to its negatively charged binding site on the carrier protein. This allows extracellular glucose to bind with a high affinity to its binding site on the carrier. For one form of the cotransport carrier, common in the kidney, there is a ratio of 1 Na^+ to 1 glucose; for a different form, found in the small intestine, the ratio is 2 Na^+ to 1 glucose. The carrier then undergoes a conformational (shape) change that transports the Na^+ and glucose to the inside of the cell (fig. 6.20). After the Na^+ and glucose are released, the carrier returns to its original conformation.

An example of countertransport is the uphill extrusion of Ca^{2+} from a cell by a type of pump that is coupled to the passive diffusion of Na^+ into the cell. Cellular energy, obtained from ATP, is not used to move Ca^{2+} directly out of the cell in this case, but energy is constantly required to maintain the steep Na^+ gradient.

An easy way to understand why examples of secondary active transport are classified as "active" is to imagine

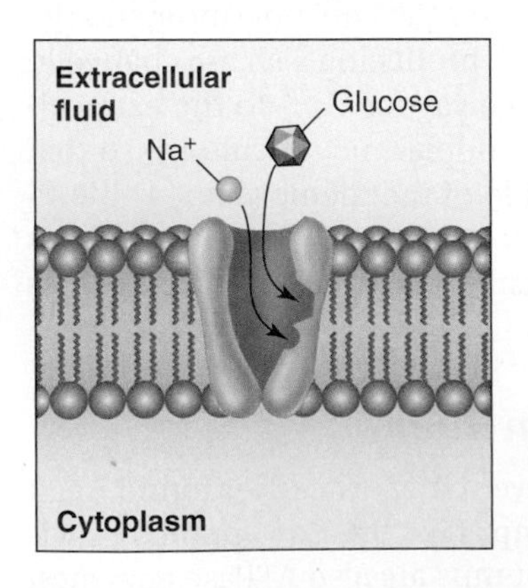

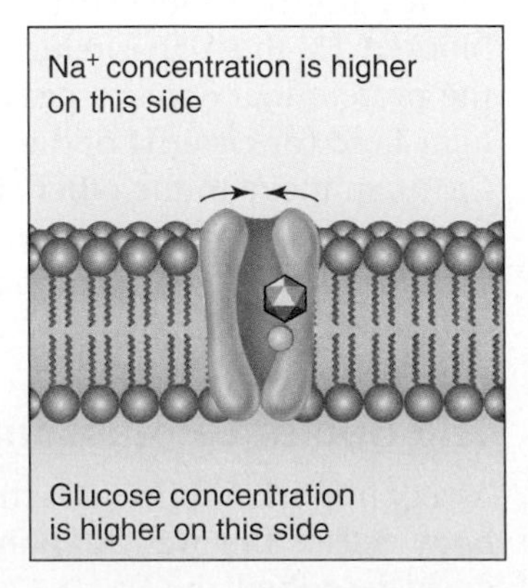

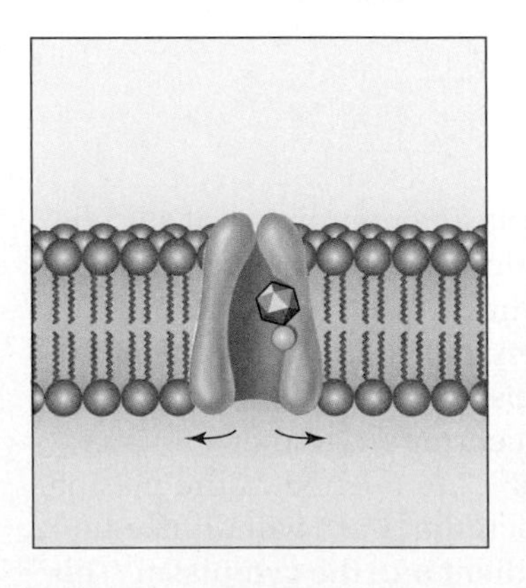

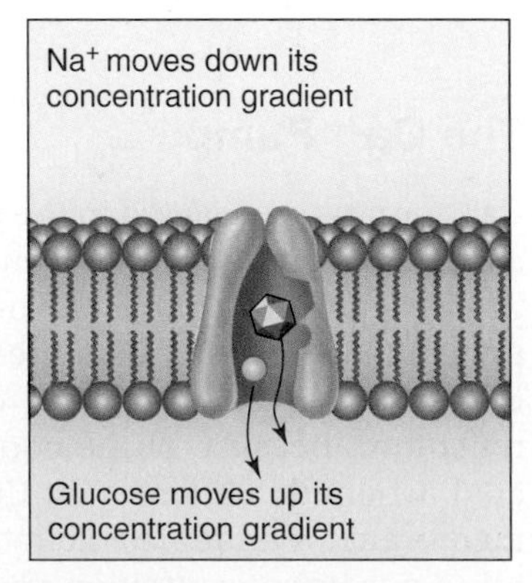

Figure 6.20 The cotransport of Na^+ and glucose. This carrier protein transports Na^+ and glucose at the same time, moving them from the lumen of the intestine and kidney tubules into the lining epithelial cells. This cotransport requires a lower intracellular concentration of Na^+, which is dependent on the action of other carriers, the Na^+/K^+ (ATPase) pumps. Because ATP is needed to power the Na^+/K^+ (ATPase) pumps, the cotransport of Na^+ and glucose depends indirectly on ATP, and so can be considered secondary active transport. The cotransport carrier shown here transports 1 Na^+ to 1 glucose, as most commonly occurs in the kidney; the carrier in the small intestine transports 2 Na^+ for 1 glucose (not shown). AP|R

what happens if a cell is poisoned with cyanide, so that it cannot produce ATP. After the primary active transport Na^+/K^+ (ATPase) pumps stop working, the concentration gradient for Na^+ gradually decreases. As this occurs, the transport of glucose from the intestinal lumen into the epithelial cells, as well as other examples of secondary active transport, likewise declines. This differs from passive transport, such as the facilitated diffusion of glucose from the blood into tissue cells, which does not depend on ATP.

Transport Across Epithelial Membranes

Epithelial membranes cover all body surfaces and line the cavities of all hollow organs (chapter 1, section 1.3). Therefore, in order for a molecule or ion to move from the external environment into the blood (and from there to the body organs), it must first pass through an epithelial membrane. The transport of digestion products (such as glucose) across the intestinal epithelium into the blood is called **absorption.** The transport of molecules out of the urinary filtrate (originally derived from blood) back into the blood is called **reabsorption.**

The cotransport of Na^+ and glucose described in the last section can serve as an example. The cotransport carriers for Na^+ and glucose are located in the apical (top) plasma membrane of the epithelial cells, which faces the lumen of the intestine or kidney tubule. The Na^+/K^+ pumps, and the carriers for the facilitated diffusion of glucose, are on the basal (bottom) plasma membrane of the epithelial cell facing the location of blood capillaries. As a result of these active and passive transport processes, glucose is moved from the lumen, through the cell, and then to the blood (fig. 6.21). Amino acids are similarly transported across the epithelial lining of the small intestine and kidney tubules. Some amino acids are cotransported by a carrier that uses the Na^+ electrochemical gradient, similar to the cotransport of glucose; however, other amino acids are transported by a carrier that uses a proton (H^+) electrochemical gradient. This H^+ gradient is created by a different carrier, a Na^+/H^+ pump, which uses the inward movement of Na^+ to transport H^+ out of the cell.

The membrane transport mechanisms described in this section move materials through the cytoplasm of the epithelial cells, a process termed **transcellular transport.** However, diffusion and osmosis may also occur to a limited extent in

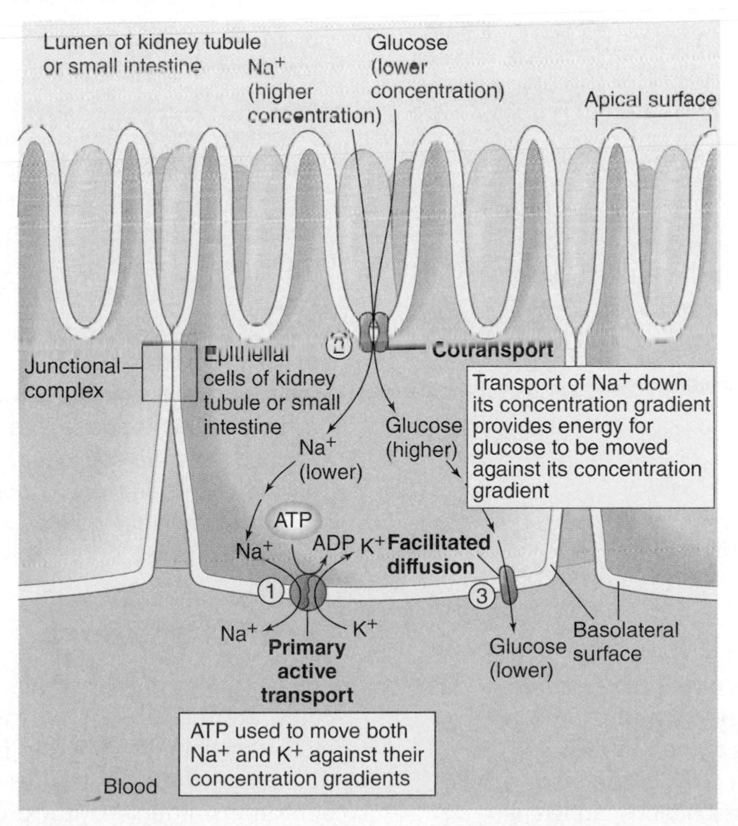

Figure 6.21 Transport processes involved in the epithelial absorption of glucose. When glucose is to be absorbed across the epithelial membranes of the kidney tubules or the small intestine, several processes are involved. (*1*) Primary active transport (the Na^+/K^+ pumps) in the basal membrane use ATP to maintain a low intracellular concentration of Na^+. (*2*) Secondary active transport uses carriers in the apical membrane to transport glucose up its concentration gradient, using the energy from the "downhill" flow of Na^+ into the cell. Finally, (*3*) facilitated diffusion of glucose using carriers in the basal membrane allows the glucose to leave the cells and enter the blood. AP|R

the very tiny spaces between epithelial cells, a process termed **paracellular transport.**

Paracellular transport between cells is limited by the **junctional complexes** that connect adjacent epithelial cells. Junctional complexes consist of three structures: (1) *tight junctions,* where the space between the two adjoining plasma membranes appears to be occluded and strands of proteins penetrate the plasma membranes to bridge the cytoskeleton actin fibers in each cell; (2) *adherens junctions,* where the plasma membranes of the two cells come very close together and are "glued" by interactions between proteins that span each membrane and connect to the cytoskeleton of each cell; and (3) *desmosomes,* where the plasma membranes of the two cells are "buttoned together" by interactions between particular desmosomal proteins (fig. 6.22).

Although the space between the plasma membranes of two cells joined by tight junctions appears to be occluded, physiological evidence suggests that the barrier can be selectively permeable. Some tight junctions are indeed tight, but some may be leaky, selectively allowing ions and molecules to pass. There are interconnected protein strands between the cells in the tight junctions, and the tight junctions that are leakier appear to have fewer of those strands. The paracellular movement through tight junctions varies in different membranes, and their leakiness may be subject to regulation.

In addition, there is bulk movement of fluid through pores in epithelial membranes. The presence of pores, and their size, depends on the extent to which junctional complexes surround each epithelial cell in the membrane. For example, the epithelial cells that compose the walls of many blood capillaries (the thinnest of blood vessels) have pores between them that can be relatively large, permitting filtration of water and dissolved molecules out of the capillaries through the paracellular route. In the capillaries of the brain, however, such filtration is prevented by tight junctions, so molecules must be transported transcellularly. This involves the cell transport mechanisms previously described, as well as the processes of endocytosis and exocytosis described next.

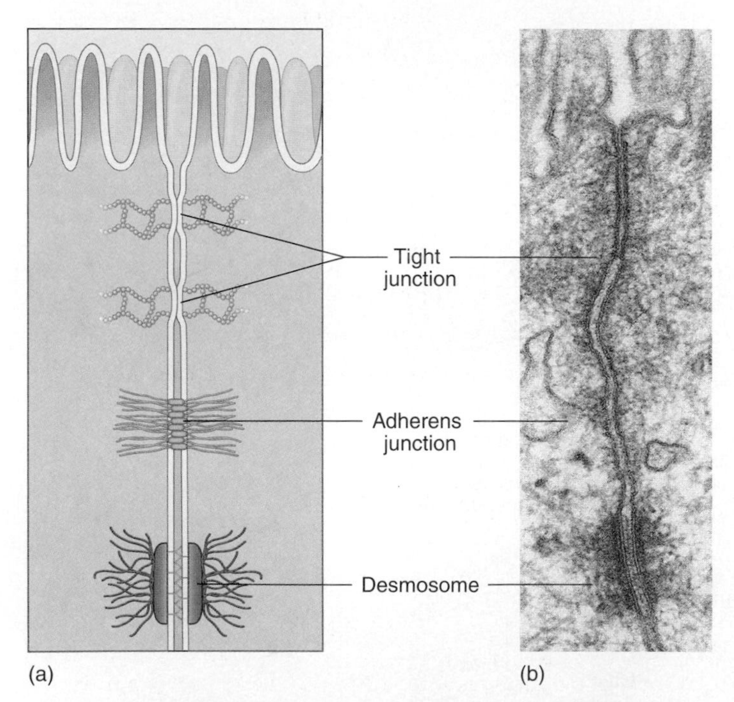

Figure 6.22 Junctional complexes provide a barrier between adjacent epithelial cells. Proteins penetrate the plasma membranes of the two cells and are joined to the cytoskeleton of each cell. Junctional complexes consist of three components: tight junctions, adherens junctions, and desmosomes. Epithelial membranes differ, however, in the number and arrangement of these components, which are illustrated in (*a*) and shown in an electron micrograph in (*b*).

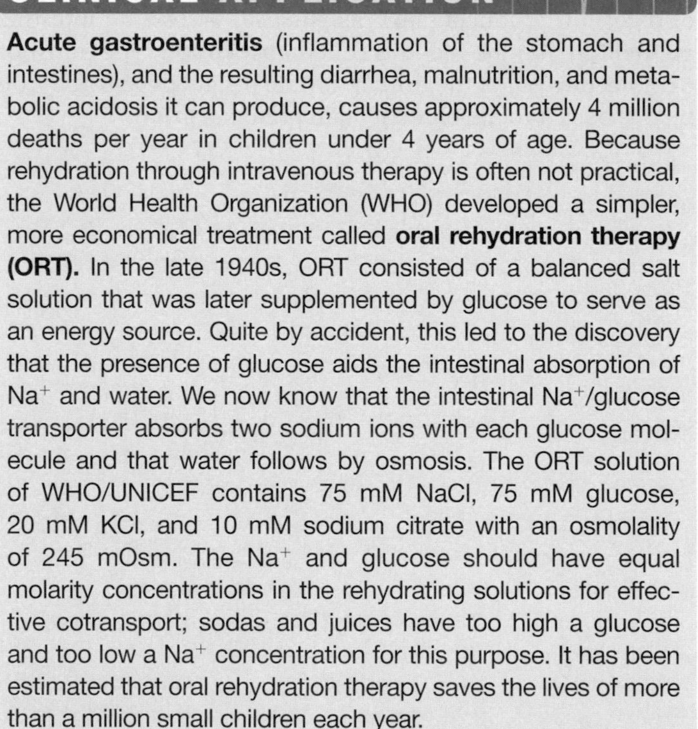

CLINICAL APPLICATION

Acute gastroenteritis (inflammation of the stomach and intestines), and the resulting diarrhea, malnutrition, and metabolic acidosis it can produce, causes approximately 4 million deaths per year in children under 4 years of age. Because rehydration through intravenous therapy is often not practical, the World Health Organization (WHO) developed a simpler, more economical treatment called **oral rehydration therapy (ORT).** In the late 1940s, ORT consisted of a balanced salt solution that was later supplemented by glucose to serve as an energy source. Quite by accident, this led to the discovery that the presence of glucose aids the intestinal absorption of Na^+ and water. We now know that the intestinal Na^+/glucose transporter absorbs two sodium ions with each glucose molecule and that water follows by osmosis. The ORT solution of WHO/UNICEF contains 75 mM NaCl, 75 mM glucose, 20 mM KCl, and 10 mM sodium citrate with an osmolality of 245 mOsm. The Na^+ and glucose should have equal molarity concentrations in the rehydrating solutions for effective cotransport; sodas and juices have too high a glucose and too low a Na^+ concentration for this purpose. It has been estimated that oral rehydration therapy saves the lives of more than a million small children each year.

Bulk Transport

Polypeptides and proteins, as well as many other molecules, are too large to be transported through a membrane by the carriers described in previous sections. Yet many cells do secrete these molecules—for example, as hormones or neurotransmitters—by the process of **exocytosis.** As described in chapter 3, this involves the fusion of a membrane-bound vesicle that contains these cellular products with the plasma membrane, so that the membranes become continuous (fig. 6.23).

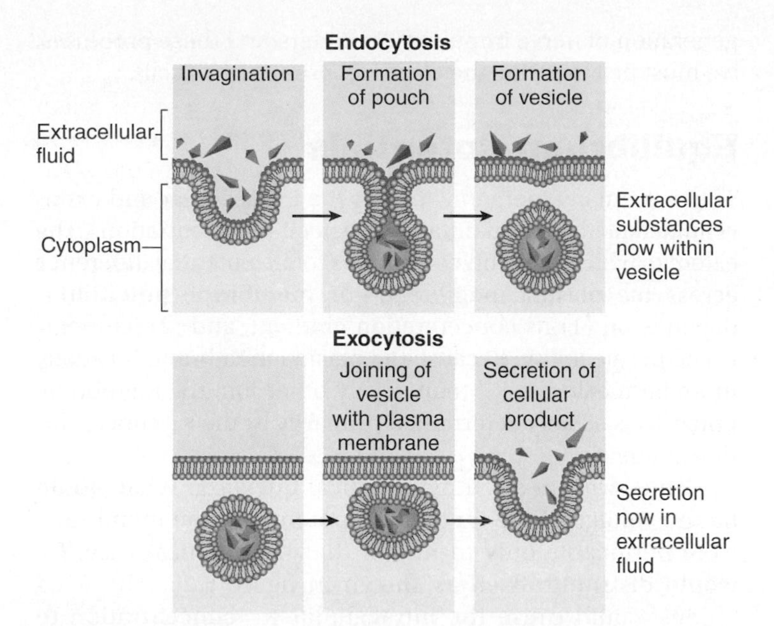

Figure 6.23 Endocytosis and exocytosis. Endocytosis and exocytosis are responsible for the bulk transport of molecules into and out of a cell. AP|R

The process of **endocytosis** resembles exocytosis in reverse. In receptor-mediated endocytosis (see fig. 3.4), specific molecules, such as protein-bound cholesterol, can be taken into the cell because of the interaction between the cholesterol transport protein and a protein receptor on the plasma membrane. Cholesterol is removed from the blood by the liver and by the walls of blood vessels through this mechanism.

Exocytosis and endocytosis together provide **bulk transport** out of and into the cell, respectively. (The term *bulk* is used because many molecules are moved at the same time.) It should be noted that molecules taken into a cell by endocytosis are still separated from the cytoplasm by the membrane of the endocytotic vesicle. Some of these molecules, such as membrane receptors, will be moved back to the plasma membrane, while the rest will end up in lysosomes.

Reference to figure 6.21 reveals that there is a definite direction, or *polarity,* to transport in epithelial cells. This figure illustrates the polarization of membrane transport processes involved in absorption and reabsorption across the epithelium lining the small intestine or kidney tubules. There is also a polarization of organelles involved in exocytosis (see fig. 3.12) and endocytosis. For example, exocytotic vesicles that bud from the Golgi complex fuse with the plasma membrane at its apical, or top, surface, while the nucleus and endoplasmic reticulum are located more toward the bottom of the cell (nearer to the basement membrane). Because of the polarity of transport processes across the plasma membrane and polarity of intracellular organelles, scientists often distinguish between the *apical surface* of epithelial cells and their *basolateral surface* (see fig. 6.21).

CHECKPOINT

7a. List the three characteristics of facilitated diffusion that distinguish it from simple diffusion.

7b. Draw a figure that illustrates two of the characteristics of carrier-mediated transport and explain how this type of movement differs from simple diffusion.

7c. Describe active transport, including primary and secondary active transport in your description. Explain how active transport differs from facilitated diffusion.

8. Discuss the physiological significance of the Na^+/K^+ pumps.

6.4 THE MEMBRANE POTENTIAL

As a result of the permeability properties of the plasma membrane, the presence of nondiffusible negatively charged molecules inside the cell, and the action of the Na^+/K^+ pumps, there is an unequal distribution of charges across the membrane. As a result, the inside of the cell is negatively charged compared to the outside. This difference in charge, or potential difference, is known as the membrane potential.

LEARNING OUTCOMES

After studying this section, you should be able to:

9. Describe the equilibrium potentials for Na^+ and K^+.

10. Describe the membrane potential and explain how it is produced.

If you understand how the membrane potential is produced, and how it is affected by the permeability of the plasma membrane to specific ions, you will be prepared to learn how neurons and muscles (including the heart muscle) produce impulses and function. Thus, this section serves as a basis for the discussion of nerve impulses that follows in chapter 7, and for discussions of muscle (chapter 12) and heart (chapter 13) function.

In section 6.3, the action of the Na^+/K^+ pumps was discussed in conjunction with the topic of active transport, and it was noted that these pumps move Na^+ and K^+ against their concentration gradients. This action alone would create and amplify a difference in the concentration of these ions across the plasma membrane. There is, however, another reason why the concentration of Na^+ and K^+ would be unequal across the membrane.

Cellular proteins and the phosphate groups of ATP and other organic molecules are negatively charged at the pH of the cell cytoplasm. These negative ions (*anions*) are "fixed" within the cell because they cannot penetrate the plasma membrane. As a result, these anions attract positively charged inorganic ions (*cations*) from the extracellular fluid that can pass through ion channels in the plasma membrane. In this way, fixed anions within the cell influence the distribution of inorganic cations (mainly K^+, Na^+, and Ca^{2+}) between the extracellular and intracellular compartments.

Because the plasma membrane is more permeable to K^+ than to any other cation, K^+ accumulates within the cell more than the others as a result of its electrical attraction for the fixed anions (fig. 6.24). So, instead of being evenly distributed between the intracellular and extracellular compartments, K^+ becomes more highly concentrated within the cell. The intracellular K^+ concentration is 150 mEq/L in the human body compared to an extracellular concentration of 5 mEq/L (mEq = milliequivalents, which is the millimolar concentration multiplied by the valence of the ion—in this case, by one).

As a result of the unequal distribution of charges between the inside and outside of cells, each cell acts as a tiny battery with the positive pole outside the plasma membrane and the negative pole inside. The magnitude of this difference in charge, or **potential difference,** is measured in *voltage.* Although the voltage of this battery is very small (less than a tenth of a volt), it is of critical importance in such physiological processes as muscle contraction, the regulation of the heartbeat, and the generation of nerve impulses. To understand these processes, we must first examine the electrical properties of cells.

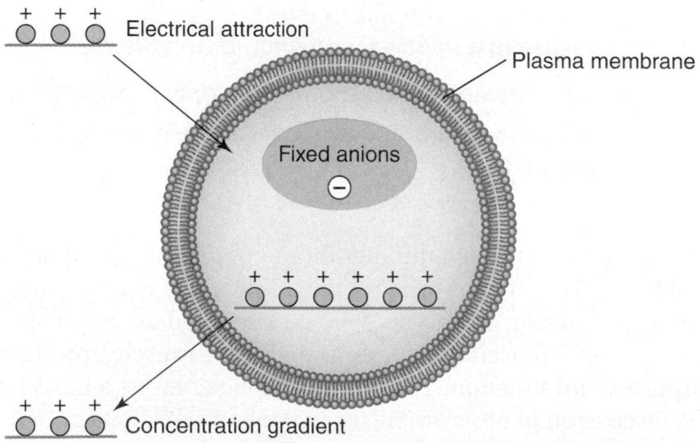

Figure 6.24 The effect of fixed anions on the distribution of cations. Proteins, organic phosphates, and other organic anions that cannot leave the cell create a fixed negative charge on the inside of the membrane. This negative charge attracts positively charged inorganic ions (cations), which therefore accumulate within the cell at a higher concentration than is found in the extracellular fluid. The amount of cations that accumulates within the cell is limited by the fact that a concentration gradient builds up, which favors the diffusion of the cations out of the cell.

Equilibrium Potentials

There are many inorganic ions in the intracellular and extracellular fluid that are maintained at specific concentrations. The extent to which each ion contributes to the potential difference across the plasma membrane—or **membrane potential**—depends on (1) its concentration gradient, and (2) its membrane permeability. Because the plasma membrane is usually more permeable to K^+ than to any other ion, the membrane potential is usually determined primarily by the K^+ concentration gradient.

Thus, we can ask a hypothetical question: What would be the voltage of the membrane potential if the membrane were permeable only to K^+? In that hypothetical case, K^+ would distribute itself as shown in figure 6.25. The fixed anions would cause the intracellular K^+ concentration to become higher than the extracellular concentration. However, once the concentration gradient (ratio of concentrations outside and inside the cell) reached a particular value, net movement of K^+ would cease. If more K^+ entered the cell because of electrical attraction, the same amount would leave the cell by net diffusion. Thus, a state of *equilibrium* would be reached where the concentrations of K^+ remained

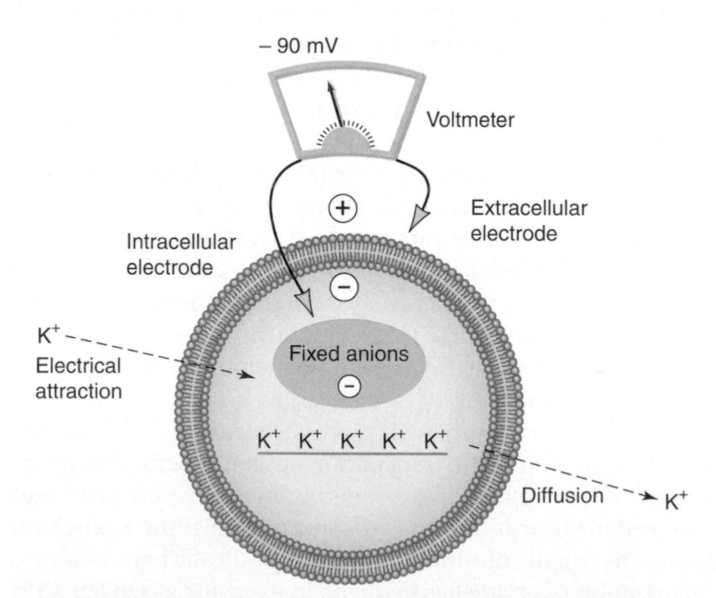

Figure 6.25 Potassium equilibrium potential. If K^+ were the only ion able to diffuse through the plasma membrane, it would distribute itself between the intracellular and extracellular compartments until an equilibrium was established. At equilibrium, the K^+ concentration within the cell would be higher than outside the cell because of the attraction of K^+ for the fixed anions. The intracellular and extracellular K^+ concentrations are normal when the inside of the cell is –90 millivolts compared to the outside of the cell. This membrane potential is the equilibrium potential (E_K) for potassium.

stable. The membrane potential that would stabilize the K^+ concentrations is known as the **K^+ equilibrium potential** (abbreviated E_K). Given the normal K^+ concentration gradient, where the concentration is 30 times higher inside than outside the cell (fig. 6.26), the value of E_K is −90 millivolts (mV). A sign (+ or −) placed in front of the voltage always indicates the polarity of the inside of the cell (this is done because when a neuron produces an impulse, the polarity briefly reverses, as discussed in chapter 7).

Expressed in a different way, a membrane potential of −90 mV is needed to produce an equilibrium in which the K^+ concentrations are 150 mM inside and 5 mM outside the cell (fig. 6.26). At −90 mV, these intracellular and extracellular concentrations are kept stable. If this value were more negative, it would draw more K^+ into the cell; if it were less negative, K^+ would diffuse out of the cell.

Now, let's ask another hypothetical question: What would the membrane potential be if the membrane were permeable only to Na^+? (This is quite different from the usual situation in which the membrane is less permeable to Na^+ than to K^+.) What membrane potential would stabilize the Na^+ concentrations at 12 mM intracellularly and 145 mM extracellularly (fig. 6.26) if Na^+ were the only ion able to cross the membrane? This is the **Na^+ equilibrium potential** (abbreviated E_{Na}). You could guess that the inside of the cell would have to be the positive pole, repelling the Na^+ and causing its concentration to be lower inside than outside the cell. The actual voltage, however, has to be calculated, as described in the next section.

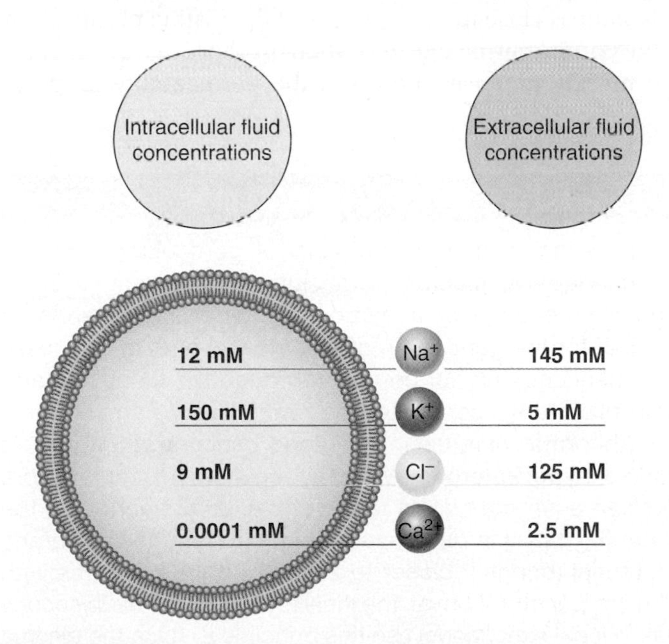

Figure 6.26 Concentrations of ions in the intracellular and extracellular fluids. This distribution of ions, and the different permeabilities of the plasma membrane to these ions, affects the membrane potential and other physiological processes.

See the *Test Your Quantitative Ability* section of the Review Activities at the end of this chapter.

This calculation reveals that an equilibrium potential of 66 mV, with the inside of the cell the positive pole, maintains the Na^+ concentration of 12 mM inside and 145 mM outside the cell. The E_{Na} is thus written as +66 mV.

Equilibrium potentials are useful to know because they tell us what happens to the membrane potential when the plasma membrane becomes highly permeable to one particular ion. The resting neuron, for example, has a membrane potential close to E_K because its membrane is most permeable to K^+. However, when it produces an impulse, it suddenly becomes highly permeable to Na^+ for a brief time, driving its membrane potential closer to E_{Na}. The resting membrane potential will be described shortly; the production of nerve impulses is explained in chapter 7, section 7.2.

Nernst Equation

The diffusion gradient depends on the difference in concentration of the ion. Therefore, the value of the equilibrium potential must depend on the ratio of the concentrations of the ion on the two sides of the membrane. The **Nernst equation** allows this theoretical equilibrium potential to be calculated for a particular ion when its concentrations are known. The following simplified form of the equation is valid at a temperature of 37° C:

$$E_x = \frac{61}{z} \log \frac{[X_o]}{[X_i]}$$

where

E_x = equilibrium potential in millivolts (mV) for ion x
X_o = concentration of the ion outside the cell
X_i = concentration of the ion inside the cell
z = valence of the ion (+1 for Na^+ or K^+)

Note that, using the Nernst equation, the equilibrium potential for a cation has a negative value when X_i is greater than X_o. If we substitute K^+ for X, this is indeed the case. As a hypothetical example, if the concentration of K^+ were 10 times higher inside compared to outside the cell, the equilibrium potential would be 61 mV (log 1/10) = 61 × (−1) = −61 mV. In reality, the concentration of K^+ inside the cell is 30 times greater than outside (150 mEq/L inside compared to 5 mEq/L outside). Thus,

$$E_K = 61 \text{ mV} \log \frac{5 \text{ mEq/L}}{150 \text{ mEq/L}} = -90 \text{ mV}$$

This means that a membrane potential of 90 mV, with the inside of the cell negative, would be required to prevent the diffusion of K^+ out of the cell. This is why the equilibrium potential for K^+ (E_K) was given as −90 mV in the earlier discussion of equilibrium potentials.

If we wish to calculate the equilibrium potential for Na^+, different values must be used. The concentration of Na^+ in the extracellular fluid is 145 mEq/L, whereas its concentration inside cells is 5 to 14 mEq/L. The diffusion gradient thus promotes the movement of Na^+ into the cell, and, in order to oppose this diffusion, the membrane potential would have to have a positive polarity on the inside of the cell. This is indeed what the Nernst equation would provide. Thus, using an intracellular Na^+ concentration of 12 mEq/L,

$$E_{Na} = 61 \text{ mV} \log \frac{145 \text{ mEq/L}}{12 \text{ mEq/L}} = +66 \text{ mV}$$

This means that a membrane potential of 66 mV, with the inside of the cell positive, would be required to prevent the diffusion of Na^+ into the cell. This is why the equilibrium potential for Na^+ (E_{Na}) was given as +66 mV in the earlier discussion of equilibrium potentials.

Resting Membrane Potential

The membrane potential of a real cell that is not producing impulses is known as the **resting membrane potential.** If the plasma membrane were only permeable to Na^+, its resting membrane potential would equal the E_{Na} of +66 mV; if it were only permeable to K^+, its resting membrane potential would equal the E_K of −90 mV. A real resting cell is more permeable to K^+ than to Na^+, but it is not completely impermeable to Na^+. As a result, its resting membrane potential is close to the E_K but somewhat less negative due to the slight inward diffusion of Na^+. Since the resting membrane potential is less negative than the E_K, there will also be a slight outward diffusion of K^+. These leakages are countered by the constant activity of the Na^+/K^+ pumps.

The actual value of the resting membrane potential depends on two factors:

1. The *ratio of the concentrations* (X_o/X_i) of each ion on the two sides of the plasma membrane.
2. The *specific permeability* of the membrane to each different ion.

Many ions—including K^+, Na^+, Ca^{2+}, and Cl^-—contribute to the resting membrane potential. Their individual contributions are determined by the differences in their concentrations across the membrane (fig. 6.27), and by their membrane permeabilities.

This has two important implications:

1. For any given ion, a change in its concentration in the extracellular fluid will change the resting membrane potential—but only to the extent that the membrane is permeable to that ion. Because *the resting membrane is most permeable to K^+*, a change in the extracellular concentration of K^+ has the greatest effect on the resting membrane potential. This is the mechanism behind the fact that "lethal injections" are of KCl (raising the extracellular K^+ concentrations and depolarizing cardiac cells).
2. A change in the membrane permeability to any given ion will change the membrane potential. This fact is central to the production of nerve and muscle impulses, as will be described in chapter 7. Most often, it is the opening and closing of Na^+ and K^+ channels that are involved, but gated channels for Ca^{2+} and Cl^- are also very important in physiology.

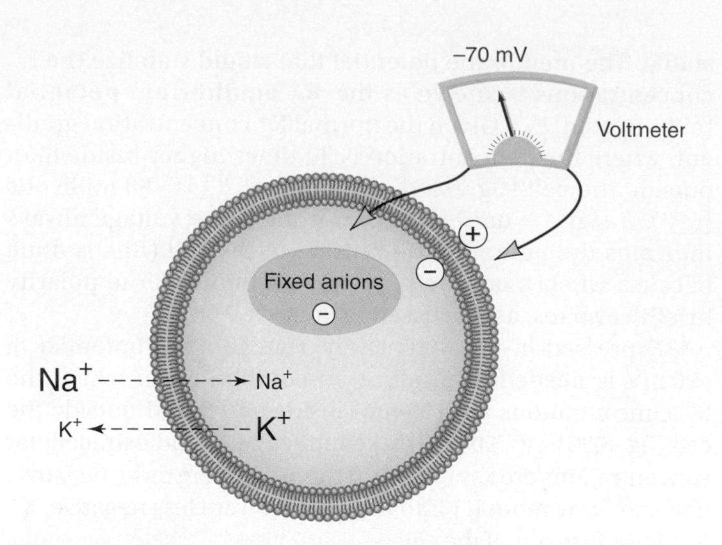

Figure 6.27 The resting membrane potential. Because some Na^+ leaks into the cell by diffusion, the actual resting membrane potential is not as negative as the K^+ equilibrium potential. As a result, some K^+ diffuses out of the cell, as indicated by the dashed lines.

The resting membrane potential of most cells in the body ranges from −65 mV to −85 mV (in neurons it averages −70 mV). This value is close to the E_K because the resting plasma membrane is more permeable to K^+ than to other ions. During nerve and muscle impulses, however, the permeability properties

CLINICAL APPLICATION

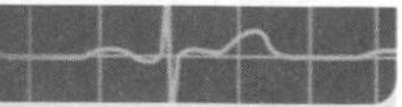

The resting membrane potential is particularly sensitive to changes in plasma potassium concentration. Since the maintenance of a particular membrane potential is critical for the generation of electrical events in the heart, mechanisms that act primarily through the kidneys maintain plasma K^+ concentrations within very narrow limits. An abnormal increase in the blood concentration of K^+ is called **hyperkalemia.** When hyperkalemia occurs, more K^+ can enter the cell. In terms of the Nernst equation, the ratio $[K_o^+]/[K_i^+]$ is decreased. This reduces the membrane potential (brings it closer to zero) and thus interferes with the proper function of the heart. Indeed, *lethal injections* (as in legal executions) use this principle to raise the plasma K^+ concentration to levels that cause cessation of the heartbeat. For these reasons, the blood electrolyte concentrations are monitored very carefully in patients with heart or kidney disease.

Clinical Investigation CLUES

Jessica's medical tests revealed that she had hyperkalemia and an abnormal electrocardiogram (ECG).

- What is hyperkalemia, and how might it influence the electrical properties of the heart, as indicated by an abnormal ECG?

change, as will be described in chapter 7. An increased membrane permeability to Na^+ drives the membrane potential toward E_{Na} (+66 mV) for a short time. This is the reason that the term *resting* is used to describe the membrane potential when it is not producing impulses.

Role of the Na^+/K^+ Pumps

Since the resting membrane potential is less negative than E_K, some K^+ leaks out of the cell (fig. 6.27). The cell is *not* at equilibrium with respect to K^+ and Na^+ concentrations. Nonetheless, the concentrations of K^+ and Na^+ are maintained constant because of the constant expenditure of energy in active transport by the Na^+/K^+ pumps. The Na^+/K^+ pumps act to counter the leaks and thus maintain the membrane potential.

Actually, the Na^+/K^+ pump does more than simply work against the ion leaks; because it transports 3 Na^+ out of the cell for every 2 K^+ that it moves in, it has the net effect of contributing to the negative intracellular charge (see fig. 6.19). This *electrogenic effect* of the pumps adds approximately 3 mV to the membrane potential. As a result of all of these activities, a real cell has (1) a relatively constant intracellular concentration of Na^+ and K^+ and (2) a constant membrane potential (in the absence of stimulation) in nerves and muscles of −65 mV to −85 mV. The processes influencing the resting membrane potential are summarized in figure 6.28.

CHECKPOINT

9a. Describe the potassium and sodium equilibrium potentials.

9b. Define *membrane potential* and explain how it is measured.

10a. Explain the relationship of the resting membrane potential to the two equilibrium potentials.

10b. What role do the Na^+/K^+ pumps play in establishing the resting membrane potential?

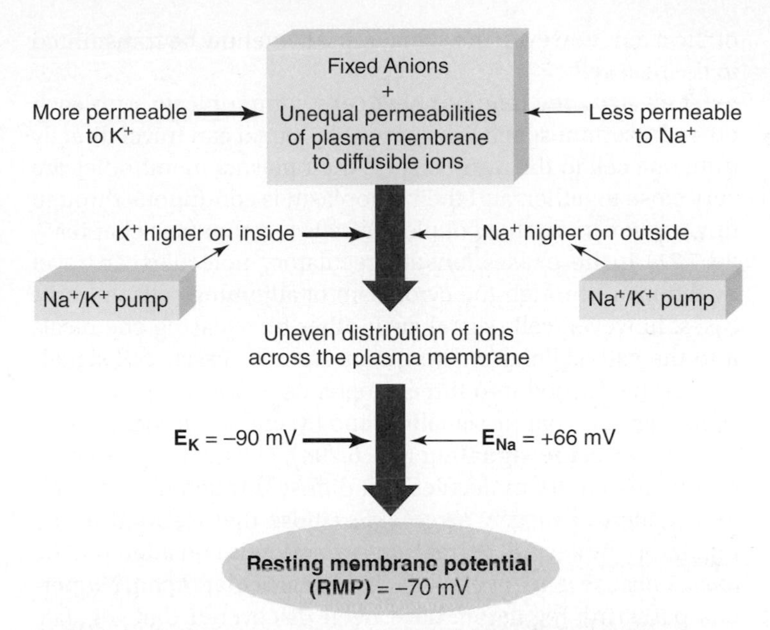

Figure 6.28 The processes that influence the resting membrane potential. As shown in this figure, the Na^+/K^+ pumps produce concentration gradients for Na^+ and K^+, and the presence of fixed anions and the different permeabilities of the plasma membrane to diffusible ions results in their unequal distribution across the plasma membrane. The greater permeability of the membrane to K^+ causes the membrane potential to be closer to the equilibrium potential for K^+ (E_K) than to Na^+ (E_{Na}). The resting membrane potential is different for different cells; a value of −70 mV is typical for mammalian neurons.

6.5 CELL SIGNALING

Cells communicate by signaling each other chemically. These chemical signals are regulatory molecules released by neurons and endocrine glands, and by different cells within an organ.

LEARNING OUTCOMES

After studying this section, you should be able to:

11. Distinguish between synaptic, endocrine, and paracrine regulation.

12. Identify where receptor proteins are located within target cells.

The membrane potential and the permeability of the plasma membrane to ions discussed in the previous section set the stage for the discussion of nerve impulses in chapter 7. Nerve impulses are a type of signal that is conducted along the axon of a neuron. When the impulses reach the end of the axon, however, the signal must somehow be transmitted to the next cell.

Cell signaling refers to how cells communicate with each other. In certain specialized cases, the signal can travel directly from one cell to the next because their plasma membranes are very close together, and their cytoplasm is continuous through tiny *gap junctions* that couple the cells together (see chapter 7, fig. 7.21). In these cases, ions and regulatory molecules can travel by diffusion through the cytoplasm of adjoining cells. In most cases, however, cells signal each other by releasing chemicals into the extracellular environment. In these cases, cell signaling can be divided into three general categories: (1) paracrine signaling; (2) synaptic signaling; and (3) endocrine signaling.

In **paracrine signaling** (fig. 6.29*a*), cells within an organ secrete regulatory molecules that diffuse through the extracellular matrix to nearby *target cells* (those that respond to the regulatory molecule). Paracrine regulation is considered to be *local,* because it involves the cells of a particular organ. Numerous paracrine regulators have been discovered that regulate organ growth and coordinate the activities of the different cells and tissues within an organ.

Synaptic signaling refers to the means by which neurons regulate their target cells. The axon of a neuron (see chapter 1, fig. 1.11) is said to *innervate* its target organ through a functional connection, or *synapse,* between the axon ending and the target cell. There is a small synaptic gap, or cleft, between the two cells, and chemical regulators called *neurotransmitters* are released by the axon endings (fig. 6.29*b*).

In **endocrine signaling,** the cells of endocrine glands secrete chemical regulators called *hormones* into the extracellular fluid. The hormones enter the blood and are carried by the blood to all the cells in the body. Only the target cells for a particular hormone, however, can respond to the hormone (fig. 6.29*c*).

In order for a target cell to respond to a hormone, neurotransmitter, or paracrine regulator, it must have specific **receptor proteins** for these molecules. A typical cell can have a few million receptor proteins. Of these, about 10,000 to 100,000 receptors can be of a given type in certain cells. Taking into account the total number of receptor genes, the alternative splicing of exons that can be produced from these genes, and the possible posttranslational modifications of proteins (chapter 3), scientists have estimated that the 200 different cell types found in the human body may have as many as 30,000 different types of receptor proteins for different regulatory molecules. This great diversity allows the many regulatory molecules in the body to exert fine control over the physiology of our tissues and organs.

These receptor proteins may be located on the outer surface of the plasma membrane of the target cells, or they may be located intracellularly in either the cytoplasm or nucleus. The location of the receptor proteins depends on whether the regulatory molecule can penetrate the plasma membrane of the target cell (fig. 6.30).

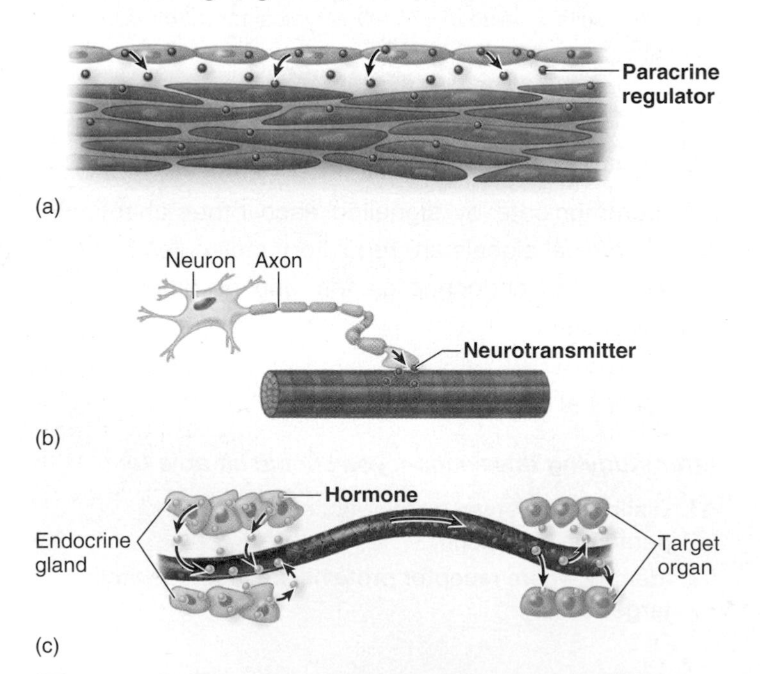

Figure 6.29 Chemical signaling between cells. (*a*) In paracrine signaling, regulatory molecules are released by the cells of an organ and target other cells in the same organ. (*b*) In synaptic signaling, the axon of a neuron releases a chemical neurotransmitter, which regulates a target cell. (*c*) In endocrine signaling, an endocrine gland secretes hormones into the blood, which carries the hormones to the target organs.

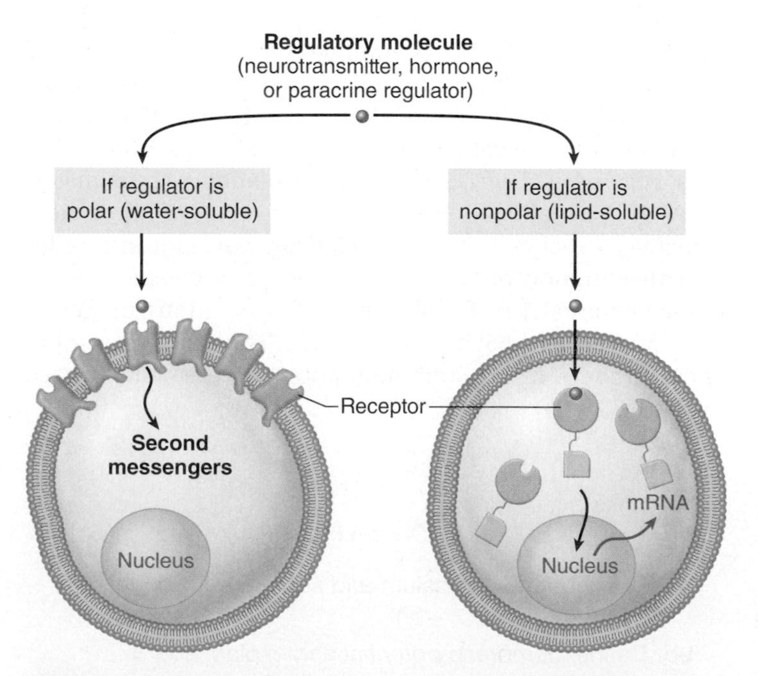

Figure 6.30 How regulatory molecules influence their target cells. Regulatory molecules that are polar bond to receptor proteins on the plasma membrane of a target cell, and the activated receptors send second messengers into the cytoplasm that mediate the actions of the hormone. Nonpolar regulatory molecules pass through the plasma membrane and bind to receptors within the cell. The activated receptors act in the nucleus to influence genetic expression. AP|R

If the regulatory molecule is nonpolar, it can diffuse through the cell membrane and enter the target cell. Such nonpolar regulatory molecules include steroid hormones, thyroid hormones, and nitric oxide gas (a paracrine regulator). In these cases, the receptor proteins are intracellular in location. Regulatory molecules that are large or polar—such as epinephrine (an amine hormone), acetylcholine (an amine neurotransmitter), and insulin (a polypeptide hormone)—cannot enter their target cells. In these cases, the receptor proteins are located on the outer surface of the plasma membrane.

Second Messengers

If a polar regulatory molecule binds to a receptor protein in the plasma membrane, how can it influence affairs deep in the cell? Even though the regulatory molecule doesn't enter the cell, it somehow has to change the activity of specific proteins, including enzyme proteins, within the cytoplasm. This feat is accomplished by means of intermediaries, known as **second messengers,** sent into the cytoplasm from the receptor proteins in the plasma membrane (fig. 6.30).

Second messengers may be ions (most commonly Ca^{2+}) that enter the cell from the extracellular fluid, or molecules produced within the cell cytoplasm in response to the binding of polar regulatory molecules to their receptors in the plasma membrane. One important second-messenger molecule is **cyclic adenosine monophosphate** (abbreviated **cyclic AMP,** or **cAMP**). The details of this regulation are described in conjunction with neural and endocrine regulation in the next several chapters (for example, see chapter 7, fig. 7.31). However, the following general sequence of events can be described here:

1. The polar regulatory molecule binds to its receptor in the plasma membrane.
2. This indirectly activates an enzyme in the plasma membrane that produces cyclic AMP from its precursor, ATP, in the cell cytoplasm.
3. Cyclic AMP concentrations increase, activating previously inactive enzymes in the cytoplasm.
4. The enzymes activated by cAMP then change the activities of the cell to produce the action of the regulatory molecule.

The polar regulatory molecule (neurotransmitter, hormone, or paracrine regulator) doesn't enter the cell, and so its actions are produced by the second messenger. For example, because the hormone epinephrine (adrenalin) uses cAMP as a second messenger in its stimulation of the heart, these effects are actually produced by cAMP within the heart cells. Cyclic AMP and several other second messengers are discussed in conjunction with the action of particular hormones in chapter 11, section 11.2.

G-Proteins

Notice that, in the second step of the previous list, the binding of the polar regulatory molecule to its receptor activates an enzyme protein in the plasma membrane *indirectly.* This is because the receptor protein and the enzyme protein are in different locations within the plasma membrane. Thus, there has to be something that travels in the plasma membrane between the receptor and the enzyme, so that the enzyme can become activated. In 1994 the Nobel Prize in Physiology or Medicine was awarded for the discovery of the **G-proteins:** three protein subunits that shuttle between receptors and different membrane effector proteins, including specific enzymes and ion channels. The three G-protein subunits are designated by the Greek letters alpha, beta, and gamma (α, β, and γ).

When the regulatory molecule reaches the plasma membrane of its target cell and binds to its receptor, the alpha subunit dissociates from the beta-gamma subunits (which stay attached to each other). The dissociation of the alpha from the beta-gamma subunits occurs because the alpha subunit releases GDP (guanosine diphosphate) and binds to GTP (guanosine triphosphate). The alpha subunit (or in some cases the beta gamma subunits) then moves through the membrane and binds to the effector protein, which is an enzyme or ion channel. This temporarily activates the enzyme or operates (opens or closes) the ion channel. Then, the alpha subunit hydrolyzes the GTP into GDP and P_i (inorganic phosphate), which causes the three subunits to reaggregate and move back to the receptor protein. This cycle is illustrated in figure 6.31.

The effector protein in figure 6.31 may be an enzyme, such as the enzyme that produces the second-messenger molecule cyclic AMP. This may be seen in the action of epinephrine and norepinephrine on the heart, shown in chapter 7, figure 7.31. Or the effector protein may be an ion channel, as can be seen in the way that acetylcholine (a neurotransmitter) causes the heart rate to slow (shown in chapter 7, fig. 7.27). Because there are an estimated 400 to 500 different G-protein-coupled receptors for neurotransmitters, hormones, and paracrine regulators (plus several hundred more G-protein-coupled receptors producing sensations of smell and taste), there is great diversity in their effects. Thus, specific cases are best considered in conjunction with the nervous, sensory, and endocrine systems in the chapters that follow.

CHECKPOINT

11. Distinguish between synaptic, endocrine, and paracrine regulation.
12. Identify the location of the receptor proteins for different regulatory molecules.

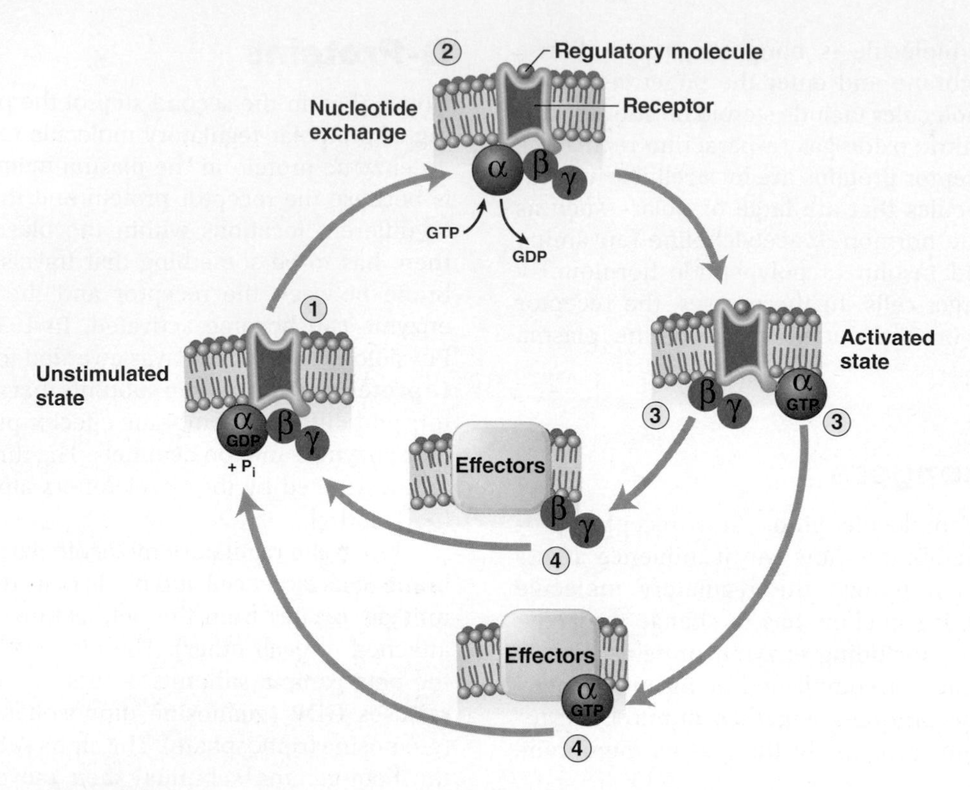

Figure 6.31 The G-protein cycle. (1) When the receptor is not bound to the regulatory molecule, the three G-protein subunits are aggregated together with the receptor, and the α subunit binds GDP. (2) When the regulatory molecule attaches to its receptor, the α subunit releases GDP and binds GTP; this allows the α subunit to dissociate from the βγ subunits. (3) Either the α subunit or the βγ complex moves through the membrane and binds to the effector protein (an enzyme or ion channel). (4) The α subunit splits GTP into GDP and P_i, causing the α and βγ subunits to reaggregate and bind to the unstimulated receptor once more. AP|R

Clinical Investigation SUMMARY

Jessica's hyperglycemia caused her renal carrier proteins to become saturated, resulting in glycosuria (glucose in the urine). The elimination of glucose in the urine and its consequent osmotic effects caused the urinary excretion of an excessive amount of water, resulting in dehydration. This raised the plasma osmolality, stimulating the thirst center in the hypothalamus. (Hyperglycemia and excessive thirst and urination are cardinal signs of diabetes mellitus.) Further, the loss of plasma water (increased plasma osmolality) caused an increase in the concentration of plasma solutes, including K^+. The resulting hyperkalemia affected the membrane potential of myocardial cells of the heart, producing electrical abnormalities that were revealed in Jessica's electrocardiogram.

See the additional chapter 6 Clinical Investigation on *Recessive Genetic Kidney Disorder* in the Connect site for this text at www.mhhe.com/Fox13.

Mc Graw Hill connect® | ANATOMY & PHYSIOLOGY

Visit this book's website at **www.mhhe.com/Fox13** for:

- Chapter quizzes, interactive learning exercises, and other study tools
- Additional clinical investigations
- Access to LearnSmart—an adaptive diagnostic tool that constantly assesses student knowledge of course material
- Ph.I.L.S. 4.0—physiology interactive lab simulations that may be used to supplement or substitute for wet labs

Interactions

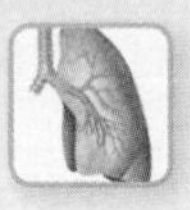
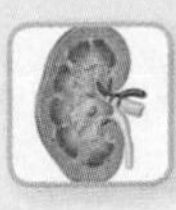

HPer Links of Membrane Transport Concepts to the Body Systems

Skeletal System

- Osteoblasts secrete Ca^{2+} and PO_4^{3-} into the extracellular matrix, forming calcium phosphate crystals that account for the hardness of bone (p. 690)

Nervous System

- Glucose enters neurons by facilitated diffusion (p. 143)
- Voltage-gated ion channels produce action potentials, or nerve impulses (p. 174)
- Ion channels in particular regions of a neuron open in response to binding to a chemical ligand known as a neurotransmitter (p. 183)
- Neurotransmitters are released by axons through the process of exocytosis (p. 182)
- Sensory stimuli generally cause the opening of ion channels and depolarization of receptor cells (p. 269)

Endocrine System

- Lipophilic hormones pass through the cell membrane of their target cells, where they then bind to receptors in the cytoplasm or nucleus (p. 323)
- Active transport Ca^+ pumps and the passive diffusion of Ca^+ are important in mediating the actions of some hormones (p. 328)
- Insulin stimulates the facilitative diffusion of glucose into skeletal muscle cells (p. 347)

Muscular System

- Exercise increases the number of carriers for the facilitative diffusion of glucose in the muscle cell membrane (p. 378)
- Ca^{2+} transport processes in the endoplasmic reticulum of skeletal muscle fibers are important in the regulation of muscle contraction (p. 372)
- Voltage-gated Ca^{2+} channels in the cell membrane of smooth muscle open in response to depolarization, producing contraction of the muscle (p. 395)

Circulatory System

- Transport processes through the capillary endothelial cells of the brain are needed in order for molecules to cross the blood-brain barrier and enter the brain (p. 171)
- Ion diffusion across the plasma membrane of myocardial cells is responsible for the electrical activity of the heart (p. 425)
- The LDL carriers for blood cholesterol are taken into arterial smooth muscle cells by receptor-mediated endocytosis (p. 437)

Immune System

- B lymphocytes secrete antibody proteins that function in humoral (antibody mediated) immunity (p. 503)
- T lymphocytes secrete polypeptides called cytokines that promote the cell mediated immune response (p. 508)
- Antigen-presenting cells engulf foreign proteins by pinocytosis, modify these proteins, and present them to T lymphocytes (p. 510)

Respiratory System

- Oxygen and carbon dioxide pass through the cells of the pulmonary alveoli (airsacs) by simple diffusion (p. 533)
- Surfactant is secreted into pulmonary alveoli by exocytosis (p. 540)

Urinary System

- Urine is produced as a filtrate of blood plasma, but most of the filtered water is reabsorbed back into the blood by osmosis (p. 590)
- Osmosis across the wall of the renal tubules is promoted by membrane pores known as aquaporins (p. 596)
- Transport of urea occurs passively across particular regions of the renal tubules (p. 595)
- Antidiuretic hormone stimulates the permeability of the renal tubule to water (p. 595)
- Aldosterone stimulates Na^+ transport in a region of the renal tubule (p. 604)
- Glucose and amino acids are reabsorbed by secondary active transport (p. 603)

Digestive System

- Cells in the stomach have a membrane H^+/K^+ ATPase active transport pump that creates an extremely acidic gastric juice (p. 625)
- Water is absorbed in the intestine by osmosis following the absorption of sodium chloride (p. 634)
- An intestinal membrane carrier protein transports dipeptides and tripeptides from the intestinal lumen into the epithelial cells (p. 651)

SUMMARY

6.1 Extracellular Environment 131

A. Body fluids are divided into an intracellular compartment and an extracellular compartment.
  1. The extracellular compartment consists of blood plasma and interstitial, or tissue, fluid.
  2. Interstitial fluid is derived from plasma and returns to plasma.

B. The extracellular matrix consists of protein fibers of collagen and elastin and an amorphorus ground substance.
  1. The collagen and elastin fibers provide structural support.
  2. The ground substance contains glycoproteins and proteoglycans forming a hydrated gel, which contains most of the interstitial fluid.

6.2 Diffusion and Osmosis 133

A. Diffusion is the net movement of molecules or ions from regions of higher to regions of lower concentration.
  1. This is a type of passive transport—energy is provided by the thermal energy of the molecules, not by cellular metabolism.
  2. Net diffusion stops when the concentration is equal on both sides of the membrane.

B. The rate of diffusion is dependent on a variety of factors.
  1. The rate of diffusion depends on the concentration difference across the two sides of the membrane.
  2. The rate depends on the permeability of the plasma membrane to the diffusing substance.
  3. The rate depends on the temperature of the solution.
  4. The rate of diffusion through a membrane is also directly proportional to the surface area of the membrane, which can be increased by such adaptations as microvilli.

C. Simple diffusion is the type of passive transport in which small molecules and inorganic ions move through the plasma membrane.
  1. Inorganic ions such as Na^+ and K^+ pass through specific channels in the membrane.
  2. Steroid hormones and other lipids can pass directly through the phospholipid layers of the membrane by simple diffusion.

D. Osmosis is the simple diffusion of solvent (water) through a membrane that is more permeable to the solvent than it is to the solute.
  1. Water moves from the solution that is more dilute to the solution that has a higher solute concentration.
  2. Osmosis depends on a difference in total solute concentration, not on the chemical nature of the solute.
    a. The concentration of total solute, in moles per kilogram (liter) of water, is measured in osmolality units.
    b. The solution with the higher osmolality has the higher osmotic pressure.
    c. Water moves by osmosis from the solution of lower osmolality and osmotic pressure to the solution of higher osmolality and osmotic pressure.
  3. Solutions containing osmotically active solutes that have the same osmotic pressure as plasma (such as 0.9% NaCl and 5% glucose) are said to be isotonic to plasma.
    a. Solutions with a lower osmotic pressure are hypotonic; those with a higher osmotic pressure are hypertonic.
    b. Cells in a hypotonic solution gain water and swell; those in a hypertonic solution lose water and shrink (crenate).
  4. The osmolality and osmotic pressure of the plasma is detected by osmoreceptors in the hypothalamus of the brain and maintained within a normal range by the action of antidiuretic hormone (ADH) released from the posterior pituitary.
    a. Increased osmolality of the blood stimulates the osmoreceptors.
    b. Stimulation of the osmoreceptors causes thirst and triggers the release of antidiuretic hormone (ADH) from the posterior pituitary.
    c. ADH promotes water retention by the kidneys, which serves to maintain a normal blood volume and osmolality.

6.3 Carrier-Mediated Transport 142

A. The passage of glucose, amino acids, and other polar molecules through the plasma membrane is mediated by carrier proteins in the cell membrane.
  1. Carrier-mediated transport exhibits the properties of specificity, competition, and saturation.
  2. The transport rate of molecules such as glucose reaches a maximum when the carriers are saturated. This maximum rate is called the transport maximum (T_m).

B. The transport of molecules such as glucose from the side of higher to the side of lower concentration by means of membrane carriers is called facilitated diffusion.
  1. Like simple diffusion, facilitated diffusion is passive transport—cellular energy is not required.
  2. Unlike simple diffusion, facilitated diffusion displays the properties of specificity, competition, and saturation.

C. The active transport of molecules and ions across a membrane requires the expenditure of cellular energy (ATP).
  1. In active transport, carriers move molecules or ions from the side of lower to the side of higher concentration.
  2. One example of active transport is the action of the Na^+/K^+ pump.
    a. Sodium is more concentrated on the outside of the cell, whereas potassium is more concentrated on the inside of the cell.
    b. The Na^+/K^+ pump helps to maintain these concentration differences by transporting Na^+ out of the cell and K^+ into the cell.

6.4 The Membrane Potential 149

A. The cytoplasm of the cell contains negatively charged organic ions (anions) that cannot leave the cell—they are "fixed" anions.
   1. These fixed anions attract K^+, which is the inorganic ion that can pass through the plasma membrane most easily.
   2. As a result of this electrical attraction, the concentration of K^+ within the cell is greater than the concentration of K^+ in the extracellular fluid.
   3. If K^+ were the only diffusible ion, the concentrations of K^+ on the inside and outside of the cell would reach an equilibrium.
      a. At this point, the rate of K^+ entry (due to electrical attraction) would equal the rate of K^+ exit (due to diffusion).
      b. At this equilibrium, there would still be a higher concentration of negative charges within the cell (because of the fixed anions) than outside the cell.
      c. At this equilibrium, the inside of the cell would be 90 millivolts negative (−90 mV) compared to the outside of the cell. This potential difference is called the K^+ equilibrium potential (E_K).
   4. The resting membrane potential is less than E_K (usually −65 mV to −85 mV) because some Na^+ can also enter the cell.
      a. Na^+ is more highly concentrated outside than inside the cell, and the inside of the cell is negative. These forces attract Na^+ into the cell.
      b. The rate of Na^+ entry is generally slow because the membrane is usually not very permeable to Na^+.

B. The slow rate of Na^+ entry is accompanied by a slow rate of K^+ leakage out of the cell.
   1. The Na^+/K^+ pump counters this leakage, thus maintaining constant concentrations and a constant resting membrane potential.
   2. Most cells in the body contain numerous Na^+/K^+ pumps that require a constant expenditure of energy.
   3. The Na^+/K^+ pump itself contributes to the membrane potential because it pumps more Na^+ out than it pumps K^+ in (by a ratio of three to two).

6.5 Cell Signaling 153

A. Cells signal each other generally by secreting regulatory molecules into the extracellular fluid.

B. There are three categories of chemical regulation between cells.
   1. Paracrine signaling refers to the release of regulatory molecules that act within the organ in which they are made.
   2. Synaptic signaling refers to the release of chemical neurotransmitters by axon endings.
   3. Endocrine signaling refers to the release of regulatory molecules called hormones, which travel in the blood to their target cells.

C. Regulatory molecules bind to receptor proteins in their target cells.
   1. The receptor proteins are specific for the regulatory molecule; there may be as many as 30,000 different types of receptor proteins for regulatory molecules in the body.
   2. If the regulatory molecule is nonpolar, it can penetrate the plasma membrane; in that case, its receptor proteins are located within the cell, in the cytoplasm or nucleus.
   3. If the regulatory molecule is polar, it cannot penetrate the plasma membrane; in that case, its receptors are located in the plasma membrane with their binding sites exposed to the extracellular fluid.
   4. When a polar regulatory molecule binds to its receptor on the plasma membrane, it stimulates the release of second messengers, which are molecules or ions that enter the cytoplasm and produce the action of the regulator within its target cell.
      a. For example, many polar regulatory molecules bind to receptors that indirectly activate an enzyme that converts ATP into cyclic AMP.
      b. The rise in cyclic AMP within the cell cytoplasm then activates enzymes, and in that way carries out the action of the regulatory molecule within the cell.
   5. Some plasma membrane receptor proteins are G-protein-coupled receptors.
      a. There are three G-protein subunits, designated alpha, beta, and gamma, which are aggregated at a plasma membrane receptor protein.
      b. When the receptor is activated by binding to its regulatory molecule, the G-proteins dissociate.
      c. Then, either the alpha subunit or the beta-gamma complex moves through the membrane to an effector protein, which is an enzyme or an ion channel.
      d. In this way, the effector protein (enzyme or ion channel) and the receptor protein can be in different locations in the plasma membrane.

REVIEW ACTIVITIES

Test Your Knowledge

1. The movement of water across a plasma membrane occurs by
   a. an active transport water pump.
   b. a facilitated diffusion carrier.
   c. simple diffusion through membrane channels.
   d. all of these.
2. Which of these statements about the facilitated diffusion of glucose is *true?*
   a. There is a net movement from the region of lower to the region of higher concentration.
   b. Carrier proteins in the cell membrane are required for this transport.
   c. This transport requires energy obtained from ATP.
   d. It is an example of cotransport.
3. If a poison such as cyanide stopped the production of ATP, which of the following transport processes would cease?
   a. The movement of Na^+ out of a cell
   b. Osmosis
   c. The movement of K^+ out of a cell
   d. All of these
4. Red blood cells crenate in
   a. a hypotonic solution.
   b. an isotonic solution.
   c. a hypertonic solution.
5. Plasma has an osmolality of about 300 mOsm. The osmolality of isotonic saline is equal to
   a. 150 mOsm.
   b. 300 mOsm.
   c. 600 mOsm.
   d. none of these.
6. Which of these statements comparing a 0.5 *m* NaCl solution and a 1.0 *m* glucose solution is *true?*
   a. They have the same osmolality.
   b. They have the same osmotic pressure.
   c. They are isotonic to each other.
   d. All of these are true.
7. The most important diffusible ion in the establishment of the membrane potential is
   a. K^+.
   b. Na^+.
   c. Ca^{2+}.
   d. Cl^-.
8. Which of these statements regarding an increase in blood osmolality is *true?*
   a. It can occur as a result of dehydration.
   b. It causes a decrease in blood osmotic pressure.
   c. It is accompanied by a decrease in ADH secretion.
   d. All of these are true.
9. In hyperkalemia, the resting membrane potential
   a. moves farther from 0 millivolts.
   b. moves closer to 0 millivolts.
   c. remains unaffected.
10. Which of these statements about the Na^+/K^+ pump is *true?*
    a. Na^+ is actively transported into the cell.
    b. K^+ is actively transported out of the cell.
    c. An equal number of Na^+ and K^+ ions are transported with each cycle of the pump.
    d. The pumps are constantly active in all cells.
11. Which of these statements about carrier-mediated facilitated diffusion is *true?*
    a. It uses cellular ATP.
    b. It is used for cellular uptake of blood glucose.
    c. It is a form of active transport.
    d. None of these are true.
12. Which of these is *not* an example of cotransport?
    a. Movement of glucose and Na^+ through the apical epithelial membrane in the intestinal epithelium
    b. Movement of Na^+ and K^+ through the action of the Na^+/K^+ pumps
    c. Movement of Na^+ and glucose across the kidney tubules
    d. Movement of Na^+ into a cell while Ca^{2+} moves out
13. The resting membrane potential of a neuron or muscle cell is
    a. equal to the potassium equilibrium potential.
    b. equal to the sodium equilibrium potential.
    c. somewhat less negative than the potassium equilibrium potential.
    d. somewhat more positive than the sodium equilibrium potential.
    e. not changed by stimulation.
14. Suppose that gated ion channels for Na^+ or Ca^{2+} opened in the plasma membrane of a muscle cell. The membrane potential of that cell would
    a. move toward the equilibrium potential for that ion.
    b. become less negative than the resting membrane potential.
    c. move farther away from the potassium equilibrium potential.
    d. all of these.
15. Which of the following questions regarding second messengers is *false?*
    a. They are needed to mediate the action of nonpolar regulatory molecules.
    b. They are released from the plasma membrane into the cytoplasm of cells.

c. They are produced in response to the binding of regulatory molecules to receptors in the plasma membrane.
d. They produce the intracellular actions of polar regulatory molecules.

Test Your Understanding

16. Describe the conditions required to produce osmosis and explain why osmosis occurs under these conditions.
17. Explain how simple diffusion can be distinguished from facilitated diffusion and how active transport can be distinguished from passive transport.
18. Compare the resting membrane potential of a neuron with the potassium and sodium equilibrium potentials. Explain how this comparison relates to the relative permeabilities of the resting plasma membrane to these two ions.
19. Describe how the Na^+/K^+ pumps contribute to the resting membrane potential. Also, describe how the membrane potential would be affected if (1) gated Na^+ channels were to open, and (2) gated K^+ channels were to open.
20. Explain how the permeability of a membrane to glucose and to water can be regulated by the insertion or removal of carrier proteins, and give examples.
21. What are the factors that influence the rate of diffusion across a plasma membrane? What structural features are often seen in epithelial membranes specialized for rapid diffusion?
22. Describe the cause-and-effect sequence whereby a genetic defect results in improper cellular transport and the symptoms of cystic fibrosis.
23. Using the principles of osmosis, explain why movement of Na^+ through a plasma membrane is followed by movement of water. Use this concept to explain the rationale on which oral rehydration therapy is based.
24. Distinguish between primary active transport and secondary active transport, and between cotransport and countertransport. Give examples of each.
25. Describe the different types of regulatory molecules found in the body. What are the target cells for each type of regulatory molecule?
26. How do nonpolar and polar regulatory molecules differ in terms of the location of their receptor proteins in the target cells and the mechanism of their actions?
27. What are G-protein-coupled receptors? Explain their function in regard to how particular regulatory molecules influence different effector proteins in the membrane.

Test Your Analytical Ability

28. Mannitol is a sugar that does not pass through the walls of blood capillaries in the brain (does not cross the "blood-brain barrier," as described in chapter 7). It also does not cross the walls of kidney tubules, the structures that transport blood filtrate to become urine (see chapter 17). Explain why mannitol can be described as osmotically active. How might its clinical administration help to prevent swelling of the brain in head trauma? Also, explain the effect it might have on the water content of urine.
29. Discuss carrier-mediated transport. How could you experimentally distinguish between the different types of carrier-mediated transport?
30. Remembering the effect of cyanide (described in chapter 5), explain how you might determine the extent to which the Na^+/K^+ pumps contribute to the resting membrane potential. Using a measurement of the resting membrane potential as your guide, how could you experimentally determine the relative permeability of the plasma membrane to Na^+ and K^+?
31. Using only the information in this chapter, explain how insulin (a polar polypeptide hormone) causes increased transport of plasma glucose into muscle cells.
32. Using only the information in this chapter, explain how antidiuretic hormone (ADH, also called vasopressin)—a polar polypeptide hormone—can stimulate epithelial cells in the kidneys to become more permeable to water.
33. Epinephrine increases the heart rate and causes the bronchioles (airways) to dilate by using cyclic AMP as a second messenger. Suppose a drug increased the cyclic AMP in heart and bronchiolar smooth muscle cells; what effects would the drug have? Could you give a person intravenous cyclic AMP and duplicate the action of epinephrine? Explain.

Test Your Quantitative Ability

Suppose a semipermeable membrane separates two solutions. One solution has 0.72 g glucose to 1.0 L of water; the other has 0.117 g NaCl to 1.0 L of water. Given that glucose has a molecular weight of 180 and NaCl has a molecular weight of 58.5, perform the following calculations.

34. Calculate the molality and osmolality of each solution (see fig. 6.10).
35. Given your answers, state whether osmosis will occur and if so, in which direction (assuming that the membrane is permeable to water but not to glucose or NaCl).

Use the Nernst equation and the ion concentration provided in figure 6.26 to perform the following calculations.

36. Calculate the equilibrium potential for K^+ (E_K) if its extracellular concentration rises from 5 mM to 10 mM. Comparing this to the normal E_K, is the change a depolarization or hyperpolarization?
37. Using the chloride (Cl^-) concentrations provided, calculate the equilibrium potential for Cl^-. Given your answer, should Cl^- enter or leave the cell if the plasma membrane suddenly becomes permeable to it (given a membrane potential of −70 mV)?

CHAPTER

7

The Nervous System

Neurons and Synapses

REFRESH YOUR MEMORY

Before you begin this chapter, you may want to review the following concepts from previous chapters:

CHAPTER OUTLINE

Clinical Investigation

Sandra's grades have been improving, and she treats herself to dinner at a seafood restaurant. However, after just beginning to eat some mussels and clams gathered from the local seashore, she complains of severe muscle weakness. Paramedics are called, and when they examine Sandra, they notice that she has a droopy eyelid and that her purse contains a prescription bottle for an MAO inhibitor. When questioned, Sandra states that she had a recent Botox treatment, and the medication was prescribed to treat her clinical depression. Further investigation reveals that the shellfish were gathered from waters at the beginning of a red tide and that Sandra's blood pressure was in the normal range.

Some of the new terms and concepts you will encounter include:

- Voltage-gated channels and the action of saxitoxin
- Neurotransmitter release and the action of botulinum toxin
- Monoamine neurotransmitters and monoamine oxidase (MAO)

7.1 NEURONS AND SUPPORTING CELLS

The nervous system is composed of neurons, which produce and conduct electrochemical impulses, and supporting cells, which assist the functions of neurons. Neurons are classified functionally and structurally; the various types of supporting cells perform specialized functions.

LEARNING OUTCOMES

After studying this section, you should be able to:

1. **Describe the different types of neurons and supporting cells, and identify their functions.**
2. **Identify the myelin sheath and describe how it is formed in the CNS and PNS.**
3. **Describe the nature and significance of the blood-brain barrier.**

The nervous system is divided into the **central nervous system (CNS),** which includes the brain and spinal cord, and the **peripheral nervous system (PNS),** which includes the *cranial nerves* arising from the brain and the *spinal nerves* arising from the spinal cord.

The nervous system is composed of only two principal types of cells—neurons and supporting cells. **Neurons** are the basic structural and functional units of the nervous system. They are specialized to respond to physical and chemical stimuli, conduct electrochemical impulses, and release chemical regulators. Through these activities, neurons enable the perception of sensory stimuli, learning, memory, and the control of muscles and glands. Most neurons cannot divide by mitosis, although many can regenerate a severed portion or sprout small new branches under certain conditions.

Supporting cells aid the functions of neurons and are about five times more abundant than neurons. In common usage, supporting cells are collectively called **neuroglia,** or simply **glial cells** (from the Middle Greek *glia* = glue). Unlike neurons, which do not divide mitotically (except for particular neural stem cells; chapter 8, section 8.1), glial cells are able to divide by mitosis. This helps to explain why brain tumors in adults are usually composed of glial cells rather than of neurons.

Neurons

Although neurons vary considerably in size and shape, they generally have three principal regions: (1) a cell body, (2) dendrites, and (3) an axon (figs. 7.1 and 7.2). Dendrites and axons can be referred to generically as *processes,* or extensions from the cell body.

The **cell body** is the enlarged portion of the neuron that contains the nucleus. It is the "nutritional center" of the neuron where macromolecules are produced. The cell body and larger dendrites (but not axons) contain *Nissl bodies,* which are seen as dark-staining granules under the microscope. Nissl bodies are composed of large stacks of rough endoplasmic reticulum that are needed for the synthesis of membrane proteins. The cell bodies within the CNS are frequently clustered into groups called *nuclei* (not to be confused with the nucleus of a cell). Cell bodies in the PNS usually occur in clusters called *ganglia* (table 7.1).

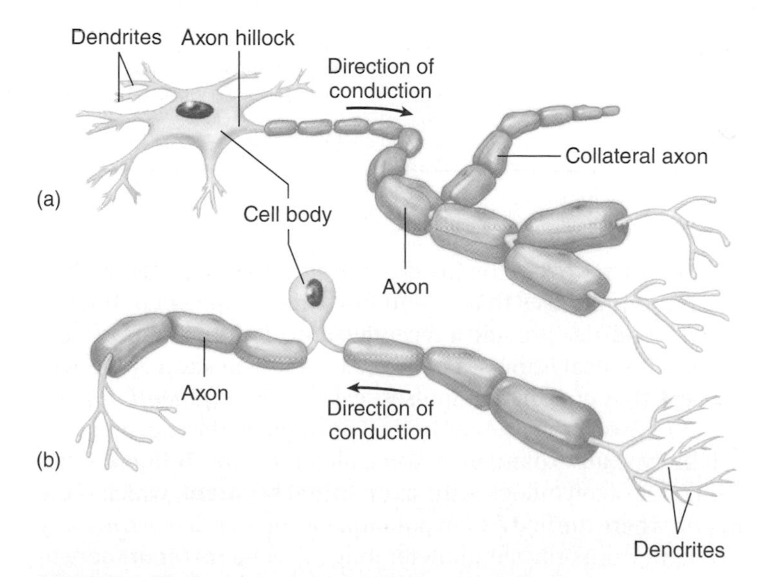

Figure 7.1 The structure of two kinds of neurons. A motor neuron (*a*) and a sensory neuron (*b*) are depicted here. AP|R

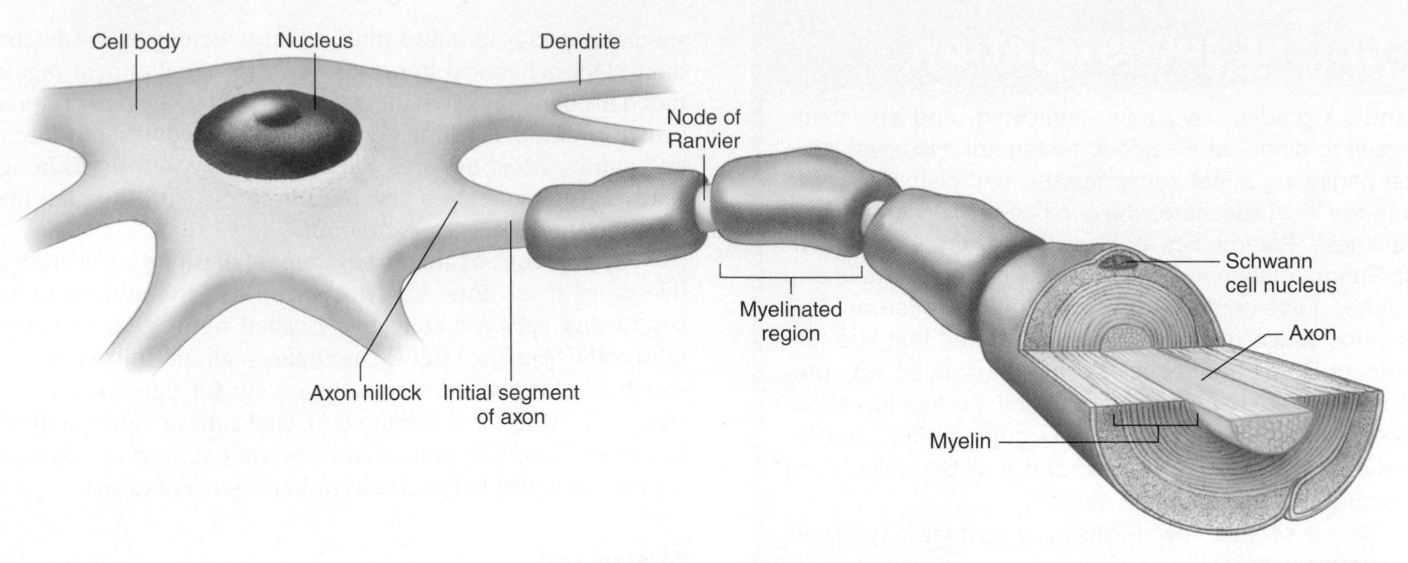

Figure 7.2 Parts of a neuron. The axon of this neuron is wrapped by Schwann cells, which form a myelin sheath. AP|R

Table 7.1 | Terminology Pertaining to the Nervous System

Term	Definition
Central nervous system (CNS)	Brain and spinal cord
Peripheral nervous system (PNS)	Nerves, ganglia, and nerve plexuses (outside of the CNS)
Association neuron (interneuron)	Multipolar neuron located entirely within the CNS
Sensory neuron (afferent neuron)	Neuron that transmits impulses from a sensory receptor into the CNS
Motor neuron (efferent neuron)	Neuron that transmits impulses from the CNS to an effector organ; for example, a muscle
Nerve	Cablelike collection of many axons in the PNS; may be "mixed" (contain both sensory and motor fibers)
Somatic motor nerve	Nerve that stimulates contraction of skeletal muscles
Autonomic motor nerve	Nerve that stimulates contraction (or inhibits contraction) of smooth muscle and cardiac muscle and that stimulates glandular secretion
Ganglion	Grouping of neuron cell bodies located outside the CNS
Nucleus	Grouping of neuron cell bodies within the CNS
Tract	Grouping of axons that interconnect regions of the CNS

Dendrites (from the Greek *dendron* = tree branch) are thin, branched processes that extend from the cytoplasm of the cell body. Dendrites provide a receptive area that transmits graded electrochemical impulses to the cell body. The **axon** is a longer process that conducts impulses, called *action potentials* (section 7.2), away from the cell body. The origin of the axon near the cell body is an expanded region called the **axon hillock.** Adjacent to the axon hillock is the **axon initial segment,** which is the region where the first action potentials are generated. Axons vary in length from only a millimeter long to over a meter or more in length (for axons that extend from the CNS to the foot). Toward their ends, axons can produce up to 200 or more branches called **axon collaterals,** and each of these can divide to synapse with many other neurons. In this way, a single CNS axon may synapse with as many as 30,000 to 60,000 other neurons.

Because axons can be quite long, special mechanisms are required to transport organelles and proteins from the cell body to the axon terminals. This **axonal transport** is energy-dependent and is often divided into a *fast component* and two *slow components.* The fast component (at 200 to 400 mm/day) mainly transports membranous vesicles (important for synaptic transmission, as discussed in section 7.3). One slow component (at 0.2 to 1 mm/day) transports microfilaments and microtubules of the cytoskeleton, while the other slow component (at 2 to 8 mm/day) transports over 200 different proteins, including those critical for synaptic function. The slow components appear to transport their cargo in fast bursts with frequent pauses, so that the overall rate of transport is much slower than that occurring in the fast component.

Axonal transport may occur from the cell body to the axon and dendrites. This direction is called **anterograde transport,**

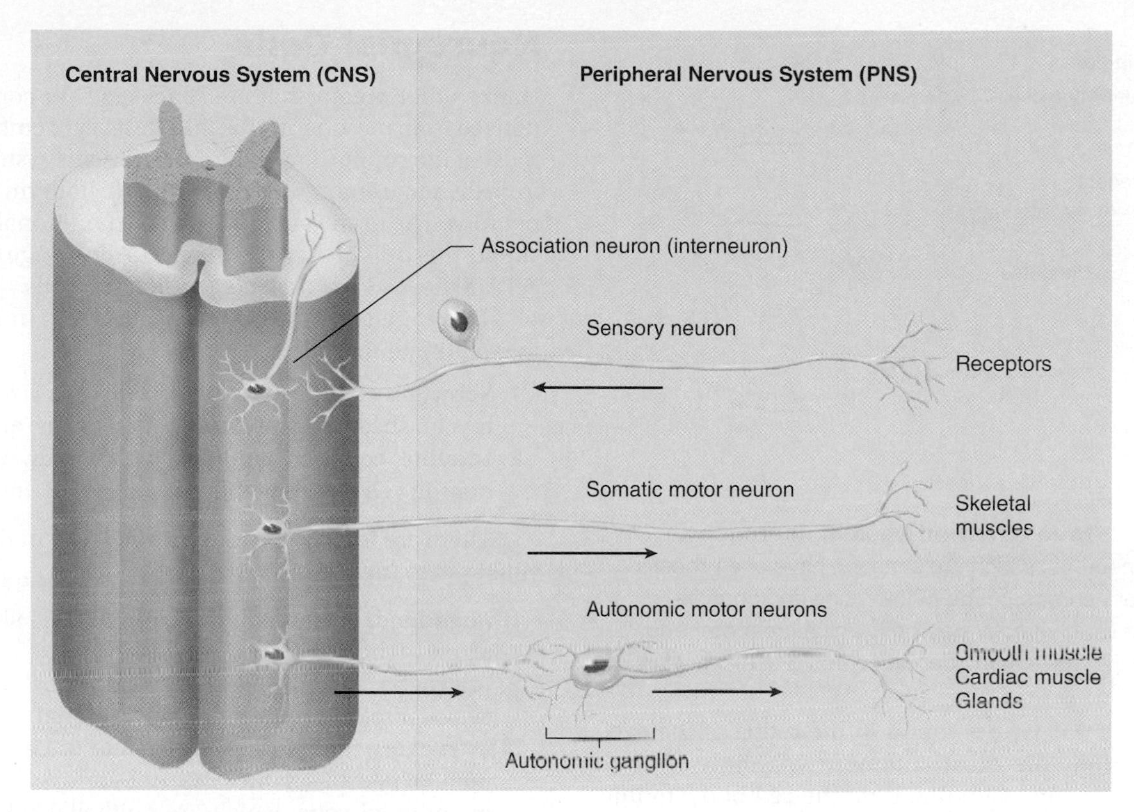

Figure 7.3 The relationship between CNS and PNS. Sensory and motor neurons of the peripheral nervous system carry information into and out of, respectively, the central nervous system (brain and spinal cord). APR

and involves molecular motors of *kinesin* proteins that move cargo along the microtubules of the cytoskeleton (chapter 3, section 3.2). For example, kinesin motors move synaptic vesicles, mitochondria, and ion channels from the cell body through the axon. Similar anterograde transport occurs in the dendrites, as kinesin moves postsynaptic receptors for neurotransmitters and ion channels along the microtubules in the dendrites.

By contrast, axonal transport in the opposite direction—that is, along the axon and dendrites toward the cell body—is known as **retrograde transport** and involves molecular motor proteins of *dyneins.* The dyneins move membranes, vesicles, and various molecules along microtubules of the cytoskeleton toward the cell body of the neuron. Retrograde transport can also be responsible for movement of herpes virus, rabies virus, and tetanus toxin from the nerve terminals into cell bodies.

Classification of Neurons and Nerves

Neurons may be classified according to their function or structure. The functional classification is based on the direction in which they conduct impulses, as indicated in figure 7.3. **Sensory,** or **afferent, neurons** conduct impulses from sensory receptors *into* the CNS. **Motor,** or **efferent, neurons** conduct impulses *out* of the CNS to effector organs (muscles and glands). **Association neurons,** or **interneurons,** are located entirely within the CNS and serve the associative, or integrative, functions of the nervous system.

There are two types of motor neurons: somatic and autonomic. **Somatic motor neurons** are responsible for both reflex and voluntary control of skeletal muscles. **Autonomic motor neurons** innervate (send axons to) the involuntary effectors—smooth muscle, cardiac muscle, and glands. The cell bodies of the autonomic neurons that innervate these organs are located outside the CNS in autonomic ganglia (fig. 7.3). There are two subdivisions of autonomic neurons: *sympathetic* and *parasympathetic.* Autonomic motor neurons, together with their central control centers, constitute the *autonomic nervous system,* the focus of chapter 9.

The structural classification of neurons is based on the number of processes that extend from the cell body of the neuron (fig. 7.4). **Pseudounipolar neurons** have a single short process that branches like a T to form a pair of longer processes. They are called pseudounipolar (from the Late Latin *pseudo* = false) because, although they originate with two processes, during early embryonic development their two processes converge and partially fuse. Sensory neurons are pseudounipolar—one of the branched processes receives sensory stimuli and produces nerve impulses; the other delivers these impulses to synapses within the brain or spinal cord. Anatomically, the part of the process that conducts impulses toward the cell body can be considered a dendrite, and the part that conducts impulses away from the cell body can be considered an axon. Functionally, however, the branched process behaves as a single, long axon that continuously conducts action potentials (nerve impulses). Only the small projections at the receptive end of the process function as typical dendrites, conducting graded electrochemical impulses rather than action potentials. **Bipolar neurons** have two processes,

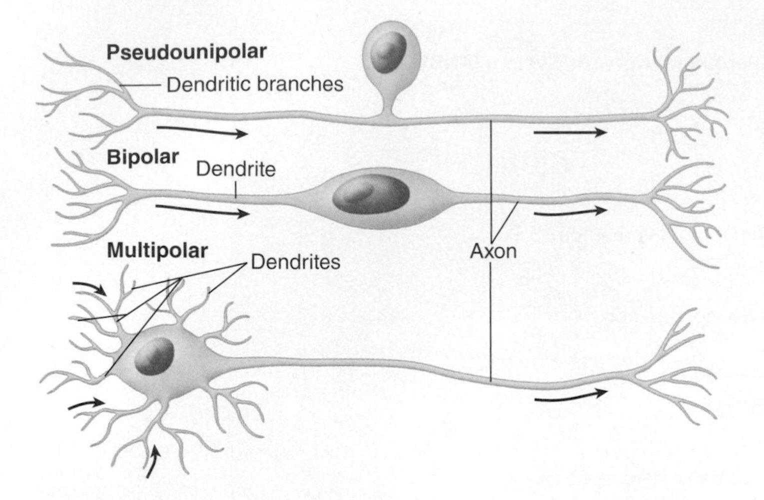

Figure 7.4 Three different types of neurons. Pseudounipolar neurons, which are sensory, have one process that splits. Bipolar neurons, found in the retina and cochlea, have two processes. Multipolar neurons, which are motor and association neurons, have many dendrites and one axon. AP|R

one at either end; this type is found in the retina of the eye. **Multipolar neurons,** the most common type, have several dendrites and one axon extending from the cell body; motor neurons are good examples of this type.

A **nerve** is a bundle of axons located outside the CNS. Most nerves are composed of both motor and sensory fibers and are thus called *mixed nerves.* Some of the cranial nerves, however, contain sensory fibers only. These are the nerves that serve the special senses of sight, hearing, taste, and smell. A bundle of axons in the CNS is called a **tract.**

Neuroglial Cells

Unlike other organs that are "packaged" in connective tissue derived from mesoderm (the middle layer of embryonic tissue), most of the supporting cells of the nervous system are derived from the same embryonic tissue layer (ectoderm) that produces neurons. The term *neuroglia* (or *glia*) traditionally refers to the supporting cells of the CNS, but in current usage the supporting cells of the PNS are often also called glial cells.

There are two types of neuroglial cells in the peripheral nervous system:

1. **Schwann cells** (also called *neurolemmocytes*), which form myelin sheaths around peripheral axons; and
2. **satellite cells,** or **ganglionic gliocytes,** which support neuron cell bodies within the ganglia of the PNS.

There are four types of neuroglial cells in the central nervous system (fig. 7.5):

1. **oligodendrocytes,** which form myelin sheaths around axons of the CNS;
2. **microglia,** which migrate through the CNS and phagocytose foreign and degenerated material;
3. **astrocytes,** which help to regulate the external environment of neurons in the CNS; and
4. **ependymal cells,** which are epithelial cells that line the ventricles (cavities) of the brain and the central canal of the spinal cord.

Microglia are unique among neuroglial cells in that they are of hematopoietic (bone marrow) origin, and indeed can be replenished by monocytes (a type of leukocyte) from the blood. Microglial cells in the "resting" state have a small cell

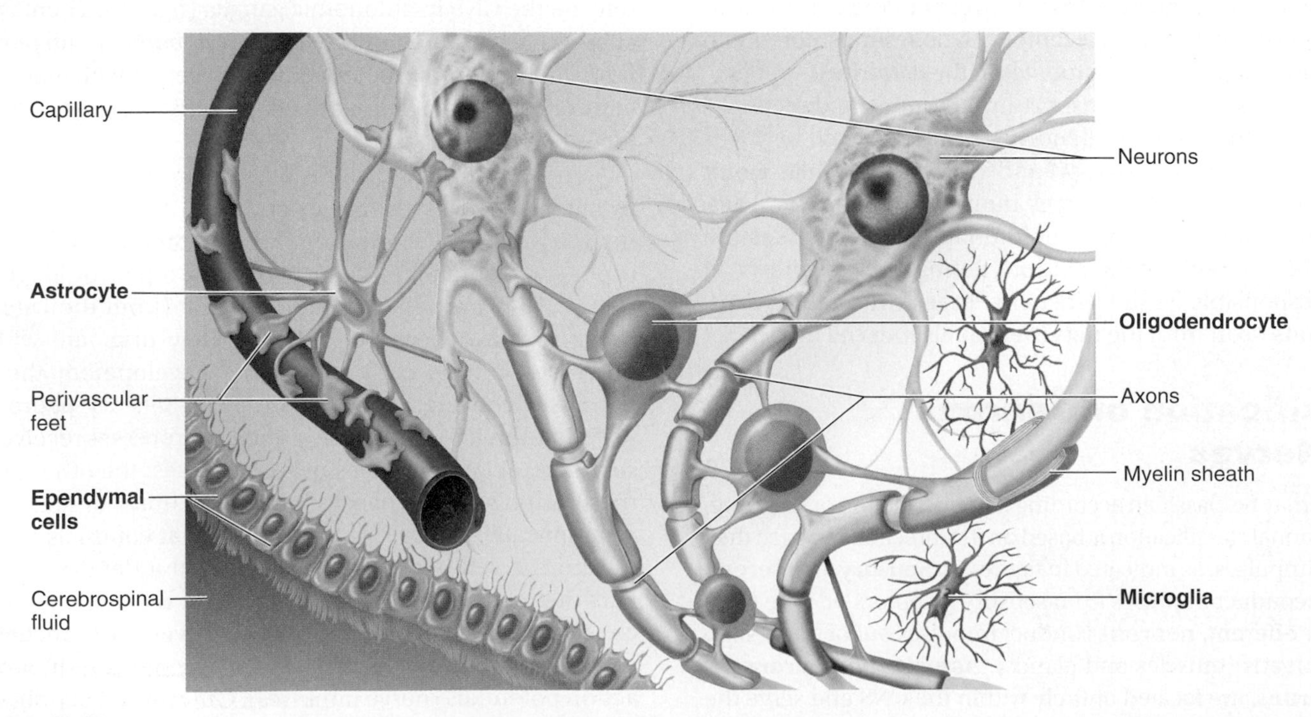

Figure 7.5 The different types of neuroglial cells. Myelin sheaths around axons are formed in the CNS by oligodendrocytes. Astrocytes have extensions that surround both blood capillaries and neurons. Microglia are phagocytic, and ependymal cells line the brain ventricles and central canal of the spinal cord. AP|R

Table 7.2 | Neuroglial Cells and Their Functions

Cell Type	Location	Functions
Schwann cells	PNS	Also called neurolemmocytes, produce the myelin sheaths around the myelinated axons of the peripheral nervous system; surround all PNS axons (myelinated and nonmyelinated) to form a neurilemmal sheath, or sheath of Schwann
Satellite cells	PNS	Support functions of neurons within sensory and autonomic ganglia; also called ganglionic gliocytes
Oligodendrocytes	CNS	Form myelin sheaths around central axons, producing "white matter" of the CNS
Microglia	CNS	Phagocytose pathogens and cellular debris in the CNS
Astrocytes	CNS	Cover capillaries of the CNS and induce the blood-brain barrier; interact metabolically with neurons and modify the extracellular environment of neurons
Ependymal cells	CNS	Form the epithelial lining of brain cavities (ventricles) and the central canal of the spinal cord; cover tufts of capillaries to form choroid plexuses—structures that produce cerebrospinal fluid

body and many fine cellular processes. These cells are not truly resting, however, because their processes are constantly waving as the cells survey their extracellular environment. Infection, trauma, or any altered state can lead to *microglial activation,* in which the cells become amoeboid in shape and are transformed into phagocytic, motile cells. They follow chemokines (chemical attractants, including ATP) to the site of the infection or damage, where they may proliferate by cell division. They can kill exogenous pathogens; remove damaged dendrites, axon terminals, myelin, and other debris within the CNS; and release anti-inflammatory chemicals. Although microglia are needed for repair, overactive microglial cells may release free radicals (chapter 19, section 19.1) that contribute to neurodegenerative diseases. The functions of the other neuroglial cells are described in detail in the next sections and summarized in table 7.2.

Neurilemma and Myelin Sheath

All axons in the PNS (myelinated and unmyelinated) are surrounded by a continuous living sheath of Schwann cells, known as the **neurilemma,** or **sheath of Schwann.** The axons of the CNS, by contrast, lack a neurilemma (Schwann cells are found only in the PNS). This is significant in terms of regeneration of damaged axons, as will be described shortly.

Some axons in the PNS and CNS are surrounded by a **myelin sheath.** In the PNS, this insulating covering is formed by successive wrappings of the cell membrane of Schwann cells; in the CNS, it is formed by oligodendrocytes. Those axons smaller than 2 micrometers (2 μm) in diameter are usually *unmyelinated* (have no myelin sheath), whereas those that are larger are likely to be *myelinated.* Myelinated axons conduct impulses more rapidly than those that are unmyelinated.

Myelin Sheath in PNS

In the process of myelin formation in the PNS, Schwann cells attach to and roll around the axon, much like a roll of electrician's tape is wrapped around a wire. Unlike electrician's tape, however, the Schwann cell wrappings are made in the same spot, so that each wrapping overlaps the previous layers. The number of times the Schwann cells wrap themselves around the axon, and thus the number of layers in the myelin sheath, is greater for thicker than for thinner axons.

The cytoplasm, meanwhile, is forced into the outer region of the Schwann cell, much as toothpaste is squeezed to the top of the tube as the bottom is rolled up (fig. 7.6). Each Schwann cell wraps only about a millimeter of axon, leaving gaps of exposed axon between the adjacent Schwann cells. These gaps in the myelin sheath are known as the **nodes of Ranvier.**

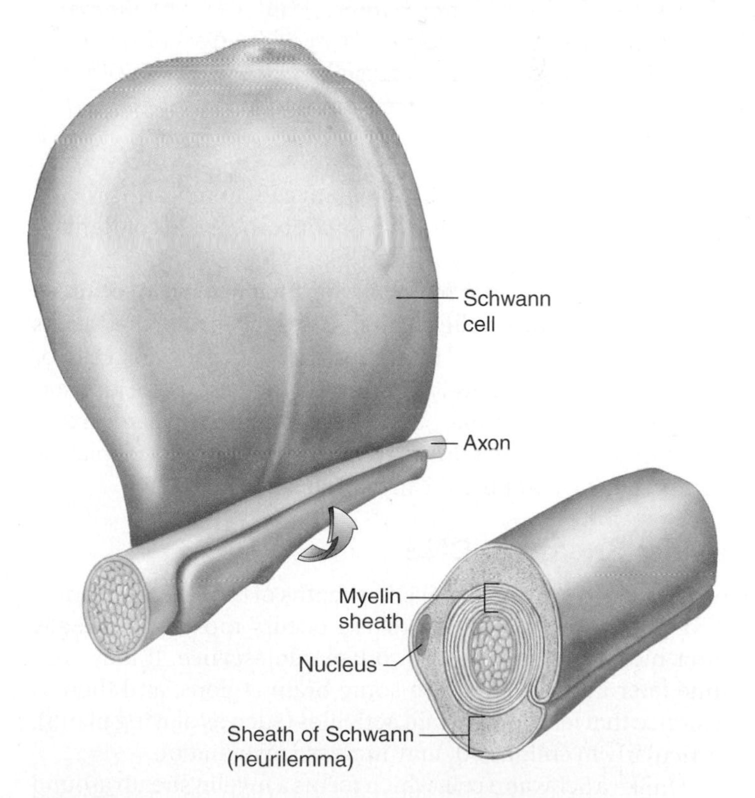

Figure 7.6 The formation of a myelin sheath around a peripheral axon. The myelin sheath is formed by successive wrappings of the Schwann cell membranes, leaving most of the Schwann cell cytoplasm outside the myelin. The sheath of Schwann is thus external to the myelin sheath. AP|R

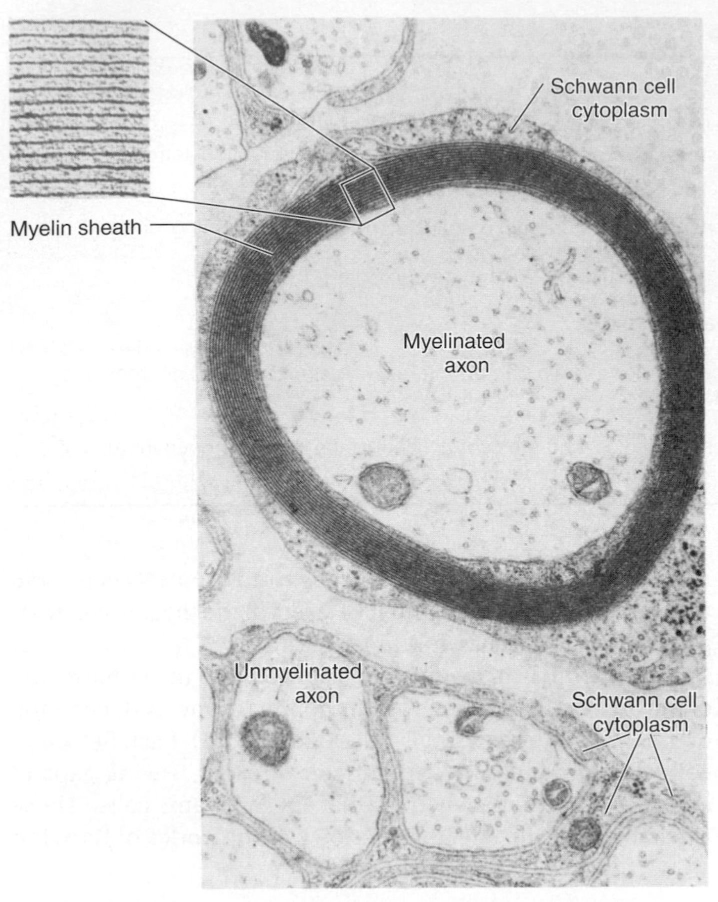

Figure 7.7 An electron micrograph of unmyelinated and myelinated axons. Notice that myelinated axons have Schwann cell cytoplasm to the outside of their myelin sheath, and that Schwann cell cytoplasm also surrounds unmyelinated axons. AP|R

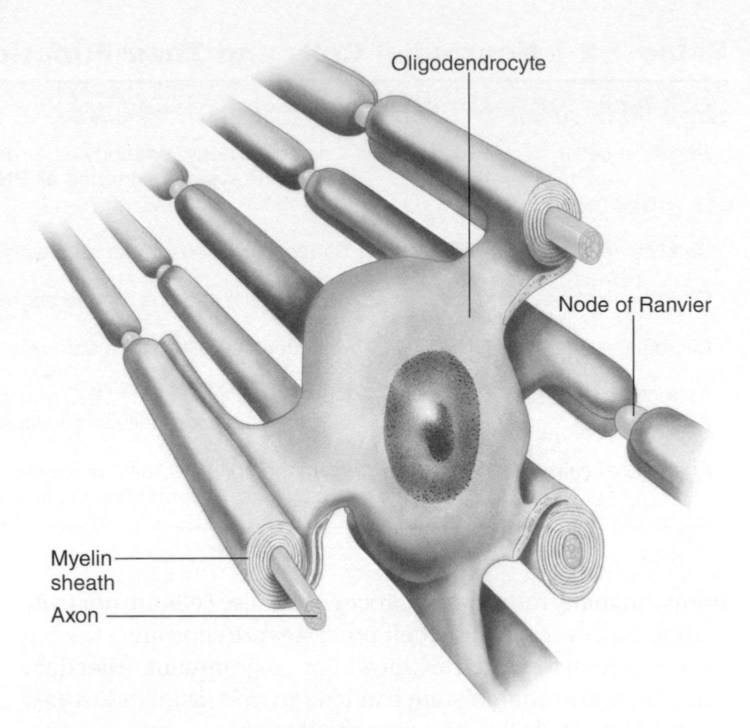

Figure 7.8 The formation of myelin sheaths in the CNS by an oligodendrocyte. One oligodendrocyte forms myelin sheaths around several axons. AP|R

The successive wrappings of Schwann cell membrane provide insulation around the axon, leaving only the nodes of Ranvier exposed to produce nerve impulses.

The Schwann cells remain alive as their cytoplasm is forced to the outside of the myelin sheath. As a result, myelinated axons of the PNS are surrounded by a living sheath of Schwann cells, or neurilemma (figs. 7.6 and 7.7). Unmyelinated axons are also surrounded by a neurilemma, but they differ from myelinated axons in that they lack the multiple wrappings of Schwann cell plasma membrane that compose the myelin sheath.

Myelin Sheath in CNS

As mentioned earlier, the myelin sheaths of the CNS are formed by oligodendrocytes. This process occurs mostly postnatally (after birth) and continues into late adolescence. It may continue later into adulthood in some brain regions, and there is evidence that learning certain activities (such as playing piano), particularly in childhood, may increase myelination.

Unlike a Schwann cell, which forms a myelin sheath around only one axon, each oligodendrocyte has extensions, like the tentacles of an octopus, that form myelin sheaths around several axons (fig. 7.8). The myelin sheaths around axons of the

CLINICAL APPLICATION

Multiple sclerosis (MS) is a common neurological disease, usually diagnosed in people (most often women) between the ages of 20 and 40. It is a chronic disease, remitting and relapsing with progressively advancing symptoms that are highly variable; these include sensory impairments, motor dysfunction and spasticity, bladder and intestinal problems, fatigue, and others. Infiltration of the CNS with lymphocytes (particularly T cells; chapter 15) and immune attack of self-antigens leads to degeneration of oligodendrocytes and myelin sheaths, which can develop hardened *scleroses,* or scars (from the Greek *sklerosis* = hardened) followed by axonal degeneration. Thus, MS is believed to be an autoimmune disease (chapter 15, section 15.6). Because this degeneration is widespread and affects different areas of the nervous system in different people, MS has a wider variety of symptoms than any other neurological disease. The causes of MS are not fully understood, but are believed to involve a number of genes that affect a person's susceptibility to environmental agents (such as viruses) that trigger an immune attack on self-antigens in the CNS. The immune cells involved are known as autoreactive T lymphocytes (chapter 15, section 15.5), which must pass from the blood to the brain to maintain the inflammation responsible for the neural destruction in MS. Treatment for MS includes drugs that reduce autoimmune activity and inflammation (such as interferon-β) and drugs that interfere with the entry of autoreactive T lymphocytes into the CNS.

CNS give this tissue a white color; areas of the CNS that contain a high concentration of axons thus form the **white matter.** The **gray matter** of the CNS is composed of high concentrations of cell bodies and dendrites, which lack myelin sheaths.

Regeneration of a Cut Axon

When an axon in a peripheral nerve is cut, the distal portion of the axon that was severed from the cell body degenerates and is phagocytosed by Schwann cells. The Schwann cells, surrounded by the basement membrane, then form a *regeneration tube* (fig. 7.9) as the part of the axon that is connected to the cell body begins to grow and exhibit amoeboid movement. The Schwann cells of the regeneration tube are believed to secrete chemicals that attract the growing axon tip, and the regeneration tube helps guide the regenerating axon to its proper destination. Even a severed major nerve may be surgically reconnected—and the function of the nerve largely reestablished—if the surgery is performed before tissue death occurs.

After spinal cord injury, some neurons die as a direct result of the trauma. However, other neurons and oligodendrocytes in the region die later because they produce "death receptors" that promote apoptosis (cell suicide; chapter 3, section 3.5). Injury in the CNS stimulates growth of axon collaterals, but central axons have a much more limited ability to regenerate than peripheral axons. Regeneration of CNS axons is prevented, in part, by inhibitory proteins in the membranes of the myelin sheaths. Also, regeneration of CNS axons is prevented by a glial scar that eventually forms from astrocytes. This glial scar physically blocks axon regeneration and induces the production of inhibitory proteins.

Three growth-inhibiting proteins, produced by oligodendrocytes, have been identified to date. These include glycoproteins that are associated with the myelin sheaths of CNS axons. These molecules inhibit the growth of a severed axon by binding to a receptor (called the *Nogo* receptor) on the axon.

Surprisingly, Schwann cells in the PNS also produce myelin proteins that can inhibit axon regeneration. However, after axon injury in the PNS, the fragments of old myelin are rapidly removed (through phagocytosis) by Schwann cells and macrophages. Also, quickly after injury the Schwann cells stop producing the inhibitory proteins. The rapid changes in Schwann cell function following injury (fig. 7.9) create an environment conducive to axon regeneration in the PNS.

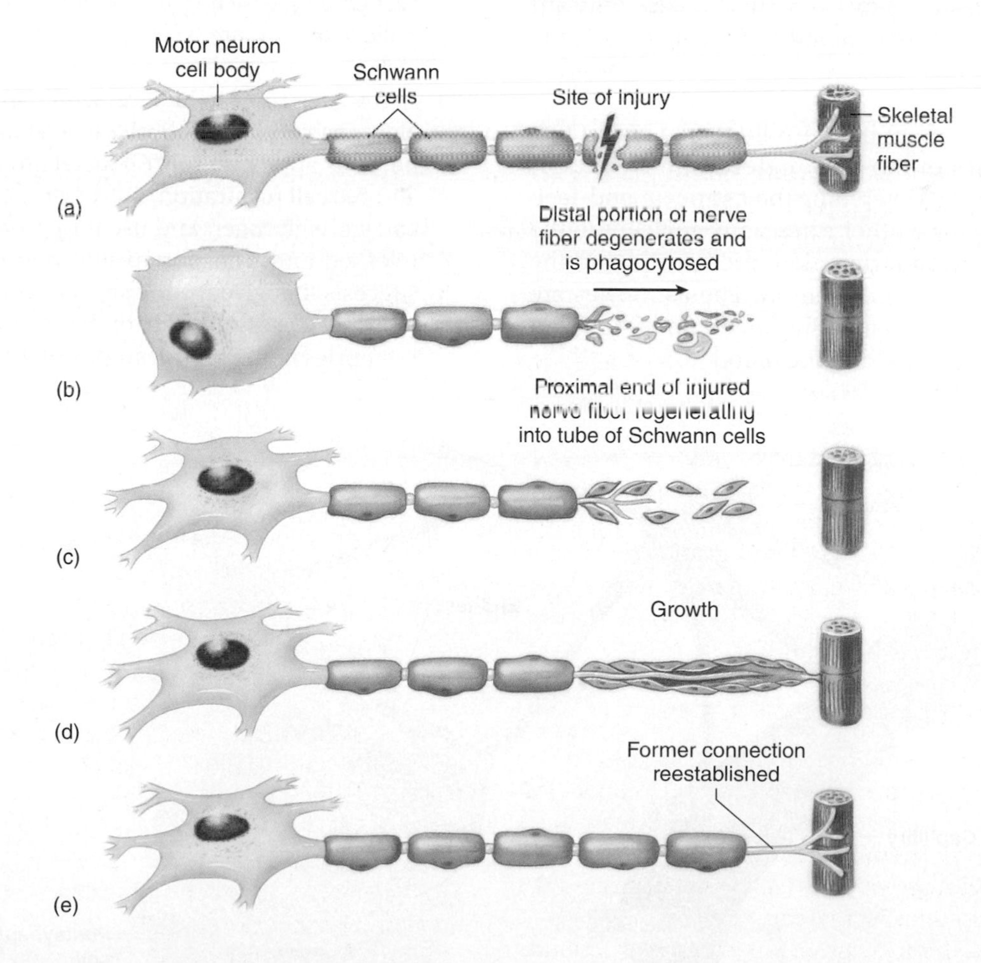

Figure 7.9 The process of peripheral neuron regeneration. (*a*) If a neuron is severed through a myelinated axon, the proximal portion may survive, but (*b*) the distal portion will degenerate through phagocytosis. The myelin sheath provides a pathway (*c*) and (*d*) for the regeneration of an axon, and (*e*) innervation is restored.

Neurotrophins

In a developing fetal brain, chemicals called **neurotrophins** promote neuron growth. *Nerve growth factor (NGF)* was the first neurotrophin to be identified; others include *brain-derived neurotrophic factor (BDNF); glial-derived neurotrophic factor (GDNF); neurotrophin-3;* and *neurotrophin-4/5* (the number depends on the animal species). NGF and neurotrophin-3 are known to be particularly important in the embryonic development of sensory neurons and sympathetic ganglia.

Neurotrophins also have important functions in the adult nervous system. NGF is required for the maintenance of sympathetic ganglia, and there is evidence that neurotrophins are required for mature sensory neurons to regenerate after injury. In addition, GDNF may be needed in the adult to maintain spinal motor neurons and to sustain neurons in the brain that use the chemical dopamine as a neurotransmitter.

Functions of Astrocytes

Astrocytes (from the Greek *aster* = star) are large stellate cells with numerous cytoplasmic processes that radiate outward (fig. 7.10). They are the most abundant of the glial cells in the CNS, constituting up to 90% of the nervous tissue in some areas of the brain.

Astrocytes have processes that terminate in *end-feet* surrounding the capillaries of the CNS; indeed, the entire surface of these capillaries is covered by the astrocyte end-feet. In addition, astrocytes have other extensions adjacent to the synapses between the axon terminal of one neuron and the dendrite or cell body of another neuron. The astrocytes are thus ideally situated to influence the interactions between neurons and between neurons and the blood.

Here are some of the proposed functions of astrocytes:

1. **Astrocytes take up K^+ from the extracellular fluid.** Because K^+ diffuses out of neurons during the production of nerve impulses (described in section 7.2), this function may be important in maintaining the proper ionic environment for neurons.
2. **Astrocytes take up some neurotransmitters released from the axon terminals of neurons.** For example, the neurotransmitter glutamate (the major excitatory neurotransmitter of the cerebral cortex) is taken into astrocytes and transformed into glutamine (fig. 7.10). The glutamine is then released back to the neurons, which can use it to reform the neurotransmitter glutamate. The glutamine from astrocytes can also be used by other neurons to produce GABA, the major inhibitory neurotransmitter in the brain.
3. **The astrocyte end-feet surrounding blood capillaries take up glucose from the blood.** The glucose is metabolized into lactic acid, or lactate (fig. 7.10). The lactate is then released and used as an energy source by neurons, which metabolize it aerobically into CO_2 and H_2O for the production of ATP. Thus, PET scans and MRI (chapter 8, section 8.2), which visualize brain locations by their metabolic activities, are based on the functions of astrocytes as well as neurons.
4. **Astrocytes release lactate, which aids neuron function.** Neurons can take in both glucose and lactate, but active neurons appear to rely on lactate to sustain a high rate of aerobic cell respiration. Astrocytes, but not most neurons, can store glycogen and use it to produce lactate, which is released and transported into neurons. Newer evidence suggests that glycogenolysis in astrocytes and subsequent lactate release is needed for the consolidation of long-term memories in the hippocampus of the brain (chapter 8).

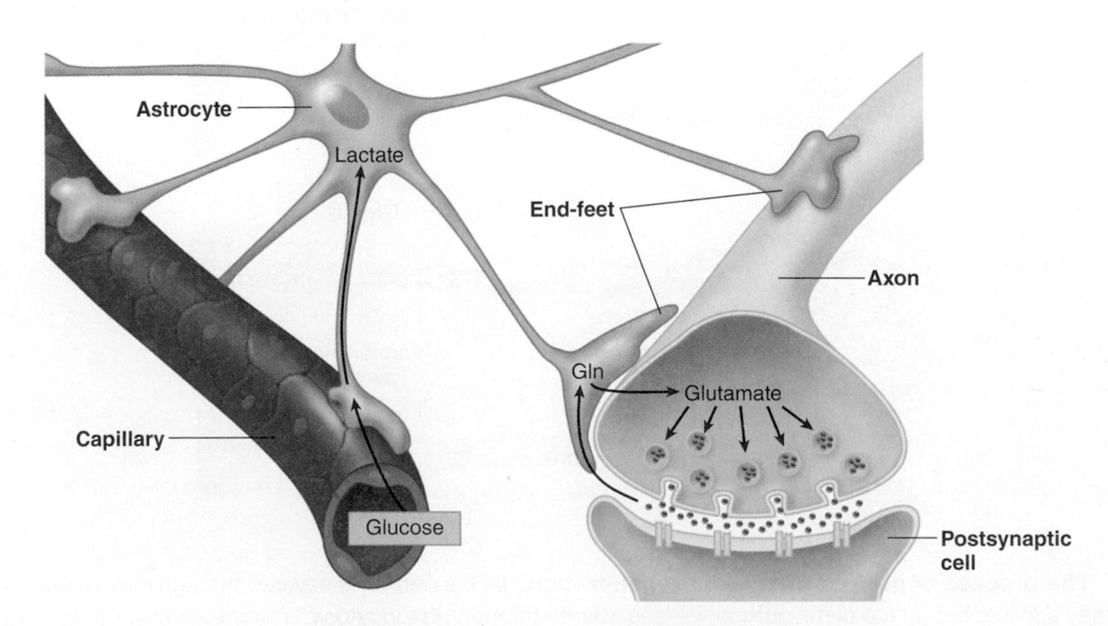

Figure 7.10 Astrocytes have processes that end on capillaries and neurons. Astrocyte end-feet take up glucose from blood capillaries and use this to help supply energy substrates for neurons. Astrocytes also take up the neurotransmitter glutamate from synapses and convert it to glutamine (Gln), which is then recycled to the neurons. AP|R

5. **Astrocytes appear to be needed for the formation of synapses in the CNS.** Few synapses form in the absence of astrocytes, and those that do are defective. Normal synapses in the CNS are ensheathed by astrocytes (fig. 7.10).
6. **Astrocytes regulate neurogenesis in the adult brain.** They appear to be needed for stem cells in the hippocampus and subventricular zone (chapter 8) to differentiate into both glial cells and neurons.
7. **Astrocytes induce the formation of the blood-brain barrier.** The nature of the blood-brain barrier is described in the next section.
8. **Astrocytes release transmitter chemicals that can stimulate or inhibit neurons.** Such so-called *gliotransmitters* include glutamate, ATP, adenosine derived from the released ATP, and D-serine. Glutamate released by astrocytes aided by D-serine stimulates a type of glutamate receptor on certain neurons (sections 7.6 and 7.7), whereas ATP or adenosine from astrocytes inhibits particular neurons.

Astrocyte physiology is also influenced by neural activity. Although astrocytes do not produce action potentials (impulses), they can be classified as excitable because they respond to stimulation by transient changes in their intracellular Ca^{2+} concentration. Action potentials in neurons can provoke a rise in Ca^{2+} within a localized region of an astrocyte, which in turn stimulates the release of ATP and other gliotransmitters that affect the synaptic transmission of neurons. When astrocytes release ATP, the adenosine derived from it by extracellular ATPase enzymes can stimulate a rise in the Ca^{2+} concentrations within nearby astrocytes. These astrocytes then also release ATP, which causes a rise in the Ca^{2+} concentrations within other astrocytes. This has been described as a *Ca^{2+} wave* that spreads among astrocytes away from the active neuron.

A rise in the Ca^{2+} concentration also can promote the production of prostaglandin E_2, which is released from the astrocyte end-feet surrounding cerebral blood vessels and stimulates vasodilation. Because this chain of events is triggered by the release of ATP from active neurons, an increase in neural activity within a brain region is thereby accompanied by an increased blood flow to that region.

Blood-Brain Barrier

Capillaries in the brain, unlike those of most other organs, do not have pores between adjacent endothelial cells (the cells that compose the walls of capillaries). Instead, all of the endothelial cells of brain capillaries are joined together by tight junctions. Unlike other organs, therefore, the brain cannot obtain molecules from the blood plasma by a nonspecific filtering process. Instead, molecules within brain capillaries must be moved through the endothelial cells by diffusion and active transport, as well as by endocytosis and exocytosis. This feature of brain capillaries imposes a very selective **blood-brain barrier.**

The structural components of the blood-brain barrier—the tight junctions between endothelial cells of brain capillaries—restricts the paracellular movement of molecules between epithelial cells (chapter 6), requiring the molecules to instead take the transcellular route and pass through the epithelial cells. Nonpolar O_2 and CO_2, as well as some organic molecules such as alcohol and barbiturates, can pass through the phospholipid components of the plasma membranes on each side of the capillary endothelial cells. Ions and polar molecules require ion channels and carrier proteins in the plasma membrane to move between the blood and brain. For example, plasma glucose can pass into the brain using specialized carrier proteins known as GLUT1. The GLUT1 glucose carriers, found in most brain regions, are always present; they do not require insulin stimulation like the GLUT4 carriers in skeletal muscles (chapter 11) or the hypothalamus (the brain region that contains hunger centers; chapters 8 and 19). There is also a metabolic component to the blood-brain barrier, including a variety of enzymes that can metabolize and inactivate potentially toxic molecules.

There is evidence that astrocytes can induce many of the characteristics of the blood-brain barrier, including the tight junctions between endothelial cells, the production of carrier proteins and ion channels, and the enzymes that destroy potentially toxic molecules. Astrocytes influence the capillary endothelial cells by secreting neurotrophins, such as glial-derived neurotrophic factor (GDNF, previously discussed). The endothelial cells, in turn, appear to secrete regulators that promote the growth and differentiation of astrocytes. This two-way communication leads to a view of the blood-brain barrier as a dynamic structure, and indeed scientists currently believe that the degree of its "tightness" and selectivity can be adjusted by a variety of regulators.

The blood-brain barrier presents difficulties in the chemotherapy of brain diseases because drugs that could enter other organs may not be able to enter the brain. In the treatment of *Parkinson's disease*, for example, patients who need a chemical called dopamine in the brain are often given a precursor molecule called levodopa (L-dopa) because L-dopa can cross the blood-brain barrier but dopamine cannot. Some antibiotics also cannot cross the blood-brain barrier; therefore, in treating infections such as meningitis, only those antibiotics that can cross the blood-brain barrier are used.

CHECKPOINT

1a. Draw a neuron, label its parts, and describe the functions of these parts.

1b. Distinguish between sensory neurons, motor neurons, and association neurons in terms of structure, location, and function.

2a. Describe the structure of the sheath of Schwann, or neurilemma, and explain how it promotes nerve regeneration. Explain how a myelin sheath is formed in the PNS.

2b. Explain how myelin sheaths are formed in the CNS. How does the presence or absence of myelin sheaths in the CNS determine the color of this tissue?

3. Explain what is meant by the blood-brain barrier. Describe its structure and discuss its clinical significance.

7.2 ELECTRICAL ACTIVITY IN AXONS

The permeability of the axon membrane to Na^+ and K^+ depends on gated channels that open in response to stimulation. Net diffusion of these ions occurs in two stages: first Na^+ moves into the axon, then K^+ moves out. This flow of ions, and the changes in the membrane potential that result, constitute an event called an action potential.

LEARNING OUTCOMES

After studying this section, you should be able to:

4. Step-by-step, explain how an action potential is produced.
5. Describe the characteristics of action potentials and explain how they are conducted by unmyelinated and myelinated axons.

All cells in the body maintain a potential difference (voltage) across the membrane, or **resting membrane potential (rmp),** in which the inside of the cell is negatively charged in comparison to the outside of the cell (for example, in neurons it is −70 mV). This potential difference is largely the result of the permeability properties of the plasma membrane (chapter 6, section 6.4). The membrane traps large, negatively charged organic molecules within the cell and permits only limited diffusion of positively charged inorganic ions. These properties result in an unequal distribution of these ions across the membrane. The action of the Na^+/K^+ pumps also helps to maintain a potential difference because they pump out 3 sodium ions (Na^+) for every 2 potassium ions (K^+) that they transport into the cell. Partly as a result of these pumps, Na^+ is more highly concentrated in the extracellular fluid than inside the cell, whereas K^+ is more highly concentrated within the cell.

Although all cells have a membrane potential, only a few types of cells have been shown to alter their membrane potential in response to stimulation. Such alterations in membrane potential are achieved by varying the membrane permeability to specific ions in response to stimulation. A central aspect of the physiology of neurons and muscle cells is their ability to produce and conduct these changes in membrane potential. Such an ability is termed *excitability* or *irritability.*

An increase in membrane permeability to a specific ion results in the diffusion of that ion down its *electrochemical gradient* (concentration and electrical gradients, considered together), either into or out of the cell. These *ion currents* occur only across limited patches of membrane where specific ion channels are located. Changes in the potential difference across the membrane at these points can be measured by the voltage developed between two microelectrodes (less than 1μm in diameter)—one placed inside the cell and the other placed outside the plasma membrane at the region being recorded. The voltage between these two recording electrodes can be visualized by connecting them to a computer or oscilloscope (fig. 7.11).

On a computer or oscilloscope screen, the voltage between the two recording electrodes over time is displayed as a line. This line deflects upward or downward in response to changes in the potential difference between the two electrodes. The display can be calibrated so that an upward deflection of the line indicates that the inside of the membrane has become less negative (or more positive) compared to the outside of the membrane. Conversely, a downward deflection of the line indicates that the inside of the cell has become more negative. The amplitude of the deflections (up or down) on the screen indicates the magnitude of the voltage changes.

If both recording electrodes are placed outside of the cell, the potential difference between the two will be zero (because there is no charge separation). When one of the two electrodes penetrates the plasma membrane, the computer will indicate that the intracellular electrode is electrically negative with respect to the extracellular electrode; a membrane potential is recorded. We will call this the *resting membrane potential (rmp)* to distinguish it from events

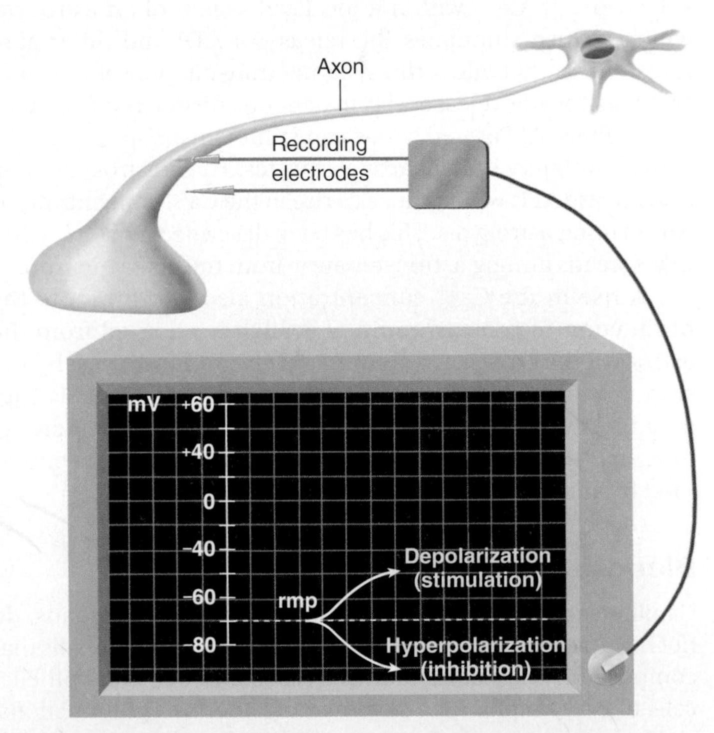

Figure 7.11 Observing depolarization and hyperpolarization. The difference in potential (in millivolts [mV]) between an intracellular and extracellular recording electrode is displayed on a computer or an oscilloscope screen. The resting membrane potential (rmp) of the axon may be reduced (depolarization) or increased (hyperpolarization). Depolarization is seen as a line deflecting upward from the rmp, and hyperpolarization by a line deflecting downward from the rmp.

described in later sections. All cells have a resting membrane potential, but its magnitude can be different in different types of cells. Neurons maintain an average rmp of −70 mV, for example, whereas heart muscle cells may have an rmp of −85 mV.

If appropriate stimulation causes positive charges to flow into the cell, the line will deflect upward. This change is called **depolarization** (or *hypopolarization*) because the potential difference between the two recording electrodes is reduced. A return to the resting membrane potential is known as **repolarization.** If stimulation causes the inside of the cell to become more negative than the resting membrane potential, the line on the oscilloscope will deflect downward. This change is called **hyperpolarization** (fig. 7.11). Hyperpolarization can be caused either by positive charges leaving the cell or by negative charges entering the cell.

Depolarization of a dendrite or cell body is *excitatory*, whereas hyperpolarization is *inhibitory*, in terms of their effects on the production of nerve impulses. The reasons for this relate to the nature of nerve impulses (action potentials), as will be explained shortly.

Ion Gating in Axons

The changes in membrane potential just described—depolarization, repolarization, and hyperpolarization—are caused by changes in the net flow of ions through ion channels in the membrane. Ions such as Na^+, K^+, and others pass through ion channels in the plasma membrane that are said to be *gated channels.* The "gates" are part of the proteins that compose the channels, and can open or close the ion channels in response to particular stimuli. When ion channels are closed, the plasma membrane is less permeable, and when the channels are open, the membrane is more permeable to an ion (fig. 7.12).

The ion channels for Na^+ and K^+ are specific for each ion. There are two types of channels for K^+. One type is gated, and the gates are closed at the resting membrane potential. The other type is not gated; these K^+ channels are thus always open and are often called *leakage channels.* Channels for Na^+, by contrast, are all gated and the gates are closed at the resting membrane potential. However, the gates of closed Na^+ channels appear to flicker open (and quickly close) occasionally, allowing some Na^+ to leak into the resting cell. As a result of these ion channel characteristics, the neuron at the resting membrane potential is much more permeable to K^+ than to Na^+, but some Na^+ does enter the cell. Because of the slight inward movement of Na^+, the resting membrane potential is a little less negative than the equilibrium potential for K^+.

Depolarization of a small region of an axon can be experimentally induced by a pair of stimulating electrodes that act as if they were injecting positive charges into the axon. If two recording electrodes are placed in the same region (one electrode within the axon and one outside), an upward deflection of the oscilloscope line will be observed as a result of this

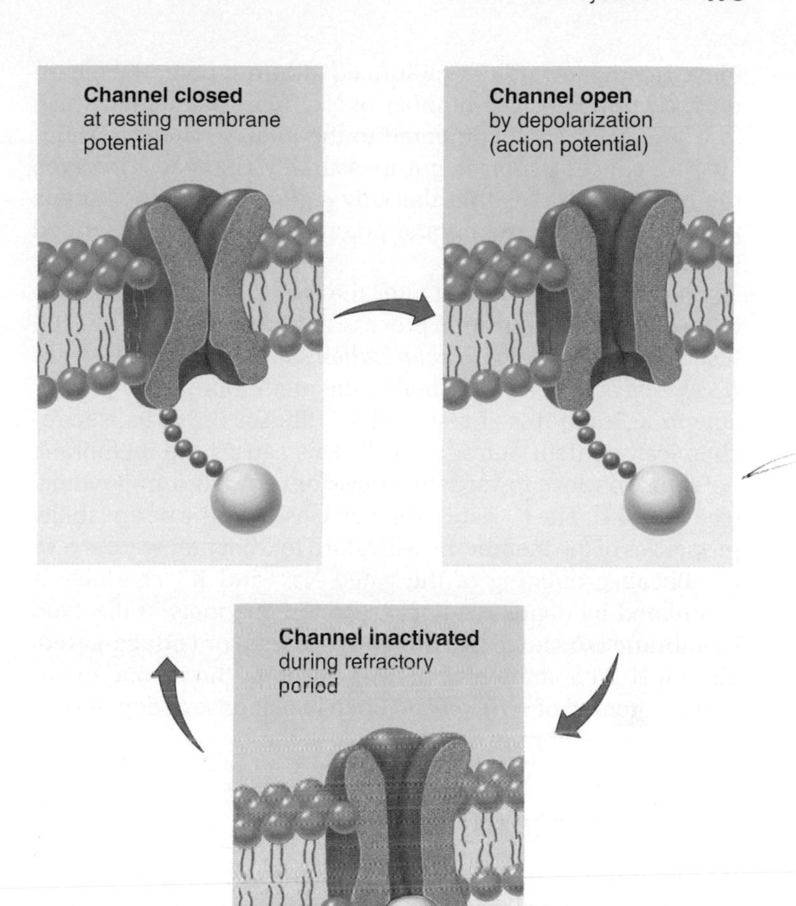

Figure 7.12 A model of a voltage-gated ion channel. The channel is closed at the resting membrane potential but opens in response to a threshold level of depolarization. This permits the diffusion of ions required for action potentials. After a brief period of time, the channel is inactivated by the "ball and chain" portion of a polypeptide chain (discussed later in the section on refractory periods).

depolarization. If the depolarization is below a certain level, it will simply decay very shortly back to the resting membrane potential (see fig. 7.18). However, if a certain level of depolarization is achieved (from −70 mV to −55 mV, for example) by this artificial stimulation, a sudden and very rapid change in the membrane potential will be observed. This is because *depolarization to a threshold level causes the Na^+ channels to open.*

Now, for an instant, the plasma membrane is freely permeable to Na^+. Because the inside of the cell is negatively charged relative to the outside, and the concentration of Na^+ is lower inside of the cell, the **electrochemical gradient** (the combined electrical and concentration gradients) for Na^+ causes Na^+ to rush into the cell. This causes the membrane potential to move rapidly toward the sodium equilibrium potential (chapter 6, section 6.4). The number of Na^+ ions that actually rush in is relatively small compared to the total, so the extracellular Na^+ concentration is not measurably changed. However, the increased Na^+ within that tiny region of axon membrane greatly affects the membrane potential, as will be described shortly.

A fraction of a second after the Na^+ channels open, they close due to an inactivation process, as illustrated in figure 7.12. Just before they do, *the depolarization stimulus causes the gated K^+ channels to open.* This makes the membrane more permeable to K^+ than it is at rest, and K^+ diffuses down its electrochemical gradient out of the cell. This causes the membrane potential to move toward the potassium equilibrium potential (see fig. 7.14). The K^+ gates will then close and the permeability properties of the membrane will return to what they were at rest.

Because opening of the gated Na^+ and K^+ channels is stimulated by depolarization, these ion channels in the axon membrane are said to be **voltage-regulated,** or **voltage-gated, channels.** The channel gates are closed at the resting membrane potential of -70 mV and open in response to depolarization of the membrane to a threshold value.

Action Potentials

We will now consider the events that occur at one point in an axon, when a small region of axon membrane is stimulated artificially and responds with changes in ion permeabilities. The resulting changes in membrane potential at this point are detected by recording electrodes placed in this region of the axon. The nature of the stimulus *in vivo* (in the body), and the manner by which electrical events are conducted to different points along the axon, will be described in later sections.

When the axon membrane has been depolarized to a threshold level—in the previous example, by stimulating electrodes—the Na^+ gates open and the membrane becomes permeable to Na^+. This permits Na^+ to enter the axon by diffusion, which further depolarizes the membrane (makes the inside less negative, or more positive). The gates for the Na^+ channels of the axon membrane are voltage regulated, and so this additional depolarization opens more Na^+ channels and makes the membrane even more permeable to Na^+. As a result, more Na^+ can enter the cell and induce a depolarization that opens even more voltage-regulated Na^+ gates. A *positive feedback loop* (fig. 7.13) is thus created, causing the rate of Na^+ entry and depolarization to accelerate in an explosive fashion.

The explosive increase in Na^+ permeability results in a rapid reversal of the membrane potential in that region from -70 mV to $+30$ mV (fig. 7.13). At that point the channels for Na^+ close (they actually become inactivated, as illustrated in fig. 7.12), causing a rapid decrease in Na^+ permeability. This is why, at the top of the action potential, the voltage does not quite reach the $+66$ mV equilibrium potential for Na^+ (chapter 6, section 6.4). Also at this time, as a result of a time-delayed effect

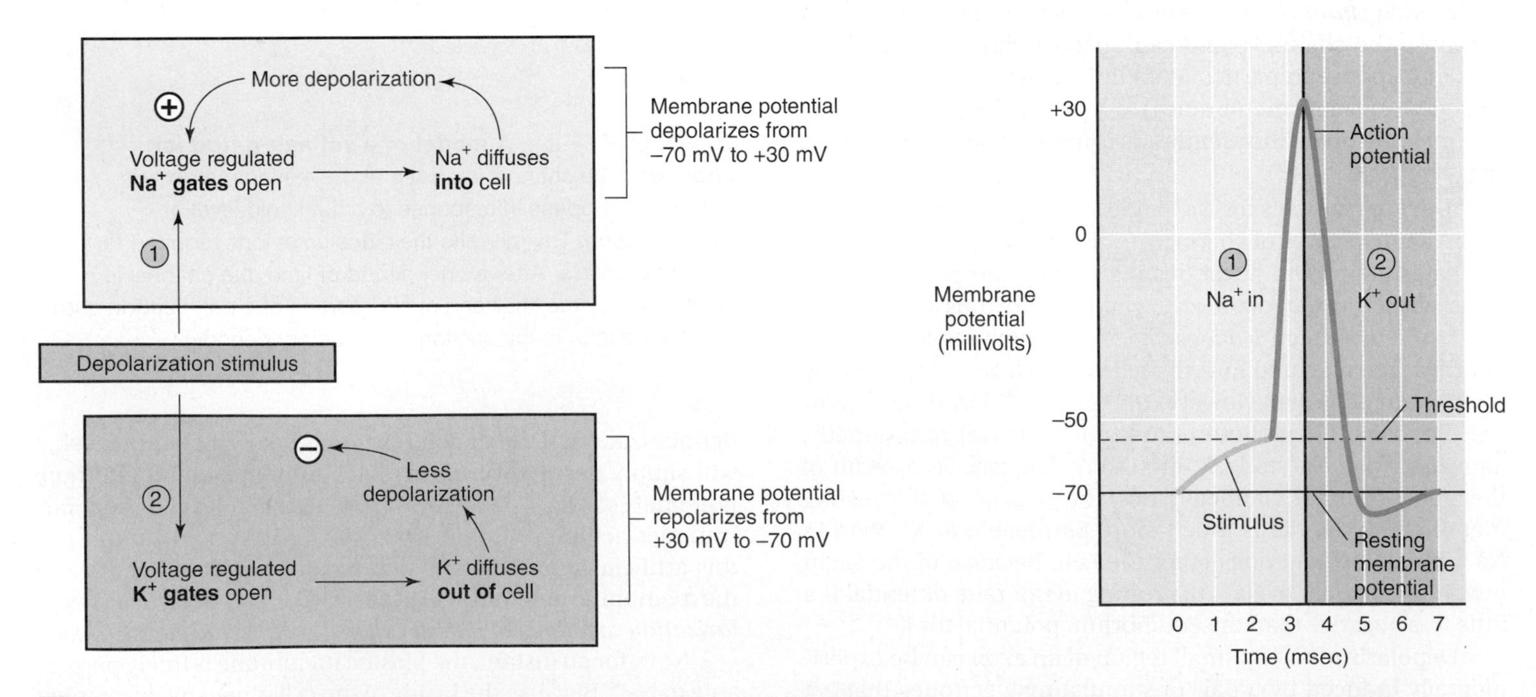

Figure 7.13 Depolarization of an axon affects Na^+ and K^+ diffusion in sequence. (1) Na^+ gates open and Na^+ diffuses into the cell. (2) After a brief period, K^+ gates open and K^+ diffuses out of the cell. An inward diffusion of Na^+ causes further depolarization, which in turn causes further opening of Na^+ gates in a positive feedback (+) fashion. The opening of K^+ gates and outward diffusion of K^+ makes the inside of the cell more negative, and thus has a negative feedback effect (–) on the initial depolarization. AP|R

CLINICAL APPLICATION

Local anesthetics block the conduction of action potentials in axons. They do this by reversibly binding to specific sites within the voltage-gated Na^+ channels, reducing the ability of membrane depolarization to produce action potentials. *Cocaine* was the first local anesthetic to be used, but because of its toxicity and potential for abuse, alternatives have been developed. The first synthetic analog of cocaine used for local anesthesia, *procaine,* was produced in 1905. Other local anesthetics of this type include *lidocaine* and *tetracaine.*

of the depolarization, voltage-gated K^+ channels open and K^+ diffuses rapidly out of the cell.

Because K^+ is positively charged, the diffusion of K^+ out of the cell makes the inside of the cell less positive, or more negative, and acts to restore the original resting membrane potential of −70 mV. This process is called **repolarization** and represents the completion of a *negative feedback loop* (fig. 7.13). These changes in Na^+ and K^+ diffusion and the resulting changes in the membrane potential they produce constitute an event called the **action potential,** or **nerve impulse.**

The correlation between ion movements and changes in membrane potential is shown in figure 7.14. The bottom portion of this figure illustrates the movement of Na^+ and K^+ through the axon membrane in response to a depolarization stimulus. Notice that the explosive increase in Na^+ diffusion causes rapid depolarization to 0 mV and then *overshoot* of the membrane potential so that the inside of the membrane actually becomes positively charged (almost +30 mV) compared to the outside (top portion of fig. 7.14). The greatly increased permeability to Na^+ thus drives the membrane potential toward the equilibrium potential for Na^+ (chapter 6, section 6.4). However, the peak action potential depolarization is less than the Na^+ equilibrium potential (+66 mV), due to inactivation of the Na^+ channels.

As the Na^+ channels are becoming inactivated, the gated K^+ channels open and the membrane potential moves toward the K^+ equilibrium potential. This outward diffusion of K^+ repolarizes the membrane. Actually, the membrane potential slightly overshoots the resting membrane potential, producing an *after-hyperpolarization* as a result of the continued outward movement of K^+ (fig. 7.14). However, the gated K^+ channels close before this after-hyperpolarization can reach the K^+ equilibrium potential (−90 mV). Then the after-hyperpolarization decays, and the resting membrane potential is reestablished.

The Na^+/K^+ pumps are constantly working in the plasma membrane. They pump out the Na^+ that entered the axon during an action potential and pump in the K^+ that had left. Remember that only a relatively small amount of Na^+ and K^+ ions move into and out of the axon during an action potential. This movement is sufficient to cause changes in the membrane potential during an action potential but does not significantly affect the concentrations of these ions. Thus, active transport (by the Na^+/K^+ pumps) is still required to move Na^+ out of the axon and to move K^+ back into the axon after an action potential.

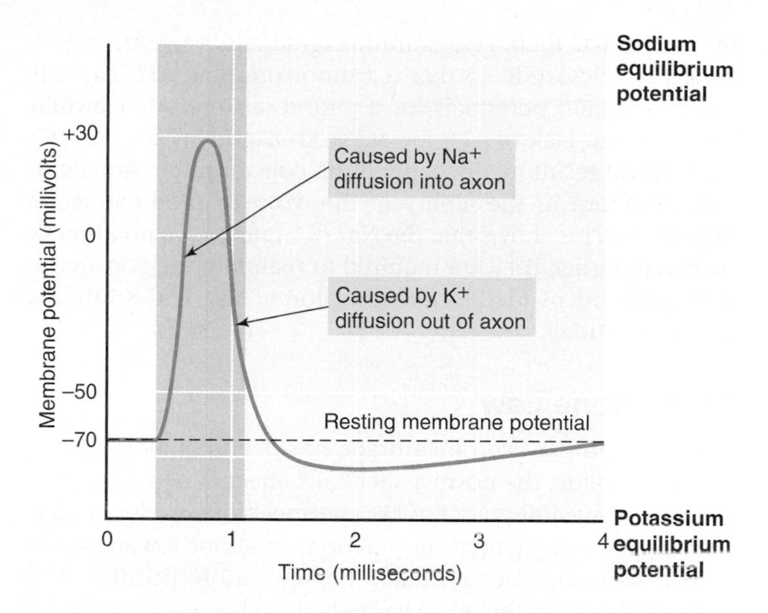

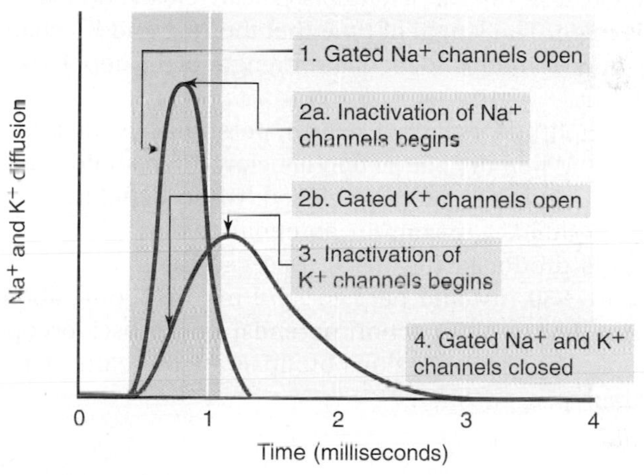

Figure 7.14 Membrane potential changes and ion movements during an action potential. The top graph depicts an action potential (blue line). The bottom graph (red lines) depicts the net diffusion of Na^+ and K^+ during the action potential. The *x*-axis for time is the same in both graphs, so that the depolarization, repolarization, and after-hyperpolarization in the top graph can be correlated with events in the Na^+ and K^+ channels and their effects on ion movements in the bottom graph. The inward movement of Na^+ drives the membrane potential toward the Na^+ equilibrium potential during the depolarization (rising) phase of the action potential, whereas the outward movement of K^+ drives the membrane potential toward the potassium equilibrium potential during the repolarization (falling) phase of the action potential. AP|R

See the *Test Your Quantitative Ability* section of the Review Activities at the end of this chapter.

Notice that active transport processes are not directly involved in the production of an action potential; both depolarization and repolarization are produced by the diffusion of ions down their concentration gradients. A neuron poisoned with cyanide so that it cannot produce ATP can still produce action potentials for a period of time. After awhile, however, the lack of ATP for active transport by the Na^+/K^+ pumps will result in a decline in the concentration gradients, and therefore in the ability of the axon to produce action potentials. This shows that the Na^+/K^+ pumps are not directly involved; rather, they are required to maintain the concentration gradients needed for the diffusion of Na^+ and K^+ during action potentials.

All-or-None Law

Once a region of axon membrane has been depolarized to a threshold value, the positive feedback effect of depolarization on Na^+ permeability and of Na^+ permeability on depolarization causes the membrane potential to shoot toward about +30 mV. It does not normally become more positive than +30 mV because the Na^+ channels quickly close and the K^+ channels open. The length of time that the Na^+ and K^+ channels stay open is independent of the strength of the depolarization stimulus.

The amplitude (size) of action potentials is therefore **all-or-none.** When depolarization is below a threshold value, the voltage-regulated gates are closed; when depolarization reaches threshold, a maximum potential change (the action potential) is produced (fig. 7.15). Because the change from −70 mV to +30 mV and back to −70 mV lasts only about 3 msec, the image of an action potential on an oscilloscope screen looks like a spike. Action potentials are therefore sometimes called *spike potentials.*

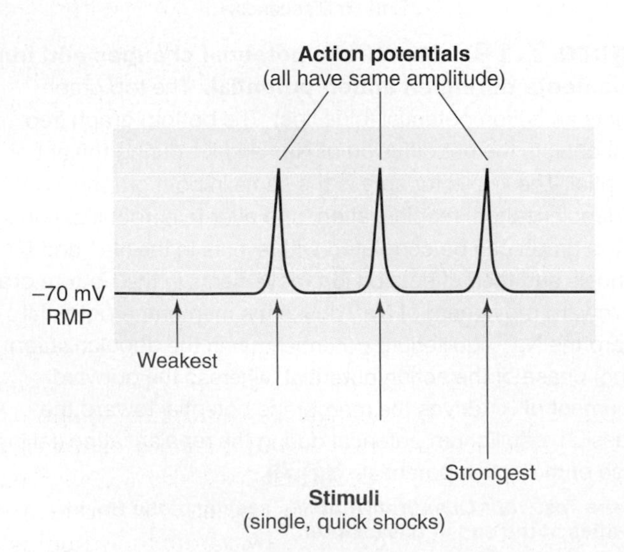

Figure 7.15 The all-or-none law of action potentials. A single, quick shock delivered to an axon can serve as a depolarizing stimulus. If the stimulus is below threshold, no action potential is produced by the axon. Once the stimulus has reached threshold, a full action potential is produced. Any greater stimulus does not produce greater action potentials. Thus, action potentials are not graded (varied); they are all-or-none.

The channels are open only for a fixed period of time because they are soon *inactivated,* a process different from simply closing the gates. Inactivation occurs automatically and lasts until the membrane has repolarized. Because of this automatic inactivation, all action potentials have about the same duration. Likewise, since the concentration gradient for Na^+ is relatively constant, the amplitudes of the action potentials are about equal in all axons at all times (from −70 mV to +30 mV, or about 100 mV in total amplitude).

Coding for Stimulus Intensity

Because action potentials are all-or-none events, a stronger stimulus cannot produce an action potential of greater amplitude. The code for stimulus strength in the nervous system is not amplitude modulated (AM). When a greater stimulus strength is applied to a neuron, identical action potentials are produced more frequently (more are produced per second). Therefore, the code for stimulus strength in the nervous system is frequency modulated (FM). This concept is illustrated in figure 7.16.

When an entire collection of axons (in a nerve) is stimulated, different axons will be stimulated at different stimulus intensities. A weak stimulus will activate only those few axons with low thresholds, whereas stronger stimuli can activate axons with higher thresholds. As the intensity of stimulation increases, more and more axons will become activated. This process, called **recruitment,** represents another mechanism by which the nervous system can code for stimulus strength.

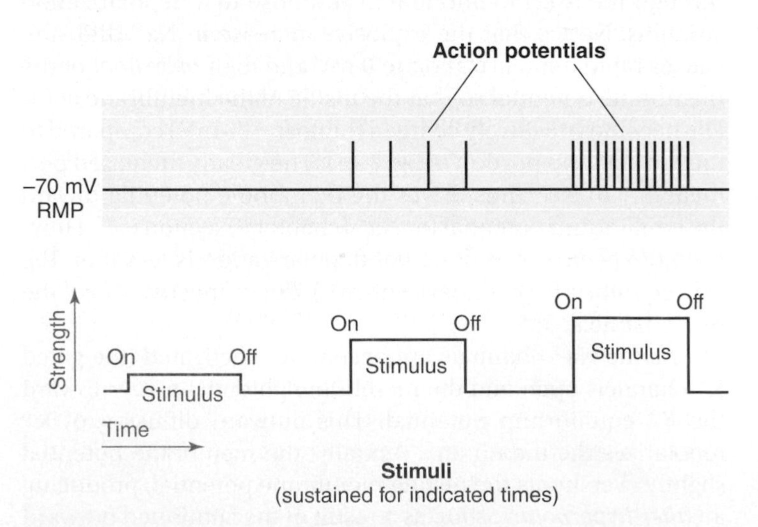

Figure 7.16 The effect of stimulus strength on action-potential frequency. Stimuli that are sustained for a period of time are given to an axon. In the first case, the stimulus is weaker than required to reach threshold, and no action potentials are produced. In the second case, a stronger stimulus is delivered, which causes the production of a few action potentials while the stimulus is sustained. In the last case, an even stronger stimulus produces a greater number of action potentials in the same time period. This demonstrates that stimulus strength is coded by the frequency (rather than the amplitude) of action potentials.

Refractory Periods

If a stimulus of a given intensity is maintained at one point of an axon and depolarizes it to threshold, action potentials will be produced at that point at a given frequency (number per second). As the stimulus strength is increased, the frequency of action potentials produced at that point will increase accordingly. As action potentials are produced with increasing frequency, the time between successive action potentials will decrease—but only up to a minimum time interval. The interval between successive action potentials will never become so short as to allow a new action potential to be produced before the preceding one has finished.

During the time that a patch of axon membrane is producing an action potential, it is incapable of responding—is *refractory*—to further stimulation. If a second stimulus is applied during most of the time that an action potential is being produced, the second stimulus will have no effect on the axon membrane. The membrane is thus said to be in an **absolute refractory period;** it cannot respond to any subsequent stimulus.

The cause of the absolute refractory period is now understood at a molecular level. In addition to the voltage-regulated gates that open and close the channel, an ion channel may have a polypeptide that functions as a "ball and chain" apparatus dangling from its cytoplasmic side (see fig. 7.12). After a voltage-regulated channel is opened by depolarization for a set time, it enters an *inactive state*. The inactivated channel cannot be opened by depolarization. The reason for its inactivation depends on the type of voltage-gated channel. In the type of channel shown in figure 7.12, the channel becomes blocked by a molecular ball attached to a chain. In a different type of voltage-gated channel, the channel shape becomes altered through molecular rearrangements. The inactivation ends after a fixed period of time in both cases, either because the ball leaves the mouth of the channel, or because molecular rearrangements restore the resting form of the channel. In the resting state, unlike the inactivated state, the channel is closed but it can be opened in response to a depolarization stimulus of sufficient strength.

The transition of the gated Na^+ channels from the inactivated to the closed state doesn't occur in all channels at the same instant. When enough Na^+ channels are in the closed rather than inactivated state, it is theoretically possible to again stimulate the axon with a sufficiently strong stimulus. However, while the K^+ channels are still open and the membrane is still in the process of repolarizing, the effects of the outward movement of K^+ must be overcome, making it even more difficult to depolarize the axon to threshold. Only a very strong depolarization stimulus will be able to overcome these obstacles and produce a second action potential. Thus, during the time that the Na^+ channels are in the process of recovering from their inactivated state and the K^+ channels are still open, the membrane is said to be in a **relative refractory period** (fig. 7.17).

Because the cell membrane is refractory when it is producing an action potential, each action potential remains a

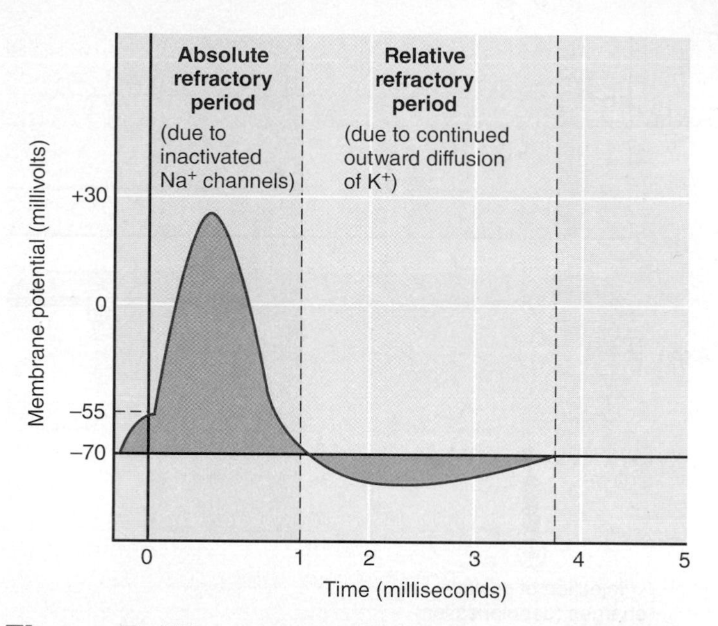

Figure 7.17 Absolute and relative refractory periods. While a segment of axon is producing an action potential, the membrane is absolutely or relatively resistant (refractory) to further stimulation. AP|R

separate, all-or-none event. In this way, as a continuously applied stimulus increases in intensity, its strength can be coded strictly by the frequency of the action potentials it produces at each point of the axon membrane.

One might think that after a large number of action potentials have been produced, the relative concentrations of Na^+ and K^+ would be changed in the extracellular and intracellular compartments. This is not the case. In a typical mammalian axon, for example, only 1 intracellular K^+ in 3,000 would be exchanged for a Na^+ to produce an action potential. Since a typical neuron has about 1 million Na^+/K^+ pumps that can transport nearly 200 million ions per second, these small changes can be quickly corrected.

Cable Properties of Neurons

If a pair of stimulating electrodes produces a depolarization that is too weak to cause the opening of voltage-regulated Na^+ gates—that is, if the depolarization is below threshold (about −55 mV)—the change in membrane potential will be *localized* to within 1 to 2 mm of the point of stimulation (fig. 7.18). For example, if the stimulus causes depolarization from −70 mV to −60 mV at one point, and the recording electrodes are placed only 3 mm away from the stimulus, the membrane potential recorded will remain at −70 mV (the resting potential). The axon is thus a very poor conductor compared to a metal wire.

The **cable properties** of neurons are their abilities to conduct charges through their cytoplasm. These cable properties are quite poor because there is a high internal resistance to the spread of charges and because many charges leak out of

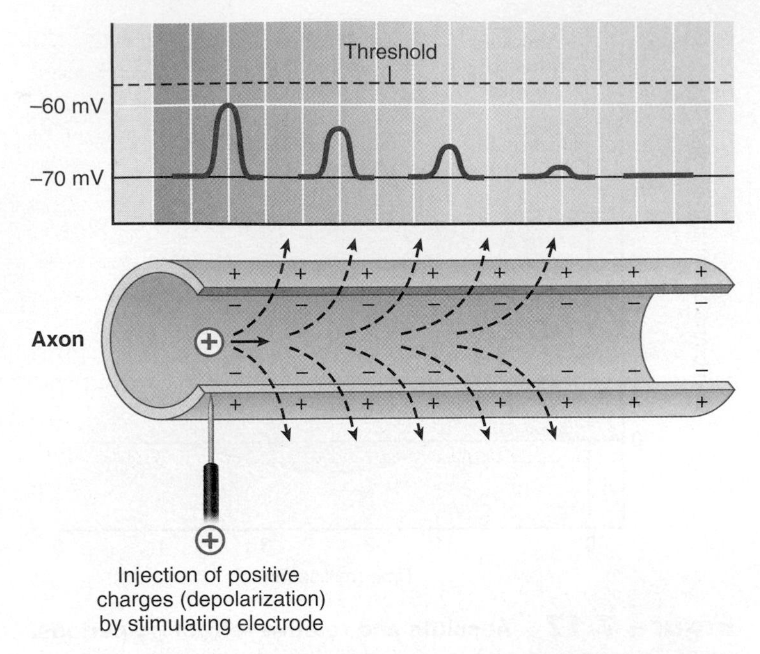

Figure 7.18 Cable properties of an axon. The cable properties of an axon are the properties that permit it to conduct potential changes over distances. If a stimulating electrode injects positive charges and produces a depolarization (*blue*) at one point in the axon, the depolarization will quickly dissipate if it doesn't trigger an action potential. The decreasing amplitude of the depolarization is due to leakage of charges through the axon membrane (*dashed arrows*). This results in a poor ability of the axon to conduct changes in potential over distances. AP|R

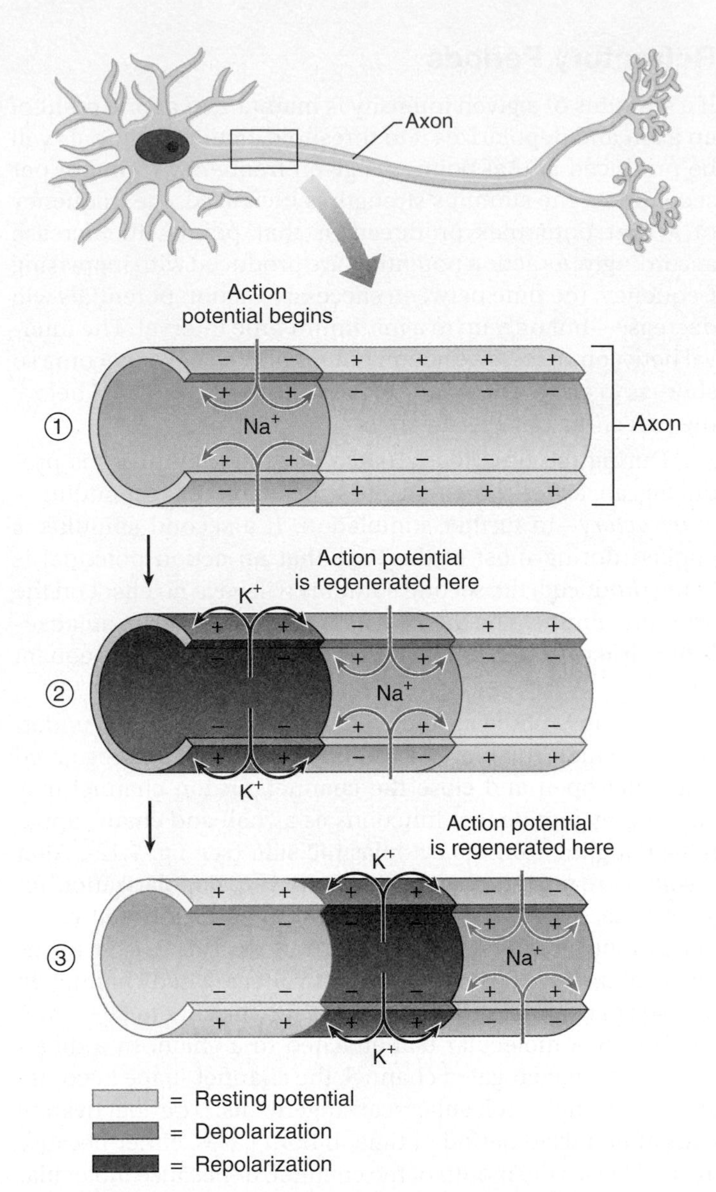

Figure 7.19 The conduction of action potentials in an unmyelinated axon. Each action potential "injects" positive charges that spread to adjacent regions. The region that has just produced an action potential is refractory. The next region, not having been stimulated previously, is partially depolarized. As a result, its voltage-regulated Na^+ gates open and the process is repeated. Successive segments of the axon thereby regenerate, or "conduct," the action potential. AP|R

the axon through its membrane (fig. 7.18). If an axon had to conduct only through its cable properties, therefore, no axon could be more than a millimeter in length. The fact that some axons are a meter or more in length suggests that the conduction of nerve impulses does not rely on the cable properties of the axon.

Conduction of Nerve Impulses

When stimulating electrodes artificially depolarize one point of an axon membrane to a threshold level, voltage-regulated channels open and an action potential is produced at that small region of axon membrane containing those channels. For about the first millisecond of the action potential, when the membrane potential changes from −70 mV to +30 mV, a current of Na^+ is entering the cell by diffusion through the open Na^+ channels. Each action potential thus "injects" positive charges (sodium ions) into the axon (fig. 7.19).

These positively charged sodium ions are conducted by the cable properties of the axon to an adjacent region that still has a membrane potential of −70 mV. Within the limits of the cable properties of the axon (1 to 2 mm), this helps to depolarize the adjacent region of axon membrane. When this adjacent region of membrane reaches a threshold level of depolarization, it too produces the action potential as its voltage-regulated gates open.

The action potential produced at the first location in the axon membrane (the initial segment of the axon) thus serves as the depolarization stimulus for the next region of the axon membrane, which can then produce the action potential. The action potential in this second region, in turn, serves as a depolarization stimulus for the production of the action potential in a third region, and so on. This explains how the action potential is produced at all regions of the axon beyond the initial segment at the axon hillock. (The depolarization stimulus for the action potential at the initial segment of the axon results from synaptic transmission, discussed in section 7.3.)

Conduction in an Unmyelinated Axon

In an unmyelinated axon, every patch of membrane that contains Na^+ and K^+ channels can produce an action potential. Action potentials are thus produced along the entire length of the axon. The cablelike spread of depolarization induced by the influx of Na^+ during one action potential helps to depolarize the adjacent regions of membrane—a process that is also aided by movements of ions on the outer surface of the axon membrane (fig. 7.19). This process would depolarize the adjacent membranes on each side of the region to produce the action potential, but the area that had previously produced one cannot produce another at this time because it is still in its refractory period.

It is important to recognize that action potentials are not really "conducted," although it is convenient to use that word. Each action potential is a separate, complete event that is repeated, or *regenerated,* along the axon's length. This is analogous to the "wave" performed by spectators in a stadium. One person after another gets up (depolarization) and then sits down (repolarization). It is thus the "wave" that travels (the repeated action potential at different locations along the axon membrane), not the people.

The action potential produced at the end of the axon is thus a completely new event that was produced in response to depolarization from the previous region of the axon membrane. The action potential produced at the last region of the axon has the same amplitude as the action potential produced at the first region. Action potentials are thus said to be **conducted without decrement** (without decreasing in amplitude).

The spread of depolarization by the cable properties of an axon is fast compared to the time it takes to produce an action potential. Thus, the more action potentials along a given stretch of axon that have to be produced, the slower the conduction. Because action potentials must be produced at every fraction of a micrometer in an unmyelinated axon, the conduction rate is relatively slow. This conduction rate is somewhat faster if the unmyelinated axon is thicker, because thicker axons have less resistance to the flow of charges (so conduction of charges by cable properties is faster). The conduction rate is substantially faster if the axon is myelinated, because fewer action potentials are produced along a given length of myelinated axon.

Conduction in a Myelinated Axon

The myelin sheath provides insulation for the axon, preventing movements of Na^+ and K^+ through the membrane. If the myelin sheath were continuous, therefore, action potentials could not be produced. The myelin thus has interruptions—the *nodes of Ranvier,* as previously described.

Because the cable properties of axons can conduct depolarizations over only a very short distance (1 to 2 mm), the nodes of Ranvier cannot be separated by more than this distance. Studies have shown that Na^+ channels are highly concentrated at the nodes (estimated at 10,000 per square micrometer) and almost absent in the regions of axon membrane between the nodes. Action potentials, therefore, occur only at the nodes of Ranvier (fig. 7.20) and seem to "leap"

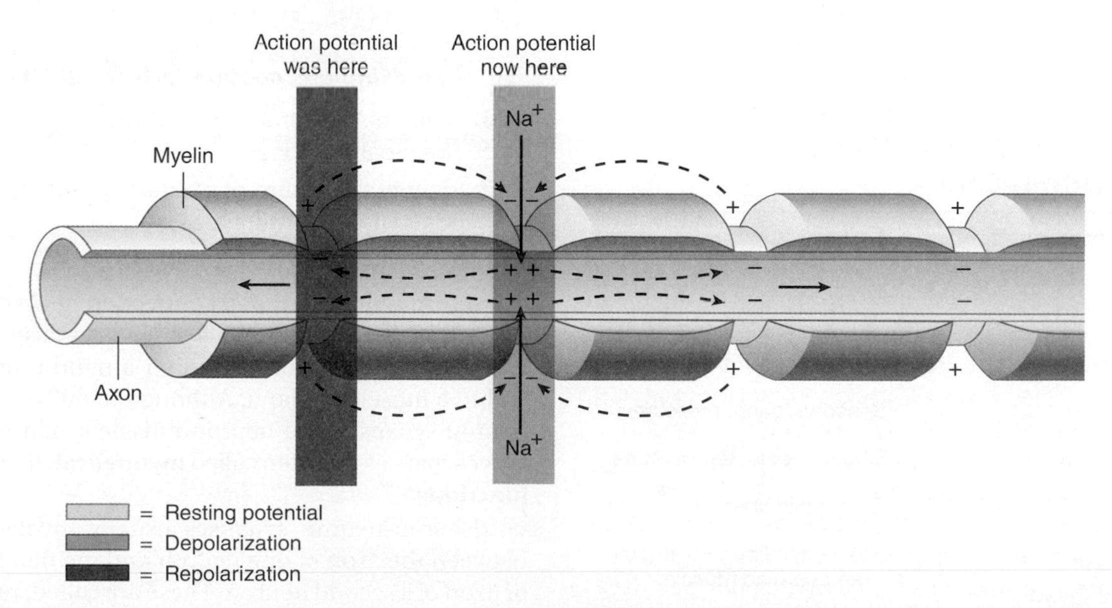

Figure 7.20 The conduction of a nerve impulse in a myelinated axon. Because the myelin sheath prevents inward Na^+ current, action potentials can be produced only at gaps in the myelin sheath called the nodes of Ranvier. This "leaping" of the action potential from node to node is known as saltatory conduction. AP|R

from node to node—a process called **saltatory conduction** (from the Latin *saltario* = leap). The leaping is, of course, just a metaphor; the action potential at one node depolarizes the membrane at the next node to threshold, so that a new action potential is produced at the next node of Ranvier.

Myelinated axons conduct the action potential faster than unmyelinated axons. This is because myelinated axons have voltage-gated channels only at the nodes of Ranvier, which are about 1 mm apart, whereas unmyelinated axons have these channels along their entire length. Because myelinated axons have more cablelike spread of depolarization (which is faster), and fewer membrane sites at which the action potential is produced (which is slower) than unmyelinated axons, the conduction is faster in a myelinated axon. Also, myelinated axons are generally thicker than unmyelinated axons, and so have less resistance to the spread of charges and a faster cable-like conduction. Conduction rates in the human nervous system vary from 1.0 m/sec—in thin, unmyelinated fibers that mediate slow, visceral responses—to faster than 100 m/sec (225 miles per hour)—in thick, myelinated fibers involved in quick stretch reflexes in skeletal muscles (table 7.3).

In summary, the speed of action potential conduction is increased by (1) increased diameter of the axon, because this reduces the resistance to the spread of charges by cable properties; and (2) myelination, because the myelin sheath results in saltatory conduction of action potentials. These methods of affecting conduction speed are generally combined in the nervous system: the thinnest axons tend to be unmyelinated and the thickest tend to be myelinated.

Table 7.3 | Conduction Velocities and Functions of Mammalian Nerves of Different Diameters*

Diameter (µm)	Conduction Velocity (m/sec)	Examples of Functions Served
12–22	70–120	Sensory: muscle position
5–13	30–90	Somatic motor fibers
3–8	15–40	Sensory: touch, pressure
1–5	12–30	Sensory: pain, temperature
1–3	3–15	Autonomic fibers to ganglia
0.3–1.3	0.7–2.2	Autonomic fibers to smooth and cardiac muscles

*See the *Test Your Quantitative Ability* section of the Review Activities in chapters 8 and 9.

CHECKPOINT

4a. Define the terms *depolarization* and *repolarization*, and illustrate these processes graphically.

4b. Describe how the permeability of the axon membrane to Na^+ and K^+ is regulated and how changes in permeability to these ions affect the membrane potential.

4c. Describe how gating of Na^+ and K^+ in the axon membrane results in the production of an action potential.

5a. Explain the all-or-none law of action potentials, and describe the effect of increased stimulus strength on action potential production. How do the refractory periods affect the frequency of action potential production?

5b. Describe how action potentials are conducted by unmyelinated nerve fibers. Why is saltatory conduction in myelinated fibers more rapid?

7.3 THE SYNAPSE

Axons end close to, or in some cases in contact with, another cell. In specialized cases, action potentials can directly pass from one cell to another. In most cases, however, the action potentials stop at the axon terminal, where they stimulate the release of a chemical neurotransmitter that affects the next cell.

LEARNING OUTCOMES

After studying this section, you should be able to:

6. Describe the structure and function of electrical and chemical synapses.

7. Identify the nature of excitatory and inhibitory postsynaptic potentials.

A **synapse** is the functional connection between a neuron and a second cell. In the CNS, this other cell is also a neuron. In the PNS, the other cell may be either a neuron or an *effector cell* within a muscle or gland. Although the physiology of neuron-neuron synapses and neuron-muscle synapses is similar, the latter synapses are often called **myoneural,** or **neuromuscular, junctions.**

Neuron-neuron synapses usually involve a connection between the axon of one neuron and the dendrites, cell body, or axon of a second neuron. These are called, respectively, *axodendritic, axosomatic,* and *axoaxonic synapses.* In almost all synapses, transmission is in one direction only—from the axon of the first (or **presynaptic**) neuron to the second (or **postsynaptic**) neuron. Most commonly, the synapse occurs between the axon of the presynaptic neuron and the dendrites or cell body of the postsynaptic neuron.

In the early part of the twentieth century, most physiologists believed that synaptic transmission was *electrical*—that is, that action potentials were conducted directly from one cell to the next. This was a logical assumption, given that nerve endings appeared to touch the postsynaptic cells and that the delay in synaptic conduction was extremely short (about 0.5 msec). Improved histological techniques, however, revealed tiny gaps in the synapses, and experiments demonstrated that the actions of autonomic nerves could be duplicated by certain chemicals. This led to the hypothesis that synaptic transmission might be *chemical*—that the presynaptic nerve endings might release chemicals called **neurotransmitters.** The neurotransmitters would then change the membrane potential of the postsynaptic cell, thereby producing action potentials if a threshold depolarization were achieved.

In 1921 a physiologist named Otto Loewi published the results of experiments suggesting that synaptic transmission was indeed chemical, at least at the junction between a branch of the vagus nerve (chapter 9; see fig. 9.6) and the heart. He had isolated the heart of a frog and, while stimulating the branch of the vagus that innervates the heart, perfused the heart with an isotonic salt solution. Stimulation of the vagus nerve was known to slow the heart rate. After stimulating the vagus nerve to this frog heart, Loewi collected the isotonic salt solution and then gave it to a second heart. The vagus nerve to this second heart was not stimulated, but the isotonic solution from the first heart caused the second heart to also slow its beat.

Loewi concluded that the nerve endings of the vagus must have released a chemical—which he called *Vagusstoff*—that inhibited the heart rate. This chemical was subsequently identified as **acetylcholine,** or **ACh.** In the decades following Loewi's discovery, many other examples of chemical synapses were discovered, and the theory of electrical synaptic transmission fell into disrepute. More recent evidence, ironically, has shown that electrical synapses do exist in the nervous system (though they are the exception), within smooth muscles, and between cardiac cells in the heart.

Electrical Synapses: Gap Junctions

In order for two cells to be electrically coupled, they must be approximately equal in size and they must be joined by areas of contact with low electrical resistance. In this way, impulses can be regenerated from one cell to the next without interruption. Adjacent cells that are electrically coupled are joined together by **gap junctions.** In gap junctions, the membranes of the two cells are separated by only 2 nanometers (1 nanometer = 10^{-9} meter). A surface view of gap junctions in the electron microscope reveals hexagonal arrays of particles that function as channels through which ions and molecules may pass from one cell to the next. Each gap junction is now known to be composed of 12 proteins known as *connexins,* which are arranged like staves of a barrel to form a water-filled pore (fig. 7.21).

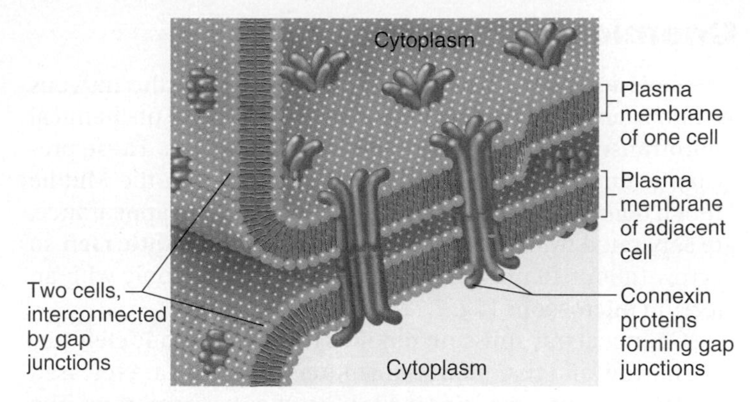

Figure 7.21 **The structure of gap junctions.** Gap junctions are water-filled channels through which ions can pass from one cell to another. This permits impulses to be conducted directly from one cell to another. Each gap junction is composed of connexin proteins. Six connexin proteins in one plasma membrane line up with six connexin proteins in the other plasma membrane to form each gap junction.

Gap junctions are present in cardiac muscle, where they allow action potentials to spread from cell to cell, so that the myocardium can contract as a unit. Similarly, gap junctions in most smooth muscles allow many cells to be stimulated and contract together, producing a stronger contraction (as in the uterus during labor). The function of gap junctions in the nervous system is less well understood; nevertheless, gap junctions are found between neurons in the brain, where they can synchronize the firing of groups of neurons. For example, gap junctions between neurons in the hippocampus, amygdala, and cerebral cortex (areas important in learning and memory; chapter 8, section 8.2) are needed for synchronized oscillations of neural activity. Gap junctions are also found between neuroglial cells, where they are believed to allow the passage of Ca^{2+} and perhaps other ions and molecules between the connected cells.

The function of gap junctions is more complex than was once thought. Neurotransmitters and other stimuli, acting through second messengers such as cAMP or Ca^{2+}, can lead to the phosphorylation or dephosphorylation of gap junction connexin proteins, causing the opening or closing of gap junction channels. There is evidence that gap junctions, like chemical synapses, undergo activity-induced changes that may be important for the formation of certain types of memory. In the retina, light causes ion conductance through the gap junctions between certain neurons to change.

Chemical Synapses

Transmission across the majority of synapses in the nervous system is one-way and occurs through the release of chemical neurotransmitters from presynaptic axon endings. These presynaptic endings, called **terminal boutons** (from the Middle French *bouton* = button) because of their swollen appearance, are separated from the postsynaptic cell by a **synaptic cleft** so narrow (about 10 nm) that it can be seen clearly only with an electron microscope (fig. 7.22).

Chemical transmission requires that the synaptic cleft stay very narrow and that neurotransmitter molecules are released near their receptor proteins in the postsynaptic membrane. The physical association of the pre- and postsynaptic membranes at the chemical synapse is stabilized by the action of particular membrane proteins. **Cell adhesion molecules (CAMs)** are proteins in the pre- and postsynaptic membranes that project from these membranes into the synaptic cleft, where they bond to each other. This Velcro-like effect ensures that the pre- and postsynaptic membranes stay in close proximity for rapid chemical transmission.

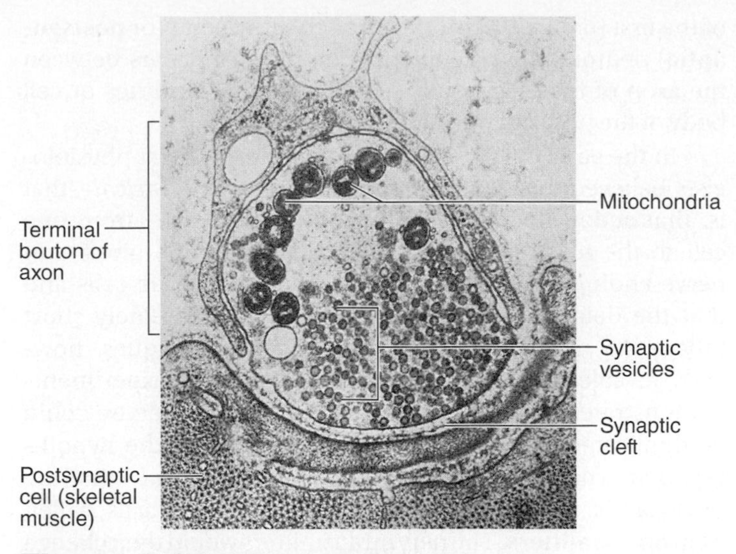

Figure 7.22 An electron micrograph of a chemical synapse. This synapse between the axon of a somatic motor neuron and a skeletal muscle cell shows the synaptic vesicles at the end of the axon and the synaptic cleft. The synaptic vesicles contain the neurotransmitter chemical. AP|R

Release of Neurotransmitter

Neurotransmitter molecules within the presynaptic neuron endings are contained within many small, membrane-enclosed **synaptic vesicles** (fig. 7.22). In order for the neurotransmitter within these vesicles to be released into the synaptic cleft, the vesicle membrane must fuse with the axon membrane in the process of *exocytosis* (chapter 3). Exocytosis of synaptic vesicles, and the consequent release of neurotransmitter molecules into the synaptic cleft, is triggered by

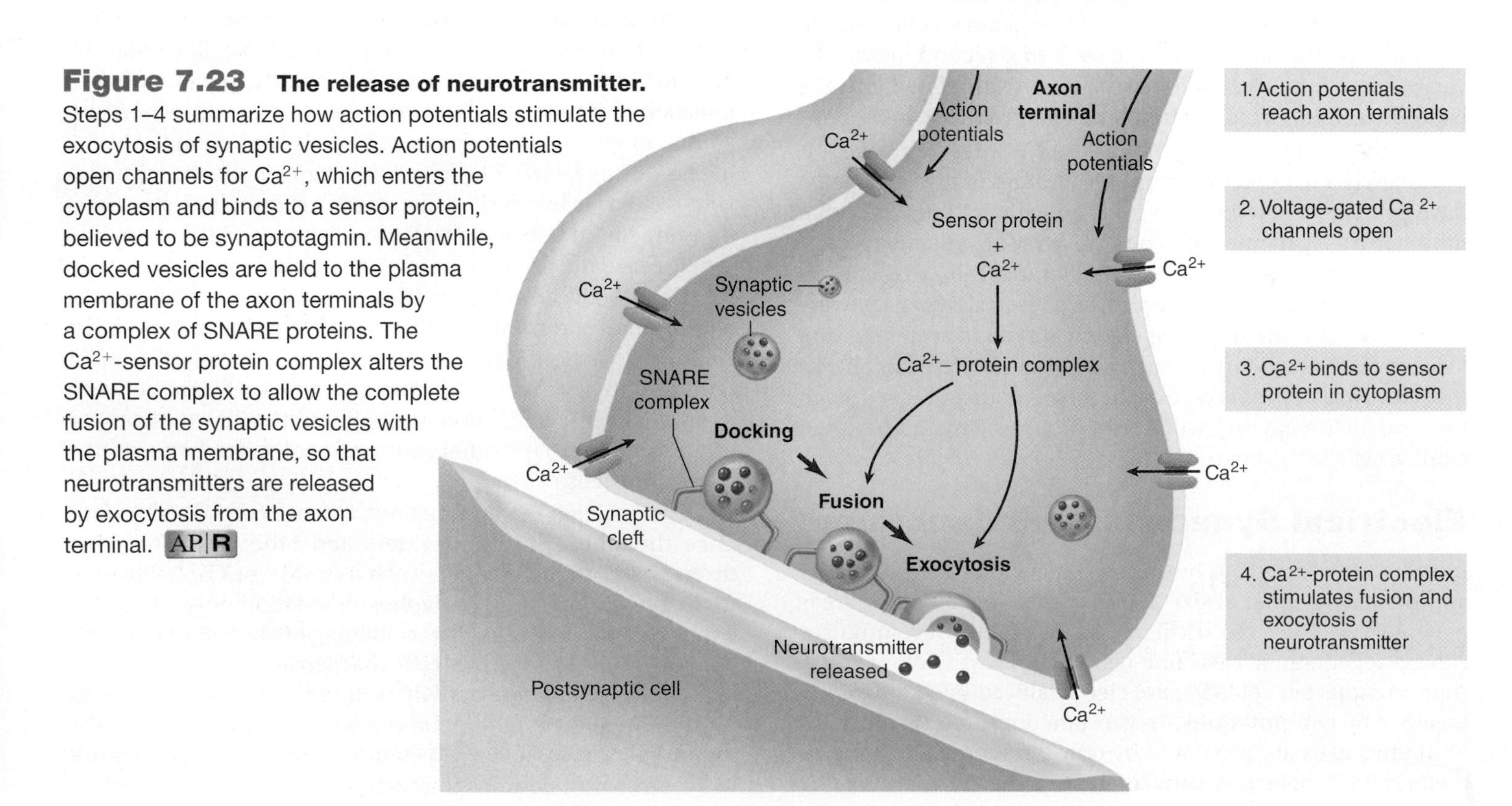

Figure 7.23 The release of neurotransmitter. Steps 1–4 summarize how action potentials stimulate the exocytosis of synaptic vesicles. Action potentials open channels for Ca^{2+}, which enters the cytoplasm and binds to a sensor protein, believed to be synaptotagmin. Meanwhile, docked vesicles are held to the plasma membrane of the axon terminals by a complex of SNARE proteins. The Ca^{2+}-sensor protein complex alters the SNARE complex to allow the complete fusion of the synaptic vesicles with the plasma membrane, so that neurotransmitters are released by exocytosis from the axon terminal. AP|R

action potentials that stimulate the entry of Ca^{2+} into the axon terminal through voltage-gated Ca^{2+} channels (fig. 7.23). When there is a greater frequency of action potentials at the axon terminal, there is a greater entry of Ca^{2+}, and thus a larger number of synaptic vesicles undergoing exocytosis and releasing neurotransmitter molecules. As a result, a greater frequency of action potentials by the presynaptic axon will result in greater stimulation of the postsynaptic neuron.

Ca^{2+} entering the axon terminal binds to a protein, believed to be *synaptotagmin,* which serves as a Ca^{2+} sensor and forms a Ca^{2+}-synaptotagmin complex in the cytoplasm. This occurs close to the location where synaptic vesicles are already *docked* (attached) to the plasma membrane of the axon terminal. At this stage, the docked vesicles are bound to the plasma membrane of the presynaptic axon by complexes of three *SNARE proteins* that bridge the vesicles and plasma membrane. The complete fusion of the vesicle membrane and plasma membrane, and the formation of a pore that allows the release of neurotransmitter, occurs when the Ca^{2+}-synaptotagmin complex displaces a component of the *SNARE,* or *fusion, complex.* This process is very rapid: exocytosis of neurotransmitter occurs less than 100 microseconds after the intracellular Ca^{2+} concentration rises.

CLINICAL APPLICATION

Tetanus toxin and **botulinum toxin** are bacterial products that cause paralysis by preventing neurotransmission. These neurotoxins function as *proteases* (protein-digesting enzymes), digesting particular components of the fusion complex and thereby inhibiting the exocytosis of synaptic vesicles. Botulinum toxin destroys members of the SNARE complex of proteins needed for exocytosis of the neurotransmitter ACh, which stimulates muscle contraction. This results in *flaccid paralysis,* where the muscles are unable to contract. Tetanus toxin acts similarly, but blocks inhibitory synapses in the CNS; this results in *spastic paralysis,* where the muscles are unable to relax.

Clinical Investigation CLUES

Botox is a preparation of botulinum toxin. Sandra had ptosis (droopy eyelid), a side effect of her Botox treatment.

- By what action does Botox exert its effects?
- How might this action be related to Sandra's ptosis?

Actions of Neurotransmitter

Once the neurotransmitter molecules have been released from the presynaptic axon terminals, they diffuse rapidly across the synaptic cleft and reach the membrane of the postsynaptic cell. The neurotransmitters then bind to specific **receptor proteins** that are part of the postsynaptic membrane. Receptor proteins have high specificity for their neurotransmitter, which is the **ligand** of the receptor protein. The term *ligand* in this case refers to a smaller molecule (the neurotransmitter) that binds to and forms a complex with a larger protein molecule (the receptor). Binding of the neurotransmitter ligand to its receptor protein causes ion channels to open in the postsynaptic membrane. The gates that regulate these channels, therefore, can be called **chemically regulated** (or **ligand-regulated**) **gates** because they open in response to the binding of a chemical ligand to its receptor in the postsynaptic plasma membrane.

Note that two broad categories of gated ion channels have been described: *voltage-regulated* and *chemically regulated.* Voltage-regulated channels are found primarily in the axons; chemically regulated channels are found in the postsynaptic membrane. Voltage-regulated channels open in response to depolarization; chemically regulated channels open in response to the binding of postsynaptic receptor proteins to their neurotransmitter ligands.

When the chemically regulated ion channels are opened, they produce a graded change in the membrane potential, also known as a **graded potential.** The opening of specific channels—particularly those that allow Na^+ or Ca^{2+} to enter the cell—produces a graded depolarization, where the inside of the postsynaptic membrane becomes less negative. This depolarization is called an **excitatory postsynaptic potential (EPSP)** because the membrane potential moves toward the threshold required for action potentials. In other cases, as when Cl^- enters the cell through specific channels, a graded hyperpolarization is produced (where the inside of the postsynaptic membrane becomes more negative). This hyperpolarization is called an **inhibitory postsynaptic potential (IPSP)** because the membrane potential moves farther from the threshold depolarization required to produce action potentials. The mechanisms by which specific neurotransmitters produce graded EPSPs and IPSPs will be described in the sections that follow.

Excitatory postsynaptic potentials, as their name implies, stimulate the postsynaptic cell to produce action potentials, and inhibitory postsynaptic potentials antagonize this effect. In synapses between the axon of one neuron and the dendrites of another, the EPSPs and IPSPs are produced at the dendrites and must propagate to the initial segment of the axon to influence action potential production (fig. 7.24).

Synaptic potentials (EPSPs and IPSPs) decrease in amplitude as they are conducted along the dendrites and cell body to the axon hillock, which is the final region where they can summate. There, the total depolarization reaching the initial

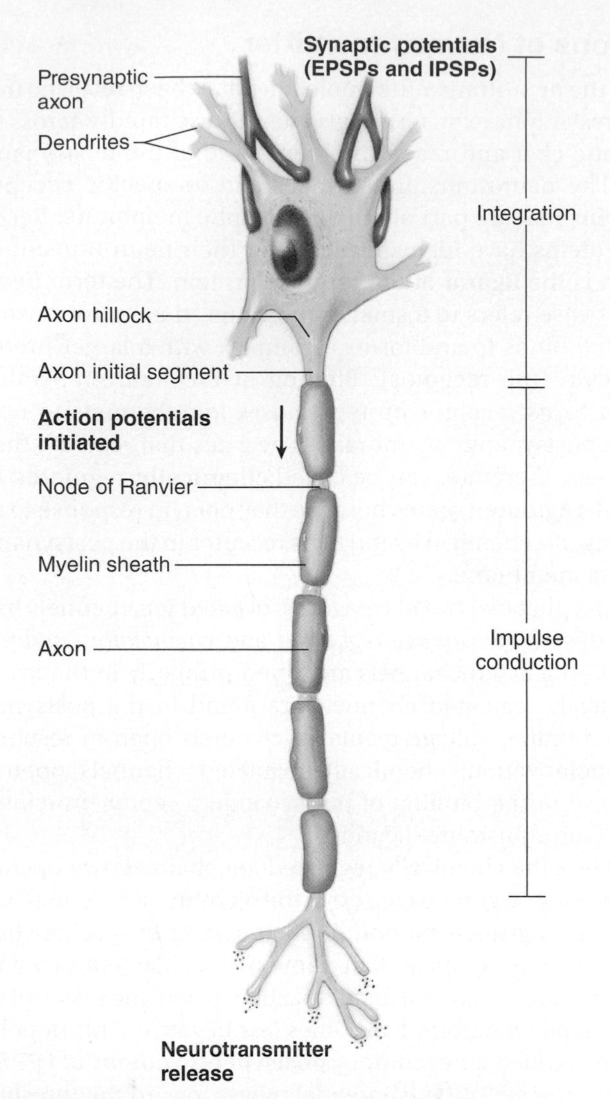

Figure 7.24 The functional specialization of different regions in a multipolar neuron. Integration of input (EPSPs and IPSPs) generally occurs in the dendrites and cell body, with the axon serving to conduct action potentials.

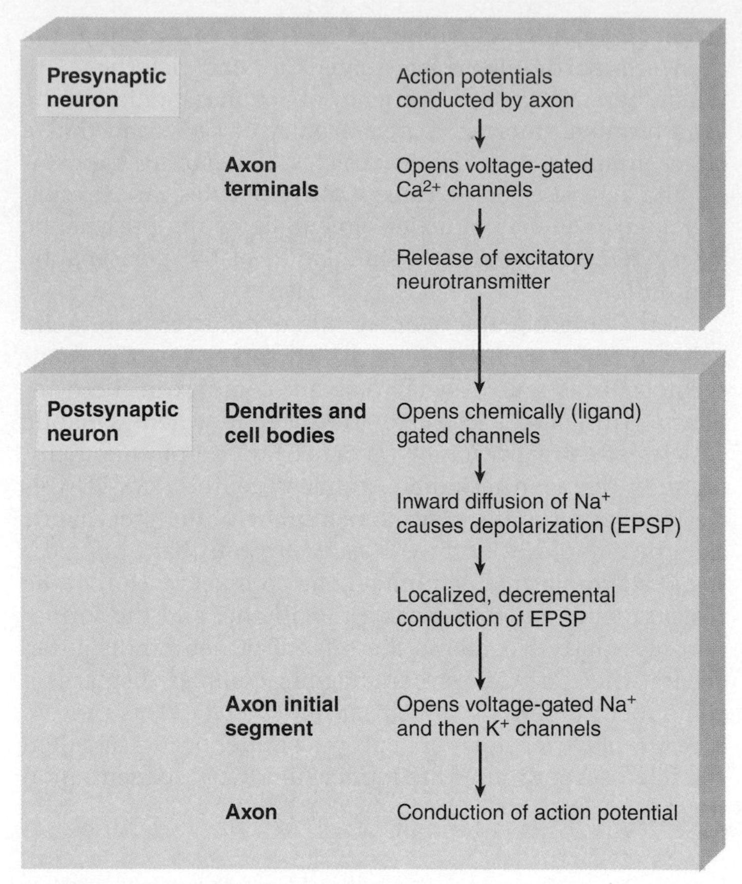

Figure 7.25 Events in excitatory synaptic transmission. The different regions of the postsynaptic neuron are specialized, with ligand-(chemically) gated channels located in the dendrites and cell body, and voltage-gated channels located in the axon.

segment of the axon determines if it will fire an action potential, and if so, the frequency of action potential production. The **initial segment of the axon,** located adjacent to the axon hillock, has a high density of voltage-gated Na^+ and K^+ channels and is where action potentials are first produced. Once action potentials are produced at the initial segment, they will regenerate themselves along the axon as previously described. These events are illustrated in figure 7.24 and summarized in figure 7.25. Additionally, in neurons of the cerebral cortex, action potentials have been shown to "back propagate" from the axon initial segment to the dendrites. This is believed to aid the synaptic events involved in learning and memory (section 7.7).

CHECKPOINT

6a. Describe the structure, locations, and functions of gap junctions.

6b. Describe the location of neurotransmitters within an axon and explain the relationship between presynaptic axon activity and the amount of neurotransmitters released.

6c. Describe the sequence of events by which action potentials stimulate the release of neurotransmitters from presynaptic axons.

7. Explain how chemically regulated channels differ from voltage-regulated channels and the nature of excitatory and inhibitory postsynaptic potentials.

7.4 ACETYLCHOLINE AS A NEUROTRANSMITTER

When acetylcholine (ACh) binds to its receptor, it directly or indirectly causes the opening of chemically regulated gates. In many cases, this produces a depolarization called an excitatory postsynaptic potential, or EPSP. In some cases, however, ACh causes a hyperpolarization known as an inhibitory postsynaptic potential, or IPSP.

LEARNING OUTCOMES

After studying this section, you should be able to:

8. Explain how ligand-gated channels produce synaptic potentials, using the nicotinic ACh receptor as an example.
9. Explain how G-protein-coupled channels produce synaptic potentials, using the muscarinic ACh receptor as an example.
10. Describe the action and significance of acetylcholinesterase.
11. Compare EPSPs and action potentials, identify where each is produced in a neuron, and explain how action potentials can be stimulated by EPSPs.

Acetylcholine (ACh) is used as an excitatory neurotransmitter by some neurons in the CNS and by somatic motor neurons at the neuromuscular junction. At autonomic nerve endings, ACh may be either excitatory or inhibitory, depending on the organ involved.

Postsynaptic cells can have varying responses to the same chemical partly because different postsynaptic cells have different subtypes of ACh receptors. These receptor subtypes can be specifically stimulated by particular toxins and are named for these toxins. The stimulatory effect of ACh on skeletal muscle cells is produced by the binding of ACh to **nicotinic ACh receptors,** so named because they can also be activated by nicotine. Effects of ACh on other cells occur when ACh binds to **muscarinic ACh receptors,** so named because these effects can also be produced by muscarine (a drug derived from certain poisonous mushrooms).

An overview of the distribution of the two types of ACh receptors demonstrates that this terminology and its associated concepts will be important in understanding the physiology of different body systems. The two types of **cholinergic receptors** (receptors for ACh) are explained in more detail in chapter 9 (see fig. 9.11).

1. *Nicotinic ACh receptors.* These are found in specific regions of the brain (chapter 8), in autonomic ganglia (chapter 9), and in skeletal muscle fibers (chapter 12). The release of ACh from somatic motor neurons and its subsequent binding to nicotinic receptors, for example, stimulates skeletal muscle contraction.
2. *Muscarinic ACh receptors.* These are found in the plasma membrane of smooth muscle cells, cardiac muscle cells, and the cells of particular glands (chapter 9). Thus, the activation of muscarinic ACh receptors by ACh released from autonomic axons is required for the regulation of the cardiovascular system (chapter 14), digestive system (chapter 18), and others. Muscarinic ACh receptors are also found in the brain.

Drugs that bind to and thereby activate receptor proteins are called **agonists,** and drugs that bind to and thereby reduce the activity of receptor proteins are **antagonists.** For example, *muscarine* (from the poisonous *Amanita muscaria* mushroom) is an agonist of muscarinic ACh receptors, whereas *atropine*—a drug derived from *Atropa belladonna,* a member of the deadly nightshade family—is an antagonist of muscarinic receptors. *Nicotine* (from tobacco plants) is an agonist for nicotinic ACh receptors; antagonists include *α-bungarotoxin* (from krait snake venom) and *curare* (see table 7.5).

Chemically Regulated Channels

The binding of a neurotransmitter to its receptor protein can cause the opening of ion channels through two different mechanisms. These two mechanisms can be illustrated by the actions of ACh on the nicotinic and muscarinic subtypes of the ACh receptors.

Ligand-Gated Channels

As previously mentioned, a neurotransmitter molecule is the *ligand* that binds to its specific receptor protein. For ion channels that are "ligand-gated," the receptor protein is also an ion channel; these are two functions of the same protein. Part of this protein has extracellular sites that bind to the neurotransmitter ligands, while part of the protein spans the plasma membrane and has a central ion channel. For example, there is a family of related ligand-gated channels that consist of five polypeptide chains surrounding an ion channel. This receptor family includes the nicotinic ACh receptors discussed here, as well as different receptors for the neurotransmitters serotonin, GABA, and glycine (discussed later in this chapter). Although there are important differences among these ligand-gated channels, all members of this family function in a similar way: when the neurotransmitter ligand binds to its membrane receptor, a central ion channel opens through the same receptor/channel protein.

The nicotinic ACh receptor can serve as an example of ligand-gated channels. Two of its five polypeptide subunits contain ACh-binding sites, and the channel opens when both

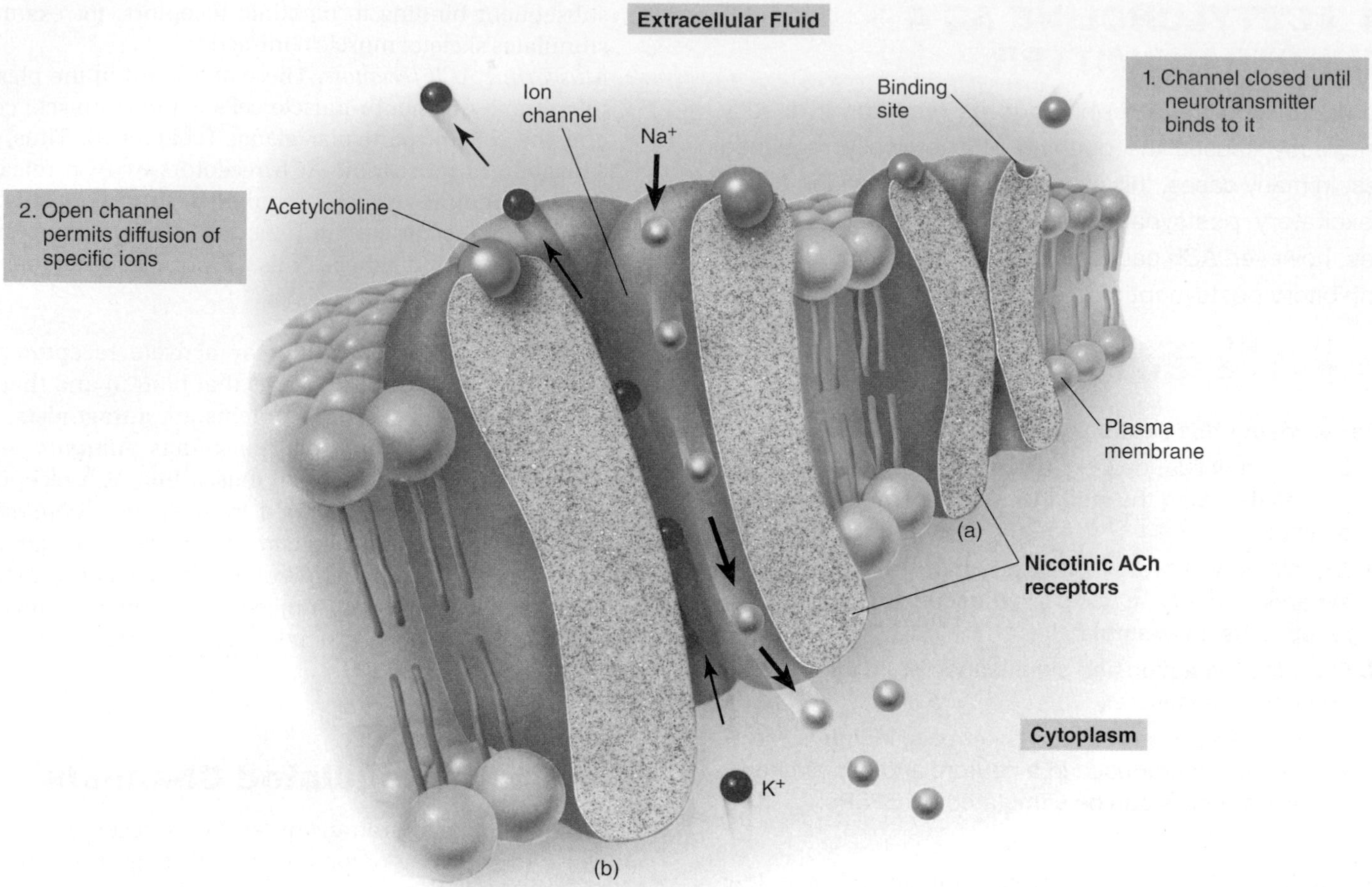

Figure 7.26 Nicotinic acetylcholine (ACh) receptors also function as ion channels. The nicotinic acetylcholine receptor contains a channel that is closed (*a*) until the receptor binds to ACh. (*b*) Na^+ and K^+ diffuse simultaneously, and in opposite directions, through the open ion channel. The electrochemical gradient for Na^+ is greater than for K^+, so that the effect of the inward diffusion of Na^+ predominates, resulting in a depolarization known as an excitatory postsynaptic potential (EPSP).

sites bind to ACh (fig. 7.26). The opening of this channel permits the simultaneous diffusion of Na^+ into and K^+ out of the postsynaptic cell. The effects of the inward flow of Na^+ predominate, however, because of its steeper electrochemical gradient. This produces the depolarization of an excitatory postsynaptic potential (EPSP).

Although the inward diffusion of Na^+ predominates in an EPSP, the simultaneous outward diffusion of K^+ prevents the depolarization from overshooting 0 mV. Therefore, the membrane polarity does not reverse in an EPSP as it does in an action potential. (Remember that action potentials are produced by separate voltage-gated channels for Na^+ and K^+, where the channel for K^+ opens only after the Na^+ channel has closed.)

A comparison of EPSPs and action potentials is provided in table 7.4. Action potentials occur in axons, where the voltage-gated channels are located, whereas EPSPs occur in the dendrites and cell body. Unlike action potentials, EPSPs have *no threshold;* the ACh released from a single synaptic vesicle produces a tiny depolarization of the postsynaptic membrane. When more vesicles are stimulated to release their ACh, the depolarization is correspondingly greater. EPSPs are therefore *graded* in magnitude, unlike all-or-none action potentials. Because EPSPs can be graded and have *no refractory period,* they are capable of *summation.* That is, the depolarizations of several different EPSPs can be added together. Action potentials are prevented from summating by their all-or-none nature and by their refractory periods.

CLINICAL APPLICATION

People with **myasthenia gravis** have muscle weakness caused by antibodies, produced by their own immune system, that bind to and block their ACh receptors. Because of this, myasthenia gravis is an autoimmune disease (discussed in chapter 15). **Paralytic shellfish poisoning,** which can be fatal, occurs when people eat shellfish containing *saxitoxin,* produced by the microorganisms in the red tide. A similar poison, *tetrodotoxin,* is produced by pufferfish. Saxitoxin and tetrodotoxin cause flaccid paralysis by blocking the voltage-gated Na^+ channels. These and other poisons that affect neuromuscular transmission are summarized in table 7.5.

Table 7.4 | Comparison of Action Potentials and Excitatory Postsynaptic Potentials (EPSPs)

Characteristic	Action Potential	Excitatory Postsynaptic Potential
Stimulus for opening of ionic gates	Depolarization	Acetylcholine (ACh) or other excitatory neurotransmitter
Initial effect of stimulus	Na^+ channels open	Common channels for Na^+ and K^+ open
Cause of repolarization	Opening of K^+ gates	Loss of intracellular positive charges with time and distance
Conduction distance	Regenerated over length of the axon	1–2 mm; a localized potential
Positive feedback between depolarization and opening of Na^+ gates	Yes	No
Maximum depolarization	+ 40 mV	Close to zero
Summation	No summation—all-or-none event	Summation of EPSPs, producing graded depolarizations
Refractory period	Yes	No
Effect of drugs	ACh effects inhibited by tetrodotoxin, not by curare	ACh effects inhibited by curare, not by tetrodotoxin

Table 7.5 | Drugs That Affect the Neural Control of Skeletal Muscles

Drug	Origin	Effects
Botulinum toxin	Produced by *Clostridium botulinum* (bacteria)	Inhibits release of acetylcholine (ACh)
Curare	Resin from a South American tree	Prevents interaction of ACh with its nicotinic receptor proteins
α-Bungarotoxin	Venom of *Bungarus* snakes	Binds to ACh receptor proteins and prevents ACh from binding
Saxitoxin	Red tide (*Gonyaulax*) algae	Blocks voltage-gated Na^+ channels
Tetrodotoxin	Pufferfish	Blocks voltage-gated Na^+ channels
Nerve gas	Artificial	Inhibits acetylcholinesterase in postsynaptic membrane
Neostigmine	Nigerian bean	Inhibits acetylcholinesterase in postsynaptic membrane
Strychnine	Seeds of an Asian tree	Prevents IPSPs in spinal cord that inhibit contraction of antagonistic muscles

Clinical Investigation CLUES

Sandra experienced severe muscle weakness after eating just a little of the local shellfish gathered at the beginning of a red tide. Mussels and clams are filter feeders that concentrate the poison (saxitoxin) in the red tide.

- How could saxitoxin produce Sandra's muscle weakness?
- Given that the diaphragm is a skeletal muscle, propose one mechanism by which paralytic shellfish poisoning could be fatal.

G-Protein-Coupled Channels

There is another group of chemically regulated ion channels that, like the ligand-gated channels, are opened by the binding of a neurotransmitter to its receptor protein. However, this group of channels differs from ligand-gated channels in that the receptors and the ion channels are different, separate membrane proteins. Thus, binding of the neurotransmitter ligand to its receptor can open the ion channel only indirectly. Such is the case with the muscarinic ACh receptors discussed in this section, as well as the receptors for dopamine and norepinephrine, discussed in section 7.5.

The muscarinic ACh receptors are formed from only a single subunit, which can bind to one ACh molecule. Unlike the nicotinic receptors, these receptors do not contain ion channels. The ion channels are separate proteins located at some distance from the muscarinic receptors. Binding of ACh (the ligand) to the muscarinic receptor causes it to activate a complex of proteins in the cell membrane known as **G-proteins**—so named because their activity is influenced by guanosine nucleotides (GDP and GTP). This topic was introduced in chapter 6, section 6.5.

There are three G-protein subunits, designated alpha, beta, and gamma. In response to the binding of ACh to its receptor, the alpha subunit dissociates from the other two subunits, which stick together to form a beta-gamma complex. Depending on the specific case, either the alpha subunit or the beta-gamma complex then diffuses through the membrane until it binds to an ion channel, causing the channel to open or close (fig. 7.27). A short time later, the G-protein alpha subunit (or beta-gamma complex) dissociates from the channel and moves back to its previous position. This causes the ion channel to close (or open). The steps of this process are summarized in table 7.6 and are illustrated in chapter 6 (see fig. 6.31).

The binding of ACh to its muscarinic receptors indirectly affects the permeability of K^+ channels. This can produce hyperpolarization in some organs (if the K^+ channels are opened) and depolarization in other organs (if the K^+ channels are closed). Specific examples should help to clarify this point.

Scientists have learned that it is the beta-gamma complex that binds to the K^+ channels in the heart muscle cells and causes these channels to open (fig. 7.27). This leads to the diffusion of K^+ out of the postsynaptic cell (because that is the direction of its concentration gradient). As a result, the cell becomes hyperpolarized, producing an inhibitory postsynaptic potential (IPSP). Such an effect is produced in the heart, for example, when autonomic nerve fibers (part of the vagus nerve) synapse with pacemaker cells and slow the rate of beat. It should be noted that inhibition also occurs in the CNS in response to other neurotransmitters, but those IPSPs are produced by a different mechanism.

There are cases in which the alpha subunit is the effector, and examples where its effects are substantially different from

Table 7.6 | Steps in the Activation and Inactivation of G-Proteins

Step 1	When the membrane receptor protein is not bound to its regulatory molecule ligand, the alpha, beta, and gamma G-protein subunits are aggregated together and attached to the receptor; the alpha subunit binds GDP.
Step 2	When the ligand (neurotransmitter or other regulatory molecule) binds to the receptor, the alpha subunit releases GDP and binds GTP; this allows the alpha subunit to dissociate from the beta-gamma subunits.
Step 3	Either the alpha subunit or the beta-gamma complex moves through the membrane and binds to a membrane effector protein (either an ion channel or an enzyme).
Step 4	Deactivation of the effector protein is caused by the alpha subunit hydrolyzing GTP to GDP.
Step 5	This allows the subunits to again reaggregate and bind to the unstimulated receptor protein (which is no longer bound to its regulatory molecule ligand).

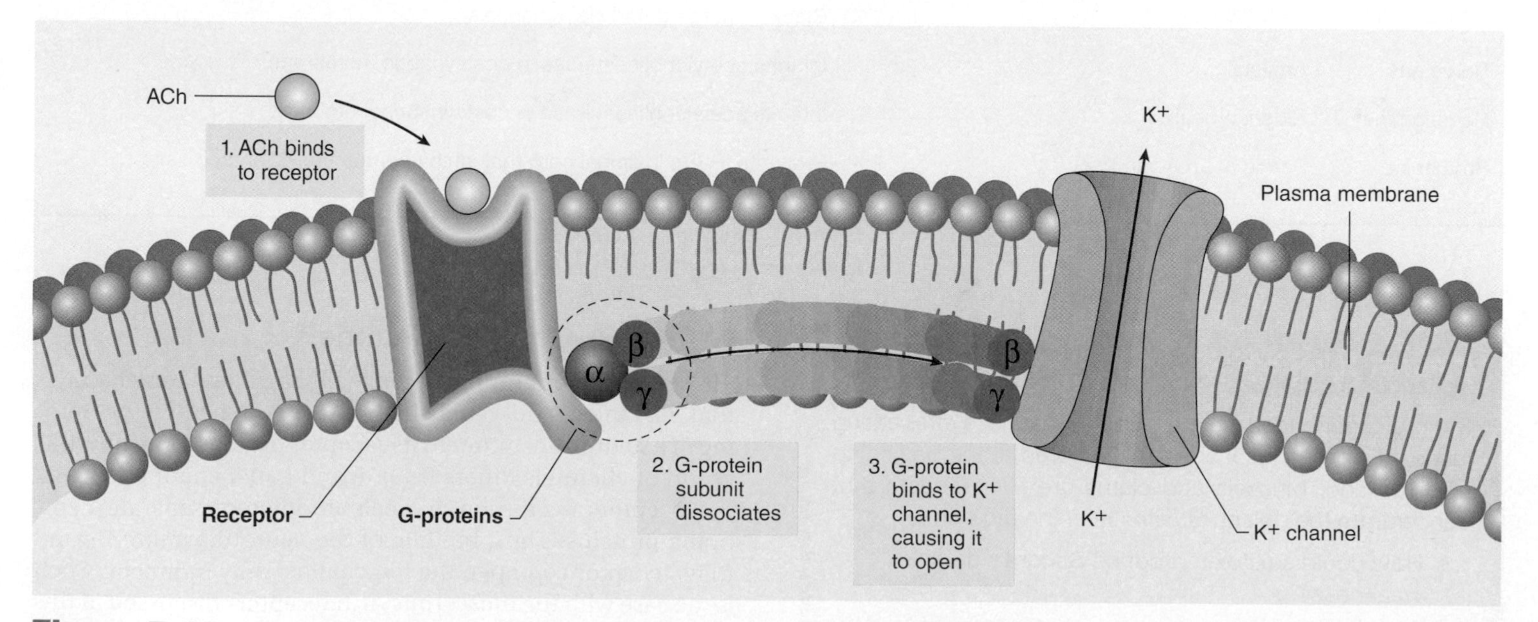

Figure 7.27 Muscarinic ACh receptors require the action of G-proteins. The figure depicts the effects of ACh on the pacemaker cells of the heart. Binding of ACh to its muscarinic receptor causes the beta-gamma subunits to dissociate from the alpha subunit. The beta-gamma complex of G-proteins then binds to a K^+ channel, causing it to open. Outward diffusion of K^+ results, slowing the heart rate.

the one shown in figure 7.27. In the smooth muscle cells of the stomach, the binding of ACh to its muscarinic receptors causes alpha subunits to dissociate and bind to gated K^+ channels. In this case, however, the binding of the G-protein subunit to the gated K^+ channels causes them to close rather than to open. As a result, the outward diffusion of K^+, which occurs at an ongoing rate in the resting cell, is reduced to below resting levels. Because the resting membrane potential is maintained by a balance between cations flowing into the cell and cations flowing out, a reduction in the outward flow of K^+ produces a depolarization. This depolarization produced in these smooth muscle cells results in contractions of the stomach (see chapter 9, fig. 9.11).

Acetylcholinesterase (AChE)

The bond between ACh and its receptor protein exists for only a brief instant. The ACh-receptor complex quickly dissociates but can be quickly re-formed as long as free ACh is in the vicinity. In order for activity in the postsynaptic cell to be stopped, free ACh must be inactivated very soon after it is released. The inactivation of ACh is achieved by means of an enzyme called **acetylcholinesterase, or AChE,** which is present on the postsynaptic membrane or immediately outside the membrane, with its active site facing the synaptic cleft (fig. 7.28). AChE hydrolyzes acetylcholine into acetate and choline, which can then reenter the presynaptic axon terminals and be resynthesized into acetylcholine (ACh).

CLINICAL APPLICATION

Nerve gas exerts its odious effects by inhibiting AChE in skeletal muscles. Since ACh is not degraded, it can continue to combine with receptor proteins and can continue to stimulate the postsynaptic cell, leading to spastic paralysis. Clinically, cholinesterase inhibitors (such as neostigmine) are used to enhance the effects of ACh on muscle contraction when neuromuscular transmission is weak, as in the disease *myasthenia gravis.*

Acetylcholine in the PNS

Somatic motor neurons form synapses with skeletal muscle cells (muscle fibers). At these synapses, or **neuromuscular junctions,** the postsynaptic membrane of the muscle fiber is known as a *motor end plate.* Therefore, the EPSPs produced by ACh in skeletal muscle fibers are often called **end-plate potentials.** This depolarization opens voltage-regulated channels that are

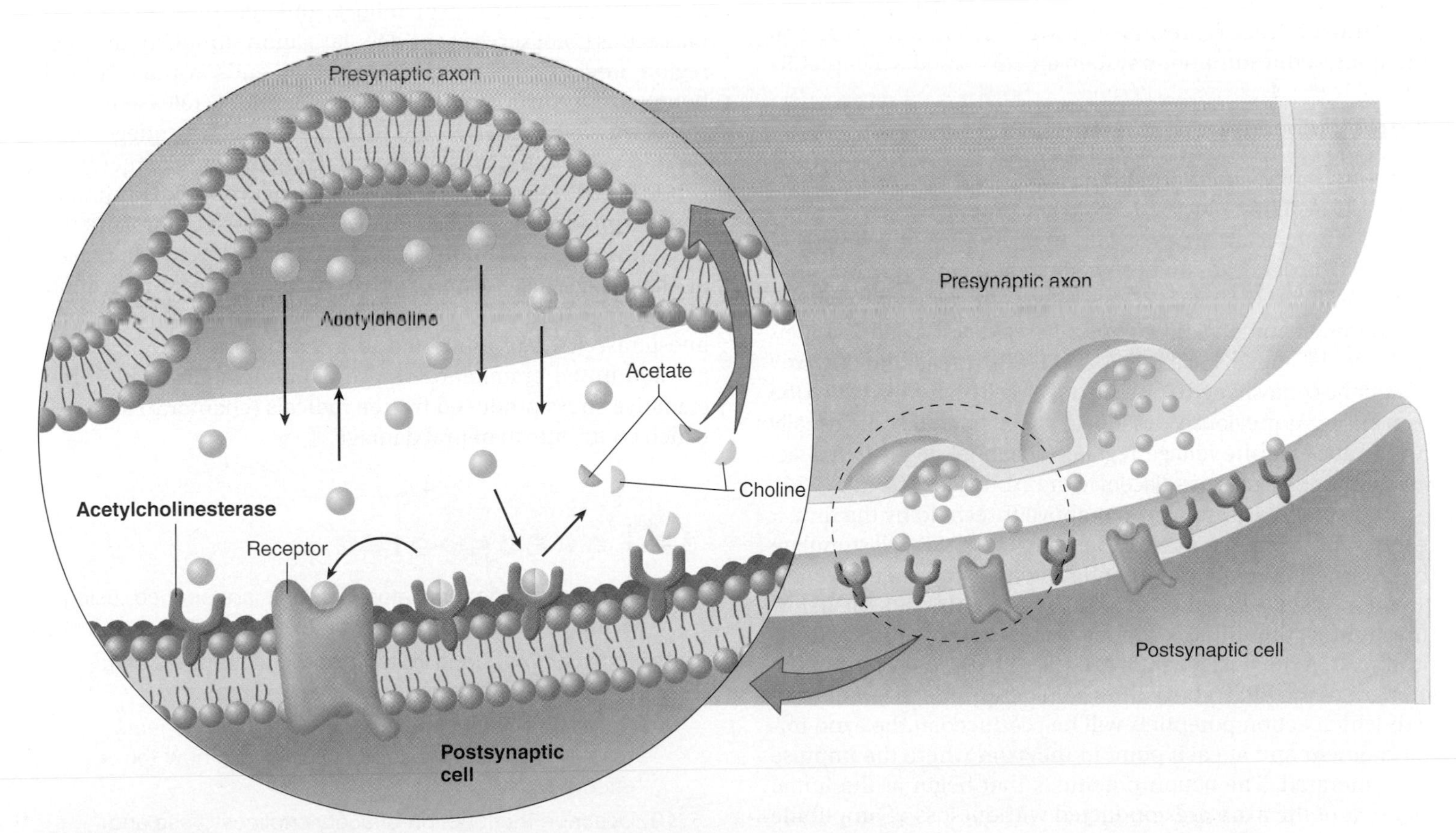

Figure 7.28 The action of acetylcholinesterase (AChE). The AChE in the postsynaptic cell membrane inactivates the ACh released into the synaptic cleft. This prevents continued stimulation of the postsynaptic cell unless more ACh is released by the axon. The acetate and choline are taken back into the presynaptic axon and used to resynthesize acetylcholine.

adjacent to the end plate. Voltage-regulated channels produce action potentials in the muscle fiber, and these are reproduced by other voltage-regulated channels along the muscle plasma membrane. This conduction is analogous to conduction of action potentials by axons; it is significant because action potentials produced by muscle fibers stimulate muscle contraction (chapter 12, section 12.2).

If any stage in the process of neuromuscular transmission is blocked, muscle weakness—sometimes leading to paralysis and death—may result. The drug *curare,* for example, competes with ACh for attachment to the nicotinic ACh receptors and thus reduces the size of the end-plate potentials (see table 7.5). This drug was first used on blow-gun darts by South American Indians because it produced flaccid paralysis in their victims. Clinically, curare is used in surgery as a muscle relaxant and in electroconvulsive shock therapy to prevent muscle damage.

Autonomic motor neurons innervate cardiac muscle, smooth muscles in blood vessels and visceral organs, and glands. There are two classifications of autonomic nerves: sympathetic and parasympathetic. Most of the parasympathetic axons that innervate the effector organs use ACh as their neurotransmitter. In some cases, these axons have an inhibitory effect on the organs they innervate through the binding of ACh to muscarinic ACh receptors. The action of the vagus nerve in slowing the heart rate is an example of this inhibitory effect. In other cases, ACh released by autonomic neurons produces stimulatory effects as previously described. The structures and functions of the autonomic system are described in chapter 9.

Acetylcholine in the CNS

There are many **cholinergic neurons** (those that use ACh as a neurotransmitter) in the CNS, where the axon terminals of one neuron typically synapse with the dendrites or cell body of another. The dendrites and cell body thus serve as the receptive area of the neuron, and it is in these regions that receptor proteins for neurotransmitters and chemically regulated gated channels are located. Graded, local EPSPs and IPSPs spread into the *axon hillock,* the cone-shaped elevation on the cell body from which the axon arises. As previously described, action potentials are usually first produced by the voltage-regulated channels in the *initial segment of the axon* located adjacent to the axon hillock (see fig. 7.24).

If the depolarization is at or above threshold by the time it reaches the initial segment of the axon, the EPSP will stimulate the production of action potentials, which can then regenerate themselves along the axon. If, however, the EPSP is below threshold at the initial segment no action potentials will be produced in the postsynaptic cell (fig. 7.29). Gradations in the strength of the EPSP above threshold determine the frequency with which action potentials will be produced at the axon initial segment and at each point in the axon where the impulse is regenerated. The action potentials that begin at the initial segment of the axon are conducted without loss of amplitude toward the axon terminals.

Earlier in this chapter, the action potential was introduced by describing the events that occurred when a depolarization stimulus was artificially produced by stimulating electrodes. Now it is apparent that EPSPs, conducted from the dendrites

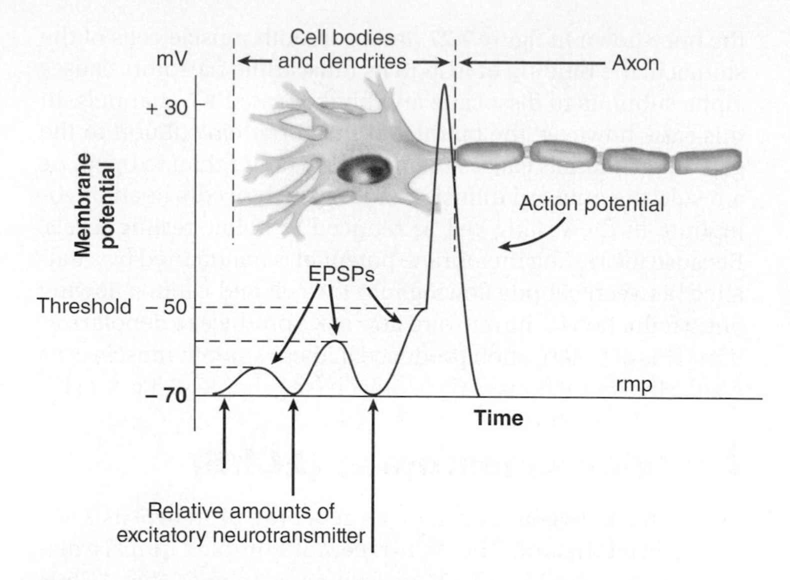

Figure 7.29 The graded nature of excitatory postsynaptic potentials (EPSPs). Stimuli of increasing strength produce increasing amounts of depolarization. When a threshold level of depolarization is produced, action potentials are generated in the axon.

and cell body, serve as the normal stimuli for the production of action potentials at the axon hillock, and that the action potentials at this point serve as the depolarization stimuli for the next region, and so on. This chain of events ends at the terminal boutons of the axon, where neurotransmitter is released.

Alzheimer's disease—the most common neurodegenerative disease and the most common cause of senile dementia—is associated with the loss of cholinergic neurons that terminate in the hippocampus and cerebral cortex, which are areas of the brain involved in memory storage (chapter 8). Possible causes of Alzheimer's disease are discussed in the Clinical Application box on page 222; treatments currently include the use of cholinesterase (AChE) inhibitors to augment cholinergic transmission in the brain, and the use of antioxidants to limit the oxidative stress produced by free radicals (chapters 5 and 19), which contribute to neural damage.

CHECKPOINT

8. Explain how ligand-gated channels are opened, using nicotinic ACh receptors as an example.

9a. Explain how ligand-gated channels operate, using muscarinic ACh receptors as an example.

9b. Describe where stimulatory and inhibitory effects of muscarinic ACh receptors occur and how these effects are produced.

10. Describe the function of acetylcholinesterase and discuss its physiological significance.

11. Compare the properties of EPSPs and action potentials, identify where in a neuron these are produced, and explain how EPSPs stimulate action potentials.

7.5 MONOAMINES AS NEUROTRANSMITTERS

A variety of chemicals in the CNS function as neurotransmitters. Among these are the monoamines, a chemical family that includes dopamine, norepinephrine, and serotonin. Although these molecules have similar mechanisms of action, they are used by different neurons for different functions.

LEARNING OUTCOMES

After studying this section, you should be able to:

12. Identify the monoamine neurotransmitters and explain how they are inactivated at the synapse.

13. Identify two neural pathways in the brain that use dopamine as a neurotransmitter, and explain their significance.

Monoamines are regulatory molecules derived from amino acids. *Dopamine, norepinephrine (noradrenalin),* and *epinephrine (adrenalin)* are derived from the amino acid tyrosine and placed in a subfamily of monoamines called **catecholamines.** The term *catechol* refers to a common six-carbon ring structure, as shown in chapter 9, figure 9.8. Dopamine is a neurotransmitter; norepinephrine is a neurotransmitter and a hormone (from the adrenal medulla); and epinephrine is the primary hormone secreted by the adrenal medulla.

Other monoamines are derived from different amino acids and so are not classified as catecholamines. *Serotonin* is derived from the amino acid tryptophan and functions as an important neurotransmitter. *Histamine* is derived from the amino acid histidine and serves as a monoamine neurotransmitter, as well as a regulator produced by nonneural tissues (discussed in later chapters). As a monoamine neurotransmitter in the brain, histamine promotes wakeful alertness; this is why some antihistamines cause drowsiness (chapter 8, section 8.4).

Like ACh, monoamine neurotransmitters are released by exocytosis from presynaptic vesicles, diffuse across the synaptic cleft, and interact with specific receptor proteins in the membrane of the postsynaptic cell. The stimulatory effects of these monoamines, like those of ACh, must be quickly inhibited so as to maintain proper neural control. The action of monoamine neurotransmitters at the synapse is stopped by (1) reuptake of the neurotransmitter molecules from the synaptic cleft into the presynaptic axon terminal, and then (2) degradation of the monoamine by an enzyme within the axon terminal called **monoamine oxidase (MAO).** This process is illustrated in figure 7.30.

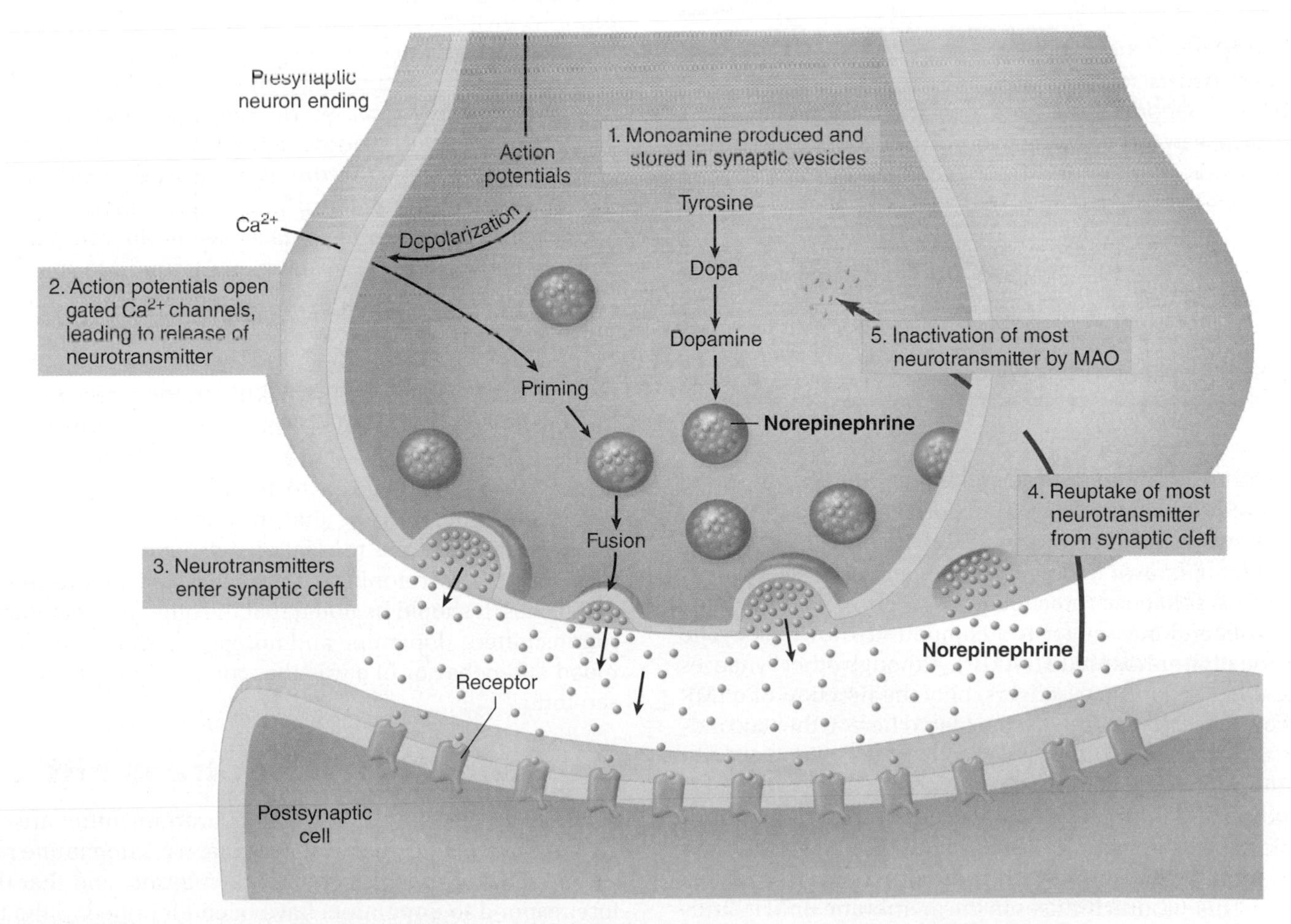

Figure 7.30 **Production, release, reuptake, and inactivation of monoamine neurotransmitters.** Most of the monoamine neurotransmitters, including dopamine, norepinephrine, and serotonin, are transported back into the presynaptic axon terminals after being released into the synaptic gap. They are then degraded and inactivated by an enzyme, monoamine oxidase (MAO). AP|R

CLINICAL APPLICATION

Monoamine oxidase (MAO) inhibitors are drugs that block monoamine oxidase, the enzyme primarily responsible for degrading the monoamine neurotransmitters. By preventing the breakdown of monoamines, these drugs increase the amount of neurotransmitters in the synaptic cleft, promoting their effects. MAO inhibitors have proven useful in the treatment of clinical depression, suggesting that a deficiency in monoamine neurotransmission contributes to this disorder. An MAO inhibitor drug is also used to treat Parkinson's disease, because it enhances the action of dopamine at the synapse. A serious potential side effect of MAO inhibitors is **hypertensive crisis** (dangerously elevated blood pressure), because the effects of epinephrine and norepinephrine (released by the sympathetic nerves and the adrenal medulla; chapter 9) act to raise the blood pressure. Hypertensive crisis may be provoked in people taking MAO inhibitors by eating foods rich in *tyramine*, a monoamine degraded by MAO that is found in cheese, preserved meats, and certain fish, among many other foods. Tyramine can also cause migraine headaches in susceptible people or those taking MAO inhibitors.

Clinical Investigation CLUES

The paramedics found a prescription bottle of an MAO inhibitor in Sandra's purse and were concerned that she may have eaten food containing high amounts of tyramine. Sandra thought the food was safe, but admitted that she needs to be more careful.

- What is MAO and what does it do?
- How might an MAO inhibitor help to treat clinical depression?
- How would food rich in tyramine be dangerous for a person taking an MAO inhibitor?

The monoamine neurotransmitters do not directly cause opening of ion channels in the postsynaptic membrane. Instead, these neurotransmitters act by means of an intermediate regulator, known as a **second messenger.** In the case of some synapses that use catecholamines for synaptic transmission, this second messenger is a compound known as **cyclic adenosine monophosphate (cAMP).** Although other synapses can use other second messengers, only the function of cAMP as a second messenger will be considered here. Other second-messenger systems are discussed in conjunction with hormone action in chapter 11, section 11.2.

Binding of norepinephrine, for example, with its receptor in the postsynaptic membrane stimulates the dissociation of the G-protein alpha subunit from the others in its complex (fig. 7.31). This subunit diffuses in the membrane until it binds to an enzyme known as *adenylate cyclase* (also called *adenylyl cyclase*). This enzyme converts ATP to cyclic AMP (cAMP) and pyrophosphate (two inorganic phosphates) within the postsynaptic cell cytoplasm. Cyclic AMP in turn activates another enzyme, *protein kinase,* which phosphorylates (adds a phosphate group to) other proteins (fig. 7.31). Through this action, ion channels are opened in the postsynaptic membrane.

Serotonin as a Neurotransmitter

Serotonin, or *5-hydroxytryptamine (5-HT),* is used as a neurotransmitter by neurons with cell bodies in what are called the *raphe nuclei* that are located along the midline of the brain stem (chapter 8). Serotonin is derived from the amino acid L-tryptophan, and variations in the amount of this amino acid in the diet (tryptophan-rich foods include milk and turkey) can affect the amount of serotonin produced by the neurons. Physiological functions attributed to serotonin include a role in the regulation of mood and behavior, appetite, and cerebral circulation.

The classical hallucinogens—LSD, mescaline, and psilocybin—exert their effects primarily by binding to and activating serotonin receptors in the cerebral cortex. Prior to 1970, when these drugs were classified as Schedule 1 drugs (and were thus highly regulated), scientists sometimes used these hallucinogens medically in attempts to treat psychiatric disorders. The role of serotonin in the regulation of mood and emotion is currently exploited by the action of the antidepressant drugs *Prozac, Paxil, Zoloft, and Luvox,* which act as **serotonin-specific reuptake inhibitors (SSRIs).** The SSRIs reduce the production of *serotonin transporter (SERT)* proteins, thereby reducing the ability of SERT proteins in the presynaptic neuron plasma membrane to clear serotonin from the synaptic cleft. This increases the ability of serotonin to stimulate its receptors in the postsynaptic membrane, an ability that aids in the treatment of depression.

Serotonin's diverse functions are related to the fact that there are many different subtypes of serotonin receptors—over a dozen are currently known. Thus, while Prozac may be given to relieve depression, another drug that promotes serotonin action is sometimes given to reduce the appetite of obese patients. A different drug that may activate a different serotonin receptor is used to treat anxiety, and yet another drug that promotes serotonin action is given to relieve migraine headaches. It should be noted that the other monoamine neurotransmitters, dopamine and norepinephrine, also influence mood and behavior in a way that complements the actions of serotonin.

Dopamine as a Neurotransmitter

Neurons that use **dopamine** as a neurotransmitter are called **dopaminergic neurons.** Neurons that have dopamine receptor proteins on the postsynaptic membrane, and that therefore respond to dopamine, have been identified in the living brain using the technique of *positron emission tomography*

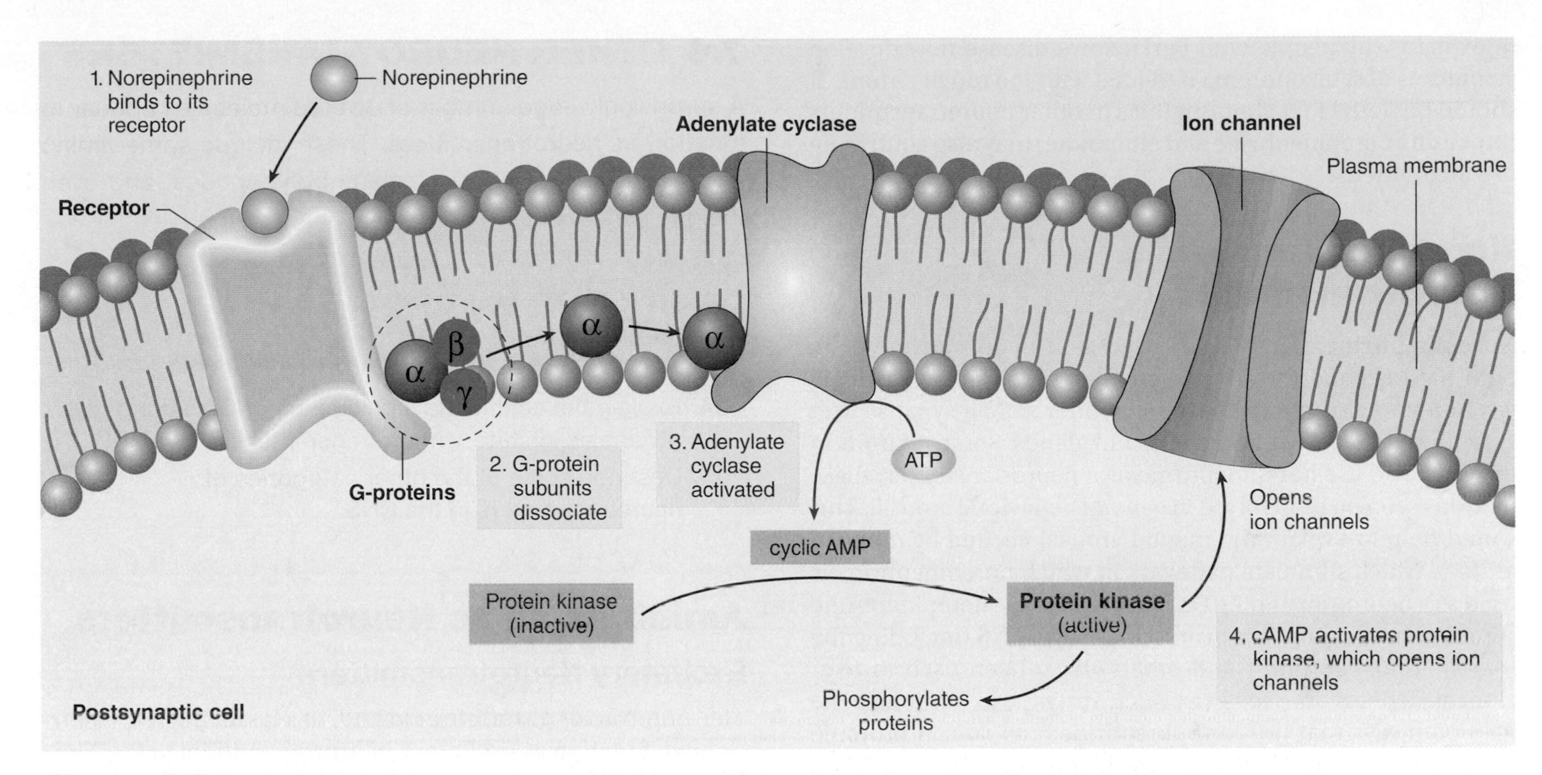

Figure 7.31 Norepinephrine action requires G-proteins. The binding of norepinephrine to its receptor causes the dissociation of G-proteins. Binding of the alpha G-protein subunit to the enzyme adenylate cyclase activates this enzyme, leading to the production of cyclic AMP. Cyclic AMP, in turn, activates protein kinase, which can open ion channels and produce other effects.

(PET) (chapter 8, section 8.2). These investigations have been spurred by the great clinical interest in the effects of dopaminergic neurons.

The cell bodies of dopaminergic neurons are highly concentrated in the midbrain. Their axons project to different parts of the brain and can be divided into two systems: the *nigrostriatal dopamine system,* involved in motor control, and the *mesolimbic dopamine system,* involved in emotional reward (see chapter 8, fig. 8.21).

Nigrostriatal Dopamine System

The cell bodies of the **nigrostriatal dopamine system** are located in a part of the midbrain called the *substantia nigra* ("dark substance") because it contains melanin pigment. Neurons in the substantia nigra send fibers to a group of nuclei known collectively as the *corpus striatum* because of its striped appearance—hence the term *nigrostriatal system.* These regions are part of the *basal nuclei*—large masses of neuron cell bodies deep in the cerebrum involved in the initiation of skeletal movements (chapter 8). **Parkinson's disease** is caused by degeneration of the dopaminergic neurons in the substantia nigra. Parkinson's disease is the second most common neurodegenerative disease (after Alzheimer's disease) and is associated with such symptoms as muscle tremors and rigidity, difficulty in initiating movements and speech, and other severe motor problems. Patients are often treated with L-dopa and MAO inhibitors in an attempt to increase dopaminergic transmission in the nigrostriatal dopamine system.

Mesolimbic Dopamine System

The **mesolimbic dopamine system** involves neurons that originate in the midbrain and send axons to structures in the forebrain that are part of the limbic system (see fig. 8.21). The dopamine released by these neurons may be involved in behavior and reward. For example, several studies involving human twins separated at birth and reared in different environments, and other studies involving the use of rats, have implicated the gene that codes for one subtype of dopamine receptor (designated D_2) in alcoholism. Other addictive drugs, including cocaine, morphine, and amphetamines, are also known to activate dopaminergic pathways.

Recent studies demonstrate that alcohol, amphetamines, cocaine, marijuana, and morphine promote the activity of dopaminergic neurons that arise in the midbrain and terminate in a particular location, the *nucleus accumbens,* of the forebrain. Interestingly, nicotine also promotes the release of dopamine by axons that terminate in this very location. This suggests that the physiological mechanism for nicotine addiction in smokers is similar to that for other abused drugs.

Drugs used to treat schizophrenia (drugs called *neuroleptics*) act as antagonists of the D_2 dopamine receptor (and can thus cause side effects resembling Parkinson's disease). This suggests that overactivity of the mesolimbic dopamine pathways contributes to schizophrenia, a concept that helps to explain why people with Parkinson's disease may develop symptoms of schizophrenia if treated with too much L-dopa. It should be noted that abnormalities in other neurotransmitters (including norepinephrine and glutamate) may also contribute to schizophrenia.

Norepinephrine as a Neurotransmitter

Norepinephrine, like ACh, is used as a neurotransmitter in both the PNS and the CNS. Sympathetic neurons of the PNS use norepinephrine as a neurotransmitter at their synapse with smooth muscles, cardiac muscle, and glands. Some neurons in the CNS also use norepinephrine as a neurotransmitter; these neurons seem to be involved in general behavioral arousal. This would help to explain the mental arousal elicited by *amphetamines,* which stimulate pathways in which norepinephrine is used as a neurotransmitter. Drugs that increase norepinephrine stimulation of synaptic transmission in the CNS (including the *tricyclic antidepressants* and others) have been used to treat clinical depression. However, such drugs also stimulate the PNS pathways that use norepinephrine, and so can promote sympathetic nerve effects that raise blood pressure.

CLINICAL APPLICATION

Cocaine—a stimulant related to the amphetamines in its action—is currently widely abused in the United States. Although early use of this drug produces feelings of euphoria and social adroitness, continued use leads to social withdrawal, depression, dependence upon ever-higher dosages, and serious cardiovascular and renal disease that can result in heart and kidney failure. The numerous effects of cocaine on the central nervous system appear to be mediated by one primary mechanism: cocaine binds to the reuptake transporters for dopamine, norepinephrine, and serotonin, and blocks their reuptake into the presynaptic axon endings. This results in overstimulation of those neural pathways that use dopamine as a neurotransmitter.

CHECKPOINT

12a. List the monoamines and indicate their chemical relationships.

12b. Explain how monoamines are inactivated at the synapse and how this process can be clinically manipulated.

13a. Describe the relationship between dopaminergic neurons, Parkinson's disease, and schizophrenia.

13b. Explain how cocaine and amphetamines produce their effects in the brain. What are the dangers of these drugs?

7.6 OTHER NEUROTRANSMITTERS

A surprisingly large number of diverse molecules appear to function as neurotransmitters. These include some amino acids and their derivatives, many polypeptides, and even the gas nitric oxide.

LEARNING OUTCOMES

After studying this section, you should be able to:

14. Explain the action and significance of GABA and glycine as inhibitory neurotransmitters.

15. Describe some of the other categories of neurotransmitters in the CNS.

Amino Acids as Neurotransmitters

Excitatory Neurotransmitters

The amino acids **glutamic acid** and, to a lesser degree, *aspartic acid,* function as excitatory neurotransmitters in the CNS. Glutamic acid (or *glutamate*), indeed, is the major excitatory neurotransmitter in the brain, producing excitatory postsynaptic potentials (EPSPs) in at least 80% of the synapses in the cerebral cortex. The energy consumed by active transport carriers needed to maintain the ionic gradients for these EPSPs constitutes the major energy requirement of the brain (action potentials produced by axons are more energy efficient than EPSPs). Astrocytes take glutamate from the synaptic cleft, as previously described, and couple this to increased glucose uptake and increased blood flow via vasodilation to the more active brain regions.

Research has revealed that each of the glutamate receptors encloses an ion channel, similar to the arrangement seen in the nicotinic ACh receptors (see fig. 7.26). Among these EPSP-producing glutamate receptors, three subtypes can be distinguished. These are named according to the molecules (other than glutamate) that they bind: (1) **NMDA receptors** (named for N-methyl-D-aspartate); (2) **AMPA receptors;** and (3) **kainate receptors.** NMDA and AMPA receptors are illustrated in chapter 8, figure 8.16.

The NMDA receptors for glutamate are involved in memory storage, as will be discussed in section 7.7 and chapter 8, section 8.2. These receptors have a channel pore that is blocked by Mg^2 and the simple binding of glutamate to these receptors cannot open the channels. Instead, two other conditions must be met at the same time: (1) the NMDA receptor must also bind to glycine (or D-serine, which is produced by astrocytes); and (2) the membrane must be partially depolarized at this time by a different neurotransmitter molecule that binds to a different receptor (for example, by glutamate binding to the AMPA receptors, as shown in fig. 8.16). Depolarization causes Mg^{2+} to be released from the NMDA channel pore, unblocking the channel and allowing the entry of Ca^{2+} and Na^{+} (and exit of K^{+}) through NMDA channels in the dendrites of the postsynaptic neuron.

Inhibitory Neurotransmitters

The neurotransmitters **GABA (gamma-aminobutyric acid),** a derivative of glutamic acid, and **glycine** (another amino acid) are inhibitory. Instead of depolarizing the postsynaptic membrane and producing an EPSP, they hyperpolarize the postsynaptic membrane and produce an IPSP. The binding of these neurotransmitters to their receptors causes the opening of Cl^{-} channels in the postsynaptic membrane (fig. 7.32). Because active transport carriers keep the concentration of Cl^{-} lower inside than outside of mature neurons, Cl^{-} will diffuse into the postsynaptic neuron. This is true as long as the postsynaptic membrane potential is less negative (more depolarized) than the chloride equilibrium potential (chapter 6, section 6.4). This may not be the case at rest, because the chloride equilibrium potential may be close to the resting membrane potential. However, if an excitatory neurotransmitter partially depolarizes the membrane, the movement of Cl^{-} through its open channels is promoted. When this occurs, the hyperpolarizing effects of Cl^{-} entering the cell make it more difficult for the postsynaptic neuron to reach the threshold depolarization required to stimulate action potentials. Thus, the opening of Cl^{-} channels by an inhibitory neurotransmitter renders excitatory input less effective.

The inhibitory effects of glycine are very important in the spinal cord, where they help in the control of skeletal movements (chapter 12; see figs. 12.30 and 12.31). Flexion of an arm, for example, involves stimulation of the flexor muscles by motor neurons in the spinal cord. The motor neurons that innervate the antagonistic extensor muscles are inhibited by IPSPs produced by glycine released from other neurons. The importance of the inhibitory actions of glycine is revealed by the deadly effects of *strychnine,* a poison that causes spastic paralysis by specifically blocking the glycine receptor proteins. Animals poisoned with strychnine die from asphyxiation because they are unable to relax the diaphragm.

GABA is the most prevalent neurotransmitter in the brain; in fact, as many as one-third of all the neurons in the brain use GABA as a neurotransmitter. Like glycine, GABA is inhibitory—it hyperpolarizes the postsynaptic membrane by opening Cl^{-} channels. Also, the effects of GABA, like those of glycine, are involved in motor control. For example, large neurons called *Purkinje cells* mediate the motor functions of the cerebellum by producing IPSPs in their postsynaptic neurons. A deficiency of GABA-releasing neurons is responsible for uncontrolled movements in people with *Huntington's disease.* Huntington's disease is a neurodegenerative disorder caused by a defect in the *huntingtin* gene (chapter 3, section 3.4).

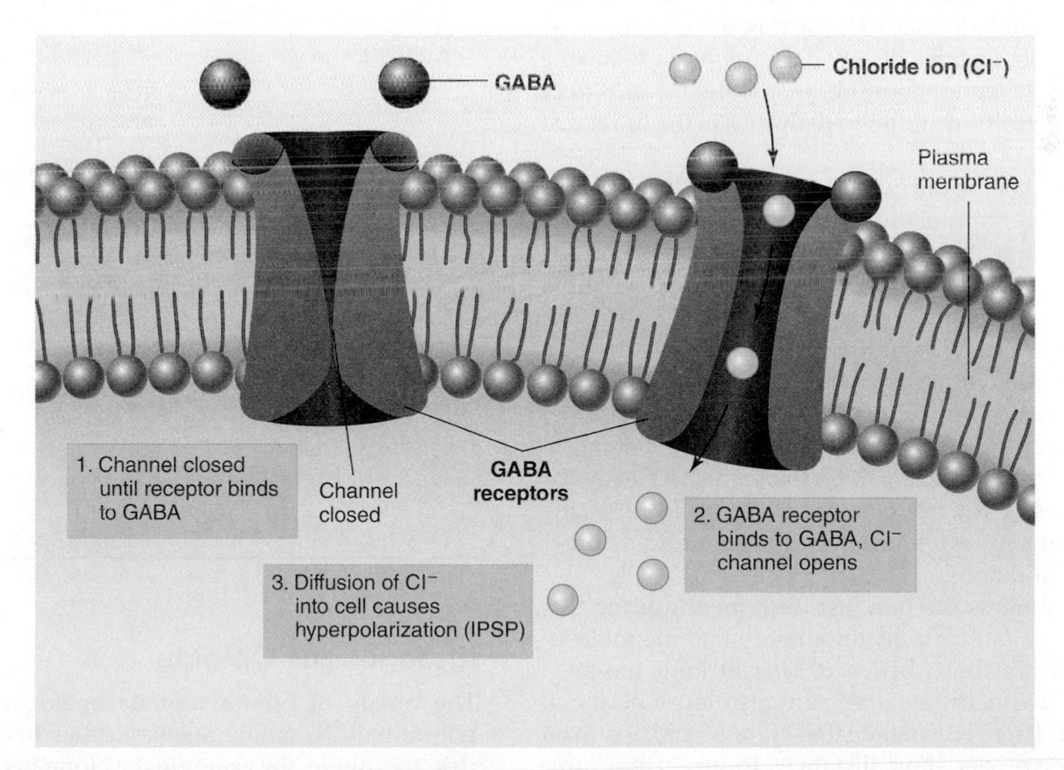

Figure 7.32 GABA receptors contain a chloride channel. When GABA (gamma-aminobutyric acid) binds to its receptor, a chloride ion (Cl^{-}) channel opens through the receptor. This permits the inward diffusion of Cl^{-}, resulting in hyperpolarization, or an IPSP.

CLINICAL APPLICATION

Benzodiazepines are drugs that act to increase the ability of GABA to activate its receptors in the brain and spinal cord. Because GABA inhibits the activity of spinal motor neurons that innervate skeletal muscles, the intravenous infusion of benzodiazepines inhibits the muscular spasms in epileptic seizures and seizures resulting from drug overdose and poisons. Probably as a result of its general inhibitory effects on the brain, GABA also functions as a neurotransmitter involved in mood and emotion. Benzodiazepines such as *Valium* are thus given orally to treat anxiety and sleeplessness.

Polypeptides as Neurotransmitters

Many polypeptides of various sizes are found in the synapses of the brain. These are often called **neuropeptides** and are believed to function as neurotransmitters. Interestingly, some of the polypeptides that function as hormones secreted by the small intestine and other endocrine glands are also produced in the brain and may function there as neurotransmitters (table 7.7). For example, *cholecystokinin (CCK),* which is secreted as a hormone from the small intestine, is also released from neurons and used as a neurotransmitter in the brain. Recent evidence suggests that CCK, acting as a neurotransmitter, may promote feelings of satiety in the brain following meals. Another polypeptide found in many organs, *substance P,* functions as a neurotransmitter in pathways in the brain that mediate sensations of pain.

Although some of the polypeptides released from neurons may function as neurotransmitters in the traditional sense (that is, by stimulating the opening of ionic gates and causing changes in the membrane potential), others may have more subtle and poorly understood effects. **Neuromodulators** has been proposed as a name for compounds with such alternative effects. Some neurons in both the PNS and the CNS produce both a classical neurotransmitter (ACh or a catecholamine) and a polypeptide neurotransmitter. These are contained in different synaptic vesicles that can be distinguished using the electron microscope. The neuron can thus release either the classical neurotransmitter or the polypeptide neurotransmitter under different conditions.

Discoveries such as the one just described indicate that synapses have a greater capacity for alteration at the molecular level than was previously believed. This attribute has been termed *synaptic plasticity.* Synapses are also more plastic at the cellular level. There is evidence that sprouting of new axon branches can occur over short distances to produce a turnover of synapses, even in the mature CNS. This breakdown and re-forming of synapses may occur within a time span of only a few hours. These events may play a role in learning and conditioning.

Table 7.7 | Examples of Chemicals That Are Either Proven or Suspected Neurotransmitters

Category	Chemicals
Amines	Histamine
	Serotonin
Catecholamines	Dopamine
	(Epinephrine—a hormone)
	Norepinephrine
Choline derivative	Acetylcholine
Amino acids	Aspartic acid
	GABA (gamma-aminobutyric acid)
	Glutamic acid
	Glycine
Polypeptides	Glucagon
	Insulin
	Somatostatin
	Substance P
	ACTH (adrenocorticotrophic hormone)
	Angiotensin II
	Endogenous opioids (enkephalins and endorphins)
	LHRH (luteinizing hormone-releasing hormone)
	TRH (thyrotrophin-releasing hormone)
	Vasopressin (antidiuretic hormone)
	CCK (cholecystokinin)
Lipids	Endocannabinoids
Gases	Nitric oxide
	Carbon monoxide
Purines	ATP

Endogenous Opioids

The ability of opium and its analogues—the **opioids**—to relieve pain (promote analgesia) has been known for centuries. Morphine, for example, has long been used for this purpose. The discovery in 1973 of opioid receptor proteins in the brain suggested that the effects of these drugs might be due to the stimulation of specific neuron pathways. This implied that opioids—along with LSD, mescaline, and other mind-altering drugs—might mimic the actions of neurotransmitters produced by the brain.

The analgesic effects of morphine are blocked in a specific manner by a drug called *naloxone.* In the same year that opioid receptor proteins were discovered, it was found that naloxone also blocked the analgesic effect of electrical brain stimulation. Subsequent evidence suggested that the analgesic effects of hypnosis and acupuncture could also be blocked by naloxone. These experiments indicate that neurons might be producing their own endogenous opioids that serve as the natural ligands of opioid receptors in the CNS. Receptor proteins for the endogenous opioids and opioid drugs have been identified and are widespread in the CNS. When the gene for one subtype of opioid receptors is knocked out in mice (chapter 3), the analgesic effect of morphine (its ability to reduce pain) is completely abolished, demonstrating the importance of opioids and their receptors in reducing pain transmission.

The endogenous opioids have been identified as a family of polypeptides produced by the brain and pituitary gland. One member is called *β-endorphin* (for "endogenously produced morphinelike compound"). Another consists of a group of five-amino-acid peptides called *enkephalins,* and a third is a polypeptide neurotransmitter called *dynorphin.*

The endogenous opioid system is inactive under normal conditions, but when activated by stressors it can block the transmission of pain. For example, a burst in β-endorphin secretion was shown to occur in pregnant women during parturition (childbirth).

Exogenous opioids such as opium and morphine can produce euphoria, and so endogenous opioids may mediate reward or positive reinforcement pathways. This is consistent with the observation that overeating in genetically obese mice can be blocked by naloxone. It has also been suggested that the feeling of well-being and reduced anxiety following exercise (the "joggers high") may be an effect of endogenous opioids. Blood levels of β-endorphin increase when exercise is performed at greater than 60% of the maximal oxygen uptake (chapter 12) and peak 15 minutes after the exercise has ended. Although obviously harder to measure, an increased level of opioids in the brain and cerebrospinal fluid has also been found to result from exercise. The opioid antagonist drug naloxone, however, does not block the exercise-induced euphoria, suggesting that the joggers high is not primarily an opioid effect. Use of naloxone, however, does demonstrate that the endogenous opioids are involved in the effects of exercise on blood pressure, and that they are responsible for the ability of exercise to raise the pain threshold.

Neuropeptide Y

Neuropeptide Y is the most abundant neuropeptide in the brain. It has been shown to have a variety of physiological effects, including a role in the response to stress, in the regulation of circadian rhythms, and in the control of the cardiovascular system. Neuropeptide Y has been shown to inhibit the release of the excitatory neurotransmitter glutamate in a part of the brain called the hippocampus. This is significant because excessive glutamate released in this area can cause convulsions. Indeed, frequent seizures were a symptom of a recently developed strain of mice with the gene for neuropeptide Y "knocked out." (Knockout strains of mice have specific genes inactivated; chapter 3, section 3.5.)

Neuropeptide Y is a powerful stimulator of appetite. When injected into a rat's brain, it can cause the rat to eat until it becomes obese. Conversely, inhibitors of neuropeptide Y that are injected into the brain inhibit eating. This research has become particularly important in light of the discovery of *leptin,* a satiety factor secreted by adipose tissue. Leptin suppresses appetite by acting, at least in part, to inhibit neuropeptide Y release. This topic is discussed in chapter 19, section 19.2.

Endocannabinoids as Neurotransmitters

In addition to producing endogenous opioids, the brain also produces compounds with effects similar to those of the active ingredient in marijuana—Δ^9-tetrahydrocannabinol (THC). These endogenous cannabinoids, or **endocannabinoids,** are neurotransmitters that bind to the same receptor proteins in the brain as does THC from marijuana.

Perhaps surprisingly, endocannabinoid receptors are abundant and widely distributed in the brain. The endocannabinoids are lipids; they are short fatty acids (*anandamide* and *2-arachidonoyl glycerol*), and the only lipids known to act as neurotransmitters. As lipids, the endocannabinoids are not stored in synaptic vessicles; rather, they are produced from the lipids of the neuron plasma membrane and released from the dendrites and cell body.

The endocannabinoids function as **retrograde neurotransmitters**—they are released from the postsynaptic neuron and diffuse backward to the axons of presynaptic neurons. Once in the presynaptic neuron, the endocannabinoids bind to their receptors and inhibit the release of neurotransmitter from the axon. This can reduce the release of either the inhibitory neurotransmitter GABA or the excitatory neurotransmitter glutamate from presynaptic axons. Endocannabinoids thereby modify the actions of a number of other neurotransmitters in the brain.

These actions may be important in strengthening synaptic transmission during learning. For example, suppose that a postsynaptic neuron receives inhibitory GABA input from one presynaptic axon and excitatory glutamate from a different presynaptic axon. If the postsynaptic neuron has just been stimulated by glutamate, the glutamate produces a depolarization that causes a rise in the cytoplasmic Ca^{2+} concentration. This promotes the release of endocannabinoids from the postsynaptic neuron, which in turn act as retrograde neurotransmitters to reduce the release of GABA from the other presynaptic axon. Such *depolarization-induced suppression of inhibition* could facilitate the use of that synapse, perhaps for learning and memory. This is a type of *long-term depression* of synaptic transmission, a form of synaptic plasticity described in section 7.7.

In contrast to the role played by endocannabinoids in learning and memory, exogenous THC obtained by smoking marijuana impairs attention, learning, and memory; some studies suggest that, in chronic marijuana users, this impairment may persist even after the drug is no longer in the body. Many studies also indicate a significantly increased risk of psychosis, particularly schizophrenia, in people who have used marijuana over an extended time.

The ability of the THC in marijuana to stimulate appetite is widely known, and it appears that the endocannabinoids may perform a similar function. The endocannabinoids stimulate overeating in experimental animals, and a drug that antagonizes the endocannabinoid receptor can block this effect. Pharmaceutical drugs in different stages of development promote the cannabinoid stimulation of appetite, while others for weight loss suppress appetite by inhibiting endocannabinoid stimulation of its receptors.

Nitric Oxide and Carbon Monoxide as Neurotransmitters

Nitric oxide (NO) was the first gas to be identified as a neurotransmitter. Produced by nitric oxide synthetase from the amino acid L-arginine in the cells of many organs, nitric oxide's actions are very different from those of the more familiar nitrous oxide (N_2O), or laughing gas, sometimes used as a mild anesthetic in dentistry.

Nitric oxide has a number of different roles in the body. Within blood vessels, it acts as a local tissue regulator that causes the smooth muscles of those vessels to relax, so that the blood vessels dilate. This role will be described in conjunction with the circulatory system in chapter 14, section 14.3. Within macrophages and other cells, nitric oxide helps to kill bacteria. This activity is described in conjunction with the immune system in chapter 15, section 15.1. In addition, nitric oxide is a neurotransmitter of certain neurons in both the PNS and the CNS. It diffuses out of the presynaptic axon and into neighboring cells by simply passing through the lipid portion of the cell membranes. In some cases, nitric oxide is also produced by the postsynaptic neuron and can diffuse back to the presynaptic neuron to act as a retrograde neurotransmitter (chapter 8; see fig. 8.16). Once in the target cells, NO exerts its effects by stimulating the production of cyclic guanosine monophosphate (cGMP), which acts as a second messenger.

In the PNS, nitric oxide is released by some neurons that innervate the gastrointestinal tract, penis, respiratory passages, and cerebral blood vessels. These are autonomic neurons that cause smooth muscle relaxation in their target organs. This can produce, for example, the engorgement of the spongy tissue of the penis with blood. In fact, scientists now believe that erection of the penis results from the action of nitric oxide, and indeed the drug *Viagra* works by increasing this action of nitric oxide (as described in chapter 20; see fig. 20.22).

In addition to nitric oxide, another gas—**carbon monoxide (CO)**—may function as a neurotransmitter. Certain neurons, including those of the cerebellum and olfactory epithelium, have been shown to produce carbon monoxide (derived from the conversion of one pigment molecule, heme, to another, biliverdin; see fig. 18.22). Also, carbon monoxide, like nitric oxide, has been shown to stimulate the production of cGMP within the neurons. Experiments suggest that carbon monoxide may promote odor adaptation in olfactory neurons, contributing to the regulation of olfactory sensitivity. Other physiological functions of neuronal carbon monoxide have also been suggested, including neuroendocrine regulation in the hypothalamus.

ATP and Adenosine as Neurotransmitters

Adenosine triphosphate (ATP) and adenosine are classified chemically as *purines* (chapter 2) and have multiple cellular functions. The plasma membrane is impermeable to organic molecules with phosphate groups, trapping ATP inside cells to serve as the universal energy carrier of cell metabolism. However, neurons and astrocytes can release ATP by exocytosis of synaptic vesicles, and this extracellular ATP, as well as adenosine produced from it by an extracellular enzyme on the outer surface of tissue cells, can function as neurotransmitters. Nonneural cells also can release ATP into the extracellular environment by different means to serve various functions. The purine neurotransmitters are released as *cotransmitters;* that is, they are released together with other neurotransmitters, such as with glutamate or GABA in the CNS. **Purinergic receptors,** designated *P1* (for ATP) and *P2* (for adenosine), are found in neurons and glial cells and have been implicated in a variety of physiological and pathological processes. For example, the dilation of cerebral blood vessels in response to ATP released by astrocytes was discussed in section 7.1.

Through the activation of different subtypes of purinergic receptors, ATP and adenosine serve as neurotransmitters when released as cotransmitters by neurons. Examples in the PNS include ATP released with norepinephrine in the stimulation of blood vessel constriction and with ACh in the stimulation of intestinal contractions. When ATP and adenosine are released by nonneural cells, they serve as paracrine regulators (chapter 6, section 6.5). Examples of ATP and adenosine acting as paracrine regulators include their roles in blood clotting (when released by blood platelets), in stimulating neurons for taste (when released by taste bud cells), and in stimulating neurons for pain (when released by damaged tissues).

14a. Explain the significance of glutamate in the brain and of NMDA receptors.
14b. Describe the mechanism of action of glycine and GABA as neurotransmitters, and discuss their significance.
15a. Give examples of endogenous opioid polypeptides, and discuss their significance.
15b. Explain how nitric acid is produced in the body, and describe its functions.

7.7 SYNAPTIC INTEGRATION

The summation of many EPSPs may be needed to produce a depolarization of sufficient magnitude to stimulate the postsynaptic cell. The net effect of EPSPs on the postsynaptic neuron is reduced by hyperpolarization (IPSPs) produced by inhibitory neurotransmitters.

LEARNING OUTCOMES

After studying this section, you should be able to:

16. Explain the nature of spatial and temporal summation at the synapse.
17. Describe long-term potentiation and depression, and explain the nature of postsynaptic and presynaptic inhibition.

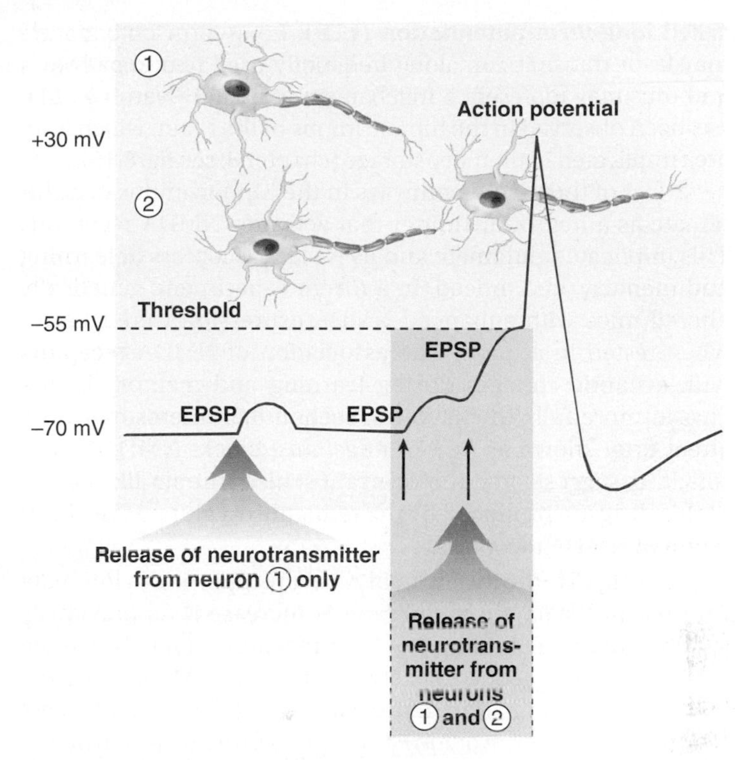

Figure 7.33 Spatial summation. When only one presynaptic neuron releases excitatory neurotransmitter, the EPSP produced may not be sufficiently strong to stimulate action potentials in the postsynaptic neuron. When more than one presynaptic neuron produces EPSPs at the same time, however, the EPSPs can summate at the axon hillock to produce action potentials. The EPSPs are recorded at the axon hillock; the action potential is recorded at the initial segment of the axon.

Because axons can have collateral branches (see fig. 7.1), **divergence** of neural pathways can occur. That is, one neuron can make synapses with a number of other neurons, and by that means either stimulate or inhibit them. By contrast, a number of axons can synapse on a single neuron, allowing **convergence** of neural pathways. Figure 7.33 shows convergence of two neurons on a single postsynaptic neuron, which can thereby integrate the input of the presynaptic neurons.

Spatial summation (fig. 7.33) occurs due to the convergence of axon terminals from different presynaptic axons (up to a thousand in some cases) on the dendrites and cell body of a postsynaptic neuron. Because of the long distances (up to hundreds of micrometers) between the distal ends of the dendrites and the initial segment of the axon, there is ample opportunity for synaptic potentials that originate in different locations to summate. In **temporal summation,** successively rapid bursts of activity of a single presynaptic axon can cause corresponding bursts of neurotransmitter release, resulting in successive waves of EPSPs (or IPSPs) that summate with each other as they travel to the initial segment of the axon. Spatial and temporal summation are important in determining the strength of the depolarization stimulus at the initial segment of the axon and thereby the frequency of action potential production.

Synaptic Plasticity

Repeated use of a particular synaptic pathway can either increase or decrease the strength of synaptic transmission for extended periods of time. This is primarily due to the insertion or removal of the AMPA type of glutamate receptors at the postsynaptic membrane. Insertion of AMPA receptors promotes *long-term potentiation,* whereas removal of AMPA receptors promotes *long-term depression* of synaptic transmission. These and other functional changes, as well as structural changes at synapses, result in **synaptic plasticity**—the ability of synapses to change in response to activity.

When a presynaptic neuron is experimentally stimulated at a high frequency, even for just a few seconds, the excitability of the synapse is enhanced—or potentiated—when this neuron pathway is subsequently stimulated. The improved efficacy of synaptic transmission may last for hours or even weeks and is called **long-term potentiation (LTP).** Long-term potentiation may favor transmission along frequently used neural pathways and thus may represent a mechanism of neural "learning." LTP has been observed in the hippocampus of the brain, which is an area implicated in memory storage (chapter 8; see fig. 8.15).

Most of the neural pathways in the hippocampus use glutamate as a neurotransmitter that activates NMDA receptors. This implicates glutamate and its NMDA receptors in learning and memory, and indeed, in a recent experiment, genetically altered mice with enhanced NMDA expression were smarter when tested in a maze. The association of NMDA receptors with synaptic changes during learning and memory is discussed more fully in chapter 8, section 8.2. Interestingly, the street drug known as *PCP* or *angel dust* blocks NMDA receptors; this suggests that the aberrant schizophrenia-like effects of this drug are produced by the reduction of glutamate stimulation of NMDA receptors.

During LTP, the insertion of AMPA receptors for glutamate into the postsynaptic membrane is increased. As previously stated, glutamate binding to AMPA receptors can produce the depolarization needed for activation of the NMDA receptors (see fig. 8.16). **Long-term depression (LTD)** is a related process involving the removal of AMPA receptors from the postsynaptic membrane. A recent report showed that, without the ability to remove AMPA receptors, LTD was impaired and learning (in rodents) was diminished.

In LTD, the postsynaptic neurons are stimulated to release endocannabinoids. The endocannabinoids then act as retrograde neurotransmitters, suppressing the release of neurotransmitters from presynaptic axons that provide either excitatory or inhibitory synapses with the postsynaptic neuron. This suppression of neurotransmitter release from the presynaptic axons can last many minutes, and has been shown to occur in several brain regions. A shorter-term form of this is *depolarization-induced suppression of inhibition (DSI).* In DSI, the depolarization of a postsynaptic neuron by excitatory input suppresses (via endocannabinoids as retrograde neurotransmitters) the release of GABA from inhibitory presynaptic axons for 20 to 40 seconds.

Long-term potentiation is produced experimentally by stimulating the presynaptic neuron with a high frequency of shocks. Long-term depression can be produced in a variety of ways, most commonly by prolonged periods of low frequency stimulation. Both LTP and LTD depend on a rise in Ca^{2+} concentration within the postsynaptic neuron. A rapid rise in the Ca^{2+} concentration causes potentiation (LTP) of the synapse, whereas a smaller but more prolonged rise in the Ca^{2+} concentration results in depression (LTD) of synaptic transmission.

Synaptic plasticity also involves structural changes in the postsynaptic neurons, including the enlargement or shrinkage of *dendritic spines* (chapter 8; see fig. 8.17). For a discussion of synaptic plasticity as it relates to memory and cerebral function, see chapter 8, section 8.2.

CLINICAL APPLICATION

The neurotransmitter glutamate is required for normal brain function and memory formation, and a certain level of glutamate is needed to sustain the health of neurons. Artificial elimination of NMDA receptor activation promotes neuron apoptosis, increasing the loss of neurons in response to injury. Paradoxically, however, the sustained exposure to glutamate and activation of NMDA receptors is toxic to neurons, a process termed **excitotoxicity.** The beneficial as well as the toxic effects of glutamate result from the entry of Ca^{2+} through the NMDA receptors. This may be because lesser amounts of Ca^{2+} promote the beneficial effect, while greater amounts promote the toxic effect. However, more recent evidence suggests that the beneficial and toxic effects may also differ in the location of the activated NMDA receptors in the dendrites and cell body. Neuronal death by excitotoxicity occurs in acute traumas such as *stroke,* as well as in neurodegenerative diseases such as *Parkinson's disease* and *Alzheimer's disease.*

Synaptic Inhibition

Although many neurotransmitters depolarize the postsynaptic membrane (produce EPSPs), some transmitters do just the opposite. The neurotransmitters glycine and GABA hyperpolarize the postsynaptic membrane; that is, they make the inside of the membrane more negative than it is at rest (fig. 7.34). Because hyperpolarization (from −70 mV to, for example, −85 mV) drives the membrane potential farther from the threshold depolarization required to stimulate action potentials, this inhibits the activity of the postsynaptic neuron. Hyperpolarizations produced by neurotransmitters are therefore called *inhibitory postsynaptic potentials (IPSPs),* as previously described. The inhibition produced in this way is called **postsynaptic inhibition.** Postsynaptic inhibition in the brain is produced by GABA; in the spinal cord it is mainly produced by glycine (although GABA is also involved).

Excitatory and inhibitory inputs (EPSPs and IPSPs) to a postsynaptic neuron can summate in an algebraic fashion. The effects of IPSPs in this way reduce, or may even eliminate, the ability of EPSPs to generate action potentials in the postsynaptic cell. Considering that a given neuron may receive as many as 1,000 presynaptic inputs, the interactions of EPSPs and IPSPs can vary greatly.

In **presynaptic inhibition,** the amount of an excitatory neurotransmitter released at the end of an axon is decreased by the effects of a second neuron, whose axon makes a synapse with the axon of the first neuron (an axoaxonic synapse). The neurotransmitter exerting this presynaptic inhibition may be GABA or excitatory neurotransmitters, such as ACh and glutamate.

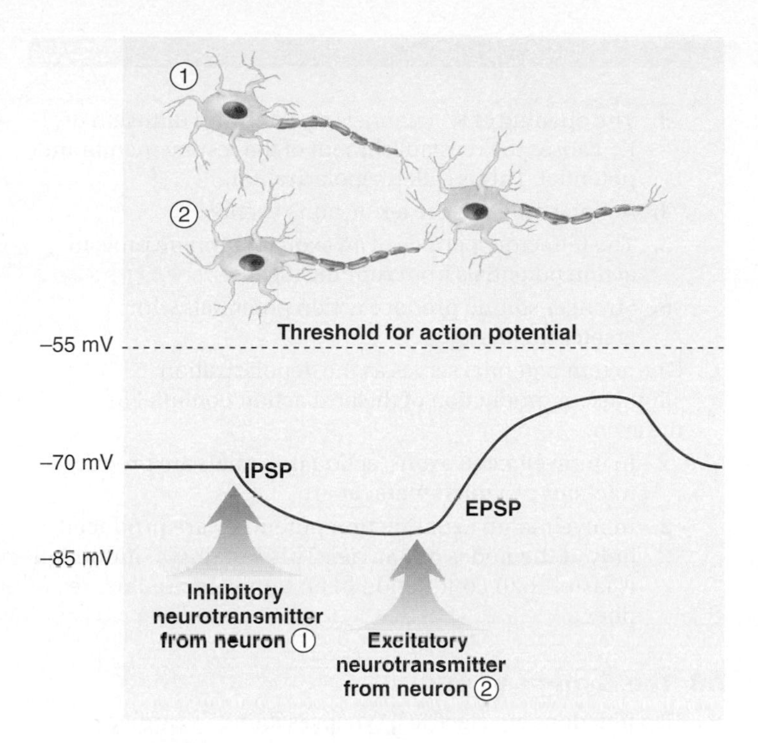

Figure 7.34 Postsynaptic inhibition. An inhibitory postsynaptic potential (IPSP) makes the inside of the postsynaptic membrane more negative than the resting potential—it hyperpolarizes the membrane. Therefore, excitatory postsynaptic potentials (EPSPs), which are depolarizations, must be stronger to reach the threshold required to generate action potentials at the axon hillock. Note that the IPSP and EPSP are recorded at the axon hillock of the postsynaptic neuron.

See the *Test Your Quantitative Ability* section of the Review Activities at the end of this chapter.

Excitatory neurotransmitters can cause presynaptic inhibition by producing depolarization of the axon terminals, leading to inactivation of Ca^{2+} channels. This decreases the inflow of Ca^{2+} into the axon terminals and thus inhibits the release of neurotransmitter. The ability of the opiates to promote analgesia (reduce pain) is an example of such presynaptic inhibition. By reducing Ca^{2+} flow into axon terminals containing substance P, the opioids inhibit the release of the neurotransmitter involved in pain transmission.

16. Define *spatial summation* and *temporal summation,* and explain their functional importance.

17a. Describe long-term potentiation, explain how it is produced, and discuss its significance.

17b. Explain how postsynaptic inhibition is produced and how IPSPs and EPSPs can interact.

17c. Describe the mechanism of presynaptic inhibition.

Clinical Investigation SUMMARY

Sandra was lucky that the shellfish were gathered when the red tide was only beginning, and that she hadn't eaten very much when she noticed her muscle weakness and stopped eating. Her muscle weakness was likely caused by saxitoxin, which blocks voltage-gated Na^+ channels and thereby interferes with the ability of excitable tissue—nerve and muscle—to produce action potentials. Her droopy eyelid (ptosis) may also have been caused by saxitoxin, but it is a potential side effect of her Botox treatment. Botox (botulinum toxin) inhibits the release of ACh at synapses between somatic motor neurons and skeletal muscles, preventing muscle contraction (relaxation of superficial facial muscles helps to smooth skin, the reason for Botox injections). The fact that her blood pressure was normal is also important, especially in view of her taking MAO inhibitor drugs and eating at a seafood restaurant. Some seafoods and associated condiments (including soy sauce) are rich in tyramine, which could have provoked a hypertensive crisis. Fortunately, Sandra was aware of the danger and was careful about her choice of foods.

See the additional chapter 7 Clinical Investigations on *Myasthenia Gravis* and *Parkinson's disease* in the Connect site for this text at www.mhhe.com/Fox13.

Visit this book's website at **www.mhhe.com/Fox13** for:

- Chapter quizzes, interactive learning exercises, and other study tools
- Additional clinical investigations
- Access to LearnSmart—An adaptive diagnostic tool that constantly assesses student knowledge of course material
- Ph.I.L.S. 4.0—physiology interactive lab simulations that may be used to supplement or substitute for wet labs

SUMMARY

7.1 Neurons and Neuroglial Cells 163

A. The nervous system is divided into the central nervous system (CNS) and the peripheral nervous system (PNS).
   1. The central nervous system includes the brain and spinal cord, which contain nuclei and tracts.
   2. The peripheral nervous system consists of nerves, ganglia, and nerve plexuses.

B. A neuron consists of dendrites, a cell body, and an axon.
   1. The cell body contains the nucleus, Nissl bodies, neurofibrils, and other organelles.
   2. Dendrites receive stimuli, and the axon conducts nerve impulses away from the cell body.

C. A nerve is a collection of axons in the PNS.
   1. A sensory, or afferent, neuron is pseudounipolar and conducts impulses from sensory receptors into the CNS.
   2. A motor, or efferent, neuron is multipolar and conducts impulses from the CNS to effector organs.
   3. Interneurons, or association neurons, are located entirely within the CNS.
   4. Somatic motor nerves innervate skeletal muscle; autonomic nerves innervate smooth muscle, cardiac muscle, and glands.

D. Neuroglial cells include Schwann cells and satellite cells in the PNS; in the CNS they include the various types of glial cells: oligodendrocytes, microglia, astrocytes, and ependymal cells.
   1. Schwann cells form a sheath of Schwann, or neurilemma, around axons of the PNS.
   2. Some neurons are surrounded by successive wrappings of supporting cell membrane called a myelin sheath. This sheath is formed by Schwann cells in the PNS and by oligodendrocytes in the CNS.
   3. Astrocytes in the CNS may contribute to the blood-brain barrier.

7.2 Electrical Activity in Axons 172

A. The permeability of the axon membrane to Na^+ and K^+ is regulated by gated ion channels.
   1. At the resting membrane potential of -70 mV, the membrane is relatively impermeable to Na^+ and only slightly permeable to K^+.
   2. The voltage-regulated Na^+ and K^+ channels open in response to the stimulus of depolarization.
   3. When the membrane is depolarized to a threshold level, the Na^+ channels open first, followed quickly by opening of the K^+ channels.

B. The opening of voltage-regulated channels produces an action potential.
   1. The opening of Na^+ channels in response to depolarization allows Na^+ to diffuse into the axon, thus further depolarizing the membrane in a positive feedback fashion.
   2. The inward diffusion of Na^+ causes a reversal of the membrane potential from -70 mV to $+30$ mV.
   3. The opening of K^+ channels and outward diffusion of K^+ causes the reestablishment of the resting membrane potential. This is called repolarization.
   4. Action potentials are all-or-none events.
   5. The refractory periods of an axon membrane prevent action potentials from running together.
   6. Stronger stimuli produce action potentials with greater frequency.

C. One action potential serves as the depolarization stimulus for production of the next action potential in the axon.
   1. In unmyelinated axons, action potentials are produced fractions of a micrometer apart.
   2. In myelinated axons, action potentials are produced only at the nodes of Ranvier. This saltatory conduction is faster than conduction in an unmyelinated nerve fiber.

7.3 The Synapse 180

A. Gap junctions are electrical synapses found in cardiac muscle, smooth muscle, and some regions of the brain.

B. In chemical synapses, neurotransmitters are packaged in synaptic vesicles and released by exocytosis into the synaptic cleft.
   1. The neurotransmitter can be called the ligand of the receptor.
   2. Binding of the neurotransmitter to the receptor causes the opening of chemically regulated gates of ion channels.

7.4 Acetylcholine as a Neurotransmitter 185

A. There are two subtypes of ACh receptors: nicotinic and muscarinic.
   1. Nicotinic receptors enclose membrane channels and open when ACh binds to the receptor. This causes a depolarization called an excitatory postsynaptic potential (EPSP).
   2. The binding of ACh to muscarinic receptors opens ion channels indirectly, through the action of G-proteins. This can cause either an EPSP or a hyperpolarization called an inhibitory postsynaptic potential (IPSP).
   3. After ACh acts at the synapse, it is inactivated by the enzyme acetylcholinesterase (AChE).

B. EPSPs are graded and capable of summation. They decrease in amplitude as they are conducted.

C. ACh is used in the PNS as the neurotransmitter of somatic motor neurons, which stimulate skeletal muscles to contract, and by some autonomic neurons.

D. ACh in the CNS produces EPSPs at synapses in the dendrites or cell body. These EPSPs travel to the axon hillock, stimulate opening of voltage-regulated channels, and generate action potentials in the axon.

7.5 Monoamines as Neurotransmitters 191

A. Monoamines include serotonin, dopamine, norepinephrine, and epinephrine. The last three are included in the subcategory known as catecholamines.

1. These neurotransmitters are inactivated after being released, primarily by reuptake into the presynaptic nerve endings.
2. Catecholamines may activate adenylate cyclase in the postsynaptic cell, which catalyzes the formation of cyclic AMP.

B. Dopaminergic neurons (those that use dopamine as a neurotransmitter) are implicated in the development of Parkinson's disease and schizophrenia. Norepinephrine is used as a neurotransmitter by sympathetic neurons in the PNS and by some neurons in the CNS.

7.6 Other Neurotransmitters 194

A. The amino acids glutamate and aspartate are excitatory in the CNS.

1. The subclass of glutamate receptor designated as NMDA receptors are implicated in learning and memory.
2. The amino acids glycine and GABA are inhibitory. They produce hyperpolarizations, causing IPSPs by opening Cl^- channels.

B. Numerous polypeptides function as neurotransmitters, including the endogenous opioids.

C. Nitric oxide functions as both a local tissue regulator and a neurotransmitter in the PNS and CNS. It promotes smooth muscle relaxation and is implicated in memory.

D. Endocannabinoids are lipids that appear to function as retrograde neurotransmitters: they are released from the postsynaptic neuron, diffuse back to the presynaptic neuron, and inhibit the release of neurotransmitters by the presynaptic neuron.

7.7 Synaptic Integration 199

A. Spatial and temporal summation of EPSPs allows a depolarization of sufficient magnitude to cause the stimulation of action potentials in the postsynaptic neuron.

1. IPSPs and EPSPs from different synaptic inputs can summate.
2. The production of IPSPs is called postsynaptic inhibition.

B. Long-term potentiation is a process that improves synaptic transmission as a result of the use of the synaptic pathway. This process thus may be a mechanism for learning.

C. Long-term depression is a process similar to long-term potentiation, but it causes depressed activity in a synapse.

REVIEW ACTIVITIES

Test Your Knowledge

1. The supporting cells that form myelin sheaths in the peripheral nervous system are
   a. oligodendrocytes.
   b. satellite cells.
   c. Schwann cells.
   d. astrocytes.
   e. microglia.
2. A collection of neuron cell bodies located outside the CNS is called
   a. a tract.
   b. a nerve.
   c. a nucleus.
   d. a ganglion.
3. Which of these neurons are pseudounipolar?
   a. Sensory neurons
   b. Somatic motor neurons
   c. Neurons in the retina
   d. Autonomic motor neurons
4. Depolarization of an axon is produced by
   a. inward diffusion of Na^+.
   b. active extrusion of K^+.
   c. outward diffusion of K^+.
   d. inward active transport of Na^+.
5. Repolarization of an axon during an action potential is produced by
   a. inward diffusion of Na^+.
   b. active extrusion of K^+.
   c. outward diffusion of K^+.
   d. inward active transport of Na^+.
6. As the strength of a depolarizing stimulus to an axon is increased,
   a. the amplitude of action potentials increases.
   b. the duration of action potentials increases.
   c. the speed with which action potentials are conducted increases.
   d. the frequency with which action potentials are produced increases.
7. The conduction of action potentials in a myelinated nerve fiber is
   a. saltatory.
   b. without decrement.
   c. faster than in an unmyelinated fiber.
   d. all of these.
8. Which of these is *not* a characteristic of synaptic potentials?
   a. They are all-or-none in amplitude.
   b. They decrease in amplitude with distance.
   c. They are produced in dendrites and cell bodies.
   d. They are graded in amplitude.
   e. They are produced by chemically regulated gates.
9. Which of these is *not* a characteristic of action potentials?
   a. They are produced by voltage-regulated gates.
   b. They are conducted without decrement.
   c. Na^+ and K^+ gates open at the same time.
   d. The membrane potential reverses polarity during depolarization.

10. A drug that inactivates acetylcholinesterase
    a. inhibits the release of ACh from presynaptic endings.
    b. inhibits the attachment of ACh to its receptor protein.
    c. increases the ability of ACh to stimulate muscle contraction.
    d. does all of these.
11. Postsynaptic inhibition is produced by
    a. depolarization of the postsynaptic membrane.
    b. hyperpolarization of the postsynaptic membrane.
    c. axoaxonic synapses.
    d. long-term potentiation.
12. Hyperpolarization of the postsynaptic membrane in response to glycine or GABA is produced by the opening of
    a. Na^+ channels.
    b. K^+ channels.
    c. Ca^{2+} channels.
    d. Cl^- channels.
13. The absolute refractory period of a neuron
    a. is due to the high negative polarity of the inside of the neuron.
    b. occurs only during the repolarization phase.
    c. occurs only during the depolarization phase.
    d. occurs during depolarization and the first part of the repolarization phase.
14. Which of these statements about catecholamines is *false?*
    a. They include norepinephrine, epinephrine, and dopamine.
    b. Their effects are increased by action of the enzyme catechol-O-methyltransferase.
    c. They are inactivated by monoamine oxidase.
    d. They are inactivated by reuptake into the presynaptic axon.
    e. They may stimulate the production of cyclic AMP in the postsynaptic axon.
15. The summation of EPSPs from numerous presynaptic nerve fibers converging onto one postsynaptic neuron is called
    a. spatial summation.
    b. long-term potentiation.
    c. temporal summation.
    d. synaptic plasticity.
16. Which of these statements about ACh receptors is *false?*
    a. Skeletal muscles contain nicotinic ACh receptors.
    b. The heart contains muscarinic ACh receptors.
    c. G-proteins are needed to open ion channels for nicotinic receptors.
    d. Stimulation of nicotinic receptors results in the production of EPSPs.
17. Hyperpolarization is caused by all of these neurotransmitters *except*
    a. glutamic acid in the CNS.
    b. ACh in the heart.
    c. glycine in the spinal cord.
    d. GABA in the brain.
18. Which of these may be produced by the action of nitric oxide?
    a. Dilation of blood vessels
    b. Erection of the penis
    c. Relaxation of smooth muscles in the digestive tract
    d. Long-term potentiation (LTP) among neighboring synapses in the brain
    e. All of these

Test Your Understanding

19. Compare the characteristics of action potentials with those of synaptic potentials.
20. In a step-by-step manner, explain how the voltage-regulated channels produce an action potential.
21. Explain how action potentials are conducted by an unmyelinated axon.
22. Explain how a myelinated axon conducts action potentials, and why this conduction is faster than in an unmyelinated axon.
23. Describe the structure of nicotinic ACh receptors, and how ACh interacts with these receptors to cause the production of an EPSP.
24. Describe the nature of muscarinic ACh receptors and the function of G-proteins in the action of these receptors. How does stimulation of these receptors cause the production of a hyperpolarization or a depolarization?
25. Once an EPSP is produced in a dendrite, how does it stimulate the production of an action potential at the axon hillock? What might prevent an EPSP from stimulating action potentials? How can an EPSP's ability to stimulate action potentials be enhanced?
26. Explain how inhibition can be produced by (a) muscarinic ACh receptors in the heart; and (b) GABA receptors in neurons of the CNS.
27. List the endogenous opioids in the brain and describe some of their proposed functions.
28. Explain what is meant by long-term potentiation and discuss the significance of this process. What may account for LTP and what role might nitric oxide play?

Test Your Analytical Ability

29. Grafting peripheral nerves onto the two parts of a cut spinal cord in rats was found to restore some function in the hind limbs. Apparently, when the white matter of the peripheral nerve was joined to the gray matter of the spinal cord, some regeneration of central neurons occurred across the two spinal cord sections. What component of the peripheral nerve probably contributed to the regeneration? Discuss the factors that promote and inhibit central neuron regeneration.
30. Discuss the different states of a voltage-gated ion channel and distinguish between these states. How has molecular biology/biochemistry aided our understanding of the physiology of the voltage-gated channels?

31. Suppose you are provided with an isolated nerve-muscle preparation in order to study synaptic transmission. In one of your experiments, you give this preparation a drug that blocks voltage-regulated Ca^{+} channels; in another, you give tetanus toxin to the preparation. How will synaptic transmission be affected in each experiment?
32. What functions do G-proteins serve in synaptic transmission? Speculate on the advantages of having G-proteins mediate the effects of a neurotransmitter.
33. Studies indicate that alcoholism may be associated with a particular allele (form of a gene) for the D_2 dopamine receptor. Suggest some scientific investigations that might further explore these possible genetic and physiological relationships.
34. Explain the nature of the endocannabinoids. Speculate about how, by acting as retrograde neurotransmitters, they might function to suppress pain in the CNS.

Test Your Quantitative Ability

Use the figure below (from figure 7.14) to answer questions 35–37:

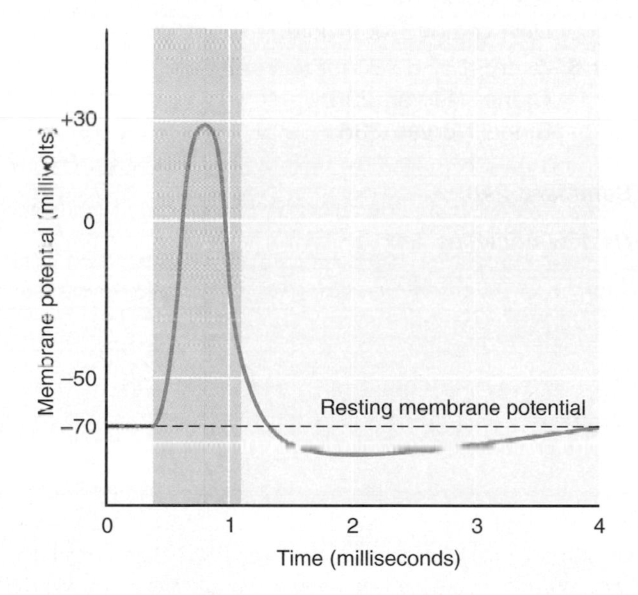

35. What is the membrane potential at 0.5 msec. after the action potential began?
36. What is the membrane potential at 1.5 msec. after the action potential began?
37. How much time was required for the membrane potential to go from the resting membrane potential to zero mV?

Use the figure below (from figure 7.34) to answer questions 38–40:

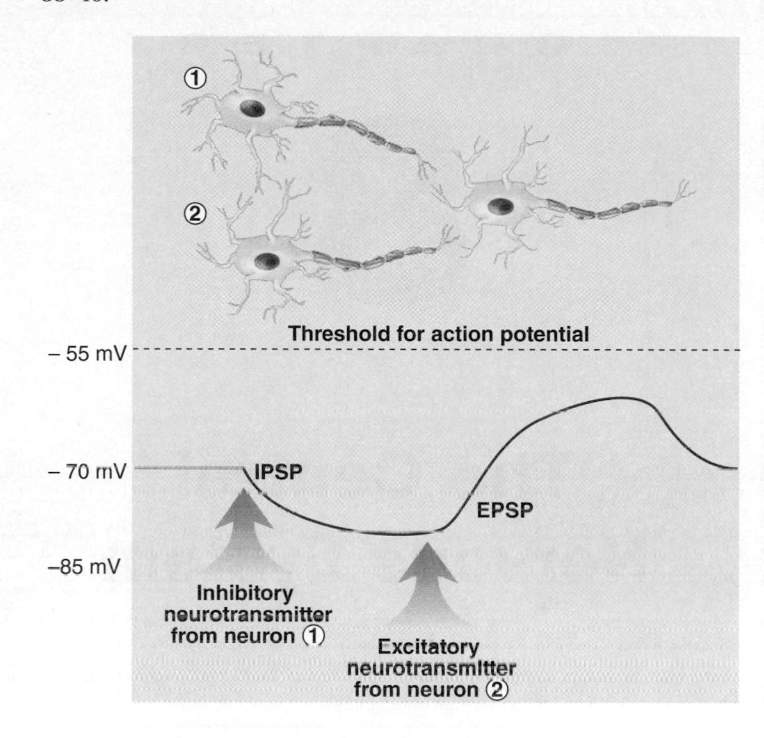

38. What is the approximate magnitude of the IPSP (how many mV)?
39. What is the approximate magnitude of the EPSP (how many mV)?
40. If the IPSP had not occurred, what would be the difference between the EPSP and the threshold required to produce an action potential?

CHAPTER

8

The Central Nervous System

REFRESH YOUR MEMORY

Before you begin this chapter, you may want to review these concepts from previous chapters:

CHAPTER OUTLINE

Clinical Investigation

Frank, a 72-year-old man, is brought to the hospital by his wife. She explains to the doctor that her husband suddenly became partially paralyzed and had difficulty speaking. In the examination, the physician determines that Frank is paralyzed on the right side of his body but is able to produce a knee-jerk reflex with either leg. When Frank is questioned, he speaks slowly and with great difficulty, but is coherent.

Some of the new terms and concepts you will encounter include:

- MRI, decussation of tracts, and cerebral lateralization
- Broca's and Wernicke's areas of the cerebral cortex, and aphasias
- Pyramidal motor system, descending tracts, and the spinal reflex arc

8.1 STRUCTURAL ORGANIZATION OF THE BRAIN

The brain is composed of an enormous number of association neurons and accompanying neuroglia, arranged in regions and subdivisions. These neurons receive sensory information, direct the activity of motor neurons, and perform such higher brain functions as learning and memory.

LEARNING OUTCOMES

After studying this section, you should be able to:

1. **Describe the embryonic origin of the CNS.**
2. **Identify the five brain regions and the major structures they contain, including the ventricles.**

The **central nervous system (CNS),** consisting of the brain and spinal cord (fig. 8.1), receives input from *sensory neurons* and directs the activity of *motor neurons* that innervate muscles and glands. The *association neurons* within the brain and spinal cord are in a position, as their name implies, to associate appropriate motor responses with sensory stimuli, and thus to maintain homeostasis in the internal environment and the continued existence of the organism in a changing external environment. Further, the central nervous systems of all vertebrates (and most invertebrates) are capable of at least rudimentary forms of learning and memory. This capability—most highly developed in the human brain—permits behavior to be modified by experience. Perceptions, learning, memory, emotions, and perhaps even the self-awareness that forms the basis of consciousness, are creations of the brain. Whimsical though it seems, the study of brain physiology is the process of the brain studying itself.

The study of the structure and function of the central nervous system requires a knowledge of its basic "plan," which is established during the course of embryonic development. The early embryo contains on its surface an embryonic tissue layer known as *ectoderm;* this will eventually form the epidermis of the skin, among other structures. As development progresses, a groove appears in this ectoderm along the dorsal midline of the embryo's body. This groove deepens, and by the twentieth day after conception it has fused to form a **neural tube.** The part of the ectoderm where the fusion occurs becomes a separate structure called the **neural crest,** which is located between the neural tube and the surface ectoderm (fig. 8.2). Eventually the neural tube will become the central nervous system, and the neural crest will become the ganglia of the peripheral nervous system, among other structures.

By the middle of the fourth week after conception, three distinct swellings are evident on the anterior end of the neural tube, which is going to form the brain: the **forebrain** *(prosencephalon),* **midbrain** *(mesencephalon),* and **hindbrain** *(rhombencephalon).* During the fifth week, these areas become modified to form five regions. The forebrain divides into the *telencephalon* and *diencephalon;* the mesencephalon

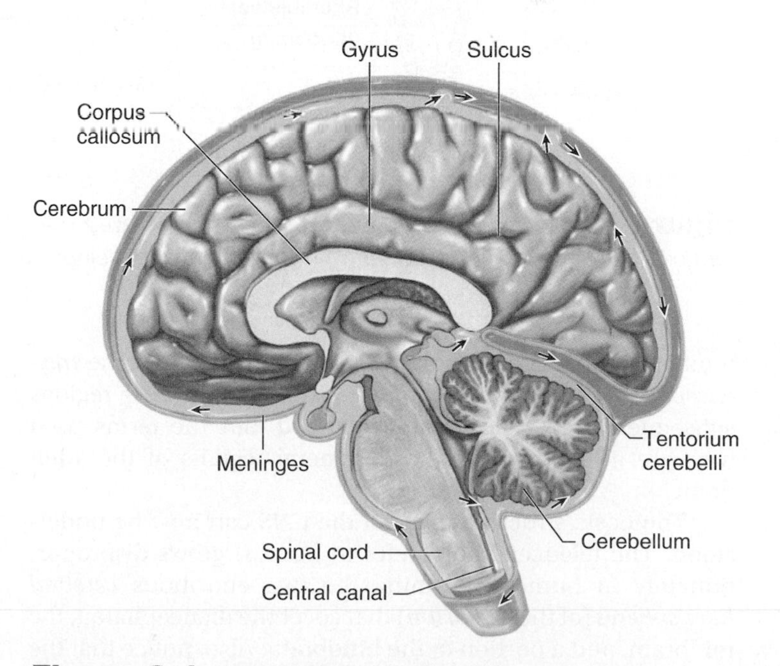

Figure 8.1 The CNS consists of the brain and the spinal cord. Both of these structures are covered with meninges and bathed in cerebrospinal fluid. AP|R

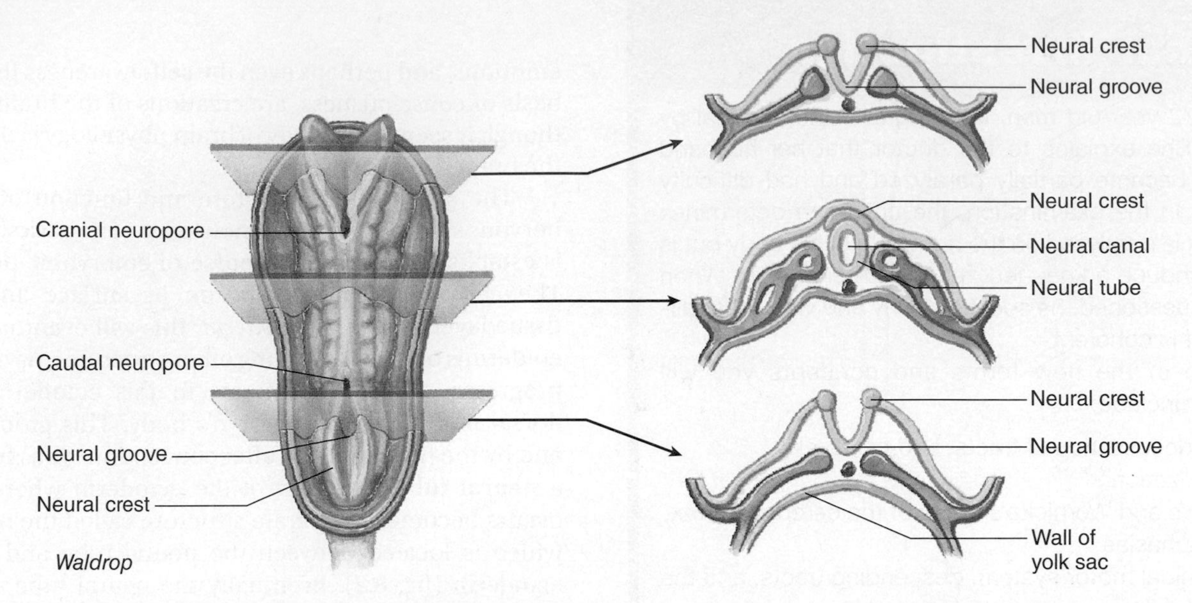

Figure 8.2 Embryonic development of the CNS. This dorsal view of a 22-day-old embryo shows transverse sections at three levels of the developing central nervous system.

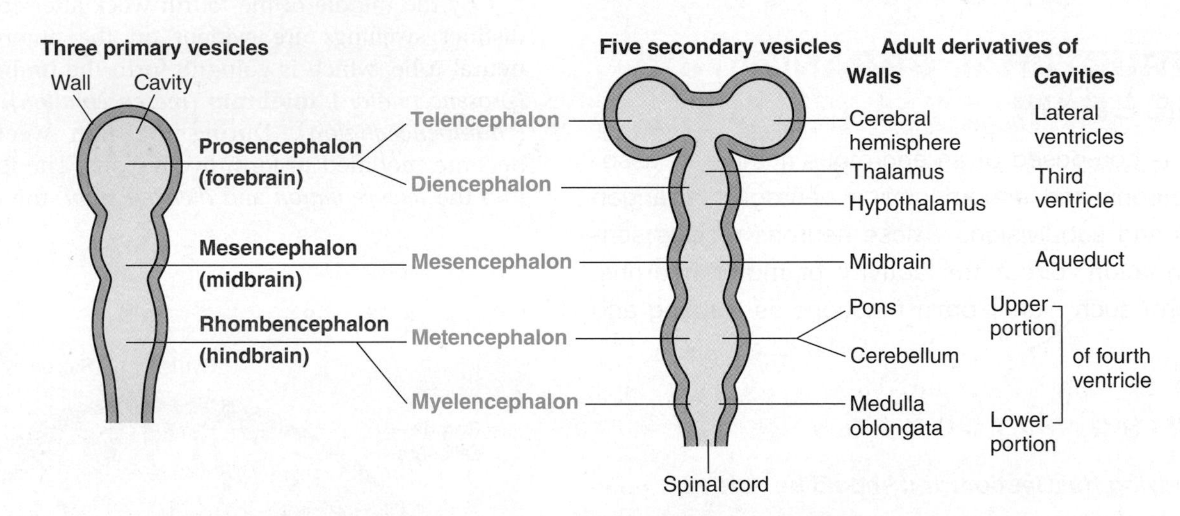

Figure 8.3 The developmental sequence of the brain. (*a*) During the fourth week, three principal regions of the brain are formed. (*b*) During the fifth week, a five-regioned brain develops and specific structures begin to form.

remains unchanged; and the hindbrain divides into the *metencephalon* and *myelencephalon* (fig. 8.3). These regions subsequently become greatly modified, but the terms used here are also used to indicate general regions of the adult brain.

The basic structural plan of the CNS can now be understood. The telencephalon (refer to fig. 8.3) grows disproportionately in humans, forming the two enormous *cerebral hemispheres* (of the *cerebrum*) that cover the diencephalon, the midbrain, and a portion of the hindbrain. Also, notice that the CNS begins as a hollow tube, and indeed remains hollow as the brain regions are formed. The cavities of the brain are known as **ventricles** and become filled with *cerebrospinal fluid (CSF)*. The cavity of the spinal cord is called the **central canal**, and is also filled with CSF (fig. 8.4).

Cerebrospinal fluid is formed by structures called the **choroid plexuses,** which are thin structures protruding into the ventricles. The choroid plexuses consist of a simple cuboidal to columnar epithelium associated with blood capillaries containing a fenestrated endothelium (chapter 13, section 13.6). These appear to form CSF by secretion, rather than by filtration of the blood plasma. As a result of secretory processes, the CSF has a remarkably constant composition, is slightly hypertonic, and differs from plasma in its concentration of various ions.

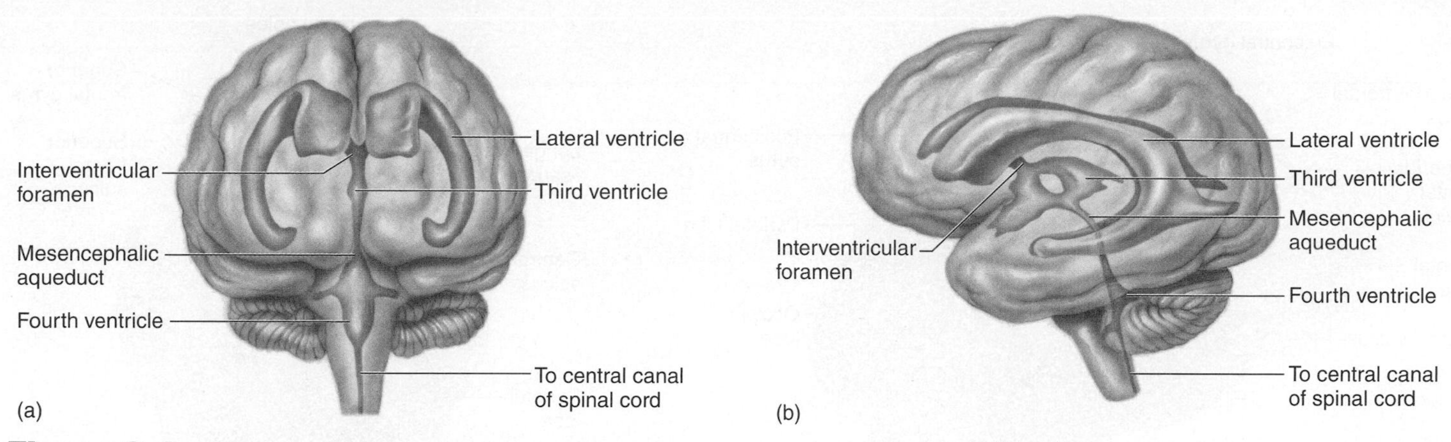

Figure 8.4 **The ventricles of the brain.** (*a*) An anterior view and (*b*) a lateral view. APIR

Constant fluid secretion by the choroid plexuses and absorption of CSF into venous blood maintain a healthy intracranial pressure of 5–15 mm Hg.

The CNS is composed of gray and white matter, as described in chapter 7. The gray matter, containing neuron cell bodies and dendrites, is found in the *cortex* (surface layer) of the brain and deeper within the brain in aggregations known as *nuclei.* White matter consists of axon tracts (the myelin sheaths produce the white color) that underlie the cortex and surround the nuclei. The adult brain contains an estimated 100 billion (10^{11}) neurons, weighs approximately 1.5 kg (3 to 3.5 lb), and receives about 15% of the total blood flow to the body per minute. This high rate of blood flow is a consequence of the high metabolic requirements of the brain; it is not, as Aristotle believed, because the brain's function is to cool the blood. (This fanciful notion—completely incorrect—is a striking example of prescientific thought, having no basis in experimental evidence.)

Scientists have demonstrated that the brains of adult mammals, including humans, contain **neural stem cells** that can develop into neurons and glial cells. **Neurogenesis**—the formation of new neurons from neural stem cells—has been demonstrated in two locations. One location is the *subventricular zone,* a thin layer of cells adjacent to the ependymal cells that line the lateral ventricles. Evidence suggests that these neural stem cells are not ependymal cells, but that ependymal cells secrete factors that promote neurogenesis in this zone. Newborn neurons migrate from the subventricular zone to the olfactory bulbs of the brain, where they become functional interneurons in a process enhanced by olfactory learning. (The olfactory receptors are themselves bipolar neurons replaced by stem cells located in the olfactory epithelium.) The other brain location in which neurogenesis has been demonstrated is an area called the *subgranular zone* within the hippocampus (see fig. 8.15). Neurogenesis within this zone results in newly formed neurons that may function within the hippocampus to aid the types of learning that depend on the hippocampus (discussed in section 8.2).

CHECKPOINT

1. Identify the three brain regions formed by the middle of the fourth week of gestation and the five brain regions formed during the fifth week.
2. Describe the embryonic origin of the brain ventricles. Where are they located and what do they contain?

8.2 CEREBRUM

The cerebrum, consisting of five paired lobes within two convoluted hemispheres, contains gray matter in its cortex and in deeper cerebral nuclei. Most of what are considered to be the higher functions of the brain are performed by the cerebrum.

LEARNING OUTCOMES

After studying this section, you should be able to:

3. **Describe the organization of the sensory and motor areas of the cerebral cortex, and the nature of the basal nuclei (basal ganglia).**
4. **Distinguish between the functions of the right and left cerebral hemispheres, and describe the significance of the limbic system.**
5. **Identify the areas of cerebral cortex involved in speech and language.**
6. **Describe the brain regions involved in memory and some synaptic events associated with learning and memory.**

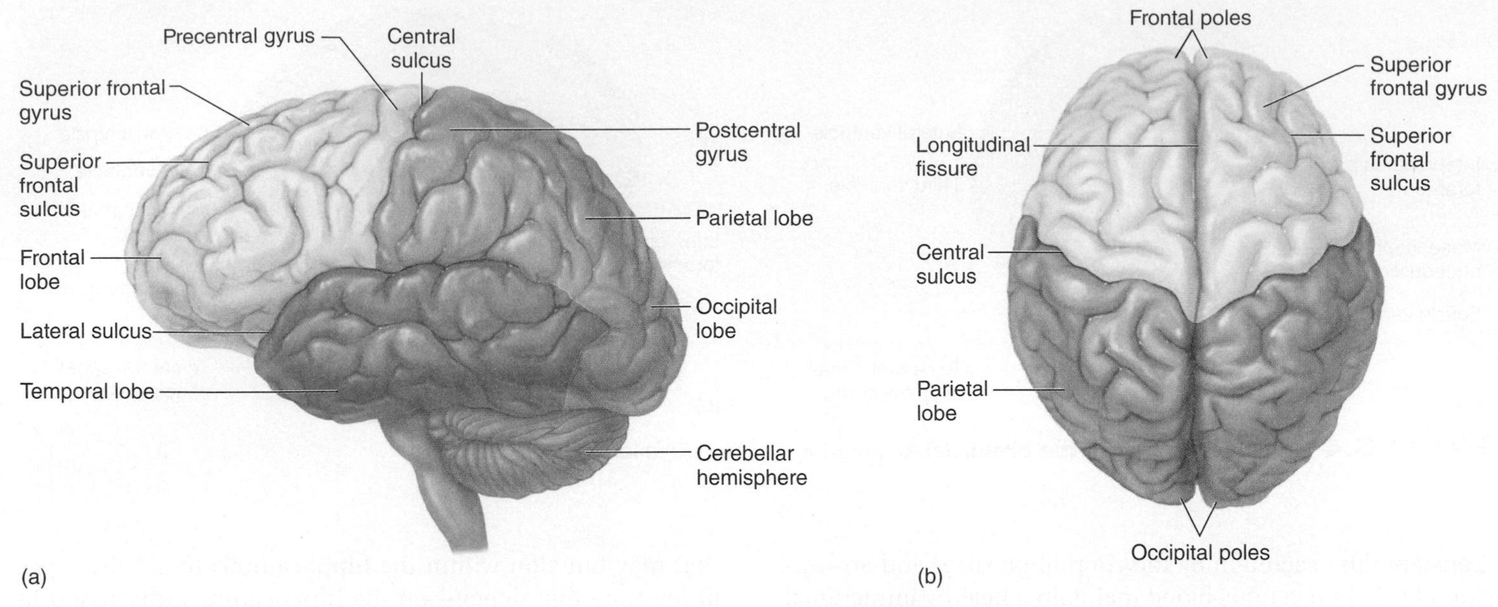

Figure 8.5 The cerebrum. (*a*) A lateral view and (*b*) a superior view. AP|R

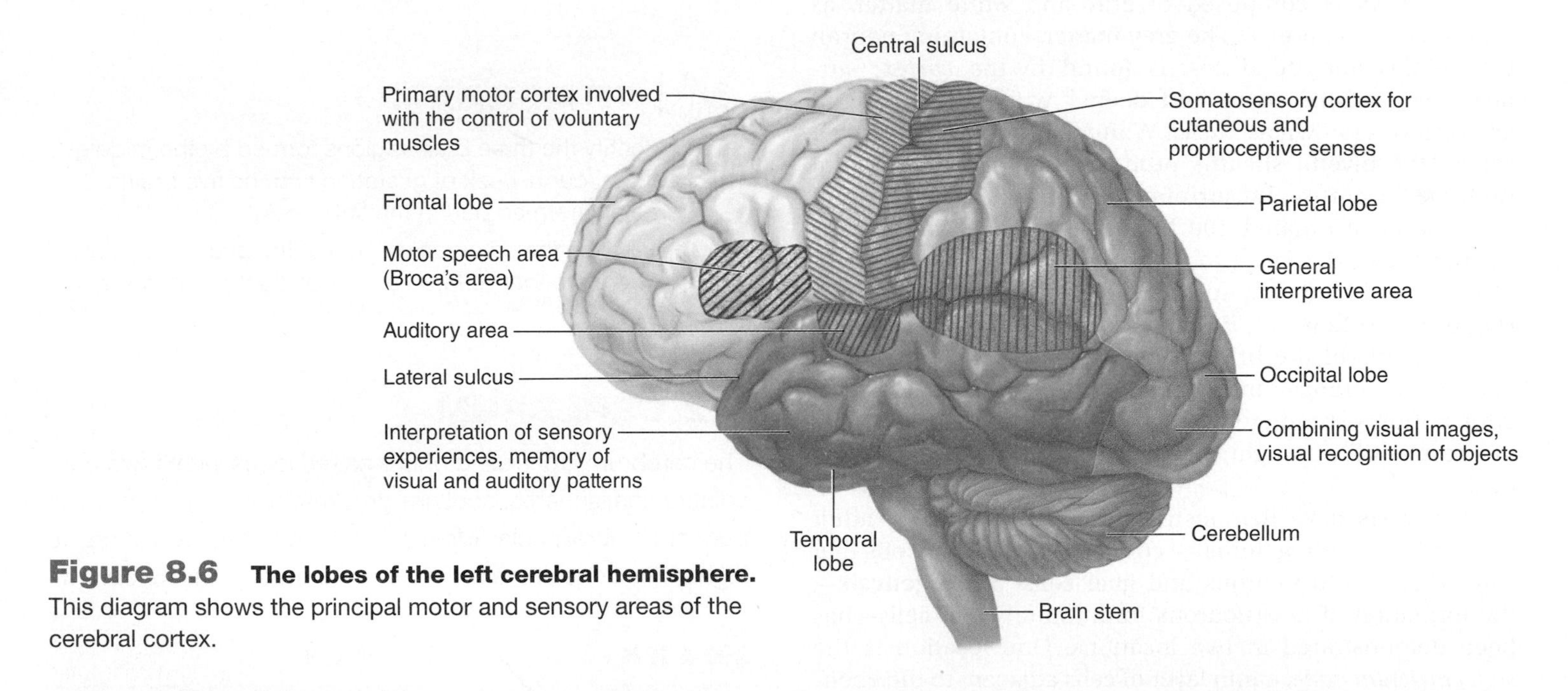

Figure 8.6 The lobes of the left cerebral hemisphere. This diagram shows the principal motor and sensory areas of the cerebral cortex.

The **cerebrum** (fig. 8.5), which is the only structure of the telencephalon, is the largest portion of the brain (accounting for about 80% of its mass) and is the brain region primarily responsible for higher mental functions. The cerebrum consists of *right* and *left hemispheres,* which are connected internally by a large fiber tract called the *corpus callosum* (see fig. 8.1). The corpus callosum is the major tract of axons that functionally interconnects the right and left cerebral hemispheres.

Cerebral Cortex

The cerebrum consists of an outer **cerebral cortex,** composed of 2 to 4 mm of gray matter and underlying white matter. The cerebral cortex is characterized by numerous folds and grooves called *convolutions.* The elevated folds of the convolutions are called *gyri,* and the depressed grooves are the *sulci.* Each cerebral hemisphere is subdivided by deep sulci, or *fissures,* into five lobes, four of which are visible from the surface (fig. 8.6). These lobes are the *frontal, parietal, temporal,* and *occipital,* which are visible from the surface, and the deep *insula* (fig. 8.7), which is covered by portions of the frontal, parietal, and temporal lobes (table 8.1).

The **frontal lobe** is the anterior portion of each cerebral hemisphere. A deep fissure, called the *central sulcus,* separates the frontal lobe from the **parietal lobe.** The *precentral gyrus* (figs. 8.5 and 8.6), involved in motor control, is located in the frontal lobe just in front of the central sulcus. The cell bodies of the interneurons located here are called *upper motor neurons* because of their role in muscle regulation (chapter 12,

Table 8.1 | Functions of the Cerebral Lobes

Lobe	Functions
Frontal	Voluntary motor control of skeletal muscles; personality; higher intellectual processes (e.g., concentration, planning, and decision making); verbal communication
Parietal	Somatesthetic interpretation (e.g., cutaneous and muscular sensations); understanding speech and formulating words to express thoughts and emotions; interpretation of textures and shapes
Temporal	Interpretation of auditory sensations; storage (memory) of auditory and visual experiences
Occipital	Integration of movements in focusing the eye; correlation of visual images with previous visual experiences and other sensory stimuli; conscious perception of vision
Insula	Memory; sensory (principally pain) and visceral integration

section 12.5). The *postcentral gyrus,* which is located behind the central sulcus in the parietal lobe of each hemisphere, contains the *somatosensory cortex.* This is the primary area responsible for the perception of *somatesthetic sensations*—sensations arising from cutaneous, muscle, tendon, and joint receptors. These neural pathways are described in chapter 10, section 10.1.

The precentral (motor) and postcentral (sensory) gyri have been mapped in conscious patients undergoing brain surgery. Electrical stimulation of specific areas of the precentral gyrus causes specific movements, and stimulation of different areas of the postcentral gyrus evokes sensations in specific parts of the body. Typical maps of these regions (fig. 8.7) show an upside-down picture of the body, with the superior regions of cortex devoted to the toes and the inferior regions devoted to the head.

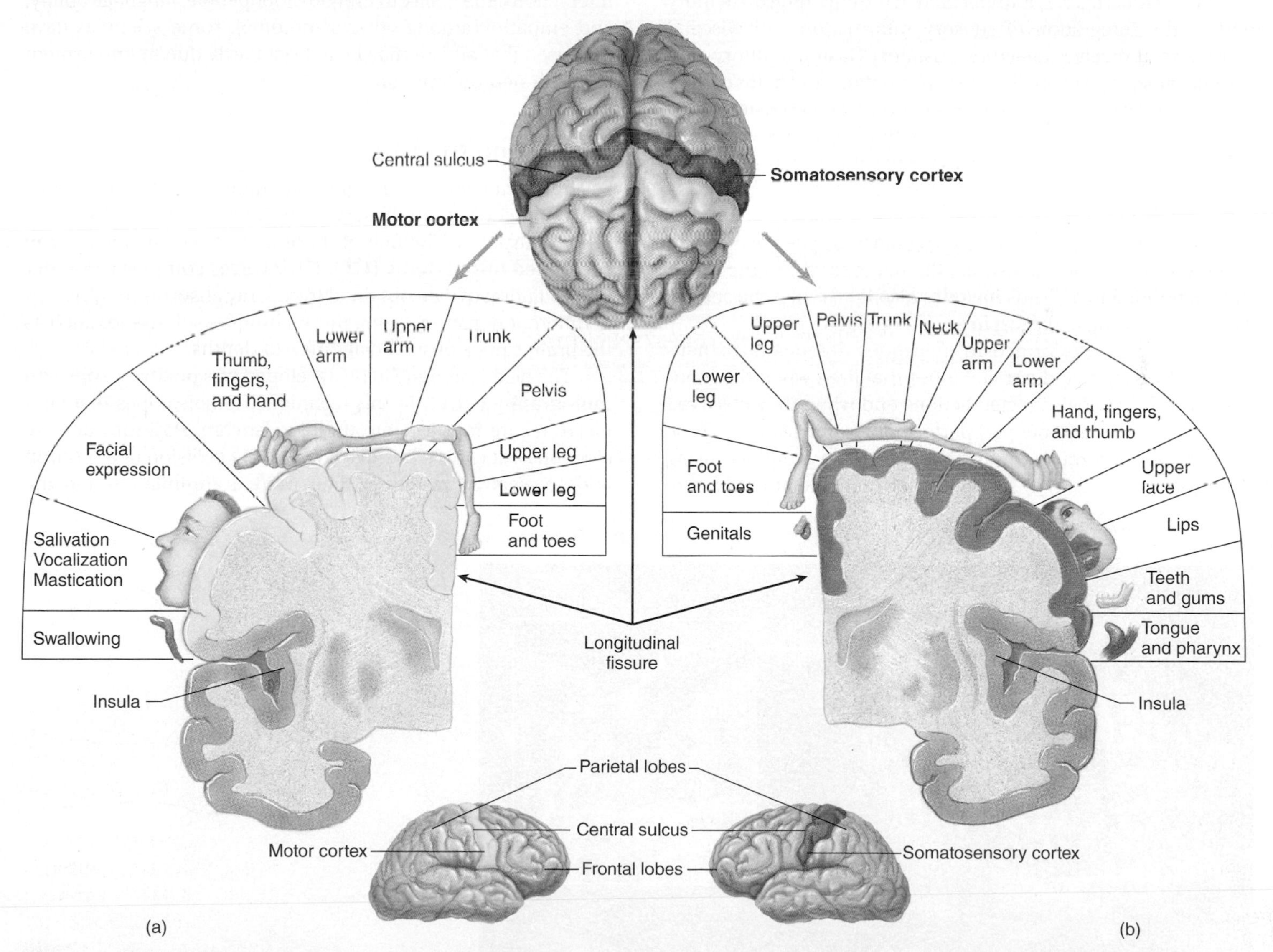

Figure 8.7 Motor and sensory areas of the cerebral cortex. (*a*) Motor areas that control skeletal muscles are shown in yellow. This region is specifically known as the *primary motor cortex* (discussed later in this chapter). (*b*) Sensory areas that receive somatesthetic sensations are shown in purple. Artistic license has been used in rendering part (*b*), because the left hemisphere receives input primarily from the right side of the body.

A striking feature of these maps is that the areas of cortex responsible for different parts of the body do not correspond to the size of the body parts being served. Instead, the body regions with the highest densities of receptors are represented by the largest areas of the sensory cortex, and the body regions with the greatest number of motor innervations are represented by the largest areas of motor cortex. The hands and face, therefore, which have a high density of sensory receptors and motor innervation, are served by larger areas of the precentral and postcentral gyri than is the rest of the body.

The **temporal lobe** contains auditory centers that receive sensory fibers from the cochlea of each ear. This lobe is also involved in the interpretation and association of auditory and visual information. The **occipital lobe** is the primary area responsible for vision and for the coordination of eye movements. The functions of the temporal and occipital lobes will be considered in more detail in chapter 10, in conjunction with the physiology of hearing and vision.

The **insula** (fig. 8.7) is implicated in the encoding of memory and in the integration of sensory information with visceral responses. It receives olfactory, gustatory (taste), auditory, and somatosensory (principally pain) information, and helps control autonomic responses to the viscera and cardiovascular system. Because it receives sensory information from the viscera, it is believed to be important in assessing the bodily states that accompany emotions. One study demonstrated that those neurons that fire in response to pain applied to the hand also fire when the subject was told that pain would be applied to the hand of a loved one; in another study, the neurons within the insula that responded to a disgusting odor also fired when the subject saw an expression of disgust in another person.

Studies first performed in macaques demonstrated neurons in the frontal and parietal lobes that fired when the monkeys performed goal-directed actions and when they observed others (monkeys and people) perform the same actions. These neurons, termed **mirror neurons,** have been identified using fMRI (discussed next) in similar locations in the human brain. Mirror neurons in the premotor areas of the frontal lobes become active when a person performs a goal-directed action or sees another individual perform the same action. Other mirror neurons involved in higher somatosensory processing in the parietal lobe respond when a person is touched in a body location or sees another individual touched in the same location. Mirror neurons also have neural connections to the insula (fig. 8.7) and cingulate gyrus (see fig. 8.18), which can provide an affective (emotional) component to the vicarious experience.

Scientists believe that mirror neurons are involved in the ability to imitate others, understand the intentions and behavior of others, and empathize with the emotions displayed by others. These abilities are required for the acquisition of social skills and perhaps also of language, a possibility supported by the observation that human mirror neurons are found in Broca's area, needed for language (see fig. 8.14). Because *autism,* better termed *autism spectrum disorder,* involves impairments in social interactions, the ability to imitate other people, language ability, and empathy (among other symptoms), some scientists have proposed that autism may be at least partly due to impairment of mirror neuron function.

Visualizing the Brain

Several relatively new imaging techniques permit the brains of living people to be observed in detail for medical and research purposes. The first of these to be developed was **x-ray computed tomography (CT).** CT involves complex computer manipulation of data obtained from x-ray absorption by tissues of different densities. Using this technique, soft tissues such as the brain can be observed at different depths.

The next technique to be developed was **positron emission tomography (PET).** In this technique, radioisotopes that emit positrons are injected into the bloodstream. Positrons are like electrons but carry a positive charge. The collision of a positron and an electron results in their mutual annihilation and the

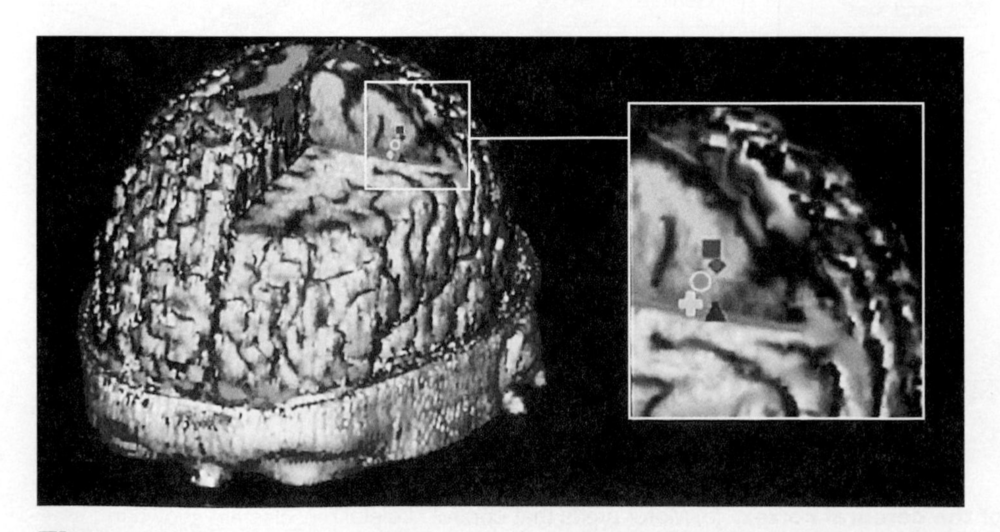

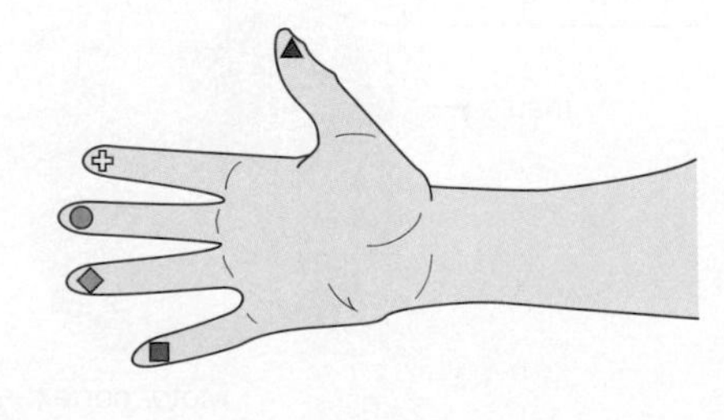

Figure 8.8 An MRI image of the brain reveals the sensory cortex. The integration of MRI and EEG information shows the location on the sensory cortex that corresponds to each of the digits of the hand.

emission of gamma rays, which can be detected and used to pinpoint brain cells that are most active. Medically, PET scans are used to determine the stage of cancer and to monitor patient responses to cancer treatments. Scientists have used PET to study brain metabolism, drug distribution in the brain, and changes in blood flow as a result of brain activity. For example, PET was used in a recent study of volunteers who were injected with (^{18}F)fluorodeoxyglucose to measure how the glucose metabolism of the brain was affected by cell phone use: metabolism was increased in regions closest to the antenna after prolonged exposure.

A newer technique for visualizing the living brain is **magnetic resonance imaging (MRI).** This technique is based on the concept that protons (H^+), because they are charged and spinning, are like little magnets. A powerful external magnet can align a proportion of the protons. Most of the protons are part of water molecules, and the chemical composition of different tissues provides differences in the responses of the aligned protons to a radio frequency pulse. This allows clear distinctions to be made between gray matter, white matter, and cerebrospinal fluid (figs. 8.8 and 8.9). In addition, exogenous chemicals known as MRI contrast agents are sometimes used to increase or decrease the signal in different tissues to improve the image.

Scientists can study the functioning brain in a living person using a technique known as **functional magnetic resonance imaging (fMRI).** This technique visualizes increased neuronal activity within a brain region indirectly, by the increased blood flow to the more active brain region (chapter 14; see fig. 14.22). This occurs because of increased release of the neurotransmitter glutamate, which causes vasodilation and increased blood flow in the more active brain regions. As a result, the active brain regions receive more oxyhemoglobin (and thus less deoxyhemoglobin, which affects the magnetic field) than they do when resting. This is known as the *BOLD response* (for "blood oxygenation level dependent contrast").

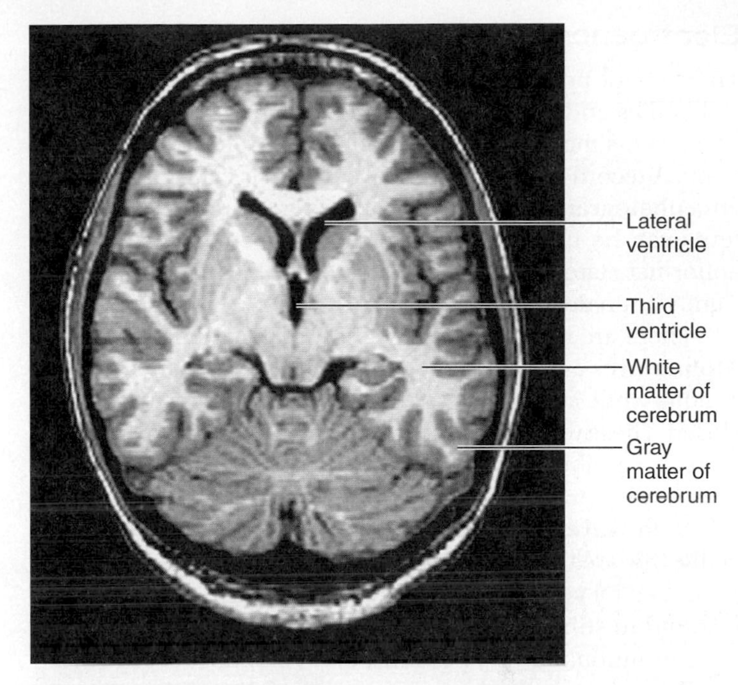

Figure 8.9 An MRI scan of the brain. Gray and white matter are easily distinguished, as are the ventricles containing cerebrospinal fluid. AP|R

Magnetoencephalogram (MEG) recordings provide images of brain activity on a millisecond time scale that can be more accurate than EEG recordings (discussed next). Because postsynaptic currents produce weak magnetic fields, thousands of these together generate magnetic fields that can be detected by sensors surrounding the head. The sensors are hundreds of SQUIDS (superconducting quantum interference devices) cooled in liquid helium to 4 degrees above absolute zero. Techniques for visualizing the functioning brain are summarized in table 8.2.

Table 8.2 | Techniques for Visualizing Brain Function

Abbreviation	Technique Name	Principle Behind Technique
EEG	Electroencephalogram	Neuronal activity is measured as maps with scalp electrodes.
fMRI	Functional magnetic resonance imaging	Increased neuronal activity increases cerebral blood flow and oxygen consumption in local areas. This is detected by effects of changes in blood oxyhemoglobin/deoxyhemoglobin ratios.
MEG	Magnetoencephalogram	Neuronal magnetic activity is measured using magnetic coils and mathematical plots.
PET	Positron emission tomography	Increased neuronal activity increases cerebral blood flow and metabolite consumption in local areas. This is measured using radioactively labeled deoxyglucose.
SPECT	Single photon emission computed tomography	Increased neuronal activity increases cerebral blood flow. This is measured using emitters of single photons, such as technetium.
CT	Computerized tomography	A number of x-ray beams are sent through the brain or other body region and are sensed by numerous detectors; a computer uses this information to produce images that appear as slices through the brain.

Source: Burkhart Bromm "Brain images of pain." *News in Physiological Sciences 16* (Feb. 2001): 244–249.

Electroencephalogram

The synaptic potentials (chapter 7, section 7.3) produced at the cell bodies and dendrites of the cerebral cortex create electrical currents that can be measured by electrodes placed on the scalp. A record of these electrical currents is called an **electroencephalogram,** or **EEG.** Deviations from normal EEG patterns can be used clinically to diagnose epilepsy and other abnormal states, and the absence of an EEG can be used to signify brain death.

There are normally four types of EEG patterns (fig. 8.10). **Alpha waves** are best recorded from the parietal and occipital regions while a person is awake and relaxed but with the eyes closed. These waves are rhythmic oscillations of 10 to 12 cycles/second. The alpha rhythm of a child under the age of 8 occurs at a slightly lower frequency of 4 to 7 cycles/second.

Beta waves are strongest from the frontal lobes, especially the area near the precentral gyrus. These waves are produced by visual stimuli and mental activity. Because they respond to stimuli from receptors and are superimposed on the continuous activity patterns, they constitute *evoked activity.* Beta waves occur at a frequency of 13 to 25 cycles per second.

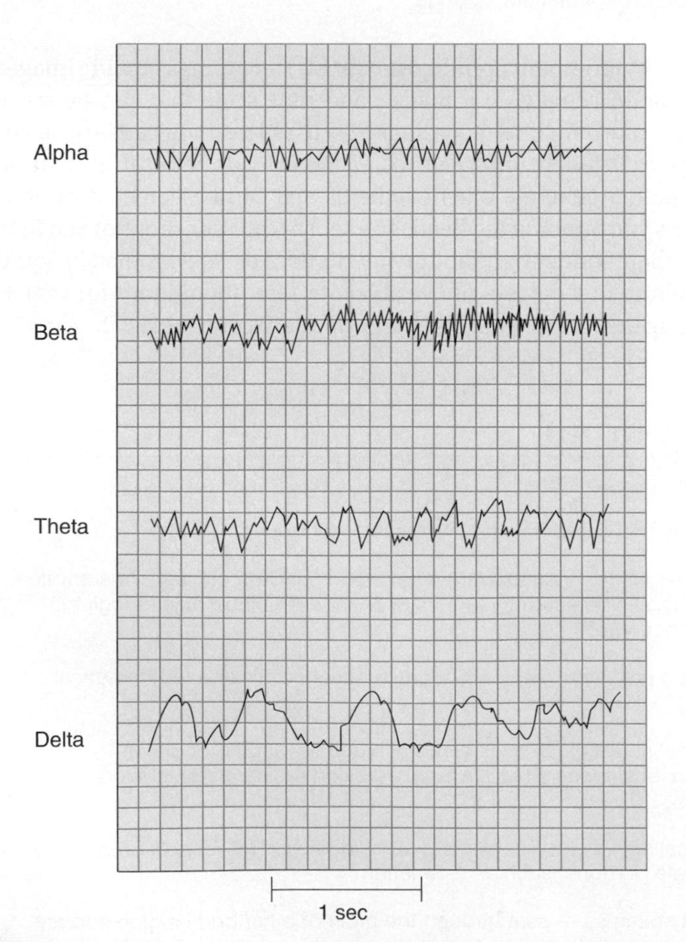

Figure 8.10 Different types of waves in an electroencephalogram (EEG). Notice that the delta waves (*bottom*) have the highest amplitude and lowest frequency.

Theta waves are emitted from the temporal and occipital lobes. They have a frequency of 5 to 8 cycles/second and are common in newborn infants and sleeping adults. The recording of theta waves in awake adults generally indicates severe emotional stress and can be a forewarning of a nervous breakdown.

Delta waves are seemingly emitted in a general pattern from the cerebral cortex. These waves have a frequency of 1 to 5 cycles/second and are common during sleep and in an awake infant. The presence of delta waves in an awake adult indicates brain damage.

Sleep

Although environmental factors affect sleep, there is evidence that sleep is genetically controlled. This is shown by sleep disorders that run in families and the heritability of sleep patterns. Histamine and several other neurotransmitters promote wakefulness, while adenosine and GABA promote sleep. The effects of serotonin are more complicated; it reduces REM sleep (discussed next) while it may stimulate non-REM sleep.

Two categories of sleep are recognized. Dreams—at least those that are vivid enough to recall upon waking—occur during **rapid eye movement (REM) sleep.** The name describes the characteristic eye movements that occur during this stage of sleep. The remainder of the time sleeping is spent in **non-REM,** or **resting, sleep.** These two stages of sleep can also be distinguished by their EEG patterns. The EEG pattern during REM sleep consists of theta waves (5 to 8 cycles per second), although the EEG is often desynchronized as in wakefulness. Non-REM sleep is divided into four stages based on the EEG patterns; stages 3 and 4 are also known as **slow-wave sleep,** because of their characteristic delta waves (1 to 5 cycles per second).

When people first fall asleep, they enter non-REM sleep of four different stages, and then ascend back through these stages to REM sleep. After REM sleep, they again descend through the stages of non-REM sleep and back up to REM sleep. Each of these cycles lasts approximately 90 minutes, and a person may typically go through about five REM-to-non-REM cycles a night. A great amount of time is spent in slow-wave sleep during the first half of a night's sleep; this gives way to mostly REM sleep during the second half of the sleep. When people are allowed to wake up naturally, they generally awaken from REM sleep.

Most neurons decrease their firing rate in the transition from waking to non-REM sleep. This correlates with a decreased energy metabolism and blood flow, as revealed by PET studies. By contrast, REM sleep is accompanied by a higher total brain metabolism and by a higher blood flow to selected brain regions than in the waking state. Interestingly, the limbic system (described shortly) is activated during REM sleep. The limbic system is involved in emotions, and part of it, the amygdala, helps to mediate fear and anxiety. Because these are common emotions during dreaming, it makes sense that the limbic system would be active during REM sleep.

During non-REM sleep, the breathing and heart rate tend to be very regular. In REM sleep, by contrast, the breathing and heart rate are as irregular as they are during waking. This may relate to dreaming and the activation of the brain regions involved in emotions during REM sleep.

Because smaller animals (which have a faster metabolism) need more sleep than bigger animals (which have a slower metabolism), some scientists believe that non-REM sleep may be needed to repair the metabolic damage to cells produced by free radicals (chapters 5 and 19). Another hypothesis regarding the importance of non-REM sleep is that it aids the neural plasticity required for learning. For example, subjects allowed to have non-REM sleep after a learning trial displayed improved performance compared to those who were not allowed to have non-REM sleep. In another study, slow-wave activity in an EEG (indicating non-REM sleep) increased in trained subjects, and the magnitude of that increase correlated with how well the subjects performed on the learned task the next morning.

These and other studies demonstrate that, although short-term memory is formed while a person is awake, the consolidation of short-term into long-term memory is promoted by sleep. Slow-wave sleep particularly benefits the consolidation of spatial and declarative memories (those that can be verbalized). REM sleep has been shown to benefit the consolidation of nondeclarative memories, but evidence suggests that both stages of sleep may participate in the consolidation of declarative and nondeclarative memories. Indeed, memory consolidation is best when slow-wave and REM sleep phases follow each other naturally.

Memory consolidation improves after a nap, but longer durations of sleep are required for maximum benefit. Evidence suggests that the time delay between the learning session and sleep is also an important consideration. Experiments indicate that a time delay of about three hours between a learning session and sleep provides better declarative memory consolidation than a delay of eight hours. These studies strongly suggest that students would improve their performance on an exam if they studied earlier and got a good night's sleep before the exam.

Basal Nuclei

The **basal nuclei** are masses of gray matter composed of neuron cell bodies located deep within the white matter of the cerebrum (fig. 8.11). Although these are more commonly called

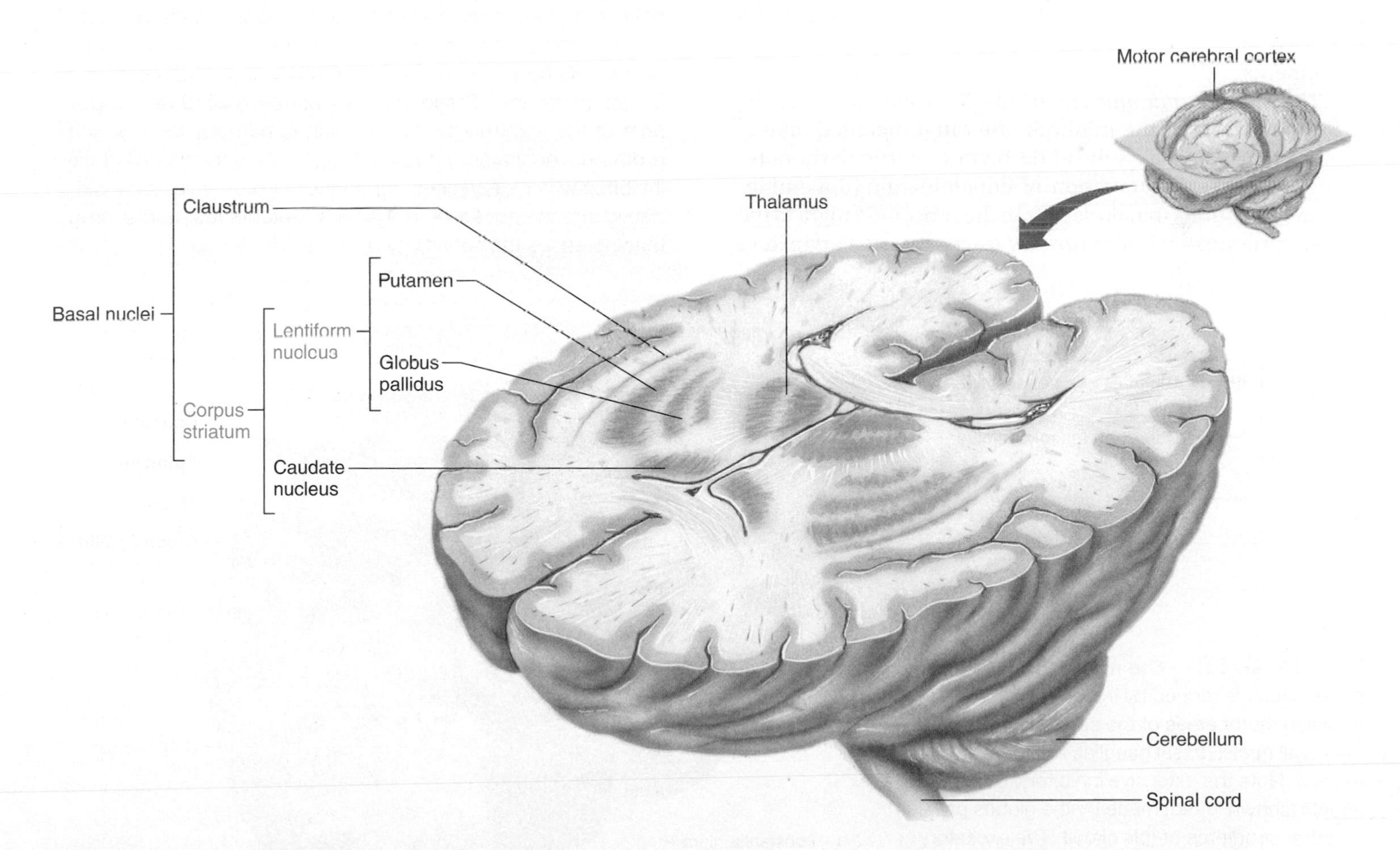

Figure 8.11 **The basal nuclei.** These are structures of the cerebrum containing neurons involved in the control of skeletal muscles (higher motor neurons).The thalamus is a relay center between the motor cerebral cortex and other brain areas. AP|R

basal ganglia, the term "basal nuclei" is more anatomically correct (nuclei and ganglia refer to clusters of cell bodies in the CNS and PNS, respectively) and will be preferred in this text.

The most prominent of the basal nuclei is the **corpus striatum,** which consists of several masses of nuclei. The upper mass, called the *caudate nucleus,* is separated from two lower masses, collectively called the *lentiform nucleus.* The lentiform nucleus consists of a lateral portion, the *putamen,* and a medial portion, the *globus pallidus.* The basal nuclei (basal ganglia) function in the control of voluntary movements.

The areas of the cerebral cortex that control movement (including the precentral motor cortex; see fig. 8.6) send axons to the basal nuclei, primarily the putamen. These cortical axons release the excitatory neurotransmitter glutamate, which stimulates neurons in the putamen. Those neurons, in turn, send axons from the putamen to other basal nuclei. These axons are inhibitory through their release of the neurotransmitter GABA. The globus pallidus and the substantia nigra (a part of the midbrain, to be discussed shortly) send GABA-releasing inhibitory axons to the thalamus. The thalamus, in turn, sends excitatory axons to the motor areas of the cerebral cortex, thereby completing a *motor circuit* (fig. 8.12). The motor circuit allows intended movements to occur while inhibiting unintended movements.

The *subthalamic nucleus* of the diencephalon and the *substantia nigra* of the midbrain are often included among the basal nuclei. The substantia nigra is particularly noteworthy because degeneration of dopaminergic (dopamine-releasing) neurons that project from the substantia nigra to the corpus striatum—the *nigrostriatal tract*—causes Parkinson's disease.

CLINICAL APPLICATION

Degeneration of the caudate nucleus (as in *Huntington's disease*) produces **chorea**—a hyperkinetic disorder characterized by rapid, uncontrolled, jerky movements. Degeneration of dopaminergic neurons to the caudate nucleus from the substantia nigra, a small nucleus in the midbrain, produces most of the symptoms of **Parkinson's disease.** The causes of Parkinson's disease are incompletely understood. An environmental agent, such as long-term exposure to a pesticide, perhaps combined with a genetic susceptibility (although Parkinson's disease rarely runs in families), is believed to provoke processes that result in neurodegeneration. These processes include defects in mitochondria leading to the release of free radicals, production of proteins called *α-synuclein,* which sticks together to form tiny fibrils of *Lewy bodies* in affected neurons, and others. Parkinson's disease is associated with rigidity, resting tremor, and difficulty in initiating voluntary movements. As discussed in chapter 7, Parkinson's disease is treated with L-dopa (the precursor of dopamine) or dopamine agonists. Unfortunately, there are complications that can limit treatment with L-dopa, including *dyskinesia*—involuntary movements that can occur when dopaminergic drugs are given. Some Parkinson's disease patients benefit from *deep brain stimulation,* where an electrode is implanted into the subthalamic nucleus or globus pallidus. Others may benefit from drugs that are in different stages of testing. These include *coenzyme Q10* (a component of the electron transport chain in mitochondria, which promotes healthier mitochondria); a monoamine oxidase inhibitor, which decreases dopamine breakdown and the oxidative stress caused by dopamine metabolites; and a drug that increases the sensitivity of dopamine receptors.

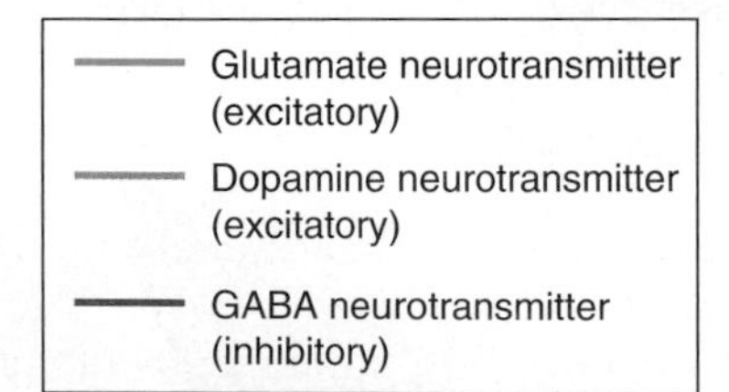

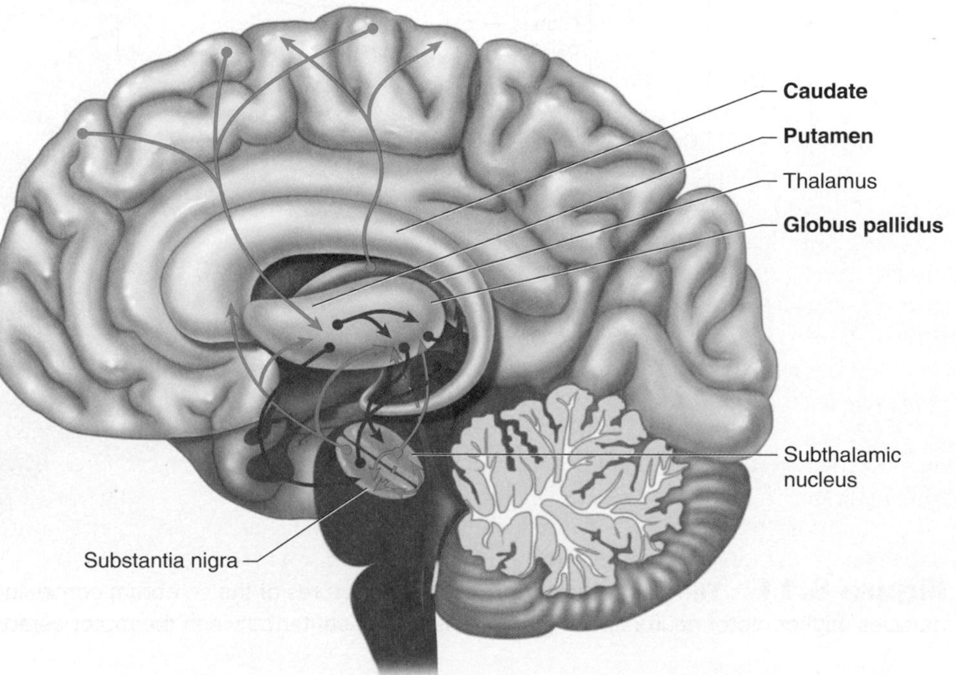

Figure 8.12 The motor circuit. The motor circuit is formed by interconnections between motor areas of the cerebral cortex, the basal nuclei (basal ganglia), and other brain regions. Note the extensive inhibitory, GABA-ergic effects (shown in red) made by the globus pallidus on other structures of this circuit. The excitatory neurotransmitters of this circuit are glutamate (green) and dopamine (blue).

Cerebral Lateralization

By way of motor fibers originating in the precentral gyrus, each cerebral cortex controls movements of the contralateral (opposite) side of the body. At the same time, somatesthetic sensation from each side of the body projects to the contralateral postcentral gyrus as a result of *decussation* (crossing over) of fibers. In a similar manner, images falling in the left half of each retina project to the right occipital lobe, and images in the right half of each retina project to the left occipital lobe. Each cerebral hemisphere, however, receives information from both sides of the body because the two hemispheres communicate with each other via the **corpus callosum,** a large tract composed of about 200 million fibers.

The corpus callosum has been surgically cut in some people with severe epilepsy as a way of alleviating their symptoms. These *split-brain procedures* isolate each hemisphere from the other, but, surprisingly, to a casual observer split-brain patients do not show evidence of disability as a result of the surgery. However, in specially designed experiments in which each hemisphere is separately presented with sensory images and the patient is asked to perform tasks (speech or writing or drawing with the contralateral hand), it has been learned that each hemisphere is good at certain categories of tasks and poor at others (fig. 8.13).

In a typical experiment, the image of an object may be presented to either the right or left hemisphere (by presenting it to either the left or right visual field only; see chapter 10, fig. 10.32) and the person may be asked to name the object. Findings indicate that, in most people, the task can be performed successfully by the left hemisphere but not by the right. Similar experiments have shown that the left hemisphere is generally the one in which most of the language and analytical abilities reside.

These findings have led to the concept of *cerebral dominance,* which is analogous to the concept of handedness—people generally have greater motor competence with one hand than with the other. Since most people are right-handed, and the right hand is also controlled by the left hemisphere, the left hemisphere was naturally considered to be the dominant hemisphere in most people. Further experiments have shown, however, that the right hemisphere is specialized along different, less obvious lines—rather than one hemisphere being dominant and the other subordinate, the two hemispheres appear to have complementary functions. The term **cerebral lateralization,** or specialization of function in one hemisphere or the other, is thus now preferred to the term **cerebral dominance,** although both terms are currently used.

Experiments have shown that the right hemisphere does have limited verbal ability; more noteworthy is the observation that the right hemisphere is most adept at *visuospatial tasks.* The right hemisphere, for example, can recognize faces better than the left, but it cannot describe facial appearances as well as the left. The right hemisphere of split-brain patients, acting through its control of the left hand, is better

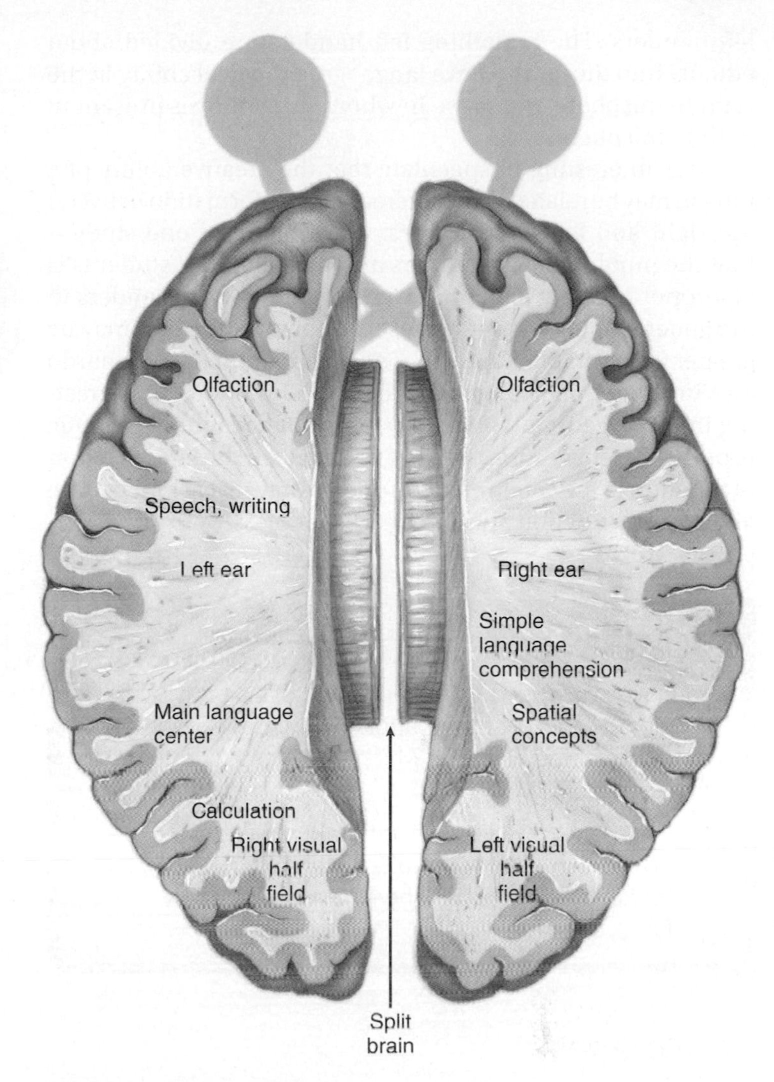

Figure 8.13 Different functions of the right and left cerebral hemispheres. These differences were revealed by experiments with people whose corpus callosum—the tract connecting the two hemispheres—was surgically split. AP|R

than the left (controlling the right hand) at arranging blocks or drawing cubes. Patients with damage to the right hemisphere, as might be predicted from the results of split-brain research, have difficulty finding their way around a house and reading maps.

Perhaps as a result of the role of the right hemisphere in the comprehension of patterns and part-whole relationships, the ability to compose music, but not to critically understand it, appears to depend on the right hemisphere. Interestingly, damage to the left hemisphere may cause severe speech problems while leaving the ability to sing unaffected.

The lateralization of functions just described—with the left hemisphere specialized for language and analytical ability, and the right hemisphere specialized for visuospatial ability—is true for 97% of all people. It is true for all right-handers (who account for 90% of all people) and for 70% of all left-handers. The remaining left-handers are divided about equally into those who have language-analytical ability in the right hemisphere and those in whom this ability is present in both hemispheres.

It is interesting to speculate that the creative ability of a person may be related to the interaction of information between the right and left hemispheres. The finding of one study—that the number of left-handers among college art students is disproportionately higher than the number of left-handers in the general population—suggests that this interaction may be greater in left-handed people. The observation that Leonardo da Vinci and Michelangelo were both left-handed is interesting in this regard, but obviously does not constitute scientific proof of this suggestion. Further research on the lateralization of function of the cerebral hemispheres may reveal much more about brain function and the creative process.

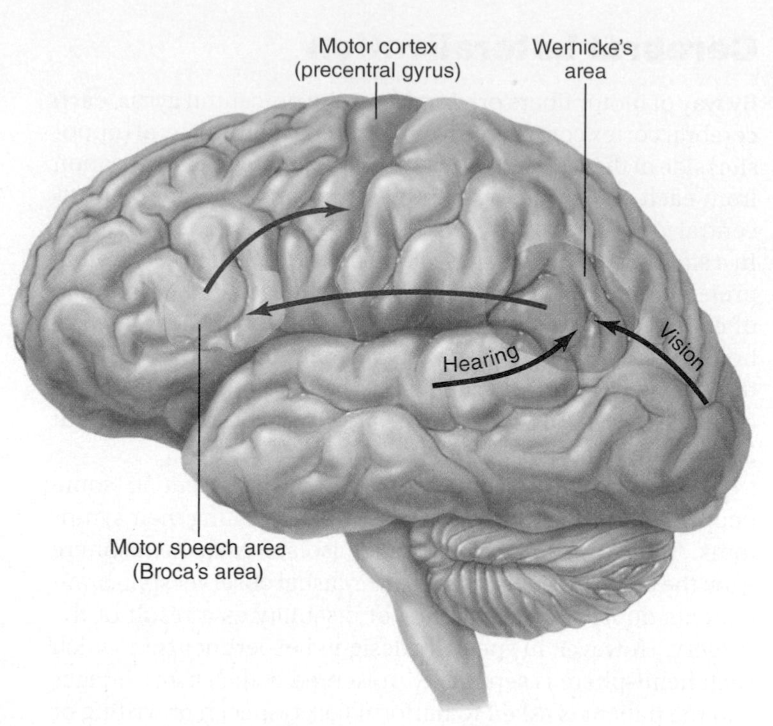

Figure 8.14 Brain areas involved in the control of speech. Damage to these areas produces speech deficits, known as aphasias. Wernicke's area, required for language comprehension, receives information from many areas of the brain, including the auditory cortex (for heard words), the visual cortex (for read words), and other brain areas. In order for a person to be able to speak intelligibly, Wernicke's area must send messages to Broca's area, which controls the motor aspects of speech by way of its input to the motor cortex.

Clinical Investigation CLUES

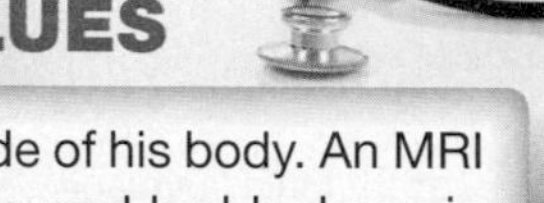

Frank was paralyzed on the right side of his body. An MRI revealed that Frank had a stroke caused by blockage in his middle cerebral artery, which provides blood to much of the brain.

- What is an MRI, and which brain regions are involved in motor control?
- Which cerebral hemisphere was damaged by Frank's stroke?

Language

Knowledge of the brain regions involved in language has been gained primarily by the study of *aphasias*—speech and language disorders caused by damage to the brain through head injury or stroke. In most people, the language areas of the brain are primarily located in the left hemisphere of the cerebral cortex, as previously described. Even in the nineteenth century, two areas of the cortex—Broca's area and Wernicke's area (fig. 8.14)—were recognized as areas of particular importance in the production of aphasias.

Broca's aphasia is the result of damage to **Broca's area,** located in the left inferior frontal gyrus and surrounding areas. Common symptoms include weakness in the right arm and the right side of the face. People with Broca's aphasia are reluctant to speak, and when they try, their speech is slow and poorly articulated. Their comprehension of speech is unimpaired, however. People with this aphasia can understand a sentence but have difficulty repeating it. It should be noted that this is not simply due to a problem in motor control, because the neural control over the musculature of the tongue, lips, larynx, and so on is unaffected.

Wernicke's aphasia is caused by damage to **Wernicke's area,** located in the superior temporal gyrus of the left hemisphere (in most people). This results in speech that is rapid and fluid but without meaning. People with Wernicke's aphasia produce speech that has been described as a "word salad." The words used may be real words that are chaotically mixed together, or they may be made-up words. Language comprehension is destroyed; people with Wernicke's aphasia cannot understand either spoken or written language.

It appears that the concept of words originates in Wernicke's area. Thus, in order to understand words that are read, information from the visual cortex (in the occipital lobe) must project to Wernicke's area. Similarly, in order to understand spoken words, the auditory cortex (in the temporal lobe) must send information to Wernicke's area.

To speak intelligibly, the concept of words originating in Wernicke's area must be communicated to Broca's area; this is accomplished by a fiber tract called the **arcuate fasciculus.** Broca's area, in turn, sends fibers to the motor cortex (precentral gyrus), which directly controls the musculature of speech. Damage to the arcuate fasciculus produces *conduction aphasia,* which is fluent but nonsensical speech as in Wernicke's aphasia, even though both Broca's and Wernicke's areas are intact.

The **angular gyrus,** located at the junction of the parietal, temporal, and occipital lobes, is believed to be a center for the integration of auditory, visual, and somatesthetic information. Damage to the angular gyrus produces aphasias, which suggests that this area projects to Wernicke's area. Some patients with damage to the left angular gyrus can speak and understand spoken language but cannot read or write. Other patients can write a sentence but cannot read it, presumably because of damage to the projections from the occipital lobe (involved in vision) to the angular gyrus.

CLINICAL APPLICATION

Recovery of language ability, by transfer to the right hemisphere after damage to the left hemisphere, is very good in children but decreases after adolescence. Recovery is reported to be faster in left-handed people, possibly because language ability is more evenly divided between the two hemispheres in left-handed people. Some recovery usually occurs after damage to Broca's area, but damage to Wernicke's area produces more severe and permanent aphasias.

Clinical Investigation CLUES

Frank spoke slowly and with great difficulty, but was coherent.

- Damage to which brain region is likely responsible for Frank's aphasia?
- If Frank's speech were fluid but nonsensical, which brain region would likely be damaged?

Limbic System and Emotion

The parts of the brain that appear to be of paramount importance in the neural basis of emotional states are the hypothalamus (in the diencephalon) and the **limbic system.** The limbic system consists of a group of forebrain nuclei and fiber tracts that form a ring around the brain stem (*limbus* = ring). Among the components of the limbic system are the *cingulate gyrus* (part of the cerebral cortex), the *amygdaloid nucleus* (or *amygdala*), the *hippocampus,* and the *septal nuclei* (figs. 8.15 and 8.18). The cingulate gyrus is a thick area of cortex that surrounds the corpus callosum and is involved in emotions (particularly negative emotions associated with pain and fear) and motivation. Studies demonstrate that the *anterior insula* is activated together with the anterior cingulate cortex during emotional experiences.

The limbic system was once called the *rhinencephalon,* or "smell brain," because it is involved in the central processing of olfactory information. This may be its primary function in lower vertebrates whose limbic system may constitute the entire forebrain. It is now known, however, that the limbic system in humans is a center for basic emotional drives. The limbic system was derived early in the course of vertebrate evolution, and its tissue is phylogenetically older than the cerebral cortex. There are thus few synaptic connections between the cerebral cortex and the structures of the limbic system, which perhaps helps explain why we have so little conscious control over our emotions.

There is a closed circuit of information flow between the limbic system and the thalamus and hypothalamus (fig. 8.15) called the *Papez circuit.* (The thalamus and hypothalamus are part of the diencephalon, described in a later section.) In the Papez circuit, a fiber tract, the *fornix,* connects the hippocampus to the mammillary bodies of the hypothalamus, which, in

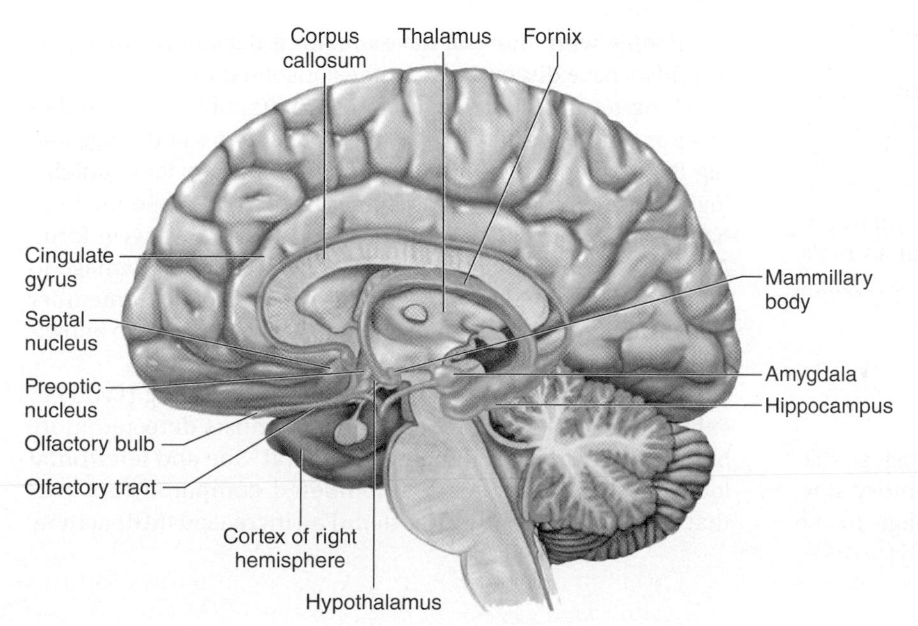

Figure 8.15 The limbic system. The left temporal lobe has been removed in this figure to show the structures of the limbic system (green). The limbic system consists of particular nuclei (aggregations of neuron cell bodies) and axon tracts of the cerebrum that cooperate in the generation of emotions. The hypothalamus, though part of the diencephalon rather than the cerebrum (telencephalon), participates with the limbic system in emotions. AP|R

turn, project to the anterior nuclei of the thalamus. The thalamic nuclei, in turn, send fibers to the cingulate gyrus, which then completes the circuit by sending fibers to the hippocampus. Through these interconnections, the limbic system and the hypothalamus appear to cooperate in the neural basis of emotional states.

Studies of the functions of these regions include electrical stimulation of specific locations, destruction of tissue (producing *lesions*) in particular sites, and surgical removal, or *ablation,* of specific structures. These studies suggest that the hypothalamus and limbic system are involved in the following feelings and behaviors:

1. **Aggression.** Stimulation of certain areas of the amygdala produces rage and aggression, and stimulation of particular areas of the hypothalamus can produce similar effects.
2. **Fear.** Fear can be produced by electrical stimulation of the amygdala and hypothalamus, and surgical removal of the limbic system can result in an absence of fear. Monkeys are normally terrified of snakes, for example, but they will handle snakes without fear if their limbic system is removed. Humans with damage to their amygdala have demonstrated an impaired ability to recognize facial expressions of fear and anger. These and other studies suggest that the amygdala is needed for fear conditioning.
3. **Feeding.** The hypothalamus contains both a *feeding center* and a *satiety center.* Electrical stimulation of the former causes overeating, and stimulation of the latter will stop feeding behavior in experimental animals.
4. **Sex.** The hypothalamus and limbic system are involved in the regulation of the sexual drive and sexual behavior, as shown by stimulation and ablation studies in experimental animals. The cerebral cortex, however, is also critically important for the sex drive in lower animals, and the role of the cerebrum is even more important for the sex drive in humans.
5. **Goal-directed behavior (reward and punishment system).** Electrodes placed in particular sites between the frontal cortex and the hypothalamus can deliver shocks that function as a reward. In rats, this reward is more powerful than food or sex in motivating behavior. Similar studies have been done in humans, who report feelings of relaxation and relief from tension, but not of ecstasy. Electrodes placed in slightly different positions apparently stimulate a punishment system in experimental animals, who stop their behavior when stimulated in these regions.

Memory

Brain Regions in Memory

Clinical studies of *amnesia* (loss of memory) suggest that several different brain regions are involved in memory storage and retrieval. Amnesia can result from damage to the temporal lobe of the cerebral cortex, the hippocampus, the head of the caudate nucleus (in Huntington's disease), or the dorsomedial thalamus (in alcoholics suffering from Korsakoff's syndrome with thiamine deficiency). A number of researchers now believe that there are several different systems of information storage in the brain. One system relates to the simple learning of stimulus-response that even invertebrates can do to some degree. This, together with skill learning and different kinds of conditioning and habits, is retained in people with amnesia.

There are different categories of memory, as revealed by patients with particular types of brain damage and by numerous scientific investigations. Scientists distinguish between **short-term memory** and **long-term memory.** Long-term memory, but not short-term memory, depends on the synthesis of new RNA and protein, so that drugs that disrupt genetic transcription or translation interfere with long-term (but not short-term) memory. People with head trauma, and patients who undergo electroconvulsive shock (ECS) therapy, may lose their memory of recent events but retain their older memories. The conversion of a short-term memory into a more stable long-term memory is called **memory consolidation.** Memory consolidation requires the activation of genes, the production of new proteins, and the formation of new synapses. There is now evidence (in rodents) that memory consolidation in the hippocampus also requires glycogenolysis and lactate production in astrocytes at about the time of training.

Long-term memory is classified as **nondeclarative** (or **implicit**) **memory** and **declarative** (or **explicit**) **memory.** Nondeclarative memory refers to memory of simple skills and conditioning (such as remembering how to tie shoelaces). Declarative memory is memory that can be verbalized; it is subdivided into **semantic** (fact) and **episodic** (event) **memory.** A semantic memory would be remembering the names of the bones; an episodic memory would be remembering the experience of taking a practical exam on the skeletal system.

People with amnesia have impaired declarative memory. Scientists have discovered that the consolidation of short-term into long-term declarative memory is a function of the **medial temporal lobe,** particularly of the *hippocampus* and *amygdala* (fig. 8.15). Although the hippocampus is important for maintaining recent memories, it is no longer needed once the memory has become consolidated into a more stable, long-term form. An amnesiac patient known as "E.P." with bilateral damage to his medial temporal lobes, for example, was able to remember well the neighborhood he left 50 years before but had no knowledge of his current neighborhood.

Using functional magnetic resonance imaging (fMRI) of subjects asked to remember words, scientists detected more brain activity in the left medial temporal lobe and left frontal lobe for words that were remembered compared to words that were subsequently forgotten. The increased fMRI activity in these brain regions seems to indicate the encoding of the memories. Indeed, lesions of the left medial temporal lobe impairs verbal memory, while lesions of the right medial temporal lobe impairs nonverbal memories, such as the ability to remember faces.

Surgical removal of the right and left medial temporal lobes was performed in one patient, designated "H.M.," in an effort to treat his epilepsy. After the surgery he was unable to consolidate any short-term memory. He could repeat a phone number and carry out a normal conversation; he could not remember the phone number if momentarily distracted, however, and if the person to whom he was talking left the room and came back a few minutes later, H.M. would have no recollection of having seen that person or of having had a conversation with that person before. Although his memory of events that occurred before the operation was intact, all subsequent events in his life seemed as if they were happening for the first time.

H.M.'s deficit was in *declarative memory.* His *nondeclarative memory*—perceptual and motor skills, such as how to drive a car—were still intact. The effects of bilateral removal of H.M.'s medial temporal lobes were due to the fact that the hippocampus and amygdaloid nucleus (fig. 8.15) were also removed in the process. Surgical removal of the left medial temporal lobe impairs the consolidation of short-term verbal memories into long-term memory, and removal of the right medial temporal lobe impairs the consolidation of nonverbal memories.

On the basis of additional clinical experience, it appears that the **hippocampus** is a critical component of the memory system. Magnetic resonance imaging (MRI) reveals that the hippocampus is often shrunken in living amnesic patients. However, the degree of memory impairment is increased when other structures, as well as the hippocampus, are damaged. The hippocampus and associated structures of the medial temporal lobe are thus needed for the acquisition of new information about facts and events, and for the consolidation of short-term into long-term memory, which is stored in the cerebral cortex. Sleep, particularly slow-wave (non-REM) sleep, but perhaps also REM sleep, is needed for optimum memory consolidation by the hippocampus. Emotional arousal, acting via the structures of the limbic system, can enhance or inhibit long-term memory storage. For example, stress has been shown to produce deficits in hippocampus-dependent learning and memory.

The **amygdala** is particularly important in the memory of fear responses. Studies demonstrate increased neural activity of the human amygdala during visual processing of fearful faces, and patients with bilateral damage to the amygdala were unable to read danger when shown threatening pictures.

The cerebral cortex is thought to store factual information, with verbal memories lateralized to the left hemisphere and visuospatial information to the right hemisphere. The neurosurgeon Wilder Penfield was the first to electrically stimulate various brain regions of awake patients, often evoking visual or auditory memories that were extremely vivid. Electrical stimulation of specific points in the temporal lobe evoked specific memories so detailed that the patients felt as if they were reliving the experience. The medial regions of the temporal lobes, however, cannot be the site where long-term memory is stored because destruction of these areas in patients treated for epilepsy did not destroy the memory of events prior to the surgery. The **inferior temporal lobes,** on the other hand, do appear to be sites for the storage of long-term visual memories.

The *left inferior frontal lobe* has recently been shown to participate in performing exact mathematical calculations. Scientists have speculated that this brain region may be involved because it stores verbally coded facts about numbers. Using fMRI, researchers have recently demonstrated that complex problem-solving involves the most anterior portion of the frontal lobes, an area called the **prefrontal cortex.** Some of the other functions ascribed to the prefrontal cortex include short term memory (as for a phone number that must be kept in mind to dial but then quickly forgotten), planning (remembering to perform sequential actions), and the inhibition of inappropriate actions (such as answering a stranger's ringing telephone). There is evidence that signals are sent from the prefrontal cortex to the inferior temporal lobes, where visual long-term memories are stored. Lesions of the prefrontal cortex interfere with memory in a less dramatic way than lesions of the medial temporal lobe.

The amount of memory destroyed by ablation (removal) of brain tissue seems to depend more on the amount of brain tissue removed than on the location of the surgery. On the basis of these observations, it was formerly believed that the memory was diffusely located in the brain; stimulation of the correct location of the cortex then retrieved the memory. According to current thinking, however, particular aspects of the memory—visual, auditory, olfactory, spatial, and so on—are stored in particular areas, and the cooperation of all of these areas is required to elicit the complete memory.

As an example of the diffuse location of memories, **working memory**—the ability to keep information in your head consciously for a short time—is stored differently depending on whether it involves keeping several numbers in your mind until you type them, or whether it involves spatial information, such as backtracking to pick up an item you skipped while browsing in a new store. However, both types of working memory require the prefrontal cortex. There are also certain generalities that can be made about long-term declarative memory and brain location. For example, the ability to recall names and categories (semantic memory) is localized to the inferior temporal lobes; different locations seem to be required for storing episodic memories. Thus, in Alzheimer's disease, episodic and semantic memory decline independently of each other.

Table 8.3 | Categories of Memories and the Major Brain Regions Involved

Memory Category	Major Brain Regions Involved	Length of Memory Storage	Examples
Episodic memory (explicit, declarative)	Medial temporal lobes, thalamus, fornix, prefrontal cortex	Minutes to years	Remembering what you had for breakfast, and what vacation you took last summer
Semantic memory (explicit, declarative)	Inferior temporal lobes	Minutes to years	Knowing facts such as what city is the capital, your mother's maiden name, and the different uses of a hammer and a saw
Procedural memory (explicit or implicit; nondeclarative)	Basal nuclei, cerebellum, supplementary motor areas	Minutes to years	Knowing how to shift gears in a car and how to tie your shoelaces
Working memory	Words and numbers: prefrontal cortex, Broca's area, Wernicke's area Spatial: prefrontal cortex, visual association areas	Seconds to minutes	Words and numbers: keeping a new phone number in your head until you dial it Spatial: mentally following a route

Source: Modified from: Budson, Andrew E. and Bruce H. Price."Memory dysfunction." *New England Journal of Medicine 352* (2005): 692–698.

Much remains to be learned about the brain locations associated with different systems of memory (table 8.3). Continued scientific investigations, including fMRI studies, patient observations, and others, will yield important new information about the relationship between different anatomical brain regions and their roles in memory storage, consolidation, and retrieval.

Alzheimer's Disease

Alzheimer's disease is the most common form of dementia, its incidence increasing with age to greater than one in three after age 85. People with Alzheimer's disease have (1) a loss of cholinergic and other neurons in the hippocampus and cerebral cortex; (2) an accumulation of extracellular protein deposits called *senile plaques;* and (3) an accumulation of intracellular proteins forming *neurofibrillary tangles.*

In Alzheimer's disease, an *amyloid precursor protein* (abbreviated *APP*) may be broken down by β-*secretase* and then γ-*secretase* into peptides called **amyloid beta (Aβ).** The Aβ peptides can associate into dimers and oligomers, and then grow into fibers in the form of β-pleated sheets (chapter 2; see fig. 2.28*c*) that compose the amyloid senile plaques. Evidence suggests that it is the soluble dimers and oligomers of the 42-amino-acid-long form of Aβ, rather than the plaques, that cause Alzheimer's disease. A small proportion (less than 1%) of people with early-onset Alzheimer's disease have mutations in the gene for APP or in *presenilin* genes that code for the catalytic portion of the γ-secretase enzyme. The vast majority of people with Alzheimer's have the "sporadic" form, caused by incompletely understood interactions between environmental and genetic influences.

Although Aβ oligomers are themselves toxic, their full ability to cause Alzheimer's disease may depend on another protein, called **tau.** Normal tau proteins bind to and stabilize microtubules in axons; in Alzheimer's disease they become excessively phosphorylated and redistributed to the neuron cell body and dendrites. There they aggregate together and become insoluble, forming the neurofibrillary tangles. These changes appear to be driven by Aβ. It is evidently not the neurofibrillary tangles, but rather the more soluble intermediate forms of tau that may produce the toxic effects. Toxic changes in Alzheimer's disease include the loss of synapses and dendritic spines, reduced ability to produce long-term potentiation (LTP), excitotoxicity (chapter 7, section 7) causing neuron apoptosis, and mitochondrial release of reactive oxygen species that produce oxidative stress and apoptosis.

For reasons not fully understood, people with a particular allele (form) of the gene for *apolipoprotein E* (a cholesterol carrier protein active in the brain) are more likely to develop Alzheimer's disease. A single copy of this allele—known as *AP0ε4*—increases the risk by a factor of 4 and two copies of this allele increase the risk by a factor of 19. More than 60% of people with Alzheimer's have at least one copy of the AP0ε4 allele.

Currently available drugs for treating Alzheimer's disease include (1) inhibitors of acetylcholinesterase (so that ACh released by the surviving cholinergic neurons can be more effective); (2) an antagonist of glutamate (to reduce its ability to promote excitotoxicity); and (3) drugs for treating depression. A variety of other medications that exploit our growing understanding of Alzheimer's disease are currently in clinical trials. Meanwhile, people are advised to engage in both mental and physical activity (to build up a "cognitive reserve" and to promote neuron health), eat a diet restricted in calories and fat (because obesity, high plasma cholesterol, and type 2 diabetes are risk factors for Alzheimer's disease), and eat a diet rich in fruits and vegetables to provide antioxidants.

Synaptic Changes in Memory

Short-term memory may involve the establishment of **recurrent** (or **reverberating**) **circuits** of neuronal activity. This is where neurons synapse with each other to form a circular path, so that the last neuron to be activated then stimulates the first neuron. A neural circuit of recurrent, or reverberating, activity may thus be maintained for a period of time. These reverberating circuits have been used to explain the neuronal basis of working memory, the ability to hold a memory (of a grocery list, for example) in mind for a relatively short period of time.

Long-term memory (unlike short-term memory) is not disrupted by electroconvulsive shock therapy, suggesting that the consolidation of memory depends on relatively permanent changes in neurons and synapses. This is supported by evidence that protein synthesis is required for the consolidation of the "memory trace." The nature of the synaptic changes involved in memory storage has been studied in the hippocampus using the processes of long-term potentiation (LTP) and long-term depression (LTD), described in chapter 7, section 7.7. The NMDA receptor for glutamate is central to these processes and to the formation of memories that depend on the hippocampus.

In *long-term potentiation (LTP),* synapses that are stimulated at high frequency exhibit subsequent increased excitability. Long-term potentiation has been studied extensively in the hippocampus, where most of the axons use glutamate as a neurotransmitter. Here, LTP is induced by the activation of the NMDA receptors for glutamate (chapter 7, section 7.6). At the resting membrane potential, the NMDA pore is blocked by a Mg^{2+} ion that does not allow the entry of Ca^{2+}, even when glutamate is present. In order for glutamate to activate its NMDA receptors, the membrane must also become partially depolarized, causing the Mg^{2+} to leave the pore. This depolarization could be produced by glutamate binding to its AMPA receptors (fig. 8.16), but a recent report suggests that D-serine released from astrocytes might also serve this function and be needed for LTP. Under these conditions, glutamate binding to its NMDA receptor causes the NMDA channel to be open so that Ca^{2+} can diffuse into the cell.

The Ca^{2+} entering through the NMDA receptors binds to **calmodulin,** a regulatory protein important for the second-messenger function of Ca^{2+}. This Ca^{2+}-calmodulin complex then activates a previously inactive enzyme called *CaMKII (calmodulin-dependent protein kinase II).* The CaMKII causes additional AMPA receptors for glutamate to move into the plasma membrane of the postsynaptic neuron. This strengthens the transmission at this synapse; then, a given amount of glutamate released from the presynaptic axon terminal produces a greater postsynaptic depolarization (EPSP).

The rise in the intracellular Ca^{2+} concentration also causes longer-term changes in the postsynaptic neuron. These more persistent changes needed for synaptic plasticity and the formation of long-term memories require Ca^{2+} to enter the nucleus and, bound to calmodulin, activate a different protein kinase. This enzyme, in turn, activates a transcription factor known as **CREB** (for "cyclic AMP response element binding protein"). CREB and other transcription factors activate genes that produce new mRNA and proteins. This may involve epigenetic changes (histone acetylation and DNA methylation; chapter 3, section 3.5) in chromatin structure, which have been shown to occur later in LTP. Scientists estimate that neurons have about 200 genes regulated by nuclear Ca^{2+}.

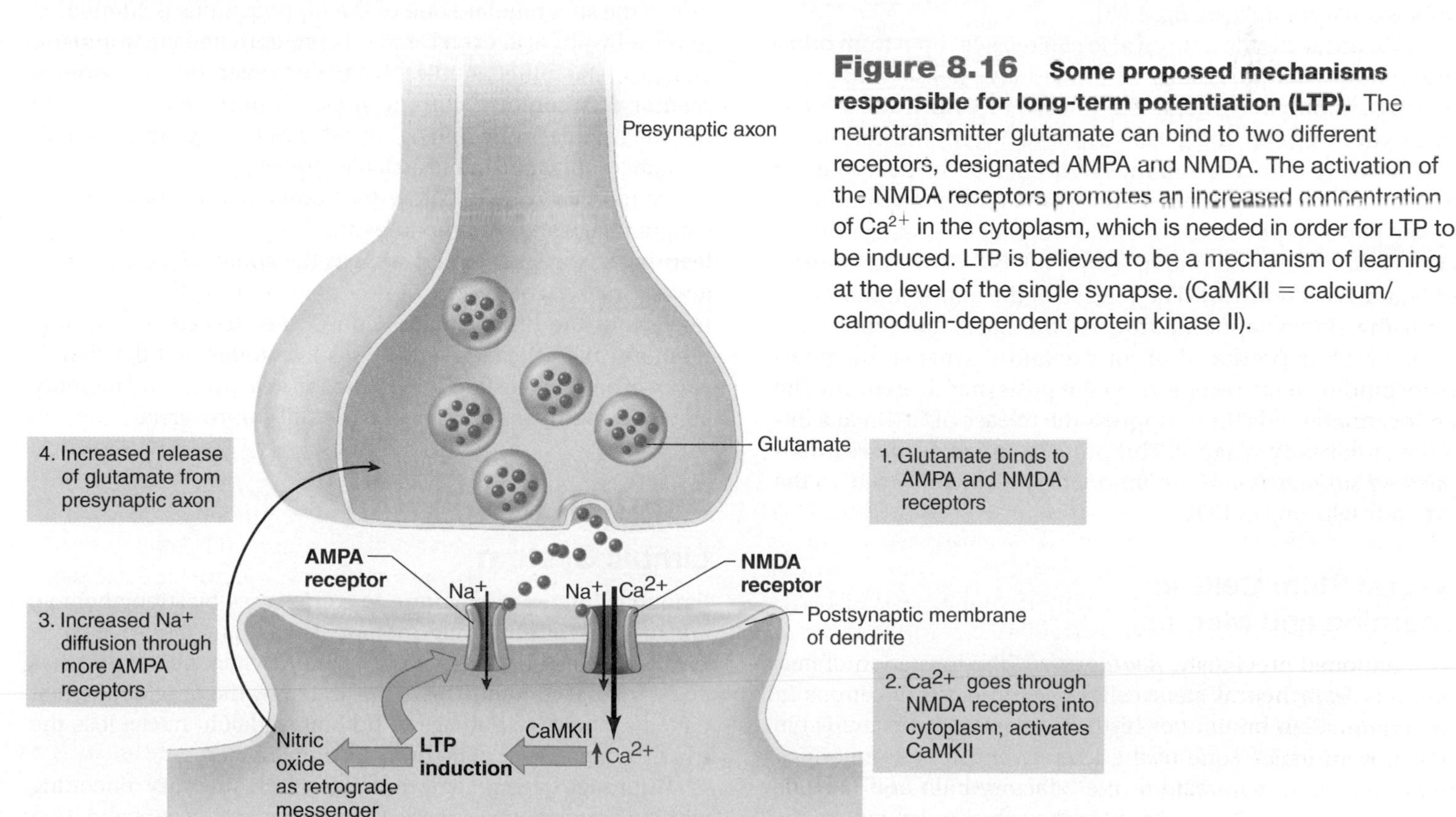
Presynaptic axon
Glutamate
1. Glutamate binds to AMPA and NMDA receptors
AMPA receptor
NMDA receptor
Na+
Na+
Ca2+
Postsynaptic membrane of dendrite
2. Ca2+ goes through NMDA receptors into cytoplasm, activates CaMKII
3. Increased Na+ diffusion through more AMPA receptors
4. Increased release of glutamate from presynaptic axon
Nitric oxide as retrograde messenger
LTP induction
CaMKII
↑Ca2+

Figure 8.16 Some proposed mechanisms responsible for long-term potentiation (LTP). The neurotransmitter glutamate can bind to two different receptors, designated AMPA and NMDA. The activation of the NMDA receptors promotes an increased concentration of Ca^{2+} in the cytoplasm, which is needed in order for LTP to be induced. LTP is believed to be a mechanism of learning at the level of the single synapse. (CaMKII = calcium/calmodulin-dependent protein kinase II).

In part, the new proteins may be needed for the production of spinelike extension from the dendrites called **dendritic spines** (fig. 8.17). *Pyramidal neurons* (a type of neuron characteristic of the cerebral cortex, hippocampus, and amygdala) have thousands of dendritic spines, where most of the EPSPs are produced in response to glutamate. Dendritic spines have been observed to enlarge and change shape during LTP, and such changes—as well as the insertion of additional AMPA receptors into the spines—may promote a prolonged improvement in synaptic transmission. During *long-term depression (LTD),* by contrast, dendritic spines have been observed to shrink or disappear. This is accompanied by the loss of AMPA receptors from the postsynaptic membrane, which primarily accounts for the reduced synaptic transmission in LTD.

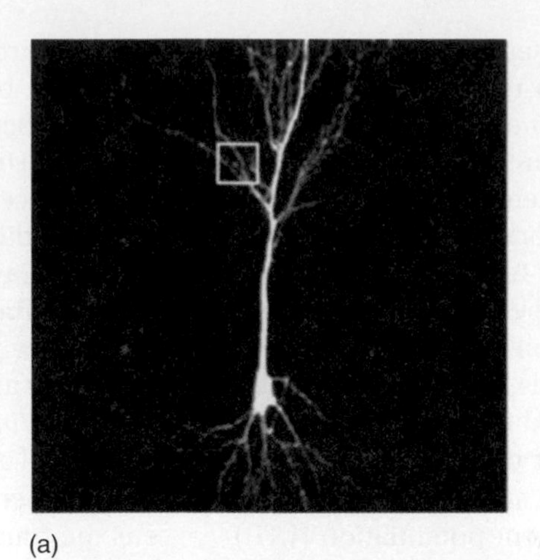

(a)

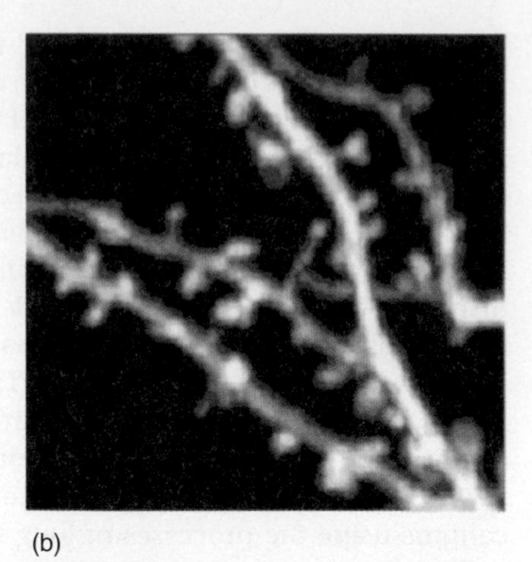

(b)

Figure 8.17 Dendritic spines. *(a)* A photomicrograph of a neuron of the human hippocampus. Dendrites are magnified in a different neuron *(b)* to show the spines. Reprinted, with permission, from the Annual Review of Physiology, Volume 64 © 2002 by Annual Reviews, www.annualreviews.org. AP|R

In some cases, LTP involves changes in the presynaptic axon as well. These changes promote an increased Ca^{2+} concentration within the axon terminals, leading to greater release of neurotransmitter by exocytosis of synaptic vesicles. The enhanced release of neurotransmitter during LTP may be produced by the release of *retrograde* messenger molecules—ones produced by dendrites that travel backward to the presynaptic axon terminals. There is evidence that nitric oxide (NO) can act as a retrograde messenger in this way, promoting LTP by increasing the amount of glutamate released from the presynaptic axon terminal (see fig. 8.16).

The postsynaptic neuron also can receive input from other presynaptic neurons, many of which may release GABA as a neurotransmitter. Through the release of GABA, these neurons would inhibit the postsynaptic neuron. The release of GABA can be reduced, and thus inhibition of the postsynaptic neuron lessened, by another retrograde messenger produced by the postsynaptic neuron. The retrograde messenger in this case is an *endocannabinoid,* a type of lipid neurotransmitter (chapter 7, section 7.6). The release of the endocannabinoids from the postsynaptic neuron is stimulated by depolarization, which is produced at an excitatory synapse by glutamate binding to its receptors on the postsynaptic neuron. The endocannabinoids then suppress the release of GABA at a different, inhibitory synapse. This process, called *depolarization-induced suppression of inhibition,* may also contribute to the synaptic learning of LTP.

Neural Stem Cells in Learning and Memory

As mentioned previously, *neurogenesis* (the formation of new neurons from neural stem cells) occurs in two locations in the mammalian brain after birth. Neurogenesis occurs (1) in the subventricular zone of the lateral ventricles, from which newborn neurons migrate to the olfactory bulb and become functional interneurons, and (2) in the subgranular zone of the hippocampus. New neurons formed in the subgranular zone develop into mature neurons and integrate into the neural networks of the dentate nucleus of the hippocampus. This is particularly exciting because the hippocampus is required for the consolidation of certain types of memory, such as episodic and spatial memory.

Neurogenesis in the subventricular zone of the lateral ventricles is implicated in olfactory memory and learning in rodents. In humans, however, there is some evidence that neurogenesis in this location may cease after 18 months of age. Neurogenesis in the subgranular zone of the hippocampus is stimulated in mice by physical exercise and by an enriched environment, changes that other studies have demonstrated to improve memory. By contrast, chronic stress reduces neurogenesis in the subgranular zone and aging reduces neurogenesis in both the subgranular and subventricular zones.

Studies in rodents show that adult neurogenesis in the subgranular zone accompanies the hippocampus-dependent learning of particular tasks, such as the ability of mice to learn a water maze. The production of new neurons in the dentate nucleus of the hippocampus contributes to spatial learning, memory, and other cognitive tasks in rodents, but the significance of neurogenesis to brain function in learning and memory generally, particularly in humans, is still controversial.

Emotion and Memory

Limbic System

Emotions influence memory, in some cases by strengthening, and in others by hindering, memory formation. The amygdala is involved in the improvement of memory when the memory has an emotional content. This is illustrated by the observation that patients who have damage to both amygdaloid nuclei lose the usual enhancement of memory by emotion.

Although strong emotions enhance memory encoding within the amygdala, stress can impair memory consolidation by the hippocampus and the cognitive functions and working memory performed by the prefrontal cortex (discussed next). As a result, stress can promote the storage of emotionally strong memories but hinder the retrieval of those memories and working memory. In this regard, researchers have demonstrated that people with *post-traumatic stress disorder* often have hippocampal atrophy. The mechanisms by which stress affects the brain are not fully understood, but it is known that during stress there is increased secretion of "stress hormones" (primarily cortisol from the adrenal cortex; chapter 11, section 11.4), and that the hippocampus and amygdala are rich in receptors for these hormones. The hippocampus and amygdala are thus targets of these hormones, and corticosteroids (including cortisol) have been shown to suppress neurogenesis in the hippocampus.

Prefrontal Cortex

As previously mentioned, the prefrontal cortex is involved in higher cognitive functions, including memory, planning, and judgment. It is also required for normal motivation and interpersonal skills and social behavior. In order to perform such varied tasks, the prefrontal cortex has numerous connections with other brain regions, and different regions of the prefrontal cortex are specialized along different lines. As revealed by patients with damage to these areas, the functions of the lateral prefrontal area can be distinguished from the functions of the orbitofacial prefrontal area.

The *orbitofrontal area* of the prefrontal cortex (fig. 8.18) seems to confer the ability to consciously experience pleasure and reward. It receives input from all of the sensory modalities—taste, smell, vision, sound, touch, and others—and has connections with many regions of the limbic system. As previously discussed, the limbic system includes several brain areas that are involved in emotion and motivation. Connections between the orbitofrontal cortex, the amygdala, and the cingulate gyrus (fig. 8.18) are notably important for the emotional reward of goal-directed behavior.

People with damage to the *lateral prefrontal area* of the precentral cortex show a lack of motivation and sexual desire, and they have deficient cognitive functions. People with damage to the orbitofrontal area of the prefrontal cortex (fig. 8.18), in contrast, have their memory and cognitive functions largely spared but experience severe impulsive behavior, verging on the sociopathic.

One famous example of damage to the orbitofrontal area of the prefrontal cortex was the first case to be described, in 1848. A 25-year-old railroad foreman named Phineas P. Gage was tamping blasting powder into a hole in a rock with a metal rod when the blasting powder exploded. The rod—3 feet 7 inches long and 1¼ inches thick—was driven through his left eye and brain and emerged through the top of his skull.

After a few minutes of convulsions, Gage got up, rode a horse three-quarters of a mile into town, and walked up a long flight of stairs to see a doctor. He recovered well, with no noticeable sensory or motor deficits. His associates, however, noted striking personality changes. Before the accident, Gage was a responsible, capable, and financially prudent man. Afterward, he appeared to have lost his social inhibitions; for example, he engaged in gross profanity, which he did not do before his accident. He was impulsive, being tossed about by seemingly blind whims. He was eventually fired from his job, and his old friends remarked that he was "no longer Gage."

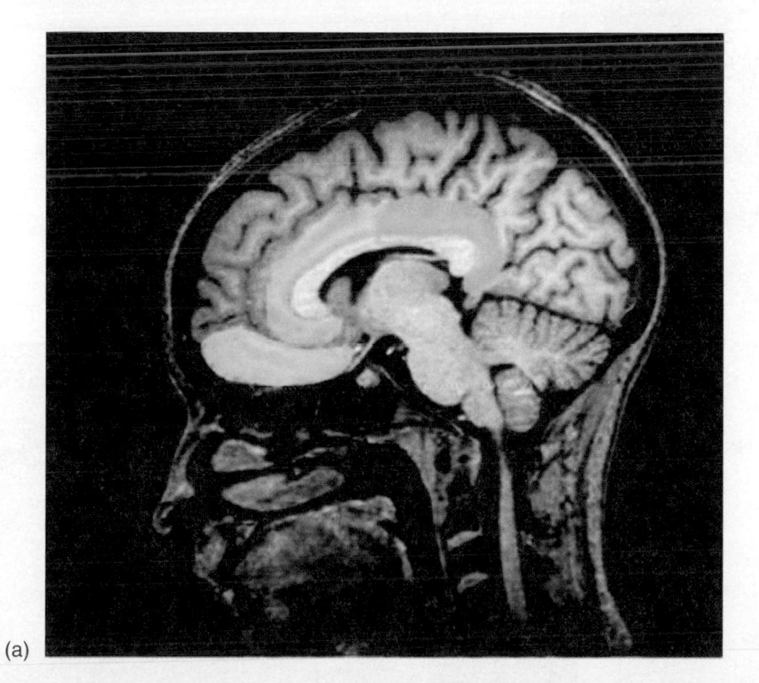

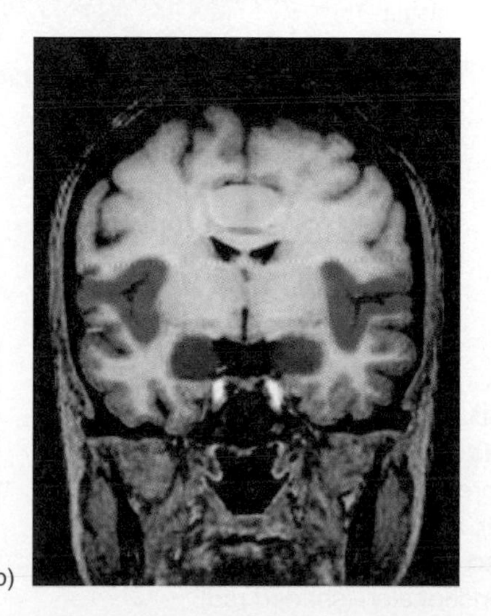

Figure 8.18 Some brain areas involved in emotion. (*a*) The orbitofrontal area of the prefrontal cortex is shown in yellow, and the cingulate gyrus of the limbic system is shown in blue-green (anterior portion) and green (posterior portion). (*b*) The insula of the cortex is shown in purple, the anterior cingulate gyrus of the limbic system in blue-green, and the amygdala in red. Reprinted Figure 2 (1st and 3rd panels) with permission from RJ Dolan, SCIENCE 298:1191–1194. Copyright 2002 AAAS.

✔ CHECKPOINT

3a. Describe the locations of the sensory and motor areas of the cerebral cortex and explain how these areas are organized.

3b. Describe the locations and functions of the basal nuclei. Of what structures are the basal nuclei composed?

4a. Identify the structures of the limbic system and explain the functional significance of this system.

4b. Explain the difference in function of the right and left cerebral hemispheres.

5. Describe the functions of the brain areas involved in speech and language comprehension.

6. Describe the brain areas implicated in memory, and their possible functions.

8.3 DIENCEPHALON

The diencephalon is the part of the forebrain that contains the epithalamus, thalamus, hypothalamus, and part of the pituitary gland. The hypothalamus performs numerous vital functions, most of which relate directly or indirectly to the regulation of visceral activities by way of other brain regions and the autonomic nervous system.

LEARNING OUTCOMES

After studying this section, you should be able to:

7. Describe the locations and functions of the thalamus and hypothalamus.

The **diencephalon,** together with the telencephalon (cerebrum) previously discussed, constitutes the forebrain and is almost completely surrounded by the cerebral hemispheres. The *third ventricle* is a narrow midline cavity within the diencephalon.

Thalamus and Epithalamus

The **thalamus** composes about four-fifths of the diencephalon and forms most of the walls of the third ventricle (fig. 8.19). It consists of paired masses of gray matter, each positioned immediately below the lateral ventricle of its respective cerebral hemisphere. The thalamus acts primarily as a relay

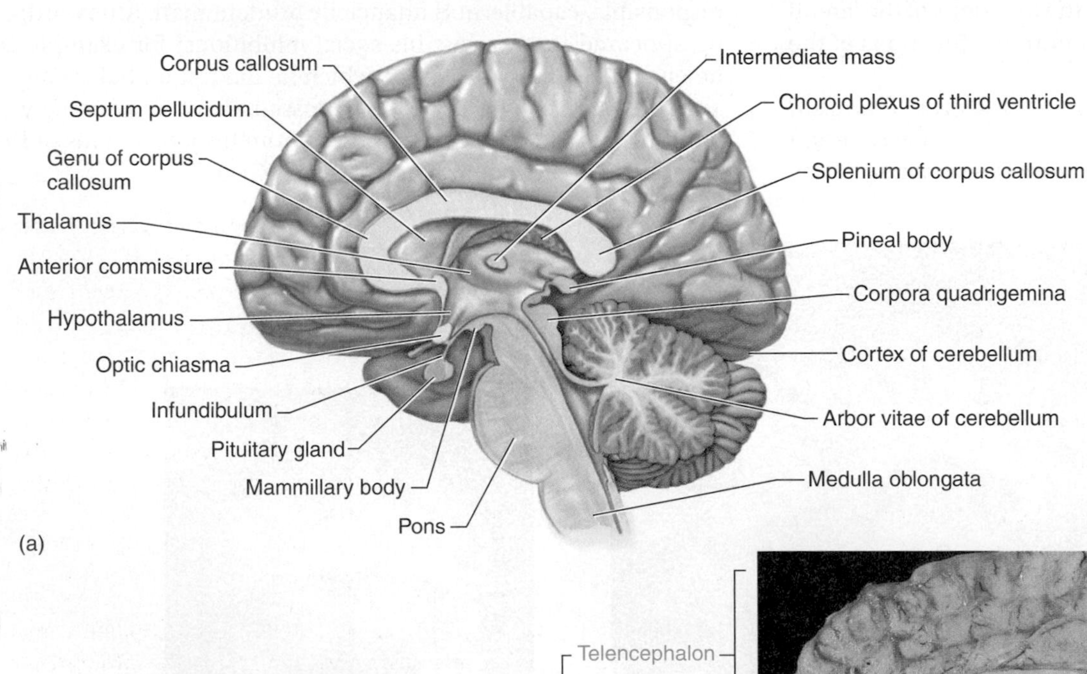

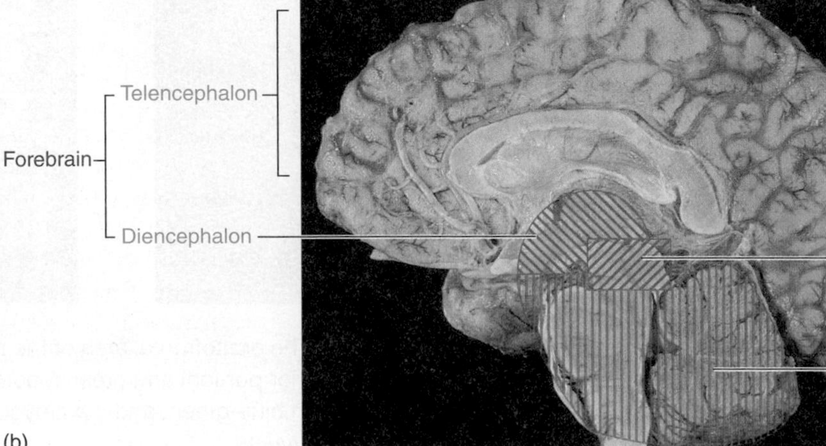

Figure 8.19 The adult brain seen in midsagittal section. The structures are labeled in the diagram shown in (*a*), and the brain regions are indicated in the photograph in (*b*). The diencephalon (shaded red) and telencephalon (unshaded area) make up the forebrain; the midbrain is shaded blue and the hindbrain is shaded green. AP|R

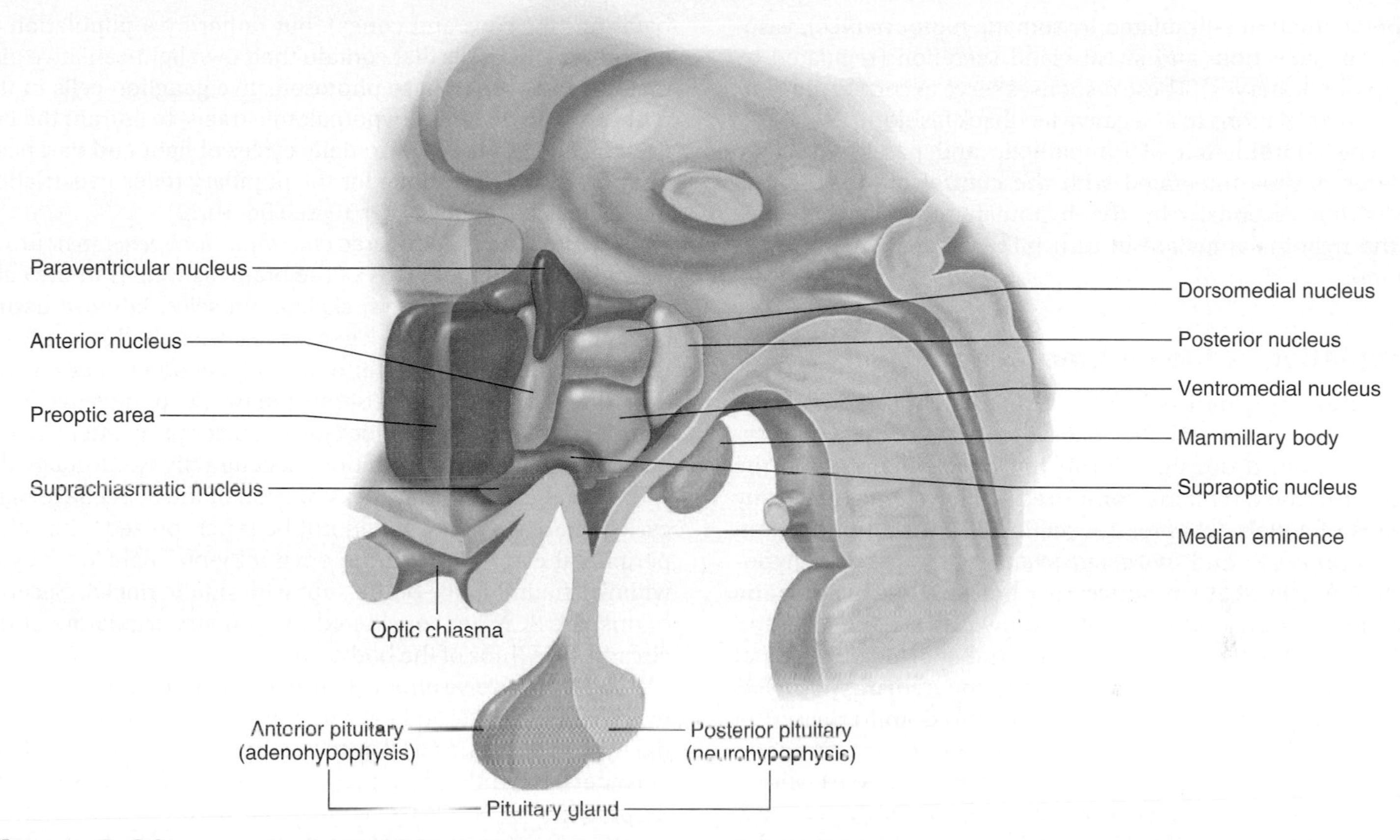

Figure 8.20 A diagram of some of the nuclei within the hypothalamus. The hypothalamic nuclei, composed of neuron cell bodies, have different functions. AP|R

center through which all sensory information (except smell) passes on the way to the cerebrum. For example, the *lateral geniculate nuclei* relay visual information, and the *medial geniculate nuclei* relay auditory information, from the thalamus to the occipital and temporal lobes, respectively, of the cerebral cortex. The *intralaminar nuclei* of the thalamus are activated by many different sensory modalities and, in turn, project to many areas of the cerebral cortex. This is part of the system that promotes a state of alertness and causes arousal from sleep in response to any sufficiently strong sensory stimulus.

The **epithalamus** is the dorsal segment of the diencephalon, containing a *choroid plexus* over the third ventricle where cerebrospinal fluid is formed. The epithalamus also contains the *pineal gland (epiphysis),* which secretes the hormone *melatonin* that helps regulate circadian (daily) rhythms (chapter 11, section 11.6).

Hypothalamus and Pituitary Gland

The **hypothalamus** is the most inferior portion of the diencephalon. Located below the thalamus, it forms the floor and part of the lateral walls of the third ventricle. This small but extremely important brain region contains neural centers for hunger and thirst; the regulation of body temperature; and hormone secretion from the pituitary gland (fig. 8.20). In addition, centers in the hypothalamus contribute to the regulation of sleep, wakefulness, sexual arousal and performance, and such emotions as anger, fear, pain, and pleasure. Acting through its connections with the medulla oblongata of the brain stem, the hypothalamus helps to evoke visceral responses to various emotional states. In its regulation of emotion, the hypothalamus works together with the limbic system.

Regulation of the Autonomic System

Experimental stimulation of different areas of the hypothalamus can evoke the autonomic responses characteristic of aggression, sexual behavior, hunger, or satiety. Chronic stimulation of the lateral hypothalamus, for example, can make an animal eat and become obese, whereas stimulation of the medial hypothalamus inhibits eating. Other areas contain osmoreceptors that stimulate thirst and the release of antidiuretic hormone (ADH) from the posterior pituitary.

The hypothalamus is also where the body's "thermostat" is located. Experimental cooling of the preoptic-anterior hypothalamus causes shivering (a somatic motor response) and nonshivering thermogenesis (a sympathetic motor response). Experimental heating of this hypothalamic area results in hyperventilation (stimulated by somatic motor nerves), vasodilation, salivation, and sweat-gland secretion (regulated by sympathetic nerves). These responses serve to correct the temperature deviations in a negative feedback fashion.

The coordination of sympathetic and parasympathetic reflexes is thus integrated with the control of somatic and endocrine responses by the hypothalamus. The activities of the hypothalamus are in turn influenced by higher brain centers.

Regulation of the Pituitary Gland

The **pituitary gland** is located immediately inferior to the hypothalamus. Indeed, the posterior pituitary derives embryonically from a downgrowth of the diencephalon, and the entire pituitary remains connected to the diencephalon by means of a stalk (chapter 11, section 11.3). Neurons within the *supraoptic* and *paraventricular nuclei* of the hypothalamus (fig. 8.20) produce two hormones—**antidiuretic hormone (ADH),** which is also known as *vasopressin,* and **oxytocin.** These two hormones are transported in axons of the *hypothalamo-hypophyseal tract* to the **neurohypophysis** (posterior pituitary), where they are stored and released in response to hypothalamic stimulation. Oxytocin stimulates contractions of the uterus during labor, and ADH stimulates the kidneys to reabsorb water and thus to excrete a smaller volume of urine. Neurons in the hypothalamus also produce hormones known as **releasing hormones** and **inhibiting hormones** that are transported by the blood to the **adenohypophysis** (anterior pituitary). These hypothalamic releasing and inhibiting hormones regulate the secretions of the anterior pituitary and, by this means, regulate the secretions of other endocrine glands (chapter 11, section 11.3).

Regulation of Circadian Rhythms

Within the anterior hypothalamus (fig. 8.20) are bilaterally located **suprachiasmatic nuclei (SCN).** These nuclei contain about 20,000 neurons that function as "clock cells," with electrical activity that oscillates automatically in a pattern that repeats about every twenty-four hours. The SCN is believed to be the major brain region involved in regulating the body's **circadian rhythms** (from the Latin *circa* = about; *dia* = day). However, for these neuron clocks to function properly, their activities must be entrained (synchronized) to each other and to the day/night cycles.

Nonmammalian vertebrates—fish, amphibians, reptiles, and birds—have photosensitive cells in their brains that can detect light passing through their skulls. In mammals, however, the daily cycles of light and darkness influence the SCN by way of tracts from the retina (the neural layer of the eyes) to the hypothalamus (see chapter 11, fig. 11.33). These *retinohypothalamic tracts* are activated not by the photoreceptors involved in vision (the rods and cones), but rather by a population of retinal ganglion cells that contain their own light-sensitive pigment, *melanopsin.* These photosensitive ganglion cells in the retina act, via the retinohypothalamic tracts, to entrain the circadian clocks of the SCN to daily cycles of light and darkness. They are also responsibile for the pupillary reflex constriction in response to light (chapter 10; see fig. 10.28).

Scientists have discovered *circadian clock genes* in neurons of the SCN and other areas of the brain, as well as in the cells of the heart, liver, kidneys, skeletal muscles, adipose tissue, and other organs. The clock genes are transcribed into mRNA, which are then translated into protein like other genes. However, there appears to be complex networks of negative feedback loops that suppress clock gene transcription after a delay, resulting in circadian oscillations of gene activity. Although the "peripheral clocks" (the clocks outside of the SCN) have daily cycles of activity, they would not be synchronized with other peripheral clocks or with the environmental light/dark cycle without the influence of the suprachiasmatic nuclei. Because of this, the SCN are considered the primary regulators of the circadian rhythms of the body.

The SCN receive photic (light) information from the retinohypothalamic tracts and have neural outputs to other nuclei of the hypothalamus, as well as to the thalamus, arcuate nucleus, amygdala, and other brain regions. By means of these neural outputs, the SCN influence circadian rhythms of body temperature, feeding, locomotor activity (movements), the autonomic nervous system, and the secretions of endocrine glands. Through autonomic nerves, the SCN can regulate circadian rhythms in the liver and other visceral organs. By indirectly influencing the secretions of the anterior pituitary, the SCN entrains the adrenal glands to produce circadian rhythms in the secretion of cortisol. The secretion of *melatonin* from the *pineal gland* is highest at night because of regulation by the SCN via sympathetic nerves (see fig. 11.33). Melatonin is a major regulator of circadian rhythms, as discussed in chapter 11, section 11.6. For example, the presence of melatonin receptors in the pancreatic islets suggests that melatonin may influence the secretion of insulin, which likewise follows a circadian rhythm.

 CHECKPOINT

7a. List the functions of the hypothalamus and indicate the other brain regions that cooperate with the hypothalamus in the performance of these functions.

7b. Explain the structural and functional relationships between the hypothalamus and the pituitary gland.

8.4 MIDBRAIN AND HINDBRAIN

The midbrain and hindbrain contain many relay centers for sensory and motor pathways, and are particularly important in the brain's control of skeletal movements. The medulla oblongata contains centers for the control of breathing and cardiovascular function.

LEARNING OUTCOMES

After studying this section, you should be able to:

8. Identify the structures and functions of the midbrain and hindbrain.
9. Describe the structure and function of the reticular activating system.

Midbrain

The *mesencephalon,* or **midbrain,** is located between the diencephalon and the pons. The **corpora quadrigemina** are four rounded elevations on the dorsal surface of the midbrain (see fig. 8.19). The two upper mounds, the *superior colliculi,* are involved in visual reflexes; the *inferior colliculi,* immediately below, are relay centers for auditory information.

The midbrain also contains the cerebral peduncles, red nucleus, substantia nigra, and other nuclei. The **cerebral peduncles** are a pair of structures composed of ascending and descending fiber tracts. The **red nucleus,** an area of gray matter deep in the midbrain, maintains connections with the cerebrum and cerebellum and is involved in motor coordination.

The midbrain has two systems of dopaminergic (dopamine-releasing) neurons that project to other areas of the brain (chapter 7, section 7.5). The *nigrostriatal system* projects from the **substantia nigra** to the corpus striatum of the basal nuclei; this system is required for motor coordination, and it is the degeneration of these fibers that produces Parkinson's disease.

Other dopaminergic neurons in the **ventral tegmental area (VTA)** of the midbrain, adjacent to the substantia nigra, are part of the *mesolimbic system* that projects dopaminergic input to the limbic system of the forebrain (fig. 8.21). This system is involved in behavioral reward (reinforcing goal-directed behavior), and has been implicated in drug addiction and psychiatric

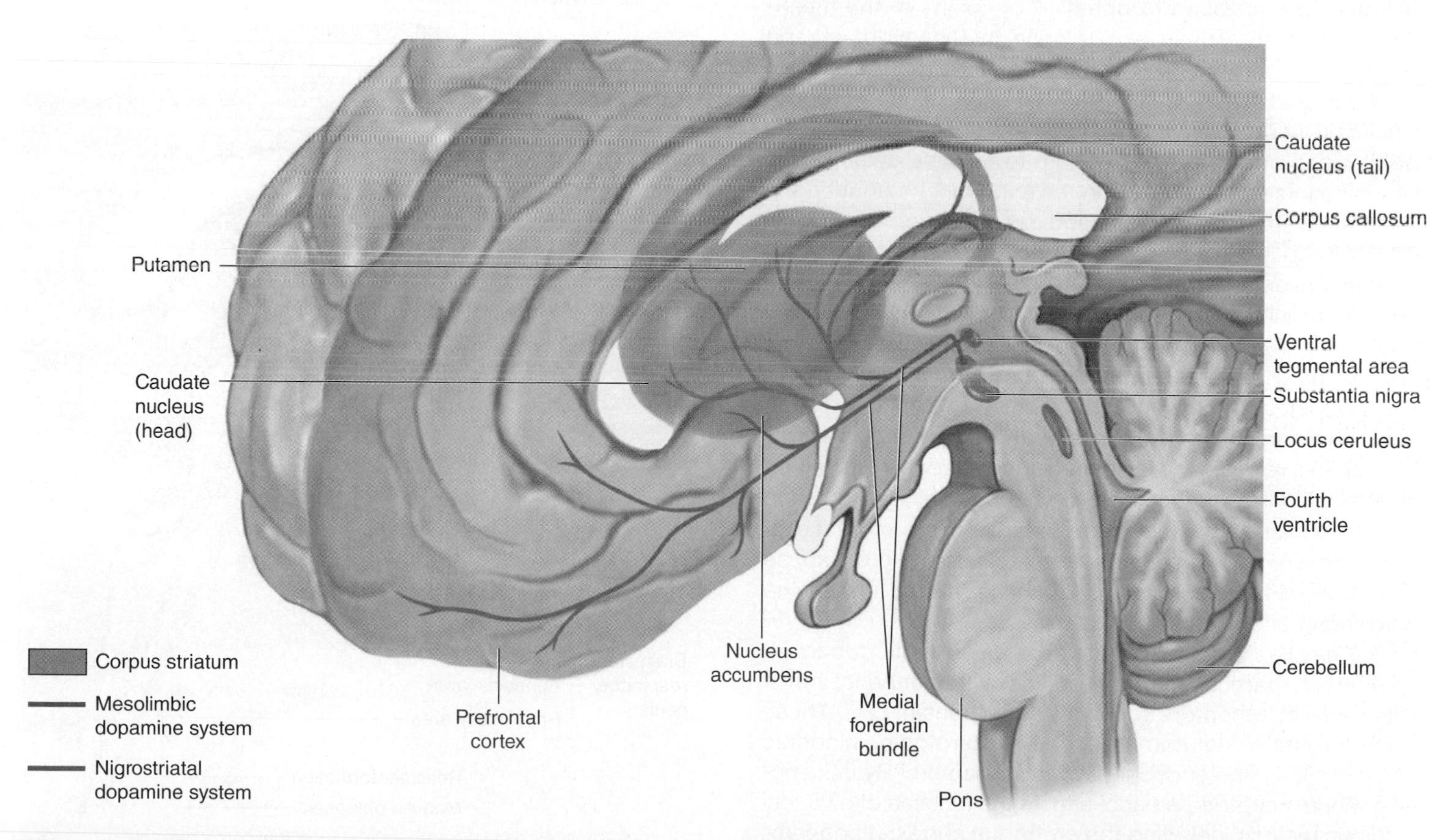

Figure 8.21 Dopaminergic pathways in the brain. Axons that use dopamine as a neurotransmitter (that are dopaminergic) leave the substantia nigra of the midbrain and synapse in the corpus striatum. This is the nigrostriatal system, used for motor control. Dopaminergic axons from the ventral tegmental area of the midbrain to the nucleus accumbens and prefrontal cortex constitute the mesolimbic system, which functions in emotional reward.

disturbances. Thus, the usual rewards for research animals in behavioral tasks become ineffective when their dopamine system is experimentally blocked. Abused drugs promote the release of dopamine in the **nucleus accumbens** in the forebrain (fig. 8.21).

The immediately rewarding effects of addictive drugs appear to be mediated by dopamine released in the nucleus accumbens, and this reward reinforces drug-seeking behavior. The nucleus accumbens receives information regarding emotions from other limbic system structures (amygdala, hippocampus, and frontal cortex) and has output to the corpus striatum (caudate nucleus and globus pallidus). This affords it an ability to relate emotions to motivated actions.

Stopping the use of an addictive drug (such as nicotine) can produce withdrawal symptoms, which cause anxiety and stress. To avoid this, the person may *relapse*—that is, resume the use of the drug despite the desire to quit and the knowledge of its negative consequences. For example, despite knowing that smoking causes 1 in 5 deaths per year in the United States (over 5 million deaths worldwide per year), each year only 3% of smokers who attempt to quit unaided are successful. Evidence suggests that relapse may result from some failure of the glutamate-releasing axons that project from the prefrontal cortex to the nucleus accumbens and other structures of the limbic system to exert control over the drug-seeking behavior.

CLINICAL APPLICATION

The positive reinforcement elicited by **abused drugs** involves the release of dopamine by axons of the mesolimbic system. These axons arise in the midbrain and terminate in the nucleus accumbens of the forebrain, deep in the frontal lobe. *Nicotine* from tobacco stimulates dopaminergic neurons in the midbrain by means of nicotinic ACh receptors. Chronic exposure to nicotine desensitizes the nicotinic ACh receptors in the midbrain, contributing to nicotine tolerance and increased dependence. The opioids (*heroin* and *morphine*) stimulate opioid receptors, and the *cannabinoids* (from marijuana) stimulate endocannabinoid receptors in the midbrain. This leads to reduced activity of GABA-releasing inhibitory neurons that synapse on the dopaminergic neurons in the ventral tegmental area. Benzodiazepines (*Valium* and *zolpidem*) may similarly reduce the inhibition of these dopaminergic neurons, increasing dopamine release by the mesolimbic dopamine system. *Cocaine* and *amphetamine* promote dopamine stimulation in the nucleus accumbens by inhibiting the reuptake of dopamine into presynaptic axons. Ironically, drug abuse can desensitize neurons to dopamine and so lessen the rewarding effects of dopamine release.

Ethanol (alcohol) stimulates the mesolimbic dopamine pathways, particularly in the nucleus accumbens, but it also affects receptors for other neurotransmitters. These include NMDA (glutamate), GABA, serotonin, nicotinic ACh, opioid, and endocannabinoid receptors. By influencing these receptors, ethanol affects the function of a variety of brain regions including the prefrontal cortex, hippocampus, amygdala, and other structures of the limbic system. Some changes in chronic alcohol abuse are permanent, perhaps because of epigenetic effects (chapter 3) that have recently been demonstrated.

Hindbrain

The *rhombencephalon,* or **hindbrain,** is composed of two regions: the metencephalon and the myelencephalon. These regions will be discussed separately.

Metencephalon

The *metencephalon* is composed of the pons and the cerebellum. The **pons** can be seen as a rounded bulge on the underside of the brain, between the midbrain and the medulla oblongata (fig. 8.22). Surface fibers in the pons connect to the cerebellum,

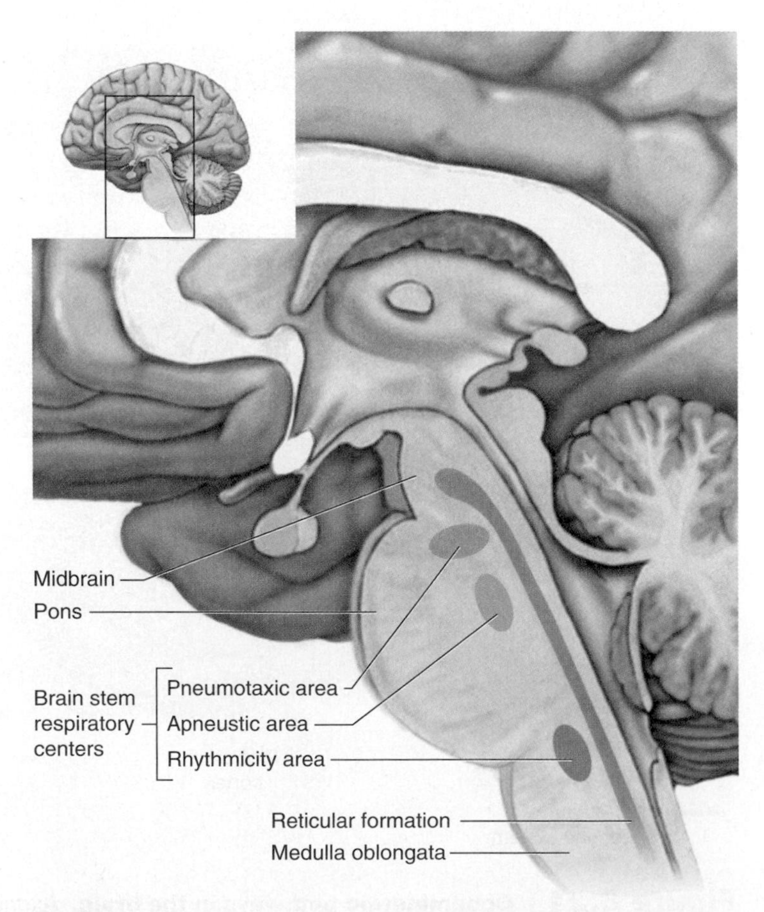

Figure 8.22 Respiratory control centers in the brain stem. These are nuclei within the pons and medulla oblongata that control the motor nerves required for breathing. The location of the reticular formation is also shown. AP|R

and deeper fibers are part of motor and sensory tracts that pass from the medulla oblongata, through the pons, and on to the midbrain. Within the pons are several nuclei associated with specific cranial nerves—the trigeminal (V), abducens (VI), facial (VII), and vestibulocochlear (VIII). Other nuclei of the pons cooperate with nuclei in the medulla oblongata to regulate breathing. The two respiratory control centers in the pons are known as the *apneustic* and the *pneumotaxic centers.* Damage to the ventral pons can produce a rare condition called *locked-in syndrome,* characterized by paralysis of almost all voluntary muscles so that communication by the aware, awake person is possible only by eye blinks.

The **cerebellum,** containing about 50 billion neurons, is the second largest structure of the brain. Like the cerebrum, it contains outer gray and inner white matter. Fibers from the cerebellum pass through the red nucleus to the thalamus, and then to the motor areas of the cerebral cortex. Other fiber tracts connect the cerebellum with the pons, medulla oblongata, and spinal cord. The cerebellum receives input from *proprioceptors* (joint, tendon, and muscle receptors) and, working together with the basal nuclei and motor areas of the cerebral cortex, participates in the coordination of movement.

The cerebellum is needed for motor learning and for coordinating the movement of different joints during a movement. It is also required for the proper timing and force required for limb movements. The cerebellum, for example, is needed to touch your nose with your finger, bring a fork of food to your mouth, or find keys by touch in your pocket or purse.

Interestingly, these functions must operate through specific cerebellar neurons known as *Purkinje cells,* which provide the only output from the cerebellum to other brain regions. Further, Purkinje cells produce only inhibitory effects on the motor areas of the cerebral cortex. Acting through this inhibition, the cerebellum aids in the coordination of complex motor skills and participates in motor learning.

Current research suggests that the cerebellum may have varied and subtle functions beyond motor coordination. Different investigations have implicated the cerebellum in the acquisition of sensory data, memory, emotion, and other higher functions. The cerebellum may also have roles in schizophrenia and autism. How these possible cerebellar functions are achieved and how they relate to the control of motor coordination are presently incompletely understood and controversial.

CLINICAL APPLICATION

Damage to the cerebellum produces **ataxia**—lack of coordination resulting from errors in the speed, force, and direction of movement. The movements and speech of people afflicted with ataxia may resemble those of someone who is intoxicated (indeed, alcohol has been shown to affect cerebellum function). This condition is also characterized by intention tremor, which differs from the resting tremor of Parkinson's disease in that it occurs only when intentional movements are made. People with cerebellar damage may reach for an object and miss it by placing their hand too far to the left or right; they will then attempt to compensate by moving their hand in the opposite direction. This back-and-forth movement can result in oscillations of the limb.

Myelencephalon

The *myelencephalon* is composed of only one structure, the **medulla oblongata,** often simply called the *medulla.* About 3 cm (1 in.) long, the medulla is continuous with the pons superiorly and the spinal cord inferiorly. All of the descending and ascending fiber tracts that provide communication between the spinal cord and the brain must pass through the medulla. Many of these fiber tracts cross to the contralateral side in elevated triangular structures in the medulla called the **pyramids.** Thus, the left side of the brain receives sensory information from the right side of the body and vice versa. Similarly, because of the decussation of fibers, the right side of the brain controls motor activity in the left side of the body and vice versa.

Many important nuclei are contained within the medulla. Several nuclei are involved in motor control, giving rise to axons within cranial nerves VIII, IX, X, XI, and XII. The *vagus nuclei* (there is one on each lateral side of the medulla), for example, give rise to the highly important vagus (X) nerves. Other nuclei relay sensory information to the thalamus and then to the cerebral cortex.

The medulla contains groupings of neurons required for the regulation of breathing and of cardiovascular responses; hence, they are known as the *vital centers.* The **vasomotor center** controls the autonomic innervation of blood vessels; the **cardiac control center,** closely associated with the vasomotor center, regulates the autonomic nerve control of the heart; and the **respiratory center** of the medulla acts together with centers in the pons to control breathing.

Reticular Activating System

In order to fall asleep, we must be able to "tune out" sensory stimulation that ascends to the cerebral cortex. Conversely, we awake rather quickly from sleep when the cerebral cortex is alerted to incoming sensory information. These abilities, and the normal cycles of sleep and wakefulness that result, depend upon the activation and inhibition of neural pathways that go from the pons through the midbrain **reticular formation,** an interconnected group of neurons (from the Latin *rete* = net). This constitutes an ascending arousal system known as the **reticular activating system (RAS).**

The RAS includes groups of cholinergic neurons (neurons that release ACh) in the brain stem that project to the thalamus; these neurons enhance the transmission of sensory information from the thalamus to the cerebral cortex. Other groups of RAS neurons located in the hypothalamus and basal forebrain

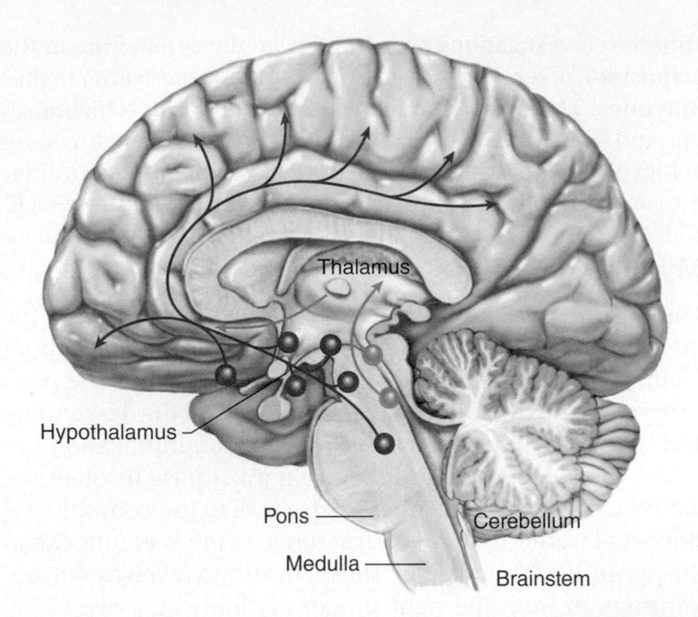

Figure 8.23 The reticular activating system (RAS). The groups of neurons shown in orange project to the thalamus, where they enhance the arousal of the cerebral cortex to sensory information relayed from the thalamus. The groups of neurons shown in red project to various locations in the cerebral cortex and more directly arouse the cerebral cortex to ascending sensory information. Activity of the RAS promotes wakefulness, and inhibition of the RAS promotes sleep.

release monoamine neurotransmitters (dopamine, norepinephrine, histamine, and serotonin) and project to various locations in the cerebral cortex (fig. 8.23). These arousal neural pathways of the RAS are inhibited by another group of neurons located in the *ventrolateral preoptic nucleus (VLPO)* of the hypothalamus, which release the inhibitory neurotransmitter GABA. The activity of VLPO and other GABA-releasing neurons is increased with the depth of sleep, and these neurons are believed to both cause and stabilize sleep. The inhibitory neurons of the VLPO and the arousal neurons that release monoamine neurotransmitters are believed to mutually inhibit each other, creating a switch that controls falling asleep and waking up.

There are other neurons of the RAS, located in the *lateral hypothalamic area* (*LHA*), that release polypeptides as neurotransmitters that promote arousal. Some of these neurons have been shown to be involved in **narcolepsy,** a neurological disorder (affecting about 1 in 2,000 people) in which the person will fall asleep inappropriately during the day despite having adequate amounts of sleep. Near the end of the twentieth century, scientists demonstrated that people with narcolepsy have a loss of LHA neurons that release a particular polypeptide neurotransmitter that promotes wakefulness. This neurotransmitter was discovered by two research groups who proposed two names for it: *orexin* and *hypocretin-1.* Narcolepsy has a genetic basis, which may promote autoimmune destruction of the approximately 70,000 orexin neurons in the lateral hypothalamus. Through extensive connections between the lateral hypothalamus and other hypothalamic areas, as well as with the limbic system and other cortical areas, orexin neurons function to promote wakefulness, the craving for food (and for abused drugs), and physical activity.

CLINICAL APPLICATION

Many drugs act on the RAS to promote either sleep or wakefulness. Amphetamines, for example, enhance dopamine action by inhibiting the dopamine reuptake transporter, thereby inhibiting the ability of presynaptic axons to remove dopamine from the synaptic cleft. This increases the effectiveness of the monoamine-releasing neurons of the RAS, enhancing arousal. The antihistamine *Benadryl,* which can cross the blood-brain barrier, causes drowsiness by inhibiting histamine-releasing neurons of the RAS. (The antihistamines that don't cause drowsiness, such as *Claritin,* cannot cross the blood-brain barrier.) Drowsiness caused by the benzodiazepines (such as *Valium*), barbiturates, alcohol, and most anesthetic gases is due to the ability of these agents to enhance the activity of GABA receptors. Increased ability of GABA to inhibit the RAS then reduces arousal and promotes sleepiness.

 **CHECKPOINT**

8a. List the structures of the midbrain and describe their functions.

8b. Describe the functions of the medulla oblongata and pons.

9. Identify the parts of the brain involved in the reticular activating system. What is the role of this system? How is it inhibited?

8.5 SPINAL CORD TRACTS

Sensory information from most of the body is relayed to the brain by means of ascending tracts of fibers that conduct impulses up the spinal cord. When the brain directs motor activities, these directions are in the form of nerve impulses that travel down the spinal cord in descending tracts of fibers.

LEARNING OUTCOMES

After studying this section, you should be able to:

10. Describe the sensory and motor pathways to and from the cerebral hemispheres.

11. Describe the structure and function of the pyramidal and extrapyramidal motor tracts.

The spinal cord extends from the level of the foramen magnum of the skull to the first lumbar vertebra. Unlike the brain, in which the gray matter forms a cortex over white matter, the gray matter of the spinal cord is located centrally, surrounded by white matter. The central gray matter of the spinal cord is arranged in the form of an H, with two *dorsal horns* and two *ventral horns* (also called posterior and anterior horns, respectively). The white matter of the spinal cord is composed of ascending and descending fiber tracts. These are arranged into six columns of white matter called *funiculi.*

The fiber tracts within the white matter of the spinal cord are named to indicate whether they are ascending (sensory) or descending (motor) tracts. The names of the ascending tracts usually start with the prefix *spino-* and end with the name of the brain region where the spinal cord fibers first synapse. The anterior spinothalamic tract, for example, carries impulses conveying the sense of touch and pressure, and synapses in the thalamus. From there it is relayed to the cerebral cortex. The names of descending motor tracts, conversely, begin with a prefix denoting the brain region that gives rise to the fibers and end with the suffix *-spinal.* The lateral corticospinal tracts, for example, begin in the cerebral cortex and descend the spinal cord.

Ascending Tracts

The ascending fiber tracts convey sensory information from cutaneous receptors, proprioceptors (muscle and joint receptors), and visceral receptors (table 8.4). Most of the sensory information that originates in the right side of the body crosses over to eventually reach the region on the left side of the brain that analyzes this information. Similarly, the information arising in the left side of the body is ultimately analyzed by the right side of the brain. For some sensory modalities, this decussation occurs in the medulla oblongata (fig. 8.24); for others, it occurs in the spinal cord. These neural pathways are discussed in more detail in chapter 10, section 10.2.

Descending Tracts

The descending fiber tracts that originate in the brain consist of two major groups: the **corticospinal,** or **pyramidal, tracts,** and the extrapyramidal tracts (table 8.5). The pyramidal tracts descend directly, without synaptic interruption, from the cerebral cortex to the spinal cord. The cell bodies that contribute fibers to these pyramidal tracts are located primarily in the *precentral gyrus,* forming the **primary motor cortex.** However, the **supplementary motor complex,** located in the superior frontal gyrus just anterior to the "leg" region of the primary motor cortex, (see fig. 8.7), contributes about 10% of the fibers in the corticospinal tracts.

From 80% to 90% of the corticospinal fibers decussate in the pyramids of the medulla oblongata (hence the name "pyramidal tracts") and descend as the *lateral corticospinal tracts.* The remaining uncrossed fibers form the *anterior corticospinal tracts,* which decussate in the spinal cord. Because of the crossing over of fibers, the right cerebral hemisphere controls

Clinical Investigation CLUES

Frank was paralyzed on the right side of his body.

- Which CNS tract, originating from which cerebral hemisphere, was likely damaged to result in Frank's paralysis?

Table 8.4 | Principal Ascending Tracts of the Spinal Cord

Tract	Origin	Termination	Function
Anterior spinothalamic	Posterior horn on one side of cord but crosses to opposite side	Thalamus, then cerebral cortex	Conducts sensory impulses for crude touch and pressure
Lateral spinothalamic	Posterior horn on one side of cord but crosses to opposite side	Thalamus, then cerebral cortex	Conducts pain and temperature impulses that are interpreted within cerebral cortex
Fasciculus gracilis and fasciculus cuneatus	Peripheral afferent neurons; ascends on ipsilateral side of spinal cord but crosses over in medulla	Nucleus gracilis and nucleus cuneatus of medulla; eventually thalamus, then cerebral cortex	Conducts sensory impulses from skin, muscles, tendons, and joints, which are interpreted as sensations of fine touch, precise pressures, and body movements
Posterior spinocerebellar	Posterior horn; does not cross over	Cerebellum	Conducts sensory impulses from one side of body to same side of cerebellum; necessary for coordinated muscular contractions
Anterior spinocerebellar	Posterior horn; some fibers cross, others do not	Cerebellum	Conducts sensory impulses from both sides of body to cerebellum; necessary for coordinated muscular contractions

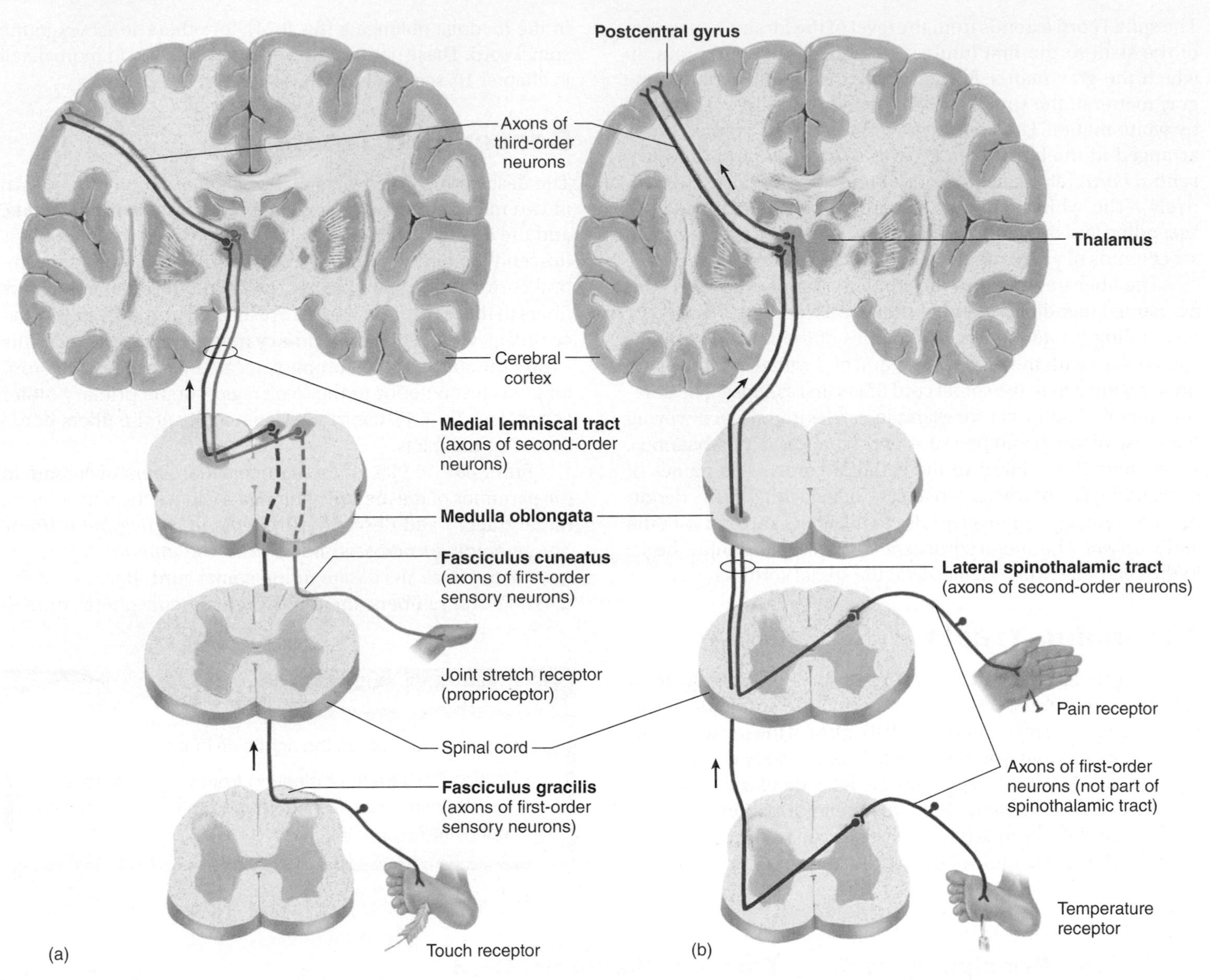

Figure 8.24 **Ascending tracts carrying sensory information.** This information is delivered by third-order neurons to the cerebral cortex. (*a*) Medial lemniscal tract; (*b*) lateral spinothalamic tract.

Table 8.5 | Descending Motor Tracts to Spinal Interneurons and Motor Neurons

Tract	Category	Origin	Crossed/Uncrossed
Lateral corticospinal	Pyramidal	Cerebral cortex	Crossed
Anterior corticospinal	Pyramidal	Cerebral cortex	Uncrossed
Rubrospinal	Extrapyramidal	Red nucleus (midbrain)	Crossed
Tectospinal	Extrapyramidal	Superior colliculus (midbrain)	Crossed
Vestibulospinal	Extrapyramidal	Vestibular nuclei (medulla oblongata)	Uncrossed
Reticulospinal	Extrapyramidal	Reticular formation (medulla and pons)	Crossed

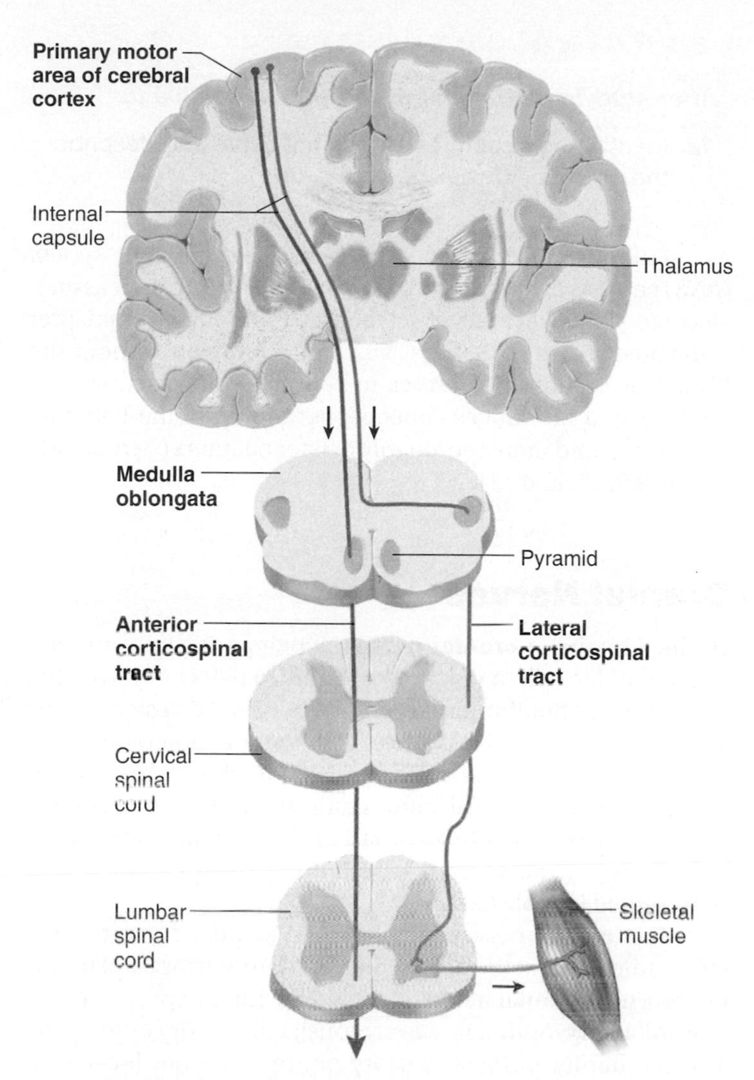

Figure 8.25 Descending corticospinal (pyramidal) motor tracts. These tracts contain axons that pass from the precentral gyrus of the cerebral cortex down the spinal cord to make synapses with spinal interneurons and lower motor neurons.

the musculature on the left side of the body (fig. 8.25), whereas the left hemisphere controls the right musculature. The corticospinal tracts are primarily concerned with the control of fine movements that require dexterity.

Because of the decussation of descending motor tracts, people who have damage to the right cerebral hemisphere (particularly of the parietal lobe) have motor deficits mostly in the left side of the body. However, patients with lesions in the left-hemisphere parietal lobe often have impaired skilled motor activity of both hands. These and other observations have led scientists to believe that the left hemisphere is specialized for skilled motor control of both hands. The left hemisphere appears to control the left hand indirectly, via projections to the right hemisphere through the corpus callosum. The right hemisphere is believed also to cross-talk with the left in the control of motor behavior, although its contributions are less well understood.

The remaining descending tracts are **extrapyramidal motor tracts.** These originate in the brain stem (table 8.5) and are largely controlled by the motor circuit structures of the corpus striatum—caudate nucleus, putamen, and globus pallidus (see figs. 8.11 and 8.12)—as well as by the substantia nigra and thalamus. This is why the symptoms of Parkinson's disease, produced by inadequate dopamine released by the nigrostriatal pathway (as previously discussed), are often referred to medically as "extrapyramidal symptoms." These symptoms demonstrate that the extrapyramidal system is needed for the initiation of body movements, maintenance of posture, control of the muscles of facial expression, and other functions.

The term *extrapyramidal* can be understood in terms of the following experiment: If the pyramidal tracts of an experimental animal are cut, electrical stimulation of the cerebral cortex, cerebellum, and basal nuclei can still produce movements. The descending fibers that produce these movements must, by definition, be extrapyramidal motor tracts. The regions of the cerebral cortex, basal nuclei, and cerebellum that participate in this motor control have numerous synaptic interconnections, and they can influence movement only indirectly by means of stimulation or inhibition of the nuclei that give rise to the extrapyramidal tracts. Notice that this motor control differs from that exerted by the neurons of the precentral gyrus, which send fibers directly down to the spinal cord in the pyramidal tracts.

The *reticulospinal tracts* are the major descending pathways of the extrapyramidal system. These tracts originate in the reticular formation of the brain stem, which receives either stimulatory or inhibitory input from the cerebrum and the cerebellum. There are no descending tracts from the cerebellum; the cerebellum can influence motor activity only indirectly by its effect on the vestibular nuclei, red nucleus, and basal nuclei (which send axons to the reticular formation). These nuclei, in

CLINICAL APPLICATION

The corticospinal tracts appear to be particularly important in voluntary movements that require complex interactions between sensory input and the motor cortex. Speech, for example, is impaired when the corticospinal tracts are damaged in the thoracic region of the spinal cord, whereas involuntary breathing continues. Damage to the pyramidal motor system can be detected clinically by the presence of **Babinski's reflex,** in which stimulation of the sole of the foot causes extension of the great toe upward and fanning of the other toes. (In normal adults such stimulation causes the plantar reflex, a downward flexion, or curling, of the toes.) Babinski's reflex is normally present in infants because neural control is not yet fully developed.

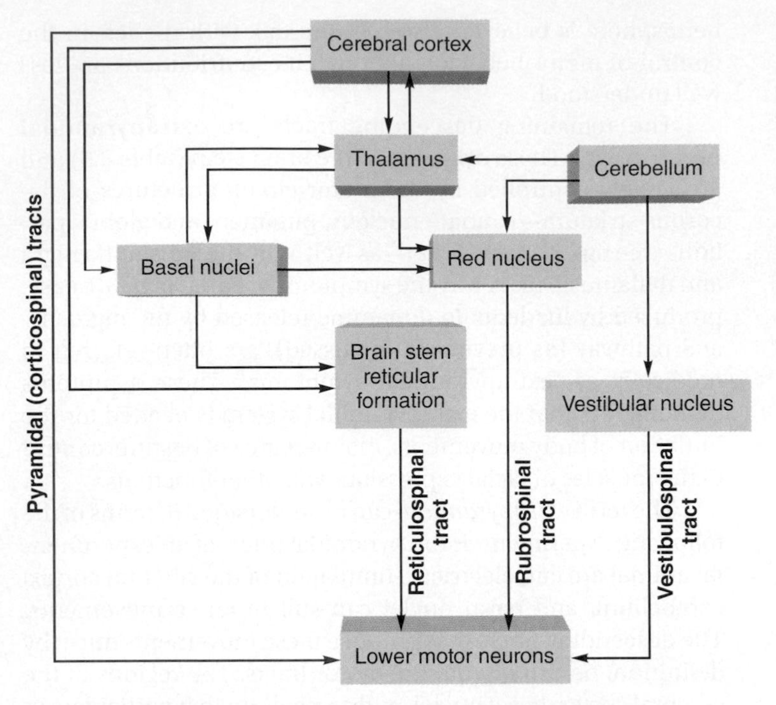

Figure 8.26 The higher motor neuron control of skeletal muscles. The pyramidal (corticospinal) tracts are shown in pink and the descending motor pathways from the brain stem that are controlled by the extrapyramidal system are shown in black.

turn, send axons down the spinal cord via the *vestibulospinal tracts, rubrospinal tracts,* and reticulospinal tracts, respectively (fig. 8.26). Neural control of skeletal muscle is explained in more detail in chapter 12.

CHECKPOINT

10. Explain why each cerebral hemisphere receives sensory input from and directs motor output to the contralateral side of the body.

11a. List the tracts of the pyramidal motor system and describe the function of the pyramidal system.

11b. List the tracts of the extrapyramidal system and explain how this system differs from the pyramidal motor system.

8.6 CRANIAL AND SPINAL NERVES

The central nervous system communicates with the body by means of nerves that exit the CNS from the brain (cranial nerves) and spinal cord (spinal nerves). These nerves, together with aggregations of cell bodies located outside the CNS, constitute the peripheral nervous system.

LEARNING OUTCOMES

After studying this section, you should be able to:

12. Identify the structures of a spinal nerve and describe the neural pathways of a reflex arc.

As mentioned in chapter 7, the *peripheral nervous system (PNS)* consists of nerves (collections of axons) and their associated ganglia (collections of cell bodies). Although this chapter is devoted to the CNS, the CNS cannot function without the PNS. This section thus serves to complete our discussion of the CNS and introduces concepts pertaining to the PNS that will be explored more thoroughly in later chapters (particularly chapters 9, 10, and 12).

Cranial Nerves

Of the 12 pairs of **cranial nerves,** 2 pairs arise from neuron cell bodies located in the forebrain and 10 pairs arise from the midbrain and hindbrain. The cranial nerves are designated by Roman numerals and by names. The Roman numerals refer to the order in which the nerves are positioned from the front of the brain to the back. The names indicate the structures innervated by these nerves (e.g., facial) or the principal function of the nerves (e.g., oculomotor). A summary of the cranial nerves is presented in table 8.6.

Most cranial nerves are classified as *mixed nerves.* This term indicates that the nerve contains both sensory and motor fibers. Those cranial nerves associated with the special senses (e.g., olfactory, optic), however, consist of sensory fibers only. The cell bodies of these sensory neurons are not located in the brain, but instead are found in ganglia near the sensory organ.

Spinal Nerves

There are 31 pairs of spinal nerves. These nerves are grouped into 8 cervical, 12 thoracic, 5 lumbar, 5 sacral, and 1 coccygeal according to the region of the vertebral column from which they arise (fig. 8.27).

Each spinal nerve is a mixed nerve composed of sensory and motor fibers. These fibers are packaged together in the nerve, but they separate near the attachment of the nerve to the spinal cord. This produces two "roots" of each nerve. The **dorsal root** is composed of sensory fibers, and the **ventral root** is composed of motor fibers (fig. 8.28). An enlargement of the dorsal root, the **dorsal root ganglion,** contains the cell bodies of the sensory neurons. The motor neuron shown in figure 8.28 is a somatic motor neuron that innervates skeletal muscles; its cell body is not located in a ganglion but instead is contained within the gray matter of the spinal cord. The cell bodies of some autonomic motor neurons (which innervate involuntary effectors), however, are located in ganglia outside

Table 8.6 | Summary of Cranial Nerves

Number and Name	Composition	Function
I Olfactory	Sensory	Olfaction
II Optic	Sensory	Vision
III Oculomotor	Somatic motor	Motor impulses to levator palpebrae superioris and extrinsic eye muscles, except superior oblique and lateral rectus; innervation to muscles that regulate amount of light entering eye and that focus the lens
	Sensory: proprioception	Proprioception from muscles innervated with motor fibers
IV Trochlear	Somatic motor	Motor impulses to superior oblique muscle of eyeball
	Sensory: proprioception	Proprioception from superior oblique muscle of eyeball
V Trigeminal		
Ophthalmic division	Sensory	Sensory impulses from cornea, skin of nose, forehead, and scalp
Maxillary division	Sensory	Sensory impulses from nasal mucosa, upper teeth and gums, palate, upper lip, and skin of cheek
Mandibular division	Sensory	Sensory impulses from temporal region, tongue, lower teeth and gums, and skin of chin and lower jaw
	Sensory: proprioception	Proprioception from muscles of mastication
	Somatic motor	Motor impulses to muscles of mastication and muscle that tenses the tympanum
VI Abducens	Somatic motor	Motor impulses to lateral rectus muscle of eyeball
	Sensory: proprioception	Proprioception from lateral rectus muscle of eyeball
VII Facial	Somatic motor	Motor impulses to muscles of facial expression and muscle that tenses the stapes
	Parasympathetic motor	Secretion of tears from lacrimal gland and salivation from sublingual and submandibular salivary glands
	Sensory	Sensory impulses from taste buds on anterior two-thirds of tongue; nasal and palatal sensation
	Sensory: proprioception	Proprioception from muscles of facial expression
VIII Vestibulocochlear	Sensory	Sensory impulses associated with equilibrium
		Sensory impulses associated with hearing
IX Glossopharyngeal	Somatic motor	Motor impulses to muscles of pharynx used in swallowing
	Sensory: proprioception	Proprioception from muscles of pharynx
	Sensory	Sensory impulses from pharynx, middle-ear cavity, carotid sinus, and taste buds on posterior one-third of tongue
	Parasympathetic motor	Salivation from parotid salivary gland
X Vagus	Somatic motor	Contraction of muscles of pharynx (swallowing) and larynx (phonation)
	Sensory: proprioception	Proprioception from visceral muscles
	Sensory	Sensory impulses from taste buds on rear of tongue; sensations from auricle of ear; general visceral sensations
	Parasympathetic motor	Regulation of many visceral functions
XI Accessory	Somatic motor	Laryngeal movement; soft palate
		Motor impulses to trapezius and sternocleidomastoid muscles for movement of head, neck, and shoulders
	Sensory: proprioception	Proprioception from muscles that move head, neck, and shoulders
XII Hypoglossal	Somatic motor	Motor impulses to intrinsic and extrinsic muscles of tongue and infrahyoid muscles
	Sensory: proprioception	Proprioception from muscles of tongue

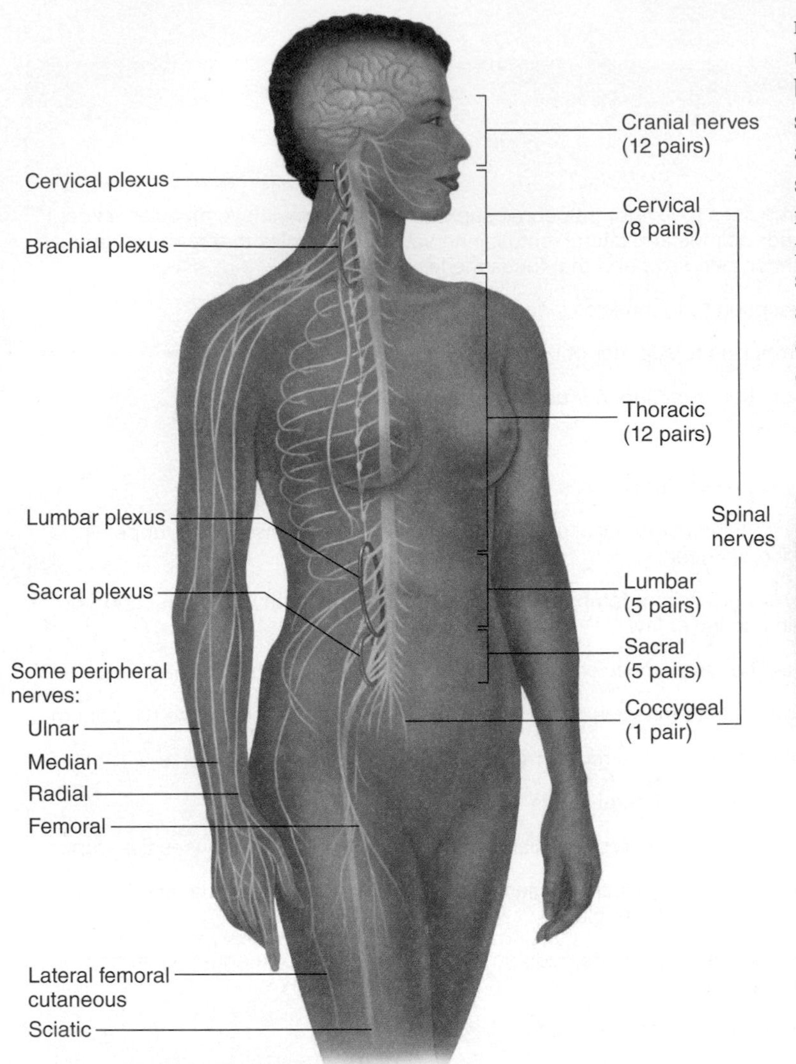

Figure 8.27 Distribution of the spinal nerves. These interconnect at plexuses (shown on the left) and form specific peripheral nerves. AP|R

the spinal cord (the autonomic system is discussed separately in chapter 9).

Reflex Arc

The functions of the sensory and motor components of a spinal nerve can be understood most easily by examining a simple reflex; that is, an unconscious motor response to a sensory stimulus. Figure 8.28 demonstrates the neural pathway involved in a **reflex arc.** Stimulation of sensory receptors evokes action potentials that are conducted into the spinal cord by sensory neurons. In the example shown, a sensory neuron synapses with an association neuron (or interneuron), which, in turn, synapses with a somatic motor neuron. The somatic motor neuron then conducts impulses out of the spinal cord to the muscle and stimulates a reflex contraction. Notice that the brain is not directly involved in this reflex response to sensory stimulation. Some reflex arcs are even simpler than this; in a muscle stretch reflex (the *knee-jerk reflex,* for example) the sensory neuron synapses directly with a motor neuron. Other reflexes are more complex, involving a number of association neurons and resulting in motor responses on both sides of the spinal cord at different levels. These skeletal muscle reflexes are described together with muscle control in chapter 12, and autonomic reflexes, involving smooth and cardiac muscle, are described in chapter 9.

Clinical Investigation CLUES

Frank produced a normal knee-jerk reflex with either leg.

- How is a knee-jerk reflex produced?
- Why would Frank produce a normal knee-jerk reflex despite his stroke?

CHECKPOINT

12a. Define the terms *dorsal root, dorsal root ganglion, ventral root,* and *mixed nerve.*

12b. Describe the neural pathways and structures involved in a reflex arc.

Clinical Investigation SUMMARY

Frank evidently suffered a cerebrovascular accident (CVA), otherwise known as a "stroke." The obstruction of blood flow in a cerebral artery damaged part of the precentral gyrus (motor cortex) in the left hemisphere. Because most corticospinal tracts decussate in the pyramids, this caused paralysis on the right side of his body. His spinal nerves were undamaged, so his knee-jerk reflex was intact. The damage to the left cerebral hemisphere apparently included damage to Broca's area, producing a characteristic aphasia that accompanied the paralysis of the right side of his body.

See the additional chapter 8 Clinical Investigation on *Ataxia* in the Connect site for this text at www.mhhe.com/Fox13.

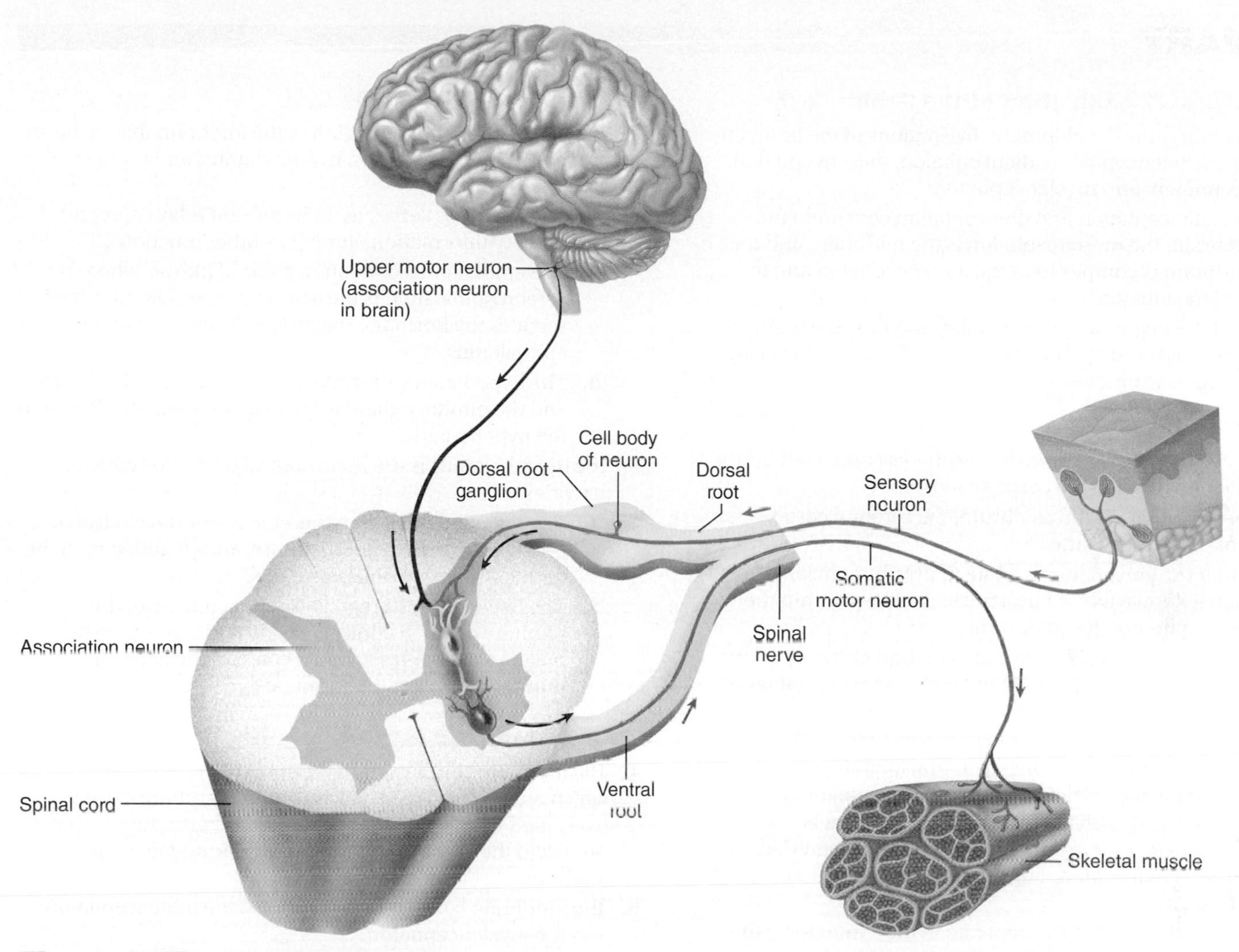

Figure 8.28 **Activation of somatic motor neurons.** Somatic motor neurons may be stimulated by spinal association neurons, as shown here, or directly by sensory neurons in a reflex arc that doesn't involve the brain. The spinal association neurons and motor neurons can also be stimulated by association neurons (called upper motor neurons) in the motor areas of the brain. This affords voluntary control of skeletal muscles. APIR

SUMMARY

8.1 Structural Organization of the Brain 207

A. During embryonic development, five regions of the brain are formed: the telencephalon, diencephalon, mesencephalon, metencephalon, and myelencephalon.

1. The telencephalon and diencephalon constitute the forebrain; the mesencephalon is the midbrain, and the hindbrain is composed of the metencephalon and the myelencephalon.
2. The CNS begins as a hollow tube, and thus the brain and spinal cord are hollow. The cavities of the brain are known as ventricles.

8.2 Cerebrum 209

A. The cerebrum consists of two hemispheres connected by a large fiber tract called the corpus callosum.

1. The outer part of the cerebrum, the cerebral cortex, consists of gray matter.
2. Under the gray matter is white matter, but nuclei of gray matter, known as the basal nuclei, lie deep within the white matter of the cerebrum.
3. Synaptic potentials within the cerebral cortex produce the electrical activity seen in an electroencephalogram (EEG).

B. The two cerebral hemispheres exhibit some specialization of function, a phenomenon called cerebral lateralization.

1. In most people, the left hemisphere is dominant in language and analytical ability, whereas the right hemisphere is more important in pattern recognition, musical composition, singing, and the recognition of faces.
2. The two hemispheres cooperate in their functions; this cooperation is aided by communication between the two via the corpus callosum.

C. Particular regions of the left cerebral cortex appear to be important in language ability; when these areas are damaged, characteristic types of aphasias result.

1. Wernicke's area is involved in speech comprehension, whereas Broca's area is required for the mechanical performance of speech.
2. Wernicke's area is believed to control Broca's area by means of the arcuate fasciculus.
3. The angular gyrus is believed to integrate different sources of sensory information and project to Wernicke's area.

D. The limbic system and hypothalamus are regions of the brain that have been implicated as centers for various emotions.

E. Memory can be divided into short-term and long-term categories.

1. The medial temporal lobes—in particular the hippocampus and perhaps the amygdaloid nucleus—appear to be required for the consolidation of short-term memory into long-term memory.
2. Particular aspects of a memory may be stored in numerous brain regions.
3. Long-term potentiation is a phenomenon that may be involved in some aspects of memory.

8.3 Diencephalon 226

A. The diencephalon is the region of the forebrain that includes the thalamus, epithalamus, hypothalamus, and pituitary gland.

1. The thalamus serves as an important relay center for sensory information, among its other functions.
2. The epithalamus contains a choroid plexus, where cerebrospinal fluid is formed. The pineal gland, which secretes the hormone melatonin, is also part of the epithalamus.
3. The hypothalamus forms the floor of the third ventricle, and the pituitary gland is located immediately inferior to the hypothalamus.

B. The hypothalamus is the main control center for visceral activities.

1. The hypothalamus contains centers for the control of thirst, hunger, body temperature, and (together with the limbic system) various emotions.
2. The hypothalamus regulates the secretions of the pituitary gland. It controls the posterior pituitary by means of a fiber tract, and it controls the anterior pituitary by means of hormones.

8.4 Midbrain and Hindbrain 229

A. The midbrain contains the superior and inferior colliculi, which are involved in visual and auditory reflexes, respectively, and nuclei that contain dopaminergic neurons that project to the corpus striatum and limbic system of the forebrain.

B. The hindbrain consists of two regions: the metencephalon and the myelencephalon.

1. The metencephalon contains the pons and cerebellum. The pons contains nuclei for four pairs of cranial nerves, and the cerebellum plays an important role in the control of skeletal movements.
2. The myelencephalon consists of only one region, the medulla oblongata. The medulla contains centers for the regulation of such vital functions as breathing and the control of the cardiovascular system.

C. The reticular activating system (RAS) is an ascending arousal system consisting of interconnected neurons of the reticular formation that extend from the pons to the midbrain.

1. Arousal is promoted by different neural tracts of the RAS that release ACh, different monoamine neurotransmitters, and a polypeptide neurotransmitter known as orexin (or hypocretin-1).
2. The activity of the RAS is inhibited by GABA-releasing neurons, and this activity is necessary for sleep.

8.5 Spinal Cord Tracts 232

A. Ascending tracts carry sensory information from sensory organs up the spinal cord to the brain.

B. Descending tracts are motor tracts and are divided into two groups: the pyramidal and the extrapyramidal systems.

1. Pyramidal tracts are the corticospinal tracts. They begin in the precentral gyrus and descend, without synapsing, into the spinal cord.
2. Most of the corticospinal fibers decussate in the pyramids of the medulla oblongata.
3. Regions of the cerebral cortex, the basal nuclei, and the cerebellum control movements indirectly by synapsing with other regions that give rise to descending extrapyramidal fiber tracts.
4. The major extrapyramidal motor tract is the reticulospinal tract, which has its origin in the reticular formation of the midbrain.

8.6 Cranial and Spinal Nerves 236

A. There are 12 pairs of cranial nerves. Most of these are mixed, but some are exclusively sensory in function.

B. There are 31 pairs of spinal nerves. Each pair contains both sensory and motor fibers.
1. The dorsal root of a spinal nerve contains sensory fibers, and the cell bodies of these neurons are contained in the dorsal root ganglion.
2. The ventral root of a spinal nerve contains motor fibers.

C. A reflex arc is a neural pathway involving a sensory neuron and a motor neuron. One or more association neurons also may be involved in some reflexes.

REVIEW ACTIVITIES

Test Your Knowledge

1. Which of these statements about the precentral gyrus is *true?*
   a. It is involved in motor control.
   b. It is involved in sensory perception.
   c. It is located in the frontal lobe.
   d. Both *a* and *c* are true.
   e. Both *b* and *c* are true.
2. In most people, the right hemisphere controls movement of
   a. the right side of the body primarily.
   b. the left side of the body primarily.
   c. both the right and left sides of the body equally.
   d. the head and neck only.
3. Which of these statements about the basal nuclei is *true?*
   a. They are located in the cerebrum.
   b. They contain the caudate nucleus.
   c. They are involved in motor control.
   d. They are part of the extrapyramidal system.
   e. All of these are true.
4. Which of these acts as a relay center for somatesthetic sensation?
   a. The thalamus
   b. The hypothalamus
   c. The red nucleus
   d. The cerebellum
5. Which of these statements about the medulla oblongata is *false?*
   a. It contains nuclei for some cranial nerves.
   b. It contains the apneustic center.
   c. It contains the vasomotor center.
   d. It contains ascending and descending fiber tracts.
6. The reticular activating system
   a. is composed of neurons that are part of the reticular formation.
   b. is a loose arrangement of neurons with many interconnecting synapses.
   c. is located in the brain stem and midbrain.
   d. functions to arouse the cerebral cortex to incoming sensory information.
   e. is described correctly by all of these.
7. In the control of emotion and motivation, the limbic system works together with
   a. the pons.
   b. the thalamus.
   c. the hypothalamus.
   d. the cerebellum.
   e. the basal nuclei.
8. Verbal ability predominates in
   a. the left hemisphere of right-handed people.
   b. the left hemisphere of most left-handed people.
   c. the right hemisphere of 97% of all people.
   d. both *a* and *b.*
   e. both *b* and *c.*
9. The consolidation of short-term memory into long-term memory appears to be a function of
   a. the substantia nigra.
   b. the hippocampus.
   c. the cerebral peduncles.
   d. the arcuate fasciculus.
   e. the precentral gyrus.

For questions 10–12, match the nature of the aphasia with its cause (choices are listed under question 12).

10. Comprehension good; can speak and write, but cannot read (although can see).
11. Comprehension good; speech is slow and difficult (but motor ability is not damaged).
12. Comprehension poor; speech is fluent but meaningless.
    a. damage to Broca's area
    b. damage to Wernicke's area
    c. damage to angular gyrus
    d. damage to precentral gyrus
13. Antidiuretic hormone (ADH) and oxytocin are synthesized by supraoptic and paraventricular nuclei, which are located in
    a. the thalamus.
    b. the pineal gland.
    c. the pituitary gland.
    d. the hypothalamus.
    e. the pons.

14. The superior colliculi are twin bodies within the corpora quadrigemina of the midbrain that are involved in
    a. visual reflexes.
    b. auditory reflexes.
    c. relaying of cutaneous information.
    d. release of pituitary hormones.
15. The consolidation of declarative memory requires the____________; working memory requires the ____________.
    a. occipital lobe; hippocampus
    b. medial temporal lobe; prefrontal cortex
    c. frontal lobe; amygdala
    d. hypothalamus; precentral gyrus

Test Your Understanding

16. Define the term *decussation,* and explain the significance of decussation in terms of the pyramidal motor system.
17. Describe the location of the hypothalamus and list its functions. Explain how it serves as a link between the nervous and endocrine systems.
18. The thalamus has been described as a "switchboard." Explain why, by describing the pathway of somatic sensory information from the receptors to the cerebral cortex.
19. Distinguish between the different types of memory and identify the brain regions that are involved in each type.
20. Describe the categories and EEG patterns of sleep, and explain the possible benefits of these categories of sleep.
21. Electrical stimulation of the basal nuclei or cerebellum can produce skeletal movements. Describe the pathways by which these brain regions control motor activity.
22. Define the term *ablation.* Give two examples of how this experimental technique has been used to learn about the function of particular brain regions.
23. Explain how "split-brain" patients have contributed to research on the function of the cerebral hemispheres. Propose some experiments that would reveal the lateralization of function in the two hemispheres.
24. What evidence do we have that Wernicke's area may control Broca's area? What evidence do we have that the angular gyrus has input to Wernicke's area?
25. State two reasons why researchers distinguish between short-term and long-term memory.
26. Describe evidence showing that the hippocampus is involved in the consolidation of short-term memory. After long-term memory is established, why may there be no need for hippocampal involvement?
27. Can we be aware of a reflex action involving our skeletal muscles? Is this awareness necessary for the response? Explain, identifying the neural pathways involved in the reflex response and the conscious awareness of a stimulus.
28. Describe the reticular activating system, and explain how amphetamines cause wakefulness and alcohol causes drowsiness.

Test Your Analytical Ability

29. Fetal alcohol syndrome, produced by excessive alcohol consumption during pregnancy, affects different aspects of embryonic development. Two brain regions known to be particularly damaged in this syndrome are the corpus callosum and the basal nuclei. Speculate on what effects damage to these areas may produce.
30. Recent studies suggest that medial temporal lobe activity is needed for memory retrieval. What is the difference between memory storage and retrieval, and what scientific evidence might allow them to be distinguished?
31. Much has been made (particularly by left-handers) of the fact that Leonardo da Vinci was left-handed. Do you think his accomplishments are in any way related to his left-handedness? Why or why not?
32. People under chronic stress can suffer atrophy of their hippocampi. How does this affect their ability to learn, and what type of learning would be most affected? What type would be less affected? Explain.
33. Which stroke victim is more likely to have impaired speech, the one with paralysis on the right side or the one with paralysis on the left side? Explain. Speculate on what changes might occur in the brain to allow gains to be made in speech recovery.
34. Neurologists have noticed that patients with lesions (damage) at the junction of the midbrain and diencephalon of the forebrain had trouble arousing themselves from sleep. In other patients, lesions in the lateral hypothalamic area produce severe sleepiness, even coma. Identify the brain system impaired by these lesions and explain how these effects could be produced.

Test Your Quantitative Ability

Table 7.3 (chapter 7), page 180, provides the axon diameters and conduction velocities required to answer the following questions.

Suppose, in a knee-jerk reflex, the sensory axon and motor axon extending between the muscle and spinal cord is each 16 inches long. The sensory axon has a diameter of 17μm, and the motor axon has a diameter of 9μm. Given that there are 2.54 cm per inch, and that a rate of 1 m/sec is equal to 2.24 miles per hour, answer the following questions.

35. What is the length of each axon in centimeters and meters?
36. What is the rate of conduction of the sensory axon in meters per second and in miles per hour?
37. What is the rate of conduction of the motor axon in meters per second and in miles per hour?
38. How long will it take, in seconds and milliseconds, for an action potential to be conducted the length of the sensory axon?
39. How long will it take, in seconds and milliseconds, for an action potential to be conducted the length of the motor axon?
40. Suppose the time from the start of action potentials in the sensory neuron and the end of action potentials in the motor neuron was measured to be 15 msec. How much time was required for synaptic transmission?

CHAPTER OUTLINE

CHAPTER 9

The Autonomic Nervous System

REFRESH YOUR MEMORY

Before you begin this chapter, you may want to review these concepts from previous chapters:

Clinical Investigation

Cathy has asthma, and had to use her inhaler before taking her physiology exam. Later, in the physiology laboratory, she measured her pulse rate and blood pressure and found them to be higher than usual. The following week, after administering some drugs (epinephrine, atropine, and others) to a frog heart, she later developed a severe headache and dry mouth. When she looked in the mirror she noticed that her pupils were dilated.

Some of the new terms and concepts you will encounter include:

- Adrenergic effects and fight-or-flight
- Alpha- and beta-adrenergic receptors and their agonists and antagonists
- Muscarinic cholinergic effects and atropine

9.1 NEURAL CONTROL OF INVOLUNTARY EFFECTORS

The autonomic nervous system helps regulate the activities of cardiac muscle, smooth muscles, and glands. In this regulation, impulses are conducted from the CNS by an axon that synapses with a second autonomic neuron. It is the axon of this second neuron in the pathway that innervates the involuntary effectors.

LEARNING OUTCOMES

After studying this section, you should be able to:

1. Describe the organization of autonomic motor neurons.
2. Describe how neural regulation of smooth and cardiac muscles differs from neural regulation of skeletal muscles.

Autonomic motor nerves innervate organs whose functions are not usually under voluntary control. The effectors that respond to autonomic regulation include **cardiac muscle** (the heart), **smooth muscles,** and **glands.** These effectors are part of the *visceral organs* (organs within the body cavities) and of blood vessels. The involuntary effects of autonomic innervation contrast with the voluntary control of skeletal muscles by way of somatic motor neurons.

Autonomic Neurons

Neurons of the peripheral nervous system (PNS) that conduct impulses away from the central nervous system (CNS) are known as *motor,* or *efferent, neurons* (chapter 7, section 7.1). There are two major categories of motor neurons: somatic and autonomic. Somatic motor neurons have their cell bodies within the CNS and send axons to skeletal muscles, which are usually under voluntary control. This was briefly described in chapter 8 (see fig. 8.28), in the section on the reflex arc. The control of skeletal muscles by somatic motor neurons is discussed in depth in chapter 12, section 12.5.

Unlike somatic motor neurons, which conduct impulses along a single axon from the spinal cord to the neuromuscular junction, autonomic motor control involves two neurons in the efferent pathway (fig. 9.1 and table 9.1). The first of these neurons has its cell body in the gray matter of the brain or spinal cord. The axon of this neuron does not directly innervate the effector organ but instead synapses with a second neuron within an *autonomic ganglion* (a ganglion is a collection of cell bodies outside the CNS). The first neuron is thus called a **preganglionic neuron.** The second neuron in this pathway, called a **postganglionic neuron,** has an axon that extends from the autonomic ganglion to an effector organ, where it synapses with its target tissue (fig. 9.1).

Preganglionic autonomic fibers originate in the midbrain and hindbrain and in the upper thoracic to the fourth sacral levels of the spinal cord. Autonomic ganglia are located in the head, neck, and abdomen; chains of autonomic ganglia also parallel the right and left sides of the spinal cord. The origin of the preganglionic

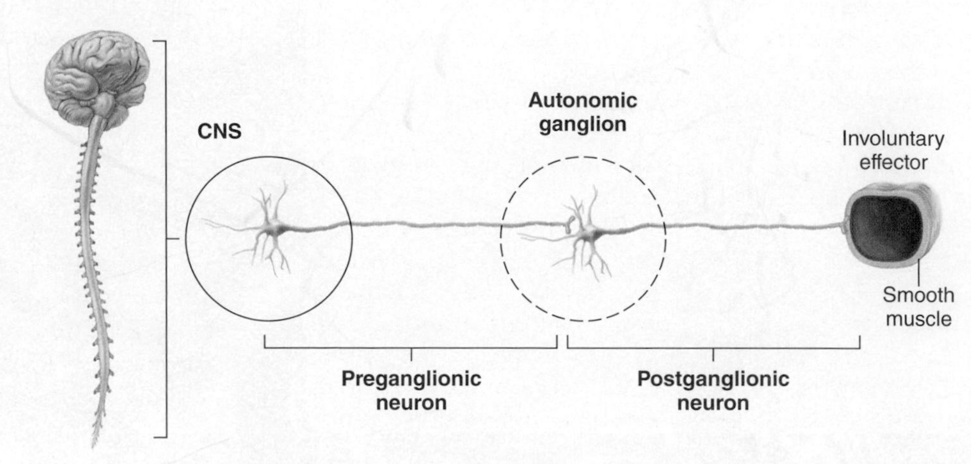

Figure 9.1 The autonomic system has preganglionic and postganglionic neurons. The preganglionic neurons of the autonomic system have cell bodies in the CNS, whereas the postganglionic neurons have cell bodies within autonomic ganglia. The sympathetic and parasympathetic divisions differ in the particular locations of their preganglionic neuron cell bodies within the CNS, and in the location of their ganglia.

Table 9.1 | Comparison of the Somatic Motor System and the Autonomic Motor System

Feature	Somatic Motor	Autonomic Motor
Effector organs	Skeletal muscles	Cardiac muscle, smooth muscle, and glands
Presence of ganglia	No ganglia	Cell bodies of postganglionic autonomic fibers located in paravertebral, prevertebral (collateral), and terminal ganglia
Number of neurons from CNS to effector	One	Two
Type of neuromuscular junction	Specialized motor end plate	No specialization of postsynaptic membrane; all areas of smooth muscle cells contain receptor proteins for neurotransmitters
Effect of nerve impulse on muscle	Excitatory only	Either excitatory or inhibitory
Type of nerve fibers	Fast-conducting, thick (9–13μm), and myelinated	Slow-conducting; preganglionic fibers lightly myelinated but thin (3μm); postganglionic fibers unmyelinated and very thin (about 1.0μm)
Effect of denervation	Flaccid paralysis and atrophy	Muscle tone and function persist; target cells show denervation hypersensitivity

fibers and the location of the autonomic ganglia help to distinguish the *sympathetic* and *parasympathetic divisions* of the autonomic system, discussed in later sections of this chapter.

The sensory neurons that conduct information from the viscera for autonomic nerve reflexes can have the same anatomy as those sensory neurons involved in somatic motor reflexes (chapter 8, fig. 8.28). That is, the sensory information enters the spinal cord on the dorsal roots of the spinal nerves. However, some important visceral sensory information can instead enter the brain in cranial nerves. For example, information about blood pressure, plasma pH, and oxygen concentration is carried into the brain by sensory axons in cranial nerves IX and X. These are mixed nerves, containing both sensory and parasympathetic motor axons.

Visceral Effector Organs

Because the autonomic nervous system helps regulate the activities of glands, smooth muscles, and cardiac muscle, autonomic control is an integral aspect of the physiology of most of the body systems. Autonomic regulation, then, plays roles in endocrine regulation (chapter 11), smooth muscle function (chapter 12), the functions of the heart and circulation (chapters 13 and 14), and, in fact, all the remaining systems to be discussed. Although the functions of the target organs of autonomic innervation are described in subsequent chapters, at this point we will consider some of the common features of autonomic regulation.

Unlike skeletal muscles, which enter a state of flaccid paralysis and atrophy when their motor nerves are severed, the involuntary effectors are somewhat independent of their innervation. Smooth muscles maintain a resting tone (tension) in the absence of nerve stimulation, for example. In fact, damage to an autonomic nerve makes its target tissue more sensitive than normal to stimulating agents. This phenomenon is called **denervation hypersensitivity.** Such compensatory changes can explain why, for example, the ability of the stomach mucosa to secrete acid may be restored after its neural supply from the vagus nerve has been severed. (This procedure is called vagotomy, and is sometimes performed as a treatment for ulcers.)

In addition to their intrinsic ("built-in") muscle tone, cardiac muscle and many smooth muscles take their autonomy a step further. These muscles can contract rhythmically, even in the absence of nerve stimulation, in response to electrical waves of depolarization initiated by the muscles themselves. Autonomic innervation simply increases or decreases this intrinsic activity. Autonomic nerves also maintain a resting tone, in the sense that they maintain a baseline firing rate that can be either increased or decreased. A decrease in the excitatory input to the heart, for example, will slow its rate of beat.

The release of acetylcholine (ACh) from somatic motor neurons always stimulates the effector organ (skeletal muscles). By contrast, some autonomic nerves release transmitters that inhibit the activity of their effectors. An increase in the activity of the vagus, a nerve that supplies inhibitory fibers to the heart, for example, will slow the heart rate, whereas a decrease in this inhibitory input will increase the heart rate.

CHECKPOINT

1. Describe the preganglionic and postganglionic neurons in the autonomic system. Use a diagram to illustrate the difference in efferent outflow between somatic and autonomic nerves.
2. Compare the control of cardiac muscle and smooth muscles with that of skeletal muscles. How is each type of muscle tissue affected by cutting its innervation?

9.2 DIVISIONS OF THE AUTONOMIC NERVOUS SYSTEM

Preganglionic neurons of the sympathetic division originate in the thoracic and lumbar levels of the spinal cord and send axons to sympathetic ganglia, which parallel the spinal cord. Preganglionic neurons of the parasympathetic division originate in the brain and in the sacral level of the spinal cord, and send axons to ganglia located in or near the effector organs.

LEARNING OUTCOMES

After studying this section, you should be able to:

3. Describe the structure of the sympathetic nervous system, locating the ganglia and the preganglionic and postganglionic neurons.
4. Explain the relationship between the sympathetic nervous system and the adrenal medulla.
5. Describe the structure and innervation pathways of the parasympathetic division of the autonomic system.

The sympathetic and parasympathetic divisions of the autonomic system have some structural features in common. Both consist of preganglionic neurons that originate in the CNS and postganglionic neurons that originate outside of the CNS in ganglia. However, the specific origin of the preganglionic fibers and the location of the ganglia differ in the two divisions of the autonomic system.

Sympathetic Division

The **sympathetic division** is also called the *thoracolumbar division* of the autonomic system because its preganglionic fibers exit the spinal cord, in the ventral roots of spinal nerves, from the first thoracic (T1) to the second lumbar (L2) levels. Most sympathetic nerve fibers separate from the somatic motor fibers and synapse with postganglionic neurons within a double row of sympathetic ganglia, called **paravertebral ganglia,** located on either side of the spinal cord (fig. 9.2). Ganglia within each row are interconnected, forming a **sympathetic chain of ganglia** that parallels the spinal cord on each lateral side.

The myelinated preganglionic sympathetic axons exit the spinal cord in the ventral roots of spinal nerves, but they soon diverge from the spinal nerves within short pathways called *white rami communicantes.* The axons within each ramus enter the sympathetic chain of ganglia, where they can travel to ganglia at different levels and synapse with postganglionic sympathetic neurons. The axons of the postganglionic sympathetic neurons are unmyelinated and form the *gray rami communicantes* as they return to the spinal nerves and travel as part of the spinal nerves to their effector organs (fig. 9.3). Because sympathetic axons form a component of spinal nerves, they are widely distributed to the skeletal muscles and skin of the body where they innervate blood vessels and other involuntary effectors.

Divergence occurs within the sympathetic chain of ganglia as preganglionic fibers branch to synapse with numerous postganglionic neurons located in ganglia at different levels in the chain. *Convergence* also occurs here when a postganglionic neuron receives synaptic input from a large number of preganglionic fibers. The divergence of impulses from the spinal

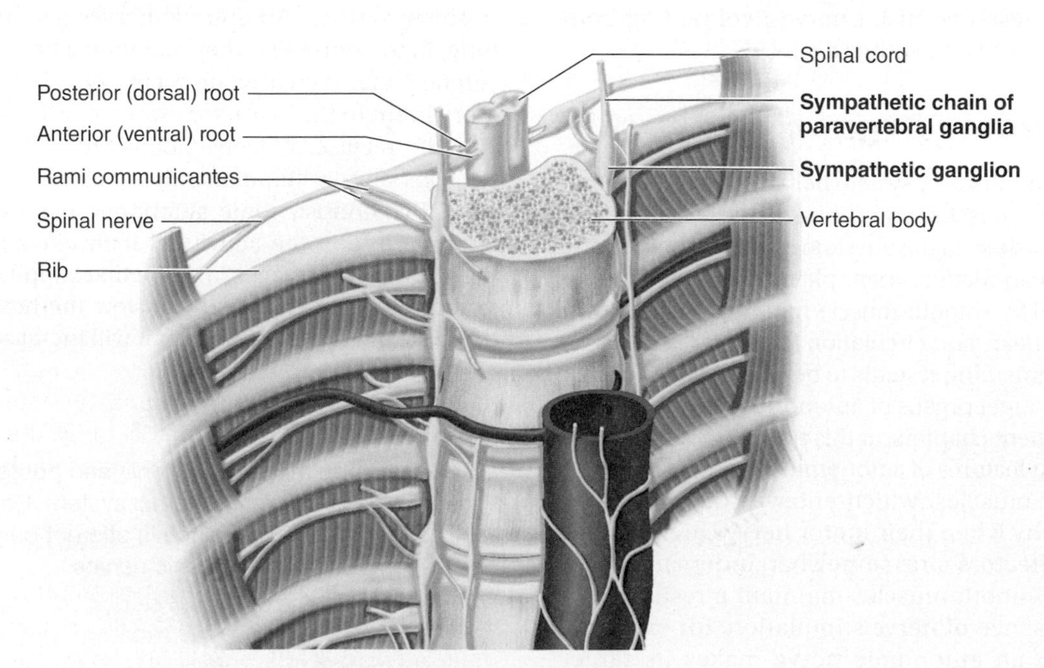

Figure 9.2 The sympathetic chain of paravertebral ganglia. This diagram shows the anatomical relationship between the sympathetic ganglia and the vertebral column and spinal cord. AP|R

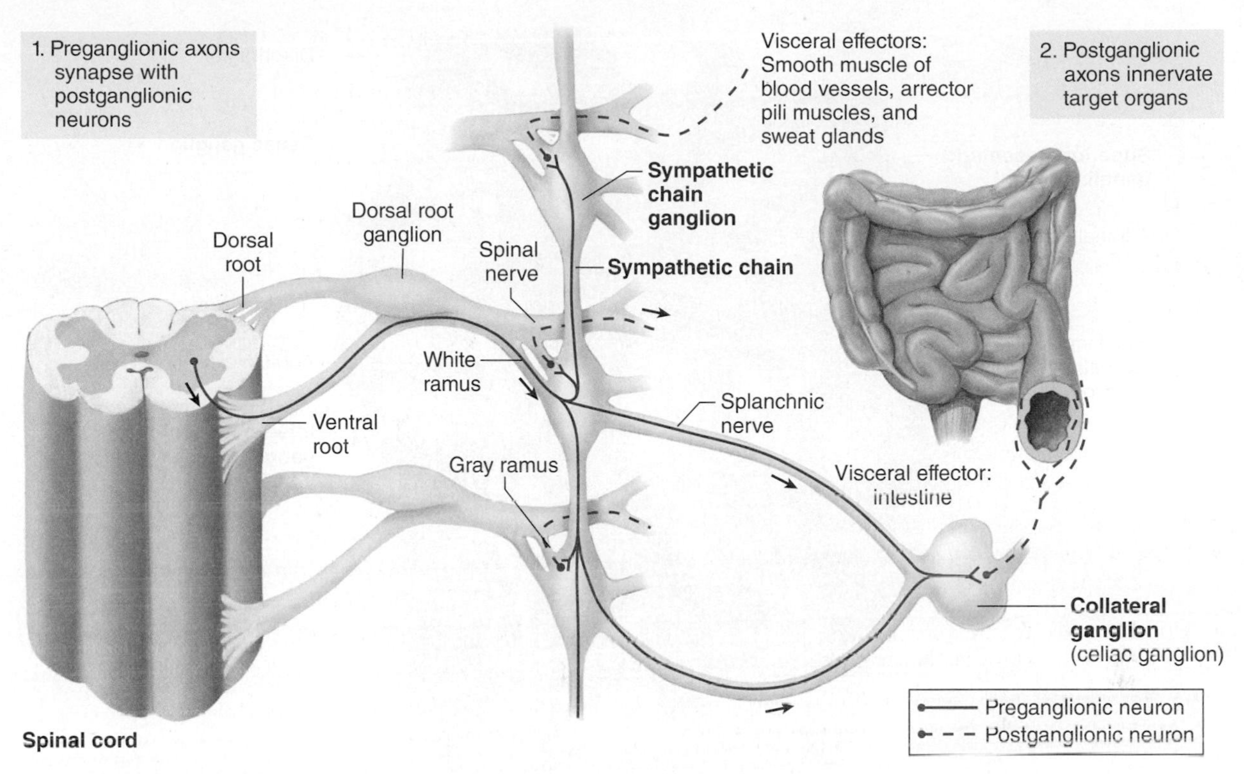

Figure 9.3 The pathway of sympathetic neurons. The preganglionic neurons enter the sympathetic chain of ganglia on the white ramus (one of the two rami communicantes). Some synapse there, and the postganglionic axon leaves on the gray ramus to rejoin a spinal nerve. Others pass through the ganglia without synapsing. These ultimately synapse in a collateral ganglion, such as the celiac ganglion.

cord to the ganglia and the convergence of impulses within the ganglia can result in the **mass activation** of almost all of the postganglionic sympathetic neurons. This mass activation allows the entire sympathetic division to be tonically (constantly) active to a certain degree and to increase its activity in response to "fight-or-flight" situations (section 9.3). However, mass activation does not always occur. In response to particular visceral stimuli (such as changes in blood pressure, blood volume, and plasma osmolality), the CNS can direct appropriate increases or decreases in the activity of postganglionic sympathetic axons to the heart and kidneys that allows these organs to compensate for the changes and maintain homeostasis.

Collateral Ganglia

Many preganglionic fibers that exit the spinal cord below the level of the diaphragm pass through the sympathetic chain of ganglia without synapsing. Beyond the sympathetic chain, these preganglionic fibers form *splanchnic nerves.* Preganglionic fibers in the splanchnic nerves synapse in **collateral,** or **prevertebral, ganglia.** These include the *celiac, superior mesenteric,* and *inferior mesenteric ganglia* (fig. 9.4). Postganglionic fibers that arise from the collateral ganglia innervate organs of the digestive, urinary, and reproductive systems.

Adrenal Glands

The paired **adrenal glands** are located above each kidney (fig. 9.4). Each adrenal is composed of two parts: an outer **cortex** and an inner **medulla.** These two parts are really two functionally different glands with different embryonic origins, different hormones, and different regulatory mechanisms. The adrenal cortex secretes steroid hormones; the adrenal medulla secretes the hormone **epinephrine** (adrenaline) and, to a lesser degree, **norepinephrine,** when it is stimulated by the sympathetic system.

The adrenal medulla can be likened to a modified sympathetic ganglion; its cells are derived from the same embryonic tissue (the neural crest, chapter 8) that forms postganglionic sympathetic neurons. Like a sympathetic ganglion, the cells of the adrenal medulla are innervated by preganglionic sympathetic fibers (fig. 9.5). The adrenal medulla secretes epinephrine into the blood in response to this neural stimulation. The effects of epinephrine are complementary to those of the neurotransmitter norepinephrine, which is released from postganglionic sympathetic nerve endings. For this reason, and because the adrenal medulla is stimulated as part of the mass activation of the sympathetic system, the two are often grouped together as a single **sympathoadrenal system.**

Parasympathetic Division

The **parasympathetic division** is also known as the *craniosacral division* of the autonomic system. This is because its preganglionic fibers originate in the brain (specifically, in the midbrain, pons, and medulla oblongata) and in the second through fourth sacral levels of the spinal column. These preganglionic

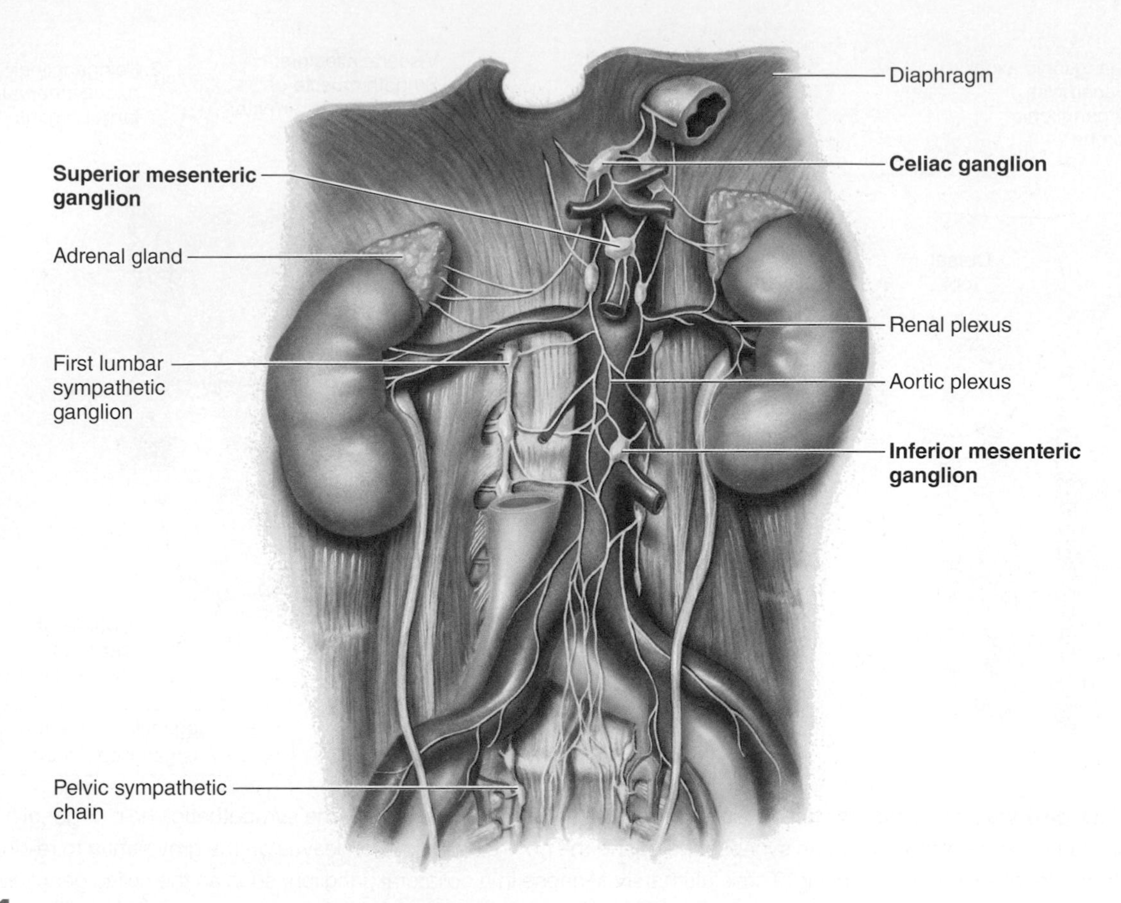

Figure 9.4 The collateral sympathetic ganglia. These include the celiac ganglion and the superior and inferior mesenteric ganglia. The paired adrenal glands are also shown. AP|R

Table 9.2 | The Sympathetic Division

Parts of Body Innervated	Spinal Origin of Preganglionic Fibers	Origin of Postganglionic Fibers
Eye	C8 and T1	Cervical ganglia
Head and neck	T1 to T4	Cervical ganglia
Heart and lungs	T1 to T5	Upper thoracic (paravertebral) ganglia
Upper extremities	T2 to T9	Lower cervical and upper thoracic (paravertebral) ganglia
Upper abdominal viscera	T4 to T9	Celiac and superior mesenteric (collateral) ganglia
Adrenal	T10 and T11	Not applicable
Urinary and reproductive systems	T12 to L2	Celiac and interior mesenteric (collateral) ganglia
Lower extremities	T9 to L2	Lumbar and upper sacral (paravertebral) ganglia

parasympathetic fibers synapse in ganglia that are located next to—or actually within—the organs innervated. These parasympathetic ganglia, called **terminal ganglia,** supply the postganglionic fibers that synapse with the effector cells.

The comparative structures of the sympathetic and parasympathetic divisions are listed in tables 9.2 and 9.3, and illustrated in figure 9.5. It should be noted that most parasympathetic fibers do not travel within spinal nerves, as do sympathetic fibers. As a result, cutaneous effectors (blood vessels, sweat glands, and arrector pili muscles) and blood vessels in skeletal muscles receive sympathetic but not parasympathetic innervation.

Four of the 12 pairs of cranial nerves (chapter 8, section 8.6) contain preganglionic parasympathetic fibers. These are the oculomotor (III), facial (VII), glossopharyngeal (IX), and vagus (X) nerves. Parasympathetic fibers within the first three of these cranial nerves synapse in ganglia located in the head; fibers in the vagus nerve synapse in terminal ganglia located in widespread regions of the body. Cranial nerves IX and X contain sensory axons as well as parasympathetic motor axons: they

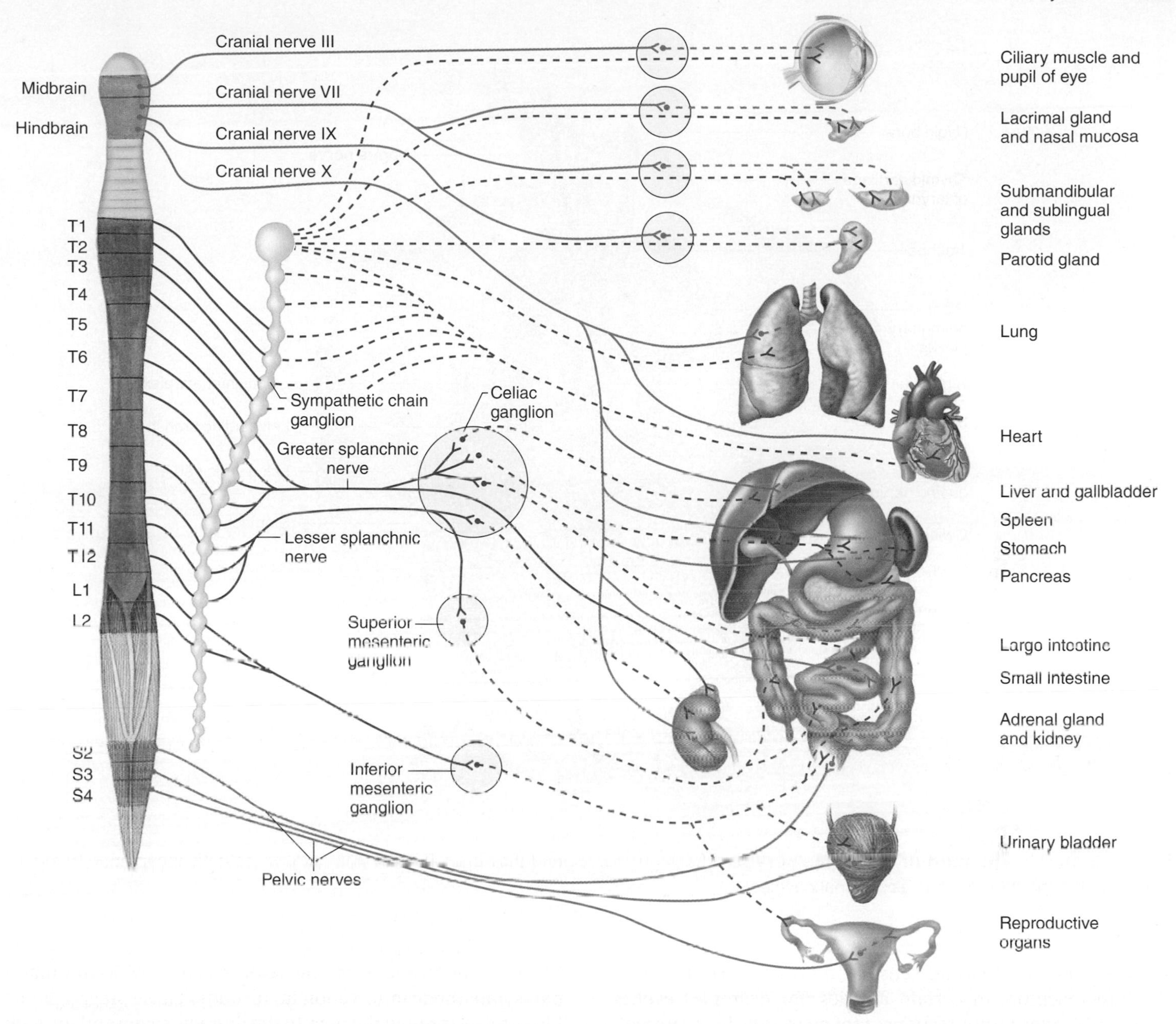

Figure 9.5 **The autonomic nervous system.** The sympathetic division is shown in red; the parasympathetic in blue. The solid lines indicate preganglionic fibers and the dashed lines indicate postganglionic fibers.

Table 9.3 | The Parasympathetic Division

Nerve	Origin of Preganglionic Fibers	Location of Terminal Ganglia	Effector Organs
Oculomotor (third cranial) nerve	Midbrain (cranial)	Ciliary ganglion	Eye (smooth muscle in iris and ciliary body)
Facial (seventh cranial)	Pons (cranial)	Pterygopalatine and submandibular ganglia	Lacrimal, mucous, and salivary glands
Glossopharyngeal (ninth cranial) nerve	Medulla oblongata (cranial)	Otic ganglion	Parotid gland
Vagus (tenth cranial) nerve	Medulla oblongata (cranial)	Terminal ganglia in or near organ	Heart, lungs, gastrointestinal tract, liver, pancreas
Pelvic spinal nerves	S2 to S4 (sacral)	Terminal ganglia near organs	Lower half of large intestine, rectum, urinary bladder, and reproductive organs

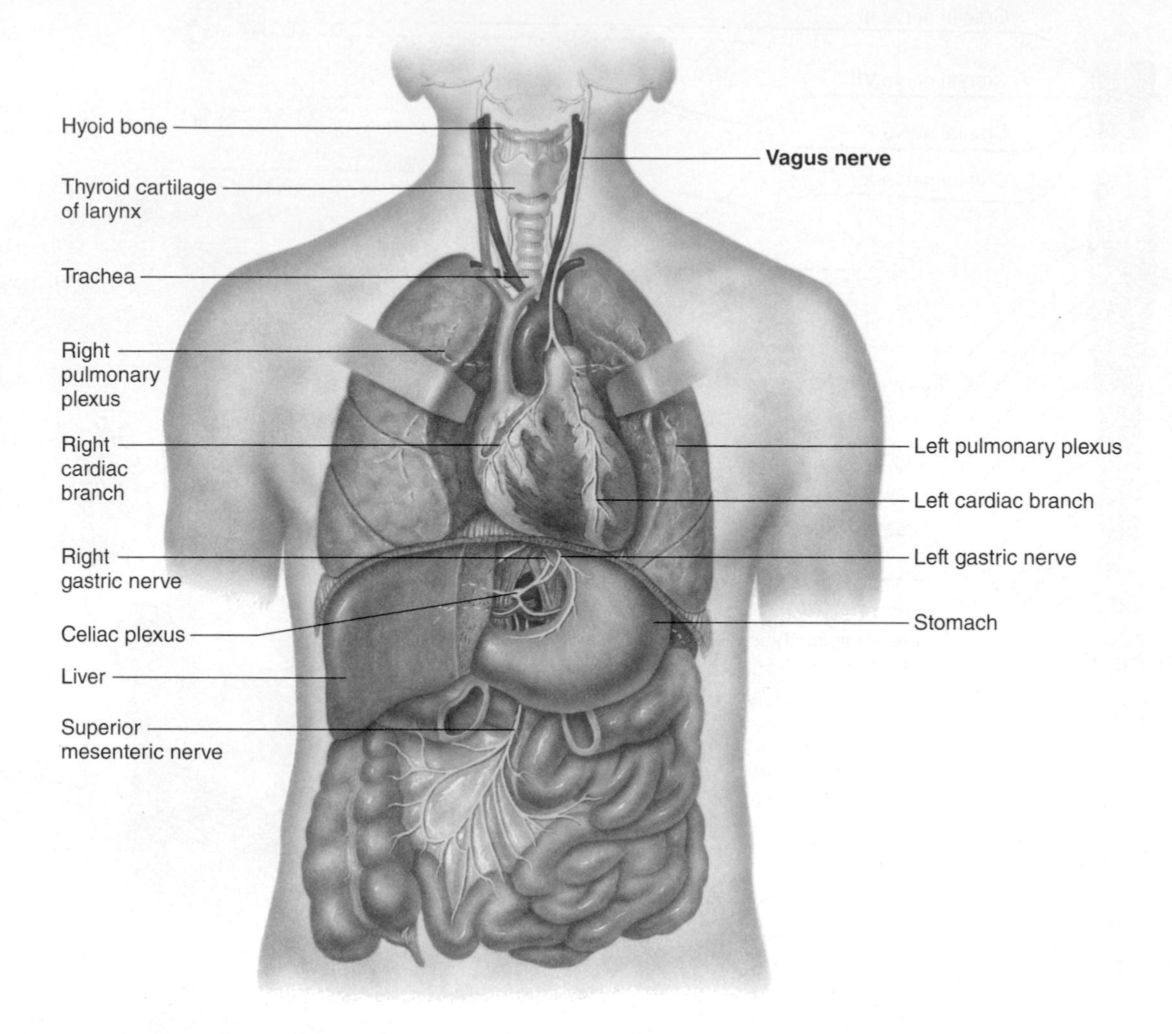

Figure 9.6 The path of the vagus nerves. The vagus nerves and their branches provide parasympathetic innervation to most organs within the thoracic and abdominal cavities.

are mixed nerves. Visceral sensory information (from blood pressure receptors in certain arteries, for example) evokes autonomic reflex motor responses (of heart rate, for example). These reflexes will be discussed in chapter 14.

The oculomotor nerve contains somatic motor and parasympathetic fibers that originate in the oculomotor nuclei of the midbrain. These parasympathetic fibers synapse in the *ciliary ganglion,* whose postganglionic fibers innervate the ciliary muscle and constrictor fibers in the iris of the eye. Preganglionic fibers that originate in the pons travel in the facial nerve to the *pterygopalatine ganglion,* which sends postganglionic fibers to the nasal mucosa, pharynx, palate, and lacrimal glands. Another group of fibers in the facial nerve terminates in the *submandibular ganglion,* which sends postganglionic fibers to the submandibular and sublingual salivary glands. Preganglionic fibers of the glossopharyngeal nerve synapse in the *otic ganglion,* which sends postganglionic fibers to innervate the parotid salivary gland.

Nuclei in the medulla oblongata contribute preganglionic fibers to the very long *tenth cranial,* or *vagus, nerves* (the "vagrant" or "wandering" nerves), which provide the major parasympathetic innervation in the body. These preganglionic fibers travel through the neck to the thoracic cavity and through the esophageal opening in the diaphragm to the abdominal cavity (fig. 9.6). In each region, some of these preganglionic fibers branch from the main trunks of the vagus nerves and synapse with postganglionic neurons located *within* the innervated organs. The preganglionic vagus fibers are thus quite long; they provide parasympathetic innervation to the heart, lungs, esophagus, stomach, pancreas, liver, small intestine, and the upper half of the large intestine. Postganglionic parasympathetic fibers arise from terminal ganglia within these organs and synapse with effector cells (smooth muscles and glands).

Preganglionic fibers from the sacral levels of the spinal cord provide parasympathetic innervation to the lower half of the large intestine, the rectum, and to the urinary and reproductive systems. These fibers, like those of the vagus, synapse with terminal ganglia located within the effector organs.

Parasympathetic nerves to the visceral organs thus consist of preganglionic fibers, whereas sympathetic nerves to these organs contain postganglionic fibers. An overall view of the autonomic nervous system, with a comparison of the sympathetic and parasympathetic divisions, can be obtained by reviewing figure 9.5.

CHECKPOINT

3. Describe the sympathetic pathway from the CNS to the (a) heart; and (b) adrenal gland.
4. Explain the functional relationship between the sympathetic division and the adrenal glands.
5. Describe the parasympathetic pathway to the eye and to the heart, identifying the neurons involved.

9.3 FUNCTIONS OF THE AUTONOMIC NERVOUS SYSTEM

The sympathetic division of the autonomic system activates the body to "fight or flight," largely through the release of norepinephrine from postganglionic neurons and the secretion of epinephrine from the adrenal medulla. The parasympathetic division often produces antagonistic effects through the release of acetylcholine from its postganglionic neurons.

LEARNING OUTCOMES

After studying this section, you should be able to:

6. Identify the neurotransmitters of the sympathetic and parasympathetic divisions, and the hormone released by the adrenal medulla.
7. Describe the effects of adrenergic stimulation on different organs, and identify the types of adrenergic receptors involved.
8. Describe the effects of parasympathetic nerve regulation, and explain how atropine and related drugs affect this regulation.
9. Describe and give examples of antagonistic, cooperative, and complementary actions of the sympathetic and parasympathetic divisions of the autonomic system.

The sympathetic and parasympathetic divisions of the autonomic system affect the visceral organs in different ways. Mass activation of the sympathetic system prepares the body for intense physical activity in emergencies; the heart rate increases, blood glucose rises, and blood is diverted to the skeletal muscles (away from the visceral organs and skin). The theme of the sympathetic system has been aptly summarized in a phrase: **"fight or flight."**

The fight-or-flight concept can give the erroneous impression that the sympathetic division is activated only in emergencies. But unlike somatic motor neurons, sympathetic neurons display tonic (continuous) activity. As a result, sympathetic nerves tonically regulate the heart, blood vessels, and other organs. Also, whereas the fight-or-flight reaction stimulates the mass activation of sympathetic nerve activity, there is also a more tailored, moment-to-moment regulation of the cardiovascular system and kidneys by the sympathetic division.

Unlike the fight-or-flight theme of the sympathetic division, there is no universally recognized phrase to describe the actions of the parasympathetic division. However, because many of its actions are opposite to those of the sympathetic division, the theme of the parasympathetic division might be described as *rest and digest,* or *repast and repose.*

The effects of parasympathetic nerve stimulation are in many ways opposite to those produced by sympathetic stimulation. The parasympathetic system, however, is not normally activated as a whole. Stimulation of separate parasympathetic nerves can result in slowing of the heart, dilation of visceral blood vessels, and increased activity of the digestive tract. Visceral organs respond differently to sympathetic and parasympathetic nerve activity because the postganglionic axons of these two divisions release different neurotransmitters.

CLINICAL APPLICATION

Cocaine blocks the reuptake of dopamine and norepinephrine into the presynaptic axon terminals. This causes an excessive amount of these neurotransmitters to remain in the synaptic cleft and stimulate their target cells. Because sympathetic nerve effects are produced mainly by the action of norepinephrine, cocaine is a *sympathomimetic drug* (a drug that promotes sympathetic nerve effects). This can result in vasoconstriction of coronary arteries, leading to heart damage (myocardial ischemia, myocardial infarction, and left ventricular hypertrophy). The combination of cocaine with alcohol is more deadly than either drug taken separately, and is a common cause of death from substance abuse.

Adrenergic and Cholinergic Synaptic Transmission

Acetylcholine (ACh) is the neurotransmitter of all preganglionic axons (both sympathetic and parasympathetic). Acetylcholine is also the transmitter released by most parasympathetic postganglionic axons at their synapses with effector cells (fig. 9.7). Transmission at these synapses is thus said to be **cholinergic.**

The neurotransmitter released by most postganglionic sympathetic nerve fibers is **norepinephrine** *(noradrenaline).* Transmission at these synapses is thus said to be **adrenergic.** There are a few exceptions, however. Some sympathetic fibers that innervate blood vessels in skeletal muscles, as well as sympathetic fibers to sweat glands, release ACh (are cholinergic).

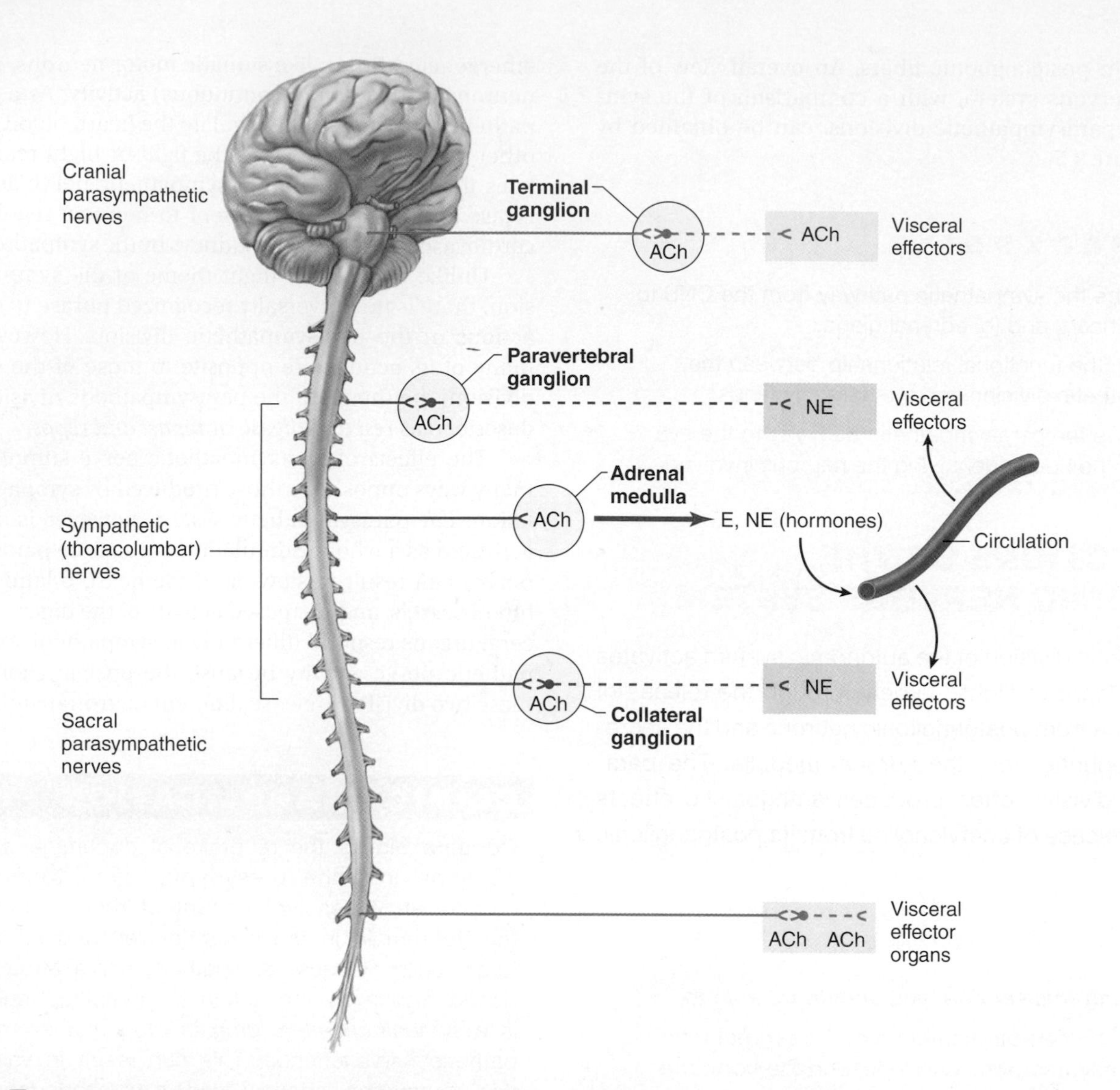

Figure 9.7 Neurotransmitters of the autonomic motor system. ACh = acetylcholine; NE = norepinephrine; E = epinephrine. Those nerves that release ACh are called cholinergic; those nerves that release NE are called adrenergic. The adrenal medulla secretes both epinephrine (85%) and norepinephrine (15%) as hormones into the blood.

Since the cells of the adrenal medulla are embryologically related to postganglionic sympathetic neurons, it is not surprising that their hormones are epinephrine (about 85%) and norepinephrine (about 15%). Epinephrine differs from norepinephrine only in that the former has an additional methyl (CH_3) group, as shown in figure 9.8. Epinephrine, norepinephrine, and dopamine (a transmitter within the CNS) are all derived from the amino acid tyrosine and are collectively termed **catecholamines** (fig. 9.8).

Where the axons of postganglionic autonomic neurons enter into their target organs, they have numerous swellings, called *varicosities,* that contain the neurotransmitter molecules. Neurotransmitters can thereby be released along a length of axon, rather than just at the axon terminal. Thus, autonomic neurons are said to form *synapses en passant* ("synapses in passing") with their target cells (fig. 9.9). Sympathetic and parasympathetic axons often innervate the same target cells, where they release different neurotransmitters that promote different (and usually antagonistic) effects.

Responses to Adrenergic Stimulation

Adrenergic stimulation—by epinephrine in the blood and by norepinephrine released from sympathetic nerve endings—has both excitatory and inhibitory effects. The heart, dilatory muscles of the iris, and the smooth muscles of many blood vessels are stimulated to contract. The smooth muscles of the bronchioles and of some blood vessels, however, are inhibited from contracting; adrenergic chemicals, therefore, cause these structures to dilate.

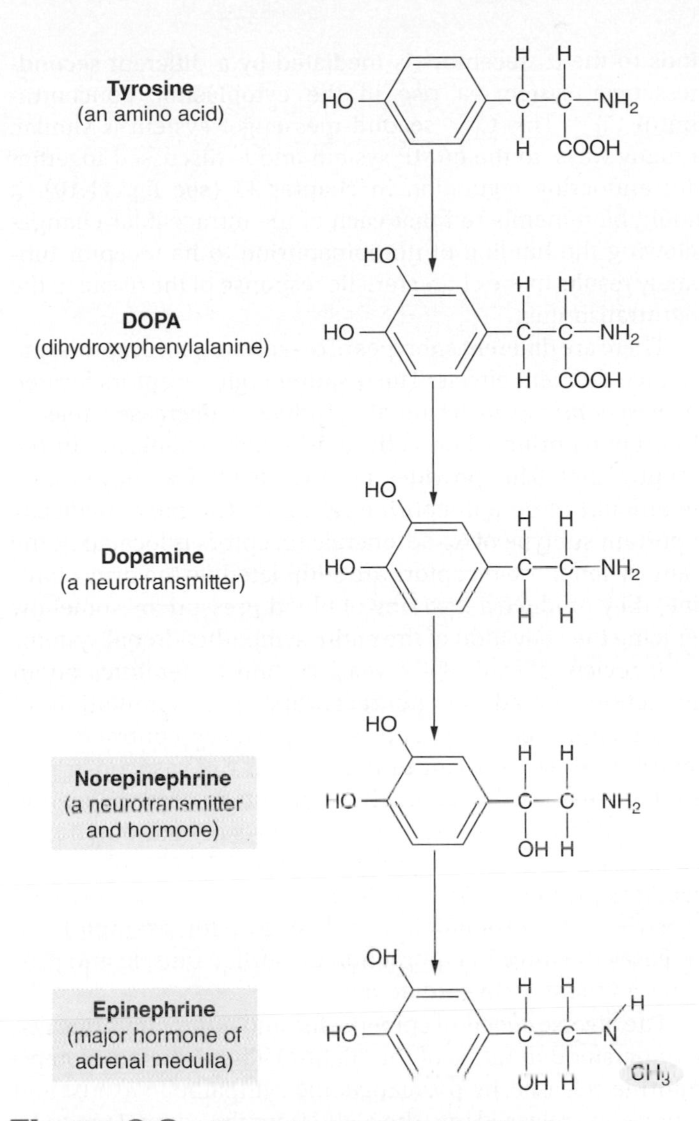

Figure 9.8 The catecholamine family of molecules. Catecholamines are derived from the amino acid tyrosine, and include both neurotransmitters (dopamine and norepinephrine) and a hormone (epinephrine). Notice that epinephrine has an additional methyl (CH_3) group compared to norepinephrine.

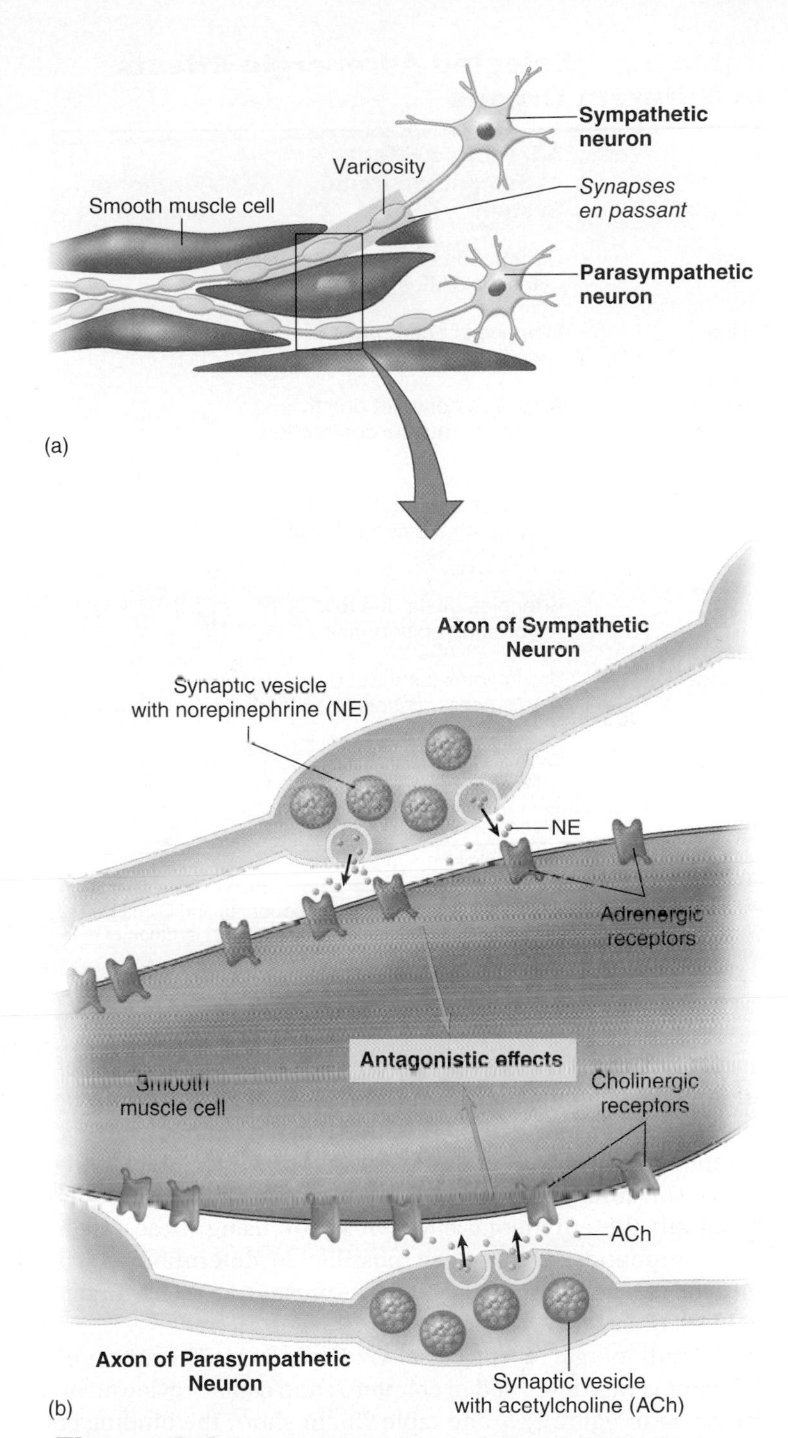

Figure 9.9 Sympathetic and parasympathetic axons release different neurotransmitters. (*a*) The axons of autonomic neurons have varicosities that form *synapses en passant* with the target cells. (*b*) In general, sympathetic axons release norepinephrine, which binds to its adrenergic receptors, while parasympathetic neurons release acetylcholine, which binds to its cholinergic receptors (discussed in chapter 7). In most cases, these two neurotransmitters elicit antagonistic responses from smooth muscles. APR

Because excitatory and inhibitory effects can be produced in different tissues by the same neurotransmitter, the responses must depend on the characteristics of the effector cells. To some degree, this is due to the presence of different membrane *receptor proteins* for the catecholamine neurotransmitters. (The interaction of neurotransmitters and receptor proteins in the postsynaptic membrane was described in chapter 7.) The two major classes of these receptor proteins are designated **alpha-** (α) and **beta-** (β) **adrenergic receptors.**

Experiments have revealed that each class of adrenergic receptor has two major subtypes. These are designated by subscripts: α_1 and α_2; β_1 and β_2. Compounds have been

Table 9.4 | Selected Adrenergic Effects in Different Organs

Organ	Adrenergic Effects of Sympathoadrenal System	Adrenergic Receptor
Eye	Contraction of radial fibers of the iris dilates the pupils	α_1
Heart	Increase in heart rate and contraction strength	β_1 primarily
Skin and visceral vessels	Arterioles constrict due to smooth muscle contraction	α_1
Skeletal muscle vessels	Arterioles constrict due to sympathetic nerve activity	α_1
	Arterioles dilate due to hormone epinephrine	β_2
Lungs	Bronchioles (airways) dilate due to smooth muscle relaxation	β_2
Stomach and intestine	Contraction of sphincters slows passage of food	α_1
Liver	Glycogenolysis and secretion of glucose	α_1, β_2

Source: Simplified from table 6-1, pp. 143–144, of Goodman and Gilman's *The Pharmacological Basis of Therapeutics.* Eleventh edition. J.E. Hardman et al., eds. 2006. McGraw-Hill.

developed that selectively bind to one or the other type of adrenergic receptor and, by this means, either promote or inhibit the normal action produced when epinephrine or norepinephrine binds to the receptor. As a result of its binding to an adrenergic receptor, a drug may either promote or inhibit the adrenergic effect. Also, by using these selective compounds, it has been possible to determine which subtype of adrenergic receptor is present in each organ (table 9.4).

All adrenergic receptors act via G-proteins. The action of G-proteins was described in chapter 7, and can be reviewed by reference to figure 7.27 and table 7.6. In short, the binding of epinephrine and norepinephrine to their receptors causes the group of three G-proteins (designated α, β, and γ) to dissociate into an α subunit and a $\beta\gamma$ complex. In different cases, either the α subunit or the $\beta\gamma$ complex causes the opening or closing of an ion channel in the plasma membrane, or the activation of an enzyme in the membrane. This begins the sequence of events that culminates in the effects of epinephrine and norepinephrine on the target cells.

All subtypes of beta receptors produce their effects by stimulating the production of cyclic AMP within the target cells. The response of a target cell when norepinephrine binds to the α_1 receptors is mediated by a different second-messenger system—a rise in the cytoplasmic concentration of Ca^{2+}. This Ca^{2+} second-messenger system is similar, in many ways, to the cAMP system and is discussed together with endocrine regulation in chapter 11 (see fig. 11.10). It should be remembered that each of the intracellular changes following the binding of norepinephrine to its receptor ultimately results in the characteristic response of the tissue to the neurotransmitter.

There are different subtypes of α_2-adrenergic receptors that produce different effects. The α_2-adrenergic receptors located on *presynaptic* axon terminals produce a decreased release of norepinephrine when activated by norepinephrine in the synaptic cleft. This provides negative feedback control over the amount of norepinephrine released. The most medically important subtype of α_2-adrenergic receptors is located in the brain. When these receptors are stimulated by the drug clonidine, they produce a lowering of blood pressure by somehow reducing the activation of the entire sympathoadrenal system.

A review of table 9.4 reveals certain generalities about the actions of adrenergic receptors. The stimulation of α_1-adrenergic receptors consistently causes contraction of smooth muscles. We can thus state that the vasoconstrictor effect of sympathetic nerves always results from the activation of alpha-adrenergic receptors. The effects of beta-adrenergic activation are more diverse; stimulation of beta-adrenergic receptors promotes the relaxation of smooth muscles (in the digestive tract, bronchioles, and uterus, for example) but increases the force of contraction of cardiac muscle and promotes an increase in cardiac rate.

The diverse effects of epinephrine and norepinephrine can be understood in terms of the "fight-or-flight" theme. Norepinephrine released by postganglionic sympathetic axons, and epinephrine released into the blood from the adrenal medulla, boost the ability of the cardiovascular system to respond to physical emergencies. The α-adrenergic receptors are more sensitive to norepinephrine, whereas the β-adrenergic receptors are more sensitive to epinephrine circulating in the blood. Stimulation of β_1- and β_2-adrenergic receptors in the heart increases heart rate and contractility. In the arterioles (small arteries) of the body, α_1-adrenergic receptors stimulate vasoconstriction and β_2-adrenergic receptors promote vasodilation in appropriate organs to prepare the body for physical exertion (fig. 9.10).

A drug that binds to the receptors for a neurotransmitter and that promotes the processes that are stimulated by that neurotransmitter is said to be an *agonist* of that neurotransmitter. A drug that blocks the action of a neurotransmitter, by contrast, is said to be an *antagonist.* The use of specific drugs that selectively stimulate or block α_1, α_2, β_1, and β_2 receptors has proven extremely useful in many medical applications. Examples of drugs that stimulate (as agonists) and block (as antagonists) adrenergic and cholinergic receptors are provided in table 9.5.

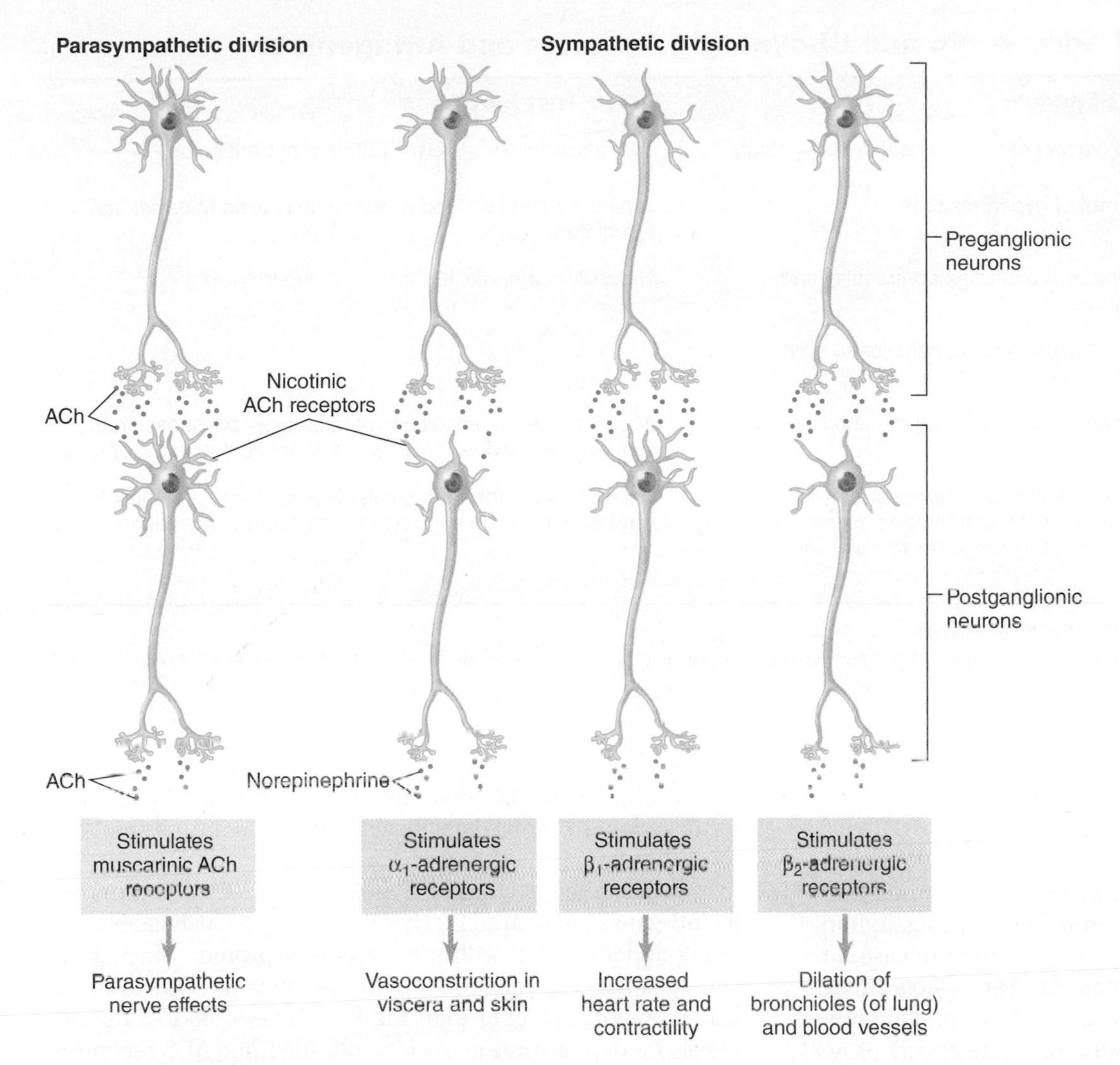

Figure 9.10 Receptors involved in autonomic regulation. Acetylcholine is released by all preganglionic neurons and stimulates postganglionic neurons by means of nicotinic ACh receptors. Postganglionic parasympathetic axons regulate their target organs through muscarinic ACh receptors. Postganglionic sympathetic axons provide adrenergic regulation of their target organs by binding of norepinephrine to α_1-, β_1-, and β_2-adrenergic receptors.

CLINICAL APPLICATION

Many people with hypertension were once treated with a beta-blocking drug known as *propranolol.* This drug blocks β_1 receptors, which are located in the heart, and thus produces the desired effect of lowering the cardiac rate and blood pressure. Propranolol, however, also blocks β_2 receptors, which are located in the bronchioles of the lungs. This reduces the bronchodilation effect of epinephrine, producing bronchoconstriction and asthma in susceptible people. A more selective β_1 antagonist, *atenolol,* is now used instead to slow the cardiac rate and lower blood pressure. At one time, asthmatics inhaled an epinephrine spray, which stimulates β_1 receptors in the heart as well as β_2 receptors in the airways. Now, drugs such as *terbutaline* that more selectively function as β_2 agonists are commonly used.

Drugs such as *phenylephrine,* which function as α_1 agonists, are often included in cold medicines because they promote vasoconstriction in the nasal mucosa. *Clonidine* is a drug that selectively stimulates α_2 receptors located on neurons in the brain. As a consequence of its action, clonidine suppresses the activation of the sympathoadrenal system and thereby helps to lower blood pressure. This drug is also helpful in treating patients with an addiction to opiates who are experiencing withdrawal symptoms.

Clinical Investigation CLUES

Kathy used her inhaler for her asthma, and had an elevated pulse rate and blood pressure in the laboratory following her physiology exam.

- How might her inhaler for asthma contribute to these elevated measurements?
- How might her exam have contributed to these elevated measurements?

Table 9.5 | Examples of Adrenergic and Cholinergic Agonists and Antagonists

Receptor Types	Drugs That Stimulate	Drugs That Block
α_1-Adrenergic	*Phenylephrine* (vasoconstrictor; nasal decongestant)	*Phentolamine* (short-term control of hypertension)
α_2-Adrenergic	*Clonidine* (control of hypertension)	*Yohimbine* (raises blood pressure; improved male sexual function)
β_1-Adrenergic	*Dobutamine* (increased cardiac contractility and cardiac output)	*Metoprolol; atenolol* (treatment of hypertension)
β_2-Adrenergic	*Terbutaline; albuterol* (dilate bronchioles to treat asthma)	
Muscarinic cholinergic	*Methacholine; pilocarpine* (pilocarpine used to constrict pupils)	*Atropine* (reduces secretions of respiratory passages; treats overactive bladder; reduces intestinal contractions; others)
Nicotinic cholinergic	*Nicotine* (no therapeutic uses; numerous toxic effects include first stimulation then depression of all autonomic ganglia and neuromuscular junctions	*D-tubocurarine* (neuromuscular blockade causing muscle relaxation during surgery and orthopedic procedures)

Selected therapeutic uses of these drugs are given in parentheses.

Modified from table 6-9, pp. 171–172, of Goodman and Gilman's *The Pharmacological Basis of Therapeutics.* Eleventh edition. 2006. McGraw-Hill.

Responses to Cholinergic Stimulation

All somatic motor neurons, all preganglionic neurons (sympathetic and parasympathetic), and most postganglionic parasympathetic neurons are cholinergic—they release acetylcholine (ACh) as a neurotransmitter. The effects of ACh released by somatic motor neurons and by preganglionic autonomic neurons are always excitatory. The effects of ACh released by postganglionic parasympathetic axons are usually excitatory, but in some cases they are inhibitory. For example, the cholinergic effect of the postganglionic parasympathetic axons innervating the heart (a part of the vagus nerve) slows the heart rate. It is useful to remember that, in general, the effects of parasympathetic innervation are opposite to the effects of sympathetic innervation.

The effects of ACh in an organ depend on the nature of the cholinergic receptor (fig. 9.11). As may be recalled from chapter 7, there are two types of cholinergic receptors—nicotinic and muscarinic. Nicotine (derived from the tobacco plant), as well as ACh, stimulates the nicotinic ACh receptors. These are located in the CNS, neuromuscular junction of skeletal muscle fibers, and in the autonomic ganglia. Nicotinic receptors are thus stimulated by ACh released by somatic motor neurons and by preganglionic autonomic neurons. Muscarine (derived from some poisonous mushrooms), as well as ACh, stimulates the ACh receptors in the visceral organs. Muscarinic receptors are thus stimulated by ACh released by postganglionic parasympathetic axons to produce the parasympathetic effects. Nicotinic and muscarinic receptors are further distinguished by the action of the drugs *curare (tubocurarine),* which specifically blocks the nicotinic ACh receptors, and *atropine* (or *belladonna*), which specifically blocks the muscarinic ACh receptors.

As described in chapter 7, the nicotinic ACh receptors are ligand-gated ion channels. That is, binding to ACh causes the ion channel to open within the receptor protein. This allows Na^+ to diffuse inward and K^+ to diffuse outward. However, the Na^+ gradient is steeper than the K^+ gradient, and so the net effect is a depolarization. As a result, nicotinic ACh receptors are always excitatory. In contrast, muscarinic ACh receptors are coupled to G-proteins, which can then close or open different membrane channels and activate different membrane enzymes. As a result, their effects can be either excitatory or inhibitory (fig. 9.11).

Scientists have identified five different subtypes of muscarinic receptors (M_1 through M_5; table 9.6). Some of these cause contraction of smooth muscles and secretion of glands, while others cause the inhibition that results in a slowing of the heart rate. These actions are mediated by second-messenger systems that will be discussed in more detail in conjunction with hormone action in chapter 11, section 11.2.

Clinical Investigation CLUES

Kathy had a headache, dry mouth, and dilated pupils following her use of various drugs on a frog heart.

- Which drug likely produced these effects?
- How did the drug work to cause these symptoms?

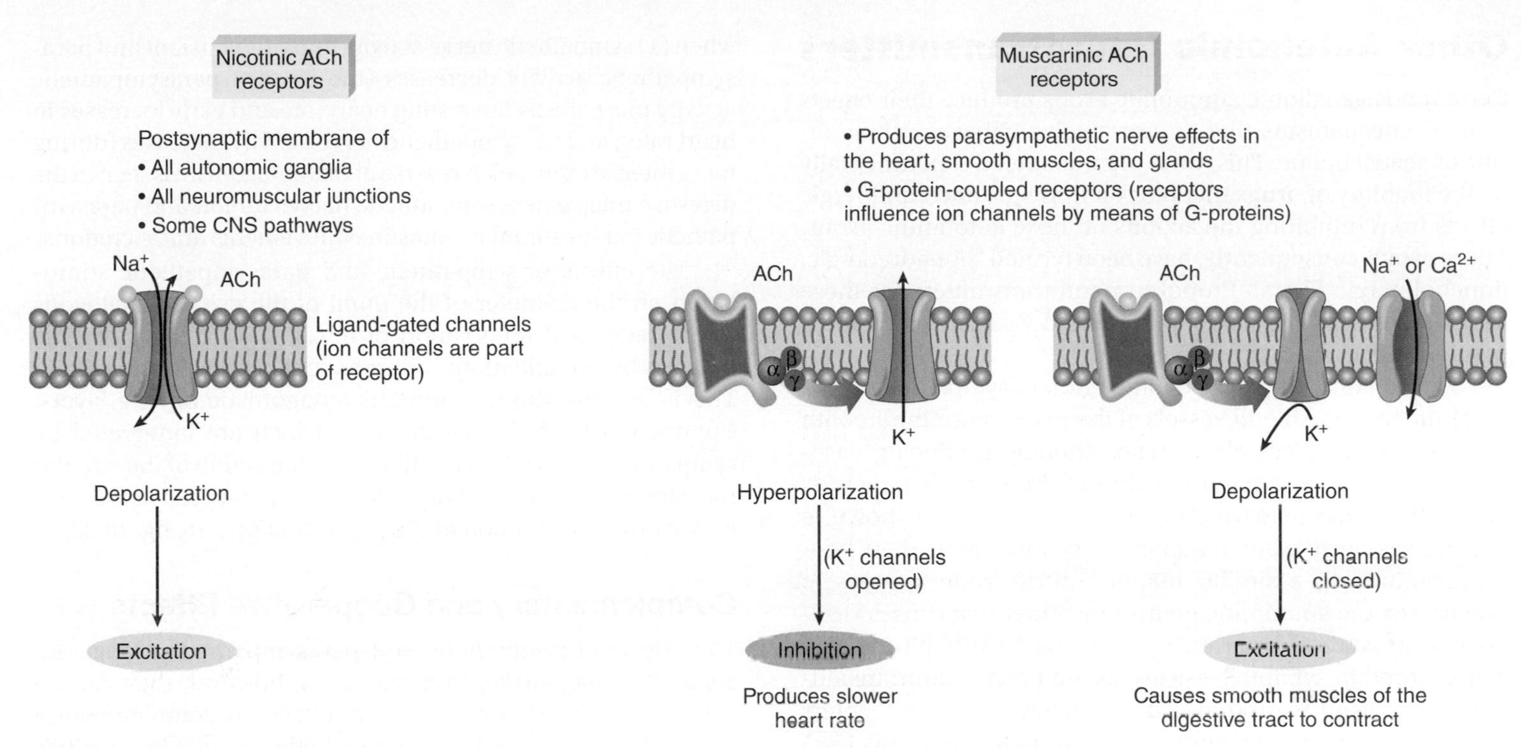

Figure 9.11 Comparison of nicotinic and muscarinic acetylcholine receptors. Nicotinic receptors are ligand-gated, meaning that the ion channel (which runs through the receptor) is opened by binding to the neurotransmitter molecule (the ligand). The muscarinic ACh receptors are G-protein-coupled receptors, meaning that the binding of ACh to its receptor indirectly opens or closes ion channels through the action of G-proteins. AP|R

Table 9.6 | Cholinergic Receptors and Responses to Acetylcholine

Receptor	Tissue	Response	Mechanisms
Nicotinic	Skeletal muscle	Depolarization, producing action potentials and muscle contraction	ACh opens cation channel in receptor
Nicotinic	Autonomic ganglia	Depolarization, causing activation of postganglionic neurons	ACh opens cation channel in receptor
Muscarinic (M_3, M_5)	Smooth muscle, glands	Depolarization and contraction of smooth muscle, secretion of glands	ACh activates G-protein coupled receptor, opening Ca^{2+} channels and increasing cytosolic Ca^{2+}
Muscarinic (M_2)	Heart	Hyperpolarization, slowing rate of spontaneous depolarization	ACh activates G-protein coupled receptor, opening channels for K^+

Source: Simplified from table 6-2, p. 119, of Goodman and Gilman's *The Pharmacological Basis of Therapeutics.* Ninth edition. J.E. Hardman et al., eds. 1996 and table 6-3, p. 156, of the Eleventh edition, 2006. McGraw-Hill.

CLINICAL APPLICATION

The muscarinic effects of ACh are specifically inhibited by the drug **atropine,** derived from the deadly nightshade plant (*Atropa belladonna*). Indeed, extracts of this plant were used by women during the Middle Ages to dilate their pupils (atropine inhibits parasympathetic stimulation of the iris). This was thought to enhance their beauty (in Italian, *bella* = beautiful, *donna* = woman). Atropine is used clinically today to dilate pupils during eye examinations, to reduce secretions of the respiratory tract prior to general anesthesia, to inhibit spasmodic contractions of the lower digestive tract, and to inhibit stomach acid secretion in a person with gastritis. Atropine is also given intramuscularly to treat exposure to *nerve gas,* which inhibits acetylcholinesterase and thereby increases synaptic transmission at both nicotinic and muscarinic ACh receptors. Atropine blocks the muscarinic effects of nerve gas, which include increased mucous secretions of the respiratory tract and muscular spasms in the pulmonary airways.

Other Autonomic Neurotransmitters

Certain postganglionic autonomic axons produce their effects through mechanisms that do not involve either norepinephrine or acetylcholine. This can be demonstrated experimentally by the inability of drugs that block adrenergic and cholinergic effects from inhibiting the actions of those autonomic axons. These axons, consequently, have been termed "nonadrenergic, noncholinergic fibers." Proposed neurotransmitters for these axons include ATP, a polypeptide called vasoactive intestinal peptide (VIP), and nitric oxide (NO).

The nonadrenergic, noncholinergic parasympathetic axons that innervate the blood vessels of the penis cause the smooth muscles of these vessels to relax, thereby producing vasodilation and a consequent erection of the penis (chapter 20, fig. 20.21). These parasympathetic axons have been shown to use the gas nitric oxide (chapter 7, section 7.6) as their neurotransmitter. In a similar manner, nitric oxide appears to function as the autonomic neurotransmitter that causes vasodilation of cerebral arteries. Studies suggest that nitric oxide is not stored in synaptic vesicles, as are other neurotransmitters, but instead is produced immediately when Ca^{2+} enters the axon terminal in response to action potentials. This Ca^{2+} indirectly activates nitric oxide synthetase, the enzyme that forms nitric oxide from the amino acid L-arginine. Nitric oxide then diffuses across the synaptic cleft and promotes relaxation of the postsynaptic smooth muscle cells.

Nitric oxide can produce relaxation of smooth muscles in many organs, including the stomach, small intestine, large intestine, and urinary bladder. There is some controversy, however, about whether the nitric oxide functions as a neurotransmitter in each case. It has been argued that, in some cases, nitric oxide could be produced in the organ itself in response to autonomic stimulation. The fact that different tissues, such as the endothelium of blood vessels, can produce nitric oxide lends support to this argument. Indeed, nitric oxide is a member of a class of local tissue regulatory molecules called *paracrine regulators* (chapter 11, section 11.7). Regulation can therefore be a complex process involving the interacting effects of different neurotransmitters, hormones, and paracrine regulators.

Organs with Dual Innervation

Most visceral organs receive **dual innervation**—they are innervated by both sympathetic and parasympathetic fibers. In this condition, the effects of the two divisions of the autonomic system may be antagonistic, complementary, or cooperative (table 9.7).

Antagonistic Effects

The effects of sympathetic and parasympathetic innervation of the pacemaker region of the heart is the best example of the antagonism of these two systems. In this case, sympathetic and parasympathetic fibers innervate the same cells. Adrenergic stimulation from sympathetic fibers increases the heart rate, whereas the release of acetylcholine from parasympathetic fibers decreases the heart rate. The heart rate is thereby increased when (1) sympathetic nerve activity remains constant and parasympathetic activity decreases (the level of parasympathetic activity most affects the resting heart rate and early increases in heart rate); and (2) sympathetic nerve activity increases (during more intense exercise). A reverse of this antagonism is seen in the digestive tract, where sympathetic nerves inhibit and parasympathetic nerves stimulate intestinal movements and secretions.

The effects of sympathetic and parasympathetic stimulation on the diameter of the pupil of the eye are analogous to the reciprocal innervation of flexor and extensor skeletal muscles by somatic motor neurons (chapter 12, section 12.5). This is because the iris contains antagonistic muscle layers. Contraction of the radial muscles, which are innervated by sympathetic nerves, causes dilation; contraction of the circular muscles, which are innervated by parasympathetic nerve endings, causes constriction of the pupils (chapter 10, fig. 10.28).

Complementary and Cooperative Effects

The effects of sympathetic and parasympathetic nerves are generally antagonistic; in a few cases, however, they can be complementary or cooperative. The effects are complementary when sympathetic and parasympathetic stimulation produce similar effects. The effects are cooperative when sympathetic and parasympathetic stimulation produce different effects that work together to promote a single action.

The effects of sympathetic and parasympathetic stimulation on salivary gland secretion are complementary. The secretion of watery saliva is stimulated by parasympathetic nerves, which also stimulate the secretion of other exocrine glands in the digestive tract. Sympathetic nerves stimulate the constriction of blood vessels throughout the digestive tract. The resultant decrease in blood flow to the salivary glands causes the production of a thicker, more viscous saliva.

The effects of sympathetic and parasympathetic stimulation on the reproductive system are cooperative. Erection of the penis, for example, is due to vasodilation resulting from parasympathetic nerve stimulation; ejaculation is due to stimulation through sympathetic nerves. The two divisions of the autonomic system thus cooperate to enable sexual function in the male. They also cooperate in the female; clitoral erection and vaginal secretions are stimulated by parasympathetic nerves, whereas orgasm is a sympathetic nerve response, as it is in the male.

There is also cooperation between the two divisions in the micturition (urination) reflex. Although the contraction of the urinary bladder is largely independent of nerve stimulation, it is promoted in part by the action of parasympathetic nerves. This is exploited clinically in helping people with incontinence (involuntary urination) caused by *overactive bladder.* In this condition, contractions of the detrusor muscle (of the urinary bladder; chapter 17, section 17.1) are stimulated by ACh released by parasympathetic axons. Newer drugs (*darifenacin* and *solifenacin*) are available to block the specific muscarinic receptor subtypes (primarily M_3) that mediate the parasympathetic stimulation of bladder contractions.

The control of micturition requires cooperation with the sympathetic division, which has antagonistic effects to the

Table 9.7 | Adrenergic and Cholinergic Effects of Sympathetic and Parasympathetic Nerves

	Effect of			
	Sympathetic		Parasympathetic	
Organ	Action	Receptor*	Action	Receptor*
Eye				
Iris				
Radial muscle	Contracts (dilates pupil)	α_1	—	—
Circular muscle	—	—	Contracts (constricts pupil)	M
Heart				
Sinoatrial node	Increases heart rate	β_1	Decreases heart rate	M
Contractility	Increases	β_1	Decreases (atria)	M
Vascular Smooth Muscle				
Skin, splanchnic vessels	Contracts (vasoconstriction)	α, β	—	—
Skeletal muscle vessels	Relaxes (vasodilation)	β_2	—	—
	Relaxes (vasodilation)	M**	—	—
Bronchiolar Smooth Muscle	Relaxes (bronchodilation)	β_2	Contracts (bronchoconstriction)	M
Gastrointestinal Tract				
Smooth muscle				
Walls	Relaxes	β_2	Contracts	M
Sphincters	Constricts	α_1	Relaxes	M
Secretion	Inhibits	α_2	Stimulates	M
Myenteric plexus	Inhibits	α	—	—
Genitourinary Smooth Muscle				
Bladder wall (detrussor m.)	Relaxes slightly	β_2	Contracts	M
Urethral sphincter	Contracts	α_1	Relaxes	M
Uterus, pregnant	Relaxes	β_2	—	—
	Contracts	α_1	—	—
Penis	Ejaculation	α_1	Erection	M
Skin				
Pilomotor smooth muscle	Contracts	α_1	—	—
Sweat glands				
Thermoregulatory	Increases	M	—	—
Apocrine (stress) in palms	Increases	α_1	—	—

*Adrenergic receptors are indicated as alpha (α) or beta (β); cholinergic receptors are indicated as muscarinic (M).

**Vascular smooth muscle in skeletal muscle has sympathetic cholinergic dilator fibers.

Source: Reproduced and modified, with permission, from Katzung, B.G.: *Basic and Clinical Pharmacology,* 6th edition, copyright Appleton & Lange, Norwalk, CT, 1995, with modifications from Goodman & Gilman's *The Pharmalogical Basis of Therapeutics,* Eleventh edition, 2006, McGraw-Hill.

parasympathetic division on the internal urethral sphincter. This smooth muscle, together with the external urethral sphincter (composed of skeletal muscle), guards the exit of the bladder to the urethra. When parasympathetic nerve activity to the detrusor muscle of the bladder increases to stimulate bladder contraction, sympathetic activity to the internal sphincter muscle must decrease to allow the sphincter to relax and the bladder to empty. Voluntary control of micturition is discussed in chapter 17, section 17.1.

Organs Without Dual Innervation

Although most organs receive dual innervation, some receive only sympathetic innervation. These include:

1. the adrenal medulla;
2. the arrector pili muscles in the skin;
3. the sweat glands in the skin; and
4. most blood vessels.

In these cases, regulation is achieved by increases or decreases in the tone (firing rate) of the sympathetic fibers. Constriction of cutaneous blood vessels, for example, is produced by increased sympathetic activity that stimulates alpha-adrenergic receptors, and vasodilation results from decreased sympathetic nerve stimulation.

The sympathoadrenal system is required for *nonshivering thermogenesis:* animals deprived of their sympathetic system and adrenals cannot tolerate cold stress. The sympathetic system itself is required for proper thermoregulatory responses to heat. In a hot room, for example, decreased sympathetic stimulation produces dilation of the blood vessels in the skin, which increases cutaneous blood flow and provides better heat radiation. During exercise, by contrast, sympathetic activity increases, causing constriction of the blood vessels in the skin of the limbs and stimulation of sweat glands in the trunk.

The sweat glands in the trunk secrete a watery fluid in response to cholinergic sympathetic stimulation. Evaporation of this dilute sweat helps to cool the body. The sweat glands also secrete a chemical called *bradykinin* in response to sympathetic stimulation. Bradykinin stimulates dilation of the surface blood vessels near the sweat glands, helping to radiate some heat despite the fact that other cutaneous blood vessels are constricted. At the conclusion of exercise, sympathetic stimulation is reduced, causing cutaneous blood vessels to dilate. This increases blood flow to the skin, which helps to eliminate metabolic heat. Notice that all of these thermoregulatory responses are achieved without the direct involvement of the parasympathetic system.

CLINICAL APPLICATION

Autonomic dysreflexia, a serious condition producing rapid elevations in blood pressure that can lead to stroke (cerebrovascular accident), occurs in 85% of people with quadriplegia and others with spinal cord lesions above the sixth thoracic level. Lesions to the spinal cord first produce the symptoms of *spinal shock,* characterized by the loss of both skeletal muscle and autonomic reflexes. After a period of time, both types of reflexes return in an exaggerated state. The skeletal muscles may become spastic in the absence of the inhibitory effects of descending tracts, and the visceral organs experience denervation hypersensitivity. Patients in this condition have difficulty emptying their urinary bladders and often must be catheterized.

Noxious stimuli, such as overdistension of the urinary bladder, can result in reflex activation of the sympathetic nerves below the spinal cord lesion. This produces goose bumps, cold skin, and vasoconstriction in the regions served by the spinal cord below the level of the lesion. The rise in blood pressure resulting from this vasoconstriction activates pressure receptors that transmit impulses along sensory nerve fibers to the medulla oblongata. In response to this sensory input, the medulla directs a reflex slowing of the heart and vasodilation. Since descending impulses are blocked by the spinal lesion, however, the skin above the lesion is warm and moist (due to vasodilation and sweat gland secretion), but it is cold (due to vasoconstriction) below the level of spinal cord damage.

Control of the Autonomic Nervous System by Higher Brain Centers

Visceral functions are largely regulated by autonomic reflexes. In most autonomic reflexes, sensory input is transmitted to brain centers that integrate this information and respond by modifying the activity of preganglionic autonomic neurons. The neural centers that directly control the activity of autonomic nerves are influenced by higher brain areas, as well as by sensory input.

The *medulla oblongata* of the brain stem controls many activities of the autonomic system. Almost all autonomic responses can be elicited by experimental stimulation of the medulla, where centers for the control of the cardiovascular, pulmonary, urinary, reproductive, and digestive systems are located. Much of the sensory input to these centers travels in the afferent fibers of the vagus nerve—a mixed nerve containing both sensory and motor fibers. The reflexes that result are listed in table 9.8.

Although it directly regulates the activity of autonomic motor fibers, the medulla itself is responsive to regulation by higher brain areas. One of these areas is the *hypothalamus,* the brain region that contains centers for the control of body temperature, hunger, and thirst; for regulation of the pituitary gland; and (together with the limbic system and cerebral cortex)

Table 9.8 | Sensory Receptors Stimulate Afferent Fibers in the Vagus, Which Transmit to the Medulla Oblongata and Cause Autonomic Reflexes

Organs	Type of Receptors	Reflex Effects
Lungs	Stretch receptors	Further inhalation inhibited; increase in cardiac rate and vasodilation stimulated
	Type J receptors	Stimulated by pulmonary congestion—produces feelings of breathlessness and causes a reflex fall in cardiac rate and blood pressure
Aorta	Chemoreceptors	Stimulated by rise in CO_2 and fall in O_2—produces increased rate of breathing, rise in heart rate, and vasoconstriction
	Baroreceptors	Stimulated by increased blood pressure—produces a reflex decrease in heart rate
Heart	Atrial stretch receptors	Antidiuretic hormone secretion inhibited, thus increasing the volume of urine excreted
	Stretch receptors in ventricles	Produces a reflex decrease in heart rate and vasodilation
Gastrointestinal tract	Stretch receptors	Feelings of satiety, discomfort, and pain

for various emotional states. Because several of these functions involve appropriate activation of sympathetic and parasympathetic nerves, many scientists consider the hypothalamus to be the major regulatory center of the autonomic system.

The *limbic system* is a group of fiber tracts and nuclei that form a ring around the brain stem (chapter 8, section 8.2). It includes the cingulate gyrus of the cerebral cortex, the hypothalamus, the fornix (a fiber tract), the hippocampus, and the amygdaloid nucleus (see fig. 8.15). The limbic system is involved in basic emotional drives, such as anger, fear, sex, and hunger. The involvement of the limbic system with the control of autonomic function is responsible for the visceral responses that are characteristic of these emotional states. Blushing, pallor, fainting, breaking out in a cold sweat, a racing heartbeat, and "butterflies in the stomach" are only some of the many visceral reactions that accompany emotions as a result of autonomic activation.

The autonomic correlates of motion sickness—nausea, sweating, and cardiovascular changes—are eliminated by cutting the motor tracts of the cerebellum. This demonstrates that impulses from the cerebellum to the medulla oblongata influence activity of the autonomic nervous system. Experimental and clinical observations have also demonstrated that the frontal and temporal lobes of the cerebral cortex influence lower brain areas as part of their involvement in emotion and personality.

Studies indicate that aging is associated with increased levels of sympathetic nervous system activity. This represents an increased level of tonic sympathetic tone in healthy adults, not an increased response to stress. It has been suggested that the higher tonic levels of sympathetic nerve activity may promote increased catabolism, generating heat and helping to combat the greater amounts of adipose tissue in the elderly. However, chronically elevated sympathetic tone may increase the risk of hypertension and cardiovascular diseases.

CLINICAL APPLICATION

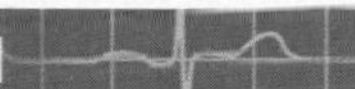

Traditionally, the distinction between the somatic system and the autonomic nervous system was drawn on the basis that the former is under conscious control whereas the latter is not. However, scientists have learned that conscious processes in the cerebrum can influence autonomic activity. In **biofeedback** techniques, data obtained from devices that detect and amplify changes in blood pressure and heart rate, for example, are "fed back" to patients in the form of light signals or audible tones. The patients can often be trained to consciously reduce the frequency of the signals and, eventually, to control visceral activities without the aid of a machine. Biofeedback has been used successfully to treat hypertension, stress, and migraine headaches.

CHECKPOINT

6. Define *adrenergic* and *cholinergic* and use these terms to describe the neurotransmitters of different autonomic nerve fibers.
7. List the effects of sympathoadrenal stimulation on different effector organs. In each case, indicate whether the effect is due to alpha- or beta-receptor stimulation.
8. Describe the effects of the drug atropine and explain these effects in terms of the actions of the parasympathetic system.
9. Explain how the effects of the sympathetic and parasympathetic systems can be antagonistic, cooperative, or complementary. Include specific examples of these different types of effects in your explanation.

Interactions

HPer Links of the Nervous System with Other Body Systems

Integumentary System

- The skin houses receptors for heat, cold, pain, pressure, and vibration (p. 270)
- Afferent neurons conduct impulses from cutaneous receptors (p. 271)
- Sympathetic neurons to the skin help to regulate cutaneous blood flow (p. 474)

Skeletal System

- The skeleton supports and protects the brain and spinal cord (p. 207)
- Bones store calcium needed for neural function (p. 690)
- Afferent neurons from sensory receptors monitor movements of joints (p. 268)

Muscular System

- Muscle contractions generate body heat to maintain constant temperature for neural function (p. 674)
- Afferent neurons from muscle spindles transmit impulses to the CNS (p. 386)
- Somatic motor neurons innervate skeletal muscles (p. 363)
- Autonomic motor neurons innervate cardiac and smooth muscles (p. 244)

Endocrine System

- Many hormones, including sex steroids, act on the brain (p. 337)
- Hormones and neurotransmitters, such as epinephrine and norepinephrine, can have synergistic actions on a target tissue (p. 321)
- Autonomic neurons innervate endocrine glands such as the pancreatic islets (p. 678)
- The brain controls anterior pituitary function (p. 334)
- The brain controls posterior pituitary function (p. 333)

Circulatory System

- The circulatory system transports O_2 and CO_2, nutrients, and fluids to and from all organs, including the brain and spinal cord (p. 405)
- Autonomic nerves help to regulate cardiac output (p. 451)
- Autonomic nerves promote constriction and dilation of blood vessels, helping to regulate blood flow and blood pressure (p. 466)

Immune System

- Chemical factors called cytokines, released by cells of the immune system, act on the brain to promote a fever (p. 497)
- Cytokines from the immune system act on the brain to modify its regulation of pituitary gland secretion (p. 514)
- The nervous system plays a role in regulating the immune response (p. 514)

Respiratory System

- The lungs provide oxygen for all body systems and eliminate carbon dioxide (p. 533)
- Neural centers within the brain control breathing (p. 553)

Urinary System

- The kidneys eliminate metabolic wastes and help to maintain homeostasis of the blood plasma (p. 582)
- The kidneys regulate plasma concentrations of Na^+, K^+, and other ions needed for the functioning of neurons (p. 604)
- The nervous system innervates organs of the urinary system to control urination (p. 584)
- Autonomic nerves help to regulate renal blood flow (p. 589)

Digestive System

- The GI tract provides nutrients for all body organs, including those of the nervous system (p. 621)
- Autonomic nerves innervate digestive organs (p. 623)
- The GI tract contains a complex neural system, called an enteric brain, that regulates its motility and secretions (p. 647)
- Secretions of gastric juice can be stimulated through activation of brain regions (p. 645)
- Hunger is controlled by centers in the hypothalamus of the brain (p. 672)

Reproductive System

- Gonads produce sex hormones that influence brain development (p. 708)
- The brain helps to regulate secretions of gonadotropic hormones from the anterior pituitary (p. 335)
- Autonomic nerves regulate blood flow into the external genitalia, contributing to the male and female sexual response (p. 711)
- The nervous and endocrine systems cooperate in the control of lactation (p. 746)

Clinical Investigation SUMMARY

Because Cathy was under stress studying for her physiology exam, the activity of her sympathoadrenal system could have been heightened. This could have raised her heart and pulse rate and elevated her blood pressure. Perhaps it also made her have an asthma attack, for which she took an inhaler containing a β_2-adrenergic agonist, causing bronchodilation. It may have had some overlapping effects on β_1- and even α_1-adrenergic receptors, which may also have contributed to the increased blood pressure, although this contribution probably was not significant. Her headache and dry mouth following her frog lab may have been produced by accidental exposure to atropine, which blocks the parasympathetic stimulation of salivary gland secretion and promotes dilation of the pupils (by blocking parasympathetic-induced pupillary constriction). The headache could have resulted from her dilated pupils.

See the additional chapter 9 Clinical Investigation on *Autoimmune Autonomic Neuropathy* in the Connect site for this text at www.mhhe.com/Fox13.

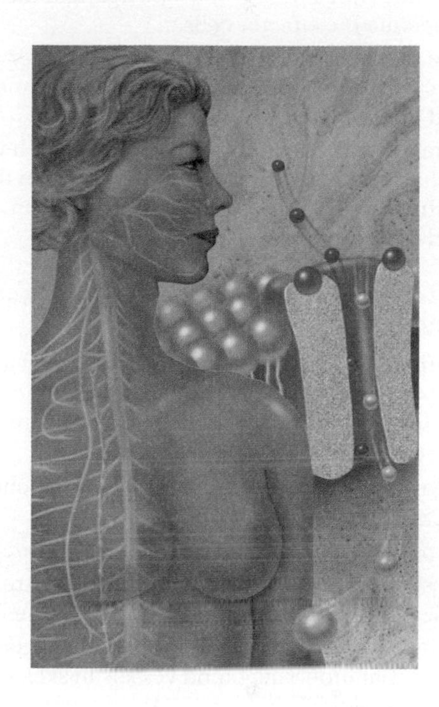

Visit this book's website at **www.mhhe.com/Fox13** for:

- Chapter quizzes, interactive learning exercises, and other study tools
- Additional clinical investigations
- Access to LearnSmart—an adaptive diagnostic tool that constantly assesses student knowledge of course material
- Ph.I.L.S. 4.0—physiology interactive lab simulations that may be used to supplement or substitute for wet labs

SUMMARY

9.1 Neural Control of Involuntary Effectors 244

A. Preganglionic autonomic neurons originate in the brain or spinal cord; postganglionic neurons originate in ganglia located outside the CNS.

B. Smooth muscle, cardiac muscle, and glands receive autonomic innervation.

1. The involuntary effectors are somewhat independent of their innervation and become hypersensitive when their innervation is removed.
2. Autonomic nerves can have either excitatory or inhibitory effects on their target organs.

9.2 Divisions of the Autonomic Nervous System 246

A. Preganglionic neurons of the sympathetic division originate in the spinal cord, between the thoracic and lumbar levels.

1. Many of these fibers synapse with postganglionic neurons whose cell bodies are located in a double chain of sympathetic (paravertebral) ganglia outside the spinal cord.
2. Some preganglionic fibers synapse in collateral (prevertebral) ganglia. These are the celiac, superior mesenteric, and inferior mesenteric ganglia.
3. Some preganglionic fibers innervate the adrenal medulla, which secretes epinephrine (and some norepinephrine) into the blood in response to stimulation.

B. Preganglionic parasympathetic fibers originate in the brain and in the sacral levels of the spinal cord.

1. Preganglionic parasympathetic fibers contribute to cranial nerves III, VII, IX, and X.
2. The long preganglionic fibers of the vagus (X) nerve synapse in terminal ganglia located next to or within

the innervated organ. Short postganglionic fibers then innervate the effector cells.

3. The vagus provides parasympathetic innervation to the heart, lungs, esophagus, stomach, liver, small intestine, and upper half of the large intestine.
4. Parasympathetic outflow from the sacral levels of the spinal cord innervates terminal ganglia in the lower half of the large intestine, in the rectum, and in the urinary and reproductive systems.

9.3 Functions of the Autonomic Nervous System 251

A. The sympathetic division of the autonomic system activates the body to "fight or flight" through adrenergic effects. The parasympathetic division often exerts antagonistic actions through cholinergic effects.

B. All preganglionic autonomic nerve fibers are cholinergic (use ACh as a neurotransmitter).
   1. All postganglionic parasympathetic fibers are cholinergic.
   2. Most postganglionic sympathetic fibers are adrenergic (use norepinephrine as a neurotransmitter).
   3. Sympathetic fibers that innervate sweat glands and those that innervate blood vessels in skeletal muscles are cholinergic.

C. Adrenergic effects include stimulation of the heart, vasoconstriction in the viscera and skin, bronchodilation, and glycogenolysis in the liver.
   1. The two main classes of adrenergic receptor proteins are alpha and beta.
   2. Some organs have only alpha or only beta receptors; other organs (such as the heart) have both types of receptors.
   3. There are two subtypes of alpha receptors (α_1 and α_2) and two subtypes of beta receptors (β_1 and β_2). These subtypes can be selectively stimulated or blocked by therapeutic drugs.

D. Cholinergic effects of parasympathetic nerves are promoted by the drug muscarine and inhibited by atropine.

E. In organs with dual innervation, the effects of the sympathetic and parasympathetic divisions can be antagonistic, complementary, or cooperative.
   1. The effects are antagonistic in the heart and pupils of the eyes.
   2. The effects are complementary in the regulation of salivary gland secretion and are cooperative in the regulation of the reproductive and urinary systems.

F. In organs without dual innervation (such as most blood vessels), regulation is achieved by variations in sympathetic nerve activity.

G. The medulla oblongata of the brain stem is the area that most directly controls the activity of the autonomic system.
   1. The medulla oblongata is, in turn, influenced by sensory input and by input from the hypothalamus.
   2. The hypothalamus is influenced by input from the limbic system, cerebellum, and cerebrum. These interconnections provide an autonomic component to some of the visceral responses that accompany emotions.

REVIEW ACTIVITIES

Test Your Knowledge

1. When a visceral organ is denervated,
   a. it ceases to function.
   b. it becomes less sensitive to subsequent stimulation by neurotransmitters.
   c. it becomes hypersensitive to subsequent stimulation.
2. Parasympathetic ganglia are located
   a. in a chain parallel to the spinal cord.
   b. in the dorsal roots of spinal nerves.
   c. next to or within the organs innervated.
   d. in the brain.
3. The neurotransmitter of preganglionic sympathetic fibers is
   a. norepinephrine.
   b. epinephrine.
   c. acetylcholine.
   d. dopamine.
4. Which of these results from stimulation of alpha-adrenergic receptors?
   a. Constriction of blood vessels
   b. Dilation of bronchioles
   c. Decreased heart rate
   d. Sweat gland secretion
5. Which of these fibers release norepinephrine?
   a. Preganglionic parasympathetic fibers
   b. Postganglionic parasympathetic fibers
   c. Postganglionic sympathetic fibers in the heart
   d. Postganglionic sympathetic fibers in sweat glands
   e. All of these
6. The effects of sympathetic and parasympathetic fibers are cooperative in
   a. the heart.
   b. the reproductive system.
   c. the digestive system.
   d. the eyes.
7. Propranolol is a beta blocker. It would therefore cause
   a. vasodilation.
   b. slowing of the heart rate.
   c. increased blood pressure.
   d. secretion of saliva.
8. Atropine blocks parasympathetic nerve effects. It would therefore cause
   a. dilation of the pupils.
   b. decreased mucous secretion.
   c. decreased movements of the digestive tract.
   d. increased heart rate.
   e. all of these.

9. Which area of the brain is most directly involved in the reflex control of the autonomic system?
   a. Hypothalamus
   b. Cerebral cortex
   c. Medulla oblongata
   d. Cerebellum
10. The two subtypes of cholinergic receptors are
    a. adrenergic and nicotinic.
    b. dopaminergic and muscarinic.
    c. nicotinic and muscarinic.
    d. nicotinic and dopaminergic.
11. A fall in cyclic AMP within the target cell occurs when norepinephrine binds to which of adrenergic receptors?
    a. α_1
    b. α_2
    c. β_1
    d. β_2
12. A drug that serves as an agonist for α_2 receptors can be used to
    a. increase the heart rate.
    b. decrease the heart rate.
    c. dilate the bronchioles.
    d. constrict the bronchioles.
    e. constrict the blood vessels.

Test Your Understanding

13. Compare the sympathetic and parasympathetic systems in terms of the location of their ganglia and the distribution of their nerves.
14. Explain the anatomical and physiological relationship between the sympathetic nervous system and the adrenal glands.
15. Compare the effects of adrenergic and cholinergic stimulation on the cardiovascular and digestive systems.
16. Explain how effectors that receive only sympathetic innervation are regulated by the autonomic system.
17. Distinguish between the different types of adrenergic receptors and state where these receptors are located in the body.
18. Give examples of drugs that selectively stimulate or block different adrenergic receptors and give examples of how these drugs are used clinically.
19. Explain what is meant by nicotinic and muscarinic ACh receptors and describe where these receptors are located in the body.
20. Give examples of drugs that selectively stimulate and block the nicotinic and muscarinic receptors and give examples of how these drugs are used clinically.

Test Your Analytical Ability

21. Shock is the medical condition that results when body tissues do not receive enough oxygen-carrying blood. It is characterized by low blood flow to the brain, leading to decreased levels of consciousness. Why would a patient with a cervical spinal cord injury be at risk of going into shock?
22. A person in shock may have pale, cold, and clammy skin and a rapid and weak pulse. What is the role of the autonomic nervous system in producing these symptoms? Discuss how drugs that influence autonomic activity might be used to treat someone in shock.
23. Imagine yourself at the starting block of the 100-meter dash of the Olympics. The gun is about to go off in the biggest race of your life. What is the autonomic nervous system doing at this point? How are your organs responding?
24. Some patients with hypertension (high blood pressure) are given beta-blocking drugs to lower their blood pressure. How does this effect occur? Explain why these drugs are not administered to patients with a history of asthma. Why might drinking coffee help asthma?
25. Why do many cold medications contain an alpha-adrenergic agonist and atropine (belladonna)? Why is there a label warning for people with hypertension? Why would a person with gastritis be given a prescription for atropine? Explain how this drug might affect the ability to digest and absorb food.

Test Your Quantitative Ability

Refer to table 7.3 (chapter 7, p. 180) to obtain axon diameters and conduction rates. Given that a rate of 1 m/sec is equal to 2.24 mph, answer the following questions.

26. What is the fastest rate (in m/sec and mph) that pain information can be transmitted?
27. How much faster is sensory information from a muscle transmitted than pain information?
28. Calculate the maximum rate of conduction (in mph) of preganglionic autonomic axons.
29. Calculate the rate of conduction (in m/sec and mph) of a postganglionic axon of average thickness.
30. How much faster does a somatic motor axon of average thickness conduct action potentials compared to a postganglionic autonomic axon of average thickness?

CHAPTER

10

Sensory Physiology

REFRESH YOUR MEMORY

Before you begin this chapter, you may want to review these concepts from previous chapters:

- **Cerebral Cortex 210**
- **Ascending Tracts 233**
- **Cranial and Spinal Nerves 236**

CHAPTER OUTLINE

Clinical Investigation

Ed, a 45-year-old man, goes to the doctor complaining of severe ear pain and reduced hearing. He has a cold, and remarks that he disembarked from an international flight the day before. The physician prescribes a decongestant and tells Ed to take an audiology test if his hearing isn't improved when his cold is better. Ed also states that he can no longer see small print clearly, even though he's never been nearsighted and his distance vision and ability to drive are not impaired. He asks if surgery might correct this.

Some of the new terms and concepts you will encounter include:

- Middle ear and auditory tube; conduction and sensorineural deafness
- Presbyopia, myopia, and hyperopia; LASIK surgery

10.1 CHARACTERISTICS OF SENSORY RECEPTORS

Each type of sensory receptor responds to a particular modality of stimulus by causing the production of action potentials in a sensory neuron. These impulses are conducted to parts of the brain that provide the proper interpretation of the sensory information when that specific neural pathway is activated.

LEARNING OUTCOMES

After studying this section, you should be able to:

1. **Explain how the stimulus modality is perceived, and how phasic receptors relate to sensory adaptation.**
2. **Describe the nature and significance of the receptor (generator) potential.**

Our perceptions of the world—its textures, colors, and sounds; its warmth, smells, and tastes—are created by the brain from electrochemical nerve impulses delivered to it from sensory receptors. These receptors **transduce** (change) different forms of energy in the "real world" into the energy of nerve impulses that are conducted into the central nervous system by sensory neurons. Different *modalities* (forms) of sensation—sound, light, pressure, and so forth—result from differences in neural pathways and synaptic connections. The brain thus interprets impulses arriving from the auditory nerve as sound and from the optic nerve as sight, even though the impulses themselves are identical in the two nerves.

We know, through the use of scientific instruments, that our senses act as energy filters that allow us to perceive only a narrow range of energy. Our vision, for example, is limited to light in a small range of electromagnetic wavelengths known as the visible spectrum. Ultraviolet and infrared light, x-rays and radio waves, which are the same type of energy as visible light, cannot normally excite the photoreceptors in our eyes. The perception of cold is entirely a product of the nervous system—there is no such thing as cold in the physical world, only varying degrees of heat. The perception of cold, however, has obvious survival value. Although filtered and distorted by the limitations of sensory function, our perceptions of the world allow us to interact effectively with the environment.

Categories of Sensory Receptors

Sensory receptors can be categorized on the basis of structure or various functional criteria. Structurally, the sensory receptors may be the dendritic endings of sensory neurons. These dendritic endings may be free, such as those that respond to pain and temperature, or encapsulated within nonneural structures, such as those that respond to pressure (see fig. 10.4). The photoreceptors in the retina of the eyes (rods and cones) are highly specialized neurons that synapse with other neurons in the retina. In the case of taste buds and of hair cells in the inner ears, modified epithelial cells respond to an environmental stimulus and activate sensory neurons.

Functional Categories

Sensory receptors can be grouped according to the type of stimulus energy they transduce. These categories include (1) **chemoreceptors,** which sense chemical stimuli in the environment or the blood (e.g., the taste buds, olfactory epithelium, and the aortic and carotid bodies); (2) **photoreceptors**—the rods and cones in the retina of the eye; (3) **thermoreceptors,** which respond to heat and cold; and (4) **mechanoreceptors,** which are stimulated by mechanical deformation of the receptor cell membrane (e.g., touch and pressure receptors in the skin and hair cells within the inner ear).

Nociceptors are pain receptors that depolarize in response to stimuli that accompany tissue damage; these stimuli include high heat or pressure and a variety of chemicals, such as adenosine, ATP, histamine, serotonin, and prostaglandin E_2. Depolarization can stimulate the production of action potentials in sensory neurons, which enter the spinal cord in the dorsal roots of spinal nerves and then relay information (via the neurotransmitters glutamate and substance P) to the brain. However, the actual perception of the pain is enhanced or reduced by a person's emotions, concepts, and expectations. This involves various brain regions that activate descending pathways in the spinal cord. Analgesia (pain reduction) depends to a large degree on the endogenous opioid neurotransmitters (including β-endorphin: chapter 7), but a nonopioid mechanism also functions to reduce the perception of pain.

Receptors also can be grouped according to the type of sensory information they deliver to the brain. **Proprioceptors** include the muscle spindles, Golgi tendon organs, and joint receptors. These provide a sense of body position and allow fine control of skeletal movements (as discussed in chapter 12). **Cutaneous** (skin) **receptors** include (1) touch and pressure receptors, (2) heat and cold receptors, and (3) pain receptors. The receptors that mediate sight, hearing, equilibrium, taste, and smell are grouped together as the **special senses.**

In addition, receptors can be grouped into **exteroceptors,** which respond to stimuli from outside of the body (such as those involved in touch, vision, and hearing), and **interoceptors,** which respond to internal stimuli. Interoceptors are found in many organs, and include mechanoreceptors and chemoreceptors. An example of mechanoreceptors are those in blood vessels that respond to stretch induced by changes in blood pressure, and chemoreceptors include those that monitor blood pH or oxygen concentration in the regulation of breathing.

Tonic and Phasic Receptors: Sensory Adaptation

Some receptors respond with a burst of activity when a stimulus is first applied, but then quickly decrease their firing rate—adapt to the stimulus—if the stimulus is maintained. Receptors with this response pattern are called *phasic receptors.* An example of a phasic receptor that responds with a pattern like that shown in figure 10.1*a* is a pacinian corpuscle (a pressure receptor—see fig. 10.4). Some other phasic receptors respond with a quick, short burst of impulses when a stimulus is first applied, and then with another quick short burst of impulses when the stimulus is removed. These phasic receptors thus provide information regarding the "on" and "off" of a stimulus. Those receptors that maintain their higher firing rate the entire time that a stimulus is applied are known as *tonic receptors* (fig. 10.1*b*).

Phasic receptors alert us to changes in sensory stimuli and are in part responsible for our ability to cease paying attention to constant stimuli. This ability is called **sensory adaptation.**

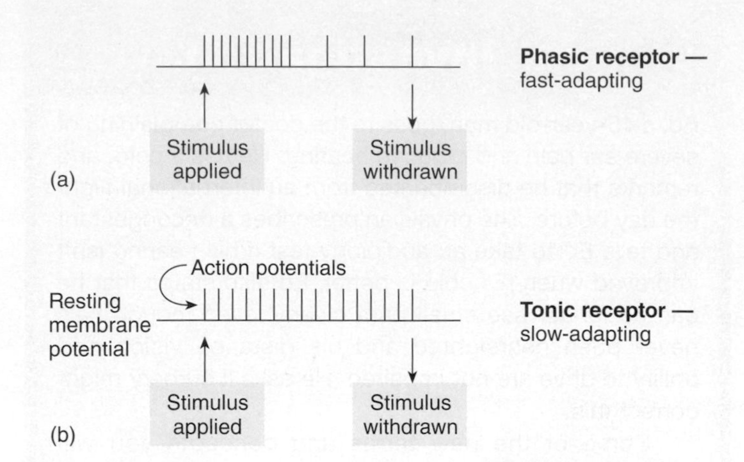

Figure 10.1 A comparison of phasic and tonic receptors. Phasic receptors (*a*) respond with a burst of action potentials when the stimulus is first applied, but then quickly reduce their firing rate if the stimulus is maintained. This produces fast-adapting sensations. Tonic receptors (*b*) continue to fire at a relatively constant rate as long as the stimulus is maintained. This produces slow-adapting sensations.

Odor, touch, and temperature, for example, adapt rapidly; bath water feels hotter when we first enter it. Sensations of pain, by contrast, adapt little, if at all.

Law of Specific Nerve Energies

Stimulation of a sensory nerve fiber produces only one sensation—touch, or cold, or pain, and so on. According to the **law of specific nerve energies,** the sensation characteristic of each sensory neuron is that produced by its normal stimulus, or *adequate stimulus* (table 10.1). Also, although a variety of different stimuli may activate a receptor, the adequate stimulus requires the least amount of energy to do so. The adequate stimulus for the photoreceptors of the eye, for example, is light, where a single photon can have a measurable effect. If these receptors are stimulated by some other means—such as by the high pressure produced by a punch to the eye—a flash of light (the adequate stimulus) may be perceived.

Table 10.1 | Classification of Receptors Based on Their Normal (or "Adequate") Stimulus

Receptor	Normal Stimulus	Mechanisms	Examples
Mechanoreceptors	Mechanical force	Deforms cell membranes of sensory dendrites or deforms hair cells that activate sensory nerve endings	Cutaneous touch and pressure receptors; vestibular apparatus and cochlea
Pain receptors	Tissue damage	Damaged tissues release chemicals that excite sensory endings	Cutaneous pain receptors
Chemoreceptors	Dissolved chemicals	Chemical interaction affects ionic permeability of sensory cells	Smell and taste (exteroceptors) osmoreceptors and carotid body chemoreceptors (interoceptors)
Photoreceptors	Light	Photochemical reaction affects ionic permeability of receptor cell	Rods and cones in retina of eye

The effect of *paradoxical cold* provides another example of the law of specific nerve energies. When the tip of a cold metal rod is touched to the skin, the perception of cold gradually disappears as the rod warms to body temperature. Then, when the tip of a rod heated to 45° C is applied to the same spot, the sensation of cold is perceived once again. This paradoxical cold is produced because the heat slightly damages receptor endings, and by this means produces an "injury current" that stimulates the receptor.

Regardless of how a sensory neuron is stimulated, therefore, only one sensory modality will be perceived. This specificity is due to the synaptic pathways within the brain that are activated by the sensory neuron. The ability of receptors to function as sensory filters so that they are stimulated by only one type of stimulus (the adequate stimulus) allows the brain to perceive the stimulus accurately under normal conditions.

Generator (Receptor) Potential

The electrical behavior of sensory nerve endings is similar to that of the dendrites of other neurons. In response to an environmental stimulus, the sensory endings produce local graded changes in the membrane potential. In most cases, these potential changes are depolarizations that are analogous to the excitatory postsynaptic potentials (EPSPs) described in chapter 7. In the sensory endings, however, these potential changes in response to stimulation are called **receptor,** or **generator, potentials** because they serve to generate action potentials in response to the sensory stimulation. Because sensory neurons are pseudounipolar (chapter 7), the action potentials produced in response to the generator potential are conducted continuously from the periphery into the CNS.

The *pacinian,* or *lamellated, corpuscle,* a cutaneous receptor for pressure (see fig. 10.4), can serve as an example of sensory transduction. When a light touch is applied to the receptor, a small depolarization (the generator potential) is produced. Increasing the pressure on the pacinian corpuscle increases the magnitude of the generator potential until it reaches the threshold depolarization required to produce an action potential (fig. 10.2). The pacinian corpuscle, however, is a phasic receptor; if the pressure is maintained, the size of the generator potential produced quickly diminishes. It is interesting to note that this phasic response is a result of the onionlike covering around the dendritic nerve ending; if the layers are peeled off and the nerve ending is stimulated directly, it will respond in a tonic fashion.

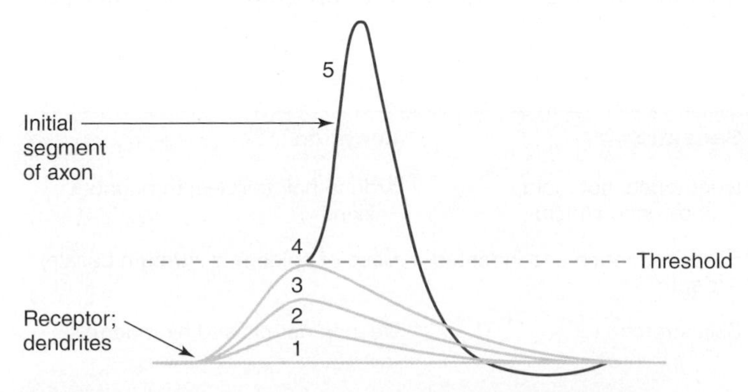

Figure 10.2 The receptor (generator) potential. Sensory stimuli result in the production of local graded potential changes known as receptor, or generator, potentials (numbers 1–4). If the receptor potential reaches a threshold value of depolarization, it generates action potentials (number 5) in the sensory neuron.

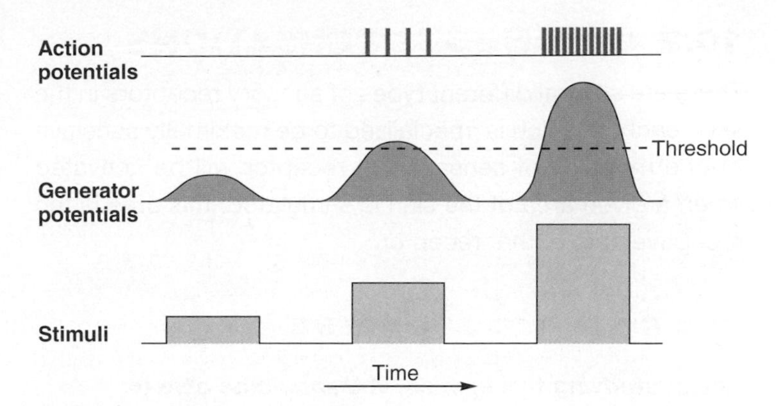

Figure 10.3 The response of tonic receptors to stimuli. Three successive stimuli of increasing strengths are delivered to a receptor. The increasing amplitude of the generator potential results in increases in the frequency of action potentials, which persist as long as the stimulus is maintained.

When a tonic receptor is stimulated, the generator potential it produces is proportional to the intensity of the stimulus. After a threshold depolarization is produced, increases in the amplitude of the generator potential result in increases in the *frequency* with which action potentials are produced (fig. 10.3). In this way, the frequency of action potentials that are conducted into the central nervous system serves as the code for the strength of the stimulus. As described in chapter 7, this frequency code is needed because the amplitude of action potentials is constant (all or none). Acting through changes in action potential frequency, tonic receptors thus provide information about the relative intensity of a stimulus.

CHECKPOINT

1a. Our perceptions are products of our brains; they relate to physical reality only indirectly and incompletely. Explain this statement, using examples of vision and the perception of cold.

1b. Explain what is meant by the law of specific nerve energies and the adequate stimulus, and relate these concepts to your answer for question 1a.

1c. Describe sensory adaptation in olfactory and pain receptors. Using a line drawing, relate sensory adaptation to the responses of phasic and tonic receptors.

2. Explain how the magnitude of a sensory stimulus is transduced into a receptor potential and how the magnitude of the receptor potential is coded in the sensory nerve fiber.

10.2 CUTANEOUS SENSATIONS

There are several different types of sensory receptors in the skin, each of which is specialized to be maximally sensitive to one modality of sensation. A receptor will be activated when a given area of the skin is stimulated; this area is the receptive field of that receptor.

LEARNING OUTCOMES

After studying this section, you should be able to:

3. Describe the sensory pathway from the skin to the postcentral gyrus.
4. Define sensory acuity and explain how it is affected by receptor density and lateral inhibition.

The **cutaneous sensations** of touch, pressure, heat and cold, and pain are mediated by the dendritic nerve endings of different sensory neurons. The receptors for heat, cold, and pain are simply the naked endings of sensory neurons. Sensations of touch are mediated by naked dendritic endings surrounding hair follicles and by expanded dendritic endings, called Ruffini endings and Merkel's discs. Merkel's discs are sensitive to the depth of skin indentation and have the highest spatial resolution of the cutaneous receptors, providing information regarding an object's texture.

The sensations of touch and pressure are also mediated by dendrites that are encapsulated within various structures (table 10.2); these include Meissner's corpuscles and pacinian (lamellated) corpuscles. In pacinian corpuscles, for example, the dendritic endings are encased within 30 to 50 onionlike layers of connective tissue (fig. 10.4). These layers absorb some of the pressure when a stimulus is maintained, which helps accentuate the phasic response of this receptor. The encapsulated touch receptors thus adapt rapidly, in contrast to the more slowly adapting Ruffini endings and Merkel's discs.

There are far more free dendritic endings that respond to cold than to warm. The receptors for cold are located in the upper region of the dermis, just below the epidermis. These receptors are stimulated by cooling and inhibited by warming. The warm receptors are located somewhat deeper in the dermis and are excited by warming and inhibited by cooling. Nociceptors are also free sensory nerve endings of either myelinated or unmyelinated fibers. The initial sharp sensation of pain, as from a pinprick, is transmitted by rapidly conducting myelinated axons of medium diameter, whereas a dull, persistent ache is transmitted by slower conducting thin unmyelinated axons. These afferent neurons synapse in the spinal cord, using substance P (an eleven-amino-acid polypeptide) and glutamate as neurotransmitters.

There is a clear distinction between the sensation of warmth and of painful heat, which activates nociceptor neurons at temperatures of about 43° C or higher. Hot temperatures produce sensations of pain through the action of a particular membrane protein in sensory dendrites. This protein, called a *capsaicin receptor,* serves as both an ion channel and a receptor for capsaicin—the molecule in chili peppers that causes sensations of heat and pain. In response to a noxiously high temperature, or to capsaicin in chili peppers, these ion channels open. This allows Ca^{2+} and Na^{+} to diffuse into the neuron, producing depolarization and resulting action potentials that are transmitted to the CNS and perceived as heat and pain.

Although the capsaicin receptor for pain is activated by intense heat, other nociceptors may be activated by mechanical stimuli that cause cellular damage. There is evidence that ATP released from damaged cells can cause pain, as can a local fall in pH produced during infection and inflammation. During an inflammation, many different cells release a wide variety of chemicals that both promote the inflammation and stimulate the nociceptor neurons. The nociceptor neurons themselves then release peptides that contribute to this "inflammatory soup" of chemicals. Some of these chemicals bind to receptor proteins on the nociceptor plasma membrane to heighten their sensitivity to temperature and touch. As a result, stimuli of temperature and touch in an inflamed area more readily elicit pain.

Table 10.2 | Cutaneous Receptors

Receptor	Structure	Sensation	Location
Free nerve endings	Unmyelinated dendrites of sensory neurons	Light touch; hot; cold; nociception (pain)	Around hair follicles; throughout skin
Merkel's discs	Expanded dendritic endings associated with 50–70 specialized cells	Sustained touch and indented depth	Base of epidermis (stratum basale)
Ruffini corpuscles (endings)	Enlarged dendritic endings with open, elongated capsule	Skin stretch	Deep in dermis and hypodermis
Meissner's corpuscles	Dendrites encapsulated in connective tissue	Changes in texture; slow vibrations	Upper dermis (papillary layer)
Pacinian corpuscles	Dendrites encapsulated by concentric lamellae of connective tissue structures	Deep pressure; fast vibrations	Deep in dermis

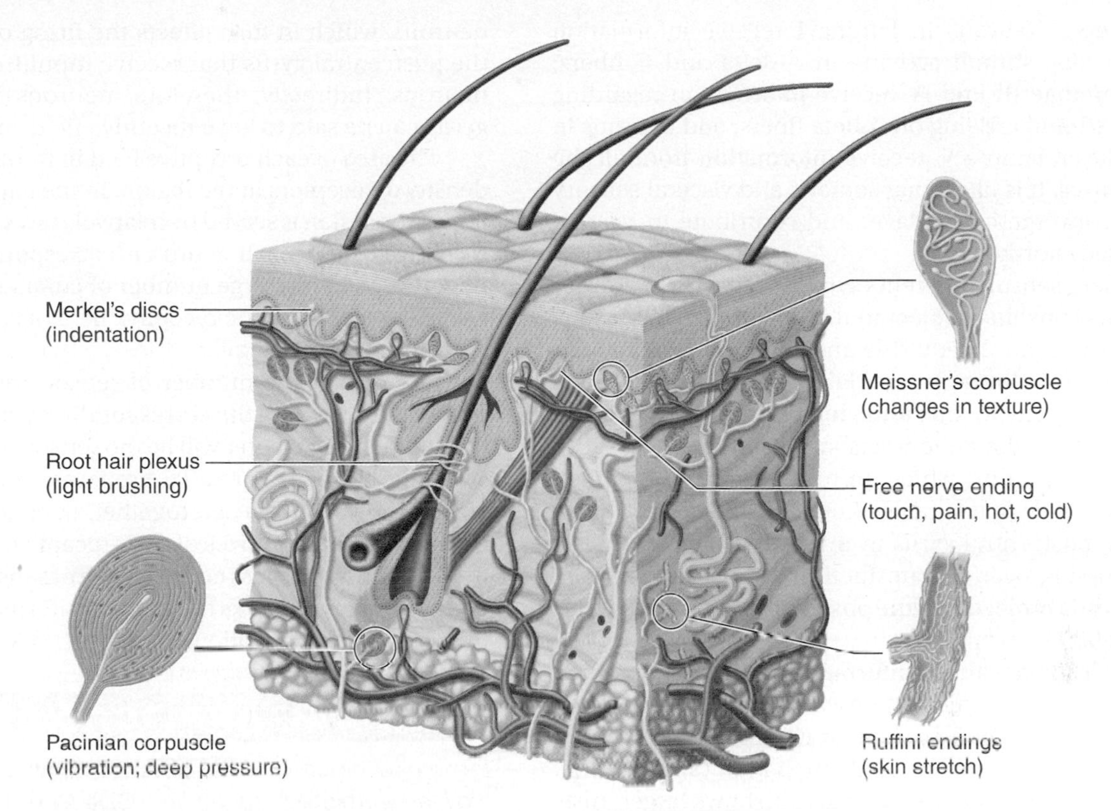

Figure 10.4 The cutaneous sensory receptors. Each of these structures is associated with a sensory (afferent) neuron. Free nerve endings are naked, dendritic branches that serve a variety of cutaneous sensations, including that of heat. Some cutaneous receptors are dendritic branches encapsulated within associated structures. Examples of this type include the pacinian (lamellated) corpuscles, which provide a sense of deep pressure, and the Meissner's corpuscles, which provide cutaneous information related to changes in texture. AP|R

Analogous to the way that capsaicin evokes the sensation of heat, menthol can produce the sensation of cold. Scientists have recently identified a membrane ion channel on sensory neurons that responds both to menthol and to cold in the 8° to 28° C range. This *cold and menthol receptor* responds to either stimulus by producing a depolarization. The cold/menthol receptors and the heat/capsaicin receptors are members of the same family of cation (Na^+ and Ca^{2+}) channels, called the *transient receptor potential (TRP)* channels. Scientists have recently identified a particular TRP channel that functions as the principal receptor for the sensation of both coolness (at temperatures below 30°C) and painful cold (at temperatures below 15°).

The sensation of *itch (pruritus)*, evoking a desire to scratch, is distinct from that of pain. Receptors for acute itch, as in a mosquito bite, are stimulated by histamine released by mast cells and basophils. Because of this, the sensation of acute itch can respond to antihistamines. The itch induced by histamine is usually accompanied by the localized redness and swelling of an inflammation (chapter 15, section 15.1). By contrast, receptors for chronic itch mostly respond to molecules other than histamine; antihistamines therefore do not relieve chronic itch. Receptors for itch stimulate unmyelinated sensory axons that conduct to the spinal cord. From there, ascending axons follow the pathways described in the next section.

Neural Pathways for Somatesthetic Sensations

The conduction pathways for the **somatesthetic senses**—which include sensations from cutaneous receptors and proprioceptors—are shown in chapter 8 (see fig. 8.24). These pathways involve three orders of neurons in series. Sensory information from proprioceptors and pressure receptors is first carried by large, myelinated axons (classified as *A-beta fibers*) that ascend in the dorsal columns of the spinal cord on the same (ipsilateral) side. These fibers do not synapse until they reach the medulla oblongata of the brain stem; hence, fibers that carry these sensations from the feet are remarkably long. After the fibers synapse in the medulla with other second-order sensory neurons, information in the latter neurons crosses over to the contralateral side as it ascends via a fiber tract, called the **medial lemniscus,** to the thalamus (chapter 8, fig. 8.24). Third-order sensory neurons in the thalamus receive this input and in turn project to the **postcentral gyrus** (the *somatosensory cortex;* see fig. 8.6).

Sensations of heat, cold, and pain are carried into the spinal cord by thin myelinated axons (classified as *A-delta fibers*) and thin unmyelinated axons (classified as *C fibers*). The sensory neurons in the dorsal roots of spinal nerves project to the dorsal horn of the spinal cord, which is organized into *laminae* (layers). Neurons in lamina I receive information regarding noxious stimuli arriving in A-delta and C fibers; neurons in laminae III and IV receive information regarding non-noxious stimuli arriving on A-beta fibers; and neurons in the deepest layer, lamina V, receive information from all the cutaneous senses. It is likely that somatic and visceral sensory information converge in this layer and contribute to *referred pain* (discussed shortly).

The primary sensory afferents synapse with second-order association neurons that project to the thalamus. Their axons cross over to the contralateral side and ascend to the brain in the **lateral spinothalamic tracts.** Fibers that mediate touch and pressure ascend in the **anterior spinothalamic tract.** Fibers of both spinothalamic tracts synapse with third-order neurons in the thalamus, which, in turn, project to the postcentral gyrus. Notice that somatesthetic information is always carried to the postcentral gyrus in third-order neurons. Also, because of crossing-over, somatesthetic information from each side of the body is projected to the postcentral gyrus of the contralateral cerebral hemisphere.

Because all somatesthetic information from the same area of the body projects to the same area of the somatosensory cortex, a "map" of the body can be drawn on the postcentral gyrus to represent sensory projection points (see fig. 8.7). This map is distorted, however, because it shows larger areas of cortex devoted to sensation in the face and hands than in other areas in the body. This disproportionately large area of the cortex devoted to the face and hands reflects the higher density of sensory receptors in these regions.

The impulses from nociceptors that are delivered to the somatosensory cortex provide information about the body location and intensity of pain. However, the emotional component of pain—the sense of "hurt"—is probably a result of impulses projecting from the thalamus to the cingulate gyrus, particularly the anterior cingulate gyrus (see chapter 8, fig. 8.18). The cingulate gyrus is a part of the limbic system, a group of brain structures involved in emotion.

Pain that is felt in a somatic location (such as the left arm) may not be the result of nociceptor stimulation in that body region, but may instead be the result of damage to an internal organ (such as the heart). This is a **referred pain** (the specific example given is known as *angina pectoris*). Another example of a referred pain is when pain in the back, under the right scapula, is caused by a gallstone when the gallbladder contracts. Referred pains are believed to result because visceral sensory and somatic sensory neurons can synapse on the same interneurons in the spinal cord. These, in turn, project to the thalamus and from there to the particular somatic location (the left arm, for example) on the somatosensory cortex.

Receptive Fields and Sensory Acuity

The **receptive field** of a neuron serving cutaneous sensation is the area of skin that, when stimulated, changes the firing rate of the neuron. Changes in the firing rate of primary sensory neurons affect the firing of second- and third-order neurons, which in turn affects the firing of those neurons in the postcentral gyrus that receive input from the third-order neurons. Indirectly, therefore, neurons in the postcentral gyrus can be said to have receptive fields in the skin.

The area of each receptive field in the skin depends on the density of receptors in the region. In the back and legs, where a large area of skin is served by relatively few sensory endings, the receptive field of each neuron is correspondingly large. In the fingertips, where a large number of cutaneous receptors serve a small area of skin, the receptive field of each sensory neuron is correspondingly small.

The greater the number of sensory receptors serving an area of the body, and the correspondingly smaller the receptive field of each, the greater will be the sensory acuity (sharpness of sensation) from that area. Two separate points of touch can be either resolved or blurred together, depending on the density of receptors and the sizes of their receptive fields. Also, resolution is improved by less convergence of sensory information on higher-order neurons as the sensory information is transmitted to the brain for perception.

CLINICAL APPLICATION

The phenomenon of the **phantom limb** was first described by a neurologist during the Civil War. In this account, a veteran with amputated legs asked for someone to massage his cramped leg muscle. It is now known that this phenomenon is common in amputees, who may experience complete sensations from the missing limbs. These sensations are sometimes useful—for example, in fitting prostheses into which the phantom has seemingly entered. However, pain in the phantom is experienced by 70% of amputees, and the pain can be severe and persistent.

One explanation for phantom limbs is that the nerves remaining in the stump can grow into nodules called neuromas, and these may generate nerve impulses that are transmitted to the brain and interpreted as arising from the missing limb. However, a phantom limb may occur in cases where the limb has not been amputated, but the nerves that normally enter from the limb have been severed. Or it may occur in individuals with spinal cord injuries above the level of the limb, so that sensations from the limb do not enter the brain. Current theories propose that the phantom may be produced by brain reorganization caused by the absence of the sensations that would normally arise from the missing limb. Such brain reorganization has been demonstrated in the thalamus and in the representational map of the body in the postcentral gyrus of the cerebral cortex.

Two-Point Touch Threshold

The approximate size of the receptive fields serving light touch can be measured by the *two-point touch threshold test.* In this procedure, the two points of a pair of calipers are lightly touched to the skin at the same time. If the distance between the points is sufficiently great, each point will stimulate a

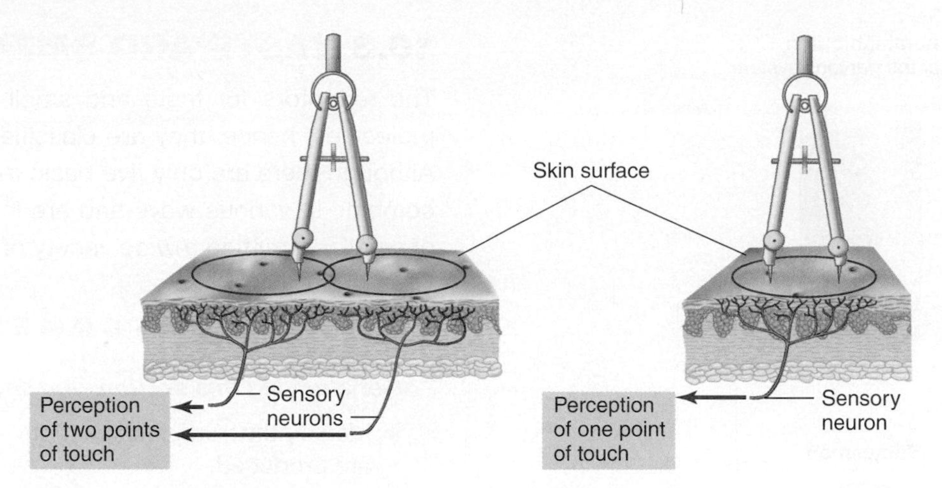

Figure 10.5 The two-point touch threshold test. If each point touches the receptive fields of different sensory neurons, two separate points of touch will be felt. If both caliper points touch the receptive field of one sensory neuron, only one point of touch will be felt.

different receptive field and a different sensory neuron—two separate points of touch will thus be felt. If the distance is sufficiently small, both points will touch the receptive field of only one sensory neuron, and only one point of touch will be felt (fig. 10.5).

The **two-point touch threshold,** which is the minimum distance at which two points of touch can be perceived as separate, is a measure of the distance between receptive fields. If the distance between the two points of the calipers is less than this minimum distance, only one "blurred" point of touch can be felt. The two-point touch threshold is thus an indication of *tactile acuity* (*acus* = needle), or the sharpness of touch perception.

The tactile acuity of the fingertips is exploited in the reading of braille. Braille symbols are formed by raised dots on the page that are separated from each other by 2.5 mm, which is slightly greater than the two-point touch threshold in the fingertips (table 10.3). Experienced braille readers can scan words at about the same speed that a sighted person can read aloud—a rate of about 100 words per minute.

Lateral Inhibition

When a blunt object touches the skin, a number of receptive fields are stimulated—some more than others. The receptive fields in the center areas where the touch is strongest will be stimulated more than those in the neighboring fields where the touch is lighter. Stimulation will gradually diminish from the point of greatest contact, without a clear, sharp boundary. What we perceive, however, is not the fuzzy sensation that might be predicted. Instead, only a single touch with well-defined borders is felt. This sharpening of sensation is due to a process called **lateral inhibition** (fig. 10.6).

Lateral inhibition and the resultant sharpening of sensation occur within the central nervous system. Those sensory neurons whose receptive fields are stimulated most strongly inhibit—via interneurons that pass "laterally" within the CNS sensory neurons that serve neighboring receptive fields.

Lateral inhibition is a common theme in sensory physiology, though the mechanisms involved are different for each sense. In hearing, lateral inhibition helps to more sharply tune the ability of the brain to distinguish sounds of different pitches. In vision, it helps the brain to more sharply distinguish borders of light and darkness; and in olfaction, it helps the brain to more clearly distinguish closely related odors.

Table 10.3 | The Two-Point Touch Threshold for Different Regions of the Body

Body Region	Two-Point Touch Threshold (mm)
Big toe	10
Sole of foot	22
Calf	48
Thigh	46
Back	42
Abdomen	36
Upper arm	47
Forehead	18
Palm of hand	13
Thumb	3
First finger	2

Source: From S. Weinstein and D.R. Kenshalo, editors, *The Skin Senses*, © 1968. Courtesy of Charles C. Thomas, Publisher, Ltd., Springfield, Illinois.

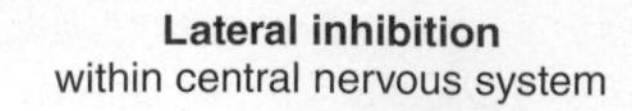

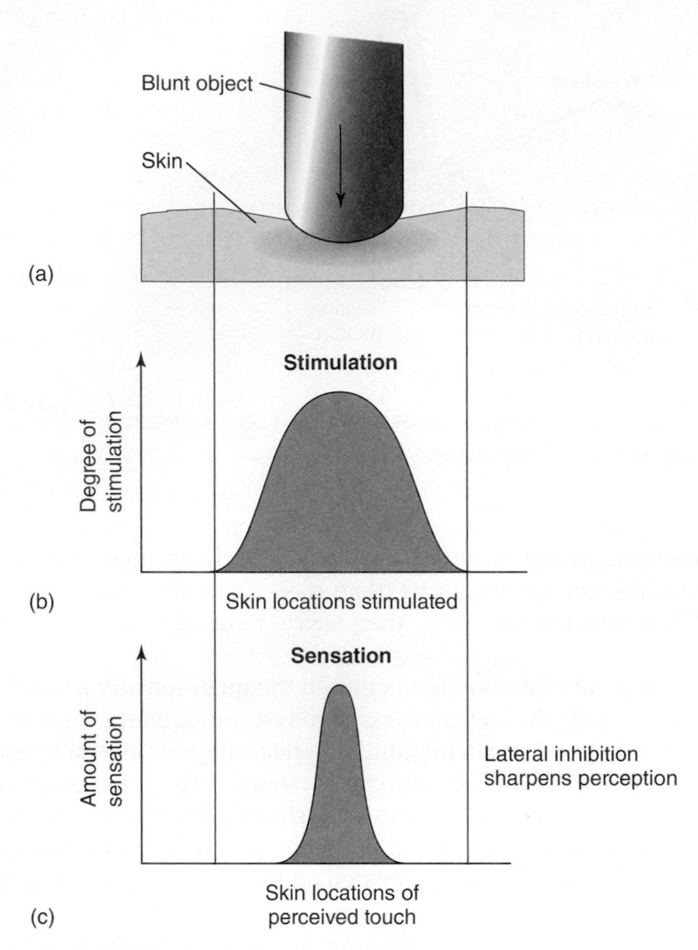

Figure 10.6 Lateral inhibition. When an object touches the skin (*a*), receptors in the central area of the touched skin are stimulated more than neighboring receptors (*b*). Lateral inhibition within the central nervous system reduces the input from these neighboring sensory neurons. Sensation, as a result, is sharpened within the area of skin that was stimulated the most (*c*).

✔ CHECKPOINT

3. Using a flow diagram, describe the neural pathways leading from cutaneous pain and pressure receptors to the postcentral gyrus. Indicate where crossing-over occurs.

4a. Define the term *sensory acuity* and explain how acuity is related to the density of receptive fields in different parts of the body.

4b. Explain the mechanism of lateral inhibition in cutaneous sensory perception and discuss its significance.

10.3 TASTE AND SMELL

The receptors for taste and smell respond to dissolved molecules; hence, they are classified as chemoreceptors. Although there are only five basic modalities of taste, they combine in various ways and are influenced by the sense of smell, permitting a wide variety of sensory experiences.

LEARNING OUTCOMES

After studying this section, you should be able to:

5. Identify the modalities of taste and explain how they are produced.

6. Explain how odorant molecules stimulate their receptors and describe how the information is conveyed to the brain.

Chemoreceptors that respond to chemical changes in the internal environment are called *interoceptors;* those that respond to chemical changes in the external environment are *exteroceptors.* Included in the latter category are *taste (gustatory) receptors,* which respond to chemicals dissolved in food or drink, and *smell (olfactory) receptors,* which respond to gaseous molecules in the air. This distinction is somewhat arbitrary, however, because odorant molecules in air must first dissolve in fluid within the olfactory mucosa before the sense of smell can be stimulated. Also, the sense of olfaction strongly influences the sense of taste, as can easily be verified by eating an onion (or almost anything else) with the nostrils pinched together.

Taste

Gustation, the sense of taste, is evoked by receptors that consist of barrel-shaped **taste buds** (fig. 10.7). Located primarily on the dorsal surface of the tongue, each taste bud consists of 50 to 100 specialized epithelial cells with long microvilli that extend through a pore in the taste bud to the external environment, where they are bathed in saliva. Although these sensory epithelial cells are not neurons, they behave like neurons; they become depolarized when stimulated appropriately, produce action potentials, and release neurotransmitters that stimulate sensory neurons associated with the taste buds. Because of this, some scientists classify the taste cells as *neuroepithelial cells.*

Taste buds are located mainly within epithelial papillae. These include *fungiform papillae* on the anterior surface of the tongue; *circumvallate papillae* on the posterior surface of the tongue; and *foliate papillae* on the sides of the tongue. Information regarding taste is transmitted from the taste buds on the fungiform papillae via the chorda tympani branch of the *facial nerve* (*VII*) and from the taste buds on the circumvallate and foliate papillae via the *glossopharyngeal nerve* (*IX*). These nerves carry taste information to a nucleus of second-order

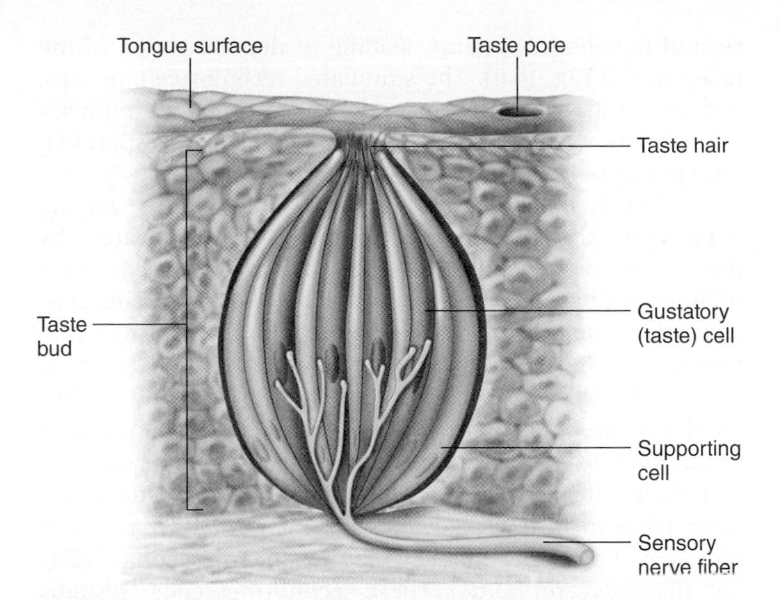

Figure 10.7 **A taste bud.** Chemicals dissolved in the fluid at the pore bind to receptor proteins in the microvilli of the sensory cells. This ultimately leads to the release of neurotransmitter, which activates the associated sensory neuron. AP|R

neurons in the medulla oblongata. From there, the second-order neurons project to the thalamus, which serves as a switchboard for directing sensory information to the cerebral cortex (chapter 8, section 8.2). Third-order neurons from the thalamus convey taste information to the *primary gustatory cortex* in the insula, and to the somatosensory cortex of the postcentral gyrus devoted to the tongue. Information is also sent to the prefrontal (orbitofrontal) cortex, which is important for taste associations and the perception of flavor.

The specialized epithelial cells of the taste bud are known as **taste cells.** The different categories of taste are produced by different chemicals that come into contact with the microvilli of these cells (fig. 10.8). Four different categories of taste are traditionally recognized: **salty, sour, sweet,** and **bitter.** There is also a more recently discovered fifth category of taste, termed **umami** (a Japanese term for "savory," related to a meaty flavor), for the amino acid glutamate (and stimulated by the flavor-enhancer monosodium glutamate). Although scientists long believed that different regions of the tongue were specialized for different tastes, this is no longer believed to be true. All areas of the tongue are able to respond to all five categories of taste. This is true even for a single taste bud, which can contain taste cells sensitive to each category of taste. However, a particular taste cell is sensitive to only one category of taste and activates a sensory neuron that transmits information regarding that specific taste to the brain.

For example, the sweet taste evoked by sugar is carried to the brain on sensory neurons devoted only to the sweet taste. Saccharin in low concentrations stimulates only the sweet receptors, but at higher concentrations can also stimulate bitter receptors and give saccharine an "aftertaste." The complex tastes we can perceive depend on the relative activities of the sensory neurons from each of the five categories of taste, together with information conveying the sense of smell from olfactory receptors. Taste is also influenced by the temperature

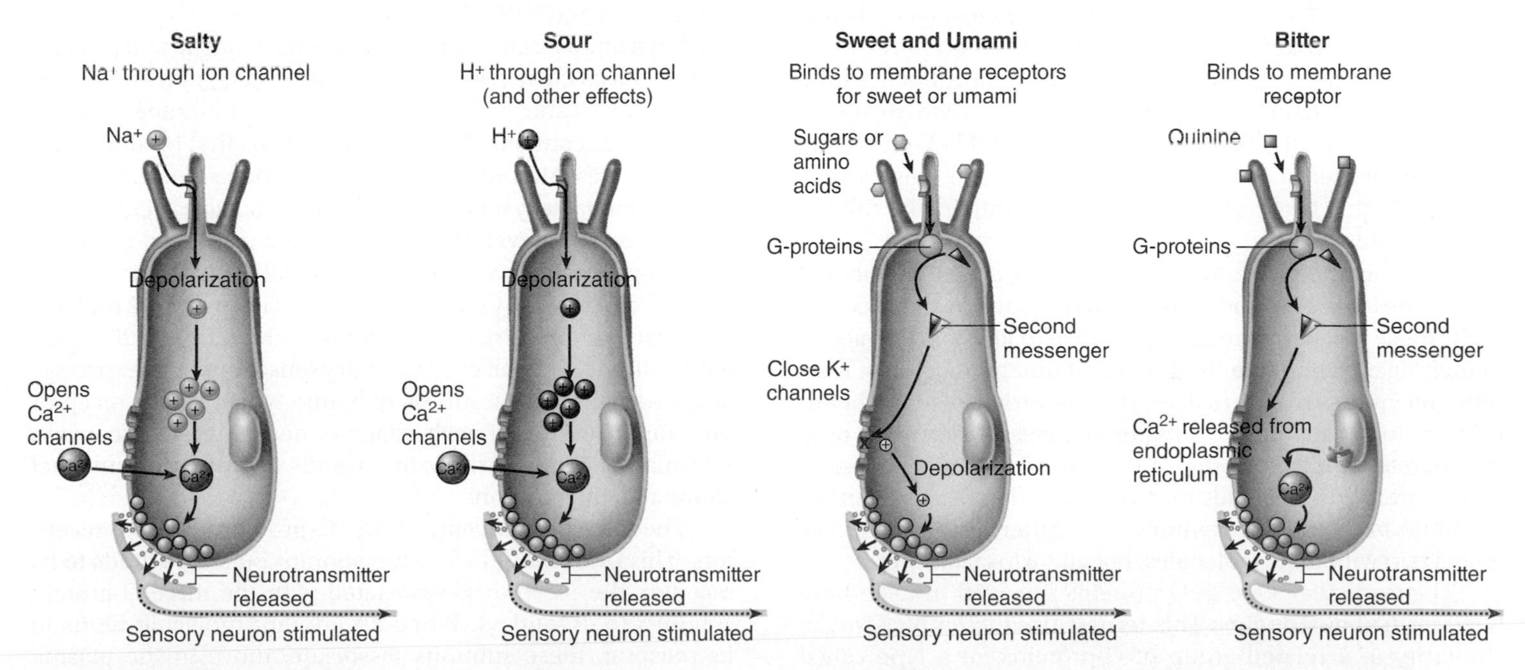

Figure 10.8 **The five major categories of taste.** Each category of taste activates specific taste cells by different means. Notice that taste cells for salty and sour are depolarized by ions (Na^+ and H^+, respectively) in the food, whereas taste cells for sweet, umami, and bitter are depolarized by sugars, the amino acids glutamate and aspartate (not shown), and quinine, respectively, by means of G-protein-coupled receptors and the actions of second messengers.

and texture of food, which stimulate receptors around the taste buds in the tongue.

The salty taste of food is due to the presence of sodium ions (Na^+), or some other cations, which activate specific receptor cells for the salty taste. Different substances taste salty to the degree that they activate these particular receptor cells. The Na^+ passes into the sensitive receptor cells through channels in the apical membranes. This depolarizes the cells, causing them to release their transmitter. The anion associated with the Na^+, however, modifies the perceived saltiness to a surprising degree: NaCl tastes much saltier than other sodium salts (such as sodium acetate). Evidence suggests that anions can pass through the tight junctions between the receptor cells and that the Cl^- anion passes through this barrier more readily than the other anions. This is presumably related to the ability of Cl^- to impart a saltier taste to the Na^+ than the other anions do.

Sour taste, like salty taste, is produced by ion movement through membrane channels. Sour taste, however, is due to the presence of hydrogen ions (H^+); all acids therefore taste sour, and the degree of sourness corresponds to the fall in pH within the taste cells. In contrast to the salty and sour tastes, the sweet and bitter tastes are produced by interaction of taste molecules with specific membrane receptor proteins.

The remaining three taste modalities—sweet, bitter, and umami—all involve interactions of the taste molecules with membrane receptors coupled to G-proteins (fig. 10.8). The ability of sweet receptors to respond to a wide variety of organic molecules is apparently due to the presence of multiple ligand binding sites in the receptor proteins. Most organic molecules, but particularly sugars, taste sweet to varying degrees when they bind to the G-protein-coupled receptors on the taste cells "tuned" to detect a sweet taste. Umami, the most recently discovered taste, evokes a savory, "meaty" sensation in response to proteins, and (along with the sweet taste) is an attractive taste modality. In humans, the G-protein-coupled umami receptors are activated only by binding of the amino acids L-glutamate and L-aspartate. Because any protein will have these amino acids in it, this is apparently sufficient to impart the umami taste modality.

Whereas the sweet and umami tastes are attractive, the bitter taste is aversive and serves to warn against toxins. Accordingly, these receptors are more sensitive to low concentrations of their ligands than are the sweet and umami receptors. Also, bitter receptors are able to detect a wide variety of toxic chemicals but do not appear to distinguish between them. Bitter taste is apparently indistinguishable if evoked by quinine or seemingly unrelated molecules that stimulate the bitter receptors. It should be noted that, although the bitter taste is generally associated with toxic molecules, not all toxins taste bitter.

The particular type of G-proteins involved in taste have been termed **gustducins.** This term is used to emphasize the similarity to a related group of G-proteins, of a type called *transducins,* associated with the photoreceptors in the eye. Dissociation of the gustducin G-protein subunit activates second-messenger systems, leading to depolarization of the receptor cell (fig. 10.8). The stimulated receptor cell, in turn, activates an associated sensory neuron that transmits impulses to the brain, where they are interpreted as the corresponding taste perception.

Although all sweet and bitter taste receptors act via G-proteins, the second-messenger systems activated by the G-proteins depend on the molecule tasted. In the case of the sweet taste of sugars, for example, the G-proteins activate adenylate cyclase, producing cyclic AMP (cAMP; see chapter 7). The cAMP, in turn, produces depolarization by closing K^+ channels that were previously open. On the other hand, the sweet taste of the amino acids phenylalanine and tryptophan, as well as of the artificial sweeteners saccharin and cyclamate, may enlist different second-messenger systems. These involve the activation of a membrane enzyme that produces the second messengers inositol triphosphate (IP_3) and diacylglycerol (DAG). These second-messenger systems are described in chapter 11, section 11.2.

Smell

The receptors responsible for **olfaction,** the sense of smell, are located in the olfactory epithelium. The olfactory apparatus consists of *receptor cells* (which are bipolar neurons), *supporting (sustentacular) cells,* and *basal stem cells.* The stem cells generate new receptor cells every one to two months to replace the neurons damaged by exposure to the environment. The supporting cells are epithelial cells rich in enzymes that oxidize hydrophobic volatile odorants, thereby making these molecules less lipid-soluble and thus less able to penetrate membranes and enter the brain.

Each bipolar sensory neuron has one dendrite that projects into the nasal cavity, where it terminates in a knob containing cilia (figs. 10.9 and 10.10). It is the plasma membrane covering the cilia that contains the receptor proteins that bind to odorant molecules. Although humans have about a thousand genes coding for olfactory receptors, most of these have accumulated mutations that prevent them from being expressed (they are "pseudogenes"), leaving an estimated 380 genes that code for 380 different olfactory receptor proteins. Through work that was awarded the Nobel Prize in Physiology or Medicine in 2004, scientists discovered that each olfactory sensory neuron expresses only one gene that produces only one type of these receptor proteins. The axon of each olfactory neuron thereby conveys information relating only to the specific odorant molecule that stimulated that neuron.

The olfactory receptors are G-protein-coupled receptors. This means that before the odorant molecule binds to its receptor, the receptor is associated with the three G-protein subunits (α, β, and γ). When an odorant molecule binds to its receptor, these subunits dissociate, move in the plasma membrane to adenylate cyclase, and activate this enzyme. Adenylate (or adenyl) cyclase catalyzes the conversion of ATP

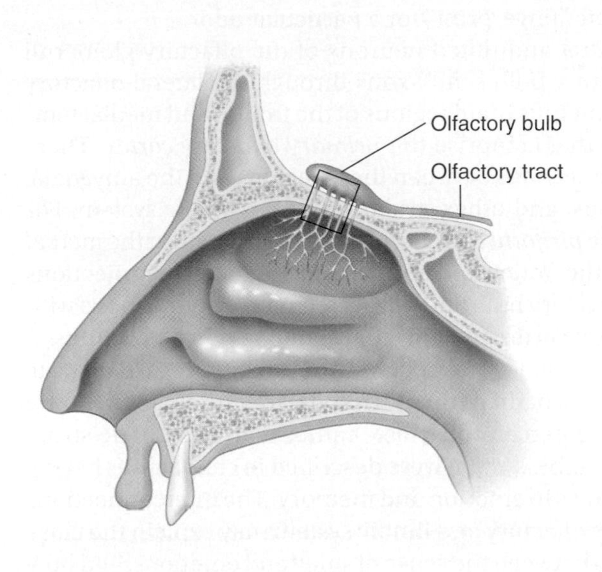

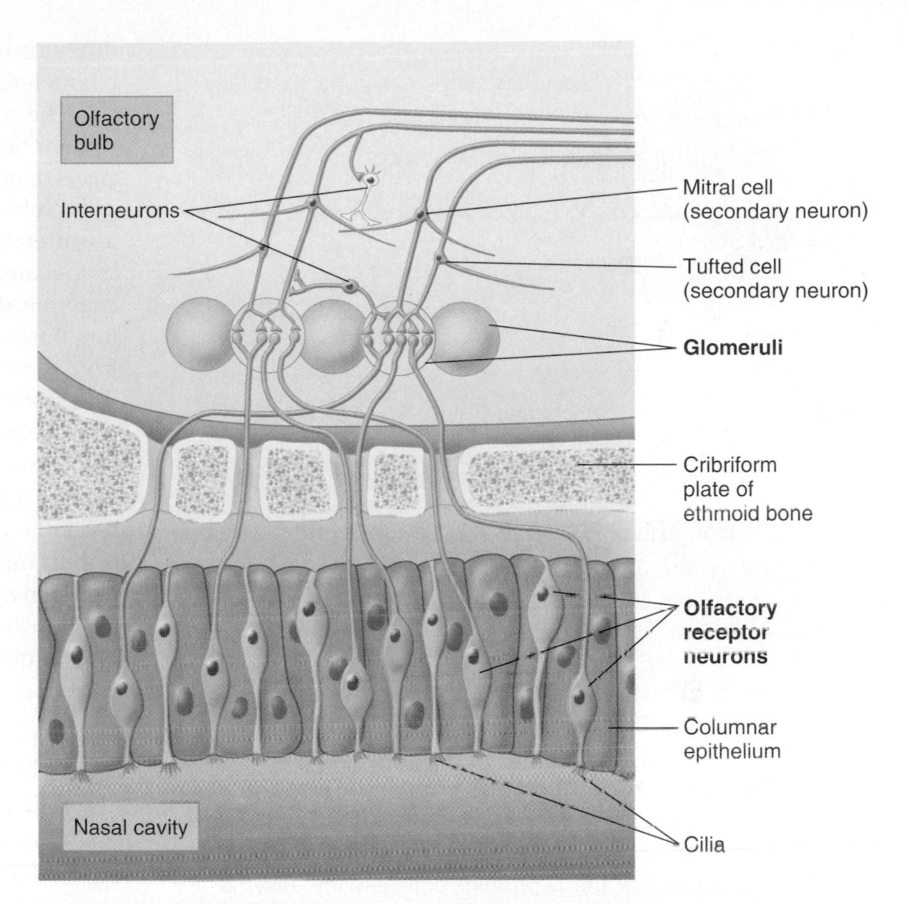

Figure 10.9 The neural pathway for olfaction. The olfactory epithelium contains receptor neurons that synapse with neurons in the olfactory bulb of the cerebral cortex. The synapses occur in rounded structures called glomeruli. Secondary neurons, known as tufted cells and mitral cells, transmit impulses from the olfactory bulb to the olfactory cortex in the medial temporal lobes. Notice that each glomerulus receives input from only one type of olfactory receptor, regardless of where those receptors are located in the olfactory epithelium. AP|R

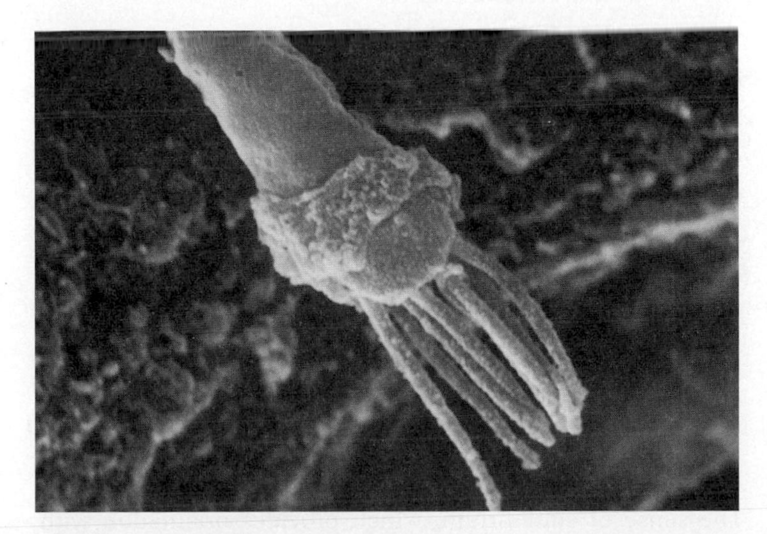

Figure 10.10 A scanning electron micrograph of an olfactory neuron. The tassel of cilia is clearly visible.

into cyclic AMP (cAMP) and PP_i (pyrophosphate). The cAMP serves as a second messenger, opening ion channels that allow inward diffusion of Na^+ and Ca^{2+} (fig. 10.11). This produces a graded depolarization, the receptor potential, which then stimulates the production of action potentials.

Up to 50 G-proteins may be associated with a single receptor protein. Dissociation of these G-proteins releases many G-protein subunits, thereby amplifying the effect many times. This amplification could account for the extreme sensitivity of the sense of smell: the human nose can detect a billionth of an ounce of perfume in air. Even at that, our sense of smell is not nearly as keen as that of many other mammals.

Once the action potential has been produced, it must be conducted into the brain to convey the olfactory sense. Each bipolar olfactory neuron has one unmyelinated axon, which projects through the holes in the cribriform plate of the ethmoid bone into the olfactory bulb of the cerebral cortex, where it synapses with second-order neurons. Therefore, unlike other

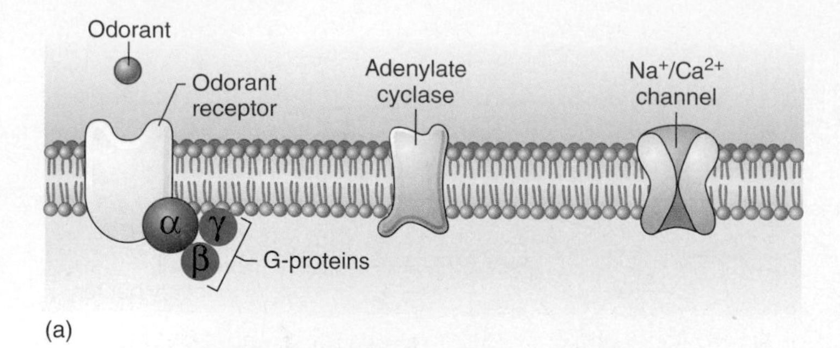

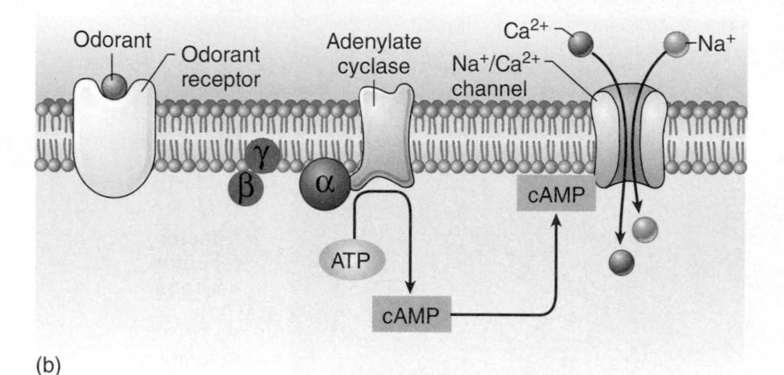

Figure 10.11 How an odorant molecule depolarizes an olfactory neuron. The olfactory receptor is coupled to many G-proteins, which dissociate upon binding of the receptor to the odorant. The α subunit of the G-proteins activates the enzyme adenylate cyclase, which catalyzes the production of cyclic AMP (cAMP). Cyclic AMP acts as a second messenger, opening cation channels. The inward diffusion of Na^+ and Ca^{2+} then produces depolarization.

sensory modalities that are first sent to the thalamus and from there relayed to the cerebral cortex, the sense of smell is transmitted directly to the cerebral cortex.

The processing of olfactory information begins in the **olfactory bulb,** where the bipolar sensory neurons synapse with neurons located in spherically shaped arrangements called *glomeruli* (see fig. 10.9). Evidence suggests that each glomerulus receives input from one type of olfactory receptor. The smell of a flower, which releases many different molecular odorants, may be identified by the pattern of excitation it produces in the glomeruli of the olfactory bulb. Identification of an odor is improved by lateral inhibition in the olfactory bulb, which appears to involve dendrodendritic synapses between neurons of adjacent glomeruli.

How can the human brain perceive as many as the estimated 10,000 different odors, if each sensory axon carries information relating to only one of about 380 olfactory receptor proteins? One reason is that a particular odorant may bind to a particular olfactory receptor protein with a high affinity, but it may also bind less avidly to other receptor proteins. In that way, a particular odorant may be perceived by the pattern of activity it produces in the glomeruli of the olfactory bulb. Evidently, the brain must somehow integrate the information from many different receptor inputs and then interpret the pattern as a characteristic "fingerprint" for a particular odor.

The mitral and tufted neurons of the olfactory glomeruli in the olfactory bulb send axons through the lateral olfactory tracts to numerous brain regions of the frontal and medial temporal lobes that comprise the *primary olfactory cortex.* There are interconnections between these regions and the amygdala, hippocampus, and other structures of the limbic system. For example, the *piriform cortex,* a pear-shaped region at the medial junction of the frontal and temporal lobes, receives projections from the olfactory bulb and makes reciprocal connections with the prefrontal cortex and amygdala, among other structures.

The prefrontal cortex receives information regarding taste as well as smell; perhaps this is why olfactory stimulation during eating can be perceived as taste rather than smell. The structures of the limbic system were described in chapter 8 as having important roles in emotion and memory. The interconnections between the olfactory and limbic system may explain the close relationship between the sense of smell and emotions, and how a particular odor can evoke emotionally charged memories.

CHECKPOINT

5. Explain how the mechanisms for sour and salty tastes are similar to each other, and how these differ from the mechanisms responsible for sweet and bitter tastes.
6. Explain how odorant molecules stimulate the olfactory receptors. Why is our sense of smell so keen?

10.4 VESTIBULAR APPARATUS AND EQUILIBRIUM

The sense of equilibrium is provided by structures in the inner ear collectively known as the vestibular apparatus. Movements of the head cause fluid within these structures to bend extensions of sensory hair cells, and this bending results in the production of action potentials.

LEARNING OUTCOMES

After studying this section, you should be able to:

7. Describe the structures of the vestibular apparatus and explain how they function to produce a sense of equilibrium.

The sense of equilibrium, which provides orientation with respect to gravity, is due to the function of an organ called the **vestibular apparatus.** The vestibular apparatus and a snail-shaped structure called the *cochlea,* which is involved in hearing, form the **inner ear** within the temporal bones

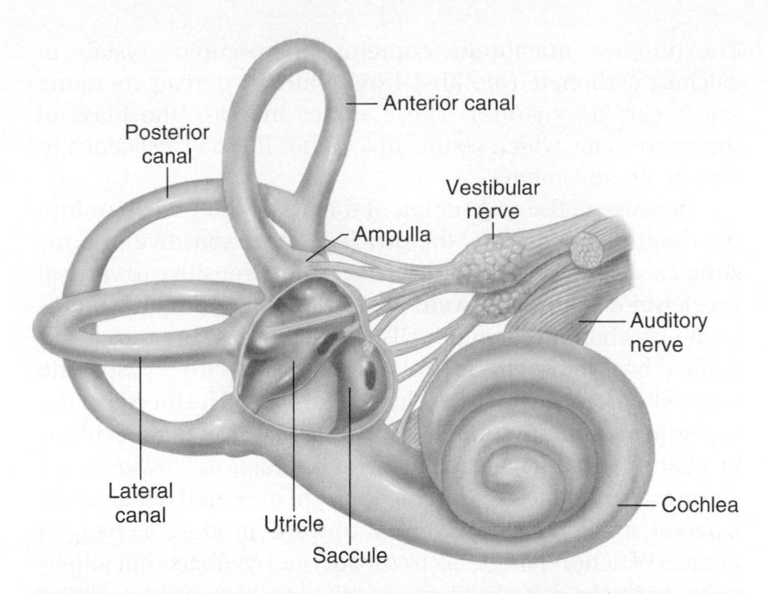

Figure 10.12 The cochlea and vestibular apparatus of the inner ear. The vestibular apparatus consists of the utricle and saccule (together called the otolith organs) and the three semicircular canals. The base of each semicircular canal is expanded into an ampulla that contains sensory hair cells. AP|R

of the skull. The vestibular apparatus consists of two parts: (1) the *otolith organs,* which include the *utricle* and *saccule,* and (2) the *semicircular canals* (fig. 10.12).

The sensory structures of the vestibular apparatus and cochlea are located within the **membranous labyrinth** (fig. 10.13), a tubular structure that is filled with a fluid called **endolymph.** Endolymph is unlike any other extracellular fluid: it has a higher K^+ concentration (higher even than in the intracellular compartment) and much lower concentrations of Na^+ and Ca^{2+} than do other extracellular fluids. Partly because of this concentration gradient, depolarization of the mechanoreceptor hair cells is produced by the passive inflow of K^+, rather than of Na^+ or Ca^{2+} as in other cells. This ion movement is also driven by the negative resting membrane potential of the hair cells, so that K^+ moves down its electrochemical gradient into the hair cells when the K^+ channels in the apical membrane of the cells are opened.

The membranous labyrinth is located within a bony cavity in the skull, the **bony labyrinth.** Within this cavity, between the membranous labyrinth and the bone, is a fluid called **perilymph.** Unlike endolymph, perilymph is fairly typical of extracellular fluids such as cerebrospinal fluid.

Sensory Hair Cells of the Vestibular Apparatus

The utricle and saccule provide information about *linear acceleration*—changes in velocity when traveling horizontally or vertically. We therefore have a sense of acceleration and deceleration when riding in a car or when skipping rope. A sense of *rotational,* or *angular, acceleration* is provided by the semicircular canals, which are oriented in three planes like the faces of a cube. This helps us maintain balance when turning the head, spinning, or tumbling.

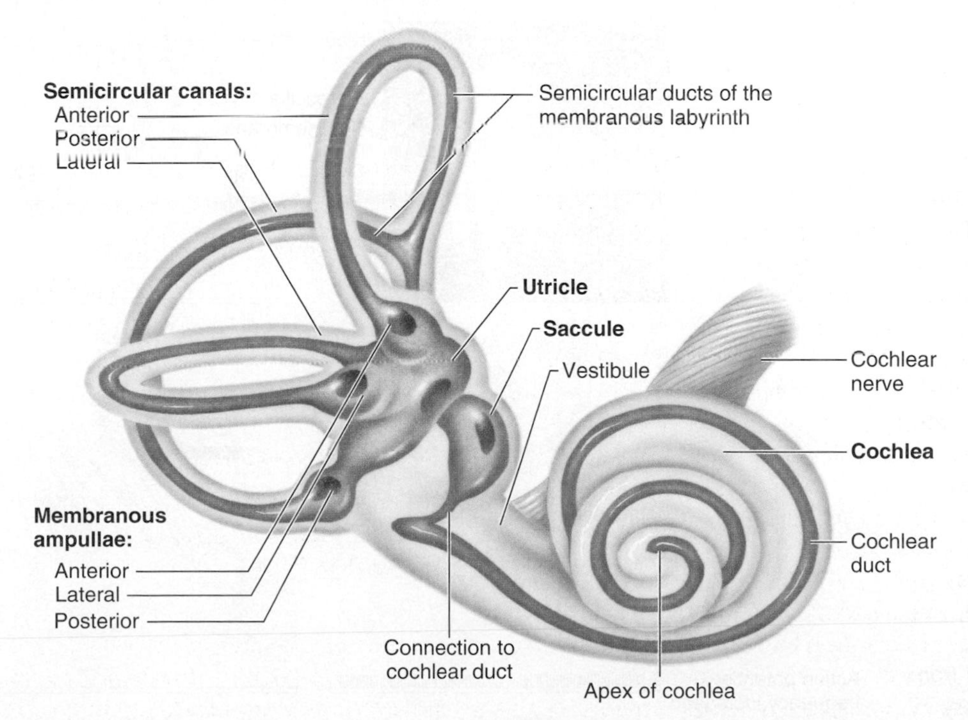

Figure 10.13 The labyrinths of the inner ear. The membranous labyrinth (darker blue) is contained within the bony labyrinth.

The receptors for equilibrium are modified epithelial cells. They are known as **hair cells** because each cell contains 20 to 50 hairlike extensions. All but one of these hairlike extensions are **stereocilia**—processes containing filaments of protein surrounded by part of the plasma membrane. One larger extension has the structure of a true cilium (chapter 3), and it is known as a **kinocilium** (fig. 10.14). When the stereocilia are bent in the direction of the kinocilium, the plasma membrane is depressed and ion channels for K^+ are opened, allowing K^+ to passively enter and depolarize the hair cell. This causes the hair cell to release a synaptic transmitter that stimulates the dendrites of sensory neurons that are part of the *vestibulocochlear nerve (VIII)*. When the stereocilia are bent in the opposite direction, the membrane of the hair cell becomes hyperpolarized (fig. 10.14) and, as a result, releases less synaptic transmitter. In this way, the frequency of action potentials in the sensory neurons that innervate the hair cells carries information about the direction of movements that cause the hair cell processes to bend.

Utricle and Saccule

The otolith organs, the **utricle** and **saccule,** each have a patch of specialized epithelium called a *macula* that consists of hair cells and supporting cells. The hair cells project into the endolymph-filled membranous labyrinth, with their hairs embedded in a gelatinous **otolithic membrane** (fig. 10.15). The otolithic membrane contains microscopic crystals of calcium carbonate (otoliths) from which it derives its name (*oto* = ear; *lith* = stone). These stones increase the mass of the membrane, which results in a higher inertia (resistance to change in movement).

Because of the orientation of their hair cell processes into the otolithic membrane, the utricle is more sensitive to horizontal acceleration and the saccule is more sensitive to vertical acceleration. During forward acceleration, the otolithic membrane lags behind the hair cells, so the hairs of the utricle are pushed backward. This is similar to the backward thrust of the body when a car quickly accelerates forward. The inertia of the otolithic membrane similarly causes the hairs of the saccule to be pushed upward when a person accelerates downward in an elevator. These effects, and the opposite ones that occur when a person accelerates backward or upward, produce a changed pattern of action potentials in sensory nerve fibers that allows us to maintain our equilibrium with respect to gravity during linear acceleration.

Semicircular Canals

The three **semicircular canals** project in three different planes at nearly right angles to each other. Each canal contains an inner extension of the membranous labyrinth called a *semicircular duct,* and at the base of each duct is an enlarged swelling called the *ampulla.* The *crista ampullaris,* an elevated area of the ampulla, is where the sensory hair cells are

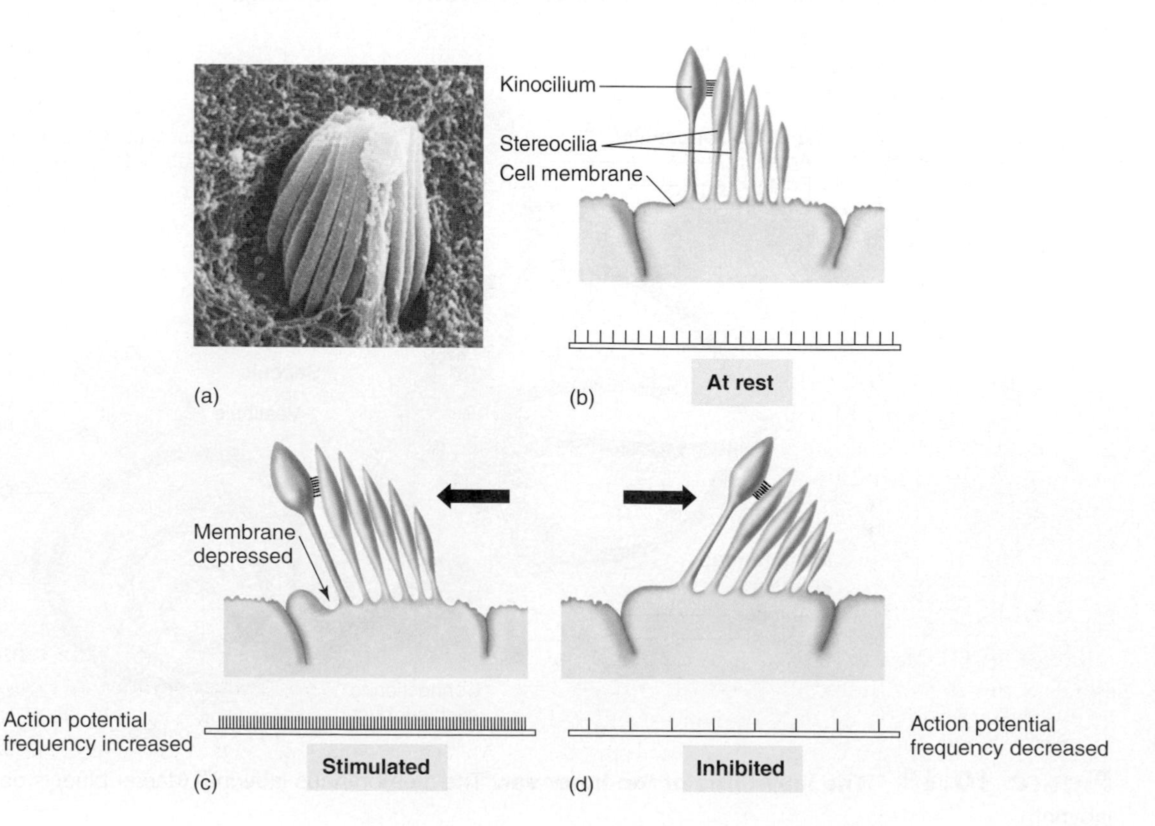

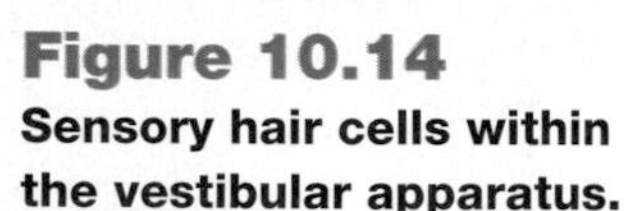

Figure 10.14
Sensory hair cells within the vestibular apparatus. (*a*) A scanning electron photograph of a kinocilium and stereocilia. (*b*) Each sensory hair cell contains a single kinocilium and several stereocilia. (*c*) When stereocilia are displaced toward the kinocilium (*arrow*), the cell membrane is depressed and the sensory neuron innervating the hair cell is stimulated. (*d*) When the stereocilia are bent in the opposite direction, away from the kinocilium, the sensory neuron is inhibited.

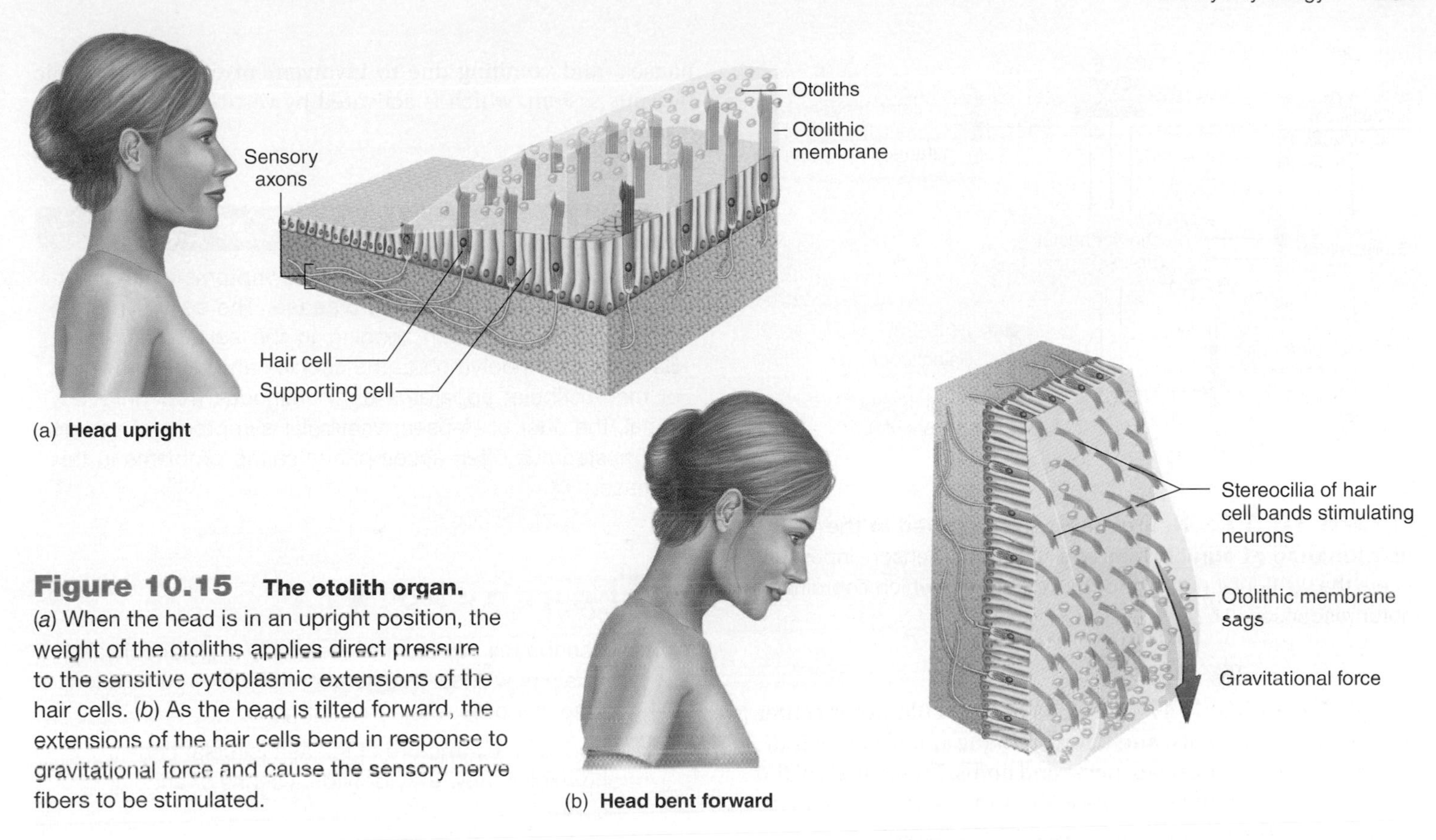

Figure 10.15 The otolith organ. (*a*) When the head is in an upright position, the weight of the otoliths applies direct pressure to the sensitive cytoplasmic extensions of the hair cells. (*b*) As the head is tilted forward, the extensions of the hair cells bend in response to gravitational force and cause the sensory nerve fibers to be stimulated.

located. The processes of these cells are embedded in a gelatinous membrane, the **cupula** (fig. 10.16), which has a higher density than that of the surrounding endolymph. Like a sail in the wind, the cupula can be pushed in one direction or the other by movements of the endolymph.

The endolymph of the semicircular canals serves a function analogous to that of the otolithic membrane—it provides inertia so that the sensory processes will be bent in a direction opposite to that of the angular acceleration. As the head rotates to the right, for example, the endolymph causes the cupula to be bent toward the left, thereby stimulating the hair cells. Hair cells in the anterior semicircular canal are stimulated when doing a somersault, those in the posterior semicircular canal are stimulated when performing a cartwheel, and those in the lateral semicircular canal are stimulated when spinning around the long axis of the body.

Neural Pathways

Stimulation of hair cells in the vestibular apparatus activates sensory neurons of the vestibulocochlear nerve (VIII). These fibers transmit impulses to the cerebellum and to the vestibular nuclei of the medulla oblongata. The vestibular nuclei, in turn, send fibers to the oculomotor center of the brain stem and to

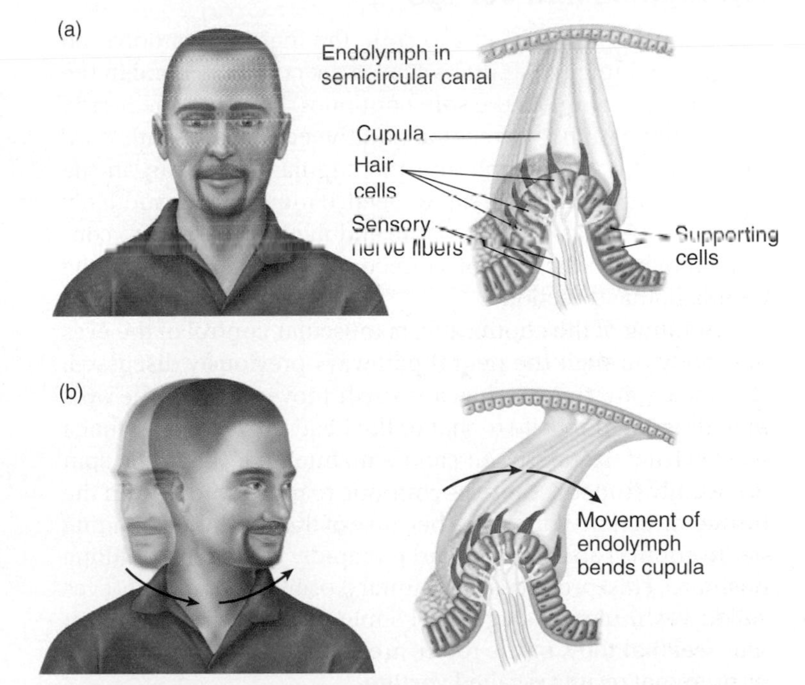

Figure 10.16 The cupula and hair cells within the semicircular canals. (*a*) Shown here, the structures are at rest or at a constant velocity. (*b*) Here, movement of the endolymph during rotation causes the cupula to bend, thus stimulating the hair cells.

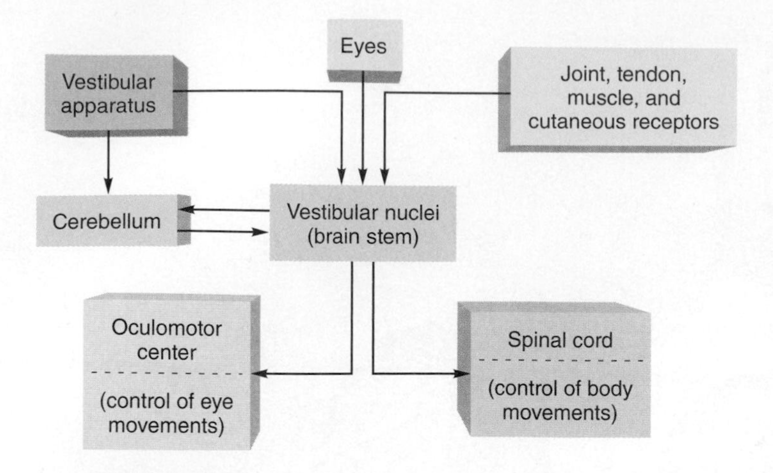

Figure 10.17 Neural pathways involved in the maintenance of equilibrium and balance. Sensory input enters the vestibular nuclei and the cerebellum, which coordinate motor responses.

the spinal cord (fig. 10.17). Neurons in the oculomotor center control eye movements, and neurons in the spinal cord stimulate movements of the head, neck, and limbs. Movements of the eyes and body produced by these pathways serve to maintain balance and "track" the visual field during rotation.

Nystagmus and Vertigo

When a person first begins to spin, the inertia of endolymph within the semicircular ducts causes the cupula to bend in the opposite direction. As the spin continues, however, the inertia of the endolymph is overcome and the cupula straightens. At this time, the endolymph and the cupula are moving in the same direction and at the same speed. If movement is suddenly stopped, the greater inertia of the endolymph causes it to continue moving in the previous direction of spin and to bend the cupula in that direction.

Bending of the cupula affects muscular control of the eyes and body through the neural pathways previously discussed. During a spin, this produces smooth movements of the eyes in a direction opposite to that of the head movement so that a stable visual fixation point can be maintained. When the spin is abruptly stopped, the eyes continue to move smoothly in the former direction of the spin (because of the continued bending of the cupula) and then are jerked rapidly back to the midline position. This produces involuntary oscillations of the eyes called **vestibular nystagmus.** People experiencing this effect may feel that they, or the room, are spinning. The loss of equilibrium that results is called **vertigo.**

Vertigo as a result of spinning is a natural response of the vestibular apparatus. Pathologically, vertigo may be caused by anything that alters the firing rate of one of the vestibulocochlear nerves (right or left) compared to the other. This is usually due to a viral infection causing vestibular neuritis. Severe vertigo is often accompanied by dizziness, pallor, sweating, nausea, and vomiting due to involvement of the autonomic nervous system, which is activated by vestibular input to the brain stem.

CLINICAL APPLICATION

Vestibular nystagmus is one of the symptoms of an inner-ear disease called **Ménière's disease.** The early symptom of this disease is often "ringing in the ears," or *tinnitus.* Because the endolymph of the cochlea and the endolymph of the vestibular apparatus are continuous through a tiny canal, the duct of Hensen, vestibular symptoms of vertigo and nystagmus often accompany hearing problems in this disease.

CHECKPOINT

7a. Describe the structure of the utricle and saccule and explain how linear acceleration results in stimulation of the hair cells within these organs.

7b. Describe the structure of the semicircular canals and explain how they provide a sense of angular acceleration.

10.5 THE EARS AND HEARING

Sound causes movements of the tympanic membrane and middle-ear ossicles that are transmitted into the fluid-filled cochlea. This produces vibrations of the basilar membrane, which is coated with hair cells. Bending of hair cell stereocilia causes the production of action potentials, which are interpreted by the brain as sound.

LEARNING OUTCOMES

After studying this section, you should be able to:

8. Explain how sound waves result in movements of the oval window and then the basilar membrane.

9. Explain how movements of the basilar membrane at different sound frequencies (pitches) affect hair cells.

10. Describe how action potentials are produced, and their neural pathways.

Sound waves are alternating zones of high and low pressure traveling in a medium, usually air or water. (Thus, sound waves cannot travel in space.) Sound waves travel in all directions from their source, like ripples in a pond where a stone has been dropped. These waves are characterized by their frequency and intensity. The **frequency** is measured in *hertz (Hz),* which is the modern designation for *cycles per second (cps).* The *pitch*

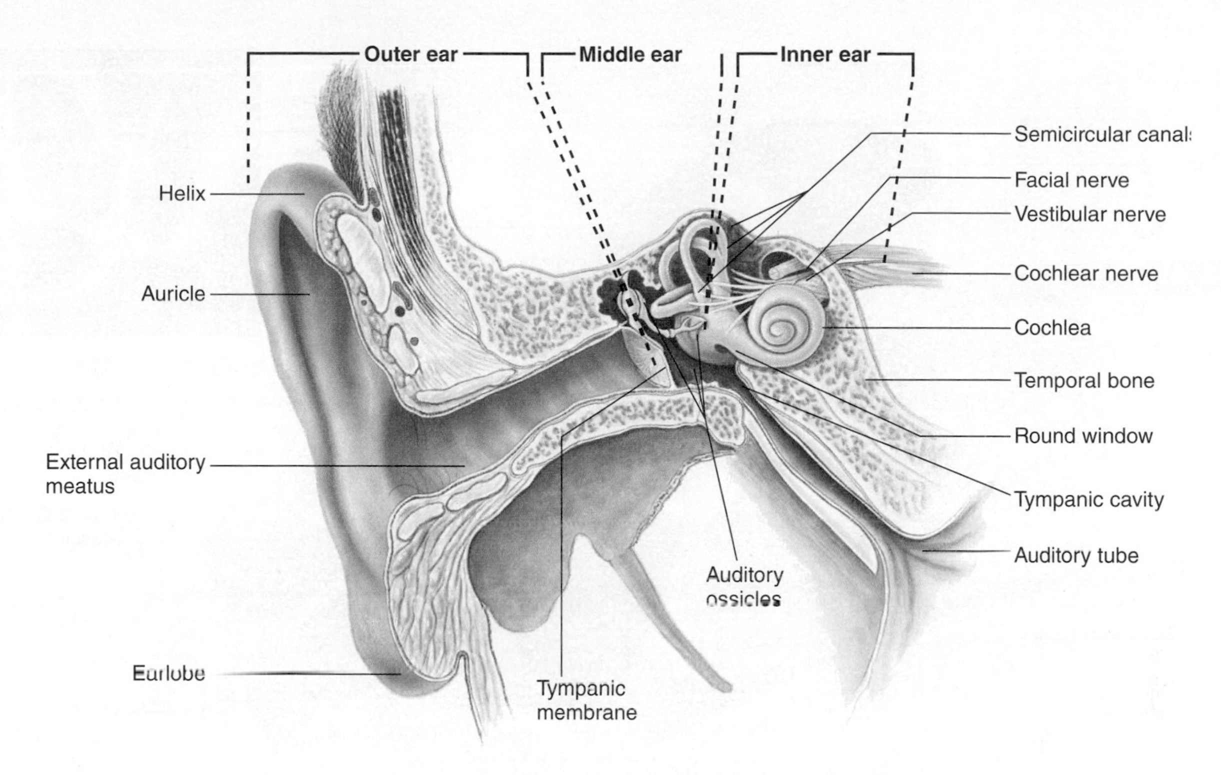

Figure 10.18 **The ear.** Note the structures of the outer, middle, and inner ear. APIR

of a sound is directly related to its frequency—the greater the frequency of a sound, the higher its pitch.

The **intensity,** or loudness, of a sound is directly related to the amplitude of the sound waves and is measured in units called *decibels (dB).* A sound that is barely audible—at the threshold of hearing—has an intensity of zero decibels. Every 10 decibels indicates a tenfold increase in sound intensity; a sound is 10 times louder than threshold at 10 dB, 100 times louder at 20 dB, a million times louder at 60 dB, and 10 billion times louder at 100 dB.

The ear of a trained, young individual can hear sound over a frequency range of 20 to 20,000 Hz, yet still can distinguish between two pitches that have only a 0.3% difference in frequency. The human ear can detect differences in sound intensities of only 0.1 to 0.5 dB, while the range of audible intensities covers 12 orders of magnitude (10^{12}), from the barely audible to the limits of painful loudness. Human hearing is optimal at sound intensities of 0 to 80 dB.

Outer Ear

Sound waves are funneled by the *pinna,* or *auricle,* into the *external auditory meatus* (fig. 10.18). These two structures form the **outer ear.** The external auditory meatus channels the sound waves to the eardrum, or **tympanic membrane.** Sound waves in the external auditory meatus produce extremely small vibrations of the tympanic membrane; movements of the eardrum during speech (with an average sound intensity of 60 dB) are estimated to be about the diameter of a molecule of hydrogen!

Middle Ear

The **middle ear** is the cavity between the tympanic membrane on the outer side and the cochlea on the inner side (fig. 10.19). Within this cavity are three **middle-ear ossicles**—the *malleus* (hammer), *incus* (anvil), and *stapes* (stirrup). The malleus is attached to the tympanic membrane, so that vibrations of this membrane are transmitted via the malleus and incus to the stapes. The stapes, in turn, is attached to a membrane in the cochlea called the *oval window,* which thus vibrates in response to vibrations of the tympanic membrane.

CLINICAL APPLICATION

Damage to the tympanic membrane or middle-ear ossicles produces **conduction deafness.** This impairment can result from a variety of causes, including otitis media and otosclerosis. In *otitis media,* which sometimes follows allergic reactions or respiratory disease, inflammation produces an excessive accumulation of fluid within the middle ear. This, in turn, can result in the excessive growth of epithelial tissue and damage to the eardrum. In *otosclerosis,* bone is resorbed and replaced by "sclerotic bone" that grows over the oval window and immobilizes the footplate of the stapes. In conduction deafness, these pathological changes hinder the transmission of sound waves from the air to the cochlea of the inner ear.

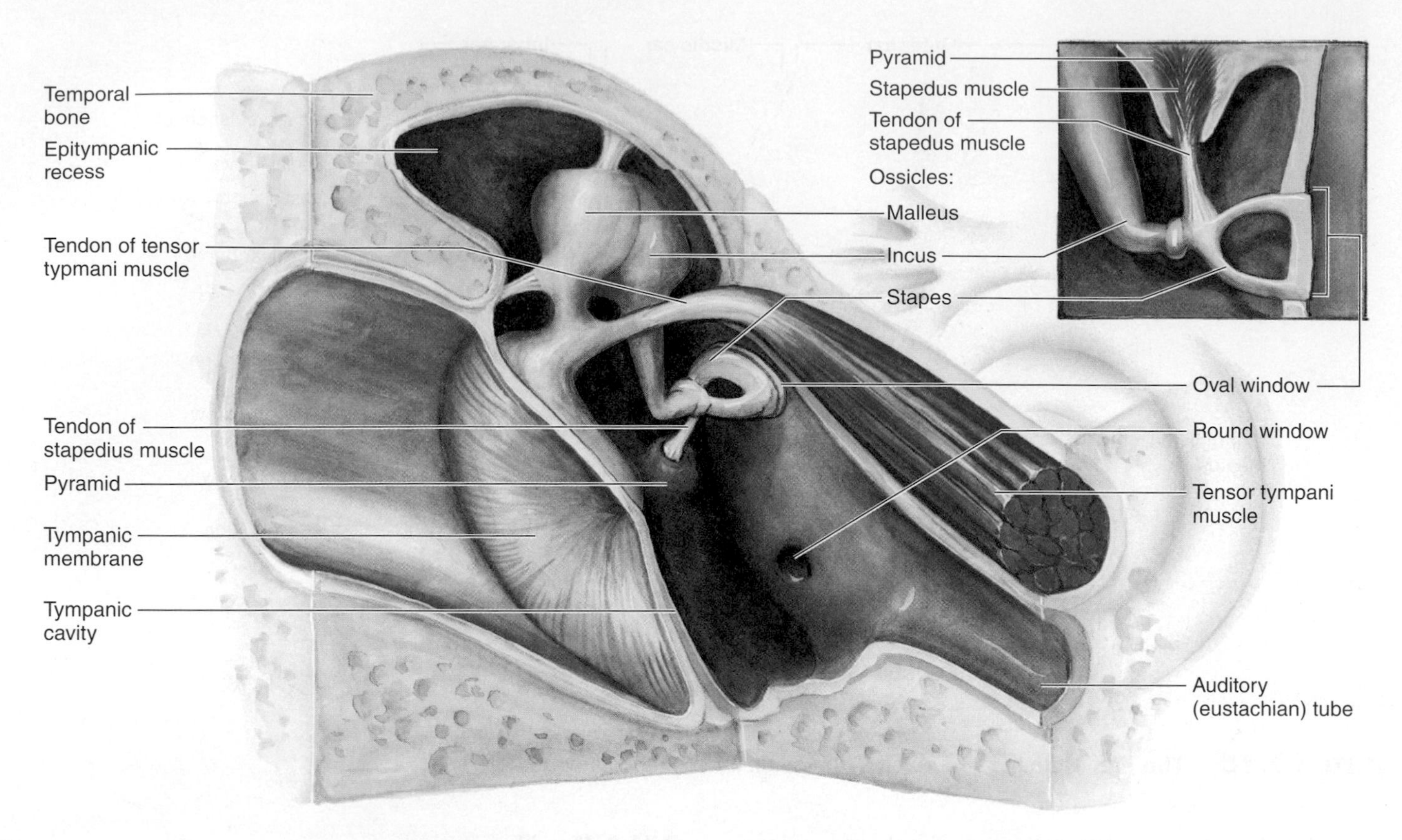

Figure 10.19 A medial view of the middle ear. The locations of auditory muscles, attached to the middle-ear ossicles, are indicated. AP|R

FITNESS APPLICATION

The **auditory (eustachian) tube** is a passageway leading from the middle ear to the nasopharynx (a cavity positioned behind the nasal cavity and extending down to the soft palate). The auditory tube is usually collapsed, so that debris and infectious agents are prevented from traveling from the oral cavity to the middle ear. In order to open the auditory tube, the *tensor tympani muscle,* attaching to the auditory tube and the malleus (fig. 10.19), must contract. This occurs during swallowing, yawning, and sneezing. People sense a "popping" sensation in their ears as they swallow when driving up a mountain because the opening of the auditory canal permits air to move from the region of higher pressure in the middle ear to the region of lower pressure in the nasopharynx.

Clinical Investigation CLUES

Ed experienced ear pain and reduced hearing when he had a cold and disembarked from an international flight.

- What may have caused Ed's ear pain and hearing impairment?
- How could this be helped by a decongestant?

The fact that vibrations of the tympanic membrane are transferred through three bones instead of just one affords protection. If the sound is too intense, the ossicles may buckle. This protection is increased by the action of the *stapedius muscle,* which attaches to the neck of the stapes (fig. 10.19). When sound becomes too loud, the stapedius muscle contracts and dampens the movements of the stapes against the oval window. This action helps to prevent nerve damage within the cochlea. If sounds reach high amplitudes very quickly, however—as in gunshots—the stapedius muscle may not respond soon enough to prevent nerve damage.

Cochlea

Encased within the dense temporal bone of the skull is an organ called the **cochlea,** about 34 mm long (the size of a pea) and shaped like the shell of a snail. Together with the vestibular apparatus (previously described), it composes the **inner ear.**

Vibrations of the stapes and oval window displace perilymph fluid within a part of the bony labyrinth known as the **scala vestibuli,** which is the upper of three chambers within the cochlea. The lower of the three chambers is also a part of the bony labyrinth and is known as the **scala tympani.** The middle chamber of the cochlea is a part of the membranous labyrinth called the **cochlear duct,** or **scala media.** Like

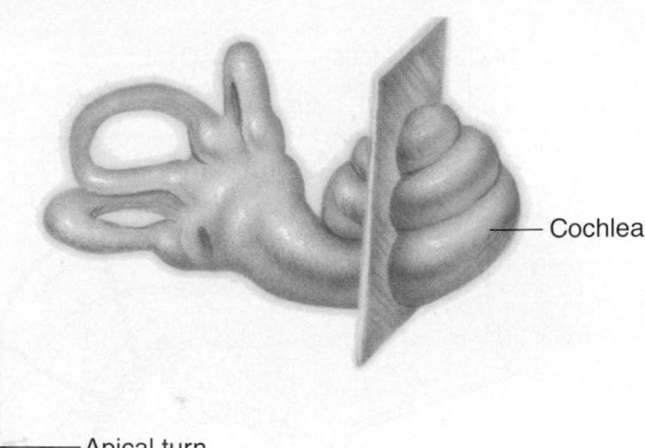

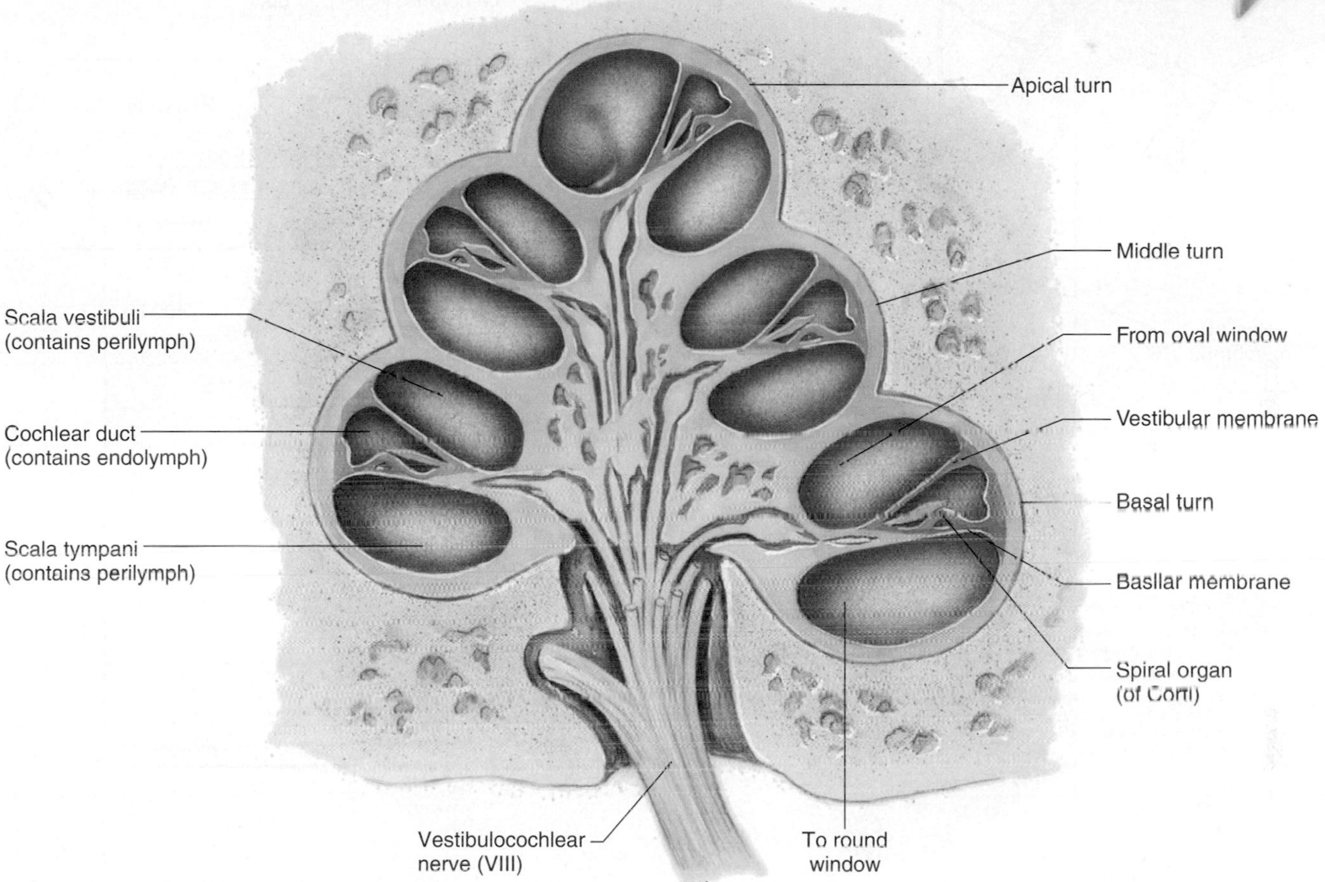

Figure 10.20 A cross section of the cochlea. In this view, its three turns and its three compartments—the scala vestibuli, cochlear duct (scala media), and scala tympani—can be seen.

the cochlea as a whole, the cochlear duct coils to form three turns (fig. 10.20), similar to the basal, middle, and apical portions of a snail shell. Because the cochlear duct is a part of the membranous labyrinth, it contains endolymph rather than perilymph.

The perilymph of the scala vestibuli and scala tympani is continuous at the apex of the cochlea because the cochlear duct ends blindly, leaving a small space called the *helicotrema* between the end of the cochlear duct and the wall of the cochlea. Vibrations of the oval window produced by movements of the stapes cause pressure waves within the scala vestibuli, which pass to the scala tympani. Movements of perilymph within the scala tympani, in turn, travel to the base of the cochlea where they cause displacement of a membrane called the *round window* into the middle-ear cavity (see fig. 10.19). This occurs because fluid, such as perilymph, cannot be compressed; an inward movement of the oval window is thus compensated for by an outward movement of the round window.

When the sound frequency (pitch) is sufficiently low, there is adequate time for the pressure waves of perilymph within the upper scala vestibuli to travel through the helicotrema to the scala tympani. As the sound frequency increases, however, pressure waves of perilymph within the scala vestibuli do not have time to travel all the way to the apex of the cochlea. Instead, they are transmitted through the *vestibular membrane,* which separates the scala vestibuli from the cochlear duct, and through the **basilar membrane,** which separates the cochlear duct from the scala tympani, to the perilymph of the scala tympani (fig. 10.20). The distance that these pressure waves travel, therefore, decreases as the sound frequency increases.

Sound waves transmitted through perilymph from the scala vestibuli to the scala tympani thus produce displacement of the vestibular membrane and the basilar membrane. Although the movement of the vestibular membrane does not directly contribute to hearing, displacement of the basilar membrane is central to pitch discrimination. Each sound

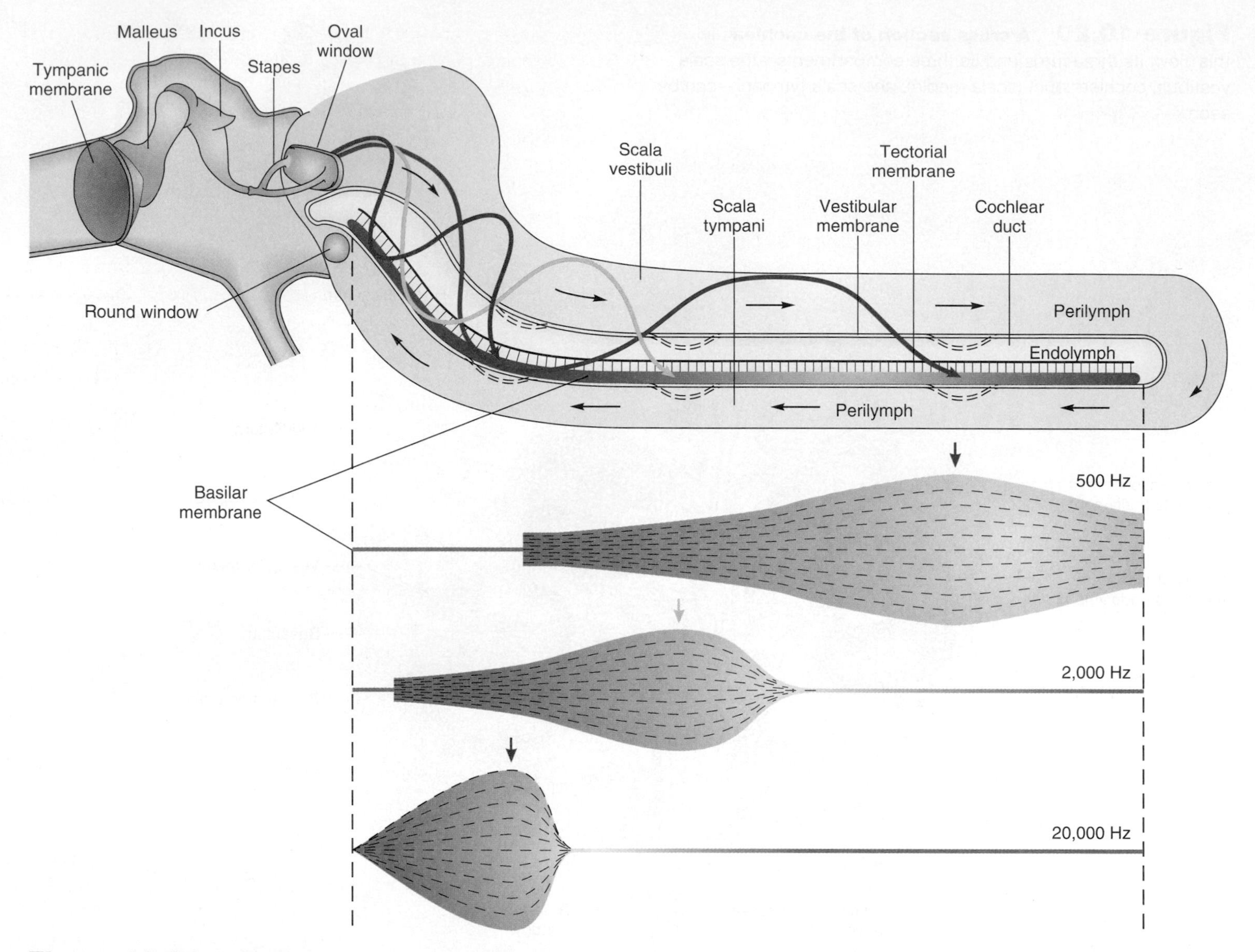

Figure 10.21 Effects of different sound pitches on the basilar membrane. Sounds of different pitch cause peak vibrations of the basilar membrane in different regions. Low-frequency (pitch) sounds, such as at 500 Hz, cause peak vibrations of the basilar membrane more toward the apex of the cochlea (to the right in this figure). High frequencies, such as 20,000 Hz, cause peak vibrations more toward the base of the cochlea (toward the left in the figure). AP|R

frequency produces maximum vibrations at a different region of the basilar membrane. Sounds of higher frequency (pitch) cause maximum vibrations of the basilar membrane closer to the stapes, as illustrated in figure 10.21.

Spiral Organ (Organ of Corti)

The sensory *hair cells* are located on the basilar membrane with their "hairs" projecting into the endolymph of the cochlear duct. The hairs are actually stereocilia, which are large, specialized microvilli arranged in bundles. The stereocilia within each bundle increase in size stepwise toward one side and are interconnected with filaments that run between the stereocilia. When the stereocilia within a bundle are bent in the direction of its tallest member, channels for K^+ open in the plasma membrane. Because the electrochemical gradient for K^+ strongly favors the passive movement of K^+ from the endolymph into the hair cells (due to the uniquely high K^+ concentration in the endolymph), the hair cells become depolarized.

There are two categories of hair cells, inner and outer. **Inner hair cells,** about 3,500 per cochlea, form one row that extends the length of the basilar membrane. Each of these inner hair cells is innervated by 10 to 20 sensory neurons in cranial nerve VIII, and these relay information regarding sound to the brain. There are also about 11,000 **outer hair cells** arranged in multiple rows: three rows in the basilar turn, four in the middle turn, and five in the apical turn of the cochlea. The outer hair cells are innervated primarily by motor axons, which cause them to shorten when they are depolarized or lengthen when they are hyperpolarized. These movements of the outer hair cells are believed to aid the sensory function of the inner hair cells, as will be described shortly.

The stereocilia of the hair cells are embedded in a gelatinous **tectorial membrane** (*tectum* = roof, covering), which overhangs the hair cells within the cochlear duct (fig. 10.22). The association of the basilar membrane, inner hair cells with sensory fibers, and tectorial membrane forms a functional unit called the **spiral organ,** or **organ of Corti** (fig. 10.22). When the cochlear duct is displaced by pressure waves of perilymph, a shearing force is created between the basilar membrane and the tectorial membrane. This causes the stereocilia to bend, and this mechanical process opens K^+ channels in the plasma membrane covering the tops of the stereocilia.

These K^+ channels face endolymph, which uniquely has a high concentration of K^+ similar to that of the intracellular compartment. Also, the endolymph of the cochlea (but not the vestibular apparatus) has an amazingly high positive potential: + 100 mV. Combined with the negative resting membrane potential of the hair cells, this produces an extremely steep electrochemical gradient favoring the entry of K^+. So, when the K^+ channels in the bent stereocilia open, K^+ moves passively down its electrochemical gradient into the hair cells. This depolarizes the hair cells and stimulates them to release glutamate, which stimulates the associated sensory neurons. The K^+ that entered the hair cells at their apical surface can then move passively out through channels in their basal surface, which face perilymph

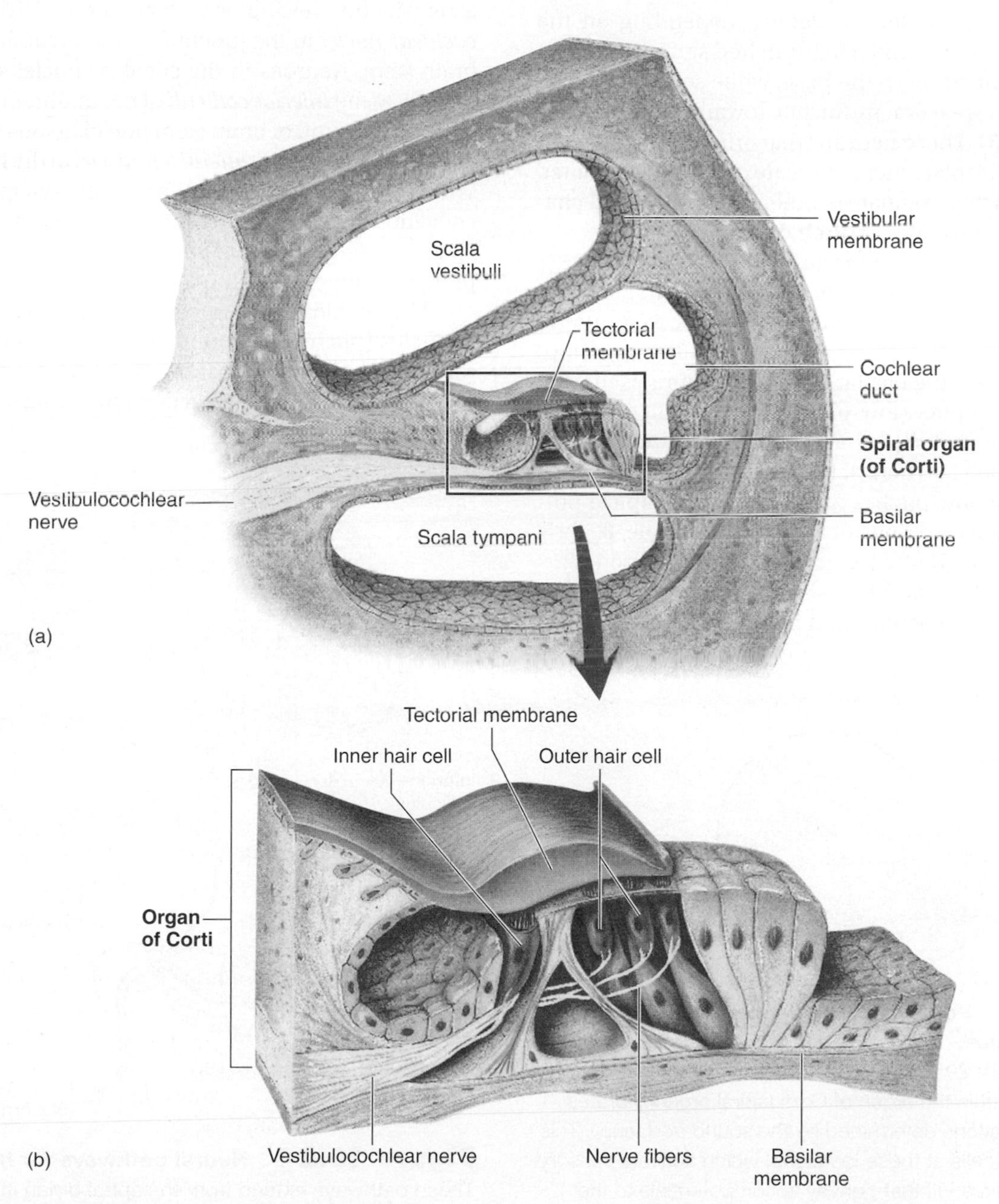

Figure 10.22 The spiral organ (organ of Corti). This functional unit of hearing is depicted (*a*) within the cochlear duct and (*b*) isolated to show greater detail.

in the scala tympani. Perilymph, as previously mentioned, has a low K^+ concentration typical of extracellular fluids.

The greater the displacement of the basilar membrane and the bending of the stereocilia, the greater the amount of transmitter released by the inner hair cell, and therefore the greater the generator potential produced in the sensory neuron. By this means, a greater bending of the stereocilia will increase the frequency of action potentials produced by the fibers of the cochlear nerve that are stimulated by the hair cells. Experiments suggest that the stereocilia need bend only 0.3 nanometers to be detected at the threshold of hearing! A greater bending will result in a higher frequency of action potentials, which will be perceived as a louder sound.

As mentioned earlier, traveling waves in the basilar membrane reach a peak in different regions, depending on the pitch (frequency) of the sound. High-pitched sounds produce a peak displacement closer to the base, while sounds of lower pitch cause peak displacement further toward the apex (see figs. 10.21 and 10.23). Those neurons that originate in hair cells located where the displacement is greatest will be stimulated more than neurons that originate in other regions. This mechanism provides a neural code for **pitch discrimination.**

Because the basilar membrane of the cochlear duct is shaped into a spiral, the base of the cochlea—its first turn—is where the basilar membrane vibrates in response to sounds of high frequency (high pitch). By contrast, the smaller apex (top) of the cochlea is where the basilar membrane vibrates most in response to low-frequency (low-pitch) sounds. In figure 10.21, a high-frequency sound that is audible to the human ear is represented as 20,000 Hz (hertz). In figure 10.23, this is shown as 20 kHz, and very low pitches are shown in the apical portions of the cochlea as fractions of a kHz; for example, 0.5 kHz is equivalent to the 500 Hz frequency depicted in figure 10.21.

However, vibrations of the basilar membrane are dampened by the fluid of the cochlea, and because of this we would be nearly deaf were it not for the actions of the outer hair cells, which act as *cochlear amplifiers.* The outer hair cells are near the center of the basilar membrane, are more than three times as numerous as the inner hair cells, and change length: they get longer when hyperpolarized and shorter when depolarized by motor neurons. These length changes magnify the effects of sound on basilar membrane vibrations and inner hair cell stimulation up to a thousand times. This allows us to hear far softer sounds than would otherwise be possible, and serves to significantly sharpen the frequency response of the basilar membrane, thereby sharpening pitch perception.

Neural Pathways for Hearing

Sensory neurons in the spiral ganglion of each ear send their axons in the vestibulocochlear nerve (VIII) to one of two *cochlear nuclei* in the junction of the medulla and pons of the brain stem. Neurons in the cochlear nuclei send axons either directly to the *inferior colliculi* of the midbrain or to the *superior olive,* a collection of brain stem nuclei. Axons from the superior olive pass through the *lateral lemniscus* to the inferior colliculus. Whatever the route, all auditory paths synapse in the inferior colliculus. Neurons in the inferior colliculus then send axons to the *medial geniculate body* of the thalamus, which in turn projects to the *auditory cortex* of the temporal lobe (fig. 10.24).

The cochlea is a frequency analyzer, in that different frequencies (pitches) of sound stimulate different sensory neurons that innervate the basilar membrane. This is because hair cells located in different places along the basilar membrane are

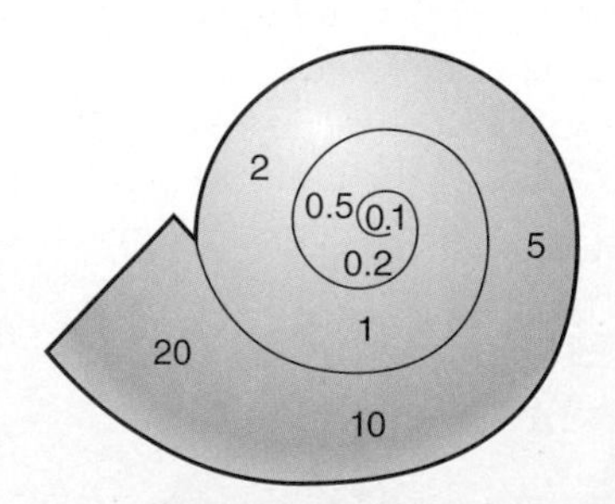

Figure 10.23 Portions of the cochlea that detect different frequencies. The numbers represent sound frequencies in kilohertz (kHz); thus, 20 = 20,000 Hz, and 0.1 = 100 Hz. The basilar membrane within the organ of Corti (spiral organ) vibrates maximally at the locations determined by the sound frequency. This stimulates inner hair cells at those locations, which activate sensory neurons of cranial nerve VIII that convey action potentials to the brain. The brain then interprets action potentials from different regions of the cochlea as sounds of different pitches.

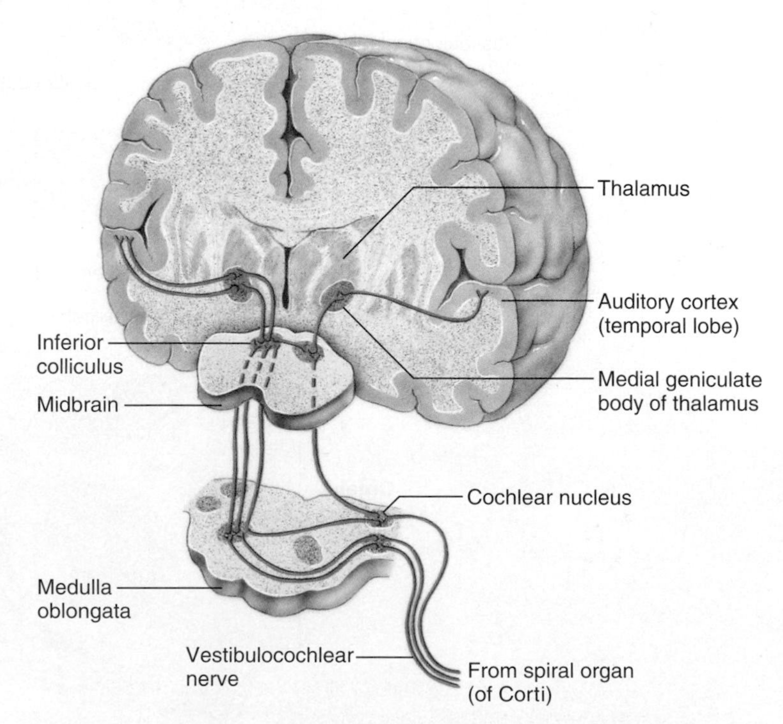

Figure 10.24 Neural pathways for hearing. These pathways extend from the spiral organ in the cochlea to the auditory cortex. The superior olive and lateral lemniscus are not shown. AP|R

most effectively stimulated by different frequencies of sound. This is known as the *place theory* of pitch, and has been previously described. Sensory neurons stimulated by low-frequency sounds, and those stimulated by high-frequency sounds, project their axons to different regions of the cochlear nucleus. The cochlear nucleus displays a **tonotopic organization,** in that different regions represent different "tones" (pitches). This separation of neurons by pitch is preserved in the tonotopic organization of the auditory cortex (fig. 10.25), which allows us to perceive the different pitches of sounds.

The analysis of pitch can be quite amazing; for example, we can recognize that a given sound frequency (such as 440 Hz) is the same regardless of whether it is played by a violin or a piano. The harmonics (multiples of a common fundamental frequency) can vary, depending on their amplitudes, and this helps produce the different characteristics of each instrument. However, if the fundamental frequency is the same, the pitch is recognized as being the same on the different instruments.

The loudness (intensity) of sounds, unlike their pitch, is coded by the frequency of action potentials. Differences in the intensity of the sounds arriving at each ear can be used to localize a sound. This *interaural intensity difference* is produced when one ear is closer to the source of the sound than the other ear and the sound frequency is above about 2000 Hz. At these higher frequencies, the wavelengths of sound are shorter than the distance between the ears. This information is supplemented by an *interaural time difference* if sound arrives at one ear before the other. The time difference is particularly important for localizing low-frequency sounds (below 2000 Hz). Humans can detect an interaural intensity difference of as little as 1–2 decibels and an interaural time difference as short as 10 microseconds. Sound localization based on intensity differences and time differences between the two ears is primarily a function of the lateral and medial superior olive, respectively.

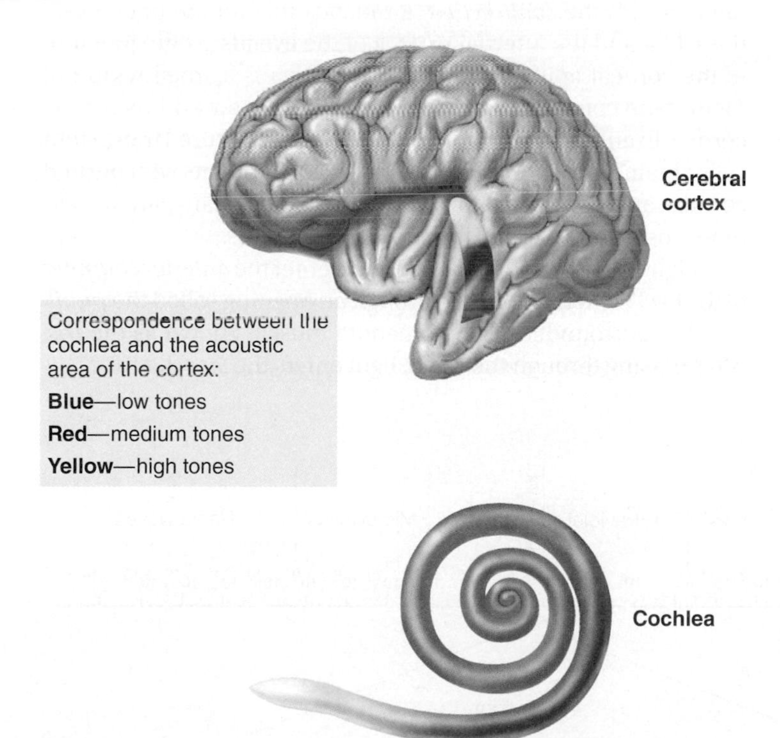

Figure 10.25 Correlation between pitch location in the cochlea and auditory cortex. Sounds of different frequencies (pitches) cause vibration of different parts of the basilar membrane, exciting different sensory neurons in the cochlea. These, in turn, send their input to different regions of the auditory cortex.

Hearing Impairments

There are two major categories of deafness: (1) **conduction deafness,** in which the transmission of sound waves through the outer and middle ear to the oval window is impaired, and (2) **sensorineural,** or **perceptive, deafness,** in which the transmission of nerve impulses anywhere from the cochlea to the auditory cortex is impaired. Conduction deafness can be caused by a variety of problems in the ability of sound waves to move through the external auditory meatus to produce vibrations of the tympanic membrane. This is most commonly due to the buildup of ear wax (cerumen), and to middle-ear damage from otitis media or otosclerosis (discussed in the previous Clinical Application box). Sensorineural deafness may result from a wide variety of pathological processes and from exposure to extremely loud sounds (as from gunshots or rock concerts). Unfortunately, the hair cells in the inner ears of mammals cannot regenerate once they are destroyed. Experiments have shown, however, that the hair cells of reptiles and birds can regenerate by cell division when they are damaged. Scientists are currently trying to determine if mammalian sensory hair cells might be made to respond in a similar fashion.

Conduction deafness impairs hearing at all sound frequencies. Sensorineural deafness, by contrast, often impairs the ability to hear some pitches more than others. This may be due to pathological processes or to changes that occur during aging. Age-related hearing impairment—called *presbycusis*—begins after age 20, when the ability to hear high frequencies (18,000 to 20,000 Hz) diminishes. Men are affected to a greater degree than women, and although the progression is variable, the deficits may gradually extend into the 4,000–8,000-Hz range. These impairments can be detected by *audiometry,* a technique in which the threshold intensity of different pitches is determined. The ability to hear speech is particularly affected by hearing loss in the higher frequencies.

People with conduction deafness can be helped by **hearing aids**—devices that amplify sounds and conduct the sound waves through bone to the inner ear. Some people with sensorineural deafness choose to have **cochlear implants.** The cochlear implant consists of electrodes threaded into the cochlea, a receiver implanted in the temporal bone, and an external microphone, processor, and transmitter. Although hair cells and most of the associated sensory dendrites have degenerated in sensorineural deafness, these devices may be effective because some dendrites survive and can be stimulated by implanted electrodes. Thus, some neurons of the spiral ganglion can be electrically stimulated to produce action potentials and convey information of low, medium, and high sound frequencies to the brain. This may restore some perception of speech in affected people.

Clinical Investigation CLUES

The physician recommended that Ed have an audiology test after recovering from his cold if his hearing was still impaired.

- What types of hearing losses are possible?
- What type of hearing impairment is most likely responsible for Ed's problem?

CHECKPOINT

8. Use a flowchart to describe how sound waves in air within the external auditory meatus are transduced into movements of the basilar membrane.
9. Explain how movements of the basilar membrane affect hair cells, and how hair cells can stimulate associated sensory neurons.
10. Explain how sounds of different intensities affect the function of the cochlea. How are different pitches of sounds distinguished by the cochlea?

10.6 THE EYES AND VISION

Light from an observed object is focused by the cornea and lens onto the photoreceptive retina at the back of the eye. The focus is maintained on the retina at different distances between the object and the eyes by muscular contractions that change the thickness and degree of curvature of the lens.

LEARNING OUTCOMES

After studying this section, you should be able to:

11. Describe the structures of the eye, and how these focus light onto the retina.
12. Explain how accommodation at different distances is accomplished.
13. Explain common disorders of refraction.

The eyes transduce energy in the electromagnetic spectrum (fig. 10.26) into nerve impulses. Only a limited part of this spectrum can excite the photoreceptors—electromagnetic energy with wavelengths between 400 and 700 nanometers (1 nm = 10^{-9} m, or one-billionth of a meter) constitutes *visible light.* Light of longer wavelengths in the infrared regions of the spectrum is felt as heat but does not have sufficient energy to excite the photoreceptors. Ultraviolet light, which has shorter wavelengths and more energy than visible light, is filtered out by the yellow color of the eye's lens. Honeybees—and people who have had their lenses removed—can see light in the ultraviolet range.

The structures of the eyeball are summarized in table 10.4. The outermost layer of the eye is a tough coat of connective tissue called the *sclera,* which can be seen externally as the white of the eyes. The tissue of the sclera is continuous with the transparent *cornea.* A clear epithelium covers the cornea and is continuous with the *conjunctiva,* a mucous membrane that covers the sclera and the internal surface of the eyelids. At the juncture of the corneal epithelium and conjunctiva is a small region of membrane containing stem cells that can renew and repair the cornea. In an exciting recent report, scientists cultured these stem cells obtained from the contralateral eyes of patients with burned corneas and used these to successfully restore transparent, self-renewing corneas in the majority of the patients.

Light passes through the cornea to enter the *anterior chamber* of the eye. Light then passes through an opening called the *pupil,* which is surrounded by a pigmented muscle known as the *iris.* After passing through the pupil, light enters the *lens* (fig. 10.27).

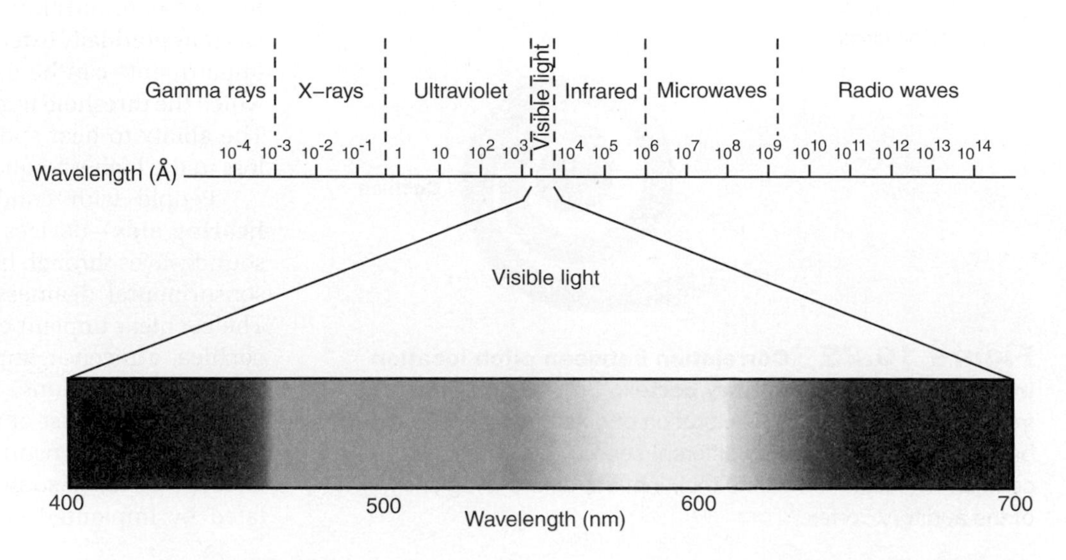

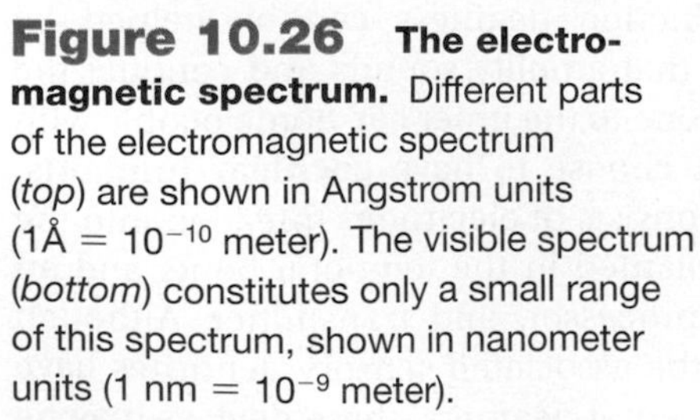

Figure 10.26 The electromagnetic spectrum. Different parts of the electromagnetic spectrum (*top*) are shown in Angstrom units (1Å = 10^{-10} meter). The visible spectrum (*bottom*) constitutes only a small range of this spectrum, shown in nanometer units (1 nm = 10^{-9} meter).
See the *Test Your Quantitative Ability* section of the Review Activities at the end of this chapter.

Table 10.4 | Structures of the Eyeball

Tunic and Structure	Location	Composition	Function
Fibrous tunic	Outer layer of eyeball	Avascular connective tissue	Gives shape to the eyeball
Sclera	Posterior outer layer; white of the eye	Tightly bound elastic and collagen fibers	Supports and protects the eyeball
Cornea	Anterior surface of eyeball	Tightly packed dense connective tissue—transparent and convex	Transmits and refracts light
Vascular tunic (uvea)	Middle layer of eyeball	Highly vascular pigmented tissue	Supplies blood; prevents reflection
Choroid	Middle layer in posterior portion of eyeball	Vascular layer	Supplies blood to eyeball
Ciliary body	Anterior portion of vascular tunic	Smooth muscle fibers and glandular epithelium	Supports the lens through suspensory ligament and determines its thickness; secretes aqueous humor
Iris	Anterior portion of vascular tunic; continuous with ciliary body	Pigment cells and smooth muscle fibers	Regulates the diameter of the pupil, and hence the amount of light entering the vitreous chamber
Internal tunic	Inner layer of eyeball	Tightly packed photoreceptors, neurons, blood vessels, and connective tissue	Provides location and support for rods and cones
Retina	Principal portion of internal tunic	Photoreceptor neurons (rods and cones), bipolar neurons, and ganglion neurons	Photoreception; transmits impulses
Lens (not part of any tunic)	Between posterior and vitreous chambers; supported by suspensory ligament of ciliary body	Tightly arranged protein fibers; transparent	Refracts light and focuses onto fovea centralis

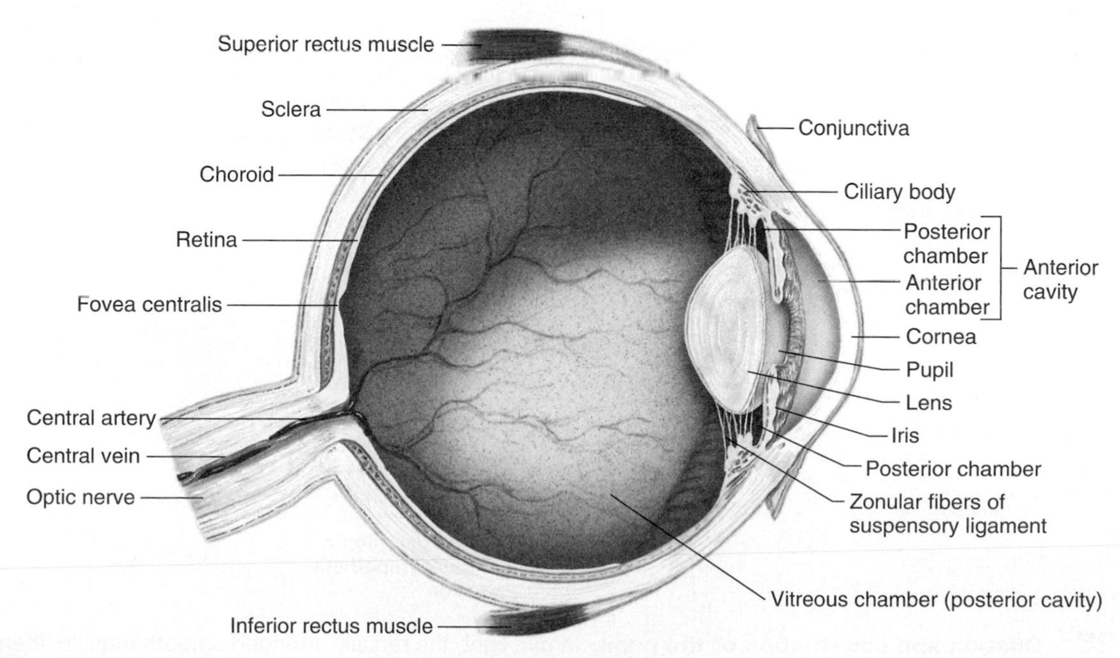

Figure 10.27 **The internal anatomy of the eyeball.** Light enters the eye from the right side of this figure and is focused on the retina. APIR

CLINICAL APPLICATION

The lens is normally completely clear, due to its very unique structure. It is composed of about a thousand layers of cells aligned in parallel and joined together tightly, so that gaps don't form as the shape of the lens changes. The lens is transparent because (1) it is avascular; (2) its cell organelles have been destroyed in a controlled process that halts before the cells die; and (3) the cell cytoplasm is filled with proteins called *crystallin.* Because of this structure, every region of the lens normally has the same refractive index. However, damage from ultraviolet light, dehydration, or oxidation may cause the crystallin proteins to change shape and aggregate to produce the cloudy patches in a person's visual field known as **cataracts.** Cataracts interfere with vision in more than half of people over the age of 65. This is generally treated by surgically replacing the lens with an artificial lens.

The iris is like the diaphragm of a camera; it can increase or decrease the diameter of its aperture (the pupil) to admit more or less light. Variations in the diameter of the pupil are similar in effect to variations in the f-stop of a camera. Constriction of the pupils is produced by contraction of circular muscles within the iris; dilation is produced by contraction of radial muscles. Constriction of the pupils results from parasympathetic stimulation through the occulomotor (III) nerve, whereas dilation results from sympathetic stimulation (fig. 10.28).

The posterior part of the iris contains a pigmented epithelium that gives the eye its color. The color of the iris of the eye is determined by the amount of pigment—blue eyes have the least pigment, brown eyes have more, and black eyes have the greatest amount of pigment. In the condition of *albinism*—a congenital absence of normal pigmentation caused by an inability to produce melanin pigment—the eyes appear pink because the absence of pigment allows blood vessels to be seen.

The lens is composed of living cells but is avascular (lacking in blood vessels), requiring its own microcirculatory system to sustain its cells. Even so, its metabolism is anaerobic and the cells near its center have low metabolic rates. The lens is transparent, composed predominately of cells called "mature fibers" that lack organelles. These cells have a flattened hexagonal shape in cross-section and are interconnected by numerous gap junctions. Interestingly, mutations in the genes that code for the connexin proteins of gap junctions (chapter 7; see fig. 7.22) are known to cause hereditary cataracts (described in the nearby Clinical Applications box).

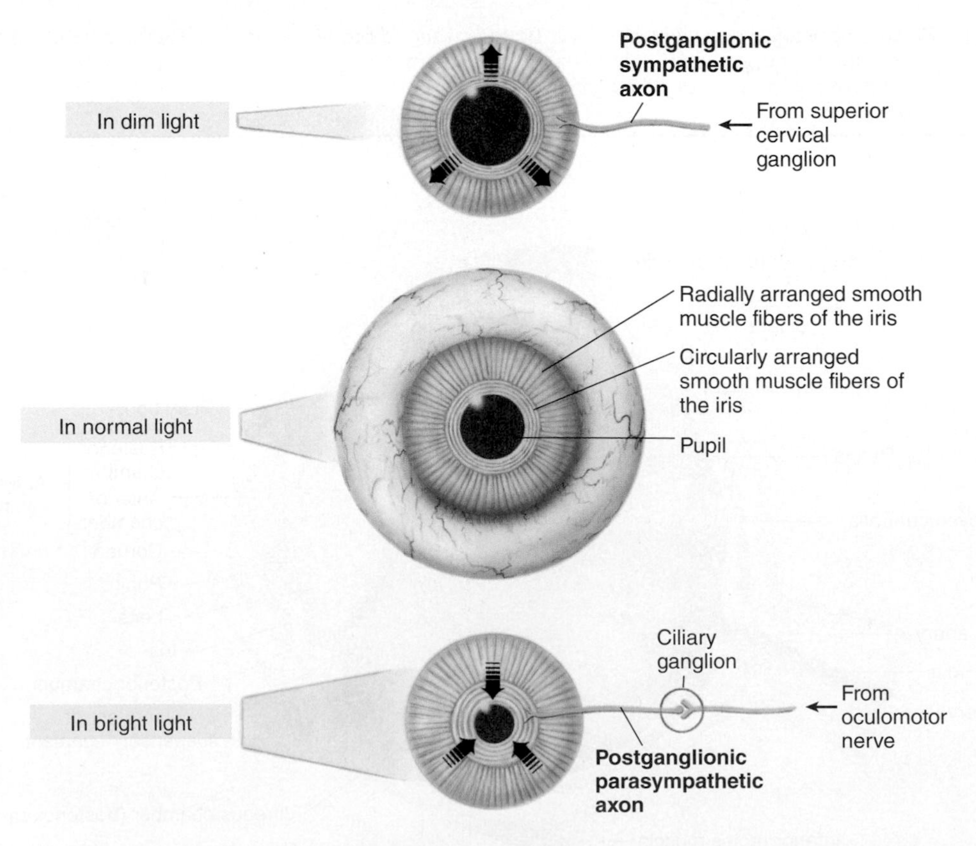

Figure 10.28 Dilation and constriction of the pupil. In dim light, the radially arranged smooth muscle fibers are stimulated to contract by sympathetic neurons, dilating the pupil. In bright light, the circularly arranged smooth muscle fibers are stimulated to contract by parasympathetic neurons, constricting the pupil.

CLINICAL APPLICATION

Glaucoma occurs when there is loss of retinal ganglion cell axons (see fig. 10.36), along with blood vessels and glia, in the optic nerve that produces characteristic changes in the appearance of the retina through an ophthalmoscope (see fig. 10.30). Glaucoma is generally caused by a problem in the flow of aqueous humor and can be classified as *closed-angle* or *open-angle.* Closed-angle glaucoma occurs when the drainage pathway for aqueous humor is blocked (by a tumor or inflammation); open-angle glaucoma occurs when the angle formed by the iris and cornea (fig. 10.29) is unobstructed, so that aqueous humor can reach the canal of Schlemm but the drainage is still inadquate. Primary open-angle glaucoma is the second leading cause of blindness in the United States, and can occur when the aqueous humor pressure is either normal (10 to 21 mmHg) or elevated. Elevated intraocular pressure, due to inadequate drainage of aqueous humor, is the most important risk factor, and the only risk factor that can be treated to prevent this disease or slow its progression.

The lens is suspended from a muscular process called the **ciliary body,** which connects to the sclera and encircles the lens. *Zonular fibers* (*zon* = girdle) suspend the lens from the ciliary body, forming a **suspensory ligament** that supports the lens. The space between the cornea and iris is the *anterior chamber,* and the space between the iris and the ciliary body and lens is the *posterior chamber* (fig. 10.29).

The anterior and posterior chambers are filled with a fluid called **aqueous humor.** This fluid is secreted by the ciliary body into the posterior chamber and passes through the pupil into the anterior chamber, where it provides nourishment to the avascular lens and cornea. The aqueous humor is drained from the anterior chamber into the *scleral venous sinus (canal of Schlemm),* which returns it to the venous blood (fig. 10.29).

The portion of the eye located behind the lens is filled with a thick, viscous substance known as the **vitreous body,** or **vitreous humor.** Light from the lens that passes through the vitreous body enters the neural layer, which contains photoreceptors, at the back of the eye. This neural layer is called the **retina.** Light that passes through the retina is absorbed by

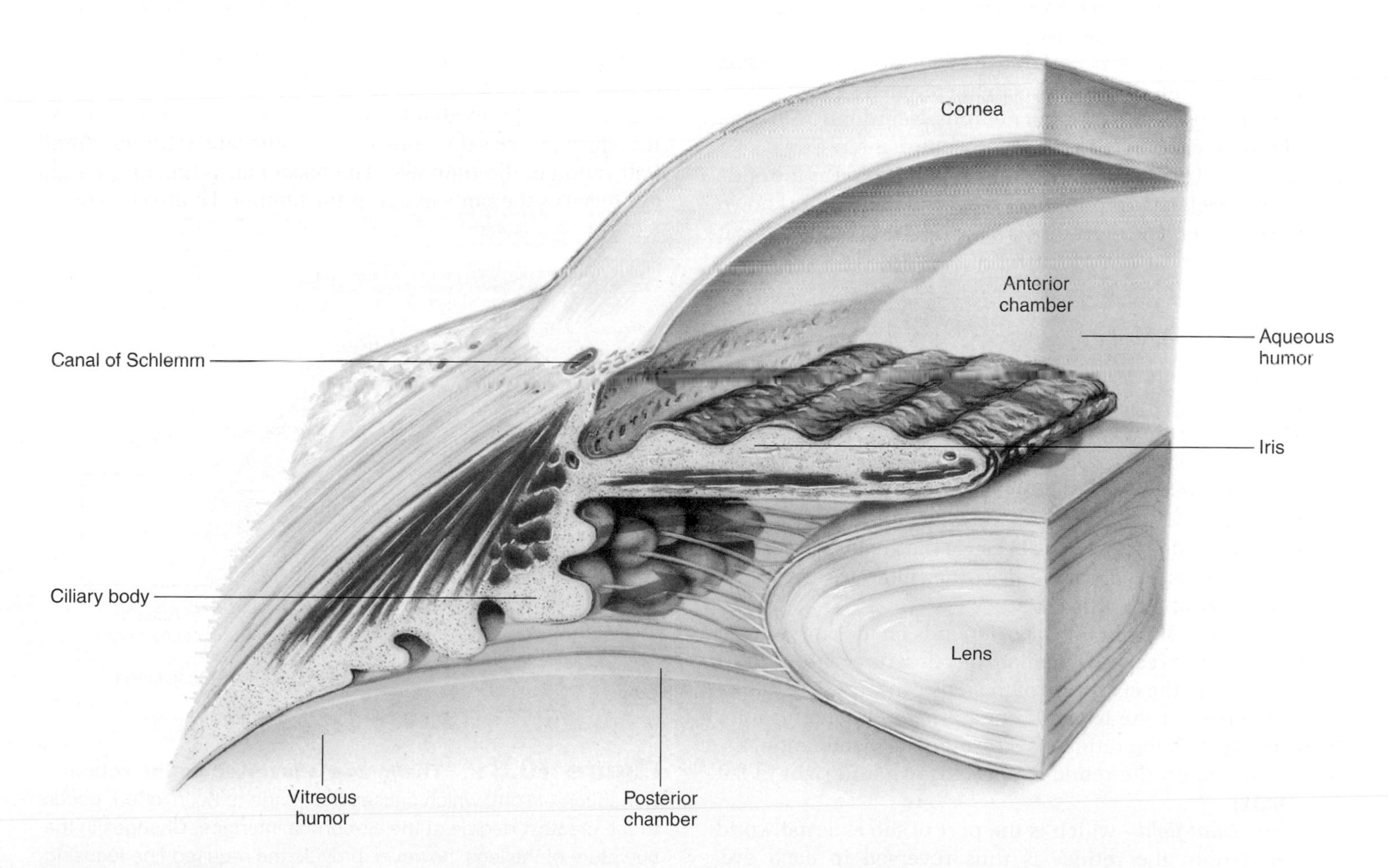

Figure 10.29 The production and drainage of aqueous humor. Aqueous humor maintains the intraocular pressure within the anterior and posterior chambers. It is secreted into the posterior chamber, flows through the pupil into the anterior chamber, and drains from the eyeball through the canal of Schlemm.

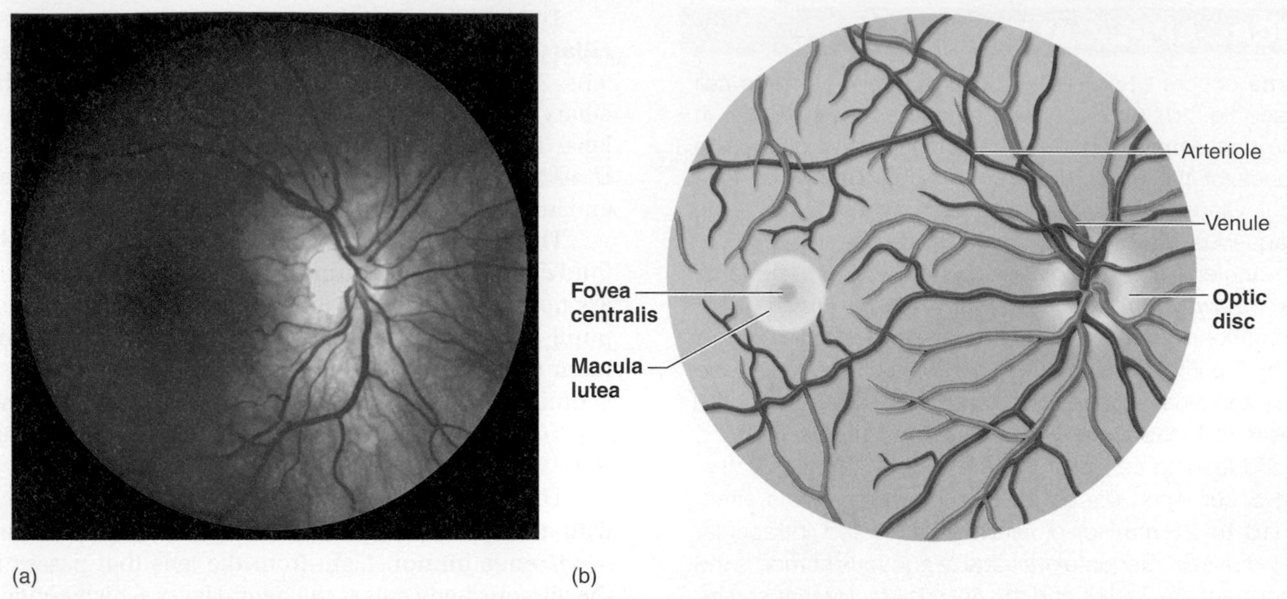

Figure 10.30 A view of the retina as seen with an ophthalmoscope. (*a*) A photograph and (*b*) an illustration of the optic fundus (back of the eye). Optic nerve fibers leave the eyeball at the optic disc to form the optic nerve. (Note the blood vessels that can be seen entering the eyeball at the optic disc.) © Steve Allen/Brand X Pictures/Getty Images

a darkly pigmented *choroid layer* underneath. While passing through the retina, some of this light stimulates photoreceptors, which in turn activate other neurons. Neurons in the retina contribute fibers that are gathered together at a region called the **optic disc** (fig. 10.30), where they exit the retina as the optic nerve. This region lacks photoreceptors and is therefore known as the *blind spot.* The optic disc is also the site of entry and exit of blood vessels.

Refraction

Light that passes from a medium of one density into a medium of a different density is *refracted,* or bent. The degree of refraction depends on the comparative densities of the two media, as indicated by their *refractive index.* The refractive index of air is set at 1.00; the refractive index of the cornea, by comparison, is 1.38; and the refractive indices of the aqueous humor and lens are 1.33 and 1.40, respectively. Because the greatest difference in refractive index occurs at the air-cornea interface, the light is refracted most at the cornea.

The degree of refraction also depends on the curvature of the interface between two media. The curvature of the cornea is constant, but the curvature of the lens can vary. The refractive properties of the lens can thus provide fine control for focusing light on the retina. As a result of light refraction, the image formed on the retina is upside down and right to left (fig. 10.31).

The *visual field*—which is the part of the external world projected onto the retina—is thus reversed in each eye. The cornea and lens focus the right part of the visual field on the left half of the retina of each eye, while the left half of the visual field is focused on the right half of each retina (fig. 10.32). The medial (or nasal) half-retina of the left eye therefore receives the same image as the lateral (or temporal) half-retina of the right eye. The nasal half-retina of the right eye receives the same image as the temporal half-retina of the left eye.

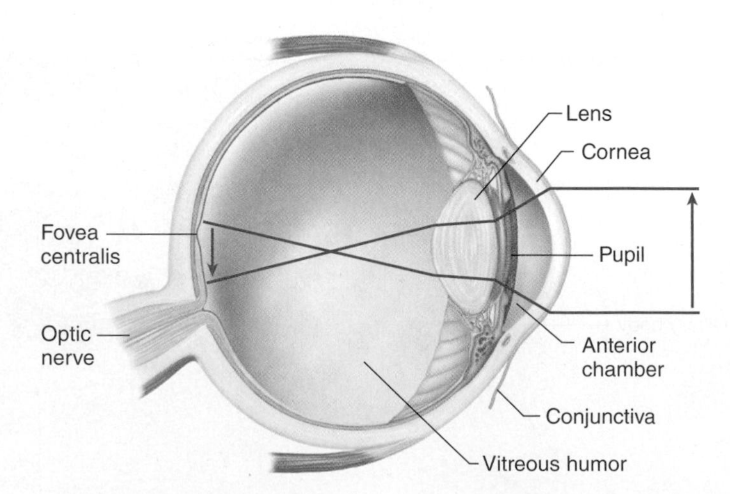

Figure 10.31 The image is inverted on the retina. Refraction of light, which causes the image to be inverted, occurs to the greatest degree at the air-cornea interface. Changes in the curvature of the lens, however, provide the required fine focusing adjustments. AP|R

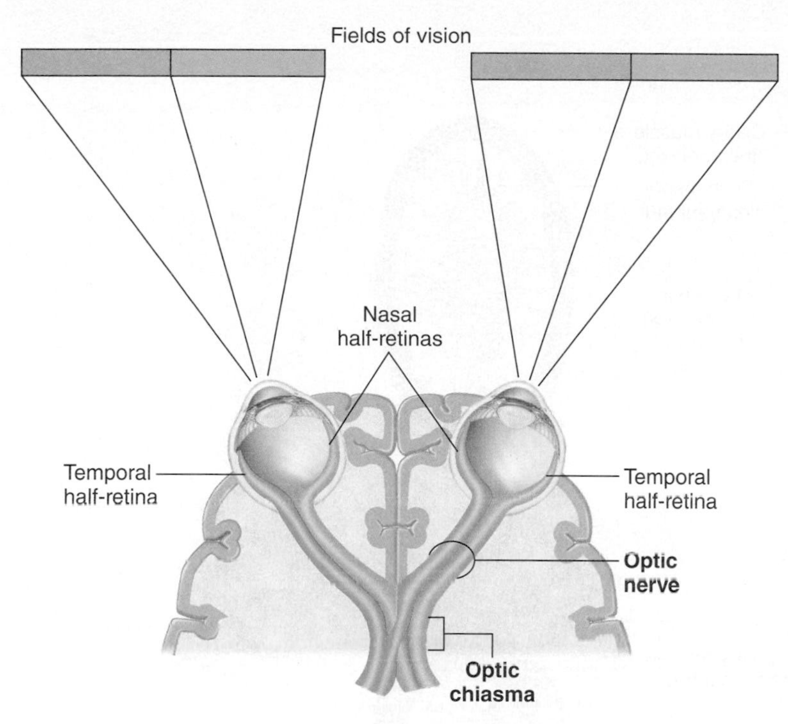

Figure 10.32 The image is switched right-to-left on the retina. The left side of the visual field is projected to the right half of each retina, while the right side of each visual field is projected to the left half of each retina. APIR

Accommodation

When a normal eye views an object, parallel rays of light are refracted to a point, or *focus*, on the retina (see fig. 10.35*a*). If the degree of refraction remained constant, movement of the object closer to or farther from the eye would cause corresponding movement of the focal point, so that the focus would either be behind or in front of the retina.

The ability of the eyes to keep the image focused on the retina as the distance between the eyes and object varies is called **accommodation.** Accommodation results from contraction of the ciliary muscle, which is like a sphincter muscle that can vary its aperture (fig. 10.33). When the ciliary muscle is relaxed, its aperture is wide. Relaxation of the ciliary muscle thus places tension on the zonular fibers of the suspensory ligament and pulls the lens taut. These are the conditions that prevail when viewing an object that is 20 feet or more from a normal eye; the image is focused on the retina and the lens is in its most flat, least convex form. As the object moves closer to the eyes, the muscles of the ciliary body contract. This muscular contraction narrows the aperture of the ciliary body and thus reduces the tension on the zonular fibers that suspend the lens. When the tension is reduced, the lens becomes more rounded and convex as a result of its inherent elasticity (fig. 10.34).

The ability of a person's eyes to accommodate can be measured by the near-point-of-vision test. The *near point of vision* is the minimum distance from the eyes at which an object can be brought into focus. This distance increases with age; indeed, accommodation in almost everyone over the age of 45 is significantly impaired. Loss of accommodating ability with age is known as **presbyopia** (*presby* = old). This loss appears to have a number of causes, including reduced flexibility of the lens and a forward movement of the attachments of the zonular fibers to the lens. As a result of these changes, the zonular fibers and lens are pulled taut even when the ciliary muscle contracts. The lens is thus not able to thicken and increase its refraction when, for example, a printed page

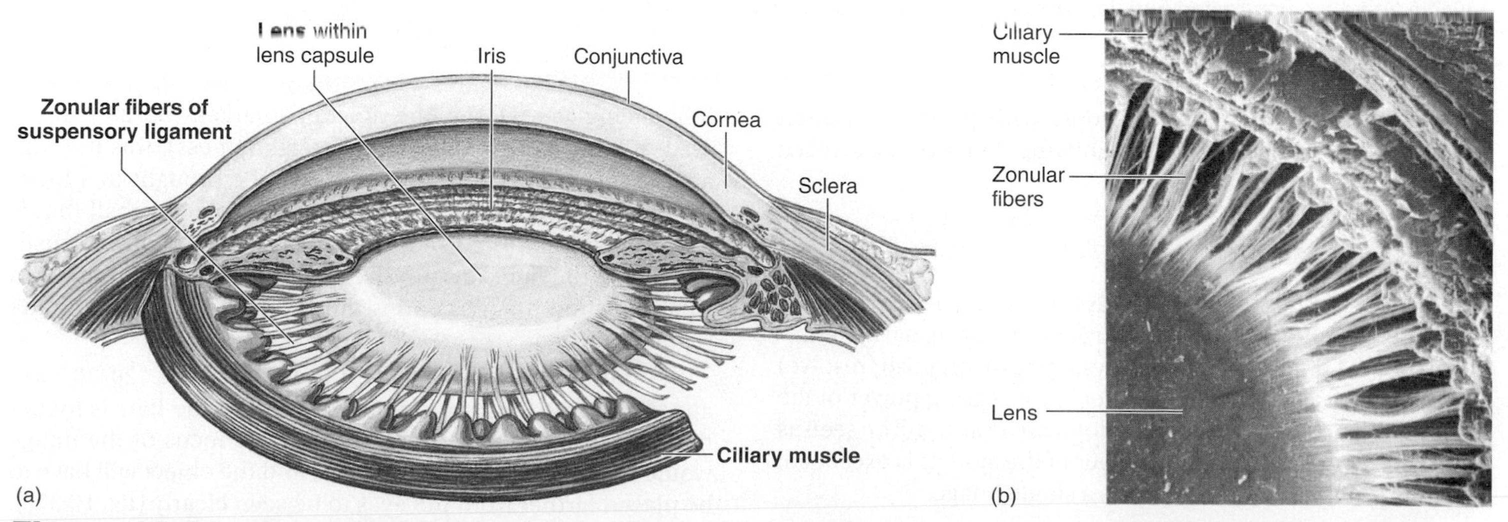

Figure 10.33 The relationship between the ciliary muscle and the lens. (*a*) A diagram, and (*b*) a scanning electron micrograph (from the eye of a 17-year-old boy) showing the relationship between the lens, zonular fibers, and ciliary muscle of the eye. Part (*b*) from "How the Eye Focuses" by James F. Koretz and George H. Handleman. APIR

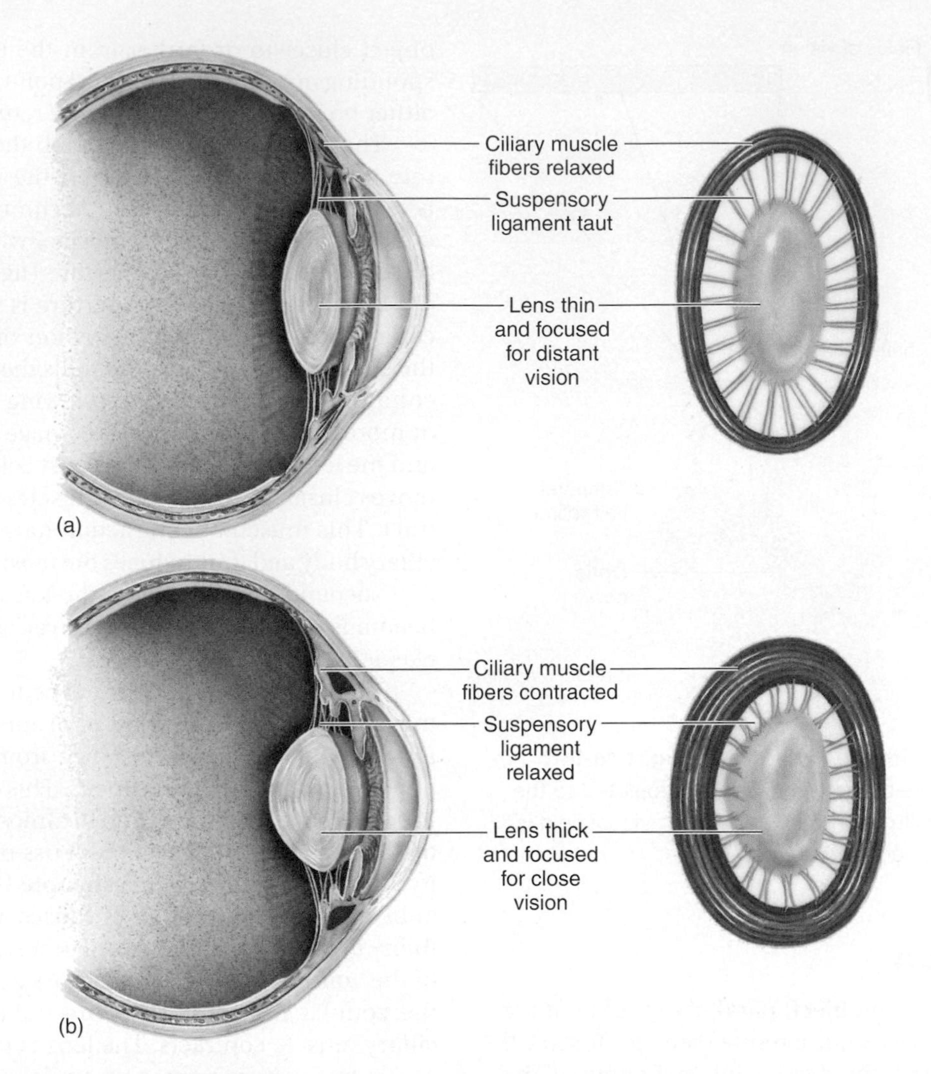

Figure 10.34 **Changes in the shape of the lens permit accommodation.** (*a*) The lens is flattened for distant vision when the ciliary muscle fibers are relaxed and the suspensory ligament is taut. (*b*) The lens is more spherical for close-up vision when the ciliary muscle fibers are contracted and the suspensory ligament is relaxed.

is brought close to the eyes. People with presbyopia often require glasses with reading (magnifying) lenses to clearly see small objects at close distances.

Visual Acuity

Visual acuity refers to the sharpness of vision. The sharpness of an image depends on the *resolving power* of the visual system—that is, on the ability of the visual system to distinguish (resolve) two closely spaced dots. The better the resolving power of the system, the closer together these dots can be and still be seen as separate. When the resolving power of the system is exceeded, the dots blur and are perceived as a single image.

Myopia and Hyperopia

When a person with normal visual acuity stands 20 feet from a *Snellen eye chart* (so that accommodation is not a factor influencing acuity), the line of letters marked "20/20" can be read. If a person has **myopia** (nearsightedness), this line will appear blurred because the image will be brought to a focus in front of the retina. This is usually due to an eyeball that is too long. Myopia is corrected by glasses with concave lenses that cause the light rays to diverge, so that the point of focus is farther from the lens and is thus pushed back to the retina (fig. 10.35*b*).

If the eyeball is too short, the line marked "20/20" will appear blurred because the focal length of the lens is longer than the distance to the retina. Thus, the focus of the image would have been behind the retina, and the object will have to be placed farther from the eyes to be seen clearly (fig. 10.35*c*). This condition is called **hyperopia** (farsightedness). Hyperopia is corrected by glasses with convex lenses that increase the convergence of light, so that the point of focus is brought closer to the lens and falls on the retina.

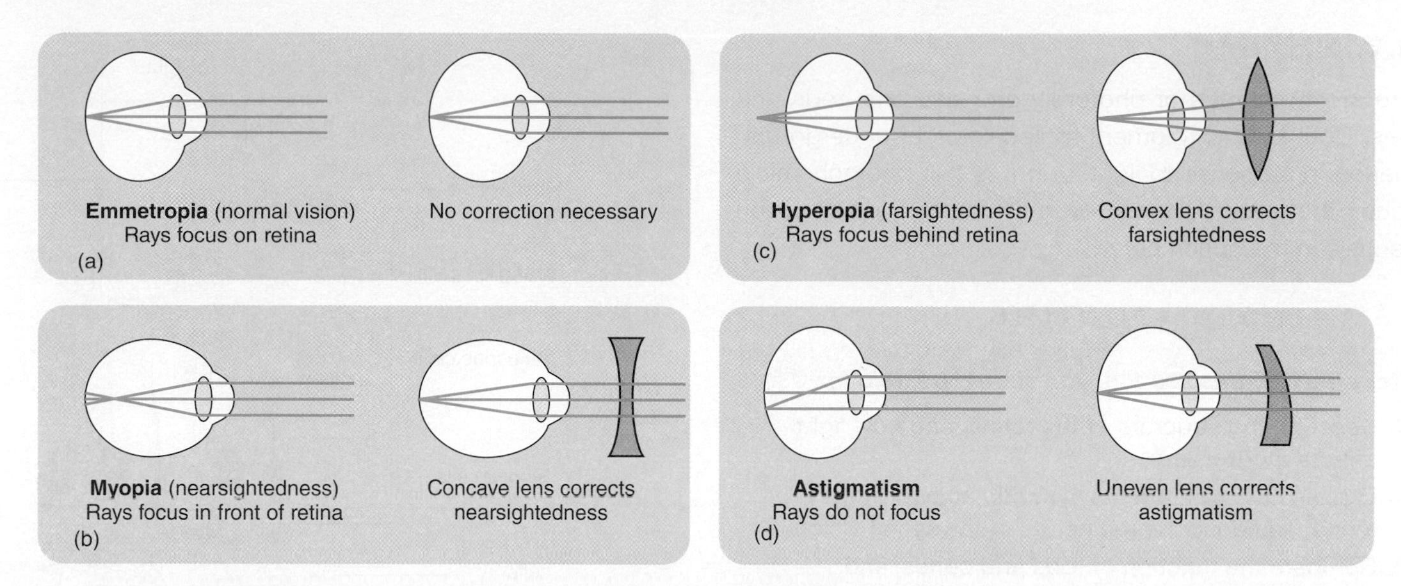

Figure 10.35 Problems of refraction and how they are corrected. In a normal eye (*a*), parallel rays of light are brought to a focus on the retina by refraction in the cornea and lens. If the eye is too long, as in myopia (*b*), the focus is in front of the retina. This can be corrected by a concave lens. If the eye is too short, as in hyperopia (*c*), the focus is behind the retina. This is corrected by a convex lens. In astigmatism (*d*), light refraction is uneven because of irregularities in the shape of the cornea or lens.

Astigmatism

Because the curvature of the cornea and lens is not perfectly symmetrical, light passing through some parts of these structures may be refracted to a different degree than light passing through other parts. When the asymmetry of the cornea and/or lens is significant, the person is said to have **astigmatism** (fig. 10.35*d*). If a person with astigmatism views a circle of lines radiating from the center, like the spokes of a wheel, the image of these lines will not appear clear in all 360 degrees. The parts of the circle that appear blurred can thus be used to map the astigmatism. This condition is corrected by cylindrical lenses that compensate for the asymmetry in the cornea or lens of the eye.

CLINICAL APPLICATIONS

Many people with refractive problems choose to have a surgical procedure know as **LASIK** (*laser-assisted in situ keratomileusis*). The surgeon first cuts a flap in the cornea, which is folded backward. Then a computer-guided laser burns corneal tissue as it reshapes the cornea. The laser reduces the curve of the cornea for myopia, reducing refraction to move the focus back to the retina; it makes the cornea more steeply curved for hyperopia, and more spherical to correct for astigmatism. LASIK cannot correct for presbyopia, so a person with this condition will need reading glasses if each eye is surgically corrected to 20/20. Alternatively, if the person also has myopia, one eye can deliberately be undercorrected for reading, while the other (dominant) eye is corrected for distance vision. This is called monovision, and is tolerated by some people better than others.

Clinical Investigation CLUES

Ed had difficulty seeing small print, even though he said he wasn't nearsighted and still had good distance vision. He asked if surgery would help.

- Which condition was most likely responsible for Ed's visual impairment?
- Would LASIK be able to help Ed?
- What kind of glasses does Ed need?

CHECKPOINT

11. Using a line diagram, explain why an inverse image is produced on the retina. Also explain how the image in one eye corresponds to the image in the other eye.

12. Using a line diagram, show how parallel rays of light are brought to a focus on the retina. Explain how this focus is maintained as the distance from the object to the eye is increased or decreased (that is, explain accommodation).

13. Explain why a blurred image is produced in each of these conditions: presbyopia, myopia, hyperopia, and astigmatism.

10.7 RETINA

There are two types of photoreceptor neurons: rods and cones. Both contain pigment molecules that undergo dissociation in response to light, and it is this photochemical reaction that eventually results in the production of action potentials in the optic nerve.

LEARNING OUTCOMES

After studying this section, you should be able to:

14. Describe the structure of the retina, and how light affects rhodopsin.
15. Explain how light affects synaptic activity in the retina, and describe the neural pathways of vision.
16. Compare the function of rods and cones, and describe the significance of the fovea centralis.
17. Describe the neural pathways required for vision.

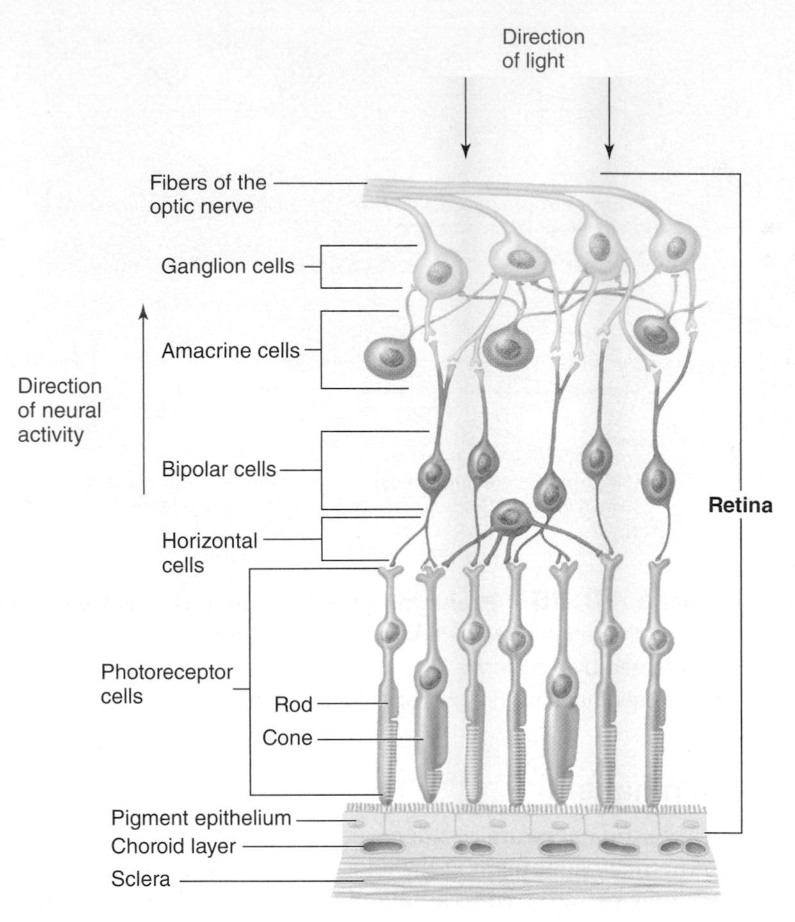

Figure 10.36 Layers of the retina. Because the retina is inverted, light must pass through various layers of nerve cells before reaching the photoreceptors (rods and cones). AP|R

The **retina** consists of a single-cell-thick pigmented epithelium, photoreceptor neurons called **rods** and **cones,** and layers of other neurons. The neural layers of the retina are actually a forward extension of the brain. In this sense, the optic nerve can be considered a tract, and indeed the myelin sheaths of its fibers are derived from oligodendrocytes (like other CNS axons) rather than from Schwann cells.

Because the retina is an extension of the brain, the neural layers face outward, toward the incoming light. Light, therefore, must pass through several neural layers before striking the photoreceptors (fig. 10.36). The photoreceptors then synapse with other neurons, so that synaptic activity flows outward in the retina.

The outer layers of neurons that contribute axons to the optic nerve are called **ganglion cells.** These neurons receive synaptic input from **bipolar cells,** which in turn receive input from rods and cones. In addition to the flow of information from photoreceptors to bipolar cells to ganglion cells, neurons called *horizontal cells* synapse with several photoreceptors (and possibly also with bipolar cells), and neurons called *amacrine cells* synapse with several ganglion cells.

Each rod and cone consists of an inner and an outer segment (fig. 10.37). The *inner segment* contains most of the cell's organelles; the *outer segment* contains hundreds of flattened membranous sacs, or **discs** (fig. 10.38), where the photopigment molecules required for vision are located. The photoreceptor cells continuously add new discs at the base of the outer segment as the tip regions are removed by the cells of the **retinal pigment epithelium** (see fig. 10.36) through a process of phagocytosis. Each retinal pigment epithelial cell is in contact with 50 to 100 photoreceptor outer segments, and daily removes the distal 10% of these outer segments by phagocytosis. This amounts to the phagocytosis of hundreds of thousands of discs over the course of a lifetime by each retinal pigment cell, making these cells the most phagocytically active cells in the body. The photoreceptors continuously produce new discs at the base of their outer segments, and these new discs migrate toward the tips to replace the lost material.

The retinal pigment epithelium is a simple, one-cell-layer-thick membrane. Microvilli project from the apical surface of the pigment epithelial cells toward the photoreceptors, aiding interactions. The basal surface of the retinal pigment epithelium contacts *Bruch's membrane,* the connective tissue basement membrane separating the pigment epithelium from the blood vessels of the choroid. Research reveals many functions of the retinal pigment epithelium important for vision, including the following:

1. Phagocytosis of the shed outer segments of the photoreceptors
2. Absorption of scattered light in the retina by melanin pigment
3. Delivery of nutrients from the blood to the photoreceptors
4. Suppression of an immune attack of the retina (thereby helping to make the retina an immunologically privileged site; chapter 15, section 15.3)
5. Conversion of visual pigment from the photoreceptors into its active form, which is recycled back to the

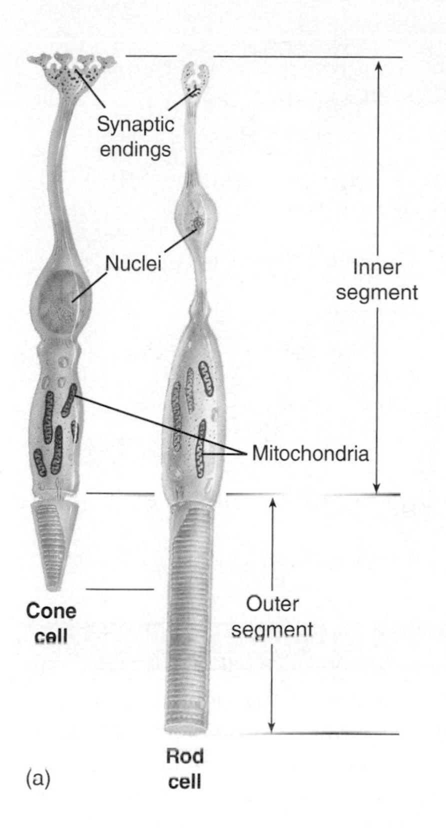

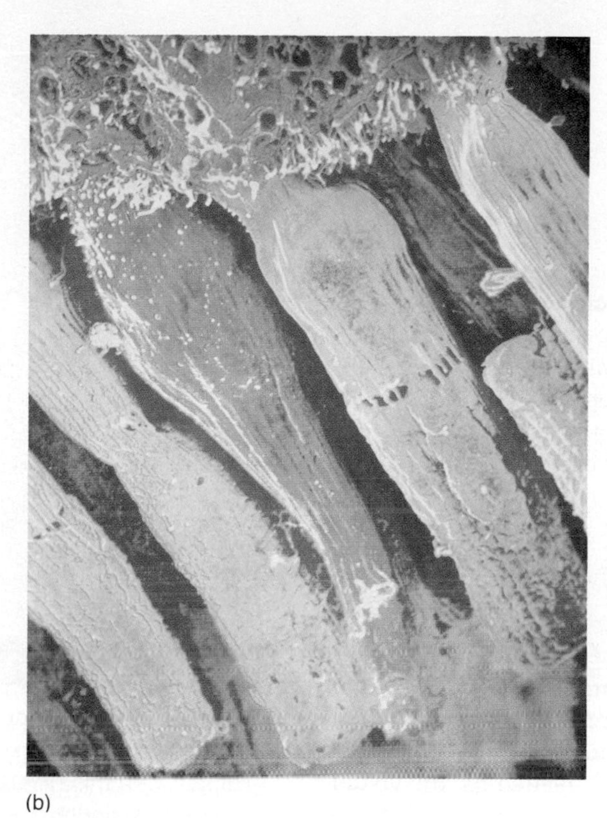

Figure 10.37 Rods and cones. (*a*) A diagram showing the structure of a rod and a cone. (*b*) A scanning electron micrograph of rods and cones. Note that each photoreceptor contains an outer and inner segment.

photoreceptors in a process called the *visual cycle of retinal* (discussed shortly)

6. Stabilization of the ion composition surrounding the photoreceptors, thereby helping them to respond appropriately to light (ion movements in the photoreceptors are shown in fig. 10.41)

Figure 10.38 Discs within rods contain rhodopsin molecules. A three-dimensional computer-generated model of the membranous discs within the outer segments of photoreceptors. In rods, each disc contains thousands of molecules of rhodopsin. Reproduced with permission of Carlos Rozas, Planeta Vivo Society, www.planetavivo.org.

Effect of Light on the Rods

The photoreceptors—rods and cones (fig. 10.37)—are activated when light produces a chemical change in molecules of pigment contained within the membranous discs of the outer segments of the receptor cells (fig. 10.38). Each rod contains thousands of molecules of a purple pigment known as **rhodopsin** in these discs. The pigment appears purple (a combination of red and blue) because it transmits light in the red and blue regions of the spectrum, while absorbing light energy in the green region. The wavelength of light that is absorbed best—the *absorption maximum*—is about 500 nm (blue-green light).

Green cars (and other green objects) are seen more easily at night—when rods are used for vision—than are red objects. This is because red light is not absorbed well by rhodopsin, and only absorbed light can produce the photochemical reaction that results in vision. In response to absorbed light, rhodopsin dissociates into its two components: the pigment **retinaldehyde** (also called **retinene** or **retinal**), which is derived from vitamin A, and a protein called **opsin.** This reaction is known as the **bleaching reaction.**

Retinal (retinene) can exist in two possible configurations (shapes)—one known as the all-*trans* form and one called the 11-*cis* form (fig. 10.39). The all-*trans* form is more stable, but only the 11-*cis* form is found attached to opsin. In response to absorbed light energy, the 11-*cis*-retinal is converted to the all-*trans* isomer, causing it to dissociate from the opsin. This dissociation reaction in response to light initiates changes in the

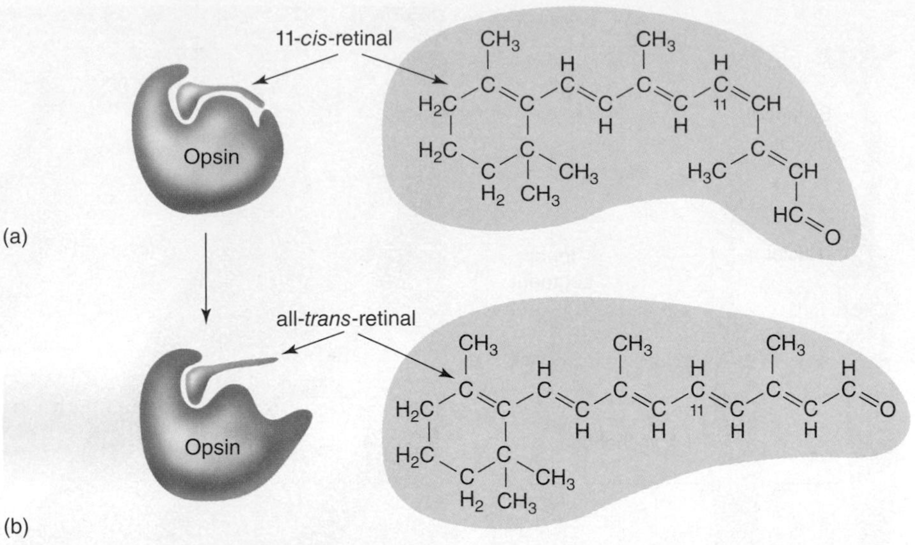

Figure 10.39 The photodissociation of rhodopsin. (*a*) The photopigment rhodopsin consists of the protein opsin combined with 11–*cis*-retinal (retinene). (*b*) Upon exposure to light, the retinal is converted to a different form, called all-*trans,* and dissociates from the opsin. This photochemical reaction induces changes in ionic permeability that ultimately result in stimulation of ganglion cells in the retina.

ionic permeability of the rod plasma membrane and ultimately results in the production of nerve impulses in the ganglion cells. As a result of these effects, rods provide black-and-white vision under conditions of low light intensity.

The retinal pigment epithelium is needed for the **visual cycle of retinal.** Photoreceptors lack the enzyme *cis-trans* isomerase, which is needed to re-isomerize (reconvert) retinal from the all-*trans* form back into the 11-*cis* form. After the absorption of light has caused the formation of the all-*trans* form of retinal, the all-*trans* retinal dissociates from the opsin and is transported from the photoreceptors into the closely associated pigment epithelial cells. There, it is re-isomerized into the 11-*cis* form and then transported back to the photoreceptors. Now, the 11-*cis* retinal can again bind to the opsin and form the active photopigment, able to respond to light. It is this recycling between the photoreceptors and the retinal pigment epithelium that is known as the visual cycle of retinal.

CLINICAL APPLICATION

Scientists have recently discovered the genetic basis for blindness in the disease **dominant retinitis pigmentosa.** People with this disease inherit a gene for the opsin protein in which a single base change in the gene (substitution of adenine for cytosine) causes the amino acid histidine to be substituted for proline at a specific point in the polypeptide chain. This abnormal opsin leads to degeneration of the photoreceptors. Rods degenerate before cones. This leads to a loss of vision that progresses from the periphery of the visual field (which contains mostly rods) to the center (which contains most of the cones). Thus, people with retinitis pigmentosa try to look directly at objects, which places the image in the center of their visual field. This direction of visual loss is reversed (from the center to the periphery of the visual field) in people with macular degeneration, who must try to see from the "corners of their eyes" (discussed in a later Clinical Application box).

Dark Adaptation

The bleaching reaction that occurs in the light results in a lowered amount of rhodopsin in the rods and lowered amounts of visual pigments in the cones. When a light-adapted person first enters a darkened room, therefore, sensitivity to light is low and vision is poor. A gradual increase in photoreceptor sensitivity, known as **dark adaptation,** then occurs, reaching maximal sensitivity in about 20 minutes. The increased sensitivity to low light intensity is partly due to increased amounts of visual pigments produced in the dark. Increased pigments in the cones produce a slight dark adaptation in the first five minutes. Increased rhodopsin in the rods produces a much greater increase in sensitivity to low light levels and is partly responsible for the adaptation that occurs after about five minutes in the dark. In addition to the increased concentration of rhodopsin, other more subtle (and less well understood) changes occur in the rods that ultimately result in a 100,000-fold increase in light sensitivity in dark-adapted as compared to light-adapted eyes.

Electrical Activity of Retinal Cells

The only neurons in the retina that produce all-or-none action potentials are ganglion cells and amacrine cells. The photoreceptors, bipolar cells, and horizontal cells instead produce only graded depolarizations or hyperpolarizations, analogous to EPSPs and IPSPs.

The transduction of light energy into nerve impulses follows a cause-and-effect sequence that is the inverse of the usual

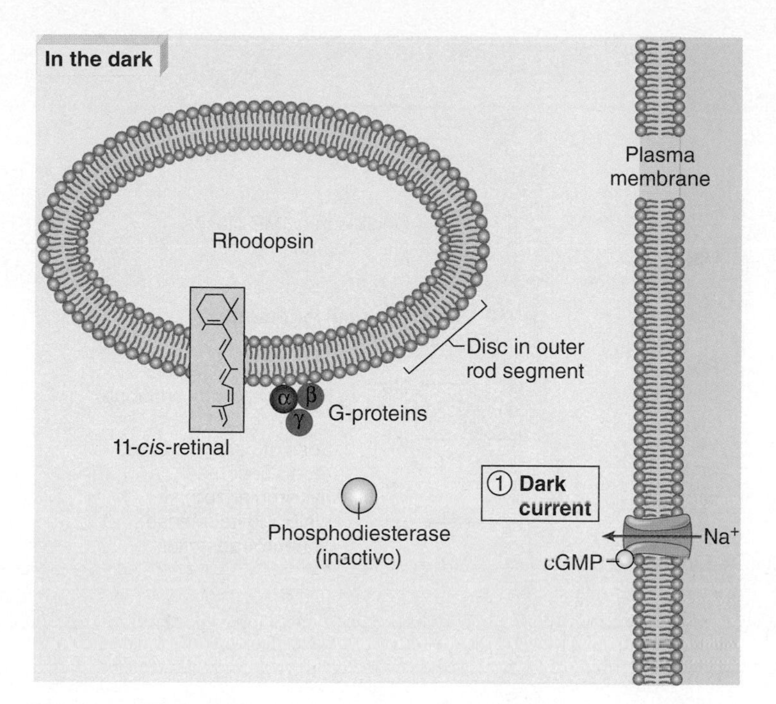

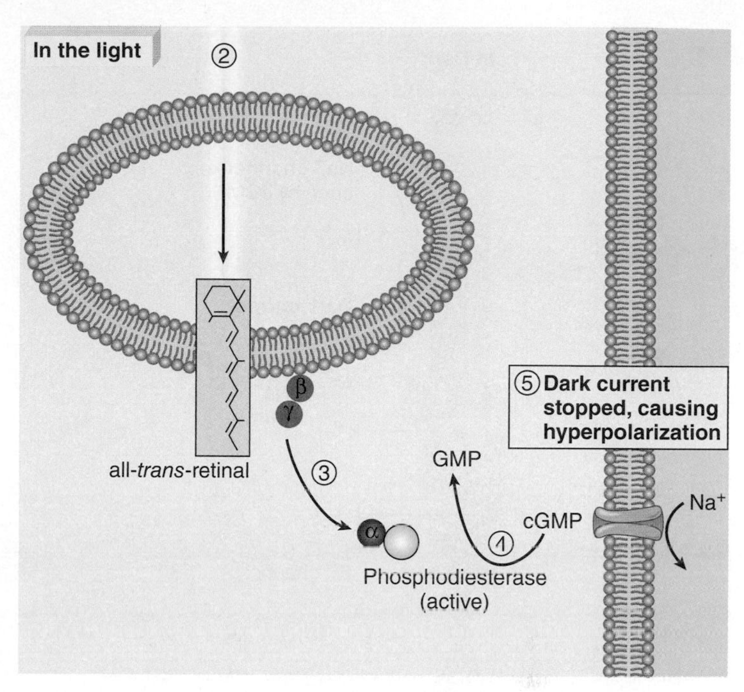

Figure 10.40 Light stops the dark current in photoreceptors. (1) In the dark, Na^+ enters the photoreceptors, producing a dark current that causes a partial depolarization. (2) In the light, 11-*cis*-retinal is converted into all-*trans*-retinal. (3) This causes G-proteins associated with the opsin to dissociate. (4) The alpha subunit binds to and activates phosphodiesterase, which converts cyclic GMP (cGMP) into GMP. (5) As a result, the Na^+ channels close, stopping the dark current and hyperpolarizing the photoreceptors. AP|R

way in which sensory stimuli are detected. This is because in the dark the photoreceptors release an inhibitory neurotransmitter that hyperpolarizes the bipolar neurons. Thus inhibited, the bipolar neurons do not release excitatory neurotransmitter to the ganglion cells. Light *inhibits* the photoreceptors from releasing their inhibitory neurotransmitter and by this means *stimulates* the bipolar cells, and thus the ganglion cells that transmit action potentials to the brain.

A rod or cone contains many Na^+ channels in the plasma membrane of its outer segment (fig. 10.40), and in the dark, many of these channels are open. As a consequence, Na^+ continuously diffuses into the outer segment and across the narrow stalk to the inner segment. This small flow of Na^+ that occurs in the absence of light stimulation is called the **dark current,** and it causes the membrane of a photoreceptor to be somewhat depolarized in the dark. The Na^+ channels in the outer segment rapidly close in response to light, reducing the dark current and causing the photoreceptor to hyperpolarize.

Cyclic GMP (cGMP) is required to keep the Na^+ channels open, and the channels will close if the cGMP is converted into GMP. Light causes this conversion and consequent closing of the Na^+ channels. When a photopigment absorbs light, 11-*cis*-retinene is converted into all-*trans*-retinene (fig. 10.40) and dissociates from the opsin, causing the opsin protein to change shape. Each opsin is associated with over a hundred regulatory G-*proteins* (chapter 6; see fig. 6.31) known as **transducins,** and the change in the opsin induced by light causes the alpha subunits of the G-proteins to dissociate. The alpha transducin (G-protein) subunits bind to and activate an equal number of the previously inactive *cGMP phosphodiesterase* enzymes, which catalyze the conversion of cGMP into GMP. This causes a very rapid fall in the concentration of cGMP within the narrow spaces of the photoreceptor outer segments, closing the cGMP-gated Na^+ channels in the plasma membrane and inhibiting the dark current (fig. 10.40). The absorption of a single photon of light can block the entry of more than a million Na^+, thereby causing the photoreceptor to hyperpolarize and release less inhibitory neurotransmitter. Freed from inhibition, the bipolar cells activate ganglion cells, and the ganglion cells transmit action potentials to the brain so that light can be perceived (fig. 10.41).

Cones and Color Vision

Cones are less sensitive than rods to light, but the cones provide color vision and greater visual acuity, as described in the next section. During the day, therefore, the high light intensity bleaches out the rods, and color vision with high acuity is provided by the cones.

Each type of cone contains retinene, as in rhodopsin, but the retinene in the cones is associated with proteins called **photopsins.** It is the three different photopsin proteins (coded by three different genes) that give each type of cone its unique light-absorbing characteristics. Each type of cone expresses

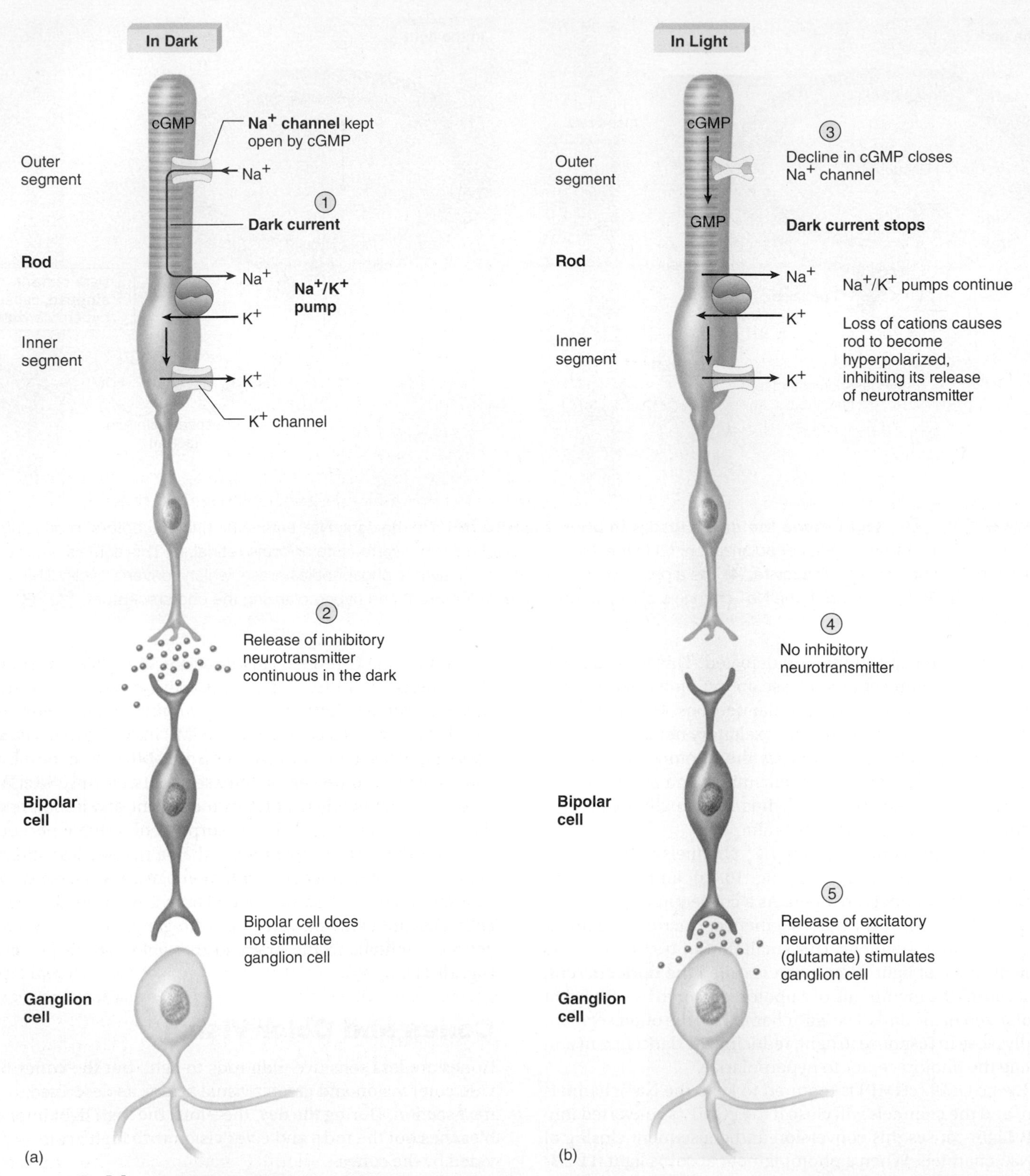

Figure 10.41 The effects of light on the retina. *(a)* In the dark, the continuous dark current (1) depolarizes the photoreceptors and causes them to (2) release inhibitory neurotransmitter at their synapses with bipolar cells. *(b)* In the light, (3) cGMP declines (due to its conversion to GMP), stopping the dark current and hyperpolarizing the photoreceptors. As a result, (4) the release of inhibitory neurotransmitter is stopped. Because they are not inhibited in the light, (5) bipolar cells release excitatory neurotransmitter at their synapses with ganglion cells so that the ganglion cell axons are stimulated to produce action potentials.

only one of these genes to produce only one of these photopsins. Humans and Old World primates (including chimpanzees, gorillas, and gibbons) have **trichromatic color vision.** We are *trichromats,* with three different types of cones. These may be designated *blue, green,* and *red,* according to the region of the visible spectrum in which each cone's pigment absorbs light best (fig. 10.42). This is each cone's *absorption maximum.*

The absorption maximum for the blue cones at 420 nanometers (nm) is in the short wavelengths, so these are also known as **S cones.** The absorption maximum for the green cones (at 530 nm) is in the middle wavelengths, and so these are called **M cones.** The red cones (with an absorption maximum of 562 nm) absorb best in the longer wavelengths, and so are **L cones.** This trichromatic vision is exploited by television and computer screens, which have only red, green, and blue pixels and yet provide us with the multitude of colors we can perceive.

The gene for the S cone pigment is located on autosomal chromosome number 7, whereas the genes for the M and L cones are located in the X chromosome. Most mammals other than humans and Old World primates have only two types of cones, M (green) and S (blue). Because of this, they are *dichromats.* Scientists believe that our trichromatic vision evolved in an ancestral species with dichromatic vision after the gene for the M cone pigment duplicated on the X chromosome. The duplicate could then have given rise to the third type of cones, the L cones able to absorb light best in the longer (red) wavelengths.

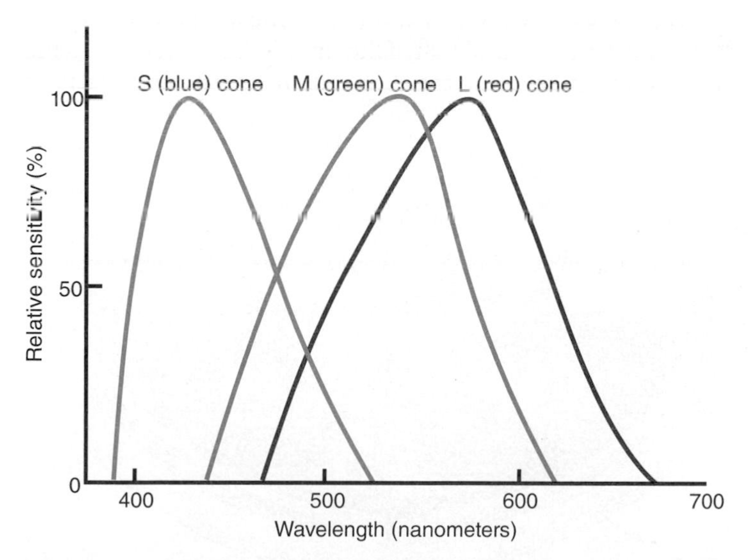

Figure 10.42 The three types of cones. Each type contains retinene, but the protein with which the retinene is combined is different in each case. Thus, each different pigment absorbs light maximally at a different wavelength. Color vision is produced by the activity of these blue cones, green cones, and red cones.

***See the *Test Your Quantitative Ability* section of the Review Activities at the end of this chapter.**

CLINICAL APPLICATION

Color blindness is caused by a congenital lack of one or more types of cones, usually the absence of either the L (red) or M (green) cones. Because such people have only two functioning types of cones, they are dichromats. The absence of functioning M cones, a condition called *deuteranopia,* is the most common form of color blindness.The absence of L cones (*protanopia*) is less common, and the absence of S cones (*tritanopia*) is the least common. People who have only one cone in the middle to long wavelength region (M or L) have difficulty distinguishing reds from greens. Because the M and L cone pigments (photopsins) are coded on the X chromosome, and because men have only one X chromosome (and therefore cannot carry the trait in a recessive state), such red-green color blindness is far more common in men (with an incidence of 8%) than in women (0.5%).

An interesting recent report demonstrated gene therapy for color blindness in male squirrel monkeys, which are all dichromats. These primates were given functioning trichromatic color vision by injecting the human gene for L-opsin (from red cones), within a virus vector (carrier), into the photoreceptor layer of the monkey's retina.

Suppose that a person has become dark adapted but wishes to be able to see (a star map, for example) without losing the dark adaptation. Because rods do not absorb red light but red cones (L cones) do, a red flashlight will allow vision due to excitation of the red cones but will not cause bleaching of the dark-adapted rods. When the red light is turned off, the rods will still be dark-adapted and the person will still be able to see.

An individual cone's response to light depends on both the wavelength (color) of the light and its intensity. For example, a green (M) cone is stimulated effectively by a weaker green light (fig. 10.42), but it can be equally stimulated by a more intense red light. The color we perceive actually depends on neural computations of the effects of a light on different types of cones. Certain ganglion cells have inputs from their receptive fields arranged into a central excitatory (or "on") region surrounded by an antagonistic "off" surround (see fig. 10.47). This allows the effects of different cones to oppose each other. There are two classes of such opposition: (1) L – M contrasts the activity of L and M cones; and (2) S – (L + M) compares the activity of S cones with the combined activity of L and M cones. This provides information about color and light intensity. The ganglion cells project to the lateral geniculate nuclei of the thalamus, which are arranged in layers that preserve this information and relay it to the primary visual cortex. The neural pathways of vision are described in more detail shortly.

Visual Acuity and Sensitivity

While reading or similarly viewing objects in daylight, each eye is oriented so that the image falls within a tiny area of the retina called the **fovea centralis.** The fovea is a pinhead-size pit (*fovea* = pit) within a yellow area of the retina called the **macula lutea** (see fig. 10.30). The pit is formed as a result of the displacement of neural layers around the periphery; therefore, light falls directly on photoreceptors in the center (fig. 10.43). Light falling on other areas, by contrast, must pass through several layers of neurons.

There are approximately 120 million rods and 6 million cones in each retina, but only about 1.2 million axons enter the optic nerve of each eye. This gives an overall convergence ratio of photoreceptors on ganglion cells of about 105 to 1. This is misleading, however, because the degree of convergence is much lower for cones than for rods. In the fovea, the ratio is 1 to 1.

The photoreceptors are distributed in such a way that the fovea contains only cones, whereas more peripheral regions of the retina contain a mixture of rods and cones. Approximately 4,000 cones in the fovea provide input to approximately 4,000 ganglion cells; each ganglion cell in this region, therefore, has a private line to the visual field. Each ganglion cell in the fovea thus receives input from an area of retina corresponding to the diameter of one cone (about 2 μm). Because of this, the only part of the visual field that is seen very clearly is the tiny part (about 1%) that falls on the fovea centralis. We are unaware of this because very rapid eye movements (called *saccadic eye movements*) continually shift different parts of the visual field onto the fovea.

Peripheral to the fovea, many rods synapse with a single bipolar cell, and many bipolar cells synapse with a single ganglion cell. A single ganglion cell outside the fovea thus may receive input from large numbers of rods, corresponding to an area of about 1 mm^2 on the retina (fig. 10.44).

Because each cone in the fovea has a private line to a ganglion cell, and each ganglion cell receives input from only a tiny region of the retina, visual acuity is greatest and sensitivity to low light is poorest when light falls on the fovea. In dim light only the rods are activated, and vision is best out of the corners of the eye so that the image falls away from the fovea. Under these conditions, the convergence of large numbers of rods on a single bipolar cell and the convergence of large numbers of bipolar cells on a single ganglion cell increase sensitivity to dim light at the expense of visual acuity. Night vision is therefore less distinct than day vision.

The difference in visual sensitivity between cones in the fovea centralis and rods in the periphery of the retina can easily be demonstrated using a technique called *averted vision.* If you go out on a clear night and stare hard at a very dim star, it will disappear. This is because the light falls on the fovea and is not sufficiently bright to activate the cones. If you then look slightly off to the side, the star will reappear because the light falls away from the fovea, onto the rods.

Neural Pathways from the Retina

As a result of light refraction by the cornea and lens, the right half of the visual field is projected to the left half of the retina of both eyes (the temporal half of the left retina and the nasal half of the right retina). The left half of the visual field is projected to the right half of the retina of both eyes. The temporal half of the left retina and the nasal half of the right retina therefore see the same image. Axons from ganglion cells in the left (temporal) half

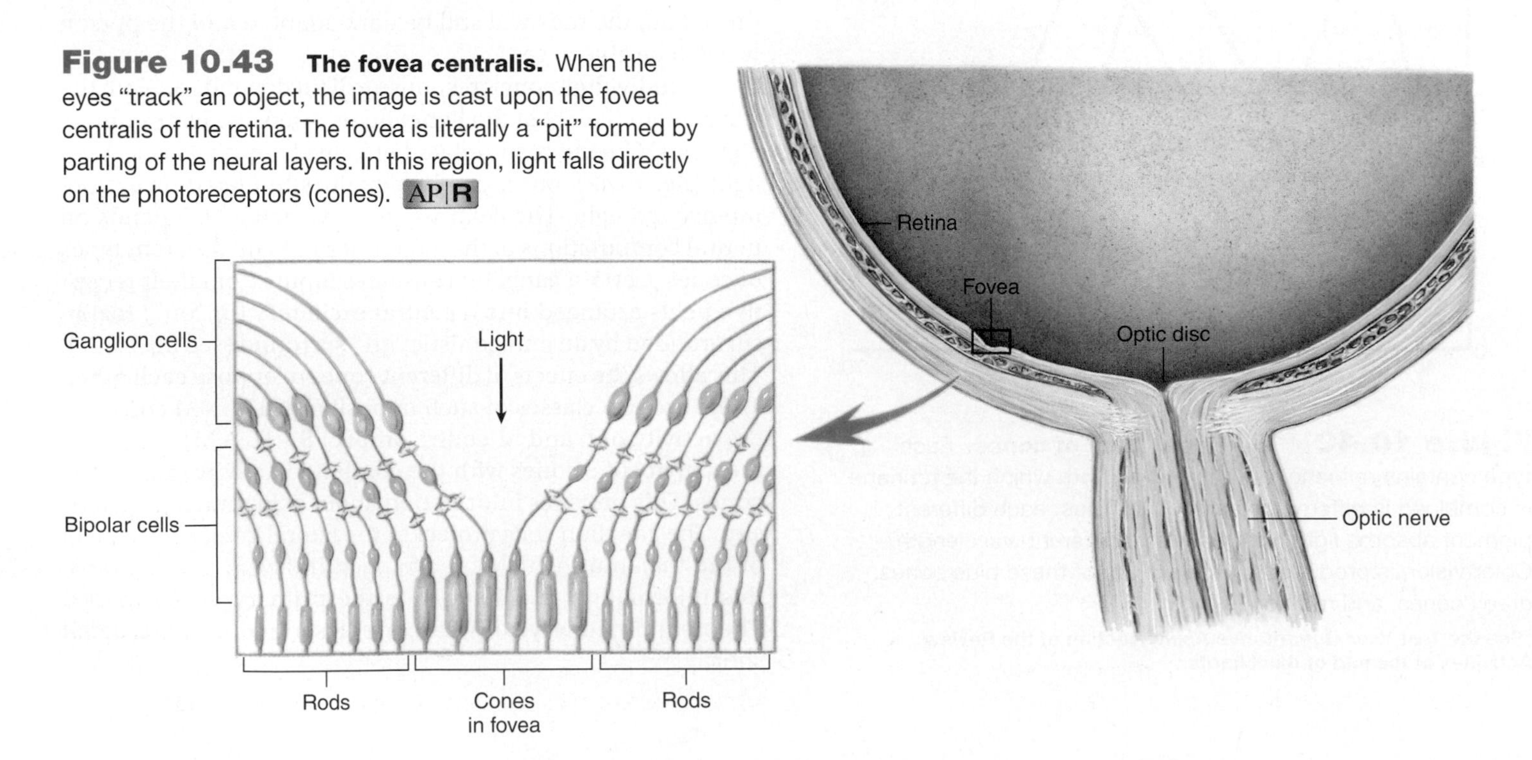

Figure 10.43 The fovea centralis. When the eyes "track" an object, the image is cast upon the fovea centralis of the retina. The fovea is literally a "pit" formed by parting of the neural layers. In this region, light falls directly on the photoreceptors (cones). AP|R

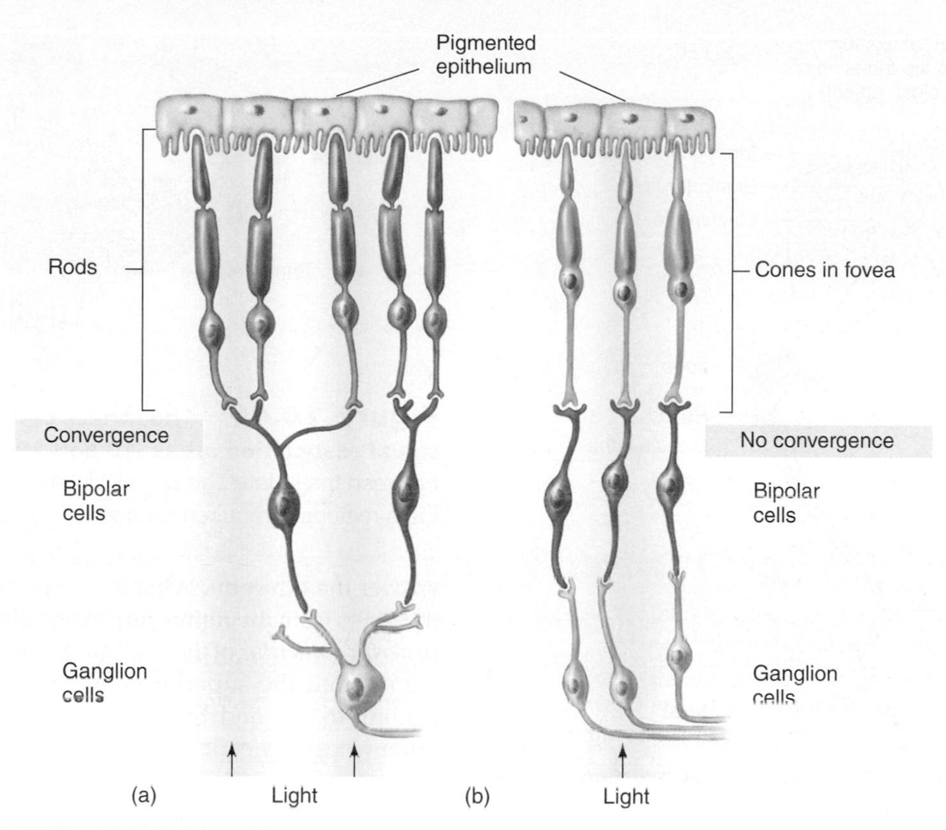

Figure 10.44 Convergence in the retina and light sensitivity. Because bipolar cells receive input from the convergence of many rods (*a*), and because a number of such bipolar cells converge on a single ganglion cell, rods maximize sensitivity to low levels of light at the expense of visual acuity. By contrast, the 1:1:1 ratio of cones to bipolar cells to ganglion cells in the fovea (*b*) provides high visual acuity, but sensitivity to light is reduced.

CLINICAL APPLICATION

Macular degeneration—involving degeneration of the macula lutea and its central fovea—affects one in three people by the age of 75 and is the leading cause of blindness in the United States. People with macular degeneration lose the clarity of vision provided by the fovea and 30% of the vision in the central region of the visual field. In most cases, the damage is believed to be related to the loss of retinal pigment epithelium in this region. This may be caused by age-related changes in the retinal pigment epithelial cells that make them more susceptible to oxidative stress (discussed in chapters 5 and 19), promoting their apoptosis (cell suicide; chapter 3, section 3.5). In its early stages, macular degeneration is detectable by the appearance of cream-colored fatty deposits called *drusen* in the macula lutea. Drusen are fatty deposits in *Bruch's membrane,* the basement membrane underlying the pigment epithelium. While drusen are found in almost everyone over the age of 50, excess drusen may cause damage to the pigment epithelium. People with this early, "dry" form of the disease have moderate vision loss, with acuity in the better eye between 20/50 and 20/100. They can usually read but have difficulty seeing well enough to drive at night.

If the disease progresses to the more serious "wet" form, there is growth of new, abnormal blood vessels (a process termed *neovascularization*) in the choroid, the layer just under the pigment epithelium and Bruch's membrane. These abnormal vessels are very leaky and cause edema. The swelling that results can increase the thickness of the macula and fovea up to three times normal and seriously disrupt vision. This neovascularization is stimulated by a paracrine regulator (chapter 6, section 6.5), *vascular endothelial growth factor (VEGF),* and is responsible for most cases of blindness in people with macular degeneration.

In addition to age, other risk factors include smoking, exposure to light, and a genetic predisposition to this disease. Genes predisposing people to macular degeneration have recently been identified, but the disease is caused by complex interactions of both genetic and environmental influences. The progression of this disease may be slowed by cessation of smoking, wearing sunglasses, and taking multivitamins with antioxidants, zinc, and perhaps lutein (found in green leafy vegetables). Treatment of the wet form currently involves injections into the eye of antibodies that bind to VEGF and prevent its stimulation of neovascularization.

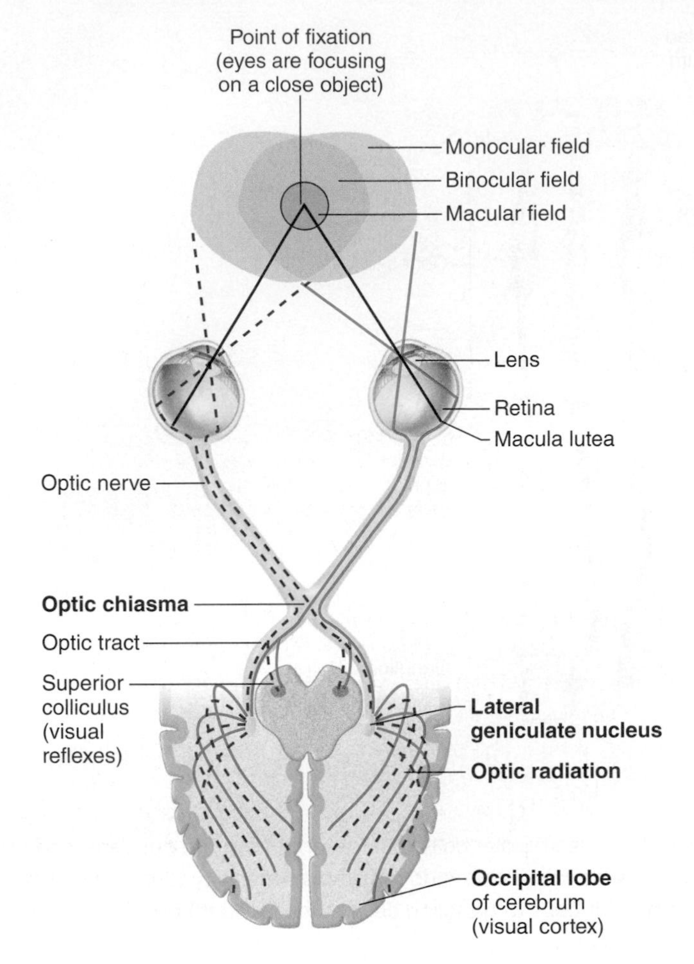

Figure 10.45 The neural pathway for vision. The neural pathway leading from the retina to the lateral geniculate body, and then to the visual cortex, is needed for visual perception. As a result of the crossing of optic fibers, the visual cortex of each cerebral hemisphere receives input from the opposite (contralateral) visual field. APR

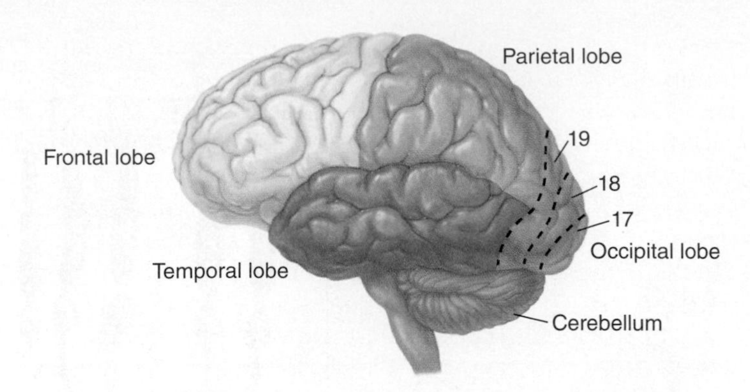

Figure 10.46 The striate cortex (area 17) and the visual association areas (18 and 19). Neural communication between the striate cortex, the visual association areas, and other brain regions is required for normal visual perception.

of the left retina pass to the left **lateral geniculate nucleus** of the thalamus. Axons from ganglion cells in the nasal half of the right retina cross over (decussate) in the X-shaped *optic chiasma* (see fig. 10.32), also to synapse in the left lateral geniculate body. The left lateral geniculate thus receives input from both eyes that relates to the right half of the visual field (fig. 10.45).

The right lateral geniculate body, similarly, receives input from both eyes relating to the left half of the visual field. Neurons in both lateral geniculate bodies of the thalamus in turn project to the **striate cortex** of the occipital lobe in the cerebral cortex (fig. 10.46). This area is also called area 17, in reference to a numbering system developed by K. Brodmann in 1906. Neurons in area 17 synapse with neurons in areas 18 and 19 of the occipital lobe (fig. 10.46).

Approximately 70% to 80% of the axons from the retina pass to the lateral geniculate bodies and to the striate cortex. This **geniculostriate system** is involved in perception of the visual field. Put another way, the geniculostriate system is needed to answer the question, What is it? Approximately 20% to 30% of the fibers from the retina, however, follow a different path to the *superior colliculus* of the midbrain (also called the *optic tectum*). Axons from the superior colliculus activate motor pathways leading to eye and body movements. The **tectal system,** in other words, is needed to answer the question, Where is it?

Neural Control of Eye Movements

Movements of the eyes are produced by contractions of the *extrinsic eye muscles,* innervated by neurons originating in the brain. For example, vertical saccadic eye movements (discussed next) are initiated by neurons in the midbrain, whereas horizontal movements are produced by activity of neurons in the pons and medulla.

There are three types of eye movements coordinated by the brain. **Saccadic eye movements** are very high-velocity movements (400° to 800° per second) of both eyes that target an image on the fovea centralis. For example, saccadic eye movements keep the images of the words you are now reading on or near the fovea, so the words at the middle and end of this sentence can be as clearly seen as those at the beginning. **Smooth pursuit movements** are slower (up to 30° per second), and match the speed of moving objects to keep their images at or near the fovea. **Vergence movements** (30° to 150° per second) cause the eyes to converge so that an image of an object is brought to the fovea of both eyes, allowing the object to be seen more clearly three-dimensionally.

Even when we fix our gaze on a stationary object, our eyes are actually moving. Such **fixational movements** are very tiny and imperceptible. These movements are required for vision, however; sight is lost when fixational movements are prevented under laboratory conditions, as would be expected by *sensory adaptation* (due to the bleaching, or photodissociation, reaction) in the stimulated photoreceptors. Sensory adaptation of vision can also be demonstrated by certain optical illusions where the image at the periphery of the visual field fades when you stare at a spot in the center.

The tectal system is also involved in the control of the intrinsic eye muscles—the iris and the muscles of the ciliary body. Shining a light into one eye stimulates the **pupillary reflex,** in which both pupils constrict. This is caused by activation of parasympathetic neurons in the superior colliculus. Postganglionic axons from the ciliary ganglia behind the eyes, in turn, stimulate constrictor fibers in the iris (see fig. 10.28). Contraction of the ciliary body during *accommodation* also involves parasympathetic stimulation by the superior colliculus.

Surprisingly, the ability to constrict the pupils maximally (by 95% in strong light) depends on light striking the ganglion cell layer, as well as the rods and cones. Scientists have discovered ganglion cells that respond to overall illumination (*luminance*) rather than to patterns and other details of seen objects. These ganglion cells make up a small population (less than 2% of the total) that contains a recently discovered photosensitive pigment called **melanopsin.** The melanopsin-containing ganglion cells depolarize and produce action potentials in response to light.

The melanopsin-containing ganglion cells seem to be uniquely responsible for the non-image-forming functions of the retina. These include: (1) the pupillary reflex (through ganglion cell projection to the optic tectum; see fig. 10.28); (2) the entrainment of circadian rhythms to the light/dark cycle (through ganglion cell projections to the suprachiasmatic nucleus, discussed in chapter 8, section 8.3); and (3) the suppression of the pineal gland secretion of melatonin by light (chapter 11, fig. 11.33; this hormone participates in the regulation of circadian rhythms). The melanopsin in these ganglion cells allows them to respond directly to light, which supplements the information they receive from rods and cones. Experiments with mice that lack melanopsin demonstrate that the input from rods and cones to the melanopsin-containing ganglion cells can produce some pupillary constriction and circadian light/dark entrainment, but pupillary constrictions greater than 50% appear to require melanopsin.

✔ CHECKPOINT

14a. Describe the layers of the retina and trace the path of light and of nerve activity through these layers.

14b. Describe the photochemical reaction in the rods and explain how dark adaptation occurs.

15. Describe the electrical state of photoreceptors in the dark. Explain how light affects the electrical activity of retinal cells.

16a. Explain what is meant by the trichromatic theory of color vision.

16b. Compare the architecture of the fovea centralis with more peripheral regions of the retina. How does this architecture relate to visual acuity and sensitivity?

17. Trace the neural pathways and explain the functions of the geniculostriate system and the tectal system.

10.8 NEURAL PROCESSING OF VISUAL INFORMATION

Electrical activity in ganglion cells of the retina and in neurons of the lateral geniculate nucleus and cerebral cortex is evoked in response to light on the retina. The way in which each type of neuron responds to light at a particular point on the retina provides information about how the brain interprets visual information.

LEARNING OUTCOMES

After studying this section, you should be able to:

18. Describe some of the higher processing of visual information.

Light cast on the retina directly affects the activity of photoreceptors and indirectly affects the neural activity in bipolar and ganglion cells. The part of the visual field that affects the activity of a particular ganglion cell can be considered its **receptive field.** As previously mentioned, each cone in the fovea has a private line to a ganglion cell, and thus the receptive fields of these ganglion cells are equal to the width of one cone (about 2 μm). By contrast, ganglion cells in more peripheral parts of the retina receive input from hundreds of photoreceptors, and are therefore influenced by a larger area of the retina (about 1 mm in diameter).

Ganglion Cell Receptive Fields

Studies of the electrical activity of ganglion cells have yielded some interesting results. In the dark, each ganglion cell discharges spontaneously at a slow rate. When the room lights are turned on, the firing rate of many (but not all) ganglion cells increases slightly. With some ganglion cells, however, a small spot of light directed at the center of their receptive fields elicits a large increase in firing rate. Surprisingly, then, a small spot of light can be a more effective stimulus than a larger area of light!

When the spot of light is moved only a short distance away from the center of the receptive field, the ganglion cell responds in the opposite manner. The ganglion cell that was stimulated with light at the center of its receptive field is inhibited by light in the periphery of its field. The responses produced by light in the center and by light in the "surround" of the visual field are *antagonistic.* Those ganglion cells that are stimulated by light at the center of their visual fields are said to have **on-center fields;** those that are inhibited by light in the center and stimulated by light in the surround have **off-center fields** (fig. 10.47).

The reason wide illumination of the retina has a weaker effect than pinpoint illumination is now clear; diffuse illumination gives the ganglion cell conflicting orders—on and off.

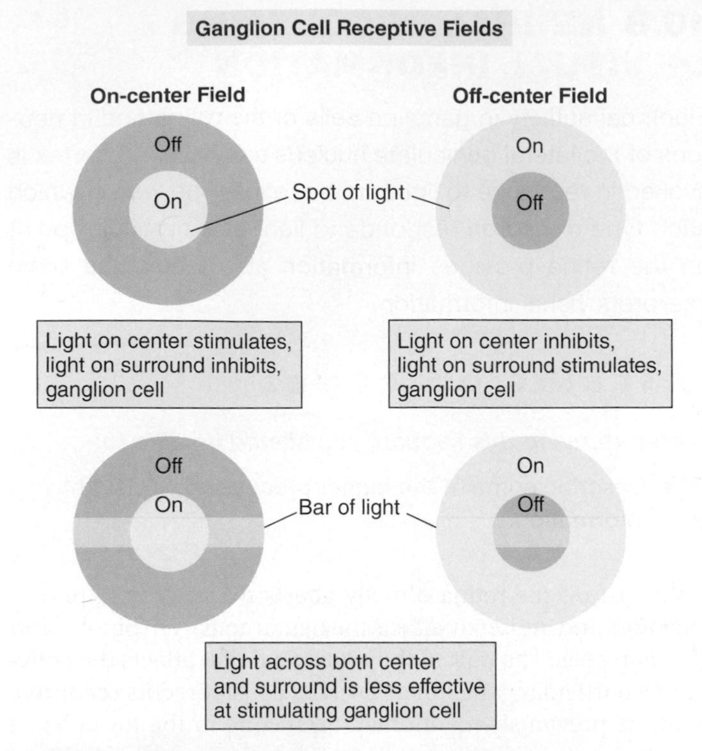

Figure 10.47 Ganglion cell receptive fields. Each ganglion cell receives input from photoreceptors in the retina that are part of the ganglion cell's receptive field. Because of the antagonism between the field's center and its surround, an image that falls across the entire field has less effect than one that excites just the center or surround. Because of this, edges of an image are enhanced, improving the clarity of vision.

Because of the antagonism between the center and surround of ganglion cell receptive fields, the activity of each ganglion cell is a result of the *difference in light intensity* between the center and surround of its visual field. This is a form of lateral inhibition that helps accentuate the contours of images and improve visual acuity.

Lateral Geniculate Nuclei

Each lateral geniculate nucleus of the thalamus receives input from ganglion cells in both eyes. The right lateral geniculate receives input from the right half of each retina (corresponding to the left half of the visual field); the left lateral geniculate receives input from the left half of each retina (corresponding to the right half of the visual field). However, within the lateral geniculate, each neuron is activated by input from only one eye.

The receptive field of each ganglion cell is the part of the retina it "sees" through its photoreceptor input. The receptive field of lateral geniculate neurons, similarly, is the part of the retina it "sees" through its ganglion cell input. Experiments in which the lateral geniculate receptive fields are mapped with a spot of light reveal that they are circular, with an antagonistic center and surround, much like the ganglion cell receptive fields.

Cerebral Cortex

Projections of axons from the lateral geniculate bodies to area 17 of the occipital lobe form the **optic radiation** (see fig. 10.45). Because these fiber projections give area 17 a striped or striated appearance, this area is also known as the *striate cortex*. As mentioned earlier, neurons in area 17 project to areas 18 and 19 of the occipital lobe. Cortical neurons in areas 17, 18, and 19 are thus stimulated indirectly by light on the retina. On the basis of their stimulus requirements, these cortical neurons are classified as *simple, complex,* and *hypercomplex*.

The receptive fields of **simple neurons** are rectangular rather than circular. This is because they receive input from lateral geniculate neurons whose receptive fields are aligned in a particular way (as illustrated in fig. 10.48). Simple cortical neurons are best stimulated by a slit or bar of light located in a precise part of the visual field (of either eye) at a precise orientation.

The striate cortex (area 17) contains simple, complex, and hypercomplex neurons. The other visual association areas, designated areas 18 and 19, contain only complex and

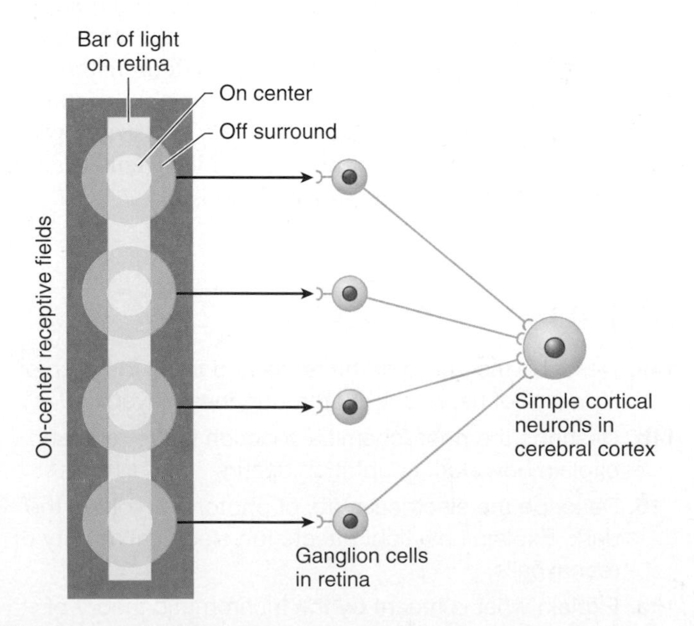

Figure 10.48 Stimulus requirements for simple cortical neurons. Cortical neurons called simple cells have rectangular receptive fields that are best stimulated by slits of light of particular orientations. This may be due to the fact that these simple cells receive input from ganglion cells that have circular receptive fields along a particular line.

hypercomplex cells. Complex neurons receive input from simple cells, and hypercomplex neurons receive input from complex cells. Complex and hypercomplex neurons have different stimulus requirements than do simple neurons. Different complex or hypercomplex neurons may be stimulated by different edges, angles, or curves; they may require that these stimuli be in particular orientations, and further require that the stimuli be moving in particular directions. These neurons must then interact with others to produce the cortical processes needed for the perception of meaningful visual information.

18a. Describe the way in which ganglion cells typically respond to light on the retina. Why may a small spot of light be a more effective stimulus than general illumination of the retina?

18b. How can the arrangement of the receptive fields of ganglion cells enhance visual acuity?

18c. Describe the stimulus requirements of simple cortical neurons.

Clinical Investigation SUMMARY

Ed was on an international flight, so he was exposed to a long flight at high altitude (even though the airplane cabin is pressurized, it still is at a lower-than-sea-level pressure). Considering his head cold, his eustachian tube may not have been able to equalize pressure on both sides of the tympanic membrane, leading to pain and reduced hearing. If this is the explanation, the symptoms should resolve with time and the aid of a decongestant. Ed's visual problem suggests that he is experiencing presbyopia, which normally begins at about Ed's age. LASIK could not compensate for the inability of the lens to perform accommodation. If Ed were nearsighted, he might try monovision LASIK surgery, but because he does not have myopia, he would still need reading glasses if he had this surgery.

See the additional chapter 10 Clinical Investigations on *Meniere's Disease* and *Transient Monocular Blindness* in the Connect site for this text at www.mhhe.com/Fox13.

Mc Graw Hill connect® |ANATOMY & PHYSIOLOGY

Visit this book's website at **www.mhhe.com/Fox13** for:

- Chapter quizzes, interactive learning exercises, and other study tools
- Additional clinical investigations
- Access to LearnSmart—an adaptive diagnostic tool that constantly assesses student knowledge of course material
- Ph.I.L.S. 4.0—physiology interactive lab simulations that may be used to supplement or substitute for wet labs

Interactions

      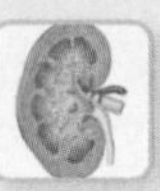 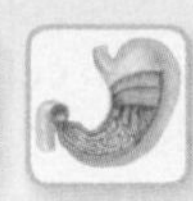 

HPer Links of the Sensory System with Other Body Systems

Integumentary System

- The skin helps protect the body from pathogens (p. 494)
- The skin helps regulate body temperature (p. 474)
- Cutaneous receptors provide sensations of touch, pressure, pain, heat, and cold (p. 270)

Skeletal System

- The skull provides protection and support for the eye and ear (p. 283)
- Proprioceptors provide sensory information about joint movement and the tension of tendons (p. 268)

Muscular System

- Sensory information from the heart helps regulate the heartbeat (p. 479)
- Sensory information from certain arteries helps regulate the blood pressure (p. 477)
- Muscle spindles within skeletal muscles monitor the length of the muscle (p. 386)

Nervous System

- Afferent neurons transduce graded receptor potentials into action potentials (p. 269)
- Afferent neurons conduct action potentials from sensory receptors into the CNS for processing (p. 165)

Endocrine System

- Stimulation of stretch receptors in the heart causes the secretion of atrial natriuretic hormone (p. 462)
- Stimulation of receptors in the GI tract causes the secretion of particular hormones (p. 645)
- Stimulation of sensory endings in the breast by the sucking action of an infant evokes the secretion of hormones involved in lactation (p. 746)

Circulatory System

- The blood delivers oxygen and nutrients to sensory organs and removes metabolic wastes (p. 405)
- Sensory stimuli from the heart provide information for neural regulation of the heartbeat (p. 479)
- Sensory stimuli from certain blood vessels provide information for the neural regulation of blood flow and blood pressure (p. 477)

Immune System

- The immune system protects against infections of sensory organs (p. 494)
- Pain sensations may arise from swollen lymph nodes, alerting us to infection (p. 502)
- The detection of particular chemicals in the brain evokes a fever, which may help to defeat infections (p. 497)

Respiratory System

- The lungs provide oxygen for the blood and provide for the elimination of carbon dioxide (p. 533)
- Chemoreceptors in the aorta, carotid arteries, and medulla oblongata provide sensory information for the regulation of breathing (p. 554)

Urinary System

- The kidneys regulate the volume, pH, and electrolyte balance of the blood and eliminate wastes (p. 582)
- Stretch receptors in the atria of the heart cause secretion of natriuretic factor, which helps regulate the kidneys (p. 608)
- Receptors in renal blood vessels contribute to the regulation of renal blood flow (p. 589)

Digestive System

- The GI tract provides nutrients for all the body organs, including those of the sensory system (p. 621)
- Stretch receptors in the GI tract participate in the reflex control of the digestive system (p. 647)
- Chemoreceptors in the GI tract contribute to the regulation of digestive activities (p. 646)

Reproductive System

- Gonads produce sex hormones that influence sensations involved in the male and female sexual response (p. 711)
- Sensory receptors provide information for erection and orgasm, as well as for other aspects of the sexual response (p. 711)

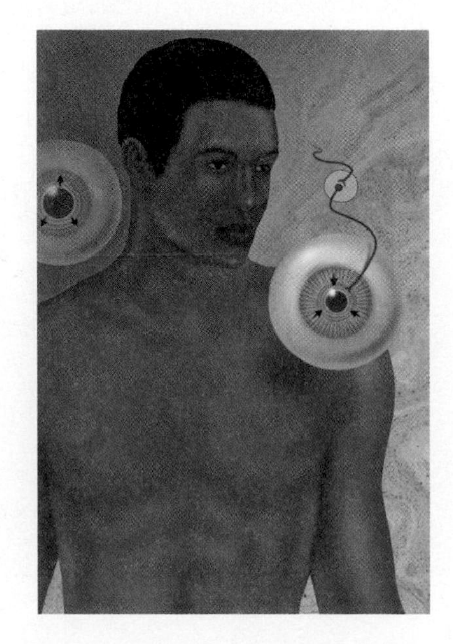

SUMMARY

10.1 Characteristics of Sensory Receptors 267

A. Sensory receptors may be categorized on the basis of their structure, the stimulus energy they transduce, or the nature of their response.

1. Receptors may be dendritic nerve endings, specialized neurons, or specialized epithelial cells associated with sensory nerve endings.
2. Receptors may be chemoreceptors, photoreceptors, thermoreceptors, mechanoreceptors, or nociceptors.
   a. Proprioceptors include receptors in the muscles, tendons, and joints.
   b. The senses of sight, hearing, taste, olfaction, and equilibrium are grouped as special senses.
3. Receptors vary in the duration of their firing in response to a constant stimulus.
   a. Tonic receptors continue to fire as long as the stimulus is maintained; they monitor the presence and intensity of a stimulus.
   b. Phasic receptors respond to stimulus changes; they do not respond to a sustained stimulus, and this partly accounts for sensory adaptation.

B. According to the law of specific nerve energies, each sensory receptor responds with lowest threshold to only one modality of sensation.

1. That stimulus modality is called the adequate stimulus.
2. Stimulation of the sensory nerve from a receptor by any means is interpreted in the brain as the adequate stimulus modality of that receptor.

C. Generator potentials are graded changes (usually depolarizations) in the membrane potential of the dendritic endings of sensory neurons.

1. The magnitude of the potential change of the generator potential is directly proportional to the strength of the stimulus applied to the receptor.
2. After the generator potential reaches a threshold value, increases in the magnitude of the depolarization result in increased frequency of action potential production in the sensory neuron.

10.2 Cutaneous Sensations 270

A. Somatesthetic information—from cutaneous receptors and proprioceptors—is carried by third-order neurons to the postcentral gyrus of the cerebrum.

1. Proprioception and pressure sensations ascend on the ipsilateral side of the spinal cord, synapse in the medulla and cross to the contralateral side, and then ascend in the medial lemniscus to the thalamus; neurons in the thalamus, in turn, project to the postcentral gyrus.
2. Sensory neurons from other cutaneous receptors synapse and cross to the contralateral side in the spinal cord and ascend in the lateral and ventral spinothalamic tracts to the thalamus; neurons in the thalamus then project to the postcentral gyrus.

B. The receptive field of a cutaneous sensory neuron is the area of skin that, when stimulated, produces responses in the neuron.

1. The receptive fields are smaller where the skin has a greater density of cutaneous receptors.
2. The two-point touch threshold test reveals that the fingertips and tip of the tongue have a greater density of touch receptors, and thus a greater sensory acuity, than other areas of the body.

C. Lateral inhibition acts to sharpen a sensation by inhibiting the activity of sensory neurons coming from areas of the skin around the area that is most greatly stimulated.

10.3 Taste and Smell 274

A. The sense of taste is mediated by taste buds.

1. There are four well-established modalities of taste (salty, sour, sweet, and bitter); a fifth, called umami, which is stimulated by glutamate, is now also recognized.
2. Salty and sour taste are produced by the movement of sodium and hydrogen ions, respectively, through membrane channels; sweet and bitter tastes are produced by binding of molecules to protein receptors that are coupled to G-proteins.

B. The olfactory receptors are neurons that synapse within the olfactory bulb of the brain.

1. Odorant molecules bind to membrane protein receptors. There may be as many as 1,000 different receptor proteins responsible for the ability to detect as many as 10,000 different odors.
2. Binding of an odorant molecule to its receptor causes the dissociation of large numbers of G-protein subunits. The effect is thereby amplified, which may contribute to the extreme sensitivity of the sense of smell.

10.4 Vestibular Apparatus and Equilibrium 278

A. The structures for equilibrium and hearing are located in the inner ear, within the membranous labyrinth.

1. The structure involved in equilibrium, known as the vestibular apparatus, consists of the otolith organs (utricle and saccule) and the semicircular canals.
2. The utricle and saccule provide information about linear acceleration, whereas the semicircular canals provide information about angular acceleration.
3. The sensory receptors for equilibrium are hair cells that support numerous stereocilia and one kinocilium.
   a. When the stereocilia are bent in the direction of the kinocilium, the cell membrane becomes depolarized.
   b. When the stereocilia are bent in the opposite direction, the membrane becomes hyperpolarized.

B. The stereocilia of the hair cells in the utricle and saccule project into the endolymph of the membranous labyrinth and are embedded in a gelatinous otolithic membrane.

1. When a person is upright, the stereocilia of the utricle are oriented vertically; those of the saccule are oriented horizontally.
2. Linear acceleration produces a shearing force between the hairs of the otolithic membrane, thus bending the stereocilia and electrically stimulating the sensory endings.

C. The three semicircular canals are oriented at nearly right angles to each other, like the faces of a cube.
1. The hair cells are embedded within a gelatinous membrane called the cupula, which projects into the endolymph.
2. Movement along one of the planes of a semicircular canal causes the endolymph to bend the cupula and stimulate the hair cells.
3. Stimulation of the hair cells in the vestibular apparatus activates the sensory neurons of the vestibulocochlear nerve (VIII), which projects to the cerebellum and to the vestibular nuclei of the medulla oblongata.
   a. The vestibular nuclei in turn send fibers to the oculomotor center, which controls eye movements.
   b. Spinning and then stopping abruptly can thus cause oscillatory movements of the eyes (nystagmus).

10.5 The Ears and Hearing 282

A. The outer ear funnels sound waves of a given frequency (measured in hertz) and intensity (measured in decibels) to the tympanic membrane, causing it to vibrate.

B. Vibrations of the tympanic membrane cause movement of the middle-ear ossicles—malleus, incus, and stapes—which in turn produces vibrations of the oval window of the cochlea.

C. Vibrations of the oval window set up a traveling wave of perilymph in the scala vestibuli.
1. This wave can pass around the helicotrema to the scala tympani, or it can reach the scala tympani by passing through the scala media (cochlear duct).
2. The scala media is filled with endolymph.
   a. The membrane of the cochlear duct that faces the scala vestibuli is called the vestibular membrane.
   b. The membrane that faces the scala tympani is called the basilar membrane.

D. The sensory structure of the cochlea is called the spiral organ or organ of Corti.
1. The organ of Corti rests on the basilar membrane and contains sensory hair cells.
   a. The stereocilia of the hair cells project upward into an overhanging tectorial membrane.
   b. The hair cells are innervated by the vestibulocochlear nerve (VIII).
2. Sounds of high frequency cause maximum displacement of the basilar membrane closer to its base, near the stapes; sounds of lower frequency produce maximum displacement of the basilar membrane closer to its apex, near the helicotrema.
   a. Displacement of the basilar membrane causes the hairs to bend against the tectorial membrane and stimulate the production of nerve impulses.
   b. Pitch discrimination is thus dependent on the region of the basilar membrane that vibrates maximally to sounds of different frequencies.
   c. Pitch discrimination is enhanced by lateral inhibition.

10.6 The Eyes and Vision 290

A. Light enters the cornea of the eye, passes through the pupil (the opening of the iris) and then through the lens, from which point it is projected to the retina in the back of the eye.
1. Light rays are bent, or refracted, by the cornea and lens.
2. Because of refraction, the image on the retina is upside down and right to left.
3. The right half of the visual field is projected to the left half of the retina in each eye, and vice versa.

B. Accommodation is the ability to maintain a focus on the retina as the distance between the object and the eyes is changed.
1. Accommodation is achieved by changes in the shape and refractive power of the lens.
2. When the muscles of the ciliary body are relaxed, the suspensory ligament is tight, and the lens is pulled to its least convex form.
   a. This gives the lens a low refractive power for distance vision.
   b. As an object is brought closer than 20 feet from the eyes, the ciliary body contracts, the suspensory ligament becomes less tight, and the lens becomes more convex and more powerful.

C. Visual acuity refers to the sharpness of the image. It depends in part on the ability of the lens to bring the image to a focus on the retina.
1. People with myopia have an eyeball that is too long, so that the image is brought to a focus in front of the retina; this is corrected by a concave lens.
2. People with hyperopia have an eyeball that is too short, so that the image is brought to a focus behind the retina; this is corrected by a convex lens.
3. Astigmatism is the condition in which asymmetry of the cornea and/or lens causes uneven refraction of light around 360 degrees of a circle, resulting in an image that is not sharply focused on the retina.

10.7 Retina 298

A. The retina contains rods and cones—photoreceptor neurons that synapse with bipolar cells.
1. When light strikes the rods, it causes the photodissociation of rhodopsin into retinene and opsin.
   a. This bleaching reaction occurs maximally with a light wavelength of 500 nm.
   b. Photodissociation is caused by the conversion of the 11-*cis* form of retinene to the all-*trans* form that cannot bind to opsin.
2. In the dark, more rhodopsin can be produced, and increased rhodopsin in the rods makes the eyes more sensitive to light. The increased concentration of rhodopsin in the rods is partly responsible for dark adaptation.

3. The rods provide black-and-white vision under conditions of low light intensity. At higher light intensity, the rods are bleached out and the cones provide color vision.
4. The pigment epithelium has many functions that are required by the photoreceptors in the retina.
   a. The pigment epithelium phagocytoses shed outer segments of the rods and cones, absorbs stray light, and has many other important functions.
   b. The photoreceptors cannot convert all-*trans* retinal back into 11-*cis* retinal; this is done by the pigment epithelium.
   c. The 11-*cis* retinal is then moved back into the photoreceptors, where it can associate with opsin to regenerate the photopigment; this is known as the visual cycle of retinal.

B. In the dark, a constant movement of Na^+ into the rods produces what is known as a "dark current."
1. When light causes the dissociation of rhodopsin, the Na^+ channels become blocked and the rods become hyperpolarized in comparison to their membrane potential in the dark.
2. When the rods are hyperpolarized, they release less neurotransmitter at their synapses with bipolar cells.
3. Neurotransmitters from rods cause depolarization of bipolar cells in some cases and hyperpolarization of bipolar cells in other cases; thus, when the rods are in light and release less neurotransmitter, these effects are inverted.

C. According to the trichromatic theory of color vision, there are three systems of cones, each of which responds to one of three colors: red, blue, or green.
1. Each type of cone contains retinene attached to a different type of protein.
2. The names for the cones signify the region of the spectrum in which the cones absorb light maximally.

D. The fovea centralis contains only cones; more peripheral parts of the retina contain both cones and rods.
1. Each cone in the fovea synapses with one bipolar cell, which in turn synapses with one ganglion cell.
   a. The ganglion cell that receives input from the fovea thus has a visual field limited to that part of the retina that activated its cone.
   b. As a result of this 1:1 ratio of cones to bipolar cells, visual acuity is high in the fovea but sensitivity to low light levels is lower than in other regions of the retina.
2. In regions of the retina where rods predominate, large numbers of rods provide input to each ganglion cell (there is a great convergence). As a result, visual acuity is impaired, but sensitivity to low light levels is improved.

E. The right half of the visual field is projected to the left half of the retina of each eye.
1. The left half of the left retina sends fibers to the left lateral geniculate body of the thalamus.
2. The left half of the right retina also sends fibers to the left lateral geniculate body. This is because these fibers decussate in the optic chiasma.
3. The left lateral geniculate body thus receives input from the left half of the retina of both eyes, corresponding to the right half of the visual field; the right lateral geniculate receives information about the left half of the visual field.
   a. Neurons in the lateral geniculate bodies send fibers to the striate cortex of the occipital lobes.
   b. The geniculostriate system is involved in providing meaning to the images that form on the retina.
4. Instead of synapsing in the geniculate bodies, some fibers from the ganglion cells of the retina synapse in the superior colliculus of the midbrain, which controls eye movement.
   a. Because this brain region is also called the optic tectum, this pathway is called the tectal system.
   b. The tectal system enables the eyes to move and track an object; it is also responsible for the pupillary reflex and the changes in lens shape that are needed for accommodation.

10.8 Neural Processing of Visual Information 307

A. The area of the retina that provides input to a ganglion cell is called the receptive field of the ganglion cell.
1. The receptive field of a ganglion cell is roughly circular, with an "on" or "off" center and an antagonistic surround.
   a. A spot of light in the center of an "on" receptive field stimulates the ganglion cell, whereas a spot of light in its surround inhibits the ganglion cell.
   b. The opposite is true for ganglion cells with "off" receptive cells.
   c. Wide illumination that stimulates both the center and the surround of a receptive field affects a ganglion cell to a lesser degree than a pinpoint of light that illuminates only the center or the surround.
2. The antagonistic center and surround of the receptive field of ganglion cells provide lateral inhibition, which enhances contours and provides better visual acuity.

B. Each lateral geniculate body receives input from both eyes relating to the same part of the visual field.
1. The neurons receiving input from each eye are arranged in layers within the lateral geniculate.
2. The receptive fields of neurons in the lateral geniculate are circular, with an antagonistic center and surround—much like the receptive field of ganglion cells.

C. Cortical neurons involved in vision may be simple, complex, or hypercomplex.
1. Simple neurons receive input from neurons in the lateral geniculate; complex neurons receive input from simple cells; and hypercomplex neurons receive input from complex cells.
2. Simple neurons are best stimulated by a slit or bar of light that is located in a precise part of the visual field and that has a precise orientation.

REVIEW ACTIVITIES

Test Your Knowledge

Match the vestibular organ on the left with its correct component on the right.

1. Utricle and saccule
2. Semicircular canals
3. Cochlea

a. Cupula
b. Ciliary body
c. Basilar membrane
d. Otolithic membrane

4. The dissociation of rhodopsin in the rods in response to light causes
   a. the Na^+ channels to become blocked.
   b. the rods to secrete less neurotransmitter.
   c. the bipolar cells to become either stimulated or inhibited.
   d. all of these.
5. Tonic receptors
   a. are fast-adapting.
   b. do not fire continuously to a sustained stimulus.
   c. produce action potentials at a greater frequency as the generator potential is increased.
   d. are described by all of these.
6. Cutaneous receptive fields are smallest in
   a. the fingertips.
   b. the back.
   c. the thighs.
   d. the arms.
7. The process of lateral inhibition
   a. increases the sensitivity of receptors.
   b. promotes sensory adaptation.
   c. increases sensory acuity.
   d. prevents adjacent receptors from being stimulated.
8. The receptors for taste are
   a. naked sensory nerve endings.
   b. encapsulated sensory nerve endings.
   c. specialized epithelial cells.
9. Which of these statements about the utricle and saccule are *true?*
   a. They are otolith organs.
   b. They are located in the middle ear.
   c. They provide a sense of linear acceleration.
   d. Both *a* and *c* are true.
   e. Both *b* and *c* are true.
10. Because fibers of the optic nerve that originate in the nasal halves of each retina cross over at the optic chiasma, each lateral geniculate receives input from
    a. both the right and left sides of the visual field of both eyes.
    b. the ipsilateral visual field of both eyes.
    c. the contralateral visual field of both eyes.
    d. the ipsilateral field of one eye and the contralateral field of the other eye.
11. When a person with normal vision views an object from a distance of at least 20 feet,
    a. the ciliary muscles are relaxed.
    b. the suspensory ligament is tight.
    c. the lens is in its most flat, least convex shape.
    d. all of these apply.
12. Glasses with concave lenses help correct
    a. presbyopia.
    b. myopia.
    c. hyperopia.
    d. astigmatism.
13. Parasympathetic nerves that stimulate constriction of the iris (in the pupillary reflex) are activated by neurons in
    a. the lateral geniculate.
    b. the superior colliculus.
    c. the inferior colliculus.
    d. the striate cortex.
14. A bar of light in a specific part of the retina, with a particular length and orientation, is the most effective stimulus for
    a. ganglion cells.
    b. lateral geniculate cells.
    c. simple cortical cells.
    d. complex cortical cells.
15. The ability of the lens to increase its curvature and maintain a focus at close distances is called
    a. convergence.
    b. accommodation.
    c. astigmatism.
    d. amblyopia.
16. Which of these sensory modalities is transmitted directly to the cerebral cortex without being relayed through the thalamus?
    a. Taste
    b. Sight
    c. Smell
    d. Hearing
    e. Touch
17. Stimulation of membrane protein receptors by binding to specific molecules is *not* responsible for
    a. the sense of smell.
    b. sweet taste sensations.
    c. sour taste sensations.
    d. bitter taste sensations.
18. Epithelial cells release transmitter chemicals that excite sensory neurons in all of these senses *except*
    a. taste.
    b. smell.
    c. equilibrium.
    d. hearing.

Test Your Understanding

19. Explain what is meant by lateral inhibition and give examples of its effects in three sensory systems.
20. Describe the nature of the generator potential and explain its relationship to stimulus intensity and to frequency of action potential production.
21. Describe the phantom limb phenomenon and give a possible explanation for its occurrence.
22. Explain the relationship between smell and taste. How are these senses similar? How do they differ?
23. Explain how the vestibular apparatus provides information about changes in the position of our body in space.
24. In a step-by-step manner, explain how vibrations of the oval window lead to the production of nerve impulses.
25. Using the ideas of the place theory of pitch and the tonotopic organization of the auditory cortex, explain how we perceive different pitches of sounds.
26. Describe the sequence of changes that occur during accommodation. Why is it more of a strain on the eyes to look at a small nearby object than at large objects far away?
27. Describe the effects of light on the photoreceptors and explain how these effects influence the bipolar cells.
28. Explain why images that fall on the fovea centralis are seen more clearly than images that fall on the periphery of the retina. Why are the "corners of the eyes" more sensitive to light than the fovea?
29. Explain why rods provide only black-and-white vision. Include a discussion of different types of color blindness in your answer.
30. Explain why green objects can be seen better at night than objects of other colors. What effect does red light in a darkroom have on a dark-adapted eye?
31. Describe the receptive fields of ganglion cells and explain how the nature of these fields helps to improve visual acuity.
32. How many genes code for the sense of color vision? How many for taste? How many for smell? What does this information say about the level of integration required by the brain for the perception of these senses?
33. Discuss the different functions of the pigment epithelium of the retina, and describe the visual cycle of retinal.
34. What makes the lens of the eye clear? What happens when cataracts form?

Test Your Analytical Ability

35. You are firing your laser cannon from your position on the bridge of your starship. You see the hostile enemy starship explode, but you hear no accompanying sound. Can you explain this? How do receptors for sight and hearing differ?
36. People with conduction deafness often speak quietly. By contrast, people with sensorineural deafness tend to speak louder than normal. Explain these differences.
37. Opioid drugs reduce the sensation of dull, persistent pain but have little effect on the initial sharp pain of a noxious stimulus (e.g., a pinprick). What do these different effects imply? What conclusion can be drawn from the fact that aspirin (a drug that inhibits the formation of prostaglandins) functions as a pain reliever?
38. Compare the role of G-proteins in the senses of taste and sight. What is the advantage of having G-proteins mediate the effect of a stimulus on a receptor cell?
39. Discuss the role that inertia plays in the physiology of the vestibular apparatus. Why is there no sensation of movement in an airplane once it has achieved cruising speed?
40. Explain why menthol feels cold on the tongue and chili sauce feels hot. What does this reveal about our perceptions of reality?
41. In order to see a dim star at night, it is better not to look directly at it. If you see it from the "corner of your eye" and then turn to look at it, it may disappear. Explain how this occurs. Also, you can stare directly at an object in daylight and continue seeing it, despite the bleaching of photoreceptors. Explain how this is possible.
42. The pigment epithelial cells may undergo apoptosis in response to the oxidative stress of free radicals (chapters 3 and 19). Explain what functions could be disturbed by a damaged pigment epithelium, and relate this to the disease of macular degeneration.

Test Your Quantitative Ability

Refer to figure 10.26 and use the figure reproduced below from figure 10.42 to answer the following questions.

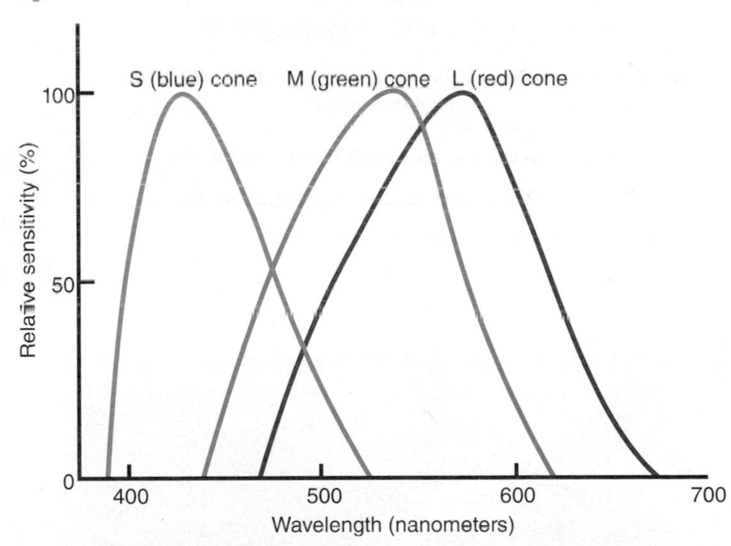

43. At which wavelength of light are the blue and green cones equally stimulated? What color is this light?
44. Which cones are most sensitive to light at 600 nm? Which are the least sensitive?
45. Which wavelength of light stimulates two cone systems to the greatest degree? What color is this light?
46. Which wavelength of light stimulates all three of the cone systems to the greatest degree? What color is this light?

CHAPTER

11

Endocrine Glands

Secretion and Action of Hormones

REFRESH YOUR MEMORY

Before you begin this chapter, you may want to review these concepts from previous chapters:

CHAPTER OUTLINE

Clinical Investigation

Rosemary, a 32-year-old office worker, goes in for a routine medical physical, where they discover that she has high blood pressure (hypertension) and high blood glucose (hyperglycemia). The physician notices that she has a generalized "puffiness." Rosemary returns another day to provide blood for several blood tests, and also to perform a test where she drinks a sweet solution and has her blood drawn periodically over the next few hours.

Some of the new terms and concepts you will encounter include:

- Thyroxine (T_4) and triiodothyronine (T_3), and myxedema
- Cortisol, ACTH, Addison's disease, and Cushing's syndrome
- Oral glucose tolerance test and diabetes mellitus

11.1 ENDOCRINE GLANDS AND HORMONES

Hormones are regulatory molecules secreted into the blood by endocrine glands. Chemical categories of hormones include steroids, amines, polypeptides, and glycoproteins. Interactions between the various hormones produce effects that may be synergistic, permissive, or antagonistic.

LEARNING OUTCOMES

After studying this section, you should be able to:

1. **Describe the chemical nature of hormones and define the terms *prehormone* and *prohormone*.**
2. **Describe the different types of hormone interactions and the significance of hormone concentrations.**

Endocrine glands lack the ducts that are present in exocrine glands (chapter 1, section 1.3). The endocrine glands secrete their products, which are biologically active molecules called **hormones,** into the blood. The blood carries the hormones to *target cells* that contain specific *receptor proteins* for the hormones, and which therefore can respond in a specific fashion to them. Many endocrine glands are organs whose primary functions are the production and secretion of hormones (fig. 11.1*a*). The pancreas functions as both an exocrine and an endocrine gland; the endocrine portion of the pancreas is composed of clusters of cells called the pancreatic islets (islets of Langerhans) (fig. 11.1*b*). The concept of the **endocrine system,** however, must be extended beyond these organs because many other organs in the body secrete hormones.

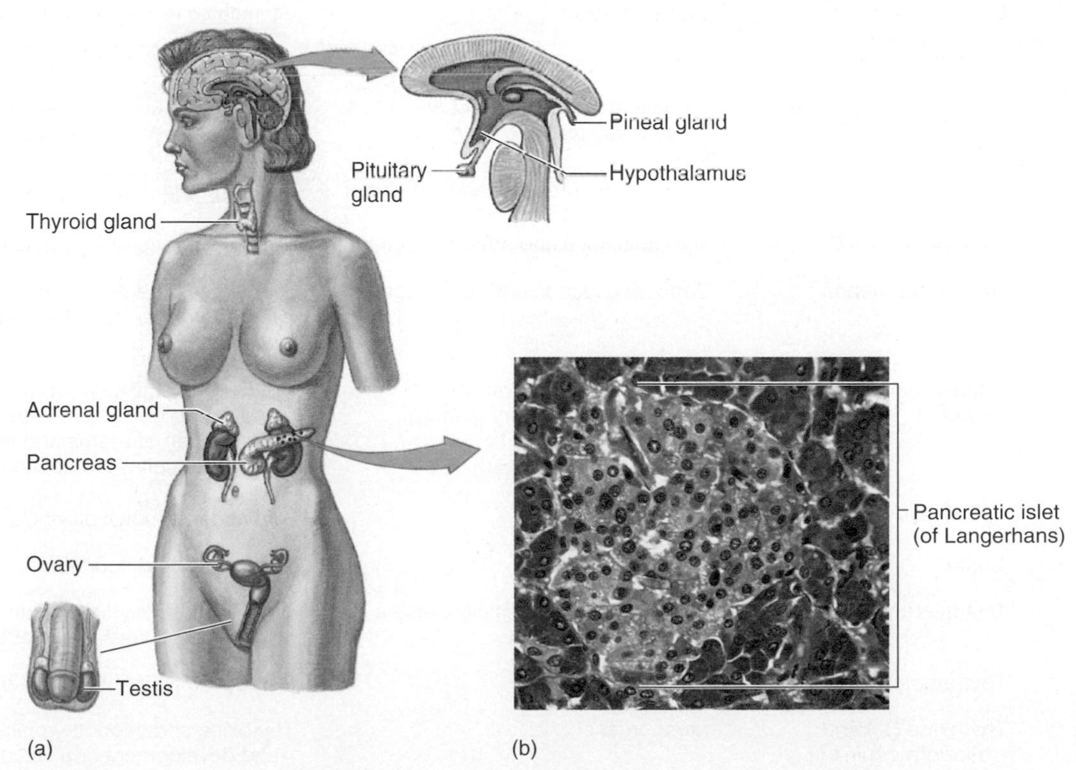

Figure 11.1 The major endocrine glands. (*a*) The anatomical location of some of the endocrine glands. (*b*) A photomicrograph of a pancreatic islet (of Langerhans) within the pancreas. APIR

These organs may be categorized as endocrine glands even though they serve other functions as well. It is appropriate, then, that a partial list of the endocrine glands (table 11.1) should include the heart, liver, adipose tissue, and kidneys.

Some specialized neurons, particularly in the hypothalamus, secrete chemical messengers into the blood rather than into a narrow synaptic cleft. In these cases, the chemical that the neurons secrete is sometimes called a *neurohormone.* In addition, a number of chemicals—norepinephrine, for example—are secreted both as a neurotransmitter and a hormone. Thus, a sharp distinction between the nervous system and the endocrine system cannot always be drawn on the basis of the chemicals they release.

Hormones affect the metabolism of their target organs and, by this means, help regulate total body metabolism, growth, and reproduction. The effects of hormones on body metabolism and growth are discussed in chapter 19; the regulation of reproductive functions by hormones is considered in chapter 20.

Table 11.1 | A Partial Listing of the Endocrine Glands

Endocrine Gland	Major Hormones	Primary Target Organs	Primary Effects
Adipose tissue	Leptin	Hypothalamus	Suppresses appetite
Adrenal cortex	Glucocorticoids (mainly cortisol) Mineralocorticoids (mainly aldosterone)	Liver and muscles Kidneys	Glucocorticoids influence glucose metabolism; mineralocorticoids promote Na^+ retention, K^+ excretion
Adrenal medulla	Epinephrine	Heart, bronchioles, and blood vessels	Causes adrenergic stimulation
Heart	Atrial natriuretic hormone	Kidneys	Promotes excretion of Na^+ and water in the urine
Hypothalamus	Releasing and inhibiting hormones	Anterior pituitary	Regulates secretion of anterior pituitary hormones
Small intestine	Secretin and cholecystokinin	Stomach, liver, and pancreas	Inhibits gastric motility and stimulates bile and pancreatic juice secretion
Islets of Langerhans (pancreas)	Insulin Glucagon	Liver, skeletal muscle, and adipose tissue primarily	Insulin promotes cellular uptake of glucose and formation of glycogen and fat; glucagon stimulates hydrolysis of glycogen and fat
Kidneys	Erythropoietin	Bone marrow	Stimulates red blood cell production
Liver	Somatomedins	Cartilage	Stimulates cell division and growth
Ovaries	Estradiol-17β and progesterone	Female reproductive tract and mammary glands	Maintains structure of reproductive tract and promotes secondary sex characteristics
Parathyroid glands	Parathyroid hormone	Bone, small intestine, and kidneys	Increases Ca^{2+} concentration in blood
Pineal gland	Melatonin	Hypothalamus and anterior pituitary	Affects secretion of gonadotrophic hormones
Pituitary, anterior	Trophic hormones	Endocrine glands and other organs	Stimulates growth and development of target organs; stimulates secretion of other hormones
Pituitary, posterior	Antidiuretic hormone Oxytocin	Kidneys and blood vessels Uterus and mammary glands	Antidiuretic hormone promotes water retention and vasoconstriction; oxytocin stimulates contraction of uterus and mammary secretory units, promoting milk ejection
Skin	1,25-Dihydroxyvitamin D_3	Small intestine	Stimulates absorption of Ca^{2+}
Stomach	Gastrin	Stomach	Stimulates acid secretion
Testes	Testosterone	Prostate, seminal vesicles, testes, and other organs	Stimulates secondary sexual development, spermatogenesis, other effects
Thymus	Thymopoietin	Lymph nodes	Stimulates white blood cell production
Thyroid gland	Thyroxine (T_4) and triiodothyronine (T_3); calcitonin	Most organs	Thyroxine and triiodothyronine promote growth and development and stimulate basal rate of cell respiration (basal metabolic rate or BMR); calcitonin may participate in the regulation of blood Ca^{2+} levels

Chemical Classification of Hormones

Hormones secreted by different endocrine glands vary widely in chemical structure. All hormones, however, can be divided into a few chemical classes.

1. **Amines.** These are hormones derived from the amino acids tyrosine and tryptophan. They include the hormones secreted by the adrenal medulla, thyroid, and pineal glands.
2. **Polypeptides and proteins.** Proteins are large polypeptides, so the distinction between the two categories is somewhat arbitrary. Antidiuretic hormone is a polypeptide with nine amino acids (table 11.2), too small to accurately be called a protein. If a polypeptide chain is larger than about 100 amino acids, such as growth hormone with 191 amino acids, it can be called a protein. Insulin blurs the two categories, because it is composed of two polypeptide chains derived from a single, larger molecule (see chapter 3, fig. 3.23).
3. **Glycoproteins.** These molecules consist of a protein bound to one or more carbohydrate groups. Examples are follicle-stimulating hormone (FSH) and luteinizing hormone (LH).
4. **Steroids.** Steroid hormones are derived from cholesterol after an enzyme cleaves off the side chain attached to the five-carbon "D" ring (fig. 11.2). Steroid hormones include testosterone, estradiol, progesterone, and cortisol.

Table 11.2 | Examples of Polypeptide and Glycoprotein Hormones

Hormone	Structure	Gland	Primary Effects
Antidiuretic hormone	9 amino acids	Posterior pituitary	Water retention and vasoconstriction
Oxytocin	9 amino acids	Posterior pituitary	Uterine and mammary contraction
Insulin	21 and 30 amino acids (double chain)	Beta cells in islets of Langerhans	Cellular glucose uptake, lipogenesis, and glycogenesis
Glucagon	29 amino acids	Alpha cells in islets of Langerhans	Hydrolysis of stored glycogen and fat
ACTH	39 amino acids	Anterior pituitary	Stimulation of adrenal cortex
Parathyroid hormone	84 amino acids	Parathyroid	Increase in blood Ca^{2+} concentration
FSH, LH, TSH	Glycoproteins	Anterior pituitary	Stimulation of growth, development, and secretory activity of target glands

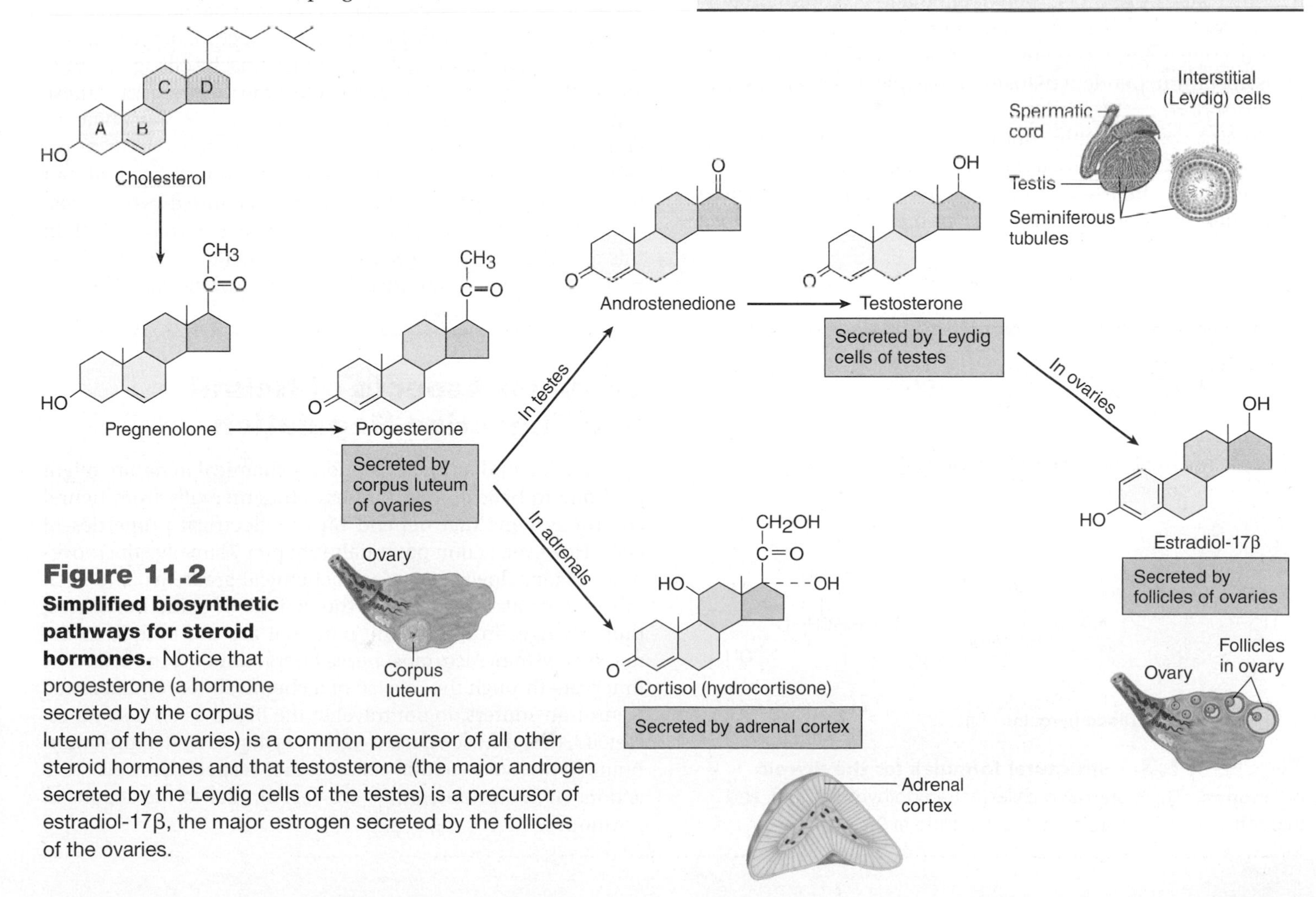

Figure 11.2 Simplified biosynthetic pathways for steroid hormones. Notice that progesterone (a hormone secreted by the corpus luteum of the ovaries) is a common precursor of all other steroid hormones and that testosterone (the major androgen secreted by the Leydig cells of the testes) is a precursor of estradiol-17β, the major estrogen secreted by the follicles of the ovaries.

In terms of their actions in target cells, hormone molecules can be divided into those that are polar, and therefore water-soluble, and those that are nonpolar, and thus insoluble in water. (For a discussion of water solubility, see chapter 2 and figure 2.6.) Because the nonpolar hormones are soluble in lipids, they are often referred to as **lipophilic hormones.** Unlike the polar hormones, which cannot pass through plasma membranes, lipophilic hormones can gain entry into their target cells. These lipophilic hormones include the steroid hormones and thyroid hormones.

Steroid hormones are secreted by only two endocrine glands: the adrenal cortex and the gonads (fig. 11.2). The gonads secrete *sex steroids;* the adrenal cortex secretes *corticosteroids* (including cortisol and aldosterone) and small amounts of sex steroids.

The major thyroid hormones are composed of two derivatives of the amino acid tyrosine bonded together (fig. 11.3). When the hormone contains 4 iodine atoms, it is called *tetraiodothyronine* (T_4), or *thyroxine.* When it contains 3 atoms of iodine, it is called *triiodothyronine* (T_3). Although these hormones are not steroids, they are like steroids in that they are relatively small, nonpolar molecules. Steroid and thyroid hormones are active when taken orally (as a pill). Sex steroids are the active agents in contraceptive pills, and thyroid hormone pills are taken by people whose thyroid is deficient (who are hypothyroid). By contrast, polypeptide and glycoprotein hormones cannot be taken orally because they would be digested into inactive fragments before being absorbed into the blood. Thus, insulin-dependent diabetics must inject themselves with this hormone.

Polar, water-soluble hormones include polypeptides, glycoproteins, and the *catecholamine* hormones secreted by the adrenal medulla, epinephrine and norepinephrine. These hormones are derived from the amino acid tyrosine (see chapter 9, fig. 9.8). Thus, like the polypeptide and glycoprotein hormones, the catecholamines are too polar to pass through the phospholipid portion of the plasma membrane. The hormone secreted by the pineal gland, melatonin, is different; derived from the nonpolar amino acid tryptophan, melatonin pills can be effective because (like steroids and thyroxine) this hormone can pass through plasma membranes. Melatonin, however, also has some similarities to the polar hormones in terms of its effects on cells.

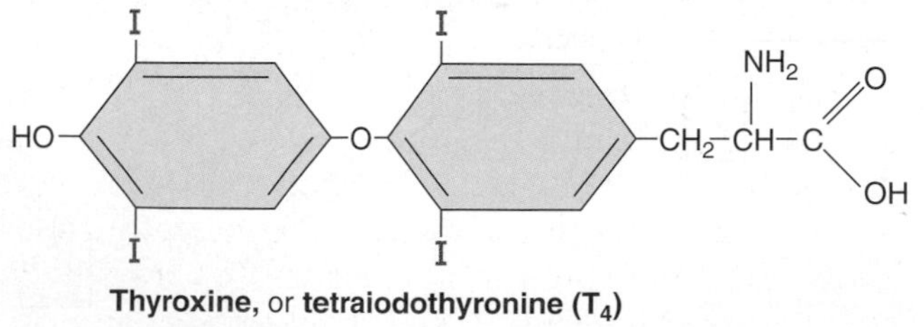

Thyroxine, or **tetraiodothyronine (T_4)**

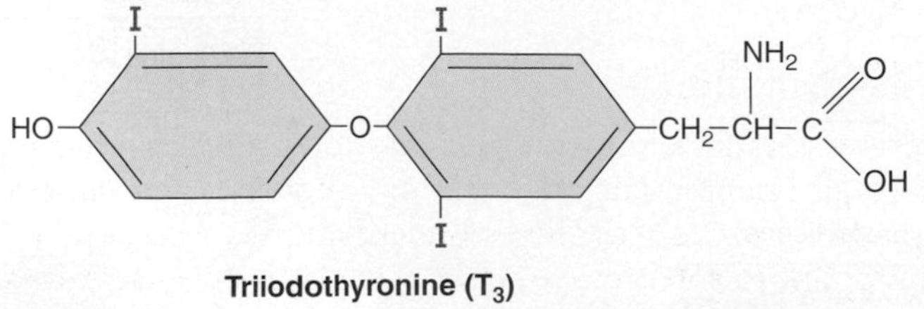

Triiodothyronine (T_3)

Figure 11.3 Structural formulas for the thyroid hormones. Thyroxine, also called tetraiodothyronine (T_4), and triiodothyronine (T_3) are secreted in a ratio of 9 to 1.

Prohormones and Prehormones

Hormone molecules that affect the metabolism of target cells are often derived from less active "parent," or **precursor,** molecules. In the case of polypeptide hormones, the precursor may be a longer chained **prohormone** that is cut and spliced together to make the hormone. Insulin, for example, is produced from **proinsulin** (see fig. 3.23) within the beta cells of the islets of Langerhans of the pancreas. In some cases, the prohormone itself is derived from an even larger precursor molecule; in the case of insulin, this molecule is called **preproinsulin.** The term **prehormone** is sometimes used to indicate such precursors of prohormones.

In some cases, the molecule secreted by the endocrine gland (and considered to be the hormone of that gland) is actually inactive in the target cells. In order to become active, the target cells must modify the chemical structure of the secreted hormone. Thyroxine (T_4), for example, must be changed into T_3 within the target cells in order to affect the metabolism of these cells. Similarly, vitamin D_3 secreted by the skin must be converted into 1,25-dihydroxyvitamin D_3 to be active in its target cells. Testosterone is an active hormone in its own right, but in some target organs it must be converted into dihydrotestosterone (DHT) or other derivatives to be active (table 11.3). In this text, the term *prehormone* will be used to designate those molecules secreted by endocrine glands that are inactive until changed by their target cells.

Common Aspects of Neural and Endocrine Regulation

The fact that endocrine regulation is chemical in nature might lead one to believe that it differs fundamentally from neural control systems that depend on the electrical properties of cells. However, action potentials (chapter 7) involve the movement of ions down their electrochemical gradients, and such movements also accompany the actions of some hormones; thus, changes in membrane potential are not unique to the nervous system. Also, most nerve fibers stimulate the cells they innervate through the release of a chemical neurotransmitter. Neurotransmitters do not travel in the blood as do hormones; instead, they diffuse across a narrow synaptic cleft to the membrane of the postsynaptic cell. In other respects, however, the actions of neurotransmitters are very similar to the actions of hormones.

Table 11.3 | Conversion of Prehormones into Biologically Active Derivatives

Endocrine Gland	Prehormone	Active Products	Comments
Skin	Vitamin D_3	1,25-Dihydroxyvitamin D_3	Conversion (through hydroxylation reactions) occurs in the liver and the kidneys.
Testes	Testosterone	Dihydrotestosterone (DHT)	DHT and other 5α-reduced androgens are formed in most androgen-dependent tissue.
		Estradiol-17β (E_2)	E_2 is formed in the brain from testosterone, where it is believed to affect both endocrine function and behavior; small amounts of E_2 are also produced in the testes.
Thyroid	Thyroxine (T_4)	Triiodothyronine (T_3)	Conversion of T_4 to T_3 occurs in almost all tissues.

Indeed, many polypeptide hormones, including those secreted by the pituitary gland and by the digestive tract, have been discovered in the brain. In certain locations in the brain, some of these compounds are produced and secreted as hormones. In other brain locations, some of these compounds apparently serve as neurotransmitters. The discovery of polypeptide hormones in unicellular organisms suggests that these regulatory molecules appeared early in evolution and were incorporated into the function of nervous and endocrine tissue as these systems evolved.

Regardless of whether a particular chemical is acting as a neurotransmitter or as a hormone, in order for it to function in physiological regulation: (1) target cells must have specific **receptor proteins** that combine with the regulatory molecule; (2) the combination of the regulatory molecule with its receptor proteins must cause a specific sequence of changes in the target cells; and (3) there must be a mechanism to turn off the action of the regulator. This mechanism, which involves rapid removal and/or chemical inactivation of the regulator molecules, is essential because without an "off-switch" physiological control would be impossible.

Hormone Interactions

A given target tissue is usually responsive to a number of different hormones. These hormones may antagonize each other or work together to produce effects that are additive or complementary. The responsiveness of a target tissue to a particular hormone is thus affected not only by the concentration of that hormone, but also by the effects of other hormones on that tissue. Terms used to describe hormone interactions include *synergistic, permissive,* and *antagonistic.*

Synergistic and Permissive Effects

When two or more hormones work together to produce a particular result, their effects are said to be **synergistic.** These effects may be additive or complementary. The action of epinephrine and norepinephrine on the heart is a good example of an additive effect. Each of these hormones separately produces an increase in cardiac rate; acting together in the same concentrations, they stimulate an even greater increase in cardiac rate. The ability of the mammary glands to produce and secrete milk (in lactation) requires the synergistic action of many hormones—estrogen, cortisol, prolactin, and oxytocin—which have complementary actions. That is, each of these hormones promotes a different aspect of mammary gland function, so that their cooperative effects are required for lactation.

A hormone is said to have a **permissive effect** on the action of a second hormone when it enhances the responsiveness of a target organ to the second hormone, or when it increases the activity of the second hormone. Prior exposure of the uterus to estradiol (the major estrogen), for example, induces the formation of receptor proteins for progesterone, which improves the response of the uterus when it is subsequently exposed to progesterone. Estradiol thus has a permissive effect on the responsiveness of the uterus to progesterone.

Vitamin D_3 is a prehormone that must be modified by enzymes in the kidneys and liver, where two hydroxyl (OH^-) groups are added to form the active hormone 1,25-dihydroxyvitamin D_3. This hormone helps to raise blood calcium levels. Parathyroid hormone (PTH) has a permissive effect on the actions of vitamin D_3 because it stimulates the production of the hydroxylating enzymes in the kidneys and liver. By this means, an increased secretion of PTH has a permissive effect on the ability of vitamin D_3 to stimulate the intestinal absorption of calcium.

Antagonistic Effects

In some situations, the actions of one hormone antagonize the effects of another. Lactation during pregnancy, for example, is inhibited because the high concentration of estrogen in the blood inhibits the secretion and action of prolactin. Another example of antagonism is the action of insulin and glucagon (two hormones from the pancreatic islets) on adipose tissue; the formation of fat is promoted by insulin, whereas glucagon promotes fat breakdown.

Effects of Hormone Concentrations on Tissue Response

The concentration of hormones in the blood primarily reflects the rate of secretion by the endocrine glands. Hormones do not generally accumulate in the blood because they are rapidly removed by target organs and by the liver. The **half-life** of a hormone—the time required for the plasma concentration of a given amount of the hormone to be reduced by half—ranges from minutes to hours for most hormones (thyroid hormone, however, has a half-life of several days). Hormones removed from the blood by the liver are converted by enzymatic reactions into less active products. Steroids, for example, are converted into more water-soluble polar derivatives that are released into the blood and excreted in the urine and bile.

The effects of hormones are very dependent on concentration. Normal tissue responses are produced only when the hormones are present within their normal, or *physiological,* range of concentrations. When some hormones are taken in abnormally high, or *pharmacological,* concentrations (as when they are taken as drugs), their effects may be different from those produced by lower, more physiological, concentrations. In part, this can be caused by the binding of the hormone at pharmacological concentrations to the receptors of different (but closely related) hormones, so that abnormal effects are produced. In the case of steroid hormones, pharmacological concentrations may cause the abnormal production of derivatives with different biological effects. For example, pharmacological amounts of androgens result in the production of abnormal amounts of estrogens, which are derived from androgens (see fig. 11.2).

Pharmacological doses of hormones, particularly of steroids, can thus have widespread and often damaging side effects. People with inflammatory diseases who are treated with high doses of cortisone over long periods of time, for example, may develop osteoporosis and characteristic changes in soft tissue structure. Contraceptive pills, which contain sex steroids, have a number of potential side effects that could not have been predicted in 1960, when "the pill" was first introduced. At that time, the concentrations of sex steroids were much higher than they are in the pills presently being marketed.

Priming Effects

Variations in hormone concentration within the normal, physiological range can affect the responsiveness of target cells. This is due in part to the effects of polypeptide and glycoprotein hormones on the number of their receptor proteins in target cells. More receptors may be formed in the target cells in response to particular hormones. Small amounts of gonadotropin-releasing hormone (GnRH) secreted by the hypothalamus, for example, increase the sensitivity of anterior pituitary cells to further GnRH stimulation. This is a *priming effect,* caused in large part by the **upregulation** of receptors. In this process, increased numbers of receptor proteins for the hormone being primed (in this case, GnRH) are inserted into the plasma membrane. Subsequent stimulation by GnRH thus causes a greater response from the anterior pituitary.

FITNESS APPLICATION

Anabolic steroids are synthetic androgens (male hormones) that promote protein synthesis in muscles and other organs. Use of these drugs by bodybuilders, weightlifters, and others is prohibited by most athletic organizations. Although administration of exogenous androgens does promote muscle growth, it can also cause a number of undesirable side effects. Because the liver and adipose tissue can change androgens into estrogens, male athletes who take exogenous androgens often develop *gynecomastia*—an abnormal growth of femalelike mammary tissue. High levels of exogenous androgens also inhibit the secretion of FSH and LH from the pituitary, causing atrophy of the testes and erectile dysfunction. The exogenous androgens also promote acne, aggressive behavior, male-pattern baldness, and premature closure of the epiphyseal discs (growth plates in bones), stunting the growth of adolescents. Female users of exogenous androgens display masculinization and antisocial behavior. In both sexes, the anabolic steroids raise blood levels of LDL cholesterol (the "bad cholesterol") and triglycerides, while lowering the levels of HDL cholesterol (the "good cholesterol"), thus predisposing users to increased risk of heart disease and stroke.

Desensitization and Downregulation

Prolonged exposure to high concentrations of polypeptide hormones has been found to *desensitize* the target cells. Subsequent exposure to the same concentration of the same hormone thus produces less of a target tissue response. This desensitization is partly due to **downregulation** of receptors—a decrease in the number of receptor proteins for a polypeptide hormone caused by continuous exposure of the target cells to high concentrations of the hormone. Such desensitization and downregulation of receptors has been shown to occur, for example, in adipose cells exposed to high concentrations of insulin and in testicular cells exposed to high concentrations of luteinizing hormone (LH).

In order to prevent desensitization from occurring under normal conditions, many polypeptide and glycoprotein hormones are secreted in spurts rather than continuously. This **pulsatile secretion** is an important aspect, for example, in the hormonal control of the reproductive system. The pulsatile secretion of GnRH and LH is needed to prevent desensitization; when these hormones are artificially presented in a continuous fashion, they produce a decrease (rather than the normal increase) in gonadal function. This effect has important clinical implications, as will be described in chapter 20, section 20.2.

1a. Compare the four chemical classes of hormones with reference to hormones within each class.

1b. Define *prohormone* and *prehormone,* and give examples of each of these molecules.

1c. Describe the common characteristics of hormones and neurotransmitters.

2a. List the terms used to describe hormone interactions and give examples of these effects.

2b. Explain how the response of the body to a given hormone can be affected by the concentration of that hormone in the blood.

11.2 MECHANISMS OF HORMONE ACTION

The mechanisms by which hormones act on their target cells depend on the chemical nature of the hormones. Nonpolar hormones can easily pass through plasma membranes and so bind to receptor proteins within their target cells. These are nuclear receptors, which work by regulating gene expression. By contrast, polar hormones do not enter their target cells but instead bind to receptors on the plasma membrane. These hormones then exert their effects through second-messenger systems.

LEARNING OUTCOMES

After studying this section, you should be able to:

3. Explain the mechanisms of action of steroid and thyroid hormones.

4. Describe the mechanisms by which other hormones exert their effects on target cells.

Although each hormone exerts its own characteristic effects on specific target cells, hormones that are in the same chemical category have similar mechanisms of action. These similarities involve the location of cellular receptor proteins and the events that occur in the target cells after the hormone has combined with its receptor protein.

Hormones are delivered by the blood to every cell in the body, but only the **target cells** are able to respond to these hormones. In order to respond to any given hormone, a target cell must have specific receptor proteins for that hormone. Receptor protein–hormone interaction is highly specific. In addition to this property of *specificity,* hormones bind to receptors with a *high affinity* (high bond strength) and a *low capacity.* The latter characteristic refers to the possibility of saturating receptors with hormone molecules because of the limited number of receptors per target cell (usually a few thousand). Notice that the characteristics of specificity and saturation that apply to receptor proteins are similar to the characteristics of enzyme and carrier proteins discussed in chapters 4 and 6.

The location of a hormone's receptor proteins in its target cells depends on the chemical nature of the hormone. Because the lipophilic hormones (steroids and thyroxine) can pass through the plasma membrane and enter their target cells, the receptor proteins for lipophilic hormones are located within the cytoplasm and nucleus. The water-soluble hormones (catecholamines, polypeptides, and glycoproteins) cannot pass through the plasma membrane, so their receptors are located on the outer surface of the membrane. In these cases, hormone action requires the activation of second messengers within the cell.

Hormones That Bind to Nuclear Receptor Proteins

Unlike the water-soluble hormones, the lipophilic steroid and thyroid hormones do not travel dissolved in the aqueous portion of the plasma; rather, they are transported to their target cells attached to plasma *carrier proteins.* These hormones must then dissociate from their carrier proteins in the blood in order to pass through the lipid component of the plasma membrane and enter the target cell, within which their receptor proteins are located (fig. 11.4).

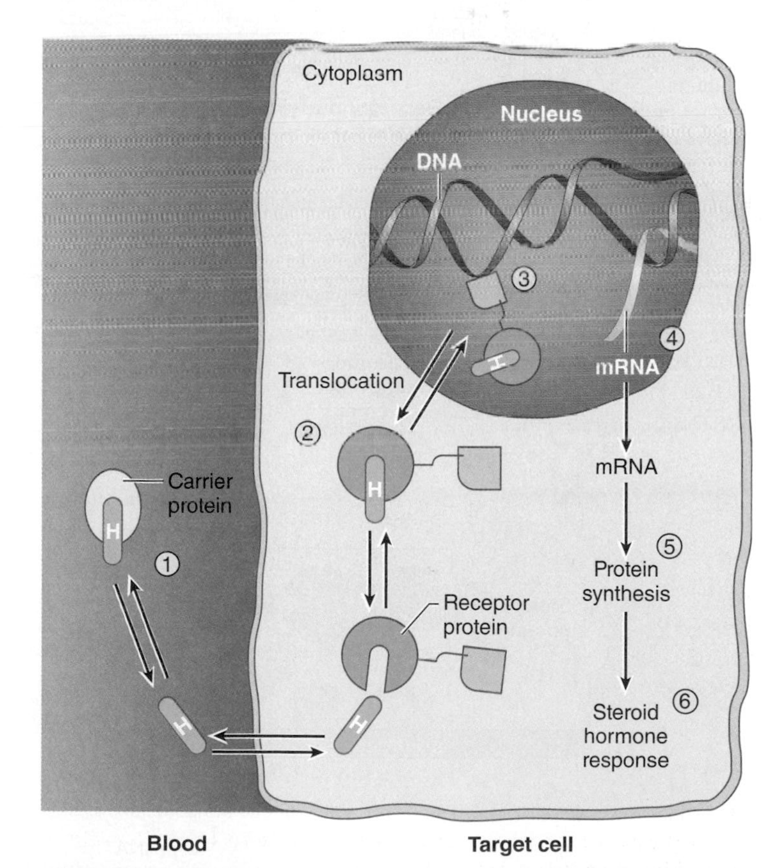

Figure 11.4 The mechanism of steroid hormone action. (1) Steroid hormones, transported bound to plasma carrier proteins, dissociate from their plasma carriers and pass through the plasma membrane of their target cell. (2) The steroid hormone binds to receptors, which may be in the cytoplasm. (3) The hormone-bound receptor translocates to the nucleus, where it binds to DNA. (4) This stimulates genetic transcription, resulting in new mRNA synthesis. (5) The newly formed mRNA codes for the production of new proteins, which (6) produce the hormonal effects in the target cell. AP|R

The receptors for the lipophilic hormones are known as **nuclear hormone receptors** because they function within the cell nucleus to activate genetic transcription (production of mRNA). The nuclear hormone receptors thus function as **transcription factors** that first must be activated by binding to their hormone ligands. The newly formed mRNA produced by the activated genes directs the synthesis of specific proteins, including enzyme proteins that change the metabolism of the target cell.

Each nuclear hormone receptor has two regions, or domains: a *ligand (hormone)-binding domain* and a *DNA-binding domain* (fig. 11.5). The receptor must be activated by binding to its hormone ligand before it can bind to a specific region of the DNA, which is called a **hormone-response element.** This is a short DNA span, composed of characteristic nucleotide bases, located adjacent to the gene that will be transcribed in response to the hormone-activated nuclear receptor.

The nuclear receptors are said to constitute a superfamily composed of two major families: the *steroid family* and the *thyroid hormone (or nonsteroid) family.* In addition to the receptor for thyroid hormone, the latter family also includes the receptors for the active form of vitamin D and for retinoic acid (derived from vitamin A, or retinol). Vitamin D and retinoic acid, like the steroid and thyroid hormones, are lipophilic molecules that play important roles in the regulation of cell function and organ physiology.

Modern molecular biology has ushered in a new era in endocrine research, where nuclear receptors can be identified and their genes cloned before their hormone ligands are known. In fact, scientists have currently identified the hormone ligand for only about half of the approximately 70 different nuclear receptors that are now known. The receptors for unknown hormone ligands are called **orphan receptors.** For example, the receptor known as the retinoid X receptor (abbreviated RXR) was an orphan until its ligand, 9-*cis*-retinoic acid (a vitamin A derivative) was discovered. The significance of this receptor will be described shortly.

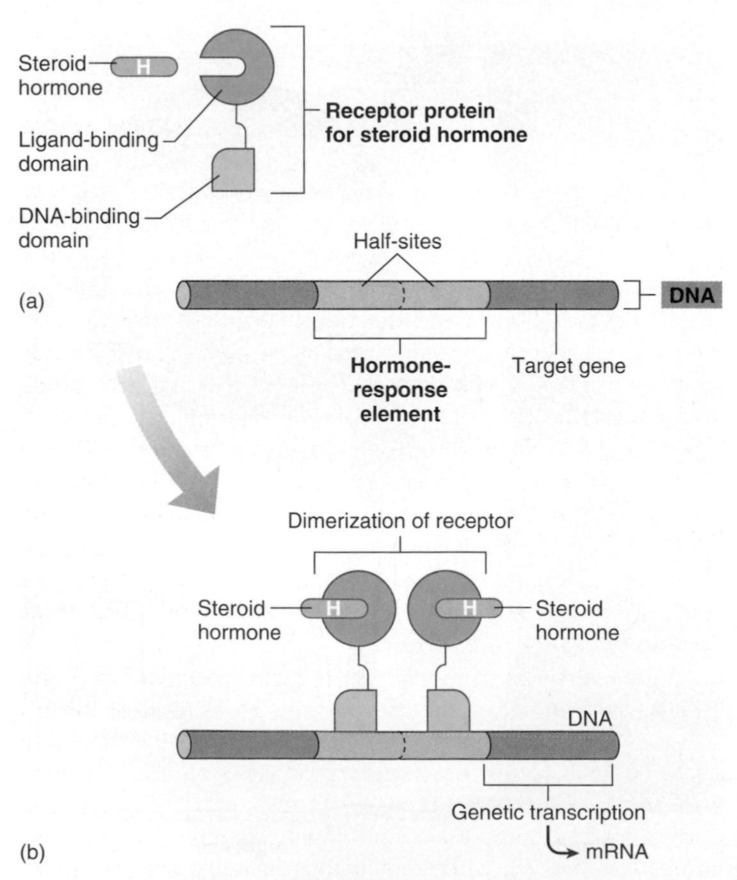

Figure 11.5 Receptors for steroid hormones. (*a*) Each nuclear hormone receptor protein has a ligand-binding domain, which binds to a hormone molecule, and a DNA-binding domain, which binds to the hormone-response element of DNA. (*b*) Binding to the hormone causes the receptor to dimerize on the half-sites of the hormone-response element. This stimulates genetic transcription (synthesis of RNA). AP|R

Mechanism of Steroid Hormone Action

In general, steroid hormones exert their effects by entering their target cells and binding to nuclear receptor proteins. Thus, they influence their target tissue by stimulating genetic transcription. This mechanism is known as the *genomic action* of steroids and requires at least 30 minutes to work. Although this is the classically described mechanism of steroid action, it doesn't explain the observation that some effects occur within seconds to minutes. These effects are too fast to be explained by binding to nuclear receptors and changes in genetic expression. This faster, *nongenomic action* of steroids may occur in the cytoplasm of the target cells and involve the activation of second-messenger systems, similar to the way polar hormones regulate their target cells (discussed shortly). The genomic actions of steroids are better established, and so only the genomic effects of steroid hormones will be described in this section.

In the classic, genomic mechanism of steroid hormone action, the receptors for the steroid hormones are located in the cytoplasm before the steroid arrives. Depending on the steroid and the tissue, however, unbound steroid receptors may be located in the nucleus as well. (The particular distribution of the receptor between the cytoplasm and nucleus varies.) When the cytoplasmic receptors bind to their specific steroid hormone ligands, they *translocate* (move) to the nucleus. Once the steroid hormone-receptor protein complex is in the nucleus, its DNA-binding domain binds to the specific hormone-response element of the DNA (see fig. 11.4).

As illustrated in figure 11.5, the hormone-response element of DNA consists of two *half-sites,* each six nucleotide bases long, separated by a three-nucleotide spacer segment. One steroid receptor, bound to one molecule of the steroid hormone, attaches as a single unit to one of the half-sites. Another steroid receptor, bound to another steroid hormone, attaches to the other half-site of the hormone-response element. The process of two receptor units coming together at the two half-sites is called **dimerization** (fig. 11.5). Because both receptor units of the pair are the same, the steroid receptor is said to form a *homodimer.* (The situation is different for the nonsteroid family of receptors, as will be described.) Once dimerization has occurred, the activated nuclear hormone receptor stimulates transcription of particular genes, and thus hormonal regulation of the target cell (see fig. 11.4).

Even this classical, genomic mechanism of steroid hormone action is an oversimplification. For example, there are drugs such as *tamoxifen* (see the next Clinical Application box) that act like estrogen in one organ while antagonizing the action of estrogen in another organ. Study of tamoxifen and other *selective estrogen receptor modulators (SERMs)* has revealed that estrogen action requires more than 20 different regulatory proteins—called **coactivators** and **corepressors**—in addition to the estrogen receptor. Coactivators and corepressors activate or repress specific transcription factors (proteins that regulate genetic transcription) but do not themselves bind to DNA. The coactivators and corepressors in this case bind to specific pockets for them in the estrogen receptor, which are separate from the hormone binding sites (ligand binding domains). By this means, the coactivators and corepressors promote or inhibit the ability of estrogen to stimulate genetic transcription. SERMs can have different effects in different organs because, even though they bind to the estrogen receptor, they may enlist coactivator proteins in one organ but not in another organ.

When a steroid hormone ligand binds to its nuclear receptor protein (at the ligand-binding domain, fig. 11.5), it changes the receptor protein structure. This causes (1) removal of a group of proteins (called *heat shock proteins*) that prevent the receptor from binding to DNA, and (2) recruitment of coactivator proteins, while corepressor proteins are prevented from binding to the receptor. The coactivator proteins form a complex that modifies the structure of the chromatin and facilitates DNA transcription (that is, RNA synthesis) at the hormone response element of DNA. As a result, the cell is stimulated to produce particular proteins by the steroid hormone.

CLINICAL APPLICATION

The ability of different tissues to respond to estrogen in different ways is important in the treatment of estrogen-sensitive cancers. For example, **breast cancer** cells are tested for estrogen receptors to see if their growth could be inhibited by anti-estrogen drugs. **Tamoxifen** and **raloxifene** are classified as **selective estrogen receptor modulators (SERMs).** Tamoxifen binds to estrogen receptors, but it has different effects in different tissues: It has an antiestrogen action in the breast, but it promotes estrogenic effects in bone and endometrial (uterine) cells. Raloxifene has an estrogenic effect on bone but not on the endometrium, and so is approved by the FDA as a treatment for osteoporosis in postmenopausal women. Scientists hope to use SERMs to learn more about estrogen's actions, and to use this knowledge to develop drugs that will selectively treat breast cancer, osteoporosis, and coronary heart disease.

Mechanisms of Thyroid Hormone Action

As previously discussed, the major hormone secreted by the thyroid gland is thyroxine, or tetraiodothyronine (T_4). Like steroid hormones, thyroxine travels in the blood attached to carrier proteins (primarily to *thyroxine-binding globulin,* or *TBG*). The thyroid also secretes a small amount of triiodothyronine, or T_3. The carrier proteins have a higher affinity for T_4 than for T_3, however, and, as a result, the amount of unbound (or "free") T_3 in the plasma is about 10 times greater than the amount of free T_4.

Approximately 99.96% of the thyroxine in the blood is attached to carrier proteins in the plasma; the rest is free. Only the free thyroxine and T_3 can enter target cells; the protein-bound thyroxine serves as a reservoir of this hormone in the blood (this is why it takes a couple of weeks after surgical removal of the thyroid for the symptoms of hypothyroidism to develop). Once the free thyroxine passes into the target cell cytoplasm, it is enzymatically converted into T_3. As discussed in section 11.1, it is the T_3 rather than T_4 that is active within the target cells.

Unlike the steroid hormone receptor proteins, the thyroid hormone receptor proteins are located in the nucleus bound to DNA (see fig. 11.7) even in the absence of their thyroid hormone ligand. The thyroid hormone response element of DNA has two half-sites, but unlike the case with the steroid receptors, the thyroid hormone receptor (for T_3) binds to only one of the half-sites. The other DNA half-site binds to the receptor for a vitamin A derivative, 9-*cis*-retinoic acid. When the thyroid hormone receptor (abbreviated *TR*) and the 9-*cis*-retinoic acid receptor (abbreviated *RXR*) bind to the two DNA half-sites of the hormone response element, the two receptors form a *heterodimer.* This term is used because these are different receptors (in contrast to the homodimer formed by two steroid hormone receptors on their DNA half-sites).

In the absence of their thyroid hormone ligand (T_3), the thyroid receptors recruit corepressor proteins that inhibit genetic transcription. Thus, although the TR and RXR are bound to DNA, the hormone response element is inhibited. When the thyroid receptor binds to its T_3 ligand, the corepressor proteins are removed and degraded by proteosomes (chapter 3), while coactivator proteins are recruited. Thus, the nuclear receptor proteins for thyroid hormone cannot stimulate genetic transcription until they bind to their hormone ligands.

The T_3 may enter the cell from the plasma, but mostly it is produced in the cell by conversion from T_4. In either case, it uses some nonspecific binding proteins in the cytoplasm as "stepping stones" to enter the nucleus, where the T_3 binds to the ligand-binding domain of the receptor (fig. 11.6). Once this occurs, the thyroid hormone receptor changes shape, enabling the removal of corepressor proteins and the recruitment of coactivator proteins that promote genetic transcription. The production of specific mRNA then codes for the synthesis of specific enzyme proteins that change the metabolism of the target cell (figs. 11.6 and 11.7).

This pattern of regulation is similar for other nuclear receptors that form heterodimers with the RXR. For example, the receptor for 1,25-dihydroxyvitamin D_3, the active form of vitamin D, also forms heterodimers with the receptor for 9-*cis*-retinoic acid (the RXR receptor) when it binds to DNA and activates genes. The RXR receptor and its vitamin A derivative ligand thus form a link between the mechanisms of action of thyroid hormone, vitamin A, and vitamin D, along with those of some other molecules that are important regulators of genetic expression.

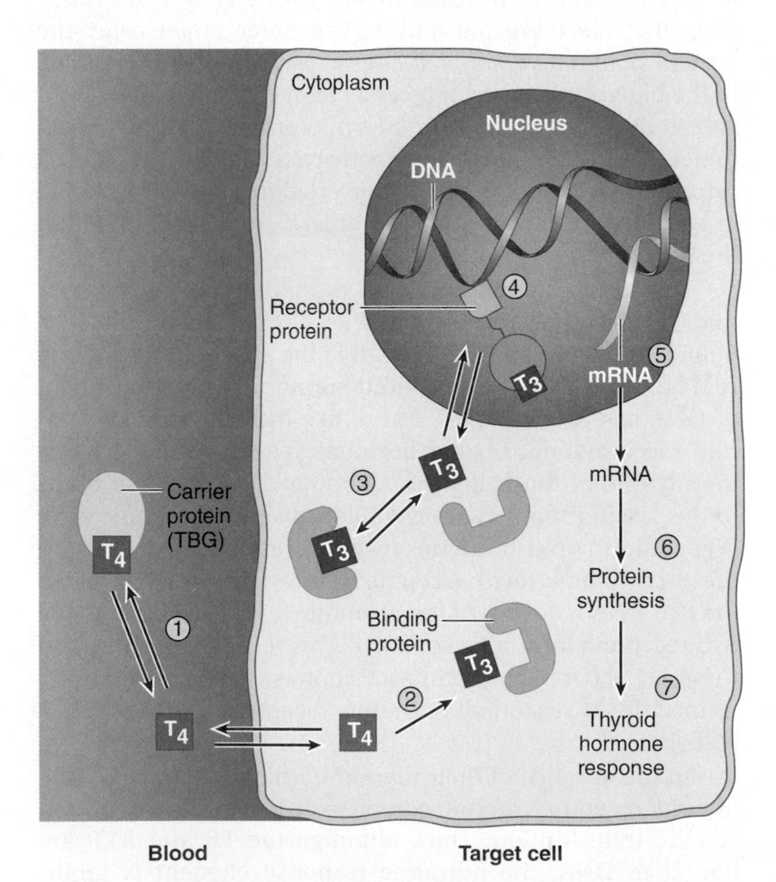

Figure 11.6 The mechanism of thyroid hormone action. (1) Thyroxine (T_4), carried to the target cell bound to its plasma carrier protein, dissociates from its carrier and passes through the plasma membrane of its target cell. (2) In the cytoplasm, T_4 is converted into T_3 (triiodothyronine), which (3) uses binding proteins to enter the nucleus. (4) The hormone-receptor complex binds to DNA, (5) stimulating the synthesis of new mRNA. (6) The newly formed mRNA codes for the synthesis of new proteins, which (7) produce the hormonal effects in the target cell. AP|R

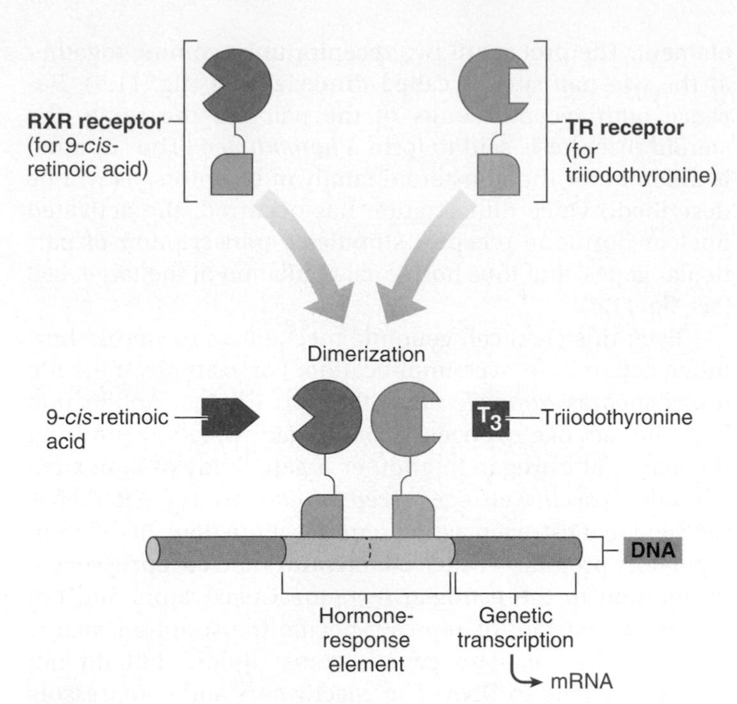

Figure 11.7 The receptor for triiodothyronine (T_3). The nuclear receptor protein for T_3 forms a dimer with the receptor protein for 9-*cis*-retinoic acid, a derivative of vitamin A. This occurs when each binds to its ligand and to the hormone-response element of DNA. Thus, 9-*cis*-retinoic acid is required for the action of T_3. The heterodimer formed on the DNA stimulates genetic transcription.

Hormones That Use Second Messengers

Hormones that are catecholamines (epinephrine and norepinephrine), polypeptides, and glycoproteins cannot pass through the lipid barrier of the target cell's plasma membrane. Although some of these hormones may enter the cell by pinocytosis, most of their effects result from their binding to receptor proteins on the outer surface of the target cell membrane. Because they exert their effects without entering the target cells, the actions of these hormones must be mediated by other molecules within the target cells. If you think of hormones as "messengers" from the endocrine glands, the intracellular mediators of the hormone's action can be called **second messengers.** (The concept of second messengers was introduced in chapter 6, section 6.5.) Second messengers are thus a component of *signal-transduction mechanisms,* because extracellular signals (hormones) are transduced into intracellular signals (second messengers).

When these hormones bind to membrane receptor proteins, they must activate specific proteins in the plasma membrane in order to produce the second messengers required to exert their effects. On the basis of the membrane enzyme activated, we can distinguish second-messenger systems that involve the activation of (1) adenylate cyclase, (2) phospholipase C, and (3) tyrosine kinase.

Adenylate Cyclase–Cyclic AMP Second-Messenger System

Cyclic adenosine monophosphate (abbreviated **cAMP**) was the first "second messenger" to be discovered and is the best understood. When epinephrine and norepinephrine bind to their β-adrenergic receptors (chapter 9), the effects of these hormones are due to cAMP production within the target cells. It was later discovered that the effects of many (but not all) polypeptide and glycoprotein hormones are also mediated by cAMP.

When one of these hormones binds to its receptor protein, it causes the dissociation of a subunit from the complex of G-proteins (discussed in chapter 7; see table 7.6). This G-protein subunit moves through the membrane until it reaches the enzyme **adenylate** (or *adenyl*) **cyclase** (fig. 11.8). The G-protein subunit then binds to and activates this enzyme, which catalyzes the following reaction within the cytoplasm of the cell:

$$\text{ATP} \rightarrow \text{cAMP} + \text{PP}_i$$

Adenosine triphosphate (ATP) is thus converted into cyclic AMP (cAMP) and two inorganic phosphates (*pyrophosphate*, abbreviated PPi). As a result of the interaction of the hormone with its receptor and the activation of adenylate cyclase, therefore, the intracellular concentration of cAMP is increased. Cyclic AMP activates a previously inactive enzyme in the cytoplasm called **protein kinase.** The inactive form of this enzyme consists of two subunits: a catalytic subunit and a regulatory subunit. The enzyme is produced in an inactive form and

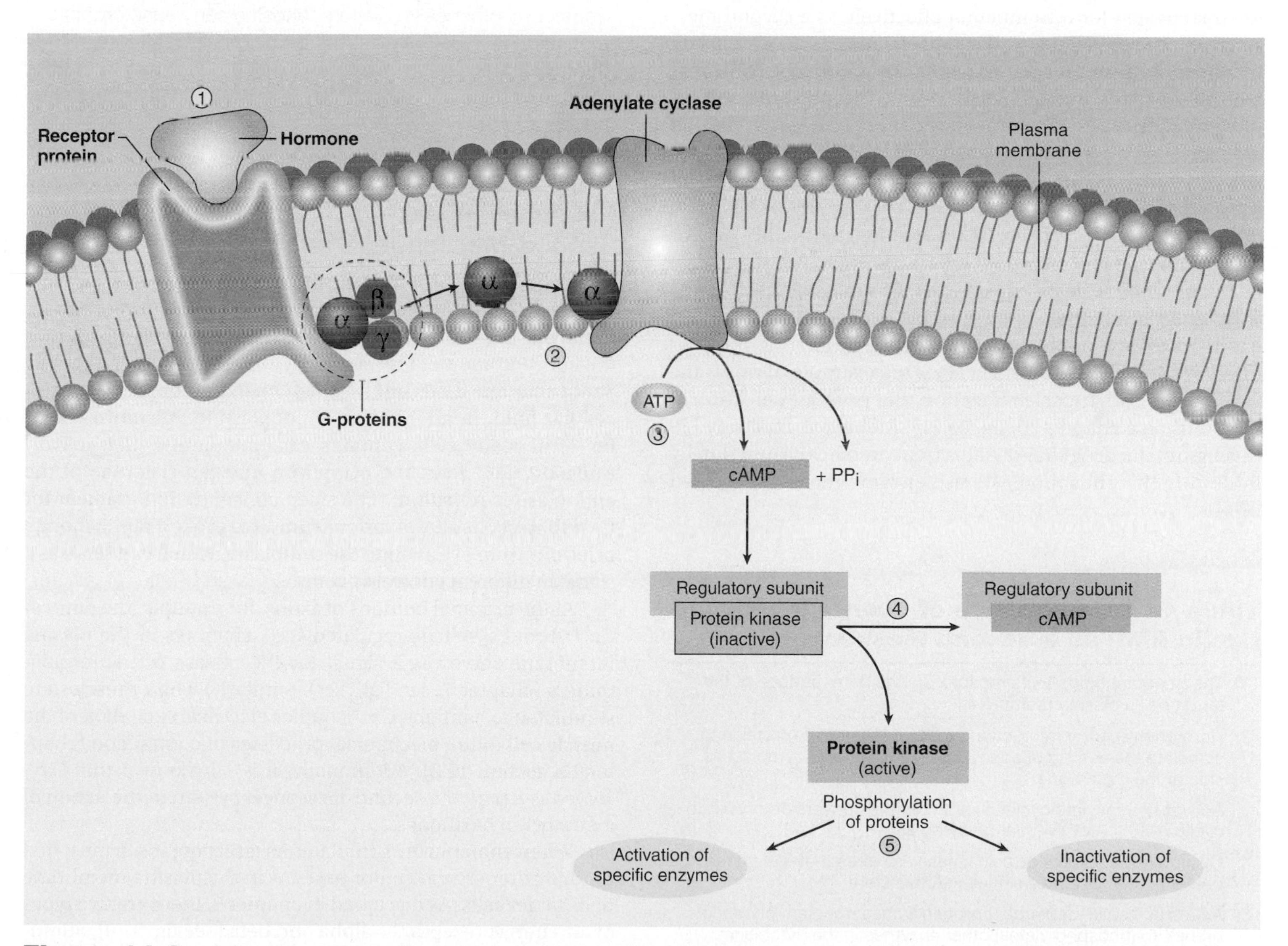

Figure 11.8 The adenylate cyclase–cyclic AMP second-messenger system. (1) The hormone binds to its receptor in the plasma membrane of the target cell. (2) This causes the dissociation of G-proteins, allowing the free α (alpha) subunit to activate adenylate cyclase. (3) This enzyme catalyzes the production of cAMP (cyclic AMP), which (4) removes the regulatory subunit from protein kinase. (5) Active protein kinase phosphorylates other enzyme proteins, activating or inactivating specific enzymes and thereby producing the hormonal effects on the target cell. AP|R

becomes active only when cAMP attaches to the regulatory subunit. Binding of cAMP to the regulatory subunit causes it to dissociate from the catalytic subunit, which then becomes active (fig. 11.8). In summary, the hormone—acting through an increase in cAMP production—causes an increase in protein kinase enzyme activity within its target cells.

Active protein kinase catalyzes the phosphorylation of (attachment of phosphate groups to) different proteins in the target cells. This causes some enzymes to become activated and others to become inactivated. Cyclic AMP, acting through protein kinase, thus modulates the activity of enzymes that are already present in the target cell. This alters the metabolism of the target tissue in a manner characteristic of the actions of that specific hormone (table 11.4).

Like all biologically active molecules, cAMP must be rapidly inactivated for it to function effectively as a second messenger in hormone action. This inactivation is accomplished by **phosphodiesterase,** an enzyme within the target cells that hydrolyzes cAMP into an inactive form. Through the action of phosphodiesterase, the stimulatory effect of a hormone that uses cAMP as a second messenger depends upon the continuous generation of new cAMP molecules, and thus depends on the level of secretion of the hormone.

In addition to cyclic AMP, **cyclic guanosine monophosphate** (cGMP) functions as a second messenger in certain cases. For example, the regulatory molecule nitric oxide (discussed in chapter 7 and in section 11.7 of this chapter) exerts its effects on smooth muscle by stimulating the production of cGMP in its target cells. One example of this is the vascular smooth muscle relaxation that produces erection of the penis (see chapter 20, fig. 20.21). Indeed, as illustrated in this figure, the drug Viagra helps treat erectile dysfunction by inhibiting the phosphodiesterase enzyme that breaks down cGMP.

Table 11.4 | Sequence of Events Involving Cyclic AMP as a Second Messenger

1. The hormone binds to its receptor on the outer surface of the target cell's plasma membrane.
2. Hormone-receptor interaction acts by means of G-proteins to stimulate the activity of adenylate cyclase on the cytoplasmic side of the membrane.
3. Activated adenylate cyclase catalyzes the conversion of ATP to cyclic AMP (cAMP) within the cytoplasm.
4. Cyclic AMP activates protein kinase enzymes that were already present in the cytoplasm in an inactive state.
5. Activated cAMP-dependent protein kinase transfers phosphate groups to (phosphorylates) other enzymes in the cytoplasm.
6. The activity of specific enzymes is either increased or inhibited by phosphorylation.
7. Altered enzyme activity mediates the target cell's response to the hormone.

CLINICAL APPLICATION

Drugs that inhibit the activity of phosphodiesterase prevent the breakdown of cAMP and thus result in increased concentrations of cAMP within the target cells. The drug **theophylline** and its derivatives, for example, are used clinically to raise cAMP levels within bronchiolar smooth muscle. This duplicates and enhances the effect of epinephrine on the bronchioles (producing dilation) in people who suffer from asthma (the causes and treatments of asthma are discussed in chapter 16).

Theophylline, along with **caffeine,** is classified as a *methylxanthine.* These drugs not only raise cAMP levels by inhibiting phosphodiesterase, they also selectively block receptor proteins for adenine. Adenine may be produced in the extracellular fluid from ATP released by autonomic neurons or by other cells. Since adenosine can cause bronchoconstriction in asthmatics, the bronchodilator effects of the methylxanthines may result from both an increase in cAMP concentrations and a decrease in adenine stimulation of its receptors.

Phospholipase C–Ca^{2+} Second-Messenger System

The concentration of Ca^{2+} in the cytoplasm is kept very low by the action of active transport carriers—calcium pumps—in the plasma membrane. Through the action of these pumps, the concentration of Ca^{2+} in the cytoplasm is reduced to only about one ten-thousandth of its concentration in the extracellular fluid. In addition, the endoplasmic reticulum (chapter 3) of many cells contains calcium pumps that actively transport Ca^{2+} from the cytoplasm into the cisternae of the endoplasmic reticulum. The steep concentration gradient for Ca^{2+} that results allows various stimuli to evoke a rapid, though brief, diffusion of Ca^{2+} into the cytoplasm, which can serve as a signal in different control systems.

At the terminal boutons of axons, for example, the entry of Ca^{2+} through voltage-regulated Ca^{2+} channels in the plasma membrane serves as a signal for the release of neurotransmitters (chapter 7; see fig. 7.23). Similarly, when muscles are stimulated to contract, Ca^{2+} couples electrical excitation of the muscle cell to the mechanical processes of contraction (chapter 12, section 12.2). Additionally, it is now known that Ca^{2+} serves as a part of a second-messenger system in the action of a number of hormones.

When epinephrine stimulates its target organs, it must first bind to adrenergic receptor proteins in the plasma membrane of its target cells. As discussed in chapter 9, there are two types of adrenergic receptors—alpha and beta (see fig. 9.10). Stimulation of the beta-adrenergic receptors by epinephrine results in activation of adenylate cyclase and the production of cAMP. Stimulation of $alpha_1$-adrenergic receptors by epinephrine, in contrast, activates the target cell via the Ca^{2+} second-messenger system (see fig. 11.10).

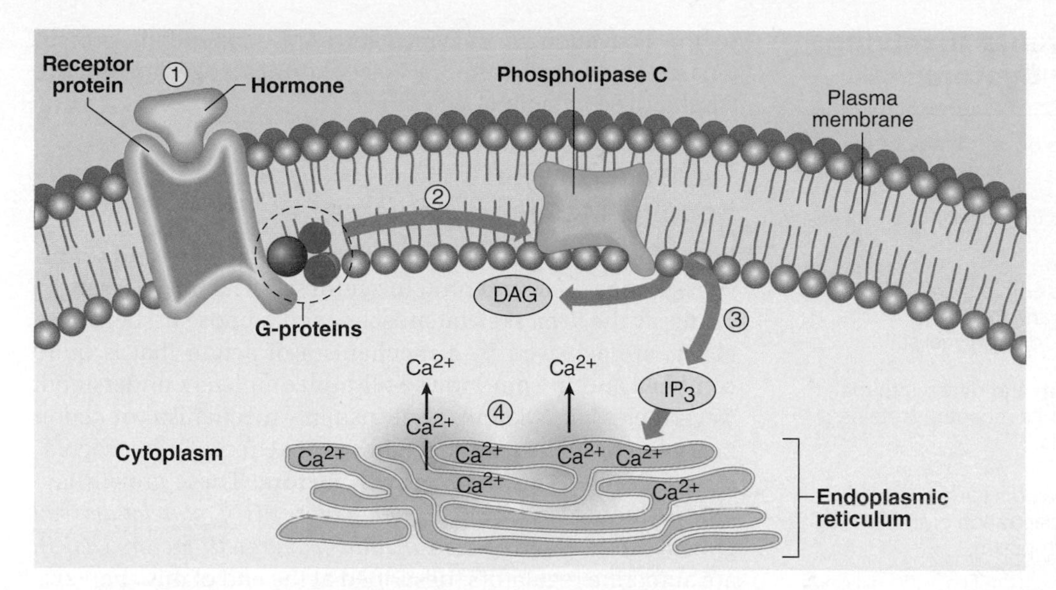

Figure 11.9 The phospholipase C–Ca^{2+} second-messenger system. (1) The hormone binds to its receptor in the plasma membrane of its target cell, (2) causing the dissociation of G-proteins. (3) A G-protein subunit travels through the plasma membrane and activates phospholipase C, which catalyzes the breakdown of a particular membrane phospholipid into DAG (diacylglycerol) and IP_3 (inositol triphosphate). (4) IP_3 enters the cytoplasm and binds to receptors in the endoplasmic reticulum, causing the release of stored Ca^{2+}. The Ca^{2+} then diffuses into the cytoplasm, where it acts as a second messenger to promote the hormonal effects in the target cell. APIR

The binding of epinephrine to its alpha-adrenergic receptor activates, via G-proteins, an enzyme in the plasma membrane known as **phospholipase C** (fig. 11.9). The substrate of this enzyme, a particular membrane phospholipid, is split by the active enzyme into **inositol triphosphate (IP_3)** and another derivative, **diacylglycerol (DAG).** Both derivatives serve as second messengers, but the action of IP_3 is somewhat better understood and will be discussed in this section.

The IP_3 leaves the plasma membrane and diffuses through the cytoplasm to the endoplasmic reticulum. The membrane of the endoplasmic reticulum contains receptor proteins for IP_3; the IP_3 is a second messenger in its own right, carrying the hormone's message from the plasma membrane to the endoplasmic reticulum. Binding of IP_3 to its receptors causes specific Ca^{2+} channels to open, so that Ca^{2+} diffuses out of the endoplasmic reticulum and into the cytoplasm (fig. 11.9).

As a result of these events, there is a rapid and transient rise in the cytoplasmic Ca^{2+} concentration. This signal is augmented, through mechanisms that are incompletely understood, by the opening of Ca^{2+} channels in the plasma membrane. This may be due to the action of yet a different (and currently unknown) messenger sent from the endoplasmic reticulum to the plasma membrane. The Ca^{2+} that enters the cytoplasm binds to a protein called **calmodulin.** Once Ca^{2+} binds to calmodulin, the now-active calmodulin in turn activates specific protein kinase enzymes (those that add phosphate groups to proteins) that modify the actions of other enzymes in the cell (fig. 11.10). Activation of specific calmodulin-dependent enzymes is analogous

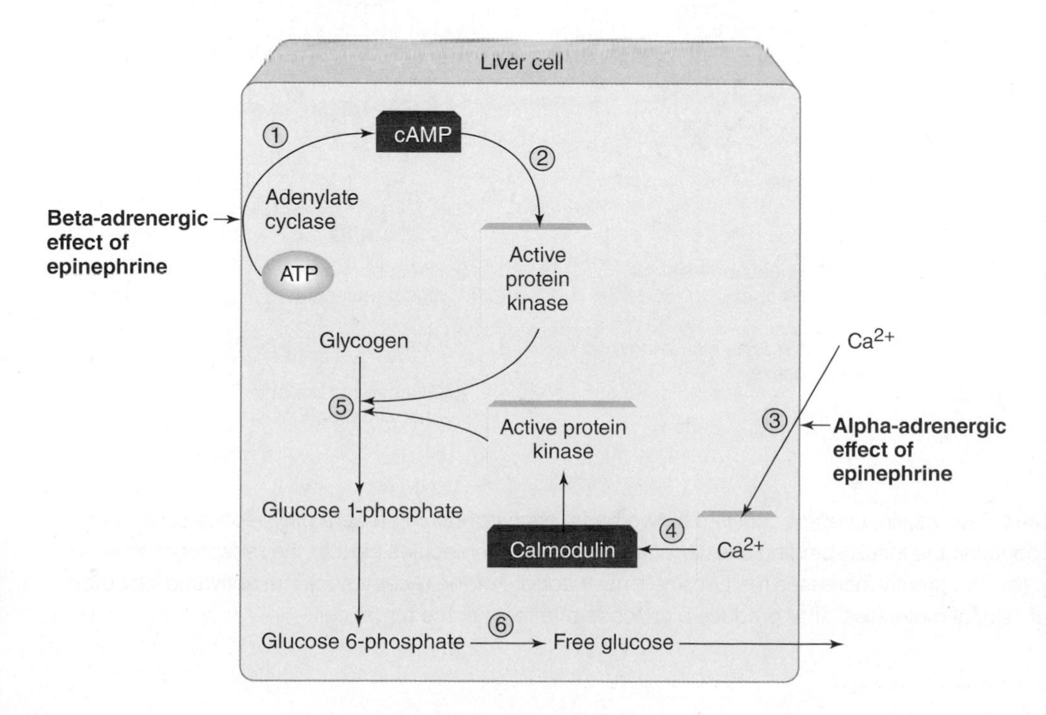

Figure 11.10 Epinephrine uses two second-messenger systems. This is shown by the action of epinephrine on a liver cell. (1) Binding of epinephrine to beta-adrenergic receptors activates adenylate cyclase and leads to the production of cAMP, which (2) activates a protein kinase. (3) Binding of epinephrine to alpha-adrenergic receptors leads to a rise in the cytoplasmic Ca^{2+} concentration, which (4) activates calmodulin. Calmodulin then activates a protein kinase, which, like the protein kinase activated by cAMP, (5) alters enzyme activity so that glycogen is converted to glucose 6-phosphate. (6) The phosphate group is removed by another enzyme, so that the liver cell secretes free glucose into the blood in response to epinephrine.

Table 11.5 | Sequence of Events Involving the Ca^{2+} Second-Messenger System

1. The hormone binds to its receptor on the outer surface of the target cell's plasma membrane.
2. Hormone-receptor interaction stimulates the activity of a membrane enzyme, phospholipase C.
3. Activated phospholipase C catalyzes the conversion of particular phospholipids in the membrane to inositol triphosphate (IP_3) and another derivative, diacylglycerol.
4. Inositol triphosphate enters the cytoplasm and diffuses to the endoplasmic reticulum, where it binds to its receptor proteins and causes the opening of Ca^{2+} channels.
5. Since the endoplasmic reticulum accumulates Ca^{2+} by active transport, there exists a steep Ca^{2+} concentration gradient favoring the diffusion of Ca^{2+} into the cytoplasm.
6. Ca^{2+} that enters the cytoplasm binds to and activates a protein called calmodulin.
7. Activated calmodulin, in turn, activates protein kinase, which phosphorylates other enzyme proteins.
8. Altered enzyme activity mediates the target cell's response to the hormone.

to the activation of enzymes by cAMP-dependent protein kinase. The steps of the Ca^{2+} second-messenger system are summarized in table 11.5.

Tyrosine Kinase Second-Messenger System

Insulin promotes glucose and amino acid transport and stimulates glycogen, fat, and protein synthesis in its target organs—primarily the liver, skeletal muscles, and adipose tissue. These effects are achieved by a mechanism of action that is quite complex, and in some ways still not completely understood. Nevertheless, it is known that insulin's mechanism of action bears similarities to the mechanism of action of other regulatory molecules known as **growth factors.** These growth factors, including *epidermal growth factor (EGF), platelet-derived growth factor (PDGF),* and *insulin-like growth factors (IGFs),* are autocrine regulators (described at the end of this chapter).

In the case of insulin and the growth factors, the receptor protein is located in the plasma membrane and is itself an enzyme known as a **tyrosine kinase.** A *kinase* is an enzyme that adds phosphate groups to proteins, and a *tyrosine* kinase specifically adds these phosphate groups to the amino acid tyrosine within the proteins. The insulin receptor consists of two alpha and two beta subunits (fig. 11.11). The beta subunits

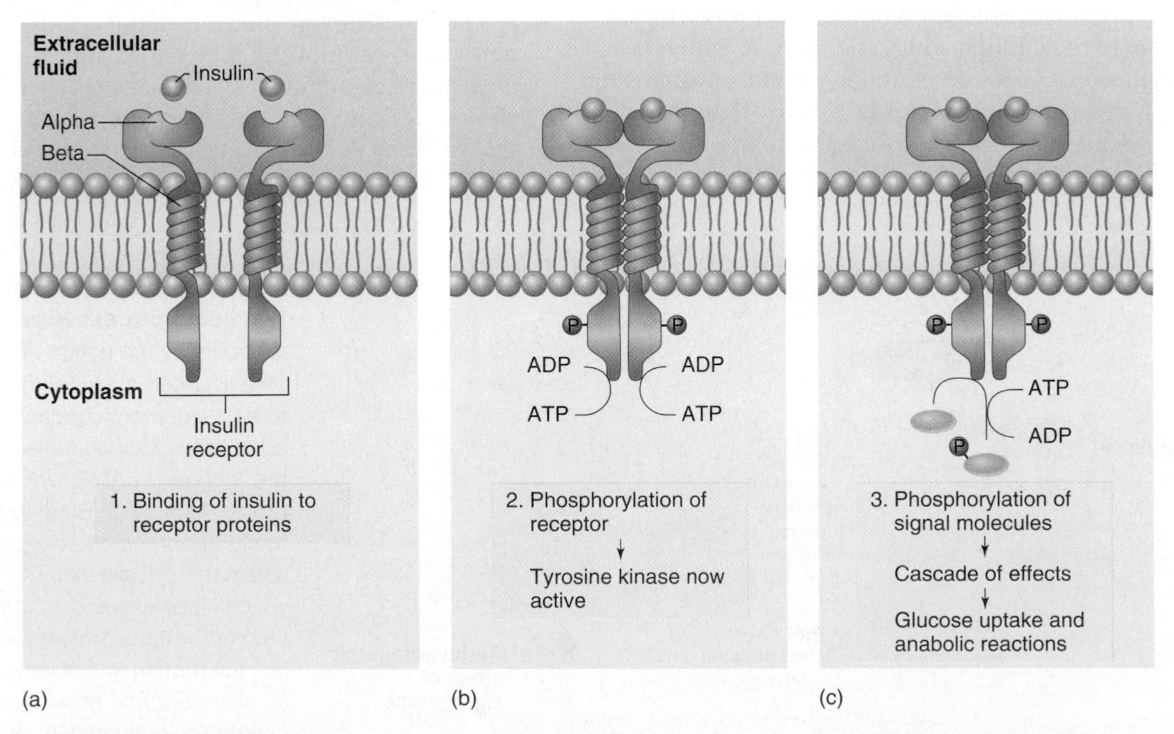

Figure 11.11 The receptor for insulin. The insulin receptor consists of two parts, each containing a beta polypeptide chain that spans the membrane, and an alpha chain that contains the insulin-binding site. (*a*) When two insulin molecules bind to the receptor, the two parts of the receptor phosphorylate each other. (*b*) This greatly increases the tyrosine kinase activity of the receptor. (*c*) The activated receptor tyrosine kinase then phosphorylates a variety of "signal molecules" that produce a cascade of effects in the target cell.

span the plasma membrane; the alpha subunits are located on the extracellular side of the plasma membrane and contain the ligand (insulin) binding sites. When insulin binds to the alpha subunits, the beta subunits are stimulated to phosphorylate each other in a process called *autophosphorylation*. This activates the tyrosine kinase activity of the insulin receptor.

The activated insulin receptor then phosphorylates *insulin receptor substrate proteins,* which provide an enzymatic docking station that activates a variety of other *signaling molecules.* These signaling molecules cause the insertion of transport carrier proteins for glucose into the plasma membrane (see fig. 11.30), and so promote the uptake of plasma glucose into tissue cells. In this way, insulin promotes the lowering of the plasma glucose concentration. Some signaling molecules activate other second-messenger systems within the target cells, allowing insulin and growth factors to regulate different aspects of the metabolism of their target cells.

The complexity of different second-messenger systems is needed so that different signaling molecules can have varying effects. For example, insulin uses the tyrosine kinase second-messenger system to stimulate glucose uptake into the liver and its synthesis into glycogen, whereas glucagon (another hormone secreted by the pancreatic islets) acts on the same cells to promote opposite effects—the hydrolysis of glycogen and secretion of glucose—by activating a different second-messenger system that involves the production of cAMP.

CHECKPOINT

3. Using diagrams, describe how steroid hormones and thyroxine exert their effects on their target cells.

4a. Use a diagram to show how cyclic AMP is produced within a target cell in response to hormone stimulation and how cAMP functions as a second messenger.

4b. Describe the sequence of events by which a hormone can cause a rise in the cytoplasmic Ca^{2+} concentration and explain how Ca^{2+} can function as a second messenger.

4c. Explain the nature and actions of the receptor proteins for insulin and the growth factors.

11.3 PITUITARY GLAND

The pituitary gland includes the anterior pituitary and posterior pituitary. The posterior pituitary stores and releases hormones that are actually produced by the hypothalamus, whereas the anterior pituitary produces and secretes its own hormones. The anterior pituitary, however, is regulated by hormones secreted by the hypothalamus, as well as by feedback from the target gland hormones.

LEARNING OUTCOMES

After studying this section, you should be able to:

5. Distinguish between the anterior and posterior pituitary, and identify the hormones secreted by each part.
6. Explain how the hypothalamus regulates both the posterior and anterior pituitary glands.
7. Describe negative feedback inhibition in the regulation of hypothalamic and anterior pituitary hormones.

The **pituitary gland,** or **hypophysis,** is located on the inferior aspect of the brain in the region of the diencephalon (chapter 8). Roughly the size of a pea—about 1.3 cm (0.5 in.) in diameter—it is attached to the hypothalamus by a stalklike structure called the *infundibulum* (fig. 11.12).

The pituitary gland is structurally and functionally divided into an anterior lobe, or **adenohypophysis,** and a posterior lobe called the **neurohypophysis.** These two parts have different embryonic origins. The adenohypophysis is

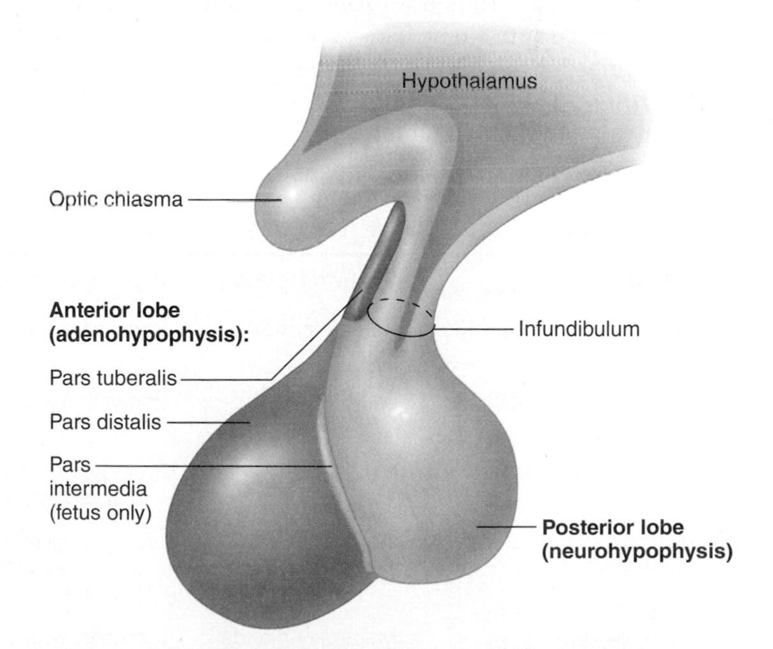

Figure 11.12 The structure of the pituitary gland. The anterior lobe is composed of glandular tissue, whereas the posterior lobe is composed largely of neuroglia and nerve fibers. APIR

derived from a pouch of epithelial tissue *(Rathke's pouch)* that migrates upward from the embryonic mouth, whereas the neurohypophysis is formed as a downgrowth of the brain. The adenohypophysis consists of two parts in adults: (1) the *pars distalis,* also known as the **anterior pituitary,** is the rounded portion and the major endocrine part of the gland, and (2) the *pars tuberalis* is a sheath of tissue that partially wraps around the infundibulum. These parts are illustrated in figure 11.12. A *pars intermedia,* a strip of tissue between the anterior and posterior lobes, exists in the fetus. During fetal development, its cells mingle with those of the anterior lobe, and in adults they no longer constitute a separate structure.

The neurohypophysis is the neural part of the pituitary gland. It consists of the *pars nervosa,* also called the **posterior pituitary,** which is in contact with the adenohypophysis, and the infundibulum. Nerve fibers extend through the infundibulum along with small neuroglia-like cells called *pituicytes.*

Pituitary Hormones

The hormones secreted by the anterior pituitary (the pars distalis of the adenohypophysis) are called **trophic hormones.** The term *trophic* means "feed." Although the anterior pituitary hormones are not food for their target organs, this term is used because high concentrations of the anterior pituitary hormones cause their target organs to hypertrophy, while low levels cause their target organs to atrophy. When names are applied to the hormones of the anterior pituitary, "trophic" (conventionally shortened to *tropic,* meaning "attracted to") is incorporated into them. This is why the shortened forms of the names for the anterior pituitary hormones end in the suffix *-tropin.* The hormones of the anterior pituitary, listed here, are summarized in table 11.6.

1. **Growth hormone (GH,** or **somatotropin).** GH promotes the movement of amino acids into cells and the incorporation of these amino acids into proteins, thus promoting overall tissue and organ growth. Some of growth hormone's actions, including growth of cartilage and bones and protein synthesis in muscles, result from a group of molecules (the somatomedins) produced by the liver under growth hormone stimulation (chapter 19, section 19.5).
2. **Thyroid-stimulating hormone (TSH,** or **thyrotropin).** TSH stimulates the thyroid gland to produce and secrete thyroxine (tetraiodothyronine, or T_4) and triiodothyronine (T_3).
3. **Adrenocorticotropic hormone (ACTH,** or **corticotropin).** ACTH stimulates the adrenal cortex to secrete the glucocorticoids, such as cortisol (hydrocortisone).
4. **Follicle-stimulating hormone (FSH,** or **folliculotropin).** FSH stimulates the growth of ovarian follicles in females and the production of sperm cells in the testes of males.
5. **Luteinizing hormone (LH,** or **luteotropin).** This hormone and FSH are collectively called **gonadotropic hormones.** In females, LH stimulates ovulation and the conversion of the ovulated ovarian follicle into an endocrine structure called a corpus luteum. In males, LH is sometimes called *interstitial cell stimulating hormone,* or *ICSH;* it stimulates the secretion of male sex hormones (mainly testosterone) from the interstitial cells (Leydig cells) in the testes.
6. **Prolactin (PRL).** This hormone is secreted in both males and females. Its best known function is the stimulation of milk production by the mammary glands of women after the birth of a baby. Prolactin plays a supporting role in the regulation of the male reproductive system by the gonadotropins (FSH and LH) and acts on the kidneys to help regulate water and electrolyte balance.

Table 11.6 | Anterior Pituitary Hormones

Hormone	Target Tissue	Principal Actions	Regulation of Secretion
ACTH (adrenocorticotropic hormone)	Adrenal cortex	Stimulates secretion of glucocorticoids	Stimulated by CRH (corticotropin-releasing hormone); inhibited by glucocorticoids
TSH (thyroid-stimulating hormone)	Thyroid gland	Stimulates secretion of thyroid hormones	Stimulated by TRH (thyrotropin-releasing hormone); inhibited by thyroid hormones
GH (growth hormone)	Most tissue	Promotes protein synthesis and growth; lipolysis and increased blood glucose	Inhibited by somatostatin; stimulated by growth hormone-releasing hormone
FSH (follicle-stimulating hormone)	Gonads	Promotes gamete production and stimulates estrogen production in females	Stimulated by GnRH (gonadotropin-releasing hormone); inhibited by sex steroids and inhibin
PRL (prolactin)	Mammary glands and other sex accessory organs	Promotes milk production in lactating females; additional actions in other organs	Inhibited by PIH (prolactin-inhibiting hormone)
LH (luteinizing hormone)	Gonads	Stimulates sex hormone secretion; ovulation and corpus luteum formation in females; stimulates testosterone secretion in males	Stimulated by GnRH; inhibited by sex steroids

The pars intermedia of the adenohypophysis ceases to exist as a separate lobe in the adult human pituitary, but it is present in the human fetus and in adults of other animals. Until recently, it was thought to secrete **melanocyte-stimulating hormone (MSH),** as it does in fish, amphibians, and reptiles, where it causes darkening of the skin. In humans, however, plasma concentrations of MSH are insignificant. Some cells of the adenohypophysis derived from the fetal pars intermedia produce a large polypeptide called *pro-opiomelanocortin (POMC).* POMC is a prohormone whose major products are beta-endorphin (chapter 7, section 7.6), MSH, and ACTH. Because part of the ACTH molecule contains the amino acid sequence of MSH, elevated secretions of ACTH (as in Addison's disease; see section 11.4) cause a marked darkening of the skin.

CLINICAL APPLICATION

Inadequate growth hormone secretion during childhood causes **pituitary dwarfism.** In an adult, by contrast, the lack of growth hormone and the other anterior pituitary hormones—a disorder called *pituitary cachexia* (*Simmonds' disease*)—produces a variety of body changes associated with premature aging. Oversecretion of growth hormone during childhood causes **gigantism.** However, excessive growth hormone secretion in an adult does not cause further growth in length because the cartilaginous epiphyseal discs have already ossified. Hypersecretion of growth hormone in an adult instead causes **acromegaly** (chapter 19; see fig. 19.17), in which the person's appearance gradually changes as a result of thickening of bones and the growth of soft tissues, particularly in the face, hands, and feet.

The posterior pituitary, or pars nervosa, stores and releases two hormones, both of which are produced in the hypothalamus.

1. **Antidiuretic hormone (ADH).** The human form of this hormone is also chemically known as **arginine vasopressin (AVP),** but the "pressor" effect (a rise in blood pressure due to vasoconstriction) is of secondary significance in humans. The "antidiuretic" effect of this hormone—its stimulation of water retention by the kidneys, so that less water is excreted in the urine—is far more significant. Because of this, the hormone will be termed antidiuretic hormone (ADH) in this text.
2. **Oxytocin.** In females, oxytocin stimulates contractions of the uterus during labor and for this reason is needed for parturition (childbirth). Oxytocin also stimulates contractions of the mammary gland alveoli and ducts, which result in the milk-ejection reflex in a lactating woman. In men, a rise in oxytocin secretion at the time of ejaculation has been measured, but the physiological significance of this hormone in males remains to be demonstrated.

CLINICAL APPLICATION

Injections of oxytocin (called **Pitocin**) may be given to a pregnant woman to induce labor if the pregnancy is prolonged or if the fetal membranes have ruptured and there is danger of infection. Labor also may be induced by injections of oxytocin in the case of severe pregnancy-induced hypertension, or *preeclampsia* (chapter 14, section 14.7). Oxytocin administration after delivery causes the uterus to regress in size and squeezes the blood vessels, thus minimizing the danger of hemorrhage.

Hypothalamic Control of the Posterior Pituitary

Both of the posterior pituitary hormones—antidiuretic hormone (ADH) and oxytocin—are actually produced in neuron cell bodies of the *supraoptic* and *paraventricular nuclei* in the hypothalamus. The ADH and oxytocin hormones produced in the hypothalamus are transported along axons of the **hypothalamo-hypophyseal tract** (fig. 11.13) to the posterior pituitary, where they are stored and later released in response to appropriate stimulation. The posterior pituitary is thus more a storage organ than a true gland.

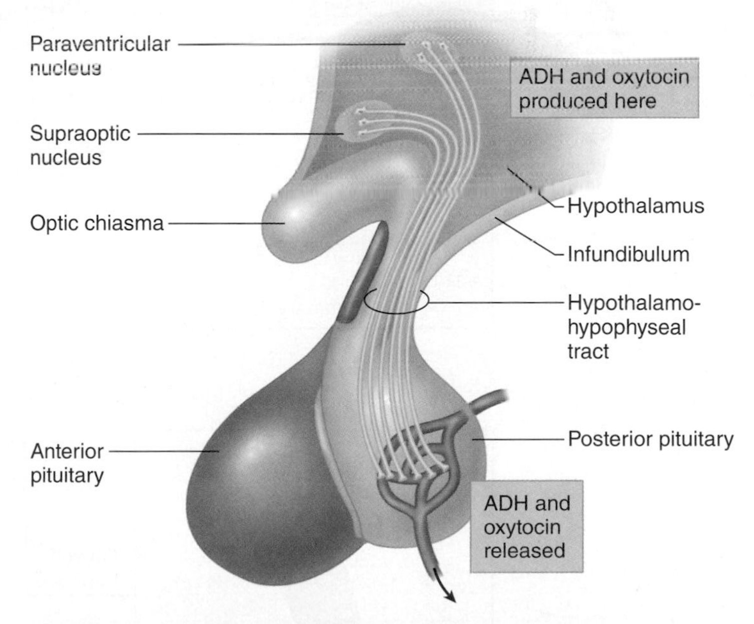

Figure 11.13 Hypothalamic control of the posterior pituitary. The posterior pituitary, or neurohypophysis, stores and releases hormones—vasopressin and oxytocin—that are actually produced in neurons within the supraoptic and paraventricular nuclei of the hypothalamus. These hormones are transported to the posterior pituitary by axons in the hypothalamo-hypophyseal tract. AP|R

The release of ADH and oxytocin from the posterior pituitary is controlled by **neuroendocrine reflexes.** In nursing mothers, for example, the mechanical stimulus of suckling acts, via sensory nerve impulses to the hypothalamus, to stimulate the reflex secretion of oxytocin (chapter 20, section 20.6). The secretion of ADH is stimulated by *osmoreceptor* neurons in the hypothalamus in response to a rise in the plasma osmolality (chapter 6, section 6.2). An increased osmolality (and osmotic pressure) stimulates an increased frequency of action potentials in the neurons that produce ADH. This causes a greater opening of voltage-gated Ca^{2+} channels at the axon terminals, which produces a greater release of ADH by exocytosis. This is similar to the way axon terminals release neurotransmitter (chapter 7; see fig. 7.23), but in this case ADH is secreted as a hormone from the posterior pituitary gland into the blood. Conversely, ADH secretion can be inhibited by sensory input from stretch receptors in the left atrium of the heart, which are stimulated when there is a rise in blood volume (chapter 14, section 14.2).

Hypothalamic Control of the Anterior Pituitary

At one time the anterior pituitary was called the "master gland" because it secretes hormones that regulate some other endocrine glands (fig. 11.14 and table 11.6). Adrenocorticotropic

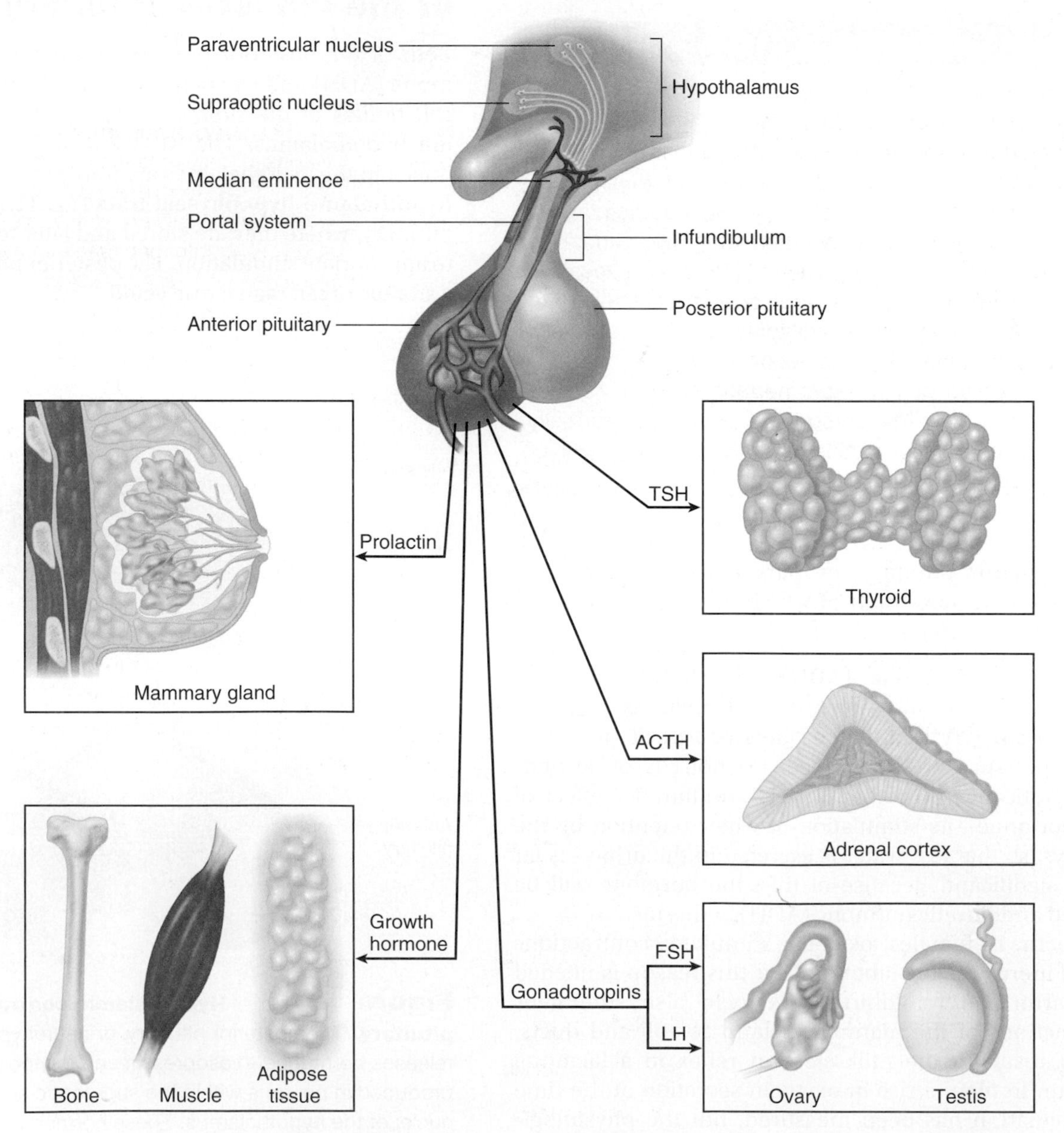

Figure 11.14 **Hormones secreted by the anterior pituitary and their target organs.** Notice that the anterior pituitary controls some (but by no means all) of the other endocrine glands.

hormone (ACTH), thyroid-stimulating hormone (TSH), and the gonadotropic hormones (FSH and LH) stimulate the adrenal cortex, thyroid, and gonads, respectively, to secrete their hormones. The anterior pituitary hormones also have a "trophic" effect on their target glands, in that the health of these glands depends on adequate stimulation by anterior pituitary hormones. The anterior pituitary, however, is not really the master gland because secretion of its hormones is controlled by hormones secreted by the hypothalamus.

Releasing and Inhibiting Hormones

Because axons do not enter the anterior pituitary, hypothalamic control of the anterior pituitary is achieved through hormonal rather than neural regulation. Releasing and inhibiting hormones, produced by neurons in the hypothalamus, are transported to axon endings in the basal portion of the hypothalamus. This region, known as the *median eminence* (fig. 11.15), contains blood capillaries that are drained by venules in the stalk of the pituitary.

The venules that drain the median eminence deliver blood to a second capillary bed in the anterior pituitary. This second capillary bed is downstream from the capillary bed in the median eminence and receives venous blood from it, so the vascular link between the median eminence and the anterior pituitary forms a *portal system.* (This is analogous to the hepatic portal system that delivers venous blood from the intestine to the liver; chapter 18, section 18.5.) The vascular link between the hypothalamus and the anterior pituitary is thus called the **hypothalamo-hypophyseal portal system.**

Regulatory hormones are secreted into the hypothalamo-hypophyseal portal system by neurons of the hypothalamus. These hormones regulate the secretions of the anterior pituitary (fig. 11.15 and table 11.7). **Thyrotropin-releasing hormone (TRH)** stimulates the secretion of TSH, and **corticotropin-releasing hormone (CRH)** stimulates the secretion of ACTH from the anterior pituitary. A single releasing hormone, **gonadotropin-releasing hormone,** or **GnRH,** stimulates the secretion of both gonadotropic hormones (FSH and LH) from the anterior pituitary.

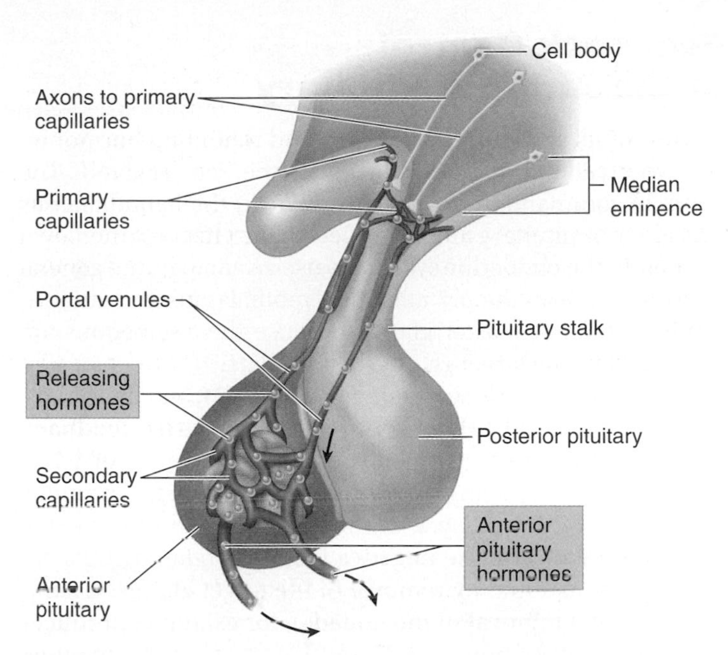

Figure 11.15 Hypothalamic control of the anterior pituitary. Neurons in the hypothalamus secrete releasing hormones (shown as green spheres) into the blood vessels of the hypothalamo-hypophyseal portal system. These releasing hormones stimulate the anterior pituitary to secrete its hormones (pink spheres) into the general circulation.

The anterior pituitary secretion of growth hormone is under the control of two polypeptide hormones from the hypothalamus. The secretion of **growth hormone–releasing hormone (GHRH)** by the hypothalamus stimulates the anterior pituitary to secrete growth hormone, whereas **somatostatin** from the hypothalamus inhibits growth hormone secretion. A **prolactin-inhibiting hormone,** identified as the neurotransmitter *dopamine,* inhibits the secretion of prolactin from the anterior pituitary. This is the most physiologically important regulator of prolactin secretion, although several factors (including oxytocin and TRH) have been shown to promote prolactin secretion when dopamine release declines.

Table 11.7 | Hypothalamic Hormones Involved in the Control of the Anterior Pituitary

Hypothalamic Hormone	Structure	Effect on Anterior Pituitary
Corticotropin-releasing hormone (CRH)	41 amino acids	Stimulates secretion of adrenocorticotropic hormone (ACTH)
Gonadotropin-releasing hormone (GnRH)	10 amino acids	Stimulates secretion of follicle-stimulating hormone (FHS) and luteinizing hormone (LH)
Prolactin-inhibiting hormone (PIH): dopamine	Catecholamine	Inhibits prolactin secretion
Somatostatin	14 amino acids	Inhibits secretion of growth hormone
Thyrotropin-releasing hormone (TRH)	3 amino acids	Stimulates secretion of thyroid-stimulating hormone (TSH)
Growth hormone-releasing hormone (GHRH)	44 amino acids	Stimulates growth hormone secretion

Feedback Control of the Anterior Pituitary

In view of its secretion of releasing and inhibiting hormones, the hypothalamus might be considered the "master gland." The chain of command, however, is not linear; the hypothalamus and anterior pituitary are controlled by the effects of their own actions. In the endocrine system, to use an analogy, the general takes orders from the private. The hypothalamus and anterior pituitary are not master glands because their secretions are controlled by the target glands they regulate.

Anterior pituitary secretion of ACTH, TSH, and the gonadotropins (FSH and LH) is controlled by **negative feedback inhibition** from the target gland hormones. Secretion of ACTH is inhibited by a rise in corticosteroid secretion, for example, and TSH is inhibited by a rise in the secretion of thyroxine from the thyroid. These negative feedback relationships are easily demonstrated by removal of the target glands. Castration (surgical removal of the gonads), for example, produces a rise in the secretion of FSH and LH. In a similar manner, removal of the adrenals or the thyroid results in an abnormal increase in ACTH or TSH secretion from the anterior pituitary.

The effects of removal of the target glands demonstrate that, under normal conditions, these glands exert an inhibitory effect on the anterior pituitary. This inhibitory effect can occur at two levels: (1) the target gland hormones can act on the hypothalamus to inhibit the secretion of releasing hormones, and (2) the target gland hormones can act on the anterior pituitary to inhibit its response to the releasing hormones. Thyroxine, for example, has long been known to inhibit the synthesis and secretion of TSH by the anterior pituitary in response to TRH stimulation (fig. 11.16). Newer evidence indicates that thyroxine also inhibits the synthesis of TRH in the paraventricular nucleus of the hypothalamus. Sex steroids inhibit both the secretion of GnRH from the hypothalamus and the ability of the anterior pituitary to secrete the gonadotropins (FSH and LH) in response to GnRH stimulation (fig. 11.17).

Evidence suggests that there are *short feedback loops* in which a particular trophic hormone inhibits the secretion of its releasing hormone from the hypothalamus. A high secretion of TSH, for example, may inhibit further secretion of TRH by this means.

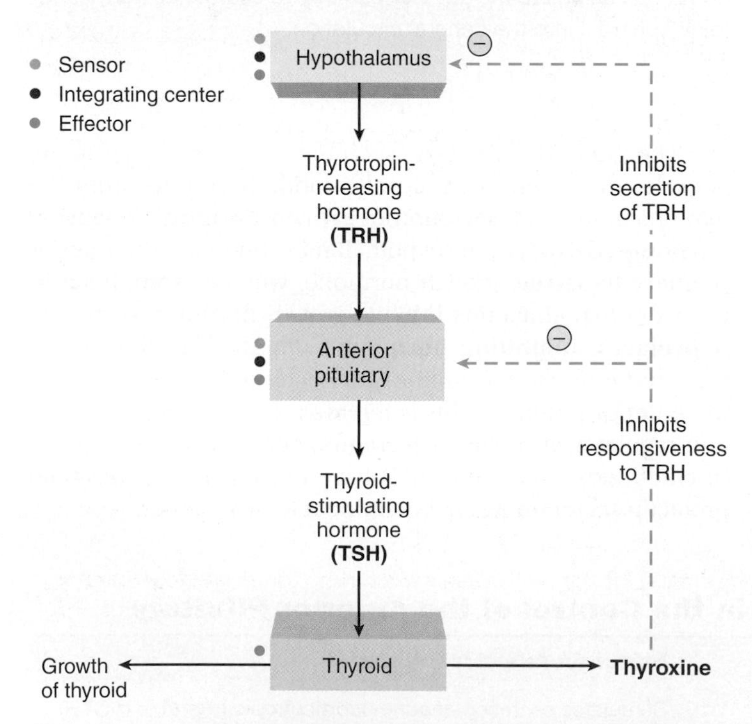

Figure 11.16 The hypothalamus-pituitary-thyroid axis (control system). The secretion of thyroxine from the thyroid is stimulated by thyroid-stimulating hormone (TSH) from the anterior pituitary. The secretion of TSH is stimulated by thyrotropin-releasing hormone (TRH) secreted from the hypothalamus. This stimulation is balanced by negative feedback inhibition (blue arrow) from thyroxine, which decreases the responsiveness of the anterior pituitary to stimulation by TRH.

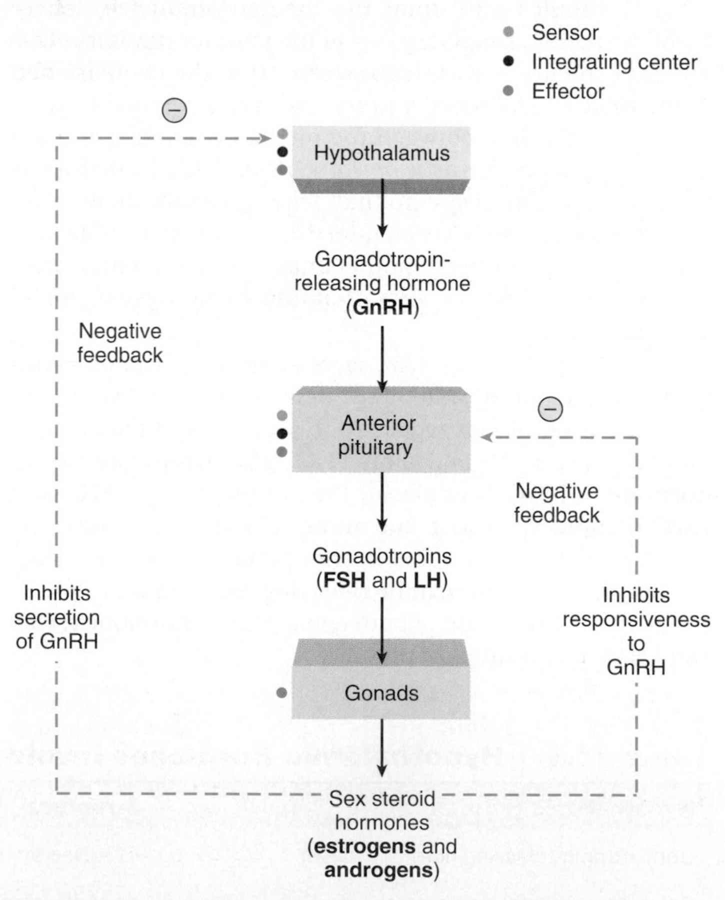

Figure 11.17 The hypothalamus-pituitary-gonad axis (control system). The hypothalamus secretes GnRH, which stimulates the anterior pituitary to secrete the gonadotropins (FSH and LH). These, in turn, stimulate the gonads to secrete the sex steroids. The secretions of the hypothalamus and anterior pituitary are themselves regulated by negative feedback inhibition (blue arrows) from the sex steroids.

Clinical Investigation CLUES 

Rosemary had a blood test in which her ACTH level was measured at one-fiftieth of normal.

- What could cause ACTH secretion to be so abnormally low?

In addition to negative feedback control of the anterior pituitary, there is one instance of a hormone from a target organ that actually stimulates the secretion of an anterior pituitary hormone. Toward the middle of the menstrual cycle, the rising secretion of estradiol from the ovaries stimulates the anterior pituitary to secrete a "surge" of LH, which results in ovulation. This is commonly described as a *positive feedback effect,* to distinguish it from the more usual negative feedback inhibition of target gland hormones on anterior pituitary secretion. Interestingly, higher levels of estradiol at a later stage of the menstrual cycle exert the opposite effect—negative feedback inhibition on LH secretion. The control of gonadotropin secretion is discussed in more detail in chapter 20, section 20.2.

Higher Brain Function and Pituitary Secretion

The relationship between the anterior pituitary and a particular target gland is described as an *axis;* the **pituitary-gonad axis,** for example, refers to the action of gonadotropic hormones on the testes and ovaries. This axis is stimulated by GnRH from the hypothalamus, as previously described. Because the hypothalamus receives neural input from "higher brain centers," it is not surprising that the pituitary-gonad axis can be affected by emotions. Indeed, the ability of intense emotions to alter the timing of ovulation or menstruation is well known.

Studies in mice show that neurons in at least 26 brain areas send axons to the GnRH-producing neurons of the hypothalamus! Considering this, it isn't surprising that various emotional states and stress can influence the menstrual cycle. These studies also show that neurons in the olfactory epithelium of the nose send information (relayed from the olfactory bulb and amygdala) to the GnRH-producing neurons of the hypothalamus. Thus, the sense of olfaction (smell) can influence the secretion of GnRH, and thereby affect the functioning of the reproductive system. This pathway is believed to be responsible for the tendency of female roommates to have synchronized menstrual cycles (described in the next Fitness Application box).

Psychological stress is known to activate the **pituitary adrenal axis,** as described more fully in section 11.4. Stressors produce an increase in CRH secretion from the hypothalamus, which in turn results in elevated ACTH and corticosteroid secretion. In addition, the influence of higher brain centers produces *circadian* ("about a day") *rhythms* in the secretion of many anterior pituitary hormones. The secretion of growth hormone, for example, is highest during sleep and decreases during wakefulness, although its secretion is also stimulated by the absorption of particular amino acids following a meal.

FITNESS APPLICATION

The projection of olfactory information from the olfactory epithelium in the nose to the GnRH-producing neurons of the hypothalamus helps to explain the **dormitory effect.** This refers to the tendency of women who room together to synchronize their menstrual cycle, an effect that can occur after only one or two cycles. This synchronization will not occur in a new roommate if her nasal cavity is plugged with cotton, suggesting that the dormitory effect is due to the action of chemicals called **pheromones.** These chemicals are excreted to the outside of the body and act through the olfactory sense to modify the physiology or behavior of another member of the same species. Pheromones are important regulatory molecules in the urine, vaginal fluid, and other secretions of most mammals, and help to regulate their reproductive cycles and behavior. The role of pheromones in humans is difficult to assess. Recently, however, scientists discovered that pheromones produced in the axillae (underarms) of women contribute to the dormitory effect.

 **CHECKPOINT**

5a. Describe the embryonic origins of the adenohypophysis and neurohypophysis, and list the parts of each. Which of these parts is also called the anterior pituitary? Which is called the posterior pituitary?

5b. List the hormones released by the posterior pituitary. Where do these hormones originate and how are their secretions regulated?

6. List the hormones secreted by the anterior pituitary and explain how the hypothalamus controls the secretion of each.

7. Draw a negative feedback loop showing the control of ACTH secretion. Explain how this system would be affected by (a) an injection of ACTH, (b) surgical removal of the pituitary, (c) an injection of corticosteroids, and (d) surgical removal of the adrenal glands.

11.4 ADRENAL GLANDS

The adrenal cortex and adrenal medulla are structurally and functionally different. The adrenal medulla secretes catecholamine hormones, which complement the sympathetic nervous system in the "fight-or-flight" reaction. The adrenal cortex secretes steroid hormones that participate in the regulation of mineral and energy balance.

LEARNING OUTCOMES

After studying this section, you should be able to:

8. Identify the hormones of the adrenal medulla, as well as the categories of corticosteroid hormones and their specific origin.

9. Describe the regulation of adrenal gland secretion, and the role of stress in adrenal secretion and function.

The **adrenal glands** are paired organs that cap the superior borders of the kidneys (fig. 11.18). Each adrenal consists of an outer cortex and inner medulla that function as separate glands. The differences in function of the adrenal cortex and medulla are related to the differences in their embryonic derivation. The adrenal medulla is derived from embryonic neural crest ectoderm (the same tissue that produces the sympathetic ganglia), whereas the adrenal cortex is derived from a different embryonic tissue (mesoderm).

As a consequence of its embryonic derivation, the adrenal medulla secretes catecholamine hormones (mainly epinephrine, with lesser amounts of norepinephrine) into the blood in response to stimulation by preganglionic sympathetic axons (chapter 9). The adrenal cortex does not receive neural innervation, and so must be stimulated hormonally (by ACTH secreted from the anterior pituitary). The cortex consists of three zones: an outer *zona glomerulosa,* a middle *zona fasciculata,* and an inner *zona reticularis* (fig. 11.18). These zones are believed to have different functions.

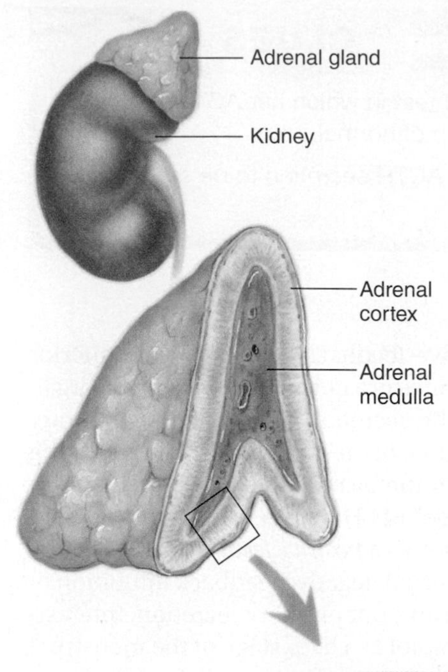

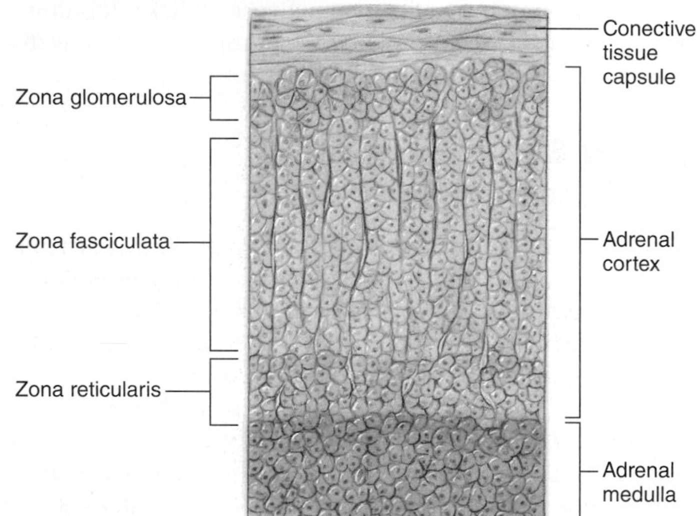

Figure 11.18 The structure of the adrenal gland, showing the three zones of the adrenal cortex. The zona glomerulosa secretes the mineralocorticoids (including aldosterone), whereas the other two zones secrete the glucocorticoids (including cortisol). AP|R

Functions of the Adrenal Cortex

The adrenal cortex secretes steroid hormones called **corticosteroids,** or **corticoids.** There are three functional categories of corticosteroids: (1) **mineralocorticoids,** which regulate Na^+ and K^+ balance; (2) **glucocorticoids,** which regulate the metabolism of glucose and other organic molecules; and (3) **adrenal androgens,** which are weak androgens (including *dehydroepiandrosterone,* or *DHEA*) that supplement the sex steroids secreted by the gonads. These three categories of steroid hormones are derived from the same precursor (parent molecule), cholesterol. The biosynthetic pathways from cholesterol diverge in the different zones of the adrenal cortex, so that a particular category of corticosteroid is produced in a particular zone of the adrenal cortex (fig. 11.19).

Aldosterone is the most potent mineralocorticoid. The mineralocorticoids are produced in the zona glomerulosa and stimulate the kidneys to retain Na^+ and water while excreting K^+ in the urine. These actions help to increase the blood volume and pressure (chapter 14, section 14.2), and to regulate blood electrolyte balance (chapter 17, section 17.5).

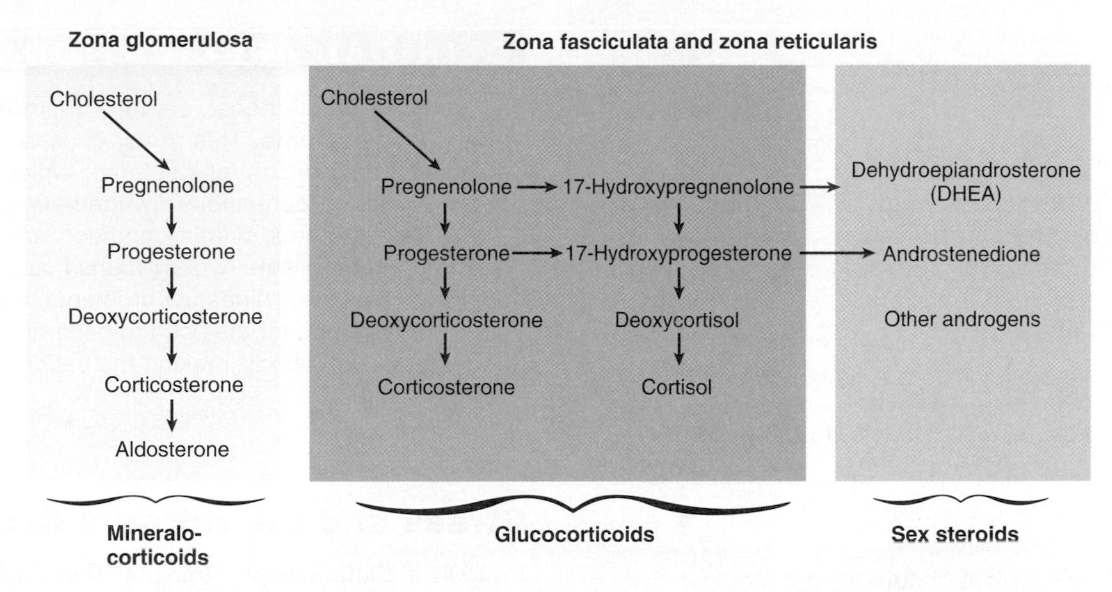

Figure 11.19 Simplified pathways for the synthesis of steroid hormones in the adrenal cortex. The adrenal cortex produces steroids that regulate Na^+ and K^+ balance (mineralocorticoids), steroids that regulate glucose balance (glucocorticoids), and small amounts of sex steroid hormones. (DHEA = dehydroepiandrosterone.) AP|R

The predominant glucocorticoid in humans is **cortisol (hydrocortisone),** which is secreted by the zona fasciculata and perhaps also by the zona reticularis. The secretion of cortisol is stimulated by ACTH from the anterior pituitary (fig. 11.20). Cortisol and other glucocorticoids have many effects on metabolism. These effects include: (1) stimulation of protein degradation; (2) stimulation of gluconeogenesis (production of glucose from amino acids and other noncarbohydrate molecules) and inhibition of glucose utilization, which help to raise the blood glucose concentration; and (3) stimulation of lipolysis (breakdown of fat) and the consequent release of free fatty acids into the blood. These effects provide more energy molecules of glucose and fatty acids in the blood, as described in chapter 19.

Exogenous glucocorticoids (taken as pills, injections, sprays, and topical creams) are used medically to suppress the immune response and inhibit inflammation. Thus, these drugs are very useful in treating inflammatory diseases such as asthma and rheumatoid arthritis. As might be predicted based on their metabolic actions, the side effects of glucocorticoids include hyperglycemia and decreased glucose tolerance. Other negative side effects include decreased synthesis of collagen and other extracellular matrix proteins and increased bone resorption, leading to osteoporosis (chapter 19, section 19.6).

CLINICAL APPLICATION

An excessively high level of corticosteroids in the blood causes **Cushing's syndrome.** This may result from the oversecretion of ACTH (usually by a tumor of the anterior pituitary) that overly stimulates the adrenal cortex to secrete corticosteroids, but it can also be produced by a tumor of the adrenal cortex that secretes excessive amounts of corticosteroids. Cushing's syndrome is characterized by changes in carbohydrate and protein metabolism, hyperglycemia, hypertension, and muscular weakness. Metabolic problems give the body a puffy appearance and can cause structural changes characterized as "buffalo hump" and "moon face."

Addison's disease is caused by inadequate secretion of both glucocorticoids and mineralocorticoids, which results in hypoglycemia, sodium and potassium imbalance, dehydration, hypotension, rapid weight loss, and generalized weakness. A person with this condition who is not treated with corticosteroids will eventually die because of severe electrolyte imbalance and dehydration. President John F. Kennedy had Addison's disease, but few knew of it because it was well controlled by corticosteroids. In the original description of the disease by Addison (1793–1860), he described a "characteristic discoloration of the skin . . . smoky appearance, or various tints or shades of deep amber or chestnut brown." This is caused by the very high secretion of ACTH, which at that concentration can stimulate melanocytes. High ACTH secretion results from inadequate negative feedback from the low glucocorticoid levels.

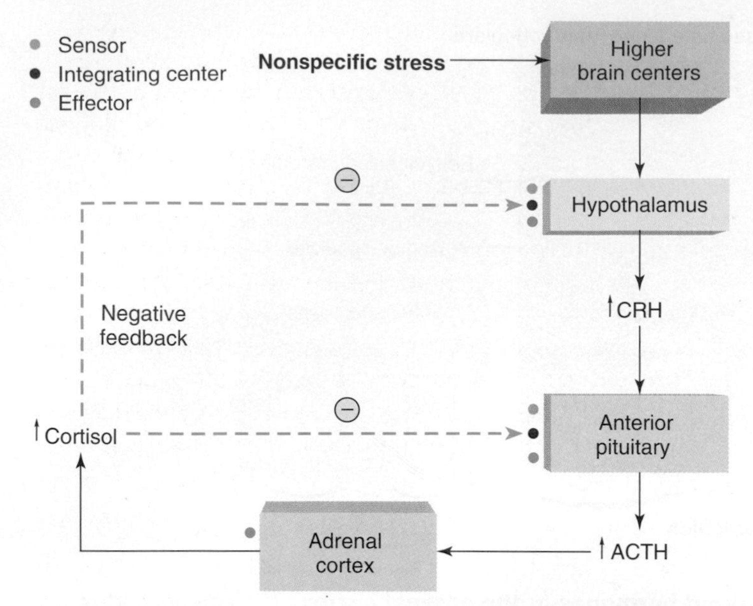

Figure 11.20 Activation of the pituitary-adrenal axis by nonspecific stress. Negative feedback control of the adrenal cortex (blue arrows) is also shown.

Clinical Investigation CLUES

Blood tests revealed that Rosemary had very high levels of cortisol together with her low levels of ACTH.

- What disease is likely responsible for these measurements, and how are these measurements related?
- How do these measurements of hormone levels relate to Rosemary's "puffiness"?

Functions of the Adrenal Medulla

The cells of the adrenal medulla secrete **epinephrine** and **norepinephrine** in an approximate ratio of 4 to 1. The effects of these catecholamine hormones are similar to those caused by stimulation of the sympathetic nervous system, except that the hormonal effect lasts about 10 times longer. The hormones from the adrenal medulla increase the cardiac output and heart rate, dilate coronary blood vessels, increase mental alertness, increase the respiratory rate, and elevate the metabolic rate.

The adrenal medulla is innervated by preganglionic sympathetic axons and secretes its hormones whenever the sympathetic nervous system is activated during "fight or flight" (chapter 9, fig. 9.7). These sympathoadrenal effects are supported by the metabolic actions of epinephrine and norepinephrine: a rise in blood glucose due to stimulation of hepatic glycogenolysis (breakdown of glycogen) and a rise in blood fatty acids due to stimulation of lipolysis (breakdown of fat). The endocrine regulation of metabolism is described more fully in chapter 19.

CLINICAL APPLICATION

A tumor of the adrenal medulla is referred to as a **pheochromocytoma.** This tumor causes hypersecretion of epinephrine and norepinephrine, which produces an effect similar to continuous sympathetic nerve stimulation. The symptoms of this condition are hypertension, elevated metabolism, hyperglycemia and sugar in the urine, nervousness, digestive problems, and sweating. It does not take long for the body to become totally fatigued under these conditions, making the patient susceptible to other diseases.

Stress and the Adrenal Gland

In 1936 a Canadian physiologist, Hans Selye, discovered that injections of a cattle ovary extract into rats (1) stimulated growth of the adrenal cortex; (2) caused atrophy of the lymphoid tissue of the spleen, lymph nodes, and thymus; and (3) produced bleeding peptic ulcers. At first he attributed these effects to the action of a specific hormone in the extract. However, subsequent experiments revealed that injections of a variety of substances—including foreign chemicals such as formaldehyde—could produce the same effects. Indeed, the same pattern occurred when Selye subjected rats to cold environments or when he dropped them into water and made them swim until they were exhausted.

The specific pattern of effects produced by these procedures suggested that the effects were due to something the procedures shared in common. Selye reasoned that all of the procedures were stressful, and that the pattern of changes he observed represented a specific response to any stressful agent. He later discovered that these effects were produced by activation of the pituitary-adrenal axis. Under stressful conditions, there is increased secretion of ACTH from the anterior pituitary, and thus there is increased secretion of glucocorticoids from the adrenal cortex.

On this basis, Selye stated that there is "a nonspecific response of the body to readjust itself following any demand made upon it." Stress causes a rise in the plasma glucocorticoid levels. Selye termed this nonspecific response the **general adaptation syndrome (GAS).** Stress, in other words, produces GAS. There are three stages in the response to stress: (1) the *alarm reaction,* when the adrenal glands are activated; (2) the *stage of resistance,* in which readjustment occurs; and (3) if the readjustment is not complete, the *stage of exhaustion,* which may lead to sickness and possibly death.

For example, when a person suffers from the stress of severe infections, trauma, burns, and surgery, the cortisol level can rise in proportion to the severity of the stress to as high as six times basal levels. There is evidence that this response of the pituitary-adrenal axis is needed for proper recovery from the illness or trauma, perhaps because cortisol and other glucocorticoids inhibit the immune response, thereby reducing damage due to inflammation. Thus, severe infections and trauma that trigger an immune response also activate mechanisms (the adrenal's secretion of cortisol) to limit that immune response. Indeed, patients who cannot secrete an adequate amount of cortisol for different reasons have an increased risk of death during an illness or trauma.

The sympathoadrenal system becomes activated, with increased secretion of epinephrine and norepinephrine, in response to stressors that challenge the organism to respond physically. This is the "fight-or-flight" reaction described in chapter 9, section 9.3. Hans Selye distinguished between neutral or positive stressors (which are "eustressful") and negative stressors (which are "distressful"), and modern research has confirmed that these differ in their neuroendocrine responses. This leads to a view that stressors include all stimuli that disrupt homeostasis. The different responses of the pituitary-adrenal axis and sympathoadrenal system to different stressors are coordinated by higher brain regions, particularly the prefrontal cortex, amygdala, and hippocampus (structures of the limbic system; chapter 8, section 8.2). These higher brain regions influence stress responses through synapses in the hypothalamus, medulla oblongata, and spinal cord. The hypothalamus-anterior pituitary-adrenal axis, with rising levels of glucocorticoids, becomes more active when the stress is of a chronic nature and when the person is more passive and feels less in control.

In rodents, the positive stress of social interactions and an enriched environment promote health and recovery from illness, whereas chronic negative stress has the opposite effects, perhaps due to glucocorticoid suppression of the immune system. Such negative stress can promote tumor growth in rodents, and similar responses have been reported in humans. There is also a dichotomy in the effects of stress on memory. Brain regions involved in memory (the prefrontal cortex, hippocampus, and amygdala) have receptor proteins for the *stress hormones* (glucocorticoids, epinephrine, and CRH), and these have been demonstrated to both promote and suppress LTP and LTD (long-term potentiation and depression, respectively; chapter 8, section 8.2). Glucocorticoids promote LTP and memory formation during stress, but severe stress and high amounts of glucocorticoids can reduce LTP and promote LTD, suppressing memory. Stress-induced chronically high cortisol secretion can even cause atrophy of the hippocampus, possibly due to inhibition of neurogenesis. The stress hormones may also act on the amygdala (important for the encoding of fearful memories) and other brain regions to contribute to the anxiety and depression associated with prolonged negative stress.

Glucocorticoids stimulate catabolism, chiefly the breakdown of muscle protein and fat. At the same time, they stimulate the liver to convert amino acids to glucose (in a process termed *gluconeogenesis*), leading to a rise in blood glucose concentration. These actions are described in detail in chapter 19, section 19.5. Through these and other effects, the glucocorticoids antagonize the actions of anabolic hormones, including growth hormone and insulin. Chronic stress, with its prolonged high secretion of glucocorticoids, can thereby aggravate *insulin resistance*—the reduced sensitivity of target tissues to insulin. Stress can thus make treatment of diabetes difficult, and can contribute to a constellation of symptoms associated with type 2 diabetes mellitus. Diabetes, and the "metabolic syndrome" associated with it, are discussed more fully in chapter 19, section 19.4.

CLINICAL APPLICATION

Because glucocorticoids such as hydrocortisone (cortisol) can inhibit the immune system and suppress inflammation, exogenous glucocorticoids—including **prednisolone** and **dexamethasone**—are medically very useful. They are given as pills or injections to treat various inflammatory conditions and to suppress the immune rejection of transplanted organs. However, as expected from the principles of negative feedback, exogenous glucocorticoids suppress the secretion of ACTH from the anterior pituitary, and thus the secretion of endogenous hydrocortisone from the adrenal cortex. Suppression of ACTH secretion can lead to atrophy of the adrenal cortex that may persist for months after the treatment with exogenous glucocorticoids.

CHECKPOINT

8a. List the categories of corticosteroids and identify the zone of the adrenal cortex that secretes the hormones within each category.

8b. Identify the hormones of the adrenal medulla and describe their effects.

9a. Explain how the secretions of the adrenal cortex and adrenal medulla are regulated.

9b. Explain how stress affects the secretions of the adrenal cortex and medulla. Why does hypersecretion of the adrenal medullary hormones make a person more susceptible to disease?

11.5 THYROID AND PARATHYROID GLANDS

The thyroid secretes thyroxine (T_4) and triiodothyronine (T_3), which are needed for proper growth and development and which are primarily responsible for determining the basal metabolic rate (BMR). The parathyroid glands secrete parathyroid hormone, which helps to raise the blood Ca^{2+} concentration.

LEARNING OUTCOMES

After studying this section, you should be able to:

10. Describe the structure of the thyroid gland, the production and actions of the thyroid hormones, and disorders of thyroid function.

11. Identify the location of the parathyroid glands and the actions of parathyroid hormone.

The **thyroid gland** is located just below the larynx (fig. 11.21). Its two lobes are positioned on either side of the trachea and are connected anteriorly by a medial mass of thyroid tissue called the *isthmus.* The thyroid is the largest of the purely endocrine glands, weighing 20 to 25 grams.

On a microscopic level, the thyroid gland consists of numerous spherical hollow sacs called **thyroid follicles** (fig. 11.22). These follicles are lined with a simple cuboidal epithelium composed of *follicular cells* that synthesize the principal thyroid hormone, *thyroxine.* The interior of the follicles contains *colloid,* a protein-rich fluid. In addition to the follicular cells that secrete thyroxine, the thyroid also contains *parafollicular cells* that secrete a hormone known as *calcitonin* (or *thyrocalcitonin*).

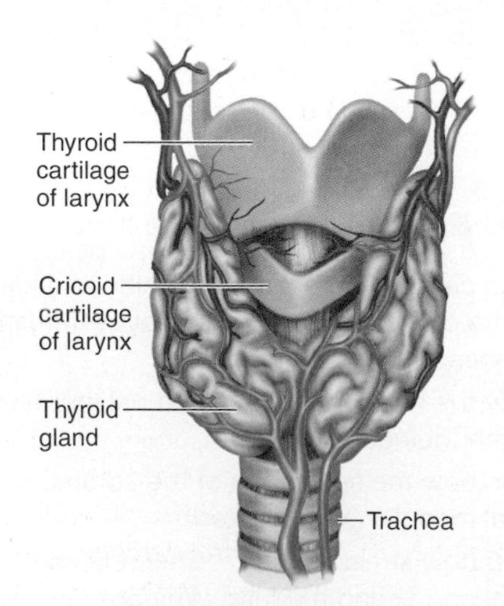

Figure 11.21 The thyroid gland. Its relationship to the larynx and trachea. APIR

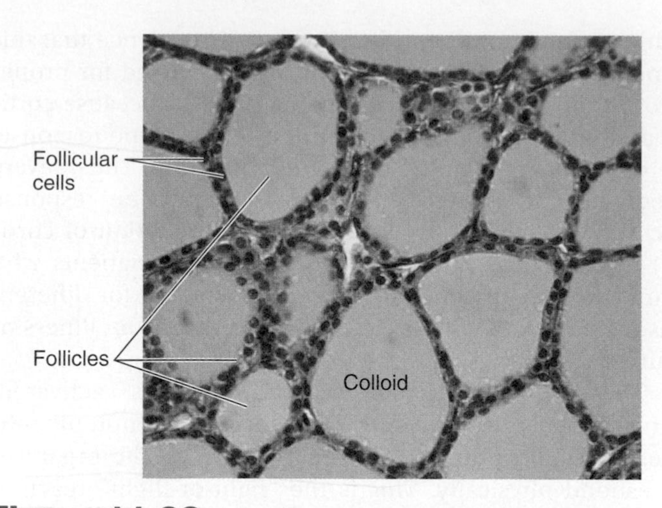

Figure 11.22 A photomicrograph (250×) of a thyroid gland. Numerous thyroid follicles are visible. Each follicle consists of follicular cells surrounding the fluid known as colloid, which contains thyroglobulin. APIR

Production and Action of Thyroid Hormones

The thyroid follicles actively accumulate iodide (I^-) from the blood and secrete it into the colloid. Once the iodide has entered the colloid, it is oxidized and attached to a specific amino acid (tyrosine) within the polypeptide chain of a protein called **thyroglobulin.** The attachment of one iodine to tyrosine produces *monoiodotyrosine (MIT);* the attachment of two iodines produces *diiodotyrosine (DIT).*

Within the colloid, enzymes modify the structure of MIT and DIT and couple them together. When two DIT molecules that are appropriately modified are coupled together, a molecule of **tetraiodothyronine** (**T_4**), or **thyroxine,** is produced (fig. 11.23). The combination of one MIT with one DIT forms **triiodothyronine (**T_3**).** Note that at this point T_4 and T_3 are still attached to thyroglobulin. Upon stimulation by TSH, the cells of the follicle take up a small volume of colloid by pinocytosis, hydrolyze the T_3 and T_4 from the thyroglobulin, and secrete the free hormones into the blood.

The transport of thyroid hormones through the blood and their mechanism of action at the cellular level was described earlier in this chapter. Through the activation of genes, thyroid hormones stimulate protein synthesis, promote maturation of the nervous system, and increase the rate of cell respiration in most tissues of the body. Through this action, thyroxine (after it is converted into T_3) elevates the *basal metabolic rate* (*BMR;* chapter 19, section 19.1), which is the resting rate of calorie expenditure by the body.

Calcitonin, secreted by the parafollicular cells of the thyroid, works in concert with parathyroid hormone (discussed

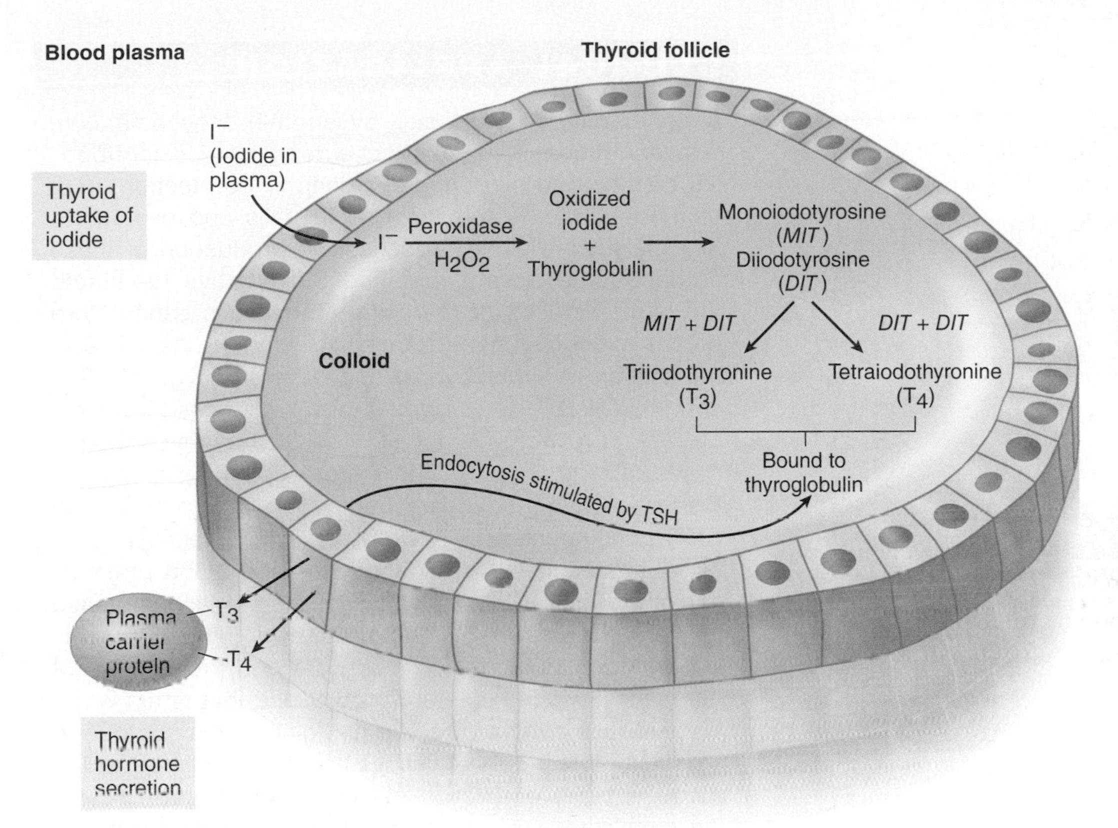

Figure 11.23 The production and storage of thyroid hormones. Iodide is actively transported into the follicular cells. In the colloid, it is converted into iodine and attached to tyrosine amino acids within the thyroglobulin protein. *MIT* (monoiodotyrosine) and *DIT* (diiodotyrosine) are used to produce T_3 and T_4 within the colloid. Upon stimulation by TSH, the thyroid hormones, bound to thyroglobulin, are taken into the follicular cells by pinocytosis. Hydrolysis reactions within the follicular cells release the free T_4 and T_3, which are secreted.

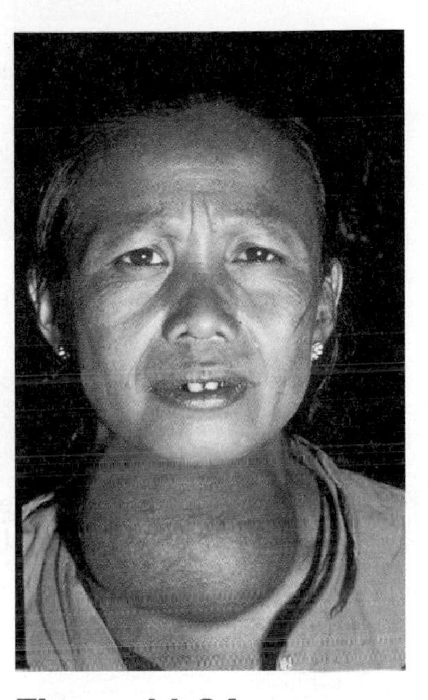

Figure 11.24 Endemic goiter is caused by insufficient iodine in the diet. A lack of iodine causes hypothyroidism, and the resulting elevation in TSH secretion stimulates the excessive growth of the thyroid.

shortly) to regulate the calcium levels of the blood. Calcitonin inhibits the dissolution of the calcium phosphate crystals of bone and stimulates the excretion of calcium in the urine by the kidneys. Both of these actions result in the lowering of blood calcium concentrations.

Diseases of the Thyroid

Thyroid-stimulating hormone (TSH) from the anterior pituitary stimulates the thyroid to secrete thyroxine; however, it also exerts a trophic (growth-stimulating) effect on the thyroid. This trophic effect is evident in people who develop an **iodine-deficiency (endemic) goiter,** or abnormal growth of the thyroid gland (fig. 11.24). In the absence of sufficient dietary iodine, the thyroid cannot produce adequate amounts of T_4 and T_3. The resulting lack of negative feedback inhibition causes abnormally high levels of TSH secretion, which in turn stimulates the abnormal growth of the thyroid. These events are summarized in figure 11.25.

People who have inadequate secretion of thyroid hormones are said to be **hypothyroid.** As might be predicted from the effects of thyroxine, people who are hypothyroid have an abnormally low basal metabolic rate and experience weight gain and lethargy. A thyroxine deficiency also decreases the ability to adapt to cold stress. Severe hypothyroidism in adults can result in **myxedema,** in which mucoproteins (glycosaminoglycans) and fluid accumulate in the subcutaneous connective tissues and viscera, producing edema that causes swelling of the hands, feet, face, and tissue around the eyes. Severe hypothyroidism produces a slowing of physical and mental activity, and can result in profound lethargy and even a myxedema coma.

Hypothyroidism can result from a thyroid gland defect or secondarily from insufficient thyrotropin-releasing hormone (TRH) secretion from the hypothalamus, insufficient TSH secretion from the anterior pituitary, or insufficient iodine in the diet. In the latter case, excessive TSH secretion stimulates abnormal thyroid growth and the development of an endemic goiter, as described previously. The hypothyroidism and goiter caused by iodine deficiency can be reversed by iodine supplements.

Because of its stimulation of protein synthesis, children need thyroxine for body growth and, most importantly, for

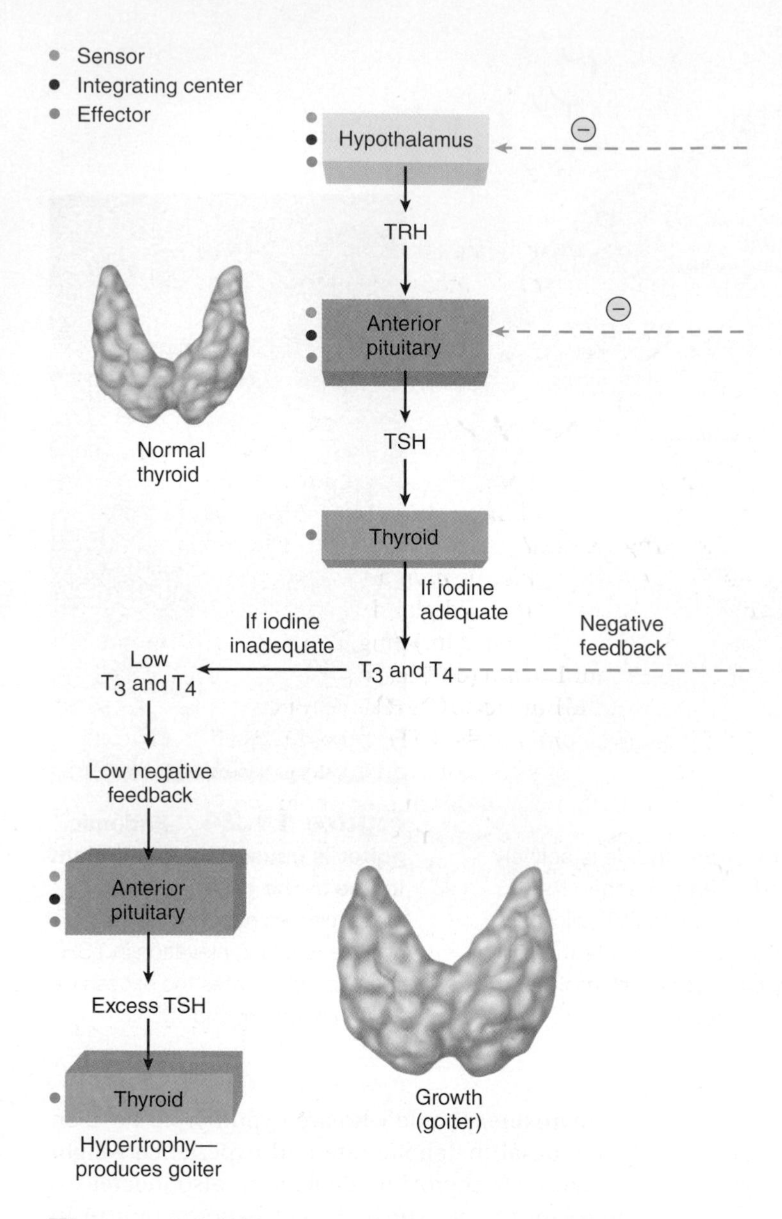

Figure 11.25 How iodine deficiency causes a goiter. Lack of adequate iodine in the diet interferes with the negative feedback control of TSH secretion, resulting in the formation of an endemic goiter.

the proper development of the central nervous system. The need for thyroxine is particularly great when the brain is undergoing its greatest rate of development—from the end of the first trimester of prenatal life to six months after birth. Hypothyroidism during this time may result in **cretinism.** Unlike people with *dwarfism,* who have inadequate secretion of growth hormone from the anterior pituitary, people with cretinism suffer severe mental retardation. Treatment with thyroxine soon after birth, particularly before one month of age, has been found to completely or almost completely restore development of intelligence as measured by IQ tests five years later.

CLINICAL APPLICATION

A goiter can be produced by another mechanism. In **Graves' disease,** circulating *autoantibodies* (antibodies that bind to self-antigens; chapter 15, section 15.6) bind to TSH receptors on thyroid cells and overstimulate the thyroid gland. Because the production of these autoantibodies is not inhibited by negative feedback, the high secretion of thyroxine that results cannot turn off the excessive stimulation of the thyroid. As a result, the person is **hyperthyroid** and develops a goiter. This condition is called *toxic goiter,* or *thyrotoxicosis.* Graves' disease is responsible for 50% to 80% of hyperthyroid cases, and is 5 to 10 times more common in women than men. Symptoms of hyperthyroidism include a high BMR accompanied by weight loss, nervousness, irritability, intolerance to heat, and increased blood pressure (table 11.8). Graves' disease is often also accompanied by *exophthalmos*, or bulging of the eyes (fig. 11.26). This is partly due to increased orbital fat and proliferation of fibroblasts, which secrete glycoproteins that retain water and produce edema. The eye pathologies may be due to TSH receptors in orbital tissues that are stimulated by the autoantibodies of Graves' disease, but cigarette smoking significantly increases the risk and severity of eye involvement.

Table 11.8 | Comparison of Hypothyroidism and Hyperthyroidism

Feature	Hypothyroid	Hyperthyroid
Growth and development	Impaired growth	Accelerated growth
Activity and sleep	Lethargy; increased sleep	Increased activity; decreased sleep
Temperature tolerance	Intolerance to cold	Intolerance to heat
Skin characteristics	Coarse, dry skin	Normal skin
Perspiration	Absent	Excessive
Pulse	Slow	Rapid
Gastrointestinal symptoms	Constipation; decreased appetite; increased weight	Frequent bowel movements; increased appetite; decreased weight
Reflexes	Slow	Rapid
Psychological aspects	Depression and apathy	Nervous, "emotional" state
Plasma T_4 levels	Decreased	Increased

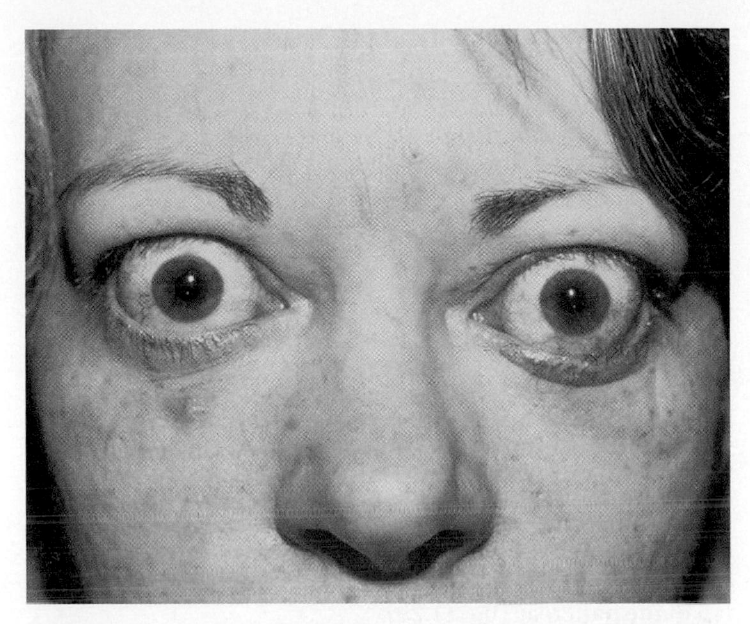

Figure 11.26 A symptom of hyperthyroidism. Hyperthyroidism is characterized by an increased metabolic rate, weight loss, muscular weakness, and nervousness. The eyes may also protrude (exophthalmos) due to edema in the orbits.

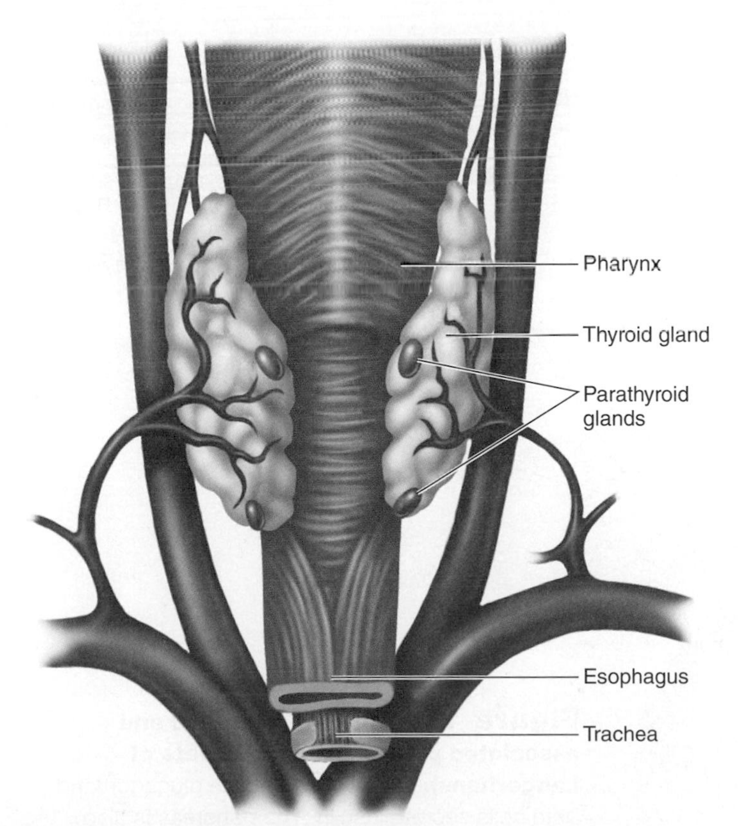

Figure 11.27 A posterior view of the parathyroid glands. The parathyroids are embedded in the tissue of the thyroid gland. APIR

Clinical Investigation CLUES

The physician stated that Rosemary's "puffiness" was not myxedema, and that her blood levels of T_4 and TSH were normal.

- What is myxedema, and what is its cause?
- How would blood measurements of T_4 and TSH have been affected if Rosemary were hypothyroid?

Parathyroid Glands

The small, flattened **parathyroid glands** are embedded in the posterior surfaces of the lateral lobes of the thyroid gland, as shown in figure 11.27. There are usually four parathyroid glands: a *superior* and an *inferior pair,* although the precise number can vary. Each parathyroid gland is a small yellowish-brown body 3 to 8 mm (0.1 to 0.3 in.) long, 2 to 5 mm (0.07 to 0.2 in.) wide, and about 1.5 mm (0.05 in.) deep.

Parathyroid hormone (PTH) is the only hormone secreted by the parathyroid glands. PTH, however, is the single most important hormone controlling the calcium concentrations in the blood. It promotes a rise in blood calcium levels by acting on the bones, kidneys, and intestine (fig. 11.28). Regulation of calcium balance is described in chapter 19, section 19.6.

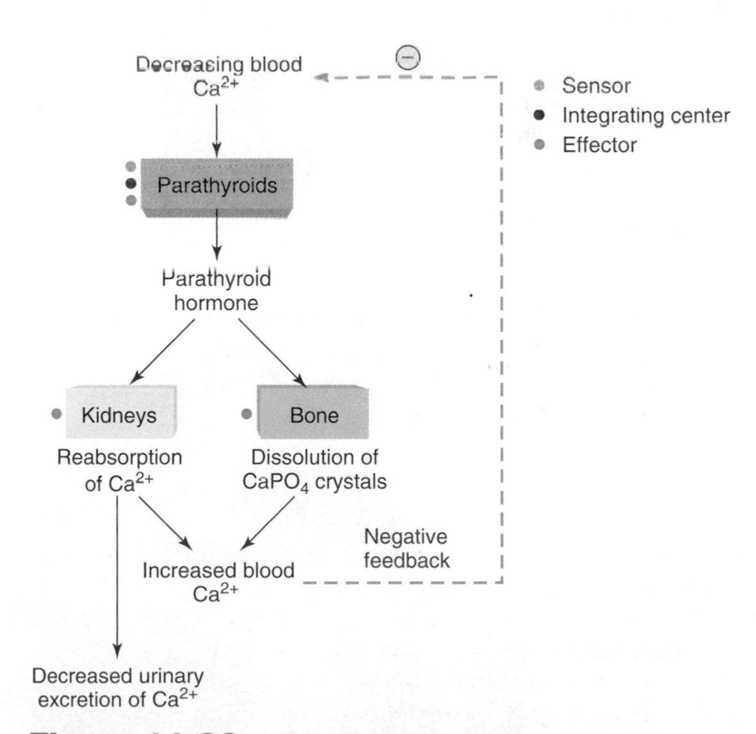

Figure 11.28 The actions of parathyroid hormone and the control of its secretion. An increased level of parathyroid hormone causes the bones to release calcium and the kidneys to conserve calcium that would otherwise be lost through the urine. A rise in blood Ca^+ can then exert negative feedback inhibition on parathyroid hormone secretion.

CHECKPOINT

10a. Describe the structure of the thyroid gland and list the effects of thyroid hormones.

10b. Describe how thyroid hormones are produced and how their secretion is regulated.

10c. Explain the consequences of an inadequate dietary intake of iodine.

11. Describe the location of the parathyroid glands and the actions of parathyroid hormone.

11.6 PANCREAS AND OTHER ENDOCRINE GLANDS

The pancreatic islets secrete two hormones, insulin and glucagon. Insulin promotes the lowering of blood glucose and the storage of energy in the form of glycogen and fat. Glucagon has antagonistic effects that raise the blood glucose concentration. Additionally, many other organs secrete hormones that help to regulate digestion, metabolism, growth, immune function, and reproduction.

LEARNING OUTCOMES

After studying this section, you should be able to:

12. Identify the endocrine cells and hormones of the pancreas, and describe the actions of these hormones.

13. Identify the pineal gland and its hormone, and describe its effect.

14. Describe the endocrine functions of the gonads and placenta.

The **pancreas** is both an endocrine and an exocrine gland. The gross structure of this gland and its exocrine functions in digestion are described in chapter 18, section 18.5. The endocrine portion of the pancreas consists of scattered clusters of cells called the **pancreatic islets** or **islets of Langerhans.** These endocrine structures are most common in the body and tail of the pancreas (fig. 11.29).

Pancreatic Islets (Islets of Langerhans)

On a microscopic level, the most conspicuous cells in the islets are the *alpha* and *beta cells* (fig. 11.29). The alpha cells secrete

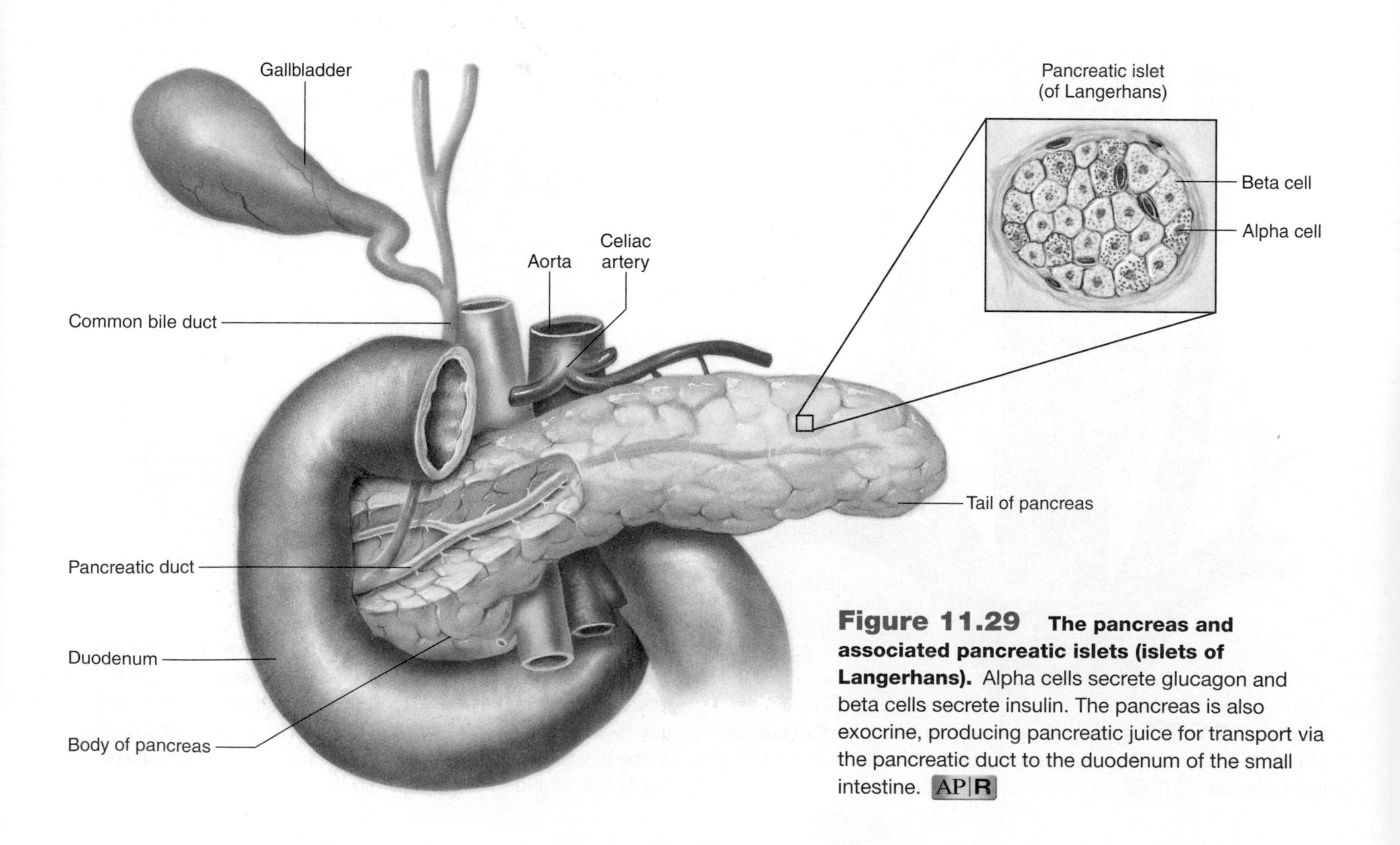

Figure 11.29 The pancreas and associated pancreatic islets (islets of Langerhans). Alpha cells secrete glucagon and beta cells secrete insulin. The pancreas is also exocrine, producing pancreatic juice for transport via the pancreatic duct to the duodenum of the small intestine. APIR

the hormone **glucagon,** and the beta cells secrete **insulin.** There are more than twice as many insulin-secreting beta cells as alpha cells in each islet.

Insulin is the primary hormone regulating the plasma glucose concentration. After a carbohydrate meal or sugary drink, the plasma glucose level rises. This rise in plasma glucose stimulates the beta cells of the islets to secrete increased amounts of insulin. Insulin then binds to its receptors in the plasma membrane of its target cells, and, through the action of signaling molecules, causes intracellular vesicles containing GLUT4 carrier proteins to translocate to the plasma membrane (fig. 11.30). These carrier proteins promote the facilitated diffusion of glucose into the cells of insulin's target organs—primarily the skeletal muscles, liver, and adipose tissue.

Also, insulin indirectly stimulates the activity of the enzyme glycogen synthetase in skeletal muscles and liver, which promotes the conversion of intracellular glucose into glycogen for storage. Insulin thereby causes glucose to leave the plasma and enter the target cells, where it is converted into the energy storage molecules of glycogen (in skeletal muscles and liver) and fat (in adipose tissue). Through these effects, insulin lowers the blood glucose concentration (fig. 11.31*a*) as it promotes anabolism (chapter 19, section 19.3). The ability of the beta cells to secrete insulin, and the action of insulin to lower the plasma glucose concentration, are tested in an oral glucose tolerance test for diabetes mellitus (chapter 19, section 19.4).

Glucagon, secreted by the alpha cells of the pancreatic islets, acts antagonistically to insulin—it promotes effects that raise the plasma glucose concentration. Glucagon secretion is stimulated

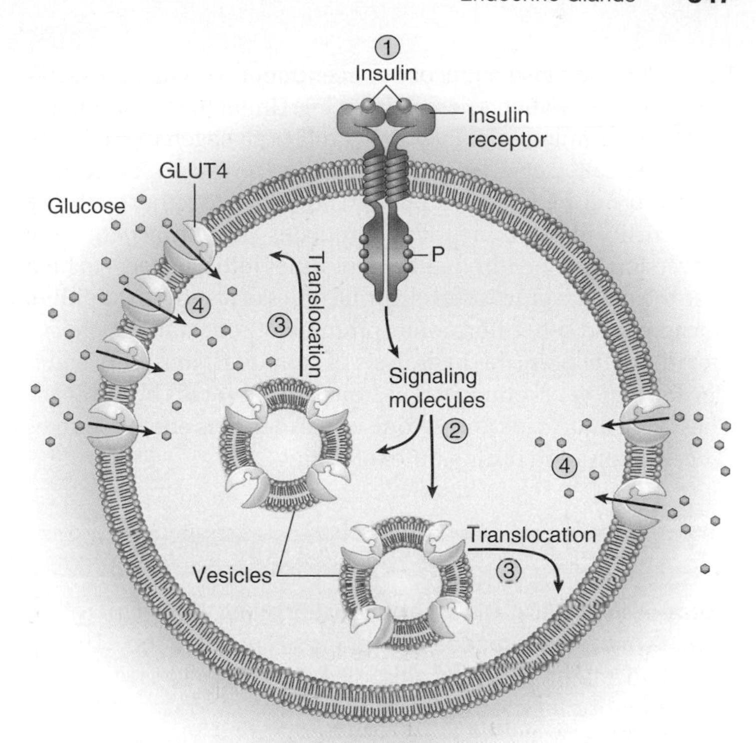

Figure 11.30 Insulin stimulates uptake of blood glucose. (1) Binding of insulin to its plasma membrane receptors causes the activation of cytoplasmic signaling molecules, which (2) act on intracellular vesicles that contain GLUT4 carrier proteins in the vesicle membrane. (3) This causes the intracellular vesicles to translocate and fuse with the plasma membrane, so that the vesicle membrane becomes part of the plasma membrane. (4) The GLUT4 proteins permit the facilitated diffusion of glucose from the extracellular fluid into the cell.

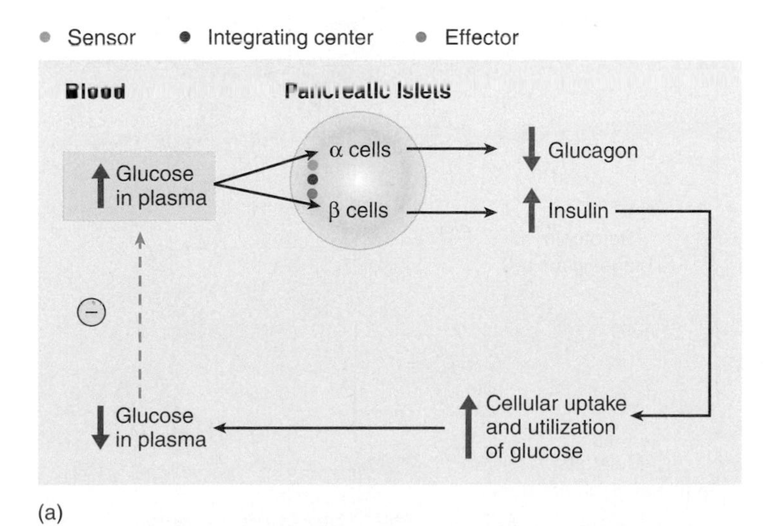

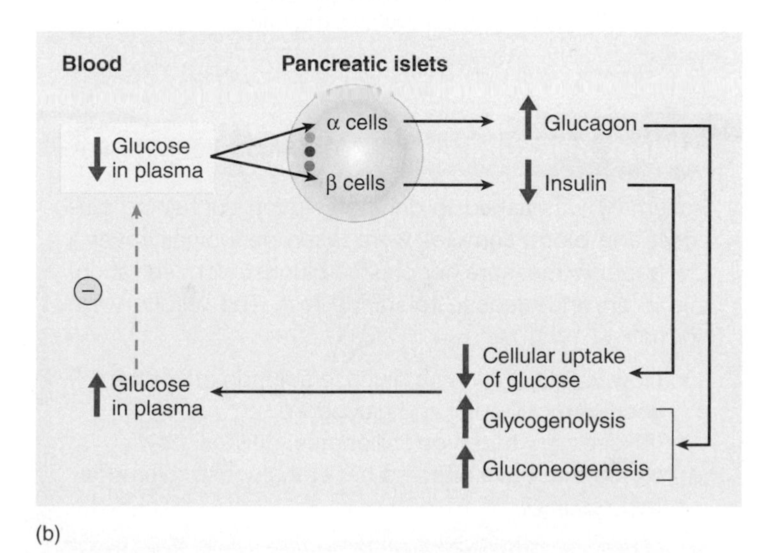

Figure 11.31 Glucose homeostasis is maintained by insulin and glucagon. (*a*) When the plasma glucose concentration rises after a meal, the beta cells secrete increased amounts of insulin (and alpha cells are inhibited from secreting glucagon). Insulin then promotes the cellular uptake of blood glucose, reducing the plasma glucose concentration so that homeostasis of blood glucose is maintained. (*b*) When the plasma glucose concentration falls, the secretion of insulin is inhibited and the secretion of glucagon is stimulated. Glucagon promotes glycogenolysis and gluconeogenesis, so that the liver can secrete glucose into the blood and maintain homeostasis of the blood glucose concentration.

by a fall in the plasma glucose concentration and insulin secretion that occurs when a person is fasting. Under these conditions, glucagon stimulates the liver to hydrolyze glycogen into glucose (a process called *glycogenolysis*), allowing the liver to secrete glucose into the blood (fig. 11.31*b*). Glucagon, together with the glucocorticoid hormones, also stimulates *gluconeogenesis*—the conversion of noncarbohydrate molecules into glucose—to help raise the plasma glucose level during times of fasting. In addition, glucagon and other hormones promote other catabolic effects, including *lipolysis* (the hydrolysis of stored fat) and *ketogenesis* (the formation of ketone bodies from free fatty acids by the liver). These free fatty acids and ketone bodies serve as energy sources for cell respiration during times of fasting.

CLINICAL APPLICATION

Diabetes mellitus is characterized by fasting hyperglycemia and the presence of glucose in the urine. There are two major forms of this disease. *Type 1,* or insulin-dependent diabetes mellitus, is caused by destruction of the beta cells and the resulting lack of insulin secretion. *Type 2,* or non-insulin-dependent diabetes mellitus (the more common form), is caused largely by decreased tissue sensitivity to the effects of insulin so that larger than normal amounts of insulin are required to produce a normal effect. The causes and symptoms of diabetes mellitus are described in more detail in chapter 19, section 19.4.

Gestational diabetes occurs in about 4% of pregnancies due to insulin secretion that is inadequate to meet the increased demand imposed by the fetus. Women who do not develop gestational diabetes have a sufficiently increased insulin secretion during pregnancy, probably due to the proliferation of beta cells in the islets.

Clinical Investigation CLUES

Rosemary was asked to drink a solution containing glucose, and blood samples were taken periodically over a few hours to measure her plasma glucose concentration; this is an oral glucose tolerance test. The results were normal.

- How would drinking a glucose solution affect the secretion of insulin and glucagon?
- If Rosemary had type 2 diabetes mellitus, how would that have affected her oral glucose tolerance test results?

Pineal Gland

The small, cone-shaped **pineal gland** is located in the roof of the third ventricle of the diencephalon (chapter 8), where it is encapsulated by the meninges covering the brain. The pineal gland of a child weighs about 0.2 g and is 5 to 8 mm (0.2 to 0.3 in.) long and 9 mm wide. The gland begins to regress in size at about age seven and in the adult appears as a thickened strand of fibrous tissue. Although the pineal gland lacks direct nervous connections to the rest of the brain, it is highly innervated by the sympathetic nervous system from the superior cervical ganglion. The pineal gland secretes the hormone **melatonin** (fig. 11.32).

The *suprachiasmatic nucleus (SCN)* of the hypothalamus (chapter 8; see fig. 8.20) regulates pineal secretion of melatonin through hypothalamic control of the sympathetic neurons that innervate the pineal gland (fig. 11.33). The SCN is also the primary center for the regulation of the body's *circadian rhythms:* rhythms of physiological activity that follow a 24-hour pattern (chapter 8, section 8.3).

Most of the neurons of the SCN produce action potentials starting at dawn. Action potential frequency increases toward the middle of the day and then decreases to become mostly silent at night. Light acts through the *retinohypothalamic tract* (fig. 11.33) to better entrain (synchronize) this spontaneous circadian rhythm to the light/dark cycle. Circadian neural activity of the SCN regulates the pineal gland via sympathetic nerves (fig. 11.33), inhibiting the pineal gland secretion of melatonin during the day. As a result, melatonin secretion begins to increase with darkness and peaks by the middle of the night.

The regulatory effect of light on the SCN, and thus the ability of light to inhibit melatonin secretion, appear to require a recently discovered retinal pigment (chapter 10, section 10.7). This pigment has been named *melanopsin,* and is found in a population of ganglion cells; thus, it is distinct from the visual

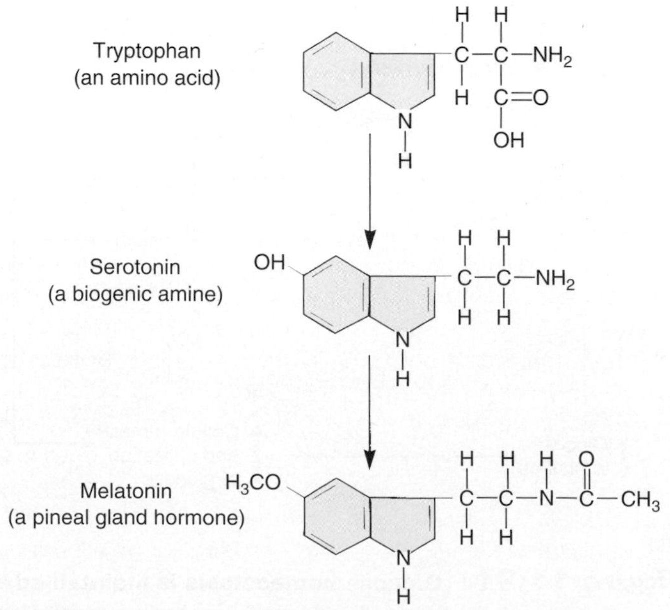

Tryptophan (an amino acid)

Serotonin (a biogenic amine)

Melatonin (a pineal gland hormone)

Figure 11.32 A simplified biosynthetic pathway for melatonin. Secretion of melatonin by the pineal gland follows a circadian (daily) rhythm tied to daily and seasonal changes in light.

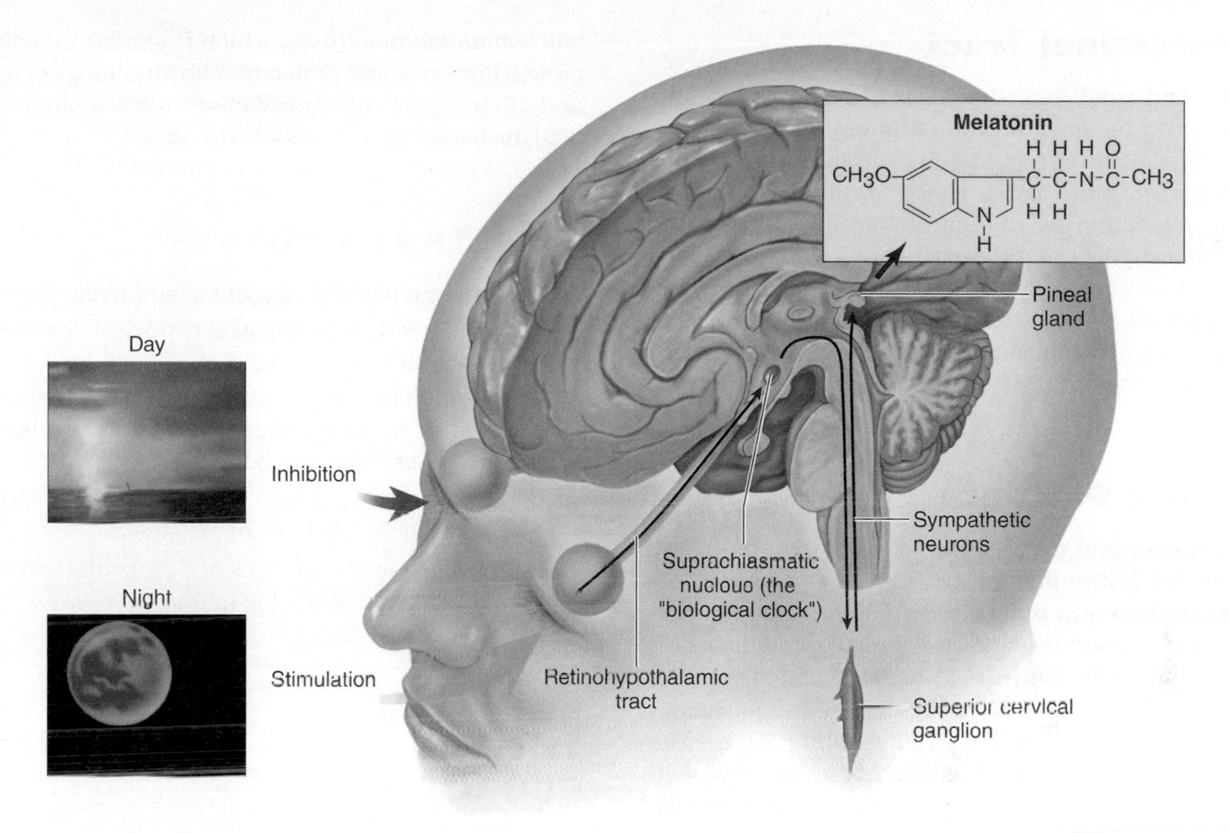

Figure 11.33 **The secretion of melatonin.** The secretion of melatonin by the pineal gland is stimulated by sympathetic axons originating in the superior cervical ganglion. Activity of these neurons is regulated by the cyclic activity of the suprachiasmatic nucleus of the hypothalamus, which sets a circadian rhythm. This rhythm is entrained to light/dark cycles by neurons in the retina. AP|R

pigments found in rods and cones. However, activation of rhodopsin and photopsins (in rods and cones, respectively) may also influence the ability of the retina to regulate circadian rhythms. The role of the SCN and the endocrine system (including the secretion of melatonin) in circadian rhythms is discussed more fully in chapter 8, section 8.3.

The pineal gland has been implicated in a variety of physiological processes. One of the most widely studied is the role of melatonin in helping to time the births of seasonally breeding animals. To do this, melatonin influences the pituitary-gonad axis; it stimulates this axis in short-day breeders such as sheep, but inhibits the axis of long-day breeders such as voles. Although there is evidence to support an antigonadotropic effect in humans, this possibility has not yet been proven. For example, excessive melatonin secretion in humans is associated with a delay in the onset of puberty. Melatonin secretion is highest in children between the ages of one and five and decreases thereafter, reaching its lowest levels at the end of puberty when concentrations are 75% lower than during early childhood. This suggests a role for melatonin in the onset of human puberty. However, because of much conflicting data, the importance of melatonin in human reproduction is still highly controversial.

The pattern of melatonin secretion is altered when a person works night shifts or flies across different time zones. There is evidence that exogenous melatonin (taken as a pill) may be beneficial in the treatment of jet lag, but the optimum dosage is not currently known. Phototherapy using bright fluorescent lamps, which act like sunlight to inhibit melatonin secretion, has been used effectively in the treatment of *seasonal affective disorder (SAD),* or "winter depression."

CLINICAL APPLICATION

Melatonin pills decrease the time required to fall asleep and increase the duration of rapid eye movement (REM) sleep; for these reasons, they may be useful in the treatment of insomnia. This is particularly significant for elderly people with insomnia, who have the lowest nighttime levels of endogenous melatonin secretion. Exogenous melatonin makes most people sleepy 30 to 120 minutes after taking the pills, although this varies widely. Melatonin pills can be used to alleviate *jet lag,* a sleep disorder produced by crossing time zones too rapidly for the circadian clock in the SCN to adjust. Endogenous melatonin secretion at night signals darkness, and so a melatonin pill taken at night before endogenous secretion begins resets the clock to an earlier time. Taking a melatonin pill at bedtime can therefore be useful after an eastward flight. After a westward flight, it may be more effective to take a lower dose of melatonin later in the night.

Gastrointestinal Tract

The stomach and small intestine secrete a number of hormones that act on the gastrointestinal tract itself and on the pancreas and gallbladder (chapter 18; the hormone actions are summarized in table 18.5). These hormones, acting in concert with regulation by the autonomic nervous system, coordinate the activities of different regions of the digestive tract and the secretions of pancreatic juice and bile. Several hormones secreted by the stomach and small intestine are also known to stimulate insulin secretion from the pancreatic islets in anticipation of a rise in blood glucose following a meal.

Gonads and Placenta

The gonads (**testes** and **ovaries**) secrete sex steroids. These include male sex hormones, or *androgens,* and female sex hormones—*estrogens* and *progesterone.* (Progesterone is produced by several endocrine glands as a precursor of their hormones, as illustrated in figure 11.2, but is secreted into the blood as a major hormone only by the ovaries and placenta.) The androgens and estrogens are families of hormones. The principal androgen secreted by the testes is *testosterone,* and the principal estrogen secreted by the ovaries is *estradiol-17β.* The principal estrogen during pregnancy, however, is a weaker estrogen called *estriol,* secreted by the placenta. After menopause, the principal estrogen is *estrone,* produced primarily by fat cells.

The testes consist of two compartments: *seminiferous tubules,* which produce sperm cells, and *interstitial tissue* between the convolutions of the tubules. Within the interstitial tissue are *Leydig cells,* which secrete testosterone (chapter 20; see fig. 20.11). Testosterone is needed for the development and maintenance of the male genitalia (penis and scrotum) and the male accessory sex organs (prostate, seminal vesicles, epididymides, and vas deferens), as well as for the development of male secondary sex characteristics.

During the first half of the menstrual cycle, estradiol-17β is secreted by small structures within the ovary called *ovarian follicles.* These follicles contain the egg cell, or *ovum,* and *granulosa cells* that secrete estrogen (chapter 20; see fig. 20.26). By about midcycle, one of these follicles grows very large and, in the process of ovulation, extrudes its ovum from the ovary. The empty follicle, under the influence of luteinizing hormone (LH) from the anterior pituitary, then becomes a new endocrine structure called a *corpus luteum* (see fig. 20.34). The corpus luteum secretes progesterone as well as estradiol-17β.

The **placenta**—the organ responsible for nutrient and waste exchange between the fetus and mother—is also an endocrine gland that secretes large amounts of estrogens and progesterone. In addition, it secretes a number of polypeptide and protein hormones that are similar to some hormones secreted by the anterior pituitary. These hormones include *human chorionic gonadotropin (hCG),* which is similar to LH, and *somatomammotropin,* which is similar in action to both growth hormone and prolactin. The physiology of the placenta and other aspects of reproductive endocrinology are considered in chapter 20.

CHECKPOINT

12a. Describe the structure of the endocrine pancreas. Which cells secrete insulin and which secrete glucagon?

12b. Describe how insulin and glucagon secretion are affected by eating and by fasting and explain the actions of these two hormones.

13. Describe the location of the pineal gland and discuss the possible functions of melatonin.

14. Explain how the gonadal and placental hormones are categorized and list the hormones secreted by each gland.

11.7 PARACRINE AND AUTOCRINE REGULATION

Many regulatory molecules produced throughout the body act within the organs that produce them. These molecules may regulate different cells within one tissue, or they may be produced within one tissue and regulate a different tissue within the same organ.

LEARNING OUTCOMES

After studying this section, you should be able to:

15. Distinguish between autocrine, paracrine, and endocrine regulation, and give examples of paracrine regulation of blood vessels.

16. Describe the production and significance of the prostaglandins.

Thus far in this text, two types of regulatory molecules have been considered—neurotransmitters in chapter 7 and hormones in the present chapter. These two classes of regulatory molecules cannot be defined simply by differences in chemical structure, because the same molecule (such as norepinephrine) could be included in both categories; rather, they must be defined by function. Neurotransmitters are released by axons, travel across a narrow synaptic cleft, and affect a postsynaptic cell. Hormones are secreted into the blood by an endocrine gland and, through transport in the blood, influence the activities of one or more target organs.

There are yet other classes of regulatory molecules. These molecules are produced in many different organs and are active

Table 11.9 | Examples of Paracrine and Autocrine Regulators

Paracrine or Autocrine Regulator	Major Sites of Production	Major Actions
Insulin-like growth factors (somatomedins)	Many organs, particularly the liver and cartilages	Growth and cell division
Nitric oxide	Endothelium of blood vessels; neurons; macrophages	Dilation of blood vessels; neural messenger; antibacterial agent
Endothelins	Endothelium of blood vessels; other organs	Constriction of blood vessels; other effects
Platelet-derived growth factor	Platelets; macrophages; vascular smooth muscle cells	Cell division within blood vessels
Epidermal growth factors	Epidermal tissues	Cell division in wound healing
Neurotrophins	Schwann cells; neurons	Regeneration of peripheral nerves
Bradykinin	Endothelium of blood vessels	Dilation of blood vessels
Interleukins (cytokines)	Macrophages; lymphocytes	Regulation of immune system
Prostaglandins	Many tissues	Wide variety (see text)
TNFα (tumor necrosis factor alpha)	Macrophages; adipocytes	Wide variety

within the organ in which they are produced. These locally acting chemicals that regulate neighboring cells within an organ are known generally as **paracrine regulators,** particularly if they are produced by one cell type and regulate neighboring cells of a different type. Regulators that act on the same cells and cell type that produce them may be called **autocrine regulators.** Some examples of these regulators are provided in table 11.9.

Examples of Paracrine and Autocrine Regulation

Many paracrine regulatory molecules are also known as **cytokines,** particularly if they regulate different cells of the immune system, and as **growth factors** if they promote growth and cell division in any organ. This distinction is somewhat blurred, however, because some cytokines may also function as growth factors. Cytokines produced by lymphocytes (the type of white blood cell involved in specific immunity—see chapter 15) are also known as *lymphokines,* and the specific molecules involved are called *interleukins.* The terminology can be confusing because new regulatory molecules, and new functions for previously named regulatory molecules, are being discovered at a rapid pace. As described in chapter 15, cytokines secreted by macrophages (phagocytic cells found in connective tissues) and lymphocytes stimulate proliferation of specific cells involved in the immune response. Neurons and neuroglia release **neurotrophins,** such as *nerve growth factor* and others (chapter 7, section 7.1), which serve as autocrine regulators in the nervous system.

The walls of blood vessels have different tissue layers (chapter 13, section 13.6), and the endothelial layer produces several paracrine regulators of the smooth muscle layer. For example, *nitric oxide*—which functions as a neurotransmitter when it is released by axon terminals (chapters 7 and 8)—is also produced by the endothelium of blood vessels. In this context it functions as a paracrine regulator when it diffuses to the smooth muscle layer of the vessels and promotes relaxation, thereby dilating the vessels. In this action, nitric oxide functions as the paracrine regulator previously known as *endothelium-derived relaxation factor.* Neural and paracrine regulation interact in this case, because autonomic axons that release acetylcholine in blood vessels cause dilation by stimulating the synthesis of nitric oxide in those vessels (see chapter 20, fig. 20.21).

The endothelium of blood vessels also produces other paracrine regulators. These include the *endothelins* (specifically *endothelin-1* in humans), which directly promote vasoconstriction, and *bradykinin,* which promotes vasodilation. These regulatory molecules are very important in the control of blood flow and blood pressure. They are also involved in the development of atherosclerosis, the leading cause of heart disease and stroke (chapter 13, section 13.7). In addition, endothelin-1 is produced by the epithelium of the airways and may be important in the embryological development and function of the respiratory system.

All paracrine/autocrine regulators control gene expression in their target cells to some degree. This is very clearly the case with the various growth factors. These include *platelet-derived growth factor, epidermal growth factor,* and the *insulin-like growth factors* that stimulate cell division and proliferation of their target cells. Regulators in the last group interact with the endocrine system in a number of ways, as will be described in chapter 19.

Prostaglandins

The most diverse group of paracrine/autocrine regulators is the **prostaglandins.** These twenty-carbon-long fatty acids contain a five-membered carbon ring. Prostaglandins are members of a family called the **eicosanoids** (from the Greek *eicosa* = twenty), which are molecules derived from the precursor *arachidonic acid.* Upon stimulation by hormones or other agents, arachidonic acid is released from phospholipids in the plasma membrane and may then enter one of two possible metabolic pathways. In one case, arachidonic acid is converted by the enzyme *cyclooxygenase* into a prostaglandin, which can then be changed by other enzymes into other prostaglandins. In the other case, arachidonic acid is converted by the enzyme *lipoxygenase* into **leukotrienes,** which are eicosanoids that are closely related to the prostaglandins (fig. 11.34). The leukotrienes are largely responsible for the symptoms of asthma.

Prostaglandins are produced in almost every organ and have been implicated in a wide variety of regulatory functions. The study of prostaglandins can be confusing because of the diversity of their actions, and because different prostaglandins may exert antagonistic effects in some tissues. For example, the smooth muscle of blood vessels relaxes (producing vasodilation) in response to prostaglandin E_2 (abbreviated PGE_2); these effects promote reddening and heat during an inflammation reaction. In the smooth muscles of the bronchioles (airways of the lungs), however, $PGF_{2\alpha}$ stimulates contraction, contributing to the symptoms of asthma.

The antagonistic effects of prostaglandins on blood clotting make good physiological sense. Blood platelets, which are required for blood clotting, produce *thromboxane* A_2. This prostaglandin promotes clotting by stimulating platelet aggregation and vasoconstriction. The endothelial cells of blood vessels, by contrast, produce a different prostaglandin, known as PGI_2 or *prostacyclin,* whose effects are the opposite—it inhibits platelet aggregation and causes vasodilation. These antagonistic effects ensure that, while clotting is promoted, the clots will not normally form on the walls of intact blood vessels (see chapter 13, fig. 13.7).

Examples of Prostaglandin Actions

Some of the regulatory functions proposed for prostaglandins in different systems of the body are:

1. **Immune system.** Prostaglandins promote many aspects of the inflammatory process, including the development of pain and fever. Drugs that inhibit prostaglandin synthesis help to alleviate these symptoms.
2. **Reproductive system.** Prostaglandins may play a role in ovulation and corpus luteum function in the ovaries and in contraction of the uterus. Excessive production of PGE_2 and PGI_2 may be involved in premature labor, endometriosis, dysmenorrhea (painful menstrual cramps), and other gynecological disorders.
3. **Digestive system.** The stomach and intestines produce prostaglandins, which are believed to inhibit gastric secretions and influence intestinal motility and fluid absorption.

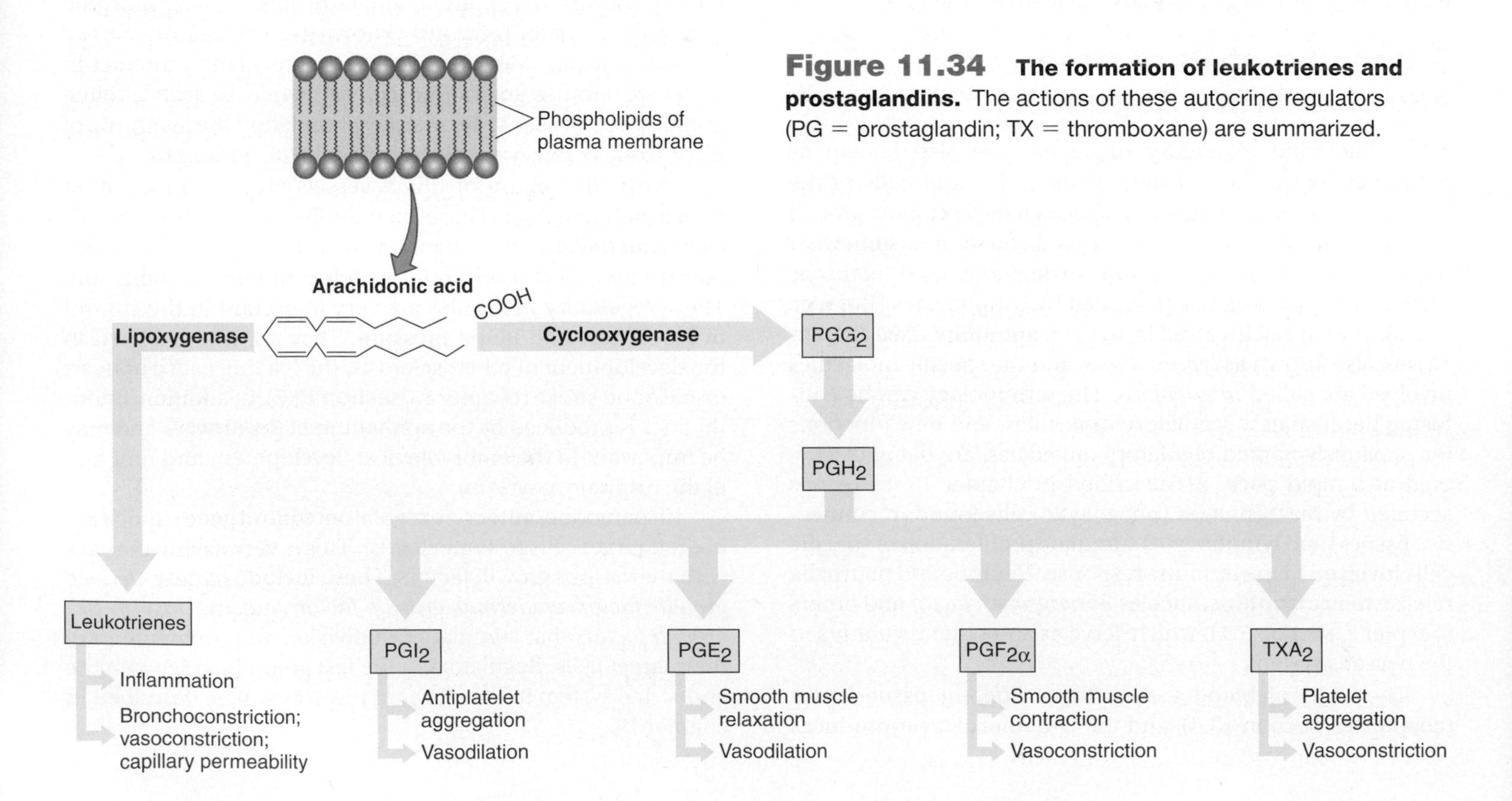

Figure 11.34 The formation of leukotrienes and prostaglandins. The actions of these autocrine regulators (PG = prostaglandin; TX = thromboxane) are summarized.

Because prostaglandins inhibit gastric secretion, drugs that suppress prostaglandin production may make a person more susceptible to peptic ulcers.

4. **Respiratory system.** Some prostaglandins cause constriction whereas others cause dilation of blood vessels in the lungs and of bronchiolar smooth muscle. The leukotrienes are potent bronchoconstrictors, and these compounds, together with $PGF_2\alpha$, may cause respiratory distress and contribute to bronchoconstriction in asthma.
5. **Circulatory system.** Some prostaglandins are vasoconstrictors and others are vasodilators. Thromboxane A_2, a vasoconstrictor, and prostacyclin, a vasodilator, play a role in blood clotting, as previously described. In a fetus, PGE_2 is believed to promote dilation of the *ductus arteriosus*—a short vessel that connects the pulmonary artery with the aorta. After birth, the ductus arteriosus normally closes as a result of a rise in blood oxygen when the baby breathes. If the ductus remains patent (open), however, it can be closed by the administration of drugs that inhibit prostaglandin synthesis.
6. **Urinary system.** Prostaglandins are produced in the renal medulla and cause vasodilation, resulting in increased renal blood flow and increased excretion of water and electrolytes in the urine.

Inhibitors of Prostaglandin Synthesis

Aspirin is the most widely used member of a class of drugs known as **nonsteroidal anti-inflammatory drugs (NSAIDs).** Other members of this class are *indomethacin* and *ibuprofen.* These drugs produce their effects because they specifically inhibit the cyclooxygenase enzyme that is needed for prostaglandin synthesis. Through this action, the drugs inhibit inflammation but produce some unwanted side effects, including gastric bleeding, possible kidney problems, and prolonged clotting time.

It is now known that there are two major isoenzyme forms of cyclooxygenase. The type I isoform (*COX-1*) is produced constitutively (that is, in a constant fashion) by cells of the stomach and kidneys and by blood platelets, which are cell fragments involved in blood clotting (chapter 13, section 13.2). The type II isoform of the enzyme (*COX-2*) is induced in a number of cells in response to cytokines involved in inflammation, and the prostaglandins produced by this isoenzyme promote the inflammatory condition.

When aspirin and indomethacin inhibit the COX-1 isoenzyme, they reduce the synthesis of prostacyclin (PGI_2) and PGE_2 in the gastric mucosa. This is believed to result in the stomach irritation caused by these NSAIDs. Indeed, inhibition of the COX-1 isoenzyme may cause serious gastrointestinal and renal toxicity in long-term use. This has spurred research into the development of next-generation NSAIDs that more selectively inhibit the COX-2 isoenzyme. These newer COX-2-selective drugs, including *celecoxib* and *rofecoxib* (for example, Celebrex and Vioxx), thus inhibit inflammation while producing fewer negative side effects in the gastric mucosa.

However, studies indicated that the COX-2 selective inhibitors produced a significant increase in the risk of myocardial infarction (heart attack) and thrombotic stroke after a year or more of treatment. This has been explained by the observation that the selective COX-2 inhibitors reduce the ability of the vascular endothelium to produce prostaglandin I_2 (which inhibits clotting and promotes vasodilation) while not inhibiting the ability of blood platelets to produce thromboxane A_2 (which promotes clotting and vasoconstriction). At the time of this writing, most of the selective COX-2 inhibitors have been withdrawn from the market. The benefits of the selective COX-2 inhibitors for gastrointestinal protection may outweigh the increased risk of cardiovascular disease in some patients, so this is a complex issue that physicians and patients may best weigh on an individual basis.

Also, inhibition of the specific COX-1 isoenzyme by aspirin can provide an important benefit. This is the isoenzyme present in blood platelets that catalyzes the production of thromboxane A_2. As mentioned previously, thromboxane A_2 is the prostaglandin produced by blood platelets that promotes platelet aggregation in the process of blood clotting (chapter 13; see fig. 13.7). While inhibition of platelet aggregation can be detrimental in certain situations, such aspirin-induced inhibition has been shown to reduce the risk of heart attacks and strokes. It should be noted that this protective effect is produced by daily doses of "baby aspirin" (81 mg), which are significantly lower than the doses used to reduce inflammation.

Acetaminophen (e.g., Tylenol) does not greatly inhibit either COX-1 or COX-2, and is not an effective anti-inflammatory drug. Yet it does reduce fever and relieve pain. It has recently been shown to work by inhibiting a newly discovered isoenzymatic form, designated *COX-3,* which is found in large amounts in the brain.

Drugs that are anti-leukotrienes have recently become available. Some (such as Zyflo) work by inhibiting the enzyme 5-lipoxygenase that forms the leukotrienes; others (such as Singulair) block the leukotriene receptors. These drugs are used for the treatment of asthma.

✔ | CHECKPOINT

15a. Explain the nature of autocrine regulation. How does it differ from regulation by hormones and neurotransmitters?

15b. List some of the paracrine regulators produced by blood vessels and describe their actions. Also, identify specific growth factors and describe their actions.

16a. Describe the chemical nature of prostaglandins. List some of the different forms of prostaglandins and describe their actions.

16b. Explain the significance of the isoenzymatic forms of cyclooxygenase in the action of nonsteroidal anti-inflammatory drugs.

Interactions

   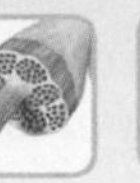   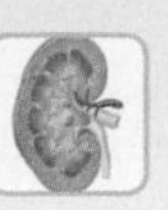  

HPer Links of the Endocrine System with Other Body Systems

Integumentary System

- The skin helps protect the body from pathogens (p. 494)
- The skin produces vitamin D_3, which acts as a prehormone (p. 693)

Skeletal System

- Bones support and protect the pituitary gland (p. 331)
- Bones store calcium, which is needed for the action of many hormones (p. 690)
- Anabolic hormones, including growth hormone, stimulate bone development (p. 689)
- Parathyroid hormone and calcitonin regulate calcium deposition and resorption in bones (p. 691)
- Sex hormones help maintain bone mass in adults (p. 692)

Muscular System

- Anabolic hormones promote muscle growth (p. 322)
- Insulin stimulates the uptake of blood glucose into muscles (p. 347)
- The catabolism of muscle glycogen and proteins is promoted by several hormones (p. 675)

Nervous System

- The hypothalamus secretes hormones that control the anterior pituitary (p. 334)
- The hypothalamus produces the hormones released by the posterior pituitary (p. 333)
- Sympathetic nerves stimulate the secretions of the adrenal medulla (p. 247)
- Parasympathetic nerves stimulate the secretions of the pancreatic islets (p. 678)
- Neurons stimulate the secretion of melatonin from the pineal gland, which in turn regulates parts of the brain (p. 348)
- Sex hormones from the gonads regulate the hypothalamus (p. 708)

Circulatory System

- The blood transports oxygen, nutrients, and regulatory molecules to endocrine glands and removes wastes (p. 405)
- The blood transports hormones from endocrine glands to target cells (p. 405)
- Epinephrine and norepinephrine from the adrenal medulla stimulate the heart (p. 451)

Immune System

- The immune system protects against infections that could damage endocrine glands (p. 494)
- Autoimmune destruction of the pancreatic islets causes type I diabetes mellitus (p. 522)
- Adrenal corticosteroids have a suppressive effect on the immune system (p. 514)

Respiratory System

- The lungs provide oxygen for transport by the blood and eliminate carbon dioxide (p. 533)
- Thyroxine and epinephrine stimulate the rate of cell respiration in the body (p. 675)
- Epinephrine promotes bronchodilation, reducing airway resistance (p. 545)

Urinary System

- The kidneys eliminate metabolic wastes produced by body organs, including endocrine glands (p. 582)
- The kidneys release renin, which participates in the renin-angiotensin-aldosterone system (p. 606)
- The kidneys secrete erythropoietin, which serves as a hormone that regulates red blood cell production (p. 560)
- Antidiuretic hormone, aldosterone, and atrial natriuretic hormone regulate kidney functions (p. 595)

Digestive System

- The GI tract provides nutrients to the body organs, including those of the endocrine system (p. 621)
- Hormones of the stomach and small intestine help to coordinate the activities of different regions of the GI tract (p. 644)
- Hormones from adipose tissue contribute to the sensation of hunger (p. 673)

Reproductive System

- Gonadal hormones help regulate the secretions of the anterior pituitary (p. 708)
- Pituitary hormones regulate the ovarian cycle (p. 724)
- Testicular androgens regulate the male accessory sex organs (p. 717)
- Ovarian hormones regulate the uterus during the menstrual cycle (p. 730)
- Oxytocin plays an essential role in labor and delivery (p. 743)
- The placenta secretes several hormones that influence the course of pregnancy (p. 742)
- Several hormones are needed for lactation in a nursing mother (p. 746)

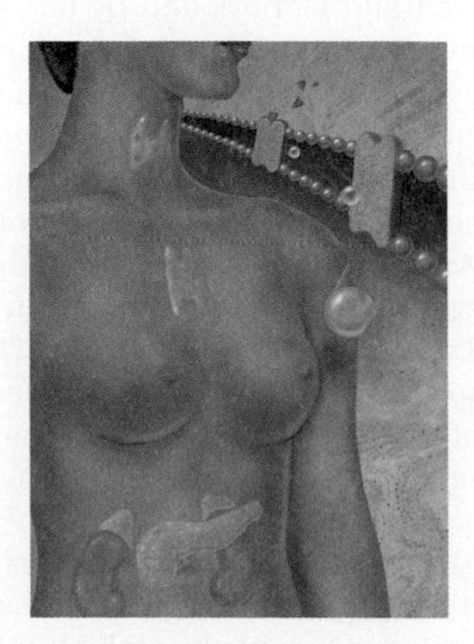

Clinical Investigation SUMMARY

Rosemary's hyperglycemia cannot be attributed to diabetes mellitus because her insulin activity is normal, as indicated by the glucose tolerance test. The symptoms might be due to hyperthyroidism, but this possibility is ruled out by the blood tests. The high blood levels of corticosteroids are not the result of ingestion of these compounds as drugs. However, the patient might have Cushing's syndrome, in which case an adrenal tumor could be responsible for the hypersecretion of corticosteroids and, as a result of negative feedback inhibition, a decrease in blood ACTH levels. This possibility is supported by Rosemary's low ACTH levels. Excessive corticosteroid levels cause the mobilization of glucose from the liver, thus increasing the blood glucose to hyperglycemic levels.

See the additional chapter 11 Clinical Investigations on *Addison's Disease* and *Hypothyroidism* in the Connect site for this text at www.mhhe.com/Fox13.

SUMMARY

11.1 Endocrine Glands and Hormones 317

A. Hormones are chemicals that are secreted into the blood by endocrine glands.
   1. The chemical classes of hormones include amines, polypeptides, glycoproteins, and steroids.
   2. Nonpolar hormones, which can pass through the plasma membrane of their target cells, are called lipophilic hormones.

B. Precursors of active hormones may be classified as either prohormones or prehormones.
   1. Prohormones are relatively inactive precursor molecules made in the endocrine cells.
   2. Prehormones are the normal secretions of an endocrine gland that must be converted to other derivatives by target cells in order to be active.

C. Hormones can interact in permissive, synergistic, or antagonistic ways.

D. The effects of a hormone in the body depend on its concentration.
   1. Abnormally high amounts of a hormone can result in atypical effects.
   2. Target tissues can become desensitized by high hormone concentrations.

11.2 Mechanisms of Hormone Action 323

A. The lipophilic hormones (steroids and thyroid hormones) bind to nuclear receptor proteins, which function as ligand-dependent transcription factors.
   1. Some steroid hormones bind to cytoplasmic receptors, which then move into the nucleus. Other steroids and thyroxine bind to receptors already in the nucleus.
   2. Each receptor binds to both the hormone and to a region of DNA called a hormone-response element.
   3. Two units of the nuclear receptor are needed to bind to the hormone-response element to activate a gene; as a result, the gene is transcribed (makes mRNA).

B. The polar hormones bind to receptors located on the outer surface of the plasma membrane. This activates enzymes that enlist second-messenger molecules.
   1. Many hormones activate adenylate cyclase when they bind to their receptors. This enzyme produces cyclic AMP (cAMP), which activates protein kinase enzymes within the cell cytoplasm.
   2. Other hormones may activate phospholipase C when they bind to their receptors. This leads to the release of inositol triphosphate (IP_3), which stimulates the endoplasmic reticulum to release Ca^{2+} into the cytoplasm, activating calmodulin.
   3. The membrane receptors for insulin and various growth factors are tyrosine kinase enzymes that are activated by binding to the hormone. Once activated, the receptor kinase phosphorylates signaling molecules in the cytoplasm that can have many effects.

11.3 Pituitary Gland 331

A. The pituitary gland secretes eight hormones.
   1. The anterior pituitary secretes growth hormone, thyroid-stimulating hormone, adrenocorticotropic hormone, follicle-stimulating hormone, luteinizing hormone, and prolactin.

2. The posterior pituitary releases antidiuretic hormone (also called vasopressin) and oxytocin, both of which are produced in the hypothalamus and transported to the posterior pituitary by the hypothalamo-hypophyseal tract.

B. The release of posterior pituitary hormones is controlled by neuroendocrine reflexes.

C. Secretions of the anterior pituitary are controlled by hypothalamic hormones that stimulate or inhibit these secretions.
1. Hypothalamic hormones include TRH, CRH, GnRH, PIH, somatostatin, and a growth hormone–releasing hormone (GHRH).
2. These hormones are carried to the anterior pituitary by the hypothalamo-hypophyseal portal system.

D. Secretions of the anterior pituitary are also regulated by the feedback (usually negative feedback) exerted by target gland hormones.

E. Higher brain centers, acting through the hypothalamus, can influence pituitary secretion.

11.4 Adrenal Glands 338

A. The adrenal cortex secretes mineralocorticoids (mainly aldosterone), glucocorticoids (mainly cortisol), and sex steroids (primarily weak androgens).
1. The glucocorticoids help to regulate energy balance. They also can inhibit inflammation and suppress immune function.
2. The pituitary-adrenal axis is stimulated by stress as part of the general adaptation syndrome.

B. The adrenal medulla secretes epinephrine and lesser amounts of norepinephrine. These hormones complement the action of the sympathetic nervous system.

11.5 Thyroid and Parathyroid Glands 342

A. The thyroid follicles secrete tetraiodothyronine (T_4, or thyroxine) and lesser amounts of triiodothyronine (T_3).
1. These hormones are formed within the colloid of the thyroid follicles.
2. The parafollicular cells of the thyroid secrete the hormone calcitonin, which may act to lower blood calcium levels.

B. The parathyroids are small structures embedded in the thyroid gland. They secrete parathyroid hormone (PTH), which promotes a rise in blood calcium levels.

11.6 Pancreas and Other Endocrine Glands 346

A. Beta cells in the islets secrete insulin, and alpha cells secrete glucagon.
1. Insulin lowers blood glucose and stimulates the production of glycogen, fat, and protein.
2. Glucagon raises blood glucose by stimulating the breakdown of liver glycogen. It also promotes lipolysis and the formation of ketone bodies.
3. The secretion of insulin is stimulated by a rise in blood glucose following meals. The secretion of glucagon is stimulated by a fall in blood glucose during periods of fasting.

B. The pineal gland, located on the roof of the third ventricle of the brain, secretes melatonin.
1. Melatonin secretion is regulated by the suprachiasmatic nucleus of the hypothalamus, which is the major center for the control of circadian rhythms.
2. Melatonin secretion is highest at night, and this hormone has a sleep-promoting effect. In many species, it also has an antigonadotropic effect and may play a role in timing the onset of puberty in humans, although this is as yet unproven.

C. The gastrointestinal tract secretes a number of hormones that help regulate digestive functions.

D. The gonads secrete sex steroid hormones.
1. Leydig cells in the interstitial tissue of the testes secrete testosterone and other androgens.
2. Granulosa cells of the ovarian follicles secrete estrogen.
3. The corpus luteum of the ovaries secretes progesterone, as well as estrogen.

E. The placenta secretes estrogen, progesterone, and a variety of polypeptide and protein hormones that have actions similar to some anterior pituitary hormones.

11.7 Paracrine and Autocrine Regulation 350

A. Autocrine regulators are produced and act within the same tissue of an organ, whereas paracrine regulators are produced within one tissue and regulate a different tissue of the same organ. Both types are local regulators—they do not travel in the blood.

B. Prostaglandins are special twenty-carbon-long fatty acids produced by many different organs. They usually have regulatory functions within the organ in which they are produced.

REVIEW ACTIVITIES

Test Your Knowledge

1. Which of these statements about hypothalamic-releasing hormones is *true?*
   a. They are secreted into capillaries in the median eminence.
   b. They are transported by portal veins to the anterior pituitary.
   c. They stimulate the secretion of specific hormones from the anterior pituitary.
   d. All of these are true.
2. The hormone primarily responsible for setting the basal metabolic rate and for promoting the maturation of the brain is
   a. cortisol.
   b. ACTH.
   c. TSH.
   d. thyroxine.

3. Which of these statements about the adrenal cortex is *true*?
   a. It is not innervated by nerve fibers.
   b. It secretes some androgens.
   c. The zona glomerulosa secretes aldosterone.
   d. The zona fasciculata is stimulated by ACTH.
   e. All of these are true.
4. Which of these statements about the hormone insulin is *true*?
   a. It is secreted by alpha cells in the islets of Langerhans.
   b. It is secreted in response to a rise in blood glucose.
   c. It stimulates the production of glycogen and fat.
   d. Both *a* and *b* are true.
   e. Both *b* and *c* are true.

Match the hormone with the primary agent that stimulates its secretion.

5. Epinephrine — a. TSH
6. Thyroxine — b. ACTH
7. Corticosteroids — c. Growth hormone
8. ACTH — d. Sympathetic nerves
   e. CRH

9. Steroid hormones are secreted by
   a. the adrenal cortex.
   b. the gonads.
   c. the thyroid.
   d. both *a* and *b*.
   e. both *b* and *c*.
10. The secretion of which of these hormones would be *increased* in a person with endemic goiter?
   a. TSH
   b. Thyroxine
   c. Triiodothyronine
   d. All of these
11. Which of these hormones uses cAMP as a second messenger?
   a. Testosterone
   b. Cortisol
   c. Insulin
   d. Epinephrine
12. Which of these terms best describes the interactions of insulin and glucagon?
   a. Synergistic
   b. Permissive
   c. Antagonistic
   d. Cooperative
13. Which of these correctly describes the role of inositol triphosphate in hormone action?
   a. It activates adenylate cyclase.
   b. It stimulates the release of Ca^{2+} from the endoplasmic reticulum.
   c. It activates protein kinase.
   d. All of these.
14. Which of these hormones may have a primary role in many circadian rhythms?
   a. Estradiol
   b. Insulin
   c. Adrenocorticotropic hormone
   d. Melatonin
15. Human chorionic gonadotropin (hCG) is secreted by
   a. the anterior pituitary.
   b. the posterior pituitary.
   c. the placenta.
   d. the thymus.
   e. the pineal gland.
16. What do insulin-like growth factors, neurotrophins, nitric oxide, and lymphokines have in common?
   a. They are hormones.
   b. They are autocrine or paracrine regulators.
   c. They are neurotransmitters.
   d. They all use cAMP as a second messenger.
   e. They all use Ca^{2+} as a second messenger.

Test Your Understanding

17. Explain how regulation of the neurohypophysis and of the adrenal medulla are related to the embryonic origins of these organs.
18. Explain the mechanism of action of steroid hormones and thyroxine.
19. Explain why polar hormones cannot regulate their target cells without using second messengers. Also explain in a step-by-step manner how cyclic AMP is used as a second messenger in hormone action.
20. Describe the sequence of events by which a hormone can cause an increase in the Ca^{2+} concentration within a target cell. How can this increased Ca^{2+} affect the metabolism of the target cell?
21. Explain the significance of the term *trophic* with respect to the actions of anterior pituitary hormones.
22. Suppose a drug blocks the conversion of T_4 to T_3. Explain what the effects of this drug would be on (a) TSH secretion, (b) thyroxine secretion, and (c) the size of the thyroid gland.
23. Explain why the anterior pituitary is sometimes referred to as the "master gland," and provide two reasons why this description is misleading.
24. Describe the role of the pituitary-adrenal axis in the response to stress. What effects on the body does this produce?
25. Describe how thyroid hormone secretion is regulated. Explain how this system is affected by (a) iodine deficiency and (b) ingestion of thyroid hormone pills.
26. Suppose a person's immune system made antibodies against insulin receptor proteins. What effect might this condition have on carbohydrate and fat metabolism?
27. Explain how light affects the function of the pineal gland. What is the relationship between pineal gland function and circadian rhythms?
28. Distinguish between endocrine and autocrine/paracrine regulation. List some of these autocrine/paracrine regulators and describe their functions.

Test Your Analytical Ability

29. Brenda, your roommate, has been having an awful time lately. She can't even muster enough energy to go out on a date. She's been putting on weight, she's always cold, and every time she pops in the workout video she complains of weakness. When she finally goes to the doctor, he finds her to have a slow pulse and a low blood pressure. Laboratory tests reveal that her T_4 is low and her TSH is high. What is the matter with Brenda? Why are her symptoms typical of this disorder, and what type of treatment will the doctor most likely prescribe?
30. Your friend Bud has the talent to be a star basketball center—if only he weren't 5 foot 8. Being well-intentioned but ignorant, you start injecting him with growth hormone as he sleeps each night. You think this is a clever strategy, but after a time you notice that he hasn't grown an inch. Instead, his jaw and forehead seem to have gotten disproportionately large and his hands and feet are swollen. Explain why the growth hormone didn't make Bud grow taller and why it had the effect it did. What disease state do these changes mimic?
31. You see your friend Joe for the first time in over a year. When you last saw him, he had been trying to bulk up by working out daily at the gym, but he was getting discouraged because his progress seemed so slow. Now, however, he's very muscular. In a frank discussion, he admits that he's been getting into trouble because he's become very aggressive. He also tells you, in strict confidence, that his testes have gotten smaller and that he's been developing breasts! What might Joe be doing to cause these changes? Explain how these changes came about.
32. Distinguish between the steroid and nonsteroid group of nuclear hormone receptors. Explain the central role of vitamin A in the actions of the nonsteroid group of receptors.
33. Suppose, in an experiment, that you incubate isolated rat testes with hCG. What would be the effect, if any, of the hCG on the testes? Explain your answer. If there was an effect, discuss its potential significance in research and clinical settings.
34. Distinguish between the genomic and nongenomic actions of steroid hormones. Which mechanism of action would be inhibited by a drug that interfered with protein synthesis? Explain.
35. People who have suffered severe stress may have difficulties remembering the stressful event. Parts of the brain, including the hippocampus, are rich in cortisol receptor proteins. Explain the relationship between these observations.
36. Diabetics who require insulin injections can experience what is called the "dawn phenomenon"—when they first wake up in the morning, they may require higher doses of insulin to control their blood glucose. People usually awaken naturally from REM sleep, which is a sleep stage that is stressful. Explain the endocrine events that are responsible for the dawn phenomenon.

Test Your Quantitative Ability

In the graph below, insulin sensitivity is plotted on the x-axis and insulin release on the y-axis, using arbitrary units where 1.0 represents average normal values for each measurement. Individuals with points anywhere in the curved graph line within the green area are considered normal. IGT = impaired glucose tolerance, indicated by the yellow area of the graph; T2DM = type 2 diabetes mellitus, indicated by the red area of the graph.

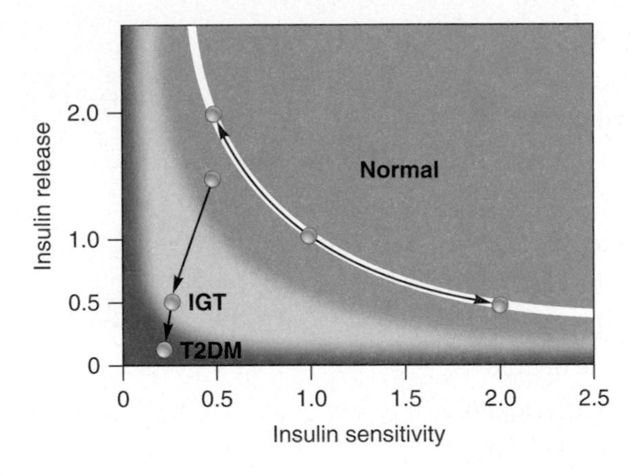

37. If a person has a 50% reduction in insulin sensitivity, what happens to insulin release for the person to remain in the normal curve?
38. If a person has a 50% increase in insulin sensitivity, what happens to insulin release for the person to remain in the normal curve?
39. A person has impaired glucose tolerance despite having an insulin secretion 50% greater than normal. What range of values must be true of the person's insulin sensitivity?
40. A person has impaired glucose tolerance despite having an insulin sensitivity 50% greater than average. What range of values must be true of the person's insulin release?

CHAPTER OUTLINE

CHAPTER

12

Muscle

Mechanisms of Contraction and Neural Control

REFRESH YOUR MEMORY

Before you begin this chapter, you may want to review these concepts from previous chapters:

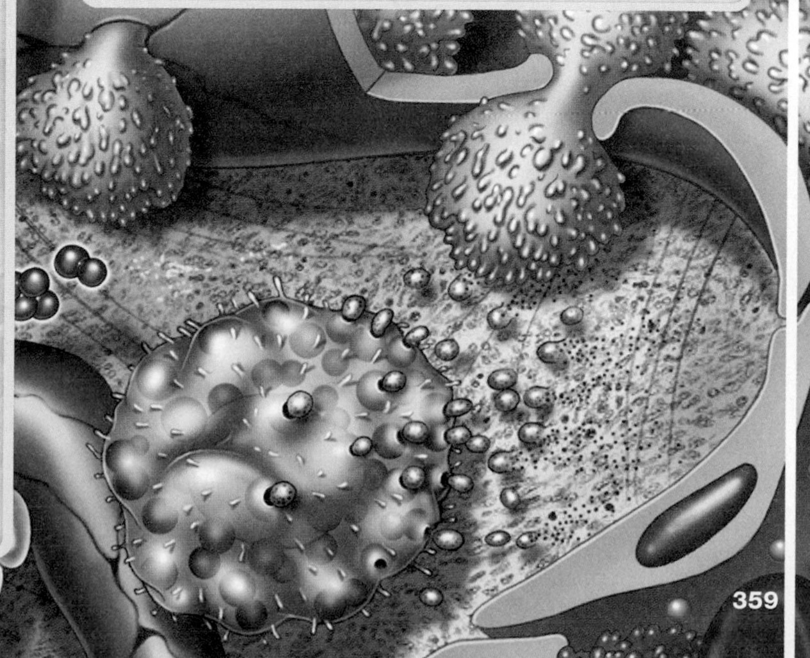

Clinical Investigation

Maria, an energetic 40-year-old who has long exercised regularly, complains to a physician that she has been experiencing unusual muscle fatigue and pain when she performs leg extensions on a machine at the gym. She states that she is taking medication for high blood pressure, and laboratory tests reveal that her plasma Ca^{2+} concentration is somewhat above the normal range.

Some of the new terms and concepts you will encounter include:

- Maximal oxygen uptake
- Ca^{2+} channel blocker
- Isotonic, concentric, and eccentric contractions
- Creatine phosphokinase and creatine phosphate

12.1 SKELETAL MUSCLES

Skeletal muscles are composed of individual muscle fibers that contract when stimulated by a somatic motor neuron. Each motor neuron branches to innervate a number of muscle fibers. Activation of varying numbers of motor neurons results in gradations in the strength of contraction of the whole muscle.

LEARNING OUTCOMES

After studying this section, you should be able to:

1. **Describe the different levels of muscle structure, and the actions of skeletal muscles.**
2. **Describe motor units, and explain the significance of recruitment of motor units.**

Skeletal muscles are usually attached to bone on each end by tough connective tissue tendons. When a muscle contracts, it places tension on its tendons and attached bones. The muscle tension causes movement of the bones at a joint, where one of the attached bones generally moves more than the other. The more movable bony attachment of the muscle, known as its *insertion,* is pulled toward its less movable attachment known as its *origin.* A variety of skeletal movements are possible, depending on the type of joint involved and the attachments of the muscles (table 12.1). When *flexor muscles* contract, for example, they decrease the angle of a joint. Contraction of *extensor muscles* increases the angle of their attached bones at the joint. The prime mover of any skeletal movement is called the **agonist muscle;** in flexion, for example, the flexor is the agonist muscle. Flexors and extensors that act on the same joint to produce opposite actions are **antagonistic muscles.**

Table 12.1 | Skeletal Muscle Actions

Category	Action
Extensor	Increases the angle at a joint
Flexor	Decreases the angle at a joint
Abductor	Moves limb away from the midline of the body
Adductor	Moves limb toward the midline of the body
Levator	Moves insertion upward
Depressor	Moves insertion downward
Rotator	Rotates a bone along its axis
Sphincter	Constricts an opening

Structure of Skeletal Muscles

The fibrous connective tissue proteins within the tendons extend around the muscle in an irregular arrangement, forming a sheath known as the *epimysium* (*epi* = above; *my* = muscle). Connective tissue from this outer sheath extends into the body of the muscle, subdividing it into columns, or *fascicles* (these are the "strings" in stringy meat). Each of these fascicles is

CLINICAL APPLICATION

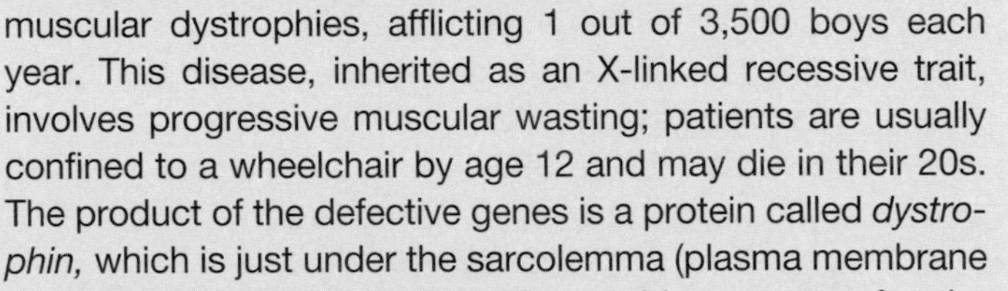

Duchenne's muscular dystrophy is the most severe of the muscular dystrophies, afflicting 1 out of 3,500 boys each year. This disease, inherited as an X-linked recessive trait, involves progressive muscular wasting; patients are usually confined to a wheelchair by age 12 and may die in their 20s. The product of the defective genes is a protein called *dystrophin,* which is just under the sarcolemma (plasma membrane of the muscle fiber). Dystrophin provides support for the muscle fiber by bridging the cytoskeleton and myofibrils in the fiber (fig. 12.1) with the extracellular matrix. Mutations that affect structures in the dystrophin complex make the sarcolemma more easily damaged, beyond the fiber's repair ability and the regenerative ability of muscle stem cells (satellite cells, discussed in section 12.4) to compensate. Defective dystrophin thereby results in muscle fiber necrosis and replacement by fibrous connective and fatty tissue.

Using this information, scientists have recently developed laboratory tests that can detect this disease in fetal cells obtained by amniocentesis. This research has been aided by the development of a strain of mice that exhibit an equivalent form of the disease. When the "good genes" for dystrophin are inserted into mouse embryos of this strain, the mice do not develop the disease. Insertion of the gene into large numbers of mature muscle cells, however, is more difficult, and so far has met with only limited success. Other potential therapies, including those involving stem cells, are also being investigated.

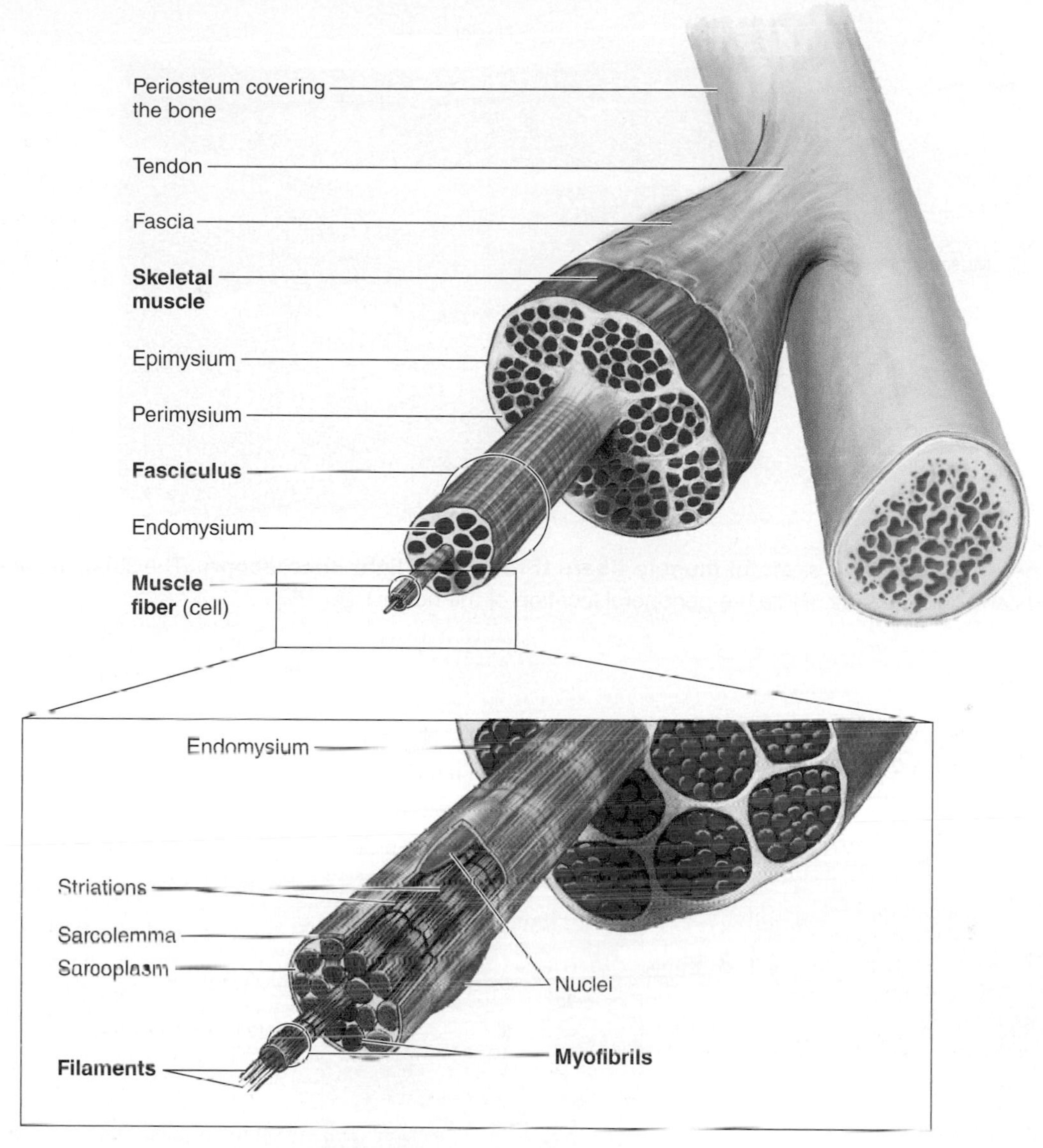

Figure 12.1 The structure of a skeletal muscle. The relationship between muscle fibers and the connective tissues of the tendon, epimysium, perimysium, and endomysium is depicted in the upper figure. Below is a close-up of a single muscle fiber. AP|R

thus surrounded by its own connective tissue sheath, which is known as the *perimysium* (*peri* = around).

Dissection of a muscle fascicle under a microscope reveals that it, in turn, is composed of many **muscle fibers,** or *myofibers.* Each is surrounded by a plasma membrane, or **sarcolemma,** enveloped by a thin connective tissue layer called an *endomysium* (fig. 12.1). Because the connective tissue of the tendons, epimysium, perimysium, and endomysium is continuous, muscle fibers do not normally pull out of the tendons when they contract.

Despite their unusual elongated shape, muscle fibers have the same organelles that are present in other cells: mitochondria, endoplasmic reticulum, glycogen granules, and others. Unlike most other cells in the body, skeletal muscle fibers are multinucleate—they contain multiple nuclei. This is because each muscle fiber is a syncytial structure (chapter 1, section 1.3). That is, each muscle fiber is formed from the union of several embryonic myoblast cells. The most distinctive feature of skeletal muscle fibers, however, is their **striated** appearance when viewed microscopically (fig. 12.2). The striations (stripes) are produced by alternating dark and light bands that appear to span the width of the fiber.

The dark bands are called **A bands,** and the light bands are called **I bands.** At high magnification in an electron microscope, thin dark lines can be seen in the middle of the I bands. These are called **Z lines** (see fig. 12.6). The labels A, I, and Z—derived in the course of early muscle research—are useful for describing the functional architecture of muscle fibers. The letters *A* and *I* stand for *anisotropic* and *isotropic,* respectively, which indicate the behavior of polarized light as it passes through these regions; the letter *Z* comes from the German word *Zwischenscheibe,* which translates to "between disc." These derivations are of historical interest only.

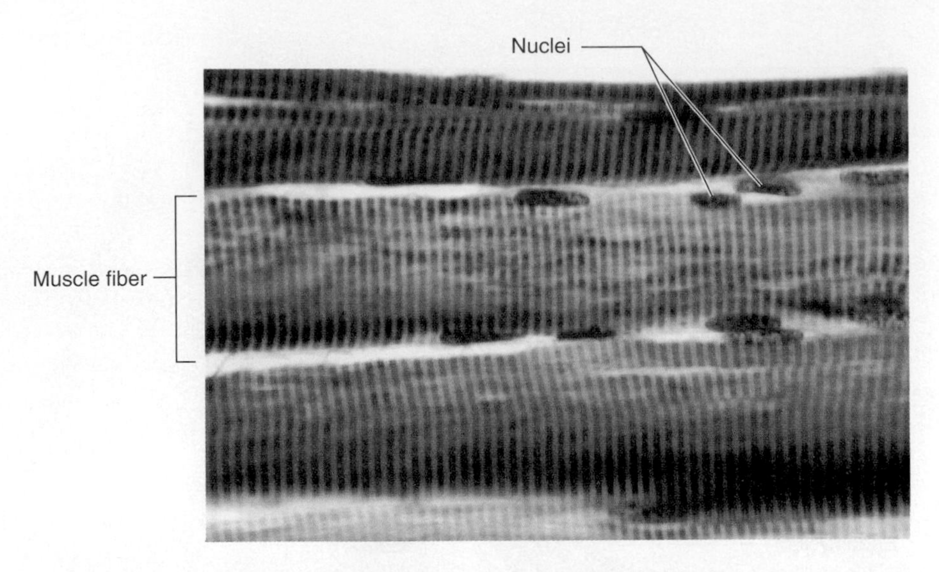

Figure 12.2 The appearance of skeletal muscle fibers through the light microscope. The striations are produced by alternating dark A bands and light I bands. (Note the peripheral location of the nuclei.) APIR

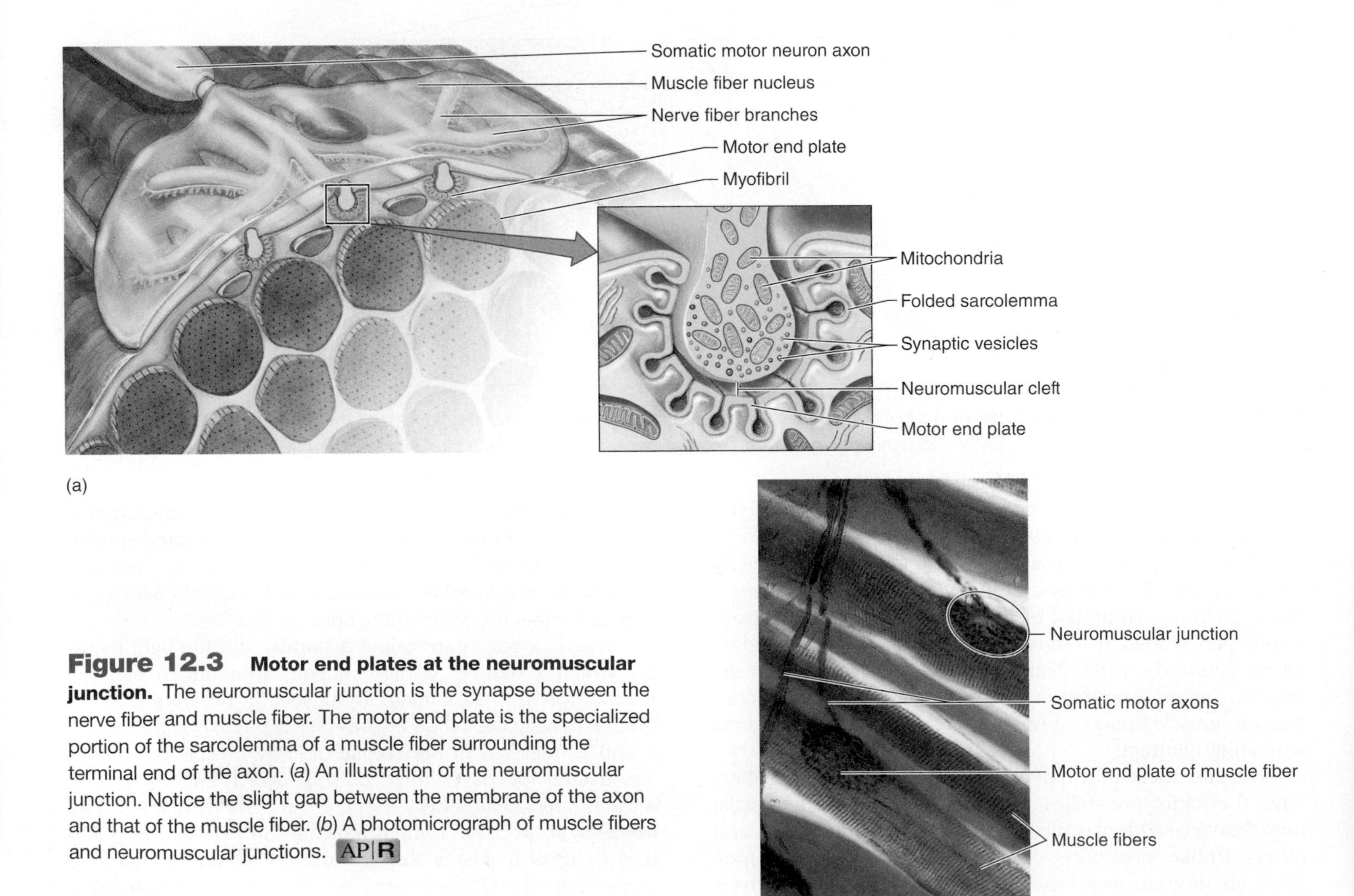

Figure 12.3 Motor end plates at the neuromuscular junction. The neuromuscular junction is the synapse between the nerve fiber and muscle fiber. The motor end plate is the specialized portion of the sarcolemma of a muscle fiber surrounding the terminal end of the axon. (*a*) An illustration of the neuromuscular junction. Notice the slight gap between the membrane of the axon and that of the muscle fiber. (*b*) A photomicrograph of muscle fibers and neuromuscular junctions. APIR

Motor Units

In vivo, each muscle fiber receives a single axon terminal from a somatic motor neuron. The motor neuron stimulates the muscle fiber to contract by liberating acetylcholine at the neuromuscular junction (chapter 7, section 7.4). The specialized region of the sarcolemma of the muscle fiber at the neuromuscular junction is known as a **motor end plate** (fig. 12.3).

The cell body of a somatic motor neuron is located in the ventral horn of the gray matter of the spinal cord and gives rise to a single axon that emerges in the ventral root of a spinal nerve (chapter 8, section 8.6). Each axon, however, can produce a number of collateral branches to innervate an equal number of muscle fibers. Each somatic motor neuron, together with all of the muscle fibers that it innervates, is known as a **motor unit** (fig. 12.4).

Whenever a somatic motor neuron is activated, all of the muscle fibers that it innervates are stimulated to contract. In vivo, *graded contractions* (where contraction strength is varied) of whole muscles are produced by variations in the number of motor units that are activated. In order for these graded contractions to be smooth and sustained, different motor units must be activated by rapid, asynchronous stimulation.

CLINICAL APPLICATION

Acetylcholine (ACh) released by the axon terminals diffuses across the synaptic cleft and binds to ACh receptors in the plasma membrane of the end plate, thereby stimulating the muscle fiber. Prior to its release, the ACh is contained in synaptic vesicles that dock and fuse with the plasma membrane of the axon terminal and undergo exocytosis (see chapter 7, fig. 7.23). The potentially deadly **botulinum toxin,** produced by the bacteria *Clostridium botulinum,* is selectively taken into cholinergic nerve endings and cleaves the proteins needed for the exocytosis of the synaptic vesicles. This blocks nerve stimulation of the muscles, producing a flaccid paralysis. Interestingly, botulinum toxin is now used medically in certain cases to relieve muscle spasms due to excessive nerve stimulation. For example, it is injected into an affected extraocular muscle in order to help correct *strabismus* (deviation of the eye). Intramuscular injections of **Botox** (a brand name for botulinum toxin), causing local muscle paralysis, are also given for the temporary cosmetic treatment of skin wrinkles.

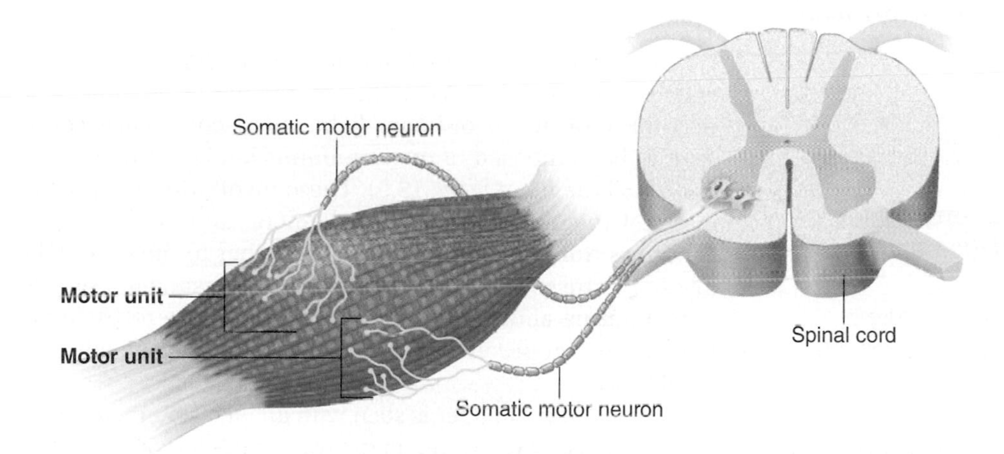

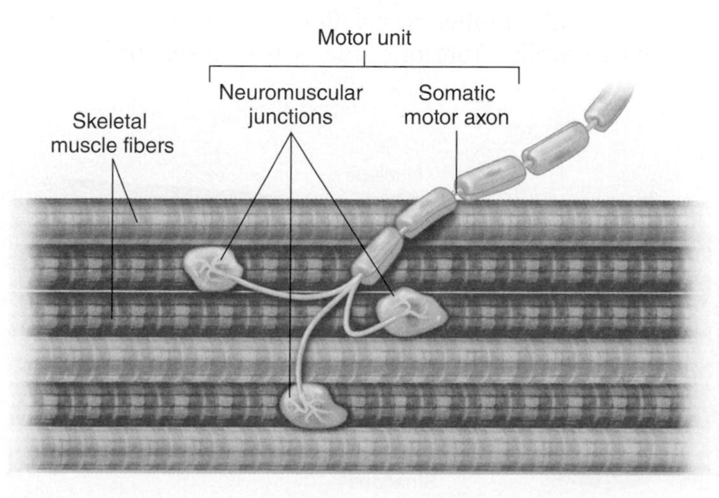

Figure 12.4 Motor units. A motor unit consists of a somatic motor neuron and the muscle fibers it innervates. (*a*) Illustration of a muscle containing two motor units. In reality, a muscle would contain many hundreds of motor units, and each motor unit would contain many more muscle fibers than are shown here. (*b*) A single motor unit consisting of a branched motor axon and the three muscle fibers it innervates (the fibers that are highlighted) is depicted. The other muscle fibers would be part of different motor units and would be innervated by different neurons (not shown).

Fine neural control over the strength of muscle contraction is optimal when there are many small motor units involved. In the extraocular muscles that position the eyes, for example, the *innervation ratio* (motor neuron:muscle fibers) of an average motor unit is one neuron per 23 muscle fibers. This affords a fine degree of control. The innervation ratio of the gastrocnemius, by contrast, averages one neuron per thousand muscle fibers. Stimulation of these motor units results in more powerful contractions at the expense of finer gradations in contraction strength.

All of the motor units controlling the gastrocnemius, however, are not the same size. Innervation ratios vary from 1:100 to 1:2,000. A neuron that innervates fewer muscle fibers has a smaller cell body and is stimulated by lower levels of excitatory input than a larger neuron that innervates a greater number of muscle fibers. The smaller motor units, as a result, are the ones that are used most often. When contractions of greater strength are required, larger and larger motor units are activated in a process known as **recruitment** of motor units.

In summary, two processes occur when you gradually increase the force of a muscle contraction. First, the motor units involved are stimulated asynchronously at greater frequency so that there is summation of contractions. The second process, which can occur at the same time, involves recruitment of additional larger motor units with more muscle fibers per motor neuron to increase the force of contraction.

CHECKPOINT

1a. Describe the actions of muscles when they contract, and define the terms *agonist* and *antagonist* in muscle action.

1b. Describe the different levels of muscle structure, explaining how the muscle and its substructures are packaged in connective tissues.

2a. Define the terms *motor unit* and *innervation ratio* as they relate to muscle function, and draw a simple diagram of a motor unit with a 1:5 innervation ratio.

2b. Using the concept of recruitment, explain how muscle contraction can be graded in its strength.

12.2 MECHANISMS OF CONTRACTION

The A bands within each muscle fiber are composed of thick filaments and the I bands contain thin filaments. Cross bridges that extend from the thick to the thin filaments cause sliding of the filaments, and thus muscle tension and shortening. The activity of the cross bridges is regulated by the availability of Ca^{2+}, which is increased by action potentials produced by the sarcolemma.

LEARNING OUTCOMES

After studying this section, you should be able to:

3. Describe the banding pattern of a myofibril, and how these bands change length during muscle contraction.

4. Explain the cross-bridge cycle and the sliding filament theory of contraction.

5. Explain excitation-contraction coupling in skeletal muscles.

When muscle cells are viewed in the electron microscope, which can produce images at several thousand times the magnification in an ordinary light microscope, each cell is seen to be composed of many subunits known as **myofibrils** (*fibrils* = little fibers) (fig. 12.5). These myofibrils are approximately 1 micrometer (1 μm) in diameter and extend in parallel rows from one end of the muscle fiber to the other. The myofibrils are so densely packed that other organelles, such as mitochondria and intracellular membranes, are restricted to the narrow cytoplasmic spaces that remain between adjacent myofibrils.

When a muscle fiber is seen with an electron microscope, its striations do not extend all the way across its width. Rather, the dark *A bands* and light *I bands* that produce the striations are seen within each myofibril (fig. 12.6). Because the dark and light bands of different myofibrils are stacked in register (aligned vertically) from one side of the muscle fiber to the

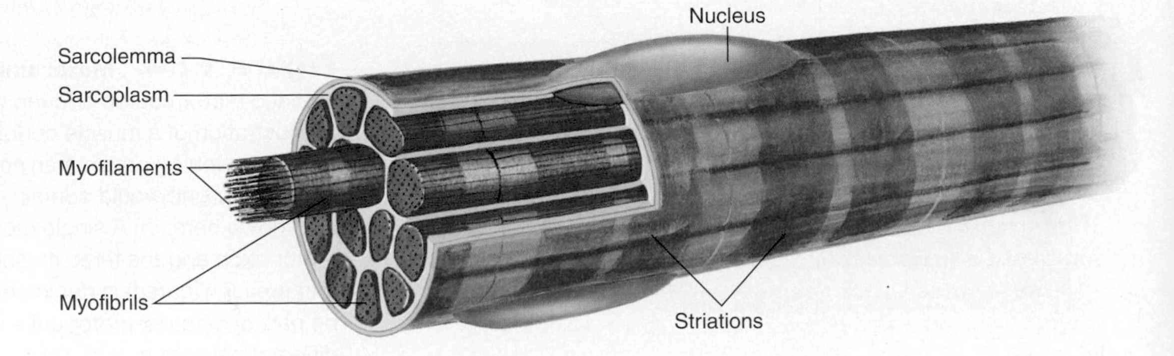

Figure 12.5 The components of a skeletal muscle fiber. A skeletal muscle fiber is composed of numerous myofibrils that contain myofilaments of actin and myosin. Overlapping of the myofilaments produces a striated appearance. Each skeletal muscle fiber is multinucleated.

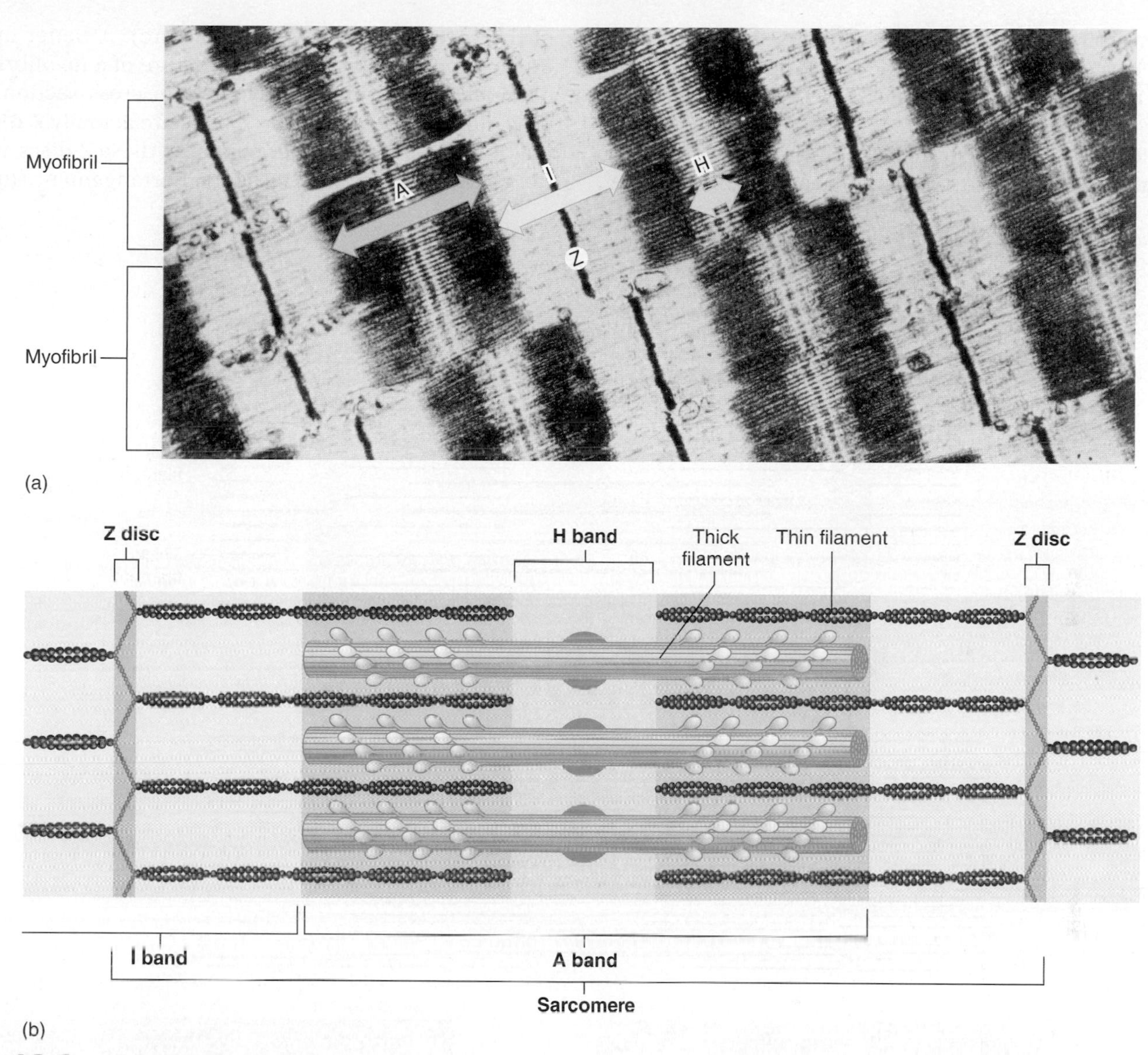

Figure 12.6 The striations of skeletal muscles are produced by thick and thin filaments. (*a*) Electron micrograph of a longitudinal section of myofibrils, showing the banding pattern characteristic of striated muscle. (*b*) Illustration of the arrangement of thick and thin filaments that produces the banding pattern. The colors used in (*a*) to depict different bands and structures correspond to the colors of (*b*). AP|R

other, and the individual myofibrils are not visible with an ordinary light microscope, the entire muscle fiber seems to be striated under a light microscope.

Each myofibril contains even smaller structures called **myofilaments.** When a myofibril is observed at high magnification in longitudinal section (side view), the A bands are seen to contain **thick filaments.** These are about 110 angstroms thick (110 Å, where 1 Å = 10^{-10} m) and are stacked in register. It is these thick filaments that give the A band its dark appearance. The lighter I band, by contrast, contains **thin filaments** (from 50 to 60 Å thick). The thick filaments are primarily composed of the protein **myosin,** and the thin filaments are primarily composed of the protein **actin.**

The I bands within a myofibril are the lighter areas that extend from the edge of one stack of thick filaments to the edge of the next stack of thick filaments. They are light in appearance because they contain only thin filaments. The thin filaments, however, do not end at the edges of the I bands. Instead, each thin filament extends partway into the A bands on each side (between the stack of thick filaments on each side of an I band). Because thick and thin filaments overlap at the edges of each A band, the edges of the A band are darker in appearance than the central region. These central lighter regions of the A bands are called the *H bands* (for *helle,* a German word meaning "bright"). The central H bands thus contain only thick filaments that are not overlapped by thin filaments.

In the center of each I band is a thin dark Z line. The arrangement of thick and thin filaments between a pair of Z lines forms a repeating pattern that serves as the basic subunit of striated muscle contraction. These subunits, from Z to Z, are known as **sarcomeres** (figs. 12.6 and 12.7*a*). A longitudinal section of a myofibril thus presents a side view of successive sarcomeres.

This side view is, in a sense, misleading; there are numerous sarcomeres within each myofibril that are out of the plane of the section (and out of the picture). A better appreciation of the three-dimensional structure of a myofibril can be obtained by viewing the myofibril in cross section. In this view, it can be seen that the Z lines are actually **Z discs,** and that the thin filaments that penetrate these Z discs surround the thick filaments in a hexagonal arrangement (fig. 12.7*b,*

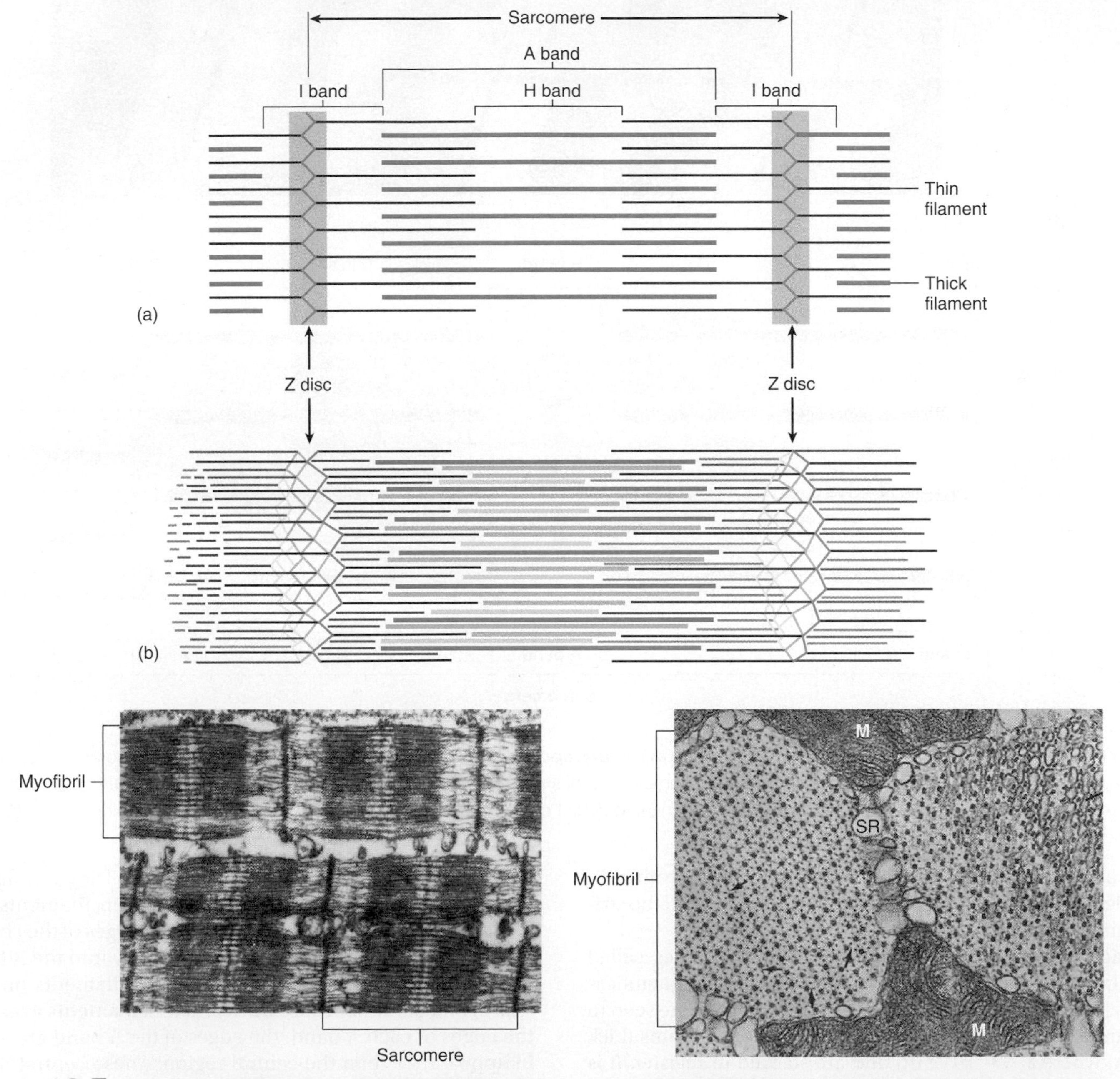

Figure 12.7 Arrangement of thick and thin filaments in a striated muscle fiber. (*a*) In a longitudinal section, the thick and thin filaments are seen to form repeating units called sarcomeres. The banding patterns of the sarcomeres are labeled I, A, and H, as shown. A corresponding electron micrograph (53,000×) is shown on the bottom left. (*b*) The three-dimensional structure of the sarcomeres is illustrated. This three-dimensional structure can be seen in a cross section of a myofibril taken through a region of overlapping thick and thin filaments (bottom right). In the electron micrograph, the arrows point to cross bridges between the thick filaments (dark dots) and thin filaments (light dots). (SR = sarcoplasmic reticulum; M = mitochondria.) Electron micrographs from R. G. Kessel and R. H. Kardon, *Tissues and Organs: A Test-Atlas of Scanning Electron Microscopy,* 1979, W. H. Freeman & Company.

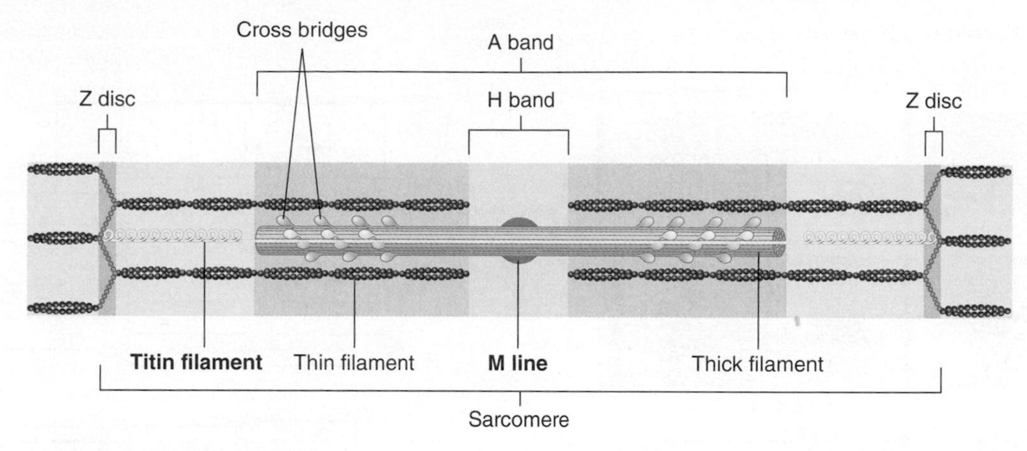

Figure 12.8 Titin filaments and M lines. The M lines are protein filaments in the middle of the A bands that join thick filaments together. Titin proteins are elastic proteins of extremely large size that run through the thick filaments, beginning at the M lines and ending at the Z discs. These stabilize the position of each thick filament within the sarcomere and serve as elastic elements that help muscles return to their resting length.

right). If we concentrate on a single row of dark thick filaments in this cross section, the alternating pattern of thick and thin filaments seen in longitudinal section becomes apparent.

Figure 12.8 indicates two structures not shown in the previous sarcomere figures. The **M lines** are produced by protein filaments located at the center of the thick filaments (and thus the A band) in a sarcomere. These serve to anchor the thick filaments, helping them to stay together during a contraction. Also shown are filaments of **titin,** the largest protein in the human body at more than 1 μm in length. Each titin protein has its amino-terminal end in a Z disc, a spring-like portion running through the I band, and a longer portion bound to the thick filament all the way to the M line. The springlike portion of titin within the I bands is highly folded when the muscle is short, but unfolds and develops passive tension when the sarcomere is stretched. Because of this, titin contributes to the elastic recoil of muscles that helps them to return to their resting lengths when they relax.

Sliding Filament Theory of Contraction

When a muscle contracts it decreases in length as a result of the shortening of its individual fibers. Shortening of the muscle fibers, in turn, is produced by shortening of their myofibrils, which occurs as a result of the shortening of the distance from Z disc to Z disc. As the sarcomeres shorten in length, however, the A bands do *not* shorten but instead move closer together. The I bands—which represent the distance between A bands of successive sarcomeres—decrease in length (table 12.2).

Table 12.2 | Summary of the Sliding Filament Theory of Contraction

1. A myofiber, together with all its myofibrils, shortens by movement of the insertion toward the origin of the muscle.
2. Shortening of the myofibrils is caused by shortening of the sarcomeres—the distance between Z lines (or discs) is reduced.
3. Shortening of the sarcomeres is accomplished by sliding of the myofilaments—the length of each filament remains the same during contraction.
4. Sliding of the filaments is produced by asynchronous power strokes of myosin cross bridges, which pull the thin filaments (actin) over the thick filaments (myosin).
5. The A bands remain the same length during contraction, but are pulled toward the origin of the muscle.
6. Adjacent A bands are pulled closer together as the I bands between them shorten.
7. The H bands shorten during contraction as the thin filaments on the sides of the sarcomeres are pulled toward the middle.

The thin filaments composing the I band, however, do not shorten. Close examination reveals that the thick and thin filaments remain the same length during muscle contraction. Shortening of the sarcomeres is produced not by shortening of the filaments, but rather by the *sliding* of thin filaments over and between the thick filaments. In the process of contraction, the thin filaments on either side of each A band slide deeper and deeper toward the center, producing increasing amounts of overlap with the thick filaments. The I bands (containing only thin filaments) and H bands (containing

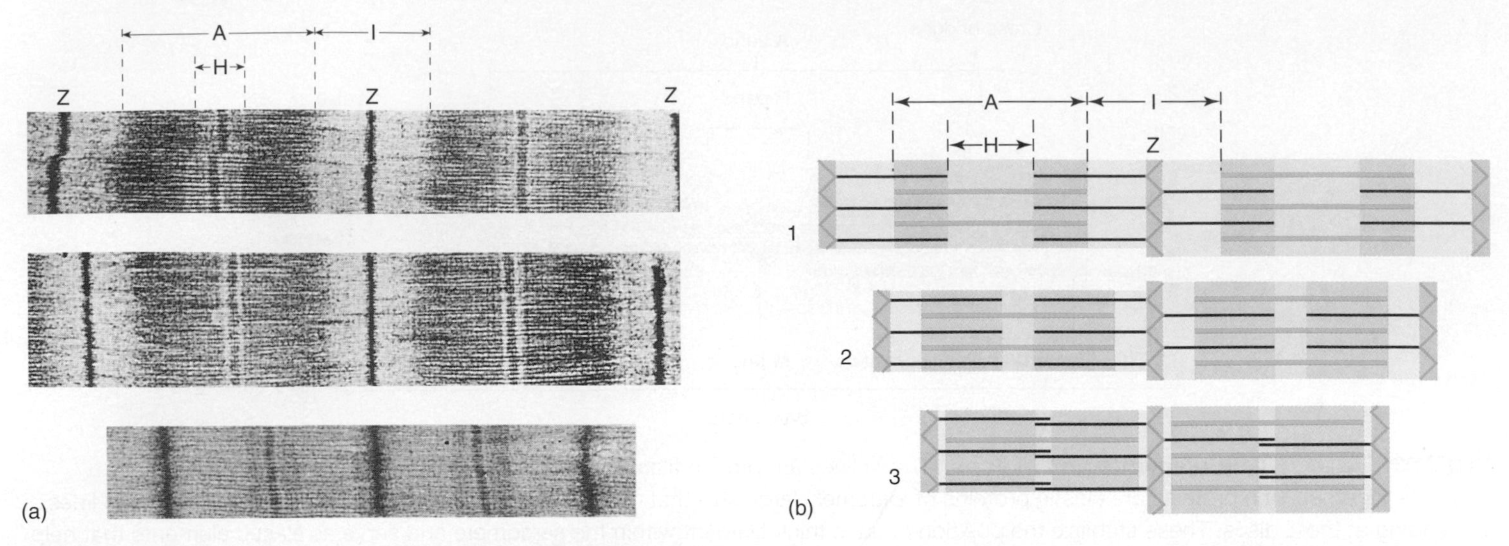

Figure 12.9 The sliding filament model of muscle contraction. (*a*) An electron micrograph and (*b*) a diagram of the sliding filament model of contraction. As the filaments slide, the Z lines are brought closer together and the sarcomeres get shorter. (1) Relaxed muscle; (2) partially contracted muscle; (3) fully contracted muscle. AP|R

only thick filaments) thus get shorter during contraction (fig. 12.9).

Cross Bridges

Sliding of the filaments is produced by the action of numerous **cross bridges** that extend out from the myosin toward the actin. These cross bridges are part of the myosin proteins that extend from the axis of the thick filaments to form "arms" that terminate in globular "heads" (fig. 12.10). A myosin protein has two globular heads that serve as cross bridges. The orientation of the myosin heads on one side of a sarcomere is opposite to that of the other side, so that, when the myosin heads form cross bridges by attaching to actin on each side of the sarcomere, they can pull the actin from each side toward the center.

Isolated muscles are easily stretched (although this is opposed in the body by the stretch reflex, described in a later section), demonstrating that the myosin heads are not attached to actin when the muscle is at rest. Each globular myosin head of a cross bridge contains an *ATP-binding site* closely associated with an *actin-binding site* (fig. 12.10, *left*). The globular heads function as **myosin ATPase** enzymes, splitting ATP into ADP and P_i.

This reaction must occur *before* the myosin heads can bind to actin. When ATP is hydrolyzed to ADP and P_i, the phosphate binds to the myosin head, phosphorylating it and causing it to change its conformation so that it becomes "cocked" (by analogy to the hammer of a gun). The position of the myosin head has changed and it now has the potential energy required for contraction. Perhaps a more apt analogy is with a bow and arrow: The energized myosin head is like a pulled bowstring; it is now in position to bind to actin (fig. 12.10, *right*) so that its stored energy can be released in the next step.

Once the myosin head binds to actin, forming a cross bridge, the bound P_i is released (the myosin head becomes dephosphorylated). This results in a conformational change in the myosin, causing the cross bridge to produce a **power stroke** (fig. 12.11). This is the force that pulls the thin filaments toward the center of the A band.

After the power stroke, with the myosin head now in its flexed position, the bound ADP is released as a new ATP molecule binds to the myosin head. This release of ADP and binding to a new ATP is required for the myosin head to break its bond with actin after the power stroke is completed. The myosin

CLINICAL APPLICATION

The detachment of a cross bridge from actin at the end of a power stroke requires that a new ATP molecule bind to the myosin ATPase. The importance of this process is illustrated by the muscular contracture called **rigor mortis** that occurs due to lack of ATP when the muscle dies. Without ATP, the ADP remains bound to the cross bridges, and the cross bridges remain tightly bound to actin. This results in the formation of "rigor complexes" between myosin and actin that cannot detach. In rigor mortis, the muscles remain stiff until the myosin and actin begin to decompose.

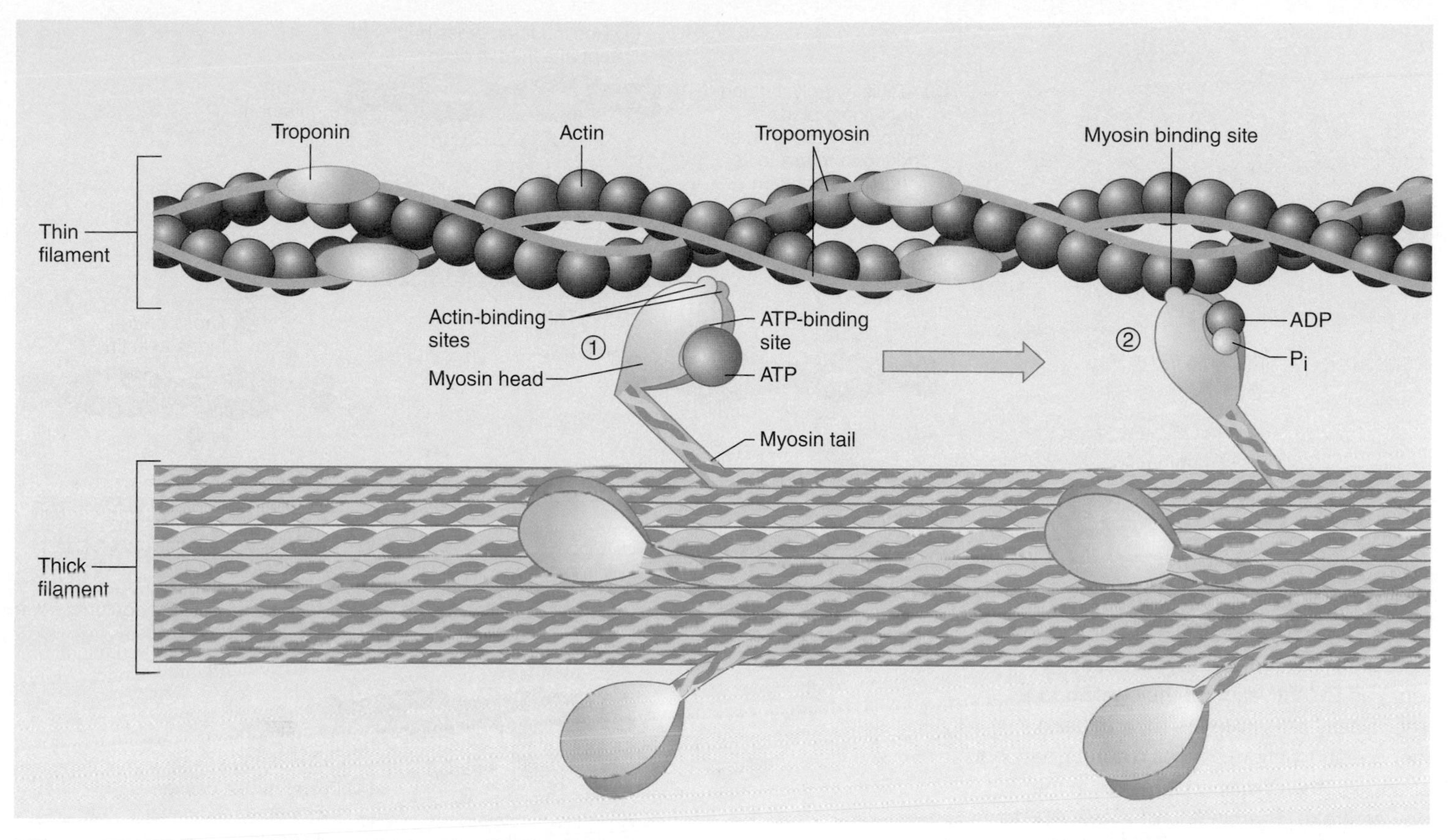

Figure 12.10 Activation of the myosin head. (1) The myosin head has an actin-binding site and an ATP-binding site, which serves as an ATPase to hydrolyze ATP. (2) When ATP is hydrolyzed into ADP and P_i, the myosin head becomes activated and changes its orientation. It is now ready to bind to the actin subunits; at this point, ADP and P_i are still attached to the myosin head.

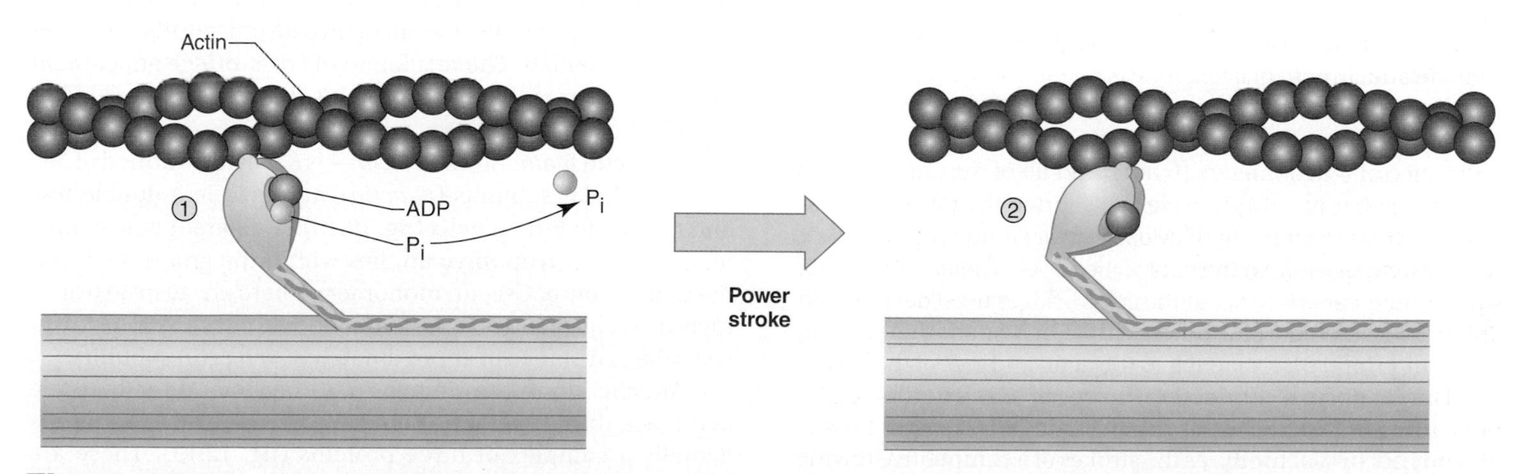

Figure 12.11 The cross-bridge power stroke. (1) The myosin head has been activated by the splitting of ATP into ADP and P_i, which remain bound. At this point, the myosin head has bonded to the actin, forming a cross bridge between the thick and thin filaments. (2) After the P_i group leaves the cross bridge, the myosin head changes its orientation, producing the power stroke that moves the actin filament. AP|R

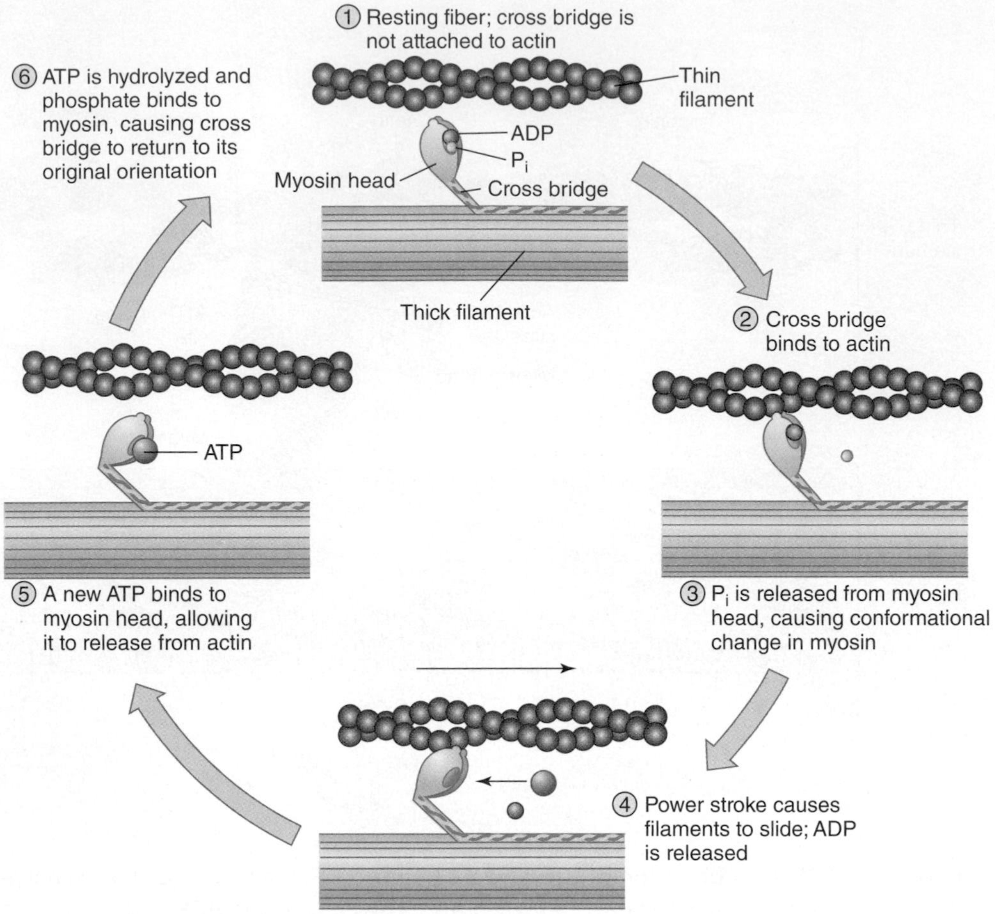

① Resting fiber; cross bridge is not attached to actin
Thin filament
ADP
P_i
Myosin head
Cross bridge
Thick filament
② Cross bridge binds to actin
③ P_i is released from myosin head, causing conformational change in myosin
④ Power stroke causes filaments to slide; ADP is released
⑤ A new ATP binds to myosin head, allowing it to release from actin
ATP
⑥ ATP is hydrolyzed and phosphate binds to myosin, causing cross bridge to return to its original orientation

Figure 12.12 The cross-bridge cycle. Hydrolysis of ATP and consequent phosphorylation of the myosin head is required for activation of the cross bridge. The release of P_i from the myosin head (dephosphorylation) causes a conformational change of the myosin that results in the power stroke. The binding of a new ATP to the myosin head allows the cross bridge to release from the actin. AP|R

head will then split ATP to ADP and P_i, and—if nothing prevents the binding of the myosin head to the actin—a new cross-bridge cycle will occur (fig. 12.12).

Note that the splitting of ATP is required *before* a cross bridge can attach to actin and undergo a power stroke, and that the attachment of a *new ATP* is needed for the cross bridge to release from actin at the end of a power stroke (fig. 12.12).

A single cross-bridge power stroke pulls the actin filament a distance of 6 nanometers (6 nm), and all of the cross bridges acting together in a single cycle will shorten the muscle by less than 1% of its resting length. Muscles can shorten up to 60% of their resting lengths, so the contraction cycles must be repeated many times. For this to occur the cross bridges must detach from the actin at the end of a power stroke, reassume their resting orientation, and then reattach to the actin and repeat the cycle.

During normal contraction, however, only a portion of the cross bridges are attached at any given time. The power strokes are thus not in synchrony, as the strokes of a competitive rowing team would be. Rather, they are like the actions of a team engaged in tug-of-war, where the pulling action of the members is asynchronous. Some cross bridges are engaged in power strokes at all times during the contraction. The force produced by each power stroke is constant, but when the muscle's load is greater, the number of cross bridges engaged in power strokes is increased to generate more force.

Regulation of Contraction

When the cross bridges attach to actin, they undergo power strokes and cause muscle contraction. In order for a muscle to relax, therefore, the attachment of myosin cross bridges to actin must be prevented. The regulation of cross-bridge attachment to actin is a function of two proteins that are associated with actin in the thin filaments.

The actin filament—or *F-actin*—is a polymer formed of 300 to 400 globular subunits (*G-actin*), arranged in a double row and twisted to form a helix (fig. 12.13). A different type of protein, known as **tropomyosin,** lies within the groove between the double row of G-actin monomers. There are 40 to 60 tropomyosin molecules per thin filament, with each tropomyosin spanning a distance of approximately seven actin subunits.

Attached to the tropomyosin, rather than directly to the actin, is a third type of protein called **troponin.** Troponin is actually a complex of three proteins (fig. 12.13). These are *troponin I* (which inhibits the binding of the cross bridges to actin), *troponin T* (which binds to tropomyosin), and *troponin C* (which binds Ca^{2+}). Troponin and tropomyosin work together to regulate the attachment of cross bridges to actin, and thus serve as a switch for muscle contraction and relaxation. In a relaxed muscle, the position of the tropomyosin in the thin filaments is such that it physically blocks the cross

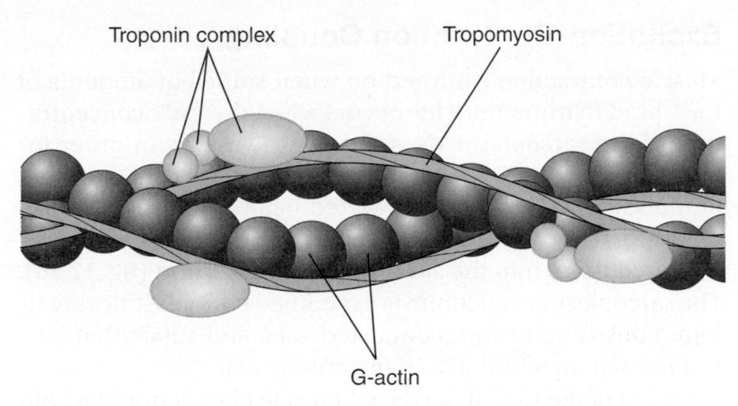

Figure 12.13 The structural relationship between troponin, tropomyosin, and actin. The tropomyosin is attached to actin, whereas the troponin complex of three subunits is attached to tropomyosin (not directly to actin).

bridges from bonding to specific attachment sites in the actin. Thus, in order for the myosin cross bridges to attach to actin, the tropomyosin must be moved. This requires the interaction of troponin with Ca^{2+}.

Role of Ca^{2+} in Muscle Contraction

Scientists long thought that Ca^{2+} only served to form the calcium phosphate crystals that hardened bone, enamel, and dentin. In 1883, Sidney Ringer published the results of a surprisingly simple experiment that changed that idea. He isolated rat hearts and found that they beat well when placed in isotonic solutions made with the hard water from a London tap. When he made the isotonic solutions with distilled water, however, the hearts gradually stopped beating. This could be reversed, he found, if he added Ca^{2+} to the solutions. This demonstrated a role for Ca^{2+} in muscle contraction, a role that scientists now understand in some detail.

In a relaxed muscle, when tropomyosin blocks the attachment of cross bridges to actin, the concentration of Ca^{2+} in the sarcoplasm (cytoplasm of muscle cells) is very low. When the muscle cell is stimulated to contract, mechanisms that will be discussed shortly cause the concentration of Ca^{2+} in the sarcoplasm to quickly rise. Some of this Ca^{2+} attaches to troponin, causing a conformational change that moves the troponin complex *and* its attached tropomyosin out of the way so that the cross bridges can attach to actin (fig. 12.14). Once the attachment sites on the actin are exposed, the cross bridges can bind to actin, undergo power strokes, and produce muscle contraction.

The position of the troponin-tropomyosin complexes in the thin filaments is thus adjustable. When Ca^{2+} is not attached to troponin, the tropomyosin is in a position that inhibits attachment of myosin heads to actin, preventing muscle contraction. When Ca^{2+} attaches to troponin, the troponin-tropomyosin complexes shift position. The myosin heads can then attach to actin, produce a power stroke, and detach from actin. These contraction cycles can continue as long as Ca^{2+} is bonded to troponin.

Relaxed muscle: tropomyosin blocks the binding site

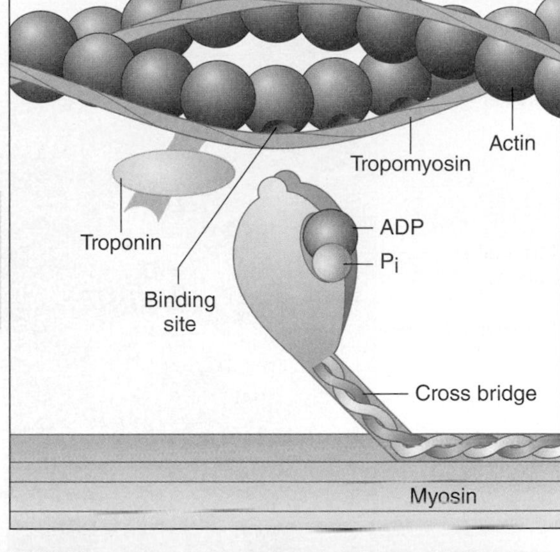

Contracting muscle: myosin head binds to actin

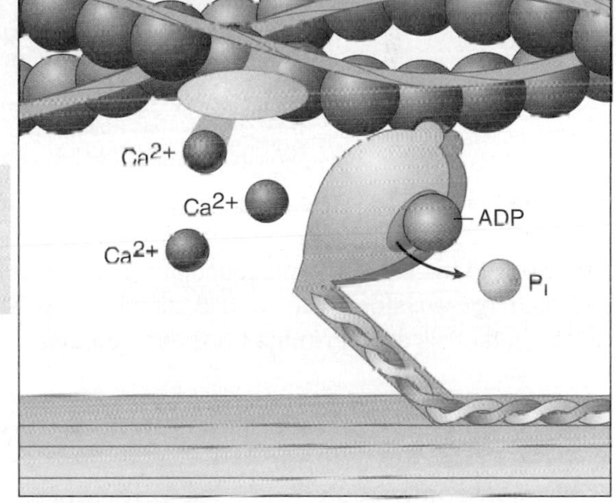

Figure 12.14 The role of Ca^{2+} in muscle contraction. The attachment of Ca^{2+} to troponin causes movement of the troponin-tropomyosin complex, which exposes binding sites on the actin. The myosin cross bridges can then attach to actin and undergo a power stroke.

Clinical Investigation CLUES

The physician notices that Maria has high muscle tone, and her lab tests reveal an elevated plasma Ca^{2+} concentration.

- How would an abnormally high plasma Ca^{2+} concentration influence skeletal muscles?
- What might cause an abnormally high plasma Ca^{2+} concentration (hint: see chapter 11)?

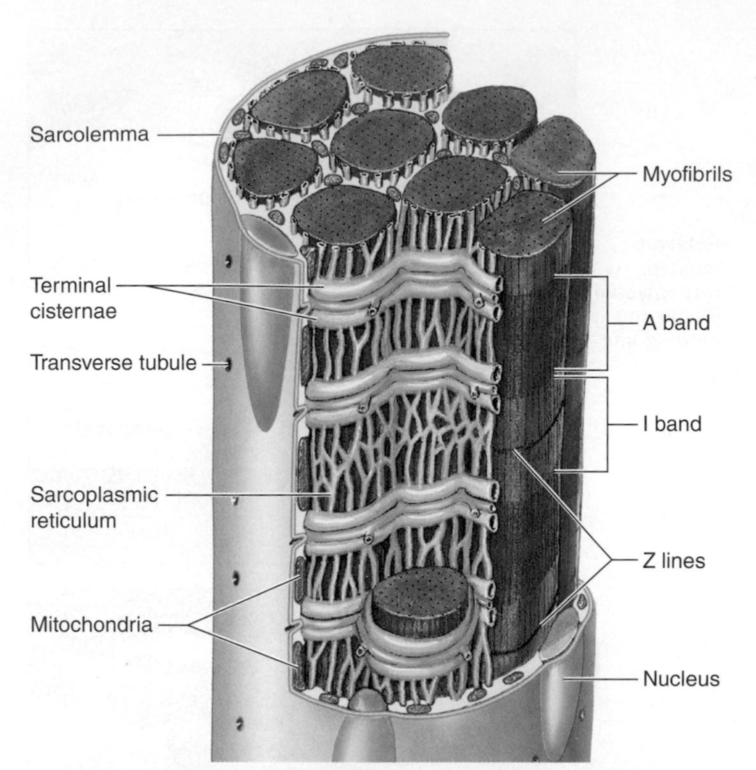

Figure 12.15 **The sarcoplasmic reticulum.** This figure depicts the relationship between myofibrils, the transverse tubules, and the sarcoplasmic reticulum. The sarcoplasmic reticulum (*green*) stores Ca^{2+} and is stimulated to release it by action potentials arriving in the transverse tubules.

Excitation-Contraction Coupling

Muscle contraction is turned on when sufficient amounts of Ca^{2+} bind to troponin. This occurs when the Ca^{2+} concentration of the sarcoplasm rises above 10^{-6} molar. In order for muscle relaxation to occur, therefore, the Ca^{2+} concentration of the sarcoplasm must be lowered below this level. Muscle relaxation is produced by the active transport of Ca^{2+} out of the sarcoplasm into the **sarcoplasmic reticulum** (fig. 12.15). The sarcoplasmic reticulum is a modified endoplasmic reticulum, consisting of interconnected sacs and tubes that surround each myofibril within the muscle cell.

Most of the Ca^{2+} in a relaxed muscle fiber is stored within expanded portions of the sarcoplasmic reticulum known as *terminal cisternae.* When a muscle fiber is stimulated to contract by either a motor neuron *in vivo* or electric shocks *in vitro,* the stored Ca^{2+} is released from the sarcoplasmic reticulum by passive diffusion through membrane channels termed **calcium release channels** (fig. 12.16); these are also called *ryanodine receptors* (after an alkaloid drug that specifically binds to them). The calcium-release channels are 10 times larger than the voltage-gated Ca^{2+} channels, permitting a very high rate of Ca^{2+} diffusion into the sarcoplasm.

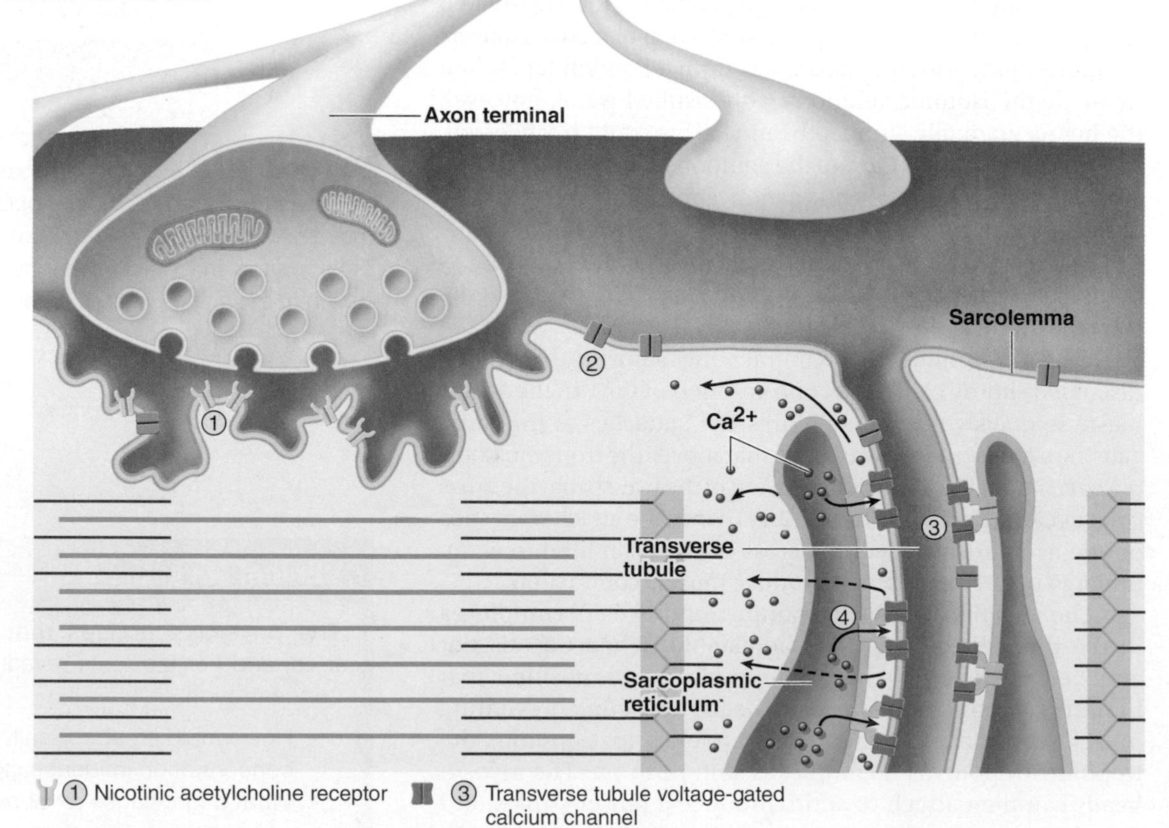

Figure 12.16 **Excitation-contraction coupling in skeletal muscle.** (1) ACh released by somatic motor neurons binds to nicotinic ACh receptors in the sarcolemma, causing a depolarization that stimulates (2) voltage-gated channels, producing action potentials. (3) The conduction of action potentials along the transverse tubules stimulates the opening of voltage-gated Ca^{2+} channels. (4) These channels in the transverse tubules are mechanically coupled to Ca^{2+} release channels in the sarcoplasmic reticulum, causing them to open. Ca^{2+} then diffuses out of the sarcoplasmic reticulum, so that it can bind to troponin and stimulate muscle contraction. AP|R

The Ca^{2+} can then bind to troponin and stimulate contraction. When a muscle fiber is no longer stimulated, the Ca^{2+} is actively transported back into the sarcoplasmic reticulum. Now, in order to understand how the release and uptake of Ca^{2+} is regulated, one more organelle within the muscle fiber must be described.

The terminal cisternae of the sarcoplasmic reticulum are separated by only a very narrow gap from **transverse tubules** (or **T tubules**). These are narrow membranous "tunnels" formed from and continuous with the sarcolemma. The transverse tubules thus open to the extracellular environment through pores in the cell surface (see fig. 12.15), and are able to conduct action potentials. The stage is now set to explain exactly how a motor neuron stimulates a muscle fiber to contract.

The release of acetylcholine from axon terminals at the neuromuscular junctions (motor end plates), as previously described, causes electrical activation of skeletal muscle fibers. *End-plate potentials* (analogous to the EPSPs described in chapter 7) are produced that generate action potentials when they reach a threshold level of depolarization. Action potentials in muscle cells, like those in nerve cells, are all-or-none events that are regenerated along the plasma membrane. This is because voltage-gated channels that produce action potentials are located immediately adjacent to the motor end plates and all along the sarcolemma and transverse tubules. It must be remembered that action potentials involve the flow of ions between the extracellular and intracellular environments across a plasma membrane that separates these two compartments. In muscle cells, therefore, action potentials can be conducted into the interior of the fiber across the membrane of the transverse tubules.

The transverse tubules contain **voltage-gated calcium channels,** also called *dihydropyridine (DHP) receptors* (after a class of drugs that specifically bind to and block these channels). These respond to membrane depolarization. When the transverse tubules conduct action potentials, the voltage-gated calcium channels undergo a conformational (shape) change. There is a direct molecular coupling between these channels on the transverse tubules and the calcium release channels (ryanodine receptors) in the sarcoplasmic reticulum. The conformational change in the voltage-gated channels in the transverse tubules directly causes the calcium release channels in the sarcoplasmic reticulum to open. This releases Ca^{2+} into the cytoplasm, raising the cytoplasmic Ca^{2+} concentration and stimulating contraction (fig. 12.16). The process by which action potentials cause contraction is termed **excitation-contraction coupling** (fig. 12.17).

This excitation-contraction coupling mechanism in skeletal muscle has been described as an *electromechanical release mechanism,* because the voltage-gated calcium channels and the calcium release channels are physically (mechanically) coupled. As a result, Ca^{2+} enters the cytoplasm from the sarcoplasmic reticulum where it is stored. However, this electromechanical release mechanism is not the full story of

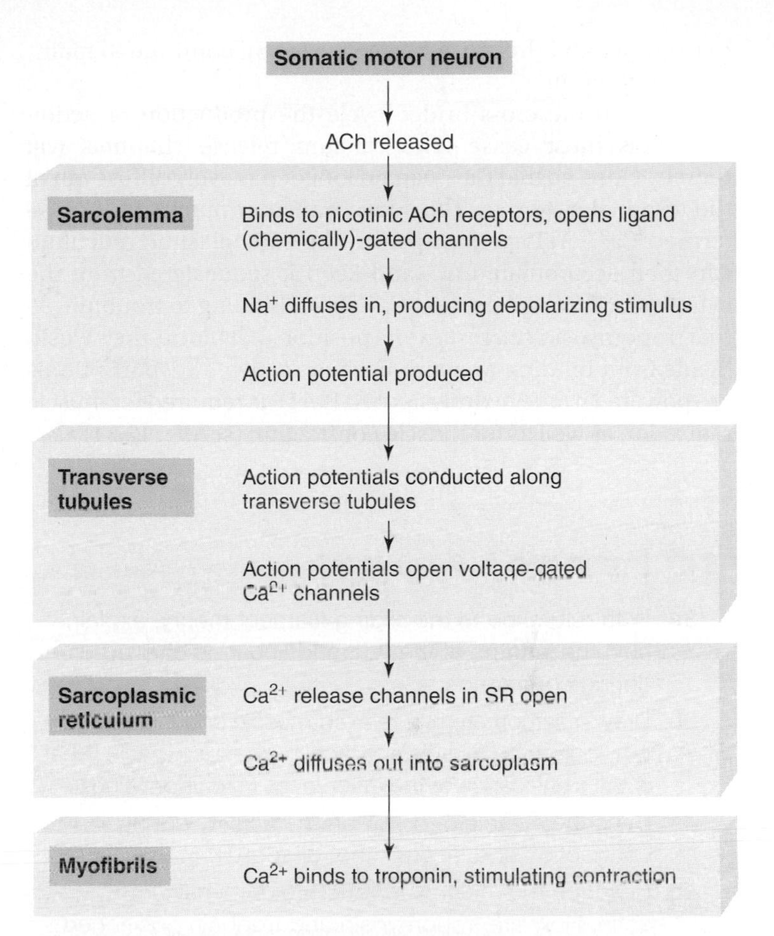

Figure 12.17 Summary of excitation-contraction coupling. Electrical excitation of the muscle fiber—that is, action potentials conducted along the sarcolemma and down the transverse tubules—triggers the release of Ca^{2+} from the sarcoplasmic reticulum. Because Ca^{2+} binding to troponin leads to contraction, the Ca^{2+} can be said to couple excitation to contraction.

how action potentials stimulate the contraction of skeletal muscles.

The membrane of the sarcoplasmic reticulum also contains a type of Ca^{2+} release channel that opens in response to a rise in the Ca^{2+} concentration in the cytoplasm. These calcium release channels are thus regulated by a *Ca^{2+}-induced Ca^{2+} release mechanism.* This mechanism contributes significantly to excitation-contraction coupling in skeletal muscle, and in cardiac muscle it is the mechanism most responsible for excitation-contraction coupling (see fig. 12.34, step 4).

Muscle Relaxation

As long as action potentials continue to be produced—which is as long as neural stimulation of the muscle is maintained—the calcium release channels in the sarcoplasmic reticulum will remain open, Ca^{2+} will passively diffuse out of the sarcoplasmic reticulum and the Ca^{2+} concentration of the sarcoplasm will remain high. Thus, Ca^{2+} will remain attached to troponin and the cross-bridge cycle will continue to maintain contraction.

To stop the cross-bridge cycle the production of action potentials must cease. The calcium release channels will thereby close, so that Ca^{2+} can no longer passively diffuse out of the terminal cisternae. The active transport pumps for Ca^{2+}—termed **Ca^{2+}-ATPase pumps**—in the sarcoplasmic reticulum can then accumulate Ca^{2+} and keep it sequestered from the cytoplasm. This will prevent Ca^{2+} from binding to troponin, so that tropomyosin can resume its position that blocks the myosin heads from binding to actin. Because active transport pumps are powered by the hydrolysis of ATP, ATP is required for muscle relaxation as well as for muscle contraction (see fig. 12.34).

CHECKPOINT

3a. With reference to the sliding filament theory, explain how the lengths of the A, I, and H bands change during contraction.

3b. Draw a sarcomere in a relaxed muscle and a sarcomere in a contracted muscle and label the bands in each. What is the significance of the differences in your drawings?

4a. Describe a cycle of cross-bridge activity during contraction and discuss the role of ATP in this cycle.

4b. Describe the molecular structure of myosin and actin. How are tropomyosin and troponin positioned in the thin filaments and how do they function in the contraction cycle?

5a. Use a flowchart to show the sequence of events from the time ACh is released from a nerve ending to the time Ca^{2+} is released from the sarcoplasmic reticulum.

5b. Explain the requirements for Ca^{2+} and ATP in muscle contraction and relaxation.

12.3 CONTRACTIONS OF SKELETAL MUSCLES

Contraction of muscles generates tension, which allows muscles to shorten and thereby perform work. The contraction strength of skeletal muscles must be sufficiently great to overcome the load on a muscle in order for that muscle to shorten.

LEARNING OUTCOMES

After studying this section, you should be able to:

6. Distinguish between the different types of muscle contractions.

7. Identify the series elastic component, and explain the length-tension relationship in striated muscles.

The contractions of skeletal muscles generally produce movements of bones at joints, which act as levers to move the loads against which the muscle's force is exerted. The contractile behavior of the muscle, however, is more easily studied *in vitro* (outside the body) than *in vivo* (within the body). When a muscle—for example, the gastrocnemius (calf muscle) of a frog—is studied *in vitro,* it is usually mounted so that one end is fixed and the other is movable. The mechanical force of the muscle contraction is transduced (changed) into an electric current, which can be amplified and displayed on a recording device (fig. 12.18). In this way the contractile behavior of the whole muscle in response to electric shocks can be studied.

Twitch, Summation, and Tetanus

Contractions of isolated muscles in response to electrical shocks mimic the behavior of muscles when they contract within the body. When the muscle is stimulated with a single electric shock of sufficient voltage, it quickly contracts and relaxes. This response is called a **twitch** (fig. 12.18*a*). The shock produces an action potential that is conducted along the sarcolemma and stimulates the release of Ca^{2+} from the sarcoplasmic reticulum. These events occur during the *latent period* between the stimulus and the contraction; the Ca^{2+} then binds to troponin and stimulates the muscle twitch. If a second electric shock is delivered before the muscle has had a chance to fully relax from the first twitch, the second twitch will "ride piggyback" on the first. This response is called **summation** (fig. 12.18*b*).

Increasing the stimulus voltage increases the frequency of action potentials and the amount of Ca^{2+} in the sarcoplasm, increasing the strength of each fiber's contraction. Increasing the stimulus voltage also activates more muscle fibers, recruiting them into the contraction. Analogous events occur *in vivo* in response to the activation of motor units. By this means, the contractions of skeletal muscles can be *graded,* or varied—a requirement for the proper control of skeletal movements. Because stimulation of muscle fibers usually results in their full contractions, stronger muscle contractions are produced mostly by *recruitment* of more muscle fibers into the contraction. Through variations in the numbers of muscle fibers participating in the contraction, skeletal muscles produce **graded contractions.**

If the stimulator is set to deliver an increasing frequency of electric shocks automatically, the relaxation time between successive twitches will get shorter and shorter as the strength of contraction increases in amplitude. This effect is known as **incomplete tetanus** (fig. 12.19). Finally, at a particular "fusion frequency" of stimulation, there is no visible relaxation between successive twitches. Contraction is smooth and sustained, as it is during normal muscle contraction *in vivo.* This smooth, sustained contraction is called **complete tetanus.** (The term *tetanus* should not be confused with the disease of the same

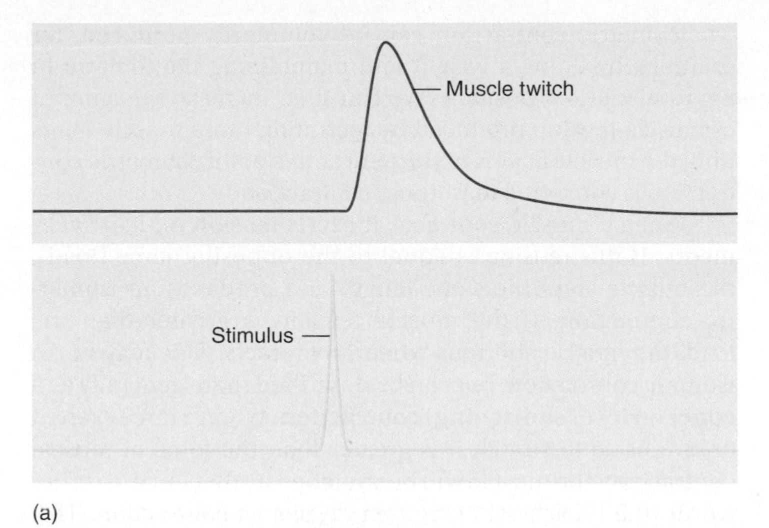

(a)

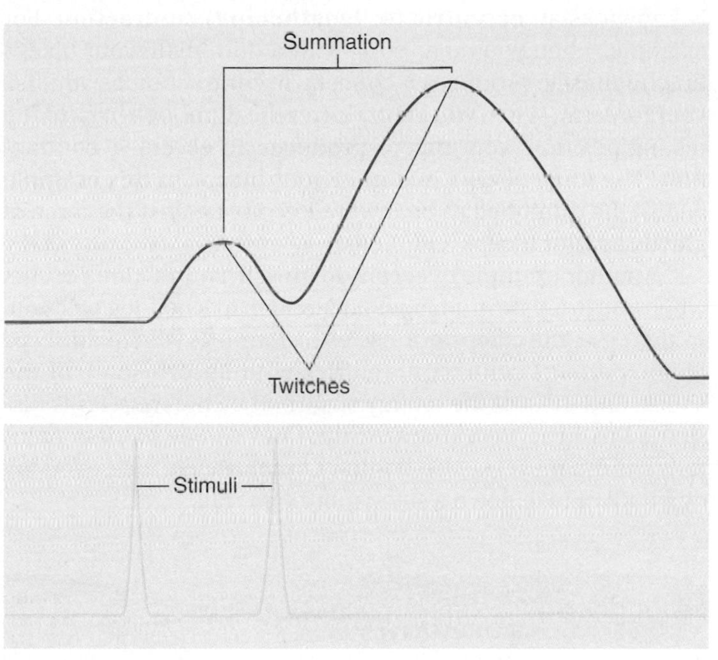

(b)

Figure 12.18 Muscle twitch and summation. (*a*) A single electrical shock to a muscle stimulates a muscle twitch. There is a latent period of a couple of milliseconds between the shock and the twitch, which can last from several to a hundred milliseconds, depending on the muscle. (*b*) A second shock delivered to the muscle before it has had a chance to relax fully from its first twitch results in a second twitch that summates with the first to produce a stronger contraction.

name, which is accompanied by a painful state of muscle contracture, or *tetany.*)

If these procedures were performed on isolated muscle fibers instead of whole muscles, similar behavior would be observed. That is, the isolated muscle fiber would twitch to a single shock and would show summation of twitches if the shocks occurred quickly one after another. This is because it only takes about 10 milliseconds for an action potential to be

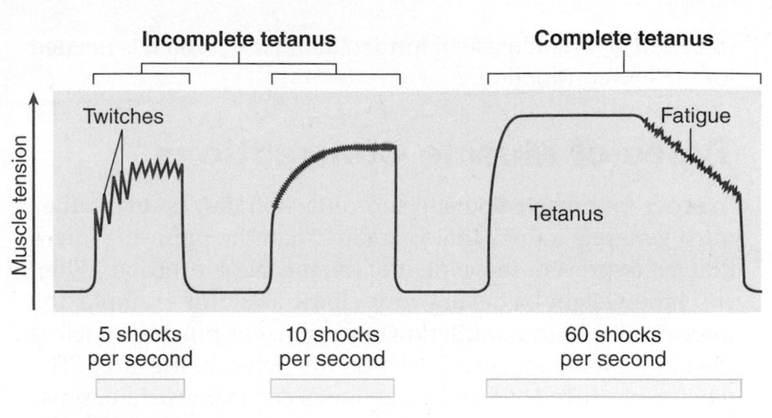

Figure 12.19 Incomplete and complete tetanus. Stimuli in the form of electric shocks are given to a muscle and the muscle twitches in response to these stimuli. When the stimuli are given in rapid succession (for example, 5 to 10 shocks per second), the twitches summate to produce an incomplete tetanus—a contraction that is sustained but "jerky." A faster frequency of stimulation (for example, at 60 shocks per second) can produce a smooth, sustained contraction known as complete tetanus. If this frequency of stimulation is maintained, the muscle gradually loses its ability to maintain the contraction; it fatigues.

conducted along the full length of a muscle fiber, whereas the contraction can last as long as 100 milliseconds. Thus, if action potentials are produced quickly in succession, Ca^{2+} will remain in the cytoplasm attached to troponin and the cross-bridge cycle will continue. In this case the muscle fiber can even be made to hold a contraction, as in complete tetanus. However, this behavior of the individual muscle fibers normally does not occur *in vivo.* As previously described, a somatic motor axon innervates a number of muscle fibers to form a *motor unit* (see fig. 12.4). When the motor axon is activated, all of the muscle fibers it innervates contract.

So, how can we produce the smooth, sustained contractions of complete tetanus *in vivo?* We do this by the **asynchronous activation** of motor units. The muscle fibers of some motor units start to twitch when those of previously activated motor units begin to relax, producing a continuous contraction of the whole muscle from the jerky contractions of the separate motor unit twitches.

Treppe

If the voltage of the electrical shocks delivered to an isolated muscle *in vitro* is gradually increased from zero, the strength of the muscle twitches will increase accordingly, up to a maximal value at which all of the muscle fibers are stimulated. This demonstrates the graded nature of the muscle contraction. If a series of electrical shocks at this maximal voltage is given to a fresh muscle so that each shock produces a separate twitch, each of the twitches evoked will be successively stronger, up to a higher maximum. This demonstrates **treppe,** or the *staircase effect.* Treppe may represent a warm-up effect, and is believed to be due to an increase in intracellular Ca^{2+}, which is needed for muscle contraction.

Types of Muscle Contractions

In order for muscle fibers to shorten when they contract, they must generate a force that is greater than the opposing forces that act to prevent movement of the muscle's insertion. When you lift a weight by flexing your elbow joint, for example, the force produced by contraction of your biceps brachii muscle is greater than the force of gravity on the object being lifted. The tension produced by the contraction of each muscle fiber separately is insufficient to overcome the opposing force, but the combined contractions of numerous muscle fibers may be sufficient to overcome the opposing force and flex your forearm. In this case, the muscle and all of its fibers shorten in length.

This process can be seen by examining the **force-velocity curve.** This graph shows the inverse relationship between the force opposing muscle contraction (the load against which the muscle must work) and the velocity of muscle shortening (fig. 12.20). The tension produced by the shortening muscle is just greater than the force (load) at each value, causing the muscle to shorten. Under these controlled experimental conditions, the contraction strength is constant at each load; this muscle contraction during shortening is thus called an **isotonic contraction** (*iso* = same; *tonic* = strength).

If the load is zero, a muscle contracts and shortens with its maximum velocity. As the load increases, the velocity of muscle shortening decreases. When the force opposing contraction (the load) becomes sufficiently great, the muscle is unable to shorten when it exerts a given tension. That is, its velocity of shortening is zero. At this point, where muscle tension does not cause muscle shortening, the contraction is called an **isometric** (literally, "same length") **contraction.**

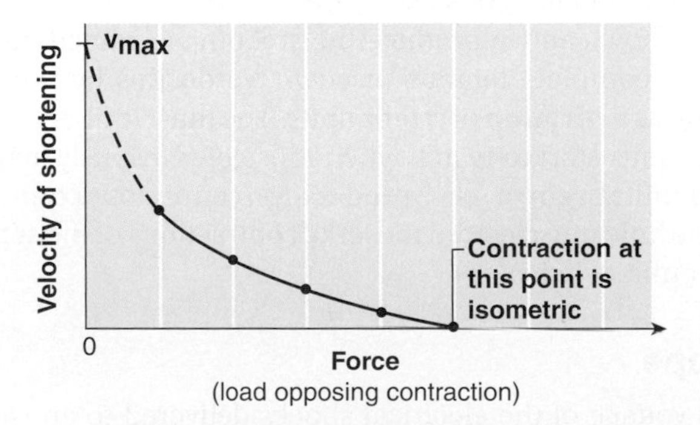

Figure 12.20 Force-velocity curve. This graph illustrates the inverse relationship between the force opposing muscle contraction (the load against which the muscle must work) and the velocity of muscle shortening. A force that is sufficiently great prevents muscle shortening, so that the contraction is isometric. If there is no force acting against the muscle contraction, the velocity of shortening is maximal (V_{max}). Since this cannot be measured (because there will always be some load), the estimated position of the curve is shown with a dashed line.

Isometric contraction can be voluntarily produced, for example, by lifting a weight and maintaining the forearm in a partially flexed position. We can then increase the amount of muscle tension produced by recruiting more muscle fibers until the muscle begins to shorten; at this point, isometric contraction is converted to isotonic contraction.

When a muscle contracts, it exerts tension on its attachments. If this tension is equal to the opposing force (load), the muscle stays the same length and produces an isometric contraction. If the muscle tension is greater than the load, the muscle shortens when it contracts. This may be an isotonic contraction, but can be described more generally as a **concentric** (or **shortening**) **contraction.** When a force exerted on a muscle to stretch it is greater than the force of muscle contraction, the muscle will be stretched by that force. In other words, the muscle will lengthen *despite* its contraction. This is known as an **eccentric** (or **lengthening**) **contraction.** For example, when you do a "curl" with a dumbbell, your biceps brachii muscle produces a concentric contraction as you flex your forearm. When you gently lower the dumbbell back to the resting position, your biceps produces an eccentric contraction. The force of contraction of your biceps in this example allows the dumbbell to be lowered gently against the force of gravity as your biceps lengthens.

Another example of eccentric muscle contractions occurs when you jump from a height and land in a flexed-leg position. In this case, the extensor muscles of your legs (the quadriceps femoris group) contract eccentrically to absorb some of the shock, and most of the energy absorbed by the muscles is dissipated as heat. Less dramatically (and somewhat less painfully), these muscles also contract eccentrically when you jog downhill or hike down a steep mountain trail.

Clinical Investigation CLUES

Maria experienced muscle fatigue and pain when doing leg extensions on a machine at the gym, where weights are lifted by contraction of the quadriceps femoris muscles and then brought back to rest by the return movement.

- What types of muscle contractions are performed when doing a leg extension and the return movement?
- How might her high blood Ca^{2+} concentration contribute to muscle fatigue and pain when performing this exercise?

Series-Elastic Component

In order for a muscle to shorten when it contracts, and thus to move its insertion toward its origin, the noncontractile parts of the muscle and the connective tissue of its tendons must first be pulled tight. These structures, particularly the collagen fibers in the tendons and connective tissues of the muscle and the molecules of titin within the sarcomeres, have *elasticity.* That is, they resist distention and spring back to their resting lengths when the distending force is released. These elastic structures provide a **series elastic component,** so called because they are in line (in series) with the force of muscle contraction. The series elastic component must be pulled tight before muscle contraction can result in muscle shortening.

When the gastrocnemius muscle was stimulated with a single electric shock as described earlier, the amplitude of the twitch was reduced because some of the force of contraction was used to stretch the series-elastic component. Quick delivery of a second shock thus produced a greater degree of muscle shortening than the first shock, culminating at the fusion frequency of stimulation with complete tetanus, in which the strength of contraction was much greater than that of individual twitches.

Some of the energy used to stretch the series-elastic component during muscle contraction is released by elastic recoil when the muscle relaxes. This elastic recoil, which helps the muscles return to their resting length, is particularly important for the muscles involved in breathing. As we will see in chapter 16, inspiration is produced by muscle contraction and expiration is produced by the elastic recoil of the thoracic structures that were stretched during inspiration.

Length-Tension Relationship

The strength of a muscle's contraction is influenced by a variety of factors. These include the number of fibers within the muscle that are stimulated to contract, the frequency of stimulation, the thickness of each muscle fiber (thicker fibers have more myofibrils and thus can exert more power), and the initial length of the muscle fibers when they are at rest.

There is an "ideal" resting length for striated muscle fibers. This is the length at which they can generate maximum force. The force that the muscle generates when it contracts is usually measured as the force required to prevent it from shortening. The muscle is made to contract isometrically, and the force required to prevent it from shortening is measured as the *tension* produced. As illustrated in figure 12.21, this tension is maximal when the sarcomeres are at a length of 2.0 to 2.25 μm. As it turns out, this is the length of the sarcomeres when muscles are at their normal resting lengths. In the body, this normal resting length is maintained by reflex contractions in response to passive stretching (described in section 12.5).

When the sarcomere lengths are greater than about 2.2 μm, the tension produced by the muscle contraction decreases with increasing sarcomere length. This is because there are fewer interactions of myosin cross bridges with actin. When the sarcomeres reach a length of about 3.6 μm, there is no overlap of thick and thin filaments, and no interactions can occur between myosin and actin. Therefore the muscle produces zero tension (fig. 12.21).

When the sarcomere length is shorter than 2.0 μm, the force generated by muscle contraction declines with decreasing

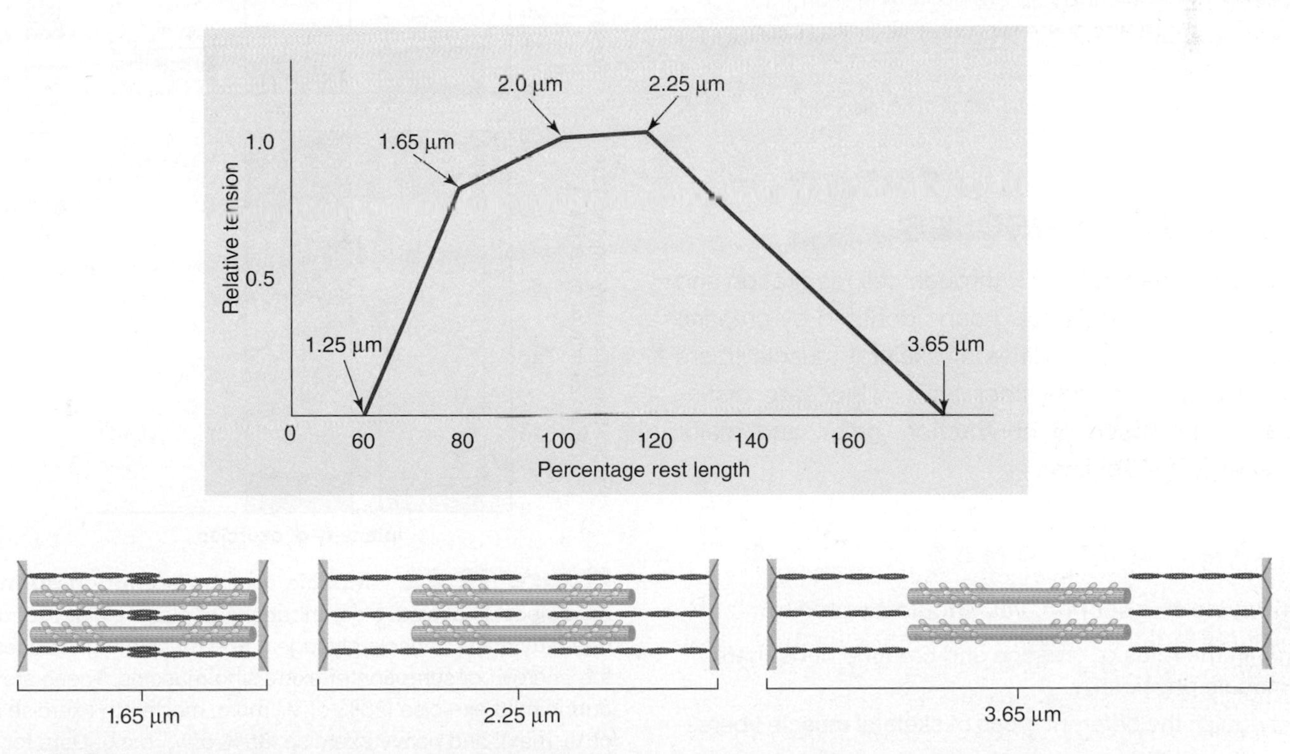

Figure 12.21 The length-tension relationship in skeletal muscles. Maximum relative tension (1.0 on the *y*-axis) is achieved when the muscle is 100% to 120% of its resting length (sarcomere lengths from 2.0 to 2.25 μm). Increases or decreases in muscle (and sarcomere) lengths result in rapid decreases in tension.

sarcomere length (fig. 12.21). This is because the cross-bridge action becomes less effective as the muscle fiber gets shorter and thicker due to: (1) the development of opposing forces (such as the fluid pressure of the sarcoplasm) as the muscle fiber gets shorter; and (2) the increasing distance between thick and thin filaments as the muscle fiber gets thicker. The double overlapping of thin filaments (see the left sarcomere in fig. 12.21) may further interfere with the action of cross bridges. The force of muscle contraction declines still further when the thick filaments abut against the Z discs at a sarcomere length of 1.7 μm. This may be due to deformation of the myosin. At a sarcomere length of 1.25 μm, the muscle produces zero force (fig. 12.21).

CHECKPOINT

6a. Explain how graded contractions and smooth, sustained contractions can be produced *in vitro* and *in vivo*.

6b. Distinguish among isotonic, isometric, concentric, and eccentric contractions, and describe what factors determine if a contraction will be isometric or isotonic.

7a. Identify the nature and physiological significance of the series-elastic component of muscle contraction.

7b. Describe the relationship between the resting muscle length and the strength of its contraction.

12.4 ENERGY REQUIREMENTS OF SKELETAL MUSCLES

Skeletal muscles generate ATP through cell respiration and through the use of phosphate groups donated by creatine phosphate. The aerobic abilities of skeletal muscle fibers differ according to muscle fiber type, which are distinguished by their speed of contraction, color, and major mode of energy metabolism.

LEARNING OUTCOMES

After studying this section, you should be able to:

8. Explain the roles of creatine and creatine phosphate in muscle physiology.

9. Distinguish the different types of skeletal muscle fibers.

10. Describe aerobic capacity, lactate threshold, and muscle fatigue.

11. Explain how exercise training affects skeletal muscles.

About 70% of the energy (ATP) consumed by muscles is used by myosin ATPase in the sarcomeres for contraction, and about 30% is used primarily for Ca^{2+} transport by the sarcoplasmic reticulum to allow muscle relaxation. Skeletal muscles at rest (not exercising) obtain most of their energy from the aerobic respiration of fatty acids. During exercise, muscle glycogen and blood glucose are also used as energy sources (fig. 12.22).

Blood glucose can be used because skeletal muscle contractions during exercise stimulate the insertion of GLUT4 carriers into the sarcolemma (chapter 6; see fig. 6.17). This occurs primarily in the transverse tubules, which comprise most of the surface area of the sarcolemma. The more intense the exercise, the greater will be the number of GLUT4 carriers inserted and thus the greater the rate of glucose uptake (fig. 12.23). This is similar to the action of insulin, which also

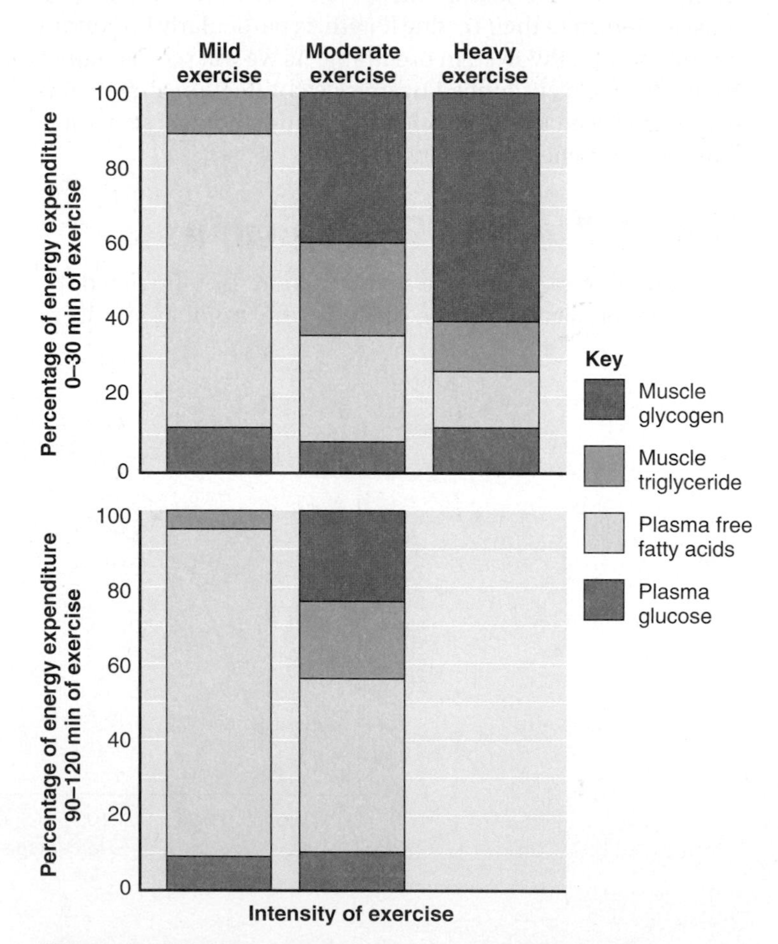

Figure 12.22 Muscle fuel consumption during exercise. The relative contributions of plasma glucose, plasma free fatty acids, muscle glycogen, and muscle triglycerides to the energy consumption of exercising muscles. These are shown during mild exercise (25% of $\dot{V}_{O_2}$ max), moderate exercise (65% of $\dot{V}_{O_2}$ max), and heavy exercise (85% of $\dot{V}_{O_2}$ max). Data for heavy exercise performed at 90 to 120 minutes are not available.

See the *Test Your Quantitative Ability* section of the Review Activities at the end of this chapter.

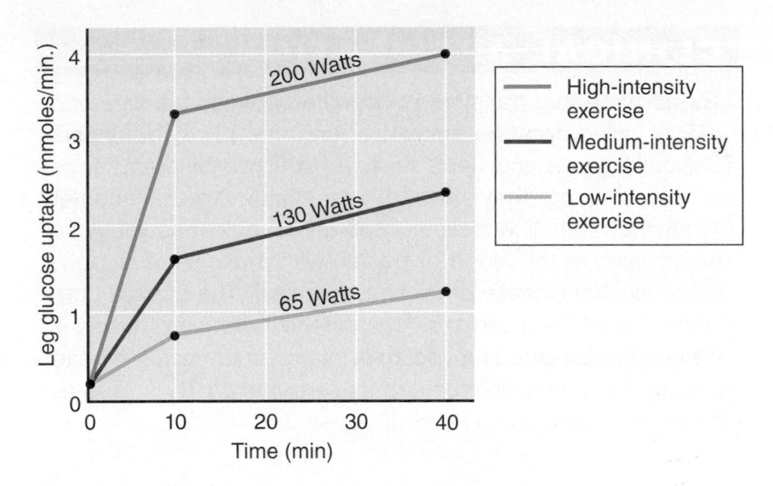

Figure 12.23 Glucose uptake in leg muscle during exercise with a cycle ergometer. Note that the uptake of blood glucose increases with the intensity of the exercise (measured in Watts) and with the exercise time. The increased uptake is largely due to the ability of muscle contraction to increase the amount of GLUT4 carriers in the sarcolemma.

stimulates the insertion of GLUT4 carriers into the sarcolemma and glucose uptake by skeletal muscles (chapter 11; see fig. 11.30). However, the signaling mechanisms by which exercise and insulin stimulate GLUT4 insertion are different, so that their effects can be additive. In addition to increased glucose uptake, exercise promotes the inhibition of glycogen synthesis and the increased uptake and oxidation of fatty acids.

Metabolism of Skeletal Muscles

Skeletal muscles metabolize anaerobically for the first 45 to 90 seconds of moderate-to-heavy exercise, because the cardiopulmonary system requires this amount of time to sufficiently increase the oxygen supply to the exercising muscles. If exercise is moderate, aerobic respiration contributes the major portion of the skeletal muscle energy requirements following the first two minutes of exercise.

Maximal Oxygen Uptake

Whether exercise is light, moderate, or heavy for a given person depends on that person's maximal capacity for aerobic exercise. The maximum rate of oxygen consumption (by aerobic respiration) in the body is called the **maximal oxygen uptake,** or the **aerobic capacity,** and is often expressed in abbreviated form as the **$\dot{V}o_2$ max.** It can be measured accurately by having a person exercise intensely on a treadmill or cycle with devices that measure the ventilation and the oxygen content of the inspired and expired air. More commonly, the $\dot{V}o_2$ max is only estmated by using equations that relate it to the heart rate and work rate during exercise.

The maximal oxygen uptake is determined primarily by a person's age, size, and sex. It is from 15% to 20% higher for males than for females and highest at age 20 for both sexes. The $\dot{V}o_2$ max ranges from about 12 ml of O_2 per minute per kilogram body weight for older, sedentary people to about 84 ml per minute per kilogram for young, elite male athletes. Some world-class athletes have maximal oxygen uptakes that are twice the average for their age and sex—this appears to be due largely to genetic factors, but training can increase the maximum oxygen uptake by about 20%.

The intensity of exercise can also be defined by the **lactate** (or **anaerobic**) **threshold.** This is the percentage of the maximal oxygen uptake at which a significant rise in blood lactate levels occurs. For average healthy people, for example, a significant amount of blood lactate appears when exercise is performed at about 50% to 70% of the $\dot{V}o_2$ max.

During light exercise (at about 25% of the $\dot{V}o_2$ max), most of the exercising muscle's energy is obtained from the aerobic respiration of fatty acids. These are derived mainly from stored fat in adipose tissue, and to a lesser extent from triglycerides stored in the muscle (see fig. 12.22). When a person exercises just below the lactate threshold, where the exercise can be described as moderately intense (at 50% to 70% of the $\dot{V}o_2$ max), the energy is derived almost equally from fatty acids and glucose (obtained from stored muscle glycogen and the blood plasma). By contrast, glucose from these sources supplies two thirds of the energy for muscles during heavy exercise above the lactate threshold.

During exercise, the carrier protein for the facilitated diffusion of glucose (GLUT4) is moved into the muscle fiber's plasma membrane, so that the cell can take up an increasing amount of blood glucose (fig. 12.23). The uptake of plasma glucose contributes 15% to 30% of the muscle's energy needs during moderate exercise and up to 40% of the energy needs during very heavy exercise. This would produce hypoglycemia if the liver failed to increase its output of glucose. The liver increases its output of glucose primarily through hydrolysis of its stored glycogen, but gluconeogenesis (the production of glucose from amino acids, lactate, and glycerol) contributes increasingly to the liver's glucose production as exercise is prolonged.

Clinical Investigation CLUES

Maria was found to have a high maximal oxygen uptake.

- What does this suggest about her lifestyle?
- Is it likely that Maria's complaint of unusual muscle fatigue and pain is simply due to her regular workout?

Oxygen Debt

When a person stops exercising, the rate of oxygen uptake does not immediately return to pre-exercise levels; it returns slowly (the person continues to breathe heavily for some time afterward). This extra oxygen is used to repay the **oxygen debt** incurred during exercise. The oxygen debt includes oxygen that was withdrawn from savings deposits—hemoglobin in blood and myoglobin in muscle (chapter 16, section 16.6); the extra oxygen required for metabolism by tissues warmed during exercise; and the oxygen needed for the metabolism of the lactic acid produced during anaerobic metabolism.

Phosphocreatine

During short, intense bouts of exercise, ATP may be used faster than it can be replenished by anaerobic metabolism and aerobic respiration. At these times the rapid renewal of ATP is extremely important for the exercise to continue. The rapid production of ATP is accomplished by combining ADP with an inorganic phosphate derived from another high-energy compound in the muscle cell known as **phosphocreatine,** or **creatine phosphate.**

Within muscle cells, the phosphocreatine concentration is more than three times the concentration of ATP and represents a ready reserve of high-energy phosphate that can be donated directly to ADP (fig. 12.24). Production of ATP from ADP and phosphocreatine is so efficient that, even though the rate of ATP breakdown rapidly increases from rest to exercise, muscle ATP concentrations decrease only slightly in aerobically adapted muscle. During times of rest, the depleted reserve of phosphocreatine can be restored by the reverse reaction—phosphorylation of creatine with phosphate derived from ATP.

Creatine is produced by the liver and kidneys, and a small amount can be obtained by eating meat and fish. In addition, some athletes take creatine monohydrate dietary supplements, which have been found to increase muscle phosphocreatine by 15% to 40%. Most studies indicate that creatine supplementation can increase muscle weight (due to increased water entry into muscle fibers), strength, and performance during short-term, high-intensity exercise; however, creatine supplementation has not been observed to improve performance during more sustained exercise. Studies of the long-term effects of creatine supplements in rodents suggest possible damaging effects to the liver and kidneys, but the health implications of these studies to long-term creatine supplementation in humans are not established.

CLINICAL APPLICATION

The enzyme that transfers phosphate between creatine and ATP is called **creatine kinase,** or **creatine phosphokinase.** Skeletal muscle and heart muscle have two different forms of this enzyme (they have different isoenzymes—chapter 4, section 4.1). The skeletal muscle isoenzyme is found to be elevated in the blood of people with muscular dystrophy (degenerative disease of skeletal muscles). The plasma concentration of the isoenzyme characteristic of heart muscle is elevated as a result of myocardial infarction (damage to heart muscle), and measurements of this enzyme are thus used as a means of diagnosing heart disease.

Clinical Investigation CLUES

Laboratory tests revealed that Maria had normal levels of creatine phosphokinase (CPK) in her blood.

- What does this test suggest about the health of her skeletal muscles and heart?

Slow- and Fast-Twitch Fibers

Skeletal muscle fibers can be divided on the basis of their contraction speed (time required to reach maximum tension) into **slow-twitch,** or **type I, fibers,** and **fast-twitch,** or **type II, fibers.** In general, the arms have more type II fibers and are accordingly faster than the muscles of the legs. These differences in contraction speed are associated with other differences, including different myosin ATPase isoenzymes that can also be designated as slow or fast. For example, researchers have measured a sixfold difference in the rate of ATP hydrolysis between the myosin of the slow soleus and the fast psoas muscles. The extraocular muscles that position the eyes have a high proportion of fast-twitch fibers and reach maximum tension in about 7.3 msec (milliseconds—thousandths of a second). The soleus muscle in the leg, by contrast, has a high proportion of slow-twitch fibers and requires about 100 msec to reach maximum tension (fig. 12.25).

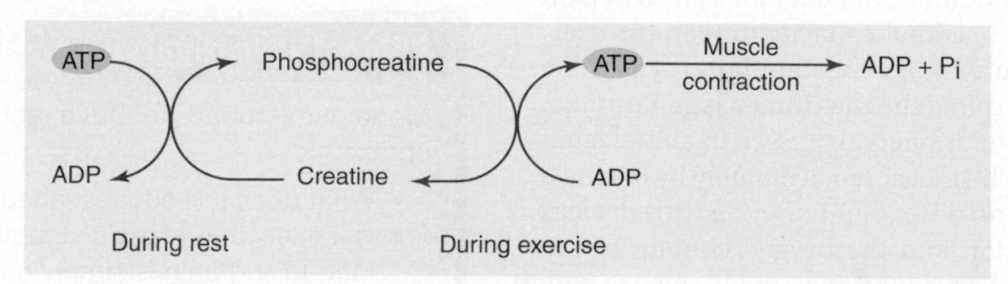

Figure 12.24 The production and use of phosphocreatine in muscles. Phosphocreatine serves as a muscle reserve of high-energy phosphate, used for the rapid formation of ATP. These reactions are catalyzed by creatine phosphokinase (CPK).

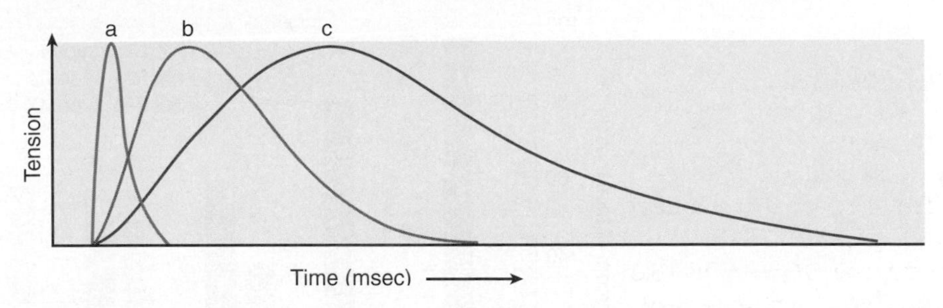

Figure 12.25 A comparison of the rates at which maximum tension is developed in three muscles. These are (*a*) the relatively fast-twitch extraocular and (*b*) gastrocnemius muscles, and (*c*) the slow-twitch soleus muscle.

Muscles like the soleus are *postural muscles;* they are able to sustain a contraction for a long period of time without fatigue. The resistance to fatigue demonstrated by these muscles is aided by other characteristics of slow-twitch (type I) fibers that endow them with a high oxidative capacity for aerobic respiration. Hence, the type I fibers are often referred to as **slow oxidative fibers.** These fibers have a rich capillary supply, numerous mitochondria and aerobic respiratory enzymes, and a high concentration of *myoglobin.* Myoglobin is a red pigment, similar to the hemoglobin in red blood cells, that improves the delivery of oxygen to the slow-twitch fibers. Because of their high myoglobin content, slow-twitch fibers are also called **red fibers.** Because slow, type I muscle fibers can obtain all of the ATP they need through aerobic respiration in mitochondria, they can contract without fatigue (discussed in the next section) longer than can other types of muscle fibers.

The thicker fast-twitch (type II) fibers have fewer capillaries and mitochondria than slow-twitch fibers and not as much myoglobin; hence, these fibers are also called **white fibers.** Fast-twitch fibers are adapted to metabolize anaerobically by a large store of glycogen and a high concentration of glycolytic enzymes.

In addition to the type I (slow-twitch) and type II (fast-twitch) fibers, human muscles have an intermediate fiber type. These intermediate fibers are fast-twitch but also have a high oxidative capacity; therefore, they are relatively resistant to fatigue. They are called type **IIA fibers,** or **fast oxidative fibers,** because of their aerobic ability. The other fast-twitch fibers are anaerobically adapted; these are called **fast glycolytic fibers** because of their high rate of glycolysis. Not all fast glycolytic fibers are alike, however. There are different fibers in this class, which vary in their contraction speeds and glycolytic abilities. In some animals, the extreme fast glycolytic fibers are of the type designated **type IIB fibers.** In humans, these fast glycolytic fibers have a different myosin protein and an even lower oxidative ability than the type IIB fibers, and so are designated as **type IIX fibers.** During maximal exercise, the type IIX fibers have the greatest rate of ATP consumption and the fastest rate of ATP and phosphocreatine depletion. The three major fiber types in humans are compared in table 12.3.

Table 12.3 | Characteristics of Muscle Fiber Types

Feature	Slow Oxidative/Type I (Red)	Fast Oxidative/Type IIA (Red)	Fast Glycolytic/Type IIX (White)
Diameter	Small	Intermediate	Large
Z-line thickness	Wide	Intermediate	Narrow
Glycogen content	Low	Intermediate	High
Resistance to fatigue	High	Intermediate	Low
Capillaries	Many	Many	Few
Myoglobin content	High	High	Low
Respiration	Aerobic	Aerobic	Anaerobic
Oxidative capacity	High	High	Low
Glycolytic ability	Low	High	High
Twitch rate	Slow	Faster	Fastest
Myosin ATPase rate	Low	Higher	Highest

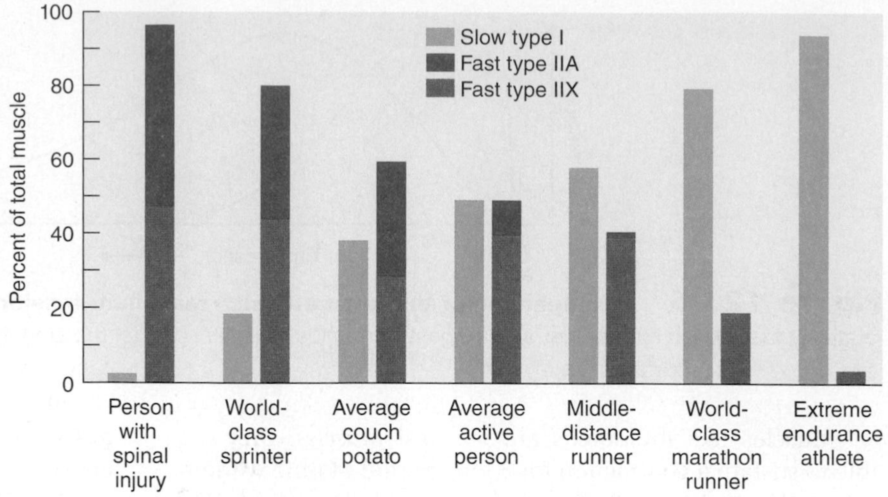

Figure 12.26 Relative abundance of different muscle fiber types in different people. The percentage of slow type I fibers, fast type IIX fibers, and intermediate fast type IIA fibers in the muscles of different people varies tremendously. This is due to differences in genetics and to the effects of physical training.

People vary tremendously in the proportion of fast- and slow-twitch fibers in their muscles (fig. 12.26). The percentage of slow-twitch, type I fibers in the quadriceps femoris muscles of the legs, for example, can vary from under 20% (in people who are excellent sprinters) to as high as 95% (in people who are good marathon runners). These differences are believed to primarily result from differences in genetics, although physical training is also an important factor. This is demonstrated by the complete disappearance of type 1 fibers from human muscle after long-term spinal cord injury (fig. 12.26).

A muscle such as the gastrocnemius contains both fast and slow-twitch fibers, although fast-twitch fibers predominate. A given somatic motor axon, however, innervates muscle fibers of one type only. The sizes of these motor units differ; the motor units composed of slow-twitch fibers tend to be smaller (have fewer fibers) than the motor units of fast-twitch fibers. As mentioned earlier, motor units are recruited from smaller to larger when increasing effort is required; thus, the smaller motor units with slow-twitch fibers would be used most often in routine activities. Larger motor units with fast-twitch fibers, which can exert a great deal of force but which metabolize anaerobically and thus fatigue quickly, would be used relatively infrequently and for only short periods of time.

Muscle Fatigue

Muscle fatigue may be defined as a reversible, exercise-induced reduction in the ability of a muscle to generate force. Fatigue during a sustained maximal contraction, when all the motor units are used and the rate of neural firing is maximal—as when lifting an extremely heavy weight—appears to be due to an accumulation of extracellular K^+. Concentrations of K^+ can become particularly high in the narrow spaces of the transverse tubules. (Remember that K^+ leaves axons and muscle fibers during the repolarization phase of action potentials.) This depolarizes the membrane potential of muscle fibers, interfering with their ability to produce action potentials. Fatigue under these circumstances lasts only a short time, and maximal tension can again be produced after less than a minute's rest.

Muscle fatigue that occurs during other types of exercise appears to have different causes. Chiefly, there is depletion of muscle glycogen and a reduced ability of the sarcoplasmic reticulum to release Ca^{2+}, leading to failure of excitation-contraction coupling. Although failure of excitation-contraction coupling is known to produce muscle fatigue, the reasons for that failure—despite nearly a century of study—are incompletely understood.

It has been known from the early twentieth century that fatigue occurs when lactate accumulates, and that restoring aerobic respiration allows muscle glycogen and contractile ability to recover. This led to the widespread belief that lowered muscle pH is caused by H^+ released from lactic acid, and that the lowered pH produced in this way causes muscle fatigue. However, ongoing research suggests that lactate production may be more coincidental with muscle fatigue than a cause of it.

Several other changes during exercise may produce muscle fatigue through their effects on Ca^{2+} release from the sarcoplasmic reticulum and Ca^{2+} stimulation of contraction. The relative contribution of each of these changes to muscle fatigue depends on the type of exercise performed. The muscle changes with exercise that may contribute to fatigue include the following:

1. Increased concentration of PO_4^{3-}, derived from the breakdown of phosphocreatine, in the cytoplasm. This is currently believed to reduce the force developed by cross bridges and to be a major contributor to muscle fatigue.
2. A decline in ATP, particularly around the junction of the transverse tubules and sarcoplasmic reticulum, that may hinder the action of the Ca^{2+} pumps. ATP declines significantly in fast-twitch fibers during high-intensity exercise, but does not decline enough to produce rigor complexes (as in rigor mortis). ATP does not measurably decline in slow-twitch fibers during exercise.
3. Depletion of muscle glycogen. The mechanisms by which this contributes to fatigue are not fully understood,

but it appears to decrease the release of Ca^{2+} from the sarcoplasmic reticulum.

4. Increased ADP in the cytoplasm. This causes a decrease in the velocity of muscle shortening during muscle fatigue.

The foregoing is a description of the reasons that muscle tissue can fatigue during exercise. When humans exercise, however, we often experience fatigue *before* our muscles themselves have fatigued sufficiently to limit exercise. Put another way, our maximum voluntary muscle force is often less than the maximum force that our muscle is itself capable of producing. This demonstrates **central fatigue**—muscle fatigue caused by changes in the CNS rather than by fatigue of the muscles themselves.

Evidence suggests that central fatigue is complex. In part, it involves a reduced ability of the "upper motoneurons" (interneurons in the brain devoted to motor control) to drive the "lower motoneurons" (in the spinal cord). Muscle fatigue thus has two major components: a peripheral component (fatigue in the muscles themselves) and a central component (fatigue in the CNS that causes reduced activation of muscles by motoneurons).

Adaptations of Muscles to Exercise Training

The maximal oxygen uptake, obtained during very strenuous exercise, averages 50 ml of O_2 per minute per kilogram body weight in males between the ages of 20 and 25 (females average 25% lower). For trained endurance athletes (such as swimmers and long-distance runners), maximal oxygen uptakes can be as high as 86 ml of O_2 per minute per kilogram. These considerable differences affect the lactate threshold, and thus the amount of exercise that can be performed before lactic acid production increases and muscles begin to fatigue. In addition to having a higher aerobic capacity, well-trained athletes also have a lactate threshold that is a higher percentage of their $\dot{V}_{O_2}$ max. The lactate threshold of an untrained person, for example, might be 60% of the $\dot{V}_{O_2}$ max, whereas the lactate threshold of a trained athlete can be up to 80% of the $\dot{V}_{O_2}$ max. These athletes thus produce less lactic acid at a given level of exercise than the average person and are less subject to fatigue.

Because the depletion of muscle glycogen places a limit on exercise, any adaptation that spares muscle glycogen will improve physical endurance. Trained athletes have a slower depletion of their stored glycogen because they derive a higher proportion of their muscle energy from the aerobic respiration of fatty acids. The greater the level of physical training, the higher the proportion of energy derived from the oxidation of fatty acids during exercise below the $\dot{V}_{O_2}$ max.

All fiber types adapt to endurance training by an increase in mitochondria, and thus in aerobic respiratory enzymes. In fact, the maximal oxygen uptake can be increased by as much as 20% through endurance training. There is a decrease in type IIX (fast glycolytic) fibers, which have a low oxidative capacity, accompanied by an increase in type IIA (fast oxidative) fibers, which have a high oxidative capacity. Although the type IIA fibers are still classified as fast-twitch, they show an increase in the slow myosin ATPase isoenzyme form, indicating that they are in a transitional state between the type II and type I fibers.

Skeletal muscles can store triglycerides both within the muscle fibers and in adipocytes between muscle fibers. The fat storage outside of the fibers is increased in obesity and type 2 diabetes mellitus (chapter 19) and reduced by aerobic exercise. The triglycerides within the muscle fibers are also increased in obesity, and in this case are associated with increased insulin resistance and type 2 diabetes. Insulin resistance (decreased insulin sensitivity) in an obese person is promoted because the skeletal muscle fibers take in fatty acids but have a reduced ability to oxidize them. Endurance-trained athletes, who have a lower risk of insulin resistance and diabetes, surprisingly also have elevated intracellular triglycerides within their skeletal muscle fibers (table 12.4). This is possible because adaptations within the muscle fibers of these athletes allow the fibers to completely oxidize fatty acids. As a result, diglycerides and long-chain fatty acids cannot accumulate where they can interfere with the insulin signaling of glucose uptake.

The beneficial effects of a moderate exercise program and a modest weight loss on insulin sensitivity are well documented. There is also evidence that regular exercise has a general anti-inflammatory effect, helping to reduce the risk of cardiovascular and pulmonary diseases, colon cancer, and other pathological conditions promoted by inflammation. The anti-inflammatory effects of exercise may be mediated by a variety of mechanisms, including reduced visceral fat (which releases pro-inflammatory molecules); increased adrenal secretion of cortisol (which suppresses inflammation); the release of interleukin-6 from exercising muscles

Table 12.4 | Effects of Endurance Training on Skeletal Muscles

1.	Improved ability to obtain ATP from oxidative phosphorylation
2.	Increased size and number of mitochondria
3.	Less lactic acid produced per given amount of exercise
4.	Increased myoglobin content
5.	Increased intramuscular triglyceride content
6.	Increased lipoprotein lipase (enzyme needed to utilize lipids from blood)
7.	Increased proportion of energy derived from fat; less from carbohydrates
8.	Lower rate of glycogen depletion during exercise
9.	Improved efficiency in extracting oxygen from blood
10.	Decreased number of type IIX (fast glycolytic) fibers; increased number of type IIA (fast oxidative) fibers

(which instigates a cascade of anti-inflammatory effects); and others.

Endurance training does not increase the size of muscles. Muscle enlargement is produced only by frequent periods of high-intensity exercise in which muscles work against a high resistance, as in weightlifting. As a result of resistance training, type II muscle fibers become thicker, and the muscle therefore grows by hypertrophy (an increase in cell size rather than number of cells). This happens first because the myofibrils within a muscle fiber thicken due to the synthesis of actin and myosin proteins and the addition of new sarcomeres. Then, after a myofibril has attained a certain thickness, it may split into two myofibrils, each of which may become thicker as a result of the addition of sarcomeres. Muscle hypertrophy, in short is associated with an increase in the size of the myofibrils and then in the number of myofibrils within the muscle fibers.

The decline in physical strength of older people is associated with a reduced muscle mass, which is due to a loss of muscle fibers and to a decrease in the size of fast-twitch muscle fibers. Aging is also associated with a reduced density of blood capillaries surrounding the muscle fibers, leading to a decrease in oxidative capacity. Resistance training can cause the surviving muscle fibers to hypertrophy and become stronger, partially compensating for the decline in the number of muscle fibers in elderly people. Endurance training can increase the density of blood capillaries in the muscles, improving the ability of the blood to deliver oxygen to the muscles. The muscle glycogen of older people can also be increased by endurance training, but it cannot be raised to the levels present in youth.

Muscle Damage and Repair

Destruction of striated muscle fibers is particularly damaging because the remaining healthy fibers cannot divide to replace the damaged ones. However, skeletal muscles have stem cells known as **satellite cells,** located between the sarcolemma and the basal lamina (the basement membrane just outside the sarcolemma). The satellite cells are activated at the site of muscle injury to differentiate into myoblasts that can fuse with the damaged muscle fibers. If the damage is more extensive, a number of satellite cells can form myoblasts that fuse to produce new muscle fibers, which can then grow thicker by the fusion of additional myoblasts. When muscles *hypertrophy,* or grow larger as a result of increased fiber thickness, the number of nuclei in each fiber must increase in proportion to the larger volume of the fiber. These new nuclei are provided by the satellite cells.

With age there is a decline in the number of satellite cells, their ability to proliferate in response to muscle damage, and their ability to form new muscle fibers. Scientists demonstrated that satellite cells from young muscles have a reduced ability to proliferate and form new muscle fibers when they are exposed to old muscle fibers, suggesting that the aged muscle fibers secrete an inhibitor of satellite cell function. One recent report suggests that this inhibitor is *transforming growth factor (TGF)-β*. Other studies suggest that elderly people with declining muscle mass produce increased amounts of a different paracrine regulator, myostatin.

Myostatin is a paracrine regulator in skeletal muscles that is able to inhibit satellite cell function and muscle growth. Lowering myostatin might thus be expected to increase muscle mass. Indeed, mice and cattle with the gene for producing myostatin "knocked out" have greatly increased muscle mass. The functions of myostatin, and the mechanisms that regulate satellite cell proliferation and formation of myotubes, have many potential health applications and are currently active areas of research.

The formation of new sarcomeres and the consequent growth of myofibrils within the muscle fiber require three gigantic muscle proteins. *Titin* (previously discussed) is an extremely long protein that spans half the length of the sarcomere, with its amino-terminal end anchored in one of the Z-discs on each side of the sarcomere and its carboxyl-terminal end in the M-band. Two other giant proteins associated with the myofibrils are *nebulin* (within the actin of the I bands) and *obscurin* (surrounding the sarcomeres primarily around the Z-discs and M-bands). These three giant proteins serve as molecular scaffolding for the formation of new sarcomeres during muscle growth and repair. For example, obscurin may help the myosin proteins assemble into the A bands, and nebulin is needed for the globular actin proteins to assemble into thin filaments of the appropriate length for the muscle. Titin serves a scaffolding function in addition to its previously described contribution to muscle elasticity, unfolding and developing passive tension when a muscle is stretched.

FITNESS APPLICATION

Muscle atrophy (reduction in size) and accompanying declines in muscle strength occur in the weight-bearing muscles of the legs when astronauts experience *microgravity* (weightlessness) for long periods. For example, reductions in muscle volume and performance were measured in the United States *Skylab* missions. However, in *Skylab 4* (which lasted 84 days), adjustments in the diet and the exercise program were able to significantly compensate for the effects of microgravity on tested muscles. Like the effects of weightlessness in astronauts, weight-bearing muscles are similarly "unloaded" in bedridden people and in people with a leg immobilized by a cast. In prolonged bed rest of two to three weeks, the calf and leg muscles experience declines in size and strength comparable to those seen in space flights. Immobilization of the leg in a cast results in more rapid declines in muscle performance and size than those observed for similar time periods in bed rest or the microgravity of space.

CHECKPOINT

8. Draw a figure illustrating the relationship between ATP and creatine phosphate, and explain the physiological significance of this relationship.
9. Describe the characteristics of slow- and fast-twitch fibers (including intermediate fibers). Explain how the fiber types are determined and list the functions of different fiber types.
10. Explain the different causes of muscle fatigue with reference to the various fiber types.
11. Describe the effects of endurance training and resistance training on the fiber characteristics of muscles.

12.5 NEURAL CONTROL OF SKELETAL MUSCLES

Skeletal muscles contain stretch receptors called muscle spindles that stimulate the production of impulses in sensory neurons when a muscle is stretched. These sensory neurons can synapse with alpha motoneurons, which stimulate the muscle to contract in response to the stretch.

LEARNING OUTCOMES

After studying this section, you should be able to:

12. Describe the components of monosynaptic muscle stretch reflexes, including the role of gamma motoneurons.
13. Describe the effects of Golgi tendon organs.
14. Explain reciprocal innervation of skeletal muscles.
15. Explain the functions of alpha and gamma motoneurons during the voluntary control of muscle contraction.

Lower motor neurons (often shortened to *motoneurons*) are somatic motor neurons with cell bodies in the brain stem and spinal cord and axons that travel within nerves to stimulate skeletal muscle contraction (table 12.5). The activity of these neurons is influenced by (1) sensory feedback from the muscles and tendons and (2) facilitory and inhibitory effects from **upper motor neurons,** which are interneurons in the brain that contribute axons to descending motor tracts. Lower motor neurons are thus said to be the *final common pathway* by which sensory stimuli and higher brain centers exert control over skeletal movements.

The cell bodies of lower motor neurons are located in the ventral horn of the gray matter of the spinal cord. Axons from these cell bodies leave the ventral side of the spinal cord to form the *ventral roots* of spinal nerves (chapter 8; see fig. 8.28).

Table 12.5 | A Partial Listing of Terms Used to Describe the Neural Control of Skeletal Muscles

Term	Description
1. Lower motoneurons	Neurons whose axons innervate skeletal muscles—also called the "final common pathway" in the control of skeletal muscles
2. Higher motoneurons	Neurons in the brain that are involved in the control of skeletal movements and that act by facilitating or inhibiting (usually by way of interneurons) the activity of the lower motoneurons
3. Alpha motoneurons	Lower motoneurons whose fibers innervate ordinary (extrafusal) muscle fibers
4. Gamma motoneurons	Lower motoneurons whose fibers innervate the muscle spindle fibers (intrafusal fibers)
5. Agonist/antagonist	A pair of muscles or muscle groups that insert on the same bone, the agonist being the muscle of reference
6. Synergist	A muscle whose action facilitates the action of the agonist
7. Ipsilateral/contralateral	Ipsilateral—located on the same side, or the side of reference; contralateral—located on the opposite side
8. Afferent/efferent	Afferent neurons—sensory; efferent neurons—motor

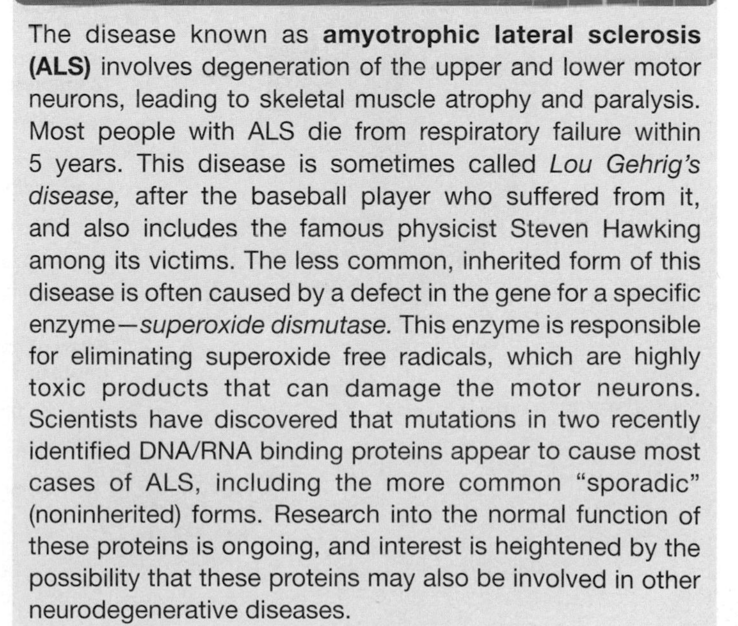

CLINICAL APPLICATION

The disease known as **amyotrophic lateral sclerosis (ALS)** involves degeneration of the upper and lower motor neurons, leading to skeletal muscle atrophy and paralysis. Most people with ALS die from respiratory failure within 5 years. This disease is sometimes called *Lou Gehrig's disease,* after the baseball player who suffered from it, and also includes the famous physicist Steven Hawking among its victims. The less common, inherited form of this disease is often caused by a defect in the gene for a specific enzyme—*superoxide dismutase.* This enzyme is responsible for eliminating superoxide free radicals, which are highly toxic products that can damage the motor neurons. Scientists have discovered that mutations in two recently identified DNA/RNA binding proteins appear to cause most cases of ALS, including the more common "sporadic" (noninherited) forms. Research into the normal function of these proteins is ongoing, and interest is heightened by the possibility that these proteins may also be involved in other neurodegenerative diseases.

The *dorsal roots* of spinal nerves contain sensory fibers whose cell bodies are located in the *dorsal root ganglia.* Both sensory (*afferent*) and motor (*efferent*) fibers join in a common connective tissue sheath to form the spinal nerves at each segment of the spinal cord. In the lumbar region there are about 12,000 sensory and 6,000 motor fibers per spinal nerve.

About 375,000 cell bodies have been counted in a lumbar segment—a number far larger than can be accounted for by the number of motor neurons. Most of these neurons do not contribute fibers to the spinal nerve. Rather, they serve as *interneurons,* whose fibers conduct impulses up, down, and across the central nervous system. Those fibers that conduct impulses to higher spinal cord segments and the brain form *ascending tracts,* and those that conduct to lower spinal segments contribute to *descending tracts.* Those fibers that cross the midline of the CNS to synapse on the opposite side are part of *commissural tracts.* Interneurons can conduct impulses up and down on the same, or *ipsilateral,* side, and can affect neurons on the opposite, or *contralateral,* side of the central nervous system.

Muscle Spindle Apparatus

In order for the nervous system to control skeletal movements properly, it must receive continuous sensory feedback concerning the effects of its actions. This sensory information includes (1) the tension that the muscle exerts on its tendons, provided by the **Golgi tendon organs,** and (2) muscle length, provided by the **muscle spindle apparatus.** The spindle apparatus, so called because it is wider in the center and tapers toward the ends, functions as a length detector. Muscles that require the finest degree of control, such as the muscles of the hand, have the highest density of spindles.

Each spindle apparatus contains several thin muscle cells called *intrafusal fibers* (*fusus* = spindle) packaged within a connective tissue sheath. Like the stronger and more numerous "ordinary" muscle fibers outside the spindles—the *extrafusal fibers*—the spindles insert into tendons on each end of the muscle. Spindles are therefore said to be in parallel with the extrafusal fibers.

Unlike the extrafusal fibers, which contain myofibrils along their entire length, the contractile apparatus is absent from the central regions of the intrafusal fibers. The central, noncontracting part of an intrafusal fiber contains nuclei. There are two types of intrafusal fibers. One type, the *nuclear bag fibers,* have their nuclei arranged in a loose aggregate in the central regions of the fibers. The other type of intrafusal fibers, called *nuclear chain fibers,* have their nuclei arranged in rows. Two types of sensory neurons serve these intrafusal fibers. **Primary,** or **annulospiral, sensory endings** wrap around the central regions of the nuclear bag and chain fibers (fig. 12.27),

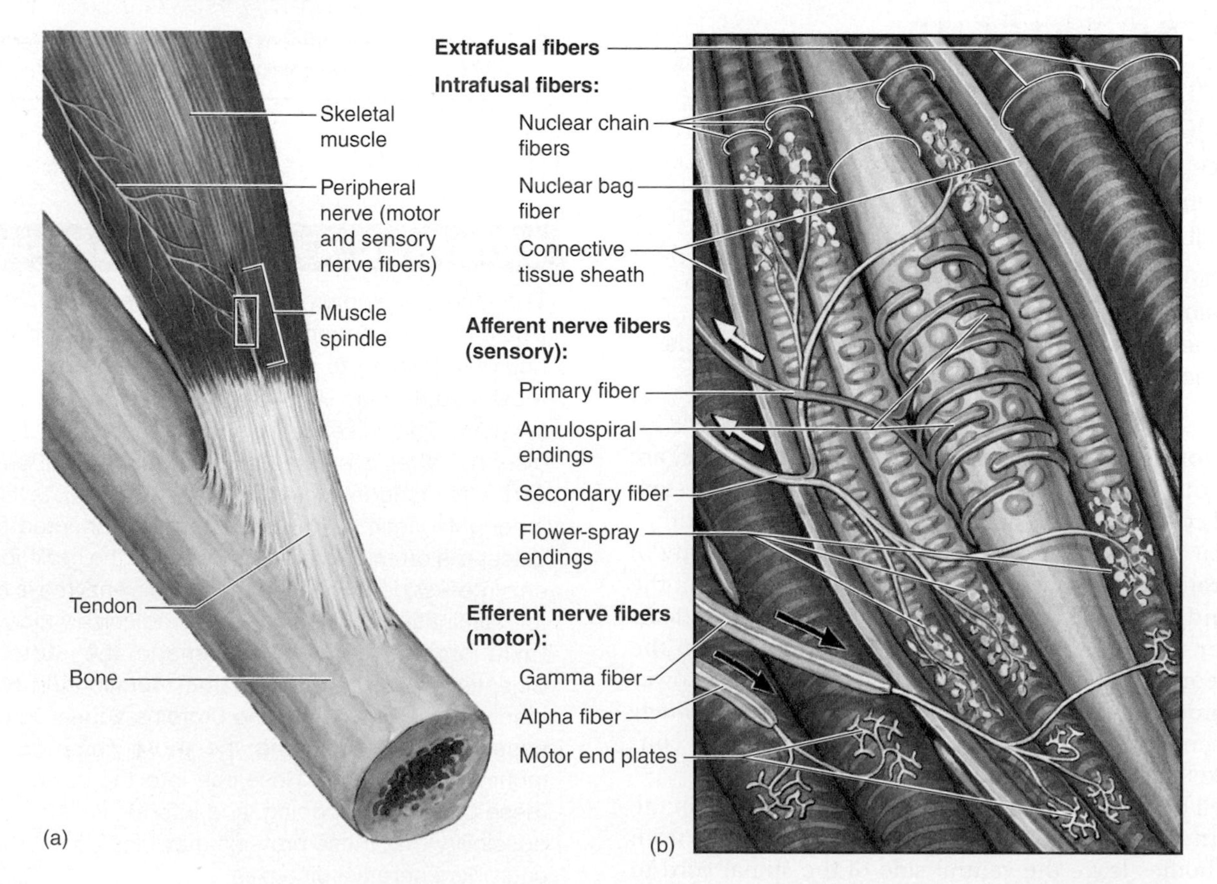

Figure 12.27 The location and structure of a muscle spindle. (*a*) A muscle spindle within a skeletal muscle. (*b*) The structure and innervation of a muscle spindle.

and **secondary,** or **flower-spray, endings** are located over the contracting poles of the nuclear chain fibers.

Because the spindles are arranged in parallel with the extrafusal muscle fibers, stretching a muscle causes its spindles to stretch. This stimulates both the primary and secondary sensory endings. The spindle apparatus thus serves as a length detector because the frequency of impulses produced in the primary and secondary endings is proportional to the length of the muscle. The primary endings, however, are most stimulated at the onset of stretch, whereas the secondary endings respond in a more tonic (sustained) fashion as stretch is maintained. Sudden, rapid stretching of a muscle activates both types of sensory endings, and is thus a more powerful stimulus for the muscle spindles than a slower, more gradual stretching that has less of an effect on the primary sensory endings. Thus, the force of this reflex contraction is greater in response to rapid stretch than to gradual stretch.

FITNESS APPLICATION

Rapid stretching of skeletal muscles produces very forceful muscle contractions as a result of the activation of primary and secondary endings in the muscle spindles and the monosynaptic stretch reflex. This can result in painful muscle spasms, as may occur, for example, when muscles are forcefully pulled in the process of setting broken bones. Painful muscle spasms may be avoided in physical exercise by stretching slowly and thereby stimulating mainly the secondary endings in the muscle spindles. A slower rate of stretch also allows time for the inhibitory Golgi tendon organ reflex to occur and promote muscle relaxation.

Alpha and Gamma Motoneurons

In the spinal cord, two types of lower motor neurons innervate skeletal muscles. The motor neurons that innervate the extrafusal muscle fibers are called **alpha motoneurons;** those that innervate the intrafusal fibers are called **gamma motoneurons** (fig. 12.27). The alpha motoneurons are faster conducting (60 to 90 meters per second) than the thinner gamma motoneurons (10 to 40 meters per second). Because only the extrafusal muscle fibers are sufficiently strong and numerous to cause a muscle to shorten, only stimulation by the alpha motoneurons can cause muscle contraction that results in skeletal movements.

The intrafusal fibers of the muscle spindle are stimulated to contract by gamma motoneurons, which represent one-third of all efferent fibers in spinal nerves. However, because the intrafusal fibers are too few in number and their contraction is too weak to cause a muscle to shorten, stimulation by gamma motoneurons results only in isometric contraction of the spindles. Because myofibrils are present in the poles but absent in the central regions of intrafusal fibers, the more distensible central region of the intrafusal fiber is pulled toward the ends in response to stimulation by gamma motoneurons. As a result, the spindle is tightened. This effect of gamma motoneurons, which is sometimes termed *active stretch* of the spindles, serves to increase the sensitivity of the spindles when the entire muscle is passively stretched by external forces. The activation of gamma neurons thereby enhances the stretch reflex. These neurons are also important in the voluntary control of skeletal muscles, as described next.

Coactivation of Alpha and Gamma Motoneurons

Most of the fibers in the descending motor tracts synapse with interneurons in the spinal cord; only about 10% of the descending fibers synapse directly with the lower motor neurons. It is likely that very rapid movements are produced by direct synapses with the lower motor neurons, whereas most other movements are produced indirectly via synapses with spinal interneurons, which in turn stimulate the motor neurons.

Upper motor neurons—interneurons in the brain that contribute fibers to descending motor tracts—usually stimulate alpha and gamma motoneurons simultaneously. Such stimulation is known as **coactivation.** Stimulation of alpha motoneurons results in muscle contraction and shortening; activation of gamma motoneurons stimulates contraction of the intrafusal fibers, and thus "takes out the slack" that would otherwise be present in the spindles as the muscles shorten. In this way, the spindles remain under tension and provide information about the length of the muscle even while the muscle is shortening.

Under normal conditions, the activity of gamma motoneurons is maintained at the level needed to keep the muscle spindles under proper tension while the muscles are relaxed. Undue relaxation of the muscles is prevented by stretch and activation of the spindles, which in turn elicits a reflex contraction (described in the next section). This mechanism produces a normal resting muscle length and state of tension, or **muscle tone.**

Skeletal Muscle Reflexes

Although skeletal muscles are often called voluntary muscles because they are controlled by descending motor pathways that are under conscious control, they often contract in an unconscious, reflex fashion in response to particular stimuli. In the simplest type of reflex, a skeletal muscle contracts in response to the stimulus of muscle stretch. More complex reflexes involve inhibition of antagonistic muscles and regulation of a number of muscles on both sides of the body.

The Monosynaptic Stretch Reflex

Reflex contraction of skeletal muscles occurs in response to sensory input and does not depend on the activation of upper motor neurons. The **reflex arc,** which describes the nerve impulse pathway from sensory to motor endings in such reflexes, involves only a few synapses within the CNS. The simplest of all reflexes—the *muscle stretch reflex*—consists of only one synapse within the CNS. The sensory neuron directly synapses with the motor neuron, without involving spinal cord interneurons. The stretch reflex is thus a **monosynaptic reflex** in terms of the individual reflex arcs (although many sensory neurons are activated at the same time, leading to the activation of many motor neurons). Stretch reflexes maintain muscles at an optimal length, as previously described under the heading "Length-Tension Relationship" in section 12.3.

The stretch reflex is present in all muscles, but it is most dramatic in the extensor muscles of the limbs. The **knee-jerk reflex (patellar tendon reflex)**—the most commonly evoked stretch reflex—is initiated by striking the patellar ligament with a rubber mallet. This stretches the entire body of the muscle, and thus passively stretches the spindles within the muscle so that sensory nerves with primary (annulospiral) endings in the spindles are activated. Axons of these sensory neurons synapse within the ventral gray matter of the spinal cord with *alpha motoneurons.* These large, fast-conducting neurons stimulate the extrafusal fibers of the quadriceps femoris muscles to produce concentric (shortening) contractions, resulting in the reflex extension of the knee joint. This is an example of negative feedback—stretching of the muscles (and spindles) stimulates shortening of the muscles (and spindles). These events are summarized in table 12.6 and illustrated in figure 12.28.

Table 12.6 | Summary of Events in a Monosynaptic Stretch Reflex

1. Passive stretch of a muscle (produced by tapping its tendon) stretches the spindle (intrafusal) fibers.
2. Stretching of a spindle distorts its central (bag or chain) region, which stimulates dendritic endings of sensory neurons.
3. Action potentials are conducted by afferent (sensory) nerve fibers into the spinal cord on the dorsal roots of spinal nerves.
4. Axons of sensory neurons synapse with dendrites and cell bodies of somatic motor neurons located in the ventral horn gray matter of the spinal cord.
5. Efferent nerve impulses in the axons of alpha motoneurons in the ventral roots of spinal nerves are conducted to the ordinary (extrafusal) muscle fibers.
6. Release of acetylcholine from the endings of alpha motoneurons stimulates the contraction of the extrafusal fibers, and thus of the whole muscle.
7. Contraction of the muscle relieves the stretch of its spindles, thus decreasing activity in the spindle afferent nerve fibers.

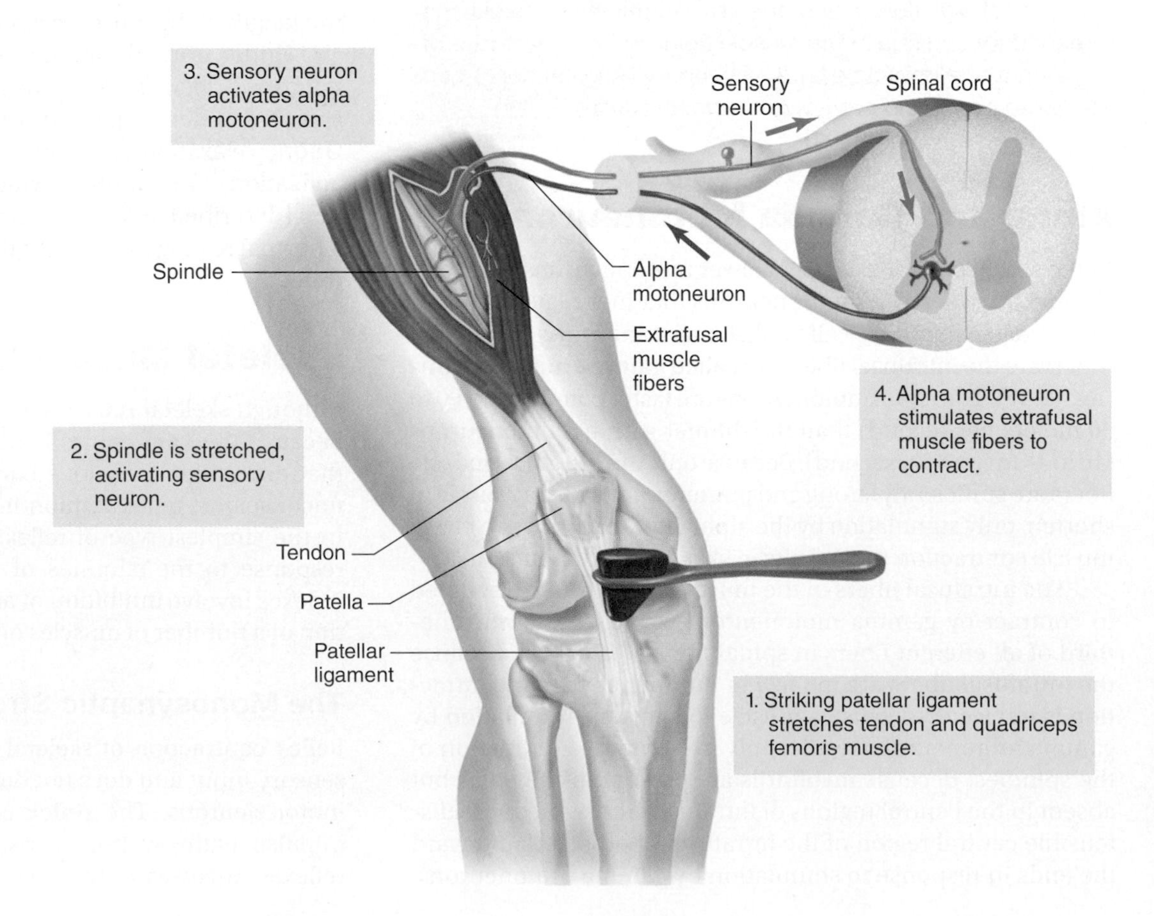
3. Sensory neuron activates alpha motoneuron.
Sensory neuron
Spinal cord
Spindle
Alpha motoneuron
Extrafusal muscle fibers
4. Alpha motoneuron stimulates extrafusal muscle fibers to contract.
2. Spindle is stretched, activating sensory neuron.
Tendon
Patella
Patellar ligament
1. Striking patellar ligament stretches tendon and quadriceps femoris muscle.

Figure 12.28 The knee-jerk reflex. This is an example of a monosynaptic stretch reflex.

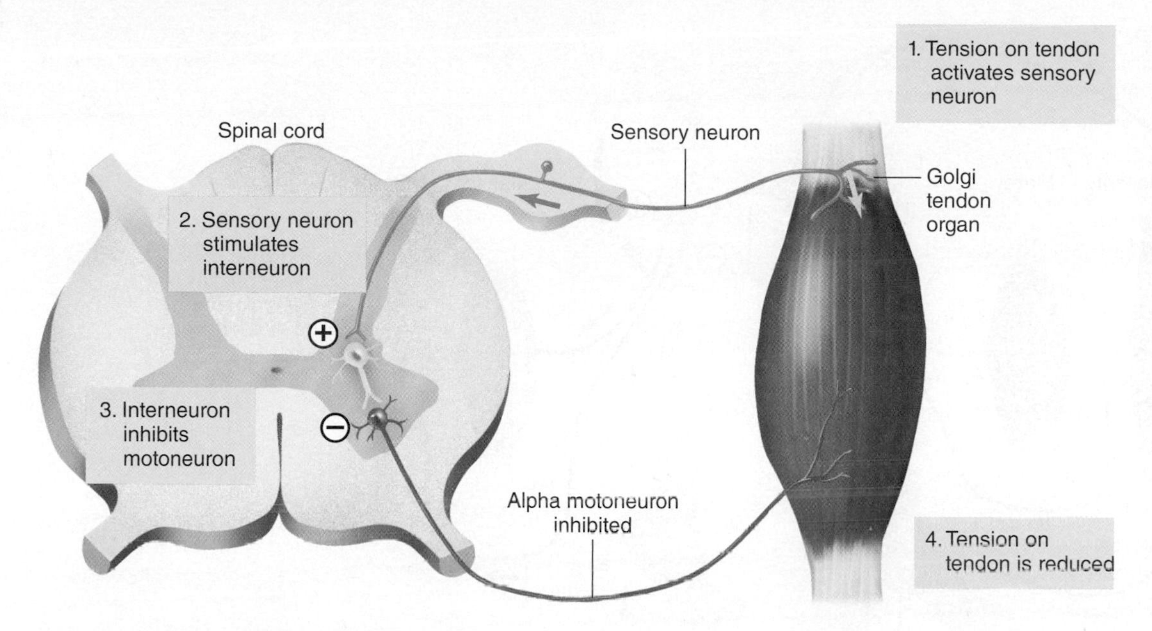
1. Tension on tendon activates sensory neuron
Spinal cord
Sensory neuron
Golgi tendon organ
2. Sensory neuron stimulates interneuron
3. Interneuron inhibits motoneuron
Alpha motoneuron inhibited
4. Tension on tendon is reduced

Figure 12.29 The action of the Golgi tendon organ. An increase in muscle tension stimulates the activity of sensory nerve endings in the Golgi tendon organ. This sensory input stimulates an interneuron, which in turn inhibits the activity of a motor neuron innervating that muscle. This is therefore a disynaptic reflex.

Golgi Tendon Organs

The **Golgi tendon organs** continuously monitor tension in the tendons produced by muscle contraction or passive stretching of a muscle. Sensory neurons from these receptors synapse with interneurons in the spinal cord; these interneurons, in turn, have *inhibitory synapses* (via IPSPs and postsynaptic inhibition—chapter 7) with motor neurons that innervate the muscle (fig. 12.29). The inhibitory Golgi tendon reflex is a **disynaptic reflex** because two synapses are crossed in the CNS. One is an excitatory synapse between a sensory neuron and a spinal interneuron, and the other is an inhibitory synapse between the spinal interneuron and the alpha motoneuron. This inhibitory reflex helps prevent dangerous tension on a tendon from excessive muscle contraction, or from muscle contraction that could add to the tension on a tendon during passive stretching of the muscle. Indeed, if a muscle is stretched extensively, it will actually relax as a result of the inhibitory effects produced by the Golgi tendon organs.

Reciprocal Innervation and the Crossed-Extensor Reflex

In the knee-jerk and other stretch reflexes, the sensory neuron that stimulates the motor neuron of a muscle also stimulates interneurons within the spinal cord via collateral branches. These interneurons inhibit the motor neurons of antagonist muscles via inhibitory postsynaptic potentials (IPSPs). This dual stimulatory and inhibitory activity is called **reciprocal innervation** (fig. 12.30).

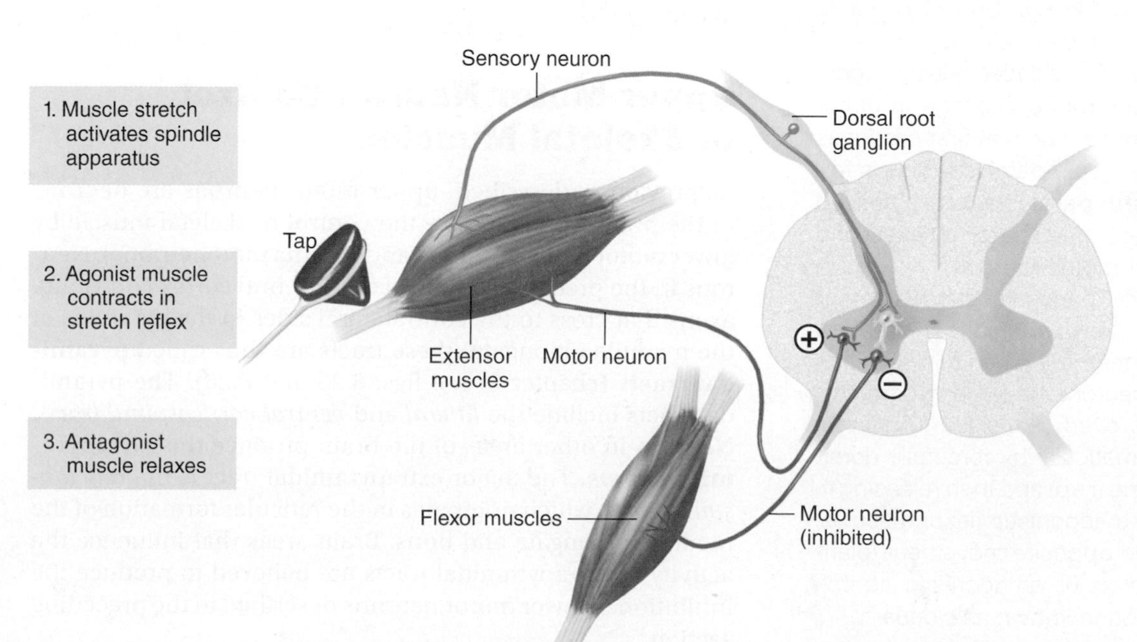
1. Muscle stretch activates spindle apparatus
2. Agonist muscle contracts in stretch reflex
3. Antagonist muscle relaxes
Sensory neuron
Dorsal root ganglion
Tap
Extensor muscles
Motor neuron
Flexor muscles
Motor neuron (inhibited)

Figure 12.30 A diagram of reciprocal innervation. Afferent impulses from muscle spindles stimulates alpha motoneurons to the agonist muscle (the extensor) directly, but (via an inhibitory interneuron) they inhibit activity in the alpha motoneuron to the antagonist muscle.

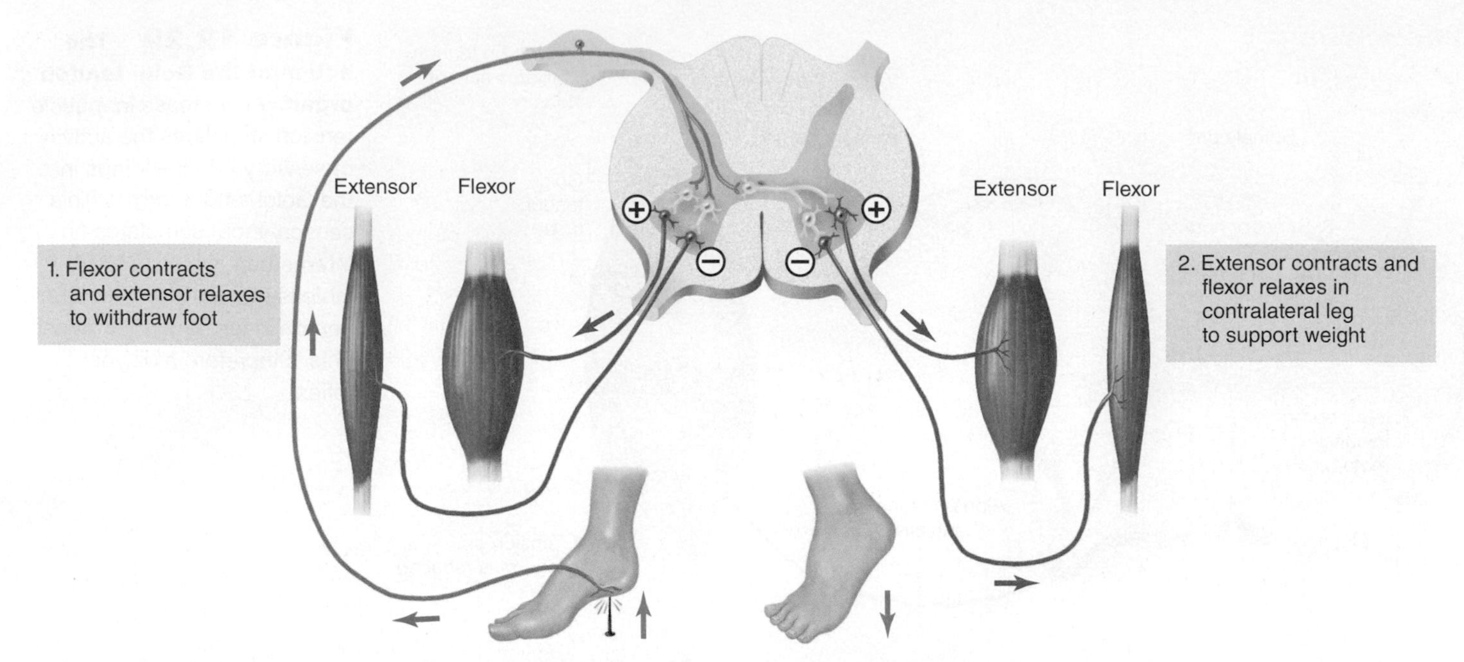
Extensor
Flexor
Extensor
Flexor
1. Flexor contracts and extensor relaxes to withdraw foot
2. Extensor contracts and flexor relaxes in contralateral leg to support weight

Figure 12.31 The crossed-extensor reflex. This complex reflex demonstrates double reciprocal innervation.

When a limb is flexed, for example, the antagonistic extensor muscles are passively stretched. Extension of a limb similarly stretches the antagonistic flexor muscles. If the monosynaptic stretch reflexes were not inhibited, reflex contraction of the antagonistic muscles would always interfere with the intended movement. Fortunately, whenever the agonist muscles are stimulated to contract, the alpha and gamma motoneurons that stimulate the antagonist muscles are inhibited.

CLINICAL APPLICATION

Damage to spinal nerves, or to the cell bodies of lower motor neurons (by poliovirus, for example), produces a **flaccid paralysis,** characterized by reduced muscle tone, depressed stretch reflexes, and atrophy. Damage to upper motor neurons or descending motor tracts at first produces spinal shock in which there is a flaccid paralysis. This is followed in a few weeks by **spastic paralysis,** characterized by increased muscle tone, exaggerated stretch reflexes, and other signs of hyperactive lower motor neurons.

The appearance of spastic paralysis suggests that upper motor neurons normally exert an inhibitory effect on lower alpha and gamma motor neurons. When this inhibition is removed, the gamma motoneurons become hyperactive and the spindles thus become overly sensitive to stretch. This can be demonstrated dramatically by forcefully dorsiflecting the patient's foot (pushing it up) and then releasing it. Forced extension stretches the antagonistic flexor muscles, which contract and produce the opposite movement (plantar flexion). Alternating activation of antagonistic stretch reflexes produces a flapping motion known as *clonus.*

The stretch reflex, with its reciprocal innervations, involves the muscles of one limb only and is controlled by only one segment of the spinal cord. More complex reflexes involve muscles controlled by numerous spinal cord segments and affect muscles on the contralateral side of the cord. Such reflexes involve **double reciprocal innervation** of muscles.

Double reciprocal innervation is illustrated by the **crossed-extensor reflex.** If you step on a tack with your right foot, for example, this foot is withdrawn by contraction of the flexors and relaxation of the extensors of your right leg. The contralateral left leg, by contrast, extends to help support your body during this withdrawal reflex. The extensors of your left leg contract while its flexors relax. These events are illustrated in figure 12.31.

Upper Motor Neuron Control of Skeletal Muscles

As previously described, upper motor neurons are neurons in the brain that influence the control of skeletal muscle by lower motor neurons (alpha and gamma motoneurons). Neurons in the precentral gyrus of the cerebral cortex contribute axons that cross to the contralateral sides in the pyramids of the medulla oblongata; these tracts are thus called **pyramidal tracts** (chapter 8, see figs. 8.25 and 8.26). The pyramidal tracts include the *lateral* and *ventral corticospinal tracts.* Neurons in other areas of the brain produce the **extrapyramidal tracts.** The major extrapyramidal tract is the *reticulospinal tract,* which originates in the reticular formation of the medulla oblongata and pons. Brain areas that influence the activity of extrapyramidal tracts are believed to produce the inhibition of lower motor neurons described in the preceding section.

Cerebellum

The **cerebellum,** like the cerebrum, receives sensory input from muscle spindles and Golgi tendon organs. It also receives fibers from areas of the cerebral cortex devoted to vision, hearing, and equilibrium.

There are no descending tracts from the cerebellum. The cerebellum can influence motor activity only indirectly, through its output to the vestibular nuclei, red nucleus, and basal nuclei (chapter 8; see fig. 8.26). These structures, in turn, affect lower motor neurons via the vestibulospinal tract, rubrospinal tract, and reticulospinal tract. It is interesting that all output from the cerebellum is inhibitory; these inhibitory effects aid motor coordination by eliminating inappropriate neural activity. Damage to the cerebellum interferes with the ability to coordinate movements with spatial judgment. Under- or overreaching for an object may occur, followed by *intention tremor,* in which the limb moves back and forth in a pendulum-like motion.

Basal Nuclei

The **basal nuclei,** also called the **basal ganglia,** include the *caudate nucleus, putamen,* and *globus pallidus* (chapter 8; see fig. 8.11). Often included in this group are other nuclei of the *thalamus, subthalamus, substantia nigra,* and *red nucleus.* Acting directly via the rubrospinal tract and indirectly via synapses in the reticular formation and thalamus, the basal nuclei have profound effects on the activity of lower motor neurons.

In particular, through their synapses in the reticular formation (see fig. 8.26), the basal nuclei exert an inhibitory influence on the activity of lower motor neurons. Damage to the basal nuclei thus results in increased muscle tone, as previously described. People with such damage display *akinesia,* lack of desire to use the affected limb, and *chorea,* sudden and uncontrolled random movements (table 12.7).

Table 12.7 | Symptoms of Upper Motor Neuron Damage

Babinski's reflex—Extension of the great toe when the sole of the foot is stroked along the lateral border
Spastic paralysis—High muscle tone and hyperactive stretch reflexes; flexion of arms and extension of legs
Hemiplegia—Paralysis of upper and lower limbs on one side—commonly produced by damage to motor tracts as they pass through internal capsule (such as by cerebrovascular accident—stroke)
Paraplegia—Paralysis of the lower limbs on both sides as a result of lower spinal cord damage
Quadriplegia—Paralysis of upper and lower limbs on both sides as a result of damage to the upper region of the spinal cord or brain
Chorea—Random uncontrolled contractions of different muscle groups (as in Saint Vitus' dance) as a result of damage to basal nuclei
Resting tremor—Shaking of limbs at rest; disappears during voluntary movements; produced by damage to basal nuclei
Intention tremor—Oscillations of the arm following voluntary reaching movements; produced by damage to cerebellum

CLINICAL APPLICATION

Parkinson's disease (or *paralysis agitans*) is a disorder of the basal nuclei involving degeneration of fibers from the substantia nigra. These fibers, which use dopamine as a neurotransmitter, are required to antagonize the effects of other fibers that use acetylcholine (ACh) as a transmitter. The relative deficiency of dopamine compared to ACh is believed to produce the symptoms of Parkinson's disease, including resting tremor. This "shaking" of the limbs tends to disappear during voluntary movements and then reappear when the limb is again at rest.

CHECKPOINT

12a. Draw a muscle spindle surrounded by a few extrafusal fibers. Indicate the location of primary and secondary sensory endings and explain how these endings respond to muscle stretch.

12b. Describe all of the events that occur from the time the patellar tendon is struck with a mallet to the time the leg kicks.

13. Explain how a Golgi tendon organ is stimulated and describe the disynaptic reflex that occurs.

14. Explain the significance of reciprocal innervation and double reciprocal innervation in muscle reflexes.

15. Describe the functions of gamma motoneurons and explain why they are stimulated at the same time as alpha motoneurons during voluntary muscle contractions.

12.6 CARDIAC AND SMOOTH MUSCLES

Cardiac muscle, like skeletal muscle, is striated and contains sarcomeres that shorten by sliding of thin and thick filaments. But while skeletal muscle requires nervous stimulation to contract, cardiac muscle can produce impulses and contract spontaneously. Smooth muscles lack sarcomeres, but they do contain actin and myosin that produce contractions in response to a unique regulatory mechanism.

LEARNING OUTCOMES

After studying this section, you should be able to:

16. **Describe the characteristics of cardiac muscle and how these compare to those of skeletal muscle.**

17. **Describe the structure of smooth muscle and explain how its contractions are regulated.**

Unlike skeletal muscles, which are voluntary effectors regulated by somatic motor neurons, cardiac and smooth muscles are involuntary effectors regulated by autonomic motor neurons. Although there are important differences between skeletal muscle and cardiac and smooth muscle, there are also significant similarities. All types of muscle are believed to contract by means of sliding of thin filaments over thick filaments. The sliding of the filaments is produced by the action of myosin cross bridges in all types of muscles, and excitation-contraction coupling in all types of muscles involves Ca^{2+}.

Cardiac Muscle

Like skeletal muscle cells, cardiac (heart) muscle cells, or **myocardial cells,** are striated; they contain actin and myosin filaments arranged in the form of sarcomeres, and they contract by means of the sliding filament mechanism. The long, fibrous skeletal muscle cells, however, are structurally and functionally separated from each other, whereas the myocardial cells are short, branched, and interconnected. Each myocardial cell is tubular in structure and joined to adjacent myocardial cells by electrical synapses, or **gap junctions** (see chapter 7, fig. 7.21).

The gap junctions are concentrated at the ends of each myocardial cell (fig. 12.32), which permits electrical impulses to be conducted primarily along the long axis from cell to cell. Gap junctions in cardiac muscle have an affinity for stain that makes them appear as dark lines between adjacent cells when viewed in the light microscope. These dark-staining lines are known as *intercalated discs* (fig. 12.33).

Action potentials that originate at any point in a mass of myocardial cells, called a **myocardium,** can spread to all cells in the mass that are joined by gap junctions. Because all cells in a myocardium are electrically joined, a myocardium behaves as a single functional unit. Thus, unlike skeletal muscles that produce contractions that are graded depending on the number of cells stimulated, a myocardium contracts to its full extent each time because all of its cells contribute to the contraction. The ability of the myocardial cells to contract, however, can be increased by the hormone epinephrine and by stretching of the heart chambers. The heart contains two distinct myocardia (atria and ventricles), as will be described in chapter 13.

Unlike skeletal muscles, which require external stimulation by somatic motor nerves before they can produce action potentials and contract, cardiac muscle is able to produce action potentials automatically. Cardiac action potentials normally originate in a specialized group of cells called the *pacemaker.* However, the rate of this spontaneous depolarization, and thus the rate of the heartbeat, are regulated by autonomic innervation. Regulation of the cardiac rate is described more fully in chapter 14, section 14.1.

CLINICAL APPLICATION

Cardiac muscle, like skeletal muscle, contains the troponin complex of three proteins (see fig. 12.13). **Troponin I** helps inhibit the binding of the myosin cross bridges to actin; **troponin T** binds to tropomyosin in the thin filaments, and **troponin C** binds to Ca^{2+} for muscle contraction. Damage to myocardial cells, which occurs in **myocardial infarction,** causes troponins to be released into the blood. Fortunately for clinical diagnosis, troponins T and I are slightly different in cardiac muscle than in skeletal muscle. Thus, troponins T and I released by damaged myocardial cells can be distinguished and measured by laboratory tests using specific antibodies. Such tests are now an important tool in the diagnosis of myocardial infarction.

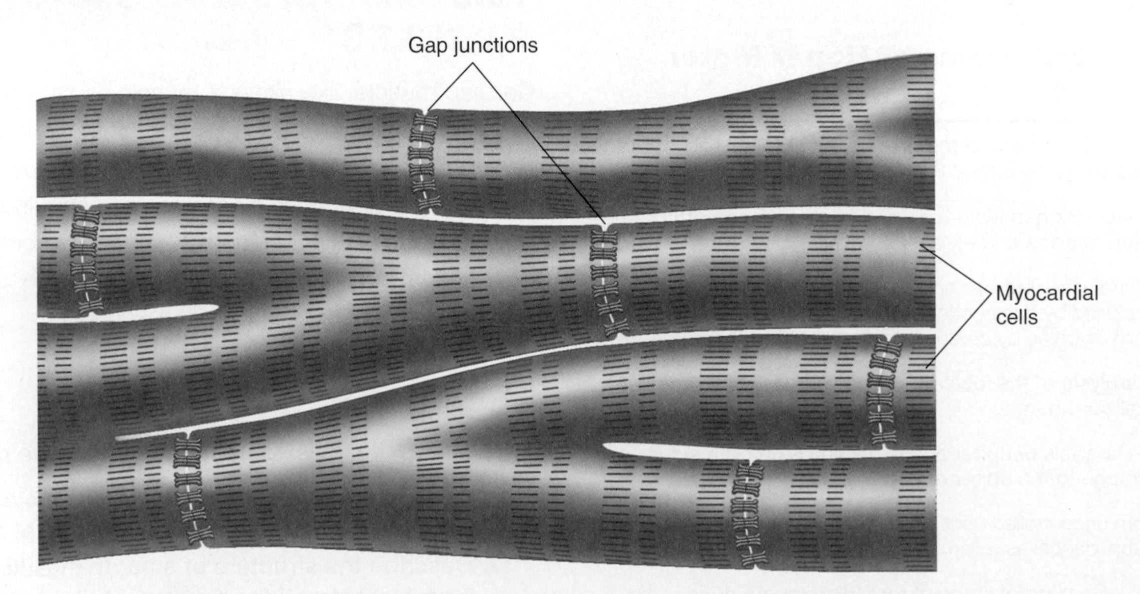

Figure 12.32 Myocardial cells are interconnected by gap junctions. The gap junctions are fluid-filled channels through the plasma membrane of adjacent cells that permit the conduction of impulses from one cell to the next. The gap junctions are concentrated at the ends of each myocardial cell, and each gap junction is composed of connexin proteins (also see chapter 7, fig. 7.21). AP|R

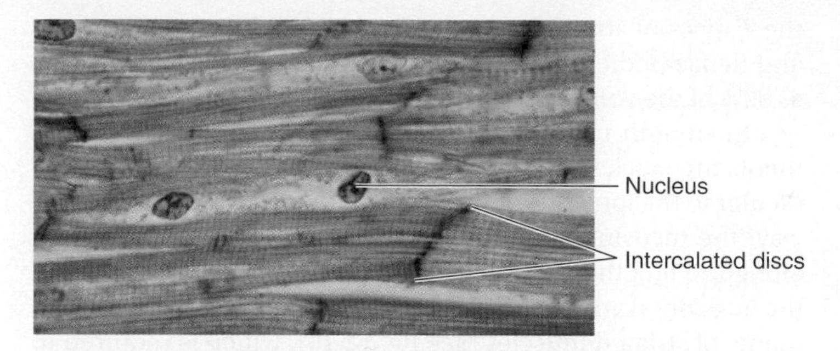

Figure 12.33 Cardiac muscle. Notice that the cells are short, branched, and striated and that they are interconnected by intercalated discs.

Also unlike skeletal muscles, where there is direct excitation-contraction coupling between the transverse tubules and sarcoplasmic reticulum (see fig. 12.16), in myocardial cells the voltage-gated Ca^{2+} channels in the plasma membrane and the Ca^{2+} release channels in the sarcoplasmic reticulum do not directly interact. Instead, the Ca^{2+} that enters the cytoplasm through the voltage-gated Ca^{2+} channels in the transverse tubules stimulates opening of the Ca^{2+} release channels of the sarcoplasmic reticulum (fig. 12.34). This process is known as **Ca^{2+}-induced Ca^{2+} release.** Thus, Ca^{2+} serves as a second messenger from the voltage-gated Ca^{2+} channels to the Ca^{2+} release channels. As a result, excitation-contraction coupling is slower in cardiac than in skeletal muscle.

Diffusion of Ca^{2+} through the plasma membrane of the transverse tubules (fig. 12.34) serves mainly to open the channels in the sarcoplasmic reticulum. The Ca^{2+} release channels in the sarcoplasmic reticulum are 10 times larger than the voltage-gated Ca^{2+} channels in the plasma membrane, and so are primarily responsible for the rapid diffusion of Ca^{2+} into the cytoplasm, which then binds to troponin and stimulates contraction. In order for the muscular chambers of the heart to relax, the Ca^{2+} in the cytoplasm must be actively transported back into the sarcoplasmic reticulum (fig. 12.34).

Smooth Muscle

Smooth (visceral) muscles are arranged in circular layers in the walls of blood vessels and bronchioles (small air passages in the lungs). Both circular and longitudinal smooth muscle layers occur in the tubular digestive tract, the ureters (which transport urine), the ductus deferentia (which transport sperm cells), and the uterine tubes (which transport ova). The alternate contraction of circular and longitudinal smooth muscle layers in the intestine produces **peristaltic waves,** which propel the contents of these tubes in one direction.

Although smooth muscle cells do not contain sarcomeres (which produce striations in skeletal and cardiac muscle), they do contain a great deal of actin and some myosin, which produces a ratio of thin to thick filaments of about 16 to 1 (in striated muscles the ratio is 2 to 1). Unlike striated muscles, in which the thin filaments are relatively short (extending from a Z disc into a sarcomere), the thin filaments of smooth muscle cells are quite long. They attach either to regions of the plasma membrane of the smooth muscle cell or to cytoplasmic protein structures called **dense bodies,** which are analogous to

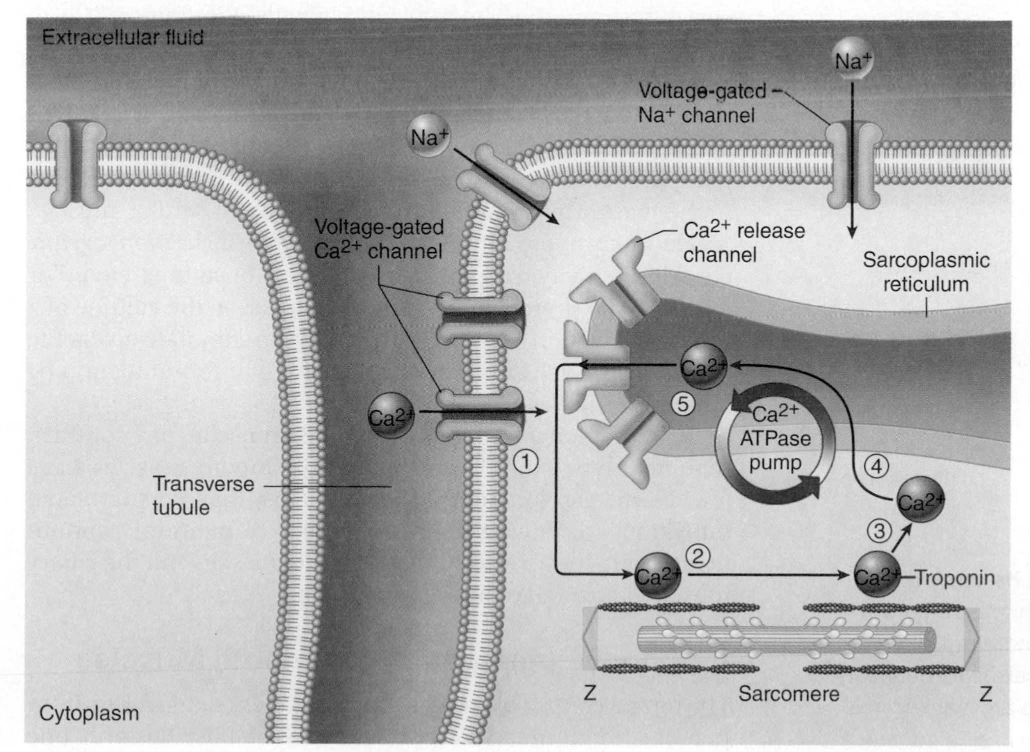

Figure 12.34 Excitation-contraction coupling in cardiac muscle. Depolarization of the plasma membrane during action potentials, when voltage-gated Na^{+} channels are opened, causes voltage-gated Ca^{2+} channels to open in the transverse tubules. (1) This allows some Ca^{2+} to diffuse from the extracellular fluid into the cytoplasm, which (2) stimulates the opening of Ca^{2+} release channels in the sarcoplasmic reticulum. This process is called Ca^{2+}-stimulated Ca^{2+} release. (3) The Ca^{2+} released from the sarcoplasmic reticulum binds to troponin and stimulates contraction. (4) A Ca^{2+} (ATPase) pump actively transports Ca^{2+} into the (5) cisternae of the sarcoplasmic reticulum, allowing relaxation of the myocardium and producing a concentration gradient favoring the outward diffusion of Ca^{2+} for the next contraction.

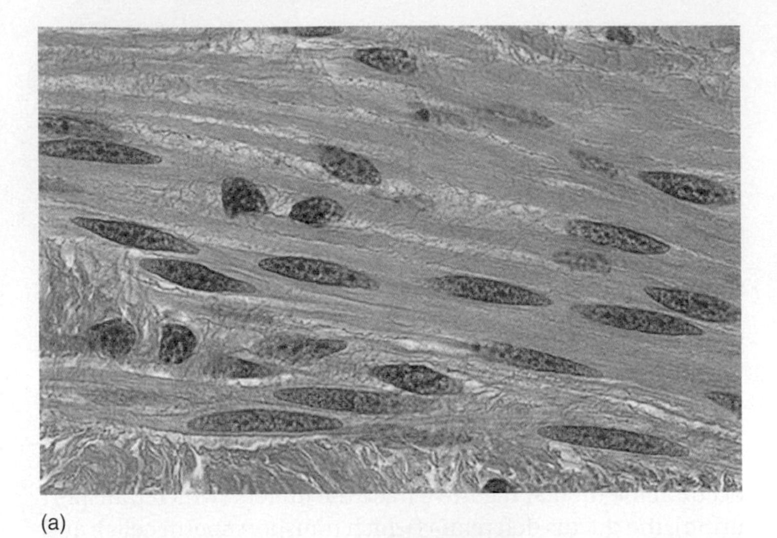

(a)

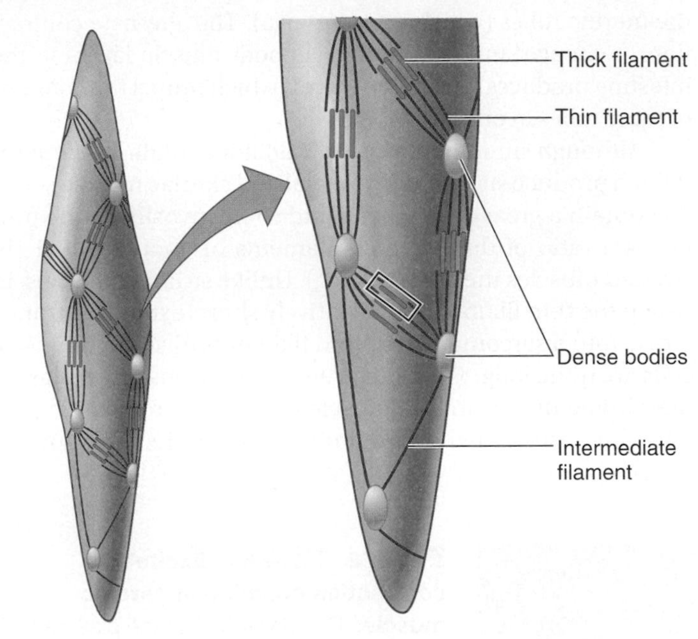

(b)

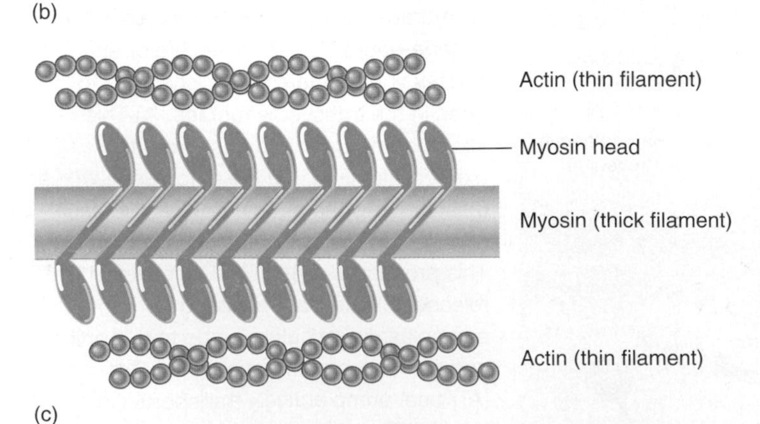

(c)

Figure 12.35 Smooth muscle and its contractile apparatus. (*a*) A photomicrograph of smooth muscle cells in the small intestine. (*b*) Arrangement of thick and thin filaments in smooth muscles. Note that dense bodies are also interconnected by intermediate fibers. (*c*) The myosin proteins are stacked in a different arrangement in smooth muscles than in striated muscles.

the Z discs of striated muscle (fig. 12.35*b*). The myofilaments and dense bodies are so numerous that they occupy as much as 90% of the volume of a smooth muscle cell.

In smooth muscle, the myosin proteins of the thick filaments are stacked vertically so that their long axis is perpendicular to the long axis of the thick filament (fig. 12.35*c*). In this way, the myosin heads can form cross bridges with actin all along the length of the thick filaments. This is different from the horizontal arrangement of myosin proteins in the thick filaments of striated muscles (see fig. 12.10), which is required to cause the shortening of sarcomeres.

The arrangement of the contractile apparatus in smooth muscle cells, and the fact that it is not organized into sarcomeres, is required for proper smooth muscle function. Smooth muscles must be able to contract even when greatly stretched—in the urinary bladder, for example, the smooth muscle cells may be stretched up to two and a half times their resting length. The smooth muscle cells of the uterus may be stretched up to eight times their original length by the end of pregnancy. Striated muscles, because of their structure, lose their ability to contract when the sarcomeres are stretched to the point where actin and myosin no longer overlap (as shown in fig. 12.21).

Single-Unit and Multiunit Smooth Muscles

Smooth muscles are often grouped into two functional categories: **single-unit** and **multiunit** (fig. 12.36). Single-unit smooth muscles have numerous gap junctions between adjacent cells that weld them together electrically; they thus behave as a single unit, much like cardiac muscle. Most smooth muscles—including those in the digestive tract and uterus—are single-unit.

Only some cells of single-unit smooth muscles receive autonomic innervation, but the ACh released by the axon can diffuse to other smooth muscle cells. Binding of ACh to its muscarinic receptors causes depolarization by closing K^+ channels, as described in chapter 9 (see fig. 9.11). Such stimulation, however, only modifies the automatic behavior of single-unit smooth muscles. Single-unit smooth muscles display *pacemaker* activity, in which certain cells stimulate others in the mass. This is similar to the situation in cardiac muscle. Single-unit smooth muscles also display intrinsic, or *myogenic,* electrical activity and contraction in response to stretch. For example, the stretch induced by an increase in the volume of a ureter or a section of the digestive tract can stimulate myogenic contraction. Such contraction does not require stimulation by autonomic nerves.

Contraction of multiunit smooth muscles, by contrast, requires nerve stimulation. Multiunit smooth muscles have few, if any, gap junctions. The cells must thus be stimulated individually by nerve fibers. Examples of multiunit smooth muscles are the arrector pili muscles in the skin and the ciliary muscles attached to the lens of the eye.

Autonomic Innervation of Smooth Muscles

The neural control of skeletal muscles differs significantly from that of smooth muscles. A skeletal muscle fiber has only one

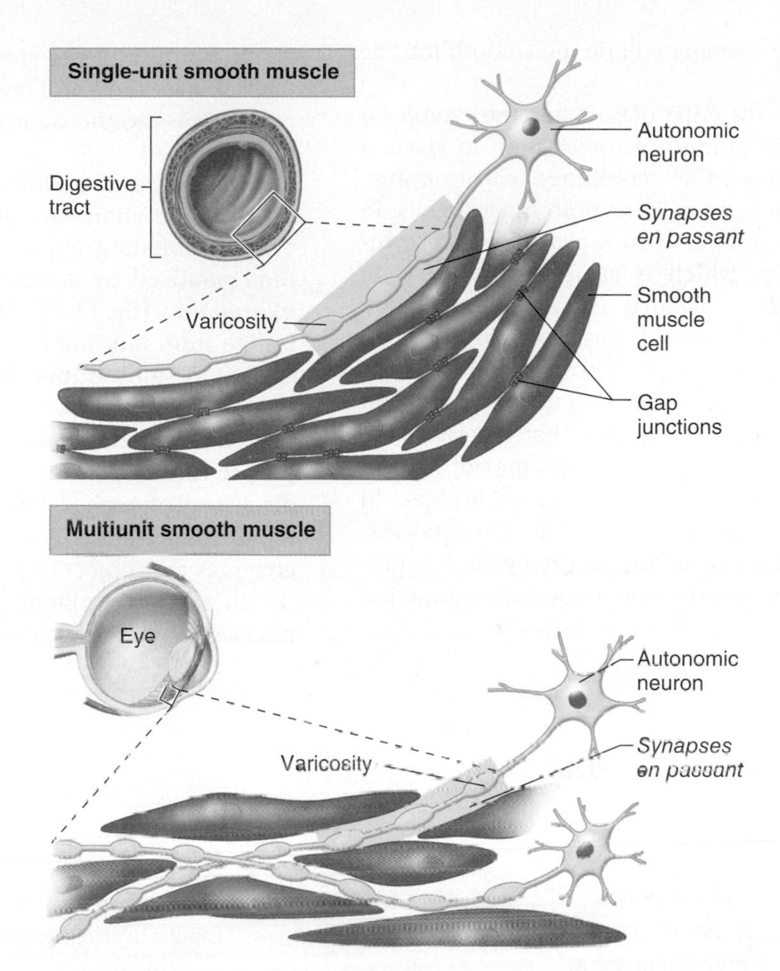

Figure 12.36 Single-unit and multiunit smooth muscle. In single-unit smooth muscle, the individual smooth muscle cells are electrically joined by gap junctions, so that depolarizations can spread from one cell to the next. In multiunit smooth muscle, each smooth muscle cell must be stimulated by an axon. The axons of autonomic neurons have varicosities, which release neurotransmitters, and which form *synapses en passant* with the smooth muscle cells.

junction with a somatic motor axon, and the receptors for the neurotransmitter are located only at the neuromuscular junction. By contrast, the entire surface of smooth muscle cells contains neurotransmitter receptor proteins. Neurotransmitter molecules are released along a stretch of an autonomic nerve fiber that is located some distance from the smooth muscle cells. The regions of the autonomic fiber that release transmitters appear as bulges, or *varicosities,* and the neurotransmitters released from these varicosities stimulate a number of smooth muscle cells. Since there are numerous varicosities along a stretch of an autonomic nerve ending, they form synapses "in passing"—or *synapses en passant*—with the smooth muscle cells. This was described in chapter 9 (see fig. 9.9) and is shown in figure 12.36.

Excitation-Contraction Coupling in Smooth Muscles

As in striated muscles, the contraction of smooth muscles is triggered by a sharp rise in the Ca^{2+} concentration within the cytoplasm of the muscle cells. However, this Ca^{2+} is *not* derived primarily from the Ca^{2+} stored in the sarcoplasmic reticulum (SR) of smooth muscle. The SR in smooth muscle is less extensive than in striated muscle, and its roles in contraction are more variable and complex. The SR of smooth muscle cells may release its stored Ca^{2+} in response to Ca^{2+} entering from the extracellular fluid (Ca^{2+}-induced Ca^{2+} release as in cardiac muscle; see fig. 12.34) and in response to other stimulators. Although the significance of Ca^{2+} released from the SR in smooth muscle contraction is controversial, there is evidence that it may play a very different role. Scientists found that Ca^{2+} released from the SR in arterial smooth muscle can activate unique K^+ channels in the plasma membrane. The resulting outward diffusion of K^+ reduces membrane excitability and promotes relaxation of the arterial smooth muscle.

Sustained smooth muscle contractions are produced in response to extracellular Ca^{2+} that diffuses through the sarcolemma into the smooth muscle cells. This Ca^{2+} enters primarily through voltage-regulated Ca^{2+} channels in the plasma membrane. The opening of these channels is graded by the amount of depolarization; the greater the depolarization, the more Ca^{2+}

will enter the cell and the stronger will be the smooth muscle contraction.

The events that follow the entry of Ca^{2+} into the cytoplasm are somewhat different in smooth muscles than in striated muscles. In striated muscles, Ca^{2+} combines with troponin. Troponin, however, is not present in smooth muscle cells. In smooth muscles, Ca^{2+} combines with a protein in the cytoplasm called **calmodulin,** which is structurally similar to troponin. Calmodulin was previously discussed in relation to the function of Ca^{2+} as a second messenger in hormone action (chapter 11, section 11.2). The calmodulin-Ca^{2+} complex thus formed combines with and activates **myosin light-chain kinase (MLCK),** an enzyme that catalyzes the phosphorylation (addition of phosphate groups) of *myosin light chains,* a component of the myosin cross bridges. In smooth muscle (unlike striated muscle), the phosphorylation of myosin cross bridges is the regulatory event that permits them to bind to actin and thereby produce a contraction (fig. 12.37).

Relaxation of the smooth muscle follows the closing of the Ca^{2+} channels and lowering of the cytoplasmic Ca^{2+} concentrations by the action of Ca^{2+}-ATPase active transport pumps.

Under these conditions, calmodulin dissociates from the myosin light-chain kinase, thereby inactivating this enzyme. The phosphate groups that were added to the myosin are then removed by a different enzyme, a *myosin light-chain phosphatase* (fig. 12.37). Dephosphorylation inhibits the cross bridge from binding to actin and producing another power stroke. In smooth muscles that maintain a contraction, the degree of contraction (muscle tone) is determined by the degree to which the myosin light chains are phosphorylated. This in turn depends on relative activity of the two enzymes discussed: phosphorylation by myosin light-chain kinase, and dephosphorylation by myosin light-chain phosphatase. Both enzymes are subject to physiological regulation.

In addition to being graded, the contractions of smooth muscle cells are slow and sustained. The slowness of contraction

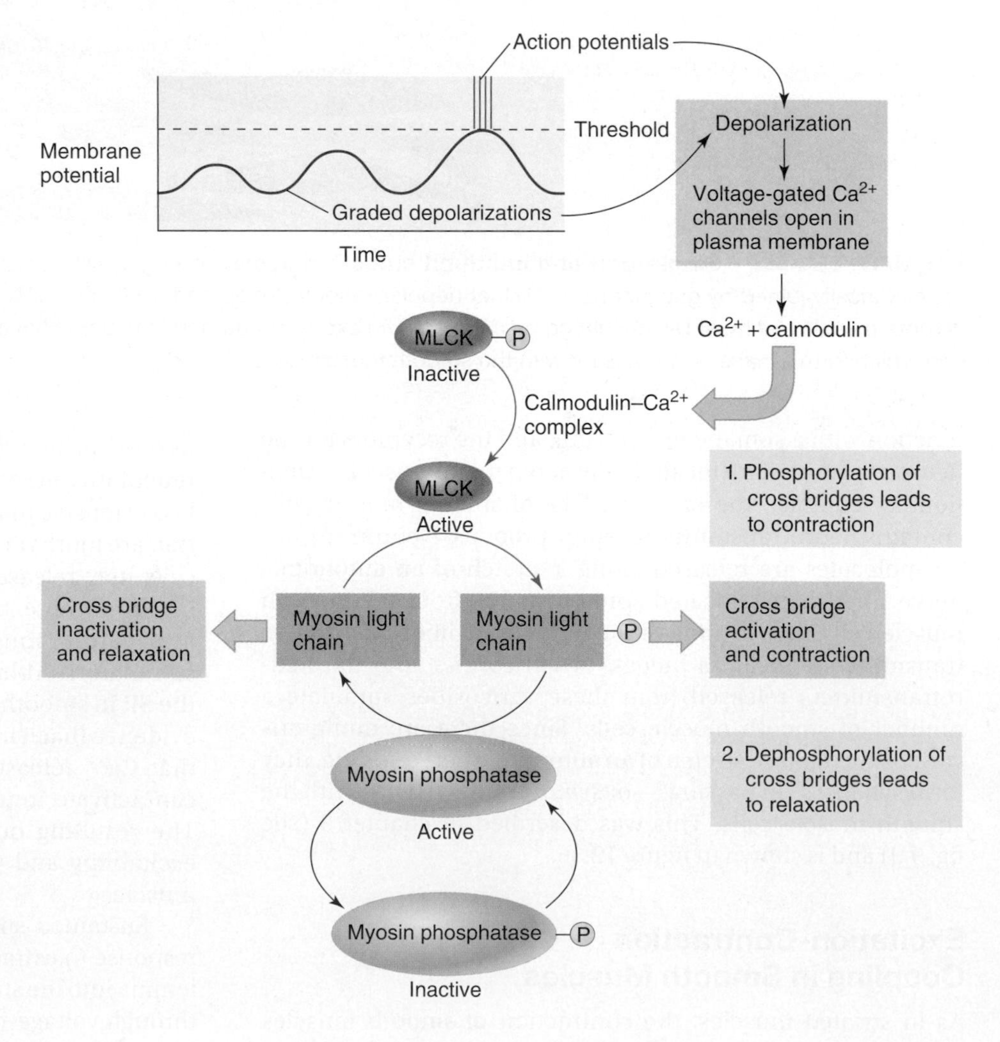

Figure 12.37 Excitation-contraction coupling in smooth muscle. When Ca^{2+} passes through voltage-gated channels in the plasma membrane it enters the cytoplasm and binds to calmodulin. The calmodulin-Ca^{2+} complex then activates myosin light-chain kinase (MLCK) by removing a phosphate group. The activated MLCK, in turn, phosphorylates the myosin light chains, thereby activating the cross bridges to cause contraction. Contraction is ended when myosin phosphatase becomes activated. Upon its activation, myosin phosphatase removes the phosphates from the myosin light chains and thereby inactivates the cross bridges.

Table 12.8 | Comparison of Skeletal, Cardiac, and Smooth Muscle

Skeletal Muscle	Cardiac Muscle	Smooth Muscle
Striated; actin and myosin arranged in sarcomeres	Striated; actin and myosin arranged in sarcomeres	Not striated; more actin than myosin; actin inserts into dense bodies and cell membrane
Well-developed sarcoplasmic reticulum and transverse tubules	Moderately developed sarcoplasmic reticulum and transverse tubules	Poorly developed sarcoplasmic reticulum; no transverse tubules
Contains troponin in the thin filaments	Contains troponin in the thin filaments	Contains calmodulin, a protein that, when bound to Ca^{2+}, activates the enzyme myosin light-chain kinase
Ca^{2+} released into cytoplasm from sarcoplasmic reticulum	Ca^{2+} enters cytoplasm from sarcoplasmic reticulum and extracellular fluid	Ca^{2+} enters cytoplasm from extracellular fluid, sarcoplasmic reticulum, and perhaps mitochondria
Cannot contract without nerve stimulation; denervation results in muscle atrophy	Can contract without nerve stimulation; action potentials originate in pacemaker cells of heart	Maintains tone in absence of nerve stimulation; visceral smooth muscle produces pacemaker potentials; denervation results in hypersensitivity to stimulation
Muscle fibers stimulated independently; no gap junctions	Gap junctions present as intercalated discs	Gap junctions present in most smooth muscles

is related to the fact that myosin ATPase in smooth muscle is slower in its action (splitting ATP for the cross-bridge cycle) than it is in striated muscle. The sustained nature of smooth muscle contraction is explained by the theory that cross bridges in smooth muscles can enter a *latch state.*

The latch state allows smooth muscle to maintain its contraction in a very energy-efficient manner, hydrolyzing less ATP than would otherwise be required. This ability is obviously important for smooth muscles, given that they encircle the walls of hollow organs and must sustain contractions for long periods of time. The mechanisms by which the latch state is produced, however, are complex and poorly understood.

The three muscle types—skeletal, cardiac, and smooth—are compared in table 12.8.

Clinical Investigation CLUES

Maria was taking a calcium-channel-blocking drug to treat her hypertension (high blood pressure).

- How does a calcium-channel-blocking drug affect the smooth muscle of blood vessels and the blood pressure?

CLINICAL APPLICATION

Drugs such as *nifedipine (Procardia)* and related newer compounds are **calcium channel blockers.** These drugs block Ca^{2+} channels in the membrane of smooth muscle cells within the walls of blood vessels, causing the muscles to relax and the vessels to dilate. This effect, called vasodilation, may be helpful in treating some cases of hypertension (high blood pressure). Calcium-channel-blocking drugs are also used when spasm of the coronary arteries (vasospasm) produces angina pectoris, which is pain caused by insufficient blood flow to the heart.

CHECKPOINT

16. Explain how cardiac muscle differs from skeletal muscle in its structure and regulation of contraction.

17a. Contrast the structure of a smooth muscle cell with that of a skeletal muscle fiber and discuss the advantages of each type of structure.

17b. Distinguish between single-unit and multiunit smooth muscles.

17c. Describe the events by which depolarization of a smooth muscle cell results in contraction and explain why smooth muscle contractions are slow and sustained.

Interactions

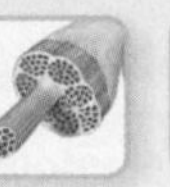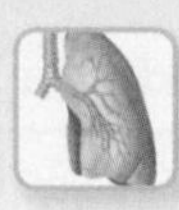

HPer Links of the Muscular System with Other Body Systems

Integumentary System

- The skin helps protect all organs of the body from invasion by pathogens (p. 494)
- The smooth muscles of cutaneous blood vessels are needed for the regulation of cutaneous blood flow (p. 474)
- The arrector pili muscles in the skin produce goose bumps (p. 20)

Skeletal System

- Bones store calcium, which is needed for the control of muscle contraction (p. 690)
- The skeleton provides attachment sites for muscles (p. 361)
- Joints of the skeleton provide levers for movement (p. 360)
- Muscle contractions maintain the health and strength of bone (p. 691)

Nervous System

- Somatic motor neurons stimulate contraction of skeletal muscles (p. 363)
- Autonomic neurons stimulate smooth muscle contraction or relaxation (p. 245)
- Autonomic nerves increase cardiac output during exercise (p. 471)
- Sensory neurons from muscles monitor muscle length and tension (p. 386)

Endocrine System

- Sex hormones promote muscle development and maintenance (p. 322)
- Parathyroid hormone and other hormones regulate blood calcium and phosphate concentrations (p. 691)
- Epinephrine and norepinephrine influence contractions of cardiac muscle and smooth muscles (p. 252)
- Insulin promotes glucose entry into skeletal muscles (p. 347)
- Adipose tissue secretes hormones that regulate the sensitivity of muscles to insulin (p. 669)

Circulatory System

- Blood transports O_2 and nutrients to muscles and removes CO_2 and lactic acid (p. 405)
- Contractions of skeletal muscles serve as a pump to assist blood movement within veins (p. 435)
- Cardiac muscle enables the heart to function as a pump (p. 392)
- Smooth muscle enables blood vessels to constrict and dilate (p. 433)

Respiratory System

- The lungs provide oxygen for muscle metabolism and eliminate carbon dioxide (p. 533)
- Respiratory muscles enable ventilation of the lungs (p. 541)

Urinary System

- The kidneys eliminate creatinine and other metabolic wastes from muscle (p. 601)
- The kidneys help regulate the blood calcium and phosphate concentrations (p. 695)
- Muscles of the urinary tract are needed for the control of urination (p. 584)

Digestive System

- The GI tract provides nutrients for all body organs, including muscles (p. 621)
- Smooth muscle contractions push digestion products along the GI tract (p. 624)
- Muscular sphincters of the GI tract help to regulate the passage of food (p. 621)

Reproductive System

- Testicular androgen promotes growth of skeletal muscle (p. 322)
- Muscle contractions contribute to orgasm in both sexes (p. 711)
- Uterine muscle contractions are required for vaginal delivery of a fetus (p. 743)

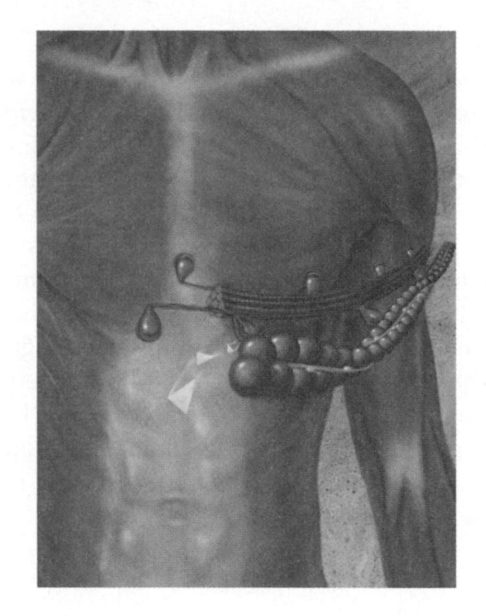

Clinical Investigation SUMMARY

Maria has a high maximal oxygen uptake, consistent with an active lifestyle, and so the muscle pain and fatigue that she reports as unusual may indicate something more serious than that caused by exercise. Her normal levels of creatine phosphokinase in the blood suggest that her cardiac and skeletal muscle cells are not damaged. However, her plasma Ca^{2+} concentration is elevated, and this could cause her high muscle tone. It may also make it difficult for her muscles to fully relax, and thereby contribute to her muscle fatigue and pain. She should be checked further to determine the cause of her high plasma Ca^{2+}; it might be due to an endocrine disturbance involving the parathyroid glands. Maria is taking a calcium channel blocker, which would relax the smooth muscle layer of blood vessels and thereby cause them to dilate, lowering her blood pressure. This drug prevents some of the blood Ca^{2+} from entering the smooth muscle cells, but is unlikely to produce a rise in the blood Ca^{2+}. Nevertheless, alternative medications to treat her hypertension should perhaps be tried.

See the additional chapter 10 Clinical Investigation on *Multifocal Motor Neuropathy* in the Connect site for this text at www.mhhe.com/Fox13.

SUMMARY

12.1 Skeletal Muscles 360

A. Skeletal muscles are attached to bones by tendons.
  1. Skeletal muscles are composed of separate cells, or fibers, that are attached in parallel to the tendons.
  2. Individual muscle fibers are covered by the endomysium; bundles of fibers, called fascicles, are covered by the perimysium; and the entire muscle is covered by the epimysium.
  3. Skeletal muscle fibers are striated.
     a. The dark striations are called A bands, and the light regions are called I bands.
     b. Z lines are located in the middle of each I band.

B. The contraction of muscle fibers *in vivo* is stimulated by somatic motor neurons.
  1. Each somatic motor axon branches to innervate numerous muscle fibers.
  2. The motor neuron and the muscle fibers it innervates are called a motor unit.
     a. When a muscle is composed of a relatively large number of motor units (such as in the hand), there is fine control of muscle contraction.
     b. The large muscles of the leg have relatively few motor units, which are correspondingly large in size.
     c. Sustained contractions are produced by the asynchronous stimulation of different motor units.

12.2 Mechanisms of Contraction 364

A. Skeletal muscle cells, or fibers, contain structures called myofibrils.
  1. Each myofibril is striated with dark (A) and light (I) bands. In the middle of each I band are Z lines.
  2. The A bands contain thick filaments, composed primarily of myosin.
     a. The edges of each A band also contain thin filaments, which overlap the thick filaments.
     b. The central regions of the A bands contain only thick filaments—these regions are the H bands.
  3. The I bands contain only thin filaments, composed primarily of actin.
  4. Thin filaments are composed of globular actin subunits known as G-actin. A protein known as tropomyosin is also located at intervals in the thin filaments. Another protein—troponin—is attached to the tropomyosin.

B. Myosin cross bridges extend out from the thick filaments to the thin filaments.
  1. At rest, the cross bridges are not attached to actin.
     a. The cross-bridge heads function as ATPase enzymes.
     b. ATP is split into ADP and P_i, activating the cross bridge.
  2. When the activated cross bridges attach to actin, they release P_i and undergo a power stroke.

3. At the end of a power stroke, the cross bridge releases the ADP and binds to a new ATP.
   a. This allows the cross bridge to detach from actin and repeat the cycle.
   b. Rigor mortis is caused by the inability of cross bridges to detach from actin because of a lack of ATP.

C. The activity of the cross bridges causes the thin filaments to slide toward the centers of the sarcomeres.
   1. The filaments slide—they do not shorten—during muscle contraction.
   2. The lengths of the H and I bands decrease, whereas the A bands stay the same length during contraction.

D. When a muscle is at rest, the Ca^{2+} concentration of the sarcoplasm is very low and cross bridges are prevented from attaching to actin.
   1. The Ca^{2+} is actively transported into the sarcoplasmic reticulum.
   2. The sarcoplasmic reticulum is a modified endoplasmic reticulum that surrounds the myofibrils.

E. Action potentials are conducted by transverse tubules into the muscle fiber.
   1. Transverse tubules are invaginations of the cell membrane that almost touch the sarcoplasmic reticulum.
   2. Action potentials in the transverse tubules stimulate the opening of Ca^{2+}-release channels in the sarcoplasmic reticulum, causing Ca^{2+} to diffuse into the sarcoplasm and stimulate contractions.

F. When action potentials cease, the Ca^{2+}-release channels in the sarcoplasmic reticulum close.
   1. This allows the active transport Ca^{2+}-ATPase pumps in the sarcoplasmic reticulum to accumulate Ca, removing it from the sarcoplasm and sarcomeres.
   2. As a result of the removal of Ca^{2+} from troponin, the muscle relaxes.

12.3 Contractions of Skeletal Muscles 374

A. Muscles *in vitro* can exhibit twitch, summation, and tetanus.
   1. The rapid contraction and relaxation of muscle fibers is called a twitch.
   2. A whole muscle also produces a twitch in response to a single electrical pulse *in vitro.*
      a. The stronger the electric shock, the stronger the muscle twitch—whole muscles can produce graded contractions.
      b. The graded contraction of whole muscles is due to different numbers of fibers participating in the contraction.
   3. The summation of fiber twitches can occur so rapidly that the muscle produces a smooth, sustained contraction known as tetanus.
   4. When a muscle exerts tension without shortening, the contraction is termed isometric; when shortening does occur, the contraction is isotonic.
   5. When a muscle contracts but, despite its contraction, is made to lengthen due to the application of an external force, the contraction is said to be eccentric.

B. The series-elastic component refers to the elastic composition of the muscle and its associated structures, which must be stretched tight before the tension exerted by the muscle can cause movement.

C. The strength of a muscle contraction is dependent upon its resting length.
   1. If the muscle is too short or too long prior to stimulation, the filaments in the sarcomeres will not have an optimum amount of overlap.
   2. At its normal resting length *in vivo,* a muscle is at its optimum length for contraction.

12.4 Energy Requirements of Skeletal Muscles 378

A. Aerobic cell respiration is ultimately required for the production of ATP needed for cross-bridge activity.
   1. Resting muscles and muscles performing light exercise obtain most of their energy from fatty acids.
   2. During moderate exercise, just below the lactate threshold, energy is obtained about equally from fatty acids and glucose.
   3. Glucose, from the muscle's stored glycogen and from blood plasma, becomes an increasingly important energy source during heavy exercise.
   4. New ATP can be quickly produced from the combination of ADP with phosphate derived from phosphocreatine.
   5. Muscle fibers are of three types.
      a. Slow-twitch red fibers are adapted for aerobic respiration and are resistant to fatigue.
      b. Fast-twitch white fibers are adapted for anaerobic respiration.
      c. Intermediate fibers are fast-twitch but adapted for aerobic respiration.

B. Muscle fatigue may be caused by a number of mechanisms.
   1. Fatigue during sustained maximal contraction may be produced by the accumulation of extracellular K^+ as a result of high levels of nerve activity.
   2. Fatigue during moderate exercise is primarily a result of anaerobic respiration by fast-twitch fibers.
      a. The production of lactic acid and consequent fall in pH, the depletion of muscle glycogen, and other metabolic changes interfere with the release of Ca^{2+} from the sarcoplasmic reticulum.
      b. Interference with excitation contraction coupling, rather than depletion of ATP, appears to be responsible for muscle fatigue.
   3. In human exercise, however, fatigue is often caused by changes in the CNS before the muscles themselves fatigue; this central fatigue reduces the force of voluntary contractions.

C. Physical training affects the characteristics of the muscle fibers.
   1. Endurance training increases the aerobic capacity of muscle fibers and their use of fatty acids for energy,

so that their reliance on glycogen and anaerobic respiration—and thus their susceptibility to fatigue—is reduced.

2. Resistance training causes hypertrophy of muscle fibers because of an increase in the size and number of myofibrils.

12.5 Neural Control of Skeletal Muscles 385

A. The somatic motor neurons that innervate the muscles are called lower motor neurons.
   1. Alpha motoneurons innervate the ordinary, or extrafusal, muscle fibers. These are the fibers that produce muscle shortening during contraction.
   2. Gamma motoneurons innervate the intrafusal fibers of the muscle spindles.

B. Muscle spindles function as length detectors in muscles.
   1. Spindles consist of several intrafusal fibers wrapped together. The spindles are in parallel with the extrafusal fibers.
   2. Stretching of the muscle stretches the spindles, which excites sensory endings in the spindle apparatus.
      a. Impulses in the sensory neurons travel into the spinal cord in the dorsal roots of spinal nerves.
      b. The sensory neuron synapses directly with an alpha motoneuron within the spinal cord, which produces a monosynaptic reflex.
      c. The alpha motoneuron stimulates the extrafusal muscle fibers to contract, thus relieving the stretch. This is called the stretch reflex.
   3. The activity of gamma motoneurons tightens the spindles, thus making them more sensitive to stretch and better able to monitor the length of the muscle, even during muscle shortening.

C. The Golgi tendon organs monitor the tension that the muscle exerts on its tendons.
   1. As the tension increases, sensory neurons from Golgi tendon organs inhibit the activity of alpha motoneurons.
   2. This is a disynaptic reflex because the sensory neurons synapse with interneurons, which in turn make inhibitory synapses with motoneurons.

D. A crossed-extensor reflex occurs when a foot steps on a tack.
   1. Sensory input from the injured foot causes stimulation of flexor muscles and inhibition of the antagonistic extensor muscles.
   2. The sensory input also crosses the spinal cord to cause stimulation of extensor and inhibition of flexor muscles in the contralateral leg.

E. Most of the fibers of descending tracts synapse with spinal interneurons, which in turn synapse with the lower motor neurons.
   1. Alpha and gamma motoneurons are usually stimulated at the same time, or coactivated.
   2. The stimulation of gamma motoneurons keeps the muscle spindles under tension and sensitive to stretch.
   3. Upper motor neurons, primarily in the basal nuclei, also exert inhibitory effects on gamma motoneurons.

F. Neurons in the brain that affect the lower motor neurons are called upper motor neurons.
   1. The fibers of neurons in the precentral gyrus, or motor cortex, descend to the lower motor neurons as the lateral and ventral corticospinal tracts.
      a. Most of these fibers cross to the contralateral side in the brain stem, forming structures called the pyramids; therefore, this system is called the pyramidal system.
      b. The left side of the brain thus controls the musculature on the right side, and vice versa.
   2. Other descending motor tracts are part of the extrapyramidal system.
      a. The neurons of the extrapyramidal system make numerous synapses in different areas of the brain, including the midbrain, brain stem, basal nuclei, and cerebellum.
      b. Damage to the cerebellum produces intention tremor, and degeneration of dopaminergic neurons in the basal nuclei produces Parkinson's disease.

12.6 Cardiac and Smooth Muscles 391

A. Cardiac muscle is striated and contains sarcomeres.
   1. In contrast to skeletal muscles, which require neural stimulation to contract, action potentials in the heart originate in myocardial cells; stimulation by neurons is not required.
   2. Also unlike the situation in skeletal muscles, action potentials can cross from one myocardial cell to another.

B. The thin and thick filaments in smooth muscles are not organized into sarcomeres.
   1. The thin filaments extend from the plasma membrane and from dense bodies in the cytoplasm.
   2. The myosin proteins are stacked perpendicular to the long axis of the thick filaments, so they can bind to actin all along the length of the thick filament.
   3. Depolarizations are graded and conducted from one smooth muscle cell to the next.
      a. The depolarizations stimulate the entry of Ca^{2+}, which binds to calmodulin; this complex then activates myosin light-chain kinase, which phosphorylates the myosin heads.
      b. Phosphorylation of the myosin heads is needed for them to be able to bind to actin and produce contractions.
   4. Smooth muscles are classified as single-unit, if they are interconnected by gap junctions, and as multiunit if they are not so connected.
   5. Autonomic neurons have varicosities that release neurotransmitters all along their length of contact with the smooth muscle cells, making *synapses en passant.*

REVIEW ACTIVITIES

Test Your Knowledge

1. A graded whole muscle contraction is produced *in vivo* primarily by variations in
   a. the strength of the fiber's contraction.
   b. the number of fibers that are contracting.
   c. both of these.
   d. neither of these.
2. The series-elastic component of muscle contraction is responsible for
   a. increased muscle shortening to successive twitches.
   b. a time delay between contraction and shortening.
   c. the lengthening of muscle after contraction has ceased.
   d. all of these.
3. Which of these muscles have motor units with the highest innervation ratio?
   a. Leg muscles
   b. Arm muscles
   c. Muscles that move the fingers
   d. Muscles of the trunk
4. The stimulation of gamma motoneurons produces
   a. isotonic contraction of intrafusal fibers.
   b. isometric contraction of intrafusal fibers.
   c. either isotonic or isometric contraction of intrafusal fibers.
   d. contraction of extrafusal fibers.
5. In a single reflex arc involved in the knee-jerk reflex, how many synapses are activated within the spinal cord?
   a. Thousands
   b. Hundreds
   c. Dozens
   d. Two
   e. One
6. Spastic paralysis may occur when there is damage to
   a. the lower motor neurons.
   b. the upper motor neurons.
   c. either the lower or the upper motor neurons.
7. When a skeletal muscle shortens during contraction, which of these statements is *false?*
   a. The A bands shorten. c. The I bands shorten.
   b. The H bands shorten. d. The sarcomeres shorten.
8. Electrical excitation of a muscle fiber *most directly* causes
   a. movement of tropomyosin.
   b. attachment of the cross bridges to action.
   c. release of Ca^{2+} from the sarcoplasmic reticulum.
   d. splitting of ATP.
9. The energy for muscle contraction is *most directly* obtained from
   a. phosphocreatine. c. anaerobic respiration.
   b. ATP. d. aerobic respiration.
10. Which of these statements about cross bridges is *false?*
    a. They are composed of myosin.
    b. They bind to ATP after they detach from actin.
    c. They contain an ATPase.
    d. They split ATP before they attach to actin.
11. When a muscle is stimulated to contract, Ca^{2+} binds to
    a. myosin. c. actin.
    b. tropomyosin. d. troponin.
12. Which of these statements about muscle fatigue is *false?*
    a. It may result when ATP is no longer available for the cross-bridge cycle.
    b. It may be caused by a loss of muscle cell Ca^{2+}.
    c. It may be caused by the accumulation of extracellular K^{+}.
    d. It may be a result of lactic acid production.
13. Which of these types of muscle cells are *not* capable of spontaneous depolarization?
    a. Single-unit smooth muscle
    b. Multiunit smooth muscle
    c. Cardiac muscle
    d. Skeletal muscle
    e. Both *b* and *d*
    f. Both *a* and *c*
14. Which of these muscle types is striated and contains gap junctions?
    a. Single-unit smooth muscle
    b. Multiunit smooth muscle
    c. Cardiac muscle
    d. Skeletal muscle
15. In an isotonic muscle contraction,
    a. the length of the muscle remains constant.
    b. the muscle tension remains constant.
    c. both muscle length and tension are changed.
    d. movement of bones does not occur.
16. Which of the following is an example of an eccentric muscle contraction?
    a. Doing a "curl" with a dumbbell
    b. Doing a breast stroke in a swimming pool
    c. Extending the arms when bench-pressing a weight
    d. Flexing the arms when bench-pressing to allow the weight to return to the chest

17. Which of the following statements about the Ca^{2+} release channels in the sarcoplasmic reticulum is *false?*
    a. They are also called ryanodine receptors.
    b. They are one-tenth the size of the voltage-gated Ca^{2+} channels.
    c. They are opened by Ca^{2+} release channels in the transverse tubules.
    d. They permit Ca^{2+} to diffuse into the sarcoplasm from the sarcoplasmic reticulum.
18. Which of the following statements is *not* characteristic of smooth muscles?
    a. Myosin phosphatase is required for contraction.
    b. They are able to conduct graded depolarizations.
    c. They can enter a latch state.
    d. They can produce graded contractions in response to graded depolarizations.

Test Your Understanding

19. Using the concept of motor units, explain how skeletal muscles *in vivo* produce graded and sustained contractions.
20. Describe how an isometric contraction can be converted into an isotonic contraction using the concepts of motor unit recruitment and the series-elastic component of muscles.
21. Explain why the myosin heads don't bind to the actin when the muscle is at rest. Then, provide a step-by-step explanation of how depolarization of the muscle fiber plasma membrane by ACh leads to the binding of the myosin heads to actin. (That is, explain excitation-contraction coupling.)
22. Using the sliding filament theory of contraction, explain why the contraction strength of a muscle is maximal at a particular muscle length.
23. Explain why muscle tone is first decreased and then increased when descending motor tracts are damaged. How is muscle tone maintained?
24. Explain the role of ATP in muscle contraction and muscle relaxation.
25. Why are all the muscle fibers of a given motor unit of the same type? Why are smaller motor units and slow-twitch muscle fibers used more frequently than larger motor units and fast-twitch fibers?
26. What changes occur in muscle metabolism as the intensity of exercise is increased? Describe the changes that occur as a result of endurance training and explain how these changes allow more strenuous exercise to be performed before the onset of muscle fatigue.
27. Compare the mechanism of excitation-coupling in striated muscle with that in smooth muscle.
28. Compare cardiac muscle, single-unit smooth muscle, and multiunit smooth muscle with respect to the regulation of their contraction.

Test Your Analytical Ability

29. Your friend eats huge helpings of pasta for two days prior to a marathon, claiming such "carbo loading" is of benefit in the race. Is he right? What are some other things he can do to improve his performance?
30. Compare muscular dystrophy and amyotrophic lateral sclerosis (ALS) in terms of their causes and their effects on muscles.
31. Why is it important to have a large amount of stored high-energy phosphates in the form of creatine phosphate for the function of muscles during exercise? What might happen to a muscle in your body if it ever ran out of ATP?
32. How is electrical excitation of a skeletal muscle fiber coupled to muscle contraction? Speculate on why the exact mechanism of this coupling has been difficult to determine.
33. How would a rise in the extracellular Ca^{2+} concentration affect the beating of a heart? Explain the mechanisms involved. Lowering the blood Ca^{2+} concentration can cause muscle spasms. What might be responsible for this effect?
34. Organs consisting mostly of highly differentiated cells appear to also contain small populations of multipotent, adult stem cells (chapters 3 and 20). Explain the significance of this statement, using skeletal muscles as an example.
35. Explain how contraction of a myocardium is analogous to a skeletal muscle twitch, and why the myocardium cannot exhibit summation and tetanus.
36. Explain how the organization of actin and myosin in smooth muscle cells differs from their organization in striated muscle cells, and the advantages of these differences.

Test Your Quantitative Ability

Refer to figure 12.22 to answer the following questions.

37. Which energy source varies the least during different intensities and durations of exercise? What is the range of its percentage of energy expenditure?
38. Which energy source varies the most during different intensities of exercise? What is the range of its percentage of energy expenditure?
39. What percentage of the energy expenditure is due to free fatty acids when a person (a) does mild exercise for 90 to 120 minutes; and (b) does heavy exercise from 0 to 30 minutes?
40. Following exercise of which intensity and duration are the muscles most depleted in stored glycogen?

CHAPTER

13

Blood, Heart, and Circulation

REFRESH YOUR MEMORY

Before you begin this chapter, you may want to review these concepts from previous chapters:

CHAPTER OUTLINE

Clinical Investigation

Jason, a 19-year-old college student, goes to the doctor complaining of chronic fatigue. The doctor palpates (feels) Jason's radial pulse and comments that it is fast and weak. He orders various tests, including an echocardiogram, an electrocardiogram, and later an angiogram. He also requests that particular blood tests be performed.

Some of the new terms and concepts you will encounter include:

- Red blood cell count, hemoglobin, and hematocrit measurements and anemia
- Ventricular septal defect and mitral stenosis
- ECG waves and sinus tachycardia
- LDL cholesterol and atherosclerosis

13.1 FUNCTIONS AND COMPONENTS OF THE CIRCULATORY SYSTEM

Blood serves numerous functions, including the transport of respiratory gases, nutritive molecules, metabolic wastes, and hormones. Blood travels through the body in a system of vessels leading from and returning to the heart.

LEARNING OUTCOMES

After studying this section, you should be able to:

1. Identify the functions and components of the circulatory system.
2. Describe the relationship between interstitial fluid, plasma, and lymph.

A unicellular organism can provide for its own maintenance and continuity by performing the wide variety of functions needed for life. By contrast, the complex human body is composed of specialized cells that depend on one another. Because most are firmly implanted in tissues, their oxygen and nutrients must be brought to them, and their waste products removed. Therefore, a highly effective means of transporting materials within the body is needed.

The blood serves this transportation function. An estimated 60,000 miles of vessels throughout the body of an adult ensure that continued sustenance reaches each of the trillions of living cells. But the blood can also transport disease-causing viruses, bacteria, and their toxins. To guard against this, the circulatory system has protective mechanisms—the white blood cells and the lymphatic system. In order to perform its various functions, the circulatory system works together with the respiratory, urinary, digestive, endocrine, and integumentary systems in maintaining homeostasis.

Functions of the Circulatory System

The functions of the circulatory system can be divided into three broad areas: transportation, regulation, and protection.

1. **Transportation.** All of the substances essential for cellular metabolism are transported by the circulatory system. These substances can be categorized as follows:
   a. *Respiratory.* Red blood cells, or *erythrocytes,* transport oxygen to the cells. In the lungs, oxygen from the inhaled air attaches to hemoglobin molecules within the erythrocytes and is transported to the cells for aerobic respiration. Carbon dioxide produced by cell respiration is carried by the blood to the lungs for elimination in the exhaled air.
   b. *Nutritive.* The digestive system is responsible for the mechanical and chemical breakdown of food so that it can be absorbed through the intestinal wall into the blood and lymphatic vessels. The blood then carries these absorbed products of digestion through the liver to the cells of the body.
   c. *Excretory.* Metabolic wastes (such as urea), excess water and ions, and other molecules not needed by the body are carried by the blood to the kidneys and excreted in the urine.
2. **Regulation.** The circulatory system contributes to both hormonal and temperature regulation.
   a. *Hormonal.* The blood carries hormones from their site of origin to distant target tissues where they perform a variety of regulatory functions.
   b. *Temperature.* Temperature regulation is aided by the diversion of blood from deeper to more superficial cutaneous vessels or vice versa. When the ambient temperature is high, diversion of blood from deep to superficial vessels helps cool the body; when the ambient temperature is low, the diversion of blood from superficial to deeper vessels helps keep the body warm.
3. **Protection.** The circulatory system protects against blood loss from injury and against pathogens, including foreign microbes and toxins introduced into the body.
   a. *Clotting.* The clotting mechanism protects against blood loss when vessels are damaged.
   b. *Immune.* The immune function of the blood is performed by the *leukocytes* (white blood cells) that protect against many disease-causing agents (pathogens).

Major Components of the Circulatory System

The **circulatory system** consists of two subdivisions: the cardiovascular system and the lymphatic system. The **cardiovascular system** consists of the heart and blood vessels, and the **lymphatic system,** which includes lymphatic vessels and lymphoid tissues within the spleen, thymus, tonsils, and lymph nodes.

The **heart** is a four-chambered double pump. Its pumping action creates the pressure head needed to push blood through the vessels to the lungs and body cells. At rest, the heart of an adult pumps about 5 liters of blood per minute. At this rate, it takes about 1 minute for blood to be circulated to the most distal extremity and back to the heart.

Blood vessels form a tubular network that permits blood to flow from the heart to all the living cells of the body and then back to the heart. *Arteries* carry blood away from the heart, whereas *veins* return blood to the heart. Arteries and veins are continuous with each other through smaller blood vessels.

Arteries branch extensively to form a "tree" of progressively smaller vessels. The smallest of the arteries are called *arterioles.* Blood passes from the arterial to the venous system in microscopic *capillaries,* which are the thinnest and most numerous of the blood vessels. All exchanges of fluid, nutrients, and wastes between the blood and tissues occur across the walls of capillaries. Blood flows through capillaries into microscopic veins called *venules,* which deliver blood into progressively larger veins that eventually return the blood to the heart.

As blood *plasma* (the fluid portion of the blood) passes through capillaries, the hydrostatic pressure of the blood forces some of this fluid out of the capillary walls. Fluid derived from plasma that passes out of capillary walls into the surrounding tissues is called *tissue fluid,* or *interstitial fluid.* Some of this fluid returns directly to capillaries, and some enters into **lymphatic vessels** located in the connective tissues around the blood vessels. Fluid in lymphatic vessels is called *lymph.* This fluid is returned to the venous blood at specific sites. **Lymph nodes,** positioned along the way, cleanse the lymph prior to its return to the venous blood. The lymphatic system is thus considered a part of the circulatory system and is discussed in section 13.8.

CHECKPOINT

1a. State the components of the circulatory system that function in oxygen transport, in the transport of nutrients from the digestive system, and in protection.

1b. Describe the functions of arteries, veins, and capillaries.

2. Define the terms *interstitial fluid* and *lymph.* How do these fluids relate to blood plasma?

13.2 COMPOSITION OF THE BLOOD

Blood consists of formed elements that are suspended and carried in a fluid called plasma. The formed elements—erythrocytes, leukocytes, and platelets—function respectively in oxygen transport, immune defense, and blood clotting.

LEARNING OUTCOMES

After studying this section, you should be able to:

3. Distinguish between the different formed elements of the blood.

4. Describe the regulation of red and white blood cell production.

5. Explain blood typing and blood clotting.

The total blood volume in the average-size adult is about 5 liters, constituting about 8% of the total body weight. Blood leaving the heart is referred to as *arterial blood.* Arterial blood, with the exception of that going to the lungs, is bright red because of a high concentration of oxyhemoglobin (the combination of oxygen and hemoglobin) in the red blood cells. *Venous blood* is blood returning to the heart. Except for the venous blood from the lungs, it contains less oxygen and is therefore a darker red than the oxygen-rich arterial blood.

Blood is composed of a cellular portion, called *formed elements,* and a fluid portion, called *plasma.* When a blood sample is centrifuged, the heavier formed elements are packed into the bottom of the tube, leaving plasma at the top (fig. 13.1). The formed elements constitute approximately 45% of the total blood volume, and the plasma accounts for the remaining 55%. Red blood cells compose most of the formed elements; the percentage of red blood cell volume to total blood volume in a centrifuged blood sample (a measurement called the *hematocrit*) is 36% to 46% in women and 41% to 53% in men (table 13.1).

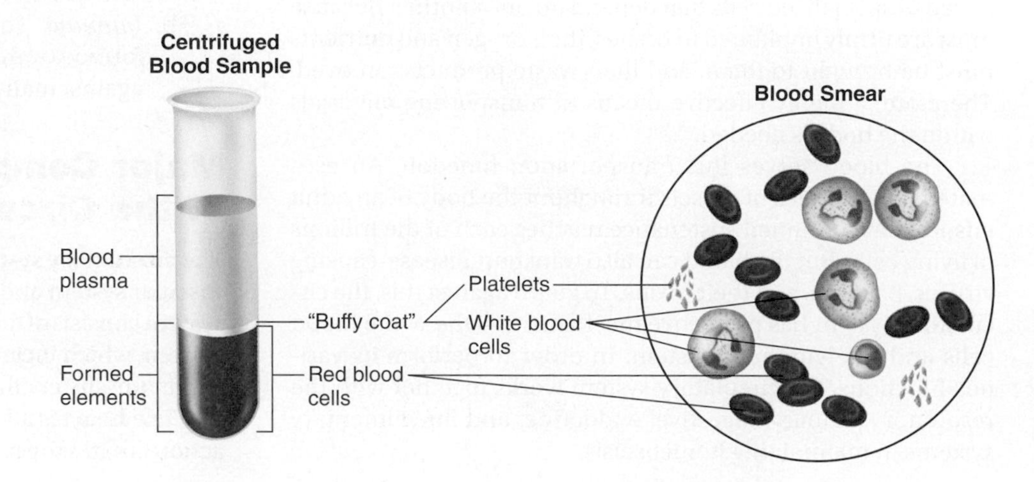

Figure 13.1 The constituents of blood. Blood cells become packed at the bottom of the test tube when whole blood is centrifuged, leaving the fluid plasma at the top of the tube. Red blood cells are the most abundant of the blood cells—white blood cells and platelets form only a thin, light-colored "buffy coat" at the interface between the packed red blood cells and the plasma. AP|R

Table 13.1 | Representative Normal Plasma Values

Measurement	Normal Range
Blood volume	80–85 ml/kg body weight
Blood osmolality	285–295 mOsm
Blood pH	7.38–7.44
Enzymes	
Creatine phosphokinase (CPK)	Female: 10–79 U/L
	Male: 17–148 U/L
Lactic dehydrogenase (LDH)	45–90 U/L
Phosphatase (acid)	Female: 0.01–0.56 Sigma U/ml
	Male: 0.13–0.63 Sigma U/ml
Hematology Values	
Hematocrit	Female: 36%–46%
	Male: 41%–53%
Hemoglobin	Female: 12–16 g/100 ml
	Male: 13.5–17.5 g/100 ml
Red blood cell count	4.50–5.90 million/mm^3
White blood cell count	4,500–11,000/mm^3
Hormones	
Testosterone	Male: 270–1,070 ng/100 ml
	Female: 6–86 ng/100 ml
Adrenocorticotrophic hormone (ACTH)	6–76 pg/ml
Growth hormone	Children: over 10 ng/ml
	Adult male: below 5 ng/ml
Insulin	2–20 µU/ml (fasting)
Ions	
Bicarbonate	24–30 mmol/l
Calcium	9.0–10.5 mg/dl
Chloride	98–106 mEq/L
Potassium	3.5–5.0 mEq/L
Sodium	135–145 mEq/L
Organic Molecules (Other)	
Cholesterol, desirable	under 200 mg/dl
Glucose	75–115 mg/dl (fasting)
Lactic acid	5–15 mg/dl
Protein (total)	5.5–8.0 g/dl
Triglyceride	under 160 mg/dl
Urea nitrogen	10–20 mg/dl
Uric acid	Male 2.5–8.0 mg/dl
	Female 1.5–6.0 mg/dl

Source: Excerpted from material appearing in *The New England Journal of Medicine*, "Case Records of the Massachusetts General Hospital," 302:37–38, 314:39–49, 351:1548–1563. 1980, 1986, 2004.

Plasma

Plasma is a straw-colored liquid consisting of water and dissolved solutes. The major solute of the plasma in terms of its concentration is Na^+. In addition to Na^+, plasma contains many other ions, as well as organic molecules such as metabolites, hormones, enzymes, antibodies, and other proteins. The concentrations of some of these plasma constituents are shown in table 13.1.

Plasma Proteins

Plasma proteins constitute 7% to 9% of the plasma. The three types of proteins are albumins, globulins, and fibrinogen. **Albumins** account for most (60% to 80%) of the plasma proteins and are the smallest in size. They are produced by the liver and provide the osmotic pressure needed to draw water from the surrounding tissue fluid into the capillaries. This action is needed to maintain blood volume and pressure. **Globulins** are grouped into three subtypes: **alpha globulins, beta globulins,** and **gamma globulins.** The alpha and beta globulins are produced by the liver and function in transporting lipids and fat-soluble vitamins. Gamma globulins are antibodies produced by lymphocytes (one of the formed elements found in blood and lymphoid tissues) and function in immunity. **Fibrinogen,** which accounts for only about 4% of the total plasma proteins, is an important clotting factor produced by the liver. During the process of clot formation (described later in this section), fibrinogen is converted into insoluble threads of *fibrin.* Thus, the fluid from clotted blood, called **serum,** does not contain fibrinogen but is otherwise identical to plasma.

Plasma Volume

A number of regulatory mechanisms in the body maintain homeostasis of the plasma volume. If the body should lose water, the remaining plasma becomes excessively concentrated—its osmolality (chapter 6) increases. This is detected by osmoreceptors in the hypothalamus, resulting in a sensation of thirst and the release of antidiuretic hormone (ADH) from the posterior pituitary (chapter 11, section 11.3). This hormone promotes water retention by the kidneys, which—together with increased intake of fluids—helps compensate for the dehydration and lowered blood volume. This regulatory mechanism, together with others that influence plasma volume, are very important in maintaining blood pressure (chapter 14, section 14.6).

The Formed Elements of Blood

The **formed elements** of blood include two types of blood cells: *erythrocytes,* or *red blood cells,* and *leukocytes,* or *white blood cells.* Erythrocytes are by far the more numerous of the two. A cubic millimeter of blood normally contains 5.1 million to 5.8 million erythrocytes in males and 4.3 million to 5.2 million erythrocytes in females. By contrast, the same volume of blood contains only 5,000 to 9,000 leukocytes.

Erythrocytes

Erythrocytes are flattened, biconcave discs about 7 μm in diameter and 2.2 μm thick. Their unique shape relates to their function of transporting oxygen; it provides an increased surface area through which gas can diffuse (fig. 13.2). Erythrocytes lack nuclei and mitochondria (they obtain energy through anaerobic metabolism). Partly because of these deficiencies, erythrocytes have a relatively short circulating life span of only about 120 days. Older erythrocytes are removed from the circulation by phagocytic cells in the liver, spleen, and bone marrow.

Each erythrocyte contains approximately 280 million **hemoglobin** molecules, which give blood its red color. Each hemoglobin molecule consists of four protein chains called *globins,* each of which is bound to one *heme,* a red-pigmented molecule that contains iron. The iron group of heme is able to combine with oxygen in the lungs and release oxygen in the tissues.

The heme iron is recycled from senescent (old) red blood cells (see chapter 18, fig. 18.22) in the liver and spleen. This iron travels in the blood to the bone marrow attached to a protein carrier called **transferrin.** This recycled heme iron supplies most of the body's need for iron. The balance of the requirement for iron, though relatively small, must be made up for in the diet. Dietary iron is absorbed mostly in the duodenum (the first part of the small intestine) and transported from the intestine bound to transferrin in the blood. The transferrin with its bound iron is taken out of the blood by cells of the bone marrow and liver by endocytosis, which is triggered by binding of transferrin to its membrane receptors.

Although the bone marrow produces about 200 billion red blood cells each day, we only need about 20 mg of iron a day for the synthesis of new hemoglobin molecules. This is a relatively small amount (because most of the iron from destroyed red blood cells is recycled, as just described), but it accounts for 80% of the daily dietary iron requirement. A dietary deficiency in iron reduces the ability of the bone marrow to produce hemoglobin, and can result in *iron-deficiency anemia.*

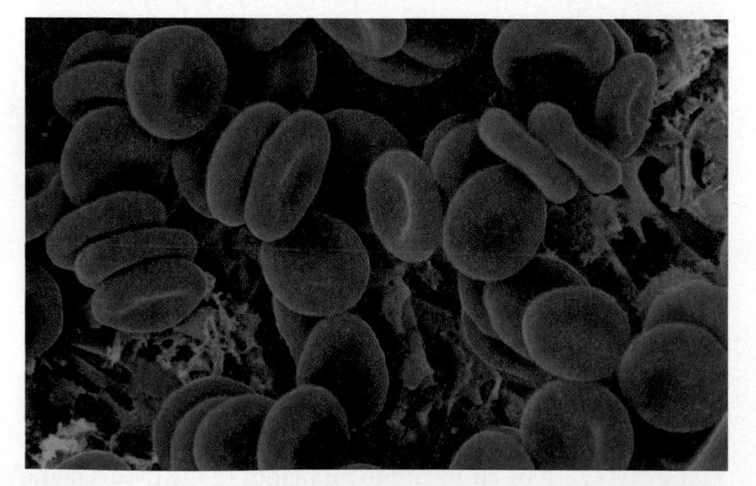

Figure 13.2 A colorized scanning electron micrograph of red blood cells. The shape of the red blood cells is described as a "biconcave disc." In reality, individual red blood cells do not look red when viewed under a microscope. AP|R

CLINICAL APPLICATION

Anemia refers to any condition in which there is an abnormally low hemoglobin concentration and/or red blood cell count. The most common type is **iron-deficiency anemia** caused by a deficiency of iron, which is an essential component of the hemoglobin molecule. In **pernicious anemia** there is an inadequate amount of vitamin B_{12}, which is needed for red blood cell production. This is usually due to atrophy of the glandular mucosa of the stomach, which normally secretes a protein called *intrinsic factor.* In the absence of intrinsic factor, the vitamin B_{12} obtained in the diet cannot be absorbed by intestinal cells. **Aplastic anemia** is anemia due to destruction of the bone marrow, which may be caused by chemicals (including benzene and arsenic) or by radiation.

Clinical Investigation CLUES

Jason's blood tests reveal that he has a low red blood cell count, hematocrit, and hemoglobin concentration.

- What condition do these tests indicate?
- How could this contribute to Jason's chronic fatigue?

Leukocytes

Leukocytes differ from erythrocytes in several respects. Leukocytes contain nuclei and mitochondria and can move in an amoeboid fashion. Because of their amoeboid ability, leukocytes can squeeze through pores in capillary walls and move to a site of infection, whereas erythrocytes usually remain confined within blood vessels. The movement of leukocytes through capillary walls is referred to as *diapedesis* or *extravasation.*

White blood cells are almost invisible under the microscope unless they are stained; therefore, they are classified according to their staining properties. Those leukocytes that have granules in their cytoplasm are called **granular leukocytes;** those without clearly visible granules are called **agranular** (or **nongranular**) **leukocytes.**

The stain used to identify white blood cells is usually a mixture of a pink-to-red stain called *eosin* and a blue-to-purple stain (methylene blue), which is called a "basic stain." Granular leukocytes with pink-staining granules are therefore called **eosinophils,** and those with blue-staining granules are called **basophils.** Those with granules that have little affinity for either stain are **neutrophils** (fig. 13.3). Neutrophils are the most abundant type of leukocyte, accounting for 50% to 70% of the leukocytes in the blood. Immature neutrophils have sausage-shaped nuclei and are called *band cells.* As the band cells mature, their nuclei become lobulated, with two to five lobes connected by thin strands. At this stage, the neutrophils are also known as *polymorphonuclear leukocytes (PMNs).*

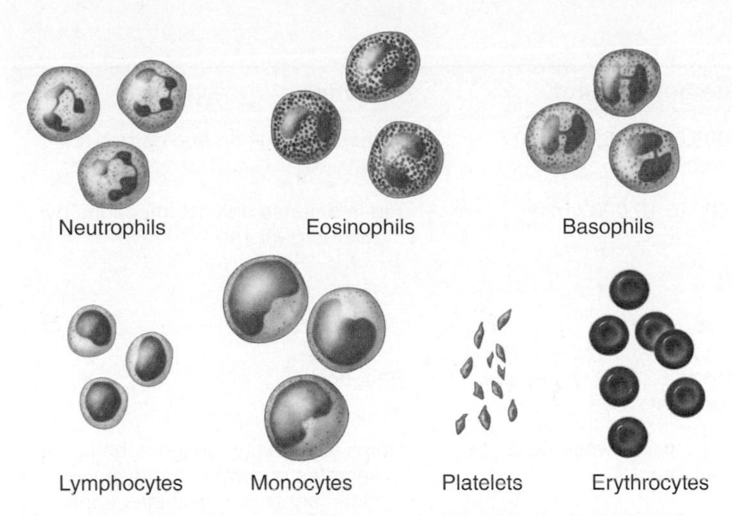

Figure 13.3 The blood cells and platelets. The white blood cells depicted above are granular leukocytes; the lymphocytes and monocytes are nongranular leukocytes. AP|R

There are two types of agranular leukocytes: lymphocytes and monocytes. **Lymphocytes** are usually the second most numerous type of leukocyte; they are small cells with round nuclei and little cytoplasm. **Monocytes,** by contrast, are the largest of the leukocytes and generally have kidney- or horseshoe-shaped nuclei. In addition to these two cell types, there are smaller numbers of *plasma cells,* which are derived from lymphocytes. Plasma cells produce and secrete large amounts of antibodies. The immune functions of the different white blood cells are described in more detail in chapter 15.

CLINICAL APPLICATION

Blood cell counts are an important source of information in assessing a person's health. An abnormal increase in erythrocytes, for example, is termed **polycythemia** and is indicative of several dysfunctions. As previously mentioned, an abnormally low red blood cell count is termed *anemia.* (Polycythemia and anemia are described in detail in chapter 16, section 16.6.) An elevated leukocyte count, called **leukocytosis,** is often associated with infection. A large number of immature leukocytes in a blood sample is diagnostic of the disease **leukemia.** A low white blood cell count, called **leukopenia,** may be due to a variety of factors; low numbers of lymphocytes, for example, may result from poor nutrition or from whole-body irradiation treatment for cancer.

Platelets

Platelets, or **thrombocytes,** are the smallest of the formed elements and are actually fragments of large cells called *megakaryocytes,* which are found in bone marrow. (This is why the term *formed elements* is used instead of *blood cells* to describe erythrocytes, leukocytes, and platelets.) The fragments that enter the circulation as platelets lack nuclei but, like leukocytes, are capable of amoeboid movement. The platelet count per cubic millimeter of blood ranges from 130,000 to 400,000, but this count can vary greatly under different physiological conditions. Platelets survive for about five to nine days before being destroyed by the spleen and liver.

Platelets play an important role in blood clotting. They constitute most of the mass of the clot, and phospholipids in their cell membranes activate the clotting factors in plasma that result in threads of fibrin, which reinforce the platelet plug. Platelets that attach together in a blood clot release *serotonin,* a chemical that stimulates constriction of blood vessels, thus reducing the flow of blood to the injured area. Platelets also secrete growth factors (autocrine regulators—chapter 11, section 11.7), which are important in maintaining the integrity of blood vessels. These regulators also may be involved in the development of atherosclerosis, as described in section 13.7.

The formed elements of the blood are illustrated in figure 13.3, and their characteristics are summarized in table 13.2.

Hematopoiesis

Blood cells are constantly formed through a process called **hematopoiesis** (also called **hemopoiesis**). The **hematopoietic stem cells**—those that give rise to blood cells—originate in the yolk sac of the human embryo and then migrate in sequence to regions around the aorta, to the placenta, and then to the liver of a fetus. The liver is the major hematopoietic organ of the fetus, but then the stem cells migrate to the bone marrow and the liver ceases to be a source of blood cell production shortly after birth. Scientists estimate that the hematopoietic tissue of the bone marrow produces about 500 billion cells each day. The hematopoietic stem cells form a population of relatively undifferentiated, multipotent adult stem cells (chapter 20, section 20.6) that give rise to all of the specialized blood cells. The hematopoietic stem cells are self-renewing, duplicating themselves by mitosis so that the parent stem cell population will not become depleted as individual stem cells differentiate into the mature blood cells. Hematopoietic stem cells are rare, but they proliferate in response to the pro-inflammatory cytokines released during infection (chapter 15, section 15.3) and in response to the depletion of leukocytes during infection. Hematopoietic stem cells are the only cells capable of restoring complete hematopoietic ability (producing all blood cell lines) upon transplantation into the depleted bone marrow of a recipient.

The term **erythropoiesis** refers to the formation of erythrocytes, and **leukopoiesis** to the formation of leukocytes. These processes occur in two classes of tissues after birth, myeloid and lymphoid. **Myeloid tissue** is the red bone marrow of the long bones, ribs, sternum, pelvis, bodies of the vertebrae, and portions of the skull. **Lymphoid tissue** includes the lymph nodes, tonsils, spleen, and thymus. The bone marrow produces all of the different types of blood cells; the lymphoid tissue produces lymphocytes derived from cells that originated in the bone marrow.

As the cells become differentiated during erythropoiesis and leukopoiesis, they develop membrane receptors for chemical

Table 13.2 | Formed Elements of the Blood

Component	Description	Number Present	Function
Erythrocyte (red blood cell)	Biconcave disc without nucleus; contains hemoglobin; survives 100 to 120 days	4,000,000 to 6,000,000 / mm^3	Transports oxygen and carbon dioxide
Leukocytes (white blood cells)		5,000 to 10,000 / mm^3	Aid in defense against infections by microorganisms
Granulocytes	About twice the size of red blood cells; cytoplasmic granules present; survive 12 hours to 3 days		
1. Neutrophil	Nucleus with 2 to 5 lobes; cytoplasmic granules stain slightly pink	54% to 62% of white cells present	Phagocytic
2. Eosinophil	Nucleus bilobed; cytoplasmic granules stain red in eosin stain	1% to 3% of white cells present	Helps to detoxify foreign substances; secretes enzymes that dissolve clots; fights parasitic infections
3. Basophil	Nucleus lobed; cytoplasmic granules stain blue in hematoxylin stain	Less than 1% of white cells present	Releases anticoagulant heparin
Agranulocytes	Cytoplasmic granules not visible; survive 100 to 300 days (some much longer)		
1. Monocyte	2 to 3 times larger than red blood cell; nuclear shape varies from round to lobed	3% to 9% of white cells present	Phagocytic
2. Lymphocyte	Only slightly larger than red blood cell; nucleus nearly fits cell	25% to 33% of white cells present	Provides specific immune response (including antibodies)
Platelet (thrombocyte)	Cytoplasmic fragment; survives 5 to 9 days	130,000 to 400,000 / mm^3	Enables clotting; releases serotonin, which causes vasoconstriction

signals that cause further development along particular lines. The earliest cells that can be distinguished under a microscope are the *erythroblasts* (which become erythrocytes), *myeloblasts* (which become granular leukocytes), *lymphoblasts* (which form lymphocytes), and *monoblasts* (which form monocytes).

Erythropoiesis is an extremely active process. It is estimated that about 2.5 million erythrocytes are produced every second in order to replace those that are continuously destroyed by the spleen and liver. The life span of an erythrocyte is approximately 120 days. Agranular leukocytes remain functional for 100 to 300 days under normal conditions. Granular leukocytes, by contrast, have an extremely short life span of 12 hours to 3 days.

The production of different subtypes of leukocytes is stimulated by chemicals called **cytokines.** These are autocrine regulators secreted by various cells of the immune system. The production of red blood cells is stimulated by the hormone **erythropoietin,** which is secreted by the kidneys. The gene for erythropoietin has been commercially cloned so that this hormone is now available for treatment of anemia, including the anemia that results from kidney disease in patients undergoing dialysis. Injections with recombinant erythropoietin significantly improve aerobic physical performance, probably because of increased hemoglobin allowing the blood to carry an increased amount of oxygen. The World Anti-Doping Code bans the use of recombinant erythropoietin for this reason, and urine from athletes is tested for erythropoietin by World Anti-Doping Agency (WADA) laboratories.

Scientists have identified a specific cytokine that stimulates proliferation of megakaryocytes and their maturation into platelets. By analogy with erythropoietin, they named this regulatory molecule **thrombopoietin.** The gene that codes for thrombopoietin also has been cloned, so that recombinant thrombopoietin is now available for medical research and applications. In clinical trials, thrombopoietin has been used to treat the *thrombocytopenia* (low platelet count) that occurs as a result of bone marrow depletion in patients undergoing chemotherapy for cancer.

CLINICAL APPLICATION 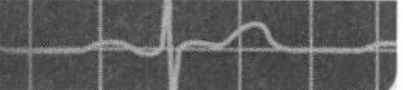

Megakaryocytes in the bone marrow and circulating platelets both have receptors that bind thrombopoietin. Thus, when there is a lowered platelet count, less thrombopoietin is bound to its receptors on the platelets and more is free to enter the bone marrow and stimulate the megakaryocytes. Conversely, when the platelet count is elevated, there is a decreased amount of unbound thrombopoietin in the plasma, and thus reduced stimulation of megakaryocytes. This is a nice homeostatic mechanism for maintaining constancy of the platelet count. However, under various medical conditions (acute blood loss, inflammation, cancer, and others), the liver may be stimulated to produce more thrombopoietin. This results in an elevated platelet count, a condition termed **thrombocytosis.**

Regulation of Leukopoiesis

A variety of cytokines stimulate different stages of leukocyte development. The cytokines known as *multipotent growth factor-1, interleukin-1,* and *interleukin-3* have general effects, stimulating the development of different types of white blood cells. *Granulocyte colony-stimulating factor (G-CSF)* acts in a highly specific manner to stimulate the development of neutrophils, whereas *granulocyte-monocyte colony-stimulating factor (GM-CSF)* stimulates the development of monocytes and eosinophils. The genes for the cytokines G-CSF and GM-CSF have been cloned, making these cytokines available for medical applications.

CLINICAL APPLICATION

Approximately 10,000 **bone marrow transplants** are performed worldwide each year. This procedure generally involves the aspiration of marrow from the iliac crest and separation of the hematopoietic stem cells, which constitute only about 1% of the nucleated cells in the marrow. Stem cells have also been isolated from peripheral blood when the donor is first injected with G-CSF and GM-CSF, which stimulate the marrow to release more stem cells. Another recent technology involves the storage, or "banking," of hematopoietic stem cells obtained from the placenta or umbilical cord blood of a neonate. These cells may then be used later in life if the person needs them for transplantation.

Regulation of Erythropoiesis

The primary regulator of erythropoiesis is *erythropoietin,* produced by the kidneys in response to tissue hypoxia when the blood oxygen levels are decreased. One of the possible causes of decreased blood oxygen levels is a decreased red blood cell count. Because of erythropoietin stimulation, the daily production of new red blood cells compensates for the daily destruction of old red blood cells, preventing a decrease in the blood oxygen content. An increased secretion of erythropoietin and production of new red blood cells occurs when a person is at a high altitude or has lung disease, which are conditions that reduce the oxygen content of the blood.

Erythropoietin acts by binding to membrane receptors on cells that will become erythroblasts (fig. 13.4). The erythropoietin-stimulated cells undergo cell division and differentiation, leading to the production of erythroblasts. These are transformed into *normoblasts,* which lose their nuclei to become *reticulocytes.* The reticulocytes then change into fully mature erythrocytes. This process takes 3 days; the reticulocyte normally stays in the bone marrow for the first 2 days and then circulates in the blood on the third day. At the end of the erythrocyte life span of 120 days, the old red blood cells are removed by the liver and by macrophages (phagocytic cells) of the spleen and bone marrow. Most of the iron contained in the hemoglobin molecules of the destroyed red blood cells is recycled back to the myeloid tissue to be used in the production of

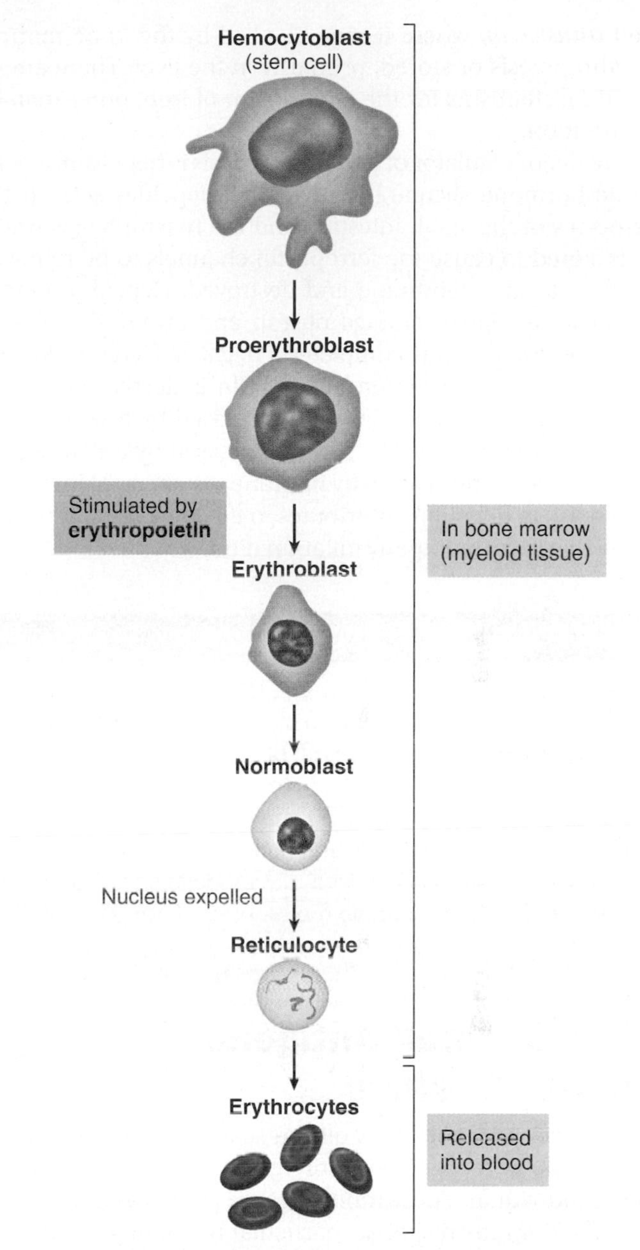

Figure 13.4 The stages of erythropoiesis. The proliferation and differentiation of cells that will become mature erythrocytes (red blood cells) occurs in the bone marrow and is stimulated by the hormone erythropoietin, secreted by the kidneys. AP|R

hemoglobin for new red blood cells (see chapter 18, fig. 18.22). The production of red blood cells and synthesis of hemoglobin depends on the supply of iron, along with that of vitamin B_{12} and folic acid.

Iron absorbed from food in the small intestine passes into intestinal epithelial cells, where it can either be stored in the cells or secreted into blood plasma through specific *ferroportin* channel proteins. Similarly, the iron derived from the heme in old red blood cells that were destroyed by macrophages can be stored in the macrophages or released into the blood through ferroportin channels. Iron travels in the blood bound to a plasma protein called *transferrin,* where it may be used by the bone marrow in erythropoiesis or stored, primarily in the liver. There are no normal mechanisms for the elimination of iron, other than by menstruation.

The major regulator of iron homeostasis is **hepcidin**, a polypeptide hormone secreted by the liver. Hepcidin acts on the enterocytes of the small intestine and the macrophages where iron is stored to cause the ferroportin channels to be removed from the plasma membrane and destroyed. Hepcidin thereby promotes the cellular storage of iron and lowers the plasma iron concentration. This completes a negative feedback loop in which the liver's production of hepcidin is decreased by iron deficiency and most anemias, and increased by excessive iron intake. Hepcidin production is also increased by inflammation (probably due to stimulation by immune system cytokines), and the increase in hepcidin contributes to the low plasma iron and anemia that can accompany inflammation.

FITNESS APPLICATION

Because of the recycling of iron, dietary requirements for iron are usually quite small. Males (and women after menopause) have a dietary iron requirement of only about 10 mg/day. Women with average menstrual blood loss need 15 mg/day, and pregnant women need 30 mg/day. Because of these relatively small dietary requirements, iron-deficiency anemia in adults is usually due not to a dietary deficiency but rather to blood loss, which reduces the amount of iron that can be recycled.

Red Blood Cell Antigens and Blood Typing

There are certain molecules on the surfaces of all cells in the body that can be recognized as foreign by the immune system of another individual. These molecules are known as *antigens.* As part of the immune response, particular lymphocytes secrete a class of proteins called *antibodies* that bond in a specific fashion with antigens. The specificity of antibodies for antigens is analogous to the specificity of enzymes for their substrates, and of receptor proteins for neurotransmitters and hormones. A complete description of antibodies and antigens is provided in chapter 15.

ABO System

The distinguishing antigens on other cells are far more varied than the antigens on red blood cells. Red blood cell antigens, however, are of extreme clinical importance because their types must be matched between donors and recipients for blood transfusions. There are several groups of red blood cell antigens, but the major group is known as the **ABO system.** In terms of the antigens present on the red blood cell surface, a person may be *type A* (with only A antigens), *type B* (with only B antigens), *type AB* (with both A and B antigens), or *type O* (with neither A nor B antigens). Each person's blood type—A, B, or O—denotes the antigens present on the red blood cell surface, which are the products of the genes (located on chromosome number 9) that code for these antigens.

Each person inherits two genes (one from each parent) that control the production of the ABO antigens. The genes for A or B antigens are dominant to the gene for O. The O gene is recessive, simply because it doesn't code for either the A or the B red blood cell antigens. The genes for A and B are often shown as I^A and I^B, and the recessive gene for O is shown as the lowercase i. A person who is type A, therefore, may have inherited the A gene from each parent (may have the genotype I^AI^A), or the A gene from one parent and the O gene from the other parent (and thus have the genotype I^Ai). Likewise, a person who is type B may have the genotype I^BI^B or I^Bi. It follows that a type O person inherited the O gene from each parent (has the genotype ii), whereas a type AB person inherited the A gene from one parent and the B gene from the other (there is no dominant-recessive relationship between A and B).

The immune system exhibits tolerance to its own red blood cell antigens. People who are type A, for example, do not produce anti-A antibodies. Surprisingly, however, they do make antibodies against the B antigen and, conversely, people with blood type B make antibodies against the A antigen (fig. 13.5). This is believed to result from the fact that antibodies made in response to some common bacteria cross-react with the A or B antigens. People who are type A, therefore, acquire antibodies that can react with B antigens by exposure to these bacteria, but they do not develop antibodies that can react with A antigens because tolerance mechanisms prevent this.

People who are type AB develop tolerance to both of these antigens, and thus do not produce either anti-A or anti-B antibodies. Those who are type O, by contrast, do not develop tolerance to either antigen; therefore, they have both anti-A and anti-B antibodies in their plasma (table 13.3).

Transfusion Reactions

Before transfusions are performed, a *major crossmatch* is made by mixing serum from the recipient with blood cells from the donor. If the types do not match—if the donor is type A, for example, and the recipient is type B—the recipient's antibodies attach to the donor's red blood cells and form bridges that cause the cells to clump together, or **agglutinate** (figs. 13.5 and 13.6). Because of this agglutination reaction, the A and B antigens are sometimes called *agglutinogens,* and the antibodies against them are called *agglutinins.* Transfusion errors that result in such agglutination can lead to blockage of small blood vessels and cause hemolysis (rupture of red blood cells), which may damage the kidneys and other organs.

In emergencies, type O blood has been given to people who are type A, B, AB, or O. Because type O red blood cells lack A and B antigens, the recipient's antibodies cannot cause agglutination of the donor red blood cells. Type O is, therefore, a *universal donor*—but only as long as the volume of plasma donated is small, since plasma from a type O person would

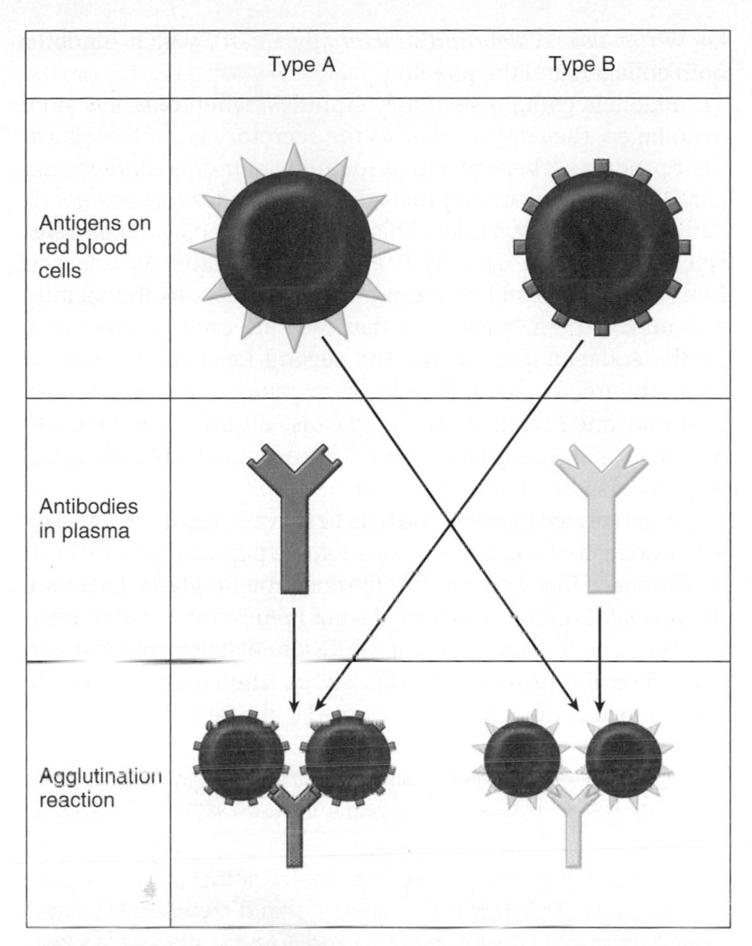

Figure 13.5 Agglutination reaction. People with type A blood have type A antigens on their red blood cells and antibodies in their plasma against the type B antigen. People with type B blood have type B antigens on their red blood cells and antibodies in their plasma against the type A antigen. Therefore, if red blood cells from one blood type are mixed with antibodies from the plasma of the other blood type, an agglutination reaction occurs. In this reaction, red blood cells stick together because of antigen-antibody binding.

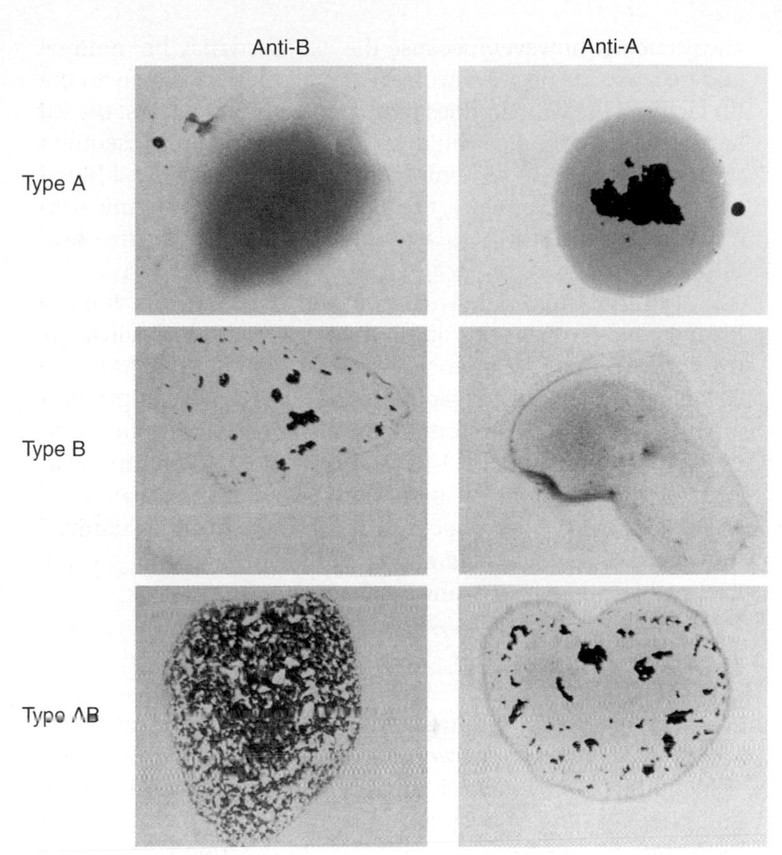

Figure 13.6 Blood typing. Agglutination (clumping) of red blood cells occurs when cells with A-type antigens are mixed with anti-A antibodies and when cells with B-type antigens are mixed with anti-B antibodies. No agglutination would occur with type O blood (not shown).

Table 13.3 | The ABO System of Red Blood Cell Antigens

Genotype	Antigen on RBCs	Antibody in Plasma
I^AI^A; I^Ai	A	Anti-B
I^BI^B; I^Bi	B	Anti-A
ii	O	Anti-A and anti-B
I^AI^B	AB	Neither anti-A nor anti-B

agglutinate type A, type B, and type AB red blood cells. Likewise, type AB people are *universal recipients* because they lack anti-A and anti-B antibodies, and thus cannot agglutinate donor red blood cells. (Donor plasma could agglutinate recipient red blood cells if the transfusion volume were too large.) Because of the dangers involved, use of the universal donor and recipient concept is strongly discouraged in practice.

Rh Factor

Another group of antigens found on the red blood cells of most people is the **Rh factor** (named for the rhesus monkey, in which these antigens were first discovered). There are a number of different antigens in this group, but one stands out because of its medical significance. This Rh antigen is termed D, and is often indicated as Rho(D). If this Rh antigen is present on a person's red blood cells, the person is **Rh positive;** if it is absent, the person is **Rh negative.** The Rh-positive condition is by far the more common (with a frequency of 85% in the Caucasian population, for example).

The Rh factor is of particular significance when Rh-negative mothers give birth to Rh-positive babies. The fetal and maternal blood are normally kept separate across the placenta (chapter 20, section 20.6), and so the Rh-negative mother is not usually exposed to the Rh antigen of the fetus during the pregnancy. At the time of birth, however, a variable degree of exposure may occur, and the mother's immune system may become sensitized and produce antibodies against the Rh antigen. This does not always occur, however, because the exposure may be minimal and because Rh-negative women vary in their sensitivity to the Rh factor. If the woman does produce antibodies against the Rh factor, these antibodies could cross the placenta in subsequent pregnancies and cause hemolysis of the Rh-positive red blood cells of the fetus. Therefore, the baby could be born anemic with a condition called *erythroblastosis fetalis,* or *hemolytic disease of the newborn.*

Erythroblastosis fetalis can be prevented by injecting the Rh-negative mother with an antibody preparation against the Rh factor (a trade name for this preparation is RhoGAM—the GAM is short for gamma globulin, the class of plasma proteins in which antibodies are found) within 72 hours after the birth of each Rh-positive baby. This is a type of passive immunization in which the injected antibodies inactivate the Rh antigens and thus prevent the mother from becoming actively immunized to them. Some physicians now give RhoGAM throughout the Rh-positive pregnancy of any Rh-negative woman.

Blood Clotting

When a blood vessel is injured, a number of physiological mechanisms are activated that promote **hemostasis,** or the cessation of bleeding (*hemo* = blood; *stasis* = standing). Breakage of the endothelial lining of a vessel exposes collagen proteins from the subendothelial connective tissue to the blood. This initiates three separate, but overlapping, hemostatic mechanisms: (1) vasoconstriction, (2) the formation of a platelet plug, and (3) the production of a web of fibrin proteins that penetrates and surrounds the platelet plug.

Platelets and Blood Vessel Walls

In the absence of blood vessel damage, platelets are repelled from each other and from the endothelium of blood vessels. The endothelium is a simple squamous epithelium that overlies connective tissue collagen and other proteins that are capable of activating platelets to begin clot formation. Thus, an intact endothelium physically separates the blood from collagen and other platelet activators in the vessel wall. In addition, the endothelial cells secrete *prostacyclin* (or *PGI_2*, a type of prostaglandin—see chapter 11, fig. 11.34) and *nitric oxide (NO),* which (1) act as vasodilators and (2) act on the platelets to inhibit platelet aggregation. In addition, the plasma membrane of endothelial cells contains an enzyme known as *CD39,* which has its active site facing the blood. The CD39 enzyme breaks down ADP in the blood to AMP and P_i (ADP is released by activated platelets and promotes platelet aggregation, as described shortly). These protective mechanisms are needed to ensure that platelets don't stick to the vessel wall and to each other, so that the flow of blood is not impeded when the endothelium is intact (fig.13.7*a*).

When a blood vessel is injured and the endothelium is broken, glycoproteins in the platelet's plasma membrane are now able to bind to the exposed collagen fibers. The force of blood flow might pull the platelets off the collagen, however, were it not for another protein produced by endothelial cells known as *von Willebrand's factor* (fig.13.7*b*), which binds to both collagen and the platelets.

Platelets contain secretory granules; when platelets stick to collagen, they *degranulate* as the secretory granules release their products. These products include *adenosine diphosphate (ADP), serotonin,* and a prostaglandin called *thromboxane A_2* (chapter 11; see fig. 11.34). This event is known as the **platelet release reaction.** The ADP and thromboxane A_2 released from activated platelets recruits new platelets to the vicinity and makes them "sticky," so that they adhere to those stuck on the collagen (fig. 13.7*b*). The second layer of platelets, in turn, undergoes a platelet release reaction, and the ADP and thromboxane A_2 that are secreted cause additional platelets to aggregate at the site of the injury. This produces a **platelet plug** (fig. 13.7*c*) in the damaged vessel.

The activated platelets also help to activate plasma clotting factors, leading to the conversion of a soluble plasma protein known as *fibrinogen* into an insoluble fibrous protein, *fibrin.* There are binding sites on the platelet's plasma membrane for fibrinogen and fibrin, so that these proteins help join platelets together and strengthen the platelet plug (fig. 13.7*c*). The clotting sequence leading to fibrin formation is discussed in the next topic.

CLINICAL APPLICATION

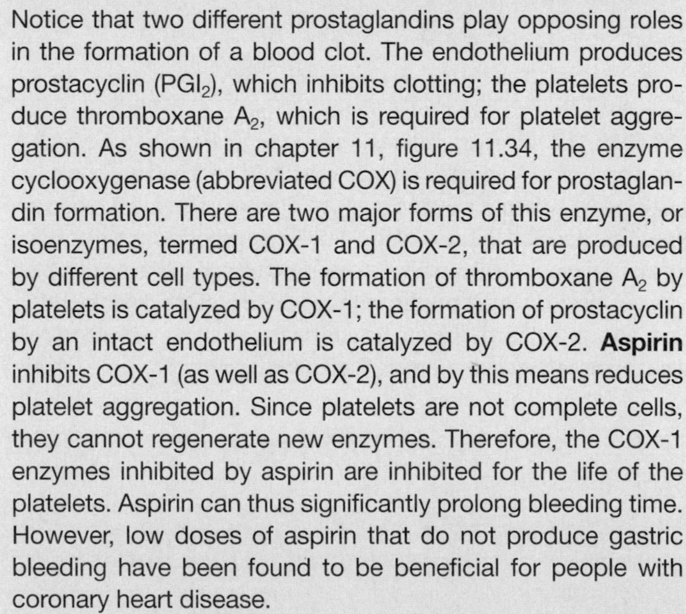

Notice that two different prostaglandins play opposing roles in the formation of a blood clot. The endothelium produces prostacyclin (PGI_2), which inhibits clotting; the platelets produce thromboxane A_2, which is required for platelet aggregation. As shown in chapter 11, figure 11.34, the enzyme cyclooxygenase (abbreviated COX) is required for prostaglandin formation. There are two major forms of this enzyme, or isoenzymes, termed COX-1 and COX-2, that are produced by different cell types. The formation of thromboxane A_2 by platelets is catalyzed by COX-1; the formation of prostacyclin by an intact endothelium is catalyzed by COX-2. **Aspirin** inhibits COX-1 (as well as COX-2), and by this means reduces platelet aggregation. Since platelets are not complete cells, they cannot regenerate new enzymes. Therefore, the COX-1 enzymes inhibited by aspirin are inhibited for the life of the platelets. Aspirin can thus significantly prolong bleeding time. However, low doses of aspirin that do not produce gastric bleeding have been found to be beneficial for people with coronary heart disease.

Drugs that inhibit platelet activation are used to help prevent *myocardial infarction* ("heart attack"; section 13.7) resulting from a **coronary thrombosis.** Aspirin does this by irreversibly acetylating (adding a two-carbon acetyl group to) COX-1 and thereby inhibiting the production of thromboxane A_2. **Plavix** (*clopidogrel*) and similar drugs inhibit platelet activation by antagonizing the platelet plasma membrane receptors for ADP. By preventing platelet aggregation caused by thromboxane A_2 and ADP, aspirin and Plavix (and others) reduce the platelet plug, which forms the bulk of thrombi in arteries.

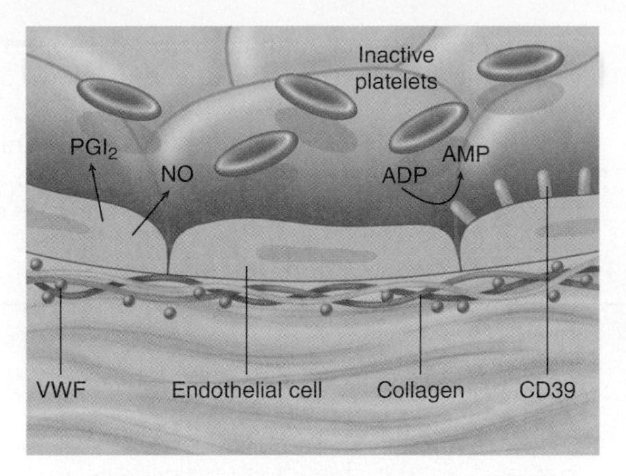

(a)

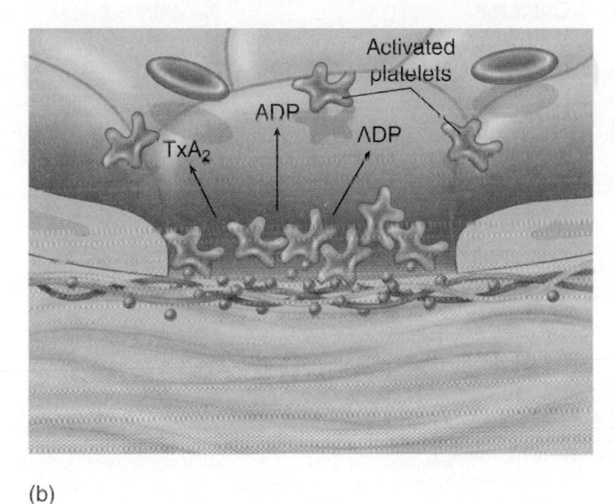

(b)

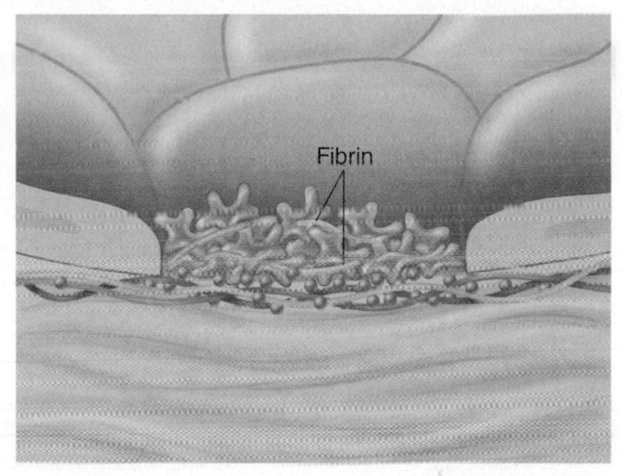

(c)

Figure 13.7 Platelet aggregation. (*a*) Platelet aggregation is prevented in an intact endothelium because it separates the blood from collagen, a potential platelet activator. Also, the endothelium secretes nitric oxide (*NO*) and prostaglandin I2 (*PGI2*), which inhibit platelet aggregation. An enzyme called *CD39* breaks down *ADP* in the blood, which would otherwise promote platelet aggregation. (*b*) When the endothelium is broken, platelets adhere to collagen and to von Willebrand's factor (*VWF*), which helps anchor the platelets that are activated by this process and by the secretion of ADP and thromboxane A2 (*TxA2*), a prostaglandin. (*c*) A platelet plug is formed and reinforced with fibrin proteins.

Clotting Factors: Formation of Fibrin

The platelet plug is strengthened by a meshwork of insoluble protein fibers known as **fibrin** (fig. 13.8). Blood clots therefore are composed of platelets and fibrin, and they usually contain trapped red blood cells that give the clot a red color (clots formed in arteries, where the blood flow is more rapid, generally lack red blood cells and thus appear gray). Finally, contraction of the platelet mass in the process of *clot retraction* forms a more compact and effective plug. Fluid squeezed from the clot as it retracts is called **serum,** which is plasma without fibrinogen, the soluble precursor of fibrin. (Serum is obtained in laboratories by allowing blood to clot in a test tube and then centrifuging the tube so that the clot and blood cells become packed at the bottom of the tube.)

The conversion of fibrinogen into fibrin may occur via either of two pathways. Blood left in a test tube will clot without the addition of any external chemicals; the pathway that produces this clot is thus called the **intrinsic pathway.** Damaged tissues, however, release a chemical that initiates a "shortcut" to the formation of fibrin. Because this chemical is not part of blood, the shorter pathway is called the **extrinsic pathway.**

The intrinsic pathway is initiated by the exposure of plasma to a negatively charged surface, such as that provided by collagen at the site of a wound or by the glass of a test tube. This *contact pathway* activates a plasma protein called factor XII (table 13.4), which is a protein-digesting enzyme (a protease). Active factor XII in turn activates another clotting factor, which activates yet another. The plasma clotting factors are numbered in order of their discovery, which does not reflect the actual sequence of reactions.

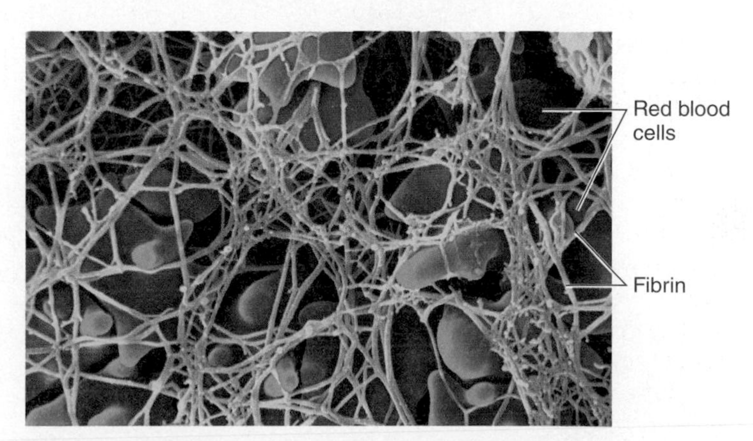

Figure 13.8 Colorized scanning electron micrograph of a blood clot. The threads of fibrin have trapped red blood cells in this image.

The next steps in the sequence require the presence of Ca^{2+} and phospholipids, the latter provided by platelets. These steps result in the conversion of an inactive glycoprotein, called **prothrombin,** into the active enzyme **thrombin.** Thrombin converts the soluble protein **fibrinogen** into **fibrin** monomers. These monomers are joined together to produce the insoluble fibrin polymers that form a meshwork supporting the platelet plug. The intrinsic clotting sequence is shown on the right side of figure 13.9.

The extrinsic pathway of clot formation is initiated by **tissue factor** (or *tissue thromboplastin,* also known as *factor III*), a membrane glycoprotein found inside the walls of blood vessels

Table 13.4 | The Plasma Clotting Factors

Factor	Name	Function	Pathway
I	Fibrinogen	Converted to fibrin	Common
II	Prothrombin	Converted to thrombin (enzyme)	Common
III	Tissue thromboplastin	Cofactor	Extrinsic
IV	Calcium ions (Ca^{2+})	Cofactor	Intrinsic, extrinsic, and common
V	Proaccelerin	Cofactor	Common
VII*	Proconvertin	Enzyme	Extrinsic
VIII	Antihemophilic factor	Cofactor	Intrinsic
IX	Plasma thromboplastin component; Christmas factor	Enzyme	Intrinsic
X	Stuart-Prower factor	Enzyme	Common
XI	Plasma thromboplastin antecedent	Enzyme	Intrinsic
XII	Hageman factor	Enzyme	Intrinsic
XIII	Fibrin stabilizing factor	Enzyme	Common

*Factor VI is no longer referenced; it is now believed to be the same substance as activated factor V.

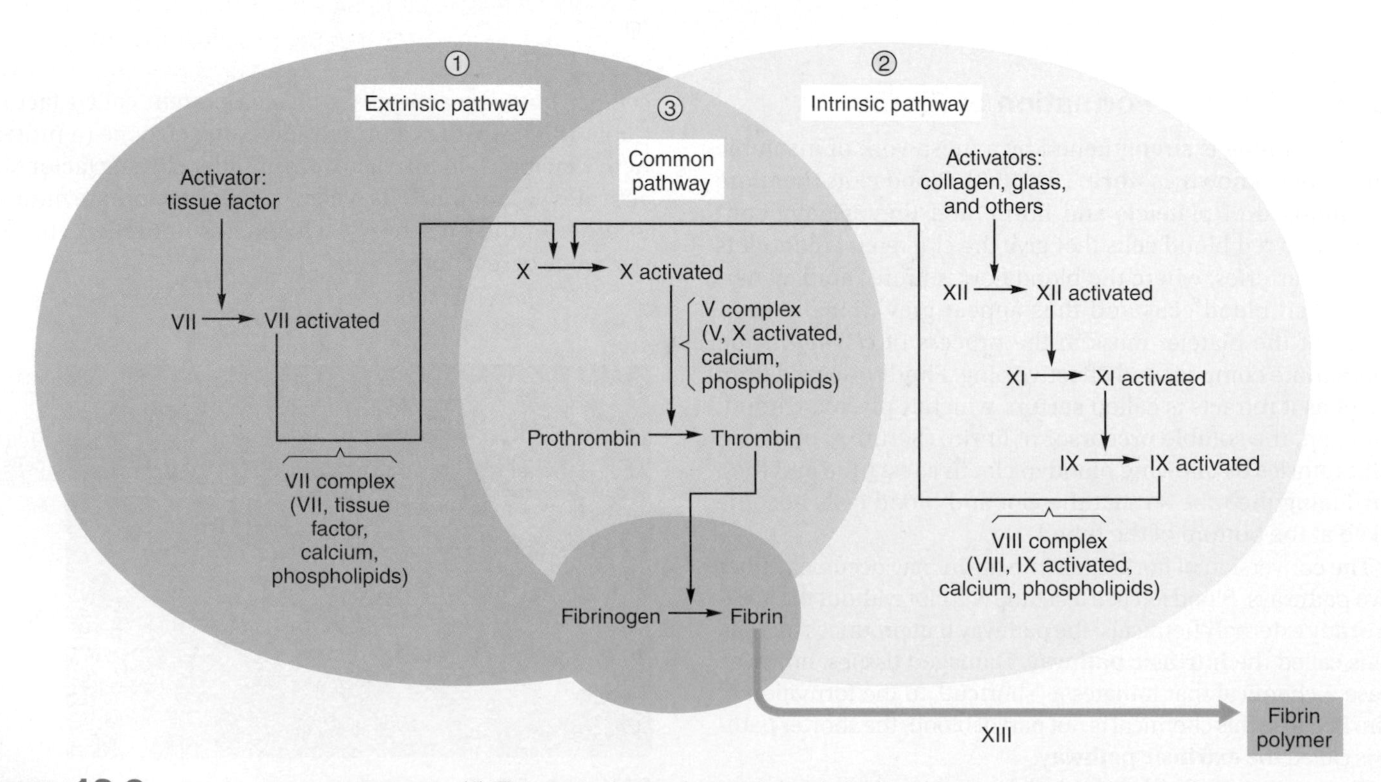

Figure 13.9 The clotting pathways. (1) The extrinsic clotting pathway is initiated by the release of tissue factor. (2) The intrinsic clotting pathway is initiated by the activation of factor XII by contact with collagen or glass. (3) Extrinsic and intrinsic clotting pathways converge when they activate factor X, eventually leading to the formation of fibrin.

(in the tunica media and tunica externa; see fig. 13.26) and the cells of the surrounding tissues. When a blood vessel is injured, tissue factor then becomes exposed to factor VII and VIIa in the blood and forms a complex with factor VIIa. By forming this complex, tissue factor greatly increases (by a factor of two million) the ability of factor VIIa to activate factor X and factor IX. This extrinsic pathway (shown on the left side of fig. 13.9) generates thrombin and fibrin more rapidly than does the intrinsic pathway and is believed to be the primary initiator of blood clotting *in vivo.*

CLINICAL APPLICATION

A number of hereditary diseases involve the clotting system. Examples of hereditary clotting disorders include two different genetic defects in factor VIII. A defect in one subunit of factor VIII prevents this factor from participating in the intrinsic clotting pathway. This genetic disease, called **hemophilia A,** is an X-linked recessive trait that is prevalent in the royal families of Europe. A defect in another subunit of factor VIII results in **von Willebrand's disease.** This is an inherited clotting disorder resulting from defective von Willebrand factor, a large glycoprotein required for rapidly circulating platelets to adhere adequately to collagen at the site of vascular injury (see fig. 13.7), and thus to form a platelet plug. In addition, von Willebrand factor protects clotting factor VIII; thus, in von Willebrand's disease there is also a deficiency in the participation of factor VIII in fibrin formation. Some acquired and inherited defects in the clotting system are summarized in table 13.5.

Dissolution of Clots

As the damaged blood vessel wall is repaired, activated factor XII promotes the conversion of an inactive molecule in plasma into the active form called *kallikrein.* Kallikrein, in turn, catalyzes the conversion of inactive *plasminogen* into the active molecule **plasmin.** Plasmin is an enzyme that digests fibrin into "split products," thus promoting dissolution of the clot.

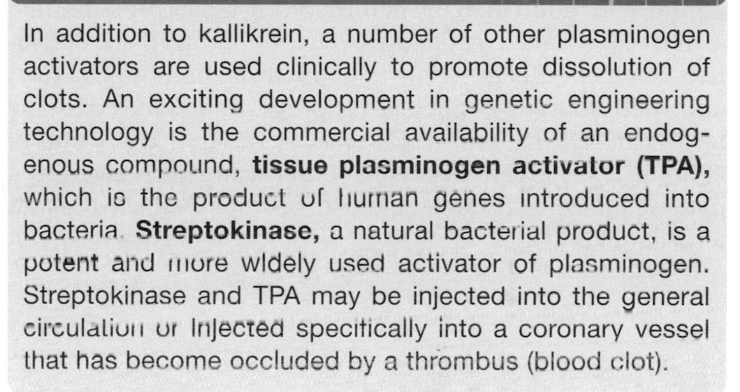
CLINICAL APPLICATION

In addition to kallikrein, a number of other plasminogen activators are used clinically to promote dissolution of clots. An exciting development in genetic engineering technology is the commercial availability of an endogenous compound, **tissue plasminogen activator (TPA),** which is the product of human genes introduced into bacteria. **Streptokinase,** a natural bacterial product, is a potent and more widely used activator of plasminogen. Streptokinase and TPA may be injected into the general circulation or injected specifically into a coronary vessel that has become occluded by a thrombus (blood clot).

Anticoagulants

Clotting of blood in test tubes can be prevented by the addition of *sodium citrate* or *ethylenediaminetetraacetic acid (EDTA),* both of which chelate (bind to) calcium. By this means, Ca^{2+} levels in the blood that can participate in the clotting sequence are lowered, and clotting is inhibited. A mucoprotein called

Table 13.5 | Some Acquired and Inherited Clotting Disorders and a Listing of Anticoagulant Drugs

Category	Cause of Disorder	Comments
Acquired clotting disorders	Vitamin K deficiency	Inadequate formation of prothrombin and other clotting factors in the liver
Inherited clotting disorders	Hemophilia A (defective factor VIII_{AHF})	Recessive trait carried on X chromosome; results in delayed formation of fibrin
	von Willebrand's disease (defective factor VIII_{VWF})	Dominant trait carried on autosomal chromosome; impaired ability of platelets to adhere to collagen in subendothelial connective tissue
	Hemophilia B (defective factor IX); also called Christmas disease	Recessive trait carried on X chromosome; results in delayed formation of fibrin
Anticoagulants		
Aspirin	Inhibits prostaglandin production, resulting in a defective platelet release reaction	
Coumarin	Inhibits activation of vitamin K	
Heparin	Inhibits activity of thrombin	
Citrate	Combines with Ca^{2+}, and thus inhibits the activity of many clotting factors	

heparin can also be added to the tube to prevent clotting. Heparin activates *antithrombin III,* a plasma protein that combines with and inactivates thrombin. Heparin is also given intravenously during certain medical procedures to prevent clotting. The *coumarin drugs* (*warfarin* and *dicumarol*) block the cellular activation of vitamin K by inhibiting the enzyme vitamin K epoxide reductase. Because activated vitamin K is required for proper blood clotting (described next), these drugs serve as anticoagulants (they are the only clinically used oral anticoagulants).

Vitamin K is needed for the conversion of glutamate, an amino acid found in many of the clotting factor proteins, into a derivative called *gamma-carboxyglutamate.* This derivative is more effective than glutamate at bonding to Ca^{2+} and such bonding is needed for proper function of clotting factors II, VII, IX, and X. Because of the indirect action of vitamin K on blood clotting, coumarin must be given to a patient for several days before it becomes effective as an anticoagulant.

CHECKPOINT

3. Distinguish between the different types of formed elements of the blood in terms of their origin, appearance, and function.
4. Describe how the rate of erythropoiesis is regulated.

5a. Explain what is meant by "type A positive" and describe what can happen in a blood transfusion if donor and recipient are not properly matched.

5b. Explain the meaning of *intrinsic* and *extrinsic* as applied to the clotting pathways. How do the two pathways differ from each other? Which steps are common to both?

13.3 STRUCTURE OF THE HEART

The heart contains four chambers: two atria, which receive venous blood, and two ventricles, which eject blood into arteries. The right ventricle pumps blood to the lungs, where the blood becomes oxygenated; the left ventricle pumps oxygenated blood to the entire body.

LEARNING OUTCOMES

After studying this section, you should be able to:

6. Distinguish between the systemic and the pulmonary circulation.
7. Describe the structure of the heart and its components.

About the size of a fist, the hollow, cone-shaped **heart** is divided into four chambers. The right and left **atria** (singular, *atrium*) receive blood from the venous system; the right and left **ventricles** pump blood into the arterial system. The right atrium and ventricle (sometimes called the *right pump*) are separated from the left atrium and ventricle (the *left pump*) by a muscular wall, or *septum.* This septum normally prevents mixture of the blood from the two sides of the heart.

Between the atria and ventricles, there is a layer of dense connective tissue known as the **fibrous skeleton** of the heart. Bundles of myocardial cells (chapter 12, section 12.6) in the atria attach to the upper margin of this fibrous skeleton and form a single functioning unit, or *myocardium.* The myocardial cell bundles of the ventricles attach to the lower margin and form a different myocardium. As a result, the myocardia of the atria and ventricles are structurally and functionally separated from each other, and special conducting tissue is needed to carry action potentials from the atria to the ventricles. The connective tissue of the fibrous skeleton also forms rings, called *annuli fibrosi,* around the four heart valves, providing a foundation for the support of the valve flaps.

Pulmonary and Systemic Circulations

Blood whose oxygen content has become partially depleted and whose carbon dioxide content has increased as a result of tissue metabolism returns to the right atrium. This blood then enters the right ventricle, which pumps it into the *pulmonary trunk* and *pulmonary arteries.* The pulmonary arteries branch to transport blood to the lungs, where gas exchange occurs between the lung capillaries and the air sacs (alveoli) of the lungs. Oxygen diffuses from the air to the capillary blood, while carbon dioxide diffuses in the opposite direction.

The blood that returns to the left atrium by way of the *pulmonary veins* is therefore enriched in oxygen and partially depleted of carbon dioxide. The path of blood from the heart (right ventricle), through the lungs, and back to the heart (left atrium) completes one circuit: the **pulmonary circulation.**

Oxygen-rich blood in the left atrium enters the left ventricle and is pumped into a very large, elastic artery—the *aorta.* The aorta ascends for a short distance, makes a U-turn, and then descends through the thoracic (chest) and abdominal cavities. Arterial branches from the aorta supply oxygen-rich blood to all of the organ systems and are thus part of the **systemic circulation.**

As a result of cellular respiration, the oxygen concentration is lower and the carbon dioxide concentration is higher in the tissues than in the capillary blood. Blood that drains from the tissues into the systemic veins is thus partially depleted of oxygen and increased in carbon dioxide content. These veins ultimately empty into two large veins—the *superior* and *inferior venae cavae*—that return the oxygen-poor blood to the

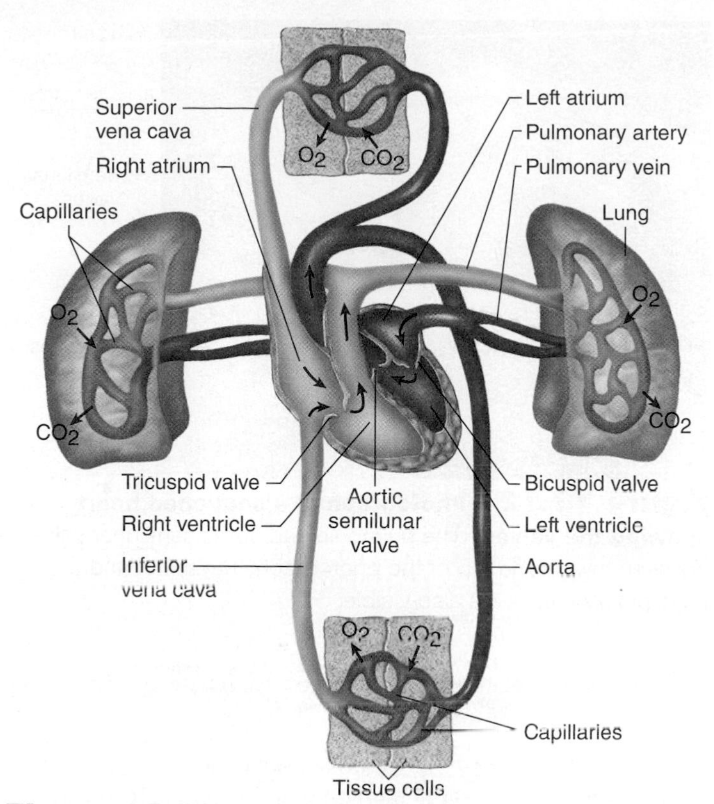

Figure 13.10 A diagram of the circulatory system. The systemic circulation includes the aorta and venae cavae; the pulmonary circulation includes the pulmonary arteries and pulmonary veins. APIR

right atrium. This completes the systemic circulation: from the heart (left ventricle), through the organ systems, and back to the heart (right atrium). The systemic and pulmonary circulations are illustrated in figure 13.10, and their characteristics are summarized in table 13.6.

The numerous small muscular arteries and arterioles of the systemic circulation present greater resistance to blood flow than that in the pulmonary circulation. Despite the differences in resistance, the rate of blood flow through the systemic circulation must be matched to the flow rate of the pulmonary circulation. Because the amount of work performed by the left ventricle is greater (by a factor of 5 to 7) than that performed by the right ventricle, it is not surprising that the muscular wall of the left ventricle is thicker (8 to 10 mm) than that of the right ventricle (2 to 3 mm).

Atrioventricular and Semilunar Valves

Although adjacent myocardial cells are joined together mechanically and electrically by intercalated discs (chapter 12; see figs. 12.32 and 12.33), the atria and ventricles are separated into two functional units by a sheet of connective tissue—the fibrous skeleton previously mentioned. Embedded within this sheet of tissue are one-way **atrioventricular (AV) valves.** The AV valve located between the right atrium and right ventricle has three flaps, and is therefore called the *tricuspid valve.* The AV valve between the left atrium and left ventricle has two flaps and is thus called the *bicuspid valve,* or, alternatively, the *mitral valve* (fig. 13.11).

The AV valves allow blood to flow from the atria to the ventricles, but they normally prevent the backflow of blood into the atria. Opening and closing of these valves occur as a result of pressure differences between the atria and ventricles. When the ventricles are relaxed, the venous return of blood to the atria causes the pressure in the atria to exceed that in the ventricles. The AV valves therefore open, allowing blood to enter the ventricles. As the ventricles contract, the intraventricular pressure rises above the pressure in the atria and pushes the AV valves closed.

There is a danger, however, that the high pressure produced by contraction of the ventricles could push the valve flaps too much and evert them. This is normally prevented by contraction of the *papillary muscles* within the ventricles, which are connected to the AV valve flaps by strong tendinous cords called the *chordae tendineae* (fig. 13.11). Contraction of the papillary muscles occurs at the same time as contraction of the muscular walls of the ventricles and serves to keep the valve flaps tightly closed.

One-way **semilunar valves** (fig. 13.12) are located at the origin of the pulmonary artery and aorta. These valves open during ventricular contraction, allowing blood to enter the pulmonary and systemic circulations. During ventricular relaxation, when the pressure in the arteries is greater than the pressure in the ventricles, the semilunar valves snap shut, preventing the backflow of blood into the ventricles.

Table 13.6 | Summary of the Pulmonary and Systemic Circulations

	Source	Arteries	O_2 Content of Arteries	Veins	O_2 Content of Veins	Termination
Pulmonary Circulation	Right ventricle	Pulmonary arteries	Low	Pulmonary veins	High	Left atrium
Systemic Circulation	Left ventricle	Aorta and its branches	High	Superior and inferior venae cavae and their branches*	Low	Right atrium

*Blood from the coronary circulation does not enter the venae cavae, but instead returns directly to the right atrium via the coronary sinus.

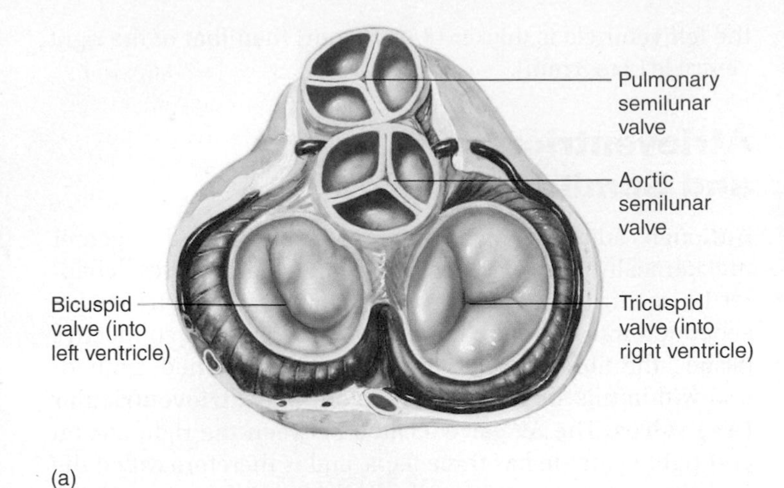

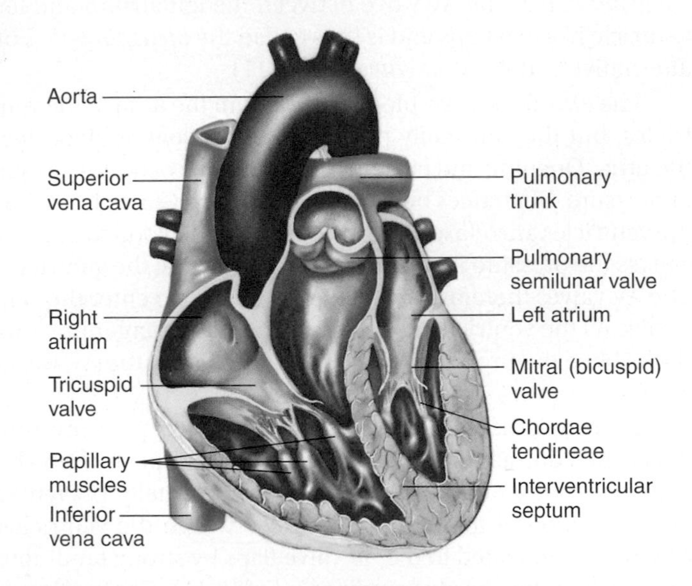

Figure 13.11 The heart valves. (*a*) A superior view of the heart valves. (*b*) A sagittal section through the heart, showing both AV valves and the pulmonary semilunar valve (the aortic semilunar valve is not visible in this view). AP|R

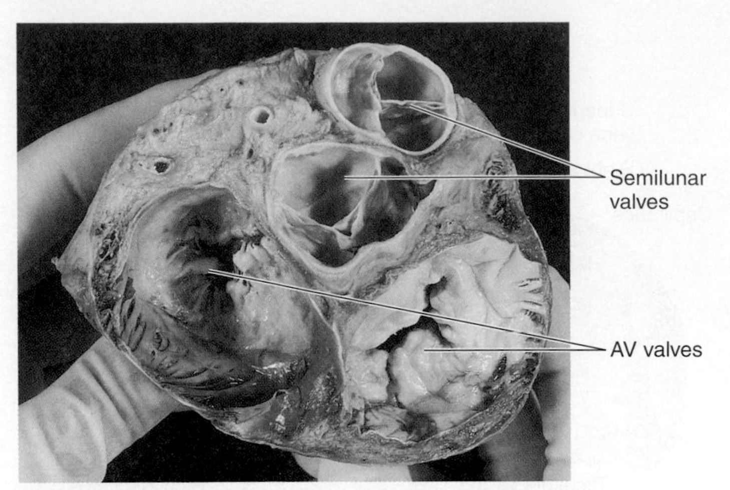

Figure 13.12 Photograph of a sectioned heart showing the valves. The pulmonic and aortic semilunar valves are seen toward the top of the photograph. The mitral and tricuspid AV valves are also visible.

Heart Sounds

Closing of the AV and semilunar valves produces sounds that can be heard by listening through a stethoscope placed on the chest. These sounds are often verbalized as "lub-dub." The "lub," or **first sound,** is produced by closing of the AV valves during isovolumetric contraction of the ventricles (section 13.4). The "dub," or **second sound,** is produced by closing of the semilunar valves when the pressure in the ventricles falls below the pressure in the arteries. The first sound is thus heard when the ventricles contract at *systole,* and the second sound is heard when the ventricles relax at the beginning of *diastole.* (Systole and diastole are discussed in section 13.4.)

CLINICAL APPLICATION

The first sound may be heard to split into tricuspid and mitral components, particularly during inhalation. Closing of the tricuspid valve is best heard at the lower sternal border, just superior to the xiphoid process on both sides of the sternum. Closing of the mitral valve is best heard at the fifth left intercostal space at the apex of the heart (fig. 13.13). The second heart sound may also be split: closing of the pulmonary and aortic semilunar valves is best heard at the second left and right intercostal spaces, respectively. However, these auscultatory positions are affected by obesity, pregnancy, ventricular hypertrophy, and other variables.

Heart Murmurs

Murmurs are abnormal heart sounds produced by abnormal patterns of blood flow in the heart. Many murmurs are caused by defective heart valves. Defective heart valves may be congenital, or they may occur as a result of *rheumatic endocarditis,* associated with rheumatic fever. In this disease, the valves become damaged by antibodies made in response to an infection caused by streptococcus bacteria (the bacteria that produce strep throat). Many people have small defects that produce detectable murmurs but do not seriously compromise the pumping ability of the heart. Larger defects, however, may have dangerous consequences and thus may require surgical correction.

In *mitral stenosis,* for example, the mitral valve becomes thickened and calcified. This can impair the blood flow from the left atrium to the left ventricle. An accumulation of blood in the left atrium may cause a rise in left atrial and pulmonary vein pressure, resulting in pulmonary hypertension. To compensate

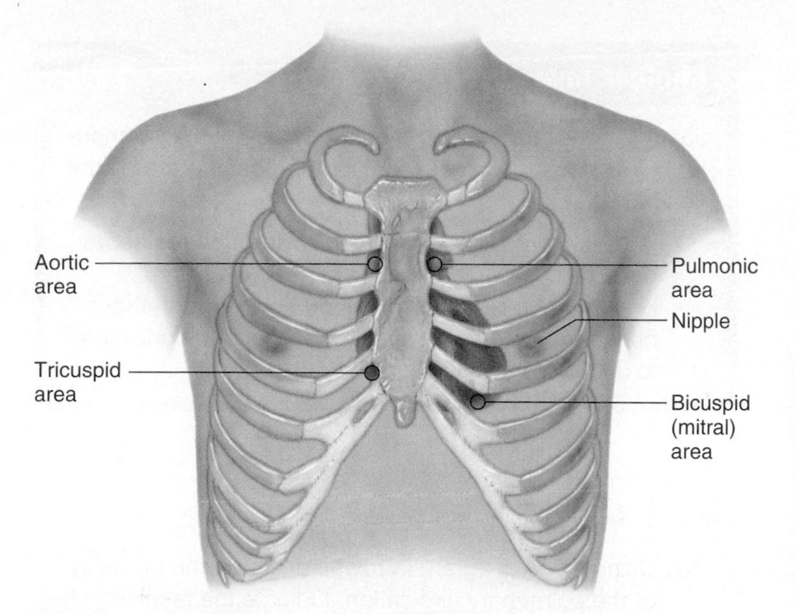

Figure 13.13 Routine stethoscope positions for listening to the heart sounds. The first heart sound is caused by closing of the AV valves; the second by closing of the semilunar valves. AP|R

for the increased pulmonary pressure, the right ventricle grows thicker and stronger.

Mitral valve prolapse (with a prevalence estimated at 2.5%) is the most common cause of chronic mitral regurgitation, where blood flows backward into the left atrium. It has both congenital and acquired forms; in younger people with mitral valve prolapse, it is usually caused by excess valve leaflet material. Although most people with this condition lack symptoms and have an apparently normal lifespan, in some people the condition can progress. Regurgitation can worsen if there is lengthening and rupture of the chordae tendinae extending from the papillary muscles to the valve flaps (see fig. 13.11). In those cases, the mitral valve may be repaired or replaced with a mechanical or biological (pig or cow) valve.

Clinical Investigation CLUES

Jason's tests revealed that he has mitral stenosis.

- What effects would this have on heart function and circulation?
- How might this contribute to Jason's chronic fatigue?

Murmurs also can be produced by the flow of blood through *septal defects*—holes in the septum between the right and left sides of the heart. These are usually congenital and may occur either in the interatrial or interventricular septum (fig. 13.14). When a septal defect is not accompanied by other abnormalities, blood will usually pass through the defect from the left to the right side, due to the higher pressure on the left side. The buildup of blood and pressure on the right side of the heart that results may lead to pulmonary hypertension and edema (fluid in the lungs).

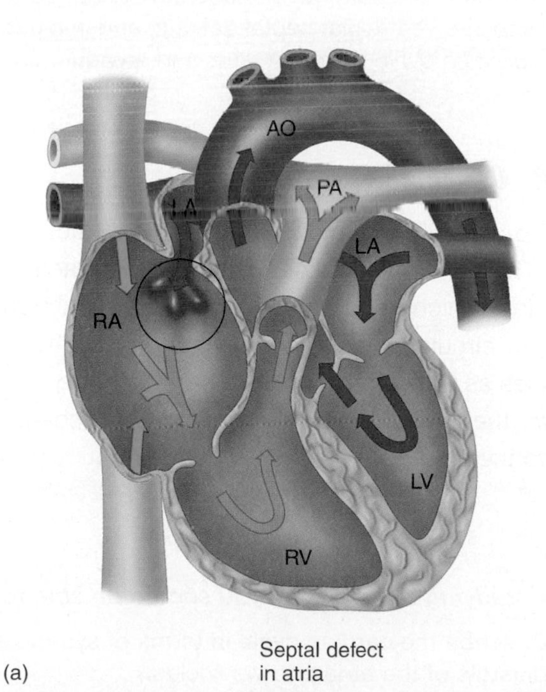

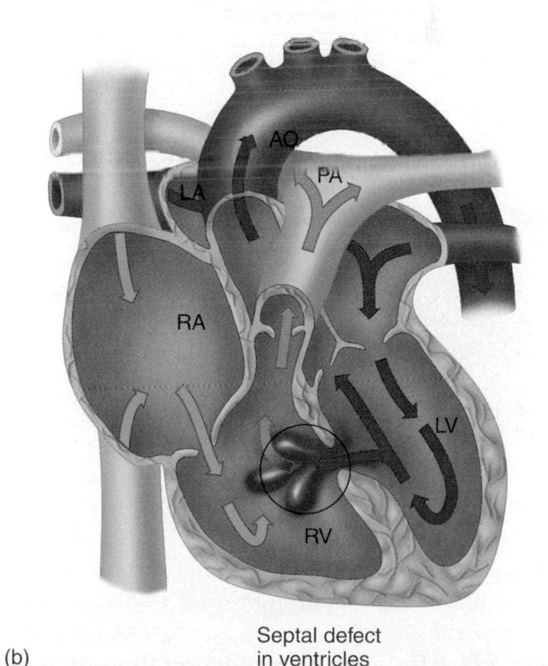

Figure 13.14 Abnormal blood flow due to septal defects. Left-to-right shunting of blood is shown (circled areas) because the left pump is at a higher pressure than the right pump in the adult heart. (*a*) Leakage of blood through a defect in the atria (a patent foramen ovale). (*b*) Leakage of blood through a defect in the interventricular septum. (RA = right atrium; RV = right ventricle; LA = left atrium; RA = right atrium; AO = aorta; PA = pulmonary artery.)

CLINICAL APPLICATION

In a fetus, the resistance to blood flow through the pulmonary circulation is quite high, because hypoxia (low oxygen) stimulates smooth muscle contraction in pulmonary arterioles, producing vasoconstriction. (This differs from systemic arterioles, which dilate in response to hypoxia.) Vasoconstriction in the pulmonary arterioles produces a high resistance to blood flow, resulting in a high pressure in the pulmonary circulation and right ventricle. In the fetal heart, unlike the postnatal heart, the pressure in the right ventricle is higher than in the left ventricle. This causes blood to go from the right to the left ventricle through an opening in the interatrial septum called the **foramen ovale** (fig. 13.14). The pressure difference also causes blood to be shunted (diverted) from the pulmonary to the systemic circulation through a connection between the pulmonary trunk and aorta called the **ductus arteriosus** (fig. 13.15). These shunts normally close after birth. When the newborn breathes, the blood oxygen levels suddenly become higher. Because of differing responses of the smooth muscle cells to increased oxygen, the rise in oxygen normally causes vasodilation and thus increased blood flow in the pulmonary vessels, but contraction (and thus closing) of the ductus arteriosus. If the foramen ovale and ductus arteriosus remain open (are *patent*) after birth, murmurs can result.

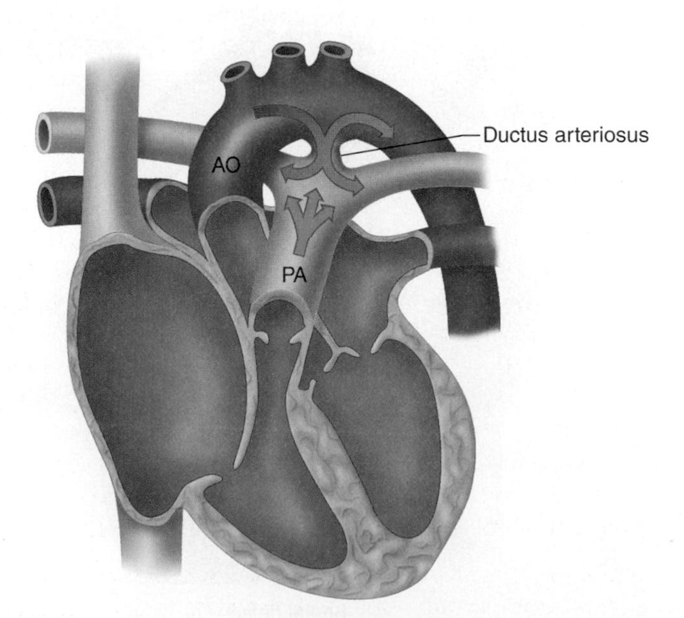

Figure 13.15 The flow of blood through a patent (open) ductus arteriosus. The ductus is normally open in a fetus but closes after birth, eventually becoming the ligamentum arteriosum. (AO = aorta; PA = pulmonary arteries.)

Clinical Investigation CLUES

Jason's echocardiogram revealed that he has a ventricular septal defect (a hole in the septum that separates the ventricles).

- In which direction will blood flow through the septal defect?
- How might this affect the blood flow through the pulmonary and systemic circulations and contribute to Jason's chronic fatigue?

CHECKPOINT

6a. Using a flow diagram (arrows), describe the pathway of the pulmonary circulation. Indicate the relative amounts of oxygen and carbon dioxide in the vessels involved.

6b. Use a flow diagram to describe the systemic circulation and indicate the relative amounts of oxygen and carbon dioxide in the blood vessels.

6c. List the AV valves and the valves of the pulmonary artery and aorta. How do these valves ensure a one-way flow of blood?

7a. Discuss how defective valves affect blood flow within the heart and produce heart murmurs.

7b. Describe the patterns of blood flow in interatrial and interventricular septal defects, and in a patent foramen ovale in both a fetus and an adult.

13.4 CARDIAC CYCLE

The two atria fill with blood and then contract simultaneously. This is followed by simultaneous contraction of both ventricles, which sends blood through the pulmonary and systemic circulations. Pressure changes in the atria and ventricles as they go through the cardiac cycle are responsible for the flow of blood through the heart chambers and out into the arteries.

LEARNING OUTCOMES

After studying this section, you should be able to:

8. Describe the cardiac cycle in terms of systole and diastole of the atria and ventricles.

9. Explain how the pressure differences within the heart chambers are responsible for blood flow during the cardiac cycle.

The **cardiac cycle** refers to the repeating pattern of contraction and relaxation of the heart. The phase of contraction is called **systole,** and the phase of relaxation is called **diastole.** When these terms are used without reference to specific chambers, they refer to contraction and relaxation of the ventricles. It should be noted, however, that the atria also contract and relax. There is an atrial systole and diastole. Atrial contraction occurs toward the end of diastole, when the ventricles are relaxed; when the ventricles contract during systole, the atria are relaxed (fig. 13.16).

The heart thus has a two-step pumping action. The right and left atria contract almost simultaneously, followed by contraction of the right and left ventricles 0.1 to 0.2 second later. During the time when both the atria and ventricles are relaxed, the venous return of blood fills the atria. The buildup of pressure that results causes the AV valves to open and blood to flow from atria to ventricles. It has been estimated that the ventricles are about 80% filled with blood even before the atria contract. Contraction of the atria adds the final 20% to the *end-diastolic volume,* which is the total volume of blood in the ventricles at the end of diastole.

Contraction of the ventricles in systole ejects about two-thirds of the blood they contain—an amount called the *stroke volume*—leaving one-third of the initial amount left in the ventricles as the *end systolic volume.* The ventricles then fill with blood during the next cycle. At an average *cardiac rate* of 75 beats per minute, each cycle lasts 0.8 second; 0.5 second is spent in diastole, and systole takes 0.3 second (fig. 13.16).

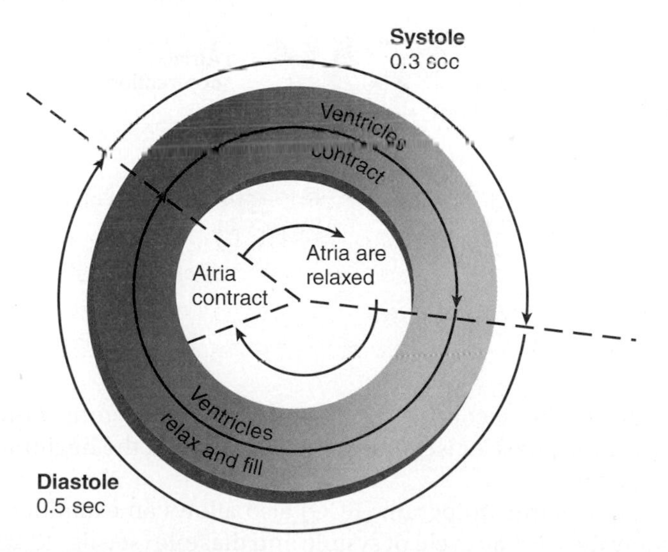

Figure 13.16 The cardiac cycle of ventricular systole and diastole. Contraction of the atria occurs in the last 0.1 second of ventricular diastole. Relaxation of the atria occurs during ventricular systole. The durations given for systole and diastole relate to a cardiac rate of 75 beats per minute.

FITNESS APPLICATION

Interestingly, the blood contributed by contraction of the atria does not appear to be essential for life. Elderly people who have **atrial fibrillation** (a condition in which the atria fail to contract) can live for many years. People with atrial fibrillation, however, become fatigued more easily during exercise because the reduced filling of the ventricles compromises the ability of the heart to sufficiently increase its output during exercise. (Cardiac output and blood flow during rest and exercise are discussed in chapter 14.)

Pressure Changes During the Cardiac Cycle

When the heart is in diastole, the pressure in the systemic arteries averages about 80 mmHg (millimeters of mercury). These events in the cardiac cycle then occur (fig. 13.17):

1. As the ventricles begin their contraction, the intraventricular pressure rises, causing the AV valves to snap shut and produce the first heart sound. At this time, the ventricles are neither being filled with blood (because the AV valves are closed) nor ejecting blood (because the intraventricular pressure has not risen sufficiently to open the semilunar valves). This is the phase of *isovolumetric contraction.*
2. When the pressure in the left ventricle becomes greater than the pressure in the aorta, the phase of *ejection* begins as the semilunar valves open. The pressure in the left ventricle and aorta rises to about 120 mmHg (fig. 13.17) when ejection begins and the ventricular volume decreases.
3. As the pressure in the ventricles falls below the pressure in the arteries, the back pressure causes the semilunar valves to snap shut and produce the second heart sound. The pressure in the aorta falls to 80 mmHg, while pressure in the left ventricle falls to 0 mmHg. During *isovolumetric relaxation,* the AV and semilunar valves are closed. This phase lasts until the pressure in the ventricles falls below the pressure in the atria.
4. When the pressure in the ventricles falls below the pressure in the atria, the AV valves open and a phase of *rapid filling* of the ventricles occurs.
5. *Atrial contraction (atrial systole)* delivers the final amount of blood into the ventricles immediately prior to the next phase of isovolumetric contraction of the ventricles.

Similar events occur in the right ventricle and pulmonary circulation, but the pressures are lower. The maximum pressure produced at systole in the right ventricle is 25 mmHg, which falls to a low of 8 mmHg at diastole.

The arterial pressure rises as a result of ventricular systole (due to blood ejected into the arterial system) and

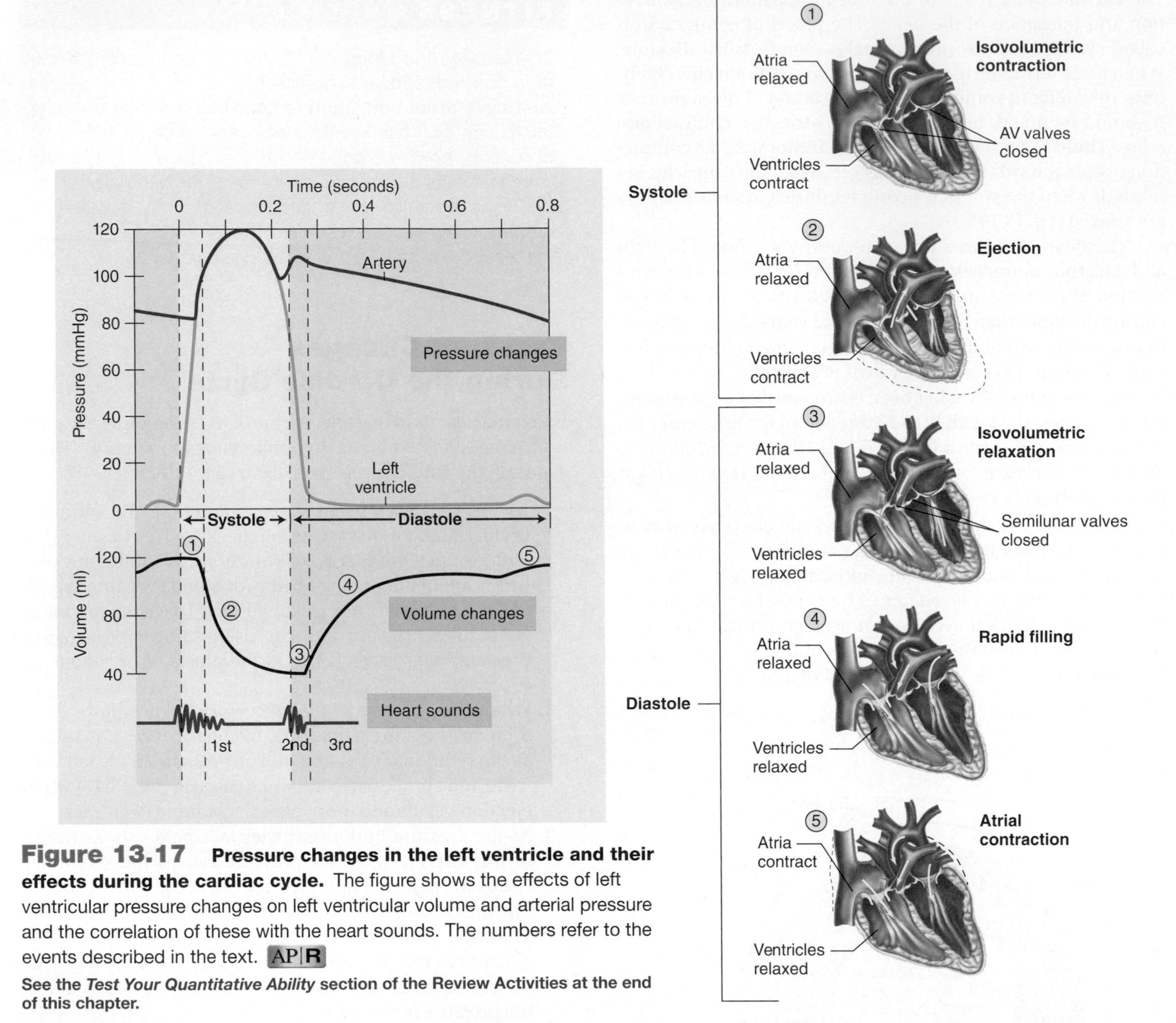

Figure 13.17 Pressure changes in the left ventricle and their effects during the cardiac cycle. The figure shows the effects of left ventricular pressure changes on left ventricular volume and arterial pressure and the correlation of these with the heart sounds. The numbers refer to the events described in the text. AP|R

See the *Test Your Quantitative Ability* section of the Review Activities at the end of this chapter.

falls during ventricular diastole (fig. 13.17). Because of this, a person's cardiac cycle can be followed by measuring the systolic and diastolic arterial pressures, and by palpating (feeling) the pulse (chapter 14, section 14.6). A pulse is felt (for example, in the radial artery of the wrist) when the arterial pressure rises from diastolic to systolic levels and pushes against the examiner's finger. Figure 13.17 reveals an inflection in the descending portion of the arterial pressure graph, which cannot be felt on palpation. This inflection is called the *dicrotic notch* and is produced by closing of the aortic and pulmonic semilunar valves. Closing of these valves produces the second heart sound and the dicrotic notch during the phase of isovolumetric relaxation at the beginning of diastole.

An electrocardiogram (ECG) also allows an examiner to follow the cardiac cycle of systole and diastole (see fig. 13.25). This is because myocardial contraction occurs in response to the depolarization stimulus of an action potential and myocardial relaxation begins during repolarization. The relationships between the electrical activity of the heart, the electrocardiogram, and the cardiac cycle are described in the next section.

CHECKPOINT

8a. Using a drawing or flow chart, describe the sequence of events that occurs during the cardiac cycle. Indicate when atrial and ventricular filling occur and when atrial and ventricular contraction occur.

8b. Describe how the pressure in the left ventricle and in the systemic arteries varies during the cardiac cycle.

9. Draw a figure to illustrate the pressure variations described in question 8b, and indicate in your figure when the AV and semilunar valves close.

13.5 ELECTRICAL ACTIVITY OF THE HEART AND THE ELECTROCARDIOGRAM

The pacemaker region of the heart (SA node) exhibits a spontaneous depolarization that causes action potentials, resulting in the automatic beating of the heart. Action potentials are conducted by myocardial cells in the atria and are transmitted to the ventricles by specialized conducting tissue. Electrocardiogram waves correspond to these events in the heart.

LEARNING OUTCOMES

After studying this section, you should be able to:

10. Describe the pacemaker potential and the myocardial action potential, and explain how the latter correlates with myocardial contraction and relaxation.

11. Describe the components of the ECG and their relationships to the cardiac cycle.

As described in chapter 12, myocardial cells are short, branched, and interconnected by gap junctions. Gap junctions function as electrical synapses, and have been described in chapter 7 (see fig. 7.21) and chapter 12 (see fig. 12.32). The entire mass of cells interconnected by gap junctions is known as a *myocardium.* A myocardium is a single functioning unit, or *functional syncytium,* because action potentials that originate in any cell in the mass can be transmitted to all the other cells. The myocardia of the atria and ventricles are separated from each other by the fibrous skeleton of the heart, as previously described. Impulses normally originate in the atria, so the atrial myocardium is excited before that of the ventricles.

Electrical Activity of the Heart

If the heart of a frog is removed from the body and all neural innervations are severed, it will still continue to beat as long as the myocardial cells remain alive. The automatic nature of the heartbeat is referred to as *automaticity.* As a result of experiments with isolated myocardial cells and of observations of patients with blocks in the conductive tissues of the heart, scientists have learned that there are three regions that can spontaneously generate action potentials and thereby function as pacemakers. In the normal heart, only one of these, the **sinoatrial node (SA node),** functions as the pacemaker. The SA node is located in the right atrium near the opening of the superior vena cava, and serves as the primary (normal) pacemaker of the heart. The two potential, or secondary, pacemaker regions—the *AV node* and *Purkinje fibers* (parts of the conduction network; see fig. 13.20)—are normally suppressed by action potentials originating in the SA node.

Pacemaker Potential

The cells of the SA node do not maintain a resting membrane potential in the manner of resting neurons or skeletal muscle cells. Instead, during the period of diastole, the SA node exhibits a slow *spontaneous depolarization* called the **pacemaker potential.** The membrane potential begins at about −60 mV and gradually depolarizes to −40 mV, which is the threshold for producing an action potential in these cells (fig. 13.18).

The spontaneous depolarization of the pacemaker potential is produced by the opening of a type of channel that, strangely, opens in response to hyperpolarization. When first discovered, the ion current through this strange channel was called a "funny current." The hyperpolarization (approaching −60 mV) stimulus that opens this channel occurs at the end of the preceding action potential. Once opened, this channel is

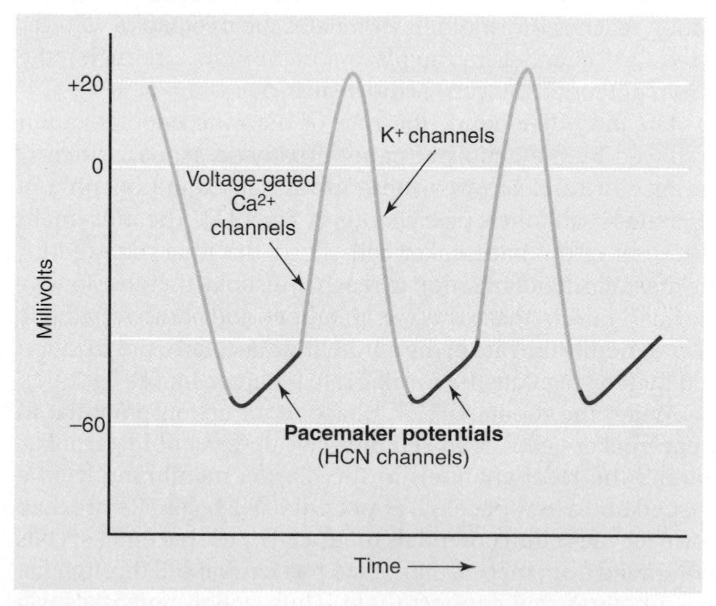

Figure 13.18 Pacemaker potentials and action potentials in the SA node. The pacemaker potentials are spontaneous depolarizations. When they reach threshold, they trigger action potentials.

permeable to both Na^+ and K^+. Because the electrochemical gradient is greater for the entry of Na^+ than for the exit of K^+, the entry of Na^+ predominates and produces a depolarization. This is similar to the way chemically gated channels produce an EPSP, as described in chapter 7. The spontaneous, automatic depolarization of the pacemaker occurs during diastole, so it can be termed a **diastolic depolarization.**

When the diastolic depolarization reaches threshold (about -40mV), it causes the opening of voltage-gated Ca^{2+} channels in the plasma membrane of the pacemaker cells. It is the inward diffusion of Ca^{2+} that produces the upward phase of the action potential (fig. 13.18). The inward current of Ca^{2+} also results in contraction of these myocardial cells (chapter 12; see fig. 12.34). Repolarization is produced by the opening of voltage-gated K^+ channels and the outward diffusion of K^+.

The diastolic depolarization occurs faster in response to epinephrine and norepinephrine. This is because these catecholamines stimulate the β_1-adrenergic receptors, causing the production of cAMP within the pacemaker cells (chapter 7; see fig. 7.31). Cyclic AMP acts like hyperpolarization to keep the pacemaker ion channels open. This is why the cardiac pacemaker channels are termed **HCN channels** (for *h*yperpolarization-activated *c*yclic *n*ucleotide-gated channels). Pacemaker HCN channels opened by cAMP (formed in response to catecholamine stimulation) allow for a faster rate of diastolic depolarization, resulting in a faster heart rate. In summary, the hyperpolarization resulting from the previous action potential causes the opening of HCN channels, and this action is augmented by cAMP as a second messenger of catecholamine stimulation. Opening of HCN channels produces a diastolic depolarization, primarily caused by an inward Na^+ current through these channels. When the diastolic depolarization reaches threshold, it stimulates the opening of voltage-gated Ca^{2+} channels in the plasma membrane, producing the action potential that causes the heartbeat.

On the other hand, the rate of diastolic depolarization is slowed by the action of parasympathetic axons, primarily because ACh released by these axons causes the opening of separate K^+ channels (see chapter 9, fig. 9.11). The movement of K^+ out of the pacemaker cells slows the time required for the diastolic depolarization to reach threshold, thereby slowing the heart rate. In these ways, sympathetic and parasympathetic nerves modify the rate of the automatic diastolic depolarization and thereby regulate the cardiac rate (chapter 14; see fig. 14.1).

When the repolarization phase of an action potential in a pacemaker cell is followed by a certain level of hyperpolarization, the HCN channels in the plasma membrane will be opened and a new pacemaker potential will begin. Before that, however, the action potentials produced by the pacemaker cells will spread from myocardial cell to myocardial cell through the gap junctions that connect them. Thus, action potentials will spread from the SA node through the atria and, by means of conducting tissue, into the ventricles.

As previously mentioned, two other regions in the heart, the AV node and Purkinje fibers, can potentially serve as pacemakers but are normally suppressed by action potentials originating in the SA node. This is because their rate of spontaneous depolarization is slower than that of the SA node. Thus, the potential pacemaker cells are stimulated by action potentials from the SA node before they can stimulate themselves through their own pacemaker potentials. If action potentials from the SA node are prevented from reaching these areas (through blockage of conduction), they will generate pacemaker potentials at their own rate and serve as sites for the origin of action potentials; they will function as pacemakers. A pacemaker other than the SA node is called an *ectopic pacemaker,* or alternatively, an *ectopic focus.* From this discussion, it is clear that the rhythm set by such an ectopic pacemaker is usually slower than that normally set by the SA node.

Myocardial Action Potential

Once another myocardial cell has been stimulated by action potentials originating in the SA node, it produces its own action potentials. The majority of myocardial cells have resting membrane potentials of about −85 mV. When stimulated by action potentials from a pacemaker region, these cells become depolarized to threshold, at which point their voltage-regulated Na^+ gates open. The upshoot phase of the action potential of nonpacemaker cells is due to the rapid inward diffusion of Na^+ through *fast Na^+ channels.* Following the rapid reversal of the membrane polarity, the membrane potential quickly declines to about −15 mV. Unlike the action potential of other cells, however, this level of depolarization is maintained for 200 to 300 msec before repolarization (fig. 13.19). This *plateau phase* results from a slow inward diffusion of Ca^{2+} through *slow Ca^{2+} channels,* which balances a slow outward diffusion of K^+. Rapid repolarization at the end of the plateau phase is achieved, as in other cells, by the opening of voltage-gated K^+ channels and the rapid outward diffusion of K^+ that results.

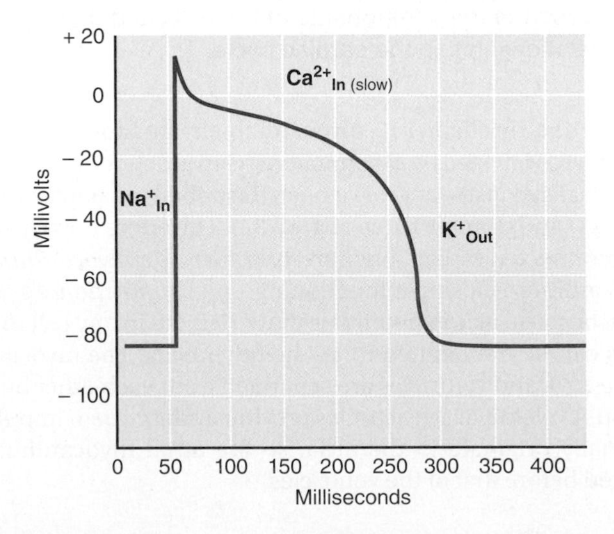

Figure 13.19 An action potential in a myocardial cell from the ventricles. The plateau phase of the action potential is maintained by a slow inward diffusion of Ca^{2+}. The cardiac action potential, as a result, is about 100 times longer in duration than the spike-like action potential in an axon.

The long plateau phase of the myocardial action potential distinguishes it from the spike-like action potentials in axons and skeletal muscle fibers. The plateau phase is accompanied by the entry of Ca^{2+}, which begins excitation-contraction coupling (as described shortly). Thus, myocardial contraction accompanies the long action potential (see fig. 13.21), and is completed before the membrane recovers from its refractory period. Summation and tetanus, as can occur in skeletal muscles (chapter 12), is thereby prevented from occurring in the myocardium by this long refractory period.

CLINICAL APPLICATION

Abnormal patterns of electrical conduction in the heart can produce abnormalities of the cardiac cycle and seriously compromise the function of the heart. These **arrhythmias** may be treated with a variety of drugs that inhibit specific aspects of the cardiac action potentials and thereby inhibit the production or conduction of impulses along abnormal pathways. Drugs used to treat arrhythmias may (1) block the fast Na^+ channel (quinidine, procainamide, lidocaine); (2) block the slow Ca^{2+} channel (verapamil); or (3) block β-adrenergic receptors (propranolol, atenolol). By this means, the latter drugs block the ability of catecholamines to stimulate the heart.

Conducting Tissues of the Heart

Action potentials that originate in the SA node spread to adjacent myocardial cells of the right and left atria through the gap junctions between these cells. Because the myocardium of the atria is separated from the myocardium of the ventricles by the fibrous skeleton of the heart, however, the impulse cannot be conducted directly from the atria to the ventricles. Specialized conducting tissue, composed of modified myocardial cells, is thus required. These specialized myocardial cells form the *AV node, bundle of His,* and *Purkinje fibers.*

Action potentials that have spread from the SA node through the atria pass into the **atrioventricular node (AV node),** which is located on the inferior portion of the interatrial septum (fig. 13.20). From here, action potentials continue through the **atrioventricular bundle,** or **bundle of His** (pronounced "hiss"), beginning at the top of the interventricular septum. This conducting tissue pierces the fibrous skeleton of the heart and continues to descend along the interventricular septum. The atrioventricular bundle divides into right and left bundle branches, which are continuous with the **Purkinje fibers** within the ventricular walls. Within the myocardium of the ventricles, the action potential spreads from the inner (endocardium) to the outer (epicardium) side. This causes both ventricles to contract simultaneously and eject blood into the pulmonary and systemic circulations.

Conduction of the Impulse

Action potentials from the SA node spread very quickly—at a rate of 0.8 to 1.0 meter per second (m/sec)—across the

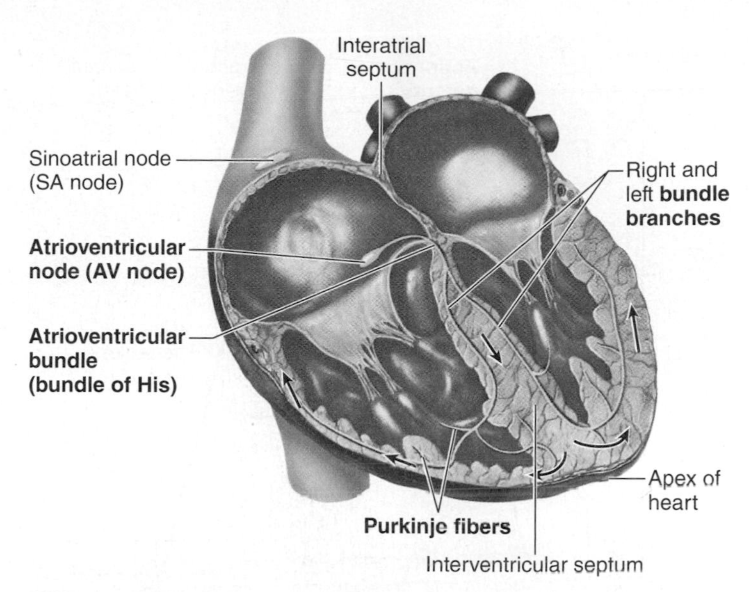

Figure 13.20 The conduction system of the heart. The conduction system consists of specialized myocardial cells that rapidly conduct the impulses from the atria into the ventricles. APIR

myocardial cells of both atria. The conduction rate then slows considerably as the impulse passes into the AV node. Slow conduction of impulses (0.03 to 0.05 m/sec) through the AV node accounts for over half of the time delay between excitation of the atria and ventricles. After the impulses spread through the AV node, the conduction rate increases greatly in the atrioventricular bundle and reaches very high velocities (5 m/sec) in the Purkinje fibers. As a result of this rapid conduction of impulses, ventricular contraction begins 0.1 to 0.2 second after the contraction of the atria.

Excitation-Contraction Coupling in Heart Muscle

The mechanism of excitation-contraction coupling in myocardial cells, involving *Ca^{2+}-stimulated Ca^{2+} release,* was discussed in chapter 12 (see fig. 12.34). In summary, action potentials conducted by the sarcolemma (chiefly along the transverse tubules) briefly open voltage-gated Ca^{2+} channels in the plasma membrane. This allows Ca^{2+} to diffuse into the cytoplasm from the extracellular fluid producing a brief "puff' of Ca^{2+} that serves to stimulate the opening of Ca^{2+} release channels in the sarcoplasmic reticulum. The amount of Ca^{2+} released from intracellular stores in the sarcoplasmic reticulum is far greater than the amount that enters from the extracellular fluid through voltage-gated channels in the sarcolemma. Thus, it is mostly the Ca^{2+} from the sarcoplasmic reticulum that binds to troponin and stimulates contraction. These events occur at *signaling complexes,* which are the regions where the sarcolemma come in very close proximity to the sarcoplasmic reticulum. There are an estimated 20,000 signaling complexes in a myocardial cell, all activated at the same time by the depolarization stimulus of the action

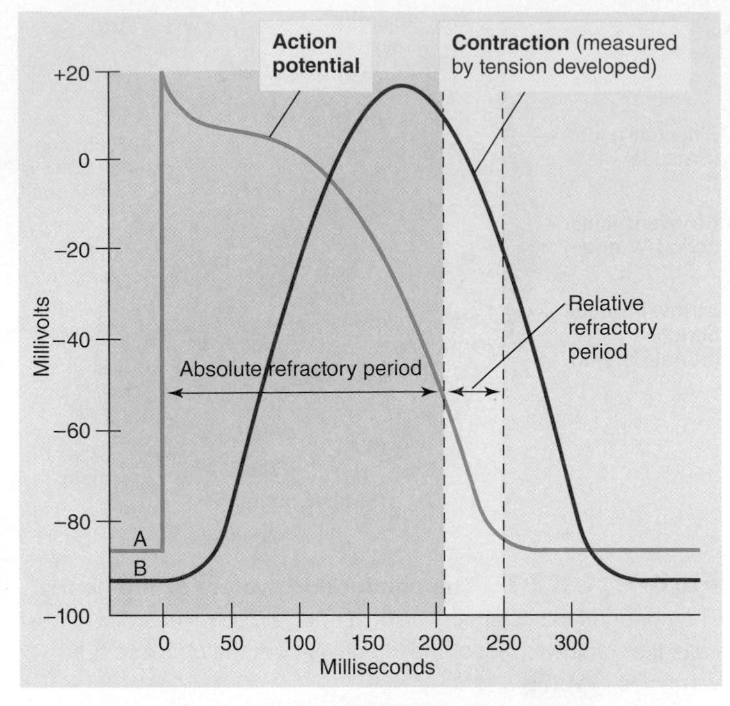

Figure 13.21 Correlation of the myocardial action potential with myocardial contraction. The time course for the myocardial action potential is compared with the duration of contraction. Notice that the long action potential results in a correspondingly long absolute refractory period and relative refractory period. These refractory periods last almost as long as the contraction, so that the myocardial cells cannot be stimulated a second time until they have completed their contraction from the first stimulus.

potential. This results in a myocardial contraction that develops during the depolarization phase of the action potential (fig. 13.21).

CLINICAL APPLICATION

People with congestive heart failure (chapter 14, section 14.7) are often treated with the drug **digitalis.** Digitalis binds to and inactivates the Na^+/K^+-ATPase pumps in the plasma membrane of myocardial cells. This reduces the ability of these pumps to extrude Na^+ from the cytoplasm, producing a rise in the cytoplasmic Na^+ concentration. This reduces the difference in Na^+ concentration between the inside and outside of the cells, so that the diffusion of Na^+ into the cells is decreased. The decreased diffusion of Na^+ into the cells then reduces the ability of the Na^+-Ca^{2+} exchangers to extrude Ca^{2+}. Thus, reduced Na^+/Ca^{2+} exchange produces a rise in the intracellular Ca^{2+} concentration, which then causes more Ca^{2+} to be taken into the sarcoplasmic reticulum. Finally, it is this increased amount of Ca^{2+} in the sarcoplasmic reticulum that is responsible for the increased strength of myocardial contraction when a person takes digitalis.

During the repolarization phase of the action potential, the concentration of Ca^{2+} within the cytoplasm is lowered by (1) active transport of Ca^{2+} back into the sarcoplasmic reticulum by a Ca^{2+}-ATPase pump, and (2) extrusion of Ca^{2+} through the plasma membrane into the extracellular fluid by a Na^+-Ca^{2+} exchanger. This exchanger is a secondary active transport carrier; it pumps Ca^{2+} out of the cytoplasm against its concentration gradient in exchange of Na^+, which moves down its concentration gradient into the cell. That is, the "downhill" movement of Na^+ into the cell powers the "uphill" extrusion of Ca^{2+}. These two mechanisms lower the cytoplasmic Ca^{2+} concentration, thereby allowing the myocardium to relax during repolarization (fig. 13.21).

Unlike skeletal muscles, the heart cannot sustain a contraction. This is because the atria and ventricles behave as if each were composed of only one muscle cell; the entire myocardium of each is electrically stimulated as a single unit and contracts as a unit. This contraction, corresponding in time to the long action potential of myocardial cells and lasting almost 300 msec, is analogous to the twitch produced by a single skeletal muscle fiber (which lasts only 20 to 100 msec in comparison). The heart normally cannot be stimulated again until after it has relaxed from its previous contraction because myocardial cells have *long refractory periods* (fig. 13.21) that correspond to the long duration of their action potentials. Summation of contractions is thus prevented, and the myocardium must relax after each contraction. By this means, the rhythmic pumping action of the heart is ensured.

The Electrocardiogram

The body is a good conductor of electricity because tissue fluids have a high concentration of ions that move (creating a current) in response to potential differences. Potential differences generated by the heart are conducted to the body surface, where they can be recorded by surface electrodes placed on the skin. The recording thus obtained is called an **electrocardiogram (ECG** or **EKG**); the recording device is called an *electrocardiograph.* Each cardiac cycle produces three distinct ECG waves, designated *P, QRS,* and *T* (fig. 13.22*a*).

Note that the ECG is not a recording of action potentials, but it does result from the production and conduction of action potentials in the heart. The correlation of an action potential produced in the ventricles to the waves of the ECG is shown in figure 13.22*b*. This figure shows that the spread of depolarization through the ventricles (indicated by the QRS, described shortly) corresponds to the action potential, and thus to contraction of the ventricles.

The spread of depolarization through the atria causes a potential difference that is indicated by an upward deflection of the ECG line. When about half the mass of the atria is depolarized, this upward deflection reaches a maximum value because the potential difference between the depolarized and unstimulated portions of the atria is at a maximum. When the entire mass of the atria is depolarized, the ECG returns to baseline because all regions of the atria have the same polarity. The spread of atrial depolarization thereby creates the **P wave** (fig. 13.23).

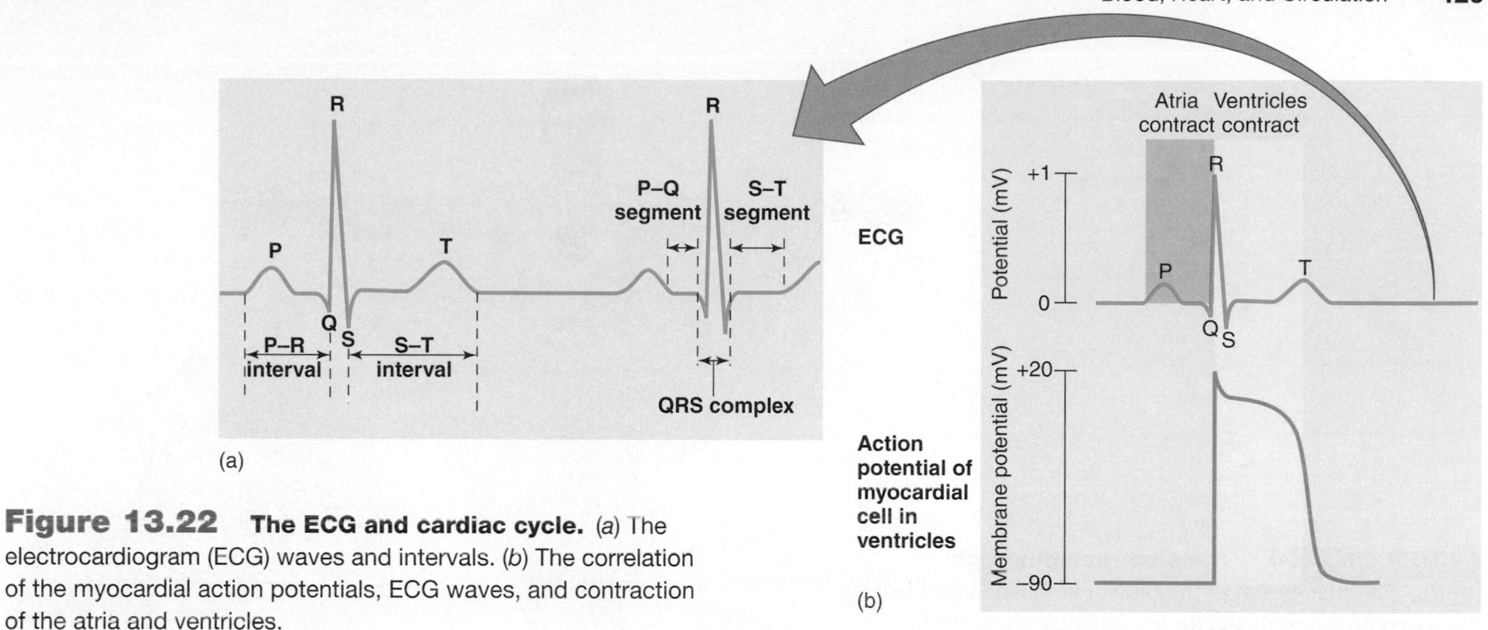

Figure 13.22 The ECG and cardiac cycle. (*a*) The electrocardiogram (ECG) waves and intervals. (*b*) The correlation of the myocardial action potentials, ECG waves, and contraction of the atria and ventricles.

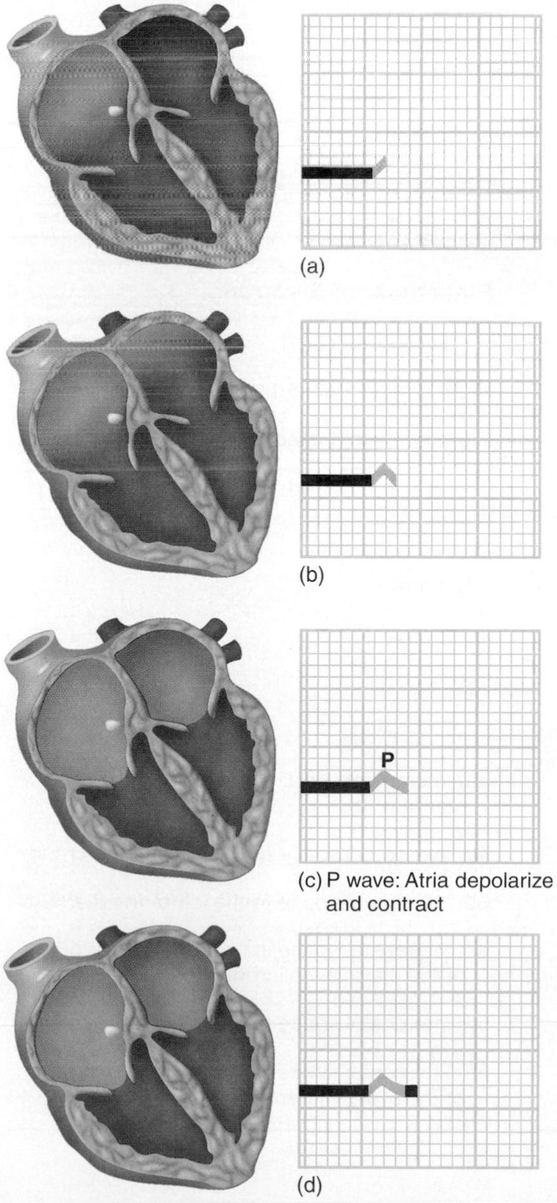

(a)

(b)

P

(c) P wave: Atria depolarize and contract

(d)

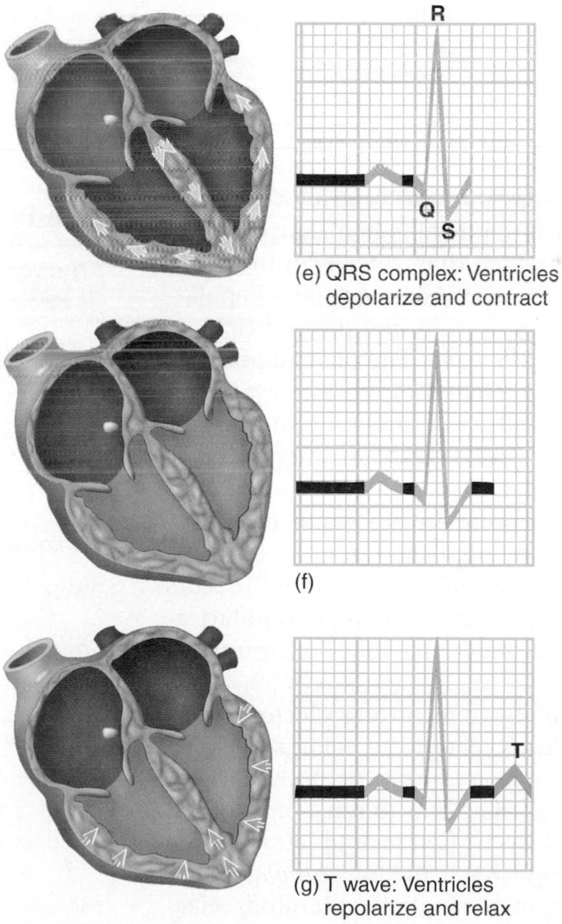

R

Q

S

(e) QRS complex: Ventricles depolarize and contract

(f)

T

(g) T wave: Ventricles repolarize and relax

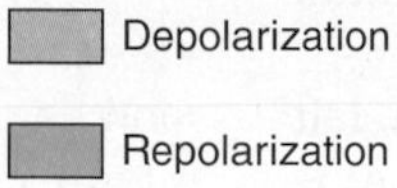

Depolarization

Repolarization

Figure 13.23 The relationship between impulse conduction in the heart and the ECG. The direction of the arrows in (*e*) indicates that depolarization of the ventricles occurs from the inside (endocardium) out (to the epicardium). The arrows in (*g*), by contrast, indicate that repolarization of the ventricles occurs in the opposite direction.

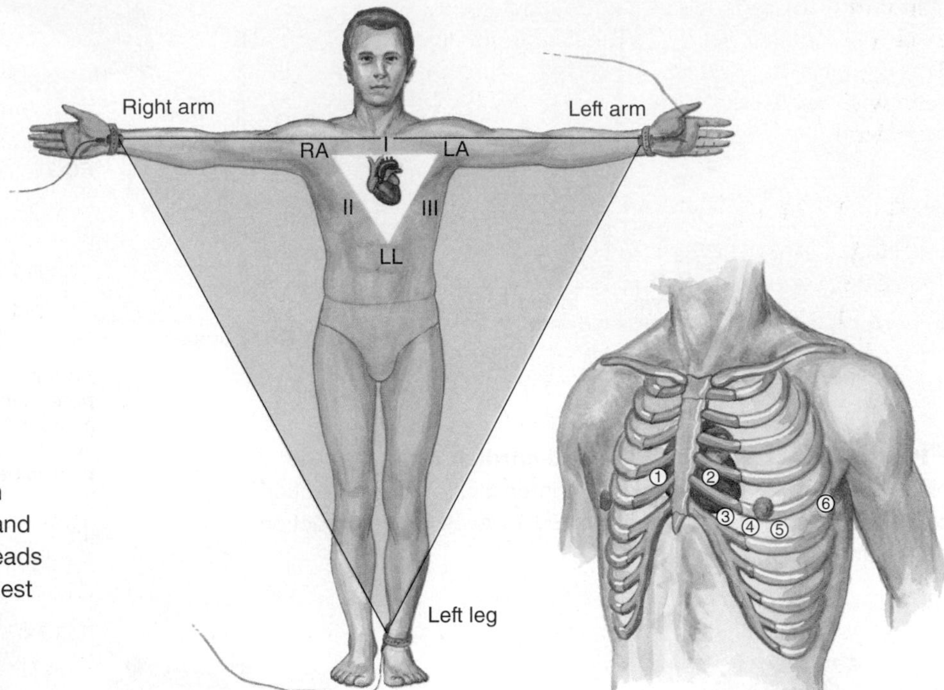

Figure 13.24 The electrocardiograph leads. The placement of the bipolar limb leads and the exploratory electrode for the unipolar chest leads in an electrocardiogram (ECG). The numbered chest positions correspond to V1 through V6, as given in table 13.7. (RA = right arm; LA = left arm; LL = left leg.)

Conduction of the impulse into the ventricles similarly creates a potential difference that results in a sharp upward deflection of the ECG line, which then returns to the baseline as the entire mass of the ventricles becomes depolarized. The spread of the depolarization into the ventricles is thereby represented by the **QRS wave.** The plateau phase of the cardiac action potential is related to the *S-T segment* of the ECG (see fig. 13.22*a*). Finally, repolarization of the ventricles produces the **T wave** (fig. 13.23). You might be surprised that ventricular depolarization (the QRS wave) and repolarization (the T wave) point in the same direction, although they are produced by opposite potential changes. This is because depolarization of the ventricles occurs from endocardium to epicardium, whereas repolarization spreads in the opposite direction, from epicardium to endocardium.

There are two types of ECG recording electrodes, or "leads." The *bipolar limb leads* record the voltage between electrodes placed on the wrists and legs (fig. 13.24). These bipolar leads include lead I (right arm to left arm), lead II (right arm to left leg), and lead III (left arm to left leg). The right leg is used as a ground lead. In the *unipolar leads,* voltage is recorded between a single "exploratory electrode" placed on the body and an electrode that is built into the electrocardiograph and maintained at zero potential (ground).

The unipolar limb leads are placed on the right arm, left arm, and left leg, and are abbreviated AVR, AVL, and AVF, respectively. The unipolar chest leads are labeled 1 through 6, starting from the midline position (fig. 13.24). Thus a total of 12

Table 13.7 | Electrocardiograph (ECG) Leads

Name of Lead	Placement of Electrodes
Bipolar Limb Leads	
I	Right arm and left arm
II	Right arm and left leg
III	Left arm and left leg
Unipolar Limb Leads	
AVR	Right arm
AVL	Left arm
AVF	Left leg
Unipolar Chest Leads	
V_1	4th intercostal space to the right of the sternum
V_2	4th intercostal space to the left of the sternum
V_3	5th intercostal space to the left of the sternum
V_4	5th intercostal space in line with the middle of the clavicle (collarbone)
V_5	5th intercostal space to the left of V_4
V_6	5th intercostal space in line with the middle of the axilla (underarm)

standard ECG leads "view" the changing pattern of the heart's electrical activity from different perspectives (table 13.7). This is important because certain abnormalities are best seen with particular leads and may not be visible at all with other leads.

Correlation of the ECG with Heart Sounds

Depolarization of the ventricles, as indicated by the QRS wave, stimulates contraction by promoting the diffusion of Ca^{2+} into the regions of the sarcomeres. The QRS wave is thus seen at the beginning of systole. The rise in intraventricular pressure that results causes the AV valves to close, so that the first heart sound (S_1, or lub) is produced immediately after the QRS wave (fig. 13.25).

Repolarization of the ventricles, as indicated by the T wave, occurs at the same time that the ventricles relax at the beginning of diastole. The resulting fall in intraventricular pressure causes the aortic and pulmonary semilunar valves to close, so that the second heart sound (S_2, or dub) is produced shortly after the T wave begins in an electrocardiogram.

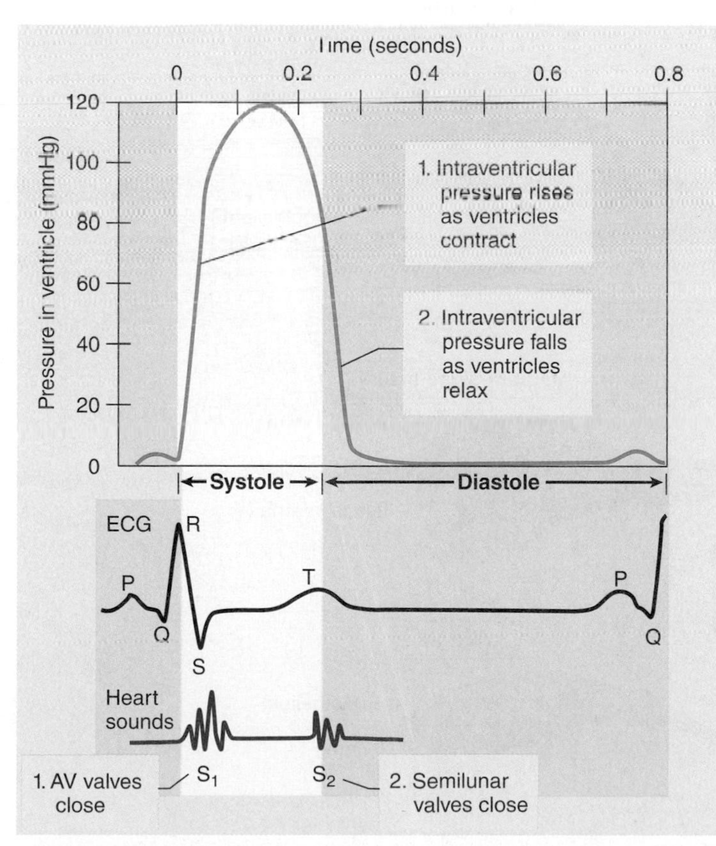

Figure 13.25 The relationship between changes in intraventricular pressure and the ECG. The QRS wave (representing depolarization of the ventricles) occurs at the beginning of systole, whereas the T wave (representing repolarization of the ventricles) occurs at the beginning of diastole. The numbered steps at the bottom of the figure correspond to the numbered steps at the top.

CHECKPOINT

10a. Describe the electrical activity of the cells of the SA node and explain how the SA node functions as the normal pacemaker.

10b. Using a line diagram, illustrate a myocardial action potential and the time course for myocardial contraction. Explain how the relationship between these two events prevents the heart from sustaining a contraction and how it normally prevents abnormal rhythms of electrical activity.

11a. Draw an ECG and label the waves. Indicate the electrical events in the heart that produce these waves.

11b. Draw a figure that shows the relationship between ECG waves and the heart sounds. Explain this relationship.

11c. Describe the pathway of electrical conduction of the heart, starting with the SA node. How does damage to the AV node affect this conduction pathway and the ECG?

13.6 BLOOD VESSELS

The thick muscle layer of the arteries allows them to transport blood ejected from the heart under high pressure. The thinner muscle layer of veins allows them to distend when an increased amount of blood enters them, and their one-way valves ensure that blood flows back to the heart. Capillaries facilitate the rapid exchange of materials between the blood and interstitial fluid.

LEARNING OUTCOMES

After studying this section, you should be able to:

12. Compare the structure and function of arteries and veins, and the significance of the skeletal muscle pumps.

13. Describe the structures and functions of different types of capillaries.

Blood vessels form a tubular network throughout the body that permits blood to flow from the heart to all the living cells of the body and then back to the heart. Blood leaving the heart passes through vessels of progressively smaller diameters, referred to as *arteries, arterioles,* and *capillaries.* Capillaries are microscopic vessels that join the arterial flow to the venous flow. Blood returning to the heart from the capillaries passes through vessels of progressively larger diameters, called *venules* and *veins.*

The walls of arteries and veins are composed of three coats, or "tunics." The outermost layer is the **tunica externa,** the middle layer is the **tunica media,** and the inner layer is the **tunica interna.** The tunica externa is composed of connective tissue, whereas the tunica media is composed primarily of smooth muscle. The tunica interna consists of three parts: (1) an innermost simple squamous epithelium, the *endothelium,* which lines the lumina of all blood vessels; (2) the basement membrane (a layer of glycoproteins) overlying some connective tissue fibers; and (3) a layer of elastic fibers, or *elastin,* forming an *internal elastic lamina.*

Although arteries and veins have the same basic structure (fig. 13.26), there are some significant differences between them. Arteries have more muscle for their diameters than do comparably sized veins. As a result, arteries appear more rounded in cross section, whereas veins are usually partially collapsed. In addition, many veins have valves, which are absent in arteries.

Arteries

In the aorta and other large arteries, there are numerous layers of elastin fibers between the smooth muscle cells of the tunica media. These large **elastic arteries** expand when the pressure of the blood rises as a result of the ventricles' contraction; they recoil like a stretched rubber band when the blood pressure falls during relaxation of the ventricles. This elastic recoil drives the blood during the diastolic phase—the longest phase of the cardiac cycle—when the heart is resting and not providing a driving pressure.

CLINICAL APPLICATION

An **aneurysm** is a localized enlargement in a large artery such as the aorta. If the aneurysm tears, which is known as a "dissection," there is often a fatal loss of blood. An aneurysm of the abdominal aorta is usually associated with advanced age and atherosclerosis (section 8.7). By contrast, an aneurysm of the thoracic aorta can occur at any age and is correlated with risk associated with mutations in a variety of genes.

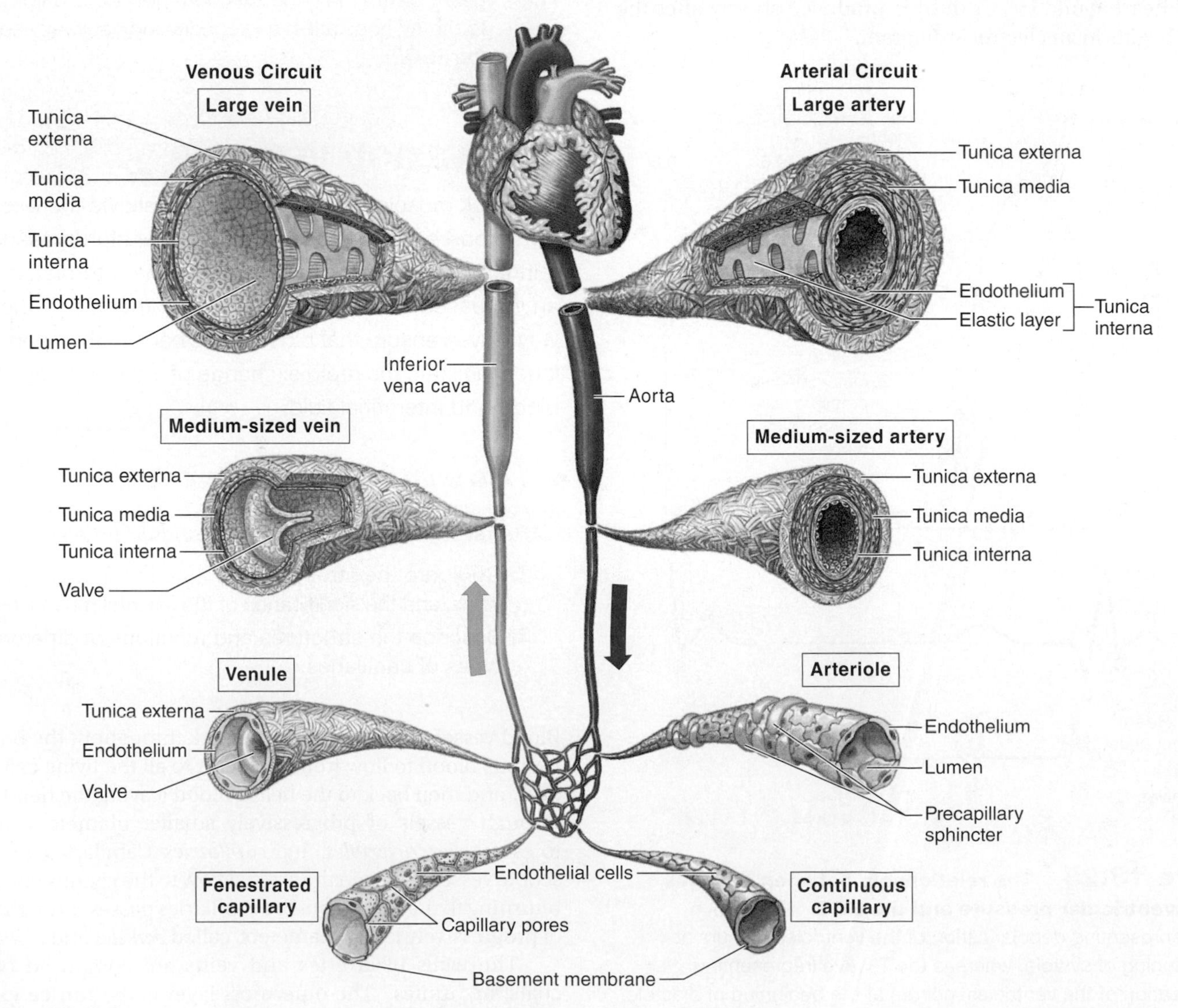

Figure 13.26 The structure of blood vessels. Notice the relative thickness and composition of the tunicas (layers) in comparable arteries and veins. APR

The small arteries and arterioles are less elastic than the larger arteries and have a thicker layer of smooth muscle for their diameters. Unlike the larger elastic arteries, therefore, the diameter of the smaller **muscular arteries** changes only slightly as the pressure of the blood rises and falls during the heart's pumping activity. Because arterioles and small muscular arteries have narrow lumina, they provide the greatest resistance to blood flow through the arterial system.

Clinical Investigation CLUES

Jason's radial pulse was fast and weak.

- What causes the pulse in the artery?
- What does the fast and weak pulse reveal about the pumping of Jason's heart?

Small muscular arteries that are 100 μm or less in diameter branch to form smaller **arterioles** (20 to 30 μm in diameter). In some tissues, blood from the arterioles can enter the venules directly through *arteriovenous anastomoses.* In most cases, however, blood from arterioles passes into capillaries (fig. 13.27). Capillaries are the narrowest of blood vessels (7 to 10 μm in diameter). They serve as the "business end" of the circulatory system, where gases and nutrients are exchanged between the blood and the tissues.

Resistance to blood flow is increased by *vasoconstriction* of arterioles (by contraction of their smooth muscle layer), which decreases the blood flow downstream in the capillaries. Conversely, *vasodilation* of arterioles (by relaxation of the smooth muscle layer) decreases the resistance and thus increases the flow through the arterioles to the capillaries. This topic is discussed in more detail in chapter 14, section 14.3. There is evidence of gap junctions between the cells of the arteriole wall in both the endothelial and smooth muscle layers. The vasoconstrictor effect of norepinephrine and the vasodilator effect of acetylcholine may be propagated for some distance along the arteriole wall by transmissions of depolarization and hyperpolarizations, respectively, through gap junctions in the vascular smooth muscle.

Capillaries

The arterial system branches extensively (table 13.8) to deliver blood to over 40 billion capillaries in the body. The number of capillary branches is so great that scarcely any cell in the body is more than 60 to 80 μm away from a blood capillary. The tiny capillaries provide a total surface area of 1,000 square miles for exchanges between blood and tissue fluid.

The amount of blood flowing through a particular capillary bed depends primarily on the resistance to blood flow in the small arteries and arterioles that supply blood to that capillary bed. Vasoconstriction in these vessels thus decreases blood flow to the capillary bed, whereas vasodilation increases blood flow. The relatively high resistance in the small arteries and arterioles in resting skeletal muscles, for example, reduces capillary blood flow to only about 5% to 10% of its maximum capacity. In some organs (such as the intestine), blood flow may also be regulated by circular muscle bands called *precapillary sphincters* at the origin of the capillaries (fig. 13.27).

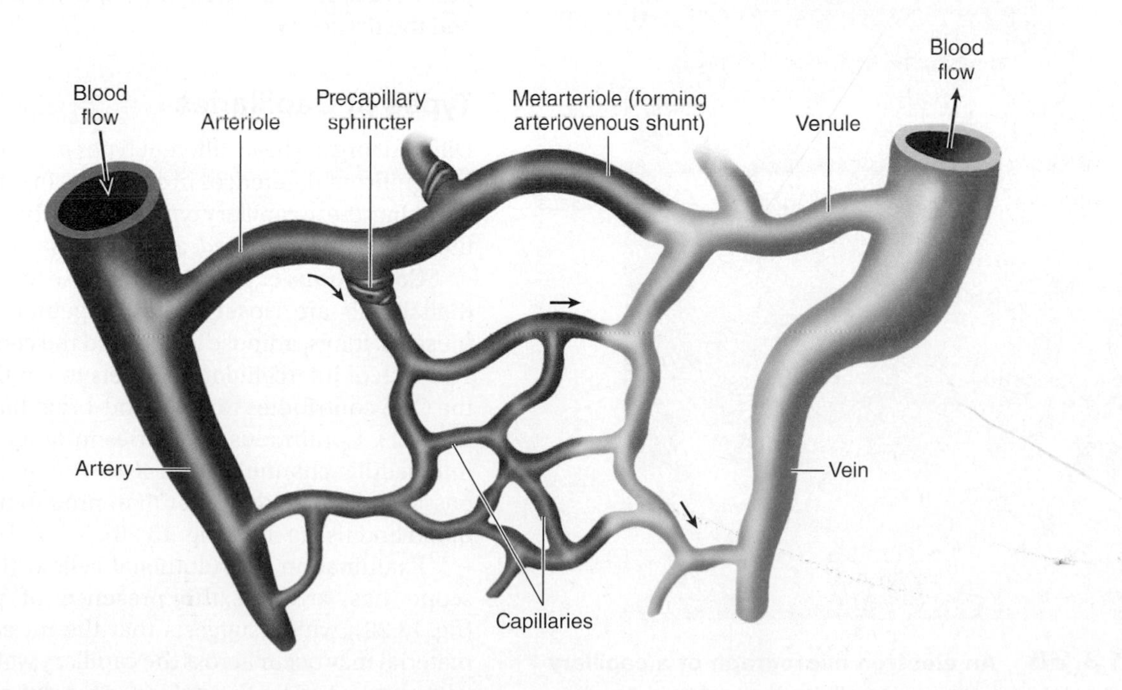

Figure 13.27 The microcirculation. Metarterioles (arteriovenous anastomoses) provide a path of least resistance between arterioles and venules. Precapillary sphincter muscles regulate the flow of blood through the capillaries.

Table 13.8 | Characteristics of the Vascular Supply to the Mesenteries in a Dog*

Kind of Vessels	Diameter (mm)	Number	Total Cross-Sectional Area (cm²)	Length (cm)	Total Volume (cm³)
Aorta	10	1	0.8	40	30
Large arteries	3	40	3.0	20	60
Main artery branches	1	600	5.0	10	50
Terminal branches	.06	1,800	5.0	1	25
Arterioles	0.02	40,000,000	125	0.2	25
Capillaries	0.008	1,200,000,000	600	0.1	60
Venules	0.03	80,000,000	570	0.2	110
Terminal veins	1.5	1,800	30	1	30
Main venous branches	2.4	600	27	10	270
Large veins	6.0	40	11	20	220
Vena cava	12.5	1	1.2	40	50
					930

*Note: The pattern of vascular supply is similar in dogs and humans.
Source: Animal Physiology, 4th ed. by Gordon et al., © 1982. Adapted by permission of Prentice-Hall, Inc., Upper Saddle River, NJ.

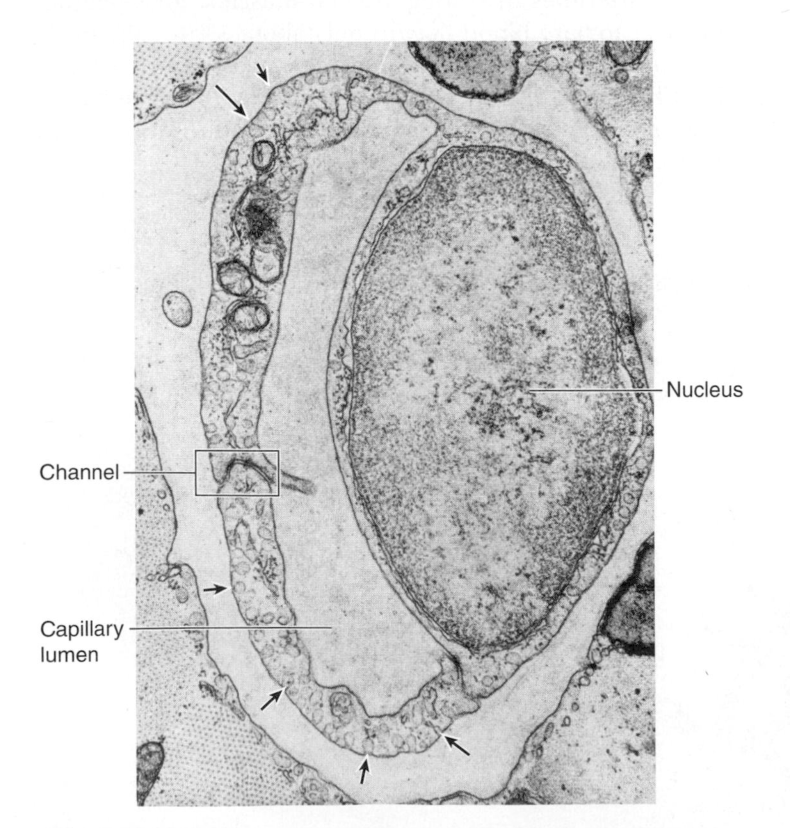

Figure 13.28 An electron micrograph of a capillary in the heart. Notice the thin intercellular channel (*middle left*) and the capillary wall, composed of only one cell layer. Arrows show some of the many pinocytotic vesicles. AP|R

Unlike the vessels of the arterial and venous systems, the walls of capillaries are composed of just one cell layer—a simple squamous epithelium, or endothelium (see fig. 13.28). The absence of smooth muscle and connective tissue layers permits a more rapid exchange of materials between the blood and the tissues.

Types of Capillaries

Different organs have different types of capillaries, distinguished by significant differences in structure. In terms of their endothelial lining, these capillary types include those that are *continuous,* those that are *fenestrated,* and those that are *discontinuous.*

Continuous capillaries are those in which adjacent endothelial cells are closely joined together. These are found in muscles, lungs, adipose tissue, and the central nervous system. The lack of intercellular channels in continuous capillaries in the CNS contributes to the blood-brain barrier (chapter 7, section 7.1). Continuous capillaries in other organs have narrow intercellular channels (from 40 to 45 Å in width) that permit the passage of molecules other than protein between the capillary blood and tissue fluid (fig. 13.28).

Examination of endothelial cells with an electron microscope has revealed the presence of pinocytotic vesicles (fig. 13.28), which suggests that the intracellular transport of material may occur across the capillary walls. This type of transport appears to be the only mechanism of capillary exchange available within the central nervous system and may account, in part, for the selective nature of the blood-brain barrier.

Fenestrated capillaries occur in the kidneys, endocrine glands, and intestines. These capillaries are characterized by wide intercellular pores (800 to 1,000 Å) that are covered by a layer of mucoprotein, which serves as a basement membrane over the capillary endothelium. This mucoprotein layer restricts the passage of certain molecules (particularly proteins) that might otherwise be able to pass through the large capillary pores. **Discontinuous capillaries** are found in the bone marrow, liver, and spleen. The distance between endothelial cells is so great that these capillaries look like little cavities (*sinusoids*) in the organ.

In a tissue that is hypoxic (has inadequate oxygen), new capillary networks are stimulated to grow. This growth is promoted by *vascular endothelial growth factor* (*VEGF,* discussed in the next Clinical Application Box). Capillary growth may additionally be promoted by *adenosine* (derived from AMP), which also stimulates vasodilation of arterioles and thereby increases blood flow to the hypoxic tissue. These changes result in a greater delivery of oxygen-carrying blood to the tissue.

CLINICAL APPLICATION

Angiogenesis refers to the formation of new blood vessels from preexisting vessels, which are usually venules. Because all living cells must be within 100 μm of a capillary, angiogenesis is required during tissue growth. Angiogenesis is thus involved in the pathogenesis of *neoplasms* (tumors) and of the blindness caused by neovascularization of the retina in *diabetic retinopathy* and *age-related macular degeneration* (the most common cause of blindness). The treatment of these diseases may therefore be improved by inhibiting angiogenesis. Treatment for *ischemic heart disease,* on the other hand, may be improved by promoting angiogenesis in the coronary circulation.

These therapies may manipulate paracrine regulators known to promote angiogenesis, including **vascular endothelial growth factor (VEGF)** and **fibroblast growth factor (FGF).** The FDA has approved the use of Avastin (bevacizumab), a monoclonal antibody (chapter 15, section 15.4) that blocks VEGF, for the treatment of metastatic colorectal cancer, nonsquamous, non-small-cell lung cancer, metastatic breast cancer, recurrent glioblastoma, and metastatic renal cell carcinoma. The FDA has also approved the use of anti-VEGF antibody injections into the eye for the treatment of the wet form of macular degeneration (discussed in the Clinical Application box on p. 305).

Veins

Most of the total blood volume is contained in the venous system. Unlike arteries, which provide resistance to the flow of blood from the heart, veins are able to expand as they accumulate additional amounts of blood. The average pressure in the veins is only 2 mmHg, compared to a much higher average arterial pressure of about 100 mmHg. These values, expressed in millimeters of mercury, represent the hydrostatic pressure that the blood exerts on the walls of the vessels.

The low venous pressure is insufficient to return blood to the heart, particularly from the lower limbs. Veins, however, pass between skeletal muscle groups that provide a massaging action as they contract (fig. 13.29). As the veins are squeezed by contracting skeletal muscles, a one-way flow of blood to the heart is ensured by the presence of **venous valves.** The ability of these valves to prevent the flow of blood away from the heart was demonstrated in the seventeenth century by William

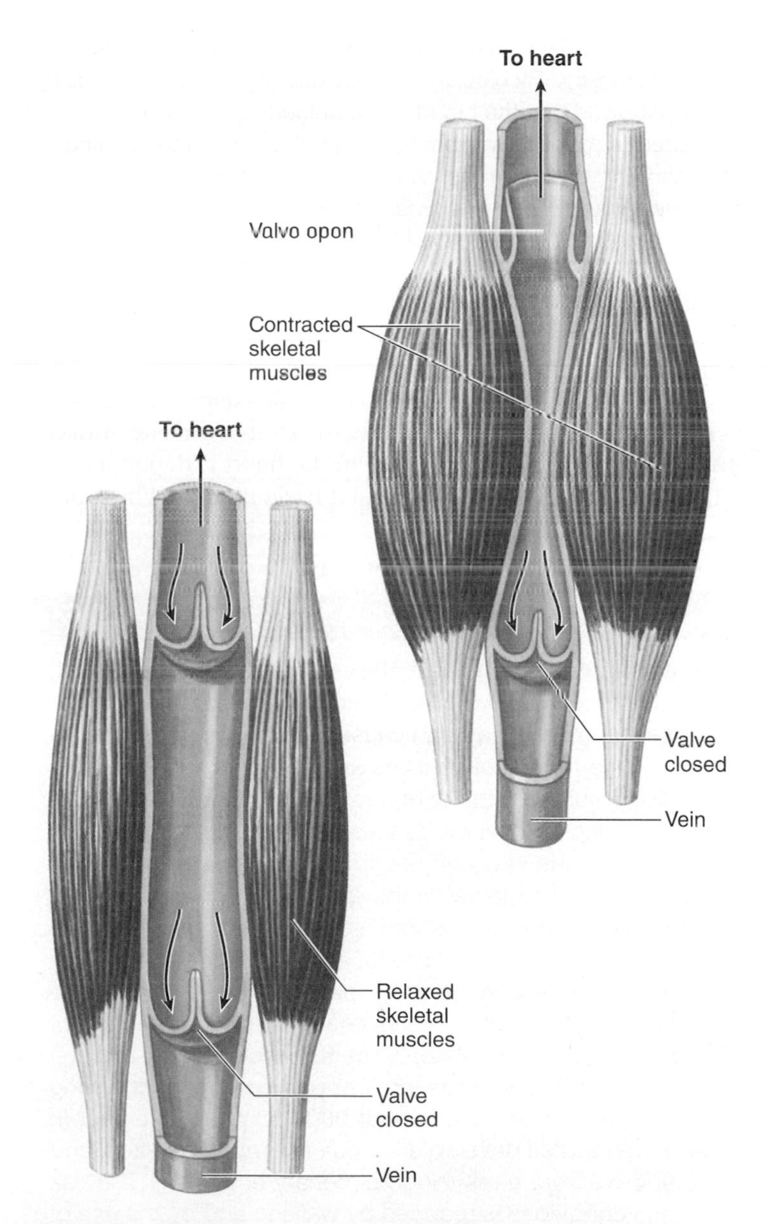

Figure 13.29 The action of the one-way venous valves. Contraction of skeletal muscles helps to pump blood toward the heart, but the flow of blood away from the heart is prevented by closure of the venous valves.

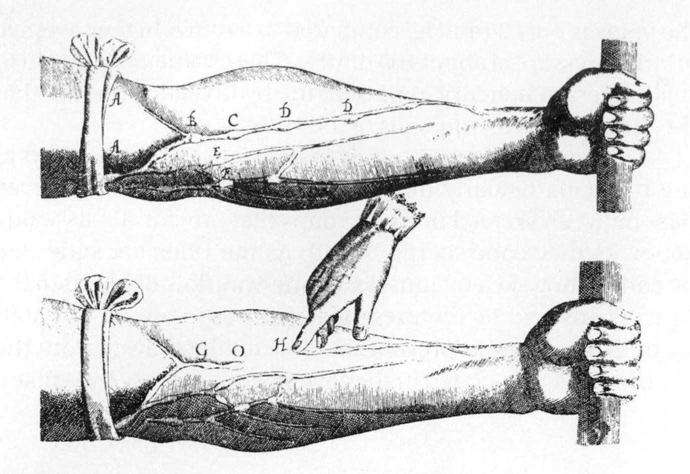

Figure 13.30 A demonstration of venous valves by William Harvey. By blocking venous drainage with a tourniquet, Harvey showed that the blood in the bulged vein was not permitted to move away from the heart, thereby demonstrating the action of venous valves. After William Harvey, *On the Motion of the Heart and Blood in Animals,* 1628.

Harvey (fig. 13.30). After applying a tourniquet to a subject's arm, Harvey found that he could push the blood in a bulging vein toward the heart, but not in the reverse direction.

The effect of the massaging action of skeletal muscles on venous blood flow is often described as the **skeletal muscle pump.** The rate of venous return to the heart is dependent, in large part, on the action of skeletal muscle pumps. When these pumps are less active, as when a person stands still or is bedridden, blood accumulates in the veins and causes them to bulge. When a person is more active, blood returns to the heart at a faster rate and less is left in the venous system.

Action of the skeletal muscle pumps aids the return of venous blood from the lower limbs to the large abdominal veins. Movement of venous blood from abdominal to thoracic veins, however, is aided by an additional mechanism—breathing. When a person inhales, the diaphragm—a muscular sheet separating the thoracic and abdominal cavities—contracts. Contraction of the dome-shaped diaphragm causes it to flatten and descend inferiorly into the abdomen. This has the dual effect of increasing the pressure in the abdomen, thus squeezing the abdominal veins, and decreasing the pressure in the thoracic cavity. The pressure difference in the veins created by this inspiratory movement of the diaphragm forces blood into the thoracic veins that return the venous blood to the heart.

CLINICAL APPLICATION

The accumulation of blood in the veins of the legs over a long period of time, as may occur in people with occupations that require standing all day, can cause the veins to stretch to the point where the venous valves are no longer efficient. This can also result from the compression of abdominal veins by a fetus during pregnancy. Venous congestion and stretching produced in this way can result in **varicose veins.** Venous congestion in the lower limbs is reduced during walking, when movements of the foot activate the soleus muscle pump. This effect can be produced in bedridden people by extending and flexing the ankle joints.

Inadequate venous blood flow in a bedridden patient increases the risk of **deep vein thrombosis,** a condition that can lead to **venous thromboembolism** (a traveling clot). It has been estimated that 900,000 patients a year in the United States develop a venous thromboembolism, and 300,000 die from a resulting pulmonary embolism. The risk of these conditions is reduced by walking and by the use of devices that compress the leg. Deep vein thrombosis most often develops in the pockets of deep venous valves, where blood flow and oxygen levels are especially low.

CHECKPOINT

12a. Describe the basic structural pattern of arteries and veins. Explain how arteries and veins differ in structure and how these differences contribute to their differences in function.

12b. Describe the functional significance of the skeletal muscle pump and illustrate the action of venous valves.

13. Explain the functions of capillaries and describe the structural differences between capillaries in different organs.

13.7 ATHEROSCLEROSIS AND CARDIAC ARRHYTHMIAS

Atherosclerosis is a disease process that can lead to obstruction of coronary blood flow. As a result, the electrical properties of the heart, and the heart's ability to function as a pump, may be seriously compromised. Abnormal cardiac rhythms, or arrhythmias, can be detected by the abnormal electrocardiogram patterns they produce.

LEARNING OUTCOMES

After studying this section, you should be able to:

14. Explain the causes and dangers of atherosclerosis.

15. Explain the cause and significance of angina pectoris.

16. Describe how different arrhythmias affect the ECG.

Atherosclerosis

Atherosclerosis is the most common form of arteriosclerosis (hardening of the arteries) and, through its contribution to heart disease and stroke, is responsible for about 50% of the deaths in the United States, Europe, and Japan. In atherosclerosis, localized **plaques,** or *atheromas,* protrude into the lumen of the artery and thus reduce blood flow. The atheromas additionally serve as sites for *thrombus* (blood clot) formation, which can further occlude the blood supply to an organ (fig. 13.31).

It is currently believed that the process of atherosclerosis begins as a result of damage, or "insult," to the endothelium. Such insults are produced by smoking, hypertension (high blood pressure), high blood cholesterol, and diabetes. The first anatomically recognized change is the appearance of *fatty streaks,* which are gray-white areas that protrude into the lumen of arteries, particularly at arterial branch points.

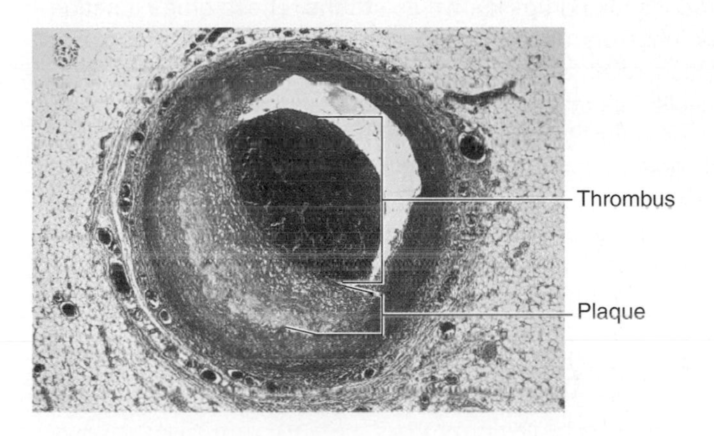

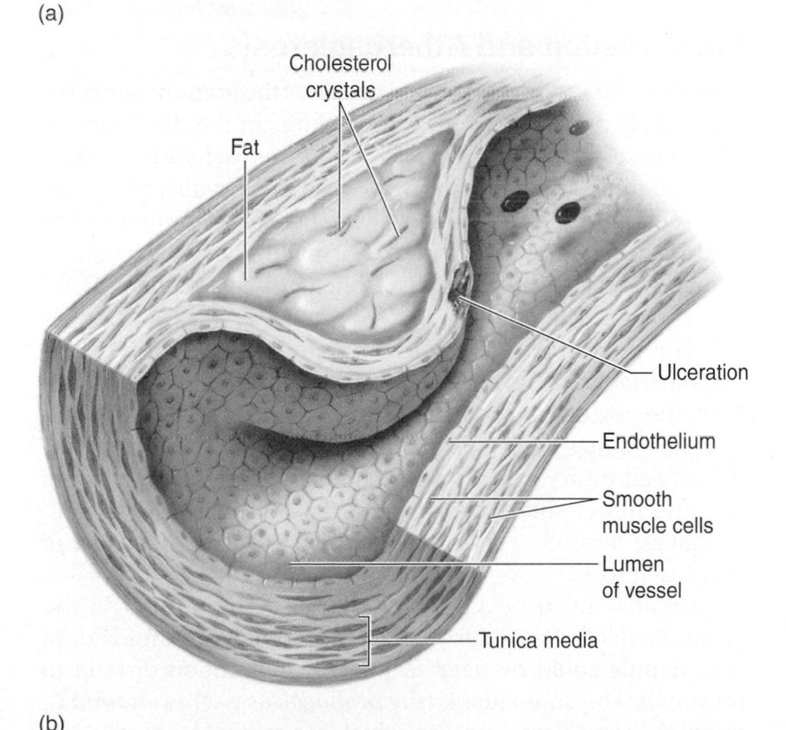

Figure 13.31 **Atherosclerosis.** (*a*) A photograph of the lumen (cavity) of a human coronary artery that is partially occluded by an atherosclerotic plaque and a thrombus. (*b*) A diagram of the structure of an atherosclerotic plaque.

These are aggregations of lipid-filled macrophages and lymphocytes within the tunica interna. They are present to a small degree in the aorta and coronary arteries of children aged 10 to 14, but progress to more advanced stages at different rates in different people. In the intermediate stage, the area contains layers of macrophages and smooth muscle cells. The more advanced lesions, called *fibrous plaques,* consist of a cap of connective tissue with smooth muscle cells over accumulated lipid and debris, macrophages that have been derived from monocytes (chapter 15), and lymphocytes. The fibrous cap of an advanced atherosclerotic lesion becomes thin and prone to rupture, promoting the formation of a thrombus.

The disease process may be instigated by damage to the endothelium, but its progression is promoted by inflammation that is stimulated by a wide variety of cytokines and other paracrine regulators secreted by the endothelium and by the other participating cells, including platelets, macrophages, and lymphocytes. Some of these regulators attract monocytes and lymphocytes to the damaged endothelium and cause them to penetrate into the tunica interna. The monocytes then become macrophages, engulf lipids, and take on the appearance of *foam cells.* Smooth muscle cells change from a contractile state to a "synthetic" state, in which they produce and secrete connective tissue matrix proteins. (This is unique; in other tissues, connective tissue matrix is secreted by cells called fibroblasts.) The changed smooth muscle cells respond to chemical attractants and migrate from the tunica media to the tunica interna, where they can proliferate.

Endothelial cells normally prevent the progression just described by presenting a physical barrier to the penetration of monocytes and lymphocytes and by producing paracrine regulators such as nitric oxide. The vasodilator action of nitric oxide helps to counter the vasoconstrictor effects of another paracrine regulator, endothelin-1, which is increased in atherosclerosis. Hypertension, smoking, and high blood cholesterol, among other risk factors, interfere with this protective function.

Cholesterol and Plasma Lipoproteins

There is considerable evidence that high blood cholesterol is associated with an increased risk of atherosclerosis. High blood cholesterol can be produced by a diet rich in cholesterol and saturated fat, or it may be the result of an inherited condition known as *familial hypercholesteremia.* This condition is inherited as a single dominant gene; individuals who inherit two of these genes have extremely high cholesterol concentrations (regardless of diet) and usually suffer heart attacks during childhood.

Lipids, including cholesterol, are carried in the blood attached to protein carriers (fig. 13.32; also see chapter 18, table 18.8). Cholesterol is carried to the arteries by plasma proteins called **low-density lipoproteins (LDLs).** LDLs are derived from *very low-density lipoproteins (VLDLs),* which

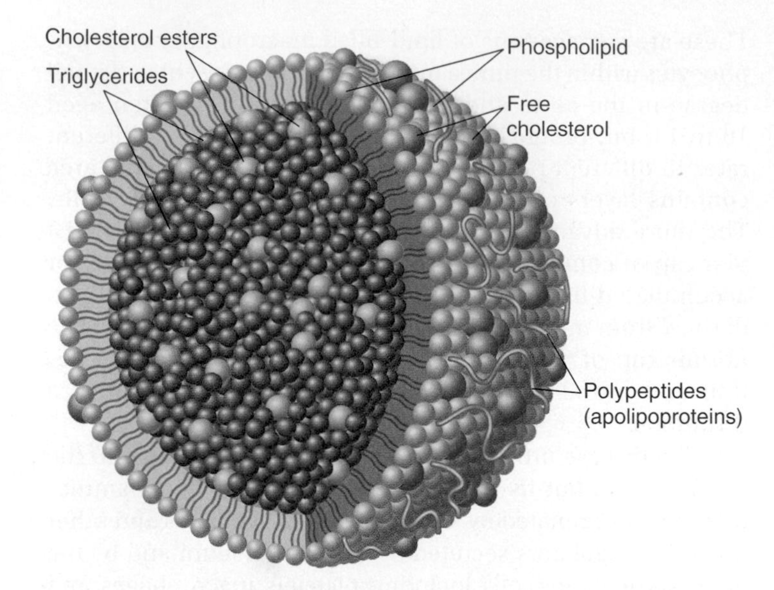

Figure 13.32 Structure of a lipoprotein. There is a core of nonpolar triglycerides and cholesterol esters coated by proteins (apolipoproteins), phospholipids, and some free cholesterol.

are small, protein-coated droplets produced by the liver and composed of cholesterol, triglycerides, free fatty acids, and phospolipids. After enzymes in various organs remove most of the triglycerides, the VLDLs become LDLs that transport cholesterol to the organs.

Cells in different organs contain receptors for the proteins (called *apolipoproteins*) in LDLs. When these apolipoproteins bind to their receptors, the cell engulfs the LDL particles by receptor-mediated endocytosis (chapter 3; see fig. 3.4). Most LDL particles are removed in this way by the liver. However, the uptake and accumulation of a particular LDL protein, *apolipoprotein B,* into the subendothelial connective tissue of an artery is believed to initiate the formation of an atherosclerotic plaque. Apolipoprotein B, enhanced by oxidation (discussed shortly), acts on the endothelium to promote the entry of monocytes into the lesion and the conversion of the monocytes into macrophages.

People who eat a diet high in cholesterol and saturated fat, and people with familial hypercholesteremia, have a high blood LDL concentration because their livers have a low number of LDL receptors. With fewer LDL receptors, the liver is less able to remove the LDL from the blood and more LDL is available to enter the endothelial cells of arteries.

High-density lipoprotein (HDL), in contrast, offers protection against atherosclerosis by carrying cholesterol away from the arterial wall. In the development of atherosclerosis, monocytes migrate through the arterial endothelium to the intima, where they become macrophages that are able to engulf oxidized LDLs (discussed shortly). The cholesterol-engorged macrophages are known as foam cells and play an important role in the development of the atherosclerotic lesion. This progress is retarded by HDL, which accepts cholesterol from the foam cells and carries it through the blood to the liver for metabolism. HDL levels are largely determined by genetics, but it is known that HDL levels are higher, and the risk of atherosclerosis is lower, in women (prior to menopause) than in men, and in people who exercise regularly. HDL levels are higher in marathon runners than in joggers, and are higher in joggers than in sedentary people. Drugs that help raise HDL levels include the *statins* (such as Lipitor), the *fibrates,* and high doses of the vitamin *niacin.*

CLINICAL APPLICATION

Many people with dangerously high LDL cholesterol concentrations take drugs known as **statins.** These drugs function as inhibitors of the enzyme *HMG-coenzyme A reductase,* which catalyzes the rate-limiting step in cholesterol synthesis. The statins therefore decrease the ability of the liver to produce its own cholesterol. The lowered intracellular cholesterol then stimulates the production of LDL receptors, allowing the liver cells to engulf more LDL-cholesterol. When a person takes a statin drug, therefore, the liver cells remove more LDL-cholesterol from the blood and thus decrease the amount of blood LDL-cholesterol that can enter the endothelial cells of arteries. Surprisingly, and for reasons not well understood, the statins also have the beneficial effect of slightly raising the blood HDL-cholesterol levels.

Inflammation and Atherosclerosis

Notice the important roles played by cells of the immune system—particularly monocytes and lymphocytes—in the development and progression of atherosclerosis. Atherosclerosis is now believed to be an inflammatory disease to a significant degree. This is emphasized by the recent evidence that measurement of blood **C-reactive protein,** a marker of inflammation, is actually a stronger predictor of atherosclerotic heart disease than the blood LDL cholesterol level.

The inflammatory process may be instigated by oxidative damage to the artery wall. When endothelial cells engulf LDL, they oxidize it to a product called *oxidized LDL.* Recent evidence suggests that oxidized LDL contributes to endothelial cell injury, migration of monocytes and lymphocytes into the tunica interna, conversion of monocytes into macrophages, and other events that occur in the progression of atherosclerosis.

Because oxidized LDL seems to be so important in the progression of atherosclerosis, it would appear that antioxidant compounds could be used to treat this condition or help to prevent it. The antioxidant drug *probucol,* as well as *vitamin C, vitamin E,* and *beta-carotene,* which are antioxidants (chapter 19, section 19.1), have decreased the formation of oxidized LDL *in vitro* but have had only limited success so far in treating atherosclerosis.

FITNESS APPLICATION

Many people can significantly lower their blood cholesterol concentration through a regimen of exercise and diet. Because saturated fat in the diet raises blood cholesterol, such foods as fatty meat, egg yolks, and internal animal organs (liver, brain, etc.) should be eaten only sparingly. The American Heart Association recommends that fat should contribute less than 30% of the total calories of a diet. By way of comparison, 40% to 50% of the calories in a fast-food meal are derived from fat. The American Heart Association also recommends that people eat at least two fish meals a week (particularly oily fish—trout, salmon, mackerel, herring, and sardines) as well as certain plant foods—including walnuts and certain plant oils such as soybean, rapeseed (canola), and linseed oils. This is because these foods are rich in omega-3 (or n-3) fatty acids, which may provide some protection against cardiovascular disease. Alternatively, fish oil capsules containing EPA and DHA (the n-3 fatty acids found in fish; chapter 19, section 19.1) may be taken. However, the single most effective action that smokers can take to lower their risk of atherosclerosis is to stop smoking.

Clinical Investigation CLUES

Jason's blood tests revealed a high total cholesterol and LDL cholesterol concentration.

- What dangers are indicated by these lab results?
- What can Jason do to reduce these dangers?

Ischemic Heart Disease

A tissue is said to be **ischemic** when its oxygen supply is deficient because of inadequate blood flow. The most common cause of myocardial ischemia is atherosclerosis of the coronary arteries. The adequacy of blood flow is relative—it depends on the tissue's metabolic requirements for oxygen. An obstruction in a coronary artery, for example, may allow sufficient coronary blood flow at rest but not when the heart is stressed by exercise or emotional conditions. In these cases, the increased activity of the sympathoadrenal system causes the heart rate and blood pressure to rise, increasing the work of the heart and raising its oxygen requirements. Recent evidence also suggests that mental stress can cause constriction of atherosclerotic coronary arteries, leading to ischemia of the heart muscle. The vasoconstriction is believed to result from abnormal function of a damaged endothelium, which normally prevents constriction (through secretion of paracrine regulators) in response to mental stress. The control of vasoconstriction and vasodilation is discussed more fully in chapter 14, section 14.3.

Myocardial ischemia is associated with increased concentrations of blood lactic acid produced by anaerobic metabolism in the ischemic tissue. This condition often causes substernal pain, which may also be referred to the left shoulder and arm, as well as to other areas. This *referred pain* (chapter 10, section 10.2) is called **angina pectoris.** People with angina frequently take nitroglycerin or related drugs that help to relieve the ischemia and pain. These drugs are effective because they produce vasodilation, which improves circulation to the heart and decreases the work that the ventricles must perform to eject blood into the arteries.

Myocardial cells are adapted for aerobic respiration and cannot metabolize anaerobically for more than a few minutes. If ischemia and anaerobic metabolism are prolonged, *necrosis* (cellular death) may occur in the areas most deprived of oxygen. A sudden, irreversible injury of this kind is called a **myocardial infarction,** or **MI.** The lay term "heart attack," though imprecise, usually refers to a myocardial infarction.

The area of dead cells is not replaced with functioning myocardial cells because mature myocardial cells can't divide. Instead, fibroblasts produce noncontractile scar tissue, which forms the infarct. The area of infarcted tissue is usually relatively small if the person is hospitalized and treated within a few hours after the onset of symptoms. However, after the heart becomes re-perfused with blood (so that it receives sufficient oxygen to resume aerobic respiration), larger numbers of myocardial cells may die. This *reperfusion injury* may be a greater threat than the initial event and is caused by apoptosis (chapter 3, section 3.5) due to the accumulation of Ca^{2+} and the production of superoxide free radicals (chapters 5 and 19) by mitochondria. Apoptosis of myocardial cells surrounding the initial lesion can greatly increase the size of the infarct and weaken the wall of the ventricle.

The infarct may thereby cause the ventricular wall to thin and distend under pressure. In recent years, scientists have investigated a variety of potential stem cell therapies for myocardial infarction. These include the use of stem cells from the bone marrow (which can secrete cytokines that promote healing); the possible differentiation of embryonic stem cells and induced pluripotent stem cells (chapter 20, section 20.6) into myocardial cells; and the transformation of fibroblasts (perhaps within an infarct) into myocardial cells. A recent report demonstrated that the mouse epicardium contains stem cells and that these could be induced to form new myocardial cells after a myocardial infarction. However, more research in these areas is required for regenerative medicine therapies to become a medical reality.

Occlusion of the coronary arteries, resulting in myocardial infarction, is the leading cause of death in the Western world. The necrosis of myocardial cells is particularly devastating because dead myocardial cells cannot be replaced by mitosis of neighboring cells. Therefore, the major medical goals are to recognize myocardial ischemia and relieve its causes before the injury becomes too great. Myocardial ischemia may be detected by changes in the S-T segment

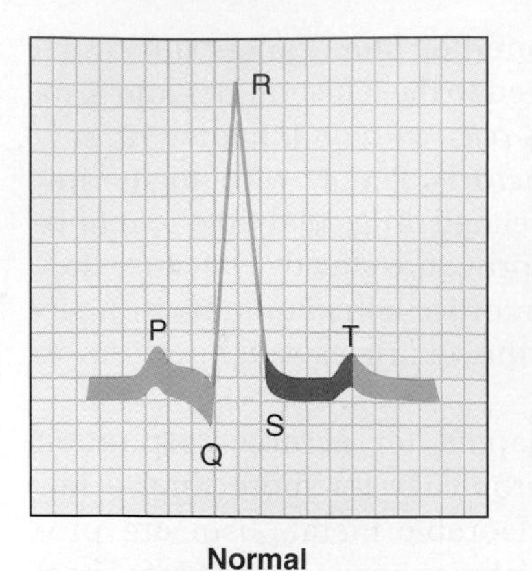

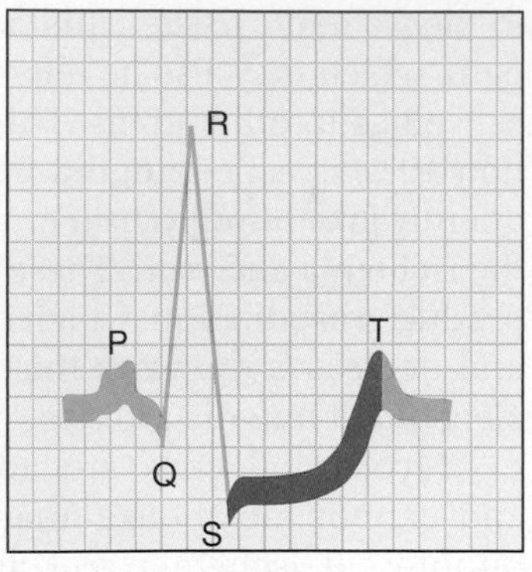

Figure 13.33 Depression of the S-T segment as a result of myocardial ischemia. This is but one of many ECG changes that alert trained personnel to the existence of heart problems.

of the electrocardiogram (fig. 13.33). Sustained occlusion of a coronary artery that produces a myocardial infarction is accompanied by an elevation of the S-T segment.

The diagnosis of myocardial infarction is aided by blood measurement of enzymes released from the infarcted tissue. Plasma concentrations of *creatine phosphokinase (CPK),* for example, increase within 3 to 6 hours after the onset of symptoms and return to normal after 3 days. Plasma levels of *lactate dehydrogenase (LDH)* reach a peak within 48 to 72 hours after the onset of symptoms and remain elevated for about 11 days. However, the plasma levels of cardiac muscle *troponin I* and *troponin T* (chapter 12) are now regarded by the medical profession as the most sensitive and specific indicators of myocardial infarction, with troponin I as particularly useful in early diagnosis. The detection and treatment of coronary thrombosis is discussed in conjunction with the coronary circulation in chapter 14, section 14.4 (see fig. 14.18).

CLINICAL APPLICATION

Stroke (cerebrovascular accident) is the third leading cause of death in the United States. Most commonly, it is produced when a region of the brain becomes *ischemic* (has inadequate blood flow) due to a thrombus (blood clot), which can be formed as a consequence of atherosclerosis. Thrombolytic drugs help, but must be administered soon after the ischemic injury to be effective. This is because neurons progressively die through a process known as *excitotoxicity* (discussed in chapter 7). Excitotoxicity occurs as a result of the ischemia-induced impairment in the removal of the excitatory neurotransmitter glutamate from synaptic clefts. Due to the accumulation of glutamate, there is excessive inflow of Ca^{2+} through NMDA receptors, particularly those located in areas of the plasma membrane away from the postsynaptic membrane. As a result, glutamate accumulation produces toxic rather than neuroprotective effects and causes death of the neurons. There is presently no effective way to prevent excitotoxicity and its consequences.

Arrhythmias Detected by the Electrocardiograph

Arrhythmias, or abnormal heart rhythms, can be detected and described by the abnormal ECG tracings they produce. Although proper clinical interpretation of electrocardiograms requires information not covered in this chapter, some knowledge of abnormal rhythms is interesting in itself and is useful in gaining an understanding of normal physiology.

A heartbeat occurs whenever a normal QRS complex is seen, and the ECG chart paper moves at a known speed, so the cardiac rate (beats per minute) can be easily obtained from an ECG recording. A cardiac rate slower than 60 beats per minute indicates **bradycardia;** a rate faster than 100 beats per minute is described as **tachycardia** (fig. 13.34).

Both bradycardia and tachycardia can occur normally. Endurance-trained athletes, for example, often have heart rates ranging from 40 to 60 beats per minute. This *athlete's bradycardia* occurs as a result of higher levels of parasympathetic inhibition of the SA node and is a beneficial adaptation. Activation of the sympathetic division of the ANS during exercise or emergencies ("fight or flight") causes a normal tachycardia.

Abnormal tachycardia occurs if the heart rate increases when the person is at rest. This may be due to abnormally fast pacing by the atria (caused, for example, by drugs), or to the development of abnormally fast *ectopic pacemakers*—cells located outside the SA node that assume a pacemaker function. This abnormal atrial tachycardia thus differs from normal, or *sinus,* (SA node) *tachycardia. Ventricular tachycardia* results when abnormally fast ectopic pacemakers in the ventricles cause them to beat rapidly and independently of the atria. This is very dangerous because it can quickly degenerate into a lethal condition known as *ventricular fibrillation.*

Clinical Investigation CLUES

Jason's ECG showed sinus tachycardia.

- How did the ECG indicate this condition?
- What could cause this condition, and how would it be related to Jason's fast pulse?

Flutter and Fibrillation

Extremely rapid rates of electrical excitation and contraction of either the atria or the ventricles may produce flutter or fibrillation. In **flutter,** the contractions are very rapid (200 to 300 per minute) but are coordinated. In **fibrillation,** contractions of different groups of myocardial cells occur at different times, so that a coordinated pumping action of the chambers is impossible.

Atrial flutter usually degenerates quickly into *atrial fibrillation,* where the disorganized production of impulses occurs very rapidly (about 600 times per minute) and contraction of the atria is ineffectual. The AV node doesn't respond to all of those impulses, but enough impulses still get through to stimulate the ventricles to beat at a rapid rate (up to 150–180 beats per minute). Since the ventricles fill to about 80% of their end-diastolic volume before even normal atrial contraction, atrial fibrillation only reduces the cardiac output by about 15%. People with atrial fibrillation can live for many years, although this condition is associated with increased mortality due to stroke and heart failure. It has been estimated that 20% to 25% of all strokes may result from thrombi promoted by atrial fibrillation. Patients with atrial fibrillation are thus often placed on antithrombotic drugs.

By contrast, people with *ventricular fibrillation* (fig. 13.34) can live for only a few minutes unless this is extended by cardiopulmonary resuscitation (CPR) techniques or the fibrillation is ended by electrical defibrillation (discussed shortly). Death is caused by the inability of the fibrillating ventricles to pump blood and thus deliver needed oxygen to the heart and brain.

Fibrillation is caused by a continuous recycling of electrical waves, known as **circus rhythms,** through the myocardium. The recycling of action potentials is normally prevented by the entire myocardium entering a refractory period as a single unit, owing to the rapid transmission of the action potential among the myocardial cells by their gap junctions and to the long duration of the action potential provided by its plateau phase (see fig. 13.21). However, if some cells emerge from their refractory periods before others, an action potential can be continuously regenerated and conducted. Recycling of electrical waves along continuously changing pathways produces uncoordinated contraction and an impotent pumping action.

Circus rhythms are thus produced whenever impulses can be conducted without interruption by nonrefractory tissue. This may occur when the conduction pathway is longer than normal, as in a dilated heart. It can also be produced by an electric shock delivered at the middle of the T wave, when different myocardial cells are in different stages of recovery from their refractory period. Finally, circus rhythms and fibrillation may be produced by damage to the myocardium, which slows the normal rate of impulse conduction.

Sudden death from cardiac arrhythmia usually progresses from ventricular tachycardia through ventricular fibrillation, culminating in *asystole* (the cessation of beating, with a straight-line ECG). Sudden death from cardiac arrhythmia is commonly a result of acute myocardial ischemia (insufficient blood flow to the heart muscle), most often due to atherosclerosis of the coronary arteries.

Fibrillation can sometimes be stopped by a strong electric shock delivered to the chest. This procedure is called **electrical defibrillation.** The electric shock depolarizes all of the myocardial cells at the same time, causing them all to enter a refractory state. Conduction of circus rhythms thus stops, and the SA node can begin to stimulate contraction in a normal fashion. This does not correct the initial problem that caused circus rhythms and fibrillation, but it does keep the person alive long enough to take other corrective measures.

A device known as an *implantable converter-defibrillator* is now available for high-risk patients. This device consists of a unit that is implanted into a subcutaneous pocket in the pectoral region, with a lead containing electrodes and a shocking coil that is threaded into the heart (usually the right ventricle). Sensors can detect when ventricular fibrillation occurs, and can distinguish between supraventricular and ventricular tachycardia (fig. 13.34). The coil can deliver defibrillating shocks if ventricular fibrillation is detected.

AV Node Block

The time interval between the beginning of atrial depolarization—indicated by the P wave—and the beginning of ventricular depolarization (as shown by the Q part of the QRS complex) is called the *P-R interval* (see fig. 13.22). In the normal heart, this time interval is 0.12 to 0.20 second in duration. Damage

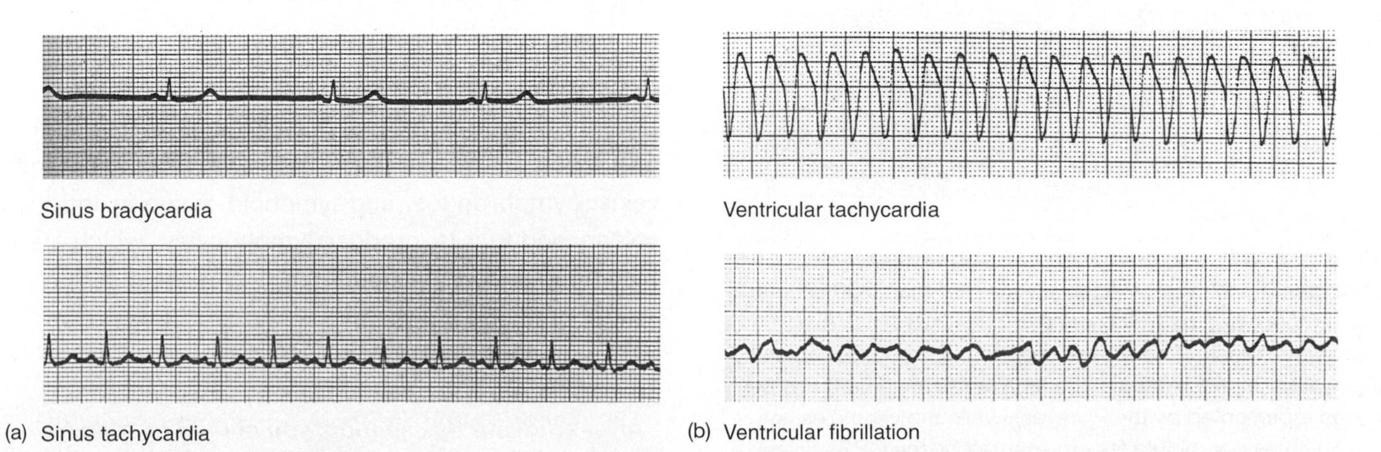

Figure 13.34 Some arrhythmias detected by the ECG. In (*a*) the heartbeat is paced by the normal pacemaker—the SA node (hence the name *sinus rhythm*). This can be abnormally slow (bradycardia—42 beats per minute in this example) or fast (tachycardia—125 beats per minute in this example). Compare the pattern of tachycardia in (*a*) with the tachycardia in (*b*). Ventricular tachycardia is produced by an ectopic pacemaker in the ventricles. This dangerous condition can quickly lead to ventricular fibrillation, also shown in (*b*).

to the AV node causes slowing of impulse conduction and is reflected by changes in the P-R interval. This condition is known as *AV node block* (fig. 13.35).

CLINICAL APPLICATION

A number of abnormal conditions, including a blockage in conduction of the impulse along the bundle of His, require the insertion of an **artificial pacemaker.** This is a battery-powered device, about the size of a locket, which may be placed in permanent position under the skin. The electrodes from the pacemaker are guided through a vein to the right atrium, through the tricuspid valve, and into the right ventricle. The electrodes are fixed to the trabeculae carnae and are in contact with the wall of the ventricle. When these electrodes deliver shocks—either at a continuous pace or on demand (when the heart's own impulse doesn't arrive on time)—both ventricles are depolarized and contract and then repolarize and relax, just as they do in response to endogenous stimulation.

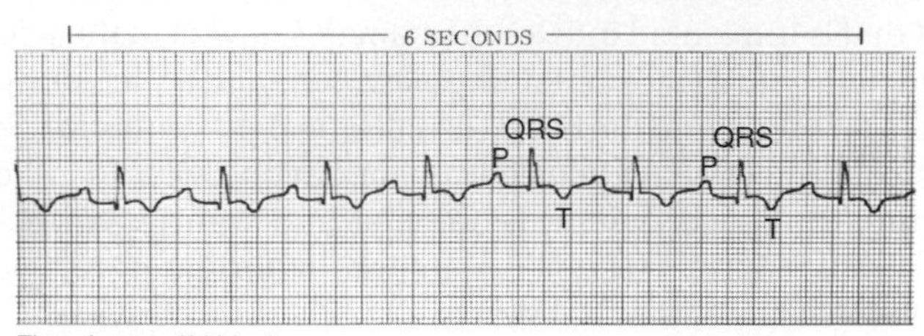

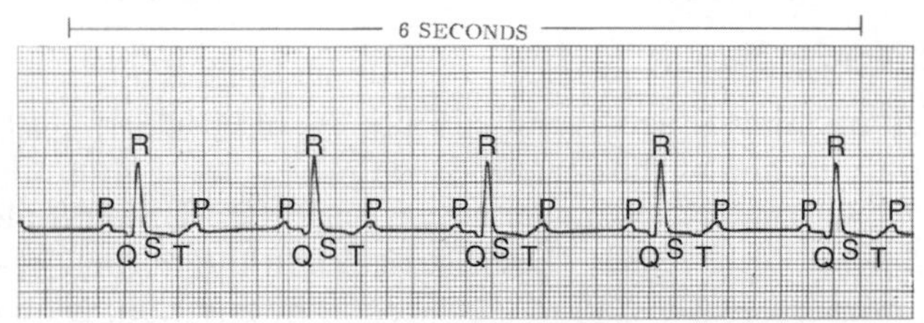

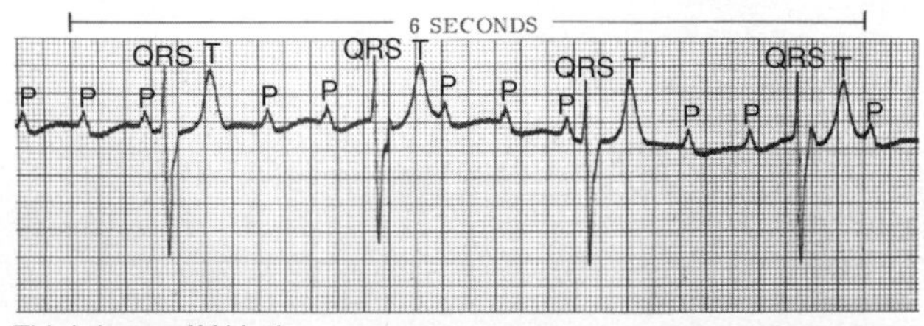

Figure 13.35 Atrioventricular (AV) node block. In first-degree block, the P-R interval is greater than 0.20 second (in the example here, the P-R interval is 0.26–0.28 second). In second-degree block, P waves are seen that are not accompanied by QRS waves. In this example, the atria are beating 90 times per minute (as represented by the P waves), while the ventricles are beating 50 times per minute (as represented by the QRS waves). In third-degree block, the ventricles are paced independently of the atria by an ectopic pacemaker. Ventricular depolarization (QRS) and repolarization (T) therefore have a variable position in the electrocardiogram relative to the P waves (atrial depolarization).

First-degree AV node block occurs when the rate of impulse conduction through the AV node (as reflected by the P-R interval) exceeds 0.20 second. **Second-degree AV node block** occurs when the AV node is damaged so severely that only one out of every two, three, or four atrial electrical waves can pass through to the ventricles. This is indicated in an ECG by the presence of P waves without associated QRS waves.

In **third-degree,** or **complete, AV node block,** none of the atrial waves can pass through the AV node to the ventricles. The atria are paced by the SA node (follow a normal "sinus rhythm"), but in complete AV node block a secondary pacemaker in the Purkinje fibers paces the ventricles. The SA node is the normal pacemaker because it has the fastest cycle of spontaneous depolarization, but in complete AV node block the action potentials from the atria cannot reach the Purkinje fibers to suppress their pacemaker activity. The pacemaker rate of the Purkinje fibers (generally about 20 to 40 beats per minute, depending on location) is abnormally slow, and the bradycardia that results is usually corrected by insertion of an artificial pacemaker.

CHECKPOINT

14. Explain how cholesterol is carried in the plasma and how the concentrations of cholesterol carriers are related to the risk for developing atherosclerosis.

15. Explain how angina pectoris is produced and discuss the significance of this symptom.

16a. Identify normal and pathological causes of bradycardia and tachycardia and describe how these affect the ECG. Also, identify flutter and fibrillation and describe how these appear in the ECG.

16b. Explain the effects of first-, second-, and third-degree AV node block on the electrocardiogram.

13.8 LYMPHATIC SYSTEM

Lymphatic vessels absorb excess interstitial fluid and transport this fluid—now called lymph—to ducts that drain into veins. Lymph nodes, and lymphoid tissue in the thymus, spleen, and tonsils, produce lymphocytes, which are white blood cells involved in immunity.

LEARNING OUTCOMES

After studying this section, you should be able to:

17. Explain how the lymph and lymphatic system relate to the blood and cardiovascular system.

18. Describe the function of lymph nodes and lymphatic organs.

The **lymphatic system** has three basic functions: (1) it transports interstitial (tissue) fluid, initially formed as a blood filtrate, back to the blood; (2) it transports absorbed fat from the small intestine to the blood; and (3) its cells—called *lymphocytes*—help provide immunological defenses against disease-causing agents (pathogens).

The smallest vessels of the lymphatic system are the **lymphatic capillaries** (fig. 13.36). Lymphatic capillaries are microscopic closed-ended tubes that form vast networks in the intercellular spaces within most organs. Because the walls of lymphatic capillaries are composed of endothelial cells with porous junctions, interstitial fluid, proteins, extravasated white blood cells, microorganisms, and absorbed fat (in the intestine) can easily enter. Once fluid enters the lymphatic capillaries, it is referred to as **lymph.**

From merging lymphatic capillaries, the lymph is carried into larger lymphatic vessels called **lymph ducts.** The walls of lymph ducts are similar to those of veins. They have the same three layers and also contain valves to prevent backflow. Fluid movement within these vessels occurs as a result of peristaltic waves of contraction (chapter 12, section 12.6). The smooth muscle within the lymph ducts contains a pacemaker that initiates action potentials associated with the entry of Ca^{2+}, which stimulates contraction. The activity of the pacemaker, and hence the peristaltic waves of contraction, are increased in response to stretch of the vessel. The lymph ducts eventually empty into one of two principal vessels: the *thoracic duct* or the *right lymphatic duct.* These ducts drain the lymph into the left and right subclavian veins, respectively. Thus interstitial fluid, which is formed by filtration of plasma out of blood capillaries (chapter 14, section 14.2), is ultimately returned to the cardiovascular system (fig. 13.37).

CLINICAL APPLICATION

Lymphedema is excessive protein and associated fluid in the interstitial tissue, caused by inadequate lymphatic drainage. At present there are no treatments for this condition, which can become progressively worse. Further, the protein-rich interstitial fluid can trigger an inflammatory reaction, leading to other degenerative changes in the surrounding connective tissues. In the tropical equatorial regions of the world, a parasitic infection of the lymphatic system by a species of nematode worms causes *elephantiasis*, a lymphedema that produces enormous swelling of the legs and scrotum (chapter 14; see fig. 14.10). Lymphedema can also be produced by radiation therapy and surgery. Breast cancer surgery, for example, is a leading cause of lymphedema.

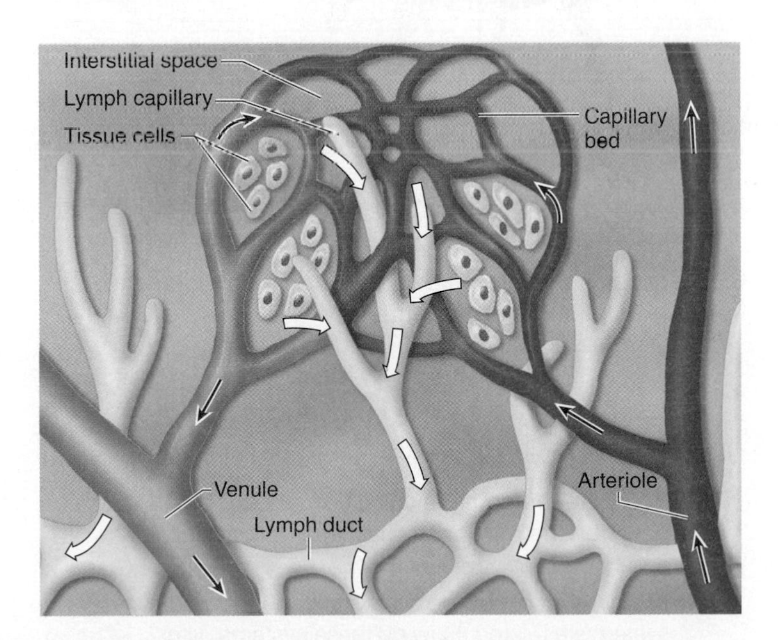

Figure 13.36 The relationship between blood capillaries and lymphatic capillaries. Notice that lymphatic capillaries are blind-ended. They are, however, highly permeable, so that excess fluid and protein within the interstitial space can drain into the lymphatic system.

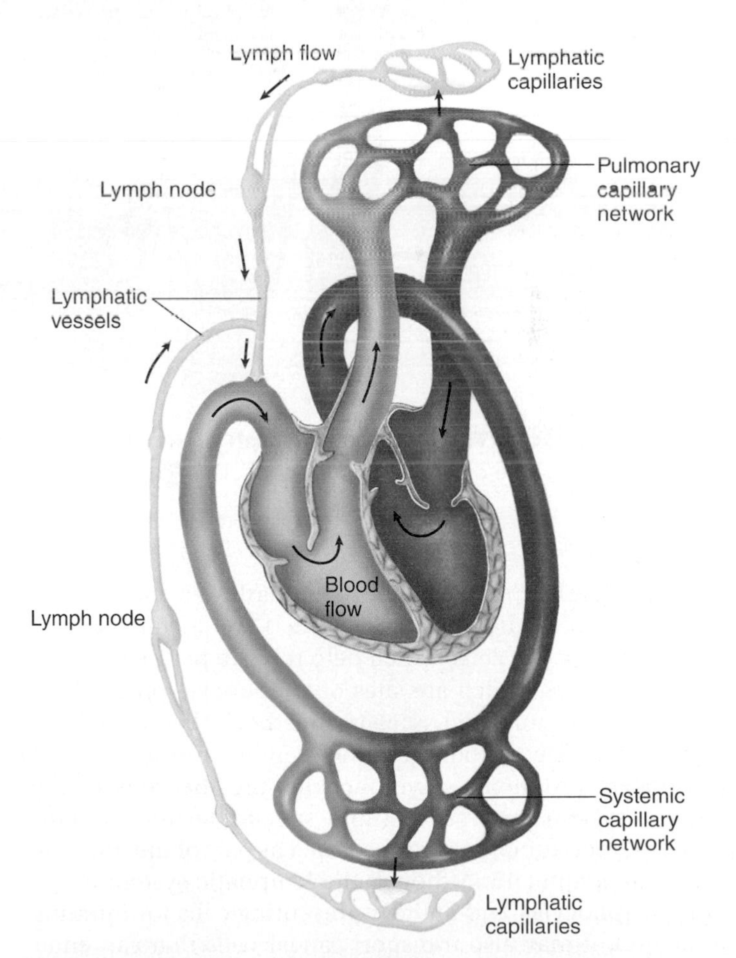

Figure 13.37 The relationship between the circulatory and lymphatic systems. This schematic illustrates that the lymphatic system transports fluid from the interstitial space back to the blood through a system of lymphatic vessels. Lymph is eventually returned to the vascular system at the subclavian veins.

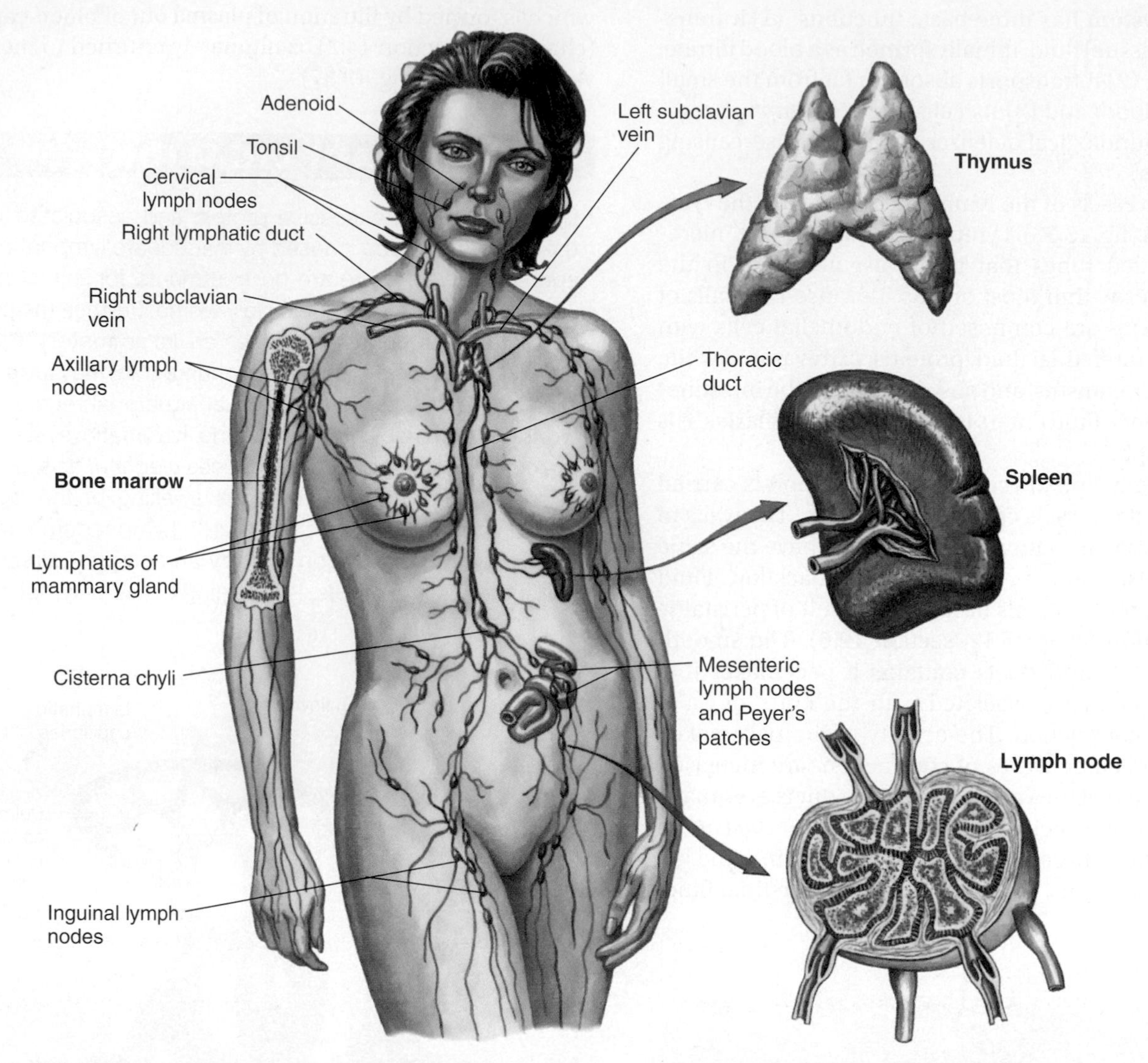

Figure 13.38 The location of lymph nodes along the lymphatic pathways. Lymph nodes are small bean-shaped bodies, enclosed within dense connective tissue capsules. AP|R

Before the lymph is returned to the cardiovascular system, it is filtered through **lymph nodes** (fig. 13.38). Lymph nodes contain phagocytic cells, which help remove pathogens, and *germinal centers,* which are sites of lymphocyte production. The tonsils, thymus, and spleen—together called **lymphoid organs**—likewise contain germinal centers and are sites of lymphocyte production. Lymphocytes are the cells of the immune system that respond in a specific fashion to antigens, and their functions are described as part of the immune system in chapter 15. Although the lymphatic system transports lymphocytes and antigen-presenting cells for immune protection, it may also transport cancer cells that can enter and later leave the porous lymphatic capillaries, thereby seeding distant organs. The lymphatic system can help cancer to spread, or *metastasize.* Metastasis to regional lymph nodes is the first step in the dissemination of tumors for cancers of the breast, colon, prostate, and others.

CHECKPOINT

17a. Compare the composition of lymph and blood, and describe the relationship between blood capillaries and lymphatic capillaries.

17b. Explain how the lymphatic system and the cardiovascular system are related. How do these systems differ?

18. Describe the functions of lymph nodes and lymphoid organs.

Clinical Investigation SUMMARY

Jason has anemia, and the reduced delivery of oxygen to his tissues probably contributed to his chronic fatigue. He also has a heart murmur due to the ventricular septal defect and mitral stenosis, which were probably congenital. These conditions could reduce the amount of blood pumped by the left ventricle through the systemic arteries and thus weaken his pulse. The reduced blood flow and consequent reduced oxygen delivery to the tissues could be the cause of his chronic fatigue. The lowered volume of blood pumped by the left ventricle could cause a reflex increase in the heart rate, as detected by his rapid pulse and the ECG tracing showing sinus tachycardia. Jason's high blood cholesterol is probably unrelated to his symptoms. This condition could be dangerous, however, as it increases his risk for atherosclerosis. Jason should therefore be placed on a special diet, and perhaps medication, to lower his blood cholesterol.

See the additional chapter 13 Clinical Investigations on *Mitral Valve Prolapse* and *AV Node Block* in the Connect site for this text at www.mhhe.com/Fox13.

SUMMARY

13.1 Functions and Components of the Circulatory System 405

A. The blood transports oxygen and nutrients to all the cells of the body and removes waste products from the tissues. It also serves a regulatory function through its transport of hormones.

1. Oxygen is carried by red blood cells, or erythrocytes.
2. White blood cells, or leukocytes, serve to protect the body from disease.

B. The circulatory system consists of the cardiovascular system (heart and blood vessels) and the lymphatic system.

13.2 Composition of the Blood 406

A. Plasma is the fluid part of the blood, containing dissolved ions and various organic molecules.

1. Hormones are found in the plasma portion of the blood.
2. The plasma proteins are albumins, globulins (alpha, beta, and gamma), and fibrinogen.

B. The formed elements of the blood are erythrocytes, leukocytes, and platelets.

1. Erythrocytes, or red blood cells, contain hemoglobin and transport oxygen.
2. Leukocytes may be granular (also called polymorphonuclear) or agranular. They function in immunity.
3. Platelets, or thrombocytes, are required for blood clotting.

C. The production of red blood cells is stimulated by the hormone erythropoietin, and the development of different kinds of white blood cells is controlled by chemicals called cytokines.

D. The major blood-typing groups are the ABO system and the Rh system.

1. Blood type refers to the kind of antigens found on the surface of red blood cells.
2. When different types of blood are mixed, antibodies against the red blood cell antigens cause the red blood cells to agglutinate.

E. When a blood vessel is damaged, platelets adhere to the exposed subendothelial collagen proteins.

1. Platelets that stick to collagen undergo a release reaction in which they secrete ADP, serotonin, and thromboxane A_2.
2. Serotonin and thromboxane A_2 cause vasoconstriction. ADP and thromboxane A_2 attract other platelets and cause them to stick to the growing mass of platelets that are stuck to the collagen in the broken vessel.

F. In the formation of a blood clot, a soluble protein called fibrinogen is converted into insoluble threads of fibrin.

1. This reaction is catalyzed by the enzyme thrombin.
2. Thrombin is derived from prothrombin, its inactive precursor, by either an intrinsic or an extrinsic pathway.
   a. The intrinsic pathway, the longer of the two, requires the activation of more clotting factors.

b. The shorter extrinsic pathway is initiated by the secretion of tissue thromboplastin.
3. The clotting sequence requires Ca^{2+} as a cofactor and phospholipids present in the platelet cell membranes.

G. Dissolution of the clot eventually occurs by the action of plasmin, which cleaves fibrin into split products.

13.3 Structure of the Heart 418

A. The right and left sides of the heart pump blood through the pulmonary and systemic circulations, respectively.
1. The right ventricle pumps blood to the lungs. This blood then returns to the left atrium.
2. The left ventricle pumps blood into the aorta and systemic arteries. This blood then returns to the right atrium.

B. The heart contains two pairs of one-way valves.
1. The atrioventricular valves allow blood to flow from the atria to the ventricles, but not in the reverse direction.
2. The semilunar valves allow blood to leave the ventricles and enter the pulmonary and systemic circulations, but they prevent blood from returning from the arteries to the ventricles.

C. Closing of the AV valves produces the first heart sound, or "lub," at systole. Closing of the semilunar valves produces the second heart sound, or "dub," at diastole. Abnormal valves can cause abnormal sounds called murmurs.

13.4 Cardiac Cycle 422

A. The heart is a two-step pump. The atria contract first, and then the ventricles.
1. During diastole, first the atria and then the ventricles fill with blood.
2. The ventricles are about 80% filled before the atria contract and add the final 20% to the end-diastolic volume.
3. Contraction of the ventricles ejects about two-thirds of their blood, leaving about one-third as the end-systolic volume.

B. When the ventricles contract at systole, the pressure within them first rises sufficiently to close the AV valves and then rises sufficiently to open the semilunar valves.
1. Blood is ejected from the ventricles until the pressure within them falls below the pressure in the arteries. At this point, the semilunar valves close and the ventricles begin relaxation.
2. When the pressure in the ventricles falls below the pressure in the atria, a phase of rapid filling of the ventricles occurs, followed by the final filling caused by contraction of the atria.

13.5 Electrical Activity of the Heart and the Electrocardiogram 425

A. In the normal heart, action potentials originate in the SA node as a result of spontaneous depolarization called the pacemaker potential.
1. When this spontaneous depolarization reaches a threshold value, opening of the voltage-regulated Na^+ gates and fast Ca^{2+} channels produces an action potential.
2. Repolarization is produced by the outward diffusion of K^+, but a stable resting membrane potential is not attained because spontaneous depolarization once again occurs.
3. Other myocardial cells are capable of spontaneous activity, but the SA node is the normal pacemaker because its rate of spontaneous depolarization is the fastest.
4. When the action potential produced by the SA node reaches other myocardial cells, they produce action potentials with a long plateau phase because of the slow inward diffusion of Ca^{2+}.
5. The long action potential and long refractory period of myocardial cells allows the entire mass of cells to be in a refractory period while it contracts. This prevents the myocardium from being stimulated again until after it relaxes.

B. The electrical impulse begins in the sinoatrial node and spreads through both atria by electrical conduction from one myocardial cell to another.
1. The impulse then excites the atrioventricular node, from which it is conducted by the bundle of His into the ventricles.
2. The Purkinje fibers transmit the impulse into the ventricular muscle and cause it to contract.

C. The regular pattern of conduction in the heart produces a changing pattern of potential differences between two points on the body surface.
1. The recording of this changing pattern caused by the heart's electrical activity is called an electrocardiogram (ECG).
2. The P wave is caused by depolarization of the atria; the QRS wave is caused by depolarization of the ventricles; and the T wave is produced by repolarization of the ventricles.

13.6 Blood Vessels 431

A. Arteries contain three layers, or tunics: the interna, media, and externa.
1. The tunica interna consists of a layer of endothelium, which is separated from the tunica media by a band of elastin fibers.
2. The tunica media consists of smooth muscle.
3. The tunica externa is the outermost layer.
4. Large arteries, containing many layers of elastin, can expand and recoil with rising and falling blood pressure. Medium and small arteries and arterioles are less distensible, and thus provide greater resistance to blood flow.

B. Capillaries are the narrowest but the most numerous of the blood vessels.
1. Capillary walls consist of just one layer of endothelial cells. They provide for the exchange of molecules between the blood and the surrounding tissues.

2. The flow of blood from arterioles to capillaries is regulated by precapillary sphincter muscles.
3. The capillary wall may be continuous, fenestrated, or discontinuous.

C. Veins have the same three tunics as arteries, but they generally have a thinner muscular layer than comparably sized arteries.
1. Veins are more distensible than arteries and can expand to hold a larger quantity of blood.
2. Many veins have venous valves that ensure a one-way flow of blood to the heart.
3. The flow of blood back to the heart is aided by contraction of the skeletal muscles that surround veins. The effect of this action is called the skeletal muscle pump.

13.7 Atherosclerosis and Cardiac Arrhythmias 436

A. Atherosclerosis of arteries can occlude blood flow to the heart and brain and is a causative factor in about 50% of all deaths in the United States, Europe, and Japan.
1. Atherosclerosis begins with injury to the endothelium, the movement of monocytes and lymphocytes into the tunica interna, and the conversion of monocytes into macrophages that engulf lipids. Smooth muscle cells then proliferate and secrete extracellular matrix.
2. Atherosclerosis is promoted by such risk factors as smoking, hypertension, and high plasma cholesterol concentration. Low-density lipoproteins (LDLs), which carry cholesterol into the artery wall, are oxidized by the endothelium and are a major contributor to atherosclerosis.

B. Occlusion of blood flow in the coronary arteries by atherosclerosis may produce ischemia of the heart muscle and angina pectoris, which may lead to myocardial infarction.

C. The ECG can be used to detect abnormal cardiac rates, abnormal conduction between the atria and ventricles, and other abnormal patterns of electrical conduction in the heart.

13.8 Lymphatic System 442

A. Lymphatic capillaries are blind-ended but highly permeable. They drain excess tissue fluid into lymph ducts.

B. Lymph passes through lymph nodes and is returned by way of the lymph ducts to the venous blood.

REVIEW ACTIVITIES

Test Your Knowledge

1. Which of these statements is *false?*
   a. Most of the total blood volume is contained in veins.
   b. Capillaries have a greater total surface area than any other type of vessel.
   c. Exchanges between blood and tissue fluid occur across the walls of venules.
   d. Small arteries and arterioles present great resistance to blood flow.
2. All arteries in the body contain oxygen-rich blood with the exception of
   a. the aorta.
   c. the renal artery.
   b. the pulmonary artery.
   d. the coronary arteries.
3. The "lub," or first heart sound, is produced by closing of
   a. the aortic semilunar valve.
   b. the pulmonary semilunar valve.
   c. the tricuspid valve.
   d. the bicuspid valve.
   e. both AV valves.
4. The first heart sound is produced at
   a. the beginning of systole.
   b. the end of systole.
   c. the beginning of diastole.
   d. the end of diastole.
5. Changes in the cardiac rate primarily reflect changes in the duration of
   a. systole.
   b. diastole.
6. The QRS wave of an ECG is produced by
   a. depolarization of the atria.
   b. repolarization of the atria.
   c. depolarization of the ventricles.
   d. repolarization of the ventricles.
7. The second heart sound immediately follows the occurrence of
   a. the P wave.
   b. the QRS wave.
   c. the T wave.
8. The cells that normally have the fastest rate of spontaneous diastolic depolarization are located in
   a. the SA node.
   c. the bundle of His.
   b. the AV node.
   d. the Purkinje fibers.
9. Which of these statements is *true?*
   a. The heart can produce a graded contraction.
   b. The heart can produce a sustained contraction.
   c. The action potentials produced at each cardiac cycle normally travel around the heart in circus rhythms.
   d. All of the myocardial cells in the ventricles are normally in a refractory period at the same time.

10. An ischemic injury to the heart that destroys myocardial cells is
    a. angina pectoris.
    b. a myocardial infarction.
    c. fibrillation.
    d. heart block.
11. The activation of factor X occurs in
    a. the intrinsic pathway only.
    b. the extrinsic pathway only.
    c. both the intrinsic and extrinsic pathways.
    d. neither the intrinsic nor extrinsic pathway.
12. Platelets
    a. form a plug by sticking to each other.
    b. release chemicals that stimulate vasoconstriction.
    c. provide phospholipids needed for the intrinsic pathway.
    d. serve all of these functions.
13. Antibodies against both type A and type B antigens are found in the plasma of a person who is
    a. type A.
    b. type B.
    c. type AB.
    d. type O.
    e. any of these types.
14. Production of which of the following blood cells is stimulated by a hormone secreted by the kidneys?
    a. Lymphocytes
    b. Monocytes
    c. Erythrocytes
    d. Neutrophils
    e. Thrombocytes
15. Which of these statements about plasmin is *true?*
    a. It is involved in the intrinsic clotting system.
    b. It is involved in the extrinsic clotting system.
    c. It functions in fibrinolysis.
    d. It promotes the formation of emboli.
16. During the phase of isovolumetric relaxation of the ventricles, the pressure in the ventricles is
    a. rising.
    b. falling.
    c. first rising, then falling.
    d. constant.
17. Peristaltic waves of contraction move fluid within which of these vessels?
    a. Arteries
    b. Veins
    c. Capillaries
    d. Lymphatic vessels
    e. All of these

Test Your Understanding

18. Describe how the pacemaker cells produce a spontaneous diastolic depolarization, and how this leads to the production of action potentials.
19. What characteristic of the SA node distinguishes it from other possible pacemaker regions and allows it to function as the normal pacemaker? How do action potentials from the SA node reach the atria and the ventricles?
20. Compare the duration of the heart's contraction with the myocardial action potential and refractory period. Explain the significance of these relationships.
21. Step by step, describe the pressure changes that occur in the ventricles during the cardiac cycle. Explain how these pressure changes relate to the occurrence of the heart sounds.
22. Can a defective valve be detected by an ECG? Can a partially damaged AV node be detected by auscultation (listening) with a stethoscope? Explain.
23. Describe the causes of the P, QRS, and T waves of an ECG, and indicate at which point in the cardiac cycle each of these waves occurs. Explain why the first heart sound occurs immediately after the QRS wave and why the second sound occurs at the time of the T wave.
24. The lungs are the only organs that receive the entire output of a ventricle. Explain this statement, and describe how this relates to the differences in structure and function between the right and left ventricles.
25. Explain the process of Ca^{2+}-induced Ca^{2+} release in the myocardium. How does this process differ from excitation-contraction coupling in skeletal muscles?
26. Explain how a cut in the skin initiates both the intrinsic and the extrinsic clotting pathways. Which pathway is shorter? Why?
27. Explain how aspirin, coumarin drugs, EDTA, and heparin function as anticoagulants. Which of these are effective when added to a test tube? Which are not? Why?
28. Explain how blood moves through arteries, capillaries, and veins. How does exercise affect this movement?
29. Explain the processes involved in the development of atherosclerosis. How might antioxidants help retard the progression of this disease? How might exercise help? What other changes in lifestyle might help prevent or reduce atherosclerotic plaques?

Test Your Analytical Ability

30. Hematopoietic stem cells account for less than 1% of the cells in the bone marrow. These cells can be separated from the others prior to bone marrow transplantation, but it is better to first inject the donor with recombinant cytokines. Identify the cytokines that might be used and describe their effects.
31. A patient has a low red blood cell count, and microscopic examination of his blood reveals an abnormally high proportion of circulating reticulocytes. Upon subsequent

examination, the patient is diagnosed with a bleeding ulcer. This is surgically corrected, and in due course his blood measurements return to normal. What was the reason for the low red blood cell count and high proportion of reticulocytes?

32. A chemical called EDTA, like citrate, binds to (or "chelates") Ca^{2+}. Suppose a person had EDTA infused into their blood. What effect would this have on the intrinsic and extrinsic clotting pathways? How would these effects differ from the effects of aspirin on blood clotting?
33. During the course of a physiology laboratory, a student finds that her PR interval is 0.24 second. Concerned, she takes her own ECG again an hour later and sees an area of the ECG strip where the PR interval becomes longer and longer. Performing an ECG measurement on herself for a third time, she sees an area of the strip where a P wave is not followed by a QRS or T; farther along in the strip, however, a normal pattern reappears. What do you think these recordings indicate?
34. A newborn baby with a patent foramen ovale or a ventricular septal defect might be cyanotic (blue). Will a two-year-old with these defects also be cyanotic? Explain your answer.
35. People with paroxysmal atrial tachycardia (commonly called "palpitations") are sometimes given drugs that block voltage-gated Ca^{2+} channels in the plasma membrane of myocardial cells in order to slow the beat. By what mechanism could these drugs help?
36. The mechanism of excitation-contraction coupling in cardiac muscle differs from that in skeletal muscle. How might these differences relate to the differences in the action potentials in cardiac muscle compared to skeletal muscle?
37. Explain how homeostasis of the circulating blood platelet count is maintained. By what mechanism would blood loss increase the platelet count?

Test Your Quantitative Ability

Refer to figure 13.17 to answer the following questions:

38. At which pressure value did blood just start to leave the left ventricle and enter the aorta?
39. How much blood was ejected by the left ventricle by the time the second heart sound was produced?
40. How much blood was in the left ventricle just before the atrial contraction?
41. How much blood was added to the left ventricle by contraction of the left atrium?
42. What is the change in intraventricular pressure between the time the first heart sound begins and the time it ends?

CHAPTER

14

Cardiac Output, Blood Flow, and Blood Pressure

REFRESH YOUR MEMORY

Before you begin this chapter, you may want to review these concepts from previous chapters:

CHAPTER OUTLINE

Clinical Investigation

Charlie, a biology student who became separated from his group on a field trip, was lost in the desert for almost two days before being found. When examined, he displayed a weak, rapid pulse, low blood pressure, and cold skin. He was hospitalized and nurses reported that he had a low output of urine. Charlie was given intravenous fluid containing albumin and soon recovered.

Some of the new terms and concepts you will encounter include:

- Baroreceptor reflex, cardiac rate, stroke volume, and cardiac output
- Colloid osmotic pressure and Starling forces; blood flow regulation
- Blood volume, ADH, and the renin-angiotensin-aldosterone system

14.1 CARDIAC OUTPUT

The pumping ability of the heart is a function of the beats per minute (cardiac rate) and the volume of blood ejected per beat (stroke volume). The cardiac rate and stroke volume are regulated by autonomic nerves and by mechanisms intrinsic to the cardiovascular system.

LEARNING OUTCOMES

After studying this section, you should be able to:

1. **Describe the extrinsic regulation of cardiac rate and contractility.**
2. **Explain the relationship between stroke volume and venous return.**
3. **Explain the Frank-Starling law of the heart.**

The **cardiac output** is the volume of blood pumped per minute by each ventricle. The average resting **cardiac rate** in an adult is 70 beats per minute; the average **stroke volume** (volume of blood pumped per beat by each ventricle) is 70 to 80 ml per beat. The product of these two variables gives an average cardiac output of 5,500 ml (5.5 L) per minute:

$$\underset{(\text{ml/min})}{\text{Cardiac output}} = \underset{(\text{ml/beat})}{\text{Stroke volume}} \times \underset{(\text{beats/min})}{\text{cardiac rate}}$$

The **total blood volume** also averages about 5.5 L. This means that each ventricle pumps the equivalent of the total blood volume each minute under resting conditions. Put another way, it takes about a minute for a drop of blood to complete the systemic and pulmonary circuits. An increase in cardiac output, as occurs during exercise, must thus be accompanied by an increased rate of blood flow through the circulation. This is accomplished by factors that regulate the cardiac rate and stroke volume.

Regulation of Cardiac Rate

In the complete absence of neural influences, the heart will continue to beat as long as the myocardial cells are alive. This automatic rhythm is a result of a spontaneous, diastolic depolarization of the pacemaker cells in the SA node. The pacemaker membrane channels are known as *HCN channels,* and these are opened in response to the hyperpolarization that occurs at the end of the preceding action potential (chapter 13, section 13.5). When the HCN channels of pacemaker cells open, they allow an inward diffusion of Na^+. This produces the depolarization stimulus for the next action potential.

Normally, however, sympathetic and parasympathetic (vagus) nerve fibers to the heart are continuously active to a greater or lesser degree. Norepinephrine from sympathetic axons and epinephrine from the adrenal medulla also open the HCN channels of the pacemaker cells, inducing a faster rate of diastolic depolarization. This causes action potentials to be produced more rapidly, resulting in a faster cardiac rate (fig. 14.1). Epinephrine and norepinephrine have this effect because (through their activation of β_1-adrenergic receptors)

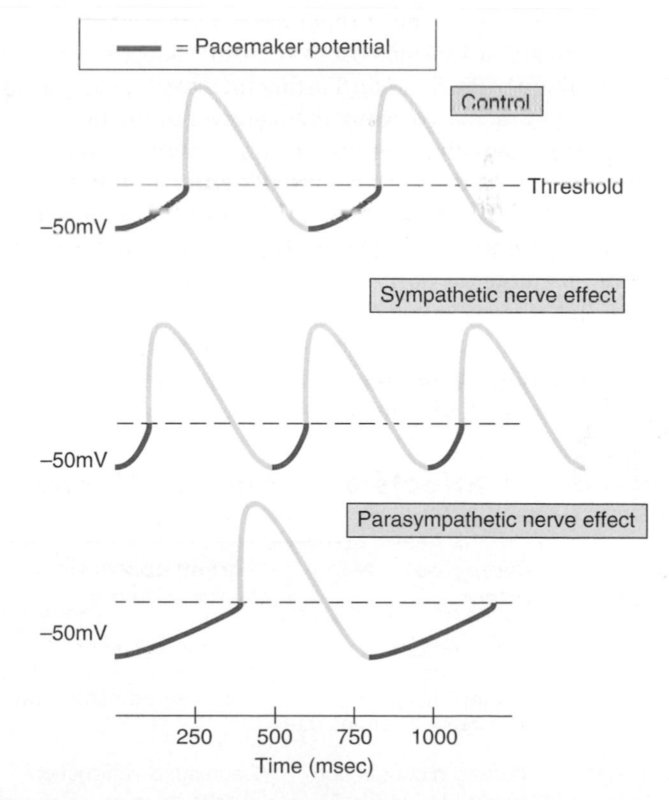

Figure 14.1 The effect of autonomic nerves on the pacemaker potentials in the SA node. The heart's rhythm is set by the rate of spontaneous depolarization in the SA node. This spontaneous depolarization is known as the pacemaker potential, and its rate is increased by sympathetic nerve stimulation and decreased by parasympathetic nerve inhibition. AP|R

they stimulate the production of cAMP in the pacemaker cells (see fig. 7.31), and the cAMP acts directly on the pacemaker HCN channels to keep them open.

Acetylcholine, released by vagus nerve endings, binds to muscarinic ACh receptors and causes the opening of separate K^+ channels in the membrane (see chapter 7, fig. 7.27; also see chapter 9, fig. 9.11). The outward diffusion of K^+ partially counters the inward diffusion of Na^+ through the HCN channels, producing a slower rate of diastolic depolarization. As a result, there is a slower rate of action potential production and thus a slower cardiac rate (fig. 14.1).

The actual pace set by the SA node at any time depends on the net effect of these antagonistic influences (see fig. 14.5). Mechanisms that affect the cardiac rate are said to have a **chronotropic effect** (*chrono* = time). Those that increase cardiac rate have a *positive chronotropic effect;* those that decrease the rate have a *negative chronotropic effect.*

Autonomic innervation of the SA node is the major means by which cardiac rate is regulated. However, other autonomic control mechanisms also affect cardiac rate to a lesser degree. Sympathetic endings in the musculature of the atria and ventricles increase the strength of contraction and cause a slight decrease in the time spent in systole when the cardiac rate is high (table 14.1).

The resting bradycardia (slow heart rate) of endurance-trained athletes is largely due to high vagus nerve activity. During exercise, the cardiac rate increases as a result of decreased vagus nerve inhibition of the SA node. Further increases in cardiac rate are achieved by increased sympathetic nerve stimulation.

The activity of the autonomic innervation of the heart is coordinated by the **cardiac control center** in the medulla oblongata of the brain stem. The cardiac control center, in turn, is affected by higher brain areas and by sensory feedback from pressure receptors, or *baroreceptors,* in the aorta and carotid arteries. In this way, a fall in blood pressure can produce a reflex increase in the heart rate (chapter 1; see fig. 1.6). This baroreceptor reflex is discussed in more detail in relation to blood pressure regulation in section 14.6.

Table 14.1 | Effects of Autonomic Nerve Activity on the Heart

Region Affected	Sympathetic Nerve Effects	Parasympathetic Nerve Effects
SA node	Increased rate of diastolic depolarization; increased cardiac rate	Decreased rate of diastolic depolarization; decreased cardiac rate
AV node	Increased conduction rate	Decreased conduction rate
Atrial muscle	Increased strength of contraction	No significant effect
Ventricular muscle	Increased strength of contraction	No significant effect

Clinical Investigation CLUES

Charlie had a weak, rapid pulse.

- What mechanism was responsible for the rapid pulse?

Regulation of Stroke Volume

The stroke volume is regulated by three variables:

1. the **end-diastolic volume (EDV),** which is the volume of blood in the ventricles at the end of diastole;
2. the **total peripheral resistance,** which is the frictional resistance, or impedance to blood flow, in the arteries; and
3. the **contractility,** or strength, of ventricular contraction.

The end-diastolic volume is the amount of blood in the ventricles immediately before they begin to contract. This is a workload imposed on the ventricles prior to contraction, and thus is sometimes called a **preload.** The stroke volume is directly proportional to the preload; an increase in EDV results in an increase in stroke volume. (This relationship is known as the *Frank-Starling law of the heart,* discussed shortly.) The stroke volume is also directly proportional to contractility; when the ventricles contract more forcefully, they pump more blood.

In order to eject blood, the pressure generated in a ventricle when it contracts must be greater than the pressure in the arteries (because blood flows only from higher pressure to lower pressure). The pressure in the arterial system before the ventricle contracts is, in turn, a function of the total peripheral resistance—the higher the peripheral resistance, the higher the pressure. As blood begins to be ejected from the ventricle, the added volume of blood in the arteries causes a rise in mean arterial pressure against the "bottleneck" presented by the peripheral resistance. Ejection of blood stops shortly after the aortic pressure becomes equal to the intraventricular pressure. The total peripheral resistance thus presents an impedance to the ejection of blood from the ventricle, or an **afterload** imposed on the ventricle after contraction has begun. This can be medically significant; a person with a high total peripheral resistance has a high arterial blood pressure, and thus a high afterload imposed on the ventricular muscle.

In summary, the stroke volume is inversely proportional to the total peripheral resistance; the greater the peripheral resistance, the lower the stroke volume. It should be noted that this lowering of stroke volume in response to a raised peripheral resistance occurs for only a few beats. Thereafter, a healthy heart is able to compensate for the increased peripheral resistance by beating more strongly. This compensation occurs by means of a mechanism described in the next section (Frank-Starling law of the heart). An inability of the heart to compensate can lead to congestive heart failure.

The proportion of the end-diastolic volume that is ejected against a given afterload depends on the strength of ventricular contraction. Normally, contraction strength is sufficient to eject 70 to 80 ml of blood out of a total end-diastolic volume of 110 to 130 ml. The **ejection fraction** is thus about 60%. The ejection fraction remains relatively constant over a range of end-diastolic volumes, so that the amount ejected per beat (stroke volume) increases as the end-diastolic volume increases. In order for this to be true, the strength of ventricular contraction must increase as the end-diastolic volume increases.

Frank-Starling Law of the Heart

Two physiologists, Otto Frank and Ernest Starling, demonstrated that the strength of ventricular contraction varies directly with the end-diastolic volume (fig. 14.2). Even in experiments where the heart is removed from the body (and is thus not subject to neural or hormonal regulation) and where the still-beating heart is filled with blood flowing from a reservoir, an increase in EDV within the physiological range results in increased contraction strength and, therefore, in increased stroke volume. This relationship between EDV, contraction strength, and stroke volume is thus a built-in, or *intrinsic,* property of heart muscle, and is known as the **Frank-Starling law of the heart.**

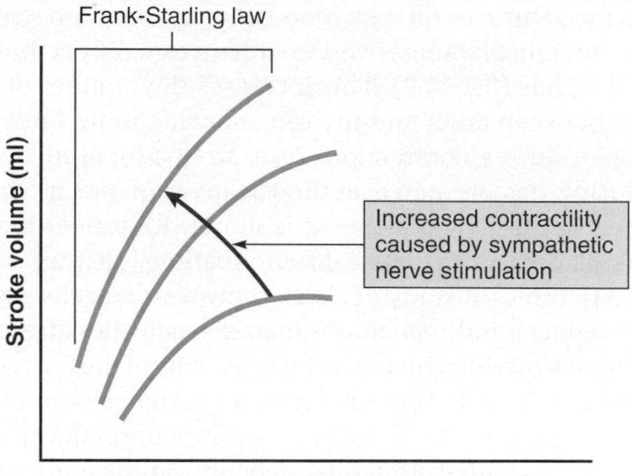

Figure 14.2 The Frank-Starling law and sympathetic nerve effects. The graphs demonstrate the Frank-Starling law: As the end-diastolic volume is increased, the stroke volume is increased. The graphs also demonstrate, by comparing the three curves, that the stroke volume is higher at any given end-diastolic volume when the ventricle is stimulated by sympathetic nerves. This is shown by the steeper curves to the left (see the red arrow).

Intrinsic Control of Contraction Strength

The intrinsic control of contraction strength and stroke volume is due to variations in the degree to which the myocardium is stretched by the end-diastolic volume. As the EDV rises within the physiological range, the myocardium is increasingly stretched and, as a result, contracts more forcefully.

As discussed in chapter 12, stretch can also increase the contraction strength of skeletal muscles (see fig. 12.21). The resting length of skeletal muscles, however, is close to ideal, so that significant stretching decreases contraction strength. This is not true of the heart. Prior to filling with blood during diastole, the sarcomere lengths of myocardial cells are only about 1.5 μm. At this length, the actin filaments from each side overlap in the middle of the sarcomeres, and the cells can contract only weakly (fig. 14.3).

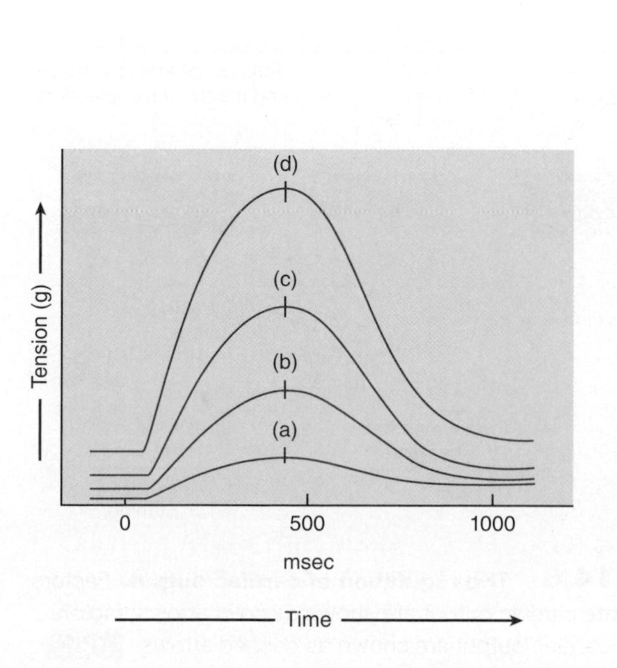

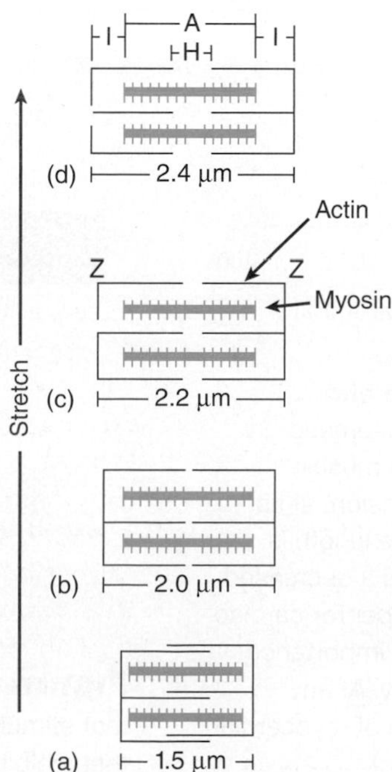

Figure 14.3 The Frank-Starling law of the heart. When the heart muscle is subjected to an increasing degree of stretch (*a* through *d*), it contracts more forcefully. The contraction strength is indicated on the *y*-axis as the tension. Notice that the time required to reach maximum contraction remains constant, regardless of the degree of stretch.

As the ventricles fill with blood, the myocardium stretches so that the actin filaments overlap with myosin only at the edges of the A bands (fig. 14.3). This increases the number of interactions between actin and myosin, allowing more force to be developed during contraction. Also, stretching of myocardial cells during diastole increases the sensitivity of the Ca^{2+}-release channels in the sarcoplasmic reticulum (SR), promoting their release of Ca^{2+} in response to stimulation (chapter 12; see fig. 12.34). A recent study of isolated myocardial cells suggests that this effect is indirect; diastolic stretch activates an enzyme, the products of which increase the sensitivity of the Ca^{2+} release channels in the SR. This results in a greater release of Ca^{2+}, which contributes to a stronger contraction. Because the degree of myocardial stretching depends on the end-diastolic volume, these mechanisms assure that an increase in end-diastolic volume intrinsically produces an increase in contraction strength and stroke volume.

As shown in figure 14.4, muscle length has a more pronounced effect on contraction strength in cardiac muscle than in skeletal muscle. That is, a particular increase in sarcomere length will stimulate contraction strength more in cardiac muscle than in skeletal muscle. This is believed to be due to an increased sensitivity of stretched cardiac muscle to the stimulatory effects of Ca^{2+}.

The Frank-Starling law explains how the heart can adjust to a rise in total peripheral resistance: (1) a rise in peripheral resistance causes a decrease in the stroke volume of the ventricle, so that (2) more blood remains in the ventricle and the end-diastolic volume is greater for the next cycle; as a result, (3) the ventricle is stretched to a greater degree in the next cycle and contracts more strongly to eject more blood. This allows a healthy ventricle to sustain a normal cardiac output when there are changes in the total peripheral resistance.

A very important consequence of these events is that the cardiac output of the left ventricle, which pumps blood into the systemic circulation with its ever-changing resistances, can be adjusted to match the output of the right ventricle (which pumps blood into the pulmonary circulation). The rate of blood flow through the pulmonary and systemic circulations must be equal to prevent fluid accumulation in the lungs and to deliver fully oxygenated blood to the body.

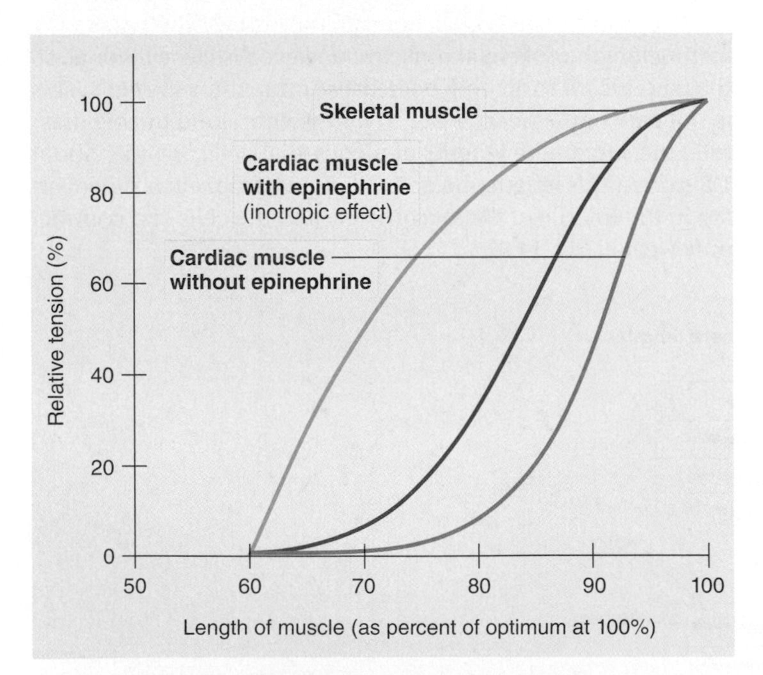

Figure 14.4 The effect of muscle length and epinephrine on contraction strength. In this schematic comparison, all three curves demonstrate that each muscle contracts with its maximum force (100% relative tension) at its own optimum length (100% optimum length). As the length is decreased from optimum, each curve demonstrates a decreased contraction strength. Notice that the decline is steeper for cardiac muscle than for skeletal muscle, demonstrating the importance of the Frank-Starling relationship in heart physiology. At any length, however, epinephrine increases the strength of myocardial contraction, demonstrating a positive inotropic effect.

Extrinsic Control of Contractility

Contractility is the strength of contraction at any given fiber length. At any given degree of stretch, the strength of ventricular contraction depends on the activity of the sympathoadrenal system. Norepinephrine from sympathetic nerve endings and epinephrine from the adrenal medulla produce an increase in contraction strength (see figs. 14.2 and 14.4). This **positive inotropic effect** results from an increase in the amount of Ca^{2+} available to the sarcomeres.

The cardiac output is thus affected in two ways by the activity of the sympathoadrenal system: (1) through a positive inotropic effect on contractility and (2) through a positive chronotropic effect on cardiac rate (fig. 14.5). Stimulation through parasympathetic nerve endings to the SA node and conducting tissue has

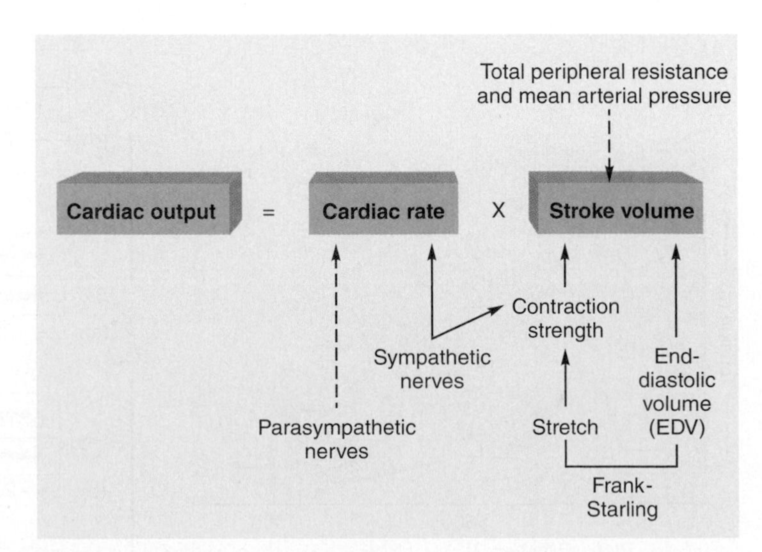

Figure 14.5 The regulation of cardiac output. Factors that stimulate cardiac output are shown as solid arrows; factors that inhibit cardiac output are shown as dashed arrows. AP|R

a negative chronotropic effect but does not directly affect the contraction strength of the ventricles. However, the increased EDV that results from a slower cardiac rate can increase contraction strength through the mechanism described by the Frank-Starling law of the heart. This increases stroke volume, but not enough to completely compensate for the slower cardiac rate. Thus, the cardiac output is decreased when the heart beats slower, a fact used by people who treat their hypertension with beta-adrenergic blocking drugs that slow the cardiac rate.

Venous Return

The end-diastolic volume—and thus the stroke volume and cardiac output—is controlled by factors that affect the **venous return,** which is the return of blood to the heart via veins. The rate at which the atria and ventricles are filled with venous blood depends on the total blood volume and the venous pressure (pressure in the veins). It is the venous pressure that serves as the driving force for the return of blood to the heart.

Veins have thinner, less muscular walls than do arteries; thus, they have a higher **compliance.** This means that a given amount of pressure will cause more distension (expansion) in veins than in arteries, so that the veins can hold more blood. Approximately two-thirds of the total blood volume is located in the veins (fig. 14.6). Veins are therefore called **capacitance vessels,** after electronic devices called capacitors that store electrical charges. Muscular arteries and arterioles expand less under pressure (are less compliant), and thus are called *resistance vessels.*

Although veins contain almost 70% of the total blood volume, the mean venous pressure is only 2 mmHg, compared to a mean arterial pressure of 90 to 100 mmHg. The lower venous pressure is due in part to a pressure drop between arteries and capillaries and in part to the high venous compliance.

The venous pressure is highest in the venules (10 mmHg) and lowest at the junction of the venae cavae with the right atrium (0 mmHg). This produces a pressure difference that promotes the return of blood to the heart. In addition, the venous return is aided by (1) sympathetic nerve activity, which stimulates smooth muscle contraction in the venous walls and thereby reduces compliance; (2) the skeletal muscle pump, which squeezes veins during muscle contraction; and (3) the pressure difference between the thoracic and abdominal cavities, which promotes the flow of venous blood back to the heart.

Contraction of the skeletal muscles functions as a "pump" by virtue of its squeezing action on veins (chapter 13; see fig. 13.29). Contraction of the diaphragm during inhalation also improves venous return. The diaphragm lowers as it contracts, increasing the thoracic volume and decreasing the abdominal volume. This creates a partial vacuum in the thoracic cavity and a higher pressure in the abdominal cavity. The pressure difference thus produced favors blood flow from abdominal to thoracic veins (fig. 14.7).

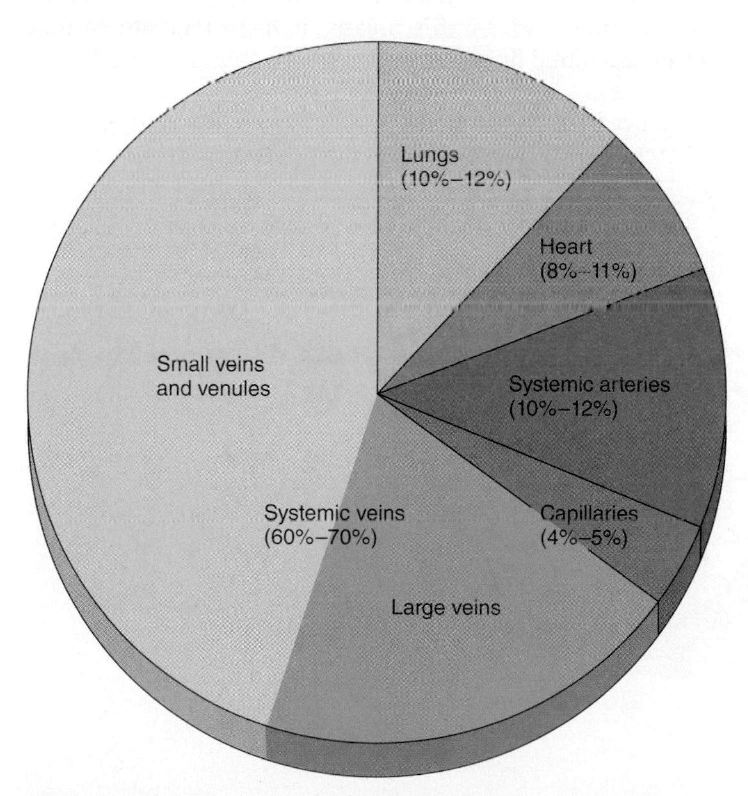

Figure 14.6 The distribution of blood within the circulatory system at rest. Notice that the venous system contains most of the blood; it functions as a reservoir from which more blood can be added to the circulation under appropriate conditions (such as exercise). APIR

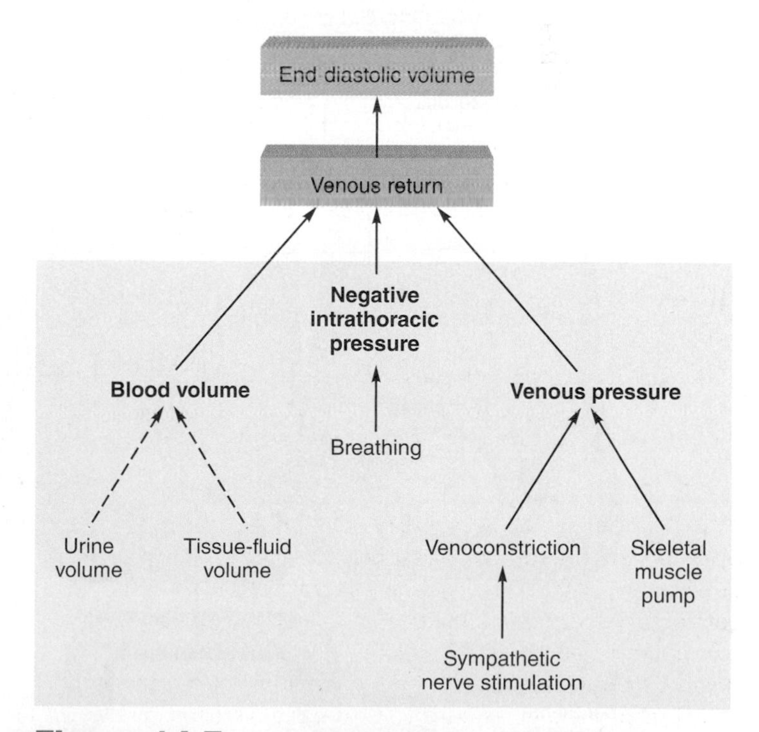

Figure 14.7 Variables that affect venous return and thus end-diastolic volume. Direct relationships are indicated by solid arrows; inverse relationships are shown with dashed arrows. APIR

CHECKPOINT

1. Describe the effects of sympathoadrenal and parasympathetic nerve activity on the cardiac rate and stroke volume.

2a. Describe the factors that regulate the venous return, and explain how the venous return is related to the end-diastolic volume and the stroke volume.

2b. Describe how the stroke volume is intrinsically regulated by the end-diastolic volume and explain the significance of this relationship.

2c. Define the terms *preload* and *afterload* and explain how they affect the cardiac output.

3. Use the Frank-Starling law of the heart to explain how an increase in venous return can result in an increase in stroke volume and cardiac output.

14.2 BLOOD VOLUME

Fluid in the extracellular environment of the body is distributed between the blood and the interstitial fluid compartments by forces acting across the walls of capillaries. The kidneys influence blood volume because urine is derived from blood plasma, and the hormones ADH and aldosterone act on the kidneys to help regulate the blood volume.

LEARNING OUTCOMES

After studying this section, you should be able to:

4. Explain the forces that act in capillaries and how edema can be produced.

5. Explain how the kidneys regulate blood volume, and the hormonal regulation of this process.

Blood volume represents one part, or compartment, of the total body water. Approximately two-thirds of the total body water is contained within cells—in the intracellular compartment. The remaining one-third is in the **extracellular compartment.** This extracellular fluid is normally distributed so that about 80% is contained in the tissues—as *tissue,* or *interstitial, fluid*—with the blood plasma accounting for the remaining 20% (fig. 14.8).

The distribution of water between the interstitial fluid and the blood plasma is determined by a balance between opposing forces acting at the capillaries. Blood pressure, for example, promotes the formation of interstitial fluid from plasma, whereas osmotic forces draw water from the tissues into the vascular system. The total volume of intracellular and extracellular fluid is normally maintained constant by a balance between water loss and water gain. Mechanisms that affect drinking, urine volume, and the distribution of water between plasma and interstitial fluid thus help to regulate blood volume and, by this means, help to regulate cardiac output and blood flow.

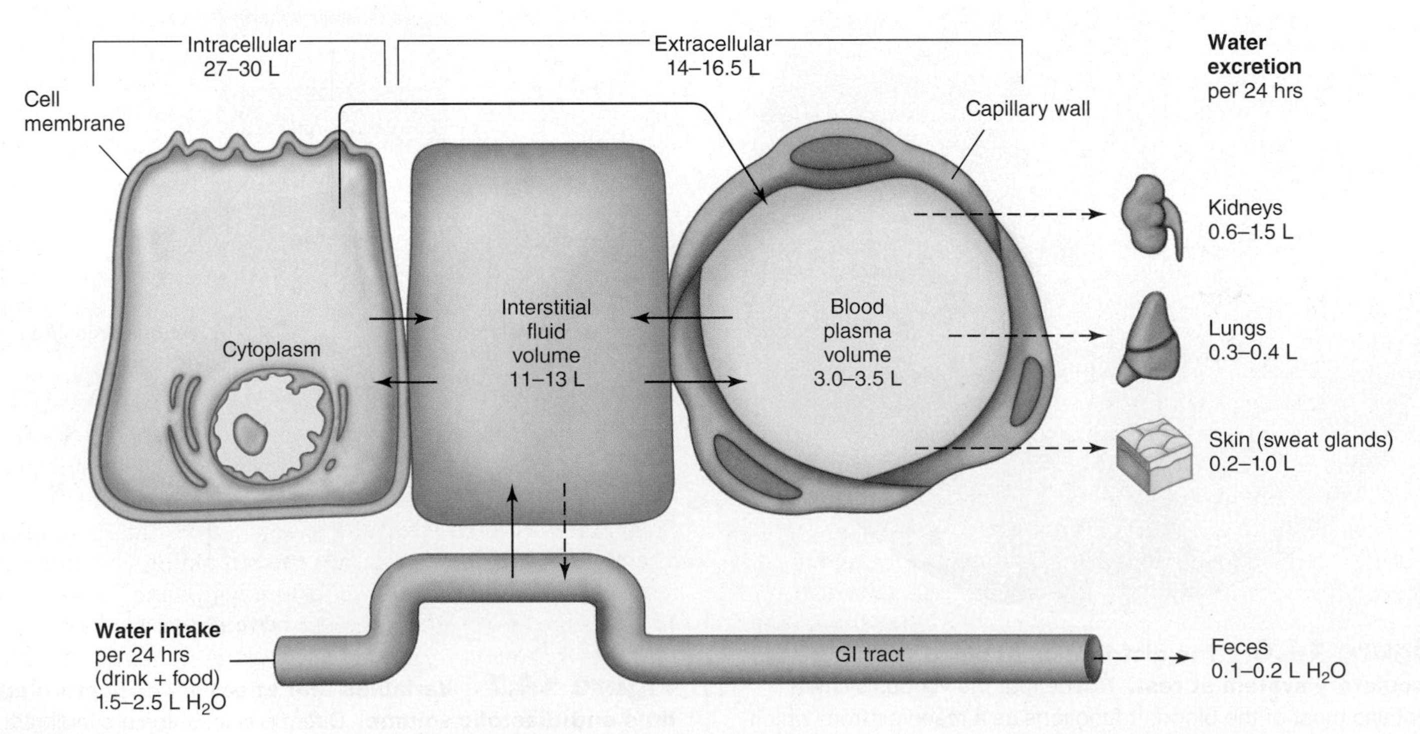

Figure 14.8 **The distribution of body water between the intracellular and extracellular compartments.** The extracellular compartment includes the blood plasma and the interstitial (tissue) fluid.

Exchange of Fluid Between Capillaries and Tissues

The distribution of extracellular fluid between the plasma and interstitial compartments is in a state of dynamic equilibrium. Tissue fluid is not normally a "stagnant pond;" rather, it is a continuously circulating medium, formed from and returning to the vascular system. In this way, the tissue cells receive a continuously fresh supply of glucose and other plasma solutes that are filtered through tiny endothelial channels in the capillary walls.

Filtration results from blood pressure within the capillaries. This hydrostatic pressure, which is exerted against the inner capillary wall, is equal to about 37 mmHg at the arteriolar end of systemic capillaries and drops to about 17 mmHg at the venular end of the capillaries. The **net filtration pressure** is equal to the hydrostatic pressure of the blood in the capillaries minus the hydrostatic pressure of tissue fluid outside the capillaries, which opposes filtration. If, as an extreme example, these two values were equal, there would be no filtration. The magnitude of the tissue hydrostatic pressure varies from organ to organ. With a hydrostatic pressure in the interstitial fluid of 1 mmHg, as it is outside the capillaries of skeletal muscles, the net filtration pressure would be 37 − 1 = 36 mmHg at the arteriolar end of the capillary and 17 − 1 = 16 mmHg at the venular end.

Glucose, comparably sized organic molecules, inorganic salts, and ions are filtered along with water through the capillary pores. The concentrations of these substances in interstitial (tissue) fluid are thus the same as in plasma. The protein concentration of interstitial fluid (2 g/100 ml), however, is less than the protein concentration of plasma (6 to 8 g/100 ml). This difference is due to the restricted filtration of proteins through the capillary pores. The osmotic pressure exerted by plasma proteins—called the **colloid osmotic pressure** of the plasma (because proteins are present as a colloidal suspension)—is therefore much greater than the colloid osmotic pressure of interstitial fluid. The difference between these two osmotic pressures is called the **oncotic pressure.** The colloid osmotic pressure of the interstitial fluid is sufficiently low to be neglected, so the oncotic pressure is essentially equal to the colloid osmotic pressure of the plasma. This value has been estimated to be 25 mmHg. Because water will move by osmosis from the solution of lower to the solution of higher osmotic pressure (chapter 6), this oncotic pressure favors the movement of water into the capillaries.

Whether fluid will move out of or into the capillary depends on the magnitude of the net filtration pressure, which varies from the arteriolar to the venular end of the capillary, and on the oncotic pressure. These opposing forces that affect the distribution of fluid across the capillary are known as **Starling forces,** and their effects can be calculated according to this relationship:

Fluid movement is proportional to:

$$\underbrace{(P_c + \pi_i)}_{\text{(Fluid out)}} - \underbrace{(P_i + \pi_p)}_{\text{(Fluid in)}}$$

where

P_c = hydrostatic pressure in the capillary
π_i = colloid osmotic pressure of the interstitial (tissue) fluid
P_i = hydrostatic pressure of interstitial fluid
π_p = colloid osmotic pressure of the blood plasma

The expression to the left of the minus sign represents the sum of forces acting to move fluid out of the capillary. The expression to the right represents the sum of forces acting to move fluid into the capillary. Figure 14.9 provides typical values for blood capillaries in skeletal muscles. Notice that the sum of the forces acting on the capillary is a positive number at the arteriolar end and a negative number at the venular end of the capillary. Examination of figure 14.9 reveals that this change is caused by the decrease in hydrostatic pressure (blood pressure) within the capillary as blood travels from the arteriolar to the venular end. The positive value at the arteriolar end indicates that the Starling forces that favor the filtration of fluid out of the capillary predominate. The negative value at the venular end indicates that the net Starling forces favor the return of fluid to the capillary. Fluid thus leaves the capillaries at the arteriolar end and returns to the capillaries at the venular end (fig. 14.9, *top*).

Clinical Investigation CLUES

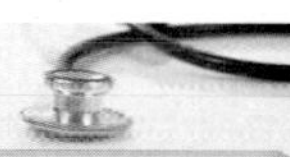

Charlie was given intravenous fluid containing albumin.

- What is the function of the albumin?
- Why was albumin used instead of dextrose (glucose)?

This "classic" view of capillary dynamics has been modified in recent years by the realization that the balance of filtration and reabsorption varies in different tissues and under different conditions in a particular capillary. For example, a capillary may be open or closed off by precapillary muscles that function as sphincters. When the capillary is open, blood flow is high and the net filtration force exceeds the force for the osmotic return of water throughout the length of the capillary. The opposite is true if the precapillary sphincter closes and the blood flow through the capillary is reduced.

Through the action of the Starling forces, plasma and interstitial fluid are continuously interchanged. The return of fluid to the vascular system at the venular ends of the capillaries, however, does not exactly equal the amount filtered at the arteriolar ends. Approximately 85% to 90% of the filtrate is returned directly to the blood capillaries; the remaining 10% to 15% is returned to the blood by way of the lymphatic system. This amounts to about 1–2 L of interstitial fluid per day, containing 20–30 g of protein per liter that enters lymphatic capillaries. As may be recalled from chapter 13 (see fig. 13.36), lymphatic capillaries are blind-ended, highly permeable vessels that drain their contents (now called lymph) into lymphatic vessels, which ultimately return this fluid to the venous system.

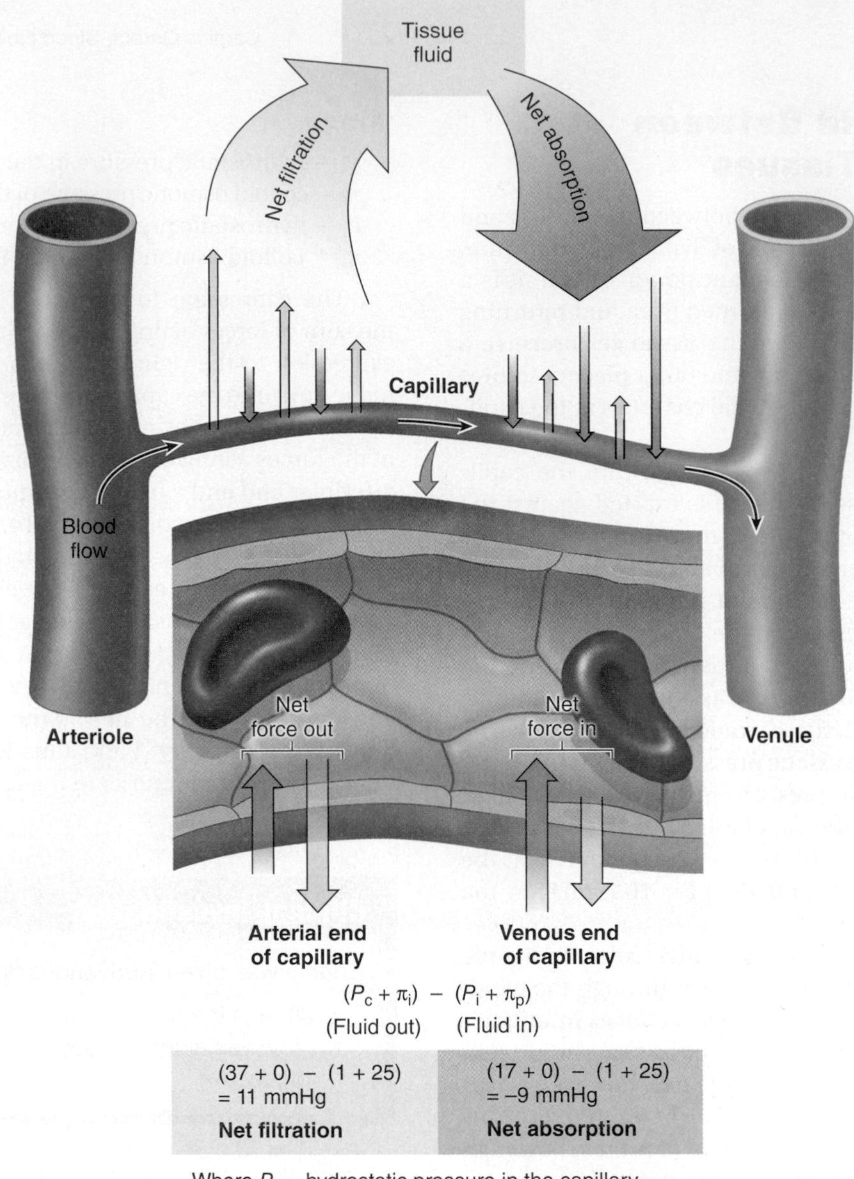

Figure 14.9 The distribution of fluid across the walls of a capillary. Tissue, or interstitial, fluid is formed by filtration (*yellow arrows*) as a result of blood pressures at the arteriolar ends of capillaries; it is returned to the venular ends of capillaries by the colloid osmotic pressure of plasma proteins (*orange arrows*). AP|R

See the *Test Your Quantitative Ability* section of the Review Activities at the end of this chapter.

Causes of Edema

Excessive accumulation of interstitial fluid is known as **edema.** This condition is normally prevented by a proper balance between capillary filtration and osmotic uptake of water and by proper lymphatic drainage. Edema may thus result from

1. *high arterial blood pressure,* which increases capillary pressure and causes excessive filtration;
2. *venous obstruction*—as in phlebitis (where a thrombus forms in a vein) or mechanical compression of veins (during pregnancy, for example)—which produces a congestive increase in capillary pressure;
3. *leakage of plasma proteins into interstitial fluid,* which causes reduced osmotic flow of water into the capillaries (this occurs during inflammation and allergic reactions as a result of increased capillary permeability);
4. *myxedema*—the excessive production of particular glycoproteins (mucin) in the extracellular matrix caused by hypothyroidism;
5. *decreased plasma protein concentration,* as a result of liver disease (the liver makes most of the plasma proteins) or kidney disease where plasma proteins are excreted in the urine;
6. *obstruction of the lymphatic drainage* due to parasitic larvae in elephantiasis (fig. 14.10 and table 14.2) or to surgery (breast surgery is a leading cause of lymphedema).

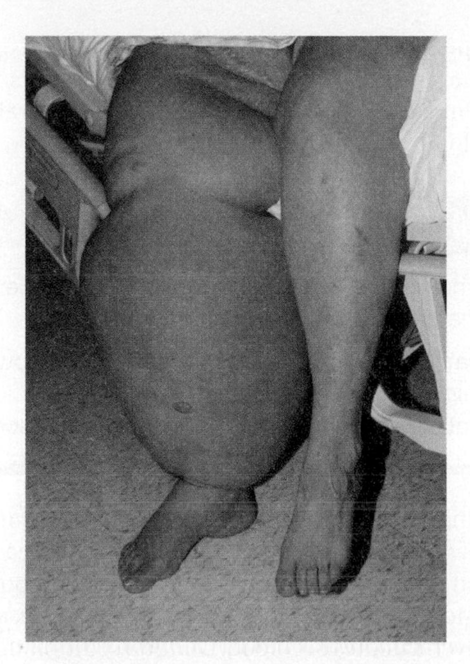

Figure 14.10 The severe edema of elephantiasis. Parasitic larvae that block lymphatic drainage produce tissue edema and the tremendous enlargement of the limbs and scrotum in elephantiasis.

CLINICAL APPLICATION

In the tropical disease *filariasis,* mosquitoes transmit a nematode worm parasite to humans. The larvae of these worms invade lymphatic vessels and block lymphatic drainage. The edema that results can be so severe that the tissues swell to produce an elephant-like appearance, with thickening and cracking of the skin. This condition is thus aptly named **elephantiasis** (fig. 14.10). The World Health Organization estimates that this disease currently affects at least 120 million people, primarily in India and Africa. A new drug regimen has been found to be 99% effective against the filariasis parasite, and a worldwide effort to eradicate this disease is now under way.

Regulation of Blood Volume by the Kidneys

The formation of urine by the kidneys begins in the same manner as the formation of interstitial fluid—by filtration of plasma through capillary pores. These capillaries are known as *glomeruli,* and the filtrate they produce enters a system of tubules that transports and modifies the filtrate (by mechanisms discussed in chapter 17). The total blood volume is only about 5.5 L, yet the kidneys produce about 180 L/day of blood filtrate; thus, most of this filtrate must be returned to the vascular system and recycled. Only about 1.5 L of urine is excreted daily; 98% to 99% of the amount filtered is **reabsorbed** back into the vascular system.

The volume of urine excreted can be varied by changes in the reabsorption of filtrate. If 99% of the filtrate is reabsorbed, for example, 1% must be excreted. Decreasing the reabsorption by only 1%—from 99% to 98%—would double the volume of urine excreted (an increase to 2% of the amount filtered). Carrying the logic further, a doubling of urine volume from, for example, 1 to 2 liters, would result in the loss of an additional liter of blood volume. The percentage of the glomerular filtrate reabsorbed—and thus the urine volume and blood volume—is adjusted according to the needs of the body by the action of specific hormones on the kidneys. Through their effects on the kidneys and the resulting changes in blood volume, these hormones serve important functions in the regulation of the cardiovascular system.

The sympathetic nervous system is also involved in the homeostasis of blood volume. An increase in blood volume is detected by stretch receptors in the atria of the heart, which selectively regulate sympathetic nerve activity. The activity of sympathetic fibers to the heart is increased, while the sympathetic nerve activity to the kidneys is reduced (there is little change in sympathetic nerve activity to other organs). Reduced sympathetic nerve stimulation of renal arteries produces vasodilation and increased blood flow, thereby promoting increased urine production to lower the blood volume and complete the negative feedback loop.

Table 14.2 | Causes of Edema

Cause	Comments
Increased blood pressure or venous obstruction	Increases capillary filtration pressure so that more tissue fluid is formed at the arteriolar ends of capillaries.
Increased tissue protein concentration	Decreases osmosis of water into the venular ends of capillaries. Usually a localized tissue edema due to leakage of plasma proteins through capillaries during inflammation and allergic reactions. Myxedema due to hypothyroidism is also in this category.
Decreased plasma protein concentration	Decreases osmosis of water into the venular ends of capillaries. May be caused by liver disease (which can be associated with insufficient plasma protein production), kidney disease (due to leakage of plasma protein into urine), or protein malnutrition.
Obstruction of lymphatic vessels	Infections by filaria roundworms (nematodes) transmitted by a certain species of mosquito block lymphatic drainage, causing edema and tremendous swelling of the affected areas.

Regulation by Antidiuretic Hormone (ADH)

One of the major hormones involved in the regulation of blood volume is **antidiuretic hormone (ADH),** also known as *vasopressin.* This hormone is produced by neurons in the hypothalamus, transported by axons into the posterior pituitary, and released from this storage gland in response to hypothalamic stimulation (chapter 11, section 11.3). The release of ADH from the posterior pituitary occurs when neurons in the hypothalamus called **osmoreceptors** detect an increase in plasma osmolality (fig. 14.11).

An increase in plasma osmolality occurs when the plasma becomes more concentrated. This can be produced either by *dehydration* or by excessive *salt intake.* Stimulation of osmoreceptors produces sensations of thirst, leading to increased water intake and an increase in the amount of ADH released from the posterior pituitary. Through mechanisms that will be discussed in conjunction with kidney physiology (chapter 17, section 17.3), ADH stimulates water reabsorption from the filtrate. A smaller volume of urine is thus excreted as a result of the action of ADH (fig. 14.11).

A person who is dehydrated or who consumes excessive amounts of salt thus drinks more and urinates less. This raises the blood volume and, in the process, dilutes the plasma to lower its previously elevated osmolality. The rise in blood volume that results from these mechanisms is extremely important in stabilizing the condition of a dehydrated person with low blood volume and pressure.

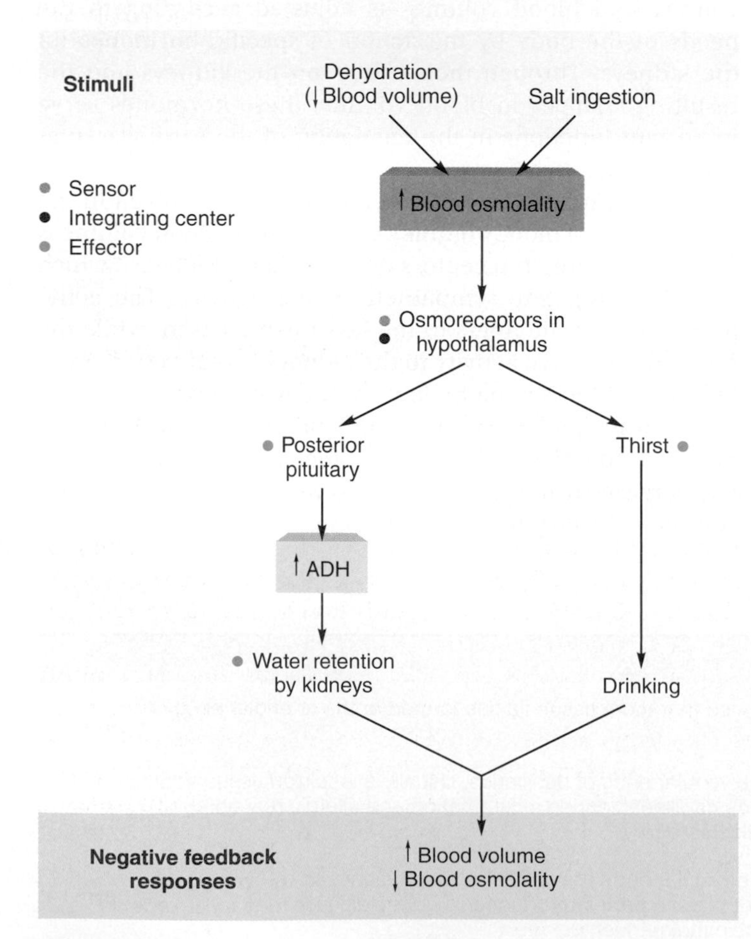

Figure 14.11 The negative feedback control of blood volume and blood osmolality. Thirst and ADH secretion are triggered by a rise in plasma osmolality. Homeostasis is maintained by countermeasures, including drinking and conservation of water by the kidneys.

Clinical Investigation CLUES

Charlie had a low output of urine, and the urine he produced had a high osmolality (concentration).

- What mechanism was responsible for his low output of highly concentrated urine?
- What benefit does Charlie derive from this mechanism?

Drinking excessive amounts of water without excessive amounts of salt does not result in a prolonged increase in blood volume and pressure. The water does enter the blood from the intestine and momentarily raises the blood volume; at the same time, however, it dilutes the blood. Dilution of the blood decreases the plasma osmolality and thus inhibits the release of ADH. With less ADH there is less reabsorption of water in the kidneys—a larger volume of more-dilute urine is excreted. Water is therefore a *diuretic*—a substance that promotes urine formation—because it inhibits the release of antidiuretic hormone.

Dilution of the blood (decreased blood osmolality) lowers ADH secretion, but a rise in blood volume itself (even in the absence of dilution) can reduce ADH secretion. This is because an increased blood volume mechanically stimulates stretch receptors in the left atrium, aortic arch, and carotid sinus, which in turn cause increased firing of sensory neurons (in cranial nerves IX and X). ADH secretion is inhibited by this sensory input, so that more water is eliminated from the blood by the kidneys.

Conversely, a lowering of blood volume by about 10% reduces stimulation of these stretch receptors, reducing the firing of their associated sensory neurons. This produces an increase in ADH secretion, which stimulates the kidneys to retain more water in the blood. These negative feedback loops thereby help to maintain homeostasis of blood volume.

FITNESS APPLICATION

During prolonged exercise, particularly on a warm day, a substantial amount of water (up to 900 ml per hour) may be lost from the body through sweating. The lowering of blood volume that results decreases the ability of the body to dissipate heat, and the consequent overheating of the body can cause ill effects and put an end to the exercise. The need for athletes to remain well hydrated is commonly recognized, but drinking pure water may not be the answer. This is because blood sodium is lost in sweat, so that a lesser amount of water is required to dilute the blood osmolality back to normal. When the blood osmolality is normal, thirst is extinguished. For these reasons, athletes performing prolonged endurance exercise should drink solutions containing sodium (as well as carbohydrates for energy), and they should drink at a predetermined rate rather than at a rate determined only by thirst.

When blood volume rises, the stimulation of stretch receptors in the atria of the heart has an additional effect: it stimulates the atria to secrete a hormone known as *atrial natriuretic peptide.* This hormone increases the excretion of salt and water in the urine, thereby working, like a decrease in ADH secretion, to lower blood volume. Atrial natriuretic peptide is discussed in a separate section.

Regulation by Aldosterone

From the preceding discussion, it is clear that a certain amount of dietary salt is required to maintain blood volume and pressure. Since Na^+ and Cl^- are easily filtered in the kidneys, a mechanism must exist to promote the reabsorption and retention of salt when the dietary salt intake is too low. **Aldosterone,** a steroid hormone secreted by the adrenal cortex, stimulates the reabsorption of salt by the kidneys. Aldosterone is thus a "salt-retaining hormone." Retention of salt indirectly promotes retention of water (in part, by the action of ADH, as previously discussed). The action of aldosterone produces an increase in blood volume, but, unlike ADH, it does not produce a change in plasma osmolality. This is because aldosterone promotes the reabsorption of salt and water in proportionate amounts, whereas ADH promotes only the reabsorption of water. Thus, unlike ADH, aldosterone does not act to dilute the blood.

The secretion of aldosterone is stimulated during salt deprivation, when the blood volume and pressure are reduced. The adrenal cortex, however, is not directly stimulated to secrete aldosterone by these conditions. Instead, a decrease in blood volume and pressure activates an intermediate mechanism, described next.

FITNESS APPLICATION

Throughout most of human history, salt was in short supply and was therefore highly valued. Moorish merchants in the sixth century traded an ounce of salt for an ounce of gold, and salt cakes were used as money in Abyssinia. Part of a Roman soldier's pay was given in salt—a practice from which the word *salary* (*sal* = salt) derives. Salt was also used to purchase slaves—hence the phrase "worth his salt."

Renin-Angiotensin-Aldosterone System

Salt deprivation results in low blood volume and pressure, as described in the previous section on ADH. This lowers the blood pressure in the renal artery and reduces the amount of NaCl and water in the renal filtrate. The *juxtaglomerular apparatus* in the kidneys (chapter 17; see fig. 17.26) senses these changes and, in response, secretes the enzyme **renin** into the blood (chapter 17, section 17.5). This enzyme cleaves a ten-amino-acid polypeptide called *angiotensin I* from a plasma protein called *angiotensinogen.* As angiotensin I passes through the capillaries of the lungs, an *angiotensin-converting enzyme (ACE)* removes two amino acids. This leaves an eight-amino-acid polypeptide called **angiotensin II** (fig. 14.12). Conditions of salt deprivation, low blood volume, and low blood pressure, in summary, cause increased production of angiotensin II in the blood. (High blood pressure, by contrast, suppresses renin secretion and thereby results in reduced production of angiotensin II.)

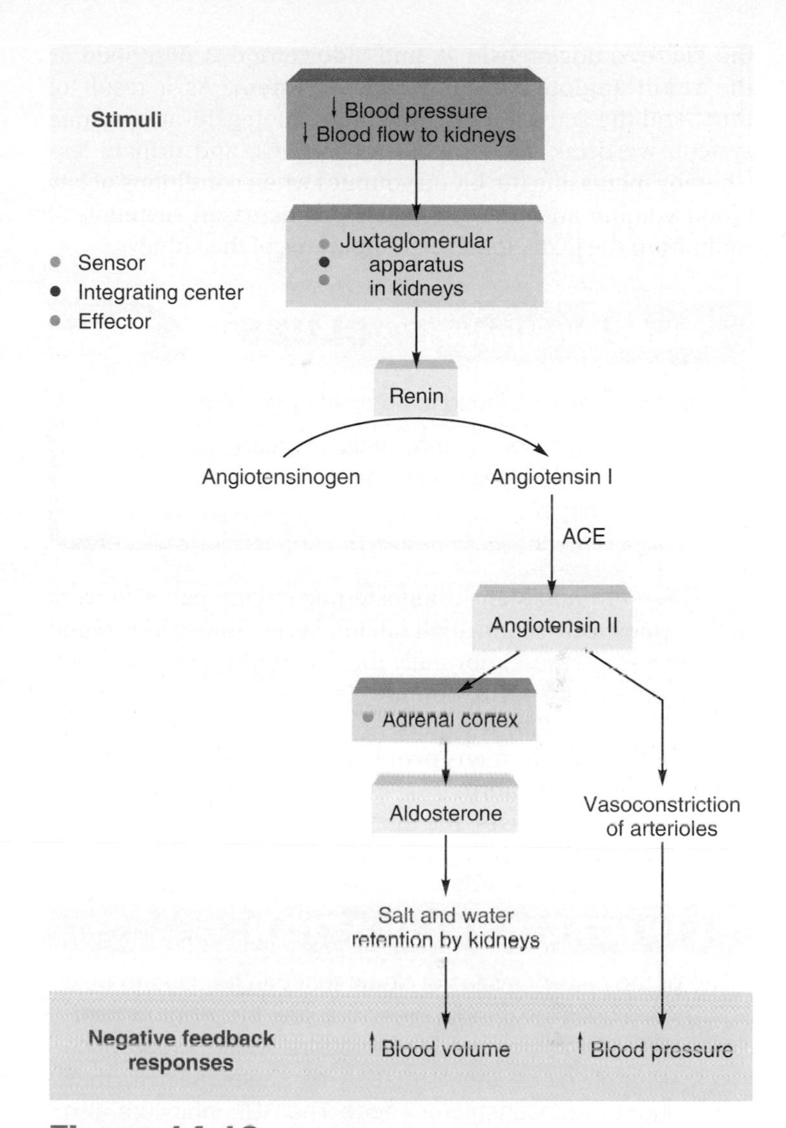

Figure 14.12 The renin-angiotensin-aldosterone system. This system helps to maintain homeostasis through the negative feedback control of blood volume and pressure. (ACE = angiotensin-converting enzyme.)

Angiotensin II exerts numerous effects that cause blood pressure to rise. Its most direct effect is to stimulate contraction of the smooth muscle layers of the small arteries and arterioles. By this means, angiotensin II is a powerful vasoconstrictor, increasing the total peripheral resistance and thus the arterial blood pressure. Angiotensin II also promotes a rise in blood volume (thereby increasing the blood pressure) by stimulating (1) the thirst center in the hypothalamus, and (2) the adrenal cortex to secrete aldosterone.

When the thirst center in the hypothalamus is stimulated, more water is taken into the intestine and then the blood. When the adrenal cortex is stimulated by angiotensin II to secrete more aldosterone, the increased aldosterone stimulates the kidneys to retain more salt and water. The relationship between the kidneys, angiotensin II, and aldosterone is described as the **renin-angiotensin-aldosterone system.** As a result of thirst and the activation of the renin-angiotensin-aldosterone system, we drink more, retain more NaCl, and urinate less (thereby increasing the blood volume) when conditions of low blood volume and pressure cause an increased secretion of renin from the juxtaglomerular apparatus of the kidneys.

Clinical Investigation CLUES

Charlie's urine was found to completely lack Na^+.

- What physiological mechanism is responsible?
- What benefit does Charlie derive from this mechanism?

The renin-angiotensin-aldosterone system can also work in the opposite direction: high salt intake, leading to high blood volume and pressure, normally inhibits renin secretion. With less angiotensin II formation and less aldosterone secretion, less salt is retained by the kidneys and more is excreted in the urine. Unfortunately, many people with chronically high blood pressure may have normal or even elevated levels of renin secretion. In these cases, the intake of salt must be lowered to match the impaired ability to excrete salt in the urine.

CLINICAL APPLICATION

One of the newer classes of drugs that can be used to treat hypertension (high blood pressure) are the **angiotensin-converting enzyme, or ACE, inhibitors.** These drugs (such as captopril) block the formation of angiotensin II, thus reducing its vasoconstrictor effect. The ACE inhibitors also increase the activity of bradykinin, a polypeptide that promotes vasodilation. The reduced formation of angiotensin II and increased action of bradykinin result in vasodilation, which decreases the total peripheral resistance. Because this reduces the afterload of the heart, the ACE inhibitors are also used to treat left ventricular hypertrophy and congestive heart failure. Another new class of antihypertensive drugs allows angiotensin II to be formed but selectively blocks the angiotensin II receptors. These drugs are called **angiotensin receptor blockers (ARBs).**

Atrial Natriuretic Peptide

Scientists have long known that a rise in blood volume, or an increased venous return for any other reason, leads to increased urine production (*diuresis*). In fact, it is a common observation that immersion in water (which increases venous return) causes increased diuresis. But how is this accomplished? Experiments suggest that the increased water excretion under conditions of high blood volume or venous return is at least partly due to an increase in the excretion of Na^+ in the urine, or *natriuresis* (*natrium* = sodium; *uresis* = making water).

Increased Na^+ excretion (natriuresis) may be produced by a decline in aldosterone secretion, but there is evidence that there is a separate hormone that stimulates natriuresis. This *natriuretic hormone* would thus be antagonistic to aldosterone and would promote Na^+ and water excretion in the urine in response to a rise in blood volume. A polypeptide hormone with these properties, identified as **atrial natriuretic peptide (ANP),** is produced by the atria of the heart.

When a person floats in water, there is an increase in venous return to the heart. This (like an increase in blood volume) stretches the atria, thereby stimulating the release of ANP. In addition, ADH secretion from the posterior pituitary is inhibited, due to sensory information traveling to the hypothalamus in the vagus nerves from stretch receptors in the left atrium. Increased ANP, together with decreased ADH, leads to a greater excretion of salt and water in the urine. This works as a negative feedback correction to lower the blood volume and thus maintain homeostasis (fig. 14.13).

CHECKPOINT

4a. Describe the composition of interstitial fluid. Using a flow diagram, explain how interstitial fluid is formed and how it is returned to the vascular system.

4b. Define the term *edema* and describe four different mechanisms that can produce this condition.

5a. Describe the effects of dehydration on blood and urine volumes. What cause-and-effect mechanism is involved?

5b. Explain why salt deprivation causes increased salt and water retention by the kidneys.

5c. Describe the actions of atrial natriuretic peptide and explain their significance.

14.3 VASCULAR RESISTANCE TO BLOOD FLOW

The rate of blood flow to an organ is related to the resistance to flow in the small arteries and arterioles. Vasodilation decreases resistance and increases flow, whereas vasoconstriction increases resistance and decreases flow. These changes occur in response to various regulatory mechanisms.

LEARNING OUTCOMES

After studying this section, you should be able to:

6. Describe the factors that affect blood flow through vessels.

7. Describe the intrinsic and extrinsic regulation of peripheral resistance.

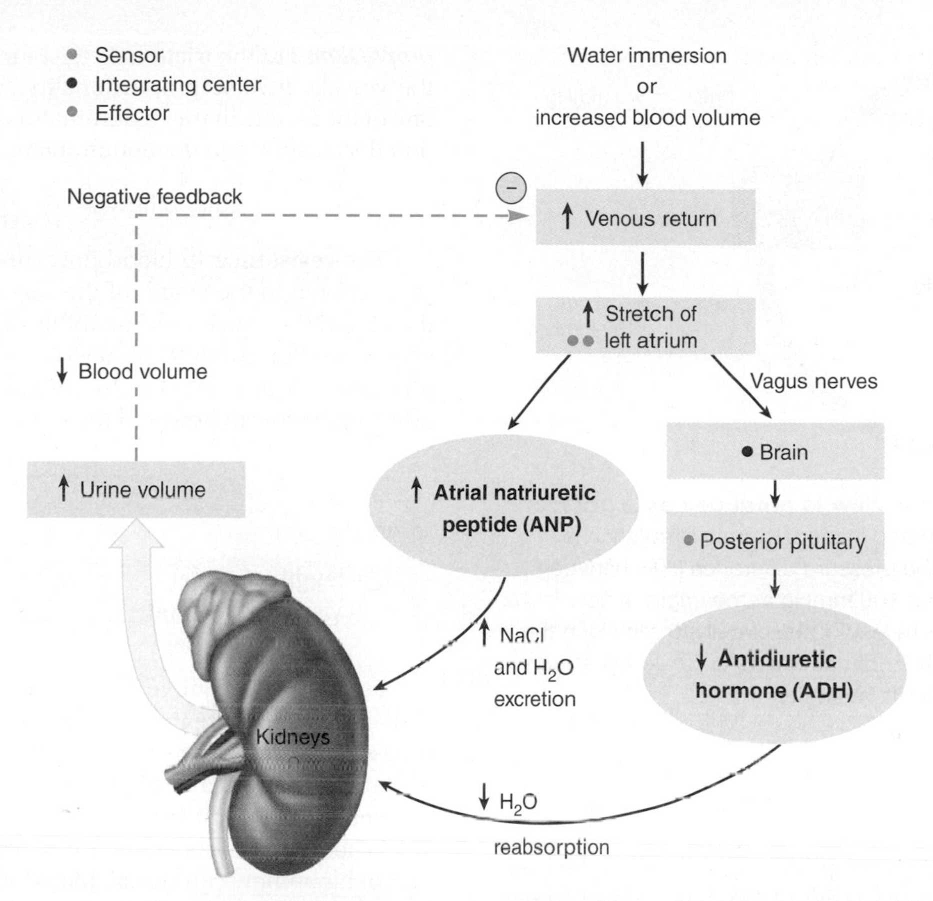

Figure 14.13 **Negative feedback correction of increased venous return.** Stimulation of stretch receptors in the left atrium causes secretion of atrial natriuretic peptide (ANP), which increases urinary excretion of salt and water. At the same time, stimulation of these stretch receptors leads to decreased antidiuretic hormone (ADH) secretion. Because ADH stimulates the kidneys to reabsorb (retain) water, a fall in ADH works together with the increased ANP to increase urine volume. This maintains homeostasis by lowering blood volume and venous return.

The amount of blood that the heart pumps per minute is equal to the rate of venous return, and thus is equal to the rate of blood flow through the entire circulation. The cardiac output is about 5 to 6 L per minute, depending upon body size and other factors. This total cardiac output is distributed unequally to the different organs because of unequal resistances to blood flow through the organs. The distribution of the cardiac output to various organs at rest, in terms of percentages and rates of blood flow, is provided in table 14.3.

Physical Laws Describing Blood Flow

The flow of blood through the vascular system, like the flow of any fluid through a tube, depends in part on the difference in pressure at the two ends of the tube. If the pressure at both ends of the tube is the same, there will be no flow. If the pressure at one end is greater than at the other, blood will flow from the region of higher to the region of lower pressure. The rate of blood flow is proportional to the pressure difference ($P_1 - P_2$) between the two ends of the tube. The term

Table 14.3 | Estimated Distribution of the Cardiac Output at Rest

Organs	Blood Flow	
	Milliliters per Minute	Percent Total
Gastrointestinal tract and liver	1,400	24
Kidneys	1,100	19
Brain	750	13
Heart	250	4
Skeletal muscles	1,200	21
Skin	500	9
Other organs	600	10
Total organs	5,800	100

Source: From O. L. Wade and J. M. Bishop, *Cardiac Output and Regional Blood Flow.* Copyright © 1962 Blackwell Science, Ltd. Used with permission.

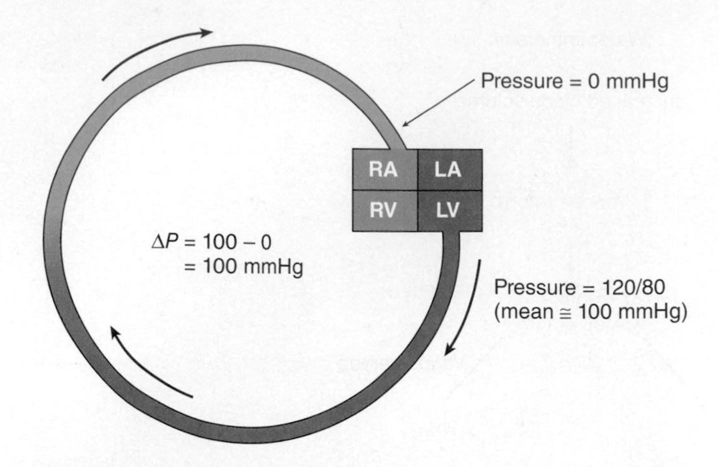

Figure 14.14 Blood flow is produced by a pressure difference. The flow of blood in the systemic circulation is ultimately dependent on the pressure difference (ΔP) between the mean pressure of about 100 mmHg at the origin of flow in the aorta and the pressure at the end of the circuit—0 mmHg in the vena cava, where it joins the right atrium (RA). (LA = left atrium; RV = right ventricle; LV = left ventricle.)

pressure difference is abbreviated ΔP, in which the Greek letter Δ (*delta*) means "change in."

If the systemic circulation is pictured as a single tube leading from and back to the heart (fig. 14.14), blood flow through this system would occur as a result of the pressure difference between the beginning of the tube (the aorta) and the end of the tube (the junction of the venae cavae with the right atrium). The average pressure, or **mean arterial pressure (MAP),** is about 100 mmHg; the pressure at the right atrium is 0 mmHg. The "pressure head," or driving force (ΔP), is therefore about $100 - 0 = 100$ mmHg.

Blood flow is directly proportional to the pressure difference between the two ends of the tube (ΔP) but is *inversely proportional* to the frictional resistance to blood flow through the vessels. Inverse proportionality is expressed by showing one of the factors in the denominator of a fraction, since a fraction decreases when the denominator increases:

$$\text{Blood flow} \propto \frac{\Delta P}{\text{resistance}}$$

The **resistance** to blood flow through a vessel is directly proportional to the length of the vessel and to the viscosity of the blood (the "thickness," or ability of molecules to "slip over" each other; for example, honey is quite viscous). Of particular physiological importance, the resistance is inversely proportional to the fourth power of the radius of the vessel:

$$\text{Resistance} \propto \frac{L\eta}{r^4}$$

where

L = length of vessel
η = viscosity of blood
r = radius of vessel

For example, if one vessel has half the radius of another and if all other factors are the same, the smaller vessel will have 16 times (2^4) the resistance of the larger vessel. Blood flow through the larger vessel, as a result, will be 16 times greater than in the smaller vessel (fig. 14.15).

When physical constants are added to this relationship, the rate of blood flow can be calculated according to **Poiseuille's** (*pwˇa-zuh'yez*) **law:**

$$\text{Blood flow} \propto \frac{\Delta P r^4 (\pi)}{\eta L (8)}$$

Vessel length (L) and blood viscosity (the Greek letter *eta,* written η) do not vary significantly in normal physiology, although blood viscosity is increased in severe dehydration and in the polycythemia (high red blood cell count) that occurs as an adaptation to life at high altitudes. The major physiological regulators of blood flow through an organ are the

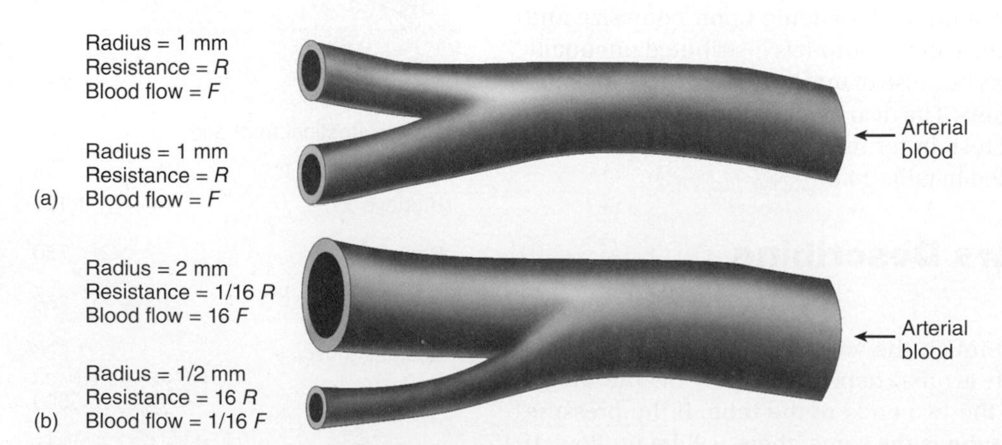

Figure 14.15 The relationships between blood flow, vessel radius, and resistance. (*a*) The resistance and blood flow are equally divided between two branches of a vessel. (*b*) A doubling of the radius of one branch and halving of the radius of the other produces a sixteenfold increase in blood flow in the former and a sixteenfold decrease of blood flow in the latter.

See the *Test Your Quantitative Ability* section of the Review Activities at the end of this chapter.

mean arterial pressure (*P*, driving the flow) and the vascular resistance to flow. At a given mean arterial pressure, blood can be diverted from one organ to another by variations in the degree of vasoconstriction and vasodilation of small arteries and arterioles (that is, by variations in vessel radius, *r*). Vasoconstriction in one organ and vasodilation in another result in a diversion, or *shunting*, of blood to the second organ. Because arterioles are the smallest arteries and can become narrower by vasoconstriction, they provide the greatest resistance to blood flow (fig. 14.16). Blood flow to an organ is thus largely determined by the degree of vasoconstriction or vasodilation of its arterioles. The rate of blood flow to an organ can be increased by dilation of its arterioles and can be decreased by constriction of its arterioles.

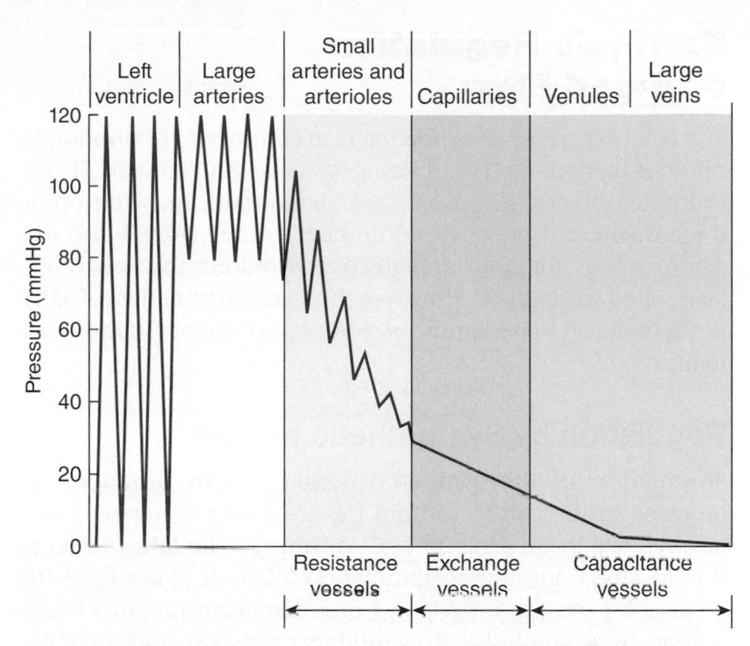

Figure 14.16 Blood pressure in different vessels of the systemic circulation. Notice that the pressure generated by the beating of the ventricles is largely dissipated by the time the blood gets into the venous system, and that this pressure drop occurs primarily as blood goes through the arterioles and capillaries.

Total Peripheral Resistance

The sum of all the vascular resistances within the systemic circulation is called the **total peripheral resistance.** The arteries that supply blood to the organs are generally in parallel rather than in series with each other. That is, arterial blood passes through only one set of resistance vessels (arterioles) before returning to the heart (fig. 14.17). Because one organ is not "downstream" from another in terms of its arterial supply, changes in resistance within one organ directly affect blood flow in that organ only.

Vasodilation in a large organ might, however, significantly decrease the total peripheral resistance and, by this means, might decrease the mean arterial pressure. In the absence of compensatory mechanisms, the driving force for blood flow through all organs might be reduced. This situation is normally prevented by an increase in the cardiac output and by vasoconstriction in other areas. During exercise of the large muscles, for example, the arterioles in the exercising muscles are dilated. This would cause a great fall in mean arterial pressure if there were no compensations. But the blood pressure actually rises during exercise, primarily because of increased cardiac output and vasoconstriction in the viscera. Also, sympathetic nerves produce cutaneous vasoconstriction at the beginning of exercise, raising blood pressure. However, when exercise is prolonged, increased metabolic heat production overrides this effect to increase the flow of warm blood to the skin for improved heat loss (see table 14.7).

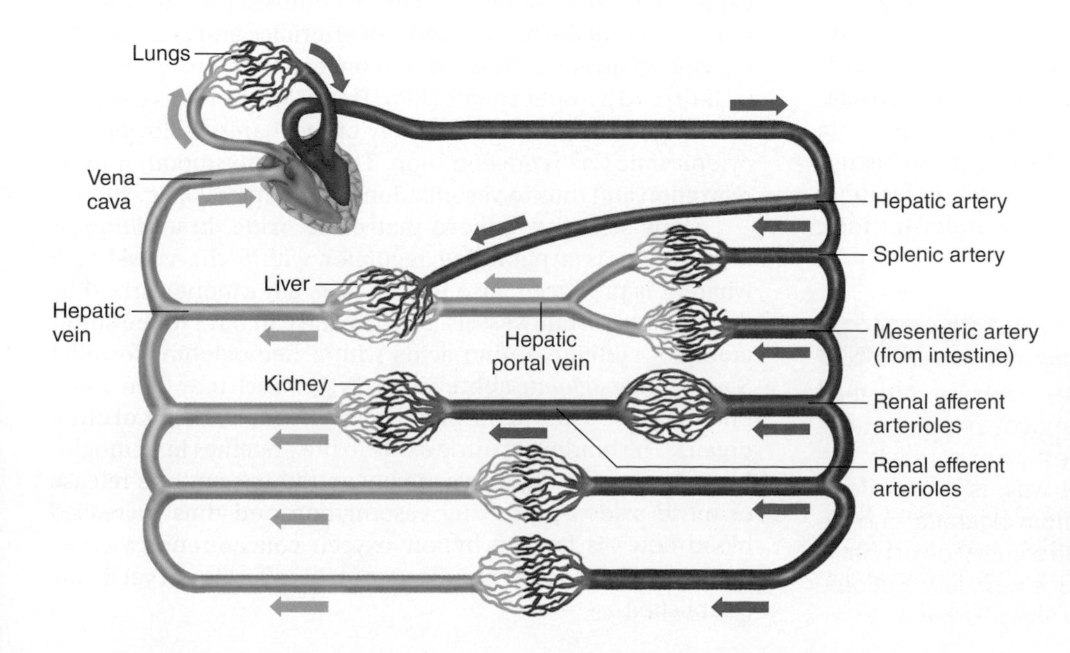

Figure 14.17 A diagram of the systemic and pulmonary circulations. Notice that with few exceptions (such as blood flow in the renal circulation) the flow of arterial blood is in parallel rather than in series (arterial blood does not usually flow from one organ to another). AP|R

Extrinsic Regulation of Blood Flow

The term *extrinsic regulation* refers to control by the autonomic nervous system and endocrine system. Angiotensin II, for example, directly stimulates vascular smooth muscle to produce vasoconstriction. Antidiuretic hormone (ADH) also has a vasoconstrictor effect at high concentrations; this is why it is also called *vasopressin.* However, this vasopressor effect of ADH is not believed to be significant under physiological conditions in humans.

Regulation by Sympathetic Nerves

Stimulation of the sympathoadrenal system produces an increase in the cardiac output (as previously discussed) and an increase in total peripheral resistance. The latter effect is due to alpha-adrenergic stimulation (chapter 9; see fig. 9.10) of vascular smooth muscle by norepinephrine and, to a lesser degree, by epinephrine. This produces vasoconstriction of the arterioles in the viscera and skin.

Even when a person is calm, the sympathoadrenal system is active to a certain degree and helps set the "tone" of vascular smooth muscles. In this case, **adrenergic sympathetic fibers** (those that release norepinephrine) activate alpha-adrenergic receptors to cause a basal level of vasoconstriction throughout the body. During the fight-or-flight reaction, an increase in the activity of adrenergic fibers produces vasoconstriction in the digestive tract, kidneys, and skin.

Arterioles in skeletal muscles receive **cholinergic sympathetic fibers,** which release acetylcholine as a neurotransmitter. During the fight-or-flight reaction, the activity of these cholinergic fibers increases. This causes vasodilation. Vasodilation in skeletal muscles is also produced by epinephrine secreted by the adrenal medulla, which stimulates beta-adrenergic receptors. During the fight-or-flight reaction, therefore, blood flow is decreased to the viscera and skin because of the alpha-adrenergic effects of vasoconstriction in these organs, whereas blood flow to the skeletal muscles is increased. This diversion of blood flow to the skeletal muscles during emergency conditions may give these muscles an "extra edge" in responding to the emergency. Once exercise begins, however, the blood flow to skeletal muscles increases far more due to other mechanisms (described shortly under Intrinsic Regulation of Blood Flow).

CLINICAL APPLICATION

Cocaine inhibits the reuptake of norepinephrine into the adrenergic axons, resulting in enhanced sympathetic-induced vasoconstriction. Chest pain, as a result of myocardial ischemia produced in this way, is a common cocaine-related problem. The nicotine from cigarette smoke acts synergistically with cocaine (through stimulation of postganglionic sympathetic neurons) to induce vasoconstriction.

Parasympathetic Control of Blood Flow

Parasympathetic endings in arterioles are always cholinergic and always promote vasodilation. Parasympathetic innervation of blood vessels, however, is limited to the digestive tract, external genitalia, and salivary glands. Because of this limited distribution, the parasympathetic system is less important than the sympathetic system in the control of total peripheral resistance.

The extrinsic control of blood flow is summarized in table 14.4.

Paracrine Regulation of Blood Flow

Paracrine regulators are molecules produced by one tissue that help to regulate another tissue of the same organ (chapter 11, section 11.7). Blood vessels are particularly subject to paracrine regulation. Specifically, the endothelium of the tunica interna produces a number of paracrine regulators that cause the smooth muscle of the tunica media to either relax or contract.

Smooth muscle relaxation results from the local effects of a number of molecules produced by the vessel endothelium, including **bradykinin, nitric oxide,** and several prostaglandins, particularly **prostaglandin I_2 (prostacyclin).** The relaxation of vascular smooth muscle produces vasodilation, which is an effect that can be medically useful. For example, the vasodilation induced by nitric oxide explains why *nitroglycerin* and related drugs (which can be converted to nitric oxide) are beneficial for the treatment of angina pectoris. As another example, people with pulmonary hypertension—a disease in which increased vascular resistance in the pulmonary circulation can lead to failure of the right ventricle—are sometimes treated by intravenous administration of prostacyclin (prostaglandin I_2).

The endothelium of arterioles contains an enzyme, *endothelial nitric oxide synthase (eNOS),* which produces nitric oxide (NO) from L-arginine. The NO diffuses into the smooth muscle cells of the tunica media of arterioles and activates the enzyme guanylate cyclase, which converts GTP into cyclic GMP (cGMP) and pyrophosphate (PP_i). The cGMP serves as a second messenger that, through a variety of mechanisms, lowers the cytoplasmic Ca^{2+} concentration. This leads to smooth muscle relaxation and thus to vasodilation (see chapter 20, fig. 20.21).

Many scientists believe that nitric oxide, in addition to functioning as a paracrine regulator within the vessel wall where it is produced, also functions as a hormone carried by the blood to distant vessels. Nitric oxide can bind to the sulfur atoms of cysteine amino acids within hemoglobin (forming *S-nitrosohemoglobin,* abbreviated *SNO*), which may then transport the nitric oxide in the blood downstream to vessels of other organs. The binding of nitric oxide to the cysteines in hemoglobin is favored by high oxygen concentrations, and the release of nitric oxide—producing vasodilation and thus increased blood flow—is favored by low oxygen concentrations. However, the physiological significance of this effect is not yet firmly established.

Table 14.4 | Extrinsic Control of Vascular Resistance and Blood Flow

Extrinsic Agent	Effect	Comments
Sympathetic nerves		
Alpha-adrenergic	Vasoconstriction	Vasoconstriction is the dominant effect of sympathetic nerve stimulation on the vascular system, and it occurs throughout the body.
Beta-adrenergic	Vasodilation	There is some activity in arterioles in skeletal muscles and in coronary vessels, but effects are masked by dominant alpha-receptor-mediated constriction.
Cholinergic	Vasodilation	Effects are localized to arterioles in skeletal muscles and are produced only during defense (fight-or-flight) reactions.
Parasympathetic nerves	Vasodilation	Effects are restricted primarily to the gastrointestinal tract, external genitalia, and salivary glands and have little effect on total peripheral resistance.
Angiotensin II	Vasoconstriction	A powerful vasoconstrictor produced as a result of secretion of renin from the kidneys; it may function to help maintain adequate filtration pressure in the kidneys when systemic blood flow and pressure are reduced.
ADH (vasopressin)	Vasoconstriction	Although the effects of this hormone on vascular resistance and blood pressure in anesthetized animals are well documented, the importance of these effects in conscious humans is controversial.
Histamine	Vasodilation	Histamine promotes localized vasodilation during inflammation and allergic reactions.
Bradykinins	Vasodilation	Bradykinins are polypeptides secreted by sweat glands and by the endothelium of blood vessels; they promote local vasodilation.
Prostaglandins	Vasodilation or vasoconstriction	Prostaglandins are cyclic fatty acids that can be produced by most tissues, including blood vessel walls. Prostaglandin I_2 is a vasodilator, whereas thromboxane A_2 is a vasoconstrictor. The physiological significance of these effects is presently controversial.

The endothelium also produces paracrine regulators that promote vasoconstriction. Notable among these is the polypeptide **endothelin-1.** This paracrine regulator stimulates vasoconstriction of arterioles, thus raising the total peripheral resistance. Endothelin-1 receptor antagonists are now medically available to block the vasoconstrictor effect of endothelin-1, and thus promote vasodilation. For example, endothelin-1 receptor antagonists may be used to help treat pulmonary hypertension. In normal physiology, the effects of endothelin-1 may be balanced by nitric oxide to help regulate blood flow and blood pressure.

Intrinsic Regulation of Blood Flow

Intrinsic, or "built-in," mechanisms within individual organs provide a localized regulation of vascular resistance and blood flow. **Autoregulation** refers to the ability of some organs—particularly the brain and kidneys—to utilize intrinsic control mechanisms to maintain a relatively constant blood flow despite wide fluctuations in blood pressure. Intrinsic mechanisms are classified as *myogenic* or *metabolic.*

Myogenic Control Mechanisms

If the arterial blood pressure and flow through an organ are inadequate—if the organ is inadequately *perfused* with blood—the metabolism of the organ cannot be maintained beyond a limited time period. However, excessively high blood pressure can also be dangerous, particularly in the brain, because this may result in the rupture of fine blood vessels (causing a cerebrovascular accident—CVA, or stroke).

Autoregulation of blood flow helps to mitigate these possibilities. Changes in systemic arterial pressure are compensated for in the brain and some other organs by the appropriate responses of vascular smooth muscle. A decrease in arterial pressure causes cerebral vessels to dilate, so that adequate blood flow can be maintained despite the decreased pressure. High blood pressure, by contrast, causes cerebral vessels to constrict, so that finer vessels downstream are protected from the elevated pressure. These responses are myogenic; they are direct responses by the vascular smooth muscle to changes in pressure.

Metabolic Control Mechanisms

Local vasodilation within an organ can occur as a result of the chemical environment created by the organ's metabolism. The localized chemical conditions that promote vasodilation include (1) *decreased oxygen concentrations* that result from increased metabolic rate; (2) *increased carbon dioxide concentrations;* (3) *decreased tissue pH* (due to CO_2, lactic acid, and other metabolic products); and (4) *release of K^+ and paracrine regulators* (such as adenosine, nitric oxide, and others) from tissue cells.

Through these chemical changes, the organ signals its blood vessels that it needs increased oxygen delivery.

The vasodilation that occurs in response to tissue metabolism can be demonstrated by constricting the blood supply to an area for a short time and then removing the constriction. The constriction allows metabolic products to accumulate by preventing venous drainage of the area. When the constriction is removed and blood flow resumes, the metabolic products that have accumulated cause vasodilation. The tissue thus appears red. This response is called **reactive hyperemia.** A similar increase in blood flow occurs in skeletal muscles and other organs as a result of increased metabolism. This is called **active hyperemia.** The increased blood flow can wash out the vasodilator metabolites, so that blood flow can fall to pre-exercise levels a few minutes after exercise ends.

CHECKPOINT

6a. Describe the relationship between blood flow, arterial blood pressure, and vascular resistance.

6b. Describe the relationship between vascular resistance and the radius of a vessel. Explain how blood flow can be diverted from one organ to another.

7a. Explain how vascular resistance and blood flow are regulated by (a) sympathetic adrenergic fibers, (b) sympathetic cholinergic fibers, and (c) parasympathetic fibers.

7b. Describe the formation and action of nitric oxide. Why is this molecule considered a paracrine regulator?

7c. Define *autoregulation* and explain how this process occurs through myogenic and metabolic mechanisms.

14.4 BLOOD FLOW TO THE HEART AND SKELETAL MUSCLES

Blood flow to the heart and skeletal muscles is regulated by both extrinsic and intrinsic mechanisms. These mechanisms provide increased blood flow when the metabolic requirements of these tissues are raised during exercise.

LEARNING OUTCOMES

After studying this section, you should be able to:

8. Explain the mechanisms that regulate blood flow to the heart and skeletal muscles.

9. Describe the circulatory changes that occur during exercise.

Survival requires that the heart and brain receive an adequate supply of blood at all times. The ability of skeletal muscles to respond quickly in emergencies and to maintain continued high levels of activity also may be critically important for survival. During such times, high rates of blood flow to the skeletal muscles must be maintained without compromising blood flow to the heart and brain. This is accomplished by mechanisms that increase the cardiac output and divert the blood away from the viscera and skin so that the heart, skeletal muscles, and brain receive a greater proportion of the total blood flow.

Clinical Investigation CLUES

When Charlie was found, he was dehydrated and his pulse was weak.

- How did dehydration affect the stroke volume and cause Charlie's pulse to be weak?

Aerobic Requirements of the Heart

The coronary arteries supply an enormous number of capillaries, which are packed within the myocardium at a density ranging from 2,500 to 4,000 per cubic millimeter of tissue. For comparison, fast-twitch skeletal muscles have a capillary density of 300 to 400 per cubic millimeter of tissue. As a consequence of its greater density of capillaries, each myocardial cell is within only 10 μm of a capillary, compared to an average distance of 70 μm in other organs. The exchange of gases by diffusion between myocardial cells and capillary blood thus occurs very quickly.

Contraction of the myocardium squeezes the coronary arteries. For this reason, unlike other organs, the blood flow in the coronary vessels is less during systole than diastole. For example, only 15% to 20% of the blood flow through the left ventricle occurs during systole when a person is not exercising. However, the myocardium contains large amounts of *myoglobin,* a pigment related to hemoglobin (the molecules in red blood cells that carry oxygen). Myoglobin in the myocardium stores oxygen during diastole and releases its oxygen during systole. In this way, the myocardial cells can receive a continuous supply of oxygen even though coronary blood flow is temporarily reduced during systole.

In addition to containing large amounts of myoglobin, heart muscle contains numerous mitochondria and aerobic respiratory enzymes. This indicates that—even more than slow-twitch skeletal muscles—the heart is extremely specialized for aerobic respiration. Almost all of the ATP produced in the heart is a result of aerobic respiration and oxidative phosphorylation within mitochondria (chapter 5, section 5.2). This is so effective that scientists estimate that the ATP within myocardial cells is completely turned over (broken down and resynthesized) every 10 seconds. The resting heart obtains 50% to 70% of this ATP from acetyl CoA produced by the β-oxidation of fatty acids (chapter 5, figure 5.14). The balance of ATP is derived almost equally from the aerobic respiration of glucose and lactate.

The normal heart always respires aerobically, even during heavy exercise when the metabolic demand for oxygen can rise to six times resting levels (largely due to increased cardiac rate). This increased oxygen requirement is met primarily by a corresponding increase in coronary blood flow, from about 80 ml at rest to about 400 ml per minute per 100 g tissue during heavy exercise.

Regulation of Coronary Blood Flow

The coronary arterioles contain both alpha- and beta-adrenergic receptors, which promote vasoconstriction and vasodilation, respectively. Norepinephrine released by sympathetic nerve fibers stimulates alpha-adrenergic receptors to raise vascular resistance at rest. Epinephrine released by the adrenal medulla can stimulate the beta-adrenergic receptors to produce vasodilation when the sympathoadrenal system is activated during the fight-or-flight reaction.

During heavy exercise, the oxygen consumption of the myocardium can increase from four to six times resting values. This involves an increased β-oxidation of fatty acids and an even greater increase in the oxidation of glucose, requiring a four- to sixfold increase in coronary blood flow due to vasodilation. The vasodilation and decreased resistance of the coronary circulation during exercise is produced partly by sympathoadrenal system changes, but mostly by intrinsic metabolic changes. As the metabolism of the myocardium increases, there are increased local tissue concentrations of carbon dioxide, K^+, and released paracrine regulators that include nitric oxide, adenosine, and prostaglandins. These act directly on the vascular smooth muscle to cause vasodilation. Exercise training (1) increases the density of coronary arterioles and capillaries, (2) increases the production of nitric oxide for promoting vasodilation, and (3) decreases the compression of the coronary vessels in systole, due to the lower cardiac rate (and thus frequency of systoles) in trained athletes.

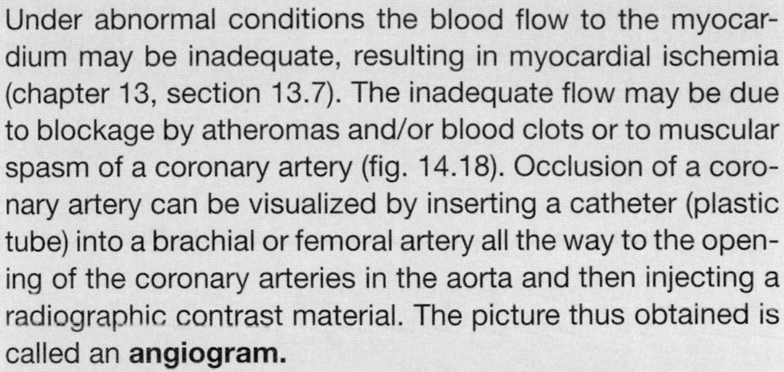

CLINICAL APPLICATION

Under abnormal conditions the blood flow to the myocardium may be inadequate, resulting in myocardial ischemia (chapter 13, section 13.7). The inadequate flow may be due to blockage by atheromas and/or blood clots or to muscular spasm of a coronary artery (fig. 14.18). Occlusion of a coronary artery can be visualized by inserting a catheter (plastic tube) into a brachial or femoral artery all the way to the opening of the coronary arteries in the aorta and then injecting a radiographic contrast material. The picture thus obtained is called an **angiogram.**

In a technique called **balloon angioplasty,** an inflatable balloon is used to open the coronary arteries. However, *restenosis* (recurrence of narrowing) often occurs. For this reason, a cylindrical support called a **stent** may be inserted to help keep the artery open. If the occlusion is sufficiently great, a **coronary bypass** may be performed. In this procedure, a length of blood vessel, usually taken from the saphenous vein in the leg, is sutured to the aorta and to the coronary artery at a location beyond the site of the occlusion (fig. 14.19).

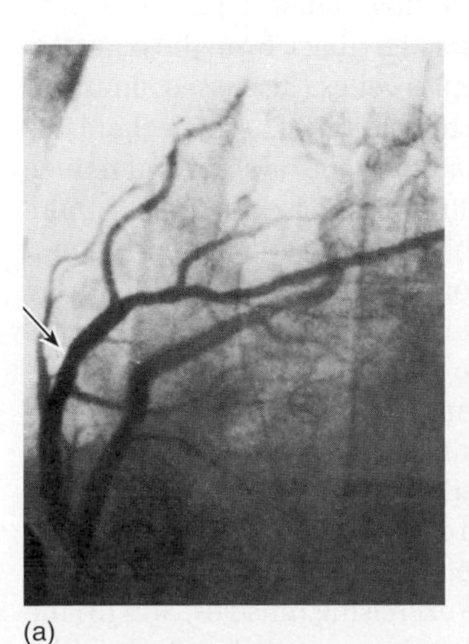

(a)

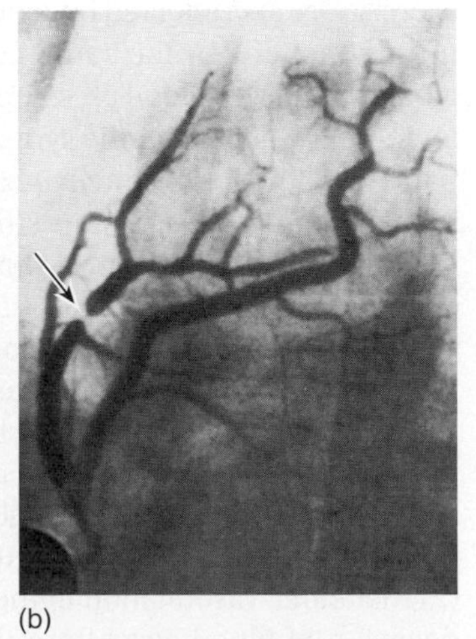

(b)

Figure 14.18 Angiograms of the left coronary artery of a heart patient. These angiograms were taken (*a*) when the patient's ECG was normal and (*b*) when the ECG showed evidence of myocardial ischemia. AP|R

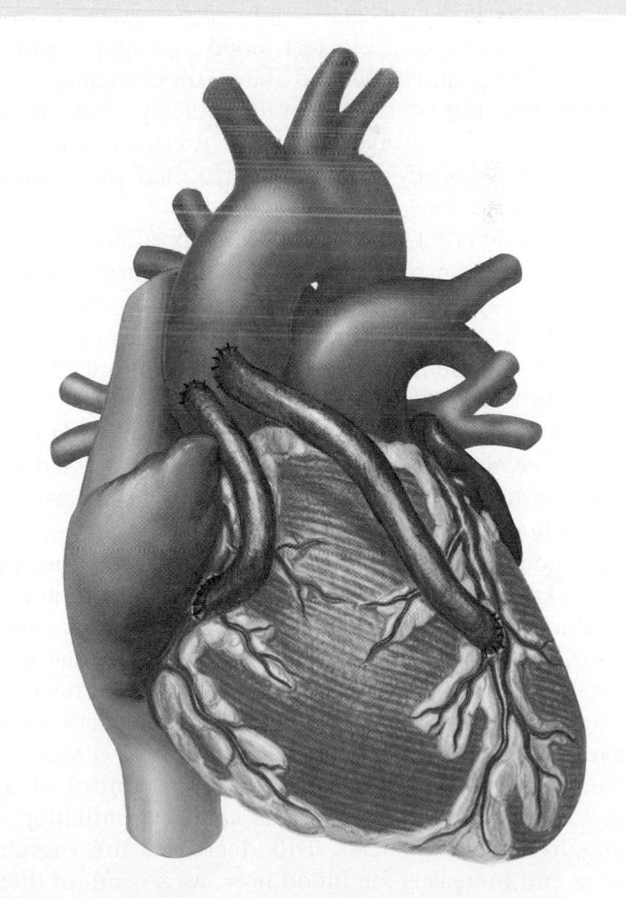

Figure 14.19 A diagram of coronary artery bypass surgery. Segments of the saphenous vein of the patient are commonly used as coronary bypass vessels.

Table 14.5 | Changes in Skeletal Muscle Blood Flow Under Conditions of Rest and Exercise

Condition	Blood Flow (ml/min)	Mechanism
Rest	1,000	High adrenergic sympathetic stimulation of vascular alpha receptors, causing vasoconstriction
Beginning exercise	Increased	Dilation of arterioles in skeletal muscles due to cholinergic sympathetic nerve activity and stimulation of beta-adrenergic receptors by the hormone epinephrine
Heavy exercise	20,000	Fall in alpha-adrenergic activity Increased cholinergic sympathetic activity Increased metabolic rate of exercising muscles, producing intrinsic vasodilation

Regulation of Blood Flow Through Skeletal Muscles

The arterioles in skeletal muscles, like those of the coronary circulation, have a high vascular resistance at rest as a result of alpha-adrenergic sympathetic stimulation. This produces a relatively low blood flow. Because muscles have such a large mass, however, they still receive from 20% to 25% of the total blood flow in the body at rest. Also, as in the heart, blood flow in a skeletal muscle decreases when the muscle contracts and squeezes its arterioles, and in fact blood flow stops entirely when the muscle contracts beyond about 70% of its maximum. Pain and fatigue thus occur much more quickly when an isometric contraction is sustained (in *static exercise*) than when rhythmic isotonic contractions are performed (in *dynamic exercise*).

In addition to adrenergic fibers, which promote vasoconstriction by stimulation of alpha-adrenergic receptors, there are also cholinergic sympathetic fibers in skeletal muscles. These cholinergic fibers, together with the stimulation of beta-adrenergic receptors by the hormone epinephrine, stimulate vasodilation as part of the fight-or-flight response to any stressful state, including that existing just prior to exercise (table 14.5). These extrinsic controls have been previously discussed and function to regulate blood flow through muscles at rest and upon the anticipation of exercise.

As dynamic exercise progresses, the vasodilation and increased skeletal muscle blood flow that occur are almost entirely due to intrinsic metabolic control. The high metabolic rate of skeletal muscles during exercise causes local changes, such as increased carbon dioxide concentrations, decreased pH (due to carbonic acid and lactic acid), decreased oxygen, increased extracellular K^+, and the secretion of adenosine. As in the intrinsic control of the coronary circulation, these changes cause vasodilation of arterioles in skeletal muscles. This decreases the vascular resistance and increases the blood flow. As a result of these changes, skeletal muscles can receive as much as 85% of the total blood flow in the body during maximal exercise (fig. 14.20).

Table 14.6 | Relationship Between Age and Average Maximum Cardiac Rate*

Age	Maximum Cardiac Rate
20–29	190 beats/min
30–39	160 beats/min
40–49	150 beats/min
50–59	140 beats/min
60+	130 beats/min

*Maximum cardiac rate can be estimated by subtracting your age from 220.

Circulatory Changes During Exercise

Both breathing and pulse rate increase within one second of exercise, suggesting that the motor cortex responsible for originating the exercise also influences the cardiovascular adjustments to exercise. However, cardiovascular changes during exercise are also affected by sensory feedback from the contracting muscles and by the baroreceptor reflex (discussed shortly in conjunction with blood pressure regulation). These mechanisms increase the activity of the sympathoadrenal system and reduce parasympathetic nerve activity during exercise. As a result, there is an increase in cardiac rate, stroke volume, and cardiac output.

The vascular resistance through muscles decreases during dynamic exercise but increases during static (isometric exercise), while the resistance to flow through the visceral organs and skin increases in both static and dynamic exercise. The increased resistance through the viscera and skin is produced by vasoconstriction due to increased activity of adrenergic sympathetic fibers. In summary, the blood flow through dynamically exercising muscles increases due to: (1) increased total blood flow (cardiac output); (2) metabolic vasodilation in the exercising muscles; and (3) the diversion of blood away from the viscera and skin. For similar reasons, coronary blood flow to cardiac muscle also increases significantly during exercise.

The total blood flow to the brain is relatively constant (fig. 14.20), but recent studies suggest that cerebral blood flow

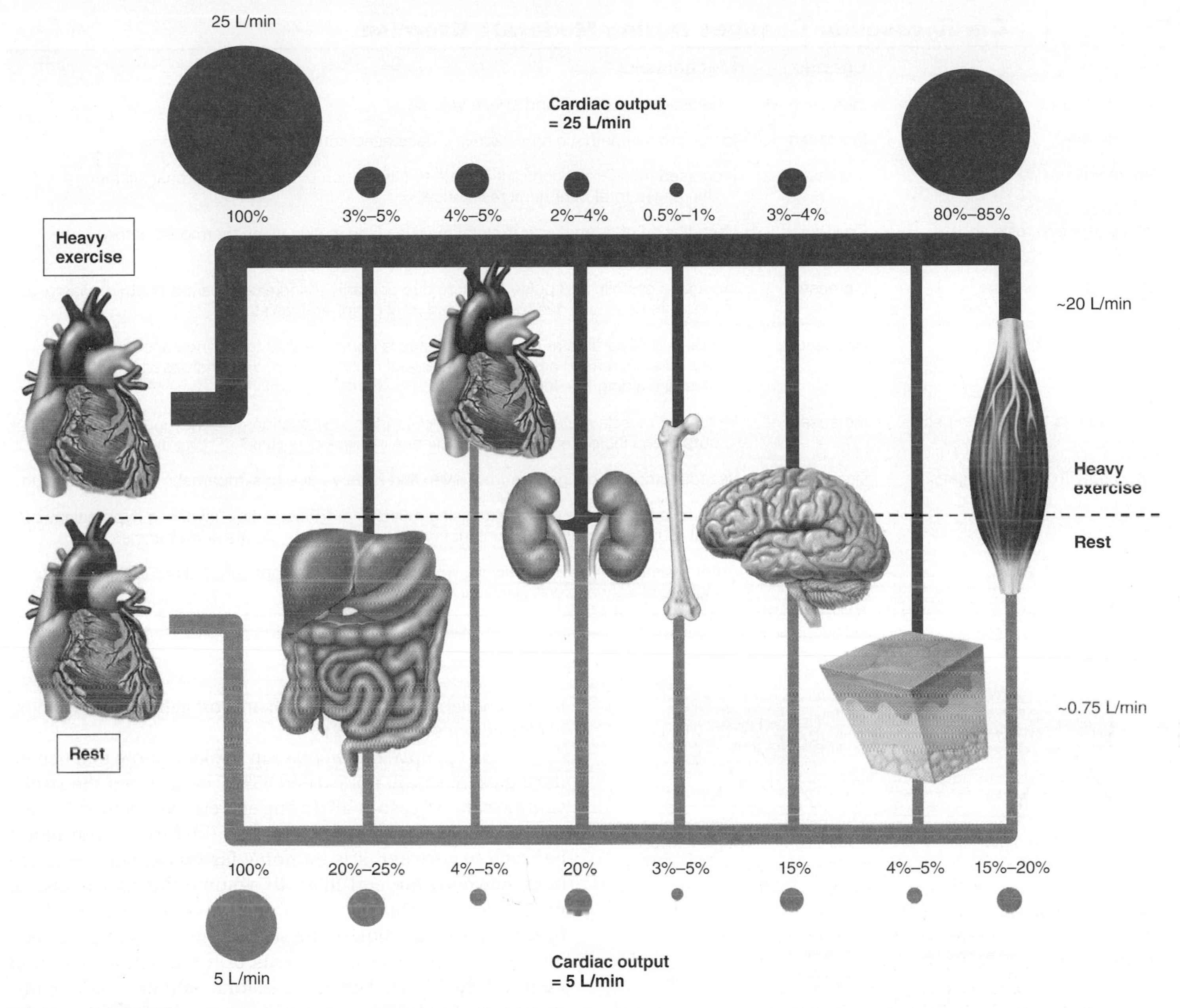

Figure 14.20 The distribution of blood flow (cardiac output) during rest and heavy exercise. At rest, the cardiac output is 5 L per minute (*bottom of figure*); during heavy exercise the cardiac output increases to 25 L per minute (*top of figure*). At rest, for example, the brain receives 15% of 5 L per minute (= 750 ml/min), whereas during exercise it receives 3% to 4% of 25 L per minute (0.03 × 25 L/min = 750 ml/min; at 4%, the brain receives 1000 ml/min). Flow to the skeletal muscles increases more than twentyfold because the total cardiac output increases (from 5 L/min to 25 L/min) and because the percentage of the total received by the muscles increases from 15% to 80%. Note that the percentage of blood flow to the skin during heavy exercise was immeasurably low, so no percentage is indicated. Also, the percentages for the blood flow to different organs at rest only add up to 95% because these percentages are estimated averages.

increases somewhat during light to moderate exercise. This is believed to result from vasodilation induced by increased metabolism of the brain regions responsible for motor control and somatosensory information. By contrast, during heavier exercise (at greater than 60% of the maximal oxygen uptake) the cerebral blood flow decreases somewhat. This is because the person hyperventilates, which lowers the blood CO_2 and thereby produces cerebral vasoconstriction. This cerebral vasoconstriction during heavy exercise may contribute to central fatigue (chapter 12, section 12.4).

During exercise, the cardiac output can increase fivefold—from about 5 L per minute to about 25 L per minute. This is primarily due to an increase in cardiac rate. The cardiac rate, however, can increase only up to a maximum value (table 14.6),

Table 14.7 | Cardiovascular Changes During Moderate Exercise

Variable	Change	Mechanisms
Cardiac output	Increased	Increased cardiac rate and stroke volume
Cardiac rate	Increased	Increased sympathetic nerve activity; decreased activity of the vagus nerve
Stroke volume	Increased	Increased myocardial contractility due to stimulation by sympathoadrenal system; decreased total peripheral resistance
Total peripheral resistance	Decreased	Vasodilation of arterioles in skeletal muscles (and in skin when thermoregulatory adjustments are needed)
Arterial blood pressure	Increased	Increased systolic and pulse pressure due primarily to increased cardiac output; diastolic pressure rises less due to decreased total peripheral resistance
End-diastolic volume	Unchanged	Decreased filling time at high cardiac rates is compensated for by increased venous pressure, increased activity of the skeletal muscle pump, and decreased intrathoracic pressure aiding the venous return
Blood flow to heart and muscles	Increased	Increased muscle metabolism produces intrinsic vasodilation; aided by increased cardiac output and increased vascular resistance in visceral organs
Blood flow to visceral organs	Decreased	Vasoconstriction in digestive tract, liver, and kidneys due to sympathetic nerve stimulation
Blood flow to skin	Increased	Metabolic heat produced by exercising muscles produces reflex (involving hypothalamus) that reduces sympathetic constriction of arteriovenous shunts and arterioles
Blood flow to brain	Unchanged*	Autoregulation of cerebral vessels, which maintains constant cerebral blood flow despite increased arterial blood pressure

*There can be slight changes in cerebral blood flow (see text), but the extent of these changes is buffered by autoregulation due to myogenic control mechanisms.

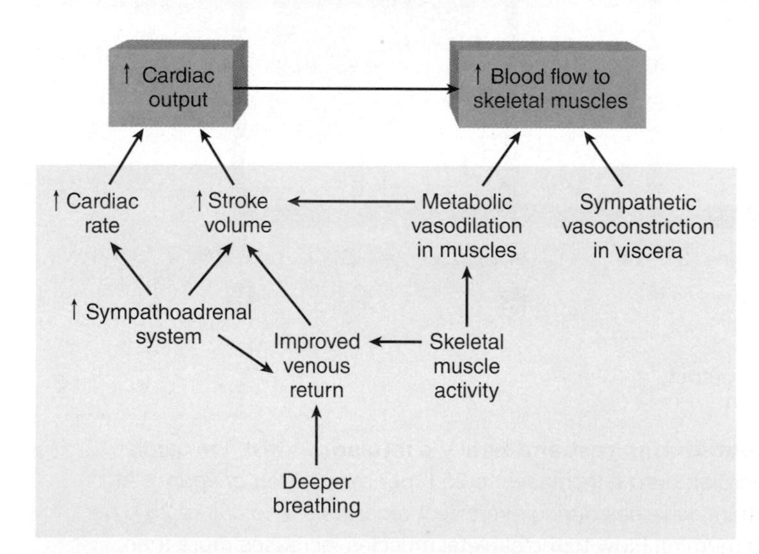

Figure 14.21 Cardiovascular adaptations to exercise. These adaptations (1) increase the cardiac output, and thus the total blood flow; and (2) cause vasodilation in the exercising muscles, thereby diverting a higher proportion of the blood flow to those muscles.

which is determined mainly by a person's age. In well-trained athletes, the stroke volume can also increase significantly, allowing these individuals to achieve cardiac outputs during strenuous exercise up to six or seven times greater than their resting values. This high cardiac output results in increased oxygen delivery to the exercising muscles; this is the major reason for the much higher than average maximal oxygen uptake ($\dot{V}_{O_2}$ max) of elite athletes (chapter 12).

In most people, the increase in stroke volume that occurs during exercise will not exceed 35%. The fact that the stroke volume can increase at all during exercise may at first be surprising, given that the heart has less time to fill with blood between beats when it is pumping faster. Despite the faster beat, however, the end-diastolic volume during exercise is not decreased. This is because the venous return is aided by the improved action of the skeletal muscle pumps and by increased respiratory movements of the diaphragm during exercise (fig. 14.21). The end-diastolic volume is not significantly changed during exercise, so any increase in stroke volume that occurs must be due to an increase in the proportion of blood ejected per stroke.

The proportion of the end-diastolic volume ejected per stroke can increase from 60% at rest to as much as 90% during heavy exercise. This increased *ejection fraction* is produced by the increased contractility that results from sympathoadrenal stimulation. There also may be a decrease in total peripheral resistance as a result of vasodilation in the exercising skeletal muscles, which decreases the afterload and thus further augments the increase in stroke volume. The cardiovascular changes that occur during exercise are summarized in table 14.7.

Endurance training often results in a lowering of the resting cardiac rate and an increase in the resting stroke volume. The lowering of the resting cardiac rate results from a greater inhibition of the SA node by the vagus nerve. The increased resting stroke volume is believed to be due to an increase in blood volume; indeed, studies have shown that the blood volume can increase by about 500 ml after only eight days of training. These adaptations enable the trained athlete to produce a larger proportionate increase in cardiac output and achieve a higher absolute cardiac output during exercise. This large cardiac output is the major reason for the improved oxygen delivery to skeletal muscles that occurs as a result of endurance training.

CHECKPOINT

8. Describe blood flow and oxygen delivery to the myocardium during systole and diastole.
9a. State how blood flow to the heart is affected by exercise. Explain how blood flow to the heart is regulated at rest and during exercise.
9b. Describe the mechanisms that produce vasodilation in exercising muscles, and list the cardiovascular changes that contribute to increased blood flow to exercising muscles.
9c. Explain how the stroke volume can increase during exercise despite a shorter time available for the heart chambers to fill with blood.

14.5 BLOOD FLOW TO THE BRAIN AND SKIN

Intrinsic control mechanisms help maintain a relatively constant blood flow to the brain. Blood flow to the skin, by contrast, can vary tremendously in response to regulation by sympathetic nerve stimulation.

LEARNING OUTCOMES

After studying this section, you should be able to:

10. Explain how blood flow to the brain is regulated.
11. Explain how blood flow to the skin is regulated.

The examination of cerebral and cutaneous blood flow is a study in contrasts. Cerebral blood flow is regulated primarily by intrinsic mechanisms; cutaneous blood flow is regulated by extrinsic mechanisms. Cerebral blood flow is relatively constant; cutaneous blood flow exhibits more variation than blood flow in any other organ. The brain is the organ that can least tolerate low rates of blood flow, whereas the skin can tolerate low rates of blood flow better than any other organ.

Cerebral Circulation

When the brain is deprived of oxygen for just a few seconds, a person loses consciousness; irreversible brain injury may occur after a few minutes. For these reasons, the cerebral blood flow is held remarkably constant at about 750 ml per minute. This amounts to about 15% of the total cardiac output at rest.

Unlike the coronary and skeletal muscle blood flow, cerebral blood flow is not much influenced by sympathetic nerve activity under normal conditions. Only when the mean arterial pressure rises to about 200 mmHg do sympathetic nerves cause a significant degree of vasoconstriction in the cerebral circulation. This vasoconstriction helps to protect small, thin-walled arterioles from bursting under the pressure, and thus helps to prevent cerebrovascular accident (stroke).

In the normal range of arterial pressures, cerebral blood flow is regulated almost exclusively by local intrinsic mechanisms—a process called *autoregulation,* as previously mentioned. These mechanisms help to ensure a relatively constant blood flow despite changes in systemic arterial pressure. The autoregulation of cerebral blood flow is achieved by both myogenic and metabolic mechanisms.

Myogenic Regulation

Myogenic regulation occurs when there is variation in systemic arterial pressure. When the blood pressure falls, the cerebral arteries automatically dilate; when the pressure rises, they constrict. This helps to maintain a constant flow rate during the normal pressure variations that occur during rest, exercise, and emotional states.

The cerebral vessels are also sensitive to the carbon dioxide concentration of arterial blood. When the carbon dioxide concentration rises as a result of inadequate ventilation (hypoventilation), the cerebral arterioles dilate. This is believed to be due to decreases in the pH of cerebrospinal fluid rather than to a direct effect of CO_2 on the cerebral vessels. Conversely, when the arterial CO_2 falls below normal during hyperventilation, the cerebral vessels constrict. The resulting decrease in cerebral blood flow is responsible for the dizziness that occurs during hyperventilation.

Metabolic Regulation

Although the mechanisms just described maintain a relatively constant total cerebral blood flow, the particular brain regions that are most active receive an increased blood flow. Indeed, active brain regions are *hyperemic*—their blood flow actually exceeds the aerobic requirements of the active neurons.

This shunting of blood between different brain regions occurs because the cerebral arterioles are exquisitely sensitive to local changes in metabolic activity, so that those brain regions with the highest metabolic activity receive the most blood. Indeed, areas of the brain that control specific processes have been mapped by the changing patterns of blood flow that

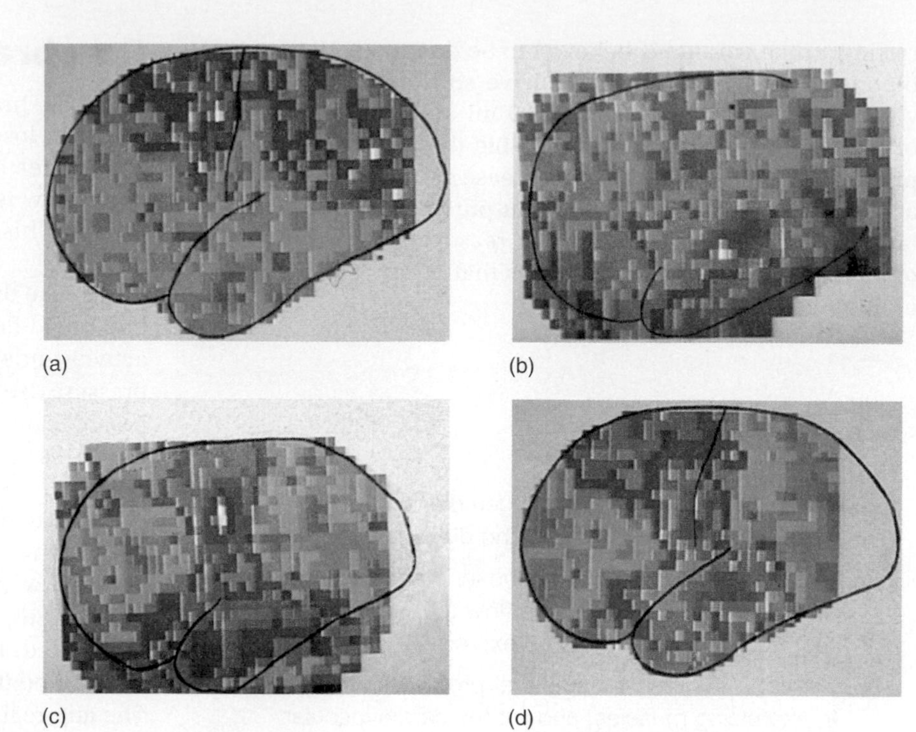

Figure 14.22 Changing patterns of blood flow in the brain. (*a*) A subject is following a moving object with his eyes; high activity is seen over the occipital lobe (visual cortex). (*b*) A subject is listening to words; high activity is seen over the temporal lobe (auditory cortex). (*c*) The subject moves his fingers on the side of the body opposite (contralateral) to the cerebral hemisphere being recorded; activity is seen over the motor cortex. (*d*) The subject counts to 20; activity is seen over the mouth area of the motor cortex, the supplementary motor area, and the auditory cortex. AP|R

result when these areas are activated. Visual and auditory stimuli, for example, increase blood flow to the appropriate sensory areas of the cerebral cortex, whereas motor activities, such as movements of the eyes, arms, and organs of speech, result in different patterns of blood flow (fig. 14.22).

The mechanisms by which increased activity within a brain region causes increased blood flow to that region are complex and not completely understood. Active neurons release many substances that stimulate vasodilation, including K^+, adenosine, nitric oxide (NO), and others. The close association of astrocytes with both neurons and cerebral vessels (chapter 7; see fig. 7.10) suggests that they may also play a role. Indeed, astrocytes have been shown to secrete vasodilator chemicals (including prostaglandin E_2 and carbon monoxide) when stimulated by the neurotransmitter glutamate, released into the synapse by neurons. Molecules released by astrocytes and active neurons could also stimulate the endothelial cells of the arterioles to produce vasodilators, including nitric oxide.

In this way, neurons, astrocytes, and endothelial cells form functioning units that couple increased neural activity to vasodilation, and thus to increased blood flow. The local vasodilation may be propagated upstream, along the wall of the arteriole into the vessels of the pia mater through gap junctions, which are located between adjacent endothelial cells and adjacent smooth muscle cells. This could serve to more effectively increase blood flow to the more active brain regions.

Cutaneous Blood Flow

The skin is the outer covering of the body and as such serves as the first line of defense against invasion by disease-causing organisms. The skin, as the interface between the internal and external environments, also helps to maintain a constant deep-body temperature despite changes in the ambient (external) temperature—a process called *thermoregulation.* The thinness and large area of the skin (1.0 to 1.5 mm thick; 1.7 to 1.8 square meters in surface area) make it an effective radiator of heat when the body temperature rises above the ambient temperature. The transfer of heat from the body to the external environment is aided by the flow of warm blood through capillary loops near the surface of the skin.

Blood flow through the skin is adjusted to maintain deep-body temperature at about 37°C (98.6°F). These adjustments are made by variations in the degree of constriction or dilation of ordinary arterioles and of unique **arteriovenous anastomoses** (fig. 14.23). These latter vessels, found predominantly in the fingertips, palms of the hands, toes, soles of the feet, ears, nose, and lips, shunt (divert) blood directly from arterioles to deep venules, thus bypassing superficial capillary loops. Both the ordinary arterioles and the arteriovenous anastomoses are innervated by sympathetic nerve fibers. When the ambient temperature is low, sympathetic nerves stimulate cutaneous vasoconstriction. Cutaneous blood flow is thus decreased, so that less heat will be lost from the body. Because the arteriovenous anastomoses also constrict, the skin may appear rosy because the blood is diverted to the superficial capillary loops. In spite of this rosy appearance, however, the total cutaneous blood flow and rate of heat loss is lower than under usual conditions.

Skin can tolerate an extremely low blood flow in cold weather because its metabolic rate decreases when the ambient temperature decreases. In cold weather, therefore, the skin requires less blood. As a result of exposure to extreme cold, however, blood flow to the skin can be so severely restricted that the tissue dies—a condition known as *frostbite.* Blood

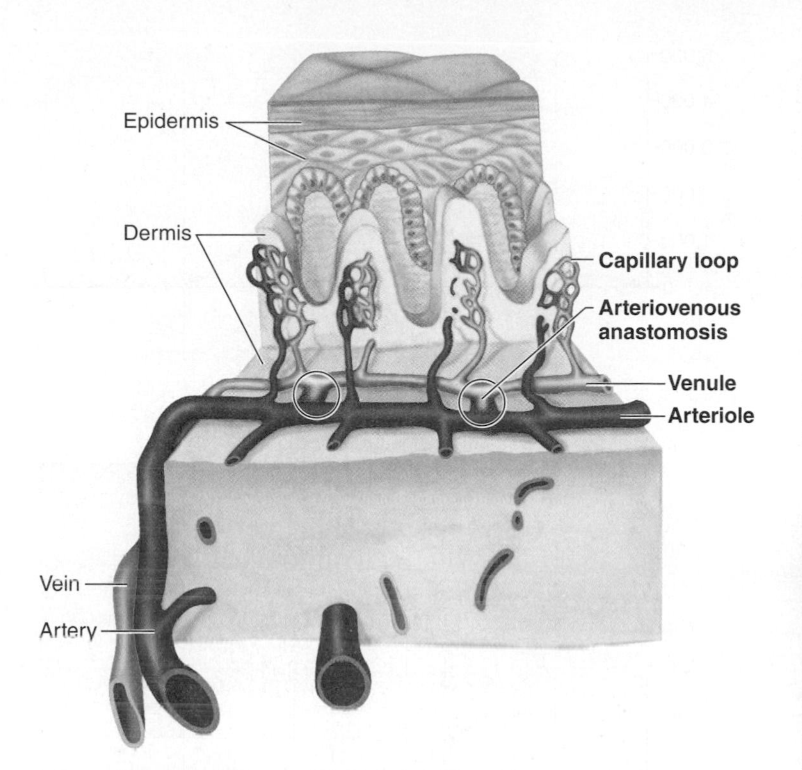

Figure 14.23 Circulation in the skin showing arteriovenous anastomoses. These vessels function as shunts, allowing blood to be diverted directly from the arteriole to the venule, and thus to bypass superficial capillary loops. 

flow to the skin can vary from less than 20 ml per minute at maximal vasoconstriction to as much as 3 to 4 L per minute at maximal vasodilation.

As the temperature warms, cutaneous arterioles in the hands and feet dilate as a result of decreased sympathetic nerve activity. Continued warming causes dilation of arterioles in other areas of the skin. If the resulting increase in cutaneous blood flow is not sufficient to cool the body, sweat gland secretion may be stimulated. Perspiration helps cool the body as it evaporates from the surface of the skin. The sweat glands also secrete **bradykinin,** a polypeptide that stimulates vasodilation.

In usual ambient temperatures, the cutaneous vascular resistance is high and the blood flow is low when a person is not exercising. In the pre-exercise state of fight or flight, sympathetic nerve activity reduces cutaneous blood flow still further. During exercise, however, the need to maintain a deep-body temperature takes precedence over the need to maintain an adequate systemic blood pressure. As the body temperature rises during exercise, vasodilation in cutaneous vessels occurs together with vasodilation in the exercising muscles.

This can cause an even greater lowering of total peripheral resistance during exercise. However, the mean arterial pressure is still high during exercise because of the increased cardiac output. If a person exercises in hot and humid weather, especially if restrictive clothing prevents adequate evaporative cooling of the skin (from perspiration), the increased skin temperature and resulting vasodilation can persist after exercise has ceased. If the total peripheral resistance remains very low as the cardiac output declines toward resting values, the blood pressure may fall precipitously; people have lost consciousness and even died as a result.

Changes in cutaneous blood flow occur as a result of changes in sympathetic nerve activity. Because the activity of the sympathetic nervous system is controlled by the brain, emotional states, acting through control centers in the medulla oblongata, can affect sympathetic activity and cutaneous blood flow. During fear reactions, for example, vasoconstriction in the skin, along with activation of the sweat glands, can produce a pallor and a "cold sweat." Other emotions may cause vasodilation and blushing.

Clinical Investigation CLUES 

When Charlie was first found, his skin was cold to the touch.

- What does this suggest about the cutaneous blood flow, and what mechanism is responsible?
- What benefit does Charlie derive from this mechanism?

CHECKPOINT

10a. Define the term *autoregulation* and describe how this process is accomplished in the cerebral circulation.

10b. Explain how hyperventilation can cause dizziness.

11. Explain how cutaneous blood flow is adjusted to maintain a constant deep-body temperature.

14.6 BLOOD PRESSURE

The pressure of the arterial blood is affected by the blood volume, total peripheral resistance, and the cardiac rate. These variables are regulated by a variety of negative feedback control mechanisms to maintain homeostasis. Arterial pressure rises and falls as the heart goes through systole and diastole.

LEARNING OUTCOMES

After studying this section, you should be able to:

12. Explain how blood pressure is regulated.

13. Describe how blood pressure is measured.

Resistance to flow in the arterial system is greatest in the arterioles because these vessels have the smallest diameters. Although the total blood flow through a system of arterioles

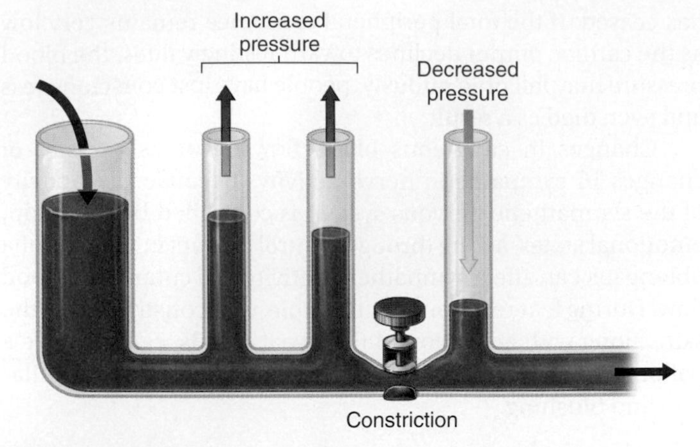

Figure 14.24 The effect of vasoconstriction on blood pressure. A constriction increases blood pressure upstream (analogous to the arterial pressure) and decreases pressure downstream (analogous to capillary and venous pressure).

must be equal to the flow in the larger vessel that gave rise to those arterioles, the narrow diameter of each arteriole reduces the flow in each according to Poiseuille's law. Blood flow and pressure are thus reduced in the capillaries, which are located downstream of the high resistance imposed by the arterioles. (The slow velocity of blood flow through capillaries enhances diffusion across the capillary wall.) The blood pressure upstream of the arterioles—in the medium and large arteries—is correspondingly increased (fig. 14.24).

The total cross-sectional area of capillaries is greater (due to their large number) than the cross-sectional areas of the arteries and arterioles (fig. 14.25), further reducing the capillary blood pressure and flow. Thus, although each capillary is much narrower than each arteriole, the capillary beds served by arterioles do not provide as great a resistance to blood flow as do the arterioles.

Variations in the diameter of arterioles as a result of vasoconstriction and vasodilation thus affect blood flow through capillaries and, simultaneously, the *arterial blood pressure* "upstream" from the capillaries. In this way, an increase in total peripheral resistance due to vasoconstriction of arterioles can raise arterial blood pressure. Blood pressure can also be raised by an increase in the cardiac output. This may be due to elevations in cardiac rate or in stroke volume, which in turn are affected by other factors. The most important variables affecting blood pressure are thus the **cardiac rate, stroke volume** (determined primarily by the **blood volume**), and **total peripheral resistance.** An increase in any of these, if not compensated for by a decrease in another variable, will result in an increased blood pressure.

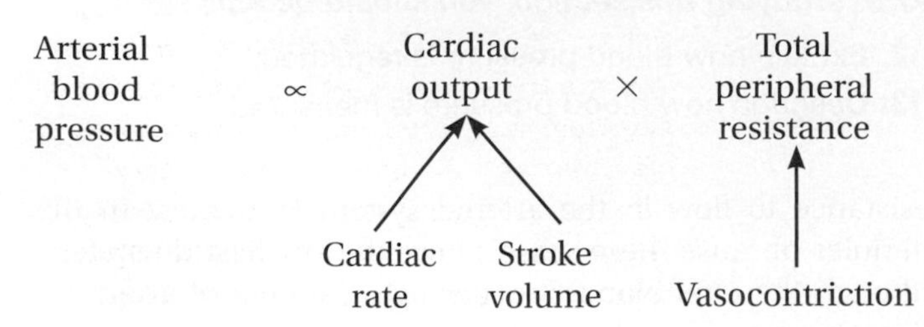

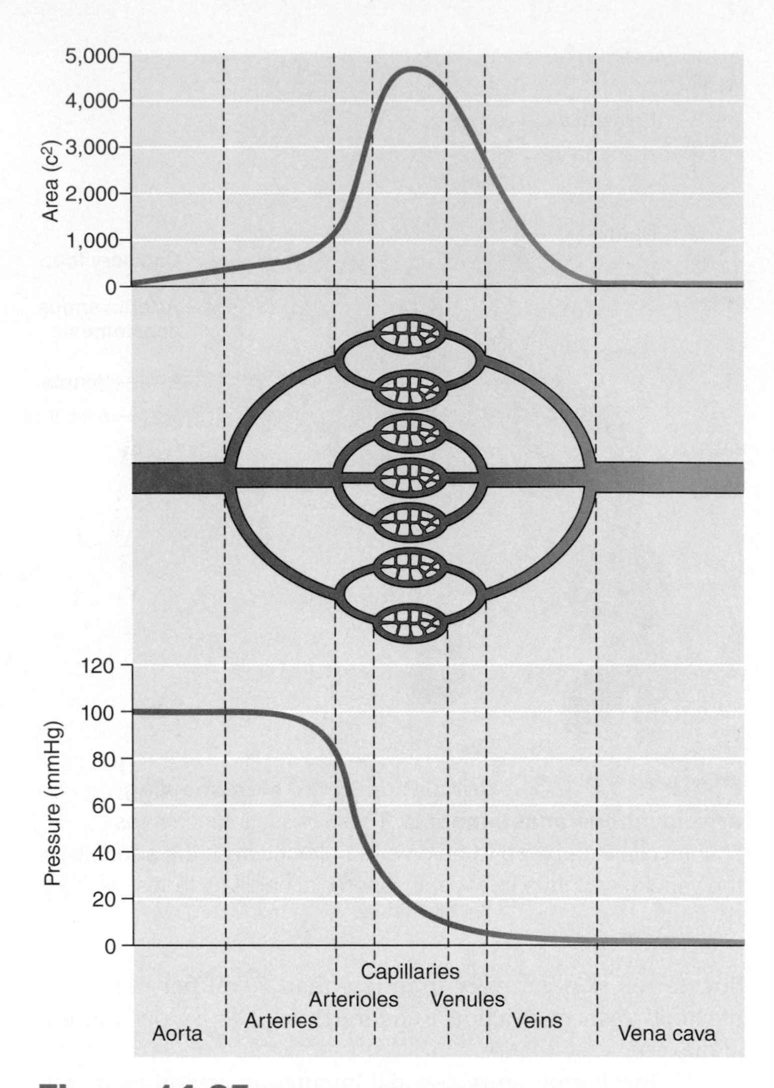

Figure 14.25 The relationship between blood pressure and the cross-sectional area of vessels. As blood passes from the aorta to the smaller arteries, arterioles, and capillaries, the cross-sectional area increases as the pressure decreases.

Blood pressure can be regulated by the kidneys, which control blood volume and thus stroke volume, and by the sympathoadrenal system. Increased activity of the sympathoadrenal system can raise blood pressure by stimulating vasoconstriction of arterioles (thereby raising total peripheral resistance) and by promoting an increased cardiac output. Sympathetic stimulation can also affect blood volume indirectly, by stimulating constriction of renal blood vessels and thus reducing urine output.

Blood pressure is measured in units of **millimeters of mercury (mmHg).** When this measurement is taken, the blood pushes on one surface of a U-shaped column of mercury while the atmosphere pushes on the other surface (see chapter 16, fig. 16.18). If the blood pressure were equal to the atmospheric pressure, the measurement would be 0 mmHg. A mean arterial pressure of 100 mmHg indicates that the blood pressure is 100 mmHg higher than the atmospheric pressure. Instruments used to measure blood pressure, called **sphygmomanometers,** contain mercury or are spring-loaded devices that are calibrated against mercurial instruments.

Baroreceptor Reflex

In order for blood pressure to be maintained within limits, specialized receptors for pressure are needed. These **baroreceptors** are stretch receptors located in the *aortic arch* and in the *carotid sinuses.* The baroreceptors are tonically (constantly) active, producing a baseline frequency of action potentials in their sensory neurons. When blood pressure is increased the walls of the aortic and carotid sinuses stretch, and this produces an increased frequency of action potentials along their sensory nerve fibers (fig. 14.26). A fall in blood pressure below the normal range, by contrast, causes a decreased frequency of action potentials in these sensory fibers.

Sensory nerve activity from the baroreceptors ascends via the vagus (X) and glossopharyngeal (IX) nerves to the medulla oblongata, which directs the autonomic system to respond appropriately. The **vasomotor control center** in the medulla regulates the degree of vasoconstriction/vasodilation, and hence helps to regulate total peripheral resistance. The **cardiac control center** in the medulla regulates the cardiac rate (fig. 14.27).

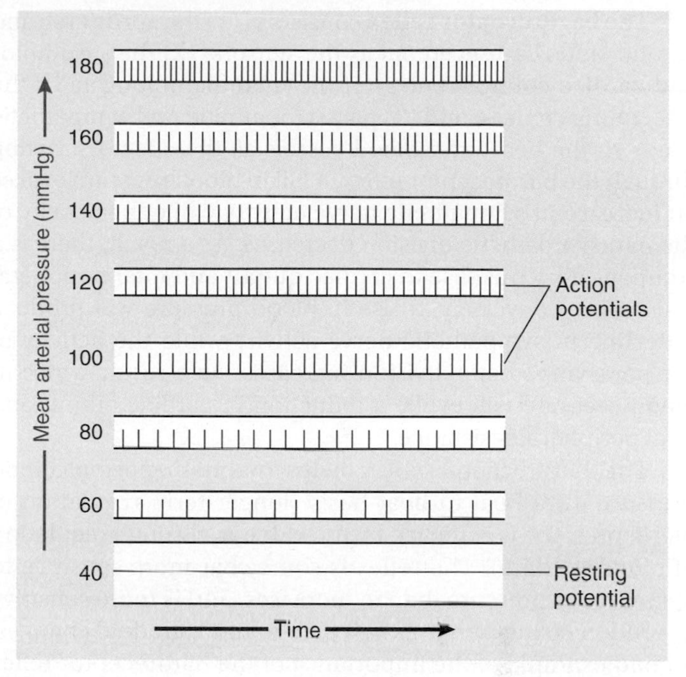

Figure 14.26 The effect of blood pressure on the baroreceptor response. This is a recording of the action potential frequency in sensory nerve fibers from baroreceptors in the carotid sinus and aortic arch. As the blood pressure increases, the baroreceptors become increasingly stretched. This results in a higher frequency of action potentials transmitted to the cardiac and vasomotor control centers in the medulla oblongata.

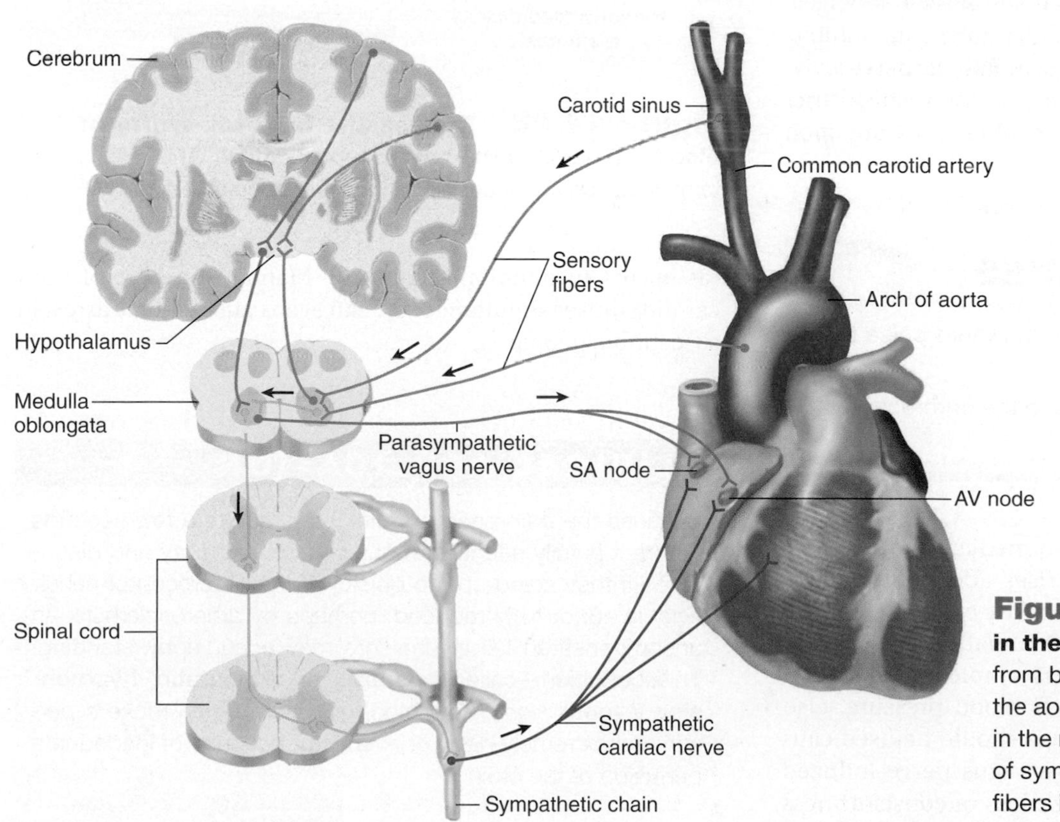

Figure 14.27 Structures involved in the baroreceptor reflex. Sensory stimuli from baroreceptors in the carotid sinus and the aortic arch, acting via control centers in the medulla oblongata, affect the activity of sympathetic and parasympathetic nerve fibers in the heart. AP|R

The **baroreceptor reflex** consists of (1) the aortic arch and carotid sinus baroreceptors as the sensors; (2) the vasomotor and cardiac control centers of the medulla oblongata as the integrating centers; and (3) parasympathetic and sympathetic axons to the heart and blood vessels as the effectors. Acting through the baroreceptor reflex, a fall in blood pressure evokes an increase in sympathetic nerve activity while the activity of the parasympathetic division decreases. As a result, there is a compensatory increase in cardiac output and total peripheral resistance. Conversely, a rise in blood pressure will produce a decline in sympathetic nerve activity while the activity of the parasympathetic division increases. As a result, a rise in blood pressure will evoke a reduction in cardiac output and total peripheral resistance.

The baroreceptor reflex helps maintain normal blood pressure on a beat-to-beat basis (longer-term regulation of blood pressure is achieved by the kidneys, through regulation of blood volume). The reflex is somewhat more sensitive to decreases in pressure than to increases, and is more sensitive to sudden changes in pressure than to more gradual changes. A good example of the importance of the baroreceptor reflex in normal physiology is its activation whenever a person goes from a lying to a standing position.

When a person goes from a lying to a standing position, there is a shift of 500 to 700 ml of blood from the veins of the thoracic cavity to veins in the lower extremities, which expand to contain the extra volume of blood. This pooling of blood in the lower extremities reduces the venous return and cardiac output, but the resulting fall in blood pressure is almost immediately compensated for by the baroreceptor reflex. A decrease in baroreceptor sensory information, traveling in the glossopharyngeal nerve and the vagus nerve to the medulla oblongata, inhibits parasympathetic activity and promotes sympathetic nerve activity. This produces an increase in cardiac rate and vasoconstriction, which help to maintain an adequate blood pressure upon standing (fig. 14.28).

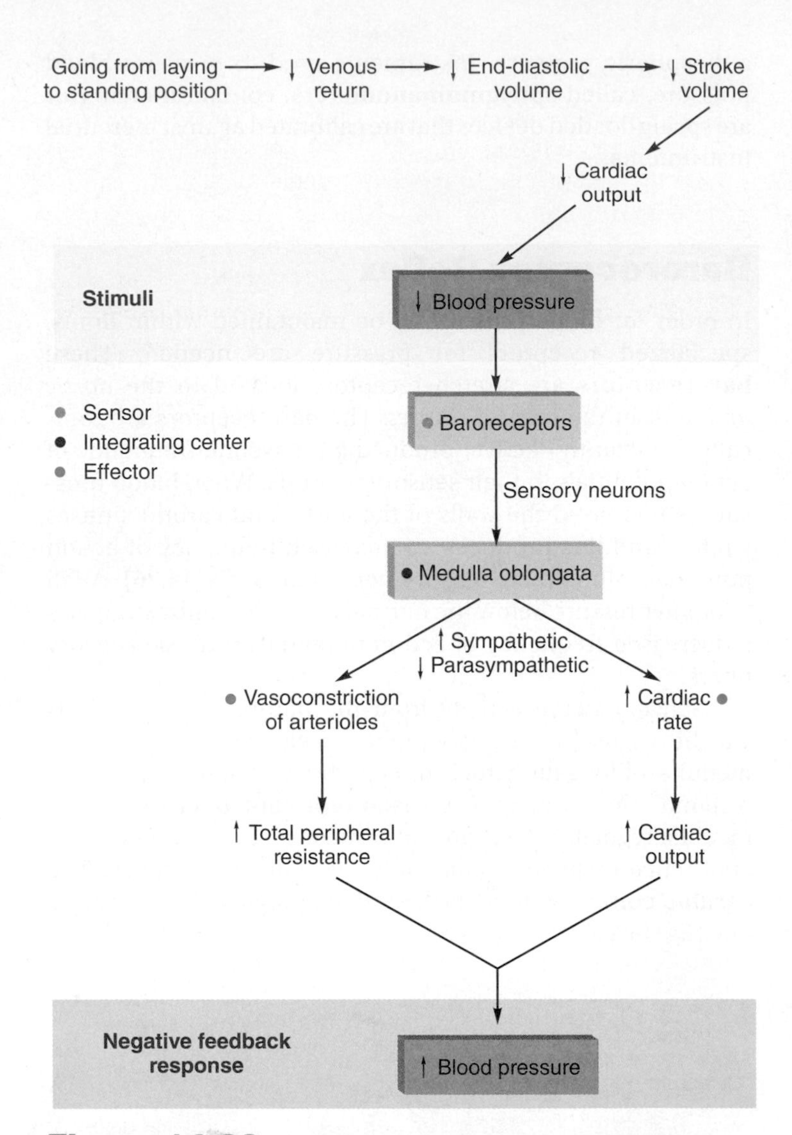

Figure 14.28 The negative feedback control of blood pressure by the baroreceptor reflex. This reflex helps to maintain an adequate blood pressure upon standing.

Clinical Investigation CLUES

When Charlie was first found, he had a rapid pulse and cold skin.

- How do these symptoms relate to the baroreceptor reflex?

Input from baroreceptors can also mediate the opposite response. When the blood pressure rises above an individual's normal range, the baroreceptor reflex causes a slowing of the cardiac rate and vasodilation. Manual massage of the carotid sinus, a procedure sometimes employed by physicians to reduce tachycardia and lower blood pressure, also evokes this reflex. Such carotid massage should be used cautiously, however, because the intense vagus-nerve-induced slowing of the cardiac rate could cause loss of consciousness (as occurs in emotional fainting). Manual massage of both carotid sinuses simultaneously can even cause cardiac arrest in susceptible people.

CLINICAL APPLICATION

Because the baroreceptor reflex may require a few seconds before it is fully effective, many people feel dizzy and disoriented if they stand up too quickly. If the baroreceptor sensitivity is abnormally reduced, perhaps by atherosclerosis, an uncompensated fall in pressure may occur upon standing. This condition—called **postural,** or **orthostatic, hypotension** (*hypotension* = low blood pressure)—can make a person feel extremely dizzy or even faint because of inadequate perfusion of the brain.

Atrial Stretch Reflexes

In addition to the baroreceptor reflex, several other reflexes help to regulate blood pressure. The reflex control of ADH release by osmoreceptors in the hypothalamus, and the control of angiotensin II production and aldosterone secretion by the juxtaglomerular apparatus of the kidneys, have been previously discussed. Antidiuretic hormone and aldosterone increase blood pressure by increasing blood volume, and angiotensin II stimulates vasoconstriction to cause an increase in blood pressure.

Other reflexes important to blood pressure regulation are initiated by **atrial stretch receptors** located in the atria of the heart. These receptors are activated by increased venous return to the heart and, in response (1) stimulate reflex tachycardia, as a result of increased sympathetic nerve activity; (2) inhibit ADH release, resulting in the excretion of larger volumes of urine and a lowering of blood volume; and (3) promote increased secretion of atrial natriuretic peptide (ANP). The ANP, as previously discussed, lowers blood volume by increasing urinary salt and water excretion.

FITNESS APPLICATION

Valsalva's maneuver is an expiratory effort against a closed glottis (which prevents the air from escaping—chapter 16, section 16.1). This maneuver, commonly performed during forceful defecation or when lifting heavy weights, increases the intrathoracic pressure. Compression of thoracic veins causes a fall in venous return and cardiac output, thus lowering arterial blood pressure. The lowering of arterial pressure then stimulates the baroreceptor reflex, resulting in tachycardia and increased total peripheral resistance. When the glottis is finally opened and the air is exhaled, the cardiac output returns to normal. The total peripheral resistance is still elevated, however, causing a rise in blood pressure. The blood pressure is then brought back to normal by the baroreceptor reflex, which causes a slowing of the heart rate. These fluctuations in cardiac output and blood pressure can be dangerous in people with cardiovascular disease. Even healthy people are advised to exhale normally when lifting weights.

Measurement of Blood Pressure

The first documented measurement of blood pressure was accomplished by Stephen Hales (1677–1761), an English clergyman and physiologist. Hales inserted a cannula into the artery of a horse and measured the heights to which blood would rise in the vertical tube. The height of this blood column bounced between the **systolic pressure** at its highest and the **diastolic pressure** at its lowest, as the heart went through its cycle of systole and diastole. Modern clinical blood pressure measurements, fortunately, are less direct. The indirect, or **auscultatory,** method is based on the correlation of blood pressure and arterial sounds first described by the Russian physician Nicolai Korotkoff in 1905.

In the auscultatory method, an inflatable rubber bladder within a cloth cuff is wrapped around the upper arm, and a stethoscope is applied over the brachial artery (fig. 14.29). The artery is normally silent before inflation of the cuff, because blood travels smoothly through the arteries. **Laminar flow** occurs when all parts of a fluid move in the same direction, parallel to the axis of the vessel. The term *laminar* means "layered"—blood in the central axial stream moves the fastest, and blood flowing closer to the artery wall moves more slowly. If flow is perfectly laminar, there is no transverse movement between these layers that would produce mixing. The blood flows smoothly and doesn't produce vibrations of the artery wall that would cause sounds. By contrast, **turbulent flow** is when some parts of the fluid move in different directions, churning and mixing the blood. Turbulent flow causes vibrations of the vessel, which may produce sounds. Before the blood pressure cuff is inflated, blood flow in the brachial artery has very little turbulence and so is silent.

When the artery is pinched, however, blood flow through the constriction becomes turbulent. This causes the artery to

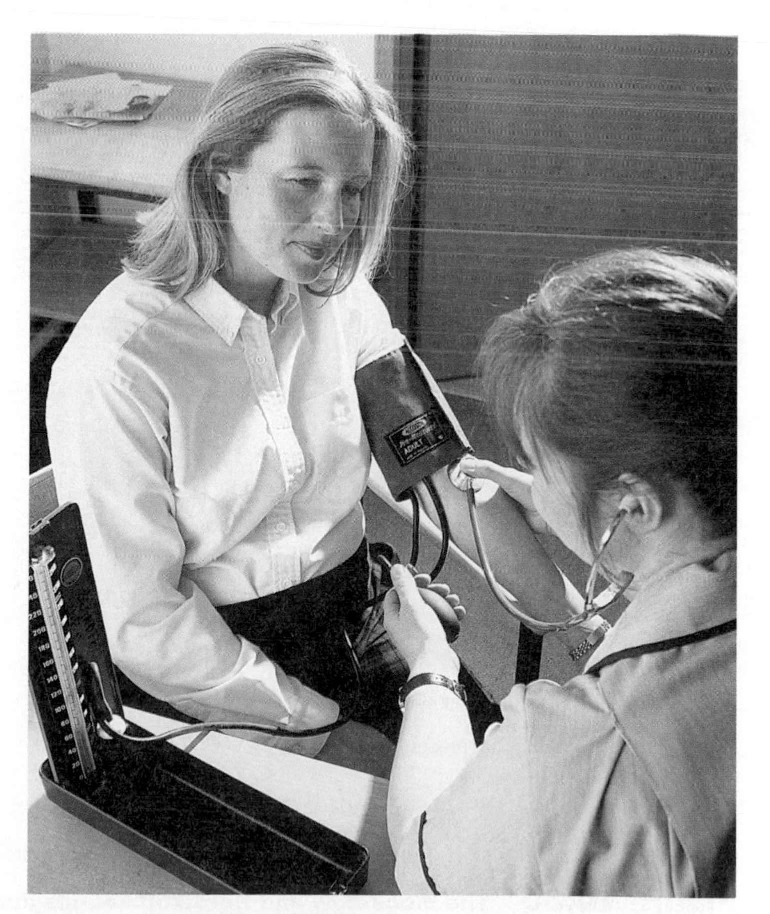

Figure 14.29 A pressure cuff and sphygmomanometer are used to measure blood pressure. The examiner is listening for the Korotkoff sounds.

produce sounds, much like the sounds produced by water flowing through a kink in a garden hose. The ability of the cuff pressure to constrict the artery is opposed by the blood pressure. So the cuff pressure must be greater than the diastolic pressure to constrict the artery during diastole. If the cuff pressure is also greater than the systolic pressure, the artery will be closed off and silent during both diastole and systole. Turbulent flow, and the sounds produced by the artery as a result of this flow, can occur only when the cuff pressure is greater than the diastolic pressure (to constrict the artery during diastole) but less than the systolic pressure. The constriction can then partially open at each systole and allow turbulent blood flow.

Let's say that a person has a systolic pressure of 120 mmHg and a diastolic pressure of 80 mmHg. When the cuff pressure is between 80 and 120 mmHg, the artery will be closed during diastole and open during systole. As the artery begins to open with every systole, turbulent flow of blood through the constriction will create sounds that are known as the **sounds of Korotkoff,** as shown in figure 14.30. These are usually "tapping" sounds because the artery becomes constricted, blood flow stops, and silence is restored with every diastole. It should be understood that the sounds of Korotkoff are *not* "lub-dub" sounds produced by closing of the heart valves (those sounds can be heard only on the chest, not on the brachial artery).

Initially, the cuff is usually inflated to produce a pressure greater than the systolic pressure, so that the artery is pinched off and silent. The pressure in the cuff is read from an attached meter called a *sphygmomanometer.* A valve is then turned to allow the release of air from the cuff, causing a gradual decrease in cuff pressure. When the cuff pressure is equal to the systolic pressure, the **first Korotkoff sound** is heard as blood passes in a turbulent flow through the constricted opening of the artery.

Korotkoff sounds will continue to be heard at every systole as long as the cuff pressure remains greater than the diastolic pressure. When the cuff pressure becomes equal to or less than the diastolic pressure, the sounds disappear because the artery remains open and laminar flow resumes (fig. 14.31). The **last Korotkoff sound** thus occurs when the cuff pressure is equal to the diastolic pressure.

Different phases in the measurement of blood pressure are identified on the basis of the quality of the Korotkoff sounds (fig. 14.32). In some people, the Korotkoff sounds do not disappear even when the cuff pressure is reduced to zero (zero pressure means that it is equal to atmospheric pressure). In these cases—and often routinely—the onset of muffling of the

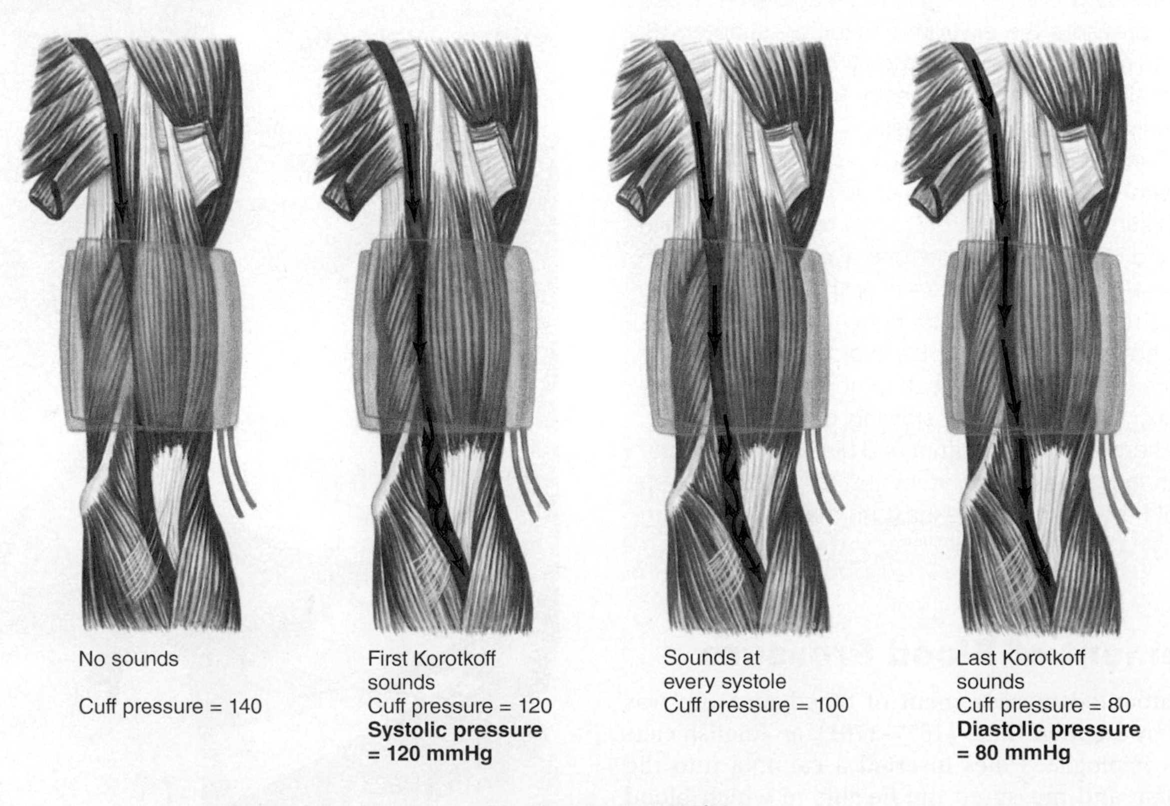

Figure 14.30 The blood flow and Korotkoff sounds during a blood pressure measurement. When the cuff pressure is above the systolic pressure, the artery is constricted. When the cuff pressure is below the diastolic pressure, the artery is open and flow is laminar. When the cuff pressure is between the diastolic and systolic pressure, blood flow is turbulent and the Korotkoff sounds are heard with each systole.

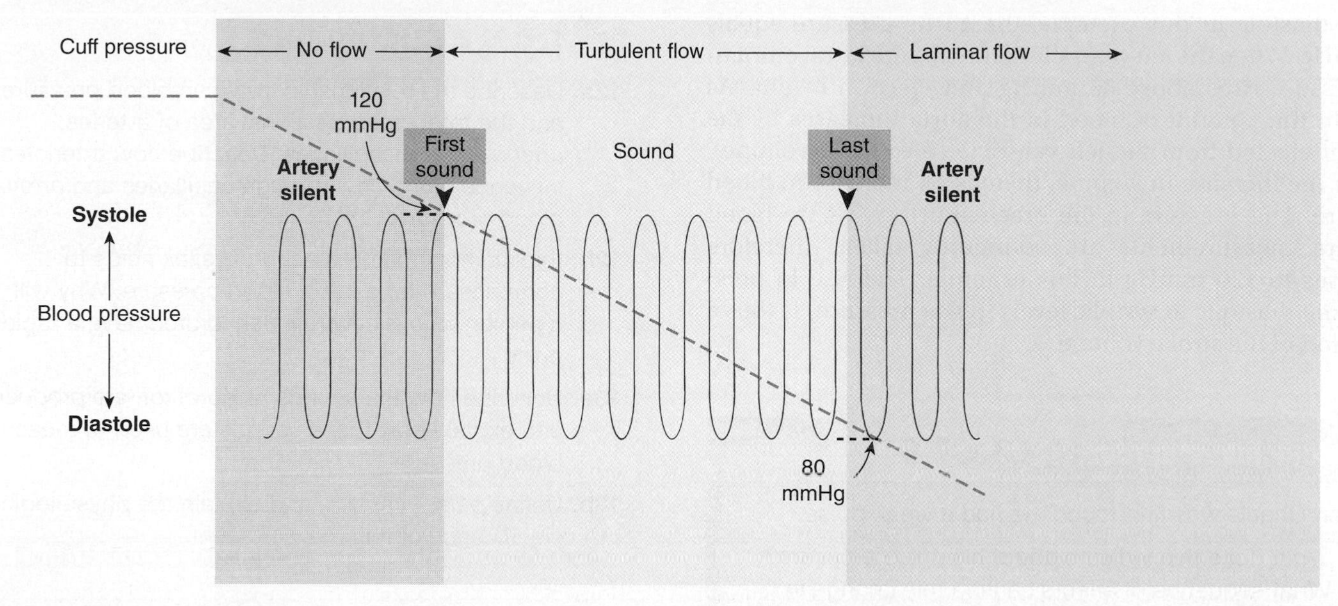

Figure 14.31 The indirect, or auscultatory, method of blood pressure measurement. The first Korotkoff sound is heard when the cuff pressure is equal to the systolic blood pressure, and the last sound is heard when the cuff pressure is equal to the diastolic pressure. The dashed line indicates the cuff pressure.

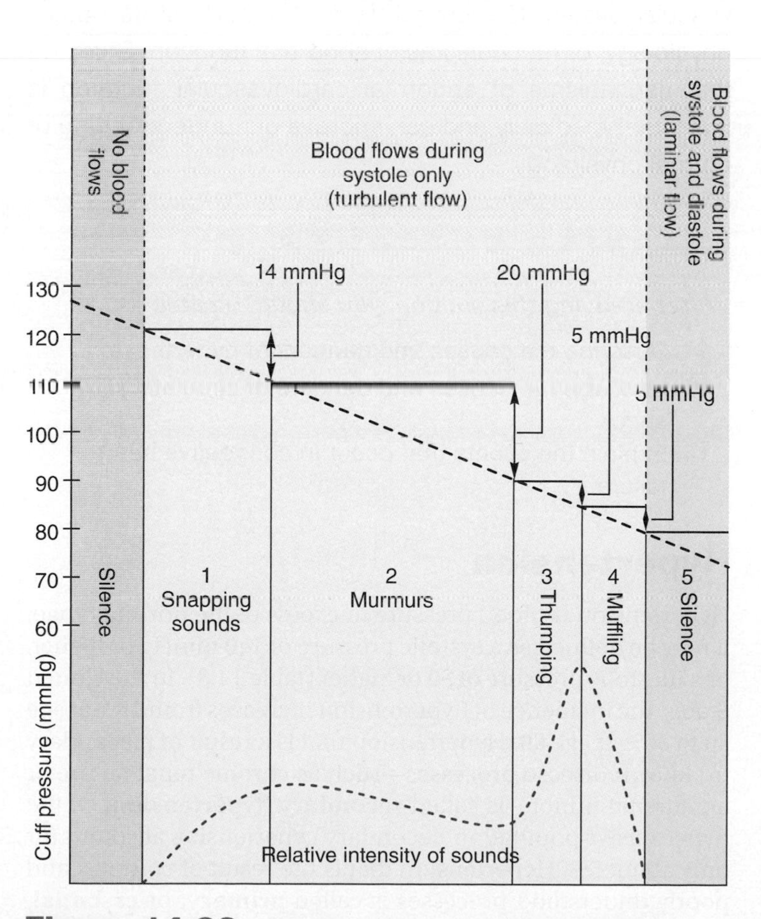

Figure 14.32 The five phases of blood pressure measurement. Not all phases are heard in all people. The cuff pressure is indicated by the falling dashed black line.

sounds (phase 4 in fig. 14.32) is used as an indication of diastolic pressure rather than the onset of silence (phase 5).

The average arterial blood pressure in the systemic circulation is 120/80 mmHg, whereas the average pulmonary arterial blood pressure is only 22/8 mmHg. Because of the Frank-Starling relationship, the cardiac output from the right ventricle into the pulmonary circulation is matched to that of the left ventricle into the systemic circulation. Since the cardiac outputs are the same, the lower pulmonary blood pressure must be caused by a lower peripheral resistance in the pulmonary circulation. Because the right ventricle pumps blood against a lower resistance, it has a lighter workload and its walls are thinner than those of the left ventricle.

Pulse Pressure and Mean Arterial Pressure

When someone "takes a pulse," he or she palpates an artery (for example, the radial artery) and feels the expansion of the artery occur in response to the beating of the heart; the pulse rate is thus a measure of the cardiac rate. The expansion of the artery with each pulse occurs as a result of the rise in blood pressure within the artery as the artery receives the volume of blood ejected from the left ventricle.

Because the pulse is produced by the rise in pressure from diastolic to systolic levels, the difference between these two pressures is known as the **pulse pressure.** A person with a blood pressure of 120/80 (systolic/diastolic) would therefore have a pulse pressure of 40 mmHg.

$$\text{Pulse pressure} = \text{systolic pressure} - \text{diastolic pressure}$$

At diastole in this example, the aortic pressure equals 80 mmHg. When the left ventricle contracts, the intraventricular pressure rises above 80 mmHg and ejection begins. As a result, the amount of blood in the aorta increases by the amount ejected from the left ventricle (the stroke volume). Due to the increase in volume, there is an increase in blood pressure. The pressure in the brachial artery, where blood pressure measurements are commonly taken, therefore increases to 120 mmHg in this example. The rise in pressure from diastolic to systolic levels (pulse pressure) is thus a reflection of the stroke volume.

Clinical Investigation CLUES 

When Charlie was first found, he had a weak pulse.

- What does this indicate about his pulse pressure?
- What sequence of effects caused this change in pulse pressure?

The **mean arterial pressure** represents the average arterial pressure during the cardiac cycle. This value is significant because it is the difference between this pressure and the venous pressure that drives blood through the capillary beds of organs. The mean arterial pressure is not a simple arithmetic average because the period of diastole is longer than the period of systole. Mean arterial pressure can be approximated by adding one-third of the pulse pressure to the diastolic pressure. For a person with a blood pressure of 120/80, the mean arterial pressure would be approximately $80 + 1/3\,(40) = 93$ mmHg.

$$\text{Mean arterial pressure} = \text{diastolic pressure} + 1/3 \text{ pulse pressure}$$

A rise in total peripheral resistance and cardiac rate increases the diastolic pressure more than it increases the systolic pressure. When the baroreceptor reflex is activated by going from a lying to a standing position, for example, the diastolic pressure usually increases by 5 to 10 mmHg, whereas the systolic pressure either remains unchanged or is slightly reduced (as a result of decreased venous return). People with hypertension (high blood pressure), who usually have elevated total peripheral resistance and cardiac rates, likewise have a greater increase in diastolic than in systolic pressure. Dehydration or blood loss results in decreased cardiac output, and thus also produces a decrease in pulse pressure.

An increase in cardiac output, by contrast, raises the systolic pressure more than it raises the diastolic pressure (although both pressures do rise). This occurs during exercise, for example, when the blood pressure may rise to values as high as 200/100 (yielding a pulse pressure of 100 mmHg).

 CHECKPOINT

12a. Describe the relationship between blood pressure and the total cross-sectional area of arteries, arterioles, and capillaries. Describe how arterioles influence blood flow through capillaries and arterial blood pressure.

12b. Explain how the baroreceptor reflex helps to compensate for a fall in blood pressure. Why will a person who is severely dehydrated have a rapid pulse?

13a. Describe how the sounds of Korotkoff are produced and explain how these sounds are used to measure blood pressure.

13b. Define *pulse pressure* and explain the physiological significance of this measurement

14.7 HYPERTENSION, SHOCK, AND CONGESTIVE HEART FAILURE

An understanding of the normal physiology of the cardiovascular system is prerequisite to the study of its pathophysiology, or mechanisms of abnormal function. Studying the mechanisms of abnormal cardiovascular function is important medically and can improve our understanding of normal physiology.

LEARNING OUTCOMES

After studying this section, you should be able to:

14. Describe the causes and dangers of hypertension.

15 Describe the causes and dangers of circulatory shock.

16. Explain the events that occur in congestive heart failure.

Hypertension

Hypertension is blood pressure in excess of the normal range; it may be defined as a systolic pressure of 140 mmHg or higher, or a diastolic pressure of 90 or higher (table 14.8). In the United States, the incidence of hypertension increases from 10% at age 30 to 50% at age 60. Hypertension that is a result of (secondary to) known disease processes—such as chronic renal failure or an adrenal tumor—is called **secondary hypertension.** Of the hypertensive population, secondary hypertension accounts for only about 5%. Hypertension that is the result of complex and poorly understood processes is called **primary,** or **essential, hypertension.** Newer evidence suggests that cardiovascular risk begins to increase when a person's systolic blood pressure

Table 14.8 | Blood Pressure Classification in Adults

Blood Pressure Classification	Systolic Blood Pressure		Diastolic Blood Pressure	Drug Therapy
Normal	Under 120 mmHg	and	Under 80 mmHg	No drug therapy
Prehypertension	120–139 mmHg	or	80–89 mmHg	Lifestyle modification;* no antihypertensive drug indicated
Stage 1 Hypertension	140–159 mmHg	or	90–99 mmHg	Lifestyle modification; antihypertensive drugs
Stage 2 Hypertension	160 mmHg or greater	or	100 mmHg or greater	Lifestyle modification; antihypertensive drugs

*Lifestyle modifications include weight reduction; reduction in dietary fat and increased consumption of vegetables and fruit; reduction in dietary sodium (salt); engaging in regular aerobic exercise, such as brisk walking for at least 30 minutes a day, most days of the week; and moderation of alcohol consumption.

Source: From the Seventh Report of the Joint National Committee on Prevention, Detection, Evaluation, and Treatment of High Blood Pressure: The JNC 7 Report. *Journal of the American Medical Association; 289* (2003): 2560–2572.

Table 14.9 | Possible Causes of Secondary Hypertension

System Involved	Examples	Mechanisms
Kidneys	Kidney disease	Decreased urine formation
	Renal artery disease	Secretion of vasoactive chemicals
Endocrine	Excess catecholamines (tumor of adrenal medulla)	Increased cardiac output and total peripheral resistance
	Excess aldosterone (Conn's syndrome)	Excess salt and water retention by the kidneys
Nervous	Increased intracranial pressure	Activation of sympathoadrenal system
	Damage to vasomotor center	Activation of sympathoadrenal system
Cardiovascular	Complete heart block; patent ductus arteriosus	Increased stroke volume
	Arteriosclerosis of aorta; coarctation of aorta	Decreased distensibility of aorta

exceeds 115 mmHg or diastolic pressure exceeds 75 mmHg. The new medical goal of maintaining a blood pressure that does not exceed 120/80 is expected to save lives by reducing the risk of heart attacks, heart failure, stroke, and kidney disease.

Diseases of the kidneys and arteriosclerosis of the renal arteries can cause secondary hypertension because of high blood volume. More commonly, the reduction of renal blood flow can raise blood pressure by stimulating the secretion of vasoactive chemicals from the kidneys. Experiments in which the renal artery is pinched, for example, produce hypertension that is associated (at least initially) with elevated renin secretion. These and other causes of secondary hypertension are summarized in table 14.9.

Essential Hypertension

The vast majority of people with hypertension have essential hypertension. Because their blood pressure is directly proportional to cardiac output and total peripheral resistance, one or both of these must be elevated. It is well established that a diet high in salt is associated with hypertension. A possible explanation for this association is that a high-salt diet causes increased plasma osmolality, which stimulates ADH secretion. Increased ADH then causes increased water reabsorption by the kidneys, increasing blood volume and thereby increasing cardiac output and blood pressure.

This sequence should be prevented by the ability of the kidneys to excrete the excess salt and water. However, the ability of the kidneys to excrete Na^+ declines with age, in part due to a gradual decline in the filtering ability of the kidneys (the glomerular filtration rate, described in chapter 17). Also, there may be inappropriately high levels of aldosterone secretion, stimulating salt and water reabsorption. This is suggested by the observation that some people with essential hypertension (who should have low renin secretion) may have normal or even elevated levels of renin, and thus increased production of angiotensin II, which stimulates aldosterone secretion.

However, there are people with essential hypertension who apparently do not have an elevated cardiac output. If that is the case, an increased total peripheral resistance must produce the high blood pressure. This could be caused by high blood volume activating stretch receptors in the atria, which in turn stimulate increased sympathetic nerve activity leading to vasoconstriction. Some scientists believe that the endothelium of arteries is at least partly responsible for the increased total peripheral resistance, through its production of paracrine regulators. For example, there may be reduced release of nitric oxide (a vasodilator) and increased release of endothelin (a vasoconstrictor). A variety of other mechanisms have also been proposed as contributing to a persistent elevation in the total peripheral resistance, and thus to essential hypertension.

The interactions between salt intake, sympathetic nerve activity, cardiovascular responses to sympathetic activity, responses to paracrine regulators from the endothelium, and kidney function make it difficult to sort out the cause-and-effect sequence that leads to essential hypertension. Some scientists believe that kidney function may be a "final common pathway" in essential hypertension, in the sense that properly functioning kidneys should be able to lower the blood volume to compensate for elevated blood pressure from any cause. It appears that a number of genes are involved in the mechanisms by which hypertension is promoted by a diet high in salt. Thus, different mechanisms may be involved to different degrees in different people. Regardless, most people with essential hypertension would benefit from a lower salt intake, and from eating more fruits and vegetables that are lower in Na^+ and higher in K^+. There is also evidence that a deficiency in calcium, and possibly vitamin D, can contribute to arterial hypertension.

Dangers of Hypertension

If other factors remain constant, blood flow increases as arterial blood pressure increases. The organs of people with hypertension are thus adequately perfused with blood until the excessively high pressure causes vascular damage. Because most patients are asymptomatic (without symptoms) until substantial vascular damage has occurred, hypertension is often referred to as a silent killer.

CLINICAL APPLICATION

Preeclampsia is a disorder specific to late pregnancy (after 20 weeks) characterized by the new onset of high blood pressure, proteinuria (the presence of protein in the urine), and edema. Only negligible amounts of proteins are normally found in the urine (chapter 17, section 17.2), and the excretion of plasma proteins in the urine can cause edema. Preeclampsia is the most common medical complication during pregnancy, affecting about 3% to 5% of pregnant women. Preeclampsia is caused by the placenta, and the symptoms almost always disappear when the placenta is eliminated at delivery. This may be related to secretions of the placenta that affect the maternal blood vessels. For example, in preeclampsia the endothelium of maternal blood vessels produces increased amounts of endothelin and thromboxane (paracrine regulators promoting vasoconstriction), and decreased amounts of nitric oxide and prostacyclin (paracrine regulators promoting vasodilation). These changes lead to vasoconstriction, resulting in decreased organ perfusion and increased blood pressure. The danger of preeclampsia is that it can quickly degenerate into a state called *eclampsia,* in which seizures occur. This can be life-threatening, and so the woman with preeclampsia is immediately treated for her symptoms and the fetus is delivered as quickly as possible.

Table 14.10 | Mechanisms of Action of Selected Antihypertensive Drugs

Category of Drugs	Examples	Mechanisms
Diuretics	Thiazide; furosemide	Increase volume of urine excreted, thus lowering blood volume
Sympathoadrenal system inhibitors	Clonidine; alpha-methyldopa	Act to decrease sympathoadrenal stimulation by bonding to $\alpha 2$-adrenergic receptors in the brain
	Guanethidine; reserpine	Deplete norepinephrine from sympathetic nerve endings
	Atenolol	Blocks beta-adrenergic receptors, decreasing cardiac output and/or renin secretion
	Phentolamine	Blocks alpha-adrenergic receptors, decreasing sympathetic vasoconstriction
Direct vasodilators	Hydralazine; minoxidil sodium nitroprusside	Cause vasodilation by acting directly on vascular smooth muscle
Calcium channel blockers	Verapamil; diltiazem	Inhibit diffusion of Ca^{2+} into vascular smooth muscle cells, causing vasodilation and reduced peripheral resistance
Angiotensin-converting enzyme (ACE) inhibitors	Captopril; enalapril	Inhibit the conversion of angiotensin I into angiotensin II
Angiotensin II–receptor antagonists	Losartan	Blocks the binding of angiotensin II to its receptor

Hypertension is dangerous for a number of reasons. One problem is that high blood pressure increases the afterload, making it more difficult for the ventricles to eject blood. The ventricles must then work harder, leading to pathological growth of their walls. This abnormal hypertrophy that can result from hypertension, or from valve defects or obesity, increases the risk of arrhythmias and heart failure. Hypertrophy due to these causes differs from the normal left ventricular hypertrophy often seen in well-trained athletes, which relieves wall stress and is believed to be beneficial.

Additionally, high pressure may damage cerebral blood vessels, leading to cerebrovascular accident, or "stroke." (Stroke is the third leading cause of death in the United States.) Finally, hypertension contributes to the development of atherosclerosis, which can itself lead to heart disease and stroke as previously described.

Treatment of Hypertension

The first form of treatment that is usually attempted is modification of lifestyle. This modification includes cessation of smoking, moderation of alcohol intake, and weight reduction, if applicable. It can also include programmed exercise and a reduction in sodium intake. People with essential hypertension may have a potassium deficiency, and there is evidence that eating food that is rich in potassium may help lower blood pressure. There is also evidence that supplementing the diet with Ca^{2+} may be of benefit, but this is more controversial.

If lifestyle modifications alone are insufficient, various drugs may be prescribed. These may include *diuretics* that increase urine volume, thus decreasing blood volume and pressure. Drugs that block β_1-adrenergic receptors (such as atenolol) lower blood pressure by decreasing the cardiac rate and are also frequently prescribed. *ACE (angiotensin-converting enzyme) inhibitors,* calcium antagonists, and various vasodilators (table 14.10) may also be used in particular situations.

Another class of drugs, the *angiotensin II-receptor blockers (ARBs),* allows angiotensin II to be formed but blocks the binding of angiotensin II to its receptors. This reduces angiotensin II–induced vasoconstriction and (via angiotensin II stimulation of aldosterone secretion) salt and water retention. ACE inhibitors and ARBs are currently the most widely prescribed drugs for the treatment of hypertension. Newer drugs include those that inhibit renin activity and in other ways reduce the activity of the renin-angiotensin-aldosterone system.

Circulatory Shock

Circulatory shock occurs when there is inadequate blood flow and/or oxygen utilization by the tissues. Some of the signs of shock (table 14.11) are a result of inadequate tissue perfusion; other signs of shock are produced by cardiovascular responses that help compensate for the poor tissue perfusion (table 14.12). When these compensations are effective, they (together with emergency medical care) are able to reestablish

Table 14.11 | Signs of Shock

	Early Sign	Late Sign
Blood pressure	Decreased pulse pressure	Decreased systolic pressure
	Increased diastolic pressure	
Urine	Decreased Na^+ concentration	Decreased volume
	Increased osmolality	
Blood pH	Increased pH (alkalosis) due to hyperventilation	Decreased pH (acidosis) due to metabolic acids
Effects of poor tissue perfusion	Slight restlessness; occasionally warm, dry skin	Cold, clammy skin; "cloudy" senses

Source: From *Principles and Techniques of Critical Care*, Vol. 1, edited by R. F. Wilson. Copyright © 1977 F. A. Davis Company, Philadelphia, PA. Used by permission.

Table 14.12 | Cardiovascular Reflexes That Help to Compensate for Circulatory Shock

Organ(s)	Compensatory Mechanisms
Heart	Sympathoadrenal stimulation produces increased cardiac rate and increased stroke volume due to a positive inotropic effect on myocardial contractility
Digestive tract and skin	Decreased blood flow due to vasoconstriction as a result of sympathetic nerve stimulation (alpha-adrenergic effect)
Kidneys	Decreased urine production as a result of sympathetic-nerve-induced constriction of renal arterioles; increased salt and water retention due to increased plasma levels of aldosterone and antidiuretic hormone (ADH)

adequate tissue perfusion. In some cases, however, and for reasons that are not clearly understood, the shock may progress to an irreversible stage and death may result.

Hypovolemic Shock

The term **hypovolemic shock** refers to circulatory shock that is due to low blood volume, as might be caused by hemorrhage (bleeding), dehydration, or burns. This is accompanied by decreased blood pressure and decreased cardiac output. In response to these changes, the sympathoadrenal system is activated by means of the baroreceptor reflex. As a result, tachycardia is produced and vasoconstriction occurs in the skin, digestive tract, kidneys, and muscles. Decreased blood flow through the kidneys stimulates renin secretion and activation of the renin-angiotensin-aldosterone system. A person in hypovolemic shock thus has low blood pressure, a rapid pulse, cold, clammy skin, and a reduced urine output.

Because the resistance in the coronary and cerebral circulations is not increased, blood is diverted to the heart and brain at the expense of other organs. Interestingly, a similar response occurs in diving mammals and, to a lesser degree, in Japanese pearl divers during prolonged submersion. These responses help deliver blood to the two organs that have the highest requirements for aerobic metabolism.

Vasoconstriction in organs other than the brain and heart raises total peripheral resistance, which helps (along with the reflex increase in cardiac rate) to compensate for the drop in blood pressure due to low blood volume. Constriction of arterioles also decreases capillary blood flow and capillary filtration pressure. As a result, less filtrate is formed. At the same time, the osmotic return of fluid to the capillaries is either unchanged or increased (during dehydration). The blood volume is thus raised at the expense of tissue fluid volume. Blood volume is also conserved by decreased urine production, which occurs as a result of vasoconstriction in the kidneys and the water-conserving effects of ADH and aldosterone, which are secreted in increased amounts during shock.

Septic Shock

Septic shock refers to a dangerously low blood pressure (hypotension) that may result from sepsis, or infection. This can occur through the action of a bacterial lipopolysaccharide called *endotoxin.* The mortality associated with septic shock is presently very high, estimated at 50% to 70%. According to recent information, endotoxin activates the enzyme nitric oxide synthase within macrophages—cells that play an important role in the immune response (chapter 15). As previously discussed, nitric oxide synthase produces nitric oxide, which promotes vasodilation and, as a result, a fall in blood pressure. Septic shock has recently been treated effectively with drugs that inhibit the production of nitric oxide.

Other Causes of Circulatory Shock

A rapid fall in blood pressure occurs in **anaphylactic shock** as a result of a severe allergic reaction (usually to bee stings or penicillin). This results from the widespread release of histamine, which causes vasodilation and thus decreases total peripheral resistance. A rapid fall in blood pressure also occurs in **neurogenic shock,** in which sympathetic tone is decreased, usually because of upper spinal cord damage or spinal anesthesia. **Cardiogenic shock** results from cardiac failure, as defined by a cardiac output inadequate to maintain tissue perfusion. This commonly results from infarction that causes the loss of a significant proportion of the myocardium. Cardiogenic shock may also result from severe cardiac arrythmias or valve damage.

Congestive Heart Failure

Cardiac failure occurs when the cardiac output is insufficient to maintain the blood flow required by the body. This may be due to heart disease—resulting from myocardial infarction or congenital defects—or to hypertension, which increases the afterload of the heart. The most common causes of left ventricular heart failure are myocardial infarction, aortic valve stenosis, and incompetence of the aortic and bicuspid (mitral) valves. This can become a vicious cycle, where a myocardial infarction causes heart failure that results in heart muscle remodeling, which can in turn promote dangerous arrythmias. Failure of the right ventricle is usually caused by prior failure of the left ventricle.

Heart failure can also result from disturbance in the electrolyte concentrations of the blood. Excessive plasma K^+ concentration decreases the resting membrane potential of myocardial cells, and low blood Ca^{2+} reduces excitation-contraction coupling. High blood K^+ and low blood Ca^{2+} can thus cause the heart to stop in diastole. Conversely, low blood K^+ and high blood Ca^{2+} can arrest the heart in systole.

The term *congestive* is often used in describing heart failure because of the increased venous volume and pressure that results. Failure of the left ventricle, for example, raises the left atrial pressure and produces pulmonary congestion and edema. This causes shortness of breath and fatigue; if severe, pulmonary edema can be fatal. Failure of the right ventricle results in increased right atrial pressure, which produces congestion and edema in the systemic circulation.

The compensatory responses that occur during congestive heart failure are similar to those that occur during hypovolemic shock. Activation of the sympathoadrenal system stimulates cardiac rate, contractility of the ventricles, and constriction of arterioles. As in hypovolemic shock, renin secretion is increased and urine output is reduced. The increased secretion of renin and consequent activation of the renin-angiotensin-aldosterone system causes salt and water retention. This occurs despite an increased secretion of atrial natriuretic peptide (which would have the compensatory effect of promoting salt and water excretion).

As a result of these compensations, chronically low cardiac output is associated with elevated blood volume and dilation and hypertrophy of the ventricles. These changes can themselves be dangerous. Elevated blood volume places a work overload on the heart, and the enlarged ventricles have a higher metabolic requirement for oxygen. These problems are often treated with drugs that increase myocardial contractility (such as digitalis, described in chapter 13), drugs that are vasodilators (such as nitroglycerin), drugs that block beta-adrenergic receptors (to reduce the strain on the heart from excessive sympathoadrenal activation), and drugs that reduce the effects of excessive activation of the renin-angiotensin system. The latter includes ACE (angiotensin converting enzyme) inhibitors and ARBs (angiotensin II receptor blockers). Diuretics—drugs that lower blood volume by increasing urine volume (chapter 17, section 17.6)—are also used to alleviate congestive heart failure.

CHECKPOINT

14. Explain how stress and a high-salt diet can contribute to hypertension. Also, explain how different drugs may act to lower blood pressure.

15a. Using a flowchart to show cause and effect, explain why a person in hypovolemic shock may have a fast pulse and cold, clammy skin.

15b. Describe the compensatory mechanisms that act to raise blood volume during cardiovascular shock.

15c. Explain how septic shock may be produced.

16. Describe congestive heart failure and explain the compensatory responses that occur during this condition.

Clinical Investigation SUMMARY

Charlie was suffering from dehydration, which lowered his blood volume and thus lowered his blood pressure. This stimulated the baroreceptor reflex, resulting in intense activation of sympathetic nerves. Sympathetic nerve activation caused vasoconstriction in cutaneous vessels—hence the cold skin—and an increase in cardiac rate (hence the high pulse rate). The intravenous albumin solution was given in the hospital to increase his colloid osmotic pressure, and thus his blood volume and pressure. His urine output was low as a result of (1) sympathetic nerve-induced vasoconstriction of arterioles in the kidneys, which decreased blood flow to the kidneys; (2) water reabsorption in response to high ADH secretion, which resulted from stimulation of osmoreceptors in the hypothalamus; and (3) water and salt retention in response to aldosterone secretion, which was stimulated by activation of the renin-angiotensin-aldosterone system. The absence of sodium in his urine resulted from the high aldosterone secretion.

See the additional chapter 14 Clinical Investigations on *Orthostatic Hypotension* and *Pheochromocytoma* in the Connect site for this text at www.mhhe.com/Fox13.

Visit this book's website at **www.mhhe.com/Fox13** for:

- Chapter quizzes, interactive learning exercises, and other study tools
- Additional clinical investigations
- Access to LearnSmart—An adaptive diagnostic tool that constantly assesses student knowledge of course material
- Ph.I.L.S. 4.0—physiology interactive lab simulations that may be used to supplement or substitute for wet labs

Interactions

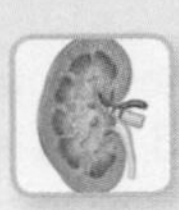

HPer Links of the Circulatory System with Other Body Systems

Integumentary System

- The skin helps protect the body from pathogens (p. 494)
- The skin provides a site for thermoregulation (p. 474)
- The circulatory system delivers blood for exchange of gases, nutrients, and wastes with all of the body organs, including the skin (p. 405)
- Blood clotting occurs if blood vessels in the skin are broken (p. 414)

Skeletal System

- Hematopoiesis occurs in the bone marrow (p. 409)
- The rib cage protects heart and thoracic vessels (p. 418)
- The blood delivers calcium and phosphate for deposition of bone and removes calcium and phosphate during bone resorption (p. 690)
- The blood delivers parathyroid hormone and other hormones that regulate bone growth and maintenance (p. 345)

Muscular System

- Cardiac muscle function is central to the activity of the heart (p. 392)
- Smooth muscle function in blood vessels regulates the blood flow and blood pressure (p. 464)
- Skeletal muscle contractions squeeze veins and thus aid venous blood flow (p. 435)
- The blood removes lactic acid and heat from active muscles (p. 382)

Nervous System

- Autonomic nerves help regulate the cardiac output (p. 451)
- Autonomic nerves help regulate the vascular resistance, blood flow, and blood pressure (p. 466)
- Cerebral capillaries participate in the blood-brain barrier (p. 171)

Endocrine System

- Epinephrine and norepinephrine from the adrenal medulla help regulate cardiac function and vascular resistance (p. 252)
- Aldosterone and other hormones influence the blood pressure (p. 482)
- The blood transports hormones to their target organs (p. 405)

Immune System

- The immune system protects against infections (p. 494)
- Lymphatic vessels drain tissue fluid and return it to the venous system (p. 443)
- Lymphocytes from the bone marrow and lymphoid organs circulate in the blood (p. 498)
- Neutrophils leave the vascular system by diapedesis to participate in aspects of the immune response (p. 486)
- The circulation carries chemical regulators of the immune response (p. 508)

Respiratory System

- The lungs provide oxygen for transport by blood and provide for elimination of carbon dioxide (p. 533)
- Ventilation helps to regulate the pH of the blood (p. 568)
- The blood transports gases between the lungs and tissue cells (p. 533)
- Breathing assists venous return (p. 455)

Urinary System

- The kidneys regulate the volume, pH, and electrolyte balance of blood (p. 582)
- The kidneys excrete waste products, derived from blood plasma, in the urine (p. 598)
- Blood pressure is required for kidney function (p. 588)

Digestive System

- Intestinal absorption of nutrients, including iron and particular B vitamins, is needed for red blood cell production (p. 408)
- The hepatic portal vein permits the enterohepatic circulation of some absorbed molecules (p. 636)
- The circulation transports nutrients from the GI tract to all the tissues in the body (p. 405)

Reproductive System

- Gonadal hormones, particularly testosterone, stimulate red blood cell production (p. 714)
- The placenta permits exchanges of gases, nutrients, and waste products between the maternal and fetal blood (p. 742)
- Erection of the penis and clitoris results from vasodilation of blood vessels (p. 711)

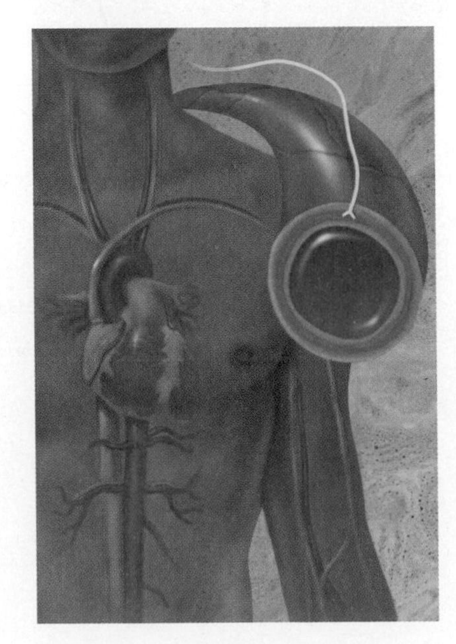

SUMMARY

14.1 Cardiac Output 451

A. Cardiac rate is increased by sympathoadrenal stimulation and decreased by the effects of parasympathetic fibers that innervate the SA node.

B. Stroke volume is regulated both extrinsically and intrinsically.
   1. The Frank-Starling law of the heart describes the way the end-diastolic volume, through various degrees of myocardial stretching, influences the contraction strength of the myocardium and thus the stroke volume.
   2. The end-diastolic volume is called the preload. The total peripheral resistance, through its effect on arterial blood pressure, provides an afterload that acts to reduce the stroke volume.
   3. At a given end-diastolic volume, the amount of blood ejected depends on contractility. Strength of contraction is increased by sympathoadrenal stimulation.

C. The venous return of blood to the heart is dependent largely on the total blood volume and mechanisms that improve the flow of blood in veins.
   1. The total blood volume is regulated by the kidneys.
   2. The venous flow of blood to the heart is aided by the action of skeletal muscle pumps and the effects of breathing.

14.2 Blood Volume 456

A. Tissue fluid is formed from and returns to the blood.
   1. The hydrostatic pressure of the blood forces fluid from the arteriolar ends of capillaries into the interstitial spaces of the tissues.
   2. Because the colloid osmotic pressure of plasma is greater than that of tissue fluid, water returns by osmosis to the venular ends of capillaries.
   3. Excess interstitial fluid is returned to the venous system by lymphatic vessels.
   4. Edema occurs when excess interstitial fluid accumulates.

B. The kidneys control the blood volume by regulating the amount of filtered fluid that will be reabsorbed.
   1. Antidiuretic hormone stimulates reabsorption of water from the kidney filtrate, and thus acts to maintain the blood volume.
   2. A decrease in blood flow through the kidneys activates the renin-angiotensin-aldosterone system.
   3. Angiotensin II stimulates vasoconstriction and the secretion of aldosterone by the adrenal cortex.
   4. Aldosterone acts on the kidneys to promote the retention of salt and water.
   5. Atrial natriuretic peptide (ANP) increases the urinary excretion of salt and water, thereby lowering the blood volume.

14.3 Vascular Resistance to Blood Flow 462

A. According to Poiseuille's law, blood flow is directly related to the pressure difference between the two ends of a vessel and inversely related to the resistance to blood flow through the vessel.
   1. Resistance to blood flow is directly proportional to the length of the vessel and the viscosity of the blood.
   2. Resistance to flow is inversely proportional to the radius, raised to the fourth power, of the vessel.

B. Extrinsic regulation of vascular resistance is provided mainly by the sympathetic nervous system, which stimulates vasoconstriction of arterioles in the viscera and skin.

C. Intrinsic control of vascular resistance allows organs to autoregulate their blood flow rates.
   1. Myogenic regulation occurs when vessels constrict or dilate as a direct response to a rise or fall in blood pressure.
   2. Metabolic regulation occurs when vessels dilate in response to the local chemical environment within the organ.

14.4 Blood Flow to the Heart and Skeletal Muscles 468

A. The heart normally respires aerobically because of its extensive capillary supply and high myoglobin and enzyme content.

B. During exercise, when the heart's metabolism increases, intrinsic metabolic mechanisms stimulate vasodilation of the coronary vessels, and thus increase coronary blood flow.

C. Just prior to exercise and at the start of exercise, blood flow through skeletal muscles increases because of vasodilation caused by the activity of cholinergic sympathetic nerve fibers. During exercise, intrinsic metabolic vasodilation occurs.

D. Since cardiac output can increase by a factor of five or more during exercise, the heart and skeletal muscles receive an increased proportion of a higher total blood flow.
   1. The cardiac rate increases because of lower activity of the vagus nerve and higher activity of sympathetic nerves.
   2. The venous return is greater because of higher activity of the skeletal muscle pumps and increased breathing.
   3. Increased contractility of the heart, combined with a decrease in total peripheral resistance, can result in a higher stroke volume.

14.5 Blood Flow to the Brain and Skin 473

A. Cerebral blood flow is regulated both myogenically and metabolically.
   1. Cerebral vessels automatically constrict if the systemic blood pressure rises too high.
   2. Metabolic products cause local vessels to dilate and supply more active areas with more blood.

B. The skin contains unique arteriovenous anastomoses that can shunt the blood away from surface capillary loops.

1. The activity of sympathetic nerve fibers causes constriction of cutaneous arterioles.
2. As a thermoregulatory response, cutaneous blood flow and blood flow through surface capillary loops increase when the body temperature rises.

14.6 Blood Pressure 475

A. Baroreceptors in the aortic arch and carotid sinuses affect the cardiac rate and the total peripheral resistance via the sympathetic nervous system.
   1. The baroreceptor reflex causes pressure to be maintained when an upright posture is assumed. This reflex can cause a lowered pressure when the carotid sinuses are massaged.
   2. Other mechanisms that affect blood volume help to regulate blood pressure.

B. Blood pressure is commonly measured indirectly by auscultation of the brachial artery when a pressure cuff is inflated and deflated.
   1. The first sound of Korotkoff, caused by turbulent flow of blood through a constriction in the artery, occurs when the cuff pressure equals the systolic pressure.
   2. The last sound of Korotkoff is heard when the cuff pressure equals the diastolic blood pressure.

C. The mean arterial pressure represents the driving force for blood flow through the arterial system.

14.7 Hypertension, Shock, and Congestive Heart Failure 482

A. Hypertension, or high blood pressure, is classified as either primary or secondary.
   1. Primary hypertension, also called essential hypertension, may result from the interaction of numerous mechanisms that raise the blood volume, cardiac output, and/or peripheral resistance.
   2. Secondary hypertension is the direct result of known specific diseases.

B. Circulatory shock occurs when delivery of oxygen to the organs of the body is inadequate.
   1. In hypovolemic shock, low blood volume causes low blood pressure that may progress to an irreversible state.
   2. The fall in blood volume and pressure stimulates various reflexes that produce a rise in cardiac rate, a shift of fluid from the tissues into the vascular system, a decrease in urine volume, and vasoconstriction.

C. Congestive heart failure occurs when the cardiac output is insufficient to supply the blood flow required by the body. The term *congestive* is used to describe the increased venous volume and pressure that result.

REVIEW ACTIVITIES

Test Your Knowledge

1. According to the Frank-Starling law of the heart, the strength of ventricular contraction is
   a. directly proportional to the end-diastolic volume.
   b. inversely proportional to the end-diastolic volume.
   c. independent of the end-diastolic volume.
2. In the absence of compensations, the stroke volume will decrease when
   a. blood volume increases.
   b. venous return increases.
   c. contractility increases.
   d. arterial blood pressure increases.
3. Which of these statements about tissue fluid is *false?*
   a. It contains the same glucose and salt concentration as plasma.
   b. It contains a lower protein concentration than plasma.
   c. Its colloid osmotic pressure is greater than that of plasma.
   d. Its hydrostatic pressure is lower than that of plasma.
4. Edema may be caused by
   a. high blood pressure.
   b. decreased plasma protein concentration.
   c. leakage of plasma protein into the interstitial fluid.
   d. blockage of lymphatic vessels.
   e. all of these.
5. Both ADH and aldosterone act to
   a. increase urine volume.
   b. increase blood volume.
   c. increase total peripheral resistance.
   d. produce all of these effects.
6. The greatest resistance to blood flow occurs in
   a. large arteries.
   b. medium-size arteries.
   c. arterioles.
   d. capillaries.
7. If a vessel were to dilate to twice its previous radius, and if pressure remained constant, blood flow through this vessel would
   a. increase by a factor of 16.
   b. increase by a factor of 4.
   c. increase by a factor of 2.
   d. decrease by a factor of 2.
8. The sounds of Korotkoff are produced by
   a. closing of the semilunar valves.
   b. closing of the AV valves.
   c. the turbulent flow of blood through an artery.
   d. elastic recoil of the aorta.
9. Vasodilation in the heart and skeletal muscles during exercise is primarily due to the effects of
   a. alpha-adrenergic stimulation.
   b. beta-adrenergic stimulation.

c. cholinergic stimulation.
d. products released by the exercising muscle cells.

10. Blood flow in the coronary circulation
    a. increases during systole.
    b. increases during diastole.
    c. remains constant throughout the cardiac cycle.
11. Total blood flow in the cerebral circulation
    a. varies with systemic arterial pressure.
    b. is regulated primarily by the sympathetic system.
    c. is maintained constant within physiological limits.
    d. increases significantly during exercise.
12. Which of these organs is able to tolerate the greatest reduction in blood flow?
    a. Brain
    b. Heart
    c. Skeletal muscles
    d. Skin
13. Which of these statements about arteriovenous shunts in the skin is *true?*
    a. They divert blood to superficial capillary loops.
    b. They are closed when the ambient temperature is very low.
    c. They are closed when the deep-body temperature rises much above 37°C.
    d. All of these are true.
14. An increase in blood volume will cause
    a. a decrease in ADH secretion.
    b. an increase in Na^+ excretion in the urine.
    c. a decrease in renin secretion.
    d. all of these.
15. The volume of blood pumped per minute by the left ventricle is
    a. greater than the volume pumped by the right ventricle.
    b. less than the volume pumped by the right ventricle.
    c. the same as the volume pumped by the right ventricle.
    d. either less or greater than the volume pumped by the right ventricle, depending on the strength of contraction.
16. Blood pressure is lowest in
    a. arteries.
    b. arterioles.
    c. capillaries.
    d. venules.
    e. veins.
17. Stretch receptors in the aortic arch and carotid sinus
    a. stimulate secretion of atrial natriuretic peptide.
    b. serve as baroreceptors that affect activity of the vagus and sympathetic nerves.
    c. serve as osmoreceptors that stimulate the release of ADH.
    d. stimulate renin secretion, thus increasing angiotensin II formation.
18. Angiotensin II
    a. stimulates vasoconstriction.
    b. stimulates the adrenal cortex to secrete aldosterone.
    c. inhibits the action of bradykinin.
    d. does all of these.
19. Which of these is a paracrine regulator that stimulates vasoconstriction?
    a. Nitric oxide
    b. Prostacyclin
    c. Bradykinin
    d. Endothelin-1
20. The pulse pressure is a measure of
    a. the number of heartbeats per minute.
    b. the sum of the diastolic and systolic pressures.
    c. the difference between the systolic and diastolic pressures.
    d. the difference between the arterial and venous pressures.

Test Your Understanding

21. Define the terms *contractility, preload,* and *afterload,* and explain how these factors affect the cardiac output.
22. Using the Frank-Starling law of the heart, explain how the stroke volume is affected by (a) bradycardia and (b) a "missed beat."
23. Which part of the cardiovascular system contains the most blood? Which part provides the greatest resistance to blood flow? Which part provides the greatest cross-sectional area? Explain.
24. Explain how the kidneys regulate blood volume.
25. A person who is dehydrated drinks more and urinates less. Explain the mechanisms involved.
26. Using Poiseuille's law, explain how arterial blood flow can be diverted from one organ system to another.
27. Describe the mechanisms that increase the cardiac output during exercise and that increase the rate of blood flow to the heart and skeletal muscles.
28. Explain why an anxious person may have cold, clammy skin and why the skin becomes hot and flushed on a hot, humid day.
29. Explain the different ways in which a drug that acts as an inhibitor of angiotensin-converting enzyme (ACE) can lower the blood pressure. Also, explain how diuretics and β_1-adrenergic-blocking drugs work to lower the blood pressure.
30. Explain how hypotension may be produced in (a) hypovolemic shock and (b) septic shock. Also, explain the mechanisms whereby people in shock have a rapid but weak pulse, cold and clammy skin, and low urine output.
31. Describe the mechanisms that have been proposed to explain how a diet high in salt could raise the blood pressure.
32. Explain how immersion of the body in water causes an increased volume of urine.

Test Your Analytical Ability

33. One consequence of the Frank-Starling law of the heart is that the outputs of the right and left ventricles are matched. Explain why this is important and how this matching is accomplished.
34. An elderly man who is taking digoxin for a weak heart complains that his feet hurt. Upon examination, his feet

are found to be swollen and discolored, with purple splotches and expanded veins. He is told to keep his feet raised and is given a prescription for Lasix, a powerful diuretic. Discuss this man's condition and the rationale for his treatment.

35. You are bicycling in a 100-mile benefit race because you want to help the cause, but you didn't count on such a hot, humid day. You've gone through both water bottles and, in the last 10 miles, you are thirsty again. Should you accept the water that one bystander offers or the sports drink offered by another? Explain your choice.
36. As the leader of a revolution to take over a large country, you direct your followers to seize the salt mines. Why is this important? When the revolution succeeds and you become president, you ask your surgeon general to wage a health campaign urging citizens to reduce their salt intake. Why?
37. Which type of exercise, isotonic contractions or isometric contractions, puts more of a "strain" on the heart? Explain.
38. As described in chapter 8, functional magnetic resonance imaging (fMRI) is based on the increased oxyhemoglobin flowing to the more active brain regions. Explain the physiological mechanisms that result in this effect when a particular brain region becomes very active.
39. Suppose you feel the pulse of a person in circulatory shock, and find that it is weak and rapid. You are also concerned that the person has a very low urine output. Explain the reasons for these observations. What treatments could be given to raise the blood pressure? Explain.
40. Athletes often have a slower resting cardiac rate than the average (a condition called *athlete's bradycardia*). What causes the slower cardiac rate? What cardiovascular adaptations allow the person to have athlete's bradycardia and yet not have a dangerously low blood pressure? What advantages might the slower resting cardiac rate have for the athlete?

Test Your Quantitative Ability

Suppose an artery divides as it enters an organ into two branches of equal radius. Use the equation for Poiseuille's law (see p. 464 and fig. 14.15) to answer the following questions:

41. If branch A is twice the length of branch B, what is true regarding the resistance in branch A compared to B?
42. If branch A now constricts to one-third its previous radius, how does this affect the resistance in branch A?
43. Given that the pressure and viscosity of the blood are constant, what will be true regarding the rate of blood flow through branch A compared to branch B?

Refer to page 457 and figure 14.9 for information regarding the calculation of Starling forces to answer the following questions, given this information:

Hydrostatic pressure inside the capillary = 34 mmHg
Colloid osmotic pressure of the blood plasma = 27 mmHg
Hydrostatic pressure in the interstitial fluid = 2 mmHg
Colloid osmotic pressure in the interstitial fluid = 4 mmHg

44. What is the total pressure favoring filtration?
45. What is the total pressure favoring absorption?
46. What is the net effect of these forces on fluid movement across the wall of the capillary?

CHAPTER OUTLINE

CHAPTER 15

The Immune System

REFRESH YOUR MEMORY

Before you begin this chapter, you may want to review these concepts from previous chapters:

Clinical Investigation

Gary, an eight-year-old child, crawled through surrounding underbrush while his parents were picnicking. He cut his thumb on a piece of rusty metal, and later, while eating his lunch, was stung by a bee for the first time in his life. The next day he developed a rash on his abdomen. When brought to the family doctor, the physician said that Gary's cut wasn't dangerous because he had received a tetanus vaccine six months earlier. He said that antihistamines wouldn't help the rash, and instead prescribed a cortisone cream. However, Gary was again stung by a bee two months later and this time the sting caused a severe swelling, for which the physician prescribed antihistamines.

Some of the new terms and concepts you will encounter include:

- Active and passive immunity and immunizations
- Inflammation; the action of cortisone
- Immediate and delayed hypersensitivity; IgE, histamine, and antihistamines

15.1 DEFENSE MECHANISMS

Nonspecific immune protection is provided by such mechanisms as phagocytosis, fever, and the release of interferons. Specific immunity, which involves the functions of lymphocytes, is directed at specific molecules or parts of molecules known as antigens.

LEARNING OUTCOMES

After studying this section, you should be able to:

1. **Describe the different elements of the innate immune system.**
2. **Describe the nature of antigens, lymphocytes, and lymphoid organs.**
3. **Explain the events that occur in a local inflammation.**

The immune system includes all of the structures and processes that provide a defense against potential *pathogens* (disease-causing agents). These defenses can be grouped in two categories: **innate** (or **nonspecific**) **immunity** and **adaptive** (or **specific**) **immunity.** Although these two categories refer to different defense mechanisms, there are areas in which they overlap.

Innate, or nonspecific, defense mechanisms are inherited as part of the structure of each organism. Epithelial membranes that cover the body surfaces, for example, restrict infection by most pathogens. The strong acidity of gastric juice (pH 1–2) also helps kill many microorganisms before they can invade the body. These external defenses are backed by internal defenses, such as phagocytosis, which function in both a specific and nonspecific manner (table 15.1).

Each individual can acquire the ability to defend against specific pathogens by prior exposure to those pathogens. This adaptive, or specific, immune response is a function of lymphocytes. Internal specific and nonspecific defense mechanisms function together to combat infection, with lymphocytes interacting in a coordinated effort with phagocytic cells.

The genes required for *innate immunity* are inherited. Since this limits the number of genes that can be devoted to this task,

Table 15.1 | Structures and Defense Mechanisms of Nonspecific (Innate) Immunity

	Structure	Mechanisms
External	Skin	Physical barrier to penetration by pathogens; secretions contain lysozyme (enzyme that destroys bacteria)
	Digestive tract	High acidity of stomach; protection by normal bacterial population of colon
	Respiratory tract	Secretion of mucus; movement of mucus by cilia; alveolar macrophages
	Genitourinary tract	Acidity of urine; vaginal lactic acid
Internal	Phagocytic cells	Ingest and destroy bacteria, cellular debris, denatured proteins, and toxins
	Interferons	Inhibit replication of viruses
	Complement proteins	Promote destruction of bacteria; enhance inflammatory response
	Endogenous pyrogen	Secreted by leukocytes and other cells; produces fever
	Natural killer (NK) cells	Destroy cells infected with viruses, tumor cells, and mismatched transplanted tissue cells
	Mast cells	Release histamine and other mediators of inflammation, and cytokines that promote adaptive immunity

innate immune mechanisms combat whole categories of pathogens. A category of bacteria called gram-negative, for example, can be recognized by the presence of particular molecules (called lipopolysaccharide) on their surfaces. In *adaptive immunity,* by contrast, specific features of pathogens are recognized. The enormous number of different genes required for this task is too large to be inherited. Instead, the variation is produced by genetic changes in lymphocytes during the life of each person after birth.

Innate (Nonspecific) Immunity

Innate immunity includes both external and internal defenses. These defenses are always present in the body and represent the first line of defense against invasion by potential pathogens.

Invading pathogens, such as bacteria, that have crossed epithelial barriers next enter connective tissues. These invaders—or chemicals, called *toxins,* secreted from them—may enter blood or lymphatic capillaries and be carried to other areas of the body. Innate immunological defenses are the first employed to counter the invasion and spread of infection. If these defenses are not sufficient to destroy the pathogens, lymphocytes may be recruited and their specific actions used to reinforce the nonspecific immune defenses.

Activation of Innate Immunity

The innate immune system distinguishes between the body's own tissue cells ("self") and invading pathogens by recognizing molecules termed **pathogen-associated molecular patterns (PAMPs)** that are unique to the invaders. The best known of these PAMPs are *lipopolysaccharides (LPS),* found in the membrane of gram-negative bacteria, and peptidoglycan from the cell walls of gram-positive bacteria.

Some cells of the innate immune system have receptor proteins—called **pathogen recognition receptors**—that recognize PAMPs. The genes that code for these receptor proteins are inherited through the germ cells (sperm and egg), which is a distinguishing characteristic of the innate immune system. By contrast, the adaptive immune system has a greater diversity of receptor protein genes because of postnatal mutation and recombination of the DNA coding for these receptors (section 15.2).

An important group of pathogen recognition receptors of the innate immune system are the **toll-like receptors,** so named because they are similar to a receptor discovered earlier in fruit flies that was named the toll receptor. (*Toll* is German for "weird"—flies that have this receptor have a weird development.) Ten different toll-like receptors have currently been identified in humans, each specific for a different type of molecule that is characteristic of invading pathogens but not of human cells. These 10 toll-like receptors enable our innate immune system to correctly identify any potential pathogen as foreign and a fit object for attack. The importance of toll-like receptors in immunity was recognized in a share of the 2011 Nobel Prize in Physiology or Medicine.

For example, exposure to LPS from bacteria stimulates one of the toll-like receptors on certain cells of the innate immune system termed *dendritic cells* and *macrophages.* These cells are stimulated to secrete **chemokines** (cell attractant molecules), which recruit other cells of the immune system, and **cytokines** (cell growth and regulatory molecules), which promote different aspects of immune responses. These include responses of both the innate immune system—including phagocytosis and a fever—and the adaptive immune system (B and T lymphocytes, discussed later).

The **complement system** helps integrate innate and adaptive immune responses. The complement system consists of proteins in plasma and other body fluids that become activated when *antibodies* (part of the adaptive immune system) bond to their molecular targets, termed *antigens* (see fig. 15.10). When this occurs, complement proteins—part of the innate immune system—promote phagocytosis, lysis (destruction) of the targeted cells, and other aspects of a local inflammation (discussed in a later section).

A local inflammation can also be produced in the absence of infection when tissue damage causes necrosis. In this case, the immmune system is exposed to **DAMPs—danger-associated molecular patterns.** DAMPs, like the PAMPs in invading microorganisms, stimulate innate immune responses and inflammation. However, when cells die by apoptosis as part of programmed cell death (chapter 3, section 3.5), they generally do not express DAMPs and thus do not provoke inflammation.

Phagocytosis

There are three major groups of phagocytic cells: (1) **neutrophils;** (2) the cells of the **mononuclear phagocyte system,** including *monocytes* in the blood and *macrophages* (derived from monocytes) and *dendritic cells* in the connective tissues; and (3) **organ-specific phagocytes** in the liver, spleen, lymph nodes, lungs, and brain (table 15.2). Organ-specific phagocytes, such as the *microglia* of the brain, are embryologically and functionally related to macrophages and may be considered part of the mononuclear phagocyte system.

The *Kupffer cells* in the liver, as well as phagocytic cells in the spleen and lymph nodes, are **fixed phagocytes.** This term refers to the fact that these cells are immobile ("fixed") in the walls of the sinusoids within these organs. As blood flows through the sinusoids of the liver and spleen, foreign chemicals and debris are removed by phagocytosis and chemically inactivated within

Table 15.2 | Phagocytic Cells and Their Locations

Phagocyte	Location
Neutrophils	Blood and all tissues
Monocytes	Blood
Tissue macrophages (histiocytes)	All tissues (including spleen, lymph nodes, bone marrow)
Kupffer cells	Liver
Alveolar macrophages	Lungs
Microglia	Central nervous system

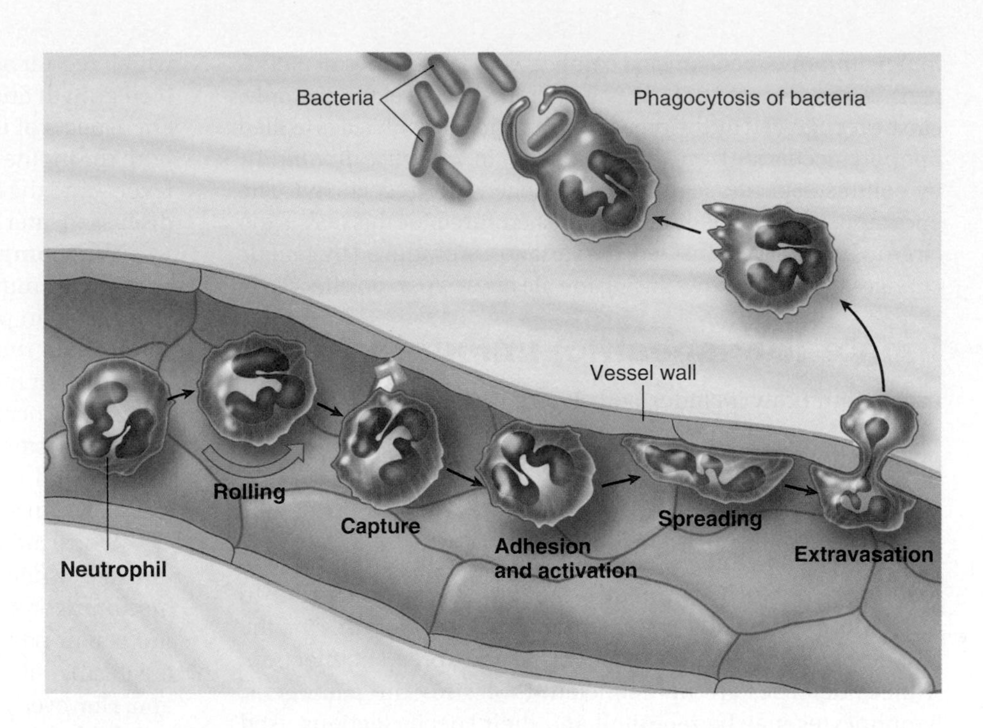

Figure 15.1 Stages involved in the migration of white blood cells from blood vessels into tissues. The figure depicts a neutrophil that goes through the stages of rolling, capture, adhesion and activation, and finally extravasation (diapedesis) through the blood vessel wall. This process is set in motion when the invading bacteria secrete certain chemicals, which attract and activate the white blood cells. The steps of extravasation require the binding of particular molecules on the white blood cell surface to receptor molecules on the surface of endothelial cells.

the phagocytic cells. Invading pathogens are very effectively removed in this manner so that blood is usually sterile after a few passes through the liver and spleen. Fixed phagocytes in lymph nodes similarly help remove foreign particles from the lymph.

Most of the time, macrophages operate separately from the immune system as they perform the simple service of clearing cellular debris, such as the remnants of cells that die normally from apoptosis. However, when they recognize pathogenic signals through the activation of their toll-like receptors, they become activated to secrete pro-inflammatory cytokines. Some of these attract resident neutrophils and monocytes within connective tissues, which are recruited to the site of an infection by a process called *chemotaxis*—movement toward chemical attractants. The chemical attractants are a subclass of cytokines known as **chemokines.** Neutrophils are the first to arrive at the site of an infection; monocytes arrive later and can be transformed into macrophages as the battle progresses.

If the infection has spread, new phagocytic cells from the blood may join those already in the connective tissue. These new neutrophils and monocytes are able to squeeze through the tiny gaps between adjacent endothelial cells in the vessel wall and enter the connective tissues. This process, called **extravasation** (or **diapedesis**), is illustrated in figure 15.1. Unlike fluid movement out of vessels, which occurs across the walls of blood capillaries (chapter 14, fig. 14.9), leukocytes leave through the walls of post-capillary venules. These narrowest venules consist only of endothelial cells and surrounding supporting cells. The leukocytes can penetrate the endothelial layer within several minutes, but may be held up by the basement membrane—a highly cross-linked network of glycoproteins supporting the endothelial cells—for up to a half hour before they can enter the surrounding connective tissue.

Phagocytic cells engulf particles in a manner similar to the way an amoeba eats. The particle becomes surrounded by cytoplasmic extensions called pseudopods, which ultimately fuse. The particle thus becomes surrounded by a membrane derived from the plasma membrane (fig. 15.2) and contained within an organelle analogous to a food vacuole in an amoeba. This vacuole then fuses with lysosomes (organelles that contain digestive enzymes), so that the ingested particle and the digestive enzymes are still separated from the cytoplasm by a continuous membrane. However, lysosomal enzymes are often released before the food vacuole has completely formed. Free lysosomal

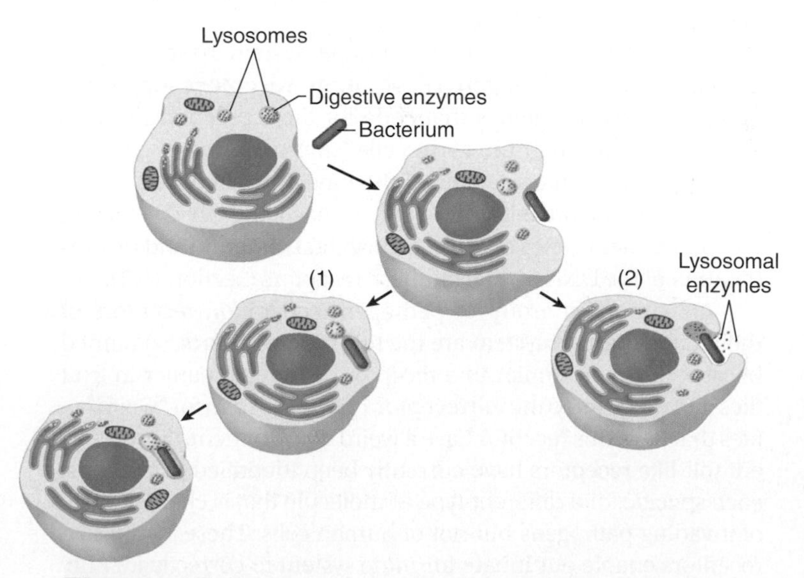

Figure 15.2 Phagocytosis by a neutrophil or macrophage. A phagocytic cell extends its pseudopods around the object to be engulfed (such as a bacterium). (Blue dots represent lysosomal enzymes.) (1) If the pseudopods fuse to form a complete food vacuole, lysosomal enzymes are restricted to the organelle formed by the lysosome and food vacuole. (2) If the lysosome fuses with the vacuole before fusion of the pseudopods is complete, lysosomal enzymes are released into the infected area of tissue. AP|R

enzymes may thereby be released into the infected area and contribute to inflammation.

Cells of our own body that commit suicide (apoptosis—chapter 3, section 3.5) signal macrophages to attack by displaying a phospholipid molecule (*phosphatidylserine*) on their surface that normally is found only on the inner layer of the plasma membrane. The phosphatidylserine provides an "eat me" signal to macrophages. However, unlike the activation of phagocytes in response to foreign pathogens, other inflammatory processes are suppressed when macrophages eat apoptotic body cells. This limits the "collateral damage" that would otherwise occur as a result of the inflammation.

Fever

Fever may be a component of the nonspecific defense system. Body temperature is regulated by the hypothalamus, which contains a thermoregulatory control center (a "thermostat") that coordinates skeletal muscle shivering and the activity of the sympathoadrenal system to maintain body temperature at about 37°C. This thermostat is reset upward in response to a chemical called **endogenous pyrogen.** In at least some infections, the endogenous pyrogen has been identified as interleukin-1β, which is first produced as a cytokine by leukocytes and is then produced by the brain itself.

The cell wall of gram-negative bacteria contains **endotoxin,** a lipopolysaccharide that stimulates monocytes and macrophages to release various cytokines. These cytokines, including *interleukin-1, interleukin-6,* and *tumor necrosis factor,* act to produce fever, increased sleepiness, and a fall in the plasma iron concentration.

Although high fevers are definitely dangerous, a mild to moderate fever may be a beneficial response that aids recovery from bacterial infections. The fall in plasma iron concentrations that accompany a fever can inhibit bacterial activity and represents one possible benefit of a fever; others include increased activity of neutrophils and increased production of interferon.

Interferons

In 1957, researchers demonstrated that cells infected with a virus produced polypeptides that interfered with the ability of a second, unrelated strain of virus to infect other cells in the same culture. These **interferons,** as they were called, thus produced a nonspecific, short-acting resistance to viral infection. Although this discovery generated a great deal of excitement, further research was hindered by the fact that human interferons could be obtained only in very small quantities; moreover, animal interferons were shown to have little effect in humans. In 1980, however, a technique called *genetic recombination* made it possible to introduce human interferon genes into bacteria, enabling the bacteria to act as interferon factories.

There are three major categories of interferons: *alpha, beta,* and *gamma interferons.* Almost all cells in the body make alpha and beta interferons. These polypeptides act as messengers that protect other cells in the vicinity from viral infection. The viruses are still able to penetrate these other cells, but the ability of the viruses to replicate and assemble new virus particles is inhibited. Viral infection, replication, and dispersal are illustrated in figure 15.3 using the HIV (human

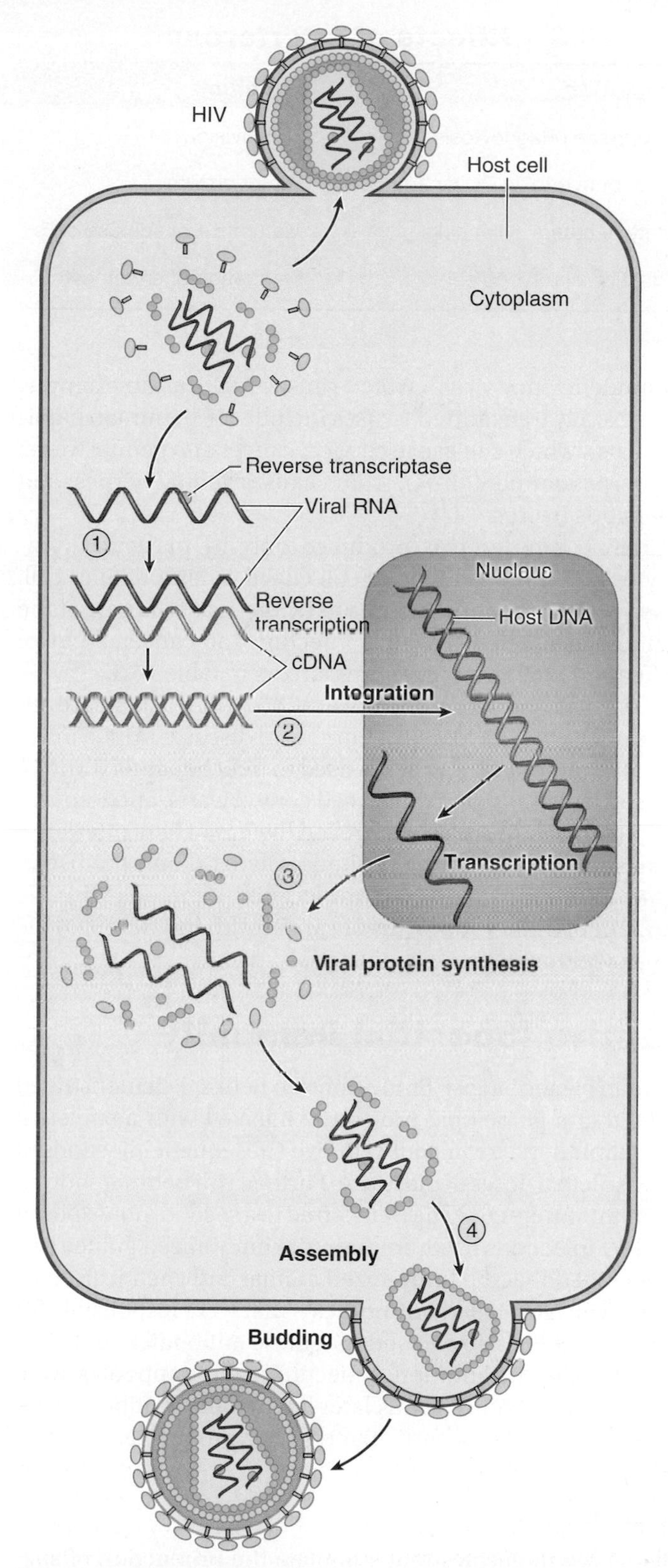

Figure 15.3 Life cycle of the human immunodeficiency virus (HIV). This virus contains RNA rather than DNA. After the viral RNA is inserted into the cell, it (1) is transcribed by reverse transcriptase into complementary DNA (cDNA), which (2) enters the nucleus and integrates into the host DNA. The cDNA then directs the synthesis of RNA, which (3) codes for viral proteins, enabling the assembly of new virus particles. (4) These new viruses are then budded from the host cell, so they can infect other cells.

Table 15.3 | Effects of Interferons

Stimulation	Inhibition
Macrophage phagocytosis	Cell division
Activity of cytotoxic ("killer") T cells	Tumor growth
Activity of natural killer cells	Maturation of adipose cells
Production of antibodies	Maturation of erythrocytes

immunodeficiency virus), which causes AIDS, as an example. Other sexually transmitted viruses include HPV (human papilloma virus), which can cause cervical cancer and genital warts; HSV (herpes simplex virus), which causes genital herpes; and the hepatitis B virus.

Gamma interferon is produced only by particular lymphocytes and a related type of cell called a natural killer cell. The secretion of gamma interferon by these cells is part of the immunological defense against infection and cancer. Some of the effects of interferons are summarized in table 15.3.

The Food and Drug Administration (FDA) has approved the use of interferons to treat a number of diseases. Alpha interferon, for example, is now being used to treat hepatitis B and C, hairy-cell leukemia, virally induced genital warts, and Kaposi's sarcoma. The FDA has also approved the use of beta interferon to treat relapsing-remitting multiple sclerosis, and the use of gamma interferon to treat chronic granulomatous disease. Treatment of numerous forms of cancer with interferon is currently in various stages of clinical trials.

Adaptive (Specific) Immunity

A German bacteriologist, Emil Adolf von Behring, demonstrated in 1890 that a guinea pig previously injected with a sublethal dose of diphtheria toxin could survive subsequent injections of otherwise lethal doses of that toxin. Further, von Behring showed that this immunity could be transferred to a second, nonexposed animal by injections of serum from the immunized guinea pig. He concluded that the immunized animal had chemicals in its serum—which he called **antibodies**—that were responsible for the immunity. He also showed that these antibodies conferred immunity only to diphtheria infections; the antibodies were *specific* in their actions. It was later learned that antibodies are proteins produced by a particular type of lymphocyte.

Antigens

Antigens are molecules that stimulate the production of specific antibodies and combine specifically with the antibodies produced. Most antigens are large molecules (such as proteins) with a molecular weight greater than about 10,000, although there are important exceptions. Also, most antigens are foreign to the blood and other body fluids. This is because the immune system can distinguish its own "self" molecules from those of any other organism ("nonself") and normally mounts an immune response only against nonself antigens. The ability of a molecule to function as an antigen depends not only on its size but also on the complexity of its structure. The plastics used in artificial implants are composed of large molecules, but they are not very antigenic because of their simple, repeating structures.

A large, complex molecule can have a number of different **antigenic determinant sites** (also called *epitopes*), which are areas of the molecule that stimulate production of, and combine with, different antibodies. Most naturally occurring antigens have many antigenic determinant sites and stimulate the production of different antibodies with specificities for these sites.

Haptens

Many small organic molecules are not antigenic by themselves but can become antigens if they bind to proteins (and thus become antigenic determinant sites on the proteins). This discovery was made by Karl Landsteiner, who also discovered the ABO blood groups (chapter 13, section 13.2). By bonding these small molecules—which Landsteiner called **haptens**—to proteins in the laboratory, new antigens could be created for research or diagnostic purposes. The bonding of foreign haptens to a person's own proteins can also occur in the body. By this means, derivatives of penicillin, for example, that would otherwise be harmless can produce fatal allergic reactions in susceptible people.

Immunoassays

When molecules act as antigens and can bind to antibodies, the antigens can be *assayed* (detected and measured) by means of the antigen-antibody reaction. If the antibodies are attached to the surface of cells, or to artificial particles such as small polystyrene beads (in commercial diagnostic tests), the antigen-antibody reaction becomes visible because the cells or particles *agglutinate* (clump). The agglutination is caused by the antigen-antibody bonds, which create bridges between the cells or particles (fig. 15.4). These agglutinated particles can be used to assay a variety of antigens, and tests that utilize this procedure are called **immunoassays.** Blood typing and modern pregnancy tests are examples of such immunoassays. In order to increase their sensitivity, modern immunoassays generally use antibodies that exhibit specificity for just one antigenic determinant site.

Lymphocytes and Lymphoid Organs

Leukocytes, erythrocytes, and blood platelets are all ultimately derived from ("stem from") unspecialized cells in the bone marrow. These *stem cells* produce the specialized blood cells, and they replace themselves by cell division so that the stem cell population is not exhausted. Lymphocytes produced in this manner seed the thymus, spleen, and lymph nodes, producing self-replacing lymphocyte colonies in these organs.

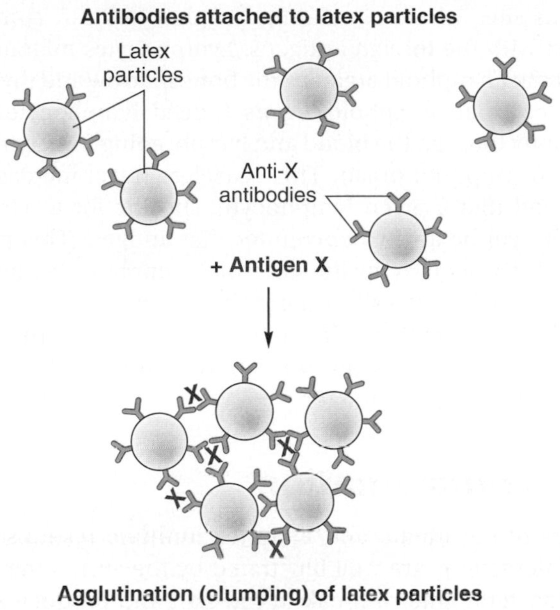

Antibodies attached to latex particles

Latex particles

Anti-X antibodies

+ Antigen X

Agglutination (clumping) of latex particles

Figure 15.4 An immunoassay using the agglutination technique. Antibodies against a particular antigen are adsorbed to latex particles. When these are mixed with a solution that contains the appropriate antigen, the formation of the antigen-antibody complexes produces clumping (agglutination) that can be seen with the unaided eye.

The lymphocytes that seed the thymus become **T lymphocytes,** or **T cells** (the letter *T* stands for *thymus-dependent*). These cells have surface characteristics and an immunological function that differ from those of other lymphocytes. The thymus, in turn, seeds other organs; about 65% to 85% of the lymphocytes in blood and most of the lymphocytes in the germinal centers of the lymph nodes and spleen are T lymphocytes. T lymphocytes, therefore, either come from or had an ancestor that came from the thymus.

Most of the lymphocytes that are not T lymphocytes are called **B lymphocytes,** or **B cells.** The letter *B* derives from immunological research performed in chickens. Chickens have an organ called the *bursa of Fabricius* that processes B lymphocytes. Since mammals do not have a bursa, the *B* is often translated as the *bursa equivalent* for humans and other mammals. It is currently believed that the B lymphocytes in mammals are processed in the bone marrow, which conveniently also begins with the letter *B*. Because the bone marrow produces B lymphocytes and the thymus produces T lymphocytes, the bone marrow and thymus are considered the **primary lymphoid organs.**

Both B and T lymphocytes function in specific immunity. The B lymphocytes combat bacterial infections, as well as some viral infections, by secreting antibodies into the blood and lymph. Because blood and lymph are body fluids (humors), the B lymphocytes are said to provide **humoral immunity,** although the term *antibody-mediated immunity* is also used. T lymphocytes attack host cells that have become infected with viruses or fungi, transplanted human cells, and cancerous cells. The T lymphocytes do not secrete antibodies; they must come in close proximity to the victim cell, or have actual physical contact with the cell, in order to destroy it. T lymphocytes are therefore said to provide **cell-mediated immunity** (table 15.4).

Table 15.4 | Comparison of B and T Lymphocytes

Characteristic	B Lymphocytes	T Lymphocytes
Site where processed	Bone marrow	Thymus
Type of immunity	Humoral (secretes antibodies)	Cell-mediated
Subpopulations	Memory cells and plasma cells	Cytotoxic (killer) T cells, helper cells, suppressor cells
Presence of surface antibodies	Yes—IgM or IgD	Not detectable
Receptors for antigens	Present—are surface antibodies	Present—are related to immunoglobulins
Life span	Short	Long
Tissue distribution	High in spleen, low in blood	High in blood and lymph
Percentage of blood lymphocytes	10%–15%	75%–80%
Transformed by antigens into	Plasma cells	Activated lymphocytes
Secretory product	Antibodies	Lymphokines
Immunity to viral infections	Enteroviruses, poliomyelitis	Most others
Immunity to bacterial infections	*Streptococcus, Staphylococcus,* many others	Tuberculosis, leprosy
Immunity to fungal infections	None known	Many
Immunity to parasitic infections	Trypanosomiasis, maybe to malaria	Most others

Thymus

The **thymus** extends from below the thyroid in the neck into the thoracic cavity. This organ grows during childhood but gradually regresses after puberty. Lymphocytes from the fetal liver and spleen, and from the bone marrow postnatally, seed the thymus and become transformed into T cells. These lymphocytes, in turn, enter the blood and seed lymph nodes and other organs where they divide to produce new T cells when stimulated by antigens.

Small T lymphocytes that have not yet been stimulated by antigens have very long life spans—months or perhaps years. Still, new T cells must be continuously produced to provide efficient cell-mediated immunity. This is particularly important following cancer chemotherapy and during HIV infection (in AIDS), when the population of T lymphocytes has been depleted. Under these conditions, the thymus can replenish the T lymphocyte population through late childhood. Repopulation of T lymphocytes occurs more slowly in adulthood, and appears to be accomplished mostly by production of T lymphocytes in the secondary lymphoid organs rather than in the thymus. This is because the thymus of adults becomes more of a fatty organ, although production of lymphocytes in the thymus has been demonstrated to occur to some extent in people over the age of 70.

Secondary Lymphoid Organs

The **secondary lymphoid organs** include the lymph nodes, spleen, tonsils, and areas called Peyer's patches under the mucosa of the intestine (see chapter 13, fig. 13.38). These organs are strategically located across epithelial membranes in areas where antigens could gain entry to the blood or lymph. The spleen filters blood, whereas the other secondary lymphoid organs filter lymph received from lymphatic vessels.

Secondary lymphoid organs capture and concentrate pathogens, present them to macrophages and other cells, and serve as sites where circulating lymphocytes can come into contact with the foreign antigens. Lymphocytes migrate from the primary lymphoid organs—the bone marrow and thymus—to the secondary lymphoid organs. Indeed, lymphocytes move constantly through the blood and lymph, going from lymphoid organ to lymphoid organ. This ceaseless travel increases the likelihood that a given lymphocyte, specific for a particular antigen, will be able to encounter that antigen. This process is aided, particularly in the case of T lymphocytes, by other cells that are known as *antigen-presenting cells* (see fig. 15.13). Secretion of chemokines (chemical attractants) by these cells increases the chances that the appropriate lymphocyte will encounter its specific antigen.

Local Inflammation

Aspects of the innate and adaptive immune responses and their interactions are well illustrated by the events that occur when bacteria enter a break in the skin and produce a **local inflammation** (table 15.5). In this case of inflammation caused by a microbial infection, the innate, nonspecific immune system becomes activated by stimulation of its toll-like pathogen recognition receptors. Macrophages and mast cells (discussed shortly), which are resident in the tissue, release a number of cytokines and chemokines that attract phagocytic neutrophils and promote the innate immune responses of phagocytosis and complement activation. (Complement proteins are also activated during adaptive immunity, as described in section 15.2.) Activated complement further increases the innate (nonspecific) responses during an inflammation by attracting new phagocytes and mast cells to the area and stimulating their activity.

Mast cells are found in most tissues, but are especially concentrated in the skin, bronchioles (airways of the lungs), and intestinal mucosa. They are identified by their content of *heparin,* a molecule that is medically important because of

Table 15.5 | Summary of Events in a Local Inflammation

Category	Events
Innate (Nonspecific) Immunity	Bacteria enter a break in the skin.
	Resident phagocytic cells—neutrophils and macrophages—engulf the bacteria.
	Nonspecific activation of complement proteins occurs.
Adaptive (Specific) Immunity	B cells are stimulated to produce specific antibodies.
	Phagocytosis is enhanced by antibodies attached to bacterial surface antigens (opsonization).
	Specific activation of complement proteins occurs, which stimulates phagocytosis, chemotaxis of new phagocytes to the infected area, and secretion of histamine from tissue mast cells.
	Extravasation (diapedesis) allows new phagocytic leukocytes (neutrophils and monocytes) to invade the infected area.
	Vasodilation and increased capillary permeability (as a result of histamine secretion) produce redness and edema.

its anticoagulant ability (chapter 13, section 13.2). However mast cells are best known for their production of **histamine,** a molecule that produces many of the symptoms of allergy (section 15.6). Histamine binds to its H_1 histamine receptors in the smooth muscle of bronchioles to stimulate bronchiolar constriction (in asthma), but produces relaxation of the smooth muscles in blood vessels (causing vasodilation). In addition, histamine promotes capiallary permeability so that more leukocytes can enter the inflamed area (fig. 15.5). Histamine and other cytokines released by mast cells are not only involved in allergic reactions, they also play roles in the immune response to infection by pathogens. Indeed, mast cells have surface receptors that recognize PAMPs on pathogens and can help activate both innate and adaptive immune responses.

Histamine and heparin, together with protease enzymes important in inflammation, are stored in *granules* within the mast cells. Either in an allergic reaction or as a normal physiological response to pathogens, mast cells are stimulated to *degranulate.* This term refers to the exocytosis of the granules, which quickly release histamine into the extracellular fluid and more slowly release the protease enzymes. With a time delay, pathogens stimulate the mast cells to produce prostaglandins and leukotrienes (chapter 11; see fig. 11.34) as well as other pro-inflammatory cytokines. These include *tumor necrosis factor,* which acts as a chemokine to recruit neutrophils to the infected site.

Leukocytes within vessels in the inflamed area stick to the endothelial cells of the vessels through interactions between

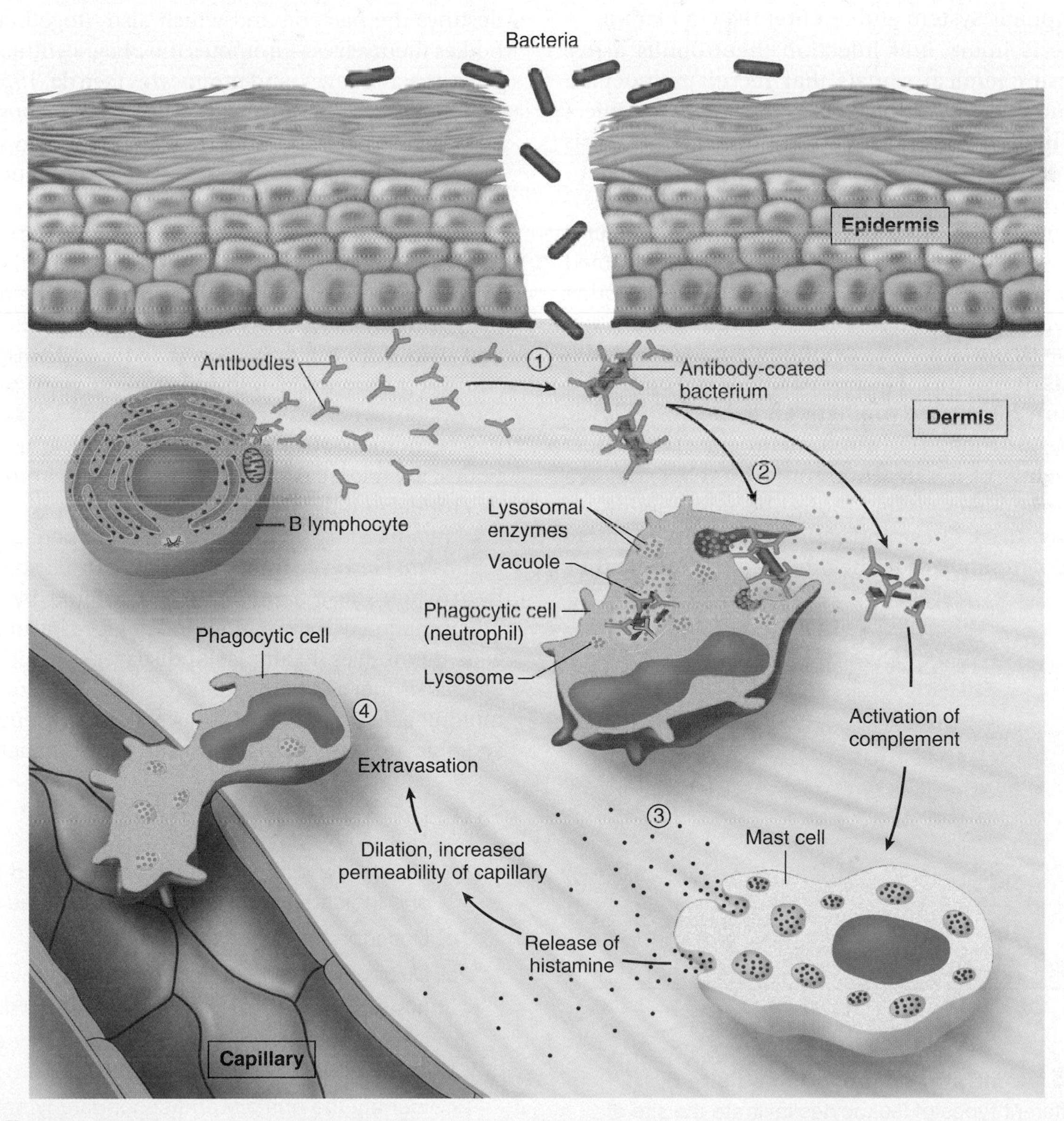

Figure 15.5 The events in a local inflammation. Antigens on the surface of bacterial cells (1) bind to antibodies, which coat the bacteria. This activates complement and (2) promotes phagocytosis by neutrophils and macrophages. Activation of complement also (3) stimulates mast cells to release histamine and other mediators of inflammation, including chemicals that promote capillary permeability and (4) extravasation (diapedesis) of leukocytes, which invade the inflamed site. AP|R

adhesion molecules on the two surfaces. The leukocytes can then roll along the wall of the vessel toward particular chemicals. As mentioned earlier, this movement, called *chemotaxis,* is produced by molecules called chemokines. Complement proteins and bacterial products may serve as chemokines, drawing the leukocytes toward the site of infection.

The leukocytes squeeze between adjacent endothelial cells (the process of extravasation, discussed earlier) and enter the subendothelial connective tissue. There, particular molecules on the leukocyte membrane interact with surrounding molecules that guide the leukocytes to the infection. The first to arrive are the neutrophils, followed by monocytes (which can change into macrophages) and T lymphocytes (fig. 15.6). Most of the phagocytic leukocytes (neutrophils and monocytes) die in the course of the infection, but lymphocytes can travel through the lymphatic system and re-enter the circulation.

When there is injury and infection, neutrophils arrive early and release chemical signals that recruit monocytes, lymphocytes, and other immune cells to the site of the infection. Neutrophils kill microoorganisms through phagocytosis and the release of enzymes and a variety of antimicrobial peptides. They also release *NETs—neutrophil extracellular traps—* composed of extracellular fibers that trap invading pathogens. In the process, the neutrophils undergo a type of programmed cell death. Through the action of proteases (protein-digesting enzymes), the neutrophils liquefy the surrounding tissues. This produces a viscous, protein-rich fluid that, together with dead neutrophils, forms **pus.** Pus can be beneficial because it produces pressure that closes lymphatic and blood capillaries, blocking the spread of bacteria away from the site of the battle.

The neutrophils also release granule proteins that are recognized by monocytes rolling along the endothelium of capillaries (see fig. 15.1). These and various cytokines promote the recruitment of monocytes to the infection site, where they can adhere to extracellular matrix proteins (chapter 6, section 6.1) and transform into macrophages. Macrophages ingest microorganisms and fragments of the extracellular matrix by phagocytosis; they also release nitric oxide, which aids in the destruction of bacteria. As neutrophils die by apoptosis, releasing more proteases and other agents that contribute to the inflammation, their remains are engulfed by macrophages. This phagocytosis of apoptotic neutrophils then causes the macrophages to release growth factors and other agents that help to end the inflammation and promote repair.

After some time, B lymphocytes are stimulated to produce antibodies against specific antigens that are part of the invading bacteria. Binding of these antibodies to antigens in the bacteria greatly amplifies the previously nonspecific response. This occurs because of greater activation of complement, which directly destroys the bacteria and which also—together with the antibodies themselves—promotes the phagocytic activity of neutrophils, macrophages, and monocytes (see fig. 15.5). The ability of antibodies to promote phagocytosis is called *opsonization.*

These effects produce the characteristic symptoms of a local inflammation: *redness* and *warmth* (due to histamine-stimulated vasodilation); *swelling* (edema) and *pus*; and *pain.* These symptoms were first described by Celsus around 40 A.D. as "rubor, calor, dolor, and tumor" (redness, heat, pain, and swelling, respectively). The pain threshold is lowered by prostaglandin E_2 (PGE_2), released as a cytokine during inflammation. PGE_2 and pain from inflammation is reduced by aspirin and other NSAIDs (nonsteroidal anti-inflammatory drugs), which inhibit the cyclooxygenase enzymes (COX-1 and COX-2) that produce prostaglandins (chapter 11, figure 11.34). If the infection continues, the release of endogenous pyrogen from leukocytes and macrophages may also produce a fever, as previously discussed.

Inflammatory processes protect the body and are required for health. However, these processes can also damage the body. Even when the inflammation is triggered by pathogens, the inflammation can sometimes inflict more harm than the pathogens themselves. Examples of diseases in which inflammation plays an important pathogenic role include Alzheimer's disease, multiple sclerosis, atherosclerosis, asthma, rheumatoid arthritis, systemic lupus erythematosus, and type 1 diabetes mellitus.

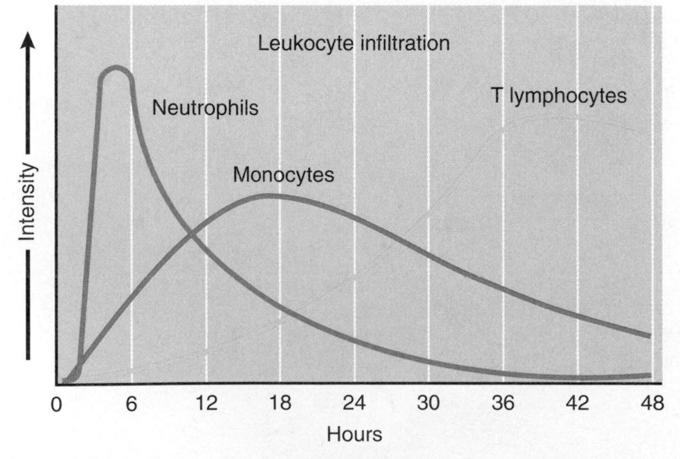

Figure 15.6 Infiltration of an inflamed site by leukocytes. Different types of leukocytes infiltrate the site of a local inflammation. Neutrophils arrive first, followed by monocytes and then lymphocytes.

See the *Test Your Quantitative Ability* section of the Review Activities at the end of this chapter.

CHECKPOINT

1a. List the phagocytic cells found in blood and lymph, and indicate which organs contain fixed phagocytes.

1b. Describe the actions of interferons.

2a. Distinguish between innate and adaptive immunity, and describe the properties of antigens.

2b. Distinguish between B and T lymphocytes in terms of their origins and immune functions.

2c. Identify the primary and secondary lymphoid organs and describe their functions.

3. Describe the events that occur during a local inflammation.

15.2 FUNCTIONS OF B LYMPHOCYTES

B lymphocytes secrete antibodies that bind to specific antigens. This bonding stimulates a cascade of reactions whereby a system of plasma proteins called complement is activated. Some of the activated complement proteins kill the cells containing the antigen; others promote phagocytosis, resulting in a more effective defense against pathogens.

LEARNING OUTCOMES

After studying this section, you should be able to:

4. Describe B lymphocytes and antibodies and explain how they function.
5. Describe the complement system and explain its functions.

A lymphocyte circulates throughout the body, going from one secondary lymphoid organ to another until it encounters the antigen that is specific to its receptors. Exposure of a B lymphocyte to the appropriate antigen activates the B cell and causes it to enter a *germinal center* of a secondary lymphoid organ, where it undergoes many cell divisions. Some of the progeny become **memory cells;** these are visually indistinguishable from the original cell and are important in active immunity. Others are transformed into **plasma cells** (fig. 15.7). Plasma cells are protein factories that produce about 2,000 antibody proteins per second.

The antibodies that are produced by plasma cells when B lymphocytes are exposed to a particular antigen react specifically with that antigen. Such antigens may be isolated molecules, as illustrated in figure 15.7, or they may be molecules at the surface of an invading foreign cell (fig. 15.8). The specific bonding of antibodies to antigens serves to identify the enemy and to activate defense mechanisms that lead to the invader's destruction.

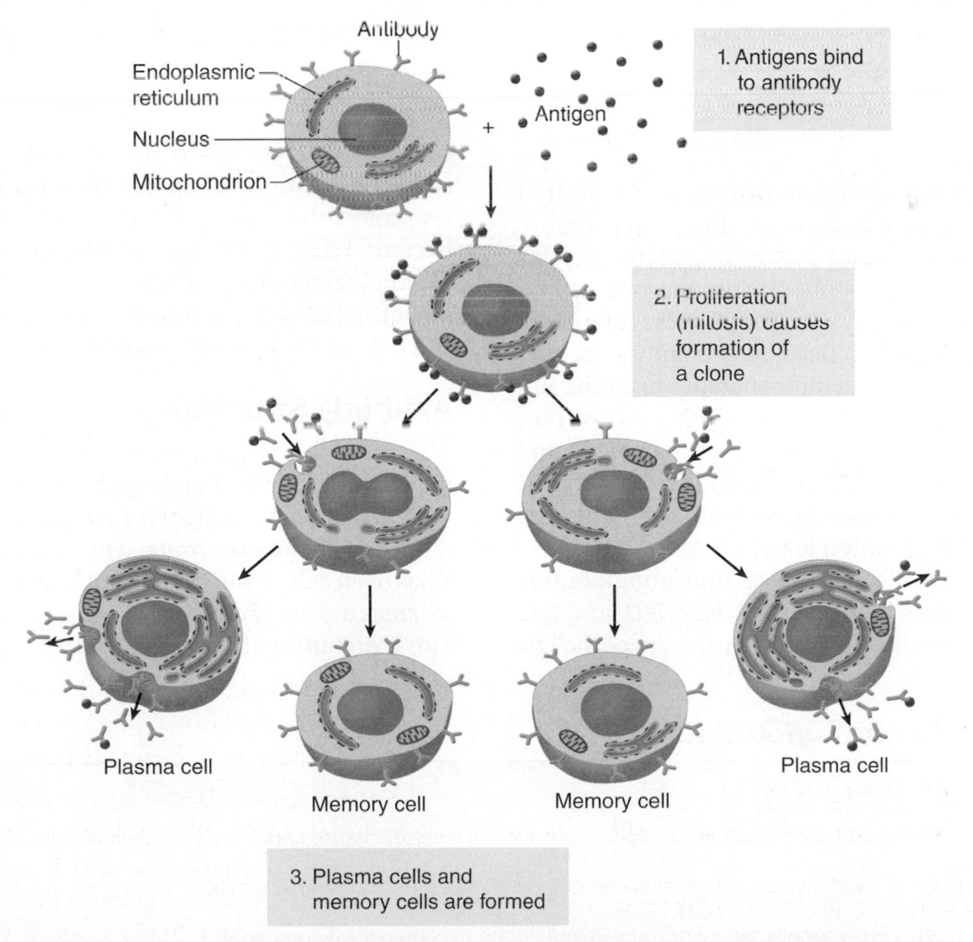

Figure 15.7 B lymphocytes are stimulated to become plasma cells and memory cells. B lymphocytes have antibodies on their surface that function as receptors for specific antigens. The interaction of antigens and antibodies on the surface stimulates cell division and the maturation of the B cell progeny into memory cells and plasma cells. Plasma cells produce and secrete large amounts of the antibody. (Note the extensive rough endoplasmic reticulum in these cells.)

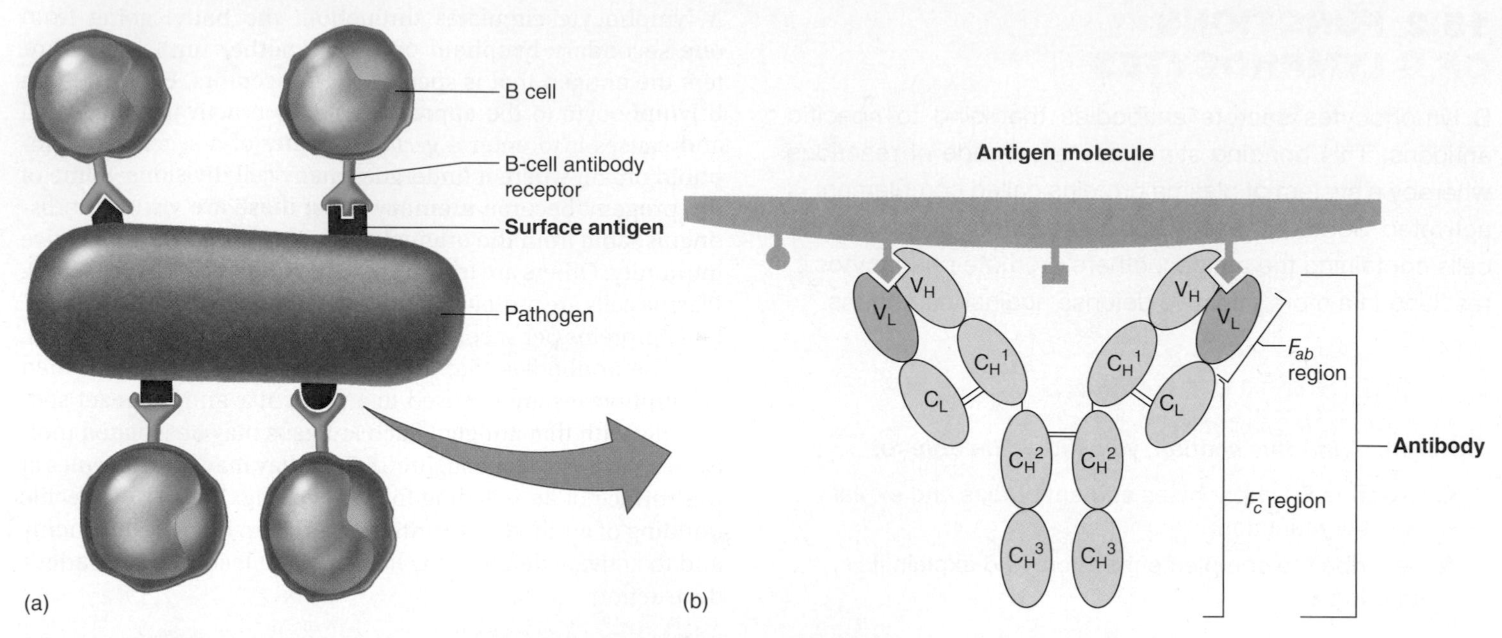

Figure 15.8 Antibodies bind to antigens on bacteria. *(a)* A pathogen such as a bacterial cell has many antigens on its surface, which can activate B cells that produce antibodies against those antigens. *(b)* Each antibody is composed of four polypeptide chains—two are heavy (H), and two are light (L). Regions that have constant amino acid sequences are abbreviated C, and those with variable amino acid sequences are abbreviated V. Antigens bind to the variable regions. Each antibody is divided into antigen-binding fragments (F_{ab}) and a crystallizable fragment (F_c). AP|R

Antibodies

Plasma proteins can be separated into five major classes by a technique called *electrophoresis,* in which the proteins move in an electric field. The five major classes of plasma proteins are albumin, alpha-1 globulin, alpha-2 globulin, beta globulin, and gamma globulin. The gamma globulin class of plasma proteins contains the antibodies. Because antibodies are specific in their actions, it follows that different types of antibodies should have different structures. An antibody against smallpox, for example, does not confer immunity to poliomyelitis and therefore must have a slightly different structure than an antibody against polio. Despite these differences, antibodies are structurally related and form only a few classes.

Antibody proteins are also known as **immunoglobulins (Ig).** There are five Ig subclasses: *IgG, IgA, IgM, IgD,* and *IgE.* Most of the antibodies circulating in the plasma are of the IgG subclass, whereas IgA is the most abundant subclass in the secretions of mucosal membranes. For example, IgA antibodies protect the mucosa of the intestine from both commensal bacteria (chapter 18) and pathogens, and are present in the external secretions of saliva and milk (table 15.6). Antibodies in the IgE subclass are involved in allergic (immediate hypersensitivity) reactions, as discussed in section 15.6.

Antibody Structure

All antibody molecules consist of four interconnected polypeptide chains. Two long, heavy chains (the *H chains*) are joined to two shorter, lighter *L chains.* Research has shown that these four chains are arranged in the form of a Y. The stalk of the Y has been called the "crystallizable fragment" (abbreviated F_c), whereas the top of the Y is the "antigen-binding fragment" (F_{ab}). This structure is shown in figure 15.8.

Table 15.6 | The Immunoglobulins

Immunoglobulin	Functions
IgG	Main form of antibodies in circulation: production increased after immunization; secreted during secondary response
IgA	Main antibody type in external secretions, such as saliva and mother's milk
IgE	Responsible for allergic symptoms in immediate hypersensitivity reactions
IgM	Function as antigen receptors on lymphocyte surface prior to immunization; secreted during primary response
IgD	Function as antigen receptors on lymphocyte surface prior to immunization; other functions unknown

The amino acid sequences of some antibodies have been determined through the analysis of antibodies sampled from people with multiple myelomas. These lymphocyte tumors arise from the division of a single B lymphocyte, forming a population of genetically identical cells (a clone) that secretes identical antibodies. Clones and the antibodies they secrete are different, however, from one patient to another. Analyses of these antibodies have shown that the F_c regions of different antibodies are the same (are constant), whereas the F_{ab} regions are variable. Variability of the antigen-binding regions is required for the specificity of antibodies for antigens. Thus, it is the F_{ab} region of an antibody that provides a specific site for bonding with a particular antigen (fig. 15.9).

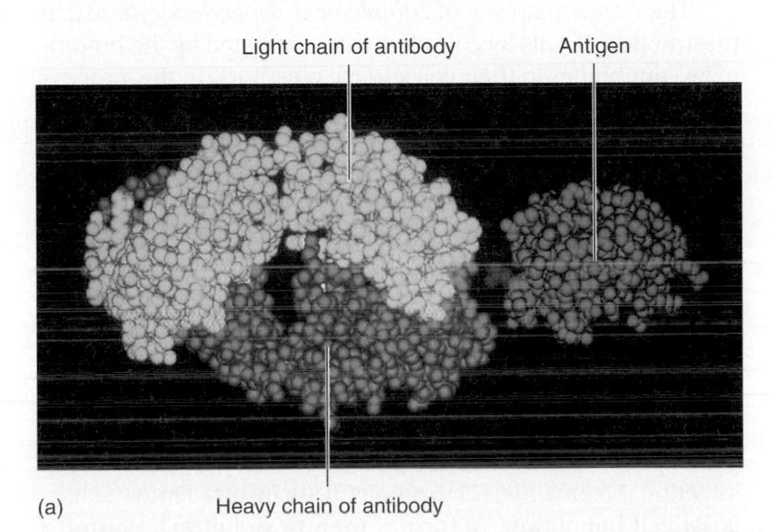

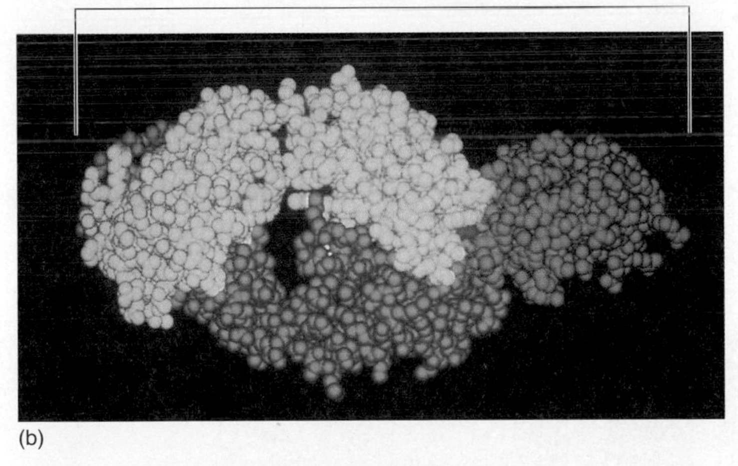

Figure 15.9 The antigen-binding site of an antibody. The structure of the F_{ab} portion of an antibody molecule and the antigen with which it combines as determined by X-ray diffraction. (*a*) The heavy and light chains of the antibody are shown in blue and yellow, respectively, and the antigen is shown in green. Note the complementary shape at the region where the two join together in (*b*). Reprinted with permission from A. G. Amit,"Three Dimensional Structure of an Antigen-Antibody Complex at 2.8 A Resolution" in *Science*, vol. 233, August 15, 1986. Copyright © 1986 American Association for the Advancement of Science.

B lymphocytes have antibodies on their plasma membranes that serve as **receptors** for antigens. Combination of antigens with these antibody receptors stimulates the B cell to divide and produce more of these antibodies, which are secreted. Exposure to a given antigen thus results in increased amounts of the specific type of antibody that can attack that antigen. This provides active immunity, as described in section 15.4.

Diversity of Antibodies

It is estimated that there are about 100 million trillion (10^{20}) antibody molecules in each individual, representing a few million different specificities for different antigens. Because antibodies that bind to particular antigens can cross-react with closely related antigens to some extent, this tremendous antibody diversity usually ensures that there will be some antibodies that can combine with almost any antigen a person might encounter. These observations evoke a question that has long fascinated scientists: How can a few million different antibodies be produced? A person cannot possibly inherit a correspondingly large number of genes devoted to antibody production.

Two mechanisms have been proposed to explain antibody diversity: (1) *antigen-independent* genetic recombination in the bone marrow; and (2) *antigen-dependent* cell division of lymphocytes in the secondary lymphoid organs.

The antigen-independent generation of antibody diversity is due to different combinations of heavy and light chains, which produce antibodies with different specificities. As a result, a person does not have to inherit a million different genes to code for a million different antibodies. If a few hundred genes code for different H chains and a few hundred code for different L chains, different combinations of these polypeptide chains could produce millions of different antibodies. The number of possible combinations is made even greater by the fact that different segments of DNA code for different segments of the heavy and light chains. Three segments in the antigen-combining region of a heavy chain and two in a light chain are coded by different segments of DNA and can be combined in different ways to make an antibody molecule. Recombination of these genes in the developing lymphocytes of the bone marrow is antigen-independent and is responsible for the initial large diversity of B cells with different antibody receptors for different antigens.

The antigen-dependent generation of antibody diversity occurs as B cells proliferate in the secondary lymphoid organs (such as the spleen and lymph nodes) in response to antigens. This diversification occurs by hypermutation (a high rate of mutations) of single base pairs in DNA. Because mutations in body cells, rather than germ cells (sperm or ova), are called somatic mutations, this mechanism is called **somatic hypermutation.** Antigen-dependent diversification of antibodies also occurs by gene recombinations. There is a switch in the constant regions of the heavy chains of the antibodies, so that the original IgM antibodies are converted into IgG, IgA, or

IgE antibodies (see fig. 15.8 and table 15.6). This is called **class switch recombination.**

Between the processes that occur during antigen-independent diversification in the bone marrow and the antigen-dependent diversification in the secondary lymphoid organs, the body can produce hundreds of billions of different antibodies to combat different pathogens.

The Complement System

The combination of antibodies with antigens does not itself cause destruction of the antigens or the pathogenic organisms that contain these antigens. Antibodies, rather, serve to identify the targets for immunological attack and to activate nonspecific immune processes that destroy the invader. Bacteria that are "buttered" with antibodies, for example, are better targets for phagocytosis by neutrophils and macrophages. The ability of antibodies to stimulate phagocytosis is termed **opsonization.** Immune destruction of bacteria is also promoted by antibody-induced activation of a system of serum proteins known as *complement.*

In the early twentieth century, scientists learned that rabbit antibodies that bind to the red blood cell antigens of sheep could not lyse (destroy) these cells unless they added certain protein components of serum. These proteins, called **complement,** constitute a nonspecific defense system that is activated by the bonding of antibodies to antigens, and by this means is directed against specific invaders that have been identified by antibodies.

The complement proteins are designated C1 through C9. These proteins are present in an inactive state within plasma and other body fluids and become activated by the attachment of antibodies to antigens. In terms of their functions, the complement proteins can be subdivided into three components: (1) recognition (C1); (2) activation (C4, C2, and C3, in that order); and (3) attack (C5 through C9). The attack phase consists of **complement fixation,** in which complement proteins attach to the cell membrane and destroy the victim cell.

There are two pathways of complement activation. The **classic pathway** is initiated by the binding of antibodies of the IgG and IgM subclasses to antigens on the invading cell's membrane. This is more rapid and efficient than the **alternative pathway,** which is initiated by the unique polysaccharides that coat bacterial cells.

The classic pathway of *complement-dependent cytotoxicity* (destruction of cells by complement) is initiated by the binding of IgG antibodies to their cell surface receptors. In this process, complement protein C1 is activated, which catalyzes the hydrolysis of C4 into two fragments, $C4_a$ and $C4_b$ (fig. 15.10). The $C4_b$ fragment binds to the cell membrane (is "fixed") and becomes an active enzyme. Then, through an intermediate step involving the splitting of C2, C3 is cleaved into $C3_a$ and $C3_b$. Acting through a different sequence of events, the alternative pathway of complement activation also results in the conversion of C3 into $C3_a$ and $C3_b$, so that the two pathways converge at this point.

The $C3_b$ converts C5 into $C5_a$ and $C5_b$. The $C3_a$ and $C5_a$ stimulate mast cells to release histamine. $C3_a$ and $C5_a$ additionally serve as powerful chemokines to attract macrophages, neutrophils, monocytes, and eosinophils to the site of the infection. Meanwhile, C5 through C9 are inserted into the bacterial cell membrane to form a **membrane attack complex** (fig. 15.11). The attack complex is a large pore that can kill the

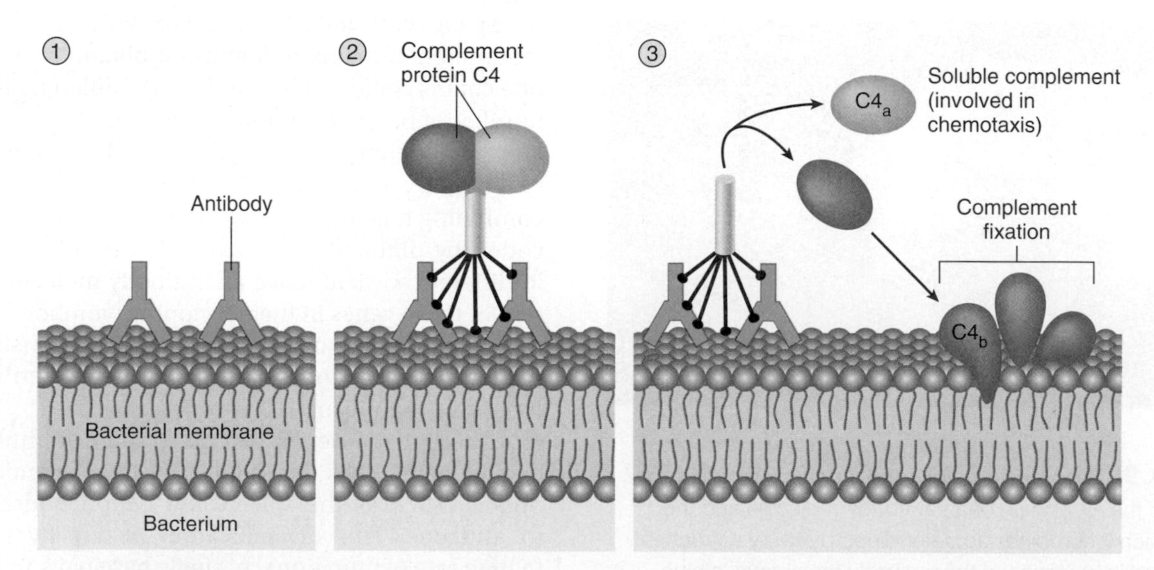

Figure 15.10 The fixation of complement proteins. (1) The formation of an antibody-antigen complex causes (2) complement protein C4 to be split (3) into two subunits—$C4_a$ and $C4_b$. The $C4_b$ subunit attaches (is fixed) to the membrane of the cell to be destroyed (such as a bacterium). This event triggers the activation of other complement proteins, some of which attach to the $C4_b$ on the membrane surface.

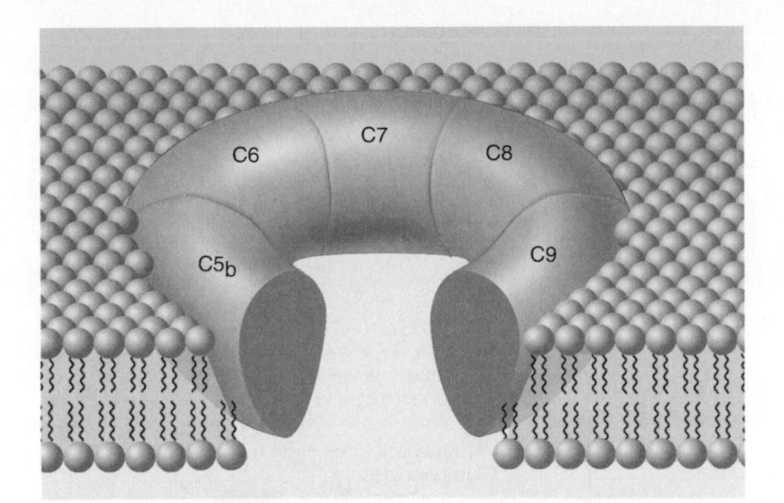

Figure 15.11 The membrane attack complex. Fixed complement proteins C5 through C9 assemble in the plasma membrane of the victim cell as a membrane attack complex. This complex forms a large pore that punctures the membrane and thereby promotes the destruction of the cell.

bacterial cell through the osmotic influx of water. Note that the complement proteins, not the antibodies directly, kill the cell; antibodies serve only as activators of this process in the classic pathway.

Complement fragments that are liberated into the surrounding fluid rather than becoming fixed have a number of effects. These effects include (1) *chemotaxis*—the liberated complement fragments attract phagocytic cells to the site of complement activation; (2) *opsonization*—phagocytic cells have receptors for $C3_b$, so that this fragment may form bridges between the phagocyte and the victim cell, thus facilitating phagocytosis; and (3) *stimulation of the release of histamine* from mast cells and basophils by fragments $C3_a$ and $C5_a$. As a result of histamine release, blood flow to the infected area is increased because of vasodilation and increased capillary permeability. This helps to bring in more phagocytic cells to combat the infection, but the increased capillary permeability can also result in edema through leakage of plasma proteins into the surrounding tissue fluid.

 CHECKPOINT

4a. Illustrate the structure of an antibody molecule. Label the constant and variable regions, the F_c and F_{ab} parts, and the heavy and light chains.

4b. Define *opsonization,* and identify two types of molecules that promote this process.

5. Describe complement fixation, and explain the roles of complement fragments that do not become fixed.

15.3 FUNCTIONS OF T LYMPHOCYTES

Killer T cells effect cell-mediated destruction of specific victim cells, and helper and suppressor T cells play supporting roles. T cells are activated only by antigens presented to them on the surface of particular antigen-presenting cells. Activated helper T cells produce lymphokines that stimulate other cells of the immune system.

LEARNING OUTCOMES

After studying this section, you should be able to:

6. Identify the different T lymphocytes and their functions.

7. Explain how T cells become activated and how they function in immunity.

The thymus processes lymphocytes in such a way that their functions become quite distinct from those of B cells. Lymphocytes residing in the thymus or originating from the thymus, or those derived from cells that came from the thymus, are all T lymphocytes. These cells can be distinguished from B cells by specialized techniques. Unlike B cells, the T lymphocytes provide specific immune protection without secreting antibodies. This is accomplished in different ways by the three subpopulations of T lymphocytes.

Killer, Helper, and Regulatory T Lymphocytes

The **killer,** or **cytotoxic, T lymphocytes** can be identified in the laboratory by a surface molecule called *CD8.* Their function is to destroy body cells that harbor foreign molecules. These are usually molecules from an invading microorganism, but they can also be molecules produced by the cell's genome because of a malignant transformation, or they may simply be body molecules that had never before been presented to the immune system.

In contrast to the action of B lymphocytes, which kill at a distance through humoral immunity (the secretion of antibodies), killer, or cytotoxic, T lymphocytes kill their victim cells by *cell-mediated destruction.* This means that they must be in actual physical contact with the victim cells. When this occurs, the killer cells secrete molecules called **perforins** and enzymes called **granzymes.** The perforins enter the plasma membrane of the victim cell and polymerize to form a very large pore. This is similar to the pore formed by the membrane attack complex of complement proteins, and results in the osmotic destruction of the victim cell. The granzymes enter the victim cell and, through the activation of caspases (enzymes involved in apoptosis—chapter 3, section 3.5), cause the destruction of the victim cell's DNA.

The killer T lymphocytes defend against viral and fungal infections and are also largely responsible for transplant rejection reactions and for immunological surveillance against cancer. Although most bacterial infections are fought by B lymphocytes, some are the targets of cell-mediated attack by killer T lymphocytes. This is the case with the tubercle bacilli that cause tuberculosis. Injections of some of these bacteria under the skin produce inflammation after a latent period of 48 to 72 hours. This *delayed hypersensitivity reaction* is cell-mediated rather than humoral, as shown by the fact that it can be induced in an unexposed guinea pig by an infusion of lymphocytes, but not of serum, from an exposed animal.

Helper T lymphocytes are identified in the laboratory by a surface molecule called *CD4.* As their name implies, these cells enhance the immune response; they improve the ability of B lymphocytes to differentiate into plasma cells and secrete specific antibodies (fig. 15.12), and they enhance the ability of killer (cytotoxic) lymphocytes to mount a cell-mediated immune response. Helper T lymphocytes aid the specific immune response of B lymphocytes and killer T lymphocytes through secretion of chemical regulators called *lymphokines,* discussed in the next section. For example, helper T lymphocytes secrete a lymphokine called *interleukin-2,* which aids the killer T lymphocyte response (see fig. 15.17).

Regulatory (previously called **suppressor**) **T lymphocytes,** abbreviated **T_{reg} lymphocytes,** provide a "brake" on the specific immune response (fig. 15.12); they inhibit the activity of killer T lymphocytes and B lymphocytes. Although there is no cell marker specific for T_{reg} lymphocytes, an important subgroup of these cells has CD4 as well as CD25 surface markers. The CD25 forms part of the receptor for IL-2 (secreted by helper T cells); people with a genetic deficiency in CD25, and thus in T_{reg} cell function, develop severe autoimmune disease and allergy. Also, T_{reg} lymphocytes have been distinguished by the activation of a gene known as *FOXP3,* which codes for a transcription factor that appears to be required for the development and maintenance of regulatory T lymphocytes.

The mechanisms by which regulatory T lymphocytes suppress immune responses are not completely understood. Immune suppression requires close proximity or even physical contact between the T_{reg} cells and their target cells. The release of cytokines, including *interleukin-10* and *TGF*β (transforming growth factor beta), from the T_{reg} cells is required in many instances of immune suppression. In other cases, the regulatory T lymphocytes promote destruction of their target cells by releasing granzymes and perforins, much like the action of killer T cells.

Because regulatory T lymphocytes suppress inappropriate specific immune responses, they guard against diseases caused by the immune system (autoimmune diseases and allergy, discussed in section 15.6). By this reasoning, people may get allergies and autoimmune diseases partly because of inadequate functioning of the regulatory T lymphocytes. Indeed, mutations in the FOXP3 gene and consequent lack of T_{reg} lymphocytes are known to cause severe autoimmune disease. On the other hand, inappropriately vigorous functioning of regulatory T lymphocytes may also promote disease. This is suggested by evidence that certain viral infections and cancers may protect themselves from immunological attack by recruiting the activity of regulatory T lymphocytes.

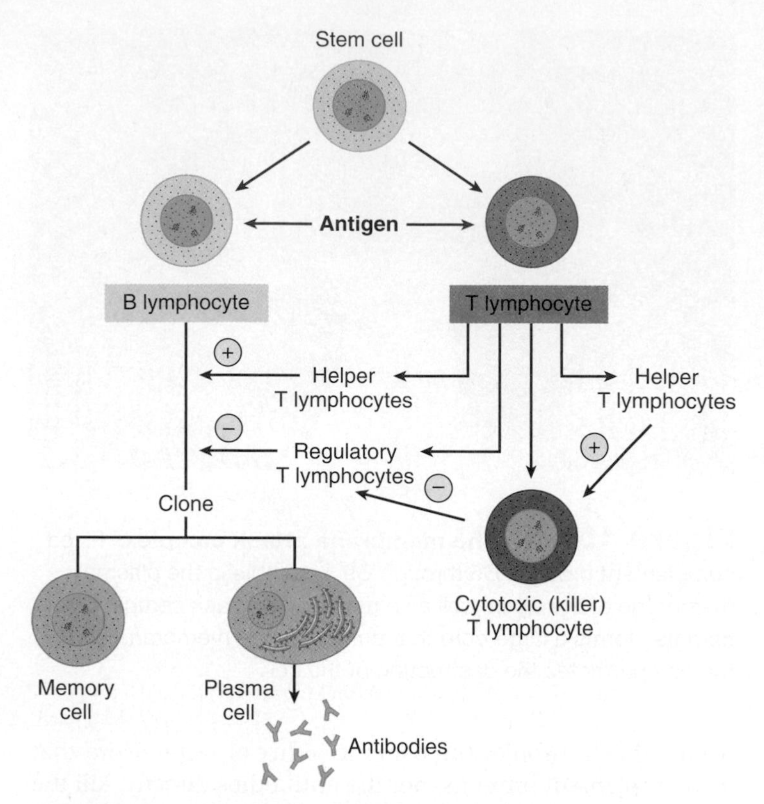

Figure 15.12 The effect of an antigen on B and T lymphocytes. A given antigen can stimulate the production of both B and T lymphocyte clones. The ability to produce B lymphocyte clones, however, is also influenced by the relative effects of helper and suppressor T lymphocytes.

Lymphokines

The T lymphocytes, as well as some other cells such as macrophages, secrete a number of polypeptides that serve as autocrine regulators (chapter 11) of the immune system. These products are generally called **cytokines;** the term **lymphokine** is often used to refer to the cytokines of lymphocytes. When a cytokine is first discovered, it is named according to its biological activity (e.g., *B-cell-stimulating factor*). Because each cytokine has many different actions (table 15.7), however, such names can be misleading. Scientists have thus agreed to use the name *interleukin,* followed by a number, to indicate a cytokine once its amino acid sequence has been determined.

Interleukin-1, a family of 11 different molecules, was the first interleukin discovered. Secreted by macrophages and other cells of the innate immune system, often in response to activation of toll-like receptors, these molecules can activate the T cell system and promote other effects (IL-1β can circulate to the brain, for example, and promote a fever). *Interleukin-2*

CLINICAL APPLICATION

Acquired immune deficiency syndrome (AIDS) has killed approximately 25 million people worldwide and orphaned 14 million children in sub-Saharan Africa. An estimated 2 million people continue to die of AIDS each year. Today, more Americans have died of AIDS than were killed in World Wars I and II combined. About 40 million people worldwide are currently infected, and since AIDS has been shown to have a latency period of approximately eight years, most will display symptoms of the disease in the near future. AIDS is caused by the **human immunodeficiency virus (HIV)** (see fig. 15.3), which specifically destroys the helper T lymphocytes. In particular, HIV targets helper T lymphocytes in the mucosa of the gastrointestinal tract, where up to 30% of all T cells reside. This results in decreased immunological function and greater susceptibility to opportunistic infections, including *Pneumocystis carinii pneumonia.* Many people with AIDS also develop a previously rare form of cancer known as *Kaposi's sarcoma.*

HIV is classified as a **retrovirus** because its genetic code is carried in its RNA. The enzyme *reverse transcriptase* is needed to transcribe this viral RNA into complementary DNA for viral replication (see fig. 15.3), and current treatments for AIDS employ drugs that inhibit this enzyme. Two different reverse transcriptase inhibitors have been combined with a protease inhibitor (protease enzymes are needed to cut viral protein into segments for assembly of the viral coat) to produce a "cocktail" that has proved to be an effective treatment. New drugs are also now available that inhibit the fusion of HIV with its victim cells by targeting the part of the glycoprotein "spokes" that protrude from the HIV particles (see fig. 15.3), which serve to anchor the virus to the plasma membrane of its victim. These and other new drugs may provide better treatment until a safe and effective vaccine is developed. This has proven to be much more difficult than expected for a number of reasons, including the extreme diversity of HIV between people and even within a single infected person, and the ability of HIV to cause massive loss of helper T cells in the intestinal mucosa within days after infection.

Hope has been raised by reports that (1) treating AIDS patients with drugs that inhibit reverse transcriptase significantly reduces the ability to infect sexual partners (by 62% to 73% in recent studies); (2) vaginal gels with antiretroviral medicines reduce the sexual transmission of AIDs to women (by 39% in the study); (3) circumcision reduces by as much as 60% the risk that a man will be infected with HIV (circumcision also reduces the risk of genital herpes by 30% and of high-risk human papillomavirus by 35%); and (4) the development of an AIDS vaccine or the use of antibodies (perhaps in a gel) that neutralize a broad range of HIV strains may be possible in the future.

Table 15.7 | Some Cytokines That Regulate the Immune System

Cytokine	Biological Functions
Interleukin-1 (IL-1)	Induces proliferation and activation of T lymphocytes
Interleukin-2 (IL-2)	Induces proliferation of activated T lymphocytes
Interleukin-3 (IL-3)	Stimulates proliferation of bone marrow stem cells and mast cells
Interleukin-4 (IL-4)	Stimulates proliferation of activated B cells; promotes production of IgE antibodies; increases activity of cytotoxic T cells
Interleukin-5 (IL-5)	Induces activation of cytotoxic T cells; promotes eosinophil differentiation and serves as chemokine for eosinophils
Interleukin-6 (IL-6)	Stimulates proliferation and activation of T and B lymphocytes
Granulocyte/monocyte-macrophage colony-stimulating factor (GM-CSF)	Stimulates proliferation and differentiation of neutrophils, eosinophils, monocytes, and macrophages

is released by helper T lymphocytes and is required for activation of killer T cells and regulatory T cells, among other functions. *Interleukin-4* is required for the proliferation and clone development of B cells. Other interleukins will be described in conjunction with their functions.

Two subtypes of helper T lymphocytes are designated $\mathbf{T_H1}$ and $\mathbf{T_H2}$. Helper T lymphocytes of the T_H1 subtype produce interleukin-2 and gamma interferon. Because they secrete these lymphokines, T_H1 cells activate killer T cells and promote cell-mediated immunity against intracellular pathogens. The lymphokines secreted by the T_H1 lymphocytes also stimulate nitric oxide production in macrophages, increasing their activity. The T_H2 lymphocytes secrete interleukin-4, interleukin-5, interleukin-13, and other lymphokines that stimulate B lymphocytes to promote humoral immunity against extracellular pathogens. Interleukin 4 and other lymphokines secreted by T_H2 cells recruit eosinophils to the site of the inflammation, help to clear parasitic infections, and induce IgE production in an allergic (immediate hypersensitivity) reaction. For example, T_H2 lymphocytes and the cytokines they secrete—IL-4, IL-5, and IL-13—play a key role in most cases of allergic asthma.

In addition to the T_H1 and T_H2 subtypes of helper T cells, there is another, more recently recognized subtype called $\mathbf{T_H17}$ cells. The T_H17 cells secrete a different group of lymphokines that include interleukin-17, which stimulates a different kind of inflammatory response dominated by neutrophils. These cells are particularly important for fighting infections in the skin, lungs, and other mucosal membranes. Cytokines released by T_H17 cells promote neutrophil recruitment and activation to combat bacteria. This is particularly important in fighting infections at mucosal surfaces (of the gut and lungs, for example) as well as in the skin. The T_H17 cells also have roles in autoimmune diseases, allergic inflammation, and immune responses that affect tumor growth.

T Cell Receptor Proteins

The antigens recognized by B lymphocytes may be either proteins or carbohydrates, but only protein antigens are recognized by most T lymphocytes. Unlike B cells, T cells do not make antibodies and thus do not have antibodies on their surfaces to serve as receptors for antigens. However, T cells do have antigen receptors on their surfaces that are closely related to the immunoglobulins. The T cell receptors differ from the antibody receptors on B cells in a very important respect: the T cell receptors *cannot bind to free antigens.* In order for T lymphocytes to respond to foreign antigens, the antigens must be presented to the T cells on the membrane of **antigen-presenting cells.**

The 2011 Nobel Prize in Physiology or Medicine was awarded to the discoverer of **dendritic cells** (fig. 15.13), which

CLINICAL APPLICATION

In response to **endotoxin,** a molecule released by bacteria, and to cytokines such as interleukin-1 and gamma interferon, macrophages are stimulated to produce *nitric oxide.* While a normal amount of nitric oxide may help macrophages destroy bacteria, an excessive release of nitric oxide may cause so much vasodilation that it results in the hypotension of *septic shock* (chapter 14, section 14.7). **Sepsis** is a leading cause of death in developed societies and kills more than 200,000 people yearly in the United States. Sepsis is caused by an infection that releases large amounts of endotoxin, which then stimulates macrophages to secrete inflammatory cytokines, including TNFα (tumor necrosis factor alpha) and interleukin-1. These cytokines produce a systemic inflammatory response, resulting in capillary leakage, tissue injury, and widespread organ failure. Scientists are trying to develop agents that prevent the release, or block the action, of the inflammatory cytokines to aid in the medical treatment of sepsis.

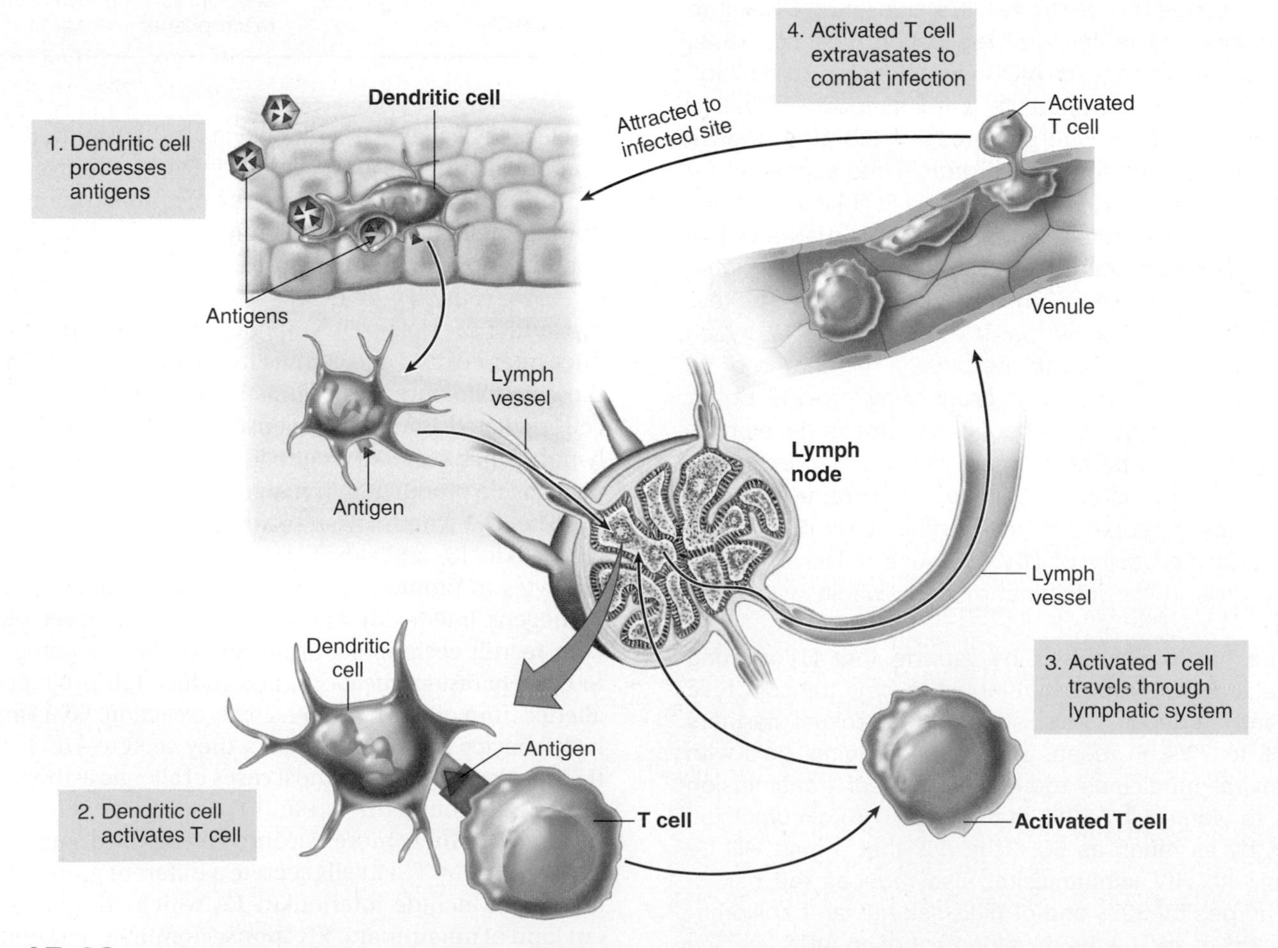

Figure 15.13 Migration of antigen-presenting dendritic cells to secondary lymphoid organs activates T cells. Once the T cells have been activated by antigens presented to them by the dendritic cells, the activated cells divide to produce a clone. Some of these cells then migrate from the lymphoid organ into the blood. Once in the blood, these activated T cells can home in on the site of the infection because of chemoattractant molecules produced during the inflammation. AP|R

together with macrophages are the chief antigen-presenting cells. Both can be derived from monocytes, which are formed in the bone marrow and migrate through blood and lymph to almost every tissue. The differentiation of monocytes into macrophages and dendritic cells occurs more in some organs than in others, and is promoted by pro-inflammatory stimuli. Antigen-presenting macrophages and dendritic cells are especially concentrated at potential sites where antigen-bearing microorganisms might enter, such as the skin, intestinal mucosa, and lungs.

For example, the epidermis contains dendritic cells called **Langerhans cells,** which are derived in the bone marrow and make up 3% to 5% of the cells in the epidermis (not counting the cornified layers), where they occupy spaces between the keratinocytes. These cells engulf protein antigens by pinocytosis, partially digest these proteins into shorter polypeptides, and then move these polypeptides to the cell surface. At the cell surface, the foreign polypeptides are associated with molecules called *histocompatibility antigens* (discussed in the next section). This allows the antigen-presenting cells to activate the T lymphocytes (fig. 15.13).

In order to interact with the correct T lymphocytes (those that have specificity for the antigen), however, the dendritic cells must migrate through lymphatic vessels to the secondary lymphoid organs, where they secrete chemokines to attract T lymphocytes. This migration affords the antigen-presenting cells the opportunity for a close encounter with the correct T lymphocytes. A T lymphocyte that does not encounter its antigen will not spend more than 24 hours in a lymph node, but this time will increase to 3 to 4 days when it becomes activated by the dendritic cells that bear its antigen. Activated T cells divide to first produce *effector T cells* (those that perform their specific functions), and then produce *memory T cells.*

Histocompatibility Antigens

Tissue that is transplanted from one person to another contains antigens that are foreign to the host. This is because all tissue cells, with the exception of mature red blood cells, are genetically marked with a characteristic combination of **histocompatibility antigens** on the membrane surface. The greater the variance in these antigens between the donor and the recipient in a transplant, the greater will be the chance of transplant rejection. Prior to organ transplantation, therefore, the "tissue type" of the recipient is matched to that of potential donors. Because the person's white blood cells are used for this purpose, histocompatibility antigens in humans are also called **human leukocyte antigens (HLAs).** They are also called *MHC molecules,* after the name of the genes that code for them.

The histocompatibility antigens are proteins that are coded by a group of genes called the **major histocompatibility complex (MHC),** located on chromosome number 6. These four genes are labeled A, B, C, and D. Each of them can code for only one protein in a given individual, but because each gene has multiple alleles (forms), this protein can be different in different people. Two people, for example, could both have antigen A3, but one might have antigen B17 and the other antigen B21. The closer two people are related, the closer the match between their histocompatibility antigens.

Interactions Between Antigen-Presenting Cells and T Lymphocytes

The major histocompatibility complex of genes produces two classes of MHC molecules, designated *class 1* and *class 2,* that are found on the cell surface. The class-1 molecules are produced by all cells in the body except red blood cells. Class-2 MHC molecules are produced only by antigen-presenting cells—macrophages, dendritic cells, and B lymphocytes. These cells present their class-2 MHC molecules together with the foreign polypeptide antigen to helper T lymphocytes. This activates the helper T lymphocytes so that they can promote the B-cell immune response.

The helper T lymphocytes can only be activated by antigens presented to them in association with class-2 MHC molecules. Killer (cytotoxic) T lymphocytes, by contrast, can be activated to destroy a victim cell only if the cell presents antigens to them in association with class-1 MHC molecules. The different requirements for class-1 or class-2 MHC molecules result from the presence of *coreceptors,* which are proteins associated with the T cell receptors. The coreceptor known as *CD8* is associated with the killer T lymphocyte receptor and interacts only with the class-1 MHC molecules; the coreceptor known as *CD4* is associated with the helper T lymphocyte receptor and interacts only with the class-2 MHC molecules. These structures are illustrated in figure 15.14.

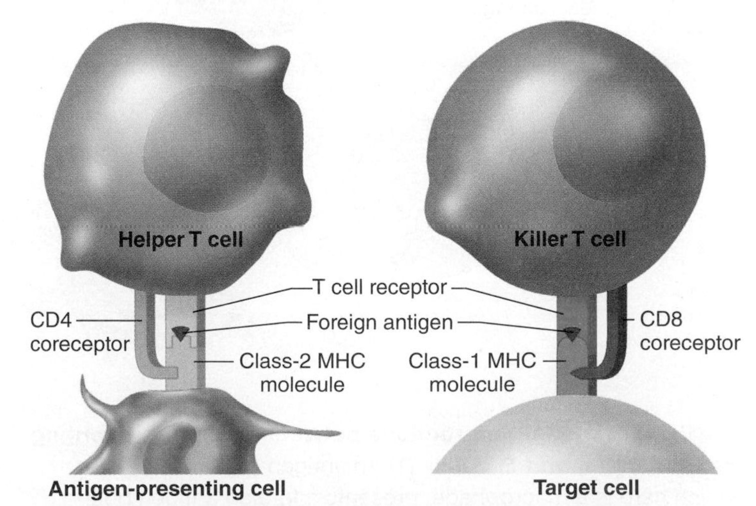

Figure 15.14 Coreceptors on helper and killer T cells. A foreign antigen is presented to T lymphocytes in association with MHC molecules. The CD4 on helper T cells, and CD8 coreceptors on killer T cells, permit each type of T cell to interact only with a specific class of MHC molecule. AP|R

T Lymphocyte Response to a Virus

When a foreign particle such as a virus infects the body, it is taken up by macrophages (or dendritic cells) via phagocytosis and partially digested. Within the macrophage, the partially digested virus particles provide foreign antigens that are moved to the surface of the plasma membrane. At the membrane, these foreign antigens form a complex with the class-2 MHC molecules. This combination of MHC molecules and foreign antigens is required for interaction with the receptors on the surface of helper T cells. The macrophages thus "present" the antigens to the helper T cells and in this way activate the helper T cells (fig. 15.15). It should be remembered that T cells are "blind" to free antigens; they can respond only to antigens presented to them by dendritic cells and macrophages in combination with class-2 MHC molecules.

The first phase of macrophage-T cell interaction then occurs: the macrophage is stimulated to secrete the cytokine known as interleukin-1. As previously discussed, interleukin-1 stimulates cell division and proliferation of T lymphocytes. The activated

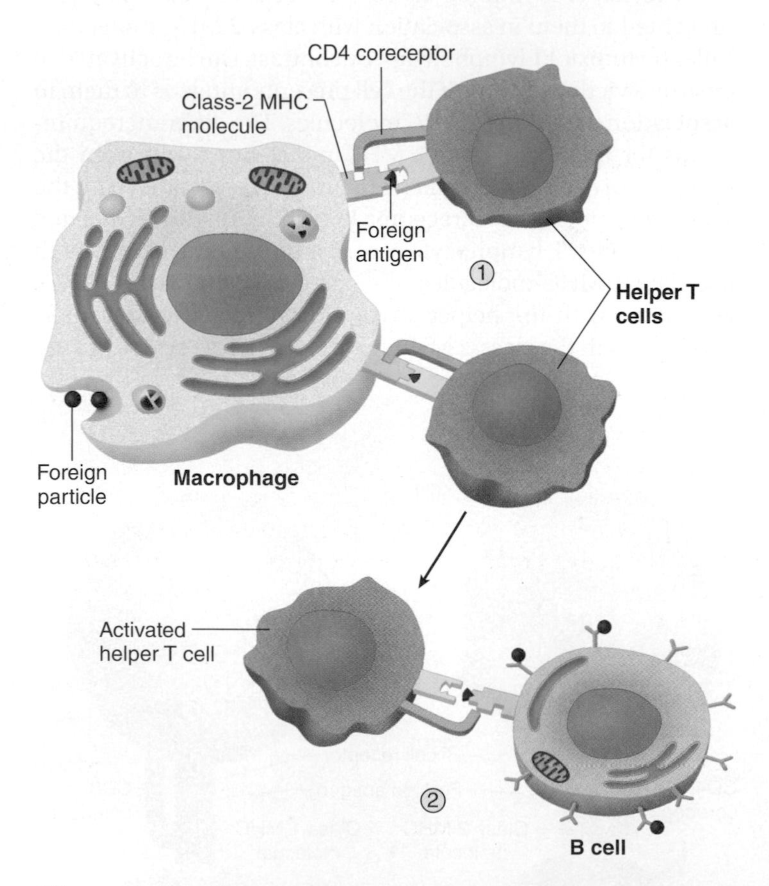

Figure 15.15 Interactions between antigen-presenting cells, T cells, and B cells. (1) An antigen-presenting cell, which here is a macrophage, presents a foreign antigen bound to a class-2 MHC molecule on its surface to a helper T cell. The T cell receptor requires that the antigen be presented in this way in order for the T cell to become activated. (2) Once activated, the helper T cell is here interacting with a B cell to improve the immune response to the foreign antigen.

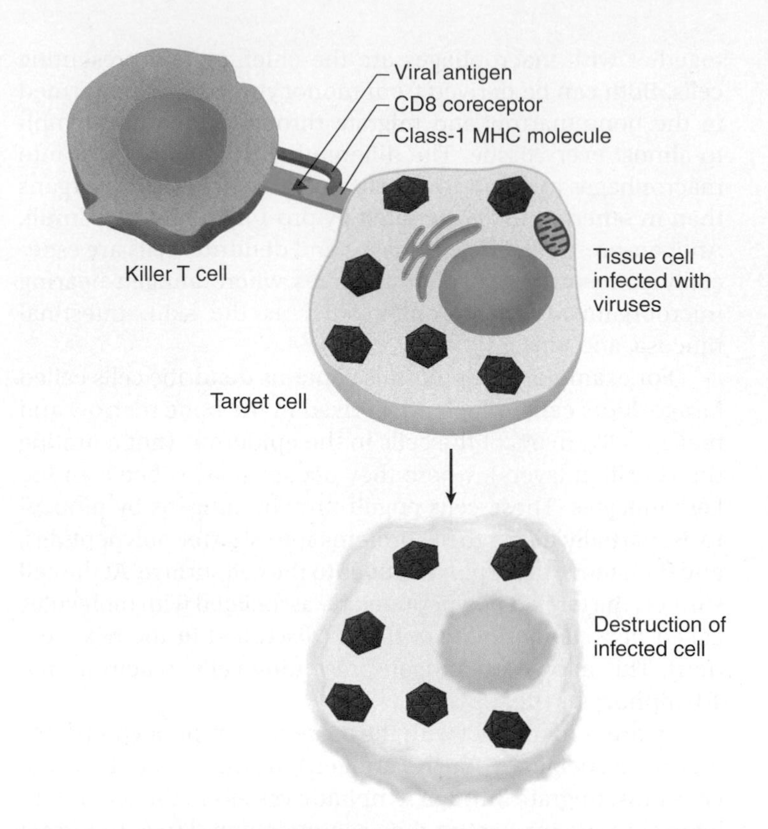

Figure 15.16 A killer T cell destroys an infected cell. In order for a killer T cell to destroy a cell infected with viruses, the T cell must interact with both the foreign antigen and the class-1 MHC molecule on the surface of the infected cell.

helper T cells, in turn, secrete macrophage colony-stimulating factor and gamma interferon, which promote the activity of macrophages. In addition, interleukin-2 is secreted by the T lymphocytes and stimulates the macrophages to secrete *tumor necrosis factor,* which is particularly effective in killing cancer cells.

Killer T cells can destroy infected cells only if those cells display the foreign antigen together with their class-1 MHC molecules (fig. 15.16). Such interaction of killer T cells with the foreign antigen-MHC class-1 complex also stimulates proliferation of those killer T cells. In addition, proliferation of the killer T lymphocytes is stimulated by interleukin-2 secreted by the helper T lymphocytes that were activated by macrophages or dendritic cells (fig. 15.17).

The network of interactions among the different cell types of the immune system now spread outward. Helper T cells can also promote the humoral immune response of B cells. It may be recalled that B cell genes for antibodies undergo somatic hypermutation and class switch recombination (section 15.2) within the germinal centers of lymph nodes. The ability of B cells to produce class-switched antibodies (from IgM to IgG, IgA, or IgE) with a high affinity for their antigens depends on helper T lymphocytes. These helper T cells home to the germinal centers of lymph nodes where they may be activated by antigen-presenting cells (fig. 15.18). There, they aid B cells to divide and develop into plasma cells and memory B cells. This

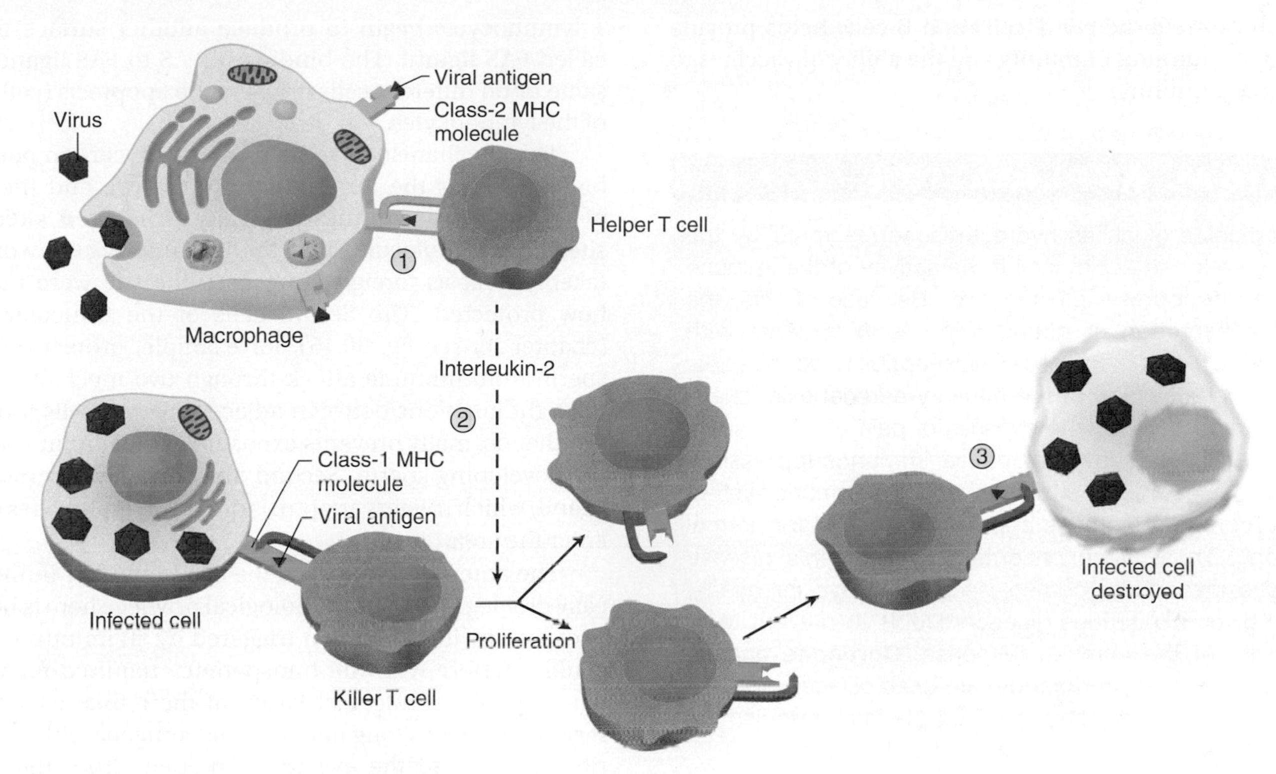

Figure 15.17 Interactions between macrophages, helper T cells, and killer T cells. (1) An antigen-presenting cell, in this case a macrophage, activates a helper T cell by presenting it with a viral antigen in association with a class 2 MHC molecule. (2) The helper T cell secretes interleukin-2, which promotes the proliferation of activated killer T cells. (3) The killer T cells destroy the cells infected with the viruses.

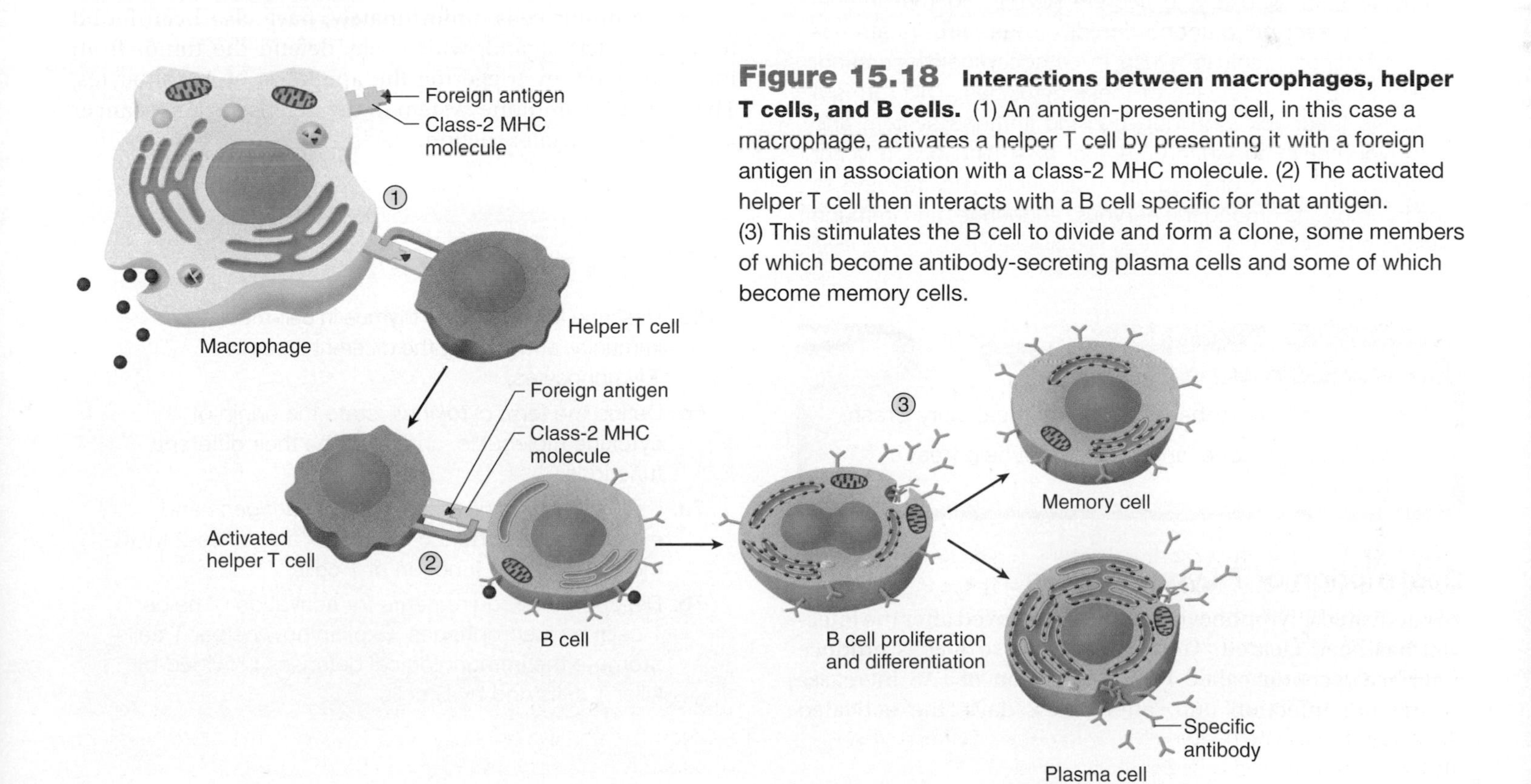

Figure 15.18 Interactions between macrophages, helper T cells, and B cells. (1) An antigen-presenting cell, in this case a macrophage, activates a helper T cell by presenting it with a foreign antigen in association with a class-2 MHC molecule. (2) The activated helper T cell then interacts with a B cell specific for that antigen. (3) This stimulates the B cell to divide and form a clone, some members of which become antibody-secreting plasma cells and some of which become memory cells.

interaction between helper T cells and B cells helps provide for long-term humoral immunity and the ability of vaccines to evoke active immunity.

CLINICAL APPLICATION

Glucocorticoids (such as hydrocortisone) secreted by the adrenal cortex can act to inhibit the activity of the immune system and suppress inflammation. Because of this, the excessive secretion of glucocorticoids in patients with *Cushing's syndrome* increases susceptibility to disease. Similarly, the activation of the pituitary-adrenal axis (chapter 11, section 11.4) by the stresses of pain and emotional trauma has been shown to cause immunosuppression. However, glucocorticoid suppression of the immune system to an appropriate degree is apparently required for normal physiology. Thus, when patients with *Addison's disease* (who have inadequate cortisol secretion) have infections, they are given exogenous glucocorticoids to prevent toxic overactivity of the immune response. **Cortisone** and its analogues (such as **prednisone**) are used clinically to treat inflammatory disorders and to inhibit the immune rejection of transplanted organs.

The immunosuppressive effects of glucocorticoids result largely from their ability to suppress the release of several pro-inflammatory cytokines. These include IL-1 and some other interleukins, interferon gamma, and TNFα. However, IL-1 and other cytokines released by microglia in the brain stimulate the hypothalamus to release CRH, which stimulates the pituitary to secrete ACTH and the adrenal cortex to secrete glucocorticoids. Thus, pro-inflammatory cytokines result in a rise in glucocorticoid secretion, which inhibits the release of these cytokines. This forms a negative feedback loop between the immune system and the hypothalamus-pituitary-adrenal axis. These and other observations have opened up a scientific field devoted to the interactions among the nervous, endocrine, and immune systems.

Clinical Investigation CLUES

The physician prescribed cortisone to treat Gary's rash.

- What is cortisone, and how does it help treat a rash?

Destruction of T Lymphocytes

The activated T lymphocytes must be destroyed after the infection has been cleared. This occurs because T cells produce a surface receptor called **FAS.** Production of FAS increases during the infection and, after a few days, the activated T lymphocytes begin to produce another surface molecule called **FAS ligand.** The binding of FAS to FAS ligand, on the same or on different cells, triggers the apoptosis (cell suicide) of the lymphocytes.

This mechanism also helps maintain certain parts of the body—such as the inner region of the eye and the tubules of the testis—as **immunologically privileged sites.** These sites harbor molecules that the immune system would mistakenly treat as foreign antigens if the site were not somehow protected. The Sertoli cells of the testicular tubules (chapter 20; see fig. 20.16), for example, protect developing sperm from immune attack through two mechanisms. First, the tight junctions between adjacent Sertoli cells form a barrier that normally prevents exposure of the immune system to the developing sperm. Second, the Sertoli cells produce FAS ligand, which triggers apoptosis of any T lymphocytes that may enter the area.

The anterior chamber of the eye is another immunologically privileged site. Immunological privilege here is beneficial because any inflammation triggered by an immune response could interfere with the transparency required for vision or damage the neurological layers of the retina, which cannot regenerate. It was long believed that antigens within the anterior chamber of the eye were "hidden" from the immune system, but more recent evidence suggests that immunological privilege here involves more active processes. These include coating of the interior of the eye with FAS ligand, which promotes apoptosis of leukocytes, and the secretion of different cytokines, which inhibit inflammation through a variety of mechanisms.

Some tumor cells, unfortunately, have also been found to produce FAS ligand, which may defend the tumor from immune attack by triggering the apoptosis of lymphocytes. The role of the immune system in the defense against cancer is discussed in a later section.

CHECKPOINT

6a. Describe the role of the thymus in cell-mediated immunity, and identify the different types of T lymphocytes.

6b. Define the term *cytokines,* state the origin of cytokine molecules, and describe their different functions.

7a. Define the term *histocompatibility antigens* and explain the importance of class-1 and class-2 MHC molecules in the function of T cells.

7b. Describe the requirements for activation of helper T cells by macrophages. Explain how helper T cells promote the immunological defenses provided by killer T cells and by B cells.

15.4 ACTIVE AND PASSIVE IMMUNITY

When a person is first exposed to a pathogen, the immune response may be insufficient to combat the disease. During the response, however, the lymphocytes that have specificity for that antigen are stimulated to divide many times and produce a clone. This is active immunity, and it can protect the person from getting the disease upon subsequent exposures.

LEARNING OUTCOMES

After studying this section, you should be able to:

8. Explain how active immunity is produced, using the clonal selection theory.

9. Explain how passive immunity is produced.

It first became known in Western Europe in the mid-eighteenth century that the fatal effects of smallpox could be prevented by inducing mild cases of the disease. This was accomplished at that time by rubbing needles into the pustules of people who had mild forms of smallpox and injecting these needles into healthy people. Understandably, this method of immunization did not gain wide acceptance.

Acting on the observation that milkmaids who contracted cowpox—a disease similar to smallpox but less *virulent* (less pathogenic)—were immune to smallpox, an English physician named Edward Jenner inoculated a healthy boy with cowpox. When the boy recovered, Jenner inoculated him with what was considered a deadly amount of smallpox, to which the boy proved to be immune. (This was fortunate for both the boy—who was an orphan—and Jenner; Jenner's fame spread, and as the boy grew into manhood he proudly gave testimonials on Jenner's behalf.) This experiment, performed in 1796, began the first widespread immunization program.

A similar but more sophisticated demonstration of the effectiveness of immunizations was performed by Louis Pasteur almost a century later. Pasteur isolated the bacteria that cause anthrax and heated them until their *virulence* (ability to cause disease) was greatly reduced (or *attenuated*), although their *antigenicity* (the nature of their antigens) was not significantly changed (fig. 15.19). He then injected these attenuated bacteria into 25 cows, leaving 25 unimmunized. Several weeks later, before a gathering of scientists, he injected all 50 cows with the completely active anthrax bacteria. All 25 of the unimmunized cows died—all 25 of the immunized animals survived.

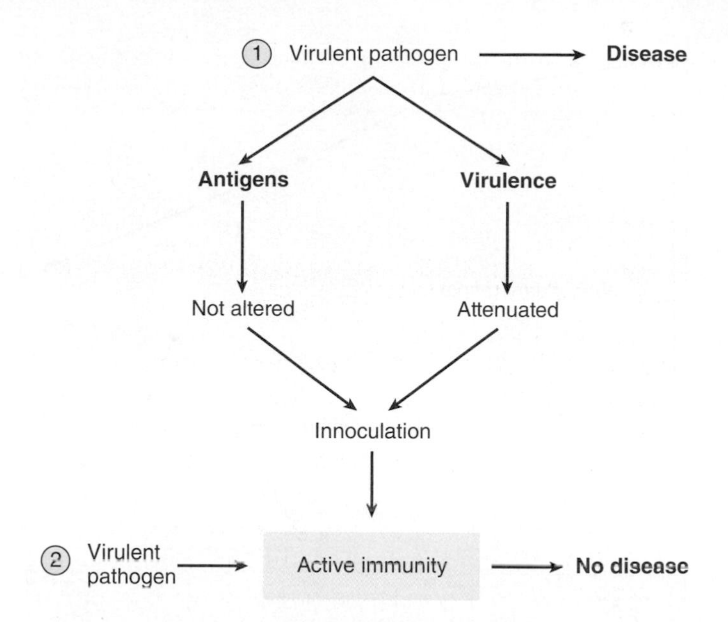

Figure 15.19 Virulence and antigenicity. (1) Active immunity to a pathogen can be gained by exposure to the fully virulent form or by inoculation with a pathogen whose virulence (ability to cause disease) has been attenuated (reduced) without altering its antigenicity (nature of its antigens). (2) Upon subsequent exposure to the same pathogen, active immunity reduces the chances of getting the disease.

Active Immunity and the Clonal Selection Theory

When a person is exposed to a particular pathogen for the first time, there is a latent period of 5 to 10 days before measurable amounts of specific antibodies appear in the blood. This sluggish **primary response** may not be sufficient to protect the person against the disease caused by the pathogen. Antibody concentrations in the blood during this primary response reach a plateau in a few days and decline after a few weeks.

A subsequent exposure of that person to the same antigen results in a **secondary response** (fig. 15.20). Compared to the primary response, antibody production during the secondary response is much more rapid. Maximum antibody concentrations in the blood are reached in less than two hours and are maintained for a longer time than in the primary response. This rapid rise in antibody production is usually sufficient to prevent the person from developing the disease.

Clonal Selection Theory

The immunization procedures of Jenner and Pasteur were effective because the people who were inoculated produced a secondary rather than a primary response when exposed to the virulent pathogens. The type of protection they were afforded does not depend on accumulations of antibodies in the blood, since secondary responses occur even after antibodies produced by the primary response have disappeared. Immunizations, therefore, seem to produce a type of "learning" in which the ability of the immune system to combat a particular pathogen is improved by prior exposure.

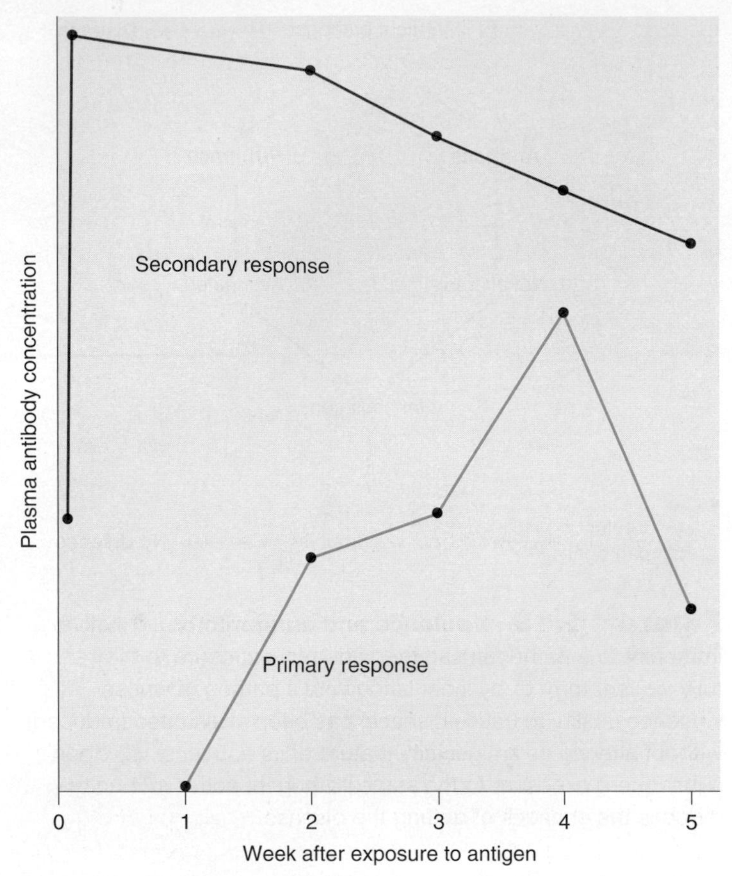

Figure 15.20 The primary and secondary immune responses. A comparison of antibody production in the primary response (upon first exposure to an antigen) to antibody production in the secondary response (upon subsequent exposure to the antigen). The more rapid production of antibodies in the secondary response is believed to be due to the development of lymphocyte clones produced during the primary response.

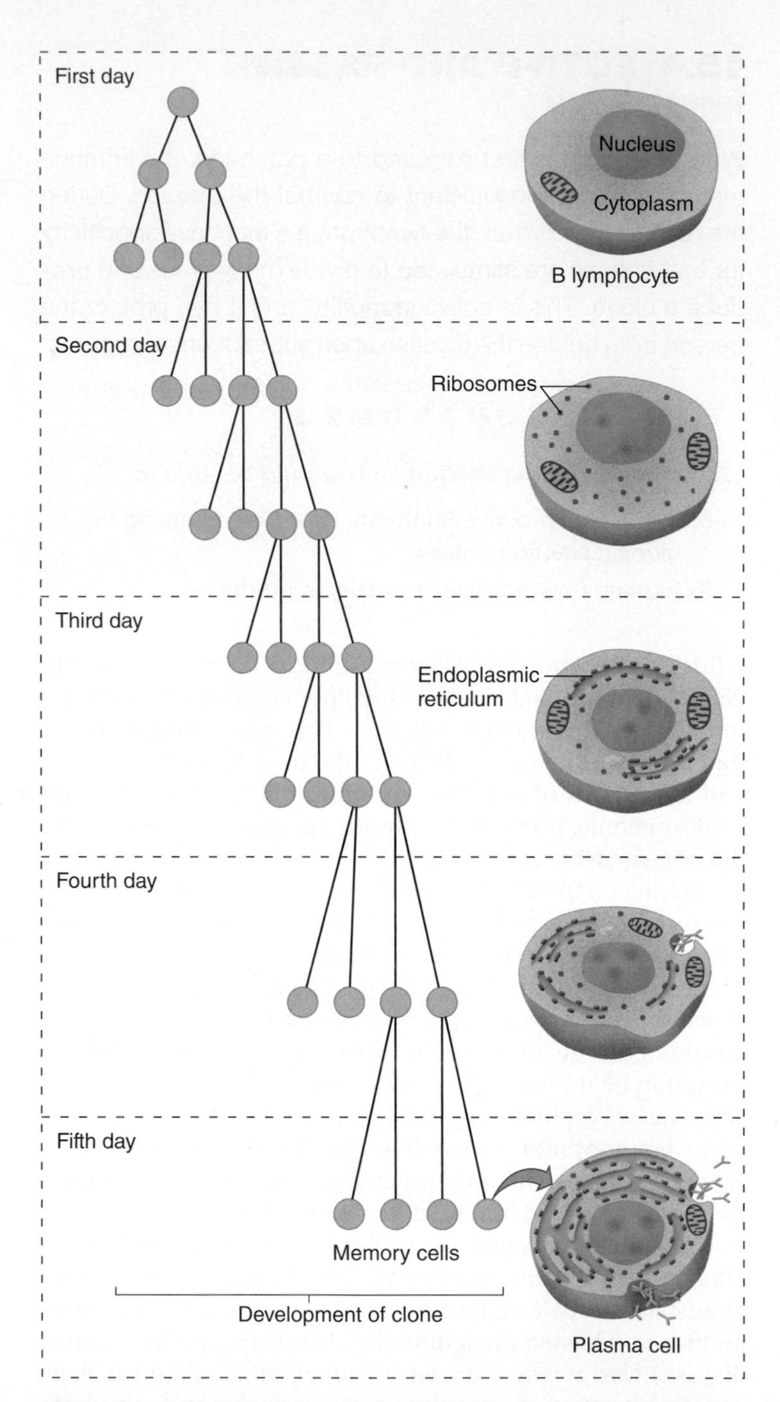

Figure 15.21 The clonal selection theory as applied to B lymphocytes. Most members of the B lymphocyte clone become memory cells, but some become antibody-secreting plasma cells. AP|R

The mechanisms by which secondary responses are produced are not completely understood; the **clonal selection theory,** however, accounts for most of the evidence. According to this theory, B lymphocytes *inherit* the ability to produce particular antibodies (and T lymphocytes inherit the ability to respond to particular antigens). A given B lymphocyte can produce only one type of antibody, with specificity for one antigen. Because this ability is genetically inherited rather than acquired, some lymphocytes can respond to smallpox, for example, and produce antibodies against it even if the person has never been previously exposed to this disease.

The inherited specificity of each lymphocyte is reflected in the antigen receptor proteins on the surface of the lymphocyte's plasma membrane. Exposure to smallpox antigens thus stimulates these specific lymphocytes to divide many times until a large population of genetically identical cells—a *clone*—is produced. Some of these cells become plasma cells that secrete antibodies for the primary response; others become memory cells that can be stimulated to secrete antibodies during the secondary response (fig. 15.21).

Notice that, according to the clonal selection theory (table 15.8), antigens do not induce lymphocytes to make the appropriate antibodies. Rather, antigens select lymphocytes (through interaction with surface receptors) that are already able

Table 15.8 | Summary of the Clonal Selection Theory (As Applied to B Cells)

Process	Results
Lymphocytes inherit the ability to produce specific antibodies.	Prior to antigen exposure, lymphocytes that can make the appropriate antibodies are already present in the body.
Antigens interact with antibody receptors on the lymphocyte surface.	Antigen-antibody interaction stimulates cell division and the development of lymphocyte clones that contain memory cells and plasma cells that secrete antibodies.
Subsequent exposure to the specific antigens produces a more efficient response.	Exposure of lymphocyte clones to specific antigens results in greater and more rapid production of specific antibodies.

to make antibodies against that antigen. This is analogous to evolution by natural selection. An environmental agent (in this case, antigens) acts on the genetic diversity already present in a population of organisms (lymphocytes) to cause an increase in number of the individuals selected.

Secondary lymphoid organs, such as the lymph nodes and spleen, contain **germinal centers** following exposure to antigens. A germinal center develops from a B cell that has been stimulated by an antigen and activated by helper T cells. This B cell then undergoes extremely rapid mitotic cell division to become the founder of a germinal center that contains a clone of memory cells and plasma cells, which secrete antibodies for the secondary immune response. These plasma cells are longer-lived than those outside of germinal centers, which produce less effective antibodies for the primary response. The proliferating B cells in the germinal center undergo somatic hypermutation (previously described), which generates a diversity of new antibodies. This could include antibodies that have a higher affinity for the stimulating antigen, thereby improving the immune response.

Active Immunity

The development of a secondary response provides **active immunity** against the specific pathogens. The development of active immunity requires prior exposure to the specific antigens, at which time the sluggishness of the primary response may cause the person to develop the disease. For example, children who get measles, chicken pox, or mumps will probably be immune to them as adults, when these diseases are potentially more serious.

Clinical immunization programs induce primary responses by inoculating people with pathogens whose virulence has been attenuated or destroyed (such as Pasteur's heat-inactivated anthrax bacteria) or by using closely related strains of microorganisms that are antigenically similar but less pathogenic (such as Jenner's cowpox inoculations). The name for these procedures—**vaccinations** (after the Latin word *vacca,* meaning "cow")—reflects the history of this technique. All of these procedures cause the development of lymphocyte clones that can combat the virulent pathogens by producing secondary responses.

The first successful polio vaccine (the Salk vaccine) was composed of viruses that had been inactivated by treatment with formaldehyde. These "killed" viruses were injected into the body, in contrast to the currently used oral (Sabin) vaccine. The oral vaccine contains "living" viruses that have attenuated virulence. These viruses invade the epithelial lining of the intestine and multiply, but do not invade nerve tissue. The immune system can, therefore, become sensitized to polio antigens and produce a secondary response if polio viruses that attack the nervous system are later encountered.

In summary, there are three ways that vaccines are currently produced. Vaccines may use:

1. "live" viruses with attenuated virulence that do not cause disease, but provoke strong immune responses against the virulent viruses. Examples include the Sabin polio vaccine and vaccinations against measles and mumps.
2. "killed" virulent viruses that do not cause the disease. Examples include the Salk polio vaccine.
3. recombinant viral proteins, produced through genetic engineering and given by themselves. Examples of this are the hepatitis B vaccine and attempted HIV vaccines.

Immunizations improved after the 1920s with the discovery that certain molecules, called **adjuvants,** could boost the immune response when delivered together with the vaccine antigens. Adjuvants are generally molecules associated with microbes, and it was later discovered that they are among the PAMPs (pathogen-associated molecular patterns) discussed earlier as activating the innate immune system through binding to toll like receptors. Adjuvants enhance the adaptive immune response by boosting the ability of antigen-presenting cells to activate B and T lymphocyte responses. For example, when adjuvants stimulate dendritic cell secretion of IL-2 and IL-12, they enhance the B cell secretion of antibodies; when adjuvants stimulate dendritic cell release of IL-4, IL-5, and IL-6, they stimulate the T cell–mediated response. Scientists continue to develop new adjuvants to bolster vaccinations against malaria, influenza, and cancer.

Clinical Investigation CLUES

Gary had previously received a tetanus vaccine, which protected him when he was cut by a piece of metal.

- What does a vaccine contain, and how would such a vaccine protect against later exposure to tetanus?

Immunological Tolerance

The ability to produce antibodies against **non-self (foreign) antigens,** while tolerating (not producing antibodies against) **self-antigens** occurs during the first month or so of postnatal life, when immunological competence is established. If a fetal mouse of one strain receives transplanted antigens from a different strain, therefore, it will not recognize tissue transplanted later in life from the other strain as foreign; consequently, it will not immunologically reject the transplant.

The ability of an individual's immune system to recognize and tolerate self-antigens requires continuous exposure of the immune system to those antigens. If this exposure begins when the immune system is weak—as in fetal and early postnatal life—tolerance is more complete and long lasting than that produced by exposure beginning later in life. Some self-antigens, however, are normally hidden from the blood, such as thyroglobulin within the thyroid gland and lens protein in the eye. An exposure to these self-antigens results in antibody production just as if these proteins were foreign. Antibodies made against self-antigens are called **autoantibodies.** Killer T cells that attack self-antigens are called **autoreactive T cells.**

The mechanisms of immunological tolerance include (1) **clonal deletion,** in which lymphocytes that recognize self-antigens are destroyed, and (2) **clonal anergy** (meaning "without working"), in which lymphocytes that recognize self-antigens are prevented from becoming activated. **Central tolerance** mechanisms are those that occur in the thymus (for T cells) and bone marrow (for B cells). Central tolerance in the thymus is achieved by the apoptosis and removal of autoreactive T cells, whereas central tolerance of B cells in the bone marrow may involve both clonal deletion and anergy. **Peripheral tolerance,** involving lymphocytes outside of the thymus and bone marrow, is due to complex mechanisms that produce anergy.

Peripheral tolerance mechanisms are needed because as lymphocytes divide, they randomly generate new antigen receptors due to gene rearrangements and somatic mutations (described previously). This is beneficial because it allows the immune system to respond to a wide variety of foreign antigens, but it also creates lymphocytes throughout life that have receptors for self-antigens. These lymphocytes would produce autoimmune diseases without mechanisms to suppress their activation. Scientists currently believe that this suppression is provided partly by *regulatory T lymphocytes* (T_{reg}). In a demonstration of this process, removal of the thymus from mice between two and four days old resulted in autoimmune disease, which was reversed by adding back T_{reg} cells.

Passive Immunity

The term **passive immunity** refers to the immune protection that can be produced by the transfer of antibodies to a recipient from a human or animal donor. The donor has been actively immunized, as explained by the clonal selection theory. The person who receives these ready-made antibodies is thus passively immunized to the same antigens. Passive immunity also occurs naturally in the transfer of immunity from mother to fetus during pregnancy and from mother to baby during nursing.

The ability to mount a specific immune response—called **immunological competence**—does not develop until about a month after birth. The fetus, therefore, cannot immunologically reject its mother. The immune system of the mother is fully competent but does not usually respond to fetal antigens for reasons that are not completely understood. Some IgG antibodies from the mother do cross the placenta and enter the fetal circulation, however, and these serve to confer passive immunity to the fetus.

The fetus and the newborn baby are thus immune to the same antigens as the mother. However, because the baby did not itself produce the lymphocyte clones needed to form these antibodies, such passive immunity gradually disappears. If the baby is breast-fed, it can receive additional antibodies of the IgA subclass in its mother's milk and *colostrum* (the secretion an infant feeds on for the first two or three days until the onset of true lactation). This provides additional passive immunity until the baby can produce its own antibodies through active immunity (see chapter 20, fig. 20.54).

Passive immunizations are used clinically to protect people who have been exposed to extremely virulent infections or toxins, such as tetanus, hepatitis, rabies, and snake venom. In these cases, the affected person is injected with *antiserum* (serum containing antibodies), also called *antitoxin,* from an animal that has been previously exposed to the pathogen. The animal develops the lymphocyte clones and active immunity, and thus has a high concentration of antibodies in its blood. Since the person who is injected with these antibodies does not develop active immunity, he or she must again be injected with antitoxin upon subsequent exposures.

In a variation on the theme of passive immunity, animals (such as mice, rabbits, or sheep) are injected with an antigen and used to obtain **monoclonal antibodies,** which are produced by an isolated, pure clone of cells. This clone is obtained by first extracting from the animal a single B lymphocyte that produces antibodies against the desired specific antigenic-determinant site. The B lymphocyte is then fused *in vitro* with a cancerous myeloma cell; this is done to form a hybrid cell that is able to divide indefinitely.

The cell divisions produce a clone called a *hybridoma* that secretes antibodies specific for a single antigenic-determinant site. This has enabled monoclonal antibodies to be produced on a commercial scale for diagnosis (of pregnancy, for example) and other laboratory tests, as well as for medical treatments. There are currently more than 25 approved therapeutic monoclonal antibodies that target cytokines (such as tumor necrosis factor), IgE, and other regulatory molecules for the treatment of many autoimmune diseases (section 15.6; see table 15.10), persistent allergic asthma, and cancers. Examples of monoclonal antibodies used in cancer treatments include *trastuzumab (Herceptin),* which targets the HER2 receptor produced by

Table 15.9 | Comparison of Active and Passive Immunity

Characteristic	Active Immunity	Passive Immunity
Injection of person with	Antigens	Antibodies
Source of antibodies	The person inoculated	Natural—the mother; artificial—injection with antibodies
Method	Injection with killed or attenuated pathogens or their toxins	Natural—transfer of antibodies across the placenta; artificial—injection with antibodies
Time to develop resistance	5 to 14 days	Immediately after injection
Duration of resistance	Long (perhaps years)	Short (days to weeks)
When used	Before exposure to pathogen	Before or after exposure to pathogen

30% of invasive breast cancers; and *bevacizumab (Avastin)*, which blocks the binding of vascular endothelial growth factor (VEGF) to its receptors for the treatment of types of colorectal, lung, kidney, and brain cancers.

Active and passive immunity are compared in table 15.9.

CHECKPOINT

8a. Describe three methods used to induce active immunity.

8b. Using graphs to illustrate your discussion, explain the characteristics of the primary and secondary immune responses.

8c. Explain the clonal selection theory and indicate how this theory accounts for the secondary response.

8d. Define *immunological tolerance,* and explain mechanisms that may be responsible for T and B lymphocytes' tolerance to self-antigens.

9. Describe passive immunity and give examples of how it may occur naturally and how it may be conferred by artificial means.

15.5 TUMOR IMMUNOLOGY

Tumor cells can reveal antigens that stimulate the destruction of the tumor. When cancers develop, this immunological surveillance system—primarily the function of T cells and natural killer cells—has failed to prevent the growth and metastasis of the tumor.

LEARNING OUTCOMES

After studying this section, you should be able to:

10. Explain the relationship between the immune system and cancer.

Oncology (the study of tumors) has revealed that tumor biology is similar to and interrelated with the functions of the immune system. Most tumors appear to be clones of single cells that have become transformed in a process similar to the development of lymphocyte clones in response to specific antigens. Lymphocyte clones, however, are under complex inhibitory control systems—such as those exerted by regulatory T lymphocytes and negative feedback by antibodies. The division of tumor cells, by contrast, is not effectively controlled by normal inhibitory mechanisms. Tumor cells are also relatively unspecialized—they *dedifferentiate,* which means that they become similar to the less specialized cells of an embryo.

Tumors are described as *benign* when they are relatively slow growing and limited to a specific location (warts, for example). *Malignant* tumors grow more rapidly and undergo **metastasis,** a term that refers to the dispersion of tumor cells and the resultant seeding of new tumors in different locations. The term **cancer,** as it is generally applied, refers to malignant tumors. The word *cancer* (Latin for "crab") stems from Hippocrates' observation of a tumor being fed by blood vessels, which resembled the claws of a crab. Cancer is now believed to result from altered expression of oncogenes (genes that promote cancer), tumor-suppressor genes, and genes that code for microRNA (miRNA; chapter 3, section 3.3).

As tumor cells dedifferentiate, they reveal surface antigens that can stimulate the immune destruction of the tumor. Consistent with the concept of dedifferentiation, some of these antigens are proteins produced in embryonic or fetal life and not normally produced postnatally. Because they are absent at the time immunological competence is established, they are treated as foreign and fit subjects for immunological attack when they are produced by cancerous cells. The release of two such antigens into the blood has provided the basis for laboratory diagnosis of some cancers. *Carcinoembryonic antigen tests* are useful in the diagnosis of colon cancer, for example, and tests for *alpha-fetoprotein* (normally produced only by the fetal liver) help in the diagnosis of liver cancer.

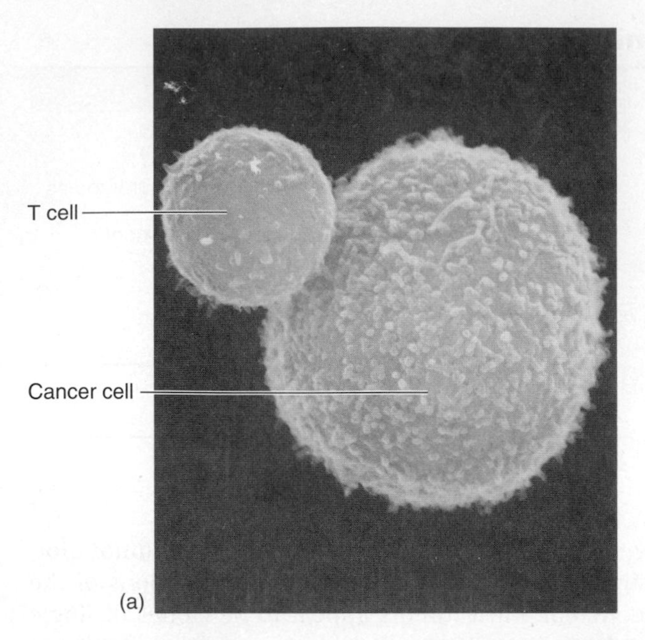

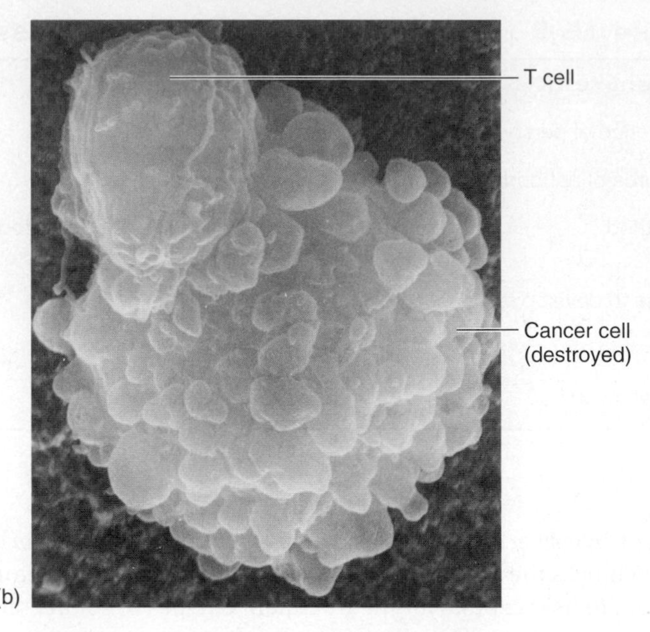

Figure 15.22 T cell destruction of a cancer cell. A killer T cell (*a*) contacts a cancer cell (the larger cell), in a manner that requires specific interaction with antigens on the cancer cell. The killer T cell releases lymphokines, including toxins that cause the death of the cancer cell, as shown in (*b*). Scanning electron micrographs © Andrejs Liepens.

The immune system can guard against cancer by recognizing antigens that are expressed in malignantly transformed cells. For example, carcinogens cause mutations, and these mutated genes can produce tumor-specific antigens. Also, some cancers are caused by viruses and so can display viral antigens that flag tumor cells. Examples include antigens of the *human papillomavirus,* which can cause cervical cancer, and antigens of the *Epstein-Barr virus,* which can cause Burkitt's lymphoma and nasopharyngeal carcinoma. Even normal self-antigens, if they are overexpressed due to gene duplication, may trigger attack by the adaptive immune system. In cases where tumor cells do not display antigens that directly stimulate T and B lymphocytes, the tumor cells may still activate the natural killer (NK) cells (discussed next) of the innate immune system. The adaptive immune response could then be triggered by molecules generated during the inflammation caused by the innate immune response to the tumor cells.

Tumor antigens activate the immune system, initiating an attack primarily by killer T lymphocytes (fig. 15.22) and natural killer cells (described in the next section). The concept of **immunological surveillance** against cancer was introduced in the early 1970s to describe the proposed role of the immune system in fighting cancer. According to this concept, tumor cells frequently appear in the body but are normally recognized and destroyed by the immune system before they can cause cancer.

However, tumors can evade immune surveillance by suppressing immunity. Tumor cells produce and secrete immunosuppressive molecules, including transforming growth factor beta (TGFβ) and FAS ligand (which binds to FAS and stimulates apoptosis of lymphocytes, as previously described). Also, the environment of the tumor can have a high population of regulatory T lymphocytes, which secrete TGFβ and interleukin-10 to suppress the immune responses to the tumor cells.

CLINICAL APPLICATION

The immune system doesn't always fight cancer; sometimes it promotes it. Chronic inflammation is associated with increased risk of cancer, particularly in the liver and intestine. For example, **hepatocellular carcinoma,** the third leading cause of cancer death worldwide, may be associated with hepatitis. The risk of hepatocellular carcinoma is about 100 times higher in people who carry the hepatitis B virus. The inflammation in hepatitis produces reactive oxygen species that impose an oxidative stress on the cells. Oxidative stress, ionizing radiation, and ultraviolet light activate a family of transcription factors called **nuclear factor-κB** (**NF-κB,** where κ is the Greek letter kappa). NF-κB stimulates the transcription of many genes that promote all aspects of inflammation. The pro-inflammatory cytokines released during the inflammation then activate NF-κB in other cells, extending the inflammation. Besides promoting inflammation, NF-κB induces genetic changes that can cause transformation into cancer. NF-κB also inhibits apoptosis (cell suicide), allowing liver tumor cells to survive, proliferate, and migrate.

Natural Killer Cells

Hairless mice of a particular strain genetically lack a thymus and T lymphocytes, yet these mice do not appear to have an especially high incidence of tumor production. This surprising observation led to the discovery of **natural killer (NK) cells,** which are lymphocytes that are related to, but distinct from, T lymphocytes. The NK cells are lymphocytes that are considered to be part of the innate immune system. Unlike the B and T lymphocytes that are part of the adaptive immune system, NK cells do not have the great receptor diversity generated by gene rearrangements to provide receptors for particular antigens. Instead, the NK cells display an array of receptors inherited through the germ cells (sperm and egg) that can target malignantly transformed cells and cells infected with intracellular pathogens, such as viruses. Also, the NK cells have inhibitory receptors that interact with MHC class 1 molecules on the person's own cells to provide tolerance to "self" and prevent autoimmune attack.

Resting NK cells release cytokines, including interferon-γ and others; they also contain intracellular granules with granzymes and perforin. Thus, like killer T lymphocytes (section 15.3), they can destroy target cells by cell-to-cell contact; unlike killer T lymphocytes, NK cells can do this efficiently without prior exposure to the foreign antigens. However, in order for NK cells to be fully effective, they must first be activated by interferon-α, interferon-β, and other pro-inflammatory cytokines.

NK cells can be activated by cytokines released by dendritic cells and by the cells of the adaptive immune system. Once activated, the NK cells release interferon-γ and other cytokines that help activate macrophages and the cells of the adaptive immune system. Because NK cells do not require prior exposure to foreign antigens, they can provide a first line of innate, cell-mediated defense that is subsequently backed up by the adaptive, specific immune responses of killer T lymphocytes. Cytokines released by both NK cells and the cells of the adaptive immune system also enlist phagocytic neutrophils and macrophages to the site of immune attack. Interestingly, recent research has shown that NK cells, like the cells of the adaptive immune system, can produce long lasting "memory" cells and mount a stronger secondary response to a subsequent challenge.

Immunotherapy for Cancer

Monoclonal antibodies have been used to combat some cancers. For example, about 20% to 25% of *breast cancer* patients have *HER2* receptors on the plasma membrane of the tumor cells. HER2 is a receptor for epidermal growth factor, a paracrine regulator that stimulates cell growth and division. For these patients, monoclonal antibodies—called *trastuzumab (Herceptin)* and *pertuzumab*—are available to block the HER2 receptors. Also, monoclonal antibodies (*rituximab,* for example) are used in the medical treatment of *non-Hodgkin's lymphoma* of B lymphocytes, the most common hematological cancer in adults. In addition to monoclonal antibodies, human interferons (from genetically engineered bacteria) are also available for medical use. These have been used for the treatment of lymphomas, renal carcinoma, and melanoma.

A team of scientists led by Dr. S. A. Rosenberg at the National Cancer Institute has pioneered the use of another lymphokine that is now available through genetic engineering techniques. This is *interleukin-2 (IL-2),* which activates both killer T lymphocytes and B lymphocytes. The investigators removed some of the blood from cancer patients who could not be successfully treated by conventional means and isolated a population of their lymphocytes. They treated these lymphocytes with IL-2 to produce *lymphokine-activated killer (LAK) cells* and then reinfused these cells, together with IL-2 and interferons, into the patients. The scientists have also identified *tumor-infiltrating lymphocytes (TIL),* which can invade solid tumors. The use of TIL cells is expensive and so far has benefited only patients with metastatic melanoma.

Besides interleukin-2 and gamma interferon, other cytokines may be useful in the treatment of cancer and are currently undergoing experimental investigations. Interleukin-12, for example, seems promising because it is needed for the changing of uncommitted helper T lymphocytes into the T_H1 subtype that bolsters cell-mediated immunity. Scientists are also attempting to identify specific antigens that may be uniquely expressed in cancer cells in an effort to help the immune system target cancer cells for destruction. In related avenues of approach, scientists have developed a vaccine of dendritic cells sensitized to tumor antigens, and have tried agents that reduce the ability of regulatory T cells to suppress the immune system's attack on cancer cells.

In another approach, scientists recently reported success treating patients with chronic lymphoid leukemia using the patient's own T cells that were previously extracted and treated with a virus containing genes for *chimeric antigen receptors.* These molecules, which function like T cell receptor proteins, contained a fragment from a monoclonal antibody against CD19, a B lymphocyte cell-surface marker. The treated T cells were infused back into the patients, where they multiplied a thousandfold and attacked both normal and leukemia B cells. Scientists are presently optimistic about this technique, but it does have toxic side effects (including depletion of B cells and low serum gamma globulin) and must be evaluated using a larger number of patients with different types of cancers.

Effects of Aging and Stress

Susceptibility to cancer varies greatly. The Epstein-Barr virus that causes Burkitt's lymphoma in some individuals (mainly in Africa), for example, can also be found in healthy people throughout the world. Most often the virus is harmless; in some cases, it causes mononucleosis (involving a limited proliferation of white blood cells). Only rarely does this virus cause the uncontrolled proliferation of leukocytes characteristic of Burkitt's lymphoma. The reasons for these differences in response to the Epstein-Barr virus, and indeed for the differing susceptibilities of people to other forms of cancer, are not well understood.

It is known that cancer risk increases with age. According to one theory, this is because aging lymphocytes gradually accumulate genetic errors that decrease their effectiveness. The functions of the thymus also decline with age in parallel with a decrease in cell-mediated immune competence. Both of these changes, and perhaps others not yet discovered, could increase susceptibility to cancer.

Numerous experiments have demonstrated that tumors grow faster in experimental animals subjected to stress than they do in unstressed control animals. This is attributed to the stress-induced increase in corticosteroid secretion (chapter 11), which suppresses the immune system; indeed, the immunosuppressive effects of cortisone are used medically for treating chronic inflammatory diseases and for reducing the immune rejection of transplanted organs. Some recent experiments, however, suggest that the stress-induced suppression of the immune system may also be due to other factors that do not involve the adrenal cortex. Future advances in cancer therapy may incorporate methods of strengthening the immune system into protocols aimed at directly destroying tumors.

CHECKPOINT

10a. Explain why cancer cells are believed to be dedifferentiated, and describe some of the clinical applications of this concept.

10b. Define the term *immunological surveillance*, and identify the cells involved in this function.

10c. Explain the possible relationship between stress and susceptibility to cancer.

15.6 DISEASES CAUSED BY THE IMMUNE SYSTEM

Immune mechanisms that normally protect the body are very complex and subject to errors that can result in diseases. Autoimmune diseases and allergies are two categories of disease that are caused not by an invading pathogen but instead by a derangement in the normal functions of the immune system.

LEARNING OUTCOMES

After studying this section, you should be able to:

11. Explain the nature of autoimmune diseases.

12. Explain immediate and delayed hypersensitivity.

The ability of the normal immune system to tolerate self-antigens while it identifies and attacks foreign antigens provides a specific defense against invading pathogens. In every individual, however, this system of defense against invaders at times commits domestic offenses. This can result in diseases that range in severity from the sniffles to sudden death.

Diseases caused by the immune system can be grouped into three interrelated categories: (1) *autoimmune diseases,* (2) *immune complex diseases,* and (3) *allergy,* or *hypersensitivity.* It is important to remember that these diseases are not caused by foreign pathogens but by abnormal responses of the immune system.

Autoimmunity

Autoimmune diseases are those produced by failure of the immune system to recognize and tolerate self-antigens. This failure results in the activation of *autoreactive T cells* and the production of *autoantibodies* by B cells, causing inflammation and organ damage (table 15.10). Cell division of B lymphocytes within germinal centers of lymph nodes increases antibody diversity through somatic hypermutation and immunoglobulin class switching (discussed in section 15.2). The production of autoreactive B cells (those that produce autoantibodies) occurs as an inevitable byproduct of these necessary processes. Autoimmune diseases result when those cells become exposed to and stimulated by the appropriate self-antigen.

Table 15.10 | Some Examples of Autoimmune Diseases

Disease	Antigen
Postvaccinal and postinfectious encephalomyelitis	Myelin, cross-reactive
Aspermatogenesis	Sperm
Sympathetic ophthalmia	Uvea
Hashimoto's disease	Thyroglobulin
Graves' disease	Receptor proteins for TSH
Autoimmune hemolytic disease	I, Rh, and others on surface of RBCs
Thrombocytopenic purpura	Hapten-platelet or hapten-absorbed antigen complex
Myasthenia gravis	Acetylcholine receptors
Rheumatic fever	Streptococcal, cross-reactive with heart valves
Glomerulonephritis	Streptococcal, cross-reactive with kidney
Rheumatoid arthritis	IgG
Systemic lupus erythematosus	DNA, nucleoprotein, RNA, etc.
Diabetes mellitus (type 1)	Beta cells in pancreatic islets
Multiple sclerosis	Components of myelin sheaths

Source: Modified from James T. Barrett, *Textbook of Immunology*, 5th ed. Copyright © 1988 Mosby-Yearbook. Reprinted by permission.

There are over 40 known or suspected autoimmune diseases that affect 5% to 7% of the population. Two-thirds of those affected are women. The most common autoimmune diseases are *rheumatoid arthritis, type 1 diabetes mellitus, multiple sclerosis, Graves' disease, glomerulonephritis, thyroiditis, pernicious anemia, psoriasis,* and *systemic lupus erythematosus.*

There are at least six reasons why self-tolerance may fail:

1. **An antigen that does not normally circulate in the blood may become exposed to the immune system.** Thyroglobulin protein that is normally trapped within the thyroid follicles, for example, can stimulate the production of autoantibodies and autoreactive T lymphocytes that cause the destruction of the thyroid; this occurs in *Hashimoto's thyroiditis.* Similarly, autoantibodies developed against lens protein in a damaged eye may cause the destruction of a healthy eye (in *sympathetic ophthalmia*).
2. **A self-antigen that is otherwise tolerated may be altered by combining with a foreign hapten.** The disease *thrombocytopenia* (low platelet count), for example, can be caused by the autoimmune destruction of thrombocytes (platelets). This occurs when drugs such as aspirin, sulfonamide, antihistamines, digoxin, and others combine with platelet proteins to produce new antigens. The symptoms of this disease usually disappear when the person stops taking these drugs.
3. **Antibodies may be produced that are directed against other antibodies.** Such interactions may be necessary for the prevention of autoimmunity, but imbalances may actually cause autoimmune diseases. *Rheumatoid arthritis,* for example, is an autoimmune disease associated with the abnormal production of one group of antibodies (of the IgM type) that attack other antibodies (of the IgG type). This contributes to an inflammation reaction of the joints characteristic of the disease.
4. **Antibodies produced against foreign antigens may cross-react with self-antigens.** Autoimmune diseases of this sort can occur, for example, as a result of *Streptococcus* bacterial infections. Antibodies produced in response to antigens in this bacterium may cross-react with self-antigens in the heart and kidneys. The inflammation induced by such autoantibodies can produce heart damage (including the valve defects characteristic of *rheumatic fever*) and damage to the glomerular capillaries in the kidneys (*glomerulonephritis*).
5. **Self-antigens, such as receptor proteins, may be presented to the helper T lymphocytes together with class-2 MHC molecules.** Normally, only antigen-presenting cells (macrophages, dendritic cells, and antigen-activated B cells) produce class-2 MHC molecules, which are associated with foreign antigens and recognized by helper T cells. Perhaps as a result of viral infection, however, cells that do not normally produce class-2 MHC molecules may start to do so and, in this way, present a self-antigen to the helper T cells. In *Graves' disease,* for example, the thyroid cells produce class-2 MHC molecules, and the immune system produces autoantibodies against the TSH receptor proteins in the thyroid cells. These autoantibodies, called *TSAb*'s for "thyroid-stimulating antibody," interact with the TSH receptors and overstimulate the thyroid gland. Similarly, in *type 1 diabetes mellitus,* the beta cells of the pancreatic islets abnormally produce class-2 MHC molecules, resulting in autoimmune destruction of the insulin-producing cells.
6. **Autoimmune diseases may result when there is inadequate activity of regulatory (suppressor) T lymphocytes.** Regulatory T lymphocytes dampen the immune response, and thus act to suppress autoimmune diseases and the chronic inflammation associated with them. There is evidence that other T lymphocytes can be converted into regulatory T lymphocytes by activation of a transcription factor known as *FOXP3.* Thus, inadequate expression of this gene may lead to insufficient activity of regulatory T lymphocytes and result in autoimmune disease. Indeed, genetically engineered mice without the gene, and people with a mutated FOXP3, suffer from severe autoimmune disease.

Immune Complex Diseases

The term *immune complexes* refers to antigen-antibody combinations that are free rather than attached to bacterial or other cells. The formation of such complexes activates complement proteins and promotes inflammation. This inflammation is normally self-limiting because the immune complexes are removed by phagocytic cells. When large numbers of immune complexes are continuously formed, however, the inflammation may be prolonged. Also, the dispersion of immune complexes to other sites can lead to widespread inflammation and organ damage. The damage produced by this inflammatory response is called immune complex disease.

Immune complex diseases can result from infections by bacteria, parasites, and viruses. In hepatitis B, for example, an immune complex that consists of viral antigens and antibodies can cause widespread inflammation of arteries (*periarteritis*). Arterial damage is caused not by the hepatitis virus itself but by the inflammatory process.

Immune complex disease can also result from the formation of complexes between self-antigens and autoantibodies. This produces systemic autoimmune disease, where the inflammation and tissue damage is not limited to a particular location.

In **rheumatoid arthritis,** there is inflammation of the synovial membranes and fluid of peripheral joints that causes damage to the cartilage and bone of the joints. Although the initial cause of this disease is unknown, there appears to be an early infiltration of the synovial membranes by helper T cells. These secrete pro-inflammatory cytokines that lead to the activation of B cells, which secrete autoantibodies. The antibodies, including *rheumatoid factors*—IgM antibodies that bind to the F_c portion of IgG antibodies, creating immune complexes—activate complement proteins. These extend the inflammation, attracting and activating phagocytic polymorphonuclear leukocytes. Rheumatoid arthritis affects joints symmetrically on both sides of the body and can have systemic symptoms (fatigue, anorexia, and weakness) owing to the release of tumor necrosis factor, IL-1, IL-6, and other pro-inflammatory cytokines into the circulation.

Systemic lupus erythematosus (SLE) is a systemic autoimmune disease involving the kidneys, joints, skin, central nervous system, and other body structures. In SLE, the affected people (90% of whom are women in their childbearing years) produce a wide range of autoantibodies. However, what most distinguishes SLE is the production of IgG antibodies against their own nuclear constituents; indeed, laboratory tests for *antinuclear antibodies (ANA)* are used to help diagnose the disease. People with SLE produce autoantibodies against their own chromatin (DNA and proteins), small nuclear ribonucleoprotein (snRNP), and others. Cells undergoing apoptosis or necrosis release these nuclear constituents, and so the immune system is always exposed to these antigens. For reasons not presently understood, people with SLE lose immune tolerance to these self-antigens. The binding of the nuclear antigens to autoantibodies can form immune complexes throughout the body, provoking inflammation that can damage organs. For example, immune complexes may enter the glomerular capillaries of the kidneys (the filtering units; chapter 17, section 17.2), provoking inflammation that can produce *glomerulonephritis.* Multiple genes may confer a genetic susceptibility to SLE, but interaction with the environment is also important. For example, exposure to ultraviolet light (in sunlight), or a variety of infections, may trigger SLE.

Allergy

The term *allergy,* often used interchangeably with *hypersensitivity,* refers to particular types of abnormal immune responses to antigens, which are called *allergens* in these cases. There are two major forms of allergy: (1) *immediate hypersensitivity,* which is due to an abnormal B lymphocyte response to an allergen that produces symptoms within seconds or minutes, and (2) *delayed hypersensitivity,* which is an abnormal T cell response that produces symptoms between 24 and 72 hours after exposure to an allergen. These two types of hypersensitivity are compared in table 15.11.

Immediate Hypersensitivity

Immediate hypersensitivity can produce *allergic rhinitis* (chronic runny or stuffy nose); *conjunctivitis* (red eyes); *allergic asthma; atopic dermatitis* (urticaria, or hives); and *food allergies.* These symptoms result from the immune response to the allergen. In people who are not allergic, the allergen stimulates one type of helper T lymphocyte, the T_H1 cells, to secrete interferon-γ and interleukin-2. In people who are allergic, dendritic cells stimulate the other type of helper T lymphocytes, the T_H2 cells, to secrete other lymphokines, including interleukin-4 and interleukin-13. These recruit eosinophils, promote mucus production from goblet cells, and stimulate smooth muscles in the bronchioles to promote the airway hyperresponsiveness of asthma (chapter 16, section 16.3). Lymphokines secreted by T_H2 cells also stimulate B lymphocytes and plasma cells to secrete antibodies of the IgE subclass instead of the normal IgG antibodies.

Table 15.11 | Allergy: Comparison of Immediate and Delayed Hypersensitivity Reactions

Characteristic	Immediate Reaction	Delayed Reaction
Time for onset of symptoms	Within several minutes	Within 1 to 3 days
Lymphocytes involved	B cells	T cells
Immune effector	IgE antibodies	Cell-mediated immunity
Allergies most commonly produced	Hay fever, asthma, and most other allergic conditions	Contact dermatitis (such as to poison ivy and poison oak)
Therapy	Antihistamines and adrenergic drugs	Corticosteroids (such as cortisone)

Unlike IgG antibodies that circulate in the blood plasma, IgE antibodies are concentrated in mucosal membranes. There, the constant fragment (F_c) portions of the antibodies bind to receptor proteins on the surface of mast cells and basophils. When a person is again exposed to the allergen, the allergen binds to these IgE antibodies. The allergen thereby cross-links the IgE bound to its receptors on the mast cells and basophils, which stimulates these cells to release **histamine** and other cytokines (including *prostaglandins* and *leukotrienes*; chapter 11, section 11.7) that produce the immediate hypersensitivity reactions (fig. 15.23).

Histamine stimulates smooth muscle contraction in the respiratory airways (producing bronchoconstriction), but causes smooth muscle relaxation in blood vessels (producing vasodilation). Histamine also increases capillary permeability, thereby promoting the exit of plasma proteins and fluid and producing a localized edema. In addition, histamine influences the specific immune response, promoting the release of inflammatory cytokines.

The symptoms of hay fever (itching, sneezing, tearing, runny nose) are produced largely by histamine and can be treated effectively by antihistamine drugs that block the H_1-histamine receptor. In asthma, the difficulty in breathing is caused by inflammation and smooth muscle constriction in the bronchioles as a result of chemicals released by mast cells and eosinophils. In particular, the bronchoconstriction in asthma is produced by leukotrienes, which are mainly secreted by activated eosinophils. Asthma is treated with epinephrine and more specific β_2-adrenergic stimulating drugs (chapter 9) that cause bronchodilation, and with corticosteroids, which inhibit inflammation and leukotriene synthesis. Asthma and its treatment are discussed more fully in chapter 16, section 16.3. Regarding food

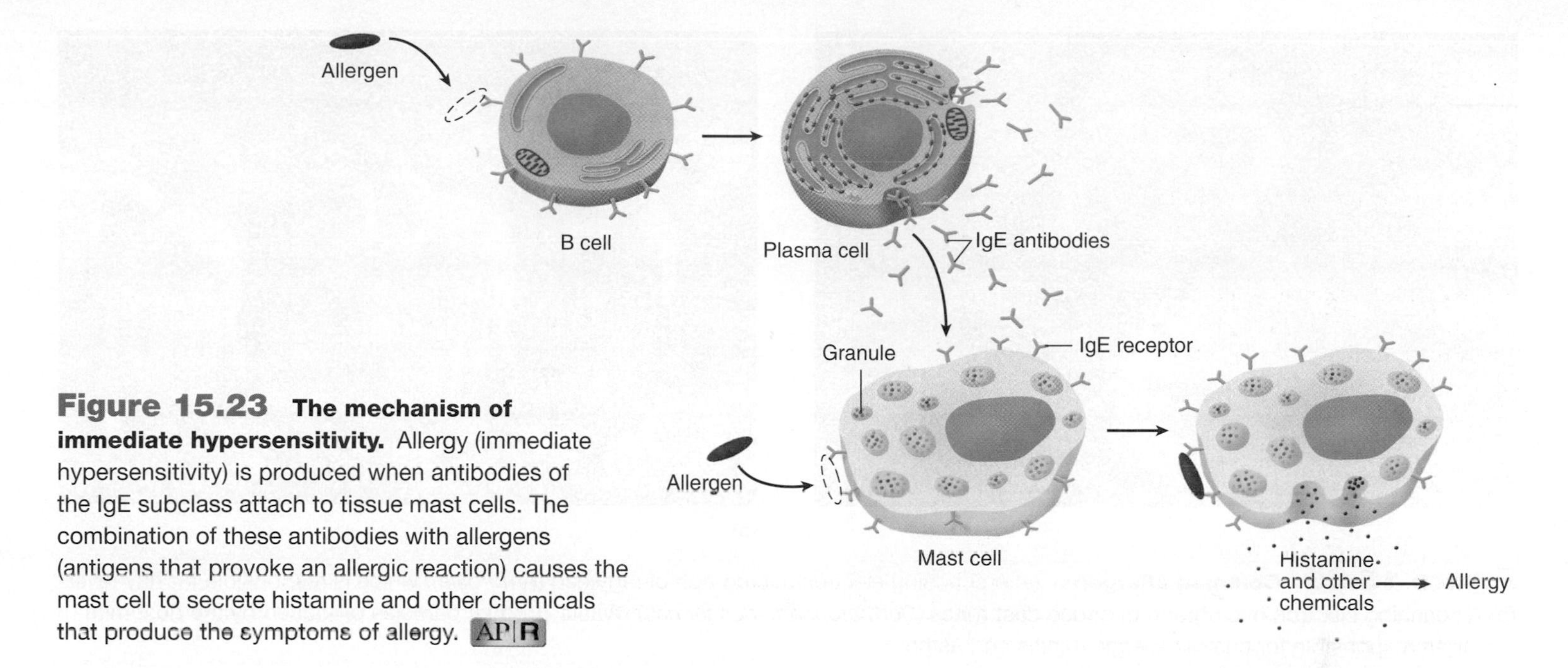

Figure 15.23 The mechanism of immediate hypersensitivity. Allergy (immediate hypersensitivity) is produced when antibodies of the IgE subclass attach to tissue mast cells. The combination of these antibodies with allergens (antigens that provoke an allergic reaction) causes the mast cell to secrete histamine and other chemicals that produce the symptoms of allergy. AP|R

allergies (to milk, eggs, peanuts, soy, wheat, and others), no specific therapy is currently available. People with a food allergy must thus be very diligent about avoiding the particular food.

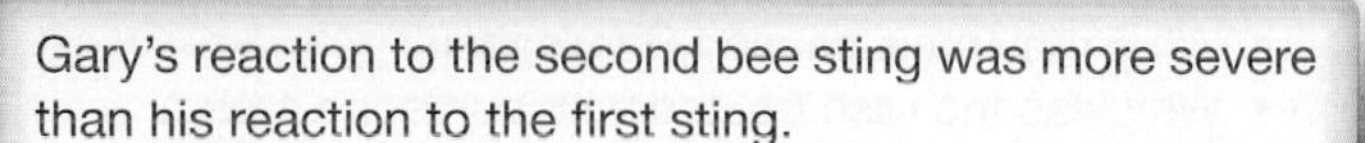

Clinical Investigation CLUES

Gary's reaction to the second bee sting was more severe than his reaction to the first sting.

- What accounts for his more severe response to the second sting?
- Why were antihistamines effective in treating Gary's response to the bee sting?

Immediate hypersensitivity to a particular antigen is commonly tested by injecting various antigens under the skin (fig. 15.24). Within a short time a *flare-and-wheal reaction* is produced if the person is allergic to that antigen. This reaction is due to the release of histamine and other chemical mediators: the flare (spreading flush) is due to vasodilation, and the wheal (elevated area) results from local edema.

Allergens that provoke immediate hypersensitivity include various foods, bee stings, and pollen grains. The most common allergy of this type is seasonal hay fever, which may be provoked by ragweed (*Ambrosia*) pollen grains (fig. 15.25*a*). People who have chronic allergic rhinitis and asthma because of an allergy to dust or feathers are usually allergic to a tiny mite (fig. 15.25*b*) that lives in dust and eats

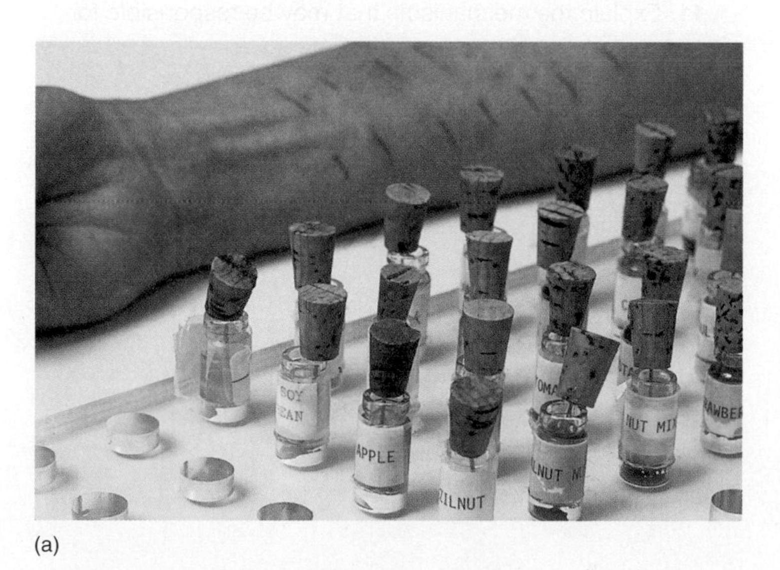

(a)

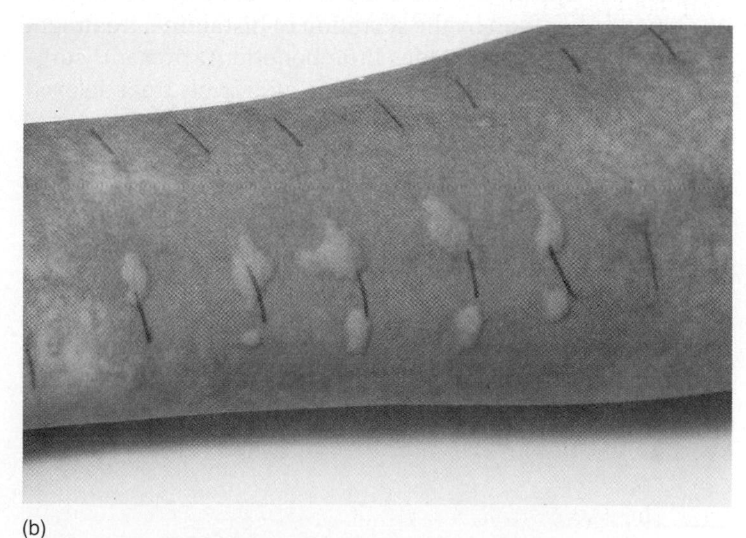

(b)

Figure 15.24 A skin test for allergy. If an allergen (*a*) is injected into the skin of a sensitive individual, a typical flare-and-wheal response (*b*) occurs within several minutes.

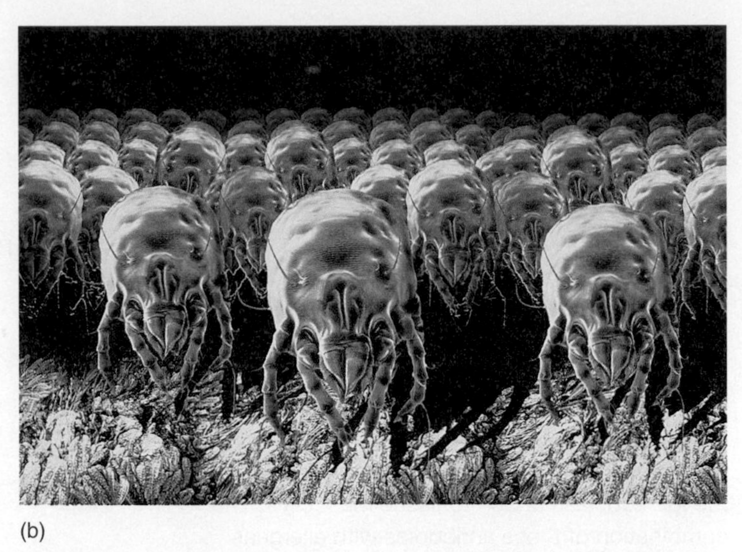

(a) (b)

Figure 15.25 Common allergens. (*a*) A scanning electron micrograph of ragweed (*Ambrosia*), which is responsible for hay fever. (*b*) A scanning electron micrograph of house dust mites (*Dermatophagoides farinae*). Waste-product particles produced by the dust mite are often responsible for chronic allergic rhinitis and asthma.

the scales of skin that are constantly shed from the body. Actually, most of the antigens from the dust mite are not in its body but rather in its feces—tiny particles that can enter the nasal mucosa, much like pollen grains. There can be more than 100,000 mite feces per gram of house dust!

Delayed Hypersensitivity

In **delayed hypersensitivity,** as the name implies, symptoms take a longer time (hours to days) to develop than in immediate hypersensitivity. This may be because immediate hypersensitivity is mediated by antibodies, whereas delayed hypersensitivity is a cell-mediated T lymphocyte response. Because the symptoms are caused by the secretion of lymphokines rather than by the secretion of histamine, treatment with antihistamines provides little benefit. At present, corticosteroids are the only drugs that can effectively treat delayed hypersensitivity.

One of the best-known examples of delayed hypersensitivity is **contact dermatitis,** caused by poison ivy, poison oak, and poison sumac. The skin tests for tuberculosis—the tine test and the Mantoux test—also rely on delayed hypersensitivity reactions. If a person has been exposed to the tubercle bacillus and consequently has developed T cell clones, skin reactions appear within a few days after the tubercle antigens are rubbed into the skin with small needles (tine test) or are injected under the skin (Mantoux test).

Clinical Investigation CLUES

Gary developed a rash on his abdomen after crawling through the underbrush.

- What may have caused Gary's rash?
- Why was the rash treated with cortisone cream rather than with antihistamines?

CHECKPOINT

11. Explain the mechanisms that may be responsible for autoimmune diseases.

12a. Distinguish between immediate and delayed hypersensitivity.

12b. Describe the sequence of events by which allergens can produce symptoms of runny nose, skin rash, and asthma.

Interactions

     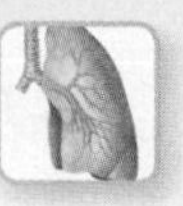 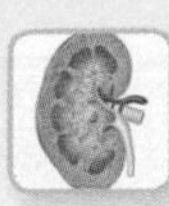  

HPer Links of the Immune System with Other Body Systems

Integumentary System

- The skin serves as first line of defense against invasion by pathogens (p. 494)
- Dendritic cells in the epidermis and macrophages in the dermis present antigens that trigger the immune response (p. 511)
- Mast cells contribute to inflammation (p. 500)

Skeletal System

- Hematopoiesis, including formation of leukocytes involved in immunity, occurs in the bone marrow (p. 409)
- The immune system protects all systems, including the skeletal system, against infection (p. 494)

Muscular System

- Cardiac muscle in the heart pumps blood to the body organs, including those of the immune system (p. 418)
- The smooth muscle of blood vessels helps regulate blood flow to areas of infection (p. 501)

Nervous System

- Neural regulation of the pituitary and adrenal glands indirectly influences activity of the immune system (p. 514)
- Nerves regulate blood flow to most organs, including the lymphatic organs (p. 466)

Endocrine System

- The pituitary and adrenal glands influence immune function (p. 339)
- The thymus regulates the production of T lymphocytes (p. 499)

Circulatory System

- The circulatory system transports neutrophils, monocytes, and lymphocytes to infected areas (p. 405)
- Hematopoiesis generates the cells required for the immune response (p. 409)

Respiratory System

- The lungs provide oxygen for transport by the blood and eliminate carbon dioxide from the blood (p. 533)

Urinary System

- The kidneys regulate the volume, pH, and electrolyte balance of the blood and eliminate wastes (p. 582)
- The immune system protects against infection of the urinary system (p. 494)

Digestive System

- The GI tract provides nutrients for all body cells, including those of the immune system (p. 621)
- Stomach acid serves as a barrier to pathogens (p. 494)
- Areas of the GI tract contain numerous lymphocytes and lymphatic nodules (p. 632)

Reproductive System

- The blood-testes barrier prevents sperm cell antigens from provoking an autoimmune response (p. 716)
- Vaginal acidity inhibits the spread of pathogens (p. 494)
- The placenta is an immunologically privileged site that is normally protected from immune attack (p. 740)
- A mother's breast milk provides antibodies that passively immunize her baby (p. 747)

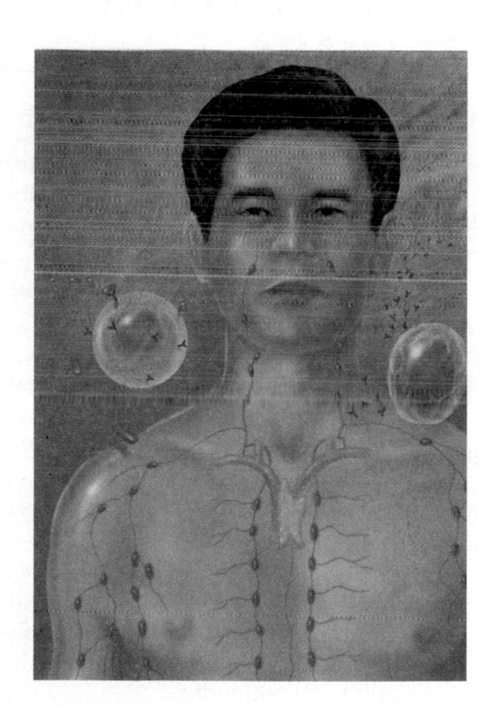

Clinical Investigation SUMMARY

While crawling through the underbrush, Gary may have been exposed to poison oak, causing a contact dermatitis. Since this is a delayed hypersensitivity response mediated by T cells, antihistamines would not have alleviated the symptoms. Cortisone helped, however, due to its immunosuppressive effect. The first bee sting did not have much of an effect, but it served to sensitize Gary (through the development of B cell clones) to the second bee sting. The second sting resulted in an immediate hypersensitivity response (mediated by IgE), which caused the release of histamine. This allergic reaction could thus be treated effectively with antihistamines. The previous tetanus vaccine that Gary had received had provided him with active immunity against tetanus.

See the additional chapter 15 Clinical Investigation on *Rheumatoid Arthritis* in the Connect site for this text at www.mhhe.com/Fox13.

SUMMARY

15.1 Defense Mechanisms 494

A. Nonspecific (innate) defense mechanisms include barriers to penetration of the body, as well as internal defenses.

1. Phagocytic cells engulf invading pathogens.
2. Interferons are polypeptides secreted by cells infected with viruses that help to protect other cells from viral infections.
3. The nonspecific (innate) immune system responds to pathogen associated molecular patterns (PAMPs) by interaction of foreign molecules with several receptors, called toll-like receptors, which recognize these uniquely foreign molecules.

B. Specific (adaptive) immune responses are directed against antigens.

1. Antigens are molecules or parts of molecules that are usually large, complex, and foreign.
2. A given molecule can have a number of antigenic determinant sites that stimulate the production of different antibodies.

C. Specific immunity is a function of lymphocytes.

1. B lymphocytes secrete antibodies and provide humoral immunity.
2. T lymphocytes provide cell-mediated immunity.
3. The thymus and bone marrow are the primary lymphoid organs, which produce lymphocytes that seed the secondary lymphoid organs.

D. Specific and nonspecific immune mechanisms cooperate in the development of local inflammation.

15.2 Functions of B Lymphocytes 503

A. There are five subclasses of antibodies, or immunoglobulins: IgG, IgA, IgM, IgD, and IgE.

1. These subclasses differ with respect to the polypeptides in the constant region of the heavy chains.
2. Each type of antibody has two variable regions that combine with specific antigens.
3. The combination of antibodies with antigens promotes phagocytosis.

B. Antigen-antibody complexes activate a system of proteins called the complement system.

1. This results in complement fixation, in which complement proteins attach to a cell membrane and promote the destruction of the cell.
2. Free complement proteins promote opsonization and chemotaxis and stimulate the release of histamine from tissue mast cells.

15.3 Functions of T Lymphocytes 507

A. The thymus processes T lymphocytes and secretes hormones that are believed to be required for an effective immune response of T lymphocytes throughout the body.

B. There are three subcategories of T lymphocytes.

1. Killer T lymphocytes kill victim cells by a mechanism that does not involve antibodies but that does require close contact between the killer T cell and the victim cell.

2. Killer T lymphocytes are responsible for transplant rejection and for the immunological defense against fungal and viral infections, as well as for defense against some bacterial infections.
3. Helper T lymphocytes secrete cytokines that stimulate activity of killer T lymphocytes and B lymphocytes.
4. Regulatory T lymphocytes (previously called suppressor T lymphocytes) inhibit the immune response, helping to prevent overactive immune responses that could cause disease.
5. The T lymphocytes secrete a family of compounds called lymphokines that promote the action of lymphocytes and macrophages.
6. Receptor proteins on the cell membrane of T lymphocytes must bind to a foreign antigen in combination with a histocompatibility antigen in order for the T cell to become activated.
7. Histocompatibility antigens, or MHC molecules, are a family of molecules on the membranes of cells that are present in different combinations in different individuals.

C. Antigen-presenting cells, such as macrophages and dendritic cells, partially digest a foreign protein, such as a virus, and present the antigens to the lymphocytes on the surface in combination with class-2 MHC antigens.

1. Helper T lymphocytes require such interaction with antigen-presenting cells in order to be activated by a foreign antigen; when activated in this way, the helper T cells secrete interleukin-2.
2. Interleukin-2 stimulates proliferation of killer T lymphocytes that are specific for the foreign antigen.
3. In order for the killer T lymphocytes to attack a victim cell, the victim cell must present the foreign antigen in combination with a class-1 MHC molecule.
4. Interleukin-2 also stimulates proliferation of B lymphocytes, and thus promotes the secretion of antibodies in response to the foreign antigen.

15.4 Active and Passive Immunity 515

A. A primary response is produced when a person is first exposed to a pathogen. A subsequent exposure results in a secondary response.

1. During the secondary response, IgM antibodies are produced slowly and the person is likely to get sick.
2. During the secondary response, IgG antibodies are produced quickly and the person is able to resist the pathogen.
3. In active immunizations, the person is exposed to pathogens of attenuated virulence that have the same antigenecity as the virulent pathogen.
4. The secondary response is believed to be due to the development of lymphocyte clones as a result of the antigen-stimulated proliferation of appropriate lymphocytes.

B. Tolerance to self-antigens occurs in prenatal development by destruction of T lymphocytes in the thymus that have specificity for the self-antigens.

1. This is called clonal deletion.
2. Clonal energy, or the suppression of lymphocytes, may also occur and may be responsible for tolerance to self-antigens by B lymphocytes.
3. When tolerance mechanisms are ineffective, the immune system may attack self-antigens to cause autoimmune diseases.

C. Passive immunity is provided by the transfer of antibodies from an immune to a nonimmune organism.

1. Passive immunity occurs naturally in the transfer of antibodies from mother to fetus.
2. Injections of antiserum provide passive immunity to some pathogenic organisms and toxins.

D. Monoclonal antibodies are made by hybridomas, which are formed artificially by the fusion of B lymphocytes and multiple myeloma cells.

15.5 Tumor Immunology 519

A. Immunological surveillance against cancer is provided mainly by killer T lymphocytes and natural killer cells.

1. Cancerous cells dedifferentiate and may produce fetal antigens. These or other antigens may be presented to lymphocytes in association with abnormally produced class-2 MHC antigens.
2. Natural killer cells are nonspecific, whereas T lymphocytes are directed against specific antigens on the cancer cell surface.
3. Immunological surveillance against cancer is weakened by stress.

15.6 Diseases Caused by the Immune System 522

A. Autoimmune diseases may be caused by the production of autoantibodies against self-antigens, or they may result from the development of autoreactive T lymphocytes.

B. Immune complex diseases are those caused by the inflammation that results when free antigens are bound to antibodies.

C. There are two types of allergic responses: immediate hypersensitivity and delayed hypersensitivity.

1. Immediate hypersensitivity results when an allergen provokes the production of antibodies in the IgE class. These antibodies attach to tissue mast cells and stimulate the release of chemicals from the mast cells.
2. Mast cells secrete histamine, leukotrienes, and prostaglandins, which are believed to produce the symptoms of allergy.
3. Delayed hypersensitivity, as in contact dermatitis, is a cell-mediated response of T lymphocytes.

REVIEW ACTIVITIES

Test Your Knowledge

1. Which of these offers a nonspecific defense against viral infection?
   a. Antibodies c. Interferon
   b. Leukotrienes d. Histamine

Match the cell type with its secretion.

2. Killer T cells a. Antibodies
3. Mast cells b. Perforins
4. Plasma cells c. Lysosomal enzymes
5. Macrophages d. Histamine

6. Which of these statements about the F_{ab} portion of antibodies is *true?*
   a. It binds to antigens.
   b. Its amino acid sequences are variable.
   c. It consists of both H and L chains.
   d. All of these are true.
7. Which of these statements about complement proteins $C3_a$ and $C5_a$ is *false?*
   a. They are released during the complement fixation process.
   b. They stimulate chemotaxis of phagocytic cells.
   c. They promote the activity of phagocytic cells.
   d. They produce pores in the victim cell membrane.
8. Mast cell secretion during an immediate hypersensitivity reaction is stimulated when antigens combine with
   a. IgG antibodies. c. IgM antibodies.
   b. IgE antibodies. d. IgA antibodies.
9. During a secondary immune response,
   a. antibodies are made quickly and in great amounts.
   b. antibody production lasts longer than in a primary response.
   c. antibodies of the IgG class are produced.
   d. lymphocyte clones are believed to develop.
   e. all of these apply.
10. Which of these cell types aids the activation of T lymphocytes by antigens?
    a. Macrophages c. Mast cells
    b. Neutrophils d. Natural killer cells
11. Which of these statements about T lymphocytes is *false?*
    a. Some T cells promote the activity of B cells.
    b. Some T cells suppress the activity of B cells.
    c. Some T cells secrete interferon.
    d. Some T cells produce antibodies.
12. Delayed hypersensitivity is mediated by
    a. T cells. c. plasma cells.
    b. B cells. d. natural killer cells.
13. Active immunity may be produced by
    a. contracting a disease.
    b. receiving a vaccine.
    c. receiving gamma globulin injections.
    d. both *a* and *b.*
    e. both *b* and *c.*
14. Which of these statements about class-2 MHC molecules is *false?*
    a. They are found on the surface of B lymphocytes.
    b. They are found on the surface of macrophages.
    c. They are required for B cell activation by a foreign antigen.
    d. They are needed for interaction of helper and killer T cells.
    e. They are presented together with foreign antigens by macrophages.

Match the cytokine with its description.

15. Interleukin-1 a. Stimulates formation of T_H1 helper T lymphocytes
16. Interleukin-2 b. Stimulates ACTH secretion
17. Interleukin-12 c. Stimulates proliferation of killer T lymphocytes
    d. Stimulates proliferation of B lymphocytes

18. Which of these statements about gamma interferon is *false?*
    a. It is a polypeptide autocrine regulator.
    b. It can be produced in response to viral infections.
    c. It stimulates the immune system to attack infected cells and tumors.
    d. It is produced by almost all cells in the body.

Test Your Understanding

19. Explain how antibodies help to destroy invading bacterial cells.
20. Identify the different types of interferons and describe their origin and actions.
21. Distinguish between the class-1 and class-2 MHC molecules in terms of their locations and functions.
22. Describe the role of macrophages in activating the specific immune response to antigens.
23. Distinguish between the two subtypes of helper T lymphocytes and explain how they may be produced.
24. Describe how plasma cells attack antigens and how they can destroy an invading foreign cell. Compare this mechanism with that by which killer T lymphocytes destroy a target cell.

25. Explain how tolerance to self-antigens may be produced. Also, give two examples of autoimmune diseases and explain their possible causes.
26. Use the clonal selection theory to explain how active immunity is produced by vaccinations.
27. Describe the nature of passive immunity and explain how antitoxins are produced and used.
28. Distinguish between immediate and delayed hypersensitivity. What drugs are used to treat immediate hypersensitivity and how do these drugs work? Why don't these compounds work in treating delayed hypersensitivity?
29. Describe regulatory T lymphocytes and their functions. What types of diseases could result from a deficiency of regulatory T lymphocytes? What type of diseases could result from inappropriate activation of regulatory T lymphocytes? Explain.
30. Explain how the innate (nonspecific) immune system is able to recognize a wide variety of different pathogens and respond to them.

Test Your Analytical Ability

31. The specific T lymphocyte immune response is usually directed against proteins, whereas nonspecific immune mechanisms are generally directed against foreign carbohydrates in the form of glycoproteins and lipopolysaccharides. How might these differences in target molecules be explained?
32. Lizards are cold-blooded; their body temperature is largely determined by the ambient temperature. Devise an experiment using lizards to test whether an elevated body temperature, as in a fever, can be beneficial to an organism with an infection.
33. Why are antibodies composed of different chains, and why are there several genes that encode the parts of a particular antibody molecule? What would happen if each antibody were coded for by only one gene?
34. As a scientist trying to cure allergy, you are elated to discover a drug that destroys all mast cells. How might this drug help to prevent allergy? What negative side effects might this drug have?
35. The part of the placenta that invades the mother's uterine lining (the endometrium) has recently been found to produce FAS ligand. What might this accomplish, and why might this action be necessary?
36. Describe the antigen-dependent and antigen-independent ways that the body generates a wide diversity of antibodies. How does this relate to the development of autoimmune diseases? Relate this to the function of regulatory T lymphocytes.
37. People with peanut allergies could go into anaphylactic shock (chapter 14, section 14.7) if they are exposed to peanuts in any form. Explain what this means and how it occurs.
38. Identify some immunologically privileged sites in the body and explain how immunological privilege is produced. How does this relate to cancer, and what does it suggest about possible treatments for cancer?

Test Your Quantitative Ability

Use figure 15.6 to answer the following questions regarding the arrival of leukocytes to the site of a local inflammation.

39. Which leukocytes are the first and last to arrive?
40. About how long does it take for the number of monocytes to peak?
41. About how long does it take for neutrophils to reach half their final number?
42. About how long does it take for T lymphocytes to peak in number?

CHAPTER

16

Respiratory Physiology

REFRESH YOUR MEMORY

Before you begin this chapter, you may want to review these concepts from previous chapters:

- **Hindbrain 230**
- **Cranial Nerves 236**
- **Characteristics of Sensory Receptors 267**

CHAPTER OUTLINE

Clinical Investigation

Harry, a taxi cab driver, was found after an assailant attacked him with a knife and punctured his chest. When he was taken to the hospital, a chest x-ray and blood tests were performed and he was treated for a "collapsed lung." Afterward, Harry revealed himself to be a chronic cigarette smoker and his physician ordered pulmonary function tests.

Some of the new terms and concepts you will encounter include:

- Pneumothorax, vital capacity, and forced vital capacity
- Blood gases, percent saturation, and carboxyhemoglobin
- Acidosis and alkalosis, obstructive and restrictive disorders

16.1 THE RESPIRATORY SYSTEM

The respiratory system is divided into a respiratory zone, which is the site of gas exchange between air and blood, and a conducting zone. The exchange of gases between air and blood occurs across the walls of respiratory alveoli, which permit rapid rates of gas diffusion.

LEARNING OUTCOMES

After studying this section, you should be able to:

1. Describe the structures and functions of the conducting and respiratory zones of the lungs.
2. Describe the location and significance of the pleural membranes.

The term *respiration* includes three separate but related functions: (1) **ventilation** (breathing); (2) **gas exchange,** which occurs between the air and blood in the lungs and between the blood and other tissues of the body; and (3) **oxygen utilization** by the tissues in the energy-liberating reactions of cell respiration. Ventilation and the exchange of gases (oxygen and carbon dioxide) between the air and blood are collectively called *external respiration.* Gas exchange between the blood and other tissues and oxygen utilization by the tissues are collectively known as *internal respiration.*

Ventilation is the mechanical process that moves air into and out of the lungs. Because the oxygen concentration of air is higher in the lungs than in the blood, oxygen diffuses from air to blood. Carbon dioxide, conversely, moves from the blood to the air within the lungs by diffusing down its concentration gradient. As a result of this gas exchange, the inspired air contains more oxygen and less carbon dioxide than the expired air. More importantly, blood leaving the lungs (in the pulmonary veins) has a higher oxygen and a lower carbon dioxide concentration than the blood delivered to the lungs in the pulmonary arteries. This is because the lungs function to bring the blood into gaseous equilibrium with the air.

Gas exchange between the air and blood occurs entirely by diffusion through lung tissue. This diffusion occurs very rapidly because of the large surface area within the lungs and the very small diffusion distance between blood and air.

Structure of the Respiratory System

Gas exchange in the lungs occurs across an estimated 300 million tiny (about 100 μm in diameter) air sacs known as **alveoli.** Their enormous number provides a large surface area (60 to 80 square meters, or about 760 square feet) for diffusion of gases. The diffusion rate is further increased by the fact that each alveolus is only one cell-layer thick, so that the total "air-blood barrier" is only two cells across (an alveolar cell and a capillary endothelial cell), or about 2 μm.

There are two types of alveolar cells, designated **type I alveolar cells** and **type II alveolar cells** (fig. 16.1). The type I alveolar cells comprise 95% to 97% of the total surface area of the lung; gas exchange with the blood thus occurs primarily through type I alveolar cells. These cells are accordingly very thin: where the basement membranes of the type I alveolar cells and capillary endothelial cells fuse, the diffusion distance

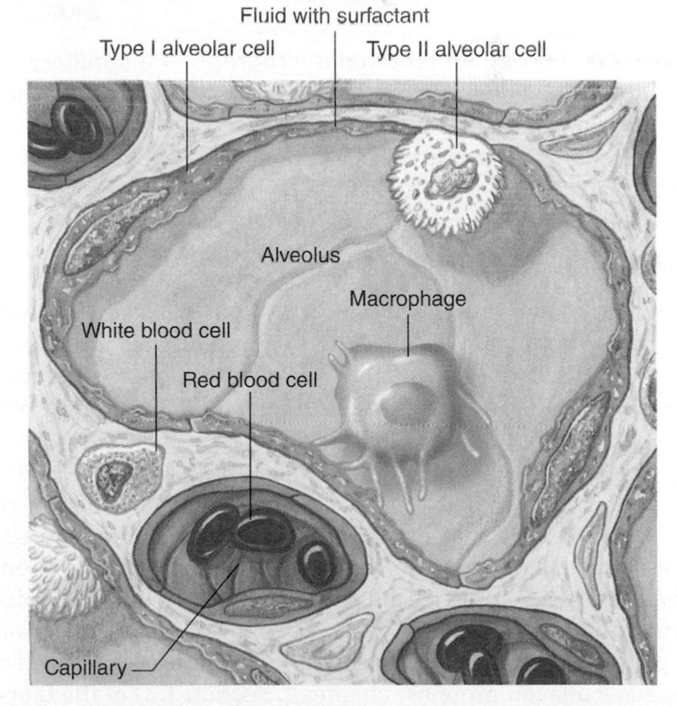

Figure 16.1 The relationship between lung alveoli and pulmonary capillaries. Notice that alveolar walls are quite narrow and lined with type I and type II alveolar cells. Pulmonary macrophages can phagocytose particles that enter the lungs. AP|R

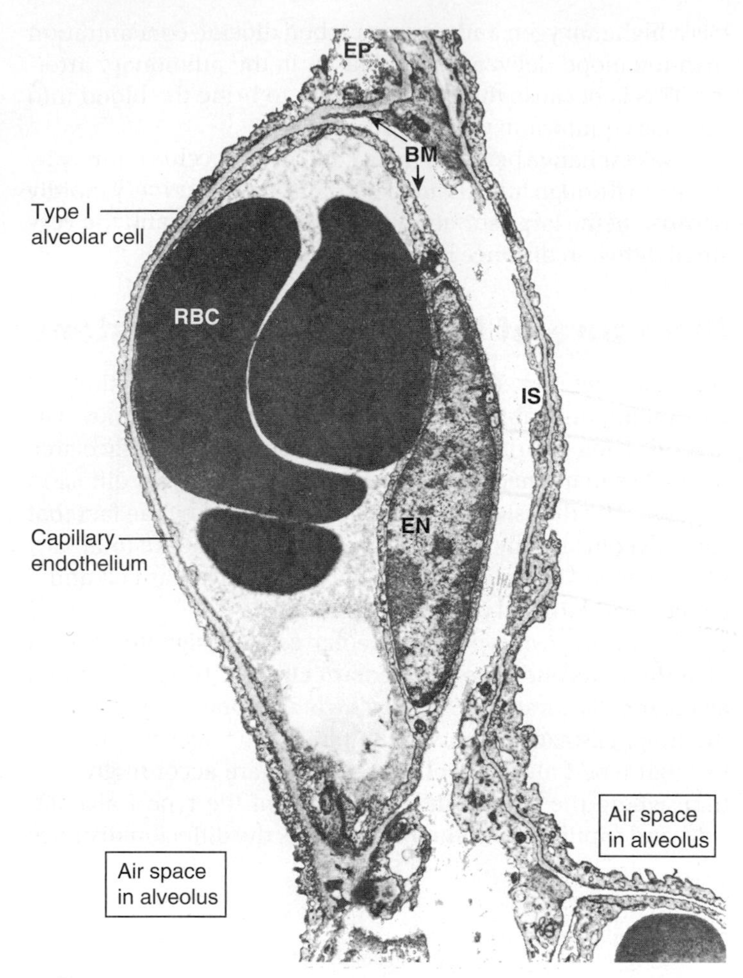

Figure 16.2 An electron micrograph of a capillary within the alveolar wall. Notice the short distance separating the alveolar space on one side (left, in this figure) from the capillary. (EP = epithelial cell of alveolus; EN = endothelial cell of capillary; RBC = red blood cell; BM = basement membrane; IS = interstitial connective tissue.) AP|R

between blood and air can be as little as 0.3 μm (fig. 16.2), which is about 1/100th the width of a human hair. The type II alveolar cells are the cells that secrete pulmonary surfactant (discussed later) and that reabsorb Na^+ and H_2O, thereby preventing fluid buildup within the alveoli.

In order to maximize the rate of gas diffusion between the air and blood, the air-blood barrier provided by the alveoli is extremely thin and has a very large surface area. Despite these characteristics, the alveolar wall isn't fragile but is strong enough to withstand high stress during heavy exercise and high lung inflation. The great tensile strength of the alveolar wall is provided by the fused basement membranes (composed of type IV collagen proteins; chapter 1, section 1.3) of the blood capillaries and the alveolar walls.

Alveoli are polyhedral in shape and are usually clustered, like the units of a honeycomb. Air within one member of a cluster can enter other members through tiny pores (fig. 16.3). These clusters of alveoli usually occur at the ends of *respiratory bronchioles,* the very thin air tubes that end blindly in alveolar sacs. Individual alveoli also occur as separate outpouchings along the length of respiratory bronchioles. Although the distance between each respiratory bronchiole and its terminal alveoli is only about 0.5 mm, these units together constitute most of the mass of the lungs.

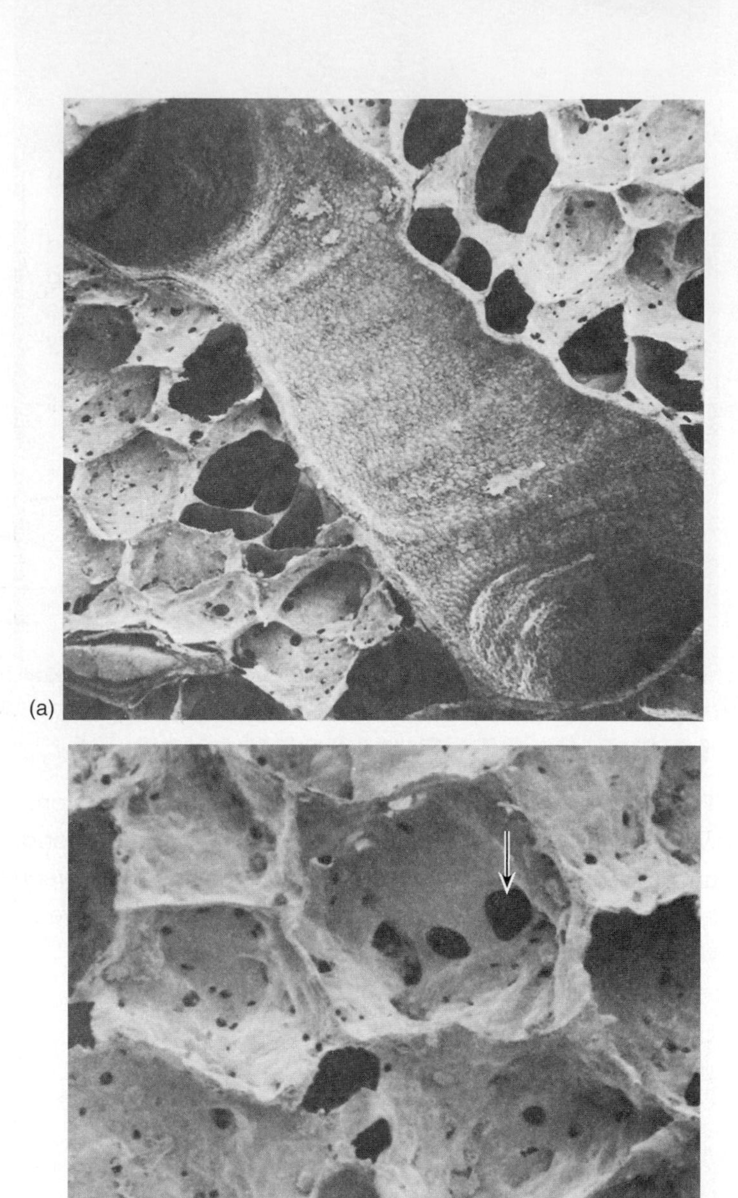

Figure 16.3 A scanning electron micrograph of lung tissue. (*a*) A small bronchiole passes between many alveoli. (*b*) The alveoli are seen under higher power, with an arrow indicating an alveolar pore through which air can pass from one alveolus to another.

The air passages of the respiratory system are divided into two functional zones. The **respiratory zone** is the region where gas exchange occurs, and it therefore includes the respiratory bronchioles (because they contain separate outpouchings of alveoli) and the terminal alveolar sacs. The **conducting zone** includes all of the anatomical structures through which air passes before reaching the respiratory zone (fig. 16.4).

Air enters the respiratory bronchioles from *terminal bronchioles,* which are the narrowest of the airways that do not have alveoli and do not contribute to gas exchange. The terminal bronchioles receive air from larger airways, which are formed from successive branchings of the *right* and *left primary bronchi.* These two large air passages, in turn, are continuous with the *trachea,* or windpipe, which is located in the neck in front of the esophagus (a muscular tube that carries food to the stomach). The trachea is a sturdy tube supported by rings of cartilage (fig. 16.5).

Air enters the trachea from the *pharynx,* which is the cavity behind the palate that receives the contents of both the oral

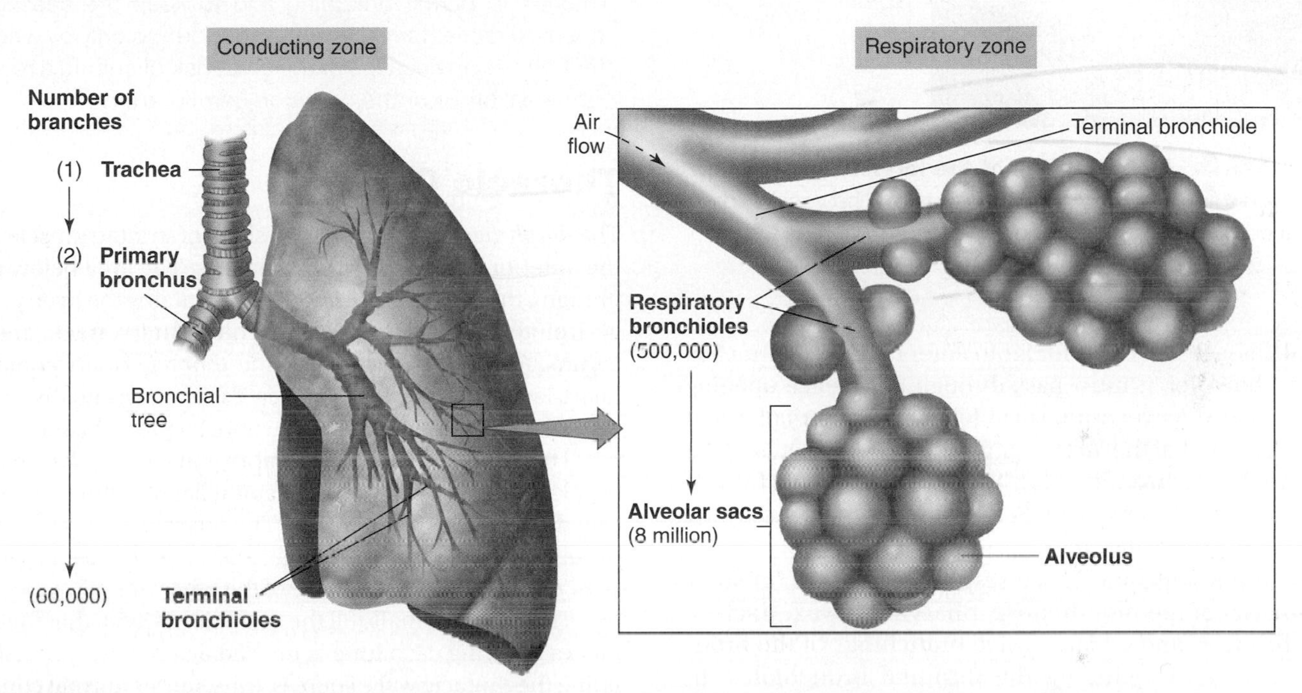

Figure 16.4 The conducting and respiratory zones of the respiratory system. The conducting zone consists of airways that conduct the air to the respiratory zone, which is the region where gas exchange occurs. The numbers of each member of the airways and the total number of alveolar sacs are shown in parentheses. AP|R

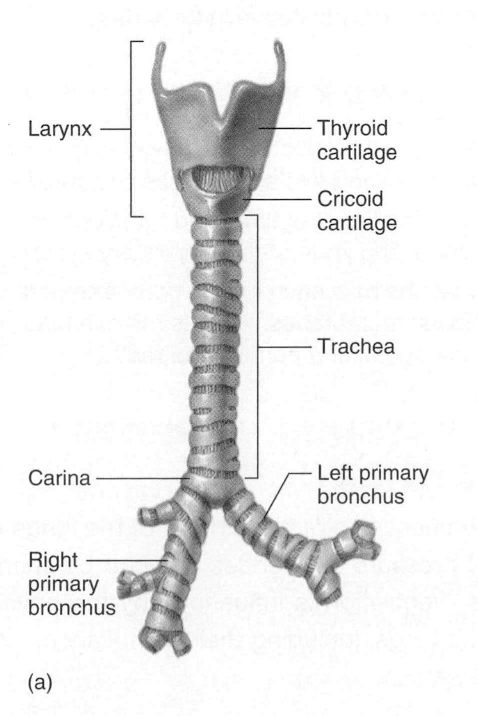

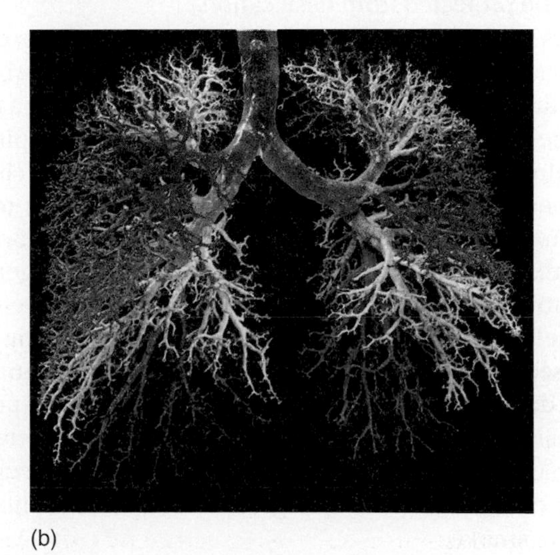

Figure 16.5 The conducting zone of the respiratory system. (*a*) An anterior view extending from the larynx to the terminal bronchi and (*b*) the airway from the trachea to the terminal bronchioles, as represented by a plastic cast. AP|R

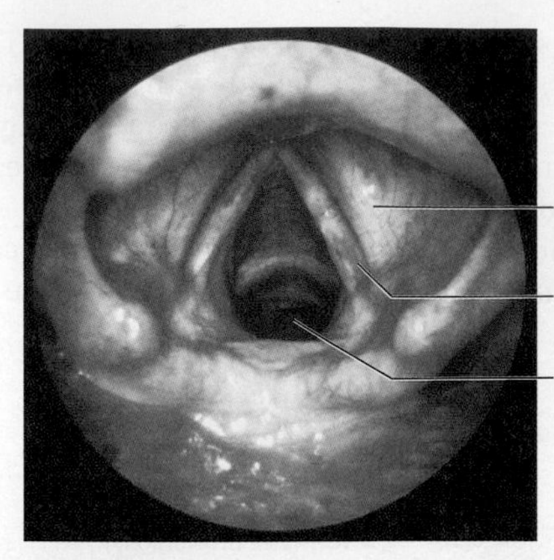

Figure 16.6 A photograph of the larynx showing the true and false vocal cords and the glottis. The vocal folds (true vocal cords) function in sound production, whereas the ventricular folds (false vocal cords) do not.

and nasal passages. In order for air to enter or leave the trachea and lungs, however, it must pass through a valvelike opening called the *glottis* between the vocal folds. The ventricular and vocal folds are part of the *larynx,* or voice box, which guards the entrance to the trachea (fig. 16.6). The projection at the front of the throat, commonly called the "Adam's apple," is formed by the largest cartilage of the larynx.

The conducting zone of the respiratory system, in summary, consists of the mouth, nose, pharynx, larynx, trachea, primary bronchi, and all successive branchings of the bronchioles up to and including the terminal bronchioles. In addition to conducting air into the respiratory zone, these structures serve additional functions: *warming* and *humidification* of the inspired air, and *filtration* and *cleaning.*

Regardless of the temperature and humidity of the ambient air, when the inspired air reaches the respiratory zone it is at a temperature of 37° C (body temperature), and it is saturated with water vapor as it flows over the warm, wet mucous membranes that line the respiratory airways. This ensures that a constant internal body temperature will be maintained and that the lung tissue will be protected from desiccation.

Mucus secreted by cells of the conducting zone structures serves to trap small particles in the inspired air and thereby performs a filtration function. This mucus is moved along at a rate of 1 to 2 cm per minute by cilia projecting from the tops of epithelial cells that line the conducting zone. There are about 300 cilia per cell that beat in a coordinated fashion to move mucus toward the pharynx, where it can either be swallowed or expectorated.

As a result of this filtration function, particles larger than about 6μm do not normally enter the respiratory zone of the lungs. The importance of this function is evidenced by *black lung,* a disease that occurs in miners who inhale large amounts of carbon dust from coal, which causes them to develop pulmonary fibrosis. The alveoli themselves are normally kept clean by the action of resident macrophages (see fig. 16.1). The cleansing action of cilia and macrophages in the lungs is diminished by cigarette smoke.

CLINICAL APPLICATION

If the trachea becomes occluded through inflammation, excessive secretion, trauma, or aspiration of a foreign object, it may be necessary to create an emergency opening into this tube so that ventilation can still occur. A **tracheotomy** is the procedure of surgically opening the trachea, and a **tracheostomy** involves the insertion of a tube into the trachea to permit breathing and to keep the passageway open. A tracheotomy should be performed only by a competent physician because of the great risk of cutting a recurrent laryngeal nerve or the common carotid artery.

Thoracic Cavity

The *diaphragm,* a dome-shaped sheet of striated muscle, divides the anterior body cavity into two parts. The area below the diaphragm, the *abdominopelvic cavity,* contains the liver, pancreas, gastrointestinal tract, spleen, genitourinary tract, and other organs. Above the diaphragm, the *thoracic cavity* contains the heart, large blood vessels, trachea, esophagus, and thymus in the central region, and is filled elsewhere by the right and left lungs.

The structures in the central region—or *mediastinum*—are enveloped by two layers of wet epithelial membrane collectively called the *pleural membranes.* The superficial layer, or *parietal pleura,* lines the inside of the thoracic wall. The deep layer, or *visceral pleura,* covers the surface of the lungs (fig. 16.7).

The lungs normally fill the thoracic cavity so that the visceral pleura covering each lung is pushed against the parietal pleura lining the thoracic wall. There is, thus, under normal conditions, little or no air between the visceral and parietal pleura. There is, however, a "potential space"—called the *intrapleural space*—that can become a real space if the visceral and parietal pleurae separate when a lung collapses. The normal position of the lungs in the thoracic cavity is shown in the radiographs in figure 16.8.

 **CHECKPOINT**

1a. Describe the structures involved in gas exchange in the lungs and explain how gas exchange occurs.

1b. Describe the structures and functions of the conducting zone of the respiratory system.

2. Describe how each lung is compartmentalized by the pleural membranes. What is the relationship between the visceral and parietal pleurae?

16.2 PHYSICAL ASPECTS OF VENTILATION

The movement of air into and out of the lungs occurs as a result of pressure differences induced by changes in lung volumes. Ventilation is influenced by the physical properties of the lungs, including their compliance, elasticity, and surface tension.

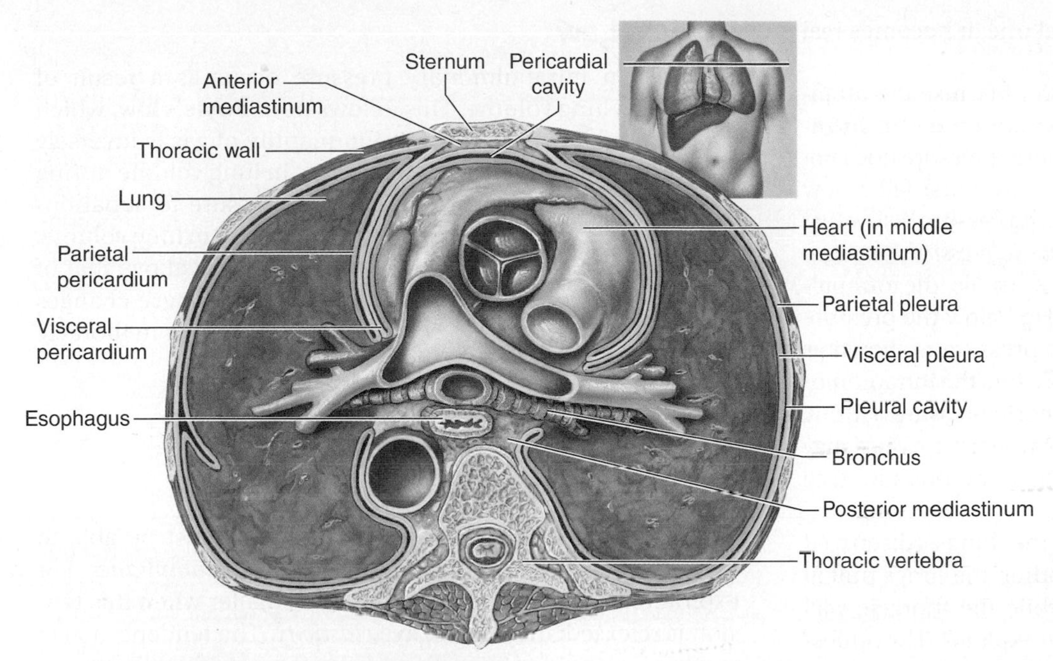

Figure 16.7 A cross section of the thoracic cavity. In addition to the lungs, the mediastinum and pleural membranes are visible. The parietal pleura is shown in green, and the visceral pleura in blue. APR

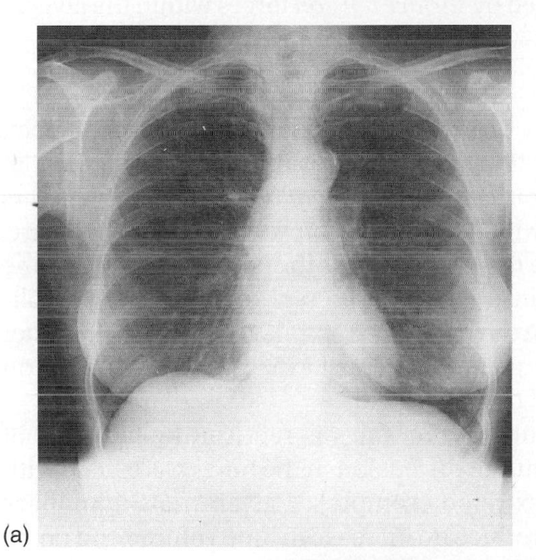

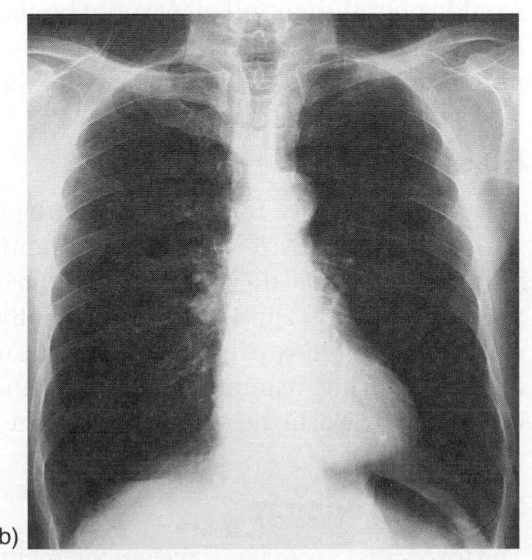

Figure 16.8 Radiographic (x-ray) views of the chest. These are x-rays (*a*) of a normal female and (*b*) of a normal male. APR

LEARNING OUTCOMES

After studying this section, you should be able to:

3. Explain how intrapleural and intrapulmonary pressures change during breathing.
4. Explain how lung compliance, elasticity, and surface tension affect breathing, and the significance of pulmonary surfactant.

Movement of air from higher to lower pressure, between the conducting zone and the terminal bronchioles, occurs as a result of the pressure difference between the two ends of the airways. Air flow through bronchioles, like blood flow through blood vessels, is directly proportional to the pressure difference and inversely proportional to the frictional resistance to flow. The pressure differences in the pulmonary system are induced by changes in lung volumes. The compliance, elasticity, and surface tension of the lungs are physical properties that affect their functioning.

Intrapulmonary and Intrapleural Pressures

The visceral and parietal pleurae are stuck to each other like two wet pieces of glass. The *intrapleural space* between them contains only a thin layer of fluid, secreted by the parietal pleura. This fluid is like the interstitial fluid in other organs; it is formed as a filtrate from blood capillaries in the parietal pleura, and it is drained into lymphatic capillaries. The major function of the liquid in the intrapleural space is to serve as a lubricant so that the lungs can slide relative to the chest during breathing. Since the lungs normally are stuck to the thoracic wall, for reasons described shortly, they expand and contract with the thoracic wall during breathing. The intrapleural space is thus more a potential space than a real one; it becomes real only if the lungs collapse.

Air enters the lungs during inspiration because the atmospheric pressure is greater than the **intrapulmonary,** or **intra-alveolar, pressure.** Because the atmospheric pressure does not usually change, the intrapulmonary pressure must fall below atmospheric pressure to cause inspiration. A pressure below that of the atmosphere is called a *subatmospheric pressure,* or *negative pressure.* During quiet inspiration, for example, the intrapulmonary pressure may decrease to 3 mmHg below the pressure of the atmosphere. This subatmospheric pressure is shown as −3 mmHg. Expiration, conversely, occurs when the intrapulmonary pressure is greater than the atmospheric pressure. During quiet expiration, for example, the intrapulmonary pressure may rise to at least +3 mmHg over the atmospheric pressure (see fig. 16.14).

Because of the elastic tension of the lungs (discussed shortly) and the thoracic wall on each other, the lungs pull in one direction (they "try" to collapse) while the thoracic wall pulls in the opposite direction (it "tries" to expand). The opposing elastic recoil of the lungs and the chest wall produces a subatmospheric pressure in the intrapleural space between these two structures. This pressure is called the **intrapleural pressure.** As indicated in table 16.1, the intrapleural pressure is lower (more negative) during inspiration because of the expansion of the thoracic cavity than it is during expiration. However, the intrapleural pressure is normally lower than the intrapulmonary pressure during both inspiration and expiration (table 16.1).

There is thus a pressure difference across the wall of the lung—called the **transpulmonary** (or **transmural**) **pressure**—which is the difference between the intrapulmonary pressure and the intrapleural pressure. Because the pressure within the lungs (intrapulmonary pressure) is greater than that outside the lungs (intrapleural pressure), the difference in pressure (transpulmonary pressure) keeps the lungs against the chest wall. The transpulmonary pressure is positive during both inspiration and expiration (table 16.1), causing the lungs to stick to the chest and thereby produce changes in lung volume as the thoracic volume changes. During inspiration, it is the transpulmonary pressure that causes the lungs to expand as the thoracic volume expands.

Table 16.1 | Pressure Changes in Normal, Quiet Breathing

	Inspiration	Expiration
Intrapulmonary pressure (mmHg)	−3	+3
Intrapleural pressure (mmHg)	−6	−3
Transpulmonary pressure (mmHg)	+3	+6

Note: Pressures indicate mmHg below or above atmospheric pressure.

Boyle's Law

Changes in intrapulmonary pressure occur as a result of changes in lung volume. This follows from **Boyle's law,** which states that the pressure of a given quantity of gas is inversely proportional to its volume. An increase in lung volume during inspiration decreases intrapulmonary pressure to subatmospheric levels; air therefore goes in. A decrease in lung volume, conversely, raises the intrapulmonary pressure above that of the atmosphere, expelling air from the lungs. These changes in lung volume occur as a consequence of changes in thoracic volume, as will be described in section 16.3.

Physical Properties of the Lungs

In order for inspiration to occur, the lungs must be able to expand when stretched; they must have high *compliance.* For expiration to occur, the lungs must get smaller when this tension is released: they must have *elasticity.* The tendency to get smaller is also aided by *surface tension* forces within the alveoli.

Compliance

The lungs are very distensible (stretchable)—they are, in fact, about a hundred times more distensible than a toy balloon. Another term for distensibility is *compliance,* which here refers to the ease with which the lungs can expand under pressure. **Lung compliance** can be defined as the change in lung volume per change in transpulmonary pressure, expressed symbolically as $\Delta V/\Delta P$. A given transpulmonary pressure, in other words, will cause greater or lesser expansion, depending on the compliance of the lungs.

The compliance of the lungs is reduced by factors that produce a resistance to distension. If the lungs were filled with concrete (as an extreme example), a given transpulmonary pressure would produce no increase in lung volume and no air would enter; the compliance would be zero. The infiltration of lung tissue with connective tissue proteins, a condition called *pulmonary fibrosis,* similarly decreases lung compliance.

Elasticity

The term **elasticity** refers to the tendency of a structure to return to its initial size after being distended. Because of their high content of elastin proteins, the lungs are very elastic and resist distension. The lungs are normally stuck to the chest wall, so they are always in a state of elastic tension. This tension increases during inspiration when the lungs are stretched and is reduced by elastic recoil during expiration.

Surface Tension

The forces that act to resist distension include elastic resistance and the **surface tension** that is exerted by fluid in the alveoli. The lungs both secrete and absorb fluid in two antagonistic

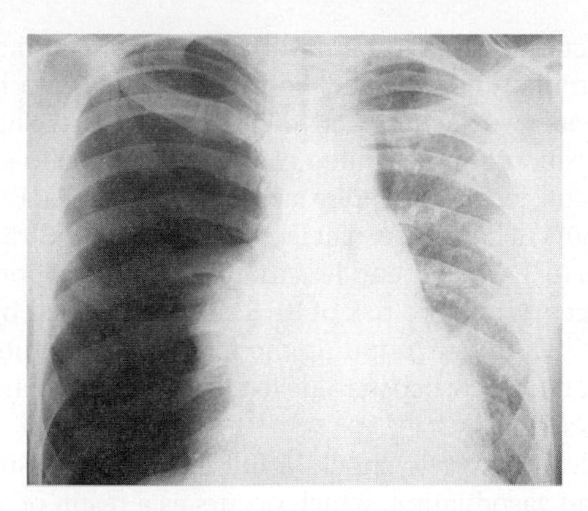

Figure 16.9 A pneumothorax of the right lung. The right side of the thorax appears uniformly dark because it is filled with air. The spaces between the ribs are also greater on the right side due to release from the elastic tension of the lungs. The left lung appears denser (less dark) because of shunting of blood from the right to the left lung.

CLINICAL APPLICATION

The elastic nature of lung tissue is revealed when air enters the intrapleural space (as a result of an open chest wound, for example). This condition, called a **pneumothorax,** is shown in figure 16.9. As air enters the intrapleural space, the intrapleural pressure rises until it is equal to the atmospheric pressure. When the intrapleural pressure is the same as the intrapulmonary pressure the lung can no longer expand. Not only does the lung not expand during inspiration, it actually collapses away from the chest wall as a result of elastic recoil, a condition called *atelectasis.* Fortunately, a pneumothorax usually causes only one lung to collapse, since each lung is contained in a separate pleural compartment.

Clinical Investigation CLUES

Harry's stab wound caused collapse of his lung.

- What condition does Harry have, and how might it look in an x-ray?
- How did the stab wound cause the lung to collapse?
- Could Harry still breathe before medical help was given? How?

processes that normally leave only a very thin film of fluid on the alveolar surface. Fluid absorption is driven (through osmosis) by the active transport of Na^+, while fluid secretion is driven by the active transport of Cl^- out of the alveolar epithelial cells. Research has demonstrated that people with cystic fibrosis have a genetic defect in one of the Cl^- carriers (called the *cystic fibrosis transmembrane regulator,* or *CFTR;* chapter 6, section 6.2). This results in an imbalance of fluid absorption and secretion, so that the airway fluid becomes excessively viscous (with a lower water content) and difficult to clear.

The thin film of fluid normally present in the alveolus has a surface tension, produced because water molecules at the surface are attracted more to other water molecules than to air. As a result, the surface water molecules are pulled tightly together by attractive forces from underneath. This surface tension acts to collapse the alveolus, and in the process increases the pressure of the air within the alveolus. As described by the **law of Laplace,** the pressure thus created is directly proportional to the surface tension and inversely proportional to the radius of the alveolus (fig. 16.10). According to this law, the pressure in a smaller alveolus would be greater than in a larger alveolus if the surface tension were the same in both. The greater pressure of the smaller alveolus would then cause it to empty its air into the larger one. This does not normally occur because, as an alveolus decreases in size, its surface tension (the numerator in the equation) is decreased at the same time that its radius

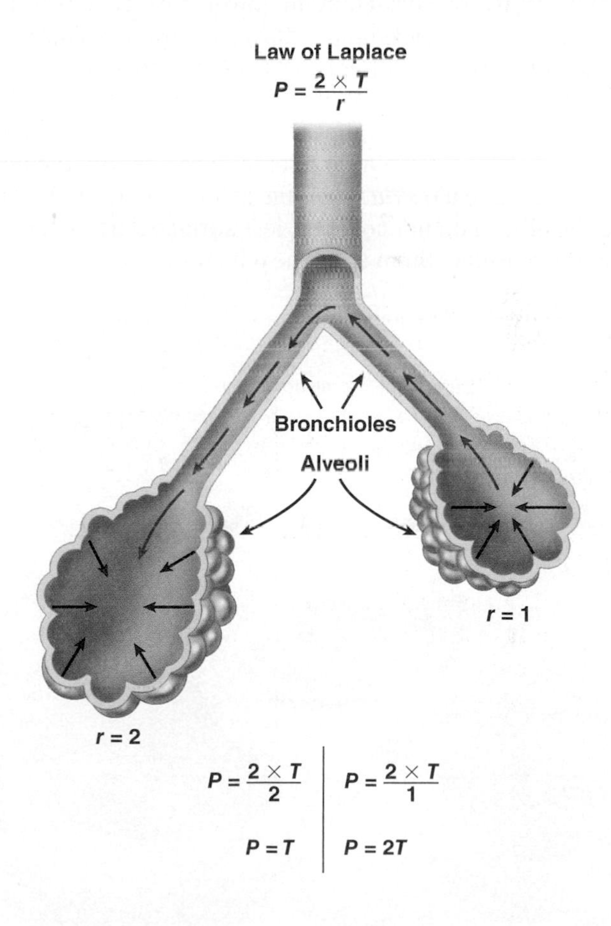

Figure 16.10 The law of Laplace. According to the law of Laplace, the pressure created by surface tension should be greater in the smaller alveolus than in the larger alveolus. This implies that (without surfactant) smaller alveoli would collapse and empty their air into larger alveoli.

(the denominator) is reduced. The reason for the decreased surface tension, which prevents the alveoli from collapsing, is described next.

Surfactant and Respiratory Distress Syndrome

Alveolar fluid contains a substance that reduces surface tension. This substance is called **surfactant**—a contraction of *surface active agent.* Surfactant is secreted into the alveoli by type II alveolar cells (fig. 16.11), and consists of phospholipids—primarily *phosphatidylcholine* and *phosphatidylglycerol*—together with hydrophobic surfactant proteins. Surfactant becomes interspersed between water molecules at the water-air interface; this reduces the hydrogen bonds between water molecules (shown in chapter 2, fig. 2.7) at the surface and thereby reduces the surface tension. As a result of this effect of pulmonary surfactant, the surface tension of the alveoli is negligible. The surfactant secreted into the alveoli by type II alveolar cells is removed by alveolar macrophages.

The ability of surfactant to lower surface tension improves as the alveoli get smaller during expiration. This may be because the surfactant molecules become more concentrated as the alveoli get smaller. Surfactant thus prevents the alveoli from collapsing during expiration, as would be predicted from the law of Laplace. Even after a forceful expiration, the alveoli remain open and a *residual volume* of air remains in the lungs. Since the alveoli do not collapse, less surface tension has to be overcome to inflate them at the next inspiration.

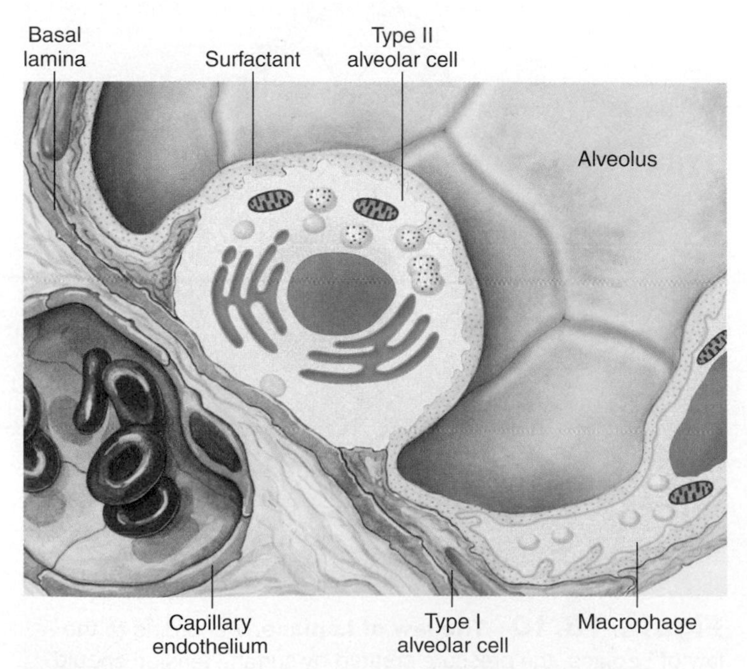

Figure 16.11 The production of pulmonary surfactant. Produced by type II alveolar cells, surfactant appears to be composed of a derivative of lecithin combined with protein.

Surfactant begins to be produced in late fetal life. For this reason, premature babies are sometimes born with lungs that lack sufficient surfactant and their alveoli are collapsed as a result. This condition is called **respiratory distress syndrome (RDS).** A full-term pregnancy lasts 37 to 42 weeks. RDS occurs in about 60% of babies born at less than 28 weeks, 30% of babies born at 28 to 34 weeks, and less than 5% of babies born after 34 weeks of gestation. The risk of RDS can be assessed by analysis of amniotic fluid (surrounding the fetus), and mothers can be given exogenous corticosteroids to accelerate the maturation of their fetus's lungs.

People with septic shock (a fall in blood pressure due to widespread vasodilation, which occurs as a result of a systemic infection) may develop a condition called **acute respiratory distress syndrome (ARDS).** In this condition, inflammation causes increased capillary and alveolar permeability that lead to the accumulation of a protein-rich fluid in the lungs. This decreases lung compliance and is accompanied by a reduced surfactant, which further lowers compliance. The blood leaving the lungs, as a result, has an abnormally low oxygen concentration (a condition called *hypoxemia*).

CLINICAL APPLICATION

Even under normal conditions, the first breath of life is difficult because the newborn must overcome great surface tension forces in order to inflate its partially collapsed alveoli. The transpulmonary pressure required for the first breath is 15 to 20 times that required for subsequent breaths, and an infant with respiratory distress syndrome must duplicate this effort with every breath. Fortunately, many babies with this condition can be saved by mechanical ventilators and by exogenous surfactant delivered to the baby's lungs by means of an endotracheal tube. The exogenous surfactant may be a synthetic mixture of phospholipids, or it may be surfactant obtained from bovine lungs. The mechanical ventilator and exogenous surfactant help keep the baby alive long enough for its lungs to mature, so that it can manufacture sufficient surfactant on its own. Unfortunately, the administration of exogenous surfactant does not help in the treatment of adults with ARDS.

✔ CHECKPOINT

3. Describe the changes in the intrapulmonary and intrapleural pressures that occur during inspiration, and use Boyle's law to explain the reasons for these changes.

4a. Explain how the compliance and elasticity of the lungs affect inspiration and expiration.

4b. Describe pulmonary surfactant and discuss its significance.

16.3 MECHANICS OF BREATHING

Normal, quiet inspiration results from muscle contraction, and normal expiration from muscle relaxation and elastic recoil. The amount of air inspired and expired can be measured in a number of ways to test pulmonary function.

LEARNING OUTCOMES

After studying this section, you should be able to:

5. Explain how inspiration and expiration are accomplished.
6. Describe lung volumes and capacities, and explain how pulmonary function tests relate to pulmonary disorders.

The thorax must be sufficiently rigid to protect vital organs and provide attachments for a number of short, powerful muscles. However, breathing, or **pulmonary ventilation,** also requires a flexible thorax that can function as a bellows during the ventilation cycle. The structure of the rib cage and associated cartilages provides continuous elastic tension, so that when stretched by muscle contraction during inspiration, the rib cage can return passively to its resting dimensions when the muscles relax. This elastic recoil is greatly aided by the elasticity of the lungs.

Pulmonary ventilation consists of two phases: *inspiration* and *expiration.* Inspiration (inhalation) and expiration (exhalation) are accomplished by alternately increasing and decreasing the volumes of the thorax and lungs (fig. 16.12).

Inspiration and Expiration

Between the bony portions of the rib cage are two layers of intercostal muscles: the **external intercostal muscles** and the **internal intercostal muscles** (fig. 16.13). Between the costal cartilages, however, there is only one muscle layer, and its fibers are oriented similar to those of the internal intercostals. These muscles are therefore called the *interchondral part* of the internal intercostals. Another name for them is the **parasternal intercostals.**

An unforced, or quiet, inspiration results primarily from contraction of the dome-shaped diaphragm, which lowers and flattens when it contracts. This increases thoracic volume in a vertical direction. Inspiration is aided by contraction of the parasternal and external intercostals, which raise the ribs when they contract and increase thoracic volume laterally. Other thoracic muscles become involved in forced (deep) inspiration. The most important of these are the *scalenes,* followed by the *pectoralis minor* and, in extreme cases, the *sternocleidomastoid muscles.* Contraction of these muscles elevates the ribs in an anteroposterior direction; at the same time, the upper rib cage is stabilized so that the intercostals become more effective. The increase in thoracic volume produced by these muscle contractions decreases intrapulmonary (intra-alveolar) pressure, thereby causing air to flow into the lungs.

Quiet expiration is a passive process. After becoming stretched by contractions of the diaphragm and thoracic muscles, the thorax and lungs recoil as a result of their elastic tension when the respiratory muscles relax. The decrease in lung volume raises the pressure within the alveoli above the atmospheric pressure and pushes the air out. During forced expiration, the internal intercostal muscles (excluding the

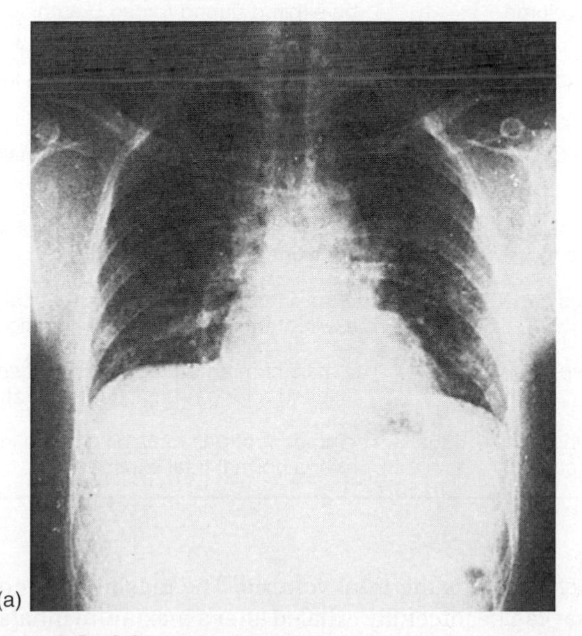
(a)

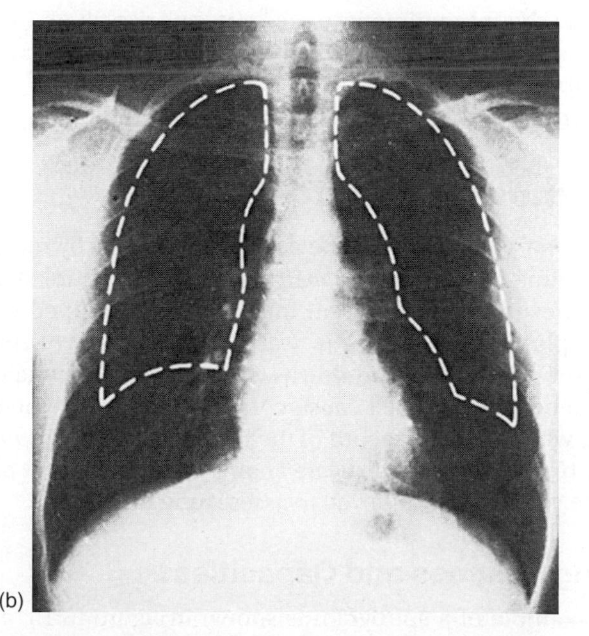
(b)

Figure 16.12 Changes in lung volume during breathing. A change in lung volume, as shown by radiographs (*a*) during expiration and (*b*) during inspiration. The increase in lung volume during full inspiration is shown by comparison with the lung volume in full expiration (*dashed lines*). AP|R

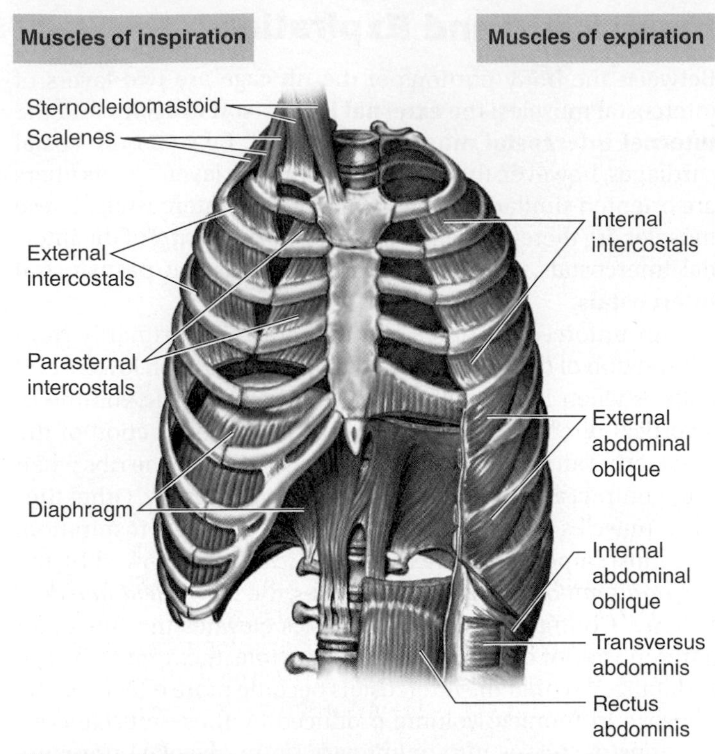

Figure 16.13 The muscles involved in breathing. The principal muscles of inspiration are shown on the left, and those of expiration are shown on the right.

interchondral part) contract and depress the rib cage. The abdominal muscles also aid expiration because, when they contract, they force abdominal organs up against the diaphragm and further decrease the volume of the thorax. By this means the intrapulmonary pressure can rise 20 or 30 mmHg above the atmospheric pressure. The events that occur during inspiration and expiration are summarized in table 16.2 and shown in figure 16.14.

Pulmonary Function Tests

Pulmonary function may be assessed clinically by means of a technique known as *spirometry.* In this procedure, a subject breathes in a closed system in which air is trapped within a light plastic bell floating in water. The bell moves up when the subject exhales and down when the subject inhales. The movements of the bell cause corresponding movements of a pen, which traces a record of the breathing called a *spirogram* (fig. 16.15). More sophisticated computerized devices are now more commonly employed to assess lung function.

Lung Volumes and Capacities

An example of a spirogram is shown in figure 16.15, and the various lung volumes and capacities are defined in table 16.3. A lung capacity is equal to the sum of two or more lung volumes. During quiet breathing, for example, the amount of air expired in each breath is the **tidal volume.** The maximum amount of air that can be forcefully exhaled after a maximum inhalation is called the **vital capacity,** which is equal to the sum of the **inspiratory reserve volume, tidal volume,** and **expiratory reserve volume** (fig. 16.15). The **residual volume** is the volume of air

Table 16.2 | Mechanisms Involved in Normal, Quiet Ventilation and Forced Ventilation

	Inspiration	Expiration
Normal, Quiet Breathing	Contraction of the diaphragm and external intercostal muscles increases the thoracic and lung volume, decreasing intrapulmonary pressure to about −3 mmHg.	Relaxation of the diaphragm and external intercostals, plus elastic recoil of lungs, decreases lung volume and increases intrapulmonary pressure to about +3 mmHg.
Forced Ventilation	Inspiration, aided by contraction of accessory muscles such as the scalenes and sternocleidomastoid, decreases intrapulmonary pressure to −20 mmHg or lower.	Expiration, aided by contraction of abdominal muscles and internal intercostal muscles, increases intrapulmonary pressure to +30 mmHg or higher.

Table 16.3 | Terms Used to Describe Lung Volumes and Capacities

Term	Definition
Lung Volumes	The four nonoverlapping components of the total lung capacity
Tidal volume	The volume of gas inspired or expired in an unforced respiratory cycle
Inspiratory reserve volume	The maximum volume of gas that can be inspired during forced breathing in addition to tidal volume
Expiratory reserve volume	The maximum volume of gas that can be expired during forced breathing in addition to tidal volume
Residual volume	The volume of gas remaining in the lungs after a maximum expiration
Lung Capacities	Measurements that are the sum of two or more lung volumes
Total lung capacity	The total amount of gas in the lungs after a maximum inspiration
Vital capacity	The maximum amount of gas that can be expired after a maximum inspiration
Inspiratory capacity	The maximum amount of gas that can be inspired after a normal tidal expiration
Functional residual capacity	The amount of gas remaining in the lungs after a normal tidal expiration

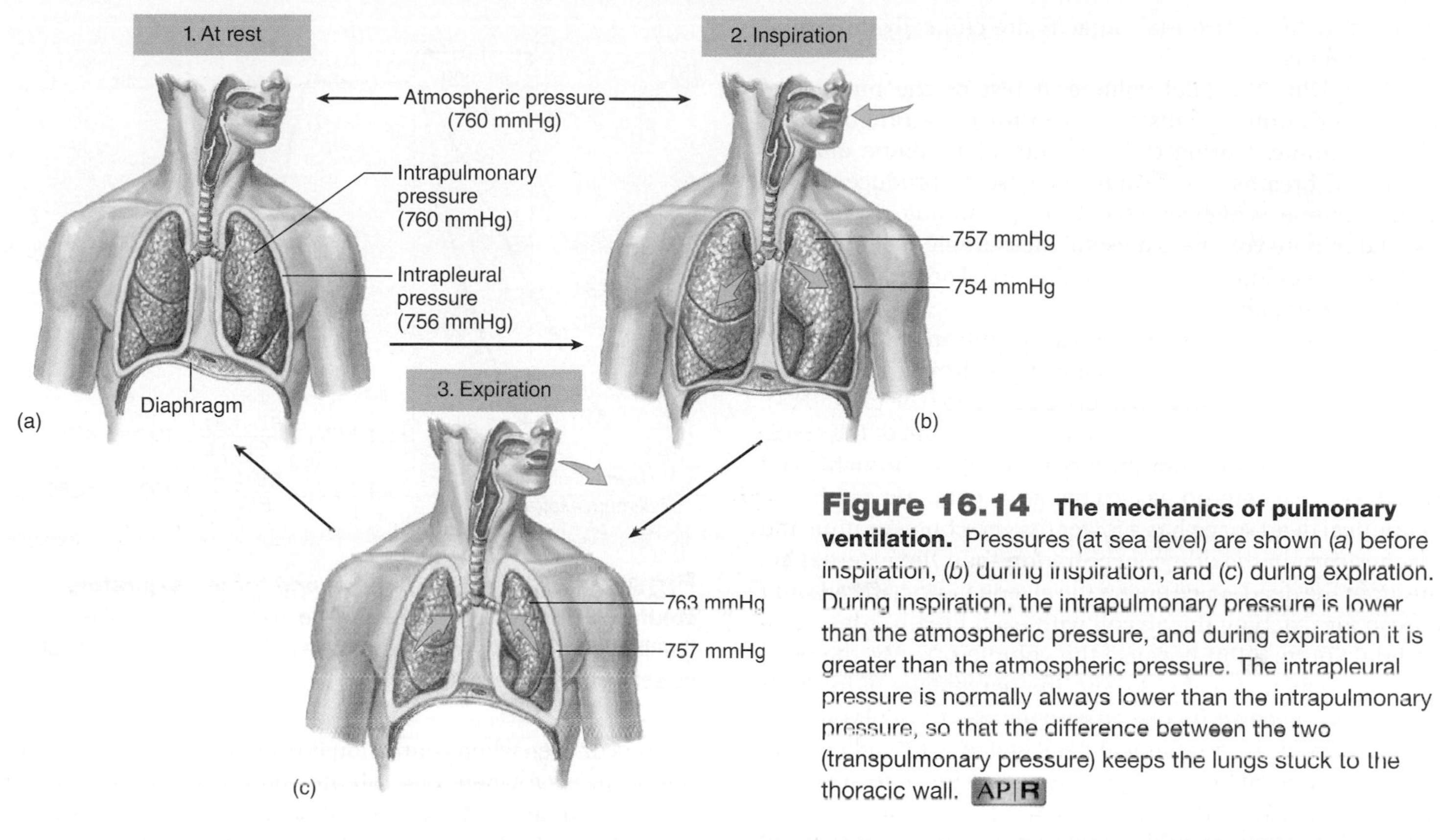

Figure 16.14 The mechanics of pulmonary ventilation. Pressures (at sea level) are shown (*a*) before inspiration, (*b*) during inspiration, and (*c*) during expiration. During inspiration, the intrapulmonary pressure is lower than the atmospheric pressure, and during expiration it is greater than the atmospheric pressure. The intrapleural pressure is normally always lower than the intrapulmonary pressure, so that the difference between the two (transpulmonary pressure) keeps the lungs stuck to the thoracic wall. AP|R

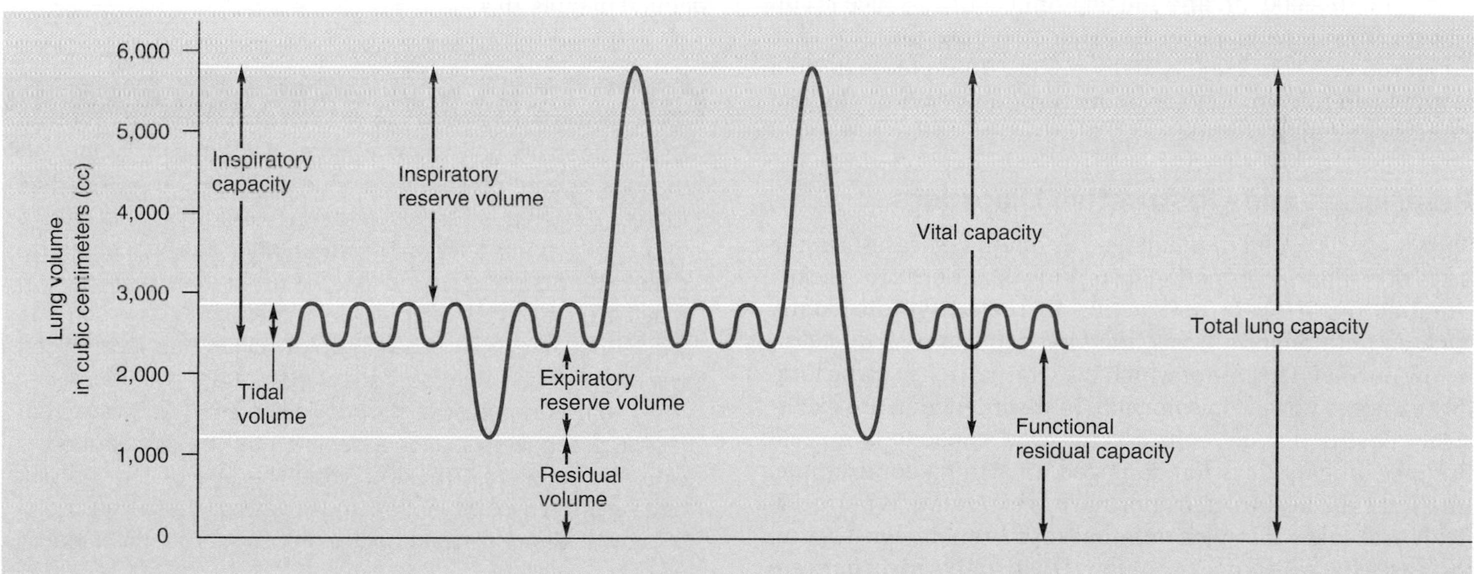

Figure 16.15 A spirogram showing lung volumes and capacities. A lung capacity is the sum of two or more lung volumes. The vital capacity, for example, is the sum of the tidal volume, the inspiratory reserve volume, and the expiratory reserve volume. Note that residual volume cannot be measured with a spirometer because it is air that cannot be exhaled. Therefore, the total lung capacity (the sum of the vital capacity and the residual volume) also cannot be measured with a spirometer.

you cannot expire, even after a maximum forced expiration. This air remains in the lungs because the alveoli and bronchioles normally do not collapse (and the larger airways are noncollapsible). The expiratory reserve volume is the additional air left in the lungs after an unforced expiration. The sum of the residual volume and expiratory reserve volume is known as the **functional residual capacity.** During quiet breathing, the tidal volume expiration ends at the functional residual capacity, and the tidal volume inspiration of the next breath begins at that level (fig. 16.15). The vital capacity and the functional residual capacity are clinically important measurements.

Multiplying the tidal volume at rest by the number of breaths per minute yields a **total minute volume** of about 6 L per minute. During exercise, the tidal volume and the number of breaths per minute increase to produce a total minute volume as high as 100 to 200 L per minute. Notice that the total minute volume is a useful measurement of breathing because it takes into account both the rate of breathing and the depth of the breaths.

It should be noted that not all of the inspired volume reaches the alveoli with each breath. As fresh air is inhaled, it is mixed with air in the **anatomical dead space** (see table 16.4). This dead space comprises the conducting zone of the respiratory system—nose, mouth, larynx, trachea, bronchi, and bronchioles—where no gas exchange occurs. Air within the anatomical dead space has a lower oxygen concentration and a higher carbon dioxide concentration than the external air. The air in the dead space enters the alveoli first, so the amount of fresh air reaching the alveoli with each breath is less than the tidal volume. But because the volume of air in the dead space is an anatomical constant, the percentage of fresh air entering the alveoli is increased with increasing tidal volumes. For example, if the anatomical dead space is 150 ml and the tidal volume is 500 ml, the percentage of fresh air reaching the alveoli is $350/500 \times 100\% = 70\%$. If the tidal volume is increased to 2,000 ml, and the anatomical dead space is still 150 ml, the percentage of fresh air reaching the alveoli is increased to $1{,}850/2{,}000 \times 100\% = 93\%$. An increase in tidal volume can thus be a factor in the respiratory adaptations to exercise and high altitude.

Restrictive and Obstructive Disorders

Spirometry is useful in the diagnosis of lung diseases. On the basis of pulmonary function tests, lung disorders can be classified as *restrictive* or *obstructive.* In **restrictive disorders,** such as pulmonary fibrosis, the vital capacity is reduced to below normal. The rate at which the vital capacity can be forcibly exhaled, however, is normal. In disorders that are exclusively obstructive, by contrast, the vital capacity is normal because lung tissue is not damaged. In asthma, for example, the vital capacity is usually normal but expiration is more difficult and takes a longer time because bronchoconstriction increases the resistance to air flow. **Obstructive disorders** are therefore diagnosed by tests that measure the rate of expiration. One such test is the **forced expiratory volume (FEV),** in which the percentage of the vital capacity that can be exhaled in the first second (FEV_1) is measured (fig. 16.16). An FEV_1 that is significantly less than 80% suggests the presence of obstructive pulmonary disease.

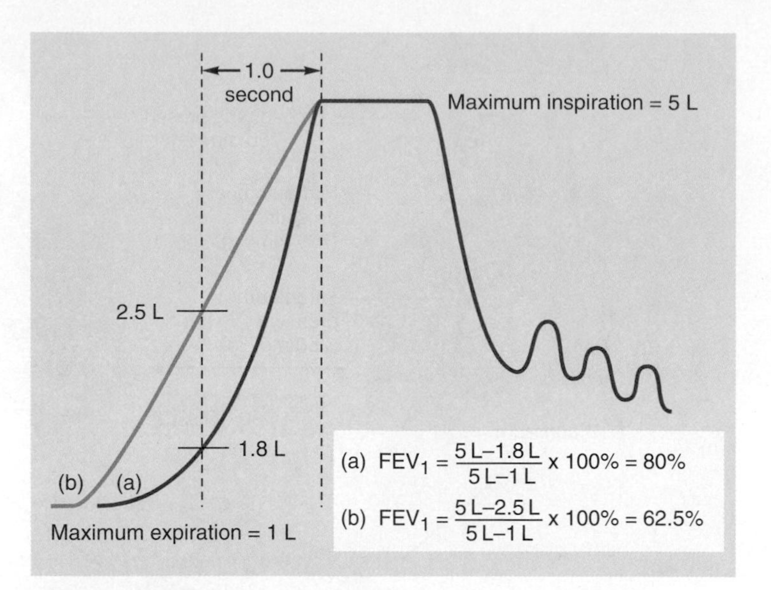

Figure 16.16 The one-second forced expiratory volume (FEV_1) test. The percentage in (*a*) is normal, whereas that in (*b*) may indicate an obstructive pulmonary disorder such as asthma or bronchitis.

Pulmonary Disorders

People with pulmonary disorders frequently complain of **dyspnea,** a subjective feeling of "shortness of breath." Dyspnea may occur even when ventilation is normal, however, and may not occur even when total minute volume is very high, as in exercise. Some of the terms associated with ventilation are defined in table 16.4.

FITNESS APPLICATION

Bronchoconstriction often occurs in response to inhaled noxious agents present in smoke or smog. The FEV_1 has therefore been used by researchers to determine the effects of various components of smog and passive cigarette smoke on pulmonary function. These studies have shown that it is unhealthy to exercise on very smoggy days and that inhaling smoke from other people's cigarettes in a closed environment can adversely affect pulmonary function.

There is normally a decline in the FEV_1 with age, but research suggests that this decline may be accelerated in cigarette smokers. Smokers under the age of 35 who quit have improved lung function; those who quit after the age of 35 slow their age-related decline in FEV_1 to normal rates.

Asthma

The dyspnea, wheezing, and other symptoms of **asthma** are produced by an obstruction of air flow through the bronchioles that occurs in episodes, or "attacks." This obstruction is caused by inflammation, mucous secretion, and bronchoconstriction. Inflammation of the airways is characteristic of asthma, and itself contributes to increased airway responsiveness to agents that promote bronchiolar constriction. Bronchoconstriction further increases airway resistance and makes breathing difficult. The increased airway resistance

Table 16.4 | Ventilation Terminology

Term	Definition
Air spaces	Alveolar ducts, alveolar sacs, and alveoli
Airways	Structures that conduct air from the mouth and nose to the respiratory bronchioles
Alveolar ventilation	Removal and replacement of gas in alveoli; equal to the tidal volume minus the volume of dead space times the breathing rate
Anatomical dead space	Volume of the conducting airways to the zone where gas exchange occurs
Apnea	Cessation of breathing
Dyspnea	Unpleasant, subjective feeling of difficult or labored breathing
Eupnea	Normal, comfortable breathing at rest
Hyperventilation	Alveolar ventilation that is excessive in relation to metabolic rate; results in abnormally low alveolar CO_2
Hypoventilation	Alveolar ventilation that is low in relation to metabolic rate; results in abnormally high alveolar CO_2
Physiological dead space	Combination of anatomical dead space and underventilated or underperfused alveoli that do not contribute normally to blood gas exchange
Pneumothorax	Presence of gas in the intrapleural space (the space between the visceral and parietal pleurae) causing lung collapse
Torr	Unit of pressure nearly equal to the millimeter of mercury (760 mmHg = 760 torr)

of asthma may be provoked by allergic reactions in which immunoglobulin E (IgE) is produced (chapter 15, section 15.6), by exercise (in exercise-induced bronchoconstriction), by breathing cold, dry air, or by aspirin (in a minority of asthmatics).

Atopic (allergic) asthma, the most common form of asthma, is a chronic inflammatory disorder of the airways characterized by *hyper-responsiveness* of the airways to inhaled allergens. Activation of the T_H2 subset of helper T lymphocytes by an allergen causes the release of several cytokines, including interleukin (IL)-4, IL-5, and IL-13 (chapter 15, section 15.3). This leads to production of IgE antibodies and pulmonary eosinophilia (a high number of eosinophils in the lung tissue). Mast cells also become more abundant in the lungs.

When the person is again exposed to the same allergen, the allergen bonds to IgE on the surface of mast cells and basophils, causing these cells to release chemicals that promote inflammation (chapter 15; see fig. 15.23). These include histamine, leukotrienes and prostaglandins (chapter 11; see fig. 11.34), and others that stimulate the bronchoconstriction and mucous secretion of asthma. With repeated exposures to allergens, there is sustained infiltration of eosinophils and basophils, increased numbers of mast cells, increased goblet cells that secrete more mucus, and increased bronchiolar smooth muscle mass. This is accompanied by hyperresponsiveness to allergens and airway irritants.

There is evidence that early exposure to diverse microbes and their products, as may occur when children grow up on farms, can protect children against later asthma. This supports the concept that the incidence of asthma has increased because of population movements to cities, causing children to be exposed to more sterile urban environments. This, combined with increased exposure to air pollutants from traffic, could be responsible for the increased prevalence of asthma observed over the last several decades.

CLINICAL APPLICATION

Asthma is often treated with glucocorticoid drugs, which inhibit inflammation. Newer drugs that block leukotriene action (such as Singulair) are also available to suppress the inflammatory response. In the past, epinephrine was used in an inhaler to stimulate beta-adrenergic receptors in bronchiolar smooth muscle and by this means promote bronchodilation, providing immediate relief during an asthma attack. However, epinephrine stimulates both the β_1-adrenergic receptors that predominate in the heart and the β_2-adrenergic receptors in the bronchioles (chapter 9, section 9.3). More selective β_2 agonist drugs, such as *albuterol*, are now used instead to promote bronchodilation without stimulating the heart to the extent that epinephrine does.

Emphysema

Alveolar tissue is destroyed in the chronic, progressive condition called **emphysema,** which results in fewer but larger alveoli (fig. 16.17). This reduces the surface area for gas exchange. Because alveoli exert a lateral tension on bronchiolar walls to keep them open, the loss of alveoli in emphysema reduces the ability of the bronchioles to remain open during expiration. Collapse of the bronchioles as a result of the compression of the lungs during expiration produces *air trapping,* which further decreases the efficiency of gas exchange in the alveoli.

The most common cause of emphysema is cigarette smoking. Cigarette smoke directly and indirectly causes the release of inflammatory cytokines, which promote inflammation by attracting and activating macrophages, neutrophils, and T lymphocytes within lung tissue. Proteinases (protein-digesting enzymes)—including matrix metalloproteinases (chapter 6) secreted from alveolar macrophages and elastase from neutrophils—cause destruction of the extracellular matrix. Proteinase degradation of the extracellular matrix is aided by the inactivation of such molecules as α_1-antitrypsin, which normally protect the lungs from destruction by proteinases. This destruction of the

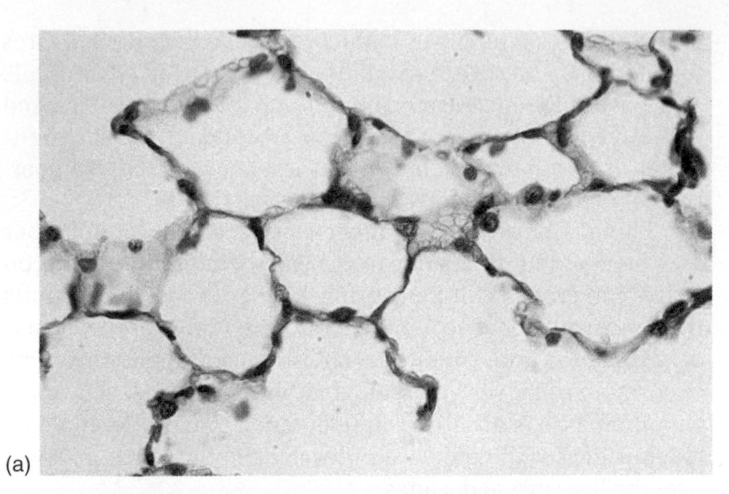

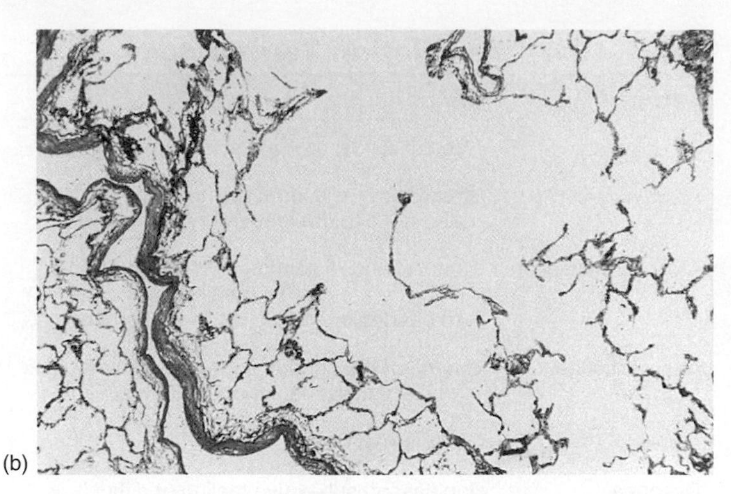

Figure 16.17 Emphysema destroys lung tissue. These are photomicrographs of tissue (*a*) from a normal lung and (*b*) from the lung of a person with emphysema. The destruction of lung tissue in emphysema results in fewer and larger alveoli.

extracellular matrix results in the loss of alveoli and enlargement of the remaining alveoli that is characteristic of emphysema.

Chronic Obstructive Pulmonary Disease (COPD)

Chronic obstructive pulmonary disease (COPD) is characterized by chronic inflammation with narrowing of the airways and destruction of alveolar walls. Included in the COPD category is *chronic obstructive bronchiolitis,* with fibrosis and obstruction of the bronchioles, and *emphysema*. This condition results in an accelerated age-related decline in the FEV_1. Although asthma is also classified as a chronic inflammatory disorder, it is distinguished from COPD in that the obstruction in asthma is largely reversible with inhalation of a bronchodilator (albuterol). Also, asthma (but not COPD) is characterized by airway hyperresponsiveness—an abnormal bronchoconstrictor response to a stimulus. The inflammatory cells characteristic of COPD are macrophages, neutrophils, and cytotoxic T lymphocytes, whereas in asthma they are helper T lymphocytes, eosinophils, and mast cells, as previously discussed.

About 90% of people with COPD are (or have been) smokers. Genetic susceptibility is also a factor; not all smokers get COPD, but a substantial proportion—10% to 20%—do. Cigarette smoke contains over 2,000 foreign compounds and many free radicals that promote inflammation and activate alveolar macrophages and neutrophils. Protein-digesting enzymes released by these activated phagocytes, together with reactive oxygen species (chapter 5, section 5.2), promote the lung damage that results in emphysema. Cigarette smoking also stimulates proliferation of the mucus-secreting goblet cells of the respiratory tract, and excessive mucus production correlates with the severity of COPD. In addition, cigarette smoking promotes remodeling of the small airways, in which additions of fibrous and muscle tissue to the bronchiolar wall narrow the lumen and contribute to the obstruction of airflow. Finally, cigarette smoke promotes remodeling in the blood vessels within the lungs, resulting in pulmonary hypertension among COPD patients. It should be noted that smoking is also the major preventable cause of **lung cancer,** which is responsible for most cancer deaths worldwide.

The vast majority of people with COPD are smokers, and cessation of smoking once COPD has begun does not seem to stop its progression. Inhaled corticosteroids, which are useful in treating the inflammation of asthma, are of limited value in the treatment of COPD. In addition to the pulmonary problems directly caused by COPD, other pathological changes may occur. These include pneumonia, pulmonary emboli (traveling blood clots), and heart failure. Patients with COPD may develop *cor pulmonale*—pulmonary hypertension with hypertrophy and eventual failure of the right ventricle. COPD is currently the fifth leading cause of death worldwide, and it has been estimated that by 2020 it will become the third leading cause of death.

Clinical Investigation CLUES

Harry's pulmonary function tests revealed that he has a slightly low vital capacity and a significantly reduced FEV_1.

- What is the significance of his low FEV_1, and what can he do to improve it?
- What is the vital capacity, and what may have caused it to be slightly low?

Pulmonary Fibrosis

Under certain conditions, for reasons that are poorly understood, lung damage leads to **pulmonary fibrosis** instead of emphysema. In this condition, the normal structure of the lungs is disrupted by the accumulation of fibrous connective tissue proteins. Fibrosis can result, for example, from the inhalation of particles less than 6 μm in size that can accumulate in the respiratory zone of the lungs. Included in this category is *anthracosis*, or black lung, which is produced by the inhalation of carbon particles from coal dust.

CHECKPOINT

5a. Describe the actions of the diaphragm and external intercostal muscles during inspiration. How is quiet expiration produced?

5b. Explain how forced inspiration and forced expiration are produced.

6a. Define the terms *tidal volume* and *vital capacity*. Explain how the total minute volume is calculated and how this value is affected by exercise.

6b. How are the vital capacity and the forced expiratory volume measurements affected by asthma and pulmonary fibrosis? Give the reasons for these effects.

16.4 GAS EXCHANGE IN THE LUNGS

Gas exchange between the alveolar air and the pulmonary capillaries results in an increased oxygen concentration and a decreased carbon dioxide concentration in the blood leaving the lungs. This blood enters the systemic arteries, where blood gas measurements are taken.

LEARNING OUTCOMES

After studying this section, you should be able to:

7. Explain how partial gas pressures are calculated, and their significance in measurements of arterial blood gases.

8. Describe the factors that influence the partial pressure of blood gases and the total content of gases in the blood.

The atmosphere is an ocean of gas that exerts pressure on all objects within it. This pressure can be measured with a glass U-tube filled with fluid. One end of the U-tube is exposed to the atmosphere, while the other side is continuous with a sealed vacuum tube. Because the atmosphere presses on the open-ended side, but not on the side connected to the vacuum tube, atmospheric pressure pushes fluid in the U-tube up on the vacuum side to a height determined by the atmospheric pressure and the density of the fluid. Water, for example, will be pushed up to a height of 33.9 feet (10,332 mm) at sea level, whereas mercury (Hg)—which is more dense—will be raised to a height of 760 mm. As a matter of convenience, therefore,

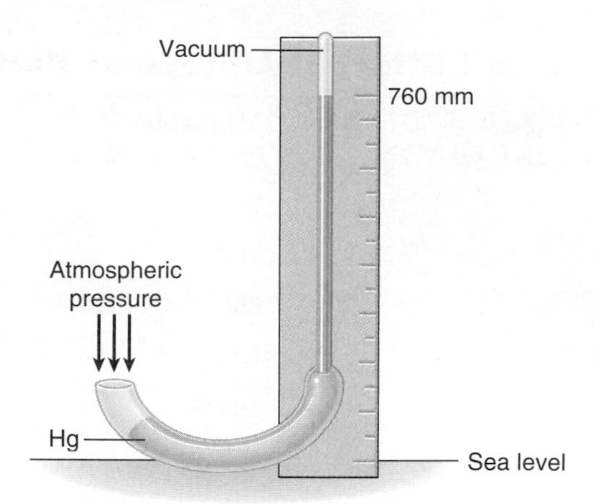

Figure 16.18 The measurement of atmospheric pressure. Atmospheric pressure at sea level can push a column of mercury to a height of 760 millimeters. This is also described as 760 torr, or one atmospheric pressure.

devices used to measure atmospheric pressure (barometers) use mercury rather than water. The atmospheric pressure at sea level is thus said to be equal to 760 mmHg (or 760 *torr*, after Evangelista Torricelli, who invented the mercury barometer in 1643), which is also described as a pressure of *one atmosphere* (fig. 16.18).

According to **Dalton's law,** the total pressure of a gas mixture (such as air) is equal to the sum of the pressures that each gas in the mixture would exert independently. The pressure that a particular gas in a mixture exerts independently is the **partial pressure** of that gas, which is equal to the product of the total pressure and the fraction of that gas in the mixture.

Dalton's law can be restated as follows: The total pressure of the gas mixture is equal to the sum of the partial pressures of the constituent gases. Because oxygen constitutes about 21% of the atmosphere, for example, its partial pressure (abbreviated P_{O_2}) is 21% of 760, or about 159 mmHg. Nitrogen constitutes about 78% of the atmosphere, so its partial pressure is equal to $0.78 \times 760 = 593$ mmHg.

These two gases thus contribute about 99% of the total pressure of 760 mmHg:

$$P_{\text{dry atmosphere}} = P_{N_2} + P_{O_2} + P_{CO_2} = 760 \text{ mmHg}$$

Calculation of P_{O_2}

With increasing altitude, the total atmospheric pressure and the partial pressure of the constituent gases decrease (table 16.5). At Denver (5,000 feet above sea level), for example, the atmospheric pressure is decreased to 619 mmHg and the P_{O_2} is therefore reduced to $619 \times 0.21 = 130$ mmHg. At the peak of Mount Everest (at 29,000 feet) the P_{O_2} is only 42 mmHg. As one descends below sea level, as in scuba diving, the total pressure increases by one atmosphere for every 33 feet. At 33 feet therefore, the pressure equals $2 \times 760 = 1{,}520$ mmHg. At 66 feet the pressure equals three atmospheres.

Table 16.5 | Effect of Altitude on Partial Oxygen Pressure (P_{O_2})

Altitude (Feet Above Sea Level)*	Atmospheric Pressure (mmHg)	P_{O_2} in Air (mmHg)	P_{O_2} in Alveoli (mmHg)	P_{O_2} in Arterial Blood (mmHg)
0	760	159	105	100
2,000	707	148	97	92
4,000	656	137	90	85
6,000	609	127	84	79
8,000	564	118	79	74
10,000	523	109	74	69
20,000	349	73	40	35
30,000	226	47	21	19

*For reference, Pike's Peak (Colorado) is 14,110 feet; Mt. Whitney (California) is 14,505 feet; Mt. Logan (Canada) is 19,524 feet; Mt. McKinley (Alaska) is 20,320 feet; and Mt. Everest (Nepal and Tibet), the tallest mountain in the world, is 29,029 feet.

Inspired air contains variable amounts of moisture. By the time the air has passed into the respiratory zone of the lungs, however, it is normally saturated with water vapor (has a relative humidity of 100%). The capacity of air to contain water vapor depends on its temperature; since the temperature of the respiratory zone is constant at 37° C, its water vapor pressure is also constant (at 47 mmHg).

Water vapor, like the other constituent gases, contributes a partial pressure to the total atmospheric pressure. The total atmospheric pressure is constant (depending only on the height of the air mass), so the water vapor "dilutes" the contribution of other gases to the total pressure:

$$P_{\text{wet atmosphere}} = P_{N_2} + P_{O_2} + P_{CO_2} + P_{H_2O}$$

When the effect of water vapor pressure is considered, the partial pressure of oxygen in the inspired air is decreased at sea level to

$$P_{O_2}\,(\text{sea level}) = 0.21\,(760 - 47) = 150 \text{ mmHg}.$$

As a result of gas exchange in the alveoli, there is an increase in the P_{CO_2}, while the P_{O_2} of alveolar air is further diminished to about 105 mmHg. The partial pressures of the inspired air and the partial pressures of alveolar air are compared in figure 16.19.

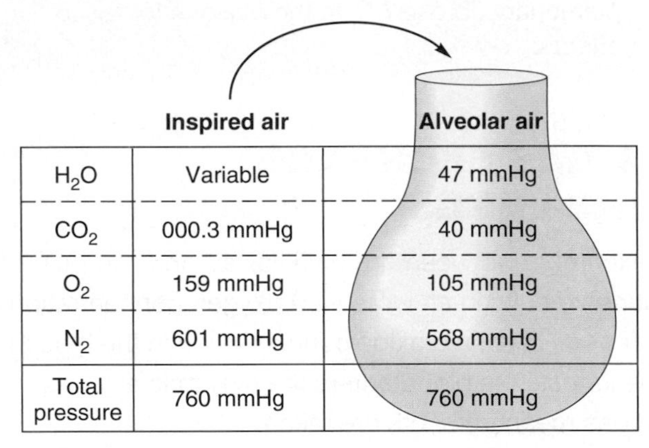

Inspired air
Alveolar air
H_2O
Variable
47 mmHg
CO_2
000.3 mmHg
40 mmHg
O_2
159 mmHg
105 mmHg
N_2
601 mmHg
568 mmHg
Total pressure
760 mmHg
760 mmHg

Figure 16.19 Partial pressures of gases in inspired air and alveolar air at sea level. Notice that as air enters the alveoli its oxygen content decreases and its carbon dioxide content increases. Also notice that air in the alveoli is saturated with water vapor (giving it a partial pressure of 47 mmHg), which dilutes the contribution of other gases to the total pressure. AP|R

See the *Test Your Quantitative Ability* section of the Review Activities at the end of this chapter.

Partial Pressures of Gases in Blood

The enormous surface area of alveoli and the short diffusion distance between alveolar air and the capillary blood quickly help to bring oxygen and carbon dioxide in the blood and air into equilibrium. This function is further aided by the tremendous number of capillaries that surround each alveolus, forming an almost continuous sheet of blood around the alveoli (fig. 16.20).

When a liquid and a gas, such as blood and alveolar air, are at equilibrium, the amount of gas dissolved in the fluid reaches a maximum value. According to **Henry's law,** this value depends on (1) the solubility of the gas in the fluid, which is a physical constant; (2) the temperature of the fluid—more gas can be dissolved in cold water than warm water; and (3) the partial pressure of the gas. Because solubility is a constant and the temperature of the blood does not vary significantly, *the concentration of a gas dissolved in a fluid (such as plasma) depends directly on its partial pressure in the gas mixture.* When water—or plasma—is brought into equilibrium with air at a P_{O_2} of 100 mmHg, for example, the fluid will contain 0.3 ml of O_2 per 100 ml fluid at 37° C. If the P_{O_2} of the gas were reduced by half, the amount of dissolved oxygen would also be reduced by half.

Blood Gas Measurements

Measurement of the oxygen content of blood (in ml O_2 per 100 ml blood) is a laborious procedure. Fortunately, an **oxygen electrode** that produces an electric current in proportion to

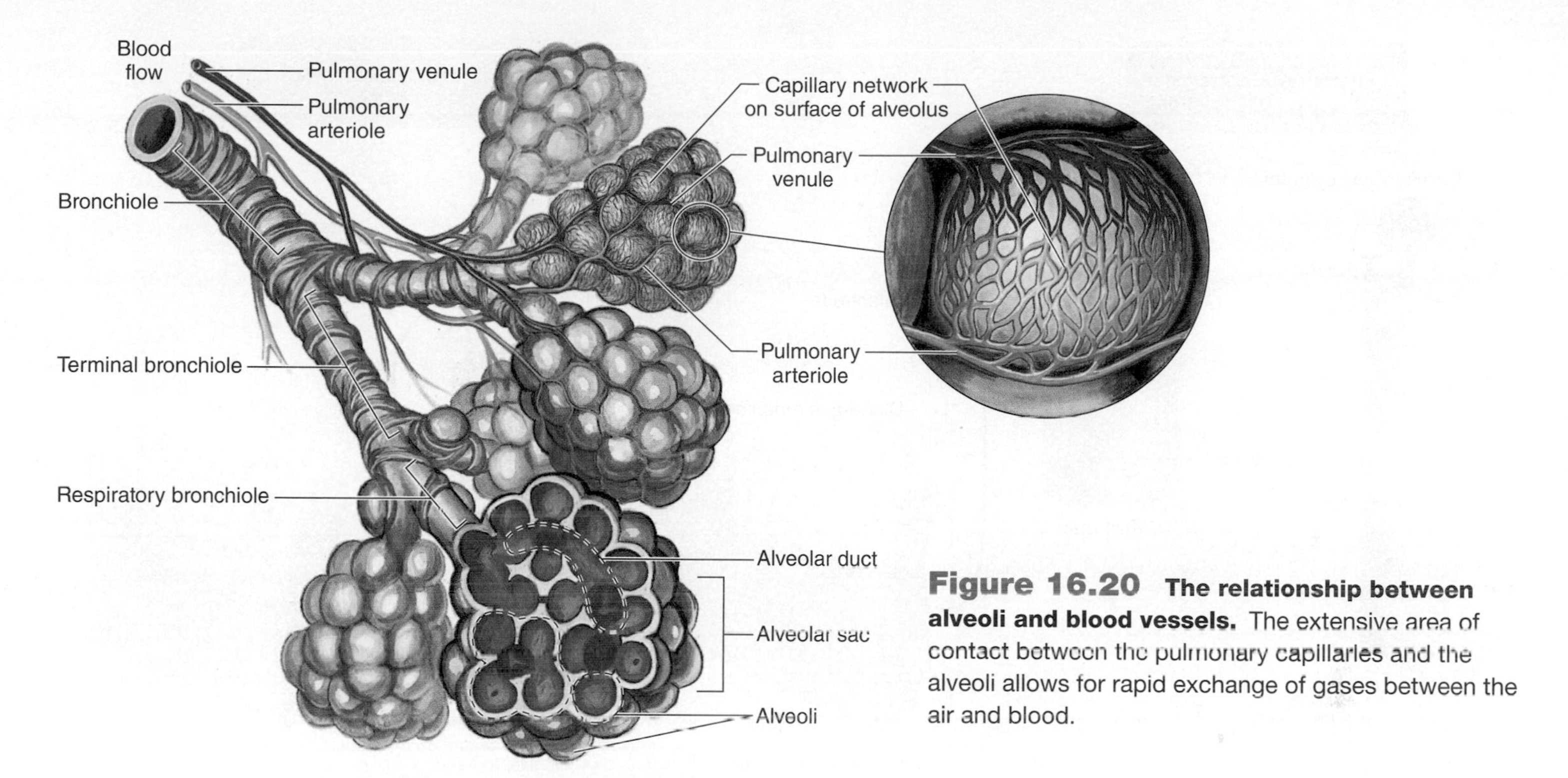

Figure 16.20 The relationship between alveoli and blood vessels. The extensive area of contact between the pulmonary capillaries and the alveoli allows for rapid exchange of gases between the air and blood.

the concentration of *dissolved oxygen* has been developed. If this electrode is placed in a fluid while oxygen is artificially bubbled into it, the current produced by the oxygen electrode will increase up to a maximum value. This maxiumum is attained when the fluid is saturated with oxygen—when it contains all of the dissolved oxygen possible (given the low solubility constant of oxygen) at a particular temperature and P_{O_2}. If the temperature is constant, the maximum amount of oxygen that can be dissolved (and thus the reading of the oxygen electrode) depends directly on the P_{O_2} of the gas.

As a matter of convenience, it can now be said that *the fluid has the same P_{O_2} as the gas.* If it is known that the gas has a P_{O_2} of 152 mmHg, for example, the deflection of a needle by the oxygen electrode can be calibrated on a scale at 152 mmHg (fig. 16.21). The actual amount of dissolved oxygen under these circumstances is not particularly important (it can be looked up in solubility tables, if desired); it is simply a linear function of the P_{O_2}. A lower P_{O_2} indicates that less oxygen is dissolved; a higher P_{O_2} indicates that more oxygen is dissolved.

If the oxygen electrode is next inserted into an unknown sample of blood, the P_{O_2} of that sample can be read directly from the previously calibrated scale. Suppose, as illustrated in figure 16.21, the blood sample has a P_{O_2} of 100 mmHg. Alveolar air has a P_{O_2} of about 105 mmHg, so this reading indicates that the blood is almost in complete equilibrium with the alveolar air.

The oxygen electrode responds only to oxygen dissolved in water or plasma; it cannot respond to oxygen that is bound to hemoglobin in red blood cells. Most of the oxygen in blood, however, is located in the red blood cells attached to hemoglobin. The oxygen content of whole blood thus depends on both its P_{O_2} and its red blood cell and hemoglobin content. At a P_{O_2} of about 100 mmHg, whole blood normally contains almost 20 ml O_2 per 100 ml blood; of this amount, only 0.3 ml of O_2 is dissolved in the plasma and 19.7 ml of O_2 is found within the red blood cells (see fig. 16.31). Because only the 0.3 ml of O_2 affects the P_{O_2} measurement, this measurement would be unchanged if the red blood cells were removed from the sample.

Significance of Blood P_{O_2} and P_{CO_2} Measurements

Because blood P_{O_2} measurements are not directly affected by the oxygen in red blood cells, the P_{O_2} does not provide a measurement of the total oxygen content of whole blood. It does, however, provide a good index of *lung function.* If the inspired air has a normal P_{O_2} but the arterial P_{O_2} is below normal, for example, you could conclude that gas exchange in the lungs is impaired. Measurements of arterial P_{O_2} thus provide valuable information in treating people with pulmonary diseases, in performing surgery (when breathing may be depressed by anesthesia), and in caring for premature babies with respiratory distress syndrome.

When the lungs are functioning properly, the P_{O_2} of systemic arterial blood is only 5 mmHg less than the P_{O_2} of alveolar air. At a normal P_{O_2} of about 100 mmHg, hemoglobin is almost completely loaded with oxygen, indicated by an *oxyhemoglobin saturation* (the percentage of oxyhemoglobin to total hemoglobin) of 97%. Given this high oxyhemoglobin saturation, an increase in blood P_{O_2}—produced, for example, by breathing 100% oxygen from a gas tank—cannot significantly increase the amount of oxygen contained in the red blood cells. It can, however, significantly increase the amount of oxygen dissolved in the plasma (because the amount dissolved is directly determined by the P_{O_2}). If the

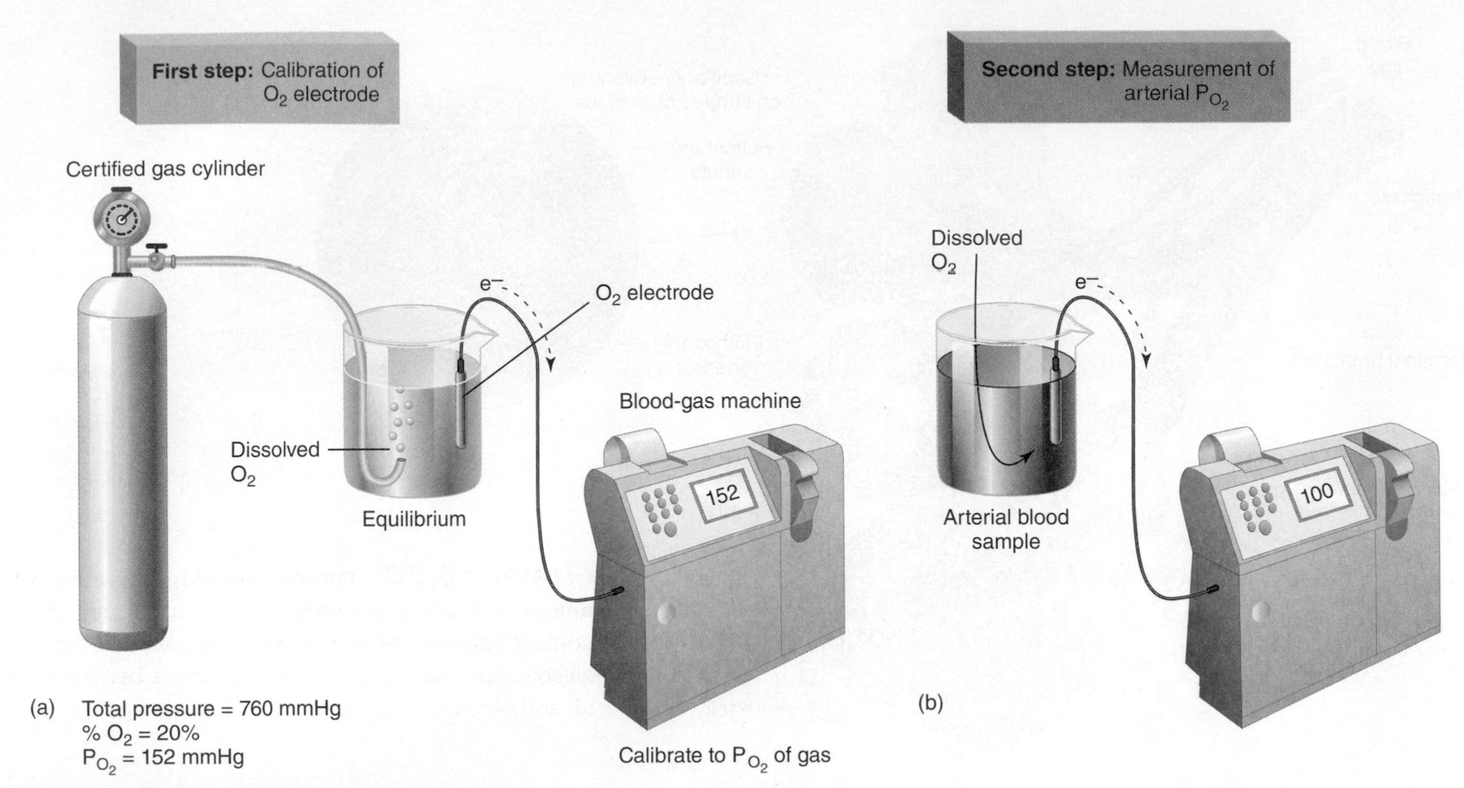

Figure 16.21 **Blood gas measurements using the P_{O_2} electrode.** (*a*) The electrical current generated by the oxygen electrode is calibrated so that the needle of the blood gas machine points to the P_{O_2} of the gas with which the fluid is in equilibrium. (*b*) Once standardized in this way, the electrode can be inserted into a fluid such as blood, and the P_{O_2} of this solution can be measured.

P_{O_2} doubles, the amount of oxygen dissolved in the plasma also doubles, but the total oxygen content of whole blood increases only slightly. This is because the plasma contains relatively little oxygen compared to the red blood cells.

Since the oxygen carried by red blood cells must first dissolve in plasma before it can diffuse to the tissue cells, however, a doubling of the blood P_{O_2} means that the *rate of oxygen diffusion* to the tissues would double under these conditions. For this reason, breathing from a tank of 100% oxygen (with a P_{O_2} of 760 mmHg) would significantly increase oxygen delivery to the tissues, although it would have little effect on the total oxygen content of blood.

An electrode that produces a current in response to dissolved carbon dioxide is also used, so that the P_{CO_2} of blood can be measured together with its P_{O_2}. Blood in the systemic veins, which is delivered to the lungs by the pulmonary arteries, usually has a P_{O_2} of 40 mmHg and a P_{CO_2} of 46 mmHg. After gas exchange in the alveoli of the lungs, blood in the pulmonary veins and systemic arteries has a P_{O_2} of about 100 mmHg and a P_{CO_2} of 40 mmHg (fig. 16.22). The values in arterial blood are relatively constant and clinically significant because they reflect lung function. Blood gas measurements of venous blood are not as useful because these values are far more variable. For example, venous P_{O_2} is much lower and P_{CO_2} much higher after exercise than at rest, whereas arterial values are not significantly affected by moderate physical activity.

CLINICAL APPLICATION

A **pulse oximeter** is commonly used in hospitals for noninvasively measuring oxyhemoglobin saturation. This device typically clips on the finger or pinna and gives readings of oxygen saturation and pulse rate within a short time, making it useful in many clinical contexts such as in emergency medicine and during anesthesia. The pulse oximeter has two light-emitting diodes (LEDs); one emits red light (wavelengths of 600–750 nm), and the other emits light in the infrared range (850–1,000 nm), both of which pass through the tissues to a sensor. Oxyhemoglobin and deoxyhemoglobin absorb light differently in these ranges, allowing microprocessors to determine the percent oxyhemoglobin saturation fairly accurately under most conditions.

Clinical Investigation CLUES

Harry's arterial P_{CO_2} was elevated when he was brought to the hospital, but after treatment it returned to normal.

- What caused the rise in his arterial P_{CO_2}?
- How did this measurement return to normal after Harry's treatment?

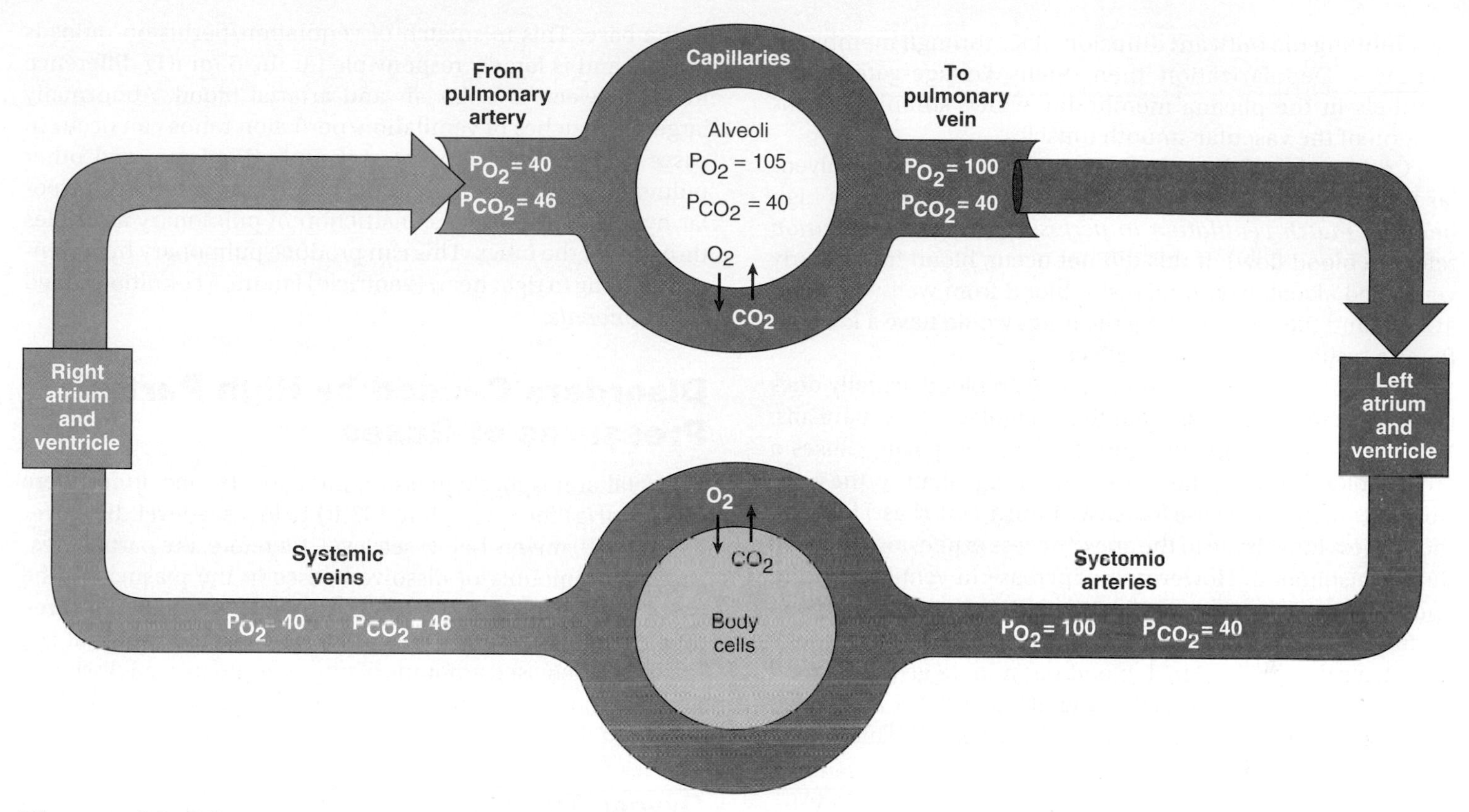

Figure 16.22 Partial pressures of gases in blood. The P_{O_2} and P_{CO_2} values of blood are a result of gas exchange in the lung alveoli and gas exchange between systemic capillaries and body cells. AP|R

Pulmonary Circulation and Ventilation/Perfusion Ratios

In a fetus, the pulmonary circulation has a high vascular resistance because the lungs are partially collapsed. This high vascular resistance helps to shunt blood from the right to the left atrium through the foramen ovale, and from the pulmonary artery to the aorta through the ductus arteriosus (chapter 13, section 13.3). After birth, the foramen ovale and ductus arteriosus close, and the vascular resistance of the pulmonary circulation falls sharply. This fall in vascular resistance at birth is due to (1) opening of the vessels as a result of the subatmospheric intrapulmonary pressure and physical stretching of the lungs during inspiration and (2) dilation of the pulmonary arterioles in response to increased alveolar P_{O_2}.

In the adult, the right ventricle (like the left) has a cardiac output of about 5.5 L per minute. The rate of blood flow through the pulmonary circulation is thus equal to the flow rate through the systemic circulation. Blood flow, as described in chapter 14, is directly proportional to the pressure difference between the two ends of a vessel and inversely proportional to the vascular resistance. In the systemic circulation, the mean arterial pressure is 90 to 100 mmHg and the pressure of the right atrium is 0 mmHg; therefore, the pressure difference is about 100 mmHg. The mean pressure of the pulmonary artery, by contrast, is only 15 mmHg and the pressure of the left atrium is 5 mmHg. The driving pressure in the pulmonary circulation is thus 15 − 5, or 10 mmHg.

Because the driving pressure in the pulmonary circulation is only one-tenth that of the systemic circulation and yet the flow rates are equal, it follows that the pulmonary vascular resistance must be one-tenth that of the systemic vascular resistance. The pulmonary circulation, in other words, is a low-resistance, low-pressure pathway. The low pulmonary blood pressure produces less filtration pressure (see chapter 14, fig. 14.9) than that produced in the systemic capillaries, and thus affords protection against *pulmonary edema.* This is a dangerous condition in which excessive fluid can enter the interstitial spaces of the lungs and then the alveoli, impeding ventilation and gas exchange. Pulmonary edema occurs when there is pulmonary hypertension, which may be produced by left ventricular heart failure.

Pulmonary arterioles constrict when the alveolar P_{O_2} is low and dilate as the alveolar P_{O_2} is raised. This response is opposite to that of systemic arterioles, which dilate in response to low tissue P_{O_2} (chapter 14, section 14.3). Dilation of the systemic arterioles when the P_{O_2} is low helps to supply more blood and oxygen to the tissues; constriction of the pulmonary arterioles when the alveolar P_{O_2} is low helps decrease blood flow to alveoli that are inadequately ventilated. The response of pulmonary arterioles to low oxygen is automatic; hypoxia causes depolarization of the vascular smooth muscle cells by inhibiting the outward diffusion of K^+ through membrane channels. Depolarization then opens voltage-gated Ca^{2+} channels in the plasma membrane, which stimulates contraction of the vascular smooth muscle.

Constriction of the pulmonary arterioles where the alveolar P_{O_2} is low and their dilation where the alveolar P_{O_2} is high helps to *match ventilation to perfusion* (the term *perfusion* refers to blood flow). If this did not occur, blood from poorly ventilated alveoli would mix with blood from well-ventilated alveoli, and the blood leaving the lungs would have a lowered P_{O_2} as a result of this dilution effect.

Dilution of the P_{O_2} of pulmonary vein blood actually does occur to some degree, despite these regulatory mechanisms. When a person stands upright, the force of gravity causes a greater blood flow to the base of the lungs than to the apex (top). Ventilation likewise increases from apex to base, because there is less lung tissue in the apex and less expansion of alveoli during inspiration. However, the increase in ventilation from apex to base is not proportionate to the increase in blood perfusion. The *ventilation/perfusion ratio* at the apex is high (0.24 L air divided by 0.07 L blood per minute gives a ratio of 3.4/1.0), while at the base of the lungs it is low (0.82 L air divided by 1.29 L blood per minute gives a ratio of 0.6/1.0). This is illustrated in figure 16.23.

Functionally, the alveoli at the apex of the lungs are overventilated (or underperfused) and they are larger than alveoli at the base. This mismatch of ventilation/perfusion ratios is normal and is largely responsible for the 5 mmHg difference in P_{O_2} between alveolar air and arterial blood. Abnormally large mismatches of ventilation/perfusion ratios can occur in cases of pneumonia, pulmonary emboli, edema, and other pulmonary disorders. In chronic pulmonary disorders, alveolar hypoxia may induce constriction of pulmonary arterioles throughout the lungs. This can produce pulmonary hypertension leading to right heart (ventricle) failure, a condition called *cor pulmonale.*

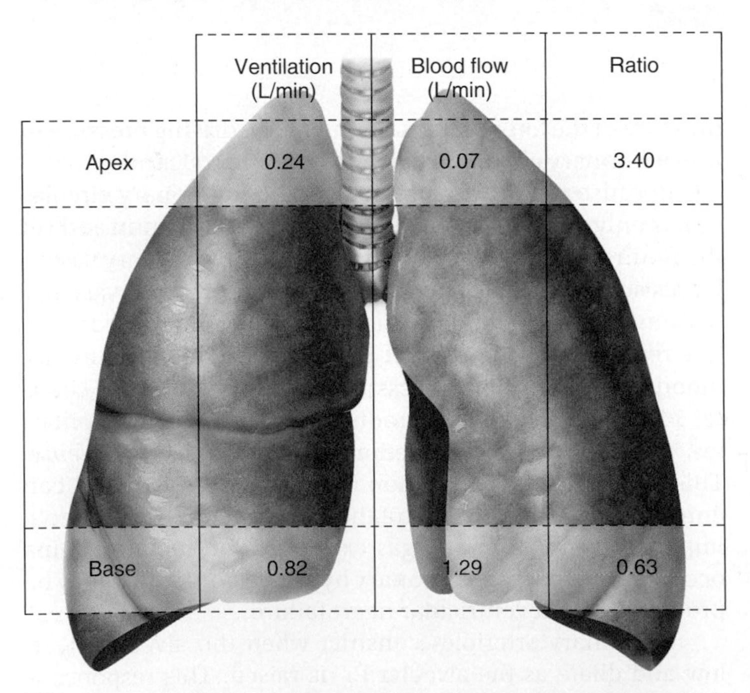

Ventilation (L/min)
Blood flow (L/min)
Ratio
Apex
0.24
0.07
3.40
Base
0.82
1.29
0.63

Figure 16.23 Lung ventilation/perfusion ratios. The ventilation, blood flow, and ventilation/perfusion ratios are indicated for the apex and base of the lungs. The ratios indicate that the apex is relatively overventilated and the base underventilated in relation to their blood flows. As a result of such uneven matching of ventilation to perfusion, the blood leaving the lungs has a P_{O_2} that is slightly lower (by about 5 mmHg) than the P_{O_2} of alveolar air.

Disorders Caused by High Partial Pressures of Gases

The total atmospheric pressure increases by one atmosphere (760 mmHg) for every 10 m (33 ft) below sea level. If a diver descends 10 meters below sea level, therefore, the partial pressures and amounts of dissolved gases in the plasma will be twice those values at sea level. At 20 meters, they are three times, and at 30 meters they are four times the values at sea level. The increased amounts of nitrogen and oxygen dissolved in the blood plasma under these conditions can have serious effects on the body.

Oxygen Toxicity

Although breathing 100% oxygen at one or two atmospheres pressure can be safely tolerated for a few hours, higher partial oxygen pressures can be very dangerous. Oxygen toxicity may develop rapidly when the P_{O_2} rises above about 2.5 atmospheres. This is apparently caused by the oxidation of enzymes and other destructive changes that can damage the nervous system and lead to coma and death. For these reasons, deep-sea divers commonly use gas mixtures in which oxygen is diluted with inert gases such as nitrogen (as in ordinary air) or helium.

CLINICAL APPLICATION

Hyperbaric oxygen therapy, in which a patient is given 100% oxygen gas at 2 to 3 atmospheres pressure to breathe for varying lengths of time, is used to treat carbon monoxide poisoning, decompression sickness, severe traumatic injury (such as crush injury), infections that could lead to gas gangrene, and other conditions. While a normal plasma oxygen concentration is 0.3 ml O_2/100 ml blood (as previously described), breathing 100% oxygen at a pressure of 3 atmospheres raises the plasma concentration to about 6 ml O_2/100 ml blood. This helps kill anaerobic bacteria such as those that cause gangrene; promote wound healing; reduce the size of gas bubbles (in the case of decompression sickness); and quickly eliminate carbon monoxide from the body. Although hyperbaric oxygen was formerly used to treat premature infants for respiratory distress, the practice was discontinued because it caused a fibrotic deterioration of the retina that frequently resulted in blindness.

Nitrogen Narcosis

Although at sea level nitrogen is physiologically inert, larger amounts of dissolved nitrogen under hyperbaric (high-pressure) conditions have deleterious effects, possibly caused by the increased amounts of nitrogen dissolved in plasma membranes at the high partial pressures. **Nitrogen narcosis** resembles alcohol intoxication; depending on the depth of the dive, the diver may experience what Jacques Cousteau termed "rapture of the deep." Dizziness and extreme drowsiness are other narcotizing effects.

Decompression Sickness

As the diver ascends to sea level, the amount of nitrogen dissolved in the plasma decreases as a result of the progressive decrease in the P_{N_2}. If the diver surfaces slowly, a large amount of nitrogen can diffuse through the alveoli and be eliminated in the expired breath. If decompression occurs too rapidly, however, bubbles of nitrogen gas (N_2) can form in the tissue fluids and enter the blood. This process is analogous to the formation of carbon dioxide bubbles in a champagne bottle when the cork is removed. The bubbles of N_2 gas in the blood can block small blood channels, producing muscle and joint pain as well as more serious damage. These effects are known as **decompression sickness,** commonly called "the bends." The primary treatment for decompression sickness is hyperbaric oxygen treatment.

Airplanes that fly long distances at high altitudes (30,000 to 40,000 ft) have pressurized cabins so that the passengers and crew do not experience the very low atmospheric pressures of these altitudes. If a cabin were to become rapidly depressurized at high altitude, much less nitrogen could remain dissolved at the greatly lowered pressure. People in this situation, like the divers that ascend too rapidly, would thus experience decompression sickness.

CHECKPOINT

7a. Explain how the P_{O_2} of air is calculated and how this value is affected by altitude, diving, and water vapor pressure.

7b. Explain how blood P_{O_2} measurements are taken, and discuss the physiological and clinical significance of these measurements.

8a. Explain how the arterial P_{O_2} and the oxygen content of whole blood are affected by (a) hyperventilation, (b) breathing from a tank containing 100% oxygen, (c) anemia (low red blood cell count and hemoglobin concentration), and (d) high altitude.

8b. Describe the ventilation/perfusion ratios of the lungs, and explain why systemic arterial blood has a slightly lower P_{O_2} than alveolar air.

8c. Explain how decompression sickness is produced in divers who ascend too rapidly.

16.5 REGULATION OF BREATHING

The motor neurons that stimulate the respiratory muscles are controlled by two major descending pathways: one that controls voluntary breathing and another that controls involuntary breathing. The unconscious rhythmic control of breathing is influenced by sensory feedback from receptors sensitive to the P_{CO_2}, pH, and P_{O_2} of arterial blood.

LEARNING OUTCOMES

After studying this section, you should be able to:

9. Explain how ventilation is regulated by the CNS.

10. Explain how blood gases and pH influence ventilation.

Inspiration and expiration are produced by the contraction and relaxation of skeletal muscles in response to activity in somatic motor neurons in the spinal cord. The activity of these motor neurons is controlled, in turn, by descending tracts from neurons in the respiratory control centers in the medulla oblongata and from neurons in the cerebral cortex.

Brain Stem Respiratory Centers

The somatic motor neurons that stimulate the respiratory muscles (see fig. 16.13) have their cell bodies in the gray matter of the spinal cord. The motoneurons of the phrenic nerve, stimulating the diaphragm, have cell bodies in the cervical level of the spinal cord; those that innervate the respiratory muscles of the rib cage and abdomen have cell bodies in the thoracolumbar region of the cord. These spinal motoneurons are regulated, either directly or via spinal interneurons, by descending axons from the brain.

The respiratory rhythm is generated by a loose aggregation of neurons in the ventrolateral region of the medulla oblongata, which forms the **rhythmicity center** for the control of automatic breathing. Within this center, four types of neurons have been identified that fire at different stages of inspiration. For simplicity, we can refer to these four types as *I neurons.* One or two types of *E neurons* have also been identified, which fire during expiration.

A large collection of inspiratory neurons forms the *dorsal respiratory group* of the medulla. These send axons that stimulate the spinal motoneurons of the phrenic nerve to the diaphragm, causing inspiration. There is also a *ventral respiratory group* of neurons in the medulla that contains both I and E neurons. The inspiratory neurons located here stimulate spinal interneurons, which in turn activate the spinal motoneurons of respiration. There is also a region of densely packed expiratory neurons in the ventral respiratory group. These E neurons inhibit the motoneurons of the phrenic nerve during expiration.

The activity of the I and E neurons varies in a reciprocal way to produce a rhythmic pattern of breathing. There is evidence that the rhythmicity of I and E neurons may be driven by the cyclic activity of particular pacemaker neurons within the rhythmicity center of the medulla. These pacemaker neurons display spontaneous, cyclic changes in the membrane potential, somewhat like the pacemaker cells of the heart (chapter 13).

The activity of the medullary rhythmicity center may be influenced by centers in the *pons.* As a result of animal research in which the brain stem is destroyed at different levels, two respiratory control centers have been identified in the pons. One area—the *apneustic center*—appears to promote inspiration by stimulating the I neurons in the medulla. The other area—the *pneumotaxic center*—seems to antagonize the apneustic center and inhibit inspiration (fig. 16.24). These are the roles of the pons in the rhythmic control of breathing in experimental animals; in humans, however, the functions of the pons in the normal regulation of breathing is currently unclear.

The brainstem respiratory centers control breathing largely via axons to the **phrenic motor nuclei** in cervical regions C3 through C6 of the spinal cord. Lower motor neurons here send axons in the *phrenic nerves* that control the diaphragm. This is why people with spinal cord injuries at about C4 often cannot breathe independently. A recent report demonstrated that rats with a comparable injury could have their breathing restored by grafting a peripheral nerve across the spinal cord lesion (forming a bridge for neuron growth) and injecting an enzyme that reduces scar tissue.

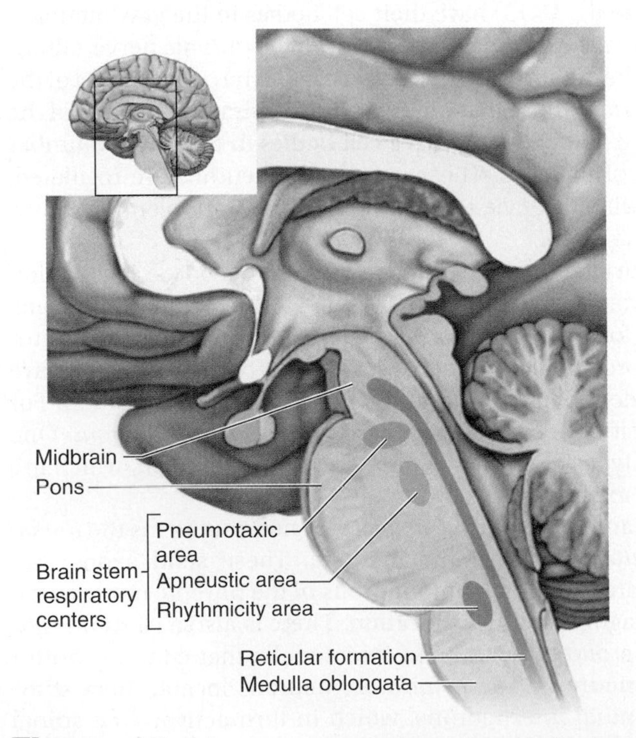

Figure 16.24 Approximate locations of the brain stem respiratory centers. The rhythmicity center in the medulla oblongata directly controls breathing, but it receives input from the control centers in the pons and from chemoreceptors. AP|R

Chemoreceptors

The automatic control of breathing is also influenced by input from *chemoreceptors,* which are collectively sensitive to changes in the pH of brain interstitial fluid and cerebrospinal fluid, and in the P_{CO_2}, pH, and P_{O_2} of the blood. There are two groups of chemoreceptors that respond to changes in P_{CO_2}, pH, and P_{O_2}. These are the **central chemoreceptors** in the medulla oblongata and the **peripheral chemoreceptors.** The peripheral chemoreceptors are contained within small nodules associated with the aorta and the carotid arteries, and they receive blood from these critical arteries via small arterial branches. The peripheral chemoreceptors include the **aortic bodies,** located around the aortic arch, and the **carotid bodies,** located in each common carotid artery at the point where it branches into the internal and external carotid arteries (fig. 16.25). The aortic

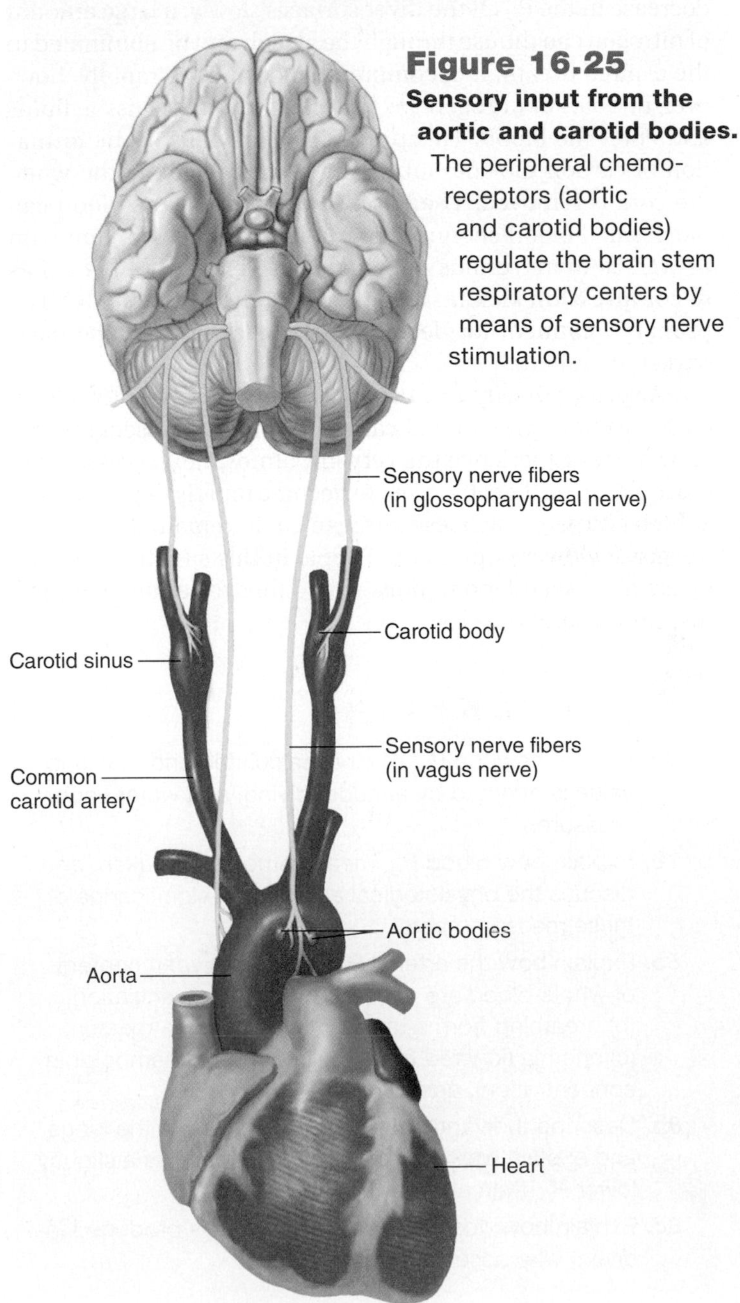

Figure 16.25 Sensory input from the aortic and carotid bodies. The peripheral chemoreceptors (aortic and carotid bodies) regulate the brain stem respiratory centers by means of sensory nerve stimulation.

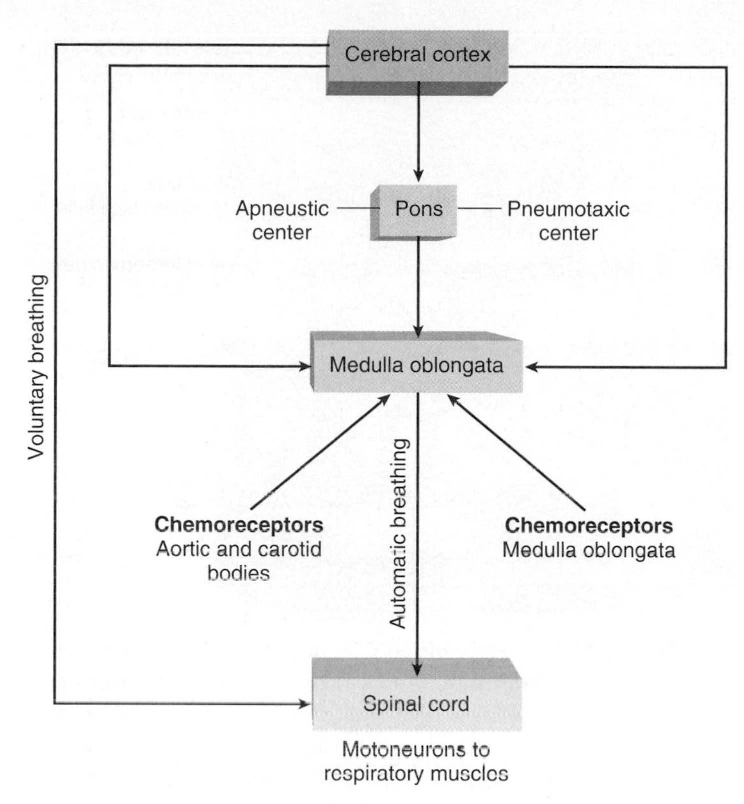

Figure 16.26 The regulation of ventilation by the central nervous system. The feedback effects of pulmonary stretch receptors and "irritant" receptors on the control of breathing are not shown in this flowchart.

and carotid bodies should not be confused with the aortic and carotid sinuses (chapter 14, section 14.6) that are located within these arteries. The aortic and carotid sinuses contain receptors that monitor the blood pressure.

The peripheral chemoreceptors control breathing indirectly via sensory nerve fibers to the medulla. The aortic bodies send sensory information to the medulla in the vagus nerve (X); the carotid bodies stimulate sensory fibers in the glossopharyngeal nerve (IX). The neural and sensory control of ventilation is summarized in figure 16.26.

CLINICAL APPLICATION

The automatic control of breathing is regulated by nerve fibers that descend in the lateral and ventral white matter of the spinal cord from the medulla oblongata. The voluntary control of breathing is a function of the cerebral cortex and involves axons that descend in the corticospinal tracts (chapter 8, section 8.5). The separation of the voluntary and involuntary pathways is dramatically illustrated in the condition called **Ondine's curse** (the term is taken from a German fairy tale). In this condition, neurological damage abolishes the automatic, but not the voluntary, control of breathing. People with Ondine's curse must remind themselves to breathe, and they cannot go to sleep without the aid of a mechanical respirator.

Effects of Blood P_{CO_2} and pH on Ventilation

Chemoreceptor input to the brain stem modifies the rate and depth of breathing so that, under normal conditions, arterial P_{CO_2}, pH, and P_{O_2} remain relatively constant. If hypoventilation (inadequate ventilation) occurs, P_{CO_2} quickly rises and pH falls. The fall in pH occurs because carbon dioxide can combine with water to form carbonic acid, which, as a weak acid, can release H^+ into the solution. This is shown in these equations:

$$CO_2 + H_2O \rightarrow H_2CO_3$$

$$H_2CO_3 \rightarrow H^+ + HCO_3^-$$

The oxygen content of the blood decreases much more slowly because of the large "reservoir" of oxygen attached to hemoglobin. During hyperventilation, conversely, blood P_{CO_2} quickly falls and pH rises because of the excessive elimination of carbonic acid. The oxygen content of blood, on the other hand, is not significantly increased by hyperventilation (because hemoglobin in arterial blood is 97% saturated with oxygen even during normal ventilation).

For these reasons, the blood P_{CO_2} and pH are more immediately affected by changes in ventilation than is the oxygen content. Indeed, changes in P_{CO_2} provide a sensitive index of ventilation, as shown in figure 16.27. Changes in plasma P_{CO_2}

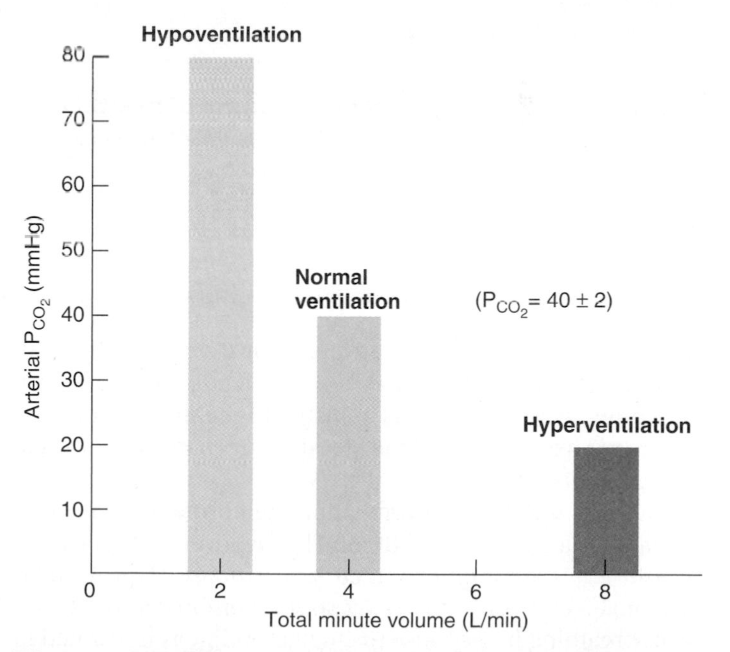

Figure 16.27 The relationship between total minute volume and arterial P_{CO_2}. These are inversely related: when the total minute volume increases by a factor of 2, the arterial P_{CO_2} decreases by half. The total minute volume measures breathing, and is equal to the amount of air in each breath (the tidal volume) multiplied by the number of breaths per minute. The P_{CO_2} measures the CO_2 concentration of arterial blood plasma.

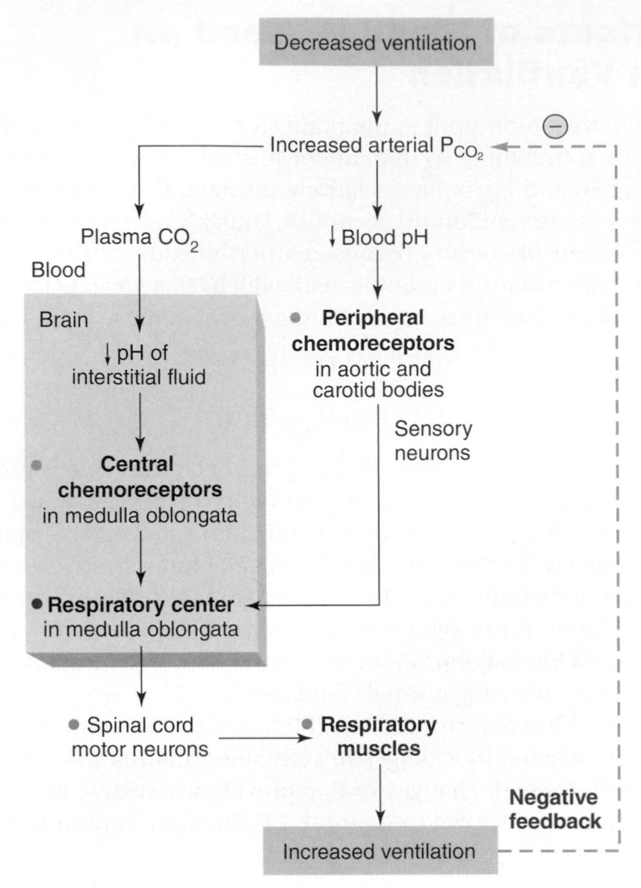

Figure 16.28 Chemoreceptor control of breathing. This figure depicts the negative feedback control of ventilation through changes in blood P_{CO_2} and pH. The blood-brain barrier, represented by the gold box, allows CO_2 to pass into the brain interstitial fluid but prevents the passage of H^+.

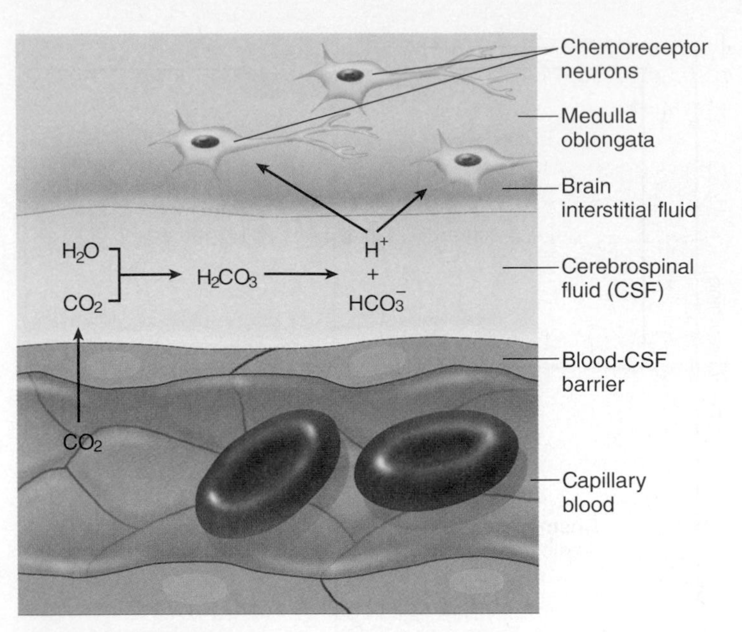

Figure 16.29 How blood CO_2 affects chemoreceptors in the medulla oblongata. An increase in blood CO_2 stimulates breathing indirectly by lowering the pH of blood and brain interstitial fluid. This figure illustrates how a rise in blood CO_2 increases the H^+ concentration (lowers the pH) of CSF and interstitial fluid and thereby stimulates chemoreceptor neurons in the medulla oblongata.

also serve as the most potent stimulus for the reflex control of ventilation, providing precise negative feedback regulation. Ventilation, in other words, is adjusted to maintain a constant P_{CO_2}; proper oxygenation of the blood occurs naturally as a side product of this reflex control.

The rate and depth of ventilation are normally adjusted to maintain an arterial P_{CO_2} of 40 mmHg. Hypoventilation causes a rise in P_{CO_2}—a condition called *hypercapnia.* Hyperventilation, conversely, results in *hypocapnia.* Chemoreceptor regulation of breathing in response to changes in P_{CO_2} is illustrated in figure 16.28.

Chemoreceptors in the Medulla

The chemoreceptors most sensitive to changes in the arterial P_{CO_2} are located on the ventrolateral surface of the medulla oblongata, near the exit of the ninth and tenth cranial nerves. These chemoreceptor neurons are anatomically separate from, but synaptically communicate with, the neurons of the rhythmicity center in the medulla.

An increase in arterial P_{CO_2} causes a rise in the H^+ concentration of the blood as a result of increased carbonic acid concentrations. The H^+ in the blood, however, cannot cross the blood-brain barrier, and thus cannot influence the medullary chemoreceptors. Carbon dioxide in the arterial blood *can* cross the blood-brain barrier and lower the pH of cerebrospinal fluid (CSF) and brain interstitial fluid (fig. 16.29). Also CO_2 produced by brain metabolism can contribute to the lower pH of the interstitial fluid. Lowered pH in the medulla oblongata stimulates the central chemoreceptors, either directly or via glial cells (which may release ATP as a transmitter in response to lowered pH).

The chemoreceptors in the medulla are ultimately responsible for 70% to 80% of the increased ventilation that occurs in response to a sustained rise in arterial P_{CO_2}. This response, however, takes several minutes. The immediate increase in ventilation that occurs when P_{CO_2} rises is produced by stimulation of the peripheral chemoreceptors.

Peripheral Chemoreceptors

The aortic and carotid bodies are not stimulated directly by blood CO_2. Instead, they are stimulated by a rise in the H^+ concentration (fall in pH) of arterial blood, which occurs when the blood CO_2 (and thus carbonic acid) is raised. In summary, the retention of CO_2 during hypoventilation rapidly stimulates the peripheral chemoreceptors through a lowering of blood pH. This is responsible for the immediate response to the elevated arterial CO_2. Then the central chemoreceptors respond to the fall in the pH of their surrounding interstitial fluid, stimulating a steady-state increase in ventilation if the blood CO_2 remains elevated.

CLINICAL APPLICATION

People who hyperventilate during psychological stress are sometimes told to breathe into a paper bag so that they rebreathe their expired air, enriched with CO_2. This procedure helps raise their blood P_{CO_2} back up to the normal range. This is needed because hypocapnia causes cerebral vasoconstriction, reducing brain perfusion and producing ischemia. The cerebral ischemia causes dizziness and can lead to an acidotic condition in the brain, which stimulates the medullary chemoreceptors to cause further hyperventilation. Breathing into a paper bag can thus relieve the hypocapnia and stop the hyperventilation.

Effects of Blood P_{O_2} on Ventilation

Under normal conditions, blood P_{O_2} affects breathing only indirectly by influencing the chemoreceptor sensitivity to changes in P_{CO_2}. Chemoreceptor sensitivity to P_{CO_2} is augmented by a low P_{O_2} (so ventilation is increased at a high altitude, for example) and is decreased by a high P_{O_2}. If the blood P_{O_2} is raised by breathing 100% oxygen, therefore, the breath can be held longer because the response to increased P_{CO_2} is blunted.

When the blood P_{CO_2} is held constant by experimental techniques, the P_{O_2} of arterial blood must fall from 100 mmHg to below 70 mmHg before ventilation is significantly stimulated (fig. 16.30). This stimulation—called a **hypoxic drive**—is apparently due to a direct effect of P_{O_2} on the carotid bodies. The carotid bodies respond to the oxygen dissolved in the plasma (as measured by the blood P_{O_2}), not to the oxygen bound by hemoglobin in the red blood cells. Because this degree of *hypoxemia*, or low blood oxygen, does not normally occur at sea level, P_{O_2} does not normally exert this direct effect on breathing.

However, there are interactions between the sensitivities of the carotid bodies to a fall in P_{O_2} and a rise in P_{CO_2}. Hypoxia (low oxygen) promotes the carotid bodies' responses to a rise in arterial P_{CO_2} and fall in pH. This effect becomes important in the regulation of breathing at higher altitudes with low P_{O_2} (section 16.9).

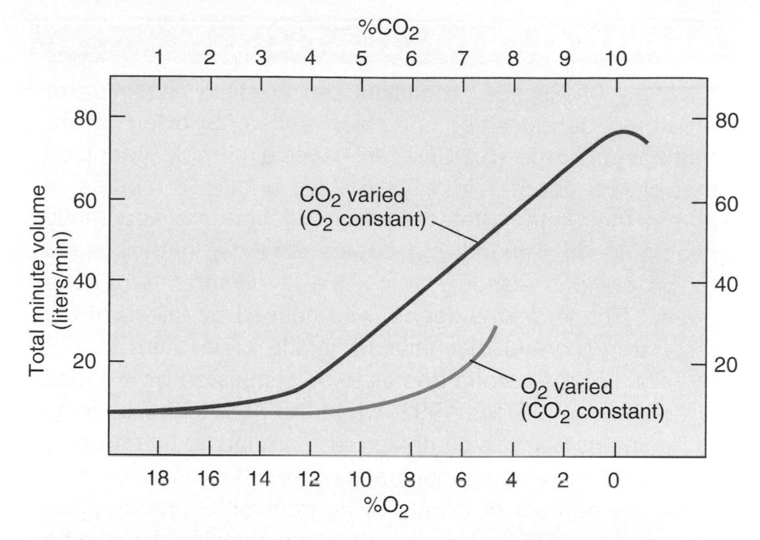

Figure 16.30 Comparing the effects of blood CO_2 and O_2 on breathing. The graph depicts the effects of increasing blood concentrations of CO_2 (see the scale at the top of the graph) on breathing, as measured by the total minute volume. The effects of decreasing concentrations of blood O_2 (see the scale at the bottom of the graph) on breathing are also shown for comparison. Notice that breathing increases linearly with increasing CO_2 concentration, whereas O_2 concentrations must decrease to half the normal value before breathing is stimulated.

Conversely, persistently elevated arterial P_{CO_2} enhances the sensitivity of the carotid bodies to a fall in P_{O_2}. Breathing in response to lowered P_{O_2} is important in emphysema, where the chemoreceptor response to chronically elevated carbon dioxide becomes blunted. This blunting results from increased secretion of bicarbonate by the choroid plexus (chapter 8), buffering the fall in pH. People with emphysema are consequently stimulated to breathe by a hypoxic drive (due to low P_{O_2}) rather than by the elevated P_{CO_2}. Over a long period of time, however, the hypoxic drive in people with emphysema can also become blunted.

The effects of changes in the blood P_{CO_2}, pH, and P_{O_2} on chemoreceptors and the regulation of ventilation are summarized in table 16.6.

Table 16.6 | Sensitivity of Chemoreceptors to Changes in Blood Gases and pH

Stimulus	Chemoreceptor	Comments
↑P_{CO_2}	Medullary chemoreceptors; aortic and carotid bodies	Medullary chemoreceptors are sensitive to the pH of cerebrospinal fluid (CSF). Diffusion of CO_2 from the blood into the brain lowers the pH of brain interstitial fluid by forming carbonic acid. Similarly, the aortic and carotid bodies are stimulated by a fall in blood pH induced by increases in blood CO_2.
↓pH	Aortic and carotid bodies	Peripheral chemoreceptors are stimulated by decreased blood pH independent of the effect of blood CO_2. Chemoreceptors in the medulla are not affected by changes in blood pH because H^+ cannot cross the blood-brain barrier.
↓P_{O_2}	Carotid bodies	Low blood P_{O_2} (hypoxemia) augments the chemoreceptor response to increases in blood P_{CO_2} and can stimulate ventilation directly when the P_{O_2} falls below 70 mmHg.

CLINICAL APPLICATION

A variety of disease processes can produce cessation of breathing during sleep, or *sleep apnea.* **Sudden infant death syndrome (SIDS)** is an especially tragic condition that claims about 1 in 1,000 babies under 12 months of age in the United States annually. Victims are apparently healthy 2- to 5-month-old babies who die in their sleep for no obvious reason—hence, the layperson's term "crib death." These deaths seem to be caused by failure of the respiratory control mechanisms in the brain stem and/or by failure of the carotid bodies to be stimulated by reduced arterial oxygen. Since 1992, when the American Academy of Pediatrics began a campaign recommending that parents put infants to sleep on their backs rather than on their stomachs, the number of infants dying from SIDS has dropped by more than 50%. However, SIDS still remains the leading cause of death in the United States of infants younger than 1 year.

Effects of Pulmonary Receptors on Ventilation

The lungs contain various types of receptors that influence the brain stem respiratory control centers via sensory fibers in the vagus nerves. **Unmyelinated C fibers** are sensory neurons in the lungs that can be stimulated by *capsaicin,* the chemical in hot peppers that creates the burning sensation. These receptors produce an initial apnea, followed by rapid, shallow breathing when a person eats these peppers or inhales pepper spray. **Irritant receptors** in the wall of the larynx, and receptors in the lungs identified as **rapidly adapting receptors,** can cause a person to cough in response to components of smoke and smog, and to inhaled particulates. The rapidly adapting receptors in the lungs are stimulated most directly by an increase in the amount of fluid in the pulmonary interstitial tissue. Because the same chemicals that stimulate the unmyelinated C fibers can cause increased pulmonary interstitial fluid (due to extravasation from pulmonary capillaries), a person may also cough after eating hot peppers.

The **Hering-Breuer reflex** is stimulated by **pulmonary stretch receptors.** The activation of these receptors during inspiration inhibits the respiratory control centers, making further inspiration increasingly difficult. This helps to prevent undue distension of the lungs and may contribute to the smoothness of the ventilation cycles. A similar inhibitory reflex may occur during expiration. The Hering-Breuer reflex appears to be important in maintaining normal ventilation in the newborn. Pulmonary stretch receptors in adults, however, are probably not active at normal resting tidal volumes (500 ml per breath) but may contribute to respiratory control at high tidal volumes, as during exercise.

CLINICAL APPLICATION

Obstructive sleep apnea is a relatively common disorder in adults in which there are at least 15 or more periods of apnea (each lasting at least 10 seconds) per hour of sleep due to temporary occlusion of the upper airways. The upper airway of the oropharynx is kept open during wakefulness by contractions of over 20 different skeletal muscles. During sleep, the pharyngeal air passage can narrow and produce obstructive sleep apnea because of obesity or other reasons. The obstruction produces periods of low arterial P_{O_2} and elevated P_{CO_2} that stimulate chemoreceptor reflexes. There are numerous other consequences, including loud snoring; partial arousal from sleep following the periods of apnea, causing sleepiness during the day; cerebral vasodilation, which produces a morning headache; metabolic alkalosis, to compensate for the respiratory acidosis produced by elevated P_{CO_2} (described later in this chapter); pulmonary hypertension; systemic hypertension; and right ventricular hypertrophy as a result of the pulmonary hypertension. People with obstructive sleep apnea often must wear *CPAP (continous positive airway pressure)* devices at night to keep the oropharynx air passage open.

CHECKPOINT

9a. Describe the roles of centers in the brain stem and cervical spinal cord in the regulation of breathing.

9b. Describe the effects of voluntary hyperventilation and breath holding on arterial P_{CO_2}, pH, and oxygen content. Indicate the relative degree of changes in these values.

10a. Using a flowchart to show a negative feedback loop, explain the relationship between ventilation and arterial P_{CO_2}.

10b. Explain the effect of increased arterial P_{CO_2} on
(a) chemoreceptors in the medulla oblongata and
(b) chemoreceptors in the aortic and carotid bodies.

10c. Explain the role of arterial P_{O_2} in the regulation of breathing. Why does ventilation increase when a person goes to a high altitude?

16.6 HEMOGLOBIN AND OXYGEN TRANSPORT

Deoxyhemoglobin loads with oxygen to form oxyhemoglobin in the pulmonary capillaries, and a portion of the oxyhemoglobin unloads its oxygen in the capillaries of the systemic circulation. The bond strength between hemoglobin and oxygen, and thus the extent of unloading, is changed under different conditions.

LEARNING OUTCOMES

After studying this section, you should be able to:

11. Describe the changes in percent oxyhemoglobin as a function of arterial P_{O_2} and explain how this relates to oxygen transport.

12. Describe the various conditions that influence the oxyhemoglobin dissociation curve and oxygen transport.

If the lungs are functioning properly, blood leaving in the pulmonary veins and traveling in the systemic arteries has a P_{O_2} of about 100 mmHg, indicating a plasma oxygen concentration of about 0.3 ml O_2 per 100 ml blood. The total oxygen content of the blood, however, cannot be derived if only the P_{O_2} of plasma is known. The total oxygen content depends not only on the P_{O_2} but also on the hemoglobin concentration. Arterial blood can carry 1.34 ml of oxygen per gram of hemoglobin. Therefore, if the P_{O_2} and hemoglobin concentration of the arterial blood are normal, this blood carries approximately 20 ml of O_2 per 100 ml of blood (fig. 16.31).

Hemoglobin

Most of the oxygen in the blood is contained within the red blood cells, where it is chemically bonded to **hemoglobin.** Each hemoglobin molecule consists of four polypeptide chains called *globins* and four iron-containing, disc-shaped organic pigment molecules called *hemes* (fig. 16.32).

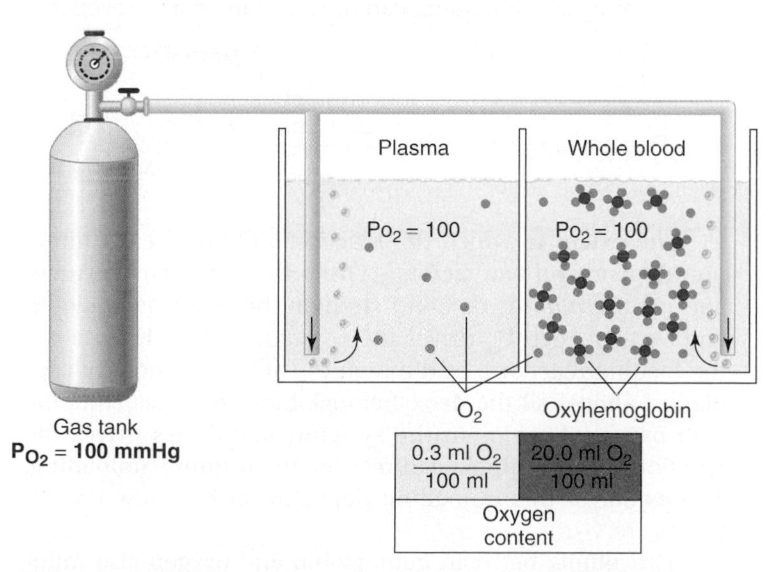

Figure 16.31 The oxygen content of blood. Plasma and whole blood that are brought into equilibrium with the same gas mixture have the same P_{O_2}, and thus the same number of dissolved oxygen molecules (shown as blue dots). The oxygen content of whole blood, however, is much higher than that of plasma because of the binding of oxygen to hemoglobin, which is normally only found in red blood cells (not shown).

The protein part of hemoglobin is composed of two identical *alpha chains,* each 141 amino acids long, and two identical *beta chains,* each 146 amino acids long. Each of the four polypeptide chains is combined with one heme group. In the center of each heme group is one atom of iron, which can combine with one molecule of oxygen. One hemoglobin molecule can thus combine with four molecules of oxygen—and since there are about 280 million hemoglobin molecules per red blood cell, each red blood cell can carry over a billion molecules of oxygen.

Normal heme contains iron in the reduced form (Fe^{2+}, or ferrous iron). In this form, the iron can share electrons and bond with oxygen to form **oxyhemoglobin.** When oxyhemoglobin dissociates to release oxygen to the tissues, the heme iron is still in the reduced (Fe^{2+}) form and the hemoglobin is called **deoxyhemoglobin,** or **reduced hemoglobin.** The term *oxyhemoglobin* is thus not equivalent to *oxidized* hemoglobin; hemoglobin does not lose an electron (and become oxidized)

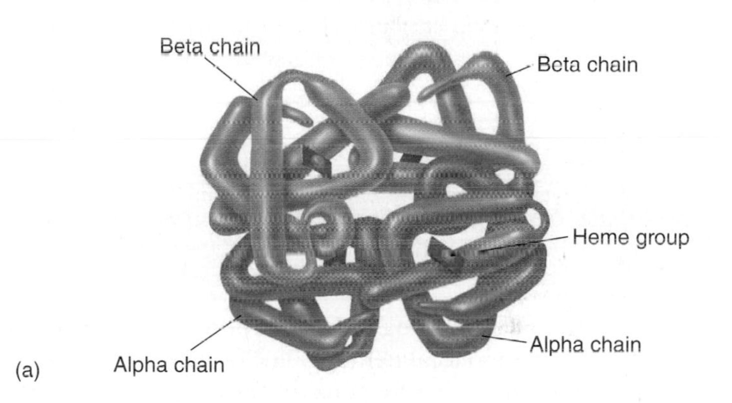

(b) Heme

Figure 16.32 The structure of hemoglobin. *(a)* An illustration of the three-dimensional structure of hemoglobin in which the two alpha and two beta polypeptide chains are shown. The four heme groups are represented as flat structures with atoms of iron (*spheres*) in the centers. *(b)* The structural formula for heme.

when it combines with oxygen. Oxidized hemoglobin, or **methemoglobin,** has iron in the oxidized (Fe^{3+}, or ferric) state. Methemoglobin thus lacks the electron it needs to form a bond with oxygen and cannot participate in oxygen transport. Blood normally contains only a small amount of methemoglobin, but certain drugs can increase this amount.

In **carboxyhemoglobin,** another abnormal form of hemoglobin, the reduced heme is combined with *carbon monoxide* instead of oxygen. Because the bond with carbon monoxide is about 210 times stronger than the bond with oxygen, carbon monoxide tends to displace oxygen in hemoglobin and remains attached to hemoglobin as the blood passes through systemic capillaries.

FITNESS APPLICATION

According to federal standards for clean air, the percentage of carboxyhemoglobin in active nonsmokers should be no higher than 1.5%. A carboxyhemoglobin concentration of greater than 3% in nonsmokers or greater than 10% in smokers indicates some degree of **carbon monoxide poisoning.** Milder cases may be caused by breathing smoggy air and smoking cigarettes; more severe cases may result from faulty furnaces with inadequate ventilation and exposure to automobile exhaust (often because of suicide attempts). Carbon monoxide poisoning can cause cardiac injury, neurological damage, inflammation, and other problems in addition to the negative effects of hypoxia. People may be treated with *normobaric oxygen* (at 1 atmosphere pressure) or with *hyperbaric oxygen therapy* (breathing 100% oxygen at more than 1.4 atmospheres pressure) within a hyperbaric chamber.

The **percent oxyhemoglobin saturation** (the percentage of oxyhemoglobin to total hemoglobin) is measured to assess how well the lungs have oxygenated the blood. The normal value for arterial blood is about 97%, with varying amounts of deoxyhemoglobin, carboxyhemoglobin, and methemoglobin composing the remainder. The oxyhemoglobin saturation and the proportion of these other forms can be measured because each hemoglobin type has a unique color; each absorbs visible light differently (has a different *absorption spectrum*). This gives oxyhemoglobin a tomato juice red color, whereas carboxyhemoglobin has a color similar to cranberry juice. The oxyhemoglobin saturation is commonly measured using a *pulse oximeter* (described on page 550), but it can be measured more precisely on a sample of arterial blood using a *blood-gas machine.*

Clinical Investigation CLUES

Harry drove a taxi and smoked cigarettes, and his blood tests showed an 18% carboxyhemoglobin saturation.

- How are these observations related?
- What is the significance of Harry's carboxyhemoglobin saturation level, particularly in light of his pulmonary function test results?

Hemoglobin Concentration

The *oxygen-carrying capacity* of whole blood is determined by its concentration of hemoglobin. If the hemoglobin concentration is below normal—in a condition called **anemia**—the oxygen content of the blood will be abnormally low. Conversely, when the hemoglobin concentration rises above the normal range—as occurs in **polycythemia** (high red blood cell count)—the oxygen-carrying capacity of blood is increased accordingly. This can occur as an adaptation to life at a high altitude.

The production of hemoglobin and red blood cells in bone marrow is controlled by the hormone *erythropoietin* (chapter 13, section 13.2), produced by the kidneys in response to tissue hypoxia. The secretion of erythropoietin—and thus the production of red blood cells—is stimulated when the amount of oxygen delivered to the kidneys is lower than normal. Red blood cell production is also promoted by androgens, which explains why the hemoglobin concentration is from 1 to 2 g per 100 ml higher in men than in women.

The Loading and Unloading Reactions

Deoxyhemoglobin and oxygen combine to form oxyhemoglobin; this is called the **loading reaction.** Oxyhemoglobin, in turn, dissociates to yield deoxyhemoglobin and free oxygen molecules; this is the **unloading reaction.** The loading reaction occurs in the lungs and the unloading reaction occurs in the systemic capillaries.

Loading and unloading can thus be shown as a reversible reaction:

$$\text{Deoxyhemoglobin} + O_2 \underset{\text{(tissues)}}{\overset{\text{(lungs)}}{\rightleftharpoons}} \text{Oxyhemoglobin}$$

The extent to which the reaction will go in each direction depends on two factors: (1) the P_{O_2} of the environment and (2) the *affinity,* or bond strength, between hemoglobin and oxygen. High P_{O_2} drives the equation to the right (favors the loading reaction); at the high P_{O_2} of the pulmonary capillaries, almost all the deoxyhemoglobin molecules combine with oxygen. Low P_{O_2} in the systemic capillaries drives the reaction in the opposite direction to promote unloading. The extent of this unloading depends on how low the P_{O_2} values are.

The affinity between hemoglobin and oxygen also influences the loading and unloading reactions. A very strong bond would favor loading but inhibit unloading; a weak bond would hinder loading but improve unloading. The bond strength between hemoglobin and oxygen is normally strong enough so that 97% of the hemoglobin leaving the lungs is in the form of oxyhemoglobin, yet the bond is sufficiently weak so that adequate amounts of oxygen are unloaded to sustain aerobic respiration in the tissues.

The Oxyhemoglobin Dissociation Curve

Blood in the systemic arteries, at a P_{O_2} of 100 mmHg, has a *percent oxyhemoglobin saturation* of 97% (which means that 97% of the hemoglobin is in the form of oxyhemoglobin). This blood is delivered to the systemic capillaries, where oxygen diffuses into the cells and is consumed in aerobic respiration. Blood leaving in the systemic veins is thus reduced in oxygen; it has a P_{O_2} of about 40 mmHg and a percent oxyhemoglobin saturation of about 75% when a person is at rest (table 16.7). Expressed another way, blood entering the tissues contains 20 ml O_2 per 100 ml blood, and blood leaving the tissues contains 15.5 ml O_2 per 100 ml blood (fig. 16.33). Thus, 22%, or 4.5 ml of O_2 out of the 20 ml of O_2 per 100 ml blood, is unloaded to the tissues.

A graphic illustration of the percent oxyhemoglobin saturation at different values of P_{O_2} is called an **oxyhemoglobin dissociation curve** (fig. 16.33). The values in this graph are obtained by subjecting samples of blood *in vitro* to different partial oxygen pressures. These percent oxyhemoglobin saturations can then be used to predict what the unloading percentages would be *in vivo* with a given difference in arterial and venous P_{O_2} values.

Figure 16.33 shows the difference between the arterial and venous P_{O_2} and the percent oxyhemoglobin saturation at rest. The relatively large amount of oxyhemoglobin remaining in the venous blood at rest serves as an oxygen reserve. If a person stops breathing, a sufficient reserve of oxygen in the blood will keep the brain and heart alive for about 4 to 5 minutes without using cardiopulmonary resuscitation (CPR) techniques. This reserve supply of oxygen can also be tapped when a tissue's requirements for oxygen are raised, as in exercising muscles.

The oxyhemoglobin dissociation curve is S-shaped, or *sigmoidal.* The fact that it is relatively flat at high P_{O_2} values indicates that changes in P_{O_2} within this range have little effect on the loading reaction. One would have to ascend as high as 10,000 feet, for example, before the oxyhemoglobin saturation of arterial blood would decrease from 97% to 93%. At more common elevations, the percent oxyhemoglobin saturation would not be significantly different from the 97% value at sea level.

At the steep part of the sigmoidal curve, however, small changes in P_{O_2} values produce large differences in percent saturation. A decrease in *venous* P_{O_2} from 40 mmHg to 30 mmHg, as might occur during mild exercise, corresponds to a change in percent saturation from 75% to 58%. Since the *arterial* percent saturation is usually still 97% during exercise, the lowered venous percent saturation indicates that more oxygen has been unloaded to the tissues. The difference between the arterial and venous percent saturations indicates the percent unloading. In the preceding example, 97% − 75% = 22% unloading at rest, and 97% − 58% = 39% unloading during mild exercise. During heavier exercise the venous P_{O_2} can drop to 20 mmHg or lower, indicating a percent unloading of about 80%.

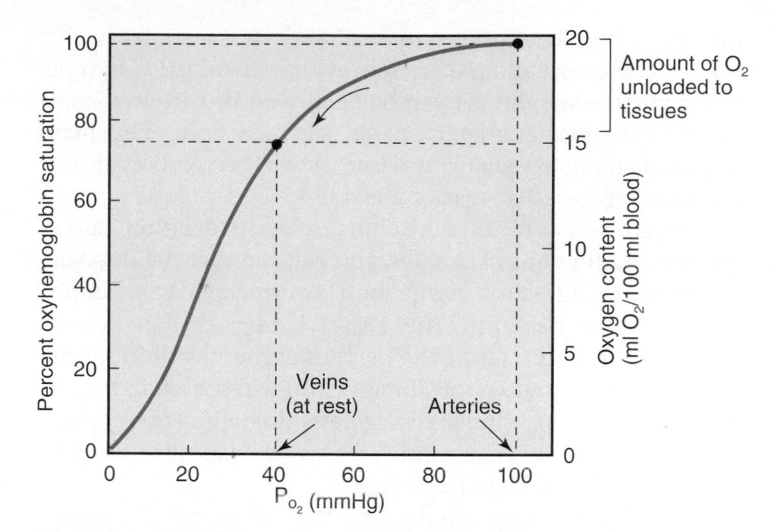

Figure 16.33 The oxyhemoglobin dissociation curve. The percentage of oxyhemoglobin saturation and the blood oxygen content are shown at different values of P_{O_2}. Notice that the percent oxyhemoglobin decreases by about 25% as the blood passes through the tissue from arteries to veins, resulting in the unloading of approximately 5 ml of O_2 per 100 ml of blood to the tissues.

Effect of pH and Temperature on Oxygen Transport

In addition to changes in P_{O_2}, the loading and unloading reactions are influenced by changes in the *affinity* (bond strength) of hemoglobin for oxygen. Such changes ensure that active skeletal muscles will receive more oxygen from the blood than they do at rest. This occurs as a result of the lowered pH and increased temperature in exercising muscles.

The affinity is decreased when the pH is lowered and increased when the pH is raised; this is called the **Bohr effect.** When the affinity of hemoglobin for oxygen is reduced, there is slightly less loading of the blood with oxygen in the lungs but

Table 16.7 | Relationship Between Percent Oxyhemoglobin Saturation and P_{O_2} (at pH of 7.40 and Temperature of 37° C)

P_{O_2} (mmHg)	100	80	61	45	40	36	30	26	23	21	19
Percent Oxyhemoglobin	97	95	90	80	75	70	60	50	40	35	30
	Arterial Blood				*Venous Blood*						

greater unloading of oxygen in the tissues. The net effect is that the tissues receive more oxygen when the blood pH is lowered (table 16.8). Since the pH can be decreased by carbon dioxide (through the formation of carbonic acid), the Bohr effect helps to provide more oxygen to the tissues when their carbon dioxide output is increased by a faster metabolism.

When you look at oxyhemoglobin dissociation curves graphed at different pH values, you can see that the dissociation curve is shifted to the right by a lowering of pH and shifted to the left by a rise in pH (fig. 16.34). If you calculate percent unloading (by subtracting the percent oxyhemoglobin saturation for arterial and venous blood), you will see that a *shift to the right* of the curve indicates a greater unloading of oxygen. A *shift to the left,* conversely, indicates less unloading but slightly more oxygen loading in the lungs.

When oxyhemoglobin dissociation curves are constructed at different temperatures, the curve moves rightward as the temperature increases. The rightward shift of the curve indicates that the affinity of hemoglobin for oxygen is decreased by a rise in temperature. An increase in temperature weakens the bond between hemoglobin and oxygen and thus has the same effect as a fall in pH. At higher temperatures, therefore, more oxygen is unloaded to the tissues than would be the case if the bond strength were constant. This effect can significantly enhance the delivery of oxygen to muscles that are warmed during exercise.

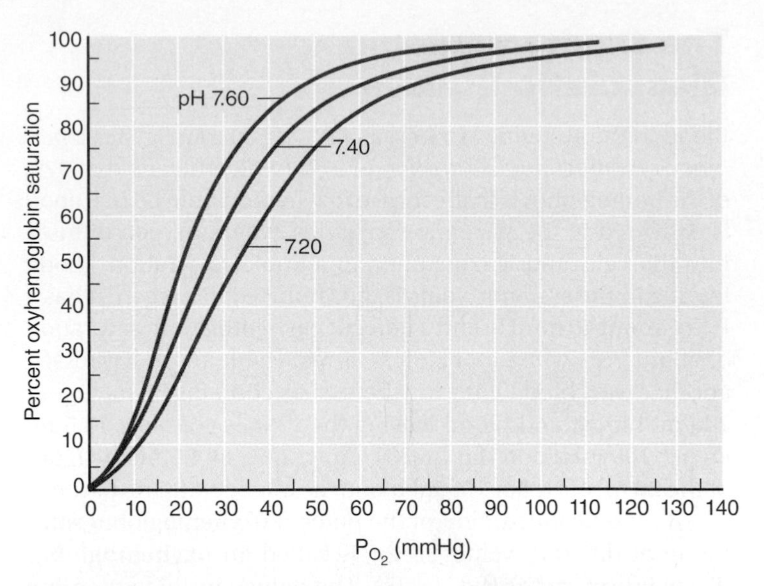

Figure 16.34 The effect of pH on the oxyhemoglobin dissociation curve. A decrease in blood pH (an increase in H^+ concentration) decreases the affinity of hemoglobin for oxygen at each P_{O_2} value, resulting in a "shift to the right" of the oxyhemoglobin dissociation curve. This is called the Bohr effect. A curve that is shifted to the right has a lower percent oxyhemoglobin saturation at each P_{O_2}.

See the *Test Your Quantitative Ability* section of the Review Activities at the end of this chapter.

Effect of 2,3-DPG on Oxygen Transport

Mature red blood cells lack both nuclei and mitochondria. Without mitochondria they cannot respire aerobically; the very cells that carry oxygen are the only cells in the body that cannot use it! Red blood cells must thus obtain energy through the anaerobic metabolism of glucose. At a certain point in the glycolytic pathway, a "side reaction" occurs in the red blood cells that results in a unique product—**2,3-diphosphoglyceric acid (2,3-DPG).**

The enzyme that produces 2,3-DPG is inhibited by oxyhemoglobin. When the oxyhemoglobin concentration is decreased, therefore, the production of 2,3-DPG is increased. This increase in 2,3-DPG production can occur when the total hemoglobin concentration is low (in anemia) or when the P_{O_2} is low (at a high altitude; fig. 16.35). The 2,3-DPG binds to deoxyhemoglobin and makes it more stable, thereby favoring the conversion of oxyhemoglobin to deoxyhemoglobin. Because of this, a greater proportion of oxyhemoglobin will unload its oxygen and be converted to deoxyhemoglobin at each P_{O_2} value. An increased concentration of 2,3-DPG in red blood cells thus increases oxygen unloading (table 16.9) and shifts the oxyhemoglobin dissociation curve to the right.

Anemia

When the total blood hemoglobin concentration falls below normal in anemia, each red blood cell produces increased amounts of 2,3-DPG. A normal hemoglobin concentration of 15 g per 100 ml unloads about 4.5 ml O_2 per 100 ml at rest, as previously described. If the hemoglobin concentration were reduced by half, you might expect that the tissues would receive

Table 16.8 | Effect of pH on Hemoglobin Affinity for Oxygen and Unloading of Oxygen to the Tissues

pH	Affinity	Arterial O_2 Content per 100 ml	Venous O_2 Content per 100 ml	O_2 Unloaded to Tissues per 100 ml
7.40	Normal	19.8 ml O_2	14.8 ml O_2	5.0 ml O_2
7.60	Increased	20.0 ml O_2	17.0 ml O_2	3.0 ml O_2
7.20	Decreased	19.2 ml O_2	12.6 ml O_2	6.6 ml O_2

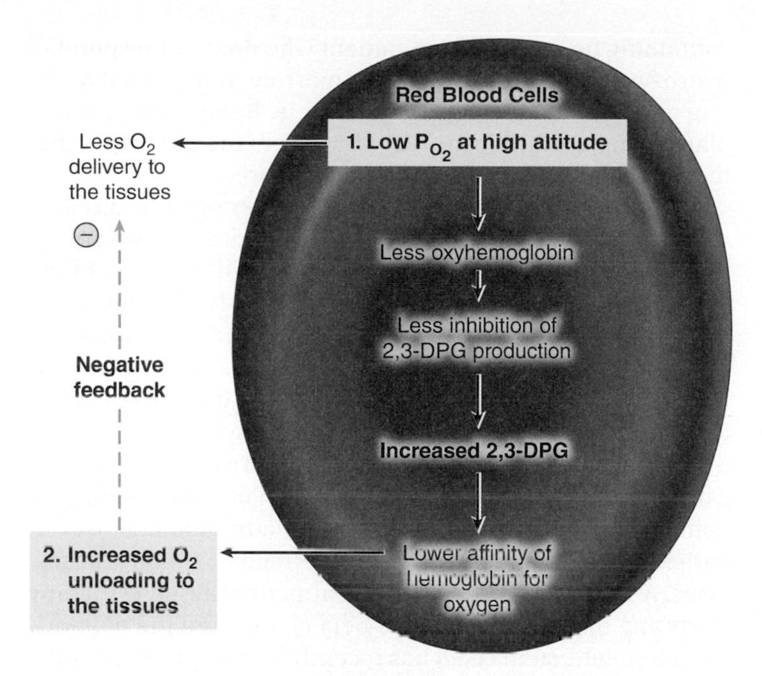

Figure 16.35 2,3-DPG promotes the unloading of oxygen to the tissues. Because production of 2,3-DPG is inhibited by oxyhemoglobin, a reduction in the red blood cell content of oxyhemoglobin (as occurs at the low P_{O_2} of high altitude) increases 2,3-DPG production. This lowers the affinity of hemoglobin for oxygen (decreases the bond strength), so that more oxygen can be unloaded. The dashed arrow and negative sign indicate the completion of a negative feedback loop.

only half the normal amount of oxygen (2.25 ml O_2 per 100 ml). However, an amount as great as 3.3 ml O_2 per 100 ml is actually unloaded to the tissues under these conditions. This occurs as a result of a rise in 2,3-DPG production that causes a decrease in the affinity of hemoglobin for oxygen.

Fetal Hemoglobin

The effects of 2,3-DPG are also important in the transfer of oxygen from maternal to fetal blood. In an adult, hemoglobin molecules are composed of two alpha and two beta chains as previously described, whereas fetal hemoglobin has two *gamma* chains in place of the beta chains (gamma chains differ from beta chains in 37 of their amino acids). Normal **adult hemoglobin** in the mother **(hemoglobin A)** is able to bind to 2,3-DPG. **Fetal hemoglobin,** or **hemoglobin F,** by contrast, cannot bind to 2,3-DPG and thus has a higher affinity for oxygen than does hemoglobin A. Because hemoglobin F can have a higher percent oxyhemoglobin than hemoglobin A at a given P_{O_2}, oxygen is transferred from the maternal to the fetal blood as these two come into close proximity in the placenta (chapter 20; see fig. 20.46).

CLINICAL APPLICATION

Hemoglobin F is the major form of hemoglobin from about 11 weeks after conception until about week 38, when hemoglobin A predominates. This involves a switch from the gene coding for the gamma chains to the gene coding for the beta chains (both genes are on chromosome 11). By late infancy, the production of gamma chains and fetal hemoglobin comprise less than 1% of the total hemoglobin. Such a turnover is possible because old red blood cells are constantly being destroyed and replaced. Heme derived from destroyed red blood cells is converted into bile pigment, or *bilirubin* (chapter 18; see fig. 18.22), which is eliminated in the fetus by transfer through the placenta. After birth, the fetal liver must eliminate the bilirubin by excreting it in the bile. Because the fetal liver is often not sufficiently developed for this task, the plasma levels of bilirubin can rise, causing a condition known as **physiological neonatal jaundice.** Putting neonates with this condition under blue light converts bilirubin to a more water-soluble form, so that their kidneys can eliminate it in the urine.

Inherited Defects in Hemoglobin Structure and Function

A number of hemoglobin diseases are produced by congenital (inherited) defects in the protein part of hemoglobin. **Sickle-cell anemia,** a disease that occurs almost exclusively in people of African heritage, is carried in a recessive state by 8% to 11% of the African American population. This disease occurs when a person inherits the affected gene from each parent and produces hemoglobin S instead of normal hemoglobin A. Hemoglobin S differs from hemoglobin A in that one amino acid is substituted for another (a valine for a glutamic acid) in the beta

Table 16.9 | Factors That Affect the Affinity of Hemoglobin for Oxygen and the Position of the Oxyhemoglobin Dissociation Curve

Factor	Affinity	Position of Curve	Comments
↓pH	Decreased	Shift to the right	Called the Bohr effect; increases oxygen delivery during hypercapnia
↑Temperature	Decreased	Shift to the right	Increases oxygen unloading during exercise and fever
↑2,3-DPG	Decreased	Shift to the right	Increases oxygen unloading when there is a decrease in total hemoglobin or total oxygen content; an adaptation to anemia and high-altitude living

chains of hemoglobin, due to a single base change in the DNA of the gene for the beta chains.

Under conditions of low P_{O_2}, when the hemoglobin is deoxygenated, hemoglobin S polymerizes into long fibers. This causes the red blood cells to have their characteristic sickle shape (fig. 16.36). It also reduces their flexibility, which hinders their ability to pass through narrow vessels and thereby reduces blood flow through organs. The long fibers of hemoglobin S also damage the plasma membrane of red blood cells and promote hemolysis, which leads to a variety of complications. Further, the damaged red blood cells can injure the vascular endothelium and cause additional symptoms of sickle-cell disease. Sickle-cell anemia is treated with the drug *hydroxyurea,* which stimulates the production of hemoglobin gamma chains instead of beta chains. As a result, the production of red blood cells containing fetal hemoglobin (hemoglobin F) is favored, with fewer red blood cells containing hemoglobin S. A patient who does not respond to hydroxyurea may receive a bone marrow transplant if a sibling or other suitable donor is available. Bone marrow transplantation has significant risks, but offers the possibility that the person's sickle-cell disease can be cured.

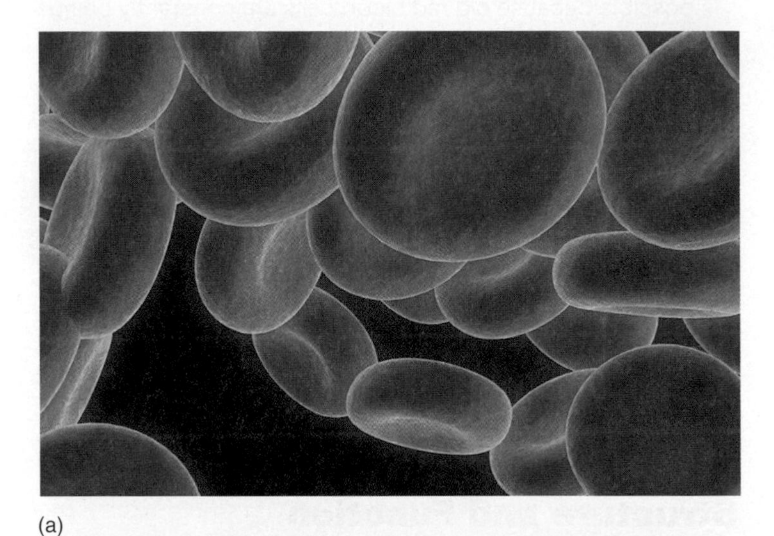

(a)

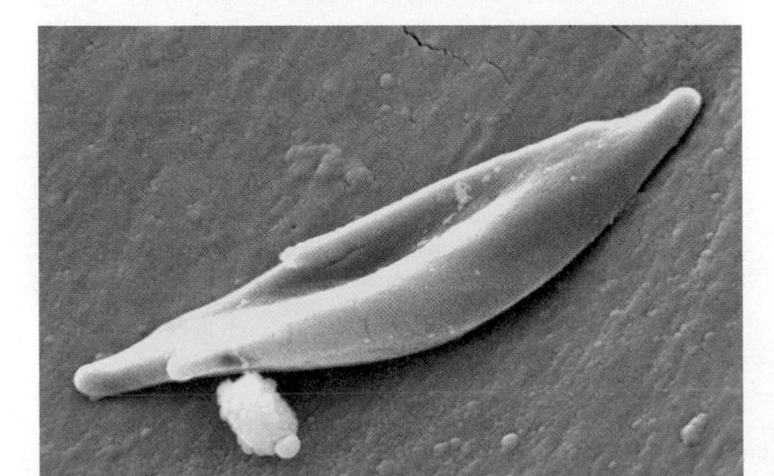

(b)

Figure 16.36 Sickle-cell anemia. (*a*) Normal red blood cells. (*b*) Sickled red blood cells as seen in the scanning electron microscope.

Thalassemia is any of a family of hemoglobin diseases found predominantly among people of Mediterranean ancestry. In *alpha thalassemia,* there is decreased synthesis of the alpha chains of hemoglobin, whereas in *beta thalassemia* the synthesis of the beta chains is impaired. Beta thalassemia can be caused by over 200 different point mutations in DNA, as well as by rare DNA deletions. This diversity of mutations produces a wide range of clinical symptoms. One of the compensations for thalassemia is increased synthesis of gamma chains, resulting in the retention of large amounts of hemoglobin F (fetal hemoglobin) into adulthood. However, patients with β-thalassemia require regular blood transfusions. An apparently successful treatment using gene therapy (inserting the gene for normal beta chains into the patient's hematopoietic stem cells) has recently been reported, but the long-term safety and efficacy of this procedure is presently undetermined.

Some types of abnormal hemoglobins have been shown to be advantageous in the environments in which they evolved. For example, a person who is a carrier for sickle-cell anemia (and who therefore has both hemoglobin A and hemoglobin S) has a high resistance to malaria. This is because the parasite that causes malaria cannot live in red blood cells that contain hemoglobin S.

Muscle Myoglobin

Myoglobin is a red pigment found exclusively in striated muscle cells (chapter 12, section 12.4). In particular, slow-twitch, aerobically respiring skeletal fibers and cardiac muscle cells are rich in myoglobin. Myoglobin is similar to hemoglobin, but it has one heme rather than four; therefore, it can combine with only one molecule of oxygen.

Myoglobin has a higher affinity for oxygen than does hemoglobin, and its dissociation curve is therefore to the left of the oxyhemoglobin dissociation curve (fig. 16.37). The shape of the myoglobin curve is also different from the oxyhemoglobin dissociation curve. The myoglobin curve is rectangular, indicating that oxygen will be released only when the P_{O_2} becomes very low.

Since the P_{O_2} in mitochondria is very low (because oxygen is incorporated into water here), myoglobin may act as a "go-between" in the transfer of oxygen from blood to the mitochondria within muscle cells. Myoglobin may also have an oxygen-storage function, which is of particular importance in the heart. During diastole, when the coronary blood flow is greatest, myoglobin can load up with oxygen. This stored oxygen can then be released during systole, when the coronary arteries are squeezed closed by the contracting myocardium.

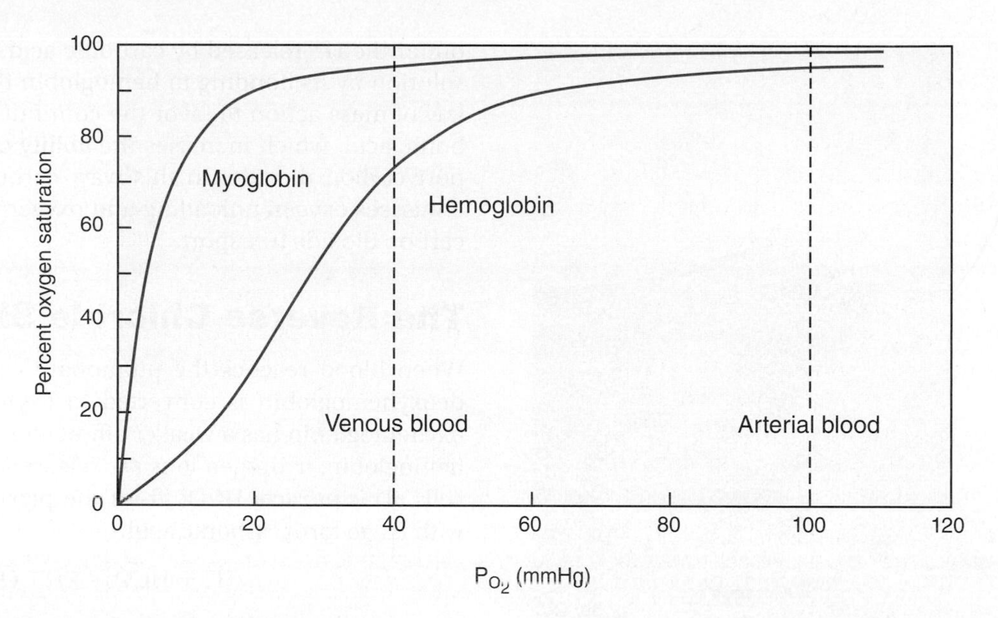

Figure 16.37 A comparison of the dissociation curves for hemoglobin and myoglobin. Myoglobin is an oxygen binding pigment in skeletal muscles. At the P_{O_2} of venous blood, the myoglobin retains almost all of its oxygen, indicating that it has a higher affinity than hemoglobin for oxygen. The myoglobin, however, does release its oxygen at the very low P_{O_2} values found inside the mitochondria.

CHECKPOINT

11a. Use a graph to illustrate the effects of P_{O_2} on the loading and unloading reactions.

11b. Draw an oxyhemoglobin dissociation curve and label the P_{O_2} values for arterial and venous blood under resting conditions. Use this graph to show the changes in unloading that occur during exercise.

12a. Explain how changes in pH and temperature affect oxygen transport, and state when such changes occur.

12b. Explain how a person who is anemic or a person at high altitude could have an increase in the percent unloading of oxygen by hemoglobin.

16.7 CARBON DIOXIDE TRANSPORT

Carbon dioxide is transported in the blood primarily in the form of bicarbonate (HCO_3^-), which is released when carbonic acid dissociates. Carbonic acid is produced mostly in the red blood cells as blood passes through systemic capillaries.

LEARNING OUTCOMES

After studying this section, you should be able to:

13. Explain how carbon dioxide is transported by the blood.

14. Explain the relationship between blood levels of carbon dioxide and the blood pH.

Carbon dioxide is carried by the blood in three forms: (1) as *dissolved* CO_2 in the plasma—carbon dioxide is about 21 times more soluble than oxygen in water, and about one-tenth of the total blood CO_2 is dissolved in plasma; (2) as *carbamino-hemoglobin*—about one-fifth of the total blood CO_2 is carried attached to an amino acid in hemoglobin (carbaminohemoglobin should not be confused with carboxyhemoglobin, formed when carbon monoxide binds to the heme groups of hemoglobin); and (3) as *bicarbonate ion,* which accounts for most of the CO_2 carried by the blood (fig. 16.38).

Carbon dioxide is able to combine with water to form carbonic acid. This reaction occurs spontaneously in the plasma at a slow rate, but it occurs much more rapidly within the red blood cells because of the catalytic action of the enzyme **carbonic anhydrase.** Since this enzyme is confined to the red blood cells, most of the carbonic acid is produced there rather than in the plasma. The formation of carbonic acid from CO_2 and water is favored by the high P_{CO_2} found in the capillaries of the systemic circulation (this is an example of the *law of mass action;* chapter 4, section 4.2).

$$CO_2 + H_2O \xrightarrow[\text{high } P_{CO_2}]{\text{carbonic anhydrase}} H_2CO_3$$

The Chloride Shift

As a result of catalysis by carbonic anhydrase within the red blood cells, large amounts of carbonic acid are produced as blood passes through the systemic capillaries. The buildup of carbonic acid concentrations within the red blood cells favors the dissociation of these molecules into hydrogen

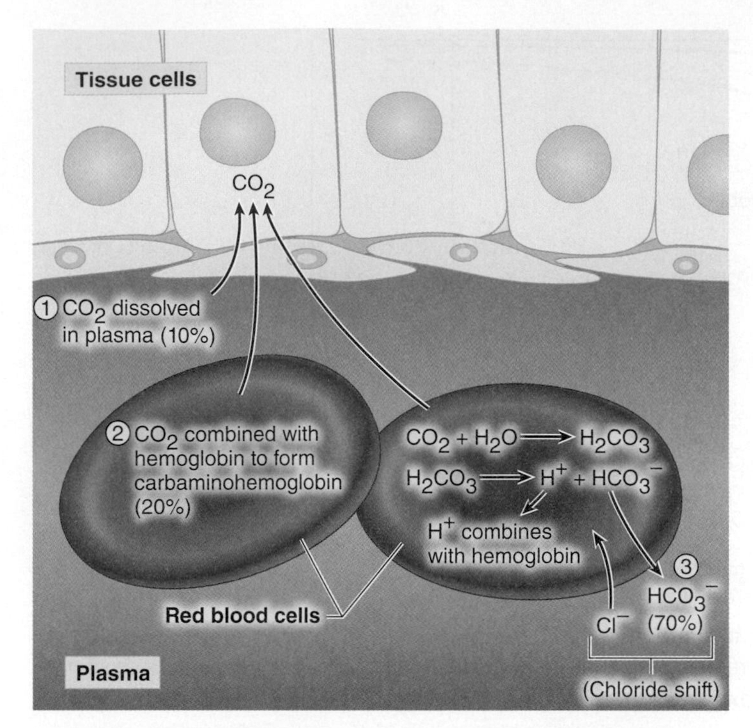

Figure 16.38 Carbon dioxide transport and the chloride shift. Carbon dioxide is transported in three forms: (1) as dissolved CO_2 gas, (2) attached to hemoglobin as carbaminohemoglobin, and (3) as carbonic acid and bicarbonate. Percentages indicate the proportion of CO_2 in each of the forms. Notice that when bicarbonate (HCO_3^-) diffuses out of the red blood cells, Cl^- diffuses in to retain electrical neutrality. This exchange is the chloride shift.

ions (protons, which contribute to the acidity of a solution) and HCO_3^- (bicarbonate), as shown by this equation:

$$H_2CO_3 \rightarrow H^+ + HCO_3^-$$

The hydrogen ions (H^+) released by the dissociation of carbonic acid are largely buffered by their combination with deoxyhemoglobin within the red blood cells. Although the unbuffered hydrogen ions are free to diffuse out of the red blood cells, more bicarbonate diffuses outward into the plasma than does H^+. As a result of the "trapping" of hydrogen ions within the red blood cells by their attachment to hemoglobin and the outward diffusion of bicarbonate, the inside of the red blood cell gains a net positive charge. This attracts chloride ions (Cl^-), which move into the red blood cells as HCO_3^- moves out. This exchange of anions as blood travels through the tissue capillaries is called the **chloride shift** (fig. 16.38).

The unloading of oxygen is increased by the bonding of H^+ (released from carbonic acid) to oxyhemoglobin. This is the *Bohr effect,* and results in increased conversion of oxyhemoglobin to deoxyhemoglobin. Now, deoxyhemoglobin bonds H^+ more strongly than does oxyhemoglobin, so the act of unloading its oxygen improves the ability of hemoglobin to buffer the H^+ released by carbonic acid. Removal of H^+ from solution by its bonding to hemoglobin then acts through the law of mass action to favor the continued production of carbonic acid, which increases the ability of the blood to transport carbon dioxide. In this way, carbon dioxide transport enhances oxygen unloading and oxygen unloading improves carbon dioxide transport.

The Reverse Chloride Shift

When blood reaches the pulmonary capillaries (fig. 16.39), deoxyhemoglobin is converted to oxyhemoglobin. Because oxyhemoglobin has a weaker affinity for H^+ than does deoxyhemoglobin, hydrogen ions are released within the red blood cells. This attracts HCO_3^- from the plasma, which combines with H^+ to form carbonic acid:

$$H^+ + HCO_3^- \rightarrow H_2CO_3$$

Under conditions of lower P_{CO_2}, as occurs in the pulmonary capillaries, carbonic anhydrase catalyzes the conversion of carbonic acid to carbon dioxide and water:

$$H_2CO_3 \xrightarrow[\text{Low } P_{CO_2}]{\text{carbonic anhydrase}} CO_2 + H_2O$$

In review, as blood goes through the systemic capillaries the carbonic anhydrase within the red blood cells converts carbon dioxide into carbonic acid. Dissociation of the carbonic acid into bicarbonate and H^+ results in the diffusion of bicarbonate out of the red blood cells into the plasma in exchange for chloride. This part of the carbon dioxide transport story is described as the chloride shift.

A **reverse chloride shift** operates in the pulmonary capillaries to convert carbonic acid to H_2O and CO_2 gas, which is eliminated in the expired breath (fig. 16.39). The P_{CO_2}, carbonic acid, H^+, and bicarbonate concentrations in the systemic arteries are thus maintained relatively constant by normal ventilation. This is required to maintain the acid-base balance of the blood (fig. 16.40), as discussed in section 16.8.

13a. List the ways in which carbon dioxide is carried by the blood, and indicate the percentage of the total carried for each.

13b. Where in the body does the chloride shift occur? Describe the steps involved in the chloride shift.

13c. Where in the body does a reverse chloride shift occur? Describe this process.

14. Using equations, show how carbonic acid and bicarbonate are formed. Explain how carbon dioxide transport influences blood pH.

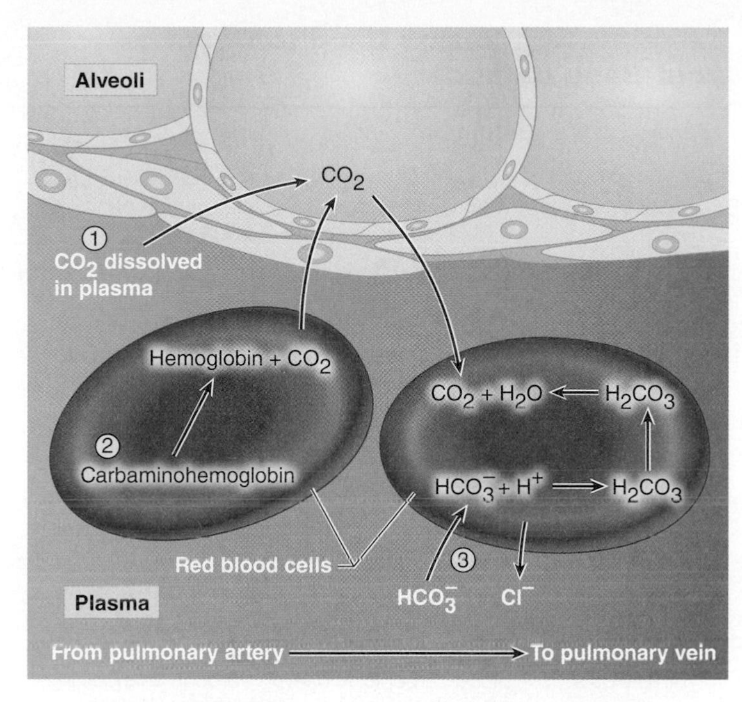

Figure 16.39 The reverse chloride shift in the lungs. Carbon dioxide is released from the blood as it travels through the pulmonary capillaries. A "reverse chloride shift" occurs during this time, and carbonic acid is transformed into CO_2 and H_2O. The CO_2 is eliminated in the exhaled air. Sources of carbon dioxide in blood include (1) dissolved CO_2, (2) carbaminohemoglobin, and (3) bicarbonate (HCO_3^-).

16.8 ACID-BASE BALANCE OF THE BLOOD

The pH of blood plasma is maintained within a narrow range of values through the functions of the lungs and kidneys. The lungs regulate the carbon dioxide concentration of the blood, and the kidneys regulate the bicarbonate concentration.

LEARNING OUTCOMES

After studying this section, you should be able to:

15. Describe the acid-base balance of the blood, and how it is influenced by the respiratory system.

Principles of Acid-Base Balance

The blood plasma within arteries normally has a pH between 7.35 and 7.45, with an average of 7.40. Using the definition of pH described in chapter 2, this means that arterial blood has a H^+ concentration of about $10^{-7.4}$ molar. Some of these hydrogen ions are derived from the ionization of carbonic acid formed from carbon dioxide and water as indicated in these equations:

$$CO_2 + H_2O \rightleftarrows H_2CO_3$$

$$H_2CO_3 \rightleftarrows H^+ + HCO_3^-$$

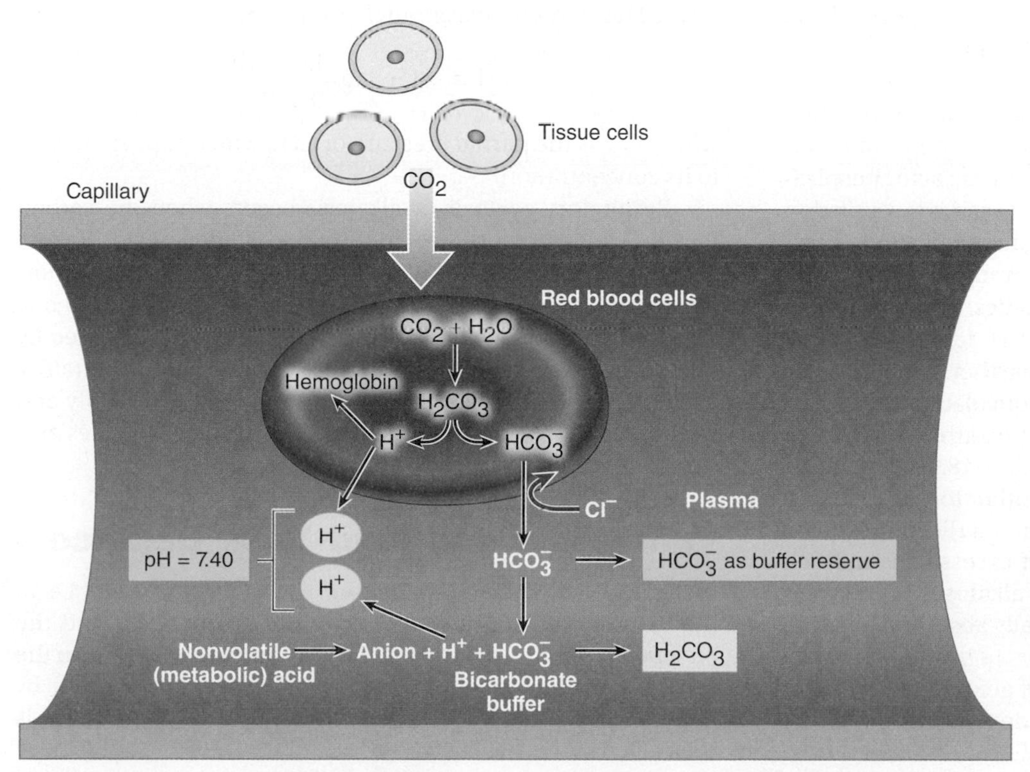

Figure 16.40 The effect of bicarbonate on blood pH. Bicarbonate released into the plasma from red blood cells buffers the H^+ produced by the ionization of metabolic acids (lactic acid, fatty acids, ketone bodies, and others). Binding of H^+ to hemoglobin also promotes the unloading of O_2.

As previously described, carbon dioxide produced by tissue cells through aerobic respiration is transported mostly as bicarbonate in the blood plasma (see fig. 16.39). During the reverse chloride shift that occurs in pulmonary capillaries, bicarbonate is converted into carbonic acid and then changed into carbon dioxide. Because CO_2 is a volatile gas released in the expired breath, carbonic acid is referred to as a *volatile acid.* This is significant because its blood concentration is uniquely regulated by breathing. All other acids in the blood—including lactic acid, fatty acids, ketone bodies, and so on—are *nonvolatile acids* that cannot be eliminated through ventilation.

Under normal conditions, the H^+ released by nonvolatile metabolic acids does not affect the blood pH because these hydrogen ions are bound to molecules that function as *buffers.* The major buffer in the plasma is the *bicarbonate* (HCO_3^-) ion, and it buffers H^+ as shown in figure 16.40 and described in this equation:

$$HCO_3^- + H^+ \rightarrow H_2CO_3$$

This buffering reaction could not go on forever because the free HCO_3^- would eventually disappear. If this were to occur, the H^+ concentration would increase and the pH of the blood would decrease. Under normal conditions, however, excessive H^+ is eliminated in the urine by the kidneys. Through this action, and through their ability to produce bicarbonate, the kidneys are responsible for maintaining a normal concentration of free bicarbonate in the plasma. The role of the kidneys in acid-base balance is described in chapter 17, section 17.5.

A fall in blood pH below 7.35 is called **acidosis** because the pH is to the acid side of normal. Acidosis does not mean acidic (pH less than 7); a blood pH of 7.2, for example, represents serious acidosis. Similarly, a rise in blood pH above 7.45 is called **alkalosis.** Both of these conditions are categorized into respiratory and metabolic components of acid-base balance (table 16.10).

Respiratory acidosis is caused by inadequate ventilation (hypoventilation), which results in a rise in the plasma concentration of carbon dioxide, and thus carbonic acid. **Respiratory alkalosis,** by contrast, is caused by excessive ventilation (hyperventilation). **Metabolic acidosis** can result from excessive production of nonvolatile acids; for example, it can result from excessive production of ketone bodies in uncontrolled diabetes mellitus (chapter 19, section 19.4). It can also result from the loss of bicarbonate, in which case there would not be sufficient free bicarbonate to buffer the nonvolatile acids. (This occurs in diarrhea because of the loss of bicarbonate derived from pancreatic juice—chapter 18, section 18.5.) **Metabolic alkalosis,** by contrast, can be caused by either too much bicarbonate (perhaps from an intravenous infusion) or inadequate nonvolatile acids (perhaps as a result of excessive vomiting). Excessive vomiting may cause metabolic alkalosis through loss of the acid in gastric juice, which is normally absorbed from the intestine into the blood.

Since the **respiratory component** of acid-base balance is represented by the plasma carbon dioxide concentration and the **metabolic component** is represented by the free bicarbonate concentration, the study of acid-base balance can be simplified. A normal arterial blood pH is obtained when there is a proper ratio of bicarbonate to carbon dioxide. The pH can be calculated given these values, and a normal pH is obtained when the ratio of these concentrations is 20 to 1. This is given by the **Henderson-Hasselbalch equation:**

$$pH = 6.1 + \log \frac{[HCO_3^-]}{0.03P_{CO_2}}$$

where P_{CO_2} is the partial pressure of CO_2, which is proportional to its concentration.

Respiratory acidosis or alkalosis occurs when the carbon dioxide concentrations are abnormal. Metabolic acidosis and alkalosis occur when the bicarbonate concentrations are abnormal (table 16.11). Often, however, a primary disturbance in one area (for example, metabolic acidosis) will be accompanied by secondary changes in another area (for example, respiratory alkalosis). It is important for hospital personnel to identify and treat the area of primary disturbance, but such analysis lies outside the scope of this discussion.

Table 16.10 | Terms Used to Describe Acid-Base Balance

Term	Definition
Acidosis, respiratory	Increased CO_2 retention (due to hypoventilation), which can result in the accumulation of carbonic acid and thus a fall in blood pH to below normal
Acidosis, metabolic	Increased production of "nonvolatile" acids, such as lactic acid, fatty acids, and ketone bodies, or loss of blood bicarbonate (such as by diarrhea), resulting in a fall in blood pH to below normal
Alkalosis, respiratory	A rise in blood pH due to loss of CO_2 and carbonic acid (through hyperventilation)
Alkalosis, metabolic	A rise in blood pH produced by loss of nonvolatile acids (such as excessive vomiting) or by excessive accumulation of bicarbonate base
Compensated acidosis or alkalosis	Metabolic acidosis or alkalosis are partially compensated for by opposite changes in blood carbonic acid levels (through changes in ventilation). Respiratory acidosis or alkalosis are partially compensated for by increased retention or excretion of bicarbonate in the urine.

Ventilation and Acid-Base Balance

In terms of acid-base regulation, the acid-base balance of the blood is divided into the *respiratory component* and the *metabolic component.* The respiratory component refers to the carbon dioxide concentration of the blood, as measured by its P_{CO_2}. As implied by its name, the respiratory component is

Table 16.11 | Classification of Metabolic and Respiratory Components of Acidosis and Alkalosis

Plasma CO_2	Plasma HCO_3^-	Condition	Causes
Normal	Low	Metabolic acidosis	Increased production of "nonvolatile" acids (lactic acids, ketone bodies, and others), or loss of HCO_3^- in diarrhea
Normal	High	Metabolic alkalosis	Vomiting of gastric acid; hypokalemia; excessive steroid administration
Low	Low	Respiratory alkalosis	Hyperventilation
High	High	Respiratory acidosis	Hypoventilation

Table 16.12 | Effect of Lung Function on Blood Acid-Base Balance

Condition	pH	P_{CO_2}	Ventilation	Cause or Compensation
Normal	7.35–7.45	39–41 mmHg	Normal	Not applicable
Respiratory acidosis	Low	High	Hypoventilation	Cause of the acidosis
Respiratory alkalosis	High	Low	Hyperventilation	Cause of the alkalosis
Metabolic acidosis	Low	Low	Hyperventilation	Compensation for acidosis
Metabolic alkalosis	High	High	Hypoventilation	Compensation for alkalosis

regulated by the respiratory system. The metabolic component is controlled by the kidneys, and is discussed in chapter 17, section 17.5.

Ventilation is normally adjusted to keep pace with the metabolic rate, so that the arterial P_{CO_2} remains in the normal range. In **hypoventilation,** the ventilation is insufficient to "blow off" carbon dioxide and maintain a normal P_{CO_2}. Indeed, hypoventilation can be operationally defined as an abnormally high arterial P_{CO_2}. Under these conditions, the concentration of carbonic acid is excessively high and *respiratory acidosis* occurs.

In **hyperventilation,** conversely, the rate of ventilation is greater than the rate of CO_2 production. Arterial P_{CO_2} therefore decreases so that less carbonic acid is formed than under normal conditions. The depletion of carbonic acid raises the pH of the blood and *respiratory alkalosis* occurs. Hyperventilation can cause dizziness because it also raises the pH of the CSF and brain interstitial fluid, which induces cerebral vasoconstriction. The resulting reduction in brain blood flow produces the dizziness.

A change in blood pH, produced by alterations in either the respiratory or metabolic component of acid-base balance, can be partially compensated for by a change in the other component. For example, a person with metabolic acidosis will hyperventilate. This is because the aortic and carotid bodies are stimulated by an increased blood H^+ concentration (fall in pH). As a result of the hyperventilation, a secondary respiratory alkalosis is produced. The person is still acidotic, but not as much so as would be the case without the compensation. People with partially compensated metabolic acidosis would thus have a low pH, which would be accompanied by a low blood P_{CO_2} as a result of the hyperventilation. Metabolic alkalosis, similarly, is partially compensated for by the retention of carbonic acid due to hypoventilation (table 16.12).

Clinical Investigation CLUES

Harry had an abnormally high arterial P_{CO_2} and an arterial pH of 7.15 when he was brought to the hospital.

- How are these measurements related?
- What caused these conditions, and how were they corrected?

CHECKPOINT

15a. Define the terms *acidosis* and *alkalosis.* Identify the two components of blood acid-base balance.

15b. Explain the roles of the lungs and kidneys in maintaining the acid-base balance of the blood.

15c. Describe the functions of bicarbonate and carbonic acid in blood.

15d. Describe the effects of hyperventilation and hypoventilation on the blood pH, and explain the mechanisms involved.

15e. Explain why a person with ketoacidosis hyperventilates. What are the potential benefits of hyperventilation under these conditions?

16.9 EFFECT OF EXERCISE AND HIGH ALTITUDE ON RESPIRATORY FUNCTION

The arterial blood gases and pH do not significantly change during moderate exercise because ventilation increases to keep pace with the increased metabolism. Adjustments are also made at high altitude in both the control of ventilation and the oxygen transport ability of the blood.

LEARNING OUTCOMES

After studying this section, you should be able to:

16. Describe the changes in the respiratory system that occur in response to exercise training and high altitude.

17. Describe the adaptations of the respiratory system to living at high altitude.

Changes in ventilation and oxygen delivery occur during exercise and during acclimatization to a high altitude. These changes help compensate for the increased metabolic rate during exercise and for the decreased arterial P_{O_2} at high altitudes.

Ventilation During Exercise

As soon as a person begins to exercise, breathing becomes deeper and more rapid to produce a total minute volume that is many times the resting value. This increased ventilation, particularly in well-trained athletes, is exquisitely matched to the simultaneous increase in oxygen consumption and carbon dioxide production by the exercising muscles. The arterial blood P_{O_2}, P_{CO_2}, and pH thus remain surprisingly constant during exercise (fig. 16.41).

It is tempting to suppose that ventilation increases during exercise as a result of the increased CO_2 production by the exercising muscles. Ventilation and CO_2 production increase simultaneously, however, so that blood measurements of P_{CO_2} during exercise are not significantly higher than at rest. The mechanisms responsible for the increased ventilation during exercise must therefore be more complex.

Two kinds of mechanisms—*neurogenic* and *humoral*— have been proposed to explain the increased ventilation that occurs during exercise. Possible neurogenic mechanisms include the following: (1) sensory nerve activity from the exercising limbs may stimulate the respiratory muscles, either through spinal reflexes or via the brain stem respiratory centers, and/or (2) input from the cerebral cortex may stimulate the brain stem centers to modify ventilation. These neurogenic theories help explain the immediate increase in ventilation that occurs as exercise begins.

Rapid and deep ventilation continues after exercise has stopped, suggesting that humoral (chemical) factors in the blood may also stimulate ventilation during exercise. Because the P_{O_2}, P_{CO_2}, and pH of the blood samples from exercising subjects are within the resting range, these humoral theories propose that (1) the P_{CO_2} and pH in the region of the chemoreceptors may be different from these values "downstream," where blood samples are taken, and/or (2) that cyclic variations in these values that cannot be detected by blood samples may stimulate the chemoreceptors. The evidence suggests that both neurogenic and humoral mechanisms are involved in the **hyperpnea,** or increased total minute volume, of exercise. (The total minute volume is increased in both hyperpnea and hyperventilation, but in hyperventilation there is also a decrease in arterial blood P_{CO_2}.)

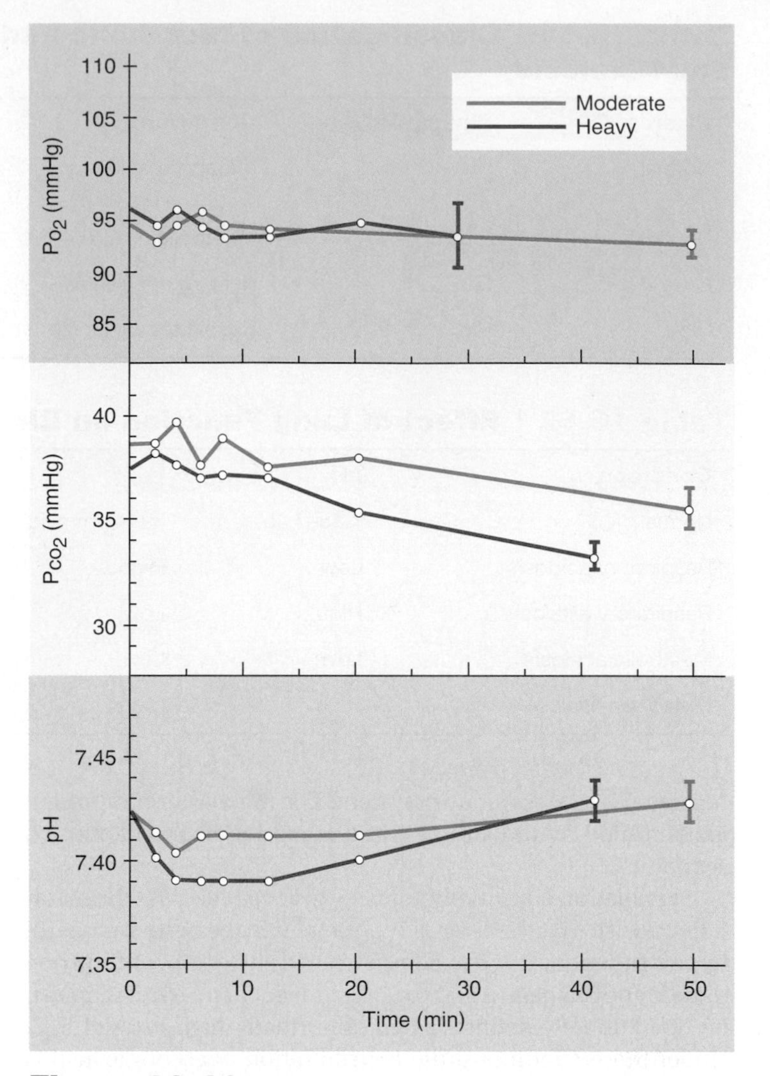

Figure 16.41 The effect of exercise on arterial blood gases and pH. Notice that there are no consistent or significant changes in these measurements during the first several minutes of moderate and heavy exercise, and that only the P_{CO_2} changes (actually decreases) during more prolonged exercise.

Lactate Threshold and Endurance Training

The cardiopulmonary system may be unable to deliver adequate amounts of oxygen to the exercising muscles at the beginning of exercise, because of the time lag required to make proper cardiovascular adjustments. During this time, therefore, the

Table 16.13 | Changes in Respiratory Function During Exercise

Variable	Change	Comments
Ventilation	Increased	In moderate exercise, ventilation is matched to increased metabolic rate. Mechanisms responsible for increased ventilation are not well understood.
Blood gases	No change	Blood gas measurements during light and moderate exercise show little change because ventilation is increased to match increased muscle oxygen consumption and carbon dioxide production.
Oxygen delivery to muscles	Increased	Although the total oxygen content and P_{O_2} do not increase during exercise, there is an increased rate of blood flow to the exercising muscles.
Oxygen extraction by muscles	Increased	Increased oxygen consumption lowers the tissue P_{O_2} and lowers the affinity of hemoglobin for oxygen (due to the effect of increased temperature). More oxygen, as a result, is unloaded so that venous blood contains a lower oxyhemoglobin saturation than at rest. This effect is enhanced by endurance training.

muscles metabolize anaerobically, and a "stitch in the side"—possibly due to hypoxia of the diaphragm—may develop. After numerous cardiovascular and pulmonary adjustments have been made, a person may experience a "second wind" when the muscles are receiving sufficient oxygen for their needs.

Continued heavy exercise can cause a person to reach the **lactate threshold,** which is the maximum rate of oxygen consumption that can be attained before anaerobic metabolism produces a rise in the blood lactate level. This generally occurs when the exercising person reaches 50% to 70% of the maximal oxygen uptake. The rise in lactic acid levels is due to the aerobic limitations of the muscles; it is not due to a malfunction of the cardiopulmonary system. Indeed, the arterial oxyhemoglobin saturation remains at 97% and venous blood draining the muscles contains unused oxygen.

The lactate threshold is higher in endurance-trained athletes than it is in other people. These athletes, because of their higher cardiac output, have a higher rate of oxygen delivery to their muscles. Endurance training also increases the skeletal muscle content of mitochondria and Krebs cycle enzymes (chapter 12, section 12.4), enabling the muscles to utilize more of the oxygen delivered to them by the arterial blood. The effects of exercise and endurance training on respiratory function are summarized in table 16.13.

Acclimatization to High Altitude

When a person from a region near sea level moves to a significantly higher elevation, several adjustments in respiratory function are made to compensate for the decreased P_{O_2} at the higher altitude. These adjustments include changes in ventilation, in hemoglobin affinity for oxygen, and in total hemoglobin concentration.

Reference to table 16.14 indicates that at an altitude of 7,500 feet, for example, the P_{O_2} of arterial blood is 69 to 74 mmHg (compared to 90 to 95 mmHg at sea level). This table also indicates that the percent oxyhemoglobin saturation at this altitude is between 92% and 93%, compared to about 97% at sea level. The amount of oxygen attached to hemoglobin, and thus the total oxygen content of blood, is therefore decreased. In addition, the rate at which oxygen can be delivered to the cells (by the plasma-derived tissue fluid) after it dissociates from oxyhemoglobin is reduced at the higher altitude. This is because the maximum concentration of oxygen that can be dissolved in the plasma decreases in a linear fashion with the fall in P_{O_2}. People may thus experience rapid fatigue even at more moderate elevations (for example, 5,000 to 6,000 feet), at which the oxyhemoglobin saturation is only slightly decreased. Compensations made by the respiratory system gradually reduce the amount of fatigue caused by a given amount of exertion at high altitudes.

Table 16.14 | Blood Gas Measurements at Different Altitudes

Altitude	Arterial P_{O_2} (mmHg)	Percent Oxyhemoglobin Saturation	Arterial P_{CO_2} (mmHg)
Sea level	90–95	97%	40
1,524 m (5,000 ft)	75–81	95%	32–33
2,286 m (7,500 ft)	69–74	92%–93%	31–33
4,572 m (15,000 ft)	48–53	86%	25
6,096 m (20,000 ft)	37–45	76%	20
7,620 m (25,000 ft)	32–39	68%	13
8,848 m (29,029 ft)	26–33	58%	9.5–13.8

Source: From P. H. Hackett et al., "High Altitude Medicine" in *Management of Wilderness and Environmental Emergencies,* 2d ed., edited by Paul S. Auerbach and Edward C. Geehr. Copyright © 1989 Mosby-Yearbook. Reprinted by permission.

Changes in Ventilation

Starting at altitudes as low as 1,500 meters (5,000 feet), the decreased arterial P_{O_2} stimulates the carotid bodies to produce an increase in ventilation. This is known as the **hypoxic ventilatory response.** The increased breathing is hyperventilation, which lowers the arterial P_{CO_2} (table 16.14) and thus produces a respiratory alkalosis. The pH of brain interstitial fluid and cerebrospinal fluid (CSF) similarly becomes alkalotic. After a couple of days, the kidneys increase their urinary excretion of bicarbonate and there is a reduced amount of bicarbonate in the CSF. This helps move the pH of blood and CSF back toward normal. However, the carotid bodies remain sensitive to the low P_{O_2}, and so the total minute volume becomes stabilized after a few days at about 2.5 L/min higher than at sea level.

An extreme example of the hypoxic ventilatory response was measured in hikers climbing Mt. Everest without supplemental oxygen. At almost 28,000 feet (near the summit at 29,029 ft), their average arterial P_{O_2} was 24.6 mmHg and the average arterial P_{CO_2} was measured at 13.3 mmHg. This P_{CO_2} is significantly lower than the normal sea-level value (about 40 mmHg), indicating hyperventilation. Hyperventilation at high altitude increases tidal volume, thus reducing the contribution of air from the anatomical dead space and increasing the proportion of fresh air brought to the alveoli. This improves the oxygenation of the blood over what it would be in the absence of the hyperventilation. Hyperventilation, however, cannot increase blood P_{O_2} above that of the inspired air. The P_{O_2} of arterial blood decreases with increasing altitude, regardless of the ventilation. In the Peruvian Andes, for example, the normal arterial P_{O_2} is reduced from about 100 mmHg (at sea level) to 45 mmHg. The loading of hemoglobin with oxygen is therefore incomplete, producing an oxyhemoglobin saturation that is decreased from 97% (at sea level) to 81%.

Nitric oxide (NO) is produced in the lungs, and a recent study demonstrated increased NO concentration in the lungs of chronically hypoxic people who live at high altitude. Because NO is a vasodilator (chapter 14), this could increase pulmonary blood flow and perhaps improve the oxygenation of the blood in these people.

Also, NO bound to sulfur atoms (and therefore abbreviated *SNOs*) in the cysteine groups of hemoglobin and other proteins may be transferred from the blood to the rhythmicity center, where it may stimulate breathing. Thus, SNOs may contribute to the hypoxic drive (increased breathing when the arterial P_{O_2} is low). These proposed mechanisms of NO action would provide partial compensations for the chronic hypoxia of life at high altitude.

The Affinity of Hemoglobin for Oxygen

Normal arterial blood at sea level unloads only about 22% of its oxygen to the tissues at rest; the percent saturation is reduced from 97% in arterial blood to 75% in venous blood. As a partial compensation for the decrease in oxygen content at high altitude, the affinity of hemoglobin for oxygen is reduced so that a higher proportion of oxygen is unloaded. This occurs because the low oxyhemoglobin content of red blood cells stimulates the production of 2,3-DPG, which in turn decreases the affinity of hemoglobin for oxygen (see fig. 16.35).

The action of 2,3-DPG to decrease the affinity of hemoglobin for oxygen thus predominates over the action of respiratory alkalosis (caused by the hyperventilation) to increase the affinity. At very high altitudes, however, the story becomes more complex. In one study, the very low arterial P_{O_2} (28 mmHg) of subjects at the summit of Mount Everest stimulated intense hyperventilation, so that the arterial P_{CO_2} was decreased to 7.5 mmHg. The resultant respiratory alkalosis (in this case, arterial pH greater than 7.7) caused the oxyhemoglobin dissociation curve to shift to the left (indicating greater affinity of hemoglobin for oxygen), despite the antagonistic effect of increased 2,3-DPG concentrations. It was suggested that the increased affinity of hemoglobin for oxygen caused by the respiratory alkalosis may have been beneficial at such a high altitude, because it increased the loading of hemoglobin with oxygen in the lungs.

FITNESS APPLICATION

Acute mountain sickness (AMS) is common in people who arrive at altitudes in excess of 5,000 feet. Cardinal symptoms of AMS are headache, malaise, anorexia, nausea, and fragmented sleep. Headache, the most common symptom, may result from changes in blood flow to the brain. Low arterial P_{O_2} stimulates vasodilation of vessels in the pia mater, increasing blood flow and pressure within the skull. The hypocapnia produced by hyperventilation, however, causes cerebral vasoconstriction. Whether there is a net cerebral vasoconstriction or vasodilation depends on the balance between these two antagonistic effects. Pulmonary edema, common at altitudes above 9,000 feet, can produce shortness of breath, coughing, and a mild fever. Cerebral edema, which generally occurs above an altitude of 10,000 feet, can produce mental confusion and even hallucinations. Pulmonary and cerebral edema are potentially dangerous and should be alleviated by descending to a lower altitude. Acute mountain sickness may be treated with a drug (*acetazolamide*) that causes the excretion of bicarbonate in the urine. This produces a metabolic acidosis that partially compensates for the respiratory alkalosis caused by hyperventilation.

Increased Hemoglobin and Red Blood Cell Production

Kidney cells sense a decreased tissue oxygen concentration (hypoxia), and in response produce and secrete erythropoietin (chapter 13, section 13.2). Erythropoietin stimulates the bone marrow to increase its production of hemoglobin and red blood cells. In the Peruvian Andes, for example, people have a total hemoglobin concentration that is increased from 15 g per 100 ml (at sea level) to 19.8 g per 100 ml. Although the percent oxyhemoglobin saturation is still lower than at sea level, the total oxygen content of the blood is actually greater—22.4 ml O_2 per 100 ml compared to a sea-level value of about 20 ml O_2 per 100 ml. These adjustments of the respiratory system to high altitude are summarized in figure 16.42 and table 16.15.

Increased red blood cell count and hemoglobin concentration at high altitude are not unalloyed benefits. The polycythemia

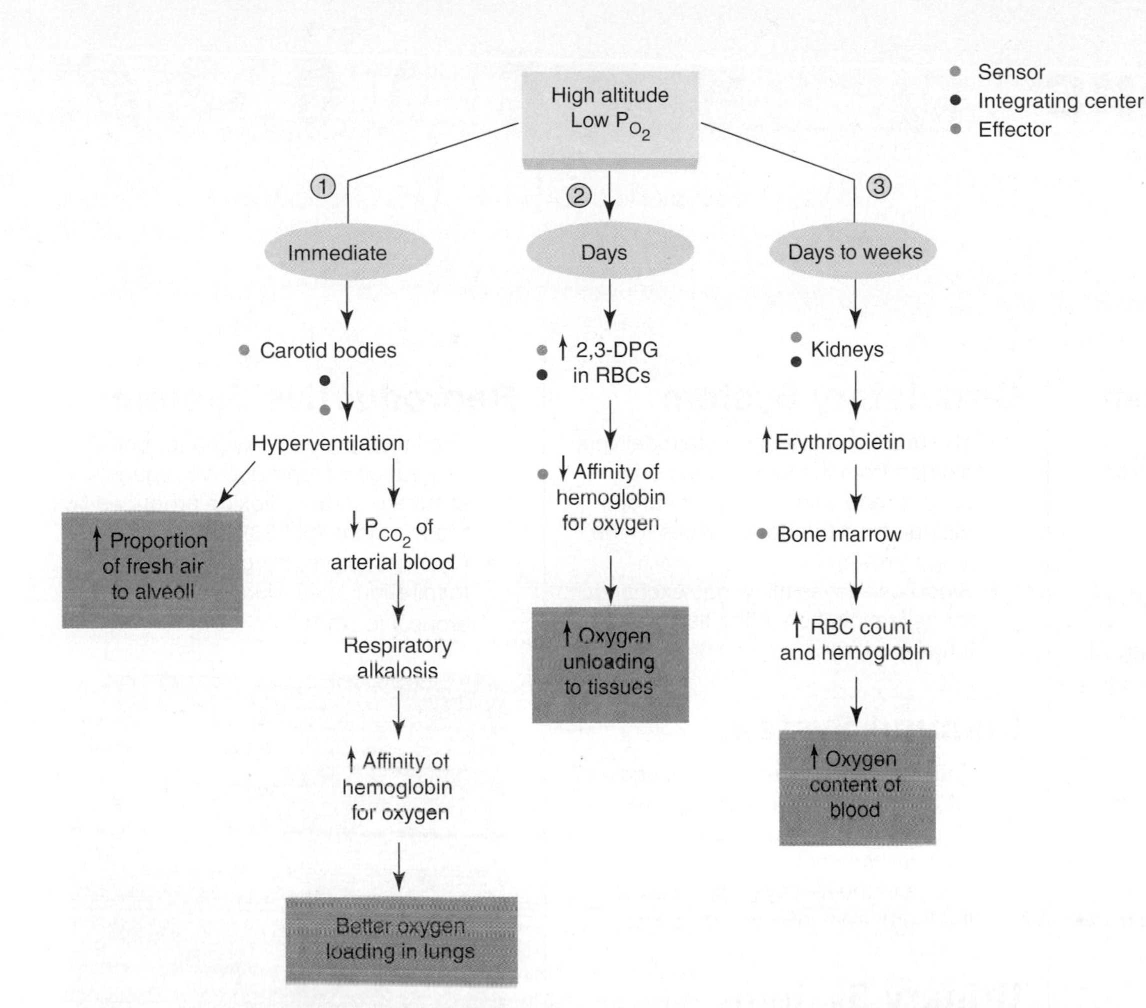

Figure 16.42 Respiratory adaptations to a high altitude. The circled numbers indicate the order of the responses, from those that occur immediately to those that require weeks to take effect. These adaptations enable people to live, work, and play at higher altitudes than would otherwise be possible.

(high red blood cell count) increases the blood viscosity, thereby increasing vascular resistance. Polycythemia can cause pulmonary hypertension, accompanied by edema and ventricular hypertrophy that can lead to heart failure. In pregnant women, polycythemia increases fetal mortality. It is interesting in this regard that Tibetan highlanders, who have lived at extreme altitudes for many thousands of years, have lower hemoglobin and red blood cell levels than other people who ascend to the same altitude.

The "ideal" hemoglobin concentration is probably close to 18 g/dl of blood. When values reach 21 to 23 g/dl, the circulation becomes abnormal and the person displays symptoms of *chronic mountain sickness.* Interestingly, the fetus attains hemoglobin concentrations that are also about 18 g/dl before birth, when blood oxygen levels are low. The hemoglobin concentrations drop rapidly after birth when blood oxygenation rises with the first breath.

CHECKPOINT

16a. Describe the effect of exercise on the P_{O_2}, P_{CO_2}, and pH blood values, and explain how ventilation might be increased during exercise.

16b. Explain why endurance-trained athletes have a higher than average anaerobic threshold.

17. Describe the changes that occur in the respiratory system during acclimatization to life at a high altitude.

Table 16.15 | Changes in Respiratory Function During Acclimatization to High Altitude

Variable	Change	Comments
Partial pressure of oxygen	Decreased	Due to decreased total atmospheric pressure
Partial pressure of carbon dioxide	Decreased	Due to hyperventilation in response to low arterial P_{O_2}
Percent oxyhemoglobin saturation	Decreased	Due to lower P_{O_2} in pulmonary capillaries
Ventilation	Increased	Due to lower P_{O_2}
Total hemoglobin	Increased	Due to stimulation by erythropoietin; raises oxygen capacity of blood to partially or completely compensate for the reduced partial pressure
Oxyhemoglobin affinity	Decreased	Due to increased DPG within the red blood cells; results in a higher percent unloading of oxygen to the tissues

Interactions

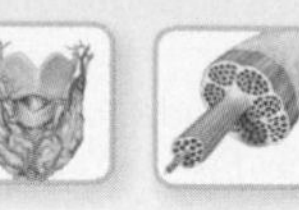

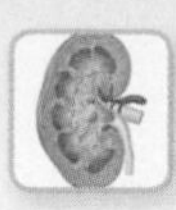

HPer Links of the Respiratory System with Other Body Systems

Integumentary System

- Nasal hairs and mucus prevent dust and other foreign material from damaging respiratory passageways (p. 536)

Skeletal System

- The lungs are protected by the rib cage, and bones of the rib cage serve as levers for the action of respiratory muscles (p. 541)
- Red blood cells, needed for oxygen transport, are produced in the bone marrow (p. 409)
- The respiratory system provides all organs, including the bones, with oxygen and eliminates carbon dioxide (p. 533)

Muscular System

- Contractions of skeletal muscles are needed for ventilation (p. 541)
- Muscles consume large amounts of oxygen and produce large amounts of carbon dioxide during exercise (p. 379)

Nervous System

- The nervous system regulates the rate and depth of breathing (p. 553)
- Autonomic nerves regulate blood flow, and hence the delivery of blood to tissues for gas exchange (p. 466)

Endocrine System

- Epinephrine dilates bronchioles, reducing airway resistance (p. 545)
- Thyroxine and epinephrine stimulate the rate of cell respiration (p. 675)

Circulatory System

- The heart and arterial system delivers oxygen from the lungs to the body tissues, and veins transport carbon dioxide from the body tissues to the lungs (p. 405)
- Blood capillaries allow gas exchange for cell respiration in the tissues and lungs (p. 433)

Immune System

- The immune system protects against infections that could damage the respiratory system (p. 494)
- Alveolar macrophages and the action of cilia in the airways help to protect the lungs from infection (p. 536)

Urinary System

- The kidneys regulate the volume and electrolyte balance of the blood (p. 582)
- The kidneys participate with the lungs in the regulation of blood pH (p. 608)

Digestive System

- The GI tract provides nutrients to be used by cells of the lungs and other organs (p. 621)
- The respiratory system provides oxygen for cell respiration of glucose and other nutrients brought into the blood by the digestive system (p. 111)

Reproductive System

- The lungs provide oxygen for cell respiration of reproductive organs and eliminate carbon dioxide produced by these organs (p. 533)
- Changes in breathing and cell respiration occur during sexual arousal (p. 711)

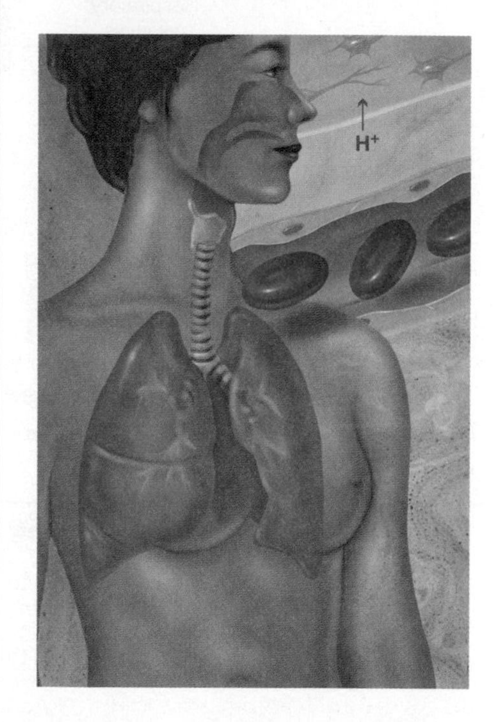

Clinical Investigation SUMMARY

The puncture wound must have admitted air into the pleural cavity (pneumothorax), raising the intrapleural pressure and causing collapse of the right lung. Since the left lung is located in a separate pleural compartment, it was unaffected by the wound. As a result of the collapse of his right lung, Harry was hypoventilating. This caused retention of CO_2, thus raising his arterial P_{CO_2} and resulting in respiratory acidosis (as indicated by an arterial pH lower than 7.35). Upon recovery, analysis of his arterial blood revealed that he was breathing adequately but that he had a carboxyhemoglobin saturation of 18%. This very high level is probably due to a combination of smoking and driving in heavily congested areas with much automobile exhaust. The high carboxyhemoglobin would reduce oxygen transport, thus aggravating any problems he might have with his cardiovascular or pulmonary system.

The significantly low FEV_1 indicates that Harry has an obstructive pulmonary problem, possibly caused by smoking and the inhalation of polluted air. A low FEV_1 could simply indicate bronchoconstriction, but the fact that Harry's vital capacity was a little low suggests that he may have early-stage lung damage, possibly emphysema. He should be strongly advised to quit smoking, and further pulmonary tests should be administered at regular intervals.

See the additional chapter 16 Clinical Investigations on *COPD* and *Hypoxemia* in the Connect site for this text at www.mhhe.com/Fox13.

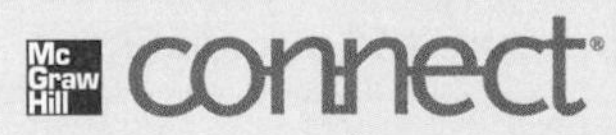

|ANATOMY & PHYSIOLOGY

Visit this book's website at **www.mhhe.com/Fox13** for:

- Chapter quizzes, interactive learning exercises, and other study tools
- Additional clinical investigations
- Access to LearnSmart—an adaptive diagnostic tool that constantly assesses student knowledge of course material
- Ph.I.L.S. 4.0—physiology interactive lab simulations that may be used to supplement or substitute for wet labs

SUMMARY

16.1 The Respiratory System 533

A. Alveoli are microscopic thin-walled air sacs that provide an enormous surface area for gas diffusion.
   1. The region of the lungs where gas exchange with the blood occurs is known as the respiratory zone.
   2. The trachea, bronchi, and bronchioles that deliver air to the respiratory zone constitute the conducting zone.

B. The thoracic cavity is delimited by the chest wall and diaphragm.
   1. The structures of the thoracic cavity are covered by thin, wet pleurae.
   2. The lungs are covered by a visceral pleura that is normally flush against the parietal pleura that lines the chest wall.
   3. The potential space between the visceral and parietal plurae is called the intrapleural space.

16.2 Physical Aspects of Ventilation 536

A. The intrapleural and intrapulmonary pressures vary during ventilation.
   1. The intrapleural pressure is always less than the intrapulmonary pressure.
   2. The intrapulmonary pressure is subatmospheric during inspiration and greater than the atmospheric pressure during expiration.
   3. Pressure changes in the lungs are produced by variations in lung volume in accordance with the inverse relationship between the volume and pressure of a gas described by Boyle's law.

B. The mechanics of ventilation are influenced by the physical properties of the lungs.
   1. The compliance of the lungs, or the ease with which they expand, refers specifically to the change in lung volume

per change in transpulmonary pressure (the difference between intrapulmonary pressure and intrapleural pressure).

2. The elasticity of the lungs refers to their tendency to recoil after distension.
3. The surface tension of the fluid in the alveoli exerts a force directed inward, which acts to resist distension.

C. On first consideration, it would seem that the surface tension in the alveoli would create a pressure that would cause small alveoli to collapse and empty their air into larger alveoli.

1. This would occur because the pressure caused by a given amount of surface tension would be greater in smaller alveoli than in larger alveoli, as described by the law of Laplace.
2. Surface tension does not normally cause the collapse of alveoli, however, because pulmonary surfactant (a combination of phospholipid and protein) lowers the surface tension sufficiently.
3. In hyaline membrane disease, the lungs of premature infants collapse because of a lack of surfactant.

16.3 Mechanics of Breathing 541

A. Inspiration and expiration are accomplished by contraction and relaxation of striated muscles.

1. During quiet inspiration, the diaphragm and external intercostal muscles contract, and thus increase the volume of the thorax.
2. During quiet expiration, these muscles relax, and the elastic recoil of the lungs and thorax causes a decrease in thoracic volume.
3. Forced inspiration and expiration are aided by contraction of the accessory respiratory muscles.

B. Spirometry aids the diagnosis of a number of pulmonary disorders.

1. In restrictive disease, such as pulmonary fibrosis, the vital capacity measurement is decreased to below normal.
2. In obstructive disease, such as asthma and bronchitis, the forced expiratory volume is reduced to below normal because of increased airway resistance to air flow.

C. Asthma results from bronchoconstriction; emphysema, asthma, and chronic bronchitis are frequently referred to collectively as chronic obstructive pulmonary disease.

16.4 Gas Exchange in the Lungs 547

A. According to Dalton's law, the total pressure of a gas mixture is equal to the sum of the pressures that each gas in the mixture would exert independently.

1. The partial pressure of a gas in a dry gas mixture is thus equal to the total pressure times the percent composition of that gas in the mixture.
2. Because the total pressure of a gas mixture decreases with altitude above sea level, the partial pressures of the constituent gases likewise decrease with altitude.
3. When the partial pressure of a gas in a wet gas mixture is calculated, the water vapor pressure must be taken into account.

B. According to Henry's law, the amount of gas that can be dissolved in a fluid is directly proportional to the partial pressure of that gas in contact with the fluid.

1. The concentrations of oxygen and carbon dioxide that are dissolved in plasma are proportional to an electric current generated by special electrodes that react with these gases.
2. Normal arterial blood has a P_{O_2} of 100 mmHg, indicating a concentration of dissolved oxygen of 0.3 ml per 100 ml of blood; the oxygen contained in red blood cells (about 19.7 ml per 100 ml of blood) does not affect the P_{O_2} measurement.

C. The P_{O_2} and P_{CO_2} measurements of arterial blood provide information about lung function.

D. In addition to proper ventilation of the lungs, blood flow (perfusion) in the lungs must be adequate and matched to air flow (ventilation) in order for adequate gas exchange to occur.

E. Abnormally high partial pressures of gases in blood can cause a variety of disorders, including oxygen toxicity, nitrogen narcosis, and decompression sickness.

16.5 Regulation of Breathing 533

A. The rhythmicity center in the medulla oblongata directly controls the muscles of respiration.

1. Activity of the inspiratory and expiratory neurons varies in a reciprocal way to produce an automatic breathing cycle.
2. Activity in the medulla is influenced by the apneustic and pneumotaxic centers in the pons, as well as by sensory feedback information.
3. Conscious breathing involves direct control by the cerebral cortex via corticospinal tracts.

B. Breathing is affected by chemoreceptors sensitive to the P_{CO_2}, pH, and P_{O_2} of the blood.

1. The P_{CO_2} of the blood and consequent changes in pH are usually of greater importance than the blood P_{O_2} in the regulation of breathing.
2. Central chemoreceptors in the medulla oblongata are sensitive to changes in blood P_{CO_2} because of the resultant changes in the pH of cerebrospinal fluid.
3. The peripheral chemoreceptors in the aortic and carotid bodies are sensitive to changes in blood P_{CO_2} indirectly, because of consequent changes in blood pH.

C. Decreases in blood P_{O_2} directly stimulate breathing only when the blood P_{O_2} is lower than 50 mmHg. A drop in P_{O_2} also stimulates breathing indirectly, by making the chemoreceptors more sensitive to changes in P_{CO_2} and pH.

D. At tidal volumes of 1 L or more, inspiration is inhibited by stretch receptors in the lungs (the Hering-Breuer reflex). A similar reflex may act to inhibit expiration.

16.6 Hemoglobin and Oxygen Transport 558

A. Hemoglobin is composed of two alpha and two beta polypeptide chains and four heme groups each containing a central atom of iron.

1. When the iron is in the reduced form and not attached to oxygen, the hemoglobin is called deoxyhemoglobin,

or reduced hemoglobin; when it is attached to oxygen, it is called oxyhemoglobin.

2. If the iron is attached to carbon monoxide, the hemoglobin is called carboxyhemoglobin. When the iron is in an oxidized state and unable to transport any gas, the hemoglobin is called methemoglobin.
3. Deoxyhemoglobin combines with oxygen in the lungs (the loading reaction) and breaks its bonds with oxygen in the tissue capillaries (the unloading reaction). The extent of each reaction is determined by the P_{O_2} and the affinity of hemoglobin for oxygen.

B. A graph of percent oxyhemoglobin saturation at different values of P_{O_2} is called an oxyhemoglobin dissociation curve.
1. At rest, the difference between arterial and venous oxyhemoglobin saturations indicates that about 22% of the oxyhemoglobin unloads its oxygen to the tissues.
2. During exercise, the venous P_{O_2} and percent oxyhemoglobin saturation are decreased, indicating that a higher percentage of the oxyhemoglobin has unloaded its oxygen to the tissues.

C. The pH and temperature of the blood influence the affinity of hemoglobin for oxygen, and thus the extent of loading and unloading.
1. A fall in pH decreases the affinity of hemoglobin for oxygen, and a rise in pH increases the affinity. This is called the Bohr effect.
2. A rise in temperature decreases the affinity of hemoglobin for oxygen.
3. When the affinity is decreased, the oxyhemoglobin dissociation curve is shifted to the right. This indicates a greater unloading percentage of oxygen to the tissues.

D. The affinity of hemoglobin for oxygen is also decreased by an organic molecule in the red blood cells called 2,3-diphosphoglyceric acid (2,3-DPG).
1. Because oxyhemoglobin inhibits 2,3-DPG production, 2,3-DPG concentrations will be higher when anemia or low P_{O_2} (as in high altitude) cause a decrease in oxyhemoglobin.
2. If a person is anemic, the lowered hemoglobin concentration is partially compensated for because a higher percentage of the oxyhemoglobin will unload its oxygen as a result of the effect of 2,3-DPG.
3. Fetal hemoglobin cannot bind to 2,3-DPG, and thus it has a higher affinity for oxygen than the mother's hemoglobin. This facilitates the transfer of oxygen to the fetus.

E. Inherited defects in the amino acid composition of hemoglobin are responsible for such diseases as sickle-cell anemia and thalassemia.

F. Striated muscles contain myoglobin, a pigment related to hemoglobin that can combine with oxygen and deliver it to the muscle cell mitochondria at low P_{O_2} values.

16.7 Carbon Dioxide Transport 565

A. Red blood cells contain an enzyme called carbonic anhydrase that catalyzes the reversible reaction whereby carbon dioxide and water are used to form carbonic acid.
1. This reaction is favored by the high P_{CO_2} in the tissue capillaries, and as a result, carbon dioxide produced by the tissues is converted into carbonic acid in the red blood cells.
2. Carbonic acid then ionizes to form H^+ and HCO_3^- (bicarbonate).
3. Because much of the H^+ is buffered by hemoglobin, but more bicarbonate is free to diffuse outward, an electrical gradient is established that draws Cl^- into the red blood cells. This is called the chloride shift.
4. A reverse chloride shift occurs in the lungs. In this process, the low P_{CO_2} favors the conversion of carbonic acid to carbon dioxide, which can be exhaled.

B. By adjusting the blood concentration of carbon dioxide, and thus of carbonic acid, the process of ventilation helps to maintain proper acid-base balance of the blood.
1. Normal arterial blood pH is 7.40. A pH below 7.35 is termed acidosis; a pH above 7.45 is termed alkalosis.
2. Hyperventilation causes respiratory alkalosis, and hypoventilation causes respiratory acidosis.
3. Metabolic acidosis stimulates hyperventilation, which can cause a respiratory alkalosis as a partial compensation.

16.8 Acid-Base Balance of the Blood 567

A. The normal pH of arterial blood is 7.40, with a range of 7.35 to 7.45.
1. Carbonic acid is formed from carbon dioxide and contributes to the blood pH. It is referred to as a volatile acid because it can be eliminated in the exhaled breath.
2. Nonvolatile acids, such as lactic acid and the ketone bodies, are buffered by bicarbonate.

B. The blood pH is maintained by a proper ratio of carbon dioxide to bicarbonate.
1. The lungs maintain the correct carbon dioxide concentration. An increase in carbon dioxide, due to inadequate ventilation, produces respiratory acidosis.
2. The kidneys maintain the free-bicarbonate concentration. An abnormally low plasma bicarbonate concentration produces metabolic acidosis.

C. Ventilation regulates the respiratory component of acid-base balance.
1. Hypoventilation increases the blood P_{CO_2}, thereby lowering the plasma pH and producing a respiratory acidosis.
2. Hyperventilation decreases the plasma P_{CO_2}, decreasing the formation of carbonic acid, and thereby increasing the plasma pH to produce a respiratory alkalosis.
3. Because of the action of the chemoreceptors, breathing is regulated to maintain a proper blood P_{CO_2} and thus a normal blood pH.

16.9 Effect of Exercise and High Altitude on Respiratory Function 570

A. During exercise there is increased ventilation, or hyperpnea, which is matched to the increased metabolic rate so that the arterial blood P_{CO_2} remains normal.

1. This hyperpnea may be caused by proprioceptor information, cerebral input, and/or changes in arterial P_{CO_2} and pH.
2. During heavy exercise, the anaerobic threshold may be reached at 50% to 70% of the maximal oxygen uptake. At this point, lactic acid is released into the blood by the muscles.
3. Endurance training enables the muscles to utilize oxygen more effectively, so that greater levels of exercise can be performed before the anaerobic threshold is reached.

B. Acclimatization to a high altitude involves changes that help to deliver oxygen more effectively to the tissues, despite reduced arterial P_{O_2}.

1. Hyperventilation occurs in response to the low P_{O_2}.
2. The red blood cells produce more 2,3-DPG, which lowers the affinity of hemoglobin for oxygen and improves the unloading reaction.
3. The kidneys produce the hormone erythropoietin, which stimulates the bone marrow to increase its production of red blood cells, so that more oxygen can be carried by the blood at given values of P_{O_2}.

REVIEW ACTIVITIES

Test Your Knowledge

1. Which of these statements about intrapulmonary pressure and intrapleural pressure is *true?*
   a. The intrapulmonary pressure is always subatmospheric.
   b. The intrapleural pressure is always greater than the intrapulmonary pressure.
   c. The intrapulmonary pressure is greater than the intrapleural pressure.
   d. The intrapleural pressure equals the atmospheric pressure.
2. If the transpulmonary pressure equals zero,
   a. a pneumothorax has probably occurred.
   b. the lungs cannot inflate.
   c. elastic recoil causes the lungs to collapse.
   d. all of these apply.
3. The maximum amount of air that can be expired after a maximum inspiration is
   a. the tidal volume.
   b. the forced expiratory volume.
   c. the vital capacity.
   d. the maximum expiratory flow rate.
4. If the blood lacked red blood cells but the lungs were functioning normally,
   a. the arterial P_{O_2} would be normal.
   b. the oxygen content of arterial blood would be normal.
   c. both *a* and *b* would apply.
   d. neither *a* nor *b* would apply.
5. If a person were to dive with scuba equipment to a depth of 66 feet, which of these statements would be *false?*
   a. The arterial P_{O_2} would be three times normal.
   b. The oxygen content of plasma would be three times normal.
   c. The oxygen content of whole blood would be three times normal.
6. Which of these would be most affected by a decrease in the affinity of hemoglobin for oxygen?
   a. Arterial P_{O_2}
   b. Arterial percent oxyhemoglobin saturation
   c. Venous oxyhemoglobin saturation
   d. Arterial P_{CO_2}
7. If a person with normal lung function were to hyperventilate for several seconds, there would be a significant
   a. increase in the arterial P_{O_2}.
   b. decrease in the arterial P_{CO_2}.
   c. increase in the arterial percent oxyhemoglobin saturation.
   d. decrease in the arterial pH.
8. Erythropoietin is produced by
   a. the kidneys. c. the lungs.
   b. the liver. d. the bone marrow.
9. The affinity of hemoglobin for oxygen is decreased under conditions of
   a. acidosis.
   b. fever.
   c. anemia.
   d. acclimatization to a high altitude.
   e. all of these.
10. Most of the carbon dioxide in the blood is carried in the form of
    a. dissolved CO_2.
    b. carbaminohemoglobin.
    c. bicarbonate.
    d. carboxyhemoglobin.
11. The bicarbonate concentration of the blood would be decreased during
    a. metabolic acidosis. c. metabolic alkalosis.
    b. respiratory acidosis. d. respiratory alkalosis.

12. The chemoreceptors in the medulla are directly stimulated by
    a. CO_2 from the blood.
    b. H^+ from the blood.
    c. H^+ in brain interstitial fluid that is derived from blood CO_2.
    d. decreased arterial P_{O_2}.
13. The rhythmic control of breathing is produced by the activity of inspiratory and expiratory neurons in
    a. the medulla oblongata.
    b. the apneustic center of the pons.
    c. the pneumotaxic center of the pons.
    d. the cerebral cortex.
14. Which of these occur(s) during hypoxemia?
    a. Increased ventilation
    b. Increased production of 2,3-DPG
    c. Increased production of erythropoietin
    d. All of these
15. During exercise, which of these statements is *true?*
    a. The arterial percent oxyhemoglobin saturation is decreased.
    b. The venous percent oxyhemoglobin saturation is decreased.
    c. The arterial P_{CO_2} is measurably increased.
    d. The arterial pH is measurably decreased.
16. All of the following can bond with hemoglobin *except*
    a. HCO_3^-.
    b. O_2.
    c. H^+.
    d. CO_2.
    e. NO.
17. Which of these statements about the partial pressure of carbon dioxide is *true?*
    a. It is higher in the alveoli than in the pulmonary arteries.
    b. It is higher in the systemic arteries than in the tissues.
    c. It is higher in the systemic veins than in the systemic arteries.
    d. It is higher in the pulmonary veins than in the pulmonary arteries.
18. The hypoxic ventilatory response occurs when low arterial P_{O_2} stimulates the
    a. aortic bodies.
    b. carotid bodies.
    c. central chemoreceptors.
    d. all of these.

Test Your Understanding

19. Using a flow diagram to show cause and effect, explain how contraction of the diaphragm produces inspiration.
20. Radiographic (x-ray) pictures show that the rib cage of a person with a pneumothorax is expanded and the ribs are farther apart. Explain why this should be so.
21. Explain, using a flowchart, how a rise in blood P_{CO_2} stimulates breathing. Include both the central and peripheral chemoreceptors in your answer.
22. Explain why a person with ketoacidosis may hyperventilate. What benefit might it provide? Also explain why this hyperventilation can be stopped by an intravenous fluid containing bicarbonate.
23. What blood measurements can be performed to detect (a) anemia, (b) carbon monoxide poisoning, and (c) poor lung function?
24. Explain how measurements of blood P_{CO_2}, bicarbonate, and pH are affected by hypoventilation and hyperventilation.
25. Describe the changes in ventilation that occur during exercise. How are these changes produced and how do they affect arterial blood gases and pH?
26. How would an increase in the red blood cell content of 2,3-DPG affect the P_{O_2} of venous blood? Explain your answer.
27. Describe how ventilation changes when a person goes from sea level to a high altitude, and explain how this change is produced. In what way is this change beneficial, and in what way might it be detrimental?
28. Explain the physiological changes in the blood's ability to transport and deliver oxygen to the tissues during the acclimatization to high altitude, and identify the time course required for these responses.
29. Compare asthma and emphysema in terms of their characteristics and the effects they have on pulmonary function tests.
30. Explain the mechanisms involved in quiet inspiration and in forced inspiration, and in quiet expiration and forced expiration. What muscles are involved in each case?
31. Describe the formation, composition, and function of pulmonary surfactant. What happens when surfactant is absent? How is this condition treated?
32. Compare and contrast asthma with chronic obstructive pulmonary disease (COPD), in terms of their causes, the structures and processes involved, and treatments.

Test Your Analytical Ability

33. The nature of the sounds produced by percussion (tapping) a patient's chest can tell a physician a great deal about the condition of the organs within the thoracic cavity. Healthy, air-filled lungs resonate, or sound hollow. How do you think the lungs of a person with emphysema would sound in comparison to healthy lungs? What kind of sounds would be produced by a collapsed lung, or one that was partially filled with fluid?

34. Explain why the first breath of a healthy neonate is more difficult than subsequent breaths and why premature infants often require respiratory assistance (a mechanical ventilator) to keep their lungs inflated. How else is this condition treated?
35. Nicotine from cigarette smoke causes the buildup of mucus and paralyzes the cilia that line the respiratory tract. How might these conditions affect pulmonary function tests? If smoking has led to emphysema, how would the pulmonary function tests change?
36. Carbon monoxide poisoning from smoke inhalation and suicide attempts is the most common cause of death from poisoning in the United States. How would carbon monoxide poisoning affect a person's coloring, particularly of the mucous membranes? How would it affect the hemoglobin concentration, hematocrit, and percent oxyhemoglobin saturation? How would chronic carbon monoxide poisoning affect the person's red blood cell content of 2,3-DPG?
37. After driving from sea level to a trailhead in the High Sierras, you get out of your car and feel dizzy. What do you suppose is causing your dizziness? How is this beneficial and how is it detrimental? What may eventually happen to help to reduce the cause of the dizziness?
38. Explain how a subatmospheric intrapleural pressure is produced, and how this relates to collapse of a lung when a person suffers an open chest wound.
39. What is the physiological advantage of the fetus having a different form of hemoglobin earlier and then switching to the adult form later in development? What is the physiological mechanism responsible for the fetus having a high blood hemoglobin concentration? How does this relate to physiological neonatal jaundice?
40. You cannot affect the oxygen delivery to your tissues by drinking "oxygenated water." However, breathing oxygen at hyperbaric pressures does increase oxygen delivery to the tissues. Explain why these two statements are true.

Test Your Quantitative Ability

Refer to page 548 and figure 16.19 and calculate the P_{O_2} of the following gas mixtures:

41. Dry air at a total pressure of 530 mmHg
42. Air saturated with water vapor at a total pressure of 600 mm Hg
43. Air saturated with water vapor at a pressure of 2.5 atmospheres

Use figure 16.34 to answer the following questions:

44. Blood at a pH of 7.40 has what percent oxyhemoglobin saturation at a P_{O_2} of (a) 70 mmHg; and (b) 20 mmHg?
45. What percentage of the oxygen carried by hemoglobin is unloaded in question 44?
46. Blood at a pH of 7.20 has what percent oxyhemoglobin saturation at a P_{O_2} of (a) 70 mmHg; and (b) 20 mmHg?
47. What percent of the oxygen carried by hemoglobin is unloaded in question 46?

CHAPTER OUTLINE

CHAPTER 17

Physiology of the Kidneys

REFRESH YOUR MEMORY

Before you begin this chapter, you may want to review these concepts from previous chapters:

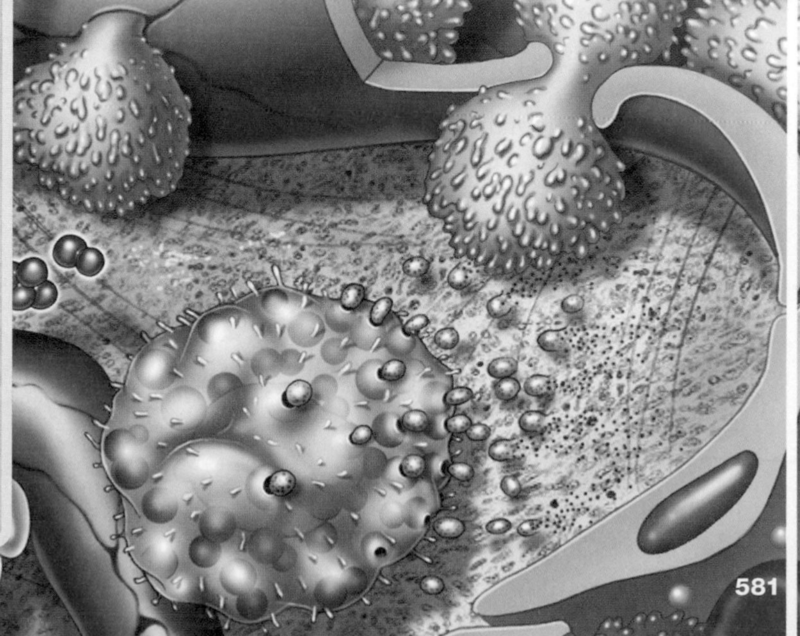

Clinical Investigation

Emily, a high school senior on the track team, went to her physician because she was concerned about the color of her urine and had a pain in her lower back. She did not experience pain when she urinated. She also admitted that she had a sore throat for a month but had continued to train with her track team. A throat culture revealed that Emily had strep throat, for which the physician prescribed an antibiotic. Emily also had mild edema, for which the physician prescribed a diuretic. Blood and urine tests were performed, and Emily's symptoms disappeared in a few weeks.

Some of the new terms and concepts you will encounter include:

- Hematuria and oliguria
- Creatinine and renal plasma clearance
- Glomerulonephritis

17.1 STRUCTURE AND FUNCTION OF THE KIDNEYS

Each kidney contains many tiny tubules that empty into a cavity drained by the ureter. Each of the tubules receives a blood filtrate from a capillary bed called the glomerulus. The filtrate is modified as it passes through different regions of the tubule and is thereby changed into urine.

LEARNING OUTCOMES

After studying this section, you should be able to:

1. Explain the functions of the kidneys.
2. Describe the gross and microscopic structure of the kidneys.
3. Trace the flow of blood and filtrate through the kidneys.

The primary function of the kidneys is regulation of the extracellular fluid (plasma and interstitial fluid) environment in the body. This is accomplished through the formation of urine, which is a modified filtrate of plasma. In the process of urine formation, the kidneys regulate:

1. the volume of blood plasma (and thus contribute significantly to the regulation of blood pressure);
2. the concentration of waste products in the plasma;
3. the concentration of electrolytes (Na^+, K^+, HCO_3^- and other ions) in the plasma; and
4. the pH of plasma.

In order to understand how the kidneys perform these functions, a knowledge of kidney structure is required.

Gross Structure of the Urinary System

The paired **kidneys** lie on either side of the vertebral column below the diaphragm and liver. Each adult kidney weighs about 160 g and is about 11 cm (4 in.) long and 5 to 7 cm (2 to 3 in.) wide—about the size of a fist. Urine produced in the kidneys is drained into a cavity known as the *renal pelvis* (= basin) and then is channeled from each kidney via long ducts—the **ureters**—to the **urinary bladder** (fig. 17.1).

A coronal section of the kidney shows two distinct regions (fig. 17.2). The outer *cortex* is reddish brown and granular in appearance because of its many capillaries. The deeper region, or *medulla,* is striped in appearance due to the presence of microscopic tubules and blood vessels. The medulla is composed of 8 to 15 conical *renal pyramids* separated by *renal columns.*

The cavity of the kidney is divided into several portions. Each pyramid projects into a small depression called a *minor calyx* (the plural form is *calyces*). Several minor calyces unite to form a *major calyx.* The major calyces then join to form the funnel-shaped renal pelvis. The renal pelvis collects urine from the calyces and transports it to the ureters and urinary bladder (fig. 17.3).

The ureter undergoes *peristalsis,* wavelike contractions similar to those that occur in the digestive tract. (This results in intense pain when a person passes a kidney stone.) Interestingly, the pacemaker of these peristaltic waves is located in the renal calyces and pelvis (see fig. 17.2), which contain smooth muscle.

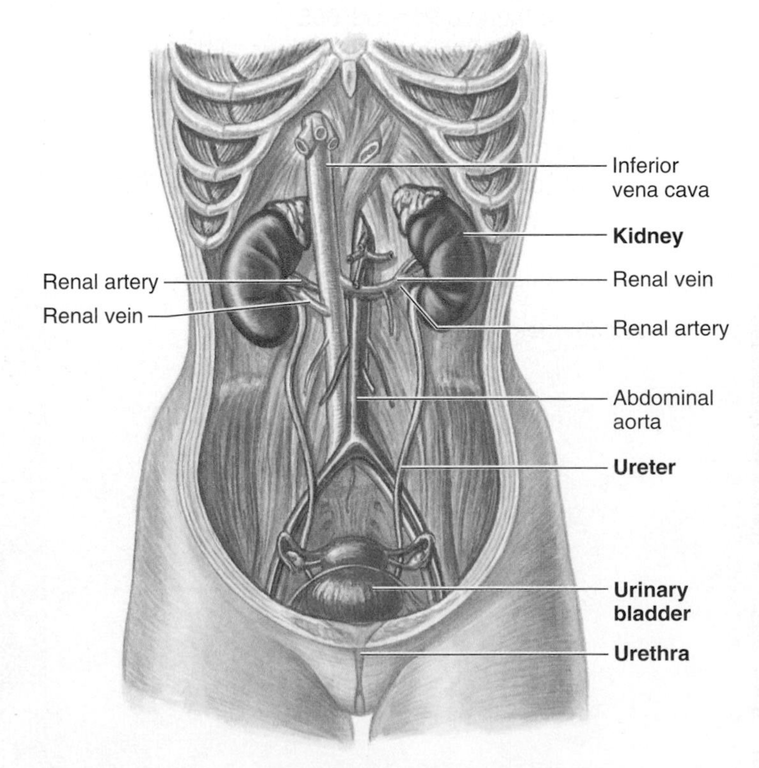

Figure 17.1 The organs of the urinary system. The urinary system of a female is shown; that of a male is the same, except that the urethra runs through the penis. AP|R

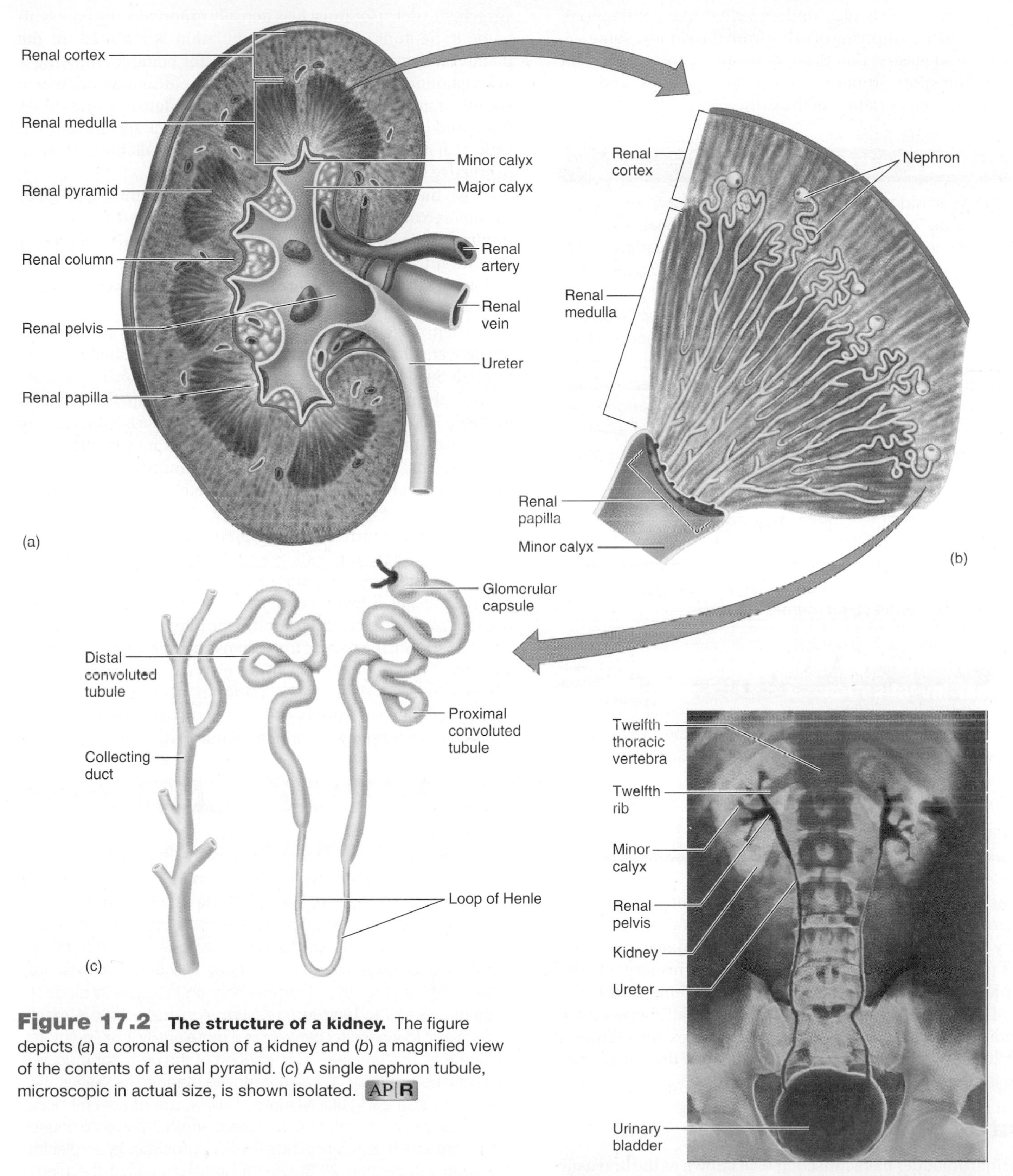

Figure 17.2 The structure of a kidney. The figure depicts (*a*) a coronal section of a kidney and (*b*) a magnified view of the contents of a renal pyramid. (*c*) A single nephron tubule, microscopic in actual size, is shown isolated. AP|R

Figure 17.3 A pseudocolor radiograph of the urinary system. In this photograph, shades of gray are assigned colors. The calyces of the kidneys, the renal pelvises, the ureters, and the urinary bladder are visible. AP|R

The calyces and pelvis also undergo rhythmic contractions, which may aid the emptying of urine from the kidney. Some scientists have suggested that these peristaltic contractions may affect the transport properties of the renal tubules, and thus influence the concentration of the urine.

CLINICAL APPLICATION

About 80% of **kidney stones** are *calcium stones,* which are composed primarily of calcium oxalate with variable traces of calcium phosphate. Other types of kidney stones are *uric acid stones, cysteine stones* (in people with cystinuria), and *struvite stones* (composed of ammonium salts of magnesium and phosphate; these are usually found in women with a urinary tract infection). Calcium kidney stones are the most common, produced when the urinary concentrations of calcium oxalate or phosphate are supersaturated—when their concentrations exceed their solubility. Large stones in the calyces or pelvis may obstruct urine flow, and smaller stones that break free and pass through a ureter can produce intense pain. Stones smaller than 5 mm usually pass through the ureter; those larger than 10 mm generally will not pass. Stones in the ureter may be removed surgically by *ureteroscopy* with *laser lithotripsy* (*litho* = stone) or be removed noninvasively by a procedure called *shock-wave lithotripsy.* Stones that are 2 cm or larger may require a surgery called *percutaneous nephrolithotomy.*

Clinical Investigation CLUES

Emily had pain in her lower back, between the twelfth rib and lumbar vertebra. Her urine showed mild hematuria (blood in the urine).

- From which organ did the pain likely originate?
- Is it likely that Emily had a kidney stone?

The **urinary bladder** is a storage sac for urine, and its shape is determined by the amount of urine it contains. An empty urinary bladder is pyramidal; as it fills, it becomes ovoid and bulges upward into the abdominal cavity. The urinary bladder is drained inferiorly by the tubular **urethra.** In females, the urethra is 4 cm (1.5 in.) long and opens into the space between the labia minora (chapter 20; see fig. 20.24). In males, the urethra is about 20 cm (8 in.) long and opens at the tip of the penis, from which it can discharge either urine or semen.

Control of Micturition

The urinary bladder has a muscular wall known as the **detrusor muscle.** Numerous gap junctions (electrical synapses; chapter 7; see fig. 7.21) interconnect its smooth muscle cells, so that action potentials can spread from cell to cell. Although action potentials can be generated automatically and in response to stretch, the detrusor muscle is densely innervated by parasympathetic neurons, and neural stimulation is required for the bladder to empty. The major stimulus for bladder emptying is acetylcholine (ACh) released by parasympathetic axons, which stimulate muscarinic ACh receptors in the detrusor muscle. As discussed in chapter 9, newer drugs that block specific muscarinic ACh receptors in the bladder are now available to treat an overactive bladder (detrusor muscle).

Two muscular sphincters surround the urethra. The upper sphincter, composed of smooth muscle, is called the *internal urethral sphincter;* the lower sphincter, composed of voluntary skeletal muscle, is called the *external urethral sphincter.* The actions of these sphincters are regulated in the process of urination, which is also known as **micturition.**

When the bladder is filling, sensory neurons in the bladder activated by stretch stimulate interneurons located in the S2 through S4 segments of the spinal cord. The spinal cord then controls the **guarding reflex,** in which parasympathetic nerves to the detrusor muscle are inhibited while the striated muscle of the external urethral sphincter is stimulated by somatic motor neurons. This prevents the involuntary emptying of the bladder. When the bladder is sufficiently stretched, sensory neuron stimulation can evoke a **voiding reflex.** During a voiding reflex, sensory information passes up the spinal cord to the pons, where a group of neurons functions as a *micturition center.* The micturition center activates the parasympathetic nerve to the detrusor muscle, causing rhythmic contractions. Inhibition of sympathetic neurons may also cause relaxation of the internal urethral sphincter. At this point, the person feels a sense of urgency but normally still has voluntary control over the external urethral sphincter, which is innervated by somatic motor neurons of the *pudendal nerve.* Incontinence would occur at a particular bladder volume unless higher brain regions inhibited the voiding reflex.

CLINICAL APPLICATION

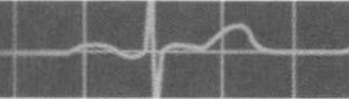

Incontinence is uncontrolled urination, a condition that can have various causes. One diagnostic tool that may be used is a *cystometrogram,* which is a graph of urinary bladder volume versus pressure. The bladder is slowly filled with normal saline solution, and the difference in pressure between the inside of the bladder and the outside (the abdominal pressure) is measured and graphed. The graph represents the compliance of the bladder—how an increase in bladder volume relates to bladder pressure. A bladder that is more compliant (distensible) can expand more easily, holding more volume at less pressure. Contraction of the detrusor muscle, as may be produced by an overactive bladder, increases the pressure at a particular volume. This is one of several possible causes of incontinence. In men, another possible cause is enlargement of the prostate (benign prostatic hyperplasia; chapter 20, section 20.3), which presses against the bladder. In women, the pelvic floor may lower with aging or after childbirth, causing the neck of the bladder to descend and increase the pressure against the urethral sphincters.

The guarding reflex permits bladder filling because higher brain regions inhibit the micturition center in the pons. These higher brain regions, including the prefrontal cortex and insula, control the switch from the guarding to the voiding reflex, thereby allowing the person to have voluntary control of micturition. When the decision to urinate is made, the micturition center in the pons becomes activated by sensory information monitoring bladder stretch. As a result, pudendal nerve activity is inhibited so that the external urethral sphincter can relax. Then the parasympathetic nerve to the detrusor muscle is activated, causing contraction of the bladder and voiding of urine. The ability to voluntarily inhibit micturition generally develops between the ages of two and three.

Microscopic Structure of the Kidney

The **nephron** (see fig. 17.2) is the functional unit of the kidney responsible for the formation of urine. Each kidney contains more than a million nephrons. A nephron consists of small tubes, or **tubules,** and associated small blood vessels. Fluid formed by capillary filtration enters the tubules and is subsequently modified by transport processes; the resulting fluid that leaves the tubules is urine.

Renal Blood Vessels

Arterial blood enters the kidney through the *renal artery,* which divides into *interlobar arteries* (fig. 17.4) that pass between the pyramids through the renal columns. *Arcuate arteries* branch from the interlobar arteries at the boundary of the cortex and medulla. A number of *interlobular arteries* radiate from the arcuate arteries into the cortex and subdivide into numerous **afferent arterioles** (fig. 17.5), which are microscopic. The

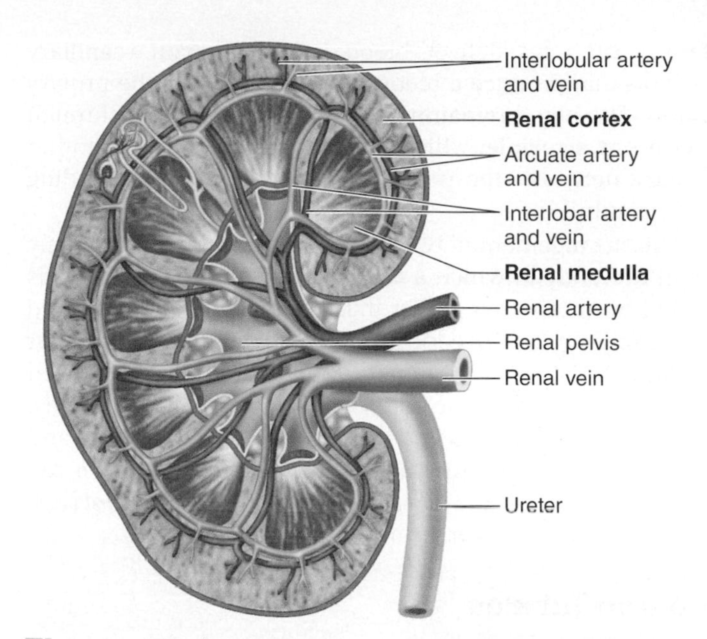

Figure 17.4 Major blood vessels of the kidney. The vessels carrying blood into the renal medulla and cortex, and those carrying blood out of the kidney, are illustrated. APIR

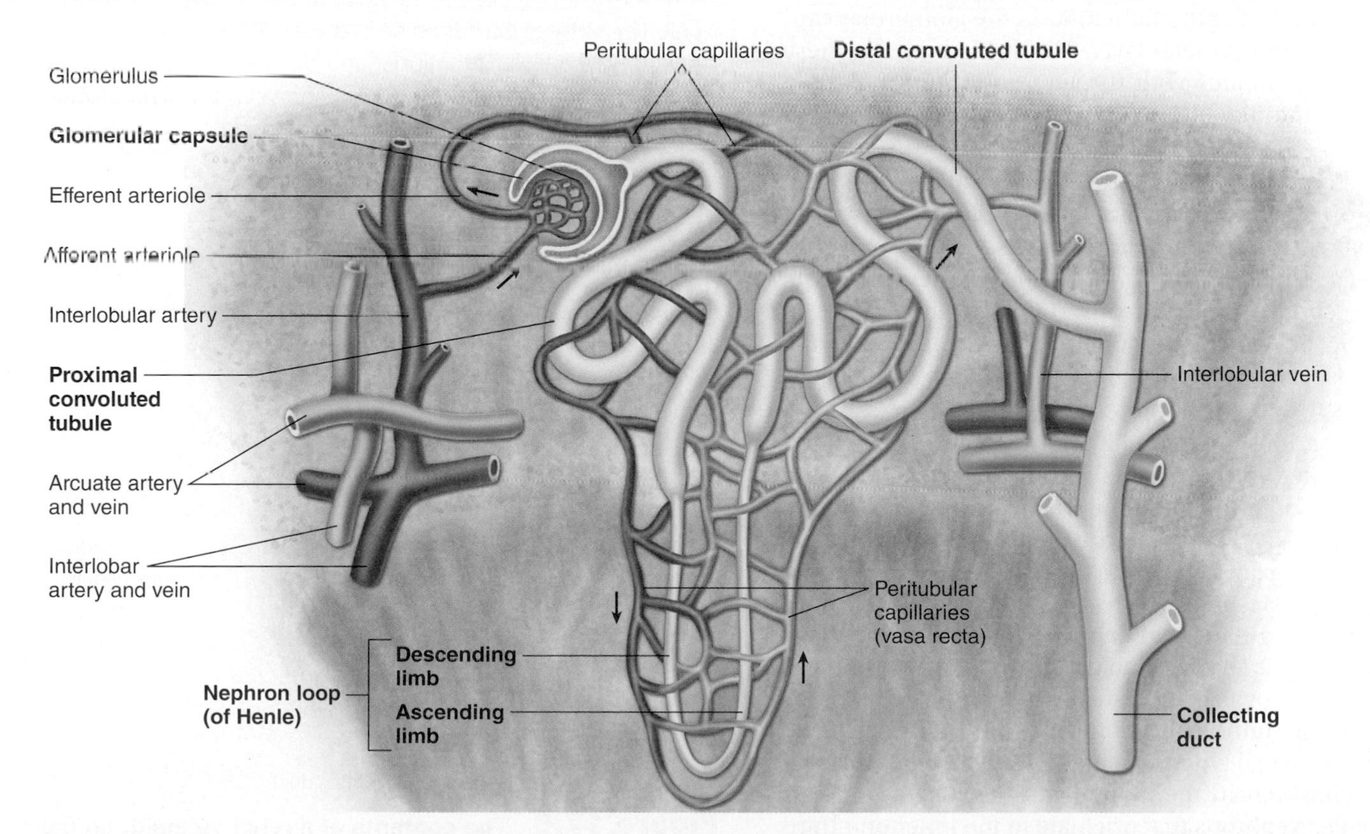

Figure 17.5 The nephron tubules and associated blood vessels. In this simplified illustration, the blood flow from a glomerulus to an efferent arteriole, to the peritubular capillaries, and to the venous drainage of the kidneys is indicated with arrows. The names for the different regions of the nephron tubules are indicated with boldface type. APIR

afferent arterioles deliver blood into **glomeruli**—capillary networks that produce a blood filtrate that enters the urinary tubules. The blood remaining in a glomerulus leaves through an **efferent arteriole,** which delivers the blood into another capillary network—the **peritubular capillaries** surrounding the renal tubules.

This arrangement of blood vessels is unique. It is the only one in the body in which a capillary bed (the glomerulus) is drained by an arteriole rather than by a venule and delivered to a second capillary bed located downstream (the peritubular capillaries). Blood from the peritubular capillaries is drained into veins that parallel the course of the arteries in the kidney. These veins are called the *interlobular veins, arcuate veins,* and *interlobar veins.* The interlobar veins descend between the pyramids, converge, and leave the kidney as a single *renal vein,* which empties into the inferior vena cava.

Nephron Tubules

The tubular portion of a nephron consists of a *glomerular capsule,* a *proximal convoluted tubule,* a *descending limb of the loop of Henle,* an *ascending limb of the loop of Henle,* and a *distal convoluted tubule* (fig. 17.5).

The **glomerular (Bowman's) capsule** surrounds the glomerulus. The glomerular capsule and its associated glomerulus are located in the cortex of the kidney and together constitute the *renal corpuscle.* The glomerular capsule contains an inner visceral layer of epithelium around the glomerular capillaries and an outer parietal layer. The space between these two layers is continuous with the lumen of the tubule and receives the glomerular filtrate, as will be described in the next section.

Filtrate that enters the glomerular capsule passes into the lumen of the **proximal convoluted tubule.** The wall of the proximal convoluted tubule consists of a single layer of cuboidal cells containing millions of microvilli; these microvilli increase the surface area for reabsorption. In the process of reabsorption, salt, water, and other molecules needed by the body are transported from the lumen, through the tubular cells and into the surrounding peritubular capillaries.

The glomerulus, glomerular capsule, and convoluted tubule are located in the renal cortex. Fluid passes from the proximal convoluted tubule to the **nephron loop,** or **loop of Henle.** This fluid is carried into the medulla in the **descending limb** of the loop and returns to the cortex in the **ascending limb** of the loop. Back in the cortex, the tubule again becomes coiled and is called the **distal convoluted tubule.** The distal convoluted tubule is shorter than the proximal tubule and has relatively few microvilli. The distal convoluted tubule terminates as it empties into a collecting duct.

The two principal types of nephrons are classified according to their position in the kidney and the lengths of their loops of Henle. Nephrons that originate in the inner one-third of the cortex—called *juxtamedullary nephrons* because they are next to the medulla—have longer nephron loops than the more numerous *cortical nephrons,* which originate in the outer two-thirds of the cortex (fig. 17.6). The juxtamedullary nephrons play an important role in the ability of the kidney to produce a concentrated urine.

CLINICAL APPLICATION

Polycystic kidney disease (PKD) is a condition usually inherited as an autosomal dominant trait that affects one in every 600 to 1,000 people. Both kidneys are greatly enlarged due to the presence of hundreds to thousands of cysts that form in all segments of the nephron tubules. The cysts eventually separate from the tubule in which they form and become filled with fluid. Over 50% of people with the autosomal dominant form of this disease (abbreviated ADPKD) develop progressive renal failure during middle age so that dialysis or kidney transplantation is required. About 85% of people with ADPKD have a mutation on chromosome 16, while the remainder have a mutation of chromosome 4. These two genes code for proteins termed *polycystin-1* and *polycystin-2,* respectively, which interact to form a complex found in different locations in the renal tubule epithelial cells. The polycystin complex is associated with the *primary cilium,* a single cilium that protrudes into the tubule lumen and bends in response to the flow of fluid through the tubule. Polycystin-1 and polycystin-2 cooperate to enable the primary cilium to function as a mechanosensor, where bending of the cilium by fluid movement in the renal tubule results in the entry of Ca^{2+} into the renal tubule cell. The entry of Ca^{2+} then serves as a second messenger for a variety of cellular functions, and disruptions of these functions by mutations affecting the polycystin proteins produce polycystic kidney disease.

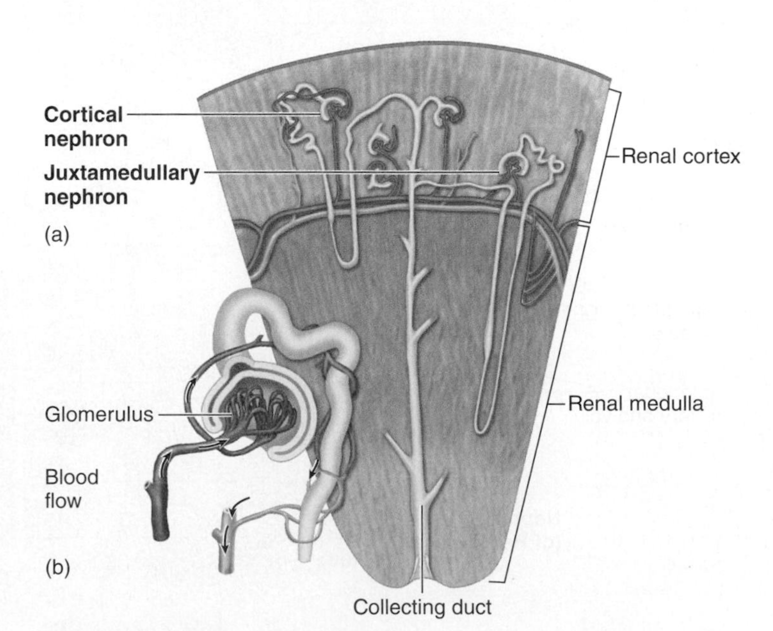

Figure 17.6 The contents of a renal pyramid. (*a*) The position of cortical and juxtamedullary nephrons is shown within the renal pyramid of the kidney. (*b*) The direction of blood flow in the vessels of the nephron is indicated with arrows. AP|R

A **collecting duct** receives fluid from the distal convoluted tubules of several nephrons. Fluid is then drained by the collecting duct from the cortex to the medulla as the collecting duct passes through a renal pyramid. This fluid, now called urine, passes into a minor calyx. Urine is then funneled through the renal pelvis and out of the kidney in the ureter.

CHECKPOINT

1. Describe the "theme" of kidney function in a single sentence and list the components of this functional theme.

2a. Draw and label the tubular components of a nephron and indicate which parts are in the cortex and which are in the medulla.

2b. Trace the course of tubular fluid from the glomerular capsules to the ureter.

3. Trace the course of blood flow through the kidney from the renal artery to the renal vein.

17.2 GLOMERULAR FILTRATION

The glomerular capillaries have large pores in their walls, and the layer of Bowman's capsule in contact with the glomerulus has filtration slits. Water, together with dissolved solutes, can thus pass from the blood plasma to the inside of the capsule and the nephron tubules.

LEARNING OUTCOMES

After studying this section, you should be able to:

4. Describe glomerular filtration and the structures and forces involved.
5. Explain the significance of the glomerular filtration rate (GFR) and how it is regulated.

Endothelial cells of the glomerular capillaries have large pores (200 to 500 Å in diameter) called fenestrae; thus, the glomerular endothelium is said to be *fenestrated.* As a result of these large pores, glomerular capillaries are 100 to 400 times more permeable to plasma water and dissolved solutes than are the capillaries of skeletal muscles. Although the pores of glomerular capillaries are large, they are still small enough to prevent the passage of red blood cells, white blood cells, and platelets into the filtrate.

Before the fluid in blood plasma can enter the interior of the glomerular capsule, it must pass through three layers that could serve as selective filters. The fluid entering the glomerular capsule is thus referred to as a *filtrate.* This is the fluid that will become modified as it passes through the different segments of the nephron tubules to become the urine.

The first potential filtration barrier is the **capillary fenestrae,** which are large enough to allow proteins to pass but are surrounded by charges that may present some barrier to plasma proteins. The second potential barrier is the **glomerular basement membrane,** a layer of collagen IV and proteoglycans (chapter 1, section 1.3) lying immediately outside the capillary endothelium. This may offer some barrier to plasma proteins, and indeed a genetic defect in collagen IV can produce inherited glomerulonephritis (*Alport's syndrome*). The glomerular basement membrane is more than five times as thick as the basement membrane of other vessels, and is the structure that most restricts the rate of fluid flow into the capsule lumen.

The filtrate must then pass through the inner (visceral) layer of the glomerular capsule, where the third potential filtration barrier is located. This layer is composed of cells called *podocytes,* shaped somewhat like an octopus with a bulbous cell body and several thick arms. Each arm has thousands of cytoplasmic extensions known as *pedicels,* or *foot processes* (fig. 17.7). These foot processes interdigitate, like the fingers of clasped hands (to mix analogies), as they wrap around the glomerular capillaries. The narrow slits between adjacent pedicles provide passageways through which the filtered molecules must pass to enter the interior of the glomerular filtrate (fig. 17.8). In some electron micrographs, a fine line may be seen between the podocyte foot processes (fig. 17.9). This is called the **slit diaphragm,** and presents the third potential filtration barrier.

All dissolved plasma solutes easily pass through all three potential filtration barriers to enter the interior of the glomerular capsule. However, plasma proteins are mostly excluded from the filtrate because of their large sizes and net negative

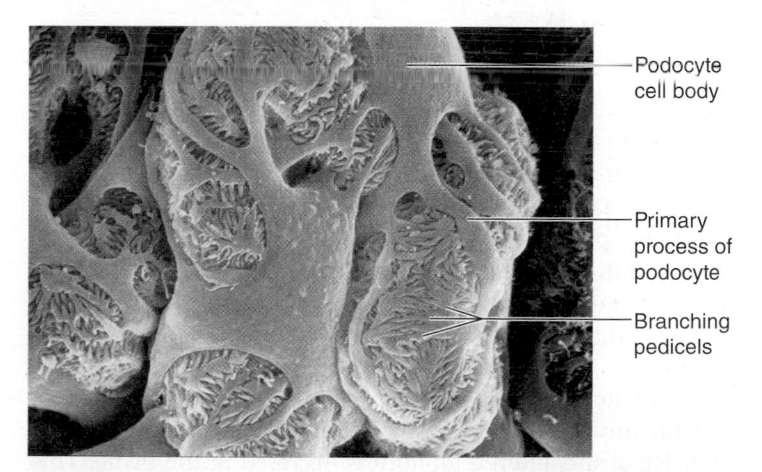

Figure 17.7 A scanning electron micrograph of the glomerular capillaries and capsule. The inner (visceral) layer of the glomerular (Bowman's) capsule is composed of podocytes, as shown in this scanning electron micrograph. Very fine extensions of these podocytes form foot processes, or pedicels, that interdigitate around the glomerular capillaries. Spaces between adjacent pedicels form the "filtration slits" (see also fig. 17.8). AP|R

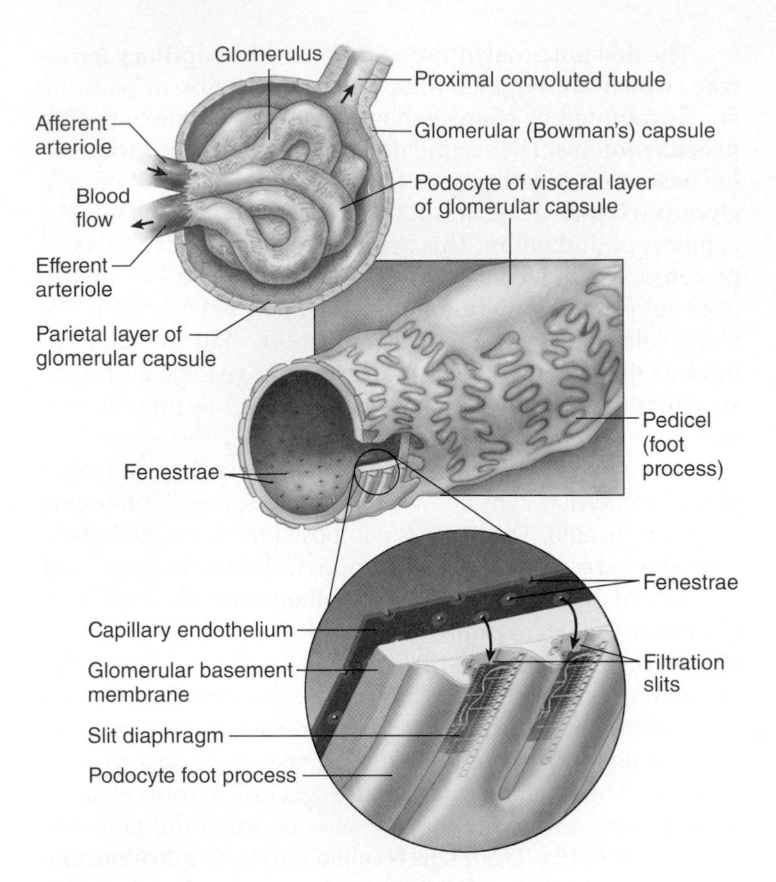

Figure 17.8 The structure of the glomerulus and capsule. An illustration of the relationship between the glomerular capillaries and the inner layer of the glomerular (Bowman's) capsule. Notice that filtered molecules pass out of the fenestrae of the capillaries and through the filtration slits to enter the cavity of the capsule. Plasma proteins are excluded from the filtrate by the glomerular basement membrane and the slit diaphragm. AP|R

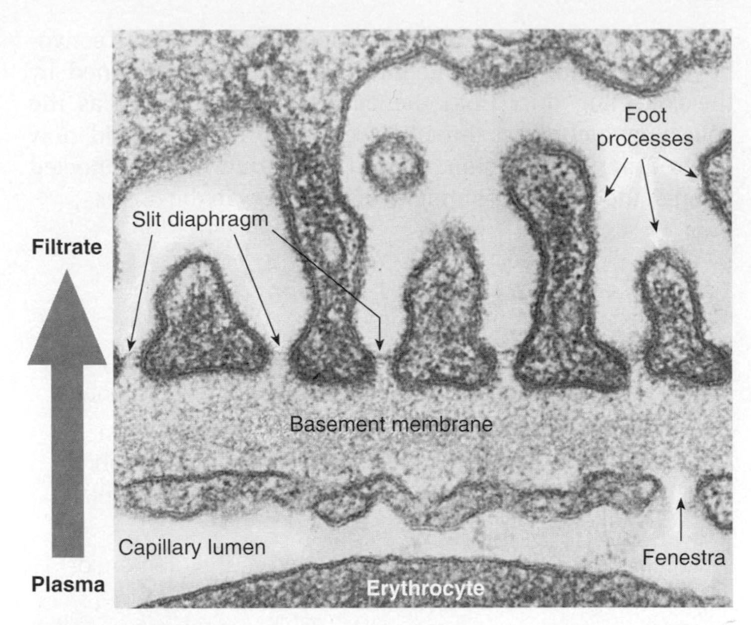

Figure 17.9 An electron micrograph of the filtration barrier. This electron micrograph shows the barrier separating the capillary lumen from the cavity of the glomerular (Bowman's) capsule.

Clinical Investigation CLUES

Emily's urine contained only trace amounts of protein.

- What prevented larger amounts of protein from entering her urine?
- If she had proteinuria (more than trace amounts of protein in her urine), what could that indicate?

charges. Until recently most scientists believed that the glomerular basement membrane was the primary filter excluding plasma proteins from the filtrate. More recent evidence indicates that the slit diaphragm poses the major barrier to the passage of plasma proteins into the filtrate. One source of evidence for this is based on the consequences of genetic defects in the proteins that compose the slit diaphragms. These defects in the slit diaphragm result in massive leakage of proteins into the filtrate, and thus in *proteinuria* (proteins in the urine).

Actually, a small amount of albumin (the major class of plasma proteins) does normally enter the filtrate, but less than 1% of this filtered amount is excreted in the urine. This is because most of the small amount of albumin that enters the filtrate is reabsorbed, or transported across the cells of the proximal tubule into the surrounding blood. In the case of filtered albumin, such reabsorption is accomplished by receptor-mediated endocytosis (chapter 3; see fig. 3.4). Proteinuria thus occurs when damage to the slit diaphragm filtration barrier causes more protein to enter the filtrate than can be reabsorbed in this way.

Glomerular Ultrafiltrate

The fluid that enters the glomerular capsule is called **filtrate,** or **ultrafiltrate** (fig. 17.10) because it is formed under pressure—the hydrostatic pressure of the blood. This process is similar to the formation of tissue fluid by other capillary beds in the body in response to Starling forces (see chapter 14; fig. 14.9). The force favoring filtration is opposed by a counterforce developed by the hydrostatic pressure of fluid in the glomerular capsule. Also, since the protein concentration of the tubular fluid is low (less than 2 to 5 mg per 100 ml) compared to that of plasma (6 to 8 g per 100 ml), the greater colloid osmotic pressure of plasma promotes the osmotic return of filtered water. When these opposing forces are subtracted from the hydrostatic pressure of the glomerular capillaries, a *net filtration pressure* of only about 10 mmHg is obtained.

Because glomerular capillaries are extremely permeable and have an extensive surface area, this modest net filtration pressure produces an extraordinarily large volume of filtrate. The **glomerular filtration rate (GFR)** is the volume of filtrate

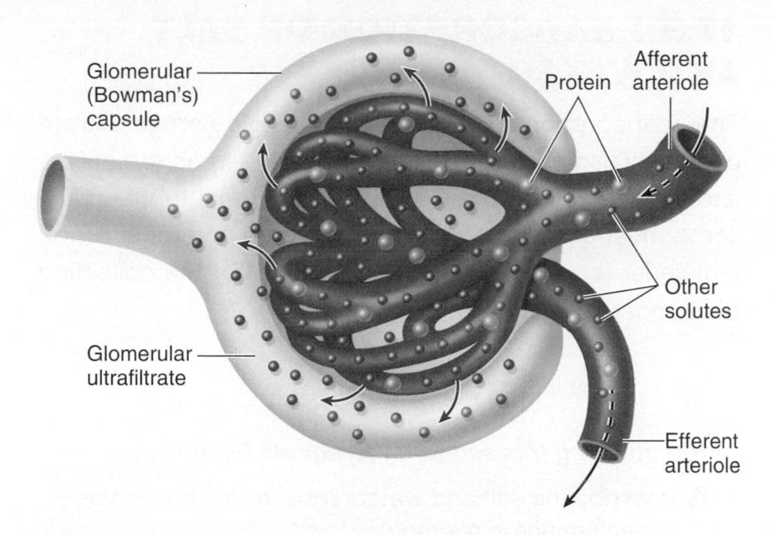

Figure 17.10 The formation of glomerular ultrafiltrate. Only a very small proportion of plasma proteins (*green spheres*) are filtered, but smaller plasma solutes (*purple spheres*) easily enter the glomerular ultrafiltrate. Arrows indicate the direction of filtration.

produced by both kidneys per minute. The GFR averages 115 ml per minute in women and 125 ml per minute in men. This is equivalent to 7.5 L per hour or 180 L per day (about 45 gallons)! Since the total blood volume averages about 5.5 L, this means that the total blood volume is filtered into the urinary tubules every 40 minutes. Most of the filtered water must obviously be returned immediately to the vascular system or a person would literally urinate to death within minutes.

Regulation of Glomerular Filtration Rate

Vasoconstriction or dilation of afferent arterioles affects the rate of blood flow to the glomerulus, and thus affects the glomerular filtration rate. Changes in the diameter of the afferent arterioles result from both extrinsic regulatory mechanisms (produced by sympathetic nerve innervation) and intrinsic regulatory mechanisms (those within the kidneys, also termed *renal autoregulation*). These mechanisms are needed to ensure that the GFR will be high enough to allow the kidneys to eliminate wastes and regulate blood pressure, but not so high as to cause excessive water loss.

Sympathetic Nerve Effects

An increase in sympathetic nerve activity, as occurs during the fight-or-flight reaction and exercise, stimulates constriction of afferent arterioles. This helps preserve blood volume and divert blood to the muscles and heart. A similar effect occurs during cardiovascular shock, when sympathetic nerve activity stimulates vasoconstriction. The decreased GFR and the resulting decreased rate of urine formation help compensate for the rapid drop in blood pressure under these circumstances (fig. 17.11).

Renal Autoregulation

When the direct effect of sympathetic stimulation is experimentally removed, the effect of systemic blood pressure on GFR can be observed. Under these conditions, surprisingly, the GFR remains relatively constant despite changes in mean arterial pressure within a range of 70 to 180 mmHg (normal mean arterial pressure is 100 mmHg). The ability of the kidneys to maintain a relatively constant GFR in the face of fluctuating blood pressures is called **renal autoregulation.**

In renal autoregulation, afferent arterioles dilate when the mean arterial pressure falls toward 70 mmHg and constrict when the mean arterial pressure rises above normal. Changes that may occur in the efferent arterioles are believed to be of secondary importance.

Blood flow to the glomeruli and GFR can thus remain relatively constant within the autoregulatory range of blood pressure values. The effects of different regulatory mechanisms on the GFR are summarized in table 17.1.

Two general mechanisms are responsible for renal autoregulation: (1) *myogenic* constriction of the afferent arteriole, due to the ability of the smooth muscle to sense and

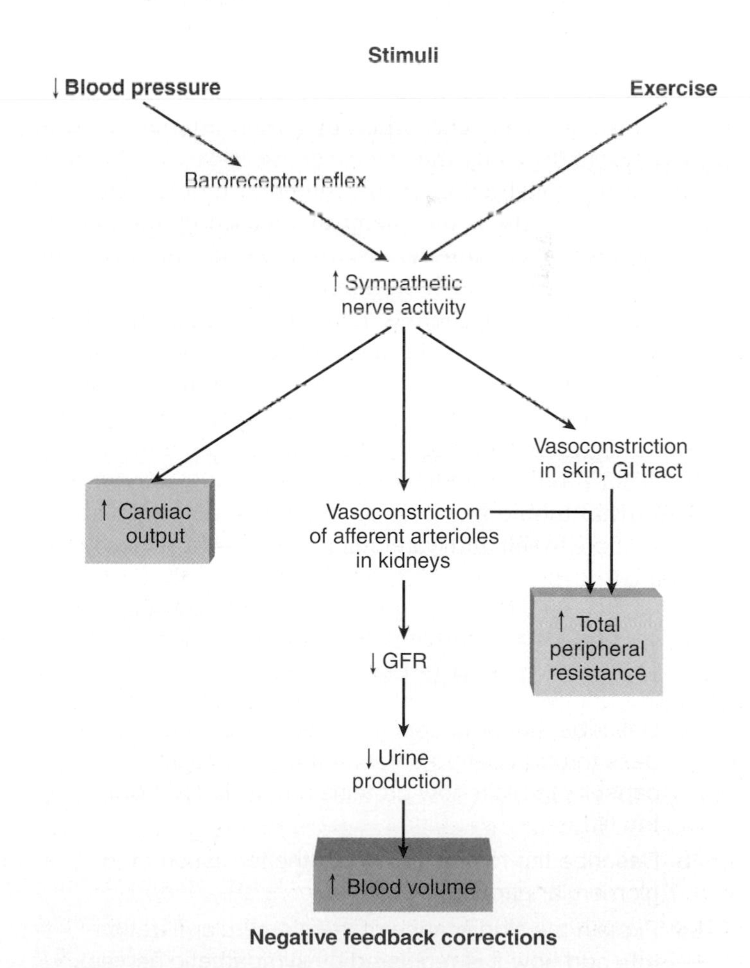

Figure 17.11 Sympathetic nerve effects. The effect of increased sympathetic nerve activity on kidney function and other physiological processes is illustrated.

Table 17.1 | Regulation of the Glomerular Filtration Rate (GFR)

Regulation	Stimulus	Afferent Arteriole	GFR
Sympathetic nerves	Activation by baroreceptor reflex or by higher brain centers	Constricts	Decreases
Autoregulation	Decreased blood pressure	Dilates	No change
Autoregulation	Increased blood pressure	Constricts	No change

respond to an increase in arterial pressure; and (2) the effects of locally produced chemicals on the afferent arteriole, which is part of a process called **tubuloglomerular feedback.** The sensor in tubuloglomerular feedback is a group of specialized cells called the **macula densa,** located in the thick portion of the ascending limb where it loops back and comes into contact with the afferent and efferent arterioles in the renal cortex. The macula densa here is part of a larger functional unit known as the *juxtaglomerular apparatus* (see fig. 17.26), which will be described in section 17.5.

When there is an increased delivery of NaCl and H_2O to the distal tubule (as occurs when increased arterial pressure causes a rise in the GFR), the macula densa releases a chemical signal causing constriction of the afferent arteriole. Scientists now believe that ATP is the chemical released by the macula densa, although adenosine derived from ATP may more directly produce vasoconstriction of the afferent arteriole.

In summary, when there is increased salt and water flowing through the distal tubule, vasoconstriction of the afferent arteriole in response to ATP (or adenosine) from the macula densa lowers the GFR. This negative feedback response reduces the salt and water entering the nephron tubule and arriving at the distal tubule. Tubuloglomerular feedback may protect the late distal tubule and cortical collecting duct—structures that contribute to salt and water reabsorption—from becoming overloaded.

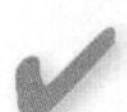

CHECKPOINT

4a. Describe the structures that plasma fluid must pass through before entering the glomerular capsule. Explain how proteins are excluded from the filtrate.

4b. Describe the forces that affect the formation of glomerular ultrafiltrate.

5a. Explain the significance of the glomerular filtration rate and how it is regulated by sympathetic nerves.

5b. Explain tubuloglomerular feedback and renal autoregulation of the GFR.

17.3 REABSORPTION OF SALT AND WATER

The reabsorption of water from the glomerular filtrate occurs by osmosis, which results from the transport of Na^+ and Cl^- across the tubule wall. The proximal tubule reabsorbs most of the filtered salt and water, and most of the remainder is reabsorbed across the wall of the collecting duct under ADH stimulation.

LEARNING OUTCOMES

After studying this section, you should be able to:

6. Describe the salt and water reabsorption properties of each nephron segment.

7. Explain the countercurrent multiplier system.

8. Explain how ADH acts to promote water reabsorption.

Although about 180 L of glomerular ultrafiltrate are produced each day, the kidneys normally excrete only 1 to 2 L of urine in a 24-hour period. Approximately 99% of the filtrate must thus be returned to the vascular system, while 1% is excreted in the urine. The urine volume, however, varies according to the needs of the body. When a well-hydrated person drinks a liter or more of water, urine production increases to 16 ml per minute (the equivalent of 23 L per day if this were to continue for 24 hours). In severe dehydration, when the body needs to conserve water, only 0.3 ml of urine per minute, or 400 ml per day, are produced. A volume of 400 ml of urine per day is the minimum needed to excrete the metabolic wastes produced by the body; this is called the **obligatory water loss.** When water in excess of this amount is excreted, the urine becomes increasingly diluted as its volume is increased.

Regardless of the body's state of hydration, it is clear that most of the filtered water must be returned to the vascular system to maintain blood volume and pressure. The return of filtered molecules from the tubules to the blood is called **reabsorption** (fig. 17.12). About 85% of the 180 L of glomerular filtrate formed per day is reabsorbed in a constant, unregulated fashion by the proximal tubules and descending limbs of the nephron loops. This reabsorption, as well as the regulated reabsorption of the remaining volume of filtrate, occurs by osmosis. A concentration gradient must thus be created between tubular filtrate and the plasma in the surrounding capillaries that promotes the osmosis of water back into the vascular system from which it originated.

Reabsorption in the Proximal Tubule

Because all plasma solutes, with the exception of proteins, are able to enter the glomerular ultrafiltrate freely, the total solute concentration (osmolality) of the filtrate is essentially the same

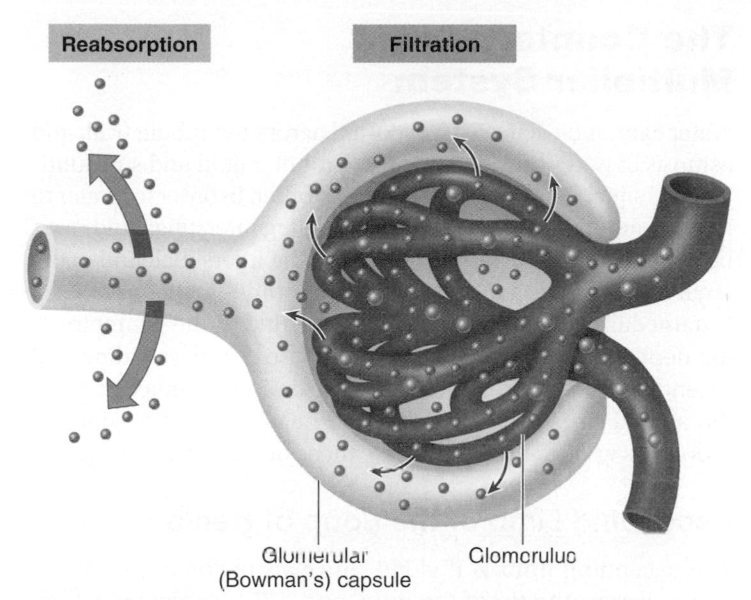

Figure 17.12 Filtration and reabsorption. Plasma water and its dissolved solutes (except proteins) enter the glomerular ultrafiltrate by filtration, but most of these filtered molecules are reabsorbed. The term *reabsorption* refers to the transport of molecules out of the tubular filtrate back into the blood.

as that of plasma. This total solute concentration is equal to 300 milliosmoles per liter, or 300 milliosmolal (300 mOsm), as described in chapter 6. The filtrate is thus said to be *isosmotic* with the plasma (chapter 6, section 6.2). Reabsorption by osmosis cannot occur unless the solute concentrations of plasma in the peritubular capillaries and the filtrate are altered by active transport processes. This is achieved by the active transport of Na^+ from the filtrate to the peritubular blood.

Active and Passive Transport

The epithelial cells that compose the wall of the proximal tubule are joined together by tight junctions only toward their apical sides—that is, the sides of each cell that are closest to the lumen of the tubule (see fig. 17.24). Each cell has four exposed surfaces: the apical side facing the lumen, which contains microvilli; the basal side facing the peritubular capillaries; and the lateral sides facing the narrow clefts between adjacent epithelial cells.

The concentration of Na^+ in the glomerular ultrafiltrate—and thus in the fluid entering the proximal tubule—is the same as in plasma. The cytoplasm in epithelial cells of the tubule, however, has a much lower Na^+ concentration. This lower Na^+ concentration is partially due to the low permeability of the plasma membrane to Na^+ and partially due to the active transport of Na^+ out of the cells by Na^+/K^+ pumps (chapter 6, section 6.3). In the cells of the proximal tubule, the Na^+/K^+ pumps are located in the basal and lateral sides of the plasma membrane but not in the apical membrane. As a result of the action of these active transport pumps, a concentration gradient is created that favors the diffusion of Na^+ from the tubular fluid across the apical plasma membranes and into the epithelial cells of the proximal tubule. The Na^+ is then extruded into the surrounding interstitial (tissue) fluid by the Na^+/K^+ pumps.

The transport of Na^+ from the tubular fluid to the interstitial fluid surrounding the proximal tubule creates a potential difference across the wall of the tubule, with the lumen as the negative pole. This electrical gradient favors the passive transport of Cl^- toward the higher Na^+ concentration in the interstitial fluid. Chloride ions, therefore, passively follow sodium ions out of the filtrate into the interstitial fluid. As a result of the accumulation of NaCl, the osmolality and osmotic pressure of the interstitial fluid surrounding the epithelial cells are increased above those of the tubular fluid. This is particularly true of the interstitial fluid between the lateral membranes of adjacent epithelial cells, where the narrow spaces permit the accumulated NaCl to achieve a higher concentration.

An osmotic gradient is thus created between the tubular fluid and the interstitial fluid surrounding the proximal tubule. Because the cells of the proximal tubule are permeable to water, water moves by osmosis from the tubular fluid into the epithelial cells and then across the basal and lateral sides of the epithelial cells into the interstitial fluid. The salt and water that were reabsorbed from the tubular fluid can then move passively into the surrounding peritubular capillaries, and in this way be returned to the blood (fig. 17.13).

Significance of Proximal Tubule Reabsorption

Approximately 65% of the salt and water in the original glomerular ultrafiltrate is reabsorbed across the proximal tubule and returned to the vascular system. The volume of tubular fluid remaining is reduced accordingly, but this fluid is still

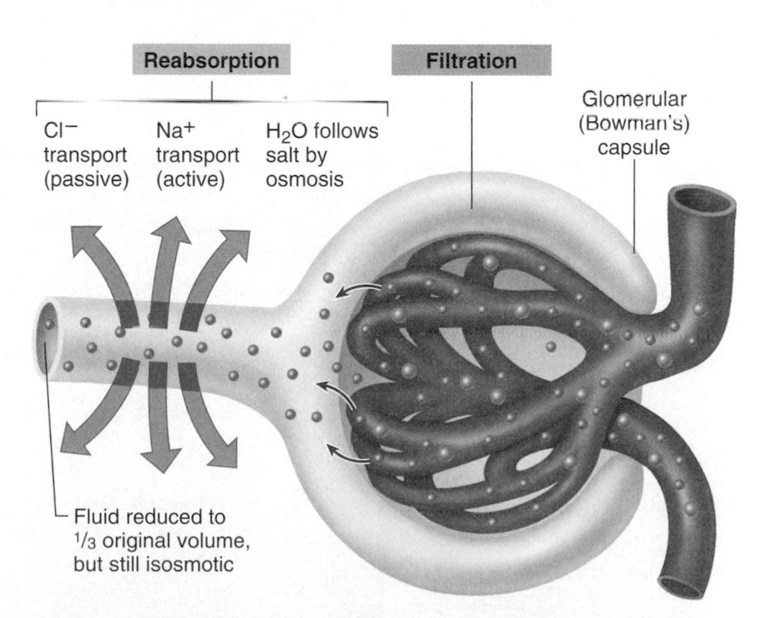

Figure 17.13 Salt and water reabsorption in the proximal tubule. Sodium is actively transported out of the filtrate (see fig. 17.24) and chloride follows passively by electrical attraction. Water follows the salt out of the tubular filtrate into the peritubular capillaries by osmosis.

isosmotic with the blood, which has a concentration of 300 mOsm. This is because the plasma membranes in the proximal tubule are freely permeable to water, so that water and salt are removed in proportionate amounts.

An additional smaller amount of salt and water (about 20%) is returned to the vascular system by reabsorption through the descending limb of the nephron loop. This reabsorption, like that in the proximal tubule, occurs constantly, regardless of the person's state of hydration. Unlike reabsorption in later regions of the nephron (distal tubule and collecting duct), it is not subject to hormonal regulation. Therefore, approximately 85% of the filtered salt and water is reabsorbed in a constant fashion in the early regions of the nephron (proximal tubule and nephron loop). This reabsorption is very costly in terms of energy expenditures, accounting for as much as 6% of the calories consumed by the body at rest.

Since 85% of the original glomerular ultrafiltrate is reabsorbed in the early regions of the nephron, only 15% of the initial filtrate remains to enter the distal convoluted tubule and collecting duct. This is still a large volume of fluid—15% × GFR (180 L per day) = 27 L per day—that must be reabsorbed to varying degrees in accordance with the body's state of hydration. This "fine tuning" of the percentage of reabsorption and urine volume is accomplished by the action of hormones on the later regions of the nephron.

The Countercurrent Multiplier System

Water cannot be actively transported across the tubule wall, and osmosis of water cannot occur if the tubular fluid and surrounding interstitial fluid are isotonic to each other. In order for water to be reabsorbed by osmosis, the surrounding interstitial fluid must be hypertonic. The osmotic pressure of the interstitial fluid in the renal medulla is raised to more than four times that of plasma by juxtamedullary nephrons. This is partly due to the geometry of the nephron loops, which bend sharply so that descending and ascending limbs are in close enough proximity to interact. Because the ascending limb is the active partner in this interaction, its properties will be described before those of the descending limb.

Ascending Limb of the Loop of Henle

The ascending limb is divided into two regions: a *thin segment,* nearest the tip of the loop, and a *thick segment,* which carries the filtrate into the distal convoluted tubule in the renal cortex. Salt (NaCl) is actively extruded from the thick segment of the ascending limb into the surrounding interstitial fluid (fig. 17.14). This is accomplished differently from the way NaCl is reabsorbed from the proximal tubule. In the cells of the thick portion of the ascending limb, the movement of Na^+

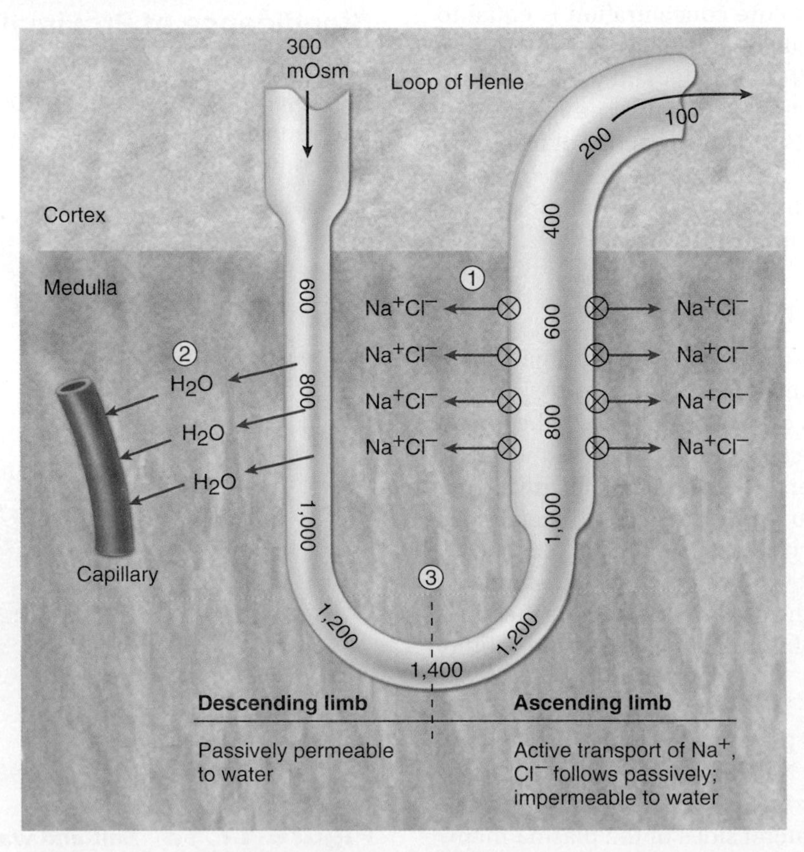

Figure 17.14 The countercurrent multiplier system. (1) The extrusion of sodium chloride from the ascending limb makes the surrounding interstitial fluid more concentrated. Multiplication of this concentration is due to the fact that (2) the descending limb is passively permeable to water, which causes its fluid to increase in concentration as the surrounding interstitial fluid becomes more concentrated. (3) The deepest region of the medulla reaches a concentration of 1,400 mOsm. (All numbers indicate milliosmolal units.)

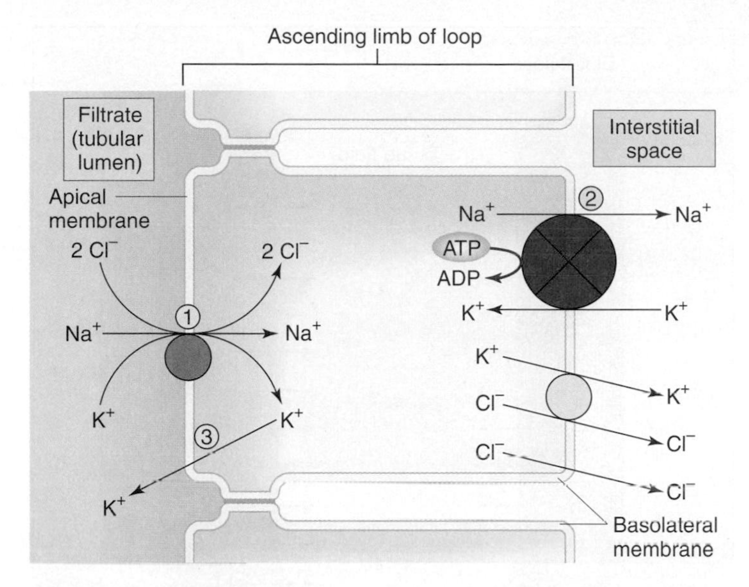

Figure 17.15 The transport of ions in the ascending limb. (1) In the thick segment of the ascending limb of the loop, Na^+ and K^+ together with two Cl^- enter the tubule cells. (2) Na^+ is then actively transported out into the interstitial space and Cl follows passively. (3) The K^+ diffuses back into the filtrate, and some also enters the interstitial space.

down its electrochemical gradient from the filtrate into the cells powers the inward secondary active transport of K^+ and Cl^-. This occurs in a ratio of 1 Na^+ to 1 K^+ to 2 Cl^-. The Na^+ is then actively transported across the basolateral membrane to the interstitial fluid by the Na^+/K^+ pumps. Cl^- follows the Na^+ passively because of electrical attraction, and K^+ passively diffuses back into the filtrate (fig. 17.15).

Although the mechanism of NaCl transport is different in the ascending limb than in the proximal tubule, the net effect is the same: salt (NaCl) is extruded into the interstitial fluid. Unlike the epithelial walls of the proximal tubule, however, the walls of the ascending limb of the loop of Henle are *not permeable to water.* The filtrate in the ascending limb thus becomes increasingly dilute as it ascends into the cortex; by contrast, the interstitial fluid surrounding the nephron loops in the medulla becomes increasingly more concentrated. By means of these processes, the tubular fluid that enters the distal tubule in the cortex is made hypotonic (with a concentration of about 100 mOsm), whereas the interstitial fluid in the medulla is made hypertonic.

Descending Limb of the Loop of Henle

The deeper regions of the medulla, around the tips of the loops of juxtamedullary nephrons, reach a concentration of 1,200 to 1,400 mOsm. In order to reach a concentration this high, the salt pumped out of the ascending limb must accumulate in the interstitial fluid of the medulla. This occurs because of the properties of the descending limb, and because blood vessels around the loop do not carry back all of the extruded salt to the general circulation. The capillaries in the medulla are uniquely arranged to trap NaCl in the interstitial fluid, as will be discussed shortly.

The descending limb does not actively transport salt, and indeed is impermeable to the passive diffusion of salt. It is, however, permeable to water. Because the surrounding interstitial fluid is hypertonic to the filtrate in the descending limb, water is drawn out of the descending limb by osmosis and enters blood capillaries. The concentration of tubular fluid is thus increased, and its volume is decreased, as it descends toward the tips of the loops.

As a result of these passive transport processes in the descending limb, the fluid that "rounds the bend" at the tip of the loop has the same osmolality as that of the surrounding interstitial fluid (1,200 to 1,400 mOsm). There is, therefore, a higher salt concentration arriving in the ascending limb than there would be if the descending limb simply delivered isotonic fluid. Salt transport by the ascending limb is increased accordingly, so that the "saltiness" (NaCl concentration) of the interstitial fluid is multiplied (see fig. 17.14).

Countercurrent Multiplication

Countercurrent flow (flow in opposite directions) in the ascending and descending limbs and the close proximity of the two limbs allow for interaction between them. Because the concentration of the tubular fluid in the descending limb reflects the concentration of surrounding interstitial fluid, and the concentration of this fluid is raised by the active extrusion of salt from the ascending limb, a *positive feedback mechanism* is created. The more salt the ascending limb extrudes, the more concentrated will be the fluid that is delivered to it from the descending limb. This positive feedback mechanism, which multiplies the concentration of interstitial fluid and descending limb fluid, is called the **countercurrent multiplier system.**

Let's imagine that fluid goes through the loop of Henle in successive steps, one following the other. Flow is really continuous, but these hypothetical steps allow us to mentally picture the countercurrent multiplication mechanism. To start with, let's suppose that the fluid that leaves the descending limb and reaches the ascending limb is at first isosmotic (300 mOsm). Through active transport, the thick ascending limb pumps out some of the NaCl. This NaCl becomes trapped in the interstitial fluid by blood vessels called the *vasa recta.* The following progression of steps will occur:

1. The interstitial fluid is now a little hypertonic due to the NaCl pumped out of the ascending limb.
2. Because of the slightly hypertonic interstitial fluid, some water leaves the descending limb by osmosis (and enters the blood) as the filtrate goes deeper into the medulla. This makes the filtrate somewhat hypertonic when it reaches the ascending limb.
3. The now higher NaCl concentration of the filtrate that enters the ascending limb allows it to pump out more NaCl than it did before, because more NaCl is now available to the carriers. The interstitial fluid now becomes yet more concentrated.

4. Because the interstitial fluid is more concentrated than it was in step 2, more water is drawn out of the descending limb by osmosis, making the filtrate even more hypertonic when it reaches the ascending limb.
5. Step 3 is repeated, but to a greater extent because of the higher NaCl concentration delivered to the ascending limb.
6. This progression continues until the maximum concentration is reached in the inner medulla. This maximum is determined by the capacity of the active transport pumps working along the lengths of the thick segments of the ascending limbs.

What does the countercurrent multiplier system accomplish? Most importantly, it increases the concentration of renal interstitial fluid from 300 mOsm in the cortex to 1,200 to 1,400 mOsm in the inner medulla. This great hypertonicity of the renal medulla is critical because it serves as the driving force for water reabsorption through the collecting ducts, which travel through the renal medulla to empty their contents of urine into the renal pelvis.

Vasa Recta

For the countercurrent multiplier system to be effective, most of the salt that is extruded from the ascending limbs must remain in the interstitial fluid of the medulla, while most of the water that leaves the descending limbs must be removed by the blood. This is accomplished by the *vasa recta*—long, thin-walled vessels that parallel the nephron loops of the juxtamedullary nephrons (see fig. 17.18). The descending vasa recta have characteristics of both capillaries and arterioles because their continuous endothelium is surrounded by smooth muscle remnants. These vessels have *urea transporters* (for facilitative diffusion) and *aquaporin proteins,* which function as water channels through the membrane (chapter 6, section 6.2). The ascending vasa recta are capillaries with a fenestrated endothelium (chapter 13, section 13.6). The wide gaps between endothelial cells in such capillaries permit rapid rates of diffusion.

The vasa recta maintain the hypertonicity of the renal medulla by means of a mechanism known as **countercurrent exchange.** Salt and other dissolved solutes (primarily urea, described in the next section) that are present at high concentrations in the interstitial fluid diffuse into the descending vasa recta. However, these same solutes then passively diffuse out of the ascending vasa recta and back into the interstitial fluid to complete the countercurrent exchange. They do this because, at each level of the medulla, the concentration of solutes is higher in the ascending vessels than in the interstitial fluid, and higher in the interstitial fluid than in the descending vessels. Solutes are thus recirculated and trapped within the medulla.

The walls of the vasa recta are permeable to water (because of aquaporin channels), NaCl, and urea, but not plasma proteins. The colloid osmotic pressure (oncotic pressure) within the vasa recta, therefore, is higher than in the surrounding tissue fluid. This results in a movement of water from the interstitial fluid into the ascending vasa recta, so that it can be removed from the renal medulla. This removal of water is needed in order to maintain the hypertonicity of the interstitial fluid.

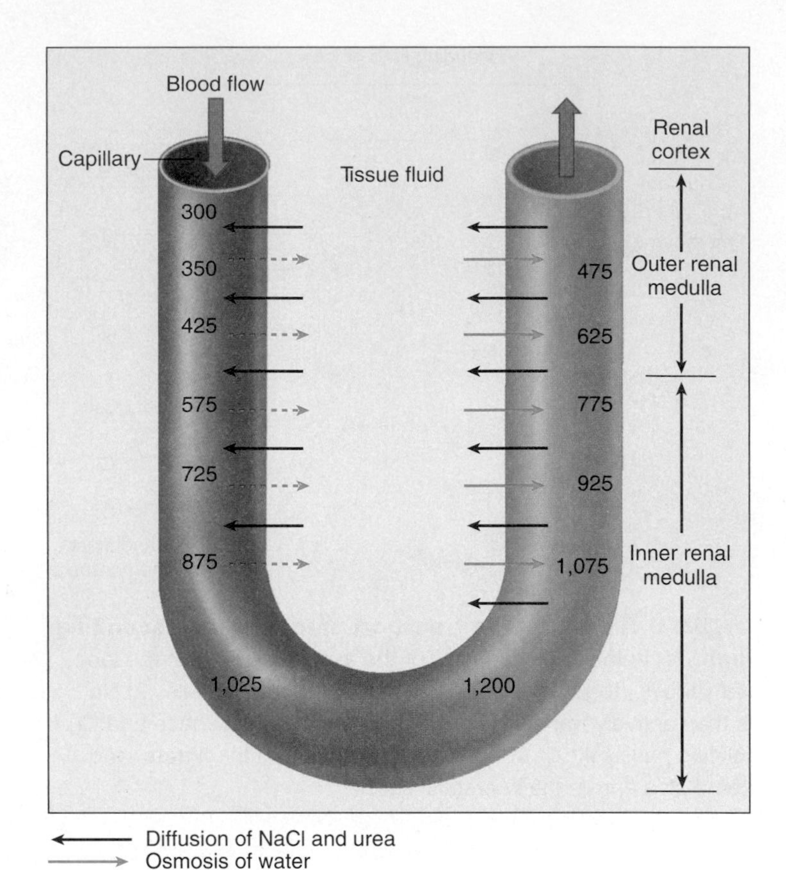

Figure 17.16 Countercurrent exchange in the vasa recta. The diffusion of salt and water first into and then out of these blood vessels helps to maintain the "saltiness" (hypertonicity) of the interstitial fluid in the renal medulla. (Numbers indicate osmolality.)

The oncotic pressure of plasma proteins in the descending vasa recta would also be expected to draw water into these vessels. However, recent evidence indicates that water actually leaves the descending vasa recta, perhaps drawn out by a higher NaCl concentration of the interstitial fluid. The amount of water removed from the descending vasa recta is less than the amount of water that enters the ascending vasa recta, so the net action of the vasa recta is to remove water from the interstitial fluid of the renal medulla (fig. 17.16).

Effects of Urea

Countercurrent multiplication of the NaCl concentration is the mechanism that contributes most to the hypertonicity of the interstitial fluid in the medulla. However, **urea,** a waste product of amino acid metabolism (chapter 5; see fig. 5.16), also contributes significantly to the total osmolality of the interstitial fluid.

The role of urea was inferred from experimental evidence showing that active transport of Na^+ occurs only in the thick

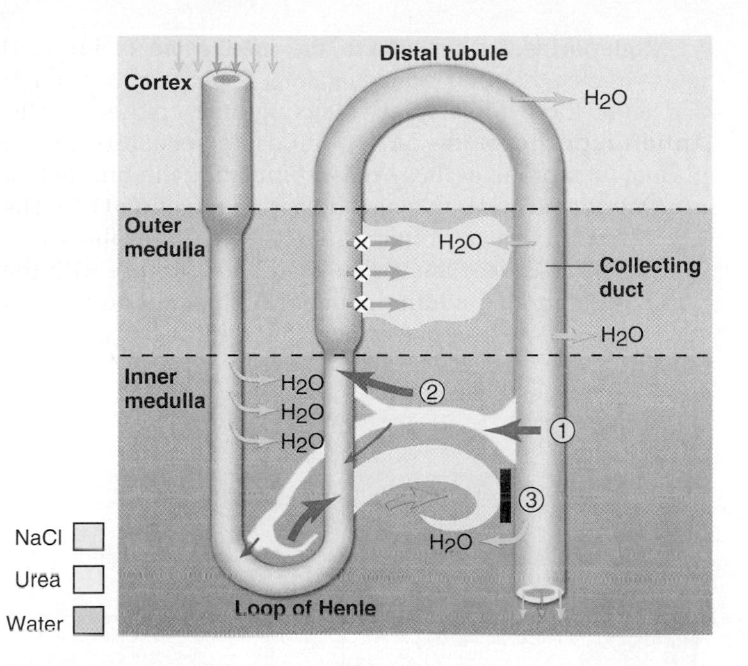

Figure 17.17 The role of urea in urine concentration. (1) Urea diffuses out of the inner collecting duct (in the renal medulla) into the interstitial fluid. (2) It can then pass into the ascending limb of the loop of Henle, so it recirculates in the interstitial fluid of the renal medulla. The urea and NaCl in the interstitial fluid of the renal medulla make it very hypertonic, so (3) water leaves the collecting duct by osmosis.

segments of the ascending limbs. The thin segments of the ascending limbs, which are located in the deeper regions of the medulla, are not able to extrude salt actively. But because salt does indeed leave the thin segments, a diffusion gradient for salt must exist, despite the fact that the surrounding interstitial fluid has the same osmolality as the tubular fluid. Investigators therefore concluded that molecules other than salt—specifically urea—contribute to the hypertonicity of the interstitial fluid.

It was later shown that the ascending limb of the loop of Henle and the terminal portion of the collecting duct in the inner medulla are permeable to urea. Indeed, the collecting duct in the inner medulla has specific urea transporters that have recently been shown to form slot-like channels, specifically allowing the polar, disc-shaped urea molecules to pass through single-file like coins through a coin slot. These channels permit a very high rate of urea diffusion into the surrounding interstitial fluid, from which it can diffuse into the ascending limb (fig. 17.17). In this way, a certain amount of urea is recycled through these two segments of the nephron. The urea is thereby trapped in the interstitial fluid where it can contribute significantly to the high osmolality of the medulla. This relates to the ability to produce a concentrated urine, as will be described next.

The transport properties of different tubule segments are summarized in table 17.2.

Collecting Duct: Effect of Antidiuretic Hormone (ADH)

As a result of active NaCl transport and countercurrent multiplication between the ascending and descending limbs and the recycling of urea between the collecting duct and the loop of Henle, the interstitial fluid is made very hypertonic. The collecting ducts must channel their fluid through this hypertonic environment in order to empty their contents of urine into the calyces. Whereas the fluid surrounding the collecting ducts in the medulla is hypertonic, the fluid that passes into the collecting ducts in the cortex is hypotonic because of the active extrusion of salt by the ascending limbs of the loops.

The collecting duct in the renal medulla is impermeable to the high concentration of NaCl that surrounds it. The wall of the collecting duct, however, is permeable to water. Because the surrounding interstitial fluid in the renal medulla is very hypertonic, water is drawn out of the collecting ducts by osmosis. This water does not dilute the surrounding interstitial fluid because it is transported by capillaries to the general circulation. In this

Table 17.2 | Transport Properties of Different Segments of the Renal Tubules and the Collecting Ducts

		Passive Transport		
Nephron Segment	**Active Transport**	**Salt**	**Water**	**Urea**
Proximal tubule	Na^+	Cl^-	Yes	Yes
Descending limb of Henle's loop	None	Maybe	Yes	No
Thin segment of ascending limb	None	NaCl	No	Yes
Thick segment of ascending limb	Na^+	Cl^-	No	No
Distal tubule	Na^+	Cl^-	No**	No
Collecting duct*	Slight Na^+	No	Yes (ADH) or slight (no ADH)	Yes

*The permeability of the collecting duct to water depends on the presence of ADH.
**The last part of the distal tubule, however, is permeable to water.

way, most of the water remaining in the filtrate is returned to the vascular system (fig. 17.18).

Note that it is the osmotic gradient created by the countercurrent multiplier system that provides the force for water reabsorption through the collecting ducts. Although this osmotic gradient is normally constant, the rate of osmosis across the walls of the collecting ducts can be varied by adjustments in their permeability to water. These adjustments are made by regulating the number of **aquaporins** (water channels) in the plasma membranes of the collecting duct epithelial cells.

Aquaporins are present in the membrane of intracellular vesicles that bud from the Golgi apparatus (chapter 3) and enter the cytoplasm of collecting duct epithelial cells. **Antidiuretic hormone (ADH)** binds to receptors in the plasma membrane of these cells, stimulating the production of cAMP as a second messenger (chapter 11, section 11.2). The cAMP then activates protein kinase, which phosphorylates proteins. This causes the vesicle membrane to fuse with the plasma membrane, adding aquaporin channels to the cell surface (fig. 17.19). Vesicle movement and fusion is similar

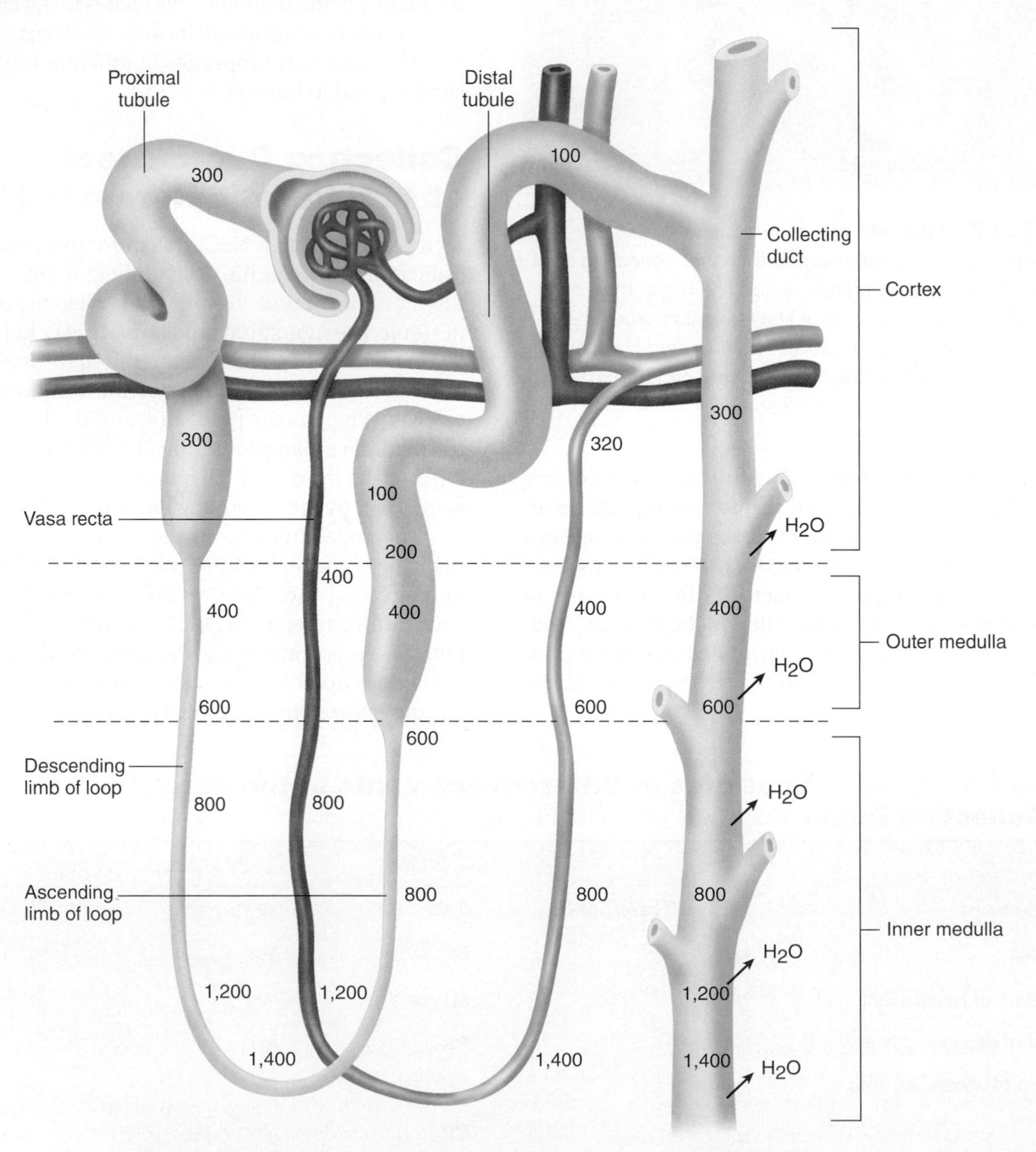

Figure 17.18 The osmolality of different regions of the kidney. The countercurrent multiplier system in the nephron loop and countercurrent exchange in the vasa recta help to create a hypertonic renal medulla. Under the influence of antidiuretic hormone (ADH), the collecting duct becomes more permeable to water, and thus more water is drawn out by osmosis into the hypertonic renal medulla and peritubular capillaries. (Numbers indicate osmolality.)

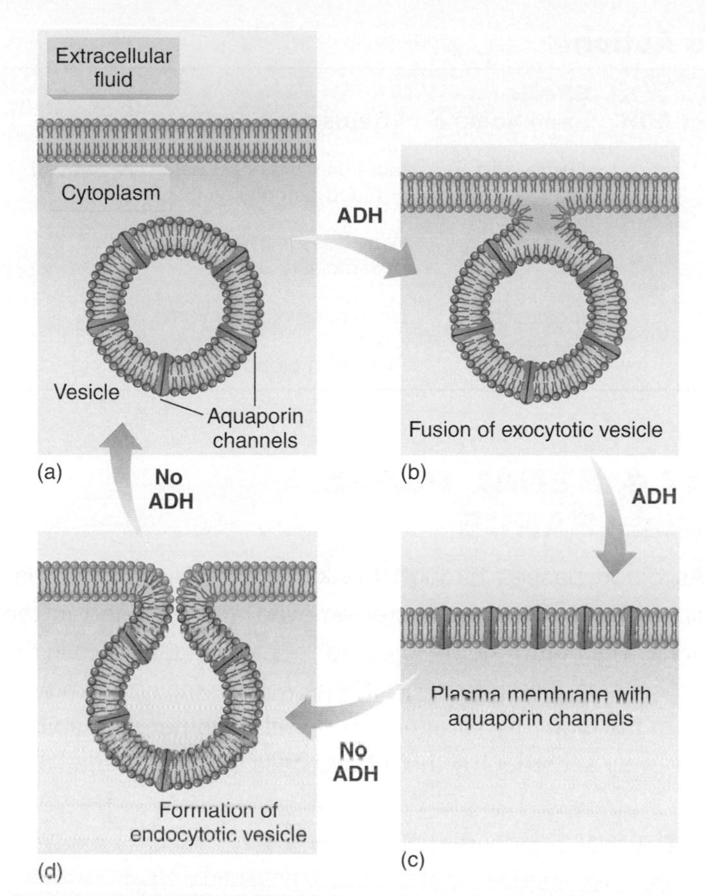

Figure 17.19 ADH stimulation of aquaporin channels. (*a*) When ADH is absent, aquaporin channels are located in the membrane of intracellular vesicles within collecting duct epithelial cells. (*b*) ADH stimulates the fusion of these vesicles with the plasma membrane and (*c*) the insertion of the aquaporin channels into the plasma membrane. (*d*) When ADH is withdrawn, the plasma membrane pinches inward (in endocytosis) to again form an intracellular vesicle and to remove the aquaporin channels from the plasma membrane.

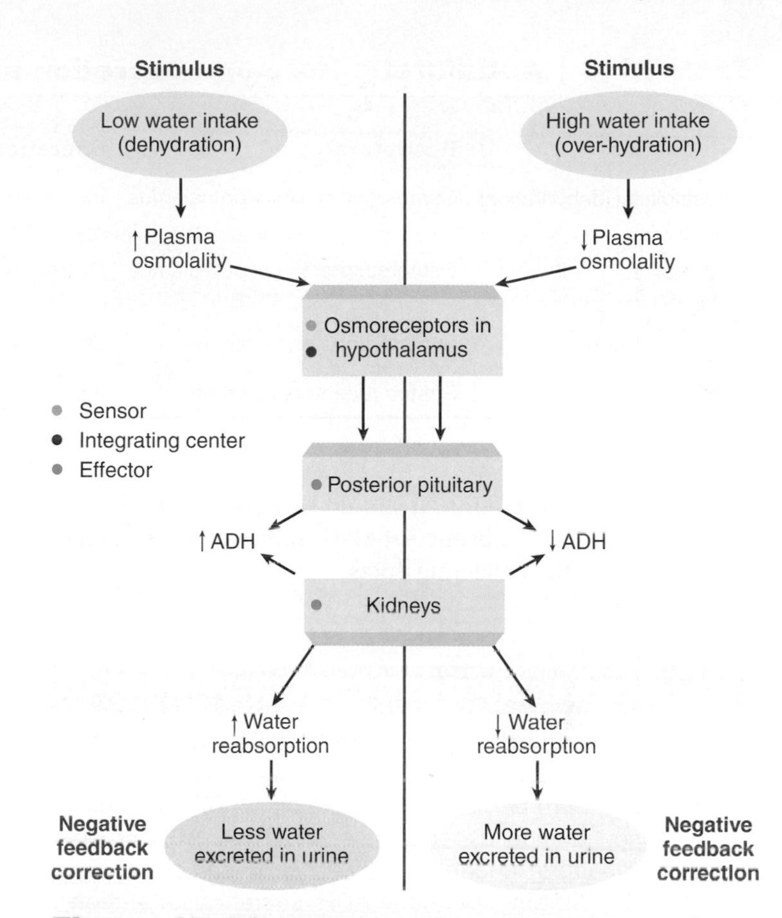

Figure 17.20 Homeostasis of plasma concentration is maintained by ADH. In dehydration (*left side of figure*), a rise in ADH secretion results in a reduction in the excretion of water in the urine. In overhydration (*right side of figure*), the excess water is eliminated through a decrease in ADH secretion. These changes provide negative feedback corrections, maintaining homeostasis of plasma osmolality and, indirectly, blood volume. AP|R

to what occurs during exocytosis, except that in this case no product is secreted.

In response to ADH, therefore, the collecting duct becomes more permeable to water. When ADH is no longer available to bind to its membrane receptors, the water channels are removed from the plasma membrane by a process of endocytosis (fig. 17.19). Endocytosis is the opposite of exocytosis; the plasma membrane invaginates to reform vesicles that again contain the water channels. Alternating exocytosis and endocytosis in response to the presence and absence of ADH, respectively, results in the recycling of water channels within the cell.

When the concentration of ADH is increased, the collecting ducts become more permeable to water and more water is reabsorbed. A decrease in ADH, conversely, results in less reabsorption of water and thus in the excretion of a larger volume of more dilute urine (fig. 17.20). ADH is produced by neurons in the hypothalamus and is released from the posterior pituitary (chapter 11; see fig. 11.13). The secretion of ADH is stimulated when osmoreceptors in the hypothalamus respond to an increase in blood osmolality above the normal range (generally 280–295 mOsm). During dehydration, therefore, when the plasma becomes more concentrated, increased secretion of ADH promotes increased permeability of the collecting ducts to water. In severe dehydration only the minimal amount of water needed to eliminate the body's wastes is excreted. This minimum, an *obligatory water loss* of about 400 ml per day, is limited by the fact that urine cannot become more concentrated than the medullary interstitial fluid surrounding the collecting ducts. Under these conditions about 99.8% of the initial glomerular ultrafiltrate is reabsorbed.

A person in a state of normal hydration excretes about 1.5 L of urine per day, indicating that 99.2% of the glomerular ultrafiltrate volume is reabsorbed. Notice that small changes in percent reabsorption translate into large changes in urine volume. Drinking more water—and thus decreasing ADH secretion (fig. 17.20 and table 17.3)—results in correspondingly larger

Table 17.3 | Antidiuretic Hormone Secretion and Action

Stimulus	Receptors	Secretion of ADH	Effects on Urine Volume	Effects on Blood
↑Osmolality (dehydration)	Osmoreceptors in hypothalamus	Increased	Decreased	Increased water retention; decreased blood osmolality
↓Osmolality	Osmoreceptors in hypothalamus	Decreased	Increased	Water loss increases blood osmolality
↑Blood volume	Stretch receptors in left atrium	Decreased	Increased	Decreased blood volume
↓Blood volume	Stretch receptors in left atrium	Increased	Decreased	Increased blood volume

volumes of urine excretion. It should be noted, however, that even in the complete absence of ADH some water is still reabsorbed through the collecting ducts.

CLINICAL APPLICATION

Diabetes insipidus may be caused by (1) drinking too much water (*polydipsia*); (2) the inadequate secretion of ADH (called *central diabetes insipidus*); or (3) inadequate ADH action due to a genetic defect in the ADH receptors or aquaporin channels (called *nephrogenic diabetes insipidus*). Without adequate ADH secretion or action, the collecting ducts are not very permeable to water. This results in the excretion of a large volume—greater than 3 L per day, and often 5 to 10 L per day—of dilute urine (with a hypotonic concentration under 300 mOsm). Excretion of so much water can cause dehydration, which produces intense thirst. If the person drinks sufficient amounts of water, symptoms of dehydration are usually absent. However, a person with diabetes insipidus may have difficulty drinking enough to compensate for the large amounts of water lost in the urine.

CHECKPOINT

6a. Describe the mechanisms for salt and water reabsorption in the proximal tubule.

6b. Compare the transport of Na^+, Cl^-, and water across the walls of the proximal tubule, ascending and descending limbs of the loop of Henle, and collecting duct.

7. Describe the interaction between the ascending and descending limbs of the loop and explain how this interaction results in a hypertonic renal medulla.

8. Explain how ADH helps the body conserve water. How do variations in ADH secretion affect the volume and concentration of urine?

17.4 RENAL PLASMA CLEARANCE

As blood passes through the kidneys, some of the constituents of the plasma are removed and excreted in the urine. The blood is thus "cleared" of particular solutes in the process of urine formation. These solutes may be removed from the blood by filtration through the glomerular capillaries or by secretion by the tubular cells into the filtrate.

LEARNING OUTCOMES

After studying this section, you should be able to:

9. **Explain how renal plasma clearance is affected by reabsorption and secretion, and how it is used to measure GFR and total renal blood flow.**

10. **Define transport maximum and renal plasma threshold, and explain their significance.**

One of the major functions of the kidneys is to eliminate excess ions and waste products from the blood. *Clearing* the blood of these substances is accomplished through their excretion in the urine. Because of renal clearance, the concentrations of these substances in the blood leaving the kidneys (in the renal vein) is lower than their concentrations in the blood entering the kidneys (in the renal artery).

Transport Process Affecting Renal Clearance

Renal clearance refers to the ability of the kidneys to remove molecules from the blood plasma by excreting them in the urine. Molecules and ions dissolved in the plasma can be filtered through the glomerular capillaries and enter the glomerular capsules. Then, those that are not reabsorbed will be eliminated in the urine; they will be "cleared" from the blood.

The process of filtration, a type of bulk transport through capillaries, promotes renal clearance. The process of reabsorption—involving membrane transport by means of carrier proteins—moves particular molecules and ions from the filtrate into the blood, and thus reduces the renal clearance of these molecules from the blood.

There is another process that affects renal clearance, a membrane transport process called **secretion** (fig. 17.21). In terms of its direction of transport, secretion is the opposite of reabsorption—secreted molecules and ions move out of the peritubular capillaries into the interstitial fluid, and then are transported across the basolateral membrane of the tubular epithelial cells and into the lumen of the nephron tubule. Molecules that are both filtered and secreted are thus eliminated in the urine more rapidly (are cleared from the blood more rapidly) than molecules that are not secreted. In summary, the process of reabsorption decreases renal clearance, while the process of secretion increases renal clearance.

By examining figure 17.21, you can see that the rate at which a substance in the plasma is excreted in the urine is equal to the rate at which it enters the filtrate (by filtration and secretion) minus the rate at which it is reabsorbed from the filtrate. This is shown in the following equation:

$$\text{Excretion rate} = (\text{filtration rate} + \text{secretion rate}) - \text{reabsorption rate}$$

It follows that if a substance in the plasma is filtered (enters the filtrate in Bowman's capsule) but is neither reabsorbed nor secreted, its excretion rate must equal its filtration rate. This fact is used to measure the volume of blood plasma filtered per minute by the kidneys, called the **glomerular filtration rate (GFR).** Measurement of the GFR is very important in assessing the health of the kidneys.

Tubular Secretion of Drugs

Many molecules foreign to the body—known generally as *xenobiotics* and including toxins and drugs—are eliminated in the urine more rapidly than would be possible by just glomerular filtration. This implies that they are secreted by membrane carriers that somehow recognize them as foreign to the body. Considering that membrane carriers are specific and that there are so many possible xenobiotic molecules, how is this accomplished?

Scientists have discovered that there is a large number of transporters whose primary function is the elimination of xenobiotics. The major group of transport proteins involved in this elimination is the **organic anion transporter (OAT)** family. These carriers mediate a sodium-independent transport that secretes some endogenous compounds—such as steroids and bile acids—as well as numerous xenobiotics, including many therapeutic and abused drugs. Relatively small xenobiotic molecules (including penicillin and PAH, discussed shortly) are eliminated by the type of OAT in the kidneys. These transporters are located in the basolateral membrane of the proximal tubule and function to secrete their transported molecules into the filtrate of the proximal tubule. Larger xenobiotics are eliminated by the type of OATs produced in the liver that transport xenobiotics into the bile (chapter 18, section 18.5).

There are also **organic cation transporters (OCTs)** that secrete particular xenobiotics such as *metformin,* a drug used to treat type 2 diabetes mellitus (chapter 19, section 19.4). Genetic studies indicate that OCT carriers vary significantly between people, suggesting that these can contribute to individual variability in the elimination of this drug—and thereby cause differences in the responsiveness to the drug.

These carriers are each specific for a broad range of molecules; they are described as being *polyspecific.* The specificity of one type of carrier overlaps with the specificity of other carriers, so that they can transport a wide variety of exogenous ("originating outside") and endogenous ("originating inside") molecules across the nephron tubules. This allows the kidneys to rapidly eliminate potentially toxic molecules from the blood. However, tubular secretion of therapeutic drugs can interfere with the ability of those drugs to work.

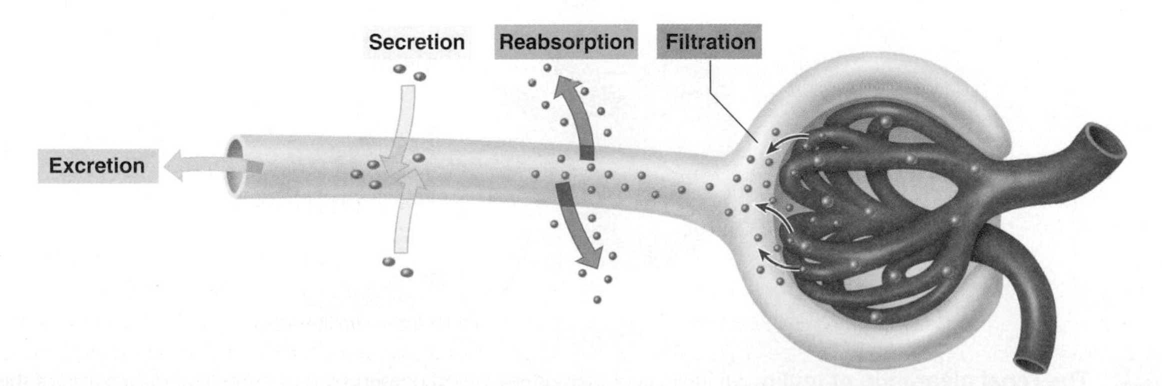

Figure 17.21 Secretion is the reverse of reabsorption. The term *secretion* refers to the active transport of substances from the peritubular capillaries into the tubular fluid. This transport is opposite in direction to that which occurs in reabsorption. In the actual nephron, most of the reabsorption and secretion occurs across the walls of the proximal tubule, although some important transport processes occur in later regions of the nephron tubule.

CLINICAL APPLICATION

Many antibiotics are secreted by the renal tubules and thus are rapidly cleared from the body. **Penicillin,** for example, is secreted into the tubular filtrate and, because of this, large amounts of the drug have to be administered. When penicillin was first used during World War II, however, it was in short supply. Scientists then discovered that a different organic anion (benzoic acid) would compete with penicillin for the carrier proteins and thus prevent penicillin from being too rapidly eliminated from the body. A newer competitor for these carriers, *probenecid,* is sometimes used to inhibit the tubular secretion of certain antibiotics. This improves the effectiveness of the antibiotic and reduces its potential toxicity to the kidneys (nephrotoxicity).

Renal Clearance of Inulin: Measurement of GFR

If a substance is neither reabsorbed nor secreted by the tubules, the amount excreted in the urine per minute will equal the amount that is filtered out of the glomeruli per minute. There does not seem to be a single substance produced by the body, however, that is not reabsorbed or secreted to some degree. Plants such as artichokes, dahlias, onions, and garlic, fortunately, do produce such a compound. This compound, a polymer of the monosaccharide fructose, is **inulin.** Once injected into the blood, inulin is filtered by the glomeruli and the amount of inulin excreted per minute is exactly equal to the amount that was filtered per minute (fig. 17.22).

If the concentration of inulin in urine is measured and the rate of urine formation is determined, the rate of inulin excretion can easily be calculated:

$$\underset{(\text{mg/min})}{\text{Quantity excreted per minute}} = \underset{\left(\frac{\text{ml}}{\text{min}}\right)}{V} \times \underset{\left(\frac{\text{mg}}{\text{ml}}\right)}{U}$$

where

V = rate of urine formation
U = inulin concentration in urine

The rate at which a substance is filtered by the glomeruli (in milligrams per minute) can be calculated by multiplying the milliliters of plasma filtered per minute (the *glomerular filtration rate,* or *GFR*) by the concentration of that substance in the plasma, as shown in this equation:

$$\underset{(\text{mg/min})}{\text{Quantity filtered per minute}} = \underset{\left(\frac{\text{ml}}{\text{min}}\right)}{GFR} \times \underset{\left(\frac{\text{mg}}{\text{ml}}\right)}{P}$$

where

P = inulin concentration in plasma

Because inulin is neither reabsorbed nor secreted, the amount filtered equals the amount excreted:

$$\underset{(\text{amount filtered})}{GFR \times P} = \underset{(\text{amount excreted})}{V \times U}$$

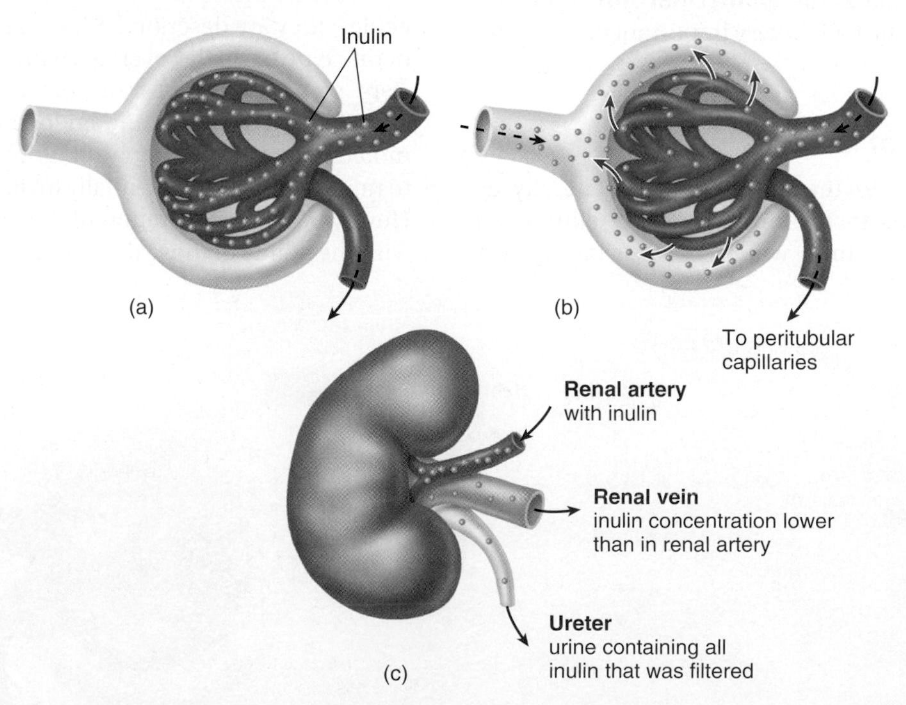

Figure 17.22 The renal clearance of inulin. (*a*) Inulin is present in the blood entering the glomeruli, and (*b*) some of this blood, together with its dissolved inulin, is filtered. All of this filtered inulin enters the urine, whereas most of the filtered water is returned to the vascular system (is reabsorbed). (*c*) The blood leaving the kidneys in the renal vein, therefore, contains less inulin than the blood that entered the kidneys in the renal artery. Because inulin is filtered but neither reabsorbed nor secreted, the inulin clearance rate equals the glomerular filtration rate (GFR). **See the *Test Your Quantitative Ability* section of the Review Activities at the end of this chapter.**

The preceding equation can now be solved for the glomerular filtration rate,

$$GFR_{(ml/min)} = \frac{V_{(ml/min)} \times U_{(mg/ml)}}{P_{(mg/ml)}}$$

Suppose, for example, that inulin is infused into a vein and its concentrations in the urine and plasma are found to be 30 mg per ml and 0.5 mg per ml, respectively. If the rate of urine formation is 2 ml per minute, the GFR can be calculated as:

$$GFR = \frac{2 \text{ ml/min} \times 30 \text{ mg/ml}}{0.5 \text{ mg/ml}} = 120 \text{ ml/min}$$

This equation states that 120 ml of plasma must have been filtered each minute in order to excrete the measured amount of inulin that appeared in the urine. The glomerular filtration rate is thus 120 ml per minute in this example.

CLINICAL APPLICATION

Measurements of the plasma concentration of **creatinine** are often used clinically as an index of kidney function. Creatinine, produced as a waste product of muscle creatine, is secreted to a slight degree by the renal tubules so that its excretion rate is a little above that of inulin. Because it is released into the blood at a constant rate, and because its excretion is closely matched to the GFR, an abnormal decrease in GFR causes the plasma creatinine concentration to rise. Thus, a simple measurement of blood creatinine concentration can indicate whether the GFR is normal and provide information about the health of the kidneys.

Clinical Investigation CLUES

Emily had oliguria (reduced urine ouput), mild edema, and an elevated plasma creatinine concentration.

- What does an elevated plasma creatinine concentration suggest?
- How might this measurement be related to Emily's oliguria and mild edema?

Clearance Calculations

The **renal plasma clearance** is the volume of plasma from which a substance is completely removed in one minute by excretion in the urine. Notice that the units for renal plasma clearance are ml/min. The simplest example is the renal plasma clearance of inulin, which is filtered but neither reabsorbed nor secreted. In that case the amount of inulin that enters the urine equals the amount that enters the glomerular filtrate. Because of this, the renal plasma clearance of inulin is equal to the glomerular filtration rate (giving a GFR of 120 ml/min in the previous example). This volume of filtered plasma, however, also contains other solutes that may be reabsorbed to varying degrees. If a portion of a filtered solute is reabsorbed, the amount of it excreted in the urine is less than the amount of it contained in the 120 ml of plasma filtered. Thus, *the renal plasma clearance of a substance that is reabsorbed must be less than the GFR* (table 17.4).

If a substance is not reabsorbed, all of the filtered amount will be cleared. If this substance is, in addition, secreted by active transport into the renal tubules from the peritubular blood, an additional amount of plasma can be cleared of that substance. Therefore, *the renal plasma clearance of a substance that is filtered and secreted is greater than the GFR* (table 17.5). In order to compare the renal "handling" of various substances in terms of their reabsorption or secretion, the renal plasma clearance is calculated using the same formula used for determining the GFR:

$$\text{Renal plasma clearance} = \frac{V \times U}{P}$$

where

V = urine volume per minute

U = concentration of substance in urine

P = concentration of substance in plasma

Clearance of Urea

Urea may be used as an example of how the clearance calculations can reveal the way the kidneys handle a molecule. Urea is a waste product of amino acid metabolism that is released by the liver into the blood and filtered into the glomerular capsules. Using the formula for renal clearance previously

Table 17.4 | Effects of Filtration, Reabsorption, and Secretion on Renal Plasma Clearance

Term	Definition	Effect on Renal Clearance
Filtration	A substance enters the glomerular ultrafiltrate.	Some or all of a filtered substance may enter the urine and be "cleared" from the blood.
Reabsorption	A substance is transported from the filtrate, through tubular cells, and into the blood.	Reabsorption decreases the rate at which a substance is cleared; clearance rate is less than the glomerular filtration rate (GFR).
Secretion	A substance is transported from peritubular blood, through tubular cells, and into the filtrate.	When a substance is secreted by the nephrons, its renal plasma clearance is greater than the GFR.

Table 17.5 | Renal "Handling" of Different Plasma Molecules

If Substance Is:	Example	Concentration in Renal Vein	Renal Clearance Rate
Not filtered	Proteins	Same as in renal artery	Zero
Filtered, not reabsorbed or secreted	Inulin	Less than in renal artery	Equal to GFR (115–125 ml/min)
Filtered, partially reabsorbed	Urea	Less than in renal artery	Less than GFR
Filtered, completely reabsorbed	Glucose	Same as in renal artery	Zero
Filtered and secreted	PAH	Less than in renal artery; approaches zero	Greater than GFR; up to total plasma flow rate (~625 ml/min)
Filtered, reabsorbed, and secreted	K^+	Variable	Variable

described and these sample values, the urea clearance can be obtained:

$$V = 2 \text{ ml/min}$$

$$U = 7.5 \text{ mg/ml of urea}$$

$$P = 0.2 \text{ mg/ml of urea}$$

$$\text{Urea clearance} = \frac{(2 \text{ ml/min})(7.5 \text{ mg/ml})}{0.2 \text{ mg/ml}} = 75 \text{ ml/min}$$

The clearance of urea in this example (75 ml/min) is less than the clearance of inulin (120 ml/min). Even though 120 ml of plasma filtrate entered the nephrons per minute, only the amount of urea contained in 75 ml of filtrate is excreted. We can conclude that the kidneys must have reaborbed some of the filtered urea. Although urea is a waste product of amino acid metabolism, a significant portion of the filtered urea (from 40% to 60%) is always reabsorbed by facilitated diffusion through the urea channels described in section 17.3. Urea diffuses out of the collecting duct and into the ascending limb, recycling in the interstitial fluid of the renal medulla and thereby contributing to its hypertonicity (see fig. 17.18).

Clearance of PAH: Measurement of Renal Blood Flow

Not all of the blood delivered to the glomeruli is filtered into the glomerular capsules; most of the glomerular blood passes through to the efferent arterioles and peritubular capillaries. The inulin and urea in this unfiltered blood are not excreted but instead return to the general circulation. Blood must therefore make many passes through the kidneys

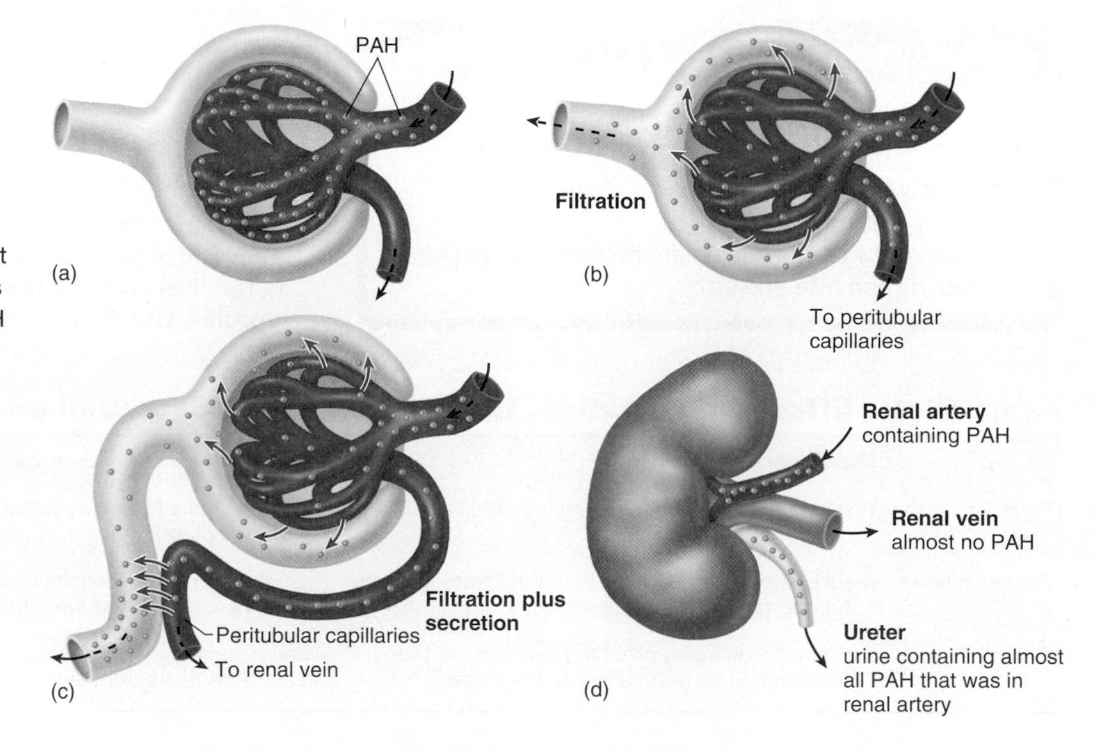

Figure 17.23 The renal clearance of PAH. Some of the para-aminohippuric acid (PAH) in glomerular blood (*a*) is filtered into the glomerular (Bowman's) capsules (*b*). The PAH present in the unfiltered blood is secreted from the peritubular capillaries into the nephron (*c*), so that all of the blood leaving the kidneys is free of PAH (*d*). The clearance of PAH therefore equals the total renal blood flow.

before it can be completely cleared of a given amount of inulin or urea.

For compounds in the unfiltered renal blood to be cleared, they must be secreted into the tubules by active transport from the peritubular capillaries. In this way, all of the blood going to the kidneys can potentially be cleared of a secreted compound in a single pass. This is the case for an exogenous molecule called **para-aminohippuric acid (PAH),** which can be infused into the blood. All of the PAH entering the peritubular capillaries is secreted by carriers of the organic anion transporter family (previously discussed) into the filtrate of the proximal tubule (fig. 17.23). Because of this, the clearance (in ml/min) of PAH can be used to measure the **total renal blood flow.** The normal PAH clearance has been found to average 625 ml/min. Since the glomerular filtration rate averages about 120 ml/min, this indicates that only about 120/625, or roughly 20%, of the renal plasma flow is filtered. The remaining 80% passes on to the efferent arterioles.

Filtration and secretion clear only the molecules dissolved in plasma, so the PAH clearance actually measures the renal plasma flow. To convert this to the total renal blood flow, the volume of blood occupied by erythrocytes must be taken into account. If the hematocrit (chapter 13) is 45, for example, erythrocytes occupy 45% of the blood volume and plasma accounts for the remaining 55%. The total renal blood flow is calculated by dividing the PAH clearance by the fractional blood volume occupied by plasma (0.55, in this example). The total renal blood flow in this example is thus 625 ml/min divided by 0.55, or 1.1 L/min.

Reabsorption of Glucose

Glucose and amino acids in the blood are easily filtered by the glomeruli into the renal tubules. However, these molecules are not present (above trace amounts) in normal urine, indicating that they must be completely reabsorbed. This occurs in the proximal tubule by secondary active transport, which is mediated by membrane carriers that cotransport glucose and Na^+ (fig. 17.24), or amino acids and Na^+.

Carrier-mediated transport displays the property of *saturation.* This means that when the transported molecule (such as glucose) is present in sufficiently high concentrations, all of the carriers become occupied and the transport rate reaches a maximal value. This is known as the **transport maximum** (abbreviated T_m). When the plasma glucose concentration is in the normal range, the glucose carriers are not saturated and the filtered glucose can be completely reabsorbed. However, when the plasma concentration is sufficiently high the filtered glucose can saturate the carriers. Then, when the rate of glucose filtration is greater than the transport maximum of the carriers, the excess glucose will continue its journey through the renal tubules and "spill over" into the urine.

The average T_m for glucose is 375 mg per minute. This is much higher than the rate at which glucose is normally delivered to the tubules. The rate of glucose filtration equals the plasma glucose concentration multiplied by the glomerular filtration rate (GFR). Since the fasting plasma glucose concentration is about 1 mg per ml, and the GFR is about 125 ml per minute, the rate of glucose filtration is about 125 mg per minute. The carriers are not saturated until 375 mg per

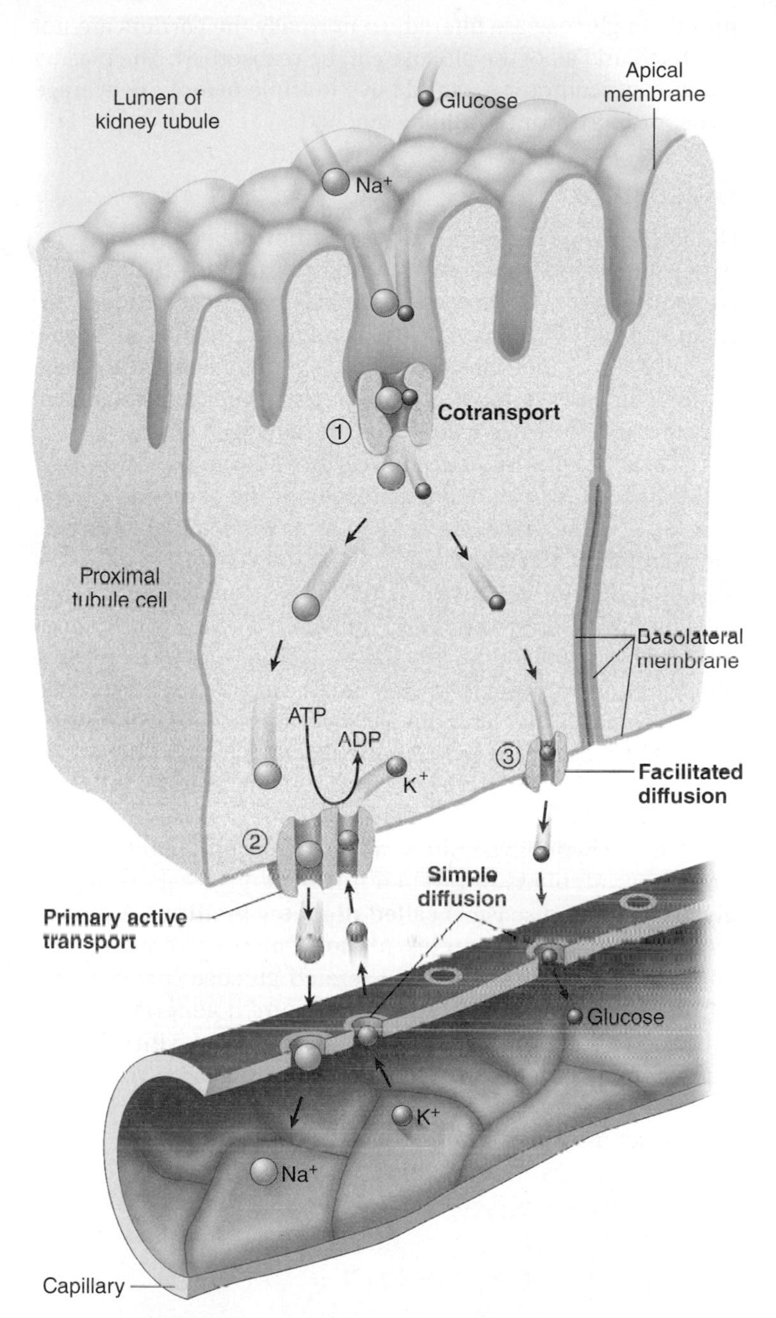

Figure 17.24 The mechanism of reabsorption in the proximal tubule. The appearance of proximal tubule cells in the electron microscope is illustrated. Molecules that are reabsorbed pass through the tubule cells from the apical membrane (*facing the filtrate*) to the basolateral membrane (*facing the blood*). (1) There is coupled transport (secondary active transport) of glucose and Na^+ into the cytoplasm, and (2) primary active transport of Na^+ across the basolateral membrane by the Na^+/K^+ pump. (3) Glucose is then transported out of the cell by facilitated diffusion and is reabsorbed into the blood.

minute of glucose are filtered, so normally the carriers are not saturated and all of the glucose can be reabsorbed. The plasma glucose concentration would have to triple before the average transport maximum would be reached.

Glycosuria

Glucose appears in the urine—a condition called **glycosuria**—when more glucose passes through the tubules than can be reabsorbed. This occurs when the plasma glucose concentration reaches 180 to 200 mg per 100 ml. Because the rate of glucose delivery under these conditions is still below the average T_m for glucose, we must conclude that some nephrons have considerably lower T_m values than the average.

The **renal plasma threshold** is the minimum plasma concentration of a substance that results in the excretion of that substance in the urine. The renal plasma threshold for glucose, for example, is 180 to 200 mg per 100 ml. Glucose is normally absent from urine because plasma glucose concentrations normally remain below this threshold value. Fasting plasma glucose is about 100 mg per 100 ml, for example, and the plasma glucose concentration following meals does not usually exceed 150 mg per 100 ml. The appearance of glucose in the urine (glycosuria) occurs only when the plasma glucose concentration is abnormally high (*hyperglycemia*) and exceeds the renal plasma threshold.

Fasting hyperglycemia is caused by the inadequate secretion or action of insulin. When this hyperglycemia results in glycosuria, the disease is called **diabetes mellitus.** A person with uncontrolled diabetes mellitus also excretes a large volume of urine because the excreted glucose carries water with it as a result of the osmotic pressure it generates in the tubules. This condition should not be confused with diabetes insipidus (discussed previously), in which a large volume of dilute urine is excreted as a result of inadequate ADH secretion or action.

CHECKPOINT

9a. Define *renal plasma clearance* and describe how this volume is measured. Explain why the glomerular filtration rate is equal to the clearance rate of inulin.

9b. Define the terms *reabsorption* and *secretion.* Using examples, describe how renal plasma clearance is affected by the processes of reabsorption and secretion.

9c. Explain why the total renal blood flow can be measured by the renal plasma clearance of PAH.

10. Define *transport maximum* and *renal plasma threshold.* Explain why people with diabetes mellitus have glycosuria.

17.5 RENAL CONTROL OF ELECTROLYTE AND ACID-BASE BALANCE

The kidneys regulate the blood concentrations of Na^+, K^+, HCO_3^- and H^+ and thereby are responsible for maintaining the homeostasis of plasma electrolytes and the acid-base balance. Renal reabsorption of Na^+ and secretion of K^+ and H^+ are stimulated by aldosterone.

LEARNING OUTCOMES

After studying this section, you should be able to:

11. Explain how the renal excretion and reabsorption of Na^+, K^+, and H^+, is regulated by the renin-angiotensin-aldosterone system.

12. Explain how the kidneys reabsorb bicarbonate, and how the kidneys contribute to the regulation of acid-base balance.

The kidneys help regulate the concentrations of plasma electrolytes—sodium, potassium, chloride, bicarbonate, sulfate, and phosphate—by matching the urinary excretion of these compounds to the amounts ingested. For example, the reabsorption of sulfate and phosphate ions across the walls of the proximal tubules is the primary determinant of their plasma concentrations. Parathyroid hormone (PTH) secretion, stimulated by a fall in plasma Ca^{2+}, acts on the kidneys to decrease the reabsorption of phosphate (chapter 19; see fig. 19.22). The control of plasma Na^+ is important in the regulation of blood volume and pressure; the control of plasma K^+ is required to maintain proper function of cardiac and skeletal muscles.

Role of Aldosterone in Na^+/K^+ Balance

Approximately 90% of the filtered Na^+ and K^+ is reabsorbed in the early part of the nephron before the filtrate reaches the distal tubule. This reabsorption occurs at a constant rate and is not subject to hormonal regulation. The final concentration of Na^+ and K^+ in the urine is varied according to the needs of the body by processes that occur in the late distal tubule and in the cortical region of the collecting duct (the portion of the collecting duct within the medulla does not participate in this regulation). Renal reabsorption of Na^+ and secretion of K^+ are regulated by **aldosterone,** the principal mineralocorticoid secreted by the adrenal cortex (chapter 11, section 11.4).

Sodium Reabsorption

Although about 90% of the filtered sodium is reabsorbed in the early region of the nephron, the amount left in the filtrate delivered to the distal convoluted tubule is still substantial. In the absence of aldosterone, 80% of this remaining amount is reabsorbed through the wall of the tubule into the peritubular blood; this represents 8% of the amount filtered. The amount of sodium excreted without aldosterone is thus 2% of the amount filtered. Although this percentage seems small, the actual amount it represents is an impressive 30 g of sodium excreted in the urine each day. When aldosterone is secreted in maximal amounts, by contrast, all of the Na^+ delivered to the distal tubule is reabsorbed. In this case urine contains no Na^+ at all.

Aldosterone stimulates Na^+ reabsorption to some degree in the *late distal convoluted tubule,* but the primary site of aldosterone action is in the **cortical collecting duct.** This is the initial portion of the collecting duct located in the renal cortex, which has different permeability properties than the terminal portion of the collecting duct in the renal medulla. Aldosterone stimulates the activity of Na^+/K^+ (ATPase) pumps in the basolateral membrane of cortical collecting duct cells. This increases the electrochemical gradient for the passive movement of Na^+ from the filtrate, through Na^+ channels in the apical membrane (facing the lumen), and into the cytoplasm.

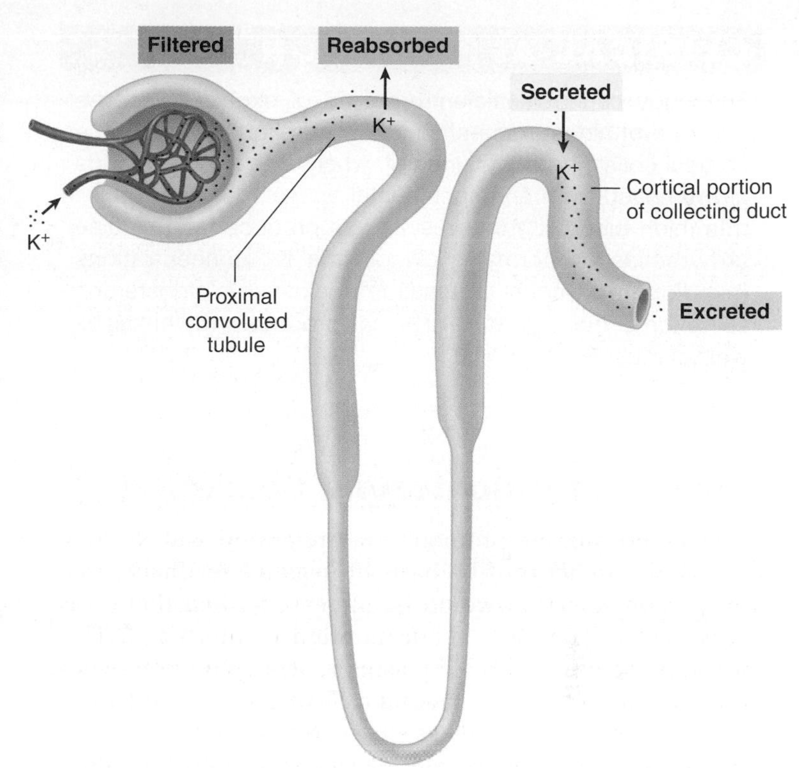

Figure 17.25 Potassium is reabsorbed and secreted. Potassium (K^+) is almost completely reabsorbed in the proximal tubule, but under aldosterone stimulation it is secreted into the cortical portion of the collecting duct. All of the K^+ in urine is derived from secretion rather than from filtration.

Potassium Secretion

About 90% of the filtered potassium is reabsorbed in the early regions of the nephron (mainly from the proximal tubule). In order for potassium to appear in the urine, it must be secreted into later regions of the nephron tubule. Secretion of potassium occurs in the parts of the nephron that are sensitive to aldosterone—that is, in the late distal tubule and cortical collecting duct (fig. 17.25).

The secretion of K^+ into the late distal tubule and cortical collecting duct matches the amount of K^+ ingested in the diet, so that the blood K^+ concentration remains in the normal range. When a person eats a K^+-rich meal, the rise in blood K^+ stimulates the adrenal cortex to secrete aldosterone. Aldosterone then stimulates an increase in the secretion of K^+ into the filtrate. In addition to this aldosterone-dependent K^+ secretion, there is also an aldosterone-independent K^+ secretion. In this process, the rise in blood K^+ directly causes additional K^+ channels to become inserted into the membrane of the cortical collecting duct. By contrast, when the blood K^+ falls, those K^+ channels are removed from the membrane by endocytosis and K^+ secretion is thereby reduced.

As Na^+ is reabsorbed in the distal convoluted tubule and cortical collecting duct, there is an increased secretion of K^+ into the filtrate. This may be a result of the potential difference created by Na^+ reabsorption, which can make the lumen negatively charged (−50 mV) compared to the basolateral side. The K^+ then enters the filtrate through a specific channel protein. Because K^+ secretion is increased when there is an increased Na^+ reabsorption, a rise in the Na^+ content of the filtrate reaching the distal tubule will cause an increased K^+ secretion. At the same time, the increased Na^+ and water in the filtrate can stimulate the juxtaglomerular apparatus to secrete renin, activating the renin-angiotensin-aldosterone system (described shortly). The increased aldosterone secretion then also stimulates more Na^+ reabsorption and K^+ secretion.

There is yet another possible mechanism by which an increased delivery of Na^+, due to increased flow of filtrate to the distal tubule, can stimulate increased K^+ secretion. Distal tubule cells contain a primary cilium (discussed in the Clinical Application box on p. 586) that protrudes into the lumen. In the distal tubules of the nephrons, the bending of the primary cilium by increased flow rates could activate K^+ channels and lead to increased K^+ secretion into the filtrate.

These mechanisms help explain how certain diuretic drugs can produce hypokalemia (low blood potassium). *Diuretics* (drugs that increase urine volume; section 17.6) that act to inhibit Na^+ transport in the nephron loop increase the delivery of Na^+ to the distal tubule. This results in increased Na^+ reabsorption and K^+ secretion in the late distal tubule and cortical collecting duct. As a result, there can be excessive urinary excretion of K^+, which may require the person who is taking the diuretic to also take potassium supplements.

CLINICAL APPLICATION

The body cannot sufficiently get rid of excess K^+ in the absence of aldosterone-stimulated secretion of K^+ into the cortical collecting ducts. Indeed, when both adrenal glands are removed from an experimental animal, the **hyperkalemia** (high blood K^+) that results can produce fatal cardiac arrhythmias. Abnormally low plasma K^+ concentrations (**hypokalemia**), as might result from excessive aldosterone secretion or from diuretic drugs, can produce arrhythmias as well as muscle weakness.

Control of Aldosterone Secretion

Because aldosterone promotes Na^+ retention and K^+ loss, one might predict (on the basis of negative feedback) that aldosterone secretion would be increased when there was a low Na^+ or a high K^+ concentration in the blood. This indeed is the case. A rise in plasma K^+ concentration depolarizes the aldosterone-secreting cells of the adrenal cortex, directly stimulating aldosterone secretion. A decrease in plasma Na^+ also promotes aldosterone secretion, but it does this indirectly. This is because decreased plasma Na^+ is accompanied by a fall in blood volume, which activates the renin-angiotensin-aldosterone system (described next).

Juxtaglomerular Apparatus

The **juxtaglomerular apparatus** is the region in each nephron where the afferent arteriole comes into contact with the last portion of the thick ascending limb of the loop (fig. 17.26). Under the microscope, the afferent arteriole and tubule in this small region have a different appearance than in other regions. *Granular cells* within the afferent arteriole secrete the enzyme **renin** into the blood. The juxtaglomerular apparatus also contains the *macula densa,* to be described shortly.

Renin catalyzes the conversion of *angiotensinogen* (a protein in the blood plasma) into *angiotensin I* (a ten-amino-acid polypeptide). Angiotensin I is converted into **angiotensin II** (an eight-amino-acid polypeptide) by *angiotensin converting enzyme (ACE).* This conversion occurs primarily as blood passes through the capillaries of the lungs, where most of the ACE is present. The secretion of renin into the plasma by the granular cells of the juxtaglomerular apparatus thereby results in the increased production of angiotensin II.

Angiotensin II, in addition to its other effects (chapter 14, section 14.2), stimulates the adrenal cortex to secrete aldosterone. Thus, secretion of renin from the granular cells of the juxtaglomerular apparatus initiates the **renin-angiotensin-aldosterone system** (chapter 14; see fig. 14.12). Conditions that result in increased renin secretion cause increased aldosterone secretion and, by this means, promote the reabsorption of Na^+ from cortical collecting duct into the blood (fig. 17.27).

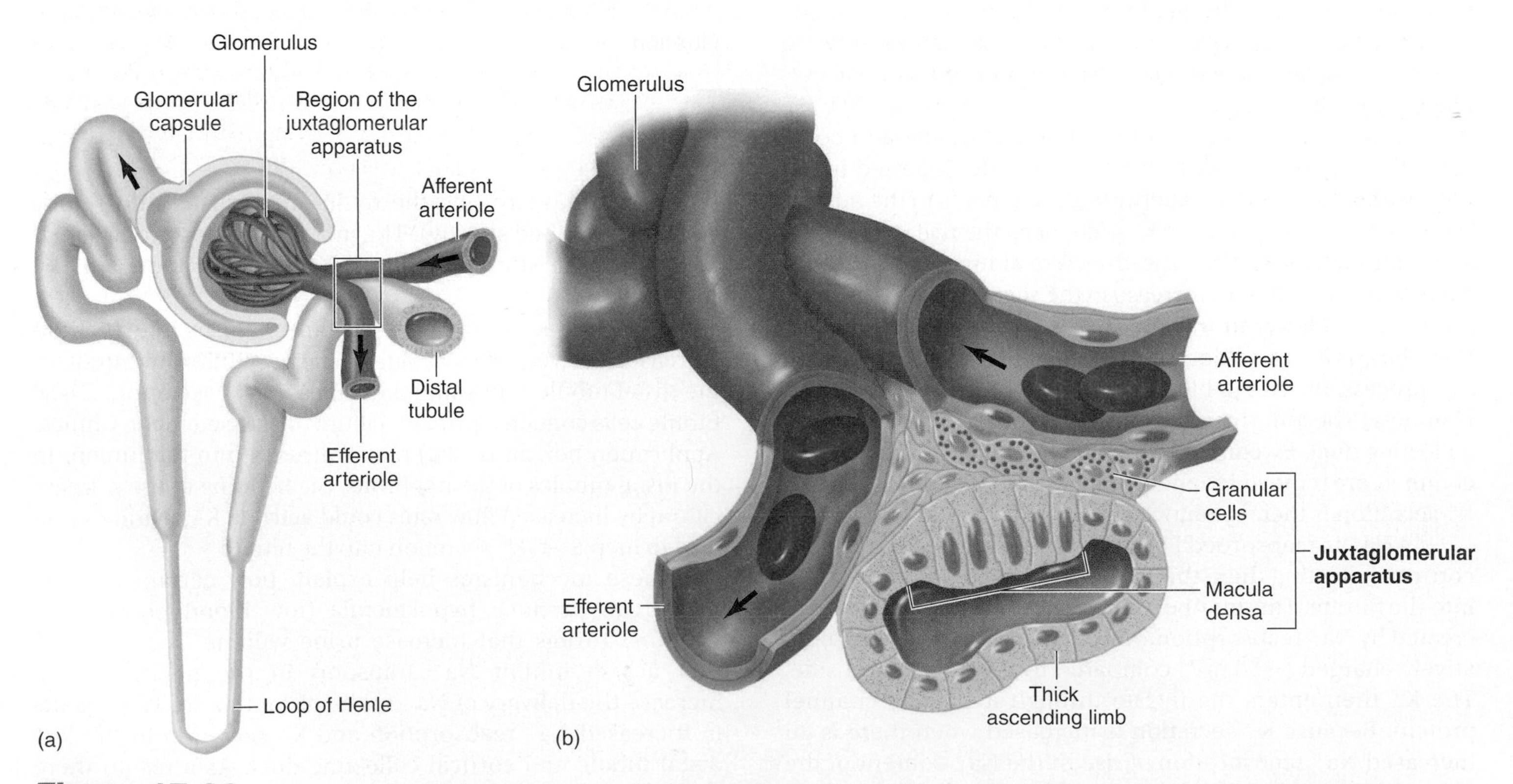

Figure 17.26 The juxtaglomerular apparatus. (*a*) The location of the juxtaglomerular apparatus. This structure includes the region where the afferent arteriole contacts the last portion of the thick ascending limb of the loop. The afferent arterioles in this region contain granular cells that secrete renin, and the tubule cells in contact with the granular cells form an area called the macula densa, seen in (*b*).

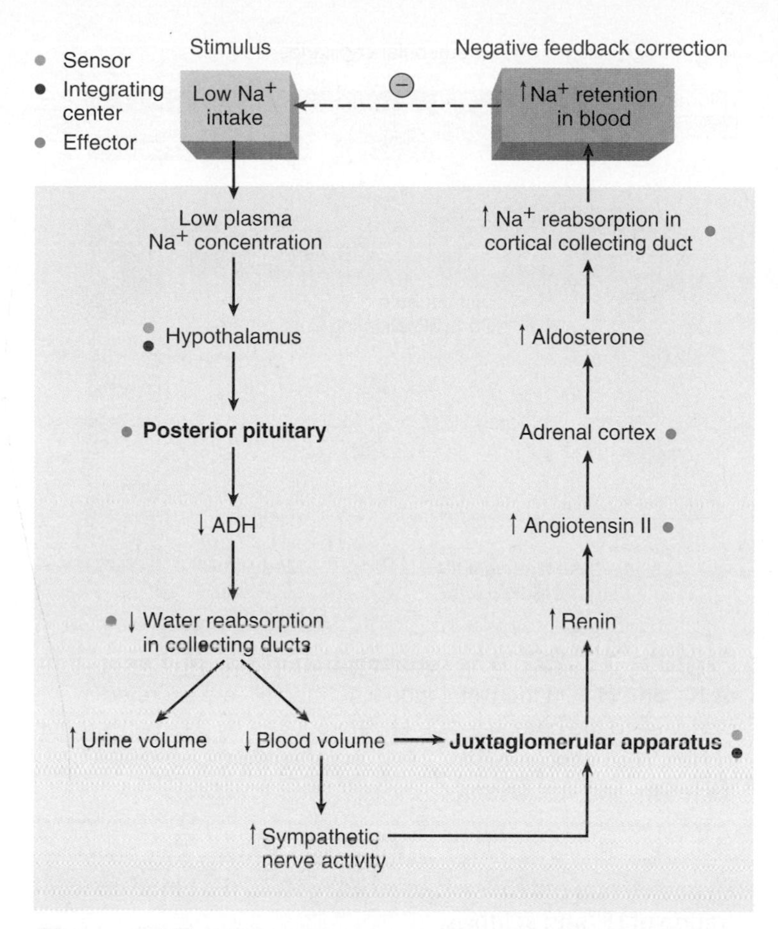

Figure 17.27 Homeostasis of plasma Na^+. This is the sequence of events by which a low sodium (salt) intake leads to increased sodium reabsorption by the kidneys. The dashed arrow and negative sign indicate the completion of the negative feedback loop.

Angiotensin II circulating through the body has long been known to raise systemic blood pressure as a result of its vasoconstrictor effects, and to stimulate aldosterone secretion. Additionally, recent evidence suggests that some angiotensin II is generated in the kidneys, where it may have important regulatory effects. These include the stimulation of embryonic kidney development and effects on reabsorption and renal blood flow.

Regulation of Renin Secretion

An inadequate dietary intake of salt (NaCl) is always accompanied by a fall in blood volume. This is because the decreased plasma concentration (osmolality) inhibits ADH secretion. With less ADH, less water is reabsorbed through the collecting ducts and more is excreted in the urine. The fall in blood volume and the fall in renal blood flow that result cause increased renin secretion. Increased renin secretion is due in part to the direct effect of blood pressure on the granular cells, which function as baroreceptors in the afferent arterioles. Renin secretion is also stimulated by sympathetic nerve activation of β_1-adrenergic receptors in the granular cells of the juxtaglomerular apparatus. This happens during the baroreceptor reflex, which increases sympathetic nerve activity when there is a fall in blood volume and pressure (chapter 14, section 14.6).

An increased secretion of renin acts, via the increased production of angiotensin II, to stimulate aldosterone secretion. Consequently, less sodium is excreted in the urine and more is retained in the blood. This negative feedback system is illustrated in figure 17.27.

Role of the Macula Densa

The region where the thick portion of the ascending limb makes contact with the granular cells of the afferent arteriole is called the *macula densa* (see fig. 17.26). When there is increased NaCl and H_2O in the filtrate, the macula densa senses this through its Na^+-K^+-$2Cl^-$ cotransporters (see fig. 17.15) and releases ATP. As described in section 17.2 on the autoregulation of the GFR, the ATP (or adenosine derived from it) stimulates the afferent arteriole to constrict. Constriction of the afferent arteriole lowers the GFR, reducing the flow of NaCl and H_2O in a negative feedback manner to complete the tubuloglomerular feedback loop.

When the plasma and filtrate concentrations of Na^+ increase, the macula densa also signals the granular cells to reduce their secretion of renin. There is thus less angiotensin II produced and less aldosterone secreted. With less aldosterone, less Na^+ is reabsorbed through the cortical collecting duct. As a result, more Na^+ (together with Cl^- and H_2O) is excreted in the urine to help restore homeostasis of blood volume. The regulation of renin and aldosterone secretion is summarized in table 17.6.

Table 17.6 | Regulation of Renin and Aldosterone Secretion

Stimulus	Effect on Renin Secretion	Angiotensin II Production	Aldosterone Secretion	Mechanisms
↓Blood volume	Increased	Increased	Increased	Low blood volume stimulates renal baroreceptors; granular cells release renin.
↑Blood volume	Decreased	Decreased	Decreased	Increased blood volume inhibits baroreceptors; increased Na^+ in distal tubule acts via macula densa to inhibit release of renin from granular cells.
↑ K^+	None	Not changed	Increased	Direct stimulation of adrenal cortex
↑Sympathetic nerve activity	Increased	Increased	Increased	α-adrenergic effect stimulates constriction of afferent arterioles; β-adrenergic effect stimulates renin secretion directly.

Atrial Natriuretic Peptide

Expansion of the blood volume causes increased salt and water excretion in the urine. This is partly due to an inhibition of aldosterone secretion, as previously described. However, it is also caused by increased secretion of a *natriuretic hormone,* a hormone that stimulates salt excretion (*natrium* is Latin for sodium)—an action opposite to that of aldosterone. The natriuretic hormone has been identified as a twenty-eight-amino-acid polypeptide called **atrial natriuretic peptide (ANP),** also called *atrial natriuretic factor* (chapter 14, section 14.2). Atrial natriuretic peptide is produced by the atria of the heart and secreted in response to the stretching of the atrial walls by increased blood volume. In response to ANP action, the kidneys lower the blood volume by excreting more of the salt and water filtered out of the blood by the glomeruli. Atrial natriuretic peptide thus functions as an endogenous diuretic.

Relationship Between Na^+, K^+, and H^+

The plasma K^+ concentration indirectly affects the plasma H^+ concentration (pH). Changes in plasma pH likewise affect the K^+ concentration of the blood. When the extracellular H^+ concentration increases, for example, some of the H^+ moves into cells and causes cellular K^+ to diffuse outward into the extracellular fluid. The plasma concentration of H^+ is thus decreased while the K^+ increases, helping to reestablish the proper ratio of these ions in the extracellular fluid. A similar effect occurs in the cells of the distal region of the nephron.

In the cells of the late distal tubule and cortical collecting duct, positively charged ions (K^+ and H^+) are secreted in response to the negative polarity produced by reabsorption of Na^+ (fig. 17.28). Acidosis (increased plasma H^+ concentration) increases the secretion of H^+ and reduces the secretion of K^+ into the filtrate. Acidosis may thus be accompanied by a rise in blood K^+. By contrast, alkalosis (lowered plasma H^+ concentration) increases the renal secretion of K^+ into the filtrate, and thus the excretion of K^+ in the urine. If, on the other hand, hyperkalemia is the primary problem, there is an increased secretion of K^+ and thus a decreased secretion of H^+.

CLINICAL APPLICATION

Aldosterone stimulation of Na^+ reabsorption can produce a potential difference between the tubule cells and the filtrate, which favors H^+ secretion. However, aldosterone can also stimulate H^+ secretion more directly by activating H^+-ATPase pumps in cortical collecting duct cells. Because aldosterone promotes the secretion of both K^+ and H^+ into the filtrate, an abnormally high aldosterone level—as occurs in **primary aldosteronism,** or **Conn's syndrome**—results in both hypokalemia and metabolic alkalosis. Conversely, abnormally low aldosterone secretion, as occurs in **Addison's disease,** can produce hyperkalemia accompanied by metabolic acidosis.

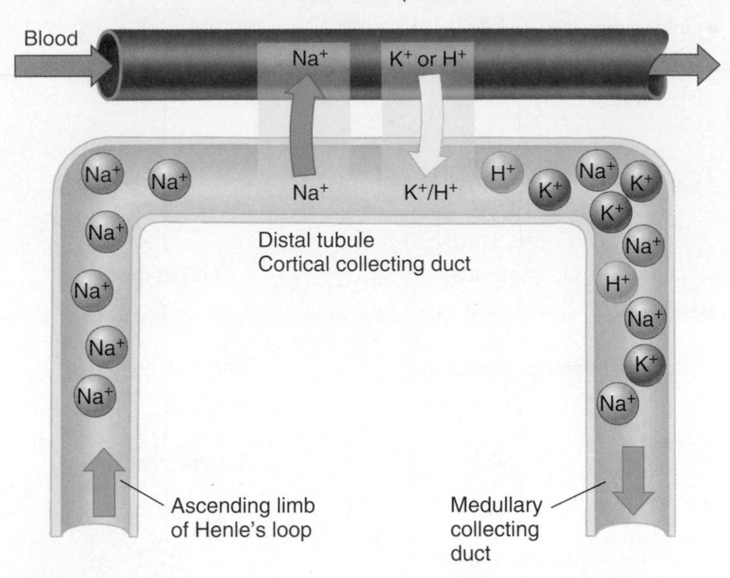

Figure 17.28 The reabsorption of Na^+ and secretion of K^+ and H^+. In the distal tubule, K^+ and H^+ are secreted in response to the potential difference produced by the reabsorption of Na^+. High concentrations of H^+ may therefore decrease K^+ secretion, and vice versa.

Hyperkalemia can thus cause an increase in the blood concentration of H^+ and acidosis.

If a person is suffering from potassium deprivation, according to recent evidence, the collecting duct may be able to partially compensate by reabsorbing some K^+. This occurs in the outer medulla, and results in the reabsorption of some of the K^+ that was secreted into the cortical collecting duct.

Renal Acid-Base Regulation

The kidneys help regulate the blood pH by excreting H^+ in the urine (mostly in buffered form, as described shortly) and by reabsorbing bicarbonate. Because the kidneys normally reabsorb almost all of the filtered bicarbonate and excrete H^+, normal urine contains little bicarbonate and is slightly acidic (with a pH range between 5 and 7). The mechanisms involved in the acidification of the urine and reabsorption of bicarbonate are summarized in figure 17.29.

Reabsorption of Bicarbonate and Secretion of H^+

The cells of the proximal tubule use Na^+/H^+ pumps to transport H^+ into the filtrate in exchange for Na^+ from the filtrate (fig. 17.29). This exchange is "antiport" cotransport (chapter 6), because the Na^+ and H^+ move in opposite directions across the apical portion of the plasma membrane (facing the tubule lumen). Antiport cotransport is a form of secondary active transport, because Na^+ diffuses down the concentration gradient maintained by primary active transport Na^+/K^+ pumps in the basolateral portion of the plasma membrane. Most of the H^+ secreted into the filtrate from the proximal tubule is used for the reabsorption of bicarbonate.

The apical membranes of the tubule cells (facing the lumen) are impermeable to bicarbonate. The reabsorption of bicarbonate must therefore occur indirectly. When the urine is acidic, HCO_3^- combines with H^+ to form carbonic acid. Carbonic acid in the filtrate is then converted to CO_2 and H_2O in a reaction catalyzed by **carbonic anhydrase.** This enzyme is located in the apical plasma membrane of the proximal tubule in contact with the filtrate. Notice that the reaction that occurs in the filtrate is the same one that occurs within red blood cells in pulmonary capillaries (chapter 16, section 16.7).

The tubule cell cytoplasm also contains carbonic anhydrase. As CO_2 concentrations increase in the filtrate, the CO_2 diffuses into the tubule cells. Within the tubule cell cytoplasm, carbonic anhydrase catalyzes the reaction in which CO_2 and H_2O form carbonic acid. The carbonic acid then dissociates to HCO_3^- and H^+ within the tubule cells. (These are the same events that occur in the red blood cells of tissue capillaries.) The bicarbonate within the tubule cell can then diffuse through the basolateral membrane and enter the blood (fig. 17.29).

Under normal condition, the proximal tubule reabsorbs 80% to 90% of the filtered bicarbonate. This process of HCO_3^- reabsorption in the proximal tubule leaves very little H^+ in the filtrate. Despite this, urine is usually more acidic than blood plasma. This is because the distal tubule secretes H^+ into the filtrate using primary active transport H^+ (ATPase) pumps (fig. 17.29), an activity that is primarily responsible for the acidification of the urine. The H^+ in the urine is mostly buffered by ammonium and phosphate buffers, as described shortly.

If a person has alkalosis, less H^+ is present in the filtrate, so that less HCO_3^- is reabsorbed; the resulting urinary excretion of HCO_3^- then helps to compensate for the alkalosis. If a person has acidosis, the proximal tubule cells can make *extra bicarbonate*—over that which is reabsorbed from the filtrate—that can enter the blood. This extra bicarbonate comes from the metabolism of the amino acid *glutamine,* derived from glutamic acid. The metabolism of one glutamine molecule yields two bicarbonate ions that are "extra" (because they were not reabsorbed from the filtrate) and two molecules of **ammonia (NH_3),** which is converted into **ammonium ion (NH_4^+)** in the filtrate. The extra bicarbonate produced by the kidneys helps compensate for acidosis, and the ammonia serves as a urinary buffer (discussed in the next section).

By these mechanisms, disturbances in acid-base balance caused by respiratory problems can be partially compensated for by changes in plasma bicarbonate concentrations. Metabolic acidosis or alkalosis—in which changes in bicarbonate concentrations occur as the primary disturbance—similarly can be partially compensated for by changes in

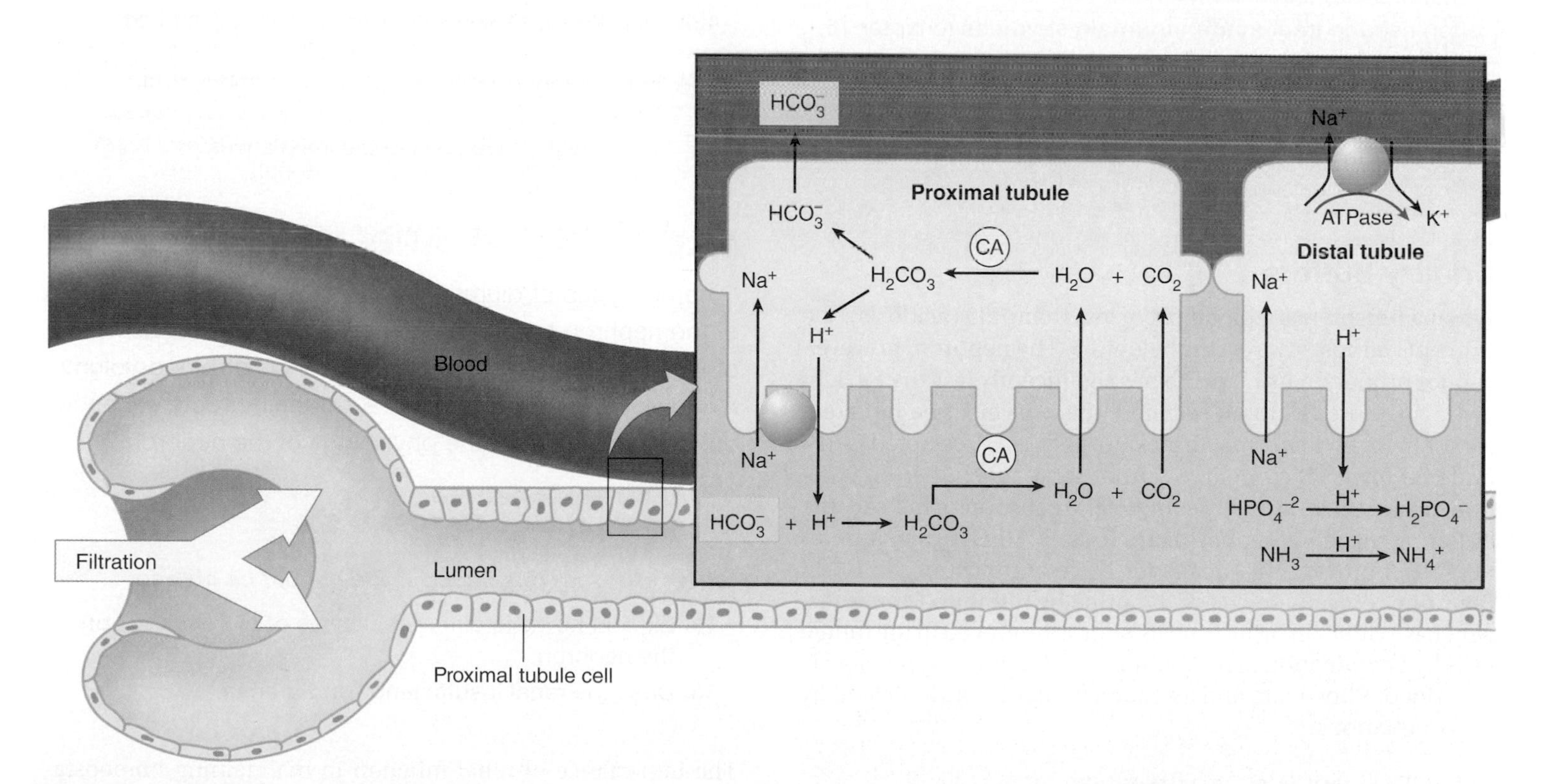

Figure 17.29 Acidification of the urine. This diagram summarizes how the urine becomes acidified and how bicarbonate is reabsorbed from the filtrate. It also depicts the buffering of the urine by phosphate and ammonium buffers. (CA = carbonic anhydrase.) The inset depicts an expanded view of proximal tubule cells. *Note:* The cells of the proximal tubule can also produce extra bicarbonate and ammonium ions from the metabolism of glutamine (not shown in this figure).

Table 17.7 | Categories of Disturbances in Acid-Base Balance

	Bicarbonate (mEq/L)*		
P_{CO_2} (mmHg)	*Less than 21*	*21–26*	*More than 26*
More than 45	Combined metabolic and respiratory acidosis	Respiratory acidosis	Metabolic alkalosis and respiratory acidosis
35–45	Metabolic acidosis	Normal	Metabolic alkalosis
Less than 35	Metabolic acidosis and respiratory alkalosis	Respiratory alkalosis	Combined metabolic and respiratory alkalosis

*mEq/L = milliequivalents per liter. This is the millimolar concentration of HCO_3^- multiplied by its valence (× 1).

ventilation. These interactions of the respiratory and metabolic components of acid-base balance are summarized in table 17.7.

CLINICAL APPLICATION

When people go to the high elevations of the mountains, they hyperventilate (as discussed in chapter 16). This lowers the arterial P_{CO_2} and produces a respiratory alkalosis. The kidneys participate in this acclimatization by excreting a larger amount of bicarbonate. This helps compensate for the alkalosis and bring the blood pH back down toward normal. It is interesting in this regard that the drug *acetazolamide,* which inhibits renal carbonic anhydrase, is often used to treat **acute mountain sickness** (chapter 16, section 16.9). The inhibition of renal carbonic anhydrase causes the loss of bicarbonate and water in the urine, producing a metabolic acidosis and diuresis that help to alleviate the symptoms.

Urinary Buffers

When a person has a blood pH of less than 7.35 (acidosis), the urine pH almost always falls below 5.5. The nephron, however, cannot produce a urine pH that is significantly less than 4.5. In order for more H^+ to be excreted, the acid must be buffered. (Actually, even in normal urine, most of the H^+ excreted is in a buffered form.) Bicarbonate cannot serve this buffering function because it is normally completely reabsorbed. Instead, the buffering reactions of phosphates (mainly HPO_4^{2-}) and ammonia (NH_3) provide the means for excreting most of the H^+ in the urine. Phosphate enters the urine by filtration. Ammonia, which is evident in urine from its odor, is produced in the tubule cells by deamination of the amino acid glutamine as previously described. Phosphate and ammonia buffer H^+ as indicated by these equations:

$$NH_3 + H^+ \rightarrow NH_4^+ \text{ (ammonium ion)}$$
$$HPO_4^{2-} + H^+ \rightarrow H_2PO_4^-$$

CHECKPOINT

11a. Describe the effects of aldosterone on the renal nephrons and explain how aldosterone secretion is regulated.
11b. Explain how changes in blood volume regulate renin secretion and how the secretion of renin helps to regulate the blood volume.
11c. Explain the mechanisms by which the cortical collecting duct secretes K^+ and H^+. How might hyperkalemia affect the blood pH?
12a. Explain how the kidneys reabsorb filtered bicarbonate and how this process is affected by acidosis and alkalosis.
12b. Suppose a person with diabetes mellitus had an arterial pH of 7.30, an abnormally low arterial P_{CO_2}, and an abnormally low bicarbonate concentration. What type of acid-base disturbance would this be? What might have caused the imbalances?

17.6 CLINICAL APPLICATIONS

Different types of diuretic drugs act on specific segments of the nephron tubule to indirectly inhibit the reabsorption of water and thus promote the lowering of blood volume. A knowledge of how diuretics exert their effects enhances our understanding of the physiology of the nephron.

LEARNING OUTCOMES

After studying this section, you should be able to:

13. Explain how the different classes of diuretics act on the nephron.
14. Describe renal insufficiency and uremia.

The importance of renal function in maintaining homeostasis, and the ease with which urine can be collected and used as a mirror of the plasma's chemical composition, make the clinical study of renal function and urine composition particularly useful. Further, the ability of the kidneys to regulate blood volume is exploited clinically in the management of high blood pressure.

Use of Diuretics

People who need to lower their blood volume because of hypertension, congestive heart failure, or edema take medications called **diuretics** that increase the volume of urine excreted. Diuretics directly lower blood volume (and hence blood pressure) by increasing the proportion of the glomerular filtrate that is excreted as urine. These drugs also decrease the interstitial fluid volume (and hence relieve edema) by a more indirect route. By lowering plasma volume, diuretic drugs increase the concentration, and thus the oncotic pressure, of the plasma within blood capillaries (see chapter 14, fig. 14.9). This promotes the osmosis of interstitial fluid into the capillary blood, helping to reduce the edema.

The various diuretic drugs act on the renal nephron in different ways (table 17.8; fig. 17.30). On the basis of their chemical structure or aspects of their actions, commonly used diuretics are categorized as *loop diuretics, thiazides, carbonic anhydrase inhibitors, osmotic diuretics,* or *potassium-sparing diuretics.*

The most powerful diuretics, which inhibit salt and water reabsorption by as much as 25%, are the drugs that inhibit active salt transport out of the ascending limb of the nephron loop. Examples of these **loop diuretics** are *furosemide (Lasix)* and *ethacrynic acid.* The **thiazide diuretics** (e.g., *hydrochlorothiazide*) inhibit salt and water reabsorption by as much as 8% through inhibition of salt transport by the first segment of the distal convoluted tubule. The **carbonic anhydrase inhibitors** (e.g., *acetazolamide*) are much weaker diuretics; they act primarily in the proximal tubule to prevent the water reabsorption that occurs when bicarbonate is reabsorbed. Largely because it also promotes the urinary excretion of bicarbonate, acetazolamide is used to treat acute mountain sickness (as previously described).

When extra solutes are present in the filtrate, they increase the osmotic pressure of the filtrate and in this way decrease the reabsorption of water by osmosis. The extra solutes thus act as **osmotic diuretics.** *Mannitol* is sometimes used clinically for this purpose. Osmotic diuresis can occur in diabetes mellitus because glucose is present in the filtrate and urine; this extra solute causes the excretion of excessive amounts of water in the urine and can result in severe dehydration of a person with uncontrolled diabetes.

All of these diuretics cause increased delivery of Na^+ to the cortical collecting ducts, which directly and indirectly stimulates increased K^+ secretion as previously described. This may cause excessive elimination of K^+ in the urine, which can dangerously lower the plasma K^+ concentration (a condition called *hypokalemia*). Hypokalemia may produce neuromuscular disorders and ECG abnormalities. People who take diuretics are usually on a low-sodium diet and must often supplement their meals with potassium chloride (KCl) to offset their loss of K^+.

For this reason, **potassium-sparing diuretics** are sometimes used. *Spironolactones (Aldactone)* are aldosterone antagonists that compete with aldosterone for cytoplasmic receptor proteins in the cells of the cortical collecting duct. These drugs thus block the aldosterone stimulation of Na^+ reabsorption and K^+ secretion. *Triamterene (Dyrenium)* is a different type of potassium-sparing diuretic that acts on the tubule more directly to block Na^+ reabsorption and K^+ secretion. Combinations of spironolactone or triamterene together with hydrochlorothiazide (*Aldactazide* and *Dyazide,* respectively) are sometimes prescribed for the diuretic treatment of hypertension.

Table 17.8 | Actions of Different Classes of Diuretics

Category of Diuretic	Example	Mechanism of Action	Major Site of Action
Loop diuretics	Furosemide	Inhibits sodium transport	Thick segments of ascending limbs
Thiazides	Hydrochlorothiazide	Inhibits sodium transport	Last part of ascending limb and first part of distal tubule
Carbonic anhydrase inhibitors	Acetazolamide	Inhibits reabsorption of bicarbonate	Proximal tubule
Osmotic diuretics	Mannitol	Reduces osmotic reabsorption of water by reducing osmotic gradient	Last part of distal tubule and cortical collecting duct
Potassium-sparing diuretics	Spironolactone	Inhibits action of aldosterone	Last part of distal tubule and cortical collecting duct
	Triamterene	Inhibits Na^+ reabsorption and K^+ secretion	Last part of distal tubule and cortical collecting duct

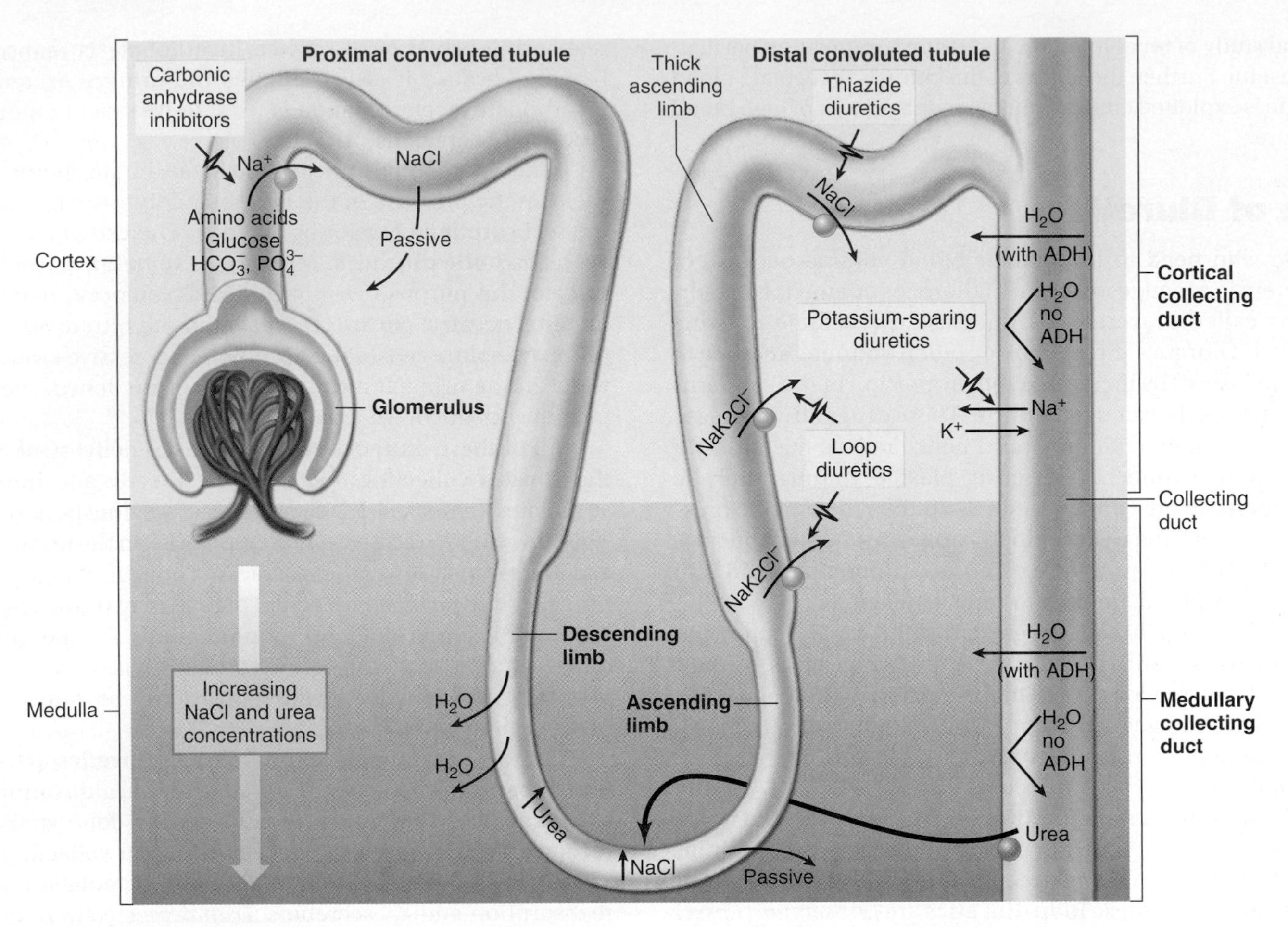

Figure 17.30 Sites of action of clinical diuretics. The different diuretic drugs act on the nephron tubules at various sites to inhibit the reabsorption of water. As a result of these actions, less water is reabsorbed into the blood and more is excreted in the urine. This lowers the blood volume and pressure.

Clinical Investigation CLUES

Emily had mild edema, for which the physician prescribed a diuretic (hydrochlorothiazide).

- How does hydrochlorothiazide act as a diuretic?
- How would this benefit Emily?

Renal Function Tests and Kidney Disease

Renal function can be tested by techniques that include the renal plasma clearance of PAH, which measures total blood flow to the kidneys, and the measurement of the GFR by the inulin clearance. The plasma creatinine concentration (see p. 601) also provides an index of renal function. These tests aid the diagnosis of kidney diseases such as glomerulonephritis and renal insufficiency. The *urinary albumin excretion rate* is a commonly performed test that can detect an excretion rate of blood albumin that is slightly above normal. This condition, called **microalbuminuria** (30–300 mg protein per day), is often the first manifestation of renal damage that may be caused by diabetes or hypertension. **Proteinuria** is present when a person excretes more than 300 mg of protein per day, and an excretion of greater than 3.5 g per day occurs in the **nephrotic syndrome**.

Acute Renal Failure

In **acute renal failure,** the ability of the kidneys to excrete wastes and regulate the homeostasis of blood volume, pH, and electrolytes deteriorates over a relatively short period of time (hours to days). There is a rise in blood creatinine concentration and a decrease in the renal plasma clearance of creatinine. This may be due to a reduced blood flow through the kidneys, perhaps as a result of atherosclerosis or inflammation of the renal tubules. The compromised kidney function may be the result of ischemia (reduced blood flow), but it may also result from excessive use of certain drugs, including nonsteroidal anti-inflammatory drugs (NSAIDs) such as phenacetin.

Glomerulonephritis

Inflammation of the glomeruli, or **glomerulonephritis,** is believed to be an *autoimmune disease*—a disease that involves the person's own antibodies (chapter 15, section 15.6). These antibodies may have been raised against the basement membrane of the glomerular capillaries. More commonly, however, they appear to have been produced in response to streptococcus infections (such as strep throat). A variable number of glomeruli are destroyed in this condition, and the remaining glomeruli become more permeable to plasma proteins. Leakage of proteins into the urine results in decreased plasma colloid osmotic pressure and can therefore lead to edema.

Clinical Investigation CLUES

Emily had a month-long streptococcus infection, and her symptoms disappeared after taking an antibiotic and a diuretic.

- How might the strep throat relate to kidney function and Emily's symptoms?

Renal Insufficiency

When nephrons are destroyed—as in chronic glomerulonephritis, infection of the renal pelvis and nephrons (*pyelonephritis*), or loss of a kidney—or when kidney function is reduced by damage caused by diabetes mellitus, arteriosclerosis, or blockage by kidney stones, a condition of **renal insufficiency** may develop. This can cause hypertension, which is due primarily to the retention of salt and water, and **uremia** (high plasma urea concentrations). The inability to excrete urea is accompanied by an elevated plasma H^+ concentration (acidosis) and an elevated K^+ concentration, which are more immediately dangerous than the high levels of urea. Uremic coma appears to result from these associated changes.

Patients with uremia or the potential for developing uremia are often placed on a *dialysis* machine. The term *dialysis* refers to the separation of molecules on the basis of their ability to diffuse through an artificial selectively permeable membrane (chapter 6; see fig. 6.4). This principle is used in the "artificial kidney machine" for **hemodialysis.** Urea and other wastes in the patient's blood can easily pass through the membrane pores, whereas plasma proteins are left behind (just as occurs across glomerular capillaries). The plasma is thus cleansed of these wastes as they pass from the blood into the dialysis fluid. Unlike the tubules, however, the dialysis membrane cannot reabsorb Na^+, K^+, glucose, and other needed molecules. These substances are kept in the blood by including them in the dialysis fluid so that there is no concentration gradient that would favor their diffusion through the membrane. By contrast, the bicarbonate concentration in the dialysate is at first higher than in the blood, favoring its diffusion into the blood. Hemodialysis is commonly performed three times a week for several hours each session.

More recent techniques include the use of the patient's own peritoneal membranes (which line the abdominal cavity) for dialysis. Dialysis fluid is introduced into the peritoneal cavity, and then after a period of time when wastes have accumulated the fluid is discarded. This procedure, called **continuous ambulatory peritoneal dialysis (CAPD),** can be performed several times a day by the patients themselves on an outpatient basis. Although CAPD is more convenient and less expensive for patients than hemodialysis, it is less efficient in removing wastes and it is more often complicated by infection.

The many dangers presented by renal insufficiency and the difficulties encountered in attempting to compensate for this condition are stark reminders of the importance of renal function in maintaining homeostasis. The ability of the kidneys to regulate blood volume and chemical composition in accordance with the body's changing needs requires great complexity of function. Homeostasis is maintained in large part by coordination of renal functions with those of the cardiovascular and pulmonary systems, as described in the preceding chapters.

CHECKPOINT

13a. List the different categories of clinical diuretics and explain how each exerts its diuretic effect.

13b. Explain why most diuretics can cause excessive loss of K^+. How is this prevented by the potassium-sparing diuretics?

14. Define *uremia* and discuss the dangers associated with this condition. Explain how uremia can be corrected through the use of renal dialysis.

Interactions

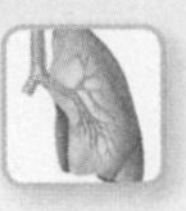

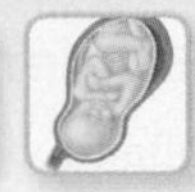

HPer Links of the Urinary System with Other Body Systems

Integumentary System

- Evaporative water loss from the skin helps control body temperature, but effects on blood volume must be compensated for by the kidneys (p. 459)
- The skin produces vitamin D_3, which is activated in the kidneys (p. 693)
- The kidneys maintain homeostasis of blood volume, pressure, and composition, which is needed for the health of the integumentary and other systems (p. 582)

Skeletal System

- The pelvic girdle supports and protects some organs of the urinary system (p. 582)
- Bones store calcium and phosphate, and thus cooperate with the kidneys to regulate the blood levels of these ions (p. 690)

Muscular System

- Muscles in the urinary tract assist the storage and voiding of urine (p. 584)
- Smooth muscles in the renal blood vessels regulate renal blood flow, and thus the glomerular filtration rate (p. 589)

Nervous System

- Autonomic nerves help regulate renal blood flow, and hence glomerular filtration (p. 589)
- The nervous system provides motor control of micturition (p. 584)

Endocrine System

- Antidiuretic hormone stimulates reabsorption of water from the renal tubules (p. 596)
- Aldosterone stimulates sodium reabsorption and potassium secretion by the kidneys (p. 604)
- Natriuretic hormone stimulates sodium excretion by the kidneys (p. 608)
- The kidneys produce the hormone erythropoietin (p. 410)
- The kidneys secrete renin, which activates the renin-angiotensin-aldosterone system (p. 606)

Circulatory System

- The blood transports oxygen and nutrients to all systems, including the urinary system, and removes wastes (p. 405)
- The heart secretes atrial natriuretic peptide, which helps to regulate the kidneys (p. 462)
- Erythropoietin from the kidneys stimulates red blood cell production (p. 410)
- The kidneys filter the blood to produce urine, while regulating blood volume, composition, and pressure (p. 582)

Immune System

- The immune system protects all systems, including the urinary system, against infections (p. 494)
- Lymphatic vessels help to maintain a balance between blood and interstitial fluid (p. 457)
- The acidity of urine provides a nonspecific defense against urinary tract infection (p. 494)

Respiratory System

- The lungs provide oxygen and eliminate carbon dioxide for all systems, including the urinary system (p. 533)
- The lungs and kidneys cooperate in the regulation of blood pH (p. 567)

Digestive System

- The GI tract provides nutrients for all tissues, including those of the urinary system (p. 621)
- The digestive system, like the urinary system, helps to eliminate waste products (p. 639)

Reproductive System

- The urethra of a male passes through the penis and can eject either urine or semen (p. 718)
- The kidneys participate in the regulation of blood volume and pressure, which is required for functioning of the reproductive system (p. 582)
- The mother's urinary system eliminates metabolic wastes from the fetus (p. 742)

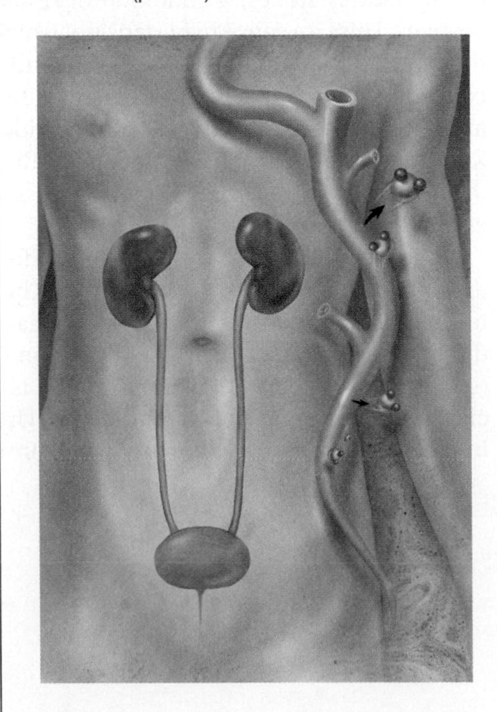

Clinical Investigation SUMMARY

The location of Emily's pain and the discoloration of her urine are indicative of a renal disorder. The elevated blood creatinine concentration could indicate a reduction in the glomerular filtration rate (GFR) as a result of glomerulonephritis, and this reduced GFR could have been responsible for the fluid retention and observed edema. The presence of only trace amounts of protein in the urine, however, was encouraging, and could be explained by her running activity (proteinuria in this case would have been an ominous sign). The streptococcus infection, acting via an autoimmune reaction, was probably responsible for the glomerulonephritis. This was confirmed when the symptoms of glomerulonephritis disappeared after treatment with an antibiotic. Hydrochlorothiazide is a diuretic that helped to alleviate the edema by (1) promoting the excretion of larger amounts of urine and (2) shifting of edematous fluid from the interstitial to the vascular compartment.

See the additional chapter 17 Clinical Investigations on *Primary Aldosteronism* and *Acute Nephritic Syndrome* in the Connect site for this text at www.mhhe.com/Fox13.

SUMMARY

17.1 Structure and Function of the Kidneys 582

A. The kidney is divided into an outer cortex and inner medulla.

1. The medulla is composed of renal pyramids, separated by renal columns.
2. The renal pyramids empty urine into the calyces that drain into the renal pelvis. From there, urine flows into the ureter and is transported to the bladder to be stored.

B. Each kidney contains more than a million microscopic functional units called nephrons. Nephrons consist of vascular and tubular components.

1. Filtration occurs in the glomerulus, which receives blood from an afferent arteriole.
2. Glomerular blood is drained by an efferent arteriole, which delivers blood to peritubular capillaries that surround the nephron tubules.
3. The glomerular (Bowman's) capsule and the proximal and distal convoluted tubules are located in the cortex.
4. The loop of Henle is located in the medulla.
5. Filtrate from the distal convoluted tubule is drained into collecting ducts, which plunge through the medulla to empty urine into the calyces.

17.2 Glomerular Filtration 587

A. A filtrate derived from plasma in the glomerulus must pass through a basement membrane of the glomerular capillaries and through slits in the processes of the podocytes—the cells that compose the inner layer of the glomerular (Bowman's) capsule.

1. The glomerular ultrafiltrate, formed under the force of blood pressure, has a low protein concentration.
2. The glomerular filtration rate (GFR) is the volume of filtrate produced by both kidneys each minute. It ranges from 115 to 125 ml/min.

B. The GFR can be regulated by constriction or dilation of the afferent arterioles.

1. Sympathetic innervation causes constriction of the afferent arterioles.
2. Intrinsic mechanisms help to autoregulate the rate of renal blood flow and the GFR.

17.3 Reabsorption of Salt and Water 590

A. Approximately 65% of the filtered salt and water is reabsorbed across the proximal convoluted tubules.

1. Sodium is actively transported, chloride follows passively by electrical attraction, and water follows the salt out of the proximal tubule.
2. Salt transport in the proximal tubules is not under hormonal regulation.

B. The reabsorption of most of the remaining water occurs as a result of the action of the countercurrent multiplier system.

1. Sodium is actively extruded from the ascending limb, followed passively by chloride.

2. Because the ascending limb is impermeable to water, the remaining filtrate becomes hypotonic.
3. Because of this salt transport and because of countercurrent exchange in the vasa recta, the interstitial fluid of the medulla becomes hypertonic.
4. The hypertonicity of the medulla is multiplied by a positive feedback mechanism involving the descending limb, which is passively permeable to water and perhaps to salt.

C. The collecting duct is permeable to water but not to salt.
1. As the collecting ducts pass through the hypertonic renal medulla, water leaves by osmosis and is carried away in surrounding capillaries.
2. The permeability of the collecting ducts to water is stimulated by antidiuretic hormone (ADH).

17.4 Renal Plasma Clearance 598

A. Inulin is filtered but neither reabsorbed nor secreted. Its clearance is thus equal to the glomerular filtration rate.

B. Some of the filtered urea is reabsorbed. Its clearance is therefore less than the glomerular filtration rate.

C. Since almost all the PAH in blood going through the kidneys is cleared by filtration and secretion, the PAH clearance is a measure of the total renal blood flow.

D. Normally all of the filtered glucose is reabsorbed. Glycosuria occurs when the transport carriers for glucose become saturated as a result of hyperglycemia.

17.5 Renal Control of Electrolyte and Acid-Base Balance 604

A. Aldosterone stimulates sodium reabsorption and potassium secretion in the late distal tubule and cortical collecting duct.

B. Aldosterone secretion is stimulated directly by a rise in blood potassium and indirectly by a fall in blood volume.
1. Decreased blood flow and pressure through the kidneys stimulates the secretion of the enzyme renin from the juxtaglomerular apparatus.
2. Renin catalyzes the formation of angiotensin I, which is then converted to angiotensin II.
3. Angiotensin II stimulates the adrenal cortex to secrete aldosterone.

C. Aldosterone stimulates the secretion of H^+, as well as potassium, into the filtrate in exchange for sodium.

D. The nephrons filter bicarbonate and reabsorb the amount required to maintain acid-base balance. Reabsorption of bicarbonate, however, is indirect.
1. Filtered bicarbonate combines with H^+ to form carbonic acid in the filtrate.
2. Carbonic anhydrase in the membranes of microvilli in the tubules catalyzes the conversion of carbonic acid to carbon dioxide and water.
3. Carbon dioxide is reabsorbed and converted in either the tubule cells or the red blood cells to carbonic acid, which dissociates to bicarbonate and H^+.
4. In addition to reabsorbing bicarbonate, the nephrons filter and secrete H^+, which is excreted in the urine buffered by ammonium and phosphate buffers.

17.6 Clinical Applications 610

A. Diuretic drugs are used clinically to increase the urine volume and thus to lower the blood volume and pressure.
1. Loop diuretics and the thiazides inhibit active Na^+ transport in the ascending limb and early portion of the distal tubule, respectively.
2. Osmotic diuretics are extra solutes in the filtrate that increase the osmotic pressure of the filtrate and inhibit the osmotic reabsorption of water.
3. The potassium-sparing diuretics act on the distal tubule to inhibit the reabsorption of Na^+ and secretion of K^+.

B. In glomerulonephritis, the glomeruli can permit the leakage of plasma proteins into the urine.

C. The technique of hemodialysis is used to treat people with renal insufficiency.

REVIEW ACTIVITIES

Test Your Knowledge

1. Which of these statements about the renal pyramids is *false*?
   a. They are located in the medulla.
   b. They contain glomeruli.
   c. They contain collecting ducts.
   d. They empty urine into the calyces.

Match the following items:

2. Active transport of sodium; water follows passively
3. Active transport of sodium; impermeable to water
4. Passively permeable to water only
5. Passively permeable to water and urea

a. Proximal tubule
b. Descending limb of loop
c. Ascending limb of loop
d. Distal tubule
e. Medullary collecting duct

6. Antidiuretic hormone promotes the retention of water by stimulating
   a. the active transport of water.
   b. the active transport of chloride.
   c. the active transport of sodium.
   d. the permeability of the collecting duct to water.
7. Aldosterone stimulates sodium reabsorption and potassium secretion in
   a. the proximal convoluted tubule.
   b. the descending limb of the loop.
   c. the ascending limb of the loop.
   d. the cortical collecting duct.
8. Substance X has a clearance greater than zero but less than that of inulin. What can you conclude about substance X?

a. It is not filtered.
b. It is filtered, but neither reabsorbed nor secreted.
c. It is filtered and partially reabsorbed.
d. It is filtered and secreted.

9. Substance Y has a clearance greater than that of inulin. What can you conclude about substance Y?
 a. It is not filtered.
 b. It is filtered, but neither reabsorbed nor secreted.
 c. It is filtered and partially reabsorbed.
 d. It is filtered and secreted.
10. About 65% of the glomerular ultrafiltrate is reabsorbed in
 a. the proximal tubule. c. the loop of Henle.
 b. the distal tubule. d. the collecting duct.
11. Diuretic drugs that act in the loop of Henle
 a. inhibit active sodium transport.
 b. cause an increased flow of filtrate to the distal convoluted tubule.
 c. cause an increased secretion of potassium into the tubule.
 d. promote the excretion of salt and water.
 e. do all of these.
12. The appearance of glucose in the urine
 a. occurs normally.
 b. indicates the presence of kidney disease.
 c. occurs only when the transport carriers for glucose become saturated.
 d. is a result of hypoglycemia.
13. Reabsorption of water through the tubules occurs by
 a. osmosis. c. facilitated diffusion.
 b. active transport. d. all of these.
14. Which of these factors oppose(s) filtration from the glomerulus?
 a. plasma oncotic pressure
 b. hydrostatic pressure in glomerular (Bowman's) capsule
 c. plasma hydrostatic pressure
 d. both *a* and *b*
 e. both *b* and *c*
15. The countercurrent exchange in the vasa recta
 a. removes Na^+ from the extracellular fluid.
 b. maintains high concentrations of NaCl in the extracellular fluid.
 c. raises the concentration of Na^+ in the blood leaving the kidneys.
 d. causes large quantities of Na^+ to enter the filtrate.
 e. does all of these.
16. The kidneys help to maintain acid-base balance by
 a. the secretion of H^+ in the distal regions of the nephron.
 b. the action of carbonic anhydrase within the apical cell membranes.
 c. the action of carbonic anhydrase within the cytoplasm of the tubule cells.
 d. the buffering action of phosphates and ammonia in the urine.
 e. all of these means.
17. Scientists currently believe that the main barrier to the filtration of proteins into the glomerular capsule is the
 a. capillary fenestrae. c. slit diaphragm.
 b. basement membrane. d. macula densa.
18. A drug that blocks the action of the organic anion transporters would
 a. increase the secretion of xenobiotics into the filtrate.
 b. keep antibiotics in the blood for a longer time.
 c. prevent glucose from being reabsorbed.
 d. cause proteinuria to occur.

Test Your Understanding

19. Explain how glomerular ultrafiltrate is produced and why it has a low protein concentration.
20. Describe the transport properties of the loop of Henle and explain the interactions between the ascending and descending limbs in the countercurrent multiplier system. What is the functional significance of this system?
21. Explain how countercurrent exchange occurs in the vasa recta and discuss the functional significance of this mechanism.
22. Explain how an increase in ADH secretion promotes increased water reabsorption and how water reabsorption decreases when ADH secretion is decreased.
23. Explain how the structure of the epithelial wall of the proximal tubule and the distribution of Na^+/K^+ pumps in the epithelial cell membranes contribute to the ability of the proximal tubule to reabsorb salt and water.
24. Describe how the thiazide diuretics, loop diuretics, and osmotic diuretics work.
25. Identify where K^+ secretion occurs in the nephron, and explain how this secretion is regulated to maintain homeostasis of blood K^+ levels. Also, explain how loop diuretics and thiazide diuretics can cause excessive K^+ secretion and hypokalemia.
26. Which diuretic drugs do not produce hypokalemia? How do these drugs function as diuretics and yet spare blood K^+?
27. Explain the mechanisms that normally prevent glycosuria. Can a person have hyperglycemia without having glycosuria? Explain.
28. Explain how filtration, secretion, and reabsorption affect the renal plasma clearance of a substance. Use this information to explain how creatinine can be used to measure the GFR.
29. What happens to urinary bicarbonate excretion when a person hyperventilates? How might this response be helpful?
30. Describe the location of the macula densa and explain its role in the regulation of renin secretion and in tubuloglomerular feedback.

Test Your Analytical Ability

31. The very high rates of urea transport in the region of the collecting duct in the inner medulla are due to the presence of specific urea transporters that are stimulated by ADH. Suppose you collect urine from two patients who have been deprived of water overnight. One has normally functioning kidneys, and the other has a genetic defect in the urea transporters. How would the two urine samples differ? Explain.
32. Two men are diagnosed with diabetes insipidus. One didn't have the disorder until he suffered a stroke. The other had withstood the condition all his life, and it had never responded to exogenous ADH despite the presence of normal ADH receptors. What might be the cause of the diabetes insipidus in the two men?
33. Suppose a woman with a family history of polycystic kidney disease develops proteinuria. She has elevated blood creatinine levels and a reduced inulin clearance. What might these lab results indicate? Explain.
34. You love to spend hours fishing in a float tube in a lake, where the lower half of your body is submerged and the upper half is supported by an inner tube. However, you always have to leave the lake sooner than you'd like because you produce urine at a faster than usual rate. Using your knowledge about the regulation of urine volume, propose an explanation as to why a person might produce more urine under these conditions.
35. You have an infection, and you see that the physician is about to inject you with millions of units of penicillin. What do you think will happen to your urine production as a result? Explain. In the hope of speeding your recovery, you gobble extra amounts of vitamin C. How will this affect your urine output?
36. Explain how the different causes of incontinence could be treated by (a) surgery, (b) a drug that blocks specific muscarinic ACh receptors, and (c) a drug that blocks the action of testosterone (by inhibiting its conversion to dihydrotestosterone; see chapter 20).
37. Potassium is both reabsorbed and secreted by the nephron. Explain this statement, and speculate about the possible benefits of the nephron handling potassium in this way.
38. What are xenobiotics, and how are we able to quickly eliminate them in the urine? Describe the carriers involved, and how they can transport many different molecules. Explain how this elimination of xenobiotics can sometimes interfere with medical treatment.

Test Your Quantitative Ability

Refer to figure 17.22 and to the renal plasma clearance formula on page 601 to calculate the answers to the following questions:

39. A woman who undergoes an inulin clearance test has an inulin concentration in her urine of 20 mg/ml; an inulin concentration in her blood of 0.70 mg/ml; and a rate of urine formation of 4 ml/min. What is her GFR?
40. If 35% of her filtered urea is reabsorbed, what would be her renal plasma clearance of urea?
41. If she is filtering exactly 20% of the total plasma flow rate to her kidneys, what would be her renal plasma clearance of PAH?
42. Suppose a substance has a renal plasma clearance of 300 ml/min. What portion of its clearance is due to its secretion across the wall of the tubules?

CHAPTER OUTLINE

CHAPTER 18

The Digestive System

REFRESH YOUR MEMORY

Before you begin this chapter, you may want to review these concepts from previous chapters:

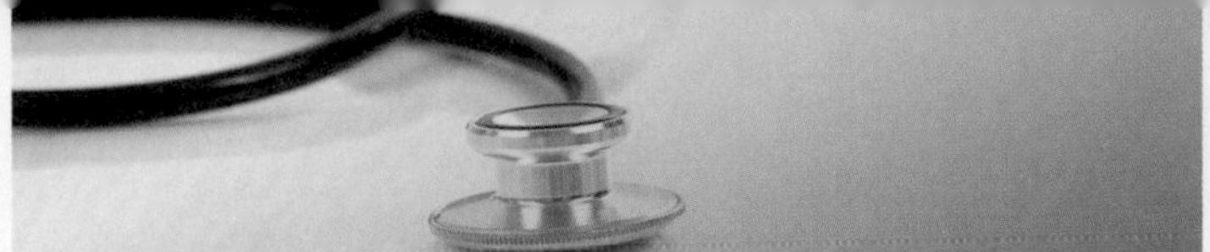

Clinical Investigation

Alan, a 23-year-old student, went to the health center complaining of severe but transient stomach pain whenever he drank wine, and pain below his right scapula whenever he ate particular foods, such as peanut butter or bacon, but not when he ate skinned chicken or fish. The physician noticed that the sclera of Alan's eyes were markedly yellow, and ordered a stool sample and a variety of blood tests to be performed.

Some of the new terms and concepts you will encounter include:

- Free and conjugated bilirubin; jaundice
- Urea, ammonia, and pancreatic amylase
- Appendicitis; gallstones; and cholecystokinin

18.1 INTRODUCTION TO THE DIGESTIVE SYSTEM

Within the lumen of the gastrointestinal tract, large food molecules are hydrolyzed into their monomers (subunits). These monomers pass through the inner layer, or mucosa, of the small intestine to enter the blood or lymph in a process called absorption. Digestion and absorption are aided by specializations of the gastrointestinal tract.

LEARNING OUTCOMES

After studying this section, you should be able to:

1. **List the functions of the digestive system.**
2. **Describe the microscopic structure of the gastrointestinal tract.**

Unlike plants, which can form organic molecules using inorganic compounds such as carbon dioxide, water, and ammonia, humans and other animals must obtain their basic organic molecules from food. Some of the ingested food molecules are needed for their energy (caloric) value—obtained by the reactions of cell respiration and used in the production of ATP—and the balance is used to make additional tissue.

Most of the organic molecules that are ingested are similar to the molecules that form human tissues. These are generally large molecules (*polymers*), which are composed of subunits (*monomers*). Within the gastrointestinal tract, the **digestion** of these large molecules into their monomers occurs by means of *hydrolysis reactions* (reviewed in fig. 18.1). The monomers thus

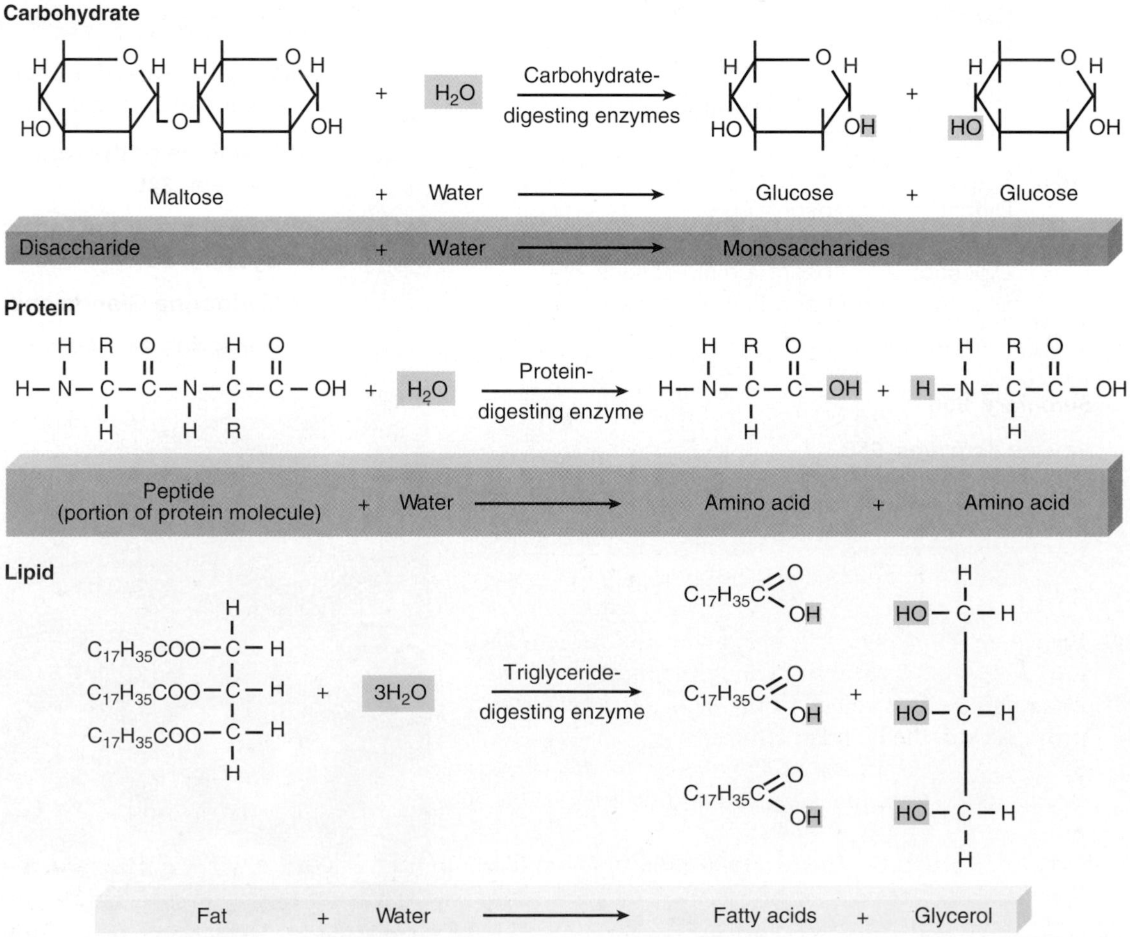

Carbohydrate

Maltose + Water → (Carbohydrate-digesting enzymes) Glucose + Glucose

Disaccharide + Water → Monosaccharides

Protein

$$H-N(H)-C(R)(H)-C(=O)-N(H)-C(H)(R)-C(=O)-OH + H_2O \xrightarrow{\text{Protein-digesting enzyme}} H-N(H)-C(R)(H)-C(=O)-OH + H-N(H)-C(R)(H)-C(=O)-OH$$

Peptide (portion of protein molecule) + Water → Amino acid + Amino acid

Lipid

$$\text{3 } C_{17}H_{35}COO-CH_x \text{ (triglyceride)} + 3H_2O \xrightarrow{\text{Triglyceride-digesting enzyme}} 3\ C_{17}H_{35}COOH + \text{glycerol } (HO-CH_2-CH(OH)-CH_2-OH)$$

Fat + Water → Fatty acids + Glycerol

Figure 18.1 **The digestion of food molecules through hydrolysis reactions.** These reactions ultimately release the subunit molecules of each food category. APR

formed are transported across the wall of the small intestine into the blood and lymph in the process of **absorption.** Digestion and absorption are the primary functions of the digestive system.

Because the composition of food is similar to the composition of body tissues, enzymes that digest food are also capable of digesting a person's own tissues. This does not normally occur, however, because a variety of protective devices inactivate digestive enzymes in the body and keep them away from the cytoplasm of the cells. The fully active digestive enzymes are normally confined to the lumen (cavity) of the gastrointestinal tract.

The lumen of the gastrointestinal tract is open at both ends (mouth and anus) and is thus continuous with the environment. In this sense, the harsh conditions required for digestion occur *outside* the body. Indigestible materials, such as cellulose from plant walls, pass from one end to the other without crossing the epithelial lining of the digestive tract; because they are not absorbed, they do not enter the body.

In *Planaria* (a type of flatworm), the gastrointestinal tract has only one opening—the mouth is also the anus. Each cell that lines the gastrointestinal tract is thus exposed to food, absorbable digestion products, and waste products. The two open ends of the digestive tract of higher organisms, by contrast, permit one-way transport, which is ensured by wavelike muscle contractions and by the action of sphincter muscles. This one-way transport allows different regions of the gastrointestinal tract to be specialized for different functions, as a "disassembly line." These functions of the digestive system include:

1. **Motility.** This refers to the movement of food through the digestive tract through the processes of
   a. *Ingestion:* Taking food into the mouth.
   b. *Mastication:* Chewing the food and mixing it with saliva.
   c. *Deglutition:* Swallowing food.
   d. *Peristalsis* and *segmentation:* Rhythmic, wavelike contractions (peristalsis), and mixing contractions in different segments (segmentation), move food through the gastrointestinal tract.
2. **Secretion.** This includes both exocrine and endocrine secretions.
   a. *Exocrine secretions:* Water, hydrochloric acid, bicarbonate, and many digestive enzymes are secreted into the lumen of the gastrointestinal tract. The stomach alone, for example, secretes 2 to 3 liters of gastric juice a day.
   b. *Endocrine secretions:* The stomach and small intestine secrete a number of hormones that help to regulate the digestive system.
3. **Digestion.** This refers to the breakdown of food molecules into their smaller subunits, which can be absorbed.
4. **Absorption.** This refers to the passage of digested end products into the blood or lymph.
5. **Storage and elimination.** This refers to the temporary storage and subsequent elimination of indigestible food molecules.
6. **Immune barrier.** The simple columnar epithelium that lines the intestine, with its tight junctions between cells, provides a physical barrier to the penetration of pathological organisms and their toxins. Also, cells of the immune system reside in the connective tissue located just under the epithelium (see fig. 18.3) to promote immune responses.

Anatomically and functionally, the digestive system can be divided into the tubular **gastrointestinal (GI) tract,** or *alimentary canal,* and **accessory digestive organs.** The GI tract is approximately 9 m (30 ft) long and extends from the mouth to the anus. It traverses the thoracic cavity and enters the abdominal cavity at the level of the diaphragm. The anus is located at the inferior portion of the pelvic cavity. The organs of the GI tract include the *oral cavity, pharynx, esophagus, stomach, small intestine,* and *large intestine* (fig. 18.2). The accessory digestive organs include the *teeth, tongue, salivary glands, liver, gallbladder,* and *pancreas.* The term *viscera* is frequently used to refer to the abdominal organs of digestion, but it can also refer to any organs in the thoracic and abdominal cavities.

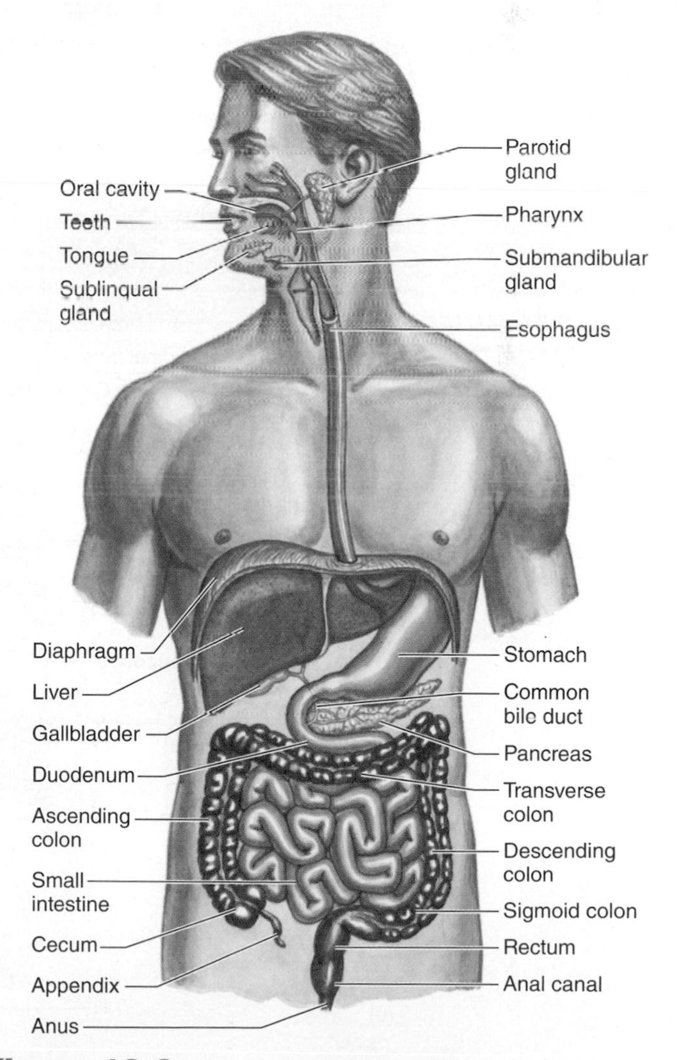

Figure 18.2 The organs of the digestive system. The digestive system includes the gastrointestinal tract and the accessory digestive organs. AP|R

Layers of the Gastrointestinal Tract

The GI tract from the esophagus to the anal canal is composed of four layers, or *tunics.* Each tunic contains a dominant tissue type that performs specific functions in the digestive process. The four tunics of the GI tract, from the inside out, are the *mucosa, submucosa, muscularis,* and *serosa* (fig. 18.3*a*).

Mucosa

The **mucosa,** which lines the lumen of the GI tract, is the absorptive and major secretory layer. It consists of a simple columnar epithelium supported by the *lamina propria,* a thin layer of areolar connective tissue containing numerous lymph nodules, which are important in protecting against disease (fig. 18.3*b*). External to the lamina propria is a thin layer of smooth muscle called the *muscularis mucosae.* This is the muscle layer responsible for the numerous small folds in certain portions of the GI tract. These folds greatly increase the absorptive surface area. Specialized goblet cells in the mucosa secrete mucus throughout most of the GI tract.

Submucosa

The relatively thick **submucosa** is a highly vascular layer of connective tissue that serves the mucosa. Absorbed molecules that pass through the columnar epithelial cells of the mucosa enter into blood and lymphatic vessels of the submucosa. In addition to blood vessels, the submucosa contains glands and nerve plexuses. The **submucosal plexus** (*Meissner's plexus*) (fig. 18.3*b*) provides a nerve supply to the muscularis mucosae of the small and large intestine.

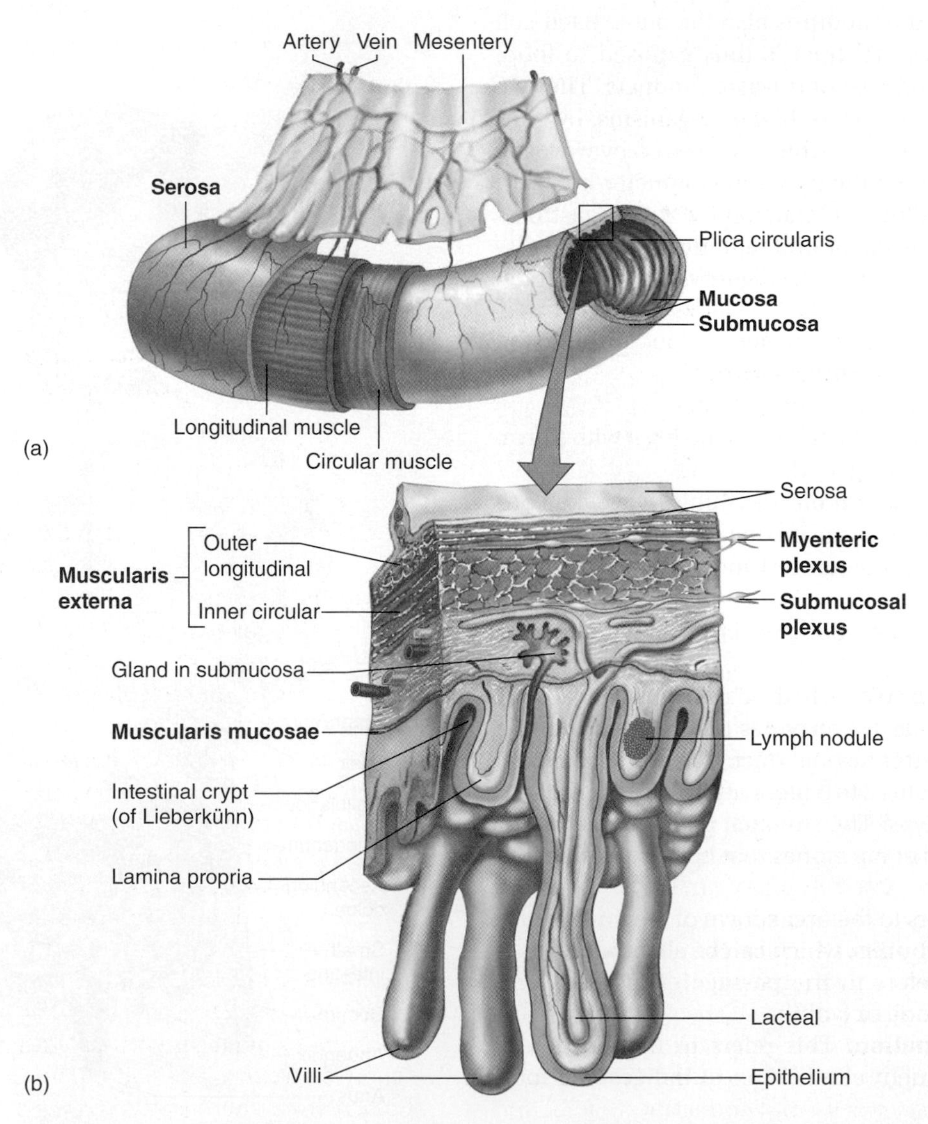

Figure 18.3 The layers of the digestive tract. (*a*) An illustration of the major tunics, or layers, of the small intestine. The inset shows how folds of mucosa form projections called villi in the small intestine. (*b*) An illustration of a cross section of the small intestine showing layers and glands. APR

Muscularis

The **muscularis** (also called the *muscularis externa*) is responsible for segmental contractions and peristaltic movement through the GI tract. The muscularis has an inner circular and an outer longitudinal layer of smooth muscle. Contractions of these layers move the food through the tract and physically pulverize and mix the food with digestive enzymes. The **myenteric plexus** (*Auerbach's plexus*), located between the two muscle layers, provides the major nerve supply to the entire GI tract. It includes fibers and ganglia from both the sympathetic and parasympathetic divisions of the autonomic nervous system.

Serosa

The outer **serosa** completes the wall of the GI tract. It is a binding and protective layer consisting of areolar connective tissue covered with a layer of simple squamous epithelium.

Regulation of the Gastrointestinal Tract

The GI tract is innervated by the sympathetic and parasympathetic divisions of the autonomic nervous system. As discussed in chapter 9, parasympathetic nerves in general stimulate motility and secretions of the gastrointestinal tract. The vagus nerve is the source of parasympathetic activity in the esophagus, stomach, pancreas, gallbladder, small intestine, and upper portion of the large intestine. The lower portion of the large intestine receives parasympathetic innervation from spinal nerves in the sacral region. The submucosal plexus and myenteric plexus are the sites where parasympathetic preganglionic fibers synapse with postganglionic neurons that innervate the smooth muscle of the GI tract.

Postganglionic sympathetic fibers pass through the submucosal and myenteric plexuses and innervate the GI tract. The effects of the sympathetic nerves reduce peristalsis and secretory activity and stimulate the contraction of sphincter muscles along the GI tract; therefore, they are antagonistic to the effects of parasympathetic nerve stimulation.

Autonomic regulation, which is "extrinsic" to the gastrointestinal tract, is superimposed on "intrinsic" modes of regulation. The gastrointestinal tract contains **intrinsic sensory neurons** that have their cell bodies within the gut wall and are not part of the autonomic system. These help in the local regulation of the digestive tract by a complex neural network within the wall of the gut called the *enteric nervous system,* or *enteric brain* (discussed in section 18.6). Regulation by the enteric nervous system complements paracrine regulation by molecules acting locally within the tissues of the GI tract, as well as hormonal regulation by hormones secreted by the mucosa.

In summary, the digestive system is regulated extrinsically by the autonomic nervous system and endocrine system, and intrinsically by the enteric nervous system and various paracrine regulators. The details of this regulation will be described in subsequent sections.

1. Define the terms *digestion* and *absorption,* describe how molecules are digested, and indicate which molecules are absorbed.

2a. Describe the structure and function of the mucosa, submucosa, and muscularis.

2b. Describe the location and composition of the submucosal and myenteric plexuses and explain the actions of autonomic nerves on the gastrointestinal tract.

18.2 FROM MOUTH TO STOMACH

Peristaltic contractions of the esophagus deliver food to the stomach, which secretes very acidic gastric juice that is mixed with the food by gastric contractions. Proteins in the resulting mixture, called chyme, are partially digested by the enzyme pepsin.

LEARNING OUTCOMES

After studying this section, you should be able to:

3. Describe the structure and functions of the esophagus and stomach.

4. Describe the composition and actions of gastric juice and explain how gastric secretion is regulated.

Mastication (chewing) of food mixes it with saliva, secreted by the salivary glands. In addition to mucus and various antimicrobial agents, saliva contains *salivary amylase,* an enzyme that can catalyze the partial digestion of starch. **Deglutition,** or swallowing, is divided into three phases: *oral, pharyngeal,* and *esophageal.* Swallowing is a complex activity that requires the coordinated contractions of 25 pairs of muscles in the mouth, pharynx, larynx, and esophagus. The muscles of the mouth, pharynx, and upper esophagus are striated and innervated by somatic motor neurons, whereas the muscles of the middle and lower esophagus are smooth and innervated by autonomic neurons. The oral phase is under voluntary control, while the pharyngeal and esophageal phases are automatic and controlled by the **swallowing center** in the brain stem.

In the oral phase, the muscles of the mouth and tongue mix the food with saliva and create a *bolus* (a mass of a size to be swallowed) of food that the tongue muscles move toward the oropharynx. Receptors in the posterior portion of the oral cavity and oropharynx stimulate the pharyngeal phase of the swallowing reflex. The soft palate lifts to close off the nasopharynx from the oropharynx (so food does not go out the nose); the vocal cords close off the opening to the larynx, and the epiglottis covers the vocal cords; the larynx is moved away from the pathway of the bolus toward the esophagus (these activities help prevent choking); and the upper esophageal sphincter relaxes. These complex activities of the pharyngeal phase take less than 1 second. In the esophageal phase of swallowing, which lasts from 5 to 6 seconds, the bolus of food is moved by peristaltic contractions toward the stomach.

Once in the stomach, the ingested material is churned and mixed with hydrochloric acid and the protein-digesting enzyme pepsin. The mixture thus produced is pushed by muscular contractions of the stomach past the pyloric sphincter (*pylorus* = gatekeeper), which guards the junction of the stomach and the duodenum of the small intestine.

Esophagus

The **esophagus** is the portion of the GI tract that connects the pharynx to the stomach. It is a muscular tube approximately 25 cm (10 in.) long, located posterior to the trachea within the mediastinum of the thorax. Before terminating in the stomach, the esophagus passes through the diaphragm by means of an opening called the *esophageal hiatus.* The esophagus is lined with a nonkeratinized stratified squamous epithelium; its walls contain either skeletal or smooth muscle, depending on the location. The upper third of the esophagus contains skeletal muscle, the middle third contains a mixture of skeletal and smooth muscle, and the terminal portion contains only smooth muscle.

Swallowed food is pushed from the oral to the anal end of the esophagus (and, afterward, of the intestine) by a wavelike muscular contraction called **peristalsis** (fig. 18.4). Movement of the bolus along the digestive tract occurs because the circular smooth muscle contracts behind, and relaxes in front of, the bolus. This is followed by shortening of the tube by longitudinal muscle contraction. These contractions progress from the superior end of the esophagus to the *gastroesophageal junction* at a rate of 2 to 4 cm per second as they empty the contents of the esophagus into the cardiac region of the stomach.

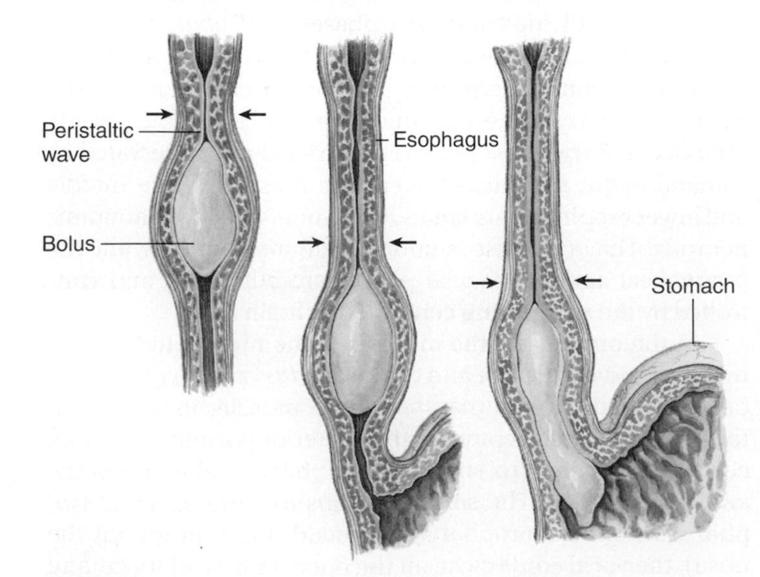

Figure 18.4 Peristalsis in the esophagus. Peristaltic contraction and movement of a bolus into the stomach.

The lumen of the terminal portion of the esophagus is slightly narrowed because of a thickening of the circular muscle fibers in its wall. This portion is referred to as the **lower esophageal (gastroesophageal) sphincter.** After food passes into the stomach, constriction of the muscle fibers of this region help prevent the stomach contents from regurgitating into the esophagus. Regurgitation would occur because the pressure in the abdominal cavity is greater than the pressure in the thoracic cavity as a result of respiratory movements. The lower esophageal sphincter must therefore remain closed until food is pushed through it by peristalsis into the stomach.

CLINICAL APPLICATION

The lower esophageal sphincter is not a true sphincter muscle that can be identified histologically, and it does at times permit the acidic contents of the stomach to enter the esophagus. This can create a burning sensation commonly called **heartburn,** although the heart is not involved. In infants under a year of age the lower esophageal sphincter may function erratically, causing them to "spit up" following meals. Certain mammals, such as rodents, have a true gastroesophageal sphincter and thus cannot regurgitate. This is why poison grains are effective in killing mice and rats.

Stomach

The J-shaped **stomach** is the most distensible part of the GI tract. It is continuous with the esophagus superiorly and empties into the duodenum of the small intestine inferiorly. The functions of the stomach are to store food, to initiate the digestion of proteins, to kill bacteria with the strong acidity of gastric juice, and to move the food into the small intestine as a pasty material called **chyme.**

Swallowed food is delivered from the esophagus to the *cardiac region* of the stomach (fig. 18.5). An imaginary

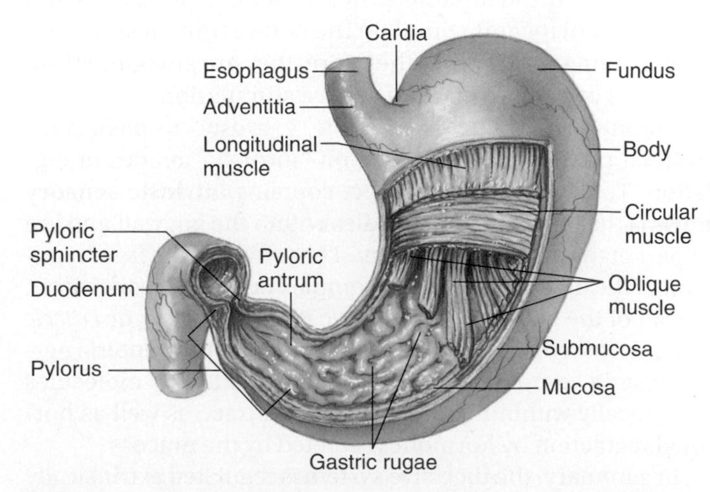

Figure 18.5 Primary regions and structures of the stomach. Notice that the pyloric region of the stomach includes the pyloric antrum (the wider portion of the pylorus) as well as the pyloric sphincter. AP|R

horizontal line drawn through the cardiac region divides the stomach into an upper *fundus* and a lower *body,* which together compose about two-thirds of the stomach. The distal portion of the stomach is called the *pyloric region.* The pyloric region begins in a somewhat widened area, the *antrum,* and ends at the *pyloric sphincter.* Contractions of the stomach churn the chyme, mixing it more thoroughly with the gastric secretions. These contractions also push partially digested food from the antrum through the pyloric sphincter and into the first part of the small intestine.

The inner surface of the stomach is thrown into long folds called *rugae,* which can be seen with the unaided eye. Microscopic examination of the gastric mucosa shows that it is likewise folded. The openings of these folds into the stomach lumen are called **gastric pits.** The cells that line the folds secrete various products into the stomach; these cells form the exocrine **gastric glands** (fig. 18.6).

Gastric glands contain several types of cells that secrete different products:

1. **mucous neck cells,** which secrete mucus (these supplement the surface mucous cells, which line the luminal surface of the stomach and the gastric pits).
2. **parietal cells,** which secrete *hydrochloric acid (HCl);*
3. **chief** (or **zymogenic**) **cells,** which secrete *pepsinogen,* an inactive form of the protein-digesting enzyme *pepsin;*
4. **enterochromaffin-like (ECL) cells,** found in the stomach and intestine, which secrete *histamine* and *5-hydroxytryptamine* (also called *serotonin*) as paracrine regulators of the GI tract;
5. **G cells,** which secrete the hormone *gastrin* into the blood; and
6. **D cells,** which secrete the hormone *somatostatin.*

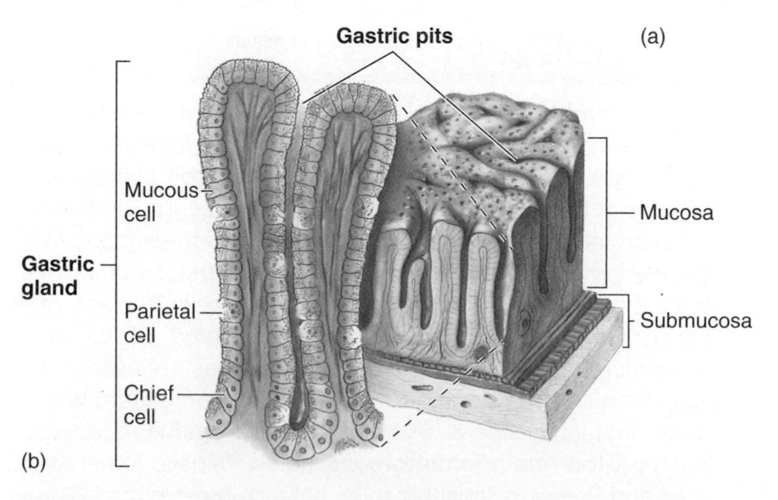

Figure 18.6 Gastric pits and gastric glands of the mucosa. (*a*) Gastric pits are the openings of the gastric glands. (*b*) Gastric glands consist of several types of cells (including mucous cells, chief cells, and parietal cells), each of which produces a specific secretion. AP|R

In addition to these products, the gastric mucosa (probably the parietal cells) secretes a polypeptide called **intrinsic factor,** which is required for the intestinal absorption of vitamin B_{12}. Vitamin B_{12} is necessary for the production of red blood cells in the bone marrow (see the next Clinical Application box). Also, the stomach has recently been shown to secrete a hormone named **ghrelin.** Secretion of this hormone rises before meals and falls after meals. This may serve as a signal from the stomach to the brain that helps regulate hunger (chapter 19, section 19.2).

The exocrine secretions of the gastric cells, together with a large amount of water (2 to 3 L/day), form a highly acidic solution known as **gastric juice.**

CLINICAL APPLICATION

The only stomach function that appears to be essential for life is the secretion of *intrinsic factor.* This polypeptide is needed for the absorption of vitamin B_{12} in the terminal portion of the ileum in the small intestine, and vitamin B_{12} is required for maturation of red blood cells in the bone marrow. Following surgical removal of the stomach (gastrectomy) a patient has to receive B_{12} injections, or take B_{12} orally together with intrinsic factor. Without vitamin B_{12}, **pernicious anemia** will develop.

Pepsin and Hydrochloric Acid Secretion

The parietal cells secrete H^+ (protons), at a pH as low as 0.8, into the gastric lumen by primary active transport (involving carriers that function as an ATPase). These carriers, known as **H^+/K^+ ATPase pumps,** transport H^+ uphill against a million-to-one concentration gradient into the lumen of the stomach while they transport K^+ in the opposite direction (fig. 18.7).

At the same time, the parietal cells' basolateral membranes (facing the blood in capillaries of the lamina propria) take in Cl^- against its electrochemical gradient by coupling its transport to the downhill movement of bicarbonate (HCO_3^-). The bicarbonate ion is produced within the parietal cells by the dissociation of carbonic acid, formed from CO_2 and H_2O by the enzyme carbonic anhydrase. Therefore, the parietal cells can secrete Cl^- (by facilitative diffusion) as well as H^+ into the gastric juice while they secrete bicarbonate into the blood (fig. 18.7).

Secretion of hydrochloric acid (HCl) by the parietal cells is stimulated by the hormone gastrin, the paracrine regulator histamine, and the neurotransmitter ACh. Gastrin is secreted by G cells of the gastric mucosa, enters the general circulation, and can stimulate parietal cells directly by binding to receptors on the parietal cell basolateral membrane. However, gastrin stimulation of HCl secretion is mostly indirect; gastrin stimulates ECL cells to secrete histamine, and histamine acts as a paracrine regulator to stimulate the parietal cells to secrete HCl (see fig. 18.30). Histamine stimulation of parietal

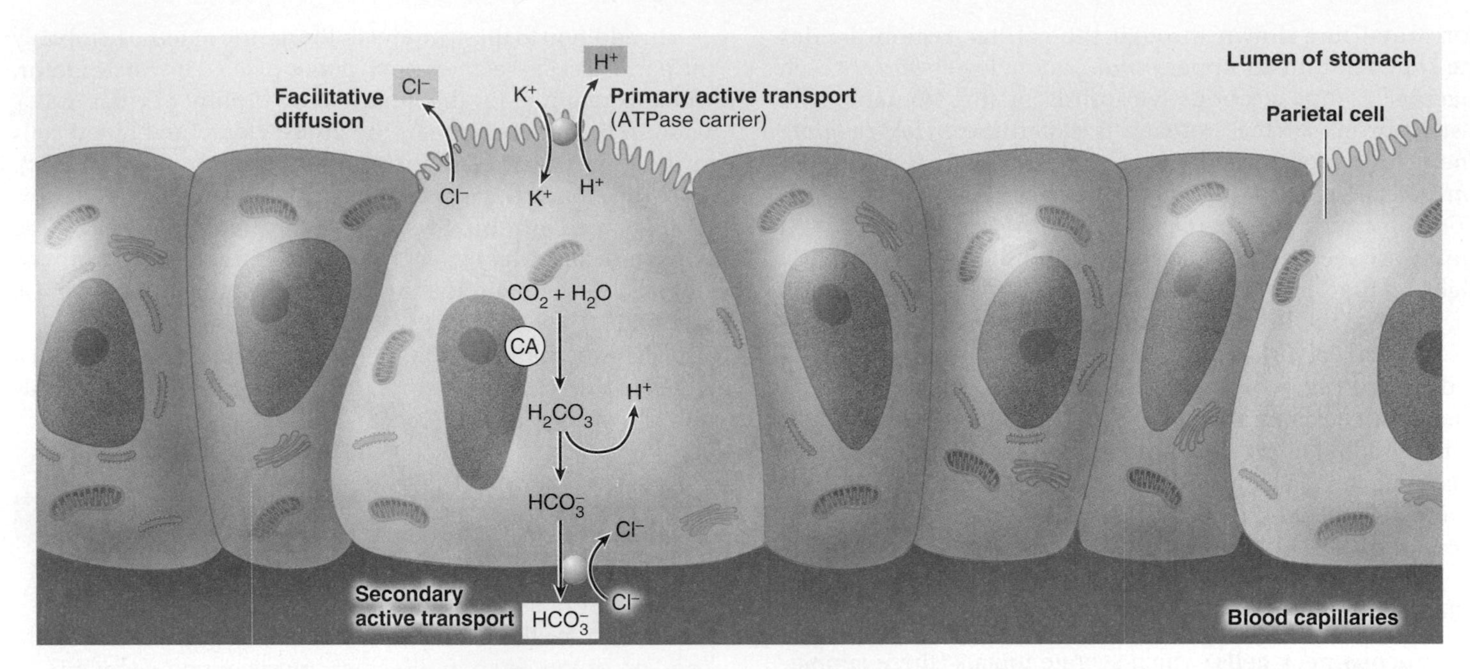

Figure 18.7 Secretion of gastric acid by parietal cells. The apical membrane (facing the lumen) secretes H^+ in exchange for K^+ using a primary active transport carrier that is powered by the hydrolysis of ATP. The basolateral membrane (facing the blood) secretes bicarbonate (HCO_3^-) in exchange for Cl^-. The Cl^- moves into the cell against its electrochemical gradient, powered by the downhill movement of HCO_3^- out of the cell. This HCO_3^- is produced by the dissociation of carbonic acid (H_2CO_3), which is formed from CO_2 and H_2O by the action of the enzyme carbonic anhydrase (abbreviated CA). The Cl^- then leaves the apical portion of the membrane by diffusion through a membrane channel. The parietal cells thus secrete HCl into the stomach lumen as they secrete HCO_3^- into the blood. AP|R

cells is mediated by the H_2 type of histamine receptor, which is different from the H_1 type of histamine receptor involved in allergic reactions (chapter 15).

Parasympathetic neurons of the vagus nerve stimulate both parietal and ECL cells, although stimulation of ECL cells is believed to be the most important effect. This is particularly true at night during sleep, when the secretion of histamine from ECL cells is most responsible for stimulating gastric HCl secretion (see fig. 18.30). This is why drugs that block H_2 histamine receptors (such as Tagamet and Zantac) are more effective at night than they are at blocking meal-stimulated HCl secretion.

The high concentration of HCl from the parietal cells makes gastric juice very acidic, with a pH of less than 2. This strong acidity serves three functions:

1. Ingested proteins are denatured at low pH—that is, their tertiary structure (chapter 2) is altered so that they become more digestible.
2. Under acidic conditions, weak pepsinogen enzymes partially digest each other—this frees the fully active pepsin enzyme as small inhibitory fragments are removed (fig. 18.8).
3. Pepsin is more active under acidic conditions—it has a pH optimum (chapter 4) of about 2.0.

As a result of the activation of pepsin under acidic conditions, the fully active pepsin is able to catalyze the hydrolysis of peptide bonds in the ingested protein. Thus, the cooperative activities of pepsin and HCl permit the partial digestion of food protein in the stomach.

The strong acid and protein-digesting action of pepsin could damage the lining of the stomach (produce a *peptic ulcer,* as described shortly). The first line of defense against such damage is a stable gel of mucus that is adherent (stuck) to the gastric epithelial surface. This **adherent layer of mucus** contains alkaline bicarbonate (HCO_3^-), secreted from the apical plasma membranes of the epithelial cells. When the stomach secretes more

CLINICAL APPLICATION

Gastroesophageal reflux disease (**GERD**) is a common disorder in which the reflux of acidic gastric juice into the esophagus causes frequent heartburn or complications. GERD can be associated with laryngitis and cough, and can produce injuries to the esophagus that include *esophagitis* (producing erosions of the mucosa and ulcers), *stricture* of the lumen, *Barrett's esophagus* (columnar cells in place of the squamous epithelium), and *adenocarcinoma*. GERD can be reduced by lifestyle modifications (such as not eating at least 3 hours before going to bed), and is often treated with drugs that inhibit the H^+/K^+ pumps in the gastric mucosa. Such **proton pump inhibitors,** including Prilosec (omeprazole) and Prevacid (lansoprazole), are also used in the treatment of peptic ulcers. Because gastric acid secretion is stimulated by histamine from the ECL cells, acid secretion can also be reduced by drugs (such as Tagamet and Zantac) that block the H_2 histamine receptors in the gastric mucosa.

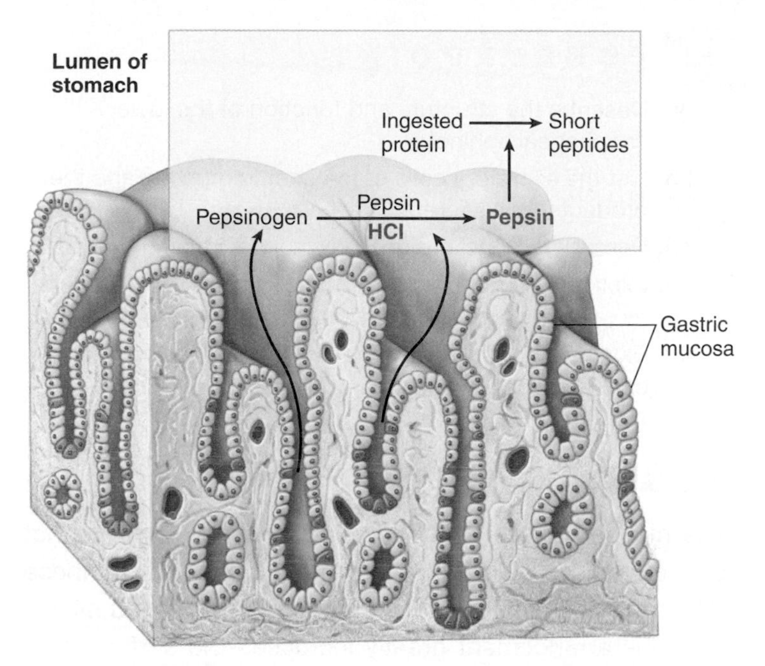

Figure 18.8 The activation of pepsin. The gastric mucosa secretes the inactive enzyme pepsinogen and hydrochloric acid (HCl). In the presence of HCl, the active enzyme pepsin is produced. Pepsin digests proteins into shorter polypeptides. AP|R

acid into the lumen, there is also more bicarbonate available to the epithelial cells for secretion into the mucus (see fig. 18.7). As a result, the pH at the epithelial surface is normally near neutral.

Also, the adherent layer of gastric mucus is the major barrier to potential damage to the stomach caused by pepsin. Little attention has historically been paid to pepsin's ability to cause damage, but there is evidence that it could pose a significant threat. Indeed, the damage to the esophagus caused by gastroesophageal reflux could be due more to pepsin than to acid. The adherent mucus layer in the stomach protects the gastric lining from pepsin by slowing its diffusion so that it doesn't normally reach the epithelial cells.

Although the adherent layer of alkaline mucus is the first line of defense against acid and pepsin damage to the stomach, there are other important protective mechanisms. These include the presence of tight junctions between adjacent epithelial cells, which prevent acid and pepsin from leaking past the epithelial barrier and destroying underlying tissues. Also, a rapid rate of epithelial cell division replaces the entire epithelium every 3 days, so that damaged cells can be rapidly replaced.

Digestion and Absorption in the Stomach

Proteins are only partially digested in the stomach by the action of pepsin, while carbohydrates and fats are not digested at all by pepsin. (Digestion of starch begins in the mouth with the action of salivary amylase and continues for a time when the food enters the stomach, but amylase soon becomes inactivated by the strong acidity of gastric juice.) The complete digestion of food molecules occurs later, when chyme enters the small intestine. Therefore, people who have had partial gastric resections—and even those who have had complete gastrectomies—can still adequately digest and absorb their food.

Almost all of the products of digestion are absorbed through the wall of the small intestine; the only commonly ingested substances that can be absorbed across the stomach wall are alcohol and aspirin. Absorption occurs as a result of the lipid solubility of these molecules. Aspirin and most other NSAIDs (nonsteroidal anti-inflammatory drugs) can promote damage to the gastric mucosa and cause bleeding, and must be avoided in people with gastric ulcers.

Gastritis and Peptic Ulcers

Peptic ulcers are erosions of the mucosa of the stomach or duodenum (produced by the action of HCl) that penetrate through the muscularis mucosa layer. In *Zollinger-Ellison syndrome,* ulcers of the duodenum are produced by excessive gastric acid secretion in response to very high levels of the hormone gastrin. Gastrin is normally secreted by the stomach, but in this case it is released by a gastrin-secreting tumor that is usually located in the duodenum or pancreas. This is a rare condition, but it does demonstrate that excessive gastric acid can cause ulcers of the duodenum. Ulcers of the stomach, however, are not believed to be due to excessive acid secretion, but rather to mechanisms that reduce the barriers of the gastric mucosa to self-digestion.

Most people who have peptic ulcers are infected with bacteria known as *Helicobacter pylori,* which have adaptations that enable them to survive in the very acidic pH of the stomach and infect almost half of the adult population worldwide. The 2005 Nobel Prize in Physiology or Medicine was awarded to two scientists who discovered that this common bacterium, rather than emotional stress or spicy food, is the cause of most cases of peptic ulcers of the stomach and duodenum. As a result of this discovery, ulcers can now be effectively treated medically with a drug regimen that consists of a *proton pump inhibitor* (a drug such as Prilosec that inhibits the K^+/H^+ pumps) combined with two different antibiotics (such as *amoxicillin* and *clarithromycin*) to suppress the *H. pylori* infection.

Eradicating the *H. pylori* infection cures duodenal ulcers in over 80% of people whose ulcers are not caused by the use of NSAIDs (nosteroidal anti-inflammatory drugs such as aspirin and ibuprofen; chapter 11, section 11.7). These drugs can damage the gastric mucosa because NSAIDs inhibit prostaglandin synthesis, and prostaglandins contribute to the mucosal barrier by stimulating mucus and bicarbonate production. This is why most cases of peptic ulcers in people who are negative for *H. pylori* are caused by the use of NSAIDs.

When the gastric barriers to self-digestion are broken down, acid can leak through the mucosa to the submucosa, causing direct damage and stimulating inflammation. The histamine released from mast cells during inflammation may stimulate further acid secretion (see fig. 18.30) and result in further damage to the mucosa. The inflammation that occurs during these events is called **acute gastritis.** This is why drugs that block the H_2 histamine receptors (such as Tagamet and Zantac) may be used to treat the gastritis.

The duodenum is normally protected from gastric acid by the adherent layer of mucus on its epithelium. Duodenal cells secrete bicarbonate into this adherent mucus layer, so that the surface epithelium is normally exposed to a neutral pH. Additional protection against gastric acid is provided through bicarbonate secreted by Brunner's glands in the submucosa, which are glands unique to the duodenum. Finally, the acidic chyme is neutralized by the buffering action of bicarbonate in alkaline pancreatic juice, which is released into the duodenum upon the arrival of the acidic chyme from the stomach. Duodenal ulcers may result from excessive gastric acid secretion and/or inadequate secretion of bicarbonate in the duodenum. Indeed, some studies have demonstrated that people with duodenal ulcers have a reduced secretion of duodenal bicarbonate in response to acid in the lumen.

People with gastritis and peptic ulcers should avoid substances that stimulate acid secretion, including coffee and alcohol, and often must take antacids (such as Tums), H_2-histamine receptor blockers (such as Zantac), or proton pump inhibitors (such as Prilosec). If the cause is excessive activity of *Helicobacter pylori,* antibiotics may also be required.

Clinical Investigation CLUES

Alan had a sharp pain in his stomach whenever he drank wine.

- What may be responsible for Alan's pain?
- What medicine might help alleviate this pain, and how would it work?

CHECKPOINT

3a. Describe the structure and function of the lower esophageal sphincter.

3b. List the secretory cells of the gastric mucosa and the products they secrete.

4a. Describe the functions of hydrochloric acid in the stomach.

4b. Explain how peptic ulcers are produced and why they are more likely to occur in the duodenum than in the stomach.

4c. Explain how gastrin and vagus nerve stimulation cause the parietal cells to secrete HCl.

18.3 SMALL INTESTINE

The mucosa of the small intestine is folded into villi that project into the lumen. In addition, the cells that line these villi have foldings of their plasma membrane called microvilli. This arrangement greatly increases the surface area for absorption and improves digestion, since digestive enzymes are embedded within the microvilli.

LEARNING OUTCOMES

After studying this section, you should be able to:

5. Describe the structure and functions of the small intestine.

6. Identify the location and describe the functions of the digestive enzymes of the small intestine.

7. Describe the muscle contractions and movements of the small intestine.

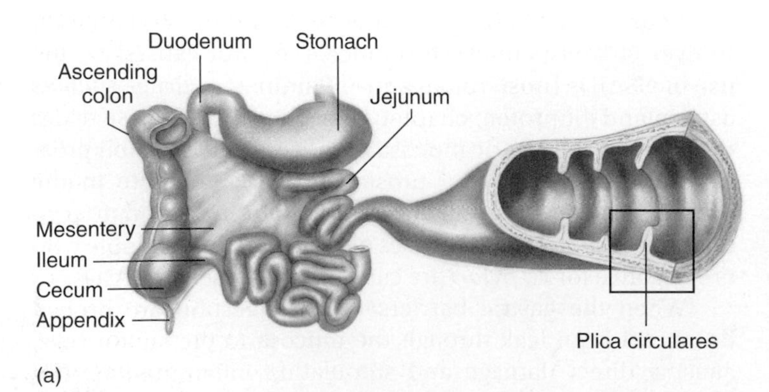

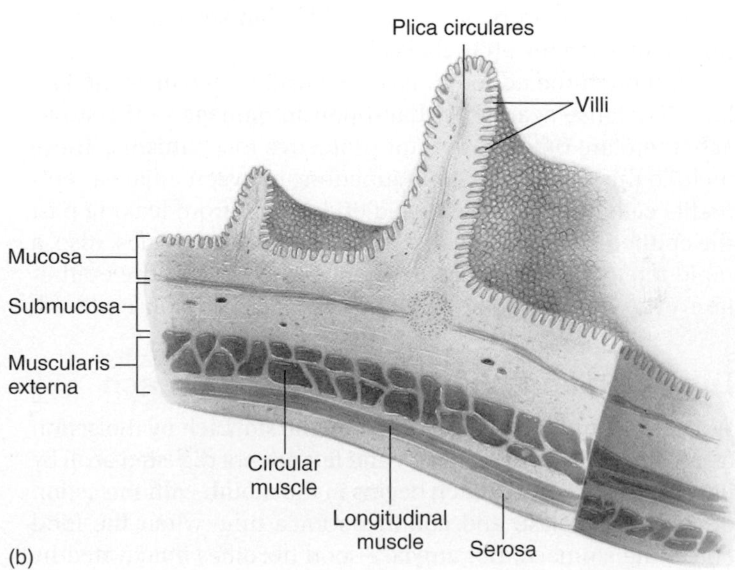

Figure 18.9 The small intestine. (*a*) The regions of the small intestine. (*b*) A section of the intestinal wall showing the tissue layers, plicae circulares, and villi. APIR

The **small intestine** (fig. 18.9) is the portion of the GI tract between the pyloric sphincter of the stomach and the ileocecal valve opening into the large intestine. It is called "small" because of its relatively small diameter compared to the large intestine. The small intestine is the longest part of the GI tract, however. It is approximately 3 m (12 ft) long in a living person, but it will measure nearly twice this length in a cadaver when the muscle wall is relaxed. The first 20 to 30 cm (10 in.) extending from the pyloric sphincter is the **duodenum.** The next two-fifths of the small intestine is the **jejunum,** and the last three-fifths is the **ileum.** The ileum empties into the large intestine through the *ileocecal valve.*

The products of digestion are absorbed across the epithelial lining of the intestinal mucosa. Absorption of carbohydrates, lipids, amino acids, calcium, and iron occurs primarily in the duodenum and jejunum. Bile salts, vitamin B_{12}, water, and electrolytes are absorbed primarily in the ileum. Absorption occurs at a rapid rate as a result of extensive foldings of the intestinal mucosa, which greatly increase its absorptive surface area. The mucosa and submucosa form large folds called *plicae circulares,* which can be observed with the unaided eye. The surface area is further increased by microscopic folds of mucosa called *villi,* and by foldings of the apical plasma membrane of epithelial cells (which can be seen only with an electron microscope) called *microvilli.*

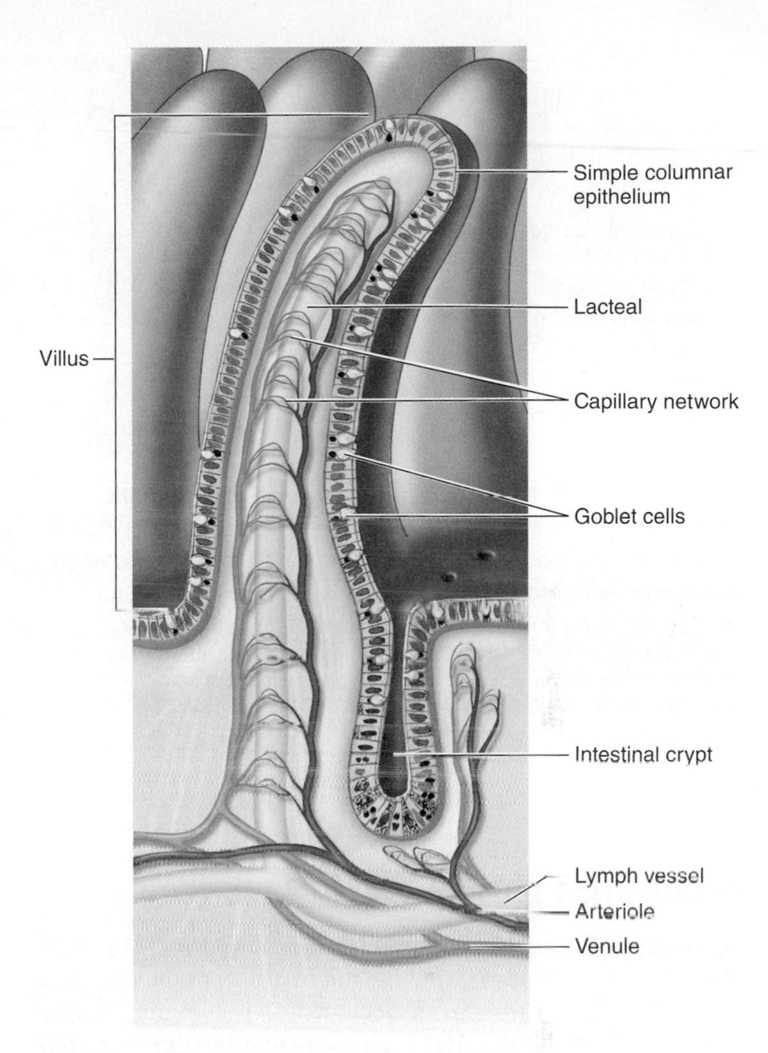

Figure 18.10 The structure of an intestinal villus. The figure also depicts an intestinal crypt (crypt of Lieberkühn), in which new epithelial cells are produced by mitosis. AP|R

Villi and Microvilli

Each **villus** is a fingerlike fold of mucosa that projects into the intestinal lumen (fig. 18.10). The villi are covered with columnar epithelial cells, among which are interspersed mucus-secreting *goblet cells.* The lamina propria, which forms the connective tissue core of each villus, contains numerous lymphocytes, blood capillaries, and a lymphatic vessel called the *central lacteal* (fig. 18.10). Absorbed monosaccharides and amino acids enter the blood capillaries; absorbed fat enters the central lacteals.

Epithelial cells at the tips of the villi are continuously exfoliated (shed) and are replaced by cells that are pushed up from the bases of the villi. The epithelium at the base of the villi invaginates downward to form narrow pouches that open through pores to the intestinal lumen. These structures are called **intestinal crypts,** or *crypts of Lieberkühn* (fig. 18.10).

At the bottom of the intestinal crypts in the small (but not the large) intestine are **Paneth cells,** which secrete antibacterial *lysozyme* and bactericidal peptides called *defensins.* The bottoms of the crypts also contain **intestinal stem cells,** which divide by mitosis to replenish themselves and to produce the specialized cells of the intestinal mucosa. Mitosis is estimated to occur twice a day in the crypts. At the top of the crypts, mitosis stops and the cells differentiate into secretory cells (Paneth cells, goblet cells, and endocrine cells) and *enterocytes* (intestinal epithelial cells). These newly formed cells migrate from the crypts to the tip of the villi, a journey of three days. Cells at the tips of the villi then undergo apoptosis and are shed into the lumen. In this way, the intestinal epithelium is renewed every four to five days.

Microvilli are formed by foldings at the apical surface of each epithelial cell membrane. These minute projections can be seen clearly only in an electron microscope. In a light microscope, the microvilli produce a somewhat vague **brush border** on the edges of the columnar epithelial cells. The terms *brush border* and *microvilli* are thus often used interchangeably in describing the small intestine (fig. 18.11).

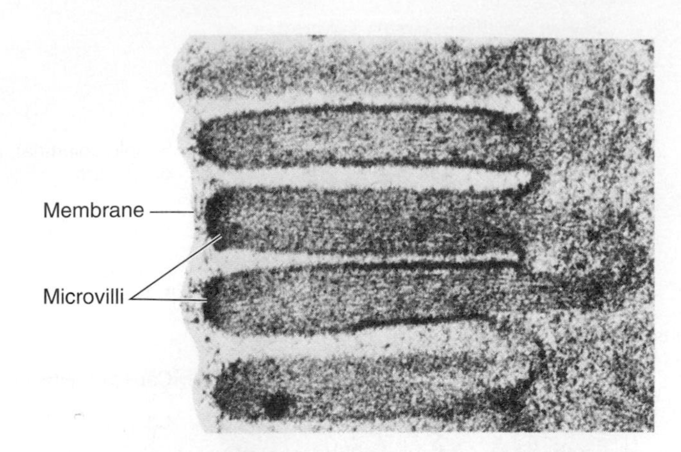

Figure 18.11 Electron micrograph of microvilli. Microvilli are evident at the apical surface of the columnar epithelial cells in the small intestine. Microvilli increase the surface area for absorption and also have the brush border digestive enzymes embedded in their plasma membranes. From Keith R. Porter, D. H. Alpers, and D. Seetharan, "Pathophysiology of Diseases Involving Intestinal Brush-Border Proteins" in *New England Journal of Medicine,* Vol. 296, 1977, p. 1047, fig. 1. Copyright © 1977 Massachusetts Medical Society. All rights reserved. APIR

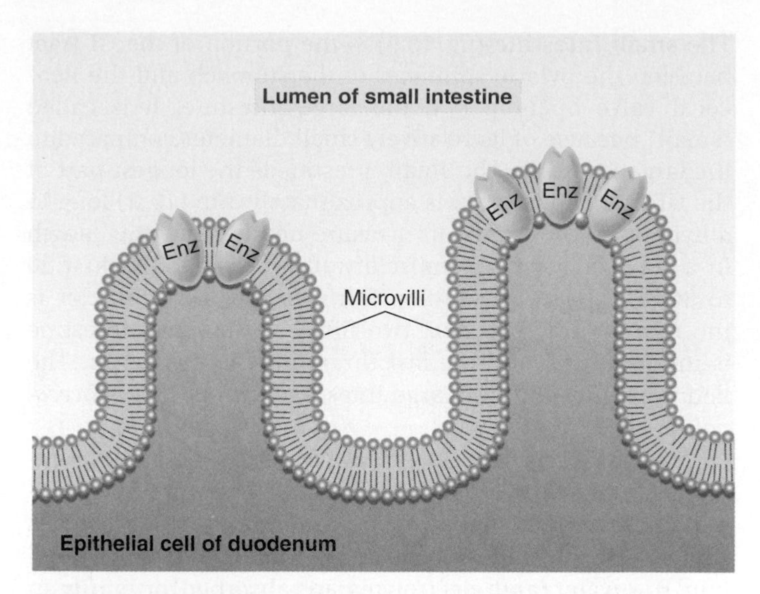

Figure 18.12 Location of brush border enzymes. The brush border enzymes (Enz) are embedded in the plasma membrane of the microvilli in the small intestine. The active sites of these enzymes face the chyme in the lumen, helping to complete the digestion of food molecules.

Intestinal Enzymes

In addition to providing a large surface area for absorption, the plasma membranes of the microvilli contain digestive enzymes that hydrolyze disaccharides, polypeptides, and other substrates (table 18.1). These **brush border enzymes** are not secreted into the lumen, but instead remain attached to the plasma membrane with their active sites exposed to the chyme (fig. 18.12). One brush border enzyme, **enterokinase** (also called **enteropeptidase**), is required for activation of the protein-digesting enzyme *trypsin,* which enters the small intestine in pancreatic juice. Activated trypsin then activates other pancreatic juice enzymes, as discussed in section 18.5 (see fig. 18.29).

CLINICAL APPLICATION

The ability to digest milk sugar, or lactose, depends on the presence of a brush border enzyme called *lactase.* This enzyme is present in all children under the age of four but becomes inactive to some degree in most adults (people of Asian or African heritage are more often lactase deficient than Caucasians). A deficiency of lactase can result in **lactose intolerance,** a condition in which too much undigested lactose in the intestine causes diarrhea, gas, cramps, and other unpleasant symptoms. Yogurt is better tolerated than milk because it contains lactase (produced by the yogurt bacteria), which becomes activated in the duodenum and digests lactose.

Table 18.1 | Brush Border Enzymes Attached to the Cell Membrane of Microvilli in the Small Intestine

Category	Enzyme	Comments
Disaccharidase	Sucrase	Digests sucrose to glucose and fructose; deficiency produces gastrointestinal disturbances
	Maltase	Digests maltose to glucose
	Lactase	Digests lactose to glucose and galactose; deficiency produces gastrointestinal disturbances (lactose intolerance)
Peptidase	Aminopeptidase	Produces free amino acids, dipeptides, and tripeptides
	Enterokinase	Activates trypsin (and indirectly other pancreatic juice enzymes); deficiency results in protein malnutrition
Phosphatase	Ca^{2+}, Mg^{2+}-ATPase	Needed for absorption of dietary calcium; enzyme activity regulated by vitamin D
	Alkaline phosphatase	Removes phosphate groups from organic molecules; enzyme activity may be regulated by vitamin D

Intestinal Contractions and Motility

Two major types of contractions occur in the small intestine: *peristalsis* and *segmentation.* Peristalsis is much weaker in the small intestine than in the esophagus and stomach. Intestinal motility—the movement of chyme through the intestine—is relatively slow and is due primarily to the greater pressure at the pyloric end of the small intestine than at the distal end.

The major contractile activity of the small intestine is **segmentation.** This term refers to muscular constrictions of the lumen, which occur simultaneously at different intestinal segments (fig. 18.13). This action serves to mix the chyme more thoroughly. Segmentation contractions occur more frequently in the proximal than in the distal end of the intestine, producing the pressure difference mentioned earlier and helping to move chyme through the small intestine.

Contractions of intestinal smooth muscles occur automatically in response to endogenous pacemaker activity, somewhat analogous to the automatic beating of the heart. In intestinal smooth muscle, however, the rhythm of contractions is paced by graded depolarizations called **slow waves** (fig. 18.14). The slow waves are produced by unique cells, often associated with autonomic nerve endings. However, these pacemaker cells are neither neurons nor smooth muscle cells; they are the cells identified histologically as the **interstitial cells of Cajal.** These compose about 5% of the cells in the muscularis layer, and have long processes that join different interstitial cells of Cajal to each other and to smooth muscle cells by gap junctions. The gap junctions permit the spread of depolarization from one cell to the next (fig. 18.15).

Slow waves spread by way of gap junctions between the interconnected cells of Cajal in the stomach and intestines. However, the slow waves can spread only a short distance

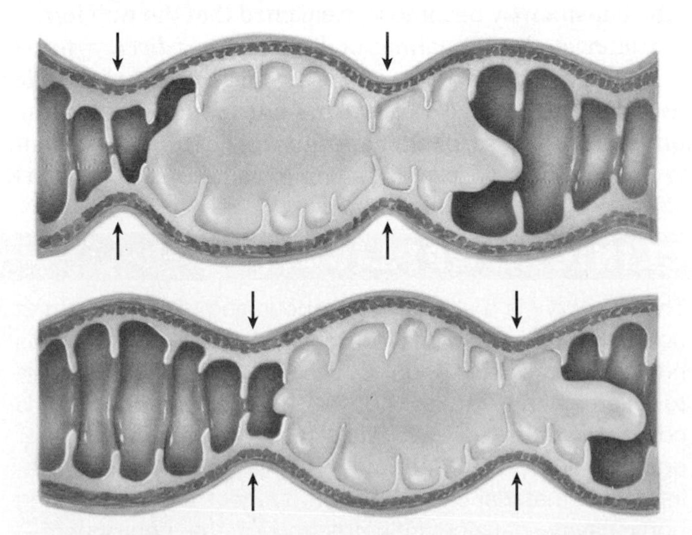

Figure 18.13 Segmentation of the small intestine. Simultaneous contractions of numerous segments of the intestine help mix the chyme with digestive enzymes and mucus.

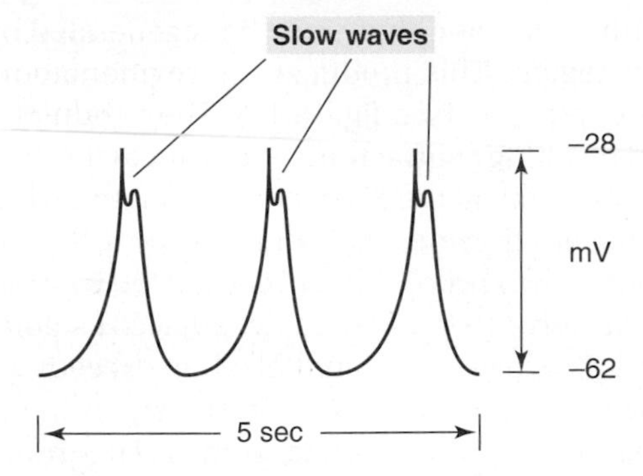

Figure 18.14 Slow waves in the intestine. The slow waves are produced by the interstitial cells of Cajal (ICC), not by smooth muscle cells, and are apparently conducted by networks of ICC that are electrically joined together within the intestinal wall. Smooth muscle cells respond to this depolarization by producing action potentials and contracting. Note that the slow waves occur much slower (with a rate measured in seconds) than do the pacemaker potentials in the heart.

See the *Test Your Quantitative Ability* section of the Review Activities at the end of this chapter.

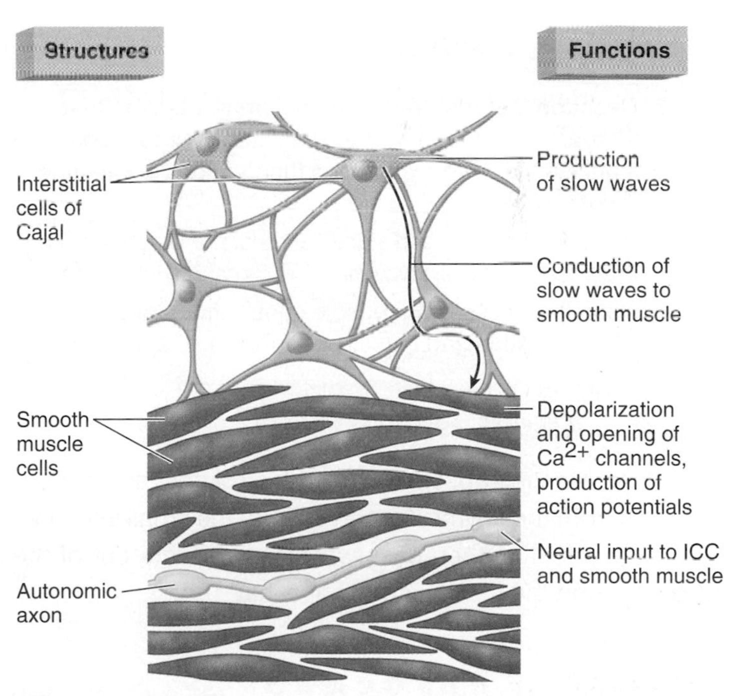

Figure 18.15 Cells responsible for the electrical events within the muscularis. The interstitial cells of Cajal (ICC) generate the slow waves, which pace the contractions of the intestine. Slow waves are conducted into the smooth muscle cells where they can stimulate opening of Ca^{2+} channels. This produces action potentials and stimulates contraction. Autonomic axons have varicosities that release neurotransmitters, which modify the inherent electrical activity of the interstitial cells of Cajal and smooth muscle cells.

(a few centimeters) and thus must be regenerated by the next pacemaker region. This produces the segmentation contractions of the intestine (see fig. 18.13). The production of slow waves, and resulting contractions, are faster at the proximal end of the intestine than at the distal end, so that there is a pressure head that pushes the intestinal contents along the GI tract.

The slow waves produced and conducted by the interstitial cells of Cajal serve to depolarize the adjacent smooth muscle cells. When the slow-wave depolarization exceeds a threshold value, it triggers action potentials in the smooth muscle cells by opening voltage-gated Ca^{2+} channels. The inward flow of Ca^{2+} has two effects: (1) it produces the upward depolarization phase of the action potential (repolarization is produced by outward flow of K^{+}); and (2) it stimulates contraction (as described in chapter 12; see fig. 12.36). Contraction may then be aided by additional calcium released from the sarcoplasmic reticulum through calcium-induced calcium release.

Autonomic nerves modify the automatic contractions of the intestine largely by influencing the enteric nervous system (section 18.6; see fig. 18.31), which in turn stimulates or inhibits the interstitial cells of Cajal. Acetylcholine, released by postganglionic axons and acting through muscarinic receptors, increases the amplitude and duration of the slow waves. Thus, it increases the production of action potentials and promotes contractions and motility of the intestine.

CHECKPOINT

5. Describe the adaptations of the small intestine that increase its surface area and explain their functional significance; also explain the function of the intestinal crypts.
6. Identify the nature and significance of the brush border enzymes and explain why many adults cannot tolerate milk.
7. Describe the smooth muscle contractions of the small intestine and explain how they are regulated.

18.4 LARGE INTESTINE

The large intestine absorbs water, electrolytes, and certain vitamins from the chyme it receives from the small intestine. The large intestine then passes waste products out of the body through the rectum and anal canal.

LEARNING OUTCOMES

After studying this section, you should be able to:

8. Describe the structure and functions of the large intestine.
9. Explain the nature and significance of the intestinal microbiota.

The **large intestine,** or **colon,** extends from the ileocecal valve to the anus, framing the small intestine on three sides. Chyme from the ileum passes into the **cecum,** which is a blind pouch (open only at one end) at the beginning of the large intestine. Waste material then passes in sequence through the **ascending colon, transverse colon, descending colon, sigmoid colon, rectum,** and **anal canal** (fig. 18.16). Waste material (feces) is excreted through the *anus,* the external opening of the anal canal.

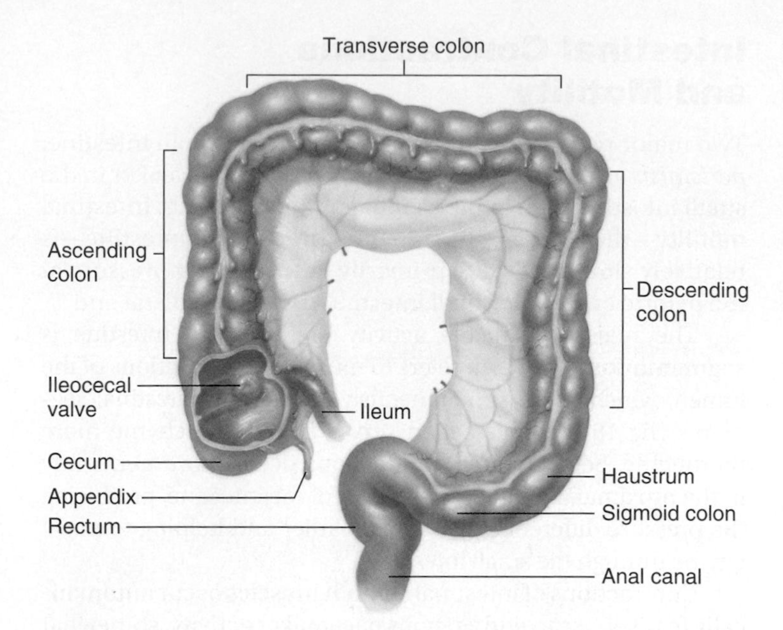

Figure 18.16 The large intestine. The different regions of the large intestine (colon) are illustrated. AP|R

The mucosa of the large intestine, like that of the small intestine, contains many scattered lymphocytes and lymphatic nodules and is covered by columnar epithelial cells and mucus-secreting goblet cells. Although this epithelium does form crypts (fig. 18.17), there are no villi in the large intestine—the intestinal mucosa therefore appears flat. The outer surface of the colon bulges outward to form pouches, or **haustra** (figs. 18.16 and 18.18). Occasionally, the muscularis externa of the haustra may become so weakened that the wall forms a more elongated outpouching, or diverticulum (*divert* = turned aside). Inflammation of one or more of these structures is called *diverticulitis.* The large intestine has little or no digestive function, but it does absorb water and electrolytes from the remaining chyme, as well as several B complex vitamins and vitamin K.

CLINICAL APPLICATION

The *appendix* is a short, thin outpouching of the cecum. It does not function in digestion, but like the tonsils, it contains numerous lymphatic nodules (see fig. 18.17) and is subject to inflammation—a condition called **appendicitis.** This is commonly detected in its later stages by pain in the lower right quadrant of the abdomen. If the appendix ruptures, infectious material can spread throughout the surrounding body cavity, causing inflammation of the peritoneum—*peritonitis.* This dangerous event may be prevented by surgical removal of the inflamed appendix (*appendectomy*).

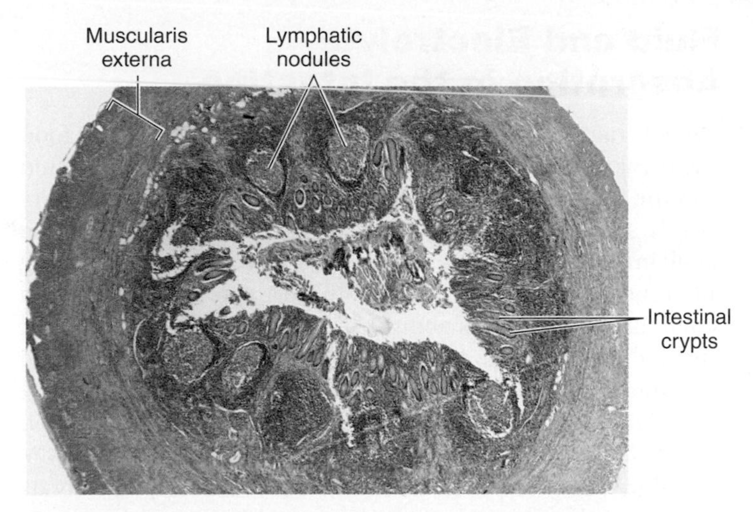

Figure 18.17 A photomicrograph of the human appendix. This cross section reveals numerous lymphatic nodules, which function in immunity.

Clinical Investigation CLUES

Alan's pains were not present unless he ate particular foods or drank wine. He did not have a fever or an elevated white blood cell count.

- Do Alan's symptoms indicate appendicitis? Why or why not?

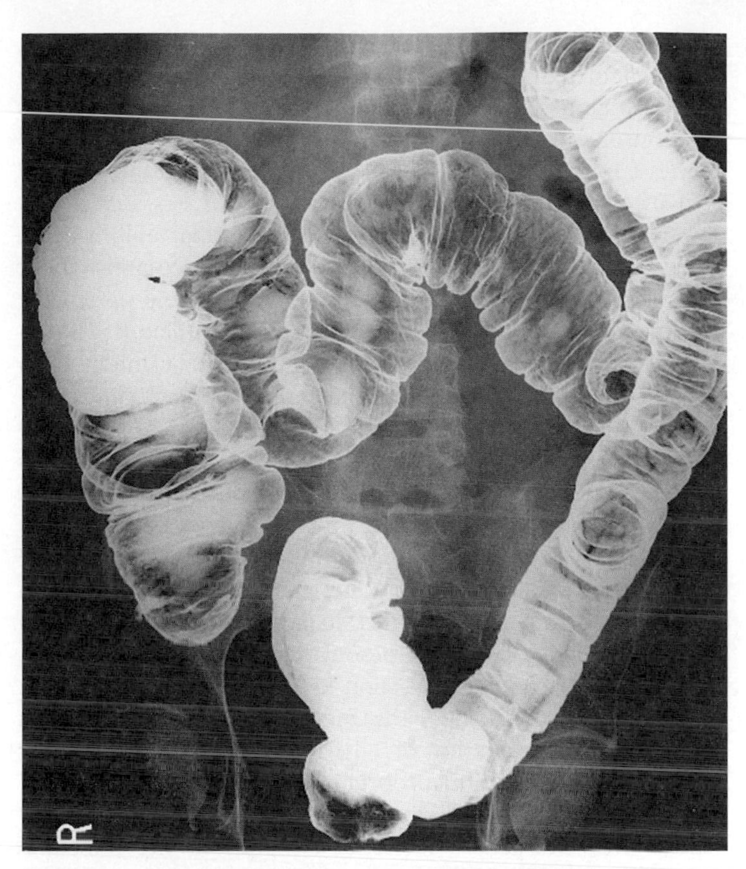

Figure 18.18 A radiograph of the large intestine. The large intestine is seen after a barium enema has been administered; the haustra are clearly visible. APIR

Intestinal Microbiota

Microorganisms, primarily bacteria, are present in relatively small numbers in the stomach and proximal portion of the small intestine. Their numbers increase in the distal ileum and are greatest in the colon, where an estimated 10^{14} reside. This enormous number is about 10 times more than the number of human cells in the body and represents several thousand different species with perhaps 100 times more genes than in human cells. These microorganisms are known collectively as the **intestinal microbiota** or **microflora.** In the colon, the intestinal microbiota are comprised mostly of anaerobic bacterial species.

The microbiota are usually described as composed of **commensal bacteria;** *commensalism* refers to a relationship where one species benefits and the other is neither benefited nor harmed. However, *mutualism* may better describe the relationship between the intestinal microbiota and their human hosts; in mutualism, both species benefit. We provide the bacteria with nutrients and an anaerobic home in our large intestine; they provide us with a variety of benefits.

Microbes from the mother invade an infant's intestine during birth, and invasion continues from the immediate environment during early infancy. Scientists believe that this initial colonization affects the composition of the microbiota later in life. This explains why the composition of the intestinal microbiota varies substantially between individuals and shows more similarity among family members than among unrelated people.

The intestinal microbiota help provide us with B vitamins and vitamin K. In addition, they ferment (through anaerobic metabolism) some indigestible molecules in the chyme that enters the large intestine. Among the most important products of commensal bacterial fermentatation are the **short-chain fatty acids**—acetate, propionate, and butyrate. These can be obtained only by bacterial fermentation of polysaccharides that resist digestion in the small intestine and enter the colon. The short-chain fatty acids are used for energy by the colonic epithelial cells and are absorbed into the blood, accounting for about 10% of the calories in the average Western diet.

In the small intestine, fluid absorption occurs primarily as a result of osmosis following glucose-stimulated Na^+ transport, but this cotransport carrier is absent from the colon. In the colon, short-chain fatty acids produced by bacterial fermentation stimulate active Na^+ and Cl^- absorption, promoting water absorption by osmosis. The short-chain fatty acids thereby help retain calories, electrolytes, and water from the colon contents. Because antibiotics reduce the population of the intestinal microbiota, scientists believe that most cases of diarrhea in people taking antibiotics are due to a reduced production of short-chain fatty acids by the commensal bacteria. By this logic, adding a form of starch that can arrive undigested to the colon should increase fluid absorption (because of increased production of short-chain fatty acids) and relieve diarrhea. This idea is currently being tested as an improved solution for *oral rehydration therapy* (chapter 6, section 6.3).

The intestine is lined by a simple columnar epithelium a mere 20 μm across that presents a huge surface area (about 200 m^2) to the 100 trillion commensal bacteria and potentially pathogenic bacteria that could provoke inflammation. The intestine is protected from bacteria in its lumen by: (1) mucus from goblet cells, which prevents most bacteria from reaching the epithelial cells; (2) secretion of antimicrobial peptides (defensins) by Paneth cells in the crypts; and (3) secretion of IgA antibodies by plasma cells. A normal intestinal microbiota helps to maintain a healthy epithelial barrier that protects deeper tissues, limits inflammatory damage, and promotes the repair of a damaged epithelium. This occurs while commensal bacteria in the colon are protected from immune attack, in part because the antigens of commensal bacteria promote production of regulatory T lymphocytes (T_{reg} cells; chapter 15, section 15.3) rather than killer T cells.

Intestinal dendritic cells are antigen-presenting cells needed for the activation of T lymphocytes. These dendritic cells reside in the lamina propria of the intestinal mucosa and send thin processes between the columnar epithelial cells, where they can engulf bacterial antigens in the intestinal lumen. Activated dendritic cells capable of presenting bacterial antigens to T cells can travel through the lymphatic vessels to distant sites, but are normally prevented from going farther than the mesenteric lymph nodes. The intestinal dendritic cells activate regulatory T cells to suppress inflammation in response to commensal bacteria while stimulating killer T cells to combat pathogenic bacteria (chapter 15, section 15.3).

CLINICAL APPLICATION

When commensal bacteria become reduced in diversity or number, protection against inflammation in the intestinal mucosa is compromised and **inflammatory bowel disease (IBD)**—including *Crohn's disease* and *ulcerative colitis*—may develop. The rising incidence of IBD in the United States may be attributed to the use of antibiotics that reduce the intestinal microbiota, as well as to the use of anti-inflammatory drugs and antimicrobial food preservatives that disturb the commensal bacteria. Many genes have been implicated in the genetic susceptibility to IBD, particularly to Crohn's disease. Bacterial antigens normally present in the intestine appear to drive the inflammation, and there is increased exposure to these bacteria because of epithelial injury, decreased production of defensin by Paneth cells, decreased goblet cells and mucus production, and changes in tight junctions that abnormally increase intestinal permeability. Erosions and ulcerations of the mucosa are accompanied by pronounced infiltration of the lamina propria by cells of the innate and adaptive immune system.

Fluid and Electrolyte Absorption in the Intestine

The GI tract receives about 1.5 L per day of water from food and drink; additionally, the GI tract secretes 8–10 L/day of fluid into the lumen. This includes contributions from the salivary glands, stomach, intestine, pancreas, liver, and gallbladder. The small intestine both secretes and absorbs water accompanying different transport processes, but these are not in balance. The small intestine secretes about 1 L per day but absorbs most of the fluid in the chyme. As a result, only about 2 L per day of fluid pass into the large intestine. The large intestine absorbs about 90% of this remaining volume, leaving less than 200 ml of fluid to be excreted in the feces.

Absorption of water in the intestine occurs passively as a result of the osmotic gradient created by the active transport of ions. The epithelial cells of the intestinal mucosa are joined together much like those of the kidney tubules and, like the kidney tubules, contain Na^+/K^+ pumps in the basolateral membrane. The analogy with kidney tubules is emphasized by the observation that aldosterone, which stimulates salt and water reabsorption in the renal tubules, also appears to stimulate salt and water absorption in the ileum.

The handling of salt and water transport in the large intestine is made more complex by the ability of the large intestine to secrete, as well as absorb, water. The secretion of water by the mucosa of the large intestine occurs by osmosis as a result of the active transport of Na^+ or Cl^- out of the epithelial cells into the intestinal lumen. Secretion in this way is normally minor compared to the far greater amount of salt and water absorption, but this balance may be altered in some disease states.

CLINICAL APPLICATION

Diarrhea is characterized by excessive fluid excretion in the feces. Three different mechanisms, illustrated by three different diseases, can cause diarrhea. In *cholera,* severe diarrhea and dehydration result from *enterotoxin,* a chemical produced by the infecting bacteria. Release of enterotoxin stimulates active NaCl transport into the lumen of the intestine, followed by the osmotic movement of water. In *celiac disease,* diarrhea is caused by damage to the intestinal mucosa induced by T lymphocytes when genetically susceptible people eat foods containing gluten (a type of protein found in wheat and other grains). In *lactose intolerance,* diarrhea is produced by the increased osmotic pressure in the intestinal lumen as a result of the presence of undigested lactose.

Defecation

After electrolytes and water have been absorbed the waste material passes to the rectum, leading to an increase in rectal pressure, relaxation of the internal anal sphincter, and the urge to defecate. If the urge to defecate is denied, feces are prevented from entering the anal canal by the external anal sphincter. In this case the feces remain in the rectum, and may even back up into the sigmoid colon. The **defecation reflex** normally occurs when the rectal pressure rises to a particular level that is determined, to a large degree, by habit. At this point the external anal sphincter relaxes to admit feces into the anal canal.

During the act of defecation the longitudinal rectal muscles contract to increase rectal pressure, and the internal and external anal sphincter muscles relax. Excretion is aided by contractions of abdominal and pelvic skeletal muscles, which raise the intra-abdominal pressure (this is part of Valsalva's maneuver; chapter 14, section 14.6). The raised pressure helps push the feces from the rectum, through the anal canal, and out of the anus.

CHECKPOINT

8a. Describe how electrolytes and water are absorbed in the large intestine, and explain how diarrhea may be produced.

8b. Describe the structures and mechanisms involved in defecation.

9. Identify the nature and significance of the intestinal microbiota.

18.5 LIVER, GALLBLADDER, AND PANCREAS

The liver regulates the chemical composition of the blood in numerous ways. In addition, the liver produces and secretes bile, which is stored and concentrated in the gallbladder prior to its discharge into the duodenum. The pancreas produces pancreatic juice, an exocrine secretion containing bicarbonate and important digestive enzymes.

LEARNING OUTCOMES

After studying this section, you should be able to:

10. Describe the structure and functions of the liver.

11. Describe the synthesis, composition, and functions of bile and explain the enterohepatic circulation.

12. Describe the composition of pancreatic juice and explain the significance of pancreatic juice enzymes.

The *liver* is positioned immediately beneath the diaphragm in the abdominal cavity. It is the largest internal organ, weighing about 1.3 kg (3.5 to 4.0 lb) in an adult. Attached to the inferior surface of the liver, between its right and quadrate lobes, is the pear-shaped *gallbladder.* This organ is approximately 7 to 10 cm (3 to 4 in.) long. The *pancreas,* which is about 12 to 15 cm (5 to 6 in.) long, is located behind the stomach along the posterior abdominal wall.

Structure of the Liver

Although the liver is the largest internal organ, it is, in a sense, only one to two cells thick. This is because the liver cells, or **hepatocytes,** form **hepatic plates** that are one to two cells thick. The plates are separated from each other by large capillary spaces called **sinusoids** (fig. 18.19).

The liver sinusoids are lined by endothelial cells with flattened processes and *fenestrae*—openings 150 to 175 nanometers in diameter that make the sinusoids very porous. Unlike the fenestrated capillaries of the kidneys and pancreas, the fenestrae of the hepatic sinusoids lack a diaphragm and a basement membrane. This makes the hepatic sinusoids much more permeable than other capillaries, even permitting the passage of plasma proteins with protein-bound nonpolar molecules such as fat and cholesterol. The sinusoids also contain phagocytic

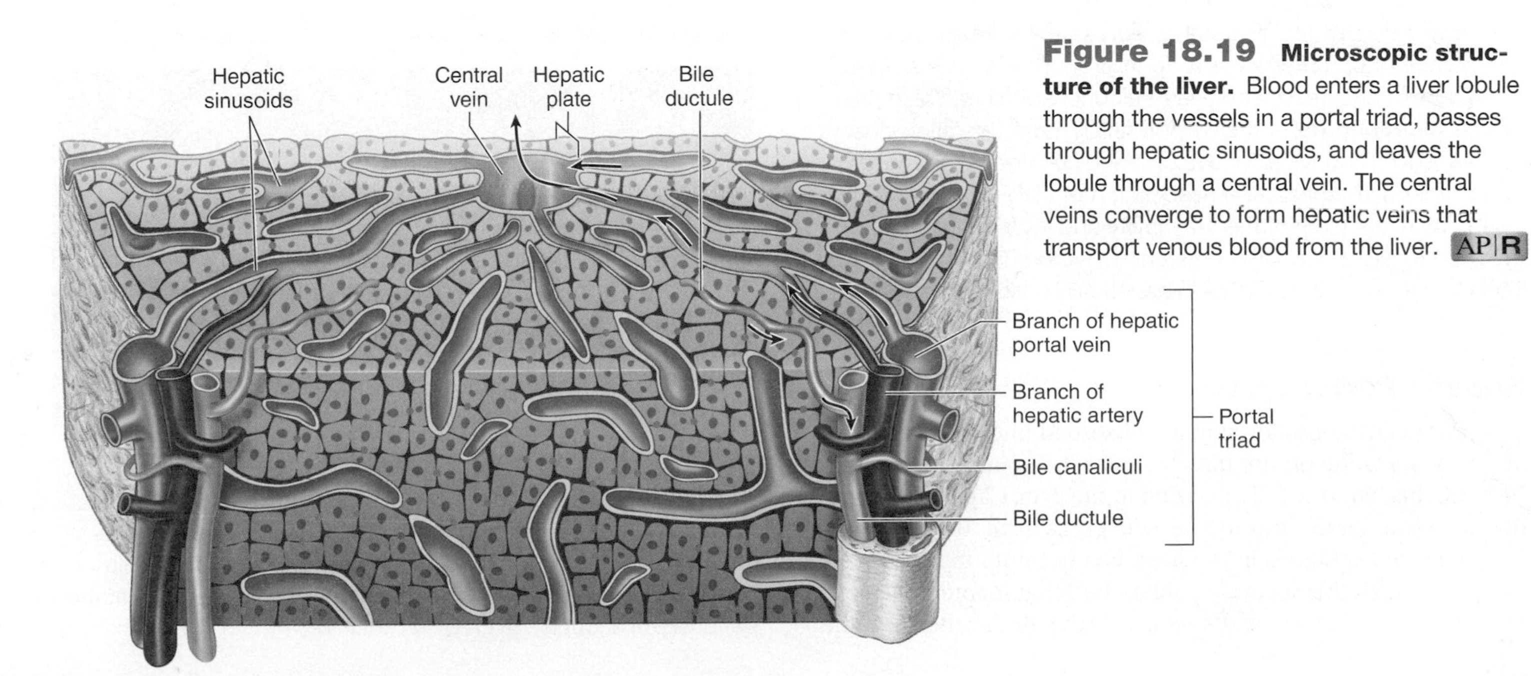

Figure 18.19 Microscopic structure of the liver. Blood enters a liver lobule through the vessels in a portal triad, passes through hepatic sinusoids, and leaves the lobule through a central vein. The central veins converge to form hepatic veins that transport venous blood from the liver. AP|R

Kupffer cells, which are part of the reticuloendothelial system (also called the mononuclear phagocyte system; chapter 15, section 15.1). The fenestrae, lack of a basement membrane, and plate structure of the liver allow intimate contact between the hepatocytes and the contents of the blood.

The liver has an amazing ability to regenerate itself. For example, if two-thirds of a rodent's liver is surgically removed, the remaining tissue will regenerate its original mass in one week. This regenerative ability is due not to stem cells, but rather to the mitotic division of the remaining hepatocytes. When the original mass is restored, cell division ceases. The same regenerative ability is seen when most toxins or infections cause the hepatocytes to die. For reasons not presently understood, hepatic damage due to alcohol abuse and viral hepatitis can cause *liver fibrosis,* where there is accumulation of collagen fibers and extracellular matrix. This can lead to *cirrhosis,* a more serious condition described in the next Clinical Application box.

CLINICAL APPLICATION

In **cirrhosis,** large numbers of liver lobules are destroyed by inflammatory processes and replaced with permanent, scar-like fibrotic connective tissue and "regenerative nodules" of hepatocytes. These regenerative nodules do not have the platelike structure of normal liver tissue and are therefore less functional. One indication of this decreased function is the entry of ammonia (produced by intestinal bacteria) from the hepatic portal blood into the general circulation. Cirrhosis may be caused by chronic alcohol abuse, biliary obstruction, viral hepatitis, or various chemicals that attack liver cells.

Chronic alcohol abuse is the third leading preventable cause of death in the United States, due to its association with liver disease, several types of cancers, deaths from accidents and violence, and other causes that result in an average of 30 lost years per alcoholic death. The regular use of alcohol can cause **steatosis** (fatty liver), where hepatocytes store large globules of fat. This can progress to cirrhosis, with the incidence of cirrhosis increasing in proportion to the amount of alcohol consumed. Long-term alcohol abuse can also produce **alcoholic hepatitis,** a potentially fatal condition characterized by the rapid onset of jaundice (discussed shortly) and other symptoms. The association of alcohol with cancer, particularly cancers of the oral cavity and esophagus, may be related to the ability of acetaldehyde—a breakdown product of alcohol—to damage DNA in genetically susceptible people.

Hepatic Portal System

The products of digestion that are absorbed into blood capillaries in the intestine do not directly enter the general circulation. Instead, this blood is delivered first to the liver. Capillaries in the digestive tract drain into the *hepatic portal vein,* which carries this blood to capillaries in the liver. It is not until the blood has passed through this second capillary bed that it enters the general circulation through the *hepatic vein* that drains the liver. The term **portal system** is used to describe this unique pattern of circulation: capillaries ⇒ vein ⇒ capillaries ⇒ vein. In addition to receiving venous blood from the intestine, the liver also receives arterial blood via the *hepatic artery.*

The hepatic portal vein drains the capillaries of the intestine, pancreas, gallbladder, omentum, and spleen, and accounts for about 75% to 80% of the blood flow to the liver. Because it contains blood coming from the intestine, the hepatic portal vein delivers nutrients and other absorbed molecules to the liver. The hepatic artery supplies the remaining 20% to 25% of the liver's incoming blood flow; however, this arterial blood flow is adjusted to compensate for changes in the blood flow through the hepatic portal vein. As a result, the total hepatic blood flow is maintained at about 25% of the cardiac output. This relatively constant hepatic blood flow is needed to maintain *hepatic clearance*—the ability of the liver to remove substances from the blood, as will be described in the section on the enterohepatic circulation.

Liver Lobules

The hepatic plates are arranged into functional units called **liver lobules** (figs. 18.19 and 18.20). In the middle of each lobule is a *central vein,* and at the periphery of each lobule are branches of the hepatic portal vein and of the hepatic artery, both of which open into the sinusoids *between* hepatic plates. Arterial blood mixes with portal venous blood containing molecules absorbed by the GI tract, and this mixed blood travels within the sinusoids from the periphery of the lobule to the central vein. The central veins of different liver lobules converge to form the hepatic vein, which carries blood from the liver to the inferior vena cava.

Bile is produced by the hepatocytes and secreted into thin channels called **bile canaliculi** *within* each hepatic plate (fig. 18.20). These bile canaliculi are drained at the periphery of each lobule by *bile ducts,* which in turn drain into *hepatic*

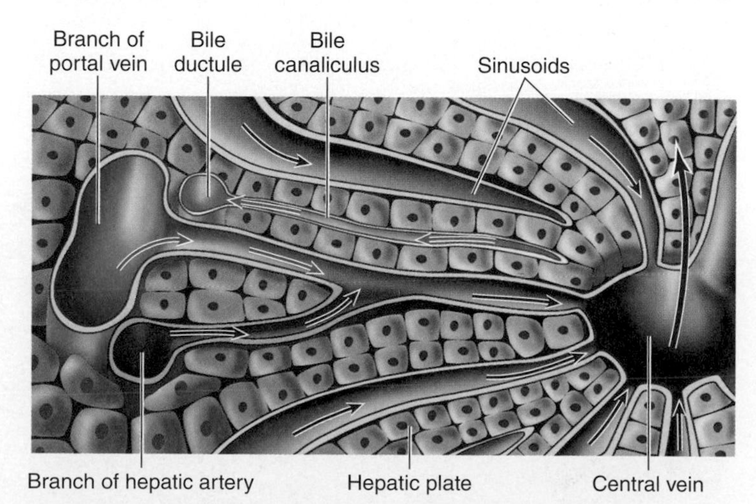

Figure 18.20 The flow of blood and bile in a liver lobule. Blood flows within sinusoids from a portal vein to the central vein (from the periphery to the center of a lobule). Bile flows within hepatic plates from the center to bile ductules at the periphery of a lobule. AP|R

Table 18.2 | Compounds Excreted by the Liver into the Bile Ducts

Category	Compound	Comments
Endogenous (Naturally occurring)	Bile salts, urobilinogen, cholesterol	High percentage reabsorbed and has an enterohepatic circulation*
	Lecithin	Small percentage reabsorbed and has an enterohepatic circulation
	Bilirubin	No enterohepatic circulation
Exogenous (Drugs)	Ampicillin, streptomycin, tetracyline	High percentage reabsorbed and has an enterohepatic circulation
	Sulfonamides, penicillin	Small percentage reabsorbed and has an enterohepatic circulation

*Compounds with an enterohepatic circulation are absorbed to some degree by the intestine and are returned to the liver in the hepatic portal vein.

ducts that carry bile away from the liver. Because blood travels in the sinusoids and bile travels in the opposite direction in bile canaliculi within the hepatic plates, blood and bile do not mix in the liver lobules.

Enterohepatic Circulation

In addition to the normal constituents of bile, a wide variety of exogenous compounds (drugs) are secreted by the liver into the bile ducts (table 18.2). The liver can thus "clear" the blood of particular compounds by removing them from the blood and excreting them into the intestine with the bile. Molecules that are cleared from the blood by secretion into the bile are eliminated in the feces; this is analogous to renal clearance of blood through excretion in the urine (chapter 17, section 17.4).

Many compounds that are released with the bile into the intestine are not eliminated with the feces, however. Some of these can be absorbed through the small intestine and enter the hepatic portal blood. These molecules are thus carried back to the liver, where they can again be secreted by hepatocytes into the bile ducts. Compounds that recirculate between the liver and intestine in this way are said to have an **enterohepatic circulation** (fig. 18.21). For example, a few grams of bile salts (discussed shortly) released into the intestine recirculate 6 to 10 times a day, with only about 0.5 g of bile salts per day excreted in the feces.

Clinical Investigation CLUES

Alan's blood tests revealed normal levels of free bilirubin, ammonia, and urea.

- What do these results suggest about the health of Alan's liver?

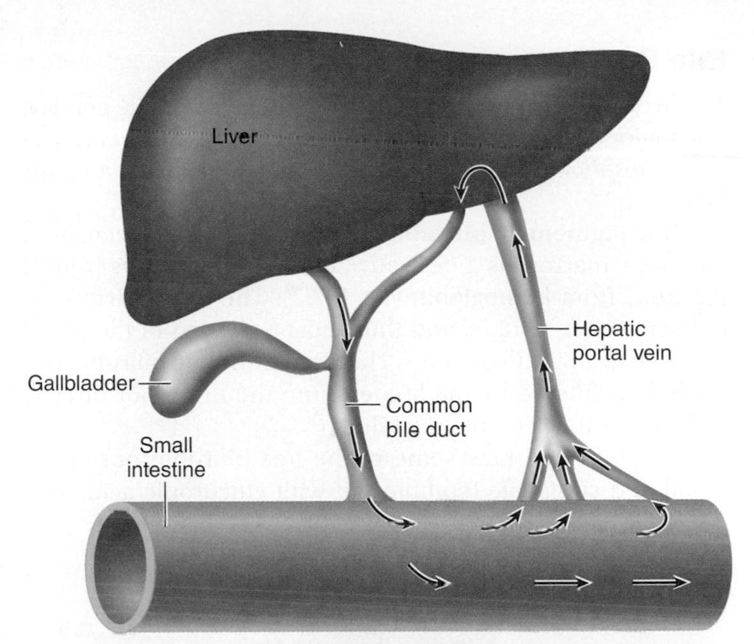

Figure 18.21 The enterohepatic circulation. Substances secreted in the bile may be absorbed by the intestinal epithelium and recycled to the liver via the hepatic portal vein.

Functions of the Liver

As a result of its large and diverse enzymatic content and its unique structure, and because it receives venous blood from the intestine, the liver has a greater variety of functions than any other organ in the body. The major categories of liver function are summarized in table 18.3.

Table 18.3 | Major Categories of Liver Function

Functional Category	Actions
Detoxication of Blood	Phagocytosis by Kupffer cells Chemical alteration of biologically active molecules (hormones and drugs) Production of urea, uric acid, and other molecules that are less toxic than parent compounds Excretion of molecules in bile
Carbohydrate Metabolism	Conversion of blood glucose to glycogen and fat Production of glucose from liver glycogen and from other molecules (amino acids, lactic acid) by gluconeogenesis Secretion of glucose into the blood
Lipid Metabolism	Synthesis of triglycerides and cholesterol Excretion of cholesterol in bile Production of ketone bodies from fatty acids
Protein Synthesis	Production of albumin Production of plasma transport proteins Production of clotting factors (fibrinogen, prothrombin, and others)
Secretion of Bile	Synthesis of bile salts Conjugation and excretion of bile pigment (bilirubin)

Bile Production and Secretion

The liver produces and secretes 250 to 1,500 ml of bile per day. The major constituents of bile are *bile pigment (bilirubin), bile salts, phospholipids* (mainly lecithin), *cholesterol,* and *inorganic ions.*

Bile pigment, or **bilirubin,** is produced in the spleen, liver, and bone marrow as a derivative of the heme groups (minus the iron) from hemoglobin (fig. 18.22). The **free bilirubin** is not very water-soluble, and thus most is carried in the blood attached to albumin proteins. This protein-bound bilirubin can neither be filtered by the kidneys into the urine nor directly excreted by the liver into the bile.

The liver can take some of the free bilirubin out of the blood and conjugate (combine) it with glucuronic acid. This **conjugated bilirubin** is water-soluble and can be secreted into the bile. Once in the bile, the conjugated bilirubin can enter the intestine where it is converted by bacteria into another pigment—**urobilinogen.** Derivatives of urobilinogen impart a brown color to the feces. About 30% to 50% of the urobilinogen, however, is absorbed by the intestine and enters the hepatic portal vein. Of the urobilinogen that enters the liver sinusoids, some is secreted into the bile and is thus returned to the intestine in an enterohepatic circulation; the rest enters the general circulation (fig. 18.23). The urobilinogen in plasma, unlike free bilirubin, is not attached to albumin. Urobilinogen is therefore easily filtered by the kidneys into the urine where its derivatives produce an amber color.

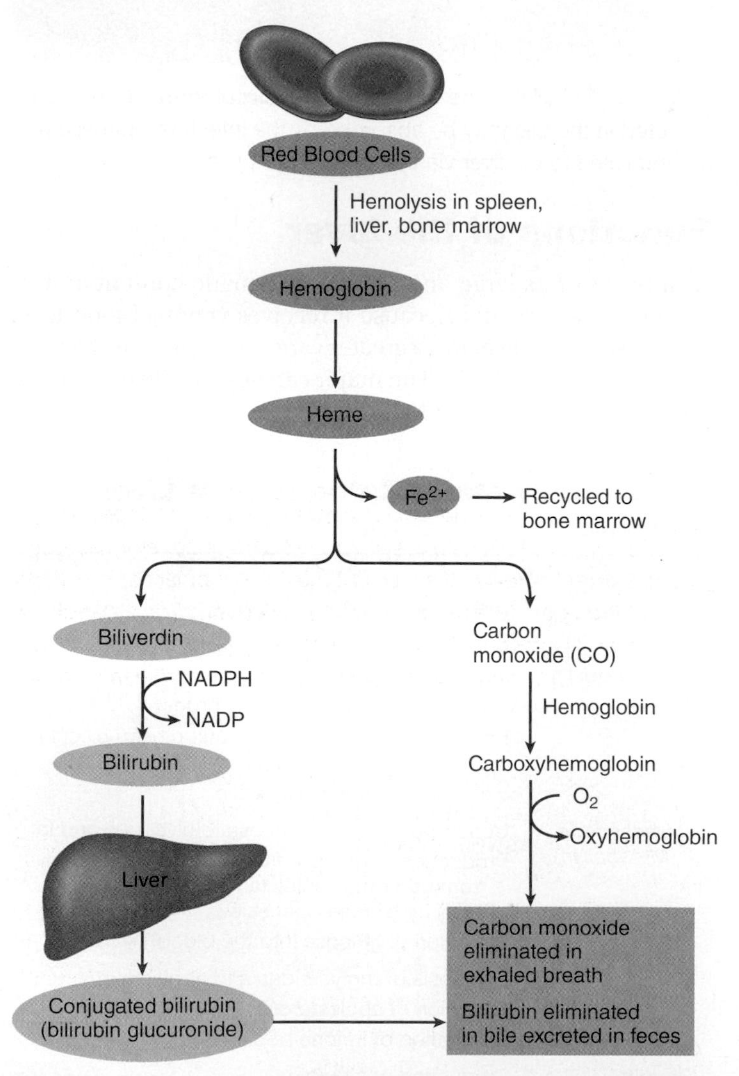

Figure 18.22 Simplified pathway for the metabolism of heme and bilirubin. Heme can be formed from the hemoglobin in red blood cells. The iron from the heme group is recycled back to the bone marrow when the heme is converted into biliverdin. Notice that carbon monoxide is produced in this process and because it is toxic, must be eliminated from the body.

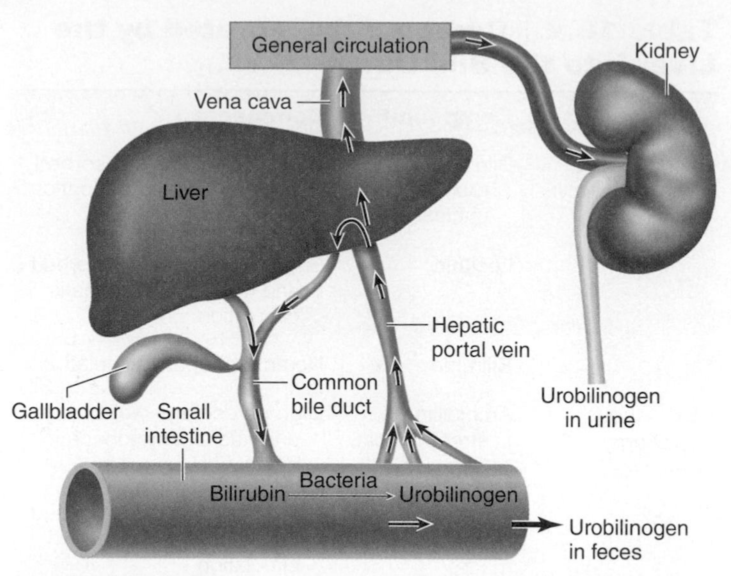

Figure 18.23 The enterohepatic circulation of urobilinogen. Bacteria in the intestine convert bilirubin (bile pigment) into urobilinogen. Some of this pigment leaves the body in the feces; some is absorbed by the intestine and is recycled through the liver. A portion of the urobilinogen that is absorbed enters the general circulation and is filtered by the kidneys into the urine.

Bile acids are derivatives of cholesterol that have two to four polar groups on each molecule. The principal bile acids in humans are *cholic acid* (fig. 18.24) and *chenodeoxycholic acid,* conjugated with the amino acids glycine or taurine to form the **bile salts.** In aqueous solutions these molecules "huddle" together to form aggregates known as **micelles** (fig. 18.24). The nonpolar parts are located in the central region of the micelle (away from water), whereas the polar groups face water around the periphery of the micelle (chapter 2; see fig. 2.22). Lecithin, cholesterol, and other lipids in the small intestine enter these micelles, and the dual nature of the bile salts (part polar, part nonpolar) allows them to emulsify fat in the chyme. *Emulsification* refers to the conversion of larger fat globules by bile acids into a finer suspension of smaller globules, which provide a greater surface area for the fat to be digested by lipase enzymes (see fig. 18.35).

The liver's production of bile acids from cholesterol is the major pathway of cholesterol breakdown in the body. This amounts to about half a gram of cholesterol converted into bile acids per day. No more than this is required, because approximately 95% of the bile acids released into the duodenum are absorbed in the ileum by means of specific carriers, and so have an enterohepatic circulation. Bile salts recirculate 6 to 10 times per day, with only about 0.5 g excreted in the feces.

CLINICAL APPLICATION

Jaundice is a yellow staining of the tissues produced by high blood concentrations of either free or conjugated bilirubin. Jaundice associated with high blood levels of conjugated bilirubin in adults may occur when bile excretion is blocked by gallstones. Because free bilirubin is derived from heme, jaundice associated with high blood levels of free bilirubin is usually caused by an excessively high rate of red blood cell destruction. This is the cause of jaundice in infants who suffer from **hemolytic disease of the newborn,** or **erythroblastosis fetalis. Physiological jaundice of the newborn** is due to high levels of free bilirubin in otherwise healthy neonates. This type of jaundice may be caused by the rapid fall in blood hemoglobin concentrations that normally occurs at birth. In premature infants, it may be caused by inadequate amounts of hepatic enzymes that are needed to conjugate bilirubin so that it can be excreted in the bile.

Newborn infants with jaundice are usually treated by exposing them to blue light in the wavelength range of 400 to 500 nm. This light is absorbed by bilirubin in cutaneous vessels and results in the conversion of the bilirubin into a more polar form that can be dissolved in plasma without having to be conjugated with glucuronic acid. This more water-soluble photoisomer of bilirubin can then be excreted in the bile and urine.

Clinical Investigation CLUES

Alan had yellowing of his sclera, and his blood tests revealed elevated levels of conjugated bilirubin.

- What does the yellowing of the sclera indicate, and what is its cause?
- What could cause the elevation in conjugated bilirubin?

Detoxication of the Blood

The liver's endothelial cells (lining the sinusoids), Kupffer cells, and dendritic cells have pathogen recognition receptors that recognize PAMPs (pathogen-associated molecular patterns; chapter 15, section 15.1), enabling them to scavenge blood-borne bacteria. The liver can also remove hormones, drugs, and other biologically active molecules from the blood by (1) excretion of these compounds in the bile as previously described; (2) phagocytosis by the Kupffer cells that line the sinusoids; and (3) chemical alteration of these molecules within the hepatocytes.

CH_3 OH CH–CH_2–CH_2–COOH HO OH

(a) **Cholic acid** (a bile acid)

Nonpolar Polar

(b) Simplified representation of bile acid

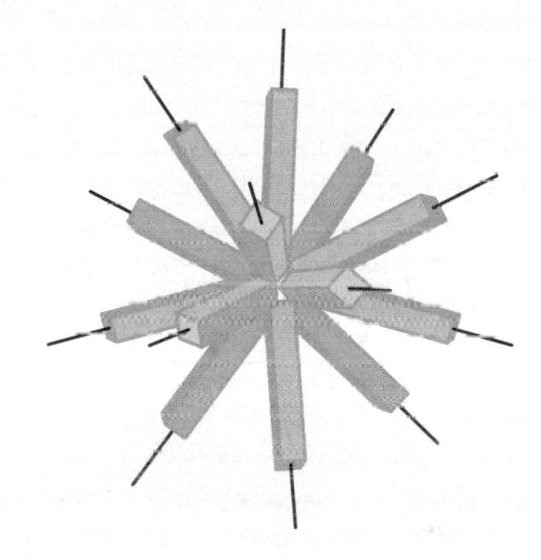

(c) **Micelle** of bile acids

Figure 18.24 Bile acids form micelles. (*a*) Cholic acid, a bile acid. (*b*) A simplified representation of a bile acid, emphasizing that part of the molecule is polar, but most is nonpolar. (*c*) Bile acids in water aggregate to form associations called micelles. Cholesterol and lecithin, being nonpolar, can enter the micelles. The bile acids in the micelles serve to emulsify triglycerides (fats and oils) in the chyme.

Ammonia, for example, is a very toxic molecule produced by deamination of amino acids in the liver and by the action of bacteria in the intestine. Because the ammonia concentration of portal vein blood is 4 to 50 times greater than that of blood in the hepatic vein, it is clear that the ammonia is removed by the liver. The liver has the enzymes needed to convert ammonia into less toxic **urea** molecules (chapter 5; see fig. 5.16), which are secreted by the liver into the blood and excreted by the kidneys in the urine. Similarly, the liver converts toxic porphyrins into **bilirubin** and toxic purines into **uric acid.**

Steroid hormones and many drugs are inactivated in their passage through the liver by modifications of their chemical structures. The liver has enzymes that convert these nonpolar molecules into more polar (more water-soluble) forms by *hydroxylation* (the addition of OH^- groups) and by *conjugation* with highly polar groups such as sulfate and glucuron

Polar derivatives of steroid hormones and drugs are less biologically active and, because of their increased water solubility, are more easily excreted by the kidneys into the urine.

Conjugation of steroid hormones and *xenobiotics* (foreign chemicals that are biologically active) makes them anionic (negatively charged) and hydrophilic (water-soluble). Thus changed, these compounds can be transported by liver cells into the bile canaliculi by multispecific **organic anion transport** carriers. These carriers have been cloned and identified as the same type that transports similar molecules into the nephron tubules (chapter 17, section 17.4). Through renal secretion and secretion into the bile, therefore, these transport carriers help the body to eliminate potentially toxic molecules.

CLINICAL APPLICATION

The liver cells contain enzymes for the metabolism of steroid hormones and other endogenous molecules, as well as for the detoxication of such exogenous toxic compounds as benzopyrene (a carcinogen from tobacco smoke and charbroiled meat), polychlorinated biphenyls (PCBs), and dioxin. The enzymes are members of a class called the **cytochrome P450 enzymes** that comprises a few dozen enzymes with varying specificities. Together, these enzymes can metabolize thousands of toxic compounds. Because people vary in their hepatic content of the different cytochrome P450 enzymes, one person's sensitivity to a drug may be greater than another's because of a relative deficiency in the appropriate cytochrome P450 enzyme needed to metabolize that drug.

The cytochrome P450 enzymes in the liver contain heme groups, like the heme in hemoglobin. Heme is classified chemically as a *porphyrin,* and excessive amounts are toxic to the bone marrow or liver. This condition is called **porphyria.** Porphyria is caused by genetic mutations that damage one of several enzymes required for the synthesis of heme, resulting in the accumulation of porphyrin heme precursors. Hepatic porphyria can produce abdominal pain, as well as neurological and psychological disturbances. Indeed, there is evidence that porphyria may have caused the "madness" of King George III, and may also have afflicted Friedrich Wilhelm I of Prussia and the artist Vincent van Gogh.

Production of the cytochrome P450 enzymes, needed for the hepatic metabolism of lipophilic compounds such as steroid hormones and drugs, is stimulated by the activation of a nuclear receptor. Nuclear receptors bind to particular molecular ligands and then activate specific genes (chapter 11; see fig. 11.5). The particular nuclear receptor that stimulates the production of cytochrome P450 enzymes is known as *SXR—for steroid and xenobiotic receptor.* A drug that activates SXR, and thereby induces the production of cytochrome P450 enzymes, would thus be expected to increase the hepatic metabolism of many other drugs. This is the mechanism responsible for many interactions among different drugs.

Secretion of Glucose, Triglycerides, and Ketone Bodies

The liver helps regulate the blood glucose concentration by either removing glucose from the blood or adding glucose to it, according to the needs of the body (chapter 5; see fig. 5.5). After a carbohydrate-rich meal, the liver can remove some glucose from the hepatic portal blood and convert it into glycogen and triglycerides through the processes of **glycogenesis** and **lipogenesis,** respectively. During fasting, the liver secretes glucose into the blood. This glucose can be derived from the breakdown of stored glycogen in a process called **glycogenolysis,** or it can be produced by the conversion of noncarbohydrate molecules (such as amino acids) into glucose in a process called **gluconeogenesis.** The liver also contains the enzymes required to convert free fatty acids into ketone bodies (**ketogenesis**), which are secreted into the blood in large amounts during fasting. These processes are controlled by hormones and are explained further in chapter 19 (see figs. 19.6 and 19.9).

Production of Plasma Proteins

Plasma albumin and most of the plasma globulins (with the exception of immunoglobulins, or antibodies) are produced by the liver. Albumin constitutes about 70% of the total plasma protein and contributes most to the colloid osmotic pressure of the blood (chapter 14, section 14.2). The globulins produced by the liver have a wide variety of functions, including transport of cholesterol and triglycerides, transport of steroid and thyroid hormones, inhibition of trypsin activity, and blood clotting. Clotting factors I (fibrinogen), II (prothrombin), III, V, VII, IX, and XI, as well as angiotensinogen, are all produced by the liver.

Gallbladder

The **gallbladder** is a saclike organ attached to the inferior surface of the liver. This organ stores and concentrates bile, which drains to it from the liver by way of the bile ducts, hepatic ducts, and *cystic duct,* respectively. A sphincter valve at the neck of the gallbladder allows a 35- to 100-ml storage capacity. When the gallbladder fills with bile, it expands to the size and shape of a small pear. Bile is a yellowish green fluid containing bile salts, bilirubin, cholesterol, and other compounds, as previously discussed. Contraction of the muscularis layer of the gallbladder ejects bile through the cystic duct into the *common bile duct,* which conveys bile into the duodenum (fig. 18.25).

Bile is continuously produced by the liver and drains through the hepatic and common bile ducts to the duodenum. When the small intestine is empty of food, the *sphincter of ampulla (sphincter of Oddi)* at the end of the common bile duct closes, and bile is forced up to the cystic duct and then to the gallbladder for storage.

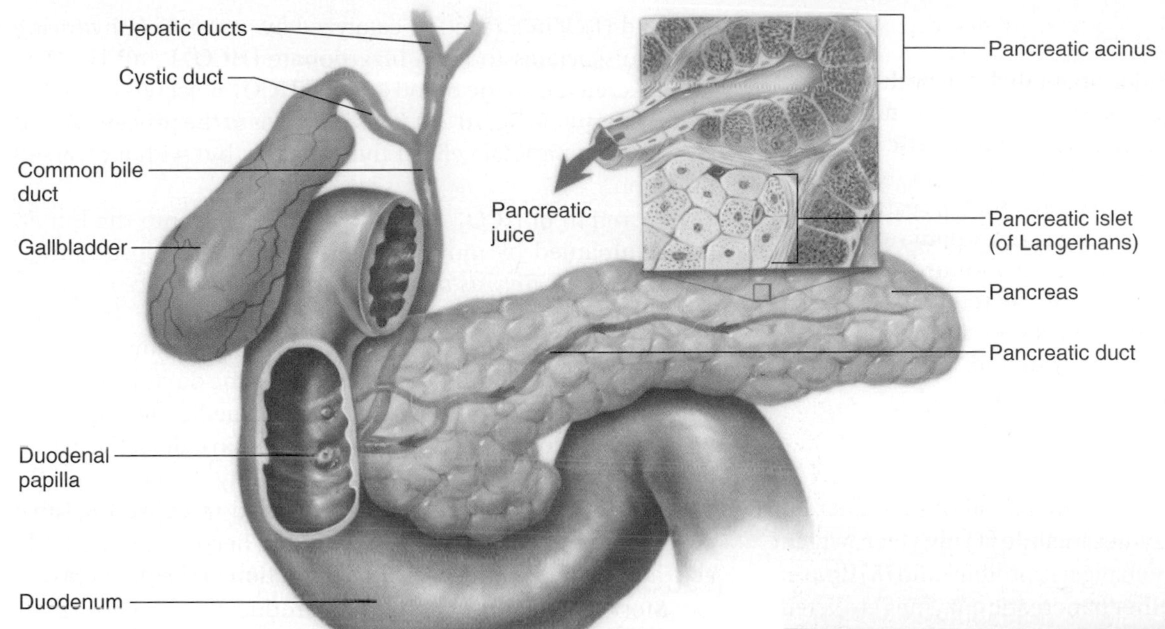

Figure 18.25 Pancreatic juice and bile are secreted into the duodenum. The pancreatic duct joins the common bile duct to empty its secretions through the duodenal papilla into the duodenum. The release of bile and pancreatic juice into the duodenum is controlled by the sphincter of ampulla (sphincter of Oddi).

CLINICAL APPLICATION

Approximately 20 million Americans have **gallstones**—small, hard mineral deposits (calculi) that can produce painful symptoms by obstructing the cystic or common bile ducts. Gallstones commonly contain cholesterol as their major component; indeed, cholesterol was discovered in 1789 when it was first isolated from gallstones. Cholesterol normally has an extremely low water solubility (20 μg/L), but it can be present in bile at 2 million times its water solubility (40 g/L) because cholesterol molecules cluster together with bile salts and lecithin in the hydrophobic centers of micelles. In order for gallstones to be produced, the liver must secrete enough cholesterol to create a supersaturated solution. The gallbladder then secretes excess mucus that serves as a nucleating agent for the formation of solid cholesterol crystals (fig. 18.26). The combination of these crystals and the mucus form a sludge that impedes the emptying of the gallbladder. In some cases, cholesterol gallstones may be dissolved by oral ingestion of bile acids. This may be combined with a treatment that involves fragmentation of the gallstones by high-energy shock waves delivered to a patient immersed in a water bath. The most common and effective treatment, however, is surgical removal of the gallbladder using a procedure called *laparoscopic cholecystectomy.*

Clinical Investigation CLUES

Alan had pain below his right scapula whenever he ate foods such as peanut butter or bacon, which are oily or fatty.

- If this pain was caused by a gallstone, how might that relate to Alan's high blood levels of conjugated bilirubin and jaundice?

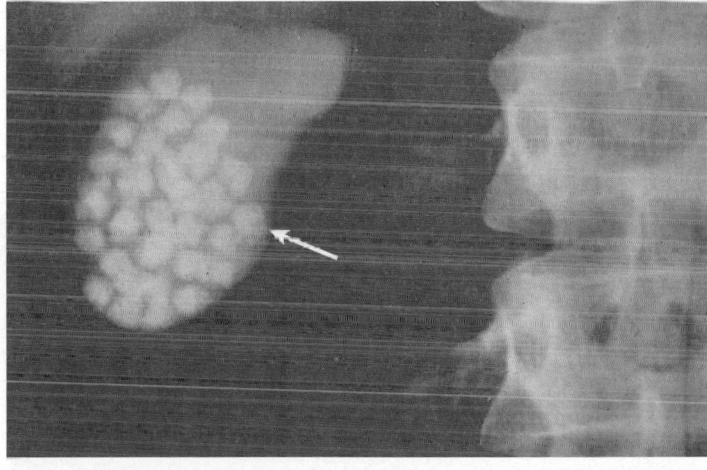

(a)

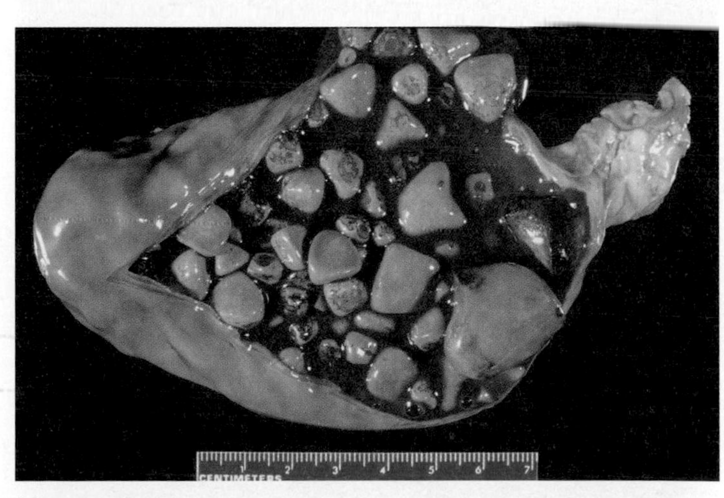

(b)

Figure 18.26 Gallstones. (*a*) A radiograph of a gallbladder that contains gallstones (biliary calculi). (*b*) A posterior view of a gallbladder that has been surgically removed (cholecystectomy) and cut open to reveal its gallstones.

Pancreas

The **pancreas** is a soft, glandular organ that has both exocrine and endocrine functions (fig. 18.27). The endocrine function is performed by clusters of cells called the **pancreatic islets,** or **islets of Langerhans** (fig. 18.27*a*), that secrete the hormones insulin and glucagon into the blood (chapter 19, section 19.3). As an exocrine gland, the pancreas secretes pancreatic juice through the pancreatic duct into the duodenum. Within the lobules of the pancreas are the exocrine secretory units, called **acini** (fig. 18.27*b*). Each acinus consists of a single layer of acinar epithelial cells surrounding a lumen, into which the constituents of pancreatic juice are secreted.

Pancreatic Juice

Pancreatic juice contains *bicarbonate* and about 20 different digestive enzymes. These enzymes include (1) **amylase,** which digests starch; (2) **trypsin,** which digests protein; and (3) **lipase,** which digests triglycerides. Other pancreatic enzymes are listed in table 18.4. It should be noted that the complete digestion of food molecules in the small intestine requires the action of both pancreatic enzymes and brush border enzymes.

Evidence suggests that bicarbonate is secreted into the pancreatic juice by the cells that line the ductules, rather than by the acinar cells (see fig. 18.27*b*). The bicarbonate is produced from CO_2 that diffuses into the cells from the blood. This occurs because of the formation of carbonic acid (from CO_2 and H_2O, in a reaction catalyzed by carbonic anhydrase), which dissociates to form bicarbonate (HCO_3^-) and H^+. The H^+ is secreted into the blood and the HCO_3^- is secreted into the pancreatic juice (fig. 18.28). This is similar to the process of acid secretion by parietal cells of the stomach, but with a reversed direction.

Secretion of HCO_3^- from the ductule cells into the lumen is accompanied by movement of Cl^- in the opposite direction. The *cystic fibrosis transmembrane conductance regulator (CFTR),* a channel for the facilitated diffusion of Cl^-, is located in the ductule cells on the membrane facing the lumen. Here, the CFTR promotes diffusion of Cl^- out of the ductule cells and back into the lumen (fig. 18.28). This is medically important because people with cystic fibrosis (who have defective CFTR function) have a greatly diminished ability to secrete HCO_3^- into the pancreatic juice. This is believed to cause digestive enzymes to build up in the pancreas and become prematurely activated, eventually leading to destruction of the pancreas.

Most pancreatic enzymes are produced as inactive molecules, or *zymogens,* so that the risk of self-digestion within the pancreas is minimized. The inactive form of trypsin, called trypsinogen, is activated within the small intestine by the catalytic action of the brush border enzyme *enterokinase* (also called *enteropeptidase*). Enterokinase converts trypsinogen to active trypsin. Trypsin, in turn, activates the other zymogens of pancreatic juice (fig. 18.29) by cleaving off polypeptide sequences that inhibit the activity of these enzymes.

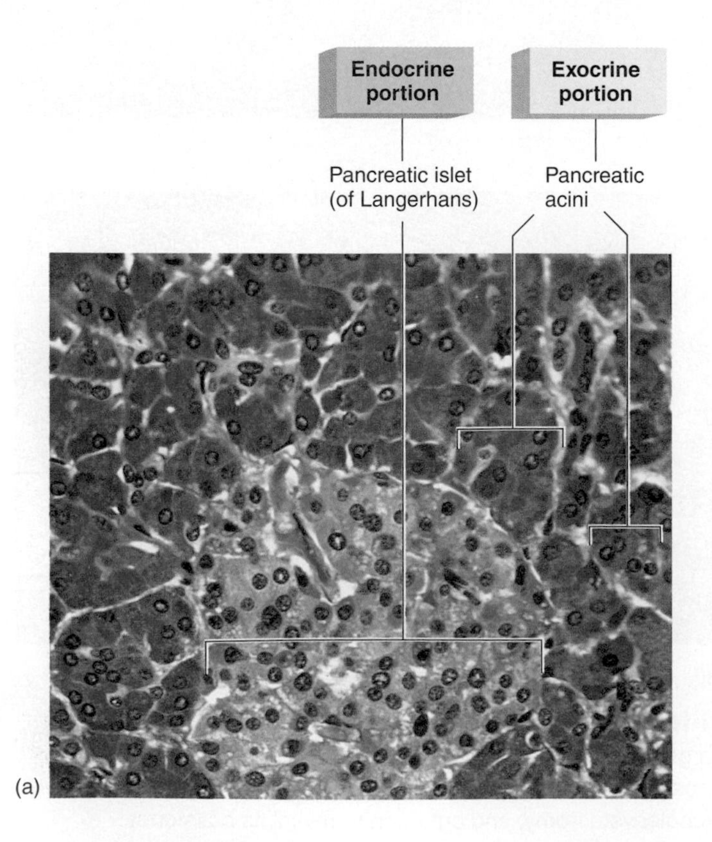

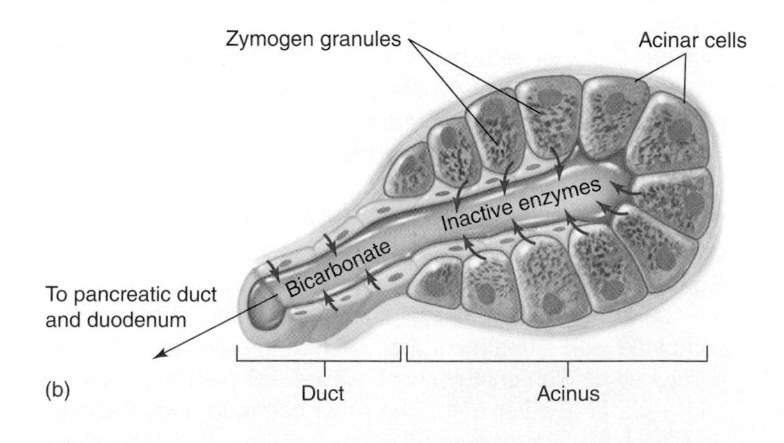

Figure 18.27 The pancreas is both an exocrine and an endocrine gland. (*a*) A photomicrograph of the endocrine and exocrine portions of the pancreas. (*b*) An illustration depicting the exocrine pancreatic acini, where the acinar cells produce inactive enzymes stored in zymogen granules. The inactive enzymes are secreted by way of a duct system into the duodenum. AP|R

Table 18.4 | Enzymes Contained in Pancreatic Juice

Enzyme	Zymogen	Activator	Action
Trypsin	Trypsinogen	Enterokinase	Cleaves internal peptide bonds
Chymotrypsin	Chymotrypsinogen	Trypsin	Cleaves internal peptide bonds
Elastase	Proelastase	Trypsin	Cleaves internal peptide bonds
Carboxypeptidase	Procarboxypeptidase	Trypsin	Cleaves last amino acid from carboxyl-terminal end of polypeptide
Phospholipase	Prophospholipase	Trypsin	Cleaves fatty acids from phospholipids such as lecithin
Lipase	None	None	Cleaves fatty acids from glycerol
Amylase	None	None	Digests starch to maltose and short chains of glucose molecules
Cholesterolesterase	None	None	Releases cholesterol from its bonds with other molecules
Ribonuclease	None	None	Cleaves RNA to form short chains
Deoxyribonuclease	None	None	Cleaves DNA to form short chains

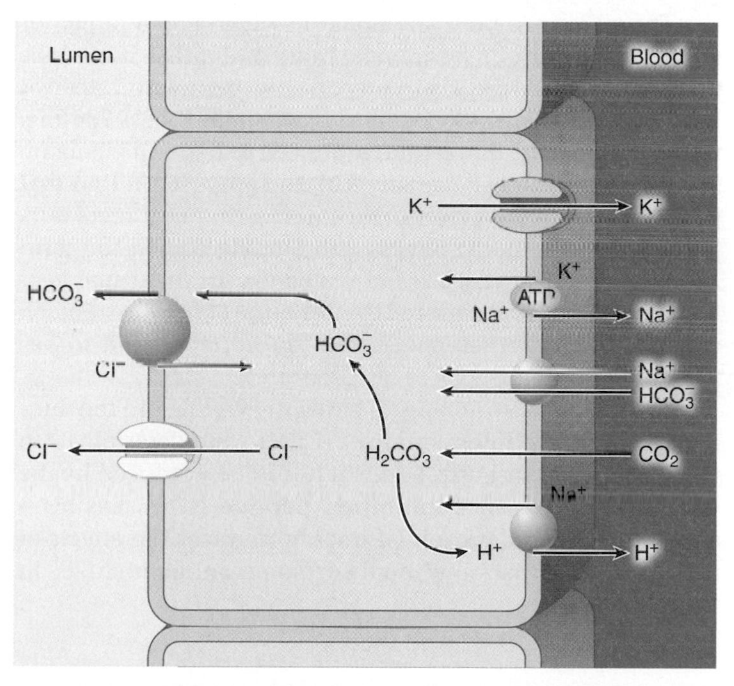

Figure 18.28 Secretion of bicarbonate into pancreatic juice. Cells of the pancreatic duct take in CO_2 from the blood and use it to generate carbonic acid (H_2CO_3). This dissociates into bicarbonate (HCO_3^-) and H^+. The HCO_3^- is secreted into the lumen of the duct by a carrier that exchanges it for Cl^-. The Cl^- then leaks passively back into the lumen through a different, CFTR chloride channel (see text for details).

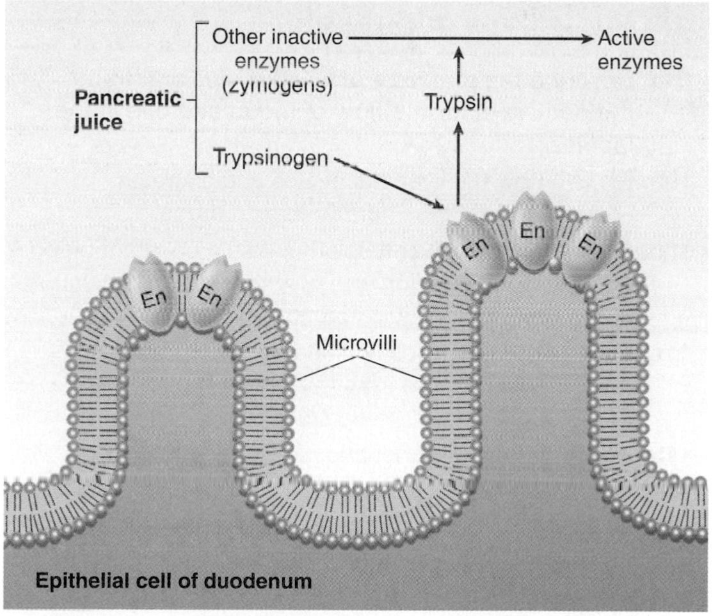

Figure 18.29 The activation of pancreatic juice enzymes. The pancreatic protein-digesting enzyme trypsin is secreted in an inactive form known as trypsinogen. This inactive enzyme (zymogen) is activated by a brush border enzyme, enterokinase (EN), located in the cell membrane of microvilli. Active trypsin in turn activates other zymogens in pancreatic juice.

The activation of trypsin, therefore, is the triggering event for the activation of other pancreatic enzymes. Actually, the pancreas does produce small amounts of active trypsin, but the other enzymes are not activated until pancreatic juice has entered the duodenum. This is because pancreatic juice also contains a small protein called *pancreatic trypsin inhibitor* that attaches to trypsin and inhibits its activity in the pancreas.

CLINICAL APPLICATION

Pancreatitis (inflammation of the pancreas) may result when conditions such as alcoholism, gallstones, traumatic injury, infections, or toxicosis from various drugs provoke activation of digestive enzymes within the pancreas. Leakage of trypsin into the blood also occurs, but trypsin is inactive in the blood because of the inhibitory action of α_1-antitrypsin and α_2-macroglobulin. (The liver secretes α_1-antitrypsin into the blood, where its primary function is to control the tissue destruction caused by the neutrophil enzyme elastase.) Pancreatic amylase may also leak into the blood, but it is not active because its substrate (starch) is not present in blood. Pancreatic amylase activity can be measured *in vitro,* however, and these measurements are commonly performed to assess the health of the pancreas.

 **CHECKPOINT**

10a. Describe the structure of liver lobules and trace the pathways for the flow of blood and bile in the lobules.

10b. Explain how the liver inactivates and excretes compounds such as hormones and drugs.

10c. Explain how the liver helps maintain a constant blood glucose concentration and how the pattern of venous blood flow enables this function.

11a. Describe the composition and function of bile and trace the flow of bile from the liver and gallbladder to the duodenum.

11b. Describe the enterohepatic circulation of bilirubin and urobilinogen.

12. Describe the endocrine and exocrine structures and functions of the pancreas and explain how the pancreas is protected against self-digestion.

18.6 REGULATION OF THE DIGESTIVE SYSTEM

The stomach begins to increase its secretion in anticipation of a meal, and further increases its activities in response to the arrival of food. The entry of chyme into the duodenum stimulates the secretion of hormones that promote contractions of the gallbladder, the secretion of pancreatic juice, and the inhibition of gastric activity.

LEARNING OUTCOMES

After studying this section, you should be able to:

13. Identify the phases and explain the mechanisms of gastric regulation.

14. Explain the regulation of pancreatic juice and bile secretion.

15. Explain the significance of the enteric nervous system.

Neural and endocrine control mechanisms modify the activity of the digestive system. The sight, smell, or taste of food, for example, can stimulate salivary and gastric secretions via activation of the vagus nerves, which help "prime" the digestive system in preparation for a meal. Stimulation of the vagus originates in the brain and is a conditioned reflex (as Pavlov demonstrated by training dogs to salivate in response to a bell). The vagus nerves are also involved in the reflex control of one part of the digestive system by another—these are "short reflexes," which do not involve the brain.

The GI tract is both an endocrine gland and a target for the action of various hormones. Indeed, the first hormones to be discovered were gastrointestinal hormones. In 1902 two English physiologists, Sir William Bayliss and Ernest Starling, discovered that the duodenum produced a chemical regulator. They named this substance **secretin,** and proposed in 1905 that it was but one of many yet undiscovered chemical regulators produced by the body. Bayliss and Starling coined the term *hormones* for this new class of regulators. In that same year other investigators discovered that an extract from the stomach antrum stimulated gastric secretion. The hormone **gastrin** was thus the second hormone to be discovered.

The chemical structures of gastrin, secretin, and the duodenal hormone **cholecystokinin (CCK)** were determined in the 1960s. More recently, a fourth hormone produced by the small intestine, **gastric inhibitory peptide (GIP),** has been added to the list of proven GI tract hormones. The effects of these and other gastrointestinal hormones are summarized in table 18.5.

Regulation of Gastric Function

Gastric motility and secretion are to some extent automatic. Waves of contraction that serve to push chyme through the pyloric sphincter, for example, are initiated spontaneously by pacesetter cells in the greater curvature of the stomach. Likewise, the secretion of HCl from parietal cells and pepsinogen from chief cells can be stimulated in the absence of neural and hormonal influences by the presence of cooked or partially digested protein in the stomach. This action involves other cells in the gastric mucosa, including the G cells, which secrete the hormone gastrin; the enterochromaffin-like (ECL) cells, which secrete histamine; and the D cells, which secrete somatostatin.

The effects of autonomic nerves and hormones are superimposed on this automatic activity. This extrinsic control of stomach function is conveniently divided into three phases: (1) the *cephalic phase;* (2) the *gastric phase;* and (3) the *intestinal phase.* These are summarized in table 18.6.

Cephalic Phase

The **cephalic phase** of gastric regulation refers to control by the brain via the vagus nerves. As previously discussed, various conditioned stimuli can evoke gastric secretion. This conditioning in humans is, of course, more subtle than that exhibited by Pavlov's dogs in response to a bell. In fact, just talking about appetizing food is sometimes a more potent stimulus for gastric acid secretion than the actual sight and smell of food.

Activation of the vagus nerves stimulates the chief cells to secrete pepsinogen. Neurotransmitters released by the vagus also stimulate the secretion of HCl by the parietal cells. This neural stimulation of HCl secretion may be partly direct, through ACh binding to muscarinic receptors on the parietal cell membrane. However, the major mechanism of neural stimulation is indirect,

Table 18.5 | Effects of Gastrointestinal Hormones

Secreted by	Hormone	Effects
Stomach	Gastrin	Stimulates parietal cells to secrete HCl Stimulates chief cells to secrete pepsinogen Maintains structure of gastric mucosa
Small intestine	Secretin	Stimulates water and bicarbonate secretion in pancreatic juice Potentiates actions of cholecystokinin on pancreas
Small intestine	Cholecystokinin (CCK)	Stimulates contraction of gallbladder Stimulates secretion of pancreatic juice enzymes Inhibits gastric motility and secretion Maintains structure of exocrine pancreas (acini)
Small intestine	Gastric inhibitory peptide or glucose-dependent insulinotropic peptide (GIP)	Inhibits gastric motility and secretion Stimulates secretion of insulin from pancreatic islets
Ileum and colon	Glucagon-like peptide-I (GLP-I)	Inhibits gastric motility and secretion Stimulates secretion of insulin from pancreatic islets
	Guanylin	Stimulates intestinal secretion of Cl^-, causing elimination of NaCl and water in the feces

Table 18.6 | The Three Phases of Gastric Secretion

Phase of Regulation	Description
Cephalic Phase	1. Sight, smell, and taste of food cause stimulation of vagus nuclei in brain 2. Vagus stimulates chief cells to secrete pepsinogen 3. Vagus stimulates gastric acid secretion mainly by stimulating ECL cells to secrete histamine, which stimulates parietal cells to secrete HCl
Gastric Phase	1. Distension of stomach stimulates vagus nerve; vagus stimulates acid secretion 2. Amino acids and peptides in stomach lumen stimulate acid secretion a. Direct stimulation of parietal cells (lesser effect) b. Stimulation of gastrin secretion; gastrin stimulates acid secretion (major effect) 3. Gastrin secretion inhibited when pH of gastric juice falls below 2.5
Intestinal Phase	1. Neural inhibition of gastric emptying and acid secretion a. Arrival of chyme in duodenum causes distension, increase in osmotic pressure b. These stimuli activate a neural reflex that inhibits gastric activity 2. In response to fat in the chyme, the duodenum secretes an enterogastrone hormone that inhibits gastric motility and secretion

through the stimulation of histamine secretion by ECL cells. The histamine secreted by the ECL cells then stimulates the parietal cells to secrete HCl (fig. 18.30).

This cephalic phase continues into the first 30 minutes of a meal, but then gradually declines in importance as the next phase becomes predominant.

Gastric Phase

The arrival of food into the stomach stimulates the **gastric phase** of regulation. Gastric secretion is stimulated in response to two factors: (1) distension of the stomach, which is determined by the amount of chyme, and (2) the chemical nature of the chyme.

Although intact proteins in the chyme have little stimulatory effect, the partial digestion of proteins into shorter polypeptides and amino acids, particularly phenylalanine and tryptophan, stimulates the chief cells to secrete pepsinogen and the G cells to secrete gastrin. Gastrin, in turn, stimulates the secretion of pepsinogen from chief cells and HCl from parietal cells, but its effect on the parietal cells is primarily indirect. Gastrin stimulates the secretion of histamine from ECL cells, and the histamine then stimulates secretion of HCl from parietal cells (fig. 18.30). A *positive feedback mechanism* thus develops. As more HCl and pepsinogen are secreted, more short polypeptides and amino acids are released from the ingested proteins. This stimulates additional secretion of gastrin and, therefore, additional secretion of HCl and pepsinogen. It should be noted that glucose in the chyme has no effect on gastric secretion, and the presence of fat inhibits acid secretion.

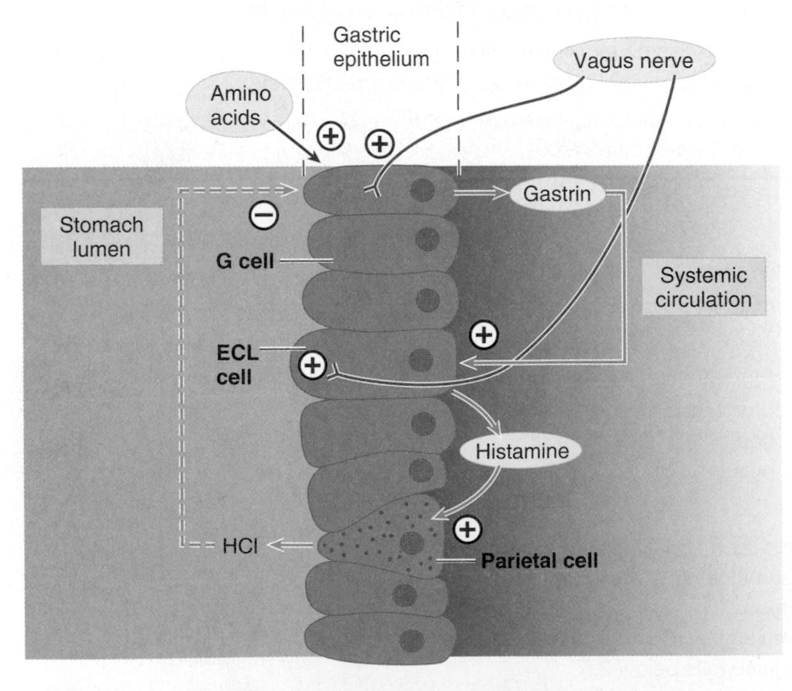

Figure 18.30 The regulation of gastric acid secretion. The presence of amino acids in the stomach lumen from partially digested proteins stimulates gastrin secretion. Gastrin secretion from G cells is also stimulated by vagus nerve activity. The secreted gastrin then acts as a hormone to stimulate histamine release from the ECL cells. The histamine, in turn, acts as a paracrine regulator to stimulate the parietal cells to secrete HCl. (⊕ = stimulation; ⊖ = inhibition.) AP|R

Secretion of HCl during the gastric phase is also regulated by a *negative feedback mechanism.* As the pH of gastric juice drops, so does the secretion of gastrin—at a pH of 2.5, gastrin secretion is reduced, and at a pH of 1.0 gastrin secretion ceases. The secretion of HCl thus declines accordingly. This effect may be mediated by the hormone somatostatin, secreted by the D cells of the gastric mucosa. As the pH of gastric juice falls, the D cells are stimulated to secrete somatostatin, which then acts as a paracrine regulator to inhibit the secretion of gastrin from the G cells.

The presence of proteins and polypeptides in the stomach helps buffer the acid and thus helps prevent a rapid fall in gastric pH. More acid can thus be secreted when proteins are present than when they are absent. In summary, arrival of protein into the stomach stimulates acid secretion in two ways—by the positive feedback mechanism previously discussed and by inhibition of the negative feedback control of acid secretion. Through these mechanisms, the amount of acid secreted is closely matched to the amount of protein ingested. As the stomach is emptied and the protein buffers exit, the pH falls, and the secretion of gastrin and HCl is accordingly inhibited.

Intestinal Phase

The **intestinal phase** of gastric regulation refers to the inhibition of gastric activity when chyme enters the small intestine. Investigators in 1886 demonstrated that the addition of olive oil to a meal inhibits gastric emptying, and in 1929 it was shown that the presence of fat inhibits gastric juice secretion. This inhibitory intestinal phase of gastric regulation is due to both a neural reflex originating from the duodenum and a chemical hormone secreted by the duodenum.

The arrival of chyme into the duodenum increases its osmolality. This stimulus, together with stretch of the duodenum and possibly other stimuli, activates sensory neurons of the vagus nerves and produces a neural reflex that inhibits gastric motility and secretion. The presence of fat in the chyme also stimulates the duodenum to secrete a hormone that inhibits gastric function. The general term for such an inhibitory hormone is an **enterogastrone.**

Several hormones secreted by the small intestine have been shown to have an enterogastrone effect. One of these hormones was even named for this action—*gastric inhibitory peptide (GIP),* secreted by the duodenum. However, subsequent research demonstrated that the major action of GIP is actually to stimulate insulin secretion (from the beta cells of the pancreatic islets) in response to glucose in food. As a consequence of this action, the acronym of the hormone was retained but it was renamed **glucose-dependent insulinotropic peptide (GIP).**

Other polypeptide hormones secreted by the small intestine that have an enterogastrone effect include **somatostatin,** produced by the stomach and intestine (as well as the brain); **cholecystokinin (CCK),** secreted by the duodenum in response to the presence of chyme; and **glucagon-like peptide-1 (GLP-1),** secreted by the ileum. These hormones help reduce gastric activity once the small intestine has received a load of chyme from the stomach, giving the intestine time to digest and absorb the food.

Interestingly, GLP-1, like GIP, is a very powerful stimulator of insulin secretion from the pancreatic islets. These two intestinal hormones stimulate the pancreatic islets to "anticipate" a rise in blood glucose by starting to secrete insulin even before the orally ingested glucose has been absorbed into the blood. Since insulin acts to lower the blood glucose concentration, this action helps maintain homeostasis of blood glucose (chapter 9, section 19.3).

FITNESS APPLICATION

The inhibitory neural and endocrine mechanisms during the intestinal phase prevent the further passage of chyme from the stomach to the duodenum. This gives the duodenum time to process the load of chyme received previously. Because the small intestine is stimulated to secrete enterogastrone by fat in the chyme, a breakfast of bacon and eggs takes longer to pass through the stomach—and makes one feel "fuller" for a longer time—than does a breakfast of oatmeal and fruit.

Regulation of Intestinal Function

Enteric Nervous System

The neurons and glial cells of the **enteric nervous system (ENS)** are organized into ganglia that are interconnected by two plexuses. The outer *myenteric (Auerbach's) plexus* is found along the entire length of the GI tract; the inner *submucosal (Meissner's) plexus* is located only in the small and large intestine. The ENS contains about 100 million neurons (roughly the same number as in the spinal cord), and has a similar diversity of neurotransmitters as the CNS. The ENS has interneurons as well as sensory and autonomic motor neurons, and its glial cells resemble the astrocytes of the brain.

Some sensory (afferent) neurons within the intestinal plexuses travel in the vagus nerves to deliver sensory information to the CNS. These are called *extrinsic afferents,* and they are involved in regulation by the autonomic nervous system. Other sensory neurons—called *intrinsic afferents*—have their cell bodies in the myenteric or submucosal plexuses and synapse with the interneurons of the enteric nervous system. An estimated 100 million intrinsic afferents greatly outnumber the 50,000 extrinsic afferents in the intestine, emphasizing the importance of local regulation of intestinal function.

Peristalsis, for example, is regulated by the enteric nervous system. A bolus of chyme stimulates intrinsic afferents (with cell bodies in the myenteric plexus) that activate enteric interneurons, which in turn stimulate motor neurons. These motor neurons innervate both smooth muscle cells and interstitial cells of Cajal, where they release excitatory and inhibitory neurotransmitters. Smooth muscle contraction is stimulated by the neurotransmitters ACh and substance P above the bolus, and smooth muscle relaxation is promoted by nitric oxide, vasoactive intestinal peptide (VIP), and ATP below the bolus (fig. 18.31).

Extrinsic afferents, together with the different peptide hormones released from the intestine, alert the brain to the conditions in the gastrointestinal tract. This information is important in the CNS regulation of digestion and in both conscious and unconscious perceptions of food intake and the state of the viscera. The brain can sometimes overpower local, enteric nervous system regulation during intense emotions of fear and anger through stimulatory parasympathetic and inhibitory sympathetic outflow from the CNS.

Paracrine Regulators of the Intestine

There is evidence that the enterochromaffin-like cells (ECL cells) of the intestinal mucosa secrete **serotonin,** or **5-hydroxytryptamine,** in response to the stimuli of pressure and various chemicals. Serotonin then stimulates intrinsic afferents, which conduct impulses into the submucosal and myenteric plexuses and there activate motor neurons. Motor neurons that terminate in the muscularis can stimulate contractions; those that terminate in the intestinal crypts can stimulate the secretion of salt and water into the lumen. The ECL cells have also been shown to produce another paracrine regulator, termed **motilin,** which stimulates contraction in the duodenum and stomach antrum.

Guanylin is a paracrine regulator produced by the ileum and colon. It derives its name from its ability to activate the enzyme guanylate cyclase, and thus to cause the production of cyclic GMP (cGMP) within the cytoplasm of intestinal epithelial cells. Acting through cGMP as a second messenger, guanylin stimulates the intestinal epithelial cells to secrete Cl^- and water and inhibits their absorption of Na^+. These actions increase the amount of salt and water lost from the body in the feces. A related polypeptide, called **uroguanylin,** has been found in the urine. This polypeptide is produced by the intestine and may function as a hormone that stimulates the kidneys to excrete salt in the urine.

CLINICAL APPLICATION

Certain *Escherichia coli* bacteria produce *heat-stable enterotoxins* that are responsible for **traveler's diarrhea.** The enterotoxins act by stimulating the same receptors on the apical membranes of the intestinal epithelial cells that are activated by guanylin. By mimicking the actions of guanylin, the enterotoxins stimulate intestinal Cl^- and water secretion to produce diarrhea.

Intestinal Reflexes

There are several intestinal reflexes that are controlled both locally, by means of the enteric nervous system and paracrine

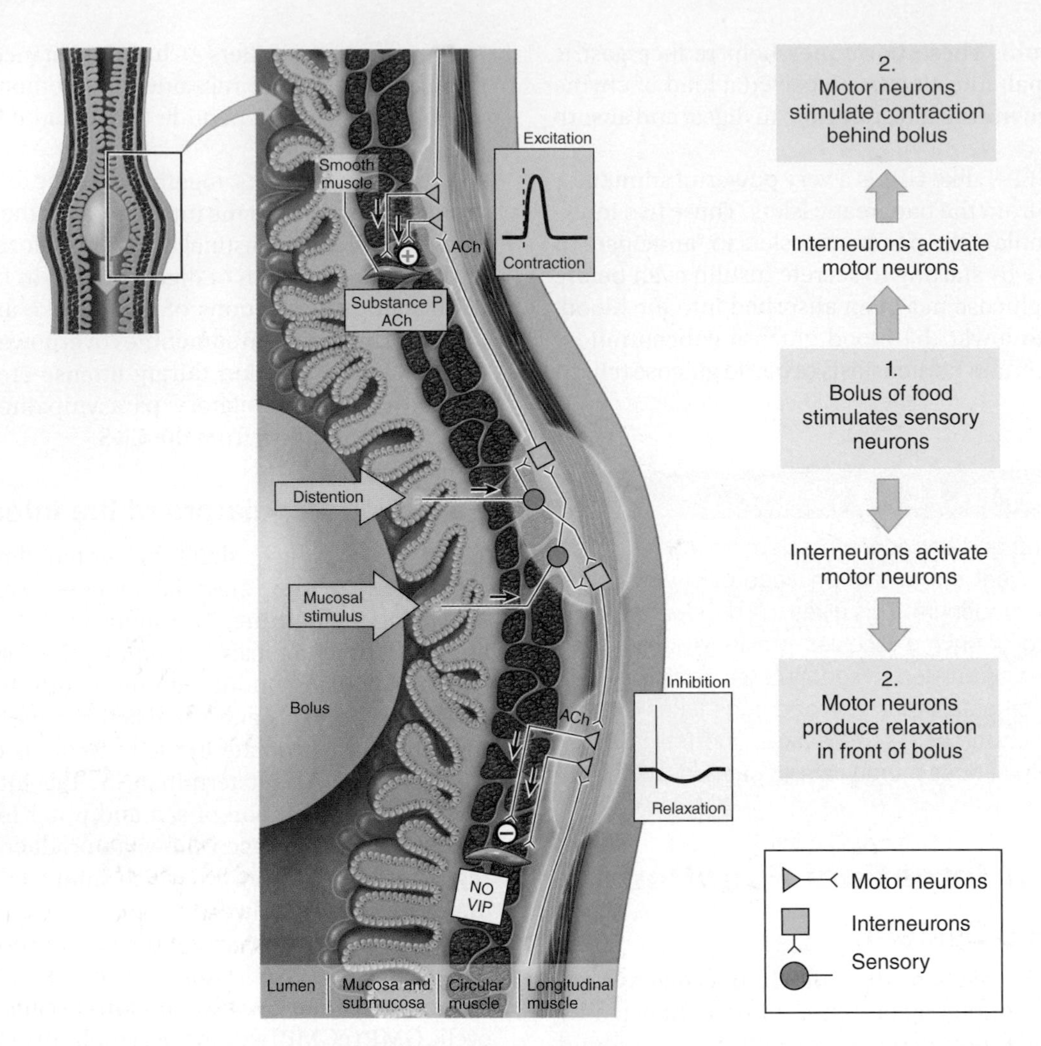

Figure 18.31 The enteric nervous system coordinates peristalsis. Peristalsis is produced by local reflexes involving the enteric nervous system. Notice that the enteric nervous system consists of motor neurons, interneurons, and sensory neurons. The neurotransmitters that stimulate smooth muscle contraction are indicated with a ⊕, while those that produce smooth muscle relaxation are indicated with a ⊖. (NO = nitric oxide; VIP = vasoactive intestinal peptide.)

regulators, and extrinsically through the actions of nerves and hormones. These reflexes include:

1. the **gastroileal reflex,** in which increased gastric activity causes increased motility of the ileum and increased movements of chyme through the ileocecal sphincter;
2. the **ileogastric reflex,** in which distension of the ileum causes a decrease in gastric motility;
3. the **intestino-intestinal reflexes,** in which overdistension of one intestinal segment causes relaxation throughout the rest of the intestine.

Regulation of Pancreatic Juice and Bile Secretion

The arrival of chyme into the duodenum stimulates the intestinal phase of gastric regulation and, at the same time, stimulates reflex secretion of pancreatic juice and bile. The entry of additional chyme into the duodenum is thereby retarded by the inhibitory effects of neural reflexes and enterogastrone, allowing time for the previous load of chyme in the duodenum to be digested with the aid of pancreatic juice enzymes and bile. The secretion of pancreatic juice and bile is stimulated both by neural reflexes initiated in the duodenum and by secretion of the duodenal hormones cholecystokinin and secretin.

Secretion of Pancreatic Juice

The secretion of pancreatic enzymes (including trypsin, lipase, and amylase) from the acinar cells is stimulated by ACh released by the vagus nerves, and by the hormone CCK secreted by the duodenum. Both ACh and CCK stimulate the acinar cells using a rise in cytoplasmic Ca^{2+} as a second messenger. Secretin, another hormone from the duodenum, can potentiate the effects of ACh and CCK on the acinar cells through the action of cyclic AMP as a second messenger.

Secretin and CCK are secreted by the duodenum in response to different stimuli and have different primary effects. Secretion of CCK is stimulated by the protein and fat content of the chyme; CCK then stimulates the secretion of the pancreatic juice enzymes that help digest these molecules. Partially digested protein and fat are the most potent stimulators of CCK secretion, which diminishes as the chyme passes out of the duodenum into the jejunum.

By contrast, the secretion of secretin by the duodenum is stimulated by a fall in duodenal pH below 4.5, caused by the arrival of acidic chyme from the stomach. This low pH is maintained for only a short time, however, because secretin then stimulates the production of bicarbonate and its secretion into pancreatic juice. Because bicarbonate in pancreatic juice neutralizes the acidic chyme in the duodenum, the low pH of the chyme is soon raised by the alkaline pancreatic juice. This helps to protect the mucosa of the duodenum and provides the pH environment for optimum activity of the pancreatic juice digestive enzymes.

Secretion of Bile

The liver secretes bile continuously, but bile secretion is increased by a meal. When bile arrives in the duodenum during a meal, the liver is stimulated to secrete more bile by the bile acids that return to the liver from the intestine via the hepatic portal vein (the enterohepatic circulation; see fig. 18.21). Endocrine and neural reflexes are also involved. Secretin stimulates the bile duct cells of the liver to secrete bicarbonate into the bile (leading to increased bile volume), and CCK enhances this effect. The secretion of CCK in response to fat in the chyme stimulates contractions of the gallbladder, allowing more bile to enter the duodenum. The bile then emulsifies the fat, aiding its digestion. Also, the arrival of chyme in the duodenum produces a neural reflex that stimulates contractions of the gallbladder.

Clinical Investigation CLUES

Alan's pain below his right scapula was triggered by eating peanut butter or bacon, but not by eating foods such as skinned chicken or fish.

- What component of the food was responsible for the painful reaction?
- How does this component of the food result in Alan's pain?

Trophic Effects of Gastrointestinal Hormones

Patients with tumors of the stomach pylorus exhibit excessive acid secretion and hyperplasia (growth) of the gastric mucosa. Surgical removal of the pylorus reduces gastric secretion and prevents growth of the gastric mucosa. Patients with peptic ulcers are sometimes treated by vagotomy—cutting of the portion of the vagus nerve that innervates the stomach. Vagotomy also reduces acid secretion but has no effect on the gastric mucosa. These observations suggest that the hormone gastrin, secreted by the pyloric mucosa, may exert stimulatory, or *trophic,* effects on the gastric mucosa. The structure of the gastric mucosa, in other words, is dependent upon the effects of gastrin.

In the same way, the structure of the acinar (exocrine) cells of the pancreas is dependent upon the trophic effects of CCK. Perhaps this explains why the pancreas, as well as the GI tract, atrophies during starvation. Since neural reflexes appear to be capable of regulating digestion, perhaps the primary function of the GI hormones is trophic—that is, maintenance of the structure of their target organs.

CHECKPOINT

13a. Describe the positive and negative feedback mechanisms that operate during the gastric phase of HCl and pepsinogen secretion.

13b. Describe the mechanisms involved in the intestinal phase of gastric regulation, and explain why a fatty meal takes longer to leave the stomach than a meal low in fat.

14. Explain the hormonal mechanisms involved in the production and release of pancreatic juice and bile.

15. Describe the enteric nervous system, and identify some of the short reflexes that regulate intestinal function.

18.7 DIGESTION AND ABSORPTION OF FOOD

Polysaccharides and polypeptides are hydrolyzed into their subunits, which are secreted into blood capillaries. Fat is emulsified by bile salts, hydrolyzed into fatty acids and monoglycerides, and absorbed into the intestinal epithelial cells. Once inside the cells, triglycerides are resynthesized, combined with proteins, and secreted into the lymphatic fluid.

LEARNING OUTCOMES

After studying this section, you should be able to:

16. Describe the processes involved in the digestion and absorption of carbohydrates and proteins.

17. Describe the processes involved in the digestion, absorption, and transport of dietary lipids.

The caloric (energy) value of food is derived mainly from its content of carbohydrates, lipids, and proteins. In the average American diet, carbohydrates account for approximately

Table 18.7 | Characteristics of the Major Digestive Enzymes

Enzyme	Site of Action	Source	Substrate	Optimum pH	Product(s)
Salivary amylase	Mouth	Saliva	Starch	6.7	Maltose
Pepsin	Stomach	Gastric glands	Protein	1.6–2.4	Shorter polypeptides
Pancreatic amylase	Duodenum	Pancreatic juice	Starch	6.7–7.0	Maltose, maltriose, and oligosaccharides
Trypsin, chymotrypsin, carboxypeptidase	Small intestine	Pancreatic juice	Polypeptides	8.0	Amino acids, dipeptides, and tripeptides
Pancreatic lipase	Small intestine	Pancreatic juice	Triglycerides	8.0	Fatty acids and monoglycerides
Maltase	Small intestine	Brush border of epithelial cells	Maltose	5.0–7.0	Glucose
Sucrase	Small intestine	Brush border of epithelial cells	Sucrose	5.0–7.0	Glucose + fructose
Lactase	Small intestine	Brush border of epithelial cells	Lactose	5.8–6.2	Glucose + galactose
Aminopeptidase	Small intestine	Brush border of epithelial cells	Polypeptides	8.0	Amino acids, dipeptides, tripeptides

50% of the total calories, protein accounts for 11% to 14%, and lipids make up the balance. These food molecules consist primarily of long combinations of subunits (monomers) that must be digested by hydrolysis reactions into free monomers before absorption can occur. The characteristics of the major digestive enzymes are summarized in table 18.7.

Digestion and Absorption of Carbohydrates

Most carbohydrates are ingested as starch, which is a long polysaccharide of glucose in the form of straight chains with occasional branchings (chapter 2; see fig. 2.15). The most commonly ingested sugars are sucrose (table sugar, a disaccharide of glucose and fructose; see fig. 2.16) and lactose (milk sugar, a disaccharide of glucose and galactose). The digestion of starch begins in the mouth with the action of **salivary amylase.** This enzyme cleaves some of the bonds between adjacent glucose molecules, but most people don't chew their food long enough for sufficient digestion to occur in the mouth. The digestive action of salivary amylase stops some time after the swallowed bolus enters the stomach because this enzyme is inactivated at the low pH of gastric juice.

The digestion of starch, therefore, occurs mainly in the duodenum as a result of the action of **pancreatic amylase.** This enzyme cleaves the straight chains of starch to produce the disaccharide *maltose* and the trisaccharide *maltriose.* Pancreatic amylase, however, cannot hydrolyze the bond between glucose molecules at the branch points in the starch. As a result, short, branched chains of glucose molecules, called *oligosaccharides,* are released together with maltose and maltriose by the activity of this enzyme (fig. 18.32).

Maltose, maltriose, and oligosaccharides are hydrolyzed to their monosaccharides by brush border enzymes located on the microvilli of the epithelial cells in the small intestine. The brush border enzymes also hydrolyze the disaccharides sucrose and lactose into their component monosaccharides. These monosaccharides are then moved across the brush border membrane by secondary active transport. One glucose molecule is cotransported with two Na^+ (chapter 6; see fig. 6.20) into the epithelial cell cytoplasm. The glucose then moves by facilitative diffusion through GLUT transporters across the basolateral membrane into the interstitial fluid and the capillary blood in the villus. In this way, the absorbed monosaccharides enter the hepatic portal vein and travel to the liver. The absorption of Na^+ that accompanies the absorption of glucose (and other nutrients) results in the simultaneous movement of Cl^- due to the potential difference created by Na^+ transport. Water follows the NaCl through the paracellular route between epithelial cells (chapter 6, section 6.3) and is absorbed into the blood along with the glucose and NaCl.

Digestion and Absorption of Proteins

Protein digestion begins in the stomach with the action of pepsin. Some amino acids are liberated in the stomach, but

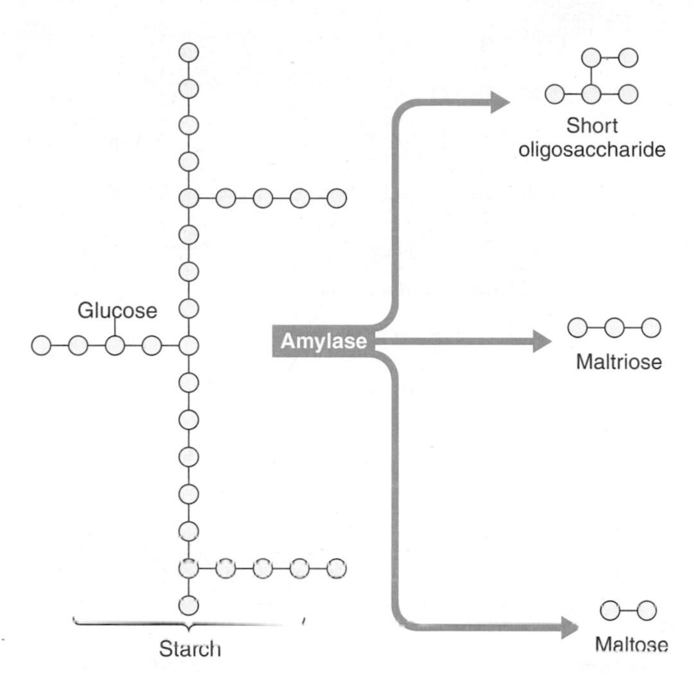

Figure 18.32 The action of pancreatic amylase. Pancreatic amylase digests starch into maltose, maltriose, and short oligosaccharides containing branch points in the chain of glucose molecules.

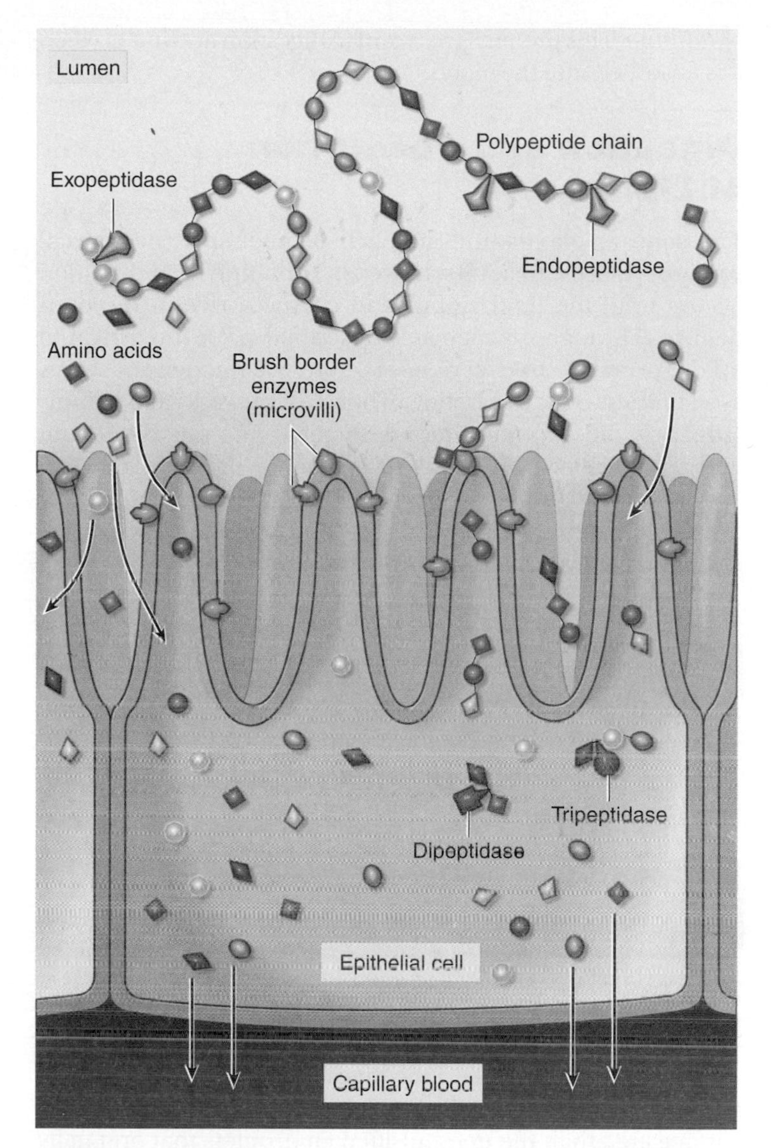

Figure 18.33 The digestion and absorption of proteins. Polypeptide chains of proteins are digested into free amino acids, peptides, and tripeptides by the action of pancreatic juice enzymes and brush border enzymes. The amino acids, dipeptides, and tripeptides enter duodenal epithelial cells. Dipeptides and tripeptides are hydrolyzed into free amino acids within the epithelial cells, and these products are secreted into interstitial fluid (not shown) and then into capillaries, which eventually drain into the hepatic portal vein.

the major products of pepsin digestion are short-chain polypeptides. Pepsin digestion helps produce a more homogeneous chyme, but it is not essential for the complete digestion of protein that occurs—even in people with total gastrectomies—in the small intestine.

Most protein digestion occurs in the duodenum and jejunum. The pancreatic juice enzymes **trypsin, chymotrypsin,** and **elastase** cleave peptide bonds in the interior of the polypeptide chains. These enzymes are thus grouped together as *endopeptidases.* Enzymes that remove amino acids from the ends of polypeptide chains, by contrast, are *exopeptidases.* These include the pancreatic juice enzyme **carboxypeptidase,** which removes amino acids from the carboxyl-terminal end of polypeptide chains, and the brush border enzyme **aminopeptidase.** Aminopeptidase cleaves amino acids from the amino-terminal end of polypeptide chains.

As a result of the action of these enzymes, polypeptide chains are digested into free amino acids, dipeptides, and tripeptides. The free amino acids are absorbed across the brush border membrane by different secondary active transport carriers, most of which cotransport the amino acids with Na^+. The dipeptides and tripeptides enter epithelial cells by the action of a single membrane carrier that has recently been characterized. This carrier functions in secondary active transport using a H^+ gradient to transport dipeptides and tripeptides into the cell cytoplasm. Within the cytoplasm, the dipeptides and tripeptides are hydrolyzed into free amino acids, which move across the basolateral membrane by facilitated diffusion to the interstitial fluid and then to the capillary blood (fig. 18.33). In this way, the absorbed amino acids enter the blood delivered to the liver by the hepatic portal vein.

Newborn babies appear to be capable of absorbing a substantial amount of undigested proteins (hence they can absorb some antibodies from their mother's first milk); in adults, however, only the free amino acids enter the portal vein. Foreign food protein, which would be very antigenic, does not normally enter the blood. An interesting exception is the protein toxin that causes botulism, produced by the bacterium *Clostridium botulinum.* This protein is resistant to digestion and intact when it is absorbed into the blood.

Digestion and Absorption of Lipids

The salivary glands and stomach of neonates (newborns) produce lipases. In adults, however, very little lipid digestion occurs until the lipid globules in chyme arrive in the duodenum. Through mechanisms described next, the arrival of lipids (primarily triglyceride, or fat) in the duodenum serves as a stimulus for the secretion of bile. In a process called **emulsification,** bile salt micelles are secreted into the duodenum and act as detergents to break up the fat droplets into tiny *emulsification droplets* of triglycerides. Note that emulsification is not chemical digestion—the bonds joining glycerol and fatty acids are not hydrolyzed by this process.

Clinical Investigation CLUES

Alan had fatty stools and a prolonged blood clotting time.

- Given what you know of Alan's condition, what could have caused his fatty stools?
- How might this relate to his prolonged clotting time? (*Hint:* vitamin K, needed for the formation of some clotting factors, is a fat-soluble vitamin.)

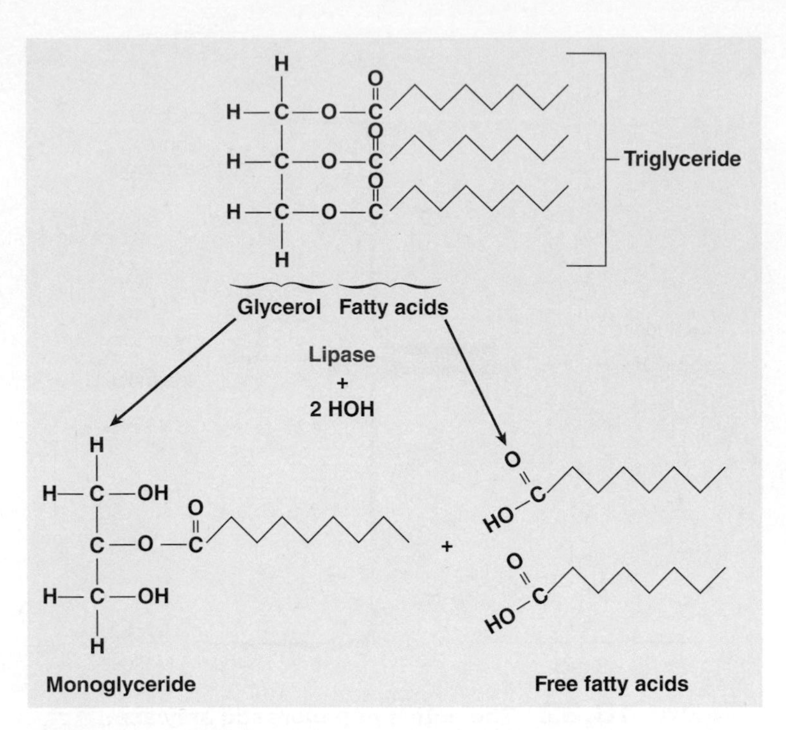

Figure 18.34 The digestion of triglycerides. Pancreatic lipase digests fat (triglycerides) by cleaving off the first and third fatty acids. This produces free fatty acids and monoglycerides. Sawtooth lines indicate hydrocarbon chains in the fatty acids.

Digestion of Lipids

The emulsification of fat aids digestion because the smaller and more numerous emulsification droplets present a greater surface area than the unemulsified fat droplets that originally entered the duodenum. Fat digestion occurs at the surface of the droplets through the enzymatic action of **pancreatic lipase,** which is aided in its action by a protein called *colipase* (also secreted by the pancreas) that coats the emulsification droplets and "anchors" the lipase enzyme to them. Through hydrolysis, lipase removes two of the three fatty acids from each triglyceride molecule and thus liberates *free fatty acids* and *monoglycerides* (fig. 18.34). **Phospholipase A** likewise

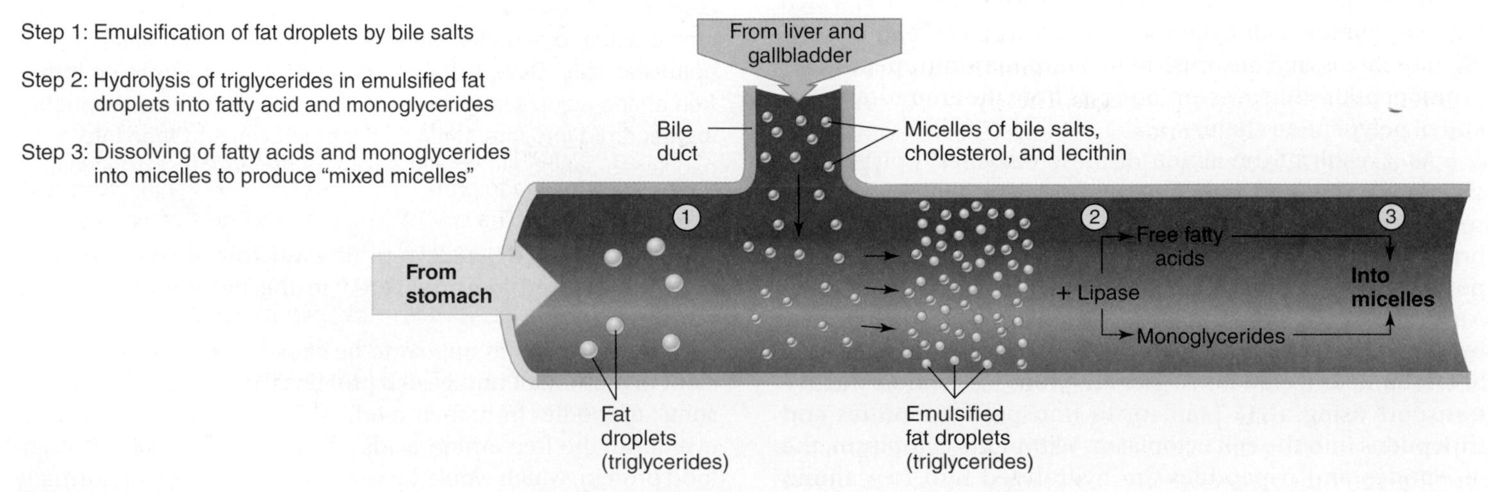

Figure 18.35 Fat digestion and emulsification. The three steps indicate the fate of fat in the small intestine. The digestion of fat (triglycerides) releases fatty acids and monoglycerides, which become associated with micelles of bile salts secreted by the liver.

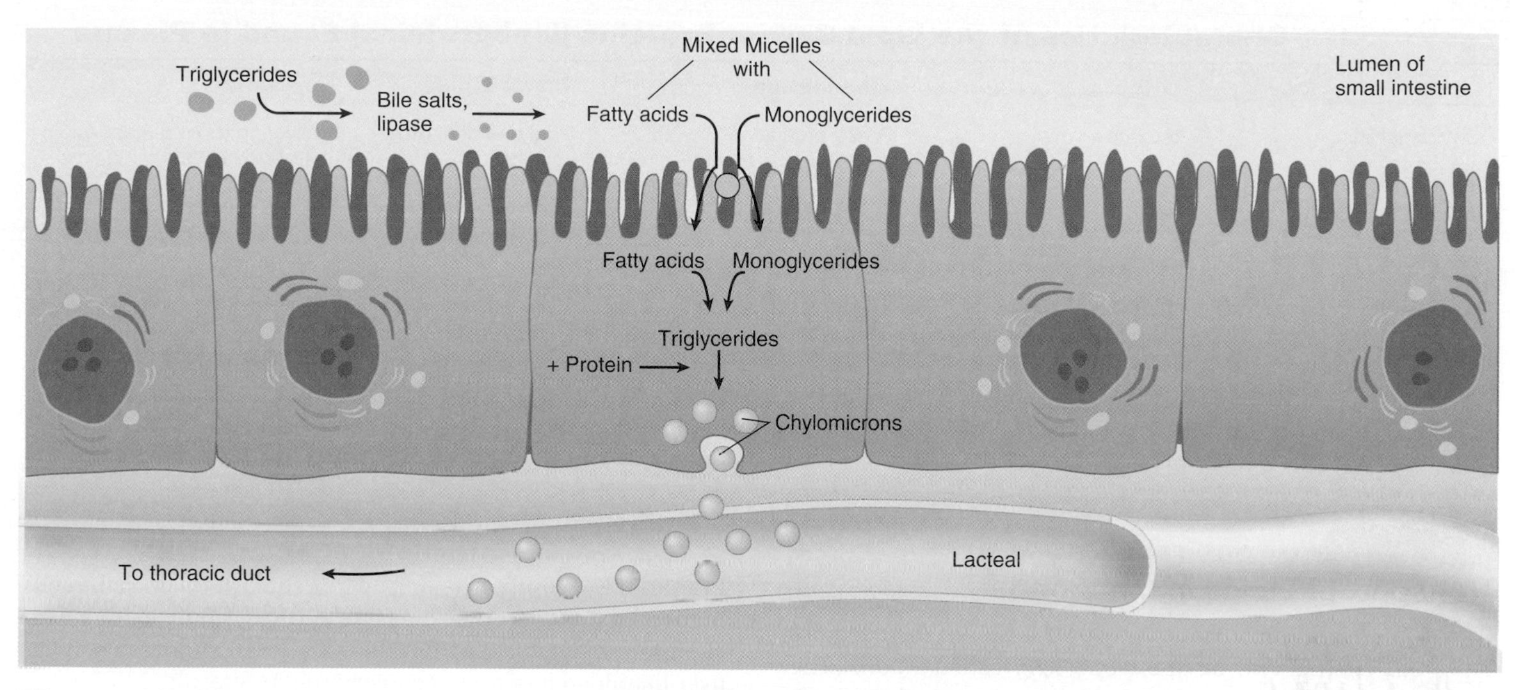

Figure 18.36 The absorption of fat. Fatty acids and monoglycerides from the micelles within the small intestine are absorbed by epithelial cells and converted intracellularly into triglycerides. These are then combined with protein to form chylomicrons, which enter the lymphatic vessels (lacteals) of the villi. These lymphatic vessels transport the chylomicrons to the thoracic duct, which empties them into the venous blood (of the left subclavian vein).

digests phospholipids such as lecithin into fatty acids and *lysolecithin* (the remainder of the lecithin molecule after the removal of two fatty acids).

The free fatty acids, monoglycerides, and lysolecithin derived from the digested lipids are more polar than the undigested lipids. They quickly enter the micelles of bile salts, lecithin, and cholesterol from the bile to form *mixed micelles* in the duodenum (fig. 18.35). The mixed micelles then move to the brush border of the intestinal epithelium where absorption occurs.

Absorption of Lipids

Free fatty acids, monoglycerides, and lysolecithin can leave the micelles and pass through the membrane of the microvilli to enter the intestinal epithelial cells. These products are then used to *resynthesize* triglycerides and phospholipids within the epithelial cells. This process is different from the absorption of amino acids and monosaccharides, which pass through the epithelial cells without being altered.

Triglycerides, phospholipids, and cholesterol are then combined with protein inside the epithelial cells to form small particles called **chylomicrons.** These tiny lipid and protein combinations are secreted by exocytosis into the central lacteals (lymphatic capillaries) of the intestinal villi (fig. 18.36). Absorbed lipids thus pass through the lymphatic system, eventually entering the venous blood by way of the thoracic duct (chapter 13, section 13.8). Chylomicrons in the blood after a fatty meal turn otherwise clear blood plasma cloudy (fig. 18.37).

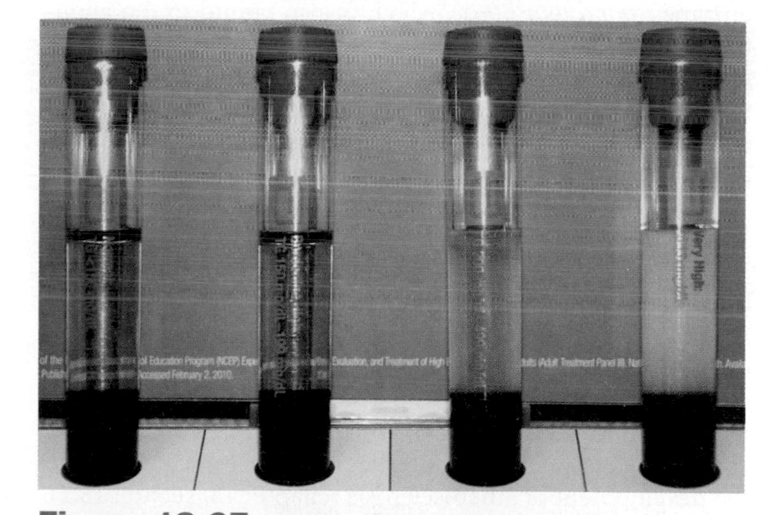

Figure 18.37 Chylomicrons can turn plasma cloudy. The turbidity (cloudiness) of the plasma is caused by triglycerides bound to protein particles such as chylomicrons and other lipoproteins. The four tubes are a simulation of increasing concentrations of lipoprotein-bound triglycerides in the plasma. The tube on the left is normal, transparent plasma, with a triglyceride concentration under 150 mg/dL. The tubes to its right have, respectively, borderline high (150–199 mg/dL); high (200–499 mg/dL); and very high (greater than 500 mg/dL) triglyceride concentration.

Table 18.8 | Characteristics of the Lipid Carrier Proteins (Lipoproteins) Found in Plasma

Lipoprotein Class	Origin	Destination	Major Lipids	Functions
Chylomicrons	Intestine	Many organs	Triglycerides, other lipids	Deliver lipids of dietary origin to body cells
Very-low-density lipoproteins (VLDLs)	Liver	Many organs	Triglycerides, cholesterol	Deliver endogenously produced triglycerides to body cells
Low-density lipoproteins (LDLs)	Intravascular removal of triglycerides from VLDLs	Blood vessels, liver	Cholesterol	Deliver endogenously produced cholesterol to various organs
High-density lipoproteins (HDLs)	Liver and intestine	Liver and steroid-hormone-producing glands	Cholesterol	Remove and degrade cholesterol

Transport of Lipids in the Blood

Once the chylomicrons have entered the blood, they acquire a protein constituent, or *apolipoprotein,* called *ApoE.* (The structure of a lipoprotein particle is illustrated in chapter 13, fig. 13.32.) This allows the chylomicrons to bind to receptor proteins for ApoE located on the plasma membrane of capillary endothelial cells in the muscles and adipose tissue. The triglyceride content of the chylomicrons can then be digested by the enzyme *lipoprotein lipase,* which is also bound to the endothelial cell plasma membrane. The hydrolysis of triglycerides releases free fatty acids that circulate in the plasma attached to albumin and are used by skeletal muscles for energy and by adipose tissue for the synthesis of stored fat. After the triglyceride content of the chylomicrons has been removed, the remaining *remnant particle* containing cholesterol is released and travels in the circulation until it is taken up by the liver.

Cholesterol and triglycerides produced by the liver are combined with other apolipoproteins and secreted into the blood as **very-low-density lipoproteins (VLDLs),** which deliver triglycerides to different organs. Once the triglycerides are removed, the VLDL particles are converted to **low-density lipoproteins (LDLs),** which transport cholesterol to various organs, including blood vessels. This can contribute to the development of atherosclerosis (chapter 13, section 13.7). Excess cholesterol is returned from these organs to the liver attached to **high-density lipoproteins (HDLs).**

The HDL particles bind to receptors in the blood vessel wall and take up phospholipids and free cholesterol. An enzyme bonds the free cholesterol to the phospholipids, producing cholesterol esters. Because the cholesterol esters are very hydrophobic they move to the center of the HDL particle, enabling the particle to continue taking up free cholesterol from the cells of the blood vessel. After the HDL particle is fully loaded with cholesterol, it detaches from the vessel wall and travels to the liver to unload its cargo of cholesterol. As a result of this activity, a high ratio of HDL cholesterol to total cholesterol affords protection against atherosclerosis. The characteristics of the different lipoproteins are summarized in table 18.8.

CHECKPOINT

16a. List the enzymes involved in carbohydrate digestion, indicating their origins, sites of action, substrates, and products.

16b. List each enzyme involved in protein digestion, indicating its origin and site of action. Also, indicate whether the enzyme is an endopeptidase or exopeptidase.

17a. Describe how bile aids both the digestion and absorption of fats. Explain how the absorption of fat differs from the absorption of amino acids and monosaccharides.

17b. Trace the pathway and fate of a molecule of triglyceride and a molecule of cholesterol in a chylomicron within an intestinal epithelial cell.

17c. Cholesterol in the blood may be attached to any of four possible lipoproteins. Distinguish among these proteins in terms of the origin and destination of the cholesterol they carry.

Interactions

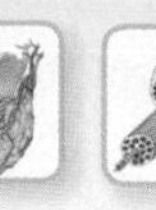

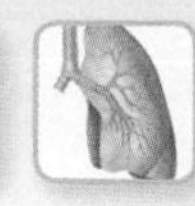
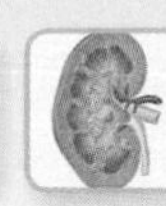

HPer Links of the Digestive System with Other Body Systems

Integumentary System

- The skin produces vitamin D, which indirectly helps to regulate the intestinal absorption of Ca^{2+} (p. 693)
- Adipose tissue in the hypodermis of the skin stores triglycerides (p. 120)
- The digestive system provides nutrients for all systems, including the integumentary system (p. 621)

Skeletal System

- The extracellular matrix of bones stores calcium phosphate (p. 690)
- The small intestine absorbs Ca^{2+} and PO_4^{3-}, which are needed for deposition of bone (p. 693)

Muscular System

- Muscle contractions are needed for chewing, swallowing, peristalsis, and segmentation (p. 621)
- Sphincter muscles help to regulate the passage of material along the GI tract (p. 621)
- The liver removes lactic acid produced by exercising skeletal muscles (p. 118)

Nervous System

- Autonomic nerves help regulate the digestive system (p. 644)
- The enteric nervous system functions like the CNS to regulate the intestine (p. 647)

Endocrine System

- Gastrin, produced by the stomach, helps regulate the secretion of gastric juice (p. 646)
- Several hormones secreted by the small intestine regulate different aspects of the digestive system (p. 644)
- A hormone produced by the intestine stimulates the pancreatic islets to secrete insulin (p. 678)
- Adipose tissue secretes leptin, which helps regulate hunger (p. 673)
- The liver removes some hormones from the blood, changes them chemically, and excretes them in the bile (p. 639)

Immune System

- The immune system protects all organs against infections, including those of the digestive system (p. 494)
- Lymphatic vessels carry absorbed fat from the small intestine to the venous system (p. 653)
- The liver aids the immune system by metabolizing certain toxins and excreting them in the bile (p. 639)
- The mucosa of the GI tract contains lymph nodules that protect against disease (p. 632)
- Acids and enzymes secreted by the GI tract provide nonspecific defense against microbes (p. 494)

Circulatory System

- The blood transports absorbed amino acids, monosaccharides, and other molecules from the intestine to the liver, and then to other organs (p. 636)
- The hepatic portal vein allows some absorbed molecules to have an enterohepatic circulation (p. 697)
- The intestinal absorption of vitamin B_{12} (needed for red blood cell production) requires intrinsic factor, secreted by the stomach (p. 625)
- Iron must be absorbed through the intestine to allow a normal rate of hemoglobin production (p. 412)
- The liver synthesizes clotting proteins, plasma albumin, and all other plasma proteins except antibodies (p. 640)

Respiratory System

- The lungs provide oxygen for the metabolism of all organs, including those of the digestive system (p. 533)
- The oxygen provided by the respiratory system is used to metabolize food molecules brought into the body by the digestive system (p. 111)

Urinary System

- The kidneys eliminate metabolic wastes from all organs, including those of the digestive system (p. 582)
- The kidneys help to convert vitamin D into the active form required for calcium absorption in the intestine (p. 693)

Reproductive System

- Sex steroids, particularly androgens, stimulate the rate of fuel consumption by the body (p. 710)
- During pregnancy, the GI tract of the mother helps to provide nutrients that pass through the placenta to the embryo and fetus (p. 742)

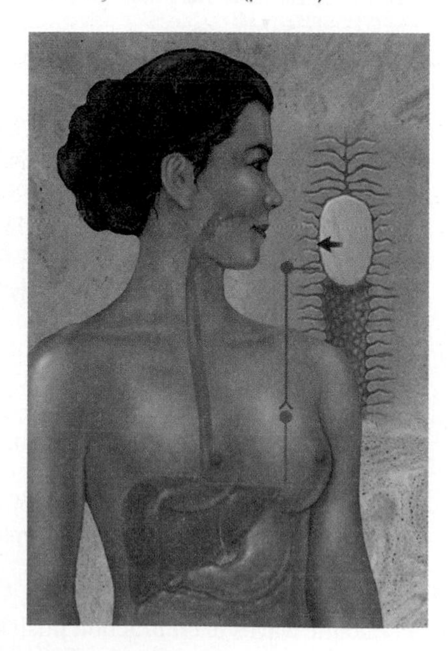

Clinical Investigation SUMMARY

Alan may have gastritis or a peptic ulcer, as suggested by the sharp pain in his stomach when he drinks wine (a stimulator of gastric acid secretion). The lack of fever and the normal white blood cell count suggest that the inflammation associated with appendicitis is absent. The yellowing of the sclera indicates jaundice, and this symptom—together with the prolonged clotting time—could be caused by liver disease. However, liver disease would elevate the blood levels of free bilirubin, which were found to be normal. The normal levels of urea and ammonia in the blood likewise suggest normal liver function. Similarly, the normal pancreatic amylase levels suggest that the pancreas is not affected.

Alan's symptoms are most likely due to the presence of gallstones. Gallstones could obstruct the normal flow of bile, and thus prevent normal fat digestion. This would explain the fatty stools. The resulting loss of dietary fat could cause a deficiency in vitamin K, which is a fat-soluble vitamin required for the production of a number of clotting factors (chapter 13)—hence, the prolonged clotting time. The pain would be provoked by oily or fatty foods (peanut butter and bacon), which trigger a reflex contraction of the gallbladder once the fat arrives in the duodenum. Contraction of the gallbladder against an obstructed cystic duct or common bile duct often produces a severe referred pain below the right scapula.

See the additional chapter 18 Clinical Investigations on *Peptic Ulcer* and *Celiac sprue* in the Connect site for this text at www.mhhe.com/Fox13.

SUMMARY

18.1 Introduction to the Digestive System 620

A. The digestion of food molecules involves the hydrolysis of these molecules into their subunits.
   1. The digestion of food occurs in the lumen of the GI tract and is catalyzed by specific enzymes.
   2. The digestion products are absorbed through the intestinal mucosa and enter the blood or lymph.

B. The layers (tunics) of the GI tract are, from the inside outward, the mucosa, submucosa, muscularis, and serosa.
   1. The mucosa consists of a simple columnar epithelium, a layer of connective tissue called the lamina propria, and a thin layer of smooth muscle called the muscularis mucosae.
   2. The submucosa is composed of connective tissue; the muscularis consists of layers of smooth muscles; the serosa is connective tissue covered by the visceral peritoneum.
   3. The submucosa contains the submucosal plexus, and the muscularis contains the myenteric plexus of autonomic nerves.

18.2 From Mouth to Stomach 623

A. Peristaltic waves of contraction push food through the lower esophageal sphincter into the stomach.

B. The stomach consists of a cardia, fundus, body, and pylorus (antrum). The pylorus terminates with the pyloric sphincter.
   1. The lining of the stomach is thrown into folds, or rugae, and the mucosal surface forms gastric pits that lead into gastric glands.
   2. The parietal cells of the gastric glands secrete HCl; the chief cells secrete pepsinogen.
   3. In the acidic environment of gastric juice, pepsinogen is converted into the active protein-digesting enzyme called pepsin.
   4. Some digestion of protein occurs in the stomach, but the most important function of the stomach is the secretion of intrinsic factor, which is needed for the absorption of vitamin B_{12} in the intestine.

18.3 Small Intestine 628

A. The small intestine is divided into the duodenum, jejunum, and ileum. The common bile duct and pancreatic duct empty into the duodenum.

B. Fingerlike extensions of mucosa called villi project into the lumen, and at the bases of the villi the mucosa forms narrow pouches called the crypts of Lieberkühn.
   1. New epithelial cells are formed in the crypts.

2. The membrane of intestinal epithelial cells is folded to form microvilli. This brush border of the mucosa increases the surface area.

C. Digestive enzymes, called brush border enzymes, are located in the membranes of the microvilli.

D. The small intestine exhibits two major types of movements—peristalsis and segmentation.

18.4 Large Intestine 632

A. The large intestine is divided into the cecum, colon, rectum, and anal canal.
   1. The appendix is attached to the inferior medial margin of the cecum.
   2. The colon consists of ascending, transverse, descending, and sigmoid portions.
   3. Bulges in the walls of the large intestine are called haustra.

B. The commensal bacteria of the intestinal microbiota are most numerous in the large intestine, where they serve a number of physiologically important functions required for health.
   1. The intestinal microbiota produce some B vitamins and vitamin K, as well as short chain fatty acids.
   2. The commensal bacteria of the intestinal microbiota help protect the GI tract from damaging inflammatory responses to pathogenic bacteria.

C. The large intestine absorbs water and electrolytes.
   1. Although most of the water that enters the GI tract is absorbed in the small intestine, 1.5 to 2.0 L pass to the large intestine each day. The large intestine absorbs about 90% of this amount.
   2. Na^+ is actively absorbed and water follows passively, in a manner analogous to the reabsorption of NaCl and water in the renal tubules.

D. Defecation occurs when the anal sphincters relax and contraction of other muscles raises the rectal pressure.

18.5 Liver, Gallbladder, and Pancreas 635

A. The liver, the largest internal organ, is composed of functional units called lobules.
   1. Liver lobules consist of plates of hepatic cells separated by capillary sinusoids.
   2. Blood flows from the periphery of each lobule, where the hepatic artery and portal vein empty, through the sinusoids and out the central vein.
   3. Bile flows within the hepatocyte plates, in canaliculi, to the bile ducts.
   4. Substances excreted in the bile can be returned to the liver in the hepatic portal blood. This is called an enterohepatic circulation.
   5. Bile consists of a pigment called bilirubin, bile salts, cholesterol, and other molecules.
   6. The liver detoxifies the blood by excreting substances in the bile, by phagocytosis, and by chemical inactivation.
   7. The liver modifies the plasma concentrations of proteins, glucose, triglycerides, and ketone bodies.

B. The gallbladder stores and concentrates the bile. It releases bile through the cystic duct and common bile duct to the duodenum.

C. The pancreas is both an exocrine and an endocrine gland.
   1. The endocrine portion, known as the islets of Langerhans, secretes the hormones insulin and glucagon.
   2. The exocrine acini of the pancreas produce pancreatic juice, which contains various digestive enzymes and bicarbonate.

18.6 Regulation of the Digestive System 644

A. The regulation of gastric function occurs in three phases.
   1. In the cephalic phase, the activity of higher brain centers, acting via the vagus nerves, stimulates gastric juice secretion.
   2. In the gastric phase, the secretion of HCl and pepsin is controlled by the gastric contents and by the hormone gastrin, secreted by the gastric mucosa.
   3. In the intestinal phase, the activity of the stomach is inhibited by neural reflexes and hormonal secretion from the duodenum.

B. Intestinal function is regulated, at least in part, by short, local reflexes coordinated by the enteric nervous system.
   1. The enteric nervous system contains interneurons, intrinsic sensory neurons, and autonomic motor neurons.
   2. Peristalis is coordinated by the enteric nervous system, which produces smooth muscle contraction above the bolus and relaxation below the bolus of chyme.
   3. Short reflexes include the gastroileal reflex, ileogastric reflex, and intestino-intestinal reflexes.

C. The secretion of the hormones secretin and cholecystokinin (CCK) regulates pancreatic juice and bile secretion.
   1. The secretion of secretin is stimulated by the arrival of acidic chyme into the duodenum.
   2. CCK secretion is stimulated by the presence of fat in the chyme arriving in the duodenum.
   3. Contraction of the gallbladder occurs in response to a neural reflex and to the secretion of CCK by the duodenum.

D. Gastrointestinal hormones may be needed for the maintenance of the GI tract and accessory digestive organs.

18.7 Digestion and Absorption of Food 649

A. The digestion of starch begins in the mouth through the action of salivary amylase.
   1. Pancreatic amylase digests starch into disaccharides and short-chain oligosaccharides.
   2. Complete digestion into monosaccharides is accomplished by brush border enzymes.

B. Protein digestion begins in the stomach through the action of pepsin.
   1. Pancreatic juice contains the protein-digesting enzymes trypsin and chymotrypsin, among others.
   2. The brush border contains digestive enzymes that help to complete the digestion of proteins into amino acids.
   3. Amino acids, like monosaccharides, are absorbed and secreted into capillary blood entering the portal vein.

C. Lipids are digested in the small intestine after being emulsified by bile salts.
   1. Free fatty acids and monoglycerides enter particles called micelles, formed in large part by bile salts, and they are absorbed in this form or as free molecules.
   2. Once inside the mucosal epithelial cells, these subunits are used to resynthesize triglycerides.
   3. Triglycerides in the epithelial cells, together with proteins, form chylomicrons, which are secreted into the central lacteals of the villi.
   4. Chylomicrons are transported by lymph to the thoracic duct and from there enter the blood.

REVIEW ACTIVITIES

Test Your Knowledge

1. Which of these statements about intrinsic factor is *true?*
   a. It is secreted by the stomach.
   b. It is a polypeptide.
   c. It promotes absorption of vitamin B_{12} in the intestine.
   d. It helps prevent pernicious anemia.
   e. All of these are true.
2. Intestinal enzymes such as lactase are
   a. secreted by the intestine into the chyme.
   b. produced by the intestinal crypts (of Lieberkühn).
   c. produced by the pancreas.
   d. attached to the cell membrane of microvilli in the epithelial cells of the mucosa.
3. Which of these statements about gastric secretion of HCl is *false?*
   a. HCl is secreted by parietal cells.
   b. HCl hydrolyzes peptide bonds.
   c. HCl is needed for the conversion of pepsinogen to pepsin.
   d. HCl is needed for maximum activity of pepsin.
4. Most digestion occurs in
   a. the mouth.
   b. the stomach.
   c. the small intestine.
   d. the large intestine.
5. Which of these statements about trypsin is *true?*
   a. Trypsin is derived from trypsinogen by the digestive action of pepsin.
   b. Active trypsin is secreted into the pancreatic acini.
   c. Trypsin is produced in the crypts (of Lieberkühn).
   d. Trypsinogen is converted to trypsin by the brush border enzyme enterokinase.
6. During the gastric phase, the secretion of HCl and pepsinogen is stimulated by
   a. vagus nerve stimulation that originates in the brain.
   b. polypeptides in the gastric lumen and by gastrin secretion.
   c. secretin and cholecystokinin from the duodenum.
   d. all of these.
7. The secretion of HCl by the stomach mucosa is inhibited by
   a. neural reflexes from the duodenum.
   b. the secretion of an enterogastrone from the duodenum.
   c. the lowering of gastric pH.
   d. all of these.
8. The first organ to receive the bloodborne products of digestion is
   a. the liver.
   b. the pancreas.
   c. the heart.
   d. the brain.
9. Which of these statements about hepatic portal blood is *true?*
   a. It contains absorbed fat.
   b. It contains ingested proteins.
   c. It is mixed with bile in the liver.
   d. It is mixed with blood from the hepatic artery in the liver.
10. Absorption of salt and water is the principal function of which region of the GI tract?
    a. Esophagus
    b. Stomach
    c. Duodenum
    d. Jejunum
    e. Large intestine
11. Cholecystokinin (CCK) is a hormone that stimulates
    a. bile production.
    b. release of pancreatic enzymes.
    c. contraction of the gallbladder.
    d. both *a* and *b.*
    e. both *b* and *c.*
12. Which of these statements about vitamin B_{12} is *false?*
    a. Lack of this vitamin can produce pernicious anemia.
    b. Intrinsic factor is needed for absorption of vitamin B_{12}.
    c. Damage to the gastric mucosa may lead to a deficiency in vitamin B_{12}.
    d. Vitamin B_{12} is absorbed primarily in the jejunum.
13. Which of these statements about starch digestion is *false?*
    a. It begins in the mouth.
    b. It occurs in the stomach.
    c. It requires the action of pancreatic amylase.
    d. It requires brush border enzymes for completion.
14. Which of these statements about fat digestion and absorption is *false?*
    a. Emulsification by bile salts increases the rate of fat digestion.
    b. Triglycerides are hydrolyzed by the action of pancreatic lipase.
    c. Triglycerides are resynthesized from monoglycerides and fatty acids in the intestinal epithelial cells.
    d. Triglycerides, as particles called chylomicrons, are absorbed into blood capillaries within the villi.
15. Which of these statements about contraction of intestinal smooth muscle is *true?*
    a. It occurs automatically.
    b. It is increased by parasympathetic nerve stimulation.
    c. It produces segmentation.
    d. All of these are true.

16. Which of the following hormones stimulates the secretion of histamine from the ECL cells?
    a. Enterogastrone
    b. Gastrin
    c. Secretin
    d. Cholecystokinin
17. Which of the following food types is most effective at provoking the intestinal phase of gastric regulation?
    a. Fat
    b. Starch
    c. Sugars
    d. Proteins
    e. Amino acids
18. Which of the following statements about the liver is *false?*
    a. Its cells cannot regenerate when damaged.
    b. It is the first organ to receive food molecules from the intestine.
    c. Its cells are in direct contact with the blood.
    d. Arterial and venous bloods mix as they travel through the sinusoids.
    e. The liver produces the bile acids and the bile pigment.

Test Your Understanding

19. Explain how the gastric secretion of HCl and pepsin is regulated during the cephalic, gastric, and intestinal phases.
20. Describe how pancreatic enzymes become activated in the lumen of the intestine. Why are these mechanisms needed?
21. Explain the function of bicarbonate in pancreatic juice. How may peptic ulcers in the duodenum be produced?
22. Describe the mechanisms that are believed to protect the gastric mucosa from self-digestion. What factors might be responsible for the development of a peptic ulcer in the stomach?
23. Explain why the pancreas is considered both an exocrine and an endocrine gland. Given this information, predict what effects tying of the pancreatic duct would have on pancreatic structure and function.
24. Explain how jaundice is produced when (a) the person has gallstones, (b) the person has a high rate of red blood cell destruction, and (c) the person has liver disease. In which case(s) would phototherapy for the jaundice be effective? Explain.
25. Describe the steps involved in the digestion and absorption of fat.
26. Distinguish between chylomicrons, very-low-density lipoproteins, low-density lipoproteins, and high-density lipoproteins.
27. Identify the different neurons present in the wall of the intestine and explain how these neurons are involved in "short reflexes." Why is the enteric nervous system sometimes described as an "enteric brain?"
28. Trace the course of blood flow through the liver and discuss the significance of this pattern in terms of the detoxication of the blood. Describe the enzymes and the reactions involved in this detoxication.
29. Drugs taken to treat gastritis and gastroesophageal reflux include proton pump inhibitors, H_2 histamine receptor blockers, and buffers. Give examples of each type of drug and explain how they may help.
30. Describe the reflexes controlling the stomach, liver, gallbladder, and pancreas that are triggered by the arrival of chyme in the duodenum.
31. Describe the intestinal microbiota, their location, and the possible benefits they may confer.

Test Your Analytical Ability

32. Which surgery do you think would have the most profound effect on digestion: (a) removal of the stomach (gastrectomy), (b) removal of the pancreas (pancreatectomy), or (c) removal of the gallbladder (cholecystectomy)? Explain your reasoning.
33. Describe the adaptations of the GI tract that make it more efficient by either increasing the surface area for absorption or increasing the contact between food particles and digestive enzymes.
34. Discuss how the ECL cells of the gastric mucosa function as a final common pathway for the neural, endocrine, and paracrine regulation of gastric acid secretion. What does this imply about the effectiveness of drug intervention to block excessive acid secretion?
35. Bacterial heat-stable enterotoxins can cause a type of diarrhea by stimulating the enzyme guanylate cyclase, which raises cyclic GMP levels within intestinal cells. Why might this be considered an example of mimicry? How does it cause diarrhea?
36. The hormone insulin is secreted by the pancreatic islets in response to a rise in blood glucose concentration. Surprisingly, however, the insulin secretion is greater in response to oral glucose than to intravenous glucose. Explain why this is so.
37. The bacteria that are part of the intestinal microbiota are usually described as commensal bacteria. What does this mean? Present arguments that the relationship is more an example of mutualism.
38. A drug swallowed as a pill or capsule may not make it into the general circulation in sufficient amounts to be effective. Explain different mechanisms that may account for this observation. Why might a drug delivered by a skin patch or nasal spray be more effective than the same drug taken orally?
39. Explain the relationship between blood and the liver's detoxication enzymes, and the possible relationship between the liver's detoxication enzymes and the American Revolution.

Test Your Quantitative Ability

Refer to figure 18.14 to answer the following questions.

40. What is the frequency of slow-wave production per minute indicated in this figure?
41. About how long does each slow wave last (what is its duration)?
42. Each slow wave and action potential represents a total depolarization of how many millivolts (mV)?
43. Each slow wave has a depolarization (not counting the action potential) of about how many millivolts (mV)?

CHAPTER

19

Regulation of Metabolism

REFRESH YOUR MEMORY

Before you begin this chapter, you may want to review these concepts from previous chapters:

CHAPTER OUTLINE

Clinical Investigation

Lisa is a moderately overweight 44-year-old who went to a physician complaining of nausea, headaches, frequent urination, and continuous thirst. She reported that both her mother and uncle are diabetics. Lisa's urine sample did not show evidence of glycosuria. She returned the next day to provide a fasting blood sample, and subsequently took an oral glucose tolerance test. The physician advised a weight reduction program and mild exercise, and mentioned that if these were not effective he would prescribe pills.

The new terms and concepts you will encounter in this chapter include:

- Insulin sensitivity and the metabolic syndrome
- Insulin resistance and impaired glucose tolerance

19.1 NUTRITIONAL REQUIREMENTS

The body's energy requirements must be met by the caloric value of food to prevent catabolism of the body's own fat, carbohydrates, and protein. Vitamins and minerals do not directly provide energy but instead are required for diverse enzymatic reactions.

LEARNING OUTCOMES

After studying this section, you should be able to:

1. Describe how various conditions affect the metabolic rate.
2. Describe the caloric and anabolic requirements of the diet and the functions of specific vitamins.
3. Identify the nature and significance of free radicals.

Living tissue is maintained by the constant expenditure of energy. This energy is obtained directly from ATP and indirectly from the cellular respiration of glucose, fatty acids, ketone bodies, amino acids, and other organic molecules. These molecules are ultimately obtained from food, but they can also be obtained from the glycogen, fat, and protein stored in the body.

The energy value of food is commonly measured in **kilocalories,** which are also called "big calories" and spelled with a capital letter (Calories). One kilocalorie (kcal) is equal to 1,000 calories; one calorie is defined as the amount of heat required to raise the temperature of one cubic centimeter of water from 14.5° to 15.5° C. The amount of energy released as heat when a quantity of food is combusted *in vitro* is equal to the amount of energy released within cells through the process of aerobic respiration (chapter 4; see fig. 4.13). This is 4 kilocalories per gram for carbohydrates or proteins and 9 kilocalories per gram for fat. When this energy is released by cell respiration, some is transferred to the high-energy bonds of ATP and some is lost as heat.

Metabolic Rate and Caloric Requirements

The total rate of body metabolism, or the **metabolic rate,** can be measured by either the amount of heat generated by the body or the amount of oxygen consumed by the body per minute. This rate is influenced by a variety of factors, discussed in section 19.2 (see fig. 19.5). For example, the metabolic rate is increased by physical activity and by eating. The increased rate of metabolism that accompanies the assimilation of food can last more than six hours after a meal.

Body temperature is also an important factor in determining metabolic rate. The reasons for this are twofold: (1) temperature itself influences the rate of chemical reactions, and (2) the hypothalamus contains *temperature control centers,* as well as temperature-sensitive cells that act as sensors for changes in body temperature. In response to deviations from a "set point" for body temperature (chapter 1), the control areas of the hypothalamus can direct physiological responses that help correct the deviations and maintain a constant body temperature. Changes in body temperature are thus accompanied by physiological responses that influence the total metabolic rate.

CLINICAL APPLICATION

Hypothermia (low body temperature)—where the core body temperature is lowered to between 26° and 32.5° C (78° and 90° F)—is often induced during open-heart or brain surgery. Compensatory responses to the lowered temperature are dampened by the general anesthetic, and the lower body temperature drastically reduces the needs of the tissues for oxygen. Under these conditions, the heart can be stopped and bleeding is significantly reduced.

The metabolic rate (measured by the rate of oxygen consumption) of an awake, relaxed person 12 to 14 hours after eating and at a comfortable temperature is known as the **basal metabolic rate (BMR).** The BMR is determined primarily by a person's age, sex, and body surface area, but it is also strongly influenced by the level of thyroid hormone secretion. A person with hyperthyroidism has an abnormally high BMR, and a person with hypothyroidism has a low BMR. The BMR may be influenced by genetic inheritance; at least some families prone to obesity may have a genetically determined low BMR.

In general, however, individual differences in energy requirements are due primarily to differences in physical activity. Daily energy expenditures may range from 1,300 to 5,000 kilocalories per day. The average values for people not engaged in heavy manual labor but active during their leisure are about 2,900 kilocalories per day for men and 2,100 kilocalories per day for women. People engaged in office work, the professions, sales, and comparable occupations consume up to 5 kilocalories per minute during work. More physically demanding occupations may require energy expenditures of 7.5 to 10 kilocalories per minute.

When the caloric intake is greater than the energy expenditures (a *positive energy balance*), excess calories are stored primarily as fat. This is true regardless of the source of the calories—carbohydrates, protein, or fat—because these molecules can be converted to fat by the metabolic pathways described in chapter 5 (see fig. 5.18).

Weight is lost when the caloric value of the food is less than the amount of calories required by cell respiration over a period of time (when there is a *negative energy balance*). Weight loss, therefore, can be achieved by dieting alone or in combination with an exercise program to raise the metabolic rate. A summary of the caloric expenditure associated with different forms of exercise is provided in table 19.1. Recent experiments, however, demonstrate why it is often so difficult to lose (or gain) weight. When subjects were maintained at 10% less than their usual weight, their metabolic rate decreased; when they were maintained at 10% greater than their usual body weight, their metabolic rate increased. The body, it seems, defends its usual weight by altering the energy expenditure as well as by regulating the food intake.

Anabolic Requirements

In addition to providing the body with energy, food also supplies the raw materials for synthesis reactions—collectively termed **anabolism**—that occur constantly within the cells of the body. Anabolic reactions include those that synthesize DNA and RNA, protein, glycogen, triglycerides, and other polymers. These anabolic reactions must occur constantly to replace the molecules that are hydrolyzed into their component monomers. These hydrolysis reactions, together with the reactions of cell respiration that ultimately break the monomers down to carbon dioxide and water, are collectively termed **catabolism.**

Acting through changes in hormonal secretion, exercise and fasting increase the catabolism of stored glycogen, fat, and body protein. These molecules are also broken down at a certain rate in a person who is neither exercising nor fasting. Some of the monomers thus formed (amino acids, glucose, and fatty acids) are used immediately to resynthesize body protein, glycogen, and fat. However, some of the glucose derived from stored glycogen, and fatty acids derived from stored triglycerides, are used to provide energy in the process of cell respiration. For this reason, new monomers must be obtained from food to prevent a continual decline in the amount of protein, glycogen, and fat in the body.

The *turnover rate* of a particular molecule is the rate at which it is broken down and resynthesized. For example, the average daily turnover for protein is 150 g/day, but because many of the amino acids derived from the catabolism of body proteins can be reused in protein synthesis, a person needs only about 35 g/day of protein in the diet. It should be noted that these are average figures and will vary in accordance with individual differences in size, sex, age, genetics, and physical activity. The average daily turnover for fat is about 100 g/day, but very little is required in the diet (other than that which supplies fat-soluble vitamins and essential fatty acids), since fat can be produced from excess carbohydrates.

The minimal amounts of dietary protein and fat required to meet the turnover rate are adequate only if they supply sufficient amounts of the essential amino acids and fatty acids. These molecules are termed *essential* because they cannot be synthesized by the body and must be obtained in the diet. The nine **essential amino acids** are lysine, tryptophan, phenylalanine, threonine, valine, methionine, leucine, isoleucine, and histidine. The **essential fatty acids** are linoleic acid and alpha-linolenic acid.

Unsaturated fatty acids—those with double bonds between the carbons—are characterized by the location of the first double bond. Linoleic acid, found in corn oil, contains 18 carbons and two double bonds. It has its first double bond on the sixth carbon from the methyl (CH_3) end, and is therefore designated as an n-6 (or omega-6) fatty acid. Alpha-linolenic acid also has 18 carbons, but it has three double bonds. More significantly for health, its first double bond is on the third carbon from the methyl end. Fatty acids that have their first double bond on the third carbon from the methyl end are described as **omega-3** (or *n-3*) **fatty acids.**

Mammals lack the enzymes needed to insert the double bond at the n-6 or n-3 positions and thus must eat linoleic and alpha-linolenic acids in their diets. Most people obtain alpha-linolenic acid derivatives (n-3 fatty acids known as EPA and DHA, described in the following Fitness Application box) primarily from fish. Studies suggest that omega-3 fatty acids may offer some protection against cardiovascular disease.

FITNESS APPLICATION

Eskimos (Inuits) who eat a traditional diet of fish and seal and whale meat have a surprisingly low blood concentration of triglycerides and cholesterol, and a low incidence of ischemic heart disease, despite the high fat and cholesterol content of their food. Several studies suggest that the omega-3 fatty acids of the cold-water fish are the source of the apparent protective effect. The omega-3 fatty acids of fish include *eicosapentaenoic acid,* or *EPA* (with 20 carbons), and *docosahexaenoic acid,* or *DHA* (with 22 carbons). The omega-3 fatty acids may help inhibit platelet function in thrombus formation, the progression of atherosclerosis, and/or ventricular arrhythmias. On the basis of the available evidence, the American Heart Association recommends that people eat at least two meals of fish—particularly oily fish (salmon, sardines, trout, mackerel, and herring)—every week.

Vitamins and Minerals

Vitamins are small organic molecules that serve as coenzymes in metabolic reactions or that have other highly specific functions. They must be obtained in the diet because the body either doesn't produce them, or produces them in insufficient quantities. (Vitamin D is produced in limited quantities by the skin, and the B vitamins and vitamin K are produced by intestinal bacteria.) There are two classes of vitamins: fat-soluble and water-soluble. The **fat-soluble vitamins** include vitamins A, D, E, and K. The **water-soluble vitamins** include thiamine (B_1), riboflavin (B_2), niacin (B_3), pyridoxine (B_6), pantothenic

Table 19.1 | Energy Consumed (in Kilocalories per Minute) in Different Types of Activities

	Weight in Pounds			
Activity	**105–115**	**127–137**	**160–170**	**182–192**
Bicycling				
10 mph	5.41	6.16	7.33	7.91
Stationary, 10 mph	5.50	6.25	7.41	8.16
Calisthenics	3.91	4.50	7.33	7.91
Dancing				
Aerobic	5.83	6.58	7.83	8.58
Square	5.50	6.25	7.41	8.00
Gardening, Weeding, and Digging	5.08	5.75	6.83	7.50
Jogging				
5.5 mph	8.58	9.75	11.50	12.66
6.5 mph	8.90	10.20	12.00	13.20
8.0 mph	10.40	11.90	14.10	15.50
9.0 mph	12.00	13.80	16.20	17.80
Rowing, Machine				
Easily	3.91	4.50	5.25	5.83
Vigorously	8.58	9.75	11.50	12.66
Skiing				
Downhill	7.75	8.83	10.41	11.50
Cross-country, 5 mph	9.16	10.41	12.25	13.33
Cross-country, 9 mph	13.08	14.83	17.58	19.33
Swimming, Crawl				
20 yards per minute	3.91	4.50	5.25	5.83
40 yards per minute	7.83	8.91	10.50	11.58
55 yards per minute	11.00	12.50	14.75	16.25
Walking				
2 mph	2.40	2.80	3.30	3.60
3 mph	3.90	4.50	6.30	6.80
4 mph	4.50	5.20	6.10	6.80

acid, biotin, folic acid, vitamin B_{12}, and vitamin C (ascorbic acid). Recommended dietary allowances for these vitamins are listed in table 19.2.

Water-Soluble Vitamins

Derivatives of water-soluble vitamins serve as coenzymes in the metabolism of carbohydrates, lipids, and proteins. **Thiamine,** for example, is needed for the activity of the enzyme that converts pyruvic acid to acetyl coenzyme A. **Riboflavin** and **niacin** are needed for the production of FAD and NAD, respectively. FAD and NAD serve as coenzymes that transfer hydrogens during cell respiration (chapter 4; see fig. 4.17). **Pyridoxine** is a cofactor for the enzymes involved in amino acid metabolism. Deficiencies of the water-soluble vitamins can thus have widespread effects in the body (table 19.3).

Table 19.2 | Recommended Dietary Allowances for Vitamins and Minerals[1]

Category	Age (Years) or Condition	Weight[2] (kg)	Weight[2] (lb)	Height[2] (cm)	Height[2] (in)	Protein (g)	Fat-Soluble Vitamins: Vitamin A (μg RE)[3]	Vitamin D (μg)[4]	Vitamin E (mg α-TE)[5]	Vitamin K (μg)
Infants	0.0–0.05	6	13	60	24	13	375	7.5	3	5
	0.5–1	9	20	71	28	14	375	10	4	10
Children	1–3	13	29	90	35	16	400	10	6	15
	4–6	20	44	112	44	24	500	10	7	20
	7–10	28	62	132	52	28	700	10	7	30
Males	11–14	45	99	157	62	45	1,000	10	10	45
	15–18	66	145	176	69	59	1,000	10	10	65
	19–24	72	160	177	70	58	1,000	10	10	70
	25–50	79	174	176	70	63	1,000	5	10	80
	51+	77	170	173	68	63	1,000	5	10	80
Females	11–14	46	101	157	62	45	800	10	8	45
	15–18	55	120	163	64	44	800	10	8	55
	19–24	58	128	164	65	46	800	10	8	60
	25–50	63	138	163	64	50	800	5	8	65
	51+	65	143	160	63	50	800	5	8	65
Pregnant						60	800	10	10	65
Lactating	1st 6 months					65	1,300	10	12	65
	2nd 6 months					62	1,200	10	11	65

[1]The allowances, expressed as average daily intakes over time, are intended to provide for individual variations among most normal persons as they live in the United States under usual environmental stresses. Diets should be based on a variety of common foods in order to provide other nutrients for which human requirements have been less well defined.

[2]Weights and heights of Reference Adults are actual medians for the U.S. population of the designated age, as reported by NHANES II. The use of these figures does not imply that the height-to-weight ratios are ideal.

[3]Retinol equivalents. 1 RE = 1 μg retinol or 6 μg β-carotene.

[4]As cholecalciferol. 10 μg cholecalciferol = 400 IU of vitamin D.

[5]α-tocopherol equivalents. 1 mg d-α-tocopherol = 1 α-TE.

Source: Reprinted with permission from *Recommended Dietary Allowances,* 10th Edition. Copyright 1989 by the National Academy of Sciences. Courtesy of the National Academy Press, Washington, D.C.

CLINICAL APPLICATION

Chronic alcohol abuse is the leading cause of thiamine (vitamin B_1) deficiency, which can produce the cardiovascular and neural impairments of the disease **beriberi.** Alcoholism with thiamine deficiency can also cause cognitive deficits. In severe form, this becomes a disorder characterized by brain damage and psychosis known as **Wernicke-Korsakoff syndrome,** or *wet brain.*

Free radicals are highly reactive molecules that carry an unpaired electron. They can damage tissues by removing an electron from, and thus oxidizing, other molecules. **Vitamin C** (a water-soluble vitamin) and vitamin E (a fat-soluble vitamin) function as *antioxidants* through their ability to inactivate free radicals. These vitamins may afford protection against some of the diseases that may be caused by free radicals.

Water-Soluble Vitamins							Minerals						
Vitamin C (mg)	**Thiamine (mg)**	**Riboflavin (mg)**	**Niacin (mg NE)[6]**	**Vitamin B_6 (mg)**	**Folate (μg)**	**Vitamin B_{12} (μg)**	**Calcium (mg)**	**Phosphorus (mg)**	**Magnesium (mg)**	**Iron (mg)**	**Zinc (mg)**	**Iodine (μg)**	**Selenium (μg)**
30	0.3	0.4	5	0.3	25	0.3	400	300	40	6	5	40	10
35	0.4	0.5	6	0.6	35	.5	600	500	60	10	5	50	15
40	0.7	0.8	9	1.0	50	0.7	800	800	80	10	10	70	20
45	0.9	1.1	12	1.1	75	1.0	800	800	120	10	10	90	20
45	1.0	1.2	13	1.4	100	1.4	800	800	170	10	10	120	30
50	1.3	1.5	17	1.7	150	2.0	1,200	1,200	270	12	15	150	40
60	1.5	1.8	20	2.0	200	2.0	1,200	1,200	400	12	15	150	50
60	1.5	1.7	19	2.0	200	2.0	1,200	1,200	350	10	15	150	70
60	1.5	1.7	19	2.0	200	2.0	800	800	350	10	15	150	70
60	1.2	1.4	15	2.0	200	2.0	800	800	350	10	15	150	70
50	1.1	1.3	15	1.4	150	2.0	1,200	1,200	280	15	12	150	45
60	1.1	1.3	15	1.5	180	2.0	1,200	1,200	300	15	12	150	50
60	1.1	1.3	15	1.6	180	2.0	1,200	1,200	280	15	12	150	55
60	1.1	1.3	15	1.6	180	2.0	800	800	280	15	12	150	55
60	1.0	1.2	13	1.6	180	2.0	800	800	280	10	12	150	55
70	1.5	1.6	17	2.2	400	2.2	1,200	1,200	300	30	15	175	65
95	1.6	1.8	20	2.1	280	2.6	1,200	1,200	355	15	19	200	75
90	1.6	1.7	20	2.1	260	2.6	1,200	1,200	340	15	16	200	75

[6]Niacin equivalents. 1 NE = 1 mg of niacin or 60 mg of dietary tryptophan.

Table 19.3 | The Major Vitamins

Vitamin	Sources	Function	Deficiency Symptom(s)
A	Yellow vegetables and fruit	Constituent of visual pigment; strengthens epithelial membranes	Night blindness; dry skin
B_1 (Thiamine)	Liver, unrefined cereal grains	Cofactor for enzymes that catalyze decarboxylation	Beriberi; neuritis
B_2 (Riboflavin)	Liver, milk	Part of flavoproteins (such as FAD)	Glossitis; cheilosis
B_6 (Pyridoxine)	Liver, corn, wheat, and yeast	Coenzyme for decarboxylase and transaminase enzymes	Convulsions
B_{12} (Cyanocobalamin)	Liver, meat, eggs, milk	Coenzyme for amino acid metabolism; needed for erythropoiesis	Pernicious anemia
Biotin	Egg yolk, liver, tomatoes	Needed for fatty acid synthesis	Dermatitis; enteritis
C	Citrus fruits, green leafy vegetables	Needed for collagen synthesis in connective tissues	Scurvy
D	Fish, liver	Needed for intestinal absorption of calcium and phosphate	Rickets; osteomalacia
E	Milk, eggs, meat, leafy vegetables	Antioxidant	Muscular dystrophy (nonhereditary)
Folate	Green leafy vegetables	Needed for reactions that transfer one carbon	Sprue; anemia
K	Green leafy vegetables	Promotes reactions needed for function of clotting factors	Hemorrhage; inability to form clot
Niacin	Liver, meat, yeast	Part of NAD and NADP	Pellagra
Pantothenic acid	Liver, eggs, yeast	Part of coenzyme A	Dermatitis; enteritis; adrenal insufficiency

Fat-Soluble Vitamins

Vitamin E has important antioxidant functions, as will be described shortly. It also acts on the immune system to block the release of pro-inflammatory cytokines and thereby reduce damage caused by inflammation. Some fat-soluble vitamins have highly specialized functions. **Vitamin K,** for example, is required for the production of prothrombin and for clotting factors VII, IX, and X. Vitamins A and D also have functions unique to each, but these two vitamins overlap in their mechanisms of action.

Vitamin A is a collective term for a number of molecules that include *retinol* (the transport form of vitamin A), *retinal* (also known as retinaldehyde, used as the photopigment in the retina), and *retinoic acid.* Most of these molecules are ultimately derived from dietary β-carotene, present in such foods as carrots, leafy vegetables, and egg yolk. The β-carotene is converted by an enzyme in the intestine into two molecules of retinal. Most of the retinal is reduced to retinol, while some is oxidized to retinoic acid. It is the retinoic acid that binds to nuclear receptor proteins (see chapter 11; fig. 11.7) and directly produces the effects of vitamin A. Retinoic acid is involved, for example, in regulating embryonic development; vitamin A deficiency interferes with embryonic development, while excessive vitamin A during pregnancy can cause birth defects. Retinoic acid also regulates epithelial membrane structure and function, and has various effects on the immune system that include the induction of regulatory T lymphocytes (chapter 15, section 15.3). These actions may account for the effectiveness of retinoids in treating acne and other skin conditions.

Vitamin D is produced by the skin under the influence of ultraviolet light, but usually it is not produced in sufficient amounts for all of the body's needs. That is why we must eat food containing additional amounts of vitamin D, and why it is classified as a vitamin even though it can be produced by the body. The vitamin D secreted by the skin or consumed in the diet is inactive in its original form; it must first be converted into a derivative (1,25-dihydroxyvitamin D_3—see fig. 19.21) by enzymes in the liver and kidneys before it can be active in the body. Once the active derivative is produced, vitamin D helps regulate calcium balance, primarily by promoting the intestinal absorption of calcium.

The nuclear receptor protein for 1,25-dihydroxyvitamin D_3 cannot activate a gene unless it forms a dimer with the RXR receptor, which binds to the active form of retinoic acid (9-*cis* retinoic acid). This is similar to the way the thyroid hormone receptor (for triiodothyronine) must dimerize with the RXR receptor to activate its target genes (chapter 11, section 11.2;

see fig. 11.7). This overlapping of receptors may permit "cross-talk" between the actions of thyroid hormone, vitamin D, and vitamin A. In view of this, it is not surprising that thyroxine, vitamin A, and vitamin D have overlapping functions—all three are involved in regulating gene expression and promoting differentiation (specialization) of tissues.

CLINICAL APPLICATION

The best-known function of vitamin D is regulation of calcium balance. Because it promotes the intestinal absorption of Ca^{2+} and PO_4^{3-}, vitamin D is needed for proper calcification of the bones. However, a vitamin D derivative called *calcipotriene* is now widely used for the treatment of **psoriasis,** a skin condition characterized by inflammation and excessive proliferation of keratinocytes (the cells of the epidermis that produce keratin). In this case, the vitamin D analogue inhibits proliferation and promotes differentiation of the keratinocytes. The beneficial effects may also be promoted by the ability of 1,25-dihydroxyvitamin D_3 to suppress inflammation through various effects on the immune system, such as inducing the formation of regulatory T lymphocytes (T_{reg} cells; chapter 15, section 15.3)

Minerals (Elements)

Minerals (elements) are needed as cofactors for specific enzymes and for a wide variety of other critical functions. Those that are required daily in relatively large amounts include sodium, potassium, magnesium, calcium, phosphorus, and chlorine (see table 19.2). In addition, the following **trace elements** are recognized as essential: iron, zinc, manganese, fluorine, copper, molybdenum, chromium, and selenium. The trace elements are required in much lower daily doses than the previously mentioned elements (minerals; table 19.2).

Free Radicals and Antioxidants

The electrons in an atom are located in *orbitals,* with each orbital containing a maximum of two electrons. When an orbital has an unpaired electron, the molecule containing the unpaired electron is called a **free radical.** Free radicals are highly reactive in the body, oxidizing (removing an electron from) other atoms, or sometimes reducing (donating their electron to) other atoms. The major free radicals are referred to as **reactive oxygen species** if they contain oxygen with an unpaired electron, or **reactive nitrogen species** if they contain nitrogen with an unpaired electron. Mitochondria are a major source of reactive oxygen species, produced by the electron transport chain as a byproduct of aerobic respiration (chapter 5, section 5.2).

The unpaired electron is symbolized with a dot superscript. Thus, reactive oxygen species include the *superoxide radical* ($O^{\bullet}_2$), the *hydroxyl radical* ($HO^{\bullet}$), and others. Reactive nitrogen species include the *nitric oxide radical* ($NO^{\bullet}$) and others. These free radicals are produced by many cells in the body and serve some important physiological functions. The superoxide radical and nitric oxide radical produced in phagocytic cells such as neutrophils and macrophages, for example, help these cells destroy bacteria. The superoxide radicals in phagocytic cells can be thought of as nonselective antibiotics, killing any infecting bacteria (as well as the neutrophils) and perhaps also injuring surrounding tissue cells, as these radicals contribute to the inflammation reaction. In addition, the superoxide radicals promote cellular proliferation (mitotic division) of fibroblasts, so that scar tissue can form. The superoxide radicals have similarly been shown to stimulate proliferation of lymphocytes in the process of clone production (chapter 15). The nitric oxide radical also has physiological actions, promoting relaxation of vascular smooth muscle and thus vasodilation (chapter 14), so that more blood can flow to the site of the inflammation. Thus, free radicals do serve useful physiological roles in the body (fig. 19.1).

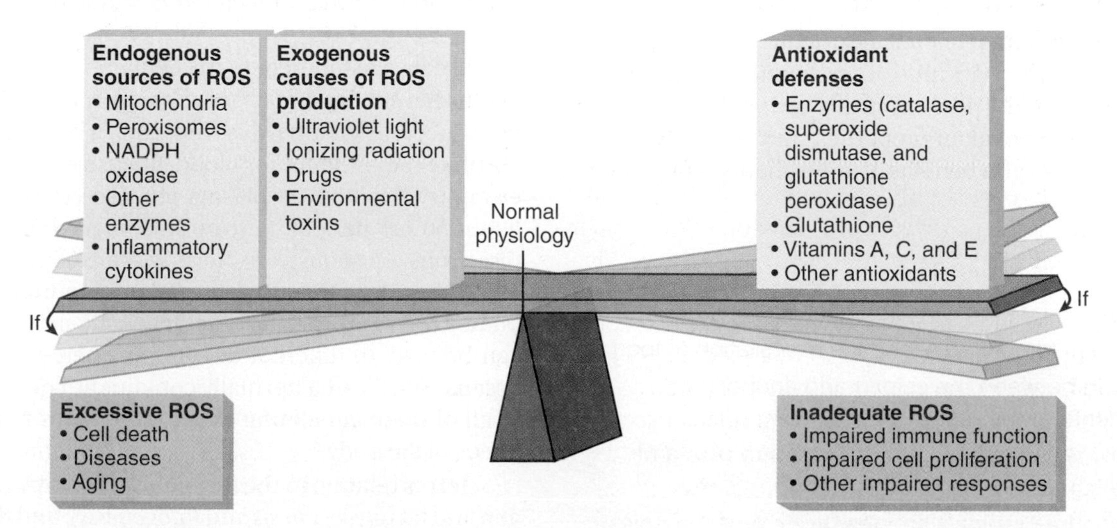

Figure 19.1 Reactive oxygen species (ROS) production and defense. Normal physiology requires that the reactive oxygen species (those that contain oxygen with an unpaired electron) be kept in balance.

Excessive production of free radicals, however, can damage lipids, proteins, and DNA, and by this means exert an **oxidative stress** on the body. Oxidative stress has wide-ranging ill effects (fig. 19.1). It promotes cell death (apoptosis), contributes to aging and degenerative diseases associated with aging, promotes the malignant growth of cancers, and contributes to all inflammatory diseases (such as glomerulonephritis, rheumatoid arthritis, and lupus erythematosus). It has been implicated in ischemic heart disease, stroke, hypertension, and a variety of neurological diseases, including multiple sclerosis, Alzheimer's disease, Parkinson's disease, and others. The wide range of diseases associated with oxidative stress stems from the widespread production of superoxide radicals in the mitochondria of all cells that undergo aerobic respiration.

The body protects itself against oxidative stress through various means, both enzymatic and nonenzymatic. The enzymes that help to prevent an excessive buildup of oxidants include *superoxide dismutase (SOD), catalase,* and *glutathione peroxidase.* The SOD enzyme catalyzes a reaction where two superoxide radicals form *hydrogen peroxide* (H_2O_2) and O_2. Hydrogen peroxide, though not a free radical, is considered a reactive oxygen species because it can generate the highly reactive hydroxyl radical. Elimination of hydrogen peroxide is accomplished by the enzyme *catalase.* In this reaction, two hydrogen peroxide molecules (H_2O_2) react to form H_2O and O_2. Also, the enzyme *glutathione peroxidase* catalyzes a reaction where H_2O_2 reacts with NADPH + H^+ to form NADP and H_2O.

The body also protects itself from oxidative stress through nonenzymatic means (fig. 19.1). One of the most important protective mechanisms is the action of a tripeptide called **glutathione.** When it is in its reduced state, glutathione can react with certain free radicals and render them harmless. Thus, glutathione is said to be the major cellular **antioxidant.** *Ascorbic acid* (vitamin C) in the aqueous phase of cells, and *α-tocopherol* (the major form of vitamin E) in the lipid phase, help in this antioxidant function by picking up unpaired electrons from free radicals. This is said to "quench" the free radicals, although in the reaction vitamins C and E themselves gain an unpaired electron and thus become free radicals. Because of their chemical structures, however, they are weaker free radicals than those they quench. Many other molecules present in foods (primarily fruits and vegetables) have been shown to have antioxidant properties, and research on the actions and potential health benefits of antioxidants is ongoing.

CHECKPOINT

1. Explain how the metabolic rate is influenced by exercise, ambient temperature, and the assimilation of food.

2a. Distinguish between the caloric and anabolic requirements of the diet.

2b. List the water-soluble and fat-soluble vitamins and describe some of their functions.

3. Identify the free radicals and reactive oxygen species, and describe their significance.

19.2 REGULATION OF ENERGY METABOLISM

The plasma contains circulating glucose, fatty acids, amino acids, and other molecules that can be used by the body tissues. The synthesis of energy reserves of glycogen and fat following a meal and the utilization of these reserves between meals are regulated by the action of hormones.

LEARNING OUTCOMES

After studying this section, you should be able to:

4. Identify the energy reserves and circulating energy substrates of the body.
5. Describe the functions of adipose tissue and the dangers of obesity.
6. Explain how different neurotransmitters and hormones regulate hunger.
7. Describe the different types of caloric expenditures and identify the hormones that regulate metabolic balance.

The molecules that can be oxidized for energy by cell respiration may be derived from the **energy reserves** of glycogen, fat, or protein. Glycogen and fat function primarily as energy reserves; for proteins, by contrast, this represents a secondary, emergency function. Although body protein can provide amino acids for energy, it can do so only through the breakdown of proteins needed for muscle contraction, structural strength, enzymatic activity, and other functions. Alternatively, the molecules used for cell respiration can be derived from the products of digestion that are absorbed through the small intestine. Since these molecules—glucose, fatty acids, amino acids, and others—are carried by the blood to the cells for use in cell respiration, they can be called **circulating energy substrates** (fig. 19.2).

Because of differences in cellular enzyme content, different organs have different *preferred energy sources.* This concept was introduced in chapter 5, section 5.3. The brain has an almost absolute requirement for blood glucose as its energy source, for example. A fall in the plasma glucose concentration to below about 50 mg per 100 ml can thus "starve" the brain and have disastrous consequences. Resting skeletal muscles, by contrast, use fatty acids as their preferred energy source. Similarly, ketone bodies (derived from fatty acids), lactic acid, and amino acids can be used to different degrees as energy sources by various organs. The plasma normally contains adequate concentrations of all of these circulating energy substrates to meet the energy needs of the body.

Terms relating to the formation of energy reserves of glycogen and fat (*glycogenesis* and *lipogenesis*), and the production of different circulating energy substrates (*glycogenolysis, lipolysis, gluconeogenesis,* and *ketogenesis*), are useful when describing

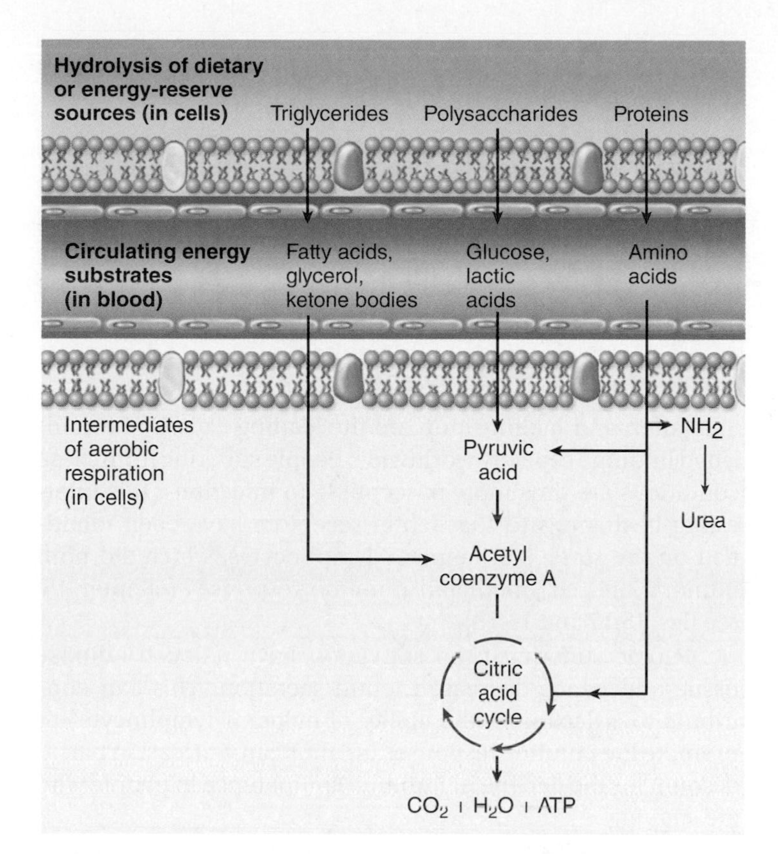

Figure 19.2 A flowchart of energy pathways in the body. The molecules indicated in the top and bottom are those found within cells, while the molecules in the middle portion of the figure are those that circulate in the blood.

metabolism. These terms and processes were introduced in chapter 5, and can be reviewed by referencing table 5.1. The hormonal regulation of metabolism is described in several sections of this chapter.

Regulatory Functions of Adipose Tissue

Many scientists believe that it is difficult for a person to lose (or gain) weight because the body has negative feedback loops that act to "defend" a particular body weight, or more accurately, the amount of adipose tissue. This regulatory system has been called an *adipostat.* When a person eats more than is needed to maintain the set point of adipose tissue, the person's metabolic rate increases and hunger decreases. Homeostasis of body weight implies negative feedback loops. Hunger and metabolism (acting through food and hormones) affect adipose cells, so in terms of negative feedback it seems logical that adipose cells should influence hunger and metabolism.

White adipose tissue (white fat) is the major site of energy storage in the body. In white fat **adipocytes** (fat cells), triglycerides are stored in a single, large droplet within each cell. The triglycerides are formed during times of plenty and are hydrolyzed (in a process termed *lipolysis*) by lipase enzymes to release free fatty acids and glycerol during times of fasting. Because the synthesis and breakdown of fat is controlled by hormones that act on the adipocytes, the adipocytes traditionally have been viewed simply as passive storage depots of fat. Recent evidence suggests quite the opposite, however; adipocytes may themselves secrete hormones that play a pivotal role in the regulation of hunger and metabolism.

Development of Adipose Tissue

Some adipocytes appear during embryonic development, but their numbers increase greatly following birth. This increased number is due to both mitotic division of the adipocytes and the conversion of preadipocytes (derived from fibroblasts) into new adipocytes. This differentiation (specialization) is promoted by a high circulating level of fatty acids, particularly of saturated fatty acids. This represents a nice example of a negative feedback loop, where a rise in circulating fatty acids promotes processes that ultimately help to convert the fatty acids into stored fat.

The differentiation of adipocytes requires the action of a nuclear receptor protein—in the same family as the receptors for thyroid hormone, vitamin A, and vitamin D—known as **PPARγ**. (*PPAR* is an acronym for *p*eroxisome *p*roliferator *a*ctivated *r*eceptor, and the γ is the Greek letter gamma, indicating the subtype of PPAR.) Just as the thyroid receptor is activated when it is bound to its ligand, PPARγ is activated when it is bound to its own specific ligand, a type of prostaglandin. When this ligand binds to the PPARγ receptor, it stimulates adipogenesis by promoting the development of preadipocytes into mature adipocytes. The PPARγ nuclear receptor protein is a transcription factor that acts as a "master switch" in the formation of adipocytes. This occurs primarily in children, because the development of new adipocytes is more limited in adults.

However, even in adults, the activation of PPARγ stimulates the formation of new adipocytes. This occurs predominately in the subcutaneous tissue, where small adipocytes are produced, rather than in the large adipocytes characteristics of deep visceral fat. This activation of PPARγ and the resulting formation of new subcutaneous adipocytes increases the sensitivity of the body to insulin, helping to lower blood glucose. How the activation of PPARγ receptors in adipocytes can produce increased sensitivity of the liver and skeletal muscles to insulin is not completely understood. However, part of the answer may involve some recently discovered hormones that are secreted by adipocytes.

Endocrine Functions of Adipose Tissue

Adipose tissue secretes regulatory molecules, collectively termed **adipokines,** which regulate hunger, metabolism, and insulin sensitivity. These include *adiponectin, leptin, resistin, tumor necrosis factor alpha,* and *retinol binding protein 4.* These adipokines are secreted into the blood and act on distant target organs, and so qualify as hormones.

In obesity, the enlarged adipocytes secrete an adipokine that attracts monocytes that can turn into macrophages. As a result, the macrophage content of the adipose tissue of an obese person can be as high as 50%. **Tumor necrosis factor-alpha (TNF_α)** is secreted by the macrophages and can reduce the ability of skeletal muscles to remove glucose from the blood in response to insulin (that is, promote *insulin resistance*). The secretion of TNF_α, as well as the release of free fatty acids from the adipose tissue (which also acts on skeletal muscles to promote insulin resistance), is increased in obesity and type 2 diabetes mellitus.

Three other adipokines that are known to be increased in obesity are **leptin, resistin,** and **retinol-binding protein 4.** There is evidence that these hormones also may contribute to the reduced sensitivity of skeletal muscles and other tissues to insulin in obesity and type 2 diabetes mellitus. By contrast, there is another adipocyte hormone—**adiponectin**—that is decreased in obesity and type 2 diabetes mellitus. Adiponectin stimulates glucose utilization and fatty acid oxidation in muscle cells. Through these actions, adiponectin has an insulin-sensitizing, antidiabetic effect.

The adipocyte hormone *leptin,* secreted in proportion to the amount of stored fat, has another important function. It acts on the hypothalamus to help regulate hunger and food intake. In other words, it acts as a signal from the adipose tissue to the brain, helping the body maintain a certain level of fat storage. This function is described in more detail shortly.

CLINICAL APPLICATION

Alcohol (ethanol) consumption can cause a **fatty liver.** The production of fat by hepatocytes results from an increased availability of fatty acids in the cell, suppressed oxidation (catabolism) of the fatty acids, and increased production of the enzymes that produce fat. Interestingly, chronic alcohol consumption in mice was found to significantly reduce *adiponectin* secretion from adipocytes. Because adiponectin stimulates fatty acid oxidation, suppressed adiponectin could increase fat synthesis in the liver, which contains adiponectin receptors and is a target of that hormone. In fact, the liver damage in these alcohol-consuming mice was largely reversed when they were given recombinant adiponectin. Because low adiponectin levels cause decreased fatty acid oxidation, low adiponectin levels are also associated with insulin resistance, type 2 diabetes, and atherosclerosis (discussed shortly).

Nonalcoholic fatty liver disease affects one-third of adults and is associated with obesity and insulin resistance. *Hepatic steatosis* is the accumulation of abnormal amounts of triglycerides in liver cells, a condition strongly correlated with insulin resistance; *steatohepatitis* refers to steatosis accompanied by inflammatory processes that injure liver cells. This can progress to *cirrhosis* and liver cancer (*hepatocellular carcinoma*) in some people, depending on genetic predisposition and other factors.

Clinical Investigation CLUES

Lisa was moderately overweight, and the physician recommended that she lose weight to help alleviate her symptoms.

- How does her weight relate to her insulin sensitivity?

Low Adiposity: Starvation

Starvation and malnutrition are the leading causes of diminished immune capacity worldwide. People suffering from these conditions are thus more susceptible to infections. It is interesting in this regard that leptin receptors have been identified on the surface of helper T lymphocytes, which aid both humoral and cell-mediated immune responses (chapter 15; see figs. 15.17 and 15.18).

People suffering from starvation have reduced adipose tissue and hence decreased leptin secretion. This can contribute to a decline in the ability of helper T lymphocytes to promote the immune response, and thus can—at least in part—account for the decline in immune competence in people who are starving.

The hypothalamus—a target of leptin action regulating appetite—is also involved in regulation of the reproductive system (chapter 20, section 20.2). There is evidence that leptin may be involved in regulating the onset of puberty and the menstrual cycle (*menarche*). Adolescent girls who are excessively thin enter puberty later than the average age, and very thin women can experience amenorrhea (cessation of menstrual cycles). Adequate amounts of adipose tissue are thus required for proper functioning of the immune and reproductive systems.

CLINICAL APPLICATION

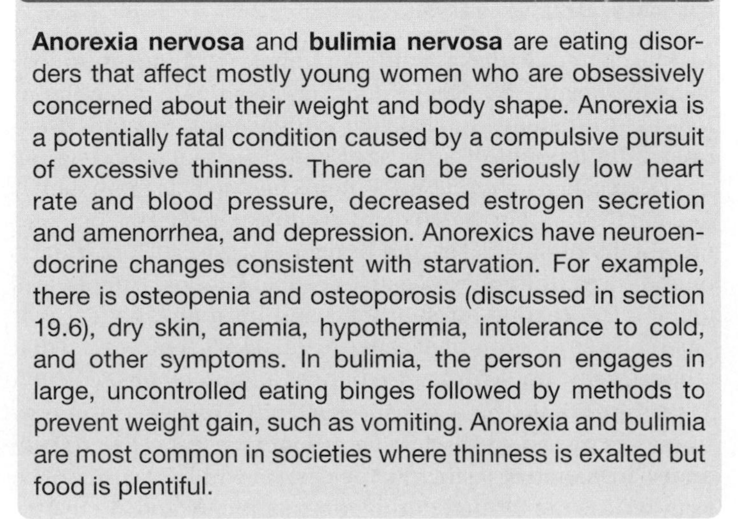

Anorexia nervosa and **bulimia nervosa** are eating disorders that affect mostly young women who are obsessively concerned about their weight and body shape. Anorexia is a potentially fatal condition caused by a compulsive pursuit of excessive thinness. There can be seriously low heart rate and blood pressure, decreased estrogen secretion and amenorrhea, and depression. Anorexics have neuroendocrine changes consistent with starvation. For example, there is osteopenia and osteoporosis (discussed in section 19.6), dry skin, anemia, hypothermia, intolerance to cold, and other symptoms. In bulimia, the person engages in large, uncontrolled eating binges followed by methods to prevent weight gain, such as vomiting. Anorexia and bulimia are most common in societies where thinness is exalted but food is plentiful.

Obesity

Obesity is a risk factor for cardiovascular diseases, diabetes mellitus, gallbladder disease, and some malignancies (particularly endometrial and breast cancer). The distribution of fat in the body is also important. White adipose tissue is divided into *visceral fat* (the fat in the mesenteries and greater omentum) and *subcutaneous fat.* There is a greater risk of cardiovascular disease when a high amount of visceral fat produces a high *waist-to-hip ratio,* or an "apple shape," as compared to a "pear shape." The larger adipocytes of visceral fat, producing the apple shape, are less sensitive to insulin than the smaller adipocytes of subcutaneous fat, and so pose a greater risk of diabetes mellitus.

The number of adipocytes increases through childhood and adolescence, but a recent study suggests that the number generally remains constant through adulthood (although there is about a 10% turnover, where new adipocytes are formed to replace those lost). Obesity in childhood results from an increase in the number of adipocytes as well as their size, whereas obesity in most adults is produced by an increase in the size of existing adipocytes. When weight is lost, the adipocytes get smaller but their number remains constant.

It is thus important to prevent further increases in weight in all overweight people, but particularly in children. Weight loss can be achieved by a calorie-restricted diet higher in proteins and lower in carbohydrates and saturated fat combined with a regular exercise program. Prolonged exercise of low to moderate intensity promotes weight loss because the skeletal muscles use fatty acids as their primary source of energy under those conditions (chapter 12; see fig. 12.22). However, once weight is lost it is often regained because of increased hunger and decreased energy expenditure by the body.

Obesity is often diagnosed using a measurement called the **body mass index (BMI).** This measurement is calculated using the following formula:

$$\text{BMI} = \frac{w}{h^2}$$

where

w = weight in kilograms (pounds divided by 2.2)
h = height in meters (inches divided by 39.4)

Obesity has been defined by health agencies in different ways. The World Health Organization classifies people with a BMI of 30 or over as being at high risk for the diseases of obesity. According to the standards set by the National Institutes of Health, a healthy weight is indicated by a BMI between 19 and 25. A BMI in the range of 25.0 to 29.9 is described as "overweight;" a BMI of over 30 is "obese." According to a recent study, however, the lowest death rates from all causes occurred in men with a BMI in the range of 22.5 to 24.9, and in women with a BMI in the range of 22.0 to 23.4. Recent surveys indicate that over 60% of adults in the United States are either overweight (with a BMI greater than 25) or obese (with a BMI greater than 30).

Obesity is strongly associated with type 2 diabetes mellitus, discussed later in this chapter. Childhood obesity and associated childhood type 2 diabetes mellitus have increased dramatically in recent years. For example, one study demonstrated a 36% increase in childhood obesity between 1988 and 1994; another showed a tenfold increase in the incidence of childhood type 2 diabetes mellitus between 1982 and 1994. According to a study done by the Rand Corporation, obesity is a greater risk factor in chronic diseases than either smoking or drinking!

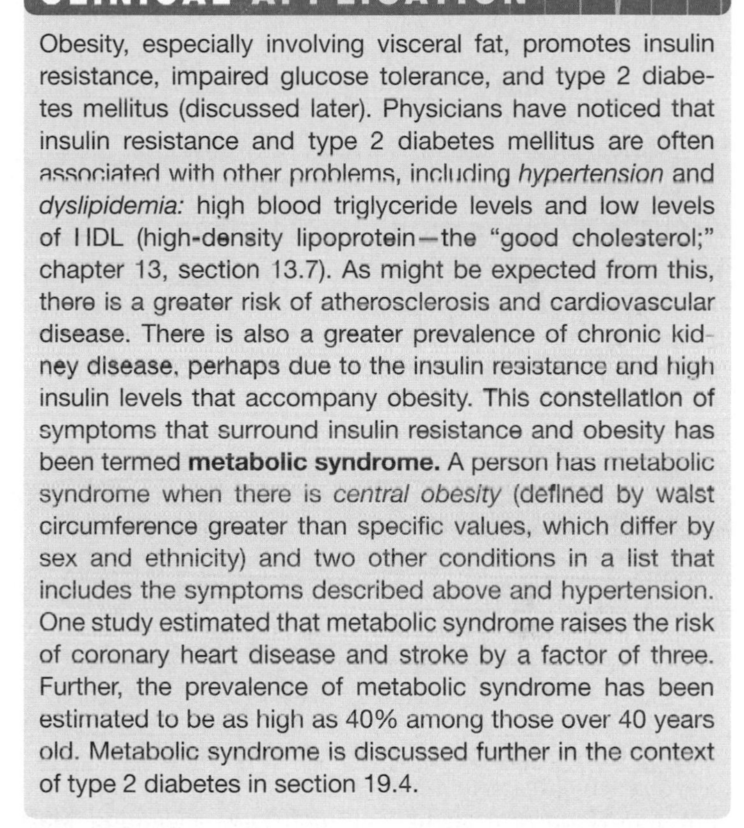

CLINICAL APPLICATION

Obesity, especially involving visceral fat, promotes insulin resistance, impaired glucose tolerance, and type 2 diabetes mellitus (discussed later). Physicians have noticed that insulin resistance and type 2 diabetes mellitus are often associated with other problems, including *hypertension* and *dyslipidemia:* high blood triglyceride levels and low levels of HDL (high-density lipoprotein—the "good cholesterol;" chapter 13, section 13.7). As might be expected from this, there is a greater risk of atherosclerosis and cardiovascular disease. There is also a greater prevalence of chronic kidney disease, perhaps due to the insulin resistance and high insulin levels that accompany obesity. This constellation of symptoms that surround insulin resistance and obesity has been termed **metabolic syndrome.** A person has metabolic syndrome when there is *central obesity* (defined by waist circumference greater than specific values, which differ by sex and ethnicity) and two other conditions in a list that includes the symptoms described above and hypertension. One study estimated that metabolic syndrome raises the risk of coronary heart disease and stroke by a factor of three. Further, the prevalence of metabolic syndrome has been estimated to be as high as 40% among those over 40 years old. Metabolic syndrome is discussed further in the context of type 2 diabetes in section 19.4.

Clinical Investigation CLUES

Lisa is moderately overweight.

- If she were obese, for which diseases would she be at an increased risk?

Regulation of Hunger and Metabolic Rate

Ideally, we should eat the kinds and amounts of foods that provide adequate vitamins, minerals, essential amino acids, and calories. Proper caloric intake maintains energy reserves (primarily fat and glycogen) and maintains body weight that is optimum for health.

Hunger and eating behavior are at least partially controlled by areas of the hypothalamus. Lesions (destruction) in the ventromedial area of the hypothalamus produce *hyperphagia,* or overeating, and obesity in experimental animals. Lesions of the lateral hypothalamus, by contrast, produce *hypophagia* and weight loss.

The neurotransmitters in the brain that may be involved in the control of eating are under investigation, and many have been implicated. Among those are the endorphins (naloxone, which blocks opioid receptors, suppresses overeating in rats); norepinephrine (injections into the brains of rats cause overeating); and serotonin (intracranial injections suppress overeating in rats). These results can be applied to humans. For example, the diet pills *Redux* (D-fen-fluramine) and *fen-phen* (L-fenfluramine) work to reduce hunger by elevating the brain levels of serotonin. However, these drugs have been removed from the market because of their association with heart valve problems.

Arcuate Nucleus of the Hypothalamus

Neurotransmitters from different brain regions, and hormonal regulators carried by the blood, are believed to regulate hunger by influencing a particular center in the hypothalamus known as the **arcuate nucleus.** Neurons from the arcuate nucleus, in turn, project to other brain areas. One of these is the *paraventricular nucleus* of the hypothalamus, an area associated with hunger (destruction of this nucleus causes overeating). Neurons in the arcuate nucleus also send axons to the *lateral hypothalamus,* an area known as a "feeding center;" axons from here project to the nucleus accumbens of the forebrain (chapter 8; see fig. 8.21), which is believed to be involved in the rewarding aspects of eating.

In terms of hunger regulation, the arcuate nucleus contains two groups of neurons. One group of neurons suppresses hunger. This group contains *POMC (proopiomelanocortin) neurons,* which secrete a large polypeptide that is converted into a *melanocortin* family of molecules including **MSH (melanocyte-stimulating hormone).** MSH acts to decrease appetite by binding to its *melanocortin receptors* in the hypothalamus. There is evidence that some cases of congenital obesity are caused by defects in one of these receptors, the melanocortin-4 receptor. In such cases MSH is unable to activate its receptors and suppress appetite. Conversely, cigarette smoking may reduce appetite because nicotine binds to nicotinic ACh receptors on POMC neurons, stimulating the release of MSH and the activation of melanocortin receptors in the hypothalamus.

Another group of neurons in the arcuate nucleus releases two neurotransmitters that promote hunger: **neuropeptide Y** and **AgRP (agouti-related protein).** Neuropeptide Y stimulates hunger directly; AgRP antagonizes MSH at the melanocortin receptor, thereby inhibiting the appetite-suppressing action of MSH. When eating results in an abundance of circulating energy substrates (fig. 19.2), the release of MSH produces a suppression of appetite and increased metabolic energy expenditure. At the same time, the neurons that release neuropeptide Y and AgRP (which would increase appetite) are inhibited. Conversely, when there is a reduction in circulating energy substrates, the neurons that release MSH are suppressed, while those that release neuropeptide Y and AgRP are activated to increase appetite (fig. 19.3).

The activation of these hypothalamic neurons and the release of their neurotransmitters must be regulated by signals from other brain areas (because psychological factors, and the smell and taste of food, influence hunger), and by signals from the body. In particular, hunger and appetite are responsive to signals from the digestive tract and from the adipose tissue. As illustrated in figure 19.4, hunger is stimulated by the hormone ghrelin and inhibited by the hormones PYY, CCK, insulin, and leptin.

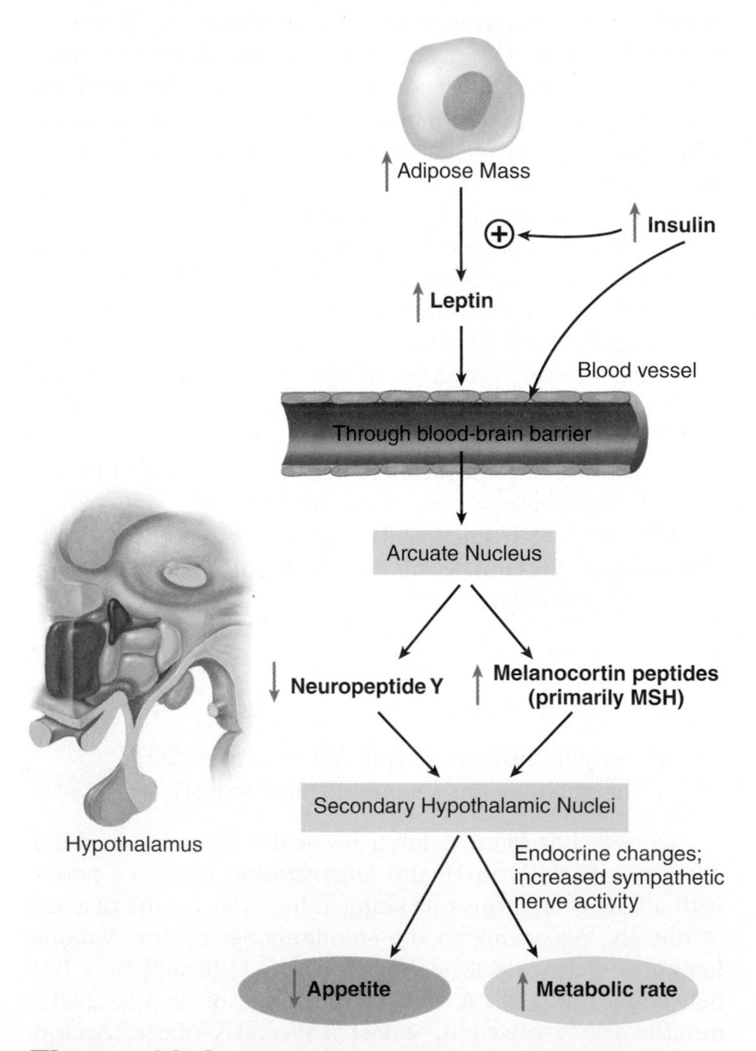

Figure 19.3 The action of leptin. Leptin crosses the blood-brain barrier to affect neurotransmitters released by neurons in the arcuate nucleus of the hypothalamus.This influences other hypothalamic nuclei, which in turn reduce appetite and increase metabolic rate. The figure also shows that insulin stimulates adipose cells to secrete leptin and is able to cross the blood-brain barrier and to act in a manner similar to leptin.

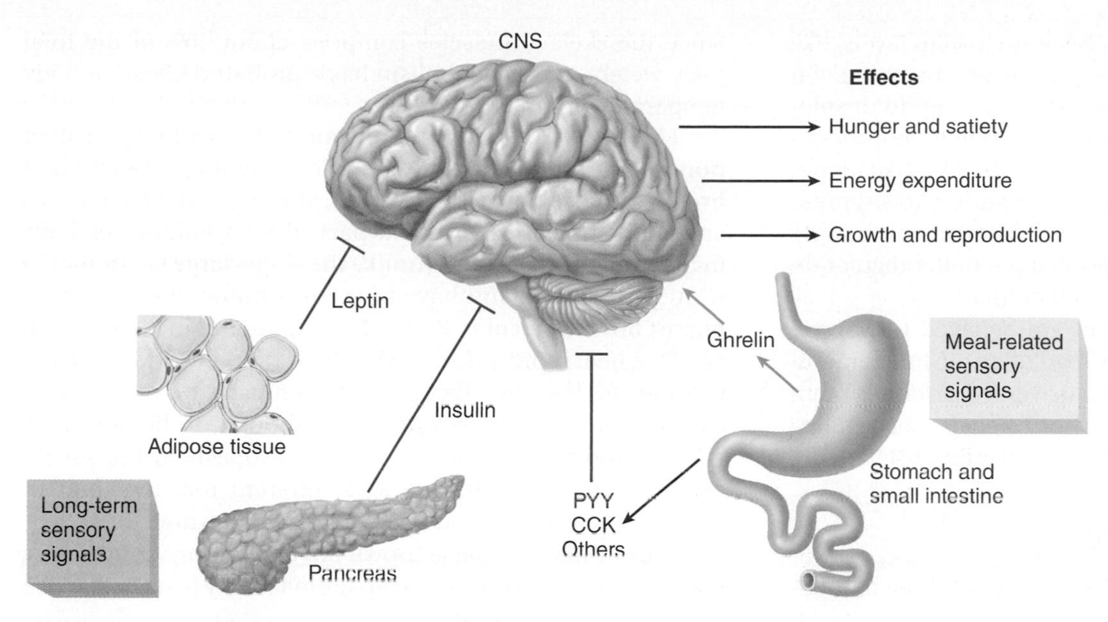

Figure 19.4 Hormonal signals that regulate feeding and energy expenditures. Ghrelin (green) is the only hormone that stimulates hunger; the other hormones (red) suppress appetite. The central nervous system (CNS) integrates this sensory information with other information (smell, taste, and psychological factors) to help regulate hunger and satiety, energy expenditures, as well as growth and reproduction.

Ghrelin, Cholecystokinin, and PYY

Signals from the stomach and small intestine regulate hunger and *satiety* (a feeling of "fullness, and thus a reduction of appetite) on a meal-to-meal basis. This signaling is performed by the secretion of several polypeptide hormones, which were earlier known to regulate digestive function by targeting different parts of the gastrointestinal tract. Later, they were shown to also regulate appetite by targeting the arcuate nucleus of the hypothalamus. Included among these are the polypeptides *ghrelin, cholecystokinin, polypeptide YY (PYY),* and several others.

Ghrelin is an important hunger signal secreted by the stomach. Ghrelin secretion rises between meals, when the stomach is empty, and stimulates hunger. This ghrelin effect results from stimulation in the arcuate nucleus of the neurons that release neuropeptide Y and AgRP. As the stomach fills during a meal, the secretion of ghrelin rapidly falls and hunger is thereby reduced. However, one study demonstrated raised levels of ghrelin in dieters who lost weight. If this raised ghrelin level enhances appetite, it may partially explain why it is so difficult for most dieters to maintain their weight loss.

Another hormone that regulates eating is the intestinal hormone **cholecystokinin (CCK).** Secretion of CCK rises during and immediately after a meal and suppresses hunger (promotes satiety). CCK thus acts antagonistically to ghrelin, helping to reduce appetite immediately after a meal.

Ghrelin and CCK are involved in the regulation of hunger on a short-term, meal-to-meal basis. Another hormone, **polypeptide YY (PYY)** secreted in greatest amounts by the ileum and colon, may reduce appetite on a more intermediate-time basis. The intestinal satiety hormones—including CCK, PYY, glucagon-like peptide-1 (GLP-1)—suppress hunger by acting in the arcuate nucleus of the hypothalamus to stimulate neurons that release MSH and inhibit neurons that release neuropeptide Y.

Leptin and Insulin

The areas of the hypothalamus that regulate hunger are also influenced by a circulating satiety factor (appetite suppressant) secreted by the adipose tissue. This satiety factor, a 167-amino-acid polypeptide, is the hormone *leptin.* Leptin is involved in more long-term hunger regulation than the hormones of the digestive tract. Leptin secretion increases as the amount of stored fat increases, suppressing appetite and thus reducing further calorie intake. Rather than regulating meal-to-meal food consumption like the hormones of the digestive tract, however, leptin helps maintain the body's usual level of *adiposity* (fat storage). In addition to suppressing appetite (promoting satiety), leptin increases the metabolic rate and calorie expenditure of the body (discussed in the next section).

Leptin secretion increases in proportion to the amount of stored fat and acts on the arcuate nucleus of the hypothalamus to suppress appetite. It does this by inhibiting neurons that release neuropeptide Y and AgRP, while stimulating neurons that release MSH (see fig. 19.3). The gene for human leptin has been cloned, and recombinant leptin is now available for the medical treatment of obesity. A small number of congenitally obese people are deficient in leptin and have been greatly helped by leptin injections; however, most obese people do not appear to have a leptin deficiency and are not much benefited by leptin injections. The levels of leptin are instead high in most obese people, suggesting that their obesity is promoted by a resistance of their hypothalamus to the appetite-suppressing effects of leptin. Resistance to leptin action can occur through different mechanisms (including reduced transport into the brain or impaired leptin signaling within neurons) that may be of varying significance in the obesity of different people.

Insulin, secreted by the β-cells of the pancreatic islets, has long been suspected of being a satiety factor. The beta cells secrete more insulin in obese than in lean people, because more insulin is required to maintain homeostasis of blood glucose as the stored fat increases. Because insulin levels, like leptin secretion, increase with increasing adiposity, insulin could act like leptin to suppress appetite. Like leptin, insulin can cross the blood-brain barrier in proportion to its concentration gradient and enter the hypothalamus. Also like leptin, insulin suppresses the release of neuropeptide Y to suppress appetite (see fig. 19.3). However, insulin's effects on hunger and satiety are complex and its significance in the regulation of eating behavior is not presently well defined.

In summary, leptin and insulin are believed to provide sensory signals for the long-term regulation of hunger, and indirectly of body weight, whereas several hormones from the gastrointestinal tract provide sensory signals that regulate hunger on a shorter term, meal-related basis (fig. 19.4).

Caloric Expenditures

The caloric energy expenditure of the body has three components (fig. 19.5):

1. **Basal metabolic rate (BMR)** is the energy expenditure of a relaxed, resting person who is at a neutral ambient temperature (about 28° C) and who has not eaten in 8 to 12 hours. This constitutes the majority (about 60%) of the total calorie expenditure in an average adult.
2. **Adaptive thermogenesis** is the heat energy expended in response to (a) changes in ambient temperature and (b) the digestion and absorption of food. This constitutes about 10% of the total calorie expenditure, although this contribution can change in response to cold and diet.
3. **Physical activity** raises the metabolic rate and energy expenditure of skeletal muscles. This contribution to the total calorie expenditure is highly variable, depending on the type and intensity of the physical activity (see table 19.1).

In adaptive thermogenesis, a cold environment evokes cutaneous vasoconstriction and shivering, which increases the metabolic rate and heat production of skeletal muscles. Since the skeletal muscles comprise about 40% of the total body weight, their metabolism has a profound effect on body temperature.

Heat production in the absence of shivering is called **nonshivering thermogenesis.** This is the major function of **brown adipose tissue (brown fat),** although skeletal muscles and other tissues also play a part. Brown adipocytes have many smaller fat droplets (unlike the single large fat droplet in white adipocytes) and have numerous mitochondria, which impart the brown color. Brown fat is abundant in infants, who can lose heat rapidly due to their high ratio of surface area to volume and have insufficient skeletal muscle for thermogenesis from shivering. Although once thought to be absent in adults, brown fat in adults—located primarily in the supraclavicular region of the neck (a different location than in infants)—was discovered by PET scans performed in nuclear medicine. This is because brown fat, like tumors, can be very metabolically active and take up radioactively labeled glucose (^{18}F-fluorodeoxyglucose).

Nonshivering thermogenesis in brown adipose tissue is due to the presence of *uncoupling protein 1 (UCP1)* in their mitochondria. These proteins provide channels in the inner mitochondrial membrane that allows protons (H^+) to move down their electrochemical gradient, from the intermembranous space to the matrix of the mitochondrion. As a result, the ability of ATP synthase to produce ATP in oxidative phosphorylation (chapter 5; see fig. 5.11) is reduced. Because ATP has an inhibitory effect on aerobic respiration (chapter 5), the reduced formation of ATP increases the metabolism of the brown adipose tissue, generating increased amounts of heat. There is some evidence that muscles may also contribute to nonshivering thermogenesis.

Recent reports demonstrated that the activity of brown adipose tissue is stimulated by exposure of the person to cold, and by the sympathoadrenal system (through the stimulation of β_3-adrenergic receptors on the brown adipocytes). There is more brown fat in women than in men, and more in lean people than in those who are overweight or obese.

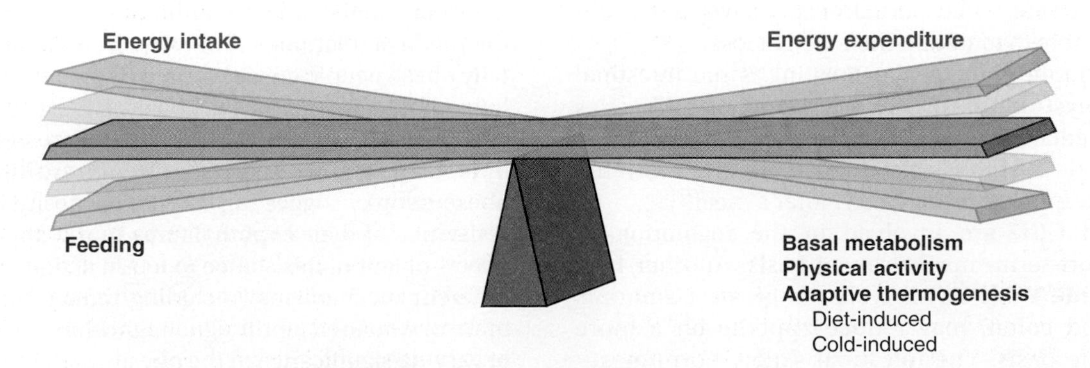

Figure 19.5 Homeostasis of body weight depends on caloric balance. Energy is taken into the body by eating, and energy is expended in basal metabolism (the basal metabolic rate, or BMR), exercise, and adaptive thermogenesis. The latter includes calories expended by the diet (the digestion and absorption of food) and metabolic heat production induced by cold.

Furthermore, lean people who have a lower body mass index (BMI) have more brown fat, and people with a higher BMI have less brown fat (although some brown fat is still present in most obese people). This suggests that the calories expended by brown fat in nonshivering thermogenesis could significantly affect body weight. Research into this relationship, and possible ways to exploit it to treat obesity and type 2 diabetes, is ongoing.

Diet is also an important regulator of adaptive thermogenesis, producing what is called the **thermic effect of food.** Starvation decreases the metabolic rate by as much as 40%, and feeding increases the metabolic rate by about 25% to 40% in average adults, with corresponding increases in heat production. This diet-induced thermogenesis is provoked most strongly by dietary carbohydrates and fats (but not proteins) through the activation of the sympathetic nervous system, which stimulates muscles and β_3-adrenergic receptors in brown adipose tissue.

The brain regulates adaptive thermogenesis, accomplished largely by activation of the *sympathoadrenal system.* Sympathetic innervation of skeletal muscles and brown fat, together with the effects of circulating epinephrine, cause increased metabolism in these tissues. Thyroxine secretion, controlled by the brain via TRH (thyrotropin-releasing hormone, which stimulates TSH secretion from the anterior pituitary—see chapter 11, fig. 11.16), is also needed for adaptive thermogenesis. Although thyroxine secretion is required for adaptive thermogenesis, the levels of thyroxine do not rise in response to cold or food, suggesting that the role of thyroid hormones is mainly a permissive one. In starvation, however, thyroxine levels do fall, suggesting that this decline may contribute to the slowdown in the metabolic rate during starvation.

During starvation, adipose tissue decreases its secretion of leptin and this fall is needed for the fall in TRH secretion that occurs in starvation. Thus, the decline in leptin may be responsible for the decline in thyroxine secretion. There is also evidence that decreasing leptin during starvation may cause a decline in the sympathetic nerve stimulation of brown fat. Through both mechanisms, the decreased leptin levels that occur during starvation could cause a slowdown in the metabolic rate. This effect would help to conserve energy during starvation. Opposite responses when leptin levels are high, conversely, would help to raise the metabolic rate and put a brake on the growth of adipose tissue (see fig. 19.3).

Hormonal Regulation of Metabolism

The absorption of energy carriers from the intestine is not continuous; it rises to high levels over a four-hour period following each meal (the **absorptive state**) and tapers toward zero between meals, after each absorptive state has ended (the **postabsorptive,** or **fasting, state**). Despite this fluctuation, the plasma concentration of glucose and other energy substrates does not remain high following periods of absorption, nor does it normally fall below a certain level during periods of fasting. Following the absorption of digestion products from the intestine, energy substrates are removed from the blood and deposited as energy reserves from which withdrawals can be made during times of fasting (fig. 19.6). This ensures an adequate plasma concentration of energy substrates to sustain tissue metabolism at all times.

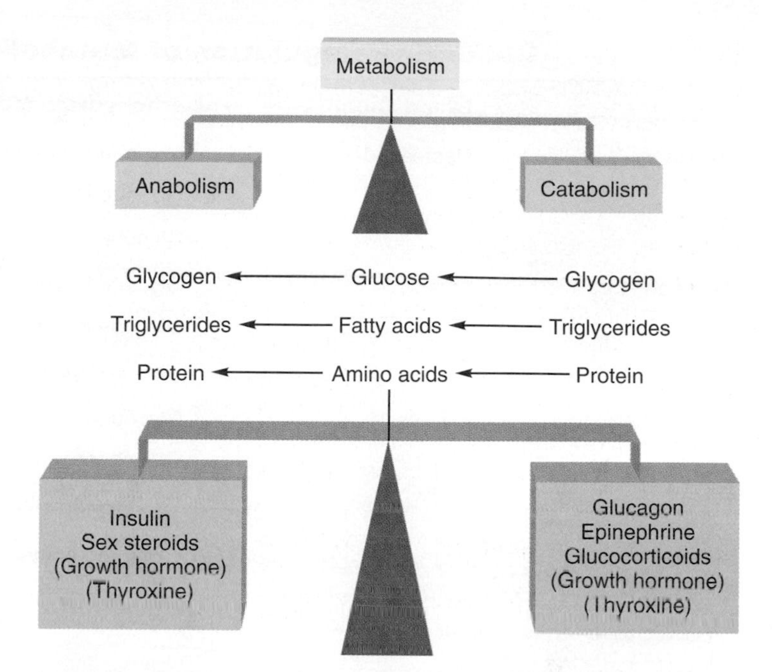

Figure 19.6 The regulation of metabolic balance. The balance of metabolism can be tilted toward anabolism (synthesis of energy reserves) or catabolism (utilization of energy reserves) by the combined actions of various hormones. Growth hormone and thyroxine have both anabolic and catabolic effects, so these hormones are shown in parentheses on both sides of the balance.

The rate of deposit and withdrawal of energy substrates into and from the energy reserves and the conversion of one type of energy substrate into another are regulated by hormones. The balance between anabolism and catabolism is determined by the antagonistic effects of insulin, glucagon, growth hormone, thyroxine, and other hormones (fig. 19.6). For example, during prolonged fasting there is increased glucagon secretion and decreased secretion of insulin (discussed in section 19.3). There is also reduced thyroxine secretion (which slows the metabolic rate); increased growth hormone secretion (which causes the release of fatty acids from adipose tissue); increased ACTH and cortisol secretion (which also promotes the release of circulating energy substrates); and decreased secretion of sex steroids (which reduces the chances of pregnancy). The specific metabolic effects of these hormones are summarized in table 19.4, and some of their actions are illustrated in figure 19.7.

Table 19.4 | Endocrine Regulation of Metabolism

Hormone	Blood Glucose	Carbohydrate Metabolism	Protein Metabolism	Lipid Metabolism
Insulin	Decreased	↑ Glycogen formation ↓ Glycogenolysis ↓ Gluconeogenesis	↑ Protein synthesis	↑ Lipogenesis ↓ Lipolysis ↓ Ketogenesis
Glucagon	Increased	↓ Glycogen formation ↑ Glycogenolysis ↑ Gluconeogenesis	No direct effect	↑ Lipolysis ↑ Ketogenesis
Growth hormone	Increased	↑ Glycogenolysis ↑ Gluconeogenesis ↓ Glucose utilization	↑ Protein synthesis	↓ Lipogenesis ↑ Lipolysis ↑ Ketogenesis
Glucocorticoids (hydrocortisone)	Increased	↑ Glycogen formation ↑ Gluconeogenesis	↓ Protein synthesis	↓ Lipogenesis ↑ Lipolysis ↑ Ketogenesis
Epinephrine	Increased	↓ Glycogen formation ↑ Glycogenolysis ↑ Gluconeogenesis	No direct effect	↑ Lipolysis ↑ Ketogenesis
Thyroid hormones	No effect	↑ Glucose utilization	↑ Protein synthesis	No direct effect

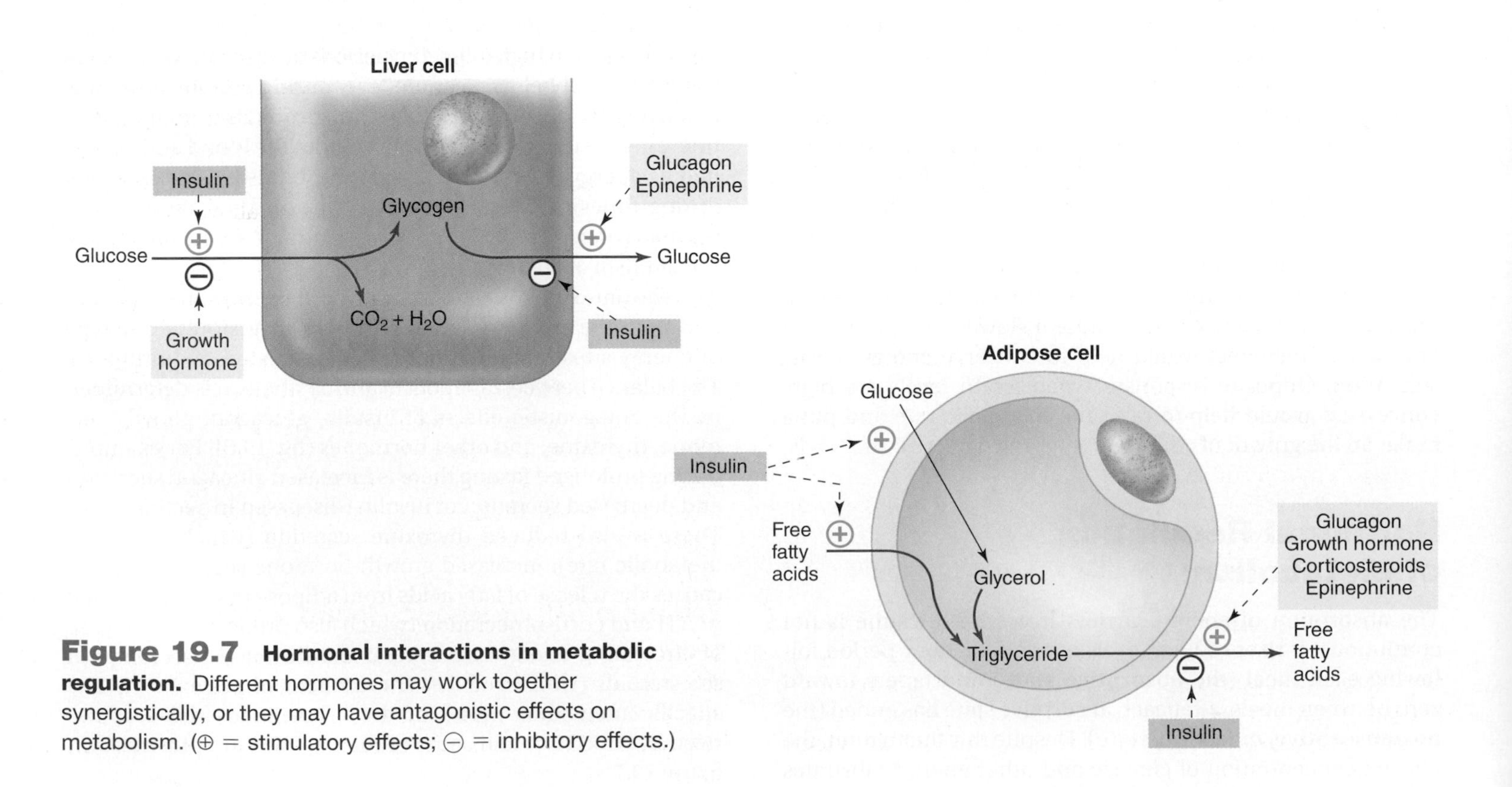

Figure 19.7 Hormonal interactions in metabolic regulation. Different hormones may work together synergistically, or they may have antagonistic effects on metabolism. (⊕ = stimulatory effects; ⊖ = inhibitory effects.)

✔ | CHECKPOINT

4. Distinguish between the energy reserves and circulating energy substrates, identifying the molecules in each category.
5. Explain the functions of adipose tissue and the dangers of obesity.
6. Identify the brain regions involved in hunger and describe the regulation of hunger by specific neurotransmitters and hormones.
7. Identify the hormones that increase or decrease blood glucose and those that stimulate lipogenesis or lipolysis.

19.3 ENERGY REGULATION BY THE PANCREATIC ISLETS

Insulin secretion is stimulated by a rise in the plasma glucose concentration, and insulin lowers the blood glucose while it promotes the synthesis of glycogen and fat. Glucagon secretion is stimulated by a fall in the plasma glucose concentration, and glucagon promotes increased blood glucose due to glycogenolysis in the liver.

LEARNING OUTCOMES

After studying this section, you should be able to:

8. Explain how the secretions of insulin and glucagon are regulated and how they change during the absorptive and postabsorptive states.
9. Explain how the liver produces glucose and why the liver but not skeletal muscles can secrete glucose into the blood.

Scattered within a "sea" of pancreatic exocrine tissue (the acini) are islands of hormone-secreting cells (chapter 18; see fig. 18.27). These pancreatic islets (islets of Langerhans) contain three distinct cell types that secrete different hormones. The most numerous are the *beta cells,* which secrete the hormone **insulin.** About 60% of each islet consists of beta cells. The *alpha cells* form about 25% of each islet and secrete the hormone **glucagon.** The least numerous cell type, the *delta cells,* produce **somatostatin,** the composition of which is identical to the somatostatin produced by the hypothalamus and the intestine.

Insulin is the major hormone that maintains homeostasis of the plasma glucose concentration, with glucagon playing an important supporting role. This homeostasis is required because the brain uses plasma glucose as its primary energy source, and indeed the brain uses about 60% of the blood glucose when the person is at rest. The plasma glucose concentration is maintained relatively constant even during exercise, when the skeletal muscle glucose metabolism can increase tenfold. This is possible because the pancreatic islet hormones regulate the ability of the liver to produce glucose (from stored glycogen and noncarbohydrate molecules) and secrete it into the blood.

Regulation of Insulin and Glucagon Secretion

The secretion of insulin and glucagon is largely regulated by the plasma concentrations of glucose and, to a lesser degree, of amino acids. These concentrations rise during the absorption of a meal (the *absorptive state*) and fall during times of fasting (the *postabsorptive state*). As a result, the secretion of insulin and glucagon changes during these conditions and helps maintain homeostasis of the plasma glucose concentration (chapter 11; see fig. 11.31).

The targets of insulin action are primarily the cells of skeletal and cardiac muscles, adipose tissue, and the liver. In these cells, intracellular vesicles containing GLUT4 carrier proteins for glucose are stimulated by insulin to fuse with the plasma membrane so that the GLUT4 carriers are at the cell surface (chapter 11; see fig. 11.30). This permits the entry of glucose into its target cells by facilitated diffusion. As a result, insulin promotes the production of the energy-storage molecules of glycogen and fat. Both actions decrease the plasma glucose concentration. Insulin also inhibits the breakdown of fat, induces the production of fat-forming enzymes, and inhibits the breakdown of muscle proteins. Thus, insulin promotes anabolism as it regulates the blood glucose concentration.

The mechanisms that regulate insulin and glucagon secretion and the actions of these hormones normally prevent the plasma glucose concentration from rising above 170 mg per 100 ml after a meal or from falling below about 50 mg per 100 ml between meals. This regulation is important because abnormally high blood glucose can damage certain tissues (as may occur in diabetes mellitus), and abnormally low blood glucose can damage the brain. This can occur because glucose enters neurons by facilitated diffusion driven by the higher glucose concentration in the plasma. When low plasma glucose concentrations reduce this diffusion gradient, the brain may not get sufficient glucose for its metabolic needs. This can result in weakness, dizziness, personality changes, and ultimately in coma and death.

Effects of Glucose and Amino Acids

The fasting plasma glucose concentration is in the range of 65 to 105 mg/dl. During the absorption of a meal, the plasma glucose concentration usually rises to a level between 140 and 150 mg/dl. This rise in the plasma glucose concentration acts on the beta cells of the islets, where it leads to closing of K^+ channels. This produces a depolarization, which opens voltage-gated Ca^{2+} channels. The resultant rise in the cytoplasmic Ca^{2+} concentration stimulates exocytosis of vesicles containing insulin (fig. 19.8). A rise in plasma glucose thus leads to a rise in insulin secretion; at the same time, it inhibits the secretion of glucagon from the alpha cells of the islets. Because insulin lowers the plasma glucose concentration (by stimulating the cellular uptake of plasma glucose), and glucagon acts antagonistically to raise the plasma glucose (by stimulating glycogenolysis in the liver), these changes in insulin and glucagon secretion help maintain homeostasis during the absorption of a carbohydrate meal.

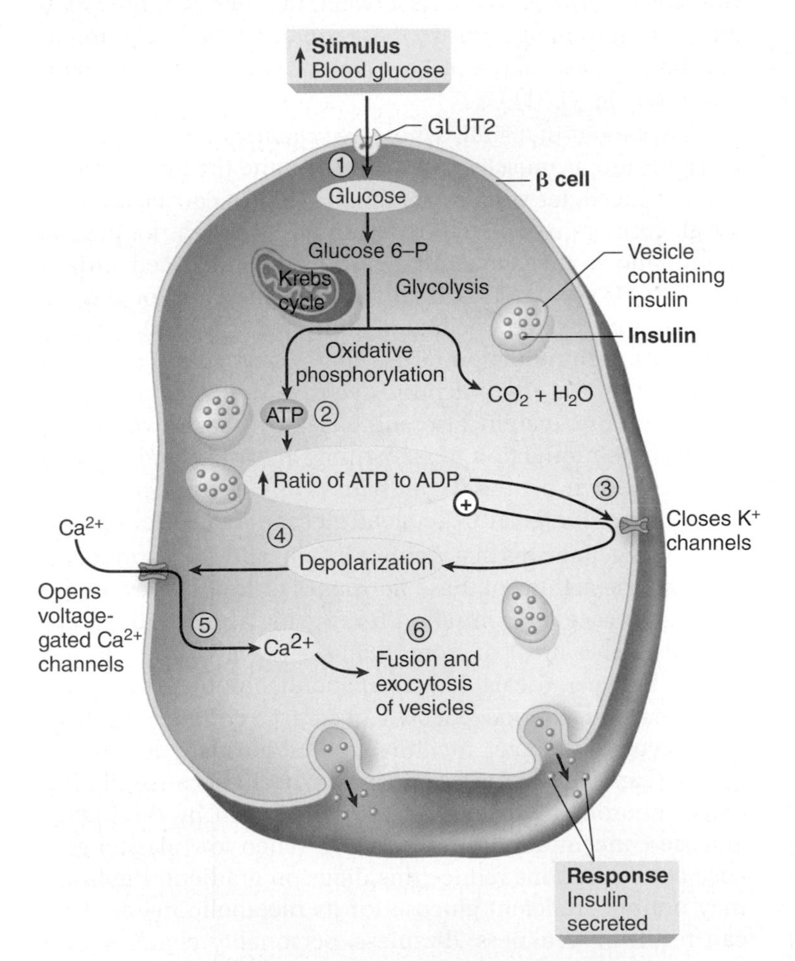

Figure 19.8 Regulation of insulin secretion. (1) A rise in blood glucose causes more glucose to enter the beta cells of the islets of Langerhans, resulting in (2) an increased production of ATP. (3) This closes K^+ channels, so that K^+ ions cannot leave the cell. (4) This produces a depolarization, which (5) opens voltage-gated Ca^{2+} channels, permitting the entry of Ca^{2+} into the cytoplasm. (6) Ca^{2+} stimulates intracellular vesicles containing insulin to fuse with the plasma membrane and release insulin by exocytosis.

During the postabsorptive (fasting) state, the plasma glucose concentration falls. As a result, insulin secretion decreases and glucagon secretion increases (chapter 11; see fig. 11.31). These changes in hormone secretion prevent the cellular uptake of blood glucose into organs such as the muscles, liver, and adipose tissue. An elevated glucagon secretion, in the presence of low insulin levels, also stimulates glycogenolysis and so promotes the release of glucose from the liver. A negative feedback loop is therefore completed, helping to retard the fall in plasma glucose concentration that occurs during fasting.

The **oral glucose tolerance test** (see fig. 19.12) is a measure of the ability of the beta cells to secrete insulin and of the ability of insulin to lower blood glucose. In this procedure, a person drinks a glucose solution and blood samples are taken periodically for plasma glucose measurements. In a normal person, the rise in blood glucose produced by drinking this solution is reversed to normal levels within two hours following glucose ingestion. In contrast, the plasma glucose concentration remains at 200 mg/dl or higher two hours after the oral glucose challenge in a person with diabetes mellitus.

Insulin secretion is also stimulated by particular amino acids derived from dietary proteins. Meals that are high in protein, therefore, stimulate the secretion of insulin; if the meal is high in protein and low in carbohydrates, glucagon secretion will be stimulated as well. The increased glucagon secretion acts to raise the blood glucose, while the increased insulin promotes the entry of amino acids into tissue cells.

Effects of Autonomic Nerves

The islets of Langerhans receive both parasympathetic and sympathetic innervation. The parasympathetic division of the autonomic system is activated during meals and stimulates both gastrointestinal function and the secretion of insulin from the beta cells of the islets. By contrast, the sympathetic division inhibits insulin secretion but stimulates glucagon secretion from the alpha cells of the islets. Glucagon and epinephrine then work together to produce a *stress hyperglycemia* (due to hydrolysis of glycogen to glucose and the secretion of glucose from the liver) when the sympathoadrenal system is activated.

Effects of Intestinal Hormones

Surprisingly, insulin secretion increases more rapidly when glucose is taken by mouth than when glucose is injected intravenously. This is because, when glucose is taken orally, the intestine secretes hormones that stimulate insulin secretion before the glucose has even been absorbed. Insulin secretion thus begins to rise "in anticipation" of a rise in blood glucose. Two intestinal hormones that are powerful stimulators of insulin secretion are *glucagon-like polypeptide 1 (GLP-1)*, secreted by the ileum, and *glucose-dependent insulinotropic polypeptide (GIP),* secreted by the duodenum (chapter 18, section 18.6). These and perhaps other intestinal hormones stimulate increased insulin secretion before there is a rise in blood glucose from the digested food, preventing an excessive rise in the blood glucose concentration following a high-carbohydrate meal.

Insulin and Glucagon: Absorptive State

The lowering of plasma glucose by insulin is, in a sense, a side effect of the primary action of this hormone. Insulin is the major hormone that promotes anabolism in the body. During absorption of the products of digestion into the blood, insulin promotes the cellular uptake of plasma glucose and its incorporation into energy-reserve molecules of glycogen in the liver and muscles, and of triglycerides in adipose cells (chapter 11; see fig. 11.31). Insulin also promotes the cellular uptake of amino acids and their incorporation into proteins. The stores of large energy-reserve molecules are thus increased while the plasma concentrations of glucose and amino acids are decreased. It should be noted that skeletal muscles are the primary targets of insulin action, and are responsible for most of the insulin-stimulated uptake of plasma glucose.

A nonobese 70-kg (155-lb) man has approximately 10 kg (about 82,500 kcal) of stored fat. Because 250 g of fat can supply the energy requirements for 1 day, this reserve fuel is sufficient for about 40 days. Glycogen is less efficient as an energy reserve, and less is stored in the body; there are about 100 g (400 kcal) of glycogen stored in the liver and 375 to 400 g (1,500 kcal) in skeletal muscles. Insulin promotes the cellular uptake of glucose into the liver and muscles and the conversion of glucose into glucose 6-phosphate. In the liver and muscles, this can be changed into glucose 1-phosphate, which is used as the precursor of glycogen. Once the stores of glycogen have been filled, the continued ingestion of excess calories results in the production of fat rather than of glycogen.

The high insulin secretion during the absorptive state, when blood glucose levels are rising, also inhibits the liver from secreting more glucose into the blood. Insulin may directly inhibit glucose production by the liver, but evidence also supports an indirect action by way of the CNS. In this, insulin acts on the hypothalamus (particularly the arcuate nucleus) to inhibit vagus nerve stimulation of the liver's glucose secretion. By contrast, during the postabsorptive state, when blood glucose and insulin secretion are falling (discussed next), the liver is freed from this inhibition and does secrete glucose into the blood.

Insulin and Glucagon: Postabsorptive State

The plasma glucose concentration is maintained surprisingly constant during the fasting, or postabsorptive, state because of the secretion of glucose from the liver. This glucose is derived from the processes of glycogenolysis and gluconeogenesis, which are promoted by a high secretion of glucagon coupled with a low secretion of insulin.

Glucagon stimulates and insulin suppresses the hydrolysis of liver glycogen, or **glycogenolysis.** Thus during times of fasting, when glucagon secretion is high and insulin secretion is low, liver glycogen is used as a source of additional blood glucose. This results in the liberation of free glucose from glucose 6-phosphate by the action of an enzyme called *glucose 6-phosphatase* (chapter 5; see fig. 5.5). Only the liver has this enzyme, and therefore only the liver can use its stored glycogen as a source of additional blood glucose. The liver has about 100 g of stored glycogen, whereas skeletal muscles have about 400 g of glycogen. However, the muscles can use their stored glycogen only for themselves, because muscles lack the glucose 6-phosphatase enzyme needed to form free glucose for secretion into the blood.

The 100 g of stored glycogen in the liver would not last long during prolonged fasting or exercise if this were the only source of glucose for the blood. However, the low levels of insulin secretion during fasting, together with elevated glucagon secretion, promote **gluconeogenesis**—the formation of glucose from noncarbohydrate molecules. Low insulin allows the release of amino acids from skeletal muscles, while glucagon and cortisol (an adrenal hormone) stimulate the production of enzymes in the liver that convert amino acids to pyruvic acid and pyruvic acid into glucose. During prolonged fasting and exercise, gluconeogenesis in the liver using amino acids from muscles may be the only source of blood glucose.

The secretion of glucose from the liver during fasting compensates for the low blood glucose concentrations and helps to provide the brain with the glucose that it needs. But because insulin secretion is low during fasting, skeletal muscles cannot utilize blood glucose as an energy source. Instead, skeletal muscles—as well as the heart, liver, and kidneys—use free fatty acids as their major source of fuel. This helps to "spare" glucose for the brain.

The free fatty acids are made available by the action of glucagon. In the presence of low insulin levels, glucagon activates an enzyme in adipose cells called *hormone-sensitive lipase.* This enzyme catalyzes the hydrolysis of stored triglycerides and the release of free fatty acids and glycerol into the blood. Glucagon also activates enzymes in the liver that convert some of these fatty acids into ketone bodies, which are secreted into the blood (fig. 19.9). Several organs in the body can use ketone bodies, as well as fatty acids, as a source of acetyl CoA in aerobic respiration.

Through the stimulation of **lipolysis** (the breakdown of fat) and **ketogenesis** (the formation of ketone bodies), the high glucagon and low insulin levels during fasting provide circulating energy substrates for use by the muscles, liver, and other organs. Through liver glycogenolysis and gluconeogenesis,

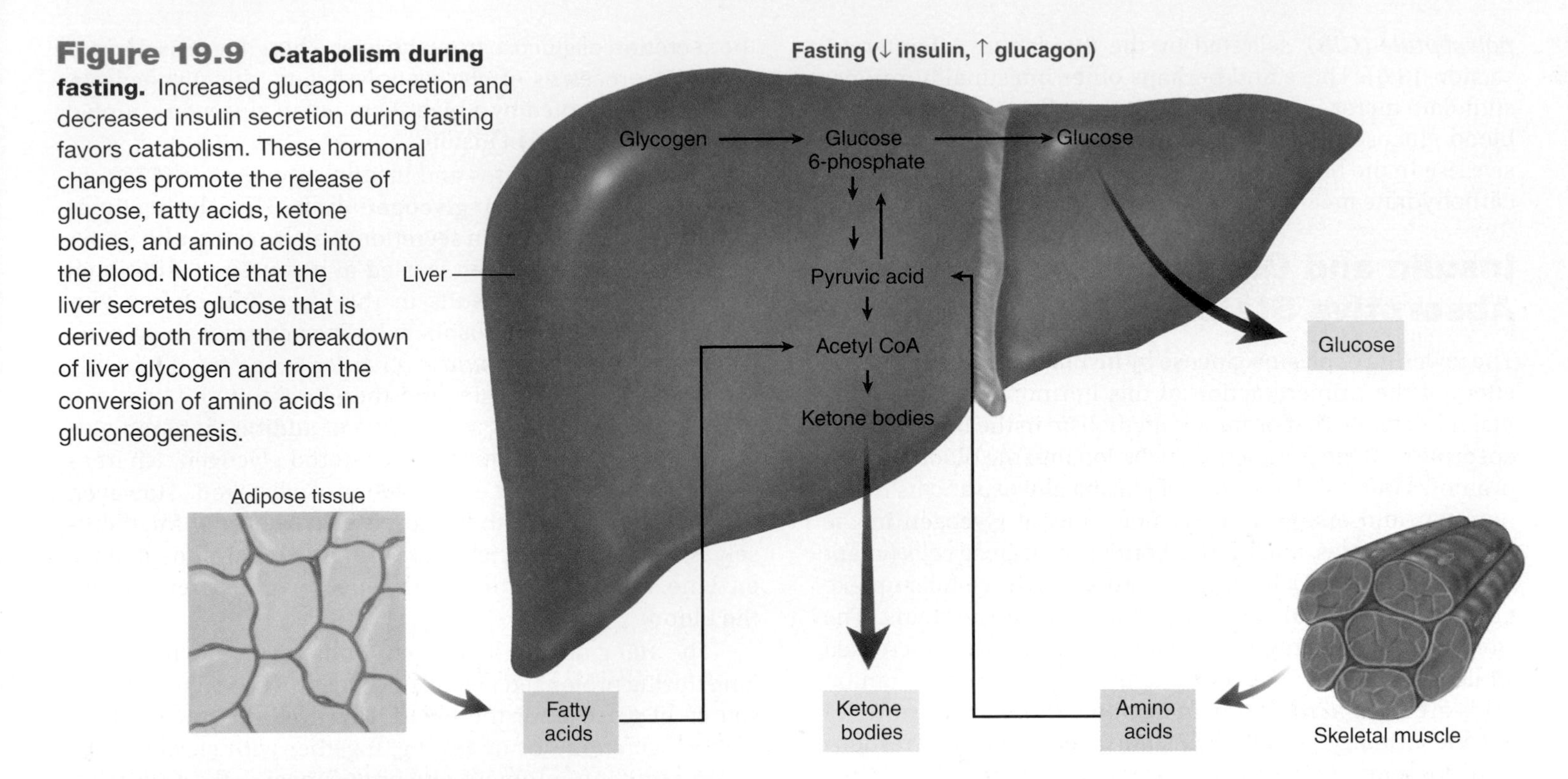

Figure 19.9 Catabolism during fasting. Increased glucagon secretion and decreased insulin secretion during fasting favors catabolism. These hormonal changes promote the release of glucose, fatty acids, ketone bodies, and amino acids into the blood. Notice that the liver secretes glucose that is derived both from the breakdown of liver glycogen and from the conversion of amino acids in gluconeogenesis.

these hormonal changes help to provide adequate levels of blood glucose to sustain the metabolism of the brain. By serving as energy substrates for the muscles and other organs, the free fatty acids and ketone bodies spare blood glucose for use by the brain. Free fatty acids also reduce the activity of glycolytic enzymes in muscles, hindering the ability of muscles to utilize glucose for energy so that more is available for the brain. These changes, together with the glucose provided by glycogenolysis and gluconeogenesis promoted by high glucagon/low insulin levels, sustain body metabolism during fasting (and exercise) conditions (fig. 19.10).

CHECKPOINT

8a. Describe how the secretions of insulin and glucagon change during periods of absorption and periods of fasting. How are these changes in hormone secretion produced?

8b. Explain how the synthesis of fat in adipose cells is regulated by insulin. Also, explain how fat metabolism is regulated by insulin and glucagon during periods of absorption and fasting.

8c. Define the following terms: *glycogenolysis, gluconeogenesis,* and *ketogenesis.* How do insulin and glucagon affect each of these processes during periods of absorption and fasting?

9. Describe two pathways used by the liver to produce glucose for secretion into the blood. Why can't skeletal muscles secrete glucose into the blood?

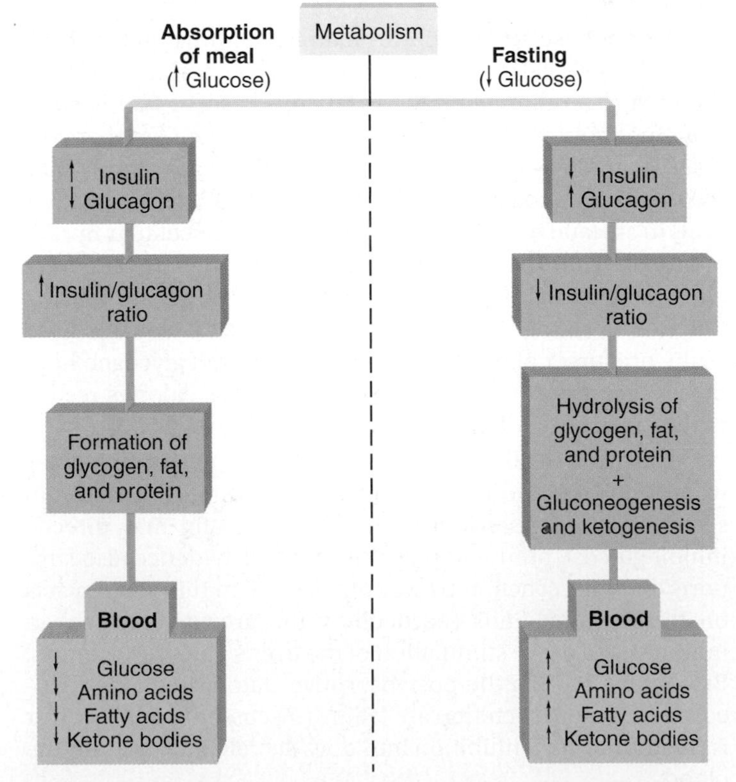

Figure 19.10 The effect of feeding and fasting on metabolism. Metabolic balance is tilted toward anabolism by feeding (absorption of a meal) and toward catabolism by fasting. This occurs because of an inverse relationship between insulin and glucagon secretion. Insulin secretion rises and glucagon secretion falls during food absorption, whereas the opposite occurs during fasting.

19.4 DIABETES MELLITUS AND HYPOGLYCEMIA

Inadequate secretion or action of insulin produces metabolic disturbances characteristic of diabetes mellitus. A person with type 1 diabetes requires injections of insulin; a person with type 2 diabetes can control this condition by other methods. A person with reactive hypoglycemia secretes excessive amounts of insulin.

LEARNING OUTCOMES

After studying this section, you should be able to:

10. Distinguish between type 1 and type 2 diabetes mellitus.

11. Explain insulin resistance and impaired glucose tolerance, and describe hypoglycemia.

Chronic high blood glucose, or hyperglycemia, is the hallmark of **diabetes mellitus.** The name of this disease is derived from the fact that glucose "spills over" into the urine when the blood glucose concentration is too high (*mellitus* is derived from a Latin word meaning "honeyed" or "sweet"). The general term *diabetes* comes from a Greek word meaning "siphon;" it refers to the frequent urination associated with this condition. The hyperglycemia of diabetes mellitus results from either the insufficient secretion of insulin by the beta cells of the pancreatic islets or the inability of secreted insulin to stimulate the cellular uptake of glucose from the blood. Diabetes mellitus, in short, results from the inadequate secretion or action of insulin. There is also elevated glucagon secretion, because insulin is less able to allow a high plasma glucose concentration to suppress the secretion of glucagon. Evidence suggests that glucagon (through its stimulation of hepatic glycogenolysis) contributes significantly to the hyperglycemia of people with type 2 diabetes mellitus (discussed shortly).

There are two major forms of diabetes mellitus. In **type 1** (or *insulin-dependent*) **diabetes,** the beta cells are progressively destroyed and secrete little or no insulin. Injections of exogenous insulin are thus required to sustain the person's life (since insulin is a polypeptide, it would be digested if taken orally; however, the FDA has approved the use of an inhaler containing a fine powder of insulin for the control of type 1 diabetes in adults). Scientists have recently succeeded in transforming some pancreatic acinar cells (chapter 18; see fig. 18.27) into insulin-secreting β-cells in rodents, a procedure that offers some hope of a possible future treatment for type 1 diabetes. Type 1 diabetes accounts for about 5% of the diabetic population.

About 95% of the people who have diabetes have **type 2** (*non-insulin-dependent*) **diabetes.** Type 1 diabetes was once known as *juvenile-onset diabetes* because this condition is usually diagnosed in people under the age of 20. Type 2 diabetes has also been called *maturity-onset diabetes,* because it is usually diagnosed in people over the age of 40. The incidence of type 2 diabetes in children is rising (due to an increase in the frequency of obesity), however, so these terms are no longer preferred. The two forms of diabetes mellitus are compared in table 19.5. (It should be noted that only the early stages of type 1 and type 2 diabetes mellitus are compared; some people with severe type 2 diabetes may also require insulin injections to control the hyperglycemia.)

Diabetes is the major cause of kidney failure and limb amputations, and is the second leading cause of blindness. Furthermore, because type 2 diabetes is associated with *metabolic syndrome* (discussed in the Clinical Application box on p. 671), in which there is high blood triglycerides and low HDL cholesterol, there is increased risk of atherosclerosis. Thus, diabetes is also a significant contributor to heart disease and stroke.

Type 1 Diabetes Mellitus

Type 1 diabetes mellitus is an autoimmune disease (chapter 15, section 15.6), and like other autoimmune diseases, the susceptibility for it is associated with genes of the major histocompatability complex (MHC) on chromosome 6. Because the concordance rate for type 1 diabetes between identical twins

Table 19.5 | Comparison of Type 1 and Type 2 Diabetes Mellitus

Feature	Type 1	Type 2
Usual age at onset	Under 20 years	Over 40 years
Development of symptoms	Rapid	Slow
Percentage of diabetic population	About 5%	About 95%
Development of ketoacidosis	Common	Rare
Association with obesity	Rare	Common
Beta cells of islets (at onset of disease)	Destroyed	Not destroyed
Insulin secretion	Decreased	Normal or increased
Autoantibodies to islet cells	Present	Absent
Associated with particular MHC antigens*	Yes	Unclear
Treatment	Insulin injections	Diet and exercise; oral stimulators of insulin sensitivity

*Discussed in chapter 15, section 15.3.

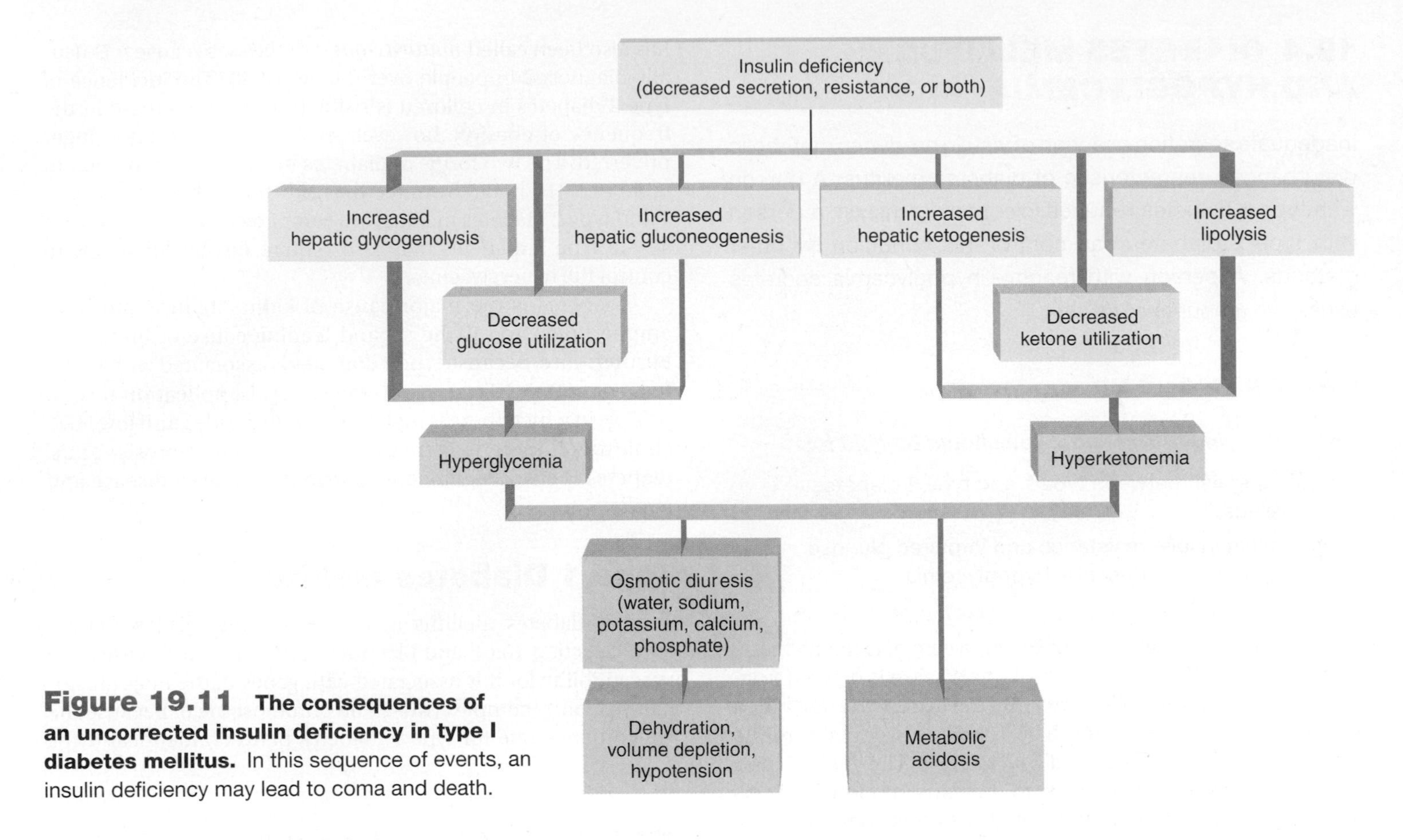

Figure 19.11 The consequences of an uncorrected insulin deficiency in type I diabetes mellitus. In this sequence of events, an insulin deficiency may lead to coma and death.

is only 40% to 60%, and because the incidence of type 1 diabetes has been increasing (by an estimated 5.3% annually in the United States), environmental factors must play a triggering role in genetically susceptible people. Prominent among the environmental factors that may trigger an autoimmune attack of the beta cells in the pancreatic islets are viruses and bacteria. Autoreactive T lymphocytes—helper and killer T cells—are believed to be most important in the progressive destruction of the insulin-secreting beta cells, although autoantibodies appear early in the course of the disease and aid diagnosis.

Removal of the insulin-secreting beta cells causes hyperglycemia and the appearance of glucose in the urine. Without insulin, glucose cannot enter the adipose cells; the rate of fat synthesis thus lags behind the rate of fat breakdown and large amounts of free fatty acids are released from the adipose cells.

In a person with uncontrolled type 1 diabetes, many of the fatty acids released from adipose cells are converted into ketone bodies in the liver. This may result in an elevated ketone body concentration in the blood (ketosis), and if the buffer reserve of bicarbonate is neutralized, it may also result in *ketoacidosis.* During this time, the glucose and excess ketone bodies that are excreted in the urine act as osmotic diuretics and cause the excessive excretion of water in the urine (chapter 17, section 17.6). This can produce severe dehydration, which, together with ketoacidosis and associated disturbances in electrolyte balance, may lead to coma and death (fig. 19.11).

In addition to the lack of insulin, people with type 1 diabetes have an abnormally high secretion of glucagon from the alpha cells of the islets. Glucagon stimulates glycogenolysis in the liver and thus helps to raise the blood glucose concentration. Glucagon also stimulates the production of enzymes in the liver that convert fatty acids into ketone bodies. The full range of symptoms of diabetes may result from high glucagon secretion as well as from the absence of insulin. The lack of insulin may be largely responsible for hyperglycemia and for the release of large amounts of fatty acids into the blood. The high glucagon secretion may contribute to the hyperglycemia and be largely responsible for the development of ketoacidosis.

Clinical Investigation CLUES

Lisa did not have glycosuria, but she complained of frequent urination and continuous thirst.

- Could Lisa have glycosuria at some times and not others?
- If so, how could that relate to her frequent urination and continuous thirst?
- Might Lisa have ketonuria, and could that contribute to her symptoms?

Type 2 Diabetes Mellitus

The effects produced by insulin, or any hormone, depend on the concentration of that hormone in the blood and on the sensitivity of the target tissue to given amounts of the hormone. Tissue responsiveness to insulin, for example, varies under normal conditions. Exercise increases insulin sensitivity and obesity decreases insulin sensitivity of the target tissues. The islets of a nondiabetic obese person must therefore secrete high amounts of insulin to maintain the blood glucose concentration in the normal range. Conversely, nondiabetic people who are thin and who exercise regularly require lower amounts of insulin to maintain the proper blood glucose concentration.

People with type 2 diabetes have abnormally low tissue sensitivity to insulin, or **insulin resistance.** In people who have type 2 diabetes, the islet beta cells fail to secrete adequate amounts of insulin to compensate for the insulin resistance. This failure seems to result from inflammatory processes in the islets. However, the concordance of identical twins with type 2 diabetes is 70% and the risk of anyone getting diabetes is also almost 70% if both parents have it. This demonstrates a strong heritable component in type 2 diabetes mellitus.

The expression of this genetic tendency toward diabetes is increased by obesity. This is particularly true if the obesity involves an "apple shape," with large adipocytes in the greater omentum (visceral fat). Insulin resistance is believed to result from increased plasma levels of free fatty acids and from adipokines (previously discussed) released by adipocytes. Insulin resistance reduces the ability of insulin to stimulate skeletal muscle, liver, and adipose tissue to take glucose out of the blood. Also, insulin is less able to inhibit the liver from producing more blood glucose. In the presence of an insulin deficiency, there is an elevated glucagon secretion that also contributes significantly to the hyperglycemia of type 2 diabetes by stimulating hepatic glycogenolysis and gluconeogenesis. Thus, the insulin resistance of type 2 diabetes raises blood glucose through increased hepatic secretion of glucose and decreased uptake of glucose into skeletal muscles and adipose tissue.

People who are prediabetic may have *impaired glucose tolerance,* which is defined (in an oral glucose tolerance test) as a plasma glucose level of 140 to 200 mg/dl at two hours following the glucose ingestion. As previously mentioned, a value here of over 200 mg/dl indicates diabetes. Because impaired glucose tolerance is accompanied by higher levels of insulin (fig. 19.12), a state of insulin resistance is suggested.

People who are obese appear to have an increased mass of beta cells in their pancreatic islets to compensate for their insulin resistance. Those who are going to become diabetic have a genetic susceptibility to β-cell failure under these conditions, where abnormal function and apoptosis of β-cells eventually leads to an inability of insulin secretion to compensate for the insulin resistance. Thus, type 2 diabetes is usually hereditary and slow to develop, and occurs most often in people who are overweight. In early, less severe cases, it can be accompanied by elevated and abnormal patterns of insulin secretion. In later, more severe cases the secretion of insulin is reduced by beta cells apoptosis and other causes of beta cell failure.

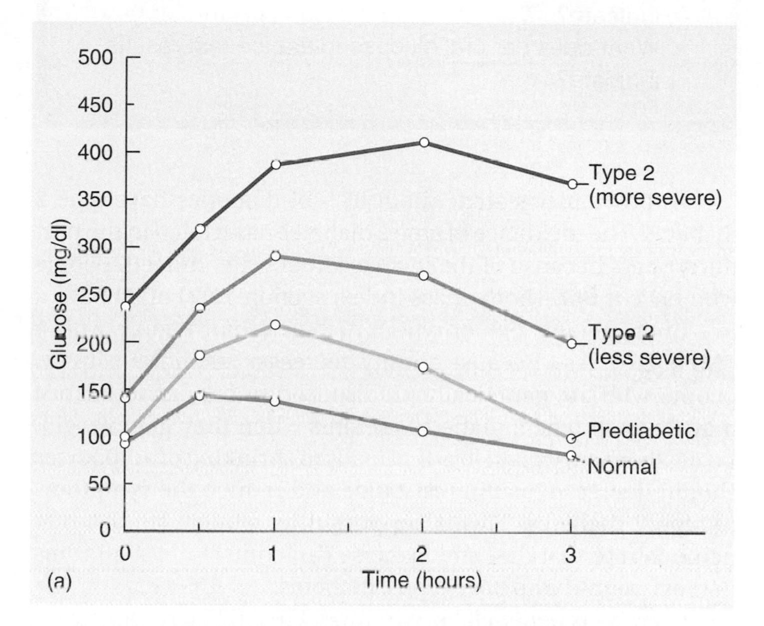

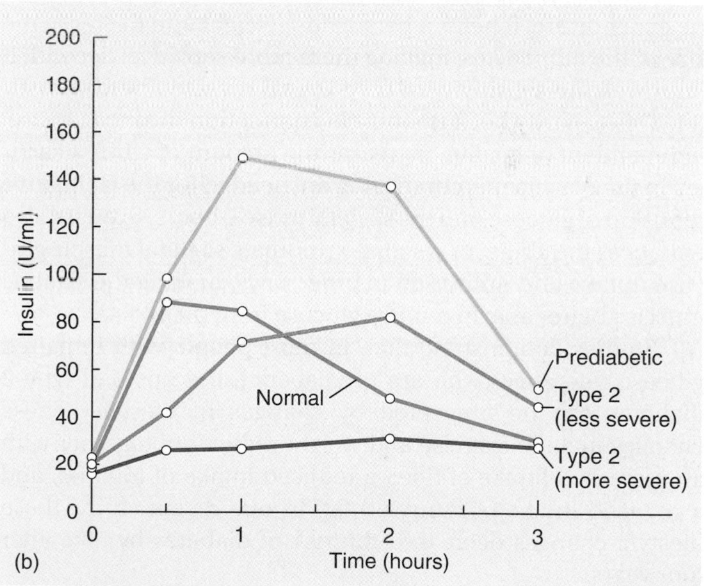

Figure 19.12 Oral glucose tolerance in prediabetes and type 2 diabetes. The oral glucose tolerance test showing (*a*) blood glucose concentrations and (*b*) insulin values following the ingestion of a glucose solution. Values are shown for people who are normal, prediabetic, and type 2 (non-insulin-dependent) diabetic. Prediabetics (those who demonstrate "insulin resistance") often show impaired glucose tolerance without fasting hyperglycemia. Data from Simeon I. Taylor et al.,"Insulin Resistance of Insulin Deficiency: Which is the Primary Cause of NIDDM?" in *Diabetes*, vol. 43, June 1994, p. 735.

See the *Test Your Quantitative Ability* section of the Review Activities at the end of this chapter.

Clinical Investigation CLUES

Lisa had a fasting blood glucose concentration of 150 mg/dl, and an oral glucose tolerance test two-hour measurement of 220 mg/dl.

- What does her fasting blood glucose measurement indicate?
- What does her oral glucose tolerance test results indicate?

As previously stated, about 95% of diabetics have type 2 diabetes. The incidence of type 2 diabetes has tripled in the past thirty years because of the increase in obesity. Indeed, people who have a BMI (body mass index; section 19.2) of 30 have a five-times greater risk of type 2 diabetes than people with a BMI of 25 or less. Because obesity decreases insulin sensitivity, people who are genetically predisposed to insulin resistance may develop type 2 diabetes mellitus when they gain weight. Logically, then, weight loss that leads to shrinking of adipocytes should decrease insulin resistance and reduce the symptoms of type 2 diabetes. This is supported by clinical studies that demonstrate that diet and exercise can control the symptoms in most people who have type 2 diabetes.

Exercise is beneficial in two ways. First, by increasing calorie expenditure, it helps a person lose weight and decrease the size of the adipocytes, making them more sensitive. Second, it improves the sensitivity of the skeletal muscle fibers to insulin. This is partly because muscle contraction during exercise, independent of insulin, increases the amount of GLUT4 carriers in the plasma membrane that are needed for the facilitative diffusion of glucose into the skeletal muscle fibers. Exercise also enhances the ability of insulin to stimulate skeletal muscle glucose uptake and utilization in other ways, making the skeletal muscles better able to remove glucose from the blood.

Studies demonstrate that, in most people with impaired glucose tolerance (who are prediabetic), the onset of type 2 diabetes can be prevented by changes in lifestyle. These changes include exercise and weight reduction, together with an increased intake of fiber, a reduced intake of total fat, and a reduced intake of saturated fat. In one recent study, these lifestyle changes decreased the risk of diabetes by 58% after four years.

In the years preceding the development of hyperglycemia in type 2 diabetes, the insulin resistance associated with obesity can result in a compensatory increase in the number of β cells in the islets. This produces an elevated secretion of insulin, but there may be impaired glucose tolerance despite the increased insulin. Insulin resistance can be associated with hypertension and dyslipidemia (and thus with increased risk of cardiovascular diseases), as discussed in the Clinical Application box on *metabolic syndrome* (see p. 671). Metabolic syndrome in obesity may be caused by inflammation; the number of macrophages in adipose tissue increases in proportion to the obesity, as do inflammation markers in the blood such as *C-reactive protein.* In obesity, adipose tissue (including adipocytes and macrophages) secretes several pro-inflammatory adipokines, including tumor necrosis factor alpha (TNF_{α}), interleukin-1, and resistin, that also reduce the insulin sensitivity of target tissues (adipose tissue, liver, and muscles). By contrast, the adipose tissue of lean people releases an anti-inflammatory adipokine—adiponectin—that increases insulin sensitivity and protects against metabolic syndrome.

People with type 2 diabetes do not usually develop ketoacidosis. The hyperglycemia itself, however, can be dangerous on a long-term basis. In the United States, diabetes is the leading cause of blindness, kidney failure, and amputation of the lower extremities. People with diabetes frequently have circulatory problems that increase the tendency to develop gangrene and increase the risk for atherosclerosis. The causes of damage to the retina and lens of the eyes and to blood vessels are not well understood. It is believed, however, that these problems result from a long-term exposure to high blood glucose, which damages tissues through a variety of mechanisms.

The **glycated hemoglobin (hemoglobin A1c)** test is a measure of the average blood glucose level over a few months and does not require overnight fasting. An Alc measurement of about 5% is normal, whereas an A1c of 5.7% to 6.4% indicates prediabetes. An A1c level above 6.5% indicates diabetes; diabetic control is considered good if the A1c measurement is maintained at 7% or less.

CLINICAL APPLICATION

If diet and exercise are not sufficiently effective at treating a person's type 2 diabetes mellitus, oral medications may be necessary. These include **sulfonylurea,** which stimulates insulin secretion, and **metformin** and the **thiazolidinediones,** which reduce the insulin resistance of the target cells. Metformin (*glucophage*) reduces hyperglycemia by stimulating glycolysis and fatty acid oxidation, thereby promoting insulin action in the liver and reducing the secretion of glucose by the liver. The thiazolidinediones include *pioglitazone (Actos)* and *rosiglitazone (Avandia);* the latter has been withdrawn because of its association with an increased risk of myocardial infarction.

Many scientists now believe that inflammatory pathways within adipose tissue, liver, and muscles cause insulin resistance. The thiazolidinediones increase the sensitivity of target tissues to insulin by activating the PPARγ nuclear receptors in both adipocytes and macrophages, thereby altering the secretion of pro-inflammatory cytokines from macrophages and adipokines from adipocytes. These drugs decrease the secretion of cytokines and adipokines that promote insulin resistance (TNFα, interleukin-1, and resistin) and increase the secretion of adiponectin, which reduces insulin resistance.

Clinical Investigation CLUES

The physician asked Lisa to diet and exercise. He said that, if necessary, he would prescribe pills.

- What condition does the physician believe is responsible for Lisa's symptoms?
- How could diet and exercise help?
- Which pills might the physician prescribe, and how do these work?

Hypoglycemia

A person with type 1 diabetes mellitus depends on insulin injections to prevent hyperglycemia and ketoacidosis. If inadequate insulin is injected, the person may enter a coma as a result of the ketoacidosis, electrolyte imbalance, and dehydration that develop. An overdose of insulin (*insulin shock*), however, can also produce a coma as a result of the hypoglycemia (abnormally low blood glucose levels) produced. The physical signs and symptoms of diabetic and hypoglycemic coma are sufficiently different to allow hospital personnel to distinguish between these two types.

Less severe symptoms of hypoglycemia are usually produced by an oversecretion of insulin from the islets of Langerhans after a carbohydrate meal. This **reactive hypoglycemia,** caused by an exaggerated response of the beta cells to a rise in blood glucose, is most commonly seen in adults who are genetically predisposed to type 2 diabetes. For this reason, people with reactive hypoglycemia must limit their intake of carbohydrates and eat small meals at frequent intervals, rather than two or three meals per day.

The symptoms of reactive hypoglycemia include tremor, hunger, weakness, blurred vision, and mental confusion. The appearance of some of these symptoms, however, does not necessarily indicate reactive hypoglycemia, and a given level of blood glucose does not always produce these symptoms. Diagnosis of reactive hypoglycemia is controversial, but an accepted criterion is a blood glucose concentration of less than 70 mg/dl when a person is experiencing hypoglycemic symptoms and a relief from these symptoms by a rise in blood glucose following a carbohydrate meal. In the oral glucose tolerance test, for example, reactive hypoglycemia is shown when the initial rise in blood glucose produced by the ingestion of a glucose solution triggers excessive insulin secretion, so that the blood glucose levels fall below normal within five hours (fig. 19.13). This test is no longer used because of the danger that it can trigger hypoglycemic symptoms.

CHECKPOINT

10a. Explain how ketoacidosis and dehydration are produced in a person with type 1 diabetes mellitus.

10b. Describe the causes of hyperglycemia in a person with type 2 diabetes. How may weight loss help to control this condition?

11. Explain the meaning of the terms *insulin resistance, impaired glucose tolerance,* and *reactive hypoglycemia.*

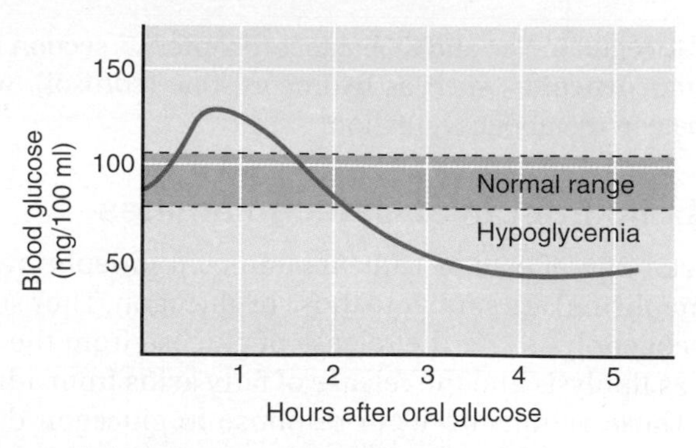

Figure 19.13 Reactive hypoglycemia. An idealized oral glucose tolerance test on a person with reactive hypoglycemia. The blood glucose concentration falls below the normal range within five hours of glucose ingestion as a result of excessive insulin secretion. Because this can be dangerous, the oral glucose tolerance test is no longer used for diagnosis of reactive hypoglycemia.

19.5 METABOLIC REGULATION BY ADRENAL HORMONES, THYROXINE, AND GROWTH HORMONE

Epinephrine, cortisol, thyroxine, and growth hormone stimulate the catabolism of carbohydrates and lipids. Thyroxine and growth hormone promote protein synthesis, body growth, and proper development of the central nervous system.

LEARNING OUTCOMES

After studying this section, you should be able to:

12. Explain how catecholamines, glucocorticoids, and thyroid hormones regulate metabolism.

13. Explain how growth hormone regulates metabolism.

The anabolic effects of insulin are antagonized by glucagon, as previously described, and by the actions of a variety of other hormones. The hormones of the adrenals, thyroid, and anterior pituitary (specifically growth hormone) antagonize the action of insulin on carbohydrate and lipid metabolism. The actions of insulin, thyroxine, and growth hormone, however, can act synergistically in the stimulation of protein synthesis.

Adrenal Hormones

The adrenal gland consists of two parts that function as separate glands, secreting different hormones and regulated by different control systems (chapter 11, section 11.4). The **adrenal medulla** secretes catecholamine hormones—epinephrine and lesser amounts of norepinephrine—in response to sympathetic nerve stimulation. The **adrenal cortex** secretes corticosteroid hormones. These are grouped into two functional categories: **mineralocorticoids,** such as aldosterone, which act on the kidneys to regulate Na^+ and K^+ balance (chapter 17, section 17.5), and **glucocorticoids,** such as hydrocortisone (cortisol), which participate in metabolic regulation.

Metabolic Effects of Catecholamines

The metabolic effects of catecholamines (epinephrine and norepinephrine) are similar to those of glucagon. They stimulate glycogenolysis and the release of glucose from the liver, as well as lipolysis and the release of fatty acids from adipose tissue. These actions occur in response to glucagon during fasting, when low blood glucose stimulates glucagon secretion, and in response to catecholamines during the fight-or-flight reaction to stress. The latter effect provides circulating energy substrates in anticipation of the need for intense physical activity. Glucagon and epinephrine have similar mechanisms of action, where both are mediated by cyclic AMP (fig. 19.14).

Sympathetic nerves, acting through the release of norepinephrine, can stimulate **β_3-adrenergic receptors** in brown adipose tissue (there appears to be few β_3 receptors in the ordinary, white fat of humans, and none in other tissues). As may be recalled from earlier in this chapter, brown fat is a specialized tissue that contains an *uncoupling protein* that dissociates electron transport from the production of ATP. As a result, brown fat can have a very high rate of energy expenditure (unchecked by negative feedback from ATP) that is stimulated by epinephrine.

Metabolic Effects of Glucocorticoids

Hydrocortisone (cortisol) and other glucocorticoids are secreted by the adrenal cortex in response to ACTH stimulation. The secretion of ACTH from the anterior pituitary occurs as part of the general adaptation syndrome in response to stress (chapter 11; section 11.4). Because prolonged fasting or prolonged exercise certainly qualify as stressors, ACTH—and thus glucocorticoid

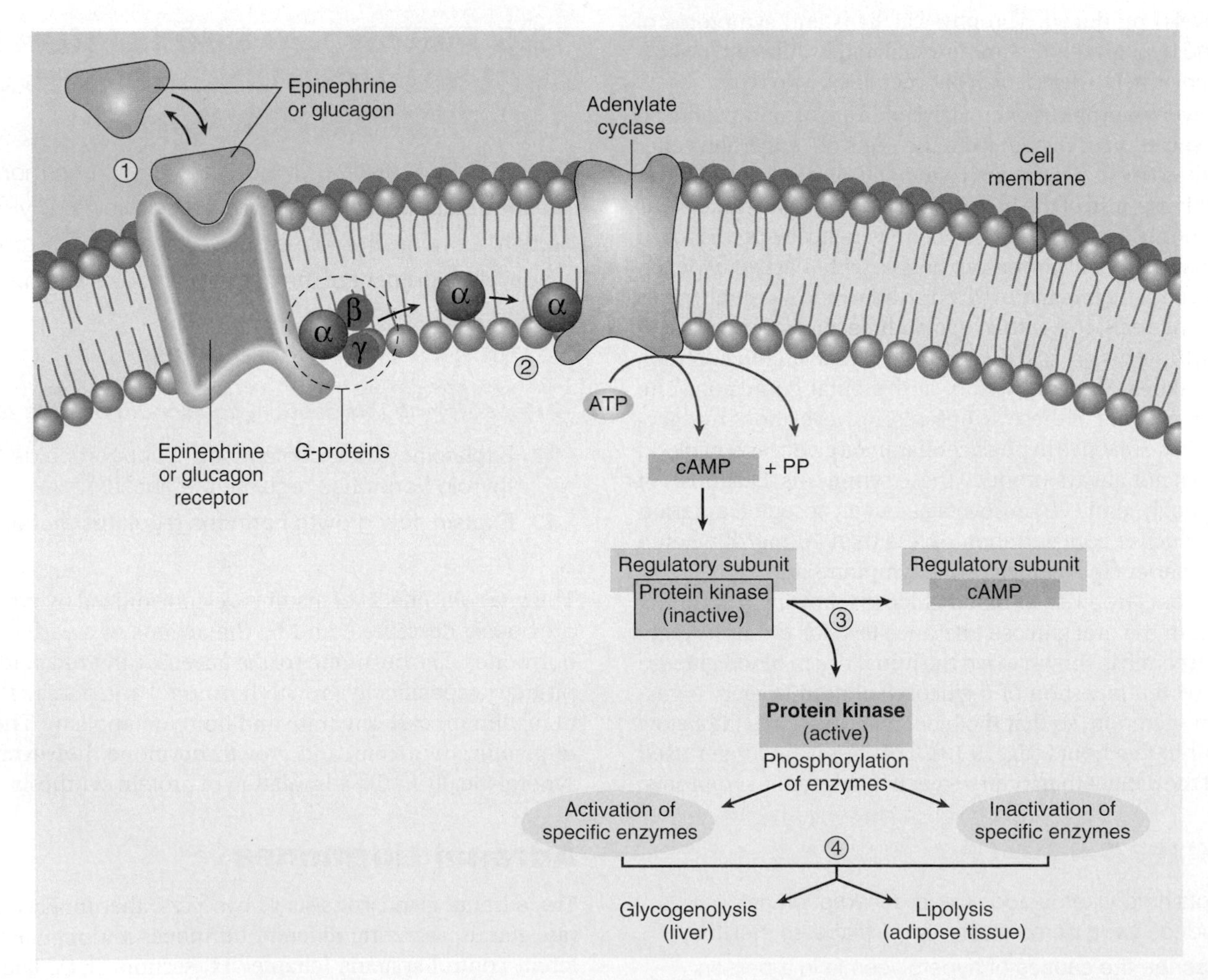

Figure 19.14 How epinephrine and glucagon affect metabolism. (1) The hormone binds to its receptor, causing G-proteins to dissociate. The alpha subunit diffuses through the membrane to activate adenylate cyclase, which catalyzes the production of cAMP as a second messenger. (3) The cAMP binds to and removes the regulatory subunit of protein kinase, activating this enzyme. (4) The activation and inactivation of different enzymes by protein kinase promotes glycogenolysis in the liver and lipolysis in adipose tissue.

secretion—is stimulated under these conditions. The increased secretion of glucocorticoids during prolonged fasting or exercise supports the effects of increased glucagon and decreased insulin secretion from the pancreatic islets.

Like glucagon, hydrocortisone promotes lipolysis and ketogenesis; it also stimulates the synthesis of hepatic enzymes that promote gluconeogenesis. Although hydrocortisone stimulates enzyme (protein) synthesis in the liver, it promotes protein breakdown in the muscles. Breakdown of muscle proteins increases the blood levels of amino acids, providing the substrates needed for gluconeogenesis by the liver. The release of circulating energy substrates—amino acids, glucose, fatty acids, and ketone bodies—into the blood in response to hydrocortisone (fig. 19.15) helps compensate for a state of prolonged fasting or exercise.

Thyroxine

The thyroid follicles secrete **thyroxine,** also called **tetraiodothyronine (T_4),** in response to stimulation by thyroid-stimulating hormone (TSH) from the anterior pituitary. The thyroid also secretes smaller amounts of **triiodothyronine (T_3)** in response to stimulation by TSH. Almost all organs in the body are targets of thyroxine action. Thyroxine itself, however, is not the active form of the hormone within the target cells; thyroxine is a prehormone that must first be converted to triiodothyronine (T_3) within the target cells to be active (chapter 11; see fig. 11.6). Acting via its conversion to T_3, thyroxine (1) regulates the rate of cell respiration and (2) contributes to proper growth and development, particularly during early childhood.

Thyroxine (via its conversion to T_3) stimulates the rate of cell respiration in almost all cells in the body—an effect believed to be due to a lowering of cellular ATP concentrations. This effect is produced by the production of uncoupling proteins (as in brown fat, discussed previously). ATP exerts an end-product inhibition (chapter 4, section 4.2) of cell respiration, so that when ATP concentrations decrease, the rate of cell respiration increases.

Much of the energy liberated during cell respiration escapes as heat, and uncoupling proteins increase the proportion of food energy that escapes as heat. Because thyroxine stimulates the production of uncoupling proteins and the rate of cell respiration, the actions of thyroxine increase the production of metabolic heat. The heat-producing, or *calorigenic* (*calor* = heat), *effects* of thyroxine are required for cold adaptation. Recent evidence suggests that thyroxine has a permissive effect on the ability of brown adipose tissue to generate heat in response to sympathetic nerve stimulation. Through this permissive effect, thyroxine contributes to adaptive thermogenesis as well as to the basal metabolic rate.

The basal metabolic rate (BMR) has two components—one that is independent of thyroxine action and one that is regulated by thyroxine. In this way, thyroxine acts to "set" the BMR. The BMR can thus be used as an index of thyroid function. Indeed, such measurements were used clinically to evaluate the condition of the thyroid prior to the development of direct chemical determinations of T_4 and T_3 in the blood. A person who is hypothyroid may have a basal O_2 consumption about 30% lower than normal, while a person who is hyperthyroid may have a basal O_2 consumption up to 50% higher than normal.

A normal level of thyroxine secretion is required for growth and proper development of the central nervous system in children. This is why hypothyroidism in children can cause cretinism. The symptoms of hypothyroidism and hyperthyroidism in adults are compared in chapter 11, table 11.8.

A normal level of thyroxine secretion is required in order to maintain a balance of anabolism and catabolism. For reasons that are incompletely understood, both hypothyroidism and hyperthyroidism cause protein breakdown and muscle wasting.

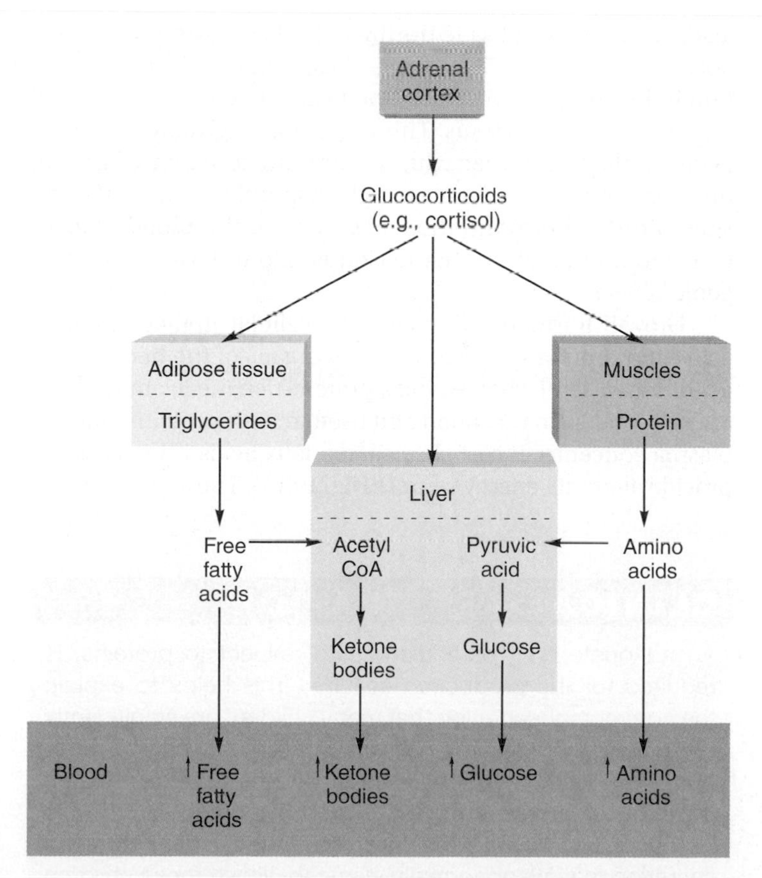

Figure 19.15 The metabolic effects of glucocorticoids. The catabolic actions of glucocorticoids help to raise the blood concentration of glucose and other energy-carrier molecules.

Growth Hormone

The anterior pituitary secretes **growth hormone,** also called **somatotropin,** in larger amounts than any other of its hormones. As its name implies, growth hormone stimulates growth in children and adolescents. The continued high secretion of growth hormone in adults, particularly under the conditions of fasting and other forms of stress, implies that this hormone can have important metabolic effects even after the growing years have ended.

Regulation of Growth Hormone Secretion

The secretion of growth hormone is inhibited by somatostatin, which is produced by the hypothalamus and secreted into the hypothalamo-hypophyseal portal system (chapter 11, section 11.3). In addition, there is also a growth hormone–releasing hormone (GHRH), which stimulates growth hormone secretion. Growth hormone thus appears to be unique among the anterior pituitary hormones in that its secretion is controlled by both a releasing and an inhibiting hormone from the hypothalamus. The secretion of growth hormone follows a circadian ("about a day") pattern, increasing during sleep and decreasing during periods of wakefulness.

Growth hormone secretion is stimulated by an increase in the plasma concentration of amino acids and by a decrease in the plasma glucose concentration. These events occur during absorption of a high-protein meal, when amino acids are absorbed. The secretion of growth hormone is also increased during prolonged fasting, when plasma glucose is low and plasma amino acid concentration is raised by the breakdown of muscle protein.

Insulin-like Growth Factors

Insulin-like growth factors (IGFs), produced by many tissues, are polypeptides that are similar in structure to proinsulin (chapter 3; see fig. 3.23). They have insulin-like effects and serve as mediators for some of growth hormone's actions. The term **somatomedins** is often used to refer to two of these factors, designated *IGF-1* and *IGF-2,* because they mediate the actions of somatotropin (growth hormone). The liver produces and secretes IGF-1 in response to growth hormone stimulation, and this secreted IGF-1 then functions as a hormone in its own right, traveling in the blood to the target tissue. A major target is cartilage, where IGF-1 stimulates cell division and growth. IGF-1 also functions as an autocrine regulator (chapter 11), because the chondrocytes (cartilage cells) themselves produce IGF-1 in response to growth hormone stimulation. The growth-promoting actions of IGF-1, acting as both a hormone and an autocrine regulator, directly mediate the effects of growth hormone on cartilage and serve as the major regulator of bone growth. These actions are supported by IGF-2, which has more insulin-like actions. The action of growth hormone in stimulating lipolysis in adipose tissue and in decreasing glucose utilization is apparently not mediated by the somatomedins (fig. 19.16).

Effects of Growth Hormone on Metabolism

The fact that growth hormone secretion is increased during fasting and also during absorption of a protein meal reflects the complex nature of this hormone's action. Growth hormone has both anabolic and catabolic effects; it promotes protein synthesis (anabolism) and in this respect is similar to insulin. It also stimulates the catabolism of fat and the release of fatty acids from adipose tissue during periods of fasting (the postabsorptive state). A rise in the plasma fatty acid concentration induced by growth hormone results in decreased rates of glycolysis in many organs. This inhibition of glycolysis by fatty acids, perhaps together with a more direct action of growth hormone, results in decreased glucose utilization by the tissues. Growth hormone thus acts to raise the blood glucose concentration, and for that reason is said to have a "diabetogenic" effect.

Growth hormone stimulates the cellular uptake of amino acids and protein synthesis in many organs of the body. These actions are useful when eating a protein-rich meal; amino acids are removed from the blood and used to form proteins, and the plasma concentration of glucose and fatty acids is increased to provide alternate energy sources (fig. 19.16). The anabolic effect

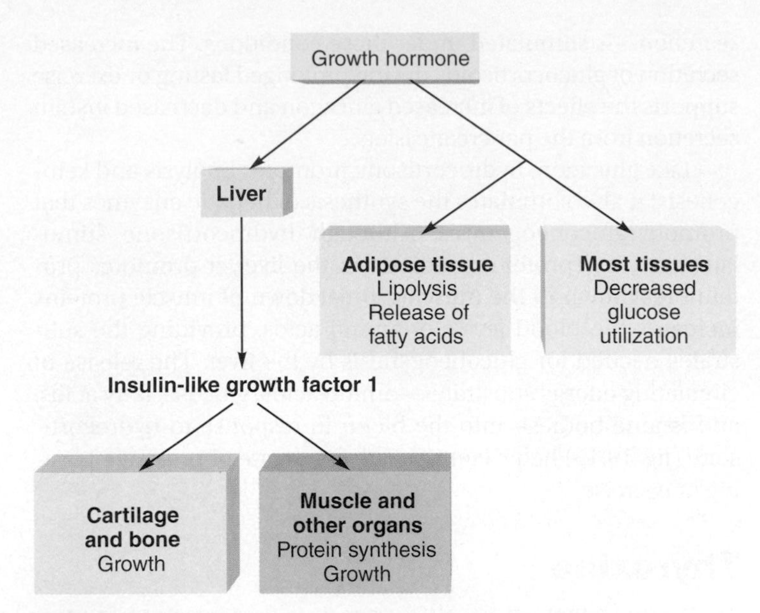

Figure 19.16 The metabolic effects of growth hormone. The growth-promoting, or anabolic, effects of growth hormone are mediated indirectly via stimulation of insulin-like growth factor 1 (also called somatomedin C) production by the liver.

CLINICAL APPLICATION

An adequate diet, particularly with respect to proteins, is required for the production of IGF-1.This helps to explain the common observation that many children are significantly taller than their parents, who may not have had an adequate diet in their youth. Children with protein malnutrition (**kwashiorkor**) have low growth rates and low levels of IGF-1 in the blood, despite the fact that their growth hormone secretion may be abnormally elevated. When these children are provided with an adequate diet, IGF-1 levels increase and growth is stimulated.

of growth hormone is particularly important during the growing years, when it contributes to increases in bone length and in the mass of many soft tissues.

Effects of Growth Hormone on Body Growth

The stimulatory effects of growth hormone on skeletal growth results from stimulation of mitosis in the epiphyseal growth plates of cartilage in the long bones of children and adolescents. This action is mediated by the somatomedins, IGF-1 and IGF-2, which stimulate the chondrocytes to divide and secrete more cartilage matrix. Part of this growing cartilage is converted to bone, enabling the bone to grow in length. This skeletal growth stops when the epiphyseal discs are converted to bone after the growth spurt during puberty, despite the continued secretion of growth hormone throughout adulthood.

An excessive secretion of growth hormone in children can produce **gigantism.** These children may grow up to 8 feet tall, at the same time maintaining normal body proportions. An excessive growth hormone secretion that occurs after the epiphyseal discs have sealed, however, cannot produce increases in height. In adults, the oversecretion of growth hormone results in an elongation of the jaw and deformities in the bones of the face, hands, and feet. This condition, called **acromegaly,** is accompanied by the growth of soft tissues and coarsening of the skin (fig. 19.17). It is interesting that athletes who take growth hormone supplements to increase their muscle mass may also experience body changes that resemble those of acromegaly.

An inadequate secretion of growth hormone during the growing years results in **dwarfism.** An interesting variant of this condition is *Laron dwarfism,* in which there is a genetic insensitivity to the effects of growth hormone. This insensitivity is associated with a reduction in the number of growth hormone receptors in the target cells. Genetic engineering has made available recombinant IGF-1, which has recently been approved by the FDA for the medical treatment of Laron dwarfism.

Age 9

Age 16

Age 33

Age 52

Figure 19.17 The progression of acromegaly in one individual. The coarsening of features and disfigurement are evident by age 33 and severe at age 52.

CLINICAL APPLICATION

Before **recombinant growth hormone** (produced by genetically engineered cells) became available, the supply of growth hormone was very limited because it could only be obtained from the pituitaries of cadavers. Now that recombinant growth hormone is available, children with idiopathic short stature (who do not have pituitary dwarfism) can receive growth hormone injections. In a recent study, such children, injected weekly for 10 years, attained an adult height that was significantly higher than predicted before the treatment. The use of growth hormone treatment for this purpose, however, is medically and ethically very controversial.

CHECKPOINT

12a. Describe the effects of epinephrine and the glucocorticoids on the metabolism of carbohydrates and lipids. What is the significance of these effects as a response to stress?

12b. Explain the actions of thyroxine on the basal metabolic rate. Why do people with hypothyroidism have a tendency to gain weight, and why are they less resistant to cold stress?

13a. Describe the effects of growth hormone on the metabolism of lipids, glucose, and amino acids.

13b. Explain how growth hormone stimulates skeletal growth.

19.6 REGULATION OF CALCIUM AND PHOSPHATE BALANCE

A normal blood Ca^{2+} concentration is critically important for contraction of muscles and maintenance of proper membrane permeability. Parathyroid hormone promotes an elevation in blood Ca^{2+} by stimulating resorption of the calcium phosphate crystals from bone and renal excretion of phosphate. 1,25-dihydroxyvitamin D_3 promotes the intestinal absorption of calcium and phosphate.

LEARNING OUTCOMES

After studying this section, you should be able to:

14. Describe the hormonal regulation of blood calcium and how this affects the skeletal system.

15. Explain how 1,25-dihydroxyvitamin D_3 is formed and describe its actions.

The calcium and phosphate concentrations of plasma are affected by bone formation and resorption, intestinal absorption of Ca^{2+} and PO_4^{3-}, and urinary excretion of these ions. These processes are regulated by parathyroid hormone, 1,25-dihydroxyvitamin D_3, and calcitonin, as summarized in table 19.6.

Bone Deposition and Resorption

The skeleton, in addition to providing support for the body, serves as a large store of calcium and phosphate in the form of crystals called *hydroxyapatite,* which has the formula $Ca_{10}(PO_4)_6(OH)_2$. The calcium phosphate in these hydroxyapatite crystals is derived from the blood by the action of bone-forming cells, or **osteoblasts.** The osteoblasts secrete an organic matrix composed largely of collagen protein, which becomes hardened by deposits of hydroxyapatite. This process is called **bone deposition. Bone resorption** (dissolution of hydroxyapatite), produced by the action of **osteoclasts** (fig. 19.18*a*), results in the return of bone calcium and phosphate to the blood.

The formation of new osteoclasts from their precursor cells is a highly regulated process. The osteoclast precursor cells produce a surface receptor protein known by its acronym, *RANK.* Osteoblasts produce the ligand for this receptor, known as *RANK ligand,* or *RANKL,* which binds to RANK and stimulates osteoclast development and activation. Although osteoblasts promote the development of osteoclasts through secretion of RANKL, they also secrete a molecule called *osteoprotegerin* that interferes with the ability of RANKL to bind to RANK and thereby blocks the development osteoclasts. Interestingly, there is a new drug currently being tested for the treatment of osteoporosis that is a monoclonal antibody to RANKL, acting (like osteoprotegerin) to prevent RANKL from binding to RANK.

Bone resorption begins when the osteoclast attaches to the bone matrix and forms a "ruffled membrane" (see fig. 19.18*b*). The bone matrix contains both an inorganic component (the calcium phosphate crystals) and an organic component (collagen and other proteins), and so the osteoclast must secrete products that both dissolve calcium phosphate and digest the proteins of the bone matrix. The dissolution of calcium phosphate is accomplished by transport of H^+ by a H^+-ATPase pump in the ruffled membrane, thereby acidifying the bone matrix (to a pH of about 4.5) immediately adjacent to the osteoclast. A channel for Cl^- allows Cl^- to follow the H^+, preserving electrical neutrality. The H^+ is derived from carbonic acid, and the Cl^- is obtained by an active transport Cl^-/HCO_3^- pump on the opposite side of the osteoclast (fig. 19.18*b*).

The protein component of the bone matrix is digested by enzymes, primarily one called cathepsin K, released by the osteoclasts. The osteoclast can then move to another site and begin the resorption process again, or be eliminated. Interestingly, there is evidence that estrogen, sometimes given to treat osteoporosis in post-menopausal women, works in part by stimulating the apoptosis (cell suicide) of osteoclasts.

The formation and resorption of bone occur constantly at rates determined by the relative activity of osteoblasts and osteoclasts. Body growth during the first two decades of life occurs because bone formation proceeds at a faster rate than

Table 19.6 | Endocrine Regulation of Calcium and Phosphate Balance

Hormone	Effect on Intestine	Effect on Kidneys	Effect on Bone	Associated Diseases
Parathyroid hormone (PTH)	No direct effect	Stimulates Ca^{2+} reabsorption; inhibits PO_4^{3-} reabsorption	Stimulates resorption	Osteitis fibrosa cystica with hypercalcemia due to excess PTH
1,25-dihydroxyvitamin D_3	Stimulates absorption of Ca^{2+} and PO_4^{3-}	Stimulates reabsorption of Ca^{2+} and PO_4^{3-}	Stimulates resorption	Osteomalacia (adults) and rickets (children) due to deficiency of 1,25-dihydroxyvitamin D_3
Calcitonin	None	Inhibits resorption of Ca^{2+} and PO_4^{3-}	Stimulates deposition	None

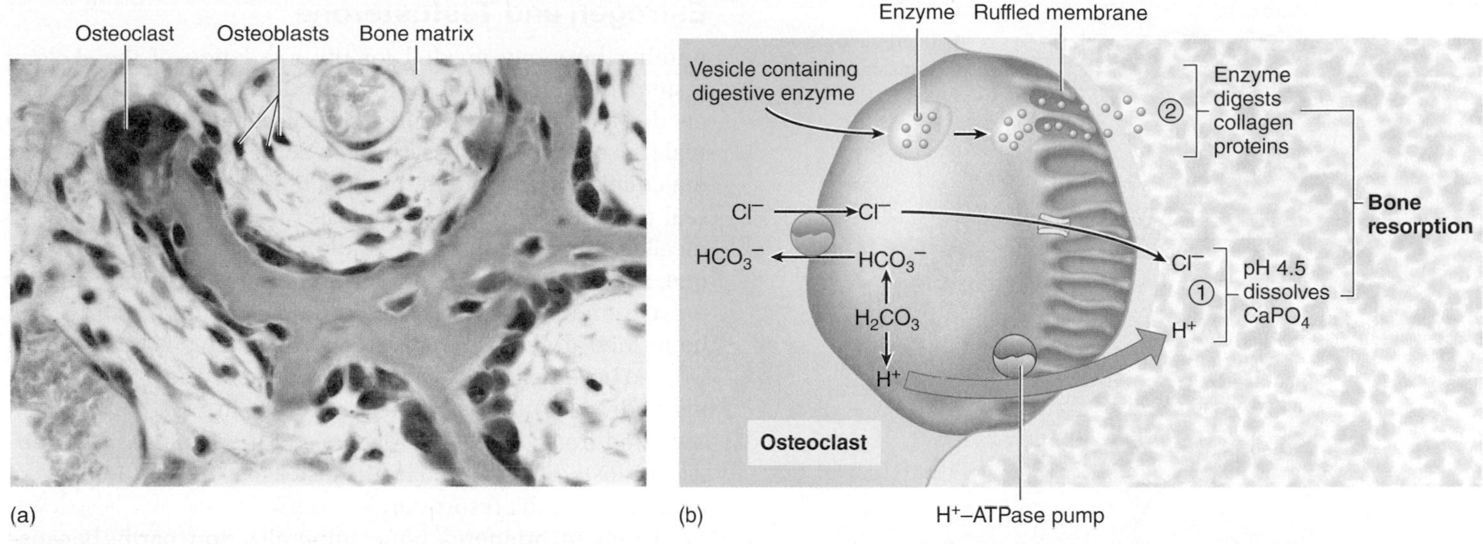

Figure 19.18 The resorption of bone by osteoclasts. (*a*) A photomicrograph showing osteoclasts and bone matrix. (*b*) Figure depicting the mechanism of bone resorption. (1) The bone is first demineralized by the dissolution of $CaPO_4$ from the matrix due to acid secretion by the osteoclast. (2) After that, the organic component of the matrix (mainly collagen) is digested by the secretion of enzyme molecules (an enzyme called cathepsin K) from the osteoclast.

bone resorption. The constant activity of osteoblasts and osteoclasts allows bone to be remodeled throughout life. The position of the teeth, for example, can be changed by orthodontic appliances (braces), which cause bone resorption on the pressure-bearing side and bone formation on the opposite side of the alveolar sockets. Peak bone mass occurs when people are in their 30s, and subsequently begins to decline. By the age of 50 or 60, the rate of bone resorption often exceeds the rate of bone deposition.

Despite the changing rates of bone formation and resorption, the plasma concentrations of calcium and phosphate are maintained by hormonal control of the intestinal absorption and urinary excretion of these ions. These hormonal control mechanisms are very effective in maintaining the plasma calcium and phosphate concentrations within narrow limits. Plasma calcium, for example, is normally maintained at about 2.5 millimolar, or 5 milliequivalents per liter (a milliequivalent equals a millimole multiplied by the valence of the ion; in this case, × 2).

The maintenance of normal plasma calcium concentrations is important because of the wide variety of effects that calcium has in the body. Calcium is needed for blood clotting, for example, and for a variety of cell signaling functions. These include the role of calcium as a second messenger of hormone action (chapter 11), as a signal for neurotransmitter release from axons in response to action potentials (chapter 7), and as the stimulus for muscle contraction in response to electrical excitation (chapter 12).

Calcium is also needed to maintain proper membrane permeability. An abnormally low plasma calcium concentration increases the permeability of the cell membranes to Na^+ and other ions. Hypocalcemia, therefore, enhances the excitability of nerves and muscles and can result in muscle spasm (tetany).

FITNESS APPLICATION

The rate of bone deposition equals the rate of bone resorption in healthy people on earth. In the **microgravity** (essentially, weightlessness) of space, however, astronauts have suffered from a slow, progressive loss of calcium from the weight-bearing bones of the legs and spine. For reasons that are not presently understood, about 100 mg of calcium are lost per day, which has reduced bone mineral density up to 20% in some astronauts who have been in space for several months. This loss cannot be countered simply by giving astronauts calcium, since hypercalcemia may cause kidney stones and other problems. The exercise machines that have been used in space have helped to prevent loss of muscle mass in astronauts, but they have not been effective in countering the problem of bone resorption.

Hormonal Regulation of Bone

Parathyroid Hormone and Calcitonin

Whenever the plasma concentration of Ca^{2+} begins to fall, the parathyroid glands are stimulated to secrete increased amounts of **parathyroid hormone (PTH)** that work to raise the blood Ca^{2+} back to normal levels. As might be predicted from this action of PTH, people who have their parathyroid glands removed (as may occur accidentally during surgical removal of the thyroid) will experience hypocalcemia. *Hypocalcemia* (low plasma calcium concentrations) is a common clinical condition with many potential causes, including inadequate parathyroid hormone secretion or activation of its receptors, insufficient vitamin D, insufficient magnesium (needed for PTH secretion and action), and others.

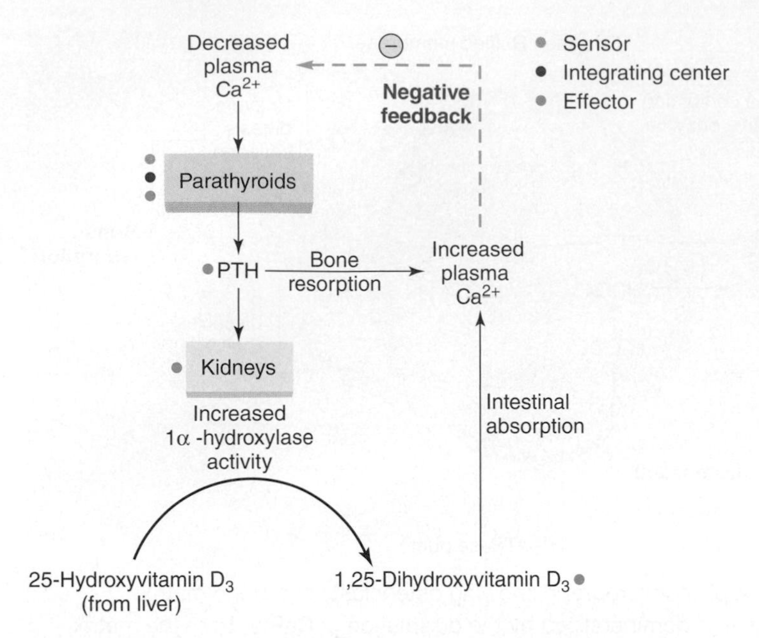

Figure 19.19 The negative feedback control of parathyroid hormone secretion. A decrease in plasma Ca^{2+} directly stimulates the secretion of parathyroid hormone (PTH). The production of 1,25-dihydroxyvitamin D_3 also rises when Ca^{2+} is low because PTH stimulates the final hydroxylation step in the formation of this compound in the kidneys.

Hypocalcemia can cause muscle twitching, spasms, severe tetany, cardiac abnormalities, and other symptoms.

Parathyroid hormone helps to raise the blood Ca^{2+} concentration through three mechanisms (fig. 19.19):

1. PTH stimulates osteoclasts to resorb bone, thereby adding Ca^{2+} and PO_4^{3-} to the blood. This is the primary mechanism of PTH action.
2. PTH stimulates the kidneys to reabsorb Ca^{2+}, but acts to decrease the renal reabsorption of PO_4^{3-}. This raises blood Ca^{2+} levels without promoting the deposition of calcium phosphate crystals in bone.
3. PTH stimulates the kidneys to produce the enzyme (1α-hydroxylase) needed to convert 25-hydroxyvitamin D_3 into the active hormone, 1,25-dihydroxyvitamin D_3 (see fig. 19.21). This active form of vitamin D then promotes the absorption of Ca^{2+} and PO_4^{3-} from food and drink across the intestinal epithelium.

Experiments in the 1960s revealed that high blood calcium in dogs could be lowered by a hormone secreted from the thyroid gland. This hormone, **calcitonin**, is secreted by the *parafollicular cells,* or *C cells,* of the thyroid, and has an effect opposite to that of parathyroid hormone and 1,25-dihydroxyvitamin D_3 (chapter 11, section 11.5). Though its physiological significance in humans is questionable, its pharmacological action (as a drug) can be useful—it inhibits the resorption of bone. People with stress fractures of vertebrae due to osteoporosis (discussed in the next Clinical Application box), may be helped by injections or nasal sprays of calcitonin.

Estrogen and Testosterone

Another hormone needed for the regulation of the skeletal system is **estrogen.** In both men and women, estrogen is needed for the epiphyseal discs (the cartilage growth plates) to seal (become bone). This estrogen comes from the ovaries via the circulation in women, but it is formed within the epiphyseal discs from circulating **testosterone** in men. (As may be recalled from chapter 11, estradiol is derived from testosterone; see fig. 11.2.) Also, proper bone mineralization, and the prevention of osteoporosis, requires the action of estrogen in bone. Men are less prone to osteoporosis than postmenopausal women because men can form estrogen (derived from circulating androgens) in their bones, whereas in women, estrogen secretion from the ovaries declines at menopause. Additionally, testosterone secreted by the testes may have a direct effect on bone to inhibit resorption.

Estrogen promotes bone mineralization partly because it stimulates the actions of osteoblasts. As mentioned previously, osteoblasts secrete new bone matrix and also produce molecules—RANK ligand (RANKL) and osteoprotegerin—that regulate osteoclast production. Estrogen affects this regulation and by that means suppresses the formation of new osteoclasts and promotes the apoptosis (cell suicide) of existing osteoclasts. In summary, estrogen promotes bone mineralization by stimulating bone deposition by osteoblasts and inhibiting bone resorption by osteoclasts.

Effects of Other Hormones

People with hyperthyroidism are more prone to osteoporosis. Since both osteoblasts and osteocytes have receptors for T_3 (triiodothyronine, derived from thyroxine—see chapter 11, figs. 11.6 and 11.7), the thyroid hormones would seem to be implicated. However, the possible role of thyroid hormones in bone physiology is poorly understood.

Leptin is a hormone discussed previously that is secreted by adipocytes and acts on the hypothalamus to reduce hunger and (through stimulation of sympathoadrenal activity) increase metabolic rate. Leptin also influences the activity of osteoblasts. Although its regulation of bone is complex, its net effect appears to be stimulation of osteoblast proliferation. This effect is believed to be indirect; leptin acts on the hypothalamus to increase sympathetic nerve activity, and sympathetic nerves stimulate osteoblasts through the binding of norepinephrine to β_2-adrenergic receptors. This leptin effect helps explain why people who are obese generally have thicker bones, while those who are malnourished (such as people with anorexia) have thinner bones.

Insulin promotes bone growth by suppressing an inhibitor of osteoblast development. Insulin also appears to stimulate osteoblasts to secrete a hormone known as *osteocalcin.* Scientists showed that when active osteocalcin is released into the blood, it stimulates the beta cells of the pancreatic islets to secrete insulin. This represents a previously unknown positive feedback mechanism regulating both bone and insulin secretion, and ties together the discussions of the regulation of metabolism and bone in this chapter.

As previously described, sex hormones promote bone growth; indeed, the loss of these hormones causes a decrease in bone mass. A recent report suggests that bone can, in turn, influence the reproductive system. Osteocalcin from osteoblasts was shown to stimulate the Leydig cells of the testes to secrete testosterone (chapter 20, section 20.3) in mice. The physiological significance of this effect in humans is not yet understood.

CLINICAL APPLICATION

The most common bone disorder in elderly people is **osteoporosis.** Osteoporosis is characterized by parallel losses of mineral and organic matrix from bone, reducing bone mass and density (fig. 19.20) and increasing the risk of fractures. Although the causes of osteoporosis are not well understood, age-related bone loss occurs more rapidly in women than in men (osteoporosis is almost 10 times more common in women after menopause than in men at comparable ages), suggesting that the fall in estrogen secretion at menopause contributes to this condition. The withdrawal of sex steroids causes increased formation of osteoclasts, producing an imbalance between bone resorption and bone deposition. Premenopausal women who have a very low percentage of body fat and amenorrhea can also have osteoporosis.

Physicians advise teenage girls, who are attaining their maximum bone mass, to eat such calcium-rich foods as milk and other dairy products. This may reduce the progression of osteoporosis when they get older. Additionally, calcium supplementation and other dietary changes are recommended for women prior to menopause. Estrogen replacement therapy is sometimes given for postmenopausal women, but this practice has been reduced in favor of alternative treatments for osteoporosis that may be safer. A selective estrogen receptor modulator (*SERM;* see chapter 11, p. 325), known as *raloxifene* (Evista), is currently approved for the treatment of osteoporosis. However, the *bisphosphonates* are the most commonly used drugs. These drugs bind to the hydroxyapatite mineral phase and work at the osteoclast's ruffled border, where they disrupt the attachment of the osteoclast to bone; they are also taken into the osteoclast and promote apoptosis. Both actions reduce bone resorption. In certain cases, the hormone calcitonin (generally derived from salmon), either as an injection or as a nasal spray, is also used to treat osteoporosis.

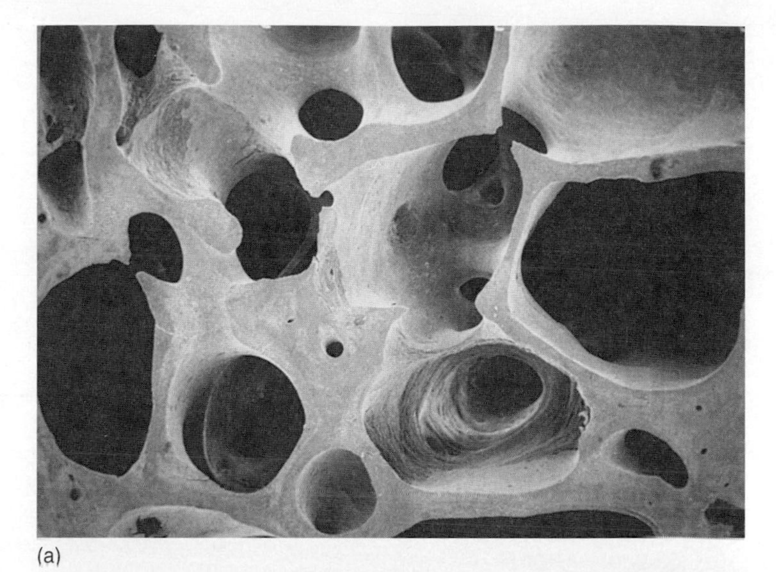

(a)

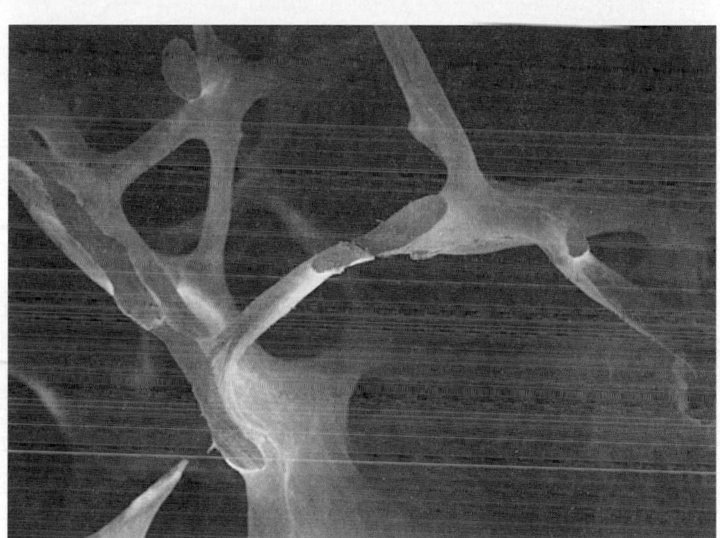

(b)

Figure 19.20 Scanning electron micrographs of bone. These biopsy specimens were taken from the iliac crest. Compare bone thickness in (*a*) a normal specimen and (*b*) a specimen from a person with osteoporosis. From L. G. Raisz, S.W. Dempster, et al., "Mechanisms of Disease" in *New England Journal of Medicine,* Vol. 218 (13):818.

1,25-Dihydroxyvitamin D_3

The production of **1,25-dihydroxyvitamin D_3** begins in the skin, where vitamin D_3 is produced from its precursor molecule (7-dehydrocholesterol) under the influence of ultraviolet B in sunlight. In equatorial regions of the globe, exposure to sunlight can allow sufficient cutaneous production of vitamin D_3. In more northerly or southerly latitudes, however, exposure to the winter sun may not allow sufficient production of vitamin D_3. When the skin does not make sufficient amounts of vitamin D_3, this compound must be ingested in the diet—that is why it is called a vitamin. Production of vitamin D in the skin provides most of a person's vitamin D; food sources of vitamin D—including fortified milk, eggs, and fish—provide only an average of 10% to 20% (in the absence of vitamin D supplementation). Whether this compound is secreted into the blood from the skin or enters the blood after being absorbed from the intestine, vitamin D_3 functions as a *prehormone;* in order to be biologically active, it must be chemically changed (chapter 11, section 11.1).

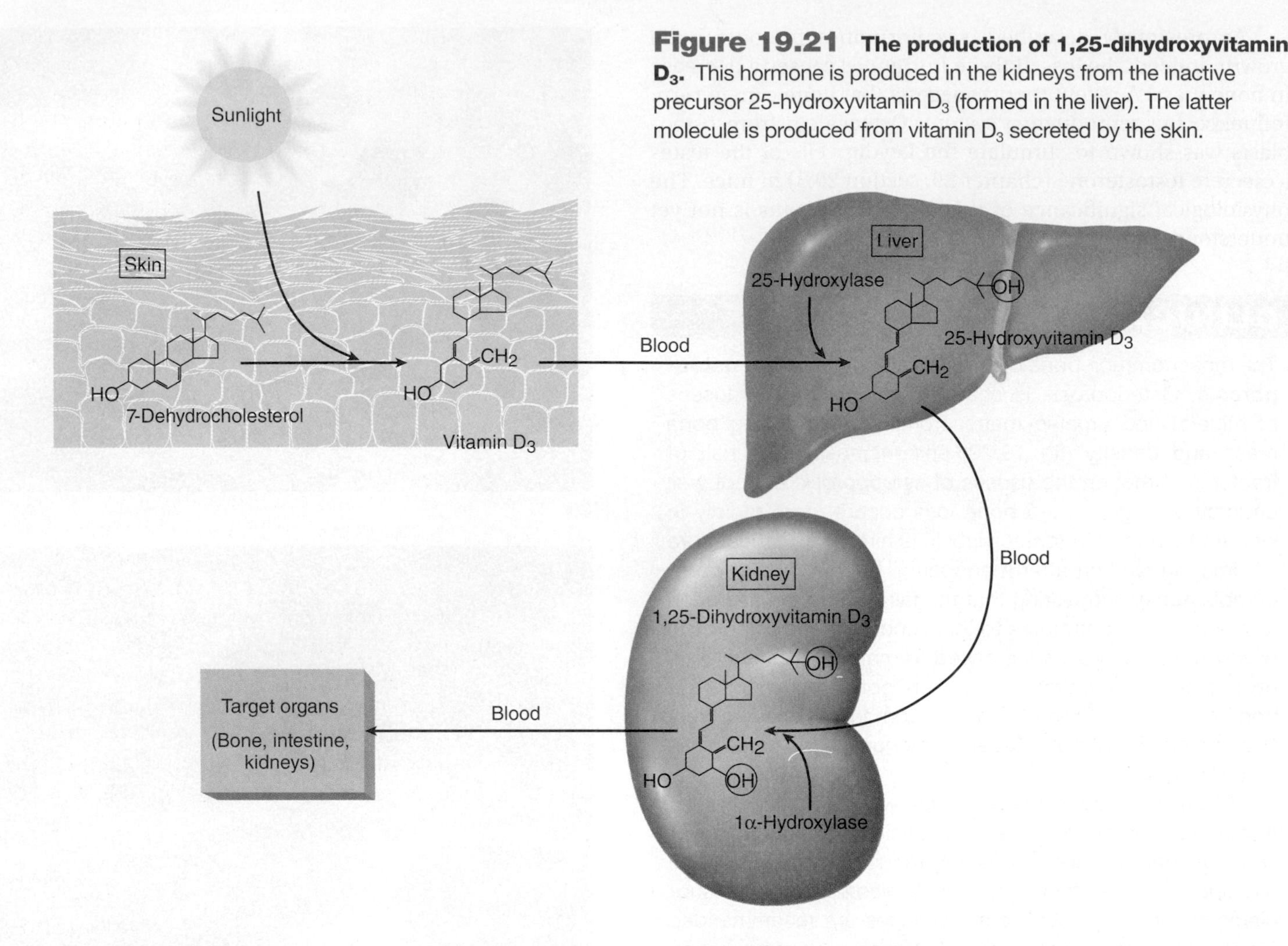

Figure 19.21 The production of 1,25-dihydroxyvitamin D_3. This hormone is produced in the kidneys from the inactive precursor 25-hydroxyvitamin D_3 (formed in the liver). The latter molecule is produced from vitamin D_3 secreted by the skin.

An enzyme in the liver adds a hydroxyl group (OH) to carbon number 25, which converts vitamin D_3 into 25-hydroxyvitamin D_3. In order to be active, however, another hydroxyl group must be added to carbon number 1. Hydroxylation of the first carbon is accomplished by an enzyme in the kidneys, which converts the molecule to 1,25-dihydroxyvitamin D_3 (fig. 19.21). The activity of this enzyme in the kidneys is stimulated by parathyroid hormone (see fig. 19.19). Increased secretion of PTH, stimulated by low blood Ca^{2+}, is thus accompanied by the increased production of 1,25-dihydroxyvitamin D_3.

The hormone 1,25-dihydroxyvitamin D_3 helps raise the plasma concentrations of calcium and phosphate by stimulating (1) the intestinal absorption of calcium and phosphate, (2) the resorption of calcium and phosphate from bones, and (3) the renal reabsorption of calcium and phosphate so that less is excreted in the urine. Notice that 1,25-dihydroxyvitamin D_3, but not parathyroid hormone, directly stimulates intestinal absorption of calcium and phosphate and promotes the reabsorption of phosphate in the kidneys. The effect of simultaneously raising the blood concentrations of Ca^{2+} and PO_4^{3-} results in the increased tendency of these two ions to precipitate as hydroxyapatite crystals in bone.

1,25-dihydroxyvitamin D_3 is needed for proper bone mineralization—indeed, inadequate amounts of this molecule cause the demineralization of osteomalacia and rickets. Yet 1,25-dihydroxyvitamin D_3 also stimulates bone resorption by promoting the formation of osteoclasts. This apparent paradox can be resolved by considering when each action predominates. When calcium intake is adequate, 1,25-dihydroxyvitamin D_3 stimulates the intestinal absorption of Ca^{2+} and PO_4^{3-} and thereby promotes bone deposition. Only when the calcium intake is inadequate does the direct effect of 1,25-dihydroxyvitamin D_3 on bone resorption become significant and raise blood Ca^{2+} to maintain homeostasis.

1,25-dihydroxyvitamin D_3 is also formed as an autocrine/paracrine regulator by the skin, breast, colon, prostate, and some of the cells of the immune system. These convert 25-hydroxyvitamin D_3 (from the liver) into the active regulator, which remains within the tissues or organ that produces it. As an autocrine or paracrine regulator within these tissues and organs, 1,25-dihydroxyvitamin D_3 promotes cell differentiation and inhibits cell proliferation (thereby protecting against cancer), and aids the function of the immune system (helping to defend against infections). There are clinical applications presently available that utilize these functions of vitamin D (such as treatment for the hyperproliferative skin disease *psoriasis,* as discussed on p. 667), and new clinical applications may emerge in the future.

CLINICAL APPLICATION

In addition to osteoporosis, a number of other bone disorders are associated with abnormal calcium and phosphate balance. In **osteomalacia** (in adults) and **rickets** (in children), inadequate intake of vitamin D results in inadequate calcification of the organic matrix in bones. Excessive secretion of parathyroid hormone results in **osteitis fibrosa cystica,** in which excessive osteoclast activity causes resorption of both the mineral and organic components of bone, which are then replaced by fibrous tissue.

Negative Feedback Control of Calcium and Phosphate Balance

The secretion of parathyroid hormone is controlled by the plasma calcium concentrations. Its secretion is stimulated by low calcium concentrations and inhibited by high calcium concentrations. Since parathyroid hormone stimulates the final hydroxylation step in the formation of 1,25-dihydroxyvitamin D_3, a rise in parathyroid hormone results in an increase in production of 1,25-dihydroxyvitamin D_3. Low blood calcium can thus be corrected by the effects of increased parathyroid hormone and 1,25-dihydroxyvitamin D_3 (fig. 19.22).

It is possible for plasma calcium levels to fall while phosphate levels remain normal. In this case, the increased secretion of parathyroid hormone and the production of 1,25-dihydroxyvitamin D_3 that result could abnormally raise phosphate levels while acting to restore normal calcium levels. This is prevented by the inhibition of phosphate reabsorption in the kidneys by parathyroid hormone, so that more phosphate is excreted in the urine (fig. 19.22). In this way, blood calcium can be raised to normal levels without excessively raising blood phosphate concentrations.

The secretion of calcitonin is stimulated by high plasma calcium levels and acts to lower blood calcium by (1) inhibiting the activity of osteoclasts, thus reducing bone resorption, and (2) stimulating the urinary excretion of calcium and phosphate by inhibiting their reabsorption in the kidneys (fig. 19.23).

Although it is attractive to think that calcium balance is regulated by the effects of antagonistic hormones, the significance of calcitonin in human physiology remains unclear. Patients who have had their thyroid gland surgically removed (as for thyroid cancer) are *not* hypercalcemic, as one might expect them to be if calcitonin were needed to lower blood calcium levels. The ability of very large pharmacological doses of calcitonin to inhibit osteoclast activity and bone resorption, however, is clinically useful in the treatment of *Paget's disease,* in which osteoclast activity causes softening of bone. It is sometimes also used to treat osteoporosis, as previously described.

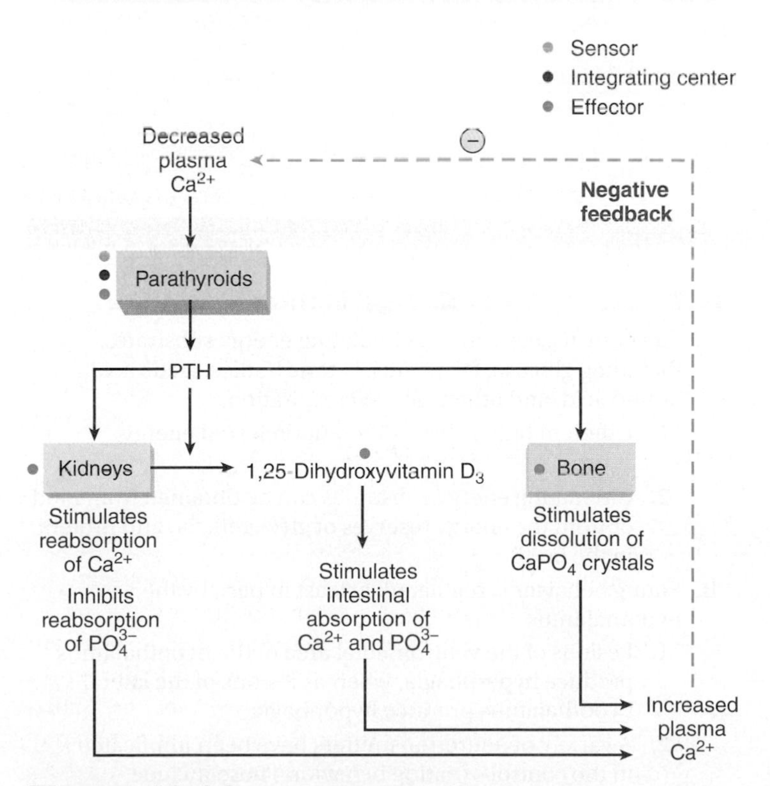

Figure 19.22 Homeostasis of plasma Ca^{2+} concentrations. A negative feedback loop returns low blood Ca^{2+} concentrations to normal without simultaneously raising blood phosphate levels above normal.

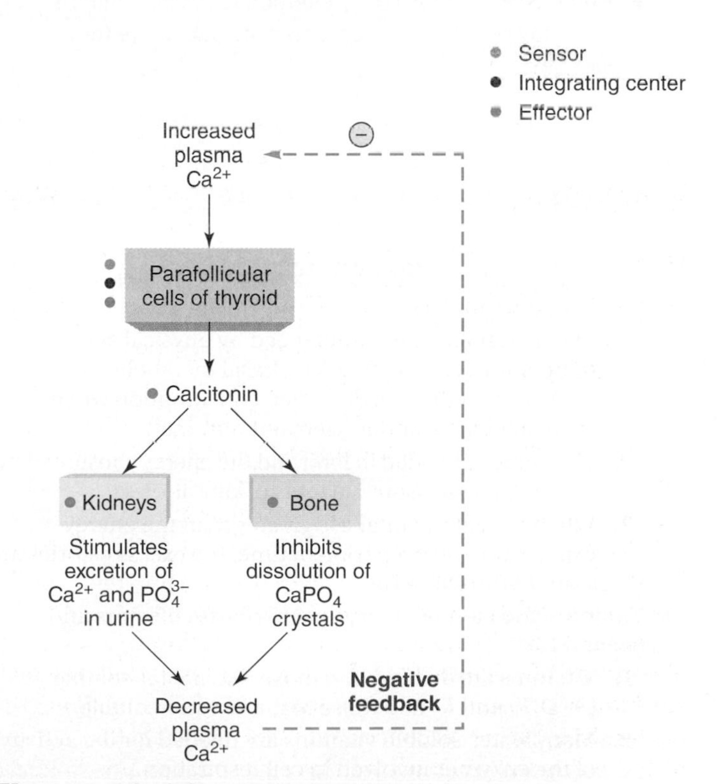

Figure 19.23 The negative feedback control of calcitonin secretion. The action of calcitonin is antagonistic to that of parathyroid hormone.

CHECKPOINT

14. Describe the mechanisms by which the secretion of parathyroid hormone and of calcitonin is regulated.

15a. List the steps involved in the formation of 1,25-dihydroxyvitamin D_3 and state how this formation is influenced by parathyroid hormone.

15b. Describe the actions of parathyroid hormone, 1,25-dihydroxyvitamin D_3, and calcitonin on the intestine, skeletal system, and kidneys, and explain how these actions affect the blood levels of calcium.

15c. Account for the different effects of 1,25-dihydroxy-vitamin D_3 on bones according to whether calcium intake is adequate or inadequate.

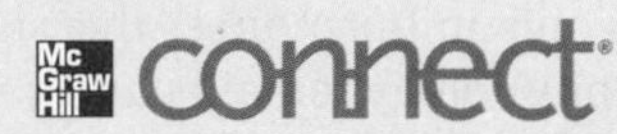

Visit this book's website at **www.mhhe.com/Fox13** for:

- Chapter quizzes, interactive learning exercises, and other study tools
- Additional clinical investigations
- Access to LearnSmart—an adaptive diagnostic tool that constantly assesses student knowledge of course material
- Ph.I.L.S. 4.0—physiology interactive lab simulations that may be used to supplement or substitute for wet labs

Clinical Investigation SUMMARY

Lisa's frequent urinations (polyuria) probably are causing her thirst and other symptoms. These symptoms, combined with her family history of a mother and an uncle with diabetes, suggest that Lisa might have diabetes mellitus. Indeed, polyuria, polyphagia (frequent eating), and polydipsia (frequent drinking)—the "three P's"—are cardinal symptoms of diabetes mellitus. The fasting hyperglycemia (blood glucose concentration of 150 mg/dl) confirm the diagnosis of diabetes mellitus. This abnormally high fasting blood glucose is too low to result in glycosuria. She could have glycosuria after meals, however, which would be responsible for her polyuria. The oral glucose tolerance test further confirms the diagnosis of diabetes mellitus, and the observations that this condition appears to have begun in middle age and that it is not accompanied by ketosis and ketonuria suggest that it is type 2 diabetes mellitus. This being the case, she could increase her tissue sensitivity to insulin by diet and exercise. If this is not sufficient, she could reduce her insulin resistance with the oral medicines metformin or the thiazolidinediones.

See the additional chapter 19 Clinical Investigations on *Sustained Hyperglycemia* and *Hypocalcemia* in the Connect site for this text at www.mhhe.com/Fox13.

SUMMARY

19.1 Nutritional Requirements 661

A. Food provides molecules used in cell respiration for energy.

1. The metabolic rate is influenced by physical activity, temperature, and eating. The basal metabolic rate is measured as the rate of oxygen consumption when such influences are standardized and minimal.
2. The energy provided in food and the energy consumed by the body are measured in units of kilocalories.
3. When the caloric intake is greater than the energy expenditure over a period of time, the excess calories are stored primarily as fat.

B. Vitamins and elements serve primarily as cofactors and coenzymes.

1. Vitamins are divided into those that are fat-soluble (A, D, E, and K) and those that are water-soluble.
2. Many water-soluble vitamins are needed for the activity of the enzymes involved in cell respiration.
3. The fat-soluble vitamins A and D have specific functions but share similar mechanisms of action, activating nuclear receptors and regulating genetic expression.

19.2 Regulation of Energy Metabolism 668

A. The body tissues can use circulating energy substrates, including glucose, fatty acids, ketone bodies, lactic acid, amino acid, and others, for cell respiration.

1. Different organs have different preferred energy sources.
2. Circulating energy substrates can be obtained from food or from the energy reserves of glycogen, fat, and protein in the body.

B. Eating behavior is regulated, at least in part, by the hypothalamus.

1. Lesions of the ventromedial area of the hypothalamus produce hyperphagia, whereas lesions of the lateral hypothalamus produce hypophagia.
2. A variety of neurotransmitters have been implicated in the control of eating behavior. These include the endorphins, norepinephrine, serotonin, cholecystokinin, and neuropeptide Y.

C. Adipose cells, or adipocytes, are both the targets of hormonal regulation and themselves endocrine in nature.

1. In children, circulating saturated fatty acids promote cell division and differentiation of new adipocytes. This activity involves the bonding of fatty acid and prostaglandin ligands with a nuclear receptor known as PPARγ.
2. Adipocytes secrete leptin, which regulates food intake and metabolism, and TNF$_\alpha$, which may help to regulate the sensitivity of skeletal muscles to insulin.

D. The control of energy balance in the body is regulated by the anabolic and catabolic effects of a variety of hormones.

19.3 Energy Regulation by the Pancreatic Islets 677

A. A rise in plasma glucose concentration stimulates insulin and inhibits glucagon secretion.
1. Amino acids stimulate the secretion of both insulin and glucagon.
2. Insulin secretion is also stimulated by parasympathetic innervation of the islets and by the action of intestinal hormones such as gastric inhibitory peptide (GIP).

B. During the intestinal absorption of a meal, insulin promotes the uptake of blood glucose into skeletal muscle and other tissues.
1. This lowers the blood glucose concentration and increases the energy reserves of glycogen, fat, and protein.
2. Skeletal muscles are the major organs that remove blood glucose in response to insulin stimulation.

C. During periods of fasting, insulin secretion decreases and glucagon secretion increases.
1. Glucagon stimulates glycogenolysis in the liver, gluconeogenesis, lipolysis, and ketogenesis.
2. These effects help to maintain adequate levels of blood glucose for the brain and provide alternate energy sources for other organs.

19.4 Diabetes Mellitus and Hypoglycemia 681

A. Diabetes mellitus and reactive hypoglycemia represent disorders of the islets of Langerhans.
1. Type 1 diabetes mellitus occurs when the beta cells are destroyed; the resulting lack of insulin and excessive glucagon secretion produce the symptoms of this disease.
2. Type 2 diabetes mellitus occurs as a result of a relative tissue insensitivity to insulin and inadequate insulin secretion; this condition is aggravated by obesity and improved by exercise.
3. Reactive hypoglycemia occurs when the islets secrete excessive amounts of insulin in response to a rise in blood glucose concentration.

19.5 Metabolic Regulation by Adrenal Hormones, Thyroxine, and Growth Hormone 685

A. The adrenal hormones involved in energy regulation include epinephrine from the adrenal medulla and glucocorticoids (mainly hydrocortisone) from the adrenal cortex.
1. The effects of epinephrine are similar to those of glucagon. Epinephrine stimulates glycogenolysis and lipolysis, and activates increased metabolism of brown fat.
2. Glucocorticoids promote the breakdown of muscle protein and the conversion of amino acids to glucose in the liver.

B. Thyroxine stimulates the rate of cell respiration in almost all cells in the body.
1. Thyroxine sets the basal metabolic rate (BMR), which is the rate at which energy (and oxygen) is consumed by the body under resting conditions.
2. Thyroxine also promotes protein synthesis and is needed for proper body growth and development, particularly of the central nervous system.

C. The secretion of growth hormone is regulated by releasing and inhibiting hormones from the hypothalamus.
1. The secretion of growth hormone is stimulated by a protein meal and by a fall in glucose, as occurs during fasting.
2. Growth hormone stimulates catabolism of lipids and inhibits glucose utilization.
3. Growth hormone also stimulates protein synthesis, and thus promotes body growth.
4. The anabolic effects of growth hormone, including the stimulation of bone growth in childhood, are produced indirectly via polypeptides called insulin-like growth factors, or somatomedins.

19.6 Regulation of Calcium and Phosphate Balance 690

A. Bone contains calcium and phosphate in the form of hydroxyapatite crystals. This serves as a reserve supply of calcium and phosphate for the blood.
1. The formation and resorption of bone are produced by the action of osteoblasts and osteoclasts, respectively.
2. The plasma concentrations of calcium and phosphate are also affected by absorption from the intestine and by the urinary excretion of these ions.

B. Parathyroid hormone stimulates bone resorption and calcium reabsorption in the kidneys. This hormone thus acts to raise the blood calcium concentration.
1. The secretion of parathyroid hormone is stimulated by a fall in blood calcium levels.
2. Parathyroid hormone also inhibits reabsorption of phosphate in the kidneys, so that more phosphate is excreted in the urine.

C. 1,25-dihydroxyvitamin D_3 is derived from vitamin D by hydroxylation reactions in the liver and kidneys.
1. The last hydroxylation step is stimulated by parathyroid hormone.
2. 1,25-dihydroxyvitamin D_3 stimulates the intestinal absorption of calcium and phosphate, resorption of bone, and renal reabsorption of phosphate.

D. A rise in parathyroid hormone, accompanied by the increased production of 1,25-dihydroxyvitamin D_3, helps to maintain proper blood levels of calcium and phosphate in response to a fall in calcium levels.

E. Calcitonin is secreted by the parafollicular cells of the thyroid gland.
1. Calcitonin secretion is stimulated by a rise in blood calcium levels.
2. Calcitonin, at least at pharmacological levels, acts to lower blood calcium by inhibiting bone resorption and stimulating the urinary excretion of calcium and phosphate.

REVIEW ACTIVITIES

Test Your Knowledge

Match these:

1. Absorption of a carbohydrate meal
2. Fasting

a. Rise in insulin; rise in glucagon
b. Fall in insulin; rise in glucagon
c. Rise in insulin; fall in glucagon
d. Fall in insulin; fall in glucagon

Match these:

3. Growth hormone
4. Thyroxine
5. Hydrocortisone

a. Increased protein synthesis; increased cell respiration
b. Protein catabolism in muscles; gluconeogenesis in liver
c. Protein synthesis in muscles; decreased glucose utilization
d. Fall in blood glucose; increased fat synthesis

6. A lowering of blood glucose concentration promotes
   a. decreased lipogenesis.
   b. increased lipolysis.
   c. increased glycogenolysis.
   d. all of these.
7. Glucose can be secreted into the blood by
   a. the liver.
   b. the muscles.
   c. the liver and muscles.
   d. the liver, muscles, and brain.
8. The basal metabolic rate is determined primarily by
   a. hydrocortisone.
   b. insulin.
   c. growth hormone.
   d. thyroxine.
9. Somatomedins are required for the anabolic effects of
   a. hydrocortisone.
   b. insulin.
   c. growth hormone.
   d. thyroxine.
10. The increased intestinal absorption of calcium is stimulated directly by
    a. parathyroid hormone.
    b. 1,25-dihydroxyvitamin D_3.
    c. calcitonin.
    d. all of these.
11. A rise in blood calcium levels directly stimulates
    a. parathyroid hormone secretion.
    b. calcitonin secretion.
    c. 1,25-dihydroxyvitamin D_3 formation.
    d. all of these.
12. At rest, about 12% of the total calories consumed are used for
    a. protein synthesis.
    b. cell transport.
    c. the Na^+/K^+ pumps.
    d. DNA replication.
13. Which of these hormones stimulates anabolism of proteins and catabolism of fat?
    a. Growth hormone
    b. Thyroxine
    c. Insulin
    d. Glucagon
    e. Epinephrine
14. If a person eats 600 kilocalories of protein in a meal, which of these statements will be *false?*
    a. Insulin secretion will be increased.
    b. The metabolic rate will be increased over basal conditions.
    c. The tissue cells will use some of the amino acids for resynthesis of body proteins.
    d. The tissue cells will obtain 600 kilocalories worth of energy.
    e. Body-heat production and oxygen consumption will be increased over basal conditions.
15. Ketoacidosis in untreated diabetes mellitus is due to
    a. excessive fluid loss.
    b. hypoventilation.
    c. excessive eating and obesity.
    d. excessive fat catabolism.
16. Which of these statements about leptin is *false?*
    a. It is secreted by adipocytes.
    b. It increases the energy expenditure of the body.
    c. It stimulates the release of neuropeptide Y in the hypothalamus.
    d. It promotes feelings of satiety, decreasing food intake.
17. A person with insulin resistance has
    a. decreased hepatic secretion of glucose and increased skeletal muscle uptake of blood glucose.
    b. decreased hepatic secretion of glucose and decreased skeletal muscle uptake of blood glucose.
    c. increased hepatic glucose secretion and increased skeletal muscle uptake of blood glucose.
    d. increased hepatic glucose secretion and decreased skeletal muscle uptake of blood glucose.
18. Which of the following is *not* an adipokine?
    a. leptin
    b. adiponectin
    c. ghrelin
    d. tumor necrosis factor alpha

Test Your Understanding

19. Compare the metabolic effects of fasting to the state of uncontrolled type 1 diabetes mellitus. Explain the hormonal similarities of these conditions.
20. Glucocorticoids stimulate the breakdown of protein in muscles but promote the synthesis of protein in the liver. Explain the significance of these different effects.
21. Describe how thyroxine affects cell respiration. Why does a person who is hypothyroid have a tendency to gain weight and less tolerance for cold?
22. Compare and contrast the metabolic effects of thyroxine and growth hormone.
23. Why is vitamin D considered both a vitamin and a prehormone? Explain why people with osteoporosis might be helped by taking controlled amounts of vitamin D.

24. Define the term *insulin resistance.* Explain the relationships between insulin resistance, obesity, exercise, and noninsulin-dependent diabetes mellitus.
25. Describe the chemical nature and origin of the somatomedins and explain the physiological significance of these growth factors.
26. Explain how insulin secretion and glucagon secretion are influenced by (a) fasting, (b) a meal that is high in carbohydrate and low in protein, and (c) a meal that is high in protein and high in carbohydrate. Also, explain how the changes in insulin and glucagon secretion under these conditions function to maintain homeostasis.
27. Using a cause-and-effect sequence, explain how an inadequate intake of dietary calcium or vitamin D can cause bone resorption. Also, describe the cause-and-effect sequence whereby an adequate intake of calcium and vitamin D may promote bone deposition.
28. Describe the conditions of gigantism, acromegaly, pituitary dwarfism, Laron dwarfism, and kwashiorkor, and explain how these conditions relate to blood levels of growth hormone and IGF-1.
29. Describe how hormones secreted by the gastrointestinal tract help to regulate hunger and satiety. Also, explain the role of adipose tissue in regulating hunger and metabolism.
30. Explain how drugs that bind to PPARγ receptors are helpful in treating type 2 diabetes mellitus.

Test Your Analytical Ability

31. Your friend is trying to lose weight and at first is very successful. After a time, however, she complains that it seems to take more exercise and a far more stringent diet to lose even one more pound. What might explain her difficulties?
32. How can a high-fat diet in childhood lead to increased numbers of adipocytes? Explain how this process may be related to the ability of adipocytes to regulate the insulin sensitivity of skeletal muscles in adults.
33. Discuss the role of GLUT4 in glucose metabolism and use this concept to explain why exercise helps to control type 2 diabetes mellitus.
34. You are running in a 10-K race and, to keep your mind occupied, you try to remember which physiological processes regulate blood glucose levels during exercise. Step by step, what are these processes?
35. Discuss the location and physiological significance of the β_3 adrenergic receptors and explain how a hypothetical β_3 adrenergic agonist drug might help in the treatment of obesity.
36. A person with type 1 diabetes mellitus accidentally overdoses on insulin. What symptoms might she experience, and why? If she remains conscious, what treatment might be offered to adjust her blood glucose level?
37. Suppose a person has a large portion of the stomach surgically removed. How might this affect the person's sense of hunger and pattern of meals? Explain.
38. Someone who drinks alcoholic beverages too often is told that her blood glucose levels are elevated. She responds, "But I don't drink anything sweet!" Explain how alcohol itself can have this effect.

Test Your Quantitative Ability

39. Calculate the BMI of a woman who is 5 feet, 3 inches tall and weighs 130 pounds, and of a man who is 5 feet, 10 inches tall and weighs 215 pounds. State whether either of these people could be classified as "overweight" or "obese."

Please refer to figure 19.12 to answer the following questions:

40. At one hour following glucose ingestion, the prediabetic graph shows what percentage of the normal one-hour level of insulin?
41. At one hour following glucose ingestion, the prediabetic graph shows what percentage of the normal one-hour level of blood glucose?
42. At two hours following glucose ingestion, the more severe type 2 diabetes graph shows what percentage of the normal two-hour level of insulin?
43. At two hours following glucose ingestion, the more severe type 2 diabetes graph shows what percentage of the normal two-hour level of blood glucose?

CHAPTER

20

Reproduction

REFRESH YOUR MEMORY

Before you begin this chapter, you may want to review these concepts from previous chapters:

CHAPTER OUTLINE

Clinical Investigation

Gloria, a second-year college student, goes to the student health service complaining that she has missed her period for several months, although she had normal periods before that. She states that she doesn't take birth control pills and a home pregnancy test was negative. Her body weight is below normal for her height, but she has normal secondary sexual characteristics. She takes thyroxine pills for her hypothyroid condition and feels fine; she even teaches aerobics and exercises at least an hour a day. The physician performs a pregnancy test and orders a variety of blood tests.

The new terms and concepts you will encounter include:

- Amenorrhea, menarche, and secondary sexual characteristics
- Endometrium, menstruation, and ovarian cycle
- Human chorionic gonadotropin (hCG) and pregnancy tests

20.1 SEXUAL REPRODUCTION

A particular gene on the Y chromosome induces the embryonic gonads to become testes. The embryonic testes secrete testosterone, which induces the development of male accessory sex organs and external genitalia. The absence of testes in a female embryo causes the development of the female accessory sex organs.

LEARNING OUTCOMES

After studying this section, you should be able to:

1. **Explain how the chromosomal sex of the zygote affects the formation of the gonads.**
2. **Describe the development of male and female accessory sex organs.**

"A chicken is an egg's way of making another egg." Phrased in more modern terms, genes are "selfish." Genes, according to this view, do not exist in order to make a well-functioning chicken (or other organism). The organism, rather, exists and functions so that the genes can survive beyond the mortal life of individual members of a species. Whether or not one accepts this rather cynical view, it is clear that reproduction is one of life's essential functions.

The incredible complexity of structure and function in living organisms could not be produced in successive generations by chance; mechanisms must exist to transmit the blueprint (genetic code) from one generation to the next. Sexual reproduction, in which genes from two individuals are combined in random and novel ways with each new generation, offers the further advantage of introducing great variability into a population. This diversity of genetic constitution helps to ensure that some members of a population will survive changes in the environment over evolutionary time.

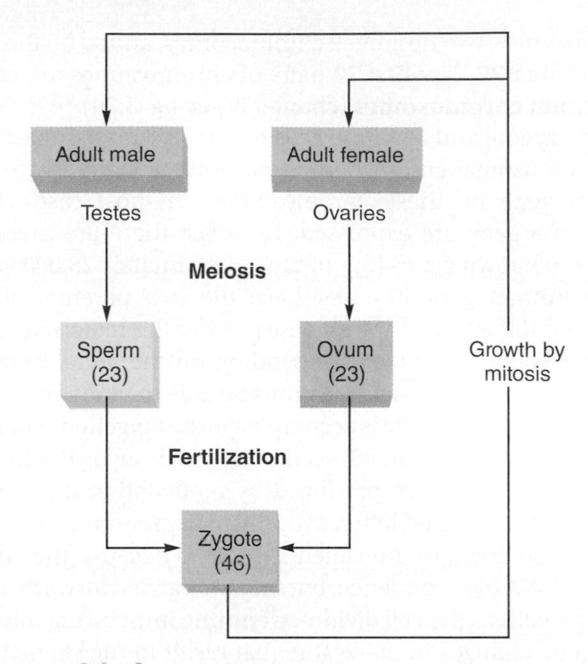

Figure 20.1 The human life cycle. Numbers in parentheses indicate the haploid state (23 chromosomes) or diploid state (46 chromosomes). AP|R

In sexual reproduction, **germ cells,** or **gametes** (sperm and ova), are formed within the *gonads* (testes and ovaries) by a process of reduction division, or *meiosis* (chapter 3; see fig. 3.30). During this type of cell division, the normal number of chromosomes in human cells—46—is halved, so that each gamete receives 23 chromosomes. Fusion of a sperm cell and ovum (egg cell) in the act of **fertilization** results in restoration of the original chromosome number of 46 in the **zygote,** or fertilized egg. Growth of the zygote into an adult member of the next generation occurs by means of mitotic cell divisions, as described in chapter 3. When this individual reaches puberty, mature sperm or ova will be formed by meiosis within the gonads so that the life cycle can be continued (fig. 20.1).

Sex Determination

Autosomal and Sex Chromosomes

Each zygote inherits 23 chromosomes from its mother and 23 chromosomes from its father. This does not produce 46 different chromosomes, but rather 23 pairs of *homologous chromosomes.* The members of a homologous pair, with the important exception of the sex chromosomes, look like each other and contain similar genes (such as those coding for eye color, height, and so on). These homologous pairs of chromosomes can be photographed and numbered. Each cell that contains 46 chromosomes (that is *diploid*) has two number 1 chromosomes, two number 2 chromosomes, and so on through pair number 22. The first 22 pairs of chromosomes are called **autosomal chromosomes** (chapter 3; see fig. 3.29).

The zygote and all of the cells it forms by mitosis have two sets of autosomal chromosomes and thus two **alleles** (forms) of each gene on these chromosomes. In most cases, both alleles of a gene are expressed. However, there are presently about 100 known genes (out of the approximately 25,000 genes in the human genome) that have the two parental alleles expressed differently. In most cases, either the maternal or the paternal allele is silenced. Depending on the gene, only the maternal or paternal allele may thus be able to function.

Silencing of an allele is accomplished by epigenetic changes to the chromatin (chapter 3, section 3.5). This involves changes in chromatin structure produced by methylation of cytosine bases in DNA and acetylation of the histone proteins in the chromatin. Such changes are called *epigenetic* because they don't alter the DNA base sequence, but they are carried forward to the daughter cells as the cell divides. **Genomic imprinting** refers to epigenetic changes in the zygote that result in the silencing of the allele derived from one parent and the expression of only the nonimprinted allele of the other parent in the offspring. Because only the maternal or paternal allele of the gene pair functions in the tissues of the offspring, these genes are particularly susceptible to mutations that can cause diseases.

The 23rd pair of chromosomes are the **sex chromosomes** (fig. 20.2). In a female, these consist of two X chromosomes, whereas in a male there is one X chromosome and one Y chromosome. The X and Y chromosomes look different and contain different genes. This is the exceptional pair of homologous chromosomes mentioned earlier.

The X chromosome has recently been sequenced and shown to have 1,090 genes; the Y chromosome, by contrast, has only about 80 genes. When meiosis occurs in the testes, the X and Y chromosomes cannot undergo recombination (chapter 3; see fig. 3.31), as can the autosomal chromosomes and the two X chromosomes in a female. Instead, only the tips of the single X chromosome in a male, which contain most of the 54 genes that are homologous in the X and Y chromosome, can recombine with the Y chromosome during meiosis. Genes outside of the tip regions are restricted to the solitary X chromosome of the male. A surprisingly large number of these *X-linked genes* are responsible for certain diseases—there are presently 168 diseases known to be caused by mutations in 113 X-linked genes. These diseases are more common in males than in females because the genes responsible, being unpaired, cannot be present in a recessive state.

Although the Y chromosome is not much to look at under the microscope (see fig. 20.2), scientists have recently determined its DNA base sequence and found that it still contains more than 23 million base pairs of euchromatin. (Euchromatin is the extended, active form of DNA; chapter 3, section 3.3.) The DNA of the Y chromosome includes X-transposed sequences almost identical to regions of the X chromosome (from which the Y chromosome is believed to have evolved), degenerate regions, and testis-specific genes. Unusually, most of the testis-specific genes (expressed by spermatogenic cells) were found to be located in huge *palindromes.* Palindromes are regions of DNA bases that read the same from either direction. The eight palindromes discovered are quite long, one up to 2.9 million base pairs! These palindromes may enable the Y chromosome to have "gene conversions," where defects in one region of the palindrome can be corrected by a corresponding region. This would substitute for the crossing-over (chapter 3; see fig. 3.31) that normally occurs between homologous chromosomes, helping protect these important genes from genetic changes and conserving the genes over evolutionary time.

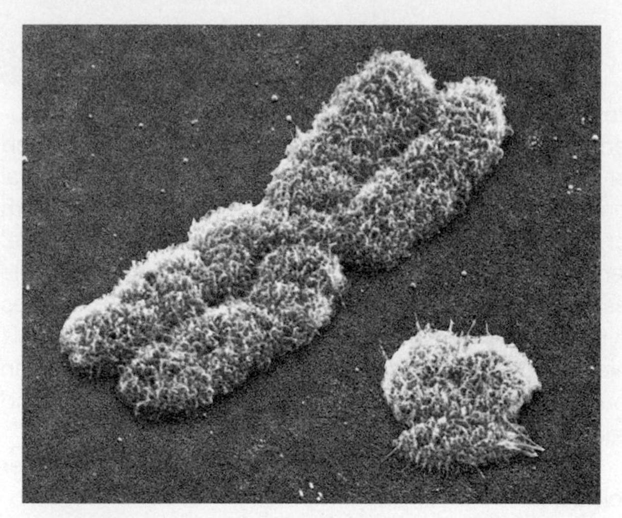

Figure 20.2 The human X and Y chromosomes. Magnified about 10,000×, these images reveal that the X and Y chromosomes are dramatically different in size and shape. Despite the small size and simple appearance of the Y chromosome, recent studies have uncovered its surprising complexity and sophistication.

When a diploid cell (with 46 chromosomes) undergoes meiotic division, its daughter cells receive only one chromosome from each homologous pair of chromosomes. The gametes are therefore said to be *haploid* (they contain only half the number of chromosomes in the diploid parent cell). Each sperm cell, for example, will receive only one chromosome of homologous pair number 5—either the one originally contributed by the mother, or the one originally contributed by the father (modified by the effects of crossing-over). Which of the two chromosomes—maternal or paternal—ends up in a given sperm cell is completely random. This is also true for the sex chromosomes, so that approximately half of the sperm produced will contain an X and approximately half will contain a Y chromosome.

The egg cells (ova) in a woman's ovary will receive a similar random assortment of maternal and paternal chromosomes. Because the body cells of females have two X chromosomes, however, all of the ova will normally contain one X chromosome. Since all ova contain one X chromosome, whereas some sperm are X-bearing and others are Y-bearing, *the chromosomal sex of the zygote is determined by the fertilizing sperm cell.* If a Y-bearing sperm cell fertilizes the ovum, the zygote will be XY and male; if an X-bearing sperm cell fertilizes the ovum, the zygote will be XX and female.

Each diploid cell in a woman's body inherits two X chromosomes, but one is inactivated so that only one X chromosome is fully active. Whether the inactivated X chromosome is the one inherited from the mother or the father is completely random, so a woman's cells are a mosaic in which the active X chromosome may be derived from either parent. The inactive X chromosome forms a clump of heterochromatin, which can be seen as a dark spot, called a *Barr body,* in cheek cells, and a "drumstick" appendage in the nucleus of some of the neutrophils in a female (fig. 20.3). These can provide a convenient microscopic test for the chromosomal sex of a person. Actually, depending on the individual woman, about 15% of the genes in the inactive X chromosome escape inactivation and another 10% are only partially inactivated.

Formation of Testes and Ovaries

Following conception, the gonads of males and females are similar in appearance for the first 40 or so days of development. During this time, cells that will give rise to sperm (called *spermatogonia*) and cells that will give rise to ova (called *oogonia*) migrate from the yolk sac to the developing embryonic gonads. At this stage, the embryonic structures have the potential to become either **testes** or **ovaries.** The hypothetical substance that promotes their conversion to testes (fig. 20.4) has been called the **testis-determining factor (TDF).**

Although it has long been recognized that male sex is determined by the presence of a Y chromosome, the genes involved have only recently been localized. In females, the Y chromosome must be absent for pathways directed by genes in the X chromosomes to result in the development of ovaries. However, in rare male babies with XX genotypes, scientists discovered that one of the X chromosomes contains a segment of the Y chromosome—the result of an error that occurred during the meiotic cell division that formed the sperm cell. Similarly, rare female babies with XY genotypes were found to be missing the same portion of the Y chromosome erroneously inserted into the X chromosome of XX males.

Through these and other observations, it has been shown that the gene for the testis-determining factor is located on the short arm of the Y chromosome. Evidence suggests that it may be a particular gene known as **SRY** (for sex-determining region of the Y). This gene is found in the Y chromosome of all mammals and is highly conserved, meaning that it shows little variation in structure over evolutionary time.

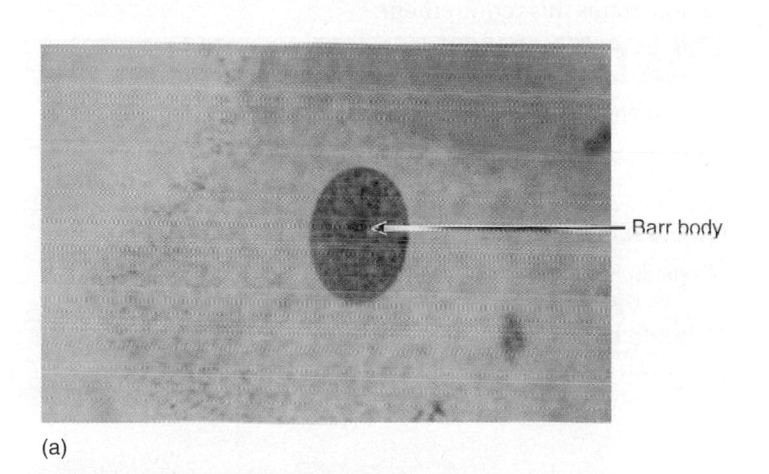

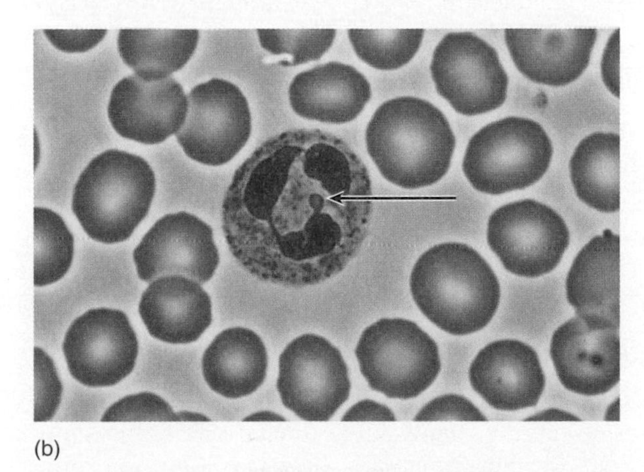

Figure 20.3 Barr bodies. The nuclei of cheek cells from females (*a*) have Barr bodies (*arrow*). These are formed from one of the X chromosomes, which is inactive. No Barr body is present in the cell obtained from a male because males have only one X chromosome, which remains active. Some neutrophils obtained from females (*b*) have a "drumstick-like" appendage (*arrow*) that is not found in the neutrophils of males.

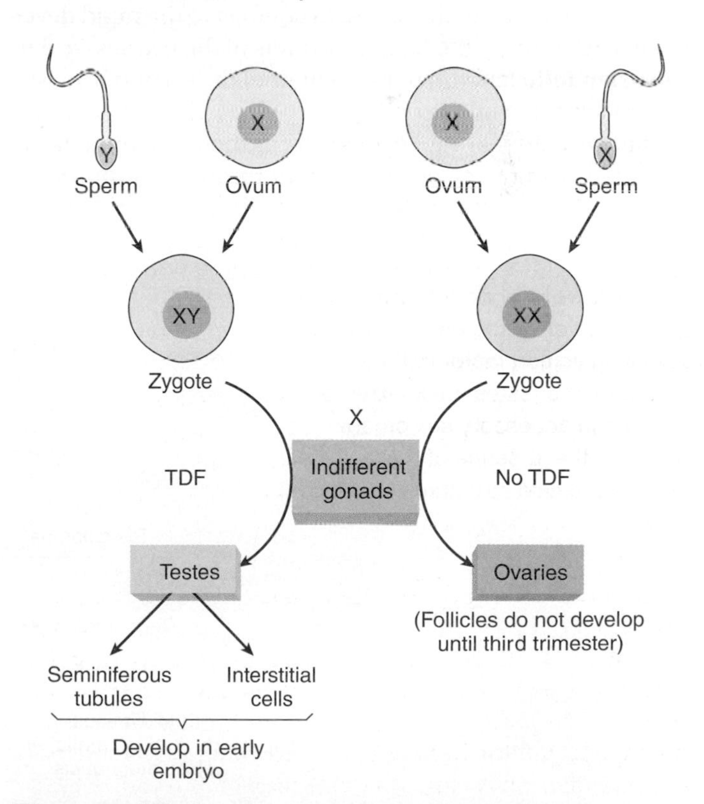

Figure 20.4 The chromosomal sex and the development of embryonic gonads. The very early embryo has "indifferent gonads" that can develop into either testes or ovaries. The testis-determining factor (TDF) is a gene located on the Y chromosome. In the absence of TDF, ovaries will develop.

CLINICAL APPLICATION

Notice that it is normally the presence or absence of the Y chromosome that determines whether the embryo will have testes or ovaries. This point is well illustrated by two genetic abnormalities. In **Klinefelter's syndrome,** the affected person has 47 instead of 46 chromosomes because of the presence of an extra X chromosome. This person, with an XXY genotype, will develop testes and have a male phenotype despite the presence of two X chromosomes. Patients with **Turner's syndrome,** who have the genotype XO (and therefore have only 45 chromosomes), have poorly developed ("streak") gonads and are phenotypically female.

The structures that will eventually produce sperm within the testes, the **seminiferous tubules,** appear very early in embryonic development—between 43 and 50 days following conception. The tubules contain two major cell types: germinal and nongerminal. The **germinal cells** are those that will eventually become sperm through meiosis and subsequent specialization. The nongerminal cells are called **Sertoli** (or **sustentacular**) **cells.** The Sertoli cells appear at about day 42. At about day 65, the **Leydig** (or **interstitial**) **cells** appear in the embryonic testes. The Leydig cells are clustered in the *interstitial tissue* that surrounds the seminiferous tubules. The interstitial Leydig cells constitute the endocrine tissue of the testes. In contrast to the rapid development of the testes, the functional units of the ovaries—called the **ovarian follicles**—do not appear until the second trimester of pregnancy (at about day 105).

The early-appearing Leydig cells in the embryonic testes secrete large amounts of male sex hormones, or *androgens* (*andro* = man; *gen* = forming). The major androgen secreted by these cells is **testosterone.** Testosterone secretion begins as early as 8 weeks after conception, reaches a peak at 12 to 14 weeks, and then declines to very low levels by the end of the second trimester (at about 21 weeks). Testosterone secretion during embryonic development in the male serves the very important function of masculinizing the embryonic structures. Testosterone levels rise again in newborn boys until the age of 3 months and then fall to almost undetectable levels by ages 7–12 months. From the age of 12 months until adolescence, the sex hormone levels are the same in both sexes.

As the testes develop, they move within the abdominal cavity and gradually descend into the *scrotum.* Descent of the testes is sometimes not complete until shortly after birth. The temperature of the scrotum is maintained at about 35° C—about 3° C below normal body temperature. This cooler temperature is needed for spermatogenesis. The fact that spermatogenesis does not occur in males with undescended testes—a condition called *cryptorchidism* (*crypt* = hidden; *orchid* = testes)—demonstrates this requirement.

FITNESS APPLICATION

Associated with each spermatic cord is a strand of skeletal muscle called the cremaster muscle. In cold weather, the cremaster muscles contract and elevate the testes, bringing them closer to the warmth of the trunk. The **cremasteric reflex** produces the same effect when the inside of a man's thigh is stroked. In a baby, however, this stimulation can cause the testes to be drawn up through the inguinal canal into the body cavity. The testes can also be drawn up into the body cavity voluntarily by trained Sumo wrestlers.

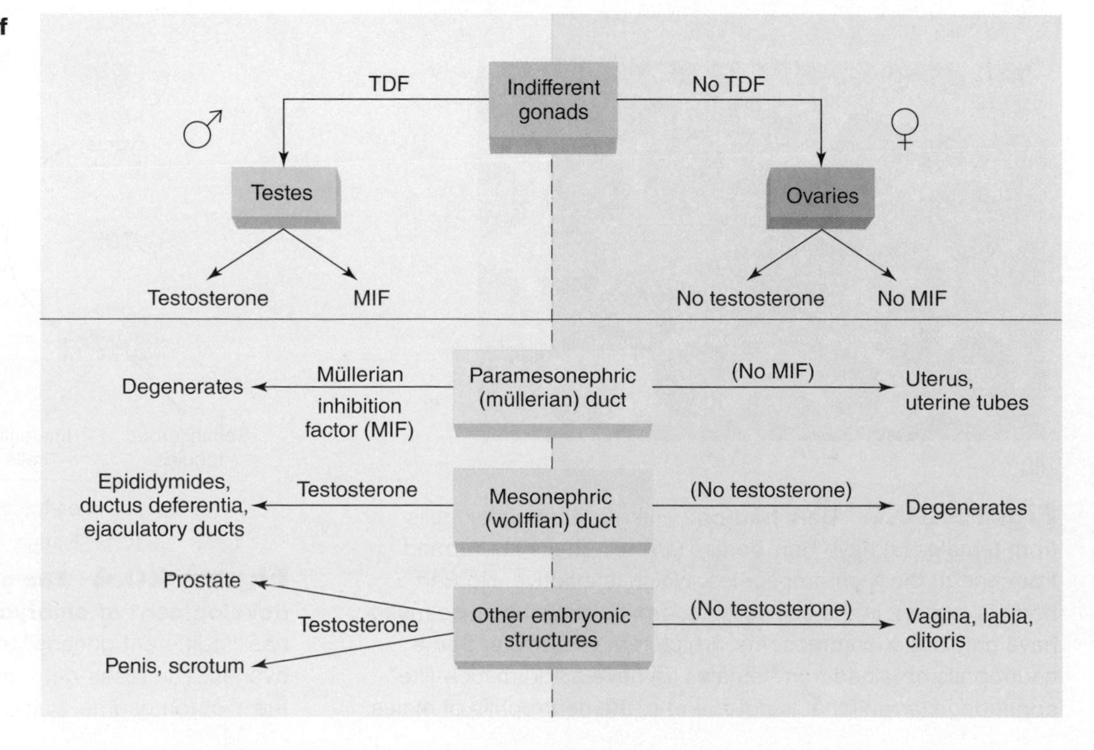

Figure 20.5 The regulation of embryonic sexual development. In the presence of testosterone and müllerian inhibition factor (MIF) secreted by the testes, male external genitalia and accessory sex organs develop. In the absence of these secretions, female structures develop.

Development of Accessory Sex Organs and External Genitalia

In addition to testes and ovaries, various internal **accessory sex organs** are needed for reproductive function. Most of these are derived from two systems of embryonic ducts. Male accessory organs are derived from the **wolffian (mesonephric) ducts,** and female accessory organs are derived from the **müllerian (paramesonephric) ducts** (fig. 20.5). Interestingly, the two duct systems are present in both male and female embryos between day 25 and day 50, and so embryos of both sexes have the potential to form the accessory organs characteristic of either sex.

Experimental removal of the testes (castration) from male embryonic animals results in regression of the wolffian ducts and development of the müllerian ducts into *female accessory organs:* the **uterus** and **uterine (fallopian) tubes.** Female accessory sex organs, therefore, develop as a result of the absence of testes rather than as a result of the presence of ovaries.

In a male, the Sertoli cells of the seminiferous tubules secrete *müllerian inhibition factor (MIF),* a polypeptide that causes regression of the müllerian ducts beginning at about day 60. The secretion of testosterone by the Leydig cells of the testes subsequently causes growth and development of the wolffian ducts into *male accessory sex organs:* the **epididymis, ductus (vas) deferens, seminal vesicles,** and **ejaculatory duct.**

The **external genitalia** of males and females are essentially identical during the first 6 weeks of development, sharing in common a *urogenital sinus, genital tubercle, urethral folds,* and a pair of *labioscrotal swellings.* The secretions of the testes masculinize these structures to form the **penis** and *spongy (penile) urethra,* **prostate,** and **scrotum.** In the absence of secreted testosterone, the genital tubercle that forms the penis in a male will become the **clitoris** in a female. The penis and clitoris are thus said to be *homologous structures.* Similarly, the labioscrotal swellings form the scrotum in a male or the **labia majora** in a female; these structures are therefore also homologous (fig. 20.6).

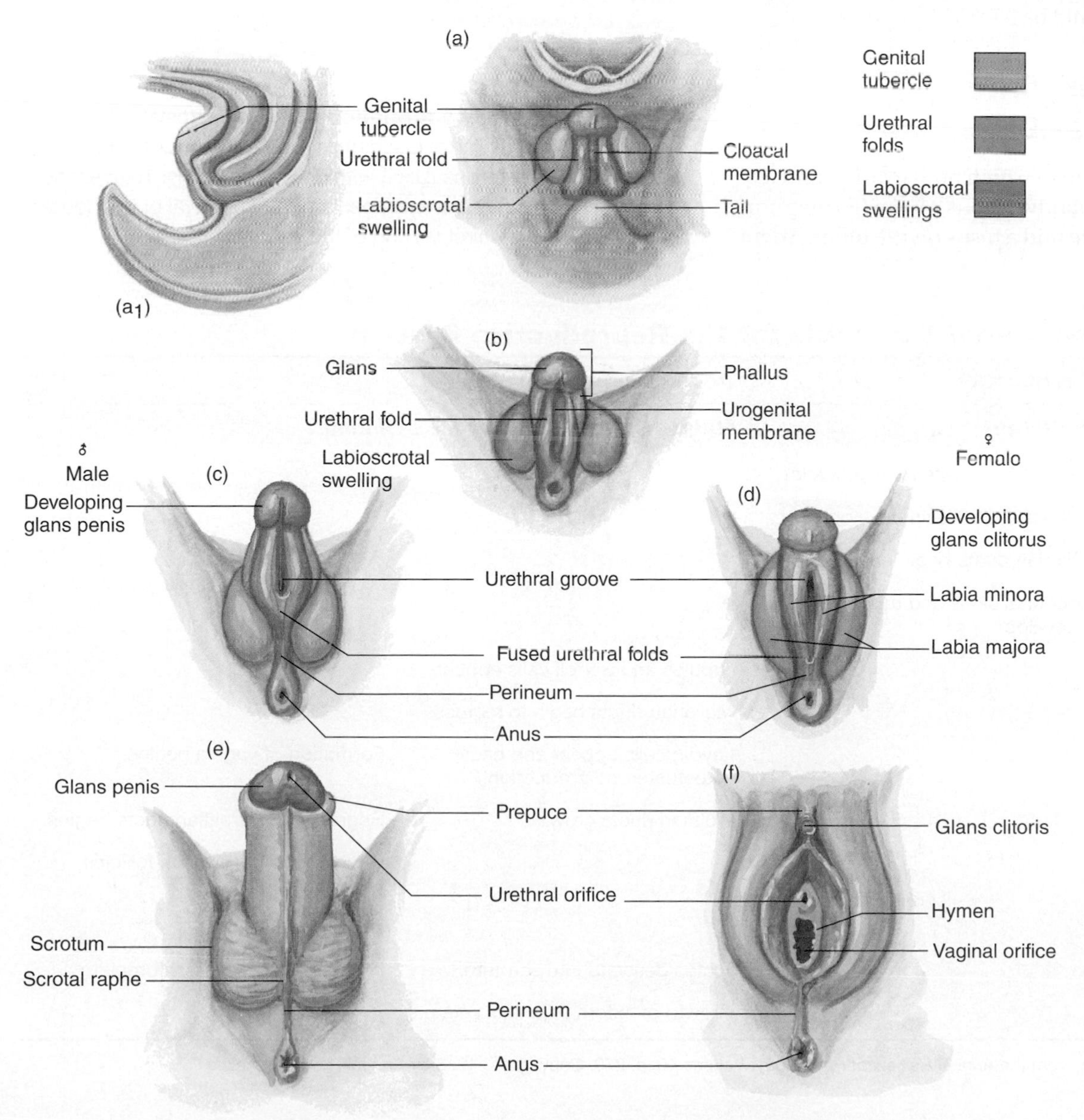

Figure 20.6 The development of the external genitalia in the male and female. (*a* [a_1, sagittal view]) At 6 weeks, the urethral fold and labioscrotal swelling have differentiated from the genital tubercle. (*b*) At 8 weeks, a distinct phallus is present during the indifferent stage. By week 12, the genitalia have become distinctly male (*c*) or female (*d*), being derived from homologous structures. (*e, f*) At 16 weeks, the genitalia are formed. **AP|R**

Masculinization of the embryonic structures occurs as a result of testosterone secreted by the embryonic testes. Testosterone itself, however, is not the active agent within all of the target organs. Once inside particular target cells, testosterone is converted by the enzyme *5α-reductase* into the active hormone known as **dihydrotestosterone (DHT)** (fig. 20.7). DHT is needed for the development and maintenance of the penis, spongy urethra, scrotum, and prostate. Evidence suggests that testosterone itself directly stimulates the wolffian duct derivatives—epididymis, ductus deferens, ejaculatory duct, and seminal vesicles.

In summary, the genetic sex is determined by whether a Y-bearing or an X-bearing sperm cell fertilizes the ovum; the presence or absence of a Y chromosome, in turn, determines whether the gonads of the embryo will be testes or ovaries. The presence or absence of testes, finally, determines whether the accessory sex organs and external genitalia will be male or female (table 20.1). This regulatory pattern of sex determination makes sense in light of the fact that both male and female embryos develop within an environment high in estrogen, which is secreted by the mother's ovaries and the placenta. If the secretions of the ovaries determined the sex, all embryos would be female.

Disorders of Embryonic Sexual Development

Hermaphroditism is a condition in which both ovarian and testicular tissue is present in the body. About 34% of hermaphrodites have an ovary on one side and a testis on the other. About

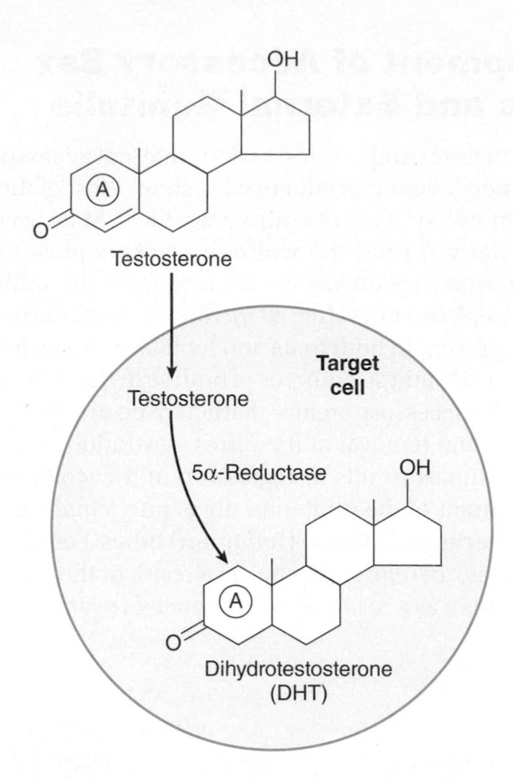

Figure 20.7 The formation of DHT. Testosterone, secreted by the interstitial (Leydig) cells of the testes, is converted into dihydrotestosterone (DHT) within the target cells. This reaction involves the addition of a hydrogen (and the removal of the double carbon bond) in the first (A) ring of the steroid.

Table 20.1 | A Developmental Timetable for the Reproductive System

Approximate Time After Fertilization			Developmental Changes	
Days	***Trimester***	***Indifferent***	***Male***	***Female***
19	First	Germ cells migrate from yolk sac.		
25–30		Wolffian ducts begin development.		
44–48		Müllerian ducts begin development.		
50–52		Urogenital sinus and tubercle develop.		
53–60			Tubules and Sertoli cells appear.	
			Müllerian ducts begin to regress.	
60–75			Leydig cells appear and begin testosterone production.	Formation of vagina begins.
			Wolffian ducts grow.	Regression of wolffian ducts begins.
105	Second			Development of ovarian follicles begins.
120				Uterus is formed.
160–260	Third		Testes descend into scrotum.	Formation of vagina is complete.
			Growth of external genitalia occurs.	

Source: Reproduced, with permission, from the *Annual Review of Physiology*, Volume 40, p. 279. Copyright © 1978 by Annual Reviews, Inc.

20% have ovotestes—part testis and part ovary—on both sides. The remaining 46% have an ovotestis on one side and an ovary or testis on the other. Hermaphroditism is extremely rare and results when some embryonic cells receive the short arm of the Y chromosome, with its SRY gene, whereas others do not. More common (though still rare) disorders of sex determination involve individuals with either testes or ovaries, but not both, who have accessory sex organs and external genitalia that are incompletely developed or that are inappropriate for their chromosomal sex. These individuals are called *pseudohermaphrodites* (*pseudo* = false).

The most common cause of female pseudohermaphroditism is *congenital adrenal hyperplasia*. This condition, which is inherited as a recessive trait, is caused by the excessive secretion of androgens from the adrenal cortex. Because the cortex does not secrete müllerian inhibition factor, a female with this condition would have müllerian duct derivatives (uterus and fallopian tubes), but she would also have wolffian duct derivatives and partially masculinized external genitalia.

An interesting cause of male pseudohermaphroditism is *testicular feminization syndrome*. Individuals with this condition have normally functioning testes but lack receptors for testosterone. Thus, although large amounts of testosterone are secreted, the embryonic tissues cannot respond to this hormone. Female genitalia therefore develop, but the vagina ends blindly (a uterus and fallopian tubes do not develop because of the secretion of müllerian inhibition factor). Male accessory sex organs likewise cannot develop because the wolffian ducts lack testosterone receptors. A child with this condition appears externally to be a normal prepubertal girl, but she has testes in her body cavity and no accessory sex organs. These testes secrete an exceedingly large amount of testosterone at puberty because of the absence of negative feedback inhibition. This abnormally large amount of testosterone is converted by the liver and adipose tissue into estrogens. As a result, the person with testicular feminization syndrome develops into a female with well-developed breasts who never menstruates (and who, of course, can never become pregnant).

Some male pseudohermaphrodites have normally functioning testes and normal testosterone receptors, but they genetically lack the ability to produce the enzyme 5α-reductase. Individuals with *5α-reductase deficiency* have normal epididymides, ductus (vasa) deferentia, seminal vesicles, and ejaculatory ducts because the development of these structures is stimulated directly by testosterone. However, the external genitalia are poorly developed and more female in appearance because DHT, which cannot be produced in the absence of 5α-reductase, is required for the development of male external genitalia.

Clinical Investigation CLUES

Gloria has normal secondary sex characteristics and used to have regular periods.

- Is it likely that she has any of the problems of sexual development discussed in this section?

CHECKPOINT

1a. Define the terms *diploid* and *haploid*, and explain how the chromosomal sex of an individual is determined.

1b. Explain how the chromosomal sex determines whether testes or ovaries will be formed.

2a. List the male and female accessory sex organs and explain how the development of one or the other set of organs is determined.

2b. Describe the abnormalities characteristic of testicular feminization syndrome and of 5α-reductase deficiency and explain how these abnormalities are produced.

20.2 ENDOCRINE REGULATION OF REPRODUCTION

The functions of the testes and ovaries are regulated by gonadotropic hormones secreted by the anterior pituitary. The gonadotropic hormones stimulate the gonads to secrete their sex steroid hormones, and these steroid hormones in turn have an inhibitory effect on the secretion of the gonadotropic hormones.

LEARNING OUTCOMES

After studying this section, you should be able to:

3. Describe the interactions among the hypothalamus, anterior pituitary, and gonads.

4. Explain mechanisms that may be responsible for puberty, and describe the male and female sexual response.

The embryonic testes during the first trimester of pregnancy are active endocrine glands, secreting the high amounts of testosterone needed to masculinize the male embryo's external genitalia and accessory sex organs. Ovaries, by contrast, do not mature until the third trimester of pregnancy. Testosterone secretion in the male fetus declines during the second trimester of pregnancy, however, so that the gonads of both sexes are relatively inactive at the time of birth.

Before puberty, there are equally low blood concentrations of *sex steroids*—androgens and estrogens—in both males and females. Apparently, this is not due to deficiencies in the ability of the gonads to produce these hormones, but rather to lack of sufficient stimulation. During *puberty*, the gonads secrete increased amounts of sex steroid hormones as a result of increased stimulation by **gonadotropic hormones** from the anterior pituitary.

Interactions Between the Hypothalamus, Pituitary Gland, and Gonads

The anterior pituitary produces and secretes two gonadotropic hormones—**FSH (follicle-stimulating hormone)** and **LH (luteinizing hormone).** Although these two hormones are named according to their actions in the female, the same hormones are secreted by the male's pituitary gland. (In the male, LH is sometimes called **interstitial cell stimulating hormone** and abbreviated **ICSH;** because this is identical to the LH in females, it is simpler and more common to refer to it as LH in both females and males.) The gonadotropic hormones of both sexes have three primary effects on the gonads: (1) stimulation of *spermatogenesis* or *oogenesis* (formation of sperm or ova); (2) stimulation of gonadal hormone secretion; and (3) maintenance of the structure of the gonads (the gonads atrophy if the pituitary gland is removed).

The secretion of both LH and FSH from the anterior pituitary is stimulated by a hormone produced by the hypothalamus and secreted into the hypothalamo-hypophyseal portal vessels (chapter 11, section 11.3). This releasing hormone is sometimes called *LHRH (luteinizing hormone-releasing hormone).* Because attempts to find a separate FSH-releasing hormone have thus far failed, and LHRH stimulates FSH as well as LH secretion, LHRH is often referred to as **gonadotropin-releasing hormone (GnRH).**

If a male or female animal is castrated (has its gonads surgically removed), the secretion of FSH and LH increases to much higher levels than in the intact animal. This demonstrates that the gonads secrete products that exert a negative feedback effect on gonadotropin secretion. This negative feedback is exerted in large part by sex steroids: estrogen and progesterone in the female, and testosterone in the male. A biosynthetic pathway for these steroids is shown in figure 20.8.

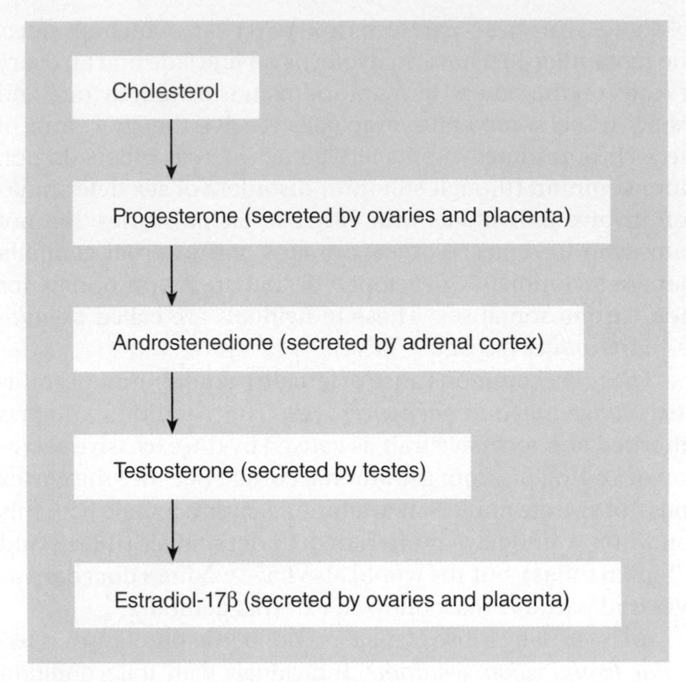

Figure 20.8 A simplified biosynthetic pathway for the steroid hormones. The sources of the sex hormones secreted in the blood are also indicated.

The negative feedback effects of steroid hormones occurs by means of two mechanisms: (1) inhibition of GnRH secretion from the hypothalamus and (2) inhibition of the pituitary's response to a given amount of GnRH. In addition to steroid hormones, the testes and ovaries secrete a polypeptide hormone called **inhibin.** Inhibin is secreted by the Sertoli cells of the seminiferous tubules in males and by the granulosa cells of the ovarian follicles in females. This hormone specifically inhibits the anterior pituitary's secretion of FSH without affecting the secretion of LH.

Figure 20.9 illustrates the process of gonadal regulation. Although hypothalamus-pituitary-gonad interactions are similar in males and females, there are important differences. Secretion of gonadotropins and sex steroids is more or less constant in adult males. Secretion of gonadotropins and sex steroids in adult females, by contrast, shows cyclic variations (during the menstrual cycle). Also, during one phase of the

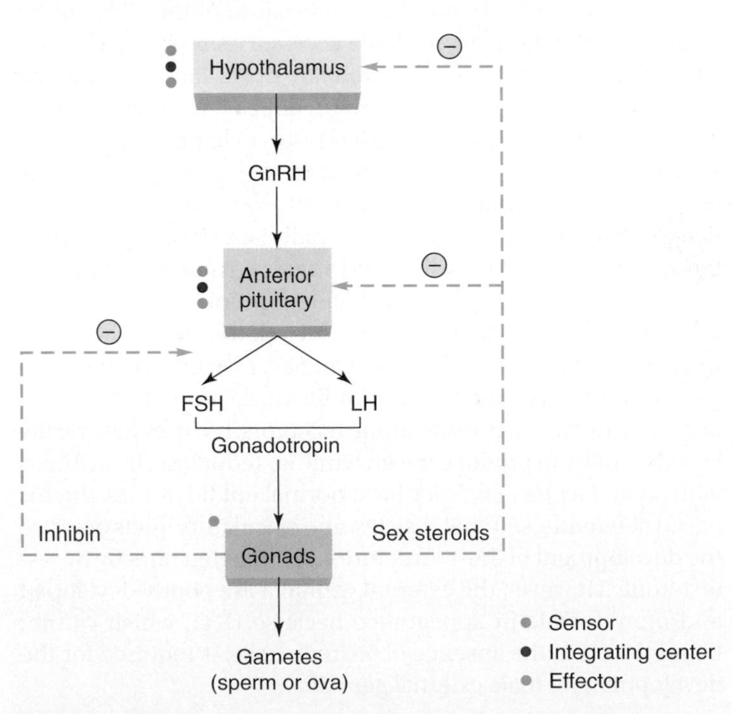

Figure 20.9 Interactions between the hypothalamus, anterior pituitary, and gonads. Sex steroids secreted by the gonads have a negative feedback effect on the secretion of GnRH (gonadotropin-releasing hormone) and on the secretion of gonadotropins. The gonads may also secrete a polypeptide hormone called inhibin that functions in the negative feedback control of FSH secretion.

female cycle—shortly before ovulation—estrogen exerts a positive feedback effect on LH secretion.

Studies have shown that secretion of GnRH from the hypothalamus is *pulsatile* rather than continuous, stimulating a similar pulsatile secretion of FSH and LH from the anterior pituitary. This **pulsatile secretion** is needed to prevent desensitization and downregulation of the target glands (chapter 11, section 11.1). It appears that the frequency of the pulses of secretion, as well as their amplitude (how much hormone is secreted per pulse), affects the target gland's response to the hormone. For example, it has been proposed that a slow frequency of GnRH pulses in women preferentially stimulates FSH secretion, while faster pulses of GnRH favor LH secretion.

CLINICAL APPLICATION

If a powerful synthetic analogue of GnRH (such as *nafarelin*) is administered, the anterior pituitary first increases and then decreases its secretion of FSH and LH. This decrease, which is contrary to the normal stimulatory action of GnRH, is due to a desensitization of the anterior pituitary evoked by continuous exposure to GnRH. The decrease in LH causes a fall in testosterone secretion from the testes, or of estradiol secretion from the ovaries. The decreased testosterone secretion is useful in the treatment of men with **benign prostatic hyperplasia.** In this condition, common in men over the age of 60, testosterone supports abnormal growth of the prostate.

The fall in estradiol secretion in women given synthetic GnRH analogues can be useful in the treatment of **endometriosis.** Endometriosis, which affects 5% to 10% of women during the reproductive years, is caused by shed endometrial tissue that moves backward and exits through the fallopian tubes to implant into the pelvic peritoneum, ovaries, or other sites. This instigates an inflammatory response, accompanied by angiogenesis (blood vessel growth), adhesions, fibrosis, and other changes that cause pelvic pain and infertility. Drugs that are GnRH analogues produce a prolonged stimulation of GnRH receptors, at first causing the anterior pituitary to release its stored gonadotropins and then causing downregulation of GnRH receptors due to their continuous (rather than pulsatile) stimulation. This results in a drastic reduction in FSH and LH secretion, which causes the ovaries to stop producing and secreting estrogen. Without stimulation by estrogen, the growth and pathological activity of the ectopic endometrial tissue is reduced.

Onset of Puberty

Secretion of FSH and LH is elevated at birth and remains relatively high for the first six months of postnatal life, but then declines to very low levels until puberty. Puberty is triggered by the increased secretion of LH. Secretion of LH is pulsatile, and both the frequency and amplitude of the LH pulses increase at puberty (secretion is greater at night than during the day).

In animals such as rats and sheep, the low secretion of LH prior to puberty is due to high sensitivity of the hypothalamus to the negative feedback effects of gonadal hormones. The rise in LH secretion at puberty is then caused by a declining sensitivity of the hypothalamus to these negative feedback effects. With less inhibition the secretion of GnRH rises, causing increased secretion of LH (and FSH). However, this does not seem to be the case for primates, including humans.

In humans, the increased secretion of GnRH occurs independently of gonadal hormones. Puberty in humans is a result of changes in the hypothalamus that allow an increased secretion of GnRH to stimulate the pulsatile secretion of LH. Studies in monkeys suggest that, prior to puberty, the secretion of GnRH is inhibited by neurons that release the neurotransmitter GABA, whereas this inhibition is reduced at puberty. Also at puberty, stimulation by the neurotransmitter glutamate is increased. Through these and other changes in the hypothalamus, increased secretion of LH (and FSH) from the anterior pituitary drive the other endocrine changes that cause puberty.

Because of the increased pulses of LH at puberty, the gonads secrete increased amounts of sex steroid hormones. Increased secretion of testosterone from the testes and of **estradiol-17β** (estradiol is the major *estrogen,* or female sex hormone) from the ovaries during puberty, in turn, produce body changes characteristic of the two sexes. Such **secondary sex characteristics** (tables 20.2 and 20.3) are the physical manifestation of the hormonal changes that occur during puberty.

Table 20.2 | Development of Secondary Sex Characteristics and Other Changes That Occur During Puberty in Girls

Characteristic	Age of First Appearance	Hormonal Stimulation
Appearance of breast buds	8–13	Estrogen, progesterone, growth hormone, thyroxine, insulin, cortisol
Pubic hair	8–14	Adrenal androgens
Menarche (first menstrual flow)	10–16	Estrogen and progesterone
Axillary (underarm) hair	About 2 years after the appearance of pubic hair	Adrenal androgens
Eccrine sweat glands and sebaceous glands; acne (from blocked sebaceous glands)	About the same time as axillary hair growth	Adrenal androgens

Table 20.3 | Development of Secondary Sex Characteristics and Other Changes That Occur During Puberty in Boys

Characteristic	Age of First Appearance	Hormonal Stimulation
Growth of testes	10–14	Testosterone, FSH, growth hormone
Pubic and axillary hair	10–15	Adrenal androgens
Body growth	11–16	Testosterone, growth hormone
Growth of penis	11–15	Testosterone
Growth of larynx (voice lowers)	Same time as growth of penis	Testosterone
Facial hair	About 2 years after the appearance of pubic hair	Testosterone
Eccrine sweat glands and sebaceous glands; acne (from blocked sebaceous glands)	About the same time as facial and axillary hair growth	Testosterone

For example, the rising secretion of estradiol in girls stimulates the epiphyseal growth plates of bones, causing a growth spurt; this is the first sign of puberty. Estradiol also stimulates breast development, and, a little later (at an average age of 11 to 13 years), **menarche**—the first menstrual flow. In boys, the rising secretion of testosterone causes growth of the penis and testes, as well as most of the other male secondary sex characteristics. Testosterone also stimulates growth in boys (via its conversion to estrogen within the epiphyseal growth plates), but the growth spurt in boys occurs at a later age than in girls (fig. 20.10). Boys gain about 35% more muscle mass at puberty as a result of testosterone-stimulated anabolism.

Estrogen produced from testosterone in the epiphyseal growth plates of males stimulates cell division of chondrocytes, synthesis of new cartilage matrix, and calcification of the bone matrix. Testosterone directly stimulates the addition of new bone under the periosteum (the membrane surrounding bones) more than estrogen does; this accounts for the larger, thicker bones in males compared to females.

Interestingly, the growth of hair in the pubic and axillary (underarm) regions in boys and girls is primarily stimulated by an increased secretion of androgens from the adrenal cortex at the time of puberty. The maturation of the pituitary-adrenal axis is separate from the pubertal changes in the pituitary-gonad axis.

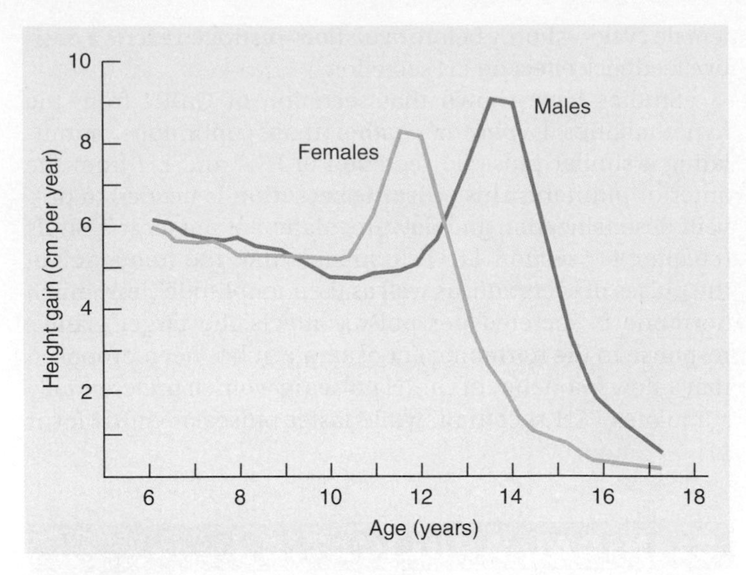

Figure 20.10 Growth as a function of sex and age. Notice that the growth spurt during puberty occurs at an earlier age in females than in males.

The age at which puberty begins is related to the amount of body fat and level of physical activity of the child. The average age of menarche is later (age 15) in girls who are very active physically than in the general population (age 12.6). This appears to be due to a requirement for a minimum percentage of body fat for menstruation to begin; this may represent a mechanism favored by natural selection to ensure that a woman can successfully complete a pregnancy and nurse the baby. Recent evidence suggests that the secretion of leptin from adipocytes (chapter 19) is required for puberty. Later in life, women who are very lean and physically active may have irregular cycles and *amenorrhea* (cessation of menstruation). This may also be related to the percentage of body fat. However, there is also evidence that physical exercise may act to inhibit GnRH and gonadotropin secretion.

Clinical Investigation CLUES

Gloria has normal secondary sex characteristics and used to have regular periods; she has low body weight and performs regular strenuous exercise.

- What are normal secondary sex characteristics?
- Could her low body weight and strenuous exercise be responsible for her missed period?

Pineal Gland

The role of the **pineal gland** in human physiology is poorly understood. It is known that the pineal, a gland located deep within the brain, secretes the hormone **melatonin** as a derivative of the amino acid tryptophan (chapter 11; see fig. 11.32) and that production of this hormone is influenced by light-dark cycles.

The pineal glands of some vertebrates have photoreceptors that are directly sensitive to environmental light. Although no such photoreceptors are present in the pineal glands of mammals, the secretion of melatonin has been shown to increase at night and decrease during daylight. The inhibitory effect of light on melatonin secretion in mammals is indirect. Pineal secretion is stimulated by postganglionic sympathetic neurons that originate in the superior cervical ganglion; activity of these neurons, in turn, is inhibited by nerve tracts that are activated by light striking the retina. The physiology of the pineal gland was discussed in chapter 11 (see fig. 11.33).

There is abundant experimental evidence that melatonin can influence the pituitary-gonad axis in seasonally breeding mammals. However, the role of melatonin in the regulation of human reproduction has not yet been clearly established.

Human Sexual Response

The sexual response, similar in both sexes, is often divided into four phases: excitation, plateau, orgasm, and resolution. The **excitation phase,** also known as **arousal,** is characterized by myotonia (increased muscle tone) and vasocongestion (the engorgement of a sexual organ with blood). This results in erection of the nipples in both sexes, although the effect is more intense and evident in females than in males. The clitoris swells (analogous to erection of the penis), and the labia minora swell to more than twice their previous size. Vasocongestion of the vagina leads to secretion of fluid, producing vaginal lubrication. Vasocongestion also causes considerable enlargement of the uterus, and in women who have not breast-fed a baby the breasts may enlarge as well.

During the **plateau phase,** the clitoris becomes partially hidden behind the labia minora because of the continued engorgement of the labia with blood. Similarly, the erected nipples become partially hidden by continued swelling of the *areolae* (pigmented areas surrounding the nipples). Pronounced engorgement of the outer third of the vagina produces what Masters and Johnson, two scientists who performed pioneering studies of the human sexual response, called the "orgasmic platform."

In **orgasm,** which lasts only a few seconds, the uterus and orgasmic platform of the vagina contract several times. This is analogous to the contractions that accompany ejaculation in a male. Orgasm is followed by the **resolution phase,** in which the body returns to preexcitation conditions. Men, but not women, immediately enter a **refractory period** following orgasm, during which time they may produce an erection but are not able to ejaculate. Women, by contrast, lack a refractory period and are thus capable of multiple orgasms.

CHECKPOINT

3a. Using a flow diagram, show the negative feedback control that the gonads exert on GnRH and gonadotropin secretion. Explain the effects of castration on FSH and LH secretion and the effects of removal of the pituitary on the structure of the gonads and accessory sex organs.

3b. Explain the significance of the pulsatile secretion of GnRH and the gonadotropic hormones.

4a. Describe the two mechanisms that have been proposed to explain the rise in sex steroid secretion that occurs at puberty. Explain the possible effects of body fat and intense exercise on the timing of puberty.

4b. Describe the effect of light on the pineal secretion of melatonin and discuss the possible role of melatonin in reproduction.

4c. Compare the phases of the sexual response in males and females.

20.3 MALE REPRODUCTIVE SYSTEM

The Leydig cells of the testes secrete testosterone, which stimulates the male accessory sex organs, promotes the development of male secondary sex characteristics, and is needed for spermatogenesis. LH stimulates the Leydig cells, whereas FSH stimulates the Sertoli cells of the seminiferous tubules.

LEARNING OUTCOMES

After studying this section, you should be able to:

5. Explain the functions of the two compartments of the testes, and how they are regulated.

6. Describe the stages of spermatogenesis, and the functions of Sertoli cells.

7. Explain how spermatogenesis is regulated.

The testes consist of two parts, or "compartments"—the seminiferous tubules, where spermatogenesis occurs, and the interstitial tissue, which contains the testosterone-secreting *Leydig cells* (fig. 20.11). The seminiferous tubules account for about 90% of the weight of an adult testis. The interstitial tissue is a thin web of connective tissue that fills the spaces between the tubules. The most abundant cells in the interstitial tissue are the Leydig cells, but the interstitial tissue is

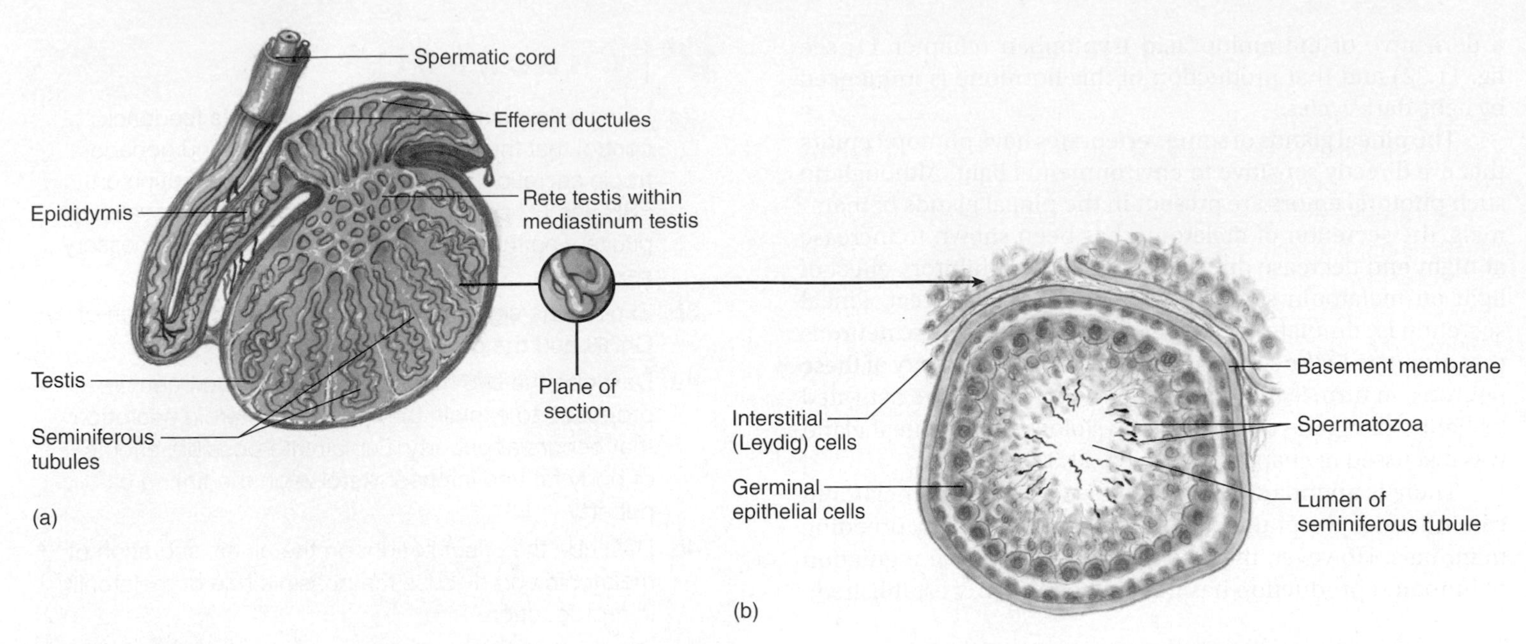

Figure 20.11 **The seminiferous tubules.** (*a*) A sagittal section of a testis and (*b*) a transverse section of a seminiferous tubule. AP|R

also rich in blood and lymphatic capillaries to transport the hormones of the testes.

With regard to gonadotropin action, the testes are strictly compartmentalized. Cellular receptor proteins for FSH are located exclusively in the seminiferous tubules, where they are confined to the *Sertoli cells.* LH receptor proteins are located exclusively in the interstitial Leydig cells. Secretion of testosterone by the Leydig cells is stimulated by LH but not by FSH. Spermatogenesis in the tubules is stimulated by FSH. The apparent simplicity of this compartmentation is an illusion, however, because the two compartments interact with each other in complex ways.

Control of Gonadotropin Secretion

Castration of a male animal results in an immediate rise in FSH and LH secretion. This demonstrates that hormones secreted by the testes exert negative feedback control of gonadotropin secretion. If testosterone is injected into the castrated animal, the secretion of LH can be returned to the previous (precastration) levels. This provides a classical example of negative feedback—LH stimulates testosterone secretion by the Leydig cells, and testosterone inhibits pituitary secretion of LH (fig. 20.12).

The amount of testosterone that is sufficient to suppress LH, however, is not sufficient to suppress the postcastration rise in FSH. In rams and bulls, a water-soluble (and, therefore, peptide rather than steroid) product of the seminiferous tubules specifically suppresses FSH secretion. This hormone, produced by the Sertoli cells, is called *inhibin.* The seminiferous tubules of the human testes also have been shown to produce inhibin, which inhibits FSH secretion in men. (There is also evidence that inhibin is produced by the ovaries, where it may function as a hormone and as a paracrine regulator of the ovaries.)

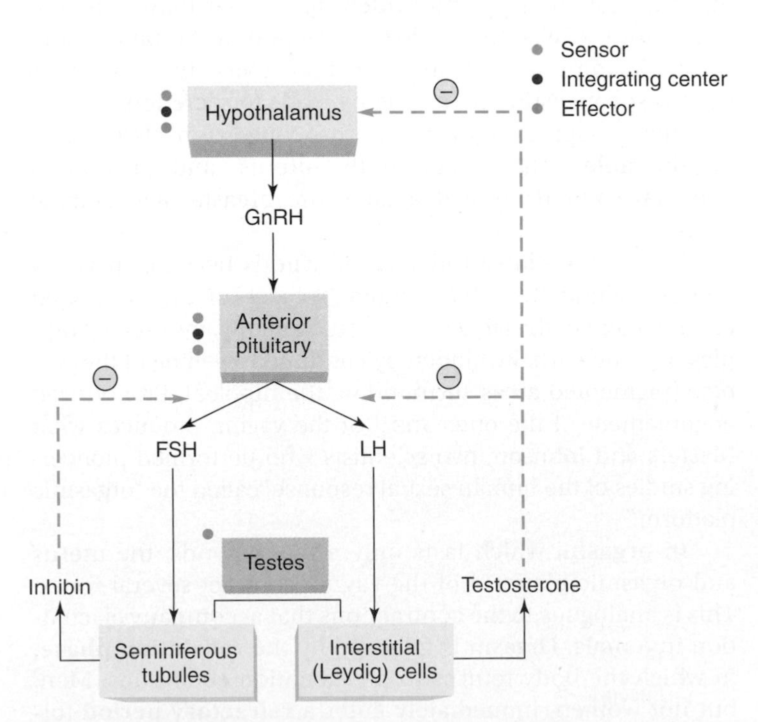

Figure 20.12 **The anterior pituitary and testes.** The seminiferous tubules are the targets of FSH action; the interstitial (Leydig) cells are targets of LH action. Testosterone secreted by the Leydig cells inhibits LH secretion; inhibin secreted by the tubules may inhibit FSH secretion.

Testosterone Derivatives in the Brain

The brain contains testosterone receptors and is a target organ for this hormone. The effects of testosterone on the brain, such as the suppression of LH secretion, are not mediated directly by testosterone, however, but rather by its derivatives that are produced within the brain cells. Testosterone may be converted by the enzyme 5α-reductase to dihydrotestosterone (DHT), as previously described. The DHT, in turn, can be changed by other enzymes into other 5α-reduced androgens—abbreviated 3α-diol and 3β-diol (fig. 20.13). Alternatively, testosterone may be converted within the brain to estradiol-17β. Although usually regarded as a female sex steroid, estradiol is therefore an active compound in normal male physiology. Estradiol is formed from testosterone by the action of an enzyme called **aromatase.** This reaction is known as *aromatization,* a term that refers to the presence of an aromatic carbon ring (chapter 2, section 2.1). The estradiol formed from testosterone in the brain is required for the negative feedback effects of testosterone on LH secretion.

There are some differences between the brains of male and female mammals that are established during fetal development and result from testosterone released from the fetal testes. Some of these differences are caused by testosterone binding to its androgen receptors in brain cells, but others are caused by the conversion of testosterone to estradiol by aromatase within certain brain neurons. In male mice, this estradiol effect is needed for later development of male sexual behavior, territoriality, and aggression.

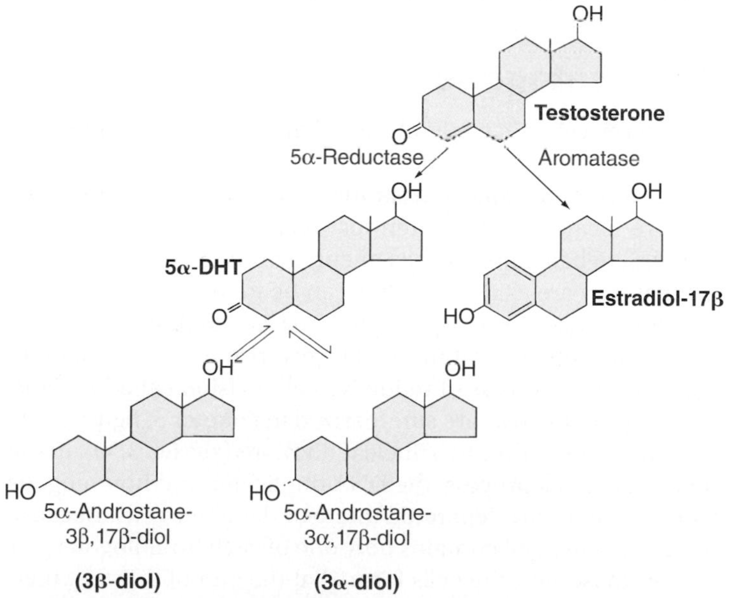

Figure 20.13 Derivatives of testosterone. Testosterone secreted by the interstitial (Leydig) cells of the testes can be converted into active metabolites in the brain and other target organs. These active metabolites include DHT and other 5α-reduced androgens and estradiol.

Testosterone Secretion and Age

The negative feedback effects of testosterone and inhibin help to maintain a relatively constant (that is, noncyclic) secretion of gonadotropins in males, resulting in relatively constant levels of androgen secretion from the testes. This contrasts with the cyclic secretion of gonadotropins and ovarian steroids in females. Women experience an abrupt cessation in sex steroid secretion during menopause. By contrast, the secretion of androgens declines only gradually and to varying degrees in men. The causes of this age-related change in testicular function are not currently known. The decline in testosterone secretion cannot be due to decreasing gonadotropin secretion, since gonadotropin levels in the blood are, in fact, elevated (because of less negative feedback) at the time that testosterone levels are declining.

Testosterone levels decline slowly in men past their 20s, commonly reaching a hypogonadal state (defined as plasma testosterone concentrations below 320 ng/dl) by age 70. Besides age, additional factors that lower plasma testosterone levels are physical inactivity, obesity, and drugs. Low testosterone levels are associated with a reduction in lean muscle and bone mass.

Endocrine Functions of the Testes

Testosterone is by far the major androgen secreted by the adult testis. This hormone and its derivatives (the 5α-reduced androgens) are responsible for initiation and maintenance of the body changes associated with puberty in males. Androgens are sometimes called **anabolic steroids** because they stimulate *anabolism* (synthesis reactions; chapter 19, section 19.1) leading to the growth of muscles and other structures (table 20.4).

Table 20.4 | Actions of Androgens in the Male

Category	Action
Sex Determination	Growth and development of wolffian ducts into epididymis, ductus deferens, seminal vesicles, and ejaculatory ducts
	Development of urogenital sinus into prostate
	Development of male external genitalia (penis and scrotum)
Spermatogenesis	At puberty: Completion of meiotic division and early maturation of spermatids
	After puberty: Maintenance of spermatogenesis
Secondary Sex Characteristics	Growth and maintenance of accessory sex organs
	Growth of penis
	Growth of facial and axillary hair
	Body growth
Anabolic Effects	Protein synthesis and muscle growth
	Growth of bones
	Growth of other organs (including larynx)
	Erythropoiesis (red blood cell formation)

Increased testosterone secretion during puberty is also required for growth of the accessory sex organs—primarily the seminal vesicles and prostate. Removal of androgens by castration results in atrophy of these organs.

Androgens stimulate growth of the larynx (causing a lowering of the voice) and promote hemoglobin synthesis (males have higher hemoglobin levels than females) and bone growth. The effect of androgens on bone growth is self-limiting, however, because they ultimately cause replacement of cartilage by bone in the epiphyseal discs, thus "sealing" the discs and preventing further lengthening of the bones.

Although androgens are by far the major endocrine products of the testes, there is evidence that Sertoli cells, Leydig cells, and developing sperm cells secrete estradiol. Further, receptors for estradiol are found in Sertoli and Leydig cells, as well as in the cells lining the male reproductive tract (efferent ductules and epididymis) and accessory sex organs (prostate and seminal vesicles). Estrogen receptors have also been located in the developing sperm cells (spermatocytes and spermatids, described in the next section) of many species, including humans. This suggests a role for estrogens in spermatogenesis, and indeed knockout mice (chapter 3) missing an estrogen receptor gene are infertile. Further, men with a congenital deficiency in aromatase—the enzyme that converts androgens to estrogens (fig. 20.13)—are also infertile.

The developing sperm cells contain aromatase for the production of estrogen (fig. 20.13), suggesting that they may regulate their environment by means of this hormone. Even mature sperm that travel through the rete testis and efferent ductules (see fig. 20.11) can produce estradiol, although their ability to do this is abolished by the time they get to the tail of the epididymis. Since the efferent ductules are rich in estrogen receptors, the estrogen delivered to them by the sperm in the lumen may help regulate their functions, including fluid reabsorption.

Estradiol produced locally as a paracrine regulator may be responsible for a number of effects in men that have previously been attributed to androgens. For example, the importance of the conversion of testosterone into estradiol in the brain for negative feedback control was described earlier. Estrogen also may be responsible for sealing of the epiphyseal plates of cartilage; this is suggested by observations that men who lack the ability to produce estrogen or who lack estrogen receptors (due to rare genetic defects only recently discovered) maintain their epiphyseal plates and continue to grow.

The two compartments of the testes interact with each other in paracrine fashion (fig. 20.14). Paracrine regulation refers to chemical regulation that occurs among tissues within an organ (chapter 11, section 11.7). Testosterone from the Leydig cells is metabolized by the tubules into other active androgens and is required for spermatogenesis, for example. The tubules also secrete paracrine regulators that may influence Leydig cell function.

Inhibin secreted by the Sertoli cells in response to FSH can facilitate the Leydig cells' response to LH, as measured by the amount of testosterone secreted. Further, it has been shown that

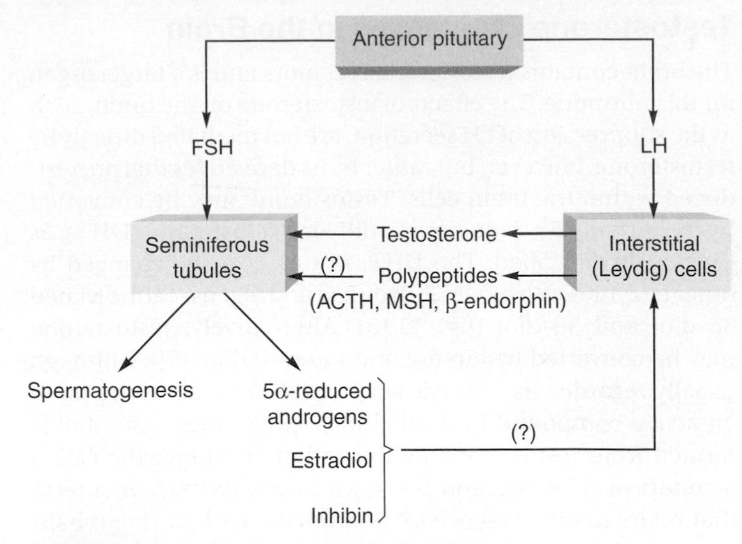

Figure 20.14 Interactions between the two compartments of the testes. Testosterone secreted by the interstitial (Leydig) cells stimulates spermatogenesis in the tubules. Leydig cells may also secrete ACTH, MSH, and β-endorphin. Secretion of inhibin by the tubules may affect the sensitivity of the Leydig cells to LH stimulation.

the Leydig cells are capable of producing a family of polypeptides previously associated with the pituitary gland and brain—ACTH, MSH, and β-endorphin. Experiments suggest that ACTH and MSH can stimulate Sertoli cell function, whereas β-endorphin can inhibit Sertoli function. The physiological significance of these fascinating paracrine interactions between the two compartments of the testes remains to be demonstrated.

Spermatogenesis

The germ cells that migrate from the yolk sac to the testes during early embryonic development become spermatogenic stem cells, called **spermatogonia.** The stem cell spermatogonia are located in the outermost region of the seminiferous tubules, right against the basement membrane (basal lamina), so that they are as close as they can be to the blood vessels in the interstitial tissue. Spermatogonia are diploid cells (with 46 chromosomes) that ultimately give rise to mature haploid gametes by a process of reductive cell division called *meiosis.* The steps of meiosis are summarized in chapter 3, figure 3.30.

Meiosis involves two nuclear divisions (see fig. 3.30). In the first part of this process, the DNA duplicates and homologous chromosomes are separated into two daughter cells. Because each daughter cell contains only one of each homologous pair of chromosomes, the cells formed at the end of this first meiotic division contain 23 chromosomes each and are haploid. Each of the 23 chromosomes at this stage, however, consists of two strands (called *chromatids*) of identical DNA. During the second meiotic division, these duplicate chromatids are separated into daughter cells. Meiosis of one diploid spermatogonium cell therefore produces four haploid cells (fig. 20.15).

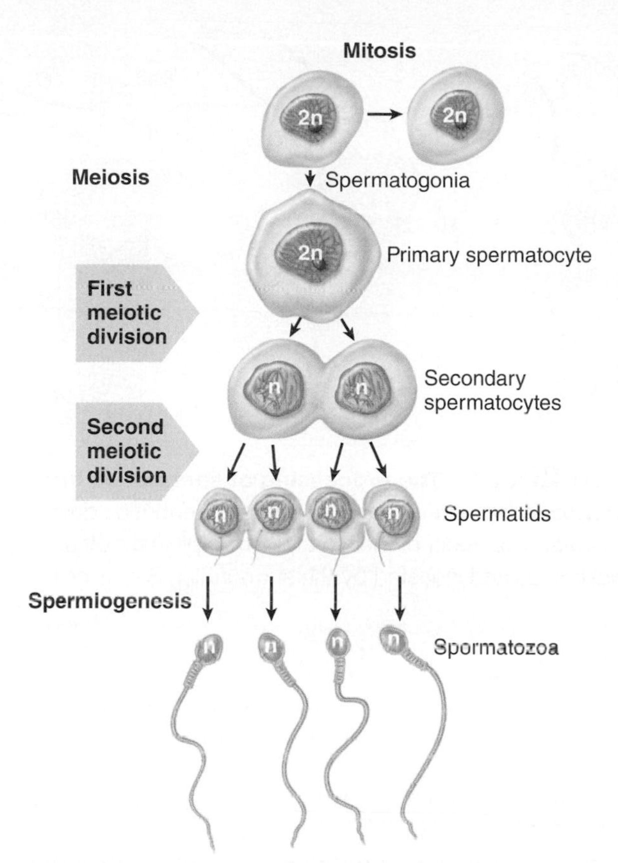

Figure 20.15 Spermatogenesis. Spermatogonia undergo mitotic division in which they replace themselves and produce a daughter cell that will undergo meiotic division. This cell is called a primary spermatocyte. Upon completion of the first meiotic division, the daughter cells are called secondary spermatocytes. Each of these completes a second meiotic division to form spermatids. Notice that the four spermatids produced by the meiosis of a primary spermatocyte are interconnected. Each spermatid forms a mature spermatozoon. AP|R

Actually, only about 1,000 to 2,000 stem cells migrate from the yolk sac into the embryonic testes. In order to produce many millions of sperm throughout adult life, these spermatogonia duplicate themselves by mitotic division and only one of the two cells—now called a **primary spermatocyte**—undergoes meiotic division (fig. 20.15). In this way, spermatogenesis can occur continuously without exhausting the number of spermatogonia.

When a diploid primary spermatocyte completes the first meiotic division (at telophase I), the two haploid cells thus produced are called **secondary spermatocytes.** At the end of the second meiotic division, each of the two secondary spermatocytes produces two haploid **spermatids.** One primary spermatocyte therefore produces four spermatids.

The sequence of events in spermatogenesis is reflected in the cellular arrangement of the wall of the seminiferous tubule. The spermatogonia and primary spermatocytes are located toward the outer side of the tubule, whereas spermatids and mature spermatozoa are located on the side of the tubule facing the lumen.

At the end of the second meiotic division, the four spermatids produced by meiosis of one primary spermatocyte are

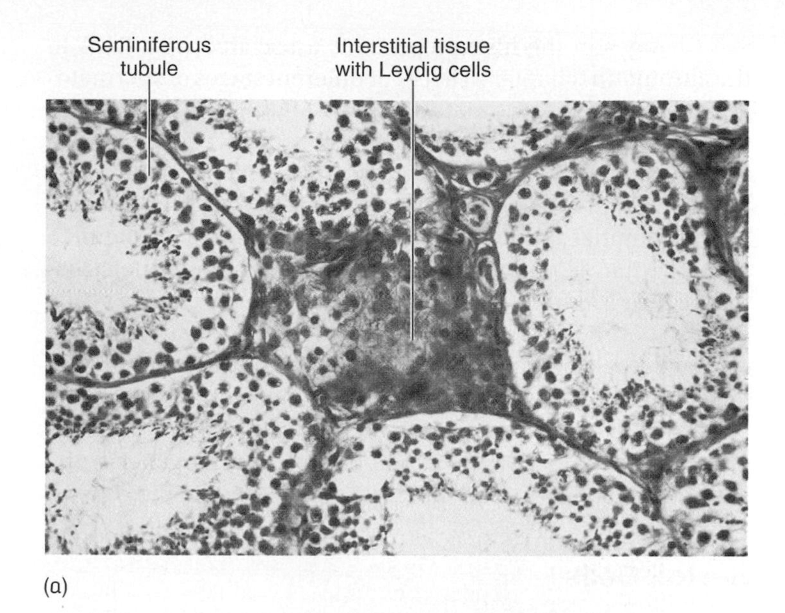

(a)

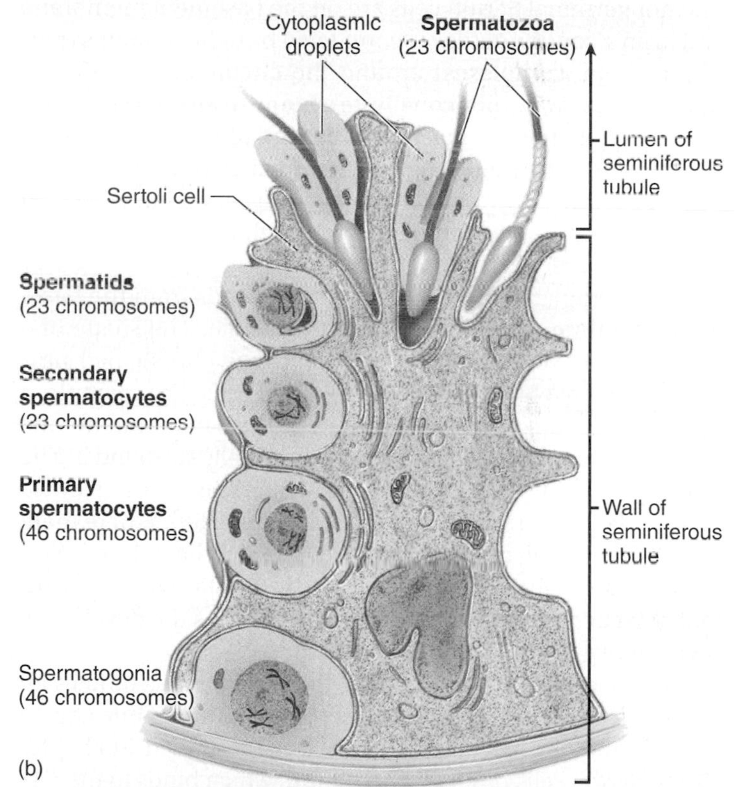

(b)

Figure 20.16 A photomicrograph and diagram of the seminiferous tubules. (*a*) A cross section of the seminiferous tubules also shows surrounding interstitial tissue. (*b*) the stages of spermatogenesis are indicated within the germinal epithelium of a seminiferous tubule. The relationship between Sertoli cells and developing spermatozoa can also be seen. AP|R

interconnected—their cytoplasm does not completely pinch off at the end of each division. Development of these interconnected spermatids into separate mature **spermatozoa** (singular, *spermatozoon*)—a process called *spermiogenesis*—requires the participation of the Sertoli cells (fig. 20.16).

Changes in the histone proteins associated with DNA in the chromatin (chapter 3) occur at different stages of spermatogenesis. Since histone modifications affect gene expression, these changes may be needed to allow proper gene expression in the future embryo. During spermiogenesis, a related type of protein called *protamines* replaces the histone proteins. The protamines induce great compaction of the chromatin, to a degree that is unique for spermatozoa. This unique structure of the chromatin then causes the nucleus to change shape during spermiogenesis. Compaction of the chromatin and the changed nuclear shape is followed by the development of the flagellum, the removal of germ cell cytoplasm by the Sertoli cells, and the appearance of the *acrosome* (a cap of digestive enzymes—see fig. 20.18). At the end of spermiogenesis, the spermatozoon is released into the lumen of the tubule.

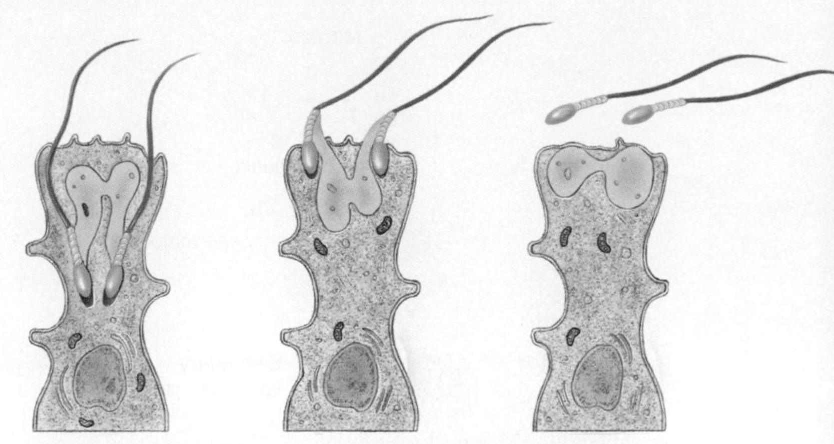

Figure 20.17 The processing of spermatids into spermatozoa (spermiogenesis). As the spermatids develop into spermatozoa, most of their cytoplasm is pinched off as residual bodies and ingested by the surrounding Sertoli cell cytoplasm.

Sertoli Cells

The nongerminal Sertoli cells are on the basement membrane and form a continuous layer connected by tight junctions (part of junctional complexes) around the circumference of each tubule. In this way, they constitute a **blood-testis barrier:** molecules from the blood must pass through the cytoplasm of the Sertoli cells before entering the germinal cells. Similarly, this barrier prevents the immune system from becoming sensitized to antigens in the developing sperm and thus prevents autoimmune destruction of the sperm.

The cytoplasm of the Sertoli cells extends from the basement membrane to the lumen of the tubule. The shape of a Sertoli cell is very complex because it has cup-shaped processes that envelop the developing germ cells. For example, a spermatogonium is located close to the basement membrane between adjacent Sertoli cells that each partially surround it. The other developing spermatocytes and spermatids are likewise surrounded by Sertoli cell cytoplasm (fig. 20.16*b*). Where there are no developing germ cells between them, the adjacent Sertoli cells are tightly joined together by junctional complexes that must be formed, broken, and re-formed as the developing sperm move toward the lumen.

The Sertoli cells help to make the seminiferous tubules an *immunologically privileged site* (protected from immune attack) through another mechanism as well. As described in chapter 15, the Sertoli cells produce **FAS ligand,** which binds to the *FAS* receptor on the surface of T lymphocytes. This triggers apoptosis (cell suicide) of the T lymphocytes and thus helps to prevent immune attack of the developing sperm.

In the process of spermiogenesis (conversion of spermatids to spermatozoa), most of the spermatid cytoplasm is eliminated. This occurs through phagocytosis by Sertoli cells of the "residual bodies" of cytoplasm from the spermatids (fig. 20.17). Phagocytosis of residual bodies may transmit regulatory molecules from germ cells to Sertoli cells. The Sertoli cells, in turn, provide molecules needed by the germ cells. It is known, for example, that the X chromosome of germ cells is inactive during meiosis. Since this chromosome contains genes needed to produce many essential molecules, it is believed that these molecules are provided by the Sertoli cells during this time.

Sertoli cells secrete a protein called **androgen-binding protein (ABP)** into the lumen of the seminiferous tubules. This protein, as its name implies, binds to testosterone and thereby concentrates it within the tubules. Production of ABP is stimulated by FSH, and FSH receptors are found only in Sertoli cells. Thus, all effects of FSH in the testes must be mediated by Sertoli cells. These effects include production of ABP, FSH-induced stimulation of spermiogenesis, and other paracrine interactions between Sertoli and Leydig cells in the testis.

At the conclusion of spermiogenesis, spermatozoa enter the lumen of the seminiferous tubules. A spermatozoon (fig. 20.18) contains a *head,* consisting mostly of a nucleus with DNA and an overlying cap known as an *acrosome* (section 20.6; see fig. 20.38), and a flagellum tail. The flagellum has a characteristic "9+2" microtubule structure (chapter 3, section 3.1) called an *axoneme.* The axoneme runs through the entire flagellum, which is divided into an upper *midpiece* that contains a fibrous sheath and mitochondria around the axoneme; a *principal piece* with only the fibrous sheath around the axoneme; and an *end piece* consisting only of the axoneme. Sperm in the seminiferous tubules are nonmotile; they become capable of flagellar movement and thus motility outside of the testis in the epididymis.

Hormonal Control of Spermatogenesis

The formation of primary spermatocytes and entry into early prophase I begins during embryonic development, but spermatogenesis is arrested at this point until puberty, when testosterone secretion rises. Testosterone is required for completion of meiotic division and for the early stages of spermatid maturation. This effect is probably not produced by testosterone

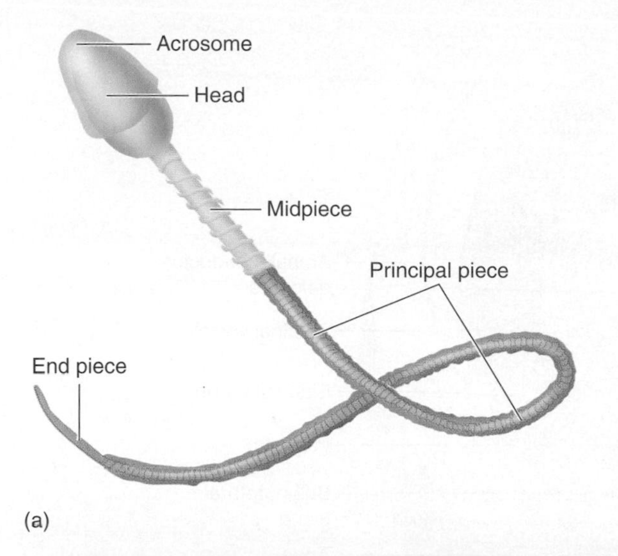

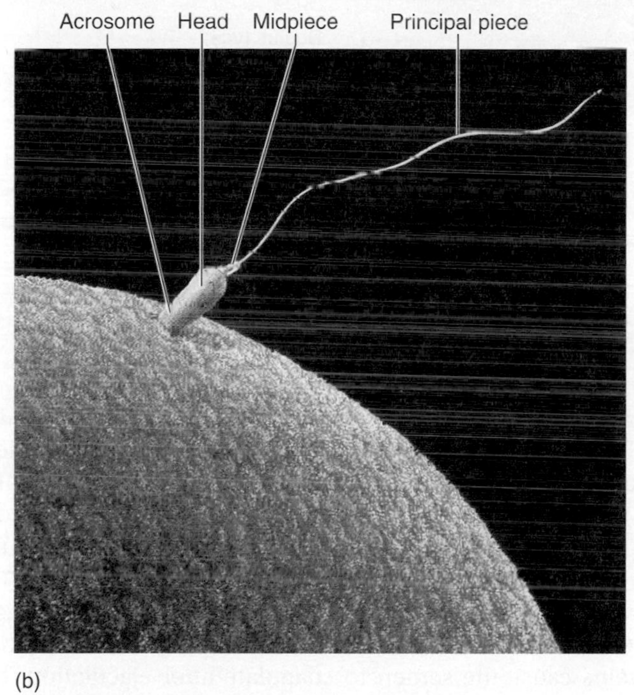

Figure 20.18 A human spermatozoon. (*a*) A diagrammatic representation and (*b*) a scanning electron micrograph in which a spermatozoon is seen in contact with an ovum.

directly, but rather by some of the molecules derived from testosterone (the 5α-reduced androgens and estrogens, described earlier) in the tubules. The testes also produce a wide variety of paracrine regulators—transforming growth factor, insulin-like growth factor-1, inhibin, and others—that may help to regulate spermatogenesis.

Testosterone, secreted by the Leydig cells under LH stimulation, acts as a paracrine regulator by stimulating spermatogenesis in the seminiferous tubules. Testosterone can initiate spermatogenesis at puberty and can maintain spermatogenesis in the adult human testis. As mentioned previously, estrogen produced in the testes (from testosterone) may also be required for spermatogenesis, although its role is not currently understood.

Surprisingly, FSH is not absolutely required for spermatogenesis, as demonstrated by men who have mutated and nonfunctional FSH receptors. LH-stimulated testosterone secretion promotes spermatogenesis, whereas FSH only enhances this effect. The FSH receptors are located in the Sertoli cells, as previously described, and FSH stimulates Sertoli cells to produce androgen-binding protein and inhibin. A newborn male has only about 10% of his adult number of Sertoli cells, and this increases to the adult number as a boy enters puberty. It appears that FSH, acting together with testosterone, promotes this proliferation of Sertoli cells. Without FSH, spermatogenesis would still occur but would commence later in puberty.

Likewise, maintenance of spermatogenesis in the adult testis requires only testosterone. However, FSH is required for maximal sperm production, and so it may be required for optimal fertility. Therefore, hypothetical male contraceptive drugs that blocked FSH might reduce, but probably would not abolish, fertility. The enhancement of testosterone-supported spermatogenesis by FSH is believed to be due to paracrine regulators secreted by the Sertoli cells.

Male Accessory Sex Organs

The seminiferous tubules are connected at both ends to a tubular network called the *rete testis* (see fig. 20.11). Spermatozoa and tubular secretions are moved to this area of the testis and are drained via the *efferent ductules* into the **epididymis** (the plural is *epididymides*). The epididymis is a tightly coiled structure, about 5 meters (16 feet) long if stretched out, that receives the tubular products. Spermatozoa enter at the "head" of the epididymis and are drained from its "tail" by a single tube, the **ductus,** or **vas, deferens.**

Spermatozoa that enter the head of the epididymis are nonmotile. This is partially due to the low pH of the fluid in the epididymis and ductus deferens, produced by the reabsorption of bicarbonate and the secretion of H^+ by active transport ATPase pumps. During their passage through the epididymis, the sperm undergo maturational changes that make them more resistant to changes in pH and temperature. The pH is neutralized by the alkaline prostatic fluid during ejaculation, so that the sperm are fully motile and become capable of fertilizing an ovum once they spend some time in the female reproductive tract. Sperm obtained from the seminiferous tubules, by contrast, cannot fertilize an ovum. The epididymis serves as a site for sperm maturation and for the storage of sperm between ejaculations.

The ductus deferens carries sperm from the epididymis out of the scrotum into the pelvic cavity. The **seminal vesicles** then add secretions that pass through their ducts; at this point, the ductus deferens becomes an **ejaculatory duct.** The ejaculatory duct is short (about 2 cm), however, because it enters

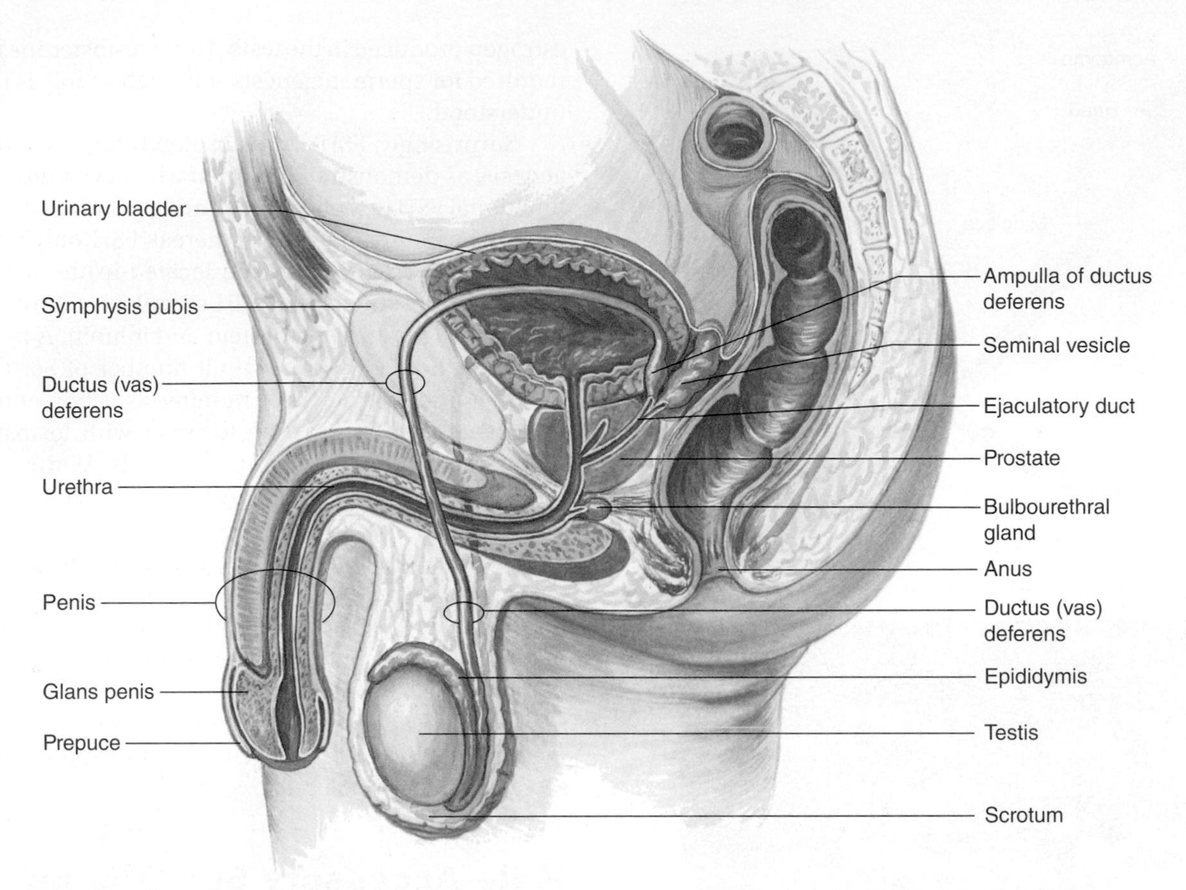

Figure 20.19 The organs of the male reproductive system. The male organs are seen here in a sagittal view. AP|R

the **prostate** and soon merges with the prostatic **urethra.** The prostate adds its secretions through numerous pores in the walls of the prostatic urethra, forming a fluid known as *semen* (fig. 20.19).

CLINICAL APPLICATION

An immunoassay for **prostate-specific antigen (PSA)** is a common laboratory test for prostate disorders, including **prostate cancer.** A more common disorder, affecting most men over 60 to a greater or lesser degree, is **benign prostatic hyperplasia (BPH).** This disorder is responsible for most symptoms of bladder outlet obstruction, where there is difficulty in urination. BPH treatment may involve a surgical procedure called *transurethral resection (TUR),* or the use of drugs. Drugs used to treat BPH include α_1*-adrenergic receptor blockers,* which decrease the muscle tone of the prostate and bladder neck (making urination easier), and 5α*-reductase inhibitors,* which inhibit the enzyme needed to convert testosterone into dihydrotestosterone (DHT). As previously described, DHT is required to maintain the structure of the prostate; therefore, a reduction in DHT may help to reduce the size of the prostate.

The seminal vesicles and prostate are androgen-dependent accessory sex organs—they will atrophy if androgen is withdrawn by castration. The seminal vesicles secrete fluid containing fructose, which serves as an energy source for the spermatozoa. This fluid secretion accounts for about 60% of the volume of the semen. The fluid contributed by the prostate contains citric acid, calcium, and coagulation proteins. Clotting proteins cause the semen to coagulate after ejaculation, but the hydrolytic action of fibrinolysin later causes the coagulated semen to again assume a more liquid form, thereby freeing the sperm.

Erection, Emission, and Ejaculation

Erection, accompanied by increases in the length and width of the penis, is achieved as a result of blood flow into the "erectile tissues" of the penis. These erectile tissues include two paired structures—the *corpora cavernosa*—located on the dorsal side of the penis, and one unpaired *corpus spongiosum* on the ventral side (fig. 20.20). The urethra runs through the center of the corpus spongiosum. The erectile tissue forms columns that extend the length of the penis, although the corpora cavernosa do not extend all the way to the tip.

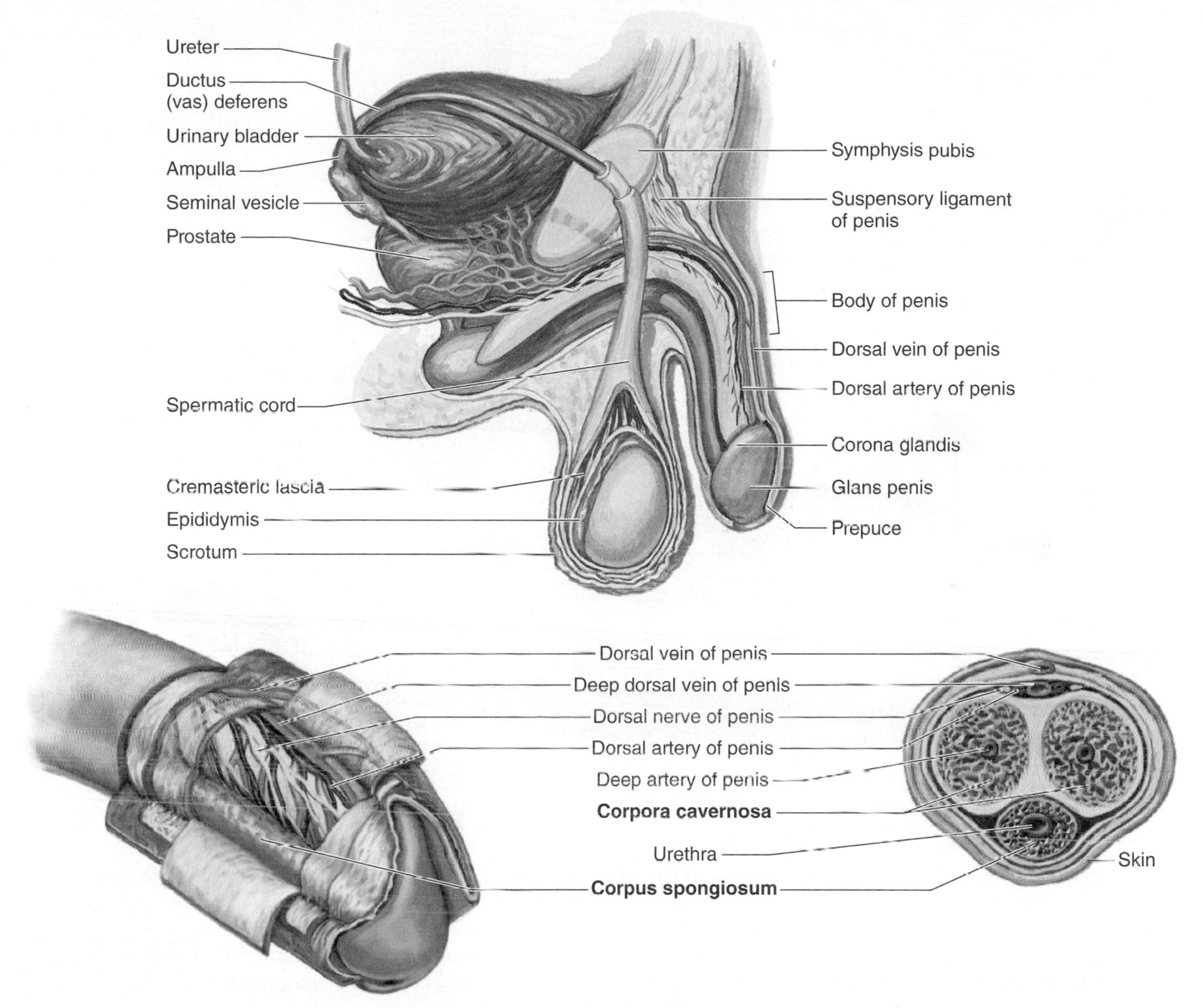

Figure 20.20 The structure of the penis. The attachment, blood and nerve supply, and arrangement of the erectile tissue are shown in both longitudinal and cross section. AP|R

Erection is achieved by parasympathetic nerve-induced vasodilation of arterioles that allows blood to flow into the corpora cavernosa of the penis. The neurotransmitter that mediates this increased blood flow is believed to be nitric oxide (fig. 20.21). Nitric oxide released by parasympathetic axons and produced by the endothelial cells of penile blood vessels activates guanylate cyclase in the vascular smooth muscle cells. Guanylate cyclase catalyzes the production of cyclic GMP (cGMP), which closes Ca^{2+} channels in the plasma membrane (fig. 20.21). This decreases the cytoplasmic Ca^{2+} concentration, causing smooth muscle relaxation (chapter 12). The penile blood vessels thereby dilate to increase the blood flow into the erectile tissue, producing an erection.

As the erectile tissues become engorged with blood and the penis becomes turgid, venous outflow of blood is partially occluded, thereby aiding erection. The term **emission** refers to the movement of semen into the urethra, and **ejaculation** refers to the forcible expulsion of semen from the urethra out of the penis. Emission and ejaculation are stimulated by sympathetic nerves, which cause peristaltic contractions of the tubular system, contractions of the seminal vesicles and

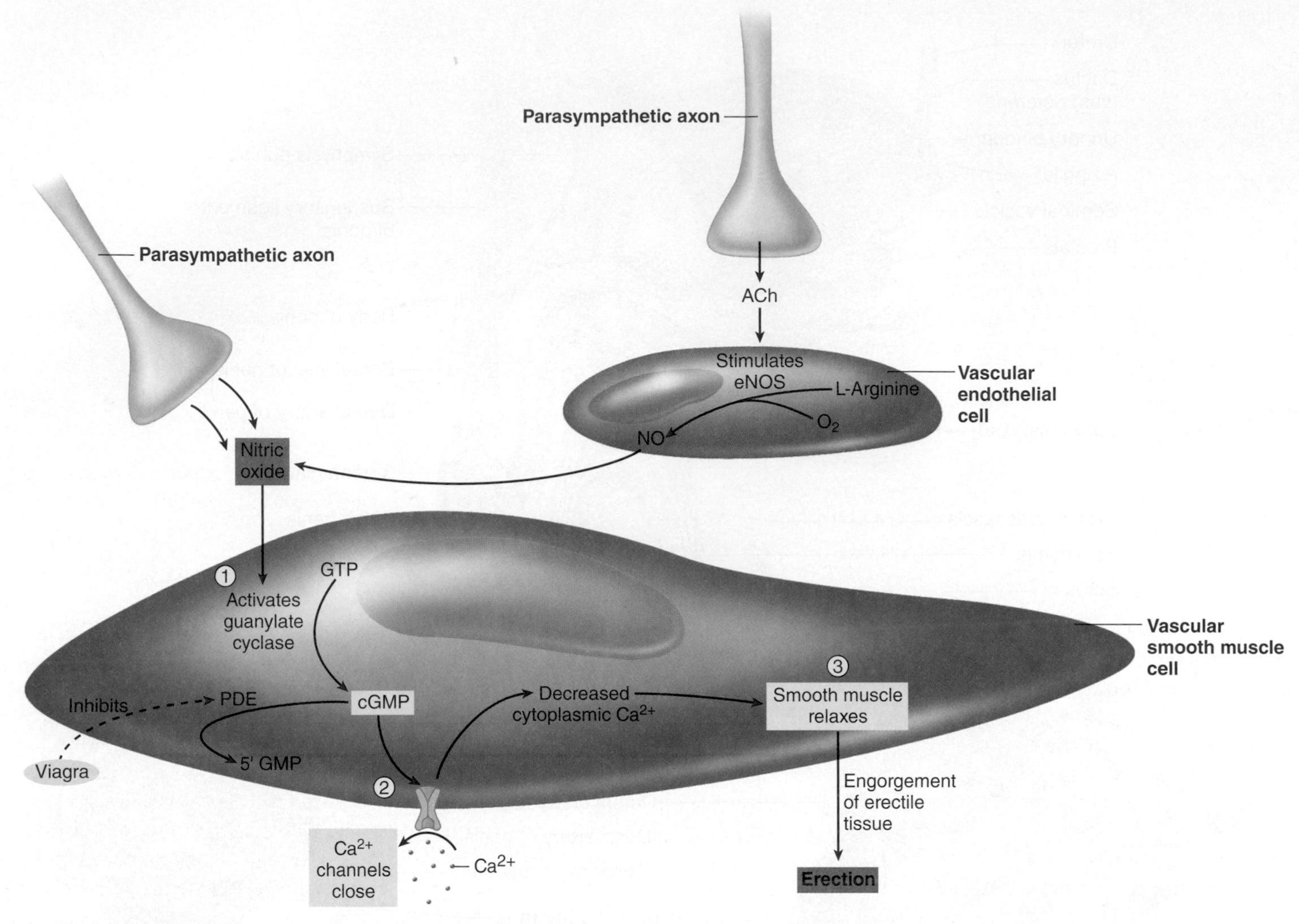

Figure 20.21 Role of nitric oxide in erection of the penis. Nitric oxide is released by parasympathetic axons in the penis, and is also produced as a paracrine regulator by endothelial cells in blood vessels of the penis. (1) Nitric oxide activates guanylate cyclase, causing the production of cyclic GMP (cGMP). (2) cGMP causes Ca^{2+} channels to close in the plasma membrane of the vascular smooth muscle cells, decreasing the cytoplasmic Ca^{2+} concentration. (3) This produces relaxation of the smooth muscle cells, causing vasodilation in the penis and engorgement of the erectile tissue. Viagra and related drugs inhibit the phosphodiesterase that catalyzes the breakdown of cGMP, thereby increasing the availability of cGMP to promote an erection.

prostate, and contractions of muscles at the base of the penis. Sexual function in the male thus requires the synergistic action (rather than antagonistic action) of the parasympathetic and sympathetic systems.

Erection is controlled by two portions of the central nervous system—the hypothalamus in the brain and the sacral portion of the spinal cord. Conscious sexual thoughts originating in the cerebral cortex act via the hypothalamus to control the sacral region, which in turn increases parasympathetic nerve activity to promote vasodilation and erection of the penis. Conscious thought is not required for erection, however, because sensory stimulation of the penis can more directly activate the sacral region of the spinal cord and cause an erection.

CLINICAL APPLICATION

Nitric oxide, released in the penis in response to parasympathetic nerve stimulation, diffuses into the smooth muscle cells of blood vessels and stimulates the production of cyclic guanosine monophosphate (cGMP). The cGMP, in turn, causes the vascular smooth muscle to relax so that blood can flow into the corpora cavernosa (fig. 20.21). This physiology is exploited by *sildenafil* (trade named *Viagra*), which can be taken as a pill to treat **erectile dysfunction.** Sildenafil blocks cGMP phosphodiesterase, an enzyme that functions to break down cGMP. This increases the concentration of cGMP and thus promotes vasodilation, leading to increased engorgement of the erectile spongy tissue with blood and consequently promoting erection (fig. 20.21).

Male Fertility

The approximate volume of semen for each ejaculation is 1.5 to 5.0 milliliters. The bulk of this fluid (45% to 80%) is produced by the seminal vesicles and 15% to 30% is contributed by the prostate. There are usually between 60 and 150 million sperm per milliliter of ejaculate. Normal human semen values are summarized in table 20.5.

A sperm concentration below about 20 million per milliliter is termed *oligospermia* (*oligo* = few) and is associated with decreased fertility. A total sperm count below about 40 million per ejaculation is clinically significant in male infertility. Oligospermia may be caused by a variety of factors. including heat from a sauna or hot tub, various pharmaceutical drugs, lead and arsenic poisoning, and such illicit drugs as marijuana, cocaine, and anabolic steroids. It may be temporary or permanent. In addition to low sperm counts as a cause of infertility, some men and women have antibodies against sperm antigens (this is very common in men with vasectomies). While such antibodies do not appear to affect health, they do reduce ferility.

Attempts have been made to develop new methods of male contraception. These have generally involved compounds that suppress gonadotropin secretion, such as testosterone or a combination of progesterone and a GnRH antagonist. Another compound, *gossypol,* which interferes with sperm development, has also been tried. These drugs can be effective but have unacceptable side effects. One of the most widely used methods of male contraception is a surgical procedure called a **vasectomy** (fig. 20.22). In this procedure, each ductus (vas) deferens is cut and tied or, in some cases, a valve or similar device is inserted. A vasectomy interferes with sperm transport but does not directly affect the secretion of androgens from Leydig cells in the interstitial tissue. Spermatogenesis continues, but the sperm cannot be drained from the testes; as a result, the sperm accumulate in "crypts" that form in the seminiferous tubules, epididymis, and ductus deferens. These crypts present sites for inflammatory reactions in which spermatozoa are phagocytosed and destroyed by the immune system. It is thus not surprising that approximately 70% of men with vasectomies develop antisperm antibodies. These antibodies do not appear to cause autoimmune damage to the testes, but they do significantly diminish the possibility of reversing a vasectomy and restoring fertility.

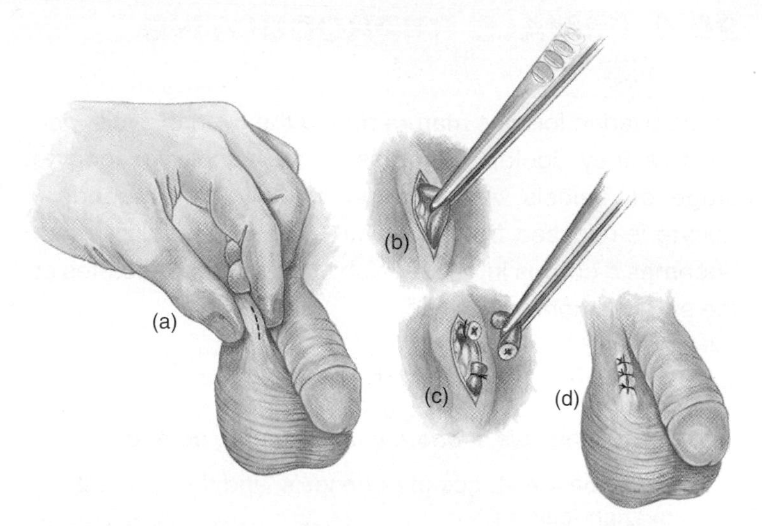

Figure 20.22 A vasectomy. In this surgical procedure a segment of the ductus (vas) deferens is removed through an incision in the scrotum.

Table 20.5 | Semen Analysis

Characteristic	Reference Value
Volume of ejaculate	1.5–5.0 ml
Sperm count	40–250 million/ml
Sperm motility	
Percentage of motile forms:	
1 hour after ejaculation	70% or more
3 hours after ejaculation	60% or more
Leukocyte count	0–2,000/ml
pH	7.2–7.8
Fructose concentration	150–600 mg/100 ml

Source: Modified from L. Glasser,"Seminal Fluid and Subfertility," *Diagnostic Medicine*, July/August 1981, p. 28. Used by permission. Other sources provide slightly different ranges.

CHECKPOINT

5a. Describe the effects of castration on FSH and LH secretion in the male. Explain the experimental evidence suggesting that the testes produce a polypeptide that specifically inhibits FSH secretion.

5b. Describe the two compartments of the testes with respect to (a) structure, (b) function, and (c) response to gonadotropin stimulation. Describe two ways in which these compartments interact.

6a. Using a diagram, describe the stages of spermatogenesis. Why can spermatogenesis continue throughout life without using up all of the spermatogonia?

6b. Describe the structure and proposed functions of the Sertoli cells in the seminiferous tubules.

7. Describe the roles of FSH and testosterone in spermatogenesis during and after puberty.

20.4 FEMALE REPRODUCTIVE SYSTEM

Some ovarian follicles mature during the ovarian cycle, and the ova they contain progress to the secondary oocyte stage of meiosis. At ovulation, usually one secondary oocyte is released from the ovary. The empty follicle then becomes a corpus luteum, which ultimately degenerates at the end of a nonfertile cycle.

LEARNING OUTCOMES

After studying this section, you should be able to:

8. Describe the stages of oogenesis, and the different ovarian follicles.

9. Describe ovulation and the formation of a corpus luteum.

The two ovaries (fig. 20.23), about the size and shape of large almonds, are suspended by means of ligaments from the pelvic girdle. Extensions called *fimbriae* of the **uterine (fallopian) tubes** partially cover each ovary. Ova that are released from the ovary—in a process called *ovulation*—are normally drawn into the uterine tubes by the action of cilia on the epithelial lining of the tubes. The lumen of each uterine tube is continuous with the **uterus** (or womb), a pear-shaped muscular organ held in place within the pelvic cavity by ligaments.

The uterus consists of three layers. The outer layer of connective tissue is the **perimetrium,** the middle layer of smooth muscle is the **myometrium,** and the inner epithelial layer is the **endometrium.** The endometrium is a stratified, squamous, nonkeratinized epithelium that consists of a *stratum basale* and a more superficial *stratum functionale.* The stratum functionale, which cyclically grows thicker as a result of estrogen and progesterone stimulation, is shed at menstruation.

The uterus narrows to form the *cervix* (= neck), which opens to the tubular **vagina.** The only physical barrier between the vagina and uterus is a plug of *cervical mucus.* These structures—the vagina, uterus, and fallopian tubes—constitute the accessory sex organs of the female (fig. 20.24). Like the accessory sex organs of the male, the female reproductive tract is affected by gonadal steroid hormones. Cyclic changes in ovarian secretion, as will be described in the next section, cause cyclic changes in the epithelial lining of the tract.

The vaginal opening is located immediately posterior to the opening of the urethra. Both openings are covered by longitudinal folds—the inner **labia minora** and outer **labia majora** (fig. 20.25). The **clitoris,** a small structure composed largely of erectile tissue, is located at the anterior margin of the labia minora.

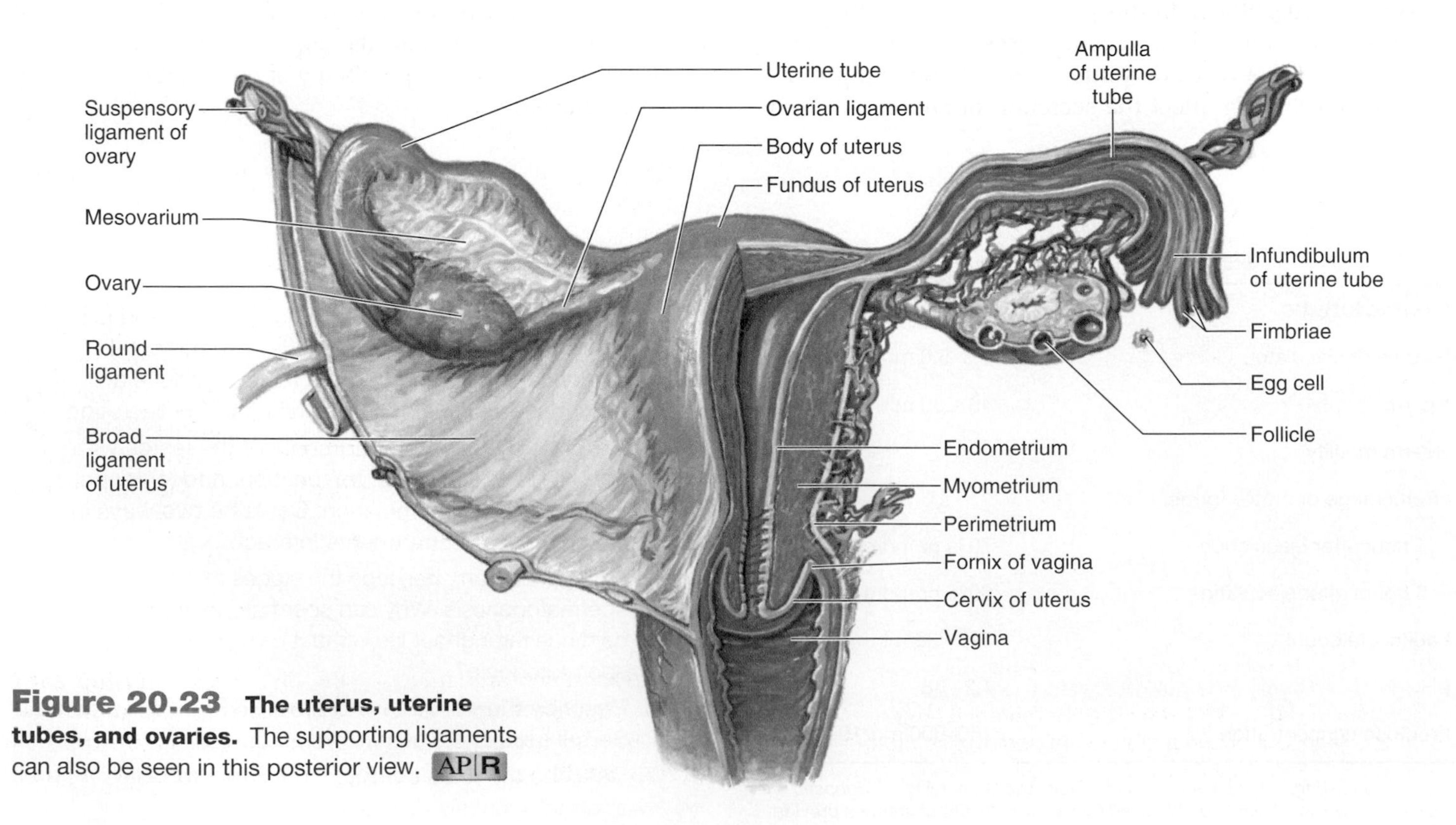

Figure 20.23 The uterus, uterine tubes, and ovaries. The supporting ligaments can also be seen in this posterior view. APIR

Figure 20.24 **The organs of the female reproductive system.** These are shown in sagittal section. AP|R

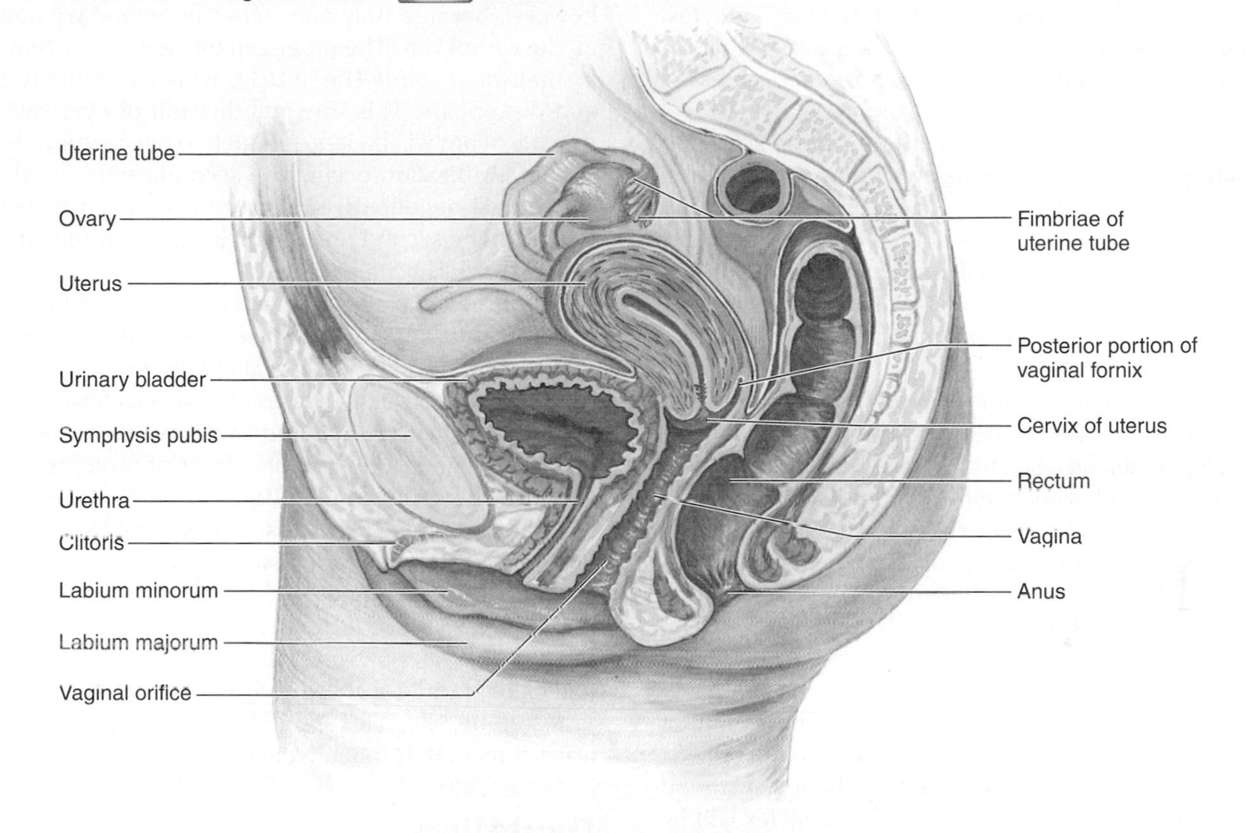

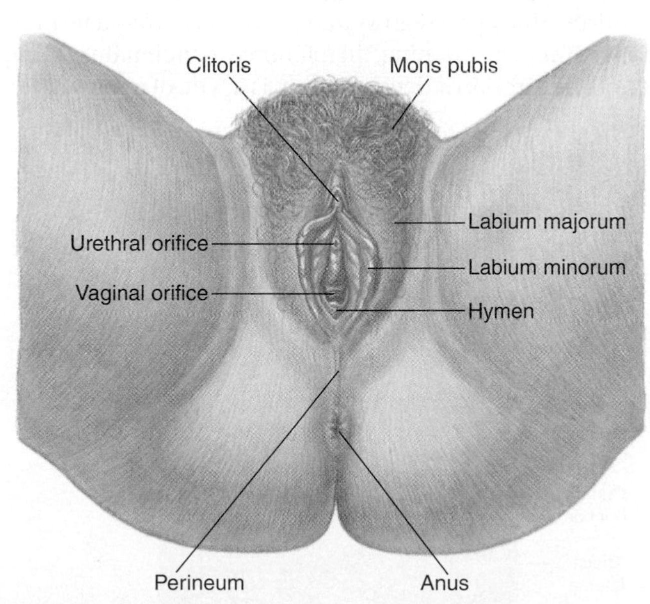

Figure 20.25 **The external female genitalia.** The labia majora and clitoris in a female are homologous to the scrotum and penis, respectively, in a male.

CLINICAL APPLICATION

Uterine fibroids (leiomyomas) are nonmalignant growths (neoplasms) that also include abundant extracellular matrix. Symptoms—including profuse menstrual bleeding and pelvic discomfort—can be produced by fibroids as small as 10 mm or as large as 20 cm. The ovarian steroid hormones estradiol and progesterone stimulate their growth, and uterine fibroids have high concentrations of estradiol and progesterone receptors. Because most fibroids occur within the wall of the uterus, surgical treatment usually requires a *hysterectomy* (surgical removal of the uterus). Indeed, the majority of hysterectomies are performed because of uterine fibroids.

Ovarian Cycle

The germ cells that migrate into the ovaries during early embryonic development multiply, so that by about five months of *gestation* (prenatal life) the ovaries contain approximately 6 million to 7 million oogonia. Most of these oogonia die prenatally through a process of apoptosis (chapter 3, section 3.5). The remaining oogonia begin meiosis toward the end of gestation, at which time they are called **primary oocytes.** Like spermatogenesis in the male, oogenesis is arrested at prophase I of the first meiotic division and the primary oocytes are still diploid.

The ovaries of a newborn girl contain about 2 million oocytes. Each is contained within its own hollow ball of cells, the *ovarian follicle.* By the time a girl reaches puberty, the number of oocytes and follicles has been reduced to 400,000. Only about 400 of these oocytes will ovulate during the woman's reproductive years, and the rest will die by apoptosis. Oogenesis ceases entirely at menopause (the time menstruation stops).

Primary oocytes that have not yet been stimulated to complete the first meiotic division are contained within tiny **primary follicles** (fig. 20.26*a*). Scientists have long believed that the primary oocytes and primary follicles could not be renewed postnatally—that a girl is born with all she will ever have, and those that are lost cannot be replaced. There has been a controversial report of the postnatal production of new oocytes and primary follicles in mice, but this observation has not yet been made in humans.

Immature primary follicles consist of only a single layer of follicle cells. In response to FSH stimulation, some of these oocytes and follicles get larger, and the follicular cells divide to produce numerous layers of **granulosa cells** that surround the oocyte and fill the follicle. Some primary follicles will be stimulated to grow still more, and they will develop a number of fluid-filled cavities called *vesicles;* at this point, they are called **secondary follicles** (fig. 20.26*a*). Continued growth of one of these follicles will be accompanied by the fusion of its vesicles to form a single fluid-filled cavity called an *antrum.* At this stage, the follicle is known as a **mature,** or **graafian, follicle** (fig. 20.26*b*).

As the follicle develops, the primary oocyte completes its first meiotic division. This does not form two complete cells, however, because only one cell—the **secondary oocyte**—gets all the cytoplasm. The other cell formed at this time becomes a small *polar body* (fig. 20.27), which eventually fragments and disappears. This unequal division of cytoplasm ensures that the ovum will be large enough to become a viable embryo should fertilization occur. The secondary oocyte then begins the second meiotic division, but meiosis is arrested at metaphase II. The second meiotic division is completed only by an oocyte that has been fertilized.

The secondary oocyte, arrested at metaphase II, is contained within a graafian follicle. Some of the granulosa cells of this follicle form a mound called the *cumulus oophorus* that supports the oocyte. Other granulosa cells form a ring around the oocyte called a *corona radiata.* Between the oocyte and the corona radiata is a thin gel-like layer of proteins and polysaccharides called the **zona pellucida** (see fig. 20.26*b*). The zona pellucida is significant because it presents a barrier to the ability of a sperm to fertilize an ovulated oocyte.

Under the stimulation of FSH from the anterior pituitary, the granulosa cells of the ovarian follicles secrete increasing amounts of estradiol (estrogen) as the follicles grow. Interestingly, the granulosa cells produce estradiol from its precursor testosterone, which is supplied by cells of the *theca interna,* the layer immediately outside the follicle (see fig. 20.26*b*).

Ovulation

Usually by the 10th to 14th day after the first day of menstruation only one follicle has continued its growth to become a fully mature graafian follicle (fig. 20.28). Other secondary follicles during that cycle regress and become *atretic*—a term that means "without an opening," in reference to their failure to rupture. Follicle atresia, or degeneration, is a type of apoptosis that

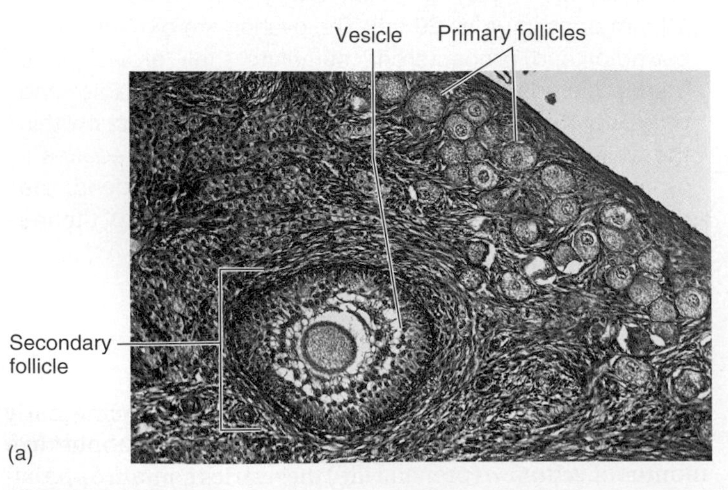

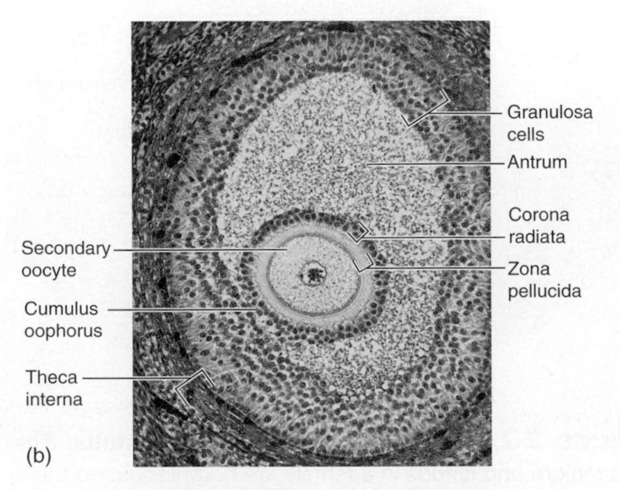

Figure 20.26 Photomicrographs of the ovary. (*a*) Primary follicles and one secondary follicle and (*b*) a graafian follicle are visible in these sections. AP|R

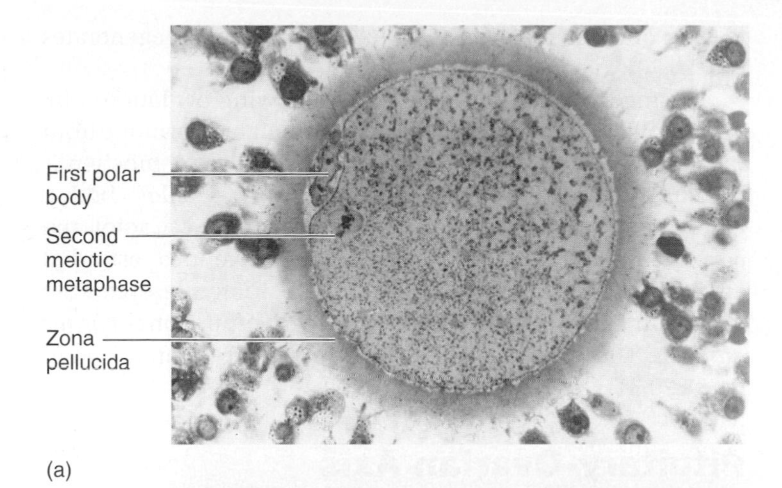

(a)

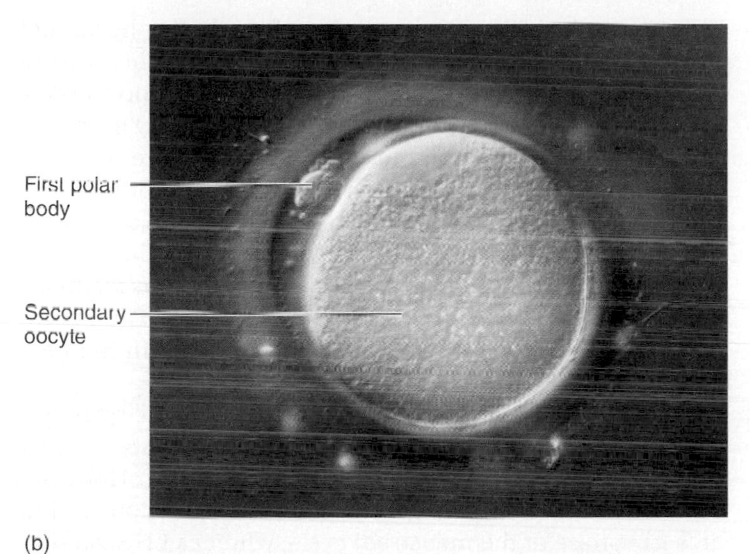

(b)

Figure 20.27 A secondary oocyte with its first polar body. The first polar body is formed by the first meiotic division. This will degenerate and a second meiotic division will not occur unless the secondary oocyte is fertilized. In that event, the second meiotic division will produce a second polar body, which also will degenerate (see fig. 20.30).

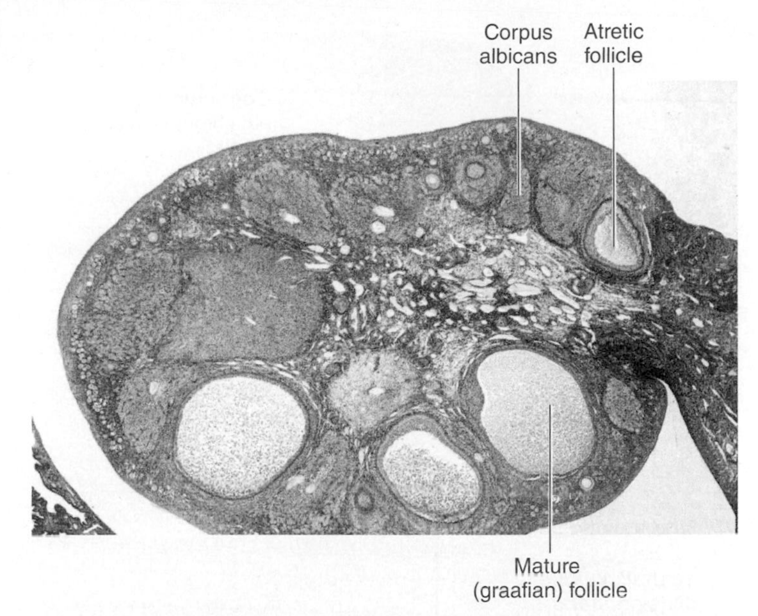

Figure 20.28 An ovary containing follicles at different stages of development. This cat ovary shows a fully mature (graafian) follicle, an atretic follicle that started to develop but then stopped, and a corpus albicans that formed from a corpus luteum of a previous cycle. APIR

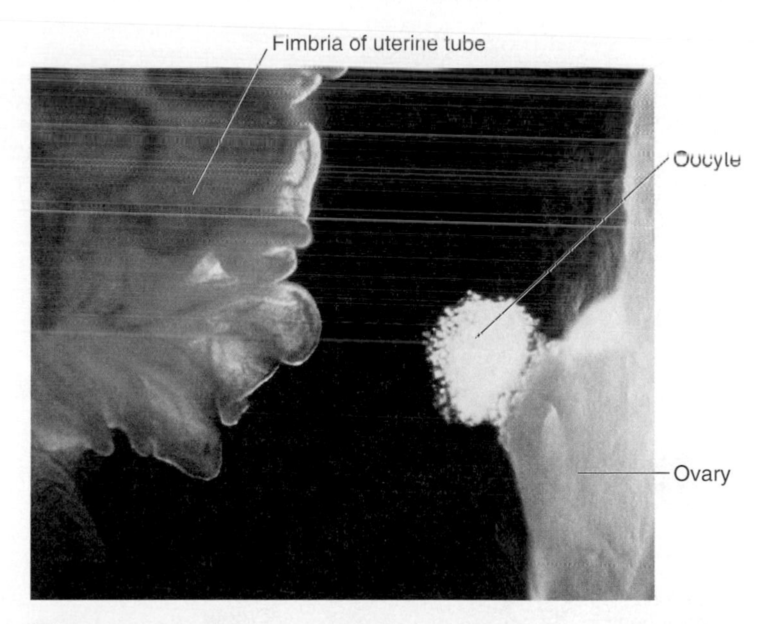

Figure 20.29 Ovulation from a human ovary. Notice the cloud of fluid and granulosa cells surrounding the ovulated oocyte.

results from a complex interplay of hormones and paracrine regulators. The gonadotropins (FSH and LH), as well as various paracrine regulators and estrogen, act to protect follicles from atresia. By contrast, paracrine regulators that include androgens and FAS ligand (chapter 3, section 3.5) promote atresia of the follicles.

The follicle that is protected from atresia and that develops into a graafian follicle becomes so large that it forms a bulge on the surface of the ovary. Under proper hormonal stimulation, this follicle will rupture—much like the popping of a blister—and extrude its oocyte into the uterine tube in the process of **ovulation** (fig. 20.29).

The released cell is a secondary oocyte surrounded by the zona pellucida and corona radiata. If it is not fertilized, it will degenerate in a couple of days. If a sperm passes through the corona radiata and zona pellucida and enters the cytoplasm of the secondary oocyte, the oocyte will then complete the second meiotic division. In this process, the cytoplasm is again not divided equally; most remains in the zygote (fertilized egg),

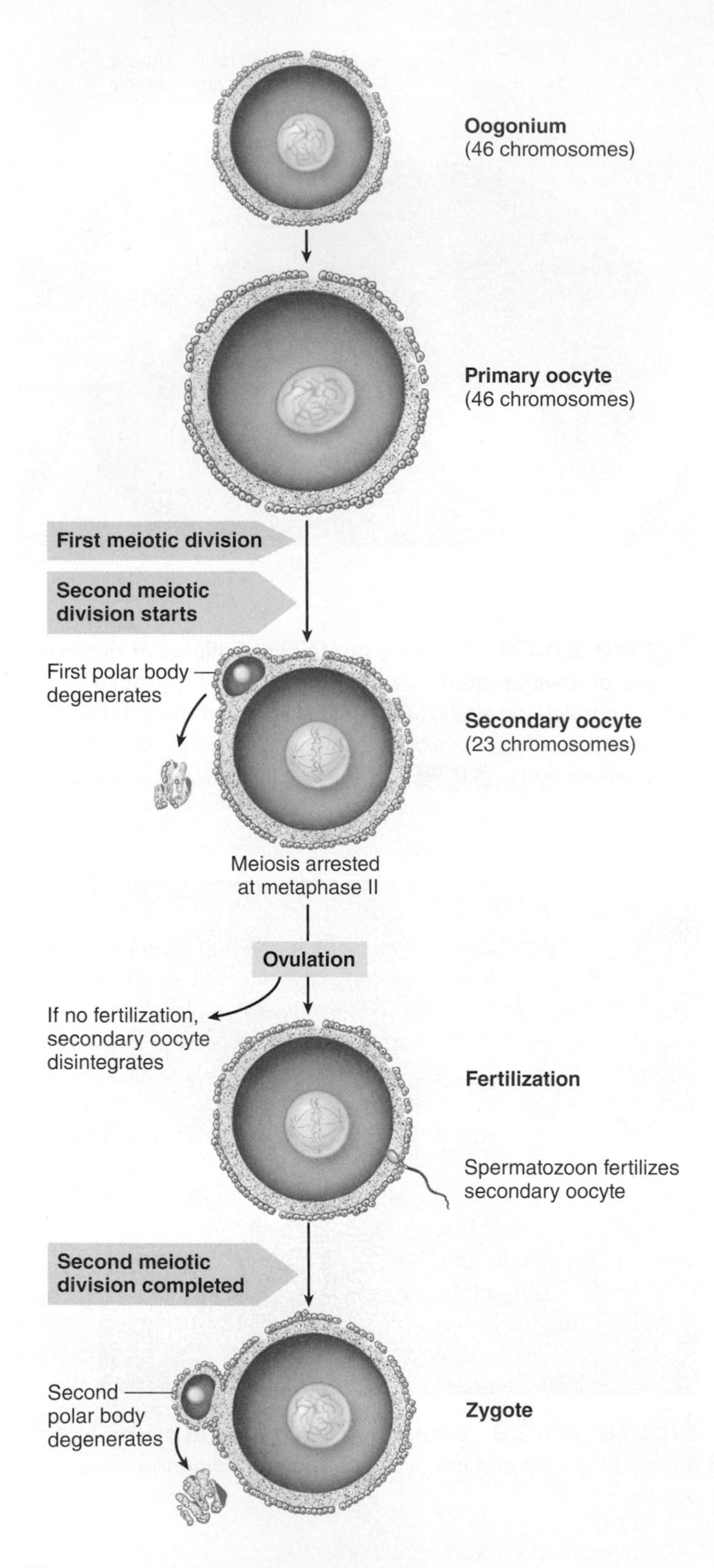

Figure 20.30 Oogenesis. During meiosis, each primary oocyte produces a single haploid gamete. If the secondary oocyte is fertilized, it forms a second polar body and its nucleus fuses with that of the sperm cell to become a zygote. Also see figure 20.40.

leaving another polar body which, like the first, degenerates (fig. 20.30).

Changes continue in the ovary following ovulation. The empty follicle, under the influence of luteinizing hormone from the anterior pituitary, undergoes structural and biochemical changes to become a **corpus luteum** (= yellow body). Unlike the ovarian follicles, which secrete only estradiol, the corpus luteum secretes two sex steroid hormones: estradiol and progesterone. Toward the end of a nonfertile cycle, the corpus luteum regresses to become a nonfunctional *corpus albicans.* These cyclic changes in the ovary are summarized in figure 20.31.

Pituitary-Ovarian Axis

The term **pituitary-ovarian axis** refers to the hormonal interactions between the anterior pituitary and the ovaries. The anterior pituitary secretes two gonadotropic hormones—follicle-stimulating hormone (FSH) and luteinizing hormone (LH)—both of which promote cyclic changes in the structure and function of the ovaries. The secretion of both gonadotropic hormones, as previously discussed, is controlled by a single releasing hormone from the hypothalamus—gonadotropin-releasing hormone (GnRH)—and by feedback effects from hormones secreted by the ovaries. The nature of these interactions will be described in detail in the next section.

Because one releasing hormone stimulates the secretion of both FSH and LH, one might expect to always see parallel changes in the secretion of these gonadotropins. However, FSH secretion is slightly greater than LH secretion during an early phase of the menstrual cycle, whereas LH secretion greatly exceeds FSH secretion just prior to ovulation. These differences are believed to result from the feedback effects of ovarian sex steroids, which can change the amount of GnRH secreted, the pulse frequency of GnRH secretion, and the ability of the anterior pituitary cells to secrete FSH and LH. These complex interactions result in a pattern of hormone secretion that regulates the phases of the menstrual cycle.

CHECKPOINT

8a. Compare the structure and contents of a primary follicle, secondary follicle, and graafian follicle.

8b. Describe oogenesis and explain why only one mature ovum is produced by this process.

9a. Define *ovulation* and describe the changes that occur in the ovary following ovulation in a nonfertile cycle.

9b. Compare the hormonal secretions of the ovarian follicles with those of a corpus luteum.

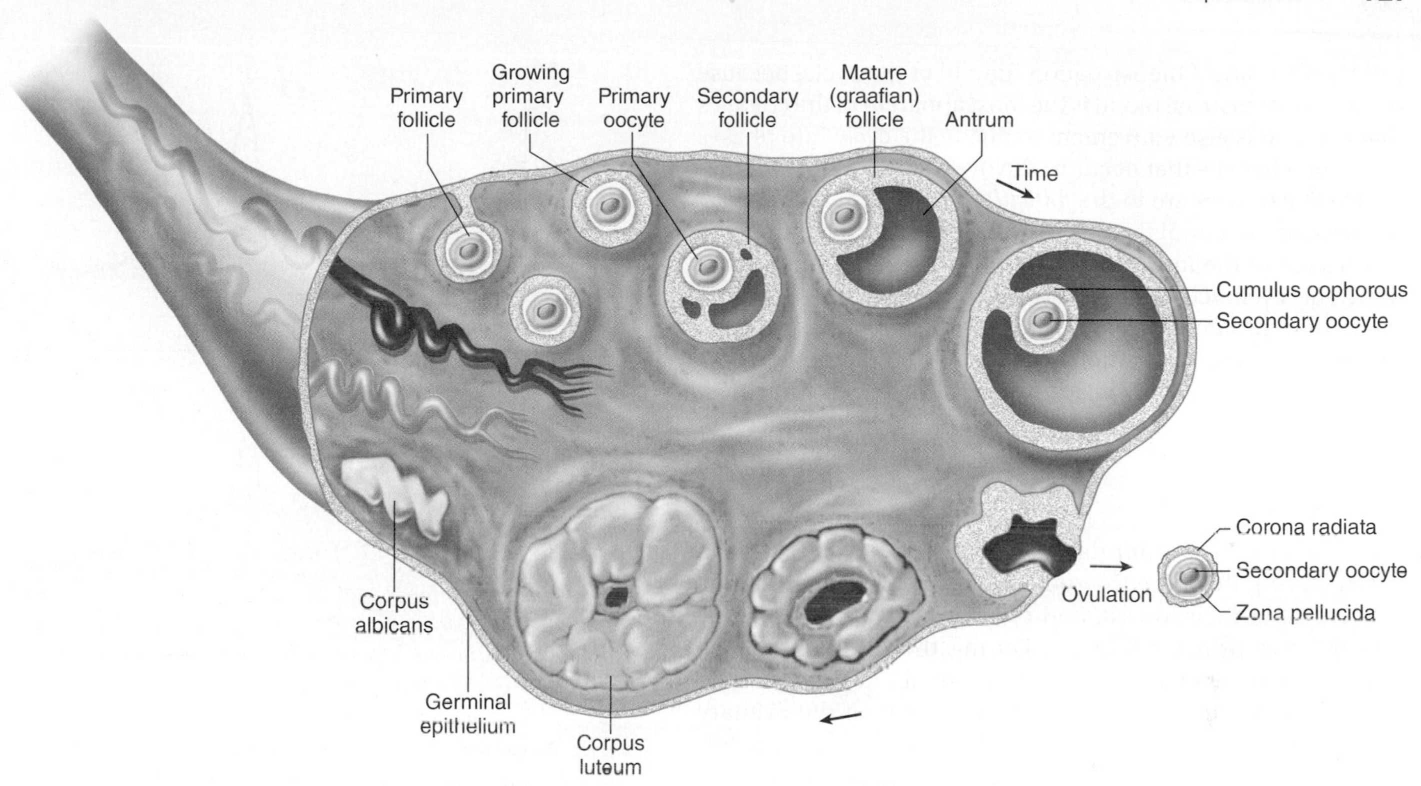

Figure 20.31 Stages of ovum and follicle development. This diagram illustrates the stages that occur in an ovary during the course of a monthly cycle. The arrows indicate changes with time.

20.5 MENSTRUAL CYCLE

Cyclic changes in the secretion of gonadotropic hormones from the anterior pituitary cause ovarian changes during a monthly cycle. The ovarian cycle is accompanied by cyclic changes in the secretion of estradiol and progesterone, which cause changes in the endometrium of the uterus during a menstrual cycle.

LEARNING OUTCOMES

After studying this section, you should be able to:

10. Describe the changes that occur in the ovaries and endometrium during each phase of the menstrual cycle.

11. Explain the hormonal regulation of ovulation, corpus luteum formation, and menstruation.

Humans, apes, and Old-World monkeys have cycles of ovarian activity that repeat at approximately one-month intervals; hence the name **menstrual cycle** (*menstru* = monthly). The term *menstruation* is used to indicate the periodic shedding of the stratum functionale of the endometrium, which becomes thickened prior to menstruation under the stimulation of ovarian steroid hormones. In primates (other than New-World monkeys) this shedding of the endometrium is accompanied by bleeding. There is no bleeding when most other mammals shed their endometrium, and therefore their cycles are not called menstrual cycles.

In human females and other primates that have menstrual cycles, coitus (sexual intercourse) may be permitted at any time of the cycle. Nonprimate female mammals, by contrast, are sexually receptive (in "heat" or "estrus") only at a particular time in their cycles, shortly before or after ovulation. These animals are therefore said to have *estrous cycles.* Bleeding occurs in some animals (such as dogs and cats) that have estrous cycles shortly before they permit coitus. This bleeding is a result of high estrogen secretion and is not associated with shedding of the endometrium. The bleeding that accompanies menstruation, by contrast, is caused by a fall in estrogen and progesterone secretion.

Phases of the Menstrual Cycle: Cyclic Changes in the Ovaries

The duration of the menstrual cycle is typically about 28 days. Because it is a cycle, there is no beginning or end and the changes are generally gradual. However, it is convenient to call the first day of menstruation "day 1" of the cycle, because the flow of menstrual blood is the most apparent of the changes that occur. It is also convenient to divide the cycle into phases based on changes that occur in the ovary and in the endometrium. The ovaries are in the *follicular phase* from the first day of menstruation until the day of ovulation. After ovulation, the ovaries are in the *luteal phase* until the first day of menstruation. The cyclic changes in the endometrium are called the menstrual, proliferative, and secretory phases. These will be discussed separately. It should be noted that the time frames used for the following discussion are only averages. Individual cycles may exhibit considerable variation.

Follicular Phase

Menstruation lasts from day 1 to day 4 or 5 of the average cycle. During this time the secretions of ovarian steroid hormones are at their lowest, and the ovaries contain only primordial and primary follicles. During the **follicular phase** of the ovaries, which lasts from day 1 to about day 13 of the cycle (this duration is highly variable), some of the primary follicles grow, develop vesicles, and become secondary follicles. Toward the end of the follicular phase, one follicle in one ovary reaches maturity and becomes a graafian follicle. As follicles grow, the granulosa cells secrete an increasing amount of **estradiol** (the principal estrogen), which reaches its highest concentration in the blood two days before ovulation at about day 12 of the cycle.

The growth of the follicles and the secretion of estradiol are stimulated by, and dependent upon, FSH secreted from the anterior pituitary. The amount of FSH secreted during the early follicular phase is believed to be slightly greater than the amount secreted in the late follicular phase (fig. 20.32), although this can vary from cycle to cycle. FSH stimulates the production of FSH receptors in the granulosa cells, so that the follicles become increasingly sensitive to a given amount of FSH. This increased sensitivity is augmented by estradiol, which also stimulates the production of new FSH receptors in the follicles. As a result, the stimulatory effect of FSH on the follicles and their secretion of estradiol increases despite the lack of an increase in FSH levels during the follicular phase. Toward the end of the follicular phase, FSH and estradiol also stimulate the production of LH receptors in the graafian follicle. This prepares the graafian follicle for the next major event in the cycle.

The rapid rise in estradiol secretion from the granulosa cells during the follicular phase acts on the hypothalamus to increase the frequency of GnRH pulses. In addition, estradiol augments the ability of the pituitary to respond to GnRH with an increase in LH secretion. As a result of this stimulatory, or **positive feedback,** effect of estradiol on the pituitary, there is an increase in LH secretion in the late follicular phase that culminates in an **LH surge** (fig. 20.32).

The LH surge begins about 24 hours before ovulation and reaches its peak about 16 hours before ovulation. It is this surge that acts to trigger ovulation. Because GnRH stimulates the anterior pituitary to secrete both FSH and LH, there is a simultaneous, smaller surge in FSH secretion. Some investigators believe that this midcycle peak in FSH acts as a stimulus for the development of new follicles for the next month's cycle.

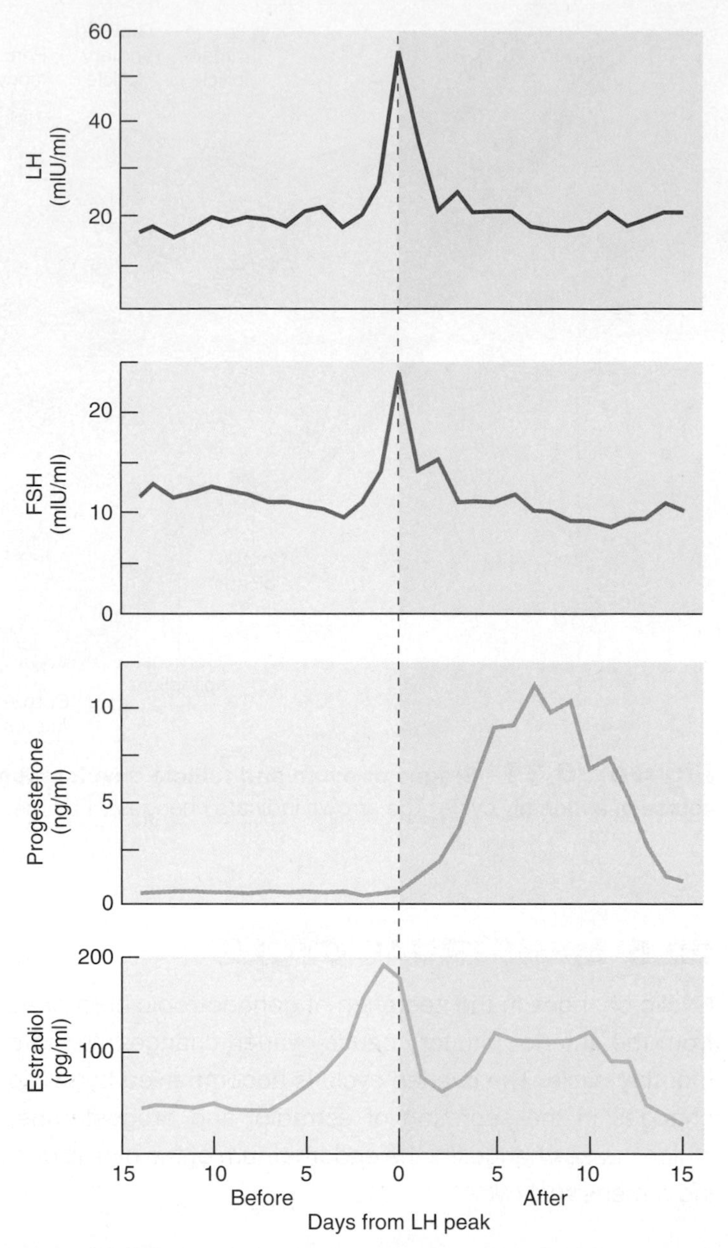

Figure 20.32 Hormonal changes during the menstrual cycle. Sample values are indicated for LH, FSH, progesterone, and estradiol during the menstrual cycle. The midcycle peak of LH is used as a reference day. (IU = international unit.) AP|R

See the *Test Your Quantitative Ability* section of the Review Activities at the end of this chapter.

Ovulation

Under the influence of FSH stimulation, the graafian follicle grows so large that it becomes a thin-walled "blister" on the surface of the ovary. The growth of the follicle is accompanied

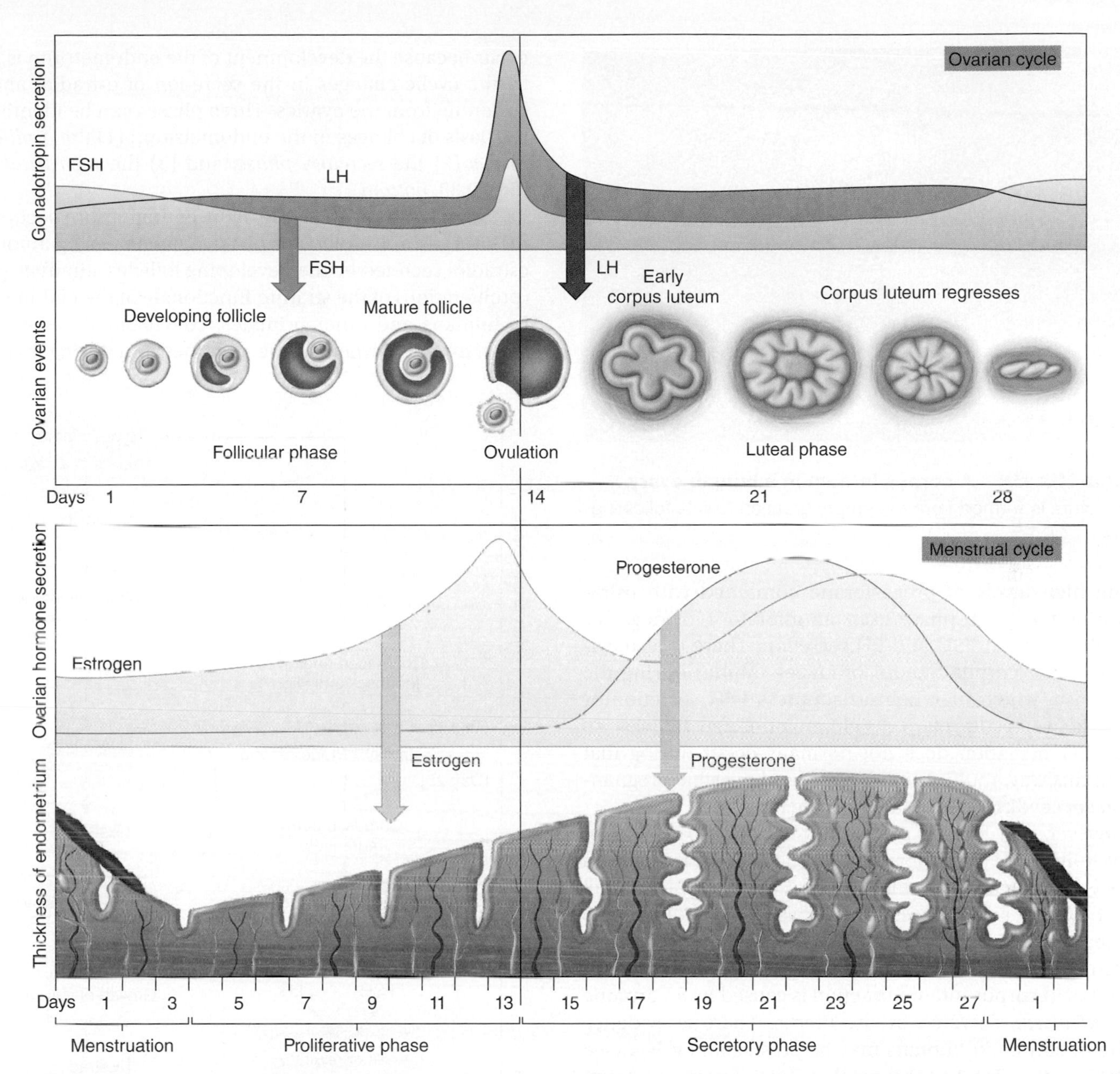

Figure 20.33 The cycle of ovulation and menstruation. The downward arrows indicate the effects of the hormones. APIR

by a rapid increase in estradiol secretion. This rapid increase in estradiol, in turn, triggers the LH surge at about day 13. Finally, the surge in LH secretion causes the wall of the graafian follicle to rupture at about day 14 (fig. 20.33, *top*). In the course of ovulation, a secondary oocyte, arrested at metaphase II of meiosis, is released from the ovary and swept by cilia into a uterine tube. The ovulated oocyte is still surrounded by a zona pellucida and corona radiata as it begins its journey to the uterus.

Ovulation occurs, therefore, as a result of the sequential effects of FSH and LH on the ovarian follicles. By means of the positive feedback effect of estradiol on LH secretion, the follicle in a sense sets the time for its own ovulation. This is because ovulation is triggered by an LH surge, and the LH surge is triggered by the increased estradiol secretion that occurs while the follicle grows. In this way, the graafian follicle is not normally ovulated until it has reached the proper size and maturity.

Luteal Phase

After ovulation, the empty follicle is stimulated by LH to become a new structure—the corpus luteum (fig. 20.34). This change in structure is accompanied by a change in function. Whereas the developing follicles secrete only estradiol, the corpus luteum secretes both estradiol and **progesterone.** Progesterone levels in the blood are negligible before ovulation but rise rapidly to a peak level during the **luteal phase,** approximately one week after ovulation (see figs. 20.32 and 20.33).

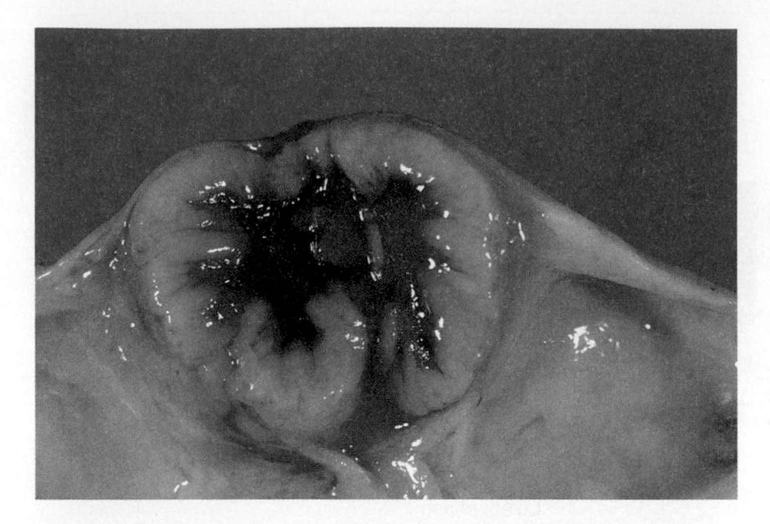

Figure 20.34 A corpus luteum in a human ovary. This structure is formed from the empty graafian follicle following ovulation. AP|R

The high levels of progesterone combined with estradiol during the luteal phase exert an inhibitory, or **negative feedback,** effect on FSH and LH secretion. There is also evidence that the corpus luteum produces inhibin during the luteal phase, which may help to suppress FSH secretion or action. This serves to retard development of new follicles, so that further ovulation does not normally occur during that cycle. In this way, multiple ovulations (and possible pregnancies) on succeeding days of the cycle are prevented.

However, new follicles start to develop toward the end of one cycle in preparation for the next. This may be due to a decreased production of inhibin toward the end of the luteal phase. Estrogen and progesterone levels also fall during the late luteal phase (starting about day 22) because the corpus luteum regresses and stops functioning. In lower mammals, the decline in corpus luteum function is caused by a hormone called *luteolysin,* secreted by the uterus. There is evidence that the luteolysin in humans may be prostaglandin $F_{2\alpha}$ (see figs. 2.24 and 11.34), but the mechanisms of corpus luteum regression in humans is still incompletely understood. Luteolysis (breakdown of the corpus luteum) can be prevented by high levels of LH, but LH levels remain low during the luteal phase as a result of negative feedback exerted by ovarian steroids. Through its secretion of estradiol and progesterone, the corpus luteum in a sense causes its own demise.

With the declining function of the corpus luteum, estrogen and progesterone fall to very low levels by day 28 of the cycle. *The withdrawal of ovarian steroids causes menstruation* and permits a new cycle of follicle development to progress.

Cyclic Changes in the Endometrium

In addition to a description of the female cycle in terms of ovarian function, the cycle can also be described in terms of the changes in the endometrium of the uterus. These changes occur because the development of the endometrium is driven by the cyclic changes in the secretion of estradiol and progesterone from the ovaries. Three phases can be identified on the basis of changes in the endometrium: (1) the *proliferative phase;* (2) the *secretory phase;* and (3) the *menstrual phase* (fig. 20.33, *bottom*).

The **proliferative phase** of the endometrium occurs while the ovary is in its follicular phase. The increasing amounts of estradiol secreted by the developing follicles stimulate growth (proliferation) of the stratum functionale of the endometrium. In humans and other primates, coiled blood vessels called *spiral arteries* develop in the endometrium during this phase.

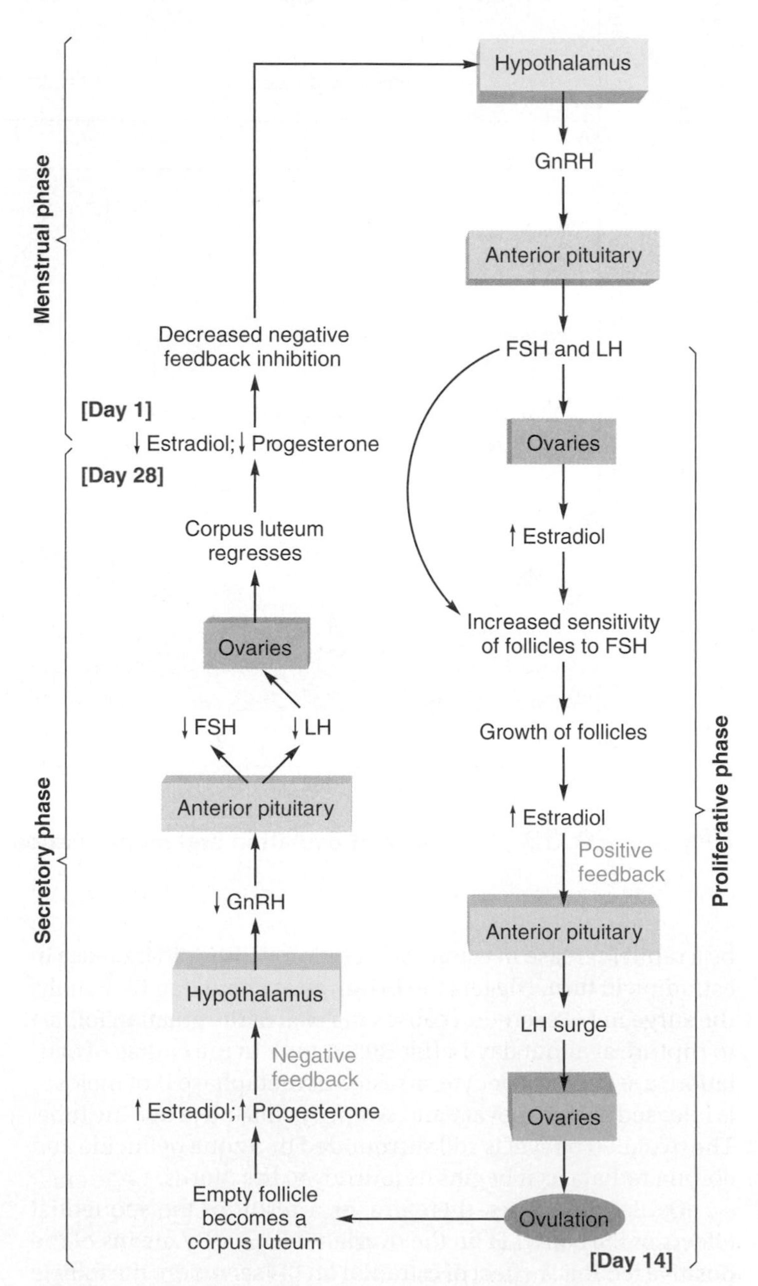

Figure 20.35 Endocrine control of the ovarian cycle. This sequence of events is shown together with the associated phases of the endometrium during the menstrual cycle.

Table 20.6 | Phases of the Menstrual Cycle

Phase of Cycle		Hormonal Changes		Tissue Changes	
Ovarian	**Endometrial**	**Pituitary**	**Ovary**	**Ovarian**	**Endometrial**
Follicular (days 1–4)	Menstrual	FSH and LH secretion low	Estradiol and progesterone remain low	Primary follicles grow	Outer two-thirds of endometrium is shed with accompanying bleeding
Follicular (days 5–13)	Proliferative	FSH slightly higher than LH secretion in early follicular phase	Estradiol secretion rises (due to FSH stimulation of follicles)	Follicles grow; graafian follicle develops (due to FSH stimulation)	Mitotic division increases thickness of endometrium; spiral arteries develop (due to estradiol stimulation)
Ovulatory (day 14)	Proliferative	LH surge (and increased FSH) stimulated by positive feedback from estradiol	Estradiol secretion falls	Graafian follicle ruptures and secondary oocyte is extruded into uterine tube	No change
Luteal (days 15–28)	Secretory	LH and FSH decrease (due to negative feedback from steroids)	Progesterone and estrogen secretion increase, then fall	Development of corpus luteum (due to LH stimulation); regression of corpus luteum	Glandular development in endometrium (due to progesterone stimulation)

Estradiol may also stimulate the production of receptor proteins for progesterone at this time, in preparation for the next phase of the cycle.

The **secretory phase** of the endometrium occurs when the ovary is in its luteal phase. In this phase, increased progesterone secretion by the corpus luteum stimulates the development of uterine glands. As a result of the combined actions of estradiol and progesterone, the endometrium becomes thick, vascular, and "spongy" in appearance, and the uterine glands become engorged with glycogen during the phase following ovulation. The endometrium is therefore well prepared to accept and nourish an embryo should fertilization occur.

The **menstrual phase** occurs as a result of the fall in ovarian hormone secretion during the late luteal phase. Necrosis (cellular death) and sloughing of the stratum functionale of the endometrium may be produced by constriction of the spiral arteries. It would seem that the spiral arteries are responsible for menstrual bleeding, since animals that lack these arteries do not bleed when they shed their endometrium. The phases of the menstrual cycle are summarized in figure 20.35 and in table 20.6.

The cyclic changes in ovarian secretion cause other cyclic changes in the female reproductive tract. High levels of estradiol secretion, for example, cause cornification of the vaginal epithelium (the upper cells die and become filled with keratin). High levels of estradiol also cause the production of a thin, watery cervical mucus that can easily be penetrated by spermatozoa. During the luteal phase of the cycle, the high levels of progesterone cause the cervical mucus to thicken and become sticky after ovulation has occurred.

CLINICAL APPLICATION

Abnormal menstruations are among the most common disorders of the female reproductive system. The term *amenorrhea* refers to the absence of menstruation. *Dysmenorrhea* refers to painful menstruation, which may be marked by severe cramping. In *menorrhagia*, menstrual flow is excessively profuse or prolonged, and in *metrorrhagia* uterine bleeding that is not associated with menstruation occurs at irregular intervals.

Effects of Pheromones, Stress, and Body Fat

Since GnRH stimulates the anterior pituitary to secrete FSH and LH, the GnRH-releasing neurons of the hypothalamus might be considered the master regulators of the reproductive system. However, the release of GnRH is itself regulated by feedback effects of ovarian hormones and by input from higher brain centers. Because of input to GnRH neurons from the olfactory system, pheromones can cause the menstrual cycle of roommates to synchronize (the *dormitory effect;* chapter 11, section 11.3). Recent evidence suggests that this pheromonal effect in humans is due to the stimulation of olfactory neurons in the nasal mucosa.

As discussed in chapter 8, the limbic system of the brain includes regions involved in emotions. Axons extend from the limbic system to the GnRH neurons of the hypothalamus. By means of these neural pathways, the secretion of GnRH, and thus of FSH and LH, can be influenced by stress and emotions.

Considering this, it is not surprising that stress can even cause a cessation of menstruation, or **amenorrhea.**

Many girls who are very thin or athletic have a delayed menarche, and women with low body fat can have irregular cycles or amenorrhea. *Functional amenorrhea* is the cessation of menstruation caused by inadequate stimulation of the ovaries by FSH and LH, which in turn is due to inadequate release of GnRH from the hypothalamus. Functional amenorrhea is most often seen in women who are thin and athletic, as well as women under prolonged stress. Intense physical exercise can suppress GnRH secretion, and reducing the exercise program can reverse the amenorrhea produced by rigorous athletic training. Leptin, secreted by adipocytes, regulates hunger and metabolism (chapter 19); it also indirectly affects the GnRH-secreting neurons of the hypothalamus. Because of this, a sufficient amount of adipose tissue and leptin secretion is required for ovulation and reproduction, and inadequate adiposity (and leptin secretion) can produce a functional amenorrhea. Treatment with exogenous leptin has been shown to benefit women with functional amenorrhea by stimulating the pulsatile secretion of GnRH and inducing menstrual periods.

Clinical Investigation CLUES

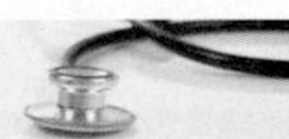

Gloria is experiencing amenorrhea (missed periods).

- What category of amenorrhea does she likely have?
- Decreased functioning of which organ is likely responsible for her amenorrhea—the endometrium, ovaries, anterior pituitary, or hypothalamus?

Contraceptive Methods

Contraceptive Pill

First sold in 1960, **oral contraceptives** are currently used by about 10 million women in the United States and 60 million women worldwide. These contraceptives usually consist of a synthetic estrogen combined with a synthetic progesterone in the form of pills that are taken once each day for three weeks after the last day of a menstrual period. This procedure causes an immediate increase in blood levels of ovarian steroids (from the pill), which is maintained for the normal duration of a monthly cycle. As a result of *negative feedback inhibition* of gonadotropin secretion, *ovulation never occurs.* The entire cycle is like a false luteal phase, with high levels of progesterone and estrogen and low levels of gonadotropins.

Because the contraceptive pills contain ovarian steroid hormones, the endometrium proliferates and becomes secretory just as it does during a normal cycle. In order to prevent an abnormal growth of the endometrium, women stop taking the steroid pills after three weeks (placebo pills are taken during the fourth week). This causes estrogen and progesterone levels to fall, permitting menstruation to occur.

The side effects of earlier versions of the birth control pill have been reduced through a decrease in the content of estrogen and through the use of newer generations of progestogens (analogues of progesterone). The newer contraceptive pills are very effective and have a number of beneficial side effects, including a reduced risk for endometrial and ovarian cancer, and a reduction in osteoporosis. However, there may be an increased risk for breast cancer, and possibly cervical cancer, with oral contraceptives.

Newer systems for delivery of contraceptive steroids are designed so that the steroids are not taken orally, and as a result do not have to pass through the liver before entering the general circulation. (All drugs taken orally pass from the hepatic portal vein to the liver before they are delivered to any other organ; chapter 18, section 18.5.) This permits lower doses of hormones to be effective. Such systems include a subcutaneous implant (Norplant), which need only be replaced after five years, and vaginal rings, which can be worn for three weeks. The long-term safety of these newer methods has not yet been established.

Clinical Investigation CLUES

Gloria stated that she is not taking birth control pills.

- If she had been taking birth control pills, how could they have caused missed periods?

Rhythm Method

Studies have demonstrated that the likelihood of a pregnancy is close to zero if coitus occurs more than six days prior to ovulation, and that the likelihood is very low if coitus occurs more than a day following ovulation. Conception is most likely to result when intercourse takes place one to two days prior to ovulation. There is no evidence for differences in the sex ratio of babies conceived at these different times.

Cyclic changes in ovarian hormone secretion also cause cyclic changes in basal body temperature. In the **rhythm method** of birth control, a woman measures her oral basal body temperature upon waking to determine when ovulation has occurred. On the day of the LH peak, when estradiol secretion begins to decline, there is a slight drop in basal body temperature. Starting about one day after the LH peak, the basal body temperature sharply rises as a result of progesterone secretion and it remains elevated throughout the luteal phase of the cycle (fig. 20.36). The day of ovulation for that month's cycle can be accurately determined by this method, making the method useful if conception is desired. However, the day of the cycle on which ovulation occurs is quite variable in many women, and so the rhythm method can be unreliable in predicting the day of ovulation for the next month's cycle, and thus unreliable for birth control. The contraceptive pill is a statistically more effective means of birth control.

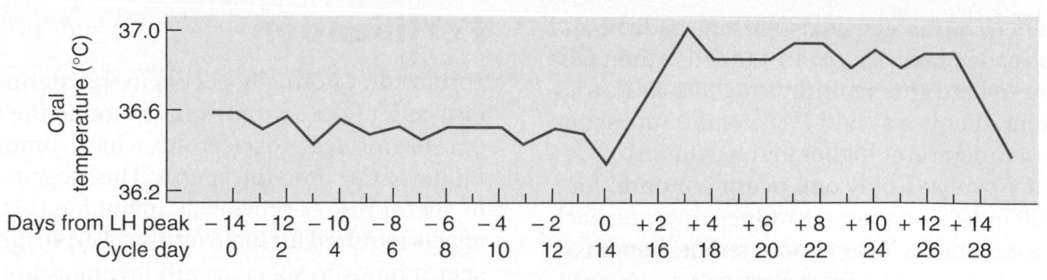

Figure 20.36 Changes in basal body temperature during the menstrual cycle. Such changes can be used in the rhythm method of birth control.

Menopause

The term **menopause** means literally "pause in the menses" and refers to the cessation of ovarian activity and menstruation that occurs at about the age of 50. During the postmenopausal years, which account for about a third of a woman's life span, the ovaries are depleted of follicles and stop secreting estradiol and inhibin. The fall in estradiol is due to changes in the ovaries, not in the pituitary; indeed, FSH and LH secretion by the pituitary is elevated because of a lack of negative feedback from estradiol and inhibin.

The only estrogen found in the blood of postmenopausal women is the weak estrogen *estrone,* formed by the mesenchymal cells in adipose tissue. Estrone is formed from weak androgens, such as *androstenedione* and *dehydroepiandrosterone (DHEA),* secreted from the adrenal cortex. Because adipose tissue is the only source of estrogen, postmenopausal women who have more adipose tissue have higher levels of estrogen and less propensity toward osteoporosis.

It is the withdrawal of estradiol secretion from the ovaries that is most responsible for the many symptoms of menopause. These include vasomotor disturbances and urogenital atrophy. Vasomotor disturbances produce the "hot flashes" of menopause, where a fall in core body temperature is followed by feelings of heat and profuse perspiration. Atrophy of the urethra, vaginal wall, and vaginal glands occurs, with loss of lubrication. There is also increased risk of atherosclerotic cardiovascular disease and increased progression of osteoporosis.

CHECKPOINT

10a. Describe the changes that occur in the ovary and endometrium during the follicular phase and explain how these changes are regulated by hormones.

10b. Describe the formation, function, and fate of the corpus luteum. Also, describe the changes that occur in the endometrium during the luteal phase.

11a. Describe the hormonal regulation of ovulation.

11b. Explain the significance of negative feedback control during the luteal phase and describe the hormonal control of menstruation.

20.6 FERTILIZATION, PREGNANCY, AND PARTURITION

Once fertilization has occurred, the secondary oocyte completes meiotic division. It then undergoes mitosis and forms an early embryonic structure called a blastocyst. Cells of the blastocyst secrete human chorionic gonadotropin, a hormone that maintains the mother's corpus luteum and prevents menstruation. Birth is dependent upon strong contractions of the uterus, which are stimulated by oxytocin.

LEARNING OUTCOMES

After studying this section, you should be able to:

12. Describe the process of fertilization and blastocyst formation, and the significance of hCG.

13. Describe the formation and functions of the placenta.

14. Explain the mechanisms that promote parturition.

15. Explain the hormonal control of lactation.

Sperm are stored in the epididymis, where they are fully developed yet incapable of fertilization. This is largely because they are kept in a slightly acidic state, with a cytoplasmic pH below 6.5. During the act of sexual intercourse, the male ejaculates an average of 300 million sperm into the vagina of the female. This tremendous number is needed because of the high sperm fatality—only about 100 survive to enter each fallopian tube. During their passage through the female reproductive tract, about 10% of the sperm gain the ability to fertilize an ovum; this ability is called **capacitation.**

In order for the sperm to become capacitated, they must be present in the female reproductive tract for at least seven hours. During this time, the pH of sperm cytoplasm is increased (made alkaline) by exposure to the increasing pH of the female reproductive tract (from pH 4 to pH 7) and by extrusion of H^+ from the sperm cytoplasm. Calcium channels unique to sperm and located in the sperm's principal piece (see fig. 20.18) are opened by alkalinization, allowing the entry of Ca^{2+}. This serves as a signal for *hyperactivation* of the flagella, which is a more vigorous, larger-amplitude beating of the flagella. Hyperactivation is required for the sperm to migrate to the site of fertilization and penetrate the egg coats (corona radiata and zona pellucida). The capacitated sperm are guided in their passage up the oviduct toward the ovum by *chemotaxis* (attraction toward particular chemicals) and *thermotaxis* (attraction toward the warmer temperatures higher in the oviduct).

A woman usually ovulates only one ovum a month, for a total of less than 450 ova during her reproductive years. Each ovulation releases a secondary oocyte arrested at metaphase of the second meiotic division. The secondary oocyte, as previously described, enters the uterine tube surrounded by its zona pellucida (a thin transparent layer of protein and polysaccharides) and corona radiata of granulosa cells (fig. 20.37).

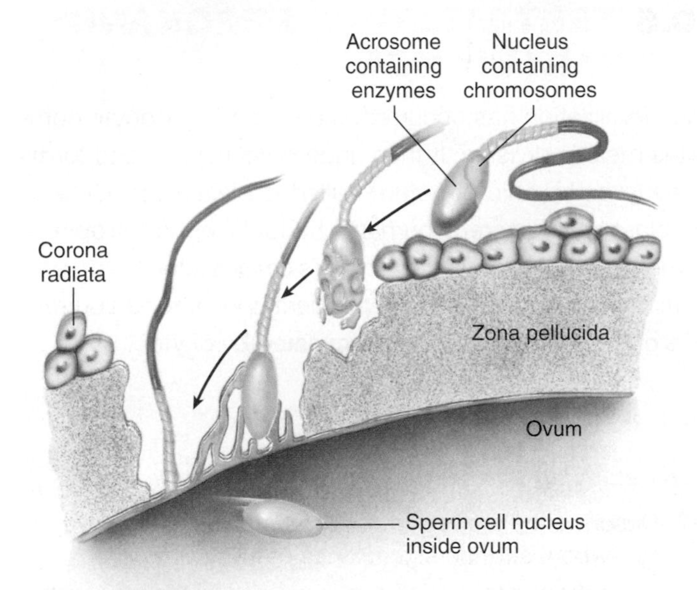

Figure 20.37 The process of fertilization. As the head of the sperm cell encounters the gelatinous corona radiata of the secondary oocyte, the acrosomal vesicle ruptures and the sperm cell digests a path for itself by the action of the enzymes released from the acrosome. When the plasma membrane of the sperm cell contacts the plasma membrane of the ovum, they become continuous, and the nucleus of the sperm cell moves into the cytoplasm of the ovum. AP|R

Fertilization

Fertilization normally occurs in the uterine tubes. The granulosa cells of the corona radiata around the ovulated secondary oocyte produce progesterone, which stimulates the immediate inflow of Ca^{2+} into the sperm. This (together with the increase in sperm pH, as previously mentioned) is needed for several effects required for male fertility. Fertilization begins when the sperm binds to specific carbohydrates in the glycoproteins of the zona pellucida.

Each sperm contains a large, enzyme-filled vesicle above its nucleus known as an **acrosome** (fig. 20.38). Binding of the sperm with the zona pellucida triggers the entry of Ca^{2+} and the **acrosome reaction.** This involves the progressive fusion of the acrosomal membrane with the plasma membrane of the sperm, creating pores through which the acrosomal enzymes can be released by exocytosis. These enzymes, including a protein-digesting enzyme and hyaluronidase (which digests hyaluronic acid, a constituent of the extracellular matrix), allow the sperm to digest a path through the zona pellucida to the oocyte.

Fertilization, acting through a second messenger (inositol triphosphate), stimulates the endoplasmic reticulum of the oocyte to release its stored Ca^{2+}. This causes a rise in cytoplasmic Ca^{2+} that spreads from the point of sperm entry to the opposite pole of the oocyte, creating a **Ca^{2+} wave.** The egg cell is so large (about 0.1 mm in diameter) that this Ca^{2+} wave takes about 2 seconds to spread from one side of the oocyte to the other. The Ca^{2+} wave activates the fertilized egg cell, causing numerous structural and metabolic changes. Some of these changes prevent other sperm from fertilizing the same oocyte. *Polyspermy* (the fertilization of an oocyte by many sperm) is thereby prevented; only one sperm can fertilize an egg cell.

When the secondary oocyte was released from the ovary at ovulation, its cell cycle was arrested at metaphase II. The Ca^{2+} wave initiated at fertilization activates proteins that allow the cell cycle to continue past this arrested stage. As a result, the secondary oocyte is stimulated to complete its second meiotic

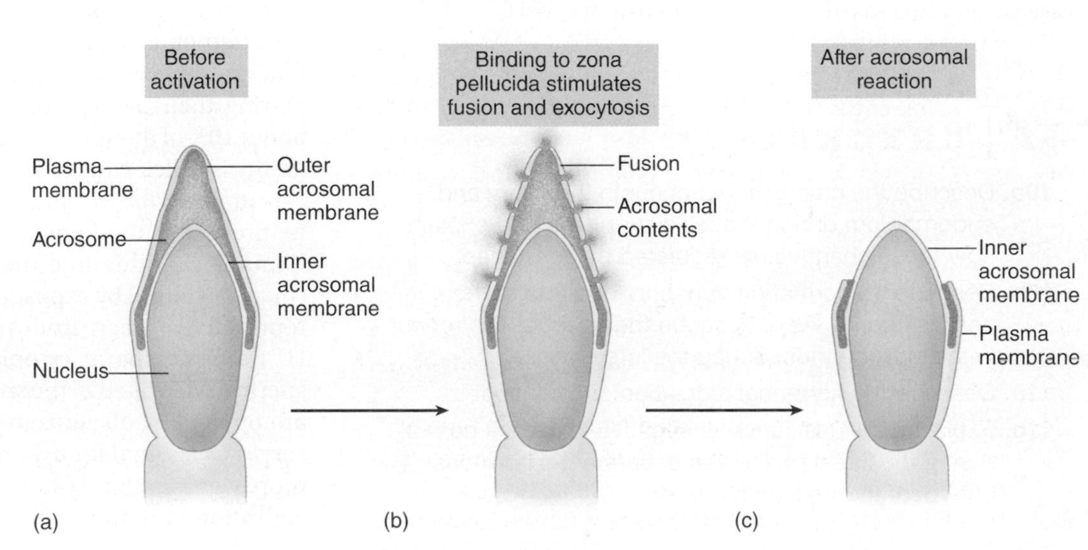

Figure 20.38 The acrosome reaction. (*a*) Prior to activation, the acrosome is a large, enzyme-containing vesicle over the sperm nucleus. (*b*) After the sperm binds to particular proteins in the zona pellucida surrounding the egg, the acrosomal membrane fuses with the plasma membrane in many locations, creating openings through which the acrosomal contents can be released by exocytosis. (*c*) When the process is complete, the inner acrosomal membrane has become continuous with the plasma membrane.

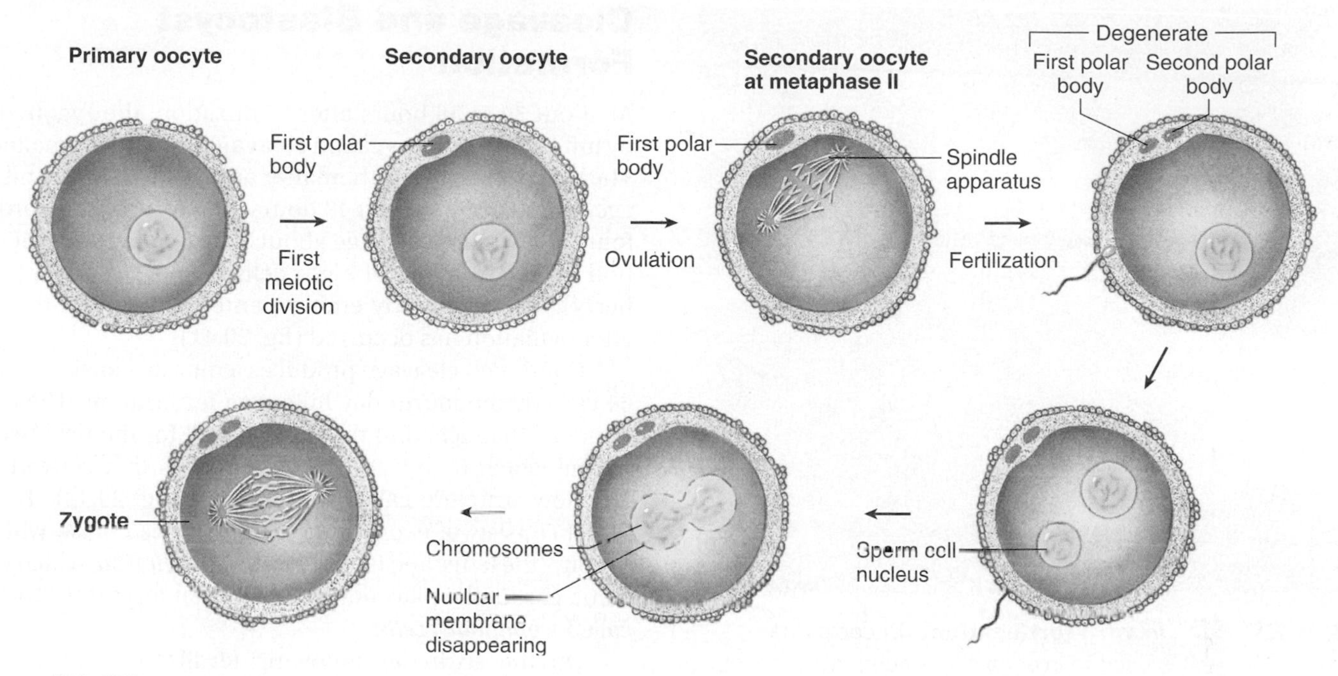

Figure 20.39 Changes in the oocyte following fertilization. A secondary oocyte, arrested at metaphase II of meiosis, is released at ovulation. If this cell is fertilized, it will complete its second meiotic division and produce a second polar body. The chromosomes of the two gametes are joined in the zygote.

division (fig. 20.39). Like the first meiotic division, the second produces one cell that contains all of the cytoplasm—the mature ovum—and one polar body. The second polar body, like the first, ultimately fragments and disappears.

Within 12 hours after fertilization the nuclear membrane in the ovum disappears and the haploid number of chromosomes (23) in the ovum is joined by the haploid number of chromosomes from the sperm cell. A fertilized egg, or **zygote,** containing the diploid number of chromosomes (46) is thereby formed (fig. 20.39). *Monozygotic twins* (identical twins) are derived from a single zygote that splits and becomes two embryos, whereas *dizygotic twins* (fraternal twins) are derived from two different zygotes produced by two ovulated oocytes fertilized by two different sperm.

It should be noted that the sperm cell contributes more than the paternal set of chromosomes to the zygote. The sperm, not the oocyte, contributes the centrosome, which is needed for the organization of microtubules into the spindle apparatus (chapter 3, section 3.5) that separates duplicated chromosomes during mitosis. The midpiece of the sperm brings some mitochondria into the oocyte during fertilization, but these do not contribute to the embryo. The paternal mitochondria, with their contents of mitochondrial DNA, are quickly eliminated in the zygote by a process of autophagy (chapter 3, section 3.2). As a result, all of the mitochondrial DNA in a person is inherited from the mother's oocyte.

A secondary oocyte that has been ovulated but not fertilized does not complete its second meiotic division, but instead disintegrates 12 to 24 hours after ovulation. Fertilization therefore cannot occur if intercourse takes place later than one day following ovulation. Sperm, by contrast, can survive up to three days in the female reproductive tract. Fertilization therefore can occur if intercourse takes place within a three-day period prior to the day of ovulation.

CLINICAL APPLICATION

The 2010 Nobel Prize in Physiology or Medicine was awarded to a British physiologist for pioneering **in vitro fertilization (IVF),** a technique in which oocytes are fertilized outside of the body and then implanted into the uterus. Gonadotropins are given to the woman to promote the development of multiple ovarian follicles, which are followed by transvaginal ultrasound. Transvaginal ultrasound is then used to guide the aspiration of follicular fluid, containing secondary oocytes, from growing follicles. Most commonly, a glass pipette is then used to inject a single, capacitated sperm through the zona pellucida and into the cytoplasm of an isolated secondary oocyte (fig. 20.40), in a technique called **intracytoplasmic sperm injection (ICSI).** ICSI is used to produce a number of embryos, which are grown in vitro for either three days (to the eight-cell stage) or five days (to the blastocyst stage). Three or more of these embryos are transferred to the woman's uterus, and the remaining embryos are preserved frozen in liquid nitrogen. This technique has allowed women who are infertile for a variety of reasons to become pregnant, resulting in the birth of about 4 million babies to date.

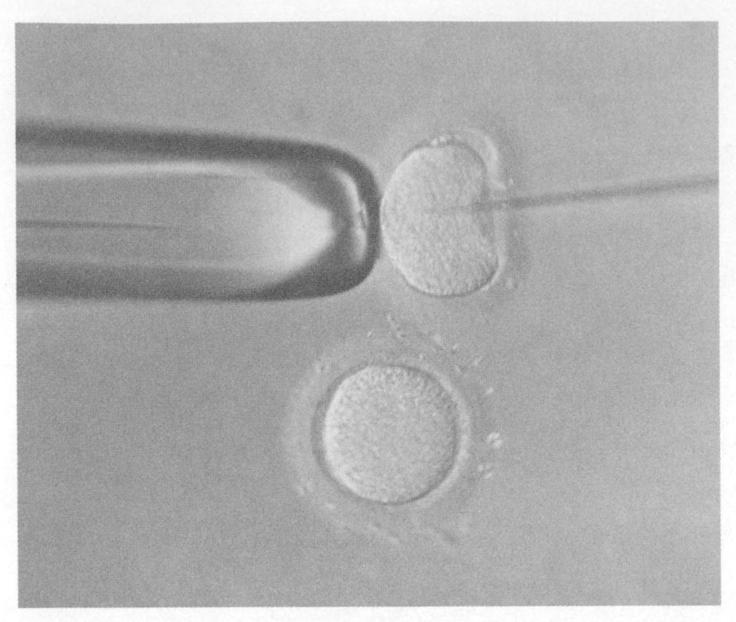

Figure 20.40 ***In vitro* fertilization.** A needle (the shadow on the right) is used to inject a single spermatozoon into a human oocyte *in vitro.*

Cleavage and Blastocyst Formation

At about 30 to 36 hours after fertilization, the zygote divides by mitosis—a process called **cleavage**—into two smaller cells. The rate of cleavage is thereafter accelerated. A second cleavage, which occurs about 40 hours after fertilization, produces four cells. A third cleavage about 50 to 60 hours after fertilization produces a ball of eight cells called a **morula** (= mulberry). This very early embryo enters the uterus three days after ovulation has occurred (fig. 20.41).

Continued cleavage produces a morula consisting of 32 to 64 cells by the fourth day following fertilization. The embryo remains unattached to the uterine wall for the next two days, during which time it undergoes changes that convert it into a hollow structure called a **blastocyst** (fig. 20.42). The blastocyst consists of two parts: (1) an *inner cell mass,* which will become the fetus, and (2) a surrounding *chorion,* which will become part of the placenta. The cells that form the chorion are called *trophoblast cells.*

On the sixth day following fertilization the blastocyst attaches to the uterine wall, with the side containing the inner

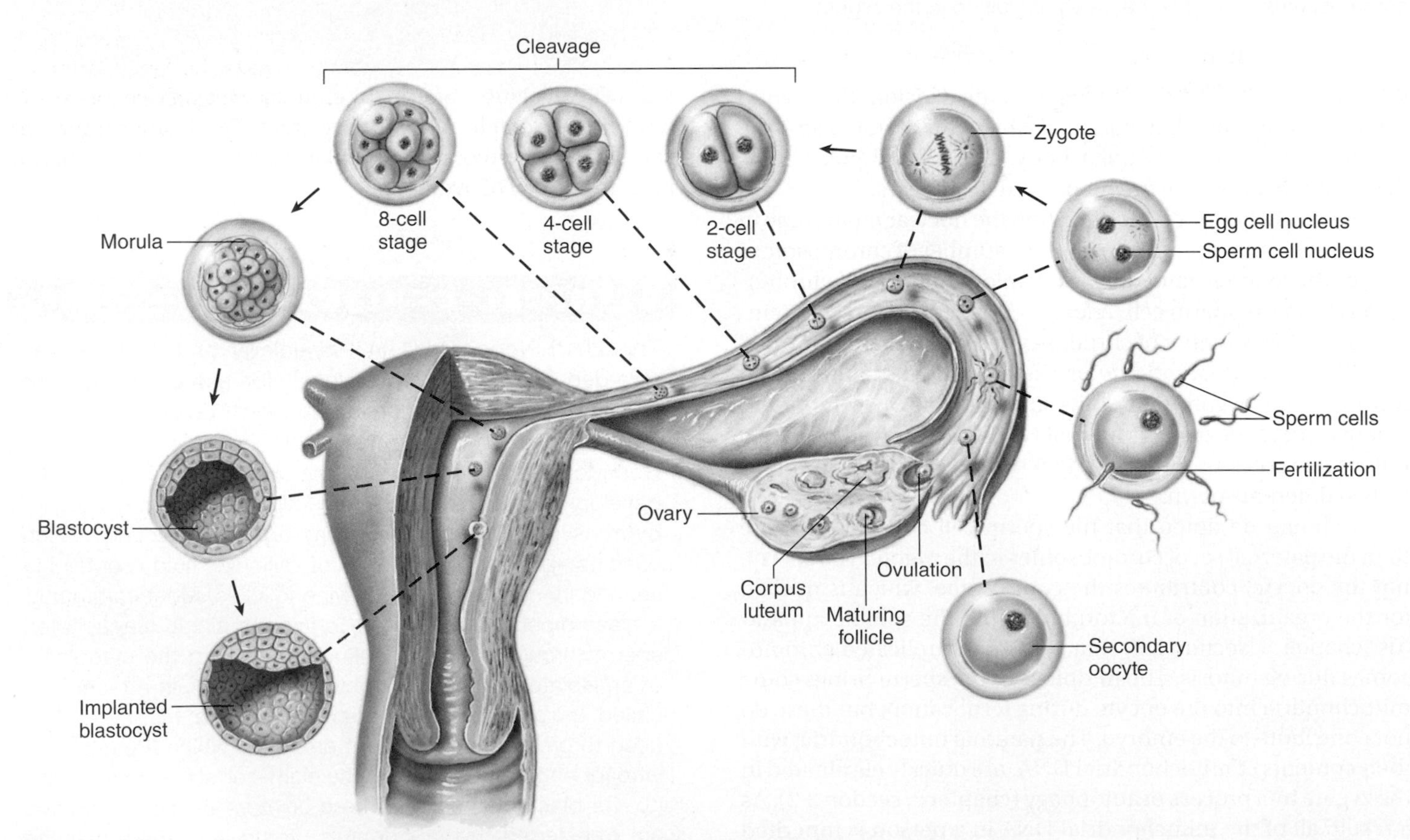

Figure 20.41 Fertilization, cleavage, and the formation of a blastocyst. A diagram showing the ovarian cycle, fertilization, and the events of the first week following fertilization. Implantation of the blastocyst begins between the fifth and seventh day and is generally complete by the tenth day. AP|R

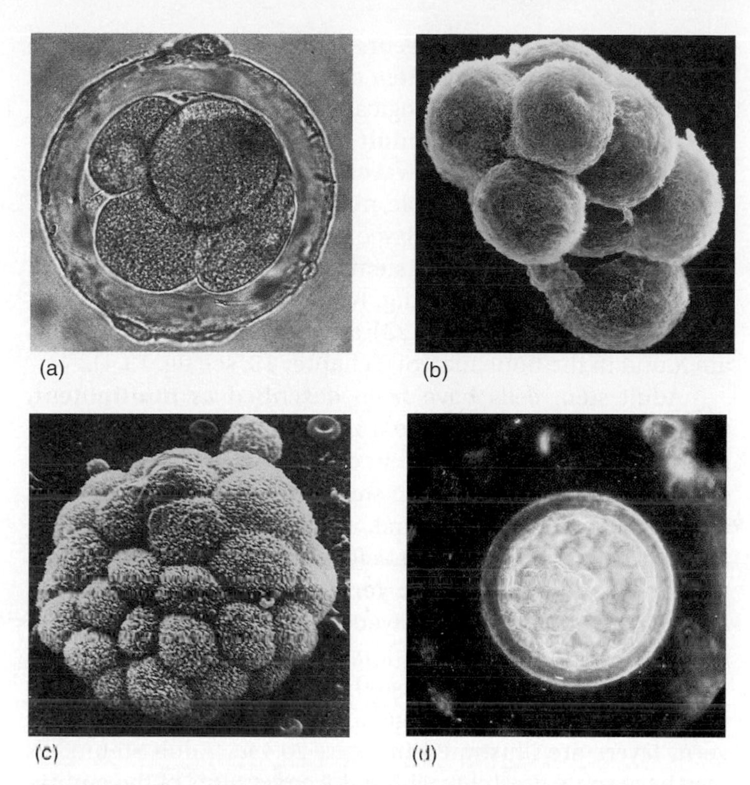

Figure 20.42 Scanning electron micrographs of preembryonic human development. A human ovum fertilized in a laboratory (*in vitro*) is seen at (*a*) the 4-cell stage. This is followed by (*b*) cleavage at the 16-cell stage and the formation of (*c*) a morula and (*d*) a blastocyst.

cell mass positioned against the endometrium. The trophoblast cells produce enzymes that allow the blastocyst to penetrate into the thick endometrium. This begins the process of **implantation,** or **nidation,** and by the seventh to tenth day the blastocyst is completely buried in the endometrium (fig. 20.43). Approximately 75% of all lost pregnancies are due to a failure of implantation, and consequently are not recognized as pregnancies.

CLINICAL APPLICATION

Progesterone, secreted from the woman's corpus luteum, is required for the endometrium to support the implanted embryo and maintain the pregnancy. A drug developed in France and approved for use in the United States promotes abortion by blocking the progesterone receptors of the endometrial cells. This drug, called **RU486,** has the generic name **mifepristone.** When combined with a small amount of a prostaglandin that stimulates contractions of the myometrium, RU486 can cause the endometrium to slough off, carrying the embryo with it. Sometimes called the "abortion pill," RU486 has generated bitter controversy in the United States. A recent study found mifepristone followed by prostaglandin treatment to be 96% to 99% effective at terminating pregnancies of 49 days or less.

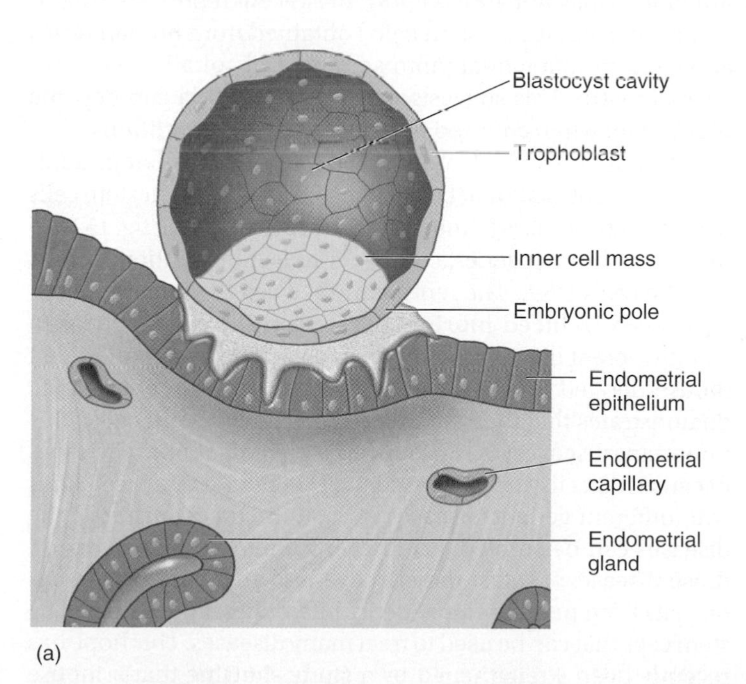

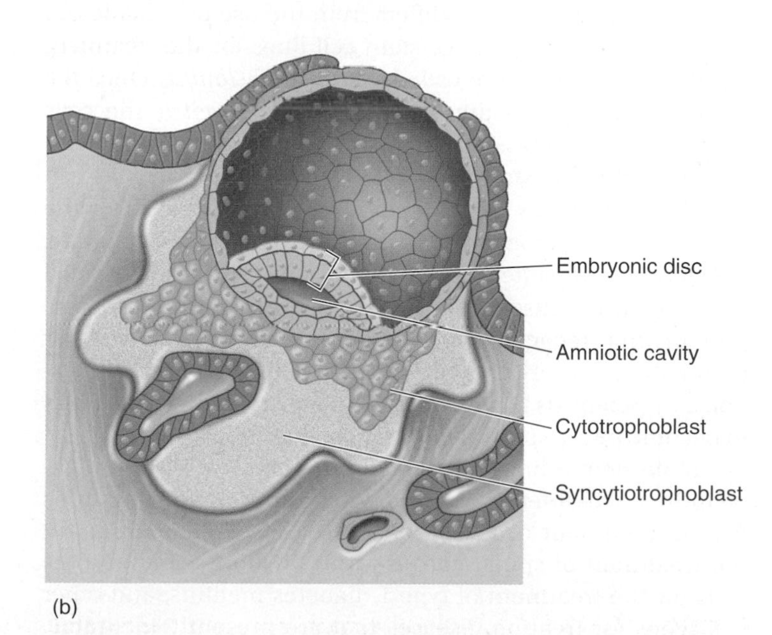

Figure 20.43 Implantation of the blastocyst. (*a*) A diagram showing the blastocyst attached to the endometrium on about day 6. (*b*) Implantation of the blastocyst at day 9 or 10.

Embryonic Stem Cells and Cloning

Only the fertilized egg cell and each of the early cleavage cells are **totipotent,** a term that refers to their ability to create the entire organism if implanted into a uterus. The nuclei of adult somatic cells, however, can be reprogrammed to become totipotent if they are transplanted into egg cell cytoplasm. Through such *somatic cell nuclear transfer,* the cloning of an entire adult organism (often called *reproductive cloning*) is possible, and indeed has been accomplished in sheep, cattle, cats, and other animals. This technique has not worked well in humans, but a recent report demonstrated that transfer of a nucleus from a diploid donor cell into an ovum that still contained its haploid chromosomes did allow a blastocyst to be produced (although all of the cells were triploid, with three sets of chromosomes). The possible use of this technique to clone humans has been widely condemned by scientists and others for many reasons, including the low probability of producing healthy children.

Part of the reason for the low probability of success in reproductive cloning relates to the processes of genomic imprinting discussed in section 20.1. In normal reproduction, the paternal chromatin (from the sperm) is "marked" by epigenetic changes, such as DNA methylation, differently than is the maternal chromatin (in the ovum). The maternal and paternal alleles of many genes are thus expressed differently in the embryo, and this differing gene expression (genomic imprinting) is needed for normal development and physiology. However, if an embryo is formed by somatic cell nuclear transfer, all of the chromatin is derived from the same somatic cell and so was subject to the same epigenetic changes. This may cause abnormalities in imprinted gene expression that may produce abnormalities in the animals that have been cloned.

Reproductive cloning differs from the use of somatic cell nuclear transfer to produce stem cell lines for the treatment of diseases, a procedure called *therapeutic cloning.* Once the embryo has reached the blastocyst stage *in vitro,* the cells of the inner cell mass are isolated and cultured. These cells, called **embryonic stem (ES) cells,** are **pluripotent:** they can form all the tissues of the body, but they cannot form the trophoblast and so cannot be used for reproductive cloning. The pluripotent state of embryonic stem cells is maintained by three core transcription factors. *Transcription factors* are proteins that recognize specific genes and either activate or inhibit their transcription (production of mRNA; chapter 3, section 3.3). Scientists have used this concept to induce adult cells to become pluripotent, as will be described shortly.

Many people hope that ES cells produced in this way might be induced to differentiate into dopamine-releasing neurons for the treatment of Parkinson's disease, other neurons for the treatment of spinal cord injuries, insulin-secreting beta cells for the treatment of type 1 diabetes mellitus, and other cell types for treating diseases that are presently incurable. A patient would not immunologically reject these cells if they were derived using the patient as the donor of the nucleus for the somatic cell nuclear transfer. However, in the first therapeutic use of ES cells, patients with spinal cord injury will receive oligodendrocyte precursor cells developed from one of the original embryonic stem cell lines, and thus will need drugs to suppress immunological rejection of these cells.

In contrast to ES cells, **adult stem cells** are found in protected locations where renewal of specialized cells is required in the adult body. For example, neural stem cells are located in the hippocampus (chapter 8; see fig. 8.15) and subventricular zone of the brain; epithelial stem cells are found in the intestinal crypts (chapter 18; see fig. 18.10) and in the bulge of hair follicles (chapter 1; see fig. 1.23); and hematopoietic stem cells are found in the bone marrow (chapter 13; see fig. 13.4).

Adult stem cells have been described as **multipotent,** because they can give rise to a number of differentiated cell types. For example, neural stem cells give rise to neurons and glial cells, and hematopoietic stem cells give rise to different types of blood cells. In general, adult stem cells are believed to differentiate into cells characteristic of their organ, and do not jump across **embryonic germ layer** (embryonic tissue) lines. The embryonic germ layers are *ectoderm* (giving rise to epidermis and neural tissues); *mesoderm* (giving rise to connective tissues and muscle tissue); and *endoderm* (giving rise to the epithelium of the lungs, gut and its derivatives). The three germ layers are illustrated in figure 20.45*a*. Adult stem cells may have some flexibility within the constraints of the embryonic germ layers, such as changing from bone marrow tissue to muscle tissue (because both are derived from mesoderm). For example, stem cells from the bone marrow can differentiate into myocardial cells, which may help repair myocardial infarction, and into skeletal muscle fibers, which may be useful in the treatment of muscular dystrophy. However, these germ layer limitations may not always apply; in a recent report, spermatogonia (spermatogenic stem cells) obtained from human testes appeared to differentiate into somatic cells of all three germ layers in vitro. This suggests that spermatogonia can become pluripotent when cultured under appropriate conditions.

Recent research demonstrated that differentiated, adult human fibroblasts could be changed into pluripotent stem cells by using retroviruses (similar to HIV; see chapter 15, fig. 15.3) to insert four human genes, coding for four transcription factors, into the cell's DNA. The reprogrammed cells, which the scientists called **induced pluripotent stem (iPS) cells,** appear to be pluripotent like embryonic stem cells. The observation that mouse-derived iPS cells can be used to generate entire mice demonstrates that they are indeed pluripotent stem cells. This important discovery has produced a great deal of excitement, because iPS cells could be used to (1) generate many cell lines with different genetic diseases, so that the mechanisms of the diseases can be studied; (2) test how different drugs work on those diseases; (3) test the effectiveness and toxicity of drugs on cells from people with genetic differences; and (4) produce stem cells that can be used to treat many diseases. This hope has recently been strengthened by a study showing that a mouse model of sickle-cell disease could be corrected using iPS cells.

A field called **regenerative medicine** involves developing future medical treatments using stem cells. Embryonic stem cells, derived using a patient's nucleus inserted into an unfertilized oocyte's cytoplasm (the technique of somatic cell nuclear transfer, previously discussed), can develop into tissues tolerated by the host's immune system. Similarly, iPS cells derived from a patient's fibroblasts may produce differentiated tissues that won't be immunologically rejected. However, there are differences between iPS cells and embryonic stem cells, including a large number of sites where the four transducing factor genes are inserted by the virus into the host DNA. These can cause genetic disruption and induce tumor formation, causing cancer. Also, iPS cells have been found to have epigenetic changes (specifically, DNA methylation; chapter 3, section 3.5) characteristic of their cells of origin rather than of embryonic stem cells. The significance of this in regenerative medicine is presently unknown.

In an exciting recent development, scientists produced iPS cells from fibroblasts derived from patients with Parkinson's disease. These scientists were then able to remove the four transducing genes from the iPS cells and induce these cells to differentiate into dopaminergic neurons. Such neurons could be used to learn about the sporadic form of Parkinson's disease (the most common form), and hopefully be the basis for future treatments. Other techniques for generating iPS cells without leaving the transforming genes in the host chromatin are also being developed. Together with the recent demonstration that iPS cells derived from the fibroblasts of a person with ALS (chapter 12) can differentiate into motor neurons, these reports and others encourage hope for the potential therapeutic benefits of iPS cells in regenerative medicine. However, scientists caution that iPS cells differ from each other and from embryonic stem cells, complicating the evaluation of their safety.

In order to get around potential problems of iPS cells, scientists recently reported that they could directly convert mouse and human skin fibroblast cells into functioning neurons. Since fibroblasts are derived from embryonic mesoderm and neurons from embryonic ectoderm, this is an example of *transdifferentiation* (conversion of one adult cell directly into another) that crosses germ layers. One group reported producing dopamine-releasing neurons derived from the fibroblasts of Parkinson's disease patients, while another group reported producing glutamine-releasing neurons from the fibroblasts of both normal and Alzheimer's disease patients. Other reports have demonstrated the ability to convert mouse fibroblasts directly into myocardial cells, neurons, hematopoietic stem cells, and others. However, as with iPS cells, more research is required to determine the safety and effectiveness of this technique.

Implantation of the Blastocyst and Formation of the Placenta

If fertilization does not take place, the corpus luteum begins to decrease its secretion of steroids about 10 days after ovulation. This withdrawal of ovarian steroids with the death of the corpus luteum causes the sloughing of the endometrium (menstruation) following day 28 of the typical cycle. If fertilization and implantation have occurred, these events must obviously be prevented to maintain the pregnancy.

Chorionic Gonadotropin

The blastocyst saves itself from being eliminated with the endometrium by secreting a hormone that indirectly prevents menstruation. Even before the sixth day, when implantation occurs, the trophoblast cells of the chorion secrete **chorionic gonadotropin,** or **hCG** (the *h* stands for "human"). This hormone is identical to LH in its effects and therefore is able to maintain the corpus luteum past the time when it would otherwise regress. The secretion of estradiol and progesterone is thus maintained and menstruation is normally prevented.

The secretion of hCG declines by the 10th week of pregnancy (fig. 20.44). Actually, this hormone is required for only the first five to six weeks of pregnancy because the placenta

CLINICAL APPLICATION

All **pregnancy tests** assay for the presence of hCG in blood or urine because this hormone is secreted by the blastocyst but not by the mother's endocrine glands. Modern pregnancy tests detect the beta subunit of hCG, which is unique to hCG and provides the least amount of cross-reaction with other hormones. Accurate and sensitive immunoassays for hCG in pregnancy tests employ antibodies that are produced by a clone of lymphocytes—termed *monoclonal antibodies* (chapter 15, section 15.4)—against the specific beta subunit of hCG. Home pregnancy kits that use these antibodies are generally accurate in the week following the first missed menstrual period.

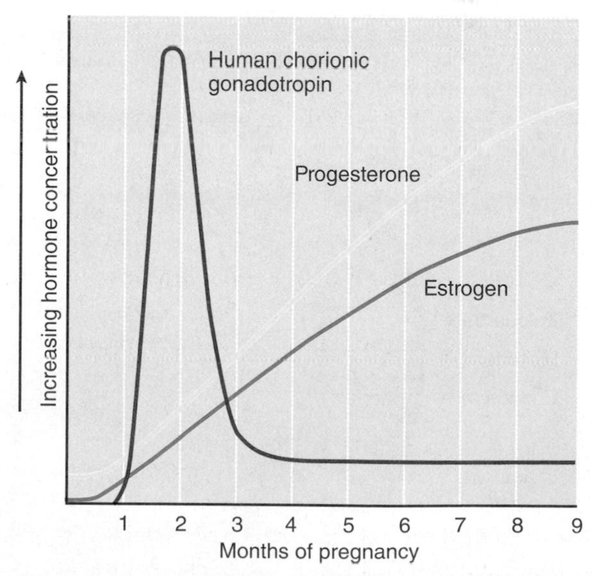

Figure 20.44 The secretion of human chorionic gonadotropin (hCG). This hormone is secreted by trophoblast cells during the first trimester of pregnancy, and it maintains the mother's corpus luteum for the first 5½ weeks. After that time, the placenta becomes the major sex-hormone-producing gland, secreting increasing amounts of estrogen and progesterone throughout pregnancy.

itself becomes an active steroid hormone-secreting gland. By the fifth to sixth week, the mother's corpus luteum begins to regress (even in the presence of hCG), but by this time the placenta is secreting more than sufficient amounts of steroids to maintain the endometrium and prevent menstruation.

Clinical Investigation CLUES

Gloria's pregnancy tests were negative.

- Which chemical does a pregnancy kit test for?
- If the pregnancy test had been positive, how would that have accounted for Gloria's amenorrhea?

Chorionic Membranes

Between days 7 and 12, as the blastocyst becomes completely embedded in the endometrium, the chorion becomes a two-cell-thick structure that consists of an inner *cytotrophoblast* layer and an outer *syncytiotrophoblast* layer (see fig. 20.43*b*). Meanwhile, the inner cell mass (which will become the fetus) also develops two cell layers. These are the **ectoderm** (which will form the nervous system and skin) and the **endoderm** (which will eventually form the gut and its derivatives). A third, middle embryonic layer—the **mesoderm**—is not yet seen at this stage. The embryo at this stage is a two-layer-thick disc separated from the cytotrophoblast of the chorion by an *amniotic cavity.*

As the syncytiotrophoblast invades the endometrium, it secretes protein-digesting enzymes that create numerous blood-filled cavities in the maternal tissue. The cytotrophoblast then forms projections, or *villi* (fig. 20.45), that grow into these pools of venous blood and produce a leafy-appearing structure called the **chorion frondosum** (*frond* = leaf). This occurs only on the side of the chorion that faces the uterine wall. As the embryonic structures grow, the other side of the chorion bulges into the cavity of the uterus, loses its villi, and takes on a smooth appearance.

Since the chorionic membrane is derived from the zygote, which inherits paternal genes that produce proteins foreign to the mother, scientists have long wondered why the mother's immune system doesn't attack the embryonic tissues. The placenta, it seems, is an "immunologically privileged site." Studies suggest that this immune protection may be due to FAS ligand, which is produced by the cytotrophoblast. T lymphocytes produce a surface receptor called FAS (chapter 15, section 15.3). The binding of FAS to FAS ligand triggers the apoptosis (cell suicide) of those lymphocytes, thereby preventing them from attacking the placenta.

Formation of the Placenta and Amniotic Sac

As the blastocyst implants in the endometrium and the chorion develops, the cells of the endometrium also undergo changes. These changes, including cellular growth and the accumulation of glycogen, are collectively called the *decidual*

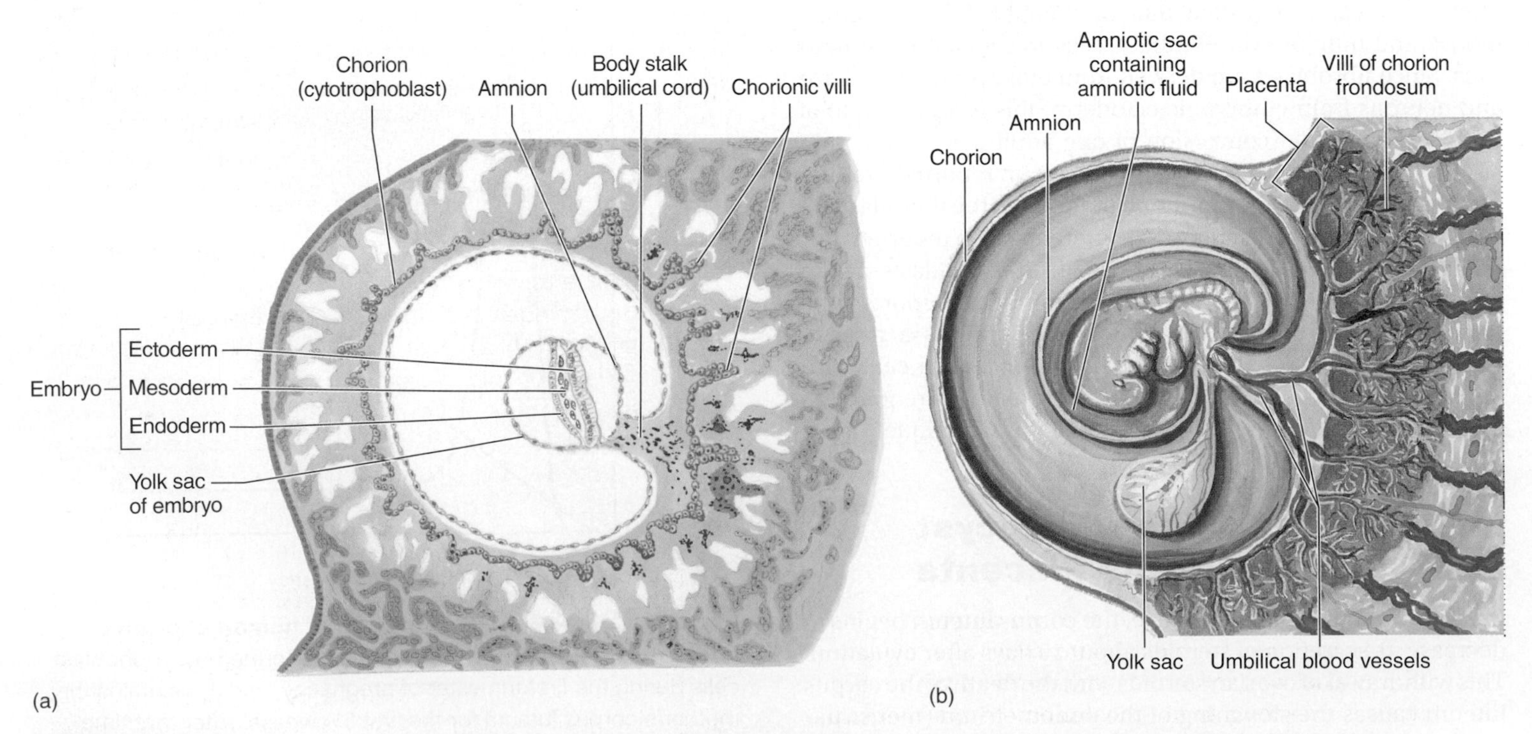

Figure 20.45 The extraembryonic membranes. After the syncytiotrophoblast has created blood-filled cavities in the endometrium, these cavities are invaded by extensions of the cytotrophoblast (*a*). These extensions, or villi, branch extensively to produce the chorion frondosum (*b*).The developing embryo is surrounded by a membrane called the amnion.

reaction. This is because the maternal tissue in contact with the chorion frondosum is called the **decidua basalis.** These two structures—chorion frondosum (fetal tissue) and decidua basalis (maternal tissue)—together form the functional unit known as the **placenta.**

Cells of the cytotrophoblast (see figs. 20.43*b* and 20.45*a*) from the chorionic villi invade the spiral arteries of the endometrium. As a result, by the end of the second trimester, the spiral arteries have been remodeled into dilated tubes lined by the cytotrophoblast. These produce a low vascular resistance, so that more maternal blood flows into the placenta.

The disc-shaped human placenta is continuous at its outer surface with the smooth part of the chorion, which bulges into the uterine cavity. Immediately beneath the chorionic membrane is the amnion, which has grown to envelop the entire embryo (fig. 20.46). The embryo, together with its umbilical cord, is therefore located within the fluid-filled **amniotic sac.**

Amniotic fluid is formed initially as an isotonic secretion. Later, the volume is increased and the concentration changed by urine from the fetus. Amniotic fluid also contains cells that are sloughed off from the fetus, placenta, and amniotic sac. Because all of these cells are derived from the same fertilized ovum, all have the same genetic composition. Many genetic abnormalities can be detected by aspiration of this fluid and examination of the cells thus obtained. This procedure is called **amniocentesis** (fig. 20.47).

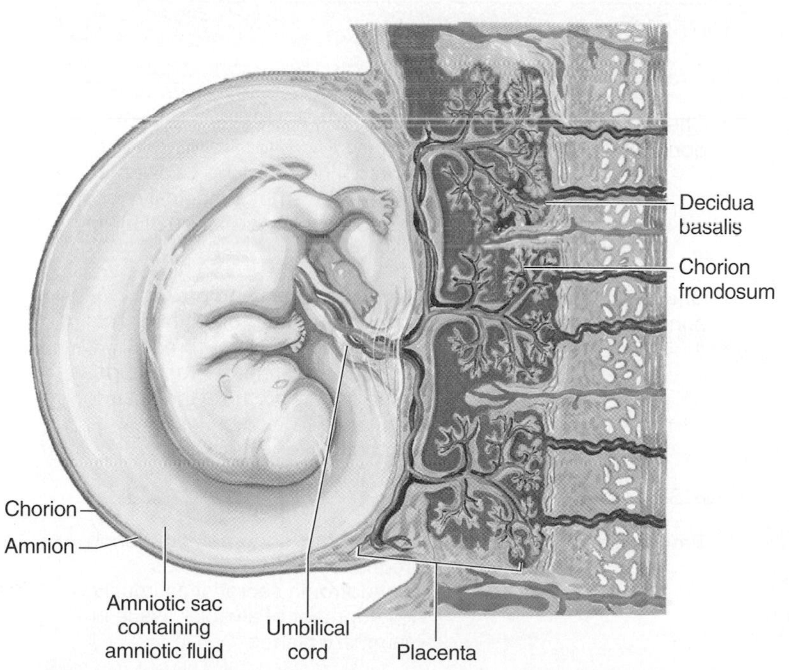

Figure 20.46 The amniotic sac and placenta. Blood from the embryo is carried to and from the chorion frondosum by umbilical arteries and veins. The maternal tissue between the chorionic villi is known as the decidua basalis; this tissue, together with the chorionic villi, forms the functioning placenta. The space between chorion and amnion is obliterated, and the fetus lies within the fluid-filled amniotic sac.

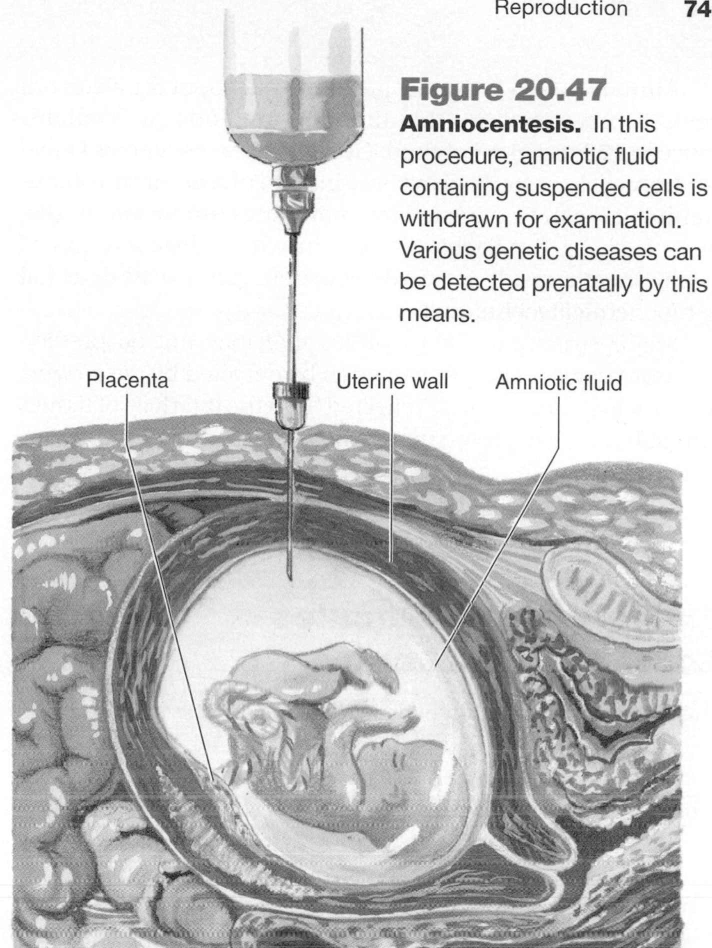

Figure 20.47 Amniocentesis. In this procedure, amniotic fluid containing suspended cells is withdrawn for examination. Various genetic diseases can be detected prenatally by this means.

CLINICAL APPLICATION

The amniotic fluid that is withdrawn contains fetal cells at a concentration too low to permit direct determination of genetic or chromosomal disorders. These cells must therefore be cultured *in vitro* for 10 to 14 days before they are present in sufficient numbers for the laboratory tests required. A newer method called **chorionic villus biopsy** is now available to detect genetic disorders earlier than permitted by amniocentesis. In chorionic villus biopsy, a catheter is inserted through the cervix to the chorion and a sample of a chorionic villus is obtained by suction or cutting. Genetic tests can be performed directly on the villus sample because it contains much larger numbers of fetal cells than does a sample of amniotic fluid. Chorionic villus biopsy can provide genetic information at 12 weeks' gestation. Amniocentesis, by contrast, cannot provide such information before about 20 weeks.

Noninvasive fetal sex determination is now possible using blood from the mother. In these tests, DNA sequences from the Y chromosome (including SRY; section 20.1) of a male fetus that enters the mother's blood is amplified and tested. These tests are apparently quite accurate after 20 weeks of gestation but inaccurate before 7 weeks.

Amniocentesis is usually performed at about the sixteenth week of pregnancy. By this time the amniotic sac contains between 175 to 225 ml of fluid. Genetic diseases such as Down syndrome (characterized by three instead of two chromosomes number 21) can be detected by examining chromosomes; diseases such as Tay-Sachs disease, in which degeneration of myelin sheaths results from a defective enzyme, can be detected by biochemical techniques.

Major structural abnormalities that may not be predictable from genetic analysis can often be detected by *ultrasound.* Sound-wave vibrations are reflected from the interface of tissues with different densities—such as the interface between the fetus and amniotic fluid—and used to produce an image. This technique is so sensitive that it can be used to detect a fetal heartbeat several weeks before it can be heard using a stethoscope.

Exchange of Molecules Across the Placenta

The *umbilical arteries* deliver fetal blood to vessels within the villi of the chorion frondosum of the placenta. This blood circulates within the villi and returns to the fetus via the *umbilical vein.* Maternal blood is delivered to and drained from the cavities within the decidua basalis that are located between the chorionic villi (fig. 20.48). Because of the structure of the placenta, only two cell layers separate molecules in the maternal blood from those in the fetal blood. In this way, maternal and fetal blood are brought close together but never mix within the placenta.

The placenta serves as a site for the exchange of gases and other molecules between the maternal and fetal blood. Oxygen diffuses from mother to fetus, and carbon dioxide diffuses in the opposite direction. Nutrient molecules and waste products likewise pass between maternal and fetal blood; the placenta is, after all, the only link between the fetus and the outside world.

But the placenta is not merely a passive conduit for exchange between maternal and fetal blood. It has a very high metabolic rate, utilizing about a third of all the oxygen and glucose supplied by the maternal blood. The rate of protein synthesis is, in fact, higher in the placenta than in the liver. Like the liver, the placenta produces a great variety of enzymes capable of converting hormones and exogenous drugs into less active molecules. In this way, potentially dangerous molecules in the maternal blood are often prevented from harming the fetus.

Mother
Intervillous pool of maternal blood
Fetus
Maternal vein
Chorion frondosum
Maternal artery
Maternal vein
Umbilical cord
Maternal artery
Umbilical artery
Umbilical vein
Uterine endometrium

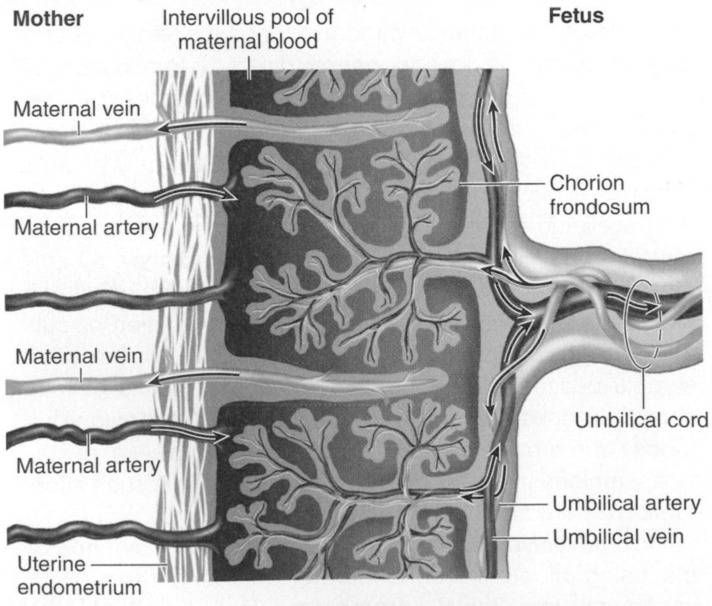

Figure 20.48 The circulation of blood within the placenta. Maternal blood is delivered to and drained from the spaces between the chorionic villi. Fetal blood is brought to blood vessels within the villi by branches of the umbilical artery and is drained by branches of the umbilical vein.

Endocrine Functions of the Placenta

The placenta secretes both steroid hormones and protein hormones. The protein hormones include **chorionic gonadotropin (hCG)** and **chorionic somatomammotropin (hCS),** both of which have actions similar to those of some anterior pituitary hormones (table 20.7). Chorionic

Table 20.7 | Hormones Secreted by the Placenta

Hormones	Effects
Pituitary-like Hormones	
Chorionic gonadotropin (hCG)	Similar to LH; maintains mother's corpus luteum for first 5½ weeks of pregnancy; may be involved in suppressing immunological rejection of embryo; also exhibits TSH-like activity
Chorionic somatomammotropin (hCS)	Similar to prolactin and growth hormone; in the mother, hCS acts to promote increased fat breakdown and fatty acid release from adipose tissue and to promote the sparing of glucose for use by the fetus ("diabetic-like" effects)
Sex Steroids	
Progesterone	Helps maintain endometrium during pregnancy; helps suppress gonadotropin secretion; stimulates development of alveolar tissue in mammary glands
Estrogens	Help maintain endometrium during pregnancy; help suppress gonadotropin secretion; help stimulate mammary gland development; inhibit prolactin secretion; promote uterine sensitivity to oxytocin; stimulate duct development in mammary glands

gonadotropin has LH-like effects, as previously described; it also has thyroid-stimulating ability, like pituitary TSH. Chorionic somatomammotropin likewise has actions that are similar to two pituitary hormones: growth hormone and prolactin. The placental hormones hCG and hCS thus duplicate the actions of four anterior pituitary hormones.

Pituitary-like Hormones from the Placenta

The importance of chorionic gonadotropin in maintaining the mother's corpus luteum for the first 5½ weeks of pregnancy has been previously discussed. There is also some evidence that hCG may in some way help to prevent immunological rejection of the implanting embryo.

Chorionic somatomammotropin acts together with growth hormone from the mother's pituitary to produce a diabetic-like effect in the pregnant woman. These two hormones promote (1) lipolysis and increased plasma fatty acid concentration; (2) glucose-sparing by maternal tissues and, therefore, increased blood glucose concentrations; and (3) polyuria (excretion of large volumes of urine), thereby producing a degree of dehydration and thirst. This diabetic-like effect in the mother helps to ensure a sufficient supply of glucose for the placenta and fetus, which (like the brain) use glucose as their primary energy source. Meanwhile, the beta cells of the mother's pancreatic islets proliferate during pregnancy to supply increased insulin, so that the development of *gestational diabetes* (chapter 11, section 11.6) is normally prevented.

Steroid Hormones from the Placenta

After the first 5½ weeks of pregnancy, when the corpus luteum regresses, the placenta becomes the major sex-steroid-producing gland. As a result of placental secretion, the blood concentration of estrogens rises to levels more than 100 times greater than those existing at the beginning of pregnancy. The placenta also secretes large amounts of progesterone, changing the estrogen/progesterone ratio in the blood from 100:1 at the beginning of pregnancy to close to 1:1 toward full-term. Progesterone "calms" the uterus during pregnancy by inhibiting genes that code for labor-promoting uterine proteins: connexin proteins for gap junctions between myometrial cells and oxytocin receptor proteins.

The placenta, however, is an "incomplete endocrine gland" because it cannot produce estrogen and progesterone without the aid of precursors supplied to it by both the mother and the fetus. For example, the placenta cannot produce cholesterol from acetate, and so it must be supplied with cholesterol from the mother's circulation. Cholesterol, which is a steroid containing 27 carbons, can then be converted by enzymes in the placenta into steroids that contain 21 carbons—such as progesterone. The placenta, however, lacks the enzymes needed to convert progesterone into androgens (which have 19 carbons). Because of this, androgens produced by the fetus (principally by the fetal adrenal

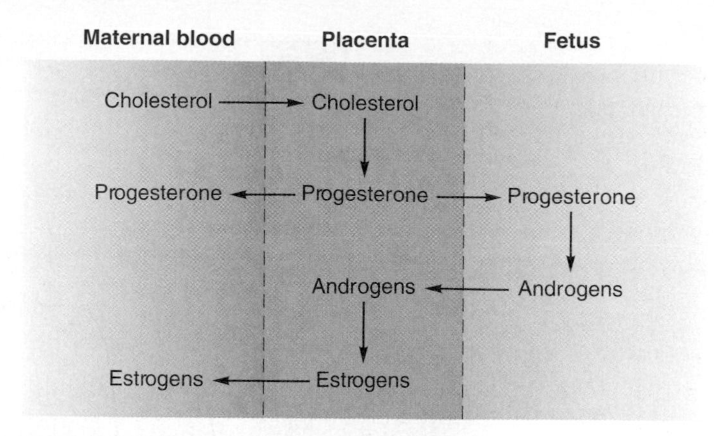

Figure 20.49 Interactions between the embryo and placenta produce the steroid hormones. The secretion of progesterone and estrogen from the placenta requires a supply of cholesterol from the mother's blood and the cooperation of fetal enzymes that convert progesterone to androgens.

cortex) are needed for the placenta to convert into estrogens (fig. 20.49), which have 18 carbons. Fetus and placenta thereby function together in the production of steroid hormones, an association termed the **fetal-placental unit.**

In addition to producing estradiol, the placenta secretes large amounts of a weak estrogen called **estriol.** The production of estriol increases tenfold during pregnancy, so that by the third trimester estriol accounts for about 90% of the estrogens excreted in the mother's urine. Because almost all of this estriol comes from the placenta (rather than from maternal tissues), measurements of urinary estriol can be used clinically to assess the health of the placenta.

Labor and Parturition

Powerful contractions of the uterus are needed to expel the fetus in the sequence of events called **labor.** These uterine contractions are known to be stimulated by two agents: (1) **oxytocin,** a polypeptide hormone produced in the hypothalamus and released by the posterior pituitary (and also produced by the uterus itself), and (2) **prostaglandins,** a class of cyclic fatty acids with paracrine functions produced within the uterus. The particular prostaglandins (PGs) involved are $PGF_{2\alpha}$ and PGE_2. Labor can indeed be induced artificially by injections of oxytocin or by insertion of prostaglandins into the vagina as a suppository.

Although labor is known to be stimulated by oxytocin and prostaglandins, the factors responsible for the initiation of labor are incompletely understood. In all mammals, labor is initiated by activation of the fetal adrenal cortex. In mammals other than primates, the fetal hypothalamus–anterior pituitary–adrenal cortex axis sets the time of labor. Corticosteroids secreted by the fetal adrenal cortex stimulate the placenta to convert progesterone into estrogens, thereby causing a fall in progesterone. This is significant because progesterone inhibits activity of the myometrium, while estrogens

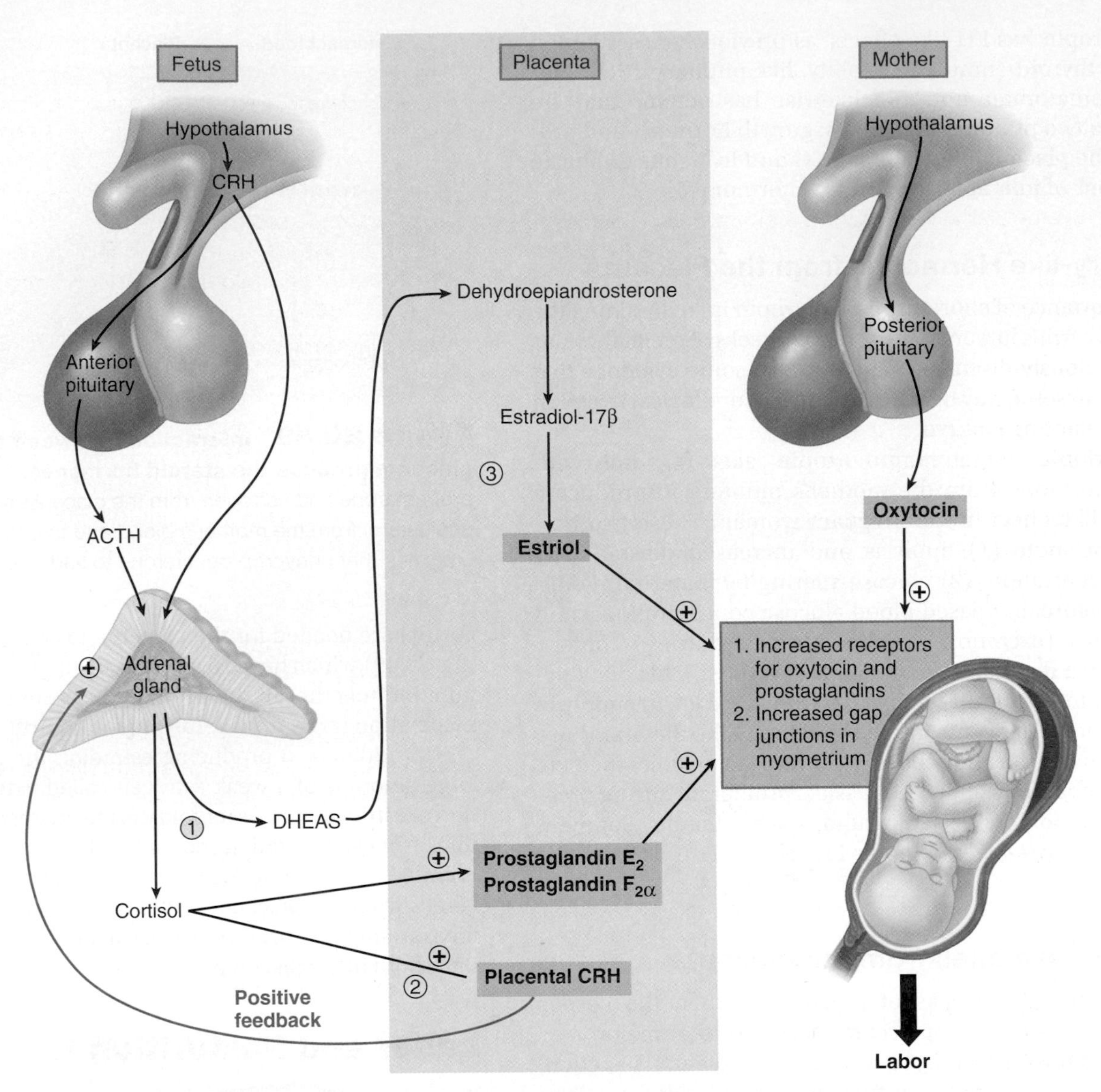

Figure 20.50 Labor in humans. (1) The fetal adrenal gland secretes dehydroepiandrosterone sulfate (DHEAS) and cortisol upon stimulation by CRH (corticotropin-releasing hormone) and ACTH (adrenocorticotropic hormone). (2) In turn, cortisol stimulates the placenta to secrete CRH, producing a positive feedback loop. (3) The DHEAS is converted by the placenta into estriol, which is needed, together with prostaglandins and oxytocin, to stimulate the myometrium of the mother's uterus to undergo changes leading to labor. The plus signs emphasize activation steps critical to this process.

stimulate the ability of the myometrium to contract. However, the initiation of labor in humans and other primates is more complex. Progesterone levels do not fall because the human placenta cannot convert progesterone into estrogens; it can only make estrogen when it is supplied with androgens from the fetus (fig. 20.49).

In primates only, the placenta produces **corticotropin-releasing hormone (CRH).** Further, it is only in humans and the great apes that the secretion of CRH rises rapidly during pregnancy. Most scientists now believe that the rate of increase of CRH secretion from the placenta is the most important determinant of when **parturition** (childbirth) will occur. CRH stimulates the secretion of ACTH from the anterior pituitary, which stimulates cortisol secretion from the adrenal cortex of both the fetus and the mother. Cortisol then acts in a positive feedback fashion to stimulate the placenta to secrete more CRH (fig. 20.50). Cortisol secreted from the fetal adrenal cortex also stimulates the maturation of the fetal lungs and their production of pulmonary surfactant, needed for lung function after the baby is born (chapter 16, section 16.2).

The fetal adrenal glands lack a medulla, but the cortex itself is composed of two parts. The outer part secretes cortisol, as does the adult adrenal cortex. The inner part, called the *fetal adrenal zone,* secretes an androgen called **dehydroepiandrosterone sulfate (DHEAS).** When ACTH secretion rises in response to placental CRH, it stimulates the fetal adrenal cortex to secrete both cortisol and DHEAS. The DHEAS travels from the fetal adrenal cortex to the placenta, where it is converted into estrogens (principally estriol). Estriol from the placenta then travels in the blood to the mother's uterus, where it increases the sensitivity of the myometrium to oxytocin and prostaglandins (fig. 20.50). This occurs because estriol stimulates the myometrium to produce (1) more receptors for oxytocin; (2) more receptors for prostaglandin $F_{2\alpha}$; and (3) more gap junctions between myometrial cells. During labor, oxytocin and prostaglandin $F_{2\alpha}$ stimulate the opening of Ca^{2+} channels in the plasma membrane for muscle contraction, and the gap junctions help to coordinate and synchronize uterine contractions.

Parturition in animals such as pigs, rats, and guinea pigs is aided by a hormone called **relaxin,** which causes softening of the pubic symphysis and relaxation of the cervix. In humans, however, this hormone does not seem to be required for parturition. Rather, relaxin secreted by the human ovary together with progesterone seem to be required during the first trimester for the decidual reaction (the formation of a decidua basalis in the endometrium). Also, relaxin promotes the growth of blood vessels into the decidua basalis, thus helping to nourish the growing embryo.

Following delivery of the baby, oxytocin is needed to maintain the muscle tone of the myometrium and to reduce hemorrhaging from uterine arteries. Oxytocin may also play a role in promoting the involution (reduction in size) of the uterus following delivery; the uterus weighs about 1 kg (2.2 lb) at term but only about 60 g (2 oz) by the sixth week following delivery.

CLINICAL APPLICATION

Genetic screening of neonates (newborns) is done in hospitals using only a drop of blood obtained by pricking the foot. Most of these blood tests do not involve DNA or chromosomal testing, yet they can detect a variety of genetic disorders, including phenylketonuria, hypothyroidism, cystic fibrosis, hemoglobin disorders such as sickle-cell anemia, and many others. Also, **umbilical cord blood banking** may be performed after birth. This is done because the umbilical cord blood contains a high concentration of hematopoietic stem cells, which can replenish the blood cell forming ability of bone marrow that has been damaged (by chemotherapy for leukemia, for example). Indeed, one unit of cord blood can reconstitute a person's entire hematopoietic system. Using banked umbilical cord blood for transplantation later in life minimizes immunological rejection.

Lactation

Each mammary gland is composed of 15 to 20 *lobes* divided by adipose tissue. The amount of adipose tissue determines the size and shape of the breast but has nothing to do with the ability of a woman to nurse. Each lobe is subdivided into *lobules,* which contain the glandular *alveoli* (fig. 20.51) that secrete the milk of a lactating female.

Figure 20.51 The structure of the breast and mammary glands. (*a*) A sagittal section and (*b*) an anterior view partially sectioned. AP|R

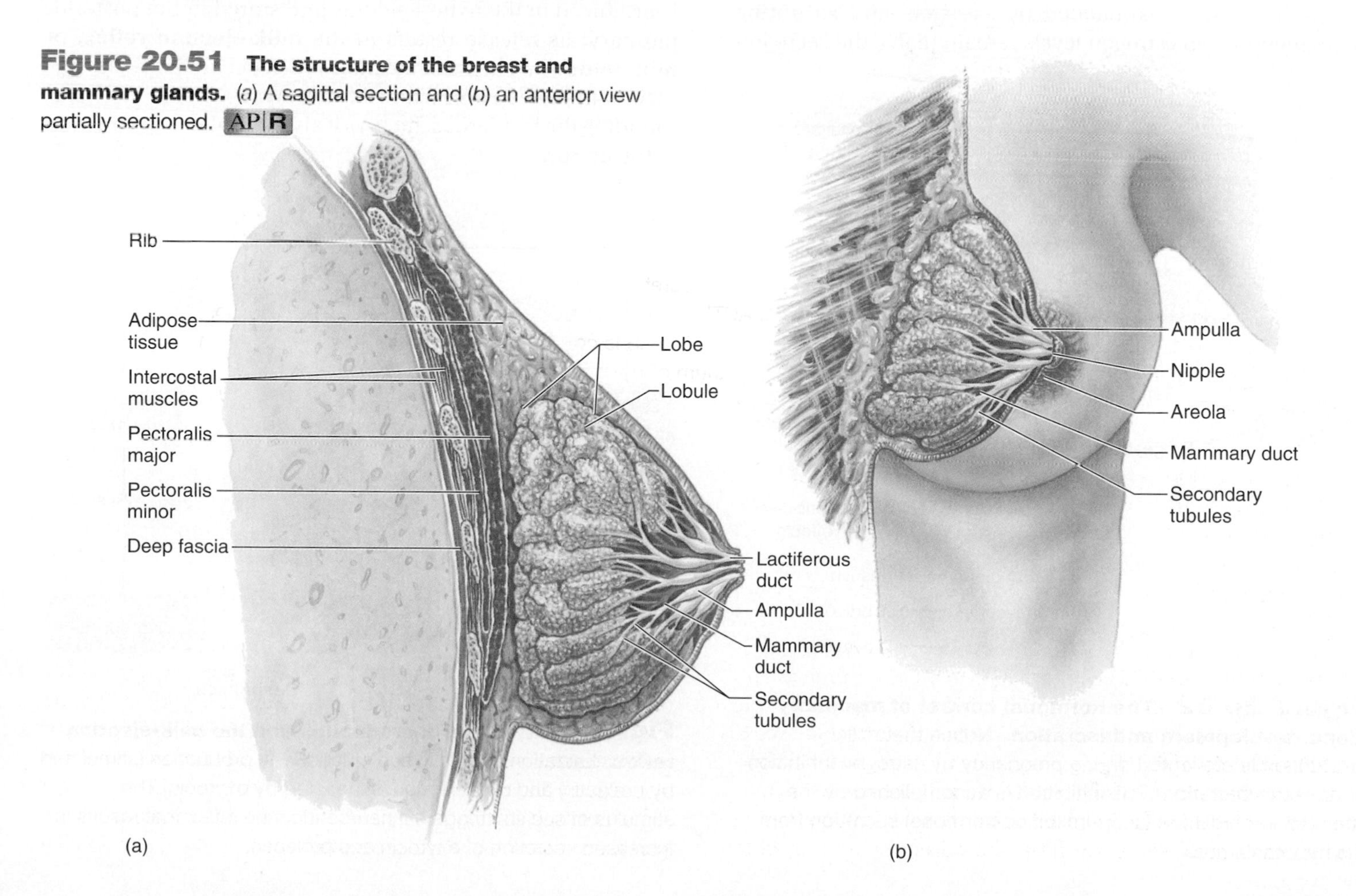

The clustered alveoli secrete milk into a series of *secondary tubules.* These tubules converge to form a series of *mammary ducts,* which in turn converge to form a *lactiferous duct* that drains at the tip of the nipple. The lumen of each lactiferous duct expands just beneath the surface of the nipple to form an *ampulla,* where milk accumulates during nursing. Epithelial cells that line the tubules and those that form the alveoli secrete water and nutrients for the lactating mammary glands. There are also specialized myoepithelial cells in the mammary glands that can contract to propel milk through its system of ducts.

The changes in the mammary glands during pregnancy and the regulation of lactation provide excellent examples of hormonal interactions and neuroendocrine regulation. Growth and development of the mammary glands during pregnancy requires the permissive actions of insulin, cortisol, and thyroid hormones. In the presence of adequate amounts of these hormones, high levels of progesterone stimulate the development of the mammary alveoli and estrogen stimulates proliferation of the tubules and ducts (fig. 20.52).

Prolactin, secreted by the anterior pituitary, stimulates the mammary glands after parturition to produce milk proteins, including casein and lactalbumin. Prolactin secretion is controlled by *prolactin-inhibiting hormone (PIH),* identified as dopamine, which is released by the hypothalamus into the hypothalamo-hypophyseal portal system of vessels. The secretion of PIH is stimulated by estrogen, and so during pregnancy—when estrogen levels remain high—the secretion of prolactin from the anterior pituitary is tonically inhibited. As a result, the high levels of estrogen during pregnancy helps prepare the mammary glands for lactation but, by inhibiting prolactin secretion, prevents milk production.

After parturition, when the placenta is expelled as the *afterbirth,* declining levels of estrogen are accompanied by an increase in the secretion of prolactin. Milk production is thereby stimulated. If a woman does not wish to breast-feed her baby she may take oral estrogens to inhibit prolactin secretion. A different drug commonly given in these circumstances, and in other conditions in which it is desirable to inhibit prolactin secretion, is *bromocriptine.* This drug binds to dopamine receptors and thus promotes the action of dopamine as the prolactin-inhibiting hormone (PIH).

The act of nursing helps to maintain high levels of prolactin secretion via a *neuroendocrine reflex* (fig. 20.53). Sensory endings in the breast, activated by the stimulus of suckling, relay impulses to the hypothalamus and inhibit the secretion of PIH. There is also indirect evidence that the stimulus of suckling may cause the secretion of a *prolactin-releasing hormone,* but this is controversial. Suckling thus results in the reflex secretion of high levels of prolactin that promotes the secretion of milk from the alveoli into the ducts. In order for the baby to get the milk, however, the action of another hormone is needed.

The stimulus of suckling also results in the reflex secretion of oxytocin from the posterior pituitary. This hormone is produced in the hypothalamus and stored in the posterior pituitary; its release results in the **milk-ejection reflex,** or **milk letdown.** The milk-ejection reflex occurs because oxytocin stimulates contraction of the myoepithelial cells surrounding the lactiferous ducts as it also stimulates contraction of the uterus.

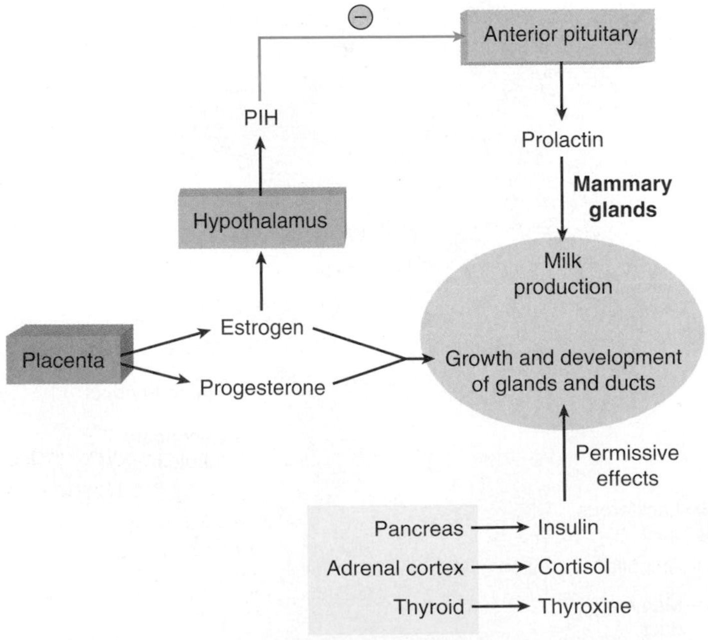

−
Anterior pituitary
PIH
Prolactin
Mammary glands
Hypothalamus
Milk production
Estrogen
Placenta
Progesterone
Growth and development of glands and ducts
Permissive effects
Pancreas → Insulin
Adrenal cortex → Cortisol
Thyroid → Thyroxine

Figure 20.52 The hormonal control of mammary gland development and lactation. Notice that milk production is prevented during pregnancy by estrogen inhibition of prolactin secretion. This inhibition is accomplished by the stimulation of PIH (prolactin-inhibiting hormone) secretion from the hypothalamus.

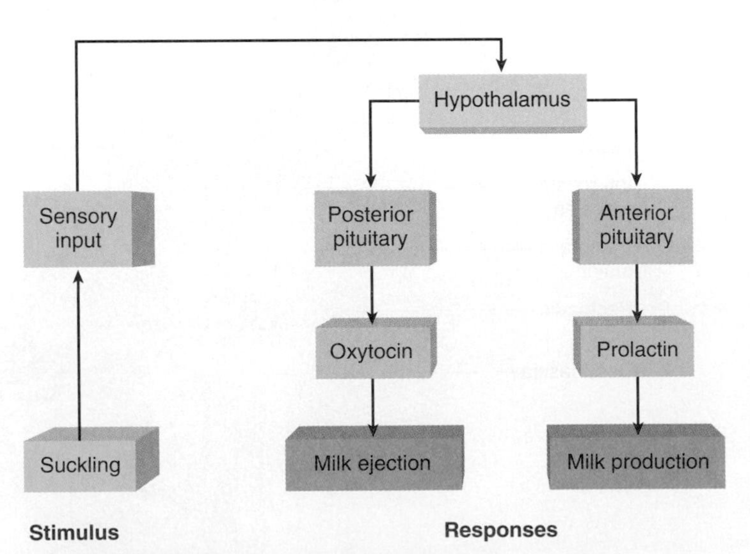

Figure 20.53 Milk production and the milk-ejection reflex. Lactation occurs in two stages: milk production (stimulated by prolactin) and milk ejection (stimulated by oxytocin).The stimulus of sucking triggers a neuroendocrine reflex that results in increased secretion of oxytocin and prolactin.

FITNESS APPLICATION

Milk letdown can become a conditioned reflex made in response to visual or auditory cues; the crying of a baby can elicit oxytocin secretion and the milk-ejection reflex. On the other hand, this reflex can be suppressed by the adrenergic effects produced in the fight-or-flight reaction. Thus, if a woman becomes nervous and anxious while breast-feeding, she will produce milk but it will not flow (there will be no milk letdown). This can cause increased pressure, intensifying her anxiety and frustration and further inhibiting the milk-ejection reflex. It is therefore important for mothers to nurse their babies in a quiet and calm environment. If needed, synthetic oxytocin can be given as a nasal spray to promote milk letdown.

Breast-feeding supplements the immune protection given to the infant by its mother. While the fetus is *in utero,* immunoglobin G (IgG—chapter 15) antibodies cross the placenta from the maternal to the fetal blood. These antibodies provide passive immune protection to the baby for the first three to twelve months after birth (fig. 20.54). Infants that are breast-fed also receive IgA antibodies from the mother's milk, which provides additional passive immune protection within the baby's intestine. In addition, the mother's milk contains cytokines, lymphocytes, and antibodies that may promote the development of the baby's system of active immunity. The ability of the baby to produce its own antibodies is not well developed for several months after birth (fig. 20.54), so the passive

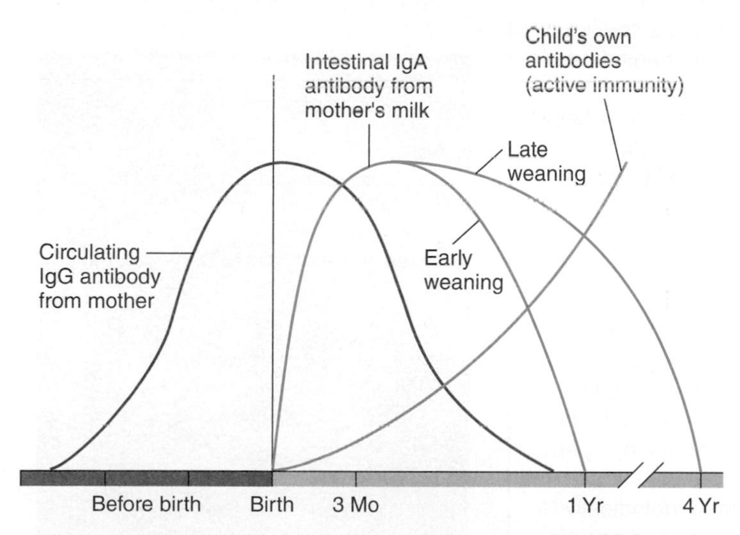

Figure 20.54 Maternal antibodies that protect the baby. Circulating IgG antibodies cross the placenta and protect the baby for 3 months to 1 year after birth. This passive immunity is supplemented by IgA antibodies in the baby's intestine obtained from the mother's milk. This protection lasts longer for babies weaned at a later age. Notice the inability of the baby to produce a large amount of its own antibodies until it is several months of age. Adapted from R. M. Zinkernagel, "Advances in immunology: Maternal antibodies, childhood infections, and autoimmune diseases." *New England Journal of Medicine, 345:18,* pp. 1331–1335.

immunity provided by maternal antibodies in breast milk may be significant in protecting the baby from a variety of infections.

Breast-feeding, acting through reflex inhibition of GnRH secretion, can also inhibit the secretion of gonadotropins from the mother's anterior pituitary and thus inhibit ovulation. Breast-feeding is thus a natural contraceptive mechanism that helps to space births. This mechanism appears to be most effective in women with limited caloric intake and in those who breast-feed their babies at frequent intervals throughout the day and night. In the traditional societies of the less industrialized nations, therefore, breast-feeding is an effective contraceptive. Breast-feeding has much less of a contraceptive effect in women who are well nourished and who breast-feed their babies at more widely spaced intervals.

CHECKPOINT

12a. Describe the changes that occur in the sperm cell and ovum during fertilization.
12b. Identify the source of hCG and explain why this hormone is needed to maintain pregnancy for the first 10 weeks.
13a. List the fetal and maternal components of the placenta and describe the circulation in these two components. Explain how fetal and maternal gas exchange occurs.
13b. List the protein hormones and sex steroids secreted by the placenta and describe their functions.
14. Identify the two agents that stimulate uterine contraction during labor and describe the proposed mechanisms that may initiate labor in humans.
15. Describe the hormonal interactions required for breast development during pregnancy and for lactation after delivery.

CONCLUDING REMARKS

We shall not cease from exploration
And the end of all our exploring
Will be to arrive where we started
And know the place for the first time.

— T.S. Eliot, *Little Gidding, V*

It may seem strange to end a textbook on physiology with the topics of pregnancy and parturition. This is done in part for practical reasons; these topics are complex, and to understand them requires a grounding in subjects covered earlier. Also, it seems appropriate to end at the beginning, at the start of a new life. Although generations of researchers have accumulated an impressive body of knowledge, the study of physiology is still young and rapidly growing. I hope that this introductory textbook will serve students' immediate practical needs as a resource for understanding current applications, and that it will provide a good foundation for a lifetime of further study.

Interactions

      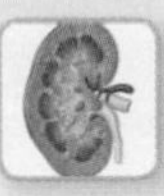  

HPer Links of the Reproductive System with Other Body Systems

Integumentary System

- The skin serves as a sexual stimulant and helps to protect the body from pathogens (p. 494)
- Sex hormones affect the distribution of body hair, deposition of subcutaneous fat, and other secondary sexual characteristics (p. 709)

Skeletal System

- The pelvic girdle supports and protects some reproductive organs (p. 722)
- Sex hormones stimulate bone growth and maintenance (p. 692)

Muscular System

- Contractions of smooth muscles aid the movement of gametes (p. 393)
- Contractions of the myometrium aid labor and delivery (p. 743)
- Cremaster muscles help to maintain proper temperature of the testes (p. 704)
- Testosterone promotes an increase in muscle mass (p. 322)

Nervous System

- Autonomic nerves innervate the organs of male reproduction to stimulate erection and ejaculation (p. 719)
- Autonomic nerves promote aspects of the human sexual response (p. 711)
- The CNS, acting through the pituitary, coordinates different aspects of reproduction (p. 337)
- The limbic system of the brain is involved in sexual drive (p. 219)
- Gonadal sex hormones influence brain activity (p. 336)

Endocrine System

- The anterior pituitary controls the activity of the gonads (p. 332)
- Testosterone secreted by the testes maintains the structure and function of the male reproductive system (p. 713)
- Estradiol and progesterone secreted by the ovaries regulates the female accessory sex organs, including the endometrium of the uterus (p. 730)
- Hormones secreted by the placenta are needed to maintain a pregnancy (p. 739)
- Prolactin and oxytocin are required for production of breast milk and the milk-ejection reflex (p. 746)

Circulatory System

- The circulatory system transports oxygen and nutrients to the reproductive organs (p. 405)
- The fetal circulation permits the fetus to obtain oxygen and nutrients from the placenta (p. 742)
- Estrogen secreted by the ovaries helps raise the level of HDL-cholesterol carriers in the blood, lowering the risk of atherosclerosis (p. 438)

Immune System

- The immune system protects the body, including the reproductive system, against infections (p. 494)
- The blood-testis barrier prevents the immune system from attacking sperm in the testes (p. 716)
- The placenta is an immunologically privileged site; it is protected against rejection by the mother's immune system (p. 740)

Respiratory System

- The lungs provide oxygen for all body systems, including the reproductive system, and provide for the elimination of carbon dioxide (p. 533)
- The red blood cells of a fetus contain hemoglobin F, which has a high affinity for oxygen (p. 563)

Urinary System

- The kidneys regulate the volume, pH, and electrolyte balance of the blood and eliminate wastes (p. 582)
- The male urethra transports urine as well as semen (p. 584)

Digestive System

- The GI tract provides nutrients for all of the organs of the body, including those of the reproductive system (p. 621)
- Nutrients obtained from the GI tract of the mother can cross the placenta to the embryo and fetus (p. 742)

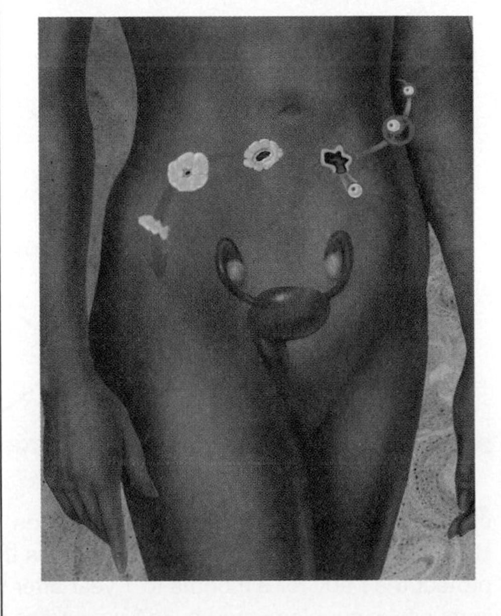

Clinical Investigation SUMMARY

Gloria's lack of menstruation was not accompanied by pain and she did not have a history of spotting or excessive menstrual bleeding. She had menstruated prior to her amenorrhea, which rules out the possibility of primary amenorrhea. Her secondary amenorrhea could have been the result of pregnancy, but this was ruled out by the negative pregnancy test. The amenorrhea could have been caused by her hypothyroidism, but she stated that she took her thyroid pills regularly and her blood test demonstrated normal thyroxine levels.

Gloria most likely has a secondary amenorrhea that is due to emotional stress, low body weight, and/or her strenuous exercise program. She should take steps to alleviate these conditions if she wants to resume her normal menstrual periods. If she refuses to gain weight and reduce her level of physical activity, her physician might recommend the use of oral contraceptives to help regulate her cycles.

See the additional chapter 20 Clinical Investigations on *Male Pituitary Adenoma* and *Polycystic Ovarian Syndrome* in the Connect site for this text at www.mhhe.com/Fox13.

SUMMARY

20.1 Sexual Reproduction 701

A. Sperm that bear X chromosomes produce XX zygotes when they fertilize an ovum; sperm that bear Y chromosomes produce XY zygotes.

  1. Embryos that have the XY genotype develop testes; those without a Y chromosome produce ovaries.

  2. The testes of a male embryo secrete testosterone and müllerian inhibition factor. MIF causes degeneration of female accessory sex organs, and testosterone promotes the formation of male accessory sex organs.

B. The male accessory sex organs are the epididymis, ductus (vas) deferens, seminal vesicles, prostate, and ejaculatory duct.

  1. The female accessory sex organs are the uterus and uterine (fallopian) tubes. They develop when testosterone and müllerian inhibition factor are absent.

  2. Testosterone indirectly (acting via conversion to dihydrotestosterone) promotes the formation of male external genitalia; female genitalia are formed when testosterone is absent.

C. Numerous disorders of embryonic sexual development can be understood in terms of the normal physiology of the developmental processes.

20.2 Endocrine Regulation of Reproduction 707

A. The gonads are stimulated by two anterior pituitary hormones: FSH (follicle-stimulating hormone) and LH (luteinizing hormone).

  1. The secretion of FSH and LH is stimulated by gonadotropin-releasing hormone (GnRH), which is secreted by the hypothalamus.

  2. The secretion of FSH and LH is also under the control of the gonads by means of negative feedback exerted by gonadal steroid hormones and by a peptide called inhibin.

B. The rise in FSH and LH secretion that occurs at puberty may be due to maturational changes in the brain and to decreased sensitivity of the hypothalamus and pituitary gland to the negative feedback effects of sex steroid hormones.

C. The pineal gland secretes melatonin. This hormone has an inhibitory effect on gonadal function in some species of mammals, but its role in human physiology is presently controversial.

D. The human sexual response is divided into four phases: excitation, orgasm, plateau, and resolution. Both sexes follow a similar pattern.

20.3 Male Reproductive System 711

A. In the male, the pituitary secretion of LH is controlled by negative feedback from testosterone, whereas the secretion of FSH is controlled by the secretion of inhibin from the testes.

  1. The negative feedback effect of testosterone is actually produced by the conversion of testosterone to 5α-reduced androgens and to estradiol.

2. The secretion of testosterone is relatively constant rather than cyclic, and it does not decline sharply at a particular age.

B. Testosterone promotes the growth of soft tissue and bones before the epiphyseal discs have sealed; thus, testosterone and related androgens are anabolic steroids.
   1. Testosterone is secreted by the interstitial Leydig cells under stimulation by LH.
   2. LH receptor proteins are located in the interstitial tissue. FSH receptors are located in the Sertoli cells within the seminiferous tubules.
   3. The Leydig cells of the interstitial compartment and the Sertoli cells of the tubular compartment of the testes secrete autocrine regulatory molecules that allow the two compartments to interact.

C. Diploid spermatogonia in the seminiferous tubules undergo meiotic cell division to produce haploid sperm.
   1. At the end of meiosis, four spermatids are formed. They develop into spermatozoa by a maturational process called spermiogenesis.
   2. Sertoli cells in the seminiferous tubules are required for spermatogenesis.
   3. At puberty, testosterone is required for the completion of meiosis, and FSH is required for spermiogenesis.

D. Spermatozoa in the seminiferous tubules are conducted to the epididymis and drained from the epididymis into the ductus deferens. The prostate and seminal vesicles add fluid to the semen.

E. Penile erection is produced by parasympathetic-induced vasodilation. Ejaculation is produced by sympathetic nerve stimulation of peristaltic contraction of the male accessory sex organs.

20.4 Female Reproductive System 722

A. Primordial follicles in the ovary contain primary oocytes that have become arrested at prophase of the first meiotic division. Their number is maximal at birth and declines thereafter.
   1. A small number of oocytes in each cycle are stimulated to complete their first meiotic division and become secondary oocytes.
   2. At the completion of the first meiotic division, the secondary oocyte is the only complete cell formed. The other product of this division is a tiny polar body, which disintegrates.

B. One of the secondary follicles grows very large, becomes a graafian follicle, and is ovulated.
   1. Upon ovulation, the secondary oocyte is extruded from the ovary. It does not complete the second meiotic division unless it becomes fertilized.
   2. After ovulation, the empty follicle becomes a new endocrine gland called a corpus luteum.
   3. The ovarian follicles secrete only estradiol, whereas the corpus luteum secretes both estradiol and progesterone.

C. The hypothalamus secretes GnRH in a pulsatile fashion, causing pulsatile secretion of gonadotropins. This is needed to prevent desensitization and downregulation of the target glands.

20.5 Menstrual Cycle 727

A. During the follicular phase of the cycle, the ovarian follicles are stimulated by FSH from the anterior pituitary.
   1. Under FSH stimulation, the follicles grow, mature, and secrete increasing amounts of estradiol.
   2. At about day 13, the rapid rise in estradiol secretion stimulates a surge of LH from the anterior pituitary. This represents positive feedback.
   3. The LH surge stimulates ovulation at about day 14.
   4. After ovulation, the empty follicle is stimulated by LH to become a corpus luteum, at which point the ovary is in a luteal phase.
   5. The secretion of progesterone and estradiol rises during the first part of the luteal phase and exerts negative feedback on FSH and LH secretion.
   6. Without continued stimulation by LH, the corpus luteum regresses at the end of the luteal phase, and the secretion of estradiol and progesterone declines. This decline results in menstruation and the beginning of a new cycle.

B. The rising estradiol concentration during the follicular phase produces the proliferative phase of the endometrium. The secretion of progesterone during the luteal phase produces the secretory phase of the endometrium.

C. Oral contraceptive pills usually contain combinations of estrogen and progesterone that exert negative feedback control of FSH and LH secretion.

20.6 Fertilization, Pregnancy, and Parturition 733

A. The sperm undergoes an acrosomal reaction, which allows it to penetrate the corona radiata and zona pellucida.
   1. Upon fertilization, the secondary oocyte completes meiotic division and produces a second polar body, which degenerates.
   2. The diploid zygote undergoes cleavage to form a morula and then a blastocyst. Implantation of the blastocyst in the endometrium begins between the fifth and seventh day.

B. The trophoblast cells of the blastocyst secrete human chorionic gonadotropin (hCG), which functions in the manner of LH and maintains the mother's corpus luteum for the first 10 weeks of pregnancy.
   1. The trophoblast cells provide the fetal contribution to the placenta. The placenta is also formed from adjacent maternal tissue in the endometrium.
   2. Oxygen, nutrients, and wastes are exchanged by diffusion between the fetal and maternal blood.

C. The placenta secretes chorionic somatomammotropin (hCS), chorionic gonadotropin (hCG), and steroid hormones.
   1. The action of hCS is similar to that of prolactin and growth hormone. The action of hCG is similar to that of LH and TSH.
   2. The major steroid hormone secreted by the placenta is estriol. The placenta and fetal glands cooperate in the production of steroid hormones.

D. Contraction of the uterus in labor is stimulated by oxytocin from the posterior pituitary and by prostaglandins, produced within the uterus.

1. Androgens, primarily DHEAS, secreted by the fetal adrenal cortex are converted into estrogen by the placenta.
2. Estrogen secreted by the placenta induces oxytocin synthesis, enhances uterine sensitivity to oxytocin, and promotes prostaglandin synthesis in the uterus. These events culminate in labor and delivery.

E. The high levels of estrogen during pregnancy, acting synergistically with other hormones, stimulate growth and development of the mammary glands.
1. Prolactin (and the prolactin-like effects of hCS) can stimulate the production of milk proteins. Prolactin secretion and action, however, are blocked during pregnancy by the high levels of estrogen secreted by the placenta.
2. After delivery, when estrogen levels fall, prolactin stimulates milk production.
3. The milk-ejection reflex is a neuroendocrine reflex. The stimulus of suckling causes reflex secretion of oxytocin. This stimulates contractions of the lactiferous ducts and the ejection of milk from the nipple.

REVIEW ACTIVITIES

Test Your Knowledge

Match these:

1. Menstrual phase
2. Follicular phase
3. Luteal phase
4. Ovulation

a. High estrogen and progesterone; low FSH and LH
b. Low estrogen and progesterone
c. LH surge
d. Increasing estrogen; low LH and low progesterone

5. A person with the genotype XO has
   a. ovaries.
   b. testes.
   c. both ovaries and testes.
   d. neither ovaries nor testes.
6. An embryo with the genotype XX develops female accessory sex organs because of
   a. androgens.
   b. estrogens.
   c. lack of androgens.
   d. lack of estrogens.
7. In the male,
   a. FSH is not secreted by the pituitary.
   b. FSH receptors are located in the Leydig cells.
   c. FSH receptors are located in the spermatogonia.
   d. FSH receptors are located in the Sertoli cells.
8. The secretion of FSH in a male is inhibited by negative feedback effects of
   a. inhibin secreted from the tubules.
   b. inhibin secreted from the Leydig cells.
   c. testosterone secreted from the tubules.
   d. testosterone secreted from the Leydig cells.
9. Which of these statements is *true?*
   a. Sperm are not motile until they pass through the epididymis.
   b. Sperm require capacitation in the female reproductive tract before they can fertilize an ovum.
   c. A secondary oocyte does not complete meiotic division until it has been fertilized.
   d. All of these are true.
10. The corpus luteum is maintained for the first 10 weeks of pregnancy by
    a. hCG.
    b. LH.
    c. estrogen.
    d. progesterone.
11. Fertilization normally occurs in
    a. the ovaries.
    b. the uterine tubes.
    c. the uterus.
    d. the vagina.
12. The placenta is formed from
    a. the fetal chorion frondosum.
    b. the maternal decidua basalis.
    c. both *a* and *b*.
    d. neither *a* nor *b*.
13. Uterine contractions are stimulated by
    a. oxytocin.
    b. prostaglandins.
    c. prolactin.
    d. both *a* and *b*.
    e. both *b* and *c*.
14. Contraction of the mammary glands and ducts during the milk-ejection reflex is stimulated by
    a. prolactin.
    b. oxytocin.
    c. estrogen.
    d. progesterone.
15. If GnRH were secreted in large amounts and at a constant rate rather than in a pulsatile fashion, which of these statements would be *true?*
    a. LH secretion will increase at first and then decrease.
    b. LH secretion will increase indefinitely.
    c. Testosterone secretion in a male will be continuously high.
    d. Estradiol secretion in a woman will be continuously high.
16. The contents of the birth control pill
    a. inhibit secretion of gonadotropins.
    b. exert negative feedback on the hypothalamus.
    c. prevent ovulation.
    d. maintain the endometrium.
    e. All of these.

17. Which of the following statements regarding spermatogenesis is *false?*
    a. Testosterone stimulates spermatogenesis.
    b. The chromatin becomes highly compacted as protamines replace histone proteins.
    c. Sertoli cells engulf spermatid cytoplasm by phagocytosis.
    d. FSH is required for spermatogenesis to occur.
18. Which of the following statements about progesterone is *false?*
    a. Progesterone is secreted by the corpus luteum
    b. Progesterone is secreted by the placenta.
    c. Progesterone decreases the basal body temperature.
    d. Progesterone secretion drops towards the end of the nonfertile luteal phase.

Test Your Understanding

19. Identify the conversion products of testosterone and describe their functions in the brain, prostate, and seminiferous tubules.
20. Explain why a testis is said to be composed of two separate compartments. Describe the interactions that may occur between these compartments.
21. Describe the roles of the Sertoli cells in the testes.
22. Describe the steps of spermatogenesis and explain its hormonal control.
23. Explain the hormonal interactions that control ovulation and cause it to occur at the proper time.
24. Compare menstrual bleeding and bleeding that occurs during the estrous cycle of a dog in terms of hormonal control mechanisms and the ovarian cycle.
25. "The [contraceptive] pill tricks the brain into thinking you're pregnant." Interpret this popularized explanation in terms of physiological mechanisms.
26. Why does menstruation normally occur? Under what conditions does menstruation not occur? Explain.
27. Explain the proposed mechanisms whereby the act of a mother nursing her baby results in lactation. By what mechanisms might the sound of a baby crying elicit the milk-ejection reflex?
28. Describe the steps of oogenesis when fertilization occurs and when it does not occur. Why are polar bodies produced?
29. Identify the hormones secreted by the placenta. Why is the placenta considered an incomplete endocrine gland?
30. Describe the endocrine changes that occur at menopause and discuss the consequences of these changes. What are the benefits and risks associated with hormone replacement therapy?
31. Explain the sequence of events by which the male accessory sex organs and external genitalia are produced. What occurs when a male embryo lacks receptor proteins for testosterone? What occurs when a male embryo lacks the enzyme 5α-reductase?
32. Describe the mechanisms that have been proposed to time the onset of parturition in sheep and humans.

Test Your Analytical Ability

33. According to your friend, there is a female birth control pill and not a male birth control pill only because the medical establishment is run by men. Do you agree with her conspiracy theory? Provide physiological support for your answer.
34. Elderly men with benign prostatic hyperplasia are sometimes given estrogen treatments. How would this help the condition? What other types of drugs may be given, and what would you predict their possible side effects to be?
35. Discuss the role of apoptosis and follicle atresia in ovarian physiology. How might this process be regulated?
36. Is it true that estrogen is an exclusively female hormone and that testosterone is an exclusively male hormone? Explain your answer.
37. Surgical removal of a woman's ovaries (ovariectomy) can precipitate menstruation. Ovariectomy in a dog or cat, however, does not cause the discharge of uterine blood. How can you explain these different responses?
38. Endometrial tissue is maintained by estrogen, and yet endometriosis is treated with a GnRH agonist drug. Explain this seeming paradox.
39. If scientists developed a new drug to block the FSH receptor in men, or to reduce a man's FSH secretion, would that lead to a male contraceptive pill? Explain, in terms of the hormonal control of spermatogenesis.
40. Describe the locations and characteristics of adult stem cells. How does their potential relate to normal embryonic tissue development? What are the potential advantages and disadvantages of using adult stem cells, compared to embryonic stem cells, for regenerative medicine (stem cell therapies)?

Test Your Quantitative Ability

Refer to figure 20.32 to answer the following questions:

41. During which range of days are the secretions of both estradiol and progesterone at their highest levels in the graph?
42. What is the percentage increase in LH secretion in the day leading up to the LH peak?
43. In the 5 days leading up to the LH peak, what is the percentage increase in estradiol secretion?
44. At 5 days after the LH peak, how many times greater is the blood concentration of progesterone compared to estradiol? (Hint: a picogram is 10^{-12} g, whereas a nanogram is 10^{-9} g.)

Appendix

Answers to Test Your Knowledge Questions

Chapter 1

1. d
2. d
3. b
4. b
5. d
6. c
7. b
8. b
9. a
10. c
11. c
12. c

Chapter 2

1. c
2. b
3. a
4. d
5. c
6. b
7. c
8. d
9. d
10. b
11. d
12. b
13. b
14. d

Chapter 3

1. d
2. b
3. a
4. c
5. d
6. a
7. a
8. c
9. a
10. b
11. e
12. e
13. b
14. a
15. b
16. a

Chapter 4

1. b
2. d
3. d
4. a
5. d
6. e
7. e
8. d
9. d
10. d

Chapter 5

1. b
2. a
3. c
4. e
5. d
6. c
7. a
8. c
9. a
10. d
11. b
12. d
13. b

Chapter 6

1. c
2. b
3. a
4. c
5. b
6. d
7. a
8. a
9. b
10. d
11. b
12. b
13. c
14. d
15. a

Chapter 7

1. c
2. d
3. a
4. a
5. c
6. d
7. d
8. a
9. c
10. c
11. b
12. d
13. d
14. b
15. a
16. c
17. a
18. e

Chapter 8

1. d
2. b
3. e
4. a
5. b
6. e
7. c
8. d
9. b
10. c
11. a
12. b
13. d
14. a
15. b

Chapter 9

1. c
2. c
3. c
4. a
5. c
6. b
7. b
8. e
9. c
10. c
11. b
12. b

Chapter 10

1. d
2. a
3. c
4. d
5. c
6. a
7. c
8. c
9. d
10. c
11. d
12. b
13. b
14. c
15. b
16. c
17. c
18. b

Chapter 11

1. d
2. d
3. e
4. e
5. d
6. a
7. b
8. c
9. d
10. a
11. d
12. c
13. b
14. d
15. c
16. b

Chapter 12

1. b
2. d
3. c
4. b
5. e
6. b
7. a
8. c
9. b
10. b
11. d
12. a
13. e
14. c
15. b
16. d
17. b
18. a

Chapter 13

1. c
2. b
3. e
4. a
5. b
6. c
7. c
8. a
9. d
10. b
11. c
12. d
13. d
14. c
15. c
16. b
17. d

Chapter 14

1. a
2. d
3. c
4. e
5. b
6. c
7. a
8. c
9. d
10. b
11. c
12. d
13. b
14. d
15. c
16. e
17. b
18. d
19. d
20. c

Chapter 15

1. c
2. b
3. d
4. a
5. c
6. d
7. d
8. b
9. e
10. a
11. d
12. a
13. d
14. c
15. b
16. c
17. a
18. d

Chapter 16

1. c
2. d
3. c
4. a
5. c
6. c
7. b
8. a
9. e
10. c
11. a
12. c
13. a
14. d
15. b
16. a
17. c
18. b

Chapter 17

1. b
2. a
3. c
4. b
5. e
6. d
7. d
8. c
9. d
10. a
11. c
12. c
13. a
14. d
15. b
16. e
17. c
18. b

Chapter 18

1. e
2. d
3. b
4. c
5. d
6. b
7. d
8. a
9. d
10. c
11. e
12. d
13. b
14. d
15. d
16. b
17. a
18. a

Chapter 19

1. c
2. b
3. c
4. a
5. b
6. d
7. a
8. d
9. c
10. b
11. b
12. c
13. a
14. d
15. d
16. c
17. d
18. c

Chapter 20

1. b
2. d
3. a
4. c
5. a
6. c
7. d
8. a
9. d
10. a
11. b
12. c
13. d
14. b
15. a
16. e
17. d
18. c

Glossary

Keys to Pronunciation

Most of the words in this glossary are followed by a phonetic spelling that serves as a guide to pronunciation. The phonetic spellings reflect standard scientific usage and can be interpreted easily following a few basic rules.

1. Any unmarked vowel that ends a syllable or that stands alone as a syllable has the long sound. For example, *ba, ma,* and *na* rhyme with *fay; be, de,* and *we* rhyme with *fee; bi, di,* and *pi* rhyme with *sigh; bo, do,* and *mo* rhyme with *go.* Any unmarked vowel that is followed by a consonant has the short sound (for example, the vowel sounds in *hat, met, pit, not,* and *but*).
2. If a long vowel appears in the middle of a syllable (followed by a consonant), it is marked with a macron (¯). Similarly, if a vowel stands alone or ends a syllable but should have a short sound, it is marked with a breve (˘).
3. Syllables that are emphasized are indicated by stress marks. A single stress mark (′) indicates the primary emphasis; a secondary emphasis is indicated by a double stress mark (″).

A

ABO system The most common system of classification for red blood cell antigens. On the basis of antigens on the red blood cell surface, individuals can be type A, type B, type AB, or type O.

absorption (*ab-sorp′shun*) The transport of molecules across epithelial membranes into the body fluids.

accommodation (*ă-kom″ŏ-da′shun*) Adjustment; specifically, the process whereby the focal length of the eye is changed by automatic adjustment of the curvature of the lens to bring images of objects from various distances into focus on the retina.

acetyl (*as′ĭ-tl, ă-sēt′l*) **CoA (Acetyl coenzyme A)** An intermediate molecule in aerobic cell respiration that, together with oxaloacetic acid, begins the Krebs cycle. Acetyl CoA is also an intermediate in the synthesis of fatty acids.

acetylcholine (*ă-sēt″l-ko′lēn*) **(ACh)** An acetic acid ester of choline—a substance that functions as a neurotransmitter chemical in somatic motor nerve and parasympathetic nerve fibers.

acetylcholinesterase (*ă-sēt″l-ko″lĭ-nes′tĕ-rās*) **(AChE)** An enzyme in the membrane of postsynaptic cells that catalyzes the conversion of ACh into choline and acetic acid. This enzymatic reaction inactivates the neurotransmitter.

acidosis (*as″ĭ-do′sis*) An abnormal increase in the H^+ concentration of the blood that lowers arterial pH below 7.35.

acromegaly (*ak″ro-meg′ă-le*) A condition caused by hypersecretion of growth hormone from the pituitary after maturity and characterized by enlargement of the extremities, such as the nose, jaws, fingers, and toes.

acrosome (*ak′ro-som*) The membranous cap on the head of a sperm that contains digestive enzymes. During fertilization, an acrosomal reaction releases these enzymes so that the sperm can tunnel through the zona pellucida and reach the plasma membrane of the oocyte.

ACTH **adrenocorticotropic** (*ă-dre″no-kor″tă-ko-trop′ik*) **hormone.** A hormone secreted by the anterior pituitary that stimulates the adrenal cortex.

actin (*ak′tin*) A structural protein of muscle that, along with myosin, is responsible for muscle contraction.

action potential An all-or-none electrical event in an axon or muscle fiber in which the polarity of the membrane potential is rapidly reversed and reestablished.

active immunity Immunity involving sensitization, in which antibody production is stimulated by prior exposure to an antigen.

active transport The movement of molecules or ions across the cell membranes of epithelial cells by membrane carriers. An expenditure of cellular energy (ATP) is required.

adaptive immunity Also called specific immunity; the ability of lymphocytes to react to specific molecular targets, or antigens, and to respond to these targets more effectively after prior exposure to them.

adaptive thermogenesis (*ther″mo-jen′e-sis*) The heat energy expended in response to changes in ambient temperature and during the digestion and absorption of food.

adenohypophysis (*ad″n-o-hi-pof′ĭ-sis*) The anterior, glandular lobe of the pituitary gland that secretes FSH (follicle-stimulating hormone), LH (luteinizing hormone), ACTH (adrenocorticotropic hormone), TSH (thyroid-stimulating hormone), GH (growth hormone), and prolactin. Secretions of the anterior pituitary are controlled by hormones secreted by the hypothalamus.

adenylate cyclase (*ă-den′l-it si′klās*) An enzyme found in cell membranes that catalyzes the conversion of ATP to cyclic AMP and pyrophosphate (PP_1). This enzyme is activated by an interaction between a specific hormone and its membrane receptor protein.

ADH **antidiuretic** (*an″te-di″yŭ-ret′ik*) **hormone,** also known as *vasopressin.* A hormone produced by the hypothalamus and released from the posterior pituitary. It acts on the kidneys to promote water reabsorption, thus decreasing the urine volume.

adipokines (*ad′i-po-kins″*) Hormones (including leptin, adiponectin, resistin, and others) secreted by adipose cells.

adiponectin (*ad-i-po-nek′tin*) A hormone secreted by adipose tissue that stimulates glucose utilization and fatty acid oxidation. Its secretion is decreased in obesity and type 2 diabetes mellitus.

adipose (*ad′ĭ-pōs*) **tissue** Fatty tissue. A type of connective tissue consisting of fat cells in a loose connective tissue matrix.

ADP **adenosine diphosphate** (*ă-den′ŏ-sēn di-fos′fāt*). A molecule that, together with inorganic phosphate, is used to make ATP (adenosine triphosphate).

adrenal cortex (*ă-dre′nal kor′teks*) The outer part of the adrenal gland. Derived from embryonic mesoderm, the adrenal cortex secretes corticosteroid hormones, including aldosterone and hydrocortisone.

adrenal medulla (*mĕdul′ă*) The inner part of the adrenal gland. Derived from embryonic postganglionic sympathetic neurons, the adrenal medulla secretes catecholamine hormones—epinephrine and (to a lesser degree) norepinephrine.

adrenergic (*ad″rĕ-ner′jik*) Denoting the actions of epinephrine, norepinephrine, or other molecules with similar activity (as in *adrenergic receptor* and *adrenergic stimulation*).

aerobic (*ă-ro′bik*) **capacity** The ability of an organ to utilize oxygen and respire aerobically to meet its energy needs.

aerobic (*ă-ro′bik*) **cell respiration** The metabolic pathways that provide electrons for the electron transport system in the mitochondrial cristae, and provide protons for the formation of ATP by oxidative phosphorylation in the mitochondria. Cells obtain most of their ATP by aerobic cell respiration.

afferent (*af′er-ent*) Conveying or transmitting inward, toward a center. Afferent neurons, for example, conduct impulses toward the central nervous system; afferent arterioles carry blood toward the glomerulus.

afterload Related to the total peripheral resistance and arterial pressure, the afterload presents an impedance to the ejection of blood from the ventricles at systole.
agglutinate (*ă-gloot´n-āt*) A clumping of cells (usually erythrocytes) as a result of specific chemical interaction between surface antigens and antibodies.
agonist (*ag´on-ist*) Regarding muscles, an agonist is a muscle that performs a particular action that is opposed by one or more antagonistic muscles. Regarding regulatory molecules and drugs, an agonist is a molecule that specifically binds to and activates a particular receptor protein; antagonistic molecules may block the ability of the agonist to activate the receptor protein.
agranular leukocytes (*a-gran´yŭ-lar loo´kŏ-sīts*) White blood cells (leukocytes) with cytoplasmic granules that are too small to be clearly visible; specifically, lymphocytes and monocytes.
albumin (*al-byoo´min*) A water-soluble protein produced in the liver; the major component of the plasma proteins.
aldosterone (*al dos´ter-ōn*) The principal corticosteroid hormone involved in the regulation of electrolyte balance (mineralocorticoid).
alkalosis (*al″kă-lo´sis*) An abnormally high alkalinity of the blood and body fluids (blood pH > 7.45).
allergen (*al´-er-jen*) An antigen that evokes an allergic response rather than a normal immune response.
allergy (*al´er-je*) A state of hypersensitivity caused by exposure to allergens. It results in the liberation of histamine and other molecules with histamine-like effects.
all-or-none law The statement that a given response will be produced to its maximum extent in response to any stimulus equal to or greater than a threshold value. Action potentials obey an all-or-none law.
allosteric (*al″o-ster´ik*) Denoting the alteration of an enzyme's activity by its combination with a regulator molecule. Allosteric inhibition by an end product represents negative feedback control of an enzyme's activity.
alpha motoneuron (*al´fă mo″tŏ-noor´on*) The type of somatic motor neuron that stimulates extrafusal skeletal muscle fibers.
alveoli (*al-ve´ŏ-li*); sing., *alveolus.* Small, saclike dilations (as in *lung alveoli*).
amniocentesis (*am″ne-o-sen-te´sis*) A procedure for obtaining amniotic fluid and fetal cells in this fluid through transabdominal perforation of the uterus.
amnion (*am´ne-on*) A developmental membrane surrounding the fetus that contains amniotic fluid; commonly called the "bag of waters."
amoeboid (*ah-me´boid*) **movement** Movement of a cell from place to place using pseudopods, the way an amoeba (a single-celled animal) moves.
amphoteric (*am-fo-ter´ik*) Having both acidic and basic characteristics; used to denote a molecule that can be positively or negatively charged, depending on the pH of its environment.
amylase (*am´il-ās*) A digestive enzyme that hydrolyzes the bonds between glucose subunits in starch and glycogen. Salivary amylase is found in saliva and pancreatic amylase is found in pancreatic juice.
an- (Gk.) Without; not.
anabolic steroids (*an″ă-bol´ik ster´oidz*) Steroids with androgen-like stimulatory effects on protein synthesis.
anabolism (*ă-nab´ŏ-liz″em*) Chemical reactions within cells that result in the production of larger molecules from smaller ones; specifically, the synthesis of protein, glycogen, and fat.
anaerobic (*an-a-ro´bik*) **metabolism** The metabolic pathway that provides energy for cells in the absence of oxygen. In animal cells, this refers to lactic acid fermentation, the formation of lactic acid from glucose. In the process, two ATP are formed per glucose molecule.
anaerobic threshold The maximum rate of oxygen consumption that can be attained before a significant amount of lactic acid is produced by the exercising skeletal muscles through anaerobic respiration. This generally occurs when about 60% of the person's total maximal oxygen uptake has been reached.
anaphylaxis (*an″ă-fĭ-lak´sis*) An unusually severe allergic reaction that can result in cardiovascular shock and death.
androgen (*an´drŏ-jen*) A steroid hormone that controls the development and maintenance of masculine characteristics; primarily testosterone secreted by the testes, although weaker androgens are secreted by the adrenal cortex.
androgen-binding protein (ABP) A protein produced by the Sertoli cells and released into the lumen of the seminiferous tubules. It is believed to concentrate testosterone in the lumen of these tubules.
anemia (*ă-ne´me-ă*) An abnormal reduction in the red blood cell count, hemoglobin concentration, or hematocrit, or any combination of these measurements. This condition is associated with a decreased ability of the blood to carry oxygen.
angina pectoris (*an-ji´nă pek´tŏ-ris*) A thoracic pain, often referred to the left pectoral and arm area, caused by myocardial ischemia.
angiogenesis (*an″je-o-jen´e-sis*) The growth of new blood vessels.
angiotensin II (*an″je-o-ten´sin*) An eight-amino-acid polypeptide formed from angiotensin I (a ten-amino-acid precursor), which in turn is formed from the cleavage of a protein (angiotensinogen) by the action of renin, an enzyme secreted by the kidneys. Angiotensin II is a powerful vasoconstrictor and a stimulator of aldosterone secretion from the adrenal cortex.
anion (*an´i-on*) An ion that is negatively charged, such as chloride, bicarbonate, or phosphate.
antagonistic effects Actions of regulators such as hormones or nerves that counteract the effects of other regulators. The actions of sympathetic and parasympathetic neurons on the heart, for example, are antagonistic.
anterior pituitary (*pĭ-too´ĭ-ter-e*) *See* adenohypophysis.
antibodies (*an´tĭ-bod″ēz*) Immunoglobulin proteins secreted by B lymphocytes that have been transformed into plasma cells. Antibodies are responsible for humoral immunity. Their synthesis is induced by specific antigens, and they combine with these specific antigens but not with unrelated antigens.
anticoagulant (*an″te-ko-ag´yŭ-lant*) A substance that inhibits blood clotting.
anticodon (*an″te-ko´don*) A base triplet provided by three nucleotides within a loop of transfer RNA that is complementary in its base-pairing properties to a triplet (the codon in mRNA). The matching of codon to anticodon provides the mechanism for translation of the genetic code into a specific sequence of amino acids.
antigen (*an´tĭ-jen*) A molecule able to induce the production of antibodies and to react in a specific manner with antibodies.
antigenic (*an-tĭ-jen´ik*) **determinant site** The region of an antigen molecule that specifically reacts with particular antibodies. A large antigen molecule may have a number of such sites.
antioxidants Molecules that scavenge free radicals, thereby relieving the oxidative stress on the body.
antiport (*an´-tĭ-port*) A form of secondary active transport (coupled transport) in which a molecule or ion is moved together with, but in the opposite direction to, Na^+ ions; that is, out of the cell; also called *countertransport.*
antiserum (*an´tĭ se″rum*) A serum containing antibodies that are specific for one or more antigens.
aphasia (*ă-fa´ze-ă*) Absent or defective speech, writing, or comprehension of written or spoken language caused by brain damage or disease. Broca's area, Wernicke's area, the arcuate fasciculus, or the angular gyrus may be involved.
apnea (*ap´ne-ă*) The temporary cessation of breathing.
apneustic (*ap-noo´stik*) **center** A collection of neurons in the brain stem that participates in the rhythmic control of breathing.
apolipoprotein (*ap´o-lip-o-pro´ten*) The protein component of plasma lipoproteins, such as chylomicrons, LDL, and HDL.
apoptosis (*ap″ŏ-to´sis*) Cellular death in which the cells show characteristic histological changes. It occurs as part of programmed cell death and other events in which cell death is a physiological response.
aquaporins (*ă-kwă-por´inz*) The protein channels in a cell (plasma) membrane that permit osmosis to occur across the membrane. In certain tissues, particularly the collecting ducts of the kidney, aquaporins are inserted into the cell membrane in response to stimulation by antidiuretic hormone.
aqueous humor (*a´kwe-us*) A fluid produced by the ciliary body that fills the anterior and posterior chambers of the eye.
arteriosclerosis (*ar-tir″e-o-sklĕ-ro´sis*) Any of a group of diseases characterized by thickening and hardening of the artery wall and narrowing of its lumen.
arteriovenous anastomosis (*ar-tir″e-o-ve´nus ă-nas″tŏ-mo´sis*) A direct connection between an artery and a vein that bypasses the capillary bed.
artery (*ar´tĕ-re*) A vessel that carries blood away from the heart.
astigmatism (*ă-stig´mă-tiz″em*) Unequal curvature of the refractive surfaces of the eye (cornea and/or lens), so that light that enters the eye

along certain meridians does not focus on the retina.

atherosclerosis (*ath″ĕ-ro-sklĕ-ro′sis*) A common type of arteriosclerosis in which raised areas, or plaques, within the tunica interna of medium and large arteries are formed from smooth muscle cells, cholesterol, and other lipids. These plaques occlude the arteries and serve as sites for the formation of thrombi.

atomic number A whole number representing the number of positively charged protons in the nucleus of an atom.

atopic dermatitis (*ă-top′ik der″mă-ti′tis*) An allergic skin reaction to agents such as poison ivy and poison oak; a type of delayed hypersensitivity.

ATP adenosine triphosphate (*ăden′ŏ-sēn tri-fos′fāt*). The universal energy carrier of the cell.

atretic (*ă-tret′ik*) Without an opening. Atretic ovarian follicles are those that fail to rupture and release an oocyte.

atrial natriuretic (*a′tre-al na″trĭ-yoo-ret′ik*) **peptide (ANP)** A chemical secreted by the atria that acts as a natriuretic hormone (a hormone that promotes the urinary excretion of sodium).

atrioventricular node (*a″tre-o-ven-trik′yŭ-lar nōd*) A specialized mass of conducting tissue located in the right atrium near the junction of the interventricular septum. It transmits the impulse into the bundle of His; also called the *AV node.*

atrioventricular valves One-way valves located between the atria and ventricles. The AV valve on the right side of the heart is the tricuspid, and the AV valve on the left side is the bicuspid, or mitral, valve.

atrophy (*at′rŏfe*) A gradual wasting away, or decrease in mass and size of an organ; the opposite of hypertrophy.

atropine (*at′rŏ-pēn*) An alkaloid drug, obtained from a plant of the species *Belladonna,* that acts as an anticholinergic agent. It is used medically to inhibit parasympathetic nerve effects, dilate the pupil of the eye, increase the heart rate, and inhibit intestinal movements.

autoantibody (*aw″to-an′tĭ-bod″e*) An antibody that is formed in response to, and that reacts with, molecules that are part of one's own body.

autocrine (*aw′tŏ-krin*) **regulation** A type of regulation in which one part of an organ releases chemicals that help to regulate another part of the same organ. Prostaglandins, for example, are autocrine regulators.

autonomic (*aw″tŏ-nom′ik*) **nervous system** The part of the nervous system that involves control of smooth muscle, cardiac muscle, and glands. The autonomic nervous system is subdivided into the sympathetic and parasympathetic divisions.

autoregulation (*aw″to-reg′yŭ la′shun*) The ability of an organ to intrinsically modify the degree of constriction or dilation of its small arteries and arterioles, and thus to regulate the rate of its own blood flow. Autoregulation may occur through myogenic or metabolic mechanisms.

autosomal chromosomes (*aw″to-so′mal kro′mŏ-sōmz*) The paired chromosomes; those other than the sex chromosomes.

axon (*ak′son*) The process of a nerve cell that conducts impulses away from the cell body.

axonal (*ak′sŏ-nal, ak-son′al*) **transport** The transport of materials through the axon of a neuron. This usually occurs from the cell body to the end of the axon, but retrograde (backward) transport can also occur.

B

baroreceptors (*bar″o-re-sep′torz*) Receptors for arterial blood pressure located in the aortic arch and the carotid sinuses.

Barr body A microscopic structure in the cell nucleus produced from an inactive X chromosome in females.

basal ganglia (*ba′sal gang′gle-ă*) Gray matter, or nuclei, within the cerebral hemispheres, forming the corpus striatum, amygdaloid nucleus, and claustrum.

basal metabolic (*ba′sal met′ă-bol′ik*) **rate (BMR)** The rate of metabolism (expressed as oxygen consumption or heat production) under resting or basal conditions 8 to 12 hours after eating.

basophil (*ba′sŏ-fil*) The rarest type of leukocyte; a granular leukocyte with an affinity for blue stain in the standard staining procedure.

B cell lymphocytes (*lim′fŏ-sīts*) Lymphocytes that can be transformed by antigens into plasma cells that secrete antibodies (and are thus responsible for humoral immunity). The *B* stands for *bursa equivalent,* which is believed to be the bone marrow.

benign (*bĭ-nīn′*) Not malignant or life threatening.

bi- (L.) Two, twice.

bile (*bīl*) Fluid produced by the liver and stored in the gallbladder that contains bile salts, bile pigments, cholesterol, and other molecules. The bile is secreted into the small intestine.

bile salts Salts of derivatives of cholesterol in bile that are polar on one end and nonpolar on the other end of the molecule. Bile salts have detergent or surfactant effects and act to emulsify fat in the lumen of the small intestine.

bilirubin (*bil″ĭ-roo′bin*) Bile pigment derived from the breakdown of the heme portion of hemoglobin.

blastocyst (*blas′tŏ-sist*) The stage of early embryonic development that consists of an inner cell mass, which will become the embryo, and surrounding trophoblast cells, which will form part of the placenta. This is the form of the embryo that implants in the endometrium of the uterus beginning at about the fifth day following fertilization.

blood-brain barrier The structures and cells that selectively prevent particular molecules in the plasma from entering the central nervous system.

blood-testis barrier The barrier formed by Sertoli cells around the seminiferous tubules, which separates the antigens in the spermatogenic cells from the immune system in the blood.

Bohr effect The effect of blood pH on the dissociation of oxyhemoglobin. Dissociation is promoted by a decrease in the pH.

Boyle's law The statement that the pressure of a given quantity of a gas is inversely proportional to its volume.

bradycardia (*brad″ĭ-kar′de-ă*) A slow cardiac rate; less than sixty beats per minute.

bradykinin (*brad″ĭ-ki′nin*) A short polypeptide that stimulates vasodilation and other cardiovascular changes.

bronchiole (*brong′ke-ōl*) the smallest of the air passages in the lungs, containing smooth muscle and cuboidal epithelial cells.

brown fat A type of fat most abundant at birth that provides a unique source of heat energy for infants, protecting them against hypothermia.

brush border enzymes Digestive enzymes that are located in the cell membrane of the microvilli of intestinal epithelial cells.

buffer A molecule that serves to prevent large changes in pH by either combining with H^+ or by releasing H^+ into solution.

bulk transport Transport of materials into a cell by endocytosis or phagocytosis, and out of a cell by exocytosis.

bundle of His (*hiss*) A band of rapidly conducting cardiac fibers originating in the AV node and extending down the atrioventricular septum to the apex of the heart. This tissue conducts action potentials from the atria into the ventricles.

C

cable properties A term that refers to the ability of neurons to conduct an electrical current. This occurs, for example, between nodes of Ranvier, where action potentials are produced in a myelinated fiber.

calcitonin (*kal″sĭ-to′nin*) Also called *thyrocalcitonin.* A polypeptide hormone produced by the parafollicular cells of the thyroid and secreted in response to hypercalcemia. It acts to lower blood calcium and phosphate concentrations and may serve as an antagonist of parathyroid hormone.

calcium release channels These are proteins in the membrane of the sarcoplasmic reticulum that allow Ca^{2+} to diffuse down its concentration gradient, out of the cisternae into the cytoplasm.

calmodulin (*kal″mod′yŭ-lin*) A receptor protein for Ca^{2+} located within the cytoplasm of target cells. It appears to mediate the effects of this ion on cellular activities.

calorie (*kal′ŏ-re*) A unit of heat equal to the amount of heat needed to raise the temperature of 1 gram of water by 1° C.

cAMP cyclic adenosine monophosphate (*ă-den′ŏ-sēn mon″o-fos′fāt*) A second messenger in the action of many hormones, including catecholamine, polypeptide, and glycoprotein hormones. It serves to mediate the effects of these hormones on their target cells.

cancer A tumor characterized by abnormally rapid cell division and the loss of specialized tissue characteristics. This term usually refers to malignant tumors.

capacitation (*kă-pas″ĭ-ta′shun*) Changes that occur within spermatozoa in the female reproductive tract that enable them to fertilize ova. Sperm that have not been capacitated in the female tract cannot fertilize ova.

capillary (*kap′ĭlar″e*) The smallest vessel in the vascular system. Capillary walls are only one cell thick, and all exchanges of molecules between the blood and tissue fluid occur across the capillary wall.

capsaicin (*kap-sa´ĭ-sin*) **receptor** Both an ion channel in cutaneous sensory dendrites and a receptor for capsaicin—the molecule in chili peppers that causes sensations of heat and pain. In response to a noxiously high temperature, or to capsaicin in chili peppers, this ion channel opens, resulting in the perception of heat and pain.

carbohydrate (*kar"bo-hi´drāt*) An organic molecule containing carbon, hydrogen, and oxygen in a ratio of 1:2:1. The carbohydrate class of molecules is subdivided into monosaccharides, disaccharides, and polysaccharides.

carbonic anhydrase (*kar-bon´ik an-hi´drās*) An enzyme that catalyzes the formation or breakdown of carbonic acid. When carbon dioxide concentrations are relatively high, this enzyme catalyzes the formation of carbonic acid from CO_2 and H_2O. When carbon dioxide concentrations are low, the breakdown of carbonic acid to CO_2 and H_2O is catalyzed. These reactions aid the transport of carbon dioxide from tissues to alveolar air.

carboxyhemoglobin (*kar-bok"se-he"mŏglo´bin*) An abnormal form of hemoglobin in which the heme is bound to carbon monoxide.

cardiac cycle The repeating pattern of contraction (systole) and relaxation (diastole) of the atria and ventricles of the heart. The chambers fill with blood during diastole and eject blood at systole.

cardiac (*kar´de-ak*) **muscle** Muscle of the heart, consisting of striated muscle cells. These cells are interconnected, forming a mass called the myocardium.

cardiac output The volume of blood pumped by either the right or the left ventricle each minute.

cardiogenic (*kar"de-o-jen´ik*) **shock** Shock that results from low cardiac output in heart disease.

carrier-mediated transport The transport of molecules or ions across a cell membrane by means of specific protein carriers. It includes both facilitated diffusion and active transport.

cast An accumulation of proteins molded from the kidney tubules that appear in urine sediment.

catabolism (*kă-tab´ŏ-liz-em*) Chemical reactions in a cell whereby larger, more complex molecules are converted into smaller molecules.

catalyst (*kata´-ă-list*) A substance that increases the rate of a chemical reaction without changing the nature of the reaction or being changed by the reaction.

catecholamines (*kat"ĕ-kol´ă-mēnz*) A group of molecules that includes epinephrine, norepinephrine, L-dopa, and related molecules. The effects of catecholamines are similar to those produced by activation of the sympathetic nervous system.

cations (*kat´i-ions*) Positively charged ions, such as sodium, potassium, calcium, and magnesium.

cell-mediated immunity Immunological defense provided by T cell lymphocytes that come into close proximity with their victim cells (as opposed to humoral immunity provided by the secretion of antibodies by plasma cells).

cellular respiration (*sel´yŭ-lar res"pĭ-ra´shun*) The energy-releasing metabolic pathways in a cell that oxidize organic molecules such as glucose and fatty acids, and that use oxygen as a final electron acceptor.

centriole (*sen´tre-ōl*) The cell organelle that forms the spindle apparatus during cell division.

centromere (*sen´trŏ-mēr*) The central region of a chromosome to which the chromosomal arms are attached.

cerebellum (*ser"ĕ-bel´um*) A part of the metencephalon of the brain that serves as a major center of control in the extrapyramidal motor system.

cerebral lateralization (*ser´ĕ-bral lat"er-al-ĭ-za´shun*) The specialization of function of each cerebral hemisphere. Language ability, for example, is lateralized to the left hemisphere in most people.

chemiosmotic (*kem-e"os-mot´ik*) **theory** The theory that oxidative phosphorylation within mitochondria is driven by the development of a H^+ gradient across the inner mitochondrial membrane.

chemoreceptor (*ke"mo-re-sep´tor*) A neural receptor that is sensitive to chemical changes in blood and other body fluids.

chemotaxis (*ke"mo-tak´sis*) The movement of an organism or a cell, such as a leukocyte, toward a chemical stimulus.

Cheyne-Stokes (*chān´stōks*) **respiration** Breathing characterized by rhythmic waxing and waning of the depth of respiration, with regularly occurring periods of apnea (failure to breathe).

chief cells The cells in the gastric glands that secrete pepsinogen. Pepsinogen is an inactive enzyme (zymogen) that is converted within the gastric lumen to the active enzyme, pepsin. Also called *zymogenic cells.*

chloride (*klor´īd*) **shift** The diffusion of Cl^- into red blood cells as HCO_3^- diffuses out of the cells. This occurs in tissue capillaries as a result of the production of carbonic acid from carbon dioxide.

cholecystokinin (*ko"lĭ-sis"to-ki´nin*) **(CCK)** A hormone secreted by the duodenum that acts to stimulate contraction of the gallbladder and to promote the secretion of pancreatic juice.

cholesterol (*kŏ-les´ter-ol*) A twenty-seven-carbon steroid that serves as the precursor for steroid hormones.

cholinergic (*ko"lĭ-ner´jik*) Denoting nerve endings that, when stimulated, release acetylcholine as a neurotransmitter, such as those of the parasympathetic system.

chondrocyte (*kon´dro-sīt*) A cartilage-forming cell.

chorea (*kŏ-re´ă*) The occurrence of a wide variety of rapid, complex, jerky movements that appear to be well coordinated but that are performed involuntarily.

chromatids (*kro´mă-tidz*) Duplicated chromosomes, joined together at the centromere, that separate during cell division.

chromatin (*kro´mă-tin*) Threadlike structures in the cell nucleus consisting primarily of DNA and protein. They represent the extended form of chromosomes during interphase.

chromosome (*kro´mŏ-sōm*) A structure in the cell nucleus, containing DNA and associated proteins, as well as RNA, that is made according to the genetic instructions in the DNA. Chromosomes are in a compact form during cell division; hence, they become visible as discrete structures in the light microscope at this time.

chylomicron (*ki"lo-mi´kron*) A particle of lipids and protein secreted by the intestinal epithelial cells into the lymph and transported by the lymphatic system to the blood.

chyme (*kīm*) A mixture of partially digested food and digestive juices that passes from the pylorus of the stomach into the duodenum.

cilia (*sil´e-ă*); sing., *cilium.* Tiny hairlike processes extending from the cell surface that beat in a coordinated fashion.

circadian (*ser"kă-de´an*) **rhythms** Physiological changes that repeat at approximately 24-hour periods. They are often synchronized to changes in the external environment, such as the day-night cycles.

cirrhosis (*sĭ-ro´sis*) Liver disease characterized by the loss of normal microscopic structure, which is replaced by fibrosis and nodular regeneration.

clonal (*klōn´al*) **selection theory** The theory in immunology that active immunity is produced by the development of clones of lymphocytes able to respond to a particular antigen.

clone (*klōn*) A group of cells derived from a single parent cell by mitotic cell division. Clone formation in lymphocytes occurs during active immunity.

CNS **central nervous system** That part of the nervous system consisting of the brain and spinal cord.

cochlea (*kok´le-ă*) The organ of hearing in the inner ear where nerve impulses are generated in response to sound waves.

codon (*ko´don*) The sequence of three nucleotide bases in mRNA that specifies a given amino acid and determines the position of that amino acid in a polypeptide chain through complementary base pairing with an anticodon in transfer RNA.

coenzyme (*ko-en´zīm*) An organic molecule, usually derived from a water-soluble vitamin, that combines with and activates specific enzyme proteins.

cofactor (*ko´fac-tor*) A substance needed for the catalytic action of an enzyme; generally used in reference to inorganic ions such as Ca^{2+} and Mg^{2+}.

colloid osmotic (*kol´oid oz-mot´ik*) **pressure** Osmotic pressure exerted by plasma proteins that are present as a colloidal suspension; also called *oncotic pressure.*

complement system A system of plasma proteins that, as part of the innate immune system, can be activated by the binding of antibodies to antigens on the membrane of a pathogenic cell. Once activated, different complement proteins serve to attract and activate phagocytic cells and to form a pore that promotes lysis of the pathogen.

compliance (*kom-pli´ans*) **1.** A measure of the ease with which a structure such as the lung expands under pressure. **2.** A measure of the change in volume as a function of pressure changes.

conducting zone The structures and airways that transmit inspired air into the respiratory zone of the lungs where gas exchange occurs. The conducting zone includes such structures as the trachea, bronchi, and larger bronchioles.

cone Photoreceptor in the retina of the eye that provides color vision and high visual acuity.

congestive (*kon-jes´tiv*) **heart failure** The inability of the heart to deliver an adequate blood flow because of heart disease or hypertension. It is associated with breathlessness, salt and water retention, and edema.

connective tissue One of the four primary tissues, characterized by an abundance of extracellular material.

Conn's syndrome Primary hyperaldosteronism in which excessive secretion of aldosterone produces electrolyte imbalances.

contractility (*kon-trak-til´i-te*) The ability of cardiac muscle to contract and shorten at any particular fiber length, independent of the preload. Contractility is increased by epinephrine.

contralateral (*kon″trā-lat´er-al*) Taking place or originating in a corresponding part on the opposite side of the body.

cornea (*kor´ne-ă*) The transparent structure forming the anterior part of the connective tissue covering of the eye.

corpora quadrigemina (*kor´por-ă kwad″rĭ-jem´ĭ-na*) A region of the mesencephalon consisting of the superior and inferior colliculi. The superior colliculi are centers for the control of visual reflexes; the inferior colliculi are centers for the control of auditory reflexes.

corpus callosum (*kor´pus kă-lo´sum*) A large transverse tract of nerve fibers connecting the cerebral hemispheres.

corpus (*kor´pus*) **luteum** (lu´teum) A yellowish gland in the ovary formed from a mature graafian follicle after it has ovulated. The corpus luteum secretes progesterone as well as estradiol.

cortex (*kor´teks*) **1.** The outer layer of an internal organ or body structure, as of the kidney or adrenal gland. **2.** The convoluted layer of gray matter that covers the surface of the cerebral hemispheres.

corticosteroid (*kor″tĭ-ko-ster´oid*) Any of a class of steroid hormones of the adrenal cortex, consisting of glucocorticoids (such as hydrocortisone) and mineralocorticoids (such as aldosterone).

cotransport Also called *coupled transport* or *secondary active transport.* Carrier-mediated transport in which a single carrier transports an ion (e.g., Na^+) down its concentration gradient while transporting a specific molecule (e.g., glucose) against its concentration gradient. The hydrolysis of ATP is indirectly required for cotransport because it is needed to maintain the steep concentration gradient of the ion.

countercurrent exchange The process that occurs in the vasa recta of the renal medulla in which blood flows in U-shaped loops. This allows sodium chloride to be trapped in the interstitial fluid while water is carried away from the kidneys.

countercurrent multiplier system The interaction that occurs between the descending limb and the ascending limb of the loop of Henle in the kidney. This interaction results in the multiplication of the solute concentration in the interstitial fluid of the renal medulla.

creatine phosphate (*kre´ă-tin fos´fāt*) An organic phosphate molecule in muscle cells that serves as a source of high-energy phosphate for the synthesis of ATP; also called *phosphocreatine.*

creatinine (*kre-at´i-nen*) A molecule formed from muscle creatine that is maintained at a normal plasma concentration by the kidneys, which filter the creatinine into the urine. Because the creatinine is not reabsorbed and is only slightly secreted, its renal plasma clearance is often used as a measure of the glomerular filtration rate (GFR).

CREB An acronym for *cyclic AMP response element binding protein.* CREB is a transcription factor that activates genes important for learning and long-term potentiation.

crenation (*krĭ-na´shun*) A notched or scalloped appearance of the red blood cell membrane caused by the osmotic loss of water from the cells.

cretinism (*krēt´n-iz″em*) A condition caused by insufficient thyroid secretion during prenatal development or the years of early childhood. It results in stunted growth and inadequate mental development.

cryptorchidism (*krip-tor´kĭ-diz″em*) A developmental defect in which the testes fail to descend into the scrotum, and instead remain in the body cavity.

curare (*koo-ră-re*) A chemical derived from plant sources that causes flaccid paralysis by blocking ACh receptor proteins in muscle cell membranes.

Cushing's syndrome Symptoms caused by hypersecretion of adrenal steroid hormones as a result of tumors of the adrenal cortex or ACTH-secreting tumors of the anterior pituitary.

cyanosis (*si´ă-no″sis*) A bluish discoloration of the skin or mucous membranes due to excessive concentration of deoxyhemoglobin; indicative of inadequate oxygen concentration in the blood.

cyclins (*si´klinz*) A group of proteins that promote different phases of the cell cycle by activating enzymes called cyclin-dependant kinases.

cytochrome (*si´tŏ-krōm*) A pigment in mitochondria that transports electrons in the process of aerobic respiration.

cytochrome P450 enzymes Enzymes of a particular kind, not related to the mitochondrial cytochromes, that metabolize a broad spectrum of biological molecules, including steroid hormones and toxic drugs. They are prominent in the liver where they help in detoxication of the blood.

cytokine (*si´to-kīn*) An autocrine or paracrine regulator secreted by various tissues.

cytokinesis (*si″to-kĭ-ne´sis*) The division of the cytoplasm that occurs in mitosis and meiosis when a parent cell divides to produce two daughter cells.

cytoplasm (*si´tŏ-plaz″em*) The semifluid part of the cell between the cell membrane and the nucleus, exclusive of membrane-bound organelles. It contains many enzymes and structural proteins.

cytoskeleton (*si″to-skel´ĕ-ton*) A latticework of structural proteins in the cytoplasm arranged in the form of microfilaments and microtubules.

D

Dalton's law The statement that the total pressure of a gas mixture is equal to the sum that each individual gas in the mixture would exert independently. The part contributed by each gas is known as the partial pressure of the gas.

dark adaptation The ability of the eyes to increase their sensitivity to low light levels over a period of time. Part of this adaptation involves increased amounts of visual pigment in the photoreceptors.

dark current The steady inward diffusion of Na^+ into the rods and cones when the photoreceptors are in the dark. Stimulation by light causes this dark current to be blocked, and thus hyperpolarizes the photoreceptors.

declarative memory The memory of factual information. This can be contrasted with nondeclarative memory, which is the memory of perceptual and motor skills.

deglutition (*de-gloo-tish´un*) The process of swallowing.

dehydration (*de″hi-dra´shun*) **synthesis** The bonding together of subunits to form a longer molecule, in a reaction that also results in the production of a molecule of water.

delayed hypersensitivity An allergic response in which the onset of symptoms may not occur until 2 or 3 days after exposure to an antigen. Produced by T cells, it is a type of cell-mediated immunity.

dendrite (*den´drīt*) A relatively short, highly branched neural process that carries electrical activity to the cell body.

dendritic (*den-drit´ik*) **cells** The most potent antigen-presenting cells for the activation of helper T lymphocytes. The dendritic cells originate in the bone marrow and migrate through the blood and lymph to lymphoid organs and to nonlymphoid organs such as the lungs and skin.

denervation (*de″ner-va´shun*) **hypersensitivity** The increased sensitivity of smooth muscles to neural stimulation after their innervation has been blocked or removed for a period of time.

deoxyhemoglobin (*de-ok″se-he″mō-glo´bin*) The form of hemoglobin in which the heme groups are in the normal reduced form but are not bound to a gas. Deoxyhemoglobin is produced when oxyhemoglobin releases oxygen.

depolarization (*de-po″lar-ĭ-za´shun*) The loss of membrane polarity in which the inside of the cell membrane becomes less negative in comparison to the outside of the membrane. The term is also used to indicate the reversal of membrane polarity that occurs during the production of action potentials in nerve and muscle cells. Also called *hypopolarization.*

deposition (*dep-ŏ-zish´on*), **bone** The formation of the extracellular matrix of bone by osteoblasts. This process includes secretion of collagen and precipitation of calcium phosphate in the form of hydroxyapatite crystals.

detoxication (*de-tok″sĭ-ka´shun*) The reduction of the toxic properties of molecules. This occurs through chemical transformation of the molecules and takes place, to a large degree, in the liver.

diabetes insipidus (*dī″ă-be′tēz in-sip′ĭ-dus*) A condition in which inadequate amounts of antidiuretic hormone (ADH) are secreted by the posterior pituitary. It results in inadequate reabsorption of water by the kidney tubules, and thus in the excretion of a large volume of dilute urine.

diabetes mellitus (*mĕ-li′tus*) The appearance of glucose in the urine due to the presence of high plasma glucose concentrations, even in the fasting state. This disease is caused by either a lack of sufficient insulin secretion or by inadequate responsiveness of the target tissues to the effects of insulin.

dialysis (*di-al′ĭ-sis*) A method of removing unwanted elements from the blood by selective diffusion through a porous membrane.

diapedesis (*dī″ă-pē-de′sis*) The migration of white blood cells through the endothelial walls of blood capillaries into the surrounding connective tissues.

diarrhea (*dī″ă re′ă*) Abnormal frequency of defecation accompanied by abnormal liquidity of the feces.

diastole (*di-as′tŏ-le*) The phase of relaxation in which the heart fills with blood. Unless accompanied by the modifier *atrial,* diastole refers to the resting phase of the ventricles.

diastolic (*dī″ă-stol′ik*) **blood pressure** The minimum pressure in the arteries that is produced during the phase of diastole of the heart. It is indicated by the last sound of Korotkoff when taking a blood pressure measurement.

diffusion (*dĭ-fyoo′zhun*) The net movement of molecules or ions from regions of higher to regions of lower concentration.

digestion The process of converting food into molecules that can be absorbed through the intestine into the blood.

dihydrotestosterone (*di-hi″dro-tes-tos′ter-on*) **(DHT)** A metabolite formed in target cells from testosterone by the enzyme 5α-reductase. DHT is more directly responsible for many of the actions of testosterone on its target organs.

1,25-dihydroxyvitamin (*di″hi-drok″se-vi′tă-min*) **D_3** The active form of vitamin D produced within the body by hydroxylation reactions in the liver and kidneys of vitamin D formed by the skin. This is a hormone that promotes the intestinal absorption of Ca^{2+}.

diploid (*dip′loid*) Denoting cells having two of each chromosome, or twice the number of chromosomes that are present in sperm or ova.

disaccharide (*di-sak′ă-rīd*) Any of a class of double sugars; carbohydrates that yield two simple sugars, or monosaccharides, upon hydrolysis.

diuretic (*di″yŭ-ret′ik*) A substance that increases the rate of urine production, thereby lowering the blood volume.

DNA **deoxyribonucleic** (*de-ok″se-ri″bo-noo-kle′ik*) **acid** A nucleic acid composed of nucleotide bases and deoxyribose sugar that contains the genetic code.

dopa (*do′pă*) **dihydroxyphenylalanine** (*di″hi-drok″se-fen″al-ă-lă-nīn*) An amino acid formed in the liver from tyrosine and converted to dopamine in the brain. L-dopa is used in the treatment of Parkinson's disease to stimulate dopamine production.

dopamine (*do′pă-mēn*) A type of neurotransmitter in the central nervous system; it is also the precursor of norepinephrine, another neurotransmitter molecule.

dopaminergic (*do″pă-mēn-er′jik*) **pathways** Neural pathways in the brain that release dopamine. The nigrostriatal pathway is involved in motor control, whereas the mesolimbic dopamine pathway is involved in mood and emotion.

2,3-DPG 2,3-diphosphoglyceric (*di-fos′fo-glis-er″ik*) acid. A product of red blood cells, 2,3-DPG bonds with the protein component of hemoglobin and increases the ability of oxyhemoglobin to dissociate and release its oxygen.

ductus arteriosus (*duk′tus ar-tir″e-o′sus*) A fetal blood vessel connecting the pulmonary artery directly to the aorta.

dwarfism A condition in which a person is undersized because of inadequate secretion of growth hormone.

dyspnea (*disp-ne′ă*) Subjective difficulty in breathing.

dystrophin (*dis-trof′in*) A protein associated with the sarcolemma of skeletal muscle cells that is produced by the defective gene of people with Duchenne's muscular dystrophy.

E

eccentric (*ek-sen′trik*) **contraction** A muscle contraction in which the muscle lengthens despite its contraction, due to a greater external stretching force applied to it. The contraction in this case can serve a shock absorbing function, as when the quadriceps muscles of the leg contract eccentrically upon landing when a person jumps from a height.

ECG **electrocardiogram** (*ĕ-lek″tro-kar′de-ŏ-gram*) (also abbreviated EKG) A recording of electrical currents produced by the heart.

E. coli (*e ko′li*) A species of bacteria normally found in the human intestine; full name is *Escherichia (esh″ĭ-rik′e-ă) coli.*

ectopic focus An area of the heart other than the SA node that assumes pacemaker activity.

ectopic pregnancy Embryonic development that occurs anywhere other than in the uterus (as in the uterine tubes or body cavity).

edema (*ĕ-de′mă*) Swelling resulting from an increase in tissue fluid.

EEG **electroencephalogram** (*ĕ-lek″tro-en-sef′ă-lŏ-gram*) A recording of the electrical activity of the brain from electrodes placed on the scalp.

effector (*ĕ-fek′tor*) **organs** A collective term for muscles and glands that are activated by motor neurons.

efferent (*ef′er-ent*) Conveying or transporting something away from a central location. Efferent nerve fibers conduct impulses away from the central nervous system, for example, and efferent arterioles transport blood away from the glomerulus.

eicosanoids (*i-ko′să-noidz*) The biologically active derivatives of arachidonic acid, a fatty acid found in cell membranes. The eicosanoids include prostaglandins and leukotrienes.

ejection fraction The ratio of the stroke volume to the end-diastolic volume of a ventricle. This is normally about 60% at rest.

elasticity (*ĕ″las-tis′ĭ-te*) The tendency of a structure to recoil to its initial dimensions after being distended (stretched).

electrolyte (*ĕ-lek′tro-līt*) An ion or molecule that is able to ionize and thus carry an electric current. The most common electrolytes in the plasma are Na^+, HCO_3^-, and K^+.

electrophoresis (*ĕ-lek″tro-fŏ-re′sis*) A biochemical technique in which different molecules can be separated and identified by their rate of movement in an electric field.

element, chemical A substance that cannot be broken down by chemical means into simpler substances. An element is composed of atoms that all have the same atomic number. An element can, however, include different forms of a given atom (isotopes) that have different numbers of neutrons, and thus different atomic weights.

elephantiasis (*el″ĕ-fan-ti′ă-sis*) A disease in which the larvae of a nematode worm block lymphatic drainage and produce edema. The lower areas of the body can become enormously swollen as a result.

EMG **electromyogram** (*ĕ-lek″tro-mi′ĕo-gram*) An electrical recording of the activity of skeletal muscles through the use of surface electrodes.

embryonic stem cells Also called ES cells, these are the cells of the inner cell mass of a blastocyst. Embryonic stem cells are pluripotent, and so are potentially capable of differentiating into all tissue types except the trophoblast cells of a placenta.

emmetropia (*em″ĭ tro′pe ă*) A condition of normal vision in which the image of objects is focused on the retina, as opposed to nearsightedness (myopia) or farsightedness (hyperopia).

emphysema (*em″fĭ-se′mă em″fĭ-ze′mă*) A lung disease in which alveoli are destroyed and the remaining alveoli become larger. It results in decreased vital capacity and increased airway resistance.

emulsification (*ĕ-mul″sĭ-fĭ-ka′shun*) The process of producing an emulsion or fine suspension. In the small intestine, fat globules are emulsified by the detergent action of bile.

end-diastolic (*dī″ă-stol′ik*) **volume** The volume of blood in each ventricle at the end of diastole, immediately before the ventricles contract at systole.

endergonic (*en″der-gon′ik*) Denoting a chemical reaction that requires the input of energy from an external source in order to proceed.

endocannabinoids (*endo-can-ab′in-oids*) Endogenous molecules that act like tetrahydrocannabinol in marijuana, and are believed to serve as retrograde neurotransmitters.

endocrine (*en′dŏ-krin*) **glands** Glands that secrete hormones into the circulation rather than into a duct; also called *ductless glands.*

endocytosis (*en″do-si-to′sis*) The cellular uptake of particles that are too large to cross the cell membrane. This occurs by invagination of the cell membrane until a membrane-enclosed vesicle is pinched off within the cytoplasm.

endoderm (*en´dŏ-derm*) The innermost of the three primary germ layers of an embryo. It gives rise to the digestive tract and associated structures, the respiratory tract, the bladder, and the urethra.

endogenous (*en-doj´ĕ-nus*) Denoting a product or process arising from within the body (as opposed to exogenous products or influences, which arise from external sources).

endolymph (*en´dŏ-limf*) The fluid contained within the membranous labyrinth of the inner ear.

endometrium (*en″do-me´tre-um*) The mucous membrane of the uterus, the thickness and structure of which vary with the phases of the menstrual cycle.

endoplasmic reticulum (*en-do-plaz´mik rĕ-tik´yŭ-lum*) **(ER)** An extensive system of membrane-enclosed cavities within the cytoplasm of the cell. Those with ribosomes on their surface are called rough endoplasmic reticulum and participate in protein synthesis.

endorphin (*en-dor´fin*) Any of a group of endogenous opioid molecules that may act as a natural analgesic.

endothelin (*en″do-the´lin*) A polypeptide secreted by the endothelium of a blood vessel that serves as a paracrine regulator, promoting contraction of the smooth muscle and constriction of the vessel.

endothelium (*en″do-the´le-um*) The simple squamous epithelium that lines blood vessels and the heart.

endotoxin (*en″do-tok´sin*) A toxin found within certain types of bacteria that is able to stimulate the release of endogenous pyrogen and produce a fever.

end-plate potential The graded depolarization produced by ACh at the neuromuscular junction. This is equivalent to the excitatory postsynaptic potential produced at neuron-neuron synapses.

end-product inhibition The inhibition of enzymatic steps of a metabolic pathway by products formed at the end of that pathway.

enkephalin (*en-kef´ă-lin*) Either of two short polypeptides, containing five amino acids, that have analgesic effects and are grouped as endorphins.

enteric (*en-ter´ik*) A term referring to the intestine.

enteric nervous system The neurons found within the wall of the gastrointestinal (GI) tract. These include sensory, association, and motor neurons. They coordinate peristalsis and other GI functions.

enterochromaffin (*en″ter-o-kro″maf´in*) **-like (ECL) cells** Cells of the gastric epithelium that secrete histamine. The ECL cells are stimulated by the hormone gastrin and by the vagus nerve; the histamine from ECL cells, in turn, stimulates gastric acid secretion from the parietal cells.

enterohepatic (*en″ter-o-hĕ-pat´ik*) **circulation** The recirculation of a compound between the liver and small intestine. The compound is present in the bile secreted by the liver into the small intestine. It is then reabsorbed and returned to the liver via the hepatic portal vein.

entropy (*en´trŏ-pe*) The energy of a system that is not available to perform work. A measure of the degree of disorder in a system, entropy increases whenever energy is transformed.

enzyme (*en´zim*) A protein catalyst that increases the rate of specific chemical reactions.

epididymis (*ep″ĭ-did´ĭ-mis*); pl., *epididymides* A tubelike structure outside the testes. Sperm pass from the seminiferous tubules into the head of the epididymis and then pass from the tail of the epididymis to the ductus (vas) deferens. The sperm mature, becoming motile, as they pass through the epididymis.

epinephrine (*ep″ĭ-nef´rin*) A catecholamine hormone secreted by the adrenal medulla in response to sympathetic nerve stimulation. It acts together with norepinephrine released from sympathetic nerve endings to prepare the organism for "fight or flight;" also known as *adrenaline.*

epithelium (*ep″ĭ-the´le-um*) One of the four primary tissue types; the type of tissue that covers and lines body surfaces and forms exocrine and endocrine glands.

EPSP **excitatory postsynaptic** (*pōst″sĭ-nap´tik*) **potential** A graded depolarization of a postsynaptic membrane in response to stimulation by a neurotransmitter chemical. EPSPs can be summated, but they can be transmitted only over short distances; they can stimulate the production of action potentials when a threshold level of depolarization is attained.

equilibrium (*e″kwĭ-lib´re-um*) **potential** The hypothetical membrane potential that would be created if only one ion were able to diffuse across a membrane and reach a stable, or equilibrium, state. In this stable state, the concentrations of the ion would remain constant inside and outside the membrane, and the membrane potential would be equal to a particular value.

erythroblastosis fetalis (*ĕ-rith″ro-blas-to´sis fe-tal´is*) Hemolytic anemia in an Rh-positive newborn caused by maternal antibodies against the Rh factor that have crossed the placenta.

erythrocyte (*ĕ-rith´rŏ-sīt*) A red blood cell. Erythrocytes are the formed elements of blood that contain hemoglobin and transport oxygen.

erythropoietin (*ĕ-rith″ro-poi´ĕ-tin*) A hormone secreted by the kidneys that stimulates the bone marrow to produce red blood cells.

essential amino acids The eight amino acids in adults or nine amino acids in children that cannot be made by the human body; therefore, they must be obtained in the diet.

estradiol (*es″tră-di´ol*) The major estrogen (female sex steroid hormone) secreted by the ovaries.

estrus (*es´trus*) **cycle** Cyclic changes in the structure and function of the ovaries and female reproductive tract, accompanied by periods of "heat" (estrus), or sexual receptivity; the lower mammalian equivalent of the menstrual cycle, but differing from the menstrual cycle in that the endometrium is not shed with accompanying bleeding.

excitation-contraction coupling The means by which electrical excitation of a muscle results in muscle contraction. This coupling is achieved by Ca^{2+}, which enters the muscle cell cytoplasm in response to electrical excitation and which stimulates the events culminating in contraction.

exergonic (*ek″ser-gon´ik*) Denoting chemical reactions that liberate energy.

exocrine (*ek´sŏ-krin*) **gland** A gland that discharges its secretion through a duct to the outside of an epithelial membrane.

exocytosis (*ek″so-si-to´sis*) The process of cellular secretion in which the secretory products are contained within a membrane-enclosed vesicle. The vesicle fuses with the cell membrane so that the lumen of the vesicle is open to the extracellular environment.

exon (*ek´son*) A nucleotide sequence in DNA that codes for the production of messenger RNA.

exteroceptor (*ek″ster-o-cep´tor*) A sensory receptor that is sensitive to changes in the external environment (as opposed to an interoceptor).

extracellular (*eks-tra-sel´u-lar*) **compartment** All of the material outside of cells, including the extracellular fluid with all of its solutes, insoluble protein fibers, and, in some cases, crystals. Also called the *extracellular matrix.*

extracellular fluid The fluid outside of cells, including the blood plasma and the interstitial fluid within the tissues. Also called the *extracellular compartment.*

extrafusal (*eks″tră-fyooz´al*) **fibers** The ordinary muscle fibers within a skeletal muscle; not found within the muscle spindles.

extraocular (*eks″tră-ok´yŭ-lar*) **muscles** The muscles that insert into the sclera of the eye. They act to change the position of the eye in its orbit (as opposed to the intraocular muscles, such as those of the iris and ciliary body within the eye).

extrapyramidal (*eks″tră-pĭ-ram´ĭ-dl*) **motor tracts** Neural pathways that are situated outside of, or that are "independent of," pyramidal tracts. The major extrapyramidal tract is the reticulospinal tract, which originates in the reticular formation of the brain stem and receives excitatory and inhibitory input from both the cerebrum and the cerebellum. The extrapyramidal tracts are thus influenced by activity in the brain involving many synapses, and they appear to be required for fine control of voluntary movements.

F

facilitated (*fă-sil´ĭ-ta″tid*) **diffusion** The carrier-mediated transport of molecules through the cell membrane along the direction of their concentration gradients. It does not require the expenditure of metabolic energy.

FAD **flavin adenine dinucleotide** (*fla´vin ad´n-ēn di-noo´kle-ō-tīd*) A coenzyme derived from riboflavin that participates in electron transport within the mitochondria.

FAS A surface receptor produced by T lymphocytes during an infection. After a few days, the activated T lymphocytes begin to produce another surface molecule, FAS ligand. The bonding of FAS with FAS ligand, on the same or on different cells, triggers apoptosis of the lymphocytes.

feces (*fe´sēz*) The excrement discharged from the large intestine.

fertilization (*fer´tĭ-lĭ-za″shun*) The fusion of an ovum and spermatozoon.

fiber, muscle A skeletal muscle cell.

fiber, nerve An axon of a motor neuron or the dendrite of a pseudounipolar sensory neuron in the PNS.
fibrillation *(fib"rĭ-la´shun)* A condition of cardiac muscle characterized electrically by random and continuously changing patterns of electrical activity and resulting in the inability of the myocardium to contract as a unit and pump blood. It can be fatal if it occurs in the ventricles.
fibrin *(fi´brin)* The insoluble protein formed from fibrinogen by the enzymatic action of thrombin during the process of blood clot formation.
fibrinogen *(fi-brin´ŏ-jen)* A soluble plasma protein that serves as the precursor of fibrin; also called *factor I.*
flaccid paralysis *(flak´sid pă-ral´ĭ-sis)* The inability to contract muscles, resulting in a loss of muscle tone. This may be due to damage to lower motor neurons or to factors that block neuromuscular transmission.
flagellum *(flă-jel´um)* A whiplike structure that provides motility for sperm.
flare-and-wheal reaction A cutaneous reaction to skin injury or to the administration of antigens produced by release of histamine and related molecules; characterized by local edema and a red flare.
flavoprotein *(fla"vo-pro´te-in)* A conjugated protein containing a flavin pigment that is involved in electron transport within the mitochondria.
follicle *(fol´ĭ-k´l)* A microscopic hollow structure within an organ. Follicles are the functional units of the thyroid gland and of the ovary.
foramen ovale *(fŏ-ra´men o-val´e)* An opening normally present in the atrial septum of the fetal heart that allows direct communication between the right and left atria.
fovea centralis *(fo´ve-ă sen-tra´lis)* A tiny pit in the macula lutea of the retina that contains slim, elongated cones. It provides the highest visual acuity (clearest vision).
Frank-Starling law of the heart The statement describing the relationship between end-diastolic volume and stroke volume of the heart. A greater amount of blood in a ventricle prior to contraction results in greater stretch of the myocardium, and by this means produces a contraction of greater strength.
free radical A molecule that contains an atom with an unpaired electron in an orbital that can contain a maximum of two electrons. Free radicals are highly reactive and function in a number of normal and pathological processes.
FSH **follicle-stimulating hormone** One of the two gonadotropic hormones secreted by the anterior pituitary. In females, FSH stimulates the development of the ovarian follicles; in males, it stimulates the production of sperm in the seminiferous tubules.

G

GABA **gamma-aminobutyric** *(gam"ă-ă-me"no-byoo-tir´ik)* **acid** An amino acid believed to function as an inhibitory neurotransmitter in the central nervous system.
gamete *(gam´ēt)* Collective term for haploid germ cells: sperm and ova.
gamma motoneuron *(gam´ă mo"tŏ-noor´on)* The type of somatic motor neuron that stimulates intrafusal fibers within the muscle spindles.
ganglion *(gang´gle-on)* A grouping of nerve cell bodies located outside the brain and spinal cord.
gap junctions Specialized regions of fusion between the cell membranes of two adjacent cells that permit the diffusion of ions and small molecules from one cell to the next. These regions serve as electrical synapses in certain areas, such as in cardiac muscle.
gas exchange The diffusion of oxygen and carbon dioxide down their concentration gradients that occurs between pulmonary capillaries and alveoli, and between systemic capillaries and the surrounding tissue cells.
gastric *(gas´trik)* **intrinsic factor** A glycoprotein secreted by the stomach that is needed for the absorption of vitamin B_{12}.
gastric juice The secretions of the gastric mucosa. Gastric juice contains water, hydrochloric acid, and pepsinogen as major components.
gastrin *(gas´trin)* A hormone secreted by the stomach that stimulates the gastric secretion of hydrochloric acid and pepsin.
gastroileal *(gas´tro-il"e-al)* **reflex** The reflex in which increased gastric activity causes increased motility of the ileum and increased movement of chyme through the ileocecal sphincter.
gates A term used to describe structures within the cell membrane that regulate the passage of ions through membrane channels. Gates may be chemically regulated (by neurotransmitters) or voltage regulated (in which case they open in response to a threshold level of depolarization).
general adaptation syndrome (GAS) The specific response of the body to nonspecific stressors, involving the activation of the hypothalamus-pituitary-adrenal axis.
generator *(jen´ĕ-ra"tor)* **potential** The graded depolarization produced by stimulation of a sensory receptor that results in the production of action potentials by a sensory neuron; also called the *receptor potential.*
genetic *(jĕ-net´ik)* **recombination** The formation of new combinations of genes, as by crossing-over between homologous chromosomes.
genetic transcription The process by which RNA is produced with a sequence of nucleotide bases that is complementary to a region of DNA.
genetic translation The process by which proteins are produced with amino acid sequences specified by the sequence of codons in messenger RNA.
genome *(je´nom)* All of the genes of an individual or in a particular species.
ghrelin *(gre´lin)* A hormone secreted by the stomach. Secretion of ghrelin rises between meals and stimulates centers in the hypothalamus to promote hunger.
gigantism *(ji-gan´tiz"em)* Abnormal body growth due to the excessive secretion of growth hormone.
glomerular *(glo-mer´yŭ-lar)* **filtration rate (GFR)** The volume of blood plasma filtered out of the glomeruli of both kidneys each minute. The GFR is measured by the renal plasma clearance of inulin.
glomerular ultrafiltrate Fluid filtered through the glomerular capillaries into the glomerular (Bowman's) capsule of the kidney tubules.
glomeruli *(glo-mer´yŭ-li)* The tufts of capillaries in the kidneys that filter fluid into the kidney tubules.
glomerulonephritis *(glo-mer"yŭ-lo-nĕ-fri´tis)* Inflammation of the renal glomeruli; associated with fluid retention, edema, hypertension, and the appearance of protein in the urine.
glucagon *(gloo´că-gon)* A polypeptide hormone secreted by the alpha cells of the islets of Langerhans in the pancreas that acts to promote glycogenolysis and raise the blood glucose levels.
glucocorticoid *(gloo"ko-kor´tĭ-koid)* Any of a class of steroid hormones secreted by the adrenal cortex (corticosteroids) that affects the metabolism of glucose, protein, and fat. These hormones also have anti-inflammatory and immunosuppressive effects. The major glucocorticoid in humans is hydrocortisone (cortisol).
gluconeogenesis *(gloo"ko-ne"ŏ-jen´ĭ-sis)* The formation of glucose from noncarbohydrate molecules, such as amino acids, lactic acid, and glycerol.
GLUT An acronym for *glucose transporters.* GLUT proteins promote the facilitated diffusion of glucose into cells. One isoform of GLUT, designated GLUT4, is inserted into the cell membranes of muscle and adipose cells in response to insulin stimulation and exercise.
glutamate *(gloo´tă-māt)* The ionized form of glutamic acid, an amino acid that serves as the major excitatory neurotransmitter of the CNS. *Glutamate* and *glutamic acid* are terms that can be used interchangeably.
glutathione *(gloo"tah-thi´on)* A tripeptide molecule that functions as the major cellular antioxidant.
glycogen *(gli´kŏ-jen)* A polysaccharide of glucose—also called *animal starch*—produced primarily in the liver and skeletal muscles. Similar to plant starch in composition, glycogen contains more highly branched chains of glucose subunits than does plant starch.
glycogenesis *(gli"kŏ-jen´ĭ-sis)* The formation of glycogen from glucose.
glycogenolysis *(gli"ko-jĕ-nol´ĭ-sis)* The hydrolysis of glycogen to glucose-1-phosphate, which can be converted to glucose-6-phosphate. The glucose-6-phosphate then may be oxidized via glycolysis or (in the liver) converted to free glucose.
glycolysis *(gli"kol´ĭ-sis)* The metabolic pathway that converts glucose to pyruvic acid. The final products are two molecules of pyruvic acid and two molecules of reduced NAD, with a net gain of two ATP molecules. In anaerobic metabolism, the reduced NAD is oxidized by the conversion of pyruvic acid to lactic acid. In aerobic respiration, pyruvic acid enters the Krebs cycle in mitochondria, and reduced NAD is ultimately oxidized by oxygen to yield water.
glycosuria *(gli"kŏ-soor´e-ă)* The excretion of an abnormal amount of glucose in the urine (urine normally contains only trace amounts of glucose).

Golgi *(gol′je)* **complex** A network of stacked, flattened membranous sacs within the cytoplasm of cells. Its major function is to concentrate and package proteins within vesicles that bud off from it. Also called *Golgi apparatus.*

Golgi tendon organ A tension receptor in the tendons of muscles that becomes activated by the pull exerted by a muscle on its tendons; also called a *neurotendinous receptor.*

gonad *(go′nad)* A collective term for testes and ovaries.

gonadotropic *(go″nad-ŏ-trop′ik)* **hormones** Hormones of the anterior pituitary that stimulate gonadal function—the formation of gametes and secretion of sex steroids. The two gonadotropins are FSH (follicle-stimulating hormone) and LH (luteinizing hormone), which are essentially the same in males and females.

G-protein An association of three membrane-associated protein subunits, designated alpha, beta, and gamma, that is regulated by guanosine nucleotides (GDP and GTP). The G-protein subunits dissociate in response to a membrane signal and, in turn, activate other proteins in the cell.

graafian *(graf′e-an)* **follicle** A mature ovarian follicle, containing a single fluid-filled cavity, with the ovum located toward one side of the follicle and perched on top of a hill of granulosa cells.

graded potential A change in the membrane potential (depolarization or hyperpolarization) with amplitudes that are varied, or graded, by gradations in the stimulus intensity. The stimuli for graded potentials in postsynaptic neurons are neurotransmitters, and the degree of depolarization or hyperpolarization produced depends on the amount of neurotransmitter molecules released by the presynaptic axons.

granular leukocytes *(loo′kŏ-sīts)* Leukocytes with granules in the cytoplasm. On the basis of the staining properties of the granules, these cells are of three types: neutrophils, eosinophils, and basophils.

Graves' disease A hyperthyroid condition believed to be caused by excessive stimulation of the thyroid gland by autoantibodies. It is associated with exophthalmos (bulging eyes), high pulse rate, high metabolic rate, and other symptoms of hyperthyroidism.

gray matter The part of the central nervous system that contains neuron cell bodies and dendrites but few myelinated axons. It forms the cortex of the cerebrum, cerebral nuclei, and the central region of the spinal cord.

growth hormone (GH) A hormone secreted by the anterior pituitary that stimulates growth of the skeleton and soft tissues during the growing years and that influences the metabolism of protein, carbohydrate, and fat throughout life.

gustducins *(gus-doo′sinz)* The G-proteins involved in the sense of taste, particularly of sweet and bitter tastes.

gyrus *(ji′rus)* A fold or convolution in the cerebrum.

H

hair cells Sensory epithelial cells with processes called hairs (cilia or stereocilia) on their apical surfaces. These are mechanoreceptors involved in equilibrium and hearing, where bending of the hairs results in stimulation of associated sensory neurons.

haploid *(hap′loid)* Denoting cells that have one of each chromosome type and therefore half the number of chromosomes present in most other body cells. Only the gametes (sperm and ova) are haploid.

hapten *(hap′ten)* A small molecule that is not antigenic by itself, but which—when combined with proteins—becomes antigenic and thus capable of stimulating the production of specific antibodies.

haversian *(hă-ver′shan)* **system** A haversian canal and its concentrically arranged layers, or lamellae, of bone. It constitutes the basic structural unit of compact bone.

hay fever A seasonal type of allergic rhinitis caused by pollen. It is characterized by itching and tearing of the eyes, swelling of the nasal mucosa, attacks of sneezing, and often by asthma.

hCG Human chorionic gonadotropin *(kor′e-on-ik gon-ad″ŏ-tro′pin)* A hormone secreted by the embryo that has LH-like actions and that is required for maintenance of the mother's corpus luteum for the first 10 weeks of pregnancy.

heart murmur An abnormal heart sound caused by an abnormal flow of blood in the heart. Murmurs are due to structural defects, usually of the valves or septum.

heart sounds The sounds produced by closing of the AV valves of the heart during systole (the first sound) and by closing of the semilunar valves of the aorta and pulmonary trunk during diastole (the second sound).

helper T cells A subpopulation of T cells (lymphocytes) that help to stimulate antibody production of B lymphocytes by antigens.

hematocrit *(he-mat′ŏ-krit)* The ratio of packed red blood cells to total blood volume in a centrifuged sample of blood, expressed as a percentage.

hematopoiesis *(he-ma-to-poy-e′sis)* The formation of new blood cells, including erythropoiesis (the formation of new red blood cells) and leukopoiesis (the formation of new white blood cells).

heme *(hēm)* The iron-containing red pigment that, together with the protein globin, forms hemoglobin.

hemoglobin *(he′mŏ-glo″bin)* The combination of heme pigment and protein within red blood cells that acts to transport oxygen and (to a lesser degree) carbon dioxide. Hemoglobin also serves as a weak buffer within red blood cells.

Henderson-Hasselbalch *(hen′der-son-has′el-balk)* **equation** A formula used to determine the blood pH produced by a given ratio of bicarbonate to carbon dioxide concentrations.

Henry's law The statement that the concentration of gas dissolved in a fluid is directly proportional to the partial pressure of that gas.

heparin *(hep′ar-in)* A mucopolysaccharide found in many tissues, but in greatest abundance in the lungs and liver. It is used medically as an anticoagulant.

hepatic *(hĕ-pat′ik)* Pertaining to the liver.

hepatitis *(hep″ă-ti′tis)* Inflammation of the liver.

Hering-Breuer reflex A reflex in which distension of the lungs stimulates stretch receptors, which in turn act to inhibit further distension of the lungs.

hermaphrodite *(her-maf′rŏ-dīt)* An organism with both testicular and ovarian tissue.

heterochromatin *(het″ĕ-ro-kro′mă-tin)* A condensed, inactive form of chromatin.

hiatal hernia *(hi-a′tal her′ ne-ă)* A protrusion of an abdominal structure through the esophageal hiatus of the diaphragm into the thoracic cavity.

high-density lipoproteins *(lip″o-pro′te-inz)* **(HDLs)** Combinations of lipids and proteins that migrate rapidly to the bottom of a test tube during centrifugation. HDLs are carrier proteins that are believed to transport cholesterol away from blood vessels to the liver, and thus to offer some protection from atherosclerosis.

histamine *(his′tă-mēn)* A compound secreted by tissue mast cells and other connective tissue cells that stimulates vasodilation and increases capillary permeability. It is responsible for many of the symptoms of inflammation and allergy.

histocompatibility *(his″to-kom-pat″ĭ-bil′ĭ-te)* **antigens** A group of cell-surface antigens found on all cells of the body except mature red blood cells. They are important for the function of T lymphocytes, and the greater their variance, the greater will be the likelihood of transplant rejection.

histone *(his′toōn)* A basic protein associated with DNA that is believed to repress genetic expression.

homeostasis *(ho″me-o-sta′sis)* The dynamic constancy of the internal environment, the maintenance of which is the principal function of physiological regulatory mechanisms. The concept of homeostasis provides a framework for understanding most physiological processes.

homologous *(hŏ-mol′-ŏ-gus)* **chromosomes** The matching pairs of chromosomes in a diploid cell.

hormone *(hor′mōn)* A regulatory chemical produced in an endocrine gland that is secreted into the blood and carried to target cells that respond to the hormone by an alteration in their metabolism.

hormone-response element A specific region of DNA that binds to a particular nuclear hormone receptor when that receptor is activated by bonding with its hormone. This stimulates genetic transcription (RNA synthesis).

humoral immunity *(hyoo′-mor-al ĭ-myoo′nĭ-te)* The form of acquired immunity in which antibody molecules are secreted in response to antigenic stimulation (as opposed to cell-mediated immunity).

hyaline *(hi′ă-līn)* **membrane disease** A disease affecting premature infants who lack pulmonary surfactant. It is characterized by collapse of the alveoli (atelectasis) and pulmonary edema; also called *respiratory distress syndrome.*

hydrocortisone *(hi″drŏ-kor′tĭ-sōn)* The principal corticosteroid hormone secreted by the adrenal cortex, with glucocorticoid action; also called *cortisol.*

hydrolysis *(hi-drol′ĭ-sis)* The splitting of a larger molecule into its subunits, in a reaction that also results in the breaking of a water molecule.

hydrophilic (*hi″drŏ-fil′ik*) Denoting a substance that readily absorbs water; literally, "water loving."

hydrophobic (*hi″drŏ-fo′bik*) Denoting a substance that repels, and that is repelled by, water; literally, "water fearing."

hyperbaric (*hi″per-bar′ik*) **oxygen** Oxygen gas present at greater than atmospheric pressure.

hypercapnia (*hi″per-kap′ne-ă*) Excessive concentration of carbon dioxide in the blood.

hyperemia (*hi″per-e′-me-ă*) Excessive blood flow to a part of the body.

hyperglycemia (*hi″per-gli-se′me-ă*) An abnormally increased concentration of glucose in the blood.

hyperkalemia (*hi″per-kă-le′me-ă*) An abnormally high concentration of potassium in the blood.

hyperopia (*hi″per-o′pe-ă*) A refractive disorder in which rays of light are brought to a focus behind the retina as a result of the eyeball being too short; also called *farsightedness.*

hyperplasia (*hi″per-pla′ze-ă*) An increase in organ size because of an increase in the number of cells as a result of mitotic cell division.

hyperpnea (*hi″perp′ne-ă*) Increased total minute volume during exercise. Unlike hyperventilation, the arterial blood carbon dioxide values are not changed during hyperpnea because the increased ventilation is matched to an increased metabolic rate.

hyperpolarization (*hi″per-po″lar-ĭ-za′shun*) An increase in the negativity of the inside of a cell membrane with respect to the resting membrane potential.

hypersensitivity (*hi″per-sen″sĭ-tiv′ĭ-te*) Another name for *allergy,* an abnormal immune response that may be immediate (due to antibodies of the IgE class) or delayed (due to cell-mediated immunity).

hypertension (*hi″per-ten′shun*) High blood pressure. Classified as either primary, or essential, hypertension of unknown cause or secondary hypertension that develops as a result of other, known disease processes.

hypertonic (*hi″per-ton′ik*) Denoting a solution with a greater solute concentration, and thus a greater osmotic pressure, than plasma.

hypertrophy (*hi-per′trŏ-fe*) Growth of an organ because of an increase in the size of its cells.

hyperventilation (*hi-per-ven″tĭ-la′shun*) A high rate and depth of breathing that results in a decrease in the blood carbon dioxide concentration to below normal.

hypotension (*hi″po-ten′shun*) Abnormally low blood pressure.

hypothalamo-hypophyseal (*hi″po-thă-lam′o-hi″po-fĭ-se′al*) **portal system** A vascular system that transports releasing and inhibiting hormones from the hypothalamus to the anterior pituitary.

hypothalamo-hypophyseal tract The tract of nerve fibers (axons) that transports antidiuretic hormone and oxytocin from the hypothalamus to the posterior pituitary.

hypothalamus (*hi″po-thal′ă-mus*) An area of the brain lying below the thalamus and above the pituitary gland. The hypothalamus regulates the pituitary gland and contributes to the regulation of the autonomic nervous system, among its many functions.

hypothermia (*hi″pŏ-ther′me-ă*) A low body temperature. This is a dangerous condition that is defended against by shivering and other physiological mechanisms that generate body heat.

hypoventilation (*hi′po-ven-ti-la′shun*) Inadequate pulmonary ventilation, such that the plasma concentration of carbon dioxide (partial pressure of carbon dioxide) is abnormally increased.

hypovolemic (*hi″po-vo-le′mik*) **shock** A rapid fall in blood pressure as a result of diminished blood volume.

hypoxemia (*hi″pok-se′me-ă*) A low oxygen concentration of the arterial blood.

hypoxic (*hi-pok′sik*) **drive** The stimulation of breathing by a fall in the plasma concentration of oxygen (partial pressure of oxygen), as may occur at high altitudes.

I

ileogastric (*il″e-o-gas′trik*) **reflex** The reflex in which distension of the ileum causes decreased gastric motility.

immediate hypersensitivity Hypersensitivity (allergy) that is mediated by antibodies of the IgE class and that results in the release of histamine and related compounds from tissue cells.

immunization (*im″yŭ-nĭ-za′shun*) The process of increasing one's resistance to pathogens. In active immunity, a person is injected with antigens that stimulate the development of clones of specific B or T lymphocytes; in passive immunity, a person is injected with antibodies made by another organism.

immunoassay (*im″yŭ-no-as′a*) Any of a number of laboratory or clinical techniques that employ specific bonding between an antigen and its homologous antibody in order to identify and quantify a substance in a sample.

immunoglobulins (*im″yŭ-no-glob′yŭ-linz*) Subclasses of the gamma globulin fraction of plasma proteins that have antibody functions, providing humoral immunity.

immunosurveillance (*im″yŭ-no-ser-va′lens*) The function of the immune system to recognize and attack malignant cells that produce antigens not recognized as "self." This function is believed to be cell mediated rather than humoral.

implantation (*im″plan-ta′shun*) The process by which a blastocyst attaches itself to and penetrates the endometrium of the uterus.

infarct (*in′farkt*) An area of necrotic (dead) tissue produced by inadequate blood flow (ischemia).

inhibin (*in-hib′in*) Believed to be a water-soluble hormone secreted by the seminiferous tubules of the testes that specifically exerts negative feedback control of FSH secretion from the anterior pituitary.

innate immunity Also called nonspecific immunity, this refers to the parts of the immune system that are inherited and can combat pathogens without prior exposure to them.

inositol triphosphate (*ĭ-no′sĭ-tol tri-fos′fāt*) (**IP_3**) A second messenger in hormone action that is produced by the cell membrane of a target cell in response to the action of a hormone. This compound is believed to stimulate the release of Ca^{2+} from the endoplasmic reticulum of the cell.

insulin (*in′sŭ-lin*) A polypeptide hormone secreted by the beta cells of the islets of Langerhans in the pancreas that promotes the anabolism of carbohydrates, fat, and protein. Insulin acts to promote the cellular uptake of blood glucose and, therefore, to lower the blood glucose concentration; insulin deficiency produces hyperglycemia and diabetes mellitus.

integrins (*in-te′grinz*) A family of glycoproteins that extend from the cytoskeleton, through the plasma membrane of cells, and into the extracellular matrix. They serve to integrate different cells of a tissue and the extracellular matrix, and to bind cells to other cells, such as neutrophils to the endothelial cells of capillaries for extravasation.

interferons (*in″ter-fer′unz*) Small proteins that inhibit the multiplication of viruses inside host cells and that also have antitumor properties.

interleukin-2 (*in″ter-loo′kin-2*) A lymphokine secreted by T lymphocytes that stimulates the proliferation of both B and T lymphocytes.

interneurons (*in″ter-noor′onz*) Those neurons within the central nervous system that do not extend into the peripheral nervous system. They are interposed between sensory (afferent) and motor (efferent) neurons; also called *association neurons.*

interoceptors (*in″ter-o-sep′torz*) Sensory receptors that respond to changes in the internal environment (as opposed to exteroceptors).

interphase The interval between successive cell divisions, during which time the chromosomes are in an extended state and are active in directing RNA synthesis.

intestinal microbiota (*mi″kro-bi-o′tah*) The microscopic organisms, principally bacteria, found in the large intestine. The bacteria are often referred to as commensal bacteria, because they do not harm the human host and are in many ways beneficial. Also called *microflora.*

interstitial (*in-ter-stish′al*) **fluid** The fluid outside of the cells within a tissue or organ. Interstitial fluid and blood plasma together compose the extracellular fluid of the body. Also called *tissue fluid.*

intestino-intestinal (*in″tes′tĭ-no-in-tes′tĭ-nal*) **reflex** The reflex in which overdistension to one region of the intestine causes relaxation throughout the rest of the intestine.

intrafusal (*in″tră-fyoo′sal*) **fibers** Modified muscle fibers that are encapsulated to form muscle spindle organs, which are muscle stretch receptors.

intrapleural (*in″tră-ploor′al*) **space** An actual or potential space between the visceral pleura covering the lungs and the parietal pleura lining the thoracic wall. Normally, this is a potential space; it can become real only in abnormal situations.

intrapulmonary (*in″tră-pul′mŏ-nar″e*) **space** The space within the air sacs and airways of the lungs.

intron (*in′tron*) A noncoding nucleotide sequence in DNA that interrupts the coding regions (exons) for mRNA.

inulin (*in′yŭ-lin*) A polysaccharide of fructose, produced by certain plants, that is filtered by the human kidneys but neither reabsorbed nor secreted. The clearance rate of injected inulin is thus used to measure the glomerular filtration rate.
in vitro (*in ve′tro*) Occuring outside the body, in a test tube or other artificial environment.
in vivo (*in ve′vo*) Occuring within the body.
ion (*i′on*) An atom or a group of atoms that has a net positive or a net negative charge because of a loss or gain of electrons.
ionization (*i″on-ĭ-za′shun*) The dissociation of a solute to form ions.
ipsilateral (*ip″sĭ-lat′er-al*) On the same side (as opposed to contralateral).
IPSP **inhibitory postsynaptic potential** A hyperpolarization of the postsynaptic membrane in response to a particular neurotransmitter chemical, which makes it more difficult for the postsynaptic cell to attain the threshold level of depolarization required to produce action potentials. IPSPs are responsible for postsynaptic inhibition.
ischemia (*ĭ-ske′me-ă*) A rate of blood flow to an organ that is inadequate to supply sufficient oxygen and maintain aerobic respiration in that organ.
islets of Langerhans (*ĭ′letz of lang′er-hanz*) Encapsulated groupings of endocrine cells within the exocrine tissue of the pancreas, including alpha cells that secrete glucagon and beta cells that secrete insulin; also called *pancreatic islets.*
isoenzymes (*i″so-en′zimz*) Enzymes, usually produced by different organs, that catalyze the same reaction but that differ from each other in amino acid composition.
isometric (*i″sŏ-met′rik*) **contraction** Muscle contraction in which there is no appreciable shortening of the muscle.
isotonic (*i″sŏ-ton′ik*) **contraction** Muscle contraction in which the muscle shortens in length and maintains approximately the same amount of tension throughout the shortening process.
isotonic solution A solution having the same total solute concentration, osmolality, and osmotic pressure as the solution with which it is compared; a solution with the same solute concentration and osmotic pressure as plasma.

J

jaundice (*jawn′dis*) A condition characterized by high blood bilirubin levels and staining of the tissues with bilirubin, which imparts a yellow color to the skin and mucous membranes.
junctional (*jungk′shun-al*) **complexes** Structures that join adjacent epithelial cells together, including the zonula occludens, zonula adherens, and macula adherens (desmosome).
juxtaglomerular (*juk″stă-glo-mer′yŭ-lar*) **apparatus** A renal structure in which regions of the nephron tubule and afferent arteriole are in contact with each other. Cells in the afferent arteriole of the juxtaglomerular apparatus secrete the enzyme renin into the blood, which activates the renin-angiotensin system.

K

keratin (*ker′ă-tin*) A protein that forms the principal component of the outer layer of the epidermis and of hair and nails.
ketoacidosis (*ke″to-ă-sĭ-do′sis*) A type of metabolic acidosis resulting from the excessive production of ketone bodies, as in diabetes mellitus.
ketogenesis (*ke″to-jen′ĭ-sis*) The production of ketone bodies.
ketone (*ke′tōn*) **bodies** The substances derived from fatty acids via acetyl coenzyme A in the liver; namely, acetone, acetoacetic acid, and β-hydroxybutyric acid. Ketone bodies are oxidized by skeletal muscles for energy.
ketosis (*ke-to′sis*) An abnormal elevation in the blood concentration of ketone bodies. This condition does not necessarily produce acidosis.
kilocalorie (*kil′ŏ-kal″ŏ-re*) A unit of measurement equal to 1,000 calories, which are units of heat. (A kilocalorie is the amount of heat required to raise the temperature of 1 kilogram of water 1° C.) In nutrition, the kilocalorie is called a big calorie (Calorie).
kinase (*ki′nās*) Any of a class of enzymes that transfer phosphate groups to organic molecules. The activity of particular protein kinases may be promoted by hormones and other regulatory molecules. These enzymes can, in turn, phosphorylate other enzymes and thereby regulate their activities.
Klinefelter's (*klīn′fel-terz*) **syndrome** The syndrome produced in a male by the presence of an extra X chromosome (genotype XXY).
knockout mice Strains of mice in which a specific targeted gene has been inactivated by developing the mice from embryos injected with specifically mutated cells.
Krebs (*krebz*) **cycle** A cyclic metabolic pathway in the matrix of mitochondria by which the acetic acid part of acetyl CoA is oxidized and substrates provided for reactions that are coupled to the formation of ATP.
Kupffer (*koop′fer*) **cells** Phagocytic cells lining the sinusoids of the liver that are part of the reticuloendothelial system.

L

lactate threshold A measurement of the intensity of exercise. It is the percentage of a person's maximal oxygen uptake at which a rise in blood lactate levels occurs. The average lactate threshold occurs when exercise is performed at 50% to 70% of the maximal oxygen uptake (aerobic capacity).
lactose (*lak′tōs*) Milk sugar; a disaccharide of glucose and galactose.
lactose intolerance The inability of many adults to digest lactose because of a deficiency of the enzyme lactase.
Laplace, law of The statement that the pressure within an alveolus is directly proportional to its surface tension and inversely proportional to its radius.
larynx (*lar′ingks*) A structure consisting of epithelial tissue, muscle, and cartilage that serves as a sphincter guarding the entrance of the trachea. It is the organ responsible for voice production.
lateral inhibition The sharpening of perception that occurs in the neural processing of sensory input. Input from those receptors that are most greatly stimulated is enhanced, while input from other receptors is reduced. This results, for example, in improved pitch discrimination in hearing.
leakage channels Ion channels in the plasma membrane that are always open because they are not gated. For example, there are leakage channels (in addition to voltage-gated channels) for K^+, that make the resting membrane more permeable to K^+ than to other ions.
leptin (*lep′tin*) A hormone secreted by adipose tissue that acts as a satiety factor to reduce appetite. It also increases the body's caloric expenditure.
lesion (*le′zhun*) **1.** A wounded or damaged area of tissue. **2.** An injury or wound. **3.** A single infected patch in a skin disease.
leukocyte (*loo′kŏ-sīt*) A white blood cell.
Leydig (*li′dig*) **cells** The interstitial cells of the testes that serve an endocrine function by secreting testosterone and other androgenic hormones.
ligament (*lig′ă-ment*) A tough cord or fibrous band of dense regular connective tissue that contains numerous parallel arrangements of collagen fibers. It connects bones or cartilages and serves to strengthen joints.
ligand (*li′gand, lig′and*) A smaller molecule that chemically binds to a larger molecule, which is usually a protein. Oxygen, for example, is the ligand for the heme in hemoglobin, and hormones or neurotransmitters can be the ligands for specific membrane proteins.
limbic (*lim′bik*) **system** A group of brain structures, including the hippocampus, cingulate gyrus, dentate gyrus, and amygdala. The limbic system appears to be important in memory, the control of autonomic function, and some aspects of emotion and behavior.
lipid (*lip′id*) An organic molecule that is nonpolar, and thus insoluble in water. Lipids include triglycerides, steroids, and phospholipids.
lipogenesis (*lip″ŏ-jen′ĕ-sis*) The formation of fat or triglycerides.
lipolysis (*li-pol′ĭ-sis*) The hydrolysis of triglycerides into free fatty acids and glycerol.
lipophilic (*lip″ŏ-fil′ik*) Pertaining to molecules that are nonpolar and thus soluble in lipids. The steroid hormones, thyroxine, and the lipid-soluble vitamins are examples of lipophilic molecules.
long-term depression (LTD) A process in which proper stimulation of a presynaptic neuron causes the postsynaptic neuron to release endocannabinoids, which suppress the release of neurotransmitters from presynaptic neurons. Depression of excitatory input from glutamate would inhibit activation of the postsynaptic neuron, but depression of inhibitory input from GABA-releasing presynaptic axons would enhance the activation of the postsynaptic neuron.
long-term potentiation (*pŏ-ten″she-a′shun*) **(LTP)** The improved ability of a presynaptic neuron that has been stimulated at high frequency to subsequently stimulate a postsynaptic neuron over a period of weeks or even months. This may represent a mechanism of neural learning.

low-density lipoproteins (*lip″o-pro′te-inz*) **(LDLs)** Plasma proteins that transport triglycerides and cholesterol to the arteries. LDLs are believed to contribute to arteriosclerosis.

lower motor neuron The motor neuron that has its cell body in the gray matter of the spinal cord and that contributes axons to peripheral nerves. This neuron innervates muscles and glands.

lumen (*loo′men*) The cavity of a tube or hollow organ.

lung surfactant (*sur-fak′tant*) A mixture of lipoproteins (containing phospholipids) secreted by type II alveolar cells into the alveoli of the lungs. It lowers surface tension and prevents collapse of the lungs, as occurs in hyaline membrane disease when surfactant is absent.

luteinizing (*loo′te-ĭ-ni″zing*) **hormone (LH)** A gonadotropic hormone secreted by the anterior pituitary. In a female, LH stimulates ovulation and the development of a corpus luteum; in a male, it stimulates the Leydig cells to secrete androgens.

lymph (*limf*) A fluid derived from tissue fluid that flows through lymphatic vessels, returning to the venous bloodstream.

lymphatic (*lim-fat′ik*) **system** The lymphatic vessels and lymph nodes.

lymphocyte (*lim′fŏ-sīt*) A type of mononuclear leukocyte; the cell responsible for humoral and cell-mediated immunity.

lymphoid (*lim′foyd*) **organs** The organs that produce lymphocytes. The primary lymphoid organs are the bone marrow and thymus, and the secondary lymphoid organs include the lymph nodes, spleen, tonsils, and Peyer's patches of the intestinal mucosa.

lymphokine (*lim′fŏ-kīn*) Any of a group of chemicals released from T cells that contribute to cell-mediated immunity.

lysosome (*li′sŏ-sōm*) An organelle containing digestive enzymes that is responsible for intracellular digestion.

M

macromolecule (*mak″rŏ-mol′ĭ-kyool*) A large molecule; a term commonly used to refer to protein, RNA, and DNA.

macrophage (*mak′rŏ-fāj*) A large phagocytic cell in connective tissue that contributes to both specific and nonspecific immunity.

macula densa (*mak′yŭ-lă den′să*) The region of the distal tubule of the renal nephron in contact with the afferent arteriole. This region functions as a sensory receptor for the amount of sodium excreted in the urine and acts to inhibit the secretion of renin from the juxtaglomerular apparatus.

macula lutea (*loo′te-ă*) A yellowish depression in the retina of the eye that contains the fovea centralis, the area of keenest vision.

malignant Denoting a structure or process that is life threatening. Of a tumor, tending to metastasize.

mast cell A type of connective tissue cell that produces and secretes histamine and heparin.

maximal oxygen uptake The maximum rate of oxygen consumption by the body per unit time during heavy exercise. Also called the *aerobic capacity,* the maximal oxygen uptake is commonly indicated with the symbol $\dot{V}_{O_2}$ max.

mean arterial pressure (MAP) An adjusted average of the systolic and diastolic blood pressures. It averages about 100 mmHg in the systemic circulation and 10 mmHg in the pulmonary circulation.

mechanoreceptor (*mek″ă-no-re-sep′tor*) A sensory receptor that is stimulated by mechanical means. Mechanoreceptors include stretch receptors, hair cells in the inner ear, and pressure receptors.

medulla oblongata (*mĕ-dul′ă ob″long-gătă*) A part of the brain stem that contains neural centers for the control of breathing and for regulation of the cardiovascular system via autonomic nerves.

megakaryocyte (*meg″ă-kar′e-o-sīt*) A bone marrow cell that gives rise to blood platelets.

meiosis (*mi-o′sis*) A type of cell division in which a diploid parent cell gives rise to haploid daughter cells. It occurs in the process of gamete production in the gonads.

melanin (*mel′ă-nin*) A dark pigment found in the skin, hair, choroid layer of the eye, and substantia nigra of the brain. It may also be present in certain tumors (melanomas).

melatonin (*mel″ă-to′nin*) A hormone secreted by the pineal gland that produces darkening of the skin in lower animals and that may contribute to the regulation of gonadal function in mammals. Secretion follows a circadian rhythm and peaks at night.

membrane potential The potential difference or voltage that exists between the two sides of a cell membrane. It exists in all cells but is capable of being changed by excitable cells (neurons and muscle cells).

membranous labyrinth (*mem′bră nus lab′ĭ-rinth*) A system of communicating sacs and ducts within the bony labyrinth of the inner ear.

menarche (*mĕ-nar′ke*) The first menstrual discharge, normally occurring during puberty.

Ménière's (*mān-yarz′*) **disease** Deafness, tinnitus, and vertigo resulting from a disease of the labyrinth.

menopause (*men′ŏ-pawz*) The cessation of menstruation, usually occurring at about age 50.

menstrual (*men′stroo-al*) **cycle** The cyclic changes in the ovaries and endometrium of the uterus that lasts about a month. It is accompanied by shedding of the endometrium, with bleeding, and occurs only in humans and the higher primates.

menstruation (*men″stroo-a′shun*) Shedding of the outer two-thirds of the endometrium with accompanying bleeding as a result of a lowering of estrogen secretion by the ovaries at the end of the monthly cycle. The first day of menstruation is taken as day 1 of the menstrual cycle.

mesoderm (*mes′ŏ-derm*) The middle embryonic tissue layer that gives rise to connective tissue (including blood, bone, and cartilage); blood vessels; muscles; the adrenal cortex; and other organs.

messenger RNA (mRNA) A type of RNA that contains a base sequence complementary to a part of the DNA that specifies the synthesis of a particular protein.

metabolic acidosis (*as″ĭ-do′sis*) **and alkalosis** (*al″kă-lo′sis*) Abnormal changes in arterial blood pH due to changes in nonvolatile acid concentration (for example, changes in lactic acid or ketone body concentrations) or to changes in blood bicarbonate concentration.

metabolic syndrome A combination of central obesity, insulin resistance, type 2 diabetes, and hypertension.

metabolism (*mĕ-tab′ŏ-liz-em*) All of the chemical reactions in the body. It includes those that result in energy storage (anabolism) and those that result in the liberation of energy (catabolism).

metastasis (*mĕ-tas′tă-sis*) A process whereby cells of a malignant tumor separate from the tumor, travel to a different site, and divide to produce a new tumor.

methemoglobin (*met-he′mŏ-glo″bin*) The abnormal form of hemoglobin in which the iron atoms in heme are oxidized to the ferrous form. Methemoglobin is incapable of bonding with oxygen.

micelle (*mi sel′*) A colloidal particle formed by the aggregation of numerous molecules.

microarray technology A technique for identifying large numbers of genes by complementary base pairing of single-stranded DNA samples to spots of known single-stranded DNA on a support medium.

microvilli (*mi″kro-vil′i*) Tiny fingerlike projections of a cell membrane. They occur on the apical (lumenal) surface of the cells of the small intestine and in the renal tubules.

micturition (*mik″tŭ-rish′un*) Urination.

milliequivalent (*mil″ĭ-e-kwiv′ă-lent*) The millimolar concentration of an ion multiplied by its number of charges.

mineralocorticoid (*min″er-al-o-kor′tĭ-koid*) Any of a class of steroid hormones of the adrenal cortex (corticosteroids) that regulate electrolyte balance.

mitochondria (*mi″to-kon′dre-ah*), sing. *mitochondrion* The organelles in cells that are responsible for the production of most of the ATP through aerobic respiration. The citric acid cycle, the electron transport system, and oxidative phosphorylation occur within the mitochondria.

mitosis (*mi-to′sis*) Cell division in which the two daughter cells receive the same number of chromosomes as the parent cell (both daughters and parent are diploid).

molal (*mo′lal*) Pertaining to the number of moles of solute per kilogram of solvent.

molar (*mo′lar*) Pertaining to the number of moles of solute per liter of solution.

mole (*mōl*) The number of grams of a chemical that is equal to its formula weight (atomic weight for an element or molecular weight for a compound).

monoamine (*mon″o-am′ēn*) Any of a class of neurotransmitter molecules containing one amino group. Examples are serotonin, dopamine, and norepinephrine.

monoamine oxidase (*mon′o-am′ēn ok′sĭ-dās*) **(MAO)** An enzyme that degrades monoamine neurotransmitters within presynaptic axon endings. Drugs that inhibit the action of this enzyme thus potentiate the pathways that use monoamines as neurotransmitters.

monoclonal (*mon-o-klo´nal*) **antibodies** Antibodies produced by a single clone of lymphocytes. These are generated medically to bind to a single antigenic determinant site of a molecule—for example, to detect hCG in a pregnancy test.

monocyte (*mon´o-sīt*) A mononuclear, nongranular, phagocytic leukocyte that can be transformed into a macrophage.

monomer (*mon´ŏ-mer*) A single molecular unit of a longer, more complex molecule. Monomers are joined together to form dimers, trimers, and polymers. The hydrolysis of polymers eventually yields separate monomers.

mononuclear leukocyte (*mon″o-noo´kle-ar loo´kŏ-sīt*) Any of a category of white blood cells that includes the lymphocytes and monocytes.

mononuclear phagocyte (*fag´ŏ-sīt*) **system** A term used to describe monocytes and tissue macrophages.

monosaccharide (*mon″ŏ-sak´ă-rīd*) The monomer of the more complex carbohydrates. Examples of monomers are glucose, fructose, and galactose. Also called a *simple sugar.*

motile (*mo´til*) Capable of self-propelled movement.

motility (*mo-til´i-te*) The property of movement. In the gastrointestinal (GI) tract, motility refers to the ability to mix its contents and move them from the oral to the anal end of the GI tract by muscular contractions.

motor cortex (*kor´teks*) The precentral gyrus of the frontal lobe of the cerebrum. Axons from this area form the descending pyramidal motor tracts.

motor neuron (*noor´on*) An efferent neuron that conducts action potentials away from the central nervous system to effector organs (muscles and glands). It forms the ventral roots of spinal nerves.

motor unit A lower motor neuron and all of the skeletal muscle fibers stimulated by branches of its axon. Larger motor units (more muscle fibers per neuron) produce more force when the unit is activated, but smaller motor units afford a finer degree of neural control over muscle contraction.

muscarinic receptors (*mus″kă-rin´ik re-sep´torz*) Receptors for acetylcholine that are stimulated by postganglionic parasympathetic neurons. Their name is derived from the fact that they are also stimulated by the chemical muscarine, derived from a mushroom.

muscle spindle (*mus´el spin´d´l*) A sensory organ within skeletal muscle that is composed of intrafusal fibers. It is sensitive to muscle stretch and provides a length detector within muscles.

myelin (*mi´ĕ-lin*) **sheath** A sheath surrounding axons formed from the cell membrane of Schwann cells in the peripheral nervous system and from oligodendrocytes in the central nervous system.

myocardial infarction (*mi″ŏ-kar´de-al in-fark´shun*) An area of necrotic tissue in the myocardium that is filled in by scar (connective) tissue.

myofibril (*mi″ŏ-fi´bril*) A subunit of striated muscle fiber that consists of successive sarcomeres. Myofibrils run parallel to the long axis of the muscle fiber, and the pattern of their filaments provides the striations characteristic of striated muscle cells.

myofilaments (*mi-o-fil´a-ments*) The thick and thin filaments in a muscle fiber. The thick filaments are composed of the protein myosin, and the thin filaments are composed primarily of the protein actin.

myogenic (*mi″ŏ-jen´ik*) Originating within muscle cells. This term is used to describe self-excitation by cardiac and smooth muscle cells.

myoglobin (*mi″ŏ-glo´bin*) A molecule composed of globin protein and heme pigment. It is related to hemoglobin but contains only one subunit (instead of the four in hemoglobin). Myoglobin is found in striated muscles, wherein it serves to store oxygen.

myoneural (*mi″ŏ-noor´al*) **junction** A synapse between a motor neuron and the muscle cell that it innervates; also called the *neuromuscular junction.*

myopia (*mi-o´pe-ă*) A condition of the eyes in which light is brought to a focus in front of the retina because the eye is too long; also called *nearsightedness.*

myosin (*mi´ŏ-sin*) The protein that forms the A bands of striated muscle cells. Together with the protein actin, myosin provides the basis for muscle contraction.

myxedema (*mik″sĭ-de´mă*) A type of edema associated with hypothyroidism. It is characterized by accumulation of mucoproteins in tissue fluid.

N

NAD **nicotinamide adenine dinucleotide** (*nik″ŏ-tină-mīd ad´nēn di-noo´kle-ŏ-tīd*) A coenzyme derived from niacin that functions to transport electrons in oxidation-reduction reactions. It helps to transport electrons to the electron-transport chain within mitochondria.

naloxone (*na″l ok-sōn, nă-lok´-sōn*) A drug that antagonizes the effects of morphine and endorphins.

natriuretic (*na″trĭ-yoo-ret´ik*) **hormone** A hormone that increases the urinary excretion of sodium. This hormone has been identified as atrial natriuretic peptide (ANP), produced by the atria of the heart.

natural killer (NK) cells These are related to killer T lymphocytes but are part of the innate immune system. NK cells attack body cells infected with viruses and tumor cells by direct contact, and also secrete cytokines that activate B and T lymphocytes as well as macrophages and neutrophils.

necrosis (*nĕ-kro´sis*) Cellular death within tissues and organs as a result of pathological conditions. Necrosis differs histologically from the physiological cell death of apoptosis.

negative feedback loop A response mechanism that serves to maintain a state of internal constancy, or homeostasis. Effectors are activated by changes in the internal environment, and the inhibitory actions of the effectors serve to counteract these changes and maintain a state of balance.

neoplasm (*ne´ŏ-plazm*) A new, abnormal growth of tissue, as in a tumor.

nephron (*nef´ron*) The functional unit of the kidneys, consisting of a system of renal tubules and a vascular component that includes capillaries of the glomerulus and the peritubular capillaries.

Nernst equation The equation used to calculate the equilibrium membrane potential for given ions when the concentrations of those ions on each side of the membrane are known.

nerve A collection of motor axons and sensory dendrites in the peripheral nervous system.

neurilemma (*noor″ĭ-lem´ă*) The sheath of Schwann and its surrounding basement membrane that encircles nerve fibers in the peripheral nervous system.

neuroglia (*noo-rog´le-ă*) The supporting cells of the central nervous system that aid the functions of neurons. In addition to providing support, they participate in the metabolic and bioelectrical processes of the nervous system, also called *glial cells.*

neurohypophysis (*noor″o-hi-pof´ĭ-sis*) The posterior part of the pituitary gland that is derived from the brain. It releases vasopression (ADH) and oxytocin, both of which are produced in the hypothalamus.

neuron (*noor´on*) A nerve cell, consisting of a cell body that contains the nucleus; short branching processes called dendrites that carry electrical charges to the cell body; and a single fiber, or axon, that conducts nerve impulses away from the cell body.

neuropeptide (*noor″o-pep´tīd*) Any of various polypeptides found in neural tissue that are believed to function as neurotransmitters and neuromodulators. Neuropeptide Y, for example, is the most abundant polypeptide in the brain and has been implicated in a variety of processes, including the stimulation of appetite.

neurotransmitter (*noor″o-trans´mit-er*) A chemical contained in synaptic vesicles in nerve endings that is released into the synaptic cleft, where it causes the production of either excitatory or inhibitory postsynaptic potentials.

neurotrophin (*noor″ŏ-trof´in*) Any of a family of autocrine regulators secreted by neurons and neuroglial cells that promote axon growth and other effects. Nerve growth factor is an example.

neutron (*noo´tron*) An electrically neutral particle that exists together with positively charged protons in the nucleus of atoms.

nexus (*nek´sus*) A bond between members of a group; the type of intercellular connection found in single-unit smooth muscles.

niacin (*ni´-ă-sin*) A water-soluble B vitamin needed for the formation of NAD, which is a coenzyme that participates in the transfer of hydrogen atoms in many of the reactions of cell respiration.

nicotinic receptors (*nik″ŏ-tin´ik re-sep´torz*) Receptors for acetylcholine located in the autonomic ganglia and in neuromuscular junctions. Their name is derived from the fact that they can also be stimulated by nicotine, derived from the tobacco plant.

nidation (*ni-da´shun*) Implantation of the blastocyst in the endometrium of the uterus.

Nissl (*nis′l*) **bodies** Granular-appearing structures in the cell bodies of neurons that have an affinity for basic stain; they correspond to ribonucleoprotein; also called *chromatophilic substances.*

nitric oxide (NO) A gas that functions as a neurotransmitter in both the central nervous system and in peripheral autonomic neurons, and as an autocrine and paracrine regulator in many organs. It promotes vasodilation, intestinal relaxation, penile erection, and aids long-term potentiation in the brain.

nociceptor (*no″sĭ-sep′tor*) A receptor for pain that is stimulated by tissue damage.

nodes of Ranvier (*ran′ve-a*) Gaps in the myelin sheath of myelinated axons, located approximately 1 mm apart. Action potentials are produced only at the nodes of Ranvier in myelinated axons.

nonpolar molecule A molecule lacking positive and negative charges and therefore not soluble in water.

norepinephrine (*nor″ep-ĭ-nef′rin*) A catecholamine released as a neurotransmitter from postganglionic sympathetic nerve endings and as a hormone (together with epinephrine) by the adrenal medulla; also called *noradrenaline.*

nuclear factor-κB Abbreviated NF-κB, where κ is the Greek letter "kappa." A transcription factor that activates genes to promote inflammation; it is also believed to contribute to cancer.

nuclear receptors Receptors that bind to both a regulatory ligand (such as a hormone) and to DNA. The nuclear receptors, when activated by their ligands, regulate genetic expression (RNA synthesis).

nucleolus (*noo-kle′ŏ-lus*) A dark-staining area within a cell nucleus; the site of production of ribosomal RNA.

nucleoplasm (*noo′kle-ŏ-plaz″em*) The protoplasm of a nucleus.

nucleosome (*noo′kle-ŏ-sōm*) A complex of DNA and histone proteins that is believed to constitute an inactive form of DNA. In the electron microscope, the histones look like beads threaded on a string of chromatin.

nucleotide (*noo′kle-ŏ-tīd*) The subunit of DNA and RNA macromolecules. Each nucleotide is composed of a nitrogenous base (adenine, guanine, cytosine, and thymine or uracil); a sugar (deoxyribose or ribose); and a phosphate group.

nucleus (*noo′klē-us*), **brain** An aggregation of neuron cell bodies within the brain. Nuclei within the brain are surrounded by white matter and are located deep to the cerebral cortex.

nucleus, cell The organelle, surrounded by a double saclike membrane called the nuclear envelope (nuclear membrane), that contains the DNA and genetic information of the cell.

nystagmus (*nī-stag′mus*) Involuntary oscillatory movements of the eye.

O

obese (*o-bēs*) Excessively fat.

oligodendrocyte (*ol″ĭ-go-den′drŏ-sīt*) A type of glial cell that forms myelin sheaths around axons in the central nervous system.

oncogene (*on′kŏ-jēn*) A gene that contributes to cancer. Oncogenes are believed to be abnormal forms of genes that participate in normal cellular regulation.

oncology (*on-kol′ŏ-je*) The study of tumors.

oncotic (*on-kot′ik*) **pressure** The colloid osmotic pressure of solutions produced by proteins. In plasma, it serves to counterbalance the outward filtration of fluid from capillaries caused by hydrostatic pressure.

oocyte (*o′ŏ-sīt*) An immature egg cell (ovum). A primary oocyte has not yet completed the first meiotic division; a secondary oocyte has begun the second meiotic division. A secondary oocyte, arrested at metaphase II, is ovulated.

oogenesis (*o″ŏ-jen′ĕ-sis*) The formation of ova in the ovaries.

opsonization (*op″sŏ-nī-za′shun*) The process by which antibodies enhance the ability of phagocytic cells to attack bacteria.

optic (*op′tik*) **disc** The area of the retina where axons from ganglion cells gather to form the optic nerve and where blood vessels enter and leave the eye. It corresponds to the blind spot in the visual field caused by the absence of photoreceptors.

organ A structure in the body composed of two or more primary tissues that performs a specific function.

organelle (*or″gă-nel′*) A structure within cells that performs specialized tasks. Organelles include mitochondria, the Golgi apparatus, endoplasmic reticulum, nuclei, and lysosomes. The term is also used for some structures not enclosed by a membrane, such as ribosomes and centrioles.

organic anion transporters (OATs) A family of transport proteins with a broad range of specificities (described as multispecific or polyspecific), able to transport many drugs and endogenous molecules across the cell membranes of the renal tubules and bile ductules. As a result of this action, the organic anion transporters are responsible for the renal secretion of many compounds (such as antibiotics) into the renal tubules, and for the secretion of many compounds into the bile, for elimination in the urine and feces, respectively.

organ of Corti (*kor′te*) The structure within the cochlea that constitutes the functional unit of hearing. It consists of hair cells and supporting cells on the basilar membrane that help to transduce sound waves into nerve impulses; also called the *spiral organ.*

osmolality (Osm) (*oz″mŏ-lal′ĭ-te*) A measure of the total concentration of a solution; the number of moles of solute per kilogram of solvent.

osmoreceptor (*oz″mŏ-re-cep′tor*) A sensory neuron that responds to changes in the osmotic pressure of the surrounding fluid.

osmosis (*oz-mo′sis*) The passage of solvent (water) from a more dilute to a more concentrated solution through a membrane that is more permeable to water than to the solute.

osmotic (*oz-mot′ik*) **pressure** A measure of the tendency for a solution to gain water by osmosis when separated by a membrane from pure water. Directly related to the osmolality of the solution, it is the pressure required to just prevent osmosis.

osteoblast (*os′te-ŏ-blast*) A bone-forming cell.

osteoclast (*os′te-ŏ-klast*) A cell that resorbs bone by promoting the dissolution of calcium phosphate crystals.

osteocyte (*os′te-ŏ-sīt*) A mature bone cell that has become entrapped within a matrix of bone. This cell remains alive because it is nourished by means of canaliculi within the extracellular material of bone.

osteomalacia (*os″te-o-mă-la′shă*) Softening of bones due to a deficiency of vitamin D and calcium.

osteoporosis (*os″te-o-pŏ-ro′sis*) Demineralization of bone, seen most commonly in postmenopausal women and patients who are inactive or paralyzed. It may be accompanied by pain, loss of stature, and other deformities and fractures.

ovary (*o′vă-re*) The gonad of a female that produces ova and secretes female sex steroids.

oviduct (*o′vĭ-dukt*) The part of the female reproductive tract that transports ova from the ovaries to the uterus. Also called the *uterine,* or *fallopian tube.*

ovulation (*ov-yŭ-la′shun*) The extrusion of a secondary oocyte from the ovary.

oxidation-reduction (*ok″sĭ-da′shun-re-duk′shun*) The transfer of electrons or hydrogen atoms from one atom or molecule to another. The atom or molecule that loses the electrons or hydrogens is oxidized; the atom or molecule that gains the electrons or hydrogens is reduced.

oxidative phosphorylation (*ok″sĭ-da′tiv fos″for-ĭ-la′shun*) The formation of ATP by using energy derived from electron transport to oxygen. It occurs in the mitochondria.

oxidative stress The damage to lipids, proteins, and DNA in the body produced by the excessive production of free radicals. Oxidative stress is believed to contribute to aging and a variety of diseases.

oxidizing (*ok′sĭ-dizing*) **agent** An atom that accepts electrons in an oxidation-reduction reaction.

oxygen (*ok′sĭ-jen*) **debt** The extra amount of oxygen required by the body after exercise to metabolize lactic acid and to supply the higher metabolic rate of muscles warmed during exercise.

oxyhemoglobin (*ok″se-he″mŏ-glo′bin*) A compound formed by the bonding of molecular oxygen with hemoglobin.

oxyhemoglobin saturation The ratio, expressed as a percentage, of the amount of oxyhemoglobin compared to the total amount of hemoglobin in blood.

oxytocin (*ok″sĭ-to′sin*) One of the two hormones produced in the hypothalamus and released from the posterior pituitary (the other being vasopressin). Oxytocin stimulates the contraction of uterine smooth muscles and promotes milk letdown in females.

P

pacemaker The group of cells that has the fastest spontaneous rate of depolarization and contraction in a mass of electrically coupled cells; in the heart, this is the sinoatrial, or SA, node.

pacemaker potential The spontaneous depolarization of the pacemaker region of the heart, the SA node, that occurs during diastole. When this depolarization reaches threshold, an action potential is produced and results in contraction (systole).

pacesetter potentials Changes in membrane potential produced spontaneously by pacemaker cells of single-unit smooth muscles.

pacinian corpuscle (*pă-sin′e-an kor′pus′l*) A cutaneous sensory receptor sensitive to pressure. It is characterized by an onionlike layering of cells around a central sensory dendrite.

PAH para-aminohippuric (*par′ă-ă-me′no-hi-pyoor″ik*) **acid** A substance used to measure total renal plasma flow because its clearance rate is equal to the total rate of plasma flow to the kidneys. PAH is filtered and secreted by the renal nephrons but not reabsorbed.

pancreatic (*pan″kre-at′ik*) **islets** *See* islets of Langerhans.

pancreatic juice The secretions of the pancreas that are transported by the pancreatic duct to the duodenum. Pancreatic juice contains bicarbonate and the digestive enzymes trypsin, lipase, and amylase.

Paneth cells Located at the bottom of intestinal crypts, Paneth cells secrete lysozyme and bactericidal peptides called defensins.

paracrine (*par′ă-krin*) **regulator** A regulatory molecule produced within one tissue that acts on a different tissue of the same organ. For example, the endothelium of blood vessels secretes a number of paracrine regulators that act on the smooth muscle layer of the vessels to cause vasoconstriction or vasodilation.

parasympathetic (*par″ă-sim″pă-thet′ik*) Pertaining to the craniosacral division of the autonomic nervous system.

parathyroid (*par″ă-thi′roid*) **hormone (PTH)** A polypeptide hormone secreted by the parathyroid glands. PTH acts to raise the blood Ca^{2+} levels primarily by stimulating resorption of bone.

parietal (*pah-ri′e-tal*) **cells** Cells in the gastric glands that secrete hydrochloric acid (HCl).

Parkinson's disease A tremor of the resting muscles and other symptoms caused by inadequate dopamine-producing neurons in the basal nuclei of the cerebrum. Also called *paralysis agitans.*

parturition (*par″tyoo-rish′un*) The process of giving birth; childbirth.

passive immunity Specific immunity granted by the administration of antibodies made by another organism.

Pasteur effect A decrease in the rate of glucose utilization and lactic acid production in tissues or organisms by their exposure to oxygen.

pathogen (*path′ŏ-jen*) Any disease-producing microorganism or substance.

pathogen recognition receptors Receptor proteins on cells of the innate immune system that recognize pathogen-associated molecular patterns (PAMPS). One important group of pathogen recognition receptors is the Toll-like receptors.

pepsin (*pep′sin*) The protein-digesting enzyme secreted in gastric juice.

peptic ulcer (*pep′tik ul′ser*) An injury to the mucosa of the esophagus, stomach, or small intestine caused by the breakdown of gastric barriers to self-digestion or by excessive amounts of gastric acid.

perfusion (*per′fyoo″zhun*) The flow of blood through an organ.

perilymph (*per′ĭ-limf*) The fluid that fills the space between the membranous and bony labyrinths of the inner ear.

perimysium (*per″ĭ-mis′e-um*) The connective tissue surrounding a fascicle of skeletal muscle fibers.

periosteum (*per″e-os′te-um*) Connective tissue covering bones. It contains osteoblasts, and is therefore capable of forming new bone.

peripheral resistance The resistance to blood flow through the arterial system. Peripheral resistance is largely a function of the radius of small arteries and arterioles. The resistance to blood flow is proportional to the fourth power of the radius of the vessel.

peristalsis (*per″ĭ-stal′sis*) Waves of smooth muscle contraction in smooth muscles of the tubular digestive tract. It involves circular and longitudinal muscle fibers at successive locations along the tract and serves to propel the contents of the tract in one direction.

permissive effect The phenomenon in which the presence of one hormone "permits" the full exertion of the effects of another hormone. This may be due to promotion of the synthesis of the active form of the second hormone, or it may be due to an increase in the sensitivity of the target tissue to the effects of the second hormone.

pH The symbol (short for potential of hydrogen) used to describe the hydrogen ion (H^+) concentration of a solution. The pH scale in common use ranges from 0 to 14. Solutions with a pH of 7 are neutral; those with a pH lower than 7 are acidic; and those with a higher pH are basic.

phagocytosis (*fag″ŏ-si-to′sis*) Cellular eating; the ability of some cells (such as white blood cells) to engulf large particles (such as bacteria) and digest these particles by merging the food vacuole in which they are contained with a lysosome containing digestive enzymes.

phenylalanine (*fen″il-al′ă-nēn*) An amino acid that also serves as the precursor for L-dopa, dopamine, norepinephrine, and epinephrine.

phenylketonuria (*fen″il-kēt″n-oor′e-ă*) **(PKU)** An inborn error of metabolism that results in the inability to convert the amino acid phenylalanine into tyrosine. This defect can cause central nervous system damage if the child is not placed on a diet low in phenylalanine.

phonocardiogram (*fo″nŏ-kar′de-ŏ-gram*) A visual display of the heart sounds.

phosphatidylcholine (*fos-fat″i-dil-ko′len*) The chemical name for the molecule also called lecithin.

phosphodiesterase (*fos″fo-di-es′ter-ās*) An enzyme that cleaves cyclic AMP into inactive products, thus inhibiting the action of cyclic AMP as a second messenger.

phospholipid (*fos″fo-lip′id*) A lipid containing a phosphate group. Phospholipid molecules (such as lecithin) are polar on one end and nonpolar on the other end. They make up a large part of the cell membrane and function in the lung alveoli as surfactants.

phosphorylation (*fos″for-ĭ-la′shun*) The addition of an inorganic phosphate group to an organic molecule; for example, the addition of a phosphate group to ADP to make ATP or the addition of a phosphate group to specific proteins as a result of the action of protein kinase enzymes.

photoreceptors (*fo″to-re-sep′torz*) Sensory cells (rods and cones) that respond electrically to light. They are located in the retina of the eyes.

pia mater (*pi′ă ma′ter*) The innermost of the connective tissue meninges that envelops the brain and spinal cord.

pineal (*pin′ e-al*) **gland** A gland within the brain that secretes the hormone melatonin. It is affected by sensory input from the photoreceptors of the eyes.

pinocytosis (*pin″ŏ-si-to′sis*) Cell drinking; invagination of the cell membrane to form narrow channels that pinch off into vacuoles. This permits cellular intake of extracellular fluid and dissolved molecules.

pituitary (*pĭ-too′ĭ-ter-e*) **gland** Also called the *hypophysis.* A small endocrine gland joined to the hypothalamus at the base of the brain. The pituitary gland is functionally divided into anterior and posterior portions. The anterior pituitary secretes ACTH, TSH, FSH, LH, growth hormone, and prolactin. The posterior pituitary releases oxytocin and antidiuretic hormone (ADH), which are produced by the hypothalamus.

plasma (*plaz′mă*) The fluid portion of the blood. Unlike serum (which lacks fibrinogen), plasma is capable of forming insoluble fibrin threads when in contact with test tubes.

plasma cells Cells derived from B lymphocytes that produce and secrete large amounts of antibodies. They are responsible for humoral immunity.

plasma membrane The selectively permeable structure that regulates the transport of materials into and out of the cell, and that separates the intracellular from the extracellular compartment. Also called the *cell membrane.*

platelet (*plāt′let*) A disc-shaped structure, 2 to 4 micrometers in diameter, derived from bone marrow cells called megakaryocytes. Platelets circulate in the blood and participate (together with fibrin) in forming blood clots.

pluripotent (*ploo-rip′ŏ-tent*) A term used to describe the ability of early embryonic cells to specialize to produce all tissues except the trophoblast cells of the placenta.

pneumotaxic (*noo″mŏ-tak′sik*) **center** A neural center in the pons that rhythmically inhibits inspiration in a manner independent of sensory input.

pneumothorax (*noo″mo-thor′aks*) An abnormal condition in which air enters the intrapleural space, either through an open chest wound or from a tear in the lungs. This can lead to the collapse of a lung (atelectasis).

PNS peripheral nervous system The nerves and ganglia.

Poiseuille's (*pwă-zū′yez*) **law** The statement that the rate of blood flow through a vessel is directly proportional to the pressure difference between the two ends of the vessel and inversely proportional to the length of the vessel, the viscosity of the blood, and the fourth power of the radius of the vessel.

polar body A small daughter cell formed by meiosis that degenerates in the process of oocyte production.

polar molecule A molecule in which the shared electrons are not evenly distributed, so that one side of the molecule is negatively (or positively) charged in comparison with the other side. Polar molecules are soluble in polar solvents such as water.

polycythemia (*pol″e-si-the´-me-ă*) An abnormally high red blood cell count.

polydipsia (*pol″e-dip´se-ă*) Excessive thirst.

polymer (*pol´ĭ-mer*) A large molecule formed by the combination of smaller subunits, or monomers.

polymorphonuclear (*pol″e-mor″fŏ-noo´kle-ar*) **leukocyte** A granular leukocyte containing a nucleus with a number of lobes connected by thin cytoplasmic strands. This term includes neutrophils, eosinophils, and basophils.

polypeptide (*pol″e-pep´tīd*) A chain of amino acids connected by covalent bonds called peptide bonds. A very large polypeptide is called a protein.

polyphagia (*pol´e-fa″je-ă*) Excessive eating.

polysaccharide (*pol″e-sak´ă-rīd*) A carbohydrate formed by covalent bonding of numerous monosaccharides. Examples are glycogen and starch.

polyuria (*pol″e-yoor´e-ă*) Excretion of an excessively large volume of urine in a given period.

portal (*por´tal*) **system** A system of vessels consisting of two capillary beds in series, where blood from the first is drained by veins into a second capillary bed, which in turn is drained by veins that return blood to the heart. The two major portal systems in the body are the hepatic portal system and the hypothalamo-hypophyseal portal system.

positive feedback A response mechanism that results in the amplification of an initial change. Positive feedback results in avalanche-like effects, as occur in the formation of a blood clot or in the production of the LH surge by the stimulatory effect of estrogen.

posterior (*pos-tēr´e-or*) At or toward the back of an organism, organ, or part; the dorsal surface.

posterior pituitary *See* neurohypophysis.

postsynaptic (*pŏst″sĭ-nap´tik*) **inhibition** The inhibition of a postsynaptic neuron by axon endings that release a neurotransmitter that induces hyperpolarization (inhibitory postsynaptic potentials).

potential (*pŏ-ten´shal*) **difference** In biology, the difference in charge between two solutions separated by a membrane. The potential difference is measured in voltage.

prehormone (*pre-hor´mōn*) An inactive form of a hormone secreted by an endocrine gland. The prehormone is converted within its target cells to the active form of the hormone.

preload The load on a muscle before it contracts. In the ventricles of the heart, the preload relates to the tension on the ventricular walls produced by their filling with the end-diastolic volume of blood.

presynaptic (*pre″sĭ-nap´tik*) **inhibition** Neural inhibition in which axoaxonic synapses inhibit the release of neurotransmitter chemicals from the presynaptic axon.

process (*pros´es, pro´ses*) **cell** Any thin cytoplasmic extension of a cell, such as the dendrites and axon of a neuron.

progesterone (*pro-jes´tĕ-rōn*) A steroid hormone secreted by the corpus luteum of the ovaries and by the placenta. Secretion of progesterone during the luteal phase of the menstrual cycle promotes the final maturation of the endometrium.

prohormone (*pro-hor´mōn*) The precursor of a polypeptide hormone that is larger and less active than the hormone. The prohormone is produced within the cells of an endocrine gland and is normally converted into the shorter, active hormone prior to secretion.

prolactin (*pro-lak´tin*) **(PRL)** A hormone secreted by the anterior pituitary that stimulates lactation (acting together with other hormones) in the postpartum female. It may also participate (along with the gonadotropins) in regulating gonadal function in some mammals.

prophylaxis (*pro″fĭ-lak´sis*) Prevention or protection.

proprioceptor (*pro″pre-o-sep″tor*) A sensory receptor that provides information about body position and movement. Examples are receptors in muscles, tendons, and joints and in the semicircular canals of the inner ear.

prostaglandin (*pros´tă-glan´din*) Any of a family of fatty acids that serve numerous autocrine regulatory functions, including the stimulation of uterine contractions and of gastric acid secretion and the promotion of inflammation.

proteasome A protease complex in the cytoplasm that digests proteins that are tagged with ubiquitin, a polypeptide that marks molecules for destruction.

protein (*pro´te-in*) The class of organic molecules composed of large polypeptides in which over a hundred amino acids are bonded together by peptide bonds.

protein kinase (*ki´nās*) The enzyme activated by cyclic AMP that catalyzes the phosphorylation of specific proteins (enzymes). Such phosphorylation may activate or inactivate enzymes.

proteinuria (*pro-te-noo´re-a*) The presence of protein in the urine in amounts greater than normal levels, as determined by specific standards.

proteome (*pro´te-om*) All of the different proteins produced by a genome.

proton (*pro´ton*) A unit of positive charge in the nucleus of atoms.

pseudohermaphrodite (*soo″dŏ-her-maf´rŏ-dīt*) An individual who has the gonads of one sex only, but some of the body features of the opposite sex. (A true hermaphrodite has both ovarian and testicular tissue.)

pseudopod (*soo″dŏ-pod*) A footlike extension of the cytoplasm that enables some cells (with amoeboid motion) to move across a substrate. Pseudopods also are used to surround food particles in the process of phagocytosis.

puberty (*pyoo´ber-te*) The period in an individual's life span when secondary sexual characteristics and fertility develop.

pulmonary (*pul´mŏ-ner″e*) **circulation** The part of the vascular system that includes the pulmonary arteries and pulmonary veins. It transports blood from the right ventricle of the heart through the lungs, and then back to the left atrium of the heart.

pupil The opening at the center of the iris of the eye.

Purkinje (*pur-kin´je*) **cells** Neurons in the cerebellum that send axons to other regions of the brain to influence motor coordination and other functions. Purkinje cells exert inhibitory effects at their synapses with other neurons.

Purkinje (*pur-kin´je*) **fibers** Specialized conducting tissue in the ventricles of the heart that carry impulses from the bundle of His to the myocardium of the ventricles.

pyramidal (*pĭ-ram´ĭ-dal*) **tracts** Motor tracts that descend without synaptic interruption from the cerebrum to the spinal cord, where they synapse either directly or indirectly (via spinal interneurons) with the lower motor neurons of the spinal cord; also called *corticospinal tracts.*

pyrogen (*pi´rŏ-jen*) A fever-producing substance.

Q

QRS complex The principal deflection of an electrocardiogram, produced by depolarization of the ventricles.

R

reabsorption (*re″ab-sorp´shun*) The transport of a substance from the lumen of the renal nephron into the peritubular capillaries.

receptive field An area of the body that, when stimulated by a sensory stimulus, activates a particular sensory receptor.

receptor proteins Proteins in target cells for regulatory molecules that are each specific for a particular regulatory molecule, and that bind to it with a high affinity and limited capacity. A regulatory molecule must bind to receptor proteins in its target cells in order to regulate those cells.

reciprocal innervation (*rĭ-sip´rŏ-kal in″er-va´shun*) The process whereby the motor neurons to an antagonistic muscle are inhibited when the motor neurons to an agonist muscle are stimulated. In this way, for example, the extensor muscle of the elbow joint is inhibited when the flexor muscles of this joint are stimulated to contract.

recruitment (*rĭ-kroot´ment*) In terms of muscle contraction, the successive stimulation of more and larger motor units in order to produce increasing strengths of muscle contraction.

recurrent (*re-kur´ent*) **circuits** Circuits of neural activity where neurons activate other neurons in a sequence that turns back on itself (also known as *reverberating circuits*). Such recurrent circuits are believed to underlie short-term memory.

reduced hemoglobin Hemoglobin with iron in the reduced ferrous state. It is able to bond with oxygen but is not combined with oxygen. Also called *deoxyhemoglobin.*

reducing agent An electron donor in a coupled oxidation-reduction reaction.

referred pain Pain originating in deep, visceral organs that is perceived to be coming from particular body surface locations. This is believed to be caused by sensory neurons from both locations synapsing on the same interneurons in the same spinal cord level.

reflex arc The neural pathway for an involuntary response to a stimulus, involving sensory information carried into the CNS in afferent neurons, motor information carried out of the CNS in efferent neurons, and in many cases also involving association neurons within the CNS. No association neurons are involved in the monosynaptic muscle stretch reflex arc.

refraction (*re-frak´shun*) The bending of light rays when light passes from a medium of one density to a medium of another density. Refraction of light by the cornea and lens acts to focus the image on the retina of the eye.

refractory (*re-frak´tŏ-re*) **period** The period of time during which a region of axon or muscle cell membrane cannot be stimulated to produce an action potential (absolute refractory period), or when it can be stimulated only by a very strong stimulus (relative refractory period).

regulatory T lymphocytes Previously called suppressor T lymphocytes, these cells inhibit the activity of B cells and helper T cells to reduce the intensity of inappropriate immune responses.

relaxin (*re-lak´sin*) A hormone secreted by the corpus luteum of the ovary that, in many mammals (but not humans), causes a softening of the pubic symphysis and relaxation of the uterus to aid parturition. In humans, however, relaxin promotes other effects that benefit the developing embryo.

releasing hormones Polypeptide hormones secreted by neurons in the hypothalamus that travel in the hypothalamo-hypophyseal portal system to the anterior pituitary and stimulate the anterior pituitary to secrete specific hormones.

REM sleep The stage of sleep in which dreaming occurs. It is associated with rapid eye movements (REMs). REM sleep occurs three to four times each night and lasts from a few minutes to over an hour.

renal (*re´nal*) Pertaining to the kidneys.

renal plasma clearance The volume of plasma from which a particular solute is cleared each minute by the excretion of that solute in the urine. If there is no reabsorption or secretion of that solute by the nephron tubules, the renal plasma clearance is equal to the glomerular filtration rate.

renal plasma threshold When a molecule in the blood plasma is filtered in the kidneys by the glomerular capillaries and reabsorbed across the nephron tubules by carrier proteins, the renal plasma threshold is the minimum plasma concentration of that molecule required to saturate the carriers and cause the molecule to appear in the urine.

renal pyramid (*pĭ´ră-mid*) One of a number of cone-shaped tissue masses that compose the renal medulla.

renin (*re´nin*) An enzyme secreted into the blood by the juxtaglomerular apparatus of the kidneys. Renin catalyzes the conversion of angiotensinogen into angiotensin II.

repolarization (*re-po″lar-ĭ-za´shun*) The reestablishment of the resting membrane potential after depolarization has occurred.

resorption (*re-sorp´shun*) **bone** The dissolution of the calcium phosphate crystals of bone by the action of osteoclasts.

respiratory acidosis (*rĭ-spīr´ă-tor-e as″ĭ-do´sis*) A lowering of the blood pH to below 7.35 as a result of the accumulation of CO_2 caused by hypoventilation.

respiratory alkalosis (*al″kă-lo´sis*) A rise in blood pH to above 7.45 as a result of the excessive elimination of blood CO_2 caused by hyperventilation.

respiratory distress syndrome **(RDS)** A lung disease of the newborn, most frequently occurring in premature infants, that is caused by abnormally high alveolar surface tension as a result of a deficiency in lung surfactant; also called *hyaline membrane disease.*

respiratory zone The region of the lungs in which gas exchange between the inspired air and pulmonary blood occurs. It includes the respiratory bronchioles, in which individual alveoli are found, and the terminal alveoli.

resting membrane potential The potential difference across a plasma membrane when the cell is in an unstimulated state. The resting potential is always negatively charged on the inside of the membrane compared to the outside.

reticular (*rĕ-tik´yŭ-lar*) **activating system (RAS)** A complex network of nuclei and fiber tracts within the brain stem that produces nonspecific arousal of the cerebrum to incoming sensory information. The RAS thus maintains a state of alert consciousness and must be depressed during sleep.

retina (*ret´-ĭ-nă*) The layer of the eye that contains neurons and photoreceptors (rods and cones).

retinal (*ret´i-nal*) Also called retinaldehyde and retinene, this is the visual pigment in rods and cones. The 11-*cis* form of retinal is combined with the protein opsin (in rods) or photopsin (in cones), and dissociates from this protein when light causes the retinal to change to its all-*trans* isomer in the bleaching reaction.

retinoic (*ret″ĭ-no´ik*) **acid** The active form of vitamin A that binds to nuclear receptor proteins and directly produces the effects of vitamin A.

retrograde neurotransmitter A neurotransmitter that is released from a postsynaptic neuron and diffuses across the synaptic cleft to the presynaptic axon terminals, to modify the release of the axon's neurotransmitter.

rhodopsin (*ro-dop´sin*) Visual purple. A pigment in rod cells that undergoes a photochemical dissociation in response to light and, in so doing, stimulates electrical activity in the photoreceptors.

rhythmicity center The area of the medulla oblongata that controls the rhythmic pattern of inspiration and expiration.

riboflavin (*ri´bo-fla″vin*) Vitamin B_2. Riboflavin is a water-soluble vitamin that is used to form the coenzyme FAD, which participates in the transfer of hydrogen atoms.

ribosome (*ri´bo-sōm*) A cytoplasmic organelle composed of protein and ribosomal RNA that is responsible for the translation of messenger RNA and protein synthesis.

ribozymes (*ri″bo-zims*) RNA molecules that have catalytic ability.

rickets (*rik´ets*) A condition caused by a deficiency of vitamin D and associated with interference of the normal ossification of bone.

rigor mortis (*rig´or mor´tis*) The stiffening of a dead body due to the depletion of ATP and the production of rigor complexes between actin and myosin in muscles.

RNA **ribonucleic** (*ri″bo-noo-kle´ik*) **acid** A nucleic acid consisting of the nitrogenous bases adenine, guanine, cytosine, and uracil; the sugar ribose; and phosphate groups. There are three types of RNA found in cytoplasm: messenger RNA (mRNA), transfer RNA (tRNA), and ribosomal RNA (rRNA).

rods One of the two categories of photoreceptors (the other being cones) in the retina of the eye. Rods are responsible for black-and-white vision under low illumination.

S

saccadic (*sa-kad´ik*) **eye movements** Very rapid, jerky eye movements that, for example, serve to maintain a focus on the fovea centralis of the retina while reading.

saltatory (*sal´tă-tor-e*) **conduction** The rapid passage of action potentials from one node of Ranvier to another in myelinated axons.

sarcolemma (*sar″cŏ-lem´ă*) The cell membrane of striated muscle cells.

sarcomere (*sar´kŏ-mēr*) The structural subunit of a myofibril in a striated muscle; equal to the distance between two successive Z lines.

sarcoplasm (*sar´kŏ-plaz″em*) The cytoplasm of striated muscle cells.

sarcoplasmic reticulum (*sar″kŏ-plaz´mik rĕ-tik´yŭ-lum*) The smooth or agranular endoplasmic reticulum of striated muscle cells. It surrounds each myofibril and serves to store Ca^{2+} when the muscle is at rest.

Schwann (*shvan*) **cell** A supporting cell of the peripheral nervous system that forms sheaths around peripheral nerve fibers. Schwann cells also direct regeneration of peripheral nerve fibers to their target cells.

second messenger A molecule or ion whose concentration within a target cell is increased by the action of a regulator molecule (e.g., a hormone or neurotransmitter) so as to stimulate the metabolism of that target cell in a way characteristic of the actions of the regulator molecule—that is, in a way that mediates the intracellular effects of the regulator molecule.

secretin (*sĕ-kre´tin*) A polypeptide hormone secreted by the small intestine in response to acidity of the intestinal lumen. Along with cholecystokinin, secretin stimulates the secretion of pancreatic juice into the small intestine.

secretion (*sĕ-kre´shun*), **renal** The transport of a substance from the blood through the wall of the nephron tubule into the urine.

semen (*se´men*) The fluid ejaculated by a male containing sperm and additives from the prostate and seminal vesicles.

semicircular canals Three canals of the bony labyrinth that contain endolymph, which is continuous with the endolymph of the membranous labyrinth of the cochlea. The semicircular canals provide a sense of equilibrium.

semilunar (*sem″e-loo´nar*) **valves** The valve flaps of the aorta and pulmonary artery at their juncture with the ventricles.

seminal vesicles (*sem´ĭ-nal ves´ĭ-k´lz*) The paired organs located on the posterior border of the urinary bladder that empty their contents into the ejaculatory duct and thus contribute to the semen.

seminiferous tubules (*sem″ĭ-nif´er-us too´byoolz*) The tubules within the testes that produce spermatozoa by meiotic division of their germinal epithelium.

selectively permeable membrane A membrane with pores of a size that permit the passage of solvent and some solute molecules, while restricting the passage of other solute molecules.

sensory neuron (*noor´on*) An afferent neuron that conducts impulses from peripheral sensory organs into the central nervous system.

serosa (*sĭ-ro´să*) An outer epithelial membrane that covers the surface of a visceral organ.

serotonin (*ser″ŏ-to´nin*) monoamine neurotransmitter, chemically known as 5-hydroxytryptamine, derived from the amino acid L-tryptophan. Serotonin released at synapses in the brain have been associated with the regulation of mood and behavior, appetite, and cerebral circulation.

serotonin transporters Transport proteins in the presynaptic membrane that take up serotonin from the synaptic cleft. Serotonin transporters are inhibited by drugs called serotonin-specific reuptake inhibitors (SSRIs).

Sertoli (*ser-to´e*) **cells** Nongerminal supporting cells in the seminiferous tubules. Sertoli cells envelop spermatids and appear to participate in the transformation of spermatids into spermatozoa; also called *sustentacular cells.*

serum (*ser´um*) The fluid squeezed out of a clot as it retracts; supernatant when a sample of blood clots in a test tube and is centrifuged. Serum is plasma from which fibrinogen and other clotting proteins have been removed as a result of clotting.

sex chromosomes The X and Y chromosomes. These are the unequal pairs of chromosomes involved in sex determination (which depends on the presence or absence of a Y chromosome). Females lack a Y chromosome and normally have the genotype XX; males have a Y chromosome and normally have the genotype XY.

shock As it relates to the cardiovascular system, a rapid, uncontrolled fall in blood pressure, which in some cases becomes irreversible and leads to death.

sickle-cell anemia A hereditary autosomal recessive trait that occurs primarily in people of African ancestry, in whom it evolved apparently as a protection (in the carrier state) against malaria. In the homozygous state, hemoglobin S is made instead of hemoglobin A, which leads to the characteristic sickling of red blood cells, hemolytic anemia, and organ damage.

sinoatrial (*si″no-a´tre-al*) **node** A mass of specialized cardiac tissue in the wall of the right atrium that initiates the cardiac cycle; the SA node, also called the *pacemaker.*

sinus (*si´nus*) A cavity.

sinusoid (*si´nŭ-soid*) A modified capillary with a relatively large diameter that connects the arterioles and venules in the liver, bone marrow, lymphoid tissues, and some endocrine organs. In the liver, sinusoids are partially lined by phagocytic cells of the reticuloendothelial system.

skeletal muscle pump A term used with reference to the effect of skeletal muscle contraction on the flow of blood in veins. As the muscles contract, they squeeze the veins and in this way help move the blood toward the heart.

sleep apnea A temporary cessation of breathing during sleep, usually lasting for several seconds.

sliding filament theory The theory that the thick and thin filaments of a myofibril slide past each other during muscle contraction, decreasing the length of the sarcomeres but maintaining their own initial length.

slow waves Pacemaker depolarizations in the intestine produced by pacemaker cells, the interstitial cells of Cajal; these produce action potentials and resulting smooth muscle contractions.

smooth muscle A specialized type of nonstriated muscle tissue composed of fusiform single nucleated fibers. It contracts in an involuntary, rhythmic fashion in the walls of visceral organs.

sodium/potassium (*so´de-um/pŏ-tas´e-um*) **pump** An active transport carrier with ATPase enzymatic activity that acts to accumulate K^+ within cells and extrude Na^+ from cells, thus maintaining gradients for these ions across the cell membrane.

somatesthetic (*so″mat-es-thet´ek*) **sensations** Sensations arising from cutaneous, muscle, tendon, and joint receptors. These sensations project to the postcentral gyrus of the cerebral cortex.

somatic (*so-mat´ik*) **motor neuron** A motor neuron in the spinal cord that innervates skeletal muscles. Somatic motor neurons are categorized as alpha and gamma motoneurons.

somatomammotropic (*so´mă-tŏ-mam″ŏ-trop´ik*) **hormone** A hormone secreted by the placenta that has actions similar to the pituitary growth hormone and prolactin; also called *chorionic somatomammotropin (hCS).*

somatomedin (*so″mă-tŏ-med´n*) Any of a group of small polypeptides that are believed to be produced in the liver in response to growth hormone stimulation and to mediate the actions of growth hormone on the skeleton and other tissues.

somatostatin (*so″mă-tŏ-stat´n*) A polypeptide produced in the hypothalamus that acts to inhibit the secretion of growth hormone from the anterior pituitary. Somatostatin is also produced in the islets of Langerhans of the pancreas, but its function there has not been established.

somatotropic (*so″mă-tŏ-trop´ik*) **hormone** Growth hormone. An anabolic hormone secreted by the anterior pituitary that stimulates skeletal growth and protein synthesis in many organs.

sounds of Korotkoff (*kŏ-rot´kof*) The sounds heard when blood pressure measurements are taken. These sounds are produced by the turbulent flow of blood through an artery that has been partially constricted by a pressure cuff.

spastic paralysis (*spas´tik pă-ral´ĭ-sis*) Paralysis in which the muscles have such a high tone that they remain in a state of contracture. This may be caused by inability to degrade ACh released at the neuromuscular junction (as caused by certain drugs) or by damage to the spinal cord.

spermatid (*sper´mă-tid*) Any of the four haploid cells formed by meiosis in the seminiferous tubules that mature to become spermatozoa without further division.

spermatocyte (*sper-mat´ŏ-sīt*) A diploid cell of the seminiferous tubules in the testes that divides by meiosis to produce spermatids.

spermatogenesis (*sper″mă-to-jen´ĭ-sis*) The formation of spermatozoa, including meiosis and maturational processes in the seminiferous tubules.

spermatozoon (*sper″mă-to-zo´on*) pl., *spermatozoa* or, loosely, *sperm.* A mature sperm cell formed from a spermatid.

spermiogenesis (*sper″me-ŏ-jen´ĭ-sis*) The maturational changes that transform spermatids into spermatozoa.

sphygmomanometer (*sfig″mo-mă-nom´ĭ-ter*) A manometer (pressure transducer) used to measure the blood pressure.

spindle fibers Filaments that extend from the poles of a cell to its equator and attach to chromosomes during the metaphase stage of cell division. Contraction of the spindle fibers pulls the chromosomes to opposite poles of the cell.

Starling forces The hydrostatic pressures and the colloid osmotic pressures of the blood and tissue fluid. The balance of these pressures determines the net movement of fluid out of or into blood capillaries.

stem cells Cells that are relatively undifferentiated (unspecialized) and able to divide and produce different specialized cells.

stereoisomers (*ster´e-o-iso-mers*) Molecules with the same atoms in the same sequence, but which differ in the three-dimensional arrangement of their atoms.

steroid (*ster´oid*) A lipid derived from cholesterol that has three six-sided carbon rings and one five-sided carbon ring. These form the steroid hormones of the adrenal cortex and gonads.

stretch reflex The monosynaptic reflex whereby stretching a muscle results in a reflex contraction. The knee-jerk reflex is an example of a stretch reflex.

striated (*stri´āt-ed*) **muscle** Skeletal and cardiac muscle, the cells of which exhibit cross banding, or striations, because of the arrangement of thin and thick filaments.

stroke volume The amount of blood ejected from each ventricle at each heartbeat.

substrate (*sub´strāt*) In enzymatic reactions, the molecules that combine with the active sites of an enzyme and that are converted to products by catalysis of the enzyme.

sulcus (*sul′kus*) A groove or furrow; a depression in the cerebrum that separates the folds, or gyri, of the cerebral cortex.

summation (*sŭ-ma′shun*) In neural physiology, the additive effects of graded synaptic potentials. In muscle physiology, the additive effects of contractions of different muscle fibers.

suppressor T cells A subpopulation of T lymphocytes that acts to inhibit the production of antibodies against specific antigens by B lymphocytes.

suprachiasmatic (*soo″pră-ki″az-mot′ik*) **nucleus (SCN)** The primary center for the regulation of circadian rhythms. Located in the hypothalamus, the SCN is believed to regulate circadian rhythms by means of its stimulation of melatonin secretion from the pineal gland.

surface tension A property of water wherein the hydrogen bonding between water molecules produces a tension at the water surface.

surfactant (*sur-fak′tant*) In the lungs, a mixture of phospholipids and proteins produced by alveolar cells that reduces the surface tension of the alveoli and contributes to the elastic properties of the lungs.

sympathoadrenal (*sim′pa-tho-a-dre′nal*) **system** The sympathetic division of the autonomic system and the adrenal medulla as they function together to promote the body changes of the "fight-or-flight" response.

symport (*sim′port*) A form of secondary active transport (coupled transport) in which a molecule or ion is moved together with, and in the same direction as, Na^+ ions; that is, into the cell.

synapse (*sin′aps*) The junction across which a nerve impulse is transmitted from an axon terminal to a neuron, a muscle cell, or a gland cell either directly or indirectly (via the release of chemical neurotransmitters).

synapses en passant The type of synapses formed by autonomic neurons with their target cells. Neurotransmitters are released into the extracellular fluid from a number of regions of the autonomic axons as they pass through the target tissue.

synapsin (*sĭ-nap′sin*) A protein within the membrane of the synaptic vesicles of axons. When activated by the arrival of action potentials, synapsins aid the fusion of the synaptic vesicles with the cell membrane so that the vesicles may undergo exocytosis and release their content of neurotransmitters.

synaptic plasticity (*sĭ-nap′tik plas-tis′ĭ-te*) The ability of synapses to change at a cellular or molecular level. At a cellular level, plasticity refers to the ability to form new synaptic associations. At a molecular level, plasticity refers to the ability of a presynaptic axon to release more than one type of neurotransmitter and to exhibit synaptic facilitation and depression.

syncytium (*sin-sish′e-um*) The merging of cells in a tissue into a single functional unit. Because the atria and ventricles of the heart have gap junctions between their cells, these myocardia behave as syncytia.

synergistic (*sin″er-jis′tik*) Pertaining to regulatory processes or molecules (such as hormones) that have complementary or additive effects.

systemic (*sis-tem′ik*) **circulation** The circulation that carries oxygenated blood from the left ventricle via arteries to the tissue cells and that carries blood depleted of oxygen via veins to the right atrium; the general circulation, as compared to the pulmonary circulation.

systole (*sis′-tŏ-le*) The phase of contraction in the cardiac cycle. Used alone, this term refers to contraction of the ventricles; the term *atrial systole* refers to contraction of the atria.

T

tachycardia (*tak″ĭ-kar′de-ă*) An excessively rapid heart rate, usually applied to rates in excess of 100 beats per minute (in contrast to bradycardia, in which the heart rate is very slow—below 60 beats per minute).

target organ The organ that is specifically affected by the action of a hormone or other regulatory process.

T cell A type of lymphocyte that provides cell-mediated immunity (in contrast to B lymphocytes, which provide humoral immunity through the secretion of antibodies). There are three subpopulations of T cells: cytotoxic (killer), helper, and suppressor.

telomere (*tel′ŏ-mēr*) A DNA sequence at the end of a chromosome that is not copied by DNA polymerase during DNA replication. This inability to copy telomeres may contribute to cell aging and death. Germinal cells (that produce gametes) and cancer cells have an additional enzyme, telomerase, which copies the telomeres.

telophase (*tel′ŏ-fāz*) The last step of mitosis and the last step of the second division of meiosis.

tendon (*ten′dun*) The dense regular connective tissue that attaches a muscle to the bones of its origin and insertion.

testes (*tes′tēz*); sing; *testis.* Male gonads. Testes are also known as *testicles.*

testis-determining factor (TDF) The product of a gene located on the short arm of the Y chromosome that causes the indeterminate embryonic gonads to develop into testes.

testosterone (*tes-tos′tĕ-rōn*) The major androgenic steroid secreted by the Leydig cells of the testes after puberty.

tetanus (*tet′n-us*) In physiology, a term used to denote a smooth, sustained contraction of a muscle, as opposed to muscle twitching.

tetraiodothyronine (*tet″rā-ī″ŏo-dŏ-thi′ro-nēn*) **(T_4)** A hormone containing four iodine atoms; also known as *thyroxine.*

thalassemia (*thal″ă-se′me-ă*) Any of a group of hemolytic anemias caused by the hereditary inability to produce either the alpha or beta chain of hemoglobin. It is found primarily among Mediterranean people.

theophylline (*the-of′ĭ-lin*) A drug found in certain tea leaves that promotes dilation of the bronchioles by increasing the intracellular concentration of cyclic AMP (cAMP) in the smooth muscle cells. This effect is due to inhibition of the enzyme phosphodiesterase, which breaks down cAMP.

thermiogenesis (*ther-me-o-jen′ĭ-sis*) The production of heat by the body through mechanisms such as increased metabolic rate.

thorax (*thor′aks*) The part of the body cavity above the diaphragm; the chest.

threshold The minimum stimulus that just produces a response.

thrombin (*throm′bin*) A protein formed in blood plasma during clotting that enzymatically converts the soluble protein fibrinogen into insoluble fibrin.

thrombocyte (*throm′bŏ-sīt*) A blood platelet; a disc-shaped structure in blood that participates in clot formation.

thrombopoietin (*throm″bo-poi-e′tin*) A cytokine that stimulates the production of thrombocytes (blood platelets) from megakaryocytes in the bone marrow.

thrombosis (*throm-bo′sis*) The development or presence of a thrombus.

thrombus (*throm′bus*) A blood clot produced by the formation of fibrin threads around a platelet plug.

thymus (*thi′mus*) A lymphoid organ located in the superior portion of the anterior mediastinum. It processes T lymphocytes and secretes hormones that regulate the immune system.

thyroglobulin (*thi-ro-glob′yŭ-lin*) An iodine-containing protein in the colloid of the thyroid follicles that serves as a precursor for the thyroid hormones.

thyroxine (*thi-rok′sin*) Also called *tetraiodothyronine,* or T_4. The major hormone secreted by the thyroid gland. It regulates the basal metabolic rate and stimulates protein synthesis in many organs. A deficiency of this hormone in early childhood produces cretinism.

tinnitus (*tĭ-ni′tus*) The spontaneous sensation of a ringing sound or other noise without sound stimuli.

tissue factor A membrane glycoprotein that initiates the extrinsic pathway of blood clotting; also called *thromboplastin* or *factor III.*

titin The largest protein the human body, titin extends from a Z disc of a sarcomere to its M line. The springlike portion of titin contributes to muscle elasticity.

tolerance, immunological The ability of the immune system to distinguish self from nonself; thus, the immune system does not normally attack those antigens that are part of one's own tissues.

tonicity (*to-nis′i-te*) The osmotic pressure of a solution in comparison to the pressure of another solution, usually blood plasma. Solutions with the same osmotic pressure are isotonic; those with a lower osmotic pressure are hypotonic; and those with a higher osmotic pressure are hypertonic.

total minute volume The product of tidal volume (ml per breath) and ventilation rate (breaths per minute).

totipotent (*tōtip′ŏ-tent*) The ability of a cell to differentiate into all tissue types, and thus to form a new organism when appropriately stimulated and placed in the correct environment (a uterus).

toxin (*tok′sin*) A poison.

toxoid (*tok′soid*) A modified bacterial endotoxin that has lost toxicity but that still has the ability to act as an antigen and stimulate antibody production.

tracts A collection of axons within the central nervous system that forms the white matter of the CNS.

transamination (*trans″am-ĭ-na′shun*) The transfer of an amino group from an amino acid to an alpha-keto acid, forming a new keto acid and a new amino acid without the appearance of free ammonia.

transferrin (*trans-fer′in*) A protein in the plasma that binds to and transports iron, so that the iron component of hemoglobin in destroyed red blood cells can be recycled back to the bone marrow for the synthesis of new hemoglobin.

transcription (*tran-skrip′shun*), **genetic** The process by which messenger RNA is synthesized from a DNA template resulting in the transfer of genetic information from the DNA molecule to the mRNA.

transducins (*trans-doo′sinz*) The G-proteins involved in vision. When light causes the photodissociation of rhodopsin, the G-protein alpha subunit dissociates from the opsin and indirectly causes a reduction in the dark current of the photoreceptors.

translation (*trans-la′shun*), **genetic** The process by which messenger RNA directs the amino acid sequence of a growing polypeptide during protein synthesis.

transplantation (*trans″plan-ta′shun*) The grafting of tissue from one part of the body to another part, or from a donor to a recipient.

transport maximum Abbreviated T_m, the maximum rate at which a substance can be transported by a carrier protein across a plasma membrane. At the transport maximum, the carriers are said to be saturated.

transpulmonary (*trans″pul′mŏ-ner″-e*) **pressure** The pressure difference across the wall of the lung; equal to the difference between intrapulmonary pressure and intrapleural pressure.

triglyceride (*tri-glis′er-id*) Fats and oils. Also known as *triacylglycerol*.

triiodothyronine (*tri″i-ŏ′dŏ-thi′ro-nēn*) **(T_3)** A hormone secreted in small amounts by the thyroid; the active hormone in target cells formed from thyroxine.

tropomyosin (*tro″pŏ-mi′ŏ-sin*) A filamentous protein that attaches to actin in the thin filaments. Together with another protein called troponin, it acts to inhibit and regulate the attachment of myosin cross bridges to actin.

troponin (*tro′pŏ-nin*) A protein found in the thin filaments of the sarcomeres of skeletal muscle. A subunit of troponin binds to Ca^{2+}, and as a result causes tropomyosin to change position in the thin filament.

trypsin (*trip′sin*) A protein-digesting enzyme in pancreatic juice that is released into the small intestine.

tryptophan (*trip′tŏ-fān*) An amino acid that also serves as the precursor for the neurotransmitter molecule serotonin.

TSH **thyroid-stimulating hormone** Also called *thyrotropin (thi″rŏ-tro′pin)*. A hormone secreted by the anterior pituitary that stimulates the thyroid gland.

tubuloglomerular (*too″byŭ-lo-glo-mer′yŭ-lar*) **feedback** A control mechanism whereby an increased flow of fluid through the nephron tubules causes a reflex reduction in the glomerular filtration rate.

tumor necrosis (*nĕ-kro′sis*) **factor (TNF)** A cytokine released by immune cells and mast cells that causes destruction of tumors and migration of neutrophils toward the site of a bacterial infection. TNF is also secreted by adipose cells and may be a paracrine regulator of insulin sensitivity.

turgid (*tur′jid*) Swollen and congested.

twitch A rapid contraction and relaxation of a muscle fiber or a group of muscle fibers.

tympanic (*tim-pan′ik*) **membrane** The eardrum; a membrane separating the external from the middle ear that transduces sound waves into movements of the middle-ear ossicles.

tyrosine kinase (*ti′rŏ-sen ki′nās*) An enzyme that adds phosphate groups to tyrosine, an amino acid present in most proteins. The membrane receptor for insulin, for example, is a tyrosine kinase. When bound to insulin, the tyrosine kinase is activated, which leads to a cascade of effects that mediate insulin's action.

U

universal donor A person with blood type O, who is able to donate blood to people with other blood types in emergency blood transfusions.

universal recipient A person with blood type AB, who can receive blood of any type in emergency transfusions.

upper motor neurons Neurons in the brain that, as part of the pyramidal or extrapyramidal system, influence the activity of the lower motor neurons in the spinal cord.

urea (*yoo-re-ă*) The chief nitrogenous waste product of protein catabolism in the urine, formed in the liver from amino acids.

uremia (*yoo-re′me-ă*) The retention of urea and other products of protein catabolism as a result of inadequate kidney function.

urobilinogen (*yoo″rŏ-bi-lin′ŏ-jen*) A compound formed from bilirubin in the intestine. Some is excreted in the feces and some is absorbed and enters the enterohepatic circulation where it may be excreted either in the bile or in the urine.

V

vaccination (*vaks″sĭ-na′shun*) The clinical induction of active immunity by introducing antigens into the body so that the immune system becomes sensitized to them. The immune system will mount a secondary response to those antigens upon subsequent exposures.

vagina (*vă-ji′nă*) The tubular organ in the female leading from the external opening of the vulva to the cervix of the uterus.

vagus (*va′gus*) **nerve** The tenth cranial nerve, composed of sensory dendrites from visceral organs and preganglionic parasympathetic nerve fibers. The vagus is the major parasympathetic nerve in the body.

Valsalva's (*val-sal′vaz*) **maneuver** Exhalation against a closed glottis, so that intrathoracic pressure rises to the point that the veins returning blood to the heart are partially constricted. This produces circulatory and blood pressure changes that could be dangerous.

vasa vasora (*va′să va-sor′ă*) Blood vessels that supply blood to the walls of large blood vessels.

vasectomy (*vă-sek′tŏ-me, va-zek′tŏ-me*) Surgical removal of a portion of the ductus (vas) deferens to induce infertility.

vasoconstriction (*va″zo-kon-strik′shun*) A narrowing of the lumen of blood vessels as a result of contraction of the smooth muscles in their walls.

vasodilation (*va″zo-di-la′shun*) A widening of the lumen of blood vessels as a result of relaxation of the smooth muscles in their walls.

vasopressin (*va″zo-pres′in*) Another name for antidiuretic hormone (ADH), released from the posterior pituitary. The name *vasopressin* is derived from the fact that this hormone can stimulate constriction of blood vessels.

vein A blood vessel that returns blood to the heart.

ventilation (*ven″tĭ-la′shun*) Breathing; the process of moving air into and out of the lungs.

vertigo (*ver′tĭ-go*) A sensation of whirling motion, either of oneself or of external objects; dizziness, or loss of equilibrium.

vestibular (*vĕ-stib′yŭ-lar*) **apparatus** The parts of the inner ear, including the semicircular canals, utricle, and saccule, that function to provide a sense of equilibrium.

villi (*vil′i*) Fingerlike folds of the mucosa of the small intestine.

virulent (*vir′yŭ-lent*) Pathogenic, or able to cause disease.

vital capacity The maximum amount of air that can be forcibly expired after a maximal inspiration.

vitamin (*vi′tă-min*) Any of various unrelated organic molecules present in foods that are required in small amounts for normal metabolic function of the body. Vitamins are classified as water-soluble or fat-soluble.

voltage-regulated channel A protein in the plasma membrane that can open to produce an aqueous channel for the passage of ions through the membrane. The channel opens in response the stimulus of a depolarization to a threshold level. Also called a *voltage-gated channel*.

W

white matter The portion of the central nervous system composed primarily of myelinated fiber tracts. This forms the region deep to the cerebral cortex in the brain and the outer portion of the spinal cord.

Z

zygote (*zi′gōt*) A fertilized ovum.

zymogen (*zi′mŏ-jen*) An inactive enzyme that becomes active when part of its structure is removed by another enzyme or by some other means.

Credits

Chapter 1

1.8, 1.9: © McGraw-Hill Companies, Inc. Al Telser, photographer; 1.10: © The McGraw-Hill Companies, Inc./ Dennis Strete, photographer; 1.11: © Ed Reschke; 1.12 (both): © Ray Simons/Photo Researchers; 1.12c, 1.13a: © Ed Reschke; 1.18: © The McGraw-Hill Companies, Inc./ Photo by Dr. Alvin Telser; 1.19a: © The McGraw-Hill Companies, Inc./Al Telser, photographer; 1.20b: © Ed Reschke; 1.21: Courtesy of Southern Illinois University, School of Medicine.

Chapter 2

2.30: © Ed Reschke.

Chapter 3

3.3 (both): © Kwang Jeon/Visuals Unlimited, Inc.; 3.4 (all): From M.M. Perry and A.B. Gilbert. Journal of Cell Science 39: 257–272, 1979. Reprinted by permission Company of Biologists, Ltd.; 3.5a: © Science Photo Library RF/Getty Images; 3.5b: © Biophoto Associates/Photo Researchers, Inc.; 3.6: © Dennis Kunkel/Phototake; 3.7: © K.G. Murin/ Visuals Unlimited, Inc.; 3.9a, 3.11a: Courtesy Keith R. Porter Endowment; 3.12a: Courtesy of Mark S. Ladinsky and Kathryn D. Howell, University of Colorado; 3.18: © E. Kiselva-D. Fawcett/Visuals Unlimited, Inc.; 3.27 (all): © Ed Reschke; 3.28a: © D.M. Phillips/Visuals Unlimited, Inc.; 3.29: © SPL/Photo Researchers.

Chapter 6

6.22b: Reprinted from CELL, Vol 112, 2003, pp. 535–548, M. Perez-Moreno et al., "Sticky business. . . .", Copyright 2003, with permission from Elsevier.

Chapter 7

7.7: From H. Webster, The Vertebrate Peripheral Nervous System, John Hubbard (ed.), © 1974 Pleanuim Publishing Corporation; 7.22: © John Heuser, Washington University School of Medicine, St. Louis, MO.

Chapter 8

8.8: Courtesy of Dr. Llinas, New York University Medical Center; 8.9: From W.T. Carpenter and R.W. Buchanan, "Medical Progress: Schizophrenia" in New England Journal of Medicine, 330:685, 1994, fig. 1A. Copyright © 1994 Massachusetts Medical Society. All rights reserved; 8.17 (both): Reprinted from Esther A. Nimchinsky, Bernardo L. Sabatini, and Karel Svoboda, "Structure and Function of Dendritic Spines," Annual Review of Physiology, Volume 64: 313–353 © 2002 by Annual Reviews, www.annualreviews.org; 8.18 (both): Reprinted Figure 2 (1st and 3rd panels) with permission from RJ Dolan, "Emotion, Cognition, and Behavior" Science 298: 1191–1194. Copyright 2002 AAAS; 8.19b: © The McGraw-Hill Company, Inc./Karl Rubin, Photographer.

Chapter 10

10.10, 10.14a: Courtesy of Dean E. Hillman; 10.30a: © Thomas Sims; 10.33b: Courtesy of Patricia N. Farnsworth, University of Medicine and Dentistry; 10.37b: © Ominkron/Photo Researchers; 10.38: Reproduced with permission of Carlos Rozas, Planeta Vivo Society, www.planetavivo.org.

Chapter 11

11.1b, 11.22: © Ed Reschke; 11.24, 11.26: © L.V. Bergman/ The Bergman Collection.

Chapter 12

12.2: © Ed Reschke; 12.3b: © Victor B. Eichler, Ph.D.; 12.6a: Dr. H.E. Huxley; 12.7b (both): Copyright by Dr. R.G. Kessel and R.G. Kardon, Tissues and Organs: A Text-Atlas of Scanning Electron Microscopy, W.H. Freeman, 1979; 12.9a: © Dr. H.E. Huxley; 12.33: © Ed Reschke; 12.35a: © The McGraw-Hill Companies, Inc./Dennis Strete, photographer.

Chapter 13

13.2: © Dr. Dennis Kunkel/Visuals Unlimited, Inc.; 13.6 (all): © Stuart Fox; 13.8: © David M. Phillips/Photo Researchers, Inc.; 13.12: © The McGraw-Hill Company, Inc./Karl Rubin, Photographer; 13.28: © Don W. Fawcett; 13.30: After William Harvey, On the Motion of the Heart and Blood in Animals, 1628; 13.31a: © Biophoto Associates/Photo Researchers.

Chapter 14

14.10: © John Greim/Photo Researchers, Inc.; 14.18a: Courtesy of Donald S. Baine; 14.18b: Courtesy of Donald S. Baine; 14.22 (all): © Niels A. Lassen, Copenhagen, Denmark; 14.29: © Ruth Jenkinson/MIDIRS/Science Photo/Photo Researchers.

Chapter 15

15.9 (both): Reprinted with permission from A. G. Amit, "Three Dimensional Structure of an Antigen-Antibody Complex at 2.8 A Resolution" in Science, vol. 233, August 15, 1986. Copyright 1986 AAAS.; 15.22 (both): © Dr. Andrejs Liepins; 15.24 (both): © SIU Biomed/Custom Medical Stock; 15.25a: © Stan Flegler/Visuals Unlimited, Inc.; 15.25b: © Dennis Kunkel/Phototake.

Chapter 16

16.2: From John J. Murray, The Normal Lung, 2nd edition, 1986, © W.B. Saunders; 16.3 (both): Courtesy American Lung Association; 16.5b: © Ralph Hutchings/Visuals Unlimited; 16.6: © Phototake; 16.8 (both), 16.9: Courtesy of Edward C. Vasquez, R.T.C.R.T., Dept. of Radiologic Technology, Los Angeles City College; 16.12 (both): From J.H. Comroe, Jr., Physiology of Respiration: An Introductory Text, 2nd editon, 1974 © Yearbook Medical Publishers, Inc. Chicago; 16.17a: © Robert Calentine/ Visuals Unlimited, Inc.; 16.17b: © Science VU/Visuals Unlimited, Inc.; 16.36a: © Ingram Publishing/SuperStock; 16.36b: CDC/Sickle Cell Foundation of Georgia: Jackie George, Beverly Sinclair/photo by Janice Haney Carr.

Chapter 17

17.3: © SPL/Photo Researchers; 17.7: © Professor P.M. Motta and M. Sastellucci/SPL/Photo Researchers; 17.9: © Donald Fawcett & D. Friend/Visuals Unlimited, Inc.

Chapter 18

18.11: From D.H. Alpers and D. Seetharan, "Pathophysiology of Diseases Involving Intestinal Brush-Border Proteins," in New England Journal of Medicine, Vol. 296, 1997, p. 1047, fig. 1. © 1977 Massachusetts Medical Society. All rights reserved; 18.17: © David Phillips/Visuals Unlimited, Inc.; 18.18: © Medical Body Scans/Photo Researchers, Inc.; 18.26a: © Dr. Sheril Burton; 18.26b: © Martin/Custom Medical Stock Photo; 18.27a: © Ed Reschke; 18.37: © Ellen Fox.

Chapter 19

19.17 (all): Reprinted from "Acromegaly Diabetes, Hypermetabolism, Proteinuria and Heart Failure, Clinicopathologic Conference" in American Journal of Medicine, Vol. 20, Issue 1, Pages 133–144, January 1956; 19.18a: © Richard Kessel/Visuals Unlimited, Inc.; 19.20 (both): © David W. Dempster, PhD, 2000.

Chapter 20

20.2: © Science Photo Library/Photo Researchers, Inc.; 20.3a: © George Wilder/Visuals Unlimited, Inc.; 20.3b: © Carolina Biological Supply Co./Visuals Unlimited, Inc.; 20.16a: © Biophoto Associates/Photo Researchers; 20.18b: © F. Leroy/SPL/Photo Researchers; 20.26 (both): © Ed Reschke; 20.27a: © Donald Fawcett/Visuals Unlimited, Inc.; 20.27b: © M. I. Walker/Photo Researchers, Inc.; 20.28: © Victor P. Eroshenko; 20.29: © Landrum B. Shettles, MD; 20.34: © Martin M. Rotker/Photo Researchers, Inc.; 20.40: Courtesy of David L. Hill; 20.42a: © Petit Format/ Photo Researchers; 20.42b: © SPL/Photo Researchers; 20.42c: © Landrum B. Shettles, MD; 20.42d: © Biophoto Associates/Photo Researchers.

Index

Note: Page references followed by the letters *f* and *t* indicate figures and tables, respectively.

A

B

C

D

E

F

H

J

K

M

N

O

P

Q

R

S

U

V

W

X

Y

Z